The Geology of Spain

Society books reviewing procedures

The Society makes every effort to ensure that the scientific and production quality of its books matches that of its journals. Since 1997, all book proposals have been refereed by specialist reviewers as well as by the Society's Books Editorial Committee. If the referees identify weaknesses in the proposal, these must be addressed before the proposal is accepted.

Once the book is accepted, the Society has a team of Book Editors (listed above) who ensure that the volume editors follow strict guidelines on refereeing and quality control. We insist that individual papers can only be accepted after satisfactory review by two independent referees. The questions on the review forms are similar to those for *Journal of the Geological Society*. The referees' forms and comments must be available to the Society's Book Editors on request.

Although many of the books result from meetings, the editors are expected to commission papers that were not presented at the meeting to ensure that the book provides a balanced coverage of the subject. Being accepted for presentation at the meeting does not guarantee inclusion in the book.

It is recommended that reference to all or parts of this book be made in one of the following ways:

GIBBONS, W. & MORENO, M. T. (eds) 2002. *The Geology of Spain*. Geological Society, London.

LUNAR, R., MORENO, M. T., LOMBARDERO, M., REGUEIRO, M., LÓPEZ-VERA, F., MARTÍNEZ DELOLMO, W., MALLO GARCÍA, J. M., SAENZ DE SANTA MARIA, J. A., GARCÍA-PALOMERO, F., HIGUERAS, P., ORTEGA, L. & CAPOTE, R. 2002. Economic and Environmental Geology. *In*: GIBBONS, W. & MORENO, M. T. (eds) *The Geology of Spain*. Geological Society, London.

The Geology of Spain

EDITED BY

WES GIBBONS

and

TERESA MORENO

Jaume Almera Institute,
CSIC, Barcelona

2002
Published by
The Geological Society
London

THE GEOLOGICAL SOCIETY

The Geological Society of London (GSL) was founded in 1807. It is the oldest national geological society in the world and the largest in Europe. It was incorporated under Royal Charter in 1825 and is Registered Charity 210161.

The Society is the UK national learned and professional society for geology with a worldwide Fellowship (FGS) of 9000. The Society has the power to confer Chartered status on suitably qualified Fellows, and about 2000 of the Fellowship carry the title (CGeol). Chartered Geologists may also obtain the equivalent European title, European Geologist (EurGeol). One fifth of the Society's fellowship resides outside the UK. To find out more about the Society, log on to www.geolsoc.org.uk.

The Geological Society Publishing House (Bath, UK) produces the Society's international journals and books, and acts as European distributor for selected publications of the American Association of Petroleum Geologists (AAPG), the American Geological Institute (AGI), the Indonesian Petroleum Association (IPA), the Geological Society of America (GSA), the Society for Sedimentary Geology (SEPM) and the Geologists' Association (GA). Joint marketing agreements ensure that GSL Fellows may purchase these societies' publications at a discount. The Society's online bookshop (accessible from www.geolsoc.org.uk) offers secure book purchasing with your credit or debit card.

To find out about joining the Society and benefiting from substantial discounts on publications of GSL and other societies worldwide, consult www.geolsoc.org.uk, or contact the Fellowship Department at: The Geological Society, Burlington House, Piccadilly, London W1J 0BG: Tel. +44 (0)20 7434 9944; Fax +44 (0)20 7439 8975; Email: enquiries@geolsoc.org.uk.

For information about the Society's meetings, consult *Events* on www.geolsoc.org.uk. To find out more about the Society's Corporate Affiliates Scheme, write to enquiries@geolsoc.org.uk.

Published by The Geological Society from:
The Geological Society Publishing House
Unit 7, Brassmill Enterprise Centre
Brassmill Lane
Bath BA1 3JN, UK

(*Orders*: Tel. +44 (0)1225 445046
       Fax +44 (0)1225 442836)
Online bookshop: http: //bookshop.geolsoc.org.uk

The publishers make no representation, express or implied, with regard to the accuracy of the information contained in this book and cannot accept any legal responsibility for any errors or omissions that may be made.

British Library Cataloguing in Publication Data
A catalogue record for this book is available from the British Library.

ISBN 978-1-86239-110-9 (hardback)
ISBN 978-1-86239-127-7 (softback)

Reprinted July 2011
Typeset by Type Study, Scarborough, UK
Printed by Hobbs the Printers, Southampton, UK.

Distributors
USA
  AAPG Bookstore
  PO Box 979
  Tulsa
  OK 74101–0979
  USA
Orders: Tel. + 1 918 584–2555
       Fax +1 918 560–2652
       E-mail *bookstore@aapg.org*

India
  Affiliated East-West Press PVT Ltd
  G-1/16 Ansari Road, Daryaganj,
  New Delhi 110 002
  India
Orders: Tel. +91 11 327–9113
       Fax +91 11 326–0538
       E-mail *affiliat@nda.vsnl.net.in*

Japan
  Kanda Book Trading Co.
  Cityhouse Tama 204
  Tsurumaki 1–3–10
  Tama-shi
  Tokyo 206–0034
  Japan
Orders: Tel. +81 (0)423 57–7650
       Fax +81 (0)423 57–7651

Contents

Acknowledgements

The idea of publishing a reference work on Spanish geology, written in English by research-active Spanish experts, was conceived in 2000. Manuscripts began arriving in January 2001, were refereed and revised during the year, and the completed work handed to the Geological Society in January 2002. Many people have worked hard to bring this project to completion. We especially wish to thank the referees who gave so much of their time to reviewing the initial manuscripts. We would also like to thank the Instituto Geológico y Minero de España for kindly allowing us to use and modify their geological map of Spain. Additionally we wish to acknowledge the support and encouragement provided by Angharad Hills and Helen Knapp at the Geological Society Publishing House, and our series editor Adrian Hartley. Most of all we thank the authors who worked constructively, courteously, and efficiently with us, despite the demands of our editorial *boli rojo. ¡Salud a todos!*

Wes Gibbons & Teresa Moreno, Cardiff 2002.

The editors gratefully acknowledge the work of the following referees:

Referee	Chapter
Aiello, L. (University College London, UK)	Quaternary
Arlegui, L. E. (Universidad de Zaragoza, Spain)	Introduction and Overview
Berry, C. (Cardiff University, UK)	Devonian
Berryman, K. (Gracefield Research Centre, New Zealand)	Alpine tectonics I
Betzler, C. (Geol.-Paleontologisches Inst. Hamburg, Germany)	Tertiary
Burgess, P. (Shell International, Netherlands)	Tertiary
Capdevila, R. (Université de Rennes, France)	Palaeozoic Magmatism
Casas, A. (Universidad de Zaragoza, Spain)	Alpine tectonics I
Clarkson, E. (Edinburgh University, UK)	Silurian
Cope, J. (Cardiff University, UK)	Jurassic
Crick, R. (University of Texas, USA)	Devonian
Dabrio, C. (Universidad Complutense, Madrid, Spain)	Tertiary
Davies, S. (Leicester University, UK)	Carboniferous
Dean, W. (National Mus. of Wales, UK)	Cambrian
Dominy, S. (Townsville University, Australia)	Economic & environmental
Fortey, R. (Nat. Hist. Mus., London, UK)	Ordovician
Friend, P. (Cambridge University, UK)	Tertiary
Fry, N. (Cardiff University, UK)	Alpine tectonics II
Gayer, R. (Cardiff University, UK)	Variscan and Pre-Variscan tectonics
Hartley, A. (Aberdeen University, UK)	Introduction and Overview
Harvey, A. (Liverpool University, UK)	Quaternary
Hertogen, J. (K.U. Leuven, Belgium)	Cenozoic volcanism I
Hesselbo, S. (Oxford University, UK)	Jurassic
Jones, S. (Durham University, UK)	Permian and Triassic
Kerr, A. (Cardiff University, UK)	Cenozoic volcanism II
Lisle, R. (Cardiff University, UK)	Alpine tectonics I
Lonergan, L. (Imperial College, UK)	Alpine tectonics II
Loydell, D. (Portsmouth University, UK)	Silurian
Martínez-Frías, J. (CSIC, Spain)	Economic & environmental
Marzo, M. (Universitat de Barcelona, Spain)	Permian and Triassic
Matte, P. (Université de Montpelier, France)	Variscan and Pre-Variscan tectonics
Millán, H. (Universidad de Zaragoza, Spain)	Alpine tectonics I
Pardo, G. (Universidad de Zaragoza, Spain)	Tertiary
Pearce, J. (Cardiff University, UK)	Cenozoic volcanism I
Pérez-Estaun, A. (CSIC, Spain)	Carboniferous
Prave, A. (St. Andrews University, UK)	Precambrian
Quesada, C. (IGME, Spain)	Precambrian
Ramsey, T. (Cardiff University, UK)	Economic & environmental
Sanders, D. (University of Innsbruck, Austria)	Cretaceous
Simó, T. (Wisconsin University, USA)	Cretaceous
Spjeldnas, N. (University of Oslo, Norway)	Ordovician
Stephens, E. (St. Andrews University, UK)	Palaeozoic Magmatism
Stillman, C. (Trinity College, Ireland)	Cenozoic volcanism II
Vera, J. A. (Universidad de Granada, Spain)	Tertiary
Wright, P. (Cardiff University, UK)	Tertiary

Contributing Authors

Benito Abalos

Departamento de Geodinámica, Universidad del País Vasco, P.O. Box 644, 48080 Bilbao. Spain. (e-mail: gppabviblg.ehu.es)

Emiliano Aguirre

Museo Nacional de Ciencias Naturales (CSIC), José Gutiérrez Abascal 2, 28006 Madrid, Spain.

Julio Aguirre

Departamento de Estratigrafía y Paleontología, Facultad de Ciencias, Universidad de Granada, Campus Fuentenueva s/n, 18002, Granada, Spain. (e-mail: jaguirre@ugr.es)

Agustina Ahijade

Departamento de Edafología y Geología, Universidad de La Laguna, Tenerife, Canary Islands, Spain. (e-mail: aahijado@ull.es)

Gaspar Alonso-Gavilán

Departamento de Geología, Universidad de Salamanca, 37071 Salamanca, Spain. (e-mail: gavilan@usal.es)

Ana M. Alonso-Zarza (Tertiary co-ordinator)

Departamento de Petrología y Geoquímica, Facultad de CC. Geológicas, Universidad Complutense, 28040 Madrid, Spain. (e-mail: Alonsoza@geo.ucm.es)

Eumenio Ancochea

Departamento de Petrología y Geoquímica, Facultad de CC. Geológicas, Universidad Complutense, 28040 Madrid, Spain.

Alfredo Arche

Instituto de Geología Económica-Departamento de Estratigrafía, CSIC-UCM. Facultad de Geología, Universidad Complutense. 28040 Madrid, Spain. (e-mail: destrati@geo.ucm.es)

Concha Arenas

Departamento de Ciencias de la Tierra, Universidad de Zaragoza, C/Pedro Cerbuna, 12. E-50009 Zaragoza, Spain. (e-mail: carenas@posta.unizar.es)

Political map of Spain showing the *Comunidades Autónomas* and the locations of contributing authors.

Consuelo Arias
Departamento Estratigrafía, Instituto de Geología Económica (CSIC-UCM), Facultad de CC. Geológicas, Universidad Complutense, 28040 Madrid, Spain. (ariasc@geo.ucm.es)

Luis E. Arlegui
Departamento de Ciencias de la Tierra, Universidad de Zaragoza, C/Pedro Cerbuna, 12. E-50009-Zaragoza, Spain. (e-mail: arlegui@posta.unizar.es)

Ildefonso Armenteros (Tertiary co-ordinator)
Departamento de Geología, Universidad de Salamanca, 37071 Salamanca, Spain. (e-mail: ilde@usal.es)

Enrique Arranz
Departamento de Ciencias de la Tierra, Universidad de Zaragoza, 50009 Zaragoza, Spain. (e-mail: earranz@posta.unizar.es)

Marc Aurell (Jurassic co-ordinator)
Departamento de Ciencias de la Tierra, Universidad de Zaragoza, E-50009 Zaragoza, Spain. (e-mail: maurell@posta.unizar.es)

Jose Miguel Azañón
Departamento de Geodinámica, Universidad de Granada, 18071-Granada, Spain & Instituto Andaluz de Ciencias de la Tierra, CSIC/Universidad de Granada, 18071-Granada, Spain. (e-mail: jazanon@ugr.es)

Juan I. Baceta
Departamento de Estratigrafía y Paleontología, Facultad de Ciencias, Universidad del País Vasco, Apartado 644, 48080 Bilbao, Spain. (e-mail: gppbacaj@lg.ehu.es)

Beatriz Bádenas
Departamento Ciencias de la Tierra, Universidad de Zaragoza, E-50009 Zaragoza, Spain.

Juan Ramón Bahamonde (Carboniferous co-ordinator)
Departamento de Geología, Universidad de Oviedo, C/Jesús Arias de Velasco s/n, 33005 Oviedo, Spain. (e-mail: jrbaham@asturias.geol.uniovi.es)

Pedro Barba
Departamento de Geología, Facultad de Ciencias, Universidad de Salamanca, 37008 Salamanca, Spain. (e-mail: barba@usal.es)

Xavier Berástegui
Servei Geologic, Institut Cartografic de Catalunya, Parc de Montjuic, 08038 Barcelona, Spain. (e-mail: xberastegui@icc.es)

Juan C. Braga (Tertiary co-ordinator)
Departamento de Estratigrafía y Paleontología, Facultad de Ciencias, Universidad de Granada, Campus Fuentenueva s/n, 18002, Granada, Spain. (e-mail: jbraga@ugr.es)

José P. Calvo
Departamento de Petrología y Geoquímica, Facultad de CC. Geológicas, Universidad Complutense de Madrid, 28040 Spain. (e-mail: jpcalvo@geo.ucm.es)

Ramón Capote
Departamento de Geodinámica Interna, Facultad de CC. Geológicas, Universidad Complutense de Madrid, 28040 Madrid, Spain. (e-mail: capote@geo.ucm.es)

Jesús E. Caracuel
Departamento de Ciencias de la Tierra, Universidad de Alicante, Carretera San Vicente del Raspeig s/n, 03690 San Vicente del Raspeig, Alicante, Spain (e-mail: Jesús.Caracuel@ua.es)

Jesús Carballeira
Departamento de Geología, Universidad de Salamanca, 37071 Salamanca, Spain. (e-mail: carba@usal.es)

Beatriz Carenas
Departamento Química Agrícola, Geología y Geoquímica, Facultad de Ciencias, Universidad Autónoma de Madrid, 28049 Madrid, Spain. (e-mail: beatrizcarenas@uam.es)

Peter Carls
Inst. F. Geowissenschaften, Techn. Universitat, Pockels-str., 3, D-38106 Braunschweig, Germany.

Juan C. Carracedo (Cenozoic volcanism, Canary Islands co-ordinator)
Estación Volcanológica de Canarias, IPNA-CSIC, P.O. Box 195, 38206 La Laguna, Tenerife, Canary Islands, Spain. (e-mail: jcarracedo@ipna.csic.es)

Jordi Carreras
Departamento de Geología, Universidad Autónoma de Barcelona, 08193 Bellaterra, Barcelona, Spain. (e-mail: jordi.carreras@uab.es)

Ramón Casillas
Departamento de Edafología y Geología, Universidad de La Laguna, Tenerife, Canary Islands, Spain.

Antonio Castro (Palaeozoic Magmatism co-ordinator)
Departamento de Geología, Universidad de Huelva, Facultad de Ciencias, Campus del Carmen, 21071 Huelva, Spain. (e-mail: dorado@uhu.es)

José Manuel Castro
Departamento de Geología, Universidad de Jaén, 23071 Jaén, Spain. (e-mail: jmcastro@ujaen.es)

Esmeralda Caus
Departament de Geología, Universitat Autonoma de Barcelona, 08193 Bellaterra, Spain. (e-mail: Esmeralda.Caus@uab.es)

José María Cebriá
Departamento de Geología, Museo Nacional de Ciencias Naturales (CSIC), 28006 Madrid, Spain. (e-mail: cebria@mncn.csic.es)

Beatriz Chacón
Departamento Estratigrafía, Instituto de Geología Económica (CSIC-UCM), Facultad de CC. Geológicas, Universidad Complutense, 28040 Madrid, Spain. (e-mail: bchacon@geo.ucm.es)

Juan R. Colmenero
Departamento de Geología, Universidad de Salamanca, Plaza de la Merced s/n, 37008 Salamanca, Spain.

L. Guillermo Corretgé
Departamento de Geología, Universidad de Oviedo, Arias de Velasco s/n, 33005 Oviedo, Spain. (e-mail: corretge@asturias.geol.uniovi.es)

Angel Corrochano
Departamento de Geología, Universidad de Salamanca, 37071 Salamanca, Spain. (e-mail: corro@usal.es)

Carmen R. Cubas
Departamento de Edafología y Geología, Universidad de La Laguna, Tenerife, Canary Islands, Spain.

Jesús De La Rosa
Departamento de Geología, Universidad de Huelva, Facultad de Ciencias, Campus del Carmen, 21071 Huelva, Spain. (e-mail: jesus@uhu.es)

Miguel Doblas
Departamento de Geología, Museo Nacional de Ciencias Naturales (CSIC), 28006 Madrid, Spain. (e-mail: doblas@mncn.csic.es)

Teodosio Donaire
Departamento de Geología, Universidad de Huelva, Facultad de Ciencias, Campus del Carmen, 21071 Huelva, Spain. (e-mail: donaire@uhu.es)

Elena Druguet
Departamento de Geología, Universidad Autónoma de Barcelona, 08193 Bellaterra, Barcelona, Spain. (e-mail: elena.druguet@uab.es)

Pere Enrique
Departamento de Geoquímica, Petrología i Prospeccío Geològica, Universitat de Barcelona, Martí i Franquès s/n, 08028 Barcelona. Spain. (e-mail: pedro@geo.ub.es)

Javier Escuder Viruete
Departamento de Petrología y Geoquímica, Universidad Complutense, 28040 Madrid, Spain. (e-mail: escudr@eucmax.sim.ucm.es)

Carlos Fernández
Departamento de Geodinámica Paleontología, Universidad de Huelva, Facultad de Ciencias, Campus del Carmen, 21071 Huelva, Spain. (e-mail: fcarlos@uhu.es)

Luis P. Fernández
Departamento de Geología, Universidad de Oviedo, C/Jesús Arias de Velasco s/n, 33005 Oviedo, Spain.

Marc Floquet
Sédimentologie-Paléontologie, Université de Provence, Place Victor Hugo, Case 67, F-1333, Marseille Cedes 3, France. (e-mail: Marc.Floquet@newsup.univ-mrs.fr)

Joan J. Fornós
Departament de Ciències de la Terra, Universitat Illes Balears, Ctra. de Valldemossa, km. 7,5. 07071, Palma de Mallorca, Spain. (e-mail: joan.fornos@uib.es)

M. Antonia Fregenal-Martínez
Departamento Estratigrafía, Instituto de Geología Económica (CSIC-UCM), Facultad de CC. Geológicas Universidad Complutense, 28040 Madrid, Spain. (e-mail: mariana@geo.ucm.es)

Carlos Galé
Departamento de Ciencias de la Tierra, Universidad de Zaragoza, 50009 Zaragoza, Spain. (e-mail: carlosgb@posta.unizar.es)

Jesus Galindo-Zaldívar
Departamento de Geodinámica, Universidad de Granada, 18071-Granada, Spain. (e-mail: jgalindo@ugr.es)

José A. Gámez Vintaned
Área de Paleontología, Facultad de Ciencias, Universidad de Extremadura, E-06071 Badajoz, Spain. (e-mail: gamez@unex.es)

Alvaro García
Departamento Estratigrafía, Instituto de Geología Económica (CSIC-UCM), Facultad de CC. Geológicas, Universidad Complutense, 28040, Madrid, Spain. (e-mail: geest01@sis.ucm.es)

Jenaro L. García-Alcalde (Devonian co-ordinator)
Departamento de Geología, Univeridad de Oviedo, C/Jesús Arias de Velasco s/n, 33005 Oviedo, Spain. (e-mail: jalcalde@asturias.geol.uniovi.es)

Victor García-Dueñas
Departamento de Geodinámica, Universidad de Granada, 18071-Granada, Spain, & Instituto Andaluz de Ciencias de la Tierra, CSIC/Universidad de Granada, 18071-Granada, Spain. (e-mail: vgarciad@ugr.es)

Manuel García-Hernández
Departamento Estratigrafía y Paleontología, Universidad de Granada, 18071 Granada, Spain.

José García-Hidalgo
Departamento de Geología, Universidad de Alcalá, 28871 Alcalá de Henares, Spain. (e-mail: jose.garciahidalgo@uah.es)

Félix García Palomero
Atlantic Cooper Holding S.A., Corta Atalaya, s/n. Minas de Riotinto, Huelva, Spain. (e-mail: Felix_Garcia@fmi.com)

José C. García-Ramos
Departamento de Geología, Universidad de Oviedo, C/Jesús Arias de Velasco s/n, 33005 Oviedo, Spain.
José M. García Ruiz
Instituto Pirenaico de Ecología, Avda. Montañana 177, Aptdo. 202, 50080 Zaragoza, Spain. (e-mail: humberto@ipe.csic.es)
Wes Gibbons (co-editor)
Earth Sciences Department, Cardiff University, Cardiff CF10 3YE, Wales, UK. (e-mail: gibbons@cardiff.ac.uk)
Javier Gil
Departamento Geología, Universidad de Alcalá, 28871 Alcalá de Henares, Spain. (e-mail: javier.gil@uah.es)
José I. Gil Ibarguchi (Variscan and Pre-Variscan Tectonics co-ordinator)
Departamento de Mineralogía y Petrología, Universidad del País Vasco, P.O. Box 644, 48080 Bilbao, Spain. (e-mail: nppgiibi@lg.ehu.es)
María Teresa Gómez Pugnaire
Departamento de Mineralogía y Petrología, Universidad de Granada, Av. de Fuentenueva, s/n. 18071 Granada, Spain. (e-mail: teresa@ugr.es)
Angel González
Departamento de Ciencias de la Tierra, Universidad de Zaragoza, C/Pedro Cerbuna, 12. E-50009 Zaragoza, Spain. (e-mail: agonzal@posta.unizar.es)
Felipe González
Departamento de Geología, Universidad de Huelva, Campus de La Rábida, 21819 La Rábida, Huelva, Spain.
Antonio Goy
Departamento de Paleontología, Universidad Complutense de Madrid, Instituto de Geología Económica (CSIC-UCM), 28040 Madrid. (e-mail: agoy@geo.ucm.es)
Jose Luis Goy
Departamento de Geología, Facultad de Ciencias, Universidad de Salamanca, 3 70 71 Salamanca, Spain. (e-mail: joselgoy@gugu.usal.es)
Rodolfo Gozalo
Departamento de Geología, Universitat de València, C/Dr. Moliner 50, 46100-Burjassot, Spain. (e-mail: rodolfo.gozalo@uv.es)
F. Javier Gracia Prieto
Departamento de Geología, Facultad de Ciencias del Mar, Universidad de Cádiz, 11510 Puerto Real, Cádiz, Spain. (e-mail: javier.gracia@uca.es)
Kai-Uwe Gräfe
Fachbereich Geowissenschaften, Universitat Bremen, Postflach 330440, D-28334 Bremen, Germany. (e-mail: ugraefe@micropal.uni-bremen.de)
Mateo Gutiérrez-Elorza (Quaternary co-ordinator)
Departamento Ciencias de la Tierra, Facultad de Ciencias, Universidad de Zaragoza, 50009 Zaragoza, Spain. (e-mail: mgelorza@posta.unizar.es)
Juan C. Gutiérrez-Marco (Ordovician co-ordinator)
Instituto de Geología Económica (CSIC-UCM), Facultad de Ciencias Geológicas, E-28040 Madrid, Spain. (e-mail: jcgrapto@geo.ucm.es)
Francisco Gutiérrez-Santolalla
Departamento Ciencias de la Tierra, Facultad de Ciencias, Universidad de Zaragoza, 50009 Zaragoza, Spain. (e-mail: fgutier@posta.unizar.es)
Nemesio Heredia
Instituto Geológico y Minero de España, Avda. República Argentina 30, 1 B, 24004 León, Spain.
Francisco Hernán
Departamento de Edafología y Geología, Universidad de La Laguna, Tenerife, Canary Islands, Spain.
Pedro Herranz Araújo
Instituto de Geología Económica (CSIC-UCM), Facultad de Ciencias Geológicas, E-28040 Madrid, Spain.
Pablo Higueras
Departamento de Ingeniería Geológica y Minera, E.U.P. Almadén, Universidad de Castilla-La Mancha, Plaza Manuel Meca 1, 13400 Almadén, Ciudad Real, Spain. (e-mail: phiguera@igem-al.uclm.es)
Antonio Jabaloy
Departamento de Geodinámica, Universidad de Granada, 180 71-Granada, Spain. (e-mail: jabaloy@ugr.es)
Marceliano Lago
Departamento de Ciencias de la Tierra, Universidad de Zaragoza, 50009 Zaragoza, Spain. (e-mail: mlago@posta.unizar.es)
Carlos L. Liesa
Departamento De Ciencias de la Tierra, Universidad de Zaragoza, C/Pedro Cerbuna 12, E-50009-Zaragoza, Spain.
Asunción Linares
Departamento de Estratigrafía y Paleontología, Universidad de Granada, Av. Fuentenueva s/n, 18005 Granada, Spain.
Eladio Liñán (Cambrian co-ordinator)
Departamento de Ciencias de la Tierra, Universidad de Zaragoza, 50007-Zaragoza, Spain. (e-mail: linan@posta.unizar.es)
Manolo Lombardero
Instituto Geológico y Minero de España, Sección de Minerales y Rocas Industriales, C/Ríos Rosas 23, 28003 Madrid, Spain. (e-mail: m.lombardero@igme.es)

Susana López
Departamento de Geología, Universidad de Huelva, Facultad de Ciencias, Campus del Carmen, 21071 Huelva, Spain. (e-mail: susana.lopez@dgeo.uhu.es)

José López-Gómez
Instituto de Geología Económica-Departamento de Estratigrafía, CSIC-UCM. Facultad de Geología, Universidad Complutense, 28040 Madrid, Spain. (e-mail: jlopez@geo.ucm.es)

José López-Ruiz (Cenozoic volcanism I co-ordinator)
Departamento de Geología, Museo Nacional de Ciencias Naturales (CSIC), 28006 Madrid, Spain. (e-mail: lopezruiz@ mncn.csic.es)

Fernando López-Vera
Facultad de Ciencias C-VI, Universidad Autónoma de Madrid, Spain. (e-mail: fernando.lopez-vera@uam.es)

Saturnino Lorenzo Alvarez
Minas de Almadén y Arrayanes. Cerco de San Teodoro s/n. 13400 Almadén, Ciudad Real, Spain. (e-mail: explorac@mayasa.es)

Rosario Lunar (Economic and Environmental Geology co-ordinator)
Departamento de Cristalografía y Mineralogía, Facultad de CC. Geológicas, Universidad Complutense de Madrid, 28040 Madrid, Spain. (e-mail: lunar@geo.ucm.es)

Aránzazu Luzón
Departamento de Ciencias de la Tierra, Universidad de Zaragoza, C/Pedro Cerbuna, 12. E-50009 Zaragoza, Spain. (e-mail: aluzon@posta.unizar.es)

Juan M. Mallo García
REPSOL-YPF. Paseo de La Castellana 280, 4ª Planta, 28046 Madrid, Spain. (e-mail: jmallog@repsol-ypf.com)

Carlos Martí
Instituto Pirenaico de Ecología, Avda. De Montañana 177, Apto. 202, 50080 Zaragoza, Spain. (e-mail: carlos@ipe.csic.es)

José M. Martín
Departamento de Estratigrafía y Paleontología, Facultad de Ciencias, Universidad de Granada, Campus Fuentenueva s/n, 18002, Granada, Spain.

Agustín Martín-Algarra
Departamento Estratigrafía y Paleontología, Universidad de Granada, 18071 Granada, Spain. (e-mail: agustin@ugr.es)

Javier Martín-Chivelet (Cretaceous co-ordinator)
Departamento Estratigrafía, Instituto de Geología Económica (CSIC-UCM), Facultad de CC. Geológicas, Universidad Complutense, 28040 Madrid, Spain. (e-mail: martinch@geo.ucm.es)

Angel Martín-Serrano
Instituto Tecnológico y Geominero de España, Ríos Rosas 23, 28003 Madrid, Spain. (e-mail: a.martinserrano@itge.mma.es)

Francisco J. Martínez
Department de Geología, Facultad de Ciencias, Universidad Autónoma de Barcelona, Bellaterra, 08193 Barcelona, Spain. (e-mail: francisco.martinez@uab.es)

Wenceslao Martínez Del Olmo
REPSOL-YPF. Paseo de La Castellana 280, 4ª Planta, 28046 Madrid, Spain. (e-mail: wmartinezo@repsol-ypf.com)

Ramón Mas
Departamento Estratigrafía, Instituto de Geología Económica (CSIC-UCM), Facultad de CC. Geológicas, Universidad Complutense, 28040 Madrid, Spain. (e-mail: ramonmas@geo.ucm.es)

Eduardo Mayoral
Departamento de Geodinámica y Paleontología, Universidad de Huelva, Avda. de las Fuerzas Armadas s/n, E-21006 Huelva, Spain. (e-mail: Mayoral@uhu.es)

Joaquín Meco
Departamento de Biología, Universidad de Las Palmas de Gran Canaria, Canary Islands, Spain.

Guillermo Meléndez (Jurassic co-ordinator)
Departamento Ciencias de la Tierra, Universidad de Zaragoza, E-50009 Zaragoza, Spain. (e-mail: gmelende@posta.unizar.es)

Nieves Meléndez
Departamento Estratigrafía, Instituto de Geología Económica (CSIC-UCM), Facultad de CC. Geológicas, Universidad Complutense, 28040 Madrid, Spain. (e-mail: nievesml@geo.ucm.es)

José Miguel Molina
Departamento de Geología, Universidad de Jaén, 23071 Jaén, Spain. (e-mail: jmmolina@ujaen.es)

Carmen Moreno
Departamento de Geología, Universidad de Huelva, Campus de La Rábida, 21819 La Rábida, Huelva, Spain.

Teresa Moreno (co-editor & Economic and Environmental Geology co-ordinator)
School of Biosciences, Cardiff University, Cardiff CF10 3US, Wales, UK (e-mail: morenot@cardiff.ac.uk)

Arsenio Muñoz (Tertiary co-ordinator)
Departamento de Ciencias de la Tierra, Universidad de Zaragoza, C/Pedro Cerbuna, 12. E-50009 Zaragoza, Spain. (e-mail: armunoz@posta.unizar.es)

Josep Anton Muñoz
Grup de Geodinamica i Analisi de Conques, Universitat de Barcelona, Spain. (e-mail: josep@natura.geo.ub.es)

Federico Oloriz (Jurassic co-ordinator)
Departamento Estratigrafía y Paleontología, Universidad de Granada, Avda. Fuentenueva s/n, 18005 Granada, Spain. (e-mail: foloriz@goliat.ugr.es)

Francisco Ortega
Unidad de Paleontología, Departamento Biología, Facultad de Ciencias, Universidad Autónoma de Madrid, 28049 Madrid, Spain. (e-mail: francisco.ortega@uam.es)

Lorena Ortega
Departamento de Cristalografía y Mineralogía, Facultad de CC. Geológicas, Universidad Complutense de Madrid, 28040 Madrid, Spain. (ortegal@eucmax.sim.ucm.es)

Teodoro Palacios
Área de Paleontología Facultad de Ciencias, Universidad de Extremadura, E-06071 Badajoz, Spain. (e-mail: medrano@unex.es)

Gonzalo Pardo
Departamento de Ciencias de la Tierra, Universidad de Zaragoza, C/Pedro Cerbuna, 12, E-50009 Zaragoza, Spain. (e-mail: gpardo@posta.unizar.es)

Miguel V. Pardo Alonso
Departamento Geología, Facultad de Biológicas, Universitat de Valencia, C/Dr. Moliner 50, 46100 Burjassot, Valencia, Spain. (e-mail: Miguel.V.Pardo@uv.es)

Emilio Pascual
Departamento de Geología, Universidad de Huelva, Facultad de Ciencias, Campus del Carmen, 21071 Huelva, Spain. (e-mail: pascual@uhu.es)

Aitor Payros
Departamento de Estratigrafía y Paleontología, Facultad de Ciencias, Universidad del Pais Vasco, Apartado 644, 48080 Bilbao, Spain. (e-mail: gpppaaga@1g.ehu.es)

Antonio Pérez
Departamento de Ciencias de la Tierra. Universidad de Zaragoza, C/Pedro Cerbuna, 12, E-50009 Zaragoza, Spain. (e-mail: anperez@posta.unizar.es)

Alfredo Pérez-González
Departamento de Geodinámica, Facultad de CC. Geológicas, Universidad Complutense de Madrid, 28040 Madrid, Spain. (e-mail: Alfredog@eucmax.sim.ucm.es)

Alberto Pérez-López
Universidad de Granada, Facultad de Ciencias, Departamento de Estratigrafía y Paleontología, Avenida Fuentenueva, 18071-Granada, Spain. (e-mail: aperezl@goliat.ugr.es)

Francisco J. Pérez Torrado
Departamento de Física-Geología, Fac. Ciencias del Mar, Universidad de Las Palmas de Gran Canaria, Canary Islands, Spain.

Agustín P. Pieren Pidal
Instituto de Geología Económica (CSIC-UCM), Facultad de Ciencias Geológicas, E-28040 Madrid, Spain. (e-mail: apieren@geo.ucm.es)

Luis Pomar
Departament de Ciències de la Terra, Universitat Illes Balears, Ctra. de Valldemossa, km. 7,5. 07071, Palma de Mallorca, Spain. (e-mail: lpomar@uib.es)

Carmen Puig
Servei Geologic, Institut Cartografic de Catalunya, Parc de Montjuic, 08038 Barcelona, Spain.

Victoriano Pujalte (Tertiary co-ordinator)
Departamento de Estratigrafía y Paleontología, Facultad de Ciencias, Universidad del Pais Vasco, Apartado 644, 48080 Bilbao, Spain. (e-mail: gpppunav@lg.ehu.es)

Cecilio Quesada
Instituto Geológico y Minero de España, Ríos Rosas, 23, 28003 Madrid, Spain. (e-mail: c.quesada@igme.es)

Santiago Quesada
Repsol-YPF, Paseo de la Castellana 280, 28046 Madrid, Spain. (e-mail: squesadag@repsol-ypf.com)

Isabel Rábano
Museo Geominero, Instituto Geológico y Minero De España, Ríos Rosas 23, E-28003 Madrid, Spain. (e-mail: i.rabano@igme.es)

Emilio Ramos (Tertiary co-ordinator)
Departament d'Estratigrafía, Paleontología i Geociències Marines, Facultat de Geologia, Universitat de Barcelona, c/Martí i Franquès, s/n. 08028, Barcelona, Spain. (e-mail: emilio@natura.geo.ub.es).

Manuel Regueiro
Instituto Geológico y Minero de España, Sección de Minerales y Rocas Industriales, C/Ríos Rosas 23, 28003 Madrid, Spain. (e-mail: m.regueiro@igme.es)

Michel Robardet (Silurian co-ordinator)
Géosciences-Rennes, UMR 6118 CNRS, Université De Rennes 1, Campus De Beaulieu, F-35042 Rennes Cedex, France. (e-mail: Michel.Robardet@univ-rennes1.fr)

Sergio Robles
Departamento Estratigrafía y Paleontología, Facultad de Ciencias, Universidad del Pais Vasco, Apto. 644, 48080 Bilbao, Spain. (e-mail: gpproors@lg.ehu.es)

Juan M. Rodríguez Salvador
Departamento de Geología, Universidad de Salamanca, 37071 Salamanca, Spain. (e-mail: bartolo@usal.es)
Eduardo Rodríguez Badiola
Museo Nacional de Ciencias Naturales, CSIC, José Gutiérrez Abascal 2, 28006 Madrid, Spain.
Luis R. Rodríguez Fernández
Instituto Geológico y Minero de España, Ríos Rosas, 23, 28003 Madrid, Spain. (e-mail: lr.rodriguez@igme.es)
Francisco Rodríguez-Tovar
Departamento de Estratigrafía y Paleontología, Universidad de Granada, Av. Fuentenueva s/n, 18005 Granada, Spain. (e-mail: fjrtovar@ugr.es)
Idoia Rosales
Departamento Estratigrafía y Paleontología, Facultad de Ciencias, Universidad del Pais Vasco, Apto. 644, 48080 Bilbao, Spain. (e-mail: gpbrofri@lg.ehu.es)
Pedro A. Ruiz-Ortiz
Departamento de Geología, Universidad de Jaén, 23071 Jaén, Spain. (e-mail: paruiz@ujaen.es)
Jose A. Saenz De Santa María
Departamento de Geología, Dirección Técnica, Hulleras del norte, S.A. (HUNOSA), Avenida de Galicia 44, 33 005 Oviedo, Spain. (e-mail: jassmb@hunosa.com)
Ramón Salas
Departament de Geoquímica, Petrología i Prospecció Geológica, Facultat de Geología, Universitat de Barcelona, 08071 Barcelona, Spain. (e-mail: ramons@natura.geo.ub.es)
Miguel Angel San José Lancha
Instituto de Geologiá Económica (CSIC-UCM), Facultad de Ciencias Geológicas, E-28040 Madrid, Spain.
José Sandoval
Departamento de Estratigrafía y Paleontología, Universidad de Granada, Av. Fuentenueva s/n, 18005 Granada, Spain. (e-mail: sandoval@goliat.ugr.es)
José Luis Sanz
Unidad de Paleontología, Departamento Biología, Facultad de Ciencias, Universidad Autónoma de Madrid, 28049 Madrid, Spain. (e-mail: josel.sanz@uam.es)
Javier Sanz López
Facultade de Ciencias de Educación, Universidade de Coruña, Paseo de Ronda 47, 15011-A Coruña, Spain. (e-mail: jasanz@udc.es)
Graciela N. Sarmiento
Instituto de Geología Económica (CSIC-UCM), Facultad de Ciencias Geológicas, E-28040 Madrid, Spain. (e-mail: villasar@eresmas.net)
Manuel Segura
Departamento de Geología, Universidad de Alcalá, 28871 Alcalá de Henares, Spain. (e-mail: manuel.segura@uah.es)
Jose Luis Simón
Departamento de Ciencias de la Tierra, Universidad de Zaragoza, C/Pedro Cerbuna 12, E-50009 Zaragoza, Spain. (e-mail: jsimon@posta.unizar.es)
Francisco Soto
Departamento de Geología, Universidad de Oviedo, C/Jesús Arias de Velasco s/n, 33005 Oviedo, Spain. (e-mail: fsoto@asturias.geol.uniovi.es)
César Suárez De Centi
Departamento de Geología, Universidad de Oviedo, C/Jesús Arias de Velasco s/n, 33005 Oviedo, Spain.
José M. Tavera
Departamento de Estratigrafía y Paleontología, Universidad de Granada, Av. Fuentenueva s/n, 18005 Granada, Spain. (e-mail: jtavera@goliat.ugr.es)
Montserrat Truyols-Massoni
Departamento de Geología, Universidad de Oviedo, C/Jesús Arias de Velasco s/n, 33005 Oviedo, Spain. (e-mail: mtruyols@asturias.geol.uniovi.es)
José M. Ugidos
Departamento de Geología, Facultad de Ciencias. Universidad de Salamanca, 37008 Salamanca, Spain. (e-mail: jugidos@usal.es)
Marta Valenzuela
Departamento de Geología, Universidad de Oviedo, C/Jesús Arias de Velasco s/n, 33005 Oviedo, Spain.
José I. Valenzuela-Ríos
Departamento Geología, Facultad de Biológicas, Universitat de Valencia, C/Dr. Moliner 50, 46100 Burjassot, Valencia, Spain. (e-mail: Jose.I.Valenzuela@uv.es)
Maria Isabel Valladares (Precambrian co-ordinator)
Departamento de Geología, Facultad de Ciencias, Universidad de Salamanca, 37008 Salamanca, Spain. (e-mail: valla@usal.es)
Juan A. Vera
Departamento Estratigrafía y Paleontología, Universidad de Granada, 18071 Granada, Spain. (e-mail: jvera@ugr.es)
Lorenzo Vilas
Departamento Estratigrafía, Instituto de Geología Económica (CSIC-UCM), Facultad de CC. Geológicas, Universidad Complutense, 28040 Madrid, Spain. (e-mail: vilasl@geo.ucm.es)

Joaquín Villena
Departamento de Ciencias de la Tierra, Universidad de Zaragoza, C/Pedro Cerbuna, 12. E-50009 Zaragoza, Spain. (e-mail: jvillena@posta.unizar.es)
Caridad Zazo
Museo Nacional de Ciencias Naturales (CSIC), José Gutiérrez Abascal 2, 28006 Madrid, Spain. (e-mail: mcnzc65@mncn.csic.es)

1 Introduction and overview

WES GIBBONS & TERESA MORENO

The geology of Spain is remarkably diverse. It includes one of the most complete Palaeozoic sedimentary successions in Europe, and an excellent record of the effects of the Variscan orogeny on the margins of the former supercontinent of Gondwana. In addition, post-Variscan Mesozoic and Cenozoic strata are widely exposed across the eastern half of Spain, from the Cantabrian and Pyrenean mountains to the Betic Cordillera and Balearic Islands (Fig. 1.1). These successions and their fauna reveal a unique Iberian palaeogeography influenced both by the widening Atlantic Ocean to the west and by events in the Tethys Ocean and Alpine–Himalayan orogen to the east. Alpine collision in Cenozoic times has created spectacular mountain belts in which the effects of both collisional and extensional processes can be observed. Neogene and Quaternary volcanism has occurred in southern, south-central and eastern mainland Spain, and the magnificent Canarian volcanoes expose one of the world's classic hot-spot-related ocean island chains (Chapters 17 and 18). Volcanism has taken place in the Canaries for over 20 Ma, and all stages of its volcanic evolution are preserved, from the early build-up of submarine seamounts to the emergence, growth and polyphase collapse of major subaerial volcanoes (Fig. 1.2a). In addition to recent volcanism, the Quaternary record in Spain (Chapter 14) encompasses many environments, from glacial to semi-desert, from Mediterranean to Atlantic, and includes the hugely important hominid site of Atapuerca (Fig. 1.1). Not only has this astonishing site yielded over 3000 middle Pleistocene human fossils from some 30 individuals living around 300 000 years ago, but there are also lower Pleistocene hominid remains more than 780 000 years old (Chapter 14).

The geological diversity of Spain is reflected in its mineral wealth (Chapter 19). The oldest evidence for organized mining goes back to the Tartessan civilization that existed in Andalusia around 3000 years ago. A key reason for Roman invasion, and subsequent close links between Iberia and the Roman Empire, was the abundance of metallic ores (especially gold) in the peninsula. Palaeozoic rocks in Spain host by far the most productive and historically important mercury mine in the world (Almadén), as well as the famous supergiant metallic deposits of the Iberian Pyrite Belt (Figs 1.1 and 1.2b). There are abundant coal deposits, the exploitation of which, although now in decline, played a key role in the industrialization of Spain over the last 150 years. The majority of mineral exploitation activity these days is concentrated on industrial minerals, especially on those used in the construction industry. In this context Spain continues to maintain its position as a world class producer of slate, marble, granite, celestite, fluorite, gypsum and many other industrial minerals, whilst still complying with increasing environmental controls on mineral exploitation. Also of importance and concern are the country's water reserves, the distribution of which is dependent on the highly varied geography and climate. Rainfall varies from over 1600 mm/year in the north, to 300 mm/year in the south (and even less in the Canary Islands), and further global warming could have dramatic consequences on an already stressed hydrological system (Chapter 19).

Spanish physical geography directly reflects the underlying geology, with the present geomorphology having been created primarily by Cenozoic events linked to the Alpine orogeny. Wedged between the African and Eurasian continental plates, Spain is bordered by two great mountain chains, namely the Pyrenees–Basque–Cantabria belt in the north, and the Betic Cordillera in the south (Fig.1.1). Both mountain ranges include peaks above 3000 m, with the highest point in mainland Spain being Mulhacén (3481 m) in the Betic Sierra Nevada. These two mountain chains broadly define the northern and southern margins of an Iberian continental plate that existed independently prior to Cenozoic orogeny. Alpine collision initially occurred in the north, where Pyrenean Iberia became partly subducted beneath the Eurasian plate. The focus of Alpine deformation then switched to the southern margin, where Ibero-African collision produced the Betics, a mountain range that subsequently underwent dramatic extensional collapse. Both of these collision zones have generated classical examples of foreland basins: the Ebro basin immediately south of the Pyrenees (Fig. 1.2c), and the Guadalquivir basin north of the Betics (Fig.1.1). The Ebro basin is bordered to the south by the high hills of the NW–SE trending Iberian Ranges (also known as 'Iberian Chains', 'Iberian Cordillera', 'Cadenas Ibéricas' or 'Sistema Ibérico'), a focus for Alpine intraplate deformation. In contrast, the Guadalquivir basin is bordered to the north by the Sierra Morena (Fig.1.1), where Alpine overprint is much less obvious.

Away from the mountainous northern and southern margins of Iberia, central Spain is dominated by two large Cenozoic basins (Fig. 1.1), drained by the Tajo river in the south, and the Duero river in the north. These basins form high, relatively flat areas known as 'mesetas', and are separated by the NE–SW trending, granite-dominated highlands of the Central Range (Sistema Central) (Fig.1.1; Chapter 15). The latter range represents another zone of Alpine intraplate deformation, and has 'popped up' basement rocks on reverse faults to produce mountain peaks that in the highest area (Sierra de Gredos) locally exceed 2500 m. Unlike the Iberian Ranges, which represent the inversion of a former Mesozoic sedimentary basin, the Central Range had little post-Variscan cover and is characterized by Palaeozoic granitic and metamorphic basement. This basement also appears west of the Duero and Tajo basins to form extensive and structurally complex outcrops from Sierra Morena in the south to the Atlantic coast of Galicia. Thus the western side of the country is sometimes referred to as 'Variscan Spain', the Palaeozoic outcrop of which forms the 'Iberian Massif', in contrast to 'Alpine Spain' to the east where the geology is dominated by Mesozoic and Cenozoic sediments and young mountain belts (Fig. 1.1).

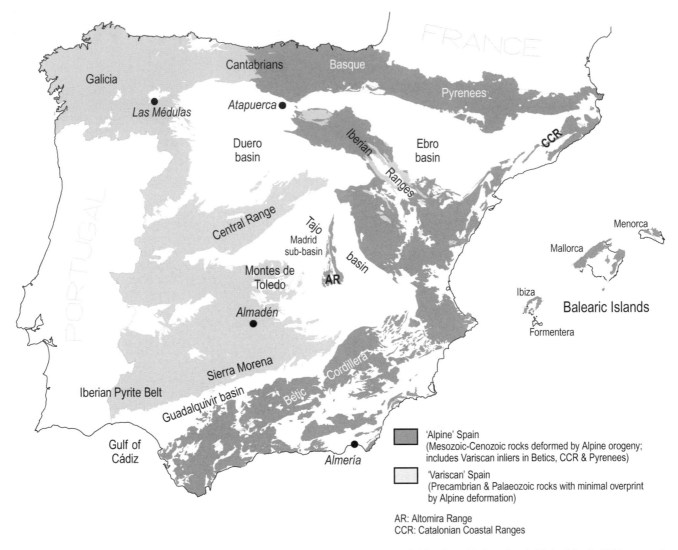

Fig. 1.1. Outline map of mainland Spain and the Balearic Islands showing the broad division into 'Variscan' and 'Alpine' Spain. White areas in Spain are Cenozoic basins.

Viewing Spain in terms of western 'Variscan' and eastern 'Alpine' areas is, of course, a simplification (see Frontispiece map). There are inliers of pre-Mesozoic Variscan rocks within the Pyrenees, Iberian Ranges and the Betic Cordillera, all overprinted by various degrees of Alpine tectonometamorphism. Similarly, far-field effects of the Alpine orogeny are detectable across the western side of Spain, well away from 'Alpine' Spain. Furthermore, pre-Variscan exposures in the west, and especially SW, include Precambrian as well as Palaeozoic rocks. These Precambrian rocks include a record of events along an ancient active plate margin, the evolution of which is commonly referred to as the Cadomian orogeny. The southwestern part of 'Variscan' Spain, in the area known as the Ossa Morena Zone, thus includes Cadomian magmatic arc-related rocks that subsequently became part of the passive Gondwanan shelf. Spain was therefore assembled from three (Cadomian, Variscan, Alpine) rather than just two orogenic cycles, although relatively little is preserved of the oldest of these.

Iberian Massif

The above-mentioned Ossa Morena Zone is one of several tectonostratigraphic zones that traditionally have been recognized in the Iberian Massif since the work of Lotze in 1945. From north to south these zones are the Cantabrian (CZ), West Asturian-Leonese (WALZ), Galician-Castilian and East Lusitanian-Alcudian (usually amalgamated under one Central Iberian Zone: CIZ), Ossa Morena (OMZ) and South Portuguese (SPZ). All these areas, except the SPZ, lay on the northern margin of the 'African' part of Gondwana during Cambrian to Devonian times and preserve thick marine successions and volcanic rocks. Cambrian sedimentary rocks are widely distributed across the Iberian Massif, as well as in northeastern and eastern Spain, and essentially comprise a diachronous carbonate succession lying between two siliciclastic sequences (Chapter 3). Ordovician sedimentary rocks (and their fossils: Fig. 1.2d) are similarly widely distributed and are characterized by siliciclastic sediments deposited on the Gondwanan margin at high latitudes (Chapter 4). By the

Fig. 1.2. 'The geology of Spain is remarkably diverse.' **(a)** El Teide: the fourth and youngest major shield volcano to have been built in central Tenerife over the last 3.5 Ma. The summit of this dormant basalt–trachyte–phonolite volcano reaches 3718 m above sea level and is the highest point in Spain (Chapter 18. Photo: Wes Gibbons). **(b)** Entredicho opencast mercury mine (Almadén) in 1996, quarrying the Silurian Criadero quartzite. The remarkable mines around Almadén have provided much of the world's mercury (Chapter 19. Photo: Wes Gibbons). **(c)** Early Miocene fluvial sedimentary rocks in the western sector of the Ebro basin, 50 km SE of Logroño. The sediments were deposited by southeastward flowing rivers within this classic example of a foreland basin lying south of the Pyrenean collisional mountain belt (Chapter 15. Photo: Arsenio Muñoz). **(d)** *Neseuretus tristani,* a trilobite representative of Iberian Ordovician rocks (Chapter 4. Photo: Juan Carlos Gutiérrez-Marco). **(e)** The 600 m-high cliffs of Cabo Ortegal in Galicia (Teresa Moreno for scale) provide spectacular exposures of exotic ultramafic mantle and high pressure lower crustal rocks emplaced over the Iberian margin during Variscan orogeny (Chapter 9. Photo: Wes Gibbons; from Moreno *et al.* 2001). **(f)** La Olla anticline, an F1 fold modified by D2 deformation during polyphase Alpine folding of Mesozoic sediments in Aliaga geological park, Iberian Ranges, Teruel province (Chapters 13 and 15. Photo: Richard Lisle).

end of the Ordovician period the Iberian area had drifted northwards to around 50°S, a trend that continued through Silurian time when the deposition of a mostly siliciclastic succession included widespread sedimentation of graptolitic black shales (Chapter 5). Devonian sedimentation took place in warm temperate to subtropical seas (c. 35°S latitude in Early Devonian), in places producing thick carbonates that include remarkable reefal deposits (Chapter 6). The deposition of Carboniferous strata (Chapter 7) was essentially coeval with Variscan deformation, when the whole Palaeozoic Gondwanan shelf sequence was variously deformed and metamorphosed during collision between Gondwana and Laurentia (Chapter 9). In order to help the reader obtain a simple overview of the Iberian Massif, each of its constituent zones is briefly considered below.

The largest Iberian Massif zone is the heterogeneous Central Iberian Zone (Fig. 1.1). The CIZ is divided into a northern part (Lotze's Galician-Castilian Zone, these days commonly referred to as the Domain of Recumbent Folds) and a southern part (Lotze's East Lusitanian-Alcudian Zone, commonly referred to as the Domain of Vertical Folds), the difference being mainly one of Variscan tectonic style, metamorphic grade, and magmatic intrusion history. CIZ rocks become increasingly deformed and metamorphosed towards the NW where they are overlain by exotic nappes stacked over the continental margin during Variscan collision (Chapter 9). This exotic belt, now exposed in NW Spain (mostly in Galicia: Fig 1.2e), is normally referred to as the Allochthonous Complexes or Galicia–Trás-os-Montes Zone, and exposes a spectacular range of metamorphic lithologies that include ophiolitic rocks, eclogites and high pressure granulites, some of which have been exhumed from depths of over 50 km.

To the north of the CIZ, the West Asturian-Leonese Zone exposes Precambrian to Devonian metasediments that were strongly folded and foliated by three Variscan (Late Carboniferous) deformation phases. The boundaries of this zone are tectonic, and, as with the CIZ, metamorphic grade and deformation intensity both increase westward along the zone. The contiguous Cantabrian Zone exposes only a relatively small area of Precambrian rocks, but shows an excellent, thick Palaeozoic succession (including the largest outcrop of Carboniferous rocks in Spain) that has undergone strong Variscan shortening. The zone is essentially a thin-skinned fold-and-thrust belt, with classic foreland-propagating deformation affecting virtually unmetamorphosed sediments. It thus represents the most external of the areas affected by Variscan collision. A remarkable feature of this zone is its strongly arcuate shape (the Ibero-Armorican arc), with a concavity facing eastwards towards the external part of the orogen.

To the south of the CIZ and OMZ, the South Portuguese Zone forms the SW margin of the Iberian Massif. This zone was an exotic terrane unrelated to Iberia prior to Variscan collision, and the Upper Palaeozoic successions in this zone (exposures include only Devonian and Carboniferous rocks) are deformed by south-vergent thrusting. The boundary between the SPZ and the OMZ includes blueschists and mélanges and represents a suture zone between Gondwanan Iberia (OMZ) and what was probably a part of Avalonia (SPZ).

Overall, the Iberian Massif can be viewed as a thick Palaeozoic Gondwanan shelf sequence of sediments and extension-related volcanic rocks, with a Proterozoic (Pan-African/Cadomian) basement. During Variscan collision this sequence was overridden by exotic nappes in the NW, and was obliquely thrust over the SPZ (Avalonia) to the south. Polyphase deformation and crustal thickening of the Iberian sequence induced widespread metamorphism, initially (D1) of compressional Barrovian type, then (D2) of lower pressure, extension-related type. Extensive partial melting during D2 produced what are now some of the largest and best preserved syntectonic and post-tectonic granitoid outcrops in the European Variscides (Chapter 8). Magmatism continued in some areas into post-Variscan (Permian) times when extensional fault-related continental sedimentary basins developed around highland areas (Chapter 10). Thus the scene became set for a long period of Mesozoic subsidence and sedimentation on and around the eroded remnants of the Variscan orogen.

Mesozoic palaeogeography and sedimentation

Encroachment of the Neotethys ocean from the east during Triassic time gradually encircled and isolated the subsiding Iberian area, creating a Pyrenean-Cantabrian basin to the north, and a Betic basin to the south. Diachronous marine transgression led to a change in many places from the deposition of continental clastic successions (Buntsandstein facies) to shallow marine carbonates separated by mudstone–evaporite units (Muschelkalk and Keuper facies). Triassic marine successions are well developed in the east Cantabrians, the Pyrenees and Catalonian Coastal Ranges, the Iberian Ranges, and the Betics (Chapter 10).

Late Triassic–Early Jurassic rifting, linked to the separation of Africa and North America, enhanced the spread of Tethyan seas. Central and western Iberia (the Iberian Massif) continued to form an emergent area through Jurassic time, but reduced in size by further encroachment of shallow epicontinental seas from the north and NE (Asturias, Basque-Cantabrian and South Pyrenean basins), east (Iberian basin), and south. The latter southern margin formed a wide Lower Jurassic carbonate shelf bordering a narrow oceanic trough that connected Tethys with the widening central Atlantic ocean, and separated Iberia from Africa. Extensional break-up of this Betic shelf created offshore pelagic and hemipelagic troughs and swells where Middle–Upper Jurassic marl–limestone successions and condensed sequences (such as the famous 'ammonitico rosso' facies) were deposited. Along the eastern side of Spain (Iberian basin and southern Catalonian Coastal Ranges), normal faults again controlled the palaeobathymetry, and a long and complex history of Jurassic sedimentation ensued, depositing mostly limestone and marl. To the north, carbonate-dominated platform sedimentation initially continued into the Basque-Cantabrian-South Pyrenean basin, where the fauna record a palaeogeography transitional between Atlantic and Tethyan waters. Upper Jurassic marine regression induced widespread erosion and continental sedimentation (Purbeck) in northern Spain, a change linked directly to the onset of rifting in the Bay of Biscay (Chapter 11).

The initiation of North Atlantic oceanic spreading at the beginning of the Cretaceous period provided the primary control over subsequent Iberian palaeogeography and sedimentation, although both climate and eustatic sea level changes were also important. The low-lying Iberian Massif remained the largest landmass (Fig. 12.1), had a subtropical, maritime climate, and was surrounded by warm seas in which carbonate deposition was mostly dominant. During periods of high relative sea level, most notably during rapid Atlantic spreading in mid-Cretaceous times, much of Iberia became flooded, and a broad Iberian basin connected the Betic area with the Bay of Biscay. As with the Jurassic record, Cretaceous marine fossils thus show a double affinity with northern and southern faunal provinces (Chapter 12).

For much of Early Cretaceous time Iberia underwent extension and transtension, with varying (but commonly rapid) subsidence rates in the various basins to the south, east and north. In the south, shallow marine sediments (Prebetic) graded into deeper water pelagic limestones and marls (Subbetic), although the Jurassic separation of the area into well-defined troughs and swells tended to decline over the Cretaceous period. Further south still, away from Iberia, turbiditic sediments derived from Africa and areas to the east produced Cretaceous 'flysch' sequences. Northwards from the Betics, the intracratonic Iberian basin covered much of eastern Spain during Cretaceous time. Initial rifting in this area produced four strongly subsident sub-basins containing thick, continental to shallow marine successions of Early Cretaceous age. During Late Cretaceous times this Iberian basin initially expanded during eustatic sea-level rise to form a wide, shallow seaway in which developed the largest expanse of carbonate platforms ever seen in Iberia. At this time of maximum flooding, the only emergent areas in Spain were the Iberian Massif in the west, and a smaller island of Variscan basement in the NE (the Ebro Massif: Fig. 12.1). Latest Cretaceous marine regression across the eastern half of Spain later led to the re-emergence and widening of land areas, and was linked to tectonic movements during the early stages of the Alpine orogeny.

Along the northern margin of Iberia, extensional break-up of the old Jurassic platform initially created several depocentres in which thick, marine Lower Cretaceous successions were deposited. This setting was replaced by more active mid-Cretaceous transtension linked to the Bay of Biscay, with strong tectonism producing pull-apart flysch basins, the emplacement of mantle rocks, and low pressure–high temperature metamorphism. As the Bay of Biscay opened, Iberia acted as a fully independent plate, rotating anticlockwise over a period of some 30 Ma before finally becoming caught between Africa and Europe during Late Cretaceous oblique convergence.

Alpine orogeny

The NW margin of Spain became a focus for convergence as early as mid-Cretaceous times, when the oceanic crust of the Bay of Biscay began to be subducted beneath Iberia, creating an accretionary prism. Continental collision, however, did not begin until Late Cretaceous times and was initiated in the eastern Pyrenean area, subsequently propagating westwards as the Iberian plate was obliquely subducted beneath the Eurasian plate. Rapidly subsiding troughs, filling with turbiditic 'flysch' sediments, became involved in synsedimentary thrusting. Nearer shore in northeastern Iberia, Pyrenean sequences show a complex evolution involving initial coastal onlap and expansion of carbonate sedimentation, followed by platform collapse and extensive olistostrome generation. The Pyrenees-Basque-Cantabrian region shows good preservation of synorogenic sedimentary successions, despite widespread Cenozoic uplift that has created unconformity between most Cretaceous and Tertiary strata. The Pyrenees is thus a mountain range formed by the inversion of a Mesozoic passive to transtensional plate margin. Convergence continued over some 60 Ma, from Late Cretaceous to middle Miocene times, with thrusting being directed northwards and southwards away from a central axial zone (where Variscan basement rocks crop out) and towards two foreland basins, the Aquitaine basin in France, and the Ebro basin in Spain (Fig. 1.2a; Chapter 13). During this time interval, northward movement of the Iberian plate induced a progressive climatic change from tropical through humid subtropical and eventually to arid subtropical conditions (Chapter 13).

Further south from the Pyrenees, in the Iberian basin, the onset of Alpine compression produced newly emergent highs separated by increasingly isolated depocentres, with the sea retreating both northwards and southwards from the intracratonic area. Intraplate stress fields, transmitted from the Iberian plate margins, reactivated formerly normal faults to form a thrust system that ultimately inverted the basin and so created the Iberian Ranges (Fig. 1.2f). Compressive deformation started in the early–middle Eocene interval and tended to progress westwards, reaching a peak in late Oligocene times but continuing throughout much of the Miocene epoch. Further west from the Iberian Ranges, intraplate stress resulted in the emergence of the Central Range and the separation of central Spain into the Tajo and Duero basins (Fig. 1.1), with the main phase of uplift taking place as recently as around 10 Ma BP.

Along the southern Iberian margin, Alpine collision occurred later than in the Iberian Ranges, so that sedimentation continued uninterrupted across the Cretaceous–Tertiary boundary in many places. The change from the transtensional setting that had characterized Iberian-African interactions since Triassic times, to one involving Cenozoic convergence, was to generate a new palaeogeography and ultimately produce the Betic Cordillera. The Betics form the westernmost Alpine mountain belt in Europe and are subdivided into an Internal Zone, generally metamorphosed and derived from the south, and an External Zone, which represents the deformed southern Iberian margin (see Frontespiece map). The Internal Zone forms a crustal wedge that was stacked over southern Iberia at the beginning of Neogene times. During the Miocene epoch deformation propagated into the Iberian margin, producing imbricate thrust stacks of Mesozoic and Palaeogene strata and creating the Guadalquivir foreland basin. Wholesale extensional collapse of the Betic mountains induced rapid exhumation of the Internal Zone and generated a series of continental and marine intermontane basins, most of which opened towards the Mediterranean Sea. Particularly noteworthy of these Neogene sedimentary successions in southern Spain is the presence of thick gypsum beds deposited in response to the Messinian salinity crisis, when the Mediterranean dried out around 5.5 Ma BP.

Neogene–Quaternary crustal extension has played an important role in generating the landscape of southern and eastern Spain. It has created not just the 'Spaghetti Western' basin-and-range topography of the internal Betic Cordillera, but has produced extensive normal faulting in the Iberian Ranges and along the eastern coast, bordering the highly subsident Valencian trough (see Frontespiece map). Cenozoic extension of Iberian crust has facilitated the extrusion of alkaline basalts in south-central (Calatrava volcanic field) and NE (around Gerona) Spain. The more dramatic extensional belt running from the Valencian trough to the Betics was linked to the eruption of calc-alkaline to ultrapotassic rocks, the best exposures of which are preserved in the Almería area of southern Spain (Frontespiece map and Fig. 1.1).

Neotectonic studies show that the primary compressive stress in Spain is oriented NW–SE, as Africa continues to push against Iberia. The focus of deformation continues to be in southern Spain where the wide, diffuse Ibero-African plate margin is converging at a rate of 4 mm/year. Seismicity associated with this convergence is concentrated in the south, but also transmitted to central and northern Spain by a network of active Quaternary faults that are currently most active in the Pyrenees and, to a much lesser extent, in the Iberian Ranges and Galicia. Although Spain is not generally recognized as a country of significant seismic risk, at least ten major, destructive earthquakes have occurred over the last 700 years (Chapter 19). The great Lisbon earthquake of 1755, for example, had its epicentre below the Gulf of Cádiz (Fig. 1.1) and was felt all over Spain, from the Portuguese border to Catalonia. Thus the Alpine orogeny continues in Spain today, a country lying across the long-active geological boundary between NW Africa and Western Europe.

2 Precambrian

M. ISABEL VALLADARES (coordinator), PEDRO BARBA & JOSÉ M. UGIDOS

The Iberian Massif provides the largest outcrop of Variscan and pre-Variscan rocks in Europe (Fig. 2.1a). It has been divided into several zones (Fig. 2.1b) and, within the Spanish part of these zones, extensive areas are occupied by Precambrian rocks (Fig. 2.1c). These Neoproterozoic successions are known under informal and local stratigraphic names so that it is not easy to provide a comprehensive synthesis of the Precambrian in Spain to a wide audience. Additional difficulties arise because different authors have focused their attention on different aspects (stratigraphy, petrology, tectonics, etc.) in widely separated geographic areas. Furthermore, widespread overprinting by Palaeozoic igneous and medium- to high-grade metamorphic processes has erased much sedimentological data and obscured correlations within and between Precambrian areas. As a consequence, many different local names have been proposed for stratigraphic units that are partially coincident with units in other localities. Thus, readers unfamiliar with geographic names may become discouraged when trying to compare and understand Precambrian stratigraphy from area to area in the Iberian Massif. To help readers,

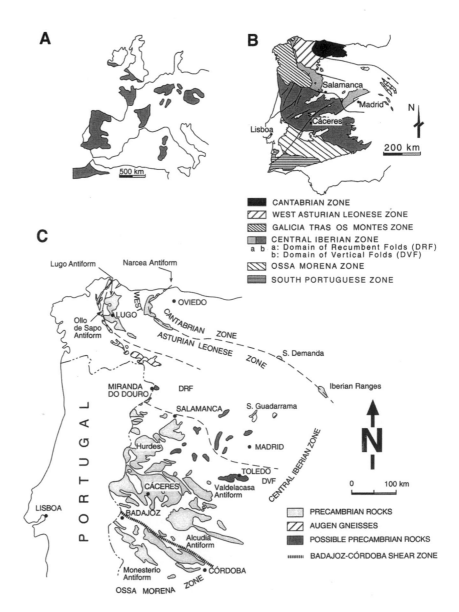

Fig. 2.1. (a) Proterozoic and Palaeozoic outcrops of the European Variscan orogen, modified from Quesada (1992). (b) Zonation of the Iberian Massif, modified from Díez Balda *et al.* (1990). (c) Outcrops of Precambrian rocks in the Spanish Iberian Massif. Key: DRF, Domain of Recumbent Folds; DVF, Domain of Vertical Folds.

CENTRAL IBERIAN ZONE

		Carrington da Costa 1950	Rodríguez Alonso 1985	Díez Balda 1986	Valladares et al. 1998, 2000	Lotze 1956	Bouyx 1970	Herranz et al. 1977; San José et al. 1990	San José 1984	Alvarez Nava et al. 1988	Robles & Alvarez Nava 1988	Palacios 1989; Vidal et al. 1994b
	CAMBRIAN	Schist-Greywacke Complex	Upper Unit	Aldeatejada Fm Monterrubio Fm	XII to V	Transition Layers	Hinojosa Series	Pusian Sequence	Pusian Gp	Valdelacasa Gp	Upper Series	Ibor-Río Huso Gp
PRECAMBRIAN					IV			Upper Alcudian	Middle Gp	Ibor Gp	Middle Series	
								Peláez et al. 1989				
			Lower Unit	Valdelacasa Series	III II I	Valdelacasa Series	Alcudia Schists	Lower Alcudian	Lower Gp	Domo Extremeño Gp — Cubilar Fm / Estomiza Fm	Lower Series — Cubilar Fm	Domo Extremeño Gp — Estenilla Fm / Cíjara Fm

Fig. 2.2. Synthesis of lithostratigraphic terminology used by previous authors for the Precambrian–Cambrian of the Central Iberian Zone.

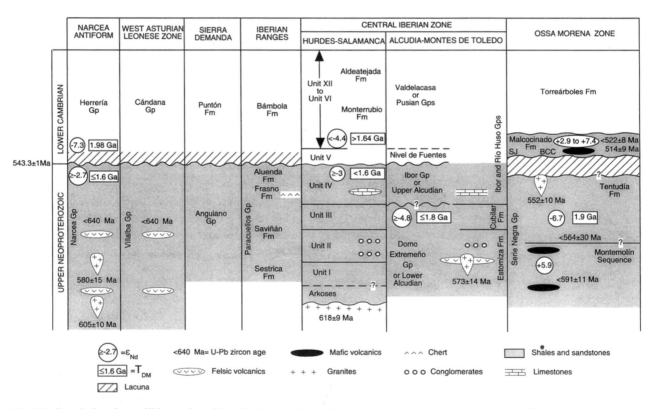

Fig. 2.3. Correlation chart of lithostratigraphic units from the Upper Neoproterozoic–Lower Cambrian in the different zones of the Spanish Iberian Massif. The Neoproterozoic–Cambrian boundary according to Bowring & Erwin (1998). U-Pb isotopic data after Lancelot *et al.* (1985), Schäfer *et al.* (1993), Ordóñez (1998), Fernández Suárez *et al.* (1998, 2000). ε_{Nd} values and T_{DM} neodymium model ages after Nägler *et al.* (1995), Ugidos *et al.* (1997b, 1999), Ordóñez (1998), Pin *et al.* (1999). Key: BBC, Bodonal–Cala Complex; SJ, San Jerónimo Fm. See Figure 2.4 for interpretation of units I–V.

the names of the stratigraphic units and correlations across the zones are summarized in Figures 2.2 and 2.3.

Current knowledge of the Precambrian sedimentary successions varies greatly between the different areas of the Spanish Iberian Massif and, in some zones, stratigraphic and sedimentological studies have not advanced significantly in the last two or three decades. This is the case in northern Spain, where the Precambrian successions are known as the Narcea schists (Lotze 1956*a*) or Mora Formation (Fm) (Compte 1959)

in the Narcea antiform (Fig. 2.1c) and as the Villalba schists (Capdevila 1969) in the Lugo antiform (West Asturian-Leonese Zone; Fig. 2.1c). In the Sierra de la Demanda (Cantabrian Zone) and the Iberian Ranges (West Asturian-Leonese Zone; Fig. 2.1c) correlative successions are known as the Anguiano schists (Colchen 1974) and Paracuellos schists (Lotze 1929) respectively. A similar situation exists in the Ossa Morena Zone, where the existence of Precambrian rocks has been known since the late nineteenth century, although the

Precambrian ages of some outcrops of generally metamorphic and unfossiliferous sequences have been a matter of intensive debate. The Precambrian rocks within the Ossa Morena Zone consist of several successions that are commonly separated by tectonic contacts and that have been affected by high-grade metamorphism and strong deformation. Thus, the primary relationships among these successions remain equivocal. The boundary between the Central Iberian and Ossa Morena zones (Fig. 2.1c) is located in the Badajoz–Córdoba shear zone (Díez Balda et al. 1990), but the interpretation of the contact is controversial. Some authors have described this contact as a Cadomian suture that was reactivated during the Variscan orogeny (Ribeiro et al. 1990c; Quesada 1990b; Ábalos et al. 1991b, 1993), while others have described it as a Variscan suture (Burg et al. 1981; Matte 1991; Azor et al. 1994).

The Central Iberian Zone is where the Precambrian successions crop out most extensively (Fig. 2.1c) and from the start these have received different names, such as the 'schist-greywacke complex' (Carrington da Costa 1950) and the Valdelacasa series and transition layers (Lotze 1956a). Subsequently, they were referred to as the Alcudian (Ovtracht & Tamain 1970) and Pusian (San José 1983) sequences and, more recently, the Domo Extremeño, Ibor and Valdelacasa (or Río Huso) groups (Vidal et al. 1994b). The Central Iberian Zone is the best known area and is where the most complete studies on the Neoproterozoic successions have been carried out. However, a major problem arises when the Neoproterozoic and Cambrian successions in the Central Iberian Zone are compared with those in the Cantabrian, West Asturian-Leonese and Ossa Morena zones. In these latter zones, an obvious angular unconformity separates the Neoproterozoic and Cambrian successions but the presence of an unconformity in the main part of the Central Iberian Zone is less clear and remains unresolved. Several reports of unconformities and disconformities in central Spain appear to have been contradicted by more recent fossil evidence (Vidal et al. 1994b), sequence stratigraphic correlation and chemostratigraphical studies (Valladares et al. 2000). The fossil record of Neoproterozoic rocks in Spain includes problematic carbonaceous fossils, ichnofossils, skeletal fossils and organic-walled phytoplankton, some of which have recently been interpreted as pseudofossils, non-fossils or preservational artifacts (Vidal et al. 1994b).

Other exposures of possible Precambrian rocks occur in the axial zone of the Pyrenees and Betics and in allochthonous units within the Iberian Massif (Galicia Tras Os Montes Zone; Fig. 2.1b). In the Spanish Pyrenees, possible Precambrian ages have been inferred for some areas of small extent in the axial zone. In these areas, high-grade metasedimentary (metagreywackes, kinzigites, migmatites and paragneisses) and metaigneous (orthogneisses and orthoamphibolites) rocks (Guitard et al. 1996) underlie a carbonate unit assigned to an Early Cambrian (Botomian) age on the basis of an association of archaeocyathans (Perejón et al. 1994). The Betics have been subdivided into three complexes, of which the lowest one (Nevado–Filábride) consists of two rock sequences. The lower of these two sequences comprises graphite-bearing metapelites with metagreywackes towards the top that contain Gloecapsomorpha sp. and Trematosphaeridium sp., suggesting a Neoproterozoic age (Gómez Pugnaire et al. 1982), while the upper sequence consists of metapelites and is considered to be a Permo-Triassic cover. In the Galicia Tras Os Montes Zone, a para-authochthonous zone known as the 'schistose domain'

lies between the Gondwanan basement of the Iberian Massif and overriding allochthonous sheets of more exotic provenance (e.g. Farias et al. 1987). Recent ion-microprobe data reported for zircons from ultramafic, mafic and metasedimentary high-pressure rocks of the allochthonous complexes (Chapter 9) suggest that the protoliths were formed in Early Ordovician times and it seems unlikely that Precambrian rocks exist (Ordóñez 1998). The lower part of the schistose domain, however, includes metagranitoid augen gneisses and the overlying Santabaia Group (Gp), which has up to 3000 m of phyllites and quartzphyllites with felsic metavolcanic rocks of pre-Silurian and possibly Neoproterozoic age (Ribeiro et al. 1990b).

Narcea antiform

In NW Spain, a long arcuate outcrop of Precambrian rocks constitutes the core of the Narcea antiform and lies along the boundary between the Cantabrian and the West Asturian-Leonese zones (Fig. 2.1c). These rocks are known as the Narcea Gp (Narcea Schists of Lotze 1956), and constitute a 1000–2000 m thick succession of alternating shales and sandstones (Fig. 2.3). Pérez Estaún (1973, 1978) defined three units: the lowest comprises interbedded porphyritic volcanic rocks, sandstones and shales; above these there are turbiditic shales and greywackes; and the uppermost unit consists predominantly of shales with interbedded sandstones.

In the northern part of the antiform, Marcos (1973) described differences between the rocks that outcrop on either flank and assigned the eastern flank to the Cantabrian Zone and the western one to the West Asturian-Leonese Zone (Fig. 2.1c). On the eastern flank, the Precambrian rocks are siltstones and shales, with thin fine-grained sublitharenites, subarkoses, litharenites, lithic arkoses and microconglomerates containing quartz, feldspars, rock fragments (metamorphic, volcanic, pelitic and chert) and detrital micas. Sedimentary structures include flute and crescent casts, slumps, normal gradation, current ripples, convoluted beds, flame structures and load marks. This succession has been interpreted as having been deposited in outer deep-sea fans. The western flank is characterized by a higher metamorphic gradient that eventually reaches the biotite isograd. Dark grey to green shales and siltstones predominate and are locally interbedded with thin layers of fine-grained sandstones. There are also medium- to coarse-grained sandstones, feldspathic or lithic arkoses and, locally, microconglomerates. Outcrops of interbedded volcanic and volcanoclastic rocks of rhyolitic, dacitic and even andesitic compositions are common (Suárez del Río & Suárez 1976; Pérez Estaún & Martínez 1978). There are also some intrusive bodies of high-K calc-alkaline metaluminous to weakly peraluminous diorites and quartzdiorites (Corretgé 1969). These rocks show moderate rare earth element (the whole REE) contents (128–142 ppm) and moderate Eu anomalies (Eu/Eu* = 0.74 to 0.77); $^{87}Sr/^{86}Sr$ ranges between 0.7047 and 0.7062 and $\varepsilon_{Nd}(600)$ ranges between −0.28 and −0.04 (Fernández Suárez et al. 1998).

The age of the Narcea Gp in the eastern flank is constrained by the presence of cyanobacterial spheromorphs (Palacios & Vidal 1992), Sphaerocongregus variabilis and Palaeogomphosphaeria caurensis; both are regarded as indicative of a middle(?)–late Vendian age (Vidal et al. 1994b). Recently, laser ablation inductively coupled plasma mass spectrometry (ICP-MS) U-Pb isotopic dating in the granitoid rocks

(Fernández Suárez *et al.* 1998) yielded intrusion ages of 605 ± 10 Ma and 580 ± 15 Ma, and the dating of detrital zircons in a greywacke of the Narcea Group in the western flank (Fernández Suárez *et al.* 2000*b*) provided ages as young as 640 Ma (Fig. 2.3).

Unconformably above the Narcea Gp lies the Cándana Gp (Lotze 1961*a*) in the West Asturian-Leonese Zone, or Herrería Gp (Compte 1959) in the Cantabrian Zone. Its fossil record indicates a Cambrian age, and the contact with the underlying Narcea Gp is an angular unconformity showing palaeoalteration due to subaerial exposure. According to Aramburu *et al.* (1992), the unconformity could be intra-Neoproterozoic since the lowest member of the Cándana Gp comprises 100–200 m of undated alternating quartzarenites (microconglomerates at the base) and shales with lenses of dolostones and limestones containing glauconite and wave ripples, and could be of Ediacaran age. However, the Herrería Gp has yielded Early Cambrian ichnofossils and acritarchs from its lowest levels (Vidal *et al.* 1994*b*). The amount of angular discordance between the Narcea Gp and Cándana/Herrería groups decreases from east to west.

West Asturian-Leonese Zone

In the West Asturian-Leonese Zone, a major outcrop of Precambrian rocks occurs in the Lugo antiform (Fig. 2.1c), and consists of metasediments and metavolcanic rocks collectively known as the Villalba schists (Capdevila 1969), and divided by Martínez Catalán (1985) into two units. In the northern limb of the Lugo antiform the lower unit contains >2000 m of grey laminated low-grade metapelites alternating with white, yellow and green thinning- and fining-upward metasandstones and metasiltstones. The metasandstones are greywackes or feldspathic subgreywackes formed by subangular medium- to fine-grained clasts of quartz and feldspar within a matrix of sericite, chlorite and muscovite. Thin layers of amphibolitic gneisses and fine-grained amphibolites are common and these represent originally carbonate-matrix sandstones and basic volcanic rocks, respectively (Capdevila 1969). The upper part of the Villalba schists is marked by alternating fine-grained metasandstones and grey-to-black shaley metapelites. The metasandstones are less abundant than in the lower part and T_{b-e} and T_{c-e} Bouma sequences have been identified. Also common are lenticular beds of metasandstones and quartzites a few metres thick. The thickness of the entire upper part is variable and may even be completely removed below the angular unconformity at the base of the Cambrian succession (Cándana Gp). The angular discordance between the Neoproterozoic and Cambrian successions is much smaller than in the Narcea antiform. In the southern limb of the antiform the lower unit comprises schists and biotitic gneisses and the upper one displays muscovitic schists with chloritoid, garnet, staurolite and aluminium silicates. The unconformity between the Neoproterozoic and Cambrian successions is less clear than in the northern limb of the antiform (Martínez Catalán 1985). Laser ablation ICP-MS isotopic dating of detrital zircons from greywackes of the Villalba Gp has provided ages of *c.* 640 Ma, similar to the ages in the Narcea Gp (Fig. 2.3; Fernández Suárez *et al.* 2000*b*).

Within the West Asturian-Leonese Zone in northeastern Spain (Iberian Ranges, Fig. 2.1c), the Precambrian succession is known as the Paracuellos Gp (Lotze 1929). Liñán & Tejero (1988) divided this group, which is 500–1500 m thick, into four formations (Sestrica, Saviñán, Frasno and Aluenda formations). The lowest two formations are alternations of laminated shale and fine-grained, graded sandstone exhibiting low-angle cross-bedding, ripple marks and flute casts. The Frasno Fm is a thin unit (4 m) of laminated chert with some lenticular shales, while the uppermost Aluenda Fm is again an alternation of sandstone and shale with planar and ripple laminations and cross-bedding in the sandstones. *Cloudina*-like shelly fossils have been recorded in a unit equivalent to the Frasno Fm, and the Saviñán Fm yielded the ichnofossil *Torrowangea rosei*, suggesting a Vendian age (Liñán & Tejero 1988; Liñán *et al.* 1994).

Central Iberian Zone

In the Central Iberian Zone (Fig. 2.1b), a Domain of Recumbent Folds (DRF) in the north and a Domain of Vertical Folds (DVF) in the south have been distinguished (Díez Balda *et al.* 1990). In the DRF Precambrian outcrops are rare and commonly metamorphosed to augen ('glandular') gneisses of controversial age and origin (Chapter 10). By contrast, in the DVF Precambrian rocks crop out extensively and preserve thick sedimentary sequences affected only by low-grade metamorphism (Fig. 2.1c).

In the western part of the DRF, Lancelot *et al.* (1985) have recognized basal orthogneisses (Miranda do Douro; Fig. 2.1c) overlain by paragneisses derived from arkoses that grade into a metamorphic complex of schists and metagreywackes which, in turn, are overlain by the augen gneisses (Ollo de Sapo Fm; Parga Pondal *et al.* 1964). The emplacement of the orthogneiss has been dated at 618 ± 9 Ma, which demonstrates the presence of a Neoproterozoic basement in western Spain (Lancelot *et al.* 1985). The schist and greywacke complex has been assigned to the Neoproterozoic or Cambrian, although evidence of depositional features is scarce given the pervasive tectonic and metamorphic overprinting. Thus, geologists working in these areas have mainly studied the petrological aspects of these rocks and the post-Precambrian events, rather than determining stratigraphical and basinal characteristics. The augen gneisses include volcanosedimentary (Navidad & Peinado 1976) and intrusive members. Previous geochronology (Lancelot *et al.* 1985; Vialette *et al.* 1987; Wildberg *et al.* 1989) indicated Late Neoproterozoic and Early Ordovician ages for these augen gneisses. Recently, Valverde-Vaquero & Dunning (2000) have provided new U-Pb dating in the Sierra de Guadarrama (Fig. 2.1c) and have reinterpreted the supposedly Late Neoproterozoic ages as Early Ordovician for the most representative types of augen gneisses of the DRF (Valverde-Vaquero & Dunning 2000).

In the DVF, Neoproterozoic sedimentary successions crop out widely, although due to lithological monotony, a scarcity of fossils and pervasive Variscan overprinting, their stratigraphy, age and correlation within the Central Iberian Zone have led to much debate. There is continuing disagreement about several key issues, notably: (a) the stratigraphic position of minor units established by different authors in the general series (Alvarez Nava *et al.* 1988; San José *et al.* 1990); (b) the total thickness attributed to the exposed Neoproterozoic succession, which ranges between 2200 and 16 500 m depending on the authors (Herranz *et al.* 1977; Díez Balda *et al.* 1990; San José *et al.* 1990; Santamaría 1995; Valladares *et al.* 1998, 2000); (c) the presence or absence of unconformities and their number, magnitude and significance. Thus, some authors favour the existence of map-scale angular unconformities due to Cadomian deformation (Bouyx 1970; Ortega & González Lodeiro 1986; San José *et al.* 1990; Díez Balda *et al.* 1990), whereas other authors have defined only sedimentary disconformities (Rodríguez Alonso 1985; Vidal *et al.* 1994*b*; Valladares *et al.* 1998, 2000).

In general terms, three thick informal units (that partially include Lower Cambrian rocks) have been recognized by most authors working in the Central Iberian Zone. These have received different names depending on the areas where the main outcrops have been studied (Fig. 2.2). The lowest of these three units comprises turbiditic deposits (Vilas & San José 1990), and contains *Gordia* ichnosp., cf. *Megagrapton*, *Neonereites* aff. *N. uniserialis*, *Phycodes* aff. *P. pedum*, *Torrowangea rosei*, *Gordia marina* and *G.* aff. *arcuata*, *Bergaueria* ? ichnosp., *Nimbia occlusa*, suggesting a late Vendian age (Vidal *et al.* 1994*a,b*). *Sphaerocongregus variabilis* was also reported from the Domo Extremeño Gp, co-occurring with *Palaeogomphosphaeria caurensis* (Liñán & Palacios 1987; Palacios 1989). The middle stratigraphic unit (Fig. 2.2) shows greater lithological diversity and has been interpreted as a progradational succession from turbidites to shelf siliciclastic and carbonate rocks, with some exposure episodes (Vilas & San José 1990). The terminal Neoproterozoic skeletal fossil *Cloudina* is abundant in situ in platform dolomites within this unit (Vidal *et al.* 1994*b*). There is also a large variety of organic ribbons of vendotaenids, similar to *Sabellidites* sp. (Vidal *et al.* 1994*b*), suggesting a late Vendian age. The palaeontological data thus suggest an uninterrupted deposition from the lower to the middle units during late Vendian times, an interpretation supported by the sequence stratigraphic framework (Valladares *et al.* 2000). The upper stratigraphic succession (Fig. 2.2) begins with an angular unconformity beneath a basal megabreccia attributed to the Cambrian and is therefore not considered in this chapter. The basal megabreccia contains limestone boulders with Neoproterozoic fossils (*Cloudina hartmannae*; Vidal *et al.* 1994*b*) and has been interpreted as recording Cambrian marine transgression (Valladares *et al.* 1998, 2000).

The lower boundary of the Neoproterozoic metasediments is not exposed in the Central Iberian Zone. However, it has been reported that in some places a metamorphosed sequence, probably equivalent to the Domo Extremeño Gp (Fig. 2.2), rests on a Precambrian gneissic basement (Lancelot *et al.* 1985). In other cases, an angular unconformity has been inferred (Herranz *et al.* 1977; Vilas & San José 1990), but because of intense Variscan deformation and metamorphism (Ribeiro 1990) the presence of this unconformity is difficult to substantiate.

San José *et al.* (1990) described the lower Alcudian succession in the southern part of the DVF (Eastern Portuguese-Alcudian Zone of Lotze 1945) as consisting of shales, sandstones and conglomerates, with rare basic and felsic volcanic rocks. The presence of pebbles of sedimentary, metamorphic and igneous rocks in the conglomerates and of igneous clasts in the sandstones suggests a mixed volcanoterrigenous provenance (San José *et al.* 1990). The estimated thickness is about 4000 m (San José *et al.* 1990), but may exceed 9000 m or possibly even 15 000 m (Santamaría 1995) in the Valdelacasa anticline (Fig. 2.1c). The Domo Extremeño Gp (Fig. 2.2) has been divided into the Estomiza and overlying Cubilar formations. The presence of complete or truncated Bouma sequences and disorganized beds in the lower of these formations has been interpreted as lobe, interlobe and channel

deposits in a submarine fan environment. In the Cubilar Fm wave- and storm-dominated offshore facies have been described and interpreted as possibly being related to distal deltaic facies (San José *et al.* 1990).

In the Alcudia–Montes de Toledo region, the lower Alcudian succession is overlain by an unconformity, although it is difficult to identify in some locations. Despite this, the unconformity is interpreted as a major stratigraphic discontinuity, implying evidence for a deformation phase (Oretanian phase; San José 1984), and marked by palaeoalteration and erosion of the substratum prior to the deposition of upper Alcudian sediments (San José *et al.* 1990). Above the unconformity is a diverse succession of mixed terrigenous–carbonate rocks (upper Alcudian) that include mudstones, sandstones, quartzites, conglomerates, limestones and dolostones. The thickness of the succession ranges from 850 m in the south and SW to 3600 m in the north and NE. Sedimentary environments change from possibly continental (alluvial fans, fluvial), fluviotidal, deltaic, siliciclastic and carbonate shorelines, to slope and submarine fans (San José *et al.* 1990).

According to San José *et al.* (1990), another important sedimentary discontinuity separates the Pusian from the upper Alcudian successions (Fig. 2.2). The discontinuity has a regional extent and has been variously interpreted as either a disconformity or an angular unconformity. There is no consensus concerning the significance of this unconformity, but for some authors (San José 1983; San José *et al.* 1990) it is undoubtedly related to an orogenic pulse (Cadomian or Assyntic phase; Herranz *et al.* 1977) followed by erosion of the substratum.

In the northern part of the DVF (Galician-Castilian Zone of Lotze 1945), detailed stratigraphic and sedimentological studies of the whole Neoproterozoic sedimentary succession have been conducted and four lithostratigraphic units (I, II, III and IV) have been distinguished (Valladares *et al.* 1998, 2000). Their most relevant features are synthesized below.

Unit I. This has an unexposed lower boundary, is >585 m thick, and is overlain by an erosion surface. It comprises massive sandstones in beds 1–3.5 m thick typified by 50–70 cm thick sets and cosets of planar or trough cross-bedded sandstone and sandstone–shale couplets as much as 20 cm thick. The sandstones are very fine- to medium-grained, with slumps, black shale clasts, complete and truncated Bouma sequences, climbing ripples and wavy and lenticular laminations. Two facies associations appear in this unit, one displaying thinning- and fining-upward cycles up to 6 m thick (interpreted as channel-fill deposits from high- and low-concentration turbidity currents), and the other comprising thickening- and coarsening-upward cycles up to 8 m thick and interpreted as lobe deposits. Such features imply that this unit was deposited at the channel–lobe transition of middle deep-sea fans (Fig. 2.4).

Unit II. This has an erosive lower boundary, a concordant, sharp upper boundary, and a thickness varying between 225 and 600 m. It is characterized by the presence of tabular conglomerate beds up to 18 m thick and laterally continuous over several kilometres that are interbedded with beds of massive sandstones, several metres thick, and rare thin sandstone–shale alternations. The conglomerates are clast- and matrix-supported, polymodal and polygenic, with a shaly or sandy black matrix that is locally laminated. The clasts are rounded to subrounded and either disorganized or with a roughly inverse-to-normal grading. The largest and most abundant clasts (pebbles and blocks) are derived from sedimentary rocks, whereas the smallest and rarest clasts are of quartz and quartzite. Clast size and the abundance of sedimentary rock clasts decrease towards the top of the unit and towards the SW, which suggests input from an inactive, unroofing source situated towards the NE. The associated fine- to medium-grained sandstones are either massive, with scattered clasts of black

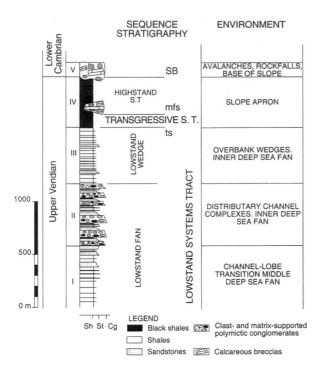

Fig. 2.4. The Neoproterozoic sedimentary succession in the central sector of the Central Iberian Zone showing an interpretative sequence stratigraphic framework. Key: Cg, conglomerate; mfs, maximum flooding surface; SB, sequence boundary; Sh, shale; St, sandstone; ST, systems tract; ts, transgressive surface. See Figure 2.3 for position of units I–V.

shale, or show planar and trough cross-bedding. This unit is made up almost exclusively of thinning- and fining-upward cycles, up to 50 m thick, which are interpreted as channel-fill deposits from debris flows and high-concentration turbidity currents in distributary channel complexes of inner deep-sea fans (Fig. 2.4).

Unit III. This shows concordant and sharp lower and upper boundaries, is around 550 m thick, and is characterized by thin sandstone–shale alternations (Fig. 2.4). The sandstones are very-fine- to fine-grained, show parallel and/or current ripple laminations, abundant slumps and slide scars and rare fining-upward cycles 1–8 m thick. This unit has been interpreted as an overbank wedge associated with an active fluviodeltaic depositional system on the adjacent shelf.

Unit IV. The lower boundary of the 40 to 500 m thick unit IV is concordant and sharp whilst the upper one is strongly erosional (Fig. 2.4). It consists of massive or laminated black shales, with phosphate nodules and numerous euhedral sulphide crystals. The shales contain dark, very fine-grained, ripple-laminated sandstone beds, up to 30 cm thick, showing flame structures and wavy tops. Commonly, within the upper half of the unit (Fig. 2.4) an intercalation of mixed rocks (siliciclastic–carbonate) appears and defines a lenticular geometry up to 3 km long and 260 m thick (Valladares 1995). This intercalation consists of alternations of limestone breccias and stratified sandstone–limestone couplets several metres thick. The breccias are clast-supported, with disorganized clasts that sometimes protrude out of the beds, and form both channelled and tabular beds up to 3.5 m thick. The stratified sandstone–limestone couplets are up to 20 cm thick with parallel and ripple laminations, low-angle planar cross-lamination, or pebbles forming inversely graded beds (Valladares & Rodríguez Alonso 1988; Valladares 1995). Throughout the entire unit, slumps, slide scars, fluid-escape and flame structures and pseudo-nodules are common. Sedimentation of the black shales occurred on a slope with high organic productivity and low-oxygen seawater contents and resulted from settling. In contrast, the deposition of the alternations of limestone breccias and sandstone–limestone couplets

resulted from debris flows and high-concentration turbidity currents derived from the adjacent platform margin and/or from the slope above a mixed slope apron (Valladares 1995). The upper boundary of unit IV with unit V is strongly erosive (Fig. 2.4). Unit V is a calcareous megabreccia with blocks derived from the lower units (IV and III), some of them showing signatures of subaerial exposure before emplacement in the megabreccia.

The stratigraphic and sedimentological data from units I–IV suggest that Neoproterozoic sedimentation in the Central Iberian Zone resulted mainly from turbidity currents, debris flows and submarine slides in slope and base-of-slope environments. Units I and II probably correspond to a type-II turbidite stage, and unit III to a type-III turbidite stage (Mutti 1985) connected directly to active deltaic systems at the shelf edge. This vertical succession resulted from a gradual reduction in the volume of mass flows, perhaps associated with a slow relative rise in sea level. Contact between units III and IV has been interpreted as a transgressive surface (Fig. 2.4), consistent with the deposition of black shales on this surface. Unit IV records the transgressive and highstand systems tracts of this depositional sequence (Fig. 2.4), the maximum flooding surface probably being marked where phosphate nodules appear within the black shales. The postulated rise in sea level would have facilitated the development of oolithic and bioclastic carbonate shoals on the shelf margin to the NE. The carbonate rocks of unit IV are suggested to represent highstand shedding of this carbonate from the platform (Valladares *et al.* 2000). The strongly erosional surface at the top of unit IV is interpreted as a type-1 sequence boundary (Fig. 2.4) since the data for the megabreccia (unit V, early cemented calcareous blocks, some of them showing signs of palaeokarst) suggest that the platform emerged following a relative fall in sea level, which probably dropped below the platform slope break. The time of non-sedimentation represented by this boundary is greater towards the platform than towards the basin, where it would be expected to pass into a correlative conformity.

Assuming the Badajoz–Córdoba shear belt to be the boundary between the Ossa Morena and Central Iberian zones, the Valle de la Serena granite (Ordóñez 1998), which intrudes acid volcanic rocks, must be considered as intruded into the Central Iberian Zone. This granite has an average age of 573 ± 14 Ma (SHRIMP zircon dating; Ordóñez 1998) and, together with the above-mentioned Miranda do Douro gneiss, is the only Neoproterozoic igneous rock dated in the Central Iberian Zone.

Ossa Morena Zone

To present a synthesis of the geology of the Ossa Morena Zone is not a simple matter since sedimentological, stratigraphical and petrological data are poorly constrained and are frequently contradictory. In recent years new data have elicited some changes in previous interpretations (see below) but some aspects of Ossa Morena Zone geology still remain unresolved. Rocks once thought to be the oldest in Spain crop out in the Badajoz–Córdoba shear zone and are known as the Azuaga Gneiss Gp (Delgado 1971). This unit consists of gneisses, amphibolites and eclogites (Ábalos 1992), with subordinate intercalations of calcsilicates and black metacherts. The rocks in this unit everywhere record medium- to high-grade metamorphic conditions and are extremely deformed. For this reason, any attempt to establish their original characteristics is unavoidably speculative. U-Pb zircon dating of orthogneisses within the Azuaga Gneiss Gp gave 509 ± 8 Ma, interpreted as the age of the formation of the igneous protolith (Ordóñez 1998).

Given the apparently Palaeozoic age of the Azuaga Gneiss Gp, the Serie Negra Gp is probably the oldest sequence seen in the Ossa Morena Zone (Fig. 2.3). Serie Negra rocks have been divided into a lower unit called the Montemolín schists and amphibolites (MSA), and an upper, apparently conformable unit of metasediments known as the Tentudía Fm. The MSA forms a 1000–2000 m-thick succession of graphite-rich pelites and psammites, with abundant amphibolites towards the top, frequent black chert lenticular interbeds and, locally, marbles. It is not clear whether the protoliths of the amphibolites were intrusive or extrusive (Ordóñez 1998), but they show a $\varepsilon_{Nd}(600)$ in the range of +5.9 to +7.41. The overlying Tentudía Fm is 500–3000 m thick (Quesada et al. 1990a) and consists of interlayered volcanogenic greywackes and slates (Eguíluz et al. 2000). The Serie Negra Gp has been dated imprecisely as mid–late Riphean on the basis of scarce (and uncertain) acritarchs (Chacón et al. 1984). However, isotopic data from detrital zircons in a biotitic schist of the Montemolín sequence and a greywacke of the Tentudía Fm have furnished an age younger than 591 ± 11 Ma and younger than 564 ± 30 Ma for the deposition of the Montemolín and Tentudía metasedimentary protoliths and the greywacke respectively (Schäfer et al. 1993; Ordóñez 1998).

The Tentudía Fm is supposedly truncated by an angular unconformity above which is a sequence of volcanosedimentary rocks known as the Bodonal–Cala complex (Hernández Enrile 1971) in the southern part of the Ossa Morena Zone, and as the Malcocinado Fm (Fricke 1941) or the San Jerónimo Fm (Liñán 1978) in the northern part of the zone (Fig. 2.3). The Malcocinado Fm is up to 1200 m thick and mainly consists of clastic rocks that include interbedded tuffs, rhyolitic flows and abundant conglomerates with pebbles of volcanic and basement rocks of slate, greywacke, granitoid rocks, quartz and black quartzite. The chemical composition of the volcanic rocks is less felsic towards the north and SE where andesite volcanics are common (Eguíluz et al. 2000). The San Jerónimo Fm contains abundant andesites with high $\varepsilon_{Nd}(T)$ values ranging from +2.9 to +7.4 (Pin et al. 1999). The Bodonal–Cala complex consists of volcanic and volcaniclastic rocks of rhyolitic to dacitic composition and includes ash layers and ignimbrites (Eguíluz et al. 2000). This succession was originally assumed to be of Vendian age on the basis of a small number of ichnofossils (Neonerites? sp. and Gordia) and acritarchs attributed to Sphaerocongregus variabilis in the San Jerónimo Fm (Fedonkin et al. 1983; Liñán & Palacios 1983). However, U-Pb zircon dating of a porphyritic rhyolite in the Bodonal-Cala complex and a reworked tuff in the Malcocinado Fm gave 514 ± 9 and 522 ± 8 respectively (Ordóñez 1998), suggesting an Early Cambrian age for at least part of this volcanosedimentary sequence (Fig. 2.3).

In the Monesterio antiform (Fig. 2.1c), two phases of Cadomian deformation and metamorphism (mainly related to the second deformation phase) have been reported (Eguíluz & Ábalos 1992). This metamorphism was of the low-pressure type and reached anatectic conditions, resulting in anatectic granodiorites and leucogranites. The age data for these granodiorites are 552 ± 16 Ma (Rb-Sr whole-rock age; cited in Dallmeyer & Quesada 1992), but conventional U-Pb monazite, xenotime and zircon data and SHRIMP dating have

given Early to Middle Cambrian ages for the anatectic granodiorite and related leucogranites (Ordóñez et al. 1997a). Hornblende from an amphibolite enclave in an anatectic granodiorite has provided an age of 616 ± 4 Ma (^{40}Ar/^{39}Ar; Dallmeyer & Quesada 1992). A Neoproterozoic granite that is unconformably covered by the Malcocinado Fm and a tonalite show intrusive ages of 552 ± 10 Ma and 544 ± 6 Ma respectively (Ordóñez 1998).

In the Badajoz–Córdoba shear zone (boundary of the Central Iberian and Ossa Morena zones; Fig. 2.1c) the metamorphic history is complex. This shear zone consists of metasedimentary rocks, orthogneisses, migmatites and amphibolites. Errorchron ages of 632 ± 103 Ma and 690 ± 134 Ma have been reported for orthogneisses (Azor et al. 1995). However, SHRIMP dating gives 460 ± 4 Ma, 465 ± 14 Ma and 511 ± 6 Ma for magmatic zircons in the orthogneisses of the shear belt. Migmatization of these orthogneisses has been dated at 335 ± 4 Ma (Ordóñez et al. 1997b). The age data for the metamorphism are 550 ± 10 Ma, 561 ± 1 Ma (^{40}Ar/^{39}Ar muscovite; Blatrix & Burg 1981; Dallmeyer & Quesada 1992) and 550 ± 3 Ma and 552 ± 3 Ma (^{40}Ar/^{39}Ar hornblende; Dallmeyer & Quesada 1992). However, zircons from a migmatite give 528 ± 6 Ma (SHRIMP dating; Ordóñez et al. 1997a). According to these latter authors, the age discrepancies are probably related to the presence of excess Ar and the Cadomian metamorphism is of Cambrian age rather than Neoproterozoic. Some amphibolites (retrogressed eclogites; Azor 1997) and a flaser gabbro have protolith ages ranging between 530 and 600 Ma (Ordóñez 1998).

Sedimentary geochemistry

Systematic studies on the geochemistry and chemostratigraphy of Neoproterozoic sediments in Spain began in the Central Iberian Zone in the early 1990s (Valladares et al. 1993) and abundant geochemical data are now available, including major and trace element and Sm-Nd isotopic data for this zone (Ugidos et al. 1997a, 1999). However, studies combining geochemical data with stratigraphic information are very scarce or lacking for other Precambrian zones in Spain. An exception is the Iberian Ranges where this kind of study has recently begun (Bauluz et al. 2000).

The compositional homogeneity of the Neoproterozoic sedimentary materials is probably their most relevant characteristic since the geochemical features of the shales do not vary significantly for hundreds of kilometres across the Central Iberian Zone (Valladares et al. 1999, 2000). Some diagrams, such as TiO_2-Zr, Al_2O_3/TiO_2-Rb/Zr and Ti/Nb-Cr/Th, help to discriminate between Neoproterozoic and Lower Cambrian shales (Fig. 2.5a, b, c). The general positive correlation of all the data in these diagrams suggests that compositions of the Neoproterozoic and Lower Cambrian shales resulted from gradually variable mixtures of two extreme compositions. The average sandstone composition shows major and trace element ratios (e.g. Al_2O_3/TiO_2, Th/Yb, Ti/Nb) comparable to those of the average shale in the Central Iberian Zone (Ugidos et al. 1997a) and the rare earth element patterns of the shales and sandstones are parallel (Fig. 2.5d). These features suggest that hydraulic sorting of minerals did not seriously affect the sedimentary materials. Moreover, the abundances of some element or ratio values (e.g. Ti, Zr, Ti/Nb) gradually decrease from unit I to unit IV while other parameters increase (e.g. Rb/Zr) similarly (Valladares et al. 2000). All

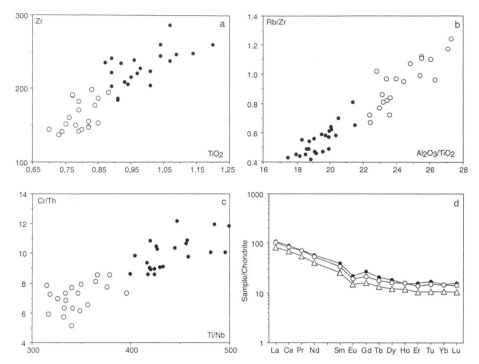

Fig. 2.5. (a, b, c) Selected major and trace element diagrams discriminating Neoproterozoic (filled circles) and Lower Cambrian (open circles) shales in the Central Iberian Zone. **(d)** Rare earth element patterns of the average Neoproterozoic shales (filled circles) from the Iberian Ranges, and the average Neoproterozoic shales and sandstones (open circles and triangles, respectively) from the Central Iberian Zone. Note the parallel patterns of the two rocks from this zone. Plotted data correspond to samples selected from more than 100 analysed shales. Samples from points close to the mapped Neoproterozoic–Cambrian boundary have been deliberately excluded for the purposes of this chapter since a relative sea-level fall favoured the removal of the uppermost Neoproterozoic sediments and their redeposition in the lowermost Lower Cambrian (Valladares *et al.* 2000). Normalizing chondrite values after Taylor & McLennan (1985).

these characteristics suggest that the source area mainly consisted of two compositions that were supplied to the basin in gradually varying proportions. Thus, a composition relatively richer in TiO_2 and Zr predominated during the sedimentation of the Neoproterozoic materials while the contribution of a composition poorer in these elements predominated during Cambrian times. All data together suggest an extensive, homogeneous and recycled source area in a passive margin tectonic setting (Ugidos *et al.* 1997*a,b*; Valladares *et al.* 2000).

The average Neoproterozoic shale of the Iberian Ranges (West Asturian-Leonese Zone; Fig. 2.1c) is similar to the average Neoproterozoic shale of the Central Iberian Zone and petrological and geochemical results in the former region also support a passive, recycled and mature source area as being the most plausible setting for Neoproterozoic sediments (Bauluz *et al.* 2000). In view of the geochemical affinities shown by both average shales, and taking into account that similar provenances have been proposed for both zones, it seems logical to conclude that both zones shared the same extensive source area. Average compositions of Cambrian shales from the Central Iberian Zone and Iberian Ranges are also similar and the same chemical parameters separate their compositions from the respective Neoproterozoic compositions in the corresponding zones (Chapter 3).

Sm-Nd isotopic results indicate $\varepsilon_{Nd}(T)$ values for the Neoproterozoic shales ranging from −4.8 to −2.9 and T_{DM} neodymium model ages from 1.55 to 2.21 Ga in the Alcudia–Montes de Toledo region in the Central Iberian Zone (Nägler *et al.* 1995). In the Hurdes–Salamanca region, $\varepsilon_{Nd}(T)$ values of the Neoproterozoic shale range between −0.4 and −2.8 and T_{DM} neodymium model ages between 1.1 and 1.69 Ga (Ugidos *et al.* 1997*b*, 1999). In the Narcea antiform data are −2.7 to −2.3 and 1.52 to 1.6 Ga (Nägler *et al.* 1995). The chemical homogeneity of the shales described above contrasts with the relatively broad range of Sm-Nd isotopic results

reported for the Central Iberian Zone. However, some ε_{Nd} values and T_{DM} neodymium model ages may be anomalous since an intensely weathered source possibly led to a redistribution of rare earth elements in certain sedimentary units of the Central Iberian Zone (Ugidos *et al.* 1999). Thus, some values of $\varepsilon_{Nd}(T)$ and neodymium T_{DM} model ages in the sedimentary succession may correspond to samples from such units and so their Sm-Nd isotopic results do not reflect the primary model ages. Samples without evidence for redistribution of rare earth elements give T_{DM} neodymium model ages younger than 1.7 Ga and have higher $\varepsilon_{Nd}(T)$ values than Cambrian rocks (Chapter 3). The Neoproterozoic sedimentary basins therefore inherited some contribution of juvenile material that was lacking or at least very reduced in the Cambrian ones. Element abundances and ratios and Sm-Nd isotopic results thus strongly support two source compositions, and discriminate Neoproterozoic from Cambrian compositions. Despite this, the geochemical data do not define a clear Neoproterozoic–Cambrian boundary because the geochemical variation through the Neoproterozoic–Lower Cambrian sedimentary succession is gradual. Furthermore, sediments from unit IV, and probably also from unit III, are interpreted to have been reworked during a relative fall in sea level and were resedimented to form at least part of Lower Cambrian units VI and VII (Valladares *et al.* 2000).

In the Ossa Morena Zone, the Tentudía greywacke is characterized by its low CaO content (CaO = 0.52%; this is not exceptional since 11 analyses of Tentudía greywackes show a range of CaO between 0.04 and 0.40%; Eguíluz 1987), a relatively low $\varepsilon_{Nd}(550)$ value (−6.7), a relatively old neodymium T_{DM} model age (1.9 Ga) and a very low initial $^{87}Sr/^{86}Sr$ ratio (0.69933), suggestive of a disturbance of the Rb-Sr system after deposition (Schäfer *et al.* 1993). All these features coincide with those shown by Cambrian shales in the Central Iberian Zone (Nägler *et al.* 1995; Ugidos *et al.* 1997*b*; Chapter 3).

Tectonic setting

The increasing acquisition of chemical and isotopic data has changed views about the antiquity of the Precambrian rocks in Spain and also about the provenance of siliciclastic sediments. Many rocks previously thought to be as old as 1350 Ma are now viewed as belonging to the latest Neoproterozoic (Schäfer et al. 1993; Ordóñez 1998), the oldest dates so far obtained being 618 ± 9 Ma (Lancelot et al. 1985), 573 ± 14 Ma (Ordóñez 1998), and 605 ± 10 and 580 ± 15 Ma (Fernández Suárez et al. 1998) for orthogneisses in the Central Iberian and West Asturian-Leonese zones, respectively. Another recent change in opinion has stemmed from the recognition that the Central Iberian and Ossa Morena zones were most likely unrelated during latest Neoproterozoic–Early Cambrian times (Valladares et al. 1993; Ugidos et al. 1997b).

Thus, previous models for the geodynamic evolution of the whole Iberian Massif (e.g. Ribeiro et al. 1990c; Quesada 1990a) suggesting a Neoproterozoic complete Wilson cycle in terms of plate rifting, drifting (middle–late Riphean), subduction (late Riphean–early Vendian) and collision (middle–late Vendian) during Cadomian or Pan-African orogeny need some review. The Ossa Morena Zone is generally interpreted as an active margin setting (but see Ordóñez 1998 for a different view), mainly on the basis of abundant basic volcanics associated with Neoproterozoic and Palaeozoic sediments (Pin et al. 1999; Eguíluz et al. 2000; Chapter 9). The Cantabrian, West Asturian-Leonese and Central Iberian zones are considered parts of an active margin of the Gondwana continent by Nägler et al. (1995), on the basis of the juvenile contribution to the Neoproterozoic supposedly synorogenic sedimentary facies. However, the geochemical maturity of the Neoproterozoic–Cambrian sediments and the lack of coeval volcanics rather suggest a Gondwanan passive margin as a more plausible tectonic setting for the Central Iberian and West Asturian-Leonese zones (Beetsma 1995; Ugidos et al. 1997a,b, 1999; Valladares et al. 1999, 2000, 2002; Bauluz et al. 2000). Fernández Suárez et al. (2000b) favour a model involving the Avalonian–Cadomian arc as the tectonic setting for the Neoproterozoic basin. In this model, the West Asturian-Leonese Zone would be located close to West Avalonia on the basis of the presence of zircons showing Grenville (0.9–1.2 Ga) ages (Fernández Suárez et al. 2000b). Thus, the proposed models are strongly contrasting and, in order to avoid a long and tedious discussion, the data most relevant to current knowledge are synthesized in Figure 2.3 and in the following three points.

(a) The Central Iberian and West Asturian-Leonese zones share similar features, such as acid volcanism in the lowermost outcropping Neoproterozoic rocks and granites intruded at c. 600 Ma BP. These rocks are overlain by uppermost Proterozoic pelite-rich successions that show the same chemical and isotopic features (key element ratios, ε_{Nd} values, neodymium T_{DM} model ages; Fig. 2.3). In both zones, Lower Cambrian shales show the same geochemical characteristics as each other, but contrast with those recorded by the Neoproterozoic shales. The chemical and isotopic (Sm-Nd) results strongly support the notion that the Neoproterozoic–Cambrian succession in the Central Iberian Zone reflects stratigraphy of the source in reverse order of age, as it was progressively unroofed (Ugidos et al. 1997b; Valladares et al. 2002).

(b) In the Cantabrian and West Asturian-Leonese zones, detrital zircons from Neoproterozoic and Lower Palaeozoic rocks have given ages of 1.8–2 Ga and 0.9–1.2 Ga. Neoproterozoic greywackes yield abundant zircons in the age range of 640–800 Ma while a Cambrian sandstone has zircons in the 550–620 Ma age range but almost completely lacks zircons of 640–800 Ma (Fernández Suárez et al. 2000b). Archean zircons were also found in both rocks. Similarly, Archean ages and ages of 1.8–2 Ga, 0.9–1.0 Ga and c. 0.6 Ga have been reported from the Ossa Morena Zone, but ages in the range of 640–800 Ma have not been found in orthogneisses or metasedimentary rocks from this zone (data in Ordóñez 1998).

(c) In the Cantabrian and West Asturian-Leonese zones, and in some areas of the Central Iberian Zone, an angular unconformity separating Neoproterozoic and Cambrian sedimentary successions has been proposed as being related to a Cadomian compressive deformation phase (Lotze 1956a; Bouyx 1970; Marcos 1973; Pérez Estaún 1978; Martínez Catalán 1985; Lancelot et al. 1985; Ortega & González Lodeiro 1986; Liñán & Tejero 1988; Wildberg et al. 1989; Díez Balda et al. 1990; San José et al. 1990). Commonly, the angle between the strata above and below the unconformity is very low so that the stratigraphic break is deduced from map interpretation (Chapter 9). However, in the central areas of the Central Iberian Zone the unconformity is a more obvious, strongly erosional surface interpreted as corresponding to a type-1 sequence boundary due to a major fall in sea level (Valladares et al. 2000). This fall in sea level was possibly eustatic since it is coeval with that seen in other basins of the world, but was also probably caused by extensional faulting and differential block subsidence related to an immature passive margin in southern and central (Valladares 1995) areas of the Central Iberian Zone during latest Neoproterozoic–Early Cambrian times (Fig. 2.6). This induced the erosive removal of the Neoproterozoic uppermost units in the upthrown blocks before the next rise in sea level at the beginning of Cambrian times. Therefore, Lower Cambrian sediments were deposited on surfaces with different slopes, depending on the position of the underlying blocks. Thus, it is suggested that the angular unconformity between Neoproterozoic and Cambrian rocks was not related to a Cadomian compressive deformation but rather to an extensional event and sea-level changes. From the authors' point of view this passive margin model for the latest Neoproterozoic–Early Cambrian times is the simplest interpretation, and one that most plausibly accounts for all the above-mentioned data from the Neoproterozoic rocks in the Cantabrian, West Asturian-Leonese and Central Iberian zones (Fig. 2.6).

The evolution of the source area and the basin for Neoproterozoic sediments in Spain were probably related to the following sequence of events.

I. Rifting of northern Gondwana into continental blocks c. 1 Ga BP. Sediments in new rift-related basins received a contribution from relatively abundant rift-related volcanics (c. 1 Ga zircons) and old continental materials (c. 1.9 Ga zircons and older).

II. Closure of the oceans, with cordilleran-type arc activity and subsequent collision of Gondwanan continental blocks (800–640 Ma zircons). Early Pan-African (Cadomian) events and development of interior orogens (Murphy & Nance 1991) were followed by late Pan-African (Cadomian) events, with extensional tectonics, acid magmatism at c. 600 Ma BP, and sedimentation of molasse deposits (Valladares et al. 2002).

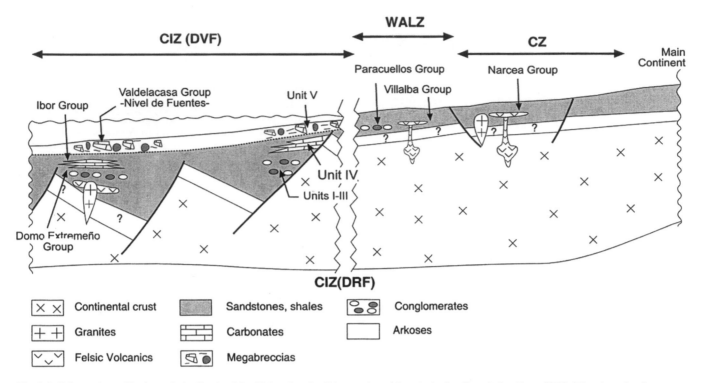

Fig. 2.6. Schematic profile through the Iberian Massif showing the lithostratigraphic units in the Cantabrian Zone (CZ), West Asturian-Leonese Zone (WALZ) and Central Iberian Zone (CIZ). The inferred environment is a passive margin and continental shelf on a cratonal basement (possibly West African Craton) in an immature passive margin across the Neoproterozoic–Cambrian boundary.

III. Development of an immature passive margin setting and sedimentation of the uppermost Neoproterozoic sediments (Narcea, Villalba, Paracuellos, Domo Extremeño and Ibor groups, units I–IV, lower and upper Alcudian). Erosion of interior orogens and related molasse deposits provided the main sources for these sediments that inherited the recycled and relatively abundant juvenile contribution, although with time there was an increasing contribution of basement materials.

IV. Extensional event, relative sea-level fall and formation of the Neoproterozoic–Cambrian type-1 sequence boundary. Basement materials would have been the main contributors to the Cambrian sedimentary basin, and there was therefore a scarcity or absence of 800–640 Ma zircons, with a contribution from 600 Ma granites occurring only after the erosion of molasse deposits had been effected.

This model is not applicable to the Ossa Morena Zone since some features do not fit in with the passive tectonic setting, such as the relative abundance of amphibolites at the top of the Montemolín sequence and the calc-alkaline volcanism in the Malcocinado Fm. Thus, it is probable that the Ossa Morena Zone was related to a volcanic arc and therefore unrelated to the other zones in Iberia. However, some problems need to be addressed before this hypothesis of an arc can be assumed. For example, the negative ε_{Nd} of the Tentudía sediments strongly contrasts with the relatively high positive values of the underlying and overlying amphibolites and andesites (Fig. 2.3). Accordingly, the Tentudía Fm probably does not belong to the same setting as the other formations. In this sense, it is relevant that metapelites (assumed to belong to the Serie Negra in the Ossa Morena Zone) in the western side of the Badajoz–Córdoba shear belt in NW Portugal show exactly the same trace element ratios and ε_{Nd} values as Tentudía metasedimentary rocks and Cambrian shales in the Central Iberian Zone (see above). It seems that more information is needed from the Ossa Morena Zone if a consistent model involving the Neoproterozoic–Cambrian sedimentary basin is to be proposed.

Constructive comments by referees T. Prave and C. Quesada and the editors greatly improved the chapter. N. Skinner is thanked for revising the English version of the chapter. Partial financial support was provided by the PB91-0321 DGICYT and PB96-1283 DGES projects of the MEC (Spain).

3 Cambrian

ELADIO LIÑÁN (coordinator), RODOLFO GOZALO, TEODORO PALACIOS,
JOSÉ ANTONIO GÁMEZ VINTANED, JOSÉ MARÍA UGIDOS & EDUARDO MAYORAL

The Iberian Peninsula has some of the most extensive Cambrian outcrops in Europe (Lotze 1961c), including a diverse, continuous record of fossils and facies, and is thus a fundamental source of biostratigraphic information for the Cambrian System and its intercontinental correlations. Most exposures of Iberian Cambrian rocks occur in the Iberian Massif, but they are also known from the Pyrenees, the Catalonian Coastal Ranges and the Iberian Ranges (Fig. 3.1).

Many exposures are geographically isolated and/or show tectonic boundaries, and facies changes are common, and these characteristics have led to a profuse stratigraphic nomenclature (see Fig. 3.2; Zamarreño 1983; Liñán *et al.* 1993a). Following Lotze (1961c), however, the Cambrian sequence can be overviewed as a diachronous Lower to Middle Cambrian carbonate sequence sandwiched by siliciclastic successions (Fig. 3.2). The lower of the siliciclastic units is entirely Lower Cambrian, whereas the upper unit ranges from upper Lower or Middle Cambrian to Upper Cambrian (Fig. 3.2). The Lower Cambrian series has been subdivided into the Corduban, Ovetian, Marianian and Bilbilian stages, and the Middle Cambrian series subdivided into the Leonian, Caesaraugustan and Languedocian stages (Fig. 3.2).

Precambrian/Cambrian boundary

The Precambrian/Cambrian boundary stratotype was erected by the International Subcommission on Cambrian Stratigraphy (ISCS) at the Fortune Head section in eastern Newfoundland (Canada) with the first appearance datum (FAD) of *Phycodes* (= *Trichophycus*) *pedum* (Landing 1994). This FAD coincides with behavioural changes, increased abundance and diversity of the benthos. Only a few Precambrian trace fossils survived and new patterns of Cambrian bioturbation were established by Arthropoda-made trace fossils and other traces, such as *Monomorphichnus*, *Psammichnites*, *Rusophycus* and *Cruziana* (Crimes 1994). The Acadobaltic Province association of *P. pedum* with *Monomorphichnus lineatus* is known from localities in northern, central and southern Spain, where the assemblage is considered as index for earliest Cambrian strata in Iberia (Fig. 3.3).

In northern Spain (Cantabrian region) the Lower Cambrian Herrería Formation (Fm) lies unconformably upon the Precambrian Narcea Group (Gp) (Chapter 2), from which organic-walled microfossils (*Palaeogomphosphaera cauriensis* and *Sphaerocongregus variabilis*) have been obtained (Palacios & Vidal 1992; Vidal *et al.* 1994b). *P. pedum* appears just above the unconformity at Irede de Luna (Palacios & Vidal 1992)

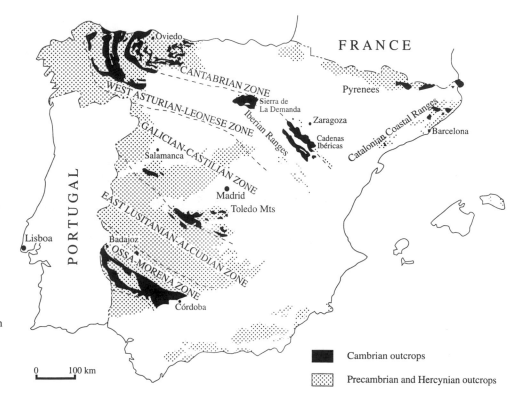

Fig. 3.1. Pre-Mesozoic geological map of Iberia showing the Cambrian outcrops. Earliest Cambrian rocks of the 'schist-greywacke complex' from Galician-Castillian and East Lusitanian-Alcudian zones are not represented.

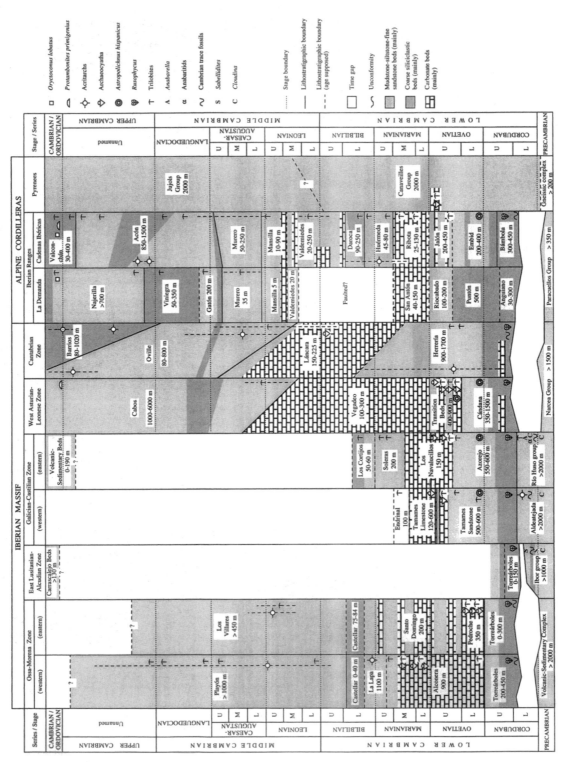

Fig. 3.2. Tentative correlation chart of Cambrian lithostratigraphic units in Spain (modified from Liñán *et al.* 1993*a*).

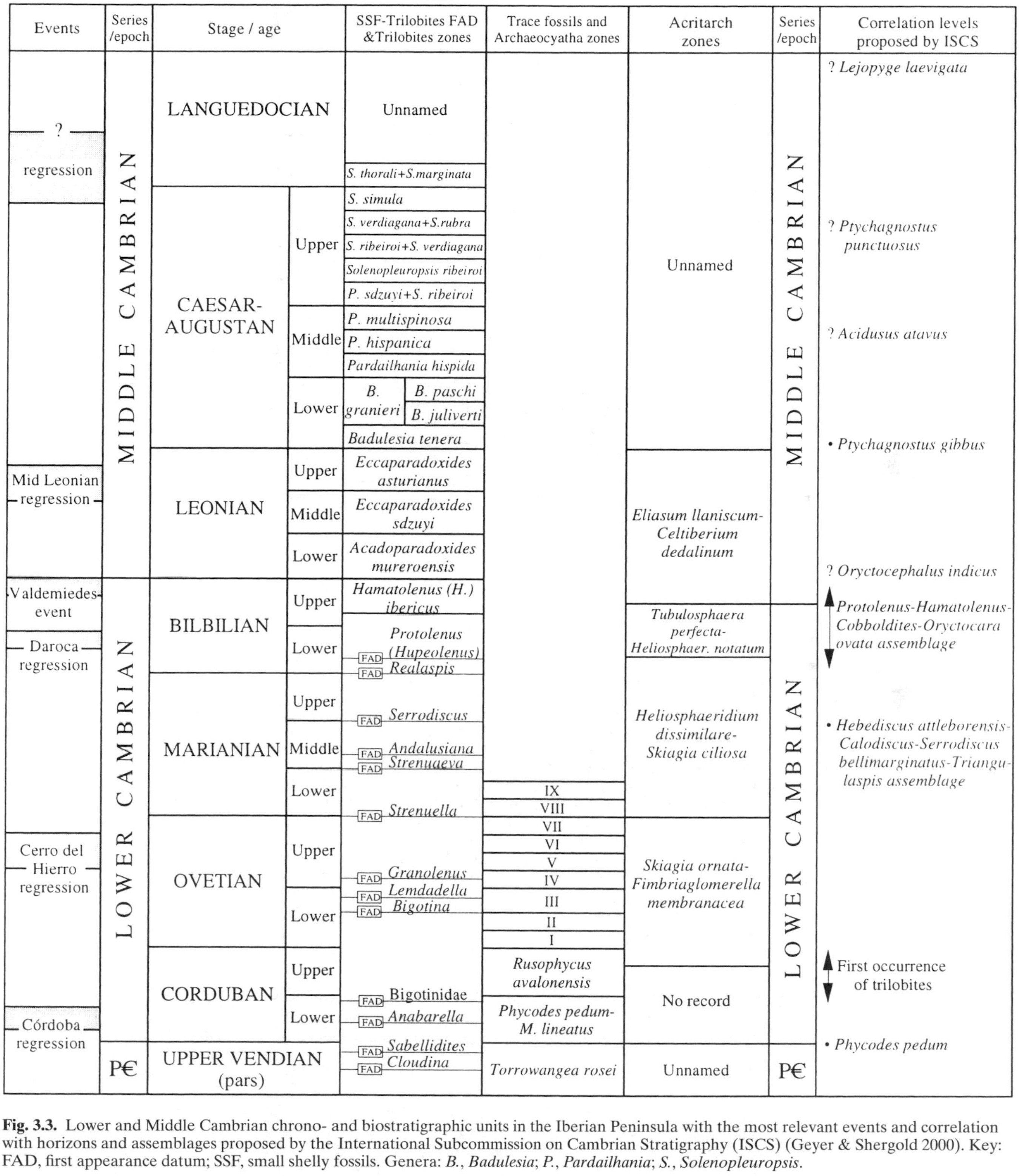

Events	Series /epoch	Stage / age		SSF-Trilobites FAD & Trilobites zones		Trace fossils and Archaeocyatha zones	Acritarch zones	Series /epoch	Correlation levels proposed by ISCS
	MIDDLE CAMBRIAN	LANGUEDOCIAN		Unnamed				MIDDLE CAMBRIAN	? *Lejopyge laevigata*
?				*S. thorali+S.marginata*					
regression									
		CAESAR-AUGUSTAN	Upper	*S. simula*			Unnamed		? *Ptychagnostus punctuosus*
				S. verdiagana+S.rubra					
				S. ribeiroi+S. verdiagana					
				Solenopleuropsis ribeiroi					
				P. sdzuyi+S. ribeiroi					
			Middle	*P. multispinosa*					? *Acidusus atavus*
				P. hispanica					
				Pardailhania hispida					
			Lower	*B. granieri*	*B. paschi*				• *Ptychagnostus gibbus*
					B. juliverti				
				Badulesia tenera					
Mid Leonian regression		LEONIAN	Upper	*Eccaparadoxides asturianus*			*Eliasum llaniscum-Celtiberium dedalinum*		
			Middle	*Eccaparadoxides sdzuyi*					
			Lower	*Acadoparadoxides mureroensis*					? *Oryctocephalus indicus*
Valdemiedes event	LOWER CAMBRIAN	BILBILIAN	Upper	*Hamatolenus (H.) ibericus*			*Tubulosphaera perfecta-Heliosphaer. notatum*	LOWER CAMBRIAN	↕ *Protolenus-Hamatolenus-Cobboldites-Oryctocara ovata assemblage*
Daroca regression			Lower	*Protolenus (Hupeolenus)* [FAD]					
				Realaspis [FAD]					
		MARIANIAN	Upper	*Serrodiscus* [FAD]			*Heliosphaeridium dissimilare-Skiagia ciliosa*		• *Hebediscus attleborensis-Calodiscus-Serrodiscus bellimarginatus-Triangulaspis assemblage*
			Middle	*Andalusiana* [FAD]					
				Strenuaeva [FAD]					
			Lower	*Strenuella* [FAD]		IX			
						VIII			
Cerro del Hierro regression		OVETIAN	Upper	*Granolenus* [FAD]		VII	*Skiagia ornata-Fimbriaglomerella membranacea*		
						VI			
						V			
			Lower	*Lemdadella* [FAD]		IV			
				Bigotina [FAD]		III			
						II			
						I			
		CORDUBAN	Upper	*Bigotinidae* [FAD]		*Rusophycus avalonensis*	No record		↕ First occurrence of trilobites
Córdoba regression			Lower	*Anabarella* [FAD]		*Phycodes pedum-M. lineatus*			• *Phycodes pedum*
	PЄ	UPPER VENDIAN (pars)		*Sabellidites* [FAD]		*Torrowangea rosei*	Unnamed	PЄ	
				Cloudina [FAD]					

Fig. 3.3. Lower and Middle Cambrian chrono- and biostratigraphic units in the Iberian Peninsula with the most relevant events and correlation with horizons and assemblages proposed by the International Subcommission on Cambrian Stratigraphy (ISCS) (Geyer & Shergold 2000). Key: FAD, first appearance datum; SSF, small shelly fossils. Genera: *B.*, *Badulesia*; *P.*, *Pardailhania*; *S.*, *Solenopleuropsis*.

followed by the FAD of *Rusophycus* and *Cruziana* (Crimes *et al.* 1977).

In northeastern Spain (Iberian Ranges), the Late Neoproterozoic Paracuellos Gp contains the FAD of *Torrowangea rosei* in the Saviñán Fm. The unconformably overlying

lowermost Cambrian rocks of the Bámbola and Embid formations have yielded suites of arthropod trace fossils: *M. lineatus*, *Rusophycus* and *Cruziana* (lower and upper Corduban).

In central Spain (western Montes de Toledo), the first

records of both *M. lineatus* and *P. pedum* are in the lower part of the Pusa shale, which overlies a Late Neoproterozoic succession with trace fossils including *T. rosei*. Both units are separated by the Fuentes olistostrome containing *Cloudina*-bearing limestone boulders. The first appearence of the *M. lineatus–P. pedum* assemblage is overlain by the FAD of *Rusophycus* within the Azorejo Fm.

In southwestern Spain (Ossa-Morena Zone), the Lower Cambrian rocks of the Torreárboles Fm also contain trace fossils including *M. lineatus* and *P. pedum*, while *Rusophycus* appears later. This unit unconformably overlies the Neoproterozoic rocks of the 'volcanic–sedimentary complex' where *T. rosei* has not been found, whereas *S. variabilis* does occur (Liñán & Palacios 1983).

Stratigraphy

The Lower Cambrian rocks of the Iberian Peninsula are represented by a thick sequence (up to more than 2000 m) of siliciclastic and carbonate materials, mostly deposited under shallow marine, generally transgressive conditions, although interspersed with several relatively minor regressive events. Middle Cambrian rocks are represented by a continuous sequence of carbonate and siliciclastic lithologies 300 to 1000 m thick, deposited under marine, sublittoral conditions in a general transgressive trend which reverses to a regressive trend at the end of Middle Cambrian times. The Upper Cambrian succession is represented by 150 to 600 m (maximum thickness is in the Iberian Ranges) of siliciclastic materials deposited under shallow marine, regressive conditions.

The detailed biostratigraphic aspects of the ten major outcrops with Cambrian rocks are analysed below, from north to south, and their correlation with the regional stages is shown in Figure 3.2.

Pyrenees and Catalonian Coastal Ranges. The Cambrian rocks here are strongly deformed and generally metamorphosed. Fossils are very rare, and the age of this sequence is based on stratigraphic position and lithological correlation. The general stratigraphy of Cambrian rocks in the axial Pyrenees was studied by Cavet (1957) and Laumonier (1988, 1996), but the strong deformation overprint has been a major impediment to elucidating the sequence in detail. The basement consists of a gneissic complex and schists that are overlain by the Canaveilles Gp, a 2000 m thick volcanosedimentary sequence with lenses of limestone assigned to the Precambrian and Cambrian on lithological grounds. The Jujols Gp consists of about 2000 m of shale, sandstone and quartzite, with lenses of limestone, and is considered Middle Cambrian to Early Ordovician in age. The Jujols Gp is unconformably overlain by Caradocian conglomerates.

Abad (1989) discovered a small outcrop of Cambrian rocks with archaeocyathans and other fossils in the Cadí thrust (southern Pyrenees), which is surrounded by Tertiary rocks. The sequence consists mainly of siliciclastic rocks interlayered with bioclastic limestones or reef mounds (Perejón *et al.* 1994). The archaeocyathans are considered typical of Zone VII of the upper Ovetian (Perejón 1986, 1994).

The pre-Silurian rocks of the Catalonian Coastal Ranges are almost unfossiliferous. Consequently, the older metasediments corresponding to siliciclastic rocks with some carbonate have often been assigned a Cambrian–Ordovician age. The lower sequence of the Les Guilleries massif has been compared with those in the Pyrenees (Julivert & Durán 1992). Recently, Sanz-López *et al.* (2000) found a small sequence, the Picamoixons unit, with ichnofossils considered as upper Corduban (Lower Cambrian).

Iberian Ranges (Cadenas Ibéricas). The Cambrian succession of the Cadenas Ibéricas (Iberian Ranges) is known through Lotze's (1929) work and was selected by Sdzuy (1971*a,b*) as the reference section for the Spanish Lower and Middle Cambrian sequences because of their trilobite record. Since Lotze's (1929) study, lithostratigraphic nomenclature for the Precambrian and Cambrian rocks has been modified (Lotze 1958, 1961; Sdzuy 1971*a*; Schmitz 1971; Álvaro 1995), and was summarized by Gozalo (1995).

The Bámbola Fm (Lotze 1929), unconformable on the Paracuellos Gp, constitutes the lowermost Cambrian in the area and is composed of 300–450 m of white quartzarenites and conglomerates with thin levels of shales interbedded in the upper part. Ichnofossils in the shales include *Monomorphichnus lineatus*, *Phycodes* and *Rusophycus* (Liñán *et al.* 1996) which suggest a Corduban age.

The Embid Fm (Lotze 1929) consists of 200–400 m of alternating brownish and greenish sandstones and shales. It yields abundant trace fossils including (in its lower part) the first record of *Cruziana*, as well as *Astropolichnus hispanicus*. An upper Corduban–upper Ovetian age is inferred, and the unit was deposited in sublittoral and littoral marine environments (Schmidt-Thomé 1973; Álvaro *et al.* 1993*a*; Sdzuy & Liñán 1993; Pillola *et al.* 1994).

The Jalón Fm (Lotze 1929) comprises 200–450 m of alternating reddish, purple and green shales and sandstones with interbedded decimetric levels of dolostones. Dolerolenid trilobites from transitional beds between the Embid and Jalón formations indicate an upper Ovetian age (Sdzuy 1987). Ichnofossils are frequent in the shales and stromatolites are common in the carbonate beds. The palaeoecological data suggest littoral to sublittoral conditions that quickly evolved to a shallow littoral environment (Schmidt-Thomé 1973; Gámez *et al.* 1991; Sdzuy & Liñán 1993). The Jalón Fm marks the beginning of Cambrian carbonate deposition in the Cadenas Ibéricas (Iberian Ranges).

The Ribota Fm (Lotze 1929) is a 25 to 130 m thick succession composed of yellow and grey dolostones, minor limestones and interbedded shales and marls containing trilobites, brachiopods, echinoderms, hyolithids and trace fossils. Gámez *et al.* (1991) cited oncolites, cryptalgal and stromatolitic laminations and gypsum pseudomorphs in dolostones of the lower part of this formation. Two successive trilobite assemblages are recorded in the shales: the lower is characterized by *Lusatiops* and *Strenuaeva*, and the upper contains *Kingaspis*, *Redlichia* and *Strenuaeva* species (Sdzuy 1971*a*). Both are typical of the Marianian stage. Acritarchs first occur in the lower horizons, with *Strenuaeva,* and continue to the top of the formation (Gámez *et al.* 1991; Liñán *et al.* 1996). Palacios & Moczydłowska (1998) cited 19 acritarch species characteristic of the *Heliosphaeridium dissimilare-Skiagia ciliosa* Zone. The formation was deposited under oscillating littoral to shallow sublittoral conditions (Schmidt-Thomé 1973; Álvaro *et al.* 1995).

The Huérmeda Fm (Lotze 1929), 45 to 80 m of green to dark blue shales with yellow dolostones, was deposited under sublittoral conditions (Álvaro *et al.* 1995). The unit contains scattered trilobites, brachiopods, echinoderms, hyolithids, acritarchs and ichnofossils. Trilobites are present at both the base and the top of the formation. The lower trilobite assemblage with *Micmacca*, *Andalusiana*, *Strenuaeva*, *Kingaspis*, *Redlichia* and *Triangulaspis* (Sdzuy 1961, 1971*a*) suggests a Marianian age. The upper trilobite assemblage with *Srenuaeva* and Protolenids (Sdzuy 1971*a*), marks the disappearance of the olenellids in Spain and indicates a Bilbilian age. Records of acritarchs by Gámez *et al.* (1991) and Liñán *et al.* (1996) have been summarized by Palacios & Moczydłowska (1998), who report 13 species characteristic of the *H. dissimilare–S. ciliosa* Zone.

The Daroca Fm (Lotze 1929) is a terrigenous sequence 90–250 m thick, including markedly heterogeneous lithologies (Sdzuy & Liñán 1993; Álvaro & Vennin 1998). The southern outcrops are mainly composed of sandstones with few conglomerates. In the central outcrops, shale forms 60–75% of the components, and the sandstone decreases towards the northwestern part of the region. In some northern localities, the trilobite *Protolenus (Hupeolenus) termierelloides* in the basal and middle parts of the Daroca Fm (Álvaro & Liñán 1997) confirms their Bilbilian age. Acritarchs occur at several levels and Palacios & Moczydłowska (1998) recognized two assemblages: the first is indicative of the *Volkovia dentifera–Liepaina plana* Zone (= *Tubulosphaera perfecta–Heliosphaeridium notatum* Zone, cf. Palacios & Delgado 1999), and the second is indicative of the *Eliasum llaniscum–Celtiberium dedalinum* Zone.

The Mesones Gp, subdivided into Valdemiedes, Mansilla and

Murero formations, is essentially composed of shales with inter-bedded carbonate nodules, dolostones and limestones. These were deposited mainly in sublittoral environments (Sdzuy & Liñán 1993).

The Valdemiedes Fm is 20–250 m thick, and has yielded trilobites, brachiopods, hyolithids, sponges, algae, annelids? and ichnofossils. Two trilobite assemblages have been recognized within the unit. The first included 11 species, characteristic of the *Hamatolenus (H.) ibericus* Zone of the upper Bilbilian. The second, at the base of the Leonian, contains 11 species which indicate the *Eoparadoxides mureroensis* Zone. Both assemblages are separated by the Valdemiedes event (Liñán et al. 1993a), interpreted as responsible for the trilobite turnover at the Lower–Middle Cambrian boundary in Iberia (Sdzuy 1961, 1971a,b, 1995; Liñán & Gozalo 1986; Liñán et al. 1993a, 1996; Álvaro et al. 1993b; Gozalo et al. 1993; Sdzuy et al. 1999). This event may be the final expression of the uppermost Lower Cambrian mass extinction, followed by trilobite diversification in the early Middle Cambrian.

The Mansilla Fm, 10–90 m thick, is made up of alternating brown dolostones and limestones and purple and violet shales containing trilobites, brachiopods, echinoderms, sponges, algae, hyolithids, monoplacophorans and ichnofossils. Trilobites in the upper part include 13 species of the late Leonian *Eccaparadoxides asturianus* Zone (Gozalo & Liñán 1995; Sdzuy et al. 1999). The highest levels contain *Badulesia tenera*, a species indicative of the earliest Caesar-augustan (Sdzuy et al. 1999).

The Murero Fm is composed of 50–250 m of green lutites with carbonate nodules and interbeds of very fine sandstone. Fossils include trilobites, brachiopods, echinoderms, annelids? (palaeo-scolecid worms), algae, hyolithids, sponges and ichnofossils of Caesaraugustan age. Liñán et al. (1996, Figs 3.25 and 3.26) reviewed the succession, which has been subdivided into 11 zones, characterized by *Badulesia, Pardailhania* and *Solenopleuropsis* species (Sdzuy 1971b, 1972; Liñán & Gozalo 1986). The lower boundary is slightly diachronous but coincides approximately with the base of the Caesar-augustan; the upper boundary is diachronous but is always upper Caesaraugustan.

The Acón Gp (Schmitz 1971; *sensu* Álvaro 1995) is a 850–1500 m thick succession of white and grey sandstones and green shales with isolated carbonate nodules. It shows important lateral facies changes, several of them named as different units, e.g. Acón Schichten (Schmitz 1971) and Almunia Schichten (Josopait 1972). Álvaro (1995) subdivided the group into five units which, in ascending order, are the Borobia, Valdeorea, Torcas, Encomienda and Valtorres formations. The Acón Gp conformably overlies the regionally diachronous top of the shaly, green Murero Fm, and fossils include trilobites, brachio-pods, echinoderms, acritarchs (Gámez et al. 1991) and ichnofossils of Caesaraugustan to Upper Cambrian age. Shergold & Sdzuy (1991) summarized the biostratigraphic data and recognized six different assemblages, ranging from Languedocian to Upper Cambrian.

The Valconchán Fm (Wolf 1980b) is 30–400 m thick and consists mainly of quartzite with interbeds of mudstone, sandstone and quartzitic sandstone, locally developed quartz conglomerate lenses and claystone breccia. In the lower part of the formation, trilobite and brachiopod faunas appear which were considered Upper Cambrian by Shergold & Sdzuy (1991); at the top is another assemblage of trilo-bites, brachiopods and echinoderms which is considered either as Upper Cambrian (Shergold & Sdzuy 1991) or as transitional between Cambrian and Ordovician (Wolf 1980b; Villas et al. 1995a).

Iberian Ranges (Sierra de La Demanda). The Cambrian stratigraphy was established by Schriel (1930) and Colchen (1974), and revised by Palacios (1982) and Shergold et al. (1983). The unconformable Anguiano conglomerate (Colchen 1974), 30–300 m thick, is composed of conglomerates with quartzite pebbles and thin levels of sandstone interbedded in the upper part. The Puntón sandstone (Colchen 1974) is 500 m thick, and composed of coarse arkosic sandstones with some interbedded conglomerate. No fossils have been found, but the units are lithologically similar to the Bámbola and Embid formations of the Cadenas Ibéricas (Iberian Ranges) and their age may not be very different (Corduban–lower Ovetian).

The overlying Riocabado beds (Colchen 1974) are 100–200 m of alternating lutite, sandstone and limestone. Liñán et al. (1993c) corre-lated this unit with the Jalón to Daroca formations, and the lithology

is comparable with the Jalón Fm in the Cadenas Ibéricas (Iberian Ranges), as proposed by Lotze (1961c) and Josopait & Schmitz (1971). However, the lack of fossils makes these correlations problematic.

The overlying San Antón dolomite (former Urbión dolomite (Josopait & Schmitz 1971) or Mansilla-San Antón dolomite (Colchen 1974)) is 40–150 m thick and has been correlated with either the Ribota dolomite (lower Marianian; see Lotze 1961c) or the Ribota, Huérmeda, Daroca and Valdemiedes formations (Josopait & Schmitz 1971), or the Valdemiedes Fm (Liñán et al. 1993c) of the Cadenas Ibéricas (Iberian Ranges). The unit exhibits different lithologies (Colchen 1974; Zamarreño 1983) and lacks fossils. Colchen (1974) commented that overlying successions to the north and south of the Demanda Mountains are different, and in general their contacts with underlying or overlying units are tectonic. For the moment, therefore, the true age and correlation remain controversial.

The Valdemiedes, Mansilla and Murero formations in Sierra de La Demanda are similar to those in the Cadenas Ibéricas (Iberian Ranges) but are less thick and may be more condensed. The first trilobite is *Badulesia tenera*, which appears in the top of the calcareous red shale of the Mansilla Fm and marks the basal Caesaraugustan. The overlying Murero Fm has several levels with trilobites, brachiopods, echinoderms and hyolithids (Josopait & Schmitz 1971; Palacios 1982) which represent successive zones of the Caesaraugustan.

The Gatón shales (Colchen 1974) are 200 m of alternating grey shale and sandstone. A massively bedded sandstone occurs at the base of the formation, but succeeding sandy layers are thinly bedded. Colchen's (1974) record of *Bailiella* cf. *levyi* in levels near Barbadillo de Herreros suggests an early Languedocian (Middle Cambrian) age.

The Viniegra Sandstone Fm (Colchen 1974) includes 300 m of sandstone, in layers several metres thick, intercalated with thin shales and lenticular beds of limestone. Brachiopods, trilobites, hyolithids and ichnofossils occur, and the trilobite species *Solenopleurina demanda* (cf. Sdzuy 1961) suggests a Languedocian age. Mergl & Liñán (1986) described the brachiopod *Westonia urbiona* from a level higher than *S. demanda* and considered its age as lower Upper Cambrian.

The Najerilla Fm (Colchen 1974), a 700 m thick unit of alternating sandstone and decalcified shale, was subdivided into three members by Shergold et al. (1983). Fossiliferous layers occur near base and top, and brachiopods, echinoderms, ichnofossils and trilobites have been described. Shergold et al. (1983) distinguished two trilobite assem-blages: one, at the base of the formation, contains *Maladioidella colcheni* and *Langyashania felixi*, and is Late Cambrian in age; the second, at the top, with *Pagodia (Wittekindtia)*, is probably Tremadocian.

Cantabrian Zone. The stratigraphy of this region was outlined by Comte (1937) and elaborated by Lotze (1961c) and Zamarreño (1972). Conglomerate, quartzite, feldspathic sandstone, rare dolomitic levels and interbedded mudstones comprise the unit known as the Herrería Fm (900–1700 m thick), which is unconformable on the Proterozoic Narcea Gp. Three members have been proposed (Aramburu et al. 1992; Aramburu & García Ramos 1993) and six palaeontological assemblages identified. The first two begin with the record of *Phycodes pedum*, followed by *Rusophycus* and *Cruziana* species (Seilacher 1970; Crimes et al. 1977; Palacios & Vidal 1992) and other Cambrian trace fossils implying a Corduban age. The following are represented by acritarchs of the *Skiagia ornata–Fimbriaglomerella membranacea*, *Skiagia ciliosa–Heliosphaeridium dissimilare* and *Volkovia dentifera–Liepaina plana* (i.e. *Tubulosphaera perfecta–Heliosphaeridium notatum* zones, typical of the *Schmidtiellus mickwitzi*, *Holmia* and *Protolenus–Proampyx* trilobite zones in Scandinavia; Palacios & Vidal 1992; Vidal et al. 1999). Another assem-blage comprises the trilobite genera *Lunolenus* and *Agraulos*, above which is a horizon with *Astropolichnus* (Lotze 1961c) and the trilo-bites *Dolerolenus*, *Lunolenus* and *Anadoxides* (Lotze 1961c; Sdzuy 1961). All the trilobite evidence suggests an upper Ovetian age (Sdzuy 1971a; Liñán et al. 1993c), but a younger age may be suggested, based on the acritarchs (Vidal et al. 1999). The Herrería Fm has been interpreted from sedimentological and palaeontological data as repre-senting transitional fluviatile–marine environments (probably braid-plain delta) where an alternation of littoral and shallow sublittoral environments evolved to sublittoral conditions (Crimes *et al.* 1977; Truyols *et al.* 1990; Aramburu *et al.* 1992).

The Láncara Fm is made up of 150–225 m of dolomite, limestone and interbedded shale. The lower member presents a persistent dolomite level, followed by a level of grey limestones and oolitic limestones, with archaeocyathans and trilobites, which is not present in the eastern Nappes region (Zamarreño 1972, 1975). Biostratigraphic events are recorded only at the top of the second level. Archaeocyathans, probably of Bilbilian age, are mentioned in the Valdoré locality (Debrenne & Zamarreño 1970; Perejón 1994) together with upper Lower Cambrian trilobites (Sdzuy 1995). The palaeontological and sedimentological data suggest supralittoral–littoral environments in the western region and shallow sublittoral conditions in the east (Zamarreño 1972, 1975; Aramburu *et al.* 1992). The upper member of the Láncara Fm consists of widespread limestone deposits followed by nodular red limestones (griotte facies) which are not present in the eastern Nappes region. Brachiopods, echinoderms, molluscs and porifera are frequent, characterizing sublittoral environments (Zamarreño & Julivert 1968; Zamarreño 1972, 1983; Sdzuy & Liñán 1993). Diachronous boundaries, from Leonian to Caesaraugustan age (Middle Cambrian), have been inferred for this member, based on trilobite records (Sdzuy 1968, 1969; Sdzuy & Liñán 1993; Gozalo *et al.* 1993).

The Oville Fm (Comte 1937), an 80–800 m siliciclastic unit, is mainly glauconitic sandstone and green shale, with many facies changes, and shows some quartzitic beds towards the top (Truyols *et al.* 1990). The unit has been divided into three members (Aramburu *et al.* 1992; Aramburu & García-Ramos 1993) all of which have diachronous boundaries (Zamarreño 1972; Sdzuy & Liñán 1993). The lowest part contain several fossiliferous levels, with trilobites (Sdzuy 1968, 1969), brachiopods and echinoderms. Diachronism between this and the preceding formation is shown by the presence of different trilobite assemblages ranging from upper Leonian to upper Caesaraugustan in age (Zamarreño 1972; Sdzuy & Liñán 1993). Above these levels is an alternation of shales, siltstones and sandstones, where Fombella (1978) found Middle Cambrian to Tremadoc acritarchs.

The Barrios Fm (Comte 1937), 80–1020 m thick, is mainly quartzite with a few intercalated shale levels. Acritarchs indicate an age between highest Middle Cambrian and Early Ordovician (Aramburu *et al.* 1992), and ichnofossils permit recognition of the Tremadoc (Crimes & Marcos 1976). The lower part of the formation may represent a lateral facies change from the upper part of the Oville Fm.

West Asturian–Leonese Zone. Historical reviews of stratigraphic studies on this area were given by Lotze (1961c), Zamarreño (1972), Pérez-Estaún (1978), Martínez Catalán (1985) and Pérez Estaún *et al.* (1990, 1992). Conglomerate, sandstone, shale and rare dolostone of the Cándana Gp (Lotze 1957a), 300–1500 m thick, lie unconformably upon the Precambrian Narcea Gp. Three biostratigraphic assemblages may be distinguished in this unit below the oldest trilobite and archaeocyathan data (Crimes *et al.* 1977). The first assemblage includes *Skolithos*, *Diplocraterion* and other selected trace fossils, and the second contains *Rusophycus*, with both assemblages suggesting a Corduban age. The ichnogenus *Astropolichnus* marks the start of assemblage three, of supposed Ovetian age. The Cándana Gp was deposited in continental, littoral and sublittoral (infralittoral) environments.

The fossiliferous Transition beds (Martínez Catalán 1985) comprise green shale and pink sandstone with lenticular carbonate beds, oncolitic levels and archaeocyathan mounds; thickness ranges from 400 to 900 m. The first palaeontological assemblage contains *Astropolichnus* (Crimes *et al.* 1977) and the trilobites *Eoredlichia* and *?Bigotinops* sp. (Liñán & Sdzuy 1978). Archaeocyathans from the lower parts of the sequence at Concha de Artedo (Debrenne & Lotze 1963; Färber & Jaritz 1964), as well as in the lower part of the Piedrafita sequence (Zamarreño & Perejón 1976), are typical of archaeocyathan Zone IV (Perejón 1986, 1994). The succeeding assemblage includes *Anadoxides* sp. and '*Wutingaspis*' in the Concha de Artedo locality (Sdzuy in Liñán & Sdzuy 1978), *Dolerolenus* in the Hermida locality (Walter 1963, 1968), and archaeocyathans (Debrenne & Lotze 1963; Walter 1963, 1968) from the middle part of the Piedrafita sequence (Zamarreño & Perejón 1976) which are typical of Perejón's Zone V. The third assemblage, containing *Dolerolenus* sp.

in the Concha de Artedo site, corresponds to an analogous assemblage from the Cantabrian region (Sdzuy in Liñán & Sdzuy 1978), and also to archaeocyathans from the upper part of the Piedrafita locality (Perejón 1986), both being included in that author's Zone VI. An upper Ovetian age for these four palaeontological levels of the Transition beds has been proposed (Perejón 1986; Liñán *et al.* 1993a).

The overlying Vegadeo Fm (Barrois 1882) is 100–300 m thick, and characterized by limestone and dolostone with microbial mats, echinoderms, trilobite and archaeocyathan remains. This suggests an essentially Early Cambrian age for the Vegadeo Fm, although it may be Middle Cambrian in its upper part (Pérez-Estaún *et al.* 1990). Littoral, shallow sublittoral and reef facies have been suggested for both formations (Zamarreño *et al.* 1975; Zamarreño & Perejón 1976; Russo & Bechstädt 1994).

The Cabos series (Lotze 1958), 1000–4500 m thick, is made up of quartzites, sandstones and slates. The basal part is composed of marly shales interbedded with sandstone and siltstone, where several fossiliferous levels have been found with trilobites, brachiopods and echinoderms of middle Leonian to upper Caesaraugustan age (Sdzuy 1969, 1971b). The remaining part of the succession contains Upper Cambrian to Arenig trilobite traces (Marcos 1973; Baldwin 1977; Pérez-Estaún *et al.* 1990, 1992). Villas *et al.* (1995a) discovered a level with *Protambonites primigenius* in the upper part of this succession, and postulated an unspecified level in the Cambro-Ordovician.

Galician-Castilian Zone (eastern part). The Precambrian/Cambrian boundary is in the lower part of the Río Huso Gp (Vidal *et al.* 1994b) above the basal Fuentes olistostrome (which bears boulders with *Cloudina*; Fig. 3.2) and within the lower Pusa shale (San José *et al.* 1974), in the uppermost part of the Río Huso Gp (Vidal *et al.* 1994b). The Pusa shale consists of more than 1000 m of lutites with intercalations of fine-grained sandstone and some slumps. The lowest part includes greenish shale; the middle part consists of black, microlaminated shale, phosphatic beds and conglomerate; and the upper unit contains very fine sandstone and a few beds of calcareous sandstone. The lower unit of the Pusa shale has yielded abundant macroscopic algae ('Vendotaenids'), *Beltanelloides?* sp. and *Sphaerocongregus variabilis* in the Río Huso section. Trace fossils from the same strata contain the *Monomorphichnus lineatus–Phycodes pedum* assemblage followed by another 13 Cambrian ichnotaxa (Brasier *et al.* 1979; Palacios 1989) suggesting that the beginning of the Phanerozoic eonothem may be placed within this lowest unit. The middle part contains phosphatized fossils, namely *Cloudina*, anabaritids, halkieriids and sponges while the upper part provided small shelly fossils (aff. *Aldanella*), hyolithids, bigotinid trilobites, possible protoconodonts (aff. *Mongolitubulus*) and chancelloriids (Palacios *et al.* 1999b). The fossils indicate sublittoral marine conditions and a lower Corduban age for the Pusa shale.

The Azorejo Fm (San José *et al.* 1974) consists of 550–600 m of coarse sandstones with alternating lutites and sandstones at the top and is interpreted as having been deposited in littoral and shallow sublittoral environments. As *Rusophycus* and *Astropolichnus hispanicus* are respectively recorded in the lower and upper parts of the Azorejo Fm (San José *et al.* 1974; Moreno *et al.* 1976; Brasier *et al.* 1979), upper Corduban–lower Ovetian ages are suggested for the whole formation according to Liñán *et al.* (1993a).

The Los Navalucillos Fm (Zamarreño *et al.* 1976) consists of limestones and dolostones, 150 m thick, alternating with shale which is more abundant in the lower part. Archaeocyathans characteristic of zones VI–VII were found at the Los Navalucillos, Urda and La Estrella localities (references in Perejón 1986; Moreno-Eiris 1987). At Los Navalucillos, trilobites were also found (Gil Cid *et al.* 1976), including *Granolenus midi* (Liñán & Gámez-Vintaned 1993), making this assemblage similar in age, upper Ovetian, to one in the Montagne Noire.

In the Soleras Fm (Zamarreño *et al.* 1976), made up of 200 m of mudstones, the trilobites *Andalusiana* and *Serrodiscus speciosus* were found (Gil Cid 1981; Liñán *et al.* 1993c), suggesting a middle-upper Marianian age for the upper part of the formation. This provides an excellent correlation with the Spanish Ossa-Morena Zone and Iberian Ranges successions, as well as Avalonia, Morocco and Germany (Álvaro *et al.* 1998). Open sublittoral conditions are suggested for this unit.

The Los Cortijos Fm (Weggen in Lotze 1961c; Walter 1977),

50–60 m thick, is a mainly arkosic sandstone unit with some quartzite and mudstone alternations. Its trilobite fauna, which represents the first discovery of Cambrian fossils in Spain (Prado, 1855), was revised by Sdzuy (1961) who identified *Realaspis, Pseudolenus* and *Kingaspis* cf. *velata* suggesting a Bilbilian age. Gil Cid & Jago (1989) enlarged the trilobite list with cf. *Latoucheia* and *Lusatiops*.

The 'volcanic–sedimentary beds' (Martín Escorza 1976) are 0–190 m thick and rest unconformably on Lower Cambrian rocks. The unit is subdivided into two parts which may be intercalated. The first is mainly acid volcanic rocks; the second is mainly quartzites with some conglomerate levels, interpreted as fluviotidal to estuarine deposits. Their age is uncertain, but Martín Escorza (1976) and San José *et al.* (1990) proposed Late Cambrian or even basal Ordovician on the basis of regional criteria.

Galician–Castilian Zone (western part). The best section is located in the Salamanca–Tamames region. There, the Aldeatejada Fm (Díez Balda 1980, 1986) has an estimated thickness of more than 2000 m and consists of siltstones and mudstones with fine intercalations of sandstones, carbonates (limestones, olistostromic breccias, conglomerates and nodules) and calcareous mudstone together with interbedded black laminated carbonaceous to grey mudstones. Palaeontological data are scarce. Tubular fossils (Vidal *et al.* 1994b) assigned to ?*Cloudina* occur in carbonate clasts within olistostromes near the base of the formation, and other phosphatic tubular fossils are in the middle part (Rodríguez Alonso 1985). The Upper Neoproterozoic–Cambrian acritarchs, *Synsphaeridium* sp. and *Micrhystridium dissimilare* (= *Heliosphaeridium* sp. *sensu* Vidal *et al.* 1994b) were found in breccias from the middle part of the formation (Díez Balda & Fournier Viñas 1981), as well as the questionable Phanerozoic trace fossil *Treptichnus*? The location of the Precambrian/Cambrian boundary within the (probably lower Corduban) Aldeatejada Fm is suggested not only by the first record of the upper Corduban trace fossil *Rusophycus* in member I of the overlying Tamames Sandstone Fm, but also by the record of indubitable Cambrian trace fossils in the upper part of the 'schist–greywacke complex' in the Hurdes-Sierra de Gata-Ciudad Rodrigo domain. Sublittoral and slope environments have been suggested for the deposition of this formation (Valladares *et al.* 1998, 2000).

The Tamames Sandstone Fm (Rölz 1975) comprises an alternating dominantly sandstone and mudstone, 500–600 m thick succession, and represents the westernmost lateral equivalent of the Azorejo Fm. In a coarse siliciclastic lower member, the Cambrian trace fossils *Skolithos, Cruziana* and *Rusophycus*, and the trilobite *Agraulos*? (Rölz 1975; Díez Balda 1986; Liñán *et al.* 1993a) suggest a Corduban age. In the mudstones and lenticular limestone intercalations of the upper member, *Astropolichnus hispanicus*, the trilobites *Giordanella*?, *Eoredlichia, Serrania* and *Granolenus*, as well as the archaeocyathans *Coscynocyathus* and *Anthomorpha* (García de Figuerola & Martínez García 1972; Rodríguez Alonso 1985) are indicative of an Ovetian age. A sublittoral (infralittoral) environment with sporadic subaerial exposure has been proposed (Díez Balda 1986; Valladares *et al.* 2000).

The Tamames Limestone Fm (García de Figuerola & Martínez García 1972) is 120–600 m thick and consists of alternating limestone and dolostone with chert, mudstone and sandstone intercalations. Archaeocyathans from Zone VII (Perejón 1972, 1986) suggest an upper Ovetian age. Changes of facies, thickness and fossils are frequent and include deposits within supralittoral, littoral, shallow sublittoral (infralittoral) and reef environments (Corrales *et al.* 1974; Valladares *et al.* 2000).

The conformable Endrinal Fm (García de Figuerola & Martínez García 1972) is 100 m thick and comprises laminated mudstone with fine sandstone interbeds. It contains *Gigantopygus*, a characteristic Marianian trilobite (Liñán *et al.* 1993c). These sediments are interpreted as deposited in littoral (Díez Balda 1986) and sublittoral environments. No Bilbilian to Upper Cambrian rocks are known below the Ordovician Armorican quartzite in this western region.

East Lusitanian-Alcudian Zone. Indubitably Cambrian rocks have only recently been discovered in this area in the upper part of the so-called 'schist–greywacke complex' (Carrington da Costa 1950) or Alcudian (Tamain 1971a) which outcrop in the flanks of Precambrian anticlines. The best fossiliferous Cambrian outcrops occur mainly in the central region (Ibor, Valdemanco, Abenójar and Alcudia anticlines) where the upper part of the 'schist–greywacke complex' was recently named the Ibor Gp (Vidal *et al.* 1994b), which is unconformably overlain in some areas by alternations of sandstone and mudstone assigned to the Torreárboles Fm (Liñán *et al.* 1993a) or to the Azorejo Fm (Monteserín *et al.* 1987).

The Ibor Gp (Vidal *et al.* 1994b) is represented by alternating sandstone and shale, more than 4000 m thick, with interbedded conglomerate, calcareous mudstone and limestone. The main characteristic is their frequent facies changes, reflected in a profuse and local lithostratigraphic nomenclature developed since the 1960s (San José *et al.* 1990). The lower part of the sequence contains *Cloudina, Sphaerocongregus variabilis* and vendotaenids (Vidal *et al.* 1994b) in the Abenójar anticline, and also *Phycodes*? in the La Serena region (Pieren *et al.* 1991), suggesting a late Vendian age. Nevertheless, in the Ibor anticline, Cambrian trace fossils such as *Phycodes pedum* (Liñán *et al.* 1984), *Palaeophycus* and *Hormosiroidea* (García-Hidalgo 1993a), as well as *Sabedillites cambriensis* (Vidal *et al.* 1994b) above *Cloudina* limestones, suggest a lower Corduban age for the middle-upper part of the Ibor Gp in this area. In the Alcudia anticline, the lower Tommotian small shelly fossil *Anabarella* cf. *A. plana* (Vidal *et al.* 1999), and *Monomorphichnus lineatus* and *Hormosiroidea* (García-Hidalgo 1993b) confirm a similar Early Cambrian age for the upper part of the Ibor Gp, where sublittoral conditions evolved to littoral during early Corduban times (Liñán & Gámez Vintaned 1993).

An alternation of quartzite, sandstone and lutite predominates in the unconformable overlying Torreárboles Fm (Liñán 1978). The trace fossils *P. pedum, Skolithos* and *Monomorphichnus* occur at the Jaraicejo locality with Palaeozoic acritarchs (Monteserín *et al.* 1987) and a Corduban age has been suggested (Liñán *et al.* 1993c) with littoral and infralittoral conditions of deposition (Liñán *et al.* 1997).

Finally, siliciclastic materials of probably Late Cambrian–early Tremadoc age are suggested by the discovery of trace fossils (García-Hidalgo 1993c) in the Carrascalejo beds (Lotze 1961c), a sequence more than 100 m thick which lies disconformably upon the underlying formation. No upper Lower to Middle Cambrian outcrops are yet known in the East Lusitanian-Alcudian Zone.

Ossa-Morena Zone: general outline. Within this southern Portuguese and Spanish region, large areas of Cambrian rocks exhibit frequent facies and thickness changes, which explains why workers on the Spanish outcrops (in Sierra Morena) have historically produced a profuse and informal stratigraphic nomenclature which makes synthesis of the geology difficult. The Cambrian outcrops have been subdivided into several fault-bounded troughs called *cubetas* (Liñán 1984a; Liñán & Quesada 1990). Lateral changes in facies and thickness of Cambrian sequences in different *cubetas*, and the bimodal volcanism existing in the whole area, have been interpreted as expressions of a rifting phase which affected the region from near the beginning of Early Cambrian times (upper lower Corduban; Mata & Munhá 1990; Liñán & Gámez-Vintaned 1993). Some authors consider that the rifting ended during Late Cambrian times (Ribeiro *et al.* 1990), whereas others believe that it continued through to Silurian times (Mata & Munhá 1990).

From a geological point of view, the best studied *cubetas* are the Sierra de Córdoba cubeta (eastern Sierra Morena) and the Alconera cubeta (western), both located on the north flank of the Precambrian Olivenza–Monesterio anticlinorium and representing the eastern and western Cambrian sequences. This flank is characterized by a Lower Cambrian sequence beginning with the Torreárboles Fm, and an upper Lower to lower Middle Cambrian sequence beginning with the Castellar Fm, these being the only units common to all sequences (for additional information see Liñán & Quesada 1990).

The Torreárboles Fm (Liñán 1978) lies unconformably upon the 'volcano-sedimentary complex' and consists of terrigenous sediments, comprising conglomerate and arkosic sandstone with interbedded shale. Its thickness is highly variable (0–450 m) due to the infilling of residual palaeorelief formed during uplift produced by the last phases of the Cadomian orogeny (Liñán & Fernández-Carrasco 1984). This erosional unconformity records the so-called Córdoba regression (Liñán & Gámez-Vintaned 1993). The Torreárboles Fm shows three trace-fossil-rich assemblages (Liñán & Palacios 1983; Liñán 1984b; Fedonkin *et al.* 1985). The first includes *Skolithos, Planolites* and *Cochlichnus*, and the second, *Phycodes pedum* and

Monomorphichnus, both assemblages suggesting a lower Corduban age. The third assemblage is defined by the lowest record of *Rusophycus*, which indicates an upper Corduban age. Ichnological and sedimentological analyses suggest that deposition occurred initially in transitional (supralittoral and littoral) environments with fluvial influences, and then in more widespread, shallow, sublittoral environments which record the first transgressive episode at the base of the Cambrian (Liñán & Fernández-Carrasco 1984). Interbedded acid volcanic rocks occur occasionally in some parts of the western and southern areas, and represent the first expression of Cambrian rifting that continued at least until Middle Cambrian times (Liñán & Gámez-Vintaned 1993).

Ossa-Morena Zone (eastern). In the Sierra de Córdoba, the formal lithostratigraphic nomenclature was defined by Liñán (1974, 1978). The Torreárboles Fm is conformably overlain by the Pedroche Fm, which comprises 350 m of limestone and shale, with occasional sandstone and dolostone which contain archaeocyathans, trilobites (*Bigotina, Lemdadella, Serrania*), brachiopods, algae *s. l.*, stromatolites, small shelly fossils (Liñán 1978; Liñán *et al.* 1982; Fernández Remolar 1999) and trace fossils. Archaeocyathan localities are numerous but the most important are Las Ermitas and Arroyo de Pedroche (Hernández Pacheco 1907; Perejón 1975*a*,*b*, 1976, 1989; Zamarreño & Debrenne 1977). The sedimentological and ecological characteristics of the Pedroche Fm suggest a sublittoral (infralittoral) environment with intermittent development of biohermal structures during early Ovetian times.

The Santo Domingo Fm, 200 m thick, overlies the Pedroche Fm and is composed of red shale, dolostone and chert-bearing limestone with stromatolites, algae and bioclastic brachiopods. No biostratigraphically significant fossils have been reported. The facies and sedimentological characteristics suggest deposition in supralittoral to restricted infralittoral environments (García Hernández & Liñán 1983). The age of the Santo Domingo Fm is interpreted as late Ovetian to early Bilbilian (Liñán *et al.* 1993*a*).

The Castellar Fm (sensu Liñán *et al.* 1995) is 75–84 m thick, lies conformably above the Santo Domingo Fm, and is composed of sandstone and conglomerate mainly deposited under littoral to shallow sublittoral conditions; it has not yet yielded biochronological fossils. The Los Villares Fm (*sensu* Liñán *et al.* 1995) comprises more than 450 m of sandstone and interbedded siltstones. The presence of Middle Cambrian trilobites at the base suggests an age extending from Leonian (*Eccaparadoxides sdzuyi* Zone) through Caesaraugustan (*Badulesia granieri* Zone) (Liñán 1978; Liñán *et al.* 1995).

Ossa-Morena Zone (western). The formal stratigraphic names of Cambrian units in the Alconera cubeta were proposed by Liñán & Perejón (1981), mainly based on the terminology of Vegas (1971). The Alconera Fm lies conformably upon the Torreárboles Fm, and consists of both calcareous and terrigenous sediments that are 900 m thick at the type section. The formation has been subdivided into a lower Sierra Gorda Member (Mb) and an upper La Hoya Mb. The former consists of 500 m of carbonates with interbedded fine-grained clastic terrigenous sediments deposited on a shallow marine platform (sublittoral). The limestones frequently show calcimicrobes. Structures of algal origin include laminations, stromatactis, stromatolites and thrombolites, which occur abundantly throughout the whole member: planar algal laminations and stromatolites predominate in the lower part, whereas stromatactis and thrombolites, occasionally associated with skeleton-bearing metazoans, characterize the upper part (Moreno-Eiris 1987). A transition towards littoral conditions in the upper part of the member is shown by limestones with chert, interpreted as the expression of the Cerro del Hierro regression (Liñán & Gámez-Vintaned 1993). In the basal Sierra Gorda Mb, the trilobite *Serrania* suggests an Ovetian age for most of the unit. An early Marianian age for the upper part of the member is demonstrated by archaeocyathids characteristic of Perejón's Zone VIII (Perejón 1986, 1994).

The lower part of the La Hoya Mb consists of shales with calcareous nodules, nodular calcilutite and limestone, deposited on a carbonate platform containing numerous superimposed reef mounds with calcimicrobes and cups of archaeocyathids (Perejón 1973; Moreno-Eiris 1987). In the upper part, the mainly carbonate succession passes gradually into one dominated by shales and sandstones about 400 m

thick. The presence of miomeroid trilobites such as *Delgadella* and *Serrodiscus* in the upper part of the La Hoya Mb indicates open marine conditions, thus reflecting the advance of the Cambrian transgression. Fossils in this member include calcimicrobes, hyolithids, sponge spicules and brachiopods. Eighteen archaeocyathid genera (Perejón 1973, 1975*a*, 1976, 1994) and nine trilobite genera are present, all of which suggest an early to late Marianian age (Liñán & Perejón 1981).

The La Lapa Fm consists of a coarsening-upward terrigenous sequence, 1100 m thick with scattered calcareous nodules near the base. It is subdivided into the Las Vegas and Vallehondo members. The Las Vegas Mb consists of 350 m of interbedded fine-grained sandstones and shales, with carbonate nodules in its lower part; fossils are scarce, being restricted to echinoderm skeletal fragments and ichnofossils. The Vallehondo Mb comprises 750 m of alternating bioturbated sandstones and shales, with the grain size of the sandstones increasing from medium to coarse towards the top of the unit, where intercalations of volcanic rocks occur. Acritarchs from this unit indicate the *Tubulosphaera perfecta–Heliosphaeridium notatum* Zone, upper Lower Cambrian (Palacios 1993; Palacios & Delgado 1999). This terrigenous unit records the beginning of a regressive episode which culminated in the deposition of the Castellar Fm and is correlated with the Daroca regression (Liñán & Gámez-Vintaned 1993).

The Castellar Fm, *sensu* Liñán *et al.* (1995), formerly the Castellar Mb (Liñán & Perejón 1981), consists of 30 m of coarse-grained sandstones and conglomerates in Alconera. The unit and its lateral equivalents crop out from the Sierra de Córdoba in the east to Vila Boim (Portugal) in the west, and represent the last expression of the Daroca regression in the Ossa-Morena Zone. An upper Bilbilian age is inferred on stratigraphic criteria.

The Playón beds are composed of shales and fine-grained sandstones with interbedded acidic and basic igneous rocks. Both types of volcanics have an alkaline character and include pyroclastics (acidic and basic ignimbrites and breccias) and lavas (rhyolites and basalts) attributed to clastic subaerial emissions and shallow submarine flows, respectively. This unit yielded several acritarch assemblages of Middle to Upper Cambrian age (Palacios 1993; Liñán *et al.* 1993*b*), as well as brachiopods and trilobites indicating the Leonian to Languedocian stages (Gozalo *et al.* 1994).

Regional stages and international correlation

Using selected trilobite assemblages as chronomarkers, Sdzuy (1971*a*) defined three regional stages (in ascending order: Ovetian, Marianian and Bilbilian) for the Lower Cambrian of the Iberian Peninsula, and Sdzuy (1971*b*) proposed three informal Middle Cambrian stages (*Acadoparadoxides*, Solenopleuropsidae and Solenopleuropsidae-free stages). For strata with Cambrian trace fossils that are overlain by rocks containing the first Ovetian fossil assemblages, Liñán (1984*b*) proposed the Corduban stage. These four regional stages were revised by Liñán *et al.* (1993*a*) using new data from trilobites, archaeocyatha and trace fossils. In the same paper, the Middle Cambrian stages were elaborated, the two first being formally defined as the Leonian and Caesaraugustan. Finally Álvaro & Vizcaïno (1998) named the highest Middle Cambrian stage the Languedocian. Recent works on acritarch biochronology permit a better definition of the stages (Gámez *et al.* 1991; Palacios & Vidal 1992; Palacios 1993; Palacios & Moczydłowska 1998) which (Fig. 3.3) have been applied to Cambrian successions in Germany (Sdzuy 1971*b*; Elicki 1997), France (Álvaro & Vizcaïno, 1998; Álvaro *et al.* 1998), Portugal (Liñán *et al.* 1997), Italy (Pillola *et al.* 1995; Perejón *et al.* 2000), Morocco (Sdzuy 1995; Sdzuy *et al.* 1999), Turkey (Dean & Monod 1997), Jordan and Israel (Rushton & Powell 1998). The rocks belonging to each of these stages, and the Upper Cambrian series, are now described in more detail.

Corduban. The Corduban stage is characterized by siliciclastic facies with abundant trace fossils. The most relevant ichnoevents are the successive first appearances of *Monomorphichnus lineatus*, *Phycodes pedum*, *Treptichnus*, *Dimorphichnus*, *Diplichnites*, *Psammmichnites*, *Rusophycus* and *Cruziana*. The lower boundary of the Corduban is defined by the first appearance of *Monomorphichnus* (Liñán *et al.* 1984, 1993*a*), which almost coincides with the first record in central Spain (Gámez-Vintaned & Liñán 1996) of *P. pedum*, index fossil for the Precambrian/Cambrian boundary at the reference stratotype in eastern Newfoundland. The appearance of the *M. lineatus–P. pedum* assemblage has been proposed as a marker for the Precambrian/Cambrian boundary in Spain and therefore corresponds to the beginning of the Corduban stage. This ichnofossil assemblage may be correlated with many other Precambrian/Cambrian sequences around the world. Upper Corduban strata contain the first records of *Rusophycus* and *Cruziana*. The Corduban stage has been correlated with the Nemakit-Daldyn and Tommotian regional stage, of Siberia (Perejón 1986).

Recently, Vidal *et al.* (1999) and Palacios *et al.* (1999*b*) have reported a number of assemblages containing small shelly fossils (*Cloudina*, *Anabarella* sp., aff. *Aldanella*, hyolithids, circothecids, orthothecids, aff. *Mongolitubulus* and chancelloriids), trace fossils and bigotinid trilobites, and interpreted these fauna as Nemakit–Daldynian to middle Tommotian in age. According to the regional chronostratigraphic framework, the above Cambrian assemblages are regarded as Corduban, including the beds containing trilobites.

Ovetian. The Ovetian stage is characterized by mixed facies, chiefly terrigenous in northern Iberia and mainly carbonate in the south. It contains the characteristic trilobite genera *Bigotina*, *Lemdadella*, *Serrania*, *Eoredlichia*, *Thoralaspis* and *Granolenus*. Its lower boundary is defined by archaeocyathids of Perejón's (1986, 1994) Zone I, and the same fossils permit a tentative correlation of the Ovetian with the Atdabanian of Siberia (Perejón 1986). On the other hand, trilobites show a good correlation between the Ovetian and the Issendalian Stage of Morocco (Geyer & Landing 1995).

Liñán *et al.* (1993*a*) considered the Herrería Fm of the Los Barrios de Luna section as the Ovetian stratotype in the Cantabrian mountains, containing dolerolenid trilobites in member III. But Palacios & Vidal (1992) and Vidal *et al.* (1999) found acritarchs of the *Skiagia ornata–Fimbriaglomerella membranacea* Zone and the *Skiagia ciliosa–Heliosphaeridium dissimilare* Zone in member I, and the *Volkovia dentifera–Liepaina plana* Zone at the top of member II, both from the Herrería Fm; these correspond to the *Schmidtiellus mickwitzi*, *Holmia* and *Protolenus–Proampyx* trilobite zones of Baltica. Only the *S. ornata–F. membranacea* Zone is correlated with Ovetian material elsewhere in Spain (Fig. 3.3). According to acritarch data, part of member I of the Herrería Fm may be Marianian in age and the top of member II may be Bilbilian, marking an important disagreement with data from trilobites *sensu* Sdzuy (1961, 1971*a*) and Liñán *et al.* (1993*a*) (Fig. 3.2).

In summary, the Ovetian is widely represented in Iberia where many well exposed sequences in north, south and central Spain are well characterized by archaeocyathid, ichnofossil (*Astropolichnus hispanicus*) and trilobite assemblages. Nevertheless, the Ovetian lower boundary is established, at present, only by means of archaeocyathids.

Marianian. The Marianian stage comprises mixed carbonate and terrigenous facies in Iberia. Its lower boundary is marked by Perejón's archaeocyathan Zone VIII and by the appearance of the trilobite *Strenuella*. It is characterized by the trilobite genera *Delgadella*, *Serrodiscus*, *Perrector*, *Eops*, *Rinconia*, *Alanisia*, *Atops*, *Hicksia*, *Termierella*, *Lusatiops*, *Triangulaspis*, *Andalusiana*, *Saukianda* and *Gigantopygus*, which permit a good correlation with the Banian Stage of the Atlas (*sensu* Geyer 1990). Marianian rocks contain trilobite genera from both the Olenellid (Occidental) and Redlichiid (Oriental) realms, as well as the cosmopolitan miomeroid genera *Delgadella*, *Calodiscus* and *Serrodiscus*, some of which are present in the Lower Cambrian successions of southeastern Newfoundland, Siberia, Sardinia and Germany (Sdzuy 1962; Geyer & Elicki 1995); they permit correlation with the Botoman Stage (Siberia) and *Callavia* Zone (Avalonia). According to Palacios & Moczydłowska (1998), Marianian rocks in the Iberian Ranges (Cadenas Ibéricas) contain

acritarchs of the *Heliosphaeridium dissimilare–Skiagia ciliosa* Zone, equivalent to the *Holmia kjerulfi* Zone in Baltica. The diversity of Marianian faunas and the presence of phytoplankton are among the most useful fossil assemblages for intercontinental correlation in the mid-Lower Cambrian.

Bilbilian. The Bilbilian stage was originally defined in mixed terrigenous and carbonate facies. Its lower boundary is marked by the appearance of the trilobites *Realaspis* and *Pseudolenus* (the so-called fauna of Los Cortijos de Malagón; Sdzuy 1961; Gil Cid & Jago 1989). Bilbilian rocks also yield the trilobites *Kingaspis*, *Alueva*, *Hamatolenus*, *Protolenus (Hupeolenus)* (the two latter characterizing the Tissafinian of Morocco *sensu* Geyer 1990), and *Onaraspis* (an Australian genus characteristic of the Ordian Stage; Öpik 1966). Toyonian archaeocyathans have been recorded from upper Bilbilian strata (Debrenne & Zamarreño 1970; Perejón 1986, 1994). Recently, Sdzuy (1995) correlated the upper Bilbilian with the *Hupeolenus* Zone of Morocco, the Toyonian of Siberia and the *Plagiura–Poliella* Zone of Laurentia.

In the Iberian Ranges (Cadenas Ibéricas), Gámez *et al.* (1991) and Palacios & Moczydłowska (1998) found several levels with acritarchs that belong the top of the *Heliosphaeridium dissimilare–Skiagia ciliosa* Zone, the *Volkovia dentifera–Liepaina plana* Zone and the basal *Eliasum llaniscum–Celtiberium dedalinum* Zone *sensu* Palacios & Moczydłowska (1998). Recently, Palacios & Delgado (1999) defined the *Tubulosphaera perfecta–Heliosphaeridium notatum* Zone as equivalent to the *V. dentifera–L. plana* Zone in the Iberian peninsula. The base of the *E. llaniscum–C. dedalinum* Zone is interpreted by Palacios & Moczydłowska (1998) and Moczydłowska (1999) as marking the Lower–Middle Cambrian boundary; this level is slightly below the FAD of *Acadoparadoxides mureroensis* (see Fig. 3.3).

Leonian. The Leonian stage was defined in a mixed carbonate and siliciclastic sequence of sublittoral facies containing the trilobites *Acadoparadoxides*, *Eoparadoxides*, *Eccaparadoxides*, *Hamatolenus (Lotzeia)*, *Protolenus (Hupeolenus)*, *Alueva*, *Ellipsocephalus*, *Conocoryphe*, *Cornucoryphe*, *Holocephalina?*, *Acadolenus*, *Asturiaspis*, *Tonkinella*, *Maccannaia*, *Dawsonia*, *Condylopyge*, *Peronopsis* and *Peronopsella*. Its lower boundary is marked by the first occurrence of *Eoparadoxides mureroensis* (Liñán *et al.* 1993*a*; Sdzuy *et al.* 1999), which is also considered as the base of the Middle Cambrian. Numerous continuous, fossiliferous sequences in the type area permit accurate correlation with the *Paradoxides oelandicus* Stage of Baltica, the *Glossopleura* and *Albertella* zones of Laurentia, the Amgan of Siberia, and the *Ornamentaspis frequens* and *Cephalopyge* zones of Morocco (Sdzuy 1971*b*, 1972, 1995; Gozalo & Liñán 1995). Recently Sdzuy *et al.* (1999) have revised the biochronology and correlation of the Leonian stage.

Caesaraugustan. The Caesaraugustan stage was also based on alternating carbonate and terrigenous facies deposited in an outer sublittoral environment. Its lower boundary is marked by the FAD of the trilobite *Badulesia tenera* (Sdzuy *et al.* 1999) and the upper boundary is located at the base of the *Solenopleuropsis thorali* + *Solenopleuropsis marginata* Zone. More than 40 trilobite species and other taxa of palaeoscolecids, graptolites, sponges, hyolithids, brachiopods, acritarchs and ichnofossils are present (Sdzuy 1961, 1968; Liñán & Gozalo, 1986; Liñán *et al.* 1995, 1996). The Caesaraugustan stage may be correlated with the *Badulesia* and *Pardailhania* zones of Morocco (Geyer & Landing 1995), and the *Paradoxides paradoxissimus* stage of Baltica (Sdzuy 1971*b*, 1972).

Languedocian. The Languedocian stage was also defined in shale and sandstone facies, its lower boundary being marked by the base of the *Solenopleuropsis thorali* Zone *sensu* Álvaro & Vizcaïno (1998). Three substages have been established in France, but cannot yet be identified in Spain. One upper Languedocian trilobitic level named by Gozalo & Liñán (1999) as the *Eccaparadoxides* group *macrocercus* permits correlation between the Iberian Ranges, Montagne Noire, Amanos Mountains and, probably, Sardinia. The assemblage is composed of *Eccaparadoxides* gr. *macrocercus*, *Chelidonocephalus* spp. and *Derikaspis* spp.; its age has been equated with the *Solenopleura brachymetopa* Biozone from Baltica.

Upper Cambrian. The Middle–Upper Cambrian boundary has not yet been recognized in the Iberian Peninsula, where Upper Cambrian successions are mainly siliciclastic and deposited under shallow marine conditions. Fossiliferous levels are scarce and, for the moment, no biostratigraphic units are defined. Upper Cambrian trilobites have been found in the La Demanda mountains and the Iberian Ranges (Shergold *et al.* 1983; Shergold & Sdzuy 1991). Acritarchs are recorded in the Cantabrian mountains (Fombella 1978), Ossa-Morena Zone (Palacios 1993) and Iberian Ranges (Palacios 1997). Figure 3.2 shows the position of assemblages near the top of the Cambrian succession which contain *Oryctoconus lobatus, Protambonites primigenius* and some trilobites; but their ages are controversial, ranging from the end of the Cambrian to the beginning of the Ordovician (see Villas *et al.* 1995a).

Sedimentary geochemistry

As in the case of Precambrian sedimentary rocks (Chapter 2) studies on the geochemistry of Cambrian sedimentary successions are relatively recent and include major and trace element and Sm-Nd isotope data in the Central Iberian Zone (CIZ) and Cantabrian Zone (Nägler *et al.* 1995; Ugidos *et al.* 1997a,b; Valladares *et al.* 1999, 2000), and whole-rock geochemical data from the Iberian Ranges (Bauluz *et al.* 2000). Chemical and isotopic results define two source compositions for Precambrian and Cambrian sedimentary shales respectively in all the zones where data are available. This suggests the same evolution of source compositions contributing to the sedimentary basins in each zone through time and suggests the same extensive, stable and recycled source area existed for all Precambrian/Cambrian sedimentary basins in the CIZ and West Asturian-Leonese Zone. Chemical results do not define a Precambrian–Cambrian boundary in the stratigraphic succession of the CIZ but record a continuous gradual variation from the oldest Neoproterozoic unit of the to the top unit of the Lower Cambrian (Valladares *et al.* 2000).

Average composition of Lower Cambrian shales from the CIZ and average shale and sandstone from the Iberian Ranges show similar major and trace element ratios (e.g. Al_2O_3/TiO_2, Rb/Zr, Th/Yb) but some differences appear when abundances and patterns of rare earth elements are compared (Fig. 3.4). Some features shown by the average shale from the CIZ, such as the almost flat pattern of heavy rare earth elements and the negative Ce anomaly (Fig. 3.4), are not evident in the average shale from the Iberian Ranges. However, these differ-

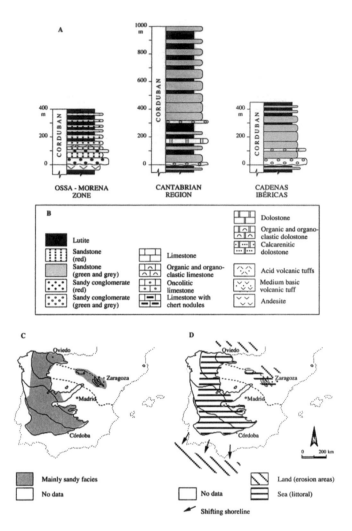

Fig. 3.5. Late early to late Corduban in Iberia. **(a)** Selected synthetic stratigraphic columns (Cadenas Ibéricas = Iberian Ranges). **(b)** Legend. **(c)** Facies distribution. **(d)** Palaeogeography (from Liñán *et al.* 1997).

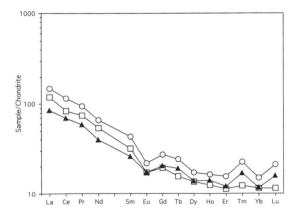

Fig. 3.4. Rare earth element patterns of average Cambrian shale (circles) and sandstone (triangles) from the Iberian Ranges, and average Cambrian shale (squares) from the Central Iberian Zone. Note the negative Ce anomaly shown by the Cambrian shale and the parallel patterns of the two other compositions. Normalizing chondrite values after Taylor & MacLennan (1985).

ences may be secondary and related to mobility of rare earth elements during weathering processes affecting source compositions as suggested for the Lower Cambrian sedimentary succession in the CIZ (Ugidos *et al.* 1999). The similar rare earth elements patterns of both the average shale and sandstone from the Iberian Ranges (Fig. 3.4) indicate the lack of a relevant sorting of heavy minerals in the sandstones relative to the shales. Thus, a recycled and homogeneous area was probably the main source supplying detrital components into the Early Cambrian basin of the Iberian Ranges. Petrological and geochemical results both from the CIZ and the Iberian Ranges indicate cratonic settings as source areas for the Cambrian sediments (Ugidos *et al.* 1997a,b; Bauluz *et al.* 2000). Therefore, and given also the geochemical affinities shown by their average shales (see above), it seems that Cambrian sedimentary successions in the CIZ and Iberian Ranges derived from the same cratonic source area. Chemical features of the Precambrian average shales from the CIZ and Iberian Ranges are also similar, but show some differences (e.g. higher TiO_2 and Zr abundances, and Ti/Nb and Cr/Th ratios; Chapter 2) to those of Early Cambrian age.

Sm-Nd isotopic results indicate relatively low $\varepsilon_{Nd}(T)$ values (around −7.5) for Lower Cambrian fine grained rocks in the Central Iberian

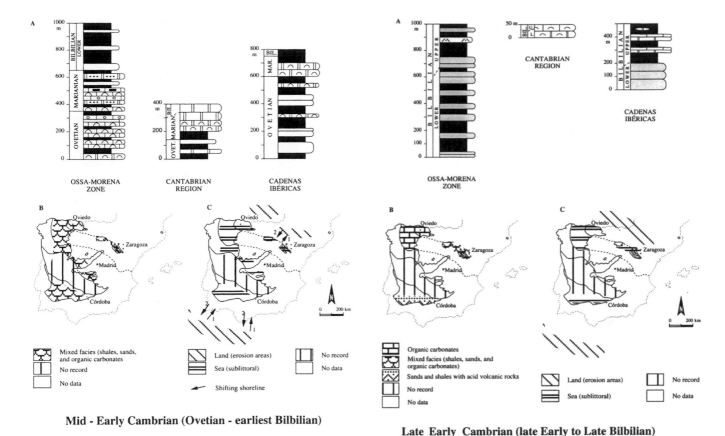

Mid - Early Cambrian (Ovetian - earliest Bilbilian)

Fig. 3.6. Ovetian to earliest Bilbilian in Iberia. (a) Selected synthetic stratigraphic columns (legend in Fig. 3.5). (b) Facies distribution. (c) Palaeogeography (from Liñán *et al.* 1997).

Late Early Cambrian (late Early to Late Bilbilian)

Fig. 3.7. Late early to late Bilbilian in Iberia. (a) Selected synthetic stratigraphic columns (legend in Fig. 3.5). (b) Facies distribution. (c) Palaeogeography (from Liñán *et al.* 1997).

Zone (Nägler *et al.* 1995; Ugidos *et al.* 1997*b*, 1999) and Cantabrian Zone (Nägler *et al.* 1995). The neodymium T_{DM} model ages range from 1.7 to 2.5 Ga in the CIZ and are around 2.0 Ga in the Cantabrian Zone. These model ages are higher than those recorded by Neoproterozoic rocks (Chapter 2) but it is unclear if all these isotopic results correspond to primary source values since there is evidence for rare earth element redistribution in some sedimentary units in the CIZ, as shown above. In spite of this uncertainty, Cambrian samples lacking evidence for rare earth element mobility (e.g. absence of Ce negative anomalies) show neodymium T_{DM} model ages around 1.7 Ga (Ugidos *et al.* 1999), which are clearly older than 1.1 Ga model ages obtained from Neoproterozoic rocks (see Chapter 2). Thus, both elemental and isotopic results strongly suggest that Cambrian sedimentary rocks from the zones considered here were derived from more felsic source areas than Precambrian ones.

Sea-level changes and palaeogeographical evolution

Cambrian sea-level changes in Iberia have been studied by Liñán (1984*a*), Liñán & Quesada (1990), Liñán *et al.* (1993*a*, 1997), Liñán & Gámez-Vintaned (1993), Álvaro *et al.* (1995), and Sdzuy *et al.* (1999). Three important regressive events (the Córdoba, Cerro del Hierro and Daroca regressions) may be recognized in Lower Cambrian strata, and another two in the Middle Cambrian succession (Fig. 3.3); the beginning and end of these sea-level changes are slightly diachronous across the Iberian Peninsula.

The Córdoba regression seems to be related to general uplift that followed the last phase of the Cadomian orogeny during the Neoproterozoic–Cambrian transition. This uplift produced hiatuses or terrigenous nearshore deposits across wide areas of the Cadomian segment of Gondwana, and more or less coincides with the Tommotian bioevent named by Brasier (1995) as CR3.

The Cerro del Hierro regression took place in late Ovetian times, but is slightly diachronous in different parts of Iberia and may reflect differential movements associated with the extensional tectonism that affected Iberia during the Cambrian period (Liñán & Gámez-Vintaned 1993). The Cerro del Hierro regression has also been recognized in Sardinia (Pillola *et al.* 1994, 1995) and may be correlated with the 'Woodlands regression' in England (Brasier 1985, 1995). In general, this event is associated with local hiatuses, carbonate littoral deposits (sometimes chert-bearing or with karstification) and red beds (sometimes with evaporite pseudomorphs) in different areas of the Northern Gondwanan margin. Another larger scale factor that probably overprinted this tectonically induced regression may be a climatic shift towards warmer conditions, as suggested by palaeoecological and sedimentological data (Gámez *et al.* 1991; Sdzuy & Liñán 1993; Álvaro *et al.* 2000).

The upper lower Bilbilian Daroca regression was first reported by Gámez *et al.* (1991) as equivalent to the

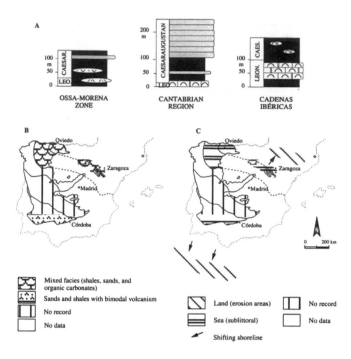

Early - mid Middle Cambrian (Leonian to Caesaraugustan)

Fig. 3.8. Leonian to Caesaraugustan in Iberia. (a) Selected synthetic stratigraphic columns (legend in Fig. 3.5). (b) Facies distribution. (c) Palaeogeography (from Liñán *et al.* 1997).

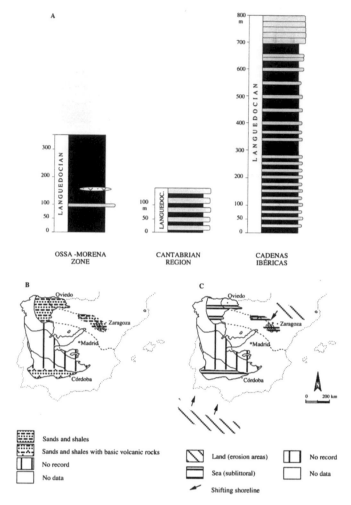

Late Middle Cambrian (Languedocian)

Fig. 3.9. Languedocian in Iberia. (a) Selected synthetic stratigraphic columns (legend in Fig. 3.5). (b) Facies distribution. (c) Palaeogeography (from Liñán *et al.* 1997).

Laurentian Hawke Bay regression (Palmer & James 1980). The Daroca regression has also been correlated with the upper Lower Cambrian regression in Baltica (Bergström & Ahlberg 1981), the end-Comleyan regression of Great Britain (Brasier 1985), the Toyonian regression of the Siberian Platform (Rowland & Gangloff 1988) and other upper Lower Cambrian regressions reported from the North China Platform, Australia and India. All these point to a global factor (probably climatic) as the cause of this geoevent (Liñán & Gámez-Vintaned 1993). Álvaro & Vennin (1998) summarized the Spanish facies produced during this regression. As pointed out by Liñán & Gámez-Vintaned (1993), all three relative sea-level drops recorded by the Lower Cambrian rocks of Iberia were relatively minor events that happened during the so-called Cambrian transgression, a long-lived global event marking the beginning of the transgressive phase of the first major Phanerozoic flooding cycle (Vail *et al.* 1991).

The Valdemiedes event (Liñán *et al.* 1993c) took place in latest Bilbilian times, thus making it younger than the widespread Daroca regression. In their reference stratotype, Liñán *et al.* (1993a) noted a change in both sedimentary mineralogy and trace fossil content, as well as the disappearance of numerous trilobite species, a dwarfing of the brachiopods, and the extinction of the Archaeocyatha (Liñán *et al.* 1993c; Sdzuy *et al.* 1999). A climatic anomaly shifting towards markedly seasonal conditions has been invoked to explain these changes. This event has been recognized in northern and southern Spain (Gozalo *et al.* 1993; Álvaro *et al.* 1993b) and in Sardinia (Pillola *et al.* 1995). The Valdemiedes event may only be recognized in areas where uppermost Lower Cambrian

sublittoral marine ecosystems were re-established after the Daroca regression or its equivalents. In areas where the Lower–Middle Cambrian transition produced coarse clastic sediments the Valdemiedes Event may be difficult to discriminate from the Daroca regression. In areas where late Early Cambrian regression produced a hiatus extending into the Middle Cambrian (as for example in Scandinavia; Ahlberg 1984), the Valdemiedes event may be included in this gap.

Trilobite palaeoecological data allow the recognition of two major Middle Cambrian regressive events. The first occurred during the Leonian stage (*Eccaparadoxides sdzuyi* Zone) and has been recognized in southern and northern Spain (Sdzuy *et al.* 1999; Liñán *et al.* 1997). From the works of Sdzuy & Liñán (1993) and Liñán *et al.* (1997), it is inferred that the second Middle Cambrian regression took place in late Caesaraugustan times (*Solenopleuropsis simula* Zone) and marked the beginning of a slow decrease in the diversity of trilobite species in Iberia.

Finally, synthetic stratigraphic columns, facies distribution

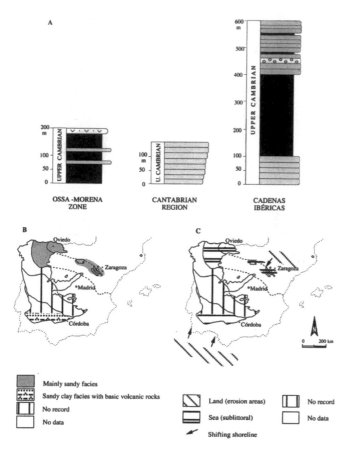

Mainly sandy facies

Sandy clay facies with basic volcanic rocks

No record

No data

Land (erosion areas)

Sea (sublittoral)

Shifting shoreline

No record

No data

Late Cambrian

Fig. 3.10. Late Cambrian in Iberia. **(a)** Selected synthetic stratigraphic columns (legend in Fig. 3.5). **(b)** Facies distribution. **(c)** Palaeogeography (from Liñán *et al.* 1997).

and palaeogeographical maps are drawn for different time spans during Cambrian times, based on maps by Liñán *et al.* (1997). Six temporal intervals have been chosen. The three Early Cambrian intervals are Corduban (Fig. 3.5), Ovetian to earliest Bilbilian (Fig. 3.6) and, finally, latest early to late Bilbilian (Fig. 3.7). This subdivision recognizes first the environment following the Córdoba regression, then gives a general view of the Early Cambrian transgression, and finally shows the distribution of ecological facies after the Daroca regression. The Middle–Late Cambrian interval is also divided into three: first, Leonian to Caesaraugustan (Fig. 3.8); second, Languedocian (after the late Caesaraugustan regressive episode; Fig. 3.9), and third, the general regression during Late Cambrian times (Fig. 3.10).

Thanks are due to W. T. Dean (National Museum of Wales, Cardiff) and the editors for their reviews of the original manuscript. We wish to thank J. Scandrett for revision of the English version. We acknowledge support from the Dirección General de Estudios Superiores (Spanish Ministerio de Ciencia y Tecnología), Projects PB96-0744, PB98-0994 and BTE2000/1145-CO2-01/02, and from the Junta de Extremadura IPR99C029 and IPR00C053.

4 Ordovician

JUAN CARLOS GUTIÉRREZ-MARCO (coordinator), MICHEL ROBARDET,
ISABEL RÁBANO, GRACIELA N. SARMIENTO, MIGUEL ÁNGEL SAN JOSÉ LANCHA,
PEDRO HERRANZ ARAÚJO & AGUSTÍN P. PIEREN PIDAL

The Iberian Peninsula comprises the most extensive outcrops of Ordovician rocks in Europe. They are mainly situated within the different 'zones' of the Variscan Iberian Massif (also referred to as the Hesperian Massif), except the South Portuguese Zone, as well as in the Palaeozoic massifs of the Iberian Cordillera (an isolated part of the Iberian Massif), the Catalonian Coastal Ranges, the Pyrenees and the Betic Cordillera (Fig. 4.1).

The first general synthesis on the Ordovician of the Iberian Peninsula was provided by Mallada (1895) and Hernández Sampelayo (1942) as part of his 'Sistema Siluriano' which also included the 'Gothlandian'. Modern stratigraphical reviews were published by Hammann (1976b), Hammann et al. (1982) and Julivert & Truyols (1983), and as different contributions included in the volumes on the Palaeozoic regional geology edited by Dallmeyer & Martínez García (1990) and Gutiérrez-Marco et al. (1992). The most detailed study was that by Hammann et al. (1982) who published a correlation chart of the Ordovician System in southwestern Europe (France, Spain and Portugal) with detailed explanatory notes on the lithology, faunal content and biostratigraphy of the different formations. Although some points must be updated, because much progress has been made during the last twenty years, the Hammann et al. (1982) review contains many details that cannot be repeated in this chapter.

Palaeogeographical setting and chronostratigraphy

Global palaeogeography

During the Ordovician Period, the Iberian Peninsula (except the South Portuguese Zone that probably belonged to Avalonia) was part of the North Gondwanan marine shelf that extended along the northern margin of the Gondwanan continent. The Iberian Ordovician consists almost entirely of terrigenous rocks (shales, siltstones and sandstones) containing low-diversity benthic assemblages of trilobites, ostracods, brachiopods, molluscs, etc., regarded as cold-water faunas indicative of high latitudes ('Mediterranean Province' of Spjeldnaes 1961 and Havlíček 1976; 'Selenopeltis Province' of Whittington & Hughes 1972; 'Calymenacean-Dalmanitacean Province' of Cocks & Fortey 1990). Limestone units are unknown, except in the upper part of the succession (Kralodvorian = approx. mid-Ashgillian).

The palaeogeographical position is also attested by lithological indicators of climate (Scotese & Barrett 1990), especially latest Ordovician glaciomarine diamictites (Robardet 1981; Robardet & Doré 1988; Brenchley et al. 1991). Although they are very few, palaeomagnetic data from Ordovician rocks of Iberia are in good agreement with a high-latitude position of the peninsula during this period (Perroud

& Bonhommet 1981; Perroud 1982, 1983; Van der Voo 1993, and references therein).

Chronostratigraphy and correlation

The general scarcity of graptolites, the almost total absence of conodonts in the Lower and Middle Ordovician, and the very distinctive shelly fossils, controlled by climate, depth and latitude, have been the source of serious difficulties with correlation between the Ordovician of the Iberian Peninsula (and more generally of all North Gondwanan regions) and the traditional Ordovician series defined in the British Isles which were frequently used, up until recent years, as the global nomenclature on regional timescales and left-hand margins of correlation charts. These difficulties, accentuated by imperfections of the regional scale based on the historical stratotypes for the British Ordovician, led several authors to propose chronostratigraphical schemes that could be used more easily and with a higher precision within the North Gondwanan regions (Havlíček & Vanek 1966; Spjeldnaes 1967; Havlíček & Marek 1973; Gutiérrez-Marco et al. 1984c, 1995; San José et al. 1992; Fig. 4.2). The range of resolution achieved by the Mediterranean scale (as for instance uppermost Dobrotivian, lower upper Oretanian, Králodvorian, etc.) is largely based on detailed biostratigraphical data from concurrent taxa of trilobites, brachiopods, echinoderms, molluscs, ostracods and graptolites (Fig. 4.3) which can hardly correlate outside this palaeogeographical area with strongly endemic faunas. Nevertheless, the North Gondwanan scheme allows direct easy internal correlations and can be regarded as of similar origin and significance to other regional scales used currently in Australasia, Baltoscandia, North America or China (Webby 1998). Although efforts have been made recently to clarify the Ordovician series and stages of the British Isles (Fortey et al. 1995, 2000; Bassett & Owens 1996), important problems still persist. The new Global Ordovician Scale (see Webby 1998; IUGS 2000), still in progress, is being constructed upon series and stage boundaries based on first appearance datum (FAD) of graptolite and conodont species that do not occur in most of the North Gondwanan regions, especially in the Iberian Peninsula. Correlation of the North Gondwanan Ordovician with the Global Ordovician Scale might thus be indirect and will necessitate use of other fossil groups, such as the chitinozoans whose biozonation (Paris 1990, 1992) should be calibrated in the global boundary stratotype sections and points (GSSP). Another independent method of correlation for the Mediterranean or North Gondwanan regional scale is provided by the sporadic record of graptolite and conodont taxa that are Baltic or Avalonian immigrants into Ibero-Armorican, Bohemian or North African regions (Gutiérrez-Marco et al. 1999b), and can

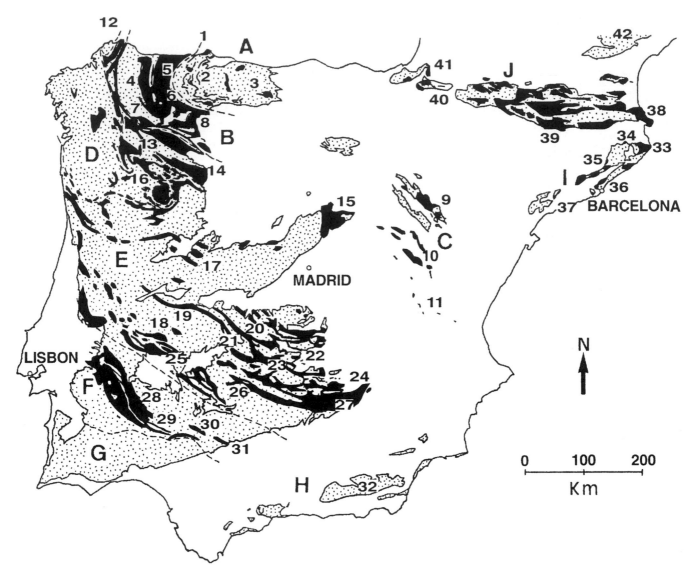

Fig. 4.1. Geological sketch map of the Iberian Peninsula showing the distribution of Ordovician rocks (in black) with reference to the main Precambrian and Palaeozoic exposures (stippled). Key: A–G, Hesperian (Iberian) Massif: A, Cantabrian Zone; B, West Asturian-Leonese Zone; C, Iberian Cordillera; D, Galicia–Trás-os-Montes Zone; E, Central Iberian Zone; F, Ossa Morena Zone; G, South Portuguese Zone (dotted lines indicate zone boundaries); H, Betic Cordilleras; I, Catalonian Coastal Ranges; J, Pyrenees. 1–42, Main Ordovician reference sections and fossil localities in Spain: 1, Cabo Peñas; 2, 'folds and nappes region'; 3, Sueve area; 4, Rececende and Villaodrid synclines (Mondoñedo Nappe); 5, Los Oscos thrust-sheet; 6, Vega de Espinareda synclinorium; 7, Caurel–Peñalba syncline; 8, Castrillo syncline; 9, Eastern Iberian Chains; 10, Albarracín anticlinorium (Western Iberian Cordillera); 11, Serranía de Cuenca anticlinorium; 12, Cabo Ortegal area; 13, Sil and Truchas synclines; 14, Alcañices synclinorium; 15, Guadarrama area (eastern 'Central System'); 16, Verín-Bragança region; 17, Tamames syncline; 18, Sierra de San Pedro and Cáceres syncline; 19, Cañaveral-Monfragüe syncline; 20, Guadarranque syncline; 21, Herrera del Duque syncline; 22, Corral de Calatrava syncline; 23, Almadén syncline; 24, Torre de Juan Abad; 25, Alange; 26, Cabeza del Buey-San Benito; 27, El Centenillo-Guadalmena; 28, Villanueva del Fresno; 29, Hinojales area; 30, Valle syncline; 31, Cerrón del Hornillo syncline; 32, Ordovician? of the Sierra Nevada; 33, Les Gavarres; 34, Les Guilleries; 35, Montseny; 36, Barcelona; 37, Serra de Miramar; 38, Eastern Pyrenees; 39, Central Pyrenees; 40, Western Pyrenees; 41, Saint-Martin-d'Arrossa (France); 42, Montagne Noire (France).

provide indirect ties for correlation of particular Spanish or Bohemian subdivisions also with reference to the Global Standard Scale. A similar procedure was adopted successfully for the Lower and Middle Cambrian chronostratigraphy of the North Gondwanan Domain, most of the regional stages defined in Spain being used already in France, Italy and Turkey (see Liñán *et al.* 1993*c*; Sdzuy *et al.* 1999; Chapter 3).

However, the Ibero-Bohemian regional scale is not completely independent from the Avalonian scheme exemplified

by the traditional British chronostratigraphy. In the earliest Ordovician, Avalonia was probably part of Gondwana, and then rifted off as a separate microcontinent and moved northwards into more temperate latitudes (Cocks 2000, 2001, and references therein). Thus the British Tremadocian and Arenigian regional chronostratigraphy can be applicable to the North Gondwanan pre-Oretanian Ordovician.

The Ordovician subdivisions recognized for the Iberian Peninsula are as follows (Fig. 4.2).

GLOBAL SERIES	GLOBAL STAGES	NORTH GONDWANA REGIONAL STAGES & SUBSTAGES		REVISED BRITISH (AVALON) REGIONAL SERIES & STAGES	
UPPER ORDOVICIAN	"Stage 6" (Ashgillian emend?)	KOSOVIAN		ASHGILL	Hirnantian
		KRALODVORIAN			Rawtheyan
	– – ? – –				Cautleyan
	– – ? – –	BEROUNIAN	Upper (Bohdalecian)	CARADOC	Pusgillian
					Streffordian
	"Stage 5" (Caradocian emend.?)		Middle (Lodenician)		Cheneyan
					Burrellian
			Lower (Chrustenician)		Aurelucian
MIDDLE ORDOVICIAN	DARRIWILIAN	DOBROTIVIAN	Upper	LLANVIRN	Llandeilian
			Lower		– – ? – –
		ORETANIAN	Upper		Abereiddian
			Lower		
	"Stage 3" (Volkhovian emend.?)	"ARENIGIAN"	Upper	ARENIG	Fennian
			Middle		Whitlandian
LOWER ORDOVICIAN	"Stage 2" (unnamed)		Lower		Moridunian
	TREMADOCIAN	TREMADOCIAN	Upper	TREMADOC	Migneintian
			Lower		Cressagian

Fig. 4.2. Standard Ordovician chronostratigraphy (Global Series and Stages: Webby 1998; IUGS 2000) and correlation with regional schemes for North Gondwana (after Havlícek 1961; Havlícek & Marek 1973; Gutiérrez-Marco *et al.* 1995) and for the British Isles (Avalonia: Fortey *et al.* 1995, 2000).

Tremadocian (global stage named after the British series; IUGS 2000).

Arenigian (British series) used here as of stage rank to avoid conflict with the former global series of the Ordovician System.

Oretanian (Mediterranean series turned into stage; San José *et al.* 1992; Gutiérrez-Marco *et al.* 1995). The Oretanian was defined in central Spain in order to obviate the modified significance of the British Llanvirn (*sensu lato*) in the 1980s and 1990s, no longer usable as the pre-Dobrotivian stage of the Mediterranean scale. The necessity of a stable conceptual framework made inadvisable the use of the Abereiddian (the present lower stage of the British Llanvirn series) which is based on a probably incomplete graptolite sequence (Maletz 1997), and has serious correlation problems for the characterization of its upper boundary, even within the British Isles. Moreover, the Abereiddian corresponds to a period of time when the faunal affinities of Gondwanan and Avalonian benthic assemblages had strongly decreased (as a consequence of the increasing separation of these two palaeogeographical units), from which serious additional problems for correlation arise. In this sense, the more recent proposal of Fortey *et al.* (2000), to correlate the base of the Dobrotivian with a level situated *within* the British Llandeilian, can give the impression that the Mediterranean scale is somewhat unsettled when, in actual fact, this proposal only reflects the uncertainty about the FAD of the graptolite *Hustedograptus teretiusculus* in the

British Isles (which defines the position of the Abereiddian/ Llandeilian boundary). The provisional FAD for the above-mentioned species (which also needs a major taxonomic revision) could vary between the lowest upper Abereiddian and the lower Llandeilian (of the British scale), even when just Eastern Avalonia and Baltica are considered (see Maletz 1997).

Dobrotivian (Mediterranean series turned into a stage; Havlícek & Marek 1973; Havlícek & Fatká 1992; Fatká *et al.* 1995; Havlícek *in* Chlupác *et al.* 1998; first usage in Spain by Gutiérrez-Marco *et al.* 1984c). This stage follows the 'pendent didymograptid period' and ends with the disappearance of *Neseuretus* in western Gondwana, that coincides precisely with the incoming of *Dalmanitina* and the *Drabovia* Fauna in the lowermost Berounian (Havlícek & Vanek 1996). The boundary between the Middle and Upper global series of the Ordovician, in the absence of any record of the key graptolite *Nemagraptus gracilis*, can be traced within the Dobrotivian by the rare occurrences in the upper third of this stage, of some graptolites (*Oepikograptus bekkeri*) and chitinozoans (*Lagenochitina ponceti* Biozone) characteristic of *N. gracilis* Biozone levels outside Gondwana (Gutiérrez-Marco *et al.* 1995; Webby 1998). This also implies that the Llanvirn–Caradoc boundary of the revised British regional chronostratigraphy would correlate with a level located within the upper Dobrotivian. Attempts at using the British series and stages in North Gondwana, for a supposedly more detailed correlation, is the source of confusing age assignations such as, for instance, the erroneous consideration as 'Aurelucian' of the Ibero-Armorican brachiopod *Apollonorthis bussacensis* (Mélou *et al.* 1999), which is an older species that occurs abundantly around the boundary between the lower and upper Dobrotivian (still Middle Ordovician).

Berounian (Mediterranean series turned into stage; Havlícek & Marek 1973; Fatká *et al.* 1995; Havlícek *in* Chlupác *et al.* 1998). This unit corresponds roughly to the Caradoc plus the basal Ashgill 'series' of the British regional scale. Some of the actual stages and substages of the British 'Caradoc' have been correlated *sensu lato* with particular beds bearing brachiopod assemblages of Anglo-Welsh influence in Spain (Villas 1985, 1992, 1995). But the permanent affinities of benthic faunas along the Upper Ordovician are expressed more clearly with Bohemia and other North Gondwanan regions (i.e. Morocco), so that the most precise correlations with the British scale do not surpass in precision those provided by the Mediterranean substages of the Bohemian Berounian (Havlícek 1961; Vanek & Vokác 1997, fig. 1).

Kralodvorian (Mediterranean series turned into stage; Havlícek & Marek 1973; Fatká *et al.* 1995; Havlícek *in* Chlupác *et al.* 1998). This unit is frequently considered as equivalent of a mid-Ashgill (pre-Hirnantian) age after the British scale. In Bohemia and SW Europe, the Kralodvorian is characterized by a distinct changeover in sedimentation, with deposition of calcareous shales and limestones, and the sudden appearance of a new shelly fauna dominated by Avalonian immigrants and cosmopolitan forms. Both phenomena were favoured by a global climate amelioration, with a decrease of latitudinal temperature gradients, and the arrival of oceanic currents coming from temperate areas of Avalonia and Baltica.

Kosovian (Mediterranean series turned into stage; Havlícek & Marek 1973; Fatká *et al.* 1995; Havlícek *in* Chlupác *et al.* 1998). This stage is coeval with the Late Ordovician glaciation on Gondwana, and correlates with the Hirnantian stage of the

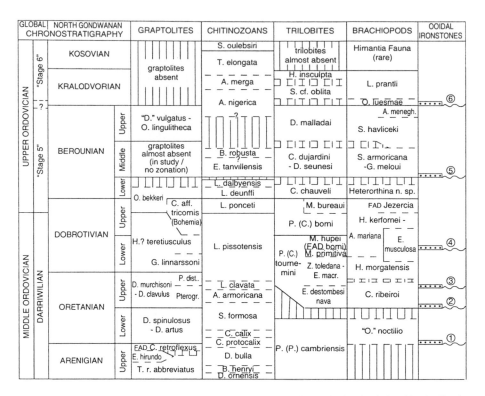

Fig. 4.3. Correlation of the main post-mid-Arenigian biostratigraphical units defined in the Iberian and SW European Ordovician, with reference to the Global Chronostratigraphy (Series and Stages) and to the North Gondwanan regional stages and substages. Biostratigraphic data compiled and adapted from Paris (1981, 1990), Gutiérrez-Marco *et al.* (1984*d*, 1995), Villas (1985), Leone *et al.* (1991) and Hammann (1992), among others, and also from unpublished work. The right-hand column indicates the position of the main ooidal ironstone beds recorded in the Hesperian Massif, where they are widely transgressive and related with discontinuities of different origin and amplitude discussed in the text (see regional successions and palaeogeography). Abbreviated palaeontological taxa are: **graptolites**: '*Diplograptus*' *vulgatus, Orthograptus lingulitheca, Oepikograptus bekkeri, Cryptograptus* aff. *tricornis, Hustedograptus*? *teretiusculus, Gymnograptus linnarssoni, Didymograptus (D.) murchisoni, D. (D.) clavulus, Pseudamplexograptus distichus, Pterograptus* spp., *Didymograptus (Jenkinsograptus) spinulosus, D. (D.) artus, Corymbograptus*? *retroflexus, Expansograptus hirundo, Tetragraptus reclinatus abbreviatus*; **chitinozoans**: *Spinachitina oulebsiri, Tanuchitina elongata, Ancyrochitina merga, Armoricochitina nigerica, Belonechitina robusta, Euconochitina tanvillensis, Lagenochitina dalbyensis, L. deunffi, L. ponceti, Linochitina pissotensis, L. clavata, Armoricochitina armoricana, Siphonochitina formosa, Cyathochitina calix, C. protocalix, Desmochitina bulla, Belonechitina henryi, Desmochitina ornensis*; **trilobites**: *Holdenia insculpta, Stenopareia* cf. *oblita, Deanaspis malladai, Crozonaspis dujardini, Deanaspis seunesi, Crozonaspis chauveli, Marrolithus bureaui, Placoparia (Coplacoparia) borni, P. (C.) tournemini, Morgatia hupei, M. primitiva, Zeliszkella toledana, Eodalmanitina macrophtalma, E. destombesi nava, Placoparia (P.) cambriensis*; **brachiopods**: *Leangella (Leptestiina) prantli, Oxoplecia luesmae, Aegiromena meneghiniana, Svobodaina havliceki, S. armoricana, Gelidorthis meloui, Heterorthina kerfornei, Aegiromena mariana, Eorhipidomella musculosa, Heterorthina morgatensis, Cacemia ribeiroi, 'Orthis' noctilio*. Macrofossil biozones mainly represent taxon-range zones named after particular taxa; except for the *Placoparia borni* trilobite biozone equivalent to the partial extent of this taxon between the disappearance of *P. tournemini* and the last record of *P. borni*. FAD corresponds to first appearance datum of some taxa. Chitinozoan biozones (Paris 1990) correspond either to total-range biozones of their index species or to partial-range biozones between the appearance of the index species of two consecutive biozones.

British Ashgill 'series'. Over many of the North Gondwanan domains, the Kosovian was a time of relative sedimentary uniformity, with rare occurrences of the worldwide *Hirnantia* Fauna.

This long introduction to the Iberian/North Gondwanan Ordovician chronostratigraphy was necessary to explain the difficulties found when considering correlations with the traditional British/Avalon regional scale or with the new Global Ordovician Standard Scale still under construction (Fig. 4.2). The boundaries proposed for the Global Series, and even for the base of the new Darriwilian International Stage or the future stages designated provisionally as 'Volkhovian', 'Caradocian' and 'Ashgillian' by Webby (1998), are based on a suite of criteria that will never be applicable to the North Gondwanan sucessions.

Ordovician successions, lithofacies and biostratigraphy

Central Iberian Zone

Despite local peculiarities, the post-Cambrian successions that occur in the different areas of the Central Iberian Zone show many similarities and indicate that all these areas had globally the same sedimentary evolution, especially during the Ordovician period. The subdivision into a southern 'East Lusitanian-Alcudian Zone' and a northern 'Galician-Castilian Zone' (Lotze 1945) corresponded mainly to differences in the subsequent Variscan tectonic style, grade of metamorphism and magmatism between a 'Domain of Vertical Folds' in the south and a 'Domain of Recumbent Folds' in the north (Díez Balda *et al.* 1990) whose boundary remains controversial.

In most regions of the Central Iberian Zone, the Ordovician

system appears as a tripartite megasequence that comprises, where it is completely preserved (Fig. 4.4):

(1) a lower transgressive part, mainly composed of sandstones, with the ubiquitous presence of the Armorican Quartzite;

(2) a middle part, composed predominantly of dark shales and siltstones with more or less important sandstone intercalations (the so-called 'Tristani' Beds' or 'Neseuretus Shales and Sandstones'); and

(3) an upper sequence (with internal unconformities) of alternating shales, siltstones and sandstones overlain by limestones and glaciomarine diamictites.

'Lower' Ordovician. The Ordovician megasequence disconformably overlies Cambrian or Upper Proterozoic rocks. The basal discontinuity

(Toledanian and/or Iberian discontinuity of Lotze 1956b) has been correlated erroneously with the Sardic Unconformity of Sardinia which is in fact intra-Ordovician (see below, Cambrian–Ordovician boundary).

The most famous and extensive lithological unit is the Armorican Quartzite, that occurs also in other regions of the Iberian Peninsula and in the Armorican Massif of western France. The Armorican Quartzite consists mainly of light-coloured thick-bedded mature sandstones and quartzites, with some shaly or silty intercalations. The thickness varies from a few metres to several hundred metres, but is generally 150–300 m. In many areas this formation overlies directly and unconformably Upper Proterozoic and/or Cambrian rocks and appears as a transgressive unit associated with the Ordovician transgression. In some areas it is underlain by commonly red coarse sediments with conglomerates, considered as precursors of the Armorican Quartzite sensu stricto. Rhyolitic to rhyodacitic volcanic rocks can occur below these sediments, and/or the Armorican

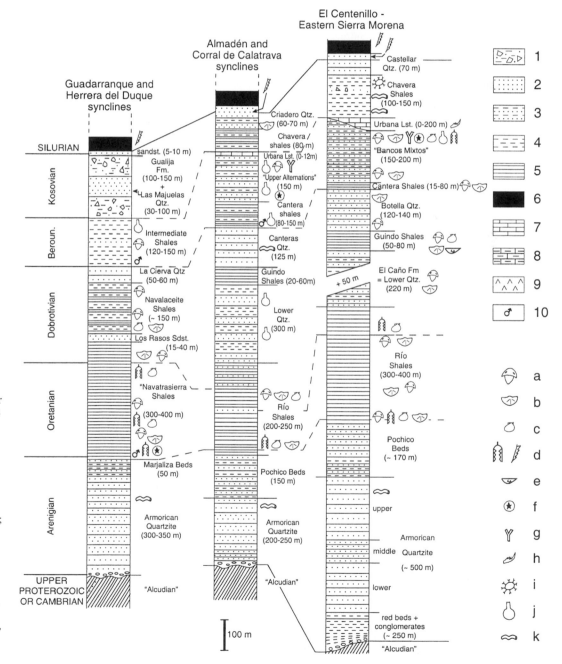

Fig. 4.4. The Ordovician successions in the southern Central Iberian Zone (after various authors, see text). Key: for all stratigraphical columns (Figs 4.4 to 4.9): Lithofacies: 1, glaciomarine diamictites; 2, sandstones and quartzites; 3, alternating sandstones, siltstones and shales; 4, siltstones and shales; 5, dark shales; 6, black shales; 7, limestones; 8, marls and calcareous shales; 9, volcanic rocks; 10, ironstones. Fossils: a, trilobites; b, brachiopods; c, bivalves; d, Ordovician (left) and Silurian (right) graptolites; e, ostracods; f, echinoderms; g, bryozoans; h, conodonts; i, acritarchs; j, chitinozoans; k, ichnofossils.

Quartzite itself. These pre-Armorican Quartzite rocks range from deformed crystal-tuff (ortho)gneissified porphyroids ('Ollo de Sapo' and related facies of the northernmost area) to volcanosedimentary units in the eastern Montes de Toledo, and even to arkose sandstones and conglomerates in the southernmost Central Iberian areas. Some of these rocks have been recently dated as 480 ± 2 Ma (Valverde Vaquero & Dunning 2000, and references therein), confirming their previously supposed Early Ordovician age (Hammann et al. 1982).

The sedimentological characteristics of the Armorican Quartzite indicate deposition in shallow water nearshore environments, under tidal, shore current and storm influence. The age cannot be determined very precisely in the Iberian Peninsula where the fossil record consists only of linguliform brachiopods (Emig & Gutiérrez-Marco 1997), rare bivalves (Gutiérrez-Marco et al. 1997b; Babin & Hammann 2001) and ichnofossils (*Skolithos, Cruziana, Daedalus, Didymaulichnus, Arthrophycus*, etc.; Moreno et al. 1976; Pickerill et al. 1984; Romano 1991) indicating an Arenigian (*sensu lato*) age, without more precision. However, in the Armorican Massif, the underlying red beds contain volcaniclastic levels dated 465 Ma (U-Pb zircon dating: Bonjour et al. 1988; Bonjour & Odin 1989; Robardet et al. 1994a), and shaly intercalations within the Armorican sandstone have yielded chitinozoans of the *Eremochitina brevis* Biozone which correlates with the early middle Arenigian and/or the upper part of the lower Arenigian (Paris 1981, 1990, 1992). The Armorican Quartzite is followed by 50 to 150 m of alternating quartzites and shales (e.g. Pochico Beds) that grade upwards into the dark shales of the middle part of the Ordovician sequence.

'Middle' Ordovician. In the middle part of the Ordovician sequence, dark shales, sometimes with nodules and a single ooidal ironstone bed of basal Oretanian age (no. 1, Fig. 4.3), predominate and correspond to the so-called '*Tristani* Beds' or '*Neseuretus* Shales and Sandstones'. The thickness varies between 150 and 1000 m, being generally about 300 m. These units are richly fossiliferous, with more than 250 species of trilobites, brachiopods, various molluscs, ostracods, echinoderms and graptolites (see Hammann 1974, 1983; Gutiérrez-Marco 1986a; Gutiérrez-Marco et al. 1984a; Rábano 1989a,b,c,d,e; Babin & Gutiérrez-Marco 1991, 1992, and references therein). The graptolites and abundant shelly fossils, especially trilobites and brachiopods, allow definition of distinct biozones from the basal Oretanian up to the lower Dobrotivian (see Gutiérrez-Marco et al. 1984c, 1995; San José et al. 1992).

The Oretanian stage corresponds to the prolongation of earlier transgressive conditions, whereas a regressive tendency appears in the Dobrotivian. The latter succession contains sandstone units, which are well developed in the southeastern Central Iberian Zone within and above the '*Neseuretus* Shales', and exemplified by the so-called 'Los Rasos Sandstones' or 'Quartzites inférieurs' and the 'Botella Quartzites' (Tamain 1967; Carré et al. 1970; Brenchley et al. 1986; San José et al. 1992).

'Upper' Ordovician. The upper part of the Ordovician tripartite sequence is much less uniform. In addition, it is incomplete in many areas, due to a small lower Berounian hiatus and also to non-deposition and/or erosion during the global lowering of the sea level induced by the late Ordovician glaciation.

The upper part of the Ordovician sequence generally begins with argillaceous units called 'Cantera Shales', 'Onnia Shales' or 'Intermediate Argilites' (Saupé 1971b; Tamain 1971b). At the base of these formations, or within their lower third, a microconglomeratic and ferruginous horizon (no. 4, Fig. 4.3) with phosphatic pebbles marks a minor stratigraphical hiatus within the lower part of the Berounian. These formations are overlain by alternating shales, siltstones and sandstones, as the so-called 'Bancos Mixtos' of the southeastern Central Iberian Zone, with abundant shelly faunas and rare graptolites (Hammann 1976a; Gutiérrez-Marco & Rábano 1987a). Then comes the Urbana limestone which contains conodonts of the Ashgillian *Amorphognathus ordovicicus* Biozone (Hafenrichter 1979, 1980; Sarmiento 1990, 1993) and rare identifiable echinoderms (Gutiérrez-Marco 2001).

The uppermost part of the Ordovician sequence is characterized by the occurrence of glaciomarine diamictite formations, e.g. Chavera, 'pelitas con fragmentos' and Gualija formations (Robardet et al. 1980; Robardet 1981; Rodríguez Núñez et al. 1989; García Palacios et al.

1996) that are also known in most areas of the Iberian Peninsula as well as in the other regions of the North Gondwanan shelf. In the Guadarranque syncline, a black shale intercalation in the middle quartzitic member of the Gualija Formation (Fm) (Las Majuelas quartzite) has yielded rare homalozoan echinoderms (Lefèbvre & Gutiérrez-Marco 2002). The glaciomarine deposits are overlain by quartzitic units (Criadero quartzite and equivalents) that were regarded generally as the base of the Silurian ('Llandovery' or 'Valentian quartzites' of the literature). In the Almadén syncline, the lower part of the Criadero quartzite has yielded a brachiopod assemblage typical of the latest Ordovician *Hirnantia* Fauna (Villas et al. 1999). In the same area, the uppermost levels, transitional into the Silurian graptolitic black shales, contain poorly preserved graptolites most probably similar to the Aeronian specimens found in the same stratigraphical position at Corral de Calatrava (García Palacios et al. 1996). It is thus most probable that the Ordovician–Silurian boundary lies within these quartzitic units.

West Asturian Leonese Zone

The West Asturian-Leonese Zone (WALZ), which comprises, from east to WSW, the Navia-Alto Sil domain, the Mondoñedo Nappe and the Caurel-Peñalba domain, is limited to the east by the Narcea antiform and to the west and the south by the Vivero and Villavieja faults and the eastern part of the Truchas and Sil synclines. Although these two synclines are considered as the northernmost structural units of the Central Iberian Zone (that continue into the eastern Sierra de Guadarrama of the Sistema Central), they are included in this section because the Ordovician shows many similarities with those of the WALZ (Fig. 4.5).

'Lower' Ordovician. In the NW of the Iberian Peninsula, the Ordovician conformably succeeds Cambrian formations. The lower limit of the Ordovician lies within the upper part of the very thick Los Cabos Group (Gp) (up to 4400 m in the Navia–Alto Sil domain) that consists of alternating shallow-water marine sandstones and shales (Marcos 1973; Marcos & Pérez-Estaún 1981; Pérez-Estaún et al. 1990, 1992, and references therein). This group extends from the Middle Cambrian in its lowermost part up into the Arenigian in its uppermost part where sandstones, closely similar to those of the Armorican Quartzite, predominate. The biostratigraphical control is very limited along the entire group that bears only ichnofossils of the Skolithos and Cruziana ichnofacies (Baldwin 1977) and a single bed with orthid brachiopods towards the middle part of the group (Villas et al. 1995a).

'Middle' Ordovician. The Los Cabos Gp passes up gradually into the Luarca Fm (or Luarca Shales). The transitional beds are generally reduced to a few tens of metres, but, in the northernmost part of the Ollo de Sapo antiform, they are thicker (Rubiana Fm: 75–150 m) and contain trilobites and graptolites of Arenigian age (Pérez-Estaún 1974b, 1978; Hammann 1983). The Luarca Fm (150–1000 m) mainly consists of dark shales that sometimes include nodule-bearing levels, volcanic intercalations and a few ironstone beds (nos 1 or 4, Fig. 4.3). A comprehensive review of the fossil content and biostratigraphy of the Luarca Fm has been published recently (Gutiérrez-Marco et al. 1999a). It gives a list of previous references and a detailed survey of the fauna that comprises about a hundred taxa, with graptolites, ostracods, trilobites, molluscs, brachiopods, bryozoans, echinoderms, acritarchs and chitinozoans. The main stratigraphical conclusion of this study is that the middle Ordovician shales are restricted to the Oretanian both in the northernmost part of the Central Iberian Zone and in the West Asturian-Leonese Zone, with a probable stratigraphical gap above the Luarca Fm.

'Upper' Ordovician. In the northernmost Central Iberian Zone and in the West Asturian-Leonese Zone, the different formations that constitute the upper part of the Ordovician succession show diverse lithologies and thicknesses, depending on their structural position. This is in clear contrast with the relative uniformity of the underlying 'middle' part of the Ordovician and of the overlying Silurian succession. The scarcity of precise biostratigraphical data in the upper part of the

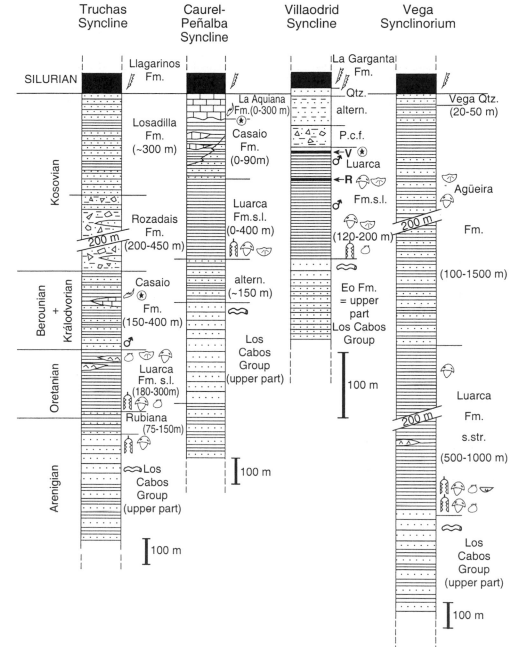

Fig. 4.5. The Ordovician successions in the northernmost Central Iberian Zone and the West Asturian-Leonese Zone. Truchas syncline after Barros (1989); Caurel-Peñalba syncline after Pérez-Estaún (1978) and Martínez Catalán *et al.* (1992*a*); Villaodrid syncline after Marcos (1973) and Gutiérrez-Marco *et al.* (1997*a*); Vega synclinorium after Marcos (1973) and Pérez-Estaún & Marcos (1981). R and V correspond respectively to Rececende and Vilargondurfe Beds. For symbols see Figure 4.4.

Ordovician sequence makes the correlations between these distinct formations difficult and uncertain.

In the Mondoñedo Nappe domain (no. 4, Fig. 4.1), it was considered that the Middle Ordovician Luarca shales were directly overlain by Silurian graptolitic black shales. However, it has been shown recently that, in the Rececende and Villaodrid synclines, the upper part of the Luarca Fm actually includes two distinct levels with Upper Ordovician faunas. The lower one, i.e. the 'Rececende beds', contains bryozoans, trilobites and brachiopods of middle Berounian age, whereas the upper one, i.e. the 'Vilargondurfe beds', has yielded echinoderm remains most probably Kralodvorian in age. Moreover these levels are overlain by alternating shales, siltstones and sandstones that include diamictite levels which can be equated with the Kosovian glaciomarine deposits and are succeeded by dark quartzites with graptolites of the basal Silurian *Parakidograptus acuminatus* Biozone (Arbizu *et al.* 1997; Gutiérrez-Marco *et al.* 1997*a*).

In the Navia–Alto Sil domain (nos 5 and 6, Fig. 4.1), the Luarca Fm

is overlain by the Agüeira Fm (Marcos 1970, 1973) that consists of a thick sequence of alternating shales, siltstones and sandstones, most of which are turbidites (Crimes *et al.* 1974; Pérez-Estaún & Marcos 1981). The thickness of the Agüeira Fm varies considerably, being 300 m in the Castrillo syncline, up to 1500 m in the Vega de Espinareda syncline (both in the Alto Sil region), and reaching its maximum (1500–3000 m) in the Navia region. The Agüeira Fm has yielded various ichnofossils and rather rare brachiopods, trilobites, bryozoans, echinoderms and gastropods from a few localities (Marcos 1973; Pérez-Estaún 1978; Pulgar *et al.* 1981, Pérez-Estaún *et al.* 1990). A recent revision of the brachiopods (M. Mélou, unpublished) has confirmed the identification of *Svobodaina havliceki* and *Aegiromena meneghiniana*, two upper Berounian species known also in the Iberian Cordillera and the southern Central Iberian Zone (Villas 1985, 1995), in Portugal (Young 1988), in Sardinia (Leone *et al.* 1991) and in the Armorican Massif (M. Mélou unpublished; Villas *et al.* 1995*b*). The Agüeira Fm is overlain by the unfossiliferous Vega de Espinareda quartzites (20–50 m) and the succeeding lower Llandovery black shales have yielded graptolites of the *Cystograptus vesiculosus* Biozone (Pérez-Estaún 1978; Gutiérrez-Marco & Robardet 1991).

In the Caurel-Peñalba domain (no. 7, Fig. 4.1), La Aquiana Fm consists of light-coloured massive limestones, frequently recrystallized. Its thickness varies greatly (0–300 m), probably due both to synsedimentary tectonics and to erosion during Late Ordovician glaciation. The La Aquiana Fm rests disconformably or para-conformably on different lithological units, generally the 'middle' Ordovician Luarca Fm, but, in some cases, the Los Cabos Gp. Locally, it rests conformably on the upper Berounian to basal Kralodvorian Casaio Fm. Due to the scarcity and poor preservation of the fossils, its age (Late Ordovician? or Early Silurian?) remained uncertain for a long time. The pelmatozoans and the conodonts found recently in the lower part of the formation clearly correlate this unit with the other Kralodvorian limestones (Sarmiento *et al.* 1999c).

In the Truchas syncline (no. 13, Fig. 4.1), the upper part of the Ordovician was first regarded as closely equivalent to the Agüeira Fm (Pérez-Estaún 1978; Pérez-Estaún & Marcos 1981; Martínez Catalán *et al.* 1992a) but more recently, it has been divided into three distinct formations (Barros 1989; Sarmiento *et al.* 1999c) which are, in ascending order, as follows. The Casaio Fm (150–400 m) comprises mostly sandstones but includes limestone levels (Trigal Limestones) with cystoids, crinoids and conodonts probably of the uppermost Berounian to Kralodvorian *Amorphognathus ordovicicus* Biozone. The overlying Rozadais Fm (400–500 m) is mainly shaly, and characterized by Kosovian diamictite levels with dispersed limestone blocks and pebbles that contain some macrofossils and conodonts typical of the *Amorphognathus ordovicicus* Biozone. Finally, the Losadilla Fm (about 300 m) comprises argillaceous slates with thin sandy intercalations and is most probably still Ordovician in age (late Kosovian), as the base of the overlying Llagarinos Fm contains basal Llandovery graptolites of the *Parakidograptus acuminatus* Biozone (Gutiérrez-Marco & Robardet 1991; Sarmiento *et al.* 1999c and references therein).

Iberian Cordillera

Towards the east, West Asturian-Leonese Zone rocks disappear below Mesozoic and Cenozoic sediments, but the Sierra de La Demanda and the Iberian Cordillera are regarded currently as the eastern prolongation of the WALZ, although the eastern part of the Iberian Ranges is sometimes considered as the prolongation of the Cantabrian Zone (Gozalo & Liñán 1988). The Sierra de La Demanda is composed mainly of Cambrian rocks and the presence of Lower Ordovician (Tremadocian) levels in the upper part of the succession (upper member of the Najerilla Fm and Brieva sandstone) is a matter of discussion (see Colchen 1974; Shergold *et al.* 1983; cf. Shergold & Sdzuy 1991). Since the first comprehensive description given by Lotze (1929) and the work by Carls (1975), the palaeontology and stratigraphy of the Ordovician of the Eastern Iberian Cordillera (Fig. 4.6; no. 9, Fig. 4.1) have been studied by numerous authors, probably because this region offers one of the most continuous and fossiliferous Lower Palaeozoic successions of the Iberian Peninsula. The Western Iberian Cordillera (nos 10 and 11, Fig. 4.1) preserves a similar, but less complete, Ordovician succession.

Eastern Iberian Cordillera. Lower Ordovician rocks conformably overlie a thick Cambrian succession and are subdivided into distinct formations that have been dated by ichnofossils, acritarchs and trilobites (Wolf 1980b, and references therein). The Cambrian–Ordovician boundary (base of the Tremadocian) was traditionally placed within the upper part of the Valconchán Fm that consists of thick-bedded quartzites with some grey to green shaly intercalations. The precise position of this limit, however, remained a matter of debate (Liñán *et al.* 1996), and the most recent data suggest that this boundary more probably lies within the upper-middle part of the Borrachón Fm. The

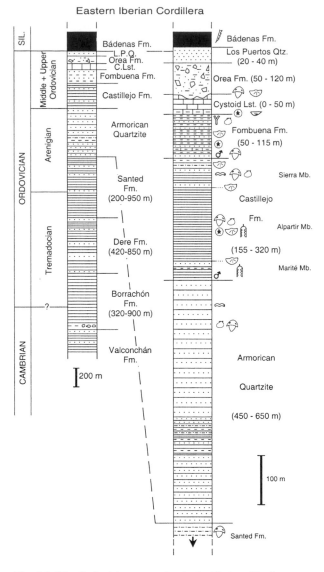

Fig. 4.6. The Ordovician succession in the Eastern Iberian Cordillera. Composite column after Wolf (1980b), Villas (1983, 1985) and Liñán *et al.* (1996). For symbols see Figure 4.4.

Borrachón Fm (320–900 m) comprises mainly green to grey laminated siltstones and shales with some sandy intercalations and has yielded acritarchs and ichnofossils. Above this are sandstones and quartzites of the Dere Fm (420–850 m) with linguliform brachiopods, trilobites and acritarchs, and finally the green to brown siltstones and shales of the Santed Fm (200–950 m). The upper part of the Santed Fm has yielded trilobites also known in the Montagne Noire of southern France, where they are dated by graptolites as lowermost Arenigian so that this formation contains the Tremadocian–Arenigian boundary. The overlying Armorican Quartzite (450–650 m) mainly consists of the typical thick-bedded quartzites with rare brachiopods, conularids, bivalves and trilobites (Babin & Hammann 2001), but shaly and silty intercalations are well developed in the middle part and allow subdivision into three members (Wolf 1980b).

The middle and upper parts of the Ordovician succession are subdivided into distinct formations that were formalized by Villas (1983, and references therein). These formations have yielded abundant and various shelly faunas of brachiopods, trilobites, molluscs, echinoderms, bryozoans and also some graptolites and conodonts; a recent review of the fossil record can be found in Liñán *et al.* (1996, and references therein).

The Castillejo Fm (155–320 m) mainly comprises dark siltstones and shales with some sandstone beds. It begins with an oolithic ironstone level (no. 2, Fig. 4.3) that probably underlines a stratigraphical gap corresponding to the lower Oretanian as the oldest graptolites found above in the formation are late Oretanian in age (*Didymograptus murchisoni* Biozone; Gutiérrez-Marco 1986a). The upper part of the Castillejo Fm has yielded *Gymnograptus linnarssoni*, *Hustedograptus*? *teretiusculus* and '*Glyptograptus raineri*' (= *Oepikograptus bekkeri*) which indicate different Dobrotivian levels (Gutiérrez-Marco 1986a), and the trilobites (Hammann 1983) and the brachiopods (Villas 1983, 1985) allow subdivision into distinct biozones. The Fombuena Fm (50–115 m) begins with a basal oolithic ironstone level (no. 5, Fig. 4.3), which is known also in the Buçaco syncline of central Portugal and in western Brittany, and marks the end of a stratigraphical gap corresponding to the lower Berounian (Villas 1992, and references therein). The first few metres consist of richly fossiliferous calcareous shales of the basal middle Berounian and the rest of the formation is made of siltstones and sandstones that yield abundant brachiopods, echinoderms, bryozoans and benthic graptolites. The uppermost levels could possibly be of upper Berounian age, probably already early Ashgillian after the British regional scale (Hammann 1983; Villas 1985, 1992). The cystoid limestone, which is the most fossiliferous lithological unit in the whole Ordovician succession, was already attributed to the 'Ashgill' by Lotze (1929). The variations in thickness (0–50 m) are due to the erosional episode related to Late Ordovician glaciation that preceded the deposition of the overlying Orea Fm. Different lithofacies have been described (Hafenrichter 1980; Hammann 1992; Vennin et al. 1998). The limestones are frequently dolomitized, and the fossils and sedimentary structures completely obliterated. However, several localities have yielded bryozoans, echinoderms, brachiopods, ostracods, conodonts, molluscs and trilobites, the most completely studied fossil groups being the echinoderms (see references in Chauvel et al. 1975; Chauvel & Le Menn 1979), the brachiopods (Villas 1983, 1985) and the trilobites (Owens & Hammann 1990; Hammann 1992). The cystoid limestone is of Kralodvorian age and, within the Iberian Peninsula, it can be correlated with the Urbana limestone of the Central Iberian Zone, the La Aquiana Fm of NW Spain, the 'pelmatozoan limestone' of the Ossa Morena Zone and the Estana Fm of the Pyrenees (see discussion with palaeogeographical considerations in Hammann 1992). The Orea Fm (50–120 m) rests unconformably on the cystoid limestone and consists mainly of poorly stratified pebble-bearing siltstones and shales that contain some sandstone beds, slump structures and spectacular dropstones in some localities. The lithofacies and the stratigraphical position are typically those of the Kosovian (or Hirnantian) glaciomarine diamictites (Carls 1975; Robardet 1981; Fortuin 1984). This formation has not yielded any macrofossils and the Early Silurian age proposed by Hafenrichter (1980), on the basis of acritarchs, cannot be maintained (see the discussion of the Ordovician–Silurian boundary below). The Orea Fm is overlain unconformably by the quartzites of the Los Puertos Fm (2 to 80 m, generally 20–40 m) the upper part of which contains Rhuddanian and Aeronian graptolites (Gutiérrez-Marco & Storch 1998). It is considered generally that the Ordovician–Silurian boundary coincides with the base of the Los Puertos Fm, but it cannot be excluded that its lower part could be Hirnantian, as is the case for the Criadero quartzite of the southern Central Iberian Zone (see above).

Western Iberian Cordillera. The Ordovician succession in the Western Iberian Cordillera (or Castilian branch: nos 10 and 11, Fig. 4.1) comprises lithostratigraphical units similar to those listed above for the Eastern Iberian Cordillera (or Aragonian branch). However, the Santed Fm is the oldest unit that crops out in this area. Between the Armorican Quartzite and the Orea Fm, the succession includes (Hammann et al. 1982; Gutiérrez-Marco et al. 1996d; and diverse explanatory notes of modern 1/50 000 scale geological maps): 160 to 200 m of shales and sandstones (upper Oretanian to uppermost Dobrotivian), with a basal oolithic ironstone (La Venta Fm, equivalent to the Castillejo Fm) (no 2, Fig. 4.3), which end locally with a quartzitic unit (Colmenarejos quartzite pro parte (p.p.); Tremedal quartzite); 250 to 270 m of middle to upper Berounian alternating shales and sandstones (Tordesilos sandstones plus Colmenarejos

quartzite p.p.; 'Bronchales Beds'); El Cabezo Fm which includes massive limestones (300–150 m) with a widespread basal horizon (2–5 m) of fossiliferous calcareous shales (Kralodvorian).

Cantabrian Zone

In the Cantabrian Zone, Ordovician rocks (Fig. 4.7) overlie the Cambrian succession in apparent conformity. However, owing to the rather poor fossil record (ichnofossils and rare acritarchs), it is considered generally that the Cambrian–Ordovician boundary could be situated either within the uppermost part of the Oville Fm in the NW and east of the Cantabrian Zone or above a stratigraphical gap (erroneously correlated with the true Sardic Unconformity) in the middle part of the Barrios Fm (Aramburu & García-Ramos 1988, 1993; Truyols et al. 1990; Aramburu et al. 1992).

The Barrios Fm (150–870 m) is mainly composed of light-coloured quartzites with minor intercalations of siltstones and shales, and some conglomeratic levels. In some localities its upper part comprises about 100 m of dark shales and finally 50 m of quartzites. The Barrios Fm contains abundant ichnofossils (*Cruziana* ichnofacies) of ?Tremadocian and Arenigian age (Crimes & Marcos 1976; Aramburu et al. 1992). In most areas of the Cantabrian Zone, younger Ordovician units do not exist and the Barrios Fm is directly overlain by Silurian, Upper Devonian or Lower Carboniferous rocks. Post-Arenigian Ordovician rocks are known along the Narcea antiform in the western part of the 'folds and nappes region', and also in the Laviana-Sueve thrust-sheet of the 'nappe region', east of the Central Coal Basin. In the first area, the Middle Ordovician fossiliferous dark shales that

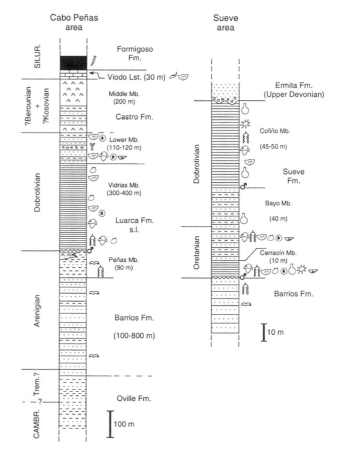

Fig. 4.7. The Ordovician successions in the Cantabrian Zone. After Gutiérrez-Marco et al. (1996a, 1999a) and Truyols et al. (1996). For symbols see Figure 4.4.

overlie the Barrios Fm have been assigned to the Luarca Fm of the West Asturian-Leonese Zone. In the Laviana–Sueve area, the name Sueve Fm has been formalized (Gutiérrez-Marco *et al.* 1996*a*).

In the Laviana–Sueve area, near Rioseco (no. 3, Fig. 4.1), the Arenigian graptolite *Azygograptus undulatus* has been found within alternating sandstones and siltstones of the uppermost part of the Barrios Fm (Gutiérrez-Marco & Rodríguez 1987). In this area, the Sueve Fm, which overlies the Barrios Fm, is a sequence (50–100 m) of dark shales that begins with an oolithic ironstone bed (no. 3, Fig. 4.3) and includes a middle member of alternating sandstones, siltstones and shales. The Sueve Fm is richly fossiferous, with trilobites, ostracods, brachiopods, molluscs and echinoderms. It has also yielded graptolites, acritarchs and relatively well preserved chitinozoans. The lower part of the formation is late Oretanian in age (upper *Didymograptus murchisoni* Biozone), which indicates that it follows a stratigraphical hiatus corresponding to the lower Oretanian and the basal upper Oretanian. The Oretanian–Dobrotivian boundary lies within the middle member, and, locally, the upper member extends up into the upper Dobrotivian (*Lagenochitina ponceti* biozone?). These levels are overlain by Upper Devonian conglomerates and sandstones.

In the western part of the Cantabrian Zone (no. 2, Fig. 4.1), Middle Ordovician fossiliferous dark shales of the Luarca Fm *sensu lato* occur as a ?tectonic slice (*c.* 10 m thick) within the Barrios Fm in its type locality at Barrios de Luna (Aramburu *et al.* 1996). A few kilometres to the ESE, at Portilla de Luna, alternating dark shales, siltstones and fine-grained sandstones (65 m thick) overlain by limestones (about 13 m thick) occur between the Barrios Fm and the Silurian Formigoso Fm. Within the limestones, brachiopods of the 'Nicolella Fauna', along with echinoderms and trilobites clearly indicate a Kralodvorian age, which suggests that the underlying poorly fossiliferous alternations are probably also Upper Ordovician (Gutiérrez-Marco *et al.* 1996*b*).

However, the main outcrops of post-Arenigian Ordovician rocks in this region are the coastal exposures of the Cabo Peñas and Cabo Vidrias (no. 1, Fig. 4.1). The most recent revision is that by Gutiérrez-Marco *et al.* (1999*a*, and references therein). In these localities, the massive quartzites of the Barrios Fm are overlain by alternating dark shales and sandstones with ichnofossils (about 90 m) that are either considered as transitional to the typical dark shales of the Luarca Fm or as its lower member (Peñas Member (Mb)). The graptolites found in these levels are most probably upper Arenigian. Then follow the typical dark shales with nodules and coquina beds of the Luarca Fm *sensu lato* (Vidrias Mb: 260 m). They begin with an oolithic ironstone level (no. 4, Fig. 4.3) that underlies a stratigraphical gap that could correspond to the whole Oretanian and the lowermost Dobrotivian, as the lowermost part of this member contains latest lower Dobrotivian trilobites (upper part of the *Placoparia tournemini* trilobite Biozone). The major part of the Vidrias Mb and the interbedded volcaniclastic rocks, siltstones and sandstones that constitute the first 100 m of the overlying Castro Fm (total thickness 340 m at Cabo Peñas; Truyols *et al.* 1996) correspond to the upper Dobrotivian (*Placoparia borni* Biozone), as attested by the fossil assemblages found in several successive levels. The middle part of the Castro Fm (*c.* 200 m at Cabo Peñas), that mainly consists of basalts with interbedded various terrigenous rocks, has not yielded any conclusive fossils and its age remains all the more uncertain (Villas *et al.* 1989; Truyols *et al.* 1996) since the upper member (Viodo limestone: 30 m) is Llandovery (Sarmiento *et al.* 1994; Villas & Cocks 1996).

Ossa Morena Zone

Ordovician rocks are known from several of the NW–SE elongated 'domains' separated by faults (Apalategui *et al.* 1990) within the Ossa Morena Zone. North of the Olivenza–Monesterio antiform they occur within the small Valle and Cerrón del Hornillo synclines, in the southeastern part of the Zafra–Alanís–Córdoba domain (nos 30 and 31, Fig. 4.1). South of this antiform, they are known in the southern part of the Elvas–Cumbres Mayores domain (south of Cañaveral de León) and in the Encinasola area of the Barrancos–Hinojales domain, in the prolongation of the Portuguese Barrancos

area units, but very little information comes from the Spanish part of the outcrop (e.g. Quesada & Cueto 1994).

In northern Seville Province, the Valle and the Cerrón del Hornillo synclines, surrounded by extensive outcrops of Cambrian rocks, are the only units in the Zafra–Alanís–Córdoba domain where post-Cambrian Palaeozoic rocks have been preserved. The complete Lower Ordovician–Lower Carboniferous succession has been established only recently (see Robardet 1976; Robardet *et al.* 1998, and references therein).

The Ordovician succession of the Valle syncline was first described by Simon (1951) and later by Gutiérrez-Marco *et al.* (1984*d*). More recent studies have investigated especially the upper part of this succession, the conodont faunas, and the Ordovician–Silurian transition, in both synclines (Jaeger & Robardet 1979; Hafenrichter 1979, 1980; Sarmiento 1993). An updated synthesis of the results is that by Robardet *et al.* (1998). The Cambrian–Ordovician transition can never be observed as the contacts are everywhere tectonized.

In the Valle syncline (Fig. 4.8; no. 30, Fig. 4.1), the lower and middle parts of the Ordovician succession crop out and comprise the following.

(1) Grey-green shales and siltstones (probably more than 200 m thick) that have yielded basal Arenigian graptolites of the *Tetragraptus phyllograptoides* Biozone. It cannot be excluded that future investigations could identify the Tremadocian–Arenigian transition in the thick shaly sequence that underlies the fossiliferous level.

(2) Dark-shales with siliceous nodules (10–15 m), with Oretanian trilobites (*Ormathops*? sp., *Selenopeltis* aff. *buchi*, *Kodymaspis puer*, *Nerudaspis* cf. *aliena*), echinoderms (*Lagenocystis pyramidalis*), brachiopods (*Euorthisina minor*), bivalves, rostroconchs, hyolitids and ostracods. The assemblage shows clear affinities with

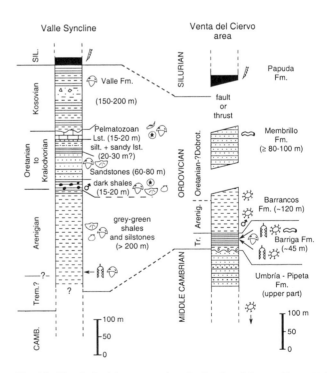

Fig. 4.8. The Ordovician successions in the Ossa Morena Zone. Valle syncline after Gutiérrez-Marco *et al.* (1984*d*), Robardet & Gutiérrez-Marco (1990*a*), and Robardet *et al.* (1998). Venta del Ciervo area after Schneider (1939), Gutiérrez-Marco (1982, 1986*a*,*b*); Mette (1989), Robardet *et al.* (1998) and Servais & Mette (2000). For symbols see Figure 4.4.

those of the Oretanian of Bohemia (Sárka Fm) and of the Ancenis area in the southern part of the French Armorican Massif.

(3) A few metres of micaceous sandstones and ferruginous oolithic levels (?no. 4, Fig. 4.3).

(4) Light-coloured micaceous sandstones (60–80 m) with brachiopods and trilobites (*Placoparia (Coplacoparia)* sp.) that suggest a Dobrotivian age.

(5) A few metres of decalcified sandy limestones that have yielded crinoid, bryozoan, trilobite and brachiopod fragments, and chitinozoans indicative of the *Belonechitina robusta* Biozone of the upper middle Berounian (Paris 1981, 1992; Gutiérrez-Marco *et al.* 1984*d*),

The overlying 'pelmatozoan limestone' and Valle formations (Fig. 4.8) occur within both synclines with the same characteristics, but in the Cerrón del Hornillo syncline the limestone is clearly transgressive (with a basal oolithic ironstone bed; no. 6, Fig. 4.3) over Lower or Middle Ordovician shales.

The 'pelmatozoan limestone' (maximum thickness 15–20 m) has been dated as Kralodvorian on the basis of conodont assemblages that indicate the *Amorphognathus ordovicicus* Biozone (Robardet 1976; Hafenrichter 1980; Sarmiento 1993). In the Cerrón del Hornillo syncline, the 'pelmatozoan limestone' has yielded gastropods (Fryda *et al.* 2001) and a rich trilobite assemblage (with *Cekovia* sp., *Symphysops* sp., *Cyclopyge* sp.; Robardet 1976; Hammann 1992, p. 40). When accessible, the uppermost part of the 'pelmatozoan limestone' shows a karstified morphology interpreted as the result of weathering and erosional processes that occurred after the sea-level fall related to growth of the latest Ordovician African ice cap. The overlying Valle Fm comprises some contorted, probably slumped beds and microconglomeratic levels, reminiscent of the uppermost Ordovician glaciomarine distal deposits. The Valle Fm is thus attributed to the Kosovian (or Hirnantian), although the fossil record is limited to a single specimen of the trilobite *Mucronaspis* sp., to probably reworked crinoid fragments in the Valle syncline and to a few modiolopsid bivalves in the Cerrón del Hornillo syncline. This interpretation is corroborated by the presence, at the base of overlying black shales, of graptolite assemblages indicative of the earliest Silurian *Akidograptus ascensus–Parakidograptus acuminatus* Biozone (Jaeger & Robardet 1979).

About 20 km north of Aracena, Cambrian to Lower Devonian rocks crop out in a complex area that corresponds to the tectonic contact between the Elvas–Cumbres Mayores and the Barrancos–Hinojales domains (Juromenha thrust fault) (no. 29, Fig. 4.1). Geological studies initiated by Schneider (1939) were continued more recently with reappraisal of the stratigraphical succession (Mette 1989). To the south of km 18 of the road between Cañaveral de León and Corteconcepción, in the Venta del Ciervo farm area, the succession (Fig. 4.8) is complicated by several tectonic contacts (Robardet *et al.* 1998). However, in the northern part of the road section, the conglomerates, feldspathic sandstones and quartzites of the Umbría-Pipeta Fm (Middle Cambrian) are overlain by the finely laminated green shales of the Barriga Fm (about 45 m). These green shales are separated from the underlying quartzites by an erosional unconformity with an irregular basal conglomeratic level. They have yielded graptolites, acritarchs, ichnofossils, some phyllocarid remains and a unique specimen of a cyclopygid trilobite. The Barriga Fm contains a remarkable graptolite fauna, discovered in this section (Schneider 1939) and later studied in more detail (Gutiérrez-Marco 1982, 1986*a,b*; Erdtmann *et al.* 1987; Gutiérrez-Marco & Aceñolaza 1987). The first graptolite assemblage (9–16.40 m above the base) contains among others *Paradelograptus onubensis*, and ends with a spectacular, 20–25 cm thick, mass-accumulation of giant rhabdosomes of *Araneograptus murrayi*. The upper assemblage (25–34 m above the base) has yielded some specimens of *Hunnegraptus* aff. *copiosus*, *Tetragraptus* and *Clonograptus*, among others. Both graptolite assemblages are clearly assignable to the latest Tremadocian (pre-*T. phyllograptoides* or pre-*T. approximatus* biozones). The acritarchs of the Barriga Fm. belong to the *messaoudensis–trifidum* acritarch assemblage' currently considered to be characteristic of the uppermost Tremadocian–lowest Arenigian (Mette 1989; Servais & Mette 2000). Dark shales and siltstones that occur to the south probably belong to the Barrancos Fm that crops out at a short distance to the east, where it has yielded upper Arenigian–lower Oretanian acritarchs (Mette

1989). The Membrillo Fm (80–100 m) mainly comprises brownish sandstones and quartzites that crop out in the Sierra Hinojales and Sierra Membrillo, to the south of the Venta del Ciervo area from which they are separated by Silurian and Lower Devonian fossiliferous rocks and by tectonic contacts. Up to now, the sandstones of the Membrillo Fm have only yielded ichnofossils of *Planolites* and *Scalarituba* types. However, they are considered to overlie the Barrancos Fm and are probably Oretanian to lower Dobrotivian in age (Mette 1989). In this area there is no evidence for younger Ordovician rocks.

In the Venta del Ciervo area, the Barrancos Fm bears in its lower third a 1–1.5 m thick ooidal ironstone bed (Gutiérrez-Marco *et al.* 1984*b*, p. 515; ?no. 1, Fig. 4.3) which is apparently absent from the same formation further SW in the Encinasola area (Terena syncline: no. 28, Fig. 4.1). There the Barrancos Fm, which is at least 450 m thick and contains ichnofossils (*Planolites, Skolithos, Rusophycus, Spirophycus, Protopaleodictyon, Megagrapton*), is traditionally regarded as a continuous sequence ranging from the basal Arenigian to the uppermost 'Caradocian' (Giese *et al.* 1994 and references therein), although this appears highly improbable to us.

Finally, it must be noted that several authors, such as Schneider (1939), Hernández Enrile & Gutiérrez Elorza (1968) and Bege (1970), considered that the conglomerates and the light-coloured quartzites that underlie the Barriga Fm in the Cañaveral–Hinojales area were representative of the Armorican Quartzite so typical of the Central Iberian Zone. This interpretation must be abandoned because these rocks are of Middle Cambrian age, as indicated by acritarchs (Mette 1989) and part of the Cambrian Umbría-Pipeta Fm. In this area, the Ordovician is represented in actual fact by the green shales of the Barriga Fm and the shales and siltstones of the Barrancos Fm.

Pyrenees and Catalonian Coastal Ranges

The oldest Ordovician fossils known up to now in the Pyrenees (nos 38 to 41, Fig. 4.1) and the Catalonian Coastal Ranges (nos 33 to 37, Fig. 4.1) are almost everywhere of middle Berounian age. The only two exceptions are: (1) poorly preserved trilobite and brachiopod fragments from pre-'Caradocian' rocks of the Les Gavarres massif in the NE Catalonian Coastal Ranges (Barnolas & García-Sansegundo 1992; no. 33, Fig. 4.1); (2) graptolites of Arenigian age from the Saint-Martin-d'Arrossa area, in the French part of the western Pyrenees (Dégardin 1979: no. 41, Fig. 4.1). In both regions, the Ordovician succession is currently subdivided into lower and upper groups.

The lower group, which up to now has not yielded any identifiable fossils, comprises a thick (up to 2000 m or more?) sequence of more or less metamorphic schists, slates and siltstones with some limestone and quartzite beds. It corresponds to the 'Sericitic Slates' of the Catalonian massifs, to the Seo Fm of the central Pyrenees, and to the Jujols Fm of the eastern Pyrenees. In the absence of any biostratigraphical data, the precise age of these formations remains uncertain; they are considered as Cambrian and Lower Ordovician (Cavet 1957; Julivert & Durán 1990*a*, 1992, among others), or mainly Cambrian and perhaps lowermost Ordovician in their uppermost part (Laumonier 1988, 1998). By comparison with the Montagne Noire of the southern French Massif Central and Sardinia, it is generally assumed that the middle part of the Ordovician is missing and that Upper Ordovician rocks lie directly and unconformably on the pre-Berounian sequence.

The upper group of Ordovician rocks comprises conglomerates, quartzites, siltstones and shales, locally with volcanic intercalations, overlain by calcareous shales and limestones ('Schistes troués', 'Grauwackes à *Orthis*'). These rocks have yielded shelly faunas with brachiopods, bryozoans, trilobites, echinoderms and bivalves, especially abundant in limestones and calcareous shales. Most of these fossils probably need taxonomic revision, but the existence of middle Berounian to Kralodvorian levels is clearly established both in the Catalonian Coastal Ranges (Barnolas *et al.* 1980; Meléndez & Chauvel 1982; Villas *et al.* 1987; Sarmiento *et al.* 1995*b*) and in the Pyrenees (Cavet 1957; Hartevelt 1970; Sanz López & Sarmiento 1995; Gil-Peña *et al.* 2000). The most precise litho- and biostratigraphical data come from the Spanish and Andorran central Pyrenees (no. 39, Fig. 4.1) where above the Jujols Fm Hartevelt (1970) described the following succession: (1) Rabassa Conglomerate (up to 100 m); (2) alternating greywackes and slates with volcanic intercalations (Cava Fm: up to 500 m); (3) calcareous shales and limestones (Estana Fm: *c.* 200 m) with Berounian to Kralodvorian brachiopods, bryozoans, cystoids and conodonts of the *Amorphognathus ordovicicus* Biozone (Sanz López & Sarmiento 1995); (4) poorly bedded dark slates of the Ansobell Fm (20 to 300 m), that contain microconglomeratic levels and locally overlie directly the Cava Fm, and are probably the equivalent of the Kosovian glaciomarine diamictites (Gutiérrez-Marco *et al.* 1998; Gil-Peña *et al.* 2000); (5) Bar Quartzite (0–20 m), a probable equivalent of the 'basal Silurian' quartzites of other Iberian regions; (6) Silurian graptolitic black shales.

The Montseny, Les Guilleries and Les Gavarres massifs of the NE Catalonian Coastal Ranges (Barnolas *et al.* 1980; Barnolas & García-Sansegundo 1992: nos 33 to 35, Fig. 4.1) comprise Upper Ordovician sequences reminiscent of that described by Hartevelt (1970). This is especially the case in the Montseny massif, where in the Riera del Avencó valley near Aiguafreda the Upper Ordovician succession comprises: slates, greywackes and siltstones with brachiopods of upper Berounian age (Villas *et al.* 1987), below siltstones and slates with limestone levels (*c.* 10–15 m) of 'Ashgill' age (Hafenrichter 1979), themselves overlain by the 'Abanco Shales' (50 m), with sedimentary breccias, that underlie Rhuddanian graptolitic black shales (Puschmann 1968*a*; Gutiérrez-Marco *et al.* 1999*d*), and are probably equivalent to the Ansobell Fm of the central Pyrenees.

The boundaries of the Ordovician System in Spain

The Cambrian–Ordovician boundary

Hammann *et al.* (1982, p. 3) and Romano (1982) were the first authors to present a generalized and updated overview of the Cambrian–Ordovician boundary in Spain. Across the whole Iberian Peninsula, the main problems for an accurate characterization of this boundary are the lack of continuous sections, and the absence of valuable fossils that could provide detailed biostratigraphical and biochronological data. Also, over large areas of central Spain, the Ordovician sequence has a transgressive character over a rather uniform basement formed by Neoproterozoic to lowest Middle Cambrian rocks. So far, neither graptolites nor conodonts of basal or lower Tremadocian age have been found in Spain, and the biostratigraphical inferences for that boundary, based on acritarch data and trace fossil distribution, need review and reappraisal.

Iberian Cordillera. The most continuous and best known Cambrian–Ordovician Spanish sequences have been found in the Eastern Iberian Cordillera (Wolf 1980*b*, and references therein). Until recently, the position of the boundary was placed at the base of a conglomeratic horizon near the top of the Valconchán Fm, which was previously related by Lotze (1956*b*) to a 'Toledanian unconformity'. The most interesting fossils coming from this horizon are the trilobite *Pagodia (Wittekindtia) alarbaensis* and some brachiopods (*Poramborthis hispanica, Protambonites primigenius*), considered of basal Tremadocian age on the basis of the occurrence of their respective generic or subgeneric taxa in rocks of this age in the north Gondwanan area and in Afghanistan. *Pagodia (Wittekindtia)* may occur also in the basal beds of the upper member of the Najerilla Fm from the Sierra de la Demanda (NW prolongation of the Iberian Cordillera), and therefore Shergold *et al.* (1983) placed the Cambrian–Ordovician boundary in coincidence with this level, although they admitted that *P. (Wittekindtia)* could initially occur in the Late Cambrian epoch with the other subgenera of the leiostegiid trilobite *Pagodia*. Later on, Shergold & Sdzuy (1991, p. 202) considered the very same horizon bearing *P. (Wittekindtia)* as terminal Late Cambrian age (Trempealeauan of the North American scale). The formal designation of the lower boundary of the Ordovician System (Cooper *et al.* 2001) led to a different perception of the position of the Cambrian–Ordovician

boundary in the Palaeozoic succession of the Iberian Cordillera, already anticipated by Shergold & Sdzuy (1991). This is because the lower-middle part of the Borrachón Fm, which clearly overlies the beds with *P. (Wittekindtia),* bears an olenid–asaphid trilobite assemblage (still undescribed, but composed of up to ten genera listed by Hammann *et al.* 1982, p. 24), which has considerable affinities with Mexican and Central Andean trilobite assemblages associated with the *Cordylodus proavus* conodont Biozone, which is unambiguously situated in the Upper Cambrian (Cooper *et al.* 2001), and provides a basis for correlation of the Cambrian–Ordovician boundary with a level placed in the upper-middle part of the Borrachón Fm of the Iberian Cordillera, and possibly also in the Brieva sandstone of the Sierra de la Demanda (which alternatively could correspond to the latest Cambrian Acerocare regressive event). It can be added that Upper Cambrian (i.e. early Franconian according to the North American biochronological scale) trilobites of Eurasiatic affinities previously occur both in the basal member of the Najerilla Fm (Sierra de la Demanda: Shergold *et al.* 1983) as well as in the lower part of the Valconchán Fm (Eastern Iberian Cordillera: Shergold & Sdzuy 1991).

West Asturian-Leonese Zone. The position of the Cambrian–Ordovician boundary within the Palaeozoic successions of the West Asturian-Leonese Zone is poorly known by comparison with its southeastern extension in the Iberian Cordillera. Baldwin (1977) tentatively situated the boundary towards the basal upper half of the Los Cabos Gp of the Navia–Alto Sil domain. However, his results are based only on the distribution of selected trace fossils (*Rusophycus* and *Cruziana* ispp.) and cannot be considered totally conclusive. The brachiopod *Protambonites primigenius*, which Villas *et al.* (1995a) suggest to be of 'a Cambro-Ordovician transition age', has been found recently below the approximate boundary proposed by Baldwin, in beds situated in the middle part of the Los Cabos Gp. However, in the Iberian Cordillera, the same species is restricted to Upper Cambrian beds (?early Franconian) and this rather suggests that the Cambrian–Ordovician boundary is most probably situated higher in the succession, possibly in the position proposed by Baldwin (1977) on the basis of ichnofossil data.

Cantabrian Zone. The location of the Cambrian–Ordovician boundary within the Cantabrian Zone is also problematic. Some palynological data (that need revision) suggest the existence of Upper Cambrian to Tremadocian acritarchs in the Upper Mb of the Oville Fm, but exclusively in sections belonging to the northwesternmost part of the zone. In the remaining areas, the Oville Fm is clearly Middle Cambrian in age, and is overlain conformably by the Barrios Fm whose lower member (La Matosa Mb) is usually considered to be of uppermost Middle to Upper Cambrian age, probably also reaching the lower Tremadocian in the east of the Cantabrian Zone (Aramburu & García-Ramos 1993). However, the discovery of an important discontinuity within the Barrios Fm (Aramburu & García-Ramos 1988) shows that the Cambrian–Ordovician boundary in the Cantabrian Zone is generally situated within a non-continuous sequence, where the middle or upper members (Ligüeria and Tanes members, respectively) of the Barrios Fm (?upper Tremadocian to basal Arenigian) transgress unconformably over the lower member of the same unit, usually considered as partly of Upper Cambrian age (Aramburu *et al.* 1992, fig. 3; Aramburu & García-Ramos 1993, fig. 4).

Central Iberian Zone. In the Central Iberian Zone, the lowermost Ordovician strata transgress over a thick sequence of Neoproterozoic to lowest Middle Cambrian rocks, with a variable angular disconformity frequently but erroneously referred to as the 'Sardic discordance' in the regional Iberian literature (for instance, Moreno *et al.* 1976; Díez Balda *et al.* 1990; Valverde-Vaquero & Dunning 2000). The same erroneously named unconformity has been claimed to exist also within the Barrios Fm of the Cantabrian Zone (Aramburu & García-Ramos 1988, 1993). A contrasting palaeotopography existed in this zone at the time when the basal Ordovician sediments were deposited and sealed the Toledanian unconformity, as they lie on both late Precambrian or Lower Cambrian rocks. The pre-Armorican Quartzite red bed successions, up to 400 m thick, occur in several grabens and half-grabens. Furthermore, associated facies also vary rapidly in small areas, from fan deltas to storm-dominated shallow-marine deposits, as seen along the Alcudia anticline.

The true Sardic unconformity in the classic sites of Sardinia, Italy (Stille 1939), is actually demonstrated to be intra-Ordovician, and separates fossiliferous rock units of Tremadocian–Arenigian and Berounian ages, respectively (see Hammann et al. 1982; Pillola & Gutiérrez-Marco 1988; Leone et al. 1991; Hammann & Leone 1997; Pillola 1998). The replacement name for the pre-Ordovician unconformity in Central Iberia, which separates two major sedimentary cycles (the Assyntic/Pan-African and the Caledono/Variscan) should be derived from one of the two unconformities recognized by Lotze (1956b) in the southern Central Iberian Zone: the so-called 'Iberian unconformity' (between the Armorican Quartzite and its basal formations of red beds and conglomerates), and the 'Toledanian unconformity' (between the basal formations of the Armorican Quartzite and the pre-Ordovician basement). As stated by San José et al. (1974) and Díez Balda et al. (1990), the lower one has a more important structural significance, and the name Toledanian unconformity is thus favoured here to replace the incorrect term of 'Sardic unconformity' in the Iberian Peninsula, where it corresponds, over most of the Iberian Massif, to a boundary between the incomplete pre-Ordovician sequence and the Lower Ordovician (also incomplete).

Other areas. Within the southernmost part of the Central Iberian Zone, formerly regarded as the northernmost Ossa Morena Zone (Lotze 1945) but recently proposed as a new Lusitan Marianic Zone (Herranz et al. 1999), a reworked assemblage of upper Tremadocian conodonts, possibly indicative of the *Paltodus deltifer* Biozone, has been found recently in a lower Middle Ordovician (c. Volkhovian) olistolith incorporated into Carboniferous mudstones (Sarmiento & Gutiérrez-Marco 1999). Sedimentary and palaeobiogeographical affinities of Ordovician and Silurian olistoliths found in the Mississippian of the Guadiato basin (northern Córdoba province) show a mixture of Central Iberian and Ossa Morena provenances together with others of unknown affinities. With regard to the Cambrian–Ordovician boundary in the Ossa Morena Zone, the only data about the oldest Ordovician sediments so far recorded are from the Barrancos–Hinojales domain (Venta del Ciervo area), where Upper Tremadocian graptolite shales of the Barriga Fm directly transgress over Middle Cambrian quartzites of the Umbría-Pipeta Fm (Robardet et al. 1998, and references therein). In the Catalonian Coastal Ranges and the Pyrenees, as noted in the text corresponding to these regions (see above), there is total uncertainty as to the position of the Cambrian–Ordovician boundary within the very thick unfossiliferous pre-Berounian (pre-Sardic) sequence (i.e. the 'Sericitic Slates', Seo, Canaveilles and Jujols formations).

The Ordovician-Silurian boundary

The existence of a Late Ordovician continental glaciation centred on the African part of the Gondwanan continent is now well established (Beuf et al. 1971; Deynoux 1980; Sutcliffe et al. 2000), although it is difficult to make precise and detailed comparisons with the Neogene–Holocene glaciation. Saharan Africa was covered by a huge ice-cap and, in most areas of the North Gondwanan Province, the Upper Ordovician–Lower Silurian transitional succession (Fig. 4.9) has registered the echoes of this major continental glaciation (Robardet & Doré 1988 and references therein).

Hirnantian glaciomarine deposits. In most areas of the Iberian Peninsula, the uppermost part of the Ordovician succession includes terrigenous deposits which consist of generally massive, poorly stratified, argillaceous siltstones and shales that include dispersed rock fragments ranging from millimetric quartz grains to pluridecimetric blocks or pebbles of various lithologies (Fig. 4.9). These diamictites (Robardet et al. 1980; Robardet 1981; Fortuin 1984) are strikingly similar to contemporaneous glaciomarine sediments deposited along the northern margin of the African ice-sheet (the so-called 'Argiles microconglomératiques' of Libya, Algeria and Morocco).

The Iberian diamictites, as well as those of other North Gondwanan European regions, are thus regarded as resulting from the same global factors and considered as glaciomarine deposits corresponding to the melting of the main ice-cap. Sediments contemporaneous with the acme of the glaciation are unknown and most probably lacking in Spain. This is due to a depositional hiatus and to the predominance of erosional processes coeval with the glacioeustatic fall of the sea level.

In some areas the apparent absence of clasts within time equivalents of the diamictites can be due to a limited amount of clast-bearing levels that can be identified only by careful bed-by-bed studies, as recently seen in the Valle Fm of the SE Ossa Morena Zone, in the Chavera Fm of the southern Central Iberian Zone, in the Ansobell Fm of the central Pyrenees and in the Abanco Shales of the Catalonian Coastal Ranges. It can also be due to the fact that the passage of floating ice in these areas was prevented either by marine currents or by any other factor controlling the iceberg or seasonal ice drift, as well as by later coastal reworking. In Spain, the majority of the macrofossils found in the diamictites are rare and poorly preserved (brachiopods, crinoids or trilobites that were most probably derived from older deposits). However, limestone pebbles from the Orea Fm of the Western Iberian Cordillera and from the Rozadais Fm of the Truchas syncline (NW Spain) have yielded Upper Ordovician conodonts of the *Amorphognathus ordovicicus* Biozone (Sarmiento 1993; Sarmiento et al. 1999c), which clearly indicates that the glaciomarine deposits cannot be older than the ?uppermost Berounian–Kralodvorian. It has been proposed, on the basis of acritarch data, that the Orea Fm of the Eastern Iberian Cordillera and the Chavera Fm of the southern Central Iberian Zone were Early Silurian in age (Hafenrichter 1980). However, this conclusion seemed uncertain because the biostratigraphical potential of these microfossils is not clearly defined, and the acritarch taxa found are long-ranging forms spanning the Ordovician–Silurian boundary.

Since then, similar glaciomarine deposits outside Iberia have yielded chitinozoans, a microfossil group whose biozonation has been precisely established (Paris 1990, 1992), and these data indicate a very high stratigraphical level within the uppermost part of the Ordovician, equivalent to the British Hirnantian (Paris et al. 1995). In several areas of the North Gondwanan Province, the diamictites are either bracketed by brachiopod *Hirnantia* faunas, as in Morocco and Algeria (Ougarta), or overlain by uppermost Ordovician graptolite assemblages of the *Normalograptus persculptus* Biozone (Robardet & Doré 1988, and references therein; Paris 1998; Underwood et al. 1998). It thus appears that the glaciomarine sediments are restricted to the uppermost part of the Ordovician and were deposited during the Kosovian (= Hirnantian). This conclusion is corroborated, in several areas of Spain, by the occurrence, above the diamictites, of basal Rhuddanian graptolite assemblages (Mondoñedo, Truchas, Iberian Cordillera, Ossa Morena Zone) or brachiopods of the *Hirnantia* Fauna (Almadén). It can be noted also that, in some areas, the glaciomarine clast-bearing beds are separated from the basal graptolite biozone of the Silurian by a certain thickness of clast-free sediments (Losadilla Fm in the Truchas syncline, upper part of the Valle Fm in the Ossa Morena Zone). These sediments represent a return to post-glacial shelf conditions, prior to the Silurian transgression, and are coeval with the upper part of the uppermost-Ordovician *Normalograptus persculptus* graptolite biozone and the *Spinachitina oulebsiri* chitinozoan biozone (Paris et al. 2000a).

Pre-diamictite stratigraphical hiatus. The diamictites apparently conformably overlie Ordovician sediments. However, depending on the region, the age of the youngest underlying rocks differs, being:

- early Kosovian in the Buçaco-Dornes area of central Portugal, as the brachiopod fauna of the Ribeira do Braçal Fm includes elements related to those of the *Hirnantia* Fauna in other regions (Brenchley et al. 1991);
- Kralodvorian (see above) in the Iberian Cordillera ('cystoid limestone'), in the Caurel-Peñalba syncline (La Aquiana Fm), in the SE Central Iberian Zone ('Urbana limestone') and in the SE Ossa Morena Zone ('pelmatozoan limestone');
- Middle Berounian in the Guadarranque and Herrera del Duque synclines of the Central Iberian Zone (Robardet et al. 1980);

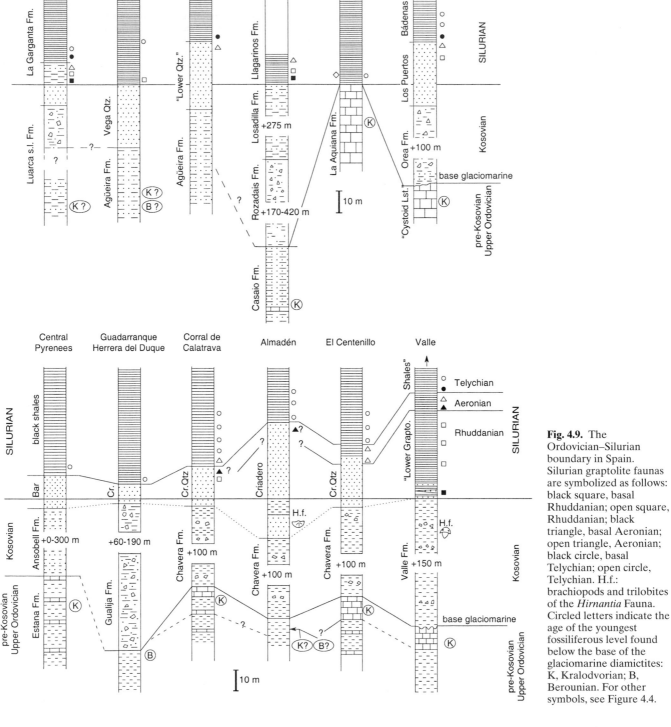

Fig. 4.9. The Ordovician–Silurian boundary in Spain. Silurian graptolite faunas are symbolized as follows: black square, basal Rhuddanian; open square, Rhuddanian; black triangle, basal Aeronian; open triangle, Aeronian; black circle, basal Telychian; open circle, Telychian. H.f.: brachiopods and trilobites of the *Hirnantia* Fauna. Circled letters indicate the age of the youngest fossiliferous level found below the base of the glaciomarine diamictites: K, Kralodvorian; B, Berounian. For other symbols, see Figure 4.4.

- most probably upper Dobrotivian (or lowermost Berounian?) in the Valongo area of north-central Portugal (Couto *et al.* 1997, and references therein).

All these data show a pre-diamictite stratigraphical hiatus of varying importance which would have resulted from erosional processes before deposition of the diamictic sediments. This hiatus most probably resulted from the global eustatic fall of the sea level that followed the development of the African continental ice-cap and

that led to emergence of a number of North Gondwanan shallow marine areas. In some areas, the top of the underlying Kralodvorian limestones seems karstified and eroded, as in the Iberian Cordillera (Carls 1975), in the southeastern Central Iberian Zone (Hafenrichter 1980), and in the SE Ossa Morena Zone (García-Ramos & Robardet 1992; Robardet *et al.* 1998). This scenario suggests that the redeposited clasts are partly of local origin and partly of remote provenance. It also implies that the deepest distal areas of the marine shelf probably were not emergent and may have registered a continuous

Ordovician–Silurian transitional sequence. This may be the case in the Truchas syncline of NW Spain and the eastern Sierra de Guadarrama, where sandstones and slates above the diamictites could be equated with the *N. persculptus* graptolite biozone.

Post-diamictite stratigraphical hiatus? In most areas, the glaciomarine Kosovian diamictites are separated from the Silurian graptolitic black shales by quartzite or sandstone units whose thickness is generally a few tens of metres (such as the 'Criadero quartzite' of the southern Central Iberian Zone and equivalent formations in other areas). Initially, these quartzitic units were considered unfossiliferous and the lowermost levels of the overlying Silurian black shales had yielded graptolite assemblages of the 'middle' part of the Llandovery (Aeronian or, more frequently, Telychian). It was therefore considered that a generalized stratigraphical hiatus corresponded to the uppermost Ordovician and to the lower Llandovery (see Truyols & Julivert 1983). Recent investigations have revealed that micaceous sandy or shaly levels within the upper part of these quartzitic units contain graptolite assemblages of Early Silurian age. In the El Centenillo area (southern Central Iberian Zone) the upper part of the 'Castellar quartzite' includes millimetric black shale interbeds that have yielded graptolites of the lower Aeronian *Pribylograptus leptotheca–Demirastrites triangulatus* biozones. At Corral de Calatrava (southern Central Iberian Zone), the uppermost 2 m of the 'Criadero quartzite', that consist of black micaceous sandstones, siltstones and shales, have yielded several successive graptolite assemblages that correspond respectively to Rhuddanian–basal Aeronian, Aeronian and basal Telychian ages (García Palacios *et al.* 1996). In the Western Iberian Cordillera of NE Spain, the upper part of the 'Los Puertos quartzite', that disconformably overlies the Kosovian glaciomarine Orea Fm, comprises several black shale intercalations and the lowest level contains Rhuddanian normalograptids (*Parakidograptus acuminatus* to *Coronograptus cyphus* biozones?; Gutiérrez-Marco & Storch 1998). In the Rececende syncline of the Mondoñedo domain (West Asturian-Leonese Zone), the Silurian black shales are underlain by a quartzitic unit, the upper part of which comprises several siltstone levels that have yielded graptolites of the Rhuddanian *Akidograptus ascensus–Parakidograptus acuminatus, Cystograptus vesiculosus* and *Coronograptus cyphus* biozones (Arbizu *et al.* 1997; Gutiérrez-Marco *et al.* 1997*a*). In the Catalonian Coastal Ranges, where the lowermost Silurian was supposedly missing or represented by unfossiliferous beds (Julivert *et al.* 1985), recent investigations have shown that the basal Rhuddanian *Akidograptus ascensus–Parakidograptus acuminatus* Biozone occurs in graptolitic black shales and cherts (lydites) from different localities of the Serra de Miramar and of the Montseny Massif (Roqué 1999).

These new data considerably restrict the extent of the early Silurian stratigraphical hiatus previously generally accepted, especially in the regions where it was supposed to correspond to the whole Rhuddanian plus the Aeronian. In many areas, the stratigraphical gap at the Ordovician–Silurian boundary seems extremely limited and, at the most, restricted to the latest Kosovian (uppermost part of the *Normalograptus persculptus* Biozone) and/or to the earliest Rhuddanian. However, a strict demonstration would necessitate a complete and continuous graptolite record, which is not available.

Within the Ordovician–Silurian transitional sequence, the most important gap is the pre-diamictite hiatus; however, it must be noted that, even when the stratigraphical succession shows the absence of an important part of the Ordovician, this hiatus was created within a very short time interval, corresponding only to a very small part of the Kosovian interval. Detailed studies of the Ordovician–Silurian transition, both in the Hodh area of SE Mauritania (Paris *et al.* 1998; Underwood *et al.* 1998) and in the northeastern Algerian Sahara (Oulebsir & Paris 1995; Paris *et al.* 2000*a*), have established a precise timetable, based on chitinozoans and graptolites, of the successive events linked to the glaciation. The maximum development of the ice-cap and the resulting erosional episodes correspond mainly to the *Normalograptus extraordinarius* graptolite biozone. The fluvioglacial and/or glaciomarine sedimentation induced by the melting of the ice corresponds to the upper part of the *Tanuchitina elongata* chitinozoan biozone (Paris 1999), i.e. the lower part of the *Normalograptus persculptus* graptolite biozone. Post-glacial conditions and sedimentation without glacial influence were re-established before the true

beginning of the Silurian period, during the upper part of the *N. persculptus* Biozone. Thus, the Iberian glaciomarine diamictites (the thickness of which is 100 m or more) were deposited in a brief time interval, probably in the order of 500 000 years. This was also the case for the pre-Silurian post-glacial sediments such as those of the Losadilla Fm (300 m) in the Truchas syncline. The unequal importance of this pre-diamictite hiatus, depending on the area considered, strongly suggests that a varied topography, possibly due to pre-Kosovian extensional tectonic basin-floor rearrangement, existed in the marine areas that emerged and were subjected to erosion when the sea level lowered during the maximum development of the African ice-sheet.

Ordovician palaeogeography of the Iberian Peninsula

To propose a precise Ordovician palaeogeographical reconstruction of the Iberian Peninsula or any other North Gondwanan region of Europe is not easy. These regions have been involved in the Variscan orogeny and, in some cases, Alpine deformation, and this has greatly obscured the evidence for Early Palaeozoic pre-Variscan palaeogeography and displaced different parts of the North Gondwanan Province. Several earlier authors have already tackled this problem (e.g. Paris & Robardet 1977; Hammann *et al.* 1982; Julivert & Truyols 1983; Robardet & Gutiérrez-Marco 1990*b*). Progress achieved in our understanding of the Spanish Ordovician successions and faunas during the last 20 years increasingly allows this problem to be considered on the basis of more complete and more precise data (see for example Paris 1998).

Central Spain: the southern Central Iberian Zone

The southern part of the Central Iberian Zone (the 'Luso-Alcudian Zone' of Lotze 1945) shows a remarkable Ordovician succession, with extensive outcrops and numerous fossil localities, that can be considered as a reference for comparison with other regions of Spain. The basal Ordovician deposits are Arenigian in age and rest transgressively and discordantly on Cambrian or Upper Proterozoic rocks. This age is clearly established for the Armorican Quartzite and is most probable for the underlying fluviotidal conglomerates and red beds. All the remaining deposits that form the Ordovician succession of central Spain show shallow stable shelf characteristics. This is attested by the sedimentological features of the Armorican Quartzite and also by those of the younger sandy units that occur within and above the Middle Ordovician open shelf '*Tristani* Beds', and in the Upper Ordovician. All these sandstones show wave ripples and hummocky cross-stratification that indicate deposition within shallow-shelf, storm-influenced environments. However, facies changes and variations in the composition of the faunal associations of trilobites and brachiopods, especially during Middle Ordovician times, show a progressive deepening gradient from south to north within the Central Iberian Zone (see Hammann & Henry 1978; Brenchley *et al.* 1986; Rábano 1989*b*). Although it is beyond the scope of the present chapter, it can be noted also that a similar deepening gradient, with a reversed north–south direction, occurs in the Mid-North Armorican regions where the Ordovician successions and benthic faunas are strongly similar to those of the Central Iberian Zone. This suggests that the two regions were closely connected during the Ordovician period, with the Armorican part in a reverse north–south orientation (Gutiérrez-Marco & Rábano 1987*b*, fig. 1; Hammann 1992) forming a direct continuation of the Central

Iberian regions (Paris & Robardet 1977; Robardet *et al.* 1990, 1994*a*).

NW Spain: northernmost Central Iberian and West Asturian-Leonese zones

In NW Spain, i.e. in the northernmost units of the Central Iberian Zone and in the West Asturian-Leonese Zone, the Lower Ordovician succession conformably follows the Cambrian rocks within the very thick Los Cabos Gp; the quartzites of the upper part of this group are rather similar to the Armorican Quartzite (Pérez Estaún 1978; Hammann *et al.* 1982; Julivert & Truyols 1983, and references therein). The dark shales of the Middle Ordovician (Oretanian) Luarca Fm contain numerous fossil taxa that also occur in the '*Tristani* Beds' of the southern Central Iberian Zone. However, the number and frequency of the fossiliferous levels, the taxonomic diversity and the abundance of individuals are lower in the West Asturian-Leonese Zone. There are also some differences in the vertical distribution of several taxa: some species known only in the lower Oretanian succession in the southern Central Iberian Zone persist in the West Asturian-Leonese Zone up to the upper Oretanian, and other taxa considered until now typical of the Dobrotivian in the southern Central Iberian Zone already occurred in the north by Oretanian times. But the most important point is the occurrence in the West Asturian-Leonese Zone, and also in the Eastern Iberian Cordillera, of taxa considered indicative of deep marine environments. This is the case for the mesopelagic graptolite genus *Pterograptus*, always found in deep-water sediments (Cooper *et al.* 1991). This genus has been identified in Upper Oretanian rocks at three localities in the Vega de Espinareda syncline and in the Castillejo Fm of the Eastern Iberian Cordillera (Gutiérrez-Marco 1986*a,b*; Gutiérrez-Marco *et al.* 1999*a*). This is also the case for a raphiophorid trilobite, found in the Villaodrid syncline, characteristic of a trilobite biofacies deeper than the *Neseuretus* biofacies developed in the southern Central Iberian Zone. These data, as well as the scarcity of bioturbation, suggest marine environments deeper than those of the southern Central Iberian Zone, with more pronounced anoxia (Gutiérrez-Marco *et al.* 1996*a*, 1999*a*). It must be added that the West Asturian-Leonese Zone is characterized by the great thickness of its Cambrian and Ordovician successions, reaching a maximum of 11 000 m in the Navia–Alto Sil domain where the Luarca and Agüeira formations are especially thick (Julivert *et al.* 1972*b*). In the palaeogeographical reconstructions proposed for the West Asturian-Leonese Zone during Early Palaeozoic times (Pérez Estaún *et al.* 1990; Martínez Catalán *et al.* 1992*a*, and references therein) this region is a continental margin affected by tectonic extension that induced the formation of fault-controlled, rapidly subsiding troughs in which thick synrift sequences accumulated (Truchas and Navia–Alto Sil domains). Among these troughs the most pronounced is the Astur–Celtiberian trough, that connects these domains with the Sierra de la Demanda and the Iberian Cordillera.

Cantabrian Zone

In the Cantabrian Zone, the massive quartzites of the upper part of the Barrios Fm were deposited during Arenigian times in tidal environments like those of the Armorican Quartzite of the southern Central Iberian Zone (Truyols *et al.* 1990;

Aramburu *et al.* 1992). This period is followed by a stratigraphical hiatus that corresponds to the Lower Oretanian and the lowermost Upper Oretanian in the east (Sueve area), and to the entire Oretanian and the lower part of the Lower Dobrotivian in the west (Cabo Peñas). The dark shales of the Sueve and Luarca formations still correspond to the neritic *Neseuretus* biofacies but, when compared with the abundant fossils of the southern Central Iberian Zone, the benthic faunas of the Cantabrian regions are rather scarce and poorly diversified. This relative paucity of the benthos, and the occurrence of epipelagic graptolites unknown in the southern Central Iberian Zone (*Pseudamplexograptus, Eoglyptograptus*), could be linked to somewhat deeper environments. These distinctive features of the Ordovician sediments of the Cantabrian Zone suggest that they were probably deposited in more distal shelf areas, a hypothesis supported by the presence of faunal immigrants (some brachiopods, graptolites and ostracods) of Baltic and Avalonian origin (Gutiérrez-Marco *et al.* 1996*a*, 1999*a*). Finally, it must be noted that, despite the significant differences shown by the Central Iberian, West Asturian-Leonese and Cantabrian zones, these three areas were part of the same continental shelf.

Ossa Morena Zone

Although undoubtedly North Gondwanan, the Lower Palaeozoic (and Lower Devonian) succession of the Ossa Morena Zone clearly differs from that of the Central Iberian Zone with which it is presently juxtaposed along the Badajoz–Córdoba Shear Zone (Robardet 1976; Robardet *et al.* 1998). With regard to the Ordovician succession, the shallow marine inner-shelf deposits that are so typical of the Central Iberian regions, such as the Armorican Quartzite and the overlying '*Tristani* Beds', do not exist in the Ossa Morena Zone *sensu stricto*. but do occur in the intervening Lusitan–Marianic Zone. The Tremadocian–Arenigian transitional beds and the Arenigian succession consist of green shales and siltstones with graptolites. Sediments deposited in outer-shelf distal environments are much more developed and some of the faunas have more affinities with those of Bohemia, especially in Oretanian rocks (Gutiérrez-Marco *et al.* 1984*d*; Robardet & Gutiérrez-Marco 1990*a,b*). The latest Tremadocian and the Kralodvorian green shales and limestones bear also some deep-water cyclopygid trilobites (Robardet 1976; Hammann 1992; Robardet *et al.* 1998). These differences and the bathymetric gradient recognized in the Ordovician of the Central Iberian Zone (deepening from south to north) indicate that the present geographical configuration does not correspond to the original positions of the two zones during Early Palaeozoic times. The two regions originally formed two separate parts of the North Gondwanan shelf (inner and outer shelf respectively) and have been juxtaposed along the Badajoz–Córdoba Shear Zone by Variscan transcurrent movements (Robardet & Gutiérrez-Marco 1990*a,b*; Eguíluz *et al.* 2000 and references therein).

Catalonian Coastal Ranges and Pyrenees

The Ordovician succession of the Catalonian Coastal Ranges and the Pyrenees remains rather poorly known, especially for its Lower and Middle series where biostratigraphical data are practically non-existent. As already mentioned, it is generally assumed (by comparison with Sardinia and the Montagne Noire of southern France) that the middle part of the

Ordovician is missing and that the Berounian directly overlies Lower Ordovician rocks transgressively and unconformably (i.e. the genuine Sardic unconformity). This highly characteristic sedimentary evolution, characterized by 'Sardic movements' between the late Arenigian and the Berounian and a 'Middle Ordovician non-sequence', is known also in the Montagne Noire and Sardinia.

Palaeogeographical summary and conclusions

Overviewing the Ordovician successions and faunas that occur in the different regions of Spain, some points of prime importance can be summarized as follows.

It is firmly established that the Ordovician successions and benthic faunas from all the Iberian regions (except the South Portuguese Zone where Ordovician rocks are not exposed) are of North Gondwanan type. Moreover, considering in detail the successions and faunas from the different regions, it is possible to distinguish distinct domains, each one showing sedimentological, faunal and stratigraphical characteristics that suggest a particular sedimentary evolution and a particular position within the North Gondwanan marine shelf.

The Tomar–Badajoz–Córdoba Shear Zone is a major Variscan transcurrent fault that has juxtaposed two distinct parts of the North Gondwanan shelf, namely the outer shelf Ossa Morena Zone with cyclopygid biofacies and the inner shelf southern Central Iberian Zone with *Neseuretus* biofacies. Thus this structural line cannot be the cryptic suture of a major Palaeozoic ocean (later obliterated by a Variscan shear zone), contrary to a widely held interpretation (e.g. Burg *et al.* 1981; Matte 1986, 2001).

The Mid-North Armorican regions of NW France were most probably in an orientation opposite to that of today, and formed the direct continuation of the southern Central Iberian regions within the shallowest part of the North Gondwanan shelf.

The overall continuity of the lithofacies and faunas, from the southernmost areas of the Central Iberian Zone towards the NNW, especially in the Lower and Middle Ordovician, most probably reflects an original palaeo- and biogeographical continuity within the North Gondwanan marine shelf with an overall SE to NW deepening trend, the deepest environments being found in the Truchas and Navia–Alto Sil areas of the northernmost Central Iberian and West Asturian-Leonese zones, respectively.

The Catalonian Coastal Ranges and the Pyrenees were most probably also situated in outer shelf areas during the Early Ordovician, but the absence of the middle part of the Ordovician is a special feature regarded by Hammann (1992, pp. 40–44) of prime importance in the palaeogeographical evolution of the North Gondwanan regions. This author considers that the regions characterized by a 'Middle Ordovician non-sequence' (Pyrenees, Catalonian Coastal Ranges, Montagne Noire, Sardinia, the Alps, parts of the Carpathians and southern Turkey) corresponding to 'unstable shelf' areas, situated along the northern border of the stable North Gondwanan shelf. The stable North Gondwanan shelf areas, that included the Central Iberian Zone, were influenced only by minor epeirogenic movements during the Ordovician period. In contrast, the 'unstable shelf' areas, characterized by continuous subsidence and very thick Upper Cambrian–Lower Ordovician sequences, were affected by Middle Ordovician uplifting movements that formed a 'Sardinian-Taurian rise'

and induced a major stratigraphical hiatus spanning the late Arenigian–early Berounian time interval. In this context, Hammann (1992) also considers that the geographical isolation of the North Gondwanan *Selenopeltis* Fauna from the Avalonian and Baltic benthic faunas was not the consequence of the opening or widening of a mid-European Rheic Ocean, but was caused by the emergence of the 'Sardinian-Taurian rise' land barrier. Another possible palaeogeographical element of a similar nature, later uplifted during Kralodvorian and Kosovian times, could be the so-called 'Cantabro-Ebroian Massif' which had an influence at a subprovincial palaeogeographic level (Vennin *et al.* 1998; Villas *et al.* 1999).

Stratigraphical breaks of varying importance also occur during the late Arenigian–early Berounian time interval within the Ordovician successions of other Iberian regions, where they are generally defined by oolithic ironstone beds that mark the return to sedimentation. Ironstone deposition is also a common and typical feature of the North Gondwanan Ordovician sequences (Gutiérrez-Marco *et al.* 1984b; Young 1989, 1992), having been studied in Spain by Lunar (1977), Lunar & Amorós (1979), Gutiérrez-Marco *et al.* (1984b), García-Ramos *et al.* (1987), García-Ramos & Robardet (1992) and Fernández *et al.* (1998), among others.

The main hiatuses (see the regional successions described above and also Fig. 4.3) are marked by ooidal ironstone beds corresponding to:

(1) the whole Dobrotivian in the northernmost Central Iberian Zone and the West Asturian-Leonese Zone (Gutiérrez-Marco *et al.* 1999a);
(2) the Lower Oretanian in the Iberian Cordillera and the Laviana–Sueve area of the Cantabrian Zone (Gutiérrez-Marco *et al.* 1984b, 1996a);
(3) the whole Oretanian and the lowermost Dobrotivian in the Cabo Peñas area of the Cantabrian Zone (Gutiérrez-Marco *et al.* 1999a);
(4) a small part of the Lower Berounian in several regions of the Iberian Peninsula (southern Central Iberian Zone, Iberian Cordillera, and possibly some parts of the West Asturian-Leonese Zone), and in the mid-North Armorican regions (Paris 1981, 1990; Young 1989, 1992; Villas 1992; Romão *et al.* 1995; Gutiérrez-Marco *et al.* 1999a).

All these stratigraphical hiatuses could be regarded as the echoes of 'Sardic movements' in the 'shelf-to-basin transitional zone' (i.e. the Cantabrian, West Asturian-Leonese and northernmost Central Iberian zones, and the Iberian Cordillera) and even in the 'stable shelf' areas (i.e. the southern Central Iberian Zone and the Mid-North Armorican regions).

Going even further, Hammann (1992, p. 43 and fig. 13) considers that extension, block faulting and rifting caused the disappearance of the 'Sardinian-Taurian rise' in early Berounian times and relate to the origin of the mid-European Rheic Ocean. The global palaeogeographical model proposed by Hammann is very attractive in many respects (see Robardet *et al.* 1994b), but this precise point seems debatable. Although a full discussion of the palaeogeographical evolution of all the North Gondwanan regions is beyond the scope of the present chapter, it can be noted that the regions characterized by the 'Sardic movements' and a 'Middle' Ordovician stratigraphical hiatus have most probably a geographical expansion smaller than supposed by Hammann (1992), Middle Ordovician rocks having been identified in the subsurface of the Aquitaine Basin (Paris & Le Pochat 1994, and references therein) and also in some regions of southern Turkey (Central Taurus: Sarmiento *et al.* 1999*b*). This is not in full accordance with the idea that their evolution during Late Ordovician times occurred during the origin of the major Rheic Ocean, the birth of which was probably earlier (Paris & Robardet 1990; Robardet *et al.* 1993, 1994*b*; Cocks & McKerrow 1993; Cocks *et al.* 1997).

Considering the North Gondwanan Ordovician palaeogeography at a wider scale, it can be noted that the Iberian Peninsula is commonly regarded as the essential part of a promontory prolonging the Moroccan part of the North Gondwanan shelf (e.g. Young 1990; Cocks & Fortey 1990; Cocks 2000). However, the Ordovician Iberian (and Armorican) benthic faunas are more similar, at the species level, to those of the NE Algerian Sahara and of Saudi Arabia than to those of Morocco. These closer faunal similarities thus suggest that, during Ordovician times, the Iberian Peninsula was not contiguous to Morocco but more probably to the Algerian Sahara (Mélou *et al.* 1999) or perhaps adjacent to Libya. This localization within the North Gondwanan shelf would explain the distribution of several Iberian and Armorican fossils that occur in Algeria and Saudi Arabia when they are unknown in Morocco. This is the case, for instance, for the late Arenigian to Oretanian problematic genus *Hanadirella* (Vannier & Babin 1995; Gutiérrez-Marco & Rábano 1996; Emig & Gutiérrez-Marco 1997), for the trilobite *Neseuretus tristani* and for brachiopods (*Cacemia ribeiroi, Apollonorthis bussacensis, Aegiromena mariana, Heterorthina kerfornei*) of Oretanian–Dobrotivian age (Fortey & Morris 1982; Mélou *et al.* 1999).

Another line of evidence suggesting that the Iberian Peninsula was possibly adjacent to Libya (and also in an intermediate position with regard to Saudi Arabia) is supplied by the resemblances between the Kralodvorian limestones of the Iberian Peninsula and the coeval carbonate mud mounds of the Djeffara Fm in the northern part of the Ghadamis basin (NE Libya: Bergström & Massa 1992; Massa & Bourrouilh 2000). The latter were deposited 'in a high latitude periglacial area' and contain a conodont fauna representative of the *Sagittodontina robusta–Scabbardella altipes* biofacies which is quite similar to coeval conodont faunas known from the Iberian Peninsula, the Armorican Massif and Thuringia, but rather different from those of the Kralodvorian limestones found in the Carnic Alps of Austria and Italy and also in Sardinia (Sweet & Bergström 1984; Sarmiento 1993).

This palaeogeographical position of the Iberian Peninsula could explain the biogeographical relations observed in the Llandovery graptolite faunas (see Chapter 5). It could also match the migration routes of the Australo-Sinian trilobites recorded in an increasing number of Upper Cambrian localities from North Africa and SW Europe (Shergold *et al.* 1983, 2000). This influx of Asiatic taxa into the Ordovician strata of SW Europe is only illustrated by the presence of some Chinese graptolite species in the Arenigian of northern Spain (Gutiérrez-Marco & Rodríguez 1987) and of SE France (J. C. Gutiérrez-Marco, unpublished). This is because the migrations operated mainly eastwards during this period (Zhou & Dean 1989; Gutiérrez-Marco *et al.* 1999*c*), except for some Australo-Sinian faunal 'incursions' in the Middle Ordovician of Saudi Arabia (El-Khayal & Romano 1985; Evans 2000).

Another problem is the proliferation of various palaeogeographical models (Linnemann 1998; Erdtmann 1998; Linnemann & Heuse 2000, and references therein) where different regions of the North Gondwanan shelf are considered as separated 'microcontinents', 'microplates' or 'terranes' that rifted away from the Gondwanan continent independently and at different moments of the Early Palaeozoic and progressively drifted northwards as autonomous microplates, to be finally amalgamated within the Late Palaeozoic Variscan Belt (c.f. 'Armorican Terrane Assemblage' (Tait *et al.* 1997) or 'Armorican microplate' (Matte 2001)). These models were based originally on palaeomagnetic data that later appeared highly questionable or erroneous. However, the microplate concept still survives and generates models that are generally totally regardless of the precise sedimentary and palaeobiogeographical affinities that existed between diverse regions; this is especially the case for the Central Iberian and mid-North Armorican regions that are frequently figured without any connection (Linnemann 1998; Erdtmann 1998; Linnemann & Heuse 2000; Matte 2001). Such models do not take into account the diversity of the environments (and the possible intrashelf basins) that existed during the Ordovician period over the huge North Gondwanan shelf.

Ordovician palaeontology of Spain

Torrubia (1754) was the first author to illustrate Spanish Ordovician fossils (some trilobites from the Western Iberian Cordillera, described as crabs). However, modern palaeontological knowledge started with Verneuil & Barrande (1855), who studied a very important collection of Central Iberian fossils obtained by Prado (1855). The latter (Spanish) author published a study on the fossils from the Armorican Quartzite north of Madrid some years later (Prado 1864).

In the first half of the twentieth century, the only important paper was a general contribution to the palaeontology of the 'Tristani Beds' by Born (1918). In the second half of the same century, and especially the last 30 years, there was a drastic renewal of interest in Ordovician fossils in Spain, mainly through the trilobite monographs published by Hammann (1974, 1983, 1992), and also many papers on macro- and microfossils of various ages and taxonomical groups.

The Ordovician fossils so far identified in Spain include about 450 taxa (excluding ichnofossils, conodonts and organic-walled microfossils), coming from more than 600 fossiliferous sections or single fossil localities distributed throughout the Spanish Variscides except the South Portuguese Zone and the Betic Cordilleras. Questionably Ordovician palynomorphs supposedly derived from high-pressure metamorphic rocks (Fombella Blanco 1989; Martínez García & Fombella 1997) are the only Ordovician fossils so far reported from the Galicia–Trás-os-Montes Zone (but see also comments by Gutiérrez-Marco *et al.* 1997*c*, p. 13).

Spanish Ordovician fossils show different patterns of extinction and recovery, mainly related to periods of great global transgressions and regressions, and with climatic crises. However, the values of the general biodiversity are strongly dependent on the sedimentary facies and environments prevailing over large areas of the high-latitude North Gondwanan Province. In the Lower Ordovician and basal Middle Ordovician successions only 33 different species have been found and the diversity is biased due to the existence of lithofacies inappropriate for the preservation of macrofossils (i.e.

the Armorican Quartzite and equivalents), also combined with the widespread stratigraphical gap corresponding to most of the Tremadocian. On the contrary, the Middle Ordovician shales and sandy tempestites ('*Tristani* Beds') show the highest values of faunal diversity, with 409 different species identified up to now. In the Upper Ordovician, the fossil record is restricted to relatively few palaeontological horizons, mostly related to coquina beds within coarse clastic sucessions, or with the Kralodvorian limestones, where in fact biodiversity is relatively high. A total of 141 species have so far been recorded in the Upper Ordovician of the Iberian Peninsula.

The most abundant and diversified invertebrate fossil groups of the Iberian Ordovician correspond to trilobites, brachiopods, echinoderms and molluscs, all of them particularly common in shallow-water shelf facies. Apart from some recent general accounts on the Middle Ordovician faunas from northwestern Spain (Gutiérrez-Marco *et al.* 1996*a*, 1999*a*) and the Upper Ordovician faunas of the entire country (Hafenrichter 1979), the main bulk of palaeontological studies (with taxonomy and systematics of the diverse Ordovician groups) come from the Central Iberian Zone and the Iberian Cordillera. This was favoured by the presence of many fossiliferous sections, in a relatively simple geological setting where fossils were not affected by important deformation or metamorphism.

The main references for Middle and Upper Ordovician **trilobites** in Spain are papers by Verneuil & Barrande (1855), Born (1918), Carré *et al.* (1970), Hammann (1974, 1976*a*, 1983, 1992), Hammann & Henry (1978), Hammann & Rábano (1987), Rábano (1981, 1983, 1984, 1989*a,b,c,d,e*), Rábano & Gutiérrez-Marco (1983), Rábano *et al.* (1985), Gutiérrez-Marco & Rábano (1987*a*) and Owens & Hammann (1990). Ordovician **brachiopods** were studied by Chauvel *et al.* (1969), Havlícek & Josopait (1972), Arbin *et al.* (1978), Mélou (1973, 1975, 1976), Villas (1983, 1985, 1992, 1995), Villas *et al.* (1987, 1989, 1995*b*, 1999) and Emig & Gutiérrez-Marco (1997). **Echinoderm** research has demonstrated enhanced interest in several groups such as diploporids, rhombiferans, crinoids, coronates, asterozoans and homalozoans (Meléndez & Hevia 1947; Meléndez 1951, 1958, 1959; Chauvel 1973; Chauvel & Truyols 1977; Chauvel *et al.* 1975; Chauvel & Meléndez 1978, 1986; Chauvel & Le Menn 1979; Meléndez & Chauvel 1982; Gutiérrez-Marco *et al.* 1984*a*, 1996*b,c*; Hammann & Schmincke 1986; Gutiérrez-Marco & Meléndez 1987; Domínguez & Gutiérrez 1990; Gutiérrez-Marco & Baeza Chico 1996; Gil Cid *et al.* 1996*a,b,c,d,e*, 1998, 1999; Aceñolaza & Gutiérrez-Marco 1998*a*; Gutiérrez-Marco & Aceñolaza 1999; Gutiérrez-Marco 2000, 2001; Lefèbvre & Gutiérrez-Marco 2002). Spanish Ordovician **molluscs** (bivalves and some rostroconchs, gastropods and cephalopods) have been described by Hernández Sampelayo (1948), Gutiérrez-Marco & Martín Sánchez (1983), Babin & Gutiérrez-Marco (1985, 1991, 1992), Fryda & Gutiérrez-Marco (1996), Gutiérrez-Marco (1997), Gutiérrez-Marco *et al.* (1997*b*, 2000), Gutiérrez-Marco & Babin (1999), Fryda *et al.* (2001) and Babin & Hammann (2001). Ordovician **graptolites** are quite rare in the Iberian Peninsula, except in some Oretanian levels, where they are numerically abundant but very poorly diversified. The main graptolite studies are from Haberfelner (1931), Hernández Sampelayo (1960), Skevington (1974), Gutiérrez-Marco (1982, 1986*a,b*), Erdtmann *et al.* (1987), Gutiérrez-Marco & Aceñolaza (1987), Gutiérrez-Marco & Rábano (1987*a*), Gutiérrez-Marco & Rodríguez (1987) and Gutiérrez-Marco *et al.* (1996*a*, 1999*a*). Other Ordovician fossils such as the organophosphatic brachiopods, hyolitids, polyplacophorans, machaeridians, conularids, bryozoans and problematica (for instance *Hanadirella*: Gutiérrez-Marco & Rábano 1996) are currently the subject of detailed taxonomy in the frame of IGCP project 410.

Ordovician **ichnofossils** from Spain, including traces, coprolites and endolithic borings were mainly presented by Radig (1964), Bouyx (1970), Seilacher (1970), Moreno *et al.* (1976), Crimes *et al.* (1974), Crimes & Marcos (1976), Moreno *et al.* (1976), Baldwin (1977), Kolb & Wolf (1979), Pickerill *et al.* (1984), Gutiérrez-Marco (1984), Romano (1991), Mayoral (1991), Mayoral *et al.* (1994) and Aceñolaza & Gutiérrez-Marco (1998*b*, 1999).

Micropalaeontological studies in the Ordovician of Spain have first focused on **palynomorphs** (Cramer-Díez *et al.* 1972; Cramer-Díez & Díez 1978; Robardet *et al.* 1980; Wolf 1980*a*; Paris 1981; Mette 1989; Albani *in* Gutiérrez-Marco *et al.* 1996*a*; Servais & Mette 2000) and **ostracods** (Vannier 1986*a,b*, 1987; Vannier *in* Gutiérrez-Marco *et al.* 1996*a*). However, in the last two decades **conodont** research is also being developed with some noticeable and promising results (Sarmiento 1990, 1993; Sanz López & Sarmiento 1995; Sarmiento & Gutiérrez-Marco 1999; Sarmiento *et al.* 1995*a*, 1999*a,b*).

The majority of the fossil assemblages so far recorded in the Ordovician rocks of Spain correspond to typical shallow-platform faunas, ranging from the subtidal shoreface zone to the upper offshore shallow outer shelf. Trilobite communities of the Middle and lowest Upper Ordovician succession are dominated by the homalonotids, and the *Neseuretus* and *Colpocoryphe* biofacies (in the sense of Rábano 1989*b* and Vidal 1998). The deeper cyclopygid biofacies is only represented in the Ossa Morena Zone (from the Upper Tremadocian to the Kralodvorian), whereas deep outer shelf and more cosmopolitan trilobite assemblages, such as the *Ovalocephalus* biofacies, occur in a few Kralodvorian localities of the Iberian Cordillera and the Cantabrian Zone (Hammann 1992; Gutiérrez-Marco *et al.* 1996*b*). The only record of a raphiophorid trilobite was in the Oretanian shales of the West Asturian-Leonese Zone, not far from other contemporaneous localities which yielded some mesopelagic graptolites (Gutiérrez-Marco *et al.* 1996*a*).

We greatly acknowledge S. C. Finney, Vicechairman of the ICS and chairperson of the Ordovician Subcommission (IUGS-UNESCO), for reading the paper and for his comments on the palaeogeographical setting and chronostratigraphy of the Iberian Ordovician; R. A. Fortey (London), N. Spjeldnaes (Oslo) and the editors for their careful and constructive reviews of the text, as well as D. Bernard (CNRS, Géosciences-Rennes) for the illustrations. Work in Spain has been supported by the projects PB96-0839 and BTE2000-1310 of the Spanish Ministry of Science and Technology. This paper is also a contribution to the IGCP Project 410 (IUGS-UNESCO) and to the Iberia project of EUROPROBE (European Science Foundation).

5 Silurian

MICHEL ROBARDET & JUAN CARLOS GUTIÉRREZ-MARCO

The Silurian rocks of Spain occur in all the zones of the Iberian (or Hesperian) Massif, except the South Portuguese Zone. Silurian rocks also crop out in other parts of the Variscan Belt that were later affected by the Alpine orogeny, in the Pyrenees, the Catalonian Coastal Ranges, the Iberian Cordillera and the Betic Cordilleras (Fig. 5.1). As in other regions of the North Gondwanan Province, the Silurian deposits of the Iberian Peninsula comprise mainly terrigenous sediments dominated by pelagic faunas. The most characteristic rocks are graptolitic black shales (the so-called 'ampelites') and the commonly mentioned uniformity of the Silurian succession mainly results from the special attention that has been paid to these richly fossiliferous rocks. Other types of rocks also occur in the Silurian sequences, however, and allow distinctions between different types of succession, as well as providing extra evidence for environmental conditions and palaeogeographical setting.

We consider that, from Cambrian to Devonian times, with the exception of the South Portuguese Zone that probably belonged to Avalonia (Oliveira & Quesada 1998, and references therein), the whole Iberian Peninsula was part of the North Gondwanan Province that extended along the northern margin of the African part of the Gondwana continent (Robardet *et al.* 2001, and references therein). The Silurian palaeolatitude of the Iberian Peninsula cannot be defined precisely, either from climatically sensitive lithofacies or from faunas, as the successions are almost entirely terrigenous and the faunas mainly pelagic. Available palaeomagnetic data are not clearly diagnostic: at Almadén, in the southern Central Iberian Zone, basaltic lavas interbedded within Llandovery graptolitic shales were remagnetized during Late Devonian or Early Carboniferous times (Perroud *et al.* 1991; Parès & Van der Voo 1992); in the Cantabrian Zone, the pre-folding magnetization (Perroud & Bonhommet 1984) of Furada/San Pedro Formation (Fm) red beds (close to the Silurian–Devonian boundary) could possibly be an Early Carboniferous remagnetization (Van der Voo 1993, p. 246). The latitudinal position of the Iberian Peninsula can thus be estimated only roughly as intermediate between the rather high or cold temperate latitudes (*c.* 50°S) of latest Ordovician times (Hirnantian glaciomarine deposits) and the warm temperate to subtropical latitudes (*c.* 35°S) of the Lower Devonian succession (limestones with local reefs).

Silurian successions, lithofacies and biostratigraphy

In this chapter, we have used the graptolite biozones (Fig. 5.2) of the generalized standard Silurian graptolite zonal sequence established and ratified by the Subcommission on Silurian Stratigraphy of the International Union of Geological Sciences (see Koren *et al.* 1996).

Cantabrian Zone

In the Cantabrian Zone (A, Fig. 5.1), Silurian rocks occur both in the 'folds and nappes region' (nos 1 and 2, Fig. 5.1; Fig. 5.3) and in the allochthonous Pisuerga–Carrión unit (no. 3, Fig. 5.1; Fig. 5.3) of the Palentian region in the SE.

In the 'folds and nappes region', the oldest Silurian deposits occur only in the Cabo Peñas area where the upper part of the El Castro Fm, i.e. the limestones of the Viodo Member (Mb) (= Viodo limestones; 30 m thick), has yielded brachiopods, bryozoans, corals, sponges, ostracods and conodonts. Biostratigraphical data derived from the conodonts (Sarmiento *et al.* 1994) and articulate brachiopods (Villas & Cocks 1996) reveal some discrepancies in suggesting respectively a pre-Telychian v. Telychian age for this limestone member.

Overlying the Viodo limestones in the Cabo Peñas area, and directly above Lower or Middle Ordovician rocks in most regions of the Cantabrian Zone, the Formigoso Fm consists of 100–300 m of black and grey siltstones and shales with a few thin sandy intercalations that show ripple laminations, hummocky cross-stratifications and trace fossils (Suarez de Centi *et al.* 1989). The Formigoso Fm has yielded abundant graptolites and palynomorphs, as well as scarce brachiopods, bivalves, cephalopods and trilobites. It extends from near the Aeronian–Telychian boundary (upper Aeronian beds have still to be confirmed) into the lower Sheinwoodian. These age assignations based on graptolite assemblages (main data compiled in Truyols *et al.* 1974) are confirmed by organic-walled microfossils, especially chitinozoans (Cramer-Díez & Díez 1978). The lowest chitinozoan assemblage was recently assigned to the *Conochitina alargada* Biozone, referred to the middle Aeronian by Verniers *et al.* (1995), on the basis of graptolites that certainly require re-evaluation.

The Formigoso Fm is overlain by the Furada (= San Pedro) Fm, which consists of 80–200 m of grey and reddish ferruginous sandstones with shaly intercalations, thin oolitic ironstone beds and, in the upper part, some sandy limestone lenses. The Furada/San Pedro Fm has yielded upper Wenlock palynomorphs from its lowermost part and Ludlow and Pridoli brachiopods, graptolites and palynomorphs from its middle and upper parts respectively. Ichnofossils are common and diverse at many levels in the succession (Suárez de Centi *et al.* 1989). Interestingly, terrestrially derived sporomorphs (miospores and cryptospores) occur abundantly within the entire formation, associated with diverse marine palynomorphs (see Rodríguez González 1983). The uppermost 30 m has yielded brachiopods, trilobites and conodonts regarded as Lochkovian and it is considered generally that the Silurian–Devonian boundary lies within the upper part of the Furada/San Pedro Fm (Truyols *et al.* 1974; Aramburu *et al.* 1992). This conclusion has been questioned, in the La Vid de Gordón section, where chitinozoans of Pridoli age occur in the uppermost part of the Furada/San Pedro Fm (Priewalder 1997). However the recent discovery and study of new important sections with a continuous record of chitinozoans has shown that the Silurian–Devonian boundary is situated between 13 m and 15 m below the top of the formation, at least in the Geras and Argojevo sections (Richardson *et al.* 2000, 2001).

In the Pisuerga–Carrión unit of the Palentian region (no. 3, Fig. 5.1), which is regarded as an allochthonous unit of West Asturian-Leonese origin, the lithological succession is somewhat different (Fig. 5.3). The lowermost part of the Silurian remains unknown. The Robledo Fm (*c.* 160 m) comprises mainly white sandstones, with shaly and silty intercalations that have provided Wenlock chitinozoans near the top. The overlying siltstones and black shales of the Las Arroyacas Fm

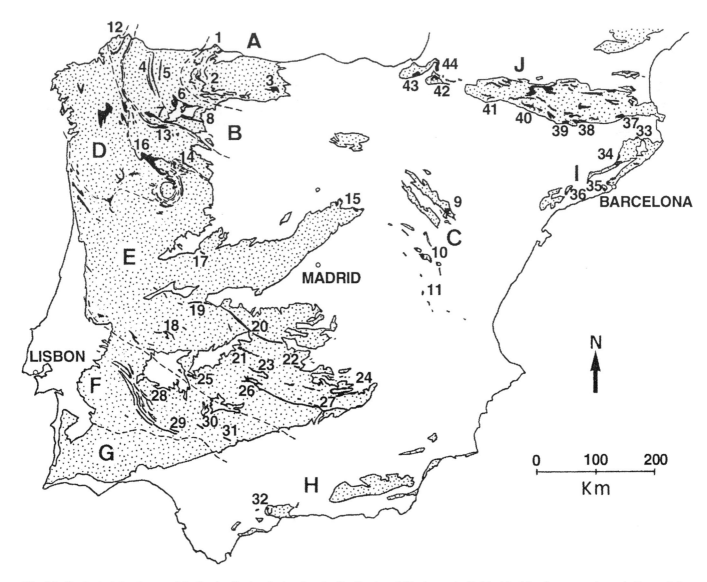

Fig. 5.1. Geological sketch map of the Iberian Peninsula showing the distribution of Silurian rocks (in black) with reference to the main Precambrian and Palaeozoic exposures (stippled). A–G, Iberian (Hesperian) Massif: A, Cantabrian Zone; B, West Asturian-Leonese Zone; C, Iberian Cordillera; D, Galicia–Trás-os-Montes Zone; E, Central Iberian Zone; F, Ossa Morena Zone; G, South Portuguese Zone; H, Betic Cordilleras; I, Catalonian Coastal Ranges; J, Pyrenees. 1 to 44: Main Silurian reference sections and fossil localities in Spain: 1, Cabo Peñas; 2, 'folds and nappes region'; 3, Palentian region; 4, Rececende and Villaodrid synclines (Mondoñedo nappe); 5, Los Oscos thrust-sheet; 6, Vega de Espinareda synclinorium; 7, Caurel-Peñalba syncline; 8, Castrillo syncline; 9, Eastern Iberian Cordillera; 10, Albarracín anticlinorium; 11, Serranía de Cuenca anticlinorium; 12, Cabo Ortegal area; 13, Sil and Truchas synclines; 14, Alcañices synclinorium; 15, Riaza and Atienza areas (Guadarrama, Central Range); 16, Verín-Bragança region; 17, Tamames syncline; 18, Sierra de San Pedro and Cáceres syncline; 19, Cañaveral-Monfragüe syncline; 20, Guadarranque syncline; 21, Herrera del Duque syncline; 22, Corral de Calatrava; 23, Almadén syncline; 24, Torre de Juan Abad; 25, Alange; 26, Cabeza del Buey-San Benito; 27, El Centenillo–Guadalmena; 28, Villanueva del Fresno; 29, Hinojales area; 30, Valle syncline; 31, Cerrón del Hornillo syncline; 32, Maláguide region; 33, Les Guilleries; 34, Montseny; 35, Barcelona; 36, Serra de Miramar; 37, Camprodon; 38, Bar-Toloriú; 39, Gerri de la Sal (Els Castells); 40, Sierra Negra; 41, Sallent de Gállego; 42, Quinto Real–Aldudes massif; 43, Cinco Villas massif; 44, Saint-Martin-d'Arrossa (France).

(300–450 m) have yielded benthic faunas, graptolites and chitinozoans of Wenlock, Ludlow and Pridoli ages. The sandstones and the carbonate beds of the Carazo Fm (*sensu stricto*: 250–380 m) have yielded chitinozoans of Pridoli age and, in the upper half, some Lochkovian brachiopods, scyphocrinoids, conodonts and chitinozoans (Jahnke *et al.* 1983; Schweineberg 1987; García-Alcalde *et al.* 1990; Gourvennec 1990; Aramburu *et al.* 1992; Rodríguez Fernández 1994).

West Asturian-Leonese Zone

The West Asturian-Leonese Zone (B, Fig. 5.1) disappears under the Cenozoic cover in the SE (Pérez-Estaún *et al.* 1990,

1992) and reappears in the Sierra de la Demanda and in the Iberian Cordillera of NE Spain (C, Fig. 5.1). The West Asturian-Leonese Zone is limited clearly to the east by the Precambrian rocks of the Narcea antiform. The western and southern limits have been a matter of discussion (Díez Balda *et al.* 1990; Martínez Catalán 1990*b*; Martínez Catalán *et al.* 1992*a*) and the Sil and Truchas synclines (no. 13, Fig. 5.1) are considered here to be the northernmost units of the Central Iberian Zone.

The West Asturian-Leonese Zone *sensu stricto* comprises, from east to west, the Navia–Alto Sil (nos 6 and 8, Fig. 5.1),

SERIES		STAGES	GRAPHTOLITES
SILURIAN	PRIDOLI		M. bouceki - I. transgrediens
			N. branikensis - N. lochkovensis
			N. parultimus - N. ultimus
	LUDLOW	LUDFORDIAN	F. formosus
			B. bohemicus tenuis - N. kozlowskii
			S. leintwardinensis
		GORSTIAN	L. scanicus
			N. nilssoni
	WENLOCK	HOMERIAN	C. ludensis
			C. praedeubeli - C. deubeli
			P. parvus - G. nassa
			C. lundgreni
		SHEINWOODIAN	C. rigidus - C. perneri
			M. riccartonensis - M. belophorus
			C. centrifugus - C. murchisoni
	LLANDOVERY	TELYCHIAN	C. lapworthi - C. insectus
			O. spiralis interval zone
			M. griestoniensis - M. crenulata
			S. turriculatus - S. crispus
			S. guerichi
		AERONIAN	S. sedgwickii
			L. convolutus
			M. argenteus
			D. triangulatus - D. pectinatus
		RHUDDANIAN	C. cyphus
			C. vesiculosus
			P. acuminatus

Fig. 5.2. Generalized Silurian graptolite zonal sequence (after Koren *et al.* 1996).

the Mondoñedo nappe (nos 4 and 5, Fig. 5.1) and the Caurel–Peñalba (no. 7, Fig. 5.1) domains. Silurian rocks occur in all three domains and these have yielded abundant graptolites, especially in the Mondoñedo nappe domain (Walter 1968; Marcos 1973; Pérez-Estaún 1978; Truyols & Julivert 1983). The Silurian graptolitic black shales overlie Ordovician rocks of different ages, including Lower Ordovician in some areas. As a whole, the Upper Silurian faunas from the West Asturian-Leonese Zone (and the northernmost part of the Central Iberian Zone) show clear palaeobiogeographical affinities with those from other North Gondwanan regions characterized by outer-shelf distal lithological successions, especially with the Barrandian area of the Czech Republic. In the same way, the Silurian of the Moncorvo and Lagoaça synclines, in the northernmost part of the Central Iberian Zone in Portugal, is a condensed succession of black shales with limestones in the upper Ludlow and Pridoli (Sarmiento *et al.* 1999*d*).

It is now established that, in the Villaodrid and Rececende synclines of the northern part of the Mondoñedo nappe, the Ordovician–Silurian transition comprises, in ascending order: (1) siltstones with dispersed clasts corresponding to the late Ashgill Hirnantian glaciomarine deposits, immediately followed by (2) thick-bedded,

dark quartzites (6–13 m) that comprise black siltstones with graptolite assemblages of Rhuddanian ages (*Parakidograptus acuminatus*, *Cystograptus vesiculosus* and *Coronograptus cyphus* biozones), overlain by (3) Aeronian and Telychian graptolitic black shales of the La Garganta Fm (Arbizu *et al.* 1997). The occurrence of Wenlock levels had been established previously (Walter 1968, 1969). Thus the black shale sequence in the Mondoñedo domain extends from the Telychian up into the Wenlock, with Sheinwoodian and Homerian faunas (Walter 1968; Romariz 1969; Marcos 1973), and is followed by chloritoid shales considered to be Upper Silurian.

In the Navia–Alto Sil domain, the Upper Ordovician Agüeira Fm (Caradoc and probably lower Ashgill) ends with the 'Vega quartzite' in the Vega de Espinareda syncline and the so-called 'lower quartzites' in the Castrillo syncline. The 'Vega quartzite' is probably uppermost Ordovician as the base of the overlying black shales has yielded graptolites of the Rhuddanian *Cystograptus vesiculosus* Biozone (Pérez-Estaún 1978; Gutiérrez-Marco & Robardet 1991). The 'lower quartzites' of the Castrillo syncline are most probably slightly younger as their upper part contains Aeronian graptolites. In the lower part of the overlying black shales three successive fossiliferous levels have yielded Telychian graptolite assemblages corresponding respectively to the *Rastrites linnaei* and *Monoclimacis griestoniensis* biozones and to the *Monoclimacis griestoniensis–Torquigraptus tullbergi* biozone interval (Gutiérrez-Marco & Storch 1997).

Within the Los Oscos thrust-sheet, at the limit between the Mondoñedo and Navia–Alto Sil domains, the black shale sequence is thicker (up to 500 m). The La Garganta Fm, that extends from basal Telychian to lower Homerian (Marcos & Philippot 1972), is overlain by the ferruginous sandstones and chloritoid shales of the so-called 'Queixoiro beds' (at least 40 m) which could be equivalent to the Furada/San Pedro Fm of the Cantabrian Zone.

In the Caurel–Peñalba domain (Fig. 5.4), Silurian black shales directly overlie the pre-Hirnantian Ashgill limestones of the La Aquiana Fm; the contact with the Peites/Valdevilla 'ferruginous level' was tectonized, karstified and mineralized in post-Palaeozoic times.

The precise stratigraphy of the black shale sequence (100–120 m) of the Peñalba syncline remains rather poorly known. However, Llandovery, Wenlock and early Ludlow graptolites have been mentioned (see references in Rábano *et al.* 1993) and the black shale sequence thus corresponds probably to the Telychian–lower Ludlow as in the Sil and Truchas synclines of the northernmost Central Iberian Zone. These black shales are overlain by about 200 m of chloritoid slates and siltstones, equivalent to the Salas Fm of the Sil syncline. In the basal part of the chloritoid slates and siltstones, a thin (0.5 m) bed of a black orthoceratid limestone has yielded cardiolid bivalves (*Cardiola docens*), conodonts (*Ozarkodina excavata excavata*, *O. confluens*, *Oulodus siluricus*, *Pseudooneotodus beckmanni*) and numerous albaillellid radiolarians, these fossils being indicative of the lower Ludfordian. Immediately above, the chloritoid slates and siltstones have yielded brachiopods, crinoids, rare solitary rugose corals and, especially, trilobites of Bohemian affinities (*Cromus* aff. *bohemicus*, *Crotalocephalus moraveci*, *Cerauroides articulatus*, *Lioharpes* sp.) that closely resemble those of the late Ludfordian *Prionopeltis archiaci* assemblage of the uppermost part of the Kopanina Fm of the Prague Basin; these levels appear equivalent to the 'Yeres quartzite' of the Sil syncline (Rábano *et al.* 1993). Above these levels the chloritoid slates contain poorly preserved graptolites (possibly *Neocolonograptus ultimus*) and nodules with orthocone nautiloids, primitive scyphocrinoid crowns and specimens of *Cardiolinka*, a bivalve genus abundant in the Pridoli. It is therefore possible that (as in the Sil syncline) most of these rocks are Pridoli in age (Piçarra *et al.* 1998) and that the Silurian–Devonian boundary lies within the upper part of the chloritoid slates. This would be in good accordance with the occurrence of Lochkovian and Pragian brachiopods, tentaculitids and conodonts in limestone levels of the overlying Peñalba Fm (Drot & Matte 1967; Pérez-Estaún 1978; Truyols-Massoni 1986; Gutiérrez-Marco *et al.* 2001).

Iberian Cordillera

In the Iberian Cordillera of NE Spain (C, Fig. 5.1), the Palaeozoic outcrops are entirely surrounded by a Mesozoic–Cenozoic

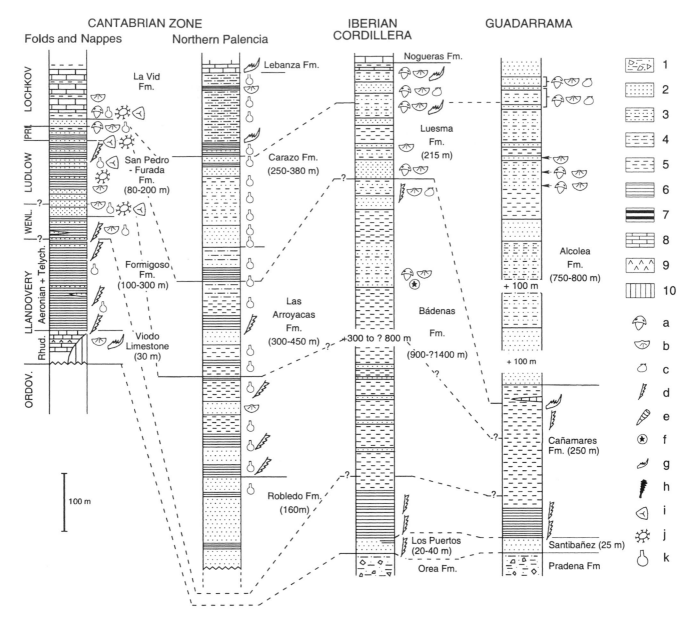

Fig. 5.3. The Silurian successions in the Cantabrian Zone, the Iberian Cordillera and the Guadarrama area of the Central Range. Cantabrian Zone, 'folds and nappes region' after Aramburu *et al.* (1992); Palencia region after Jahnke *et al.* (1983) and Schweineberg (1987); Iberian Cordillera after Carls (1977) and Gutiérrez-Marco & Storch (1998); Guadarrama area after Bultynck & Soers (1971) and Carls (1977). Key: for all stratigraphical columns (Figs 5.3 to 5.9): Lithofacies: 1, glaciomarine diamictites; 2, sandstones and quartzites; 3, alternating sandstones, siltstones and shales; 4, siltstones; 5, shales; 6, black shales; 7, black cherts ('lydites'); 8, limestones; 9, volcanic rocks; 10, gap. Fossils: a, trilobites; b, brachiopods; c, bivalves; d, graptolites; e, orthocone nautiloids; f, crinoids; g, conodonts; h, tentaculitids; i, spores; j, acritarchs; k, chitinozoans.

cover and correspond most probably to the eastern prolongation of the West Asturian-Leonese Zone. They are currently subdivided into eastern (or Aragonese) and western (or Castilian) branches (respectively nos 9 and 10, Fig. 5.1). The Silurian successions are rather similar overall to those of the Central Range and the southern part of the Central Iberian Zone.

In both branches of the Iberian Cordillera, the Silurian succession (Fig. 5.3; Carls 1977, 1983; Truyols & Julivert 1983; Gutiérrez-Marco & Storch 1998; Storch & Gutiérrez-Marco 1998) begins with the quartzites of the Los Puertos Fm (2–80 m, generally 20–40 m) that overlie the clast-bearing silty shales of the Orea Fm (Hirnantian

glaciomarine diamictites; Robardet & Doré 1988). Then follows the Bádenas Fm that consists mainly of black shales with nodules and sandstone intercalations that are especially abundant in the upper part. The Bádenas Fm is 300–400 m thick (but incomplete) in the Castilian Branch (Western Iberian Cordillera) and much thicker (900–1400 m) in the Aragonese Branch (Eastern Iberian Cordillera), where it is overlain by the sandstones of the Luesma Fm (about 200 m; Fig. 5.3).

The oldest Silurian fossils come from shaly intercalations in the upper part of the Los Puertos Fm (Gutiérrez-Marco & Storch 1998; Storch & Gutiérrez-Marco 1998). These graptolite assemblages are of Rhuddanian and Aeronian age (*Parakidograptus acuminatus*, *Coronograptus cyphus*, *Demirastrites triangulatus* and *Lituigraptus convolutus* biozones).

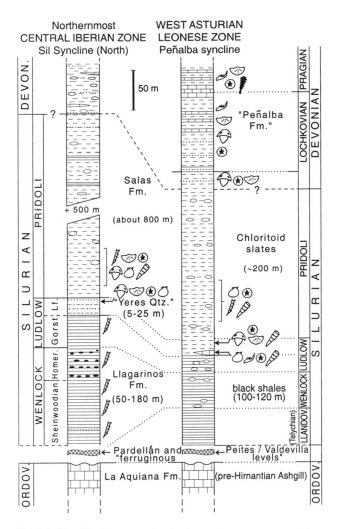

Fig. 5.4. The Silurian successions in the northernmost part of the Central Iberian Zone and the West Asturian-Leonese Zone. After Rábano *et al.* (1993), Arbizu *et al.* (1997) and Gutiérrez-Marco *et al.* (2001). For symbols see Figure 5.3.

The black shales of the Bádenas Fm have yielded abundant graptolite faunas indicating that the formation extends from the late Aeronian to basal Telychian *Rastrites linnaei* Biozone up into the basal Ludfordian *Saetograptus leintwardinensis* Biozone (Gutiérrez-Marco & Storch 1998). The black shales and nodules also contain brachiopods, bivalves, cephalopods, eurypterids, phyllocarids, cornulitids, trilobites and conodonts. In the Aragonese branch, sandstone members of the upper part of the Bádenas Fm yield shallow-water brachiopods, echinoderms, molluscs, conodonts and trilobites.

Finally, the upper part of the Luesma Fm has provided successive assemblages of Pridoli brachiopods and Lochkovian conodonts and brachiopods (Carls 1977; Truyols & Julivert 1983).

Central Iberian Zone

The Central Iberian Zone (E, Fig. 5.1) is currently subdivided into the Domain of Recumbent Folds (DRF) in the north and the Domain of Vertical Folds (DVF) in the south, these corresponding roughly to Lotze's (1945) original distinction of the Galician-Castilian (= DRF) and East Lusitanian-Alcudian zones (= DVF; Díez Balda *et al.* 1990). Both domains show common stratigraphical characteristics that reflect a common

sedimentary and palaeogeographical evolution but several separate areas of Silurian rocks can be distinguished (Truyols & Julivert 1983; Gutiérrez-Marco *et al.* 1990; San José *et al.* 1992).

Southern and southeastern Central Iberian Zone. Exposures of Lower Palaeozoic rocks in different synclines in Spain and Portugal (nos 18 to 27, Fig. 5.1) contain Silurian successions that have much in common (Fig. 5.5). The uppermost Ordovician glaciomarine formations (Gualija Fm of the Guadarranque and Herrera del Duque synclines, Chavera Fm of the Campo de Calatrava and the eastern Sierra Morena) are conformably overlain by a ubiquitous quartzitic formation, whose thickness varies between a few metres and 70 m. These quartzites have various local names (e.g. 'Castellar quartzites', 'Upper quartzites', 'Criadero quartzite') and correspond to the so-called 'Llandovery or Valentian quartzite' of previous publications. The best known is the 'Criadero quartzite' that hosts the famous mercury deposit of the Almadén mining district (Saupé 1990, and references therein; Chapter 19). These quartzites are overlain by graptolitic black shales with nodules that also contain sandstone beds, especially in their upper part. Then follows a sequence of alternating shales, siltstones and sandstones that progressively passes up into Lower Devonian sandstones where this part of the Palaeozoic succession has been preserved (Figs 5.5 and 5.9).

The age (Late Ordovician or Early Silurian) of the basal quartzitic units has long been debated and the discussion was somewhat confused by erroneous lithological correlations between quartzitic units of different ages (Robardet *et al.* 1980; García Palacios *et al.* 1996). Recent studies have shown that the Ordovician–Silurian boundary lies within the basal quartzitic units, as indicated by the following data:

(1) in the northern flank of the Almadén syncline (no. 23, Fig. 5.1), the lower part of the 'Criadero quartzite' contains *Hirnantia sagittifera* and *Plectothyrella crassicosta chauveli*, two brachiopods typical of the latest Ordovician *Hirnantia* fauna (Villas *et al.* 1999);

(2) in a nearby locality, its uppermost beds have yielded poorly preserved biserial graptolites that could be Rhuddanian or Aeronian;

(3) the uppermost part of the same formation contains graptolites of basal Aeronian age (Gutiérrez-Marco & Pineda Velasco 1988) at El Centenillo (no. 27, Fig. 5.1) and most probably Rhuddanian or basal Aeronian assemblages at Corral de Calatrava (no. 22, Fig. 5.1; García Palacios *et al.* 1996; Storch *et al.* 1998).

Numerous authors have mentioned graptolite faunas from the black shale sequence that overlies the Criadero Fm (see references in Truyols & Julivert 1983). Recent studies in the Guadarranque (no. 20, Fig. 5.1; Robardet *et al.* 1980; Rodríguez Núñez *et al.* 1989), Herrera del Duque (no. 21, Fig. 5.1; Pieren Pidal & Gutiérrez-Marco 1990), El Centenillo (no. 27, Fig. 5.1; Gutiérrez-Marco & Pineda Velasco 1988) and Corral de Calatrava (no. 22, Fig. 5.1; García Palacios *et al.* 1996; Storch 1998; Storch *et al.* 1998) synclines have resulted in more precise stratigraphical attributions. The total thickness of the black shales is about 20–35 m, except in the Almadén syncline where they are interbedded with basaltic lavas and pyroclastic rocks with a total thickness of about 400 m (Saupé 1971*a,b*). In addition to graptolites, the black shales have yielded conodonts (Sarmiento & Rodríguez Núñez 1991), brachiopods, orthoconic nautiloids, trilobites (García Palacios & Rábano 1996), hyolithids, phyllocarids, cornulitids and large fragments of eurypterids.

Although the basal horizons of the black shales have been frequently tectonized at the contact with the underlying quartzites, the graptolite assemblages found in the lowermost black shale levels indicate the basal Telychian (*Rastrites linnaei* Biozone). Generally, the black shales extend over the whole Telychian and reach possibly the lowermost part of the Wenlock. Sheinwoodian and doubtful lowermost Homerian graptolite assemblages occur at Corral de Calatrava (Storch *et al.* 1998).

The overlying 150–400 m of alternating shales, siltstones and sandstones ('San Pablo Schichten' of Butenweg 1968; 'Übergangsschichten Silur-Devon' of Puschmann 1970; 'Cerro Escudero Group' of Pardo Alonso & García-Alcalde 1996) contain only rare or poorly preserved graptolites, orthoconic nautiloids and bivalves. Precise stratigraphical

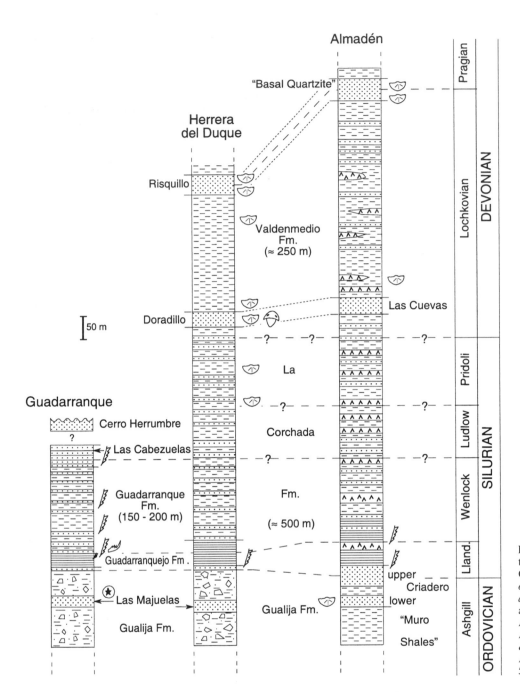

Fig. 5.5. The Silurian successions in the southern Central Iberian Zone. Guadarranque syncline after Robardet *et al.* (1980) and Rodríguez-Núñez *et al.* (1989); Herrera del Duque syncline after Puschmann (1970) and Pardo Alonso & García Alcalde (1996); Almadén syncline after Pardo Alonso & García Alcalde (1996) and Pardo Alonso (1998). For symbols see Figure 5.3.

assignments are therefore generally problematical. However, Sheinwoodian graptolites occur in their lowermost part in the Herrera del Duque syncline (Pieren Pidal & Gutiérrez-Marco 1990); lower Ludlow graptolites of the *Neodiversograptus nilssoni* Biozone have been found in the Guadarranque syncline in the upper part of the Las Cabezuelas sandstone member of the Guadarranque Fm (Rodríguez Núñez *et al.* 1989); basal Homerian (*Cyrtograptus lundgreni* Biozone) and basal Ludfordian (*Saetograptus leintwardinensis* Biozone) are known from the Alange syncline (no. 25, Fig. 5.1; Gutiérrez-Marco *et al.* 1997).

In the synclines where Devonian rocks have not been eroded, such as those of Herrera del Duque, Almadén and El Centenillo (respectively nos 21, 23 and 27, Fig. 5.1), the position of the Silurian–Devonian boundary long remained very uncertain. This was due to the absence of any precisely dated fossil assemblage (of late Ludlow or Pridoli age) between the levels with Wenlock or lower Ludlow graptolites and the fossiliferous sandstones supposedly corresponding

to the base of the Devonian ('basal quartzite' of previous authors). However, these are now known to contain brachiopods of late Lochkovian–early Pragian age (Pardo & García-Alcalde 1984). Recent studies in the Herrera del Duque syncline (Pardo Alonso & García-Alcalde 1996; Pardo Alonso 1998) have resulted in a better assessment of the real thickness of the alternating shales, siltstones and sandstones and subdivision into distinct lithological units (Fig. 5.5) with fossiliferous levels. It now appears that this sequence comprises, in ascending order:

(1) about 500 m of alternating shales, quartzites and micaceous sandstones, the 'Alternancias de La Corchada', most probably equivalent to the Guadarranque Fm of the Guadarranque syncline;
(2) the Doradillo Fm (40 m of quartzites and sandstones), equivalent to the Las Cuevas quartzites of the Almadén syncline;
(3) the Valdenmedio Fm, mainly composed of grey and mauve shales and siltstones (250 m);

(4) the quartzites (25–30 m) of the Cerro Risquillo Fm that corresponds to the 'Siegen-Quarzit' of Puschmann (1970) and the 'Basal quartzite' of other authors.

The upper part of the 'Alternancias de La Corchada' has yielded crinoid fragments, tentaculitids, trilobites and brachiopods with *Ancillotoechia* aff. *ancillans* and *Microsphaeridiorhynchus*? cf. *nucula*. The Doradilla Fm contains *Ancillotoechia* aff. *ancillans* and the trilobite *Trimerus* cf. *acuminatus* that could indicate the lower Lochkovian. The lowermost part of the Valdenmedio Fm has yielded *Ancillotoechia* aff. *ancillans* in the Herrera del Duque syncline and equivalent levels in the Almadén syncline contain the brachiopods *Mesodouvillina* sp. and *Howellella*? sp. that suggest a Lochkovian age. Finally, the Cerro Risquillo Fm bears Lochkovian fossils in its lower part and Pragian ones in its upper part.

The Silurian–Devonian boundary lies most probably either within the upper part of the 'Alternancias de La Corchada' or at the upper limit of this formation (Pardo Alonso & García-Alcalde 1996; Pardo Alonso 1998).

Northeast Central Range. In the eastern Sierra de Guadarrama (NE Central Range), specifically in the Atienza and Riaza area (no. 15, Fig. 5.1), the Silurian succession (Fig. 5.3) is similar to that of the southern part of the Central Iberian Zone. This succession begins with the quartzites of the Santibáñez Fm (20–30 m) that overlie conformably Hirnantian shales and siltstones and could be equivalent to the Criadero Fm. The overlying Cañamares Fm (190–250 m) consists mainly of black shales with quartzite intercalations in the upper part. Graptolite assemblages indicate that the black shales extend from undetermined Aeronian levels up to the basal Gorstian *Neodiversograptus nilssoni* Biozone. In the uppermost part of the Cañamares Fm, an argillaceous limestone has yielded conodonts of the Pridoli *Ozarkodina eosteinhornensis* Biozone. The overlying sandstones of the Alcolea Fm (750–800 m) contain brachiopods and trilobites of Pridoli and Lochkovian ages. The Silurian–Devonian boundary lies approximately in the middle part of this formation (Bultynck 1971; Bultynck & Soers 1971; Soers 1971; Bischoff 1974; Carls 1977; Truyols & Julivert 1983; Fernández Casals & Gutiérrez-Marco 1985; Azor et al. 1992). As a whole, this succession closely resembles that of the eastern Iberian Cordillera situated immediately to the east.

West and south of the Ollo de Sapo anticlinorium. Fossiliferous Silurian rocks have been recognized in the El Barquero synform and along the Viveiro–Begonte–Guntín band (Parga Pondal & Gómez de Llarena 1963; Matte 1968; Iglesias & Robardet 1980). Silurian rocks and faunas are also known in the Alcañices synform (no. 14, Fig. 5.1), a complex structure in which different structural units with different stratigraphical successions are juxtaposed (Díez-Balda et al. 1990; González-Clavijo 1997). The Silurian and Devonian successions are not very clearly elucidated, but recent studies have shown that most of the Silurian graptolite localities are within the Manzanal del Barco Fm (up to ?500 m) that comprises black shales, mudstones, quartzites, conglomerates and limestones. This formation has yielded graptolites of the Aeronian *Demirastrites triangulatus* and *Lituigraptus convolutus* biozones, most of the Telychian biozones, the lower Wenlock *Cyrtograptus centrifugus* or *C. murchisoni* Biozone, and probably of the Homerian and the lower Ludlow. In the eastern part of the synform, the overlying Almendra Fm (300 m of turbiditic grey shales and limestones) has yielded conodonts that indicate ages from the Pridoli up to the late Pragian or earliest Emsian. In the western part of the structure, a possible equivalent of the Almendra Fm could be the lower part of the San Vitero Fm (more than 1500 m), a terrigenous flyschoid unit of greywackes, shales and conglomerates with olistholiths of various lithologies.

Truchas and Sil synclines. This area exposes the northernmost units of the Spanish part of the Central Iberian Zone (no. 13, Fig. 5.1), situated east and north of the Ollo de Sapo anticlinorium.

In the Truchas syncline, the uppermost part of the Ordovician sequence comprises the fragment-bearing siltstones of the Rozadais Fm and the laminated shales, siltstones and sandstones of the overlying Losadilla Fm (Sarmiento et al. 1999c). In the central part of the syncline, the base of the Llagarinos Fm contains Rhuddanian graptolites of the *Parakidograptus acuminatus* Biozone at La Baña and of the lower *Cystograptus vesiculosus* Biozone at Monte Llagarinos.

However, there is clear diachronism of the base of the Silurian sequence in this area, as the lowermost black shales that directly overlie the pre-Hirnantian Ashgill limestones of the La Aquiana Fm are Telychian in the southern part of the Sil syncline and lowermost Wenlock in the northern part of the same unit (Gutiérrez-Marco & Robardet 1991). The contact is marked by the so-called Pardellán ferruginous level which has been considered to be an Upper Ordovician hardground; however, in several localities, this ferruginous level crosses obliquely the limit between the two formations and probably corresponds to the weathering of tectonized beds (Gutiérrez-Marco & Rábano 1997, and references therein).

In the Sil syncline the Silurian sequence (Fig. 5.4) comprises a lower unit of graptolite black shales (Llagarinos Fm: 150–180 m) and an upper unit of chloritoid shales and slates with nodules and limestone lenses in the uppermost part (Salas Fm: about 800 m), apparently continuous with the overlying Devonian. Separating these two lithological units, a fairly continuous intercalation of chloritoid shales and siltstones alternating with quartzitic beds (the so-called 'Yeres quartzite', up to 25–30 m) is commonly present.

The black shales of the Llagarinos Fm contain abundant graptolite faunas (Gutiérrez-Marco & Rábano 1997). Sheinwoodian beds, with diagnostic assemblages of the *Monograptus flexilis*, *Cyrtograptus rigidus* and the *Cyrtograptus ramosus–C. perneri* biozones, are overlain by richly pyritic beds of Homerian age that extend up into the *Colonograptus ludensis* Biozone and include, in the upper Homerian, a level with abundant synrhabdosomes of *Colonograptus deubeli* (Gutiérrez-Marco & Lenz 1998). Above these beds, in the uppermost part of the black shale sequence, there are lower Ludlow graptolites of the *Neodiversograptus nilssoni* Biozone.

The 'Yeres quartzite' has yielded some brachiopods, echinoderms, solitary rugose corals, bivalves, cephalopods and trilobites of Bohemian affinities that allow correlation with the upper Ludfordian faunas of the Prague area (Rábano et al. 1993; Gutiérrez-Marco et al. 2001).

The lower third of the Salas Fm contains the Pridoli graptolite *Neocolonograptus ultimus* (Piçarra et al. 1998). The same beds have yielded trilobites, brachiopods, bivalves, orthocone nautiloids and scyphocrinoid remains; the trilobite assemblage includes *Crotalocephalus transiens*, *Struszia*? cf. *concomitans*, *Cromus* aff. *leirion*, which are known from the Pridoli Fm of Bohemia (Gutiérrez-Marco et al. 2001), which is in good accordance with the occurrence of *Neocolonograptus ultimus*.

Galicia–Trás-os-Montes Zone

In the northwestern part of the Iberian Peninsula (North Portugal and NW Spain), the 'Galicia–Trás-os-Montes Zone' (GTOMZ; D, Fig. 5.1) corresponds to a pile of thrust units superimposed on top of the autochthonous Central Iberian Zone during the Variscan orogeny (Ribeiro et al. 1990b; Martínez Catalán et al. 1996). Rare and poorly preserved graptolites have been found locally in the very thick succession of metasediments (probably 8000 m; Farias Arquer 1992) of the so-called 'Parautochthonous Thrust Complex' (Romariz 1962, 1969; Matte 1968). These are of Telychian age in the lower part of the Nogueira Group (Gp) (900–1000 m) and of Llandovery and Wenlock ages in the black slates of the lower part of the overlying Paraño Gp (2900–3200 m), especially around the Bragança Massif.

Ossa Morena Zone

Within the Ossa Morena Zone (F, Fig. 5.1), the best-documented Silurian successions occur in the Valle and Cerrón del Hornillo synclines, in the southeastern part of this zone (Seville province; no. 30 and 31, Fig. 5.1), and conformably overlie the late Ashgill Valle Fm. The latter comprises dark shales and siltstones that include matrix-supported microconglomeratic levels, and is regarded as representing glaciomarine sediments of Hirnantian age. The Silurian–Lochkovian succession, 130 to 150 m thick (Fig. 5.6), consists almost entirely of 'ampelitic' black shales with some intercalations of siliceous slates and black cherts. The Silurian succession in the Ossa Morena Zone appears clearly different from those which occur in the contiguous Central Iberian Zone immediately to the north (Robardet 1976; Robardet & Gutiérrez-Marco 1990b).

The lowermost part of the black shale succession ('lower graptolitic shales') mostly comprises sandy shales, with a thin (0.50–0.80 m) black limestone level occurring in the Ludlow series (Fig. 5.6). The most important lithological change occurs in the Pridoli with the 'Scyphocrinites limestone' (10–15 m), a lithological unit of alternating limestones and shales, that separates the 'lower graptolitic shales' from the 'upper graptolitic shales' (Fig. 5.6). The entire Silurian to Lochkovian succession is fossiliferous, with a fauna consisting mainly of graptolites. Most of the graptolite biozones have been identified, indicating a continuous succession extending from the base of the Rhuddanian (Akidograptus ascensus–Parakidograptus acuminatus Biozone) up to the uppermost Pridoli (Istrograptus transgrediens Biozone) and into the Lochkovian (Jaeger & Robardet 1979; Robardet & Gutiérrez-Marco 1990a; Gutiérrez-Marco et al. 1996d; Robardet et al. 1998). Orthoconic nautiloids (Michelinoceras michelini, Arionoceras cf. arion), bivalves (Cardiola docens) and the graptolite Bohemograptus bohemicus cf. tenuis are known in the Ludlow

limestone. The 'Scyphocrinites limestone' has yielded scyphocrinoids, trilobites (Cromus cf. krolmusi, C. aff. leirion, Crotalocephalus cf. transiens, Bohemoharpes sp., Denckmannites sp., Leonaspis sp.), bivalves (Joachimia impatiens, Snoopya insolita, Patrocardia evolvens, Dualina aff. secunda), cephalopods (Cycloceras bohemicum), conodonts (seven species of Oulodus, Pseudoonetodus and Ozarkodina), ostracods (Bolbozoe sp.), graptolites (Neocolonograptus parultimus, N. ultimus, Istrograptus transgrediens), rare solitary corals and rare brachiopods (Robardet et al. 2000). Anoxic conditions persisted during the whole Rhuddanian–late Lochkovian interval with the exception of a large part of the Pridoli when bottom anoxia was probably weaker during deposition of the 'Scyphocrinites Limestone'.

The tripartite succession of the southeastern Ossa Morena Zone compares closely, in terms of both lithology and faunas, with the typical 'Thuringian triad' (Jaeger 1976, 1977): the 'Scyphocrinites limestone' is a precise equivalent of the 'Ockerkalk' (Robardet 1982; Robardet & Gutiérrez-Marco 1990a) and the end of the anoxic sedimentation occurred in both areas in earliest Devonian times, with green-brown shales and siltstones yielding Pragian trilobites, ostracods and brachiopods in the Cerrón del Hornillo and Valle synclines (Racheboeuf & Robardet 1986; Robardet et al. 1991).

Westwards, Silurian rocks occur also in the Hinojales and Villanueva del Fresno areas (nos 28 and 29, Fig. 5.1) where the succession is poorly known but apparently does not include any limestone member similar to the 'Scyphocrinites limestone'. The black shale sequence contains graptolite assemblages corresponding to the

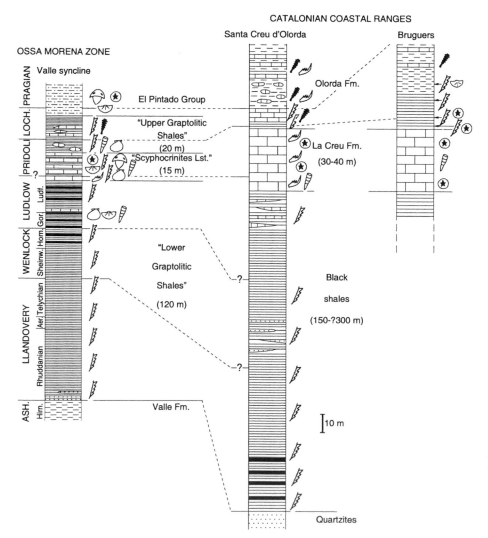

Fig. 5.6. The Silurian successions in the SE Ossa Morena Zone and the Catalonian Coastal Ranges. SE Ossa Morena after Jaeger & Robardet (1979) and Robardet et al. (1998, 2000); Catalonian Coastal Ranges after Julivert et al. (1985), García-López et al. (1990), Ferrer et al. (1992) and unpublished data by Gutiérrez-Marco & Robardet. For symbols see Figure 5.3.

interval between the Aeronian *Lituigraptus convolutus* Biozone and the Telychian *Oktavites spiralis* Biozone. The overlying alternating shales, siltstones and quartzites have yielded acritarchs of Wenlock and Ludlow age (Mette 1989; Piçarra *et al.* 1992).

Catalonian Coastal Ranges

In the Catalonian Coastal Ranges of NE Spain (I, Fig. 5.1), the occurrence of Silurian and Devonian rocks and fossils in several Palaeozoic massifs (nos 33 to 36, Fig. 5.1) has been known for more than 100 years, especially in the Barcelona area. The Silurian sequence overlies Upper Ordovician quartzites, siltstones and dark shales that have yielded upper Caradoc to middle Ashgill faunas (Villas *et al.* 1987). Hirnantian glaciomarine sediments have not been recognized, but the upper Ashgill may be represented by several tens of metres of unfossiliferous dark shales (Julivert & Durán 1992). The Silurian succession and the Silurian–Devonian transition in the Catalonian Coastal Ranges correlate very precisely with the successions known in the Ossa Morena Zone, in SE Sardinia and in Thuringia, and the La Creu Fm corresponds to the 'carbonate intermezzo' represented by the 'Ockerkalk' in SE Sardinia and Thuringia, and by the '*Scyphocrinites* limestone' in the Ossa Morena Zone.

The Silurian succession (Fig. 5.6) comprises a black shale sequence (150 to ?300 m) with some chert intercalations in its lower part, sandy levels in the middle, and limestone beds and lenses in the upper part. These graptolitic black shales begin with the basal Llandovery *Parakidograptus acuminatus* Biozone, documented both in the Montseny massif (Puschmann 1968*a,b*) and in the Sierra de Miramar (Roqué 1999), and extend throughout the Llandovery, the Wenlock and part of the Ludlow, up to the ?*Saetograptus leintwardinensis* Biozone of the basal Ludfordian (Julivert *et al.* 1985; Julivert & Durán 1990*a*).

The carbonate sequence that overlies the graptolitic black shales comprises two formations. The lower one, the La Creu Fm, consists of 30–40 m of dark grey massive nodular limestones with thin shale intercalations. The overlying Olorda Fm (35–40 m) consists mainly of light-coloured marls and nodular limestones.

The La Creu Fm has yielded crinoids, orthoconic nautiloids, bivalves and conodonts indicating that it is mainly Pridoli in age (*Ozarkodina eosteinhornensis* Biozone), the lowermost beds still being in the Ludlow. It was generally considered that the upper part was Lochkovian (García López *et al.* 1990, and references therein), but graptolites recently found in the lowermost part of the overlying Olorda Fm at Bruguers are still Pridoli in age and show that the Silurian–Devonian boundary lies within the lowermost part of the Olorda Fm (Piçarra *et al.* 1998; Gutiérrez-Marco *et al.* 1998). The rest of the Olorda Fm (35–40 m) has yielded brachiopods, dacryoconarid tentaculitids and conodonts indicating that it corresponds to the Lochkovian (with graptolites of the *Monograptus uniformis, M. praehercynicus* and *M. hercynicus* biozones), Pragian and part of the Emsian (Julivert *et al.* 1985; García López *et al.* 1990; Ferrer *et al.* 1992; Racheboeuf *et al.* 1993; Lenz *et al.* 1997; Chlupac *et al.* 1997; Gutiérrez-Marco *et al.* 1999*d*).

The Silurian–Lower Devonian succession of the Gavarres Massif (in NE Catalonia) is closely similar, with Silurian graptolitic black shales overlain by the limestones of the Can Riera and the Montnegre formations that correspond respectively to the La Creu and Olorda formations of the Barcelona area (Sanz-López 1995; Sanz-López *et al.* 1998).

Pyrenees

Within the Pyrenees (J, Fig. 5.1), Palaeozoic rocks were involved in both the Variscan and Alpine orogenies. However, reasonably complete Silurian successions have been preserved in several areas of the Spanish eastern and central Pyrenees, such as those of Camprodon (eastern Pyrenees: no. 37, Fig. 5.1), Bar-Toloriú (south of Andorra: no. 38, Fig. 5.1), Sierra Negra (SW of the Maladeta granitic massif: no. 40, Fig. 5.1), and Sallent de Gállego (western central Pyrenees: no. 41, Fig. 5.1). In the western Pyrenees, Silurian black shales are known from the Cinco Villas and the Aldudes–Quinto Real 'massifs' (no. 42 to 44, Fig. 5.1; Juch & Schäfer 1974; Requadt 1974; Heddebaut 1975; Dégardin 1988, 1990, 1995, and references therein).

The Silurian succession consists of black shales that include limestone beds and nodules in the upper part, the total estimated thickness being 120–200 m in the eastern and central Pyrenees (Fig. 5.7) and apparently much greater in the western Pyrenees (up to ? 850 m). The chronostratigraphical assignments are based on graptolites in the black shales and on conodonts in the limestones. Other fossil groups are much rarer, but orthoconic nautiloids, bivalves, crinoids and ostracods also occur in the carbonates and a few trilobites of Bohemian affinities have been found in black shales of various ages (Gaertner 1930; Dégardin & Pillet 1983; Dégardin 1995, and references therein).

The Ordovician–Silurian transition is not defined clearly: Hirnantian glaciomarine sediments have not been identified and the lowermost part of the black shales is frequently tectonized, which most probably explains why the Rhuddanian is so rarely identified. However, in the Spanish and Andorran central Pyrenees (Hartevelt 1970), the Ansobell Fm could be the equivalent of the Hirnantian glaciomarine sediments and the Bar quartzite could correspond to the 'basal Silurian' quartzites of other Iberian regions (Gutiérrez-Marco *et al.* 1998, pp. 25, 29; Gil-Peña *et al.* 2000).

In most areas, the lower part of the black shales has yielded graptolite assemblages of Aeronian and Telychian ages. This fossiliferous lithofacies extends into the Wenlock (Sheinwoodian and Homerian) and a few localities have also yielded Gorstian and Ludfordian graptolites. In the upper part of the Silurian succession, the only record of Pridoli graptolites comes from the La Seo d'Urgell area (Haude 1992) and the stratigraphy is generally based on conodonts found in limestone beds and carbonate nodules (Dégardin 1988, 1990, 1995; García-López *et al.* 1996). A few occurrences of early Llandovery and Wenlock conodonts have also been reported from the central Pyrenees (Sanz López & Sarmiento 1995; Valenzuela Ríos 1996).

In the eastern and central Pyrenees, the position of the Silurian–Devonian boundary can usually be rather precisely defined between conodont assemblages of the Pridoli *Ozarkodina eosteinhornensis* and the Lochkovian *Icriodus woschmidti* biozones (Dégardin 1988; García-López *et al.* 1996). The most precise study of the Pridoli–Lochkovian boundary has been done in the Els Castells unit of the southern central Pyrenees (near Gerri de la Sal; no. 39, Fig. 5.1) that has yielded graptolites, scyphocrinoids and conodonts (Llopis Lladó & Rosell 1968; Sanz-López *et al.* 1999, and references therein).

In the western Pyrenees, the black shales of the Arneguy and Anzabal formations have yielded graptolites of Llandovery, Wenlock and Ludlow ages at several localities of the Aldudes–Quinto Real and the Cinco Villas 'massifs' (nos 42 and 43, Fig. 5.1; Juch & Schäfer 1974; Requadt 1974; Heddebaut 1975; Dégardin 1995). However, the precise biostratigraphy of these thick black shale units remains poorly known, due to the scarcity and tectonic deformation of the fossils. Locally, in the upper part of the black shales, limestone lenses have yielded conodonts of Pridoli age. They are overlain by limestones and dolomites (Ondarolle and Inzulegui formations) with brachiopods and conodonts of 'Gedinnian' (Lochkovian) age. In the Saint-Martin-d'Arrossa area, in the French part of the Aldudes Massif (no. 44, Fig. 5.1), the graptolitic black shales (about 120–150 m) extend from the Aeronian up into the early Gorstian and their upper part consists of 30–40 m of alternating sandstones and siltstones that probably correspond to the upper Ludlow and the Pridoli, as Lochkovian brachiopods have been found immediately above. This area seems to have been characterized by a coarser terrigenous influx during Late Silurian times (Heddebaut 1975).

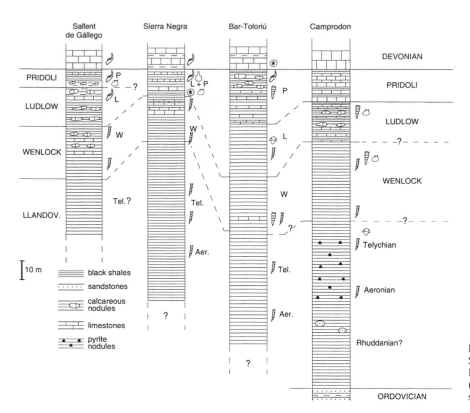

Fig. 5.7. The Silurian successions in the Spanish central and eastern Pyrenees. After Dégardin (1988, 1995), Llopis Lladó & Rosell (1968) and Sanz López et al. (1999). For symbols see Figure 5.3.

Betic Cordilleras

Within the Betic Cordilleras of SE Spain (H, Fig. 5.1), the so-called 'Maláguide Complex' (= 'Betic' of Málaga, no. 32, Fig. 5.1) comprises Palaeozoic rocks. The Morales Fm (more than 200 m thick) extends from the Upper Ordovician into lowermost Devonian. Its upper part consists of shales and silt-stones with chert and limestone intercalations (with some tintinnids and conodonts) and begins with a discontinuous unit of conglomerates (10–30 m) that marks approximately the Ordovician–Silurian boundary; this upper part probably corresponds to the Llandovery–lower Ludlow. The uppermost levels of the Morales Fm (probably upper Ludlow to Lochkovian) consist of limestones with tentaculitids, cephalopods and conodonts (Gómez Pugnaire 1992). Conodonts of Wenlock, Ludlow and Pridoli to Lochkovian ages have also been recorded from limestone olistoliths within the younger (Devonian to Carboniferous) Sancti Petri and Almogía formations (Rodríguez Cañero 1993; García-López et al. 1996; Sarmiento et al. 1998).

The boundaries of the Silurian System in the Iberian Peninsula

Ordovician–Silurian boundary

The problem of accurate placement of the Ordovician–Silurian boundary in Spain has been fully discussed in Chapter 4. However, it can be repeated here that this boundary can be fixed clearly in regions in which graptolite assemblages of the *Akidograptus ascensus–Parakidograptus acuminatus* Biozone occur. Such basal Rhuddanian assemblages have been found in several regions of the Iberian Peninsula, either within the

lowermost part of the Silurian black shale sequence (cf. Valle and Truchas synclines, Catalonian Coastal Ranges) or within the upper part of the quartzites that occur frequently at the Ordovician–Silurian transition (cf. Mondoñedo nappe, Iberian Cordillera and southern Central Iberian Zone). The hypothesis of an important and generalized stratigraphical hiatus including the lower part (Rhuddanian and Aeronian) of the Llandovery series (Truyols & Julivert 1983) can no longer be accepted.

The base of the black shale sequence, i.e. the beginning of the euxinic sedimentation of black, organic-rich deposits, is somewhat diachronous across the Iberian Peninsula. Low-oxygen conditions appeared during early Rhuddanian times in the Ossa Morena Zone, in the Catalonian Coastal Ranges, at some localities in the West Asturian-Leonese Zone and in the northernmost Central Iberian Zone, and possibly also in the eastern and central Pyrenees. In other regions, such as in the Peñalba and Mondoñedo domains of the West Asturian-Leonese Zone, in the Iberian Cordillera and in most areas of the Central Iberian Zone, widespread and long-lived anoxia or dysoxia was delayed until Aeronian and, more commonly, Telychian times.

Silurian–Devonian boundary

The precise position of the Silurian–Devonian boundary is more or less easily defined, depending of the nature of the succession. This boundary can be identified rather precisely where the fossil record is continuous and comprises representatives of key fossil groups whose vertical distribution is clearly established in reference sections of the Pridoli and the Lochkovian. This is the case in the Ossa Morena Zone, in the Catalonian Coastal Ranges and in the eastern and central

Pyrenees, where successive assemblages of graptolites, cono-donts and/or chitinozoans occur within condensed successions of black shales and limestones (Fig. 5.8).

Although the first appearance of the conodont *Icriodus woschmidti* is just below the true base of the Lochkovian, the limit between the *Ozarkodina eosteinhornensis* and the *Icriodus woschmidti* biozones can be considered practically equivalent to the Silurian–Devonian boundary (Klapper 1977; Kríz *et al.* 1986). In the same way, the chitinozoan *Urnochitina urna* is characteristic of the Pridoli (although its total range includes some decimetres of the basal Lochkovian in the global stratotype section) and the base of the *Eisenackitina bohemica* Biozone, although slightly younger than the precise Silurian–Devonian boundary, can be considered practically indicative of this limit (Verniers *et al.* 1995; Paris *et al.* 2000*b*).

The solution of the problem is less easy in areas where the fossil record consists of shelly faunas (brachiopods, trilobites and bivalves) that generally occur episodically within thick ter-rigenous sequences. This is the case in the Cantabrian Zone, in the Iberian Cordillera and in most areas of the Central Iberian Zone, where the Upper Silurian–Lower Devonian transition rocks consist of thick sequences of alternating sand-stones, siltstones and shales (Fig. 5.9). The stratigraphical range of the shelly fossils in these regions is not very precisely established, although they are generally regarded to be of 'Gedinnian' age. More recent studies, in the French Armori-can Massif (Morzadec *et al.* 1988) and in Spain (Carls 1988), have shown that some brachiopods (e.g. *Howellella mercuri* and *Platyorthis* ex gr. *monnieri*) and trilobites (*Acastella heberti*) appear to be characteristic of the Lochkovian. Micro-fossils, such as chitinozoans, can also be of great help in these terrigenous successions.

Ossa Morena Zone. The Valle and Cerrón del Hornillo synclines of the SE Ossa Morena Zone (Fig. 5.8) show a fairly complete Pridoli

and Lochkovian graptolitic sequence. In the Valle syncline, the Silurian–Devonian boundary occurs within the 'upper graptolitic shales' (15–20 m). Their lower part (4–5 m) comprises black shales with calcareous nodules which have yielded the upper Pridoli grapto-lite *Istrograptus transgrediens* (Piçarra *et al.* 1998). Their middle part consists of black shales with limestone nodules and lenses in which two successive Lochkovian graptolite assemblages have been found, a lower one with *Monograptus praehercynicus*, in black shales, and an upper one with *M. hercynicus*, within limestone lenses that also contain the dacryoconarid tentaculitoid *Homoctenowakia bohemica bohemica* (Gessa *et al.* 1994; Lenz *et al.* 1997).

Catalonian Coastal Ranges. In the Catalonian Coastal Ranges, the Silurian–Devonian boundary (Fig. 5.8) was generally drawn within the upper part of the La Creu Fm. This conclusion, based on conodont faunas (Walliser 1964; García López *et al.* 1990) at Santa Creu d'Olorda, seemed confirmed in the nearby locality of Bruguers by the occurrence of Lochkovian graptolite assemblages within shales of the lower part of the Olorda Fm, corresponding to the *Monograptus uniformis*, *M. praehercynicus* and *M. hercynicus* biozones (Ferrer *et al.* 1992; Lenz *et al.* 1997; Chlupac *et al.* 1997; Gutiérrez-Marco *et al.* 1999*d*). However, the base of the Olorda Fm has recently yielded graptolites of Pridoli age at Bruguers (Piçarra *et al.* 1998). Consider-ing these graptolites and the contrasting opinions concerning the evol-utionary lineages and the stratigraphical significance of the conodonts found in this area, it seems most probable that the uppermost part of the La Creu Fm and the basal levels of the Olorda Fm are still Pridoli in age and that the Silurian–Devonian boundary lies within the lowermost member A of the Olorda Fm.

Pyrenees. In the eastern and central Spanish Pyrenees (Fig. 5.8), the upper part of the Silurian black shales contains carbonate nodules and limestone intercalations that precede the more massive lime-stones of the Lower Devonian succession. In the Silurian–Devonian transitional sequence, biostratigraphy is based on conodont faunas and the position of the Silurian–Devonian boundary can be localized quite precisely at Camprodon, at Bar-Toloriú, in the Sierra Negra and at Sallent de Gállego, between the late Pridoli *Ozarkodina eosteinhornensis* Biozone and the Lochkovian *Icriodus woschmidti* Biozone (Dégardin 1988, pp. 203–224). Gerri de la Sal, in the Els Castells unit, is the only locality where Lochkovian graptolites,

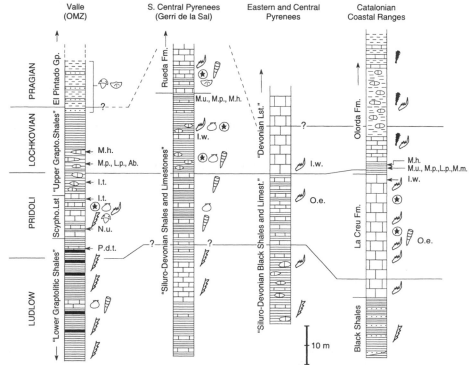

Fig. 5.8. The Silurian–Devonian boundary in the condensed successions of the Ossa Morena Zone, the central and eastern Pyrenees and the Catalonian Coastal Ranges. In Figures 5.8 and 5.9, diagnostic fossil names are abbreviated as follows. Conodonts: O.e., *Ozarkodina eosteinhornensis*; I.w., *Icriodus woschmidti*. Chitinozoans: U.u., *Urnochitina urna*; E.b., *Eisenackitina bohemica*. Brachiopods: P.m., *Platyorthis* ex gr. *monnieri*; H.m., *Howellella mercuri*. Trilobites: A.h., *Acastella heberti*; A.t., *Acastella tiro*. Graptolites: Ab., *Abiesgraptus* sp.; I.t., *Istrograptus transgrediens*; N.u., *Neocolonograptus ultimus*; L.p., *Linograptus posthumus*; M.b., *Monograptus bouceki*; M.h., *Monograptus hercynicus*; M.m., *Monograptus microdon*; M.p., *Monograptus praehercynicus*; M.u., *Monograptus uniformis*; P.d.t., *Pristiograptus dubius thuringicus*. For symbols see Figure 5.3.

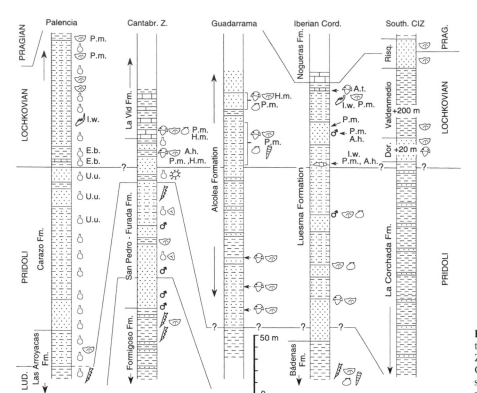

Fig. 5.9. The Silurian–Devonian boundary in the terrigenous successions of the Cantabrian Zone, the Guadarrama area, the Iberian Cordillera and the Central Iberian Zone. For symbols see Figure 5.3 and abbreviated fossil names, Figure 5.8.

namely *Monograptus uniformis*, *M. praehercynicus* and *M. hercynicus*, have been found in a conodont-bearing succession (Llopis Lladó & Rosell 1968; Sanz-López *et al.* 1999).

In the Saint-Martin-d'Arrossa area of the French Western Pyrenees, the Silurian–Devonian boundary can be placed approximately within the upper part of poorly fossiliferous siltstones and sandstones that have yielded the Lower Devonian ('Gedinnian') brachiopod *Howellella mercuri* in their uppermost levels (Heddebaut 1975; Dégardin 1988).

Cantabrian Zone. In the 'folds and nappes domain' (Fig. 5.9), the upper part of the Furada/San Pedro Fm is almost unanimously considered to be of Lochkovian age, on the basis of both macro- and microfaunal data. Bivalves, trilobites (e.g. *Acastella heberti*) and brachiopods (e.g. *Howellella mercuri*, *Platyorthis* ex gr. *monnieri*), found in the uppermost 20–50 m are regarded as indicative of a 'Gedinnian', i.e. Lochkovian age (see Truyols *et al.* 1974, 1990; Julivert *et al.* 1983a; Aramburu *et al.* 1992). This is also supported by micro-fossil data, especially chitinozoans (Cramer-Díez & Díez 1978; Richardson *et al.* 2000) and sporomorphs (Rodríguez González 1983; Richardson *et al.* 2001). However, a recent re-examination of the chiti-nozoan assemblage from the uppermost shaly intercalation of the San Pedro Fm in the La Vid de Gordón section (10 m below the base of the overlying La Vid Fm) has shown that this level corresponds to the middle part of the Pridoli and, thus, that the San Pedro Fm in this section is probably entirely Silurian (Priewalder 1997).

In the Palentine area (Fig. 5.9), where the succession is somewhat different, the Silurian–Devonian boundary lies within the upper part of the Carazo Fm, somewhere below the levels that contain brachio-pods (*Platyorthis* ex gr. *monnieri*) and conodonts of the *Icriodus woschmidti* Biozone (Jahnke *et al.* 1983). The chitinozoan sequence allows a more precise location of the boundary, between the Pridoli assemblages with *Urnochitina urna* and those of the basal Lochkovian *Eisenackitina bohemica* Biozone (Schweineberg 1987).

West Asturian-Leonese Zone. In the Peñalba syncline of the south-ernmost West Asturian-Leonese Zone, as well as in the Sil syncline of the northernmost Central Iberian Zone, the upper part of the Silurian succession consists of chloritoid shales (Fig. 5.4). A large part of the Salas Fm of the Sil syncline (and of the corresponding chloritoid slates of the Peñalba syncline) is certainly Pridoli, as attested by trilobites and bivalves that correlate with lower Pridoli assemblages of Bohemia and by the graptolite *Neocolonograptus ultimus* (Gutiérrez-Marco *et al.* 2001). The Silurian–Devonian boundary is most probably situated within the upper part of the Salas Fm because, in the Peñalba syncline, the chloritoid shales are overlain by alternating limestones and shales that pass up into limestones with Lower Devonian brachiopods, dacry-oconarids and conodonts (Piçarra *et al.* 1998; Gutiérrez-Marco *et al.* 2001). However, the precise position of the boundary cannot be fixed easily, because an abundance of chloritoid crystals impedes graptolite taxonomic identifications.

Iberian Cordillera and Guadarrama. In the Aragonese branch of the Iberian Cordillera and in the Guadarrama region of the Central Range (Fig. 5.9), the Silurian–Devonian boundary occurs within a thick succession of alternating sandstones, siltstones and shales in which fossiliferous levels are rather rare.

In the eastern Iberian Cordillera (Carls 1977, 1988), several fossil-iferous levels within the lower third of the Luesma Fm (about 200 m) have yielded bivalves and brachiopods probably of latest Silurian (Pridoli?) age. About 150 m above the base, bivalves, trilobites (*Acastella heberti*), brachiopods (*Howellella mercuri*, *Platyorthis* ex gr. *monnieri*) and conodonts of the *Icriodus woschmidti* Biozone indicate a Lochkovian age. The Silurian–Devonian boundary thus lies within the middle part of the Luesma Fm.

The situation is almost the same in the Guadarrama area of the Central Range (Carls 1977, 1988, and references therein) where the Cañamares Fm (300 m) extends up into the uppermost Silurian as attested by the occurrence of conodont assemblages of the Pridoli *Ozarkodina eosteinhornensis* Biozone in argillaceous limestones. The overlying sandstones and shales of the Alcolea Fm (about 850 m) have yielded, at 64 m, 84 m and 110 m above the base, brachiopods that are most probably still of latest Silurian age. Higher fossiliferous levels, between 190 and 230 m above the base, contain *Platyorthis* ex gr. *monnieri*, other brachiopods and bivalves that suggest a Lochkovian age.

Central Iberian Zone. In the Sil syncline (Fig. 5.4), the position of the boundary has been discussed above (West Asturian-Leonese Peñalba syncline). In the Alcañices synform, in the northern part of the Central Iberian Zone, studies (Aldaya *et al.* 1976; Quiroga 1982) have shown the existence of distinct conodont assemblages of Pridoli to Emsian ages. However, at the present state of knowledge, the structural complexity of this unit excludes any clear discussion of the precise location of the Silurian–Devonian boundary.

In the southern Central Iberian Zone (Fig. 5.9), recent studies in the Herrera del Duque and Almadén synclines (Pardo Alonso & García Alcalde 1984, 1996; Pardo Alonso 1998) allow a greater precision in placing the Silurian–Devonian boundary. This can now be drawn either within the upper part of the 'Alternancias de La Corchada' or at the upper limit of this formation, and not at the base of the so-called 'basal quartzite' of previous authors (now Cerro Risquillo Fm) which is of late Lochkovian–early Pragian age.

Overview and palaeogeography

Regional analyses show that across the different regions of the Iberian Peninsula during Silurian times, differences in sedimentary successions and faunas (Fig. 5.10) probably corresponded to differences in palaeoenvironmental conditions and palaeogeographical positions.

The first type of succession is characterized by continuous euxinic black shale or black shale–black limestone sedimentation during the whole Silurian Period (and also Lochkovian), without any important clastic influx of coarser terrigenous sediment. The reappearance of oxygenated conditions and sediments occurred only during Early Devonian times. These euxinic sequences are generally rather condensed and the total thickness of the Silurian succession does not exceed 200 m. The most characteristic succession of this type is found in the Ossa Morena Zone and is almost identical to that of Thuringia in Germany, both for lithofacies

and faunas, the '*Scyphocrinites* limestone' being the precise stratigraphical equivalent of the 'Ockerkalk'. A similar succession occurs also in the Moncorvo syncline and in the nearby autochthonous units of the northernmost part of the Central Iberian Zone in Portugal (Sarmiento *et al.* 1999*d*). The Silurian succession of the Catalonian Coastal Ranges consists also of a condensed sequence of black shales and black cherts, with the limestones of the La Creu Fm (late Ludlow and Pridoli) being equivalent to the '*Scyphocrinites* limestone' of the SE Ossa Morena Zone. In the central and eastern Pyrenees, the rather condensed Silurian succession (<200 m thick) appears to be of the same type, with limestones in the upper part, mainly in the Ludlow and the Pridoli. Finally, the thicker Silurian successions in the Peñalba and Sil synclines of NW Spain (at the limit between the Central Iberian and West Asturian-Leonese zones) also record long-lasting euxinic argillaceous sedimentation, including a large part of the Pridoli.

These regions are the only areas of the Iberian Peninsula where Pridoli and Lochkovian graptolites have been found (Lenz *et al.* 1997; Piçarra *et al.* 1998). The non-graptolitic faunas, such as the trilobites and bivalves of the Peñalba and Sil synclines, the Ossa Morena Zone and the Pyrenees, show clear Bohemian affinities (Rábano *et al.* 1993; Robardet *et al.* 2000; Gutiérrez-Marco *et al.* 2001). It can be presumed that all these regions, where the terrigenous influx was permanently weak and where the faunas were almost exclusively pelagic, were situated at some distance from the land, on the outer, distal part of the North Gondwanan marine shelf.

The second type of succession begins with sandstone units at the Ordovician–Silurian transition. Euxinic black shale sedimentation started generally during Telychian, and persisted during Wenlock and, in some regions, early Ludlow times. This black shale sequence is overlain by thick units of

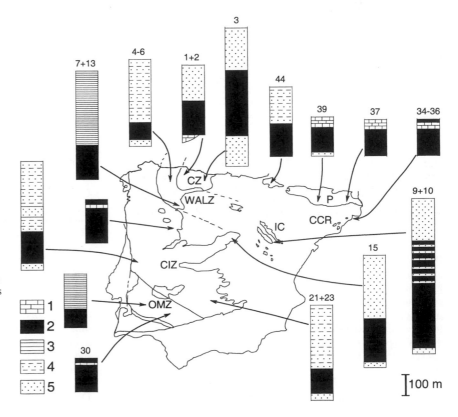

Fig. 5.10. The different types of Silurian successions in the Iberian Peninsula (after Gutiérrez-Marco *et al.* 1998). Dominant lithofacies: 1, limestones; 2, black shales; 3, shales and siltstones; 4, alternating sandstones, siltstones and shales; 5, sandstones. CCR, Catalonian Coastal Ranges; CIZ, Central Iberian Zone; CZ, Cantabrian Zone; IC, Iberian Cordillera; OMZ, Ossa Morena Zone; P, Pyrenees; WALZ, West Asturian-Leonese Zone. Numbers at the top of columns correspond to the localities and sections listed in Figure 5.1. The three columns from Portugal are, from north to south, Moncorvo, Dornes-Mação and Barrancos.

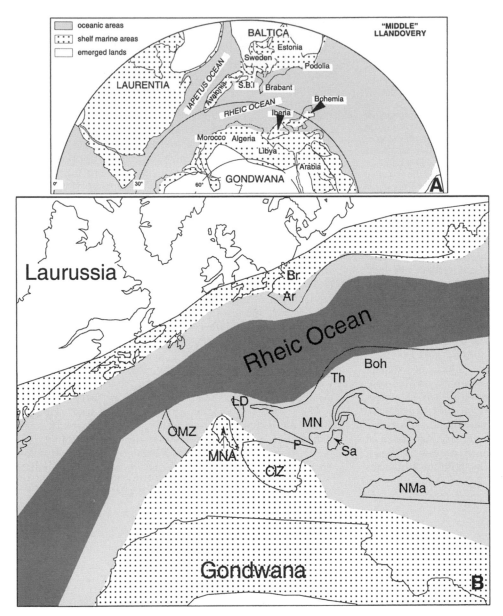

Fig. 5.11. (a) Global Silurian palaeogeography of Atlantic bordering regions (after Paris 1993). (b) Tentative palaeogeographical reconstruction of the North Gondwanan regions during the Silurian period (after Robardet *et al.* 1994; Gutiérrez-Marco *et al.* 1998). Key: white, land areas; light grey, inner shelf; medium grey, outer shelf; dark grey, oceanic areas. Ar, Ardenne; Boh, Bohemia; Br, Brabant; CIZ, Central Iberian Zone; LD, Ligerian Domain of the Armorican Massif; MN, Montagne Noire; MNA, Middle-North Armorican Domain; NMa, northern Maghreb; OMZ, Ossa Morena Zone; P, Pyrenees; Sa, Sardinia; Th, Thuringia.

alternating sandstones, siltstones and shales of Ludlow and Pridoli age, that herald the sandstone units of the lowest Devonian successions. The coarser clastic sediments appeared progressively; thin sandstone beds are commonly intercalated with the upper part of the black shale sequence, and increase in abundance upwards. However, the sandstone–siltstone–shale alternations developed fully during late Ludlow and Pridoli times. Within this upper part of the succession, fossils are scarce and consist mainly of shelly faunas. The absence of graptolites is probably due to environmental conditions that controlled their distribution in the marine areas and not to post-depositional destruction, because organic-walled microfossils (chitinozoans) have been preserved in these rocks.

This type of succession occurs across most of the Central Iberian Zone, in the Iberian Cordillera, in some units of the West Asturian-Leonese Zone (Mondoñedo?), in the Cantabrian Zone and in the western Pyrenees. On a larger scale, they also characterize the mid- and north Armorican

regions in NW France that were closely linked with the Central Iberian Zone (Paris & Robardet 1977). These areas, where the terrigenous influx was important and continuous during Late Silurian and Early Devonian times, were situated a short distance from terrestrial source areas, within the inner proximal part of the North Gondwanan shelf.

The geographical distribution of these two types of succession in the Iberian Peninsula (Fig. 5.10) is not random and some organization can be discerned, especially when the effects of the Variscan orogeny are eliminated and the probable pre-orogenic palaeogeography is restored (Fig. 5.11). A general north–south trend can be seen, from shallow-water inner-shelf environments in the southern Central Iberian Zone towards more distal and deeper outer-shelf conditions in the northernmost Central Iberian and West Asturian-Leonese zones. If it is confirmed that the flyschoid San Vitero Fm of the western part of the Alcañices synform is actually Siluro-Devonian and not a younger (Carboniferous?) unit, its occurrence in the NW of the Central Iberian Zone would provide a supplementary argument in favour of deeper environments in this direction. The condensed successions of the Catalonian Coastal Ranges and Pyrenees, deposited in outer-shelf environments, fit rather well in this scheme. It is the same for the Ossa Morena Zone, the present position of which is the result of sinistral strike-slip movement along the Tomar–Badajoz–Córdoba shear zone and does not correspond to its original pre-Variscan position which was most probably somewhere to the NW.

There are several good biogeographical markers for the shallowest areas of the Silurian marine shelf in the Iberian Peninsula: the graptolites *Metaclimacograptus asejradi* (from the *Rastrites linnaei* and *Spirograptus turriculatus* biozones), *M. flamandi* (from the *Torquigraptus arcuatus* to lower *T. tullbergi* biozones) and *Parapetalolithus meridionalis* (from the *T. arcuatus* to *Monoclimacis griestoniensis* biozones). These species occur abundantly in Telychian black shales of the southern Central Iberian Zone, the Iberian Cordillera, the Guadarrama area and the Castrillo syncline of the West Asturian-Leonese Zone. They have never been found in the Ossa Morena Zone, the Pyrenees, the Catalonian Coastal Ranges or in the boundary area between the West Asturian-Leonese and Central Iberian zones, and their presence in the former areas links the shallow Iberian platforms with those inshore environments of the North African pericratonic and intracratonic basins of North Gondwana (Algeria, Libya). *Metaclimacograptus flamandi* was not distinguished from *Paraclimacograptus? brasiliensis* in previous Iberian studies, the latter being a slightly older species (Rhuddanian–Aeronian), closely allied to *P.? libycus* (Aeronian), but also indicative of similar shallow shelf areas in the South American and North African parts of Gondwana, respectively. A typical graptolite that encompasses the Silurian transgression that extended over North Gondwana is *Neodiplograptus fezzanensis*, characteristic of the upper Rhuddanian (*Coronograptus cyphus* Biozone) of Libya, Algeria and Niger, and also found in peri-Gondwanan Europe (Bohemia; Storch 1983). We report here the recent and still unpublished identification of the species from southern Spain (Valle syncline).

However, the global palaeogeography proposed (Fig. 5.11) is obviously oversimplified and the pre-Variscan arrangement of the distinct regions that constitute the Iberian Peninsula was certainly more complicated. This is attested by the occurrence, in the Cantabrian Zone and in the Western Pyrenees, of Silurian successions which imply the existence of nearby emergent land areas. Previous authors had already postulated the existence of land areas in the north and NNW of the Iberian peninsula ('Ebroia': Llopis Lladó 1965; Heddebaut 1975; 'Cantabro-Ebroian Massif': Carls 1983, 1988). This indicates that a rather complicated topography was probably superimposed on the general south-to-north deepening trend outlined above.

Similar difficulties arise when the Silurian palaeogeography is considered, on a larger scale, for the whole North Gondwanan shelf. The North African regions at present situated to the south of the Iberian Peninsula, in Morocco, do not show Silurian successions and faunas suggestive of proximal shallow water conditions. This could result either from late or post-Palaeozoic modifications of the original palaeogeography or from a complicated original picture.

Silurian palaeontology of Spain

The most common Silurian fossils in Spain are the graptolites, generally abundant in the black shale units (5–40 m) of the lower part (Telychian–Sheinwoodian) of many Silurian successions, and of 'iberotypical' shallow shelf type. Most of the deeper Silurian sequences of 'Mediterranean' type bear a more complete and more continuous record of Silurian graptolites in somewhat condensed successions of black shales, which frequently include cephalopod limestone intercalations that provide remarkable benthic faunas of Bohemian type.

Although the common presence of Silurian fossils has been reported in many regional studies and in geological map explanatory notes, Silurian palaeontology in Spain has, until relatively recently, been limited mostly to taxonomic identifications. However, there has been an increasing interest in Spanish Silurian faunas and their contribution to global correlation and palaeobiogeography, so that new taxonomic and detailed biostratigraphical studies have been undertaken. Several outstanding Silurian sections in Sierra Morena, Campo de Calatrava and western Iberian Cordillera have revealed, in terms of graptolite biostratigraphy, the potential for high resolution correlation, equivalent to that in the best Silurian sections known in other regions of Europe (Bohemia, Thuringia, Wales, Scotland, Sardinia), Asia and North America.

The main references for Silurian **graptolites** in Spain are the papers by Almera (1891*b*), Haberfelner (1931), Suñer Coma (1957), Hernández Sampelayo (1926, 1960), Quintero (1962), Romariz (1963, 1969), Jaeger & Robardet (1979), Julivert *et al.* (1985), Dégardin (1984, 1988), Rodríguez Núñez *et al.* (1989), Gutiérrez-Marco & Robardet (1991), Gutiérrez-Marco *et al.* (1996*d*, 1997, 1999*d*), Gutiérrez-Marco & Storch (1997, 1998), Roqué (1999), Gutiérrez-Marco & Lenz (1998), Piçarra *et al.* (1998), Storch *et al.* (1998), Storch (1998) and Gutiérrez-Marco (1999). Silurian **trilobites** have been studied by Kegel (1929), Gaertner (1930), Hernández Sampelayo (1944), Gandl (1972), Dégardin & Pillet (1983), Rábano *et al.* (1993), García Palacios & Rábano (1996), Robardet *et al.* (2000), and Gutiérrez-Marco *et al.* (2001). Other macrofossil groups have been studied only locally: **brachiopods** (Kegel 1929; Carls 1974; Gourvennec 1990; Villas & Cocks 1996), **molluscs** (Vidal 1914; Kegel 1929; Cano Alonso *et al.* 1958; Bogolepova *et al.* 1998; Robardet *et al.* 2000; Gutiérrez-Marco *et al.* 2001), **tentaculitoids** (Jaeger 1986), **crinoids** (Haude 1992; Gutiérrez-Marco *et al.* 2001) and **sponges** (Rigby *et al.* 1997), as well as some **ichnofossils** (Suárez de Centi *et al.* 1989).

Micropalaeontological studies in the Silurian of Spain have focused mainly on **chitinozoans**, **acritarchs** and terrestrially derived **sporomorphs**, identified and described in about 30 papers from the Cantabrian Zone (Cramer-Díez & Díez 1978; Rodríguez González 1983, and references therein; Schweineberg 1987; Priewalder 1997; Richardson *et al.* 2000, 2001), the Pyrenees (Dégardin & Paris 1978) and the Ossa Morena Zone (Mette 1989). **Conodonts** have been studied by, for example, Bultynck (1971), Dégardin (1988), Sarmiento & Rodríguez Núñez (1991), Rodríguez Cañero (1993), Sarmiento *et al.* (1994, 1998), Sanz López (1995), Sanz López & Sarmiento (1995), Valenzuela-Ríos (1996), Sanz López *et al.* (1998, 1999) and Gutiérrez-Marco *et al.* (2001). The study of other microfossil groups, such as **radiolarians** (Dégardin & De Wever 1985; Gutiérrez-Marco *et al.* 2001) and **ostracods** (Dégardin & Léthiers 1982), is still in its early stages.

In the immediate future, palaeontological studies will focus mainly on microfossils, graptolites and trilobites, for refining biostratigraphical and geochronological correlation of many Silurian sections in Spain, as well as on detailed studies of bivalves, cornulitids, eurypterids, machaeridians, ostracods, bryozoans and muellerisphaerids for pure palaeontological and palaeobiogeographical purposes.

This chapter is a completely revised and updated version of a previous synthesis concerning the Silurian of the Iberian Peninsula (Gutiérrez-Marco *et al.* 1998). We are very grateful for the permission given by Emilio Custodio Gimena, Director of the Spanish Geological Survey, to make a new adaptation of text and figures. Illustrations have been prepared by D. Bernard (CNRS, Géosciences-Rennes). E. Clarkson (Edinburgh) and D. Loydell (Portsmouth) are thanked for their careful and constructive reviews. Work in Spain by J. C. G.-M. has been supported by the projects PB96-0839 and BTE2000-1310 of the Spanish Ministry of Science and Technology. This work is also a contribution to the IGCP Project 421 (IUGS-UNESCO) and to the Iberia project of EUROPROBE (European Science Foundation).

6 Devonian

JENARO L. GARCÍA-ALCALDE (coordinator), PETER CARLS,
MIGUEL V. PARDO ALONSO, JAVIER SANZ LÓPEZ, FRANCISCO SOTO,
MONTSERRAT TRUYOLS-MASSONI & JOSÉ I. VALENZUELA-RÍOS

The Devonian was one of the first Palaeozoic periods to be intensively studied in Spain. A few years after the formal definition of the Devonian by A. Sedgwick and R. I. Murchison in Devon, the French naturalists E. de Verneuil and A. d'Archiac (1845) noticed the occurrence of Devonian shelly fossil faunas in Asturias (north Spain). Later on, Prado & Verneuil (1850) enlarged the known Devonian outcrop area to the neighbouring province of Leon, and Prado (1856) extended this to Palencia province. Verneuil & Collomb (1853), Verneuil & Lorière (1854) and Verneuil & Lartet (1863) demonstrated Devonian rocks in the Iberian Ranges, and both Almera (1891c) and Barrois (1892) were pioneers in the study of Devonian rocks in the Catalonian Coastal Ranges. In southern Spain the seminal work on the system belongs to E. de Verneuil and J. Barrande (Prado *et al.* 1855), and in the Balearic Islands Hermite (1879) discovered the Devonian succession of Minorca. The history of Devonian research in other Spanish areas is in general much more recent, and was mainly developed in the twentieth century (Julivert *et al.* 1983).

Devonian rocks everywhere in Spain were deposited in marine conditions, although in varied settings ranging from supratidal to subtidal environments. The thickest and most complete Devonian succession in Spain is found in the Cantabrian and WestAsturo-Leonian zones and in the Basque Pyrenees (a–f and w, Fig. 6.1). There are also important but incomplete and discontinuous Devonian exposures in the Central Iberian and Ossa Morena zones (q–s, Fig. 6.1). The Devonian of the South Portuguese Zone consists mainly of Famennian rocks, and other incomplete Devonian successions occur in the Catalonian Coastal Ranges and in the Betic Cordillera (including Minorca, in the Balearic Islands) (Fig. 6.1). Maximum thicknesses have been recorded in the Iberian Ranges (*c.* 4000 m) and Basque Pyrenees (*c.* 3500 m) successions, as compared with around 2000 m in the Asturo-Leonian domain, and <1000 m in the Palentian domain.

Spanish Devonian facies are mostly calcareous (with remarkable reefal developments) in the Asturo-Leonian domain but are mostly siliciclastic in other areas, although significant thicknesses of Lower Devonian limestones occur in the Palentian domain and eastern Guadarrama, the Pyrenees and the Catalonian Coastal Ranges. Volcanic and plutonic rocks are frequent in the Devonian successions of the Central Iberian, Ossa Morena and South Portuguese zones, and a regional metamorphic overprint affects much of the West-Asturo-Leonian, northern Central Iberian, Pyrenean and Catalonian outcrops.

Many well-preserved fossil faunas allow good regional correlations across the so-called Ibarmaghian domain that extended from Meguma in Nova Scotia to North Africa via the Armorican Massif and Iberian Peninsula. Relatively frequent intercalations of nearshore and offshore facies provide pelagic

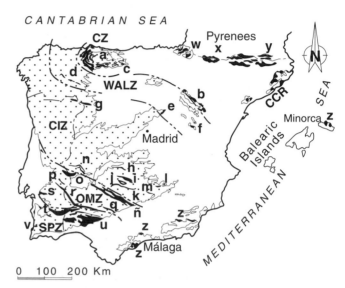

Fig. 6.1. Distribution of Devonian outcrops (sedimentary rocks) in the Iberian Peninsula and French Pyrenees.

CANTABRIAN SEA

0   100   200 Km

—··— Zone boundary   IBERIAN MASSIF
—·— Country border   Devonian outcrops

CZ: CANTABRIAN ZONE. Asturo-Leonian Domain (a) and Iberian Ranges (b)
WALZ: WESTASTURO-LEONIAN ZONE. Palentian domain (c) Peñalba-Caurel (d), Eastern Guadarrama (e) & Henarejos (f)
CIZ: CENTRAL-IBERIAN ZONE. Alcañices-Duero (g), Herrera del Duque (h), Almadén (i), Guadalmez (j), Cabeza del Buey-Fuencaliente (k), Santa Cruz de Mudela (l), El Centenillo (m), Sierra de San Pedro (n)
OMZ: OSSA-MORENA ZONE. Obejo-Valsequillo-Puebla de La Reina domain (ñ), La Codosera-Portalegre (o), Estremoz (p), Cerrón del Hornillo-Valle (q), Barrancos-Venta del Ciervo (r), Montemor-o-Novo (s)
SPZ: SOUTH PORTUGUESE ZONE. Pulo do Lobo (t), Iberian Pyrite Belt (u), SW Portugal (v)
Basque (w), Central (x) and Eastern Pyrenees (y)
Betic Cordillera (z) including the Balearic Islands (z)
CCR: Catalonian Coastal Ranges

fossils useful for long-distance and even worldwide correlations. Such biostratigraphical data reveal a palaeogeographic coherence to the Iberian Massif during Devonian times, with the area forming a north Gondwanan region close to Africa with no large ocean developed to the south (Carls 1988). Carls has argued against mobilist theories that saw Iberia as a mosaic of terranes along the northern margin of Gondwana, colliding during Early Devonian to Carboniferous times (e.g. Paris & Robardet 1977). The idea of palaeobiogeographic coherence to the whole Ibarmaghian domain was further supported by

Gourvennec *et al.* (1997) who used brachiopod data to show close linkage from at least Silurian to early Emsian times. Modern faunal studies also show that early in the Devonian period the north Gondwanan–Ibero-Armorican region and the south Baltic–Rhenohercynicum region were in close proximity, with no intervening 'Rheic Ocean' (Carls & Valenzuela-Ríos 1998).

Cantabrian mountains (JLG-A, FS, MT-M)

Two well-differentiated Devonian facies are represented in the Cantabrian mountains in the so-called Asturo-Leonian and Palentian domains (Brouwer 1964; García-Alcalde *et al.* 1990). The Asturo-Leonian domain extends across most of Asturias and León and the NW corner of Palencia, and is represented by rocks mainly of nearshore facies. In contrast the Palentian domain is composed of rocks mainly representing offshore facies and occurs chiefly in northern Palencia, with smaller outcrops also found in northeastern León and in western Santander (Fig. 6.2). Both of these facies can be interpreted as different bathymetric components of a single marine platform.

Asturo-Leonian domain

The source area for most of the Asturo-Leonian Devonian sediments was the western prolongation of an older massif (the Cantabro-Ebroian massif of Carls 1983, 1988) hidden at present below Variscan synorogenic and post-orogenic rocks of the Central Coal Basin, Ponga and Picos de Europa areas (Fig. 6.3). However, the eastward progression of the Variscan orogenesis across north Spain produced significant topographical relief near the western and southern margins of the Asturo-Leonian domain. Consequently, from late Frasnian times onwards, clastic supplies to the Cantabrian Zone (CZ) basin were increasingly derived from the Central Iberian area (Frankenfeld 1982).

Vertical movements during the initiation of the Variscan orogeny, acting together with a marked global early Famennian sea-level fall, resulted in the emergence of much of the marine platform and an ensuing peneplanation that affected even the deeply eroded remnants of the Cantabro-Ebroian massif (Julivert *et al.* 1983). In late Famennian time a slight eastward tilting of the platform allowed the progradation of a thin unit of clastic marine sediments across progressively older strata (Figs 6.4 and 6.5).

A Cantabrian Frasnian transgression has sometimes been invoked, a hypothesis based rather weakly on the occurrence of supposedly Frasnian fossils at the top of Ordovician sandstones in eastern Asturias (W. Struve *in* Radig 1966). However, this undescribed and unfigured fauna consists of badly preserved brachiopod moulds of genera that can be found during the entire Late Devonian interval. Moreover, other brachiopod faunas in similar stratigraphical situations in the same region have proven to be of latest Famennian to early Tournaisian age (J. L. García-Alcalde, unpublished data). Another controversy concerns the stratigraphical gaps cited by Buggisch *et al.* (1982), based on the occurrence of supposed palaeokarst features in Middle and Upper Devonian limestones.

The Variscan thrust sheet system, directed towards the centre of the Asturian arc, resulted in close juxtaposition of different parts of the ancient Devonian marine platform. Thus, a reasonably complete and complex picture of the platform through time can be achieved within a relatively reduced geographical area, with seven broad facies associations being recognized (Fig. 6.5). The Devonian successions usually show shallowing and/or coastal trends northwards and eastwards from a deeper region broadly represented today by the Devonian outcrops located at the SW curve of the Asturian arc (Figs 6.1 and 6.3; García-Alcalde 1995, and references therein). This shallowing-upward trend reflects the fact that the pre-Famennian Asturo-Leonian Devonian succession developed in a regressive context (in contrast to a worldwide transgressive trend) due to regional vertical movements. The most complete succession occurs in the outer part of the Asturian arc (Somiedo–Correcilla unit, Fig. 6.3) and consists of *c.* 2000 m of alternating terrigenous and calcareous rocks, sometimes with reef-building episodes.

Three main lithostratigraphical classifications related to geographic areas have been proposed (Figs 6.2 and 6.6). The Asturian terminology was introduced by Barrois (1882) and refined further by other authors (García-Alcalde 1992, and references therein). The Leonian terminology was essentially due to Comte (several articles published between 1936 and 1959 e.g. Comte 1936). Credit for the Palentian terminology (Valsurvio dome) falls upon Koopmans (1962) (Compuerto area) and Kanis (1956) (San Martín-Ventanilla area).The age and correlation of the Asturo-Leonian units are shown in Fig. 6.6. The mostly nearshore facies of the Asturo-Leonian Devonian are frequently unsuitable for the occurrence of biochronologically relevant fossils such as conodonts and ammonoids. Therefore a precise identification of the formal Devonian chronostratigraphic boundaries has not yet been possible. Dating and correlation have usually been achieved by means of brachiopod-based informal biostratigraphical units (Faunal Intervals; Fig. 6.6). Extensive fossil lists can be consulted in García-Alcalde (1995, 1996). Selected Cantabrian fossils useful for biostratigraphical purposes are included in Fig. 6.7.

Other correlative techniques subsidiary to biostratigraphical ones, both geochemical and geophysical, have been attempted on the Devonian successions of the Cantabrian mountains in recent years. Carbon and oxygen stable isotope techniques have until now given no results because of the diagenetic and tectonic overprint. Magnetosusceptibility event and cyclostratigraphy (MSEC) techniques have given promising results in the definition of a worldwide lower Emsian–upper Emsian boundary (Ellwood *et al.* 2001) and they may prove useful in the identification and correlation of other important Devonian chronostratigraphic boundaries. The exposed formations in the Valsurvio dome are closely similar to neighbouring Leonian units (Fig. 6.3) but are variably metamorphosed and always cleaved. On the other hand exposures in the San Martín–Ventanilla area (Fig. 6.6) cannot be directly correlated with other CZ units (Koopmans 1962; Julivert *et al.* 1983).

The Silurian–Devonian Furada and San Pedro formations (Fig. 6.4) usually comprise *c.* 200–250 m thick, red, ferruginous sandstones, with thin shaly beds and sandy limestone with dolostone lenses, the latter being more abundant in the Devonian, upper part of both formations. Strong bioturbation is frequent at the bottom of the sandy layers. Brachiopods, molluscs, trilobites, icriodid conodonts and palynomorphs are relatively abundant in some coquinas. The facies can be interpreted as corresponding to a nearshore, high energy environment battered by frequent storms.

The abundant sandy supply of the Furada and San Pedro formations ended abruptly, an event followed by a 400–600 m thick, predominantly calcareous series, the so-called Rañeces or La Vid Gp (Fig. 6.2). The Nieva Formation (Lochkovian/Pragian) is composed of up to

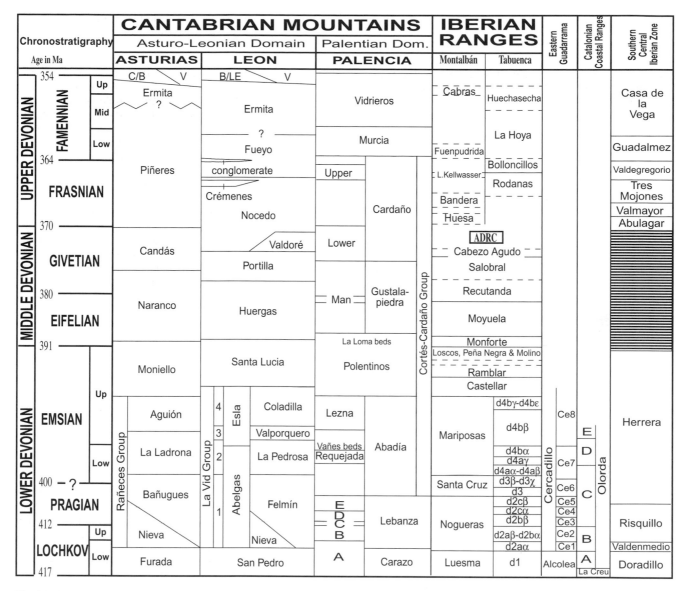

Fig. 6.2. Age and correlation of the Devonian lithological units of several Spanish areas. The formal SDS Pragian/Emsian boundary is not characterizable in Spain from the first occurrence of the index-fossil *Polygnathus kitabicus*. The present authors have not reached agreement over whether the *Kitabicus* boundary is the best Pragian–Emsian boundary level because *Polygnathus kitabicus* may be a junior synonym of *P. pyreneae*. Horizontal hatching: stratigraphical gap. Uneven line: sedimentary break (stratigraphic gap?). Abbreviations for uppermost Famennian units as in Figure 6.3. Absolute ages after Gradstein & Ogg (1996).

200 m of thick-bedded, sometimes dolomitized, bioclastic limestones, alternating with dark, argillaceous bioturbated limestones and thin shale layers, and siltstone lenses with hummocky cross-stratification. Occasional lag deposits composed of phosphate pebbles and fish remains occur. At the top of the formation limited reef-building episodes represented by small patch reef and biostromes presage major Cantabrian reefal development in overlying formations. Brachiopods are the most abundant fossils. The formation was deposited in the deeper, distal setting of the Asturo-Leonian area and laterally replaced eastwards, northwards and southwards by *c.* 150–200 m (exceptionally up to 400 m thick, west of Cape Peñas) of dolostones, dolomitized bioclastic limestones and thick sandstone intercalations of the Bañugues/Felmin Fm (Lochkovian–earliest Emsian). The Bañugues/Felmin dolostones show abundant inorganic and organic laminations, cryptalgal structures, stromatoliths, birdseyes, mudcracks, intraformational breccias and gypsum pseudomorphs. West of Cape Peñas, reefal episodes abound at the top of the formation but in general the dolostones are poorly fossiliferous. La Ladrona Fm (*ex* Ferroñes Fm *auctorum*) (Emsian) and Leonian equivalents form a *c.* 130–140 m thick succession of very fossiliferous, grey, argillaceous limestones alternating with dark shales that are more important both at the bottom and at the top of the unit. Two superposed rhythmic sequences can be perceived: decimetric to metric scale sequences composed of successive hummocky stratified storm layers, and thicker, decametric scale sequences possibly related to eustatic pulses like the rhythmothems described by Carls (1988) (see below). Brachiopods, crinoids and trilobites are abundant in the limestones, and dacryoconarids and ostracods are common in the relatively distal dark shales. The Aguión Fm (late Emsian) and equivalent units (Fig. 6.6) are rather heterogeneous lithological units. They usually consist of *c.* 200 m thick, very fossiliferous, pink- to red-coloured crinoidal limestones, dark marls and grey to brown shales. Bioturbated, argillaceous limestones with mudcracks and thin dolostone lenses occur in the upper part of the formation. Biostromal and patch reef episodes are rather frequent. Crinoids, brachiopods, corals and bryozoans are the most important fossils.

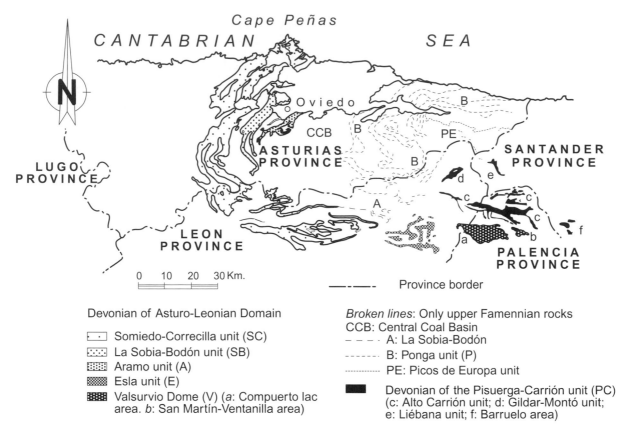

Fig. 6.3. Devonian outcrops in the Cantabrian mountains.

The calcareous sedimentation of the Rañeces/La Vid Group (Gp) and the occurrence of small reefal build-ups illustrate the warm, probably tropical, setting of the sedimentary basin during Early Devonian times. The origin of the Bañugues/Felmin dolostones may be analogous to the tidal flats occurring in present-day dry, tropical areas (Zamarreño 1976). The Nieva limestones originated on an open marine platform battered by storms; sometimes the storms were so frequent that amalgamation of storm layers resulted, forming thick coquinas. The Ladrona and the other Asturo-Leonian equivalent formations formed in open marine conditions relatively far from the coast. Initially, the sediments were deposited above storm wave base, but continued deepening later produced mostly fine siliciclastic deposits in calm and quiet conditions interrupted only rarely by hurricanes. The Aguión and equivalent formations were also deposited in tropical, open marine conditions, relatively far from the coast. The development of frequent coralline biostromes and patch reefs suggests greater bottom winnowing and oxygenation, related to shallowing conditions consistent with the occurrence of mudcracks and dolostones. The shallowing trend was accentuated in the Valsurvio dome, and perhaps in the Asturo-Leonian area, leading to partial emergence with formation of thin bauxite layers representing lateritic palaeosoils formed by weathering under tropical conditions (Koopmans 1962; Closas & Font-Altaba 1960).

The Moniello and equivalent formations (Fig. 6.6) comprise *c.* 250–260 m thick, argillaceous limestones with thin intervening bioturbated shales at the bottom and top of the formation, and a very fossiliferous reefal middle member. However, towards the Asturian arc core the reefal signature weakens, locally even disappearing completely, and fossils become rarer and rarer. This evolution is tied to a marked lithological change, with increasing occurrence of mudcracks, stromatolites, red shales, siltstones and sandstones (Coo *et al.* 1971; Méndez-Bedia 1976). Corals, stromatoporoids, bryozoans and brachiopods are the dominant fossils. The Moniello and equivalent formations exhibit mainly sublittoral facies formed on a shallow,

open marine platform, although near the Cantabrian massif peritidal facies also existed.

The Naranco and equivalent formations (Fig. 6.6) (early Eifelian–Givetian) are siliciclastic units reaching thicknesses of more than 500 m in the northwestern Asturo-Leonian domain and less than 50 m near Ventanilla in the San Martín–Ventanilla area of the Valsurvio dome. These formations display two successions, the lower of which is characterized by coarser and more abundant clastic supply (García-Ramos 1978). Ferruginous red sandstones are common in the lower half and in the uppermost part of the Naranco and Hornalejo formations, and characterize the whole San Martín Fm. The facies distribution is irregular with rapid variations occurring not only towards the inside of the Asturian arc, but even parallel to it. In general the Asturian and Palentian facies are more sandy and bioturbated than the Leonian ones which, especially in the southern distal part of the Asturo-Leonian domain, comprise pelagic-like, euxinic, nodular black shales. Brachiopods, molluscs and bryozoans dominate in the sandstones as opposed to dacryoconarids, ostracods and ammonoids in the distal black shales. The wide range of facies displayed by the Naranco and equivalent formations indicates varied depositional marine and transitional environments linked to large south-sloping deltas.

A return to reefal conditions is recorded by the Candás Fm and equivalent units (Fig. 6.6) (Givetian–earlier Frasnian). These formations consist of *c.* 100–200 m thick, very fossiliferous limestones, as well as argillaceous and sandy limestones with minor shaly intercalations. In the San Martin–Ventanilla area this succession (the Rivera Fm) is overlain with erosional unconformity by a 1–2 m thick, quartz sandstone layer (Fig 6.6; Koopmans 1962; Wagner & Winkler-Prins 1999). Reefal episodes with abundant corals, stromatoporoids, crinoids and brachiopods occur everywhere in the Asturo-Leonian formations, mostly in their upper parts. The Candás and equivalent formations originated on a shallow marine platform in sublittoral conditions although peritidal facies are also found (Reijers 1972).

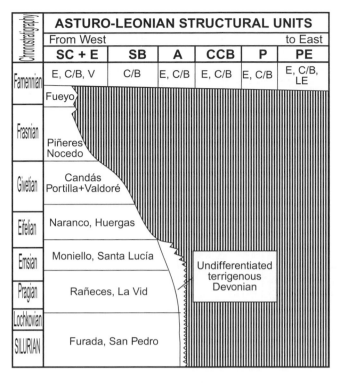

Chronostratigraphy	ASTURO-LEONIAN STRUCTURAL UNITS					
	From West					to East
	SC + E	SB	A	CCB	P	PE
Famennian	E, C/B, V	C/B	E, C/B	E, C/B	E, C/B	E, C/B, LE
Frasnian	Fueyo Piñeres Nocedo					
Givetian	Candás Portilla + Valdoré					
Eifelian	Naranco, Huergas					
Emsian	Moniello, Santa Lucía					
Pragian	Rañeces, La Vid		Undifferentiated terrigenous Devonian			
Lochkovian	Furada, San Pedro					
SILURIAN						

Fig. 6.4. The pre-upper Famennian sedimentation gap across a west-to-east transect in the northern branch of the Asturo-Leonian domain. Vertical hatching: rocks not represented. Abbreviations for structural units as in Figure 6.2. Abbreviations for uppermost Famennian units: B, Baleas; C, Candamo; E, Ermita; LE, Las Ermitas; V, Vegamián.

The development of a carbonate platform and reefal build-ups reflect the tropical setting of the Asturo-Leonian domain during Middle Devonian times. The remarkable clastic input (Naranco and equivalent formations) recorded between two main reefal stages (Moniello, Candás and equivalent formations) could have an epeirogenic significance detached from the climate. Abundant storm deposits and hummocky cross-bedding probably record hurricane events.

The Candás Fm is overlain in Asturias by a c. 400 m thick, fairly siliciclastic succession, the Piñeres and Ermita formations (early Frasnian–Famennian). The Piñeres Fm consists of calcareous, ferruginous sandstones, very fossiliferous marls and lenses of sandy and argillaceous limestones that occasionally develop small reefal bodies. Some levels are strongly bioturbated (*Zoophycos* and others), and brachiopods and bryozoans are common. The Ermita Fm comprises a rather thin succession of cross-bedded, ferruginous, calcareous and microconglomeratic quartz sandstones, with intervening layers of siltstones and shales. Occasionally badly preserved brachiopod and mollusc faunas occur. In the Valsurvio dome (Compuerto area) the Camporredondo Fm is composed of c. 300–500 m thick, almost unfossiliferous, yellow and white quartz sandstones with brown shales at the base and c. 1–2 m thick, coarse to microconglomeratic, quartz sandstones at the top. In the San Martín–Ventanilla area, the Rivera Fm (Fig. 6.6) is directly overlain by quartz sandstones closely similar in facies and thickness to those of the upper part of the Camporredondo Fm.

In Leon, the Portilla Fm (*sensu stricto*) is overlain by the Nocedo Fm, which consists of c. 300 m thick alternations of shales, siltstones, calcareous sandstones and limestones, usually arranged in two regressive–transgressive sequences. The lower part is in general more calcareous than the upper one, and it contains several minor biostromal reefal horizons that can pass laterally into true reef limestones (Valdoré limestone; Fig. 6.6). The uppermost part of the Nocedo Fm

exhibits a similar lateral facies shift from calcareous sandstones to coralline biostromal limestones, especially in the Esla unit (Crémenes limestone). Brachiopods and crinoids are relatively abundant. In a very restricted region, SW of the Asturo-Leonian domain (Somiedo–Correcilla unit; Fig. 6.3), the Nocedo Fm is overlain by c. 130 m of black nodular shales with pelagic faunas (mostly nautiloids, bivalves and ostracods), with a thin, polymict conglomerate in the lower part, and an alternation of shales and sandstones in the upper one (Fueyo Fm). Above the Fueyo Fm, the Ermita Fm comprises up to 60 m of cross-bedded sandstones and quartz sandstones, siltstones, thin shale layers and sandy limestone lenses, with abundant microconglomeratic sandstone layers. In most of Asturias and Leon provinces, the top of the Ermita Fm grades both upwards and laterally into 1–15 m thick, sandy crinoidal limestone and/or pure grainstone and packstone limestones with abundant conodonts (Baleas or Candamo formations), or occasionally to black, poorly fossiliferous shales with phosphate nodules and radiolarites (Vegamián Fm). The Baleas and Candamo formations, Vegamián Fm and perhaps the uppermost part of the Ermita Fm overlap the Devonian/Carboniferous boundary (Fig. 6.6).

The upper part of the Camporredondo Fm, and the quartz sandstones that terminate the Devonian succession in the San Martín–Ventanilla area, are here interpreted as corresponding to the transgressive Ermita Fm (Fig. 6.6; Wagner & Winkler-Prins 1999). The occurrence of a stratigraphical gap below this formation is probable because of the sharp erosive contact with the preceding beds, but the actual amount of time missing at this unconformity is unknown due to the lack of unambiguous fossil information. In general the thin clastic Ermita Fm is of latest Famennian age in Asturias and León.

The increasing clastic input during latest Devonian times triggered the destruction of reefs in the Asturo-Leonian domain (Frankenfeld 1982), although tropical conditions still prevailed because in periods of weakened siliciclastic sedimentation reefal build-ups tended to reappear. Despite such attempts to re-establish calcareous ecosystems, however, Famennian deposits remained completely devoid of coralline reefal elements. The Piñeres and Ermita formations and equivalent units originated in subtidal to supratidal, shallow to very shallow water environments. The Fueyo Fm represents an important intermediate transgressive event with some authors interpreting it as turbiditic (García-Ramos & Colmenero 1981). However, the lack of proper sedimentological and palaeogeographical arguments seems to indicate a sublittoral, open marine platform deposit formed on the distal part of a deltaic system. The Baleas/Candamo limestones resulted from submarine shoals separated by slightly deeper channels occupied by more or less coarse clastic sediments (Ermita sandstones and Vegamián shales).

Palentian domain

The Palentian domain is composed of north-directed Namurian thrust sheets containing Upper Silurian to Lower Carboniferous rocks (Rodríguez Fernández 1994, and references therein). By the end of Namurian–Westphalian A times these Palentian nappes were thrust over an antiformal stack formed in the Valsurvio region (Valsurvio dome) and underwent gravitational collapse to form rootless units in the Pisuerga–Carrión area, resting on relatively autochthonous synorogenic Carboniferous rocks. Later on, they were overprinted by an intense Variscan deformation due to the successive piling of other Cantabrian nappes (the Esla, Ponga and Picos de Europa units; Fig. 6.3).

The Palentian nappes extend across the provinces of Palencia, León and Santander and form four main outcrop areas, separated by important thrusts and faults: the Alto Carrión, Barruelo, Gildar-Montó and Liébana units (Fig. 6.3). In general, the lithology and thickness of the Palentian Devonian rocks are rather homogeneous although some of them display minor facies variations. The most complete, although still very discontinuous, Devonian successions are found in the

Facies association / Reference localities → Lithostratigraphic units ↓	A — LUANCO, CORNELLANA, VEGA LOS VIEJOS, ALBA, ARGOVEJO, COMPUERTO PEÑA CORADA	B — PERLORA, VENEROS, SOMIEDO, POLA DE GORDÓN, VALDORÉ	C — FERROÑES, CANDAMO, EL TORNO, TAMEZA	D — OVIEDO, LA SOBIA, CALDAS DE LUNA, VEGACERVERA, CORRECILLA	E — PEDROVEYA, LOREDO, TUIZA	F — EL SUEVE, LAVIANA, LLANES, PONGA REGION, ARMADA KLIPPEN	G — PICOS DE EUROPA
CAMPORREDONDO — ERMITA, FUEYO	Well developed coarse to microconglomeratic sandstones. Nodular shale(Fueyo Fm.) present only in the more internal CZ unit (southern Somiello-Correcillas unit: Alba syncline)	Thin to very thin, microconglomeratic sandstones or lacking	Very thin, microconglomeratic sandstones or lacking	Very thin, microconglomeratic sandstones or lacking	Thin, microconglomeratic sandstones	More or less thick microconglomeratic sandstones or lacking	More or less thick microconglomeratic sandstones or lacking
BALEAS, CANDAMO, or LAS PORTILLAS	Usually no Baleas/Candamo limestones	Thin, rather well developed Baleas/Candamo limestones	Thin, rather well developed Baleas/Candamo limestones or lacking	Thin Baleas/Candamo limestones or lacking	Very thin Baleas/Candamo limestones	Sometimes very thin Baleas/Candamo limestones or lacking	Thin Las Portillas limestones or lacking
PIÑERES or NOCEDO	Thick calcareous sandstones with well developed interbedded limestone levels	Calcareous sandstones with bad developed interbedded limestone levels, disappearing northwards and eastwards in the southern branch of the CZ	LACKING	LACKING	LACKING	LACKING (only a doubtful reference to Frasnian sandstones in the Asturian coastal region; Radig 1966)	LACKING
RIVERA — CANDÁS, PORTILLA (including Valdoré lst.) or VALCOVERO	Well developed reefal limestones with interbedded calcareous sandstones	Well developed reefal limestones with rare interbedded calcareous sandstones	Thin reefal limestones disappearing towards the Asturian Arc core as well eastwards in the southern branch of the CZ	LACKING	LACKING	LACKING	LACKING
S. MARTINI — NARANCO, HUERGAS or HORNALEJO	In Asturias sandstones and shales. In León, shales, sandstones and thin interbedded limestones. In Palencia shales, sandstones, and thin interbedded limestones (Hornalejo Fm.)	Sandstones, shales and thin interbedded limestones. Shales more important than sandstones in León	Thick sandstone series with interbedded shales and limestones (Naranco and upper Ventanilla + San Martín Fms.)	Thin sandstone series with very few shales (Naranco Fm.) disappearing towards the Asturian Arc core as well eastwards in the southern branch of the CZ	LACKING	LACKING	LACKING
VENTANILLA — MONIELLO, STA. LUCÍA or OTERO	Very fossiliferous reefal limestones. No birdseyes nor mud-cracks. ("Ensenada de Moniello" type, Méndez-Bedia 1976)	Very fossiliferous reefal limestones. Few or no birdseyes nor mud-cracks. ("Ensenada de Moniello" type, Méndez-Bedia 1976)	Fossiliferous limestones with birdseyes and interbedded reefal siltstones and sporadic reefal episodes. Mudcracks present or lacking. ("San Pedro" type, Méndez-Bedia 1976)	Non reefal, poorly fossiliferous limestones with interbedded red siltstone and abundant birdseyes and mudcracks. ("S. Pedro" and "Las Ventas" types, Méndez-Bedia 1976), disappearing eastwards in the southern branch of the CZ	Non reefal, poorly fossiliferous limestones ("Las Ventas" type, Méndez-Bedia 1976), laterally replaced by shales, limestones, dolostones and sandstones or disappearing towards the Asturian Arc core.	LACKING	LACKING
RAÑECES Gr. (Aguión, La Ladrona, Bañugues & Nieva Fms.) LA VID Gr. (Coladilla, Valporquero, La Pedrosa & Felmín Fms.) COMPUERTO	Very fossiliferous, limestones, marls & shales at the top; limestones, dolostones and shales at the bottom. Sporadic reefal episodes beginning in the Nieva Fm.	Very fossiliferous limestones, marls & shales at the top; poorly fossiliferous dolostones, shales and sporadic interbedded limestones at the bottom.	Fossiliferous limestones, marls & shales at the top; non fossiliferous dolostones at the bottom	Fossiliferous limestones, marlstones & shales at the top; non fossiliferous dolostones, and sometimes a thick intercalation of sandstones at the bottom; disappearing eastwards in the southern branch of the CZ	Very poorly fossiliferous limestones, shales and sandstones.	LACKING	LACKING

Fig. 6.5. Devonian facies associations in the Cantabrian Zone, recording the overall shallowing and/or coastal trend that characterizes the Asturo-Leonian Devonian successions. Names included below the seven facies associations (A–G) are either localities or areas where the successions are most typically developed. The term 'birdseyes' refers to small pores produced in micrite, by processes such as desiccation and gas escape, that can be subsequently infilled with calcite spar or sediment.

Alto Carrión and Barruelo units. In Gildar-Montó, Lochkovian to upper Emsian rocks are lacking and in Liébana only Famennian rocks occur.

According to Henn & Jahnke (1984) the Palentian domain can be interpreted as the southern continuation of the Asturo-Leonian platform exposed in the WestAsturo-Leonian Zone. During Lochkovian to Pragian time the depth of the Palentian basin was only slightly greater than in the Asturo-Leonian domain and the successions of both domains are quite similar. However, from Emsian times onwards a stratigraphic differentiation is clearly noticeable. The Palentian domain developed mainly mudstone facies with carbonates formed in anoxic or semi-anoxic environments that were unsuitable for benthic faunas. The long distance to the Cantabro-Ebroian massif, the main source area in the Cantabrian region, is reflected in a low average sedimentation rate, not greater than 1.5 cm per 1000 years (García-Alcalde *et al.* 1988).

The stratigraphical classification of the Palentian Devonian succession is from Veen (1965) who used names previously introduced by other authors (Alvarado & Hernández Sampelayo 1945; Binnekamp 1965) and also created new names. Jahnke *et al.* (1983) and Montesinos & Truyols-Massoni (1987) proposed later minor modifications. The age and correlation of the Palentian units are shown in Fig. 6.2. The Palentian offshore facies have yielded many fossil faunas useful for worldwide correlation (Fig. 6.7; comprehensive lists are given in García-Alcalde *et al.* 1988, 1990). An abridged description of the Palentian formations is given below.

The Carazo Fm (Pridolian? – early Lochkovian) consists of *c.* 200–300 m thick, bioturbated sandstones and shales with frequent storm layers. Reliable palaeontological information on the position of the Silurian–Devonian boundary is so far lacking. The lower part is composed of thick quartzitic and ferruginous sandstone episodes while the upper part is more shaly. The ratio of calcareous rocks increases upwards, and, near the top of the formation, coquinas yield rich Lochkovian faunas that include brachiopods, crinoids, bivalves, gastropods, tentaculitids, ostracods and trilobites. Early Lochkovian sedimentation took place in a rather stagnant, possibly anoxic basin. It was punctuated by short periods of biotic recovery interpreted as due to storms or hurricanes. The occurrence of haematite, phosphorite and sedimentary features like cross-bedding, ripples, grading and

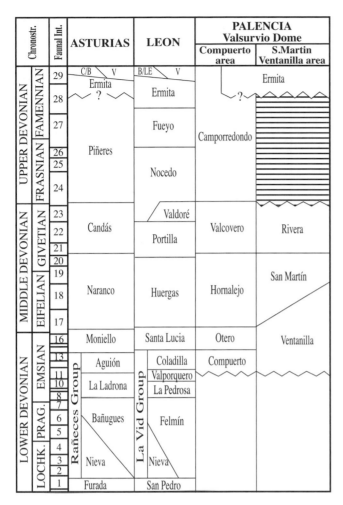

Fig. 6.6. Age and correlation of the Asturo-Leonian lithostratigraphical units. Uneven line: sedimentary break (stratigraphic gap?) or lowest outcropping level. Horizontal hatching: stratigraphic gap. Abbreviations: Chronostr., chronostratigraphy; Lochk., Lochkovian; Prag., Pragian; Faunal Int., faunal intervals. Abbreviations for uppermost Famennian units as in Figure 6.3.

lamination in the lower quartzitic member could indicate a nearshore, high-energy environment (Krans *et al.* 1982). Even allowing for the frequent tectonic repetition and sometimes loss of parts of the Carazo succession that remained unnoticed until recent times, the number and thickness of the lower quartzitic sequences seems to have decreased northwards (Rodríguez Fernández 1994) suggesting that the clastic input was from the south (Krans *et al.* 1982; Wagner *et al.* 1984). The occurrence of hurricanes and the increasing deposition of calcareous sediments reflect the tropical situation of the basin.

As in the Asturo-Leonian domain, the abundant sand supply of the Carazo Fm ended abruptly and a wide and shallow carbonate platform came into being. The Lebanza Fm (Lochkovian–Pragian) is composed of *c.* 160 m thick (decreasing northwards to less than 80 m), very fossiliferous (mostly brachiopods and crinoids) limestones with very thin interbedded shales. At the base of the formation cross-bedding and minor slumping occur. In the upper part, the shaly component increases gradually up into the overlying Abadía Fm. In the upper part some reefal tracts with disphyllid biostromes and stromatoporoid boundstones develop. Above this coralline member a poorly fossiliferous sequence of calcareous mudstones with pellets, hardgrounds and thin layers of laminated dolostones occurs. The Lebanza Fm formed in a sublittoral, shallowing-upward, tropical environment.

Above the Lebanza Fm is an alternation of very tectonized, fine siliciclastic rocks and limestones known as the Cortés-Cardaño Gp (Figs 6.2 and 6.7). The lowermost formation of the group, the Abadía Fm (Pragian to late Emsian) consists of *c.* 150–200 m thick, dark grey to brown, fissile shales and thin siltstone layers with a *c.* 16 m thick, lenticular–stratified (hummocky and ripple cross-bedded) limestone member in the lower part (Requejada Member (Mb)). Below the Requejada Mb, dark grey mudstones, wackestones and bioclastic packstones with hummocky cross-bedding and interbedded, very bioturbated (*Zoophycos,* and others), calcareous siltstones yield an important fauna of trilobites, dacryoconarids and solitary corals and pleurodictyiids. The Requejada Mb itself is poorly fossiliferous, but immediately above it are very fossiliferous, yellowish weathered, nodular to pseudonodular, sometimes bioturbated mudstones, wackestones and bioclastic packstones (with hummocky cross-bedding at the base, and ripple cross-bedding upwards) and interbedded grey, calcareous siltstones and shales (Vañes beds). In the Vañes beds the oldest Cantabrian ammonoid fauna, with *Erbenoceras filalense* (Fig. 6.7), has been found. In the uppermost part of the Abadía Fm, *c.* 10–20 m thick, argillaceous limestone and shales (Lezna Mb) yield mixed offshore/nearshore faunas. As a whole the Abadía Fm was deposited in very quiet water with a slow rate of sedimentation and long intervals of abiotic bottom conditions.

The Polentinos Fm (late Emsian–early Eifelian) is a *c.* 60 m thick (decreasing northward to less than 30 m), alternation of bioturbated (*Zoophycos,* and others) platy to nodular, grey-bluish limestones and dark grey, sandy shales and siltstones, the latter more frequent in the middle and upper part (La Loma beds) of the formation. Poor neritic (trilobites, solitary corals, tabulates and rare brachiopods) and pelagic (ammonoids, dacryoconarids and ostracods) faunas occur and both icriodid and polygnathid conodonts have also been recorded.

The Gustalapiedra Fm (early Eifelian–late Givetian) consists of *c.* 50–60 m thick, black shales and scattered lenses of dark mudstones and wackestones. The upper part of the formation comprises around 8 m of thick-bedded, ferruginous, fine-grained, micaceous, bioturbated (*Skolithos*) sandstones and siltstones with intervening thin silty to calcareous mudstones (Man Mb). *Pinacites jugleri* and other important ammonoids occur near the base of the Gustalapiedra Fm. Below the Man sandstones, in *c.* 4 m thick, nodular, yellowish to reddish mudstones and calcareous shales, ammonoids (*Subanarcestes macrocephalus* and others) and dacryoconarids (*Nowakia otomari* and others) abound, and brachiopods (*Imatrypa infima, Quasidavidsonia vicina* and others) are also present. Dacryoconarids, ammonoids and conodonts of the *Polygnathus ensensis* Biozone are relatively abundant at the top of the formation (Fig. 6.7). The Gustalapiedra Fm was deposited in quiet water but in relatively shallow conditions with a lowered sedimentary input. The sandstones of the Man Mb record the arrival of an important clastic supply related to the Kacak-Otomari event (see below).

The Cardaño Fm (late Givetian–earliest Famennian) is a *c.* 75 m thick alternation of nodular mudstones and dark shales, the former usually concentrated in two calcareous members (Fig. 6.2) at the base and near the top of the formation. The calcareous content decreases northwards so that limestone layers grade into shales with calcareous nodules. Pharciceratid ammonoids, dacryoconarids and homoctenids are frequent in the lower part of the unit while manticoceratids and beloceratids are the more conspicuous ammonoid faunas in the upper part. The Cardaño Fm represents a continuation of the quiet conditions and low sedimentation rates prevalent in the underlying Gustalapiedra Fm.

Above the Cardaño Fm a marked lithological change is recorded by the Murcia Fm (early Famennian) (Fig. 6.2) which comprises *c.* 100–150 m (decreasing northwards to less than 80 m) thick-bedded, usually very tectonized, light- to dark-coloured, graded and cross-bedded quartz sandstones and quartzites with thin intervening black shales. Load and groove casts are frequent at the bottom of the clastic layers. Coarse-grained to microconglomeratic quartz sandstones develop upwards through the formation. Plant remains, phacopid trilobites (*Trimerocephalus* and others) and pelagic fossils such as *Guerichia* and cheiloceratid ammonoids occur in the shales (Fig. 6.7) and there is a complete absence of tentaculitoids and corals. The Murcia Fm has been interpreted as a turbiditic deposit in an otherwise quiet sublittoral environment.

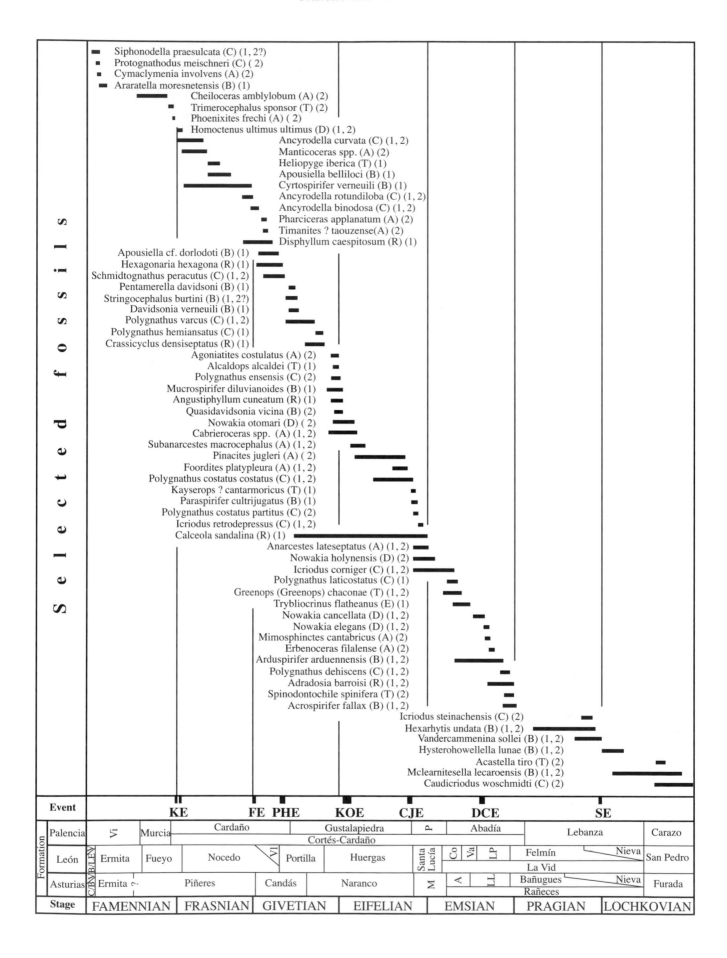

The Vidrieros Fm (early Famennian–earliest Carboniferous?) is composed of *c.* 50 m thick, reddish, nodular to lenticular limestones interbedded with dark shales. Very abundant and varied ammonoid and conodont faunas occur. The lowermost part of the formation yields early Famennian forms such as *Cheiloceras* and conodonts of the lower *Palmatolepis marginifera* zone, but most of the formation is middle and late Famennian in age, with *Sporadoceras* and clymeniids and conodonts of the *Scaphignathus velifer* to *Bispathodus costatus* zones. The uppermost part may overlap the Devonian–Carboniferous boundary, as suggested by the occurrence of conodonts of the *Siphonodella sulcata* zone.

Global events in the Cantabrian mountains

Devonian global events (Walliser 1996) have usually been defined and recognized in pelagic facies rocks, although of course labelling such events as 'global' requires their recognition in both nearshore as well as offshore facies. The tectonic juxtaposition of bathymetrically distant parts of the same platform in the Cantabrian mountains (see above) allows the identification of well-documented Devonian events in both offshore (Palentian domain) and nearshore (Asturo-Leonian domain) facies.

Many global events have been recognized in Devonian times. Here only the most clear-cut events identifiable in the Cantabrian mountains are briefly described (Fig. 6.7). More complete information can be found in García-Alcalde (1997, 1998, and references therein). Most critical index fossils for the Sulcatus event (SE, Fig. 6.7) are lacking in the Cantabrian mountains. The SE can be recognized however, as in the type-region (Bohemia), by a marked colour change that occurs in the middle part of the Lebanza Fm (Palentian domain) and in the upper part of the Nieva Fm (Asturo-Leonian domain). This colour change coincides with a rather distinct faunal turnover at Faunal Intervals 4/5 (Fig. 6.6). Late Lochkovian spiriferids such as *Howellella (Hysterohowellella)* and *H. (Iberohowellella)* evolve, respectively, the first *Hysterolites* and *Vandercammenina*. Several new stropheodontid genera, such as *Boucotstrophia*, *Fascistropheodonta* and *Plicostropheodonta*, occur at the event interval. Shortly afterwards small coralline biohermal mounds develop for the first time in the region.

The Daleje–Cancellata event (DCE) is readily traceable everywhere in the Cantabrian mountains, being marked by the gradual replacement of dark to black shales by limestones and marly rocks at the transition between the *Nowakia elegans* and *N. cancellata* biozones (Fig. 6.7). The faunal change at this level was quite gradual, and coincides with a marked magnetic susceptibility shift that has been identified worldwide (Ellwood *et al.* 2001). In the Palentian domain the loosely coiled ammonoids *Erbenoceras* and *Mimosphinctes*, and the complex-sutured *Celaeceras*, end at the base of the *N. cancellata* biozone.

The Chotec–Jugleri event (CJE) is also easy to recognize in both nearshore and offshore facies. It is represented by a lithologic change from light-coloured limestones of the upper part of the Moniello and equivalent formations (Figs 6.5, 6.6 and 6.7) to dark to black shaly rocks, with interbedded marls, siltstones or sandstones in the Asturo-Leonian domain. The CJE occurs roughly at the base of the *Polygnathus costatus costatus* Biozone, just above the Emsian–Eifelian boundary. The ammonoid turnover related to the CJE is recorded in the Palentian domain by the demise of *Anarcestes lateseptatus* and the first occurrence of *Foordites platypleura*. Throughout the Cantabrian mountains, the trilobite faunas, especially the Asteropyginae and Proetidae, suffered losses close to 100%. In the Asturo-Leonian domain the Rhenish OCA (characterized by brachiopods such as *U. orbignyanus*, *P. cultrijugatus* and *A. alatiformis*) fauna typical of pre-CJE levels, as well many other important brachiopod genera of Faunal Interval 17 (Fig. 6.6) were practically wiped out.

The Kacak–Otomari event (KOE) is recorded in the Palentian domain at the top of the Man Mb (Gustalapiedra Fm), and in the Asturo-Leonian area at the start of the upper sedimentary cycle recorded in the Naranco, Huergas and Hornalejo formations, when siliciclastic supply was reduced (see above). In the Palentian domain *Pinacites jugleri*, *Subanarcestes macrocephalus* and other ammonoids end at, or prior to, the KOE level, while *Agoniatites* and *Cabrieroceras*, as well as *Nowakia otomari*, overlap it. In the Asturo-Leonian domain a very characteristic fauna with *Mucrospirifer* and other brachiopods of Givetian affinities, bizarre solitary rugose corals (*Angustiphyllum*, *Crassicyclus*), bivalves (*Gosseletia devonica*) and trilobites (*Alcaldops*) occur above the KOE.

The Pharciceras event (PHE) cannot be matched with any obvious lithological change. In the Palentian domain it lies in the Cardaño Fm somewhere between the demise of all previously known ammonoids (except *Tornoceras*) in the *P. varcus* Biozone, and the first occurrence of multilobate pharciceratids, *Mesobeloceras* and *Timanites* in the *Schmidtognathus hermanni-P. cristatus* Biozone (Fig. 6.7). In the Asturo-Leonian domain the PHE occurred at the uppermost part of the Portilla Fm and correlative formations, and was coupled with the most marked brachiopod extinction event (*Stringocephalus* and others) recorded in the region at Faunal Interval transition 22–23 (Fig. 6.6). The important coral–stromatoporoid biostromes and bioherms of the second great phase of reefal development in the Cantabrian mountains were also wiped out. On the other hand, the opportunities arising from this extinction event led to the appearance of the important cosmopolitan genera *Douvillina* and *Apousiella* and the rise of Cyrtospiriferidae (although some tenticospiriferids are known from the Lower *P. varcus* Biozone).

General agreement on the exact date of the Frasnian event (FE) has not yet been reached. Here it is considered as immediately preceding the Givetian–Frasnian boundary. In the Palentian domain the event is recorded by the disappearance of most pharciceratids and of *Timanites? taouzense* just prior to the occurrence of conodonts of the Lower *P. asymmetricus* Biozone, at the top of the Lower Limestone Mb of the Cardaño Fm (Figs 6.2 and 6.7). In the Asturo-Leonian domain the event marked the demise of the coral–stromatoporoid reefal tracts at the topmost Valdoré limestone and Candás Fm (Fig. 6.7). Thereafter, the Cantabrian Devonian reefs came virtually to an end and never recovered their past splendour. The brachiopods and trilobites experienced a

Fig. 6.7. Devonian biostratigraphy of the Cantabrian mountains. Abbreviations: formations: A, Aguión; B, Baleas; C, Candamo; Co, Coladilla; LE, Las Ermitas; LL, La Ladrona; LP, La Pedrosa; M, Moniello; P, Polentinos; V, Vegamián; Va, Valporquero; Vl, Valdoré; Vi, Vidrieros. Events: CJE, Chotec event; DCE, Daleje-Cancellata event; FE, Frasnes event; KE, Kellwasser events; KOE, Kacak-Otomari event; PHE, Pharciceras event; SE, Sulcatus event. Selected taxa: A, ammonoid: B, brachiopod; C, conodont; D, dacryoconarid or homoctenid; E, crinoid; R, rugose coral; T, trilobite. 1, occurrence in the Asturo-Leonian domain; 2, occurrence in the Palentian domain.

limited radiation pulse after the FE, producing several new species of *Apousiella, Heliopyge* and others (Fig. 6.7).

The Lower and Upper Kellwasser events (LKE, UKE) are widely distinguished by the onset of black shales and/or black limestones (Kelwasserkalk facies) at the base of the upper *Palmatolepis rhenana* Biozone and at the uppermost *Palmatolepis linguiformis* Biozone. The UKE was one of the most important events in Earth history and was marked by a drastic faunal and floral turnover in most species. In the Palentian domain the LKE and UKE are recorded in the upper part of the Cardaño Fm (Figs 6.2 and 6.7) by the disappearance of all previously known ammonoids, bivalves, tentaculitoids (dacryoconarids and homoctenids) and brachiopods. Not far above the extinction level, new ammonoid, bivalve and trilobite forms occur (*Phoenixites frechi, Falcitornoceras falciculum, Guerichia* spp. and *Trimerocephalus* spp.) (Fig. 6.7). In the Asturo-Leonian domain it is not possible at present to identify the LKE and UKE because of unsuitable facies and a widespread pre-upper Famennian gap. The polymict conglomerate found in the lower part of the Fueyo Fm could represent a major eustatic fall in sea level (UKE; Fig. 6.2).

Iberian Ranges and eastern Guadarrama (PC)

In the Iberian Ranges and eastern Guadarrama, NE central Spain (an area sometimes referred to as Celtiberia; Carls 1988), small outcrops of deformed Devonian strata are found near both margins of the southeastern continuation of the WestAsturo-Leonian Zone and perhaps of the Cantabrian Zone (according to Gozalo & Liñán 1988) (Figs 6.1 and 6.8). Variscan deformation is moderate to weak but there is an incompleteness of exposure due to both Variscan and Alpine

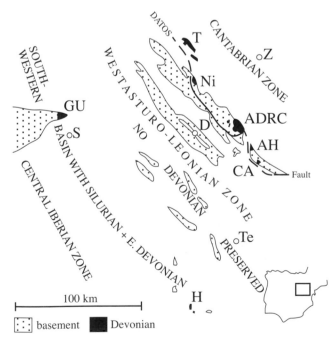

Fig. 6.8. Outcrop areas of Devonian strata in Celtiberia. Abbreviations: ADRC, axial depression of Río Cámaras; AH, Anadón-Huesa del Común; CA, Cabezos Altos; GU, Eastern Guadarrama; H, Henarejos; N, Nogueras; Ni, Nigüella; T, Tabuenca. Towns: D, Daroca; S, Sigüenza; Te, Teruel; Z, Zaragoza.

faulting. This area is interpreted as part of an Ibero-Armorican intraplate basin that subsided strongly in Devonian times. On its southwestern (Castilian) side, it was bordered by the nascent northern Central Iberian (Galician-Castilian) Zone, whose early uplift provided abundant supplies of sand. Only Lower Devonian strata are preserved on this side and they are deformed by Variscan folds (Guadarrama, Henarejos; Sommer 1965; Carls 1969, 1988; Bultynck & Soers 1971). Near the northeastern margin of the basin, alongside the Cantabro-Ebroian massif, in southern Aragón, the delimitation of the basin versus the massif is a matter of discussion, depending on the period considered. The sediments commonly exhibit platformal characteristics, which might suggest deposition directly on the stable massif. However, the thickness of most formations exceed those of their equivalents in the more stable Cantabrian Zone of Asturias and León (Fig. 6.1), indicating a transition to greater basinal subsidence in southern Aragón.

The axial depression of Río Cámaras (ADRC = Nogueras area) (Figs 6.2 and 6.9) is the best outcrop area of Lower Devonian to middle Givetian strata (Carls 1988) because it has escaped strong compression, and so its sediments and faunas are well preserved (Luesma Fm to Cabezo Agudo Fm; Fig. 6.2). The small Nigüella area (Fig. 6.8), although cleaved and folded, provides additional Lochkovian to lower Emsian faunas (Valenzuela-Ríos 1984; Carls & Valenzuela-Ríos 1998). In the northern outcrop area of the Montalbán anticline (i.e. area between Anadón-Huesa del Común; Figs 6.2 and 6.8; Marin & Plusquellec 1973; Quarch 1975; Carls & Lages 1983), some mainly upper Emsian, Givetian and lower Frasnian strata occur, although they are rather strongly fractured. In the Montalbán (Cabezos Altos) area (Fig. 6.8) are exposures of Emsian and Eifelian, but mainly Givetian (and possibly Upper Devonian) strata, but they are cleaved and still incompletely dated (Quarch 1975). Finally, a Cantabrian Zone middle Frasnian to Famennian succession is well known from four inliers surrounded by Mesozoic rocks near Tabuenca (Gozalo 1994) (Figs 6.2 and 6.8).

Caledonian deformation is absent from Celtiberia so that there is no gap or obvious facies change between the Silurian and Devonian periods and sedimentation was probably continuous at least up to middle Famennian times. No evidence remains to link the Devonian succession with that of the Lower Carboniferous, because uppermost Visean and Namurian beds rest unconformably on the former (Quarch 1975).

Close to 4000 m of Devonian strata are known, these being mostly shales, fine-grained sandstones and quartzites, although shelly limestones and marls are common, but thinner. Almost all beds are laterally continuous and record fully marine, mostly shallow-water neritic environments. Variations in subsidence and water depth, and of siliciclastic supply, have produced numerous rhythmothems (Carls 1988, fig. 4). The siliciclastic rocks are rather mature (arkoses and true greywackes do not occur) and apart from a few microconglomerates (Late Devonian of Tabuenca) there are no true conglomerates. Flysch-type graded bedding, reefs, evaporites, red beds, cherts and volcanic rocks are also absent. Numerous horizons with phosphate nodules and the lack of colonial Rugosa and Stromatoporida hint at upwelling of cool water from deep sea to the east (Palaeotethys).

Most lithic units are fossiliferous and there is a particular wealth of neritic, turbid water brachiopod lineages. In some

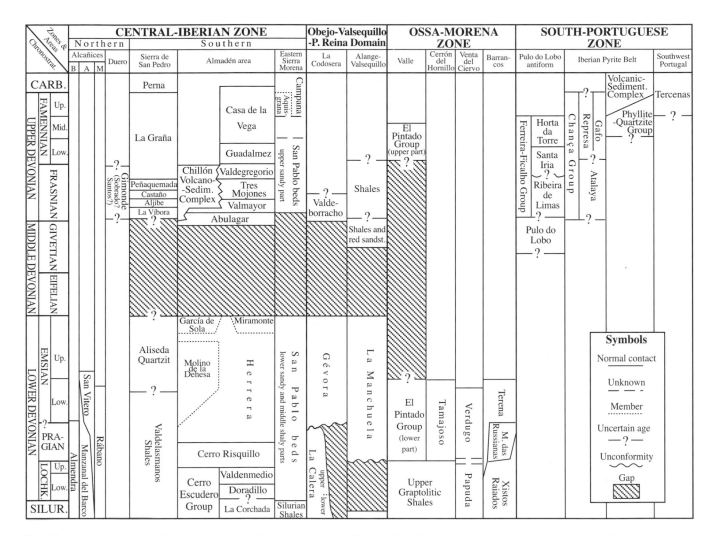

Fig. 6.9. Correlation chart of Devonian stratigraphical units in Central Iberian, Ossa Morena and South Portuguese zones, and in Obejo–Valsequillo–Puebla de la Reina domain. Tectonic units in Alcañices area: B, Bajo Esla; A, Río Aliste; M, Río Manzanas.

rhythmothems, pelagic faunas of black shales and limestones provide a high potential for correlations between neritic and pelagic environments. Up to now, such correlations have been based on brachiopods, trilobites, conodonts, ostracods, dacryoconarids, ammonoids and micro-ichthyoliths.

Most Lower Devonian successions in the area retain a rather uniform character, even across the (now eroded) central zone of the basin, with differences between ADRC, Nigüella and eastern Guadarrama being only transient. In contrast, both Givetian and Late Devonian facies varied along the Ebroian basin margin.

Early Devonian. Devonian sedimentation began with the deposition of the Alcolea Fm in the eastern Guadarrama and the Luesma Fm in the ADRC (Figs 6.2 and 6.8; Carls 1977). Both formations were deposited during Pridolian to early Lochkovian times and are essentially alternations of quartzites with mostly dark shales (like those of the underlying Silurian succession). The Alcolea Fm is 760 m thick, of which about 400 m are basal Devonian (member Al 4). Of the 225 m thick Luesma Fm in the ADRC, only the upper 75 m are Devonian, as shown by the appearance of the first Devonian index trilobite *Acastella heberti* and conodonts (primitive *Icriodus woschmidti* ssp.).

Each of three members within the Luesma Fm begins with white,

thick-bedded quartzites and fines upward. The frequency and diversity of their faunas increase upward, as does the lithological variation due to the appearance of coquinites, very thin local quartz microconglomerates, concentrations of phosphate nodules, and iron oolites, which develop a little above the quartzites. In the ADRC, there is an iron oolite 30 m below the Silurian–Devonian boundary; a lenticular iron oolite with abundant phosphate exists locally at the end of *A. heberti*, and a third one follows immediately above the top of the Luesma Fm, within the range of *Acastella tiro*.

The abundant supply of quartz sand to the eastern Guadarrama (and to the Carazo Fm of Palencia; Fig. 6.2) came from the nascent Galician-Castilian Zone of Central Iberia and deposited the Alcolea Fm. Brachiopod faunas of the ADRC can also be found in the Alcolea Fm (Carls & Valenzuela-Ríos 1999) and some are traceable as far as the eastern Armorican Massif (Pridolian Plougastel Fm and traditional 'Grès à *Orthis monnieri*' of early Lochkovian age).

Together with the iron oolites and trace fossils *Skolithos* (only in a few beds) and *Rusophycus/Cruziana* (scarce), the associations of brachiopods and trilobites recall the depositional environments of the Ordovician quartzites of Mauro-Ibero-Armorica. Thus, these widespread supplies of mature quartz sands are not peak effects of final movements of the Caledonian collision. They are instead the final examples of a specific combination (repeatedly occurring since the Cambrian period) of climatic conditions and anorogenic source areas shedding great amounts of quartz sand throughout Ibero-Armorica.

The abrupt cessation of sand supply during early Lochkovian times (early in the *Acastella tiro* Zone) led to the formation of shelly limestones and shale interbeds of the Nogueras Fm (Fig. 6.2) on the slowly subsiding shallow shelf along the Ebroian side of the basin. There, middle and upper Lochkovian strata measure only 55 m, whereas the equivalents in the Guadarrama are about 200 m thick (units Ce1–2 *partim* = MS 1–10 *partim*). Initially, masses of silt still diluted the carbonates in the Guadarrama area, but these diminished during middle Lochkovian and early late Lochkovian times and a slight deepening allowed the transient invasion of pelagic organisms such as cosmopolitan conodonts and the oldest nowakiids, and favoured clearer-water benthos. During this interval, the more turbid water taxa typical of the shallow northeastern part of the basin did not invade the eastern Guadarrama area so that the faunas on both sides of the basin differed (even the conodonts). Across the Lochkovian–Pragian boundary, renewed Galician-Castilian uplift once again shed sand to produce nearly 80 m of sediments in the eastern Guadarrama area, whereas only a few thin beds of sandstones were deposited on the opposite side of the basin.

Early within Pragian times, units d2cα-2–6 of the Aragonian Nogueras Fm and the equivalent units MS 12b–g in the Cercadillo Fm of eastern Guadarrama (Fig. 6.2) were deposited as rhythmothems of shales, siltstones and brachiopod limestones, commonly in extremely shallow waters. The succession is nearly identical on both sides of the basin except for the higher input of silt and fine sand from the Galician-Castilian uplift in the Guadarrama, where the thickness reaches 230 m compared to only some 30 m in the ADRC. This succession is similar to the Fm de l'Armorique in the west of the Armorican Massif, offering some of the best evidence for the continuity of the Ibero-Armorican trough (Carls & Valenzuela-Ríos 1999).

A synchronous change of facies, interpreted as corresponding to a transgressive pulse, is traceable from Celtiberia through Armorica just after the entry of the conodont *Icriodus simulator*. In Aragón moderate deepening resulted in the deposition of about 30 m of alternating marls and shales, the top unit of the Nogueras Fm, with diverse brachiopods that are mostly known also in Armorica and northwestern Africa. The appearance of clearer-water brachiopods suggests a rough correlation with the Konjeprusy limestone of Bohemia. Conodont correlations at this horizon are still tentative (Carls &Valenzuela-Ríos 2000), but its closeness to the recently redefined beginning of the Emsian (*kitabicus* boundary) is certain.

In the ADRC, the overlying Santa Cruz Fm measures 300 m and ranges into the Emsian (according to its formal definition; Fig. 6.2). A number of rhythmothems are well preserved in the ADRC and are comparable to those analysed by Guillocheau (1991) in the Rade de Brest. Each begins with a few beds of shelly limestones at the start of a deepening pulse which culminates with fine shales deposited from calm waters. These grade into arenaceous alternations that coarsen and shallow upwards, until the sand supply ends abruptly and the next rhythmothem begins. In the shelly limestones and sandstones, brachiopods are abundant. A similar succession can also be recognized at Henarejos, but it is not developed coevally in the eastern Guadarrama where there is a similar thickness of sediment, although only fine shale was deposited, with planktonic *Styliolina* and hardly any benthos.

High in the Santa Cruz Fm at Nigüella, sandstones bear the essential brachiopods of the classic middle Siegenian in the Rhenohercynicum (approximately Pragian in Fig. 6.2), but with an ancestral *Acrospirifer* aff. *primaevus*. Only a little later, this association invaded the Rhenohercynicum during a marine transgression after a long interval of deltaic deposition (Carls & Valenzuela-Ríos 1999). This easy movement of turbid-water fauna argues against any presence of a 'Rheic Ocean' between Ibero-Armorica and the Rhenohercynicum, and therefore also against the elaborate scenario offered by Franke & Oncken (1995) concerning its closure.

The Santa Cruz Fm of the ADRC ends with 100 m of arenaceous deltaic facies sediments, deposited in a local syntectonic depression. The abrupt end of this sand supply by deepening occurred within the range of *Polygnathus excavatus* n. ssp. 114. This favoured the invasion of the so-called western Armorican 'faune de monstres' during the deposition of the basal shelly limestones of the Mariposas Fm (Fig. 6.2) in the ADRC. These limestones form the base of 200 m of shaly and calcareous sediments with abundant faunas, that grade from shallow-water biofacies to hemipelagic ones. Brachiopods, conodonts, trilobites, dacryoconarids and, finally, primitive ammonoids enable correlation of the following boundaries: the beginning of the Emsian (in the classic German version) just above the base of the formation (acme of the dacryoconarid *Guerichina*); the top of the Pragian (in the classic Bohemian version) near the base of limestone unit d4aγ (entry of *Icriodus bil. bilatericrescens*); and the early–late Emsian boundary (and Bohemian equivalent) in unit d4bβ of the Mariposas Fm (Gandl 1972; Carls *et al.* 1972). Across the latter level, the global transgressive Daleje 'event' is recognizable as a gradual, long-term process.

The youngest preserved obvious input from the Galician-Castilian source area in the Guadarrama is represented by a few metres of arenaceous sediments with a shallow-water brachiopod fauna of early Emsian age, this being the equivalent of shales in early unit d4aβ1 of the Mariposas Fm in the ADRC (Fig. 6.2). In the succeeding units of the Guadarrama section (Bultynck & Soers 1971), 10 m of limestones and marls and about 50 m of shales with marls contain conodonts and other elements of pelagic biofacies with ammonoids from *Mimagoniatites* up to *Anarcestes lateseptatus*, reaching possibly to above the middle of the late Emsian interval. The contrast with the coeval thick and largely arenaceous upper Emsian strata of the ADRC indicates inactivity of the former Galician-Castilian source area. No further sedimentary evidence of Devonian uplift of that zone is as yet proven.

Shallow neritic facies returned gradually to the ADRC during the middle of late Emsian times (within the *Polygnathus serotinus* conodont Zone) to produce the Castellar Fm (160 m) and Ramblar Fm (42 m known) (Fig. 6.2). Following this, over six unconnected partial sections represent a total of possibly 500 m of sediment in which formations have not yet been defined because their succession is unknown. The final 180 m of the upper Emsian succession shows sandstones with typically Rhenish faunas (Loscos Fm) followed by increasingly shaly and slightly ferruginous sediments (Peña Negra and Molino formations). Thus late Emsian sediments in the ADRC total about 1000 m in thickness. Above the (hemi)pelagic shales (125 m) influenced by the Daleje 'event', the neritic succession is mostly arranged in rhythmothems like those of the Santa Cruz Fm (see above), consisting mostly of shales, with sandstones and some thin limestone horizons. By contrast with the Lochkovian–early Emsian thickness of 590 m in the same area, there was an increase of the overall deposition rate by a factor of 5, reaching 0.2 km/Ma.

Middle Devonian. At the beginning of Eifelian times terrigenous input to the ADRC decreased during the deposition of the mainly shaly and silty Monforte Fm (60 m), which is in part the equivalent of the Upper Mb of the calcareous Santa Lucía Fm of León (Fig. 6.2) and has a rich neritic brachiopod fauna resembling that of the Lauch unit in the Eifel Hills in Germany. Just after the start of the *Polygnathus costatus costatus* conodont Zone, the Moyuela Fm begins with shales that grade upwards into a very thin unit of black shales with a few levels of nodular limestone and pelagic fauna. About 5 m of thickness seem to represent the *Polygnathus australis* Zone, and the final bed contains conodonts of the *Polygnathus kockelianus* Zone (Wang 1991). The latter is the oldest (pre-Givetian) *pumilio* Horizon known and has furnished placoderm and crossopterygian fish. Above it, 4 m of black shales full of epiplanktic pelecypods and *Styliolina* represent the Kacak event.

Near the base of the overlying 125 m thick Recutanda Fm (Fig. 6.2), conodonts indicate a transition into the Givetian stage. Black shales with pelagic fauna continue, but coarsen slightly upwards; then fine-grained sandstone beds up to tens of metres thick, interpreted as distal turbidites, record the resumption of increased clastic input.

In the ADRC, the Anadón–Huesa area, and at Cabezos Altos (Montalbán area), several unconnected sections are or may be of early Givetian age (Quarch 1975; Carls & Lages 1983). Their thickness may total several hundred metres and they are dominated by neritic alternations of shale and sandstone, although there are also units of black shales with *Styliolina*. One of the latter, the Barreras section (90 m; Carls 1988), is interrupted by a packet of black marls and wavy limestones with *Styliolina*, minute crinoid remains, asteropyginid trilobites (including *Greenops* cf. *boothi*, an Appalachian species) and a huge placoderm (unstudied). The conodont *Icriodus obliquimarginatus* delimits it to an early part of the Givetian, probably not younger than the lower *Polygnathus varcus* Zone.

The Salobral Fm and Cabezo Agudo Fm, totalling probably 350 m in thickness, constitute a continuous succession of alternating shales and sandstones with shallow neritic brachiopod faunas and also some limestone beds (Fig. 6.2). The latter help the subdivision of these sediments into rhythmothems, like the Lower Devonian examples, and at their boundary, both formations have yielded conodonts of the middle *varcus* Zone. The orthid brachiopod *Tropidoleptus* in the Salobral Fm, and the homanolotid trilobite *Dipleura* in the upper part of the Cabezo Agudo Fm, are emigrants from South America. *Dipleura lanvoiensis* correlates with the late Givetian Lanvoy Fm of western Armorica.

It appears that the Givetian sediment thickness reached about 1000 m, most of which is siliciclastic fine-grained sand, siltstone and mudstone. This is in contrast with both the Middle Devonian gap in southern Central Iberia as well as with the largely biohermal middle and late Givetian limestones in León and Asturias. Whether the sand inputs came from the Ebroian Massif or from the Galician-Castilian Zone remains unknown, but their maturity indicates that they are not synorogenic.

Late Devonian. The Givetian–Frasnian transition is obscured by Variscan faulting and the oldest recognized Upper Devonian strata are represented by the Huesa Fm (Fig. 6.2) which, in the Anadón–Huesa area (Montalbán area) reach a thickness of 600–800 m. The succession is mainly made up of neritic shales and sandstones with occasional *Schizophoria*, stropheodontids and atrypids. There are repeated intercalations of black shales with *Styliolina*, *Buchiola* and poorly preserved ammonoids, as well as some thin lenticular limestones with *Homoctenus*, conodonts (no *Palmatolepis*) and *Manticoceras*. Because the boundary between the *Palmatolepis transitans* and *Pa. punctata* zones (formerly lower and middle *Polygnathus asymmetricus* Zone) is in the younger Bandera Fm (Fig. 6.2), the Huesa Fm was deposited during only part of the *Pa. transitans* Zone (obsolete Lower *P. asymmetricus* Zone), probably over less than 0.5 Ma, so it was presumably deposited at an average rate of over 1 km/Ma (>1 mm/year).

In the ADRC and the Anadón–Huesa area, the Bandera Fm (c. 200 m) is a fine shale unit with epiplanktic *Buchiola*, a few small orthoconic cephalopods and rare ammonoids (*Beloceras*), but no benthos. The conodont age of its lower portion lies on the boundary of the *Pa. transitans* and *Pa. punctata* zones. Other fragmentary exposures in these areas include 5 m of black shale dated as Lower Kellwasser Horizon by means of conodonts from limestone concretions with fish remains (Fig. 6.2). Below and above this level, there are possibly 100 m of arenaceous and shaly alternations, but no section has as yet been measured and correlated.

Part of Famennian time is represented in the Anadón–Huesa area by the Fuenpudrida Fm that shows more than 150 m of fine shale with *Guerichia* 'venusta'. It has a tectonic lower contact, but grades upwards into a quartzite unit (Quarch 1975) (Fig. 6.2).

A continuous succession of over 1300 m of Frasnian–Famennian siliciclastic strata has been reconstructed by Gozalo (1994) in the Tabuenca outcrops and has been dated mainly by means of planktonic as well as benthonic ostracods and some ammonoids and conodonts. Based on the correlation of conodont and ostracod zonations, the shaly base of this succession is close to the shaly Bandera Fm of the ADRC and the early Frasnian *Pa. punctata* Zone, whereas equivalents of the Huesa Fm do not crop out. The Tabuenca succession probably ends in late Famennian times. Essentially, the background sediment of this succession is fine shale with pelagic faunas of dacryoconarids, ostracods, a few ammonoids and scarce conodonts on bedding planes (including the only *Palmatolepis* of Celtiberia), *Buchiola* (Frasnian) and *Guerichia* (Famennian), as well as benthic ostracods and a few phacopid trilobites. The shale is frequently interrupted by the sudden intrusion of masses of mostly mature quartz sand that commonly forms the dominant lithology in beds up to several metres thick and sometimes capped by ferruginous hardgrounds. Channelling, soft pebbles, ripples, cross-bedding, swash marks and microconglomerates commonly accompany the coarser sands. After the succession of the older Palaeozoic orthoquartzite formations, which ended in early Lochkovian times, this is a very remarkable recurrence of comparatively coarse and commonly quite mature terrigenous input. Although there are occasional plant remains, the facies is essentially infralittoral, but not deltaic, and nor is it flysch-like (despite the interpretation by Ziegler 1988). Gozalo (1994) found three major, essentially transgressive intervals that are ended by very sudden regressive pulses. The middle to late Frasnian accumulation rate near Tabuenca may have reached nearly 0.2 km/Ma (0.2 mm/year) which is less than one-fifth of the rate in the preceding Huesa Fm of the Anadón–Huesa area, despite the fact that the amount of quartz sand is much higher in the Tabuenca area. Additionally, when Famennian strata of the Anadón–Huesa area are compared to those of the Tabuenca succession, it appears that less of the quartz sand supply reached the southeastern area. The quartz sand formations of Tabuenca suggest Late Devonian uplift of parts of Ebroia which provided a large source area with long-distance transport. Overall, the Upper Devonian succession of northeastern Celtiberia displays similar trends to those along the southern margin of Cantabria (compare Raven 1983).

Plate kinematic aspects

The Palaeozoic, especially Upper Silurian and Lower Devonian, faunas of Celtiberia make an important contribution to the zoogeographic database for plate kinematic scenarios. In particular, the remarkable lithostratigraphic similarities with Armorica provide robust constraints on any model of Devonian palaeogeography (Carls 1988; Carls & Valenzuela-Rios 1999).

As Lower and Middle Devonian rocks (and their faunas) in Celtiberia are closely similar to their age equivalents in the middle and northern parts of the Armorican Massif, all these belonged to the same sedimentary basin, the Ibero-Armorican trough (Carls 1971; Carls & Valenzuela-Ríos 1999). Close faunal relations, especially regarding the Early Devonian neritic faunas, exist likewise with northern Africa and make Celtiberia the central link in the Ibarmaghian faunal subprovince (Plusquellec 1987; i.e. the Mauro-Ibero-Armorican subprovince; Carls 1988). This precludes the existence, during Early Devonian times, of wide oceans between Africa and Iberia, as well as between Iberia and Armorica.

The autochthonous Devonian facies developed in the Cantabrian regions of northern Spain (Asturian and Leonian; Figs 6.2 and 6.5), have lower total thicknesses, but higher proportions of carbonates. The Cantabrian biostromal and biohermal limestones and faunas in particular have no Celtiberian counterparts, and neither stromatoporoids nor colonial rugose corals could grow in Celtiberia. This may have been due not only to the abundant siliciclastics and the turbid waters in Celtiberia, but perhaps also to cooler water temperatures in Celtiberia, where diagenetic dolomitization is also absent. In combination with numerous horizons in the ADRC that contain phosphate nodules, these conditions suggest upwelling of cool oceanic waters in the Celtiberian area which was perhaps located at the western end of Palaeotethys.

Late Silurian neritic brachiopod faunas are rare in Celtiberia, but they share a great part of their taxa with the southern British Isles (*Craniops*, *Baturria*, *Salopina*, *Shaleria*, *Nucleospira*, *Howellella*), which indicates close palaeogeographic linkage. In early Lochkovian times, Rhenish trilobites (Homalonotinae, Acastavinae) and the turbid-water brachiopods *Platyorthis*, *Dalejina*, *Fulcriphoria*, *Iridistrophia*, *Protathyris*, *Howellella*, *Podolella* and *Mutationella*, mostly on species level, provide zoogeographic links between Ibero-Armorica and the Rhenohercynicum (plus Podolia) and also militate against oceanic separation.

During the late Lochkovian to early Pragian interval (about 8 Ma), deltaic facies spread across the Rhenohercynicum and marine shelly faunas disappeared there, while numerous and diverse Ibero-Armorican faunas developed. Their lack in the Rhenohercynicum area supports the hypothesis of the 'Rheic Ocean' as providing a wide zoogeographic separation. However, as soon as marine facies returned to the Ardenno-Rhenish middle Pragian, it was invaded by a fauna of brachiopods that had just developed as an ecologic community in Celtiberia (Carls & Valenzuela-Ríos 1998). In context with Pridolian to Lower Devonian faunas in the Meguma Terrane and other regions of eastern North America, which reveal relations with both the Rhenohercynicum and Mauro-Ibero-Armorica (Bouyx *et al.* 1992, 1997; Gourvennec *et al.* 1997), these faunal connections argue against any wide ocean between all of these regions.

During the predominance of (hemi)pelagic faunas in the lower Emsian and in early parts of the upper Emsian sucessions, the zoogeographic tool of neritic assemblages was blunted, and possible plate kinematic effects on faunas would not have been especially noticeable. However, Emsian species of *Subcuspidella* and middle to uppermost Emsian and lowest Eifelian species of *Paraspirifer* and *Alatiformia*, living in turbid neritic waters, were common in both Celtiberia and the Basque Massifs of the western Pyrenees, and link both regions very closely with the Rhenohercynicum area.

Celtiberian brachiopod faunas younger than early Eifelian have not yet been evaluated sufficiently to help with detailed zoogeographic interpretations. Nevertheless, the presence of the trilobites *Greenops (G.)* cf. *boothi* (Asteropyginae; Gandl 1972) and *Dipleura* (Homalonotinae), and of the orthid brachiopod *Tropidoleptus* in the Givetian of the ADRC constitute neritic faunistic links that extend to Pennsylvania and Ontario. These links do not support the suggestions made by Ziegler (1988) for 'positive areas' with 'active fold belts' and an intervening 'anorogenic, cratonic' core between Iberia and Laurussia during Givetian time. Frasnian benthic ostracods indicate that Celtiberia remained an integral part of the European–Maghrebian domain during Late Devonian times (Lethiers 1983; Gozalo 1994), with no intervening oceanic separation.

Caurel–Peñalba syncline (JS-L)

The Caurel–Peñalba syncline is located in the western West-Asturo-Leonian Zone (WALZ) (Fig. 6.1). In this area the Silurian–Devonian transition is included in the upper part of the Salas Fm, which comprises *c.* 200 m of shales with interbedded siltstones and sandstones, developed distally from the Iberian Late Silurian siliciclastic systems (Chapter 5). The overlying Peñalba Fm is <200 m thick and subdivided into the Seceda and Carucedo members. The basal unit of the Seceda Mb consists of 8–20 m thick, crinoidal carbonate bars, locally with sandy bioclastic limestones and quartzites and horizons rich in brachiopods. Conodonts indicate an early Lochkovian age, equivalent to the uppermost part of the Luesma Fm in Aragon (Fig. 6.2). Above it are 75 m of shales with bioclastic carbonate beds interpreted as storm deposits, and containing middle to late Lochkovian conodonts. The highest Devonian beds comprise at least 75 m of bioturbated nodular and crinoidal limestones (Carucedo Mb). This Carucedo Mb has yielded Pragian conodonts such as *Icriodus angustoides castilianus* and *Icriodus* cf. *simulator*, and some brachiopod faunas of imprecise position dated as Pragian or Emsian (Drot & Matte 1967; García-Alcalde *in* Pérez-Estaún

1978). The Peñalba Fm, except its basal limestones, could represent a more distal sedimentary equivalent to units such as the Lebanza Fm (Palentian domain) or the Nogueras Fm (Celtiberia) (see above).

Central Iberian Zone (MV, PA)

The Central Iberian Zone (CIZ) Devonian series can be subdivided into unrelated northern (Alcañices–Carbajales and Duero valley successions) and southern sequences (Fig. 6.1). The former are closer both geographically and geologically to the eastern Guadarrama and WALZ (including the Palentian domain) sequences than to the southern CIZ ones.

Northern CIZ Devonian

An accurate stratigraphical reconstruction has not so far been achieved in the Alcañices–Carbajales synform because of a complex tectonic overprint. Devonian outcrops in different nappes, and the relationships among them, are difficult to study and correlate. In the Bajo Esla unit (Fig. 6.9), the so-called Almendra Fm consists of *c.* 300 m of shales with interbedded limestones, calcareous shales and sandstones of Ludlovian to Devonian age. The upper part of the Rábano Fm (sandstone and shale alternations) in the Río Manzanas unit is considered to be of Early Devonian age. The Manzanal de Barco Fm in the Río Aliste unit comprises shales, siltstones, shaly limestones, greywackes, sandstones and microconglomerates, and has recently been dated as also being of Early Devonian age. Overlying this formation is the San Vitero Fm which comprises flyschoid greywackes, sandstones and shales of Devonian age (Quiroga 1982). All of these formations everywhere overlie Upper Silurian shales and limestones (Fig. 6.9).

Calcareous sedimentation was more important in the northern CIZ than in eastern Guadarrama, and the quartzitic Pridolian successions typical of the latter are lacking. Rare Lochkovian conodont faunas similar to those in the eastern Guadarrama have been found in the northern series, i.e. *Icriodus rectangularis, Ozarkodina excavata wurmi* (cited by Carls, in Aldaya *et al.* 1976). Other localities have yielded Pragian to upper Emsian conodonts such as *Polygnathus pireneae, P. gronbergi, P. laticostatus, Ozarkodina miae* and *Pandorinellina steinhornensis*, and dacryoconarids such as *Nowakia cancellata* (Truyols-Massoni & Quiroga 1981). The last-mentioned faunas are also known in Guadarrama, but the lack of icriodids and the occurrence of *P. pireneae* could indicate a deeper setting of the Alcañices Devonian sedimentary basin. In fact, the lithofacies where the conodonts occur are usually interpreted as turbidites.

Other northern Devonian outcrops have been reported in the Portuguese area of the Duero valley. They comprise grey sandstones, greywackes and shales probably deposited by turbidity currents. Several different formation names (Sobrado, Santos and Gimonde formations) have been applied to this succession (Fig. 6.9). Pereira *et al.* (1999) reported Frasnian spores associated with probably reworked Early Devonian spores.

The relationships between the Duero valley and the Spanish outcrops referred to above are unclear, because the former exhibits features similar to those of the allochthonous units of the Galicia–Tras-os-Montes Zone (see Chapter 9) and are strongly tectonized.

Southern CIZ Devonian

The first documented reference to the CIZ Devonian in the Almadén area belongs to Prado *et al.* (1855), although precise stratigraphic and palaeontologic studies were achieved much later (references in Pardo Alonso & Gozalo Gutiérrez 1999). Pardo Alonso & García-Alcalde (1996) provided the first modern overview of the Devonian series and faunas. In the southern CIZ three areas each with slightly different geological features can be recognized, these being from west to east, the Sierra de San Pedro and Cáceres synclines, the Almadén area, and the eastern Sierra Morena area (Figs 6.1 and 6.9). The best and most complete Devonian successions occur in the Almadén area (Herrera del Duque, Almadén and Guadalmez synclines, and northern limb of the Pedroches syncline, near Cabeza del Buey; Fig. 6.1). All Devonian stages except the Eifelian and a large part of the Givetian can be recognized from abundant fossil faunas. A Middle Devonian stratigraphical gap was first identified by Puschmann (1967) and it seems commonplace everywhere across the southern CIZ as well as southwards in the Ossa Morena Zone (and in the Obejo–Valsequillo–Puebla de la Reina domain; Fig. 6.9).

The Lochkovian to Frasnian southern CIZ formations probably developed in a shallow marine platform battered by frequent storms with important siliciclastic inputs that became coarser during recurrent shallowing pulses. Ultimately this trend produced tidal flats, estuarine deposits and exposure and non-deposition during intra-Devonian unconformity. From early Famennian times onwards a deepening phase developed in the central area and resulted, in Guadalmez, in the deposition of dark, nodular shales with pelagic faunas, and outer shelf *Palmatolepis–Bispathodus* conodont biofacies (García-López *et al.* 1999). The occurrence of nearshore faunas in coarser terrigenous sediments both westwards and eastwards from Guadalmez, however, indicates maintenance of shallower settings closer to the basin margin.

The Almadén Devonian successions are composed mainly of siliciclastic rocks, including thick formations of mudstones and sandstones at the hundred-metre scale, alternating with thinner formations of sandstones and orthoquartzites at the decametric scale. Limestones are restricted to the Molino de la Dehesa Mb (Emsian), within the Herrera Fm, and the Casa de la Vega Fm (Famennian–Carboniferous). Thick volcanic intercalations are common, mainly in the Frasnian rocks of the Almadén syncline.

Thicknesses vary largely across the area, with the thickest successions occurring in Almadén (>1750 m of Lochkovian to late Frasnian age) and thinning towards both north and south (730 m of Lochkovian to upper Frasnian rocks in Herrera del Duque, and 1400 m of Lochkovian to uppermost Famennian rocks, in Guadalmez; see Fig. 6.1). The Silurian/Devonian boundary has not been documented so far but it could be located at the 'Alternancia de La Corchada' (Fig. 6.9). The top of the Doradillo Fm would be of Lochkovian age according to the occurrence of *Trimerus* cf. *acuminatus* (Pardo Alonso & García-Alcalde 1996; Pardo Alonso 1998). Fossils are scarce in the Lower Devonian rocks of the Guadalmez syncline and lacking in the Pedroches syncline. At the base of the Cerro Risquillo Fm ('Cuarcita de Base', *auctorum*) rare Lochkovian brachiopods occur (Pardo Alonso & García-Alcalde 1996). Therefore the unfossiliferous Valdenmedio Fm, located between the Doradillo and Cerro Risquillo formations, could also be of Lochkovian age. In Herrera del Duque and Almadén, the top of the Cerro Risquillo Fm yields Pragian brachiopods, such as *Hysterolites* aff. *korneri*, *H.* cf. *gandli*, and *Paulinella* cf. *guerangeri*.

The lowermost part of the overlying Herrera Fm may also be Pragian, but most of the unit belongs in the Emsian stage. The Herrera Fm comprises *c.* 250–500 m of sandy shales with abundant interbedded alternations of sandstones and quartzites. In Herrera del

Duque and Almadén a middle calcareous member, the Molino de la Dehesa Mb, comprises locally very fossiliferous, bioclastic, sandy limestones sometimes with intervening micritic and stromatolithic layers. The strongly diachronous base of this member could locally be of latest Pragian age but in general it is of Emsian, even late Emsian age, based on the occurrence of conodonts such as *Icriodus* cf. *fusiformis*, *Caudicriodus celtibericus* and *Ozarkodina steinhornensis steinhornensis* (Pardo Alonso & García-Alcalde 1996; García-López *et al.* 1999).

In Herrera del Duque, latest Emsian brachiopods occur in ferruginous quartzite layers (García de Sola Mb; Fig. 6.9) immediately below a Middle Devonian unconformity (Pardo Alonso 1999*a*). In an equivalent stratigraphical position, thick ferruginous sandstones and quartz conglomerates occur in Guadalmez (Miramonte Mb; Fig. 6.9). These deposits were formed in very shallow (locally even fluvial) conditions. The first formation above the Middle Devonian unconformity is the Abulagar Fm and is also proximal in origin. Therefore the unconformity has been interpreted as probably due to a regional tectonic uplift (Pardo Alonso & García-Alcalde 1996) that developed into a folding phase southwards in the Obejo–Valsequillo domain (Herranz Araújo 1994).

The base of the Abulagar Fm is composed of late Givetian or Frasnian, white orthoquartzites yielding bivalves. Above this, brachiopods become the dominant fossil faunas higher in the Abulagar and overlying Frasnian formations (Valmayor, Tres Mojones and Valdegregorio), although other shelly faunas, especially bivalves, gastropods and rare trilobites, also occur. The abundance of fossils has allowed the definition of a pluritaxonic biozonation (bivalves and brachiopods) which has proved to be very useful everywhere in the southern CIZ (Pardo & García-Alcalde 1984). This biozonation comprises a basal late Givetian? Bivalve Cenozone and successive Frasnian, brachiopod-based, *Cyphoterorhynchus*, *Douvillina alvarezi/Eoschuchertella jordani*, *Apousiella almadenensis* and *Pradochonetes muelleri* biozones. The upper part of the Valdegregorio Fm continues to yield Frasnian faunas (*Nowakia* sp. and *Spinulicosta*? sp.). In the Almadén syncline most of the Frasnian sequence consists of a thick, mixed, sedimentary–volcanic succession (Chillón volcanosedimentary complex; Fig. 6.9) recording a resumption of the Silurian volcanism that sourced the famous Almadén mercury ores.

The Frasnian–Famennian boundary occurs at the base of the Guadalmez Fm. Nodular black shales yield early Famennian epiplanktic bivalves such as *Buchiola* and *Guerichia*, and ammonoids of the *Phoenixites frechi* and *Falcitornoceras falciculum* zones (Montesinos & Sanz-López 1999*a*). The characteristic lower Guadalmez Fm facies are largely developed in other coeval Iberian, French and German settings, and may be related to the onset of anoxic conditions during the development of the global Kellwasser (Lower and Upper) events. Late Famennian black shales and limestone lenses and nodules are sparsely distributed throughout the Guadalmez and Pedroches areas. Sedimentation during this period was condensed and probably included important hiatuses because in just a few metres of the lower third of the Casa de la Vega Fm, there are conodonts of early Famennian *Pa. Crepida*, late Famennian middle *Siphonodella praesulcata* and late Tournaisian *Scaliognathus anchoralis* zones (García-López *et al.* 1999).

In eastern Sierra Morena (mainly in Torre de Juan Abad and in the Santa Cruz de Mudela syncline; Fig. 6.1) the Devonian succession exhibits a more clastic facies than in Almadén, and has been called the 'San Pablo beds' (Butenweg 1968) (Fig. 6.9). This unit is composed of >450 m sandstones, quartzites and shales, the former being more abundant at the top and bottom, and the shales dominating in the middle part of the unit. The lower, ferruginous, sandy unit also includes quartz conglomerates and some limestone layers that are comparable to the Molino de la Dehesa Mb of the Herrera Fm in the Almadén area. These limestones have yielded Pragian to late Emsian brachiopods, conodonts and dacryoconarids (Pardo Alonso & García-Alcalde 1996). The upper sandy unit usually yields Frasnian brachiopods of the *Apousiella almadenensis* Biozone, including the type species. However, in the easternmost outcrops (Torre de Juan Abad) early Famennian brachiopods such as *Eoparaphorhynchus* also occur (Pardo Alonso & García-Alcalde 1996; Pardo Alonso 2000). Late Famennian ostracods have been described in the 'Aquisgrana Shales' (base of the Campana Fm) in the El Centenillo–La Carolina area

(Charpentier *et al.* 1976). The Campana Fm (Famennian–Early Carboniferous) has been interpreted as unconformably overlying Silurian beds (Gutiérrez Marco *et al.* 1990), but the anomalous contact may actually be tectonic (Pardo Alonso & García-Alcalde 1996; Pardo Alonso 1999*b*).

In the western area, the Devonian is well represented in the Cáceres and Sierra de San Pedro synclines (Fig. 6.1). The most complete succession (*c.* 1500 m thick) occurs in the Sierra de San Pedro syncline and is very close to other CIZ successions. Soldevila Bartolí (1992*b*) recorded Emsian brachiopod faunas in the so-called 'Aliseda quartzite' (Fig. 6.9) with abundant *Brachyspirifer*, *Euryspirifer*, *Arduspirifer*, *Uncinulus*, *Schizophoria*, *Plebejochonetes* and other forms. The Valdelasmanos shale includes Upper Silurian to lower Emsian rocks. Frasnian brachiopod faunas corresponding to the *A. almadenensis* and *Pradochonetes muelleri* biozones are widely distributed across the area (La Víbora, Aljibe, Castaño and Peñaquemada formations). To date, the latest Frasnian interval is represented by a single locality where the succession comprises light grey to pink shales yielding articulate and inarticulate brachiopods, homoctenids and entomozoan ostracods. Late Famennian brachiopods are abundant at the top of the Devonian series in La Graña Fm, which is composed of greywacke, shaly and microconglomeratic facies (Pardo Alonso & García-Alcalde 1996).

Obejo–Valsequillo–Puebla de la Reina domain and Ossa Morena Zone (MV, PA)

The Obejo–Valsequillo–Puebla de la Reina domain (OVPD) (Delgado *et al.* 1977) extends between the so-called 'Badajoz–Córdoba shear zone' to the south and the Pedroches syncline to the north, embracing on its western side the Spanish La Codosera–Puebla de Obando area and the Portuguese Portalegre–Dornes area (Figs 6.1 and 6.9). It has usually been included in the Ossa Morena Zone (OMZ) but sometimes also in the CIZ (see discussions in Pardo Alonso & García-Alcalde 1996; Eguíluz *et al.* 1999; Pardo Alonso 1999*b*). For the moment we prefer to consider the OVPD as a separate domain of unclear affinities (Herranz Araújo *et al.* (1999) named it the 'Lusitano–Mariánica' Zone). It differs from the CIZ because it shows good development of Lower Devonian carbonate successions yielding abundant and characteristic Lochkovian–Pragian faunas, a wider intra-Devonian gap that embraces sometimes Emsian to upper Famennian rocks, and a lack of pelagic Famennian facies. The thick Lower Devonian OVPD carbonate facies rocks are coeval with the shaly, even graptolitic, OMZ facies rocks, and this has been considered as one of the main differences between the zones. However, the significance of this difference is now greatly reduced due to the recent dating of the thick OMZ Estremoz marble as Upper Silurian?–Lower Devonian.

Obejo–Valsequillo–Puebla de la Reina domain

The best known Devonian successions in the OVPD occur in La Codosera syncline and in its Portuguese prolongation in the Portalegre–Serra de Sao Mamede syncline, as well as in other synclines cropping out further NW (Amendoa and Dornes, in Portugal; Fig. 6.1). La Codosera–Portalegre syncline Devonian outcrops have been subdivided by Soldevila Bartolí (1992*a*) into the La Calera, Gévora and Valdeborracho formations (Fig. 6.9), the first two extending into Portugal. The Portuguese outcrops are in general more fossiliferous than the Spanish ones. The Silurian–Devonian transition occurs in a shaly and sandy, ferruginous unit that has received different formation names (i.e. 'La Calera unit', 'Sao Mamede upper

sandstones', 'Serra do Luaçao Fm'). In Portalegre and Dornes, Late Silurian to Pragian shelly faunas have been reported (Perdigao 1967, 1973, 1974, 1979; Cooper *et al.* 2000), including brachiopods such as '*Camarotoechia*' *aequicosta*, *Schizophoria provulvaria*, *Platyorthis circularis*, *Brachyspirifer rousseaui*, *Hysterolites hystericus* and *Mauispirifer gosseletti*, and bivalves such as *Pterinatella laevis* and *L. (Leiopteria) pseudolaevis*.

The Gévora unit is a >1000 m thick succession of shales and interbedded dolostones and limestones with a basal, thin, widely distributed 0.5–1 m thick conglomeratic horizon. The equivalent Portuguese units are the Xistos de Sao Juliao-Calcários do Escusa Fm (in Portalegre) and the Dornes Fm in Dornes. Sometimes the Gévora Fm unconformably overlies Caradocian to Wenlockian rocks, indicating that La Calera Fm is missing. Emsian brachiopods have been recorded in Portalegre (Perdigao 1967, 1974) and Pragian to Emsian? faunas in Dornes. Perdigao (1967, 1974, 1979) dated the uppermost levels (Calcários do Escusa) as Eifelian, but Racheboeuf & Robardet (1986) deduce that they are not younger than late Emsian. In fact the brachiopods determined by Perdigao are very close to the late Emsian forms found at the top of the Herrera Fm in Almadén.

The Spanish succession culminates in the Valdeborracho Fm that lacks an equivalent Portuguese correlative unit. The Valdeborracho Fm comprises *c.* 300 m of shales, siltstones and quartzites with rare Frasnian brachiopods such as *Apousiella* cf. *almadenensis* and *Cyphoterorhynchus* cf. *domenechae*.

In the Alange–Valsequillo area (Fig. 6.1), the Devonian succession lies unconformably over Silurian rocks. The basal unit is the La Manchuela Fm (Racheboeuf *et al.* 1986; Fig. 6.9) which consists of *c.* 200 m (thickening eastwards toward the Sierra del Pedroso) of dark shales with limestone lenses developed in interbedded ferruginous, sometimes oolithic, calcareous sandstones, with Pragian to late Emsian brachiopods, corals and trilobites (Rodríguez 1978; Rodríguez & Soto 1979). Above the La Manchuela Fm are *c.* 30 m of grey to red, very fine-grained, sericitic shales, with white to red sandstones in the lower part. Brachiopods such as *Eostrophalosia* sp., *Devonochonetes* aff. *kerfornei*, *Dagnachonetes* sp., *Longispina* sp., *Ilmenia* aff. *subhians*, and *Carpinaria* sp. and corals indicate a Givetian, probably late Givetian, age (Pardo Alonso & García-Alcalde 1996; Rodríguez & Soto 1979). Trilobites, tentaculitids and bivalves are also conspicuous components of the shelly fauna. A shaly overlying succession yields early Frasnian conodonts of the middle to upper *Polygnathus asymmetricus* Zone, late Frasnian brachiopods, entomozoan ostracods and tentaculitids (Racheboeuf *et al.* 1986) and other possibly early Famennian fossils. Similar facies and faunas occur east-southeastward in the central and eastern parts of the Cordoba outcrop (Julivert *et al.* 1983). There, Lochkovian, basal, dark quartzites overlain by locally thick Pragian to lower Emsian, biohermal limestones occur. Higher up are *c.* 100 m of green shales and white fossiliferous deltaic orthoquartzites with interbedded sandstones and calcareous shales (Pérez Lorente 1979). These quartzites have been previously dated as Givetian but they must correspond instead to the latest Devonian (García-Alcalde, pers. comm.). In several isolated and discontinuous outcrops Frasnian fossils have also been reported. Oliveira *et al.* (1986*b*) and Herranz Araújo (1994), among others, support the idea that Upper Devonian rocks unconformably overlie a Lower Devonian erosion surface, and suggest that the intra-Devonian gap is related in this area to tectonic compression and folding. Both the geometry of reef outcrops and palaeocurrent directions measured on the orthoquartzites indicate that the source area was situated to the north. The Alange Frasnian facies and faunas are similar to the Sierra de San Pedro ones, although this is an exception because in general the OVPD and southern CIZ Devonian successions are quite different.

Ossa Morena Zone

The best known OMZ (*sensu stricto*) Devonian successions occur in the Spanish Valle, Cerrón del Hornillo and Venta del Ciervo synclines (Fig. 6.1). Equivalent successions are also developed in Portugal near Barrancos. Robardet (1976),

Jaeger & Robardet (1979), Racheboeuf & Robardet (1986), Weyant *et al.* (1988), Robardet *et al.* (1988, 1991), Gessa *et al.* (1994) and Lenz *et al.* (1996) described the most complete stratigraphic sequence and their faunas in the core of the Valle syncline (Fig. 6.9).

The Silurian–Devonian transition occurs in graptolite-bearing black shales with Pridolian (*Monograptus transgrediens*) to late Lochkovian (*M. praehercynicus*, *Linograptus posthumus*, *M. hercynicus* and the dacryoconarid *Homoctenowakia bohemica bohemica*) faunas. Above the graptolitic shales the El Pintado Gp crops out and comprises *c.* 180 m of green, nodular shales with pebbles and some griotte-like limestone layers in its lower part, and sandy limestones full of spiriferid brachiopods and dark shales with interbedded limestone nodules and lenses in the upper part. Pragian to early Emsian brachiopods, ostracods and phacopid trilobites occur in the lower part of this group (Racheboeuf & Robardet 1986; Robardet & Gutiérrez-Marco 1990a; Robardet *et al.* 1991). In the upper part of the El Pintado Gp, the limestone nodules have yielded Famennian conodonts of the Lower *Pa. crepida* to Upper *Pa. marginifera* zones. Worthy of mention is the lack of documented Eifelian to early Famennian rocks that might record the greatest extension of the intra-Devonian unconformity in the region. In Cerrón del Hornillo a similar Lower Devonian succession, the Tamajoso Fm, occurs but it is unconformably overlain by Tournaisian beds (Racheboeuf & Robardet 1986; Robardet *et al.* 1991; Fig. 6.9).

In Venta del Ciervo (north Aracena; Fig. 6.1) Lower Devonian rocks form the Papuda and Verdugo formations (Schneider 1951) (Fig. 6.9). The 60–70 m thick, unfossiliferous Papuda Fm may span the Silurian–Devonian boundary, as suggested by stratigraphical similarities with the Silurian–Devonian successions of Valle and Cerrón del Hornillo synclines. The Verdugo Fm has been subdivided into a basal Verdugo greywacke (15 m thick) and an overlying Verdugo shaly unit ('tonschiefer') 80–100 m thick. The latter shaly unit has yielded late Pragian to early Emsian, offshore faunas with brachiopods belonging to the *Andalucinetes hastatus* and *Ctenochonetes robardeti* zones, and trilobites, ostracods and graptolites of the *Monograptus yukonensis* zone (Racheboeuf & Robardet 1986; Robardet *et al.* 1991; Lenz *et al.* 1997). The relationships of the Verdugo and other OMZ formations are as yet unknown due to a strong tectonic overprint. The Portuguese continuation of the above formations in the Barrancos area is represented by the Xistos Raiados Fm, which comprises thin alternations of shales and siltstones showing a metamorphic overprint of chloritoid grade. The upper part of the Xistos Raiados Fm passes laterally westwards into the turbiditic Lochkovian to Emsian Terena Fm and eastwards to the Pragian Monte das Russianas Fm which comprises green to grey shales with crinoidal limestones (Fig. 6.9).

South Portuguese Zone

The South Portuguese Zone (SPZ) Devonian is developed both in Spain and Portugal (Fig. 6.1) but it is much better represented in the latter than in the former. Five different domains have been recognized (Oliveira 1990), these being the Beja-Acebuche ophiolite complex, the Baixo Alentejo Flysch Gp, the Pulo do Lobo antiform, the Iberian Pyrite Belt and the Southwest Portugal domain, but definitely dated Devonian rocks occur only in the last three of these.

Pulo do Lobo antiform

The Pulo do Lobo antiform (PLA) core (Fig. 6.1) exposes the Pulo do Lobo Fm which comprises highly tectonized phyllites and orthoquartzites with interbedded metabasalts in the lower part. The formation is of unknown thickness and is interpreted as composed of oceanic metasediments and metabasic rocks of Givetian–Frasnian? age.

In the PLA northern limb, the Pulo do Lobo Fm is overlain

by the Ferreira-Ficalho Gp (*c.* 500 m thick; Fig. 6.9), which consists of the Ribeira de Limas (strongly tectonized phyllites, metagreywackes and some tuffites), Santa Iria (a flysch-like succession of shales, siltstones and greywackes) and Horta da Torre (shales, sandstones and siltstones, with orthoquartzites in the topmost part) formations. The Santa Iria and Horta da Torre formations are less tectonized than the Ribeira de Limas Fm, and it is possible that the Santa Iria Fm may unconformably overlie the Ribeira de Limas Fm. Frasnian to middle Famennian palynomorphs and acritarchs have been found in the Santa Iria and Horta da Torre formations (Oliveira *et al.* 1986b; Eden 1991).

In the southern limb of the PLA occurs the flysch-like Chança Gp (*c.* 1100 m thick) which has traditionally been considered as equivalent to the Ferreira-Ficalho Gp (*auctorum*). Quesada (1998) has recently shown it would correlate better with the Iberian Pyrite Belt domain.

Iberian Pyrite Belt

In the Iberian Pyrite Belt (IPB) area (Fig. 6.1) different formations of unknown thicknesses have been placed within a Chança Gp, which comprises the Atalaya Fm (phyllites and sandstones) overlain by the Represa Fm (siltstones and greywackes) and Gafo Fm (greywackes, siltstones and shales with interbedded basic and felsic volcanics and dykes). The two last formations possibly replace one another laterally. Late Famennian spores and acritarchs have been found at the upper part of the group.

Oliveira (1990) has distinguished a Phyllite–Quartzite Gp and the 'volcanic-sedimentary complex' in the IPB (Fig. 6.9). The Phyllite–Quartzite Gp is a metric to decametric alternation of shales and sandstones with intervening conglomerates, of unknown total thickness. Sedimentary structures and fossils (trilobites and brachiopods) indicate a shallow-water depositional environment in an epicontinental sea. In the upper part of the unit, lenses and nodules of bioclastic limestones have yielded abundant Famennian conodonts of the lower *Marginifera* to lower *Praesulcata* zones (Boogaard 1963; Boogaard & Schermerhorn 1980, 1981). At the topmost part of the group Strunian (latest Famennian) palynomorphs occur.

The 'volcanic-sedimentary complex' consists of *c.* 100–600 m thick acid and basic volcanic rocks with massive sulphide mineralizations alternating with dark to black shales and siltstones. The base of the complex has yielded in different localities either latest Strunian spores (Pereira *et al.* 1996) or early Carboniferous spores. The unit was deposited in extensional conditions during bimodal volcanism.

Southwest Portugal domain

The Southwest Portugal (SWP) domain is characterized by sparse outcrops near the Portuguese Atlantic coast (Fig. 6.1) surrounded by Carboniferous rocks of the Baixo Alentejo Flysch Gp. The SWP Devonian succession is represented by the Tercenas Fm, the base of which remains unknown (Fig. 6.9). The formation is 100 m thick and consists of two superposed sequences comprising heterolithic sandstone and shales, and sandstones respectively, the former topped by sandstones and the latter by conglomerates. Rare spores, clymeniid ammonoids, brachiopods and corals indicate a late Famennian–earlier Tournaisian age for the formation (Korn 1997; Pereira 1999). Sedimentary structures suggest the Tercenas

Fm was deposited in a sublittoral environment. The Tercenas Fm is similar to the IPB clastic formations and so it is suggested that both the SWP and IPB could belong to the same sedimentary basin and a single suspect terrane (Ribeiro *et al.* 1990) that was closer to Avalonia than to Gondwana during Late Devonian times (Oliveira & Quesada 1998).

Catalonian Coastal Ranges (MT-M, JL, G-A, FS)

The Catalonian Coastal Ranges (CCR; Figs. 6.1 and 6.10) run parallel to the Mediterranean coast in NE Spain, and form two mountainous areas separated by the Vallés-Penedés depression (pre-littoral Catalonian graben). The current relief results from Palaeogene compression during the Alpine orogeny, followed by an extensional phase at the beginning of Neogene times. The CCR comprise Palaeozoic rocks variably deformed and metamorphosed during the Variscan orogeny, cropping out discontinuously due to the occurrence of numerous plutonic bodies, and a partial cover of both Mesozoic and Cenozoic rocks.

The best documented, though incomplete, Devonian successions occur in the northern area of the CCR, close to Barcelona (such as at Santa Creu d'Olorda, Bruguers, El Papiol, Montcada, Molins de Rei), at St. Martí de Llémana (Las Guilleries, Girona), El Figaró (Montseny, NNE of Barcelona) and at Montnegre and Can Lliure (Les Gavarres, Girona), although some outcrops have also been cited further at Vilella Alta (Priorat, Tarragona; Fig. 6.10). Devonian sediments were deposited in hemipelagic to pelagic conditions and fossils are both diverse and relatively abundant. The Silurian–Devonian transition, as well as the Lochkovian, Pragian and Emsian stages, are well represented (Fig. 6.2). The youngest exposed rocks are usually of Emsian age but Famennian and Upper Devonian rocks have also been cited in El Papiol (Puschmann 1968) and in Vilella Alta, respectively.

The best studied series crops out at Santa Creu d'Olorda (Barcelona) where Julivert *et al.* (1985) have defined two units that can also be identified at other localities: La Creu Fm (late Ludlovian to Lochkovian) and the overlying Olorda Fm (late Lochkovian to late Emsian; Fig. 6.2). The occurrence of Devonian rocks in the CCR was first reported by Almera (1891c) and Barrois (1892). Almera (1914) divided the Silurian–Devonian sequence into a Silurian, massive limestone lower member and a Devonian limestone–marl upper member. This division broadly accords with the La Creu and Olorda formations. In addition, both units are broadly equivalent to Puschmann's (1968) 'Orthoceraten Kalke' and 'Tentaculiten Schiefer', respectively.

La Creu Fm comprises *c.* 30–40 m of massive, nodular, stylotized limestones (cephalopod/crinoidal wackestones) with thin intervening pelites. Crinoid–brachiopod coquinas interpreted as storm layers are common. The development of hardground records important cementation and dissolution events, presumably related to low sedimentation rates in the basin. These lithofacies have been interpreted as hemipelagites to pelagites developed on the inner part of a carbonate ramp (García-López *et al.* 1990). Most of the formation is of Silurian age but conodonts and graptolites recovered from the upper part (Walliser 1964; García-López *et al.* 1990) have allowed the identification of the Silurian–Devonian boundary *c.* 5 m below the top of the unit. *Ozarkodina excavata excavata* and *O. excavata wurmi* as well as *Scyphocrinites* sp. have been found *c.* 10 m below the top, while in the last few metres of the formation, graptolites of the *Monograptus hercynicus* Zone occur (Julivert *et al.* 1985). In Montnegre and Can Lliure (Les Gavarres) the so-called 'Calizas de Can Riera' consist of grey limestones (wackestones) with Lochkovian conodonts (Sanz-López *et al.* 1998) and are equivalent to the La Creu limestones, and contain many fossil fragments interpreted as storm layers (Barnolas & García-Sansegundo 1992).

The *c.* 40 m thick Olorda Fm has been subdivided at its type locality into 'A' to 'E' members (Fig. 6.2). Graptolites, dacryoconarids, conodonts and brachiopods are abundant and trilobites, corals, eurypterids and crinoids also occur. 'A' Mb consists of *c.* 2–3 m of dark shales and fine black sandstones, with chert nodules and polymetallic oxides. 'B' Mb comprises *c.* 5 m of bioturbated hemipelagites to pelagites, composed of yellowish limestones (dacryoconarid-wackestones) and reddish shales sedimented on the middle or upper part of the carbonate ramp. 'C' Mb comprises *c.* 15 m of nodular limestones (dacryoconarid-wackestones/packstones) with intercalated green to bluish marls with the marl/limestone ratio increasing upwards. These facies have been interpreted as belonging to the middle part of the carbonate ramp. 'D' Mb comprises 13 m of stratified limestones (dacryoconarid-packstones) and marls. The limestone–marls alternations suggest a distal part of the carbonate ramp environment receiving cyclic terrigenous inputs. 'E' Mb comprises 4 m of green shales that form the highest preserved part of the Olorda Fm (García-López *et al.* 1990).

In the type locality of the Olorda Fm the 'A' Mb belongs to the *Homoctenowakia bohemica* Zone due to the occurrence of the zone fossil. Moreover it yields conodonts mainly belonging to the *Ozarkodina delta* zone, and graptolites of the *M. hercynicus* Zone (*Monograptus hercynicus* and *Linograptus posthumus*) (Greiling & Puschmann 1965; Julivert *et al.* 1985). *L. posthumus* has also been cited in other CCR localities in the *M. praehercynicus* Zone (Gutiérrez-Marco *et al.* 1999b). Julivert *et al.* (1985) have recorded faunas of the *M. uniformis* zone at Can Castany (Cervelló, SW Molins de Rei). Eurypterids (*Pterygotus* cf. *barrandei*, *Acutiramus* sp. and *A. perneri*) have also been described at the *M. uniformis* Zone in the *Scyphocrinites* beds (Chlupac *et al.* 1997).

'B' Mb is of Lochkovian age and contains conodonts such as *Ozarkodina remscheidensis remscheidensis*, *O. stygia*, *O. excavata tuma*, *Ancyrodelloides omus*, *A. transitans* and *A. trigonicus*? which also belong within the *O. delta* Zone (García-López *et al.* 1990). At the top of the unit uppermost Lochkovian faunas occur (*Paranowakia* cf. *intermedia* and *N. acuaria*) (Chlupac *et al.* 1985). Close to Bruguers, in grey, dacryoconarid-rich shales most likely belonging to the 'B' Mb, a Lochkovian brachiopod fauna with *Orbiculoidea* sp., *Plectodonta* (*Plectodonta*) *mimica*, *Jacetanella brugesensis* as well as strophochonetids and lissatrypids has been described (Racheboeuf *et al.* 1993).

Even though the index-fossil for the earlier Pragian *Eognathodus sulcatus* has not yet been found in the CCR, the Lochkovian–Pragian boundary would be situated at the base of the 'C' Mb on account of the abundant occurrence of *N. acuaria*. 'C' Mb yields abundant dacryoconarid and rare conodont (*Pandorinellina steinhornensis miae* and *Caudicriodus* sp.) faunas of Pragian age although at the top of the unit

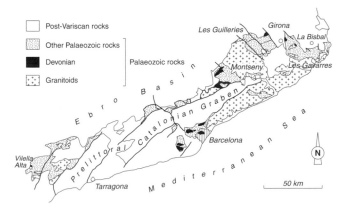

Fig. 6.10. Simplified geological map of the Catalonian Coastal Ranges with the location of the main Palaeozoic massifs and Devonian localities.

early Emsian faunas (*Polygnathus dehiscens* and *Nowakia praecursor*) occur (García-López *et al.* 1990). The 'C' Mb has also been recognized near St. Martí de Llémana (Les Guilleries) after the occurrence of *Nowakia* cf. *acuaria* in a predominantly calcareous sequence (Durán 1990).

'D' Mb yields Emsian conodonts of the *Polygnathus gronbergi* and *P. inversus/laticostatus* zones (García-López *et al.* 1990). At Montnegre and Can Lliure (Les Gavarres) is a Devonian succession probably equivalent to the Olorda Fm, the so-called 'Calizas y Margas de Montnegre'. At Can Lliure (near Girona) marl and limestone beds equivalent to 'B' Mb yielded Lochkovian–Pragian conodonts (*Ancyrodelloides omus*, *Ozarkodina pandora*, *O. remscheidensis*, *Icriodus* cf. *claudiae*, *Pelekysgnathus serratus* and other). At Montnegre (near La Bisbal) the most calcareous succession equivalent to 'D' Mb yielded conodonts of the *Polygnathus dehiscens*, *P. gronbergi*, *P. inversus*, *P. serotinus* and *P. costatus patulus* zones, and a probably late Emsian brachiopod fauna (*Prokopia* and *Dalejodiscus*) (Barnolas & García-Sansegundo 1992, Sanz-López *et al.* 1998). The 'E' Mb has not yielded any fauna at all.

Betic Cordillera and Minorca (MT-M, FS, JL, G-A)

Betic Cordillera

Devonian rocks have been preserved in places within the Nevado-Filábride and Maláguide complexes of the internal Betic Cordillera (Fig. 6.11; see Chapter 16). The Nevado-Filábride complex represents the deepest tectonic unit exposed in the Betics and has been identified in several tectonic windows in the Sierra Nevada and the Sierra de los Filabres. It has been further subdivided into the so-called 'Lower Series' (Palaeozoic I, or Veleta nappe and Palaeozoic II, or Mulhacén Nappe) and 'Upper Series' (Gómez Pugnaire & Franz 1988). Devonian rocks have been recorded in the highly metamorphosed Veleta nappe. Lafuste & Pavillon (1976) described supposedly Eifelian *Chaetetes* cf. *salairicus* and undetermined crinoids, brachiopods, gastropods and orthoceratids in marbles with intervening black schists at La Yegua Blanca, near Cartagena (Murcia). However, the best palaeontologically documented Devonian exposures occur in the Maláguide Complex in slightly metamorphosed to unmetamorphosed rocks near Málaga (between Estepona and Fuengirola), and along a narrow strip between Sierra Arana (Granada) and Vélez Rubio (Murcia) (Fig. 6.11).

The Malaguide Palaeozoic succession has been described as the Piar Gp which contains the Morales, Santi Petri, Falcoña, Almogía and Marbella formations (Martin Algarra 1987; Gómez Pugnaire 1992; Fig. 6.12). Mon (1971) found conglomerates towards the top of the Morales Fm, overlain by unfossiliferous carbonate beds that by comparison with the Moroccan Rif were interpreted as Silurian. However Devonian fossils such as *Spathognathodus frankenwaldensis*, *Nowakia*, *Styliolina*, trilobite fragments, crinoids and other echinoderms were subsequently reported by Geel (1973) from the 'Tentaculites Limestone Member' in the upper part of the formation. Both the Santi Petri and the overlying Carboniferous Almogía Fm (Fig. 6.12) contain calcareous olistoliths of unknown provenance but with conodonts representing most of the Devonian stages (Rodríguez-Cañero 1993; Herbig 1983). The Santi Petri Fm is interpreted as a Devonian unit (Mon 1971; Herbig 1983; Martín Algarra 1987), but fossils have so far been found only in olistoliths. Near Málaga and between Estepona and Fuengirola, *Ozarkodina excavata excavata* and *O. remscheidensis remscheidensis* associated with *Scyphocrinites* could represent the Silurian–Devonian boundary (Rodríguez Cañero *et al.* 1997). Rodríguez Cañero (1993) has also found conodonts belonging to the Eifelian–Givetian (*Polygnathus pseudofoliatus* to *Polygnathus hemiansatus* lineage), Givetian–Frasnian (*Mesotaxis falsiovalis* to *M. asymmetrica* lineage) and Frasnian–Famennian (*Palmatolepis praetriangularis* to *Palmatolepis triangularis* lineage) boundaries. Several conodont biozones have been identified (Rodríguez Cañero 1993), as follows. Early Devonian: *Ancyrodelloides delta*, the upper part of the *Pedavis pesavis*, the lower part of the *Eognathodus sulcatus*, and the lower part of the *Polygnathus gronbergi* zones. Middle Devonian: *Tortodus kockelianus*, *P. xylus ensensis*, Lower and Middle *P. varcus* and *Klapperina disparilis* zones. Late Devonian: *Palmatolepis transitans*, Upper *Pa. rhenana*, lower part of the *Pa. linguiformis*, Lower, Middle and Upper *Pa. triangularis*, Lower, Middle, Upper and Uppermost *Pa. crepida* and Lower *Pa. rhomboidea* zones. The Almogía IIIA section (Málaga) shows an important shift in conodont biofacies, associated with sedimentological changes and sedimentary interruptions that can be correlated with the latest Frasnian Kellwasser events (Rodríguez Cañero 1993). A similar conodont turnover has been reported by Herbig (1985) in Arroyo de la Cruz, near Marbella, related to a stratigraphical gap embracing the uppermost *Pa. gigas* to lower *Pa. triangularis* zones. Near Xiquena (Vélez Rubio), McGuillavry *et al.* (1960) recorded Devonian *Nowakia* sp., *Styliolina* and trilobites. Finally Foucault & Paquet (1970) cited *Tentaculites* sp. in Sierra Arana (Fig. 6.11).

Minorca

In the Balearic Islands the Devonian is represented only in the Tramuntana area on the northern coastal region of Minorca (Fig. 6.13). The Palaeozoic sequence here comprises Lochkovian to Namurian syntectonic marine sediments unconformably overlain by Permo-Triassic sedimentary cover (Bourrouilh

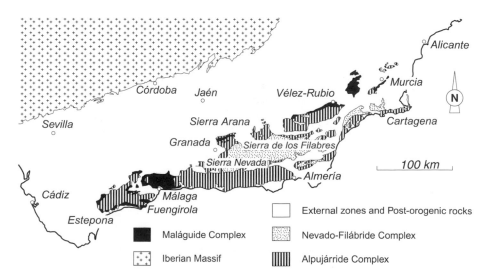

Fig. 6.11. Simplified geological map of the Betic Cordillera showing the three defined units of the Internal Zones (Maláguide, Nevado-Filábride and Alpujárride complexes).

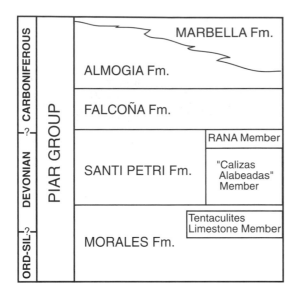

Fig. 6.12. Palaeozoic lithostratigraphic units of the Maláguide Complex.

1983). The occurrence of Devonian rocks in Minorca was first reported by Hermite (1879). They outcrop mainly between Fornells Bay and Mica inlet, although a smaller eastern outcrop occurs close to Addaia (Fig. 6.13).

Lochkovian graptolites of the *Monograptus praehercynicus* to *M. hercynicus* zones occur (Bourrouilh 1983), with the latter also yielding late Lochkovian conodonts (*Oneotodus? beckmanni* and *Spathognathodus wurmi*) and Pragian *Nowakia acuaria*. Late Pragian and Emsian conodonts (*Icriodus huddlei celtibericus, Ozarkodina typica denckmanni, Spathognathodus steinhornensis steinhornensis, Icriodus sigmoidalis, Polygnathus linguiformis foveolatus,* and others) and dacryoconarids (*Viriatellina* sp., *Styliolina* sp. and *Nowakia* sp.) have been also found. From Eifelian times onwards the faunas are clearly reworked. Some turbiditic beds have yielded mixed late Emsian to Eifelian faunas (conodonts such as *Icriodus* cf. *corniger, Polygnathus webbi* y *P. linguiformis,* tentaculitids such as *Tentaculites scalariformis, T. bellulus, Heteroctenus* cf. *mosolovicus* and *Uniconus* sp., dacryoconarids such as *Nowakia cancellata* and *N.* cf. *sulcata,* ostracods and bryozoans). The Givetian and Frasnian stages have been identified by the occurrence of conodonts of the *P. varcus* and *P. asymmetricus* zones, respectively. The uppermost part of the Devonian series

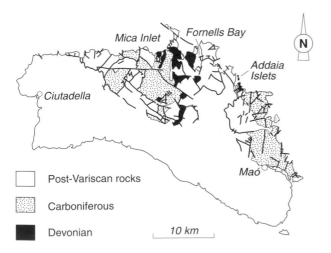

Fig. 6.13. Simplified geological map of Minorca showing Palaeozoic outcrops and localities where Devonian rocks occur.

comprises debris flows including turbiditic blocks (Escull d'es Francès and equivalent beds). The flow probably formed during late Frasnian times or later (Bourrouilh 1983) because the boulders yield Givetian to late Frasnian conodonts (*Ancyrodella rotundiloba, A. gigas, A. lobata, A. buckeyensis, Polygnathus asymmetricus asymmetricus, P. a. ovalis, P. linguiformis, P. ancyrognathoideus, Palmatolepis punctata, Ancyrognathus triangularis,* and others), corals (*Macgeea multizonata, Phillipsastrea chenouensis, Peneckinella minima, Dysphyllum caespitosum, Haplothecia* sp. 1, and so on), brachiopods (*Cyrtospirifer tenticulum*), trilobites, tentaculites and crinoids. In this part of the succession extrusive mafic igneous rocks are very common.

Pyrenees (JIV-R, JS-L)

Devonian rocks are exposed in the Pyrenees within six tectonic units that preserve the Palaeozoic basement to the Mesozoic and Cenozoic cover (Fig. 6.14; Chapter 15). These tectonic subdivisions are the Benasque unit, Las Nogueras unit, Cadí nappe, Gavarnie nappe, North Pyrenean massifs and Basque massifs. The distribution of Devonian facies and benthic faunas, as well as the occurrence of a Pragian siliciclastic wedge (the San Silvestre quartzite, see below), reefal limestones and several sedimentary gaps, indicate a shallow sedimentary setting for the south-central, western and Basque Pyrenees (Carls 1988). Deeper settings with hemipelagic rocks, although with local sedimentary highs (Sanz López 1995), are represented in the south-central, eastern and northern Pyrenees (Fig. 6.14). The maximum lithological diversity across the area occurred from late Givetian to Frasnian times, when relatively thick sedimentary formations were deposited; there was a prominent development of reef complexes, and a significant influx of early Frasnian siliciclastic detritus in the western, central and Basque Pyrenees. In contrast, very monotonous sedimentary conditions resulting in the formation of condensed cephalopod limestones characterize the whole Pyrenean region during middle to late Famennian time.

The Silurian–Devonian transition facies and faunas are rather homogeneous and show alternating black shales and limestones, a 10–15 m thick, 'Orthoceras' black limestone, and black shales with limestones nodules and interbedded lenticular limestones (the so-called 'late Silurian carbonate episode'; García-López *et al.* 1996). Several Pyrenean localities (Fig. 6.15) have yielded latest Pridolian to earlier Lochkovian conodont faunas. In the Las Nogueras/Cadí nappe unit (Fig. 6.14) fragmented *Scyphocrinites* abound. At Gerri de la Sal, *Scyphocrinites* loboliths associated with *Icriodus woschmidti woschmidti* and other conodonts provide a very good approximation to the Silurian–Devonian boundary (Valenzuela-Ríos 1994) and the same association has been found in the nearby section of Sarroca (Sanz-López *et al.* 1999).

Devonian facies and faunas vary greatly from the Lochkovian onwards across the different Pyrenean structural units and this has led to a complex and sometimes confusing use of local lithostratigraphic terminology. An attempt to integrate this terminology on a correlation diagram is provided by Fig. 6.15.

Benasque Unit

Early Devonian. Above black limestones yielding conodonts of earliest Devonian age, a range of subsequent Early Devonian facies can be observed in the Benasque unit. In Sierra Negra, for example, there are 50–80 m of sandy slates and impure limestones (Rueda Fm). In contrast, at Baliera the succession comprises a thick series of slates and argillaceous limestones (Aneto Fm) (Fig. 6.15). Valenzuela-Ríos (1994) distinguished two members in this Aneto Fm, the lower one

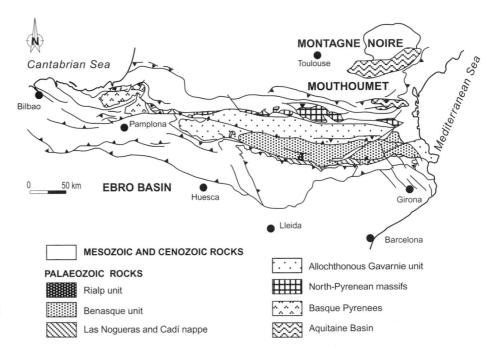

Fig. 6.14. Simplified geological map of the Pyrenees showing the situation of the main structural Alpine units displaying Palaeozoic successions.

being more calcareous than an upper one dominated by shales. *Icriodus angustoides* ssp. and '*Ozarkodina*' *eladioi* occur in a limestone bed within the lower member indicating an early Lochkovian age.

Above the Aneto Fm is the Gelada Fm which comprises 100–120 m of alternating sandy, calcareous and slightly carbonaceous shales with a few interbedded sandy and argillaceous limestone layers, and a prominent 25 m thick sequence of impure dark blue limestones in the lower part. This overall rather sandy formation is conformably overlain in Sierra Negra by massive limestones (Castanesa Fm) and in Baliera by limestones, dolostones and quartzites of greatly variable thickness (Basibé Fm). The Castanesa Fm comprises 30–60 m of dark grey limestones, with the lower part locally dolomitized. Poor and badly preserved fossils prevent the accurate dating of this unit in Sierra Negra. The Basibé Fm forms a prominent topographic feature across its outcrop and has been subdivided into the Ponferrat (mainly nodular carbonate), San Silvestre (quartzites) and Llaviero (dark blue to black platy limestones) members. It decreases in thickness eastwards mainly due to the wedging out of the quartzitic San Silvestre Mb. The maximum thickness thus occurs in the west (175 m at Barranco de Urmella where the San Silvestre Mb is 60 m thick; Mey 1967). The Ponferrat Mb comprises 20–60 m of pink-yellow, nodular limestones and grey limestones and dolostones alternating with calcareous shales. It has yielded the earliest *Polygnathus pireneae* and *Pedavis* sp., indicating an early Pragian age (Valenzuela-Ríos 1994, and unpublished data). The San Silvestre Mb consists of alternating white, light brown or bluish quartzites and dolostones, with the quartzite/dolostone ratio increasing upwards. It has yielded Pragian conodonts (*I. curvicauda*, *I.* aff. *vinearum*, at the lower part, and *P. pireneae* in the upper one; Sanz-López 1995). Finally, the Llaviero Mb comprises 20–50 m of dark bluish, platy limestones yielding abundant late Pragian to early Emsian conodonts (*I. curvicauda*, *I. sigmoidalis*, *I. celtibericus*, *P. kitabicus*, *P. excavatus*, *I. bilatericrescens bilatericrescens* and '*Pandorinellina*' *steinhornensis* ssp.) (Valenzuela-Ríos 1994, and unpublished data).

In the Sierra Negra and at Baliera both the Castanesa Fm and the Basibé Fm are overlain by the Fonchanina Fm (Fig. 6.15) which comprises *c.* 15–60 m of dark shales with a middle intercalation, 5–10 m thick, of dark platy limestone. The age of the Fonchanina Fm can only be inferred from its stratigraphic position because diagnostic fossils are so far lacking.

The overlying Mañanet Fm is made up of nodular variegated limestones with interbedded red and green, argillaceous limestones and shales. The thickness of the Mañanet Fm varies greatly from Sierra Negra (35–70 m thick) to Baliera (80–280 m thick). In the latter area the thickness decreases both westwards (to 100–120 m including a few beds of fine-grained sandstones), and northwards (to 53 m in north Benasque). In Denúy, García-López *et al.* (1990) found tabulate favositids and thamnoporids, the brachiopod *Uncinulus pila*, and several forms of *Pandorinellina steinhornensis* of the *Polygnathus gronbergi* Zone, close to the base of the formation, and *P. cooperi cooperi* and *P. costatus* spp. (Emsian or earlier Eifelian) in the topmost limestone layers.

In the Segre area (Fig. 6.15), the earlier Lochkovian is represented by black shales with intervening layers of black limestones ('A' Lithosome; Valenzuela-Ríos 1994) followed by light orange, well layered limestones with interbedded marly shales ('B' Lithosome; Valenzuela-Ríos 1994). Both lithosomes and the correlative sections at Gerri de la Sal (in the Las Nogueras/Cadí nappe unit) yield a rich and diverse Lochkovian conodont fauna invaluable for worldwide correlations. Among this fauna are complete phylogenic sequences of the genera *Ancyrodelloides*, *Flajsella*, *Lanea*, and others almost complete of the genus *Pedavis* and of *Criteriognathus pandora*. The occurrence of more 'endemic' genera as *Icriodus* and *Pelekysgnathus* enables correlations with shallower sequences. The overlying Rueda Fm shows an increased pure limestone content yielding earlier Pragian conodonts such as *I. steinachensis* and *Pelekysgnathus serratus brunsvicensis* (Valenzuela-Ríos 1994). Above this, the 20–120 m thick Castanesa Fm (Hartevelt 1970) has yielded at its basal part Pragian conodonts such as *P. mariannae* and *P. pireneae* (Valenzuela-Ríos 1997). The overlying Villech Fm comprises *c.* 85–125 m of red shales and limestones with intercalated carbonaceous green shales. Misunderstandings concerning the dating of the Villech Fm have been due to the erroneous correlation of the lower part of this formation with the underlying Castanesa Fm (i.e. Boersma 1973; see discussion in Sanz-López 1995). The Villech Fm has yielded ammonoids in the so-called '*Anarcestes*' level (first described by Dalloni 1930).

Middle and Late Devonian. In Sierra Negra and Baliera (Fig. 6.15), the Vilaller Fm overlies the Mañanet Fm and comprises *c.* 100–400 m of brown and green shales with some interbedded limestone and lenticular calcareous sandstones with *I.* cf. *culicellus* of late Emsian or early Eifelian age (Sanz-López 1995). Near the Esera springs, these shales and sandstones have been named the Cerler or Eriste shales, and yield *Paraspirifer cultrijugatus* (Ríos *et al.* 1979).

The base of the next formation (Renanué Limestone Fm) has

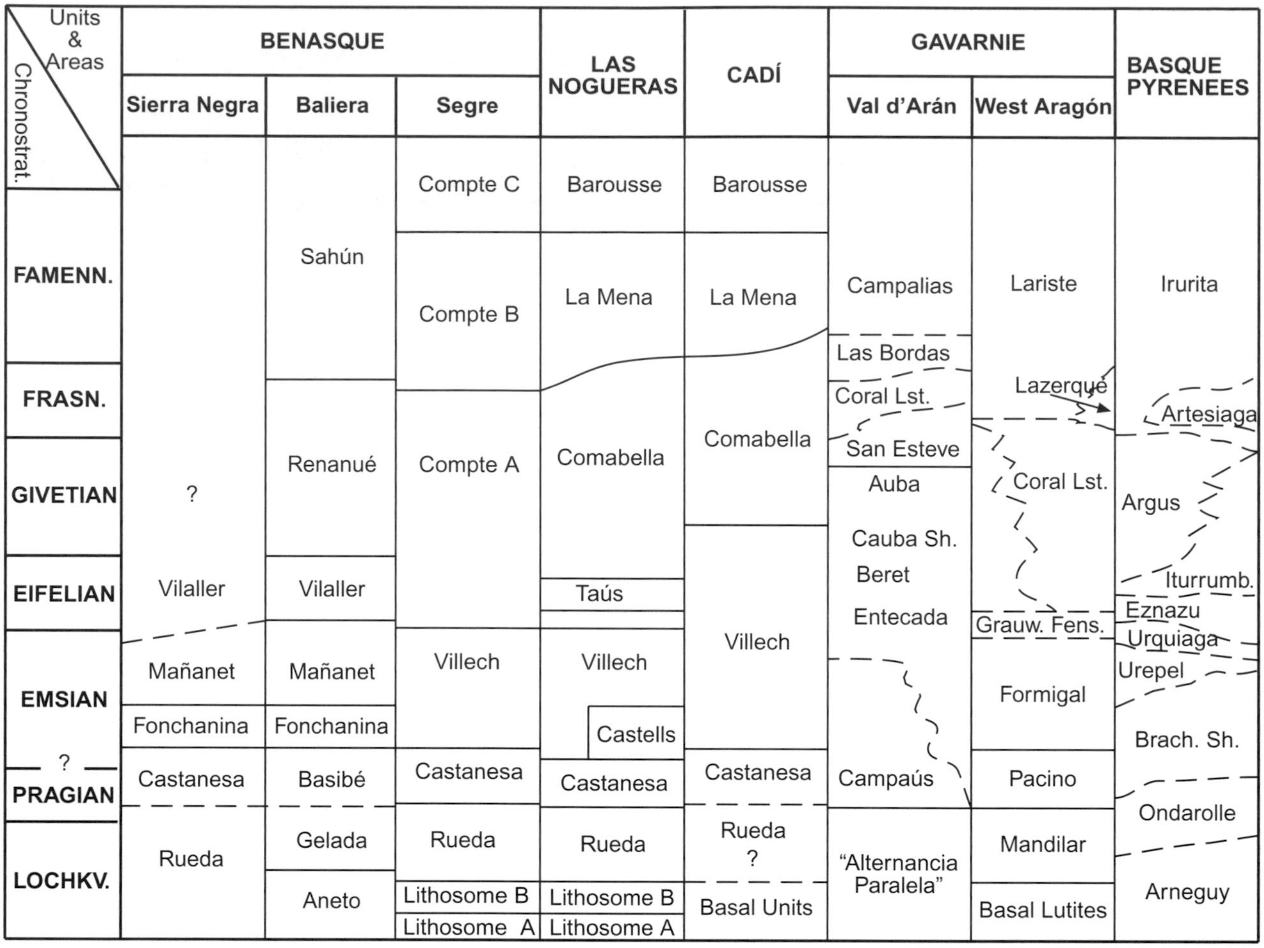

Chronostrat.	BENASQUE			LAS NOGUERAS	CADÍ	GAVARNIE		BASQUE PYRENEES
	Sierra Negra	Baliera	Segre			Val d'Arán	West Aragón	
		Sahún	Compte C	Barousse	Barousse			
FAMENN.		Sahún	Compte B	La Mena	La Mena	Campalias	Lariste	Irurita
FRASN.	?	Renanué	Compte A	Comabella	Comabella	Las Bordas / Coral Lst. / San Esteve / Auba	Lazerqué / Coral Lst.	Artesiaga / Argus
GIVETIAN	?	Renanué	Compte A	Comabella	Comabella	Cauba Sh. / Beret	Coral Lst.	Argus / Iturrumb.
EIFELIAN	Vilaller	Vilaller		Taús	Villech	Entecada	Grauw. Fens.	Eznazu / Urquiaga
EMSIAN	Mañanet / Fonchanina	Mañanet / Fonchanina	Villech	Villech / Castells	Villech		Formigal	Urepel / Brach. Sh.
PRAGIAN ?	Castanesa	Basibé	Castanesa	Castanesa	Castanesa	Campaús	Pacino	Ondarolle
LOCHKV.	Rueda	Gelada / Aneto	Rueda / Lithosome B / Lithosome A	Rueda / Lithosome B / Lithosome A	Rueda ? / Basal Units	"Alternancia Paralela"	Mandilar / Basal Lutites	Arneguy

Fig. 6.15. Correlation chart of the Pyrenean Devonian stratigraphical units. Abbreviations: Brach. Sh, brachiopod shale; Chronostrat, chronostratigraphic units; Famenn, Famennian; Frasn, Frasnian; Grauw. Fens, 'Fenestella greywacke'; Iturrumb, Iturrumburu; Lochkv, Lochkovian; Lst, limestone.

yielded latest Eifelian to earliest Givetian conodonts (*P. pseudofoliatus, P. eiflius, P. l. linguiformis* and icriodids). Earliest Givetian to early Frasnian–upper *Mesotaxis falsiovalis* zone conodonts have been reported in a continuous section *c.* 50 m thick. In overlying limestones that form discontinuous outcrops totalling *c.* 120–130 m, Boersma (1973) cited Frasnian conodonts of the lower *Palmatolepis hassi* to the *Pa. jamieae* zones.

The uppermost Devonian rocks of the region are the so-called 'Esquistos de Sahún', which comprise *c.* 20–50 m of limestones and shales with, in their lower part, conodonts of the upper *Pa. rhenana* zone (Sanz-López 1995). Dalloni (1910) also found brachiopods such as *Productella productoides, P. subaculeata* and *Cyrtospirifer verneuili*. These youngest Devonian exposures are overlain unconformably by Permo-Triassic rocks.

Las Nogueras/Cadí nappe unit

Early Devonian. In the Las Nogueras area (Fig. 6.15), the Devonian succession is comparable to that of Segre (Benasque unit), and has been subdivided into A and B units, and the Rueda and Castanesa formations. A and B units reach a thickness of about 25 m, whereas the Rueda Fm is *c.* 95–135 m thick and yields trilobite (Villalta & Rosell 1969) and dacryoconarid faunas (*Nowakia acuaria*) (Llopis Lladó 1966). The Castanesa Fm near San Sebastià-Sarroca (Llopis Lladó & Rosell 1968) comprises *c.* 50 m of grey, bluish or red coloured, light to dark limestones and rare calcareous shales. Early

Pragian to early Emsian conodonts such as *Pelekysgnathus serratus brunsvicensis, P. pireneae, I.* aff. *simulator, I. curvicauda, I. celtibericus, P. excavatus, I. sigmoidalis* and '*O.*' *steinhornensis miae* have been reported at Castells (Boersma 1973; Sanz-López 1995; Valenzuela-Ríos 1997, and unpublished data).

Sanz-López (in Montesinos & Sanz-López 1999*b*) redefined the boundaries of the Villech Fm. Thus the uppermost part of the Castanesa Fm *sensu* Boersma (1973) would correspond to the Villech Fm at Guardia d'Ares. In the latter locality the formation is *c.* 65 m thick and yields conodonts of the lower *P. gronbergi* Zone (Sanz López 1995), as well as ammonoids such as *Mimagoniatites* and others (Montesinos & Sanz López 1999*b*). At the top of both the Guardia d'Ares and Comte sections, Boersma (1973) reported conodonts of the late Emsian *P. serotinus* Zone. In Espaén, Llopis Lladó (1966) cited dacryoconarids of the lower–upper Emsian transition such as *Nowakia barrandei* and *N. cancellata*. In Castells the upper part of the Villech Fm (70 m thick) has yielded late Emsian ammonoids such as *Latanarcestes noeggerathi* and *Sellanarcestes* cf. *tenuior* (Montesinos & Sanz-López 1999*b*).

In Castells (Fig. 6.15), above the Castanesa Fm, Sanz-López (in Montesinos & Sanz-López 1999*b*) described a local unit, the 'Castell beds' (included in the Fonchanina Fm by Boersma 1973) comprising *c.* 30 m of shales and marls with interbedded brown shaly limestones. The upper part of this unit is more calcareous and consists of light grey marls with pink, green or brown interbedded limestones with early Emsian conodonts.

In the Cadí nappe (Fig. 6.15) area the so-called 'basal units'

(Sánz-López 1995) could partially correspond to the A and B units and yield an important Lochkovian conodont fauna including forms of *Ancyrodelloides*, *Lanea* and *Flajsella*. The lower part of the overlying Rueda Fm comprises *c.* 15–50 m of limestones, calcareous shales and marls with calcareous nodules, and yields early Pragian conodonts such as *P. serratus brunsvicensis* and *I. steinachensis*. The Castanesa Fm comprises *c.* 10–80 m of grey and red, sometimes nodular limestones with scattered intervening marly beds and crinoidal limestones developed at the top of the thicker sections. Pragian (*P. serratus brunsvicensis*) to early Emsian conodonts have been reported (Sanz-López 1995). The thickness of the overlying Villech Fm varies between 25 and 50 m, and a calcareous breccia formed by collapse of the 'Castanesa platform' occurs locally in the lower part of the formation. The Villech Fm yields a relatively rich Emsian ammonoid fauna (Montesinos & Sanz-López 1999b). In the lower part of the formation *Erbenoceras* e.g. *filalense* is framed above and below it, by early Emsian conodonts of the *P. gronbergi* and *P. inversus* zones, respectively. From nearby beds, Kullmann & Calzada (1982) reported *Mimosphinctes* cf. *cantabricus* and *Mimagoniatites* sp. The middle part of the Villech Fm has yielded several late Emsian *Mimagoniatites* forms whereas in the upper part *Sellanarcestes* sp., *Latanarcestes* sp. and *Anarcestes* sp. associated with late Emsian conodonts of the *P. serotinus* and *P. patulus* zones have been reported.

Middle and Late Devonian. In both the LNC and Benasque (Segre area) units (Figs 6.14 and 6.15) the Villech Fm is overlain by variegated nodular limestones of the Compte Fm which has been subdivided into three members (A to C) (see summary and correlations in Hartevelt 1970 and Boersma 1973). These members probably correspond to the Comabella Fm (A), La Mena Fm (B; both defined by Sanz López 1995; see Montesinos & Sanz López 1999b) and Barousse Fm (C; Perret 1993).

The Comabella Fm mostly comprises *c.* 20–370 m of nodular, hemipelagic limestones, although in some parts of the Cadí and Segre areas these are interleaved with carbonate slope–apron systems and reefal platform limestones. Conodont faunas belonging to latest Emsian *P. patulus*, late Frasnian upper *Pa. rhenana* or early Famennian upper *Pa. triangularis* zones occur at different localities (Sanz López 1995). The lower part of the Comabella Fm has yielded the so-called '*Anarcestes* fauna', and Eifelian *Pinacites* cf. *eminens* has also been reported. In Castells (LNC), Sanz López (in Montesinos & Sanz López 1999b) defined the 'Taús beds' resting on the lower part of the Comabella Fm which comprises *c.* 17 m of nodular light grey limestones. In contrast, the Taús beds comprise *c.* 40 m of black to grey-bluish shales and green marls with dacryoconarids, trilobites, conodonts and *Anarcestes plebeius* in their lower part. Overlying the Taús beds are *c.* 25 m of light grey, nodular limestones with latest Eifelian to early Famennian conodonts (Sanz-López 1995, and unpublished data) and Frasnian ammonoids such as *Beloceras* cf. *tenuistriatum*, *B.* cf. *subacutum* and *Manticoceras intumescens* (Llopis Lladó 1966; Llopis Lladó & Rosell 1968).

La Mena Fm ('griotte limestones' *auctorum*) comprises *c.* 12–30 m of dark red, nodular limestones with abundant early Famennian ammonoids such as *Cheiloceras verneuili*, *Ch. subpartitum* and *Ch. amblylobum* (Dalloni 1930; Schmidt 1931; Llopis Lladó 1966). In the Cadí nappe, the La Mena Fm shows carbonate bars and storm layers with brachiopods. At the top of the unit, late Famennian conodonts of the lower *Pa. marginifera* zone have been found (Sanz López 1995).

The Barousse Fm comprises 25–66 m of light grey, often nodular, cephalopod limestones of Famennian to early Tournaisian age. Schmidt (1931) reported *Platyclymenia annulata* from Baró and several taxa of the '*Gonioclymenia* Stufe' in the Segre valley. Llopis Lladó (1966) cited *Platyclymenia* sp., *Kosmoclymenia undulata* and *Gonioclymenia speciosa* in the Segre valley, and *Sporadoceras latilobatum*, *S. biferum*, *S. münsteri*, *Kosmoclymenia* sp. and *Prionoceras* sp. in Las Nogueras. The upper part of the Barousse Fm includes a shaly 'B' and a nodular limestone 'C' level (Boyer *et al.* 1974; Perret 1993; Sanz López 1995), the former representing the Hangenberg event and the latter yielding the first Carboniferous conodonts of the *Siphonodella sulcata* Zone.

The Gavarnie Unit

Early Devonian. In the Marimaña massif (Val d'Arán; Fig. 6.15), the so-called 'Alternancia Paralela' (Palau & Sanz 1989) comprises 40–60 m of grey limestones and black shales correlated with the Rueda Fm and overlain by the 'Campaús limestones' which are equivalent to the Castanesa, Fonchanina and Mañanet formations of the Benasque unit (Figs 6.14 and 6.15). The Campaús Limestones yield the Emsian conodont *I. bilatericrescens*.

In Val d'Arán, the Basal limestone, Pala Megdia shales and Cauba limestones correlate with the Campaús limestones and are overlain by a predominantly shaly sequence known as the Entecada or Beret shales (Fig. 6.15).

In the western central Pyrenees, the Lower Devonian sequence starts with the so-called 'Basal Lutites' that are overlain by a *c.* 100–200 m thick alternation of shales and limestones (Mandilar Fm). On purely lithological grounds the Mandilar Fm may be correlated with the Rueda Fm (Fig. 6.15).

The overlying Pacino Fm correlates with the Castanesa Fm and comprises 40–60 m of dolomitic limestones, dolostones, shales and limestones with Emsian brachiopods, trilobites, corals and conodonts (Bixel *et al.* 1985). Above this is the Formigal Fm which is composed of *c.* 250–600 m thick shales with interbedded limestones, sandstones and greywackes. Limestones high in this succession have furnished Emsian to lower Eifelian fossils such as *Fenestella plebeia*, *Phacops potieri*, *Euryspirifer pellicoi* and *Paraspirifer cultrijugatus* (Dalloni 1910; Mirouse 1966). The overlying red and green limestones have yielded in some localities *Anarcestes subnautilinus* and *Agoniatites* sp.

Middle and Late Devonian. In Val d'Arán (Fig. 6.15), a 90–150 m thick succession of Eifelian shales form the Entecada, Cauba and Beret formations, and these are overlain by a sandy unit, the Auba sandstones (130 m thick; García-Sansegundo 1992). Above this the 55 m thick Sant Esteve limestone yields middle Givetian conodonts. There then follows the so-called 'coral limestone' which is 40 m thick and has conodonts of the Frasnian middle *P. asymmetricus* Zone (García-López *et al.* 1991), although northwards it has yielded conodonts of the lower *P. asymmetricus* Zone (Sanz López 1995) in the so-called 'Dacryoconarid limestone' (Palau & Sanz 1989). These limestones are overlain by a thick (up to 600 m) Frasnian siliciclastic succession (Las Bordas sandstone). Southwards these clastic rocks pass into La Tuca shales and limestones, 170 m thick (García Sansegundo 1992) with middle to late Frasnian conodonts. The youngest Devonian strata are preserved in synclinal cores where early Famennian black shales and limestones (Campalias shale) are sometimes exposed (Palau & Sanz 1989).

In the southwestern Pyrenees (West Aragón on Fig. 6.15), brachiopods of the Emsian–Eifelian transition, such as *P. cultrijugatus* and *Alatiformia alatiformis*, have been reported (e.g. Mirouse 1966) in shales with interbedded limestones and sandstones. These beds could partially correspond to the so-called '*Fenestella* Greywacke' that yields a rich earliest Eifelian fauna with brachiopods, conodonts, crinoids and fish remains (Valenzuela-Ríos & Carls 1996).

Overlying the '*Fenestella* Greywacke', are *c.* 150–400 m of limestones with Eifelian conodonts and reefal limestones ('coral limestones'; Mirouse 1966; Joseph *et al.* 1980). These limestones become thin and nodular northwards where they are overlain by a Frasnian siliciclastic sequence. The uppermost part of the coralline limestones consists of marls with interbedded nodular limestone with brachiopods, corals and early Frasnian conodonts (Mirouse 1966).

The Lazerque series (Fig. 6.15) in the Subordán valley comprises 120 m of sandy limestones and shales with brachiopods and corals and includes allochthonous limestone blocks (Joseph *et al.* 1980) with late Frasnian conodonts. The Lazerque series passes northwards into the Lariste unit (Mirouse 1966) which is up to *c.* 300–350 m thick, and consists of shales with sandstones and scarce limestones yielding Frasnian *Phillipsastrea*, brachiopods, and conodonts (Joseph *et al.* 1980). Mirouse (1966), based on the occurrence of *I. cornutus* and *P. semicostatus*, suggested that the deposition of these limestones continued into lower Famennian times.

Basque Pyrenees (PC, JS-L, JIV-R)

In the Basque massifs (Cinco Villas and Aldudes–Quinto Real massifs, in Spain; Fig. 6.14) there are thick Devonian successions that differ strongly from the other Pyrenean units. This region was first studied by German and French geologists (e.g. Requadt 1972, 1979; Heddebaut 1975). The complex interfingering of siliciclastic deposits derived from the Cantabro-Ebroian massif (Carls 1983) and carbonates, has resulted in a complex nomenclature, and complete elucidation of the Devonian stratigraphy has not yet been achieved due to faulting, folding, cleavage and poor fossil preservation.

In Aldudes–Quinto Real (Fig. 6.15), the Arneguy Shale Fm (up to 850 m thick) starts with Silurian black shales and dark limestones (30 m thick), but it is mainly of Lochkovian age, reaching locally up into earliest Pragian times. In early Lochkovian times, immature sand was deposited in the basin and forms beds alternating with dolomitic carbonates. Most of the Pragian succession is also characterized by interfingering dolostones and quartzitic sandstones, including the Ondarolle Fm (up to 450 m thick) and the Aldudes quartzite (up to 500 m thick) that may be equivalent to the San Silvestre quartzite of the Benasque unit.

The arenaceous-dolomitic sedimentation ceased during early Emsian times, and gave way to the deposition of the so-called 'Brachiopod Shale' Fm (100–200 m thick). Brachiopods such as *Acrospirifer*, large *Rhenostrophia* and *Leptostrophiella explanata* have been reported in the 'Brachiopod Shale'. The overlying Urepel Fm (500–600 m thick) is another quartzite/dolostone complex of earliest to early late Emsian age. Shales near the base of this formation contain trilobites such as *Kayserops obsoletus*.

The Urquiaga Fm comprises *c.* 200 m of dolostones and 300 m of banded limestones with interbedded shales. In the upper part of the formation there are brachiopods of the Emsian–Eifelian transition such as *Plicathyris alejensis* and *Paraspirifer cultrijugatus*. The overlying Eznazu Fm comprises >200 m of dark shales with limestone lenses that have yielded Eifelian conodonts, and ferruginous shales full of brachiopods (mainly *Mucrospirifer* species). Above this, the laterally interfingering Argus shale and Iturrumburu limestone formations yield Givetian conodonts. The Iturrumburu Fm (*c.* 250 m thick) develops only locally and could extend into early Frasnian times. Elsewhere in the area, dark shales (800–1000 m thick) represent the Middle Devonian epoch.

The Irurita Gp comprises 1200–1500 m of quartzites and shales with Frasnian cephalopods and conodonts. The highest accumulation rate in the Aldudes–Quinto Real basin is recorded by the early Frasnian succession, which includes tuffitic and conglomeratic layers. The Artesiaga shales (500 m thick) record fine-grained siliciclastic sedimentation that gave way during late Frasnian times to the deposition of brachiopod–coral limestones and sandstones. These are in turn overlain by black shales and bioclastic limestones with early Famennian conodonts, and nodular limestones ranging from middle Famennian to earliest Carboniferous in age.

The Irurita Gp thins abruptly westwards, and in the westernmost Pyrenees (Cinco Villas massif) Famennian rocks rest unconformably on Ordovician rocks. This situation parallels that of Celtiberia (Iberian Ranges) and the Cantabrian mountains in the southwestern margin of the Cantabro-Ebroian massif during Frasnian times, when fault activity led to renewed erosion and increased sedimentation rates.

The Devonian succession of the Cinco Villas massif is even more poorly known than that exposed at Aldudes–Quinto Real. Requadt (1972) recorded middle upper Emsian to lower Eifelian faunas (*Alatiformia, Subcuspidella, Anathyris, Plicathyris*) from the shaly Kalforro Fm (300–500 m thick) and the overlying Marquesenea limestone.

Devonian reef development in Spain (FS, MT-M, JLG-A)

Devonian reef limestones have been known in Spain since the second half of the nineteenth century. Earliest publications on reefal facies focused basically on classical stratigraphical and palaeontological descriptions, mainly of reef-building organisms (stromatoporoids and corals). Palaeoecological and sedimentological studies began in the latter part of the twentieth century.

Even though the Devonian was a period of intensive reefal development worldwide and reef buildings of this age are abundantly represented in the European Variscan Massif, an overall picture of Spanish Devonian reef carbonates has not yet been established. This is due to the poor availability of some data, the lack of work in some areas, and the overprint of the Variscan orogeny. This section summarizes current information from reefal carbonate facies of the most studied areas (Cantabrian Zone plus Palentian domain, western Pyrenees and northern Ossa Morena Zone; Fig. 6. 1).

Main reefal episodes

Devonian successions in southern and eastern Spain are mainly incomplete and predominantly consist of clastic sediments, which represented unfavourable reefal environments. On the contrary, Devonian successions are more complete in northern Spain (Cantabrian Zone and western Pyrenees) and show more varied sedimentary environments that included different reefal carbonate facies.

The first hints of Devonian reefal facies in Iberia developed in early Pragian times, both in the Nieva Fm (Cantabrian Zone) and in the Lebanza Fm, C Mb (Palentian domain; Fig. 6.2). They are represented by abundant fasciculate rugose corals (mainly Disphyllidae) and, occasionally, by stromatoporoids forming locally biostromal patch reefs of limited extent in quiet lagoon environments (Krans *et al.* 1982). In the Asturo-Leonian domain the remaining part of the Pragian and the early Emsian ages are represented by deposits of tidal-flat environments generally unsuitable for reef-building organisms. However, in early Emsian times (lowermost part of La Ladrona Fm; Fig. 6.2), brief reefal episodes featuring colonial coral biostromes occur again (García-Alcalde 1996).

Other brief and localized reefal episodes developed with the Rañeces Gp and equivalent units, especially in the late Emsian Aguión and equivalent formations (Fig. 6.2). One of these can be observed in the lowermost part of the Coladilla Fm (Colle, Leon Province), and consists of *c.* 1 m thick, biostromal levels, built mainly by fasciculate rugose corals (Soto 1982). According to Stel (1975) the growth of these reef structures would be favoured by temporal oxygenation of otherwise anoxic waters by hurricanes and strong storms. Another reef deposit, in the Aguión Fm, occurs in the so called 'Arnao Platform' (Asturian Coast) and consists of a 5 m thick biostrome mainly built by bryozoans and tabulate corals. This patch reef was probably built in relatively quiet water on a shallow platform (Alvarez Nava & Arbizu 1986).

In areas far from the Cantabrian Zone, reef development started later (probably in the late Emsian at the same time as the deposition of limestones and dolostones in the eastern and northeastern Pyrenees. The eastern domain of the Basque Pyrenees (Adarza area, NE of Aldudes–Quinto Real Massif; Figs 6.1 and 6.14) developed a large carbonate platform with reef structures built by corals, and preserved in the uppermost part of the Urquiaga Fm (Heddebaut 1973; Fig. 6.15).

In the northern Ossa Morena Zone (Sierra del Pedroso; Fig. 6.1), a continuous band of reef limestones made up of stromatoporoid and coral biostromes and bioherms has been reported (Rodriguez 1978; Rodriguez & Soto 1979). In this region the carbonate platform overlies lowermost Devonian detrital deposits and slopes smoothly and progressively northeastward away from a southwestern located emergent area (Perez-Lorente 1979; Herranz Araújo 1985).

The most important reefal episode in the Early Devonian history of the Cantabrian Zone took place during latest Emsian times, coinciding with the deposition of the Moniello and equivalent Asturo-Leonian formations (Figs 6.2 and 6.6). Reef structures occur mainly in the lower and middle parts of the formations, although there are also some minor developments in the upper part. They consist of biostromes of variable thickness, built by hemispheric, laminar and tabular stromatoporoids and massive tabulate corals (Favositida), as well as by fasciculate rugose corals. There are also bioherms built by stromatoporoids and massive as well as branching corals, which exhibit sedimentological and palaeoecological features characteristic of organic build-ups (Méndez-Bedia 1976, 1984; Méndez-Bedia & Soto 1984). In general, these reef frameworks have been formed on a large carbonate platform perhaps close to its distal margin (Méndez-Bedia et al. 1994).

During Middle Devonian times, throughout the late Eifelian interval, several carbonate levels in the western Pyrenees (from the Gavarnie mountains to near the Anie peak; Fig. 6.14) have lithological and faunistic characteristics of reef deposits. Such is the case of the 'Canau limestone' (east Gavarnie), formed by bioherms built by stromatoporoids and corals (Joseph 1973), the 'Caillou de Soques series' (Ossau valley, Brousset stream), with reefal limestones (Mirouse 1966), the 'Peña Foradada limestone' (Gallego valley, north Sallent), formed by 300 m thick reefal limestones (Joseph et al. 1980; Galera-Fernández 1987), and a 200–300 m thick limestone series located in western Aragón (Lazerque valley, SW Lescun) and built by bioherms of rugose and tabulate corals (Joseph et al. 1980).

During Givetian times, reefal development in the Cantabrian Zone resumes. Important reefal episodes are related to the deposition of the Candás and equivalent formations (Figs 6.2 and 6.7). Reef structures are characterized by the occurrence of tabulate coral, rugose coral, tabulate coral/stromatoporoid biostromes, as well as bioherms formed by massive, irregular tabulate corals and/or stromatoporoids, with a variable presence of massive rugose corals (Méndez-Bedia et al. 1994). Detailed facies distribution and sequential analysis, carried out mainly on good outcrops of the Candás Fm, in Asturias, indicate a carbonate ramp model of reef evolution (Fernández et al. 1997). In the western Pyrenees, Givetian reefal carbonate facies have been mentioned in many localities. In Peyreget (Ossau valley; Mirouse 1966; Joseph et al. 1980), there is a 40 m thick, level of reefal limestones. In the southern part of Monte Tobazo (SW Somport pass; Joseph 1973), a c. 40 m thick argillaceous limestone band is composed of stromatoporoid biostromes and massive, lamellar and branching coral biostromes (Joseph 1973). Last but not least, in the Canfranc area (western Aragón), sandy limestones rich in corals occur (Joseph et al. 1980). All the mentioned reef structures correspond mainly to a quiet-water, carbonate platform, in a relatively uniform sedimentary environment (Andrews et al. 1996).

Upper Devonian reef carbonates have been found only in northern Spain (Cantabrian Zone and western Pyrenees). Two locally restricted Frasnian reefal episodes of minor importance have been cited in the Cantabrian Zone, at the top of the Piñeres Fm and in the upper part of the Nocedo Fm in the so-called 'Crémenes limestone' (Fig. 6.2). The upper part of the mainly siliciclastic Piñeres Fm comprises calcareous sandstones, limestones and argillaceous limestones, displaying thin branching coral (rugose and tabulate) biostromes (Fernández et al. 1995). The latest Frasnian Crémenes limestone is a biostromal unit formed by laminar stromatoporoids and branching as well as laminar corals (rugose and tabulate). According to Loevezijn (1989) this reefal episode developed in a general transgressive sequence, on the distal part of the platform in high-energy, well-oxygenated water.

The demise of the Cantabrian reefal tracts took place progressively from late Givetian to late Frasnian times. Two global transgressive events, the *Pharciceras* (late Givetian) and the Frasnes (Givetian–Frasnian transition) events (Fig. 6.7), together with local epeirogenic movements, greatly increased the rates of extinction of reef-building organisms (Raven 1983; Loevezijn 1989; García-Alcalde 1998, and unpublished data). However, the final collapse was probably due to the Kellwasser events (latest Frasnian) acting synergistically with regional pre-Variscan epeirogenic pulses that introduced large amounts of terrigenous material into the basin (García-Alcalde 1998, and unpublished data).

In the western Pyrenees, the Frasnian facies shows carbonates alternating with detrital layers rich in brachiopods. Remarkably well-developed reefal limestones occur in the SW Pyrenees (western Aragón; Figs 6.14 and 6.15) in such places as:

(1) 'Saint André Peak unit' (west Gavarnie), in a 100 m thick, massive limestone level built by stromatoporoids and corals (Joseph 1973);
(2) the upper part of the 'Lariste Peak series' (western Aragón, south Lescun), made up of thin banks of biostromal coral limestones (Mirouse 1966);
(3) Lazerque valley (SW Lescun), in a shaly series with interbedded limestones formed by branching rugose coral meadows in life position, stromatoporoids and tabulate corals (Joseph et al. 1980);
(4) south of Monte Tobazo (SW Somport pass) in a massive limestone sequence, built by corals (rugose and tabulate) and scarce stromatoporoids (Joseph 1973);
(5) Canfranc area, in sandy limestone beds, rich in corals (Galera-Fernández 1987).

Any effects of the *Pharciceras*, Frasnes and Kellwasser events on the western Pyrenees reef structures remain as yet unrecognized. However, just as in the Cantabrian Zone, Frasnian reef development was closely coupled with the general evolution of the sea floor that was characterized by positive Variscan movements in several Pyrenean areas (Andrews et al. 1996). Late Devonian (Famennian) Pyrenean successions are completely devoid of reef-building organisms.

J. L. G.-A., M. V. P.-A., F.S. and M. T.-M. acknowledge economic and scientific support from research projects PB 98/1542 of the Direccion General Enseñanza Superior e Investigation Cientifica de España, and UNESCO IGCP 421.

7 Carboniferous

JUAN R. COLMENERO, LUIS P. FERNÁNDEZ, CARMEN MORENO,
JUAN R. BAHAMONDE (coordinator), PEDRO BARBA, NEMESIO HEREDIA &
FELIPE GONZÁLEZ

The Carboniferous rocks of Spain crop out mainly in the Iberian Massif, which occupies almost half of the Iberian Peninsula. There are additionally a number of smaller Carboniferous inliers, separated by a Mesozoic and Tertiary cover, exposed in the Iberian Ranges, Pyrenees, Catalonian Coastal Ranges, Minorca and the Betic Cordillera (Fig. 7.1). Variscan and sometimes Alpine tectonism has variously overprinted these Carboniferous outcrops, commonly obscuring their original relationships.

During the Carboniferous period, sedimentation was coeval with the Variscan orogeny, in contrast to earlier Palaeozoic sedimentation in a rift to passive margin setting. The strong tectonic control on Variscan sedimentation resulted in mobile, unstable basins, with sedimentary successions that show rapid temporal and spatial changes in lithofacies and thickness. This has promoted a proliferation of stratigraphic units of only local importance, making regional correlations difficult.

The Carboniferous successions in Spain are generally dominated by siliciclastic rocks that vary from deep water turbidite successions to shallow marine, coal-bearing coastal and fully continental formations. Deep water turbidite successions are in many places referred to as the 'Culm', a term coined by Fiege (1936), although the Culm successions of the Iberian Variscan Belt are not always Early Carboniferous in age as in the original definition. In some areas limestones are locally important or even dominant and thin, condensed units of limestones or shales are widespread during Tournaisian and Visean times. An exception to the dominantly sedimentary record is provided by volcanic and volcaniclastic rocks which are abundant in the southern part of the Iberian Massif, mainly in the South Portuguese Zone. Finally, thick coal-bearing outcrops of economic importance occur in some sectors, where a coal mining industry has thrived since the nineteenth century.

This chapter presents a comprehensive review of these Carboniferous successions, but avoids detailed local information where possible. For more detail the reader is referred to four key publications (and references therein) that wholly or partly deal with the Carboniferous of Spain: Comba *et al.* (1983), Lemos de Sousa & Oliveira (1983), Martínez Díaz (1983), Dallmeyer & Martínez García (1990). In addition, there is a wealth of monographs, papers and PhD theses, and 1:50 000 and 1:200 000 geological maps of Spain (MAGNA Series, edited by the Spanish Geological Survey or IGME/ITGE).

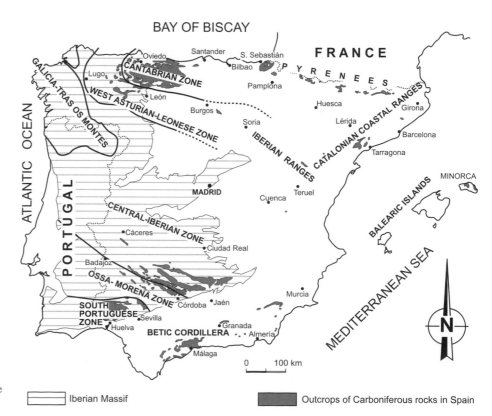

Fig. 7.1. Distribution of the Carboniferous outcrops in Spain. The zones that constitute the Iberian Massif are also shown.

Iberian Massif

The Iberian Massif constitutes the southwestern part of the European Variscan orogen and formed during Middle Devonian–Early Permian times as a consequence of the collision between Laurentia and Gondwana (Bard *et al.* 1980; Matte 1991; Martínez-Catalán *et al.* 1997, 1999). Strong tectonic shortening and an arcuate geometry (Ibero-Armorican arc) are its most characteristic features. On the basis of structural style, petrology and stratigraphy Lotze (1945), Julivert *et al.* (1972) and Farias *et al.* (1987) divided the Palaeozoic rocks of the Iberian Massif into six zones or domains. From north to south these are the Cantabrian, West Asturian-Leonese, Central Iberian, Galicia–Tras-os-Montes, Ossa Morena and South Portuguese zones (Fig. 7.1; see Chapter 9). The largest volume of Carboniferous sediments occurs in the two external zones, the Cantabrian and South Portuguese zones. The preserved Carboniferous record is described in two sections. One is devoted to the northern part of the Iberian Massif, and includes the Cantabrian and West Asturian-Leonese zones. The second section addresses the southern part of the Iberian Massif comprising the South Portuguese and Ossa Morena zones, and the southern part of the Central Iberian Zone.

The Cantabrian Zone occurs in the core of the Ibero-Armorican arc and represents the foreland thrust and fold belt in the north of the Iberian Massif. The zone is characterized by a thin-skinned tectonic style, weak to non-existent metamorphism and a virtual absence of igneous rocks (Julivert 1971; Marcos & Pulgar 1982). As a result, a relatively thick Palaeozoic succession was deformed into a set of imbricate thrusts and related folds. The strong tectonic shortening has largely modified the original palaeogeography, juxtaposing originally distinct and disparate successions. Julivert (1971)

divided the Cantabrian Zone into five tectonostratigraphic domains. From the most internal to the most external, these five domains are the Fold and Nappe, Central Asturian Coalfield, Ponga nappe, Picos de Europa and Pisuerga-Carrión provinces (Fig. 7.2). These were later subdivided by Pérez-Estaún *et al.* (1988) into major nappe units emplaced as a foreland-propagating sequence, although with the displacement vectors changing from one unit to the next (see Chapter 9). This process profoundly influenced the location and temporal evolution of syntectonic sedimentary basins. The strong tectonic shortening, in conjunction with uncertainties over the timing of nappe emplacement, are significant obstacles to the stratigraphical and sedimentological reconstruction of the Cantabrian Zone.

The West Asturian-Leonese Zone contains only scattered Stephanian B and C successions that unconformably overlie a deformed Precambrian to Lower Palaeozoic basement overprinted by low to medium grade metamorphism. Further out from the core of the Ibero-Armorican arc, Carboniferous sedimentary rocks are very scarce in the Central Iberian Zone and occur mainly in its southern part, close to the boundary with the Ossa Morena Zone. In the northern part of the Central Iberian Zone, there are two small outcrops of siliciclastic rocks of possible but unproven Carboniferous age, San Clodio in Lugo province and San Vitero in Zamora province (Pérez-Estaún 1974a), but these will not be dealt with here.

The Ossa Morena Zone is bounded by the Badajoz–Córdoba shear zone in the north, and by the Aracena (or Beja–Acebuches) ophiolite in the south. The zone has been viewed as a wide shear zone (Quesada 1983), with a Carboniferous succession distributed across several scattered outcrops oriented parallel to strike slip faults.

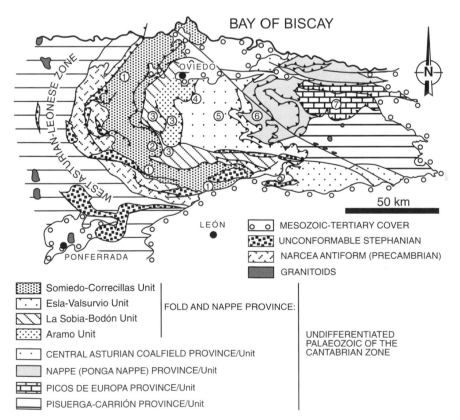

Fig. 7.2. Geological sketch map of the Cantabrian Zone showing the tectonostratigraphic provinces and tectonic units (based on Julivert *et al.* 1972; Pérez-Estaún *et al.* 1988). Numbers refer to sections of Figure 7.3B.

Classically, the South Portuguese Zone has been subdivided into three geological domains (Pulo do Lobo, Iberian Pyrite Belt, and SW Portugal), but of these only the Iberian Pyrite Belt exposes Carboniferous rocks in Spain. These three domains have been interpreted as exotic terranes which collided with the Iberian block at the onset of the Variscan orogeny (Quesada 1990*b*; Quesada *et al.* 1994) to produce south-verging, thin-skinned thrust sheets. Carboniferous sediments in the Iberian Pyrite Belt are restricted to Tournaisian and Visean units which are widely distributed and thicken to the south.

Northern Iberian Massif

In the Cantabrian Zone, the Carboniferous is represented by several thousand metres of rocks spanning the Tournaisian–Stephanian interval (Figs 7.3 and 7.4). Some interpretations of the Cantabrian Zone argued for an extensional to strike slip setting (e.g. Heward & Reading 1980; Bowman 1985; Wagner & Winkler Prins 1985; Nijman & Savage 1989), whereas most workers have envisaged sedimentation taking place in an evolving foreland basin (e.g. Marcos & Pulgar 1982; Agueda *et al.* 1991). Due to its stratigraphic complexity, the Carboniferous rocks of the Cantabrian Zone have been subdivided into a large number of predominantly local stratigraphic units. However, rising above this local detail, we argue that Carboniferous tectonostratigraphic evolution can be overviewed in terms of six informally defined sequences (*sensu* Sloss 1963). These sequences can be recognized on a regional scale and are bound by unconformities or their correlative conformities. Each of these six major sequences is thought to represent a stratigraphic unit deposited during the emplacement of major nappe units, although some of them seem locally to be composite. The bounding unconformities are related primarily to the uplift of these tectonic units and to the subsequent forward shift of associated foredeeps. The role of changes in eustatic sea level is thought generally to have been less important, although interaction between eustacy and regional tectonics must have had some influence. Generally, these six sequences have a progradational shallowing-upward character and their bounding surfaces are thought to record deepening events. In some cases, second-order cycles of the same character can be detected (see Salvador 1989, 1993; Bahamonde 1990; Barba 1991; Barba & Colmenero 1994). For shallow marine to coastal deposits, these second-order cycles are composed of cyclothems tens to hundreds of metres in thickness (see also, for example, Sánchez de la Torre *et al.* 1981; Fernández *et al.* 1988; Águeda *et al.* 1991).

The older sequences only preserve the marine and transitional marine strata deposited in the foreland basin in front of thrust units, because their more proximal correlative units were completely removed by later erosion. The younger sequences are more completely preserved and also include continental sediments deposited in intermontane basins developed on the already deformed orogen. In some cases, these continental strata are still connected with their marine correlatives. The outermost part of the foreland basin (Pisuerga-Carrión province) has preserved the most complete Carboniferous record in the Cantabrian Zone (Rodríguez Fernández & Heredia 1987).

In the northern part of the Iberian Massif coals formed from late Namurian–Westphalian to Stephanian times in coastal and alluvial settings. Two main occurrences are distinguished, one Westphalian, the other Stephanian in age. The Westphalian coals (mainly Bolsovian and Westphalian D), are paralic and best developed in the Central Asturian Coalfield (see below). The Stephanian coals are mainly limnic and formed in intermontane basins found in the Cantabrian and West Asturian-Leonese zones, and to a lesser extent in the coeval foreland basin-fill successions in the Pisuerga-Carrión province (see below). Coal rank varies from high-volatile bituminous coals to anthracite/meta-anthracite. Rank increases from the Westphalian to the Stephanian coals, and from the Cantabrian Zone to the West Asturian-Leonese Zone (Colmenero & Prado 1993). Irrespective of age, the highest rank coals occur in metamorphic belts, parallel to the Variscan structural grain, and due to the intrusion of igneous bodies during late Stephanian to Early Permian times.

Sequence 1. This sequence (upper Famennian/lower Tournaisian–Arnsbergian) disconformably overlies Upper Devonian terrigenous deposits and its top is generally conformable with the base of sequence 2. The sequence records the inversion from a passive continental margin to an active margin and reflects Variscan deformation of the inner parts of the orogen. Sequence 1 comprises three condensed units: the laterally equivalent Baleas and Vegamián formations and the overlying Alba Formation (Fm) (see Figs 7.3 and 7.4).

The Baleas and Vegamián formations were deposited in two different domains of the Cantabrian Zone and both reflect slow deposition in a shallow marine setting following transgression over the Devonian continental margin. This late Famennian–Early Carboniferous transgression event has been recognized elsewhere in the world (e.g. Savoy 1992; Alekseev *et al.* 1996) and could be the result of either tectonic or glacio-eustatic processes. The Baleas Fm (upper Famennian–lower Tournaisian; see Sánchez de Posada *et al.* 1990) occurs in the western part of the Cantabrian Zone. The formation is a thin (5–10 m thick) white limestone unit formed of skeletal grainstones to wackestones with a variable proportion of quartz-sand grains. The Vegamián Fm (middle Tournaisian–lower Visean; see Sánchez de Posada *et al.* 1990) is the equivalent formation in the eastern part of the Cantabrian Zone, and comprises a 10–60 m thick succession of black shales and mudstones with chert lenses and phosphate nodules (Fig. 7.5a). These two formations represent different shelf settings (Fig. 7.6a). The Baleas Fm limestones were deposited in shallow-water, well oxygenated shelf seas with benthic communities, whereas the Vegamián Fm records deeper water sedimentation on the shelf edge (Sánchez de la Torre *et al.* 1983; Wendt & Aigner 1985).

Maximum transgression occurred during Visean times and led to the deposition of the Tournaisian–Arnsbergian Alba Fm (also called Genicera Fm or Carboniferous griotte limestones). This 20–30 m thick, laterally uniform unit gradationally overlies the Baleas and Vegamián formations (Fig. 7.5b), and comprises red nodular limestones with red shale partings and beds, and a red radiolarite horizon. Pelagic fauna are dominant, including cephalopods, with subordinate benthic faunal debris. The Alba Fm is interpreted as a condensed unit deposited on a relatively well oxygenated pelagic platform with gentle slopes in several hundred metres of water (Sánchez de la Torre *et al.* 1983; Wendt & Aigner 1985; Fig. 7.6b).

Sequence 2. Sequence 2 is Namurian A (Serpukhovian) in age and comprises the Barcaliente and Olleros formations, and the lower part of the Prioro Group (Gp) (Figs 7.3 and 7.4). This sequence generally has a gradational contact with sequence 1, but in many places its upper boundary is a disconformity, displaying breccias, karst surfaces and depositional or erosional hiatuses. The sequence records the onset of terrigenous supply derived from the orogen in the Cantabrian Zone. During this period, the sedimentary basin had two main elements. A foredeep located in the west and south was the

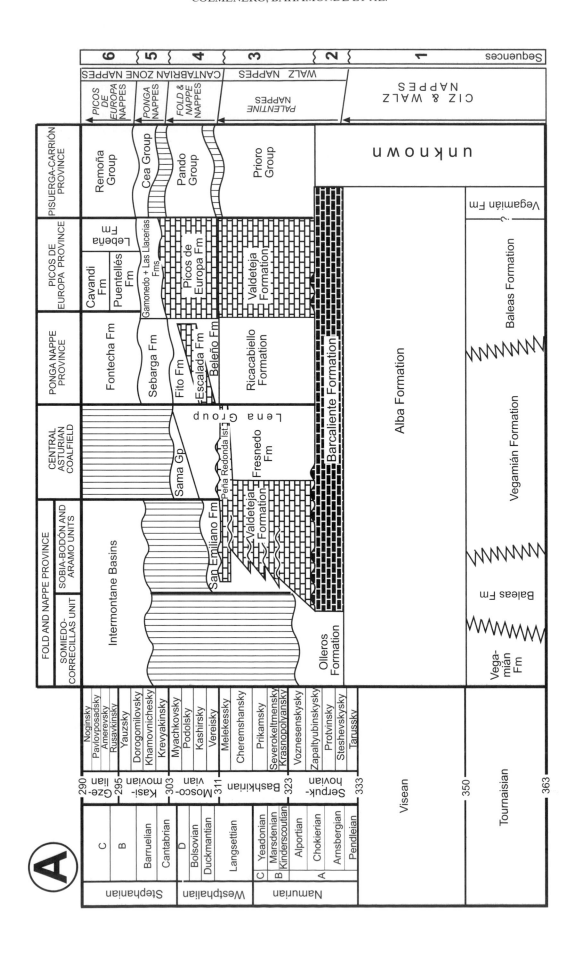

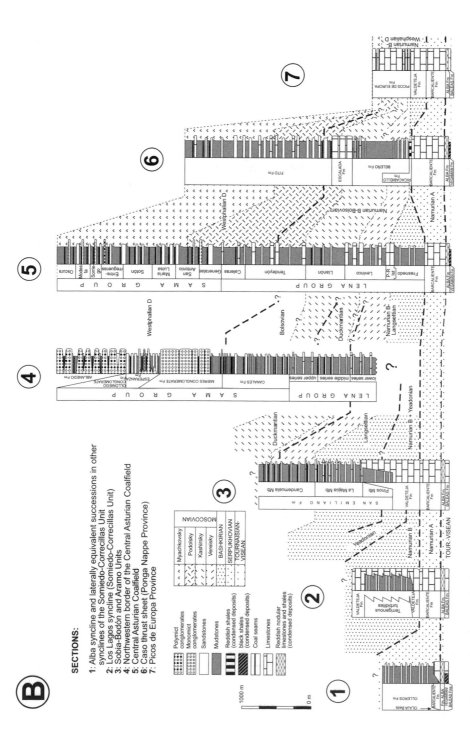

Fig. 7.3. **(a)** Stratigraphic correlation chart of the Cantabrian Zone showing the distribution and relationships of the main lithostratigraphic units discussed in the text. The timing of emplacement of the main groups of nappes and the six sequences discussed in the text are in columns to the right of the chart (CIZ, Central-Iberian Zone; WALZ, West Asturian-Leonese Zone). Absolute ages are after Harland *et al.* (1990). Correlation between the West European and Russian scales based on Villa (1988, and references therein), Harland *et al.* (1990) and Jones (1995). **(b)** Selected synthetic stratigraphic sections of the synorogenic, Tournaisian–Westphalian D (sequences 1 to 4) succession in the Cantabrian Zone (the Pisuerga-Carrión Province has been excluded, because its succession is very complex and difficult to summarize). Notice the wedge shape of the deposits as revealed by the isochrones. Updated from Fernández (1995); data from Fernández & Barba (unpublished) for sections 1 and 2; Fernández (1990, 1993) for sections 3 and 4; Fernández (1990, 1993), Villa & Heredia (1991), Bahamonde & Colmenero (1993) and Salvador (1993) for sections 5 and 6; Marquínez (1978) and Farias (1982) for section 7. No horizontal scale involved. See location of sections in Figure 7.2.

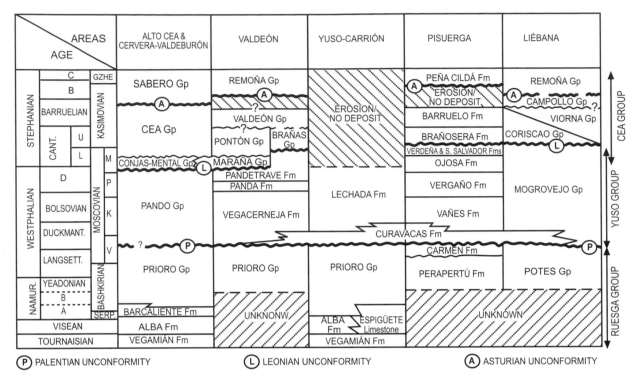

Fig. 7.4. Stratigraphic correlation chart of the Pisuerga-Carrión Province. The Ruesga, Yuso and Cea groups correspond to nomenclature used in the Dutch literature. The areas shown are located on Figure 7.7a.

site of predominantly turbidite deposition (Olleros Fm and lower part of Prioro Gp). Distally, towards the east and north, a larger, shallower water realm extended over the remainder of the Cantabrian Zone and was the site of uniform carbonate sedimentation (Barcaliente Fm; Figs 7.5c and 7.6c).

The Barcaliente Fm averages 300–350 m in thickness and thins towards the east. The formation is mainly composed of laminated, thinly bedded, dark grey limestones and mudstones, and contains few fossils. The upper part of the formation is dominated by calcareous mudstones with pseudomorphs of anhydrite and gypsum (González Lastra 1978; Hemleben & Reuther 1980; Sánchez de la Torre *et al.* 1983). In the Fold and Nappe province the top of the formation is frequently formed of extremely poorly sorted breccias (Porma breccia; Reuther 1977).

The depositional setting for the Barcaliente Fm has been a matter of discussion. González Lastra (1978) and Sánchez de la Torre *et al.* (1983) suggested a restricted, highly stressed shallow water environment, prone to evaporite precipitation. In contrast, Hemleben & Reuther (1980), who studied the formation in the inner parts of the Cantabrian Zone, interpreted the limestone beds as turbidites and concluded a deeper, slope setting. Eichmüller & Seibert (1984) considered that both interpretations were valid and envisaged a restricted shallow water platform, passing to the west and south into a slope setting, and eventually into the deepest water foredeep. The top of the formation shows signs of emergence and instability; the Porma breccia has been interpreted as a collapse breccia related to dissolution of evaporites (González Lastra 1978). Additionally, in the eastern end of the Fold and Nappe province, Rodríguez Fernández (1994) and Wagner & Winkler Prins (1999) describe an important discontinuity displaying karst and subaerial exposure features.

In the innermost units of the Fold and Nappe province, the Barcaliente Fm limestones are replaced by coeval siliciclastic turbidites (Fig. 7.3b). These turbidites are mainly preserved in the Alba syncline (Correcillas unit) where the Alba Fm is gradationally overlain by a 14 m thick interval of red and greenish shales (Olaja beds). The Olaja beds pass upwards into a 400–500 m thick, siliciclastic turbidite succession with some interleaved limestones, similar to those of the Barcaliente Fm (Olleros Fm, or lower part of the Cuevas Fm). The Olleros Fm consists of alternating shales and sandstone beds with some conglomerates organized into several stacked turbidite systems. In the Pisuerga-Carrión province, the foredeep fill is represented by a terrigenous succession similar to the Olleros Fm, in which several stacked turbidite systems have also been recognized (Rodríguez Fernández 1994).

Sequence 3. Sequence 3 was deposited during Bashkirian times (Namurian B–Langsettian) and consists of both siliciclastic and carbonate lithologies. Thick limestone deposits (the Valdeteja Fm) extend mainly over the external part of the Fold and Nappe province, the northeastern parts of the Central Asturian coalfield and Ponga nappe provinces, and the Picos de Europa province (Fig. 7.5d, e). Other, more siliciclastic parts of the succession extend across the Fold and Nappe province (some unnamed units, the upper part of the Cuevas Fm, and the Pinos and La Majúa members of the San Emiliano Fm), the Central Asturian coalfield (the Fresnedo Fm and the Peña Redonda limestone of the Lena Gp), Ponga nappe (Ricacabiello Fm), and the Pisuerga-Carrión province (Prioro Gp; *sensu* Rodríguez Fernández 1994; see Fig. 7.3). Sequence 3 reflects a change in the basin geometry driven by

Fig. 7.5. Palaeogeographic sketch maps showing the distribution of the terrigenous and carbonate deposits in the Cantabrian Zone over the Carboniferous period. Map H represents a maximum transgression that took place during part of Podolsky times. The Pisuerga-Carrión Province has been left blank due to the complexity of the distribution of its deposits. No palinspastic restoration has been allowed for.

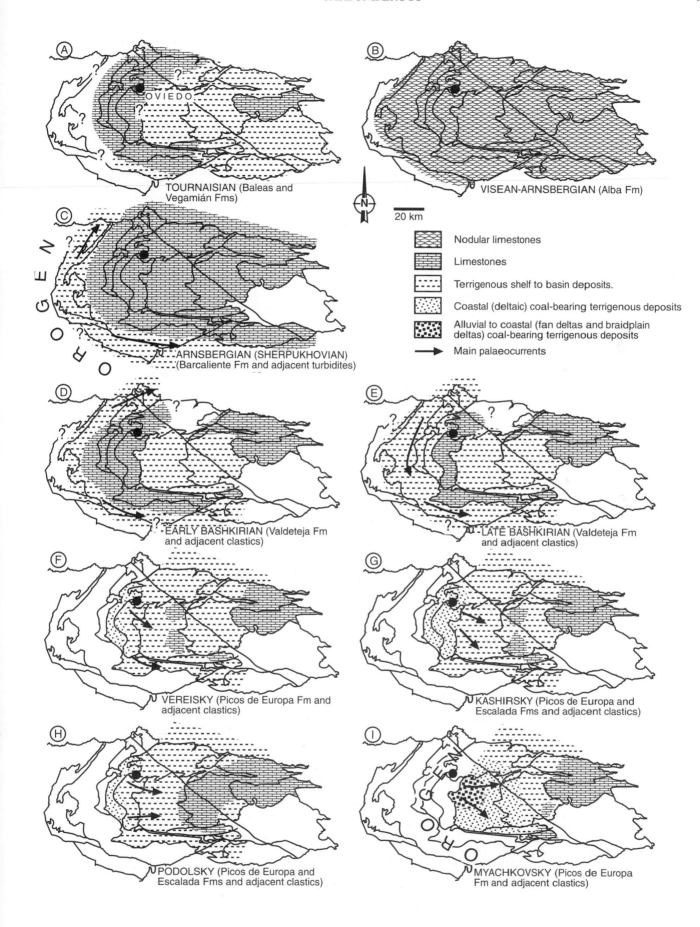

A. TOURNAISIAN (Baleas and Vegamián Fms)

B. VISEAN-ARNSBERGIAN (Alba Fm)

N

20 km

Nodular limestones

Limestones

Terrigenous shelf to basin deposits.

Coastal (deltaic) coal-bearing terrigenous deposits

Alluvial to coastal (fan deltas and braidplain deltas) coal-bearing terrigenous deposits

Main palaeocurrents

C. ARNSBERGIAN (SHERPUKHOVIAN) (Barcaliente Fm and adjacent turbidites)

D. EARLY BASHKIRIAN (Valdeteja Fm and adjacent clastics)

E. LATE BASHKIRIAN (Valdeteja Fm and adjacent clastics)

F. VEREISKY (Picos de Europa Fm and adjacent clastics)

G. KASHIRSKY (Picos de Europa and Escalada Fms and adjacent clastics)

H. PODOLSKY (Picos de Europa and Escalada Fms and adjacent clastics)

I. MYACHKOVSKY (Picos de Europa Fm and adjacent clastics)

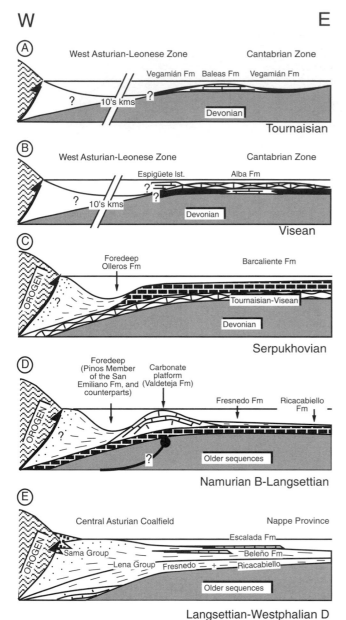

Fig. 7.6. Conceptual, simplified sketches showing the main stages of foreland basin evolution in the Cantabrian Zone during the Carboniferous period. Note that neither the Pisuerga-Carrión nor the Picos de Europa provinces have been included, due to palaeogeographic uncertainties and correlation problems. (**c–e**) are based on Fernández (1990).

a more advanced stage of nappe emplacement (Fig. 7.6d) which caused the foredeep to migrate over the Barcaliente Fm carbonates. Shallower marine carbonate deposits (Valdeteja Fm) occupied progressively less areal extent in the Fold and Nappe province (Fig. 7.5d, e). This narrow and shallow carbonate belt passed into deeper water dominated by siliciclastic sedimentation to produce the Fresnedo Fm (Central Asturian Coalfield) and the Ricacabiello Fm (Ponga nappe) (Fig. 7.6d).

The Valdeteja Fm limestones were deposited on raised platforms, in which platform top, platform margin, steep slope and basin facies can be distinguished (Eichmüller 1985; Bahamonde *et al.* 1997; Kenter *et al.* 2002). An extensive carbonate platform, up to 1000 m thick, developed in the northeastern part of the Ponga nappe (Sierra del Cuera area) during Bashkirian times (Bahamonde *et al.* 1997; Kenter *et al.* 2002). Initially, a low-angle ramp with microbial mud deposits nucleated above the Barcaliente Fm. Combined aggradation and progradation produced a flat-topped shallow water platform characterized by a steep microbial boundstone margin (shelf system). To the SE, in the Picos de Europa province, the Valdeteja Fm represents a ramp system that thins to approximately 100 m along its southern margin. According to Fernández (1990, 1993), the Valdeteja Fm in the Fold and Nappe province was deposited on submarine highs produced by the incipient emplacement of new tectonic units (blind thrusts). The formation comprises up to five vertically stacked platforms, separated by discontinuity surfaces, that are locally draped by siliciclastic sediment tongues (see below). This architecture indicates a more unstable and subsidence-prone setting closer to the orogen than the single platform present in the eastern outcrops.

The southern and western flanks of the Valdeteja platforms in the Fold and Nappe province interfinger with orogen-derived terrigenous sediments that filled the foredeep. The best preserved of these successions is the 1500 m thick, shallowing-upward succession that comprises the Pinos and La Majúa members of the San Emiliano Fm (Sobia-Bodón and Aramo units; Fig. 7.3b). The Pinos Member (Mb) comprises shales with interleaved turbidite sandstones, and locally abundant calcareous breccias and olistoliths derived from the Valdeteja platforms which bound the turbidite successions. The basin and slope deposits of the Pinos Mb are overlain by the La Majúa Mb, which is characterized by alternations of shales and sandstones with some coal seams and several limestone bands. This member was deposited in a coastal, shallow water shelf setting by minor prograding deltas (Bowman 1985; Fernández 1990, 1993). To the north, this succession passes into shallow water to coal-bearing coastal deposits (e.g. La Camocha succession), which are mostly concealed below the post-Variscan cover. To the SE and east, the deep water Pinos Mb sediments are the lateral equivalents of La Majua Mb and are capped by a single limestone unit (the 'caliza masiva') that is coeval with the youngest Valdeteja Fm platform sediments and the Peña Redonda limestone of the Central Asturian Coalfield (see below). Thus, at least in this case, it seems clear that the main source of orogen-derived sediments was located in the north, from where the depositional systems axially prograded to the south.

The Valdeteja Fm carbonate platforms represented a barrier that prevented the sandy input from the hinterland to enter the more distal parts of the sedimentary basin (mainly the Central Asturian Coalfield and Ponga nappe; Fig. 7.6d). In the Central Asturian Coalfield to the east is a 400 m thick shaly unit (the Fresnedo Fm) capped by the Peña Redonda limestone which has conspicuous karstified subaerial exposure surfaces (Salvador 1989, 1993) and represents the top of the sequence (Fig. 7.3b). Boulders and olistoliths derived from the back-flank of the Valdeteja platform and turbiditic sandstone beds occur interbedded with the Fresnedo Fm shales (Salvador 1993). Eastward, in the Ponga nappe province, the Fresnedo shale wedge pinches out giving way to a condensed succession, up to 60 m thick, formed of red shales with chert and manganese nodules (Ricacabiello Fm; Sjerp 1967; Bahamonde & Colmenero 1993). The Central Asturian coalfield and Ponga nappe represent a broad, sediment-starved marine basin, the depth of which increased eastward to about 500–600 m (Sierra del Cuera outcrop; Bahamonde *et al.* 1997; Kenter *et al.* 2002).

The Pisuerga-Carrión province represents a more unstable area of the basin where a 1500–2000 m thick terrigenous turbidite succession accumulated. The succession is dominated by black shales and sandstones, but polymict conglomerates, calcareous breccias and olistoliths, derived from the Barcaliente and Valdeteja formations, are also common or very abundant in some areas. This succession has been given different names depending on the location (see Figs 7.4 and 7.7a): Prioro Gp in the Esla area (Alonso 1985), Perapertú and Carmen formations (Wagner & Wagner-Gentis 1963) and Cervera Fm in the north of Palencia province; Potes Gp and Piedras Luengas limestones (Graaff 1971) in La Liébana area. Following Rodríguez

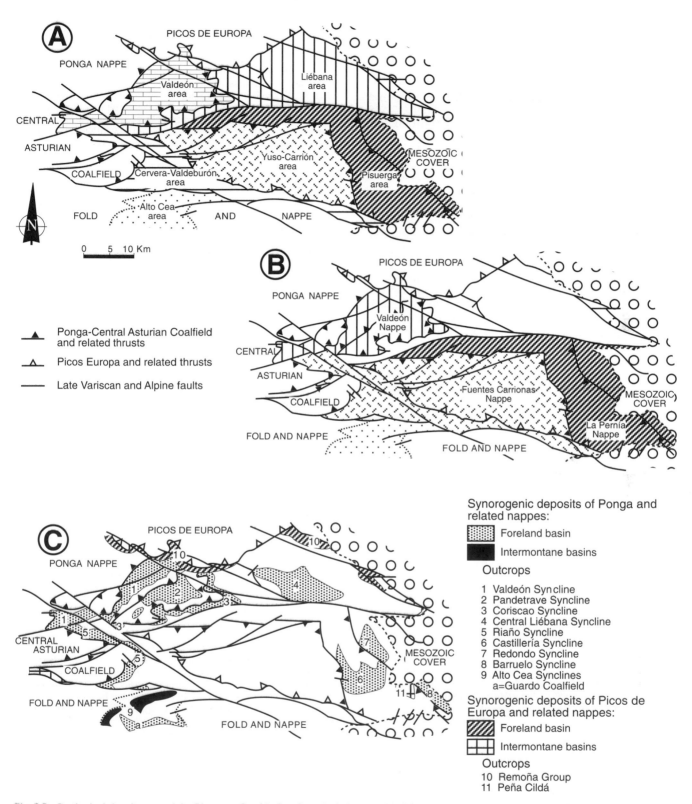

Fig. 7.7. Geological sketch maps of the Pisuerga-Carrión Province depicting: **(a)** the different areas discussed in the text; **(b)** the thrust sheets and nappes; **(c)** the location of the several synorogenic deposits related to the emplacement of the Ponga-Central Coalfield and Picos de Europa units.

Fernández (1994), in this paper we place all these units within the Prioro Gp. This group represents the synorogenic wedge related to the emplacement of the Palentine nappes which ended up entering the basin as gravitationally driven tectonic units (Frankenfeld 1983;

Marquínez & Marcos 1984; Rodríguez Fernández & Heredia 1987). In this province, the top of sequence 3 is represented by a regional unconformity, the Palentian unconformity (Wagner 1960).

Sequence 4. This sequence developed during much of Moscovian times (from late Langsettian to mid-late Westphalian D; Fig. 7.3). In the Fold and Nappe province the top of the sequence is only Kashirsky (Duckmantian) in age due to later erosion. Sequence 4 records the burial of the barrier represented by the Valdeteja Fm in the Fold and Nappe province and the subsequent spread of siliciclastic sedimentation across the entire Cantabrian Zone. Carbonate platforms only continued to develop in the Picos de Europa province and the northeastern part of the Ponga nappe province (Figs 7.5f–i and 7.6e). The succession is dominated by terrigenous deposits, which are frequently coal-bearing, and limestones are generally subordinate, although some conspicuous and thick carbonate units occur. Águeda *et al.* (1991) proposed a depositional model for this sequence in which five characteristic facies associations are represented: (a) dominantly conglomeratic alluvial and fan delta; (b) deltaic lobes; (c) shallow marine clastics; (d) carbonate shelves; (e) basinal shales with turbidites and olistoliths. Marcos & Pulgar (1982), Alonso (1985), Rodríguez Fernández & Heredia (1987) and Fernández (1990) envisaged these siliciclastic successions as clastic wedges fed from the west and SW, and related to the emplacement of several nappes of the Fold and Nappe province (Somiedo-Correcilla, Sobia-Bodón, Aramo, Esla and Valsurvio nappes).

The units comprising this sequence are the Candemuela Mb and the Villamanín beds of the San Emiliano Fm, and their laterally equivalent units (Fold and Nappe province); the portion of the Lena Gp overlying the Peña Redonda limestone and the Sama Gp (Central Asturian Coalfield); the Beleño, Escalada and Fito formations (Ponga nappe province); and the Picos de Europa Fm (Picos de Europa and northeastern part of Ponga nappe provinces) (Fig. 7.3). In the Pisuerga-Carrión province, the succession is bounded by the Palentian (see above) and Leonian unconformities, and is referred to here as the Pando Gp (see Figs 7.3a and 7.4).

In the inner parts of the basin (Fold and Nappe province), the Candemuela Mb (Fig. 7.3b) of the San Emiliano Fm and the laterally equivalent units to the north are formed of coal-bearing sandstone–mudstone alternations, deposited in a coastal, deltaic environment with minor episodes of both shallow marine and alluvial deposits (Bowman 1985; Fernández 1990, 1993). Towards the SE, the coeval Villamanín beds of the San Emiliano Fm record a more distal setting with mainly siliciclastic shelf sediments and some transient carbonate platforms (see also Wagner *et al.* 1971). Both the Villamanín and Candemuela units display a gradational base, although locally a disconformity is recognized, with unconformity becoming clearer eastwards.

In the Central Asturian Coalfield, sequence 4 includes the upper part of the Lena Gp above the Peña Redonda limestone and the Sama Gp (Fig. 7.3). The Lena Gp consists of sandstone–mudstone alternations with some conspicuous limestone units and rare coal seams. In the Sama Gp, limestones are virtually absent, economic coal seams are common and some conglomeratic intervals occur towards the upper part of the succession. On the northwestern border of the Central Asturian Coalfield, these conglomerates form two mappable units (Mieres and Olloniego conglomerates, Fig. 7.3b), each *c.* 1000 m thick, that thin rapidly towards the east and SE.

The gradational boundary between the Lena and Sama groups is diachronous across the Central Asturian Coalfield and youngs from Bolsovian, in the northwestern part, to Westphalian D, in the central-eastern part. This coarsening- and shallowing-upward cycle reaches a thickness of *c.* 5500 m and evolves from shallow-marine deposits with minor deltaic episodes (Lena Gp) to an interval of coal-bearing deltaic and fan-deltaic sediments, with both marine and alluvial episodes (Sama Gp). Barba (1991), Salvador (1993) and Barba & Colmenero

(1994) have further distinguished second-order cycles of the same character, composed of cyclothems tens to hundreds of metres in thickness (see also Sánchez de la Torre *et al.* 1981; Fernández *et al.* 1988; Águeda *et al.* 1991).

In the Ponga nappe province, sequence 4 consists of three units. From base to top, these are the Beleño, Escalada and Fito formations (Fig. 7.3b). The Beleño Fm is an 800 m thick shaly succession with a prominent interval of thick channelized turbidite sandstones at the base, and limestones, rooted horizons and coal seams at the top. The overlying Escalada Fm is a diachronous unit, 200 to 300 m thick, made up of massive and irregular beds of micritic and skeletal limestones. Finally, the Fito Fm comprises shales and sandstones with some intercalated coal seams and limestones. The Beleño and Fito formations represent two eastward-thinning clastic wedges which filled the marine basin (Bahamonde 1990; Bahamonde & Colmenero 1993).

In the Picos de Europa province and the adjacent areas of the northeastern part of the Ponga nappe province, sequence 4 comprises a 600–900 m thick carbonate unit, the Picos de Europa Fm (Fig. 7.3b), which overlies the Valdeteja Fm. The Picos de Europa Fm is a steep-fronted carbonate shelf, whose margin was onlapped by the detrital units of the Ponga nappe. Whether this large (at least 3500 km² wide) new carbonate platform was isolated or connected with a continental area to the east remains unknown, because its eastern border is buried beneath a thick Mesozoic cover.

This platform consists of several facies belts (Bahamonde *et al.* 1997, 2000). Internal areas are dominated by monotonous, non-stratified deposits that represent a shallow, open marine lagoon. External areas, several kilometres wide, show three domains characterized by distinctive patterns: (1) proximal horizontally bedded shallow platform deposits with a weakly rhythmic character; (2) mud mounds passing downslope into margin-derived breccias, representing margin and slope clinoforms with a relief of up to 500 m and a maximum depositional dip of 32°; (3) a low-angle toe-of-slope, where slope beds interfinger with a cyclic alternation of basinal sediments. In the NW part of the Ponga nappe similar architecture and lithofacies have been described for both the Valdeteja and Picos de Europa formations (Bahamonde *et al.* 1997; Kenter *et al.* 2002), with well-developed shallowing-upward cycles at the top of the platformal successions.

In the Pisuerga-Carrión province, turbidite and olistolithic successions of sequence 4 continue to reflect a deep, unstable and strongly subsident basin. Several local stratigraphic units have been defined: the Pando Gp (Alto Cea area), Mogrovejo Gp (La Liébana), Vegacerneja, Panda and Pandetrave formations (Valdeón area), Curavacas and Lechada formations (Yuso-Carrión area), and Vañes, Vergaño, Ojosa, Verdeña and San Salvador formations (Pisuerga area) (Figs 7.4 and 7.7a; see Heredia *et al.* 1990). In this chapter all these units will be referred to as the Pando Gp (*sensu lato*).

A very thick (up to 1000 m) and widespread quartzite conglomerate unit (Curavacas Conglomerate Fm and laterally equivalent conglomerates) overlies the Palentian unconformity. This Duckmantian unit (Wagner 1960) represents the cannibalization of uplifted sectors of the basin and comprises alluvial to deep water, fan-deltaic sediments fed from the south (Colmenero *et al.* 1988). The remainder of the Pando Gp (*sensu lato*) generally forms a monotonous, 1000–1500 m thick, deep water succession comprising shales, sandy turbidites, conglomerates and chaotic deposits, including olistoliths of Devonian limestones (Alonso 1985; Rodríguez-Fernández 1994). In some localities in the Pisuerga area, e.g. the Casavegas syncline, an upper unit of the Pando Gp (*sensu lato*), the Vergaño Fm, consists of coal-bearing sandstone–mudstone alternations with two interleaved algal limestone episodes. Graaff (1971) interpreted these rocks as the deposits of deltas that prograded towards the northeastern, deeper parts of the basin.

In the Valdeón area, the Pando Gp (*sensu lato*) consists of the Vegacerneja, Panda and Pandetrave formations (Heredia *et al.* 1990). The Vegacerneja Fm is a fining- and shallowing-upward sequence formed of sandstone–mudstone interbeds. The overlying Panda Fm is a widely distributed carbonate unit which thins to the west and consists of skeletal limestones and algal mounds in the lower part and of calcareous breccias in the upper part (Maas 1974; Heredia *et al.* 1990). Finally, the overlying Pandetrave Fm consists of two members. The lower member comprises sandstone–mudstone interbeds that pass upwards into a prominent sandstone interval. The upper member is dominated by a shale succession with interleaved terrigenous and calcareous turbidites, conglomerates, limestone breccias and olistoliths. This upper member records a deepening of the basins and may actually represent the most distal parts of the Maraña Gp (sequence 5).

Sequence 5. This succession spans the upper Westphalian D to Stephanian B interval (Kasimovian–Khamovnichesky) and is bounded by the Leonian and Asturian unconformities (Wagner 1959). In contrast to sequences 1–4, it only crops out in the Pisuerga-Carrión, Picos de Europa and Ponga nappe provinces (see Fig. 7.3a), and in the most external part of the Esla nappe (Fold and Nappe province). This reduction in size of the sedimentary basin was due to the advancement of the orogenic front. The inner sectors of the former sedimentary basin, including the Central Asturian Coalfield and Ponga nappe units, were thrust to the east and NE (Chapter 10) and uplifted. Deformation also created high reliefs in the already emplaced Esla nappe unit (Iwaniw 1985; Alonso 1985). Finally, thrust propagation into the remnant foreland basin (Pisuerga-Carrión province) gave rise successively to the Valdeón, Fuentes Carrionas and La Pernía nappes (Fig. 7.7b; Heredia 1991; Rodríguez-Fernández 1994). The successions of sequence 5 can be divided into those which accumulated in the foreland basin (Pisuerga-Carrión province) and those that were deposited in intermontane basins developed in the already deformed orogen. The foreland basin units group consists of the Maraña, Pontón and Valdeón groups (Valdeón and Cervera-Valdeburón area); the Coriscao, Viorna and Campollo groups (La Liébana area); and the Brañosera and Barruelo formations (Pisuerga area) (Figs 7.4 and 7.7a, c). The intermontane units comprise the Conjas-Mental and Cea groups (Alto Cea area in the Esla nappe; Figs 7.4 and 7.7a, c); Gamonedo, Demúes and Dobros formations and Las Llacerias beds (Picos de Europa province); and the Sebarga beds (Ponga nappe province) (Fig. 7.8).

The Maraña, Pontón and Valdeón groups represent the synorogenic wedge related both to the emplacement of the Ponga nappe, and (in part) to movements of the Gildar thrust of the Valdeón nappe (Fig. 7.9a, b; Heredia 1991). These three units are bound by minor unconformities and form a 3000 m thick shallowing-upward cycle. The lower unit (Maraña Gp) is dominated by shale, with some calcareous breccias and large olistoliths. The overlying Pontón Gp is also shaly with turbidite sandstones, conglomerates and breccias. Finally, the upper unit (Valdeón Gp) is conglomeratic and includes some coal seams. Towards the east, in La Liébana area, these three units laterally change to the so-called Coriscao Gp. This group comprises a lower, 500 m thick olistostromal interval containing limestone olistoliths 10–100 m in size (Rodríguez Fernández 1994). Above this a middle turbidite interval comprises mudstone–sandstone alternations, and an upper interval is formed of shallow water mudstones and sandstones, and contains some coal seams. According to Heredia (1991), the Coriscao Gp filled a small basin created by the loading associated with Valdeón nappe emplacement (Fig. 7.9b, c). Further to the east, this sequence is represented by the Viorna and Campollo groups (La Liébana central syncline in La Liébana area) and the Brañosera and Barruelo formations (Casavegas, Castillería, Redondo and Barruelo synclines in the Pisuerga area; Figs 7.4 and 7.7). All these units constitute the synorogenic sediments related to the eastward emplacement of the Fuentes Carrionas nappe (Rodríguez Fernández & Heredia 1987). In the more proximal settings, the Viorna Gp and the overlying Campollo Gp (Rodríguez-Fernández & Heredia 1987; Rodríguez-Fernández 1994) consist of deep water deposits including olistoliths and fan-delta sediments comprising both polymict and calcareous conglomerates and breccias, turbidites and slumped units. Both groups record an intense deformation event and are bound by an unconformity formed during the tightening of La Liébana central syncline. The Brañosera and the overlying Barruelo formations

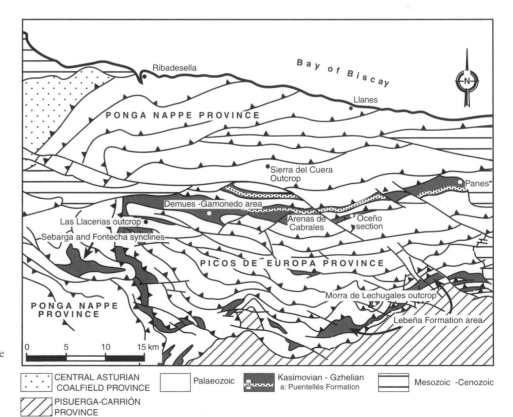

Fig. 7.8. Kasimovian and Gzelian outcrops in the northeastern part of the Cantabrian Zone showing the main localities of the studied successions. Modified from Farias (1982), Marquínez (1989) and Sánchez de Posada *et al.* (1999).

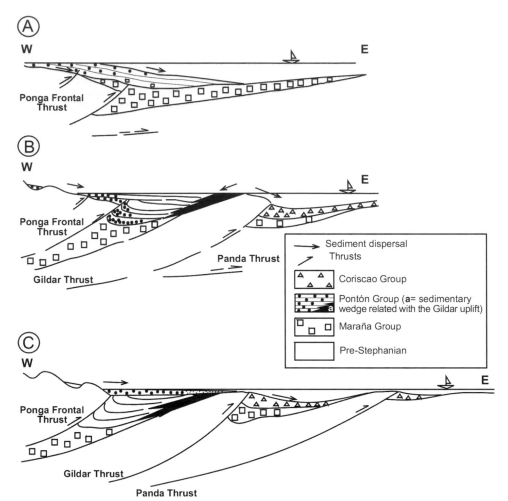

Fig. 7.9. Tectonosedimentary evolution of the Valdeón area (Pisuerga-Carrión Province; see location in Fig. 7.7a) during Cantabrian–Barruelian times: **(a)** during the emplacement of the Ponga frontal thrust; **(b)** during the emplacement of the Gildar thrust of the Valdeón nappe; **(c)** during the emplacement of the Panda thrust of the Valdeón nappe.

preserve a shallowing-upward succession, up to 2000 m thick, which occurs in synclines of the Pisuerga area (northern part of Palencia province). In the Redondo syncline, the Brañosera Fm averages 750 m in thickness and comprises a tripartite succession of megaturbidite deposits overlain by thinly bedded and fine-grained overbank turbidite sediments, finally capped by thick-bedded and coarse-grained turbidite deposits (Bahamonde & Nuño 1991). The upper interval grades into the Barruelo Fm which is some 1250 m thick and consists of fossiliferous, marine and lacustrine shales, rooted shales, channelized sandstone bodies and coal seams (Wagner & Winkler Prins 1985; Bahamonde & Nuño 1991).

In the Alto Cea area (Fig. 7.7), sequence 5 buries palaeoreliefs generated in the Esla nappe, and it is represented by two conglomerate-rich units (Conjas-Mental and Cea groups; Fig. 7.4; Alonso 1985). The Conjas-Mental Gp is several hundred metres thick and consists of quartzitic conglomerates, sandstones and shales, with sharp lateral and vertical facies changes. The Cea Gp unconformably overlies the Conjas-Mental Gp or the underlying Pando Gp, and is a 1000 m thick succession of basal conglomerates, mainly quartzitic and sub-ordinately calcareous, passing upwards into sandstones, shales and coal seams. This group was deposited in alluvial fan and fan-delta systems (Helmig 1965; Loon 1972; Iwaniw 1984; Alonso 1985), and to the east, it enters the foreland basin where it forms the Guardo coalfield (Fig. 7.7c). In this coalfield, the Cea Gp consists of a 2000 m thick, coal-bearing succession of sandstones, siltstones, and shales, deposited in coastal, mainly deltaic, environments (Saldaña 1993) during late Westphalian D to early Cantabrian times (Wagner & Winkler Prins 1985).

The deposits of this sequence also crop out along a belt 25 km long by 4 km wide, located between the northeastern sector of the Ponga

nappe and Picos de Europa provinces (Gamonedo-Cabrales area; Fig. 7.8) forming the Gamonedo Fm (Fig. 7.3a) and the laterally equivalent Demués Fm and Dobros beds (see Villa & Martínez García 1989; Sánchez de Posada *et al.* 1999). This formation is Cantabrian to Barruelian in age (upper Myachkovsky to upper Khamonichesky; Wagner & Martínez García 1998; Martínez García & Villa 1998), and unconformably overlies older Carboniferous limestones. The palaeotopography of the underlying carbonate-rimmed shelves led to an irregular distribution of later sediments, which display rapid lateral changes in thickness and lithofacies. In the east (Arenas de Cabrales outcrops), where the Gamonedo Fm overlies the carbonate shelf, it is relatively thin and comprises coal-bearing shallow water deposits that thin out towards the easternmost outcrops (Oceño-Panes area). In the west, the Gamonedo Fm overlies the basinal areas adjacent to the carbonate shelf, and shows a siliciclastic succession that dramatically thickens up to 1000 m and forms a shallowing-upward sequence from deep to shallow water deposits. The latter include fan-deltas located adjacent to emergent palaeoreliefs (see Bruner *et al.* 1998; Bruner & Smosna 2000).

Towards the southern part of the Picos de Europa province (Las Llacerias and Morra de Lechugales outcrops; Fig. 7.8; see Ginkel & Villa 1991), the Picos de Europa Fm is conformably overlain by a 100 m thick unit (Las Llacerias Fm; Ginkel & Villa 1999) consisting of an alternation of dark marls and marly limestones, together with rudstone and sandy grainstone tempestites (Marquínez *et al.* 1982; Bahamonde *et al.* 2000).

In the Ponga nappe, the 250 m thick Sebarga beds (lower Kasimovian) unconformably overlie older Carboniferous units in the core of a narrow syncline (Sebarga syncline; Figs 7.3 and 7.8). The Sebarga beds are characterized by basal coarse-grained clastic deposits and overlying fossiliferous marine shales. The interval records the progradation of fan-delta systems into a semi-closed marine basin (Colmenero & Bahamonde 1986; Colmenero *et al.* 1988).

Sequence 6. This sequence comprises the youngest Carboniferous sediment, deposited during Dorogomilovsky (mid-Kasimovian) to Gzelian times (from Stephanian B to C). It is underlain by the Asturian unconformity, and overlain by the pre-Permian unconformity which represents the onset of the Alpine cycle. The foreland basin-fill successions of this sequence crop out in the Pisuerga-Carrión and Picos de Europa provinces, where they were deposited during the southward emplacement of the Picos de Europa unit and the latest thrust episodes of the Ponga nappe unit. Sequence 6 also occurs in a series of intermontane basin-fill successions scattered through the Fold and Nappe province, the Narcea antiform and the West Asturian-Leonese Zone (Figs 7.2 and 7.10).

The foreland basin successions consist of the Puentellés, Cavandi and Lebeña formations (Picos de Europa province; see Fig. 7.3a), the Remoña Gp (Pisuerga-Carrión province; see Fig. 7.4), and some unnamed terrestrial units. In the northern part of the Picos de Europa province an erosional surface cuts into the older Kasimovian strata, and above this surface cross-bedded fluviatile sandstones were deposited along with shales, marls, calcareous mudstones and rudstone intervals (Las Llacerias outcrops; Bahamonde *et al.* 2000). This succession is connected with conglomeratic deposits that crop out to the west, near the boundary between the Ponga nappe and Picos de Europa provinces. Eastward, in the Oceño-Panes area, sequence 5 is absent and the top of the Picos de Europa Fm displays a palaeokarst surface with lateritic crust development (O. Merino, pers. comm.) overlain by thin, sandy and conglomeratic, terrestrial deposits (basal beds of the Puentellés Fm) that infill the palaeorelief. In the Gamonedo-Cabrales-Panes area, these mainly terrigenous deposits are conformably overlain by a 250–325 m thick calcareous unit, the Dorogomilovsky Puentellés Fm (see Fig. 7.8), which has been divided into two members (Villa & Bahamonde 2001). The lower member is

100–125 m thick and consists of cyclic alternations of calcareous breccias and conglomerates, pebbly sandstones, graded and laminated sandy grainstones and bioturbated marls. This member is interpreted as flood-dominated alluvial to shelfal lobes entering a restricted marine carbonate ramp. The upper member is 150–200 m thick and consists of dark pseudonodular mudstones with metre-scale intercalations of boundstones. It represents the gradual backstepping of the detrital lobes and the establishment of normal carbonate sedimentation (Villa & Bahamonde 2001).

The Puentellés Fm is gradationally overlain by the Gzelian Cavandi Fm (see Fig. 7.3a). This formation is at least 170 m thick and mainly comprises sandstone–mudstone alternations deposited from turbidity currents. In the middle part a 30 m thick carbonate megaturbidite, which represents the collapse of the Puentellés carbonate ramp, occurs, and in the upper part the succession contains some intervals of calcarenites, conglomerates and breccias. In the southern areas of the Picos de Europa province (Fig. 7.8) this sequence is represented by a unit several hundred metres thick (Lebeña Fm) that consists of a shaly succession with interleaved sandy and calcareous turbidites and olistoliths.

In the Pisuerga-Carrión province, the Stephanian B Remoña Gp is a 900 m thick succession that occurs in the northern part adjacent to the overthrusting Picos de Europa province (see Fig. 7.7c). This unit rests on an angular unconformity and mainly consists of shales with interleaved calcareous and quartzitic conglomerates, lithic sandstones, calcareous olistoliths and olistostromes.

The intermontane basin-fill successions on the Fold and Nappe province, Narcea antiform and West Asturian-Leonese Zone unconformably overlie basement units ranging in age from Carboniferous to Precambrian. These successions crop out in a series of synclines of variable size that are arranged more or less parallel to the Variscan structures (Fig. 7.10). It is uncertain if these isolated outcrops are the remnants of a smaller number of large basins or if, in some cases, they record individual basins. They are the El Bierzo coalfield (on the West Asturian-Leonese Zone), La Magdalena, Villablino, Rengos, Carballo, Cangas and Tineo coalfields (mainly on the Narcea antiform), and the Fontecha, Canseco-Salamón, Sabero and Ciñera-Matallana coalfields (on the Cantabrian Zone). The age of these successions, established by flora (e.g. Wagner 1971*a*), youngs westwards and evolves from Barruelian–Stephanian B (Sabero

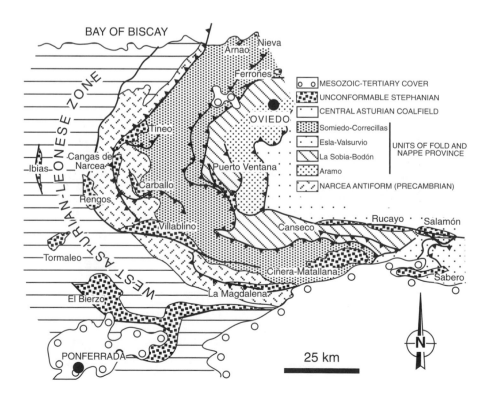

Fig. 7.10. Schematic geological map of part of the Cantabrian Zone showing the outcrops of the uncomformable Stephanian intermontane basin-fill successions.

coalfield), through Stephanian B (Canseco-Rucayo, Ciñera-Matallana and La Magdalena coalfields) to Stephanian B–C (Villablino, El Bierzo, Tineo, Cangas, Carballo and Rengos coalfields). These basins are thought to have formed during late Variscan tectonism, either in an extensional context (West Asturian-Leonese Zone) or linked with reactivated pre-existing thrusts and wrench faults (Cantabrian Zone). During this stage (Stephanian and Early Permian) important igneous activity occurred in the West Asturian-Leonese Zone and, to a lesser extent, in the Pisuerga-Carrión province. This magmatism resulted in numerous granite stocks, dykes and sills, and induced low pressure–high temperature metamorphic bands orientated parallel to tectonic structures (Fernández-Suárez 1994; Dallmeyer *et al.* 1997). This localized metamorphism increased coal rank or even produced natural cokes, and some tonstein marker beds are found in these successions (e.g. see Knight *et al.* 2000).

The continental intermontane basin-fill of each of these successions typically forms a large-scale (1500 to 3000 m thick) fining-upward sequence (Heward 1978*b*; Colmenero *et al.* 1996) that can be divided into three units. The lower unit fills palaeoreliefs cut into the basement and is made up of laterally discontinuous breccias, whose clasts derive from the immediately underlying succession, suggesting limited transport. The middle unit comprises quartzitic and polymictic conglomerates, interbedded with sandstones, minor shales and occasional coal seams. The upper unit is characterized by a thick cyclic alternation of lithic sandstones, shales and coal seams, and includes some intervals of fossiliferous lacustrine shales. In the smallest basins (e.g. Canseco-Salamón, Tineo, Cangas) only breccias and conglomerates are preserved, whereas in the larger basins (Sabero, Ciñera-Matallana, Villablino, El Bierzo) sandstones, shales and coal seams are predominant.

The successions deposited in these intermontane basins have been interpreted as recording alluvial fan-lake environments (e.g. Corrales 1971; Wagner 1971*b*; Heward 1978*a*; Colmenero *et al.* 1996), with occasional marine connections (Knight 1971). According to these models (see Fig. 7.11), four distinctive facies belts can be distinguished: proximal-fan coarse-grained debrites, mid-fan gravelly and sandy braided-stream deposits, distal-fan and floodplain deposits with coals, and lacustrine fossiliferous shales. The overall, composite fining-upward trend of these successions has been related to back-faulting at the basin margins and to the progressive denudation of the highlands (Heward 1978*b*).

Southern Iberian Massif

We include in the southern sector of the Iberian Massif all those Carboniferous outcrops located in Spain within a zone bounded by the Pedroches batholith in the north and the Tertiary cover of the Guadalquivir basin in the south. The Puertollano basin, located to the north of the Pedroches batholith, is also included. This area comprises the southern part of the Central Iberian Zone, as well as the Ossa Morena and South Portuguese zones (Figs 7.1 and 7.12).

The South Portuguese Zone and southern part of the Central Iberian Zone are characterized by a thin-skinned deformation style with stacked nappes and thrust sheets together with associated folds. The structures mostly diverge from a common décollement level and display a symmetric facing with respect to the Ossa Morena Zone in which a complex network of décollement faults occurs. Carboniferous rocks are abundant in the South Portuguese Zone, are less well represented in the southern part of the Central Iberian Zone, and occur only patchily in the Ossa Morena Zone (Fig. 7.12) where they are distributed along strike-slip faults. The successions tend to be lithologically monotonous and poor in fossils in the South Portuguese and southern Central Iberian zones, but are more varied and fossiliferous (although more deformed) in the Ossa Morena Zone. Overall, the southern part of the Iberian Massif displays quite a complete Lower Carboniferous record, whereas Upper Carboniferous rocks are poorly represented, although Stephanian successions are better represented, patchily distributed, and show relatively uniform stratigraphic and sedimentological features.

The chronostratigraphic chart in Figure 7.13 depicts the age and the main stratigraphic and sedimentological features of the Carboniferous successions in each of the above-mentioned zones. These data have been collected from several sources (Strauss 1970; Schemerhorn 1971; Lécolle 1977; Routhier *et al.* 1978; Pérez Lorente 1979; Oliveira *et. al.* 1979; Boogaard & Vázquez Guzmán 1981; Broutin 1981; Oliveira 1983, 1990; Andreis &Wagner 1983; Gabaldón *et al.* 1983; Quesada 1983; Simancas 1983; Robardet *et al.* 1986; Gabaldón 1989; Gutiérrez Marco *et al.* 1990; Quesada *et al.* 1990*b*; Rodríguez 1992, 1996; Giese *et al.* 1994; Sáez *et al.* 1996). The South Portuguese Zone is represented by the succession of the Iberian Pyrite Belt, whereas the southern part of the Central Iberian Zone is represented by the Pedroches and Guadiato successions (Fig.

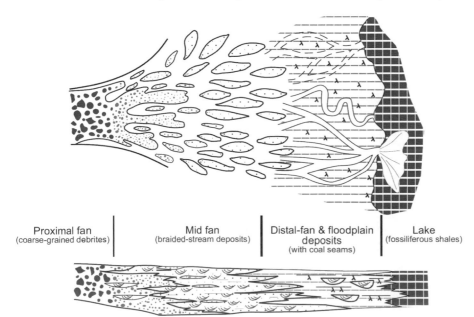

Proximal fan
(coarse-grained debrites)

Mid fan
(braided-stream deposits)

Distal-fan & floodplain deposits
(with coal seams)

Lake
(fossiliferous shales)

Fig. 7.11. Schematic alluvial fan-lake depositional model applied to the Stephanian coal-bearing successions, showing the inferred distribution of facies associations. Adapted from Colmenero *et al.* (1996).

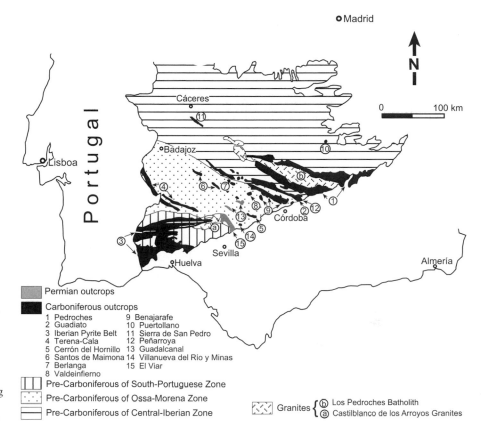

Fig. 7.12. Schematic geological map showing the Carboniferous outcrops in the southern part of the Iberian Massif (see also Fig. 7.1).

Legend:
- Permian outcrops
- Carboniferous outcrops
  1 Pedroches
  2 Guadiato
  3 Iberian Pyrite Belt
  4 Terena-Cala
  5 Cerrón del Hornillo
  6 Santos de Maimona
  7 Berlanga
  8 Valdeinfierno
  9 Benajarafe
  10 Puertollano
  11 Sierra de San Pedro
  12 Peñarroya
  13 Guadalcanal
  14 Villanueva del Río y Minas
  15 El Viar
- Pre-Carboniferous of South-Portuguese Zone
- Pre-Carboniferous of Ossa-Morena Zone
- Pre-Carboniferous of Central-Iberian Zone
- Granites { ⓑ Los Pedroches Batholith / ⓐ Castilblanco de los Arroyos Granites

7.13). The sedimentary record of the Ossa Morena Zone is more difficult to summarize due to a lack of detailed correlations between the different outcrops, which have classically been considered as individual basins. Six stratigraphic sections are presented for this zone and correspond to the Terena-Cala, Los Santos de Maimona, Valdeinfierno, Cerrón del Hornillo, Berlanga and Benajarafe successions. The chronostratigraphic chart in Figure 7.13 illustrates the following points.

(1) Carboniferous sedimentation was initiated in late Tournaisian times (except for the Iberian Pyrite Belt), became dominated by turbidites during late Visean times, and finally gave way to terrestrial conditions during Stephanian C times.

(2) The Lower Carboniferous sedimentary record is much better represented and is more varied than that of the Upper Carboniferous. Marine sediments are restricted to the Lower Carboniferous, whereas terrestrial sediments occur in the upper Lower Carboniferous and Upper Carboniferous successions.

(3) The relationships of the Carboniferous successions to their basement varies between areas. In the Iberian Pyrite Belt of the South Portuguese Zone, sedimentation was continuous across the Devonian–Carboniferous boundary. In the Ossa Morena Zone, the Carboniferous succession unconformably overlies basement rocks as old as Precambrian. Finally, in the southern part of the Central Iberian Zone, Carboniferous rocks overlie Devonian successions, both conformably and, in places, unconformably.

(4) The sedimentary successions of the Iberian Pyrite Belt and the southern part of the Central Iberian Zone are very similar. In contrast, the Ossa Morena Zone displays a more heterogeneous and irregular distribution of deposits. This reflects terrestrial and shallow marine sedimentation in the Ossa Morena Zone during Early Carboniferous times, in contrast to prevailing marine sedimentation in the other two zones.

(5) The timing of the volcanic activity differs between the Iberian Pyrite Belt and southern Central Iberian Zone (older than late Visean), on one side, and the Ossa Morena Zone (late Visean) on the other.

(6) There are no coal deposits older than Westphalian age, with the exception of the Lower Carboniferous coals in the Ossa Morena Zone.

During Early Carboniferous times, two large marine basins were located to the north (Pedroches basin, southern part of the Central Iberian Zone) and south (Iberian Pyrite Belt basin, South Portuguese Zone) of a mainly emergent landmass (Fig. 7.14) represented by the Ossa Morena Zone, and the northernmost domain of the South Portuguese Zone (Pulo do Lobo domain). During Late Carboniferous times, coal-bearing successions were deposited in isolated continental basins. These basin were mainly post-orogenic, although some, such as the Peñarroya basin, developed in a synorogenic context. As a result, there is a stratigraphic gap between Namurian and lower Stephanian deposits in most areas of the southern part of the Iberian Massif, except within the continental basins in the Ossa Morena Zone (Fig. 7.13).

Marine basins

The Lower Carboniferous successions of the two marine basins belonging to the South Portuguese Zone and the

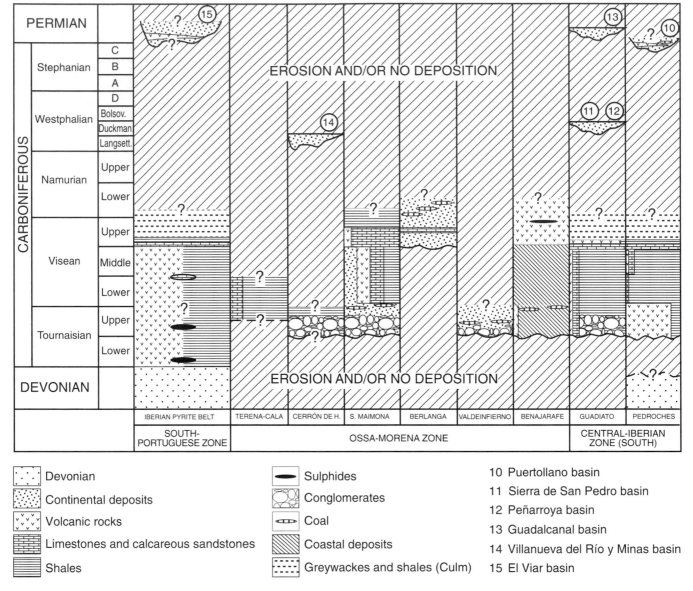

Fig. 7.13. Chronostratigraphic chart showing the distribution of the Carboniferous rock types and deposits in the southern part of the Iberian Massif. The width of the rock types in each column represents their relative abundance.

southern part of the Central Iberian Zone are quite similar. In both areas, two stratigraphic units can be distinguished: a lower one (Tournaisian–middle Visean) consisting of sedimentary and volcanic rocks, and an upper one (upper Visean to possibly lower Namurian) mainly formed of turbidites. In the Iberian Pyrite Belt these two units are known as the 'volcano-sedimentary complex' and Culm Gp respectively. In the southern part of the Central Iberian Zone, both units are collectively referred to as the Culm (Pérez Lorente 1979; Gabaldón *et al.* 1983; Quesada 1983; Gabaldón 1989). In the Pyrite Belt, the 'volcano-sedimentary complex' has been related to an extensional, crustal thinning stage (Munhá 1983; Simancas 1983; Sawkins 1990; Sáez *et al.* 1996). The overlying Culm Gp is thought to record the inversion stage of the basin as a response to the onset of the Variscan orogeny (Leistel *et al.* 1998a; Oliveira *et al.* 1979; Moreno 1987).

The southern marine basin: Iberian Pyrite Belt. The Iberian Pyrite Belt marine basin has some specific features that distinguish it from the basins of the Cantabrian Zone. The onset of Carboniferous sedimentation is characterized by bimodal subaerial and submarine volcanic activity, which is unevenly distributed in space and time. The volcanic and sedimentary rocks display sharp thickness and facies changes and alternate in a very complex way, not yet fully understood, to form the most characteristic stratigraphic unit of the Iberian Pyrite Belt. There is no established stratigraphic framework valid for the whole area, although three major stratigraphic units occur (Schermerhorn 1971). From base to top, these are the Devonian Phyllite-Quartzite Gp, the Tournaisian–middle Visean 'volcano-sedimentary complex' and the middle Visean–Namurian Culm Gp. Figure 7.15 displays a generalized lithological column that includes the subvolcanic rocks that cross-cut the succession and is valid for the entire area. The column is based on that from Sáez *et al.* (1996) and summarizes those from Strauss (1970) for the western part of the Spanish Iberian Pyrite Belt, and from Lecolle (1977) and Routhier *et al.* (1978) for the eastern and central part of the domain.

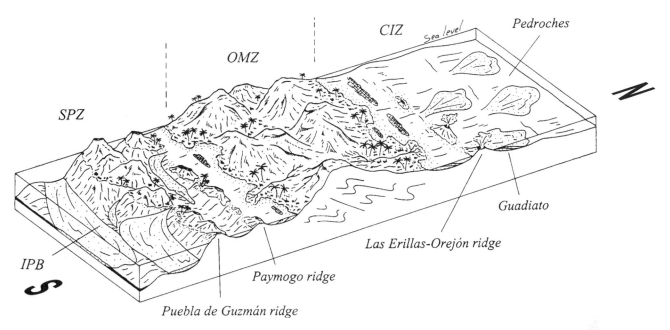

Fig. 7.14. Block diagram showing the idealized palaeogeography in the southern part of the Iberian Massif for the Early Carboniferous epoch (after Gabaldón *et al.* 1983; Moreno 1987, 1993). CIZ, Central Iberian Zone; IPB, Iberian Pyrite Belt; OMZ, Ossa-Morena Zone; SPZ, South Portuguese Zone.

At the Devonian–Carboniferous boundary, the first Variscan tectonism was extensional and compartmentalized the previously existing homogeneous Devonian shallow shelf (see Chapter 6) into a series of horsts and grabens (Moreno *et al.* 1996). These tectonic events were accompanied by intense magmatic activity that included catastrophic eruption of rhyolitic pyroclastic rocks. The fault-controlled sub-basins became filled or partially filled by volcanic and volcaniclastic rocks, creating an extremely complex palaeogeography (Moreno *et al.* 1995, 1996). The resulting 'volcano-sedimentary complex' is thus characterized by variable proportions of clastic sedimentary rocks and volcanics. The two end-members of the succession are sedimentary rocks virtually devoid of volcanic lithologies, and volcanic and subvolcanic rocks with minor siliciclastic sediments, in places showing peperitic textures recording wet sediment–magma interaction (Boulter 1993; Almodóvar *et al.* 1998; Donaire *et al.* 1998). However, the succession most commonly displays an intermediate character between both end-members typified by a mixture of siliciclastic sedimentary rocks, felsic volcaniclastic rocks and subvolcanic intrusions. Discontinuous intervals of siliceous rocks and limestones, and massive sulphide bodies also occur. The 'volcano-sedimentary complex' of the Iberian Pyrite Belt hosts some of the largest deposits of massive sulphides in the world (Chapter 19), including the famous Riotinto mining district which has been exploited for 5000 years (Leblanc *et al.* 2000). These deposits have been interpreted as Iberian-type sulphides (Sáez *et al.* 1999), a type of deposit that is intermediate between the exhalative-sedimentary (SEDEX) and volcano-sedimentary (VMS) types.

Five volcanic episodes interfingering with shales can be distinguished in the 'volcano-sedimentary complex' (Fig. 7.15), three formed of felsic rocks and two of mafic rocks (Routhier *et al.* 1978). Concretion-rich black shales occur at the base of the 'volcano-sedimentary complex' below the first felsic volcanic episode (Sáez & Moreno 1997). This initial volcanic interval consists mainly of shales with interleaved pumice-rich felsic tuffs derived from erosion of pyroclastic deposits. This succession forms one of the most important sulphide-bearing deposits in the Iberian Pyrite Belt. Commonly, these shaly horizons formed the conduits for magmas of the first mafic volcanic episode which produced large (up to 500 m thick and several tens of square kilometres in area) basaltic sills and sometimes pillow-lavas. The second felsic episode is mainly subvolcanic and consists of thin, peperitic sills intruded into terrigenous deposits (Boulter 1993). The second mafic episode, also represented by subvolcanic rocks,

was minor and has only a local distribution. It is capped by discontinuous jasper horizons, sometimes with manganese mineralizations and distinctive purple shales that form a regional marker horizon (IGME 1982). Overlying the purple shales, the third felsic episode forms the top of the 'volcano-sedimentary complex'. This latest episode is made up of volcano-detrital rocks, siliceous horizons and some pyroclastic intervals, all deposited in a coastal environment (Moreno & Sequeiros 1989). Gradually, these sediments pass into Posidonia- and goniatite-rich shales (the Basal Shaly Fm of Moreno 1987; Moreno & Sequeiros 1989) which represents the base of the Culm Gp.

Both biostratigraphic and radiometric data on the 'volcano-sedimentary complex' are scarce. Recently, palynomorphs collected from shales interleaved in the first felsic episode have been dated as belonging to the Devonian–Carboniferous boundary (the LN palynomorph biozone; Lake *et al.* 1988; Lake 1991; Pereira *et al.* 1996; Oliveira *et al.* 1979; González *et al.* 2000). U/Pb dating on zircons has confirmed this age (Nesbitt *et al.* 1999). The basal shales of the Culm Gp were dated as being of middle-late Visean age (Lake *et al.* 1988; Lake 1991; Oliveira & Wagner-Gentis 1983), so that a late Famennian–middle Visean age can be assigned to the 'volcano-sedimentary complex'.

The Culm Gp includes all the post-volcanic sedimentary rocks and consists of three units (Fig. 7.15): the Basal Shaly unit, the overlying Culm facies Turbidite Fm, which is in thrust contact with the overlying Shallow-Shelf Sandy Fm (Moreno 1987, 1993). The Basal Shaly Fm is a mixed volcanic and sedimentary succession, and includes volcaniclastic sediments of the upper part of the third felsic volcanic episode, lenticular limestone and calcareous sandstone horizons, and an overlying interval of Posidonia- and goniatite-rich shales. This unit represents the end of the volcanism in the Iberian Pyrite Belt, the reworking of the volcanic products in shallow marine environments and the onset of pelagic sedimentation during quiescent periods on ramp-type shelves (Moreno & Sequeiros 1989).

The Culm facies Turbidite Fm is a monotonous turbidite succession of shales, lithic sandstones and some conglomerate intervals, which forms a southward-thickening, coarsening-upward sequence up to several thousand metres thick (Moreno 1987, 1988, 1993; Moreno & Sáez 1989). Sandstone petrography, turbidite-facies distribution and palaeocurrent patterns suggest the existence of two coeval turbidite systems fed from two different sources: the Ossa Morena Zone, and the Iberian Pyrite Belt (Fig. 7.14). The larger of the two turbidite

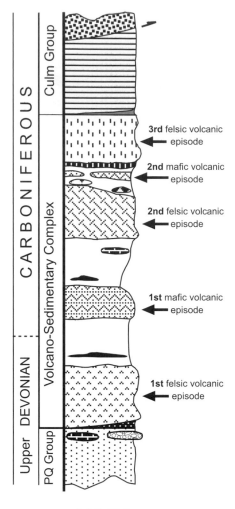

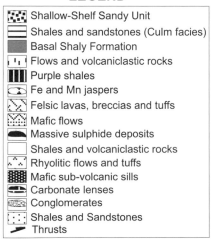

LEGEND

- Shallow-Shelf Sandy Unit
- Shales and sandstones (Culm facies)
- Basal Shaly Formation
- Flows and volcaniclastic rocks
- Purple shales
- Fe and Mn jaspers
- Felsic lavas, breccias and tuffs
- Mafic flows
- Massive sulphide deposits
- Shales and volcaniclastic rocks
- Rhyolitic flows and tuffs
- Mafic sub-volcanic sills
- Carbonate lenses
- Conglomerates
- Shales and Sandstones
- Thrusts

Fig. 7.15. Synthetic stratigraphic column of the Iberian Pyrite Belt. No vertical scale is shown due to the highly variable thickness of the stratigraphic units.

systems was sourced from the Iberian Pyrite Belt and displays a sediment dispersal pattern subparallel to the major WNW–ESE trending Variscan structures. The other, smaller turbidite system was fed from both source areas and ran NE–SW, parallel to pre-Variscan fractures and volcanic alignments (Rambaud Pérez 1969).

The Shallow-Shelf Sandy Fm forms a coarsening-upward sequence of alternating shales and quartzose sandstones. This unit was deposited in a coastal to shelf setting, with a sediment source derived from the cannibalization of an emergent 'volcano-sedimentary complex'. It is suggested that the shoreline was located to the north, although its exact position is unknown due to thrusting (Moreno 1993).

The northern marine basin: Pedroches basin. The northern marine basin includes the southern part of the Central Iberian Zone and the northernmost part of the Ossa Morena Zone (Fig. 7.14). The southernmost outcrops, the Benajarafe and Berlanga sections, display lagoonal and deltaic features respectively (Gabaldón *et al.* 1983) and consist of mudstones, sandstones, coal seams and carbonate rocks (Fig. 7.13). Volcanic and sedimentary rocks are also present in these successions and crop out along the major Variscan lineations and parallel to the palaeoshoreline. These rocks form two successive volcanic cycles: a lower mafic (La Campana) and an upper felsic (Erillas-Paredón) cycle (Apalategui *et al.* 1985). The sediments resulted from the erosion of the volcanic edifices and were deposited in wave-influenced, nearshore environments. Polymetallic sulphide deposits are also associated with these volcanic cycles, and are similar to those of the Iberian Pyrite belt, although of far less economic importance.

Northwards, in the Guadiato area, the succession consists of marine deposits, the lower part of which may correlate with the 'volcano-sedimentary complex' of the Iberian Pyrite Belt and consists of mudstones, sandstones, calcareous sandstones, minor limestones and volcanic rocks (Fig. 7.13). These rocks were deposited on a mixed, terrigenous and calcareous, shallow marine shelf. The upper part of the succession comprises Culm turbidites and reflects a deepening of the basin to the north. In the Guadiato area, storm-driven shallow water turbidity currents resulted in the deposition of turbidites with hummocky cross-stratification (Gabaldón 1989), whereas the Pedroches area succession represents a deeper water turbidite system (Fig. 7.14). This upper interval forms a shallowing- and coarsening-upward sequence which reflects the gradual filling of the basin as the depositional systems prograded northwards along with the migration of Variscan deformation.

Despite the similarities between the northern and southern marine basins of the southern part of the Iberian Massif, several important differences exist. Firstly, volcanism is less important in the northern basin where it is mainly mafic, in contrast to the dominantly felsic volcanism of the southern basin (Iberian Pyrite Belt). Secondly, limestones and calcareous sandstones are common in the northern marine basin, but rare in the southern basin. Thirdly, a wide diversity of coastal deposits in the northern basin suggests that it was shallower and bathymetrically more even than the southern basin. The southern basin appears to have been characterized by deep troughs bordered by narrow coastal belts (Fig. 7.14).

There are two additional isolated outcrops of Carboniferous marine rocks in the Ossa Morena Zone: the Terena-Cala and Los Santos de Maimona outcrops (see Fig. 7.12). Their exact palaeogeographical location is problematical, but given their present location within the Ossa Morena Zone, the Los Santos de Maimona outcrop was probably linked with the northern basin of Los Pedroches, whereas the Terena-Cala outcrop connected with the southern basin of the Pyrite Belt (Quesada 1983; Quesada *et al.* 1990*b*). In the western Terena-Cala outcrop, the Terena Fm (Ruiz-López *et al.* 1979; Oliveira 1983) is a *c.* 1000 m thick turbidite succession of shales with interbedded sandstones and conglomerates, which was deposited in a subsiding trough (Schermerhorn 1971). On top of the Terena Fm are calcareous sandstone lenses that contain upper Tournaisian–early Visean conodonts (Boogaard & Vázquez Guzmán 1981). In the eastern part, around Cala, the turbidite succession is absent and the calcareous sandstone lenses directly overlie a Silurian to basal Devonian succession (Ruiz-López *et al.* 1979). These lenses are similar to the calcareous sandstones of the southern basin (Iberian Pyrite Belt; see above), but they are thicker and represent shallower water environments interpreted as inner shelf bars, and possibly toe-of-slope deposits formed after destruction of organic build-ups by storm events (Boogaard & Vazquez Guzmán 1981).

The Terena Fm has classically been considered to record the onset of the main Variscan deformation phase in the southern part of the

Iberian Massif. Although this formation has been described both as Late Devonian (Perdigao *et al.* 1982) and Early Carboniferous (Pfefferkorn 1968; Schermerhorn 1971; Giese *et al.* 1994), recent discoveries of graptolites and palynomorphs suggest instead an Early Devonian age, at least for the base of the formation.

In the Los Santos de Maimona outcrop, the Early Carboniferous interval is represented by a 1500 m thick succession of both sedimentary and volcanic rocks deposited in fault-controlled continental and shallow marine basins (Rodríguez 1992, 1996; Giese *et al.* 1994). Conglomerates, sandstones, shales, limestones, coal seams, mafic lavas and pyroclastic felsic rocks are the prevailing deposits, and show an irregular distribution with sharp lateral facies changes.

Continental basins

During Early Carboniferous times there was widespread land across the southern Iberian Massif, the erosion of which shed a large volume of sediments into the adjacent marine basins. However, some fault-controlled, rapidly subsiding basins formed in these continental areas and became filled by thick piles of sediment over short periods of time. The only preserved example of these basins is the Tournaisian Valdeinfierno basin, which is located in the Ossa Morena Zone (see Figs 7.12 and 7.13). This pull-apart basin (Gabaldón & Quesada 1986) is characterized by thick marginal breccias and conglomerates, surrounding a central coal-bearing lacustrine setting. Upper Carboniferous (Westphalian) continental basin-fill successions are much more common in the area, and include those of the Villanueva del Río y Minas, Sierra de San Pedro and Peñarroya basins (see Figs 7.12 and 7.13). Their stratigraphical and sedimentological signatures suggest an alluvial-lacustrine setting similar to that of the Stephanian basins from the Cantabrian and West Asturian-Leonese zones (see above). The Westphalian coal-bearing successions have historically supported an important coal mining industry, which in the case of the Peñarroya basin is still active. The Westphalian fill of the Sierra de San Pedro and Villanueva del Río y Minas basins unconformably overlies a pre-Carboniferous substratum, whereas the Peñarroya succession (Andreis & Wagner 1983) is thrust over Lower Carboniferous marine strata of the Guadiato basin. The geodynamic interpretation of these syn- to late-orogenic continental basins is still under discussion. Wagner (1999) envisages the Peñarroya basin as a strike-slip basin unrelated to the earlier marine basin stage, whereas Martínez-Poyatos *et al.* (1998) interpret it as a late stage piggy-back basin, formed as thrust emplacement advanced to the north.

The youngest Carboniferous successions occur in the Puertollano (Central Iberian Zone) and El Viar (South Portuguese Zone) basins (see Figs 7.12 and 7.13). A Stephanian–Autunian age was assigned to the Puertollano basin infill, whereas the El Viar basin is predominantly Permian in age, although the onset of sedimentation in this basin possibly took place at the end of Stephanian times (Broutin 1981). Both are post-orogenic intermontane basins related with a late Variscan activity and their successions include volcanic rocks. The detailed stratigraphy of both basins has been described by several authors (Broutin 1981; Simancas 1983; Wallis 1983; Gabaldón & Quesada 1986; Sierra & Moreno 1997, 1998; Sierra *et al.* 1999, 2000).

Iberian Ranges

The Iberian Ranges represent the eastern continuation of the West Asturian-Leonese Zone, and Palaeozoic rocks are isolated within Mesozoic and Cenozoic strata. They include some small Carboniferous outcrops that are located in two main areas: the Sierra de la Demanda and the Montalbán massif (Fig. 7.16).

In the Sierra de la Demanda, an old coal mining district, a 600 m thick Carboniferous succession (Duckmantian to lower Stephanian; Colchen 1971, 1974) unconformably overlies Cambro-Ordovician rocks (Fig. 7.16b). It consists of five fining-upward megasequences, which in turn can be grouped into two major lithostratigraphic units (Villena & Pardo 1983). The lower lithostratigraphic unit comprises three conglomerate intervals separated by sandstone–mudstone alternations, which locally contain coal seams and flora-rich beds towards the top. The upper unit consists of thinly bedded sandstones and fossiliferous marine mudstones, including some dolomitic lenses in the upper part.

The Montalbán Carboniferous outcrop is a small (80 km^2), NW–SE oriented area, located in NE Teruel province (Fig. 7.16c). This Carboniferous succession, up to 1000 m thick, unconformably overlies Devonian strata and was dated as Visean to Namurian/lower Westphalian (Sacher 1966; Berger *et al.* 1968) based on analysis of the flora, or as Namurian to Westphalian (Langsettian) using goniatite stratigraphy (Quarch 1975). More recently, Meléndez *et al.* (1983) suggested a possible Stephanian age for the uppermost part of the succession, grading up into Permian strata. Several units have been distinguished in this succession (Quarch 1975). The lower unit, or Segura Fm, is a 200–250 m thick, sandstone-dominated interval, interpreted as deposited on a wave-dominated shelf. The middle unit (La Hoz Fm), is a shale succession several hundred metres thick with some interbedded hemipelagic limestones and thin sandstone beds (Villena & Pardo 1983, and references therein). These sediments represent a distal submarine fan, which prograded towards the north and NW in a deep marine basin generated by faulting of the previous shelf. The upper part of the Carboniferous succession consists of the Armillas, Peñarroyas, Montalbán and Torre formations. These formations comprise sandstones with horizons of breccias and olistoliths which might represent proximal submarine fan, slope and prograding shelf facies associations (Fig. 7.16c; Villena & Pardo 1983).

Close to the north, in the Puig Moreno area, three small (<1 km^2) outcrops of Carboniferous rocks record a 250 m thick turbidite succession with both siliciclastic and calcareous beds, which is quite similar to the La Hoz Fm of the Montalbán massif (Villena & Pardo 1983). This succession has been classically considered as Namurian in age (Gross 1966), although Villa *et al.* (1996) deduced an upper Kasimovian age on the basis of fusulinids. This latest age date contradicts all previous data for the Carboniferous successions of the Iberian Ranges.

Finally, in the core of an eroded and faulted anticline near Henarejos village, Cuenca province (Meléndez *et al.* 1983), another small (<1 km^2) outcrop of Carboniferous strata unconformably overlies Silurian shales (see Fig. 7.16a). This outcrop, dated as Stephanian B (Fonollá *et al.* 1972) or C (Wagner *et al.* 1985), consists of a 200 m thick coal-bearing succession of conglomerates, sandstones and shales deposited in

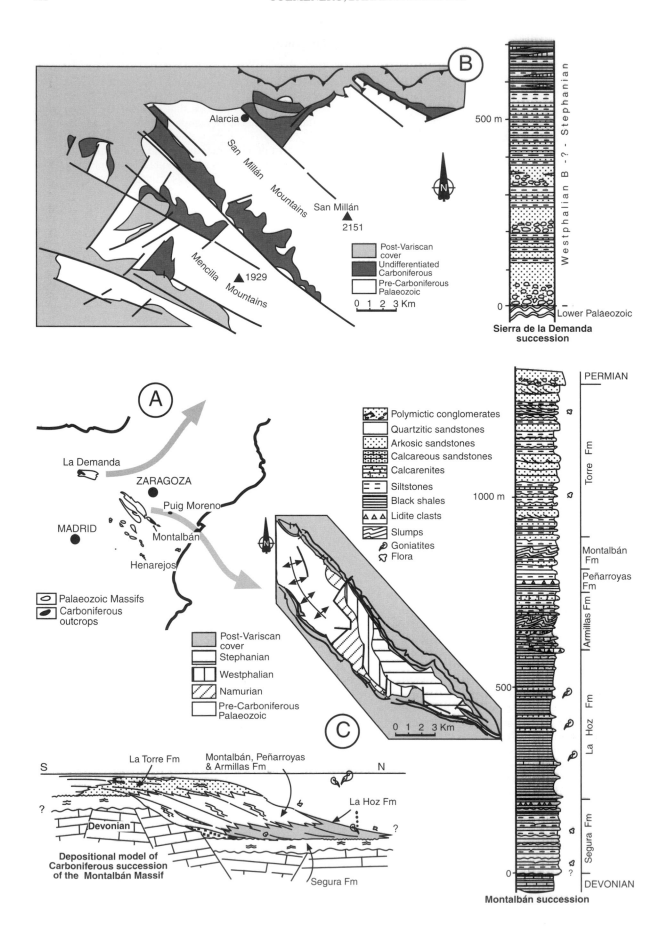

Sierra de la Demanda succession

Montalbán succession

Polymictic conglomerates
Quartzitic sandstones
Arkosic sandstones
Calcareous sandstones
Calcarenites
Siltstones
Black shales
Lidite clasts
Slumps
Goniatites
Flora

Post-Variscan cover
Stephanian
Westphalian
Namurian
Pre-Carboniferous Palaeozoic

Post-Variscan cover
Undifferentiated Carboniferous
Pre-Carboniferous Palaeozoic

Depositional model of Carboniferous succession of the Montalbán Massif

Fig. 7.17. Sketch map of the northeastern Iberian Peninsula showing the distribution of Carboniferous outcrops in the Palaeozoic massifs of both the Pyrenees (axial zone and Basque massifs) and the Catalonian Coastal Ranges. The Mouthoumet and Montagne Noire massifs are also included (see text for details). Reproduced by permission of the BRGM (France) and IGME (Spain).

an intermontane basin. Fining-upward cyclothems tens of metres thick, with conglomerates or sandstones at the base and capped by carbonaceous shales or a coal seam, characterize the succession. The presence of a 1.6 m thick coal seam has enabled coal mining since the nineteenth century.

Pyrenees

In the Pyrenees, Carboniferous rocks occur in the axial zone and (further west) in the Basque massifs, areas that are separated from each other by Mesozoic and Tertiary rocks (Fig. 7.17). Both areas are overprinted by intense Alpine deformation and their original mutual relationships remain unknown. As a consequence, the Palaeozoic successions of the Pyrenees have not yet been placed within a coherent palaeogeographic scheme. However, the polarity of the Variscan structures and the timing of onset of the synorogenic turbidite sedimentation suggest that most of the outcrops belong to the northern branch of the Ibero-Armorican arc, so that the foreland-propagating Variscan orogeny advanced southwards across the area. Based on the Palaeozoic stratigraphy, the axial zone can be divided into the eastern, central and western Pyrenees, and the Basque massifs divided into an eastern and a western domain, the latter forming the Cinco Villas massif (Fig. 7.17).

Pre-, syn- and late-orogenic successions are recognized in the Pyrenean Carboniferous (see Waterlot 1983; Carreras & Debat 1996; Delvolvé 1995; Lucas & Gisbert-Aguilar 1995; Majesté-Menjoulas & Ríos 1995, and references therein). The pre-orogenic succession consists of an interval of limestones, cherts and shales (Tournaisian–Visean), a few tens of metres thick, locally overlain by thinly bedded, dark grey limestones

with shale partings (mainly Namurian A). This succession displays a remarkable lateral uniformity of facies and has been interpreted as deposited during the inversion of the basin following Devonian extension (Majesté-Menjoulas and Ríos 1995). The conformably overlying synorogenic succession is formed of a turbidite interval (the so-called Culm) which is Namurian to Westphalian in age towards the west and SW but is as old as Visean in the Montagne Noire massif, to the NE of the Pyrenees (Fig. 7.18). Finally, the late-orogenic succession occurs as scattered Permo-Stephanian outcrops comprising terrestrial, coal-bearing deposits and volcanic rocks that rest unconformably on older rocks. Although the Pyrenean Carboniferous successions are rather incomplete, they are broadly similar to those of the Cantabrian Zone. The so-called pre- and synorogenic successions parallel their counterparts in the Cantabrian Zone where condensed carbonate and shale deposits (Tournaisian–Visean) are overlain either by Namurian limestones (Barcaliente and Valdeteja formations) or by diachronous terrigenous turbidites.

Pre-orogenic succession. In the axial zone, the pre-orogenic succession varies and bears different relationships to underlying Devonian rocks from east to west. In the eastern and central Pyrenees, Carboniferous rocks conformably overlie Upper Devonian (Fammenian) deposits. These youngest Devonian rocks generally consist of lower Famennian pink to reddish, nodular limestones (griotte limestones) overlain by upper Famennian grey to pinkish, nodular limestones (supra-griotte limestones). The base of the Carboniferous succession is placed in the upper part of the supra-griotte limestones and, in some localities, is marked by a thin interval of shales. The next unit, which may directly overlie the Devonian portion of the supra-griotte limestone unit, is formed of black cherts with black shale intercalations and phosphatic nodules (lower cherts). The lower cherts are generally overlain by grey, sometimes nodular, limestones (intervening limestones), above which occurs a unit of grey or green cherts

Fig. 7.16. (a) Palaeozoic massifs and Carboniferous outcrops in the Iberian Ranges. Schematic geological maps showing the distribution of the Carboniferous outcrops in the Sierra de la Demanda (b) and Montalbán (c) massifs, their respective general stratigraphic successions, and a depositional model for the Montalbán Massif. Based on Colchen (1971, 1974), Quarch (1975), Meléndez *et al.* (1983) and Villena & Pardo (1983).

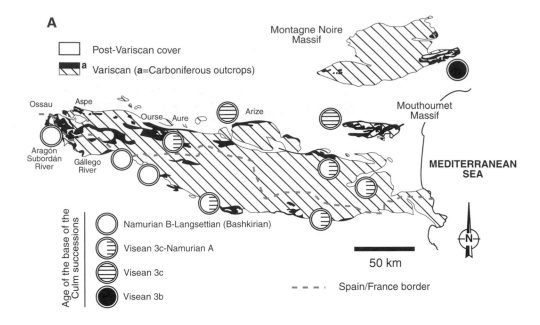

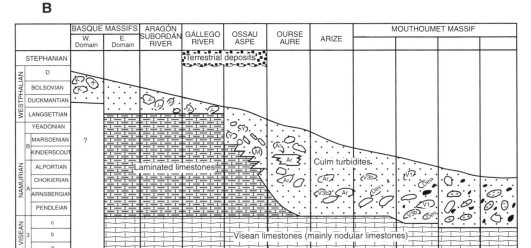

Fig. 7.18. (a) Simplified sketch map of the Pyrenean axial zone and the Montagne Noire and Mouthoumet massifs (see also Fig. 7.17) showing the southwestward younging of the synorogenic turbidite sedimentation. **(b)** Chronostratigraphic chart showing the diachroneity of the synorogenic turbidite succession (Culm) from the Mouthoumet massif to the Cinco Villas massif (western domain of the Basque massifs). Simplified from Carreras & Debat (1996) and Delvolvé (1995), respectively. Reproduced by permission of the BRGM (France) and IGME (Spain).

commonly containing pyroclastic layers (upper cherts). Finally, the top of the pre-orogenic succession is formed of grey nodular limestones (post-chert limestones), which, either directly or by means of an intervening interval of green and reddish shales, are gradationally overlain by the synorogenic turbidite succession (Culm). Biostratigraphical dating, mainly based on conodonts, places the Tournaisian–Visean boundary in the upper part or at the top of the intervening limestones while the post-chert limestones are late Visean in age.

In the western Pyrenees, the pre-orogenic succession more commonly shows an angular unconformity with the underlying Devonian succession, which can be as old as Early Devonian (e.g. García Sansegundo 1992). Also, the Tournaisian–Visean succession is less complete, lacking lower and even upper Tournaisian deposits, and is more laterally variable than in the eastern and central Pyrenees. The pre-orogenic succession consists of various types of limestones (grey or pinkish nodular limestones, grey fine-grained laminated limestones and bioclastic, sometimes sandy, limestones) and chert units, whose mutual relationships have not yet been deciphered.

In most of the western Pyrenees outcrops, the Visean limestones are not overlain by the turbidites of the synorogenic succession, but by a distinctive unit mainly composed of dark grey, thinly bedded or massive, laminated limestones with shale interbeds, and locally with interleaved breccia beds. The clasts of these breccias are of the host limestones and show little or no signs of transport. This unit is in turn gradationally overlain by the turbidites of the synorogenic succession. This distinctive unit is generally Namurian A in age, although its base can be upper Visean and, in some localities, its top is dated as Namurian B or even Namurian C.

In the Basque massifs, there is generally a continuous succession from Devonian to Carboniferous as seen in the eastern and central Pyrenees, only some localities in the western domain showing a basal unconformity (see below). In the eastern domain, the Tournaisian record comprises the upper part of the supra-griotte limestones overlain by dark grey and black cherts. Visean dark grey limestones grade up into dark shales which can be reddish in the lower part. The top of the pre-orogenic succession is formed of fine-grained, dark limestones (Namurian A), or of nodular limestones alternating with variegated shales (Upper Visean-Namurian A), which are overlain by grey, thinly bedded laminated limestones, with breccia beds in the upper part (Namurian B).

In the western domain, the supra-griotte limestones are replaced by dark grey, thinly bedded limestones. Overlying these limestones, the remainder of the Tournaisian and Visean record consists of greenish shales overlain by black cherts above which Culm turbidites occur. At some localities in the western domain the pre-orogenic succession is absent and the synorogenic succession directly overlies the basement.

Synorogenic succession. The synorogenic succession is formed of mainly terrigenous turbidite deposits which are up to around 1000 m thick and whose top is not preserved. These deposits are generally informally referred to as 'the Culm', although local formations have been defined. In general, the Culm consists of interbedded sandstones and dark shales. In some localities, thin beds of dark, fine-grained, laminated limestones occur interleaved with the shales. These limestones are similar to the Namurian limestones of the pre-orogenic succession and have been interpreted as mainly hemipelagites. Subordinate, locally conspicuous, conglomerates and carbonate breccias and olistoliths occur at several different levels, but are most common in the upper part of the successions where they lie within a generally muddier succession. White quartz and black chert are the most abundant clasts present in all conglomeratic layers, although igneous and metamorphic clasts can be very common at some localities, whereas limestone clasts dominate in others. The limestone clasts include Silurian and Devonian lithologies but are most commonly Carboniferous in age (Fig. 7.18b).

The age of the Culm has been mainly determined by dating the limestone clasts of breccias and olistoliths. The succession youngs towards the SW, with the oldest strata being upper Visean–Namurian A, and the youngest, which occur in the western domain of the Basque Massifs, being lower Moscovian (Kashirsky) in age (Fig. 7.18b).

The synorogenic succession of the Pyrenees has been interpreted (Schulze 1982; Delvolvé 1987) as slope, canyon and submarine-fan sediments deposited in a southwestward-migrating foredeep. This foredeep would have been bound by the Variscan orogen to the NNE and by a carbonate platform margin in the SSW. Carbonate platform development is considered to have been coeval with turbidite sedimentation in the foredeep (Fig. 7.18b). Sediment dispersal was from the north, but once flows entered the foredeep they deflected along the basin axis, flowing either towards the west or the east.

Late-orogenic succession. Late-orogenic, Permo-Stephanian successions occur in scattered outcrops, resting with angular unconformity on older rock, and only slightly affected by the Variscan orogeny. Three regional lithostratigraphic units, bounded by unconformable to conformable surfaces, have been distinguished on the basis of colour, forming grey (basal), transitional, and red (top) units. The two oldest units consist mainly of coal-bearing deposits with a volcanic component and anthracitic coals have been locally exploited at some of these outcrops (e.g. Campo de la Troya). The red unit is entirely Permian in age and thus will not be dealt with here.

The grey unit (Stephanian B) consists of sedimentary and acid volcanic rocks, and forms a fining-upward sequence from basal breccias and conglomerates to sandstones and shales with some coal seams. Andesitic volcanic rocks also occur, and in some cases almost form the entire grey unit. The transitional unit (Stephanian C–Autunian) paraconformably to conformably overlies the grey unit and in turn is unconformably overlain by the Permian red unit. The transitional unit consists of a single fining-upward sequence of conglomerates, sandstones, mudstones and coal seams that evolves upwards into mudstones with interleaved tuff beds. The top of this unit commonly contains lacustrine limestones with stromatolites, charophytes and ostracods. In some localities, acid to intermediate volcanic rocks, including rhyodacitic lavas and pyroclastic deposits, can be found interleaved with the sediments, and can in places even constitute the entire unit.

These Permo-Stephanian successions were deposited in alluvial and lacustrine environments within transtensional basins that were bordered by volcanic cones. Small alluvial fans in the lower part of the grey unit, and ephemeral streams in the transitional unit, account for the coarser clastic sedimentation. During deposition of the grey and transitional units the warm, humid climate probably became progressively more arid (Lucas & Gisbert-Aguilar 1995).

Catalonian Coastal Ranges

In the Catalonian Coastal Ranges Carboniferous rocks form a small number of isolated outcrops, the most important of which is located in the southern part (El Priorato region, Tarragona province; Fig. 7.17). These Variscan inliers seem to have had the same palaeogeographic position within the Variscan orogen as the Pyrenean massifs, and belonged to the northern branch of the Ibero-Armorican arc (Carreras & Debat 1996). These successions are comparable to those of the eastern Pyrenees (see above) and two intervals, comparable to the Pyrenean pre- and synorogenic successions, can be distinguished (Anadón et al. 1983b).

Although the pre-orogenic succession conformably overlies Devonian rocks in the El Priorato section, an unconformity with Silurian to Devonian rocks exists elsewhere. The succession shows a thin, condensed interval with two units, each just 10–30 m thick. The lower unit (probably Tournaisian) is generally formed of black cherts with phosphate nodules. The upper unit (uppermost Tournaisian–lower Visean) consists of nodular limestones that laterally pass into green and purple shales with some interleaved, thin limestone beds. Either unit can be absent at some localities, and this is interpreted as due to later erosion.

The second interval represents a Culm turbidite succession and is best preserved in El Priorato, where it reaches a maximum thickness of 2000 m. It consists mainly of sandstone and mudstone with interleaved, subordinate conglomerates. In El Priorato, the base of this second interval is formed of megabreccia beds containing clasts of black cherts and nodular limestones. Limestone horizons also occur towards the top of this Culm succession and have been dated as middle-upper Visean to Namurian in age, although it is also suggested that the upper part could be Westphalian (Anadón et al. 1983b). Several turbidite systems fed from the NW have been identified, whereas the basal megabreccias show a south or SE provenance (Maestro-Maideu et al. 1998).

Minorca

The Carboniferous succession is mainly exposed in the north-eastern part of Minorca (Fig. 7.19) and represents a small remnant of sediments once deposited across a wide basin. The succession is turbiditic, and unconformably overlies Upper Devonian rocks that are also turbidites. Rosell & Elízaga (1989) and Arribas et al. (1990) have separated two unconformable units based on facies associations. The lower unit (Tournaisian–Lower Visean) is a 1000 to 2000 m thick succession, made up of shales interbedded with lithic sandstones and interpreted as channel and overbank turbidites. This unit also contains interleaved calcareous turbidites and an olistostromic body (megaturbidite) containing fragments of Upper Devonian–Lower Carboniferous condensed successions (radiolarites, red limestones and dark shales) as well as olistoliths of basic and acid volcanic rocks. The upper unit (upper Visean–Namurian Culm facies) is a several thousand metre thick turbidite succession of thick, coarse-grained channel deposits interleaved with thin-bedded, fine-grained overbank deposits. Post-Namurian basic volcanism has been described in Minorca by Bourrouilt (1973). These Carboniferous Minorca turbidites have a northern provenance (Henningsen 1982), and thus may be comparable with those from the Pyrenees and Catalonian Coastal Ranges (see above).

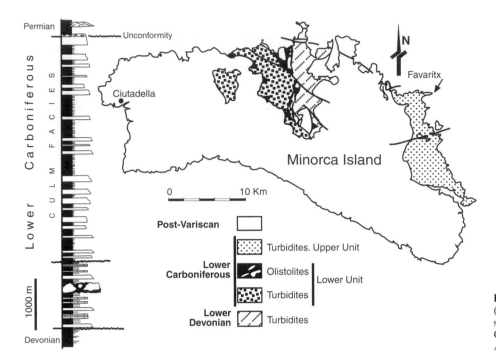

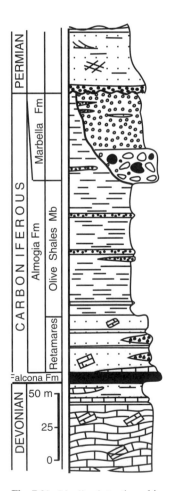

Fig. 7.19. Palaeozoic outcrops in Minorca (modified from Rosell & Elízaga 1989) and schematic stratigraphic section of the Carboniferous rocks (modified from Arribas *et al.* 1990).

Betic Cordillera

A poorly exposed and intensely deformed Palaeozoic succession crops out in the internal areas of the Betic Cordillera. Only in the Maláguide unit does a relatively low metamorphic overprint allow recognition of original sedimentary textures, although it is still difficult to identify the Devonian–Carboniferous boundary. In general terms, the Carboniferous succession has a turbiditic character, comprises shales with sandstone interbeds and rare marine fossils, and was deposited on basin floor and continental slope areas (Herbig 1983). Three stratigraphic units have been differentiated in the Carboniferous succession (Fig. 7.20). The lowest of these (Tournasian Falcona Fm) conformably overlies Devonian limestones and consists of a 5–10 m thick interval of black cherts capped by a thin interval of pseudonodular pelagic limestones. The middle unit is the Visean–lower Bashkirian Almogia Fm (Mon 1971). This middle unit is divided into a 40–60 m thick lower part (Retamares Mb; Kockel & Stoppel 1963) comprising lithic sandstones with some polymictic conglomerate beds interleaved in the basal part, and a 150 m thick upper part of green shales with scattered greywackes (Olive Shales Mb; Blumenthal 1949). Finally, the upper unit (Bashkirian Marbella Fm; Blumenthal 1949) consists of amalgamated, calcareous and polymictic conglomerate debrites eroding the top of the Olive Shales Mb and passing upwards into lithic sandstones. The Carboniferous succession is unconformably overlain by red continental Permian deposits (Geel 1973). During Late Devonian to Visean times, doleritic (and sometime aplitic) sills and dykes intrude the Palaeozoic strata (Mollat 1968; Buntfuss 1970; Herbig 1983).

R. Sáez and S. Sierra are acknowledged for their invaluable help. Referees S. Davies and A. Pérez-Estaun, and the editors, are thanked for helping to improve the chapter. Thanks also go to C. Quesada and F. Simancas for the information provided. I. Carmena (IGME, León) is thanked for the computer drafting of figures. Financial support has been provided by the Search Project PB98–0960 from the PAI (RNM 0173 Gp) and the PPI of the Universidad de Huelva, and by the Search Project MCT-00-BTE-0580 from the DGICYT.

Fig. 7.20. Idealized stratigraphic section of the Carboniferous succession in the Maláguide Unit of the Betic Cordillera showing the main lithostratigraphic units (modified from Herbig 1983).

8 Palaeozoic Magmatism

ANTONIO CASTRO, L. GUILLERMO CORRETGÉ, JESÚS DE LA ROSA,
PERE ENRIQUE, FRANCISCO J. MARTÍNEZ, EMILIO PASCUAL, MARCELIANO LAGO,
ENRIQUE ARRANZ, CARLOS GALÉ, CARLOS FERNÁNDEZ, TEODOSIO DONAIRE
& SUSANA LÓPEZ

Most Palaeozoic magmatic rocks in Spain were produced during the Variscan orogeny, and there are excellent and abundant examples of both volcanic and plutonic lithologies. Volcanic units include those in the world-famous Iberian Pyrite Belt, and plutonic rocks exposed in the Iberian Massif include some of the largest and best granite outcrops in the European Variscides. Magmatic rocks are present in all the Iberian tectonostratigraphic zones into which the Variscan orogen in Spain has been classically divided. In addition to these Variscan igneous rocks, there is also evidence for earlier magmatism, including widespread exposures of Neoproterozoic–Cambrian (Cadomian) age, and the diatreme-like breccias linked to the origin of the remarkable mercury mineralization at Almaden.

In this chapter we deal initially with Palaeozoic volcanic rocks, with special emphasis on the volcanism related to the generation of the Iberian Pyrite Belt. With regard to the Variscan granitoid rocks we have grouped these according to compositional features and relative age, rather than by tectonostratigraphic zones (the latter approach does not contribute to a better understanding of the magmatism because the emphasis is on differences and not on similarities). However, Variscan granites of the Iberian Massif are described separately from other granitic massifs in the Pyrenees and Catalonian Coastal ranges, because of their geographic separation and the lack of obvious direct links between them.

Outcrops of distinctive mafic and ultramafic rocks, mostly related to granitoids of the appinite–granodiorite association (cf. Pitcher 1997), are treated separately, not least because of their importance in international debates over the links between this mantle-derived magmatism and the generation of large granite batholiths. The close connection between Variscan metamorphism and magmatism also justifies separate analysis, as do the petrology and spatial distribution of latest-Variscan dyke swarms. Together with the Permian volcanics of the Viar basin and other similar Stephanian–Permian basins in the Pyrenees, these dykes represent the last Palaeozoic magmatic event in Spain. Finally, the chapter ends with an overview of the contribution made by studies of Spanish granitoid rocks to our understanding of granite petrogenesis.

Volcanic rocks (TD, EP, AC, ML, EA, CG)

Pre-Mesozoic volcanic rocks in the Iberian Massif show very different ages, styles and tectonic settings (Fig. 8.1), with the oldest rocks being erupted in latest Neoproterozoic times. The Palaeozoic–Neoproterozoic transition is not a major boundary in terms of Iberian magmatism because Precambrian calc-alkaline volcanism overlaps with that of Early Cambrian age, with little or no petrographic or geochemical difference. These ancient rocks, which are mostly exposed in the Ossa Morena Zone, the Pyrenees, the Catalonian Coastal Ranges and the West Asturian-Leonese Zone, record different stages of ongoing Cadomian subduction (Eguiluz et al. 2000). Upper Cambrian and younger volcanic rocks can be subdivided into post Cadomian/pre-Variscan (Upper Cambrian–Lower Devonian), and Variscan (Upper Devonian to Permian) groups. The former group comprises small volumes of very diverse volcanic products that include Cambrian alkaline rocks in the Ossa Morena Zone, Cambrian to Silurian rocks in the Cantabrian Zone, plus a range of Palaeozoic volcanic lithologies in the Central Iberian and West Asturian-Leonese zones, the Iberian Ranges and the Pyrenees. With regard to the Variscan volcanic rocks, these are found in the Central Iberian, Ossa Morena and South Portuguese zones. Finally, late orogenic (mostly Permian) lavas and volcaniclastic successions are preserved in the South Portuguese and Cantabrian zones, the Iberian Ranges and the Pyrenees.

Palaeozoic volcanic rocks show very different degrees of textural, structural and chemical preservation, due to hydrothermal alteration, deformation and/or low-grade metamorphism. Wherever possible, we use the classification by McPhie et al. (1993) for volcanic facies descriptions, but the reader should be aware of the many uncertainties that still remain, not only due to alteration, but also to the use of genetic terms. We have included in this chapter a brief description of the rocks commonly grouped as Ollo de Sapo Formation (Fm) that form the pre-Tremadocian basement exposed across large areas of the central and northern Iberian Massif.

Cadomian volcanism (TD, EP, ML, EA, CG)

Abundant volcanic and volcaniclastic rocks occur interbedded as thick successions exposed on both limbs of the Monasterio antiform of the Ossa Morena Zone (Chapters 2 and 10). The successions show some differences between each of the antiformal limbs, so they have been described as the Malcocinado Fm to the north and the Bodonal–Cala complex to the south (Fig. 8.2). Both successions lie unconformably on Precambrian rocks belonging to the Serie Negra Group (Gp), and are themselves unconformably overlain by slightly deformed, Lower Cambrian terrigenous and carbonate sequences (Eguiluz et al. 2000). Volcanic rocks in the Malcocinado Fm include both coherent rocks, described as flows, and volcaniclastic rocks ('tuffs') interbedded with clastic sediments that include abundant conglomerate horizons containing pebbles of volcanic and basement rocks. Chemical composition is mostly rhyolitic in the westernmost part of the formation, but becomes progressively more andesitic to the SE (e.g. San Jerónimo Fm). Sediments with reworked volcanic material are abundant

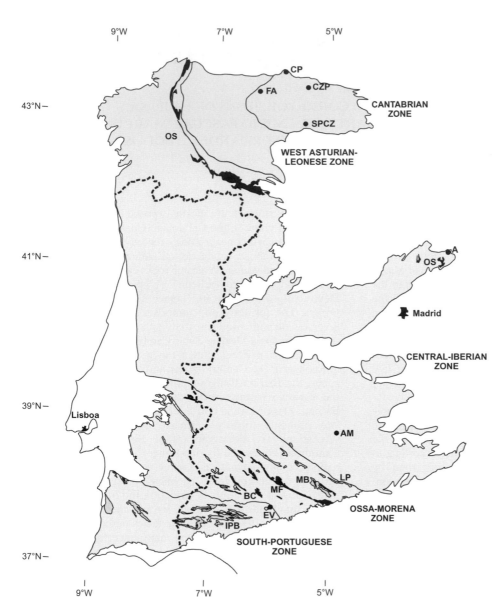

Fig. 8.1. Location of the main outcrops of volcanic rocks in the Iberian Massif. Key: A, Atienza area Permian volcanic rocks; AM, Almadén district; BC, Bodonal–Cala complex; CP, Cabo Peñas area pre-Permian tholeiitic association; CZP, Permian volcanic rocks from the Cantabrian Zone; EV, El Villar basin; FA, Pre-Permian alkaline association of the Faradón area; IPB, Iberian Pyrite Belt; LP, Los Pedroches basin; MB, Matachel-Benajarafe basin; MF, Malcocinado Fm; OS, Ollo de Sapo domain; SPCZ, southern part of the Cantabrian Zone pre-Permian alkaline association.

close to the top of the formation and, in places, Lower Cambrian arkosic sandstones (Torreárboles Fm) occur between the Malcocinado Fm and overlying younger Cambrian rocks (Eguiluz *et al.* 2000). The Bodonal–Cala complex consists of volcanic and volcaniclastic rocks (ash layers, ignimbrites and related epiclastic arkose) of mainly rhyolitic to dacitic composition. Crystal-rich tuffs are more abundant close to the base of the complex, whereas finer-grained volcaniclastic rocks predominate towards the top, above which there is a gradual transition to Lower Cambrian sediments (Cañuelo limestone; Eguiluz *et al.* 2000).

Radiometric age constraints on these rocks remain poor. U-Pb zircon dates range from *c.* 522 Ma BP for the Malcocinado Fm to *c.* 514 Ma BP for the Bodonal–Cala complex (Ordóñez *et al.* 1998). Palaeontological dating of the San Jerónimo Fm yields significantly older ages, as this formation contains middle to upper Vendian fossils (Quesada *et al.* 1990*a*). Given that the San Jerónimo rocks represent only the top of a thick volcanosedimentary pile in the Córdoba area, this could indicate that the formation is unrelated to the Cambrian units.

However, their common calc-alkaline geochemical affinity suggests that all of the units considered were formed during different stages of Cadomian arc evolution. The Malcocinado Fm may correspond to an earlier or juvenile stage because it includes tholeiitic rocks, whereas the Bodonal–Cala complex may represent more evolved arc materials as it includes K-rich or even shoshonitic rocks (Sánchez-Carretero *et al.* 1989*a*, 1990; Eguiluz *et al.* 2000). In the eastern part of the unit, the San Jerónimo Fm has been interpreted in terms of an active margin environment located on relatively juvenile crust.

In addition to the Cadomian rocks exposed in the Ossa Morena Zone, we include here some porphyroid and orthogneiss units found in the Cudillero and Tineo areas within the Narcea antiform of the West Asturian-Leonese Zone. These rocks are thought to derive from interlayered volcanics and tuffs within the slates and turbiditic greywackes of the Narcea Precambrian sequence. Volcanic facies also include sills, as well as extrusions or shallow intrusions. Rhyolites and dacites predominate, but andesitic and gabbroic rocks have also been recognized (Corretgé *et al.* 1990).

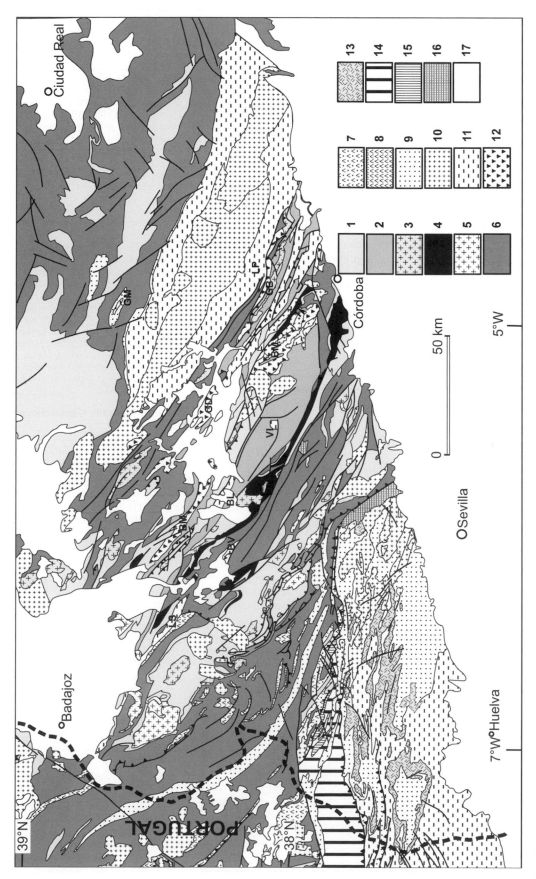

Fig. 8.2. Geological map of the South Portuguese and Ossa Morena Zone. Key: 1, Proterozoic rocks; 2, undifferentiated Proterozoic–Lower Palaeozoic rocks; 3, Neoproterozoic plutonic rocks; 4, Proterozoic volcanic and sedimentary rocks (Malcocinado Fm and Bodonal–Cala complex); 5, Cambrian to Ordovician plutonic rocks; 6, Cambrian to Silurian rocks; 7, Lower Cambrian volcaniclastic rocks; 8, Cambrian volcanic and volcaniclastic rocks; 9, Devonian sedimentary and volcaniclastic rocks; 10, Upper Palaeozoic plutonic rocks; 11, Carboniferous sedimentary rocks; 12, Carboniferous volcanic rocks: GB, Guadalbarbo basin; BM, Benajarafe–Matachel basin; GD, Guadiato basin; LP, Los Pedroches basin; 13, Iberian Pyrite Belt volcanic and subvolcanic rocks; 14, Pulo do Lobo domain; 15, Beja–Acebuches ophiolitic complex; 16, Permian basins; 17, Mesozoic and Cenozoic rocks.

Although both age and chemical constraints are very poor, these rocks are similar to those described above in that both are unconformably overlain by Lower Cambrian sediments. In addition, the lower part of the Lower Cambrian sequence contains rhyolite pebbles, indicating the existence of Precambrian volcanism. Metamorphosed porphyritic volcanic rocks within the Precambrian Villalba schists in the Lugo dome also record Precambrian volcanism (Bastida & Pulgar 1978).

Pre-Caradoc volcanic rocks are widely exposed across the eastern Pyrenees whereas only minor expressions of this magmatism outcrop in the central and western Pyrenees. The volcanic units defined in the eastern Pyrenees can be grouped into a lower unit, composed of three main rock types – quartz-tholeiites, metarhyolites and alkali-transitional orthogneisses – and an upper unit, mainly composed of calc-alkaline metavolcanites and porphyritic rhyolites. A synthesis of similar volcanic units in the French Pyrenees is provided by Navidad (in Barnolas & Chiron 1996).

In the Catalonian Coastal Ranges, basic rocks crop out in the Cap de Creus and Guilleries massifs (Fig. 8.3). A Variscan metamorphic overprint has transformed the Cap de Creus outcrops to mid-amphibolite facies metabasites and meta-andesites. Their geochemical composition (Navidad & Carreras 1995) suggests a tholeiitic affinity, and they lie within the Cadaqués-Cap de Creus series (Carreras & Los Santos 1982), which is correlated with the Cambrian Canaveilles and Cabrils formations (Laumonier 1988). These formations also contain Cambrian felsic metavolcanic rocks of rhyodacitic and rhyolitic composition (including ignimbrites) and calc-alkaline affinity (Navidad & Carreras 1995).

Cambrian volcanism in the Ossa-Morena Zone (TD, EP)

Cambrian volcanic rocks in the southern Ossa Morena Zone are commonly referred to as the Umbria–Pipeta volcanics (Sánchez-Carretero *et al.* 1990). The best outcrops are located in the Alconera syncline, on the southern flank of the Monesterio antiform, as well as within the Aracena massif (Fig. 8.2). In both Spain and Portugal, recent descriptions of these rocks tend to subdivide them into two different groups, roughly corresponding to Early and Middle Cambrian ages (Sagredo & Peinado 1992).

Lower Cambrian volcanic rocks are intensely altered, so they have been described as spilites and keratophyres (Dupont 1979). The rock association has been broadly described as tholeiitic, mostly by comparison with similar rocks in Portugal (Sagredo & Peinado 1992), although major-element geochemical abundances correspond to basalts, trachybasalts and rhyolites. In any case, the available data

are insufficient for chemical characterization. Middle Cambrian volcanism produced basic, intermediate and acid rocks that crop out on both flanks of the Monesterio antiform. Among these, acid rocks are abundant in the northern flank of this structure, whereas they are not known on the southern flank (Dupont & Vegas 1978). The Middle Cambrian age is relatively well constrained, especially in the Zafra area, where volcanic rocks overlie the upper part of the Lower Cambrian succession and are interbedded with Middle Cambrian slates and sandstones (Liñán & Perejón 1981). Basic volcanic rocks include pillow lavas and hyaloclastites (Bard 1969; Sagredo & Peinado 1992), and intermediate and felsic rocks are dominated by volcani-clastic lithologies. The best descriptions correspond to exposures on the northern flank, where ignimbrites have been recorded. Basic rocks are plagioclase basalts, in places trachybasalts, whereas intermediate rocks are trachytes with plagioclase and alkali feldspar phenocrysts. Acid rocks contain alkali feldspar phenocrysts, sometimes with quartz phenocrysts, classified as K-feldspar-phyric trachytes and rhyolites (Sagredo & Peinado 1992).

Intense hydrothermal alteration makes geochemical data difficult to interpret. However, immobile element geochemistry, including Nb/Y v. Zr/TiO_2 data, on rocks from the northern limb of the Monesterio Antiform, indicates that basic rocks are alkaline basalts, intermediate rocks are mainly trachyandesitic, and acid rocks correspond either to rhyolites (but close to trachyandesites) or to a comendite–pantellerite series. However, the occurrence of pink (titaniferous?) augite in basic rocks of the southern antiformal limb (Bard 1969) suggests that the whole Middle Cambrian volcanic suite is dominantly alkaline. In all, the available data have led to the interpretation of Lower Cambrian basic volcanic rocks as related to the onset of a rifting process, whereas Middle Cambrian volcanic lithologies are considered to represent a further stage of rifting, subsequently stopped by migration of the rifting axis.

Cambrian to Silurian volcanism in the Cantabrian and West Asturian-Leonese zones (TD, EP)

Volcanic rocks crop out mainly in the southern part of the Cantabrian Zone, with less abundant outcrops also occurring in the NW (Corretgé & Suárez 1990). In the south, they mainly crop out within the so-called Oville and Barrios formations, which roughly correspond to the whole Cambrian and Ordovician succession. Volcanic rocks commonly form necks that cut across bedding surfaces. The upper contacts of these volcanic masses are either interbedded with, or overlain by Cambro-Ordovician quartzites. The phase of maximum volcanic activity seems to correlate with the deposition of the Upper Cambrian/Lower Ordovician Barrios Fm (Aramburu 1989), although some activity occurred after this (Heinz *et al.* 1985).

In general, the volcanic necks show relatively simple conical shapes enlarging upwards, up to 1 km long by 500 m wide. Volcanic facies have been classified in four major groups: (1) 1–10 m thick massive horizons of lapilli tuffs and massive tuffaceous sandstones; (2) up to

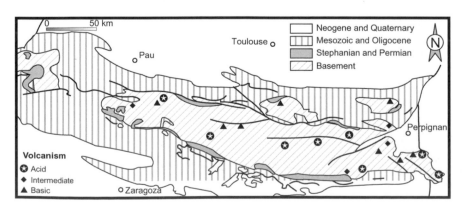

Fig. 8.3. Geological sketch map of the Pyrenees, showing the location of pre-Stephanian volcanic outcrops.

1 m thick tuffaceous sandstones showing conspicuous reverse bedding; (3) 0.1–1 m thick tuffaceous sandstones exhibiting normal bedding; and (4) up to 0.1 m thick tuffaceous sandstones with parallel bedding and planar to cross-bedding. All of these volcaniclastic rocks include country rock xenoliths, bombs and fragments of subvolcanic rocks that have been correlated with sills intruding the Oville Fm (see below). Apart from volcaniclastic rocks, dykes and lava flows occur rarely, as does an alkali basalt plug. Lapilli tuffs consist largely of vesicular pyroclasts, with minor porphyritic fragments. Tuffaceous and other volcanogenic sandstones are volcanic wackes and sublitharenites laterally grading to quartzarenites.

According to the composition of volcanic lapilli and ash, it has been proposed that all of these volcaniclastic rocks originated from a volatile-rich trachybasaltic parental magma (Heinz $et\ al.$ 1985). Again, most of the rocks are intensely altered, so most geochemical evidence is obtained from immobile elements. For instance, bombs are rich in alkalis, P_2O_5 and TiO_2, corresponding to hydrothermally altered trachybasalts and olivine alkali basalts. On a TAS diagram, basic volcaniclastic rocks plot as basalts, andesitic basalts and andesites, whereas bombs correspond to trachybasalts and trachyandesites. Subvolcanic rocks intrusive into the Oville Fm should be named tephrites/basanites and basaltic trachyandesites and andesites.

Volcaniclastic rocks are spatially and compositionally related to discontinuous, alkali-rich dolerite sills in the Oville Fm (Loeschke & Zeidler 1982; Heinz $et\ al.$ 1985). These are up to 80 m thick and can be traced across wide areas. They consist of plagioclase, diopsidic to titanaugite, titaniferous biotite, olivine, feldspathoids and rare K-feldspar and, although distinctly richer in K, Ti, P and Zr, their average composition is similar to trachybasalts. It is generally agreed that the sills were emplaced from Middle Cambrian to Lower Ordovician times, although Loeschke (1983) has obtained a Rb-Sr date of 399 ± 5 Ma BP for the Cerrecedo sill.

Minor volcaniclastic rocks have also been described in the Furada-San Pedro Fm, which overlies the Barrios Fm and has been dated as Late Silurian. Gallastegui $et\ al.$ (1992) consider these lithologies to be similar to the above-described volcaniclastic rocks, except in that volcanogenic sandstones are commoner than tuffs. Heinz (1984) has suggested that erosion of a palaeovolcanic landscape could have occurred together with deposition of the Furada-San Pedro Fm, but a reactivation of volcanic activity during the Silurian period cannot be excluded.

In the central and northern parts of the West Asturian-Leonese Zone, volcanic rocks are both rare and poorly known. Most of them are Middle to Late Cambrian (perhaps including lower Tremadoc) in age and included within the Oville Fm. They include discontinuous horizons of olivine basalts, trachytes and basaltic agglomerates, these latter grading to sandstones, quartzites and shales with little volcanic material. In addition, volcanic and subvolcanic rocks crop out in the Luarca (Llanvirn–Llandeilo) and Castro (Caradoc–Ashgill) formations of the Cabo de Peñas region. The Luarca Fm includes a 26 m thick concordant sill with finer-grained border zones (chilled margins?), described as an andesitic basalt. The Castro Fm comprises vitric and crystal tuffs, together with basaltic to andesitic basalt flows. Corretgé & Suárez (1990) have described these rocks as andesitic basalts, andesites, dacites and quartz-latites. Correlation between them and the southern Cantabrian zone is not evident, but Gallastegui $et\ al.$ (1992) have pointed out the volcanologic, petrographic and geochemical similarities of volcanism throughout the Cantabrian segment, which suggests a linked evolution, perhaps with a waning of volcanic activity after Middle Ordovician times.

Evidence for Palaeozoic igneous activity in the West Asturian-Leonese Zone is scarce and obscured by Variscan tectonometamorphism. Most basic and ultrabasic rocks are metamorphosed sills interlayered with Lower Cambrian quartzites near Villalba in the Lugo Dome. Basic rocks, considered to be tholeiitic, consist of amphibolites interbedded with actinolite or chlorite schists, together with some olivine-rich metaperidotites (cumulates?) (Capdevila 1966, 1969). Rhyolitic metavolcanics also occur here and consist of quartz, microcline and muscovite. In the Truchas area acidic rocks are represented by albitic tuffs and lavas, which are interpreted to have formed from magmas differentiated from the tholeiites, and therefore related to them in time. Finally, several felsic, porphyritic dykes post-date F1 Variscan structures (Martínez-Álvarez $et\ al.$ 1975).

The Ollo de Sapo Formation (AC)

One of the most significant geological units of the northern and central Iberian massif is the Ollo de Sapo Fm (Hernández Sampelayo 1922; Parga Pondal $et\ al.$ 1964). Most of the Ollo de Sapo outcrops form part of a great antiform more than 600 km in length that extends from the coast of Galicia towards the Central Range (Central System). This antiform separates two important zones of the Variscan massif, the West Asturian-Leonese Zone and the Central Iberian Zone (Lotze 1945; Julivert $et\ al.$ 1974). The Ollo de Sapo Fm has received special attention by geologists for two main reasons. Firstly, it represents major Early Ordovician acidic, calc-alkaline magmatic activity, and, secondly, it is a plausible candidate for a protolith for the generation of granitic batholiths during the Variscan orogeny (Capdevila 1969; Ortega & Gil Ibarguchi 1990; Holtz 1987; Castro $et\ al.$ 2000). Different lithologies can be distinguished within this formation. The dominant rock is composed of a mica–quartz–plagioclase matrix that encloses large (up to 10 cm) K-feldspar megacrysts, creating a texture sometimes referred to as a 'glandular gneiss'. Subordinate lithologies include metapelites and finer-grained felsic rocks that include volcanic and subvolcanic rhyolites. The silicic nature of the lithologies and their calc-alkaline geochemistry clearly suggest a volcanosedimentary origin resulting from subaerial explosive ignimbritic volcanism (Navidad $et\ al.$ 1992). Such an interpretation implies oceanic plate subduction along an Ordovician active plate margin. In some areas of the northern Iberian Massif, the Ollo de Sapo Fm has been affected by ultrametamorphism and anatexis giving rise to a more pronounced meta-igneous character. Many radiometric U-Pb ages on zircons from these apparently igneous-derived rocks have been carried out in the last few years, and most indicate an Ordovician age (e.g. Valverde-Vaquero & Dunning 2000). These are mostly in agreement with previous Rb-Sr whole-rock ages on the same rocks (Vialette $et\ al.$ 1987) in the Central Range (Central System) (Table 8.1). Although the meaning of the radiometric data is still somewhat controversial, a volcaniclastic origin for most of the Ollo de Sapo lithologies is clear from petrological and field relationships. However, it is equally clear that the rocks were affected by Variscan metamorphic episodes involving anatexis and the production of many leucogranites (Castro $et\ al.$ 2000). Although there are rare, metre-scale dykes and sills within the Ollo de Sapo rocks (e.g. in Hiendelaencina, Guadalajara), there is no direct evidence for the existence of large Ordovician granite intrusions in the Central Range (Central System) and adjacent areas of the Central Iberian zone. The coincidence of zircon ages in both low grade and anatectic grade Ollo de Sapo rocks indicates a protolith, rather than metamorphic, age. The granitic appearance of many of the Ollo de Sapo rocks is the result of high grade Variscan metamorphism and anatexis affecting an Ordovician volcaniclastic protolith. The age of the migmatization is constrained to around 337 ± 3 Ma and 330 ± 2 Ma, according to data from monacites in granite veins within the anatectic gneisses in Buitrago (Valverde-Vaquero & Dunning 2000). Information about the internal zoning of the dated zircons, which were mechanically abraded before analysis (Valverde-Vaquero & Dunning 2000) would be required for a more precise geological interpretation of the reported ages.

The Ordovician U-Pb zircon ages from the low grade facies

Table 8.1. Geochronological data of Pre-Variscan rocks of the northern Iberian Massif

Pluton	Rock type	Age (Ma)	Method	$^{87}Sr/^{86}Sr_T$	Source
Galicia Allochthonous Complex					
Ortegal					
Cariño	Gneiss	355 ± 8	Rb-Sr Bt	0.7076	1
Cariño	Gneiss	380 ± 4	Ar-Ar (Hb)		1
Chimparra	Gneiss	422 ± 4	U-Pb Zr		1
Chimparra	Gneiss	349 ± 7	Rb-Sr WR-Ms	0.71034	1
Chimparra	Gneiss	271 ± 8	Rb-Sr WR-Bt	0.71034	1
Chimparra	Gneiss	540	SHRIMP U-Pb Zr		2
Chimparra	Gneiss	382 ± 3	U-Pb	U-Pb Rt	3
Ortegal	Gneiss	417 ± 3	U-Pb Zr		1
Ordenes					
Corredoiras	Augengneiss	500 ± 2	U-Pb Zr		4
Curtis	Orthogneiss	450 ± 25	Rb-Sr WR	0.7085	5
Curtis	Orthogneiss	283	U-Pb Zr		5
Mellid	Orthogneiss	409 ± 24	Rb-Sr WR	0.7100	6
Mellid	Orthogneiss	444 ± 40	Rb-Sr WR	0.7090	5
Mellid	Orthogneiss	481	U-Pb Zr		5
Mellid	Orthogneiss	459	U-Pb Zr		5
Mellid	Orthogneiss	482	U-Pb Zr		5
Monte Castelo	Bt granitoids	500 ± 2	U-Pb Mz-Zr		4
Monte Castelo	Gabbro	499 ± 2	U-Pb Zr		4
Malpica-Tuy					
La Guía	Riebeckite orthogneiss	486 ± 24	Rb-Sr WR		7
Malpica a Noya	Orthogneiss	452	Rb-Sr WR	0.708	5
Malpica a Noya	Orthogneiss	459	Rb-Sr WR	0.709	5
Noya	Orthogneiss	378	U-Pb Zr	0.708	5
Noya	Orthogneiss	466 ± 11	Rb-Sr WR	0.7078	8
Noya	Orthogneiss	471 ± 6	Rb-Sr WR	0.7073	8
Vigo	Orthogneiss	469 ± 8	Rb-Sr WR	0.7100	5
Vigo	Orthogneiss	550	Rb-Sr WR		5
Tormes-Sanabria					
Miranda do Douro	Orthogneisses	618 ± 9	U-Pb Zr		9
Porto	Augengneiss	1797 ± 30	U-Pb Zr		9
Porto	Augengneiss	325 ± 3	U-Pb Zr		9
San Sebastian	Leucocratic orthogneiss	465 ± 10	U-Pb Zr		10
Sanabria	Augengneiss	488	U-Pb Zr SHRIMP		11
Viana del Bollo	Orthogneiss	465 ± 10	U-Pb Zr		10
Central System					
Gredos					
Peña Negra	Gneiss	528 ± 14	Rb-Sr WR	0.7071	12
Guadarrama					
Abantos	Metagranitic augengneiss	474 ± 7	Rb-Sr WR	0.707	13
Buitrago	Bt gneiss	2000 to 370	U-Pb Zr		14
Buitrago	Aplite	482 +14/−11	U-Pb Zr		15
Buitrago	Foliated leucogranite	482 +9/-8	U-Pb Zr		15
Buitrago	Foliated megacrystic granite	488 +10/−8	U-Pb Zr		15
El Cardoso	Gneiss	480–510	U-Pb Zr		14
El Cardoso	Gneiss	480 ± 2	U-Pb Zr		15
El Cardoso	Gneiss	500–550	U-Pb Zr		14
Hiendelaencina	Orthogneiss	2000 to 380	U-Pb Zr		14
La Morcuera	Augengneiss	477 ± 4	U-Pb Zr		15
Otero de Herreros	Augengneiss	494 ± 10	Rb-Sr WR	0.7087	13
Pedrezuela	Gneiss	476 ± 10	Rb-Sr WR	0.7106	16
Riaza	Mylonitic megacrystic granite	469 +16/−8	U-Pb Zr		15
Villar de Pradena	Leucogneiss	471 ± 12	Rb-Sr WR	0.7093	14

Sources: 1, Peucat *et al.* (1990); 2, Schäfer *et al.* (1993); 3, Valverde-Vaquero & Fernández (1996); 4, Abati *et al.* (1999); 5, Kuijper *et al.* (1982); 6, Van Calsteren (1977); 7, Priem *et al.* (1966); 8, García Garzón *et al.* (1981); 9, 10, Lancelot *et al.* (1985); 11, Gebauer & Martínez-García (1993); 12, Pereira *et al.* (1992); 13, Vialette *et al.* (1987); 14, Wildberg *et al.* (1989); 15, Valverde-Vaquero & Dunning (2000); 16, Vialette *et al.* (1986).

of the Ollo de Sapo Fm presumably records the age of the volcanism. Zircons from anatectic gneisses at Bercimuelle (SE Salamanca) have yielded a bimodal age distribution of 500 Ma and 350 Ma (J. D. De la Rosa, unpublished laser probe data). The older ages may record an early Ordovician protolith, and the younger ages may be recording the Variscan migmatization. An even older, Proterozoic, age for the terrigenous component of the Ollo de Sapo Fm is indicated by Nd model ages that cluster around 1.5 Ga (Beetsma 1995; Castro *et al.* 1999*a*). Low Sr initial ratios (calculated at 480 Ma) of around 0.706–0.710 indicate a Rb-depleted source (either mantle or crust) was involved in their genesis. Another example of complex age inheritance is provided by the Porto de Sanabria gneiss (Zamora), which is similar to many Ollo de Sapo 'glandular gneisses' of the Central Range (Central System). Lancelot *et al.* (1985) produced two U-Pb zircon age populations (1.8 Ga and 325 Ma; Table 8.1). The younger age records the Variscan metamorphism, whereas the older age is interpreted as the age of the protolith, possibly inherited from the terrigenous component of the volcaniclastic material from which the gneiss was formed. From these observations, it seems very likely that a wide spectrum of inherited ages may be identified from different zircon populations or even from different zones of a single zoned grain. The Ordovician volcanism supplied one of these ages, but Variscan and Precambrian ages are also recorded in these complex rocks.

Ordovician–Silurian volcanism in the Pyrenees and Catalonian Coastal Ranges (ML, EA, CG)

Upper Ordovician to Lower Silurian alkaline igneous rocks are widely exposed in the Spanish Pyrenees, especially in the Roc de Frausa (Liesa 1988) and Ribes de Fresser (Martí *et al.* 1986) massifs (Fig. 8.3). In the latter, Late Ordovician bimodal calk-alkaline magmatism produced andesitic rocks in the lower part of the succession, and rhyolitic rocks, including ignimbrites, in its upper part (Martí *et al.* 1986). Outcrops in the Catalonian Coastal Ranges correspond to two different volcanic events. The first of these produced Late Ordovician volcaniclastic materials and massive rhyolites that crop out in the Gavarres and Guilleries massifs (Barnolas *et al.* 1980; Duran *et al.* 1984; Navidad & Barnolas 1991; Fig. 8.3) as well as in other smaller areas (Serra 1990; Ubach 1990). According to the available geochemical data (Duran *et al.* 1984; Navidad & Barnolas 1991; Navidad 1996), these rocks record dacitic-rhyolitic calc-alkaline volcanism probably related to anatectic crustal melting in an extensional tectonic setting. The second volcanic episode is represented by small outcrops of Upper Ordovician–Lower Silurian (Llandovery) rocks in the Sierra de Collcerola (Barcelona; Julivert *et al.* 1987; Gil-Ibarguchi & Julivert 1988). This volcanism produced basic volcanic and subvolcanic rocks, whose geochemical composition suggests an alkaline affinity.

Ordovician–Carboniferous volcanism in the Central Iberian Zone (TD, EP)

Lower Palaeozoic successions in the Central Iberian Zone (CIZ) are characterized by large platform-related, mostly sili-ciclastic sediments with well-dated and very continuous marker horizons so that the ages of volcanic and subvolcanic rocks are also reasonably well constrained. During Late Ordovician times up to 46 m of volcaniclastic rocks were deposited in El Centenillo area (Jaén), close to the SE corner of the CIZ, and have been indirectly dated as lower Ashgillian (Gutiérrez-Marco & Rábano 1987*a*; Gutiérrez-Marco *et al.* 1990). These volcaniclastic rocks occur interbedded with the Urbana limestone and comprise mostly volcanic mudstones and sandstone, locally interpreted as pyroclastic, but exhibiting in most cases reworked, sedimentary textures (Pineda 1987). No geochemical data are available, but the volcanic rocks have been described as bimodal (rhyolitic with minor basalts) and may therefore be related to (proto)-rift stages preceding Silurian extension. These rocks can be correlated with others described in Portugal (Porto de Santa Ana and Vimioso formations).

Apart from the Almadén area, described below, Silurian igneous rocks are most abundant in the northern part of the CIZ, where they include widespread (but rare) subvolcanic dolerites and abundant acid volcanic rocks. The dolerites form concordant or subconcordant sills up to 10 m thick and several hundred metres long, and include pyroxene-bearing olivine dolerites, pyroxene dolerites and quartz-dolerites with amphibole. They are usually very altered and dominated by albite, epidote, actinolite and chlorite. According to their chemical composition, they have been tentatively interpreted as tholeiitic (Ancochea *et al.* 1988), but in other areas similar rocks include alkaline-transitional as well as tholeiitic varieties (Ribeiro 1986).

Regarding the acid volcanism, this is restricted to the northern CIZ, mainly occurring in the Cabo Ortegal area, close to the Ollo de Sapo Fm, as well as in the Alcañices syncline. In Cabo Ortegal, volcanic rocks occur both at the top and bottom of the Silurian sequence. They consist of metamorphosed dacites and rhyolites, forming continuous outcrops up to 30 km in length with a maximum thickness of 150 m. In the Alcañices syncline, volcanics occur close to the base of the Silurian sequence. They are mainly metamorphosed dacites and rhyolites up to 50 m in thickness. In both cases, acid rocks correspond to high-K calc-alkaline and shoshonitic series (Arenas 1985; Ancochea *et al.* 1988).

Upper Devonian rocks are known only from the Almadén and Sierra de San Pedro areas. In the latter locality subvolcanic sills have been reported in the upper part of the Frasnian succession, where both dolerite and rhyolite sills occur close to the top of a 700 m thick siliciclastic sequence that includes andesitic tuffs and breccias (Gutiérrez Marco *et al.* 1990). The age of the volcanism is similar to that of other volcanic areas in the southern Iberian Massif, and all this magmatism is likely to be linked to that in the Ossa Morena Zone, described below. With regard to the Almadén area, the total volume of volcanic and related rocks is relatively small, as is the case in Sierra de San Pedro. However, they are particularly interesting not only because they developed over a considerably longer time span but also because they are related to the world-class Almadén mercury deposits, which have produced up to one-third of total world mercury consumption (Higueras *et al.* 2000). The magmatic rocks include brecciated diatreme-like bodies of altered olivine basalt (known as 'frailesca'), sills and flows ranging in composition from basanite-nephelinite through olivine basalt, pyroxene basalt, trachybasalt, trachytes and rare rhyolites, and ultramafic xenoliths and breccia clasts (Chapter 19).

Variscan volcanism in the Ossa Morena Zone (TD, EP)

It is generally accepted that during the Variscan orogeny the Ossa Morena Zone was partly a shallow marine or emergent area, compartmentalized into a number of small basins. In each of these basins, Carboniferous deposition differs in detail, both over time and in the type of related volcanic activity.

The northernmost sedimentary depocentre containing volcanic sequences is the Early Carboniferous Los Pedroches basin, in which both acid and basic volcanics crop out (Pérez Lorente 1979). Pillowed metabasalts, consisting of albite, clinopyroxene, epidote, prehnite and opaque minerals, occur near the base of the succession, are probably tholeiitic, and are interbedded with volcaniclastic rocks and shales. In addition to these basic rocks, red mudstone horizons (commonly named 'polvos de hematites') have been interpreted as very fine grained ash. Basic volcanic rocks are locally overlain by felsic volcaniclastic lithologies comprising tuffs and cinerites (Pérez Lorente 1979). Age constraints are poor, although K-Ar ages of 348 ± 18 Ma to 334 ± 16 Ma have been reported.

The Guadiato basin, located to the south and separated from the above by a narrow band of Neoproterozoic or Lower Palaeozoic rocks, comprises siliciclastic and carbonate sediments interbedded with minor volumes of basic and acid volcanic rocks. The basic rocks are pillowed basalts, mainly cropping out in the La Alhondiguilla area, whereas the acid rocks are rhyolites which locally form necks or plugs in carbonate host rock (Sánchez-Carretero *et al.* 1990).

South of the Guadiato basin, the SE part of the Benajarafe-Matachel Carboniferous basin contains large volumes of volcanic rocks cross-cut by younger, high level plutonic rocks. This complex sequence of plutonic and volcanic rocks was first described by Delgado-Quesada (1971) as Los Ojuelos–La Coronada igneous complex, although this name is probably better applied to the basic part of the plutonic series (Burgos & Pascual 1976; Pascual 1981). The term Villaviciosa–La Coronada igneous alignment has also been used loosely to designate the SE part of the Benajarafe–Matachel basin (Pascual & Pérez-Lorente 1987) where igneous rocks (both volcanic and plutonic) predominate. We prefer to deal with the volcanic and plutonic rocks separately because no genetic link has been demonstrated between the two groups (a partly similar approach has been adopted by Sánchez-Carretero *et al.* 1990).

The succession in the Benajarafe–Matachel basin has two major groups, named La Campana and Erillas, in both of which volcanic rocks are generally more abundant than sediments (Sánchez-Carretero *et al.* 1990). The La Campana Gp consists of lava flows concordantly overlying Lower Carboniferous rocks and passing laterally into blocky epiclastic rocks interbedded with Carboniferous sediments. Rock compositions have been interpreted to correspond to andesites and dacites (Ceperuela dacites; Delgado-Quesada *et al.* 1985), although some varieties could be basaltic, in view of the occurrence of pseudomorphs of olivine phenocrysts (Pascual & Pérez-Lorente 1987). In all, the paucity of chemical data and the intense hydrothermal alteration make classification unsafe, although petrographic features suggest that most volcanics correspond to types intermediate between dacite and andesite (Sánchez-Carretero *et al.* 1990). The Erillas group consists of dacites and rhyolites, also interbedded with Carboniferous sediments.

These volcanic successions are generally thought to be related to extensional tectonism (Sánchez-Carretero *et al.* 1989a; Pascual & Pérez-Lorente 1987) although again, given the paucity of chemical data, this interpretation must be viewed with caution. It is worth noting that the Erillas volcanic group is related in time and space to massive sulphide mineralization (Baeza-Rojano *et al.* 1981). Spore studies have yielded a late Tournaisian–early Visean age for the sediments interbedded with these volcanic sequences.

Finally, some volcanic rocks also occur within the Carboniferous Bienvenida, Los Santos de Maimona and Valverde de Leganés basins. In all cases these rocks comprise small volumes of lava flows and subvolcanic bodies which have been described as ranging from basaltic andesites to rhyolites, the latter becoming dominant towards the top of the series. Although all of these small basins (as well as those previously described) are not entirely coeval, they are usually accepted as ranging from Dinantian to late Visean in age. Given their similarities with other Lower Carboniferous volcanic successions in larger basins, it is postulated that they were also formed in a (trans?)tensional environment.

Devonian–Carboniferous volcanism in the South Portuguese zone (Iberian Pyrite Belt) (TD, EP)

In contrast to other areas in the Spanish Iberian Massif, where volcanic rocks are scarce, Devonian–Carboniferous volcanism is a major, distinctive feature of the South Portuguese Zone. This is due not only to the abundance of volcanic rocks, but also to their complexity, the tectonic setting in which they occur, and the close spatial/temporal relationship between volcanic rocks and massive sulphide deposits. The latter form a world-class group of volcanic-hosted (VHMS) deposits with more than 80 known sulphide masses, among which are several especially remarkable supergiant deposits, such as Riotinto (Leistel *et al.* 1998b).

The northern contact of the South Portuguese Zone is currently interpreted to correspond to a major suture in the Iberian Variscan orogen, marking the subduction of a South Portuguese (?Avalonian) plate beneath the Ossa Morena Zone (Quesada *et al.* 1994; Castro *et al.* 1999b). It is subdivided into five tectonostratigraphic units, which have been interpreted to correspond to terranes accreted to the Ossa Morena Zone (Quesada 1990b). Volcanic rocks (as well as massive sulphide deposits) are found only in the central terrane, named the Iberian Pyrite Belt (IPB), which to the north is in contact with the Pulo do Lobo Gp, commonly interpreted as part of a Middle to Late Devonian accretionary prism. Three major stratigraphic groups are recognized in the IPB, but volcanic rocks and massive sulphide deposits occur only within the so-called volcanic-siliceous complex (VSC), which overlies the upper Fammenian Phyllite-Quartzite (PQ) Gp and is covered by Lower Carboniferous sediments of the Culm Gp (Schermerhorn 1971).

The VSC comprises upper Fammenian to Visean detrital and chemical sediments interlayered with volcanic and subvolcanic rocks, with rapid lateral and vertical facies changes. Estimates of the thickness of the VSC vary widely from tens to hundreds of metres. Published correlations within the VSC (e.g. Routhier *et al.* 1978) should be considered with caution (see Leistel *et al.* 1998b). Difficulties have arisen over whether units are intrusive or shallow extrusive (Boulter 1993; Soriano & Martí 1999), and intense Variscan deformation (Silva *et al.* 1990; Quesada 1998) can disrupt the original stratigraphy to make correlations even more complicated. Recently published local stratigraphic sequences can be found in Sáez *et al.* (1996), Leistel *et al.* (1998b) and Soriano & Martí (1999).

Until recently it has been commonly accepted that the thick volcanic piles in the VSC consisted mainly of coherent lavas and volcaniclastic facies such as pyroclastic flows (Strauss & Madel 1974; Salpeteur 1976; Lécolle 1977; Routhier *et al.* 1978; Soler 1980). However, some authors have recently claimed that volcanic rocks in the VSC are minor compared to subvolcanic rocks, and that many of the supposedly volcanic breccias described in previous works were actually produced by peperitic intrusive sill–sediment interaction. Infilling of in situ fractured subvolcanic bodies with unconsolidated, wet sediments to produce peperites and other brecciated rocks could explain the observed textural relationships (Boulter 1993; Soriano & Martí 1999). Despite this, however, most authors accept that true volcaniclastic rocks do also occur in the IPB and that they include hot pyroclastic flows (Pascual *et al.* 1994; Sáez *et al.* 1996; Leistel *et al.* 1998b; Almodóvar *et al.* 1998; Onézime 2001).

Volcanic rocks in the IPB are mostly either basaltic or felsic (dacite to rhyolite), although some andesites do occur (Munhà 1983), so that magmatism is generally considered as bimodal. Despite their wide outcrop and broad time span, volcanic rocks in the VSC show no major spatial or temporal variations in chemical composition (Sáez *et al.* 1996; Mitjavila *et al.* 1997; Almodóvar *et al.* 1998). The chemical and mineralogical composition of both basic and felsic rocks in the VSC is strongly influenced by post-magmatic hydrothermal alteration, with at least two distinct alteration stages (Sáez *et al.* 1999a). The first stage is pervasive on a regional scale, leaves few relict igneous minerals, and has been described as 'hydrothermal metamorphism' (Munhà & Kerrich 1980) or 'regional hydrothermal alteration' (Barriga & Kerrich 1984). The second alteration stage is superimposed on the former in areas close to sites of mineralization, and produced chloritization, sericitization, silicification and other alteration processes that completely change the original mineralogy, texture and chemistry of the protolith. In these circumstances, sampling of completely fresh igneous rocks is not feasible, even if performed far away from ore-related alteration haloes (Thiéblemont *et al.* 1998). Typically, rocks with relatively minor regional alteration are considered as fresh, but even these rocks exhibit intense alteration. Despite this metamorphic overprint, there is general agreement that at least some of the basaltic rocks in the VSC are mildly alkaline, suggesting a Late Devonian–Early Carboniferous extensional tectonic setting. The evidence for this includes high TiO_2 and P_2O_5 contents, high TiO_2 content in clinopyroxenes and high Nb contents with regard to La and Y (Munhà 1983; Mitjavila *et al.* 1997; Thiéblemont *et al.* 1998; Fernández-Martín *et al.* 1996). However, most of the basaltic rocks in the VSC analysed to date exhibit tholeiitic character, both in whole-rock and relict pyroxene chemistry (Fig. 8.4). $(La/Lu)_n$ and $(Eu/Sm)_n$ values found in tholeiitic basalts in the VSC correspond to continental tholeiites (Mitjavila *et al.* 1997). This has been interpreted as an indication of the existence of more than one type of basic volcanic rocks (Thiéblemont *et al.* 1998), or in terms of mixing of basalts generated from different sources (Mitjavila *et al.* 1997).

The most abundant felsic rocks in the VSC are subalkaline, ranging from dacite to rhyolite (Mitjavila *et al.* 1997; Thiéblemont *et al.* 1998). Dacites exhibit plagioclase and quartz phenocrysts, together with amphibole, biotite and rare clinopyroxene, whereas rhyolites show more abundant quartz and minor plagioclase or biotite phenocrysts, or can be aphyric. Chemically, felsic rocks have been classified as low-Al, high-Yb *sensu* Arth (1979) (Thiéblemont *et al.* 1998). Even admitting that alteration may have modified rare earth element (REE) and other trace element distributions, felsic rocks have been claimed to exhibit roughly parallel REE and spider distribution patterns, and so have been interpreted as having formed separately, by variable degrees of partial melting. Accordingly, Mitjavila *et al.* (1997) concluded that dacites, rhyolites and andesites cannot be mutually related by fractional crystallization. This conclusion, however, has been questioned by Almodóvar *et al.* (1998) after detailed studies on closely related rocks. In both cases REE patterns (some of them showing positive Eu anomaly) are not parallel, but gradually evolve from flat to progressively steeper, with an increasingly more pronounced negative Eu anomaly. From these and other major and trace element data, they concluded that felsic rocks formed a single rock series, mostly evolving by fractional crystallization. Thiéblemont *et al.* (1998) have generalized and quantitatively modelled this process in terms of plagioclase + amphibole + accessory mineral separation from a parental dacitic magma. Melt inclusion evidence additionally supports dacite–rhyolite fractional crystallization.

This fractional crystallization controversy also involves the distinctly less abundant andesites. These are quartz-poor rocks, exhibiting plagioclase and clinopyroxene phenocrysts in a plagioclase-rich microcrystalline groundmass (Munhà 1983). Mitjavila *et al.* (1997) have shown that in the Spanish VSC, a relatively large number of rocks covers the field of andesitic rocks on the Zr/TiO_2 v. Nb/Y diagram, roughly depicting a continuum with basalts and the more

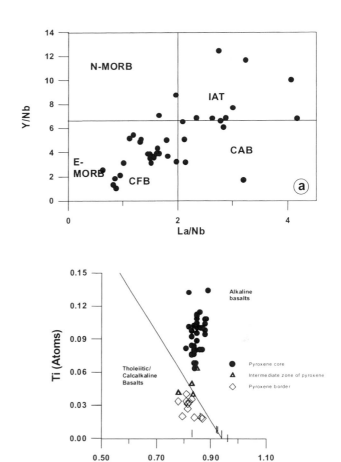

Fig. 8.4. (a) Y/Nb v. La/Nb diagram for the most basic lavas ($SiO_2 <$ 55%) of the Iberian Pyrite Belt (from Thieblemont *et al.* 1998). (b) Pyroxene compositions of basic rocks of the Iberian Pyrite Belt (modified from Fernández Martín *et al.* 1996).

evolved dacites and rhyolites. However, these authors claim that andesites cannot be related to basalts or dacites because of their mutual isotopic relationships and other chemical features (although andesites and dacites completely overlap on some trace element plots; Mitjavila *et al.* 1997). If a fractional crystallization model is alternatively accepted, then most andesites should be considered primitive or cumulate rocks (those showing a positive Eu anomaly), related to the dacite–rhyolite series (Almodóvar *et al.* 1998: Thiéblemont *et al.* 1998). Nevertheless, an origin of *some* of the andesitic rocks by mixing/contamination of basalts (Mitjavila *et al.* 1997; Thiéblemont *et al.* 1998) cannot be excluded with the available data. Whatever the case, there is a general consensus that basalts and the more evolved rocks belonging to the (andesite)–dacite–rhyolite series were generated from different sources, so that geochemical evidence available to date largely confirms the initial views pointing out that VSC magmatism is bimodal (Munhà 1983).

Considering the IPB petrology and geochemistry, another point of controversy concerns the magma source(s) for felsic magmatism. Although undoubtedly crustal, some authors suggest that the dacite–rhyolite rocks were formed from different segments of a thick, mature continental crust (Mitjavila *et al.* 1997). Andesites would have formed in a different way, by crustal contamination of basaltic rocks. Other authors have stressed that geochemical evidence, in particular regarding REE distribution, is not consistent with magma segregation from a garnet-bearing source, so indicating a possibly more shallow and/or immature crustal source for felsic rocks (Thiéblemont *et al.* 1998). Some additional evidence, however, seems to favour a relatively mature, thick crust as the most probable source for felsic volcanism. Firstly, model ages for zircons in granitoids from the South

Portuguese Zone yield dates of around 1100 Ma (De la Rosa *et al.* 1999). Secondly, Sn-W mineralizations in the IPB, which are linked to late granitoids (Sáez *et al.* 1988), were most probably related to an evolved crust. The predominance of dacitic and rhyolitic rocks, as well as other geochemical considerations, tends to exclude a purely island arc environment (Mitjavila *et al.* 1997; Thiéblemont *et al.* 1998). With all these uncertainties in mind, a consensus has been reached that the VSC complex in the South Portuguese Zone was generated during crustal extension (Munhà 1983; Sáez *et al.* 1996; Mitjavila *et al.* 1997; Leistel *et al.* 1998; Thiéblemont *et al.* 1998*b*; Onézime 2001). Beyond this, however, there remain serious differences of opinion regarding the regional significance and plate tectonic setting of this extensional area.

Stephanian–Permian volcanism in the Pyrenees (ML, EA, CG)

Several magmatic events, ranging in age from Stephanian to Late Permian (Saxonian or even Thuringian), are exceptionally well recorded in three areas of the Pyrenees (Fig. 8.5). These areas are: (a) western Pyrenees (Cinco Villas massif); (b) central Pyrenees (Ossau, Anayet, Aragón-Subordán areas and Las Paules basin); and (c) eastern Pyrenees (Erill-Castell and Sierra del Cadí areas; Gisbert 1981; Bixel 1984; Martí *et al.* 1986; Cabanís & Le Fur-Balouet 1989; Martí 1991; Valero 1991; Briqueu & Innocent 1993; Gilbert *et al.* 1994; Innocent *et al.* 1994; Barnolas & Chiron 1996). Five different magmatic episodes and two tectonomagmatic cycles can be identified (Bixel 1984, 1988). The first cycle shows an initial bimodal episode, dated as late Stephanian (using fossil plants), with peraluminous dacites and rhyolites at the base, and low-K calc-alkaline andesites at the top. These rocks are well represented in the Midi d'Ossau cauldron complex and the Sierra del Cadí outcrops. Briqueu & Innocent (1993) produced zircon U-Pb ages for both the peraluminous (278 ± 5 Ma) and calc-alkaline (272 ± 3 Ma) phases in the Midi d'Ossau complex. These ages indicate an Early Permian age for this episode, although whole-rock and garnet Sm-Nd ages in the Sierra del Cadí range between 320 ± 2 Ma and 313 ± 14 Ma (Namurian; Gilbert *et al.* 1994). The second magmatic episode is represented by high-K calc-alkaline andesites and dacites. No age data are available for this episode, but an Autunian age seems probable. Finally, the third episode comprises peraluminous potassic rhyolites that crop out only in the Sierra del Cadí area and are also probably Autunian in age. Innocent *et al.* (1994) proposed a petrogenetic model involving mixing of anatectic crustal magmas and lithospheric mantle magmas to explain the generation of different calc-alkaline magmas of the three episodes in the Ossau complex (although melting of lower crustal mafic granulites could also account for the

magmatism). Gilbert *et al.* (1994) obtained very similar isotopic ratios for the Sierra del Cadí volcanic rocks, but proposed a purely crustal source model in which melting of the Pyrenean crust was triggered by the intrusion of basic magmas derived from the upper mantle and subsequently modified by assimilation processes during their ascent.

The second cycle (Late Permian) developed under late- and post-orogenic (initially transpressive then extensional) conditions (Carreras & Capellá 1994), and shows two magmatic phases. The first of these is Autunian–Saxonian in age and represented by transitional to alkaline andesites which crop out only in the Anayet, Aragón-Subordan and Seo d'Urgell areas (Bixel 1984; Martí 1986; Valero 1991). Isotopic data are only available for rocks of the Anayet crop out (Innocent *et al.* 1994) and giving an $\varepsilon^i_{Nd} = +3.7$ value for the Anayet andesites, which suggests a mantle origin, although with significant contamination. On the contrary, Cabanis (in Barnolas & Chiron 1996), suggests that these rocks were derived from an enriched or even strongly metasomatized mantle source. The second, and last, magmatic phase is Saxonian or even Thuringian in age, and produced alkali (in some cases transitional) basalts and dolerites that crop out as dykes and sills in the Cinco Villas massif (western Pyrenees; 1 in Fig. 8.5; Bixel 1988; Cabanís & Le Fur-Balouet 1989; Innocent *et al.* 1994; Lasheras *et al.* 1999*a,b,c*), in the Anayet massif (Bixel 1984; Valero 1991; Fig. 8.5), and in the eastern French Pyrenees. Isotopic ratios (Innocent *et al.* 1994) point to a clearly mantle-derived origin for these basalts, as they have $\varepsilon^i_{Nd} = +7.4$ and a $^{87}Sr/^{86}Sr$ value of 0.70752. These authors also note the strong compositional similarity of these basalts to those generated in Ocean Island Basalt (OIB)-type environments. This episode clearly developed under an extensional tectonic setting, and is the first expression of post-Variscan crustal thinning, which was to continue during early Mesozoic time, as attested by Triassic tholeiitic dolerites (the so-called 'ophites') of the Pyrenean margins.

Stephanian-Permian volcanism of the Iberian Ranges and Atienza (ML, EA, CG)

Stephanian–Permian calc-alkaline magmatism is widely represented (over 1000 mapped outcrops) in the Iberian Ranges, which can be divided into eastern (Aragonese) and western (Castilian) branches (Fig. 8.6). Exposures are mostly sills (with some dykes) and, less commonly, volcaniclastic rocks. The intrusions commonly show a geometry controlled by pre-existing Variscan structures, and are especially common in the

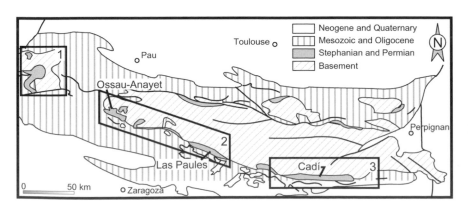

Fig. 8.5. Geological sketch of the Pyrenees with location of the three main areas with Stephanian–Permian volcanism: 1, Western Pyrenees (Cinco Villas Massif); 2, Central Pyrenees (Ossau-Anayet, and Las Paules areas); and 3, Eastern Pyrenees (Erill-Castell – Estac and Sierra del Cadí areas).

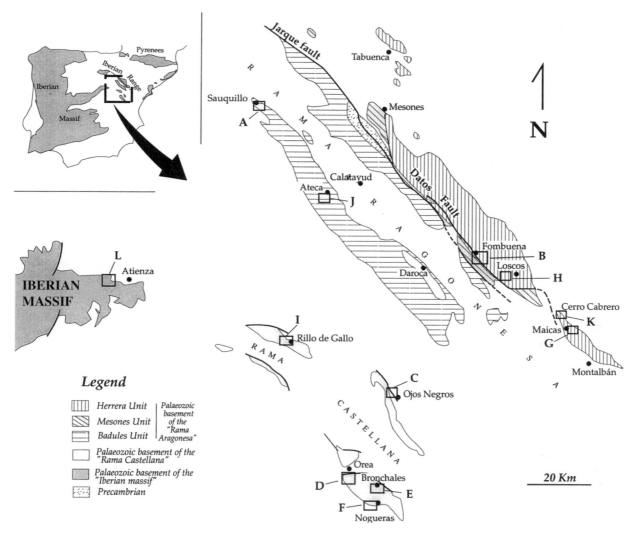

Fig. 8.6. Geological sketch map of the Iberian Ranges and easternmost Iberian Massif, showing the location of the main Stephanian–Permian volcanic outcrops.

southeastern areas of both Aragonese and Castilian branches (Sierra de Albarracín; E and F in Fig. 8.6; Lago *et al.* 1993, 1996). They are typically andesitic, with both basalts (Ojos Negros; C in Fig. 8.6) and rhyolites (J and K in Fig. 8.6) being rare. Volcaniclastic outcrops (Lago *et al.* 1991, 1994, 1995; Conte *et al.* 1987; Hernando 1973; Ramos *et al.* 1976) are also widely exposed in many areas of the Iberian Ranges (A, B, C, D, I in Fig. 8.6) as well as in the Atienza area (L in Fig. 8.6). Volcaniclastic materials are interbedded within sedimentary levels containing fossil plants which gave an Autunian age (Conte *et al.* 1987; Lago *et al.* 1991, 1995; Lago & Pocoví 1991), which agrees with those obtained by the K-Ar method on biotite in the Fombuena (292 Ma; Conte *et al.* 1987) and Atienza (287 ± 12 Ma; Hernando *et al.* 1980) andesites.

The magmatism was related to the development of sedimentary basins, being controlled by normal dip-slip faults reactivating late Variscan strike-slip structures (Capote 1983b). A wide variety of enclaves (sedimentary, medium- and high-grade metamorphic and even granitoid rocks) included within these rocks link this magmatism with deep basement

fracturing, crustal assimilation, and rapid magma ascent (Lago *et al.* 1991). The Maicas outcrop (G in Fig. 8.6) is exceptionally rich in granitoid, metapelitic and sedimentary enclaves (up to 30% of the volume of the dyke) that are commonly bordered by a hybrid reaction rim, and aligned parallel to the base of the dyke. Metapelitic enclaves are variably assimilated, from well preserved enclaves bordered by a narrow reaction rim to fully assimilated ones, only evidenced by garnet (Alm_{70-60}) xenocrysts. The Maicas and Noguera de Albarracín outcrops (F and G in Fig. 8.6) show the best examples of such high-grade enclaves which contain biotite, sillimanite, spinel, plagioclase (An_{60-33}), garnet (Alm_{72-63}) and minor corundum. Xenocrystic garnet (Alm_{70-61}) is widespread in all the hypabyssal and volcaniclastic rocks. Their compositional range and textural features support their metamorphic origin and so they are interpreted as a relict of assimilated metapelitic enclaves.

Most rocks show pervasive alteration, although a few intrusions preserve the original, relatively unmodified, composition, notably the Ojos Negros basalt (Lago *et al.* 1994; C in Fig. 8.6), the Loscos microgabbro (Lago & Pocoví 1991; H in Fig. 8.6), and some of the pyroxenitic and amphibolitic andesites of the Montalbán anticline (G and K in Fig. 8.6). Basalts are porphyritic with zoned olivine (Fo$_{73-68}$), clinopyroxene (Fs$_{8.8-11.8}$) and plagioclase (An$_{83-72}$) crystals. Pyroxenic (basaltic) andesites show rare orthopyroxene (En$_{78}$Wo$_3$Fs$_{19}$), clinopyroxene (En$_{50}$Wo$_{40}$Fs$_{10}$), plagioclase (An$_{81-27}$), minor hornblendic amphibole, and abundant quartz xenocrysts. Amphibolitic andesites show plagioclase (An$_{70-20}$), hornblendic amphibole, rare clinopyroxene crystals and garnet xenocrysts (Alm$_{70-60}$), commonly with reaction rims. Andesitic dacites and dacites consist of plagioclase, alkali feldspar, biotite, quartz and minor amphibole, whereas rhyolites (sometimes flow banded) show alkali feldspar, albite, quartz and minor tourmaline crystals. Tourmalinized rhyolites occur in the Bronchales outcrop (D in Fig. 8.6; Lago *et al.* 1993, 1995) and are related to pre-Triassic B-rich hydrothermal activity.

Data points define a continuous trend from basalts to rhyolites (Fig. 8.7), and the chondrite-normalized (Sun & McDonough 1989) spider diagram (Fig. 8.8) shows Nb, K and P depletions whereas Ba, Zr, Th, U and light REEE (LREE; La, Ce, Sm and Nd) display positive anomalies which are probably related to assimilation of crustal materials. The chondrite-normalized (Boynton 1984) REE plot (Fig. 8.9) reveals a differentiation trend from basalts to andesites and the relatively steep pattern related to LREE enrichment. These geochemical features are typical of calc-alkaline associations and are in agreement with the Ni and Cr contents.

The Atienza area, in the easternmost part of the Central Range (Central System) (Fig. 8.6), is characterized by Permian calc-alkaline magmatism which is very similar to that of the Iberian Ranges. A genetic relationship between the Atienza magmatism and the coeval granitoids of the Central Range (Central System) seems probable. Up to 15 volcanic bodies crop out, composed mainly of lava flows and volcaniclastic units produced by subaerial activity with a minimum K-Ar (whole-rock) age of 287 ± 12 Ma (Hernando *et al.* 1980). The base of the volcanic sequence comprises andesite (both pyroxenic and amphibolitic varieties) and dacite lava flows, which unconformably overlie a post-Silurian basement. The upper part of the sequence shows andesitic volcaniclastic rocks and lavas coeval with Permian sedimentation (Hernán *et al.* 1981). Whole-rock compositions of these rocks (squares in Fig. 8.7; Ancochea *et al.* 1980) are very similar to those of the andesitic rocks of the Iberian Ranges. Enclaves are again common, and include microdiorite, high grade metapelites, garnet xenocrysts, and low grade metasediments (e.g. quartzites).

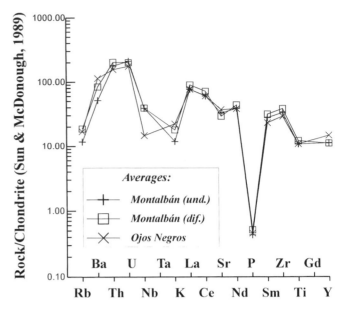

Fig. 8.8. Chondrite-normalized spider plot for the most representative Stephanian–Permian volcanic rocks of the Iberian Ranges (data given in Table 8.2). The lines represent the composition of selected compositions from the Ojos Negros basaltic unit and the Montalbán anticline andesites.

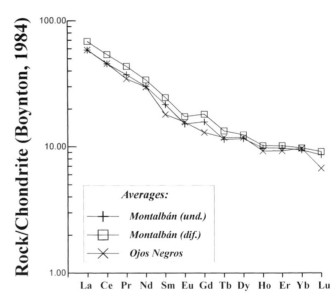

Fig. 8.9. Chondrite-normalized REE patterns for the same compositions as in Figure 8.8 (data given in Table 8.2). The lines represent the composition of the same selected compositions (Ojos Negros basaltic unit (source 12 in Table 8.2) and Montalbán andesites (sources 1 and 6 in Table 8.2).

Permian volcanism in the Cantabrian and South Portuguese zones (TD, EP)

Late Variscan volcanic rocks of Permian age are known only in the north (Cantabrian Zone) and SW (South Portuguese Zone). In the Cantabrian Zone, they comprise rare, hydrothermally altered volcaniclastic rocks and lava flows, with pseudomorphed plagioclase, clinopyroxene, olivine and

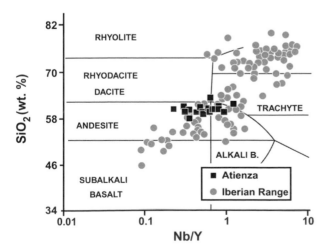

Fig. 8.7. SiO$_2$ v. Nb/Y plot for the Iberian Ranges and Atienza Stephanian–Permian volcanic rocks.

biotite phenocrysts (Corretgé & Suárez 1990). Their chemistry suggests either an alkaline or high-K calc-alkaline character, but whether they formed in a compressional environment or within a rifting zone is still an open question.

Exposures of latest Palaeozoic volcanic rocks in the South Portuguese Zone are restricted to the El Viar basin (Fig. 8.2) where earliest Permian basalt flows are interbedded with alluvial fan deposits. Dominant textures in vesiculated zones towards the top of individual flows are microporphyritic, with either a fluidal or microcrystalline groundmass. Most phenocrysts are olivine, but samples are intensely altered, with several zeolite minerals, calcite and prehnite. Non-vesiculated rocks are better preserved, consisting of plagioclase, clinopyroxene, olivine, opaques and very rare quartz (Simancas 1983). Chemical data are scarce and strongly suggest oxidization of iron and other alteration processes that have probably changed the original, igneous chemical composition, although Simancas & Rodríguez Gordillo (1980) describe them as olivine tholeiites. Finally, in addition to these basaltic flows, Sierra et al. (2000) have reported primary pyroclastic and resedimented volcaniclastic deposits near to the top of the Permian sequence ('grey volcaniclastic sequence'), of probable middle Autunian age.

Intrusive rocks of the northern and central Iberian Massif (AC, LGC, JDR, FJM)

To allow a better understanding of plutonism in the Iberian massif, the region has been divided into northern and southern domains. The northern area lies to the north of (and includes) the Los Pedroches batholith (Fig. 8.10), and is characterized by plutons of granite–monzogranite–granodiorite composition. In the southern area, south of the Los Pedroches batholith, Variscan and pre-Variscan intrusives show a more varied compositional spectrum from gabbros and diorites to tonalites of obviously calc-alkaline affinity. The source materials are also different in each area and some granitoid types, such as cordierite monzogranites, are scarce in the southern domain. The Los Pedroches batholith itself, together with the tonalite plutons of the Central Extremadura batholith, form a granodiorite–tonalite line running parallel to the Variscan structures and reflecting some underlying tectonic control on granite emplacement and generation.

Across much of NW and central Spain, intense Variscan tectonometamorphism, together with the intrusion of large granite–granodiorite batholiths, have obliterated evidence for previous orogenic events. Indirect evidence from Nd isotopic signatures (e.g. Moreno-Ventas et al. 1995; Villaseca et al. 1998; Castro et al. 1999a) point to the presence of an old basement of Middle to Late Proterozoic age, for which an older Pan-African terrigenous component has been proposed (Schäfer et al. 1993). Some of the large high-grade anatectic zones that appear around granodiorite batholiths in the Central Range (Central System) are derived, at least in part, from igneous protoliths of granite to tonalite composition. Thus, a crustal basement with granites and metamorphic rocks presumably existed, at least in the internal zones of the Variscan belt, prior to the Variscan orogeny.

With regard to pre-Variscan Palaeozoic intrusions, these are common within the thrust-stacked nappes of the NW Iberian massif (Chapter 10) (Arenas et al. 1986; Martínez Catalán et al. 1997, 1999). Recent radiometric dates of these magmatic rocks have confirmed Palaeozoic ages (e.g. Peucat

et al. 1990; Schäfer et al. 1993; Valverde & Fernández 1996; Abati et al. 1999; Valverde-Vaquero & Dunning 2000). Three units have been distinguished in these allochthonous complexes: (1) a basal unit; (2) an ophiolitic unit; and (3) an upper unit. The basal unit is characterized by the presence of alkaline and peralkaline granite intrusions (Ribeiro & Floor 1987) emplaced during Ordovician times (460–480 Ma BP). This anorogenic Ordovician magmatism is attributed to an episode of rifting occurring along the northern margin of Gondwana, similar to the Ordovician alkaline complexes of the Ossa Morena Zone in the southern domain. These intrusions were strongly deformed during the emplacement of the complexes. Overlying the basal unit there is a 395 Ma old ophiolitic sequence that includes metagabbros, amphibolites and plagiogranites.

Due to the absence of significant Ordovician tectonometamorphic activity in the autochthonous part of the Iberian massif, there has been a tendency to consider any pre-Variscan orthogneiss as derived from some Precambrian igneous protolith. Recent radiometric U-Pb dates on zircons from these rocks have, however, generally produced Palaeozoic ages (e.g. Valverde-Vaquero & Dunning 2000), mostly in agreement with previous Rb-Sr whole-rock dating of the same rocks (Vialette et al. 1987). Many of these intrusions, however, show complex isotopic patterns, with Variscan, Lower Palaeozoic and Precambrian ages reflecting both polyphase rock histories and inheritance from older protoliths (e.g. Lancelot et al. 1985; Castro et al. 1999a).

Variscan granites (AC, LGC)

Initial subdivision of Variscan granitoid rocks into syntectonic and post-tectonic types (Oen 1958, 1970) was later refined by Ferreira et al. (1987) and López-Plaza & Martínez-Catalán (1987), who distinguished three categories of intrusion: (1) syntectonic massifs, deformed by F1 or F2; (2) syntectonic massifs, deformed by F3; (3) late and post-tectonic granites. Although in some areas two generations of anatectic granite melts can be distinguished, one pre-D2 and the other syn- or late D2 (Escuder et al. 1996), the largest production of granite magmas occurred late with respect to the D2 deformation phase in the Central Iberian Zone and was probably related to crustal-scale extension (Valverde-Vaquero & Dunning 2000). U-Pb ages of monazite in metamorphic rocks indicate that this D2 extensional event took place towards the end of Early Carboniferous times (337 to 326 Ma BP; Table 8.2).

Pinto (1983) and Pinto et al. (1987) distinguished several groups of Variscan granitoids based on their ages, which range from Devonian to Permian. This magmatic activity lasted around 120 Ma and appears to have taken place in pulses separated by 10 to 15 Ma (Table 8.2). However, many age determinations have considerable errors and an accurate estimation of the number and regional location of pulses is yet not possible. If isotopic ages and structural information are considered together, the intrusions can be viewed as just two groups, one showing ages older than 300–305 Ma, and the other displaying ages younger than 290 Ma (sometimes referred to as the 'older' and 'younger' granites). Nevertheless, this simple distinction should be tempered with the realization that Variscan tectonism was diachronous, at times allowing coeval extension and compression at different places in the orogen.

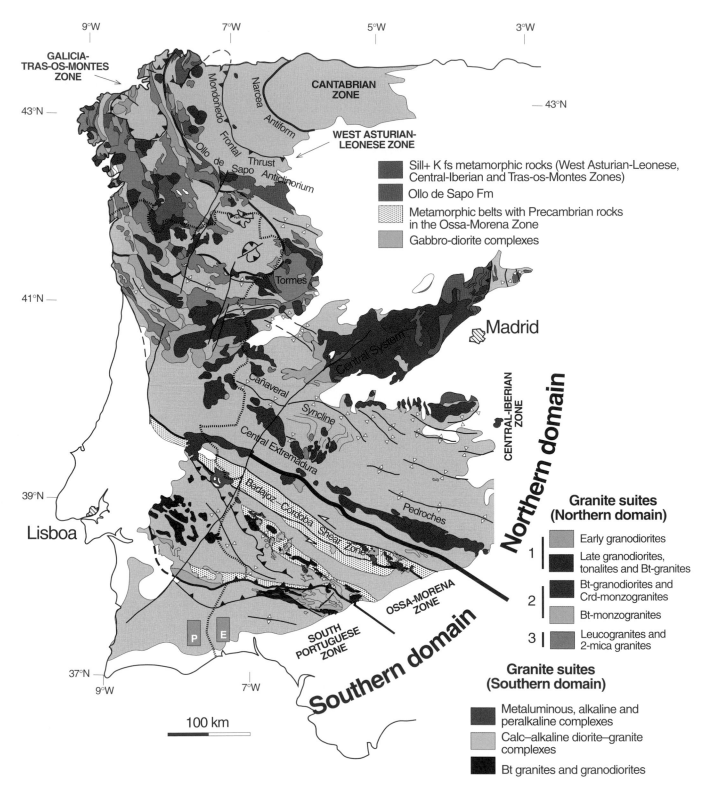

Fig. 8.10. Geological map of the Iberian Massif showing the distribution of intrusive rocks. These have been grouped in suites according to compositional and petrographic criteria. The distinction between early and late granodiorites is restricted to Galicia. The relationships between granitoids of this suite (suite 1) and the Variscan tectonic phases are not uniform along the massif. In the Central Range (Central System) most granodiorites and monzogranites seem to be late with respect to the main deformation phases. Details of the intrusive rocks of the southern domain are depicted in the map of Figure 8.16.

Table 8.2. Geochronological data of granite suites of the northern Iberian Massif

Pluton	Suite	Rock type	Age (Ma)	Method	$^{87}Sr/^{86}Sr_T$	Source
Galicia						
Bayo	1	Granodiorite	275	Rb-Sr Bt		1, 2
Bayo-Vigo	1	Bt granodiorite	348 ± 26	Rb-Sr WR		3
Bayo-Vigo	1	Vaugnerite	348 ± 48	Sm-Nd		3
Bayo-Vigo	1	Bt granodiorite	349.5 ± 14.5	U-Pb Zr		3
Betanzos	3	Granite	273	K-Ar Bt		4
Betanzos	3	Granite	301	K-Ar Ms		4
Boal	3	Monzogranite	292 ± 3	U-Pb Mz		5
Boiro	3	Granite	286	K-Ar Bt		1, 2
Boiro	3	Granite	293	K-Ar Ms		1, 2
Caldas de Reyes	1	Bt granite	287 ± 10	Rb-Sr WR	0.7084	6
Chantada	1	Granite	287	K-Ar Bt		4
Courio	1	Q-monzodiorite	297 ± 6	U-Pb Zr		7
El Barquero	3	Granite	286	K-Ar Bt		4
El Ferrol	3	Granite	292	K-Ar Bt		4
El Pato	1	Bt granodiorite	284 ± 8	Rb-Sr Ms		8
El Portomouro	3	Granite	292	K-Ar Bt		4
Estrada	3	Granite	277	K-Ar Bt		4
Finisterre	3	Granite	279	Rb-Sr Bt		1, 2
Finisterre	3	Granite	280	K-Ar Bt		4
Finisterre	3	Granite	283	Rb-Sr Bt		1, 2
Forgoselo	3	Leucogranite	317 ± 6	Rb-Sr WR	0.7092	9, 10
Friol	3	Two mica granite	330 ± 10	Rb-Sr WR	0.7172	9
Guitiriz	1	Bt granite	303	K-Ar Bt		4
La Coruña	1	Bt granite	308	K-Ar Ms		4
La Estrada	3	Granite	316	K-Ar Ms		4
La Guardia	3	Two mica granite	311 ± 21	Rb-Sr WR		11
La Tojiza	1	Granodiorite	283.8 ± 0.7	$^{40}Ar/^{39}Ar$		12
La Tojiza	1	Granodiorite	295 ± 2	U-Pb Mz		5
Lage	3	Granite	278	K-Ar Bt		4
Lalin	3	Granite	293	K-Ar Ms		4
Los Ancares	3	Leucogranite	289 ± 3	U-Pb Mz		5
Lugo	3	Granite	284	K-Ar Bt		4
Negreira	3	Granite	278	K-Ar Bt		4
Noya	3	Granite	287	K-Ar Bt		4
Padrón	3	Granite	294	K-Ar Bt		4
Padrón	3	Granites	319	K-Ar Ms		4
Penedo Gordo	3	Leucogranite	317 +9/−5	U-Pb Zr		5
Peña Prieta	1	Granodiorite	292 +2/−3	U-Pb Zr Sph		7
Porcia	1	Gabbro-diorite	295 ± 13	U-Pb Zr		5
Porriño	1	Bt granite	292	K-Ar Bt		4
Puebla de Parga	1	Bt granodiorite	358 ± 20	Rb-Sr WR	0.713	9
Puebla de Parga	3	Leucogranite	286 ± 2	U-Pb Mz		5
San Ciprián-Burela	3	Leucogranite	274.1 ± 0.7	$^{40}Ar/^{39}Ar$		12
Sarria	3	Leucogranite	313 ± 2	U-Pb Mz		5
Sarria	3	Two mica granite	282.2 ± 0.8	$^{40}Ar/^{39}Ar$		12
Tapia-Salave	1	Bt grandiorite	287 ± 8	Rb-Sr WR		8
Teijeiro	3	Granite	308	K-Ar Bt		4
Veiga	2	Bt Crd Monzogranite	256 ± 6	Rb-Sr WR	0.71026	14
Vivero		Horneblendite	293 +3/−2	U-Pb Zr		5
Vivero	1	Granodiorite	323 +9/−5	U-Pb Zr		5
Vivero	1	Bt granite	274	K-Ar Bt		4
Central System						
Béjar-Plasencia						
Béjar	2	Bt granite	295 ± 4	Rb-Sr WR		15
Béjar	2	Crd monzogranite	307 ± 5	Rb-Sr SM		15
Béjar	2	Granodiorite, Tonalite	318 ± 17	Rb-Sr WR	0.7078	15
Linares de Riofrío	2	Bt granite	271 ± 6	K-Ar Bt		16
Montemayor del Río	2	Bt granite	269 ± 6	K-Ar Bt		16
Montemayor del Río	2	Bt granite	270 ± 6	K-Ar Bt		16
Sequeros	2	Bt granite	281 ± 6	K-Ar Bt		16
Sequeros	2	Bt granite	280 ± 6	K-Ar Bt		16
Gredos						
Avila	3	Two mica granite	295 ± 13	Rb-Sr WR	0.71221	17
Avila	3	Two mica granite	297 ± 26	Rb-Sr WR	0.71144	17
Avila	3	Leucogranites	305 ± 16	Rb-Sr WR	0.71486	17
Avila	2	Monzogranite	306 ± 8	Rb-Sr WR	0.70833	17

Table 8.2. Continued

Pluton	Suite	Rock type	Age (Ma)	Method	$^{87}Sr/^{86}Sr_T$	Source
Gredos (continued)						
Avila	2	Granodiorite	310 ± 9	Rb-Sr WR	0.70775	17
Avila	1	Granodiorite	317 ± 13	Rb-Sr WR	0.70834	17
Avila	1	Granodiorite	327 ± 8	Rb-Sr WR	0.70941	17
Avila		Appinite	340 ± 18	Rb-Sr WR	0.70582	17
Avila	3	Leucogranite	344 ± 5	Rb-Sr WR	0.72299	17
Gredos	1	Granodiorite-Appinite	295	Rb-Sr WR	0.7077	18
Peña Negra	1	Granodiorite	310 ± 6	Rb-Sr WR	0.7096	19
Prado de las Pozas		Appinite	416 ± 21	Rb-Sr WR	0.7051	19
Guadarrama						
Alpedrete	2	Granite	261 ± 13	Rb-Sr WR	0.71117	20
Atalaya Real	1	Bt-Granite	284 ± 13	Rb-Sr WR	0.7129	20
Buitrago	3	Leucogranite	330 ± 2	U-Pb Mz		21
Buitrago	3	Leucogranite	337 ± 3	U-Pb Mz		21
Cabeza Mediana	3	Granite	291 ± 3	Rb-Sr WR	0.7113	22
El Tiemblo	1	Granodiorite	322 ± 5	Rb-Sr WR		23
Hiendelaencina	3	Granite	297 ± 10	Rb-Sr WR		24
Hoyo de Pinares	2	Granite	320 ± 11	Rb-Sr WR		23
Hoyo de Pinares	2	Granite	301 ± 15	Rb-Sr WR		23
La Granja	1	Bt-Monzogranite	275 ± 11	Rb-Sr WR	0.7129	22
La Granja	1	Bt-Monzogranite	299 ± 55	Rb-Sr WR	0.71212	20
La Pedriza	3	Leucogranite	305 ± 6	Rb-Sr WR	0.7073	22
Las Cabezas	1	Granodiorite	330 ± 17	K-Ar		25
Moralzarzal-Colmenar Viejo	1	Bt-Monzogranite	327 ± 4	Rb-Sr WR	0.7084	22
Navas	3	Leucogranite	284 ± 4	Rb-Sr WR		23
Navas	3	Leucogranite	290 ± 6	Rb-Sr WR		23
Navas del Marqués	3	Granite	302 ± 4	Rb-Sr WR		23
Villacastín	1	Granodiorite	323 ± 47	Rb-Sr WR	0.70701	20
Villacastín El Espinar	1	Bt-Monzogranite	344 ± 8	Rb-Sr WR	0.7061	22
Toledo						
Mora	2	Granite	320 ± 8	Rb-Sr WR	0.7103	26
BEX-Pedroches						
Albalá	2	Monzogranite	313 ± 10	K-Ar		27
Albuquerque-Nisa	2	Bt-granite	281 ± 10	Rb-Sr WR		28
Albuquerque-Nisa	2	Bt-granite	290 ± 8	Rb-Sr WR		28
Andujar	1	Granodiorite	291 ± 15	K-Ar		27
Campanario	1	Monzogranite	305 ± 10	K-Ar		27
Campanario	1	Monzogranite		Rb-Sr WR		29
Fontanosa	1	Granodiorite	302 ± 10	Rb-Sr WR		30
Logrosán	2	Monzonitic granite	337 ± 17	K-Ar		25
Quintana	1	Granodiorite	295 ± 15	K-Ar		25

Sources: 1, Priem *et al.* (1970); 2, Serrano Pinto & Gil Ibarguchi (1987); 3, Gallastegui (1993); 4, Ries (1979); 5, Fernández-Suárez *et al.* (2000); 6, Fourcade *et al.* (1991); 7, Valverde *et al.* (1999); 8, Suárez *et al.* (1978); 9, Capdevilla & Vialette (1970); 10, Cocherie (1978); 11, Van Calsteren (1977); 12, Dallmeyer *et al.* (1997); 13, Valverde *et al.* (1999); 14, Ortega *et al.* (2000); 15, Pinarelli & Rottura (1995); 16, Yenes *et al.* (1996); 17, Bea *et al.* (1999); 18, Moreno-Ventas *et al.* (1995); 19, Pereira *et al.* (1992); 20, Villaseca *et al.* (1995); 21, Escuder *et al.* (1998); 22, Ibarrola *et al.* (1987); 23, Casillas *et al.* (1991); 24, Bischoff *et al.* (1978) in Vialette *et al.* (1987); 25, Bellon *et al.* (1979); 26, Andonaegui (1990); 27, Penha & Arribas (1974); 28, Mendes (1968); 29, Alonso *et al.* (1999); 30, Leutwein *et al.* (1970). Suite 1, early and late granodiorites; Suite 2, biotite granodiorite and cordierite monzogranite; Suite 3, leucogranites and two-mica granites.

With regard to chemical composition, early studies recognized that northern Iberian granitoid rocks can be broadly grouped into peraluminous leucogranites (generally associated with high grade, anatectic zones) and a suite of granodiorites, normally also peraluminous (e.g. Schemerhorn 1959; Oen 1958, 1970; Capdevila 1969; Corretgé 1971). An important contribution to this granite classification scheme was the recognition by Capdevila *et al.* (1973) and Corretgé *et al.* (1977) of a third suite called 'séries des charactéres mixtes ou intermediaires'. Typical of these 'mixed-feature granites' are cordierite-rich monzogranites forming large epizonal plutons in many different areas of the Iberian massif. These

Fig. 8.11. Multicationic diagrams (Debon & Le Fort 1983) showing representative granite complexes from the northern domain **(a)** and southern domain **(b)** of the Iberian Massif. Trends are of Al-cafemic type in the northern domain and cafemic in the southern. Data sources are as follows. Northern domain: Espenuca (Ortega 1998); Gredos (Moreno-Ventas 1991); Toledo (Barbero *et al.* 1995); Sayago (López-Moro 2000); Puente del Congosto (A. Castro, unpublished data); Sanabria (A. Castro *et al.* (unpublished data)); Vivero (Galán *et al.* 1996); Pedroches (Donaire *et al.* 1999); Bayo–Vigo (Gallastegui 1993). Southern domain: Badajoz–Córdoba (Azor *et al.* 1995); Castillo and Monesterio (Eguiluz 1988); Ojuelos–La Coronada–Villaviciosa de Córdoba (Sánchez Carretero *et al.* 1989b); Barcarrota (Castro 1981); Burguillos del Cerro–Valencia del Ventoso-Brovales (Pons 1982); Santa Olalla (Bateman *et al.* 1992); Sierra de Sevilla batholith (De la Rosa 1992).

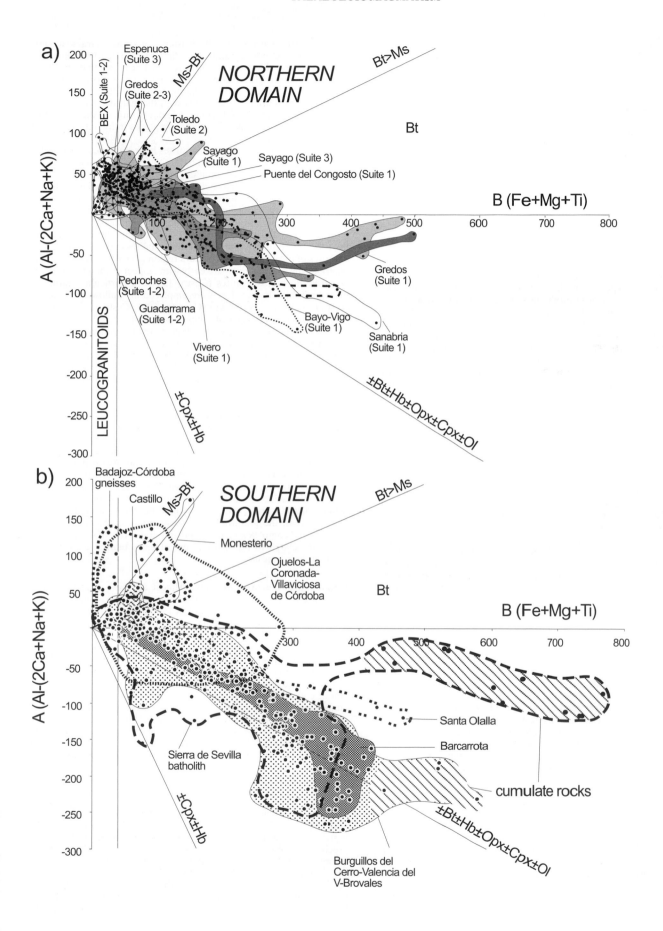

three intrusive suites – granodiorite, monzogranite and leucogranite – are shown on Figure 8.10, and described below. Figure 8.11a shows the general chemical trends displayed by these granitoid rocks (Bea *et al.* 1987).

Granodiorite suite. There are both early and late generations of granodiorites. The early suite is syntectonic, was emplaced along Variscan structures, and is associated in space and time with anatectic granites developed during the Variscan D2 deformation episode. Intrusion of the late granodiorite suite was linked with crustal extension, and produced large batholiths that cross-cut D2 structures in the country rocks. Radiometric ages show the early suite to have intruded around 345 Ma BP, and the late suite at around 300–280 Ma BP (Bea *et al.* 1999; Fernández-Suárez *et al.* 2000a; Valverde-Vaquero & Dunning 2000, and references therein). There are similarities in rock types between each suite that suggest they may record part of a single episode of granitoid generation that was diachronous along the orogenic belt.

The granodiorite suite is characterized by a wide compositional range from quartz-gabbro to granite *sensu stricto*, although the dominant rock type is a medium grained, biotite granodiorite. The commonest features of granodiorites assigned to this suite are as follows. (1) The dominant mafic mineral is biotite, but hornblende and cordierite can occasionally be present, and muscovite and Al-silicates may be present only in local zones of contamination. (2) Granodiorites normally show a close time and space relationship with mafic and ultramafic rocks, with quartz-gabbros and quartz-diorites being the dominant rock types, although hornblendites, pyroxenites, and many other mafic lithologies occur (see below). (3) Plagioclase shows complex zoning and calcic (andesine-labradorite) cores that are in part relicts of skeletal cores and in part relicts of resorbed calcic crystals (Castro 2001). (4) Zircon and allanite are typical accessory minerals. (5) Geochemically, they are characterized by a metaluminous to slightly peraluminous composition, with CaO >1.5 wt% and K_2O *c.* 4 wt%. (6) They display an alumino-cafemic trend in the A–B diagram (Fig. 8.11a). (7) They are characterized by a low $^{87}Sr/^{86}Sr$ initial ratio, normally between 0.702 and 0.708, and a very variable $^{143}Nd/^{144}Nd$ initial ratio.

The origin of these granodiorites is controversial. A purely crustal origin has been suggested, involving partial melting of igneous protoliths (e.g. Villaseca *et al.* 1998; Bea *et al.* 1999; Donaire *et al.* 1999). Although the isotopic features may be accounted for by more than a single petrogenetic model, the combination of field, geochemical and isotopic relationships and the use of experimental constraints point to a petrogenetic model based on an open-system process involving assimilation and melting of continental protoliths by mantle-derived fluids and/or magmas (e.g. Moreno-Ventas *et al.* 1995; Castro *et al.* 1999a).

In addition to the overall similarities listed above, different parts of the Iberian Massif show distinctive features, as illustrated by a consideration of the three areas Galicia–Zamora, Central Range (Central System), and Extremadura-Pedroches (Fig. 8.10). The first of these, Galicia–Zamora, mostly clearly shows the distinction between the early and late granodiorite suites. Early granodiorites, such as those forming the Bayo-Vigo and Vivero massifs, were deformed by Variscan D2 and D3 tectonic phases of the Variscan orogeny (López-Plaza & Martínez Catalán 1987), and form elongated massifs running subparallel to the main Variscan structures. The late granodiorite suite is well represented in Galicia by several massifs ranging in composition from tonalite to granodiorite and biotite granite. These form nearly rounded plutons cross-cutting the main regional structures. They were not deformed by the main regional phases but may show magmatic foliations produced during emplacement.

In the Central Range (Central System), most of the granitoids of the granodiorite suite are late with respect to the main deformation phases. They were emplaced as a large batholith (Fig. 8.12) built up by means of multiple laminar intrusions of granitic magma (Fernández & Castro 1999) favoured by a regional north–south crustal extension that post-dates the main deformation phases of the Variscan orogeny (Doblas 1991). These granodiorites are peraluminous, with biotite dominant, and contain mafic microgranular enclaves of tonalitic to quartz-dioritic composition (Moreno Ventas *et al.* 1995; Villaseca *et al.* 1998; Bea *et al.* 1999). A particular feature in the Central Range

(Central System) intrusions is the existence of a gradual transition into granitoids of the monzogranite ± cordierite suite (Bea 1985; Bea *et al.* 1999). This transition occurs on a regional scale so that the limits traced on Figure 8.12 are only approximate. On a more local scale, kilometre-sized patches of cordierite monzogranite either appear surrounded by biotite granodiorites, or at the margins of the batholith near the contacts with high grade, migmatitic areas. Also shown on Figure 8.12 are mafic and ultramafic complexes intimately associated with the granodiorites.

Early porphyritic biotite granodiorites have been described in the Central Range (Central System) batholith (Bea 1985; Bea & Pereira 1990), in Piedrahita and surrounding areas, and form small, kilometre-sized, lensoidal or tabular bodies. They show a marked foliation subparallel to the main foliation of the country rock migmatites and gneisses (Bea 1985; Bea & Pereira 1990). Further south from the Central Range (Central System), the same intrusive suite is well represented by the medium grained, biotite granodiorite of the Los Pedroches batholith, and by the biotite tonalites and associated granodiorites of the central Extremadura region (Fig. 8.10). In both areas, granodiorites and tonalites are early with respect to the emplacement of cordierite monzogranites (Castro 1984). A particular feature of these batholiths is that they were emplaced at shallow levels of the crust at depths of about 6 km (Corretgé 1971; Castro 1984; Donaire 1995), and along crustal-scale shear zones that created the room for plutons as either tension gashes oblique to the main direction of shear (east–west to NW–SE) or linked to megakink structures induced by the same shear zones (Castro 1984, 1987; Castro & Fernández 1998; Fernández & Castro 1999). Despite this link with D2 tectonism, only a few plutons in central Extremadura show a well-developed solid-state foliation. In these deformed granitoid rocks, shear bands and Schistosité-Cisaillement (SC) fabrics indicate a sense of movement compatible with antithetic shear zones associated with the main deformation event that accommodated the intrusions (Castro 1987).

Locally, the granodiorites of Los Pedroches show subvolcanic facies characterized by the presence of orthopyroxene. These appear at the NW end of the batholith (Zalamea area) and also forming minor stocks such as the Garlitos intrusion at the north of Los Pedroches main body. In the Extramadura area, the rocks of this suite lack subvolcanic facies and are more tonalitic, more aluminous (locally with cordierite), and more potassic than the Pedroches granodiorite. In both areas, the granitoid intrusions contain mafic microgranular enclaves and complex zoned plagioclases, indicative of disequilibrium processes related to the involvement of more than a single magma batch in their genesis.

Peraluminous leucogranite suite. Almost all these granitoid rocks were intruded between 310 and 330 Ma BP, and have abundant pelitic inclusions and alumina-rich minerals (andalusite, sillimanite and, in the allochthonous massifs, abundant cordierite as well). They show very low CaO content, low $^{87}Sr/^{86}Sr$, widely variable $^{147}Sm/^{144}Nd$ ratios (0.1196–0.1762; Beetsma 1995), an absence of heavy REE (HREE) depletion, and display limited geochemical variation in comparison with the biotite calc-alkaline granitoids. The group includes all anatectic autochthonous, para-autochthonous and allochthonous granites, and is related in space and time to a Variscan regional D2 plutonometamorphism that overprints a Barrovian D1 phase (Escuder *et al.* 1996; Valverde & Fernández 1996). A good example of this association between anatexites and granites is found in the El Tormes dome (Fig. 8.13). The interdependence between low P, high T metamorphism and granitoid development is widely accepted (e.g. Lux *et al.* 1986) and advective heating should be considered as an important cause–effect mechanism in the general pattern of metamorphic domes and metamorphic belts in the Iberian massif. Although the evidence points to purely thermal crustal processes (decompression melting; Castro *et al.* 2000) in the genesis of two mica granitoids, the genesis of cordierite-rich two-mica granitoids may represent more complex processes. Barbero *et al.* (1995) have precisely established the thermobaric conditions in the anatectic complex of Toledo situated at the core of the Central Iberian Zone. This complex displays basic rocks and calc-alkaline granites with almost coeval granulite facies metamorphism which yielded important volumes of synorogenic (D3), strongly peraluminous, cordierite-bearing granitoids. The peak conditions of this metamorphism

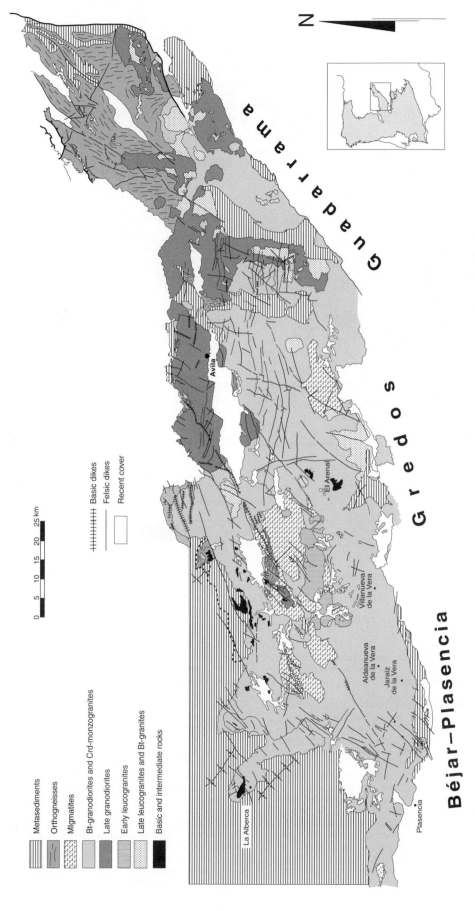

Fig. 8.12. Geological map showing the distribution of granitic and metamorphic rocks in the Central Range (Central System). The most significant mafic complexes are also depicted. This map has been compiled since the following sources: Mapa Geológico y Minero de Extremadura and Mapa Geológico y Minero de Castilla–León for the Béjar–Plasencia and Gredos sectors, and compilation by Villaseca *et al.* (1993) for the Guadarrama sector.

Metasediments

Orthogneisses

Mlgmatites

Bt-granodiorites and Crd-monzogranites

Late granodiorites

Early leucogranites

Late leucogranites and Bt-granites

Basic and intermediate rocks

Basic dikes

Felsic dikes

Recent cover

0   5   10   15   20   25 km

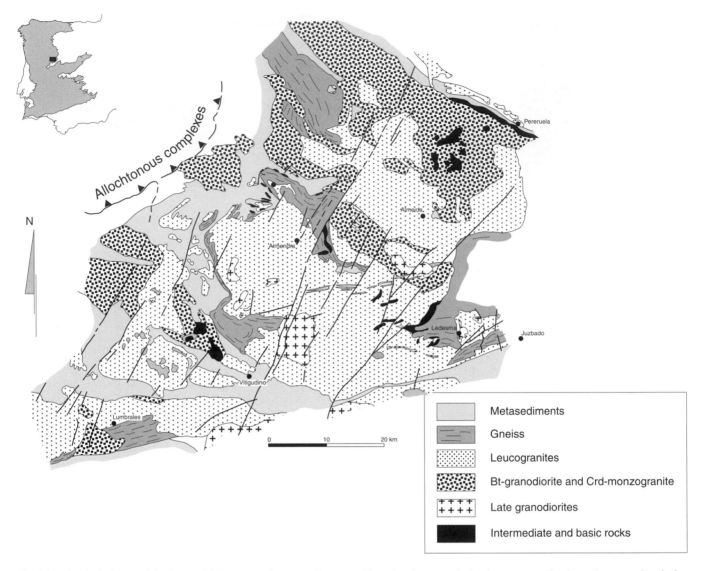

Fig. 8.13. Geological map of the Domo del Tormes and surrounding areas. Note the close association between peraluminous leucogranites (suite 3) and glandular gneisses. Note also the association of intermediate rocks and cordierite–granodiorites and monzogranites (suite 2). Based mainly on the mapping of Martínez (1974), Carnicero (1980) and López Plaza (1982), and the synthesis by López Plaza & Carnicero (1987).

(800 ± 50°C and 4–6 kbar) are higher than elsewhere in the Central Iberian Zone, and can be related to the ascent of mantle-derived magmas produced during crustal extension.

Monzogranite ± cordierite suite. Granites of the monzogranite ± cordierite suite (known as 'Serie Mixta' among Spanish geologists) crop out both in epizonal, zoned plutons as well as irregular patches with transitional contacts into biotite-rich granodiorites. The suite is mostly younger than the granodiorites, with the exception of the very late granodiorites in the northern part of the Iberian Massif (Corretgé *et al.* 1990), although at deep crustal levels, as the case of the Gredos massif in central Spain, they show transitions to granodiorites. The most typical Iberian representatives of the suite are found in zoned plutons of the Central Extremadura and Los Pedroches batholiths (Fig. 8.14). Cordierite monzogranites are typical of this suite, as is a concentric zoning with transitions into two-mica granites and late aplitic granites. Grain size, amount of K-feldspar megacrysts, and cordierite content all vary from one pluton to another, with the coarsest facies (cordierite crystals up to 6 cm long) being found in the Cabeza de Araya massif. In the Alcuéscar and Trujillo plutons

(Fig. 8.14), the cordierite monzogranites locally show transitions to biotite granodiorites in which microgranular enclaves of tonalite composition are normally present. These enclaves are also present, although very rarely, in the cordierite monzogranites.

Many workers have emphasized the petrographic peculiarities of these complex granites which contain quartz, plagioclase, K-feldspar, biotite ± muscovite ± cordierite ± andalusite ± sillimanite ± garnet (Corretgé 1971; Capdevila *et al.* 1973; Ugidos 1973, 1990; Corretgé *et al.* 1977, 1985; Barrera *et al* 1982; Bea 1982; Brandebourger *et al.* 1983; Bea & Moreno Ventas 1985). Their distinctive features include the common presence of magmatic K-feldspar megacrysts, complex zoned plagioclase crystals with multiple resorption surfaces and dendritic Ca-rich cores (Castro 2001), and sometimes conspicuous amounts of andalusite. The main geochemical features, in terms of major elements, of these peraluminous and phosphorus-rich monzogranites (Bea *et al.* 1992) are revealed by the A–B multicationic diagram (De la Roche 1964; Debon & Le Fort 1983). They plot in a position intermediate between peraluminous leucogranites and granodiorites (Fig. 8.11a), always in the peraluminous (A > 0) domain.

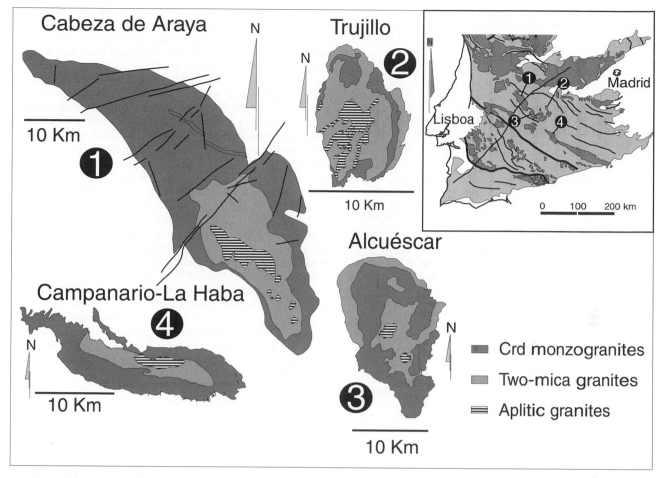

Fig. 8.14. Geological sketch of four zoned plutons from the Central Extremadura and Pedroches batholiths. These plutons have in common the zonal distribution of granitic facies and the presence of cordierite in the coarse-grained monzogranites (pluton 1 from Corretgé (1971); plutons 2 and 3 from Castro (1984); pluton 4 from Alonso Olazabal *et al.* (1999)).

Mafic and ultramafic rocks associated with the Variscan granitoids (JDR, AC, LGC)

The association of mafic and ultramafic igneous rocks with granodiorite batholiths is a feature of many orogenic belts (Pitcher 1997). In the Variscan orogen in Spain this association is well represented by small but common gabbro–granodiorite complexes in the internal and deeper level parts of the Central Iberian Zone and in the axial zone of the Pyrenees. These massifs record the input of mantle magmas with possible thermal and geochemical consequences for the generation of the granitoids, and have commonly been referred to as 'basic precursors' (Capdevila 1969; Bea *et al.* 1999). They show an anomalous chemistry with high MgO, K_2O and water contents, low silica, and a crustal Nd isotope signature, together with distinctive textures, part cumulate and part pegmatitic. Here, we follow the proposal by Pitcher (1997) in which all textural varieties, including cumulates and pegmatitic hornblendites, are grouped as an 'appinite suite' in which cortlandtites (noritic cumulates), hornblendites and vaugnerites (biotite-rich diorites) are included. In our experience, all these types, including transitions between them, can appear together in breccia-like complexes. A good example is the Sanabria complex, where ultramafic rocks form large angular fragments

enclosed by biotite diorites and cross-cut by tonalite and granodiorite dykes. These rocks are interpreted as hybrid andesitic melts coming from an enriched mantle source or as hybrid restites left after the partial extraction of a granodiorite melt produced by assimilation of pelitic gneisses by basalt magmas (Patiño Douce 1995; Castro *et al.* 1999*a*). Hornblendites of the appinite suite are the hydrated equivalent of residual noritic mafic granulites in the lower crust. Remnants of these residual rocks may be transported upwards in the crust by granodiorite magmas, becoming hydrated and transformed into appinites. In any case, they represent an important mantle input for heat and fluids that were involved in the generation of granodiorite batholiths.

The Sanabria complex is representative of a type of mafic complex associated with anatectic zones. Another type of intrusive group, typified by the Puente del Congosto complex in the Central Range (Central System) (Fig. 8.12), is instead characterized by complex structures that are synplutonic with, and located peripherally around, the batholithic granodiorites. They can be partly intrusive into the metasedimentary country rocks, developing large contact aureoles, and typically form roughly concentrically zoned complexes with cortlandtites and appinites at the core surrounded by diorites and tonalites. Localized zones of magma mixing and mingling are common, and mafic lithologies show lobate contacts with enclave-rich coeval granodiorite which, in the same place, can backvein and intrude the mafic bodies. Such relations form during cooling as the mafic system reaches the solidus before the consolidation of the granodiorite melt. Magmatic flow in the hosting granodiorite may induce brecciation and backveining of the consolidated mafic rocks.

The Sanabria-type complexes are also zoned, with the more mafic lithologies appearing in the core and rimmed by diorite and tonalite. The rocks at the core are normally brecciated and intruded by diorites and granodiorites. Partially disaggregated fragments of pelitic gneisses may appear within these breccia-like complexes, and the diorites and tonalites surrounding the mafic and ultramafic core are strongly foliated and interleaved with migmatized gneisses. Some of the best examples of the Sanabria type are the Finisterre, Vivero (Galán 1984) and Bayo-Vigo (Gallastegui 1993) complexes in Galicia. Another similar intrusive unit is the Sayago complex in Zamora, in which three magmatic association trends have been distinguished and were again probably linked to contamination processes (López Moro 2000). These three associations are dioritic, monzonitic-melanosyenitic, and granodioritic with abundant vaugnerites and amphibole-rich intermediate rocks (appinites).

In the Central Range (Central System) (Fig. 8.12), appinite complexes are mostly of the peribatholithic type (Puente type). However, some remnants of dismembered complexes of Sanabria type appear in close association with migmatites that form large, kilometre-sized inclusions within the granodiorites. The Puente del Congosto complex (Fernández *et al.* 1997*b*) is one of the most studied in this sector, with others including El Arenal in Gredos (Moreno Ventas 1991), El Mirón in Avila (Franco & García de Figuerola 1986), and El Tiemblo (Casillas 1989) in Guadarrama.

Plutonism and metamorphism (FJM)

There is a close link between plutonism and post-D1 metamorphism across much of the Iberian massif. This connection is not only confined to obvious contact aureoles, but can be seen across wide areas overprinted by 'regional contact' metamorphism. In the Boal-Los Ancares dome, for example, a prograde zoning with biotite and andalusite-cordierite not only develops in close relationship with some granitic stocks but is widespread across a relatively large area. A logical deduction is that underlying this area there is a granite volume larger than that outcropping. In the central and western Iberian Massif, elongated domes coalesce or are linked one to another and form part of a large, medium to high grade metamorphic area running from Brittany across Galicia and northern Portugal to the Central Range (Central System) (Fig. 8.15; Martínez & Rolet 1988; Martínez *et al.* 1988). Within this area synforms containing less metamorphosed, younger Siluro-Devonian materials are preserved as narrow strips or elongated patches, the most prominent of these being the klippen of allochthonous terranes (Chapter 9).

Metamorphic zoning in the elongated domes or belts tends to show a symmetrical pattern away from an axial zone containing granodiorite and (more commonly) peraluminous leucogranite plutons. The older of these granites tend to be lens-shaped and syntectonic with the dominant foliation which subsequently becomes folded over the antiformal dome as syntectonic granites continue to intrude. Earlier Barrovian assemblages produced during Variscan crustal thickening, become disequilibrated in migmatitic pelititic restites as these late domes develop. The rocks thus record a syndoming clockwise decompression path leading from peak pressures of around 8 kbar to low pressure conditions recorded by cordierite–K-feldspar–biotite–sillimanite (± garnet) mineral assemblages. The low-P conditions are connected with abundant granite magmatism, probably recording decompression melting in deeper domains during the doming stage, when garnet becomes resorbed into plagioclase and cordierite

coronas in pelitic restites at conditions of 650–700°C and 3–4.5 kbar (as seen in the El Tormes dome; Martinez *et al.* 1988).

In some areas, such as the Portuguese Moncorvo-Vila Real belt (Fig. 8.15), or in the above-mentioned Boal-Los Ancares belt, the Barrovian episode is lacking due to the absence of enough crustal thickening. In these areas the metamorphic history records only a single low-P, high-T event connected with granite intrusion, i.e. it is a pure contact metamorphism, with a prograde zoning from regional chlorite zone through biotite–andalusite/cordierite–sillimanite zones in metapelites. Chloritoid and staurolite occur in the Fe-richer protolith compositions, and, in these cases, staurolite can become resorbed into prograde andalusite–biotite aggregates.

Some authors have interpreted the low-P-high-T metamorphism as induced by extensional orogenic collapse (Escuder-Viruete 2000), comparing the associated structures with those in the Tertiary metamorphic core complexes of Arizona and New Mexico (Crittenden *et al.* 1980). Such a model requires rapid exhumation for the high grade domes, and remains controversial (e.g. Druguet 1997). The same antiformally folded foliation, which is synkinematic with the growth of staurolite–andalusite/fibrolite–biotite, has been interpreted differently in different areas. For example, whereas a model involving synextensional collapse has been proposed for the El Tormes dome (Escuder-Viruete 2000), one invoking syncompressional thrusting has been suggested for the Lugo dome (Aller & Bastida 1993). Further work should help to clarify the teconometamorphic mechanism responsible for these pluton-related low-P-high-T belts in Iberia.

Intrusive rocks of the Southern Iberian Massif (AC, JDR)

In contrast with the northern part of the Iberian Massif, further SW the Ossa Morena Zone preserves abundant evidence for pre-Variscan (mostly Cambro-Ordovician) plutonism. Four main compositional suites (calc-alkaline, metaluminous to peralkaline, anatectic peraluminous, and gabbro–diorite–tonalite) can be distinguished across an area subdivided into five tectonic domains (Fig. 8.16) which, from north to south, are (1) Mérida-Peraleda, (2) Badajoz–Córdoba, (3) Olivenza–Monesterio–Lora, (4) Aroche–Aracena and (5) Sierra Norte de Sevilla. The main major-element geochemical features of these southern Iberian intrusive rocks are summarized in the A–B and TAS diagrams (Figs 8.11b and 8.17). Table 8.3 shows the available age determinations of these intrusive complexes.

Calc-alkaline suite. This is represented by composite intrusions of granodiorites, tonalites and diorites, ranging in age from 544 Ma to 305 Ma. Typical examples include Burguillos del Cerro, Valencia del Ventoso, Brovales and Santa Olalla intrusions in the Olivenza–Monesterio alignment. These are concentrically zoned plutons with gabbros and diorites at the cores and granodiorites and tonalites at the margins. In Santa Olalla, which contains cordierite-bearing granitoids (Bateman *et al.* 1992), gabbros are concentrated to the north of the complex (Casquet 1980) and are intermingled with Bt-tonalites.

Metaluminous to peralkaline suite. This is represented by alkaline meta-aluminous granites and syenites, diorites and alkaline gabbroic complexes. These are mostly Ordovician in age, although some Carboniferous intrusions are present in the Badajoz–Córdoba alignment. The alkaline suite was related to intracontinental rifting, some complexes being emplaced by cauldron subsidence (e.g. Barcarrota; Carnicero & Castro 1982; Fig. 8.18), showing a structure identical to the annular complexes of the Niger–Nigeria and Oslo rifts. This

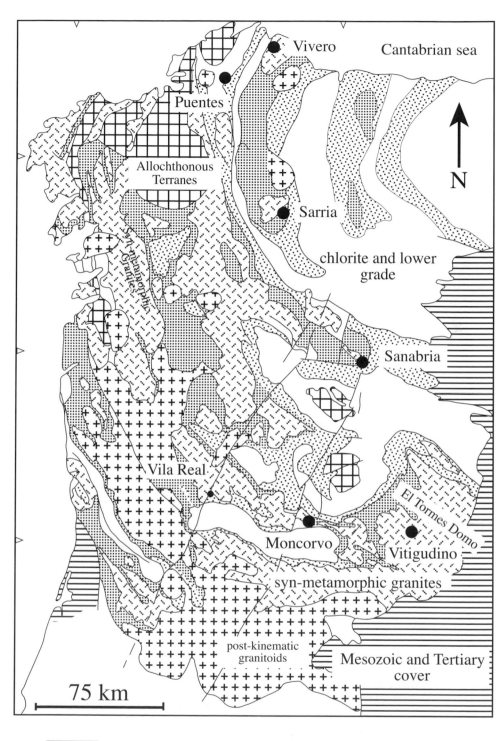

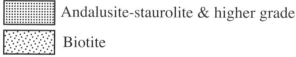

Fig. 8.15. Distribution of Variscan metamorphic belts or elongated domes in the NW Iberian peninsula. Also shown are synmetamorphic and post-metamorphic granites, the latter cross-cutting the belts and inducing a local contact metamorphic overprint.

Andalusite-staurolite & higher grade

Biotite

Ordovician rifting episode (503–470 Ma BP) produced the complexes of Barcarrota, El Castillo and Almendral, which are located at the core of the Olivenza–Monesterio antiform and escaped Variscan tectonic overprint. In contrast, alkaline, rift-related, Ordovician complexes located within the Badajoz–Córdoba shear belt, were intensively deformed and metamorphosed during the Variscan orogeny (the gneisses of Aceuchal and Almendralejo; Fig. 8.16).

Anatectic peraluminous suite. This comprises mostly Cambrian granites and migmatitic rocks in high grade cores of Cadomian

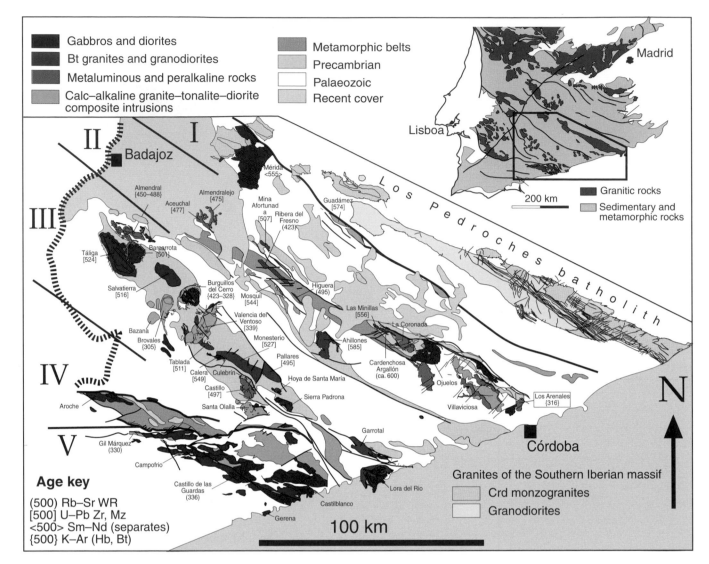

Fig. 8.16. Geological sketch of the southern Iberian Massif showing the main intrusive rocks and magmatic alignments. Reported ages (in Ma) on this figure (labels) are the most representative as ages of intrusion. A complete list of age determinations, errors and methods is given in Table 8.3. The map has been compiled and synthesized from the 1:50 000 ITGE geological maps and complemented with descriptions from the following sources. Sector I (Mérida-Peraleda): Mérida (Gonzalo 1988; Bandrés *et al.* 1999); Guadámez–Peraleda (Castro 1987; Bandrés *et al.* 1999). Sector II (Badajoz–Córdoba): Almendralejo–Aceuchal, Mina Afortunada, Ribera del Fresno, Higuera de Llerena, Mosquil (Abalos 1990; Abalos *et al.* 1991*a*); Ahillones (Sánchez Carretero *et al.* 1989); Cardenchosa, Azuaga and Riscal (Azor 1997); Villaviciosa–Ojuelos–Coronada (Pascual 1981; Pascual & Pérez Lorente 1987; Sánchez Carretero *et al.* 1989*a,b*). Sector III (Olivenza–Monesterio–Lora): Almendral (Sánchez Carretero *et al.* 1999); Barcarrota (Castro 1981; Carnicero & Castro 1982); Táliga (Carnicero & Castro 1982); Brovales, Burguillos del Cerro, Valencia del Ventoso (Pons 1982); Tablada, Calera, Monesterio, Pallares, Castillo, Culebrín, Hoya de Sta. María, Sierra Padrona (Eguíluz 1987); Santa Olalla (Casquet 1980; Eguíluz *et al.* 1989); Garrotal, Lora del Río (Apraiz 1998). Sector IV (Aroche–Aracena): Aroche (Bard 1969); Los Molares (Castro *et al.* 1996*a*; El–Hmidi 2000). Sector V (Sierra Norte of Seville): Simancas (1983), De la Rosa (1992).

metamorphic belts (Monesterio, Badajoz–Córdoba and Aracena; e.g. Eguiluz *et al.* 2000). Late Proterozoic examples also exist, such as those found within the Badajoz–Córdoda alignment (Cardenchosa–Argallón, *c.* 600 Ma BP: Azor *et al.* 1995; and Azuaga, 611 Ma BP: Schäfer 1990) and in the Mérida–Peraleda alignment (Guadamez, 574 Ma; Ordóñez 1998).

Gabbro–diorite–tonalite suite. This can in places be associated with calc-alkaline granitoids and with metaluminous and peralkaline granitoids (Fig. 8.16). Rocks of this suite have been dated at 555 Ma BP (Bandrés *et al.* 2000) in Mérida and at 503 Ma BP in Barcarrota (Ochsner 1993). Gabbroic rocks of the Variscan orogenic cycle are well represented in the Badajoz–Córdoba alignment (Ojuelos–La

Coronada; Pascual 1981) and in the southernmost part of the Iberian massif in the South Portuguese Zone (Castillo de las Guardas and Castilblanco; De la Rosa *et al.* 1993). Noritic rocks of boninite affinity have been described within the Aroche–Aracena alignment (Los Molares, 340–328 Ma BP; Castro *et al.* 1999*b*).

The calc-alkaline suite has the typical signature of subduction-related magmatism such as is seen in the Andes and Sierra Nevada batholiths of America (Figs 8.11b and 8.17). The generation of these calc-alkaline granitoids implies the existence of a Precambrian continental basement in the Ossa Morena Zone. Inherited zircons in intrusions record ages between 3.0 and 0.6 Ga (De la Rosa *et al.* 1999), similar to zircons in the Serie Negra metasediments (2.1 Ga and older; Shäfer *et al.* 1993). Nd model ages reported for the granitoids of the

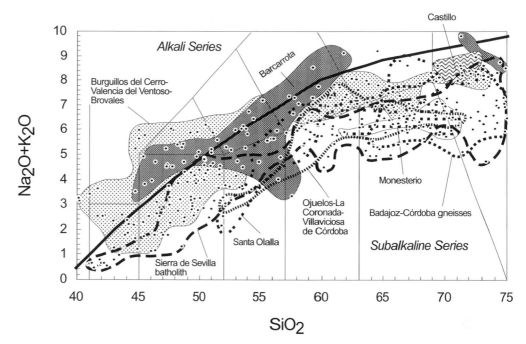

Fig. 8.17. Total alkalis–silica (TAS) diagram showing representative intrusive complexes of the southern domain of the Iberian Massif. Basic and intermediate rocks of the Ossa Morena Zone are richer in alkalis in comparison with similar rocks of the South Portuguese Zone (Seville batholith). Field divisions are according to Le Bas *et al.* (1986) and alkaline–subalkaline division line according to Irvine & Baragar (1971).

Sierra Norte de Sevilla are within the range 1.3–1.1 Ga (J. D. De la Rosa, unpublished data), but more work is necessary to define more clearly the nature of this basement.

Intrusive rocks of the Pyrenees and Catalonian Coastal Ranges (PE)

Plutonic masses in the northeastern corner of the Iberian peninsula occur in the Alpine-controlled Variscan outcrops of the Pyrenees and Catalonian Coastal Ranges (Fig. 8.19). The largest of these outcrops is in the east–west striking Pyrenean axial zone, which is up to 400 km long and 100 km wide. Pyrenean plutons crop out over an area of about 2600 km² and form about 30 intrusions more than 10 km², eight of which have batholithic dimensions (more than 100 km²). Batholiths such as those of Cauterets Panticosa, Andorra-Mont Lluís, Querigut-Millas, and the St. Laurent–La Jonquera massifs (Fig. 8.19) record a multiple intrusion history, with several

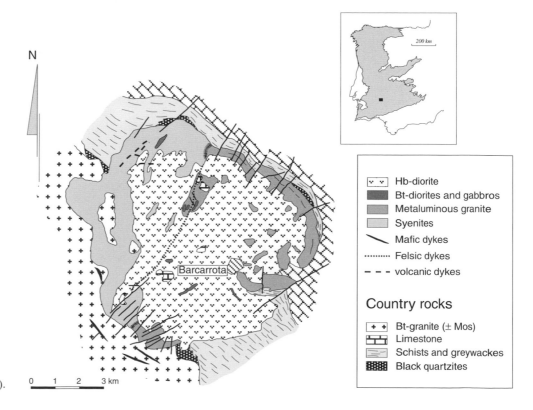

Fig. 8.18. Geological map of the alkaline–anorogenic Barcarrota zoned complex (from Castro 1981).

Legend:
- Hb-diorite
- Bt-diorites and gabbros
- Metaluminous granite
- Syenites
- Mafic dykes
- Felsic dykes
- volcanic dykes

Country rocks
- Bt-granite (± Mos)
- Limestone
- Schists and greywackes
- Black quartzites

0 1 2 3 km

Table 8.3. Geochronological data of intrusive suites in the southern Iberian Massif

Pluton	Suite	Rock type	Age (Ma)	Method	$^{87}Sr/^{86}Sr_T$	Source
Sector I. Mérida–Peraleda						
Mérida	1	Gabbro	554.7 ± 3.4	Sm-Nd Grn-WR	0.512024	1
Valle de la Serena	4	Granite	574	U-Pb Zr SHRIMP		2
Sector II. Badajoz–Córdoba						
Aceuchal	3	Bt-Hast orthogneiss	477.1 +4.8/−4.1	U-Pb Zr		3
Almendralejo	3	Riec-Aeg orthogneiss	474 +9.5/−6.3	U-Pb Zr		3
Almendralejo	3	Riec-Aeg orthogneiss	475 ± 11	Rb-Sr WR		4
Almendralejo	3	Riec-Aeg orthogneiss	473 ± 22	Rb-Sr WR		4
Mina Afortunada	2	Anatectic gneiss	506.7 +8.8/−6.6	U-Pb Zr		3
Ribera del Fresno	2	Gneiss	423 ± 38	Rb-Sr WR		4
Mosquil	4	Amp-Px tonalite	544 +6/-5	U-Pb Zr		3
Ahillones	2	Fine grained granite	585 ± 5	U-Pb Zr		5
Las Minillas	2	Bt orthogneiss	556 +159/−67	U-Pb Zr		3
Las Minillas	2	Bt orthogneiss	474	U-Pb SHRIMP		6
Cardenchosa	2	Orthogneiss	690 ± 134	Rb-Sr WR	0.7031	7
Azuaga	2	Gneiss	611 +17/−11	U-Pb Zr		5
Arroyo Argallón	2	Orthogneiss	632 ± 103	Rb-Sr WR	0.7052	7
Riscal & Higuera de Llerena	2	Gneiss	495 ± 13	Rb-Sr WR	0.7109	7
Villaviciosa C-La Coronada	3	Granite	316 ± 16	K-Ar WR		8
Villaviciosa C-La Coronada	3	Granite	332 ± 17	K-Ar WR-Bt		8
Obejo	4	Diorite	327 ± 16	K-Ar		8
Venta de Azuel (Córdoba)	4	Granodiorite	342 ± 17	K-Ar		8
Sector III. Barcarrota–Monesterio–Lora del Rio						
Almendral	3	Syenite-granite	450 ± 12	K-Ar		9
Almendral	3	Syenite-granite	481 ± 10	K-Ar		9
Taliga	2	Bt granite	541 ± 42	Rb-Sr WR		10
Taliga	2	Bt granite	530 ± 32	Rb-Sr WR		10
Taliga	2	Bt granite	474 ± 8	K-Ar Ms		11
Taliga	2	Bt granite	385 ± 11	K-Ar Ms		11
Taliga	2	Bt granite	369 ± 10	K-Ar Ms		11
Taliga	2	Bt granite	524 +1.3/-1.1	U-Pb Mz		3
Taliga	2	Bt granite	482 +37/-54	U-Pb Zr		3
Barcarrota	1	Gabbro-diorite	503.1 +4.5/-2.1	U-Pb Zr		3
Barcarrota	3	Amp-granite	500.8 +3.6/-3.2	U-Pb Zr		3
Barcarrota	3	Q-monzonite	482 ± 10	Rb-Sr WR		10
Barcarrota	3	Q-monzonite	593 ± 11	K-Ar WR		11
Barcarrota	3	Q-monzonite	471 ± 8	K-Ar WR Hb		11
Barcarrota	3	Q-monzonite	475 ± 10	K-Ar WR Bt		11
Salvatierra de los Barros	2	Bt granite	564 ± 160	Rb-Sr		12
Salvatierra de los Barros	2	Bt granite	516 +8.8/-2.7	U-Pb Mz		3
Salvatierra de los Barros	2	Bt granite	431 ± 15	K-Ar Bt		13
Brovales	4	Diorite-tonalite	305 ± 11	K-Ar Bt		13
Burguillos del Cerro	4	Monzonite	423 ± 30	K-Ar Bt		13
Burguillos del Cerro	4	Granodioritic dykes	328 ± 10	K-Ar Bt		13
San Guillermo	4	Granite	279 ± 10	K-Ar Ms		13
Valencia del Ventoso	4	Monzodiorite	339 ± 50	K-Ar Bt-Hb		13
Valencia del Ventoso	4	Dolerite	478 ± 100	K-Ar Bt		13
Tablada	2	Granite	494 ± 61	Rb-Sr WR		12
Tablada	2	Granite	494 ± 18	Rb-Sr WR		12
Tablada	2	Granite	511 ± 8	U-Pb Mz		3
Tablada	2	Granite	512 +7.8/-5.0	U-Pb Xe		3
Valuengo	2	Two mica granite	159 ± 10	K-Ar Ms		13
Valuengo	2	Two mica granite	102 ± 50	K-Ar Ms		13
Calera	2	Granite	549 ± 16	U-Pb Zr SHRIMP		2
Monesterio	2	Granodiorite	528 ± 100	Rb-Sr WR		12
Monesterio	2	Granodiorite	552 ± 72	Rb-Sr WR		12
Monesterio	2	Granodiorite	526.8 +9.9/-7.0	U-Pb Xe		3
Monesterio	2	Granodiorite	476 +13/-17	U-Pb Zr		3
Monesterio	2	Granodiorite	530	U-Pb Zr SHRIMP		3
Monesterio	2	Granodiorite	565	U-Pb Zr SHRIMP		14
Pallares	2	Granodiorite	495 +7/-8	U-Pb Zn		5
Pallares	2	Granodiorite	507 ± 21	Sm-Nd Wr-Ap		5
Pallares	2	Granodiorites	571 ± 190	Rb-Sr WR		15
Castillo	3	Hastss-Monzogranites	497.6 +9.5/-7.1	U-Pb Zr		3
Castillo	3	Hastss-Monzogranites				

Table 8.3. Continued

Pluton	Suite	Rock type	Age (Ma)	Method	$^{87}Sr/^{86}Sr_T$	Source
Sector IV. Aroche–Aracena						
Los Molares	1	Metanorites	340 ± 23	Sm-Nd WR		16
Los Molares	1	Metanorites	328 ± 4	Rb-Sr WR	0.708094	16
Los Romeros	2	Migmatites	331 ± 27	Rb-Sr WR	0.717293	16
Los Molares	1	Nebulites	232 ± 4	Rb-Sr WR		16
Cortegna	2	Migmatites	351 ± 58	Rb-Sr WR	0.712059	16
Sector V. Sierra Norte de Sevilla						
Castillo de las Guardas-Castilblanco des los Arroyos	1	Cumulate rocks, Hb-gabbros	336 ± 98	Rb-Sr WR	0.70416	17

Sources: 1, Bandrés *et al.* (2000); 2, Ordóñez (1998); 3, Ochsner (1993); 4, García-Casquero *et al.* (1985); 5, Schäfer (1990); 6, H. J. Schäfer in Azor *et al.* (1995); 7, Azor *et al.* (1995); 8, Bellon *et al.* (1979); 9, Galindo & Portugal Ferreira (1988); 10, Galindo *et al.* (1990); 11, Galindo *et al.* (1988). 12, Quesada, in Ochsner (1993); 13, Dupont *et al.* (1981); 14, Schäfer *et al.* (1993); 15, Cueto *et al.* (1983); 16, Castro *et al.* (1999); 17, De la Rosa *et al.* (1993). Suite 1, gabbros and diorites; Suite 2, biotite-granites and granodiorites; Suite 3, Metaluminous and peralkaline rocks; Suite 4, Calc-alkaline granite–tonalite–diorite composite intrusions.

clearly distinguishable plutonic bodies. In the Catalonian Coastal Ranges there are a number of Variscan intrusions of various shapes and sizes and distributed along a NE–SW trending belt about 270 km long and up to 50 km wide.

Following Autran *et al.* (1970) the Pyrenean granitoids can be divided into upper, middle and lower massifs, depending upon their level of emplacement, these three categories corresponding closely to the epizonal, mesozonal and catazonal granitoids of Buddington (1959). Granitoids belonging to the upper massif are the most abundant and form large multiple-

intrusion batholithic masses with their upper levels emplaced into a country rock envelope of very low metamorphic grade. The middle massifs are much smaller and essentially consist of numerous stocks and plutons of peraluminous leucogranites linked to zones of amphibolite facies regional metamorphism (Autran *et al.* 1970). Finally, small intrusive bodies are associated with very high grade rocks of the lower massifs, the best example being the Ansignan charnockitic complex emplaced into granulitic gneisses of the Agly massif (Autran *et al.* 1970; Fonteilles 1970).

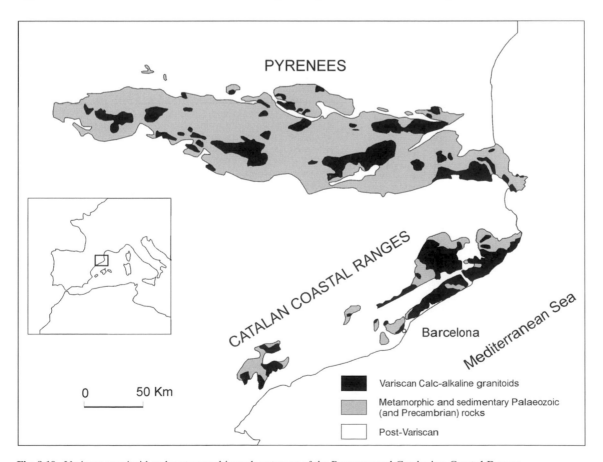

Fig. 8.19. Variscan granitoid and metamorphic rock outcrops of the Pyrenees and Catalonian Coastal Ranges.

In the Variscan outcrops of the Catalonian Coastal Ranges plutonic rocks are dominant and most are essentially similar to the Pyrenean upper massif intrusions, with the largest intrusion being the huge Montnegre–Montseny–Guilleries–Costa Brava–Gavarres batholith (Fig. 8.20). A few high-level leucogranitic intrusions of alkaline affinity have also been recognized (Ferrés 1998; Ferrés & Enrique 1994, 1996). The geochemical affinities, in terms of major elements, of these plutonic associations are depicted in the A–B multicationic diagram of Figure 8.21. Table 8.4 shows the available radiometric dating of these main plutonic complexes from Pyrenees and Catalonian Coastal Ranges.

Pyrenean upper massif intrusions. The average depth of emplacement of these plutons is 6–8 km (Pouget *et al.* 1989), although the roof of some intrusions seems to have reached levels close to the surface (intrusive contacts in the uppermost pre-Stephanian Palaeozoic country rock are present in several plutons such as Cauterets–Panticosa, Maladeta and Quérigut; Zwart 1968; Leterrier 1972; Debon 1975; Charlet 1979). Hypabyssal or subvolcanic minor intrusions or stocks directly related to the plutonic rocks are relatively rare (Debon & Autran 1996). In the upper levels of intrusions the contacts are very sharp and the country rock Variscan structures are clearly cross-cut by, and enclosed as xenoliths within, the granitoid rocks. Intrusion-related deformation of the epizonal country rocks on a small scale is generally unimportant, although deformation on a large scale occurs in several massifs (Pesquera 1985; Bouchez & Gleizes 1995; Leblanc *et al.* 1996; Palau 1995), giving the plutons a syntectonic character.

Well developed contact metamorphic aureoles surround granitic plutons in their upper levels, superimposed on a previous regional metamorphism, as is well seen along the eastern margins of the Mont-Lluís and Saint Arnac plutons (Guitard *et al.*1984; Autran 1996a). In contrast, at lower intrusive levels, high-grade regional metamorphism, deformation and contact metamorphism can be almost synchronous with the emplacement of the magmatic bodies, e.g. Mont-Lluís (Autran 1996a; Autran *et al.* 1970), La Jonquera (Liesa 1988), Northern Cap de Creus (Carreras & Druguet 1994; Enrique 1995), Trois Seigneurs (Leblanc *et al.* 1996).

Many of the intrusions show a stratiform, laccolithic or domal morphology, several kilometres thick, such as is seen at Lys (Clin 1959), Andorra–Mont Lluís and St. Laurent (Autran *et al.* 1970), and Cauterets–Panticosa (Debon 1975). Dyke-like feeders to these structures can crop out in their most eroded parts (Autran *et al.* 1970; Autran 1996a; Autran & Cocherie 1996). Other shapes (such as funnel, balloon, oil drop or mushroom) have been inferred for some of the plutons. Intrusion mechanisms proposed for different plutons include magmatic stoping (e.g. Costabona pluton; Autran 1996b) and diapirism (e.g. Lesponne, Néouvielle and Quérigut: Soula *et al.* 1986; Pouget *et al.* 1989).

Concentric structures in the intrusive bodies are common, with the most basic lithologies (gabbroids and diorites) generally located in the outermost parts, and the most acidic ones (granites, leucogranites and aplites) in the intrusion core. Good examples of this zonation are displayed by the massifs of Bordères (Forghani 1965), Cauterets-Panticosa (Debon 1975, 1980), Quérigut (Leterrier 1972; Marre 1973), Maladeta (Charlet 1979, 1982), Marimanya (Palau 1995, 1998) and Mont Lluís (Autran *et al.* 1970; Autran 1980). Reverse concentric structures with peripheral granites and inner gabbro-dioritic rocks, as seen in the Aya pluton (Pesquera 1985), are only rarely present. Zoning in Pyrenean batholiths and plutons can be sharp or indistinct. Sometimes the change in composition or texture from one zone to another is due to successive magmatic pulses clearly separated in time, producing sharp and irregular contacts between the different intrusions, such as at Quérigut (Leterrier 1972) and Mont Lluís (Autran 1996a). However, continuous and progressive changes in composition without intrusive contacts are more typical (Bassiés massif; Debon & Guitard 1996; and Marimanya; Palau 1995). Many of the Pyrenean intrusive complexes also display internal magmatic fabrics, as revealed both by alignment of mineral sand enclaves, such as is seen at Quérigut and Marimanya (Palau 1995), and by magnetic susceptibility anisotropy (Gleizes *et al.* 1993, 1997; Bouchez & Gleizes 1995; Leblanc *et al.* 1996).

The Pyrenean intrusions mostly comprise medium grained biotite granodiorite (sometimes with small amounts of hornblende), although tonalite and monzogranite also occur. In addition, some muscovite- and cordierite-bearing monzogranitic facies exist, e.g. Maladeta's Aneto granodiorites (Charlet 1979; Arranz 1997) and Cauterets porphyritic monzogranite (Debon 1975). Moreover, porphyritic granodiorites and monzogranites with alkali-feldspar megacrysts up to 5 cm are not uncommon, e.g. Quérigut's porphyroid monzogranite from Bains d'Escouloubre (Leterrier 1972), Maladeta's hornblende–biotite porphyritic granite (Charlet 1979) and Saint Laurent's pink porphyritic granite (Cocherie 1985). There is a good correlation between the mafic content (expressed as the B parameter from De la Roche (1964): $B = (Fe + Mg + Ti)/5.55$) and the Quartz, Alkali feldspar, Plagioclase (QAP)-modal composition of the main plutonic units of the Pyrenees. Following Debon & Guitard (1996), monzogranites can be grouped into those with B-values lower than 6% (two-mica or biotitic leucogranites) and those with B-values between 6 and 12% (biotitic monzogranites, sometimes porphyritic and, occasionally, peraluminous containing cordierite or muscovite). Biotitic granodiorites are the most widespread plutonic rocks. Generally they have a mafic content ranging between 12 and 19%. Some types contain small amounts of hornblende. Tonalites have B-values between 19 and 29%, and still more mafic lithologies (B > 29%) include several intermediate, basic and ultrabasic rocks, all of which are clearly subordinate to the main body of the felsic pluton. Fine to medium grained biotite–hornblende quartzdiorites and diorites are widespread and often form small stocks or heterogeneous mixed zones in association with tonalites or granodiorites, e.g. Quérigut massif (Leterrier 1972) and Cauterets-Panticosa massif (Debon 1975, 1980). They are usually also present as microgranular enclaves in most granodiorites and tonalites.

Another interesting group includes hornblende–biotite gabbroids and ultramafic rocks. Generally, pyroxene is rare or absent in these rocks, such as Quérigut (Leterrier 1972), Bielsa, Millares and Lis-Caillaouas (Enrique 1989) but in some plutons it is present in significant amounts, e.g. quartzgabbros and gabbronorites from Bordères (Forghani 1965), Taüll gabbros from the Maladeta massif (Charlet 1979, 1982; Vitrac-Michard *et al.* 1980) and noritic gabbros from Saint-Laurent de Cerdans (Autran & Cocherie 1996). Ultramafic rocks are mainly represented by coarse-grained hornblendites containing variable quantities of olivine, phlogopite and pyroxenes (cortlandtites). Large poikilitic amphiboles (1–2 cm) typically include small rounded or automorphic olivine, giving these rocks a characteristic texture. They form very localized small masses spatially related to hornblende gabbros. Sometimes a gradual transition from hornblende gabbros, through mela-hornblende gabbros and plagioclase–hornblendites to cortlandtites is found. This association is similar to the appinitic suite described in Donegal (Pitcher & Berger 1972). Ultramafic rocks such as these are relatively abundant in the Quérigut massif (Leterrier 1972; Marre 1973) and in the Saint Laurent-Albera (Autran *et al.* 1970; Cocherie 1985; Autran & Cocherie 1996).

Pyrenean middle massif intrusions. The granitoid rocks of the middle massifs form a group of numerous small stocks and plutons (rarely larger than 5 km²) which are always located in medium-grade metamorphic areas. They are almost entirely composed of peraluminous leucogranites and pegmatites. The main intrusions, from west to east, are: (1) Tramezaygues and Bossost (Kleinsmiede 1960; Zwart 1962, 1963a, 1979; Pouget *et al.* 1988); (2) La Ruse and Estibat (Trois Seigneurs massif) (Kriegsman 1989; Wickham 1987; Mercier *et al.* 1996); (3) Ax-les-Thermes (Zwart 1965; Raguin 1977; Autran 1996c); (4) Canigou (Guitard 1970; Autran 1980, 1996c). These intrusions were typically emplaced into sillimanite grade Lower Palaeozoic metasediments and Precambrian gneisses. They always lie close to (or in) the anatectic domain and rarely intrude lower grade metamorphic zones, a fact strongly suggesting an anatectic origin involving migmatization of pelitic and gneissic rocks in metamorphic cores.

The shape of these intrusions is generally complex and may be poorly defined. Sheet-like morphologies follow the main regional foliation and are separated by metamorphic screens. Only a weak

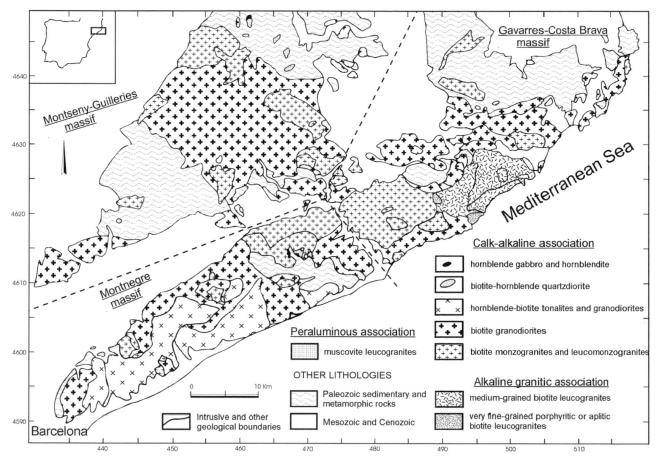

Fig. 8.20. Main plutonic units of the Catalonian coastal batholith.

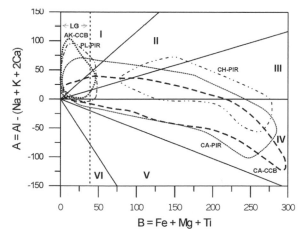

Fig. 8.21. Plot of the main plutonic associations from the Pyrenees and Catalonian coastal batholith on the A–B diagram (Debon & Le Fort 1983). Pyrenees: CA-PIR, calc-alkaline association; CH-PIR, charnockitic association; PL-PIR, peraluminous leucocratic association. Catalonian coastal batholith: AK-CCB, alkaline leucogranitic association; CA-CCB, calc-alkaline association.

foliation is usually observed in the granites and for this reason they are generally considered syntectonic with the late deformation phases (e.g. Canigou and Ax-les-Thermes; Autran 1996c). Good examples are provided by the Tramezaygues and Bossost leucogranites which form several groups of intrusions with different morphologies and

sizes (cupola, concordant or cross-cutting dykes and sills, irregular stocks), all emplaced in sillimanite grade Cambro-Ordovician metasediments. They are associated with numerous pegmatites and aplites and usually contain screens of the enclosing schists.

The laccolithic Ax leucogranite (15 km²) develops over the top of the Aston-massif orthogneisses and several apical apophyses intrude into the Palaeozoic andalusite-bearing micaschists (Autran 1996c). In the lower part of the pluton the leucogranite is locally in contact with migmatitic orthogneiss. Biotite and muscovite defines a marked foliation in this granite, which has contacts that commonly cross-cut the country rock foliation. The intrusion has been interpreted as syn-kinematic with the doming of the Aston metamorphic massif.

Another example is the Canigou leucogranite (10 km²), which is located at the core of a late antiformal structure. A great number of vertical intrusions feed a series of sheet-like concordant bodies of peraluminous leucogranites and pegmatites which occasionally cross-cut folds related to the doming of the Canigou massif (Guitard 1970; Guitard et al. 1984; Autran 1996c). Pegmatites are also common in most of the other Pyrenean metamorphic areas, such as Bosost, Trois Seigneurs, Albera and Cap de Creus (Zwart 1962, 1979; Autran et al. 1970; Carreras 1975). These pegmatites form dyke-like intrusions, sills or irregular stocks ranging from a few centimetres to several tens of metres thick. They are always located inside the sillimanite or andalusite–cordierite zones and are variably deformed by late-Variscan tectonic phases.

QAP compositions of plutonic rocks from the Middle massifs are almost totally restricted to monzogranites that closely correspond to cotectic melts in the Q-Ab-Or system. Essentially these rocks are composed of quartz, oligoclase and microcline, but they may contain several types of mafic or accessory minerals. Textures can also be very variable ranging from aplitic to coarse grained (pegmatitic). The largest plutons are usually medium grained and have a homogeneous

Table 8.4. Isotopic dating of plutonic rocks from the Pyrenees and Catalonian coastal batholith

	Intrusive complex	Method	Age (Ma)	$(^{87}Sr/^{86}Sr)_i$	Reference
Pyrenees					
Calc-alkaline					
Quérigut	Rb-Sr	(a)	303 ± 10	0.7091 ± 2	12
	Rb-Sr	(a)	282 ± 5	0.7095 ± 2	2
St Laurent	Rb-Sr	(a)	282 ± 5	0.7106 ± 2	5
Montlouis-Andorra	U-Pb	(b)	303 ± 3		14
Bassiès	Rb-Sr	(a)	276 ± 16	0.7115 ± 3	8
	Rb-Sr	(a)	290 ± 17	0.7112 ± 7	9
Trois Seigneurs	Rb-Sr	(a)	310 ± 6	0.7106 ± 2	7
Maladeta	Rb-Sr	(a)	277 ± 7	0.7117 ± 3	1
Néouvielle	Rb-Sr	(a)	300 ± 20	0.7111 ± 5	6
Peraluminous					
Canigou	Rb-Sr	(a)	295 ± 12	0.71503	11
	U-Pb	(c)	298 & 305		13
Soulcem	Rb-Sr	(a)	292 ± 13	0.7171 ± 18	9
Tramezaygues	Rb-Sr	(a)	301 ± 15	0.7142 ± 13	9
Ax-les-Thermes	Rb-Sr	(a)	301 ± 15	0.7142 ± 13	9
Trois-Seigneurs	Rb-Sr	(a)	309 ± 34	0.7149 ± 13	7
Charnockitic					
Ansignan	U-Pb	(d)	309 ± 5		3
Ansignan	U-Pb	(d)	315 ± 5		3
Ansignan	U-Pb	(c)	314 ± 7		4
Treilles	U-Pb	(d)	293 ± 14		10
Catalan Batholith					
Calc-alkaline					
Montnegre	Rb-Sr	(a)	296 ± 4	0.7107 ± 1	15
Montnegre	Rb-Sr	(a)	284 ± 7	0.7103 ± 1	16
Montnegre	Rb-Sr	(a)	287 ± 3	0.7102 ± 1	17
Montnegre	K-Ar	(e)	284 ± 4		18
Montnegre	Ar/Ar	(e)	286 ± 1		19
Llafranc	Ar/Ar	(e)	288 ± 1		20
		(f)	295 ± 2		21
Tossa	Ar/Ar	(e)	285 ± 2		22
		(e)	286 ± 2		22
		(f)	285 ± 3		22
Aiguablava	Ar/Ar	(e)	288 ± 3		22
Alkaline granites					
Tossa	Ar/Ar	(e)	286 ± 2		22
Tossa	Rb-Sr	(a)	275 ± 6	0.7085 ± 15	22

References: 1, Vitrac-Michard *et al.* (1980); 2, Fourcade & Allègre (1981); 3, Postaire (1982); 4, Respaut & Lancelot (1983); 5, Cocherie (1985); 6, Alibert *et al.* (1988); 7, Bickle *et al.* (1988); 8, Debon & Zimmermann (1988); 9, Majoor (1988); 10, Pin (1989); 11, Gibson (1989); 12, Fourcade & Javoy (1991); 13, Vitrac-Michard & Allègre (1975); 14, Romer & Soler (1995); 15, Enrique & Debon (1987); 16, Del Moro & Enrique (1996); 17, Enrique *et al.* (1999); 18, Solé *et al.* (1998); 19, Solé (1993); 20, Pérez *et al.* (1997); 21, Ferrés *et al.* (1997); 22, Ferrés (1998); 23, Cocherie *et al.* (1996). Rocks and minerals used: (a) whole rocks; (b) titanite; (c) monazite; (d) zircon; (e) biotite; (f) amphibole.

texture (e.g. Ax-les-Thermes granite). Muscovite or two-mica leucogranites are the most widespread petrographic type, and they commonly contain tourmaline and garnet and sometimes cordierite or sillimanite. Associated pegmatites may also contain a great diversity of accessory minerals such as biotite, muscovite, tourmaline, garnet, sillimanite, andalusite, cordierite and, in a few cases, rare-element-bearing minerals such as cassiterite, columbite-tantalite, ambligonite, beryl, zircon and apatite (Autran *et al.* 1970; Druguet *et al.* 1995; Malló *et al.* 1995; Alfonso *et al.* 1995; Alfonso & Melgarejo 2000; Clin & Debon 1996; Mercier *et al.* 1996).

Pyrenean lower massif intrusions. Catazonal outcrops of rocks belonging to the granulite facies (Vielzeuf 1984) occur only as sparse massifs and tectonic slices in the northern Pyrenean massifs, e.g. Agly massif (Fonteilles 1970; Respaut & Lancelot 1983), Port de Saleix pyriclasites (Azambre & Ravier 1978), Castillon massif (Roux 1977), Treilles norite (Pin 1989), Ursuya massif (Boissonnas 1974). Small plutons and stocks of granitoids, and more basic rocks, are occasionally found in them. The most noteworthy feature of these plutonic

rocks is that they are mineralogically in equilibrium with their granulitic envelope that has an anhydrous character. A good example is provided by the Ansignan charnockitic pluton (Autran 1996*d*), a small intrusion (about 10 km²) first described by Guitard (1960). It was emplaced into low-pressure granulitic paragneisses (hypersthene zone) of the Agly massif (eastern Pyrenees). Cartographic data suggest that it may form a laccolithic intrusion about 800 m thick (Fonteilles 1970). To the east the intrusion divides into several sheets which locally cross-cut the lithological banding of the gneisses. A magmatic foliation parallel to that of the gneisses is generally present. The main outcrop of the pluton consists of a dark-coloured highly porphyritic granodiorite which is relatively rich in biotite and hypersthene. Near the top of the intrusion there are numerous sills and heterogeneous bands of leucogranites containing garnet and biotite. In contrast, near the base of the laccolith are several layers of hypersthene biotitic diorites up to 200 m thick.

Catalonian Coastal Ranges. Variscan plutonic rocks in the Catalonian Coastal Ranges form a batholith about 1500 km² in size, 95% of which comprises calc-alkaline intrusions, with the remainder being more alkaline in character (Figs 8.19 and 8.20). This Catalan coastal batholith is made up of a large group of small- to medium-sized plutons intruded into a Palaeozoic sedimentary sequence ranging from Cambro-Ordovician to Early Carboniferous in age. The emplacement of the granitoid plutons was one involving brittle fracture mechanisms without distortion of the country rocks (i.e. magmatic stoping). The form of the plutons displays more or less flat roofs and steep walls giving way to complex morphologies. Some of these have probably been formed by cauldron subsidence. This type of emplacement, coupled with the occasional occurrence of explosive breccias (pebble dykes) in the southern massifs, suggests that the intrusion of magma took place at very high levels in the crust. Generally (but not always), intrusive contacts between different intrusions are also of brittle type following fault planes. In these cases chilled margins are not normally found, a fact that records a small thermal difference between the magmatic intrusion and its country rock.

All the plutonic intrusions show simple magmatic textures and sharp contacts with their host rocks, cross-cutting the main Variscan structures. Lower Triassic sediments lie unconformably on an eroded surface that includes both folded Lower Carboniferous sediments and these late Variscan plutonic rocks (Ashauer & Teichmuller 1935). The envelope of the granitoid intrusions is generally affected by a syntectonic low-grade regional metamorphism, although locally (e.g. Montseny-Guilleries massif) the metamorphism can reach amphibolite facies. A static contact metamorphism, overprinting the regional metamorphism, surrounds all the plutonic intrusions and gave rise to pelitic hornfels, quartzites, calc-silicate hornfels and marbles. P-T conditions of 1–1.5 kbar and 700°C have been estimated in the aureole from the central part of the batholith (Gil-Ibarguchi & Julivert 1988).

The Catalan coastal batholith is dominated (in decreasing order of abundance) by calc-alkaline granodiorites, monzogranites and tonalites, with minor amounts of quartzdiorites and some small outcrops of gabbroic rocks (Enrique 1984, 1985, 1990). Biotitic granodiorites (sometimes hornblende-bearing) are the most widespread plutonic rocks. They are generally equigranular, medium- or coarse-grained, but fine-grained types are also rather common. In some areas, mainly in the northern Gavarres–Costa Brava massif, porphyritic facies containing large K-feldspars are frequent. The monzogranites (usually with <5% biotite) are abundant in the northern part of the batholith but relatively rare elsewhere. They are typically equigranular (rarely porphyritic) but are highly variable in grain size, mineralogy and intrusion size and shape. The tonalites are similar in texture and form medium to large intrusions, but they are restricted to the central (Montnegre) and southern (Alforja, Riudecanyes) parts of the batholith. Biotite and hornblende are the normal mafic minerals in these rocks, although hypersthene joins the mafic assemblage in many of the southernmost outcrops. Quartzdiorites form numerous small plutons and stocks scattered throughout the batholith. In some cases dioritic magma was emplaced synchronously into larger granodioritic or tonalitic magma bodies giving rise to complex synplutonic relationships. Broken and dispersed dioritic magmatic masses supplied most of the microgranular enclaves found in granodiorites and tonalites of the Catalan coastal batholith. Two principal types of quartzdiorites may be distinguished: (i) hornblende- and biotite-rich (sometimes containing accessory orthopyroxene) fine- to medium-grained quartzdiorites; (ii) biotite-rich fine-grained quartzdiorites with little or no amphibole.

There are two kinds of hornblende gabbros (bojites of Streckeisen 1979) associated with the calc-alkaline rocks of the Catalan coastal batholith. The first is a group of biotitic hornblende quartzgabbros which represent a gradual transition from less mafic biotite–hornblende quartzdiorites by increasing anorthite in plagioclase and amount of mafic minerals. The average anorthite content in these rocks is higher than 50% and, as in the more felsic lithologies, the plagioclase is intensely zoned. The second group of gabbroic rocks usually does not contain biotite and shows a wide range in mafic content, grading from leucohornblende gabbros to melahornblende gabbros which may grade to ultramafic hornblendites. Their mineralogy consists of very calcic plagioclase (bytownite) (Solé 1993) and amphibole. Quartz, clinopyroxene and opaque minerals can also be present as minor phases. Amphibole is usually present as large,

anhedral grains, slightly brown in thin section, surrounding euhedral plagioclase crystals (ophitic texture).

Closely associated with the second group of hornblende gabbros are small bodies of ultramafic rocks dominated by amphibole without plagioclase. Sometimes these hornblendites (cortlandtites: Ratcliffe et al. 1982) contain up to 20% olivine (Fo_{80-75}) and minor amounts of phlogopite, clinopyroxene and orthopyroxene (and occasionally green spinel). Typically these rocks consist of coarse-grained (up to 3 cm in grain size) anhedral amphibole (tschermakite to magnesio-hornblende in the cores) which include poikilitically numerous rounded olivine crystals. As in the Pyrenees these rocks have similar features with the appinitic series from Donegal (Pitcher & Berger 1972), and they are only known from three occurrences in the Catalan coastal batholith: Montnegre, Susqueda (Enrique 1984, 1990) and Tossa.

A noteworthy feature of the Catalan coastal batholith is the presence of prominent dyke swarms trending mostly in NE–SW and ENE–WSW directions (Almera 1914; San Miguel de la Cámara 1936; San Miguel 1966; Enrique 1985; Durán 1985). The dykes tend to be concentrated heterogeneously in different parts of the batholith (e.g. Costa Brava, Susqueda, Montnegre, Alforja; Fig. 8.20). The compositions of the main dykes are closely linked to those of the various plutonic calc-alkaline rock types described above. The dominant compositions are granodioritic and granitic, although leucogranitic, tonalitic, dioritic and gabbrodioritic are also usually present. The dykes range from a few centimetres to more than 100 m in width, and from a few metres up to 5 or 6 km in length (Enrique 1990).

In addition to the main calc-alkaline plutonic intrusions there is a prominent 65 km² alkaline leucogranitic intrusion known as the Cadiretes alkaline complex (Ferrés 1998), located in the Gavarres–Costa Brava massif to the NE (Ferrés & Enrique 1996). It comprises several highly acidic ($SiO_2 > 74\%$) leucogranitic (mafic minerals <5%) intrusions. Intrusive contacts are suggestive of a ring complex structure about 10 km in diameter, built by successive cauldron subsidence processes. The main coarse- and medium-grained intrusions would represent the cupolas of two major ringdykes. In the borders of these intrusions, minor, discontinuous stocks may represent a later, uncompleted ring-dyke. This is cross-cut by genetically related felsic, granophyric and porphyritic microleucogranitic dykes. The stocks are very fine grained, show micrographic textures and contain miarolitic cavities, suggestive of a very shallow level of emplacement. All these intrusive rocks show a very homogeneous mineralogy, essentially composed of quartz, K-feldspar, sodic plagioclase (sodic oligoclase to albite) and biotite. Biotite is the only mafic mineral and is enriched in annite and in halogens, relative to biotite from calc-alkaline granites.

Late Variscan dyke complexes (SL, LGC, AC)

Late Variscan (Late Carboniferous–Permian) dyke swarms of various compositions are widespread across the Iberian Massif, and post-date most of the plutonic complexes. Their intrusion is generally related in space and time with the igneous rocks into which they intrude and normally linked to late Variscan faulting. In general, there is a close correlation between the geochemical affinity of the dykes and that of the hosting plutonic rocks. Camptonitic lamprophyres are abundant in regions where K-rich, mafic and ultramafic plutonic rocks are present, especially in the Central Range (Central System) batholith. Porphyritic felsites are abundant in the large granite batholiths of the Central Range (Central System) and Pedroches batholiths (Fig. 8.22). Dolerites form dense dyke swarms in the Sierra de Sevilla batholith where diorites dominate the hosting plutons. In some cases, such as that of some monzonitic dykes in the Pedroches batholith, the mineral assemblages of dykes and hosting granite are nearly coincident, with cordierite as the dominant mafic mineral in both. Overall these dykes can be grouped into mafic (lamprophyres, dolerites, microdiorites) and felsic (granitic, monzonitic and granodioritic porphyritic felsites) (Fig. 8.22). Although

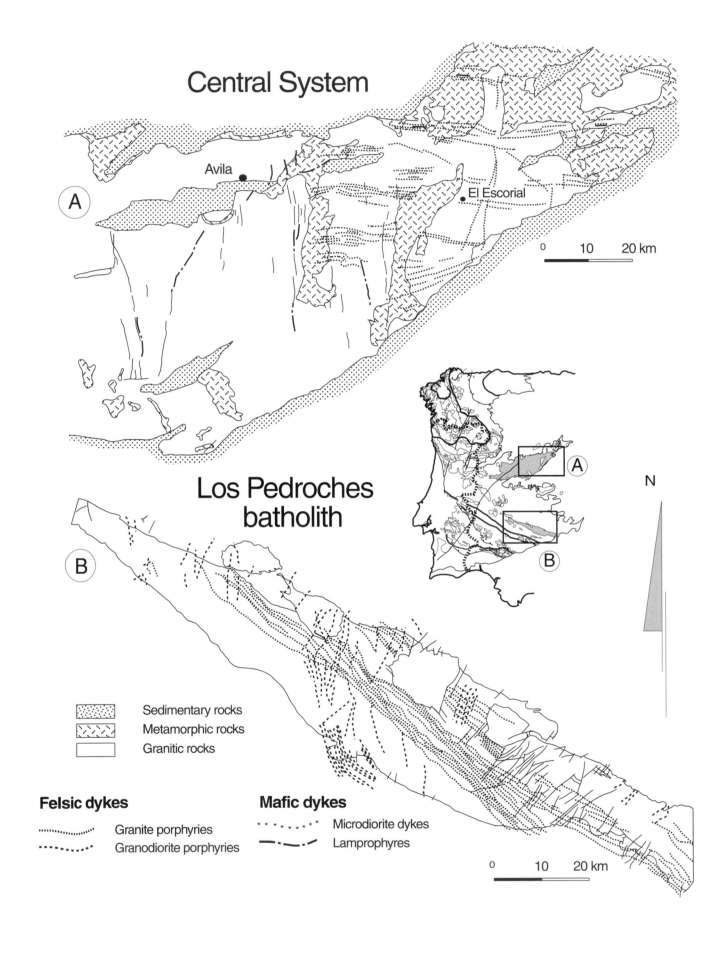

Central System

Avila

A

El Escorial

0 10 20 km

Los Pedroches
batholith

B

N

Sedimentary rocks

Metamorphic rocks

Granitic rocks

Felsic dykes

.... Granite porphyries

– – – – Granodiorite porphyries

Mafic dykes

* * * * * * * Microdiorite dykes

–··–··– Lamprophyres

0 10 20 km

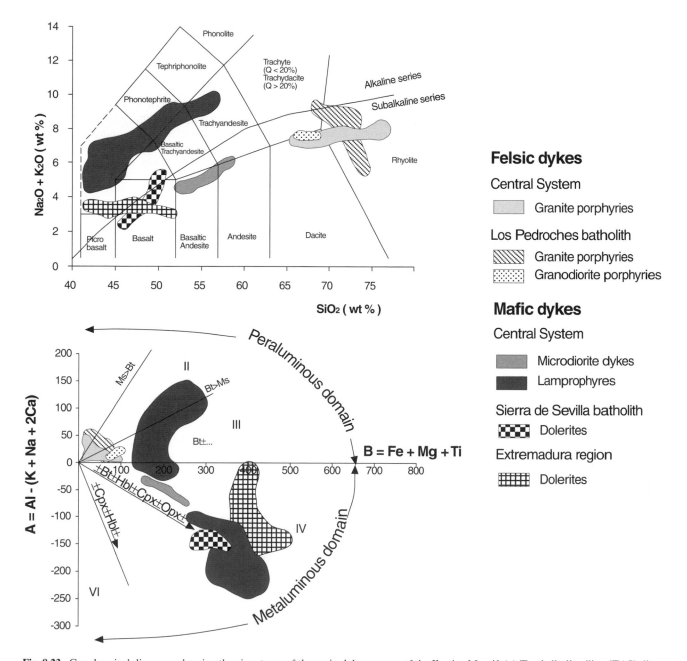

Fig. 8.23. Geochemical diagrams showing the signatures of the main dyke swarms of the Iberian Massif. **(a)** Total alkalis–silica (TAS) diagram with field divisions according to Le Bas *et al.* (1986) and alkaline–subalkaline division line according to Irvine & Baragar (1971). **(b)** A–B diagram of Debon & Le Fort (1983).

isolated dykes occur throughout plutonic exposures across the Iberian Massif, dyke swarms are particularly abundant in the eastern Central System, in the Los Pedroches batholith, and in the Sierra de Sevilla area in the South Portuguese Zone.

Central Range (Central System). This area shows abundant mafic and felsic dykes in various orientations (Fig. 8.22). Lamprophyres

(dominantly camptonites with subordinated kersantites) form long (15–40 km), steeply dipping, and narrow (1–10 m), north–south orientated dykes with Ti–augite and kaersutitic amphibole phenocrysts and feldspathic and carbonated ocelli (Villaseca & De la Nuez 1986; Bea *et al.* 1999). They plot in the region of alkali basalt series in the total alkalis–silica diagram (Fig. 8.23), and their age is constrained at 283 ± 30 Ma (Rb–Sr; Bea *et al.* 1999). Another common group of dykes in

Fig. 8.22. Geological sketches of representative areas of the Iberian Massif showing the distribution and orientations of the main dyke swarms of post-Variscan age (Late Carboniferous and Permian). Lamprophyres (mostly camptonites) are well represented in the Central Range (Central System) and post-date other dyke swarms such as the felsic porphyries and dolerites. Granitic porphyries are especially abundant in Los Pedroches batholith. Dolerites (diabases) dominate in the southernmost Variscan massif in the Sierra de Sevilla batholith. Central Range (Central System) swarms according to Villaseca & De la Nuez (1986) and Huertas & Villaseca (1994). Pedroches dyke swarms according to Donaire (1995).

the Central Range (Central System) occurs as long (>1 km), narrow (< 5 m), east–west orientated microdiorites (Casillas 1989; Huertas & Villaseca 1994). They have a fine-grained mafic matrix with phenocrysts of plagioclase, amphibole and biotite, and plot in the fields of basaltic andesites and andesites in the TAS diagram (Fig. 8.23). They have chemical features resembling those of the vaugnerites (diorites and tonalites) of the appinite suite that appear in association with the granodiorite intrusions.

Porphyritic felsites ranging in composition from granodiorite to monzogranite and granite are the most abundant dykes in the Central Range (Central System) batholith. They have variable widths from a few centimetres to more than 20 m, and on occasions up to 300 m. They are >1 km long (sometimes up to 15 km), mostly orientated east–west, and have a steep dip (Ubanell 1981). They show a quartzofeldspathic matrix and a phenocryst assemblage dominated by quartz, perthitic orthoclase, plagioclase ($An_{4–56}$), biotite and, in some dykes, cordierite and muscovite (Casillas 1989). A lower age limit for the dykes is constrained by the youngest granite cross-cut by porphyritic felsites (278 ± 16 Ma; Mendes *et al.* 1972). An upper limit is provided by the Triassic age of the Plasencia dyke that cross-cuts all Variscan pluton-related dykes in the Central Range (Central System) and Extremadura. An interesting aspect of these felsic dykes (Ubanell 1981) is that they are preferentially concentrated in the biotite-rich granites and granodiorites and absent in the two-mica granites. The compositional and age relationships suggest that microdiorites and felsic dykes may have the same origin as the vaugnerites (tonalites and enclaves) and granodiorites into which they intruded.

Los Pedroches and Extremadura regions. The dominant dyke composition in the Los Pedroches batholith is granitic to granodioritic, with mafic dykes (dolerites and lamprophyres) being very rare and younger than the felsic types. The width of these felsic dykes varies from a few metres to more than 100 m, and they can exceed 100 km in length (Fig. 8.22). Two main groups of felsic dykes, each chemically similar to their hosting pluton, can be distinguished in Los Pedroches. The first is a swarm of granodiorite–monzonite porphyritic felsites with a microcrystalline matrix and a phenocryst (0.4–2 cm) assemblage dominated by plagioclase ($An_{28–42}$), biotite and quartz. It has been proposed that the broadly north–south orientation of this dyke swarm is coincident with the Riedel fractures that fed the elongated granodiorite body (Aranguren *et al.* 1997). The second group of dykes forms a dense, younger swarm of granitic porphyritic felsites with a microcrystalline, quartzofeldspathic matrix and large phenocrysts of K-feldspar (>5 cm), quartz (0.1–0.5 cm), albitic plagioclase (0.1–0.7 cm) and biotite. This dyke swarm runs subparallel to the length of the batholith and, in some places, is abundant enough to fill half of the space occupied by the hosting granodiorite. Their emplacement implies an important extension perpendicular to the main axis of the batholith.

In Central Extremadura the dyke swarms are dominated by dolerites (with minor lamprophyres and felsites) in the western sector (Alcántara–Brozas; Garcia de Figuerola *et al.* 1974), with only rare dykes of dolerites, lamprophyres and felsic porphyries occurring to the east (cross-cutting the granites and metasediments of the Montánchez–Trujillo area and the Zújar dome). The Alcántara–Brozas dolerites form narrow (0.4–10 m) and long (>1 km) east–west orientated dykes with a subophitic matrix, plagioclase (An_{46}) and clinopyroxene phenocrysts, and a tholeiitic basaltic and picrobasaltic geochemistry (Fig. 8.23).

Sierra de Sevilla batholith. One of the most dense dyke swarms in the Iberian massif is found in the South Portuguese Zone, in the Sierra de Sevilla batholith (De la Rosa 1992). Around Castilblanco de los Arroyos this steeply dipping dolerite dyke swarm forms more than half of the outcrop. Individual dykes range up to several metres wide, but commonly they coalesce to form dyke-in-dyke structures with a chilled margin on only one side. They are orientated in three main directions, NW–SE, NNE–SSW and E–W (García Navarro 2001), with dykes of the latter group being the most abundant and thickest. They are geochemically similar to those in Extremadura, with a clear tholeiitic affinity (Fig. 8.23). The age of these dolerites is poorly constrained, but they are younger than any granite in the area (336 Ma; De la Rosa *et al.* 1993) and older than the continental Permian sediments of the El Viar basin (Chapter 10).

The Spanish granitoids: an overview (AC, LGC, CF)

Spanish granitoid rocks are remarkably varied, abundant and interesting, and show a wide range of segregation, ascent and emplacement mechanisms, so that their study can make a significant contribution to our understanding of granite petrogenesis. Recent geochemical studies on Spanish granitoids (e.g. Bea *et al.* 1992, 1999; Ortega & Gil Ibarguchi 1990; Moreno Ventas *et al.*. 1995; Villaseca *et al.* 1999), combined with experimental work (Castro *et al.* 1999a, 2000) and new field maps and detailed descriptions, have led to the development of new hypotheses concerning their origins and emplacement.

Before the introduction of the Australian (I-type and S-type) classification scheme (Chappell & White 1974), a similar classification had been proposed for the Iberian granitoids by Capdevila (1969) and Capdevila *et al.* (1973). The calc-alkaline and alkaline granite series of Capdevila *et al.* (1973) broadly coincides with the I-types and S-types respectively. In addition, in Spain a transitional type dominated by cordierite–monzogranites (Corretgé *et al.* 1977) was also recognized, and thus highlighted early the problem with adopting an oversimplistic approach to IS-type classification schemes. In this chapter we have preferred to adopt a more descriptive, and non-genetic classification by using the concept of granitoid suites in a broad sense.

Leaving aside old controversies, there now exists a general consensus in accepting a hybrid origin for the granodiorite suite, based mainly on their isotope signatures and experimental data, and an anatectic origin for the leucogranite suite (see below). The origin of the monzogranite suite, however, remains more controversial but some kind of hybrid nature again seems most likely. This genetic scheme is very similar to that originally proposed by Capdevila *et al.* (1973).

Petrogenesis: the problem of source (AC)

One of the most commonly applied geochemical constraints is Sr-Nd isotope systematics. However, such data have failed to produce unambiguous answers to fundamental questions about the nature of the crustal protolith involved in the generation of granite magmas. The interpretation of the trends displayed by granites and associated basic and intermediate rocks in initial ratio diagrams (Fig. 8.24) are normally complex and explicable by more than a single mechanism. This complexity is well understood if we consider the fact that crustal protoliths are very commonly mixtures of old, recycled clastic components and fresh volcanic components, with various crustal and/or mantle origins. Knowledge of the history of the crustal sources is prerequisite for the correct interpretations of granite isotope geochemistry (Castro *et al.* 1999a). An additional complexity may arise if mixing between crustal-derived granites and mantle melts occurred at the time of granite generation and emplacment. Only the combined use of geochemical relationships and experimentally determined phase relationships may yield reasonable interpretations about the origin of granite magmas.

Experimental constraints (Castro *et al.* 1999a, 2000) have confirmed the Ollo de Sapo gneiss as a potential source for many Variscan granites, as suggested previously by Ortega & Gil Ibarguchi (1990). The greywackes of the thick Neoproterozoic–Lower Cambrian (Complejo esquisto–grauváquico) successions in the Iberian Massif have also proven to be a

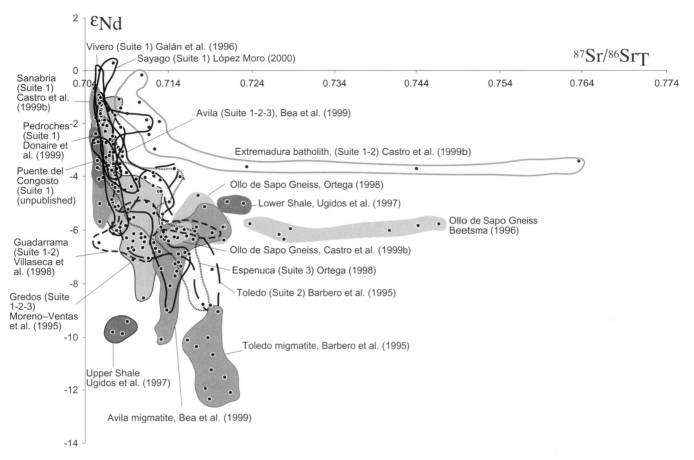

Fig. 8.24. Diagram of Sr and Nd initial isotope ratios of intrusive rocks (granitoids and basic rocks) and potential protoliths of the northern Iberian Massif. Three main crustal sources can be distinguished in the diagram according to differences in the crustal residence age as recorded by the Nd initial ratios (calculated at 300 Ma). The youngest source materials have an εNd around −4 (Tdm = 1.0 Ga), the oldest source material has values of εNd around −10 (Tdm = 1.9) and the intermediate sources have εNd of about −6 (Tdm = 1.5 Ga). The latter correspond to the Ollo de Sapo volcaniclastic gneisses and other similar glandular gneisses. These ages do not correspond to the stratigraphic ages. They are inverted (cf. Castro *et al.* 1999*a*) in such a way that the more negative values correspond to the stratigraphically younger rocks (mostly metasediments and volcaniclastic greywackes). Migmatites, leucosomes and anatectic leucogranites mirror the isotope ratios of the sources. Accordingly, less radiogenic (Nd) sources are found in the granitoids of Central Extremadura compared with the more radiogenic (Nd) sources for the granitoids of the Central Range (Central System). Most of the leucogranites (suite 3) from Galicia plot in the region of the Ollo de Sapo Fm. Granitoids of suite 1, mostly granodiorites, are plotted along curvilinear trends linking crustal sources and basic and intermediate rocks (appinites, not distinguished in the diagram). In many massifs, granodiorites and mafic (appinitic) rocks are plotted in the same region, sharing common isotope signatures. Rocks of suite 2 (monzogranites) occupy intermediate regions of the diagram between granodiorites and anatectic granites.

potential protolith for the leucogranites and two-mica granites. Both sources have in common a volcaniclastic provenance, and isotope ratios that would be transferred to the granite melts may be at least as complex as those of the source. For the dominant plutons (granodiorites and monzogranites), however, the problem is even more complex due to the absence of a recognizable source material able to produce granodiorite melts. This is a global-scale problem, because granodiorites are too calcic to be derived from a metasedimentary (e.g. pelitic) source and too potassic to be derived from an amphibolitic protolith. The possibility of an igneous protolith (e.g. a K-rich andesite, or a granodiorite) may be ruled out on the basis of phase relationships (Castro *et al.* 1999*a*; Patiño Douce 1999). A reaction between water-rich andesitic melts and crustal rocks provides a plausible explanation for the generation of granodiorite batholiths. This implies the formation of a composite source and further

melting during lithosphere extension. The mechanism is presently under study and the preliminary results combining experimental work and geochemistry are very favourable. Consequently, a hybrid origin for the granodiorites and monzogranites is therefore the most plausible hypothesis. cordierite–monzogranites are characterized by their low Sr isotope initial ratios (0.706–0.710). Good Rb-Sr errorchrons and isochrons are obtained plotting granodiorites and mafic enclaves together with the Cordierite–monzogranites (e.g. Pinarelli & Rotura 1995; Moreno Ventas *et al.* 1995; Ibarrola *et al.* 1987; Castro *et al.* 1999*a*). These relationships, as well as Sr-Nd isotope curvilinear trends, strongly suggest that these suites have mixed signatures of both fresh mantle components and old recycled crustal materials. The close association of granodiorites and basic rocks of the so-called 'basic precursors' is fundamental to an understanding of the origin of granodiorite batholiths.

Appinite–granodiorite association: the crust–mantle connection (AC, LGC)

Pitcher (1997) defined the appinite–granodiorite association based on observations in the Caledonian granitoids and associated basic rocks of Scotland and Ireland. This association has been also noted by Sabatier (1991) for the Vaugnerites of the Massif Central (France) and in the Cortlandt and Rosetown complexes in New York (Bender *et al.* 1982). The same rock association is found in the Spanish granitoids, especially in the Central Iberian Zone and the Pyrenees. The most relevant examples of appinite–granodiorite associations in Spanish geology have been mentioned (e.g. Vivero, Puente del Congosto, Gredos). The origin of these atypical mafic and ultramafic rocks (appinites, hornblendites, vaugnerites, cortlandtites) has remained enigmatic and is still a matter of debate. The crustal signatures in terms of isotope ratios (Sr-Nd-O) in rocks of mafic and ultramafic composition have puzzled petrologists. The observation of zircon and monazite in the appinitic rocks of Gredos led Bea *et al.* (1999) to

consider a contaminated mantle petrogenesis involving tectonically interleaved crustal metasediments. The observation of a compositional gap between mafic and intermediate lithologies led Bender *et al.* (1982) to propose an origin by liquid inmiscibility for the rocks of Cortlandt and Rosetown. Similarly, in the Variscan belt of Spain the mafic and ultramafic rocks are normally considered as 'basic precursors', like remnants of an early basaltic magmatism that contributed to heat transport from the mantle and to the partial melting of crustal protoliths that gave rise to the granitoids. However, a more modern view of the appinite–granodiorite association is that these mafic rocks may represent hybrid magmas coming from melting of subducted mélanges within the mantle. Assimilation of crustal rocks by fluids released from these mafic melts may produce granodiorite melts and ultramafic residues. The products of the reaction are a hornblende–plagioclase solid assemblage crystallized in equilibrium with a granodiorite melt. These model predictions are comparable with the real compositions of natural granodiorites and associated mafic rocks (Castro *et al.* 1999a).

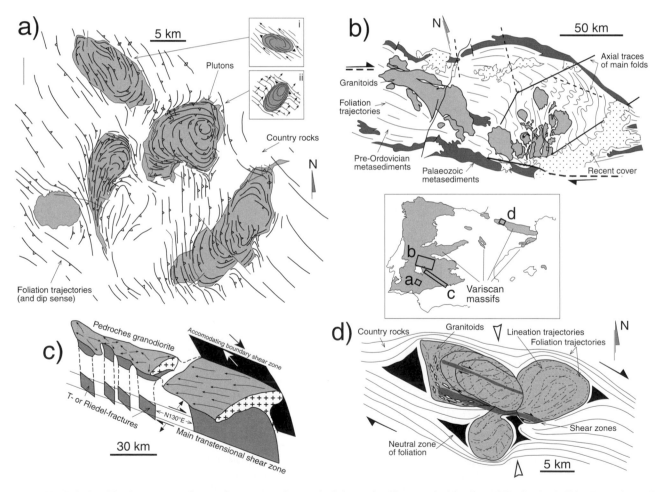

Fig. 8.25. Relationships between granite emplacement and tectonic deformation illustrated with selected Iberian batholiths. Inset shows the location of these plutons. See text for an explanation. **(a)** Sierra Morena plutons (modified from Brun & Pons 1981). **(b)** The Central Extremadura batholith (simplified from Fernández & Castro 1999). **(c)** The Los Pedroches batholith (modified from Aranguren *et al.* 1997). Arrows are inferred magma flow trajectories, according to magnetic susceptiblity data. **(d)** Cauterets-Panticosa granite complex (simplified from Gleizes *et al.* 1998b). Foliation and lineation trajectories within plutons are the result of interpolation of measured magnetic fabrics.

Granite emplacement: the space problem and the role of tectonics (CF, AC)

Granite magmatism plays a central role in the evolution of the continental crust. The processes by which granite magmas are segregated from their source region and vertically transported through the crust have been intensively studied and debated during the last decades (e.g. Petford *et al.* 2000). Of particular interest is the long-lived controversy regarding the mechanisms by which granite plutons grow, the so-called space problem. An important contribution to this problem has come from the study of natural examples from the Iberian Variscan belt, such as the Central Extremadura batholith, the Sierra Morena plutons and Los Pedroches batholith. These are nowadays extensively cited cases for specific models of emplacement mechanisms.

The importance of tectonic activity in making space for magma, and the interference between regional (tectonic) and local (magmatic) strain fields were topics addressed by Brun & Pons (1981) in their study of the Sierra Morena plutons. Their map of foliation trajectories in the Burguillos anticline (Fig. 8.25) may be considered as a classic example of perturbation of the strain field by the presence of granite bodies. Brun & Pons (1981) described the presence of triple points in the foliation trend, close to the plutons, and local obliquities of foliation at the pluton boundaries. They concluded that two distinct finite strain patterns are consistent with the observed foliation (and stretching lineation) map. The first pattern is due to a combination of ballooning and transcurrent shear (Fig. 8.25; Burguillos pluton), and the second one is a result of interference between ballooning and thrust shearing (Fig. 8.25; Salvatierra de los Barros pluton). The close association between pluton emplacement and activity of shear zones is also apparent in the Central Extremadura batholith (Fig. 8.25). The work by Castro (1986) in this area clearly demonstrated the role played by strike-slip shear zones in making space for granite magma. Plutons in the Central Extremadura batholith are synkinematic with respect to the second regional phase of deformation, and intruded a low-grade pelitic metasedimentary series of Neoproterozoic–Early Cambrian age. Using structural and petrological evidence, Castro (1986) suggested that the batholith was emplaced within a broad, east–west trending, crustal-scale dextral shear zone. Emplacement of large masses of granodiorites linked to extensional tectonics was proposed by Oen (1970) for the Central Range

(Central System) batholith (central Spain), and later corroborated by Doblas (1991).

Detailed geophysical data, especially from gravity and seismic surveys, can be very important in the analysis of emplacement mechanisms. Use of these techniques revealed the three-dimensional shape of the Cabeza de Araya pluton in the Central Extremadura batholith (Vigneresse & Bouchez 1997). Cabeza de Araya is frequently cited as a typical example of a wedge-shaped pluton, characteristic of regions dominated by transcurrent movements (Améglio *et al.* 1997). Combined gravity and magnetic susceptibility data revealed two root or feeder zones in the Cabeza de Araya pluton (Vigneresse & Bouchez 1997). Ascent of magma towards the Cabeza de Araya pluton was favoured by a double-gash opening. These gashes are probably the result of transtension in a linked system of two overstepping shear zones (Fernández & Castro 1999). A detailed magnetic susceptibility study of the Los Pedroches batholith (Fig. 8.25) indicates that it can be considered as an elongate, composite laccolith fed by a 130° east-trending transtensional shear zone (Aranguren *et al.* 1997). Alternative magma conduits are secondary T- or Riedel fractures associated with the main structure (Fig. 8.25). The sigmoidal pattern of the magnetic lineations in the Cauterets–Panticosa granite complex (Fig. 8.25) also suggests a syntectonic emplacement for these plutons, linked to a dextral transpressive shear zone (Gleizes *et al.* 1998b).

An important modern controversy about granite emplacement concerns the timescales of pluton growth. Interpretation of data from the Central Extremadura batholith suggests that most plutons were filled in a period of 1000 years or less. Inferred strain rates in the local structures governing the emplacement of plutons in this batholith are 10^{-10} to 10^{-11} s^{-1} (Fernández & Castro 1999), which is several orders of magnitude greater than the average crustal strain rate. Evidence coming from a few plutons around the world, including the Central Extremadura batholith, is challenging the view of granitic magma evolution as a slow process requiring millions of years to proceed (Petford *et al.* 2000).

We are in debt to so many colleagues for comments and discussions on diverse aspects of Spanish geology. We especially recognize careful and constructive reviews by R. Capdevila and E. Stephens which have helped to improve the chapter. We also recognize the hard editorial work performed by W. Gibbons and T. Moreno.

9 Variscan and Pre-Variscan Tectonics

BENITO ÁBALOS, JORDI CARRERAS, ELENA DRUGUET,
JAVIER ESCUDER VIRUETE, MARÍA TERESA GÓMEZ PUGNAIRE,
SATURNINO LORENZO ALVAREZ, CECILIO QUESADA,
LUIS ROBERTO RODRÍGUEZ FERNÁNDEZ &
JOSÉ IGNACIO GIL-IBARGUCHI (coordinator)

Outcrops of pre-Mesozoic rocks in Spain form various massifs that relate to both Variscan (Late Palaeozoic) and pre-Variscan tectonic settings (Fig. 9.1). The largest one of these is the so-called Iberian Massif, an autochthonous massif across which an almost complete, undisturbed geotraverse of the European Variscan orogen has been preserved. Other massifs occur as variably reworked basement complexes in Alpine chains. These are: (i) the various pre-Mesozoic massifs of the Iberian and Catalonian Coastal ranges, that can basically be considered autochthonous with respect to the Iberian Massif; (ii) the basement massifs of the axial zone of the Pyrenees; and (iii) parts of the internal zones of the Betics. The latter two are essentially exotic with respect to the Iberian Massif. Several tectonic syntheses have been published so far on this orogen (e.g. Matte 1986, 1991; Julivert & Martínez 1987; Dallmeyer & Martínez-García 1990; Martínez-Catalán 1990a; Ribeiro et al.

1990c; Quesada 1990b, 1992; Quesada et al. 1991; Shelley & Bossière 2000).

It is agreed that the European Variscan belt resulted from the oblique collision and interaction between Palaeozoic supercontinents (Gondwana, Laurentia and Baltica) and a number of continental microplates during Neoproterozoic through Palaeozoic times. These microcontinents included fragments of magmatic arcs formed previously during a process of continental convergence at the margins of the major Neoproterozoic continental masses. Such a process resulted in the so-called Cadomian, Avalonian or Pan-African orogeny, developed during Late Proterozoic to earliest Palaeozoic time. Subsequent Palaeozoic disruption and amalgamation of continental fragments and their accretion to Baltica and Laurentia gave rise to Palaeozoic foldbelts around the northern Atlantic area: Caledonian and Variscan along the eastern

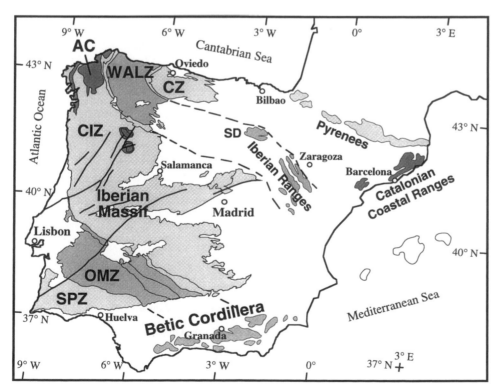

Fig. 9.1. Geological sketch map of the Iberian Peninsula showing the location of the principal pre-Mesozoic massifs. The zones into which the Iberian Massif has traditionally been subdivided are: CIZ, Central Iberian Zone; CZ, Cantabrian Zone; OMZ, Ossa Morena Zone; SPZ, South Portuguese Zone; WALZ, West Asturian-Leonese Zone. Other abbreviations: AC, allochthonous complexes; SD, Sierra de la Demanda.

boundaries, and Salinic, Acadian and Alleghenian in western sectors. Application of the tectonostratigraphic terrane concept (Coney *et al.* 1980) to the Spanish segment of the Variscan orogen has provided new tools with which to decipher the origins and tectonic evolution of the different units that form the Iberian Massif and its correlatives. A terrane is a fault-bounded tectonic unit with specific stratigraphical, petrological and structural characteristics and, on a lithospheric scale, major terrane boundaries are typically transcurrent faults and sutures.

The Iberian Massif has been traditionally subdivided into a number of tectonostratigraphic zones that in the context of the European Variscan foldbelt generally show great lateral continuity (Kossmat 1927; Lotze 1945; Julivert *et al.* 1972a). These are, from north to south, the Cantabrian, West Asturian-Leonese, Central Iberian, Ossa Morena and South Portuguese zones (Fig. 9.1). Whilst some of these are true tectonostratigraphic terranes, others are not, or even consist of various terranes themselves. In the sections that follow, the tectonic features of pre-Mesozoic massifs in Spain are described taking into account their relationship to pre-Variscan or Variscan orogeny, and their geological context within the different zones of the Iberian Massif or within the Spanish Alpine belts.

Pre-Variscan tectonics: the Cadomian orogen in Spain

The Cadomian orogen developed as an Andean-type destructive plate margin between *c.* 645 and 540 Ma BP in the northern Armorican massif type area (Chantraine *et al.* 1988; Dissler *et al.* 1988; Miller *et al.* 1999; Samson & D'Lemos, 1999). Its evolution was similar to the coeval Avalonian orogen of eastern North America and the British Isles (Lefort & Haworth 1979; Lefort 1983; Keppie & Dallmeyer 1987; Murphy *et al.* 1990; Nance *et al.* 1991; Keppie 1993; Gibbons & Harris 1994). An Andean type of geodynamic evolution is widely accepted for this orogen. Subsequent extensional events largely controlled the unconformable deposition of platformal Palaeozoic sequences on Neoproterozoic rocks at the onset of Palaeozoic basin opening. The effects of the Cadomian orogeny in Spain are recognizable in a number of areas of the Iberian Massif (most notably in the Ossa Morena Zone) and in the Iberian Ranges (Lancelot *et al.* 1985; López Díaz 1995; Álvaro & Vennin 1998; Fernández-Suárez *et al.* 1998).

Central Iberian Zone

Pre-Variscan deformation is manifested in the Central Iberian Zone (CIZ) by unconformities in the pre-Ordovician stratigraphic sequences and by the existence of pre-Variscan structures. In the CIZ, different groups of Neoproterozoic to Lower Cambrian rocks have been defined (Chapter 2), and are limited by angular unconformities (Brasier *et al.* 1979; Liñán *et al.* 1984; Liñán & Palacios 1987; San José *et al.* 1990; Díez Balda *et al.* 1990; Vidal *et al.* 1994b; Santamaría 1995). Pre-Variscan deformation did not produce visible internal strain and metamorphism (Díez Balda *et al.* 1990). North–south and NNW–SSE trending Cadomian folds have been described deforming the Precambrian basement in several outcrops of the central and southern CIZ (Fig. 9.2), such as in Las Hurdes dome (Rodríguez Alonso 1985), the Valdelacasa, Navalpino (López Díaz 1992, 1995), Ibor (Nozal Martín *et al.* 1988) and Alcudia anticlines (Palero 1993), and in the 'Domo

Extremeño' area around Cáceres (Pieren *et al.* 1991). In the Trás-os-Montes area of northern Portugal, Ribeiro (1974) described a box-fold geometry for pre-Variscan deformation, with long subhorizontal and short subvertical limbs, related to east–west to NE–SW trending extensional faults.

The Cadomian deformation was followed in the CIZ by a tectonomagmatic event, characterized by strong crustal thinning and associated melting processes, which produced the extensive granitic magmatism of the Ollo de Sapo Formation (Fm) (Navidad *et al.* 1992; Ortega *et al.* 1996; Valverde Vaquero 1997; Valverde Vaquero & Dunning 2000). Tectonically, this gave rise to the 'Sardic' unconformity (Julivert *et al.* 1972a) and the subsequent development of a passive margin sequence during Ordovician–Early Devonian times (Pérez Estaún *et al.* 1991; Azor *et al.* 1992). Extensional tectonism (Martínez Catalán *et al.* 1996) also produced an intense volcanic activity concentrated in some areas of the southern CIZ, such as the Almadén syncline.

Ossa Morena Zone

The evolution of the Ossa Morena Zone (OMZ) during Neoproterozoic–earliest Palaeozoic times is characterized by the occurrence of two main geodynamic settings: (i) an active margin in the northern part, which includes a volcanic arc in northernmost sectors (the so-called Obejo–Valsequillo–Puebla de la Reina domain; see inset map on Fig. 9.3) and a back-arc basin now lying towards the south in present-day coordinates (the so-called Badajoz–Córdoba blastomylonitic belt); and (ii) an intraplate domain in the southern part. The volcanic arc was active around 600 Ma BP as a result of subduction of the outboard Cadomian ocean beneath the western border of Gondwana (cf. Eguíluz *et al.* 1995, 2000). Extension behind the arc between 600 and 550 Ma BP led to the development of a back-arc basin, followed by tectonic inversion of the basin during contractional crustal thickening between 560 and 520 Ma BP, and finally cratonic development, relaxation, erosion and thinning of the previously thickened crust between 510 and 480 Ma BP. The Badajoz–Córdoba blastomylonitic belt cited above (BCSZ on Fig. 9.2) marks the Variscan boundary between the Ossa Morena and Central Iberian zones. The pre-Variscan boundary between these zones occurs along the northern side of the Obejo–Valsequillo–Puebla de la Reina domain (inset map on Fig. 9.3). This boundary is a fault (the Peraleda fault on Figs 9.3 and 9.11a), now largely concealed by an unconformable Palaeozoic cover so that its position is commonly inferred.

The Neoproterozoic to earliest Palaeozoic sedimentary and volcanic rocks of the southern (intraplate) part of the OMZ crop out in antiformal cores (Fig. 9.3) unconformably overlain by Lower Cambrian rocks. Within the Monesterio area (Fig. 9.3) these rocks were only weakly overprinted by Variscan events, whereas to the north and south Neoproterozoic exposures are generally less abundant and strongly overprinted by Variscan events. The Precambrian rocks belong to two main groups: (i) a lower group, mainly sedimentary, older than 565 Ma (Schäfer *et al.* 1993) and known as Serie Negra; and (ii) an upper group, mainly volcaniclastic, of latest Neoproterozoic to earliest Palaeozoic age. A number of medium- to high-grade metamorphic units and various granitoid intrusions additionally constitute this zone.

The Serie Negra rocks in the Olivenza–Monesterio area on Fig. 9.3 record two phases of Cadomian deformation (D1 and D2), and coeval metamorphism. Mesoscopic D1 structures are observed only rarely as east–west trending folds facing southward, and S_1 occurs as a relict schistosity in microlithons of the D2 foliation. D2 produced

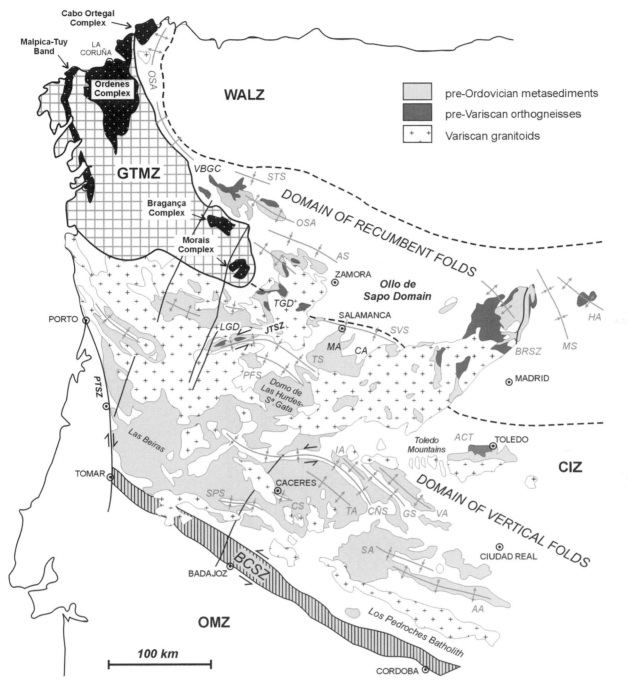

Fig. 9.2. Structural sketch map of the NW and central sectors of the Iberian Massif (modified after Díez Balda *et al.* 1990). BCSZ, Badajoz–Córdoba shear zone (note that BCSZ is presented in this figure as a Variscan limit between the OMZ and Central Iberain zones, see inset map of Fig. 9.3 for further details); CIZ, Central Iberian Zone; GTMZ, Galicia–Trás-os-Montes Zone; OMZ, Ossa Morena Zone; WALZ, West Asturian-Leonese Zone. In the Domain of Recumbent Folds abbreviations refer to the following structures: AS, Alcañices synform; HA, Hiendelaencina antiform; MS, Majaelrayo synform; OSA, Ollo de Sapo antiform; STS, Sil–Truchas synform. In the Domain of Vertical Folds abbreviations refer to the following structures: AA, Alcudia antiform; CÑS, Cañaveral synform; CS, Cáceres synform; GS, Guadarranque synform; IA, Ibor antiform; PFS, Peña de Francia synform; SA, Almadén synform; SPS, Sierra de San Pedro synform; SVS, Salamanca–Villarmayor synform; TA, Tirteafuera antiform; TA, Trujillo antiform; TS, Tamames synform; VA, Valdelacasa antiform. Other abbreviations refer to relevant Variscan structures and gneiss complexes cited in the text: ACT, anatectic complex of Toledo; BRSZ, Berzosa–Riaza shear zone; CA, Castellanos antiform; JTSZ, Juzbado–Traguntia shear zone; LGD, Lumbrales gneiss dome; MA, Martinamor antiform; PTSZ, Porto-Tomar shear zone; TGD, Tormes gneissic dome; VBGC, Viana do Bolo gneiss complex.

southward-facing, E–ESE trending, recumbent folds, and a sub-horizontal axial plane cleavage with mineral and stretching lineations trending NW–SE. A penetrative tectonic banding/schistosity occurs in the lower structural levels. Map-scale fold interference patterns

resulted from D1 and D2 deformation phase superposition (Fig. 9.4). Progressive D2 regional metamorphism was of the low-pressure type and shows a decrease in pressure–temperature conditions towards higher structural levels. An anatectic core in lowermost structural

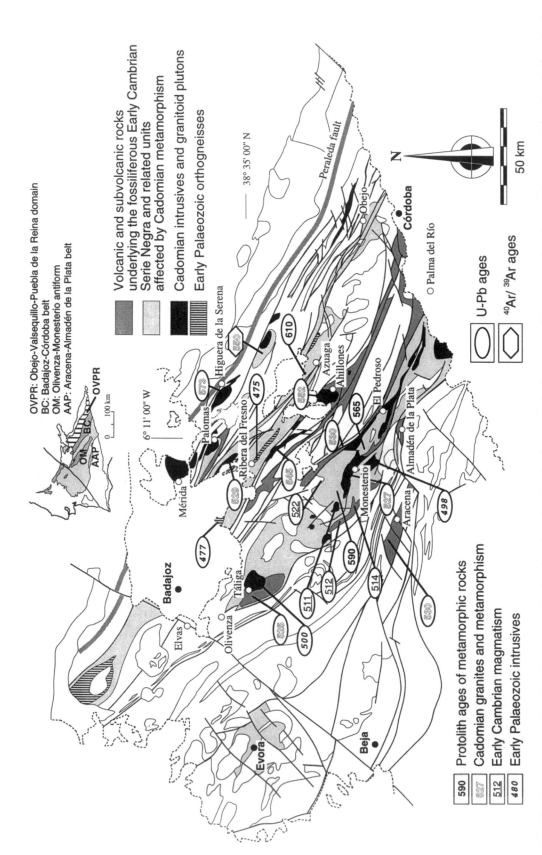

Fig. 9.3. Geological map of the Ossa Morena zone showing the principal units related to the Cadomian orogeny (Neoproterozoic to Lower Palaeozoic rocks) and a selection of radiometric datings yielding Cadomian ages (adapted from Eguíluz *et al.* 2000).

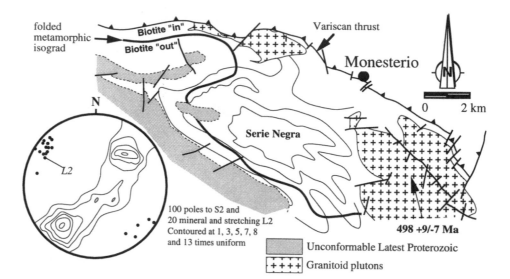

Fig. 9.4. Interpretative structural map after Eguíluz (1988) showing a reconstruction of Cadomian superimposed fold geometries. Folds affect latest Proterozoic bedding surfaces (thin solid lines in the Serie Negra) and metamorphic isograds (thick line). They are post-dated by a post-Cadomian discordance, and are cross-cut by a Cambrian granitoid pluton.

levels of the Monesterio area and emplacement of associated igneous rocks (granodiorites to leucogranites) has been related to the thermomechanical effects induced by tectonic doubling of a thin (c. 15–20 km) immature continental crust (Eguíluz & Ábalos 1992). The uppermost lithospheric mantle might have been involved too, as indicated by thrust slices of serpentinized spinel peridotite emplaced at upper structural levels prior to being unconformably overlain by early Cambrian sedimentary sequences (Ábalos & Díaz Cusí 1995). Age data for Cadomian metamorphism (Fig. 9.3) range from 524 (^{40}Ar/^{39}Ar amphibole; cf. Dallmeyer & Quesada 1992) to 562 Ma (U-Pb zircon; Ordóñez et al. 1997a). Similar ages obtained for the tectonothermal events that affected the Serie Negra in northern sectors of the OMZ range from 510–530 Ma (U-Pb zircon; Ochsner 1993; Ordóñez et al. 1997a) up to c. 540–560 Ma (^{40}Ar/^{39}Ar muscovite; Blatrix & Burg 1981). In the southern sectors data are scarce but include a c. 510 Ma U-Pb zircon age (De la Rosa et al. 1998).

In the northern sector of the OMZ, between Badajoz and Córdoba (Fig. 9.3), a number of allochthonous metamorphic units structurally underlying the Serie Negra occur in the so-called Badajoz–Córdoba blastomylonitic belt, which is a part of a much wider structure, the Badajoz–Córdoba shear zone. These metamorphic units contain schists, gneisses and amphibolites with relics of high pressure and high temperature metamorphism. The age of deformation and metamorphism in them is uncertain due to contradictory age data (e.g. Schäfer 1990) and to the intensity of the Variscan ductile deformation. D1 ductile thrusts and inverted high-pressure metamorphism are post-dated by the intrusion c. 480 Ma BP of granitoid plutons (Ochsner 1993) which cut across them (Ábalos 1992) or were emplaced at thrust contacts that are delineated locally by kilometre-size serpentinite outcrops (probably uppermost subcontinental lithospheric mantle remnants; cf. Ábalos & Díaz Cusí 1995). Relevant structures associated with thrusts in the high-grade allochthons and in their lower-grade para-autochthon include SSW-facing recumbent folds trending SSE (Fig. 9.5b) and foliations that contain broadly north–south trending stretching lineations in which shear-sense criteria provide a consistent pattern of south-directed shear (Ábalos et al. 1991b; Ábalos 1992). Although there is clear petrographic evidence for these units having been affected by metamorphic conditions typical of subduction zones (Mata & Munhá 1986; Eguíluz et al. 1990), radiometric studies provide few constraints. These include: (i) Neoproterozoic ages of gneisses and retrogressed eclogites, interpreted as protolith ages (U-Pb zircon, Rb-Sr whole-rock, e.g. Schäfer 1990; Azor et al. 1995); (ii) Cambro-Ordovician ages of orthogneiss units, interpreted as the time of emplacement of granitoid plutons subsequently transformed into orthogneisses (U-Pb zircon, Rb-Sr whole-rock, e.g. García Casquero et al. 1985; Lancelot et al. 1985); (iii) Silurian ages of eclogite and calc-silicate rock samples, interpreted as reflecting

eo-Variscan high-pressure metamorphism (U-Pb zircon, Sm-Nd garnet whole-rock; Schäfer et al. 1991); and (iv) Carboniferous ages of amphibolites, interpreted as post-peak metamorphic cooling ages (^{40}Ar/^{39}Ar amphibole; Dallmeyer & Quesada 1992). The geometry and kinematics of Cadomian deformation and the nature of metamorphism in the northern Ossa Morena Zone outside the Badajoz–Córdoba blastomylonitic belt (around Mérida and Higuera de la Serena in the Fig. 9.3) are poorly known.

Cadomian plutonic rocks are spatially and genetically related to the two groups of Neoproterozoic to earliest Palaeozoic rocks cited earlier in this section. The northern intrusives occur principally to the north of the Badajoz–Córdoba blastomylonitic belt and are mainly diorite–tonalite with minor gabbro and ultramafic rocks. They have yielded ages between c. 545 Ma (U-Pb zircon; Ochsner 1993) and 585 Ma (U-Pb zircon; Schäfer 1990) and intruded both the Serie Negra and the unconformably overlying Neoproterozoic to Lower Cambrian volcanic successions (Bellon et al. 1979; Blatrix & Burg 1981; Schäfer 1990; Dallmeyer & Quesada 1992; Ochsner 1993; Ordóñez 1998; Bandrés et al. 2000). These plutonic rocks were themselves unconformably overlain by Upper Cambrian to Lower Ordovician successions or, more rarely, by fossiliferous Lower Cambrian sedimentary rocks. The granitoids exhibit a calc-alkaline affinity and a mixture of crust and mantle components. They have been considered as early orogenic and related to the formation of a Cadomian magmatic arc (Eguíluz et al. 1997, 2000; Bandrés et al. 2000). In central sectors of the Ossa Morena Zone (Olivenza–Monesterio area in the Fig. 9.3) Cadomian plutonic rocks occur in a linear belt extending over 200 km. They are essentially anatectic granites and granodiorites emplaced into medium- to high-grade metamorphic areas (Sánchez-Carretero et al. 1990). They are generally peraluminous and have a calc-alkaline affinity although intermediate and mafic rocks are exceedingly rare. Partial melting models suggest a source area similar in composition to upper levels of the Serie Negra (Eguíluz et al. 2000). The Early Cambrian age of this magmatism is constrained between c. 525 and 530 Ma (U-Pb zircon; Ochsner 1993; Ordóñez et al. 1997a). More rarely, high-level albite-rich granitoids (dated at 511 Ma by U-Pb zircon; Ochsner 1993) were emplaced into low-grade rocks and intrude Lower Cambrian carbonate beds. In the southernmost area of the OMZ (Aracena-Almadén de la Plata area of Fig. 9.3) no evidence has been reported so far to suggest the occurrence of Cadomian plutonic rocks.

West Asturian-Leonese and Cantabrian zones

An unconformable contact beneath the earliest preserved Cambrian rocks was identified in these areas by Lotze (1956a) and De Sitter (1961). Along the eastern limb of the Narcea

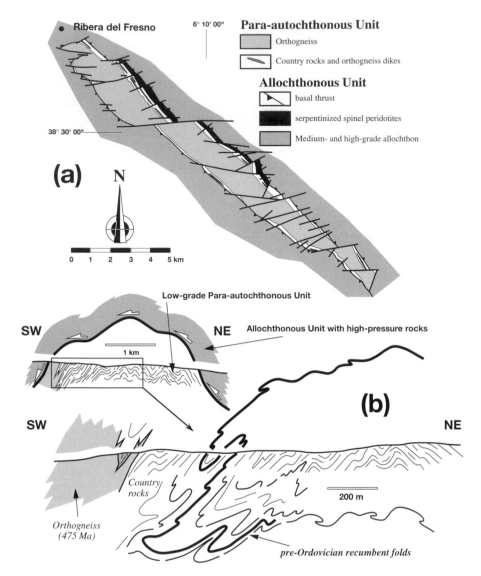

Fig. 9.5. (a) Geological map of the serpentinized spinel-facies orogenic lherzolite massif along the contact between the allochthonous and the para-autochthonous units in the Badajoz–Córdoba shear belt. **(b)** Cross-section situated immediately to the NE of the area presented in (a) showing the inverted metamorphism between the allochthonous and para-autochthonous units, and the geometrical relationships between thrust and fold structures and an early Ordovician granitoid intrusion mylonitized during the Variscan orogeny. Adapted from Ábalos (1992).

antiform, where the Variscan tectonometamorphic overprint is in places relatively mild, this unconformity is sharp and angular (Figs 9.6 and 9.8; Julivert & Martínez García 1967; Pérez Estaún 1973). In the western limb of this antiform regional foliation development obscures the nature of the contact between Precambrian and Cambrian rocks, although the geometry of intersection lineations above and below the contact again demonstrates it to be unconformable (Matte 1967; Marcos 1973). The same Precambrian–Cambrian unconformity is recognizable on a regional scale in the Lugo dome area (Fig. 9.9) of the West Asturian-Leonese Zone (Lotze 1957*b*; Martínez Catalán 1985) and has been interpreted as the result of Cadomian tectonism. Reconstruction of Precambrian structures in this area is, however, very difficult due to the intensity of Variscan deformation. The angle between the strata above and below the unconformity rarely exceeds 10–15° and so it is assumed that the Precambrian folds were either large-scale, rounded and gentle, or small-scale and localized monoclines.

It is assumed that the Neoproterozoic rocks of the

Cantabrian and the West Asturian-Leonese zones record Cadomian deformation events but no associated metamorphism. Furthermore, exposures of the igneous or metamorphic basement of this orogenic belt are generally lacking. The existence of remnants of Cadomian magmatism in these areas has been revealed recently by Fernández-Suárez *et al.* (1998), who dated intrusive, arc-derived magmatic rocks of the basal parts of the Precambrian Narcea Group (Gp) at 580–600 Ma. Subsequent radiometric studies (Fernández-Suárez *et al.* 1999, 2000) have enabled the reconstruction of the geodynamic evolution of this sector of NW Iberia, involving phases of Cadomian subduction and arc construction (800–640 Ma BP) and back-arc basin development (640–600 Ma BP). The subduction was probably long-lived and terminated at *c.* 570–580 Ma BP without a geological record of collisional events. The basin within which all the Precambrian successions cited above were deposited formed during an extensional event closely following Cadomian tectonism (Vidal *et al.* 1994*b*). Subsequent extension of this orogenic edifice would explain the occurrence of later magmatic events

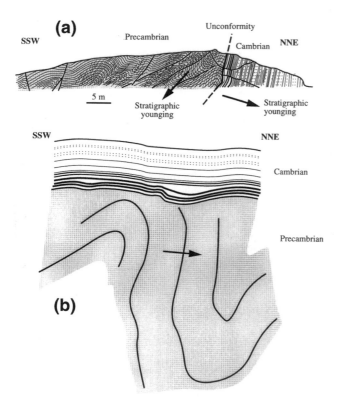

Fig. 9.6. (a) Cross-section of the eastern limb of the Narcea antiform near Barrios de Luna showing the geometrical relationships between the lowermost Cambrian and Precambrian successions (after Julivert & Martínez García 1967) and their stratigraphical polarity (indicated by the arrows; after Pérez Estaún 1973). Thrust faults, minor folds and axial planar foliations in the Precambrian rocks are interpreted as Variscan structures. **(b)** Reconstruction of Precambrian folds geometry after back-rotation of the basal unconformity of the Lower Cambrian succession to a subhorizontal attitude and removal of Variscan structures. The arrow indicates the direction of stratigraphic younging.

during Early Ordovician times (*c.* 480 Ma BP) and of regional unconformities (the 'Sardic' unconformity) recognized throughout the Iberian Massif.

Iberian Ranges

Precambrian rocks crop out in restricted areas of the Sierra de la Demanda (Colchen 1974) and SW of Zaragoza (near Calatayud; Lotze 1961*b*; Figs 9.1 and 9.7). Unravelling the geometrical relationships with the overlying Lower Cambrian successions is hampered by a lack of suitable outcrops. Thus, Lotze (1961*b*) suggested an unconformable contact, whereas Colchen (1974) noticed that the contact is always tectonized. However, taking into account the palaeogeographical and palaeotectonic correlation of these Precambrian outcrops with their correlatives in the West Asturian-Leonese Zone (Fig. 9.1), the interpretation of the original basal Cambrian contact as unconformable can be accepted as reasonable (e.g. Schmidt-Thomé 1973; Vidal *et al.* 1994*b*) and is known in the regional geological literature as the 'Asyntic' unconformity. The unconformable character of the Lower Cambrian on slightly metamorphosed Neoproterozoic successions has been identified by Liñán & Tejero (1988) SW of Zaragoza (Fig. 9.7).

Precambrian rocks in the Sierra de la Demanda area, south of Anguiano (Fig. 9.7), bear a subvertical foliation not displayed by overlying Lower Cambrian conglomerates. The latter contain rounded polycrystalline clasts of quartzite containing microstructural evidence of ductile deformation prior to their erosion and Cambrian deposition (Ábalos 2001). Pretectonic clasts akin to the former have also been reported by Álvaro & Vennin (1998) in the Neoproterozoic outcrops located to the SW of Zaragoza. This evidence argues in favour of a record of Cadomian tectonism in the basement of the Iberian Ranges (Vidal *et al.* 1994*b*).

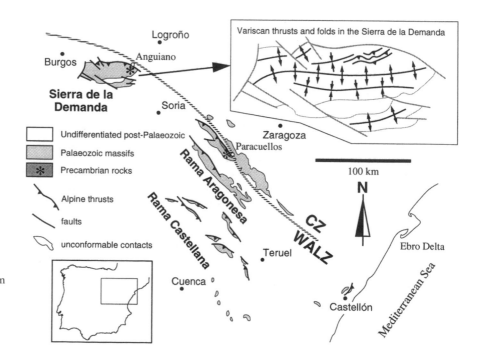

Fig. 9.7. Geological sketch map showing the pre-Mesozoic massifs of the Iberian Ranges and the outcrops of Precambrian rocks (near Anguiano and Paracuellos). Trends of Variscan structures in the Sierra de la Demanda inset traced after Colchen (1974). CZ, Cantabrian Zone; WALZ, West Asturian-Leonese Zone (after Gozalo & Liñán 1988). See text for further details.

Variscan tectonics: the Variscan orogen in the Iberian Massif

The term Variscan refers to the orogenic episode developed between Late Devonian and Late Carboniferous times, although used in a broader sense it can also include 'Eovariscan' tectonothermal events of Early Devonian and Late Silurian age. In central and western Europe, rocks affected by the Variscan orogeny crop out within a number of pre-Mesozoic massifs separated by large Mesozoic basins. The extent of this sector of the Variscan foldbelt in Europe is close to 3000 km long and 700–800 km wide (Matte 1991). The development of the Variscan foldbelt was the result of the interaction of a number of fragments of continental and oceanic lithosphere between Laurasia and Gondwana.

The different tectonostratigraphic zones that define the European Variscan foldbelt generally show great lateral continuity (Kossmat 1927; Lotze 1945; Julivert *et al.* 1972*a*). This has been classically interpreted in terms of a bilateral symmetry (Franke & Engel 1986; Matte 1991). Hence, in spite of the many uncertainties that exist, the tectonostratigraphic zones distinguished in the Iberian Peninsula (Fig. 9.1) can be correlated with confidence with equivalent zones in other countries of central and western Europe. For example, the South Portuguese Zone has been correlated with areas from SW England (the Lizard) and the Rhenohercynian zone of central Europe, characterized by the occurrence of autochthonous Devonian to Carboniferous passive continental margin sequences and allochthonous units including oceanic and active margin remnants. Also, the Ossa Morena Zone has been correlated in part with the Armorican Massif of France and with the Saxothuringian zone, which represents a Cambro-Ordovician rift basin (Franke *et al.* 1995) and includes para-autochthonous sequences (Thuringian facies) and eclogite-bearing allochthonous units (Bavarian facies). Finally, the Central Iberian Zone is correlated with central Brittany, the French Massif Central and the Moldanubian zone of central Europe (Edel & Weber 1995), and consists of Cambrian to Devonian sequences that unconformably overlie low- to high-grade Cadomian basement (Tepla-Barrandian) and high-grade metamorphic units (including granulites, eclogites and metaperidotites) that are intruded by Variscan granites (Moldanubikum).

As will be shown in detail, the effects of the Variscan orogeny in Spain are clearly recognizable in the different tectonostratigraphic zones of the Iberian Massif and in the basement massifs of the Iberian Ranges and the Catalonian Coastal Ranges. In the internal zones of the Pyrenees and the Betic Cordillera Variscan features have been variably overprinted by Alpine tectonism and so are difficult to unravel.

Cantabrian Zone

The Cantabrian Zone (CZ; Fig. 9.1) is located in the core of the arc described by the Variscan orogen (Ibero-Armorican arc or Asturian arc) and represents an easterly directed foreland thrust and fold belt. Comprehensive descriptions of the structure of the CZ, including a historical background, can be found in Julivert (1971, 1978, 1979, 1981, 1983*a*), Savage (1979, 1981), Pérez Estaún *et al.* (1988), Pérez Estaún & Bastida (1990) and Alonso *et al.* (1992).

Two remarkable structural features of the CZ are, firstly, an arcuate shape with its concavity pointing towards the external zones and, secondly, a well developed thin-skinned structure with abundant Carboniferous synorogenic deposits (Fig. 9.8). The relatively thick Palaeozoic succession was deformed by a set of imbricate thrusts and cogenetic folds, and by late high-angle faults. Strong tectonic shortening intensely modified the original palaeogeography, bringing together successions that originally lay far apart. Julivert (1971) and Pérez Estaún *et al.* (1988) divided the Cantabrian Zone into five tectonostratigraphic units: the Folds and Nappes, Central Asturian coalfield, Ponga nappe, Picos de Europa and Pisuerga–Carrión domains (Fig. 9.8). The Folds and Nappes domain has been further subdivided into a number of thrust nappe systems: Somiedo-Correcilla, La Sobia-Bodón, Aramo, Esla nappe and Valsurvio. Each of these consists of a distinctive stratigraphic succession within a principal thrust-bounded allochthon whose internal structure is complicated by secondary décollements, thrusts and duplexes, and by two sets of folds (parallel and normal to the cartographic traces of thrusts) of irregular areal distribution.

The thin-skinned general structural pattern of the CZ is dominated by a principal décollement located within the Lower–Middle Cambrian limestones of the Láncara Fm (Julivert 1971), situated close to the base of a 3–4 km thick Cambrian to Silurian pre-tectonic sedimentary succession (Fig. 9.8). Some minor décollements exist within upper stratigraphic horizons. In the western part of the CZ thrusts also cut Neoproterozoic rocks, whereas to the east several thrust surfaces diverge upwards from the Láncara décollement cutting across younger formations (here forming ramps) and along the contacts between them (forming flats or décollements). Thus, in detail, various footwall and hanging wall flat and ramp structures and fault bend folds can be recognized. Strain within the thrust surfaces depends upon the rheological nature of the rocks involved, and often affects only thin units (a few metres thick) of carbonate formations that undergo recrystallization, cataclasis and brecciation (Arboleya 1981, 1989). Deformation within the main body of the thrust slices took place at shallow crustal levels and in general did not involve foliations or metamorphism. The latter reached the transition from anchimetamorphism to very low grade metamorphism in the innermost thrust units, according to García López *et al.* (1997) and Bastida *et al.* (1999). Tectonic shortening was, in spite of this, very important: it is currently estimated at 70%, and the accumulated displacement shown in the cross-section of Figure 9.8 is *c.* 150 km (Perez Estaún *et al.* 1988).

Two major thrust systems can be distinguished in the cross-section of Figure 9.8: an earlier one exposed in the west (the Folds and Nappes domain), and a later one developed underneath, with an incomplete and much thinner Lower Palaeozoic succession that permitted renewed deformation migration towards the foreland. The second thrust system affected the Central Asturian coalfield (containing an up to 5 km thick synorogenic Carboniferous sequence), the Picos de Europa (made of a large number of thin, imbricated thrust slices involving only Carboniferous massive limestones) and the relatively autochthonous Pisuerga–Carrión domain. Nappe emplacement generally followed a forward-propagating sequence and thrusts propagated progressively, first from west and south to east and NE, then from north to south, through the Central Asturian coalfield, Ponga nappe, Picos de Europa and Pisuerga–Carrión domains. As the thrust slices were emplaced, they sequentially changed the displacement vectors from one unit to the next. Structures with orthogonal and even

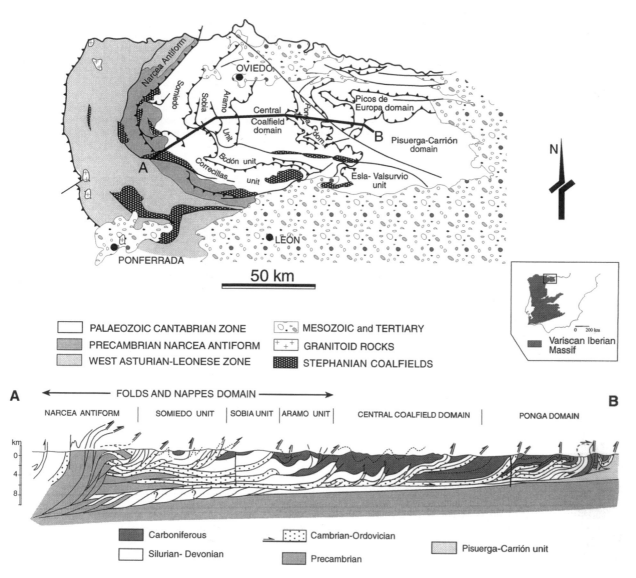

Fig. 9.8. Geological sketch map and structural cross-section of the Cantabrian Zone. See text for further details.

opposite tectonic transport directions can be recognized (Pérez Estaún *et al.* 1988): the first emplaced tectonic units (Esla nappe and Somiedo–Correcillas nappe of the Folds and Nappes domain) were directed to the north and NE, whereas the units emplaced last (Picos de Europa) were transported to the south. This gave rise to the passive rotation of older structural units and determined, at least in part, their arcuate geometry, as supported by structural (Julivert & Arboleya 1984*a,b*, 1986; Perez Estaún *et al.* 1988), palaeomagnetic (Hirt *et al.* 1992; Parés *et al.* 1994; Stewart 1995; Van der Voo *et al.* 1997) and palaeostress studies (Kollmeier *et al.* 2000). As a result, the oldest thrust sheets were not only folded by their associated lateral and frontal structures (fault bend folds or fault propagation folds), but also by subsequent reactivations, suggesting that frontal structures were reactivated as lateral and oblique ones and vice versa. This explains why nappes in the southern branch of the arc, which are the oldest, are much more deformed than younger nappes of the northern branch. All these features combined to produce a complex structural

pattern in the Cantabrian Zone, in which Carboniferous deposition was coeval, and genetically connected with structural events.

On the basis of their orientation with respect to the trend of the Asturian Arc, two sets of folds can be distinguished in the CZ: the so-called transverse (or radial) and arched (or longitudinal) folds (Julivert & Marcos 1973). The arched set runs parallel to thrust traces, whereas the radial set is transverse to them. Folds of the arched set are related to the thrust sheet geometry and can be interpreted as either leading edge folds, dorsal or frontal culmination walls (Pérez Estaún *et al.* 1988). Concomitantly, the folds of the transverse set can be related to lateral structures of the thrusts. The existence of transverse folds permits study of the deep geometry of the thrust sheets, since down-plunge perspectives of them can be observed in the eroded cores of the antiforms. Both fold sets underwent important tightening and reactivation during and after the emplacement of the unit or units to which they are related. This has given rise to several types of cartographic interference patterns both in the pre-tectonic (e.g. Julivert 1983*a*) and syntectonic successions (e.g. Alonso 1989).

The Narcea antiform (Fig. 9.8) is an antiformal stack with Precambrian rocks exposed in its core, and provides a structural transition to the internal zones of the orogen. It is located at the western border of the CZ and its antiformal structure is constrained by two principal thrusts (the Trones and La Espina thrusts) that run parallel to its trace, and by a number of out-of-sequence thrusts superimposed on them. Some geological features of this first-order structure have been described in a previous section dealing with pre-Variscan tectonics and will be studied in further detail in the section that deals with the West Asturian-Leonese Zone.

The deep seismic profile ESCICANTABRICA-1 has imaged the crustal structure of the CZ (Pérez Estaún *et al.* 1994, 1995; Gallastegui *et al.* 1997). The seismic profile illustrates the thin-skinned tectonic style of the deformation in the CZ and exhibits both a gently west-dipping basal detachment and the branching of thrusts toward the foreland. The seismic fabric of the CZ differs from that of the transition to the hinterland areas located beneath the Narcea antiform. This transition zone corresponds at depth with a crustal scale ramp that joins the foreland thrust belt detachment with structures at lower crustal levels in the hinterland (West Asturian-Leonese Zone). The existence of a lower crustal duplex is suspected there, and the presence of 'seismic crocodiles' (Meissner 1989) resembling flake tectonic structures has been suggested by Pérez Estaún *et al.* (1994, 1995).

The structural features of each of the major allochthons that constitute the CZ have been previously described in detail in several research and review articles (e.g. Julivert 1983*a*; Pérez Estaún & Bastida 1990; Alonso *et al.* 1992, and references therein). They will not be summarized here except for the case of the unit that occupies the core of the Cantabrian arc (Pisuerga–Carrión domain) because it contains Carboniferous synorogenic deposits that record parts of the tectonic evolution of the other domains of the CZ that have not been preserved in the structures themselves.

The relatively autochthonous Pisuerga–Carrión domain (Rodríguez Fernández & Heredia 1987) that crops out in the southwestern sector of the CZ has been overthrust by three separate thrust nappes coming from the south, west and north (the Esla thrust of the Folds and Nappes domain, the Ponga nappe, and the Picos de Europa domain, respectively). This unit displays some features that make it different from other units of the CZ. These are, namely, the presence of a number of allochthonous sheets containing Siluro-Devonian rocks that were affected by some degree of metamorphism, the abundance of synorogenic, unconformity-bounded Carboniferous deposits, and the occurrence of one or two penetrative foliations. The structural units that constitute the Pisuerga–Carrión domain (the so-called Palentine nappes) have been regarded in the geological literature as gravitational nappes (Wagner 1971*c*), that either rooted to the north (Ambrose 1972; Savage 1979) or formed olistoliths that originated in the south or SE (Frankenfeld 1983; Marcos & Marquínez 1984). Rodríguez Fernández & Heredia (1987) have suggested a structural and tectonosedimentary evolution model for this domain in which the centripetal vergences of the other CZ thrust domains gave rise to structures with equally centripetal vergences in the Pisuerga–Carrión domain. Additionally, the emplacement of each allochthonous unit led to the deposition of a synorogenic clastic wedge with a basal unconformity that evolves into stratigraphic continuity away from the thrust fronts. The centripetal character of synsedimentary thrust emplacement would also account for the fact that some units overthrust with variable directions and generated different clastic wedges whose depocentres were located in changing positions through time.

The pre-orogenic tectonosedimentary evolution of the CZ is characterized by a Cambrian–Devonian sedimentary wedge deposited on a continental shelf in a passive-margin-type geodynamic setting. This is consistent with other zones of the Iberian Massif, namely the Central Iberian and West Asturian-Leonese zones. Initially, intracratonic rifting occurred during the Cambrian, Ordovician and earliest Silurian periods and was followed by a more or less constant marine transgression from east to west and SW during Late Silurian and earliest Devonian times. These conditions were dramatically altered in Late Devonian times, when sediments onlapped the CZ basin eastwards. The origin of this transgressive event, together with the influx of sediment from the west at that time (Frankenfeld 1982; Rodríguez Fernández *et al.* 1985), is thought to be related to the onset of crustal thickening in areas of the Iberian Variscan orogen situated to the west of the CZ. The isostatic response to this outboard loading resulted in down-flexure of the CZ basin, which became, during Early Carboniferous times, the site of marine deposition showing typical 'synorogenic' features such as dramatic facies and thickness variations and frequent internal 'syntectonic' unconformities.

The palaeogeographic reconstruction of the CZ during the Tournaisian, Visean and lower Namurian stages suggests a shallow marine environment typical of a restricted platform basin with subtidal highs and internal low-angle slopes. This restricted shelf area was rimmed to the west and south by a forebulge that impeded the arrival of clastic deposits sourced from the emergent hinterland. The forebulge possibly was a site of reef growth. The foredeep, probably located in the West Asturian-Leonese Zone as witnessed by the Carboniferous San Clodio succession, was probably the result of crustal thickening in areas located nearer to the hinterland (Central Iberian Zone, Galicia–Trás-os-Montes Zone). The clastic wedge of this foredeep was probably eroded by cannibalism during orogenic shortening of the West Asturian-Leonese Zone, with the eroded deposits being recycled and incorporated into the infill of younger foredeeps in the CZ.

The first symptoms of transformation of the CZ itself into a foredeep appeared during early Namurian times (Sepukhovian). At this time, coeval deposition of terrigenous and carbonate turbidites took place. These deposits are partially preserved in the western and southern sectors of the Somiedo–Correcillas unit of the Folds and Nappes domain and in the Palentine nappes of the Pisuerga–Carrión domain. Toward the east and/or north, a wider carbonate shelf extended over the remainder of the CZ that formed part of the foreland passive margin.

During the late Namurian–early Westphalian (Bashkirian-Langsettian) interval a dramatic compartmentalization of the CZ sedimentary basin took place, the change in basin geometry being driven by thrust nappe emplacement. Thick limestone units in the Somiedo–Correcillas, La Sobia–Bodón, Aramo and Valsurvio thrust units of the Folds and Nappes domain were deposited in blind thrust-related submarine highs. The emplacement of these thrust units is registered as discontinuity surfaces draped by terrigenous tongues (Fernández 1990, 1993). Coeval successions in the synorogenic Pisuerga–Carrión domain formed clastic wedges in front of allochthonous units situated to the south (Rodríguez Fernández & Heredia 1990). The carbonate shelf deposits of the Picos de Europa domain, situated to the north, formed the foreland passive margin.

During the middle-late Westphalian interval (latest Langsettian to Moscovian), the remainder of the sedimentary basin was infilled by clastic wedges related to two distinct foredeeps. The middle Westphalian (Kashirsky–Podolsky) deposits in the Central Asturian coalfield and Ponga nappe domains constitute the clastic wedge developed in the front of the advancing Aramo and La Sobia–Bodón thrusts of the Folds and Nappes domain. Early Westphalian fan-delta deposits in the Aramo and La Sobia–Bodón units represent the final continental terrigenous infill of piggy-back basins. Lower-middle Westphalian successions in the Pisuerga–Carrión domain formed another clastic wedge developed in front of simultaneously emplacing southern units (Valsurvio). At the time, the carbonate shelf deposits of the Picos de Europa domain formed a foreland passive margin whose size was being progressively reduced.

During latest Westphalian and Stephanian times, basin compartmentalization increased. Synorogenic foredeeps were formed in frontal parts of the nappes. These were markedly smaller than those described above, suggesting the existence of a thickened crust by that time. This fact became particularly evident during the middle

Stephanian interval, as demonstrated by the existence of a foredeep in front of the Picos de Europa domain. This foredeep was coeval with intermontane basins in the southern CZ containing a terrigenous, continental infill. The origin of the latter basins was related to activity along vertical faults (León, Sabero–Gordón, etc.) affecting the previously thickened crust.

Summarizing, the clastic wedges associated with different tangential structures show that they had variable directions of sediment supply and that depocentre migration through time occurred within them (Rodríguez Fernández & Heredia 1987). The pattern of nappe emplacement resulted in a central area at the core of the arc (Pisuerga–Carrión domain) in which successive clastic wedges and the highest structural and stratigraphic complexity of the CZ have been preserved. The several unconformities and unconformity-bounded synorogenic deposits preserved here demonstrate that in the Pisuerga–Carrión domain synorogenic conditions persisted throughout the Namurian–Stephanian time interval. This record has permitted a reconstruction of the entire sequence of tectonosedimentary events that occurred in the CZ during the Carboniferous period.

The late Stephanian (and perhaps Permian) evolution of the CZ registers a change in tectonosedimentary style. Thin-skinned tectonics were no longer active at this time due to the significant crustal thickening and structural blocking produced by progressive nappe stacking. Deformation became localized along vertical faults, which might have been of lithospheric importance as suggested locally by the existence of mantle-derived magmatism (Gallastegui et al. 1990). Contemporaneous deposits are exclusively alluvial fan systems draining into intermontane basins in the West Asturian-Leonese and Cantabrian zones.

West Asturian-Leonese Zone

A number of comprehensive review papers exist on Variscan tectonics of the West Asturian-Leonese Zone (WALZ) upon which the report that follows is based. These are notably the contributions by Julivert (1983b), Martínez-Catalán et al. (1990, 1992b) and Pérez-Estaún et al. (1991), which include abundant reference sources of previous papers and incorporate the results of several Theses devoted to the area. The Variscan structure of the WALZ (Fig. 9.9) is characterized by the widespread association of fold nappes, ductile thrusts and foliations. They were formed as a result of three principal deformation phases broadly coeval with various stages of regional metamorphism and syn- and post-tectonic granitoid plutonic intrusions. Broadly, the intensity of deformation and metamorphism increases from the east (where the foreland Cantabrian Zone crops out) to the west (the internal zone of the Variscan orogen: Fig. 9.1). Based upon stratigraphical evidence, the structure of this zone was initiated before the Early Carboniferous epoch, which is the age of the oldest syntectonic successions (Pérez Estaún 1974a). Regional correlations with palaeogeographically equivalent domains of the Iberian Ranges also support a pre-Westphalian B age of Variscan deformation. Radiometric dating of relevant mineral geochronometers (e.g. Dallmeyer et al. 1997, and references therein) also indicates that tectonic events occurred in the middle to Late Carboniferous interval (c. 290–310 Ma BP).

Three phases of deformation (D1, D2 and D3) are regionally recognized (Martínez Catalán 1985; Pérez Estaún et al. 1991, and references therein). Deformation phase D1 gave rise to large recumbent folds with an associated axial-planar tectonic foliation (S_1) and numerous minor structures (Fig. 9.9). The geometry of these folds varies from the west, where they are clearly recumbent, to the east, where they turn progressively into inclined folds (Bastida 1980). Their axes are subhorizontal and parallel to the curvature of the Cantabrian Arc, and their vergence is towards the foreland. The Mondoñedo fold nappe is the most significant of these structures in the western area of the WALZ, with an inverted limb of up to 20 km in length. In detail, this nappe represents an anticlinorium with a number of kilometre-scale second-order anticlines and intervening synclines (e.g. the Villaodriz syncline; Fig. 9.9). Other first-order recumbent anticlines and synclines exist, as well as smaller folds from map-scale to microscopic in size. The geometry of these structures and the magnitude of the associated tectonic shortening vary longitudinally, from the northern to the southern areas (Matte 1968). A large-scale, subhorizontal, ductile shearing mechanism has been proposed to explain the generation of D1 structures (Bastida et al. 1986), and was most likely related to active crustal-scale shear zones.

Deformation phase D2 gave rise to ductile thrusts in the western sectors (Martínez Catalán 1980) and to brittle–ductile thrusts to the east (Marcos 1973). Thrusts are subparallel to D1 fold axial surfaces or cut across them at low angles (Fig. 9.9). A wealth of minor structures formed during D2 among which crenulation foliations (S_2), mineral and stretching lineations, folds with curved hinges and sheath folds are typical. The maximum elongation deduced from the minor structures shows a general west-to-east trend and a top-to-the-east vergence, though in detail a radiating pattern has been recognized (Martínez Catalán 1980). Since these are localized structures they are not always present in the outcrops, and in spite of the overprinting geometrical relationships observed in places they should not be considered completely independent of the D1 structures (Julivert 1983b) with which they share common kinematics. The principal ductile thrusts, notably the basal thrust of the Mondoñedo nappe, were originally subhorizontal shear zones formed in the inverted limbs of D1 recumbent folds. To the east this ductile thrust is connected to the Los Oscos thrust zone (Fig. 9.9), which according to Marcos (1973) is a major tectonic boundary within the West Asturian-Leonese Zone. All these structures can be considered as the natural continuation of the horizontal shearing related to D1, though now localized within discrete bands (Martínez Catalán et al. 1990).

The third deformation phase (D3) gave rise to open, map-scale upright folds. They are homoaxial with D1 folds, though their axial surfaces are subvertical, and are associated with a crenulation foliation (S_3). In the eastern domains of the zone, bands with widespread S_3 foliations overprinting gently dipping earlier foliations (S_1 and S_2) alternate with bands within which a subvertical principal foliation is present. The former commonly coincide with the gently dipping normal limbs of asymmetric D1 folds, whereas the latter relate to the steeply inclined limbs (Pulgar 1980). In the western sector of the WALZ this phase caused the folding of the basal thrust of the Mondoñedo nappe and led to the formation of the Lugo dome, in which the thrust and the innermost parts of the nappe crop out defining tectonic windows (e.g. the Xistral window; Fig. 9.9).

A later phase of deformation can be proposed for the formation of the so-called radial or transversal folds. These are gentle, upright folds whose axes strike perpendicular to the trend of the structures formed during the previous phases of deformation.

The limits of the WALZ are tectonic. The western limit corresponds to the Vivero transtensional ductile fault, whereas the eastern one is marked by the La Espina and Trones thrusts (Fig. 9.9). The Vivero fault is arcuate, running for approximately 150 km parallel to the trend of the Cantabrian arc. It dips 30–60° to the west, displaces the hanging wall block dextrally and down to the west, and cuts the basal thrust of the Mondoñedo nappe. The fault is associated with a kyanite belt (Martínez et al. 1996; Reche et al. 1998) genetically related to a ductile medium- to high-grade shear zone up to 2–3 km thick and with a throw 10–12 km down-to-the-west in the northern part (Martínez Catalán 1985). To the south and SE the throw decreases and the fault itself bifurcates and disappears as it enters the Caurel syncline (Martínez Catalán et al. 1992b).

The La Espina and Trones thrusts separate a non-metamorphic footwall and a highly strained and metamorphosed allochthonous hanging wall (Gutiérrez Alonso

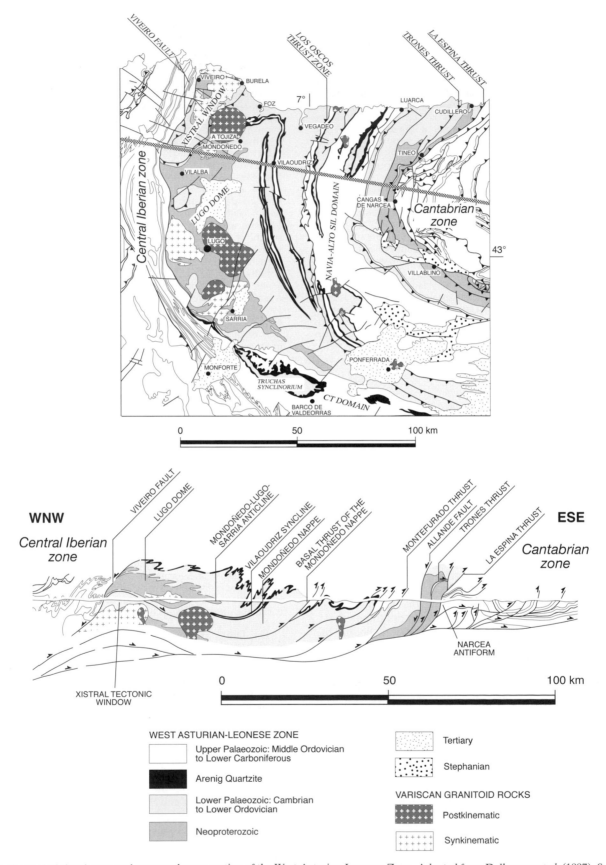

Fig. 9.9. Geological sketch map and structural cross-section of the West Asturian-Leonese Zone. Adapted from Dallmeyer *et al.* (1997). See text for further details.

1996). The thrusts are steeply dipping and trend parallel to the curvature of the Cantabrian arc. They are associated with ductile shear zones (340–400°C, 3–4 kbar) in the hanging wall blocks containing foliations parallel to both bedding and to the thrusts. The minimum shortening accommodated by both thrusting and internal deformation has been estimated at c. 75 km. The overall geometry of the La Espina and Trones thrusts forms a complex antiformal culmination due to duplex formation and late out-of-sequence thrusting. The Lugo dome and the Vivero fault to the west are also interpreted as structural culminations above a deeper duplex of shear zones that have been recently imaged with seismic reflection profiling techniques (Pérez Estaún et al. 1994; Martínez Catalán et al. 1995; Gutiérrez Alonso 1996).

The WALZ has been subdivided into three domains (Fig. 9.9) on the basis of tectonostratigraphy and plutonometamorphism (Matte 1968; Marcos 1973): (i) the Mondoñedo nappe domain, bounded below by the Los Oscos thrust zone and in the west by the Vivero fault; (ii) the Navia–Alto Sil domain, bounded above by the Los Oscos thrust zone and below by the La Espina and Trones thrusts; and (iii) the Caurel–Truchas domain, a complex fold association that represents the lateral continuation of the Mondoñedo nappe and thrust structures, together with the Vivero fault.

The internal structure of the Mondoñedo nappe domain is dominated by the D1 Mondoñedo anticlinorial structure, its basal D2 thrust, and the effects of D3 folding on both, as has been outlined above. The internal structure of the Navia–Alto Sil domain is characterized by a combination of east- to NE-facing (towards the Cantabrian arc concavity) inclined (inverse limbs of up to 2–4 km) to upright folds formed during D1, and by a series of thrusts kinematically related to the folds but formed during D2 (Marcos 1973; Bastida 1980). Broadly speaking, the folds define a synclinorium with second-order anticlines and synclines whose subhorizontal axes can be traced for up to 50 km. In the Caurel–Truchas domain east–west trending structures replace the Mondoñedo nappe as the latter narrows towards the south (Matte 1968; Pérez Estaún 1978). The Caurel structure is a north-facing recumbent anticline associated with a pair of recumbent synclines. The anticline occupies a position structurally in the rear of the Mondoñedo nappe, forming a narrow band that disappears northwards into the Vivero fault. To the SE these folds become smaller and upright, and the uppermost syncline broadens and forms the so-called Truchas synclinorium (Fig. 9.9).

Regional metamorphism in the WALZ is essentially synkinematic (Capdevila 1969) with medium pressure characteristics, whereas thermal metamorphism is local and related to the intrusion of granitoid plutons. The intensity of metamorphism increases from the east to the west, that is towards the internal parts of the Mondoñedo nappe and its autochthon, where high-temperature amphibolite facies conditions developed (Suárez et al. 1990). To the east greenschist facies and lower-grade metamorphism predominated, with the biotite zone being reached only within shear zones in the hanging walls of the La Espina and Trones thrusts. The distribution of metamorphic mineral isograds indicates that greenschist and higher-grade facies affect only the westernmost quarter of the WALZ. In several medium-grade areas andalusite- and sillimanite-bearing parageneses are superimposed on earlier staurolite- and kyanite-bearing parageneses. Late, epizonal retrograde metamorphism is associated with the basal thrust of the Mondoñedo nappe and to D3 structures and shear zones.

The higher-grade mineral parageneses in the WALZ are genetically related to D1 and D2 regional deformations, since crystals of sillimanite, kyanite and andalusite are often aligned within S_1 and S_2 foliations. In the vicinity of the Vivero fault hanging wall, replacement of andalusite parageneses by kyanite in relation to D3 suggests downdragging and burial of the hanging wall (Martínez et al. 1996; Reche et al. 1998). At this time, metamorphic isograd distribution in the Vivero fault footwall (the Lugo dome) was being modified and telescoped (Reche et al. 1998) due to high-temperature and low-pressure retrogression during structural doming and uplift.

The portion of the WALZ affected by higher-grade metamorphism coincides with a structural and thermal dome in which most of the granitoid massifs of the zone were intruded (Martínez et al. 1988). Some of them cut across the isograd pattern of regional metamorphism, whilst others are closely related to the highest-grade parts. Thermal aureoles are generally small, and commonly are more evident where the country rocks are of lower grade (Bellido et al. 1987).

Granitoid pluton emplacement is also closely related to the phases of regional deformation described earlier in this section. Four broad groups of granitoid rocks have been recognized: (1) synkinematic with D2 (G1 granites); (2) intruded between D2 and D3 (G2 granites); (3) intruded syn-late-post D3 (G3 granites); (4) clearly post-D3 granites (G4 granites) (Capdevila 1969; Capdevila et al. 1973; Bellido et al. 1987, 1992). Recent study of relevant granitoid plutons concerning their internal structure, three-dimensional shape, and structural relationships with their country rocks using geological and geophysical techniques (Aranguren & Tubía 1992; Aranguren 1997) has shown that some plutons were sheared under submagmatic to high-grade subsolidus conditions by both the basal thrust of the Mondoñedo nappe and the Vivero fault. This implies syntectonic emplacement during D2 and an overlap between D2 thrust movements and transtensional fault activity (supposedly D3). Other plutons, considered as post-D3 intrusions, exhibit internal geometries that suggest some continued control by the regional D3 stress/strain fields during their emplacement.

Central Iberian Zone

The Central Iberian Zone (CIZ) is a rather heterogeneous zone that comprises areas affected by high-grade metamorphism with abundant granitoids, as well as areas affected by very low-grade metamorphism (Fig. 9.2). It characteristically shows Arenig rocks ('Armorican quartzite') lying unconformably on an irregularly distributed basement of latest Proterozoic to Cambrian age (Julivert et al. 1972a; Julivert & Martínez 1987; Díez Balda et al. 1990): the so-called 'Sardic' unconformity. A distinctive feature of the CIZ is the occurrence in its northwestern area of a number of high-grade metamorphic massifs containing mafic and ultramafic rocks as well as the imprint of high-pressure metamorphism (Figs 9.2 and 9.10).

The CIZ is bounded to the north and east by the so-called 'Ollo de Sapo Anticlinorium' (Julivert et al. 1972a), that separates it from the West Asturian-Leonese Zone, and to the south by a series of thrust boundaries with the Badajoz–Cordoba shear zone, which separate the CIZ from the Ossa Morena Zone (Quesada 1990b; Ábalos 1992).

Arenas et al. (1986) distinguished within the CIZ an autochthonous area and the so-called Galicia–Trás-os-Montes Zone (GTMZ), which is allochthonous over the CIZ and occupies most of its northwestern part (Fig. 9.10). The autochthon of the CIZ is made of Neoproterozoic to

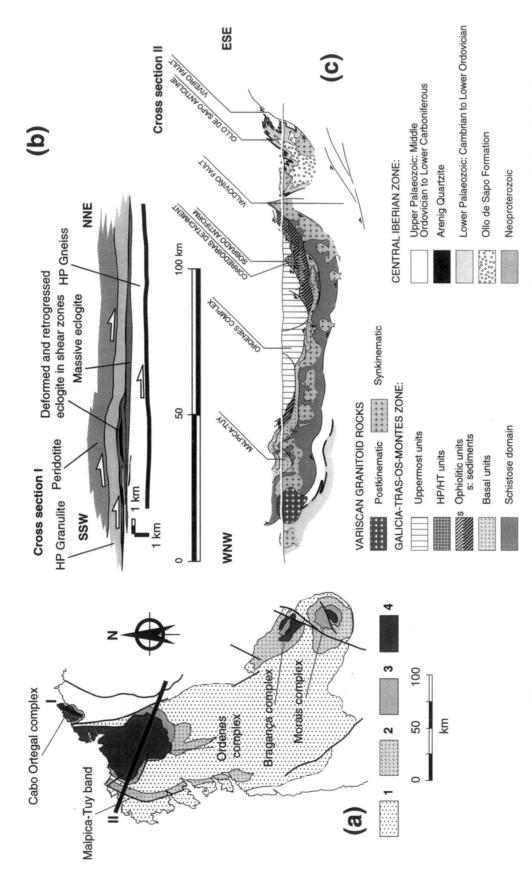

Fig. 9.10. (a) Geological sketch map of NW Iberia showing the outcrop of the units that constitute the 'allochthonous complexes' (2, basal unit or lower allochthon; 3, ophiolitic unit; 4, upper allochthon) and of their para-autochthon (1, schistose domain of Galicia–Trás-os-Montes). The cross-sections I and II show the simplified structure of the Cabo Ortegal Complex in a section parallel to the high-pressure mineral and stretching lineations **(b)** and the combined structure of the Malpica-Tuy and Órdenes Complexes and of their para-autochthon **(c)** in a section near-perpendicular to the trend of the orogen. Adapted from Dallmeyer *et al.* (1997).

Palaeozoic metasediments and igneous rocks that were deformed by folds and thrusts. The metasediments were deposited over the same basement upon which they now rest. This unit formed part of the continental margin of Gondwana and records both the Cadomian orogeny and Cambrian–Ordovician rifting. Synorogenic Devonian and Carboniferous sequences have been preserved in its external zones. The GTMZ comprises two domains or terranes with contrasting palaeogeographic and tectonometamorphic histories separated by a major thrust boundary: the para-autochthonous 'schistose domain' of Galicia–Trás-os-Montes (GTMSD) and the domain of the 'allochthonous complexes'.

The 'allochthonous complexes' of NW Spain

The 'allochthonous complexes' crop out in NW Iberia (Spain and Portugal) as megaklippen in five structural synforms (the Cabo Ortegal, Malpica-Tuy and Órdenes complexes in Spain and the Bragança and Morais complexes in Portugal; Fig. 9.10). They consist of a pile of Proterozoic to Lower Palaeozoic metasedimentary and metaigneous units that preserve evidence of both Variscan and pre-Variscan deformation and metamorphism.

It is well accepted nowadays that these complexes were overthrust onto the Gondwanan margin during the Variscan orogeny (e.g. Ries & Shackleton 1971; Bard et al. 1980; Iglesias et al. 1983; Bastida et al. 1984; Vogel 1984; Arenas et al. 1986; Matte 1991). Also, it is accepted by most that these complexes represent fragments of variably subducted continental and oceanic Palaeozoic lithosphere that were subsequently obducted and incorporated into the orogenic belt during Variscan collision (Bernard-Griffiths et al. 1985; Peucat et al. 1990; Santos Zalduegui et al. 1997; Martínez Catalán et al. 1997). Alternative hypotheses on the origin and tectonothermal evolution of this regional allochthon propose a Precambrian age of the subduction-related high-pressure events (e.g. Vogel & Abdel Monem 1971; Marques et al. 1992, 1996; Santos et al. 1995, 1997).

The lithologic and tectonometamorphic characteristics of the complexes have been described and synthesized in several papers among which those of Arenas et al. (1986), Gil Ibarguchi & Arenas (1990) and Ribeiro et al. (1990b) will normally be followed here. Arenas et al. (1986) published the first comprehensive lithostratigraphical synthesis. They distinguished a number of nappe units that might represent dismembered and duplicated terrane relics. Currently, taken as a whole, three main structural elements (each one is composed of a small number of lithostrigraphical units) may be recognized within the complexes that may be correlated at a regional scale. From bottom to top, these are: a basal unit or lower allochthon, an ophiolitic unit, and an upper allochthon.

The basal unit consist of schists, paragneisses, and alternating felsic and mafic igneous rocks, and crops out most extensively in the Malpica-Tuy complex. The magmatism recorded in this unit reflects an Ordovician rifting episode (460–480 Ma BP; Van Calsteren et al. 1979; Ribeiro & Floor 1987; Pin et al. 1992; Santos Zalduegui et al. 1996). No ophiolite units separate this unit from the underlying para-autochthon, and thus it is considered to be a part of the continental margin of Gondwana. Notwithstanding, its allochthonous character can be easily demonstrated as it records initial high-pressure regional metamorphism (Martínez Catalán et al. 1996) related to subduction within an accretionary wedge 380–375 Ma ago and subsequently uplifted through 370–355 Ma BP (Arenas et al. 1995, 1997; Rodríguez Aller et al. 1997). Younger ages, around 340–330 Ma, have been reported for blueschist facies rocks within southern sectors of the basal units (Gil Ibarguchi & Dallmeyer 1991).

The ophiolitic units consist of basalts, pillow breccias, dolerites, metagabbros, plagiogranites, amphibolites and pervasively serpentinized ultramafic rocks containing chromite pods. They are repeated by several thrust faults (Díaz García et al. 1999b) and represent oceanic sequences outboard from Gondwana. The ophiolite thrust sheets record a wide variety of metamorphic conditions (from intermediate-pressure to exceptionally high-pressure, and from greenschist to granulite facies), the higher-grade ones being situated in the highest structural positions. Radiometric evidence supports the contemporaneity of metamorphism and Variscan compressive events and continental accretion 390–375 Ma ago (Van Calsteren et al. 1979; Peucat et al. 1990; Dallmeyer & Gil Ibarguchi 1990; Dallmeyer et al. 1991, 1997) occurring soon after oceanic crust generation 395 Ma ago (Díaz García et al. 1999a).

The lowermost units of the upper allochthon (best exposed in the Cabo Ortegal region) are mostly high-pressure rocks, including granulites and gneisses, eclogites and chromite-bearing ultramafic rocks (e.g. Moreno et al. 2001; Santos et al. 2002; Fig. 9.10; Chapter 19). These have been called 'high temperature/high pressure allochthonous units'. They are the most distinctive element of the complexes and the ones that first suggested an interpretation in terms of an allochthonous origin.

The structurally uppermost units of the upper allochthon (best exposed in the Ordenes complex) are made of a thick sequence of terrigenous metasediments, large orthogneiss massifs, and variably recrystallized mafic intrusives that record greenschist- to high-grade-amphibolite- or granulite-facies medium-pressure metamorphism. The protoliths of the orthogneisses and metagabbros have yielded Late Cambrian to Early Ordovician ages by conventional and SHRIMP U-Pb zircon dating, while a first tectonothermal event was nearly coeval as deduced from the age of monazite in metasediments, and a second prograde metamorphic episode occurred at c. 390 Ma BP (Kuijper et al. 1982; Dallmeyer & Tucker 1993; Ordóñez 1998; Abati et al. 1999; Abati 2000). These units could be related to Cambro-Ordovician orogenic and rift-related events comparable to those recorded in Pan-African, Cadomian and Avalonian areas of Circum-Atlantic borderlands (Quesada 1990c), or to those of the Meguma terrane (Martínez Catalán et al. 1997) or, more locally, to Cadomian areas from other parts of the Iberian Massif (Eguíluz et al. 2000).

The available geochronological data and metamorphic textural relationships permit the distinction of three main episodes of regional metamorphism, in addition to episodes of contact metamorphism related to the emplacement of pre-Variscan (biotitic granitoids and gabbros emplaced c. 460–480 Ma ago; e.g. Kuijper et al. 1982) and Variscan intrusives (granites and granodiorites emplaced c. 340–280 Ma ago; e.g. Serrano Pinto & Gil Ibarguchi 1987).

The first metamorphic episode is recorded by high-pressure granulite and eclogite assemblages in variably deformed granulite (mafic to felsic), N-MORB eclogite, metagabbro, metaserpentinite, metaperidotite and ortho- and paragneisses (Vogel 1967; Gil Ibarguchi et al. 1990, 1999; Girardeau & Gil Ibarguchi 1991; Mendia 2000). Precambrian (Ries & Shackleton 1971; Ribeiro et al. 1987; Santos et al. 1995, 1997), Early Palaeozoic (Kuijper et al. 1982; Bernard-Griffiths et al. 1985; Peucat et al. 1990) and Middle–Late Palaeozoic ages (Santos Zalduegui et al. 1996; Ordóñez et al. 2001) have all been proposed for this metamorphism, which is particularly evident in the Cabo Ortegal, Órdenes and Bragança complexes. The peak of this metamorphism attained temperatures of c. 800°C at minimum pressures of 17 kbar (Gil Ibarguchi et al. 1990), reached during the first regional deformation phase. In eclogites, high-pressure granulites and metaperidotites on Cabo Ortegal there exist remnants of these first-phase structures and metamorphism at all scales (from textures and microstructures to map-scale duplexes and sheath folds; cf. Ábalos et al. 1994; Ábalos 1997; Gil Ibarguchi et al. 1999, Azcárraga 2000). The exhumation and emplacement of these subducted lithologies has been recorded in their mineral chemistry and pressure–temperature paths and involved strong syntectonic decompressions occurring early during Variscan collision (Gil Ibarguchi & Arenas 1990).

Amphibolite facies retrogression of the previous parageneses gave rise to a great number of amphibolite-facies rocks and, locally, to intermediate-temperature eclogites. This was simultaneous with a second regional deformation phase, and its age has been estimated at *c.* 390–400 Ma (Van Calsteren *et al.* 1979; Peucat *et al.* 1990) or slightly younger (Santos Zalduegui *et al.* 1996). The recognition and interpretation of the structures associated with this deformation phase vary among different authors and among each allochthonous complex. In the Cabo Ortegal complex, for example, some authors have proposed that large recumbent folds formed during this episode (Bastida *et al.* 1984, and references therein), whereas others suggest that they relate to the first deformation phase (e.g. Azcárraga 2000) and recognize discrete, low-angle, amphibolite-facies shear zones as D2 structures. In the Órdenes complex, amphibolite-facies ductile deformation appears to be pervasive in several units and has been interpreted as related to extensional deformation during orogenic collapse after lithospheric thickening. The Malpica-Tuy band was affected by prograde regional metamorphism during this period, producing amphibolites, C-type eclogites and high-pressure low- to intermediate-temperature parageneses in acid rocks and metapelites. The lithologies eclogitized during this episode formed part of the margin of Gondwana and were subducted later than the highest-grade rocks.

The final emplacement of the allochthonous structure as a whole was coeval with medium-pressure greenschist-facies metamorphism of variable intensity. This episode presents different characteristics depending on the nature of the protoliths. Thus, it represents retrogressive metamorphism for some lithologies (e.g. eclogites and granulites), or progressive metamorphism for previously unmetamorphosed rocks. This episode is related to the final emplacement of the principal allochthonous units and as such, generates intense mylonitic deformations along the basal thrusts (Arenas 1985; Díaz García 1986). Pressure–temperature (P–T) conditions for this episode have been calculated at *c.* 370–425°C and 2.5–3.5 kbar (Arenas 1985) and its age has been constrained around 360 Ma (Dallmeyer *et al.* 1997). The metamorphic evolution of the previously unmetamorphosed units reflects the compression and heating that took place when they were overridden by hotter allochthonous units (Gil Ibarguchi & Arenas 1990).

Early theories on the origin of these complexes suggested that they either represented the relics of an old Proterozoic basement, or were related to an early Palaeozoic rift (Matte & Ribeiro 1967; Vogel 1967; Van Calsteren 1977; Van Calsteren & Den Tex 1978). The interpretation of gravimetric surveying by Van Overmeeren (1975) and Keasberry (1979) supported these theories. Alternatively, Ries & Shackleton (1971) suggested that the complexes might be the remnants or klippen from a large metamorphic nappe dismembered and amalgamated with the Variscan foldbelt. Bayer & Matte (1979) reinterpreted the geophysical data cited above and built a coherent geological model involving nappes rooted to the west of the current western Iberian coast. The emplacement of these nappes would have required west-to-east tectonic translation of at least 100 km during Siluro-Devonian collision between Gondwana and Laurentia (Lefort & Ribeiro 1980; Burg *et al.* 1981, 1987).

Most workers agree that tangential tectonics dominate the structural style of NW Iberia. The trends of earlier Variscan structures image the curvature of the Ibero-Armorican arc, and the apparent vergence of fold structures toward the arc centre have all been interpreted in the context of a west-to-east sense of tectonic emplacement. However, when the ductile strain recorded by both the nappe units and their contacts is considered, a puzzling kinematic picture emerges. Most metamorphic nappes and recumbent folds bear stretching lineations parallel to the fold axes that also trace the curvature of the Ibero-Armorican arc. In the nappe contacts, stretching lineations with the same direction are parallel to the apical

axes of sheath folds, indicating the direction of tectonic transport (e.g. Iglesias *et al.* 1983). These directions can vary between the allochthonous complexes: NNW trending and SSE vergent in the Bragança and Morais complexes (Anthonioz 1972; Marques *et al.* 1992), WNW–ESE trending and east vergent in parts of the Órdenes complex (Díaz García *et al.* 1999*a,b*), and north- to NE trending and NNE facing in eastern (Martínez Catalán & Arenas 1992) and western (Van Zuuren 1969) parts of the Órdenes complex, in the Cabo Ortegal complex (Girardeau & Gil Ibarguchi 1991; Ábalos *et al.* 1994) and in the Malpica-Tuy band (Gil Ibarguchi & Ortega Gironés 1985). Commonly, they are parallel to the arc curvature and have been recently interpreted in the context of large orogen-parallel tectonic displacements (Vauchez & Nicolas 1991; Shelley & Bossière 2000) due to oblique plate convergence.

It is agreed in the recent literature that the regional allochthon contains remnants of oceanic, lower crustal, and upper mantle units related to various geodynamic settings that include oceanic, arc- and other subduction-related environments. However, there is still some disagreement over the timing of events. A Precambrian age of the high-pressure events was postulated by Marques *et al.* (1992, 1996), Santos *et al.* (1995, 1997) and others, whereas Early Palaeozoic extension followed by almost coeval subduction events were proposed by Bernard-Griffiths *et al.* (1985), Peucat *et al.* (1990), Santos Zalduegui *et al.* (1997) and Martínez Catalán *et al.* (1997).

Para-autochthonous domain of NW Spain

The schistose domain of Galicia–Trás-os-Montes (GTMSD) constitutes the relative autochthon of the allochthonous complexes. It comprises a thick wedge of Palaeozoic siliciclastic rocks with abundant volcaniclastic inclusions, and was deformed and metamorphosed during the Variscan orogeny. It is bounded by a basal thrust and its internal structure is imbricated (Ribeiro *et al.* 1990*c*). This domain was part of the same continental margin as the CIZ autochthon, stratigraphic correlations between them being possible. However, although no significant metamorphic differences have been reported, the GTMSD was thrust over the autochthon, and it exhibits a higher degree of deformation. Structural analysis of the metasediments that constitute this domain reveals the occurrence of three principal deformation phases accompanied by metamorphism and plutonism during the Variscan orogeny (Barrera *et al.* 1989; Farias 1992). The metamorphism reflects compression and heating at the same time as the overriding allochthonous nappes were undergoing decompression and cooling. Medium and relatively high-pressure low-grade parageneses formed (Capdevila 1969; Bastida *et al.* 1984; Arenas 1985) suggesting that the thickness of the allochthonous pile was of the order of 8–10 km in the case of Cabo Ortegal and Ordenes complexes, and up to 22 km beneath the southern complexes of Bragança and Morais.

The first deformation phase, D1, gave rise to a penetrative foliation, which in most areas has been transposed by subsequent deformation. D1 folds are rare throughout the GTMSD, in contrast with the neighbouring zones of the Iberian Massif, where they are common at all scales.

D2 represents a widespread non-coaxial deformation episode, and gave rise to a variety of structures and microstructures such as different tectonic foliations, folds and shear zone rocks. The latter are associated with ductile shear zones and two thrust zones: one that emplaced the allochthonous complexes onto the GTMSD (up to 8 km thick according to Marquínez 1984), and the other being the basal thrust of this domain separating it from the CIZ (recognized first by Ribeiro 1974). D2 folds are often small-scale (micro- and mesofolds), and exhibit low interlimb angles, subhorizontal axial surfaces, and large axial dispersion (curved hinges; cf. Farias 1992). S_2 foliations of several types (depending upon the lithology of the materials affected) form the dominant foliation in most of the GTMSD. They grade into mylonitic foliations close to the thrust zones cited above. In the highest-grade areas S_2 is a schistosity preserving ample evidence of relic S_1 fabrics. The basal thrust of the GTMSD is generally sub-parallel to lithological contacts of the allochthon, although it cuts across previous map-scale folds of the autochthon. The trace of this basal thrust is discontinuous as at several places it is overprinted by granitoid intrusions.

D3 gave rise to map-scale upright folds associated with abundant higher-order minor folds with subvertical crenulations. D3 synforms commonly preserve outcrops of GTMSD rocks, whereas D3 antiforms are rare since such areas are typically occupied by granitoid batholith intrusions. The intrusions appear to have influenced the morphology of coeval D3 folds and foliations (Marquínez 1984; Farias 1989, 1992), and are themselves cut by NW–SE trending, subvertical D3 shear zones. Finally, later Variscan deformations gave rise to various systems of normal and wrench faults, to kink-band folding related to relaxation of the orogen, and to folds with steeply plunging axes.

The metamorphic grade increases from the core of D3 synforms towards their limbs. In these areas the chlorite, biotite, garnet, staurolite, andalusite, sillimanite and sillimanite–K-feldspar isograds of progressive regional metamorphism can be mapped. The isograd pattern in relation to D2 and D3 structures as well as the microstructural observations indicate that this Barrovian-type regional metamorphism formed simultaneously with D2. The highest-grade parageneses relate to the D2–D3 interval, whereas the granitic intrusions and the concomitant increase in the regional thermal gradient (with a progressive change towards low-pressure metamorphism) relates to D3 (Barrera et al. 1989; Farias 1992). Late D3 intrusions are associated with metamorphic contact aureoles within which pyroxene hornfels facies were locally reached.

Autochthon of NW Spain

Located in the footwall of the Variscan allochthonous and para-autochthonous units described above, the autochthonous CIZ is characterized by extensive granitic magmatism and by the presence of high-grade metamorphic complexes of regional extent (Martínez et al. 1988). Its Variscan tectonic evolution began in Middle Devonian times, during nappe stacking (Matte 1986; Díez Balda et al. 1990; Martínez Catalán et al. 1997), which is consistent with a Lower-Devonian age for the youngest rocks of the pre-orogenic sequence and with the development of synorogenic Upper Devonian–Lower Carboniferous flysch (Pérez Estaún et al. 1991). Radiometric ages of c. 359 Ma have been obtained for the oldest Variscan deformations in the CIZ (Dallmeyer et al. 1997).

The CIZ has undergone a polyphase Variscan deformational history, with four deformation phases (D1–D4) having been recognized (González Lodeiro 1980; Díez Balda et al. 1990; Macaya et al. 1991; Azor et al. 1992; Díez Balda & Vegas 1992; Escuder Viruete et al. 1994; Escuder Viruete 1998). The two principal phases, D1 and D2, are characterized by heterogeneous ductile deformation, produced the present structure of the CIZ, and are related to the main metamorphic evolution. The late deformation phases, D3 and D4, are more localized but are responsible for the broad macrostructural outcrop patterns of the CIZ. The development of each deformation phase is geographically heterogeneous.

Differences in the geometry of D1 folds allowed Díez Balda et al. (1990) to distinguish two structural domains within the CIZ: a so-called Domain of Recumbent Folds (DRF), that extends along a c. 200 km wide belt close to the West Asturian-Leonese Zone, and a Domain of Vertical Folds (DVF), that occupies most of the central and southern domains of the CIZ (Fig. 9.2). The DRF is also known as the Ollo de Sapo domain (Parga Pondal et al. 1983; Chapter 8). D2 structures are more conspicuous in the DRF, where the Variscan collision gave rise to a thicker continental crust, higher-grade regional metamorphism, and to widespread synkinematic leucogranitic magmatism. The DRF/DVF division reflects significant differences in the pre-Ordovician stratigraphy (San José et al. 1990) and the distribution of a Tremadoc–Arenig felsic magmatic belt in the CIZ (Valverde Vaquero & Dunning 2000). In the DVF, Ordovician rocks rest unconformably on areally extensive, low-grade Precambrian and Lower Cambrian rocks, and locally on undated Cambro-Ordovician felsic, volcanosedimentary rocks (Toledo Ranges; e.g. San José et al. 1990). In the DRF, by contrast, there are no proven Cambrian rocks (Gutiérrez Marco et al. 1990) and Lower Ordovician rocks rest over areally extensive low-grade pre-Variscan augen-gneisses (Ollo de Sapo Fm: see Chapter 8), or tectonically on top of medium- to high-grade metamorphic complexes with abundant granitic orthogneisses (Julivert & Martínez 1987; Azor et al. 1992; Navidad et al. 1992; Ortega et al. 1996; Escuder Viruete 1998; Valverde Vaquero & Dunning 2000).

D1 structures relate to crustal shortening. In the DRF they are dominated by vergent/recumbent folds and thrusts in the upper structural levels, and by fold–nappe stacks in the lower ones. They all originated by NE to E directed, subhorizontal heterogeneous shearing (Vacas & Martínez Catalán 1987; Macaya et al. 1991; Escuder Viruete 1998). In this domain, the main D1 macrostructures are the Ollo de Sapo and Hiendelaencina anticlinoriums, and the Sil-Truchas, Alcañices, Riba de Santiuste and Majaelrayo synforms (Parga Pondal et al. 1964; González Lodeiro 1980; Díez Balda et al. 1990; Macaya et al. 1991; Azor et al. 1992; Fig. 9.2). In the upper structural levels of the DRF, D1 deformation produced NW–SE to north–south trending asymmetric isoclinal folds that are defined by the structural trend of the Armorican quartzite. Associated S_1 foliations are axial planar to these folds and dip towards the SSW. The intersection lineation (L_1) between S_1 and the bedding plunges gently towards either the NW or the SE. In the lower structural levels of the DRF, large-scale D1 structures include a series of kilometre-scale orthogneiss wedges imbricated with thin metasedimentary sheets. A NE directed sense of thrusting is indicated in these rocks by numerous shear criteria parallel to subhorizontal L_1 stretching lineations. The stacking of orthogneiss lobes has been interpreted by Macaya et al. (1991) as a basement-cover imbrication produced by NE vergent lower to mid-crustal thrusting synchronous with the overturned folding and more brittle thrust movements in the upper crust.

Crustal thickening during D1 induced M_1 prograde metamorphism. In pelitic rocks, the syn-D1 to inter-D1–D2 growth of chlorite, biotite, chloritoid, garnet, staurolite, kyanite and sillimanite porphyroblasts records a Barrovian-type metamorphic sequence. Isograds are well preserved in the Ollo de Sapo anticlinorium (Martínez et al. 1988; Arenas 1991), in the Salamanca area (Díez Balda et al. 1995; Escuder Viruete et al. 1995) and in the sector of the Central Range located to the north and NE of Madrid (Arenas et al. 1980; Casquet & Navidad 1985; González Casado 1987; Escuder Viruete et al. 1998). However, preservation of this M_1 sequence is commonly incomplete in the lower structural levels as a consequence of subsequent D2 deformation. In greenschist facies metapelites pressure greater than 4 kbar and temperatures (based on appearance of biotite) near 425°C have been estimated for this event (Escuder Viruete 1998). Amphibolite-facies assemblages such as Grt + Bt + St + Ky developed regionally in metapelites under temperature conditions between 570 and 625°C and pressures of over 5 kbar. In upper amphibolite facies metapelites, an intermediate-pressure prograde evolution (from 6 ± 0.5 kbar, 600 ± 25°C to 8.5–9 kbar, 725 ± 25°C) has been calculated (Escuder Viruete *et al.* 2000). These P–T estimates agree with the replacement in metabasites of Hbl + Pl + Qtz by Grt ± Cpx assemblages, associated with an ilmenite S_1 fabric, which establishes a syn-D1 temperature rising to upper amphibolite facies conditions. Recently, M_1 eclogite facies relics (14 kbar, 725–775°C) have been described in metabasites from deep crustal realms cropping out in the Central Range (Barbero & Villaseca 2000).

By contrast, D1 structures in the DVF are characterized by subvertical folds formed by a simple coaxial subhorizontal shortening normal to the trend of the belt (Díez Balda *et al.* 1990; Díez Balda & Vegas 1992). The macrostructures formed during D1 in this domain are narrow NW–SE trending synclines limited by the Armorican quartzites and broad anticlines (with wavelengths of several kilometres) cored by pre-Ordovician rocks. These structures are very regular and continuous in the NW sector of the DVF (traceable for more than 250 km). However, in the SE sector they exhibit more irregular geometries as a result of dome and basin interference patterns produced by folding during the subsequent deformation phases (Julivert *et al.* 1983*b*; Díez Balda *et al.* 1990). The main D1 macrostructures in the DVF are the Salamanca–Villarmayor, Tamames, Peña de Francia, Cañaveral, Cáceres, Sierra de San Pedro and Almadén synforms, and the Valdelacasa, Ibor, Urda, Esteras, Trujillo, Navalpino, Tirteafuera and Alcudia antiforms (Fig. 9.2). These folds exhibit a subvertical axial-planar S_1 foliation and a subhorizontal L_1 intersection lineation. Excluding the Anatectic complex of Toledo (Barbero 1995), the most frequent metamorphic conditions registered in the DVF during D1 are those of the prehnite–pumpellyite and low-T greenschist facies.

Variscan contractional structures are variably overprinted by a major D2 extensional event (Díez Balda *et al.* 1990, 1995; Doblas *et al.* 1994*a*; Escuder Viruete *et al.* 1994), attributed to large-scale collapse of the thickened continental crust. D2 and D1 fabrics are generally similar, and are distinguished by D2 having different SE-facing geometries, top-to-the-SE general kinematics, overprinting relationships, and metamorphic conditions. In the DRF, D2 extensional structures conform to a regional linked system of kilometre-scale subhorizontal extensional shear zones exposed in several NW–SE culminations. In these shear zones hanging wall low-grade (or even anchizonal) metamorphic rock units up to 4 km thick are juxtaposed against footwall high-grade gneissic complexes. Examples of these D2 macrostructures are the Viana do Bolo gneiss complex, the Tormes gneissic dome, the Lumbrales, Martinamor and Castellanos antiforms, the Berzosa–Riaza shear zone and the Anatectic complex of Toledo (Díez Balda *et al.* 1990, 1995; Hernández Enrile 1991; Díez Balda & Vegas 1992; Escuder Viruete *et al.* 1994, 1995, 1998; Barbero 1995; Valverde Vaquero 1997). Early ductile shearing evolved to more brittle deformations. A wide variety of kinematic indicators are associated with S_2 mylonitic fabrics, including S-C structures, shear bands, garnets with helicitic and sigmoidal inclusion trails, asymmetric tails and pressure shadows of porphyroclasts, oblique grain-shape fabrics and quartz crystallographic fabrics. The general shear sense, parallel to a mesoscopic L_2 stretching lineation, is regionally consistent across more than 300 km and indicates that upper structural levels moved towards the SE (subparallel to the structural trends of the CIZ) during D2.

The metamorphic event associated with D2 crustal extension (M_2) overprinted earlier compressional M_1 Barrovian metamorphic signatures. M_2 metamorphism is of a different nature in the footwall and hanging wall parts of each extensional system. The basal part of hanging wall blocks normally displays a condensed sequence of M_2 isograds (cordierite, andalusite and sillimanite), parallel to regional S_2 fabrics and to the extensional contact with the footwall units. Low-P upper amphibolite facies conditions (T > 600°C) are common. The high metamorphic gradient points to high vertical heat flows and heating during extension (Escuder Viruete *et al.* 1994). In the Tormes gneissic dome, the paragenetic relationships in sillimanite-bearing metapelites from 100 m above the basal contact, indicate a P–T path during D2 for the upper unit passing under the triple point of the aluminium silicate polymorphs and close to an isobaric heating. This path is compatible with the sequence of M_2 isograds and metamorphic zones observed in the field. In upper levels, thermobarometric calculations in conjunction with the observed phase compatibilities, constrain the P–T conditions to low-P/high-T re-equilibration for Bt

+ And + Crd-bearing rocks at 2.6–3.2 kbar and 540–570°C (Martínez *et al.* 1988; Escuder Viruete *et al.* 1994). In the footwall, the syn-D2 path deduced from the sequence of reaction textures and thermobarometry is characterized by: (i) an initial (near)-isothermal decompression phase of several kilobars (8–9 kbar, 700–750°C to 3.3–4.7 kbar, 640–670°C; Escuder Viruete *et al.* 1997), related to their tectonic exhumation and juxtaposition with the upper unit; and (ii) subsequent quasi-isobaric cooling associated with thermal re-equilibration of the unit in its new structural position, which is recorded by low-P assemblages and retrogression from amphibolite to greenschist facies conditions in late S_2 mylonitic fabrics.

During D2 deformation, the deepest structural realms of each extensional system passed through the metamorphic peak, reaching high-T amphibolite to lower-P granulite facies conditions (Gil Ibarguchi & Martínez 1982; Martínez *et al.* 1988; Barbero 1995; Escuder Viruete *et al.* 1997; Barbero & Villaseca 2000), and undergoing extensive partial melting (López Plaza 1982; Castro *et al.* 2000). The age of D2–M_2 is constrained by the 325 to 318 Ma ages of emplacement of syn- to post-S_2 granitoid intrusion (Serrano Pinto & Gil Ibarguchi 1987; Macedo 1988). However, the age of the beginning of the extensional deformation could be slightly older, as indicated by the 337 ± 2 Ma U-Pb age obtained by Valverde Vaquero (1997).

D2 structures in the DVF are very rare, but can be found in the Guadarranque syncline (Moreno 1977), the Yébenes area (Vázquez *et al.* 1992) and the southern limb of the Extremadura dome. These structures form normal ductile–brittle shear zones (up to several centimetres thick and with associated chloritization), subhorizontal folds (with a crenulation axial-planar S_2 fabric), and low-angle elongated quartz veins parallel to a NW–SE trending stretching lineation. The syn-D2 P–T path of footwall units (e.g. in the Tormes gneiss dome) is consistent with the tectonic exhumation of deep crustal levels along an extensional system composed of a crustal-scale shear zone and later low-grade normal detachments. In the upper unit, the parallelism of the M_2 isograds, S_2 fabrics and the low-grade detachments suggests a relationship between the D2 extensional event and the low-P/mid-high-T mineral metamorphic growth. According to the U-Pb data the D2 tectonothermal event is Visean to Namurian in age and coincides with initial thrusting in the foreland of the Variscan Iberian Massif (Cantabrian and South Portuguese zones), confirming the synorogenic character of the D2 extensional process.

The present-day geometry of the CIZ resulted from the superposition of later structures, including a set of open antiforms and synforms related to a D3 contractional event, a system of syn- to post-D3, ENE–WSW trending strike-slip sinistral shear zones, and a system of ductile–brittle and brittle D4 normal detachments and faults (Díez Balda *et al.* 1990; Doblas 1991; Díez Balda & Vegas 1992; Doblas *et al.* 1994*a*; Escuder Viruete 1998). D3 and D4 deformations occurred under greenschist facies retrograde conditions at higher crustal levels. In lower structural levels high temperatures locally persisted as a consequence of the high thermal gradients inherited from D2 and the intrusion of synkinematic granitoid rocks. The mineral assemblages present in mylonites of strike-slip shear zones indicate deformation under lower greenschist facies conditions.

D3 was a discontinuous, localized deformation. In the DRF, the structures formed during D3 are upright, open to tight folds, centimetric to kilometric in size with NW–SE to east–west trending axes, and essentially homoaxial with mylonitic L_2 (Fig. 9.2). At the outcrop scale, D3 folds are characterized by a fold-axis-parallel crenulation lineation (L_3), sometimes associated with non-pervasive, subvertical spaced cleavage (S_3). On the regional scale, the major D3 folds define a set of subvertical or moderately NE-vergent antiforms and synforms, which fold the extensional D2 shear zones, M_2 metamorphic isograds and the late-D2 extensional detachments. The sigmoidal appearance of major D1 structures in map view is due to syn- to post-D3 sinistral strike-slip shears trending ENE–WSW (Díez Balda & Vegas 1992), such as the Juzbado–Traguntia shear zone (Jiménez Ontiveros & Hernández Enrile 1983), which displaces the D1 Tamames syncline (Fig. 9.2). In the Montes de Toledo, D3 shears have a NW–SE trend and again a sinistral strike-slip movement. There they gave rise to the sigmoidal appearance of the Herrera del Duque syncline and the Esteras anticline, as well as to dome and basin fold interference patterns. During D3, a conjugate system of NW–SE to NNW–SSE trending strike-slip dextral shear zones also developed.

The Porto–Tomar shear zone might be related to this system (Fig. 9.2), as well as the regional inflections of S_1 cleavage produced by dextral folds with subvertical axes (Castro 1986; Díez Balda & Vegas 1992). In summary, D3 folding and syn- to post-D3, ENE–WSW sinistral and NNW–SSE conjugate dextral subvertical shear zones are compatible with a large-scale NE–SW subhorizontal shortening.

D4 structures are ductile–brittle shears in lower structural levels and medium- to high-angle normal faults in the upper ones. The ductile–brittle shears occur in several main families. NW–SE to east–west shears have low to moderate angles of dip of 10–50° and associated normal displacements. NE–SW to ENE–WSW conjugate sets of shears exhibit steep dips (>60°). Other brittle normal faults trend NW–SE to east–west and dip between 45 and 75° towards either the north or the south. D4 faults are infilled by breccias and subparallel bands of slightly foliated cataclastic rocks. Given the great variety of orientations of D4 shears and D4 normal faults described in different sectors of the CIZ, the late D4 extension seems to have been radial. However, in several localities these structures are related to a NE–SW to north–south extension direction, opposite to the previous D3 shortening (Doblas 1991; Doblas et al. 1994a).

Ossa Morena Zone

The Variscan cycle in the Ossa Morena Zone (OMZ) began with the fragmentation of the Cadomian orogen during an extensional event that led to the formation of a Cambrian marine platform in a southern domain while areas to the north probably remained emergent. Within the southern basin, pelitic rocks of Middle to Late Cambrian age were deposited, whereas in the emergent, northern sector such deposits are commonly lacking. Progressive rifting during Late Cambrian to Early Ordovician times led to the development of troughs and the formation of a continental platform and an ocean basin. The early stages of rifting produced bimodal alkaline rocks that overlie the Lower Cambrian carbonate platform rocks. The younger rocks include basalts, spilites and quartz-keratophyres that are especially thick (several hundreds of metres) in southern areas. Emplacement ages for coeval bimodal intrusions cluster around 480 Ma (Priem 1970; Lancelot & Allegret 1982; García Casquero et al. 1985), although older ages of c. 500 Ma have been obtained for some alkaline granitoids (U-Pb zircon; Ochsner 1993). While the extensional process might not have been uniform, the development of the basin appears to be continuous from mid-latest Cambrian up to Silurian times.

Variscan deformation within the OMZ was controlled strongly by the heterogeneity resulting from the existence of a Cadomian basement and a Palaeozoic sedimentary cover, as a result of which large-scale strain partitioning took place. Within the basement, the first Variscan deformation phase gave rise to important ductile to brittle–ductile shear zones (Fig. 9.11) including the Monesterio thrust, which has an estimated tectonic displacement vector in excess of 20 km towards the SSW (Fig. 9.11; Eguíluz 1988). Large-scale fold–nappes with axial-planar cleavage (up to 20 km inverted limbs) and brittle–ductile low-angle thrusts originated during the same episode in the Palaeozoic cover (Vauchez 1975), and are rooted in basement shear zones. Mylonitic bands and axial-planar foliation developed in the core of the fold and nappe structures and the general south to SSE direction of movement is slightly oblique to the regional NW–SE trend. The pre-Variscan intrusive rocks also underwent deformation with intensities proportional to the distance from associated main structures.

The second phase of Variscan deformation gave rise, in both the basement and cover, to long-wavelength upright folds trending NW–SE along with brittle subvertical sinistral transcurrent shear zones with the same strike (Fig. 9.11). A crenulation foliation is developed locally in the innermost zones of fold hinges. Two important shear zones, Southern Iberian and Badajoz–Córdoba, recorded an intense ductile deformation with a generalized component of sinistral oblique convergence. Both shear zones show tectonic features typical of suture zones (Apalategui et al. 1990; Simancas et al. 2001), and this indeed is the case for the Southern Iberian shear zone (Fig. 9.11) which is clearly a Variscan suture (Quesada 1990b; Ábalos et al. 1991a; Crespo-Blanc & Orozco 1991). The Badajoz–Córdoba shear zone, however, shows regional transcurrent deformation (characterized by penetrative subvertical foliations containing subhorizontal NW–SE stretching and mineral lineations) and associated amphibolite- to greenschist-facies metamorphism related to an intracontinental sinistral ductile shear zone (Burg et al. 1981) which reactivated an older suture (Quesada 1990b,c, 1997; Ábalos 1992).

With the exception of southern sectors of the OMZ, regional metamorphism associated with Variscan crustal thickening generally was unimportant, reaching at most greenschist or lower amphibolite facies conditions (Fig. 9.11). However, along a band near the southern boundary of the OMZ, in the Aracena massif and lateral correlatives of the Almadén de la Plata and Beja areas (Fig. 9.3), Variscan regional metamorphism involved a prograde history that reached anatectic conditions under low-pressure conditions (Bard 1969). This P–T evolution was apparently related to mantle upwelling, either within an extensional context (Ábalos et al. 1991a) or in connection with the subduction of an oceanic ridge (Castro et al. 1996b, 1999a). Eclogitic and other high-pressure rocks, which are likely remnants of Variscan subduction, are well exposed in westernmost sections of the OMZ (Fonseca & Ribeiro 1993). The effect of the Variscan low-pressure metamorphism on these rocks is unclear. Subsequent retrogression in the OMZ appears to be related to thrust movements along ductile shear zones that emplaced the Ossa Morena units (Aracena massif and correlatives) on to volcaniclastic rocks of the South Portuguese Zone. Finally, extensional structures were generated through reactivation and tectonic inversion of major thrusts and shear zones formed during the previous stages of crustal thickening. Detailed structural studies of some of the domes with anatectic cores (Fig. 9.11) show an irregular reactivation of former structures, which might in some cases be inherited from the Cadomian orogenic stage (Apraiz 1998; Eguíluz et al. 2000).

Variscan calc-alkaline plutonic rocks show complex and commonly inversely zoned structures with mafic rocks at the core and granitic rocks near the margins. Gabbro and monzogranite compositions predominate, showing abundant cumulate structures and magmatic foliations. The petrological data suggest an important contribution of mantle components to the geochemistry of these magmas (Sánchez-Carretero et al. 1990). The intrusions had originally been interpreted as synkinematic with respect to the first Variscan deformation phase (Brun & Pons 1981). Most recent studies, however, suggest that pluton emplacement was controlled by major Variscan structures and stress fields and that the intrusions are only affected by late-Variscan brittle structures (Eguíluz et al. 1989).

Synorogenic foreland basins formed adjacent to the rising foldbelt, and became filled with Carboniferous sediments (Apalategui et al. 1990). One of the basins was located over

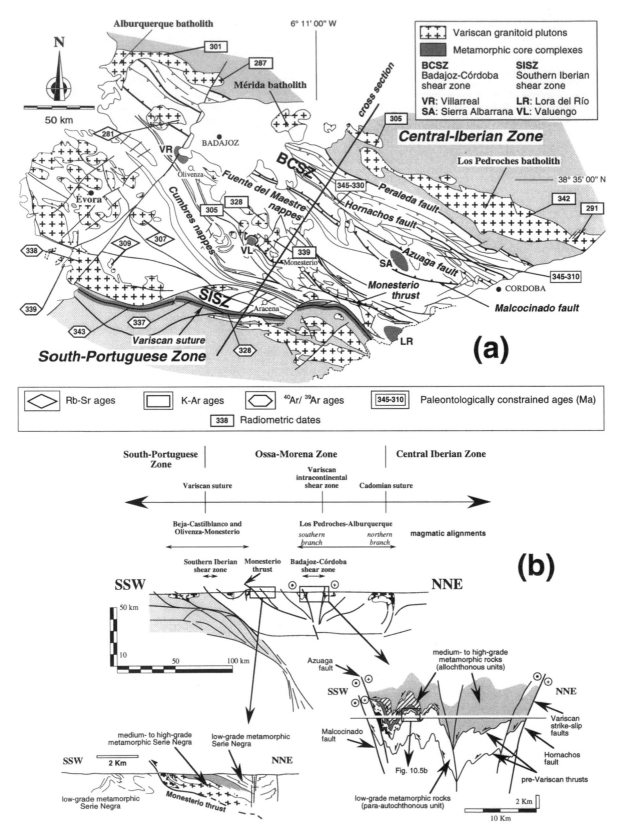

Fig. 9.11. (a) Geological map of the Ossa Morena Zone showing the principal tectonic features related to the Variscan orogeny and a selection of radiometric determinations yielding Variscan ages (adapted from Eguíluz *et al.* 2000; Castro 1999*a*). (b) Sketch cross-section of the Ossa Morena Zone (after Ábalos & Díaz Cusí 1995) showing the close relation between major strike-slip faults, sutures and magmatic alignments. Detailed cross-sections of the geometry of the Monesterio thrust and of the Badajoz–Córdoba shear zone adapted from Eguíluz & Ábalos (1992) and Ábalos (1992).

the northernmost regional shear zone (Los Pedroches basin), while a larger one formed to the south of the subduction zone along the boundary of the OMZ (South Portuguese deposits). Intramontane basins formed during Late Carboniferous times (Stephanian) and were filled with unconformable continental deposits attesting to the end of the orogeny. Late Variscan brittle intracontinental deformation completed the cycle.

South Portuguese Zone

The South Portuguese Zone (SPZ) constitutes the southern-most division of the Iberian Massif. It is bounded to the west and south by Mesozoic and Cenozoic sedimentary successions related to passive margins of the Iberian peninsula, to the east by the Cenozoic infill of the foreland basin of the Alpine Betic orogen, and to the north it exhibits complex tectonic relationships with other units of the Iberian Massif. A substantial part of the SPZ crops out in Portugal.

A complicating feature of the SPZ (Fig. 9.12) is that no basement or Early Palaeozoic rocks crop out, and only Upper Devonian through lower Westphalian metasedimentary and metavolcanic rocks are exposed. These rocks show clear evidence for an origin within a platformal continental margin environment that was progressively affected by tectonic events related to its proximity to, and docking with, the active southern margin of the Iberian Massif (Munhá 1983; Oliveira 1983, 1990; Quesada et al. 1991; Quesada 1990b, 1992, 1996, 1998). The rocks were penetratively deformed and intruded by the Carboniferous Sierra Norte batholith complex (Fig. 9.12). In the northeastern area of the SPZ Carboniferous granitoid rocks were affected by Stephanian strike-slip faulting during which diabase and microgranite dyke swarms were emplaced and flood basalts extruded. The Variscan complex is unconformably overlain by the remnants of a late Autunian terrestrial intermontane basin (Viar basin; Broutin 1981; Simancas 1985).

Traditionally, four tectonic domains have been recognized in the SPZ, each showing distinctive stratigraphical, structural and metamorphic records (Schermerhorn 1971; Routhier et al. 1978; Oliveira 1983). These are, from north to south (Fig. 9.12): the Pulo do Lobo, Pyrite Belt, Baixo Alentejo flysch and Southwest Portugal domains. They are separated by south-facing thrusts with a notable sinistral strike-slip component (Silva 1989; Silva et al. 1990; Quesada 1996, 1998). Three further divisions have been recently identified in the Spanish segment of the SPZ, separated by oblique fault zones: the so-called western, central and eastern blocks (Quesada 1996, 1998). All these domains exhibit significant differences in their stratigraphic record and in their relationships between volcanic rocks and ore deposits (Rambaud 1978; Routhier et al. 1978; Leistel et al. 1994, 1998).

Recently, oceanic affinities have been recognized in metasedimentary and metaigneous successions that crop out discontinuously or as dismembered klippen in the northernmost structural domain of the SPZ (the Pulo do Lobo unit or terrane; Munhá et al. 1986b; Silva 1989; Eden 1991; Quesada et al. 1994) and in southernmost units of the Ossa Morena Zone (the so-called Beja–Acebuches ophiolite). This has led some authors to propose that such rocks might constitute a zone (the 'Pulo do Lobo Zone') of comparable status, from a tectonostratigraphic point of view, to the South Portuguese and Ossa Morena zones (Quesada 1990b, 1992; Quesada et al. 1991, 1994; Fonseca 1996).

Structurally, the SPZ constitutes a south-facing, thin-skinned fold and thrust arcuate belt in which the dominant structures change progressively from a north–south direction in the westernmost parts, to east–west in the eastern sectors. Within this large-scale imbricate fan, divided along strike by oblique lateral ramps, Variscan deformation gave rise notably to thrust systems closely associated with a penetrative foliation (Fig. 9.12). The intensity of the deformation varies both parallel to the arcuate belt of the SPZ from the west (where it is more intense) to the east, and normal to the belt from the north to the south (less severe).

A number of deformation phases have been recognized in the area (e.g. Simancas 1983) although the structural history was essentially progressive. Thrust propagation followed a forward (southward) and downward sequence by footwall collapse in front of the advancing orogenic wedge. Upper crustal imbrication was accompanied throughout the entire deformation process by a component of sinistral wrenching, indicating a transpressional event. In detail, a great variety of thrust systems (Boyer & Elliot 1982) and fault-related folds can be recognized in the field (Silva 1989; Quesada 1996, 1998) ranging in size from the millimetre to the kilometre scale. Duplexes are the most common thrust structures in the region (Fig. 9.12), and in particular are closely associated with the ore deposits of the Pyrite Belt (Leistel et al. 1998), where large rheological contrasts occurred between extremely competent massive pyrite and phyllosilicate-rich softened alteration zones. Fault-related folds include detachment, fault-propagation, break-thrust and fault-bend types. These exhibit south-facing asymmetric fold geometries whose genetic and temporal association with thrust systems and lateral ramps is clearly indicated. Tectonic foliations also developed in association with the folds in all but the most competent lithologies at the presently exposed crustal level. They are roughly axial planar to the folds in cross-section, but usually show a remarkable transection on a horizontal plane, therefore providing evidence for the transpressional nature of deformation (Ribeiro & Silva 1983; Simancas 1983; Silva et al. 1990; Quesada 1998).

Many thrust sheets have a basal detachment located within a suitably weak horizon (frequently carbonate-rich layers within the Famennian Phyllite-Quartzite Gp) and this is the reason why no older rocks are seen at the present exposure level in the SPZ. The basal detachment of the whole fold and thrust belt coincides with the brittle–ductile transition within the crust of the SPZ (Ribeiro & Silva 1983; Silva 1989; Quesada 1996, 1998). It is interpreted to be located at a depth of c. 10 km, as indicated by a low seismic velocity layer at this depth (Ribeiro et al. 1990b). A 5–8 km thick layer exists below the detachment, and could represent a Palaeozoic cover deposited on a middle to deep crustal segment (a Precambrian basement?) that extends to the Moho at a depth of 30–35 km. Whether or not the crust beneath the detachment reacted congruently in geometric and kinematic terms with the upper crust is impossible to ascertain with the structural and geophysical data available so far.

A quantitative approach to the calculation of the amount of shortening, crustal thickening and thrust displacement has been attempted by Quesada (1998). In gross terms, the detached upper crust of the SPZ was shortened to at least half its original length and has almost tripled its pre-deformation thickness. Average displacements of the major thrusts are of the order of a few tens of kilometres, which is an order of magnitude smaller than the estimated displacements of the foreland nappes of NW Iberia. However, similar or even larger left-lateral strike-slip translations might have taken place along the northern suture boundary of the SPZ and within other major fault zones.

Many of the thrusts and lateral ramps represent the reactivation of former normal and transfer faults, respectively. Indirect evidence for this is clear in the case of the Pyrite Belt. Here, a submarine bimodal volcanism was widespread during most of Early Carboniferous time, giving rise to the so-called 'volcano-sedimentary complex' (Schermerhorn 1971) and included both mantle- and crustal-derived melts (Munhá 1983). The Iberian Pyrite Belt is a world-class metallogenic province (Chapters 8 and 19) associated with the volcano-sedimentary complex. The belt consists of a large number of base metal stockworks and massive sulphide and manganese ore deposits (and

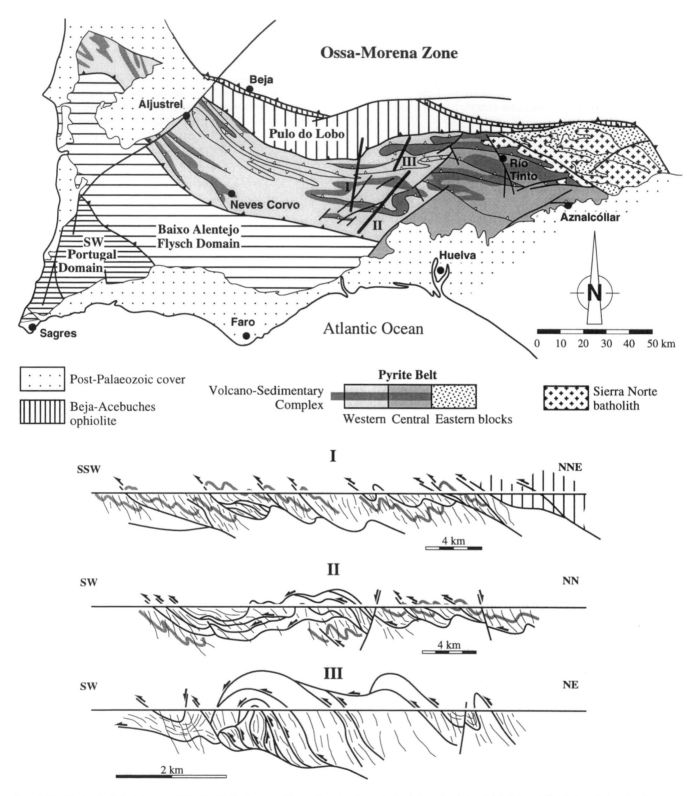

Fig. 9.12. Geological sketch map of the South Portuguese Zone showing the principal domains into which it is subdivided, and structural cross-sections of the central part of the Pyrite Belt showing the geometrical relationships between thrusts, folds (depicted by thick grey lines) and foliations (thin solid lines). Adapted from Quesada (1998). See text for further details.

associated hydrothermal alteration belts), some of them of huge dimensions. Both magmatic ascent and hydrothermal circulation would have been greatly favoured by coeval generation of extensional fissures of crustal/lithospheric scale. This is supported by the fact that many volcanic centres, ore deposits and alteration belts are spatially associated with current fault zones notably with south-facing, east–west striking oblique thrusts, and with NE–SW and subordinate, conjugate NW–SE striking tear faults or lateral ramps (cf. Rambaud

1978; Routhier *et al.* 1978; Leistel *et al.* 1994, 1998; Quesada 1996, 1998). This can also be the origin of the NE–SW faults that define the boundaries of the western, central and eastern blocks cited above. Direct evidence for this extensional event in other domains of the SPZ outside the Pyrite Belt is debatable due to the absence of volcanic and ore deposits. However, the striking geometrical and kinematic similarity of the transpressional Variscan structures throughout the SPZ argues in favour of a comparable tectonic setting. The thermal anomaly responsible for the magmatic and metallogenic evolution of the Pyrite Belt remains a subject open to speculation. The coexistence of a mantle thermal anomaly and associated igneous and hydrothermal processes together with lithospheric thinning may have had a feedback influence on one another, that can explain the singular characteristics of the Pyrite Belt both in a regional and a global perspective.

The northern part of the SPZ, the Late Devonian Pulo do Lobo domain, displays some tectonic features that make it special. Its exposure is restricted to a 150 km long, 20 km wide, east–west orientated, south-facing, sigmoidal strike-slip duplex sandwiched between SPZ units (to the south and below) and Ossa Morena units (to the north and above). Its internal structure is extremely complicated and still poorly known due to a combination of intense internal deformation, a densely imbricated set of fractures and shear zones, a general metamorphic overprint, a poor knowledge of the stratigraphy, and a paucity of chronologic data (both palaeontologic and radiometric). Although its overall structure defines a broad antiform, the existence of notable differences in the northern and southern limbs (Silva 1989; Oliveira 1990) suggests that it is most likely an antiformal stack of oblique thrust sheets. The rocks that constitute the Pulo do Lobo unit contain ample evidence for an oceanic origin. First, they are dominated by a pelite–quartzwacke succession in which N-MORB and minor acid volcanics and dykes occur, and locally this succession overlies MORB-type metabasalts (Crespo Blanc 1989). This suggests that it could correspond to a sedimentary cover laid down on ocean floor and/or thinned, outermost continental margin environments (Quesada *et al.* 1994). Second, they contain greywacke–pelite turbiditic successions (Santa Iría Fm; Carvalho *et al.* 1976a), and sedimentary mélanges and olistostromes, including ophiolitic (Peramora Mélange Fm; Eden 1991) and non-ophiolitic formations (Alájar Fm; Eden 1991; and Horta da Torre Fm; Oliveira 1990) that could represent synorogenic deposition in fore-arc or trench environments. Finally, the Pulo do Lobo rocks relate to MORB-type igneous sequences in which a complete ophiolitic stratigraphy can be reconstructed despite their present extreme dismemberment (Beja-Acebuches Ophiolite; Munhá *et al.* 1986b; Quesada *et al.* 1994; Castro *et al.* 1996b).

Variscan metamorphism in the SPZ is complex and reflects the superposition of regional metamorphism on an earlier, areally restricted (notably to the Pyrite Belt) episode of hydrothermal metamorphism related to the emplacement and cooling of volcanic complexes and massive ore deposits. Both types of metamorphism were post-dated by contact metamorphism associated with the aureoles of Carboniferous intrusives that reached the pyroxene hornfels facies close to the contact with the plutons.

Hydrothermal metamorphism caused the metasomatic alteration of basaltic and rhyolitic precursors (hydration, carbonation and oxidation reactions and Na-K mobilization) through the convective circulation of marine waters and hot brines. The metamorphic grade changes abruptly at the scale of a few metres from greenschist to zeolite facies, through chlorite–epidote assemblages. Geothermal gradients were elevated and the ambient temperatures likely oscillated between 50–100°C and 350–400°C.

Except for the cases of the Pulo do Lobo and the Pyrite Belt structural domains, regional metamorphism in the SPZ is of very low grade or absent. Based upon the mineral assemblages present in metabasic protoliths and on the variation of illite crystallinity, four different zones have been distinguished in the metamorphosed areas of the SPZ (Munhá 1990). The lowest grade, anchizonal zeolite-facies assemblages, occur at the southern sectors of the Pyrite Belt (zone 1).

Slightly higher-grade prehnite–pumpellyite assemblages are widespread in the main body of the Pyrite Belt (zone 2). The transition between these metamorphic zones represents temperatures of 200–290°C. Towards the contact with the Pulo do Lobo, situated to the north, the metamorphic grade reaches the transition to the greenschist facies at temperatures of 300–350°C (zone 3). Finally, greenschist facies metamorphism affected the Pulo do Lobo (zone 4), where temperatures of 400–450°C were reached. The associated pressures were probably up to 2 kbar, thus reflecting geothermal gradients of 40–50°C/km.

The presence of high-pressure metamorphic rocks (eclogites and blueschists; de Jong *et al.* 1991; Pedro *et al.* 1995; Fonseca 1996; Leal *et al.* 1997) and Lower Devonian–lower Visean arc-related igneous rocks in southern sectors of the Ossa Morena Zone (Santos *et al.* 1987, 1990) is clearly indicative of the existence of a Variscan suture contact and proves the northward polarity of subduction in the SPZ (Ribeiro *et al.* 1990b; Quesada 1990b, 1992; Quesada *et al.* 1991, 1994). According to Quesada (1990b, 1992) and Quesada *et al.* (1994), the Pulo de Lobo rocks (Fig. 9.12) would represent an ancient accretionary prism developed at the southern margin of the Ossa Morena Zone as a result of oblique, north-directed subduction of the SPZ. Late Devonian palaeontological age determinations for greywacke–pelite turbiditic successions and sedimentary mélanges of the Pulo do Lobo (Oliveira *et al.* 1986; Giese *et al.* 1986a; Oliveira 1990) are interpreted as roughly the time when the accretion took place. No age control exists for the ophiolitic rocks. This process of subduction/accretion (and eventual north-directed obduction; Fonseca 1996) lasted until the intervening oceanic tract was entirely consumed (by late Visean times), leading then to oblique continent–continent collision that brought this unit and the adjacent Ossa Morena Zone to obduct over the underlying SPZ. This culminated in a process of continent–continent collision between the Ossa Morena Zone (an outboard part of Gondwanan Iberia) and the SPZ (probably a part of Avalonia) during Late Carboniferous times. This interpretation (Munhá *et al.* 1986b; Ribeiro *et al.* 1990c; Quesada 1990b, 1992; Quesada *et al.* 1991, 1994; Fonseca 1996) has gained wide acceptance and has helped to improve significantly the current understanding of the Variscan orogeny in the whole Iberian Massif and its correlation with other segments of the European Variscides.

By latest Carboniferous times a relaxation in the convergence vector caused a return to transtensional conditions in the vicinity of the suture allowing emplacement of a large, WNW–ESE orientated, composite batholith made of diorite–gabbro and granite that transects the suture zone proper (the so-called Sierra Norte batholith in the east and Beja batholith in the west of the SPZ; Figs 9.11b and 9.12). On the basis of geochemical similarity, it has been speculated (Bellido 1996; Quesada 1998) that this batholith might have been fed by the same deep magma source that generated the Pyrite Belt volcano-sedimentary complex. If this were true, it would imply persistence of the mantle thermal anomaly throughout the entire Variscan orogeny.

The basement massifs of Alpine foldbelts

Iberian Ranges

Outcrops of pre-Mesozoic rocks in the Iberian Ranges occur in the core of alpine anticlines or anticlinorial structures, often bounded by high-angle reverse faults. They are regarded as

'Palaeozoic massifs' in the regional geological literature. The most important of these massifs are the Sierra de la Demanda, located in the NW edge of the Iberian Ranges, and two parallel, elongated massifs located in the central segment of the Iberian Ranges SW of Zaragoza (Fig. 9.7). Other numerous but much smaller (a few kilometres wide) Palaeozoic massifs exist throughout. The principal stratigraphical and structural characteristics of these inliers were first recognized by Lotze (1929) and Brinkmann (1931). A wealth of geological studies (mostly stratigraphical and palaeontological) have been published since then, notably (from a tectonic perspective) Colchen (1974) and Capote & González Lodeiro (1983), on which the following account is based. With respect to the Spanish segment of the Variscan orogenic belt, these massifs are located in the southern branch of the Ibero-Armorican arc and exhibit important stratigraphical and structural affinities with the West Asturian-Leonese Zone of the Iberian Massif. Gozalo & Liñán (1988) have correlated the western parts of the pre-Mesozoic Iberian Ranges with the West Asturian-Leonese Zone, and the easternmost parts with the Cantabrian Zone, the limit between them being delineated by outcrops of Precambrian rocks that are correlated with the Precambrian of the Narcea antiform. As regards the age of Variscan deformation, the youngest rocks affected by D1 and D2 structures are of Namurian and lower Westphalian age (Sacher 1966). The oldest post-tectonic rocks crop out in the Sierra de la Demanda, and are Westphalian B–D sediments that unconformably overlie much older folded and cleaved Palaeozoic rocks (Colchen 1974).

Variscan and Alpine structural trends in the Iberian Ranges are subparallel, so that the massifs contain the record of two orogenies whose effects are sometimes difficult to separate (e.g. Liesa & Casas Sainz 1994), although study of the unconformity beneath the Mesozoic succession commonly resolves the issue. The dominant structural trends in the Variscan Iberian Ranges change from east–west to WNW–ESE in the Sierra de la Demanda, to NW–SE and north–south in the central segments. The principal structures are folds and thrusts whose geometry and throw can vary along strike. Both were generated in the course of two principal deformation phases (D1 and D2) accompanied by very low-grade regional metamorphism. Regional metamorphism during D1 and D2 gave rise to chlorite growth across wide areas, but the biotite isograd was reached only in western sectors of the Sierra de la Demanda, where the metamorphic grade increases from east to west.

The folds are map-scale cylindrical structures and exhibit either parallel morphologies or slightly thickened hinges. Variscan folds commonly show a subvertical, axial planar foliation in the less competent units. Folds and foliations are regarded as D1 Variscan structures. Locally, the foliation exhibits steep to moderate dips to the SW, which denote structural vergence toward the east or NE, with gently plunging intersection lineations. In the Sierra de la Demanda a progressive variation in fold style can be recognized from west to east. In the west folds are clearly inclined, sometimes exhibit inverted limbs, and are associated with north–NNE directed thrusts and NNE–SSW stretching lineations (Colchen 1974). Toward the east, folds are larger scale (minor folds are rare) and fold axial surfaces are steeper. This structural variation appears to reflect a strain gradient from west to east probably related to progressive lithological changes in the Palaeozoic succession.

Thrusts are clearly younger structures than the folds, with which they share a common strike. Thrust surfaces often form small angles with D1 fold axial surfaces. Usually they are high-angle isolated reverse faults or wedge-like thrust associations with slips of some

hundreds of metres, directed toward the NE. They are considered as D2 Variscan structures, and both fold interference patterns and incipient crenulation foliation development related to D2 have also been recognized locally. D3 Variscan deformation corresponds to very restricted, NW–SE trending, subvertical angular folds that deform D1 and D2 structures.

Catalonian Coastal Ranges

Outcrops of pre-Mesozoic rocks in the Catalonian Coastal Ranges (CCR) form two NE–SW trending bands, separated by a Neogene graben subparallel to the Mediterranean coast in NE Spain (Fig. 9.13). These outcrops form massifs of variable size (e.g. the so-called Gavarres, Guilleries, Montseny, Montnegre and Collcerola massifs) that consist of Palaeozoic rocks intruded by large volumes of Variscan granitoids. The CCR are located in a transitional area between the northern branch of the Ibero-Armorican arc (to which the neighbouring Pyrenees clearly belong) and the core of the arc. They exhibit palaeogeographical affinities with parts of the Cantabrian Zone (notably the Pisuerga–Carrión units; Fig. 9.8). Comprehensive reviews of the principal stratigraphical and structural characteristics of the area, including the historical background, have been carried out by Julivert & Martínez (1980, 1983) and more recently by Julivert & Durán (1990*b*, 1992). The report that follows is based upon them.

The Variscan structure is difficult to unravel due to the scattered disposition of the massifs, the scarcity of major structures, the monotony of the stratigraphical sequence and the lack of stratigraphical key units. A northeastern segment of the ranges can be distinguished where the oldest and deepest rocks crop out, as can a southern one where Carboniferous rocks predominate. Broadly, a dominant first phase (D1) and later minor phases of Variscan deformation (D2 and D3) can be distinguished whose geometrical expression depends upon the structural level considered. The age of these Variscan deformations is difficult to elucidate. Visean–Namurian conglomerate units overlie the pre-Carboniferous successions unconformably and contain granitic and metamorphic clasts, but the age of the Catalonian granites is not well known and thus the possibility exists that the age of the earliest deformations and metamorphism may be pre-Carboniferous.

In the structurally shallower, lower-grade areas, the trend of D1 structures is rather constant, broadly east–west, and oblique to the trend of Alpine structures. Apart from some rare mappable structures such as kilometre-scale folds, one can recognize metric to decametric, east–west trending inclined folds and a regional D1 foliation (S_1). The folds are evident when they deform competent layers. In the structurally highest and less metamorphic realms they are south-vergent structures. The regional D1 foliation also trends roughly east–west and grades between a continuous foliation in metapelites and a poor, spaced cleavage in quartzites. It usually transposes the bedding, a gently plunging intersection lineation being observed in the foliation planes orientated parallel to the mesoscopic fold axes. Also in the less metamorphosed areas, penetrative pencil structures result from the intersection between the D1 foliation and bedding. The dip attitude of S_1 forms alternating belts of north- and south-dipping foliations at all scales.

In structurally deeper, higher-grade metamorphic areas, no major D1 structures can be observed, though both major tight folds and thrusts possibly exist. Minor structures, by contrast, are widespread and include a D1 foliation (S_1) subparallel to bedding. Minor isoclinal, intrafolial folds are common in schists and marbles. Their axial surfaces are parallel to S_1, whose orientation does not differ significantly in dip and strike from that of lower-grade areas in the same massifs. Subhorizontal, east–west stretching lineations on the foliation

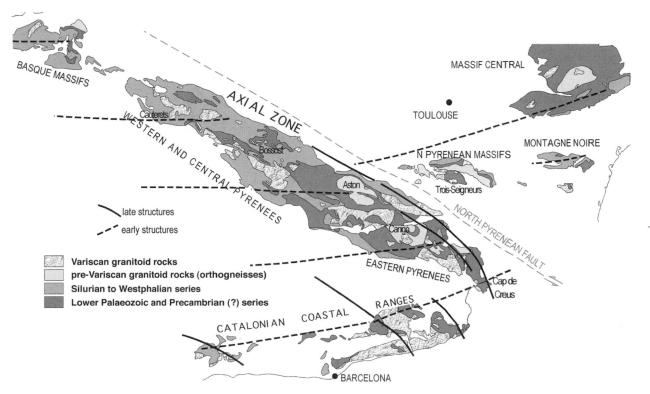

Fig. 9.13. Sketch map of the outcrops of pre-Mesozoic rocks in the Pyrenees, the Catalonian Coastal Ranges and neighbouring French massifs in a reconstruction of the pre-Alpine geotectonic setting of NE Iberia. Restoration of the North Pyrenean Fault and Alpine contraction are based on Barnolas & Chiron (1996).

planes have been reported by Julivert & Durán (1990b). Alternating bands of north- and south-dipping foliations also occur in the lower structural levels.

Later phases of deformation (D2 and D3) gave rise to both gently and steeply dipping crenulations and to open chevron and kink folds. The former are discontinuous foliations that show attitudes that vary from subvertical to subhorizontal. By contrast, the orientation of the associated crenulation lineations is rather constant. Though S_2/S_3 usually cut across D1 foliations at a high angle, sometimes they are difficult to distinguish from S_1. In the higher-grade areas, these crenulations can transpose the S_1 and become the principal foliations. These deformations are considered to be responsible for the formation of the alternating bands of north- and south-dipping foliations, in which case they might be represented by both major upright south-facing folds or by folds with subhorizontal axial surfaces.

Contact and regional low- to medium-grade metamorphism are widespread (Sebastián et al. 1990). Regional metamorphism appears to be syn- to late-tectonic with respect to D2 and of low-pressure type, as evidenced by the sequence of metamorphic index minerals biotite–andalusite/cordierite–sillimanite. Metamorphic isograds are closely spaced, suggesting elevated geothermal gradients and the likely existence of thermal domes. The duration of metamorphism is poorly known, although microtextural relationships and the occurrence of pre- and syn-D2 cordierite and of post-D2 biotite and andalusite suggest a relatively long time span. Granitoid intrusion commonly post-dated the regional metamorphism and induced contact metamorphism in the country rocks, reaching the hornblende and pyroxene hornfels facies (Gil Ibarguchi & Julivert 1988). However, contact metamorphism is locally syn-tectonic with respect to D3.

Pyrenees

Variscan basement in the Pyrenees crops out mainly along the so-called axial zone (Fig. 9.13). North of the axial zone, the North Pyrenean massifs comprise Variscan basement surrounded by Mesozoic cover rocks. The Pyrenean axial zone and the North Pyrenean domain are separated by the North Pyrenean fault (Fig. 9.13). This fault played its most important role during rotation and sinistral strike-slip of Iberia prior to Pyrenean (Alpine) compression (Le Pichon et al. 1970). Consequently, the North Pyrenean massifs occupied a more easterly position than their present-day setting with regard to the axial zone. This must be taken into account in any attempt to reconstruct Variscan structure. South of the axial zone, small outcrops of Variscan basement are involved, together with their cover rocks, in Pyrenean tectonics.

A question that frequently arises when dealing with the Variscan structure of the Pyrenees refers to the degree of Alpine overprinting superimposed on the basement. Although there is no doubt that basement rocks were involved in the Alpine fold and thrust belt, it is generally agreed that penetrative ductile structures, metamorphism and igneous activity in the basement are related to the Variscan orogeny. Post-Variscan events would only have been responsible for localized deformation along thrust zones (brittle and ductile), general doming and tilting of the Variscan structures, and for localized metamorphism along the North Pyrenean fault. Thus, the degree of reworking of Variscan structures by Alpine events is sufficiently slight to enable reconstruction of the Variscan structures.

The study of the Variscan structure of the Pyrenees has been the target of many groups of geologists (Barnolas & Chiron 1996), knowledge of Variscan structure having increased significantly as a result of the work of Dutch and French schools. However, there have been many interpretations, and critical questions relating to the structural evolution and tectonic regimes involved in the Variscan Pyrenees are still under debate and controversial. Thus, it is currently impossible to give a picture of the Pyrenean Variscan structure with which all authors would be comfortable (Carreras & Debat 1996, and references therein). The first explanations (e.g. Matte 1976; Zwart 1963b) proposed that crustal shortening dominated throughout the Variscan orogeny. Such hypotheses were gradually replaced by tectonic models that included crustal extension at a given stage of the orogeny (e.g. Soula 1982; Verhoeff *et al.* 1984; Soula *et al.* 1986; Van den Eeckhout & Zwart 1988). Models considering extension only, though commonly quoted in the literature (e.g. Wickham & Oxburgh 1986; Vissers 1992), are not consistent with field evidence and thus can be ruled out. Other recent interpretations consider that crustal shortening was associated with dextral wrenching, and thence deformation, metamorphism and magmatism would have taken place in a transpressive tectonic setting (Carreras & Druguet 1994; Gleizes *et al.* 1997, 1998a). Given the debates over Pyrenean structure and tectonic history, it is especially important to focus on a number of key features that are outlined below.

The Variscan rocks of the Pyrenees exhibit a polyphase structural evolution that in some domains is contemporaneous with magmatism and an associated low pressure–high temperature metamorphism. This tectonic setting is analogous to the one observed in rocks of the Variscan massifs outcropping to the south (Catalonian Coastal Ranges) and to the north (Montagne Noire; see Fig. 9.13), where they are unaffected by the Alpine orogeny. The tectonic evolution of basement rocks in the Pyrenees thus needs to be considered in a broader regional frame in which, in addition to the above-mentioned massifs, other Variscan exposures should be included (e.g. Sardinia and the Iberian Ranges).

The regional structural trend is WNW–ESE, with mappable folds showing WNW–ESE trending axes and associated south to SW vergences. Fold axial surfaces and axial-planar foliations usually dip to the north. Their attitude is subvertical close to the North Pyrenean fault, but their dip decreases gradually towards the southern border of the axial zone. The most characteristic and striking structural feature of the Variscan Pyrenees on a regional scale is the occurrence of broad domes and pinched synforms. Domes are cored by medium- to high-grade orthogneisses and/or Cambro-Ordovician metasediments, whereas the synforms contain non-metamorphic to low-grade younger rocks (Silurian, Devonian or Carboniferous).

Another outstanding feature is the existence of a vertical variation in tectonic style. In the highest structural levels, the regional foliation exhibits a steeply dipping, homoclinal attitude rather constant over large areas. Regional foliations are axial-planar with respect to folds defined by bedding anisotropy and/or a weak, bedding-parallel older foliation. Commonly, these foliations are slaty cleavages or penetrative crenulation cleavages. In contrast, in the lowest structural levels several penetrative foliations can be recognized, the older ones displaying complex folding and transposition patterns. This structural zonation led to the definition of two domains: the so-called supra- and infrastructure (Zwart 1963b). Chronological correlation of structures observed in the infrastructure with those observed in the suprastructure is a matter of discussion.

At all structural levels there exists ample evidence for polyphase tectonism, and minor structures (i.e. minor folds and foliations) abound. However, there is no clear relationship between them and the larger-scale, mappable structures. Additionally, the number of deformation phases recognized can vary considerably from place to place and from one author to another. As a consequence, a widely accepted and reliable deformational history valid for large areas of the Variscan basement is not at present available.

Variscan metamorphism is characterized by its distribution in confined medium- to high-grade areas that are bounded by much larger low- or very-low-grade metamorphic domains. Usually, the transition belts between the higher- and lower-grade tracts exhibit high thermal gradients. Metamorphic zones are cored either by orthogneisses or by migmatitic schists. Variscan magmatism is widespread in the Pyrenean basement and is mainly represented by sheeted granitoid batholiths and stocks. These are preferentially emplaced in the shallower structural domains so that granitoid rocks are rare in medium- to high-grade metamorphic domains. There is a tendency for the plutons to be aligned parallel to the dominant Variscan trend. Though there is no general agreement between different authors, compelling evidence exists supporting the syntectonic emplacement of most of these granitoids (Gleizes *et al.* 1997, 1998a) and the contemporaneity of deformation and metamorphism.

A widespread mylonite belt with associated shear zones runs roughly parallel to the Variscan and Pyrenean east–west dominant trend. The shear zones affect orthogneisses, metasediments and granitoids. Concomitant mylonitization took place under retrograde, greenschist-facies metamorphic conditions. Whether the age of this mylonite belt is Variscan and the nature of its tectonic significance are matters of conjecture and controversy.

Divergent views on Pyrenean tectonics emphasize the problem of relating minor structures with regional- and crustal-scale ones. This is especially so when regional foliations in the deeper structural levels are interpreted. Most authors accept that these foliations were originally flat-lying. However, gently dipping foliations can form both in extensional and compressional settings, as they both can be related to subhorizontal heterogeneous shearing and recumbent folding. Some authors (e.g. Guitard 1964) suggested that the sheet-like shape of orthogneiss bodies (likely Lower Palaeozoic granite sills; Guitard *et al.* 1996) was the result of recumbent folding during the early stages of Variscan orogeny. This interpretation was based upon the acceptance of stratigraphic equivalence between the metasediments overlying and underlying the gneiss sheets (Guitard 1970). In the Canigó massif, however, the sequences overlying the Canigó orthogneisses (the Canavelles series) are considerably younger than the sequences situated below (Balaig schists; Navidad & Carreras 1995). Therefore, although early stages of compressional tectonics cannot be excluded, undisputable large-scale mappable structures which would support this assumption have not been described in deep-seated structural domains. It is also commonly accepted that flat-lying foliations are structural features characteristic of deep-seated tectonic realms. However, this generalization is commonly incorrect. What are thought to be originally flat-lying foliations form currently asymmetrical domes and antiforms. Such a disposition could have been achieved even if the initial foliation had been in a moderately dipping homoclinal attitude. Moreover, it appears that in some massifs the regional foliation was initially steeply inclined, which argues against flat-lying foliations being indicative of extensional structures.

As stated above, the complex structures of deep-seated structural levels contrast with the simpler deformation patterns observed in shallower domains. Here, no signs are found of later doming or large-scale folding except for the case of some rather open domal structures. Structural evidence for the existence of a folding and thrusting event prior to the development of the regional foliation in the suprastructure has been presented by Bodin & Ledru (1986) and Losantos *et al.* (1986). It thus seems that while in deep-seated realms there was a synschistose deformation event and a later one which involved folding and doming of earlier structures, in shallow-seated domains regional folding pre-dated at least in part foliation development. If this reflects diachronism during tectonism, this could be the reason for the many discrepancies among proposed interpretations due to the fact that foliation development in deep-seated levels was initially considered to be coeval with that of shallower domains. Alternative explanations consider that there was no ubiquitous foliation development either along or across the Pyrenean segment of the Variscides (Carreras & Capellà 1994). According to the latter hypothesis, early foliations formed with variable attitudes and degree of penetration in deep-seated domains, whereas shallower levels deformed by folding and thrusting with minor or only local development of associated cleavages (Fig. 9.14). Progressive deformation and the migration of deformation fronts within the orogen would have led to development of penetrative foliations in shallow structural levels (the first regional foliation here) at the time the early penetrative structures of the deep-seated realms were being refolded and transposed by new ones. In this framework, the orthogneiss massifs constitute stretched and refolded sill-like pre-Variscan granite intrusions (Fig. 9.14).

The occurrence in the Pyrenees of low-pressure, high-temperature metamorphism with associated high thermal gradients has been varyingly interpreted in relation to Variscan tectonics. One interpretation is the so-called 'effet de socle' model. This refers to the controlling effect of gneissic domes on the distribution of metamorphism and metamorphic isograds (Guitard 1970). However, the model does not account for the existence of metamorphic domes that are not cored by orthogneissic rocks (for example the Bossost and Cap de Creus massifs). Other models couple metamorphism and extensional tectonism (e.g. Wickham & Oxburgh 1986). These are difficult to apply since no compelling structural evidence exists which supports the occurrence of a regional extensional tectonic phase in the Variscan Pyrenees. Alternative models consider that high thermal gradients are related to strong horizontal strain gradients associated with transpressional deformation (Druguet 2001). In these latter models, syntectonic magmatism in the deep-seated domains (characterized by the intrusion of diorite and granitoid bodies) could have enhanced, if not caused, the metamorphism. Relevant examples of the intimate association between the three processes (transpression, magmatism and metamorphism) have been described in the Cap de Creus and in the Trois-Seigneurs massifs.

Summarizing, and regardless of discrepancies in the detailed reconstruction of deformation phases, it appears that in the Pyrenean segment of the Variscan orogen Variscan tectonic evolution overall was characterized by (i) a broadly transpressional regime which started with contraction and ended

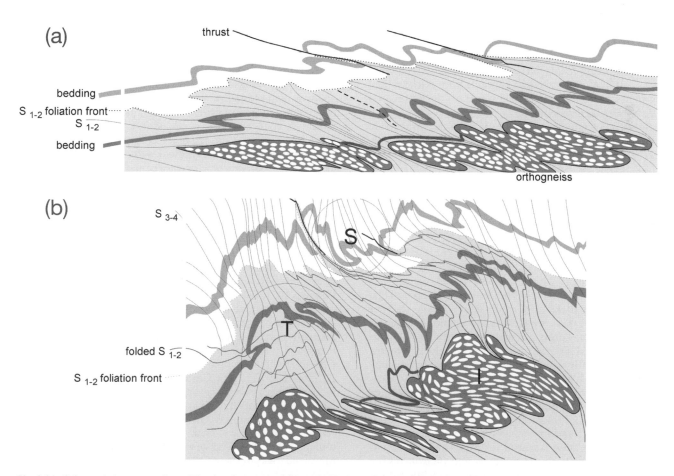

Fig. 9.14. Schematic interpretation of the development of the main Variscan structural features in the basement of the Pyrenees. (a) Early to main stage. (b) Main to late stage (see text for further details). Key: S, suprastructure; I, infrastructure; T, transitional zone. For the sake of simplicity, the evolution of metamorphic zones and granitoid emplacement, which are related to the early stages of orogeny and would also account for the narrowing of the metamorphic zones, have not been included in the drawings.

with wrench-dominated transpression, (ii) a clockwise rotation of structures to NW–SE trends, and (iii) a progressive
localization of deformation along narrow shear zones
(mylonite belts) during post-peak-metamorphic cooling of the
orogen. An interpretation of the development of the main
Variscan structural features in the basement of the Pyrenees is
presented in the Figure 9.14. Heterogeneous shearing in the
early to main stage led to the development of penetrative foliations (composite S_{1-2} foliation) in deep-seated levels (Fig.
9.14a), while shortening at shallow levels was accommodated
by folds and thrusts without (or with only locally developed)
foliations. Progression of deformation during the main to late
stages increased horizontal shortening leading to the development of a ubiquitous foliation in the shallower domains
(suprastructure, S in Fig. 9.14b). In deeper levels (infrastructure, I) horizontal shortening induced the folding and
crenulation of older foliations, and a crenulation foliation
(S_{3-4}) formed with a variable degree of penetration. This
shortening event also caused doming of gneissic sheets (Fig.
9.14b). The main features in the infrastructure (I) are a generally less steep attitude of foliations, which show a variable
degree of superimposed folding, with late foliations becoming
locally prevalent. In the passage between these domains, a
transitional zone (T in Fig. 9.14b) is characterized by a very
complex structural pattern in which the dominant foliation can
change from one domain to another depending on the relative
intensity of the earlier, main or late deformation events.

The Betic Cordillera

The Betic Cordillera forms part of a collisional mountain belt
generated during convergence between the African and
Iberian plates in Tertiary times (Dewey *et al.* 1989; Comas *et
al.* 1992, 1999). The three metamorphic complexes that can be
distinguished in its internal zones, the Maláguide, Alpujárride
and Nevado–Filábride complexes (Chapter 16), consist mainly
of metasedimentary Mesozoic and pre-Mesozoic rocks (Fallot
et al. 1960; Egeler & Simon 1969; Lafuste & Pavillon 1976;
Gómez Pugnaire *et al.* 1982). These complexes underwent pervasive Alpine deformation and metamorphism that obliterated most of the effects of the pre-Alpine deformational
events, whose relics are described and discussed below.

The Veleta and Mulhacén nappes of the Nevado–Filábride
complex (the structurally lowermost complex) are classically
considered as formed by a thick pre-Mesozoic basement and
a pre-Alpine basement with an Alpine cover, respectively
(Puga *et al.* 1974). However, it has been recently proposed that
all of the Mulhacén rocks have a pre-Mesozoic age (Gómez
Pugnaire *et al.* 2000). The principal foliation and the regional
high- to intermediate-pressure metamorphism are Alpine.
Pre-Alpine (regional D1) structures are only locally recognized, most frequently as inclusions within the Alpine metamorphic porphyroblasts.

Orientated graphite inclusions (within chloritoid, garnet,
plagioclase, white mica) that define an axial-planar foliation
(S_1) related to isoclinal folds can be recognized in the Veleta
nappe. Low-pressure and low-temperature conditions (400°C
and *c.* 4 kbar) have been deduced for pre-Alpine metamorphism, accepting that both the minerals and their inclusions
formed during a pre-Alpine (Variscan or older) metamorphic
episode (Gómez Pugnaire & Franz 1988). Some authors,
however, argue that this metamorphism is also Alpine (Puga
& Díaz de Federico 1976; Vissers 1981; De Jong 1991), in spite

of the great contrast with the pressures and temperatures
prevalent during the Alpine metamorphism elsewhere.

Relics of pre-Alpine structures and minerals in the Mulhacén nappe appear very locally in the basement rock units
(Puga & Díaz de Federico 1976; Gómez Pugnaire & Sassi
1983). Hornfels-like metapelites contain a low-pressure and
high-temperature assemblage (3–4 kbar and 650°C) consisting
of exceptionally large crystals of andalusite, staurolite, garnet,
chloritoid and biotite. This assemblage is partially replaced by
a higher-pressure paragenesis (kyanite after andalusite,
garnet after biotite, kyanite and chloritoid replacing staurolite,
etc.) clearly related to the first Alpine metamorphic event.
Deformational pre-Alpine features are rare and include orientated inclusions in some porphyroblasts (Puga & Díaz de Federico 1976; Vissers 1981; Gómez Pugnaire 1981) and foliations,
folds and lineations described in several outcrops and interpreted as older than the first-phase Alpine structures by
Vissers (1981). Magmatic events clearly related to the latest
stages of the Variscan orogeny can be found both in the basement and cover series of the Mulhacén nappe. These are meta-
granitic bodies cropping out in the basement sequences that
preserve the magmatic mineralogy and texture as well as minerals formed by contact metamorphism in the country rocks
(Nijhuis 1964b). In the cover sequences, the metagranites
occur concordantly with the metasediments and their Rb/Sr
and Sm/Nd radiometric signatures also indicate a Palaeozoic
age for its emplacement (Nieto *et al.* 1997; Gómez Pugnaire
et al. 2000).

In the Alpujárride complex a pre-Mesozoic basement and a
Triassic cover can be distinguished (Kozur & Simon 1972;
Martín & Braga 1987a). In contrast with the Nevado–Filábride
complex, the basement series of the Alpujárride show progressive high-temperature metamorphism (from greenschist
to granulite facies) and contain large allochthonous sheets of
ultramafic rocks (the Ronda peridotites) emplaced onto basement metasedimentary sequences (Torné *et al.* 1992; Tubía
et al. 1997). In the earliest studies on the Alpujárride complex
most of the metamorphic and deformational features of the
basement were ascribed to the Variscan orogeny (e.g. Copponex
1958; De Vries & Zwaan 1967; Boulin 1968; Fernex 1968;
Aldaya 1969; Egeler & Simon 1969; Kornprobst 1971) due to
the very low-grade metamorphism of the Alpine orogeny in
the cover series. In more recent studies, the high-grade metamorphism in the basement has been interpreted as the result
of the hot emplacement of the Ronda peridotites (Lundeen
1978; Torres Roldán 1979; Obata 1980; Tubía & Cuevas 1986;
Balanyá 1991; De Jong 1991; García Casco 1993). The same
conclusion was reached for the central and eastern part of the
Alpujárride complex because of the progressive evolution of
the metamorphism from the basement to the cover as well as
the similar deformational evolution (Navarro Vilá 1976;
Cuevas 1990). Recognition of high-pressure (*c.* 10 kbar;
Azañón & Goffé 1997; García Casco & Torres Roldán 1996)
metamorphic assemblages in the cover and the occurrence of
eclogite lenses interlayered within the high-grade sequence of
the basement (Tubía & Gil Ibarguchi 1991) suggest a similar
Alpine evolution for both. However, the *c.* 300 Ma age
obtained for the high-grade metamorphism in the pre-
Mesozoic rocks of the western Alpujárride is clearly related to
the Variscan orogeny (Acosta 1998; Sánchez Rodríguez 1998).
It is now clear therefore that these rocks underwent Variscan
metamorphism prior to the emplacement of the Ronda peridotites at *c.* 20 ± 0.7 Ma BP (Sánchez Rodríguez 1998) and,

consequently, prior to the Alpine orogeny. In the central and eastern part of the Alpujárride complex the existence of a Palaeozoic metamorphic crust is evidenced by inherited Rb/Sr and Sm/Nd isotopic data found in the calc-alkaline Neogene volcanic province, which originated as a consequence of the opening of the Alborán basin (Zeck & Whitehouse 1999; Zeck *et al.* 1999).

The Maláguide complex (the structurally uppermost complex) is made of a low-grade metamorphic sequence partially dated as Palaeozoic but with a Mesozoic–Cenozoic sedimentary cover (Geel 1973; Paquet 1974; Navarro Vilá 1976; Mäckel 1985; Lonergan 1993). The lower part of the Palaeozoic sequence consists of andalusite-bearing mica-schists overlain by phyllites and carbonates. In this case, the age of the main deformation and metamorphism, whether Variscan (Chalouan 1986; Tubía 1988) or Alpine (Torres Roldán 1979; Platzman *et al.* 2000), is a matter of debate.

Radiometric data obtained by Chalouan & Michard (1990) indicate a possible overprint by Alpine metamorphism. However, the lack of zircon recrystallization from the basement sequences suggests the absence of post-Variscan overprint or, alternatively, a very low temperature for Alpine metamorphism (Sánchez Rodríguez 1998).

This study was financially supported by the Spanish DGICYT (grants PB97-617 and PB98-0143 to B. A. and J. I. G. I., PB91-0192-CO2 and PB94-1396-CO2 to J. E. V), the European Union FEDER R&D Program (grant 1FD97-1177 to C. Q.), the MAGNA (IGME) mapping program (J. E. V.), and the Grupo de Investigación de la Junta de Andalucía (grant RNM-0145 to M. T. G. P). R. Gayer and Ph. Matte are thanked for careful and constructive reviews of the manuscript. W. Gibbons and T. Moreno are thanked for the editorial handling. This is a contribution to IGCP Project 453 'Ancient and recent orogens'.

10 Permian and Triassic

JOSÉ LÓPEZ-GÓMEZ, ALFREDO ARCHE & ALBERTO PÉREZ-LÓPEZ

The Permo-Triassic succession in Spain records the change from a Pangaea configuration and compressive tectonic regime inherited from the Variscan orogeny, to an extensional tectonic setting accompanied by continental break-up and westward expansion of the Neotethyan ocean (Fig. 10.1). During latest Carboniferous–Early Permian times, latest Variscan orogenic extension associated with andesitic volcanism produced small continental basins in the Pyrenees, the east and central Iberian Ranges, and along the southern margin of the Iberian Massif. Extension continued into and during Late Permian times, creating half-graben and graben continental sedimentary basins bounded by Palaeozoic highs. Subsequent westward expansion of Neotethys led to successive marine transgression–regression cycles along the eastern and southern margins of the Iberian plate, which drowned the Palaeozoic highs during Middle Triassic (Ladinian) times. Although the Atlantic Ocean did not open during Triassic times, slow subsidence allowed the Neotethys to prograde westwards around the Iberian Massif, so that the Pyrenean-Cantabrian, Betic and Lusitanian basins became interconnected and extended towards the Grand Banks area (Fig. 10.1; Jansa *et al.* 1980).

Permo-Triassic southward propagation of pre-existing Norwegian-Greenland sea rift systems and westward propagation of the Tethys rift system represented the initial phase of post-Variscan plate reorganization, spanning some 90 Ma (Fig. 10.1a). The development of different rift systems during this initial break-up of Pangaea was related to a series of strike-slip faults that dissected the Variscan foldbelt and its associated foreland areas. The development of the different rift systems of central and western Europe was not coeval. During Stephanian–Autunian times, minor faulting accompanied the development of basins caused by the gradual decay of thermal anomalies induced by transtensional deformation (Ziegler 1988*b*). However, a major graben that had already started to form during the deposition of the Rotliegendes conglomerates and sandy series occupied the area from Poland to the Black Sea, while Iberia, as well as part of NW Africa, formed a coherent positive area during this time (Fig. 10.1a). Permian seas rapidly advanced westward along the axes of these pre-existing rift systems, aided by continental rifting combined with a glacio-eustatic rise in sea level (Ziegler 1990). Thus, while carbonate shelves of the Tethys realm covered part of western and central Europe at the end of the Permian period, basins in Iberia were filled by continental clastics (Fig. 10.1b). Many of these Iberian basins, initiated during latest Carboniferous–Early Permian times (Sopeña *et al.* 1988), were to become at least temporarily connected to the Tethys sea by the end of the Permian period. The development of Permian to Middle Triassic graben systems of Iberia and other areas in central and southern Europe was intimately related both to the rapid southward propagation of the northern Europe–Greenland sea rift and to the development and westward propagation of the Tethys sea (Fig. 10.1c) (Arche & López-Gómez 1996). Finally, by Late Triassic times, Tethys had advanced westward beyond western Poland via the Bay of Biscay rift and the Iberian basin to encircle and isolate the Iberian meseta (Iberian Massif). Shallow platform carbonates were deposited on the Iberian plate and are comparable with similar successions elsewhere in Europe.

It may be concluded that the general development of Late Permian–Triassic basins in the Iberian plate was similar to the development of those in the rest of Europe, but not coeval with them. Thus, similar facies ('Saxonian', Buntsandstein, Muschelkalk and Keuper) are found in different European basins, but the age of the sediments is different because the rift systems propagating from the east and NE took several million years to evolve and so created diachronous facies changes.

The Permian–Triassic boundary in Spain lies somewhere in a continental red bed basal succession, and so there are few biostratigraphic markers to allow precise dating around this boundary. Thus it is very difficult to separate the two systems in the Iberian peninsula and both Permian and Triassic successions are described together in this chapter. The Permo-Triassic record corresponding to the Iberian peninsula can be broadly subdivided into the classic tripartite, Germanic-type Permo-Triassic – Buntsandstein, Muschelkalk and Keuper – using these terms as facies descriptions and not time intervals. The continental facies have been subdivided into three unconformity-bounded units: Autunian (Early Permian), 'Saxonian' (Late Permian) and Buntsandstein (Late Permian–Early Triassic).

Continental deposits have been dated mainly by means of palynological assemblages as well as a few freshwater conchostraceae, macroflora remains and isotope absolute age determinations obtained on basal volcanic complexes. Marine deposits have been dated mainly using ammonites, foraminifera, conodonts, bivalves and palynological assemblages. Absolute ages of stages and system boundaries quoted in this chapter are those of Gradstein *et al.* (1995). The degree of precision of the available regional studies on which this chapter is based is, however, very variable and depends on the available outcrops. The Catalonian Coastal Ranges and the Iberian Ranges are better known than areas such as the Cantabrian Zone or southern Iberian Massif. Thus, regional correlations are still open to debate, especially for the early, continental sedimentary record. In contrast, the Middle–Upper Triassic sedimentary record can be correlated in much more detail (Sopeña *et al.* 1988; López-Gómez *et al.* 1998).

The following sections firstly offer an overview of the tectonic setting, indicating some of the controlling factors influencing the origin and development of the Permo-Triassic basins. This is followed by a description of Permo-Triassic stratigraphy within the various basins through time.

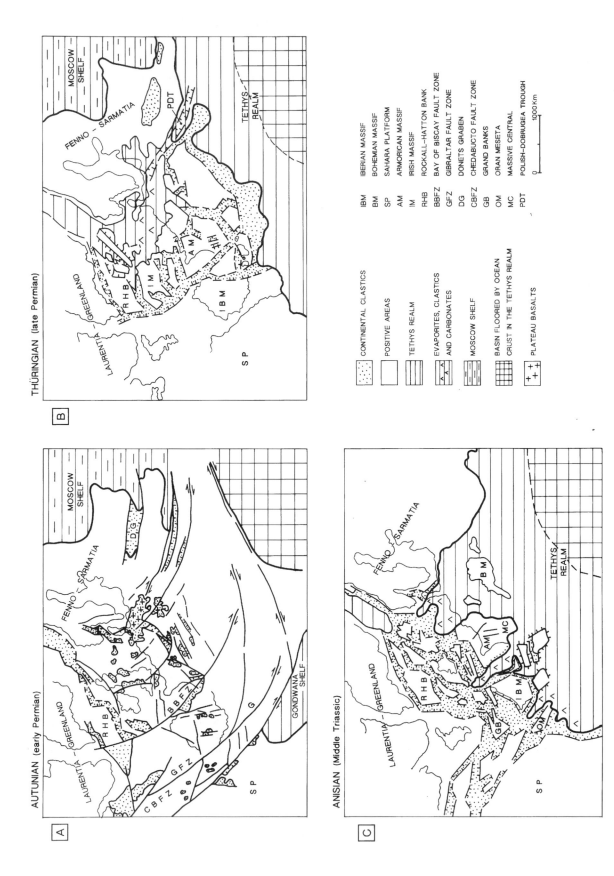

Fig. 10.1. Location and paleogeographical evolution of the Iberian peninsula and western Tethys area during Early Permian–Middle Triassic times (modified from Arche & López-Gómez 1996).

Tectonic setting

The final, extensional collapse phase of the Variscan orogeny took place during Late Carboniferous (Stephanian B–C) to Early Permian time across most of the Iberian plate. Basin development at that time was coeval with granitic and monzo-granitic magmatism and high heat flow which led to marked uplift and extension, perhaps in part as decollements along former thrust planes. This created intermontane continental basins along the margins of the Iberian plate along the northern (e.g. Demués and San Tirso formations in the Cantabrians), eastern (e.g. Minas de Henarejos basin), and southern Iberian margins (e.g. Río Viar basin in Andalusia). Despite their synorogenic character, none of these basins contain volcanic rocks, a fact that can be explained by low-angle basin boundary faults not reaching the asthenosphere (i.e. detachments along former thrusts). Denudation following uplift eliminated most of the Stephanian basins in the Iberian Ranges and along the southern margin of the Iberian Massif, and the tectonic regime became dominated by strike-slip faulting.

The evolution of the Iberian area during latest Carboniferous to Late Triassic times can be divided into three successive periods, each with differing tectonic, magmatic and sedimentological characteristics: latest Carboniferous to Early Permian, Late Permian, and latest Permian to Late Triassic.

Latest Carboniferous–Early Permian rift basins

The first sign of an extensional regime in the Iberian plate is recorded by a series of small, elongated basins, each a few kilometres long, along the Asturian-Pyrenean margin, the Iberian Ranges and southern margin of the Iberian Massif. This period of time was one of tectonic readjustment of plates by transtensional faulting in the Chedabucto–Gibraltar and Bay of Biscay areas, and a conjugate wrench zone, originating in the Iberian basin, following an ancient suture running across the microplate (Salas & Casas 1993; Doblas et al. 1994b; Arche & López-Gómez 1996; Valero 1993). This extension began in the central Pyrenees area during late Stephanian times and shortly afterwards in the other areas. During Early Permian (Autunian) times, a series of half-graben, intermontane basins became infilled by alluvial fans, slope breccias and lacustrine deposits associated in their lower part with volcaniclastic rocks of calc-alkaline affinities, such as andesites with subordinate rhyolites and basalts (Muñoz et al. 1983; Lago et al. 1992). Episyenites were intruded along tensional structures in the adjacent Central Range (González-Casado et al. 1996) around the same time (280–277 Ma BP).

These basins originated under a transtensional regime resembling the onset of the Red Sea rift (Bonnatti 1985) and the East African rift (Ebinger 1989), in which regularly spaced hot mantle plumes precede extension and propagate along the strike of the future rift system. Tensional stress varied along the wrench faults and transtensional half-grabens were formed at the weakest points. Unlike in the Stephanian basins, the presence of thick volcaniclastic deposits at the base of the basin fill indicates that the basin boundary faults were initially deep enough to reach the asthenosphere. Lacustrine deposits at the top of the volcaniclastic succession and the lower part of the siliciclastic succession record a change from freshwater to saline lakes, indicating progressive aridity in Iberia, ending with a dominance of floodplain–fluvial channel sandstone sequences. This development is clearly illustrated in the Pyrenean basins (Gisbert 1981; Valero 1993). Finally, transtensional faulting ceased during the late (?) Autunian and this sedimentary phase ended with a period of uplifting, tilting and erosion.

Late Permian rift basins

Widespread extension began during Late Permian (Thüringian) times in the Iberian plate, and basins were created in the Cantabrian-Pyrenean Zone the Catalonian Coastal Ranges and the Iberian Ranges (Fig. 10.1), where thick red bed sequences, usually called 'Saxonian' facies, accumulated in grabens and halfgrabens. Strain propagated along strike and the basins widened and became interconnected laterally over time, although the extension rate was not uniform so that rapidly extending areas became basins whereas less rapidly extending areas became intervening highs or narrowings. Overall vertical development shows a fining-upward trend in most of the basins, starting with transverse alluvial fan deposits and followed by longitudinal braided river deposits, with alluvial plain and lacustrine deposits at the top (Virgili et al. 1976; Sopeña et al. 1988; Valero & Gisbert 1992; Arche & López-Gómez 1996). The commonly cyclic nature of the 'Saxonian' facies can be explained by pulses of rapid extension and rift shoulder uplifts followed by subdued extension, reduced sediment supply and probable onlap over the hanging wall of the grabens and half-grabens. Rapid lateral migration of the depocentres over time indicates the asymmetric nature of the extension.

There was no volcanic activity during this period in the Iberian Ranges, where basin boundary faults show shallow listric geometries at depths of only 13–14 km, thus remaining in the brittle domain of the lithosphere (Arche & López-Gómez 1996, 1999). However, basaltic flows and dykes do appear in the western and central Pyrenees (Valero & Gisbert 1992; Lucas & Gisbert 1995), indicating deep-seated, vertical basin boundary faults reaching the asthenosphere in these areas. Subsidence rates were high in the Cantabrian-Pyrenean basins, where more than 1500 m of red beds accumulated, but more subdued in the Iberian basin, where no more than 300 m of sediments are found. This period of sedimentation was terminated by a phase of uplift, tilting and erosion.

Late Permian–Early Triassic basins

During latest Thüringian–Anisian times there was a drastic reorganization of sedimentary basins and palaeoenvironments. An extensional tectonic regime prevailed right across the eastern half of the Iberian plate, leading to the development of symmetric basins bounded by listric, conjugate faults in the Iberian Ranges and the Cantabrian-Pyrenees areas, and the creation of the Ebro and Tajo basins. This extension also induced southward-propagating rifting along the eastern margin of the Iberian plate to create a Catalan–Valencia–Betic basin (Marzo & Calvet 1985; Sopeña et al. 1988; Arche & López-Gómez 1996). Extension was not accompanied by volcanic activity, although monzogranitic dykes were intruded in the Central Range (González-Casado et al. 1996).

The subsidence rate was moderate to low, and decreased with time in all the basins. Sedimentary infilling (Buntsandstein facies) was very similar in each basin: alluvial fan conglomerates followed by sandy braided river deposits, with amalgamated channelled sand bodies and a total lack of alluvial plain deposits. Subsidence was episodic, punctuated by brief quiescent periods marked by erosion and gentle regional tilting, although, at the beginning of Anisian times, there was a more general interruption of sedimentation with widespread unconformities. A final short-lived but intense rifting episode took place in the Iberian Ranges, causing graben polarity inversion and leading to the accumulation of up to 650 m of coarsening-upward fluvial sequences (Ramos 1979; García-Royo & Arche 1987; Pérez-Arlucea 1985; Arche & López-Gómez 1999).

Extension produced basins with east–west (Pyrenees), NW–SE (Iberian Ranges) and NE–SW (Catalan–Valencia–Prebetic) orientations. Arche & López-Gómez (1996) therefore proposed the coexistence of two different strain fields during this period, one related to the final stages of development of the Bay of Biscay and Gibraltar transform faults, and the other being a pure extensional field propagating towards the SW with time, from the Boreal basin (the Hesse–Burgundy rift) to the Catalan–Prebetic basin.

Middle–Upper Triassic mature extensional basins

Extension, subsidence and sedimentation prior to Middle Triassic times can be interpreted as fault-related synrift events according to the model proposed by Allen & Allen (1990). The subsequent, thermally controlled post-rift period of development of the Iberian plate commenced during Anisian times and continued until Late Jurassic times. The Tethys sea prograded over the eastern margin of the plate successively during at least five major transgression–regression cycles, each one onlapping further westward with time. These five cycles are: (a) Röt-lower Muschelkalk (Anisian); (b) middle–upper Muschelkalk (latest Anisian–Ladinian); (c) lower Keuper (Carnian); (d) upper Keuper (late Carnian–Norian); (e) Imón or Isábena carbonates (upper Norian). The detailed stratigraphy for each of these events is presented later.

Palaeozoic highs that bounded Late Permian–Early Triassic continental rift basins were drowned during late Ladinian times (Ortí et al. 1996; López-Gómez et al. 1998) and a single basin was established (Fig. 10.10b). Sedimentation was not, however, uniform across this basin: in some subsiding areas hundreds of metres of evaporites accumulated, and relative highs with predominantly siliciclastic sedimentation were formed during both the (c) and (d) transgressive cycles. The shallow marine calcareous and evaporitic sediments passed laterally to coastal and continental siliciclastic deposits, leading to many misinterpretations of the sedimentary record (see Ortí & Salvany 1990; López-Gómez et al. 1998, for references and details).

Subsiding areas containing evaporites (Fig. 10.12) were identified by Castillo-Herrador (1974) and their shape and extent were described by Ortí (1990) and Jurado (1990). They are oval in shape and may have arisen from areas of high heat flow during earlier stages so that subsequently (in the Late Triassic) they subsided faster and deeper than surrounding regions. It should be noted that these are not fault-bounded basins but rather warps of 100–300 km in size.

Keuper facies rocks contain numerous subvolcanic bodies and sills of alkaline basalts of uncertain age (Lago & Pocovi 1984; Navidad & Álvaro 1985; Mitjavilla & Martí 1986; Bastida et al. 1989). Their absolute age has not yet been established due to the degree of subsequent alteration they present, but some authors (Schermerholm et al. 1978; Ferreira 1983) have linked this magmatism to early phases of North Atlantic opening, and the establishment of the Messejana–Plasencia dolerite dyke system. If this theory proves correct, these volcanic rocks herald a new, major tectonic event that was to influence Iberia for the rest of Phanerozoic time.

Early Permian sediments

Early Permian basins in Iberia are located along the Asturias-Pyrenean Zone, the Iberian Ranges and the southern margin of Iberia. There were several isolated rift basins with very different subsidence rates and probably the infilling was not coeval in all of them. Usually these basins contain a lower succession of volcanic and volcaniclastic rocks, and an upper succession of continental red beds without volcanic rocks.

Cantabrian Zone

Permian rocks of the Cantabrian Zone crop out in Asturias, Santander and in the north of Palencia province (Fig. 10.2). Permian sedimentation took place in a region of very irregular topography with a relief of up to 500 m between troughs and swells (García-Mondejar et al. 1989). Three main east–west aligned outcrops (Unquera, Cangas de Onís and Cabuérniga) in the eastern and central areas, and a NW–SE trending outcrop (Ventaniella) in the western area, clearly subdivide zones with different stratigraphic characteristics (Fig. 10.2). Broadly speaking, the NW–SE alignment separates the Carboniferous central coal basin to the west (devoid of Permian sediments) from a Picos de Europa area to the east (with Early Permian sediments).

The presence of Permian sediments in the Cantabrian Zone was first described by Patac (1920) after his discovery of flora *Callipteris conferta*, *Walchia piniformis*, *Walchia hipnoides*, *Pecopteris arborescens* and *Pecopteris pluckeneti* in the Pola de

Fig. 10.2. The Cantabrian Zone. Location of the Permo-Triassic outcrops and lithostratigraphy in the main sections of the three study areas. Units: bcu (Autunian), basal conglomerate unit; bcu ('Saxonian'), basal conglomerate unit; 'lb', limestone breccia; rmc, red marls with cavities; usu, upper siltstone unit; umc, upper marls and clays; V mb, Valbuena Mb; T mb, Torazo Mb; C mb, El Cayo Mb; A mb, Arboleya Mb; V mb, Villaescusa Mb; l mb, Lower Mb; m mb, Middle Mb; u mb, Upper Mb; Bu, Buntsandstein facies; Mu, Muschelkalk facies; Ke, Keuper facies; M. as., brown clays and black slates with anhydrite and gypsum. References: 1, Martínez García (1991); 2, Pieren *et al.* (1995); 3, Martínez García (1999); 4, Wagner & Martínez García (1982); 5, Martínez García *et al.* (1994); 6, Suarez Vega (1974); 7, Valenzuela (1988); 8, Martínez García (1983a); 9, Martínez García *et al.* (1998); 10, Gutierrez-Claverol & Manjón (1984); 11, Carreras *et al.* (1979); 12, Martínez García (1981); 13, Gervilla *et al.* (1978); 14, Martínez García & Villa (1998); 15, Manjón *et al.* (1992); 16, Mamet & Martínez García (1995); 17, Gand *et al.* (1997); 18, Maas (1974); 19, Robles *et al.* (1987); García-Mondéjar *et al.* (1989); 21, Llopis (1961); 22, Dubar *et al.* (1963). VL, Ventaniella lineament. Localities: A, Arenas de Cabrales; AV, Avilés; Bap, Bárcena de Pié de Concha; CO, Cangas de Onís; CP, Cervera de Pisuerga; CV, Caravia; DO, Dobros; G, Gijón; GA, Gamonedo; HO, Hozarco; L, Lastres; LC, La Camocha; LCB, Los Corrales de Buelna; LE, Lebeña; LL, Llanes; M, Merodio; MC, Mestas de Con; O, Oviedo; PA, Peña Sagra; PAR, Pico Paraes; PD, Pandébano; PE, Pen; PL, Peña Labra; PN, Panes; PO, Potes; PS, Pola de Siero; PSa, Peña Sagra; R, Ribadesella; Re, Reinosa; S, Salinas; SL, Sama de Langreo; SO, Sotres; ST, San Tirso; T, Tresviso; Un, Unquera; V, Villaviciosa.

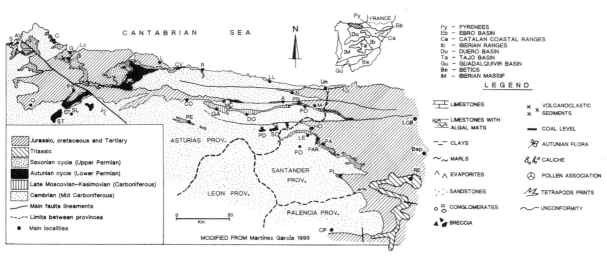

MODIFIED FROM Martínez García 1999

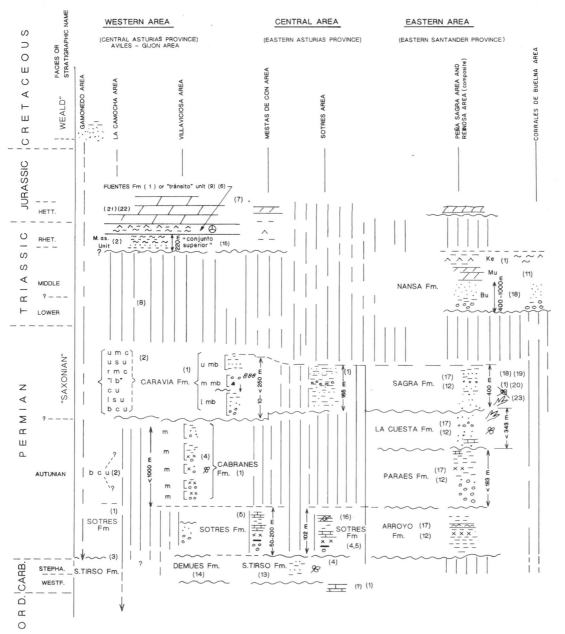

Siero area, Asturias province. However, for many subsequent decades these sediments, as well as those immediately underlying Lower Jurassic dolomites, were simply referred to as 'post-Variscan' or 'Permo-Triassic' sediments. Initial detailed stratigraphic and sedimentological studies (De Jong 1971; Prado 1972; Maas 1974; Sánchez de la Torre *et al.* 1977) only briefly described these sediments, until new evidence of Early Permian (Autunian) flora (Martínez García 1981; Wagner & Martínez García 1982) emerged in the 1980s and allowed the differentiation of two lithological groups (Martínez García 1983*a*): a lower or Viñón Group (Gp) comprising conglomerates, marls, limestones and volcaniclastic rocks, and an upper or Villaviciosa Gp, comprising mainly red detrital sediments with no volcaniclastic rocks. More recently, this nomenclature was abandoned by the same author who proposed four new formations: Sotres, Cabranes, Caravia and Fuentes, from base to top respectively (Martínez García 1991). The first and latter two formations are equivalent to the Viñón Gp and Villaviciosa Gp respectively (Martínez García 1991).

A summary of the stratigraphic nomenclature of the Cantabrian Zone is shown in Figure 10.2, which shows a subdivision into western, central and eastern areas that mostly lie within Asturias province but also include part of western Santander and northern Palencia provinces. Lower Permian rocks are exposed in all three areas, and when well developed are represented by the Sotres and Cabranes formations. The Sotres Formation (Fm) lies unconformably on Lower Palaeozoic basement or on the marine Carboniferous Demués or San Tirso formations (Gervilla *et al.* 1978; Wagner & Martínez García 1982; Martínez García *et al.* 1991). The Sotres Fm is dated as Early Permian in age using both macrofauna (Wagner & Martínez García 1982) and pollen and spore associations (Martínez García 1981). It is up to 102 m thick in the Sotres area and comprises conglomerates, volcaniclastic sediments and isolated coal measures at the base, and limestones with algal mats intercalated with black siltstones that represent most of the upper part of the formation (Martínez García 1991). In the east, the equivalent formation has been defined as 'unit 1' (Martínez García *et al.* 1994) or the Arroyo Fm (Gand *et al.* 1997) in the Peña Sagra area (Santander province; Fig. 10.2.)

The Cabranes Fm apparently lies unconformably on the Sotres Fm although, based on the megaflora data of Wagner & Martínez García (1982), it is also of Early Permian age. The formation reaches 600 m in thickness in the Villaviciosa area, where it has been divided into five members (Suárez Rodríguez 1988). It is only 70 m in the Avilés zone to the west, and it disappears altogether across the central area, or is reduced to 1 m in thickness, as seen in La Camocha zone where it was described as 'the basal conglomeratic unit' by Pieren *et al.* (1995). It mostly comprises volcanic and volcano-sedimentary successions with interbedded shales, although siliceous conglomerates, sandstones and siltstones are also present. The Cabranes Fm is represented in the Peña Sagra

area by 'unit 2 and unit 3' (Martínez García *et al.* 1994) or by their equivalents, the Paraes Fm and overlying La Cuesta Fm (Gand *et al.* 1997). The Paraes Fm can reach a thickness of 183 m and is composed of white and grey sandstones with intercalated conglomeratic lenses.

Pyrenean basin

The present-day configuration of the Pyrenees is the result of intense compressive Alpine tectonics (Chapter 15), so that attempts to restore Permian and Triassic palaeogeography in this area must be considered only provisional. The Permian sediments of the Pyrenees were deposited in several small, isolated continental basins of highly variable age and infilling history. In spite of their differences, two sedimentary successions, each bounded by unconformities (Fig. 10.3), may be discerned in most basins. The early studies of Nagtegaal (1969) and Gisbert (1983, 1984), and a recent synthesis by Lucas & Gisbert (1995), and Debon *et al.* (1995), offer a plethora of data and references.

The Early Permian (Stephanian–Autunian) succession in the Pyrenean area is well developed in the Erill-Castell and Seo d'Urgell–Adrahent basins, among others (Fig. 10.3), and in the Somport area. Four essentially conformable sedimentary units may be identified (Aguiró Fm, Erill-Castell Fm, Malpás Fm and Peranera Fm; Nagtegaal 1969), these being roughly equivalent (Fig. 10.3) to the Punta de la Garganta Fm of Roger (1970) and the 'grey unit' (UG), 'transition unit' (UT) and 'lower red unit' (URI) of Gisbert (1981, 1984) and Lucas & Gisbert (1995).

The lowermost unit (Aguiró Fm, Punta de la Garganta Fm or the lower part of the grey unit) crops out only in the central and western Pyrenees, and probably along the northern margin of the Basque Pyrenees (Fig. 10.3), and consists of conglomerates, slope breccias, tuffs and coal measures. It lies unconformably on a Variscan basement and contains plant remains of *Mixoneura ovata, Linopteris brongniarti* and *Alethopteris gandini* (Nagtegaal 1969; Gisbert *et al.* 1985) (Stephanian B age). However, Briqueu & Innocent (1993) proposed an absolute age for the volcanic rocks in the lower part of the grey unit of 278 ± 5 Ma (Autunian). This is a recurrent problem in the study of continental basins of this age, since it is very difficult to distinguish the latest Carboniferous from Early Permian ages based only on plant remains.

The next unit (Erill-Castell Fm, Lower Campo de Troya Fm or middle grey unit) crops out in the central and western Pyrenees and NE Basque Pyrenees, has been dated as 272 ± 3 Ma (Autunian; Briqueu & Innocent 1993), and comprises volcanic and volcaniclastic rocks of andesitic and rhyodacitic composition, in places reaching over 500 m in thickness (Nagtegaal 1969; Lucas & Gisbert 1995). These volcanic rocks lie conformably on the Aguiró Fm and uncomformably on the Variscan basement, and the upper part of the succession shows a transition to fluvial red beds, sometimes marked by palaeosol horizons.

The Malpás Fm lies conformably on the Erill-Castell Fm and is equivalent to the top of the grey unit and transitional unit and to the top of the Campo de Troya Fm. It locally exceeds 300 m in thickness, and crops out in every Permian basin preserved in the Pyrenees. It consists of fining-upward sequences of sandstone, variagated siltstones and coal measures, interpreted as deposited by alluvial fans and

---

Fig. 10.3. The Pyrenees. Location of the Permian and Triassic outcrops and scheme of the described lithological units in the main sections of the three subdivided study areas. Units: UG, grey unit (including the Aguiró, Punta de la Garganta and Erill-Castell formations, as well as part of the Malpás Fm); UT, transitional unit (including the Campo de Troja Fm and part of the Malpás Fm); URI, lower red unit (including the Peranera and Somport formations); URS, upper red unit (including the Pic de Baralet and Peña de Macantón formations); ISA, Isábena Fm. Because of space limitations the following localities cited in the text are not shown in the figure, but are close to the numbered localities indicated: Somport (8), Gavarnie (between 8 and 9), Cotiella (9), Les Nogueres (10), Erill-Castell (11), Montsec (12), Seo de Urgell-Adrahert (13), Cadí (14), Fresser-Pedraforca (15), Pobla de Lillet, Bac Grillera and Figueres-Montgrí (to the south of 16–17). Tectonic units: GTS, Gabarnie thrust sheet; No, Noguera; SM, Sierras Marginales; PTS, Pedraforca thrust sheet; AZ, Axial Zone; LU, lowermost units. South Pyrenean Zone (SPZ) lies south of the Axial Zone.

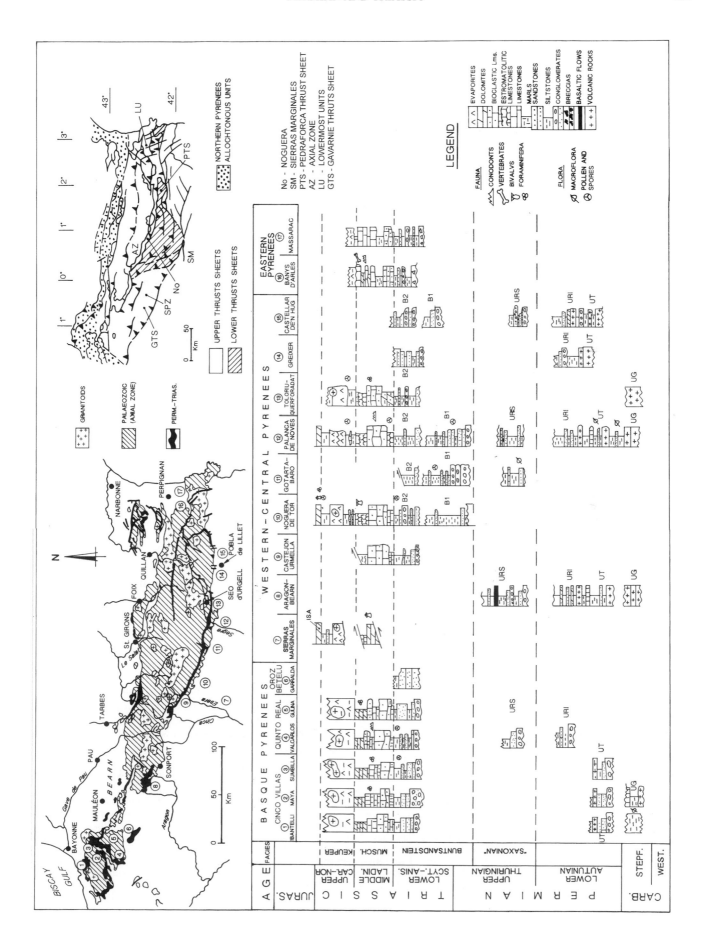

meandering rivers, intercalated with limestones with stromatolites and ostracodes, interpreted as shallow lacustrine deposits. The formation also includes dacitic-rhyolitic volcanic rocks (lava flows and pyroclastic complexes), and the coal measure facies has yielded a rich macroflora (Álvarez *et al.* 1969; Robert 1980)

The Early Permian succession ends with the Peranera Fm, equivalent to the lower red unit and the Somport Fm. It comprises up to 1000 m of red conglomerates and sandstone, red siltstones with macroflora (Dalloni 1930; Gisbert *et al.* 1985), and rhyolitic tuffs and ignimbritic flows, and is truncated by an overlying erosive unconformity. This unit is interpreted as braided river, debris flow and ephemeral lake deposits associated with volcanic cones. The great thickness and extensive outcrop indicates a very high subsidence rate.

Iberian basin and Central Range

The Iberian basin is currently geographically divided into the two NW–SE oriented Aragonese and Castilian branches of the Iberian Ranges, separated by the Tertiary Calatayud Teruel basin (Figs 10.4 and 10.5). The northwesternmost part of the Castilian branch is geographically linked to the northeasternmost part of the Central Range (Fig. 10.4).

Early Permian sediments of the Castilian Branch were deposited in a series of intramontane, transtensional, half-graben basins that have been identified from Noviales–Atienza in the NW to Boniches in the centre and Eslida–Desierto de Las Palmas in the SE (Fig. 10.4). Subsidence rates varied in different basins so that sedimentary succession ranges from 150 to 2000 m in thickness (Hernando 1977; Sopeña 1979; Ramos 1979; Pérez-Arlucea & Sopeña 1985; López-Gómez & Arche 1986; Sopeña *et al.* 1988; Van Wees *et al.* 1998; Arche & López-Gómez 1999).

Castilian branch. The Noviales–Atienza basin contains 50–150 m thick andesitic lava flows and pyroclastic rocks lying unconformably on the Lower Palaeozoic basement and dated as 282 ± 12 Ma (Hernando *et al.* 1980). They are overlain by 200 m of red mudstones and sandstones and a second volcanic episode followed conformably by more than 1000 m of coarsening-upward red beds of alluvial fan–fluvial origin. The Palmaces basin (Fig. 10.4) contains breccias and volcaniclastic rocks at the base, followed by a 700 m thick coarsening upward succession of siltstones to conglomerates that contain *Estheria tenella* near the base (Sopeña 1979). In the nearby Valdesotos basin, the volcaniclastic lower part of the sedimentary infilling contains macroflora including *Callipteris conferta* and *C. raimondii* and palynomorphs such as *Potonieisporites novicus* and *Vittatina costabilis* (Autunian age; Sopeña 1979; Broutin *et al.* 2000).

The Molina de Aragón basin shows the same initial volcaniclastic interval, but above this there is a grey-black succession of mudstones, ash beds and dolomites, deposited in a shallow lake environment (Ramos 1979) showing a subdued subsidence rate and lack of siliciclastic alluvial fan input into the basin (Fig. 10.4). This grey-black mudstone interval in this succession contains a rich palynomorph assemblage including *Potonieisporites novicus*, *Vittatina costabilis* and *Pytiosporites westfaliensis*.

To the SE, volcanic rocks bounded by angular unconformities are found in the Orea–Orihuela del Tremedal area (Pérez-Arlucea &

Sopeña 1985), and these clearly correlate in composition and stratigraphic position with the volcanic rocks of the previously described basins. Volcanic rocks have also been identified at the base of the Eslida section (Fig. 10.4). Finally, in the Boniches area, red breccias (Tabarreña breccias unit; López-Gómez & Arche 1993) unconformably overlying the Lower Palaeozoic basement, are unconformably covered by Late Permian conglomerates, and could represent the local correlatives of the Autunian deposits of other basins (TB; Fig. 10.4).

Early Permian sediments are scarce in the Aragonese branch of the Iberian Ranges and restricted to reduced, isolated outcrops in the Moncayo area (Reznos and Paniza sections) and Montalbán area (Fombuena section; Fig. 10.5). The age of these sediments has been derived from palaeoflora and isotopic studies as well as by comparison with better-known areas of the Castilian branch (Ramos 1979).

Aragonese branch. The Reznos section described by Rey & Ramos (1991) is completed with information from a well-log described by De la Peña *et al.* (1977) in Paniza. The succession consists of about 100 m of alternating mudstones, conglomerates and sandstones with some intercalated thin (4 m) levels of volcaniclastic sediments that Rey & Ramos (1991) called the Arroyo Riduero conglomerates, sandstones and mudstones unit (ARCSM). Palaeoflora discovered in the middle part of the section include *Callipteris conferta*, *Cathaysiopteris whitei*, *Dizeufiotheca* sp., *Equisetites elongatus*, *Gamophyllites* sp., *Gigantonoclea largrelli*, *Koretrophyllites*, *Labachia piniformis* and *Umbellaphyllites annularioides*, among others, and were considered Autunian in age (Early Permian) by De la Peña *et al.* (1977). A similar outcrop, described 2 km SE of the previous section (Desparmet *et al.* 1972), may also be of Early Permian age, although it is devoid of palaeontological markers. The Early Permian sediments from the Paniza section were described by Del Olmo *et al.* (1983), and comprise up to 230 m of conglomerates, with quartzite clasts derived from the basement and rhyolite fragments and mudstones with thin intercalated sandstones and dolomites. The Fombuena section (Fig. 10.5) shows Autunian sediments composed of volcanic tuffs with *lapilli* and small bombs (Conte *et al.* 1987). Isotope analysis and palaeoflora (Lendínez *et al.* 1989) dated these sediments as near the Carboniferous–Permian boundary, although they could represent only the uppermost Carboniferous.

Southern margin of the Iberian Massif intermontane basins

The rifting phase that began during Early Permian times (Sopeña *et al.* 1988) gave rise to a series of intermontane basins in southern Iberia. The deposits which infilled these basins now crop out only in a restricted area of the southern edge of the Iberian Massif and within the 'internal zone' of the Betic Cordillera where they are principally schists and gneisses which are difficult to distinguish and study in detail. Outside the Betic Cordillera, southern Iberian Permian deposits are all continental and restricted to small basins in northern Sevilla province, the most important of which is that of the Viar river (Fig. 10.6).

Fig. 10.4. The Castilian branch of the Iberian Ranges. Location of the Permian and Triassic outcrops and scheme of the described lithological units for the main sections of the subdivided study areas. Autunian units: PLC, Pálmaces lower conglomerates; PMS, Pálmaces mudstones and sandstones; PS, Pálmaces sandstones; PUC, Pálmaces upper conglomerates; TB, Tabarreña breccias; VSC, volcano-sedimentary complex. 'Saxonian' units: AMS, Alcotas mudstones and sandstones; HGC, Hoz de Gallo conglomerates; MB, Montesoro beds; RSC, Riba de Santiuste conglomerates. Buntsandstein facies units: AS, Arandilla sandstones; CS, Cañizar sandstones; CSM, Cercadillo sandstones and mudstones; ESS, Eslida siltstones and sandstones; FM, Fraguas mudstones; MCMM, Marines clays, marls and mudstones; PB, Prados beds; PSM, Pedro sandstones and mudstones; RGS, Rillo de Gallo sandstone; RMS, Rillo mudstones and sandstones; RSS, Riba de Santiuste sandstones; TMS, Torete mudstones and sandstones. Muschelkalk facies units: ADM, Albarracín dolostones and marls; AMG, Alcotas marls and gypsum; CDL, Cañete dolomites and limestones; LD, Landete dolomites; RDML, Royuela dolostones, marls and limestones; TDs, Tramacastilla dolostones. Keuper facies units: GCG, Gavilanes clays and gypsum. Other units: ID, Imón dolomites; MSMG, Mas sandstones, marls and gypsum are lateral equivalent of the middle Muschelkalk; TMG, Tramacastilla marls and gypsum.

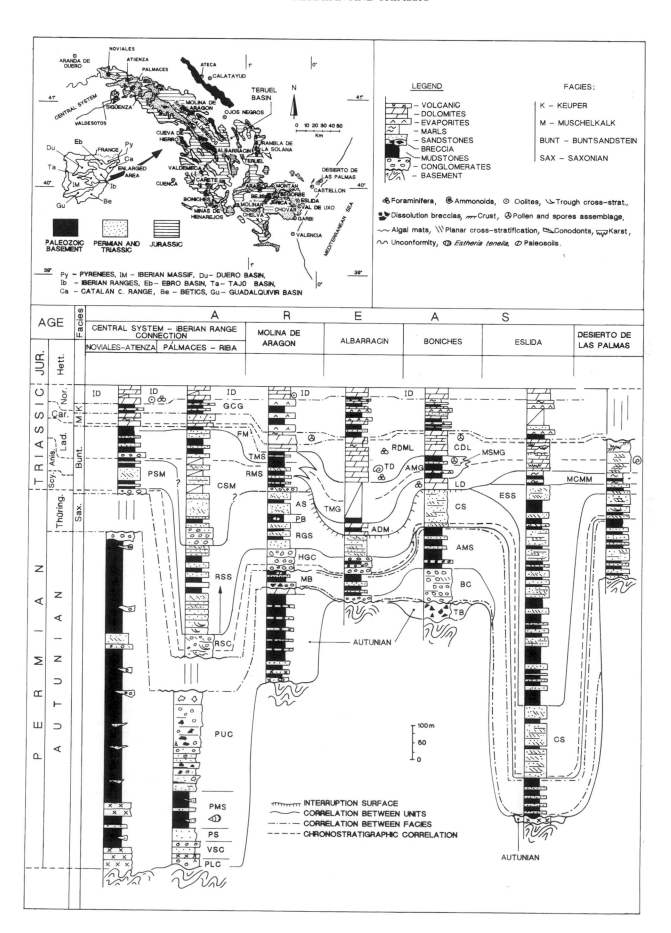

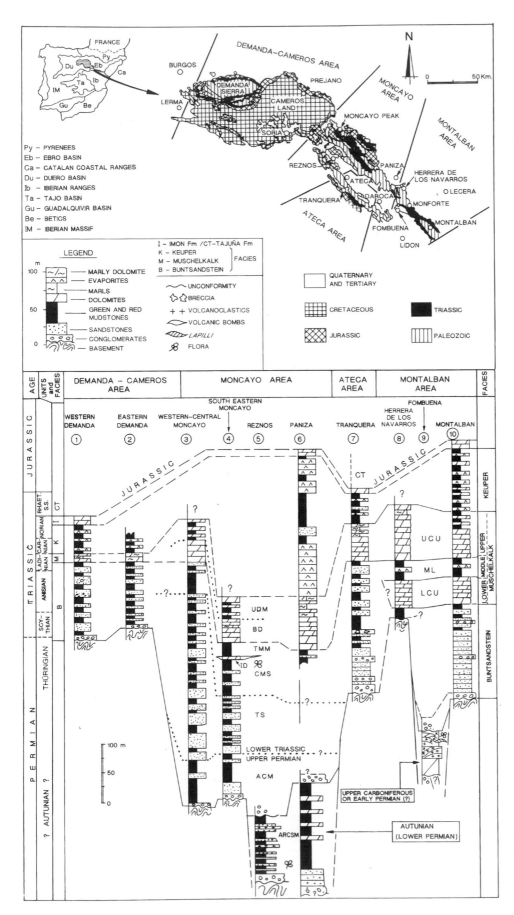

Fig. 10.5. The Aragonese branch of the Iberian Ranges. Location of the Permian–Triassic outcrops and scheme of the described lithological units for the main sections of the subdivided study areas. Permian units: ARCSM, Arroyo Ridruero conglomerates, sandstones and mudstones; ACM, Araviana conglomerates and sandstones. Triassic units: TS, Tierga sandstones; CMS, Cálcena mudstones and sandstones; TMM, Trasobares mudstones and marls; LCU, lower carbonate unit; ML, middle level; UCU, upper carbonate unit. Sections: 1, 2, 8, 10 from Sopeña *et al.* (1988); 3,4 from Arribas (1984); 5 from De la Peña *et al.* (1977) and Meléndez *et al.* (1995); 7 from García-Royo & Arche (1987); 9 from Lendínez *et al.* (1989).

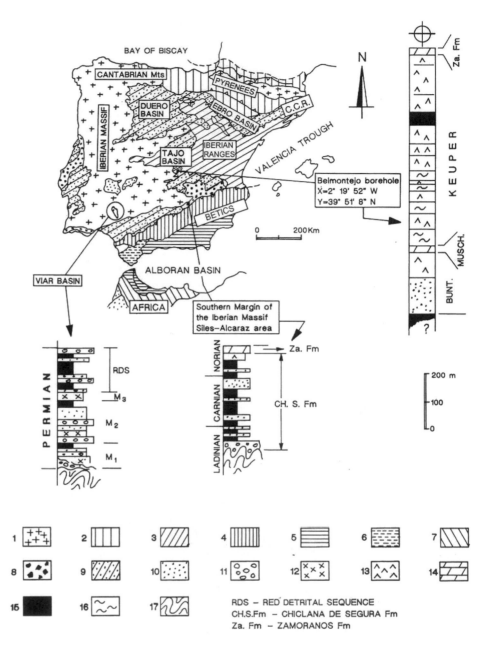

Fig. 10.6. Present-day location of the Tajo basin, the southern margin of the Iberian Massif and representative sections from each one. Sections: Viar basin, from Sierra & Moreno (1997); southern margin of the Iberian Massif, composite from Fernández (1984), Fernández & Dabrio (1985), Fernández & Gil (1989); Tajo basin, from borehole in Belmontejo area, Cuenca province (IGME 1987). The palaeogeographical location of the southern margin of the Iberian Massif is represented by the number 7 in Figure 10.13. Legend: 1, Variscan basement of the Iberian Massif; 2, deformed Mesozoic cover of the Pyrenees; 3, Mesozoic of the Iberian Ranges; 4, Mesozoic external units of the Betic Cordillera; 5, internal units of the Betic–Rif realm; 6, flysch units of the Gibraltar arc; 7, Mesozoic of the African margin; 8, undeformed Mesozoic cover; 9, Tertiary onshore basin; 10, sandstones; 11, conglomerates; 12, volcaniclastics; 13, gypsum; 14, dolomites; 15, mudstones; 16, marls; 17, Palaeozoic basement; CCR, Catalonian Coastal Ranges. Geographical scheme modified from Banda (1996).

RDS – RED DETRITAL SEQUENCE
CH.S.Fm – CHICLANA DE SEGURA Fm
Za. Fm – ZAMORANOS Fm

Viar basin. The Viar basin is a narrow, NNW–SSE oriented intermontane, post-Variscan Permian basin (Simancas 1980) bounded by extensional faults that were active during sedimentation (Sierra & Moreno 1997). It was filled with over 300 m of continental detrital sediments, volcaniclastic deposits and volcanic rocks. Five lithostratigraphic units have been identified (Simancas 1983). The three lower units, located near Guadalcanal, comprise volcanic rocks erupted from fissures and interbedded with layers of conglomerates, sandstones and mudstones interpreted as alluvial fan deposits of Early Permian age (Broutin 1977, 1981). The two upper units constitute the 'red detrital succession' of Simancas (1983; Fig. 10.6) comprising red conglomerates, sandstones and mudstones, interpreted as braided river deposits and of uncertain age (possibly Triassic) as an unconformity separates them from the three underlying units (Sopeña et al. 1985).

Late Permian–Early Triassic sediments

The sedimentary record corresponding to this time interval has been traditionally divided into a 'Saxonian' facies, of late

Late Permian (Thüringian) age, and a 'Buntsandstein' facies, of latest Permian (late Thüringian)–Early Triassic age. 'Saxonian' facies is a loose term for late Permian red beds derived from central Europe, whereas 'Buntsandstein' facies traditionally refers to latest Permian–Early Triassic red beds.

During this period of time, red beds were deposited in rift basins along the northern and eastern margins of the Iberian Massif and the related Catalan, Ebro and Pyrenean basins. They are roughly equivalent to the classic Buntsandstein facies of northern Europe. As in most of western Europe, these red beds can be subdivided into an upper part (or Buntsandstein *sensu stricto*) overlying unconformably a lower one ('Saxonian' facies, not always present). Both of these units can lie unconformably on Lower Palaeozoic basement or on Lower Permian (Autunian) deposits. A major difference between the Iberian peninsula and the rest of Europe, however, is that the angular unconformity that separates the

Buntsandstein and 'Saxonian' facies lies within the Upper Permian succession in Spain and therefore does not mark the Permian–Triassic boundary as is the case elsewhere.

Cantabrian Zone

It is difficult to differentiate Late Permian and Early Triassic rocks in the Cantabrian Zone due to the scarcity of outcrops and lack of data on the age of most rocks. Late Permian ('Saxonian') sediments are represented by the Caravia Fm (Fig. 10.2) (Martínez García 1991) which rests unconformably on the Carboniferous Cabranes, Sotres, Demués or San Tirso formations, or even directly on the Ordovician basement. The main sedimentary difference between the Caravia Fm and the Cabranes Fm is the lack of volcaniclastic sediments and the general red colour of the Caravia Fm. This latter formation is of economic interest due to its mineral resources such as F, Pb, Zn and Ba.

Caravia Formation. This formation is divided into lower, middle and upper members that broadly correspond to the seven units described for the La Camocha area by Pieren *et al.* (1995) (Fig. 10.2). The lower member comprises conglomerates, red sandstones and red siltstones and attains a thickness of over 100 m in the Villaviciosa area. The middle member sometimes lies unconformably on the lower one, can reach tens of metres in thickness, and comprises red breccias and calcareous conglomerates with some caliche horizons. The upper member consists of a succession of red and green siltstones with some intercalated sandstone beds.

The maximum thickness of the Caravia Fm is found in the Villaviciosa, Gijón and Caravia areas, where it reaches hundreds of meters (Martínez García 1991, 1999). It is also well developed in the Sotres-Tresviso and Peña Sagra areas (Martínez García *et al.* 1994; Gand *et al.* 1997) and in the central and eastern Cantabrian Zone (Fig. 10.2), although the thickness and lateral continuity of the outcrops are less pronounced compared to the western area. Also in this western area (Paraes peak) different *in situ* fossil tracks of *Hyloidichnus major* and *Limnopus* cf. *zeilleri* tetrapods have been described, as well as the invertebrates *Scoyenia gracilis* and *Isopodichnus* cf. *minutus* (Martínez García *et al.* 1994; Gand *et al.* 1997). In the Gijón area (Fig. 10.2), a succession of red beds equivalent to the Caravia Fm consists of marls (5–50 m) and dolomites, together with calcareous and siliceous conglomerates in the lower part, red sandstones and pelites (20–40 m) and marly limestones (5–80 m) in the middle part, and alternating red and grey sandstones and pelites (60–120 m) in the upper part (Martínez García 1991). In the Sama de Langreo area (western Cantabrians; Fig. 10.2), Martínez García (1991) describes the presence of the Caravia Fm close to Sotres, where it reaches 165 m in thickness. Although already attributed to the Permian, the age of the Caravia Fm has not yet been proven. However, based on the fact that it overlies Autunian sediments and underlies the Buntsandstein facies in the eastern zone (west of Santander province), Martínez García (1991) considered the sediments of the Caravia Fm as most likely to be Late Permian in age (Fig. 10.2).

Peña Sagra and Peña Labra areas. The Peña Sagra and Peña Labra areas expose the most complete succession of both Early and Late Permian sediments, together attaining a thickness of 1000 m (Fig. 10.2). Robles *et al.* (1987) and García-Mondéjar *et al.* (1989) performed a detailed sedimentological study of these Early–Late Permian sediments, interpreting the sedimentary environment in terms of semi-arid alluvial fan complexes with clearly differentiated proximal and distal areas. Proximal successions are characterized by coarse-grained, channelled sandstones and a general fining-upward tendency, whereas distal successions show finer-grained sediments representing wide floodplains with isolated lakes and fluvial systems that finally covered most of the previous volcanic palaeorelief. Palaeocurrents in both proximal and distal successions indicate a north to NW transport direction, and provenance studies on conglomerate reveals a siliceous and volcanic origin from a nearby source area to the SW, probably controlled by the Ventaniella fault system (Fig. 10.2).

The absence of Late Permian sediments in the Cantabrian basin (Fig. 10.2) has been attributed to intense erosion that produced palaeorelief overlain by Triassic sediments (Martínez García 1999), well shown at Puente Pumar, SE of Peña Sagra. This overlying Triassic siliciclastic succession, resting unconformably on Late Permian rocks in the areas of Peña Sagra, Peña Labra, Liébana and Reinosa (western Santander province) and Cervera de Pisuerga (northern Palencia province), was first described during the 1960s (Papa 1964; Smit 1966). It was defined as the Nansa Fm by Maas (1974), related to the Buntsandstein facies by different authors (e.g. Maas 1974; Carreras *et al.* 1979; García-Mondejar *et al.* 1989; Martínez García 1991, 1999) and is considered by some authors to be of Early Triassic age (Martínez García 1991, 1999). It ranges in thickness from around 400 m east of Lebeña, to more than 1000 m in the Nansa river valley, east of Peña Sagra. The succession, when complete, begins with homogeneous and well-rounded quartzitic conglomerates with a quartzite–sandstone matrix, interbedded in places with cross-bedded sandstone and shale. This conglomeratic facies reaches 90 m in thickness in the Peña Labra area, but this thickness decreases rapidly to 10 m in the north and SE. These conglomerates grade upward into cross-bedded, coarse-grained sandstones interspersed with thin shale intercalations. A thinner and laterally less extensive conglomerate locally develops about 100 m above the basal conglomerates and the upper part of the succession consists of red and purple silts and shaly mudstones.

Pyrenean basin

The Late Permian succession in the Pyrenees is represented by the Pic de Baralet Fm, equivalent to the upper red unit or the Peña de Macantón Fm, and can be correlated with the 'Saxonian' facies in the Iberian Ranges (Ramos 1979) and the Caravia and Sagra formations of the Cantabrian Zone (Fig. 10.3). It is bounded by unconformities (Mirouse 1962; Gisbert *et al.* 1985) and consists of red breccias, conglomerates, sandstones and siltstones and associated thin limestone, nodular dolomites and caliche horizons. These deposits are interpreted as alluvial fans, braided rivers, alluvial plains and permanent shallow lakes, organized in at least three fining-upward successions up to 800 m thick (Valero 1993). Intercalated alkaline basalt, sills and lava flows occur at several levels and have been dated as 268 ± 7 Ma (early Thüringian; Bixel 1984).

The Late Permian–Early Triassic units of the Pyrenean basin are markedly irregular in their distribution (Fig. 10.3). The lower part (B1; Gisbert 1983), of Thüringian–Early Triassic age, is only developed in the western-central area, whereas the upper part (B2; Gisbert 1983) is found along the entire chain. In the Banys d'Arles and Massarac areas of the eastern Pyrenees, these Buntsandstein facies sediments of probable Early Triassic age (B1; Fig. 10.3), consist of red siltstones, sandstones and conglomerates lying unconformably on the deeply weathered granitic basement with a thin basal breccia. These exposures are found within a tectonic unit referred to as 'the lowermost unit' (see Chapter 16 and Muñoz *et al.* 1986). Conglomeratic beds are organized into fining-upward sequences and are thicker in the upper third of the sections (Souquet 1986b). A 8–18 m thick unit of variegated marls and thin stromatolitic layers marks the transition to the Muschelkalk facies.

The Buntsandstein facies succession of the central and western Pyrenees shows a lower unit that is not always present and lies uncomformably on the Palaeozoic basement or on different Early Permian units, and an upper unit that crops out all along the chain. The latter unit lies uncomformably on the lower within the western part of the central Pyrenees, but the contact is conformable further east. The presence or absence of the lower unit suggests a structural high separating two east–west trending basins, but a precise palaeogeographic reconstruction is still not available. The lower unit initially comprises up to 60 m of fining-upward cycles of conglomerates, sandstones and red siltstones at the base, with a palynomorph assemblage (Palanca de Noves section; Fig. 10.3) (Broutin *et al.* 1988) that contains *Falcisporites, Limitisporites, Gardenasporites, Jugasporites, Lueckisporites, Endosporites* and *Densoisporites* (Thüringian). Above this are 20–65 m of sandstone and siltstone containing several palynomorph assemblages (Broutin *et al.* 1988; Calvet *et al.* 1993) including *Triadispora staplini, Alisporites grauvogeli, Illinites kosankei* and *Stellapollenites thiergartii* (lower Anisian). All across the Pyrenees, evidence may be found of a rapid, normal vertical transition to calcareous Muschelkalk facies and palaeocurrent data consistently indicate a SSW flow direction.

Catalonian basin

Late Permian and Early Triassic sediments of the Catalonian Coastal Ranges (CCR) were first described by Bauza (1876), Gombau (1877), Mallada (1890), Almera (1891*a*), Bataller (1933) and Elias (1934) among others. However, detailed studies were published only from 1950 onwards, notably byVirgili (1958) but also by Virgili & Julivert (1954) and Solé Sabarís *et al.* (1956). In this area Buntsandstein facies rocks lie directly on a Palaeozoic basement (Fig. 10.7) and are considered to be of Late Permian–Early Triassic age, based mainly on pollen and spore assemblages (Marzo & Calvet 1985; Solé de Porta *et al.* 1987; Calvet & Marzo 1994) and by comparisons with the Iberian Ranges successions. An Anisian age for carbonate sediments (Muschelkalk facies) immediately above the Buntsandstein facies is well established (see also discussion in Marzo 1980).

Detailed sedimentological studies (Marzo & Anadón 1977; Anadón *et al.* 1979; Marzo 1980, 1986; Calvet & Marzo 1994; Roca *et al.* 1999) show that the beginning of the deposition of continental Buntsandstein facies rocks sedimentation in the CCR was controlled by NE–SW and NW–SE trending fault systems. This network of lineaments produced three main grabens (Montseny–Llobregat, Garraf and Miramar–Priorat), each one initially evolving independently with its own source areas (Gómez-Gras 1993*a*). Connections between these basins occurred near the end of the Buntsandstein facies sedimentation phase (Fig. 10.7). Lithostratigraphy has been defined for each of the three individual grabens (Marzo 1980; Calvet & Marzo 1994), with only the uppermost unit (the 'upper evaporitic-silty-sandy complex'; Röt facies) extending beyond these three areas to cover the whole Catalonian Coastal Ranges.

The Montseny–Llobregat area shows four lithological units starting with a basal breccia unit about 10 m thick, lying unconformably above the Palaeozoic basement and interpreted as local scree deposits. Above this unit, or directly on the Paleozoic basement, lies the Riera de St. Jaume unit (up to 25 m thick) which comprises conglomerates, red-pink sandstones and red siltstones with palaeosols, and represents a fining-upward succession of alluvial origin. The white-red pebbly sandstones of the overlying Caldés unit (up to 5 m thick) lie above the Riera de St. Jaume unit, or directly on the Palaeozoic basement; this unit is interpreted as distal braided fluvial, mixed load sediments. The overlying Figaró unit comprises a thick (up to 235 m) succession of fining-upward cycles of pink sandstones and red siltstones, with palaeocurrents flowing towards the WSW. It is interpreted as recording high sinuosity fluvial systems of mixed or suspended load, probably during relatively rapid subsidence.

The Garraf area also shows four lithological units. The lowest unit is the Garraf conglomerates which rest directly on the Palaeozoic basement, reach up to 53 m in thickness, and comprise red conglomerates with intercalated levels of sandstones up to 9 m thick. This unit is only found in the western part of the Garraf area, is clearly related to its faulted border, and is interpreted as gravelly to mixed, proximal to medial, alluvial braided stream deposits. The Bruguers conglomerates and sandstones unit is laterally continuous with the previous unit and is interpreted as its distal equivalent. The Erampruyá sandstones unit conformably overlies the Bruguer unit, reaches 50 m in thickness, contains red sandstones and is interpreted as sandy braided river deposits. Above, and again in conformable contact with this latter unit, is found the Aragall unit, comprising red siltstones with intercalated pink sandstone levels up to 2 m in thickness, and interpreted as meandering fluvial systems.

The Miramar–Prades–Priorat area shows three lithological units. The lowest unit is the Bellmunt de Siurana conglomerates, sandstones and siltstones, which reaches a thickness of 40 m, and only crops out along the SW border of this area. This unit shows a clear fining-upward tendency that grades from white conglomerates to red siltstones with intercalated thin sandstone levels and red conglomeratic sandstones, and is interpreted as deposited by fluvial bedload braided streams with an intercalated episode of floodplain deposits. The Prades unit can reach 25 m in thickness, comprises quartz conglomerates and sandstones, and lies on the previous unit or unconformably on the Palaeozoic basement. It grades upwards to the upper Prades sandstones unit, the thickest unit in this area (up to 85 m), which is composed of red, coarse to fine subarkosic sandstones. The two latter units are interpreted as recording a change from proximal to more distal braided river systems.

The uppermost Buntsandstein facies unit is the 'upper evaporitic-silty-sandy complex' that can reach a thickness of 35 m and corresponds to the 'limit clays' unit (Virgili 1958) and the Röt facies of central Europe (Schriel 1929). This is the only unit of the Buntsandstein facies represented across the whole Catalonian Coastal Ranges and records the transition to the Muschelkalk facies, and the final quiescence of the basement faults that had previously controlled sedimentation (Calvet & Marzo 1994). It is mainly composed of alternating siltstones and thin carbonate and evaporite beds, although isolated thin sandy or conglomeratic levels may also appear. It is interpreted as representing the change from mudflats to evaporitic sabkhas during the transition from alluvial to marine environments.

Ebro basin

The Ebro basin has been a sedimentary depocentre since Late Permian (?) and Triassic times and can easily be differentiated from the Pyrenean, Catalan and Iberian basins at least until the deposition of middle Ladinian sediments. Ebro basin Permian–Triassic deposits contain well-developed Buntsandstein, Muschelkalk and Keuper facies, and since both lower and upper Muschelkalk carbonate units are present in the area, they form part of the so-called Mediterranean Triassic (Jurado 1990; López-Gómez *et al.* 1998). Most of the basin presently lies under a thick Mesozoic and Cenozoic succession and the only direct information available has been derived from commercial oil wells (Fig. 10.8). The rocks actually crop out only in part of the Aragonese branch of the Iberian Ranges and in the transition to the southwestern Catalonian Coastal Ranges (i.e. the Llavería section; Fig. 10.8). The basin was bounded by emergent older Palaeozoic rocks to the south (the Ateca–Castellón high), the east (Lleida and Girona highs that separate the Ebro basin from the Catalonian basin, except for the connecting Maestrat depocentre in the SE corner), and the north (an ill-defined Ebro high, separating the Ebro basin from the Pyrenean-Cantabrian basin).

The Buntsandstein facies is always thick, sometimes in excess of 400 m (Fig. 10.8; Fraga and Mayals sections), but no volcanic rocks have been identified at the base, so it seems that the Lower Permian (Autunian) is not represented in this basin. Conglomeratic units at the

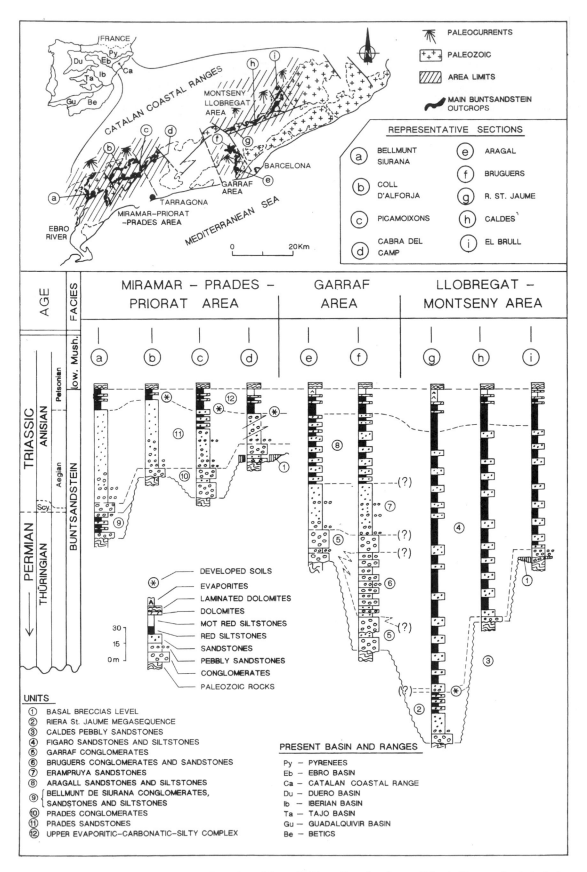

Fig. 10.7. The Catalonian Coastal Ranges. Location of the main Upper Permian–Lower Triassic (Buntsandstein facies) outcrops and stratigraphic columns for main sections. Modified from Calvet & Marzo (1994).

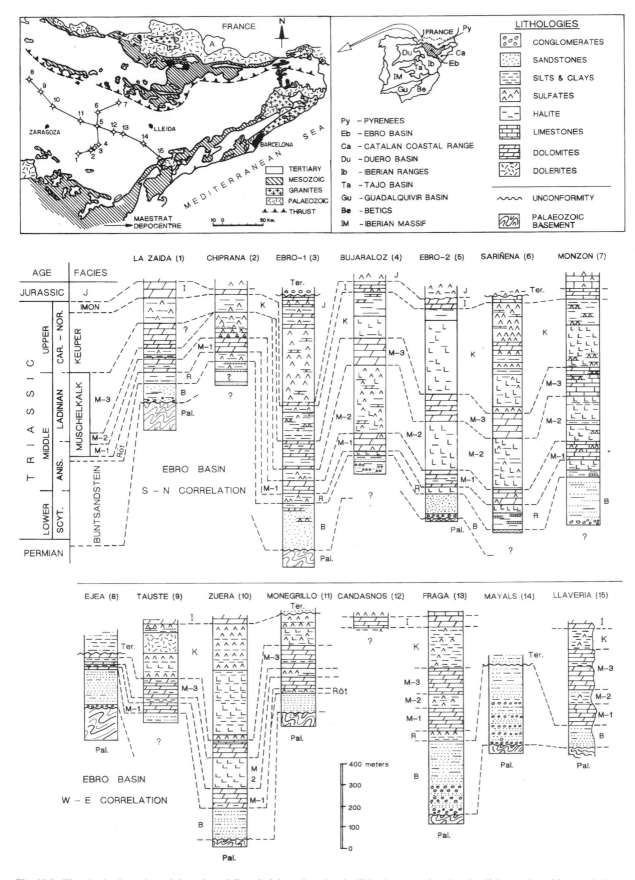

Fig. 10.8. Ebro basin. Location of the selected (borehole) sections for the Triassic succession showing lithostratigraphic correlations and facies.

base of many of the boreholes can be correlated with the Late Permian (Thüringian) Hoz de Gallo Conglomerates Fm in the Iberian Ranges (Fig. 10.4) and Garraf Conglomerates Fm in the Catalonian Coastal Ranges (Fig. 10.7), although they are absent west of a line drawn between Bujaraloz and Sariñena (Fig. 10.8). The upper Buntsandstein facies comprises sandstones and alternations of sandstones and silt-stones, comparable to the Cañizar and Eslida formations in the Iberian Ranges (Fig. 10.4) and Eramprunyá and Figaró formations in the Catalonian Coastal Ranges (Fig. 10.7). These deposits are assumed to be of Early Triassic age, and thin to about 150–200 m towards the west. A transition to marine sedimentation (Röt facies; Fig. 10.8) conformably follows the Buntsandstein facies in the east and central parts of the basin, pinching out west of Zuera (Fig. 10.8). It consists of red siltstones followed by evaporites showing halite in the central part of the basin and anhydrite in marginal regions. This facies can be correlated with the Marines Fm in the Iberian basin (Fig. 10.4) and the 'upper evaporitic-calcareous-silty complex' in the Catalonian Coastal Ranges (Fig. 10.7), and an early Middle Triassic (Anisian) age is inferred.

Iberian Ranges

Late Permian ('Saxonian') and latest Permian–Early Triassic ('Buntsandstein') rocks are found in both the Castilian and Aragonese branches of the Iberian Ranges. The Late Permian ('Saxonian') successions comprise continental siliciclastic red beds lying unconformably upon Variscan basement or Lower Permian (Autunian) sediments and volcaniclastic rocks. The overlying Buntsandstein facies rocks usually rest uncon-formably upon 'Saxonian' and older rocks (Figs 10.4 and 10.5), expanding the area of the sedimentation to create the Tajo basin (Castillo-Herrador 1974). The succession again com-prises continental siliciclastic rocks with abundant conglomer-ates and sandstones.

Castilian branch. Late Permian ('Saxonian') red beds were deposited in two isolated basins in the Castilian branch of the Iberian area. In one of the basins, in the Molina de Aragón area (Montesoro Fm), about 120 m of breccias, sandstones and red siltstones with prominent caliche profiles accumulated (Ramos 1979; Arche et al. 1983; Pérez-Arlucea & Sopeña 1985; López-Gómez & Arche 1993). The other basin extended southeastwards at least to the present-day Mediter-ranean coastline (Fig. 10.4), and preserves a succession of alluvial fan conglomerates (Boniches Fm, up to 270 m thick) overlain by at least three fining-upward sequences of conglomerates, sandstones, red silt-stones and, occasionally, dolomites (Alcotas Fm and Tormón Fm, together up to 150 m thick and roughly equivalent to the Montesoro Fm; Arche et al. 1983; López-Gómez 1985; López-Gómez & Arche 1993). These rocks are interpreted as being of fluvial channel, alluvial plain and shallow lake origin, and caliche palaeosol horizons are common, especially in the upper part of the formation. The two basins were separated by the Cueva de Hierro–Tremedal Palaeozoic high (Pérez-Arlucea & Sopeña 1985; López-Gómez 1985). The age of these formations is well constrained, as a rich palynoflora association has been found both in the conglomerates and in overlying red beds (Boulouard & Viallard 1982; Arche et al. 1983; Doubinger et al. 1990; Sopeña et al. 1995) including Lueckiesporites virkiae, Nuskoisporites dulhunti and Falcisporites schaubergeri of Late Permian (Thüringian) age. In sharp contrast to their time equivalents in the Pyrenees, these Late Permian formations lack volcanic rocks.

The Buntsandstein facies rocks record sedimentation that progres-sively expanded over the Palaeozoic basement, and the two sub-basins of the previous cycle became a single, complex graben (the early Tajo basin). The sediments can be divided into two major successions that are not always both present in the Iberian area: a lower one, compris-ing laterally restricted conglomerates (Hoz de Gallo Fm (HGC)) and more widespread sandstones (Rillo de Gallo Fm (RGS) or the equivalent Cañizar Fm (CS)), and an upper one of irregular distri-bution, comprising sandstones (Arandilla Fm (AS) or the equivalent Eslida Fm (ESS)), sandstones and siltstones (Prados Fm (PB) and Rillo Fm (RMS)) and siltstones (Torete Fm (TMS), equivalent to the

Marines Fm (MCMM); Fig. 10.4). Where the upper succession is missing, the top of the lower one is a hiatus marked by soil horizons and a sharp unconformable contact with the lower Muschelkalk facies.

The lower succession shows a marked fining-upward succession, with the Hoz de Gallo conglomerates interpreted as being of alluvial fan origin and forming an apron derived from an elevated SW footwall margin of the rift basin identified from Pálmaces to Albarracín. The conglomerates contain a palynomorph assemblage comprising Lueck-iesporites virkkiae, Nuskoisporites dulhuntyi and Klausipollenites schaubergeri (Thüringian age; Sopeña et al. 1995). The overlying Rillo de Gallo (or its equivalent Cañizar) Fm, consists of pink to red cross-bedded sandstones of sandy braided river origin with palaeocurrents parallel to the basin axis. These sandstones consist of amalgamated channel deposits totally devoid of fine grained floodplain sediments and are interpreted as having been deposited during a very slow subsidence rate. The top of these deposits contains a palynomorph assemblage of Triadispora staplini, T. falcata and Illinites sp. (Anisian age; Doubinger et al. 1990).

The second major succession was not deposited in some parts of the central Iberian basin, where the Muschelkalk facies lies uncon-formably on the Cañizar Fm, but regional reactivation of the basin boundary faults in the SE led to the deposition of more than 700 m of fluvial sediments (Eslida Fm (ESS); Arche & López-Gómez 1999) overlain by red siltstones (Marines Fm (MCMM)) which record a transition to marine deposits (Fig. 10.4). A coeval but different succes-sion of mainly sandstones and mudstones was deposited to the NW of Molina de Aragón (Fig. 10.4), and comprises the Prados, Arandilla, Rillo and Torete formations. As the latter is the lateral equivalent of the middle Muschelkalk sabkha deposits (El Mas Fm (MSMG)) or the equivalent Tramacastilla Fm (TMG)) to the SE, the Rillo Fm must be the time-equivalent of the lower Muschelkalk facies (Sopeña et al. 1988, 1995). Further NW (Alcolea-Sigüenza area), the lower part of the upper Muschelkalk facies passes laterally into fluviodeltaic sand-stones and mudstones (García-Gil 1991, 1995).

Aragonese branch. Both 'Saxonian' and Buntsandstein facies show considerable lateral change southeastwards across the Aragonese branch of the Iberian Ranges, from Demanda and Cameros to the Montalbán area (Fig. 10.5). The lack of palaeontological data for the oldest sediments prevents determination of their age, so some could be Triassic rather than Permian (Arribas 1984). Early publications on these sediments include those of Dereims (1898), Würm (1911), Lotze (1929), Richter (1930) and Vilas et al. (1977). More recently, detailed studies of this facies in the Aragonian branch were performed in the Moncayo area (Arribas 1984, 1985, 1987; Arribas & Soriano 1984; Arribas et al. 1985; Rey & Ramos 1991). Arribas (1984, 1985) described four units from base to top – Araviana conglomerates and mudstones (ACM), Tierga sandstones (TS), Cálcena mudstones and sandstones (CMS) and Trasobares mudstones and marls (TMM) – and a tentative correlation of these units is shown in Figure 10.5.

The Araviana unit is attributed to the latest Permian, is about 60 m thick, and shows a lower part which rests with erosive unconformity on basement or Lower Permian (Autunian) sediments. This part of the succession is mainly composed of conglomerates with planar cross-stratification, interpreted as recording well-developed alluvial systems with clear proximal–distal evolution from base to top. The Araviana unit consists of red mudstones with intercalated sandstones and dolomite levels and is interpreted as deposited by sheet floods and on terminal fans with well-developed floodplains. The Tierga unit is about 58 m thick and consists of red and white arkoses and subarkoses, with planar and trough cross-stratification and ripples, and red mudstones. It is considered to be Lower–Middle Triassic in age (Arribas 1985, 1987; Arribas et al. 1985; Rey & Ramos 1991), and interpreted as deposited during the migration of bars in different sized channels under flow regime oscillations (as indicated by erosive surfaces) during sporadic flash-floods. The Cálcena unit is considered Anisian in age according to palaeoflora associations found close to Ródanas, 34 km NE of Reznos, in the Moncayo area (Díez et al. 1994, 1996). This unit consists of about 14 m of red mudstones and inter-calated decimetric pink sandstone levels, interpreted as deposited on a mixed carbonate siliciclastic intertidal flat (Arribas & Soriano 1984; Arribas 1985). This unit shows a dolomitic level up to 7 m in thickness, the Illueca dolomites (ID) (Arribas 1984, 1985), comprising yellow dolomites with ripples, algal lamination, mud-cracks and evaporite crystals. The Trasobares unit consists of about 40 m of red mudstones, clays and marls with intercalated decimetre-scale dolomites and gypsum levels and is interpreted as deposited on a supratidal flat (Arribas & Soriano 1984; Arribas 1985).

Betic Cordillera

Two main tectonic domains may be distinguished in the Betic Cordillera (see Chapter 16): the 'external zone', which includes sediments deposited in the southern Iberian continental margin; and the 'internal zone', which corresponds to the so-called Alboran domain (García-Dueñas & Balanyá 1986; Fig. 10.9). During the Triassic period, this latter domain was situated further east, but it approached the Iberian Massif during the Alpine orogeny. Most Permian and Triassic units within the Betics are strongly deformed, and some are highly metamorphosed, which, combined with a general absence of index fossils, inhibits correlation and palaeogeographical interpretation (Pérez-López & Sanz de Galdeano 1994; Sanz de Galdeano 1997; Pérez-López 1998).

Rocks attributed to the Permo-Triassic have been described from all three of the internal Betic tectonic units, namely the Nevado–Filábride, Alpujárride and Malaguide complexes. Within the Nevado–Filábride complex several schistose lithostratigraphic units have been described (Puga 1976), one of the most representative being the Tahal Fm (Nijhuis 1964a). In addition to several hundred metres of schists, some Permo-Triassic units in this complex include quartzite, marble and gneiss, and metaconglomerate in their lower part (Vissers 1981; Martínez-Martínez 1984–85; De Jong 1991; Fig. 10.9). Within the Alpujárride complex there are metamorphosed Permian and Early Triassic phyllites and quartzites, with some metacarbonate intercalations (Soediono 1971; Akkerman et al. 1980; Delgado et al. 1981). Finally, within the Malaguide complex, rocks resting on a Palaeozoic basement have been traditionally ascribed to the Permo-Triassic, due to their red siliciclastic facies and to the presence of plant remains (Ansted 1860). Upper–Middle Triassic pollen within these rocks (Martín-Algarra 1995), however, supports previous interpretations of a relatively young age (Kozur et al. 1985) and it seems unlikely that the Malaguide Complex includes Permian or Early Triassic red beds.

Balearic Islands

Betic structures continue northeastwards into the Balearic Islands where there are a number of outcrops of Late Permian–Early Triassic rocks (Fig. 10.9). The islands are the emergent part of the Balearic promontory, a submarine structural high that forms the NE part of the Betic Cordillera (Rodriguez-Perea et al. 1987). Permian and Triassic rocks crop out in the three major islands, Majorca, Minorca and Ibiza (Fig. 10.9). The structural development of the Balearic Islands during Mesozoic and Tertiary time is a very complex sequence of mainly extensional events and short compressive phases (e.g. Banda & Santanach 1992; Dañobeitia et al. 1992; Gelabert et al. 1992).

The Late Permian–Middle Triassic (pro parte) succession in the Balearic Islands is represented by the 'lower clastic unit' of Rodriguez-Perea et al. (1987). This unit, roughly equivalent to the Buntsandstein facies, crops out in Majorca and Minorca

and was subdivided in Majorca by Ramos & Doubinger (1989) and Ramos (1995) into three formations: Port des Canonge (oldest), Asá and Son Serralta formations (youngest).

The Port des Canange Fm crops out only along the northern coast of Majorca and the central-northern part of Minorca, and lies unconformably on the Palaeozoic (Carboniferous?) in Majorca (Ramos & Rodriguez-Perea 1985). The base consists of up to 20 m of clast-supported breccias with metamorphic and sandstone clasts in a red matrix, overlain by up to 200 m of red mudstones and sandstones (Ramos & Rodriguez-Perea 1985; Rodriguez-Perea et al. 1987; Ramos & Doubinger 1989; Arribas et al. 1990; Gómez-Gras 1993b; Ramos, 1995). Siltstone units, 7–12 m thick, contain parallel lamination, ripple cross-laminae and horizons of calcareous nodules, probably of pedogenic origin. Thin tabular sandstone bodies, with sharp bases and fining-upward character, are also found, with common amalgamation of individual sandstone units. Sandstone levels are of tabular geometry and fining-upward character, erosive bases are covered by a clay-chip lag and the main internal structure shows trough and tabular cross-stratification, with current ripples capping the sequence. This unit is interpreted as a distal meandering fluvial system with extensive floodplain deposits with crevasse-splay, suspension deposits and pedogenic horizons. The formation has been dated in Minorca by means of pollen assemblages consisting of *Lueckisporites virkkiae*, *Falcisporites stabilis*, *Klausipollenites schaubergerii* and *Illinites unicus* among others, indicating a Late Permian (Thüringian) age (Broutin et al. 1992). This formation can be correlated with the 'Saxonian' facies of the Iberian Ranges, i.e. the 'Capas de Montesoro' (Ramos 1979), the Alcotas Mudstones and Sandstones Fm or the Tormón Mudstones and Sandstones Fm (Pérez-Arlucea & Sopeña 1985).

The Asá Fm lies on the previous formation by means of a gentle unconformity. It consists of amalgamated sandstone sheets, with parallel lamination, low-angle planar cross-stratification and climbing ripples, with some metric red mudstone intercalations. The formation is about 120 m thick (Rodriguez-Perea et al. 1987; Ramos 1995) and is interpreted as upper-flow regime deposits with flashy discharge through narrow channels evolving vertically to wide shallow channels. Floodplain deposits are scarce, probably due to repeated reworking. The age of the Asá Fm is well-constrained by several pollen findings (Ramos & Doubinger 1989) near the middle part of the section. They contain, among others, *Lueckisporites virkkiae*, *Nuskoisporites dulhunti*, *Falcisporites zapfei* and *Klausipollenites schaubergerii* of Late Permian (Thüringian) age. The top of the formation is so far not dated. The Asá Fm can be correlated with the Cañizar Fm (López-Gómez 1985) and the Hoz de Gallo conglomerates and Rillo de Gallo sandstones formations in the Iberian Ranges (Ramos 1979) (Fig. 10.4).

The uppermost siliciclastic unit is the Son Serralta Fm (Ramos 1995) which lies conformably on the Asá Fm and is conformably overlain by Muschelkalk facies rocks. It is about 65 m thick and can be subdivided into a lower interval, 35 m thick, of sandstones, and an upper interval, 30 m thick, of red mudstones, solution breccias and carbonate concretion horizons. Tabular cross-stratification and parallel lamination are the main internal structures in the sandstone beds and are usually organized in coarsening-upward successions with wave ripples and flaser-bedding in the transition to the muddy upper interval. These sandstones were deposited in a distal braidplain environment (Ramos 1995) evolving into coastal shallow deltaic or beach-prograding complexes. The mudstones with wave ripples, carbonate beds and solution breccias of the upper half can be interpreted as coastal mudflats with shallow evaporitic ponds, preceding the installation of the marine carbonate platform of the Muschelkalk facies.

The lower part of the formation has been dated in Majorca by means of pollen and spores, among others *Porcellispora longdonensis*, *Sulcosaccispora minuta*, *Triadispora staplini* and *Alisporites grauvogeli*, of Anisian (middle to late?) age. These deposits can be correlated with the Eslida and Marines formations (López-Gómez et al. 1993) and the Nivel de Prados and Areniscas del Río Arandilla formations (Ramos 1979) of the Iberian Ranges (Fig. 10.4).

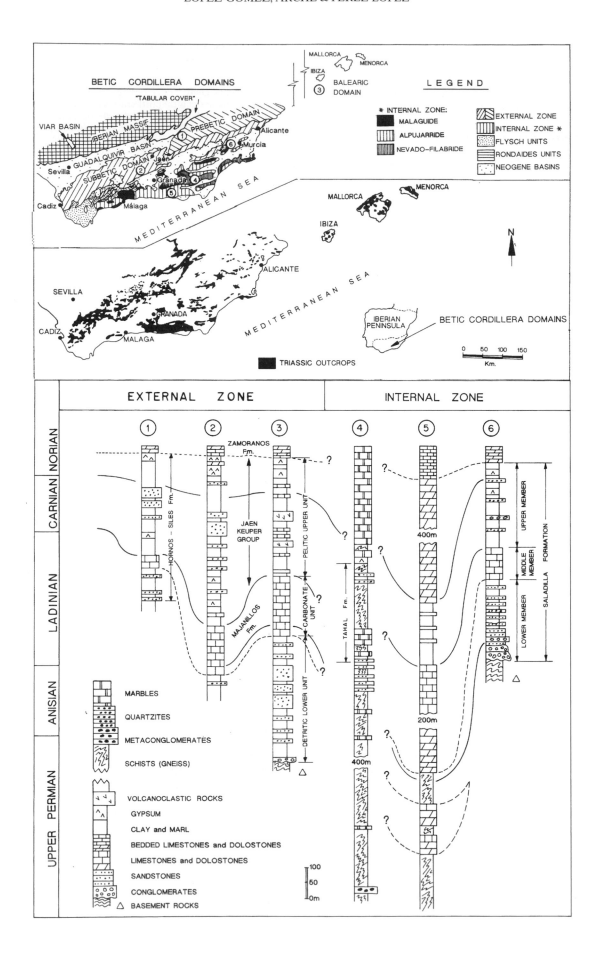

Middle–Late Triassic sediments

Westward propagation of the Tethys sea along the entire European plate meant that it reached the Iberian plate during Anisian times. Thus, the Triassic epicontinental shallow marine sediments of the Iberian peninsula form part of the westernmost area of the Tethys margin. These sediments consist of three carbonate units separated by two mudstone–evaporite units and represents three transgressive–regressive main cycles that onlapped the western and southern margins of the Palaeozoic Iberian Massif and related blocks (López-Gómez et al. 1998; Fig. 10.10). The dating of these units in both the Iberian peninsula and Balearic Islands is based on ammonoids, foraminifera, conodonts and pollen and spore associations (López-Gómez et al. 1998).

The two lower carbonate units correspond to the Muschel-kalk facies. Although in the Iberian peninsula this term is used as a facies term and does not correlate directly with classic Germanotype nomenclature, these units probably correspond to comparable periods of transgression and regression in central Europe. Dated as lower (Anisian) and upper (Ladinian) Muschelkalk, these units represent a single carbonate sequence close to what is now the Mediterranean coastline, where the intermediate mudstone–evaporite unit gave way eastwards to carbonates. The third carbonate unit is the youngest and most extensive of the three; it is called the Imón or Isábena Fm, and is Norian in age.

The limits of westward transgressions or propagation of the three carbonate units in the Iberian peninsula are clearly different and are shown in Figure 10.10. Four terms have been recently introduced, based on which (if any) of the three carbonate units are present (López-Gómez et al. 1998): 'Hesperian Triassic', devoid of carbonate Muschelkalk units; 'Iberian Triassic', showing the presence of the upper Muschel-kalk unit; 'Mediterranean Triassic', showing the presence of both carbonate Muschelkalk units separated by a mudstone–evaporite unit; 'Levantine-Balearic Triassic', showing the presence of a single carbonate unit that represents the two carbonate Muschelkalk units without the intermediate mudstone–evaporite unit that laterally passed to carbonates (Fig. 10.10).

Cantabrian Zone

Throughout the Cantabrian Zone only a few outcrops are found preserving sediments between well-dated Permian and Jurassic rocks. These sediments have been traditionally considered to be of Triassic age (e.g. Maas 1974; García-Mondejar et al. 1989; Martínez García 1991; Gand et al. 1997) but dating was not based on palaeontological criteria. Recently, two pollen and spore assemblages found SW of La Camocha in the so-called 'Tránsito' unit (Suárez Vega 1974) and Fuentes Fm (Martínez García 1991; Fig. 10.2) have been interpreted as Upper Triassic (Rhetian; Martínez García et al. 1998).

Fuentes Fm or 'Tránsito' unit. The Fuentes Fm mainly crops out in the Villaviciosa area (Gutierrez-Caverol & Manjón 1984; Martínez García 1999) in central Asturias province, and comprises a succession of alternating red and green mudstones, siltstones, and thin sandstones with gypsum beds that may be locally abundant. In this area, and in the Avilés region, this succession passes conformably up into well-dated Lower Jurassic (Hettangian) limestones.

The 'Tránsito' unit close to Gijón attains some 10 m in thickness and overlies a 220 m thick succession of red and green marls, siltstones and sandstones that corresponds to the 'Conjunto Superior' unit (Manjón et al. 1992) or 'brown clays and black slates with anhydrite and gypsum' unit (Pieren et al. 1995). Martínez García et al. (1998) consider that this latter succession could also be of Rhetian age, i.e. similar in age to the overlying 'Transito' unit. The lower contact of the 'Conjunto Superior' unit always lies on the Caravia Fm of Permian age and is probably tectonic (Martínez García 1991). These Upper Triassic sediments show significant lateral changes in thickness but in many places are completely eroded, such as at Langreo (20 km east of Oviedo) or Cangas de Onís (15 km SW of Ribadesella), where Cretaceous sediments are deposited directly on the Permian Caravia Fm (Martínez García 1991).

Only the higher of the two Muschelkalk carbonate units is found in the Cantabrian Zone, and even this is restricted to a few outcrops to the east (mainly in the Bárcena de Pié de Concha area, and north of Corrales de Buelna and Reinosa; Fig. 10.2). These sediments shows 25 m of alternating dolomites and limestones with remains of bivalves and ostracods in the Bárcena de Pié de Concha area (Carreras et al. 1978, 1979). Towards the north, in Corrales de Buelna, this alternating succession passes into calcareous sandstones indicating the pinching out of this carbonate platform.

Keuper facies rocks are commonly tectonically deformed and/or show diapiric doming, which makes it difficult to discern its thickness or lithology. Only a few general works have described this facies in the Cantabrian Zone (Carreras et al. 1979; Salvany 1990) and reveal the succession to be mainly composed of red clays with intercalated levels of dark gypsum and sporadic green basalts.

The third and last Triassic marine carbonate unit (equivalent to the Isábena Fm in the Pyrenees, and the Imón Fm in the Iberian Ranges) is nowhere seen over the entire Cantabrian Zone. However, in Santander province, isolated limestone or dolomitic levels normally thought to be of Early Jurassic age (but without reliable palaeontological criteria) might prove to be older. Another possibility is that the Fuentes Fm and the lower 'Conjunto Superior' unit could represent a lateral equivalent of the Keuper facies and Isábena Fm.

Pyrenean Basin

The Middle–Upper Triassic succession of the Pyrenean Basin crops out in the Basque Pyrenees, the southern margin of the western-central Pyrenees, and in the eastern Pyrenees (Fig. 10.3).

Western-central and eastern Pyrenees. The Triassic of the lowermost units (Chapter 15) is always incompletely represented, because most of the Keuper facies and the Isábena Fm (the uppermost Triassic carbonate unit) are eroded (Fig. 10.3). Muschelkalk facies rocks are 60 m thick and consist of bioclastic dolomites, thin-bedded dolomites containing *Pseudofurnishius murcianus* and *Lingula tenuisima* (Ladinian age; Destombes 1950), and grey marls overlain by alternating massive and thin-bedded dolomites with *Pachipleurosaurus* sp., *Colobodus* sp., *Perleidus* sp. and *Birgeria* sp. (Ladinian age; Mazin & Martín 1983). These deposits can, therefore, be correlated with the upper carbonate unit of the Muschelkalk facies or the M-3 of the Catalonian Coastal Ranges.

The Triassic of the lower thrust sheets (Chapter 15) is complete, with good exposure of Buntsandstein, Muschelkalk, Keuper and Isábena Fm rocks (Fig. 10.3), and has been studied by Virgili (1958), Nagtegaal (1969), Gisbert (1981, 1984), Lucas et al. (1980), Broutin

---

Fig. 10.9. The Betic Cordillera Triassic outcrop map and stratigraphic columns for main sections.

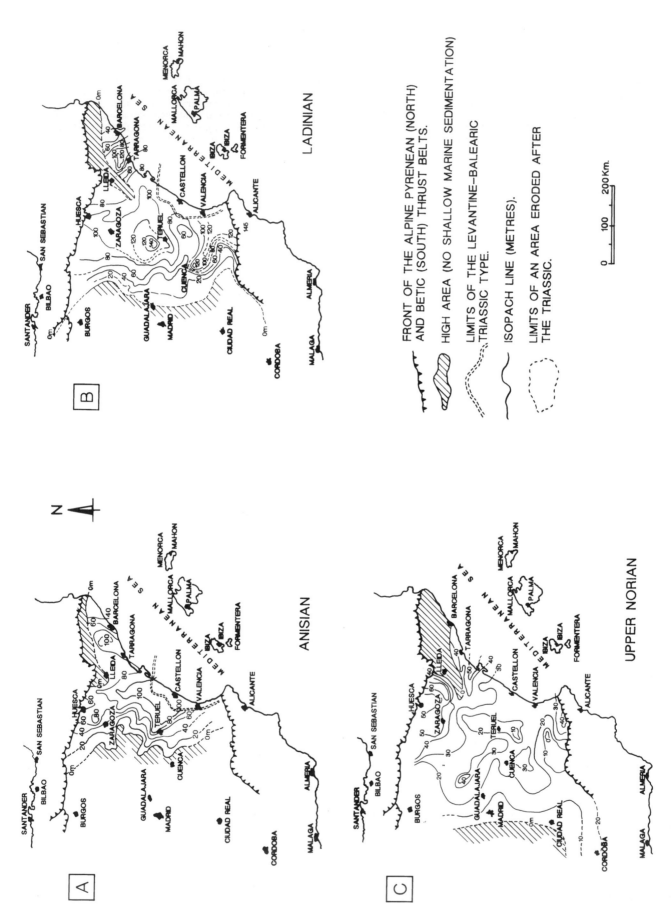

Fig. 10.10. Isopach maps of the three Triassic platforms of the Iberian Peninsula: Anisian, Ladinian and upper Norian. The two lower platforms correspond to the lower and upper Muschelkalk facies respectively, and the upper one corresponds to the Imón–Isábena formations. Modified from López-Gómez et al. (1998).

et al. (1988), Frèchenges & Peybernes (1991) and Calvet *et al.* (1993), among others. Muschelkalk facies rocks range in thickness from 50 to 85 m and have been subdivided into three lithological units which are, from base to top, dolomitic marls, bioclastic limestones and sandstones, and thin-bedded dolomites, interpreted as recording a complete transgressive–regressive cycle. A palynological assemblage found in the Cadí area (Calvet *et al.* 1993) in the dolomitic marls unit contains *Triadispora suspecta, Ovalipollis ovalis, Praecirculina granifer, Duplicisporites scurrilis* and *Camerosporites secatus* (middle–late Ladinian age). The overlying bioclastic unit contains foraminifera such as *Aulotortus praegasei, Lamelliconus procerus, Triadodiscus eomesozoicus* and *Nodosaridae,* as well as the conodont *Pseudofurnishius murcianus* (late Ladinian age; Freychengues & Peybernes 1991). In the eastern part of the central Pyrenees, as well as in the Cadí area, the overlying thin-bedded unit contains palynological assemblages comprising *Ovallipollis ovalis, Staurosaccites quadrifidus, Triadispora* sp., *Praecirculina granifer, Camerosporites secatus, C. densus* and *Patinasporites quadruplicis* (late Ladinian–?basal Carnian age; Calvet *et al.* 1993). Clearly, these rocks are the time equivalent of the upper Muschelkalk (M-3) facies of the Catalonian Coastal Ranges (Virgili 1958). A gradual transition marks the passage to overlying Keuper facies rocks.

Keuper facies rocks are usually strongly tectonized, but despite this their overall thickness can be estimated at 180–250 m (Salvany 1990). The succession has been subdivided into five lithological units which are, from base to top, grey siltstones and limestones, red clays, variegated gypsums, white gypsums and variegated clays and limestones. The lowermost unit contains a palynological assemblage found in several localities formed by *Classopollis* sp., *Granuloperculatipollis rudis, Triadispora suspecta, T. crassa* and *Ovalipollis cultus* (lower–middle Norian age). The uppermost unit, the Isábena Fm, harbours an assemblage of *Classopollis* sp., *Rhaetipollis germanicus, Suessia swabiana* and *Deltoidospora* (late Norian–Rhaetian age).

The Isábena Fm is the youngest Triassic deposit in the lower thrust sheets (Chapter 15), and comprises 20–30 m of oolitic limestones followed by homogeneous dolomitic mudstones of remarkable uniformity and lateral extent (Fig. 10.3). Its age is well defined by the presence of *Rhaetavicula contorta* (Virgili 1963), palynological assemblages containing *Cerebropollenites pseudomassulae* and *Corollina zwolinskae* (Baudelot & Taugourdeau-Lanz 1986; Calvet *et al.* 1993), and foraminifera such as *Glomospirella rossetta, G? apenninica, Aulotortus friedli* and *Agathammina inconstans* (Márquez & Trifonova 1990; Rhetian age). The upper boundary is an erosional unconformity.

The Triassic of the upper thrust sheets is found in the Marginal Sierras and associated thrust units, and shows only partial exposure of Muschelkalk and Keuper facies rocks and Isábena Fm (Pocovi 1978a; Martínez 1982). The Muschelkalk facies never exceeds 50 m in thickness and invariably presents faulted upper and lower limits. It consists of bioclastic limestones and thin-bedded dolomites containing undetermined *Nodosaridae* and *Lingula* sp. (Ladinian age) in the Montsec area. It is equivalent to the upper carbonate unit of the Muschelkalk facies or the upper Muschelkalk, or M-3 (from Virgili 1958) of the Catalonian Coastal Ranges.

The Keuper facies, some 150 m thick, comprises three units which, from base to top, are variegated siltstones, white gypsums, and variegated siltstones and limestones. Baudelot & Taugourdeau-Lanz (1986) found a palynological assemblage with *Classopollis* sp., *Ovallipollis pseudoalata, Rhaetepollis germanicus* and *Schwassia schwabiana* (late Norian–Rethian age) in the uppermost unit, near Pobla de Lillet. Finally, the Isábena Fm is up to 30 m thick, and has yielded a palynological assemblage of *Cerebropollenites pseudomassulae* and *Corollina zwolinskae* (Rethian age) at its base in the Bac Grillera area (Vachard *et al.* 1990).

The Basque Pyrenees. The Basque Pyrenees are separated from the central-eastern Pyrenees by a major fault zone called the Pamplona or Estella–Velate fault (Muñoz *et al.* 1983). Triassic sediments in this area have been described by Muller (1969), Solé & Villalobos (1974), Villalobos (1975) and Campos (1979), as well as in palaeontological studies by Calvet *et al.* (1993) and Lucas *et al.* (1980). Buntsandstein, Muschelkalk and Keuper facies rocks and the Isábena Fm can be found all over the area. The structural continuity of the Basque Pyrenees with the Cantabrian mountains is now

clear (Cámara 1989), but the absence of outcrops between the eastern Basque Country and the western Cantabrian Zone as yet precludes precise correlations.

Buntsandstein facies rocks lie unconformably on Early Permian sediments or the Palaeozoic basement (Fig. 10.3). It can be subdivided into two major sedimentary successions in sharp but conformable contact. A lower succession consists of a basal interval, 0 to 40 m in thickness, of fining-upward conglomeratic sequences overlain by 40–200 m of large-scale, cross-stratified sandstone which, despite a lack of reliable dating markers, may be correlated with the 'lower Buntsandstein' succession of the central-eastern Pyrenees (Thüringian–Scytian? age, Late Permian–Early Triassic). The second succession consists of fining-upward sandstone–siltstone alternations, with variegated siltstones at the top (120–200 m thick), where macroflora including *Yuccites vogesiacus, Neocalamites* sp. and *Schizoneura* sp. (Muller 1969, 1973) were found along with a palynological assemblage of *Stellapollenites thiergartii, Strioabietites aytugii, Triadispora* sp. and *Alisporites* sp. (Anisian age). This succession can be correlated with the second Buntsandstein succession seen in the central-eastern Pyrenees. There is a transitional upper contact with the overlying Muschelkalk facies rocks.

The carbonates of the Muschelkalk facies, up to 70 m thick (Fig. 10.3), consist of marls, bioclastic limestones and thin-bedded dolomites and dolomicrites. In its central part, the facies contains *Sephardiella mungoensis, Nodosaria ordinata* and *Aulotortus* sp. (Ladinian age). These levels can be correlated with the upper carbonate unit of the Muschelkalk facies (upper Muschelkalk or M-3) of the Catalonian Coastal Ranges. Keuper facies rocks are poorly exposed and were heavily deformed during Alpine compression. This succession can reach 120 m (Muller 1969) and comprises red and green siltstones, subordinate gypsum horizons, and common subvolcanic basaltic rocks (ophites).

Catalonian Basin

The Middle–Upper Triassic succession in the Catalonian Coastal Ranges is represented by all three previously described carbonate units (lower and upper Muschelkalk and overlying Imón Fm) separated by two mudstone–evaporite units (Fig. 10.11). These sediments have been described since the nineteenth century, with the early workers offering only general descriptions (Almera 1891a; Würm 1919; Vilaseca 1920; Bataller 1933; Schmidt 1933; Sos 1933). More specific description of localized areas typified mid-twentieth century publications (e.g. Ríos & Almela 1954; Virgili & Juliver 1954; Solé Sabarís *et al.* 1956), with the exception of Virgili (1958) who included all the Triassic sediments of the Catalonian Coastal Ranges in a very detailed report. More recent works deal mainly with sedimentology, palaeoecology, sequence stratigraphy and tectonics (Esteban *et al.* 1977; Ortí & Bayo 1977; Via Boada *et al.* 1977; Anadón *et al.* 1979; Calvet & Ramón 1987; Salvany & Ortí 1987; Solé de Porta *et al.* 1987; Santiesteban & Taberner 1987; Calvet *et al.* 1990; Calvet & Marzo 1994; Calvet & Tucker 1995; Morad *et al.* 1995; Roca *et al.* 1999).

The lowermost carbonate unit of the Triassic, or lower Muschelkalk (M-1, according to Virgili 1958), succession is 70 m thick in the northeastern area yet reaches 120 m in the SW (Fig. 10.11). The lower boundary shows a gradual transition from the 'upper evaporitic-silty-sandy complex' of the upper Buntsandstein facies (Fig. 10.11; Marzo 1980). The upper boundary is a sharp contact with the overlying siliciclastic-evaporitic deposits of the middle Muschelkalk. The lower Muschelkalk has been divided into four members which are, from base to top, El Brull, Olesa, Vilella Baixa and Colldejou (Calvet & Ramón 1987; Calvet *et al.* 1990; Calvet & Marzo 1994). The whole carbonate succession represents a transgressive–regressive cycle developing from intertidal stromatolitic layers (El Brull) to bioclastic shoals (Olesa), shallowing-upward cycles of subtidal carbonate mudstones, with black

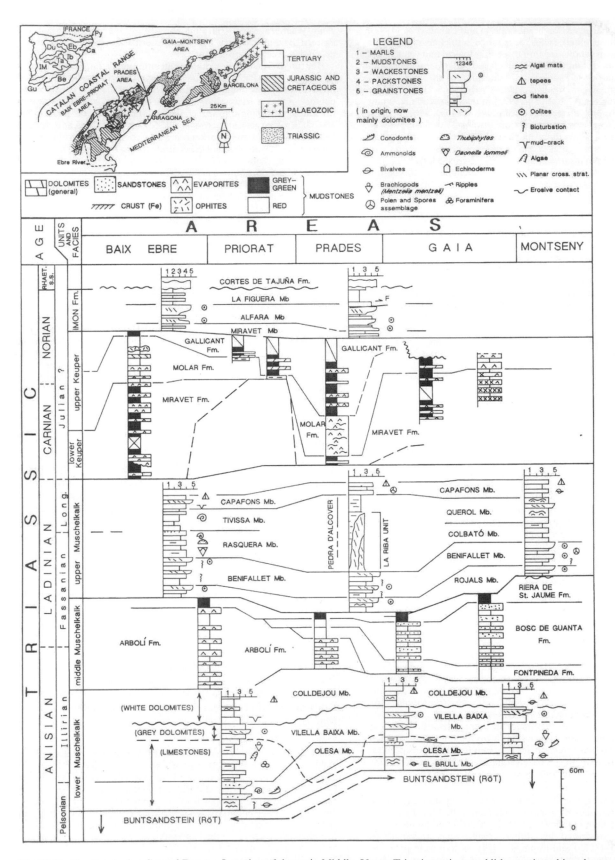

Fig. 10.11. The Catalonian Coastal Ranges. Location of the main Middle–Upper Triassic sections and lithostratigraphic columns for the five selected areas. Sections modified from Calvet & Marzo (1994), Calvet & Tucker (1995), Castelltort (1986), Salvany (1986) and Salvany & Ortí (1987). Basins in the location map: 1, Pyrenees; 2, Ebro; 3, Catalan; 4, Iberian; 5, Duero; 6, Tajo; 7, Guadalquivir; 8, Betics.

paper shales indicating anoxic conditions (Vilella Baixa), to sabkha deposits (Colldejou). The age of this carbonate unit is Anisian based on the presence of ammonoids (Virgili 1958; Marzo & Calvet 1985; Ramón & Calvet 1987), foraminifera (Budurov *et al.* 1993) and conodonts (Calvet *et al.* 1987; Ramón & Calvet 1987; March 1991; Calvet & Marzo 1994). In general, it is possible to correlate these deposits with coeval deposits of the Iberian Ranges using biostratigraphical and sedimentological data (López-Gómez *et al.* 1998).

The middle siliciclastic-evaporitic Muschelkalk unit (M-2, according to Virgili 1958) has been poorly defined in the Iberian peninsula due to the poor quality of outcrops that are sometimes mistaken for those of the Keuper facies. However, the Catalonian Coastal Ranges show the best continuous outcrops of this unit in the Iberian peninsula, possibly due to the fact that it contains less evaporites (Fig. 10.11). The first detailed description of this unit, which is 50–115 m thick, was made by Virgili (1955, 1958). Later, Castelltort (1986) and Morad *et al.* (1995) reported marked synsedimentary tectonics for this unit, controlling the transport of siliciclastic sediments from elevated areas in the NE towards the central part of the basin where evaporite deposits had accumulated. The dominant lithologies are sands, evaporites and mudstones with smaller percentages of dolomites that vary depending on the area. Sedimentation was accompanied by several episodes of syntectonic alkaline volcanism in the southern areas (Castillo-Herrador 1974; Calvet & Marzo 1994). The middle Muschelkalk was divided by Castelltort (1986) into four units which are, from base to top and from east to west: (1) Fontpineda mudstones and gypsum, representing supralittoral, evaporitic plain deposits, 10 to 25 m thick; (2) Bosc de Guanta sandstones and mudstones, developed towards the north and generated by ephemeral fluvial flows within a terminal alluvial fan system that passes laterally to; (3) L'Arbolí mudstones and gypsum towards the central and southern areas, interpreted as a playa lake or sabkha deposits; (4) the Riera de Sant Jaume Fm, which attains a thickness of 10 m and is interpreted as supralittoral, muddy intertidal flat deposits and represents the transition to the upper Muschelkalk.

The upper Muschelkalk of the Catalonian Coastal Ranges (M-3, according to Virgili 1958) is up to 130 m thick and can be subdivided into several units equivalent to members (Calvet *et al.* 1987, 1990; Calvet & Marzo 1994; Calvet & Tucker 1988, 1995). These members are not always easy to correlate across the whole Catalonian Coastal Ranges and the following have been defined for each of the differentiated areas (Fig. 10.11): Gaiá-Montseny (Rojals, Benifallet, Colbató, Querol and Capafons members), Prades (Rojals, Benifallet, La Riba, Pedra d'Alcover and Capafons members) and Priorat-Baix Ebre (Rojals, Benifallet, Rasquera, Tivissa and Capafons members). Most of the upper Muschelkalk of the Catalonian Coastal Ranges records a major transgressive–regressive cycle with a significant break in the lower part seen in the Baix Ebre–Priorat areas. The Rojals and Benifallet members represent the transgressive stage, from carbonate mudstones of restricted lagoonal origin to the sandy rim of a shallow ramp. The Collbató Member (Mb) represents a regressive period from a deep proximal ramp to a sandy bioclastic rim of a shallow ramp, and the Querol Mb represents an intersupratidal environment with sporadic bioclastic lenses that indicate high-energy subtidal environments. La Riba Mb represents a reef complex that clearly shows evidence for shallowing upwards (Calvet & Tucker 1995), and is onlapped by the La Pedra d'Alcover Mb that infills the inter-reef. The Rasquera Mb, which contains *Daonella*, ammonoids and *Tubiphytes*, is interpreted as a distal to proximal shallowing-upward deep ramp deposit. Finally, the Tivissa and Capafons members are interpreted as outer to shallower inner ramp and intertidal to hypersaline supratidal deposits respectively. The age of the upper Muschelkalk is Longobardian (late Ladinian) based on conodonts (Hirsch 1966; March 1991), foraminifera (Márquez & Trifonova 1990; Budurov *et al.* 1993) and ammonoids (Virgili 1958; Marzo & Calvet 1985; Calvet & Marzo 1994; Goy 1995).

The Keuper facies in the Catalonian Coastal Ranges was first studied in detail by Virgili (1958) and subsequently by Ortí & Bayo (1977), Salvany & Ortí (1987), and particularly by Salvany (1986, 1990). There are two clear superposed evaporitic cycles in the Catalonian Coastal Ranges differentiated by some of the above authors into lower and upper evaporitic units (Fig. 10.11). The lower Keuper, well-represented in the NE and SW zones, is formed by the

c. 100 m thick Miravet Clays and Gypsum Fm (Salvany 1986). This formation comprises well-stratified gypsum, grey clays and thin intercalated dolomite levels, and is interpreted as a regressive depositional sequence developing from mud flats to sabkhas or an evaporitic restricted lagoon. The upper Keuper succession represents a transgressive depositional environment, is about 50–100 m thick and comprises the Molar Fm (clays and gypsum) overlain by the Gallicant Fm (clays and carbonates; Fig. 10.11). The Molar Fm shows intercalated volcano-sedimentary rocks (Mitjavila & Martí 1986), and is interpreted as sabkhas or restricted ponds with sporadic alluvial incursions. The Gallicant Fm is interpreted as recording small lakes within which very thin and ephemeral algal-laminated sediments were deposited and later affected by early diagenetic dolomitization. These calcareous sediments indicate the transition to the upper carbonate unit or Imón Fm. The different units of the Keuper of the Catalonian Coastal Ranges have been correlated with equivalents in the Levantine and Pyrenean Iberian basins (Salvany & Ortí 1987; Salvany 1990), and their age is Carnian–Norian (Calvet & Marzo 1994) based on pollen and spore associations.

The Imón Fm (Goy *et al.* 1976) represents the youngest epicontinental Triassic carbonate formation and shows the widest distribution. It consists of well-stratified dolomites that can be over 50 m in the Catalonian Coastal Ranges. A gradual transition from the Keuper may be observed at the base, while the top presents an erosion surface with the overlying Cortes de Tajuña Fm (Goy *et al.* 1976). The Imón Fm is subdivided into the Miravet (oldest), Alfara and La Figuera (youngest) members (Fig. 10.11). The Miravet Mb (approximately 10 m thick) consists of dolomudstones with parallel lamination and abundant evaporite pseudomorphs, marly dolomite and shales, and is interpreted as having been deposited in carbonate sabkhas. The Alfara Mb (15–25 m thick) consists of laminated dolomudstones–dolowackestones–dolopackstones with oolites, current ripples and cross-stratification showing coarsening and thickening-upward cycles, and is interpreted as lagoonal deposits developing into the bioclastic rim of a ramp. The La Figuera Fm (10–15 m thick) lies in sharp contact with the Alfara Mb and consists of laminated dolomudstones containing calcite-infilled evaporite pseudomorphs and some local thin bivalve shell horizons, and has been interpreted as a coastal sabkha. The age of the Imón Fm is latest Alaunian–middle Sevatian (latest Norian *sensu stricto* lower Rhaetian *sensu lato*) based on foraminifera associations (see discussion in López-Gómez *et al.* 1998).

Iberian Ranges

As seen in the Catalonian Coastal Ranges described above, Middle and Upper Triassic sediments in the Castilian branch of the Iberian Ranges mostly show the same tripartite carbonate succession (lower and upper Muschelkalk and Imón Fm) separated by two mudstone–evaporite horizons (Fig. 10.4). In the Aragonese branch of the Iberian Ranges, however, the lower Muschelkalk carbonates are only found in the extreme SE corner, around Montalbán (Fig. 10.5). As with other areas in Spain, sediments of this age record successive Neotethys sea transgressive–regressive cycles that onlapped the western and southern margins of the Palaeozoic Iberian Massif.

Castilian branch. The study of the Muschelkalk facies in the Castilian branch of the Iberian Ranges has yielded important results over the last 30 years (e.g Hinkelbein 1969; Boulouard & Viallard 1971; Viallard 1973; Marin 1974; Hernando 1977; Doubinger *et al.* 1977; Sopeña 1979; Ramos 1979; Visscher *et al.* 1982; Virgili *et al.* 1983; López-Gómez 1985; Pérez-Arlucea 1985; Márquez-Aliaga 1985; Sopeña *et al.* 1988, 1995; García-Gil 1991; Sánchez-Moya 1992; López-Gómez & Arche 1993; López-Gómez *et al.* 1993, 1998). The lower Muschelkalk (which pinches out in the area between Molina de Aragón and Cueva de Hierro) was defined as the Albarracín Dolostones and Marls Fm (ADM) in the Albarracín area (Pérez-Arlucea & Sopeña 1985) and as the Landete Dolomites Fm (LD) in the Boniches area (López-Gómez & Arche 1986; Fig. 10.4). Correlation between the different members of these formations was proposed in López-Gómez *et al.* (1998). The unit has been dated as Anisian

(Pelsonian–Illyrian; Pérez-Arlucea & Trifonova 1993; Márquez *et al.* 1994; Sopeña *et al.* 1995; López-Gómez *et al.* 1998) based on pollen and spore associations, foraminifera and ammonite criteria. Reaching a thickness of 85 m in the Molinar section, the unit lies unconformably on the Cañizar Fm in the central and NW areas or on the Rillo de Gallo (RGS) and Prado Beds (PB) in the SE, and conformably on the Röt facies (Marines Fm, MCMM) near the present Mediterranean coastline. This unit comprises mainly dolomites that record a general transgressive–regressive cycle developing from bioclastic bars to restricted lagoon carbonate mudstones (representing the deepest sediments), and then to intertidal stromatolites and sabkha deposits.

The overlying middle Muschelkalk mudstone–evaporite unit was rheologically weak enough to provide a suitable detachment surface during Alpine compressive tectonics, so that its detailed stratigraphy remains poorly known. It was defined as the Mas sandstones, marls and gypsum (MSMG) in the Boniches area (López-Gómez & Arche 1986), and Tramacastilla mudstones and gypsum (TMG) in Tramacastilla, close to the Albarracín area (Pérez-Arlucea & Sopeña 1985; Fig. 10.4). Many of the logs have been obtained from well-log cross-sections (Castillo-Herrador 1974; Ortí *et al.* 1996; IGME 1987). Correlations between well-logs and surface outcrops are almost impossible due to dissolution processes affecting the evaporite sediments. Mudstones, and even sandstones, dominate most of the outcrops of the central and western areas, while evaporites are common towards the east but grade into carbonates that constitute part of the single carbonate Muschelkalk unit between the Boniches and Eslida sections (Fig. 10.4). In general terms, from bottom to top, this unit represents a regressive–transgressive stage which starts at the end of the carbonate unit of the lower Muschelkalk and shows sedimentary continuity with the carbonate unit of the upper Muschelkalk. To date, however, its general palaeogeography has been difficult to establish, although there is a clear continental influence towards the west. The age of this unit, determined using pollen and spore assemblages, is upper Anisian to lower Ladinian (Doubinger *et al.* 1990; Sopeña *et al.* 1995).

The carbonates of the upper Muschelkalk were deposited right across what are now the Iberian Ranges (Fig. 10.4). This unit has been defined as the Tramacastilla Dolomites Fm (TD) and Royuela Dolomites, Marls and Limestones Fm (RDML) for the lower and upper part respectively, close to the Albarracín area (Pérez-Arlucea & Sopeña 1985), and the Cañete Dolomites and Limestones Fm (CDL) close to the Boniches area (López-Gómez & Arche 1986). Correlation between the different members of these formations was proposed by López-Gómez *et al.* (1998). It is of Ladinian age (upper Fassanian–Longobardian) based on ammonites, foraminifera, conodonts and pollen and spore assemblages (Doubinger *et al.* 1977, 1990; Boulouard & Viallard 1982; Márquez-Aliaga 1985; Pérez-Arlucea & Trifonova 1993; Goy 1995; Sopeña *et al.* 1995; López-Gómez *et al.* 1998). This unit reaches 120 m in thickness in the Albarracín area but is reduced to 46 m in the Atienza area in the NW (Fig. 10.4). The lower contact oversteps Buntsandstein rocks in the NW, but elsewhere there is a normal vertical transition from the middle Muschelkalk. The upper contact is represented by a short sedimentary hiatus below the overlying Keuper facies. These carbonate sediments broadly represent a transgressive–regressive cycle, which developed initially from intertidal environments to shoals and protected lagoons in the middle part of the unit, back to intertidal–supratidal environments at the top.

The overlying Keuper facies rocks form a succession of mudstones, marls and gypsum. As they are rheologically weak, they have (like the mudstone–evaporite unit of the middle Muschelkalk) proven susceptible to Alpine deformation, so that their thickness is difficult to assess (probably in the range 60 to 250 m). These sediments represent an advanced stage of basin infilling, marking the development of a sabkha environment with continental siliciclastic influence from the NW. The age of these sediments is lower Carnian–upper Norian based on pollen and spore assemblages (Doubinger *et al.* 1990; Sopeña *et al.* 1995; López-Gómez *et al.* 1998). This unit was defined for the Pálmaces area as Los Gavilanes clays and gypsum (GCG; Fig. 10.4; Sopeña 1979).

The uppermost epicontinental marine carbonate formation is referred to as the Imón Fm (Goy *et al.* 1976) in both the Iberian and Catalonian Coastal Ranges, and is the equivalent of the Isábena Fm previously described for the Pyrenees. It consists of up to 35 m of well-stratified dolomites, with gradational contacts both at the base (Keuper) and top (Cortes de Tajuña Fm; Goy *et al.* 1976), and is subdivided into three unnamed members. The lower member consists of alternating dolomudstones and dolomitic breccias interpreted as coastal sabkha or carbonate–evaporite tidal flat deposits. The middle member consists of laminated dolomudstones–dolowackestones, and dolowackestones–dolopackestones with oolites and peloids, and is interpreted as recording the transition from a confined lagoon to the bioclastic rim of a ramp. The upper member comprises dolomudstones with evaporite pseudomorphs and stromatolitic domes, and is interpreted as sabkha deposits. The age of the Imón Fm is upper Norian (Goy & Márquez-Aliaga 1998; Gómez & Goy 1997b; López-Gómez *et al.* 1998).

Aragonese branch. The Muschelkalk facies succession in the Aragonian branch shows two carbonate units separated by a siliciclastic-evaporitic unit towards the east in the Montalbán area (Fig. 10.5), whereas elsewhere it is solely represented by the upper carbonate unit. Early general descriptions of the Muschelkalk of this area include Donayre (1873), Dereims (1898), Würm (1911), Tricalinos (1928), Lotze (1929), Richter (1930), Richter & Teichmüller (1933) and Vilas *et al.* (1977). The lower carbonate unit (LCU) was first described in detail by Meléndez *et al.* (1995) and was placed in what they called the 'lower sequence', which also includes the upper part of the Buntsandstein succession (Röt facies) which is separated from the LCU by a a ferruginous surface. The LCU can be well-correlated across the Montalbán area (Fig. 10.5) reaching about 58 m in thickness and pinching out towards the west. It consists of well-stratified dolomites that show bioturbation, parallel lamination, planar cross-stratification, stromatolites and tepees, and mudcracks at the top of the unit. Close to the Herrera de Los Navarros section, this unit is separated from the Palaeozoic basement by only a few metres of Buntsandstein facies sediments. The LCU unit is interpreted as recording the transgressive–regressive development of a carbonate ramp with different subenvironments changing from subtidal to supratidal (Meléndez *et al.* 1995).

The siliciclastic-evaporitic unit that separates both Muschelkalk carbonate units, termed the 'middle level' (ML) by Meléndez *et al.* (1995), corresponds to the Trasobares Mudstones and Marls (TMM; Arribas 1985) in the Moncayo area, and lies directly on the Palaeozoic basement close to Ateca village (Fig. 10.5). This unit is correlated with the Mas Fm (López-Gómez & Arche 1993) in the Castilian branch of the Iberian Ranges (Fig. 10.4), and with the 'middle red interval' (Virgili 1955) of the Catalonian Coastal Ranges. This unit is mainly formed of red marls and clays, although intercalated thin levels of sandstones, gypsum and dolomites can also appear. Its thickness varies from about 45 m in the Montalbán area to only 7 m in the Paniza section of the Moncayo area (Fig. 10.5). It is interpreted as recording transitional marine–continental deposition with increasing marine dominance towards the east.

Deposition of the upper carbonate unit (UCU; Meléndez *et al.* 1995) extended over the whole Aragonese branch (Fig. 10.5). This unit corresponds to the basal dolomites (BD) and upper dolomites and marls (UDM) defined by Arribas (1985) in the Moncayo area (López-Gómez *et al.* 1998). The unit consists of nodular, well-stratified and laminated dolomites with alternating centimetric to metric levels of grey and green marls in the upper half of the unit. Planar and trough cross-stratification and bioclastic levels are frequent in the lower part of the unit, and algal laminations, tepee structures, mudcracks and bioturbation are very common in the upper part. Its thickness can reach 120 m in the Anadón section, located between the villages of Montalbán and Monforte, but decreases to 40 m, 4 km west of Ateca (Fig. 10.5). This unit is interpreted as recording the transgressive–regressive development of a carbonate ramp with different subenvironments that range from subintertidal in the lower part to intersupratidal in the upper part (García-Royo & Arche 1987; Meléndez *et al.* 1995). The age of this unit is Ladinian, based on ammonoids, pollen and spore associations, foraminifera and conodonts taken mainly from areas close to the Castilian branch (see review in López-Gómez *et al.* 1998).

Keuper facies rocks show very poor outcrops in the Aragonese branch so that, in comparison to the adjacent Ebro basin, there have been few detailed investigations. Only general works, such as those of Del Olmo *et al.* (1983), García-Royo & Arche (1987), Sopeña *et al.* (1988), Lendínez *et al.* (1989) and Salvany (1990), have described the succession. Salvany (1990) considered that most of the Aragonese branch represented an elevated area during the deposition of the Keuper sediments, as previously did Marin (1974). The thickness of this unit shows substantial lateral change, varying from <45 m in the Ateca area to about 190 m in the Moncayo area (Del Olmo *et al.* 1983). It is mainly composed of grey, green and yellow mudstones with intercalated evaporite levels and sporadic sandstone and dolomite levels. Close to the upper contact of this facies, Bastida *et al.* (1989) described a 13 m thick level of alkaline basalts in Arándiga, 38 km north of Ateca, related to intracontinental rifting. The Keuper facies is mainly interpreted as representing sabkha deposits with more continental influence towards the west in the Demanda–Cameros area (Fig. 10.5). The age of these sediments is Carnian to Norian (Castillo-Herrador 1974; Torres 1990; Gómez & Goy 1997*b*) according to pollen and spore association criteria.

The uppermost carbonate unit was defined as the Imón Fm by Goy *et al.* (1976) (also present in the Catalonian Coastal Ranges, and equivalent to the Isábena Fm of the Pyrenees). The Imón Fm comprises about 18 m of well-stratified dolomites showing a gradual transition from Keuper facies rocks at the base. This unit is laterally widespread, covering almost all the eastern part of the Iberian peninsula (López-Gómez *et al.* 1998), and is divided into three members: lower, middle and upper. The lower member comprises alternating dolomudstones and dolomitic breccias and is interpreted as representing carbonate–evaporite tidal flat deposits. The middle member consists of dolomudstones–dolopackstones with cross-stratification and oolites and is interpreted as recording the development from a restricted lagoon to a bioclastic ramp of a shallow ramp. The upper member consists of dolomudstones with evaporite pseudomorphs and is interpreted as sabkha deposits. The age of the Imón Fm is latest Alautunian–middle Sevatian (latest Norian *sensu stricto*–lower Rhaetian *sensu lato*; López-Gómez *et al.* 1988), and it grades upwards into the overlying Jurassic Cortes de Tajuña Fm (CT) or Lécera Fm (Gómez & Goy 1999).

The Tajo basin

The Tajo basin (sometimes referred to as the Cuenca basin) was formed during Anisian times to the SE of a narrow, linear NW–SE trending Palaeozoic high separating it from the Iberian basin. This high was formed by the elevated Palaeozoic footwall block of the Iberian Basin boundary fault, and became drowned by shallow marine carbonates during Ladinian times so that a single basin was established in eastern Iberia during the Late Triassic epoch. The basin was bounded to the NW by the Iberian Massif, opened eastwards to the Tethys, and was linked with the Betic basin to the south.

The oldest deposits of the basin are found at its northern edge, where 100–140 m of Buntsandstein sandstones lie unconformably on the Palaeozoic basement. These are followed by 9–65 m of shallow marine carbonates, that are easily correlated with the upper part of the siliciclastic continental (Buntsandstein) deposits of the Iberian basin (Arandilla to Torete formations) and the upper Muschelkalk (Cañete or Albarracín-Royuela Fms; Fig. 10.4) in the NW part of this basin.

During Triassic times, the Tajo basin became divided into a more rapidly subsiding area to the east (Cuenca depocentre), and more slowly subsiding areas to the west and south (Fig. 10.12). The Cuenca depocentre contains Keuper facies rocks with well-developed halite and anhydrite intervals that together may exceed 1000 m in thickness and can be subdivided into two major successions (Ortí 1990) separated by a sandstone interval of fluvial origin (K2 unit or Manuel Fm; Ortí 1974). Triassic sedimentation came to an end with the

deposition of dolomites of the Imón Fm, as also seen in the Iberian, Ebro and Catalan basins.

The present-day configuration of Triassic deposits along the western edge of the basin does not reflect their original extension, but rather the effect of partial erosion during Early–Middle Cretaceous times. This western margin exposes only siliciclastic sediments of Carnian–Norian age and continental origin. However, correlation with the classic Keuper deposits to the east of the Iberian peninsula can be made due to the presence of the Manuel Fm that acts as a marker bed (Fig. 10.9; Arche *et al.* 2001).

The southeastern margin of the Tajo basin was of irregular shape and red beds and evaporites infilled a marked palaeorelief. These sediments were not deformed during the Alpine orogeny and have been referred to as the 'tabular cover' of the Meseta or Chiclana de Segura Fm (López-Garrido 1969). According to palynological data, the age of the red beds lies somewhere between Ladinian and latest Norian times (Besems 1981).

Triassic sedimentation began during Ladinian times with conglomeratic deposits of alluvial fans and braided rivers (Fernández 1984; Fernández & Dabrio 1985). Above these conglomerates, a succession of mostly sandstones and clays is interpreted as recording sandy river and isolated playa-lake environments that were laterally related to alluvial environments (Fernández & Gil 1989). Beds of pedogenic carbonates and horizons with abundant iron oxides are also present in these successions (Fernández & Dabrio 1985). The fluvial sequence grades upwards into an upper unit consisting of evaporites and mudstones that is irregularly distributed across this area and is interpreted as sabkha or lagoonal deposits of Norian age. A shallow-marine Norian carbonate unit occurs at the top of some sections, and could be correlated with coeval carbonate units elsewhere in eastern Spain (Fernández & Gil 1989).

Betic domain

Middle–Upper Triassic rocks are well represented in southern mainland Spain and have been intensively studied. They show different facies related to differences in subsidence history between and within the Betic external and internal zones (Fig. 10.9). In general, they record shallow epicontinental facies (mostly tidal flats, shallow platform and ramp), but in some cases they are continental in origin within the Betic external zones. Buntsandstein facies rocks are rare; Muschelkalk facies rocks are mostly limestones and marls, and are overlain by a late Triassic (Norian) carbonate unit, equivalent to the Imón and Isábena formations described above. Within the Betic internal zones, Middle–Upper Triassic rocks are known from all three main tectonic units, and are typified by continental red beds in the Malaguide Complex, marine carbonates in the Alpujarride Complex, and marbles, schists and quartzites in the Nevado–Filábride Complex.

The palaeogeography of the Betic basin during Early Triassic times was the result of rifting and the formation of graben systems that marked the beginning of a long period of regional extensional deformation. Depocentres with different subsidence rates developed on the SE Iberian Massif, and different transgressive–regressive cycles also strongly influenced sedimentation. An initial transgressive stage (Anisian) deposited carbonates in the Alpujárride and the Nevado–Filábride (?) domains (Fig. 10.13). A Ladinian transgressive

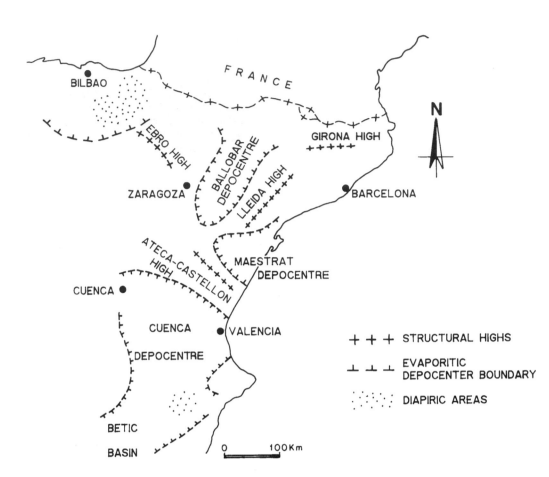

Fig. 10.12. Main subsiding areas during the Middle–Upper Triassic extensional basins in the Iberian peninsula. Modified from Castillo-Herrador (1974), Ortí (1990) and Jurado (1990).

stage can be identified right across the Betic Cordillera, with all areas showing carbonate units of greater or lesser thickness (from 20 m to hundreds of metres). Regression during Carnian times particularly affected domains nearest to the foreland areas (external zone and Malaguide domain of the internal zone). Shallow platform and coastal zone deposition continued in the Alpujarride domain througout this regressive phase. At the end of the Triassic period, during Norian times, the Betic basin reached its maximum extent, and marine conditions spread to the marginal facies at edges of the depocentre. In the areas of greatest subsidence, such as the internal zone, carbonates continued to be deposited until Liassic times.

The marine carbonates of the Betic basin were deposited on a shallow platform and tidal flats of an epicontinental sea which was connected to the Tethys. The fauna of the external zone carbonates is related to the Sephardi Bioprovince (Hirsch et al. 1987), although there may have been some influence of Tethys fauna during Anisian times (Goy & Pérez-López 1996). The Ladinian carbonates from the internal zone show fauna from the Sephardi Bioprovince (López-Garrido et al. 1997) as well as from the Tethys domain, whereas the fauna of the Norian carbonates is mostly derived from the Tethys (Braga & Martín 1987a; López-Gómez et al. 1998). The influence of the Tethys is most obvious in the areas of greatest subsidence during transgressive phases.

Betic external zone. The epicontinental external zone Triassic rocks deposited on the southern margin of Iberia have long been differentiated into Buntsandstein, Muschelkalk and Keuper facies (Bertrand

& Kilian 1889) that linked them to the Triassic in the central European basin (Blumenthal 1927). In the central area of the external zone, one can identify a carbonate formation of the Muschelkalk facies (Majanillos Fm), five detrital and evaporitic formations which constitute the Jaén Keuper Gp (Pérez-López 1991), and one upper carbonate formation of Norian age (Zamoranos Fm; Fig. 10.7; Pérez-López et al. 1992). All these formations show significant variation in thickness and lithofacies depending on local palaeogeography (Martín-Algarra 1987; Martín-Algarra et al. 1995).

Rocks attributed to probable Buntsandstein facies crop out in very few places, and have been definitively identified below the Muschelkalk carbonates only in the northernmost sector of the basin (external Prebetic; Fig. 10.7) near the Meseta border, and in several places in Murcia province (southern internal Prebetic; Pérez-Valera et al. 2000). The unit consists of red and grey mudstones, thin layers of gypsum, and sandstones with mudcracks together interpreted as coastal plain deposits. These rocks are probably equivalent to the upper part of the Buntsandstein, although the possibility exists that they are instead coeval with the middle Muschelkalk of the Iberian Ranges, this being especially likely for the southern Prebetic outcrops.

The Muschelkalk deposits consist of a carbonate succession increasingly dominated by marls towards the upper part. Three main members may be identified in the Majanillos Fm in the Subbetic domain. The lower member has been broadly interpreted as represent in a deepening-upward sequence in which subtidal and ramp deposits can be identified. The middle member presents a predominance of marly facies deposited on a shallow platform on which there are frequent storm deposits. The facies belonging to the higher member shows lagoonal and tidal flat deposits that correspond to a muddy shallowing-upward sequence. The Keuper deposits are characterized by variegated shales, sandstones, gypsum and sometimes by basic intrusive rocks (Pérez-López 1991). Although none of the outcrops presents a complete section, five different lithostratigraphic units have

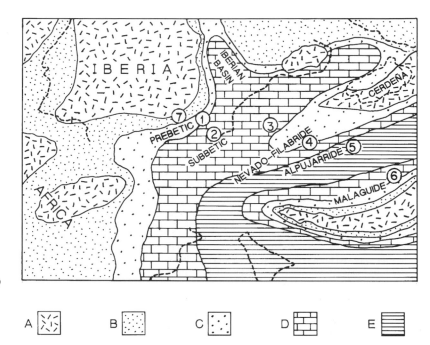

Fig. 10.13. Palaeogeographical reconstruction of the Betic basin for Early Triassic times showing the location of the different domains and their deposits. Key: A, positive areas; B, continental siliciclastic sediments; C, coastal deposits; D, shallow marine deposits; E, deep marine deposits. See also Figure 10.9 for the domain locations. Number 7 indicates the location of the southern margin of the Iberian Massif (see section in Fig. 10.6). Modified from Martin–Algarra *et al.* (1995).

been distinguished in the Subbetic domain (Jaén Keuper Gp). The thickness and facies of these units varies from the internal Subbetic to the Prebetic domain. The units can be correlated with the Carnian–Norian formations that crop out in the eastern Iberian peninsula (units K1 to K5; Ortí 1974). The lower one (unit K1) is mainly formed by clays and evaporites, although sandstone beds are sometimes common. The rocks of this unit can be attributed to a fluvial-coastal environment, with widespread development of lakes and salt pans. The second unit (unit K2) consists mainly of sandstones, interpreted as the deposits of a fluvial environment and terminal alluvial system. The third unit (K3) consists of red clays with gypsum, interpreted as a saline mud flat with environments of sabkha and lagoon deposits (Pérez-López 1996). The last two units are mainly composed of gypsum, the fourth one (K4) comprising clay with nodular gypsum, and the upper one (K5) characterized by laminated gypsum with dolomite beds.

A Norian carbonate unit (known as the Zamoranos Fm in the Subbetics) has been identified above the K5 unit and consists of bedded dolomites and limestones with red siliciclastic intercalations. This carbonate unit, correlated with the Imón Fm and Isabena formations elsewere in Spain (López-Gómez *et al.* 1998), is of variable thickness, and records facies ranging from shallow platform to tidal flat with thin continental intercalations as one moves from the Subbetic to Prebetic areas (Pérez-López *et al.* 1992). This facies is interpreted as shallow platform and tidal flat deposits with thin continental facies intercalations.

The Carcelen anhydrite unit (Ortí 1987) is situated above the Zamoranos Fm. It is an evaporitic unit which shows only gypsum, carbonate rocks and shales in its exposures, but in boreholes is revealed to contain anhydrite, shales and dolomite beds with important intercalations of halite. This unit crops out in several places in the central part of the Subbetic zone and in the eastern part of the Prebetic zone, although it is very difficult to recognize since some of its facies are similar to those of the K5 unit. The Carcelén unit is interpreted as recording lagoonal and coastal sabkha sedimentation in a palaeogeographical location similar to that of the K5 unit.

Betic internal zone. Malaguide complex Middle–Upper Triassic rocks are mostly continental red siliciclastic sediments lying unconformably on a Palaeozoic basement. Soediono (1971), Roep (1972) and Geel (1973) called these deposits the Saladilla Fm (Fig. 10.9), and Martín-Algarra (1995) extended recognition of this formation over the entire Malaguide domain and into its equivalent Rifian domain (Gomaride)

in North Africa. Three members may be differentiated, with the lowest showing a fining-upward sequence of conglomerates, sandstones and clays. The second member comprises carbonates (limestones and dolostones) which are variable in thickness and can be attributed to the Ladinian. The upper member consists of marls and thin gypsum beds, and above this a dolostone formation overlain by Liassic limestones has been described (Roep 1972). These dolostones show good stratification in their lower part where thin layers of sandstones and conglomerates have been attributed to the Rhetian (Martín-Algarra 1995).

The Middle–Upper Triassic rocks of the Alpujarride complex crop out extensively and are mainly represented by carbonate marine facies of variable thickness (Delgado *et al.* 1981; Braga & Martín 1987a). The lower part of this carbonate succession comprises Anisian–Ladinian interbedded dolostones, marly limestones and limestones with phyllites, and are interpreted as lagoonal deposits with patch reefs mostly of dasycladacean facies (Braga & Martín 1987a,b; Martín & Braga 1987b). The marls and clays of the middle part are interpreted as coastal flats and represent a transition to the dolostones of the upper (Norian) part of the succession, which are considered as having been deposited in tidal, lagoon and barrier environments. Overlying Rhaetian sediments are represented by limestones deposited on a shallow platform. A sedimentary transition can be observed between successions of the continental facies of the Malaguide complex and the marine successions of the Alpujarride complex. For this reason, intermediate units have been defined between the two complexes in some outcrops (Sanz de Galdeano 1997; Sanz de Galdeano *et al.* 2001). Finally, the metamorphosed middle–upper Triassic rocks of the Nevado–Filábride complex are marbles with schists and frequent quartzites, and intrusions of metabasic bodies, but their lithostratigraphy is poorly known.

Balearic Islands

The Balearic Islands show a Middle–Upper Triassic stratigraphy similar to the external zone Triassic of the Betic Cordillera, although deeper marine facies are more common (Fig. 10.9). During Middle–Late Triassic times, shallow marine carbonates, marls and evaporites were deposited around the eastern and southern margins of Iberia in a series of transgressive–regressive cycles. Brief periods of continental, alluvial sedimentation, took place during early Ladinian times in Catalonia and early Carnian times in SE Iberia. As seen over most of Europe, the Balearic succession can be sub-divided into a lower carbonate unit or Muschelkalk facies and an upper evaporitic-siliciclastic unit or Keuper facies, capped by shallow water marine carbonates of late Norian age.

The Middle Triassic (*pro parte*) succession of the Balearic Islands (the middle carbonate unit of Rodriguez-Perea *et al.* 1987; ?late Anisian–Ladinian) consists of an epicontinental, marine carbonate interval of Levantine–Balearic Type (López-Gómez *et al.* 1998) described by Darder (1914), Virgili (1952), Bourrouilh (1973), Colom (1975), Rodriguez-Perea *et al.* (1987) and Llompart *et al.* (1987) among others. Some authors (Bourrouilh 1973; Colom 1975; Rodriguez-Perea *et al.* 1987) describe a thin middle level of marls and red mudstones in parts of the Muschelkalk of Majorca, but some of these outcrops could be Keuper slices in between thrust sheets because of the structural complexities in the northern sierra (Gelabert *et al.* 1992).

The Muschelkalk facies rocks of Minorca have to date yielded the richest ammonoid faunas in Spain. The lower part lies conformably on the Buntsandstein and consists of 32 m of grey micritic limestones (Llompart *et al.* 1987). Bourrouilh (1973) found *Pseudomonotis schmidti* in these levels, of possible Anisian age. The unit is capped by a highly bioturbated, micritic limestone rich in chert nodules.

The intermediate unit consists of micritic limestones and marls horizons, 76 m thick, of intertidal to inner shelf origin. Four ammonite and different bivalve levels have been found in this unit (Llompart *et al.* 1987) ranging from *Israelites* cf. *ramonensis* (early Ladinian) and *Eoprotrachyceras curionii* and the first *Daonella lommelli* (also early Ladinian) to the classic *Protrachiceras hispanicum–Daonella lomelli* (late Ladinian) and *Trachiceras aon–Costatoria goldfussi, C. decussata* association (late Ladinian). Foraminifera such as *Lamelliconus procerus, Aulotortus praegaschei* and *Triadodiscus eomesozoicus*, of late Ladinian age (Vachard *et al.* 1989), have been found associated with *Daonella lomelli*, and pollen like *Triadispora, Ovalipollis, Camerosporites* and *Paracirculina* (late Ladinian).

The upper unit consists of yellowish dolomites 10–20 m thick, that have not yielded fossils up to now, but probably correspond to an uppermost Ladinian age. The intermediate and upper units can be correlated with the M-3 or upper Muschelkalk of the Catalonian Coastal Ranges and the eastern Iberian Ranges (López-Gómez *et al.* 1998), but doubt remains over the age of the lower unit.

The Muschelkalk facies rocks of Majorca are comparable to those of Minorca but a marl, dolomitic breccia and red mudstone level is found at the top of the lower unit in the latter island. This facies has been described in the classic papers of Darder (1914) and Virgili (1952), and Colom (1975) provides a good compilation of these papers and many others.

Modern studies are very few, but Rodriguez Perea *et al.* (1987) review the available sections and propose correlations with the Catalonian Coastal Ranges, and Álvaro & Del Olmo (1992) and Álvaro *et al.* (1992) present a detailed geological map of Majorca and a revision of Muschelkalk facies rocks stratigraphy. These latter authors subdivided the Middle Triassic carbonates in Majorca into four units that, in a broad way, can be correlated with the Minorca sections. A basal dolomite level, 80 m thick, is followed by solution breccias, cryptalgal laminated dolomites and mudstones that together may be equivalent to the middle Muschelkalk (or M-2 from Virgili 1958) of Catalonia (Fig. 10.11). The succession continues with two upper units, the first one showing a shallowing-upward sequence, equivalent to the intermediate unit of Minorca, where Virgili (1952) found ammonites of late Ladinian age (*hispanicum* zone) and *Daonela lomelli*. The uppermost unit comprises marls, grey dolomites, gypsum levels and some foraminifera like *Frandicularia woodwardii*.

The Middle Triassic carbonates of Ibiza are poorly known. They crop out at the base of the Alpine thrust units and their age is not proven. Rangheard (1972) and Rodriguez-Perea *et al.* (1987) subdivide these facies into two units: the base is never exposed and the sections begin with 35 m of dark grey dolomites with fenestral porosity and cryptalgal lamination, followed by 80 m of dark grey, bioturbated dolomites with 'zebra' horizons. The upper part of this section consists of *c.* 70 m of laminated limestones, nodular bio-turbated limestones and marls.

The Late Triassic sediments of the Balearic Islands, as also seen all over the Iberian peninsula, comprise red mudstones, evaporites (or upper silty unit, from Rodriguez-Perea *et al.* 1987) and, locally, volcanic rocks capped by dolomites, solution breccias and marls (Keuper facies rocks and Imón Fm, respectively; Rodriguez-Perea *et al.* 1987). They crop out all over the islands, but are heavily distorted by alpine compressional events, and can be distinguished from the Keuper facies elsewhere in the Betic domain due to their lack of sandstone beds.

The Majorca outcrops, both in the Serra de la Tramuntana and the Serra de Levant, show a thick Keuper facies succession up to 300 m thick (Álvaro 1987). It consists of red mudstones and marls lying conformably on the Muschelkalk facies, with fine-grained sandstones, dolomites and solution breccias, and massive gypsum horizons in the uppermost 40 m. There are numerous sills and lava flows intercalated in the series (Navidad & Álvaro 1985), with picritic basalts at the base evolving into olivine basalts towards the top. Both submarine lava flows and pyroclastic horizons are found, and secondary spilite horizons are present in the fine-grained facies. They are interpreted as alkaline basalts emplaced in an intraplate geodynamic regime.

The upper carbonate part of the Upper Triassic succession (Imón Fm) consists of 20 m of laminated limestones and dolostones, and massive dark grey dolostones. The laminated interval has been dated as Norian by Boutet *et al.* (1982) on the basis of a pollen assemblage containing *Corollina meyeriana, Triadispora* sp., *Tsugaepollenites pseudomassulae* and *Ovalipollis ovalis*. Finally, the Keuper facies rocks of Minorca and Ibiza are much thinner (Rangheard 1972; Rodriguez-Perea *et al.* 1987) and consist of red mudstones, gypsum horizons and rare basaltic sills, evolving towards the top into the same dolomites as seen in Majorca.

This work was funded by Projects PB 98-0488 and PB 97-1201 of the DGICYT, Ministerio de Ciencia y Tecnología, Spain, and Grupo de Investigación de Junta de Andalucía RNM 0163. We thank F. Calvet, A. Márquez-Aliaga, L. Márquez and A. Goy for constructive comments on the paper and field support and A. Burton for the English revision. We are indebted to W. Gibbons, M. Marzo and S. J. Jones for helpful comments while reviewing the manuscript. We express sincere gratitude to T. Moreno for her patience and cordiality.

11 Jurassic

MARC AURELL, GUILLERMO MELÉNDEZ, FEDERICO OLÓRIZ
(coordinators), BEATRIZ BÁDENAS, JESÚS E. CARACUEL,
JOSÉ CARLOS GARCÍA-RAMOS, ANTONIO GOY, ASUNCIÓN LINARES,
SANTIAGO QUESADA, SERGIO ROBLES,
FRANCISCO J. RODRÍGUEZ-TOVAR, IDOIA ROSALES, JOSÉ SANDOVAL,
CESAR SUÁREZ DE CENTI, JOSÉ M. TAVERA & MARTA VALENZUELA

At the beginning of the Jurassic period, southern European areas formed a single continental mass open to the east (western Tethys), and the Iberian plate lay between latitude 25°N and 35°N. It was separated from the larger European plate to the north by a narrow trough corresponding to the early rifting of the Bay of Biscay. To the NW it was separated from the Laurentia–Greenland Plate by an epicontinental sea showing a typical horst and graben structure, which would eventually become the palaeogeographical connection between the northern and central Atlantic. The opening of the Bay of Biscay took place between latest Jurassic and early Campanian times, giving rise to SE-directed movement and anti-clockwise rotation of the Iberian plate (e.g. Ziegler 1988b; Osete et al. 2000).

Jurassic palaeogeography was characterized by a large part of the central and western Iberian plate forming an emergent massif (the so-called Iberian Massif), whilst the surrounding areas were occupied by intracratonic basins that formed shallow epicontinental seas, predominantly filled with marine carbonate deposits (Fig. 11.1). Those areas, located to the north and NE of the Iberian Massif, correspond from west to east to Asturias, the Basque-Cantabrian basin, and the South Pyrenean basin. To the east extended the Iberian basin, whereas the southern margin of the Iberian Massif was occupied by a wide carbonate platform parallel to a narrow oceanic trough connecting Tethys with the central Atlantic Ocean. These areas together comprised the south Iberian margin basin, whose proximal area, close to the Iberian Massif, formed the Prebetic shelf system, whereas further south the Subbetic basin and Balearic shelf and basin were developed. The principal stratigraphic features and sedimentary evolution of Iberian Jurassic basins are described in this chapter starting in the north with Asturias and then moving clockwise around east Iberia to the Betics. Also described are Jurassic deposits belonging to southernmost terranes (the Alboran domain) which collided with the Iberian margin during Alpine convergence between Africa and Iberia.

Asturias (GM, JCG-R, MV, CS, MA)

Jurassic outcrops in the region of Asturias (Cantabrian coast, north Iberia) form a patchy pattern across the northern sector of the province, north of the Palaeozoic core of the Cantabrian mountains. According to Ramírez del Pozo (1969), Mesozoic outcrops define three main sedimentary domains, which from north to south are: the Gijón–Villaviciosa basin, along the present coast, the so-called 'intermediate mobile fringe', and the Oviedo trough. The classical outcrops are those of the

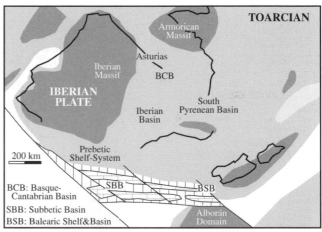

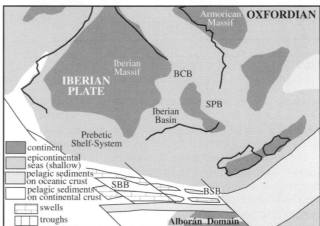

Fig. 11.1. Palaeogeographic and palinspastic reconstructions of the Iberian Plate during the Toarcian and during the Oxfordian, indicating the different sedimentary domains of east and south Iberia. Modified from Vera (1998).

Gijón-Villaviciosa basin and form a near-continuous band along the Cantabrian coast, between the localities of Gijon and Ribadesella (Fig. 11.2). The basic lithostratigraphic framework for the Jurassic units was first established by Suárez-Vega (1974) and subsequently modified by Valenzuela et al. (1986). More specific and detailed works were published by Fernández-López & Suárez-Vega (1979), Valenzuela et al. (1985, 1989, 1992) and García-Ramos et al. (1992). An updated

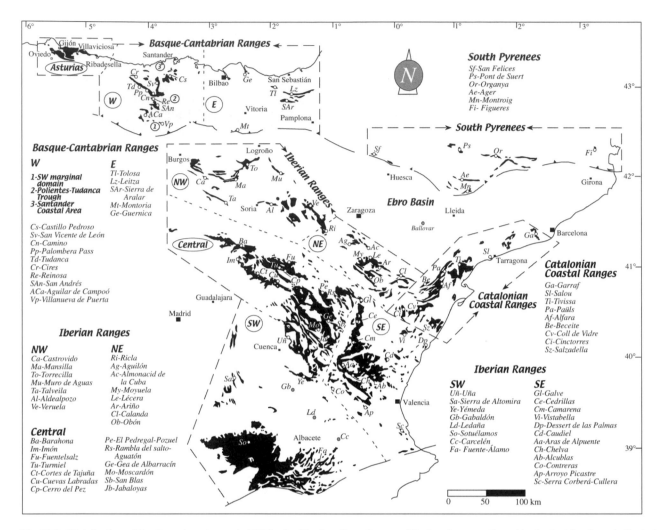

Fig. 11.2. Distribution of the Jurassic outcrops in NE Spain. The location of some of the key logs mentioned in the text has been indicated.

synthesis of Jurassic stratigraphy has been presented by García-Ramos & Gutiérrez-Claverol (1995), and the entire succession can be viewed as comprising two important depositional megasequences bounded by major unconformities.

The lower megasequence corresponds to the Villaviciosa Group (Gp), which ranges in age from Hettangian to early Bajocian, is dominated by marls and limestones (Fig. 11.3). Tectonic factors played an important role in the distribution of facies and thickness in this megasequence from middle to late Toarcian times, when extensional faulting became increasingly evident, especially during the Aalenian to early Bajocian interval. This led to the break-up of the platform into blocks, producing important irregularities in the basin, as is evident from the important lateral variations of facies and thickness in the Aalenian to lower Bajocian deposits (Fernández-López & Suárez-Vega 1979). This process was probably connected with an early opening phase of the Bay of Biscay and is also reflected in the central North Sea by a general process of doming and subsequent development of a rift basin.

A tectonic episode, probably around the Middle–Late Jurassic boundary, was responsible for doming, uplift, and sub-aerial exposure of the whole basin followed by an initial process of rifting. This led to the development of an irregular horst and

graben relief, and the intense erosion of Lower and Middle Jurassic deposits. The sedimentary infill of the newly created continental basin with siliciclastic sediments marks the beginning of the upper megasequence, which corresponds to the Ribadesella Gp. These littoral and continental environments were quickly colonized by numerous vertebrate groups, such as dinosaurs, turtles, crocodiles, flying reptiles and fish. The Ribadesella Gp is separated from the underlying units by an erosional disconformity (Fig. 11.3). It comprises an essentially terrigenous succession, which may be subdivided into four units: the La Ñora, Vega, Tereñes, and Lastres formations.

Early–Middle Jurassic. The lowest unit, the Gijón Formation (Fm), comprises a evaporitic, dolomitic, carbonate complex up to 100 to 150 m thick. The recorded facies associations indicate a sabkha to hypersaline coastal lagoon environment, gradually evolving towards a barrier-lagoon and microtidal flat system. Marine influence is indicated very early by the rare presence of ammonites. A single specimen of *Psiloceras (Caloceras) pirondi* (Reynes) from the lower Hettangian, Planorbis Zone, has been recorded (Suárez-Vega 1974). Above this point, no further biostratigraphic interval has been characterized by ammonites in this unit, until the upper Sinemurian part of the succession within the overlying Rodiles Fm.

The Rodiles Fm comprises a marine, rhythmically bedded marl and limestone successsion, deposited on a carbonate ramp at various depths, from above fairweather wave base to below storm wave base.

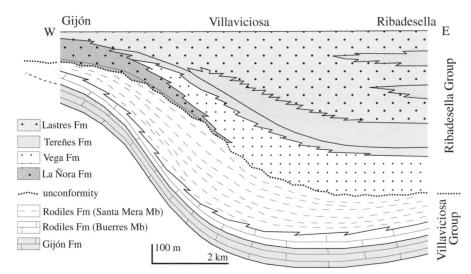

Fig. 11.3. Facies distribution of the Jurassic of Asturias (approximate thickness scale). See Figure 11.2 for location of reference logs.

Ammonites are more common, witnessing open marine conditions between late Sinemurian and early Bajocian times. All ammonite zones from this interval have been identified (Suárez-Vega 1974; Fernández-López & Suárez-Vega 1979), from Obtusum to Sauzei zones. Two members have been defined: a lower Buerres Member (Mb), overlain by the Santa Mera Mb. The Buerres Mb (upper lower Sinemurian–upper Sinemurian) comprises 30 m of alternating marl and limestone interbedding ordered in metric- to decametric-scale shallowing-upward cycles and showing a typical nodular aspect. Tempestitic structures, such as microhummocky and swaley cross-bedding, are common. The Santa Mera Mb (Pliensbachian to lower Bajocian) comprises up to 130 m of regular, well-bedded marls and limestones ordered in metric- to decametric-scale, deepening- or shallowing-upward cycles (Gómez & Goy 2000). Black shale episodes are also recorded in this unit, at the lower Pliensbachian Jamesoni Zone and at upper Pliensbachian Margaritatus Zone respectively (Borrego *et al.* 1996).

Late Jurassic. The La Ñora Fm is mostly composed of siliceous conglomerates derived from the SW, resulting from the erosion of Lower Palaeozoic rocks in a source area located in the West-Asturian-Leonese Zone. These deposits represent an initial phase of infill of a wide palaeovalley orientated SW–NE, leading to the growth of a complex of alluvial fan sedimentary systems.

The Vega Fm, a series of fluviatile deposits up to 160 m thick, constitutes the lateral equivalent of the La Ñora Fm to the NE. It shows complex interbedding of siliceous conglomerates, sandstones and mudstones, forming a series of metre-scale fining-upward cycles, locally with carbonate lacustrine intervals. This unit was deposited on an alluvial plain crossed by ephemeral, high sinuosity rivers under semi-arid climatic conditions favouring the development of calcrete deposits on the floodplains.

The overlying unit, the Tereñes Fm, comprises a fine-grained succession, up to 130 m at the type locality, of dark grey to black mudstones and marls including common carbonate nodules and thin limestone intercalations, as well as calcareous shell beds of gastropods and bivalves and sands in the lower part. To the west, it becomes progressively richer in sand, and can be difficult to differentiate from the overlying Lastres Fm. This unit is interpreted as the infill of a marginal, restricted sub-basin (half-graben) isolated from the open sea by a tectonic threshold that would have prevented the arrival of drifted shells, and/or the biogeographic dispersal of macrofauna.

Finally, the Lastres Fm comprises over 500 m of interbedded sandstone and mudstone with sporadic intercalations of carbonate shell beds and marls similar to those of the Tereñes Fm. Within the Lastres Fm it is worth noting the presence of common wood and other plant remains (Valenzuela *et al.* 1998). Some of these have given rise to exceptionally high quality jet deposits (Blanco *et al.* 1996). This unit resulted from the vertical piling up of small deltaic, fluvially dominated systems leading to the same restricted basin as for the

Tereñes Fm. However, occasional communication with the open sea allowed the incursion of rare ammonites as drifted shells, which have proven to be of great value for dating the series as upper Kimmeridgian, Eudoxus Zone. So far, only three specimens of ammonites have been reported from this interval; the first two, *Aulacostephanus* aff. *eudoxus* (d'Orbigny) and *Aspidoceras* cf. *longispinum* (J. de C. Sowerby), were quoted and illustrated by Suárez-Vega (1974) whilst a third one, belonging to the genus *Eurasenia*, was described by Olóriz *et al.* (1988).

Basque-Cantabrian basin (SR, SQ, IR, MA, GM, BB)

Jurassic rocks of the Basque-Cantabrian basin crop out in two main areas located in the eastern and western parts of the basin, and separated by a large central area occupied by outcrops of Cretaceous and Tertiary sediments. The Jurassic deposits of the eastern area form a series of east–west trending outcrops extending to the north of Pamplona and south of San Sebastián. Some other small outcrops extend south of Vitoria (Montoria) and east of Bilbao (Guernica) (Fig. 11.2). In the western area the outcrops are more abundant and generally better exposed, with the largest being located north and south around Reinosa, as well as in the valleys of the Besaya (San Vicente de León) and Pas (Castillo Pedroso) rivers. Some other scattered outcrops appear in the SW marginal domain (related to the southern frontal thrust system of the basin) and in the Santander coastal area (Fig. 11.2).

The Jurassic succession is represented by two stratigraphic units of different age and sedimentary environment (Fig. 11.4), which are referred to in the regional geological literature as 'marine Jurassic' and 'continental Jurassic' or Purbeck (Rat 1962; Soler y José 1971, 1972*a*,*b*; Robles *et al.* 1989). The lower (marine) Jurassic unit comprises the main part of the succession (Dahm 1966). The upper continental-transitional unit was mainly developed during Early Cretaceous times, although its basal deposits belong to the Jurassic (late Tithonian), at least in the western domain of the basin (Hernández *et al.* 1999).

The lower unit is limited by regional unconformities and is referred to here as the marine Jurassic megasequence (MJM). Both unconformities are related to extensional tectonism, with the first stage occurring at the Triassic–Jurassic transition and linked to Triassic rifting (Quesada & Robles 1995). The second extensional stage, developed during latest

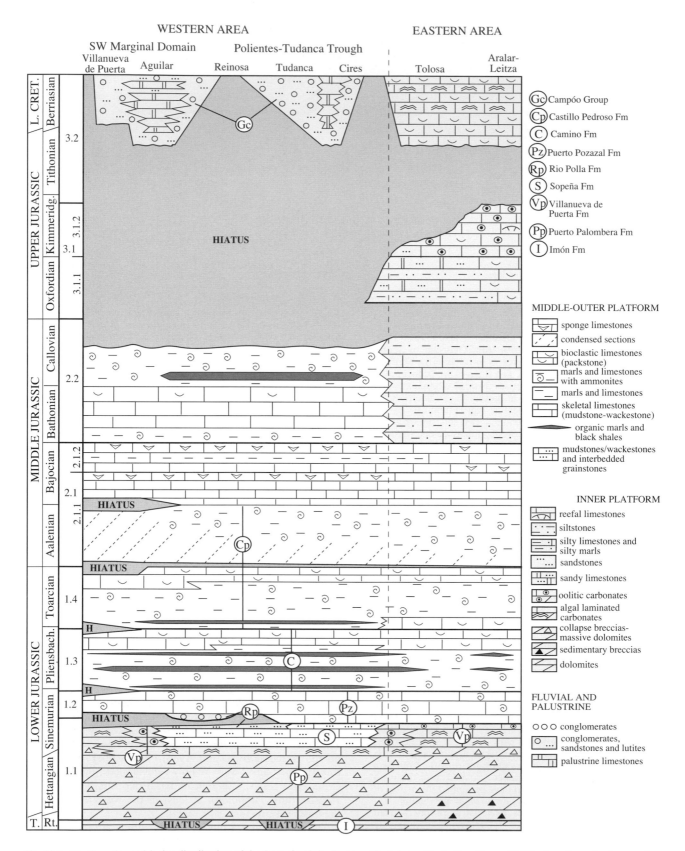

Fig. 11.4. Stratigraphy and facies distribution of the Jurassic of the Basque-Cantabrian Basin. See Figure 11.2 for location of reference logs.

Jurassic–Early Cretaceous times, has been related to early phases of the Bay of Biscay Rift (Robles *et al.* 1989, 1996; Pujalte *et al.* 1996; Hernández 2000). This episode affected not only the evolution of northern Iberia, but also influenced development of the Iberian basin to the east (Salas & Casas 1993). The erosional gap involved in the upper unconformity shows a marked lateral variation from west to east. As a consequence, in the western domain only Lower and Middle Jurassic rocks occur, whereas in the eastern area the Upper Jurassic succession is still partly preserved (Fig. 11.4).

During sedimentation of the MJM the Basque-Cantabrian basin formed a part of a large epeiric sea, bounded by the Iberian Massif to the south and by the Armorican Massif to the north, and occupying an intermediate position between the Tethyan and Boreal realms. Palaeontological data indicate a clear Boreal affinity, although incursions of Tethyan fauna have been recorded at different stratigraphic intervals. Sedimentation of this megasequence took place in a tectonic setting of relative quiescence ('inter-rift stage' of Quesada & Robles 1995), but recorded abrupt thickness variations (450–1000 m) indicating differential subsidence, most probably related to reactivation of extensional faults in the basement (Quesada *et al.* 1991, 1993). The analysis of unconformities and facies cycles has allowed the definition of several unconformity-bounded units of different scales within the MJM. Such units can be generally correlated with those defined in other basins of northern and eastern Iberia. In addition, analysis of the vertical evolution of the type and content of organic matter has allowed the identification of smaller scale transgressive–regressive cycles in the middle and upper part of the Lower Jurassic succession (Quesada & Robles 1995; Quesada *et al.* 1997).

The tectonic extension initiated during Late Jurassic times caused a dramatic change in the tectonostratigraphic and palaeogeographic setting of the basin. This resulted in subaerial exposure and widespread erosion of the MJM and the subsequent deposition of the uppermost Jurassic–Lower Cretaceous megasequence (Purbeck megasequence). The transition from relative tectonic quiescence during the development of the MJM, to the extensional and active tectonic context that characterizes the Purbeck megasequence, reflects the onset of the earliest stage of the Bay of Biscay Rift. The continental and brackish deposits that characterize the Purbeck megasequence were developed in separated basins controlled by the activity of both ancient basement faults and younger listric faults that formed in the Mesozoic cover (Ramírez del Pozo 1971; Soler y José 1971; Pujalte 1989*a*; Pujalte & Robles 1989; Pujalte *et al.* 1996; Robles *et al.* 1996; Hernández *et al.* 1999; Hernández 2000).

Early Jurassic

At the end of Triassic times a wide epicontinental platform was formed along the northeastern margin of Iberia. Late Triassic deposits, attributed to the Rhaetian, are represented by peritidal thin-bedded limestone and dolostones. As seen further east in the Iberian basin, Late Triassic tectonic extension caused the break-up of this extensive epicontinental platform. For this reason in some places (e.g. San Andrés and Camesa) the thin-bedded Rhaetian dolomites are completely eroded (Robles *et al.* 1989). In other areas (e.g. Tudanca) breccias and 'carniolas' at the base of the MJM unconformably overlie early Triassic (Buntsandstein) or even Permian

deposits (García-Mondéjar *et al.* 1986). However, the basal unconformity of the MJM is recorded in many places only by an irregular contact between the lower, Rhaetian dolomite unit and overlying breccia unit. The Triassic–Jurassic boundary has been generally located within the basal breccia ('carniolas') on the basis of lithostratigraphic correlation with its lateral equivalent, the Cortes de Tajuña Fm of the Iberian Ranges (see below).

Sedimentation during the Early Jurassic epoch took place on a wide carbonate platform, connected to the east with the Iberian platform. This platform suffered a progressive drowning, evolving from shallow carbonate ramp environments (inner–middle platform) during the Hettangian–early Sinemurian interval, to a hemipelagic ramp system with development of suboxic environments in the trough areas (middle–outer platform) from early Sinemurian to early Toarcian times.

Early Jurassic differential subsidence was very important, with subsidence rate increasing from early Sinemurian onwards and reaching a maximum during the late Sinemurian–early Pliensbachian interval. This phenomenon generated an arrangement of the basin into troughs, such as the Polientes–Tudanca trough, and swells which show a general orientation from NW to SE. These palaeogeographic elements have been identified in both cross-sections and isopach maps, constructed from surface and well data for the Pliensbachian–Toarcian interval (Quesada *et al.* 1991; Quesada & Robles 1995). In the western part of the basin, troughs and swells coincide with main basement fault systems which were active both during Early Triassic extension (García Mondéjar *et al.* 1986; García-Espina 1997), as well as during subsequent rifting of the Bay of Biscay which took place during the Late Jurassic–Early Cretaceous interval (Hernández 2000). This suggests that intense differential subsidence during middle to late Early Jurassic times might have been linked into the same extensional basement fault system (Quesada *et al.* 1991, 1993).

The upper Sinemurian–Toarcian succession can be considered as a long-term transgressive–regressive cycle. One of the most noticeable phenomena of this period was the development of oxygen-deficient troughs where extensive sedimentation of organic-rich facies took place. They are concentrated in two time intervals, early-middle Pliensbachian and early Toarcian (Quesada & Robles 1995). The lower succession is thicker (up to 100 m in the troughs of the western domain) and more widespread than the upper one, which only occurs in the western domain. In the other basins of eastern and northern Iberia, except for the Asturias basin (García-Ramos *et al.* 1992), sedimentation of similar Early Jurassic organic facies did not occur. However, the early Toarcian anoxic interval has been widely recognized in other Boreal and Tethyan areas (Jenkyns & Clayton 1986; Jiménez *et al.* 1996). Characterization of organic matter has been carried out through total organic carbon (TOC) and visual kerogen analyses (in immature samples), gas chromatography–mass spectrometry, and Rock-Eval pyrolysis. These analyses revealed an organic matter of type II, with TOC values up to 25 wt%, HI (Hydrogen Index) values up to 816, and an organic assemblage dominated by amorphous and algal material, with minor amounts of humic kerogens (Suárez-Ruiz 1987; Quesada *et al.* 1996, 1997). Carbonate stable-isotope analyses of the late Sinemurian–Toarcian cycle show a close coincidence in timing between black-shale deposition, $d^{18}O_{carb}$ minimum and $d^{13}C_{carb}$ maximum, suggesting that warmer sea-surface palaeotemperatures (greenhouse effect) probably enhanced primary organic productivity and water-column stratification (Rosales *et al.* 2001). These palaeoenvironmental conditions, along with an irregular sea-floor palaeotopography, caused stagnation and bottom water anoxia, favouring the preservation of large amounts of organic matter. Early Jurassic organic deposits constitute a hydrocarbon source rock of basinwide extension (Quesada *et al.* 1991, 1993; Garmendia & Robles 1991; Quesada & Robles 1995). They have been genetically correlated with the petroleum of the Ayoluengo oil field, within the Polientes trough. This is the only oil field discovered so far onshore Spain (Quesada *et al.* 1996, 1997).

The upper Sinemurian–Toarcian succession can be divided into three transgressive–regressive cycles dated respectively as late Sinemurian, Pliensbachian and Toarcian. Each cycle is bounded by unconformities, which are better developed in the marginal areas (SW marginal domain) and in the depositional swells, where they are usually represented by paraconformities related to transgressive events.

Hettangian–early Sinemurian. The oldest Jurassic succession, ranging from Hettangian to early Sinemurian in age, was deposited on a shallow water carbonate ramp. It shows a general transgressive trend, evolving from a coastal sabkha environment in the Hettangian stage ('carniolas', or massive dolomites: Puerto de la Palombera Fm) to shallow tidal flats during early Sinemurian times (dolostones and limestones defined as Villanueva de Puerta Fm). These deposits constitute the lateral equivalent of the unit J1 of Soler y José (1971), units L1–L4 of Garmendia & Robles (1991) and units I–III of Bádenas *et al.* (1997) and Gallego & Meléndez (1997) for the eastern part of the basin. The transgression is more evident in northern and central areas of the western domain. In these areas, early Sinemurian limestones (the Sopeña Fm) and sandstones (the Rio Polla Fm), characteristic of a large storm-dominated carbonate ramp (middle platform), overlie the lower dolomitic units (Fig. 11.4; Robles & Quesada 1995). In other basins of northern and eastern Iberia similar middle-platform environments were not developed during early Sinemurian times.

The unconformity at the base of the sequence is represented by an erosional surface related to the Late Triassic extensional stage. This erosional surface was responsible for local removal of Rhaetian dolomites (Imón Fm) and the presence, at the base of the sequence, of polymictic breccias including clasts up to 2 m across (e.g. in Leitza; Gallego *et al.* 1994). The upper boundary shows marked lateral variations: in the SW marginal domain it is represented by an angular unconformity, whereas in the Polientes trough (San Andrés-Reinosa) it grades into a disconformity, and in the eastern and northern areas of the basin the upper boundary is represented by a paraconformity (Fig. 11.4). This unconformity is related to tectonism, occurring at the end of early Sinemurian times, which produced the subaerial exposure and erosion of the SW part of the basin (forced regression), followed by a generalized transgressive event. In the eastern and northern areas the boundary is a transgressive surface, developed on top of the middle–inner platform facies of early Sinemurian age.

Deposits of the lower part of the sequence constitute the Puerto Palombera Fm, and show both facies and thickness variations between outcrop and subsurface. In subsurface the unit comprises dolostones and limestones interbedded with anhydrite beds, dolomites with anhydrite nodules and minor dark marl intervals, generally organized in metric to decametric sequences. In outcrop, however, evaporites are absent, and the unit comprises massive vuggy dolostones and breccias with angular clasts of laminated limestones and dolostones (dissolution collapse breccias). Breccias alternate with limestone and dolomite beds that include laminar and hemispheric stromatolites, which predominate in the base and top of the unit. The unit is interpreted as having formed in an internal platform environment with episodic development of coastal sabkhas and intertidal flats. Thickness in outcrop normally ranges from 80 to 125 m (with the remarkable exception of the Leitza area, where it reaches 300 m). However, in subsurface the unit is thicker, exceeding 250 m in the Polientes Trough.

The transition from the Puerto Palombera Fm to the overlying Villanueva de Puerta Fm (dolomites and limestones) is gradual and diachronous. This suggests that the early Sinemurian transgression was progressively expanding towards the margin, eventually reaching the marginal areas of the basin (the SW marginal domain and the swells of the eastern area; Garmendia & Robles 1991). The Villanueva de Puerta Fm comprises well-bedded limestones in which three main facies associations, commonly organized in shallowing-upward sequences, have been recognized: (a) limestones and dolostones with laminated and hemispherical stromatolites; (b) calcisiltites and mudstones or wave-rippled skeletal wackestones; (c) thin- to thick-bedded oolitic and skeletal grainstones with current ripples and megaripples or hummocky cross-lamination. These facies associations represent inner ramp environments, from muddy intertidal flats to subtidal shoals. The thickness of the unit ranges from 15 to 100 m, the highest values being broadly coincident with the maximum development of the oolitic shoals, as opposed to more northern outcrops, where the thickness is strongly reduced and it grades laterally and vertically into the limestones of the Sopeña Fm (Fig. 11.4).

The Sopeña Fm is a mainly carbonate unit formed in a storm-dominated ramp. The formation represents the transgressive maximum event of the Hettangian–early Sinemurian sequence, and is only developed in the deepest areas of the basin (Polientes–Tudanca trough and Santander coastal domain). It is formed by a 100 to 130 m thick sequence of tabular to nodular limestones (wackestone to packstone) including thin intercalations of skeletal and oolitic grainstones. These intervals show typical structures of storm-current deposits formed in a large ramp located between the fairweather and storm wave base (Robles & Quesada 1995). This ramp opened northwards and was more than 80 km wide, although the absence of open marine fossil remains suggests a partly confined environment.

By the end of early Sinemurian times an initial tectonic uplift pulse in the SW margin of the basin probably induced rapid regression, accompanied by the input of siliciclastic sand deposits which now form the lower part of the Rio Polla Fm sandstones. This unit, like the preceding Sopeña Fm, was only deposited in the western zone (Fig. 11.4) and reaches a maximum thickness of 15 m. It constitutes a complex northward-prograding sequence and three main facies groups can be recognized: (a) sandy limestones showing hummocky cross-stratification and superimposed symmetrical ripples; (b) coarse-grained siliciclastic and bioclastic sandstones showing bimodal cross-lamination and superimposed orthogonal ripples; and (c) limestones with algal lamination. This succession represents a regressive sequence on a storm-dominated tidal flat, recording both subtidal and intertidal environments (Robles & Quesada 1995). In southern areas this unit is truncated by the middle Sinemurian unconformity. Laterally it shows a gradual northward transition to the carbonate facies of the Sopeña Fm (Fig. 11.4).

Late Sinemurian. Late Sinemurian deposition began with a transgressive event that followed an unconformity formed at the early–late Sinemurian boundary. In the SW part of the basin the first transgressive deposits occurred above an uneven subaerial erosion surface and comprise a thin (up to 5 m thick) and irregular unit of siliciclastic conglomerates and sandstones that have calcareous intraclasts. This unit, which comprises the upper part of the Rio Polla Fm sandstones (Fig. 11.4), has been interpreted as the result of the filling of an incised valley excavated on top of the early Sinemurian shallow carbonate platform. Outcrops are limited to the area of the Polientes–Tudanca trough. In the rest of the basin the upper Sinemurian succession is represented by a single carbonate unit made of micritic (mudstone to bioclastic wackestone) limestones and marls known as the Puerto de Pozazal Fm, a unit equivalent to the UTS III of Braga *et al.* (1988), and Unit IV of Bádenas *et al.* (1997) and Gallego & Meléndez (1997). This unit was formed in middle to outer platform areas and has yielded the earliest marine Jurassic fossils, including ammonites, belemnites, brachiopods, crinoids, gastropods and other benthic organisms. The unit shows a highly variable thickness (8–90 m in the west and 5–55 m in the east). These differences may be partly related to the progressive onlap of the base of the unit from more basinal areas (Polientes–Tudanca trough) toward the SW marginal domain (Braga *et al.* 1988; Pujalte *et al.* 1988; Robles *et al.* 1989; Quesada & Robles 1995).

Pliensbachian. Lower and upper limits of the Pliensbachian succession correspond to highly bioturbated, iron-encrusted surfaces suggesting deepening events. The basal unconformity lies in the terminal uppermost Sinemurian (intra-Raricostatum Zone) and is seen right across the basin, whilst the upper limit, located in the uppermost Pliensbachian (intra-Spinatum Zone), is poorly developed in the east. Both unconformities are better developed in the SW marginal domain, where the depositional hiatus is probably longer. The sequence comprises a set of limestones and organic-rich marls that form the Camino Fm (Fig. 11.4). It is equivalent to the UTS IV–VI of Braga et al. (1988), and to the Unit V of Bádenas et al. (1997) and Gallego & Meléndez (1997). It comprises interbedded limestones and marls and includes three organic matter-rich intervals (1–4 wt% TOC; Quesada & Robles 1995). The number of carbonate beds increases upwards: mudstones and wackestones grade upward into bioclastic packstones, and even grainstones. Bioturbation is generally absent in the horizons displaying maximum organic content, but intense *Chondrites* and *Planolites* traces occur in certain limestone beds, increasing upward in the sequence. Fossil remains include belemnites, brachiopods, pectinids, echinoderms, foraminifera (nodosarids and other benthic groups), ostracods and most especially ammonites, which have allowed the recognition of the Pliensbachian zones within this unit (Braga et al. 1985, 1988; Gallego & Meléndez 1997).

The thickness of the sequence ranges from 30 m in the SW margin of the basin to 130 m in the depocentre (Polientes–Tudanca trough), and is less than 80 m in the east. In the west the succession includes up to four black shale horizons located in the early–middle Jamesoni Zone, Ibex to early Davoei Zone, Celebratum Subzone, and Subnodosus Subzone (Braga et al. 1988; Quesada & Robles 1995). Laminated black shales and organic-rich marls are characterized by scarce to absent burrowing and benthic fauna, high-quality hydrogen-rich type I/II kerogens, and an organic assemblage mainly composed of algae/amorphous material with minor amounts of humic kerogens (Quesada & Robles 1995). In the east only the middle early Pliensbachian and upper Pliensbachian black shale horizons were deposited (Garmendia & Robles 1991; Gallego & Meléndez 1997). The thickness of the black shale horizons varies from a few centimetres in the SW marginal domain to 15 m in the depocentre (Polientes–Tudanca trough). The unit was deposited in an outer ramp environment with development of intraplatform troughs where black shales preferentially developed. The outer ramp, located below the storm wave base, shallowed in latest Pliensbachian times to above the storm wave base. As a whole, the unit constitutes a transgressive–regressive cycle. Organic matter-rich intervals were formed under anoxic conditions during successive transgressive events in earliest Pliensbachian, middle early Pliensbachian, and middle late Pliensbachian times. Bioclastic limestones on top of the sequence were deposited during the final, latest Pliensbachian, regressive phase of the cycle (Braga et al. 1988; Quesada & Robles 1995).

Toarcian. This sequence was deposited from the Pliensbachian–Toarcian boundary to the end of the Toarcian stage, and represents a deepening event following the deposition of bioclastic limestones on top of the Camino Fm. It is bounded by unconformities, better developed in the west and particularly in the SW marginal domain (Fig. 11.4). The lower unconformity is marked by a ferruginous crust followed by a rapid increase in the amount of marly facies. This change of facies is particularly sharp in the SW (Pujalte et al. 1988). The upper unconformity is located at the base of the Aalenian (Opalinum Zone) and has been detected in the Polientes–Tudanca trough and the SW marginal domain (Fernández-López et al. 1988b; Pujalte et al. 1988; Canales et al. 1993). The unit is formed by limestones and marly limestones interbedded with grey to black marls, known as the Castillo Pedroso Fm (Fig. 11.4). It is equivalent to Unit VI of Bádenas et al. (1997) and Gallego & Meléndez (1997). In the west these deposits form a carbonate sequence showing dominantly marl at the base, and an upward increase in the proportion and thickness of limestone beds. In the eastern area these deposits show a cyclic trend, with the marly intervals mainly developed in the central part of the unit (Fig. 11.4). Calcareous facies basically comprise mudstone to wackestone, showing common traces of bioturbation (*Chondrites*), with sporadic bioclastic packstone and grainstone intervals, generally at the upper part of the unit. Fossil content comprises ammonites, belemnites, brachiopods, bivalves and nodosariid foraminifera. Detailed palaeontological and biostratigraphical studies based on ammonites in the Polientes–Tudanca trough have allowed identification of all Toarcian zones within the sequence (Fernández-López et al. 1988b). They have also shown the presence of small internal unconformities (Goy et al. 1994) which can be correlated with those defined by Yébenes et al. (1988) in the Iberian Ranges. Finally, a black shale horizon has been identified at the base of the sequence (upper part of the Tenuicostatum Zone), and correlates with an early Toarcian global anoxic event (Jenkyns & Clayton 1986; Rosales et al. 2001). This level has only been identified in the Polientes–Tudanca trough, where it reaches its maximum thickness (1 m), and pinches out towards the SW marginal domain (Quesada & Robles 1995).

The Toarcian succession as a whole shows strong thickness differences in different areas: 40–125 m in the east, less than 25 m in the SW marginal domain, and up to 80 m in the Polientes–Tudanca trough. Condensed sequences in the SW include common redeposited levels which may contain reworked ammonites from late Toarcian times onwards (Fernández-López et al. 1988b). Deposition of the Castillo Pedroso Fm occurred in a middle–outer platform that was temporarily subjected to restricted water circulation and dysaerobic conditions on the sea bottom, as suggested by the fossil record, bioturbation structures and a high content of organic matter (Fernández-López et al. 1988b; Quesada & Robles 1995). Conditions became more oxygenated and shallower in the upper part of the sequence, as well as in the SW marginal domain.

Middle Jurassic

The carbonate ramp of the Basque-Cantabrian basin records Middle Jurassic tectonic instability. The Aalenian sequence in the Basque-Cantabrian basin is generally thicker and more complete than in the other basins, although sedimentary discontinuities, unconformities and condensed or redeposited levels containing reworked fossils are also common, especially in the SW marginal domain (Pujalte et al. 1988; Fernández-López et al. 1988b). In some cases redeposited beds with reworked fragments of vertebrate bones have been reported from inner areas of the basin such as the Polientes–Tudanca trough (Canales et al. 1993). The Bajocian was a more stable interval for the development of carbonate ramps, and deepening and shallowing did not generate large differences in separate areas of the basin. Differential subsidence was progressively attenuated, reaching minimum values in latest Bajocian times, when a spongiolithic shallow platform interval developed, providing an excellent regional correlation level (Robles et al. 1989) that has also been detected in well logs (Quesada et al. 1991, 1993).

During the Bathonian and Callovian stages the western and eastern areas of the basin differed in their sedimentary evolution. The east was characterized by the development of shallow carbonate ramps subjected to the input of fine-grained terrigenous materials, whereas to the west, outer carbonate ramp sedimentation persisted for longer. Differential subsidence between the SW marginal domain and the Polientes–Tudanca trough increased during the Callovian stage. In the western area four deepening–shallowing cycles have been identified, corresponding to the Aalenian–early Bajocian, late Bajocian, Bathonian, and Callovian intervals respectively. Jurassic deposits younger than early Callovian in age are not preserved in this area, most probably because of removal by pre-late Tithonian erosion during the early stages of the Bay of Biscay rifting (Hernández et al. 1999). Marine sedimentation probably continued at least during the rest of the Middle Jurassic, as is suggested by the pelagic characteristics of the youngest facies, and by the more complete Middle Jurassic marine sedimentary record in the eastern area of the basin.

Aalenian–Bajocian. Aalenian to Bajocian deposits are represented by a predominantly carbonate interval showing remarkable differences between the western and eastern areas. In the east this carbonate sequence (Unit VII of Bádenas *et al.* 1997) shows no evidence of unconformity at the Early–Middle Jurassic boundary, so that in the Montoria sector ammonites of the Aalenian Murchisonae and Concavum zones have been reported (Gallego & Meléndez 1997). The thickness of the unit ranges from 100 to 200 m and its lower part comprises interbedded limestone and marly limestone forming a general thickening-upward succession. Minor sequences are capped by thick, bioturbated limestone beds (wackestone to fine-grained packstone with filaments and peloids) containing common traces of *Zoophycos*. In the Aralar range, sponge limestones, locally forming metre-thick bioherms, are common in the upper Bajocian Niortense Zone (Fontana *et al.* 1994*b*). In the west two main sequences are recognized (Fig. 11.4) which respectively correspond to the Aalenian–early Bajocian (sequence 2.1.1) and late Bajocian (sequence 2.1.2).

The base of the first sequence is an unconformity, located at the Early–Middle Jurassic boundary. In the SW marginal domain this boundary is represented by a hiatus affecting the uppermost Toarcian Aalensis Zone and the base of the Aalenian Opalinum Zone (Fernández-López *et al.* 1988*b*; Pujalte *et al.* 1988; Canales *et al.* 1993). Aalenian deposits at the base of this unit comprise a regular succession of mudstone to wackestone facies, interbedded with dark grey marls (upper part of the Castillo Pedroso Fm). Fossil content includes ammonites, belemnites, brachiopods, nodosariid foraminifera and other fossil remains. This unit shows a maximun thickness of 18 m in the Poliente s–Tudanca trough, where all Aalenian zones appear to be represented in spite of numerous small unconformities recorded by ferruginous surfaces, condensed and redeposited levels (Fernández-López *et al.* 1988*b*; Canales *et al.* 1993). Along the SW margin the number of unconformities increases as the thickness of the succession is reduced to 4 m. The upper part of the unit comprises a thick-bedded mudstone to wackestone sequence containing common *Zoophycos*, bioclasts and filaments (i.e. sections of shells of the pelagic bivalve *Bositra*). Sponge mounds are recorded in the Sauzei and Humphriesianum zones (Fernández-López *et al.* 1988*b*). The total thickness of the sequence ranges from 70 m in the Polientes–Tudanca trough to 17 m in the SW, reflecting differential subsidence that waned upwards. The unit evolved from hemipelagic, poorly oxygenated middle–outer platform environments (Aalenian) to a pelagic platform (early Bajocian), culminating in a sharp increase of depositional energy and obvious shallowing at the top of the sequence (Humphriesianum Zone).

The late Bajocian sequence constitutes a deepening–shallowing cycle formed by a well-bedded succession of mudstone to bioclastic and filament wackestone beds, interbedded with thin marls. Marly intervals increase progressively to become predominant in the middle part of the unit. Fossil content comprises ammonites, belemnites, brachiopods, bivalves and trace fossils (*Zoophycos*). The thickness of limestone beds increases progressively upwards (Parkinsoni Zone), where bioclastic wackestone and even packstone levels were dominant. Also noteworthy is the widespread development of bioherms formed by siliceous sponges and algae, with single sponge mounds reaching up to 2–3 m high. Redeposited sponge debris along with other fossil remains (ammonites, bivalves, belemnites and serpulids among other) are also common. This unit was formed in an open carbonate platform under pelagic conditions for most of the time. The deeper facies of the succession corresponds to the Garantiana Zone, whereas at the top of the sequence (Parkinsoni Zone) higher energy facies are widespread, indicating shallow platform conditions above storm wave base by the end of Bajocian times (Fernández-López *et al.* 1988*b*; Robles *et al.* 1989; Quesada *et al.* 1990, 1993). The thickness of the unit ranges from 50 m in the Polientes–Tudanca trough to 27 m in the SW, reflecting a decrease of differential subsidence compared to former stages. At a more general, basin scale, the top of the unit shows a concentration of reworked ammonites from the upper Parkinsoni Zone, preserved as glauconitic and phosphatic internal moulds.

Bathonian–Callovian. The Bathonian–Callovian interval is recorded by sedimentary successions showing marked differences between separate domains of the basin. In the eastern area sedimentation took place in a shallow carbonate platform crossed by swells and troughs produced by extensional tectonics, whereas in the west open-platform conditions were more typical (Fig. 11.4). In the east this interval corresponds to several stratigraphic units defined by former authors: Unit J3 (p.p. (part)) of Soler y José (1971), Unit D of Garmendia & Robles (1991), Unit VIII of Bádenas *et al.* (1997) and Units VII (p.p.) and VIII of Gallego & Meléndez (1997). Between Aralar and Leitza, the succession comprises 20–30 m of micritic and silty limestones organized into beds 0.5 to 1 m thick. In the lower part, bioclastic silty limestones (packstone) with filaments predominate, whereas in the upper part calcareous mudstones prevail. Fossil content includes bivalves, ammonites, belemnites, brachiopods and echinoderms. In Montoria the Bathonian and Callovian sequences are separated by a non-sequence represented by a level containing reworked ammonites and iron ooids (Gallego & Meléndez 1997). The Callovian succession forms a regular interbedding of bioturbated, fossiliferous micritic limestones and silty marls. Ammonites are common, which allows the recognition of the lower Callovian Bullatus and Gracilis zones and, probably, the base of the middle Callovian Anceps Zone (Gallego & Meléndez 1997).

In the western area two main sedimentary sequences are distinguished, corresponding to the Bathonian and early Callovian intervals respectively (Robles *et al.* 1989; Quesada *et al.* 1993). The Bathonian sequence shows a rapid deepening trend in the Zigzag Zone (Fernández-López *et al.* 1988*b*), which resulted in a general drowning of the upper Bajocian shallow spongiolitic ramp, and deposition of marls at the base of the sequence. The unit is formed by metric to decametric sequences of marls and limestones; mudstone to wackestone facies with bioclasts and filaments, as well as ammonites, belemnites, brachiopods and trace fossils (*Zoophycos*). Sedimentation took place in an open, outer carbonate platform. Both sedimentary areas show only slight thickness differences, i.e. 70 m in the Polientes–Tudanca trough and 45 m in the SW marginal domain suggesting differential subsidence was partly attenuated during this time.

The upper sequence is early Callovian in age (Dahm 1966) and comprises a rhythmic succession of yellowish-grey marls and limestones (mudstone to wackestone facies with filaments) containing common ammonites and belemnites, and scarcer bivalves and brachiopods. This sequence was deposited on an outer carbonate platform environment showing an upward trend to more restricted conditions. During deposition differential subsidence increased again and the system of sedimentary swells and troughs acquired similar characteristics to those of the middle-upper Lower Jurassic succession. This palaeogeography favoured the development of a marly level rich in organic matter in the Bullatus Zone (Fig. 11.4), providing another potential Jurassic hydrocarbon source rock in the western area (Quesada *et al.* 1993). The organic-rich layer reaches a thickness of 25 m in the Polientes–Tudanca trough (where the whole sequence reaches its maximum thickness of 180 m) although it wedges out to the SW (Robles *et al.* 1989; Quesada *et al.* 1991). The upper boundary to the whole sequence is an intra-Jurassic erosional surface.

Upper Jurassic

Following deposition of marine sediments in Oxfordian and Kimmeridgian times, subsequent tectonism related to the Bay of Biscay Rift terminated deposition of the MJM and began a new sedimentary cycle of continental and paralic sediments (Purbeck megasequence). The base of this megasequence is marked by a major erosional unconformity seen across many southern European sedimentary basins. The stratigraphic gap represented by this unconformity is wider in the western area, where the youngest recorded marine rocks are middle Callovian in age, so that marine Upper Jurassic units are recorded only in the east (Fig. 11.4). Equivalent marine Upper Jurassic rocks were probably initially deposited in the west, but subsequently eroded during pre-late Tithonian times.

Oxfordian–Kimmeridgian. This succession, which is restricted to the eastern part of the Basque-Cantabrian Basin, has been the subject of detailed analysis in the Aralar range (Floquet & Rat 1975; Bádenas

1996). The lower boundary is a non-sequence involving a remarkable facies change, marked by the disappearance of the pelagic bivalve limestone unit (*Bositra*), a characteristic facies of the underlying sequence. In the Aralar range the existence of a stratigraphic gap between the Middle and Upper Jurassic successions spanning at least the late Callovian to early middle Oxfordian interval has also been proposed (Bulard *et al.* 1979). Bádenas (1996) has differentiated two unconformity-bound sequences, broadly corresponding to the Oxfordian and Kimmeridgian sequences (3.1.1 and 3.1.2, in Fig. 11.4). The Oxfordian sequence (Unit IX of Bádenas *et al.* 1997) is formed by a regular, 50 m thick succession of cherty limestones (pelloidal and bioclastic packstone) stratified in decimetric beds and containing scarce echinoderms, benthic foraminifera and sponge spicules. Farther north, in the Tolosa and Leitza areas, the lateral equivalent of the unit (Unit M1 of Garmendia & Robles 1991) is formed by a 140 m thick interval of calcareous siltstone and sandstone. Such lateral thickness and facies variations in the unit indicate the persistence of swells and troughs (Garmendia & Robles 1991).

In the Aralar range the Kimmeridgian sequence (i.e. Unit X of Bádenas *et al.* 1997) comprises a 90–140 m thick limestone succession including oolitic packstone to grainstone facies with variable quantities of other components such as peloids, intraclasts and bioclasts. Coral and chaetetid bioherms are also locally recorded, enclosing abundant microbial crusts, and form both tabular beds and isolated patches up to 1 m thick. Laterally, this unit grades into a sandstone succession of deltaic origin, showing thickening- and coarsening-upward sequences. Oolitic and bioclastic limestones have been reported farther north (Tolosa; Unit M2 of Garmendia & Robles 1991) and west (Montoria; Unit XI of Gallego & Meléndez 1997). The upper boundary is an angular unconformity surface showing evidence of subaerial erosion and overlain by sediments of the Purbeck megasequence (Soler y José 1971, 1972*b*; Garmendia & Robles 1991).

Late Tithonian-late Berriasian. The rifting initiated at the Jurassic–Cretaceous boundary resulted in the deposition of the Purbeck megasequence from late Tithonian to Valanginian times. This megasequence comprises sediments which have been the subject of numerous stratigraphical, sedimentological and palaeontological studies (e.g. Rat 1962; Ramírez del Pozo 1971; Schudack 1987; Soler y José 1972*a,b*; Pujalte 1982*a*, 1989*a*; Pujalte *et al.* 1996; Robles *et al.* 1996; Hernández *et al.* 1999; Hernández 2000). Most debates concerning these rocks have traditionally focused on their age and stratigraphic correlation.

In the western area several tectonostratigraphic units bounded by regional unconformities have been identified within this megasequence, but only the lower one (Campóo Gp) belongs partly within the Jurassic (Pujalte *et al.* 1996; Hernández *et al.* 1999). The Campóo Gp comprises the oldest tectonostratigraphic unit of the rifting stage, and has been dated as late Tithonian to early late Berriasian on the basis of charophyte associations (Hernández *et al.* 1999). The lower boundary of the unit is the unconformity overlying the MJM. This surface may be locally covered by a thin, *in situ* or slightly reworked horizon of laminate calcrete (Hernández 2000).

The succession comprises a thick package of palustrine carbonates within fluviatile and alluvial siliceous and carbonate detrital units (Fig. 11.4). It has a highly variable thickness, reaching over 500 m at the depocentre, in the region of Aguilar de Campóo where two main units can be distinguished, the Aguilar and Frontada formations (Hernández *et al.* 1999). Only the lower part of the Aguilar Fm belongs within the uppermost Jurassic succession, which represents the initiation of a palustrine basin equivalent to the Cameros basin in the NW Iberian Ranges. These deposits are interpreted as the infill of a complex tectonic graben created by the reactivation of Variscan faults resulting from north–south extensional stress. This graben was split into smaller sub-basins bounded by synsedimentary listric faults (Hernández 2000), extending across the former Polientes–Tudanca trough and, partly, on to the SW marginal domain (Fig. 11.4).

In the eastern area (Guernica, Tolosa, Leitza, Aralar) deposits corresponding to the Purbeck megasequence (uppermost Jurassic to lowermost Cretaceous) are represented by the 'Serpulid Limestone Unit' of Soler y José (1972*a,b*), equivalent to the lower PW Unit of Garmendia & Robles (1991). This unit lies unconformably on Upper Jurassic rocks and forms a regular 90–120 m thick succession of

dark bioclastic limestones containing serpulids, gastropods, bivalves (oysters and other groups) and ostracods (Fig. 11.4). Beds are commonly organized in fining-upward sequences showing laminar stromatolites on top, interpreted as recording shallowing-upward cycles developed on an intertidal flat (Garmendia & Robles 1991).

South Pyrenean Basin (MA, GM)

Jurassic outcrops in the southern Pyrenees are allochthonous, strongly tectonized and widely scattered (Fig. 11.2). Furthermore, Jurassic successions are commonly incomplete, due to a major phase of erosion during the Early Cretaceous opening of the Bay of Biscay, north Pyrenean rifting and the uplift of the south Pyrenean margin. Jurassic sequences are best preserved in the northern and central part of the range (Organyá, Pont de Suert), but are markedly less complete in the southernmost areas, indicating that erosion was more intense southwards. Both east and west from this central area there are only a few scattered remnant outliers, such as at Figueras-Montgrí (to the east) and San Felices (to the west), preserving evidence for the former continuity of Jurassic outcrops (Fig. 11.2).

South Pyrenean Jurassic rocks record the existence of an epicontinental platform developed on the northern margin of the Iberian plate. Sedimentation on this platform was controlled by synsedimentary extensional tectonics which produced important palaeogeographic differences between separate areas during the evolution of the basin (Peybernés 1976; Peybernés & Pelissiée 1985; Souquet 1986*a*). At the beginning of Early Jurassic times a wide epicontinental platform extended throughout the south Pyrenean area, connected with both the Atlantic and Tethys oceans. The platform underwent a progressive flooding until the Early–Middle Jurassic boundary, after which a reactivation of Atlantic rifting in the Bay of Biscay led to the separation of the basin into a western area open to the Atlantic (Boreal), and an eastern area (Mesogean) linked to the Tethys (Peybernés & Pelissié 1985; Souquet 1986*a*). These two areas were separated by a wide, emergent (or shallow marine) area in the central part of what is now the south Pyrenees. During Late Jurassic times this central area was open to the east (the expanding Ligurian ocean), whereas the western area became progressively uplifted and exposed.

Early Jurassic. Basal ferruginous breccias (5–8 m thick) rest unconformably on Upper Triassic dolomites and older rocks, locally containing intraclasts and some occasional basaltic debris. They are overlain by 8–12 m of laminated dolomitic limestones containing ostracods, bivalves and echinoids (the *Diademopsis* Limestones unit of Fauré 1984), in turn overlain by an unconformity. Both the breccias and the limestones have been assigned to the lowermost Jurassic by some authors (Fauré 1984; Ramón 1989), although from a tectonostratigraphic point of view they represent the final stage of a dominantly Triassic cycle (Souquet 1986*a*). Above them, four depositional sequences can be recognized within the Lower Jurassic succession (Fig. 11.5). They are very similar to those defined in the eastern Iberian sedimentary basins, so it may be concluded that both areas were probably connected.

The first major sedimentary sequence, Hettangian to Sinemurian in age (1.1 in Fig. 11.5), includes two main lithological units interpreted as a deepening-upward sequence within an inner platform environment. The lower unit, 40–80 m thick, comprises dolomites and massive dolomitic breccias with abundant dissolution structures. This unit grades southwards (Montroig) into a 100–250 m thick succession of anhydrites with interbedded laminated dolostone. The upper unit, 25–70 m thick, is formed of well-bedded dolostone and limestones, and it is overlain by a ferruginous hardground. Laminated dolostones

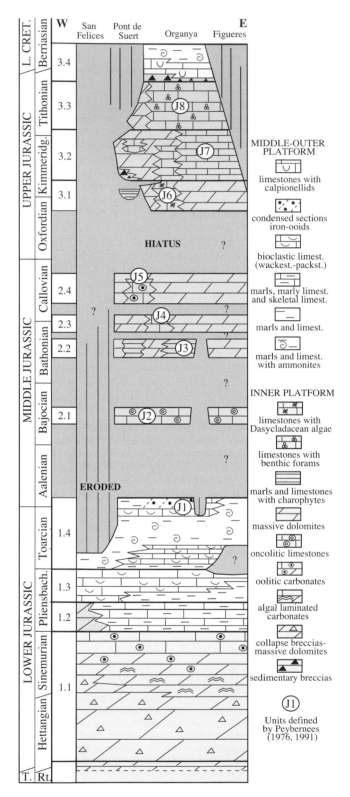

Fig. 11.5. Stratigraphy and facies distribution of the Jurassic of the South Pyrenean Basin. The Middle and Upper Jurassic stratigraphy is largely based on Souquet (1986*a*) and Peybernés (1991). See Figure 11.2 for location of reference logs.

are dominant in the lower part, whilst oolitic limestones are more common in the upper part. A Sinemurian age has been inferred for this unit on the basis of the algal content, mainly dasycladaceans and codiaceans.

The second major sequence, early Pliensbachian in age, records a sudden deepening of the platform, under open marine, middle to outer platform conditions, except in the west (San Felices) where inner platform facies persisted. In the central areas, this second sequence (1.2 in Fig. 11.5) comprises 10–30 m of fossiliferous marls with thin interbeds of bioclastic limestones. Ammonites from the lower interval of the sequence belong to the lower Pliensbachian Jamesoni, Ibex and Davoei zones. An upper, mostly bioclastic interval contains abundant brachiopods and belemnites and rare ammonites typical of the upper Pliensbachian, Stokesi Zone. The sequence as a whole forms a thickening-, shallowing-upward cycle overlain by a ferruginous hardground.

The third major sequence, some 15 m thick, comprises two carbonate units that form a new thickening- and shallowing-upward cycle (1.3 in Fig. 11.5). The lower interval is a marly unit containing ammonites from the Stokesi Zone, whereas the upper unit comprises bioclastic limestones containing ammonites from the uppermost Pliensbachian, Spinatum Zone. In the westernmost areas (San Felices) the sequence is mainly formed of bioclastic limestones with thin, fossiliferous marly interbeds containing abundant bivalves (pectinids; pholadomyids), brachiopods, ammonites, belemnites and crinoids (Comas-Rengifo *et al.* 1989). The upper boundary is a sharp discontinuity surface marked by a ferruginous hardground, and represents a stratigraphic gap omitting the lowermost Toarcian. In the easternmost areas (Figueras–Montgrí) bioclastic limestones from the upper Pliensbachian, Stokesi Zone are bounded by an important discontinuity, omitting the uppermost Pliensbachian and lower Toarcian. Above this discontinuity surface a reworked level contains abundant ammonites from the middle Toarcian Thouarsense and Insigne zones (Llompart *et al.* 1984).

The fourth major sequence (1.4 in Fig. 11.5) forms a thick interval (up to 80 m in the central areas) ranging from lower to upper Toarcian in age. It comprises three main carbonate units (limestone and marl) typical of middle to outer carbonate platform, which together form a deepening–shallowing cycle. The lower unit shows 2–6 m of fossiliferous bioclastic rocks containing abundant brachiopods, and is assigned to the lower-middle Toarcian, Tenuicostatum to Variabilis zones. It is capped by a sharp encrusted, ferruginous discontinuity surface. This interval appears to be absent in the easternmost areas (Figueras) whilst towards the western areas (San Felices) it grades laterally into marls with ammonite-rich marly limestone interbeds over 20 m thick. The second unit is a 30–50 m thick marly interval, with abundant ammonites from the middle-upper Toarcian Thouarsense, Insigne and Pseudoradiosa zones. The third unit, a 3–12 m thick interval known as the *Gryphaea* Lumachelic limestones and marls, corresponds to the uppermost Toarcian, Aalensis Zone. Fossil content includes abundant crinoids, bivalves, ostracods, gastropods, serpulids, foraminifera and bryozoans. On top of this sequence, a thin marly limestone interval has locally yielded rare ammonites characterizing the basal Aalenian, Opalinum Zone (Peybernes 1976).

Middle Jurassic. The appearance at the Lower–Middle Jurassic boundary of a central sedimentary high bounded by two wide outer platform areas to the west (Basque-Cantabrian basin) and to the east (Bas-Languedoc basin), resulted in a large part of the South Pyrenean Basin becoming temporarily emergent or covered by only a shallow sea. Consequently, the stratigraphic record of the Middle Jurassic is highly discontinuous and shows major facies changes and gaps (Fig. 11.5). This record preserves a 100–200 m thick, mostly dolomitized succession, traditionally known as the Lower Dolomite unit. It is formed by thick dolomitic intervals with thinner interbeddings of algal limestones in which benthic foraminifera are common. Both litho- and biofacies indicate deposition under shallow marine platform conditions. A precise dating of the series has been long hampered by the mostly dolomitic character of the sequence and biochronological inadequacy of benthic foraminifera recorded in non-dolomitized intervals. According to some authors (Peybernés 1991) this south Pyrenean shallow platform would have been flooded only during transgressive 'peaks'. Four depositional sequences are recognized within this whole dolomitic interval, with large stratigraphic gaps probably existing between them (Fig. 11.5), and correspond to stratigraphic Units J2 to J5 of Peybernés (1976).

The first major sequence is represented by the 'oncolitic dolostones and limestones' (Unit J2). This unit is well-represented in most areas of the South Pyrenean Basin, and comprises a 30 m thick, partly dolomitized, oolitic and oncolitic carbonate interval containing common algal remains (dasycladaceans) as well as bryozoans, gastropods and corals. Laterally equivalent deposits in middle to outer platform facies appear in the southernmost Pyrenees, around Montroig. These mainly comprise fossiliferous marly limestones containing common brachiopods, and are late early and late Bajocian in age. A basal discontinuity separates this unit from the underlying upper Toarcian–lower Aalenian (Opalinum Zone) limestone and marl unit (the *Gryphaea* Limestones). Widespread emergence is inferred for this non-sequence interval, which would have locally produced erosion of part of the previously deposited Lower Jurassic successions.

The second major sequence corresponds to Unit J3 of Peybernés (1976), and comprises a 8–10 m thick sequence of lignitic marls and limestones with charophytes and ostracods. The fossil content suggests a probable middle Bathonian age for this unit, which grades laterally into dolomites, and is separated from the underlying sequence by an important discontinuity that corresponds to the Bajocian–Bathonian boundary.

The third sequence comprises a 10 m thick unit of white, massive limestone (Unit J4a of Peybernés 1976). Microfossil content suggests a roughly late Bathonian to Callovian age. The unit is interpreted to have been deposited in a shallow inner platform under restricted brackish water, undergoing episodic subaerial exposure. This white limestone, as well as its dolomitic lateral equivalent unit, oversteps underlying units and has been described as transitional from a synrift to a post-rift stage in the evolution of the Pyrenean basin (Peybernés 1991).

The fourth major sequence comprises the lithological unit known as *Trocholine* limestones (Unit J5). It forms a thick (over 100 m) carbonate interval of mostly oolitic limestones containing abundant coral remains and foraminifera (large-size specimens of *Trocholina*) suggesting a general Callovian age. These rocks crop out in the central northern area (Pont de Suert) and again grade laterally into massive dolomitized facies. The sequence is bounded on top by an important Middle–Late Jurassic discontinuity that is locally difficult to recognize due to dolomitization of the sequence (Ramón *et al.* 1992).

Upper Jurassic. Upper Jurassic deposits comprise up to 700 m of carbonate rocks, mainly recording inner to middle platform facies conditions, and range in age from late Oxfordian to earliest Cretaceous (Berriasian). The development of this cycle was controlled by an important tilting of the Pyrenean basin, involving uplift and subaerial exposure in the west and subsidence in the east, where a more or less continuous sedimentary record is preserved. Therefore two sedimentary realms (western and eastern) can be recognized (Peybernés 1976). To the west, between the Esera and Noguera Pallaresa rivers (e.g. Pont de Suert), the Upper Jurassic sequence extends only from lower Kimmeridgian to the Kimmeridgian–Tithonian boundary. The succession includes a lower breccia, presumably Kimmeridgian in age, which lies erosionally across the underlying Callovian *Trocholina* limestones, and an upper, thick (up to 350 m) sequence of partly dolomitized argillaceous limestones containing common foraminifera (lituolids). To the east, between the Noguera Pallaresa and Segre rivers (around Organyá), above the Middle–Late Jurassic boundary discontinuity a more complete series includes *Anchispirocyclina* massive limestones, and the Calpionellid marls and limestones the age of which continues up into earliest Cretaceous times (Berriasian).

Deposits of this Upper Jurassic succession are classified into four major depositional sequences, which show a progressive retreat of the coastline towards the SE (Fig. 11.5). As a general rule, each sequence begins with the exposure and erosion of the western area, followed by transgression towards the west (Peybernés 1976; Souquet 1986a).

The first sequence, well represented in the east, comprises some 50 m of oolitic and bioclastic limestone containing algal remains (dasycladaceans) and foraminifera (lituolids) including *Alveosepta jaccardi*, which indicate an upper Oxfordian to lower Kimmeridgian age (Unit

J6). To the west it corresponds to a thinner unit of limestones deposited in a lagoonal environment and containing charophytes and coal remains. This unit directly overlies the Callovian *Trocholina* limestone unit of the preceding sequence. According to the inferred age for this first Upper Jurassic sequence, the stratigraphic gap associated with the Middle–Late Jurassic boundary ranges at least across the late Callovian to late Oxfordian interval.

The second sequence shows a wider geographic distribution and, in the west, it is again represented by breccias and micritic limestones, with a basal 25 m thick polymict breccia resulting from the erosion of the unconformably underlying units. This is overlain by successive breccia units interbedded with micritic, argillaceous limestone containing foraminifera (Lituolids), bivalves (Ostraeidae), serpulids and crinoids that, as a whole, indicate deposition in a restricted, inner platform environment. Above this breccia-dominated succession is a thick (up to 350 m), locally dolomitized, micritic limestone unit. The occurrence of lituolids such as *Everitcyclammina virguliana* and *Alveosepta jaccardi* allows this sequence to be assigned to the late Kimmeridgian–early Tithonian interval. To the east, this sequence is thinner (<200 m) and shows micritic–dolomitic rocks grading into laminated micritic limestones (Unit J7) deposited in a restricted lagoonal environment, and then into massive limestones containing coprolites, charophytes and large foraminifera (lituolids), typical of a somewhat more open marine environment.

The third sequence, well represented in the eastern area (Organyá), comprises a thick (up to 250 m) carbonate succession including a lower calcareous breccia unit followed by massive, mostly oolitic limestones containing coprolites and foraminifera, and fine-grained dolomites (Unit J8). The scattered record of *Anchispirocyclina lusitanica* suggests a latest Jurassic (Tithonian) age for this unit. To the west this succession wedges out and grades into massive dolostones that include collapse-dissolution breccias.

The fourth sequence is represented only in the central-eastern area, in the valley of the river Segre. It begins with a dolomitic breccia unit known as the 'boundary breccia' unit. At Organyá this unit is up to 200 m thick and formed by angular clasts resulting from the erosion of the underlying Jurassic units. The breccia is overlain by a 20–40 m thick carbonate unit of oncolitic and calpionellid-bearing micrites (*Calpionella alpina, Crassicollaria parvula*) indicating a Berriasian age, with overlying marls containing ammonites of the genus *Pseudosubplanites*. This sequence includes rare beds of micritic limestone with calpionellids, ammonites, benthic foraminifera (*Trocholina*) and ostracods. This fossil content indicates that the Pyrenean basin was more open to deeper marine conditions towards the east, where it was linked to the Bas-Languédoc basin.

East Iberian basins (MA, GM, BB)

Jurassic deposits in the Iberian Ranges and southern Catalonian Coastal Ranges crop out widely across the eastern half of the Iberian Peninsula (Fig. 11.2). Excellent field exposures combined with the existence of scattered well-log data from intervening areas (Castillo Herrador 1974; Morillo & Meléndez 1979; Fontana *et al.* 1994a) allow a detailed reconstruction of facies and thickness variations of units (Figs 11.6–11.7). Jurassic deposits in these eastern Iberian basins are bounded overall by important discontinuities of regional extent and can therefore be viewed as a simple, Jurassic supersequence. The bounding discontinuities are linked to major phases of extensional fault reactivation (Salas & Casas 1993). The unconformity at the Triassic–Jurassic transition is related to extensional movements that led to the westward extension of Tethys. In contrast, extensional tectonism at the latest Jurassic–Early Cretaceous boundary was additionally linked to the opening of the central Atlantic and the early stages of the Bay of Biscay rift. Lowermost Cretaceous deposits (lower to midupper Berriasian) have been traditionally included within the Jurassic supersequence.

Normal faults controlled Jurassic sedimentation in eastern

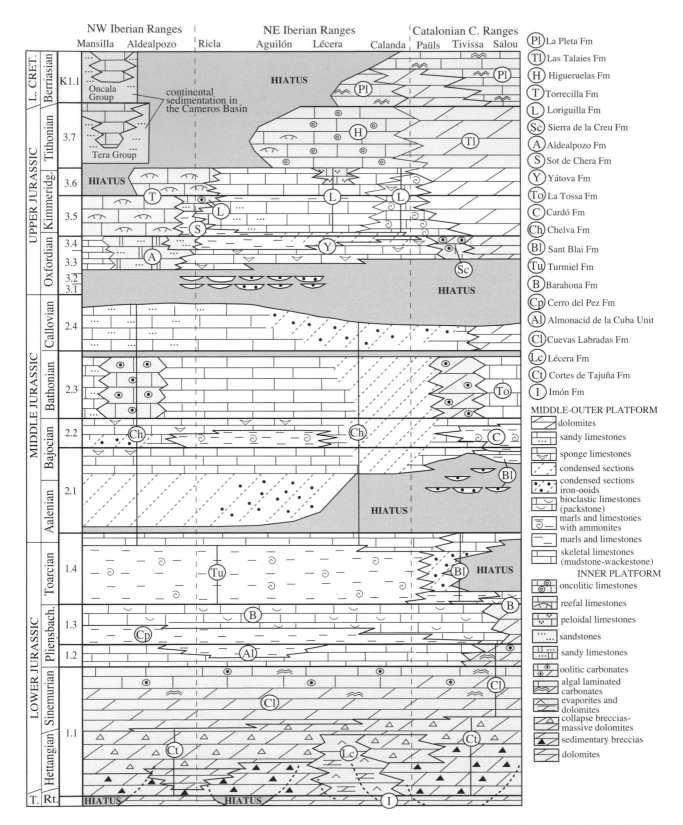

Fig. 11.6. Stratigraphy and facies distribution of the Jurassic of the northern part of the Iberian basin. See Figure 11.2 for location of reference logs.

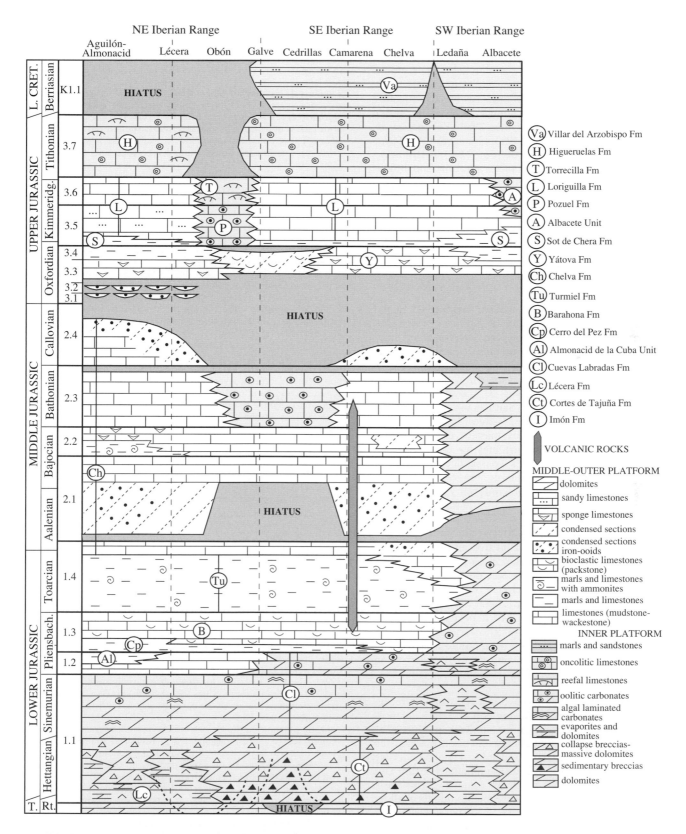

Fig. 11.7. Stratigraphy and facies distribution of the Jurassic of the central part of the Iberian Basin. See Figure 11.2 for location of reference logs.

Iberia, and differential movements produced substantial variations in both thickness and facies over space and time. Additionally, widespread transgressive and regressive events recorded across the basin suggest a certain degree of regional relative sea-level change controlling stratigraphic discontinuities and facies distribution. On the basis of such discontinuities, a series of unconformity-bounded units (third-order sequences) can be recognized (Giner 1980; Yébenes *et al.* 1988; Aurell 1991; Salas & Casas 1993; Gómez & Goy 1997*a*, 2000; Fernández-López 1997; Aurell *et al.* 2000). Distribution of depositional sequences, main facies patterns and lithostratigraphic units for the East Iberian Jurassic, are shown across two general sections, from west to east (ranging from northern Iberian Ranges to the Catalonian Coastal Ranges) and from north to south (Figs. 11.5 and 11.6). The sequences generally consist of deepening–shallowing cycles bounded by flooding surfaces.

Early Jurassic

By the end of Triassic time extensional faulting led to the break-up the Rhaetian epicontinental platform, which was represented across wide areas by bedded dolostones and limestones deposited in peritidal environments (Imón Fm; Goy *et al.* 1976). This locally produced a basal Jurassic angular unconformity, such as seen in the Garraf massif, in the Catalonian Coastal Ranges (Esteban & Juliá 1973), the surroundings of Sierra del Moncayo, in Zaragoza (San Román & Aurell 1992) and at Mansilla, in the NW Iberian Ranges. In all these areas the basal Jurassic unit, formed by massive dolostones and breccias (the Cortes de Tajuña Fm), may directly overlie the Buntsandstein facies of the Lower Triassic deposits, or even Palaeozoic rocks. However, across large areas of the basin, this basal discontinuity is in fact a paraconformity between Late Triassic dolostones (Imón Fm) and the basal Jurassic unit (Cortes de Tajuña Fm). The Triassic–Jurassic boundary has been tentatively placed within the lower part of the Cortes de Tajuña Fm on the basis of fossil bivalve and palynomorph associations (Goy *et al.* 1976; Gómez & Goy 1998).

After deposition of a Lower Jurassic dolomitic-evaporitic-breccia unit, Sinemurian to Toarcian sedimentation took place on a wide carbonate platform open to the north. From a general point of view, sediments deposited during the Early Jurassic epoch record the progressive deepening of a carbonate platform that reached a maximum during the late early Toarcian interval. A shallowing event then led to an important stratigraphic discontinuity around the Toarcian–Aalenian boundary.

Within the Lower Jurassic succession it is possible to recognize a series of deepening–shallowing depositional sequences bounded by encrusted, ferruginous surfaces produced by transgressive events. The first such event took place at the early–late Sinemurian boundary, and led to a general flooding of the northern (Basque-Cantabrian) sedimentary realm. This first event did not, however, reach the eastern Iberian basin where inner platform conditions were maintained throughout the Sinemurian stage. Here, in turn, a further transgressive event at the end of Sinemurian times produced middle and outer ramp facies in the northern parts of the Iberian basin (Fig. 11.8). Further transgression at the early–late Pliensbachian boundary and into early Toarcian times progressively reached more southerly areas. However, in marginal parts of the basin, inner platform conditions were maintained during

the whole of Early Jurassic time. These four transgressive events allow the distinction of four depositional sequences that are most clearly recognizable in the most open (i.e. central to northern) areas of the basin (Figs 11.6 and 11.7).

Sequence 1: Hettangian–Sinemurian. The first Jurassic depositional sequence in the Iberian basin ranges from latest Rhaetian to Sinemurian in age. It includes the Cortes de Tajuña Fm and a lateral, evaporitic facies equivalent (the anhydrite and carbonate unit known as the Lécera Fm; Gómez & Goy 1998), as well as the lower part of an overlying well-bedded dolostone and limestone unit known as the Cuevas Labradas Fm (Goy *et al.* 1976). As a whole, this sequence represents a general transgressive cycle, from evaporitic coastal sabkha to subtidal to intertidal plain environments (peritidal cycles).

The thickness and facies distribution of the Cortes de Tajuña Fm were controlled by extensional tectonics. Break-up of the Late Triassic ramp led to the development of a series of half-graben basins which were filled with facies typical of sabkha to shallow marine environments, including carbonate breccia, evaporites, 'carnioles' (i.e. massive dolomitic limestones and evaporites) as well as bedded dolomites and limestones. The total thickness of this unit ranges widely from 30 to 400 m. Highest thickness values are recorded in the northeastern Iberian Ranges, in the Ebro valley (Ebro sub-basin), and at the SW margin of the Iberian Ranges (Cuenca sub-basin). The facies can be grouped into three main categories, changing laterally into each other (Fig. 11.8).

1. *Sedimentary breccia.* This facies is well developed in the northern Iberian Ranges and the Catalonian Coastal Ranges, and varies in thickness from 50 and 300 m. The facies comprises a set of polymictic, poorly sorted breccia levels, with clasts ranging in size from a few centimetres to large olistolite dimensions. They are formed by subangular to slightly rounded carbonate clasts, including dolostones, laminated mudstone and oolitic grainstone, embedded within a micritic carbonate or sandy matrix. This unit developed upon the break-up of the pre-Jurassic carbonate platform, by the combined effect of synsedimentary block tectonics and early dissolution of interbedded evaporitic levels. Gravitational (slump) structures and dramatic thickness variations help define the blocks tilted by normal faults (Giner 1978, 1980; San Román & Aurell 1992; Campos *et al.* 1996; Bordonaba *et al.* 1999).

2. *Evaporite.* A thick unit (150–400 m) formed by gypsum (anhydrite in borehole samples) with interbedded carbonates occurs in both the Ebro and the Cuenca sub-basins. Both the facies association and the mineralogical and isotopic composition indicate formation in a coastal sabkha sedimentary environment, under episodic marine influence (Castillo Herrador 1974; Ortí 1987).

3. *Collapse breccias and massive dolostones.* This is the most typical facies at the base of the sequence, and is widespread across the basin. It is formed by massive, highly porous dolostones, interbedded with breccia levels that include angular dolostone clasts. The origin of these breccias and their high porosity have been related to evaporite dissolution and subsequent collapse under supratidal conditions, either just after their deposition (Giner 1980; Bordonaba & Aurell 2001), or else during a later burial stage (Morillo & Meléndez 1979; Gómez & Goy 1998). The total thickness for this unit generally ranges between 50 and 150 m, but minimum values (below 50 m) are locally recorded at the westernmost margin of the basin, as well as in the east, on the so-called Maestrazgo High.

From Sinemurian times onwards a period of relative tectonic stability led to the establishment of a wide and shallow carbonate platform, episodically exposed to supratidal conditions. Deposits of this stage correspond to the lower part of the Cuevas Labradas Fm. The total thickness of this unit ranges from 50 to 150 m, the highest values being recorded in the central part of the basin along several sedimentary troughs following a general north–south direction. However, over the Tarragona and Maestrazgo highs, the thickness of this unit, which spans the Sinemurian to early Pliensbachian interval, does not commonly exceed 50 m (Fig. 11.8). The formation comprises well-bedded, micritic laminated and oolitic limestones in a series of shallowing-upward sequences that terminate in intertidal to supratidal facies. As a general rule, muddy-type sequences dominate in the lower part whilst grainy sequences are dominant in the upper part. The thickness and number of sequences vary from one place to another,

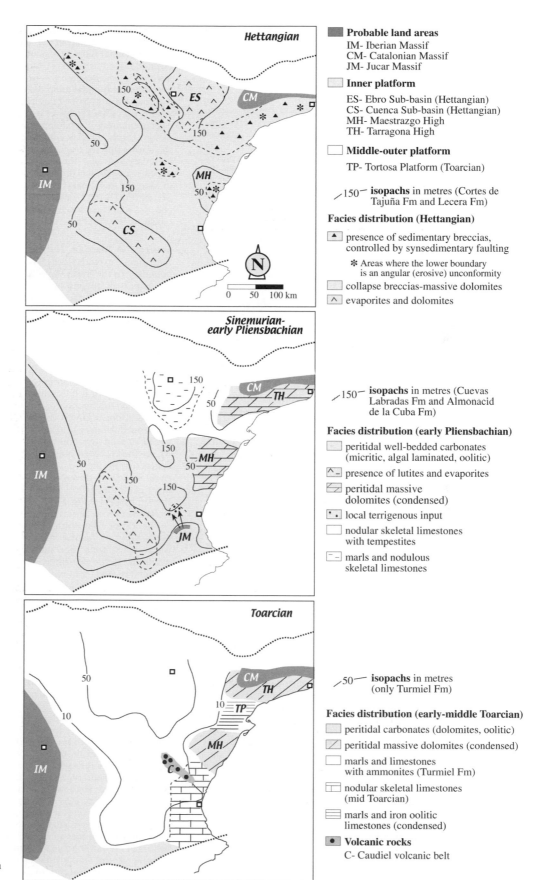

Fig. 11.8. Palaeogeography and facies distribution of east Iberia during Hettangian, Sinemurian–Early Pliensbachian and Toarcian.

reflecting local variations of subsidence. West from Valencia a coeval unit comprising carbonate breccias and conglomerates (including centrimetre-sized clasts and pebbles), interbedded with internal platform facies, has been described by Gómez (1979). This has led to the hypothesis of an uplifted block in this area (the Júcar massif) during Sinemurian times (Fig. 11.8).

Sequence 2: Early Pliensbachian. The deposition of this second sequence begins with a sharp discontinuity around the Sinemurian–Pliensbachian boundary. This discontinuity is marked by a ferruginous hardground that heralds the establishment of middle and outer platform facies (limestones and marls) in the central-northern part of the basin (the Almonacid de la Cuba unit). Farther south, along the western and eastern margins of the basin, sedimentary conditions typical of an inner ramp still persisted, leading to the development of a thick carbonate and dolomitic succession (the upper part of the Cuevas Labradas Fm) forming a series of shallowing-upward sequences typical of subtidal to inter- and supratidal environments. In the SW (Cuenca–La Mancha), green marls and clays interbedded with dolomitic facies are typical (Azéma *et al.* 1971; Morillo & Meléndez 1979). These facies reveal the presence of sabkha environment conditions receiving episodic terrigenous supply (Fig. 11.8).

In the northern part of the basin this unit is represented by middle to outer platform facies, i.e. well-bedded nodular fossiliferous limestones stratified in decimetric banks, interbedded with thin marly layers, and containing common brachiopods, bivalves, gastropods, foraminifera, echinoderms and ammonites. The thickness of this unit ranges between 10 and 120 m and it usually forms a series of shallowing-upward sequences 1 to 3 m thick beginning with a marly interval and capped by a ferruginous, encrusted hardground. Tempestite levels commonly can be recognized within these sequences, indicating deposition on a middle carbonate ramp above storm wave base level. Ammonite records from this interval have allowed recognition of the lower Pliensbachian Jamesoni (p.p.), Ibex, and Davoei (p.p.) zones (Comas-Rengifo 1985; Comas-Rengifo *et al* 1988, 1999).

In the central and northern areas, the upper part of this sequence shows a higher content of marls and marly limestones. These deposits form the Almonacid de la Cuba unit, the total thickness of which ranges from 40 m at the type locality (Comas-Rengifo 1985; Comas-Rengifo *et al.* 1999) to 10–20 m farther SE, in the region of Obón (Mouterde *et al.* 1978). This unit has also been recorded farther east within the Catalonian Coastal Ranges, at Alfara, near Tortosa (Comas-Rengifo *et al.* 1998). It comprises interbedded marls and limestones (fossiliferous mudstone to packstone) showing a series of metric-scale thickening- and shallowing-upward sequences with abundant bivalves, brachiopods, gastropods, echinoderms, and rarer ammonites. Locally, at some points (e.g. Castel de Cabra, near Obón), the upper levels show a bioclastic limestone interval containing common hermatypic coral remains and calcareous algae (*Solenopora, Cayeuxia, Uragiella, Palaeodasycladus*) and providing evidence for a local shallowing of the platform at the end of this cycle (Bordonaba *et al.* 2000).

Sequence 3: Late Pliensbachian. The depositional sequence corresponding to this interval is bounded by two ferruginous encrusted surfaces each interpreted as recording deepening events. The lower surface was developed in the upper Davoei Zone (late early Pliensbachian), and the upper surface was developed at the Pliensbachian–Toarcian transition. Across much of the basin this sequence comprises two lithostratigraphic units, the Cerro del Pez Fm (a mostly marly unit) and the Barahona Fm (a bioclastic limestone unit). Both units were deposited on a wide epicontinental platform interpreted as a carbonate ramp. Marls belonging to the Cerro del Pez Fm were deposited below the normal wave base, in the middle and outer areas of the platform. They form a 5–15 m thick sequence of marls with thin limestone and marly limestone interbeddings (fossiliferous mudstone) containing common brachiopods, bivalves, gastropods, echinoderms and ammonites. In contrast, the Barahona Fm comprises bioclastic limestones deposited under strong wave influence on the middle part of the ramp. They form a 10–30 m thick sequence of bioclastic limestones stratified in discontinuous, uneven banks (fossiliferous wackestone to packstone containing abundant brachiopods, bivalves, foraminifera, belemnites, echinoderms, and rare ammonites).

Towards the margins of the basin, in the Tarragona and Maestrazgo highs (to the NE), and in the Altomira areas (to the SW), both units grade into inner ramp carbonate and dolomite facies. Such facies have been described by Giner (1980) in the section of Salou, as representing a set of shallowing-upward sequences of muddy and grainy type developed in subtidal and tidal environments.

These two units as a whole are interpreted as a complete deepening–shallowing cycle. The lower part of the marly Cerro del Pez Fm forms a set of thinning-, deepening-upward sequences, whilst the upper part along with the bioclastic limestones of the Barahona Fm displays a shallowing-upward trend (Comas-Rengifo *et al.* 1999). Generally, the boundary between the Cerro del Pez and Barahona formations appears roughly synchronous, within the Stokesi Zone or at the base of Margaritatus Zone (Comas-Rengifo *et al.* 1988, 1999). Towards the SE (eastern sector; Gómez 1979) the lower, Cerro del Pez Fm is absent and the whole sequence is represented only by the shallower water bioclastic limestones of the Barahona Fm.

Sequence 4: Toarcian. The final sequence within the Lower Jurassic succession extends roughly from the Pliensbachian–Toarcian boundary to the lower Aalenian, Opalinum Zone. It is bounded by two important discontinuities of regional extent. The lower discontinuity (flooding surface) corresponds to a widespread ferruginous encrusted surface on top of the Barahona Fm. Above this discontinuity comes a thick marly unit, the Turmiel Fm, which comprises interbedded marls and micritic, fossiliferous limestones containing common ammonites, brachiopods, bivalves, belemnites, foraminifera and other groups. The boundary between the Barahona and Turmiel formations falls generally within the Spinatum Zone (upper Pliensbachian), but in some areas (e.g. in the Tortosa Platform, Catalonian Coastal Ranges), it falls within the lower Toarcian succession. As a general rule this boundary is slightly diachronous, being progressively younger from north to south (Gómez 1991). This means that the deeper sedimentary areas in the platform were in the north, which would explain the dominantly mid-European character of faunas within the recorded ammonite associations (Elmi *et al.* 1989). The connection during the Toarcian with the northernmost Iberian basin could have been through the Basque-Cantabrian basin. The shallow, inner platform environments, located in the southernmost part of the Iberian basin (see Fig. 11.8), impeded direct connections between the Betic and Iberian basins (Sandoval *et al.* 2001).

The Turmiel Fm forms a 20–60 m thick, monotonous interval of marls with rare regular limestone interbeds (mudstone to bioclastic wackestone), which become more common upwards. The unit as a whole is highly fossiliferous: ammonites, brachiopods, bivalves, belemnites and foraminifera are dominant, whereas echinoids, gastropods and crinoids are less common. The biostratigraphic value of the rich nannoplankton content of this unit has only recently been recognized (Perilli 1999). This unit has been the subject of numerous detailed palaeontological, biostratigraphical and sedimentological studies throughout the Iberian Ranges, the main studied sections being those of Castrovido (Comas-Rengifo *et al.* 1988), Ricla and La Almunia (Goy & Martínez 1990), Obón (Mouterde *et al.* 1978), Almonacid de la Cuba (Comas-Rengifo 1985, Comas-Rengifo *et al.* 1999), Rambla del Salto-Aguatón (Arche *et al.* 1977; Barrón *et al.* 1999) and the classic section of Fuentelsaz (Goy & Ureta 1986, 1990; Perilli 1999), recently designated as the Aalenian GSSP (global boundary stratotype section and point).

The Turmiel Fm was deposited on an open epicontinental, relatively deep platform, which favoured colonization by ammonite populations during early Toarcian times (Gómez & Goy 1997*a*). Marls and limestones form a succession of metre-thick, thinning- and thickening-upward sedimentary sequences, which have been interpreted as deepening and shallowing cycles. Recent sequence stratigraphic analysis (Gómez & Goy 2000) has led to the recognition of four main deepening–shallowing sequences, with maximum depths reached during late early Toarcian times (Serpentinus-Bifrons zones). The four sequences can be recognized throughout the whole Iberian basin as well as in the Asturian and Basque-Cantabrian basins.

In the central and northern part of the Iberian basin the Toarcian cycle ends with a series of thickening, shallowing-upward minor sequences, ranging from Pseudoradiosa to Aalensis Zone (upper Toarcian) or even lower Aalenian, Opalinum Zone. The greatest thickness of the Aalensis and Opalinum zones is found in the central Iberian Ranges (more than 35 m in Fuentelsalz; 14.5 m in Moyuela). In the NW the thickness is notably less (4–13 m) (Ureta 1985; Goy & Ureta 1990; Sandoval *et al.* 2001), and in the south stratigraphic gaps are frequent. As a general rule, the lower Aalenian Opalinum Zone and the uppermost Toarcian, Pseudoradiosa and Aalensis zones are absent southwards of a palaeogeographic line defined by the Noguera–Aguatón fault (Fernández-López & Gómez 1990*a*). This synsedimentary lineament trended WSW–ENE, separating the area of Sierra de Albarracín and Sierra Palomera from the northern part of the Castillian branch (Sierra Menera), and would have played a major role in the delineation of the platform at the junction of the lower–middle Aalenian boundary (see below). In the Sierra de Albarracín area, between Gea de Albarracín and San Blas (Embalse del Arquillo section) near Teruel, this sequence is highly incomplete, with upper Toarcian and Aalenian strata being partly absent. The Early–Middle Jurassic boundary is here marked by a wide stratigraphic gap (Fernández-López & Gómez 1990*b*).

In the marginal areas of the basin the Toarcian ammonite-rich, marly unit (Turmiel Fm) is poorly developed or even absent (Fig. 11.8). The middle–upper Toarcian succession comprises nodular limestones with a thickness of over 30 m in the Valencia area (SE Iberian Ranges), at localities such as Alcublas and Sot de Chera (Gómez 1979). In the southwestern areas (Cuenca province, Sierra de Altomira; south of Albacete) it is represented by dolomitic or oolitic limestone facies (Morillo & Meléndez 1979). In the northeastern, Catalonian margins, across palaeogeographically high areas such as Maestrazgo (eastern Iberian Ranges) and Tarragona (Catalonian Coastal Ranges), the Turmiel Fm is either absent or else represented by a thin dolomitic sequence. Between these shallow areas a carbonate, outer platform area was developed during early Toarcian times (the Tortosa platform; Fernández-López *et al.* 1996, 1998*a*). This Tortosa platform formed a rather shallow, low-subsidence area during Toarcian times, resulting in deposits of this age commonly being condensed or incomplete. Pliensbachian to early Toarcian deposits (bioclastic limestones) of the Barahona Fm are overlain by the Sant Blai Fm (Fernández-López *et al.* 1996). This unit ranges widely from early Toarcian to early Bajocian in age, varies from a few metres to over 60 m thick, and comprises regularly bedded limestones (mudstone to bioclastic wackestone) with thin intercalations of marls and marly limestones. In the area of Cap Salou, the lower Toarcian Tenuicostatum and Serpentinus zones, are still represented by the underlying, bioclastic limestones of the Barahona Fm (Fig. 11.6). A large stratigraphic gap separates these deposits from the overlying Sant Blai Fm limestones, which are of early Bajocian age here. To the south, Toarcian deposits are represented by the three lower members of the Sant Blai Fm (Fernández-López *et al.* 1998*b*). The Tenuicostatum and Serpentinus zones are represented by the lower unit, the El Caragol Mb, which comprises a variable thickness (0–9 m) of marls and limestones. The late early Toarcian is represented by the Alfara Mb, comprising iron-ooid limestones up to 7 m thick. Finally, the late Toarcian–early Aalenian (Opalinum Zone) part of the succession (Paüls Mb), comprises a bioclastic wackestone interval (0–30 m thick) with flint nodules and thin marly intercalations.

Middle Jurassic

The important discontinuity that roughly coincides with the Lower–Middle Jurassic boundary has been related to the reactivation of normal faults, which resulted in the break-up of the wide Lower Jurassic carbonate platform of eastern Iberia into small NE–SW orientated sub-basins and platforms (Fig. 11.9). As a result of this synsedimentary tectonic activity, the Middle Jurassic stratigraphic record is irregular and discontinuous, showing considerable thickness and facies variations, as well as common non-sequences of variable duration and extent. In the southeastern areas volcanic rocks and pyroclastic rocks of the NW–SE trending Caudiel volcanic belt are interbedded with upper Pliensbachian and Toarcian sediments (Fig. 11.8). This volcanic activity, which extended further

south during Middle Jurassic times to form the Alcublas pyroclastic belt (Fig. 11.9), has been related to large NW–SE faults reactivated at the Early–Middle Jurassic transition (Gómez 1979; Ortí & Vaquer 1980).

In northeastern areas, south from the Tarragona high, the Tortosa platform was still active as a relatively subsident and deep sedimentary basin (Fernández-López *et al.* 1996, 1998*a*). Farther south, in the palaeogeographically high areas of Maestrazgo, Middle Jurassic sequences are normally dolomitized or poorly represented and record inner platform facies (Canerot 1971; Giner 1980). In the central part of the basin there existed two open sedimentary areas, known as the Aragonese platform and the Castillian platform. These platforms were separated by the Montalbán–Ejulve high which was dominated by oolitic facies from late Bajocian times onwards and connected to the south with the Maestrazgo high (Bulard 1972; Gómez 1979; Aurell *et al.* 1999*b*). The Aragonese and Tortosa platforms were connected through a narrow palaeogeographic corridor, the Beceite seaway, in which Middle Jurassic sequences are strongly condensed, the total thickness rarely exceeding 25 m. Further SW from this central area Middle Jurassic deposits are typical of inner platform environments. Data from well-logs at Cuenca and Albacete have revealed the local presence of oolitic limestones and dolostones with interbedded evaporites (Morillo & Meléndez 1979). Finally, to the NW, in the provinces of Soria and Rioja, the Iberian basin was connected with the Basque-Cantabrian basin via the Soria Seaway, a narrow corridor bounded to the north by the uplifted Ebro massif. This emergent area became active from the Bajocian–Bathonian transition onwards, as can be inferred by the increase in siliciclastic supply (Bulard 1972; Wilde 1990).

From a lithostratigraphic point of view, Middle Jurassic units across wide areas in the central part of the basin have all been placed within the Chelva Fm, a generally thick, well-bedded carbonate unit (Gómez 1979; Gómez & Goy 1979). Several different lithological intervals can be recognized within this important unit, which can be subdivided into two main facies: an open, relatively deep platform biomicrite facies association, typical of middle to outer carbonate platform environments, and a shallow, oolitic to dolomitic facies, typical of inner platform conditions. Both facies groups may interfinger and develop unequally, with the oolitic–dolomitic facies reaching its maximal lateral extent from latest Bajocian to early Bathonian times (Fig. 11.9). A different lithostratigraphy exists on the Tortosa platform (Catalonian Coastal Ranges) where the Middle Jurassic succession includes the upper part of the Sant Blai Fm, the marly Cardó Fm, and the thick limestones and dolostones of La Tossa Fm, together ranging in age from latest Bajocian to middle Callovian (Fig. 11.6; Fernández-López *et al.* 1996, 1998*a,b*).

The Middle Jurassic sequence in the Iberian basin has been the subject of numerous studies in many sections across the Iberian Ranges where it is represented by ammonite-rich, outer platform carbonate facies. In the northwestern part, the Castrovido (Ureta 1985), Talveila (Fernández-López *et al.* 1988*a*) and Muro de Aguas (Ureta 1988) sections are especially noteworthy. In the Aragonese platform the most relevant sections are Ricla (Sequeiros & Cariou 1984; Cariou *et al.* 1988; Meléndez & Lardiés 1988), Aguilón (Sequeiros & Meléndez 1979; Martínez *et al.* 1997), Moyuela (Ureta *et al.* 1999), Moneva (Lardiés 1990; Aurell *et al.* 1999*a*), Obón (Mouterde *et al.* 1978, Fernández-López 1985) and Ariño-Oliete

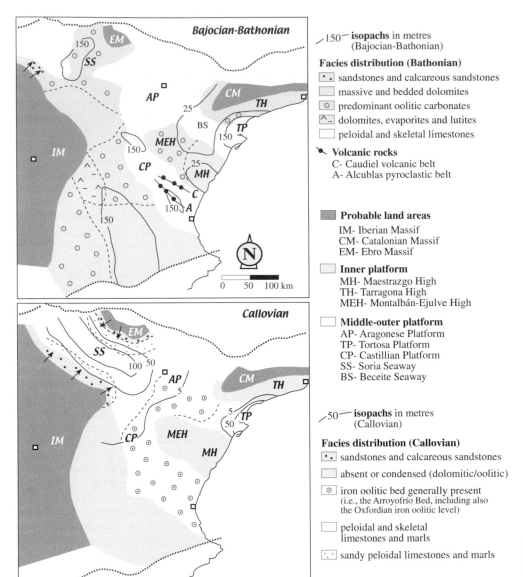

/150— **isopachs** in metres
      (Bajocian–Bathonian)

Facies distribution (Bathonian)

⌞•˙⌟ sandstones and calcareous sandstones

☐ massive and bedded dolomites

⊙ predominant oolitic carbonates

⌐˄‿⌐ dolomites, evaporites and lutites

☐ peloidal and skeletal limestones

Volcanic rocks
   C- Caudiel volcanic belt
   A- Alcublas pyroclastic belt

▇ **Probable land areas**
   IM- Iberian Massif
   CM- Catalonian Massif
   EM- Ebro Massif

▨ **Inner platform**
   MH- Maestrazgo High
   TH- Tarragona High
   MEH- Montalbán-Ejulve High

☐ **Middle-outer platform**
   AP- Aragonese Platform
   TP- Tortosa Platform
   CP- Castillian Platform
   SS- Soria Seaway
   BS- Beceite Seaway

/50— **isopachs** in metres
      (Callovian)

Facies distribution (Callovian)

⌞•˙⌟ sandstones and calcareous sandstones

☐ absent or condensed (dolomitic/oolitic)

⊙ iron oolitic bed generally present
   (i.e., the Arroyofrío Bed, including also
   the Oxfordian iron oolitic level)

☐ peloidal and skeletal
   limestones and marls

⌞˙˙⌟ sandy peloidal limestones and marls

Fig. 11.9. Palaeogeography and facies distribution of east Iberia during Bajocian–Bathonian and Callovian.

(Meléndez *et al.* 1997). In the Castillian platform the Middle Jurassic succession has been studied in detail at Fuentelsaz (Goy & Ureta 1986, 1990; Perilli 1999), El Pedregal-Pozuel (Villena *et al.* 1971; Fernández-López 1985), Aguatón-Sierra Palomera (Fernández-López *et al.* 1985), Moscardón (Fernández-López *et al.* 1978; Giner & Barnolas 1980; Fernández-López 1985), Gea de Albarracín-San Blas (Fernández-López 1985; Fernández-López & Gómez 1990*b*) and Chelva-Domeño, in the southeastern Iberian Ranges (Gómez 1979).

Sequence stratigraphic analyses of the Middle Jurassic successions have been carried out by Fernández-López (1997). The combined analysis of facies and taphonomic features of ammonite assemblages in the outer platform areas has led to the recognition of four main sedimentary deepening–shallowing-upward cycles bounded by important, widespread discontinuities which can be recognized across the east Iberian basin. These four sequences are discussed in more detail below.

Sequence 1: Aalenian to lower Bajocian. This first Middle Jurassic sequence rests directly on the discontinuity that approximates to the Lower–Middle Jurassic boundary (see above). According to Fernández-López (1997) this sequence overall represents a progressive deepening process followed by a shallowing event at the upper Humphriesianum Zone. The stratigraphic record of this interval is mostly irregular, sparse and discontinuous. In particular, the Aalenian–earliest Bajocian (Discites and Leviuscula zones) record contains stratigraphic gaps that show very different ranges in different parts of the basin. These discontinuities may be associated with condensed levels containing phosphatic and ferruginous ooids and may also represent subaerial exposure events, as has been demonstrated in Sierra de Albarracín, west from Teruel (Fernández-López & Gómez 1990*a,b*). Average values of sediment thickness for this interval range from 5 to 20 m.

Northwards from the Noguera–Aguatón fault (Fernández-López & Gómez 1990*a*; see above), the Opalinum and Murchisonae (p.p.) zones are usually well developed whilst the Bradfordensis and Concavum zones, as well as the lowermost Bajocian Discites and Laeviuscula zones, are usually incomplete or even absent. In Sierra Menera (i.e. at Pozuel-El Pedregal section) and eastern areas

(Obón–Montalbán) middle and upper Aalenian sediments are virtually absent. In other areas, however, such as Aguilón–Belchite (south from Zaragoza), Aguatón (in Sierra Palomera), Moscardón (Sierra de Albarracín) and around Valencia (at Chelva and Domeño), the Aalenian stage is represented by one, or several, irregular reworked horizons containing phosphatic and ferruginous ooids. Similar levels containing reworked ammonites derived from the lowermost Bajocian succession have also been reported from the northwestern Iberian Ranges (Ricla) and the Soria seaway area (Talveila).

The range of this late Aalenian–early Bajocian hiatus is progressively reduced northwards on the Aragonese platform. In the Lécera–Moyuela area, south from Zaragoza, middle Aalenian, Murchisonae and Bradfordensis zones are represented by a condensed sequence (1.8 m). In the NW Iberian Ranges, the Sierra de la Demanda (Burgos province) and Sierra de Cameros (Muro de Aguas), all Aalenian zones are more or less well-developed in carbonate facies (Ureta 1988; Ureta et al. 1999).

The late early Bajocian interval (Propinquans and Humphriesianum zones) comprises a generally thicker sequence displaying values from 20 to 50 m and a mainly continuous record. Sponge facies and bioherm developments are locally recorded, as seen at Talveila (NW Iberian Ranges). Fossiliferous biomicrite (mudstone to packstone) facies are generally dominant, containing brachiopods, bivalves, ammonites, belemnites, sponges, echinoids and crinoids, as well as bioturbation traces, such as Zoophycos and Thalassinoides. They are developed as several metre-thick, coarsening and thickening, shallowing-upward carbonate sequences. Maximum sediment thicknesses occur in the central-western part of the Castillian platform (El Pedregal–Pozuel section), whereas minimum thicknesses are found farther NE in the Aragonese platform (Obón section; c. 5 m thick).

In the Beceite seaway, lower Bajocian deposits are irregular and discontinuous, and are represented by outer platform carbonate successions, generally less than 10 m thick. Farther east, on the Tortosa platform, this interval includes the upper two members of the Sant Blai Fm (see above), the Tivenys and Salou members. The Tivenys Mb shows a patchy distribution, being recorded in some areas only, where it is represented by up to 2 m of phosphatic to ferruginous-ooid wackestones, usually associated with reworked levels including common litho- and bioclasts and condensed ammonite associations characterizing the Concavum, Discites and Laeviuscula zones (uppermost Aalenian to lower Bajocian). The Salou Mb, in turn, comprises interbedded mudstone and marl, and corresponds to the early Bajocian, Propinquans and Humphriesianum zones, although its upper boundary may be diachronous at a zonal scale. The whole sequence shows a wide range in thickness, from less than 1 m to over 60 m (Fernández-López et al. 1998b).

Sequence 2: Late Bajocian. At the early–late Bajocian boundary a general deepening trend is recorded throughout the Iberian basin. Maximum depth values were reached in the Niortense–lower Garantiana zones, interrupted by a shallowing trend that began at the Niortense–Garantiana zone boundary and reached a maximum at the end of the Bajocian stage (upper Parkinsoni Zone), when regional discontinuity developed (Fernández-López 1985, 1997; Fernández-López & Meléndez 1995). Mean thicknesses for this upper Bajocian sequence range between 20 and 60 m, although in the south (in the Chelva-Domeño sections of Valencia province) thicknesses never exceed 5 m. Biomicritic, mudstone to wackestone facies are generally dominant forming thickening, shallowing-upward successions, and sponge facies are also important, locally developing as bioherms of variable size. The age of such bioherms ranges from south to north: from Niortense Zone in Sierra de Albarracín (Moscardón), Garantiana Biochron in Ricla (SW from Zaragoza) to Parkinsoni Biochron in Talveila (NW Iberian Ranges).

To the east, on the Tortosa platform, this sequence is represented by a thick marly unit, the Cardó Fm (Cadillac et al. 1981; Fernández-López et al. 1996) and by a further thick carbonate unit, La Tossa Fm. The Cardó Fm is formed of bioturbated mudstones and marly limestones with common traces of Zoophycos, interbedded with marls. Ammonites are common along with pelagic bivalves (Bositra) and more scarce belemnites, brachiopods and benthic bivalves. Deposits of this unit generally form metric, thickening and coarsening,

shallowing-upward sequences. The thickness of this unit may exceed 100 m although it thins out abruptly towards the northern and western margins of the basin to less than 5 m. The age of this unit also shows important variations: in the central part of the basin the lower boundary is early Bajocian in age, whilst in the western areas it is late Bajocian (Niortense Zone). Towards the top, the Cardó Fm passes gradually into the overlying La Tossa Fm, a thick carbonate unit formed of limestones and dolostones. Three intervals are recognized within the La Tossa Fm, only the lowest of which (5–35 m thick) still belongs within the upper Bajocian succession. This latter unit comprises well-bedded limestones (mudstone to wackestone) arranged in metric, thickening, shallowing-upward sequences, containing abundant ammonites and bivalves (Bositra) as well as trace fossils such as Zoophycos and Thalassinoides (Fernández-López et al. 1996, 1998a).

Sequence 3: Bathonian. The Bajocian sequence ends with a general, widespread discontinuity of regional extent which partly affects the upper Bajocian deposits (Parkinsoni Zone) and, in some areas, the lowermost Bathonian, Zigzag Zone (Mensink & Mertmann 1984; Fernández-López 1985, 1997; Wilde 1990). According to Fernández-López (1997) Zigzag to Progracilis zones record a deepening phase in the basin whereas during late Bathonian times a progressive shallowing led to the development of a general stratigraphic discontinuity at the Bathonian–Callovian boundary (Discus Zone).

As in the preceding sequences, several different facies groups can be distinguished in this interval, which can be classified as showing a biomicritic facies typical of an outer, somewhat deep platform, and a dolomitic and grain-supported facies (bioclastic, peloidal and oolitic grainstone to packstone) more typical of shallower, proximal areas of the platform. From the Bajocian–Bathonian boundary onwards the shallower water facies became widely established across the basin (Fig. 11.9), especially north of the Maestrazgo region, on the Montalbán–Ejulve high (Bulard 1972; Gómez 1991; Aurell et al. 1999b) and in the Soria seaway, where terrigenous supply from the Iberian Massif also gave rise to siliciclastic facies (Wilde 1990).

Middle to outer platform facies rocks, located on both margins of the Soria seaway, are represented by a unit of silty, filament-rich peloidal limestones up to 100 m thick. Farther SE, in the Aragonese platform, siliciclastic influence is lower, and the Bathonian succession comprises 20–40 m of mostly wackestone to packstone, filament-rich, peloidal limestones. An exception occurs in the deposits of the Beceite seaway, at the eastern end of the Aragonese platform (Fig. 11.9), where Bathonian deposits show the irregular, discontinuous sedimentation of typically outer platform peloidal limestones (filament-rich wackestone to packstone), and the total thickness rarely exceeds 10 m (Bulard 1972; Meléndez et al. 1999).

Farther east, on the Tortosa platform (Fig. 11.9), Bathonian deposits correspond to the middle part of the La Tossa Fm. They are represented by 40–70 m of well-bedded limestones (mudstone to wackestone), which gradually pass westwards into massive and cross-bedded dolomitized oolitic grainstone to packstone. Dolostones are likewise the dominant lithology across large areas on the Maestrazgo and Tarragona highs, where they show variable thickness (40–250 m; Canerot 1971; Cadillac et al. 1981; Fernández-López et al. 1996, 1998a).

Sequence 4: Callovian. The Bathonian–Callovian boundary is marked by a sharp, widespread stratigraphic discontinuity of variable range (Gómez 1979; Mensink & Mertmann 1984; Lardiés 1990; Wilde 1990; Fernández-López 1997), with the most extreme taphonomic and stratigraphic condensation being reached during late Bathonian times, equivalent to the Subcontractus and Discus zones (Fernández-López 1997). As a general rule, the lower Callovian succession was deposited during a deepening interval in the basin, whilst middle and upper Callovian sediments provide evidence of shallowing, which reached a maximum at the Callovian–Oxfordian boundary. With the notable exception of some localities in the northwestern sector of the basin (e.g. Ricla), upper Callovian, Athleta and Lamberti (p.p.) zone deposits are either absent or else extremely reduced as a reworked level. In the central and southern part of the basin, this condensed section shows ferruginous ooids (Bulard 1972; Sequeiros & Meléndez 1979; Sequeiros & Cariou 1984; Lardiés 1990).

The Callovian sequence is thickest in the western part of the

Castillian and Aragonese platforms (20–30 m) and in the central and southern part of the Soria seaway (Fig. 11.9) where thicknesses of over 100 m have been recorded (Bulard 1972; Wilde 1990; Goy *et al.* 1979). The succession is typically characterized by a regular alternation of marls and bioclastic–peloidal limestones, which at the margins of the Soria seaway contain a high proportion of siliciclastic grains. In contrast, in the middle–outer areas of the Castillian and Aragonese platforms, as well as in the Beceite seaway, the Callovian succession (which spans the lower Callovian, Bullatus and Gracilis (p.p.) zones) is markedly incomplete, condensed and only rarely thicker than 5 m. It includes a thin lithological interval formed by well-bedded to undulating yellowish bioclastic (biomicritic) limestones with ferruginous ooids, the latter facies belonging to the lower part of the Arroyofrío Bed. This ferruginous ooidal horizon might partly represent the middle Callovian, Anceps Zone (Lardiés 1990), although intense reworking affecting the ammonites recorded in this bed makes it difficult precisely to assess its age, as most of the lower to middle Callovian ammonite specimens it contains are reworked internal moulds (Lardiés 1990; Aurell *et al.* 1994a; Meléndez *et al.* 1997).

Callovian deposits appear to be completely absent in the wide area represented by the Montalbán–Ejulve and Maestrazgo highs (Fig. 11.9). Across this area, middle Oxfordian fossiliferous limestones (carbonate outer platform facies) overlie an erosional uneven surface that cuts a Bathonian oolitic–dolomitic limestone sequence. In the southwestern part of the Iberian basin (Serranía de Cuenca) Bathonian deposits are mostly represented by inner platform oolitic and dolomitic facies (Meléndez & Ramírez del Pozo 1972). However, in this area, as in the central part of the Soria seaway, Callovian outer platform transgressive deposits overlie Bathonian inner platform sediments (Fig. 11.9). Further evidence for a regional transgressive trend during early Callovian times is reflected in the record of fining- and thinning-upward sequences of the Gracilis Zone in western areas. However, tilting of blocks during this interval in central and eastern areas led to an early shallowing of the platform at the early–middle Callovian boundary, and the development of condensed iron–oolitic Callovian to middle Oxfordian sequences, thus producing a reversed sedimentary trend (Fernández-López & Meléndez 1995). This long interval of subaerial to shallow marine conditions was punctuated by ephemeral flooding episodes marked by sedimentation of the iron-ooid Arroyofrío Bed, which contains fossils derived from deeper water environments, most remarkably phylloceratid ammonoids.

In the NE of the Tortosa platform (Fig. 11.9), Callovian deposits reach maximum thicknesses of *c.* 45 m in outer platform carbonate facies. They constitute the upper half of the La Tossa Fm: fossiliferous limestones (filament mudstone to wackestone) and marly limestones, ranging in age from early to middle Callovian, Bullatus to Anceps Zone. This unit thins to the west, at the Beceite seaway, where it is absent or very poorly represented (Fernández-López *et al.* 1996, 1998a).

Late Jurassic

The Middle–Upper Jurassic boundary is marked by an important sedimentary discontinuity across the eastern Iberian basins. This transition has been the subject of numerous detailed studies, which have demonstrated the stratigraphic gap to be of variable range (Bulard 1972; Bulard *et al,.* 1974; Gómez 1979; Giner 1980; Meléndez *et al.* 1983, 1990; Aurell *et al.* 1994b; Ramajo & Aurell 1997). The gap is minimal in the area of the Soria seaway and northwestern Aragonese platform (Ricla–Aguilón) where it spans at least the latest Callovian to earliest Oxfordian interval (i.e. Lamberti and Mariae zones). In contrast, the maximum range is reached in shallow uplifted areas to the east, such as the Montalbán–Ejulve high (Fig. 11.9), where a middle Oxfordian sponge limestone unit (Yátova Fm) lies with erosional unconformity on a presumed Bathonian succession of oolitic limestones and dolostones typical of inner platform conditions.

Upper Jurassic palaeogeographic reorganization of eastern Iberia led to the development of a wide, homogeneous carbonate ramp showing a much less pronounced relief than previously, and dipping more or less uniformly eastwards. Coeval sinking of the Maestrazgo high led to the development of open platform facies from middle Oxfordian times onwards (Canerot 1971; Fig. 11.10). Fault reactivation then induced facies change and a sudden increase in sedimentary thickness from latest Oxfordian times (Salas 1989). Farther north, on the Tarragona high (in the Garraf massif) although Lower and Middle Jurassic deposits are under-represented or locally absent, Kimmeridgian to lower Tithonian deposits form a thick dolomitic succession generally deposited in shallow platform environments (Esteban & Juliá 1973; Giner 1980). Middle Oxfordian to Kimmeridgian outer platform carbonate facies reached the southwestern areas of the basin (Cuenca province), an area occupied by shallow, inner platform facies during Early and Middle Jurassic times. To the NW, during the Oxfordian to early Kimmeridgian interval, the central part of the Soria seaway was occupied by inner platform facies, and it was not until the end of Kimmeridgian times that uplift of the western margin of the basin (Iberian Massif) induced the eventual emergence of the Soria seaway and the displacement of the coastline to the east (Fig. 11.10).

The upper boundary of the dominantly Jurassic supersequence is normally an angular unconformity linked to a phase of tectonic activity and the deposition of continental facies sediments in fault-bounded basins. In the eastern Iberian Ranges, this unconformity is located above the lowermost Cretaceous deposits (lower to mid-upper Berriasian) which are therefore included within the supersequence (Aurell & Meléndez 1993; Salas & Casas 1993; Aurell *et al.* 1994b). A stratigraphic gap of variable range is associated with this unconformity.

Sequence stratigraphic studies of the Upper Jurassic succession in eastern Iberian basins carried out by various authors over the last twenty years have led to the recognition of eight depositional sequences (Giner 1980; Salas 1989; Aurell 1991; Aurell & Meléndez 1993; Aurell *et al.* 2000; Bádenas & Aurell in press). Sequences J3.1 to J3.4 are Oxfordian, sequences J3.5 and J3.6 are Kimmeridgian, and sequences J3.7 and K1.1 are Tithonian–Berriasian in age (Figs 11.6 and 11.7).

Oxfordian. Sedimentation during early and early middle Oxfordian times in the Iberian basin was discontinuous and irregular, and the sedimentary record is only preserved in the middle and outer areas of the platform. In these areas a single level, up to 1 m thick, contains abundant upper lower to lower middle Oxfordian ammonites. Quite commonly, lower Oxfordian specimens are associated with reworked internal moulds of Callovian ammonites. Among other common fossil groups are belemnites, echinoderms, bivalves, foraminifera, sponges and brachiopods. Ammonites normally show consistent evidence for taphonomic dispersal (post-mortem shell drifting) and allochthony (Fernández-López & Meléndez 1995). Also in central and eastern areas this level contains abundant ferruginous ooids and pisoids, and corresponds to the upper part of the Arroyofrío Bed (Fig. 11.9). This condensed bed is generally interpreted as having been deposited on a shallow carbonate platform during a low relative sea level, which would have been responsible for widespread emergence of marginal areas of the basin and the stratigraphic gap that exists at the Callovian–Oxfordian boundary in platform areas. Carbonate sedimentation in the outer platform areas was reduced or non-existent, hence leading to repeated reworking of sediments by current action. Ammonites in these beds are most commonly preserved as fragmented internal moulds displaying clear traces of reworking such as abrasion or a sharp difference between the infill and the sedimentary matrix (Aurell *et al.* 1994b). Such abrasion surfaces are interpreted as developed by the action of currents on an

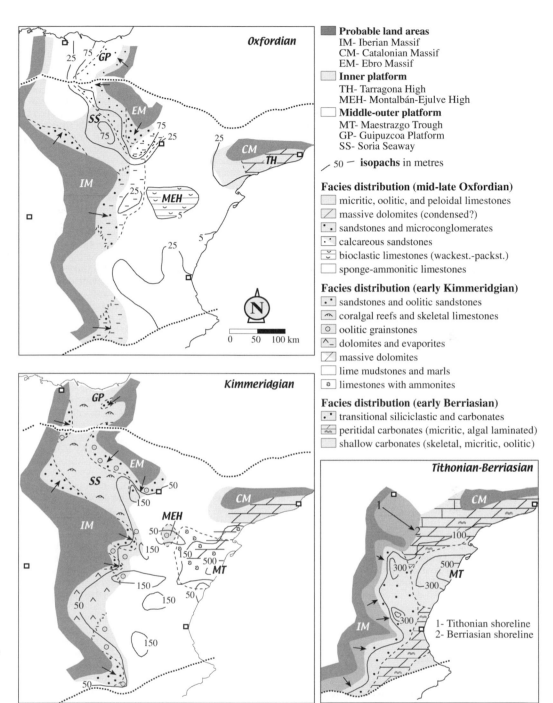

Fig. 11.10. Palaeogeography and facies distribution of east Iberia during Oxfordian, Kimmeridgian and Tithonian.

exhumed fossil ammonite under shallow-water conditions (Fernández López & Meléndez 1994). The origin of ferruginous ooids and pisoids has been the subject of discussion. They might have been formed within the platform under shallow subtidal conditions or, alternatively, resedimented and transported from adjacent emergent areas (Aurell & Meléndez 1993; Aurell *et al.* 1994a; Ramajo & Aurell 1997).

Ammonites from the iron–oolitic limestone Arroyofrío Bed, or from the laterally equivalent biomicrite level at Ricla and adjacent areas, record the existence across the platform of two ephemeral flooding episodes, leading to the deposition of sequences J3.1 and J3.2. The lower sequence (J3.1) is early Oxfordian in age and produced by widespread flooding during the lower Claromontanus (or Cordatum)

Zone (Meléndez *et al.* 1983; Meléndez 1989). Its upper boundary is marked by a stratigraphic gap, comprising the middle and upper part of Cordatum Zone (the Costicardia and Cordatum subzones). The upper sequence (J3.2) is early middle Oxfordian in age and is interpreted as corresponding to a second, short flooding event. It is represented by an upper ferruginous ooid band, containing ammonites belonging to the lower Plicatilis Zone (i.e. Paturattensis or Vertebrale Subzone). A further stratigraphic gap on top omits the upper part of the Plicatilis Zone (Antecedens Subzone) and, in the more eastern areas, the lowermost Transversarium Zone (Parandieri to lower Luciaeformis Subzone).

During the middle Oxfordian Transversarium Zone, another important transgressive event produced a general submergence of the basin, favouring the colonization of the platform by sponges and other benthic and nektoplanktonic communities (Aurell & Meléndez 1993; Ramajo *et al.* 1999). The homogeneity of the basin, inherited from early Oxfordian times, and general tectonic stability during this period, led to the development of a wide, carbonate ramp. Sponge limestone facies (Yátova Fm) typical of an outer ramp environment are widespread across the central and eastern part of the basin and show no significant facies variation. Mean thickness values range from 10 to 20 m although reducing to less than 5 m across shallower areas such as the Montalbán–Ejulve high, where the Yátova Fm is represented by bioclastic packstones (Fig. 11.10). Farther south, in areas formerly occupied by thick inner platform Lower and Middle Jurassic deposits (Cuenca, La Mancha), middle and upper Oxfordian sediments reach maximum thickness values (up to 45 m). The Yátova Fm exhibits great fossil diversity, including sponges, crinoids, brachiopods, bivalves, belemnites, echinoids and serpulids. Especially noteworthy is the record of abundant ammonites that have allowed a precise recognition of the successive zones of the middle and late Oxfordian (Meléndez 1989; Meléndez & Fontana 1993; Meléndez *et al.* 1997).

More diversified facies, typical of inner ramp environments, are developed in the marginal areas of the basin. Southwards from the Tarragona high (Fig. 11.10) this interval is occupied by oncolitic, oolitic and peloidal facies, corresponding to the Serra de la Creu Fm (Salas 1989; Fernández-López *et al.* 1996; Aurell *et al.* 1999*a*). In the central part of the Soria seaway, the Oxfordian sequence comprises a 25 to 75 m thick siliciclastic and calcareous unit, the Aldealpozo Fm, deposited in a peritidal environment and organized in shallowing-upward sequences. The facies shows a locally high terrigenous, siliciclastic supply from uplifted areas to the NE (Ebro massif) and NW (Iberian Massif; Alonso & Mas 1990). To the north, beyond this narrow area, these inner ramp carbonates grade laterally into deeper facies connecting the Iberian basin with the Basque-Cantabrian basin.

In the northwestern part of the Iberian basin, a stratigraphic discontinuity is recorded between the upper Oxfordian Hypselum and Bimammatum zones. This discontinuity has been recently proposed as the sequence boundary between sequences J3.3 and J3.4, the latter comprising marls with bioclastic, locally glauconitic intercalations. In the western, proximal areas (the area of the Sierra Menera, in Teruel province) a sudden increase in terrigenous supply is recorded as early as Bifurcatus Zone by a thick marly unit (over 60 m) with sandy limestone interbeds (Corbalán & Meléndez 1986). The discontinuity between sequences J3.3 and J3.4 is located within this unit, and marked by a sharp hardground surface. Towards the more distal areas of the platform these facies grade into the more typical sponge limestone facies, with abundant ammonites. In such areas the sequence boundary J3.3–J3.4 can be more difficult to recognize.

Kimmeridgian. During the Kimmeridgian stage the same sedimentary conditions persisted as for the Oxfordian, i.e. a wide homogeneous carbonate ramp open to the east (Fig. 11.10). Nevertheless at the Oxfordian–Kimmeridgian boundary a sharp sedimentary change takes place. Thick marl and mudstone sequences were deposited in the middle and outer ramp areas whilst reefal and oolitic facies were developed in the more proximal and shallow areas of the platform.

In the outer platform areas an important sedimentary discontinuity is evident in the lower part of the Planula Zone, and is marked by a bored hardground surface locally rich in glauconite. The associated stratigraphic gap may partly affect the underlying, Hauffianum Zone and the lower part of the Planula Zone (i.e. the Tonnerrense or Proteron Subzone: Aurell *et al.* 1999*a*). In the central part of the Soria seaway this boundary is an irregular surface showing traces of subaerial exposure and a stratigraphic gap of variable range (Alonso & Mas 1990). This discontinuity is related to an increase of tectonic activity detected at the Oxfordian–Kimmeridgian boundary, and which led to the development of sedimentary troughs showing markedly different subsidence rates, including the local presence of angular unconformities (Aurell *et al.* 1997; Bádenas 2001). In the central part of the basin a series of subsident troughs was formed, causing the development of thick sedimentary sequences during Kimmeridgian times. Mean thickness values normally range from 150 to 225 m, whilst maximum values, up to 600 m, have been recorded in

the Maestrazgo trough (Salas 1989; Salas & Casas 1993; Bádenas & Aurell in press).

In the areas of the Soria seaway coral and algal reef facies corresponding to the Torrecilla Fm developed from lower Kimmeridgian times onwards. These facies show higher siliciclastic content in the marginal areas (Benke *et al.* 1981; Alonso & Mas 1990). Shallow carbonate and siliciclastic facies extend to the eastern Basque-Cantabrian basin, to the so-called Guipuzcoa platform (Pujalte 1989*a*; Bádenas 1996), and in the western margin of the Iberian basin (Bádenas *et al.* 1993; Aurell & Bádenas 1997; Bádenas & Aurell 1997, 2001). In certain areas (such as in the SW: the Buenache de la Sierra Fm) a local record of dolostones and evaporitic facies records the development of coastal sabkha environments in the innermost parts of the basin (limestones and brecciated dolomites: Meléndez & Ramírez del Pozo 1972; Morillo & Meléndez 1979). In the Montalbán–Ejulve high, Kimmeridgian deposits comprise mainly oolitic and reefal facies (Aurell *et al.* 1999*b*; Bádenas & Aurell 2001).

In the more outer platform areas, lower Kimmeridgian deposits are represented by the Sot de Chera Fm comprising up to 50–100 m of marls and followed by the Loriguilla Fm, a thick (50–150 m) rhythmic alternation of marls and mudstones (Gómez & Goy 1979). Siliciclastic supply is more evident in the basin margins, but becomes increasingly weak eastwards where the marls of the Sot de Chera Fm grade progressively into carbonate micrite facies of the Loriguilla Fm, typical of outer ramp environments. The origin of the carbonate mud in this unit has been related to the oolitic and reefal facies of marginal areas characterized by a high carbonate production (Aurell *et al.* 1995, 1998; Bádenas & Aurell 2001). In the outer, eastern areas of the basin (Calanda region) the Loriguilla Fm contains abundant ammonites which have allowed recognition of the successive zones of the Kimmeridgian to lower Tithonian succession (Geyer & Pelleduhn 1979; Meléndez *et al.* 1990). Some of these zones have been also reported in the southern edge of the basin, in the Fuentealamo section (Behmel 1970).

Two depositional sequences are defined within the Kimmeridgian succession, separated by a stratigraphic discontinuity that is recognizable as a transgressive surface in the marginal areas of the basin. The first sequence, J3.5, extends from early Kimmeridgian (Planula Zone) to early late Kimmeridgan (upper part of the Acanthicum Zone) in age. The sedimentary evolution of this sequence reflects a progressive sea-level rise during early Kimmeridgian times, followed by a highstand at the beginning of the late Kimmeridgian interval. In the Soria seaway reefal facies accumulated during this time. Farther southwards oolitic and siliciclastic facies of the so-called Pozuel Fm were developed (Bádenas *et al.* 1993; Bádenas & Aurell 2001). In contrast, in the easternmost block (the Maestrazgo area), thick micritic successions were formed, showing local development of sponge bioherms (the Polpis Fm; Salas 1989).

The second Kimmeridgian depositional sequence (i.e. sequence J3.6) ranges from late Kimmeridgian (Eudoxus Zone) to early Tithonian (Hybonotum Zone) in age. During this interval, uplift of the western margin of the basin led to the closure of the Soria seaway and retreat of the sea from the Basque-Cantabrian basin. During the initial transgressive stage a widespread development of algal and coral reef formations took place along the marginal areas of the basin, most remarkably in the Sierra de Albarracín, around Jabaloyas (Giner & Barnolas 1979; Fezer 1988; Errenst 1990; Leinfelder 1993; Aurell & Bádenas 1997). In the outer areas of the platform, reef facies grade into micrites (Loriguilla Fm) which commonly contain sponge-rich intervals (Geyer & Pelleduhn 1979). Farther east, towards the basin depocentre (Maestrazgo trough), the sequence is represented by a thick succession (up to 300 m) of marls and limestones rich in organic matter, which included anoxic episodes (the Mas d'Ascla Fm; Salas 1989).

Tithonian to mid-upper Berriasian. Uplift of the western margin of the basin, combined with a sea-level fall, produced an important regression at the onset of the Tithonian stage (Aurell & Meléndez 1993; Aurell *et al.* 1994*b*). Marine sedimentation became restricted to the eastern Iberian basin, taking place on a shallow platform (Fig. 11.10). In contrast, in the area of the former Soria seaway, a rapidly subsiding continental basin was developed: the Cameros basin (e.g. Martín-Closas 1989). Farther north, in the western Basque-Cantabrian basin, a new continental depocentre was formed during the Jurassic–Cretaceous transition (Campoó Gp, see above).

Two sedimentary episodes, corresponding to two depositional sequences, are currently distinguished within this stratigraphic interval. The first is the Tithonian sequence (J3.7), represented across most of the basin by a 40–80 m thick succession of oolitic, oncolitic, biohermal and bioclastic limestones forming the Higueruelas Fm (Gómez & Goy 1979; Aurell & Meléndez 1993; Aurell et al. 1994b). The lateral equivalent in the eastern, more open area is the Bovalar Fm, best represented in the Maestrazgo trough where it may attain thicknesses close to 850 m (Salas 1989). This formation comprises fossiliferous bioturbated mudstone to wackestone containing abundant remains of algae, foraminifera and bivalves, with oolitic and bioclastic (packstone to grainstone) intervals. These sets of facies form successive coarsening- and thickening-upward sedimentary sequences. Salas (1989) has reported the occasional presence of calpionellid and tintinid wackestone facies, which would represent the most open, deeper areas of the platform. In the northwestern margin of the basin these carbonate deposits are intensely dolomitized, and they are known as the Les Talaies Fm.

The upper sedimentary sequence is lower and mid-upper Berriasian in age (Sequence K1.1). In the Maestrazgo trough, it comprises the upper part of the above-mentioned Bovalar Fm. To the north, the time-equivalent unit is a 40–70 m thick dolomitic and limestone unit called La Pleta Fm, deposited under peritidal environments. It comprises a set of muddy carbonate intervals showing shallowing-upward sequences. This formation includes well-developed laminated algal intervals and marly intercalations containing charophytes which have allowed precise dating of the unit (Martín-Closas 1989). In the southwestern margin of the basin this unit is represented by the Villar del Arzobispo Fm, which is up to 400–500 m thick in the Galve-Cedrillas and Villar del Arzobispo areas (Fig. 11.10). This formation comprises mixed carbonate and siliciclastic (clays and sandstones) lithologies, with local indications of tidal flat conditions (Mas et al. 1984).

The Betic Cordillera and Balearic Islands (FO, AL, AG, JS, JEC, FJR-T, JKT)

Betic Cordillera (FO)

Jurassic rocks in the Betic Cordillera of SSE Spain crop out over an area of 600 × 200 km (Fig. 11.11A). A NW–SE transect across the external part of this orogen moves from the autochthonous–parauthochtonous Prebetic to allochthonous Intermediate Units and Subbetic rocks. The Subbetic region has been subdivided from NW to SE into the External, Median, and Internal Subbetic (Fig. 11.11C), the latter area showing lateral equivalents further SW (Penibetic). Less well understood are the Jurassic rocks known from the allochthonous Internal Zones of the Betics (the Rondaides and Malaguides), and the so-called Betic Dorsal successions which typically crop out at the boundary between the External and the Internal Zones (Martín-Algarra 1987; Vera 1988; Sanz de Galdeano 1997; also see Chapter 16 for Betic tectonostratigraphy). Even more arcane are fragmentary data from Jurassic rocks included as exotic blocks in sandy deposits from the Trough Flysch Units. These blocks are probably derived from palaeoslopes in the Internal Zones (Alboran Domain) and/or the African margin.

Since the early 1970s, the Jurassic evolution of the SSE Iberian palaeomargin has been linked to the spreading history in the central-north Atlantic. During the Jurassic period the complex plate boundary system related to the Maghrebian–Gibraltar Transform Zone (Fig. 11.11B) involved sinistral transcurrent tectonics and the generation of extensional, and locally compressional, strike-slip basins across an area

measuring 300 km or so from NW to SE (e.g. Andrieux et al. 1971; Dewey et al. 1973; Borrouilh & Gorsline 1979, 1980; Bourgois 1980; Mascle 1980; Wildi 1983; Torres Roldan et al. 1986; Martín-Algarra 1987; Guerrera et al. 1993). Within this general tectonic setting, different models have variously emphasized the roles of extension and transtension in the Jurassic evolution of the SSE palaeomargin of Iberia: passive extensional Atlantic-type (Puga 1977; Azéma et al. 1979; García-Hernández et al. 1980; Vera 1988; Ogg et al. 1983; with an increased role for listric faults, half-graben tectonics and tilting in Vera 1998 and Molina et al. 1999b), transtensional (Martín-Algarra 1987; Ziegler 1988b; García-Hernández et al. 1989) and transform–transtensive (Andrieux et al. 1989; Dercourt et al. 1994; Sanz de Galdeano 1997).

Major advances in our understanding of Betic Jurassic geology have mostly been based on facies analysis (Fig. 11.12A) and biochronostratigraphy (mainly ammonites; Fig. 11.12B). The earliest Jurassic transgression resulted in the onset of generalized carbonate shelf conditions that favoured colonization by a neritic–benthic biota (bivalves, crinoids, brachiopods, algae and others). The break-up and drowning of this neritic shelf system occurred diachronously during the Early Jurassic epoch, and hemipelagic to pelagic deposition (occasionally interrupted by resedimentation events) became overwhelmingly dominant from the middle–late Early Jurassic interval (Upper Pliensbachian; i.e. Domerian–Toarcian in peri-Mediterranean Alpine stratigraphy, west Tethys) offshore from the Prebetic area, which persisted as an epicontinental environment throughout Jurassic times (Figs 11.11C, 11.12A). Offshore hemipelagic and pelagic conditions led to the deposition of marly limestone rhythmites in troughs and 'ammonitico rosso' and related facies on underwater swells. The deepest troughs accumulated siliceous marls, and volcanic rocks (central and eastern Median Subbetic) and calcareous turbidites were locally present. Highly condensed deposition on submarine swells, and even local emergence, were typical of late Middle and earliest Late Jurassic times. The differentiation of the area into swells and troughs seaward from the epicontinental (neritic) shelf system (Prebetic) was completed during Late Jurassic times. The principal physiographic features southeastwards across the area were the slope and northern trough adjacent to the Prebetic shelf (= Intermediate units), the northern swells (= External Subbetic), the median trough (= Median Subbetic), the southern swells (= Internal Subbetic and Penibetic) and the southernmost slope of the SSE paleomargin of Iberia (Fig. 11.11C). This increased differentiation into swells and troughs during Late Jurassic time was accompanied by resedimentation, slumping and increased volcanic activity in the median trough, which became topographically subdivided by the growth of a mid-Subbetic volcanic ridge (Comas 1978). Peak transgressive conditions during middle Oxfordian times resulted in major flooding of the Middle Jurassic Prebetic carbonate shelf system (Chorro Fm in Fig. 11.12A), with initial deposition of condensed and variously reworked sediments followed by hemipelagites. Subsequent regression from the early middle Kimmeridgian interval onwards was ultimately responsible for the emergence of land and the development of tidal environments within the Prebetic area towards the end of the Jurassic period. Offshore from the shelf, in the Subbetic area and its lateral equivalents, neither shallowing nor emergence were widespread (although they have been locally identified) during the Late Jurassic epoch. Jurassic lithostratigraphy in the Subbetic Zone (Fig.

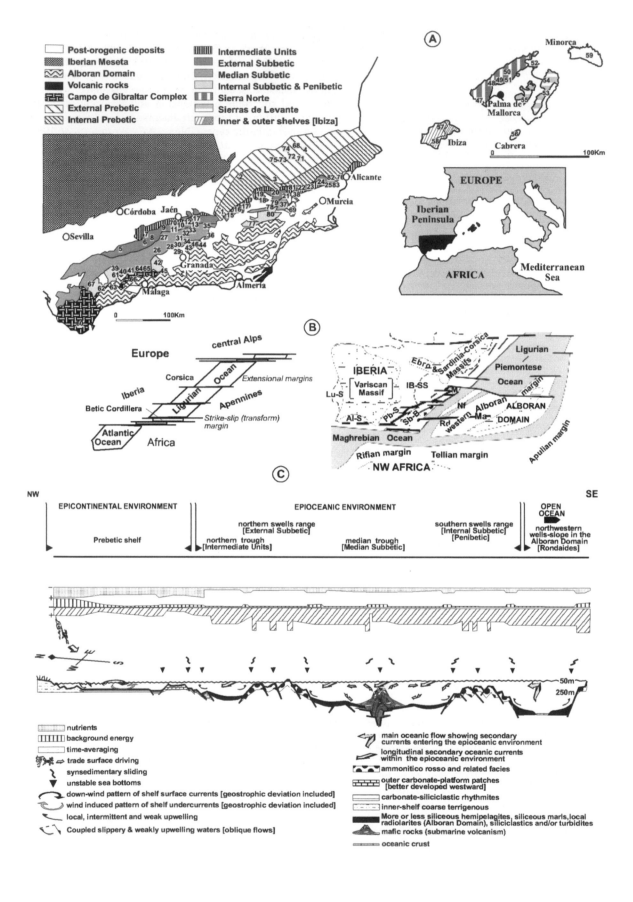

11.12A) has recently been summarized by Rey (1993), Nieto (1996) and O'Dogherty *et al.* (1997).

In the SE palaeomargin of Iberia, two major ecosedimentary environments (Olóriz 2000) existed during the Jurassic period. These were based on the persistence of epicontinental conditions in the Prebetic and the northward irregular progression of epioceanic conditions (ecologically oceanic environments but developed on the continental crust) which reached their northernmost advance early during the Late Jurassic epoch (see above).

Jurassic oceanic crust, and therefore true oceanic environments (Fig. 11.11B), probably existed between Iberia and Africa southwards from the Betic External Zones (e.g. Bernoulli 1972; Puga 1977, 1980; Vera 1988; Wildi 1983; Dercourt *et al.* 1985; Martín-Algarra 1987; Ziegler 1988*b*; Puga *et al.* 1993, 1995; although see Seyfried 1979; Vera & Molina 1998). The age postulated for the appearance of this oceanic crust varies from Early Jurassic (Argyriadis *et al.* 1980; Puga 1980; Puga *et al.* 1993, 1995) to mid-Early Jurassic (Azéma *et al.* 1979), to around the Early–Middle Jurassic boundary (Laubscher & Bernoulli 1977 connecting the Atlantic and the Ligurian-Piedmont and Vardar oceans; Bourgois 1980), and finally to the late Middle and/or earliest Late Jurassic (Vera 1981, 1988; Wildi 1983; Lemoine *et al.* 1987; Ziegler 1988*b*; Andrieux *et al.* 1989; Martín-Algarra *et al.* 1998). Whatever the case, oceanic crust in this region is widely assumed to have been present during the Late Jurassic epoch (references not cited above include e.g. Bernoulli & Lemoine 1980; Dercourt *et al.* 1985; Martín-Algarra 1987; Ricou 1987; Guerrera *et al.* 1993; Vera 1998). Oceanic crust and related oceanic to quasi-oceanic basins of Middle Jurassic age are here assumed to have separated the External Zones of the SE palaeomargin of Iberia from the Internal Zones in the Alboran Domain (Fig. 11.11B), in which intricate swell–trough physiography is envisaged (stepped slopes, perched basins and tilted/rotated blocks), including sea floor with oceanic crust within what was later to become the Nevado–Filabrides Complex (Guerrera *et al.* 1993; Puga *et al.* 1993).

Unconformities linked to major events in the structural evolution of the southeastern palaeomargin of Iberia during the Jurassic period (Fig. 11.12A) have been identified and interpreted mainly in terms of tectonic–eustatic interactions (García-Hernández *et al.* 1989; Marques *et al.* 1991; Rey 1993; Nieto 1996; Vera 1998; Molina *et al.* 1999*b*). More specifically they have been related, mainly, to: (1) the break-up of an Early Jurassic carbonate shelf (a well-known event in western Tethys; e.g. Bernoulli & Jenkyns 1974); (2) incipient oceanic spreading during late to latest Middle Jurassic times, coeval with spreading in the central North Atlantic (Ogg *et al.* 1983); (3) the beginning of deep siliciclastic deposition within the flysch trough bordering northwestern Africa (Wildi 1983); (4) the opening of the basin between the External Zones of the Betic Cordillera and the Alboran Domain (Dercourt *et al.* 1985), including volcanism (Martín-Algarra 1987; Martín-Algarra *et al.* 1998); (5) regional latest Jurassic–earliest Cretaceous events affecting palaeogeography and palaeoceanography all around western Tethys (e.g. Azéma *et al.* 1979; Martín-Algarra 1987; Vera 1988; García-Hernández *et al.* 1989; Marques *et al.* 1991; Cecca *et al.* 1992). On an even broader scale, discontinuities and major ecosedimentary events during the Late Jurassic epoch have been identified in epicontinental shelves surrounding Iberia and northwestern Africa (Marques *et al.* 1991) and correlated with major disconformities in the Gulf of Mexico Basin and related areas (Olóriz *et al.* in press). For further references and an overview of present knowledge and perspectives for future research in the Jurassic of the Betic Cordillera, see Vera (1998).

Balearic Islands (FO)

The Balearic archipelago is bounded by the Valencia trough (NW), the Provence basin (NE), and the North Africa basin (south) (Fig. 11.11A; M in Fig. 11.11B for palaeogeography). Majorca, the largest island, shows an overall horst–graben structure produced by late Miocene–Quaternary listric faults superimposed on an Alpine thrust system that had previously induced a shortening of around 50–56% (Ramos-Guerrero *et al.* 1989). Geomorphologically, Majorca has traditionally been subdivided into three areas: Sierra Norte (Serra de Tramuntana; NW mountain range or northern horst), Llanos Centrales (Central Plains or intermediate graben, filled by Late Miocene–Quaternary deposits) and the Sierras de

Fig. 11.11. (A) Location and geological sketch with selected Jurassic outcrops (1 to 83) in the Betic Cordillera and the Balearic archipelago. 1, Puerto Lorente; 2, Segura de la Sierra; 3, Elche de la Sierra; 4, Fuente Alamo-Cerrón; 5, Sierra de Estepa; 6, Sierra de Gaena-Carcabuey; 7, Sierra de Cabra; 8, Sierra de los Judios; 9, Sierra del Ahillo; 10, La Coronilla-Collado de Gracia; 11, Cornicabra-Ventisquero; 12, Otiñar-Rio Frío; 13, Cárcel-Las Pilas, Puerto Rico; 14, La Cerradura; 15, Puente Duda; 16, Cortijo Mazagrán-Sierra de la Sagra; 17, Sierra de Jorquera; 18, Sierra de Mojantes; 19, Cortijo de Majarazán; 20, Cerro de Mai Valera; 21, Sierra de Quipar; 22, Sierra de Lugar; 23, Sierra del Corque; 24, Sierra de Crevillente; 25, Sierra del Reclot; 26, Sierra de Chanzas; 27, Sierra de San Pedro; 28, Illora-Sierra Pelada; 29, Sierra Elvira; 30, Colomera; 31, Puerto del Cegrí-Sierra de las Cabras; 32, Sierra de Montillana; 33, Sierra de Alta Coloma; 34, Iznalloz; 35, Cerro Méndez; 36, Río Fardes; 37, Sierra de Ponce; 38, Sierra de Ricote; 39, Sierra de Cañete; 40, Sierra de Huma; 41, Torcal de Antequera; 42, Sierra Gorda; 43–44, northern Sierra Harana (Subbetic); 45, Baños de Alhama; 46, southern Sierra Harana (Alboran Domain); 47, southern Sierra Norte (Andraitx area); 48–51, Central Sierra Norte (Son Patx, Sierra de Cuber, Sierra de Alfabia, Aumedrá); 52, northern Sierra Norte (Alcudia area); 53, central Sierras de Levante; 54, northern Sierras de Levante (Artá-Betlem); 55, Randa-Petra; 56, Sa Cova Blava; 57, NW Ibiza (Eubarca-Fournou-Rey Unit); 58, SE and NE Ibiza (Ibiza Unit); 59, Minorca; 60, El Chivalo; 61, Castillones; 62, La Almola; 63, Convento Nieves; 64, Gallo Vilo; 65, Sierra Prieta; 66, Alpandeire-Ardales area (Nieves Unit); 67, Corredor del Boyar; 68, Chinchilla de Monte Aragón area; 69, Sierra Espuña; 70, Sierra de Jabalcuz–Sierra de La Grana–Cerro de San Cristóbal (Jabalcuz Unit); 71, Cancárix area; 72, Hellín area; 73, Elche-Férez area; 74, Pozo Cañada area; 75, El Entredicho area; 76, Sierra de Fontcalent; 77, Sierra de María; 78, Collado del Lentiscar–Rambla del Gasón; 79, Sierra del Tornajo; 80, Sierra del Almirez; 81, Almorchón-Sierra del Oro (Rameles-Oro Unit); 82, Novelda area (La Mola Unit); 83, Zarzadilla de Totana–Cejo de la Grieta (Canteras Unit). **(B)** Palaeogeographic overview. Left: adapted from Lemoine *et al.* (1987); right: author's interpretation based on Guerrera *et al.* (1993) and Puga *et al.* (1993) with modifications and the assumption of the southeasternmost slope of the Subbetic Basin as being the most distal component of the southeastern palaeomargin of Iberia. Key: Al-S, Algarve shelf; IB-SS, Iberian shelf system; Lu-S, Lusitanian shelf; M, Majorca and related areas, except Minorca; Ma, Malaguide complex; Nf, Nevado-Filabride complex; Pb-S, Prebetic shelf; Rd, Rondaide complex; Sb-B, Subbetic basin. **(C)** Cross section profile synthesis showing mainly Late Jurassic palaeomargins in SE Iberia and SW Alboran Domain (based on Olóriz (2000) with additions by the author).

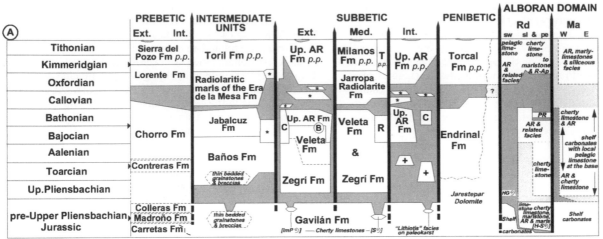

(A) Lithostratigraphic correlation chart of the Betic domains (Prebetic, Intermediate Units, Subbetic, Penibetic, Alboran Domain).

(B) Ammonite biostratigraphic zonation.

STAGE	SUBSTAGE	AMMONITE ZONES	Subzones
TOARCIAN	LATE	AALENSIS	Buckmani / Aalensis / Mactra
		REYNESI	
		FALLACIOSUM	Fallaciosum / Mediterraneum
	MIDDLE	GRADATA	Alticarinatus / Gemma
		BIFRONS	Bifrons / Sublevisoni
	EARLY	SERPENTINUM	Levisoni / Striatus
		POLYMORPHUM	
PLIENSBACHIAN	LATE	EMACIATUM	Elisa / Solare
		ALGOVIANUM	Lavidorsatum / Meneghinii / Accuratum / Bertrandi / Ragazzonii
		LAVINIANUM	Cornacaldense / Portisi
	EARLY	DILECTUM	
		DEMONENSE	*Metaderoceras* / *Tropidoceras*
		AENIGMATICUM	Circumcrispatum / Aenigmaticum
SINEMURIAN	LATE	RARICOSTATUM	
		OBTUSUM	
	EARLY	TURNERI	
		SEMICELATUM	
		BUCKLANDI	
HETTANG.	LATE	ANGULATA	
		LIASICUS	
	EAR.	PLANORBIS	

STAGE	SUBSTAGE	AMMONITE ZONES	Subzones
CALLOVIAN	LATE	LAMBERTI	
		ATHLETA	
	MIDDLE	CORONATUM	
		ANCEPS	
	EARLY	GRACILIS	
		BULLATUS	
BATHONIAN	LATE	DISCUS	
		ORBIS	
	MIDDLE	COSTATUS	Suspensum / Bullatimorphus
		SOFANUS	
	EARLY	AURIGERUS	Postpollubrum / Yeovilensis
		ZIGZAG	Macrescens / Dimorphitiformis
BAJOCIAN	LATE	PARKINSONI	Dimorphus / Daubenyi
		GARANTIANA	
		NIORTENSE	Sauzeanum / Phaulus
	EARLY	HUMPHRIESIANUM	Blagdeni / Humphriesianum / Romani
		PROPINQUANS	Hebridica / Patella
		LAEVIUSCULA	Laeviuscula / Ovalis
		DISCITES	Subsectum / Walkeri
AALENIAN	MIDDLE LA.	CONCAVUM	Limitatum / Concavum
		BRADFORDENSIS	Gigantea / Bradfordensis
		MURCHISONAE	Murchisonae / Haugi
	EA.	OPALINUM	Comptum / Opalinum

STAGE	SUBSTAGE	AMMONITE ZONES	Subzones
TITHONIAN	LATE	DURANGITES	
		TRANSITORIUS	
		SIMPLISPHINCTES	
	EARLY	BURCKHARD-TICERAS	
		ADMIRANDUM	
		RICHTERI	
		VERRUCIFERUM	
		ALBERTINUM	
		HYBONOTUM	
KIMMERIDGIAN	LATE	BECKERI	Pressulum ✳
		CAVOURI	
		COMPSUM	
		DIVISUM	Uhlandi ✳ / Divisum
	EARLY	STROMBECKI HYPSELOCYCLUM	
		PLATYNOTA	Betica
		PLANULA	Galar / Planula ✳
OXFORDIAN	LATE	BIMAMMATUM	Hauffianum/Tiziani / Bimammatum ✳ / Hypselum ✳
		BIFURCATUS	Grossouvrei / Bifurcatoides/Stenocycloides
	MIDDLE	RIAZI (TRANSVERSARIUM)	
		PLICATILIS	Antecedens
	EARLY	RENNGERI	

Levante (Serres de Llevant; SE mountain range or southern horst). Particular topics of Majorcan geology have been recently revisited in Fornos (1998).

Except for Minorca, the Balearic Islands form the emergent part of the Balearic promontory, which is usually viewed as the northeastern prolongation of the Betic Cordillera. However, Majorca and Minorca were probably more closely linked to the Catalonian region in Palaeozoic and Triassic times (Meléndez et al. 1988), and ammonite assemblages were closely related to those in NW Europe during much of the Early Jurassic epoch (Alvaro et al. 1989). The distinctive character of Balearic Jurassic geology is further emphasized by: (1) a prominent siliciclastic horizon of early Pliensbachian age (i.e. Carixian in peri-Mediterranean Alpine stratigraphy, west-Tethys) in Majorca and Cabrera (Arbona et al. 1984–85); (2) the prominent role of gravitational resedimentation (first emphasized by Pomar 1976) in these areas; (3) a stratigraphical gap affecting the upper Lower, the Middle and the lower Upper Jurassic succession in Ibiza (Fourcade et al. 1982). Recognition of a distinctive Balearic palaeogeographic identity has been emphasized by Bourrouilh & Gorsline (1979), Andrieux et al. (1989) and Dewey et al. (1989), with alternative views preferring closer links with the Betic area presented by Fourcade et al. (1977) and Guerrera et al. (1993).

Jurassic evolution in the Balearic area was influenced by reactivated Variscan fractures that largely determined depositional environments. Additional palaeogeographic influence on sedimentation resulted from the proximity to an emergent mainland to the west (Massif Valencien of Fourcade et al. 1977), and from links with the epicontinental shelves of eastern and southeastern Iberia. Away from this epicontinental platform system (Platform Citrabétique of Fourcade et al. 1977, 1982) increasingly open marine conditions, although with an irregular sea bottom, existed to the ESE (Sillon Citrabétique of Fourcade et al. 1977, 1982) indicating an adjacent epioceanic fringe (Olóriz et al. 1993; Olóriz 2000). Depositional models for the Jurassic succession in the Balearic Islands have emphasized gravitational-sliding events, commonly triggered by tectonic and tectonic–eustatic pulses affecting carbonate slopes and plateaux (e.g. Pomar 1976; Barnolas & Simó 1984). Evidence for such Jurassic events is well displayed on the islands of Majorca (Barnolas & Simó 1984) and Cabrera (Arbona et al. 1984–85), and precise data exist for interpreting Late Jurassic sedimentation in epioceanic swell–slope, shelf–slope and slope–basin conditions.

Another characteristic of Balearic geology is evidence for widespread unconformities (Rangheard 1970, 1971; Bourrouilh 1973; Fourcade et al. 1977, 1982; Alvaro et al. 1984a, 1989; Arbona et al. 1984–85; Caracuel & Olóriz 1998), the most significant being identified between the lower Pliensbachian and the lower Toarcian (see Bernoulli & Jenkyns 1974 for its

diachronous record in Tethyan palaeomargins), as well as at both the base and top of the Upper Jurassic–lowermost Cretaceous succession. Many additional, but less prominent, discontinuities have been identified throughout the Jurassic succession (Alvaro et al. 1989; Caracuel & Olóriz 1998).

A model interpreting Jurassic evolution in Majorca has been proposed by Barnolas & Simó (1984), complemented by Alvaro et al. (1989), and correlated with data from Cabrera by Arbona et al. (1984–85). In Majorca, initial depositional environments changed from ramp (Hettangian–Sinemurian) to barred shelf with a clayey intrashelf basin (Pliensbachian). Close to the Pliensbachian–Toarcian boundary, the arrival of siliciclastic detritus from the NW was followed by the Toarcian break-up of the shelf, the differentiation of the Sierra Norte horst, a low, muddy depocentre in the Randa area of central Majorca (Fig. 11.11), and a carbonate shelf deepening southwards (slope) in the Sierras de Levante. This configuration persisted during Middle and Late Jurassic times, with evidence for palaeogeographic reorganization close to both the Middle–Late Jurassic and latest Jurassic–earliest Cretaceous boundaries. South from the Sierras de Levante source areas, the island of Cabrera records a distinctive depositional evolution until the Late Jurassic epoch (including earlier break-up during Early Jurassic times, and accentuated bottom instability during the Middle Bathonian to Middle Kimmeridgian interval). Late Jurassic deposits in the island of Cabrera show sedimentation patterns similar to epioceanic swell–slope conditions in the Sierra Norte, but resedimentation and the latest Jurassic–earliest Cretaceous instability were less accentuated. Both Ibiza and Minorca maintained carbonate shelf environments, although the former evolved to more open mid-outer shelf conditions with increasing siliciclastic influence during Late Jurassic times.

Lower Jurassic: Betic Cordillera (AL, FO)

Early Jurassic events began with a generalized transgression that resulted in a broad carbonate Bahamian-type shelf which later broke up and became differentiated into inner-mid neritic and more pelagic areas of deposition. The initial shelf conditions produced a thick carbonate succession (>100 m of dolomites and limestones, sometimes overlain by cherty limestones, known as the Gavilán Fm in the Subbetic Zone and Carretas Fm in the Prebetic; Figs 11.12A and 11.13), which typifies pre-Lower Pliensbachian Jurassic rocks across most of the area. The ecosedimentary conditions corresponded to tidal and shallow neritic environments mainly colonized by benthic organisms. Oolitic and oncolitic grainstones to packstones, algal wackestones (dasycladaceans, rodophiceans, cianophiceans), wackestones with forams (lituolids, textulariids, nodosarids), and peloidal mudstones to packestones contain remains of brachiopods (terebratulids, rhynchonellids), echinoderms (crinoids), bivalves ('Lithiotis' assemblage) and gastropods. Facies assemblages mainly record an

Fig. 11.12. (A) Jurassic lithostratigraphy in the Betic Cordillera based on the synthesized interpretation of significant units made by the authors (see text). Key: AR, ammonitico rosso; B, Burete Fm, a brecciated massive limestone interpreted by Rey (1993) as the local expression of a probable salt tectonics event of late Bajocian to earliest Bathonian age; C, Camarena Fm; HG, hardground with ammonites in the Almola section; H-S, Hettangian-Sinemurian ammonites; lmP, lowermost Pliensbachian ammonites; Ma, Malaguides; PR, Parauta Radiolarite Fm; R, Ricote Fm; R-Ap, Rosso ad Aptychi; Rd, Rondaide complex (pe, perched basins; sl, slope; sw, swell areas); S, Sinemurian ammonites; T, Tornajo marly limestone Fm; Up. AR, Upper Ammonitico Rosso; indicated by asterisks in small boxes; dark and pale grey shading represents, respectively, well known and less well known hiatuses related to significant and usually overimposed unconformities; ‣, approximate place for comparatively minor unconformities without precise indication of hiatuses; +, local incomplete record of the Zegrí Fm; small boxes. **(B)** Biostratigraphic scale for Jurassic ammonites in the Betic Cordillera and the Balearic archipelago (mainly Majorca): left, Lower Jurassic; centre, Middle Jurassic; right, Upper Jurassic. Lower and Middle Jurassic compiled by J. Sandoval, Upper Jurassic compiled by the author. Grey shading (Discus and Lamberti ammonite zones which are not identified in the Middle Jurassic of the Betic Cordillera and Majorca). Asterisks in the Upper Jurassic indicate biohorizons described in Olóriz (1978), Olóriz et al. (1999a,b) and Caracuel et al. (2000).

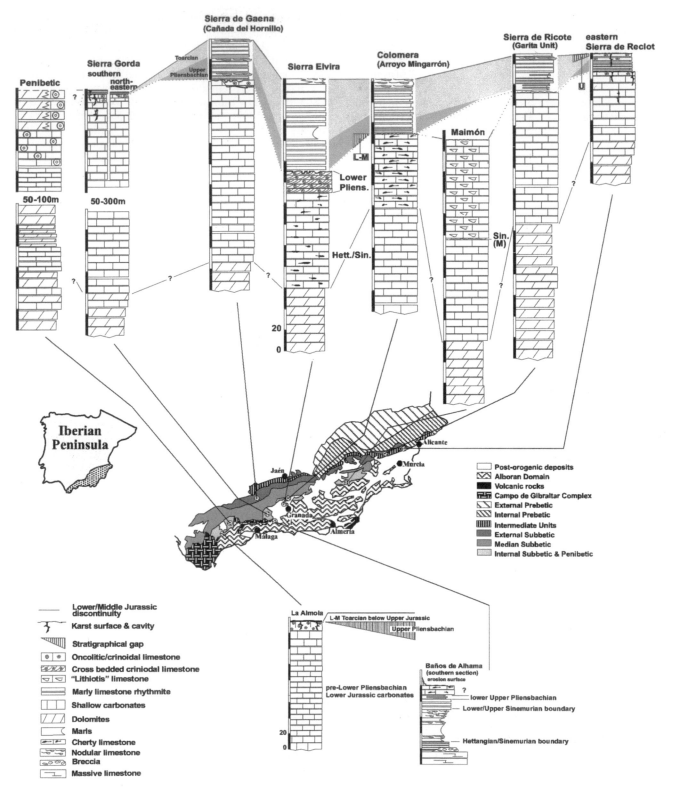

Fig. 11.13. Geological sketch map with location of selected Lower Jurassic sections in the Betic Cordillera. Upper columns belong to the southeastern palaeomargin of Iberia. Lower columns belong to the southwestern palaeomargin of the Alboran Domain (Rondaide Complex). Key: L, lower; M, middle; Pliens.; Pliensbachian; Sin., Sinemurian; U, Upper. Data for sections in the SSE palaeomargin of Iberia: Penibetic (Martín-Algarra 1987); Sierra Gorda (García-Hernández *et al.* 1986–87); Sierra de Gaena-Cañada del Hornillo (Braga 1983; Molina 1987); Sierra Elvira (García-Dueñas 1967; Linares *et al.* 1971; Rivas 1973); Colomera-Arroyo Mingarrón (Rivas 1973); Maimón (Rey *et al.* 1990; Rey 1993); Sierra de Ricote-Garita Unit and Sierra de Reclot (Nieto 1996). Data for sections in the Alboran Domain: La Almola (Martín-Algarra 1987); Baños de Alhama southern section (Busnardo *et al.* 1966; Braga *et al.* 1984*b*). Numbers in scale-bars indicate metres. Pre-lower Pliensbachian Jurassic rocks (shelf carbonates) are referred to in text as the Gavilán Fm.

irregular distribution of tidal (algal mats and stromatolites, fenestral and laminated mudstones, black pebbles) and inner-shelf environments under variable depth and energy. Relatively high energy conditions are recorded by oncolitic–oolitic limestones, whereas middle to low energy environments produced 'Lithiotis' patches (best developed eastward in the Subbetic area) and wackestones with forams (Azéma *et al.* 1979; García Hernández *et al.* 1989; Rey *et al.* 1990). Occasional records of Sinemurian ammonites (*Arnioceras*; Linares *et al.* 1971) are locally known from cherty limestones far southeastward from the mainland shoreline of the Iberian (or Variscan) Massif, in areas that were to evolve into troughs (southern Median Subbetic) during mid-Early Jurassic times (Sierra Elvira section in Fig. 11.13).

Early indications of tectonic instability of the shelf are provided by localized Sinemurian submarine volcanism (García-Hernández *et al.* 1989). Above the pre-Lower Pleinsbachian shelf carbonates are crinoidal grainstones and packstones, showing parallel lamination, cross-bedding, hiatuses, condensed deposition and hardgrounds, recording an environment found widely around the western Tethys at this time. These sediments have been interpreted as tidal and inner-shelf deposits that accumulated during the diachronous break-up of the former shelf. An exception to this environment is provided by the so-called Penibetic area to the south (Figs 11.11 and 11.12A) where oolitic facies conditions (Endrinal Fm; Martín-Algarra 1987) persisted during a large part of the Middle Jurassic.

The earliest major intra-Jurassic unconformity in the External Zones is Early Pliensbachian in age and occurs on top of the crinoidal limestones. This unconformity has been interpreted as recording carbonate shelf break-up, and is associated with karstification, neptunian dykes and breccias in the Subbetic region, and influxes of continental detritus in the Prebetics (García-Hernández *et al.* 1986–87, 1989). Further SE, in the Internal Zones (the Rondaides in the southwestern Alboran Domain; Figs 11.12A and 11.13), carbonate shelf break-up began earlier (Martín-Algarra 1987). This is recorded by mid-Hettangian to mid-Sinemurian ammonites from deeper water marly limestone rhythmites and cherty, red nodular limestones (Baños de Alhama section in Fig. 11.13), Hettangian conglomerates, and breccias and karstification close to the Hettangian–Sinemurian boundary. However, as seen in the External Zones, carbonate shelf conditions persisted on blocks that became karstified close to the Early–Middle Jurassic boundary (swell areas in the Rondaide Complex in Fig. 11.12A; La Almola in Fig. 11.13). Eastward in the Internal Zones (the Malaguides area; Figs 11.12A and 11.13), the earliest Jurassic carbonate shelf received pulses of continental detritus, while cherty limestones accumulated in deeper water troughs during late Early Jurassic times. Shelf break-up in the Malaguides was coeval with the generalized intra-lower Pliensbachian event in the southeastern palaeomargin of Iberia. Block structuring followed Early Jurassic shelf break-up in the Malaguides, and the Malaguides shelf was irregularly karstified before generalized transgression during mid-Early Jurassic times, when further palaeogeographical changes began to take effect in the Betic External Zones.

Neritic inner shelf facies showing episodes of continental influence persisted in the Prebetic Zone (Madroño, Colleras and Contreras formations; Fig. 11.12A), but in the Intermediate units (Northern trough: Figs 11.11, 11.12A and 11.13) 100 m thick breccias, which are the lateral equivalent to hemipelagites of the Baños Fm, are interpreted as linked to tectonic instability (Ruíz-Ortiz 1980). Probably coeval conglomerates and breccias have been identified further south on slopes in the Penibetic area (Fig. 11.11; Martín-Algarra 1987). In the Subbetic area, increased but varying subsidence rates produced undersea swells and troughs, and a change to lower energy, middle to outer shelf environments of low to moderate depth, colonized by ammonites (Zegrí Fm; Fig. 11.12A). During Late Pliensbachian and Toarcian times deposition in the deeper parts of this Subbetic area produced limestone, cherty limestone and marly limestone rhythmites, episodically affected by storms, whereas 'ammonitico rosso' facies developed on gentle swells and their slopes (Fig. 11.13). Differential subsidence controlled the thicknesses of local successions, and faulting induced episodes of slumping and the resedimentation of shallow carbonate debris. In both the External and Median Subbetic areas, rhythmic marls, marly limestones and limestones (Zegrí Fm, 0–250 m thick) dominated, with more restricted marly 'ammonitico

rosso' facies first appearing during the Late Pliensbachian and becoming widespread under a more calcareous subfacies during the Toarcian in central areas of the External Subbetic (Braga *et al.* 1981). Large areas in the eastern External Subbetic show stratigraphical gaps in the Late Pliensbachian to Late Toarcian record (Nieto 1996), and in some of them the Zegrí Fm is completely absent (Rey 1993). In the central sector of the Median Subbetic, late Early Jurassic synsedimentary sliding was common and submarine volcanism was locally present during latest Toarcian times (Comas 1978). Southward, the Internal Subbetic and the Penibetic (two left-hand columns on Figs 11.12A and 11.13), as well as the so-called High Chain locally, were areas comparatively resistant to sinking and they persisted as open sea isolated, locally emergent platforms, which provided a source for carbonate detritus subsequently resedimented in surrounding deeper water areas receiving sediments that resulted in siliceous marly limestone rhythmites.

Southeastward, in the Internal Zones (southwestern Alboran Domain), variably marly and cherty limestones with nodular intercalations upwards characterized deposition in slope areas of the Malaguides, and more outlying areas such as the Rondaides (Figs 11.12A and 11.13) during the mid-late Early Jurassic, away from shallow blocks on which shelf carbonate conditions persisted.

Lower Jurassic biostratigraphy in the Betics is mainly based on ammonites. Cosmopolitan Psiloceratidae (*Psiloceras*, *Tmaegoceras*), Schlotheimiidae (*Schlotheimia*, *Waehneroceras*), Arietitidae (*Arnioceras*, *Asteroceras*, *Coroniceras*, *Epophioceras*, *Vermiceras*) and Echioceratidae (*Paltechioceras*) with a Mediterranean aspect characterize the pre-Lower Pliensbachian assemblages, while Mediterranean Hildoceratoidea, Psiloceratoidea and Eoderoceratoidea allow post-Sinemurian biostratigraphic correlation (Rivas 1973; Braga *et al.* 1981, 1984a,b; Braga 1983; Jiménez 1986). Post-Sinemurian stage boundaries in the External Zones are based on the first appearance of *Gemmellaroceras* (lowermost Pliensbachian), the turnover between *Protogrammoceras dilectum* and *Fuciniceras portisi* and *monestieri* (Lower–Upper Pliensbachian boundary), the first appearance of *Dactylioceras* (Pliensbachian–Toarcian boundary) and the first appearance of *Leioceras lineatum* above the last appearance of *Pleydellia* sp. gr *buckmani* (Toarcian–Aalenian boundary).

Lower Jurassic: Balearic Islands (AG)

Although the Lower Jurassic successions in the islands of Majorca, Cabrera, Ibiza and Minorca have been studied since the early nineteenth century (see Colom & Escandell 1960–62, Colom 1975, 1980; Bourrouilh 1973; Llompart 1979, 1980), only in the last two decades have formally defined lithostratigraphic units been established (Alvaro *et al.* 1984a, 1989), and elucidated by sedimentological, geochemical and biostratigraphical studies (Arbona *et al.* 1984–85, Barnolas & Simó 1984; Prescott 1988; Goy & Ureta 1988; Goy *et al.* 1995).

Jurassic sedimentation across the Balearic area was initially relatively homogeneous, although on Majorca palaeogeographic separation into sub-basins has been detected in the western Sierra Norte and areas close to Alcudia and Petra. A mid-Pliensbachian discontinuity has been recognized, and interpreted as linked to transgression, but a more significant unconformity developed later in the Early Jurassic epoch (Alvaro *et al.* 1989). This latter stratigraphic gap separates Pliensbachian from Toarcian rocks in Majorca, and is recorded by a prominent hardground which marks a change in facies from shallow water carbonates to open shelf hemipelagic marls and limestones (Fig 11.14). The diachronous nature of this change is illustrated by the fact that in Majorca open shelf facies became established by the Toarcian–Aalenian boundary, initially depositing locally ferruginous sediments containing reworked fossils from several biochronozones, whereas renewed sedimentation in Cabrera was delayed until late Bajocian times.

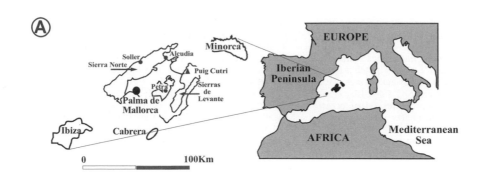

SIERRA NORTE · SIERRAS DE LEVANTE

				SIERRA NORTE	SIERRAS DE LEVANTE
JURASSIC	MIDDLE	BAJOCIAN	UPPER	CÚBER Fm	CÚBER Fm
			LOWER	GORG BLAU Fm	GORG BLAU Fm
		AALENIAN		GORG BLAU Fm	GORG BLAU Fm
	LOWER	TOARCIAN	UPPER		
			LOWER		
		PLIENSBACHIAN	UPPER	ES COSCONAR Fm	ES COSCONAR Fm
			LOWER	SA MOLETA Mb · ES RACÒ Mb	ES RACÒ Mb
		SINEMURIAN		SOLLER Fm · ES BARRACA Mb	SOLLER Fm · ES BARRACA Mb
		HETTANGIAN		MAL PAS Fm	MAL PAS Fm
UPPER TRIASSIC				FELANITX Fm	

			Iberic Cordillera	Betic Cordillera	Balearic Islands		
					Majorca		Cabrera (1) & Minorca (2)
					Sierra Norte	S. de Levante	
M.J.	AAL	L.A.	Opalinum	Opalinum	Opalinum	*Opalinum	
LOWER JURASSIC	TOARCIAN	UPPER	Aalensis	Aalensis	Aalensis	*Aalensis	*Aalensis (1)
			Pseudoradiosa	Reynesi	Pseudoradiosa	*Pseudoradiosa	
			Insigne	"Fallaciosum"	*Insigne/ *Thouarsense	*Insigne	
			Thouarsense				
		LOWER	Variabilis	Gradata	*Variabilis		Variabilis (2)
			Bifrons	Bifrons	Bifrons		Bifrons (2)
			Serpentinus	Serpentinus	Serpentinus		Serpentinus (2)
			Tenuicostatum	Polymorphum			
	PLIENSBACHIAN	UPPER	Spinatum	Emaciatum			
			Margaritatus	Algovianum	Algovianum		
			Stokesi	Lavinianum			*Stokesi (1)
		LOWER	Davoei	Dilectum			*Davoei (1)
			Ibex	Demonense	Ibex		*Ibex (1)
			Jamesoni	Aenigmaticum	Jamesoni		*Jamesoni (1)

With regard to the Lower Jurassic fossil record in the Balearic Islands, ammonite assemblages in Majorca are especially interesting because they inhabited areas situated between Alpine–Mediterranean seas and epicontinental seas belonging to northwestern Europe. Available data confirm the occurrence of NW European taxa (these continuing into Middle Jurassic times). Uppermost Toarcian rocks contain the oldest evidence for some typically Mediterranean ammonoids: Phylloceratidae, Lytoceratidae and others, such as *Osperlioceras sourensis* (Perrot), which have a restricted distribution from North Africa to southern Europe.

In the Sierra Norte, Early Pliensbachian ammonoids and brachiopods belong to the NW European Province, while Mediterranean forms are rare. Early Toarcian ammonite assemblages include Dactylioceratidae and Hildoceratidae, which are known in the Betic Cordillera, the Iberian Ranges and NW Europe. Late Toarcian ammonite assemblages, especially those from the Pseudoradiosa and Aalensis zones, show strong similarities with those of NW Europe, including the same succession of species of *Pleydellia* (Goy & Ureta 1988). The mixed provenance of Early Jurassic ammonite assemblages is well recorded in the Sierras de Levante where relatively common *Polyplectus discoides* (Zieten) (which occurred in both bioprovinces but was more abundant in Mediterranean seas) occurs together with typical species from NW Europe such as *Cotteswoldia costulata* (Zieten) or *Pleydellia leura* Buckman, as well as *Oxyparoniceras,* which is unknown or very scarce in the Mediterranean province.

In Cabrera, Mediterranean influence is evident earlier than in Majorca, as shown by the frequent presence of *Fuciniceras* and *Reynesocoeloceras* in the Upper Pliensbachian succession.

Majorca. Lower Jurassic successions are lithologically homogeneous with only minor local differences recorded during the Pliensbachian stage, and are similar to those described elsewhere in Tethyan areas (Alvaro *et al.* 1984a, 1989). The 50–150 m thick unfossiliferous Mal Pas Dolomites and Breccias Fm (Fig. 11.14) shows dolomitic breccias (carniolas) and a Hettangian age is assumed from regional correlation. Overlying it is the Sóller Carbonate Fm which shows a shallowing succession in its lowermost part, known as the Es Barraca Limestones Mb (200–300 m), and dated as Sinemurian–Pliensbachian in age from microfossils. Yellow marls of the Sa Moleta Mb (0–30 m) are recorded locally, reflecting the localized palaeogeographic separation into sub-basins at this time, and contain Lower Pliensbachian fossils from the Jamesoni and Ibex Zones. The highest member in this formation comprises siliciclastic deposits of the Es Racó Quartzitic Sandstones Mb (5–30 m), which probably represent the upper part of early Pliensbachian time. Directly above the Sóller Carbonate Fm are marls and crinoidal limestones (the Es Cosconar Fm, 5–50 m thick) containing upper Pliensbachian fossils and capped by a hardground at the top. Colom (1975) cited also Toarcian ammonites in the Es Cosconar area. Directly above the hardground is the Gorg Blau Marly and Nodular Limestones Fm (20–50 m in Sierra Norte and more than 30 m in Sierras de Levante), which shows from bottom to top: (a) ferruginous horizons (Barnolas & Simó 1984; Prescott 1988) indicating reworking with reelaborated ammonites (post-Variablis Zone in Sierra Norte, and pre-uppermost Pseudoradiosa Zone in Sierras de Levante; Goy & Ureta 1988 and Goy *et al.* 1995, respectively); (b) marly limestones with ammonites belonging to the Pseudoradiosa and Aalensis zones (Upper Toarcian) and to the Opalinum, Murchisonae, Bradfordensis and Concavum zones (Aalenian); (c) nodular and well

bedded limestones with Lower Bajocian ammonites. The top of this formation is a reworked horizon with local boring and a ferruginous coating.

Cabrera. In the island of Cabrera, Arbona *et al.* (1984–85) identified 240 m of more or less dolomitized limestones, composed dominantly of lower Lower Jurassic pelsparites and biomicrites (Colom 1980). The whole sedimentary package can be correlated with the Soller Fm and, probably, to the lowermost part of the Mal Pas Fm in Majorca. Over a large part of Cabrera there is crinoidal limestone with ferruginous surfaces (locally a hardground), which constitutes the top of the Lower Jurassic sequence. Arbona *et al.* (1984–85) identified packstones with detrital quartz, crinoids and forams, and then grainstones with crinoids and lower Pleinsbachian ammonoids. Locally, the hardground at the top shows lowermost upper Pliensbachian ammonites. Overlying the Lower Jurassic succession there is a reworked horizon of 0.5 m average thickness that contains Toarcian to upper Bajocian ammonites.

Ibiza. Up to around 120 m of massive to thick bedded dolomites and dolomitic limestones of undetermined age overlie Keuper facies rocks. A previous interpretation by Colom & Escandell (1960–62) envisaged emergence during Early and Middle Jurassic time, and Rangheard (1971) interpreted the succession as comprising highest Triassic–lowermost Lower Jurassic 'well bedded dolomitic limestones' that overlie Keuper facies, with overlying 'dolomites and dolomitic limestones' of Early and, perhaps, Middle Jurassic age.

Minorca. Lower Jurassic deposits are mainly dolomites, with locally abundant algae, and marly limestones with Toarcian fossils. Bourrouilh (1973) and Llompart (1979) described: (a) grey dolomites with medium to thick bedding (150–200 m); (b) calcareous marls with crinoids and ferruginous concretions (4–5 m); (c) marly limestones with limonite concretions and lower Toarcian brachiopods (2 m); and (d) grey marls (5–6 m) containing Toarcian brachiopods and which are capped by a ferruginous surface (hardground).

Biostratigraphy and biochronology. Biostratigraphic and biochronologic correlation in the Balearic Islands is based on ammonite assemblages (see Fig. 11.4 for correlations between the Balearic Islands, Betic and Iberian basins). At present, there is no evidence for Hettangian and Sinemurian ammonites. In Majorca, the oldest Pliensbachian ammonites are known from western Sierra Norte within the Sa Moleta Mb close to Sóller. *Uptonia jamesoni* and *Polymorphites* belong to the lower Pliensbachian (Jamesoni Zone) and *Tropidoceras* may be lower to mid-lower Pliensbachian (Fig. 11.14). Alvaro *et al.* (1989) mentioned old citations of *Seguenziceras algovianum* Oppel, an upper Pliensbachian species, by Fallot, although Dilectum Zone ammonites are unknown.

According to Arbona *et al.* (1984–85), crinoidal limestones in Cabrera contain the early–middle lower Pliensbachian *Gemmellaroceras aenigmaticum* (Gemmellaro), '*Miltoniceras*' gr. *sellae* (Gemmellaro) and *Tropidoceras* gr. *flandrini* (Dumortier) and, locally, the middle-upper lower Pliensbachian *Tropidoceras desmonense* (Gemmellaro) and *Fuciniceras detractum* (Fucini). Within the upper hardground, these authors found lowermost upper Pliensbachian *Fuciniceras portisi* (Fucini), *Calaiceras*? sp., and *Reynesocoeloceras psiloceroides*? (Fucini). In addition, Gómez-Llueca (1920) mentioned lower Pliensbachian *Tropidoceras masseanus* (D'Orbigny), *T.* cf. *flandrini* (Dumortier), *Aegoceras nautiliformis* Buckman and *Racophyllites* cf. *mimatensis* (D'Orbigny).

Toarcian ammonites in Majorca are relatively abundant and permit the identification of all standard zones except the Tenuicostatum/Polimorphum Zone. The oldest specimen known in the Sierra Norte is probably *Murleyiceras murleyi* (Buckman), figured in Colom (1975, pl. 4, fig. 6) and mentioned by Alvaro *et al.* (1989) as *Hildaites murleyi*, which is a typical species from the lower part of the Serpentinus Zone. The Bifrons Zone has been identified in marly deposits of

---

Fig. 11.14. (A) Location. **(B)** Lower Jurassic biochronostratigraphy and lithostratigraphy in Majorca (modified from Alvaro *et al.* 1989). **(C)** Biostratigraphic scales and correlation in the Balearic archipelago based on Bourrouilh (1973), Colom (1975), Llompart (1979), Arbona *et al.* (1984–85), Goy & Ureta (1988), Alvaro *et al.* (1989), Goy *et al.* (1995), and references therein. Reference biostratigraphy in the Betic Cordillera and the Iberian Ranges based on Braga *et al.* (1984a) and Goy *et al.* (1988). Asterisks indicate that typical taxo-records from these zones correspond to reworked reelaborated ammonites.

the Es Cosconar Fm, which according to Colom contain *Porpoceras*, *Hildoceras* and *Mercaticeras*. Other Toarcian ammonites, namely Hildoceratinae, Phymatoceratinae, Grammoceratinae and Dactylioceratidae have been reworked in ferruginous horizons at the bottom of the Gorg Blau Fm in several outcrops of the Sierra Norte. The upper Toarcian succession contains typical ammonites of the Pseudoradiosa Zone (*Dumortieria* and *Geczyceras*) and Aalensis Zone (*Pleydellia* and *Cotteswoldia*) assembled with Phylloceratidae and rare Lytoceratidae (Goy & Ureta 1988). Upper Toarcian ammonites such as *Pseudogrammoceras, Pseudolillia, Gruneria, Catulloceras, Dumortieria, Pleydellia, Cotteswoldia, Oxyparoniceras, Polyplectus, Osperlioceras, Geczyceras* and *Hammatoceras*, which are reworked, were found in a ferruginous horizon at the bottom of the Gorg Blau Fm in Puig Cutri, northern Sierras de Levante (Alvaro *et al.* 1989; Goy *et al.*1995). The Aalenian Opalinum Zone is documented by *Leioceras* and *Tmetoceras* associated with rare *Vacekia*. In Cutri, *Leioceras* appears within a reworked horizon that contains reworked (reelaborated) ammonites belonging to several Toarcian zones.

In Cabrera, the reworked horizon contains upper Toarcian to upper Bajocian ammonites. Thus, Gómez-Llueca (1920) mentioned *Pleydellia* gr. *aalensis* (Zieten), *Euedmetoceras* (*Rhodaniceras*) *infernensis* (Roman) and several species of *Stephanoceras, Teloceras, Normannites* and *Oppelia*. In addition, in Gómez-Llueca's collection there is a specimen of *Leioceras lineatum* Buckman obtained from the small island of Conejera.

In Minorca, Bourrouilh (1973) and Llompart (1979) identified *Soaresirhynchia bouchardi* (Davidson), a typical species from the Toarcian, Serpentinus Zone, directly above the 'calcareous marls with crinoids' package. Overlying this are *Harpoceras* sp., *Hildoceras lusitanicum* Meister, *H.* gr. *bifrons* (Bruguière) together with a typical brachiopod assemblage belonging to the Spanish bioprovince (García Joral & Goy 2000) that shows *Homeorhynchia batalleri* (Dubar), *Homeorhynchia meridionalis* (Deslongchamps), *Telothyris jauberti* (Deslongchamps), *Sphaeroidothyris perfida* (Choffat) and *Lobothyris hispanica* (Dubar). These taxa characterize the lower Toarcian, Serpentinus and Bifrons Zones. In addition, *Crassiceras* sp. is known from the lowermost upper Toarcian (Variabilis Zone).

Middle Jurassic: Betic Cordillera (JS)

During Middle Jurassic times shallow neritic shelf conditions continued to prevail in the Prebetic area, with coastal and even continental facies occurring locally. In the Intermediate units (northern trough) Middle Jurassic rocks show a pelagic or hemipelagic character and are red nodular limestones and oolitic limestones with minor intercalations of limestones with chert and marl beds. Further offshore, deeper waters in the Subbetic area allowed more or less continuous pelagic or hemipelagic sedimentation throughout the Middle Jurassic epoch. Basin physiography was, however, very irregular due to severe intracontinental rifting, which caused two high swell areas (Northern or External Subbetic and Southern or Internal Subbetic) and one relatively deep trough (southern trough or Median Subbetic; Fig. 11.11). Further south, in the Rondaides (Betic Internal Zones), pelagic limestones, marls and radiolarites accumulated.

Prebetic. The most complete Middle Jurassic sections show *c.* 150 m thick dolomites and limestones in the southern part of the Sierra de Cazorla (Fig.11.15C; Acosta & García-Hernández 1988). In the upper part of the sections, 45–50 m thick limestones constitute a

transgressive depositional sequence (upper Middle Jurassic) containing minor shallowing-upward sequences with beds 0.5 to 1 m thick and the dominant microfacies comprising oolitic limestones with benthic foraminifers. Thin beds of micritic limestones with abundant ostracods and charophytes, and some levels of packstones and wackestones (with ostracods, fragments of charophytes and ooliths) occur near the top of the succession.

Northern trough (intermediate units). During the Middle Jurassic epoch this area was well differentiated in the central part of the Betic Cordillera (provinces of Jaén and Granada) whereas to east and west the intermediate units succession grades into that typical of the External Subbetic area (see below). In the Jaén area (Jabalcuz-San Cristobal unit) Middle Jurassic sediments start with red nodular marly limestones (6 to 8 m thick) containing some Aalenian ammonites, which correspond to the upper part of the Los Baños Fm (Ruíz-Ortíz 1980). The rest of the Middle Jurassic sediments correspond to the Jabalcuz Fm (Ruíz-Ortíz 1980), which comprises approximately 150 m of well-stratified oolitic limestones, limestones with chert and some marl beds. The dominant microfacies is grainstone with rare filaments, crinoids and benthic foraminifers, and packstone with faecal pellets and *Bositra* filaments deposited in a shallow pelagic environment. Laterally (Carceles-Carluco unit; Ruíz-Ortíz 1980) Middle Jurassic sediments gradually pass into laminated limestones with intercalated marly limestones and marls (packstones to wackestones of filaments and pellets).

Northern swells (External Subbetic). Middle Jurassic sediments are especially well developed in the central and eastern sectors of the External Subbetic and their main stratigraphic features are summarized in Fig. 11.15A. Aalenian sediments are generally very reduced and condensed, with red nodular limestones or dark red ferruginous nodular limestones (limonitic crust) predominating. The uppermost Lower Aalenian (Opalinum Zone, Comptum Subzone) and/or the uppermost Aalenian (Concavum Zone, Limitatum Subzone) are recognized locally (Linares & Sandoval 1993). Stratigraphical gaps of variable amplitude affect Aalenian deposits (locally omitting the entire section), especially for the Lower and/or Middle Aalenian part of the succession (Linares & Sandoval 1993).

Bajocian sediments are, in contrast, generally well represented, with a thickness ranging from 4–5 m to more than 240 m in some sections (Sandoval 1983; Molina 1987; Rey 1993; Nieto 1996). A typical succession is as follows: (a) well-stratified oolitic limestones (Camarena Fm; Molina, 1987) with individual bed thicknesses ranging from 0.2 to 0.5 m, and a predominant microfacies of grainstones with pellets, abundant ooliths and varied benthic foraminifers; (b) light grey micritic limestones (packstones and wackestones) alternating with grey marls and marly limestones with cherty nodules (Veleta Fm; Molina 1987), with bed thicknesses between 0.2 and 1.25 m, and radiolarians, crinoids, and filaments of *Bositra* and trace fossils (*Zoophycos, Thalassinoides* and *Chondrites*) being locally abundant; (c) grey to pinkish nodular limestones (wackestones to mudstones) with bed thicknesses 0.15–1.2 m, and containing filaments, fragments of echinoderms, radiolarians, benthic and planktonic foraminifers (conoglobigerinids), and small ammonites; (d) yellowish-red marls, nodular marly limestones and red nodular limestones from the lower part of the upper Ammonitico Rosso Fm (Molina 1987) and typified by mainly upper Bajocian wackestones to packstones (bed thickness again 0.15–1.2 m) with abundant *Bositra* filaments, benthic and planktonic (conoglobigerinids) foraminifers, echinoderms, radiolarians, and ammonite embryos. Abundant and diversified ammonite assemblages, especially in the 'ammonitico rosso' facies, enable the recognition of the standard Mediterranean zones: Discites, Laeviuscula, Propinquans, Humphriesianum, Niortense (Leptosphinctes Zone in

Fig. 11.15. Some of the most representative Middle Jurassic sections of the Betic Cordillera and Majorca. **(A)** External Subbetic (pelagic swell): Puerto Escaño, Carcabuey, province of Córdoba (left) and Sierra de Lugar, Fortuna, province of Murcia (right). **(B)** Median Subbetic (trough), Agua Larga section, Campillo de Arenas, province of Jaén. **(C)** Prebetic (epicontinental shallow platform), Sierra de Cazorla, province of Jaén (modified from Acosta & García-Hernández 1988). **(D)** Internal Subbetic (pelagic swell): Sierra Harana, Cogollos Vega, province of Granada (left) and Sierra Gorda, Loja, province of Granada (right). **(E)** Internal Zones, Nieves unit (Rondaide complex), La Encina section, south of Ronda, province of Málaga (modified from O'Dogherty *et al.* 2001). **(F)** Island of Majorca: Serra de Tramuntana, Alfabia-Cúber area (left) and Serra de Llevant, Puig Cutri-Puig de Ses Fites (right).

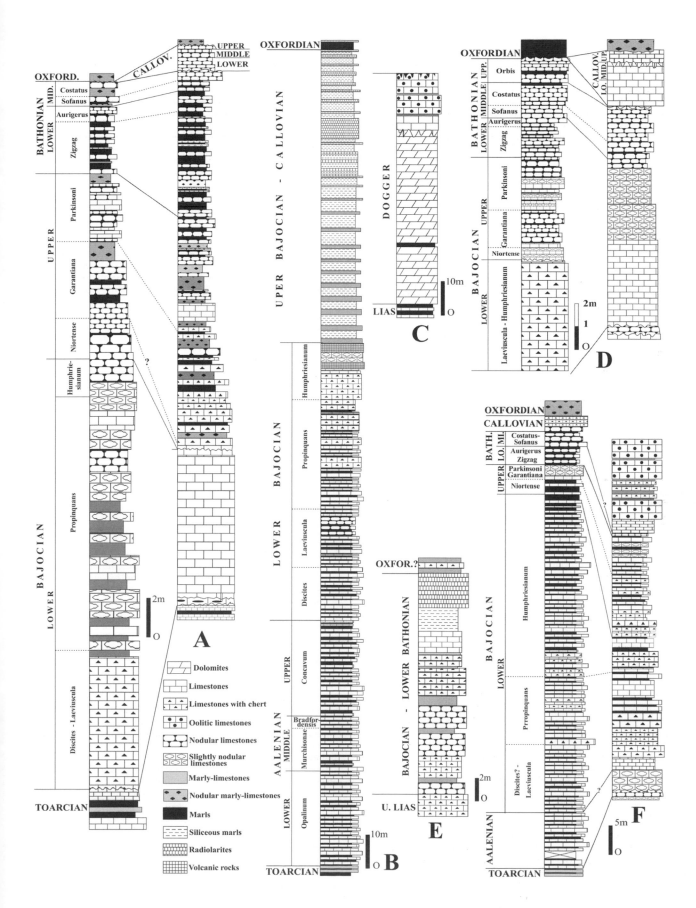

Sandoval 1983), Garantiana (Annulatum in Sandoval 1983) and Parkinsoni.

Bathonian rocks are well represented in the External Subbetic (Fig. 11.15A), their thickness being variable but generally less than 12 m, with bed thicknesses between 0.5 and 0.9 m. Violet-yellowish or red marls and nodular marly limestones predominate in the lower Bathonian succession, whereas red nodular limestones ('ammonitico rosso') dominate in the middle and the upper Bathonian. Microfacies consist of wackestones and packstones with *Bositra* filaments, radiolarians, benthic foraminifers, planktonic foraminifers (conoglobigerinids), and ammonite embryos. Macrofossils are mainly represented by ammonites, among which typically Mediterranean phylloceratids and lytoceratids predominate. Abundant ammonitids provide valuable data for biostratigraphy: the Zigzag, Aurigerus (lower Bathonian), Sofanus, Costatus (middle Bathonian) and Orbis (upper Bathonian) zones are usually recognizable.

Callovian sediments are poorly represented (Sierras de Estepa and Cabra, La Cornicabra, and Sierras de Quípar, Lugar, Corque and Crevillente). Thicknesses range from a few centimetres (limonitic, ferruginous phosphate crusts) to 6 m (Sequeiros 1974, 1987). The dominant lithology consists of well-stratified red, sometimes yellowish, nodular limestones with abundant ferruginous phosphate concretions, which sometimes contain ammonites. Microfacies are wackestones to packstones with abundant filaments (*Bositra*), benthic and planktonic (conoglobigerinids) foraminifers, fragments of crinoids and ammonite embryos. The Bullatus, Gracilis, Anceps (Substeinmanni zone in Sequeiros 1974), Coronatum and Athleta (Orionoides zone in Sequeiros 1974) ammonite zones can be recognized.

In the External Subbetic major stratigraphical gaps and discontinuities occur at the Lower–Middle Jurassic, lower Bajocian/upper Bajocian, Bathonian/Callovian and Callovian/Oxfordian boundaries (Sequeiros 1974, 1987; Seyfried 1978; Sandoval 1983, 1990; O'Dogherty *et al.* 2000). The entire Callovian and the lower Oxfordian intervals are missing in places.

Southern trough (Median Subbetic). Middle Jurassic sediments are especially well developed in the central and eastern sectors of the Median Subbetic (Fig. 11.15B), and their main stratigraphic features can be summarized as follows. The Aalenian, Bajocian, Bathonian and the Callovian successions present varied features, as in some sectors the sediments corresponding to these stages are partially lacking and/or show notable variations in thickness (Sandoval 1983, 1990; Linares & Sandoval 1993). Predominant lithofacies are marls, marly limestones (mudstones), interbedded limestones with chert and red nodular limestones ('ammonitico rosso' facies). Oolitic limestones (Ricote Fm) and siliceous marlstones and radiolarites (lower part of the Jarropa Radiolarite Fm) appear locally (Sandoval 1983; Linares & Sandoval 1993; Nieto 1996; O'Dogherty *et al.* 1997).

The thickness of the generally well represented Aalenian successions varies from less than 5 m ('ammonitico rosso' sections) to more than 70 m (marls and marly limestone rhythmites; Linares & Sandoval 1993). Marly limestones are predominantly mudstones with radiolarians, some *Bositra* filaments and benthic foraminifers. Trace fossils (especially *Chondrites* and *Zoophycos*) can be very abundant. Ammonite assemblages are abundant and diverse, and enable us to recognize lower (Opalinum Zone: Opalinum and Comptum Subzones), middle (Murchisonae Zone: Haugi and Murchisonae Subzones; and Bradfordensis Zone: Bradfordensis and Gigantea Subzones) and upper (Concavum Zone: Concavum and Limitatum Subzones) Aalenian strata.

The Bajocian succession can exceed 200 m and provides abundant exposures in the central part of the Betic Cordillera (especially in the Campillo de Arenas/Montejícar area; Sandoval 1983) as well as in the eastern part (Seyfried 1978; Sandoval 1983, 1990; Nieto 1996). Bajocian rocks are generally well stratified alternations of grey, yellowish and cream pelagic limestones (0.1 to 1 m thick mudstones to wackestones with *Bositra* filaments and radiolarians) and marls. Interbedding of limestones with chert and red, whitish or violet marly nodular limestones are also relatively common in the Humphriesianum, Niortense and Parkinsoni zones. Trace fossils (especially *Chondrites* and *Zoophycos*) may be abundant locally. Oolitic limestones (Ricote Fm; Nieto 1996), calcareous radiolarites,

radiolaritic limestones (mudstones to wackestones with *Bositra* filaments and radiolarians) and green marls are also present (Sandoval 1983; Nieto 1996; Molina *et al.* 1999a). Ammonites are by far the predominant macrofauna with biochronological interest. The standard ammonite zones of Discites, Laeviuscula, Propinquans, Humphriesianum and Niortense have been recognized in many sections. Ammonites are scarcer in the Garantiana and Parkinsoni Zones, but they are recognized locally.

Bathonian and Callovian sediments are composed fundamentally of calcareous radiolarites, radiolarian limestones and green marls (Sandoval 1983) corresponding to the lower member of the Jarropa Radiolarite Fm (O'Dogherty *et al.* 1997; Molina *et al.* 1999a). Red, white or yellow siliceous nodular limestones and limestones with chert also occur (Sandoval 1983). Intercalations of pillow lavas and pyroclastic rocks, sometimes of great thickness, appear in the Middle Jurassic succession of many Median Subbetic localities.

Southern swells (Internal Subbetic). Outcrops with Middle Jurassic sediments are common in the central part of the Betic Cordillera (Sandoval 1983) and in some areas further west (Martín-Algarra 1987), with the best exposures being located in Sierra Harana and Sierra Gorda (Fig. 11.15D). The upper Aalenian is recognized in Sierra Gorda by a only few centimetres of micritic grey limestones, which overlie Lower Jurassic sediments. In contrast, the Bajocian succession varies from 3 to 20 m in thickness, comprising mostly grey limestones with nodules of chert, grey-yellow limestones and violet or red nodular marly limestones (wackestones and packstones with abundant *Bositra* filaments, radiolarians, and benthic foraminifers). Ammonites are abundant in some sections and the Laeviuscula?, Propinquans, Humphriesianum, Niortense, Garantiana and Parkinsoni zones have been recognized (Sandoval 1983, 1990).

Bathonian sediments comprise 4 to 8 m of grey, yellow, violet or red marls and nodular marly limestones. Microfacies show wackestones and packstones with abundant *Bositra* filaments, radiolarians, benthic foraminifers and planktonic foraminifers (conoglobigerinids). Ammonites are abundant in many localities and the Zigzag, Aurigerus, Sofanus, Costatus and Orbis Zones can be recognized (Sandoval 1983).

In the Internal Subbetic, Callovian sediments have only been recognized in the Sierra Gorda (Sequeiros 1974), where Callovian rocks vary between a few centimetres and 2 m in thickness. The dominant lithology comprises well-stratified grey micritic limestones with dendrites of pyrolusite. Microfacies are wackestones to packstones with abundant filaments of *Bositra*, benthic and planktonic foraminifers (conoglobigerinids), fragments of crinoids and small ammonites. Ammonite assemblages enable the recognition of the Gracilis, Anceps, Coronatum and Athleta Zones (Sequeiros 1974). As seen in the northern swells, there are major stratigraphical gaps and discontinuities at the Lower–Middle Jurassic, lower Bajocian–upper Bajocian, upper Bathonian–lower Callovian, and Callovian–Oxfordian boundaries (Sequeiros 1974; Sandoval 1983; O'Dogherty *et al.* 2000).

The Middle Jurassic succession is well developed as the Endrinal Fm throughout the area known as the Penibetic (Fig. 11.12; for example at Teba, Torcal de Antequera, Hacho de Montejaque, Sierra Blanquilla, Jarastepar, Benaocaz, El Chorro), but the rocks lack age-diagnostic faunas (Martín-Algarra 1987). The formation ranges from 200 m to >300 m in thickness and comprises: (a) finely stratified grey, yellowish or cream-coloured limestones (mudstones to wackestones) with gastropods, scarce bivalves, fragments of corals, algae, benthic foraminifers and crinoids; (b) well stratified pelagic oolitic limestones (grainstones) with bioclasts of corals, algae, echinoderms, some bivalves and benthic foraminifers; (c) limestones with corals (rudstones, grainstones, packstones) with algae, gastropods, benthic foraminifers, etc., which locally constitute bioherms; (d) grey or cream-coloured limestones (rudstones, grainstones, packstones) with cherty nodules.

Rondaides. In the Internal Zones, the Middle Jurassic succession is well represented in the Nieves unit (Rondaides complex, south of Ronda) where it is included in the Parauta Radiolarite Fm and associated limestones (O'Dogherty *et al.* 2001). The Middle Jurassic succession (Fig. 11.15E) starts with red, subordinately greenish and strongly stylolitized limestones with red or green marly intercalations, and chert nodules. These limestones are up to 10 m thick and generally overlie a well-bedded succession of greyish limestones and marls. In places, nodular limestones may alternate with platy marly limestones and marls with chert nodules. Microfacies show wackestones–packstones with pelagic bivalves, radiolarians, benthic and planktonic (conoglobigerinids) foraminifers, and small ammonites. Radiolarites show finely and rhythmically stratified alternations (decimetre to centimetre thick horizons) of brownish-red, carbonate-free radiolarian chert, and grey calcareous radiolarian chert that grades into radiolarian-bearing siliceous limestones (O'Dogherty et al. 2001).

Middle Jurassic: Balearic Islands (JS)

Middle Jurassic rocks in the Balearic Islands are well developed in Majorca where they are exposed in both the NW (Serra de Tramuntana) and SE of the island, as well as in Cabrera (Fig. 11.11A). Each of these areas is described below.

Serra de Tramuntana (Sierra Norte). The Middle Jurassic (Aalenian–Callovian) succession of the Serra de Tramuntana area has many features in common with that of the External Subbetic. Pelagic sediments are dominant, with marls and marly limestones alternating with limestones containing chert and nodular limestones (Fig. 11.15F, left), and some radiolarite levels also occurring (Alvaro et al. 1989; Sandoval 1994). Middle Jurassic sediments have been subdivided into the Gorg Blau, Cúber and Puig d'en Paré formations (Figs 11.14 and 11.17; Alvaro et al. 1989). Aalenian sediments are included in the marly limestone interval of the Gorg Blau Fm where the standard ammonite zones of Opalinum, Murchisonae, Bradfordensis and Concavum have been recognized (Alvaro et al. 1989).

Bajocian rocks are widespread in the Serra de Tramuntana, but the best outcrops are located in the Alfabia–Cúber, Son Vidal, and Bianamar areas (Alvaro et al. 1989; Sandoval 1994). In the Alfabia–Cúber area, Bajocian sediments comprise >50 m of alternating grey marls and marly limestones (wackestones and mudstones) with bed thicknesses of 0.1–0.5 m, and radiolarians, filaments (Bositra) and benthic foraminifers (Cúber Fm). Abundant nodules of chert and scarcely nodular limestones with limonitic concretions occur in the Humphriesianum Zone. Conglomeratic nodular limestones (wackestones and packstones) with abundant filaments (Bositra), radiolarians, and benthic foraminifers (Puig d'en Paré Fm) occur in upper Bajocian rocks. Although not abundant, ammonite assemblages are significant for recognition of standard ammonite zones (Alvaro et al. 1989; Caracuel et al. 1994; Sandoval 1994): Discites, Laeviuscula, Propinquans and Humphriesianum (Lower Bajocian) and Niortense, Garantiana? and Parkinsoni (Upper Bajocian).

The thickness of Bathonian–Callovian sediments of Serra de Tramuntana varies from 2 to 7 m, and they are included in the Puig d'en Paré Fm (Alvaro et al. 1989). The predominant facies consists of grey to red conglomeratic nodular limestones (wackestones and packstones with abundant Bositra filaments, radiolarians, and benthic and planktonic (conoglobigerinids) foraminifers), with bed thickness varying from 0.1 to 0.3 m. Ammonite assemblages enable the differentiation of the Zigzag and Aurigerus zones (lower Bathonian) and the Sofanus and Costatus Zones (middle Bathonian). The lower Callovian (Gracilis Zone) is recognized in a few localities, where it directly overlies a middle Bathonian stratigraphical gap.

Serres de Llevant (Sierras de Levante). The best representative Middle Jurassic successions crop out in Puig Cutri-Puig de ses Fites (Fig. 11.15F, right), Betlem and Son Amoixa (Alvaro et al. 1989; Sandoval 1994). The lower boundary is marked by a ferruginous horizon separating the Es Corconar and Gorg Blau formations, and containing a mixed upper Toarcian–lower Aalenian ammonite assemblage (Alvaro et al. 1989; Goy et al. 1995). The middle–upper Aalenian succession consists of approximately 5 m of partially nodular red limestones with radiolarians, filaments of pelagic bivalves, benthic foraminifers and fragments of crinoids, plus 30–40 m of well stratified grey marly limestone and marls with radiolarians, filaments, and benthic foraminifers (Gorg Blau Fm; Alvaro et al. 1989; Sandoval 1994). Bajocian sediments (including the upper Gorg Blau, the Cúber and lower Puig d'en Paré formations) are c. 40–50 m thick, consist of well

bedded (bed thicknesses 0.1–0.7 m), grey marly limestones, limestones with chert (and limonitic concretions), partially nodular limestones, and thin interbedded grey marls. Trace fossils, especially Zoophycos and Chondrites, are locally abundant. Dominant microfacies are mudstones and wackestones with filaments (Bositra), radiolarians, benthic foraminifers and pellets. The ammonite fauna is abundant at some levels and all the standard Bajocian zones have been recognized. The early Bathonian interval is represented by a few centimetres of nodular limestones with abundant filaments (Bositra) and scarce radiolarians (upper part of the Puig d'en Paré Fm). Middle Bathonian–Callovian sediments are oolitic limestones and marly limestones and limestones with chert intercalations (40 to 150 m thick) of the Cutri Fm (Alvaro et al. 1989; Sandoval 1994).

Cabrera. On the island of Cabrera Middle Jurassic rocks are of pelagic or hemipelagic character and similar to those of the Serres de Llevant of Majorca. They are well represented in the Es Ca des Vaixell, Sa Croveta and (especially) in the Sa Cova Blava sections (Arbona et al. 1984–85). In the latter locality Middle Jurassic sediments are as follows: (a) 0.5 m of yellowish limestones (packstones with crinoids, ostracods, filaments of Bositra, and benthic and planktonic foraminifers) which contain uppermost Toarcian–lower Bajocian ammonoids (Arbona et al. 1984–85); (b) 30–35 m of green to red nodular marly limestones and nodular limestones with chert (Puig d'en Paré Fm). Microfacies are wackestones and packstones with abundant radiolarians and filaments (Bositra) and rarer ostracods, crinoids and benthic and planktonic (conoglobigerinids) foraminifers. Ammonites are abundant and representative, enabling the recognition of the late Bajocian (Niortense, Garantiana and Parkinsoni zones) and the early Bathonian intervals; (c) 25–35 m of grey oolitic limestones (grainstones with benthic foraminifers), which end with 3 m of well bedded (0.3 m) alternations of oolitic limestones and wackestones (Cutri Fm); (d) 9–10 m of well bedded slightly nodular limestones (bed thickness 0.2–0.4 m) containing some beds of compact limestones with chert. Microfacies show wackestones and packstones with radiolarians, filaments and some fragments of crinoids. The ammonite fauna (scarce) enable the recognition of middle–late Bathonian ages. In Cabrera, Middle Jurassic rocks are topped by a discontinuous horizon of breccias with upwardly thinning decimetric intraclasts and an erosive base.

Upper Jurassic: Betic Cordillera (FO, JEC, FJR-T, JMT)

By Late Jurassic times there was a clear separation between epicontinental (neritic) and epioceanic ecosedimentary environments (Olóriz 2000) along the SSE palaeomargin of Iberia. An accentuation of the trough and swell geomorphology is reflected by the widespread existence of 'ammonitico rosso' and related facies on swells (Upper Ammonitico Rosso Fm; Fig. 11.12A), and to siliceous limestone, marly limestone and marl deposits in troughs (Figs 11.12A and 11.16). The 'ammonitico rosso' is a key facies for palaeoenvironmental interpretation (e.g. Seyfried (1978, 1979) for an early seminal approach in the area; see below for more recent interpretations) and for biostratigraphic research. Discontinuous sedimentation and variable reworking characterized sediments corresponding to ammonitico rosso facies, the mature sections of which typically contain numerous discontinuities with hiatuses below the resolution of ammonite biostratigraphy (i.e. mean rates of deposition largely lower than those corresponding to episodes of sedimentation). Precise biostratigraphy based on ammonites (and calpionellids) has been worked from sections in both the External and Internal Subbetic areas (Fig. 11.16), mainly from the central sector (Casa Blanca and El Cardador sections) but also from western (Lentejuela and Teba sections; La Almola section in the Alboran Domain) and eastern areas (Sierra de Quipar section). Ammonite biostratigraphy has been broadly understood since the 1970s (Barthel et al. 1966; Sequeiros 1974; Olóriz 1978; Tavera 1985;

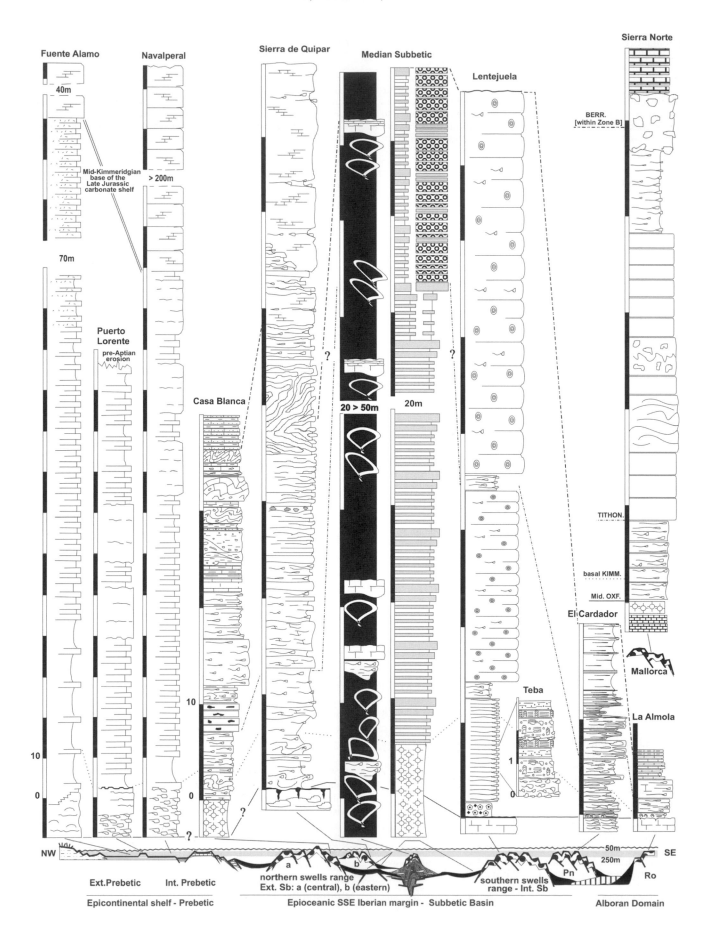

Caracuel 1996) but continues to be refined at the level of sub-zones and horizons (e.g. Caracuel *et al.* 1999, 2000; Olóriz *et al.* 1999*a*), as well as used in integrated macro-microfossil biostratigraphy (e.g. Tavera *et al.* 1994). Ecostratigraphic interpretations in the Late Jurassic ammonitico rosso facies, based on the combined analysis of facies and fossil assemblages (Olóriz *et al.* 1991, 1995*a,b*, 1996; Caracuel 1996), can help interpret depositional sequences and aspects of geobiological evolution. Non-condensed epioceanic deposits are poor in macrofossils and palaeontological information based on microfossils is still incomplete.

Epicontinental marly limestones and nodular-like limestones are poorer in ammonites and other Upper Jurassic guide fossils (Lorente Fm in Fig. 11.13A; three columns on left in Fig. 11.16). Updated neritic ammonite biostratigraphy as well as ecostratigraphy are progressing in the Prebetic (see below), as is microfossil biostratigraphy, supported by benthic forams and calcareous algae, and by a small amount of data on calpionellids (see below). For recent progress in ecostratigraphy, palaeoichnology and cyclostratigraphy, see Rodríguez-Tovar (1993), Olóriz *et al.* (1992, 1994) and Olóriz & Rodríguez-Tovar (1999, 2000).

Upper Jurassic unconformities and tectonosedimentary units identified on swells in the External and Internal Subbetic (García-Hernández *et al.* 1986–87, 1989; Marques *et al.* 1991; Rey 1993; Nieto 1996) have been interpreted in terms of tectonic–eustatic interactions of second and third order. Higher-order cycles, system tracts and/or deviations from expected patterns in the relationship between fossil assemblages and facies have been elucidated through ecostratigraphic interpretations (Olóriz *et al.* 1993, 1995*b*, 1996; Caracuel 1996; Caracuel *et al.* 1998). The effects of localized fault block movements were superimposed upon the regional transtensional setting of SSE Iberia, and a Late Jurassic–Early Cretaceous long-term transgressive–regressive cycle, forcing changes in ecosedimentary conditions and hence local variations in the sedimentary and fossil record (e.g. Olóriz *et al.* 1996).

Prebetic. The northern shelf (Prebetic) of the Betic Cordillera in Late Jurassic times corresponded to an epicontinental (neritic) environment. (As in the preceding text, the terms External and Internal Prebetic (Jerez-Mir 1973) refer to northern, comparatively proximal, and southern, comparatively distal, shelf environments, respectively.)

Late Jurassic deposits (middle Oxfordian, Antecedens to Riazi zones) rest on Early–Middle Jurassic shallow carbonates capped by a hardground. The Oxfordian succession (the lower part of the Lorente Fm; Fig. 11.12A) is up to 20 m thick, comprises nodular-like rubbly (lumpy) limestones (mainly wackestones) and related facies, bioclastic wackestones to occasional packestones with ferruginous pisolites, rhythmic limestones (mudstones to wackestones) and marly limestones with sponges, forams (Conoglobigerinidae, nodosariids), ammonites, bivalves, brachiopods and others, and subordinate but locally typical oncolitic and stromatolitic horizons (e.g. Behmel 1970; Foucault 1971; Azéma *et al.* 1979; García-Hernández *et al.* 1979, 1981; Acosta 1989; Olónz *et al.* 2000). Topmost Oxfordian deposits (Bimammatum Zone) indicate the change to increased inflows in fine siliciclastics and to less favourable life conditions for the most typical late Oxfordian benthos (sponges). Whatever the local lithology, the Oxfordian–Kimmeridgian boundary is only identifiable through ammonite biostratigraphy.

The lowermost Kimmeridgian (Planula Zone) comprises marls, marly limestone rhythmites, or nodular-like limestones below a fossiliferous hardground (García-Hernández *et al.* 1981; García-Hernández & López-Garrido 1988; Acosta 1989, and Rodríguez-Tovar 1993; Olóriz & Rodríguez-Tovar 1993*b*) overlain by marls to clays up to 30 m thick. An overlying marly limestone rhythmite shows increasing carbonate upwards and is thicker (up to 100 m; middle to upper parts of the Lorente Fm; Fig. 11.12A) and more detrital eastwards, where it incorporates sandstones containing quartz and plant remains (Behmel 1970; Fourcade 1970; García-Hernández et al. 1979). On the whole, depositional conditions during the middle Kimmeridgian interval shifted to inner carbonate environments (Sierra del Pozo Fm in Fig. 11.12A) to produce up to 200 m of massive dolomites passing locally into mudstones and oolitic–oncolitic limestones, and into bioclast-rich sandy limestones eastwards and northeastwards (Behmel 1970; Fourcade 1970; García-Hernández 1978; Azéma et al. 1979; García-Hernández & López-Garrido 1979).

With regard to the youngest part of the Jurassic succession in the Prebetics, the record of middle–upper Kimmeridgian deposits is variable and becomes progressively poorer northward due to shallowing, although sandy marls and sandstones deposited in shallow marine and transitional environments (brackish waters with

Fig. 11.16. Selected Upper Jurassic sections in the Betic Cordillera and Majorca. Betic sections on the SSE palaeomargin of Iberia and the southwestern palaeomargin of the Alboran Domain (Rondaide Complex). Complementary information for the single section in Majorca is given in Figure 11.17. Note biostratigraphic boundaries at the stage level depicted on the Sierra Norte column and correlated among other sections. See legend on Figure 11.11. Key: Ext. Sb, External Subbetic; Int. Sb, Internal Subbetic; Pn, Penibetic; Ro, Rondaides. Data for sections in the SSE palaeomargin of Iberia: Prebetic sections at Fuente Alamo, Puerto Lorente and Navalperal (Marques *et al.* 1991); Subbetic sections at Casa Blanca (Comas *et al.* 1981); Sierra de Quipar (Caracuel 1996); Median Subbetic (Comas 1978; Rey 1993; O'Dogherty *et al.* 1997; Vera & Molina 1998); El Cardador (Comas *et al.* 1981; Caracuel 1996); Penibetic sections at Lentejuela (Cruz-Sanjulián *et al.* 1973; Olóriz *et al.* 1979) and Teba (Comas *et al.* 1981). Data for La Almola section in the Alboran Domain (Martín-Algarra *et al.* 1983; Martín-Algarra 1987). Data for the Sierra Norte section in Majorca (Caracuel *et al.* 1995; Olóriz *et al.* 1995*a*; Caracuel 1996; Caracuel & Olóriz 1998). Numbers in scale bars indicate metres. Legend for lithology given below.

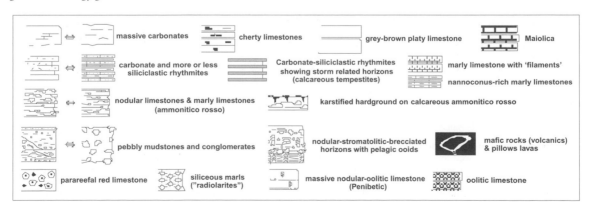

charophytes) are known towards the NE. Marine carbonates accumulated at this time in the southern part of the External Prebetic area (Fourcade 1970; Azéma *et al.* 1979; García-Hernández & López-Garrido 1979), but no Portlandian–Berriasian sediments are known from the External Prebetic. Southwards, into the Internal Prebetic area, a succession of more than 300 m of upper Kimmeridgian to Portlandian carbonates shows, from base to top: dolomites, limestones with intercalated conglomerates, and alternating limestones and marls containing conglomerate horizons with ferruginous coating of clasts (Fourcade 1970; Foucault 1971; Azéma *et al.* 1979; García-Hernández & López-Garrido 1979).

Across the Prebetic shelf, Oxfordian to lower-Middle Kimmeridgian biostratigraphy is based on epicontinental (Submediterranean-type, i.e. neritic) ammonites (Behmel 1970; García-Hernández *et al.* 1979; Rodríguez-Tovar 1993; Olóriz & Rodríguez-Tovar 1993*a,b*; Olóriz *et al.* 1999*b*). From the middle Kimmeridgian interval onwards, Upper Jurassic to lowermost Cretaceous biostratigraphy has been based on benthic foraminifera and Dasycladaceaeans (e.g. Fourcade 1970; García-Hernández 1978; García-Hernández & López-Garrido 1979). Biostratigraphic control on discontinuity surfaces has been used to support a sequence stratigraphic interpretation in which two major discontinuities bound an Upper Jurassic supercycle (García-Hernández & López-Garrido 1988; Acosta 1989; García-Hernández *et al.* 1989; Marques *et al.* 1991; Rodríguez-Tovar 1993). Lower order tectonic–eustatic interactions produced a widely recognized middle Oxfordian–lower upper Kimmeridgian megasequence. Similar megasequence interpretations have been proposed for the lower-upper Kimmeridgian–Berriasian (Marques *et al.* 1991) and middle Kimmeridgian–lower Valanginian (García-Hernández & López-Garrido 1988; García-Hernández *et al.* 1989) successions. On an even finer scale, ecostratigraphic analysis in middle Oxfordian to lower upper Kimmeridgian (Acanthicun Zone) deposits have supported the recognition of third-order sequences, in which the influence of tectonic pulses has been identified (Rodríguez-Tovar 1993; Olóriz *et al.* 1994; Olóriz & Rodríguez-Tovar 1999, 2000).

Northern trough (intermediate units). In the northern trough (Fig. 11.11), 20 to 60 m of greenish to reddish siliceous clays, and siliceous marls and marly limestones (Radiolarite marls of the Era de la Mesa Fm) were deposited during the Oxfordian and probably the mid-late Callovian (Fig. 11.12A). These deposits contain reworked oolitic grainstones westward, and become more siliceous and thinner eastward (Ruíz-Ortiz 1980). The radiolarite marls of the Era de la Mesa Fm have been considered to correlate with the more siliceous Jarropa Radiolarite Fm, which was deposited southward in the median trough (Median Subbetic) at depths interpreted as between a few hundred to less than 200 m (O'Dogherty *et al.* 1997).

Latest Oxfordian, Kimmeridgian and early Tithonian deposition in the northern trough comprises up to 200 m of thick marly limestone, limestone and marl rhythmite (Toril Fm), which included significant resedimentation events (conglomerates and calcareous turbidites) beginning in the mid-Kimmeridgian in the west and around the Kimmeridgian–Tithonian boundary in the east (Ruíz-Ortiz 1980; Olóriz & Ruíz-Ortiz 1980). Depositional environments corresponded to small submarine fans, which distributed carbonate-rich inflows southward (but secondarily also to the east and west) from surrounding highs and neritic shelves that retrograded northward during this period (Ruíz-Ortiz 1980).

During late Tithonian to late early to early mid-Berriasian times, distal fan calcareous turbidites (including neritic shelf remains and black pebbles) resulted in the generalized deposition of cherty limestones with calpionellids (Ruíz-Ortiz 1980). Updated reinterpretation of these calpionellids indicates that mass sliding of condensed ammonitico rosso from nearby swells was mainly an intra-Berriasian event, probably coeval with, or slightly earlier than that causing the monotonous deposition of marly limestone rhythmites of the Carretero Fm (Comas *et al.* 1982) and lateral equivalents (e.g. Miravetes Fm) throughout the SSE palaeomargin of Iberia.

Subbetic (northern and southern swells, and median trough). Across both the northern and southern swell areas (External Subbetic and Internar Subbetic and Penibetic, respectively) ammonitico rosso facies rocks overlie a widespread and locally complex late Middle Jurassic unconformity (García-Hernández *et al.* 1986–87; Molina 1987;

Marques *et al.* 1991; Rey 1993; Nieto 1996). These ammonitico rosso rocks, from a few metres to more than 70 m thick, range in age from middle Oxfordian to early mid-Berriasian (Upper Ammonitico Rosso Fm; Fig. 11.12A), although some sections show stratigraphical gaps (e.g. missing Oxfordian and less frequently part of the Kimmeridgian). A common unconformity clearly identifiable on swell deposits has a maximal range of early–mid-late Bathonian to early mid-Oxfordian, but larger ones are known locally (Late Jurassic resting on Early Jurassic carbonates; Fig. 11.12A). Very local records exist of lower Oxfordian deposits, these either being related to karstic cavities on Middle Jurassic carbonates, or representing the earliest 'normal' ammonitico rosso sedimentation overlying late Callovian hardgrounds (Castro *et al.* 1990; Caracuel *et al.* 2000). Karstic cavities up to tens of metres deep and excavated in older Jurassic deposits show ammonitico rosso infilling of Late Jurassic age in both the Internal and External Subbetic (e.g. Castro *et al.* 1990; Rey 1993; Nieto 1996). Directly above such karstic surfaces, red mudstones–wackestones with Conoglobigerinidae, breccias, and oolitic–crinoidal packstones are the oldest deposits, which laterally pass to extensive hardgrounds. More 'complete' ammonitico rosso successions typically show wackestones with Conoglobigerinidae, then *Saccocoma* and finally with calpionellids, during Oxfordian, Kimmeridgian–early Tithonian and late Tithonian to early mid-Berriasian sedimentation (e.g. Sierra de Quipar and El Cardador sections in Fig. 11.16). Other components in this microfacies include radiolarians, dinoflagellate cysts, ostracods, and young bivalves, gastropods and ammonites, all of which may be locally abundant. The succession is sometimes interbedded with grainstones to packstones containing *Saccocoma* and peloids, as well as stromatolitic horizons and 'eventites' (storm- and/or seismic-related resedimentation). Bottom instability is well documented by slumping, pebbly mudstones and breccias (e.g. Casa Blanca and Sierra de Quipar sections in Fig. 11.16) resulting from tilting and sliding, as well as by structured palaeosurfaces (faulting) and associated neptunian dykes. Increasing instability was to characterize the transition from Jurassic to earliest Cretaceous time when swell areas became locally emergent (García-Hernández *et al.* 1986–87; Molina & Ruíz-Ortiz 1990; Rey 1993; Nieto 1996).

Photic zone depths have been envisaged for calcareous ammonitico rosso facies, and it is unlikely that more marly ammonitico rosso facies were deposited under significantly deeper conditions (up to around 200 m depth) as they grade laterally into oolitic limestones (Martín-Algarra 1987; Rey 1993) and siliceous deposits (Jarropa Radiolarite Fm in the median trough (= Median Subbetic); Figs 11.11, 11.16), especially after the reinterpretation of 'deep' rhythmites (Milanos Fm; Molina & Vera 1996; Nieto 1996; O'Dogherty *et al.* 1997). The gradual change from a shallow rhythmite (Tornajo Marly-Limestone Fm) to ammonitico rosso facies is also known (Rey 1993).

Deposition on the slopes of both the northern and southern swell areas lying in the transition into the median trough (i.e. the sides of the intervening median trough: Median Subbetic on Fig. 11.11), produced comparatively thicker ammonitico rosso successions with increased marly and/or siliceous content, as well as resedimented deposits (Casa Blanca section in Fig. 11.16). The latter were produced by slumping which episodically produced significant debris flows, that became especially common during late Tithonian and early in Berriasian times (Comas *et al.* 1981; Molina 1987; Rey 1993; Nieto 1996), when swell instability peaked (Casa Blanca and Sierra de Quipar sections in Fig. 11.16). In swells of the eastern External Subbetic areas lying in the transition into the northern trough (intermediate units) received post-Oxfordian limestones that included frequent calcareous turbidites and debris flows (Nieto 1996).

Within the southern swell areas in the External Zones of the Betic Cordillera, westernmost and, locally, eastern outcrops show calcareous Jurassic rocks with a dominance of oolitic facies and a missing lowermost Cretaceous succession (Martín-Algarra 1987; Nieto 1993). Up to 170 m of Upper Jurassic oolitic packstones–grainstones with intercalated ammonitico rosso containing stromatolitic horizons (Lentejuela and Teba sections in Fig. 11.16) are found in the Penibetic, where this sedimentary package represents the upper member of the Torcal Fm (Martín-Algarra 1987; Fig. 11.12A).

Upper Jurassic deposits in the trough between the northern (External Subbetic) and southern (Internal Subbetic) swells (Figs 11.11 and 11.12A) are best known in the central and eastern Subbetic area (see Median Subbetic in Fig. 11.16). Deposition during Oxfordian and early Kimmeridgian ? times produced siliceous clays and marls (upper member of the Jarropa Fm; O'Dogherty *et al.* 1997), with overlying cherty, marly limestone rhythmites intercalated with calcareous tempestite horizons (Milanos Fm; Molina & Vera 1996) deposited during the Kimmeridgian–Tithonian interval (Fig. 11.16). These successions can be more or less clearly identified elsewhere in the Median Subbetic (Fig. 11.12A), and have been interpreted as representing hemipelagic–pelagic conditions evolving from a depth of several hundred metres, and locally less than 200 m, for the Jarropa Fm (O'Dogherty *et al.* 1997) to around 200 m and less for the Milanos Fm (Molina & Vera 1996). Due to an irregular bottom topography, these sediments are interleaved with marly, nodular and cherty limestones, especially where there was a transition to upper slope conditions. In contrast to the Jurassic deepening identified eastward in the Median Subbetic (e.g. Nieto 1996), shallowing has been interpreted in the more central Subbetic areas (Vera & Molina 1998) during deposition of the Kimmeridgian-Tithonian Milanos Fm. As elsewhere in the Subbetic, late Tithonian–earliest Berriasian instability induced resedimentation events, producing breccias and calcareous turbidites in the Median Subbetic trough (e.g. Comas 1978). Late Jurassic–earliest Cretaceous instability was accompanied by intermittent submarine volcanism identified in the central and eastern sectors of the Median Subbetic (Comas 1978; Rey 1993), with mafic rocks becoming intercalated with red, nodular and variably fossiliferous pelagic limestones (Median Subbetic in Fig. 11.16). A pile of pillow lavas grew several hundred metres thick to form the 5–10 km wide mid-Subbetic volcanic ridge which topographically subdivided the Median Subbetic trough (Comas 1978). This volcanic ridge reached shallow waters, as deduced from oolitic deposition, and episodically received the ammonitico rosso and related facies (Comas *et al.* 1981; Nieto 1993) that occur persistently in other Subbetic seamounts.

Rondaides and Malaguides (Internal Betics). Upper Jurassic sedimentary rocks in the Internal Zones of the Betic Cordillera (summarized in Martín-Algarra 1987) are less well known than in the External Zones, but they show similar sedimentary facies and pelagic fossil assemblages, and record both late-Middle and latest Jurassic–earliest Cretaceous emergences on swells seen elsewhere, as well as providing further evidence for the Late Jurassic tectonic instability discussed above. In the Rondaides, on the western flank of the Alboran Domain (Figs 11.11 and 11.12A), siliceous clays and marls, and reddish, nodular, marly and cherty wackestones were deposited away from local swells (sl and pe in Fig. 11.12A), on which more calcareous ammonitico rosso and other condensed facies dominated (sw in Fig. 11.12A; La Almola section in Fig. 11.16). The sedimentary environment for these sediments has been interpreted as a starved slope close to a deep basin (Marin-Algarra 1987). In the Malaguides (Figs 11.11 and 11.12A), Late Jurassic deposition changed from marly and more or less nodular and cherty grey to reddish wackestones, locally grading into reddish greenish siliceous marls, to locally thin (but generally missing) Late Jurassic ammonitico rosso facies that can fill erosion surfaces. Martín-Algarra (1987) interpreted these sediments as recording Late Jurassic deepening producing starved outer slope environments swept by marine currents, and more stable inner slope environments under shallow pelagic shelf (locally emergent?) conditions.

Upper Jurassic: Balearic Islands (JEC, FO)

Upper Jurassic rocks are exposed on the islands of Majorca, Cabrera, Ibiza and Minorca. Majorcan successions preserve a range of depositional settings from NW to SE: epioceanic swells (1 in Fig. 11.17) and slopes (2, 5, 6 in Fig. 11.17), slope–basin (3, 4 in Fig. 11.17), and shallow carbonate shelf (7 in Fig. 11.17). The thickness of these successions is equally varied, ranging from 60 m in ammonitico rosso facies deposited on epioceanic swells (Sierra Norte) to more than 300 m for slope–basinal areas receiving frequent gravity flows (3 in Fig. 11.17) from shallow shelves (7 in Fig. 11.17). Updated Upper Jurassic biostratigraphy (ammonites,

tintinnids) and ecostratigraphic interpretations of macrofossil assemblages are provided in Olóriz *et al.* (1995a, 1996, 1998), Caracuel *et al.* (1995), Caracuel (1996) and Caracuel & Olóriz (1999), based on bed-by-bed sampling in the Sierra Norte (Alfabia and Son Torrelles formations in Fig. 11.17). Outside the Sierra Norte, Upper Jurassic biostratigraphy needs to be improved.

Underlying Majorcan Upper Jurassic deposits there is an unconformity with a maximum time gap of late Bathonian–middle Oxfordian (Caracuel *et al.* 1995), as recorded from the swells and surrounding areas. An exception to this is the local preservation of Callovian siliceous marls and imprecisely dated conglomeratic and/or oolitic limestones of the Puig d´en Parè and Cutri formations (Fig. 11.17). There are also reports of isolated Callovian ammonites (e.g. Arbona *et al.* 1984–85), but no precise section and biostratigraphy have been reported. Barnolas & Simó (1984) interpreted a significant palaeogeographical change in Majorca close to the Middle–Upper Jurassic boundary, as seen elsewhere in western Tethys.

The Upper Jurassic rocks of Cabrera, although close to southern Majorca and aligned with the Sierras de Levante, actually show a closer palaeogeographic affinity with the Sierra Norte. Upper Jurassic rocks in Ibiza are mostly nodular-like limestones with Conoglobigerinidae (Oxfordian), thin bedded limestones with *Saccocoma* (Kimmeridgian), and sandy limestones and dolomites to the north and marly limestones southward (Portlandian), deposited under shelf environments. Finally, Upper Jurassic rocks on Minorca are mostly shallow shelf dolostones thought to have been deposited some distance away from the other islands (Minorca is separated from them by a fault zone).

Majorca. Upper Jurassic palaeoenvironments can be subdivided into swells (and their adjacent upper slopes), troughs (and their adjacent lower slopes), and carbonate shelf. Rocks deposited on swells mainly crop out in the Sierra Norte (Figs 11.16 and 11.17). As recorded in the Betic Cordillera and elsewhere in western Tethys, ammonite-bearing, marly to calcareous, red-grey, and locally cherty ammonitico rosso and wackestones to packstones with dominant Conoglobigerinidae, *Saccocoma*, radiolaria and dinoflagellate cysts (Alfabia Fm; Alvaro *et al.* 1989) developed in the middle Oxfordian to late Kimmeridgian interval (1 in Fig. 11.17). Latest Kimmeridgian to earliest Tithonian well stratified but fossil-poor, 30–40 m thick, brownish mudstones (Aumedrà Fm; Alvaro *et al.* 1984a) show a brief departure from ammonitico rosso facies in the Sierra Norte (less than two ammonite chronozones; Caracuel & Olóriz 1998, 1999; Fig. 11.17). Above this, mainly marly ammonitico rosso (Son Torrelles Fm; Alvaro *et al.* 1989) were re-established on epioceanic swells in Sierra Norte from the early (but not earliest) Tithonian to early-middle Berriasian p.p., as dated by ammonites and calpionellids. These ammonitico rosso deposits contain horizons of calcareous turbidites/tempestites?, slumps, and pebbly mudstones of late early Tithonian to early mid-Berriasian age (Caracuel & Olóriz 1998), consistent with a phase of tectonic instability. Northeastward in this range, thinner successions with a locally significant contribution of algae exist (2, 3 in Fig. 11.17).

In the Cap Pinar-Son Fe area (2 in Fig. 11.17), which geographically belongs to the Sierra Norte but shows Lower and Middle Jurassic rocks with affinity to the Sierras de Levante (Alvaro *et al.* 1989), red and marly ammonitico rosso rocks of early late Oxfordian age (Bifurcatus Chron) show slumps revealing epioceanic swell–slope conditions later replaced by the influence of lower slope deposits supplied from carbonate shelves (Puig d'en Borràs Fm; see below) until a return to swell conditions during the Tithonian stage. A similar change during the Oxfordian stage occurred in the Betlem–Camp d'en Porrassa area (5 in Fig. 11.17), in the northern Sierras de Levante, where lower slope conditions persisted throughout the Late Jurassic epoch, supplied from carbonate shelves.

Barnolas & Simó (1984) assumed subphotic zone conditions for the

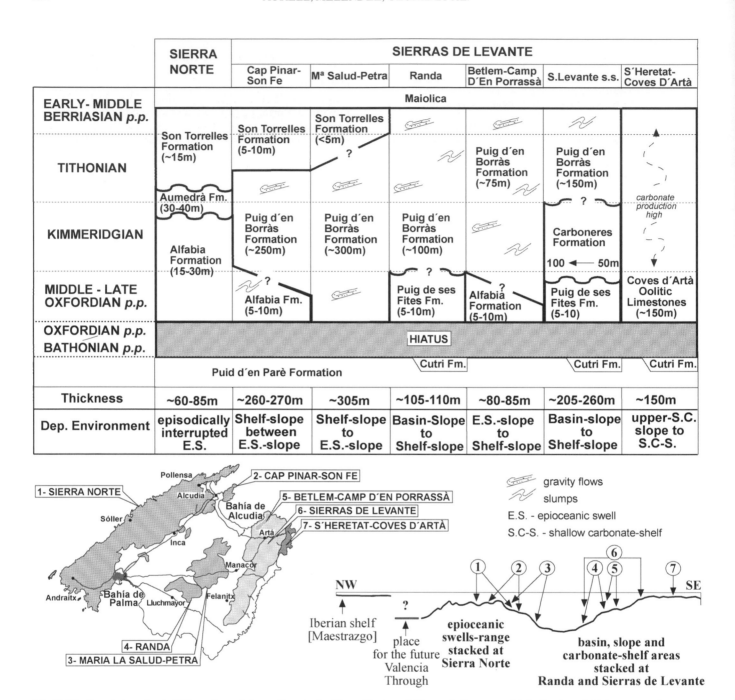

Fig. 11.17. Correlation chart of Upper Jurassic formations in the Sierra Norte and Sierras de Levante, with indication of thicknesses and depositional environments in Majorca (top). Map (lower left) and cross sections (lower right) showing Upper Jurassic palaeography of Majorca.

Sierra Norte during Late Jurassic times, although epioceanic equivalent swell areas in the Betic Cordillera have usually been interpreted as shallower depositional settings (see above). Bioclastic packstones in the Alfabia and Son Torrelles formations have been identified and related to storm activity (Alvaro *et al.* 1984a; Caracuel & Olóriz 1998), although their differentiation from distal turbidites might be investigated. On the basis of available data, we argue that Late Jurassic epioceanic plateaux, as identified in the Sierra Norte (Caracuel & Olóriz 1998), were not confined to subphotic depths in western Tethys. This hypothesis accords with depth interpretations for carbonate-poor pelagic sediments related to the deepest episodes in the 'pelagic carbonate platforms' in Umbria–Marche (Santantonio 1993), as well as with those recently proposed for classically 'deep' deposits in the Betic Cordillera (see above).

Away from the epioceanic swells, lower slope to slope–basin areas crop out in the Sierras de Levante (6 in Fig. 11.17), and Maria La Salud-Petra and Randa (3, 4 in Fig. 11.17), the latter two belonging to the subsiding Randa Sector (Alvaro *et al.* 1984a) identified in the Llanos Centrales (central plains). Carbonate shelf and related upper to lower slope settings crop out in the Sierras de Levante. Red siliceous clays, marls and marly limestones of assumed Oxfordian age accumulated in slope basin areas (Puig de ses Fites Fm; Alvaro *et al.* 1984a). In the Maria La Salud-Petra area (3 in Fig. 11.17), persistent lower slope facies showing slumped cherty mudstones with pelagic microfossils (as in ammonitico rosso facies; see above), conglomerates and bioclastic grainstones with resedimented oolites, shallow-water forams and algae, fragmented corals and crinoids (Puig d´en Borrás Fm; Álvaro *et al.* 1984a), were deposited throughout the Late Jurassic epoch, except for imprecisely dated late Tithonian to early mid-Berriasian ammonitico rosso (Son Torrelles Fm; Fig. 11.17) below Maiolica facies. In contrast to the erosional and highly diachronous base of the Puig d´en Borrás Fm, the top of this unit in Randa, Betlem-Camp d'en Porrassà and Sierras de Levante *sensu stricto* (4, 5, 6 in Fig. 11.17) seems to be isochronous (or only slightly diachronous) within the early Berriasian interval.

Carbonate shelf deposits crop out in the northern Sierras de Levante (S'Hererat-Coves D'Artà) throughout a poorly known part of the Late Jurassic succession (7 in Fig. 11.17). These rocks, which include beds proven to be Oxfordian in age, are oolitic and skeletal grainstones, and dolomitized limestones (Fornos *et al.* 1988). Southward in Sierras de Levante *sensu stricto* (6 in Fig. 11.17), middle-late Oxfordian–Kimmeridgian deposition on a shelf-related upper slope took place over older Oxfordian slope–basin deposits (see above). This upper slope deposition resulted in cherty, well-stratified and crinoid-rich turbiditic grainstones intercalated with autochthonous, laminated mudstones (Carboneres Fm; Alvaro *et al.* 1984a) that contain less abundant pelagic microfossils than ammonitico rosso. During Tithonian and Berriasian times, the latter before Maiolica deposition, lower-slope conditions with shelf influence continued, producing the Puig d'en Borràs Fm deposits mentioned above.

Depositional areas corresponding to Upper Jurassic outcrops in Majorca received nannoconid-rich oozes (Maiolica mudstones to wackestones) that ended the Late Jurassic–earliest Cretaceous supercycle. As noted for the Betic Cordillera (see above), Barnolas & Simó (1984) identified in Majorca latest Jurassic–earliest Cretaceous tectonic instability that forced significant palaeogeographic change. Olóriz *et al.* (1995a) dated this episode as a non-basal Zone B and/or Zone C (upper lower–lower middle Berriasian) event in central Sierra Norte (Sierra Norte section in Fig. 11.16), which was confirmed by Caracuel (1996) and Caracuel & Olóriz (1998) who identified small-range variation of stratigraphical gaps in the area using calpionellids.

Cabrera. Available Upper Jurassic bio- and lithostratigraphy are limited, but permit general comparisons between Cabrera and Majorca. Arbona *et al.* (1984–85) interpreted Late Jurassic slope conditions to have existed in Cabrera, with a comparatively distant location from coeval source areas supplying Majorcan basins, suggested by reduced evidence for resedimentation. On the basis of Majorcan data (see above), updated interpretation of information in Arbona *et al.* (1984–85) indicates: (a) depositional evolution during the Late Jurassic epoch in Cabrera closer to that in the Sierra Norte than to the Sierras de Levante (Majorca), as shown by epioceanic swell–slope conditions (ammonitico rosso facies of the Alfabia and Son Torrelles formations) interrupted by distal and presumably shelf-related slope deposits (brownish mudstones) during late/latest Kimmeridgian? to early Tithonian times (Aumedrà Fm); (b) thinner Upper Jurassic supercycle deposits in Cabrera, compared with Majorca, a feature that we relate to significant instability (8 m thick debris flow deposits) in Cabrera between the middle Bathonian and the middle Kimmeridgian; (c) less latest Jurassic–earliest Cretaceous instability in Cabrera due to the absence of thick debris flow deposits, which are typical of the Majorcan succession (especially in Sierra Norte); (d) cherty horizons (skeletal packstones) in the mid-Kimmeridgian ammonitico rosso in Cabrera accord with the absence of typical slope–basin deposits in the area, reinforce our interpretation of an epioceanic swell–slope depositional environment (point a), and contrast with the record of slope–basinal siliceous clays and marls (Puig de ses Fites Fm) in the Sierras de Levante (Majorca).

Arbona *et al.* (1984–85) compared the Lower–Middle Jurassic succession in Cabrera and Majorca and emphasized an earlier (early late Pliensbachian) break-up unconformity during Early Jurassic times in Cabrera, the record of early Pliensbachian siliciclastics closer to those of the Sierras de Levante than in Sierra Norte, the probability of a wider stratigraphical gap between the Lower and the Middle

Jurassic, and significant resedimentation between the mid-Bathonian and the Kimmeridgian stage. The different affinities between Cabrera and different parts of the Majorcan succession through Jurassic time emphasize the reality of block tectonics within some segments of the northwestern palaeomargins of the Tethys. This is a fact that should be taken into account before accepting the classical and oversimplified idea of the Balearic archipelago (except Minorca) as representing the NE expression of Betic External Zones (see the comparison of Cabrera and External Zones of the Betic Cordillera in Arbona *et al.* 1984–85).

Ibiza and Minorca. Palaeogeographic interpretation of Ibiza and Minorca has long been discussed on the basis of stratigraphic and structural data (e.g. Fourcade *et al.* 1977, 1982), the former being limited by the fossil record mainly inner-neritic biotas that impede precise biostratigraphy and correlations.

Ibiza has proved especially controversial, with Jurassic stratigraphy being considered as representing entirely shelf (Prebetic, e.g. Rangheard 1970; Azéma *et al.* 1979) or shelf-to-basin affinity (Prebetic to Intermediate units and Subbetic, included in the Plate-Forme and Sillon Citrabétiques, e.g. Fourcade *et al.* 1977, 1982). Upper Jurassic rocks in Ibiza overlie shallow water dolostones and are mostly nodular-like limestones with Conoglobigerinidae (Oxfordian), thin bedded limestones with *Saccocoma* (Kimmeridgian), and sandy limestones and dolomites (to the north) and marly limestones to the south (Portlandian), deposited under shelf environments. We argue that the existence of a NW–SE Jurassic shelf-to-basin trend in Ibiza is real, but alternative palaeogeographic interpretations (hypotheses a and b below) are possible depending on whether (a) a Middle Jurassic carbonate shelf existed (Rangheard 1970; Fourcade *et al.* 1977) or (b) there is instead a stratigraphical gap between the mid-Lower Jurassic and middle Oxfordian successions (Fourcade *et al.* 1982). Hypothesis (a) supports a Prebetic-type pelagic shelf (i.e. no pelagic influences during the Middle Jurassic) showing outer shelf environments southeastward, which were inhabited by calpionellids during the latest Jurassic–earliest Cretaceous interval. Therefore, the Ibiza series (*sensu* Rangheard 1970) should be equivalent to the Internal Prebetic (outer shelf), although the absence of cherty limestones and green marly limestones during the middle to early Late Jurassic interval leads us to discard any reference to basin-type deposits (Sillon Citrabétique, either Intermediate Units or External Subbetic; see above). Hypothesis (b) reinforces the palaeogeographic character of the Ibiza shelf system due to a significant stratigraphical gap in the Jurassic series that separates this area from westward epicontinental shelves (Prebetic), which do not show this feature. This hypothesis accords with a differentiation of segments within northwestern Tethyan palaeomargins during the Jurassic period, and supports the notion that the Balearic islands (except Minorca) conformed to one of these segments in the east Iberian palaeomargin (e.g. Meléndez Hevia *et al.* 1988), which is consistent with major fracture zones envisaged as having existed in this area (e.g. Fig. 11.11B).

Minorca is separated from Majorca by a fault zone inherited from the Variscan basement (e.g. Bourrouilh & Gorsline 1979) and shows Palaeozoic affinity with the Corso-Sardinian Block and NE Iberia (Catalanides) (e.g. Fourcade *et al.* 1977, 1982). Minorca was located *c.* 70 km NW from Majorca during the Jurassic, and its rock records shows persistent shallow shelf conditions, the deposition of dolomitic carbonates throughout the Jurassic period, and a lack of open sea influence in the latest Jurassic–earliest Cretaceous interval.

Research in the Basque-Cantabrian Ranges was funded by project UE-1999/8 of the Basc Government. Projects P35/97 (Aragon Government), PB1998-1260 and BTE2000-1148 (M.C.T., Spain) supported research in the Iberian and Catalalonian Coastal Ranges. Research in the Betic Cordillera has been supported by CICYT and DGICYT Projects and Concerted Actions (MCyT) included in the research programme of the EMMI Group (KNK-178, funta de Andalucía, Spain).

12 Cretaceous

JAVIER MARTÍN-CHIVELET (coordinator), XAVIER BERÁSTEGUI,
IDOIA ROSALES, LORENZO VILAS, JUAN ANTONIO VERA,
ESMERALDA CAUS, KAI-UWE GRÄFE, RAMÓN MAS, CARMEN PUIG,
MANUEL SEGURA, SERGIO ROBLES, MARC FLOQUET,
SANTIAGO QUESADA, PEDRO A. RUIZ-ORTIZ,
M. ANTONIA FREGENAL-MARTÍNEZ, RAMÓN SALAS,
CONSUELO ARIAS, ALVARO GARCÍA, AGUSTÍN MARTÍN-ALGARRA,
M. NIEVES MELÉNDEZ, BEATRIZ CHACÓN, JOSÉ MIGUEL MOLINA,
JOSÉ LUIS SANZ, JOSÉ MANUEL CASTRO, MANUEL GARCÍA-HERNÁNDEZ,
BEATRIZ CARENAS, JOSÉ GARCÍA-HIDALGO, JAVIER GIL &
FRANCISCO ORTEGA

Cretaceous rocks crop out extensively in the three main Alpine orogenic belts of Spain: the Betic Cordillera, the Pyrenees and the Iberian Ranges. These rocks, deformed during Cenozoic Alpine convergence, are almost entirely sedimentary (with the exception of rare volcanic and metamorphic rocks) and were deposited in an enormous variety of environments ranging from alluvial fans to pelagic seas. There were four main basins – Betic, Pyrenean, Basque-Cantabrian and Iberian (Fig. 12.1) – each of which originated in Triassic and Jurassic times in response to continental break-up at the start of the Alpine cycle. They subsequently underwent a polyphase evolution in the Cretaceous period when palaeogeography and sedimentation in Iberia were strongly influenced by the relative movements of the contiguous Eurasian and African plates. Initiation of the North Atlantic spreading in earliest Cretaceous time led to a decrease in relative

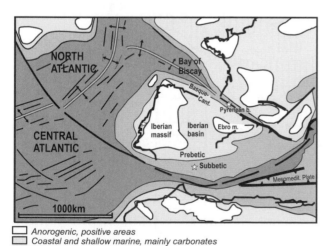

Fig. 12.1. Palaeotectonic–palaeogeographic map of Iberia and surrounding areas during the Cretaceous (Santonian), showing the main tectonosedimentary domains for that time. Based on Ziegler (1988a).

sinistral motion between Iberia and Africa (e.g. Ziegler 1988a). This was followed by a phase of rapid counterclockwise rotation of Iberia relative to Europe and the progressive opening of the Bay of Biscay, which lasted from late Aptian to early Campanian times (e.g. Olivet 1996). Finally, a third phase of basin evolution was heralded by the onset of Late Cretaceous oblique convergence between Africa and Europe (e.g. Savostin et al. 1986; Reicherter & Pletsch 2000).

In addition to this changing tectonic setting, other factors, such as climate and eustasy, were also important controlling influences. The Cretaceous climate of Iberia was largely subtropical, semi-arid to humid (e.g. Alvárez-Ramis 1980; Floquet 1991; Giménez et al. 1993). With a palaeolatitude of 25–30°N (Dercourt et al. 2000), Iberia would have lain within the subtropical anticyclonic belt, with NE trade winds probably being dominant (e.g. Barron & Moore 1994). Its palaeogeographic configuration (a group of islands with low topography; Fig. 12.1) suggests that the regional climate was largely maritime, and strongly influenced by the warm, north-equatorial ocean current (e.g. Martín-Chivelet 1999). Climatic conditions favoured the development of thick sequences of carbonates in both shallow and deep marine conditions, as well as in lacustrine environments of the continental interior.

Added to the effects of tectonism and climate, eustasy had also an important role in determining the palaeogeography. During the Cretaceous period, it is well known that unusually rapid sea-floor spreading, mainly due to the opening of the Atlantic ocean, produced high mid-ocean ridges which displaced water out of the ocean basins to cause the largest transgressive episodes since Early Palaeozoic times. High mid-Cretaceous sea levels in particular resulted in flooding of wide areas within the Iberian interior. Only the Iberian and the Ebro Variscan massifs remained positive features (Fig. 12.1). Between them, during part of the Cretaceous period, the Iberian intracontinental basin operated as a seaway connecting the Bay of Biscay with the Betic basin. This palaeogeographic configuration, in addition to the location of Iberia within the Tethys realm, was of crucial biogeographical importance. Assemblages of marine fossils indicate a double affinity with European and African faunal provinces (e.g. Masse et al. 1992; Martín Chivelet 1992). Furthermore, specific assemblages within the Pyrenees have allowed the definition of a faunal subprovince for this area (e.g. Philip 1985; Hottinger et al. 1989).

This chapter has been prepared by integrating data from different research teams across Spain, with the basic objective of offering a broad and modern perspective of the Cretaceous record in that country. It has been separated into four main sections, each of which deals with a major tectonosedimentary domain (Betic, Pyrenean, Basque-Cantabrian and Iberian basins). Each main section has been further divided into several parts that refer to the three main episodes in the Cretaceous evolution of Iberia and its sedimentary basins. These episodes start with a prolonged interval, lasting nearly all of early Cretaceous time, which was characterized by repeated transtensional or extensional deformation events and rapid differential subsidence in all the sedimentary basins. This was followed by a second interval (approximately late Albian to late Santonian) characterized by relative tectonic quiescence, thermal subsidence and the development of the highest carbonate platforms of the Mesozoic era, which progressively overstep the Early Cretaceous basin margins. Finally, a third interval (Campanian–Maastrichtian) saw the onset of crustal shortening in the Pyrenees, as well as, to a lesser extent, in the other basins.

Betics (JAV, LV, PAR-O, JM-Ch, CA, JMM, JMC, BCH, MG-H)

The Betic Cordillera is an Alpine mountain belt located in southern Spain. It is characterized by fold and thrust structures that formed in response to considerable shortening during mainly early-mid Miocene compression and transpression. Three areas are usually distinguished (Fig. 12.2): the Betic External Zones and the Betic Internal Zones, separated by the Campo de Gibraltar Complex (Azéma et al. 1979; García-Hernández et al. 1980). Previous syntheses of the Cretaceous system of the Betic Cordillera were published by Fallot (1943) and Vera et al. (1982).

The Betic External Zones form a complex of thrust sheets comprising thick successions of Triassic to Lower Miocene sedimentary rocks detached from a Palaeozoic basement that corresponds to the southern prolongation of the Iberian Variscan massif (Fig. 12.2). These rocks, which include thick Cretaceous sequences, were deposited on the southern Iberian continental margin (SICM; or Betic Margin in some papers). They record a complex evolution during Cretaceous times, including the later part of a spreading and transform stage (Oxfordian–early Albian) and a complete post-rifted margin stage (Late Cretaceous; Vera 2001).

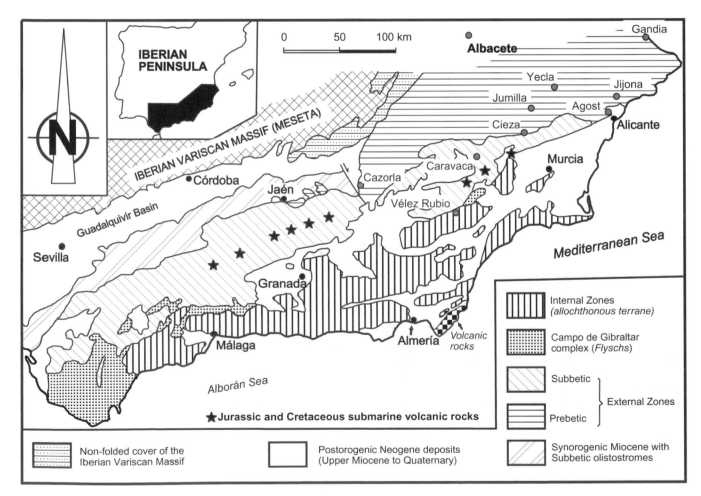

Fig. 12.2. Map of the Betic Cordillera, showing the External and Internal Zones, and the Campo de Gibraltar complex. In the Betic External Zones the Prebetic and the Subbetic are also differentiated. The main outcrops of the Subbetic Jurassic and Cretaceous submarine volcanic rocks are marked. After Vera & Molina (1999).

The External Zones are subdivided into two tectonic zones (Fig. 12.2) that are roughly based on the two great palaeogeographic domains of the SICM: a platform area to the north, the Prebetic; and a mainly pelagic area to the south, the Subbetic. Cretaceous sediments from the Prebetic share features with those of other Iberian platforms (e.g. Iberian Ranges), in contrast to the Subbetic sediments, which are similar to those of the pelagic domains from other Alpine chains (Alps, Apennines, Carpathians, etc.).

The Betic Internal Zones, together with the Internal Zones of the Rif (Morocco) and Kabylias (Algeria), formed a crustal unit called the Mesomediterranean microplate, then located south of Sardinia. A fragment of that Mesomediterranean microplate (the Alborán domain) subsequently moved westward during early (and middle) Miocene times, forming the allochthonous terrane that today forms the Betic Internal Zones. The Alboran Domain collided with the SICM towards the end of middle Miocene times (Sanz de Galdeano & Vera 1992). In the Betic Internal Zones, Cretaceous sediments are, however, very reduced, discontinuous, and restricted to the Malaguide Complex (Chapter 16).

The Campo de Gibraltar complex is made up of allochthonous units of sediments deposited in the so-called 'flysch troughs', which once formed the northwestern margin of the Mesomediterranean microplate. In these troughs, Cretaceous hemipelagites and turbidites were deposited on oceanic crust and/or highly thinned continental crust. Turbidite sedimentation continued during the Palaeogene interval and reached its maximum sedimentation rate in late Oligocene and Aquitanian times. Later, the collision of the Alboran Domain with the SICM caused the destruction of the flysch troughs and the expulsion of their turbidite sediments westwards (Campo de Gibraltar Complex).

Lower Cretaceous of the Prebetic Zone (LV, CA, MG-H, PAR-O, JMC)

The Prebetic Zone (Fig. 12.2) is a broad Alpine tectonic unit that corresponds to the most external portion of the foreland fold-and-thrust belt of the Betic Cordillera. It consists of a para-autochtonous sedimentary cover of Mesozoic–Cenozoic age, which is detached from the Variscan basement along the upper Triassic evaporites. These Mesozoic–Cenozoic rocks include thick successions (up to 3 km) of Cretaceous carbonates and clastics that formed in the shallow areas of the ancient continental margin of Iberia (the southern Iberian continental margin), in environments which ranged from continental to outer platform and hemipelagic settings.

In the Prebetic Zone, two main subareas are traditionally recognized on the basis of stratigraphic and tectonic criteria: the External and the Internal Prebetic. The External Prebetic was closer to the emergent Iberian Massif (Meseta), and was characterized during Cretaceous times by the development of shallow carbonate (and mixed carbonate–siliciclastic) platform environments. The Internal Prebetic, located basinwards of the former (towards the SE), represented coeval but more distal areas, including mid- and outer platform, slope and hemipelagic settings, the latter marking a transition into the deep marine environments of the Subbetic.

From a palaeogeographic point of view, and based exclusively on Cretaceous tectonosedimentary features, a more specific domain has been proposed and broadly adopted in recent papers dealing with the External Prebetic. This is the

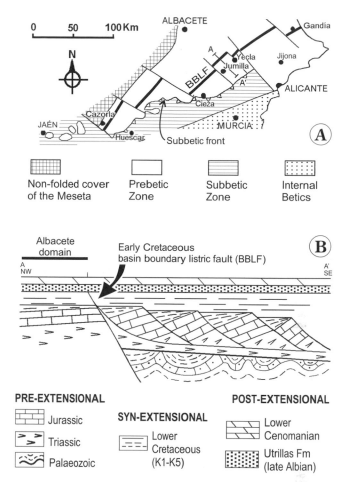

Fig. 12.3. (A) Simplified map of the present-day location, within the Prebetic Zone, of a major listric fault that acted as the basin boundary (BBLF) during the Early Cretaceous extensional phases. The fault separated the less subsident Albacete domain (northwards) from the rest of the Prebetic Zone. Based on Vilas & Querol (1999). (B) Schematic interpretative model of the Early Cretaceous extensional structure of the Prebetic, showing the Albacete domain and the listric boundary fault. Note that the Palaeozoic basement does not seem to be affected by extension. Triassic salts acted as the main detachment level for listric faults. Not to scale.

Albacete domain (cf. Vilas *et al.* 1982; Martín-Chivelet 1992; Fig. 12.3A), defined as a 'cratonic like' slowly subsiding area that marks the transition between the Meseta and the more subsident areas of the rest of the External Prebetic. This domain, characterized by very discontinuous, continental to very shallow marine Cretaceous sedimentation, became separated from the rest of the Prebetic area during Early Cretaceous time by a prominent SW–NE striking, SE dipping extensional fault (Fig. 12.3). This Early Cretaceous basin boundary fault was the largest of a family of faults produced by regional extension associated with the opening of the North Atlantic. The crustal extension gave rise to new accommodation space towards the interior of the Iberian plate, in areas of relatively shallow basement that until that time had shown generally low subsidence rates. Within this extensional basin framework thick successions of shallow marine and continental carbonates and clastics were deposited (e.g. Jerez-Mir 1973; Rodríguez Estrella 1977, 1979; Arias 1978; García-Hernández

1978; Vilas *et al.*1993; Castro 1998; Ruiz-Ortiz *et al.* 1996), these being noticeably less complete and much thinner in the Albacete domain. At present, these Prebetic sediments occur both as outcrops, exposed in the northeastern Betic Cordillera, and in the subsurface, underlying Subbetic materials.

Basin analysis (Vilas *et al.* 1993) has allowed characterization of the Early Cretaceous tectonosedimentary evolution of the Prebetic area, starting with a general increase in the regional subsidence already seen from Oxfordian times onwards. Late Berriasian break-up and subsequent fragmentation of the former Jurassic carbonate platform was followed by over 45 Ma of continued, multiepisodic, extensional tectonics, until late Albian times, when tectonic subsidence gave way to a generalized, complex, thermal subsidence.

Within this long time interval, stratigraphic analyses (Vilas *et al.*1993; Castro 1998; Ruiz-Ortiz & Castro 1998) allow recognition of five main tectonosedimentary episodes (K1–K5 on Fig. 12.4). These are each separated by significant changes in basin palaeogeography, mainly controlled by extensional tectonic events. In fact each episode is represented by a stratigraphic unit bounded by tectonically induced unconformities. These units internally show a relatively high degree of

sedimentary continuity, and record in many cases several third-order accommodation changes that allow the definition of depositional sequences. For this reason, these major unconformity-bounded units have been named sequence sets and details of each one are summarized below. It should be noted that the northern extension of sequence sets corresponding to episodes K1 and K2 was limited by the listric margin fault, whereas for those of episodes K3, K4 and K5, the area of sedimentation spread landward of this fault and partially covered the Albacete domain.

As should be expected for such an extensional setting, most accommodation was generated at the start of each of the five episodes, when tectonic subsidence reached its highest values. It become attenuated towards the end of each episode, when the development of carbonate platforms became dominant. Furthermore, on a broader timescale, subsidence rates generally diminished from late Berriasian to Albian times, a trend that coincided with an increase in the development of carbonate platforms.

Lower Cretaceous Prebetic sequence sets. *Episode K1.* The boundaries of this episode have been ascribed to the base of the late Berriasian and the boundary between early and late Valanginian times (Fig. 12.4). The episode represents the sedimentary response to

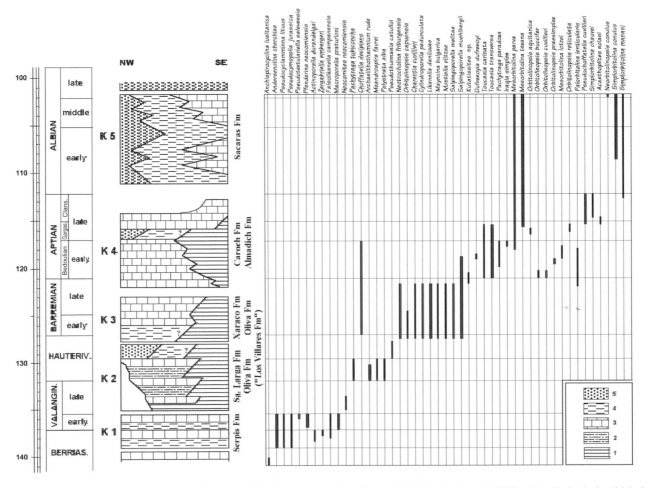

Fig. 12.4. Chronostratigraphic chart for the Lower Cretaceous succession (timescale by Gradstein *et al.* 1995) of the Prebetic, in which the five major genetic units or sequence sets (K1 to K5) are shown, together with lithostratigraphic units, depositional systems, and the vertical distribution of most indicative fossils. Key: l, distal outer platform facies; 2, proximal outer platform facies; 3, inner platform facies; 4, littoral facies; 5, continental facies. Based on data from García-Hernández (1978, 1981), Masse *et al.* (1992, 1993, 1998), Company (1987), Castro (1998) and Aguado *et al.* (1999).

fracturing of the Jurassic carbonate platforms, with the first substantial deposition of terrigenous material. Sedimentary facies varied, from tidal in the northern zone to open platform facies in the southern zone, and subsidence is spatially highly variable.

Episode K2. The age of this episode ranges from the early late Valanginian transition to the latest early Hauterivian interval. Its base corresponds to regional drowning with the development of outer platform sedimentation with ammonites and, in the Alicante region, condensed distal and hemipelagic facies became established. Dominant facies are outer platform, with some reef levels towards the end of the episode. It ends with a substantial terrigenous supply that covers the carbonate platform.

Episode K3. Unequally represented in the area, this episode represents the top of late Hauterivian to late Barremian times (Fig. 12.4). At its maximum extent it was characterized by a shallow carbonate platform with dasyclads limestones as the dominant facies. In the Albacete area, this episode is represented by lacustrine marls, limestones, and evaporites with carophytes, infilling half-grabens. Hemipelagic and distal platform sedimentation continued in the Alicante region.

Episode K4. This episode starts in late Barremian times and finishes at the Aptian–Albian boundary, with sedimentation extending landwards beyond the area in which, up until now, deposition had taken place. The most representative facies of this episode is the classical 'Urgonian', characterized by abundant rudist and orbitoline limestones. Towards the south, south of the Alicante region, these grade into hemipelagic facies.

Episode K5. Covering the early to late Albian interval, this represents the last extensional episode with a high degree of facies variation, from palustrine terrigenous sedimentation to an outer platform in the south of Alicante.

Post-rift sequences in the Prebetic Zone (late Albian–Santonian) (JM-Ch)

In the shallow areas of the southern Iberian continental margin (SICM), the multiepisodic extensional interval that had characterized Early Cretaceous time gave way to a tectonically more quiescent interval during which the thermal cooling of the lithosphere controlled basin subsidence. This fact, together with a mid-Cretaceous sea-level highstand, allowed the spread of shallow marine waters tens of kilometres landwards, resulting in the development of the widest carbonate platforms ever known in the area.

The long-term thermal subsidence that characterized the basin during the late Albian–Santonian interval was, however, punctuated by a series of regional tectonic events that caused block movements and that substantially modified basin depocentres and distribution of sedimentary systems within the SICM. For this reason, we have differentiated within the late Albian to late Santonian interval, three main episodes of basin evolution, each of them characterized by specific palaeogeographical patterns, subsidence trends and vertical stacking of sedimentary facies. The first of these episodes is late Albian–middle Cenomanian in age and is characterized by a remarkable absence of tectonic activity and a corresponding homogeneity of facies. The second, middle Cenomanian–early Coniacian in age, was much more unstable, being characterized by fault block movements and by a strong tectonic differentiation of several palaeogeographic domains. Finally, the third episode, late Coniacian to Santonian in age, was, like the first one, an interval of relative tectonic quiescence. All mentioned ages for units and unconformities are based on published biostratigraphic data (Martín-Chivelet *et al.* 1990; Martín-Chivelet 1992, 1995; De Ruig 1992; Giménez *et al.* 1993).

Late Albian–middle Cenomanian carbonate platform. The late Albian–middle Cenomanian interval is characterized in the Prebetic area by the development of the widest carbonate platforms seen in the Cretaceous period. Shallow seas covered the entire area, including the Albacete domain, resulting in a remarkably uniform lateral distribution of facies. Deposits corresponding to this interval form a major unconformity-bounded unit (sequence set K6 of Fig. 12.5) within which one major transgressive–regressive cycle, containing five minor transgressive–regressive events, has been recognized (Martín-Chivelet 1992; Giménez *et al.* 1993).

Sedimentation started in late Albian times with the installation of a vast, low-sinuosity fluvial system (Elízaga 1980; Martín-Chivelet 1992) which is represented by the Utrillas Formation (Fm). Basinwards and upwards, these continental clastics grade into the Jumilla and Chera formations which are heterolithic and consist of alternating peritidal/lagoonal sandy marls and shallow marine carbonates (Martín-Chivelet 1994; Castro 1998). Further basinwards, in the deepest areas of the Internal Prebetic, fluvial and coastal siliciclastic rocks of the Utrillas Fm and mixed carbonate–siliciclastic inner platform deposits of the Jumilla and Chera formations grade into sandy marls and carbonates (e.g. Paquet 1969; Hoedemaeker 1973). Those deposits, interpreted as prodelta and outer shelf facies (Leret *et al.* 1982), yield abundant ammonites pertaining to the *Inflata*, *Dispar* and *Mantelli* zones (Cremades & Linares 1982; García-Hernández *et al.* 1982; Linares & Cremades 1988).

The mixed deposits of the Jumilla and Chera formations and their outer platform equivalents prevented the installation of a huge carbonate platform of early Cenomanian age. This new platform is represented by three laterally equivalent units (Fig 12.5): the Villa de Ves Fm (thick peritidal cyclic successions with supratidal sabkha deposits), the Alatoz Fm (inner-shelf to shelf-edge deposits: massive or mega-cross-bedded orbitolinid grainstones and packstones, bioturbated wackestones, and rudist biostromes, all of them strongly dolomitized) and a part of the Jaén Fm (inner and outer platform limestones, partially dolomitized). This huge platform system grades basinwards into pithonella-rich hemipelagic carbonates.

The relationship between these lithological units is summarized in Figure 12.5. The basal facies of Alatoz (and their equivalents in the Jaén Fm) recorded a rapid and extensive transgression that marked the establishment of a wide carbonate ramp (Martín-Chivelet 1995). Later, the ramp evolved into a flat-topped platform in response to the rapid accretion of the shelf margin. A vast tidal-flat complex (Villa de Ves Fm) developed landward of the platform edge, covering all the External Prebetic, including the low subsiding Albacete domain. The platform persisted until the beginning of middle Cenomanian times, when it ended abruptly in response to a major palaeogeographic change caused by a broad regional tectonic event. This event marked the onset of a prolonged period of tectonic instability in the margin that lasted until early Coniacian times (Martín-Chivelet 1996).

Middle Cenomanian–early Coniacian tectonism. The middle Cenomanian to early Coniacian interval was characterized by a succession of tectonic events that controlled facies distribution, subsidence patterns and basin geometry. This tectonic activity has been recognized in different areas of the Prebetic (e.g. Hoedemaeker 1973; De Ruig 1992; Martín-Chivelet 1992, 1995, 1996), and caused a complex structure of topographic highs and troughs in response to a multiphase reactivation of listric faults. This fault-dominated structure is well known in the Prebetic of Jumilla-Yecla region, in the northern part of Murcia province (Martín-Chivelet 1995). Here mid-late Cenomanian reactivation of listric faults in the Betics caused the development of an WSW–ENE elongated trough with shallow marine carbonate sedimentation until the end of the Cenomanian stage (Carada, Cuchillo and Moratillas formations) and very reduced peritidal deposits and palaeosols during the Turonian and early Coniacian (Alarcón Fm; Figs 12.5 and 12.6A). This trough was bounded by two elevated and emergent areas with no sedimentation. The elevated area to the NE roughly coincides with the Albacete domain, and the area located to the SE was a narrow (3–5 km wide), ENE–WSW trending zone which has been named Franja Anómala ('anomalous fringe'; Martínez del Olmo *et al.* 1982). The new palaeotectonic feature that formed the Franja Anómala extended over several hundreds of kilometres along the Prebetic (Martínez del Olmo *et al.* 1982; Fig. 12.6) and for much

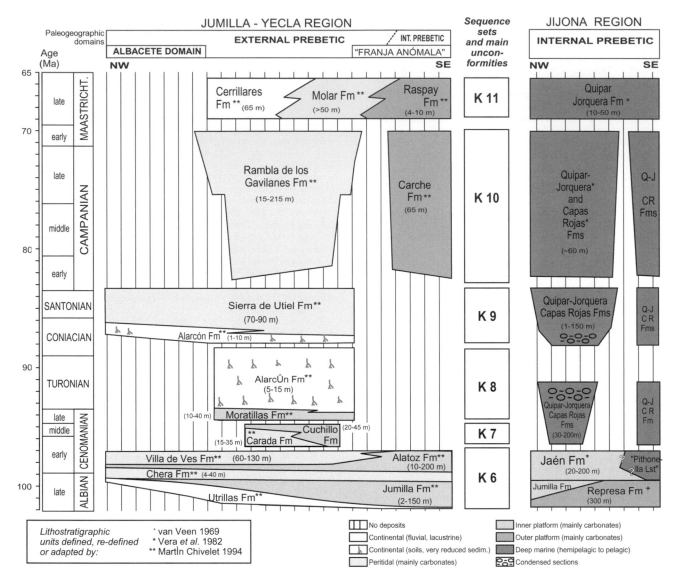

Fig. 12.5. Chronostratigraphic chart (timescale by Gradstein *et al.* 1995) for the Upper Cretaceous succession in two representative sectors of the Prebetic Zone: The Jumilla–Yecla region and the Jijona region (see Fig. 12.2). The chart summarizes the distribution of litostratigraphic units and main depositional systems, grouped into major unconformity-bounded units or sequence sets (K6 to K11; numbers continue from those from the Lower Cretaceous succession, see Fig. 12.4). Partly based on data form Martín Chivelet (1992, 1995), De Ruig (1992), Castro (1998) and Chacón & Martín-Chivelet (2001*b*).

of Late Cretaceous time it formed a barrier separating the shallow marine facies in the NW (External Prebetic) from deep marine facies in the SE (Internal Prebetic).

Basinwards of the Franja Anómala, in the region of Alicante, middle Cenomanian to lower Coniacian deposits correspond to open marine facies, and unconformably overlie rocks of Albian to early Cenomanian age (Fig. 12.6B). These younger rocks record deeper water conditions than the underlying sediments, and consist of mudstones–wackestones bearing calcispherulidae and planktonic foraminifera, forming successions that vary greatly in their thickness (Baena & Jerez 1982), and which are commonly condensed. In areas of higher accumulation rates, upper Cenomanian deposits can locally show huge slump folds and olistoliths consisting of blocks of upper Albian and lower Cenomanian limestone (Leclerc 1971; Hoedemaeker 1973) suggesting contemporaneous faulting in the Prebetic shallow marine areas. This idea is also supported by the presence of several important hiatuses and abrupt lateral changes in thickness within the middle Cenomanian to lower Coniacian successions (Leclerc 1971; Leclerc & Azéma 1976), which probably resulted from

synsedimentary tilting of fault blocks. De Ruig (1992) describes preserved listric faults in the Jijona area (Alicante), that were active during the late Cenomanian–Turonian interval (Fig. 12.6B). These faults were sealed by early Senonian deposits, suggesting that fault activity ceased during latest Turonian or early Coniacian times.

The middle-late Cenomanian tectonism has been interpreted as related to the generalized tilting of Iberia towards the NW (Martín-Chivelet 1990; Martín-Chivelet & Giménez 1993). This tilting possibly resulted from an acceleration in the oceanic spreading rate in the Bay of Biscay (Floquet 1991) combined with the effects of transpressional tectonics occurring along at the Iberia–Africa boundary (cf. De Jong 1990; Kuhnt & Obert 1991). Plate tilting induced an increase in the subsidence rate of the Basque-Cantabrian and North Iberian basins (Floquet 1991; Alonso *et al.* 1993; Gräfe & Wiedmann 1993) and a deceleration or inversion of the subsidence in the southern areas of the Meseta. In the SICM, the response was complex, with the External Prebetic (including the Albacete domain) recording mid-Cenomanian fault movements, overall end-Cenomanian uplift and only very low Turonian–Coniacian subsidence, whereas in the Internal Prebetic

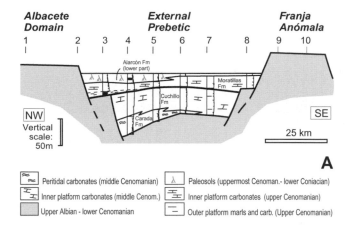

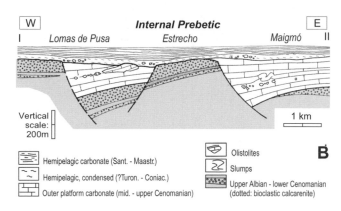

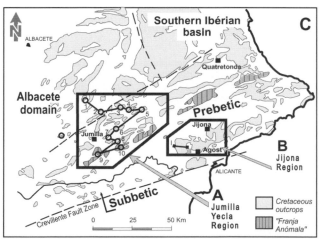

Fig. 12.6. Schematic reconstruction of middle to late Cenomanian fault-block geometries in two representative areas of the Prebetic Zone. **(A)** The Jumilla–Yecla region (based on Martín Chivelet 1992). **(B)** The Jijona region in the Sierra del Cid–Maigmó area (based on De Ruig 1992). **(C)** Location of the cross-sections in the western part of the Prebetic. Note the different scale on the two schemes.

(basinwards from the Franja Anómala) the block movements were accompanied by a generalized deepening and deposition of hemipelagic successions. This palaeogeography, with the external areas nearly completely emergent and the internal areas drowned, together with the complex pattern of highs and troughs that resulted from tectonic movements, explains the poor Turonian record in the

Prebetic region. Turonian platform successions are thin, with amalgamated palaeosols, and hemipelagic deposits are typically condensed (e.g. Baena & Jerez 1982; Leret *et al.* 1982; Martín-Chivelet 1990, 1992; De Ruig 1992; Martín-Chivelet & Giménez 1993).

Coniacian-Santonian tectonic stability. Coniacian relaxation of the intraplate stresses that had controlled subsidence since middle Cenomanian times (Martín-Chivelet 1996) ushered in a new, prolonged interval of thermal subsidence during which carbonate sedimentation returned to nearly the whole basin. The Franja Anómala, however, still continued to form a structural high (indicating that sedimentation was in part controlled by inherited structures) which acted as an emergent palaeogeographic barrier that separated shallow marine (to the NW) and open marine environments (to the SE; Martín-Chivelet 1995).

Shallow marine (mainly peritidal) environments extended over the External Prebetic area, including the Albacete domain (Martín-Chivelet & Giménez 1992), and are represented by the Sierra de Utiel Fm (Fig. 12.5), which yields rich algal and benthic foraminiferal assemblages including, towards the top, abundant specimens of the *Lacazina* group (e.g. Fourcade 1970; Azéma *et al.* 1979; Ramírez del Pozo & Martín-Chivelet 1994). Southwards of the Franja Anómala, in the Internal Prebetic, hemipelagic sedimentation took place, depositing Coniacian to Santonian thin-bedded carbonates unconformably on erosional surfaces affecting Turonian or older sediments. At the base of these carbonates, near Jijona (Alicante), De Ruig (1992) described the presence of a conglomerate bed with included clasts of Albian to Turonian age.

Latest Cretaceous evolution of the Prebetic Zone (JM-Ch, BCh)

During latest Cretaceous time, the SICM experienced a complex evolution in response to dramatic geodynamic changes in the western Tethys. These changes were related to the replacement of the divergent (transtensional) motions between Africa and Europe that had controlled the evolution of the basin since the Triassic period, by a generally convergent setting that was to culminate in Tertiary shortening and destruction of the basin. The SICM thus evolved from a passive margin into a convergent one in which the formerly wide carbonate platforms that had characterized this domain gave way to narrow (tens of kilometres wide) and mixed (carbonate–siliciclastic) platforms, located between the emergent areas to the north (roughly coinciding with the Albacete domain) and extensive deep marine zones to the south (cf. Azéma *et al.* 1979; Martín-Chivelet 1996). Recent detailed tectonostratigraphic and palaeogeographic analyses have shown that changes affecting the continental margin were much more intense than previously suspected. Three strong tectonic episodes, dated as latest Santonian–earliest Campanian, middle Maastrichtian and Maastrichtian–Palaeocene boundary, have been recognized throughout the Prebetic area (Martín-Chivelet 1995, 1996; Chacón & Martín-Chivelet 2001a,b). These episodes caused major changes in basin geometry, palaeogeography, subsidence patterns and sedimentation.

Latest Santonian–earliest Campanian palaeogeographic change. Latest Santonian to earliest Campanian tectonism marked the onset of the transition from a passive to convergent margin (Martín-Chivelet *et al.* 1997). It caused a complete restructuring of the former carbonate platform (Sierra de Utiel Fm) creating a complex mosaic of environments of different bathymetries, terrigenous inputs and ecological conditions. Within that mosaic shallow marine waters occupied a very narrow (10–30 km wide), ENE–WSW elongated belt, adjacent to the emergent areas of the Albacete domain, on which coastal lakes, tidal flats and inner platform settings developed (Rambla de los Gavilanes Fm; Figs 12.5 and 12.7). In that shallow fringe, sedimentation was dominated by carbonates in the western and

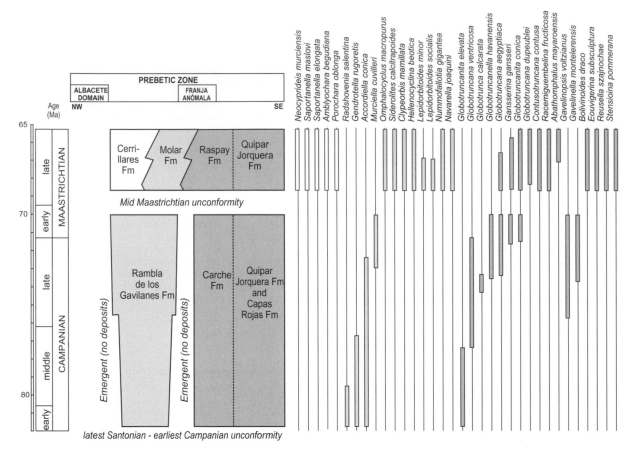

Fig. 12.7. Biochronostratigraphic summary chart of the Campanian and Maastrichtian rocks in the Prebetic, showing the vertical distribution of the most indicative taxa for each main depositional system (from coastal to deep marine). Based partly on Fourcade (1970), Company *et al.* (1982), Martín-Chivelet (1992), Pons *et al.* (1994), Martín Chivelet *et al.* (1995) and Chacón & Martín Chivelet (2001*a,b*).

central areas of the Prebetic (e.g. Fourcade 1970), with an increasing terrigenous influx towards the east probably deposited by longshore currents and derived from a large deltaic system in the Quatretonda area (Champetier 1972; Philip 1983).

In deeper water environments, facies are much more homogeneous and consist of marly limestones and few marls which correspond to the Carche Fm and to part of the Quipar-Jorquera and Capas Rojas formations (Fig 12.5). All these units have yielded abundant planktonic foraminifera which allow the recognition of the biozones *Gansserina elevata, G. ventricosa, G. calcarata, G. falsostuarti,* and the lower and middle part of the biozone of *G. gansseri* (cf. Company *et al.* 1982; Ramírez del Pozo & Martín-Chivelet 1994; Chacón & Martín-Chivelet 2001*a,b*). The Carche Fm corresponds to outer platform deposits, and developed on most of the Franja Anómala and adjacent areas basinwards (e.g. Azéma 1977; Azéma *et al.* 1979; Martín-Chivelet 1992; Chacón & Martín-Chivelet 1999). This unit grades basinwards into the Quipar-Jorquera and Capas Rojas formations (Figs 12.5 and 12.7). These latter units, defined in the Subbetic, comprise a white or reddish, homogeneous series of rhythmic carbonates and marls, deposited under hemipelagic to pelagic conditions.

Middle Maastrichtian compression. A new rapid tectonic episode took place during middle Maastrichtian times, leading to a drastic reorganization of the palaeogeography, so that the former mosaic of environments gave way to a more homogeneous mixed carbonate–siliciclastic platform. The unconformity that marks that event (Fig. 12.7) implies some erosion in the shallow areas of the basin. In deeper environments, Chacón & Martín-Chivelet (2001*b*) recognize synsedimentary reverse faulting and development of olistholiths and massive slumps associated with the unconformity, indicating the compressional character of the tectonic episode.

Within the new palaeogeographical configuration sedimentation took place in narrow belts trending ENE, getting deeper to the SE (Kenter *et al.* 1990; Martín-Chivelet *et al.* 1995; Martín-Chivelet 1996). In the innermost areas of the platform, adjacent to the emergent areas (Albacete domain), shallow coastal lakes became established, and mudstones, marls and a few limestones (Cerrillares Fm; Martín-Chivelet 1994) were deposited under mostly freshwater conditions although with some influx of saline waters (Martín-Chivelet *et al.* 1995). The Cerrillares Fm passes laterally and basinward into the mixed shallow marine deposits of the Molar Fm (Martín-Chivelet 1994), which includes tidal, lagoonal and rudist-coral reefal facies (Martín-Chivelet 1992; Pons *et al.* 1994; Martín-Chivelet *et al.* 1995; Fig. 12.7). These facies occupied a narrow belt that rapidly grades into outer shelf and deeper envionments towards the SE.

Deeper water facies consist of hemipelagic, green to grey marls with some limestone beds, and by marl–limestone alternations, in which stratification is often masked by syndepositional slumps and other gravitational structures. Turbiditic beds can be also present, showing calcarenitic textures and containing abundant reworked fossils from the inner platform environments (Hoedemaeker 1973; Vera *et al.* 1982; Chacón & Martín-Chivelet 2001*a,b*). All these facies belongs to the Raspay Fm and the equivalent part of the Quipar-Jorquera Fm (deeper facies than the former). Both units yield abundant planktonic foraminifera that characterize the upper part of the *Gansserina gansseri* biozone and the *Abathomphalus mayaroensis* biozone (Company *et al.* 1982; Ramírez del Pozo & Martín-Chivelet 1994; Chacón & Martín-Chivelet 2001*a,b*).

K–T boundary. The Cretaceous–Tertiary transition has a very complete record in some areas of the Prebetic as well as in those of the Subbetic. Excellent stratigraphic sections such as those from Agost (Internal Prebetic) and Caravaca (Subbetic), which show sedimentary continuity across the K–T boundary, have become key sections for studying the K–T global event (e.g. DePaolo *et al.* 1983; Groot *et al.* 1989; Smit 1990; Canudo *et al.* 1991; Martínez-Ruiz 1994; Arz & Arenillas 1996; Molina *et al.* 1996; Pardo *et al.* 1996; Kaiho & Lamolda 1999; Martínez-Ruiz *et al.* 1999; Stoll & Schrag 2000). However, a gradual transition across the K–T boundary is actually an exception within the continental margin, and in most areas of the Prebetic, the Maastrichtian deposits are separated from Palaeocene or younger rocks by a regional unconformity of variable size. In the distal, deeper areas of the Prebetic, the unconformity can be marked by prominent, intensely burrowed and bored, mineralized hardgrounds, that sometimes show pelagic stromatolite laminae (Vera & Martín-Algarra 1994), or by a level of catastrophic sedimentation, represented by megabreccias, megaslumps or debris flows. In proximal areas, late Maastrichtian shallow marine and lacustrine deposits (Molar and Cerrillares formations) are usually capped by an erosional surface overlain by Eocene or younger deposits. At some points, however, lacustrine marls and evaporites of possible early Palaeocene age overlie

late Maastrichtian rocks (e.g. Fourcade 1970; Champetier 1972; Martín-Chivelet 1991). Further work on these deposits could reveal interesting data on the K–T passage in continental environments.

Cretaceous of the pelagic domains: Subbetic, flysch troughs and Betic Internal Zones (JAV, AM-A, JMM, PAR-O)

A very wide palaeogeographic domain (the Subbetic), characterized by the abundance of pelagic facies, has been recognized in the southern Iberian continental margin (Fig. 12.8). The later northward thrusting of the Subbetic over the Prebetic produced a boundary easily recognizable on a geological map, and is the main criterion for differentiating the two tectonic units. Cretaceous sediments also occur in the Campo de Gibraltar Complex (Fig. 12.2). These sediments, originally deposited in the deep flysch troughs, now occupy

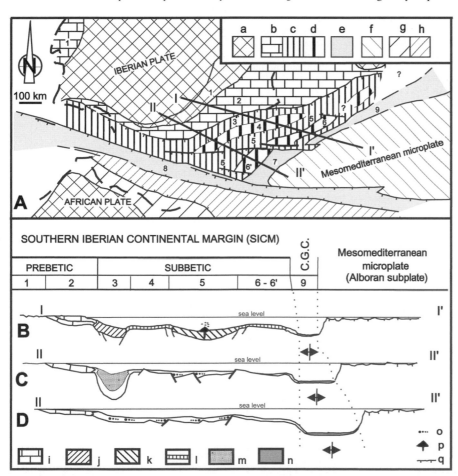

Fig. 12.8. Palaeogeographic and palinspastic reconstruction of the southern Iberian continental margin (SICM). **(A)** Earliest Cretaceous palinspastic restored palaeogeographic map (modified after Vera & Martín-Algarra 1994). Legend: a, continental areas (in Iberian and African plates); b, c, d, SICM (b, shelf areas in Prebetic; c, slope, trough and basin areas in Subbetic; d, plagic swells in Subbetic); e, oceanic basins, including the Campo de Gibraltar complex and related flysch basins; f, Mesomediterranean microplate; g, h, North African margin (g, shelf areas; h, slope and basinal areas). **(B)** Latest Jurassic palinspastic cross-section (location I–I' in A; modified after Vera & Martín-Algarra 1994). **(C)** Palinspastic reconstruction corresponding to the Early–Late Cretaceous boundary (location I–I' in A; modified after Vera & Molina 1999). **(D)** Palinspastic reconstruction of the latest Senonian (location II–II' in A; modified after Vera & Molina 1999). Key (A–D): tectonostratigraphic units in SICM 1 to 6: 1, unfolded sedimentary cover in Iberian Variscan Massif and the externalmost Prebetic; 2, Prebetic with Lower Cretaceous deposition; 3, intermediate domain trough; 4, External Subbetic swell; 5, Median Subbetic trough; 6, Internal Subetic swell; 6', Penibetic. Deep, oceanic and semi-oceanic basins 7 to 9: 7, Campo de Gibraltar flysch basin; 8, North African flysch basin; 9, westernmost Ligurian Ocean). C.G.C. Basin of the Campo de Gibraltar complex. Key (B–D): i, peritidal, lagoonal, platform and reefal carbonates, locally also fluvial, coastal and deltaic sands and sandstones; j, hemipelagic marl–limestone rhythmites, with calcareous turbidites; k, pelagic marl–limestone rhythmites, with interbedded cherty calcisiltite tempestites; l, 'rosso ammonitico' facies and related condensed facies; m, hemipelagic marl–limestone rhythmites with cherty siliciclastic turbidites; n, pelagic marl–limestone rhythmites; o, pelagic marls and clays, with carbonate turbidites; p, volcanics; q, carbonate turbidite levels; r, non-depositional surfaces and condensed sedimentation.

allochthonous positions. Finally, Cretaceous sediments have been observed in part of the sedimentary cover of the Betic Internal Zones, in the Malaguide Complex.

Subbetic. Four palaeogeographic domains can be distinguished in the Subbetic area (Fig. 12.8) and formed well-defined troughs and swells during Middle and Late Jurassic times (Azéma *et al.* 1979; García-Hernández *et al.* 1980). In the troughs, the Intermediate Domain (3 in Fig. 12.8) and the Median Subbetic (5 in Fig. 12.8), thick successions of pelagic sediments were deposited. In contrast, reduced and/or condensed sequences were deposited in the swells, the External and Internal Subbetic (4 and 6 in Fig. 12.8, respectively). The differences between these Subbetic troughs and swells progressively decreased during the Cretaceous period. The Intermediate domain was a rapidly subsident pelagic trough in which a very thick succession of Cretaceous siliciclastic and carbonate turbidites was deposited. In contrast, Cretaceous rocks of the adjacent External Subbetic swell mainly comprise pelagic limestones and marls, with ammonites, radiolaria, planktonic foraminifera and calcareous nannoplankton. The Cretaceous of the Median Subbetic shows similar features to that of the External Subbetic, with two main differences: the presence of black shale facies and submarine volcanic rocks interbedded with pelagic facies sediments. In the Internal Subbetic the Cretaceous is also characterized by pelagic limestones and marls, locally with black shale facies, with a prominent hiatus in the southwesternmost sector, an area sometimes referred to as the Penibetic (6' in Fig. 12.8). This unconformity omits the great part of the Lower Cretaceous and part of the Upper Cretaceous succession (Fig. 12.9).

In the Subbetic Lower Cretaceous succession, four formations have been differentiated. Three of them (Villares, Carretero and Cerrajón formations) are completely included in the Lower Cretaceous and the fourth (Fardes Fm and coeval formations of more local extent) is Lower Cretaceous but locally passes into the Upper Cretaceous (Fig. 12.9). All these formations are characterized by a dominance of pelagic facies (marls and marly limestones), with abundant ammonites (Company 1987), radiolaria (O'Dogherty 1994) and calcareous nannoplankton (Aguado 1992). The differences among them arise from the presence or absence of turbidite intercalations, the nature of the turbidite beds, and their ages.

One of the most relevant features is the thickness of the Lower Cretaceous in the Intermediate Domain, which can exceed 2000 m, the maximum known thickness for this time interval anywhere in the SICM. This remarkably high value indicates that the Intermediate Domain was the most rapidly subsiding area along the margin during the Early Cretaceous epoch (as well as during Late Jurassic time). The greatest thickness and highest average sedimentation rates are recorded by the Los Villares and Cerrajón formations. The former is late Berriasian–Hauterivian age, locally >1000 m thick, and has an average sedimentation rate of 0.08 mm/year. The Cerrajón Fm (Barremian–Albian) comprises mostly siliciclastic turbidites, is up to 1200 m thick, and has a mean sedimentation rate of 50 m/Ma. The presence of disconformities (Fig. 12.9) and evidence for erosion within the Cerrajón Fm (Aguado *et al.* 1996), reduces the significance of the averaged sedimentation rate. In the remaining palaeogeographic domains, farther from the continent (External Subbetic, Median Subbetic and Internal Subbetic), the Lower Cretaceous succession is represented mainly (or exclusively) by the Carretero Fm. This formation shows frequent variations in thickness, generally being thicker towards the north and thinner towards the south, although with sudden local changes. This variability in thickness is explained by the presence of listric faults bounding half-grabens that were filled by this formation (Aguado *et al.* 1991). The additional presence of abundant slumps and breccias indicates the existence of slopes on the bottom as well as remarkable tectonic instability.

Subbetic Lower Cretaceous. Four main features are particularly characteristic of the Subbetic Lower Cretaceous succession (Fig. 12.9). The first of these is the local presence of submarine volcanic lava (with pillow lavas) in the Median Subbetic. Secondly, there is a great abundance of siliciclastic turbidites in the Barremian–Albian part of this succession in the Intermediate Domain (Cerrajón Fm), with large, canyon-like erosive features, channelled sequences of amalgamated beds, thickening- and thinning-upward cycles, and beds with redeposited remains of plants and orbitolinids. All this evidence points to a genesis related to submarine canyons cut in the Prebetic platform nearer to the continent (Ruiz-Ortiz 1980). Boulders of rudist limestone and other shallow water lithologies, up to 1 m³ in volume, were derived from the erosion of the platform, and are now found in

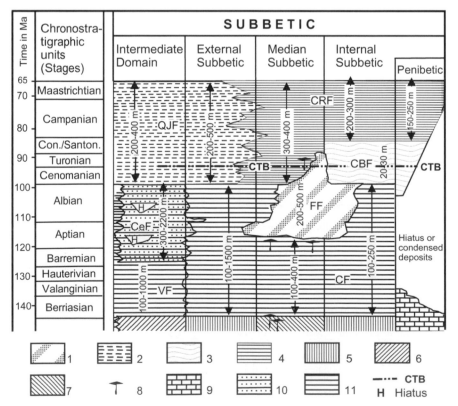

Fig. 12.9. Chrono-lithostratigraphic diagram illustrating the evolution of the different pelagic domains (simplified from Vera 2001). Legend: 1, black shales and marls, locally with calcareous turbidites and olistoliths; 2, marls and marly limestones with planktonic foraminifera and calcareous nannoplankton, locally with calcareous turbidites; 3, white marls and marly limestones with planktonic foraminifera and calcareous nannoplankton (lower member) and cherty limestones (upper member); 4, pink marls and marly limestones with planktonic foraminifera and calcareous nannoplankton, rarely with rudists; 5, condensed limestones, mainly 'ammonitico rosso' facies; 6, limestone–marl rhythmites with interbedded calcareous turbidites (Toril Fm); 7, limestone–marl rhythmites with interbedded calcareous tempestites (Milanos Fm); 8, volcanic submarine rocks; 9, pelagic oolitic limestones and 'ammonitico rosso' facies (Torcal Fm); 10, marls and marly limestones with siliciclastic turbidites; 11, pelagic limestone–marl rhythmites with ammonites. CTB, Cenomanian–Turonian boundary anoxic event. Names of the formations: CBF, Capas Blancas Fm; CeF, Cerrajón Fm; CF, Carretero Fm; CRF, Capas Rojas Fm; FF, Fardes Fm; QJF, Quípar-Jorquera Fm; VF, Villares Fm. Most of these units were defined or redefined by Vera *et al.* (1982). Absolute timescale from Gradstein *et al.* (1995).

submarine canyon infill (Ruiz-Ortiz *et al.* 1996, 2001). The third characteristic feature is the presence of the Fardes Fm across wide sectors of the margin. This formation is characterized by black anoxic facies, similar to those recognized in the Atlantic (Reicherter *et al.* 1996). In addition, significant volumes of calcareous turbidites and even large Jurassic olistoliths (Hernández-Molina *et al.* 1991) are locally found in the Fardes Fm. These latter deposits are linked to areas adjacent to fault escarpments, revealing the local importance of synsedimentary tectonics (Rey 1993). Other formations, similar to the Fardes Fm but containing radiolarites and of more limited lateral extent, have also been recognized (e.g. the Carbonero Fm; Molina 1987; Aguado *et al.* 1993) and are here included under the general term Fardes Fm. The fourth characteristic feature is the local presence in the Penibetic area of a hiatus that entirely omits Lower Cretaceous strata, so that Upper Cretaceous sediments directly overlie Jurassic–lowermost Cretaceous limestones (Martín-Algarra 1987; Martín-Algarra *et al.* 1992). Locally even Lower Cretaceous sediments can be partially or totally absent or, alternatively, they can be affected by extreme condensation (Molina 1987; Castro & Ruiz-Ortiz 1991).

Subbetic Upper Cretaceous. In the Subbetic Upper Cretaceous succession, another four formations are differentiated (Fig. 12.9). These are the Capas Blancas Fm (Upper Cretaceous only), the Quípar-Jorquera and the Capas Rojas formations, which continue into the Palaeogene, and the Fardes Fm, which is mainly Lower Cretaceous, as mentioned above, but locally extends up into the Cenomanian–Turonian interval. All the formations are characterized by a prevalence of pelagic facies (marls, marly limestones and limestones) with abundant radiolaria, calcareous nannoplankton and planktonic foraminifera. They differ in their dominant colour and in the presence or absence of turbidites intercalated with the pelagic facies. In general, these successions have very low sedimentation rates, always less than 0.01 mm/year, and typically show pelagic cyclicity on a decimetric scale, probably related to astronomically induced climate cycles (Milankovitch cycles).

The most representative lithostratigraphic unit is the Capas Rojas Fm (Fig. 12.9), which began to be deposited in Cenomanian (or Turonian) times and continued until the beginning or the end of the middle Eocene, depending on the sector. This formation crops out extensively in the Subbetic area and has an equivalent in other Alpine domains (Alps and Apennines), where it is known as the Scaglia Rossa. Recently, Vera & Molina (1999) reviewed the Capas Rojas Fm and drew attention to the presence of isolated rudists in life position in numerous Subbetic localities, suggesting a deposition depth of a few hundred metres. Locally, submarine volcanism continued until the Santonian (Molina *et al.* 1998), as evidenced by the volcanic flows with pillow lavas in this formation. The Quípar-Jorquera Fm is a lateral equivalent of the Capas Rojas Fm, with the differences between the two formations being limited to colour (white and red respectively), and to the presence in the former of calcareous turbidites, which are more abundant in the areas closer to the Prebetic platform (Nieto 1996).

The uppermost part of the Fardes Fm (Fig. 12.9) is characterized by the presence of black shales deposited in response to the oceanic anoxic event of the Cenomanian–Turonian boundary (CTB in Fig. 12.9). The Capas Blancas Fm, of limited thickness and deposited in the adjacent Internal Subbetic area (Fig.12.9), shows two members separated by a thin level of black shales, also linked to the Cenomanian–Turonian anoxic event. The lower member is composed of micritic limestones and marls, whereas the upper member comprises cherty limestones.

Campo de Gibraltar Cretaceous. Cretaceous sediments have been recognized in only a few exposures and comprise shales and marls with siliciclastic and carbonate turbidite intercalations (see Vera *et al.* 1982; Martín-Algarra 1987; Martín-Algarra *et al.* 1992). The turbidites were derived from the North African margin (Massylian flysch) or from the Mesomediterranean microplate (Mauritanian flysch) and are generically known as 'Cretaceous flysch'. Martín-Algarra *et al.* (1992) pointed out the good chronostratigraphic correlation between the Lower Cretaceous siliciclastic turbidites of the

SICM and those of the Campo de Gibraltar complex. These authors associate the genesis of this coeval deposition to three tectonic–eustatic sea-level falls, during the Hauterivian, Aptian and Albian stages.

The Berriasian–Hauterivian stages in the Mauritanian flysch are represented in some of the units by red and green clays with marly limestone and calcareous *Aptychus*-rich turbidite (*Aptychus* microbreccia) intercalations. The Barremian–Aptian–Albian successions are mainly composed of siliciclastic micaceous sandstone turbidites with intercalated mudstones, organized in a thickening-upward sequence several hundred metres thick. The Cenomanian and Turonian stages are represented by radiolarites associated with dark marls dated by radiolaria. The youngest part of the Massylian Cretaceous succession consists of green clays with calcareous turbidite beds (microbreccias), whereas the Mauritanian equivalent shows similar features, but very locally also has siliciclastic turbidite beds intercalated with Campanian–Maastrichtian mudstones.

Maláguide Complex Cretaceous. The Maláguide Complex is the only unit of the Betic Internal Zones containing clearly dated Cretaceous sediments. There is a sharp contrast between the widespread outcrop of this complex and only extremely rare exposure of its Mesozoic–Tertiary cover, especially of the Cretaceous outcrops. Only in three areas (Málaga, Vélez-Rubio and Sierra Espuña) have Cretaceous rocks been clearly identified (Vera *et al.* 1982). Overall, Cretaceous sedimentation in the Mesomediterranean (Alboran) microplate was very reduced, frequently discontinuous, and predominantly pelagic.

In the Málaga area, the oldest Cretaceous sediments are Neocomian white micritic limestones and Albian red limestones filling cavities in a palaeokarst surface. Overlying these sediments are several metres of *Globotruncana*-bearing brecciated limestones and then 15–20 m of pinkish marls, also with *Globotruncana* of Senonian age in the base. In the Vélez-Rubio region (Almería province), micritic limestones crop out in strongly condensed stratigraphic sections. In some of these sections, no more than 20 m thick, almost all the stages of the Cretaceous can be recognized by means of planktonic foraminifera (Roep 1980). In Sierra Espuña (Murcia province), Cretaceous sections up to 100 m thick have been described in one of the Maláguide Complex units. The section is made up of 50 m of Berriasian–Valanginian *Calpionella*-bearing limestones, overlain by Aptian–Albian pelagic marls that locally change to phosphorites filling karstic cavities in Jurassic limestones. The Upper Cretaceous is formed of limestones with planktonic foraminifera and white marls with *Globotruncana* of Senonian age, *Inoceramus* fragments and sponge spicules.

Pyrenees (XB, EC, CP)

The Pyrenean mountain belt was formed as a result of broadly north–south crustal contraction due to the collision of the European and Iberian plates from latest Cretaceous to late Oligocene times. From south to north the main structural units are: the southern foreland basin (Ebro basin), the upper thrust sheets (involving the cover succession), the lower thrust sheets (involving basement), the north Pyrenean fault zone, the north Pyrenean thrust sheets (or North Pyrenean Zone), and the northern foreland basin (Aquitaine basin) (Muñoz 1992; Berástegui *et al.* 1993; Chapter 13). The upper and lower thrust sheets form the South Pyrenean Zone, which shows east–west trending folds and south verging thrusts (Séguret 1972; Garrido 1973; Choukroune 1976; Figs 12.10, 12.11 and

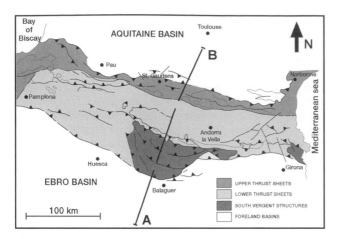

Fig. 12.10. Geographic and geological setting of the Pyrenees. A–B: ECORS Pyrenees cross-section (Fig. 12.11).

12.12). The north Pyrenean fault zone acted as the boundary between both plates during Mesozoic and Palaeogene times. An interpretation of the present-day internal structure of the central Pyrenees (Losantos *et al.* 1988) is shown in Figure 12.11a.

Cretaceous rocks crop out in several different structural units (Fig. 12.12). During Mesozoic times the North Pyrenean Zone (NPZ) was a part of the south European continental margin, while the South Pyrenean Zone (SPZ) was a part of the north Iberian continental margin. Crustal-scale restored cross-sections (Muñoz 1992; Teixell 1992, 1996; Berástegui *et al.* 1993; Vergés 1999) provide an insight into the geometry of the Cretaceous basins, and help in understanding the factors controlling sedimentation and the main events that took place in the Pyrenean area. Figure 12.11b is a reproduction of the restored ECORS Pyrenees cross-section to middle Cenomanian times (Berástegui *et al.* 1993).

The general situation of the northern Iberian margin at the beginning of the Cretaceous period was inherited from older major events. After a rifting episode (Late Triassic–Early Jurassic) related to the separation of North America and Africa, a more quiet phase allowed the development of extensive carbonate platforms during Jurassic times. Renewed extension (Late Jurassic–Early Cretaceous) then produced the opening of the Bay of Biscay during Aptian–Albian times, and late Albian–early Cenomanian emplacement of lherzolite and thermal metamorphism along the north Pyrenean fault zone (Ravier 1959; Albarede & Michard-Vitrac 1978).

Upper Cretaceous sedimentation extended beyond the rift structures. For about 10 Ma (middle Cenomanian to middle Santonian) there was relative tectonic stability in the southern Pyrenean domains, although with continued high activity

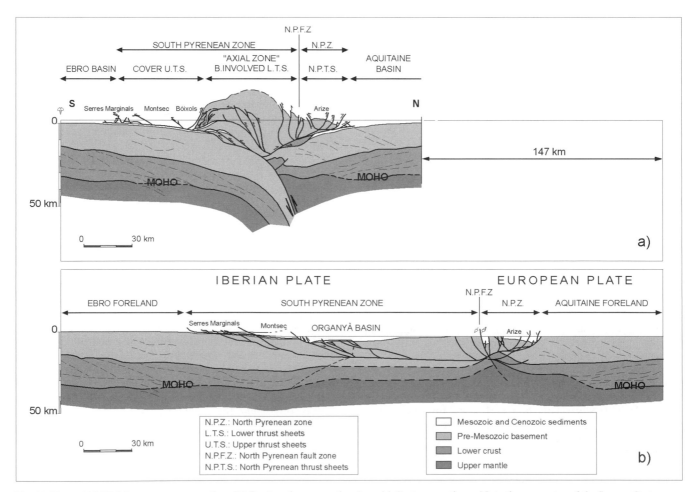

Fig. 12.11. (**a**) ECORS Pyrenees cross-section. (**b**) Restored cross-section for mid-Cretaceous times. Note the geometry of the Lower Cretaceous basins (simplified from Berástegui *et al.* 1993).

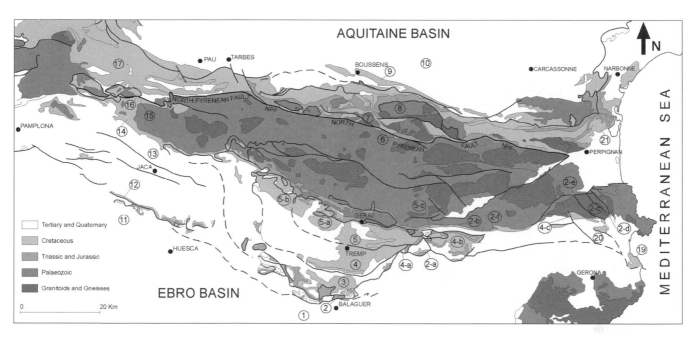

Fig. 12.12. Geological sketch map of the Pyrenees. Numbers refer to sections in Figures 12.13 and 12.14.

along the north Pyrenean fault zone. Subsequent plate convergence between Iberia and Europe initiated basin inversion during late Santonian times, a process completed during latest Cretaceous and early Cenozoic collision and foreland basin growth (Puigdefàbregas & Souquet 1986). The geometry and sequential arrangement of the sedimentary infill in the Pyrenean Cretaceous basins reflect the record of the above mentioned major tectonic events (Figs 12.13 and 12.14; note that Fig. 12.14 shows only palaeogeographic reconstructions for the southern margin of the basin).

Lower Cretaceous of the Pyrenees (XB, EC, CP)

The Lower Cretaceous succession in Spain developed in response to an incipient triple junction between the Grand Banks of Newfoundland, Iberia and western Europe that later evolved into active sea-floor spreading (Hubbard *et al.* 1985). In the Pyrenean area the early Berriasian–middle Cenomanian record can be divided into the following four main stages: early extension, carbonate platform, late extension and thermal flow periods.

Early extensional phase. In response to the early Cretaceous extensional event, the old Jurassic platforms were fragmented and a number of small sedimentary basins were formed along the edges of the Iberian and European plates. One of these basins, at present completely inverted, is the so-called Organyà Basin (Berástegui *et al.* 1990) in the south-central Pyrenees. Its dimensions and the exceptional exposures of its infill allow it to be used here as an example to illustrate the sedimentary record of the Early Cretaceous rifting (see Figs 12.13 and 12.14).

The onset of the rifting phase is recorded in the Organyà Basin (Fig 12.13, chart I–I', section 5, and Fig. 12.14, chart IV–IV', sections 5a, 5b and 5c) by the Late Jurassic unconformity. The first sediment related to this early extensional period is a calcareous-dolomitic breccia which was probably deposited at the toe of submarine fault escarpments. On this breccia or, in its absence, directly overlying Jurassic strata, lies a succession of marls, limestones, silty sandstones and carbonaceous silty limestones. As a whole this series shows a shallowing-up evolution, from lime mudstones with calpionellids and marls with ammonites at the base, to siliciclastic and carbonaceous sediments with charophytes deposited in shallow delta plain environments at the top. The middle parts of the series consist of bioclastic limestones with trocholines and larger agglutinated foraminifera. Fossils in these sediments indicate an age of Berriasian to Valanginian for the first extensional events.

Carbonate platform phase. The above-mentioned early extensional events were followed in the Organyà Basin (Fig. 12.13, chart I–I', section 5, and Fig. 12.14, chart IV–IV', sections 5a, 5b, 5c) by a high subsidence period characterized by the accumulation of an important thickness of Barremian–early Aptian shallow platform carbonates (Urgonian facies). These rest unconformably on the sediments deposited during early extension, marking a hiatus extending over the late Valanginian to Hauterivian interval (Caus *et al.* 1990). The Urgonian carbonates consist of an alternation of bioclastic limestones, including fragmented rudists and bryozoa, deposited in open marine, high-energy environments, and muddy limestones rich in agglutinated foraminifera (orbitolinids) and charophytes deposited, respectively, in inner platform and fresh water (ponds or lakes) environments. A high subsidence rate produced very thick depositional sequences (up to 1000 m) and gradual changes in facies (Bernaus 2000). Towards the end of this period and prior to the demise of the carbonate platform, an anoxic event was recorded by the sedimentation of black, rich organic matter, which can be correlated with the first worldwide Cretaceous oceanic anoxic event (OAE 1a).

Late extensional phase. Onset of the second extensional phase is recorded in the Organyà Basin (Fig. 12.13, chart I–I', section

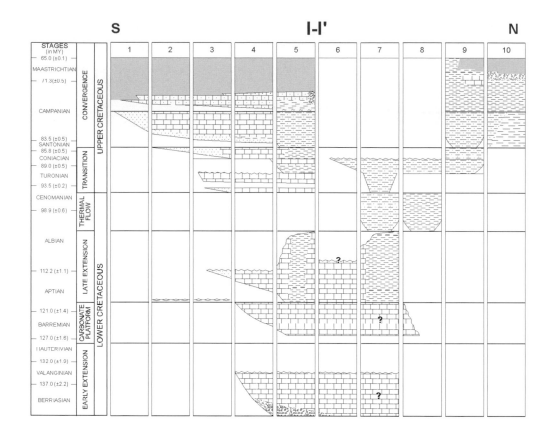

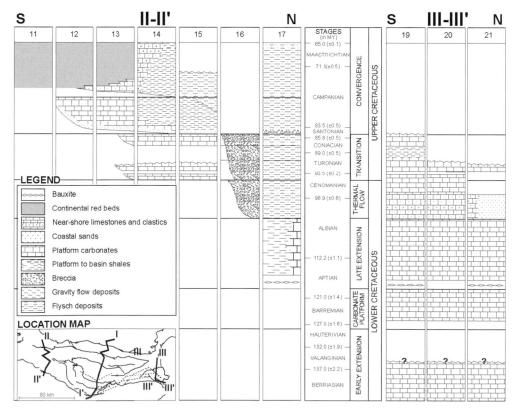

Fig. 12.13. Stratigraphic charts in the Pyrenees showing the distribution of sedimentary facies in the successive structural units, from south to north. Partially based on Berástegui *et al.* (1993), Ducasse and Velasque (1988), Teixell (1996) and Institut Cartografic de Caralunya (1994–97). See Figure 12.12 for location. Chart I–I', central Pyrenees: section 1, Ebro foreland basin; sections 2 and 3, Serres Marginals thrust sheet; section 4, Montsec thrust sheet; section 5, Boixols thrust sheet; section 6, cover of the axial zone; section 7, north Pyrenean fault zone basins; section 8, North Pyrenean Zone; sections 9 and 10, Aquitane foreland basin. Chart II–II', western central Pyrenees: section 11, Ebro foreland; section 12, Sierras Marginales Aragonesas thrust sheet; section 13, Jaca basin; sections 14 and 15, cover of the axial zone; section 16, Lakora thrust sheet; section 17, north Pyrenean Zone (Mauleon and Arzaq basins). Chart III–III', eastern Pyrenees: sections 19 and 20, Empordà thrust sheet; section 21, Nappe des Corbières.

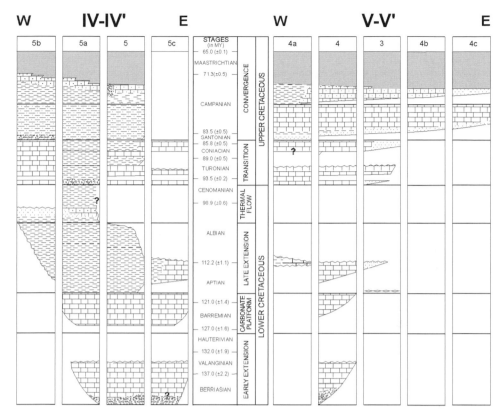

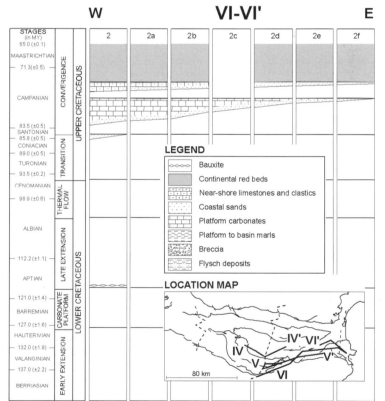

Fig. 12.14. Stratigraphic charts illustrating the facies distribution from west to east along the south Pyrenean margin. See Figure 12.12 for location. Chart IV–IV', central areas of the south Pyrenean margin. Note that the trend of these figures is roughly parallel to the basin axis. The section corresponds to palaeogeographical distal settings in relation to those shown on cross-sections V–V' and VI–VI'. Sections 5b, 5a and 5 are in Boixols thrust sheet; section 5c is in the Upper Pedraforca thrust sheet. Chart V–V', central and eastern areas of the south Pyrenean margin. Although the sections are on different structural units, they have been positioned on the chart considering their respective palaeogeographical settings. This section represents intermediate positions between sections IV–IV' and VI–VI'. Sections 4a and 4 are on the Montsec thrust sheet; section 3 is on the Serres Marginals thrust sheet; section 4b is on the Lower Pedraforca thrust sheet; section 4c is on the Bac Grillera thrust sheet. Chart VI–VI', proximal areas of south Pyrenean margin. The sections have been placed on the chart from palaeogeographical criteria. Section 2, Serres Marginals thrust sheet; section 2a, Port del Comte thrust sheet; section 2b, Cadí thrust sheet (western area); section 2c, Empordà thrust sheet (Biure); sections 2d and 2e, autochthonous; section 2f, Cadí thrust sheets (eastern area).

5, and Fig. 12.14, chart IV–IV', sections 5a, 5b, 5c) by a gentle angular unconformity, marking an early Aptian hiatus in sedimentation on top of the Urgonian carbonates. The old carbonate platform became covered by lower Aptian basinal shales (Martínez 1982). During Aptian–earliest middle Albian times, rapid subsidence characterized the central areas of the basin, where almost 3000 m of hemipelagic sediments were deposited, while on its margins narrow carbonate platforms with massive coral-bearing bodies developed. Two main lithologies are vertically and laterally related: clays and grey to black marls with ammonites and sponge spicules (deposited in offshore, slope and basin environments), and orbitolinid and bioconstructed limestones and marls (related to the narrow carbonate platform complexes). Whereas in the Organyá area the middle Albian strata are unconformably overlain by upper Cretaceous sediments, further west the remaining Lower Cretaceous succession is well exposed in the Sopeira basin (Fig. 12.14, chart IV–IV', sections 5a and 5b). Elsewhere, similar late extension phase successions occur in other Pyrenean basins (Fig. 12.13, chart III–III', sections 19 to 21). In the emergent areas, lateritic and bauxitic soils (Combes 1969) were developed over the old, subaerially exposed carbonate platforms.

Thermal flow phase. This phase spanned from middle Albian to middle Cenomanian times (about 8 Ma), and was characterized by a change from regional extension to localization of tectonic movements along the north Pyrenean fault zone, where transtension produced pull-apart flysch basins (Souquet *et al.* 1985). The period is referred to as the thermal flow phase because subsequent crustal thinning in the north Pyrenean fault zone area resulted in low pressure–high temperature metamorphism. Granulitic and ultrabasic upper mantle (lherzolite) rocks were emplaced at upper crustal levels in this area synchronously, or immediately after, Albo-Cenomanian flysch basin formation (Ravier 1959; Albarede & Michard-Vitrac 1978). Platform and slope sedimentation shifted to the western areas of the Pyrenean domain, where upper Albian sediments are found onlapping older deposits.

In the South Pyrenean Zone, the main extensional basin-forming event ended around middle Albian times. Late Albian uplift and deep erosion occurred across much of the southern Pyrenean domain, although localized extensional troughs were still generated in some places. The Sopeira basin (Fig. 12.14, chart IV–IV', sections 5a and 5b) is used here to illustrate this episode, it being a narrow, small depocentre that was generated at the start (late Albian) of the thermal flow stage (Caus *et al.* 1997). From a structural point of view, the Sopeira basin in its early stages appeared as a small half-graben, bounded to the south by a low-angle extensional fault. Basal sediments were deposited in the downthrown extensional basin above a Jurassic or reduced lower Cretaceous succession, and sedimentation was initially confined to the hanging-wall of the extensional fault. The basin shows high sedimentation rates as the result of a combination of fault-induced subsidence and a high terrigenous supply. In the centre of the basin sedimentation was continuous, beginning with marly limestones, followed by ferruginous bioclastic limestones and quartz-bearing calcarenites. Along the margins, sedimentation was interrupted by unconformities and dominated by clastic deposits. In the Bonansa area (Fig. 12.14, chart IV–IV', section 5b), a condensed red hardground known as La Selva (Souquet 1967) coincides with this unit.

The palaeontological content of these sediments, according to Peybernès (1976), suggests a late Albian to early Cenomanian age.

This previous episode was followed by more rapid subsidence which led to the sedimentation of lower to middle Cenomanian marls and marly limestones within the basinal area during a progressive deepening which is well registered by the fossil record (Caus *et al.* 1993, 1997). The marls overlapped the previous depositional area, overlying the Lower Cretaceous substratum in the basin margins. They reflect, therefore, a more homogeneous subsidence over a broader area, thus indicating thermal subsidence after the main extensional faulting episodes in the Pyrenees. The main extensional fault bounding the so-called Sopeira basin was progressively onlapped by marly sediments as a vast shallow-water limestone platform area developed. A similar type of sedimentation to that described in the Sopeira basin is also recorded in eastern Pyrenean basins (Fig. 12.13, chart III–III').

As has been mentioned above, during late Albian–early Cenomanian times, strike-slip fault activity along the north Pyrenean fault zone (Figs 12.10 and 12.11) generated small and deep basins which were rapidly infilled by flysch deposits. In this area severe crustal thinning occurred synchronously with, or immediately after, basin formation. This induced thermal metamorphism in the more deeply buried sediments ('flysch ardoisier'; Fig. 12.13, chart I–I', section 7). Lower crustal granulite rocks, as well as ultrabasic upper mantle rocks (lherzolites) were emplaced along the north Pyrenean fault zone. Further north continuous, or almost continuous, sedimentation was restricted to the main basin axis of the present North Pyrenean Zone, where non-metamorphic flysch-type materials ('flysch noir') were deposited (Fig. 12.13, chart I–I', section 8 and chart II–II', section 17).

Towards the western areas of the present-day Pyrenees, upper Albian to lower Cenomanian sediments are found in the North Pyrenean Zone (Teixell 1992, 1996), directly overlying Palaeozoic rocks (Fig. 12.13, chart II–II', sections 16 and 17). The geometry and distribution of these deposits indicate a source area located to the south, and an opening of the basin to the north. In the southern areas of this basin, the deposits mainly consist of conglomerates (e.g. Mendibelza 'breccia'; Fig. 12.13, chart II–II', section 16), while to the north, they show an evolution to flysch-type deposits (Fig. 12.13, chart II–II', section 17), which overlie Lower Cretaceous limestones and shales (Ducasse & Velasque 1988).

To summarize, the flysch-type sediments are interpreted as syntectonic deposits in a narrow and deep trough system developed along the north Pyrenean fault zone. They represent slope and basin facies, and grade, in the direction of the present-day Aquitaine basin, into black marls with ammonites and planktonic foraminifera, orbitolinid-bearing carbonates and finally to terrigenous deposits, in a similar way as that described for the Sopeira basin. The Pyrenean upper Albian–lower Cenomanian flysch-type deposits are largely known as 'flysch noir' ('black flysch') with the 'flysch ardoisier' being its metamorphic equivalent in age.

Upper Cretaceous of the Pyrenees (XB, EC, CP)

A second major, mostly marine Cretaceous succession records the period following Early Cretaceous to middle Cenomanian extension (and transtension), and lasted until the onset of plate collision in Early Cenozoic times. Post-extension Upper

Cretaceous sequences in the Pyrenean area extended to the south towards the Ebro basin area, and to the north towards the Aquitaine basin area (Fig 12.13, charts I–I' and II–II'). From a tectonostratigraphic point of view, this Upper Cretaceous succession can be viewed as being divided into two units representing transition and convergence. A middle Cenomanian unconformity marks the onset of the transition phase that lasted for about 10 Ma until middle Santonian times. Similarly, a middle Santonian unconformity marks the onset of the convergence phase which extended until late Palaeocene times.

Transition phase. During late Cenomanian–early Santonian times, both basin orientation and depositional profile were partially inherited from previous events, although sediments overlapped the former basin margins and unconformably overstepped Lower Cretaceous and older rocks (Fig. 12.11b). During this episode, deposits reflected homogeneous and moderate subsidence related to lithospheric thermal contraction after the initial extensional episodes. In shallow areas of the Pyrenean domain, stable carbonate platforms developed and retreated landwards (Bilotte 1985; Simó 1986), while shales, limestones and breccias characterized deposition in deep parts of the basin. As a general example to illustrate the sedimentary record of platform and slope domains, we will use the sediments cropping out in the present-day South Pyrenean Zone (Fig. 12.14, charts IV–IV' and V–V'). For flysch deposits we will refer to the North Pyrenean Zone (Fig. 12.13, chart I–I', sections 6 to 9, and chart II–II', section 17).

Upper Cenomanian events in the Pyrenean area were dominated by an important marine transgression, which can also be identified in several parts of the world (Hankock & Kauffman 1979). In the platformal areas of in the South Pyrenean Zone, shallow carbonate deposits (*Praealveolina* limestones) covered previously exposed older sediments, whereas in deeper domains, sedimentation was almost continuous and was characterized by the deposition of marly limestones and marls with planktonic foraminifera (rotaliporids) and ammonites.

Near the Cenomanian–Turonian boundary, open marine sediments contain an unusual amount of calcisphaerulids, indicating an intense primary productivity in the surface waters due to a relative increase in the richness of nutrients at that time. Mineral components would have run off into the basin during erosion of the surrounding uplifted land in response to Albian–early Cenomanian uplift, and accumulated in relatively deep sea water. The supply of nutrients from below to the surface layers of the ocean (euphotic zone) was assured by upwelling currents. Part of the organic matter, generated by primary productivity in the water surface, was transported to the sea bottom. The fraction not used by bathyal organisms was accumulated and decomposed by microbial organisms, acting on the organic debris without contact with the atmosphere for a long period of time. In this way, the concentration of oxygen in the water column was reduced, registering an oxygen minimum zone that could have expanded upwards. The accumulation of decomposed organic debris together with fine-grained sediments, was responsible for the deposition of organic-rich sediments (black shales). These black shale layers in the Pyrenees can be correlated with those from other areas of the Atlantic margin (OAE-2). Eutrophication phenomena during the Cenomanian–Turonian interval not only involved the open marine realm, but also had a drastic effect on the organisms living in shallow water

platform areas and, by extension, on the sedimentation itself. The disappearance of the main carbonate producers stopped the production of skeletal grains, and a hiatus in the platform sedimentary record was subsequently generated (Caus *et al.* 1993, 1997).

During middle and late Turonian times, pelagic sedimentation continued in the basin. In the platform areas, eutrophic conditions prevailing during the previous interval decreased progressively with time. Benthic organisms were then able to recolonize the neritic realm and so carbonate sedimentation was re-established. The carbonate platform development ended in middle Santonian times because the basin margins emerged at that time.

Flysch-type sedimentation during this transition period was confined to the northern edge of the current axial zone and to the North Pyrenean Zone, where Cenomanian to Coniacian turbidites overlie the old black flysch ('flysch noir' and 'flysch ardoisier') deposits (Fig. 12.13, chart I–I', sections 6, 7, 8 and 9), and have been given several local names (e.g. 'flysch gris' and 'flysch à fucoides'; Bilotte 1985; Ducasse & Velasque 1988). Polygenic breccia beds are sometimes included within the turbiditic fine-grained deposits, with clasts of Palaeozoic rocks and/or Upper Cretaceous limestones, and are interpreted as debris flow deposits or megaturbidites.

Convergence phase. During late Santonian to late Maastrichtian times, the northern edge of the Iberian plate, and therefore the Pyrenean area, behaved as a convergent margin. This convergent zone was superimposed on structures of the former divergent margin related to the opening of the Bay of Biscay, and on the sediments deposited during the thermal-driven subsidence that characterized the transition period. At the beginning of the convergence period, sedimentation patterns in the Pyrenean domain initially did not change much, but then they became increasingly affected by westward-propagating compression (Choukroune 1976) which changed both basin geometry and the composition of the sedimentary infill.

An early development during this phase (late Santonian and early Campanian) was the generation of a rapidly subsiding, terrigenous flysch trough in the Southern Pyrenean Zone (Rosell 1967). This turbiditic trough (Fig. 12.13, chart I–I', section 5) had an elongated shape, with a NE–SW trending axis, roughly parallel to the axis of the present mountain chain. The trough was coeval with the growth of hanging-wall anticlines related to the emplacement of early thrust sheets (e.g. Sant Corneli anticline and Boixols thrust sheet, north of Tremp; Fig. 12.12). In the shallower areas of the basin, the sedimentary response started with a coastal onlap of platform carbonates, which then developed into a shallow carbonate platform system that extended landwards beyond the former basin margins. Thus, the lower and middle Campanian sediments record the maximum extension of the Late Cretaceous sea in the Pyrenean domain (Fig. 12.13, chart I–I', sections 1, 2, 3 and 4, and chart II–II', sections 12 and 13; Fig. 12.14, charts V–V' and VI–VI'). In general these shallow water carbonate platform sediments mainly consist of bioclastic limestones. Open marine platform deposits are represented by nodular marls and shales containing planktonic foraminifera and ammonites (Gallemí *et al.* 1983; Gómez-Garrido 1989), and basin facies in the trough consist of siliciclastic turbidites, which were deposited above irregularly eroded older sediments.

A change from this early convergence to a more widespread compression took place in the Pyrenean Basin during late Campanian times and initially resulted in the collapse of the platform margin (Simó 1986). This event was recorded in the Southern Pyrenean Zone by olistostromes containing clasts including Cretaceous carbonates, Triassic shales and various basement rocks, all transported as sediment gravity flows from the platform margin to the toe of the slope (Fig. 12.13, chart I–I', section 5). Coeval correlative shallow platform environments show emergence and erosion (Fig. 12.14, charts V–V' and VI–VI'). Subsequently, the south Pyrenean basin records a shallowing-upward, regressive tendency, from turbidites at the bottom to offshore shales and nearshore siliciclastics at the top (Arbués *et al*. 1996). In the southern and eastern Pyrenean domains, continental red beds (Garumnian facies: late Campanian to middle Palaeocene) are preserved on the very top of these sediments.

To the north of the western end of the axial zone, south of the north Pyrenean fault zone (Figs 12.12 and 12.13, chart II–II', sections 14 and 15), a rapidly deepening trough formed, producing terrigenous deep marine sedimentation during late Santonian to Maastrichtian times (Teixell 1996). Unlike further east, in this area, the Cretaceous–Tertiary boundary is recorded within marine sediments, but palaeontological evidence suggests the existence of a gap in between the two periods.

Convergence in the north Pyrenean domain (Fig. 12.13, chart I–I', sections 9 and 10; and chart II–II', section 17) is recorded by shallow marine and coastal deposits in the eastern areas (Fig. 12.13, chart I–I') grading into deep flysch basin sediments in the west. The Cretaceous–Tertiary boundary again lies within Garumnian red beds in the eastern parts of the Pyrenees, and within marine deposits to the west. The cessation of subsidence in the north Pyrenean Cretaceous flysch basin is recorded by the deposition of limestone conglomerates at the beginning of Danian times.

The Basque-Cantabrian Basin (IR, K-UG, SR, SQ, MF)

The Basque-Cantabrian Basin (BCB) is a large sedimentary basin that developed on thinned continental crust between the European and Iberian plates during the Cretaceous period. The basin constitutes the western extension of the Pyrenean folded belt along extensive onshore areas of central-northern Spain and offshore areas of the Bay of Biscay. Nowadays, the basin lies between the Tertiary Ebro and Duero Basins, the Palaeozoic Asturian and Basque massifs, and the Bay of Biscay (Fig. 12.15A). The Cretaceous succession of the BCB was initiated by latest Jurassic rifting of the Iberian and European plates to create the Bay of Biscay (Fig. 12.15B). This rifting resulted in the break-up of previously widespread and relatively homogeneous Jurassic carbonate ramps (Chapter 11), and the emergence and erosion of former marine areas. During Cretaceous time the BCB was bounded by the land area of the Iberian Massif to the west and SW, and connected toward the east and SE with the Pyrenean and Iberian basins respectively. It was partially separated from the Pyrenean basin by the palaeogeographic highs of the Landes (now underthrusted and buried by Cretaceous sediments in the offshore platform of the Bay of Biscay) and Basque massifs.

The Cretaceous evolution of the BCB can thus be related to the kinematic relationship between the European and Iberian plates, and closely linked to the opening of the North Atlantic Ocean and the Bay of Biscay in Cretaceous times (Malod & Mauffret 1990; Olivet 1996). According to several authors (Montadert *et al*. 1979; Grimaud *et al*. 1982; Boillot & Malod 1988; Malod & Mauffret 1990; Olivet 1996), the BCB initially underwent a NNE–SSW direction of simple stretching during an Early Cretaceous rifting episode, perpendicular to the axes of the main NW–SE structural trends. Later, it changed to a NW–SE direction of stretching as the initial simple extension was replaced by left-lateral strike-slip motion along the main NW–SE faults. Transtensional movement along these latter faults produced oblique slip and pull-apart sub-basins in which great thicknesses of sediments accumulated. This change accompanied the anticlockwise rotation of Iberia with respect to Europe, as spreading started in the western Bay of Biscay (Montadert *et al*. 1979; Boillot & Malod 1988). In the eastern Bay of Biscay, no oceanic crust was formed, and the plate motion between Iberia and Europe was relieved instead by sinistral strike-slip faulting along the plate boundary in the Pyrenean area. Palaeomagnetic data suggest that Iberia separated from Europe during late Aptian to Santonian times (Montadert *et al*. 1979; Boillot & Malod 1988; Olivet 1996). Recent palaeostress reconstructions, however, based on measurement of striations preserved on lower and middle Albian synsedimentary fault scarps, suggest instead that rift-related NE–SW extensional tectonism persisted and predominated in the BCB at least until middle Albian times (Lepvrier *et al*. 1992). Extension between both plates reached its maximum during early Santonian times with a 100 to 150 km wide gap (Mathey 1987; Olivet 1996). During the Campanian to Maastrichtian interval, Iberia rotated back northwards and collided with Europe, a plate movement that culminated in the Palaeogene Pyrenean orogeny.

The boundary between the Iberian and European plates probably lies in the faulted area of the Biscay synclinorium and its associated submarine volcanism from the late Albian onwards (Fig. 12.15A), which has been considered as the western continuation of the north Pyrenean fault zone (Choukroune 1992) via the Leiza fault (with associated metamorphism in the Nappes des Marbres; Rat 1988; García-Mondéjar 1996; García-Mondéjar *et al*. 1996). To the north of this fault zone, the area around Gernika, Zumaya and San Sebastián has a European palaeomagnetic signature and belonged to Europe in the Cretaceous period (Van der Voo 1969; Vandenberg 1980; Schott & Peres 1987). To the south of this fault zone, the area around the Bilbao anticlinorium belonged to Iberia, as does the Le Danois Block (underthrust by oceanic crust in the Bay of Biscay; Malod *et al*. 1982; Olivet 1996) which forms the northwestward extension of the BCB. To the east, the area of Nappes des Marbres and the Dépression Intermédiaire (Fig. 12.15A) was situated at the plate boundary. In the Basque massifs the plate boundary can be traced between the Cinco Villas massif (European plate) and the Aldudes massif (Iberian plate) (Schwentke 1990; Schwentke & Kuhnt 1992).

It is difficult to define palaeogeographical domains valid for the entire Cretaceous succession because sedimentary systems

Fig. 12.15. (**A**) Simplified geological map of the Basque Cantabrian basin (BCB) showing the distribution of the Cretaceous outcrops in the area. The present-day location of some of the faults, sections and wells referred to in the text and figures has been indicated. (**B**) Map illustrating the main Albian palaeogeographic domains and palaeotectonic features of the Basque-Cantabrian basin (BCB).

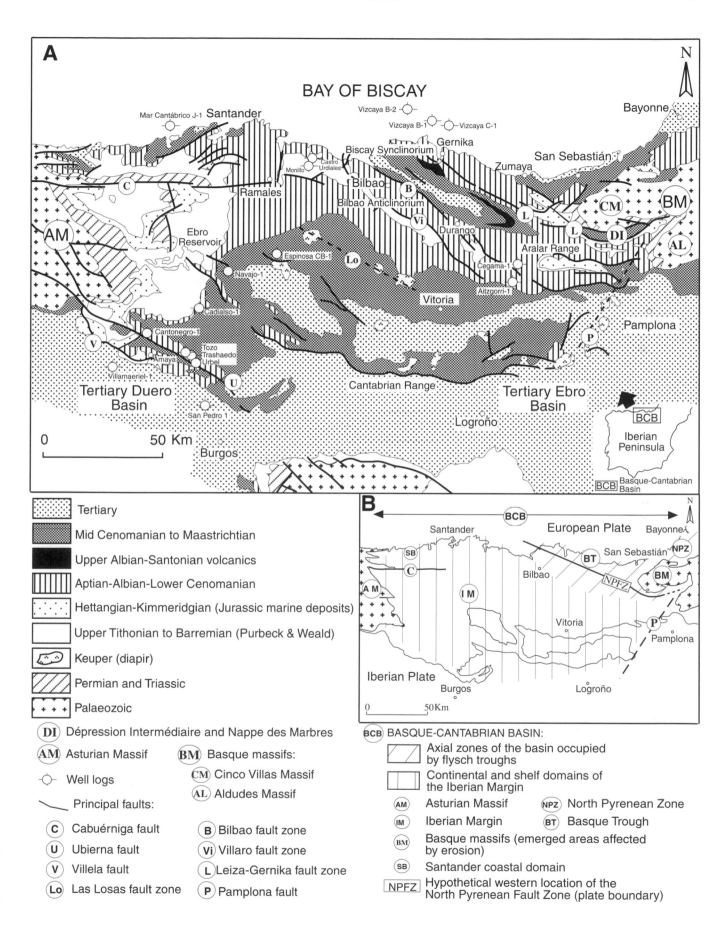

A

BAY OF BISCAY

N

Mar Cantábrico J-1 Santander

Vizcaya B-2

Vizcaya B-1 Vizcaya C-1

Gernika

Bayonne

Biscay Synclinorium

San Sebastián

Castro Urdiales

Zumaya

Monillo

Bilbao

B

Ramales

Bilbao Anticlinorium

Vi

CM

BM

AM

Ebro Reservoir

Durango

L

L

DI

Espinosa CB-1

Lo

Aralar Range

AL

Navajo-1

Cegama-1

Vitoria

Aitzgorri-1

Cadialso-1

Pamplona

Cantonegro-1

P

V

Amaya

Tozo Trashaedo Urbel

U

Cantabrian Range

Tertiary Ebro Basin

Villamaeriel-1

Tertiary Duero Basin

San Pedro 1

Logroño

0   50 Km

Burgos

BCB

Iberian Peninsula

BCB Basque-Cantabrian Basin

Tertiary

Mid Cenomanian to Maastrichtian

Upper Albian-Santonian volcanics

Aptian-Albian-Lower Cenomanian

Hettangian-Kimmeridgian (Jurassic marine deposits)

Upper Tithonian to Barremian (Purbeck & Weald)

Keuper (diapir)

Permian and Triassic

Palaeozoic

DI Dépression Intermédiaire and Nappe des Marbres

AM Asturian Massif

BM Basque massifs:

◇ Well logs

CM Cinco Villas Massif

Principal faults:

AL Aldudes Massif

C Cabuérniga fault

B Bilbao fault zone

U Ubierna fault

Vi Villaro fault zone

V Villela fault

L Leiza-Gernika fault zone

Lo Las Losas fault zone

P Pamplona fault

B

N

Santander

BCB

European Plate

Bayonne

SB

NPZ

C

BT

San Sebastián

AM

IM

Bilbao

BM

Vitoria

P

Iberian Plate

Pamplona

Burgos

Logroño

0   50Km

BCB BASQUE-CANTABRIAN BASIN:

Axial zones of the basin occupied by flysch troughs

Continental and shelf domains of the Iberian Margin

AM Asturian Massif

NPZ North Pyrenean Zone

IM Iberian Margin

BT Basque Trough

BM Basque massifs (emerged areas affected by erosion)

SB Santander coastal domain

NPFZ Hypothetical western location of the North Pyrenean Fault Zone (plate boundary) and depositional depocentres shifted over time, but from Albian times three distinct sedimentary areas became established (Fig. 12.15B). These comprise the following, from SW to NE. (a) The Iberian margin (including the Santander coastal domain, to the north of the Cabuérniga fault, which developed as an independent structural unit for most of Cretaceous time), characterized by continental and shelf sedimentation. To the SE, the Pamplona fault (Fig. 12.15A, B) is an old and deep fracture in the Iberian basement, an interpretation supported by seismic profiles from the area (Gräfe 1994). It separated the subsiding BCB in the west from a more stable block in the east (Ebro massif); (b) The Basque trough in the axial zones of the basin was located across the plate boundary area (north Pyrenean fault zone; NPFZ on Fig. 12.15B) and characterized by the development of Albian and Late Cretaceous flysch troughs with turbiditic and pelagic deposition. (c) The high blocks of the European margin (Landes and Cinco Villas massifs) constitute the third area.

Geological and subsurface well studies have enabled the subdivision of the Cretaceous tectonosedimentary evolution of the basin into a multiphase rifting stage (mainly early Cretaceous), and a post-rift stage (Late Cretaceous). The latter phase was accompanied by intensive alkaline within-plate volcanism (late Albian–Santonian) occurring on thinned continental crust along the fault systems that formed the plate boundary (Azambre & Rossy 1976; Lamolda *et al.* 1983; Wiedmann & Boess 1984; Meschede 1985; Cabanis & Le Fur-Balquet 1990; Castañares *et al.* 2001). This magmatism started in the NW before extending SE in the Biscay synclinorium.

Spreading ceased in the Bay of Biscay in early Campanian times, and convergence began. This compression is associated with Campanian–Maastrichtian siliciclastic deposits in shallow marine (coastal sandstones) and deep marine (flysch) environments. It culminated in the Palaeogene Pyrenean orogeny and northward subduction of Iberian crust in the Pyrenees (Engeser & Schwentke 1986; Choukroune *et al.* 1990). In the area of Le Danois Bank, southwardly directed underthrusting or subduction has been reported (Malod *et al.* 1982). This latter observation has led to the hypothesis that a reversal in the subduction direction occurred between the Basque-Cantabrian and Pyrenean areas (Engeser & Schwentke 1986; Turner 1996), probably across the NE–SW striking Pamplona fault system.

Early Cretaceous of the Basque-Cantabrian Basin (IR, SR, SQ)

The Early Cretaceous succession of the BCB can be related to the different stages of the Bay of Biscay rifting, because this played an important role in controlling the distribution and evolution of the sedimentary environments. The onset of rifting ranged between the end of Kimmeridgian and Tithonian times, probably starting earlier in the west (Asturian basin; Lepvrier & Martínez-García 1990) and propagating progressively eastwards (BCB). In the BCB pre-rift Jurassic marine deposition thus persisted at least until Kimmeridgian times in the eastern part of the basin (Guipuzcoa area), and the first synrift deposits are late Tithonian in age (Hernández *et al.* 1999; Hernández 2000; Chapter 11). Syndepositional tectonics in the area are indicated by strong differential subsidence, which controlled abrupt changes in both thickness and facies across NW–SE, N–S, E–W and NE–SW trending faults, most

of which were reactivated during the Tertiary tectonic inversion. The rifting has been divided into early, middle and late stages (Fig. 12.16), each separated by major regional unconformities.

Early rift stage (late Tithonian–Barremian): graben phase. Late Jurassic rift initiation caused the reactivation of E–W and NW–SE trending basement faults within the northern Iberian continental margin, and the development of elongated grabens in which huge amounts of synrift deposits began to accumulate. These initially isolated grabens progressively expanded, with onlap and overstep of the graben margins by successively younger synrift deposits sealing the older active faults. This evolution is recorded by the successive development of two major depositional units or facies groups bounded by regional unconformities (Fig. 12.16): the Purbeck facies group (late Tithonian–early Valanginian) and the Weald facies group (late Valanginian–Barremian) (Pujalte 1977, 1981, 1989*b*; Hernández *et al.* 1999).

During this time sedimentary basin geometry in the west and SW part of the BCB was controlled partly by the activity of E–W and NW–SE ancient basement faults and other related normal synsedimentary faults, which created a system of structural grabens filled by thick successions of mainly continental (fluvial to palustrine) and transitional marine deposits (Pujalte & Robles 1989; Pujalte *et al.* 1996; Robles *et al.* 1996; Hernández 2000). An additional influence on sedimentary facies was a variation from continental to transitional marine environments connected to the sea toward the NE and SE of the BCB (Soler y José 1972*a,b*; Pujalte 1982*a*, 1985; Garmendia & Robles 1991). Given this background, four palaeogeographic areas can be distinguished in the BCB during this early rift stage: the Cantabrian and Bilbao grabens to the west and central part of the basin, the Guipuzcoa platform to the east, and the emergent Cantabrian-Obarenes domain to the south.

The Cantabrian graben lies in the west, within the provinces of Cantabria, northern Palencia and northern Burgos. It is more than 50 km wide and contains up to 4 km of continental and transitional marine sediments. Its southern limit was the Iberian Massif, now buried below the Tertiary Duero basin, and both field and subsurface data suggest that the Ubierna and Villela faults acted during this time as basin-bounding faults (Figs 12.15 and 12.17). The northern limit moved northwards through time from a position south of the Cabuérniga ridge to one just off the present coastline. Within the graben two more subsident depocentres, the Polientes depocentre to the south and the Toranzo depocentre to the north, were separated by an east–west trending diapiric high, roughly aligned with the Ebro reservoir.

The Bilbao domain was an important trough in the central part of the basin, broadly coinciding with the position of the present-day Bilbao anticlinorium. It was filled by restricted marine shales, sandstones and minor limestones and dolostones, but its relationship with the Cantabrian graben to the west is still poorly known. The Guipuzcoa platform domain lay to the east, including the Gernika area and parts of the Guipuzcoa and Navarra provinces, and was the site of calcareous and terrigenous sediments deposited in transitional-marine environments. Further south, the Cantabria-Obarenes domain (Cantabrian ranges) was mainly a site of non-deposition or even erosion during the early rift stage (Pujalte 1989*b*).

The 'Purbeck' facies group is represented in the Cantabrian graben by two successive units: the Campóo and Cabuérniga groups (Fig.12.16; Pujalte 1977, 1985; García de Cortazar & Pujalte 1982; Robles *et al.* 1989, 1996; Pujalte *et al.* 1996; Hernández *et al.* 1999; Hernández 2000). The older synrift sediments (Campóo Group) were deposited discontinuously in several small, isolated, mainly alluvial-palustrine basins controlled by active faults. They accumulated in two principal half-graben depocentres: the Aguilar trough in the Polientes area and the Cires trough in the Toranzo area (Robles *et al.* 1996; Hernández *et al.* 1999). Both of these troughs lay in the active hanging-wall of the two more important bounding faults of the Cantabrian graben (the Villela and Cabuérniga faults to the south and north respectively). Their sedimentary infill consists of up to 1000 m of alluvial and fluvial siliciclastic rocks and carbonate-rich palustrine sediments, tufa deposits and evaporites (Tozo, Trashaedo, Urbel wells) associated with perennial, shallow lakes formed on seasonal floodplain wetlands. These troughs were separated by an intervening less subsident area where up to 100 m of fluvial conglomerates, sandstones and calcareous mudstones were deposited in east–west trending palaeovalleys.

The succeeding Cabuérniga Group (Gp) (Fig. 12.16; Pujalte 1977, 1985; García de Cortazar & Pujalte 1982; Hernández *et al.* 1999) has a more widespread area of deposition, overstepping the older synrift sediments (Campóo Gp), and sealing the previous troughs. The succession (up to 1000 m thick) largely consists of dark calcareous mudstones, siliciclastic and calcareous sandstones and sandy limestones with fresh-brackish to transitional marine fossils including stromatolitic structures, bivalves, gastropods, bryozoans, sponges, crinoids, coralline algae and sparse corals, recording a marine flooding of the rift basin (Pujalte 1985). These deposits show an increasing marine influence toward the north and NE, while their southern equivalent deposits, in the vicinity of the Ubierna fault (Fig. 12.17), comprise fluvial conglomerates and sandstones and red mudstones with palaeosoils. In the Bilbao region the Castro Urdiales and Monillo wells discovered shallow marine limestones in an equivalent stratigraphic position, suggesting that the Campóo and Cabuérniga groups entered the open sea in this area. During this time the sector to the north of the Cabuérniga ridge (Santander coastal domain) was subjected to erosion or non-deposition. In the Guipuzcoa platform domain the boundary between the underlying marine Jurassic deposits and the Purbeck–Weald succession is locally represented (e.g. Aralar range) by an angular unconformity with associated karstification ('postkimmeridgian unconformity'; Floquet & Rat 1975) that separates Kimmeridgian shallow water limestones below from latest Jurassic–Early Cretaceous transitional marine Purbeckian facies above (Soler y José 1971; Floquet & Rat 1975; Bulard *et al.* 1979; Pujalte 1989b; Garmendia & Robles 1991; Badenas 1999; Chapter 11). The succeeding sequence consists of about 100 m of dark limestones with gastropods, ostracods, charophytes and serpulid bioherms ('serpulid limestone'), deposited in fresh to brackish water transitional-marine environments. In the southern sector of the Guipuzcoa domain the Aitzgorri-1 and Cegama-1 wells drilled thick series of continental evaporitic deposits tentatively attributed to Late Jurassic–Neocomian times (Sánchez Ferrer 1991). This may indicate the presence of internally draining basins in the south part of the Guipuzcoa domain, similar of those described for the Cantabrian graben (Pujalte *et al.* 1996).

The 'Weald' facies group of the Cantabrian graben is separated from the Purbeck by Los Llares unconformity (Pujalte 1981, 1982b). This unconformity is related to a major tectonic pulse which reactivated the basin-bounding faults and thus rejuvenated the relief of the source areas, causing the influx of large volumes of siliciclastic sediments into the basin. The 'Weald' is represented by the Pas Group (Pujalte 1977, 1981, 1982b, 1985), a unit comprising 200–2000 m of siliciclastic channel-fill sandstones with associated red overbank mudstones and lacustrine black organic mudstones and sandstones (Barcena Mayor and Vega de Pas formations; Pujalte 1977; Fig. 12.16). These sediments extend in places beyond the limits of the underlying synrift deposits, resting directly on Jurassic (Santander coastal domain) or Triassic (Ebro reservoir high) rocks. In the Bilbao domain the 'Weald' is represented by the Villaro Fm (Pujalte 1982a; García Garmilla 1987) which consists of up to 2000 m of black shales, sandstones and limestones, interpreted as brackish deposits. Similarly, in the Guipuzcoa platform domain, the Weald is represented by up to 100 m of sandstones, black shales, marls and bioclastic limestones resting on Purbeckian serpulid limestones, together interpreted as recording brackish transitional to coastal marine environments.

Middle rift stage (Aptian–early Albian): carbonate platform phase. The Aptian-early Albian middle rifting episode is characterized by less active tectonism and a progressive basin

downwarping, which combined with a global rise in sea level to cause a generalized marine transgression leading to the deposition of Urgonian-type carbonate facies in a shallow epeiric sea (Rat 1959; Pascal 1985; García-Mondéjar 1990). The transgression opened marine connections with the Tethys ocean, so that warm currents favoured the widespread development of reefal, shallow-water carbonate platforms on shelf areas.

The base of the Urgonian succession shows an angular unconformity in some localities, and is characterized by sudden encroachment of marine sandstones and marls (García-Mondéjar 1990). In those localities where terrestrial facies prevailed (e.g. Polientes area), the basal unconformity is marked by an erosional surface between two fluvial units that separate Urgonian sediments from underlying Weald deposits (Fig. 12.16; García-Mondéjar & Pujalte 1975). During Aptian–Albian times, the Asturian, Duero, Ebro and the Landes and Basque massifs were still emergent areas. However, the depositional area of the Urgonian sediments was enlarged compared to that of the underlying early rift deposits (Purbeck–Weald), expanding over previously exposed areas. As a result, in places the Urgonian sediments may overlie Palaeozoic rocks, as seen in the Santander coastal domain or in the Mar Cantabrico J-1 well. Regional correlations and seismic data indicate a northeastward migration of Urgonian depocentres with respect to the previous ones (Fig. 12.17).

At the beginning of the Aptian stage sedimentation across the area ranged from fluvial sandstones and mudstones in the southern and western sectors (Rio Yera Fm; García-Mondéjar 1982), into a northerly shallow marine muddy shelf (Ernaga Fm; García-Mondéjar 1982). In the Bilbao area, a large sandy delta-estuary occurred (Ereza Fm; García-Mondéjar 1982) with the local development of oyster and rudist carbonate patches, recording a progressive but rapid onset of carbonate deposition. Shortly after, carbonate sedimentation expanded over large areas of the basin during the early Aptian, creating broad, shallow-water carbonate banks. Thereafter (from late Aptian times), a SW–NE restored section across the basin (Fig. 12.17) shows the establishment of two opposite shallow-water areas in the SW and NE, separated by an intervening deeper area with marly deposition (Bilbao trough; Figs 12.16 and 12.17), that was later to become a flysch trough. At the same time, alluvial plain and fluvial facies of the Escucha Fm (late Aptian–middle Albian) began to accumulate further south (Figs 12.16 and 12.17), depositing siliciclastic sediments containing lignites. These 'Escucha' facies sediments pass to the north and NE into transitional-marine environments and deltaic deposits, and then into carbonate platforms (Ramales limestones and lateral equivalents) and associated slope deposits, with a central intrashelf deeper embayment around the Bilbao area. This larger intrashelf trough formed as a result of differential subsidence and was filled by a huge pile of basinal lutites, marls and hemipelagic limestones with scattered siliciclastic turbidites and resedimented deposits (Bilbao Marls; García-Mondéjar 1982). An equivalent (though narrower) depositional scenario developed on the northeastern side of the Bilbao trough (Gernika area), where thick carbonate banks and reefal limestone interdigitated with siliciclastic rocks connected to fan-deltas and Gilbert-type deltas sourced from the emerging Landes massif to the north (Agirrezabala 1996) and graded to the south into the basinal deposits of the Bilbao trough.

Shallow shelf Urgonian carbonate platforms developed across relatively slowly subsiding areas and on top of tilted blocks, typically producing around 1000 m of sediment thickness, although up to 4000 m accumulated in places (Rio Ason-Sía area; García-Mondéjar 1985). This can be compared to >3000 m of contemporaneous deeper water sediments locally preserved in the Bilbao trough (Rosales *et al.* 1989; García-Mondéjar 1990). Platform facies typically consist of micritic limestones with rudist and coral bioherms, along with miliolids, orbitolinids and other benthic foraminifera, oyster-like

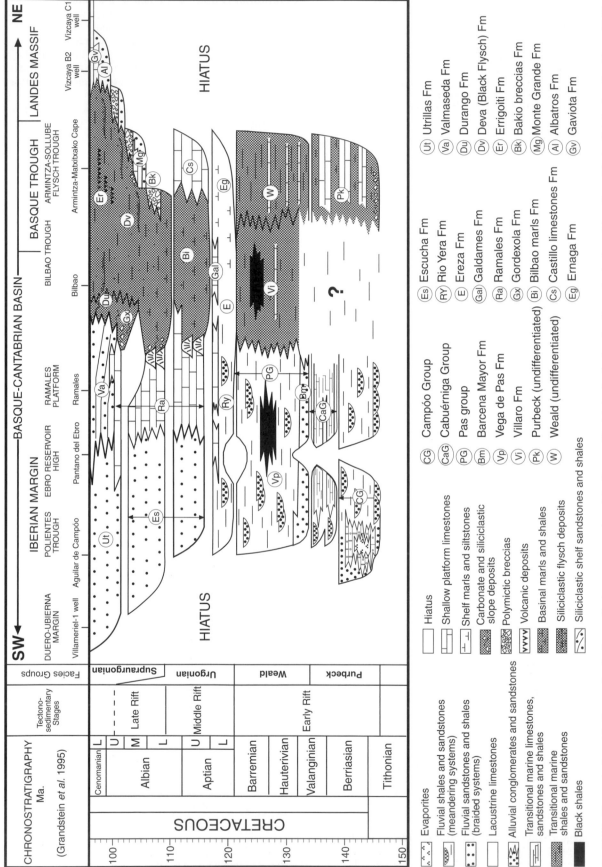

Fig. 12.16. Chronostratigraphic diagram for the Lower Cretaceous succession of the Basque-Cantabrian basin showing the main stratigraphic units and facies distribution. See Figure 12.15 for location of reference sections and wells and Figure 12.17 for equivalent lithostratigraphy. Timescale after Gradstein *et al.* (1995).

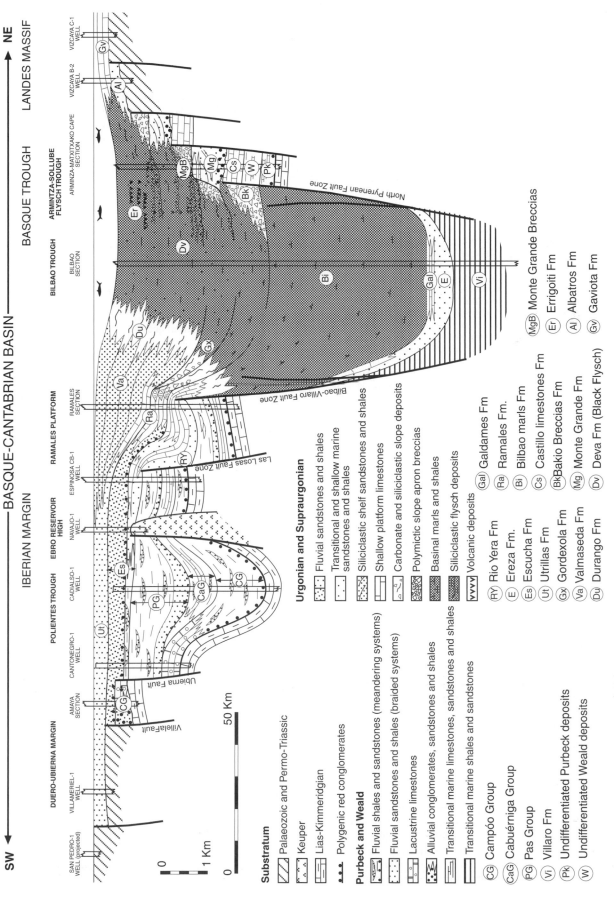

Fig. 12.17. SW–NE transverse cross-section of the Basque-Cantabrian basin restored for the early Cenomanian. It shows the development of a series of fault-bounded sedimentary grabens where synsedimentary faults are overlain and sealed by progressively younger synrift units. See Figure 12.15 for location of reference sections and wells and Figure 12.16 for equivalent chronostratigraphy.

Substratum

- Palaeozoic and Permo-Triassic
- Keuper
- Lias-Kimmeridgian
- Polygenic red conglomerates

Purbeck and Weald

- Fluvial shales and sandstones (meandering systems)
- Fluvial sandstones and shales (braided systems)
- Lacustrine limestones
- Alluvial conglomerates, sandstones and shales
- Transitional marine limestones, sandstones and shales
- Transitional marine shales and sandstones

Urgonian and Supraurgonian

- Fluvial sandstones and shales
- Transitional and shallow marine sandstones and shales
- Siliciclastic shelf sandstones and shales
- Shallow platform limestones
- Carbonate and siliciclastic slope deposits
- Polymictic slope apron breccias
- Basinal marls and shales
- Siliciclastic flysch deposits
- Volcanic deposits

CG Campóo Group
CaG Cabuérniga Group
PG Pas Group
Vi Villaro Fm
Pk Undifferentiated Purbeck deposits
W Undifferentiated Weald deposits

RY Rio Yera Fm
E Ereza Fm.
Es Escucha Fm
Ut Utrillas Fm
Gx Gordexola Fm
Va Valmaseda Fm
Du Durango Fm

Gal Galdames Fm
Ra Ramales Fm.
Bi Bilbao marls Fm
Cs Castillo limestones Fm
Bk Bakio Breccias Fm
Mg Monte Grande Fm
Dv Deva Fm (Black Flysch)

MgB Monte Grande Breccias
Er Errigoiti Fm
Al Albatros Fm
Gv Gaviota Fm

Chondrodonta, gastropods and calcareous algae. Micritic mud mounds as large as 250 m wide and 150 m thick proliferated in platform slope and margin environments (García-Mondéjar & Fernández-Mendiola 1995; Neuweiler *et al.* 1999). These micritic build-ups are usually made of algal (*Bacinella irregularis*) and microbialite/sponge frames (Neuweiler 1993; Rosales *et al.* 1995). Four major stages of Urgonian carbonate platform development are seen in the BCB (Fig. 12.16): early Aptian, late Aptian–earliest Albian, early–middle Albian, and late Albian (García-Mondéjar 1990; Fernández-Mendiola *et al.* 1993; Rosales 1999). These stacked platform limestone sequences are separated by three regional unconformities (Fig. 12.16), and can be internally subdivided into smaller depositional sequences. The main unconformities are usually accompanied by an influx of siliciclastic sediments.

During the development of the first Urgonian carbonate sequence (early Aptian, 30–200 m thick) flat rudist and coral carbonate banks (Galdames Fm and lateral equivalents; García-Mondéjar 1982) spread extensively across a shallow epeiric shelf that expanded to the south as far as the Ebro Reservoir and Alava. Carbonate banks were separated by siliciclastic sediments trapped in intervening shallow troughs, although bathymetrical differences between carbonate platforms and surrounding terrigenous areas remained low. During the second Urgonian carbonate stage (late Aptian–earliest Albian) there was a significant reduction in the area of shallow-water carbonate production compared to the first stage. Differential subsidence increased and bathymetrical differences between platforms and basins began to become more accentuated, leading to the development of a deeper intraplatform trough in the Bilbao synclinorium area, filled by basinal deposits (Bilbao trough). Platform carbonate deposits developed surrounding the Bilbao basinal area (Bilbao trough). During this stage platforms displayed a preferential progradational, distally steepened ramp geometry. The two youngest Urgonian carbonate phases (early–middle Albian and late Albian respectively) formed during the next late rift stage, principally as aggrading isolated carbonate build-ups and residual platforms. The replacement of carbonate sedimentation by siliciclastic deposits (Supraurgonian episode; Fig. 12.16) occurred progressively from early–middle Albian times, but is not until the late Albian (late Vraconian) interval when siliciclastic sediments invaded most of the shallow platform areas and the final shutdown of carbonate deposition occurred (García-Mondéjar 1990; López-Horgue *et al.* 1994; Rosales 1999).

Late rift stage (early–middle Albian) and transition to post-rift stage (late Albian–early Cenomanian): flysch phase. Successive pulses of renewed rifting during early–middle Albian times marked a turning point in the evolution of the BCB. This late rifting episode was characterized by maximum differential subsidence with formation of differentially foundering fault blocks and by the development of the first flysch troughs ('black flysch'). From Albian times onwards the BCB comprised an Iberian margin, a European margin (the latter marked by the southern edge of the Landes massif and the Cinco Villas massif), and an intervening WNW–ESE trending deep-sea trough (Fig. 12.18) in which thick piles of turbiditic, pelagic and volcanic deposits accumulated (García-Mondéjar 1982; Feuillée 1983; Robles *et al.* 1988*b*; Vicente-Bravo & Robles 1995). Tectonic features during this stage included NW–SE trending faults parallel to the rift axis, with transverse E–W, N–S and SW–NE striking faults forming complex patterns of different depocentres (Vicente-Bravo & Robles 1995; García-Mondéjar *et al.* 1996).

Sedimentary systems initiated during the last stages of the rifting (Albian) persisted until early Cenomanian times (transition to post-rift stage), and show a marked onlap onto previously marginal areas (Fig. 12.19). Albian to early Cenomanian deposits variously rest on the Asturias, Ebro and Duero basement massifs in the Iberian margin, and on the Landes and Cinco Villas massifs in the European margin. In

the latter area, Albian sequences rest on Buntsandstein sandstones and Palaeozoic rocks as seen in the Biscay offshore wells (e.g. Vizcaya C-1, Vizcaya B-1, Vizcaya B-2) and around the Basque massifs (Fig. 12.19). In the axial areas of the basin, occupied by the flysch trough, slope apron and turbiditic deposits of the 'black flysch' successively expanded across the continental and shallow marine deposits that fringed the emergent massif of Landes (today buried below the Basque allochthonous pile) and Cinco Villas. This progressively enlarging flysch trough area constituted the Basque trough (Fig. 12.18), which was connected to the NE with the North Pyrenean Zone. Therefore, during Albian–early Cenomanian times the BCB can be subdivided into two large depositional areas, the Iberian margin and the Basque trough.

The Iberian margin. In the Iberian margin, as the result of rifting during early Albian times, many of the previous large Urgonian carbonate platforms underwent fragmentation into fault blocks, uplift, exposure, and finally collapse and drowning (Aranburu *et al.* 1992; Rosales *et al.* 1994; Rosales 1995, 1999). Coeval volcanic activity in the Durango area was linked to the activity of the Bilbao fault (Fernández-Mendiola & García-Mondéjar 1995). In some half-grabens and other rapidly foundering sites there was erosion of submarine canyons and deposition of syntectonic terrigenous and mass-transport deposits (calcareous and siliciclastic turbidites, slumps and carbonate breccias; García-Mondéjar *et al.* 1993; Rosales *et al.* 1994). In the southwestern BCB (Polientes area), this tectonic event resulted in uplift and development of incised valleys on top of meandering fluvial sandstones and mudstones (García-Mondéjar & Fernández-Mendiola 1992). These incised valleys became filled by coarse fluvial quartz-conglomerates of the Escucha Fm (Fig. 12.17).

In the central BCB a reorganization of the former carbonate platform areas took place, leading to the development of a new style of carbonate deposition (third Urgonian carbonate stage, early–middle Albian). Differential vertical movements on basement blocks related to tectonic activity along NW–SE and SW–NE trending faults and caused the formation of shallow-water carbonates on top of some uplifted blocks. Previous carbonate ramps evolved into thick carbonate build-ups (up to 500 m thick) with steeper slopes that aggraded to keep up with increasing relative sea level. Platform aggradation was interrupted at times by periods of emergence, related to vertical movements of individual fault blocks (Rosales *et al.* 1994; Gómez-Pérez *et al.* 1998). Tectonic tilting caused fracturing and gravity sliding of platform margin carbonates, leading to the emplacement of slumps, megabreccias and large olistoliths in slope and basinal areas. A hiatus with extensive meteoric diagenesis and palaeokarst development occurred throughout the mid-Albian interval in many platform areas (Gómez-Pérez 1994; Rosales 1995, 1999; Aranburu 1998; López-Horgue *et al.* 2000). This middle Albian stage of platform exposure was followed at the beginning of late Albian times by a general collapse of the platform margins. In the Bilbao embayment these events produced megabreccia horizons (containing discontinuous but aligned olistoliths up to 500 m long), and a thick, rapidly deposited wedge (up to 2 km) of mostly mudstones, siltstones and sandstone turbidites (Gordexola Fm; Zuluaga *et al.* 1992; Gómez-Pérez *et al.* 1994) that rapidly filled up the Bilbao trough, levelling the submarine relief between the basin and surrounding platforms. During deposition of this siliciclastic wedge there was no sedimentation on the carbonate palaeohighs, which remained exposed.

Following this early–middle Albian evolution, three major depositional domains can be defined for the late Albian–early Cenomanian palaeogeography of the Iberian Margin: (a) a low subsiding continental area to the south, in which the fluvial Utrillas Fm was deposited; (b) an intermediate shelf area of faster subsidence in the central part of the basin, northeastward of the Las Losas fault area (Fig. 12.17), in which huge piles of deltaic and platform siliciclastic rocks of the Valmaseda Fm and equivalent units were deposited; (c) a central basin where deposition of the 'black flysch' took place.

The Utrillas Fm (late Albian–early Cenomanian, 50–300 m thick) consists of fluvial sandstones and conglomerates with thin intercalations of mudstones and lignites (Arnaiz *et al.* 1991). It rests

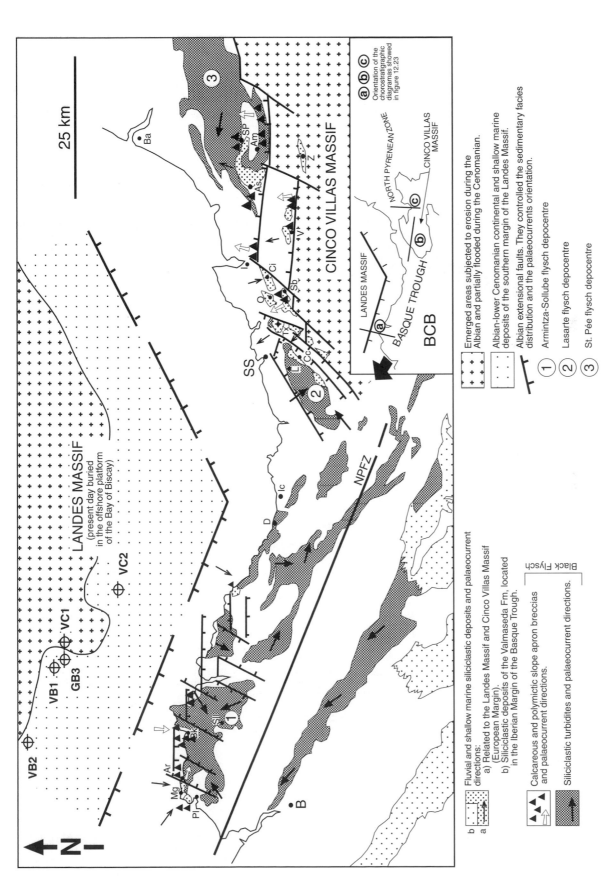

Fig. 12.18. Palaeotectonic and facies map of the Basque trough (a deep sub-basin developed in the axial zone of the BCB) and western end of the North Pyrenean Zone. The map superimposes the outcrops the main palaeofaults and the palaeocurrents of the Albian–early Cenomanian continental, shallow marine and turbiditic siliciclastic sediments. Also shown is a palinspastic restoration of the Landes massif, later overthrust by the deposits of the Basque trough during Alpine compression (modified from Vicente Bravo & Robles 1995). Offshore wells: VB1, Vizcaya B1; VB2, Vizcaya B2; VC1, Vizcaya C1; VC2, Vizcaya C2; GB3, Gaviota B3. Localities: Am, Amotz; Ar, Armintza; As, Ascain; B, Bilbao; Ba, Bayonne; Bk, Bakio; Cc, Cuatro Caminos; Ci, Castillo del Inglés; D, Deva; I, Irún; Ic, Iciar; L, Lasarte; Mg, Monte Grande; O, Oyarzun; P, Plencia; Sb, Sarobe; Sll, Sollube; SP, St. Pée; SS, San Sebastian; Tx, Txoritoquieta; V, Vera; Z, Zugarramurdi. NPFZ, western location of the North Pyrenean Fault Zone.

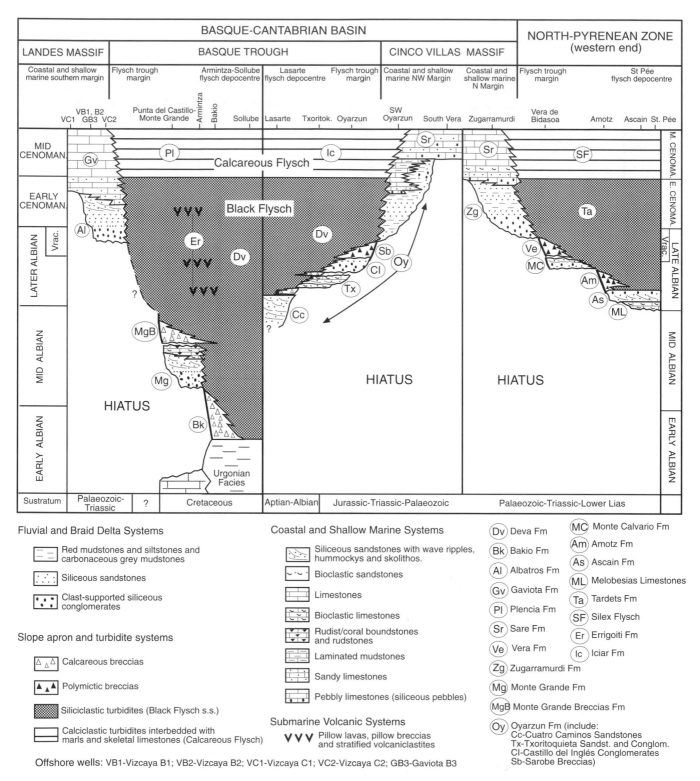

Fig. 12.19. Chronostratigraphic chart for the Albian–Middle Cenomanian deposits of the Basque trough and western end of the North Pyrenean Zone. Note the diachroneity of the sedimentary facies and the progressive expansion of the flysch deposits of the Basque trough, onlapping the previous marginal areas. The flysch trough margins show a stepped configuration caused by active synsedimentary back faulting.

unconformably on a Variscan basement (Iberian and Asturian massifs) and on older Mesozoic sediments, showing a discordance with the underlying Escucha Fm. Northward, the Utrillas Fm grades into the Valmaseda Fm (García-Mondéjar 1982) which comprises an extremely thick wedge (up to 3500 m thick) of prograding siliciclastic,

storm-dominated shelf sediments (Fig. 12.17; Pujalte & Monge 1985; Pérez-García *et al.* 1993, 1997) over former carbonate platforms and basinal areas. During late Albian times platform siliciclastic rocks were dominant over carbonate facies (fourth Urgonian carbonate stage). The carbonates became restricted to residual platforms on top of the formerly more widespread carbonate areas, and localized build-ups on some terrigenous-free structural palaeohighs (e.g. Albeniz-Eguino limestones) as well as on active Triassic salt diapirs (e.g. Caniego limestones on the Mena diapir; Reitner 1982, 1986; Engeser *et al.* 1984). To the SE (Estella area), lateral equivalents of the Valmaseda Fm consist of mudstones and siltstones, with minor sandstones and limestones, deposited in a shallow marine platform extending from shoreface to offshore environments (Zufia Fm; García-Mondéjar 1982).

Basque trough. The Basque trough was an irregular graben-like feature developed in the axial zone of the BCB and limited by the Iberian margin to the south and the emergent Landes and Cinco Villas massifs to the north (Fig. 12.18). It was internally segmented by NE–SW and NW–SE trending faults that formed a series of step-like features bounding many sub-basins or depocentres (e.g. Armintza-Sollube depocentre in the southern edge of the Landes block, and Lasarte depocentre in the western margin of the Cinco Villas massif) which extended northwards into the North Pyrenean Zone (e.g. St. Pée depocentre in the northern margin of the Cinco Villas massif; Vicente-Bravo & Robles 1995). This created a complex basin topography, as recorded by facies changes, palaeocurrent distribution and dramatic thickness variations, from a few hundred metres at the margins of the sub-basins to >3000 m in some depocentres (Figs 12.18 and 12.19).

During the Albian–early Cenomanian interval three main types of depositional systems existed across the area: (1) fluvial and deltaic; (2) coastal and shallow marine carbonate and siliciclastic; (3) basinal slope apron and turbiditic (Deva Fm). The fluvial-alluvial and shallow marine sediments unconformably overlie older Mesozoic or Palaeozoic rocks and developed along narrow shelves fringing the southern edge of the Landes (Monte Grande, Albatros and Gaviota formations) and Cinco Villas emergent massifs (e.g. Oyarzun, Zugarramurdi and Ascain formations; Figs 12.17 and 12.19), whereas the central and deeper parts of the basin were occupied by siliciclastic turbiditic systems of the Deva Fm. Palaeocurrent analyses indicate a dual, broadly northeastern and southwestern provenance for these sediments (Fig. 12.18; Vicente-Bravo & Robles 1991, 1995). Tectonic instability gave rise to slope aprons where thick successions (up to 500 m) of carbonate and polymictic megabreccias were deposited on hanging-wall depocentres attached to faulted palaeoescarpments bordering emergent massifs. Examples include the Bakio megabreccias on the southern edge of the Landes massif (García-Mondéjar & Robador 1986–87; Robles *et al.* 1988a), and the Amotz and Vera breccias on the northern margin of Cinco Villas-La Rhune (Razin 1989; Figs 12.18 and 12.19).

Progressive widening and deepening of the flysch basin from the central areas to the outer marginal areas created successively younger marginal fault-bounded steps, at the base of which accumulated highly diachronous slope apron deposits, eroded from uplifted faulted blocks, and siliciclastic turbiditic deposits. During a succeeding relative sea-level rise previously emergent blocks were flooded and overlain by coastal and shallow marine sediments (e.g. Monte Grande Fm; Pujalte *et al.* 1986–87; Robles *et al.* 1988a,b; Fig. 12.19). This pattern of sedimentation occurred repeatedly from early Albian to early Cenomanian times (Fig. 12.19). Flysch sedimentation was initiated earlier in the NW (Armintza-Sollube flysch depocentre; Fig. 12.19) and propagated progressively eastwards (Lasarte and St. Pée flysch depocentres). In the Armintza-Sollube sub-basin the older flysch sequences of the Deva Fm are early Albian in age (Vicente-Bravo & Robles 1995) and were deposited at the foot of subaerially exposed uplifted blocks of older Urgonian platforms (e.g. Punta del Castillo limestones; García-Mondéjar & Pujalte 1983). Further east (Gernika sector), carbonate platform deposits persisted during the early Albian–earliest middle Albian interval, and were bounded by active NW–SE and NE–SW fault escarpments with associated rockfall megabreccias and olistoliths (Agirrezabala & García-Mondéjar 1989). Subsequently, 'black flysch' was deposited during late Albian times, after a basal hiatus spanning most of the middle Albian interval (Agirrezabala 1996), a time gap that increases northwards (Landes massif) and eastwards (Cinco Villas massif), where the basement can be as old as Palaeozoic and the 'black flysch' sequences are younger (Fig. 12.19). In the eastern area, surrounding the Cinco Villas massif (Lasarte and St. Pée depocentres), 'black flysch' deposits are late

Albian in age (EVE 1988, 1991; Razin 1989; Souquet & Peybernés 1991). The Deva Fm contains thin intercalations of volcaniclastic sediments and thick pillow lavas (Errigoity Fm; Castañares *et al.* 1997, 2001), which precede the huge piles of Late Cretaceous volcanic rocks that were to be erupted in the same context (Mathey 1986).

Late Cretaceous of the Basque-Cantabrian Basin (K-UG, MF, IR)

During Late Cretaceous times, plate motion produced sinistral transtension which in turn created a series of en-echelon pull-apart sub-basins within the Basque-Cantabrian region (Fig.12.20). Palaeogeographically, two main domains can be

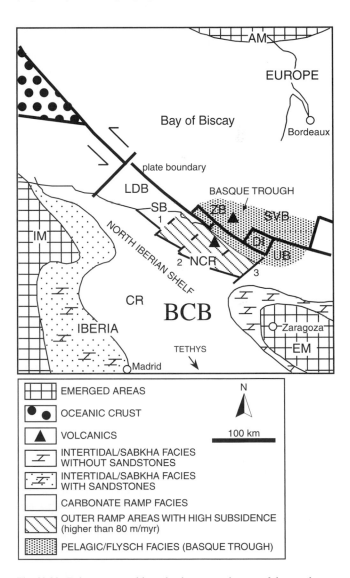

Fig. 12.20. Palaeogeographic and palaeotectonic map of the north Spanish continental margin and BCB during Turonian–Santonian. Modified after Gräfe (1999) with additional data from Schwentke (1990). Key: AM, Armorican massif; CR, Castilian ramp; DI, Dépression Intermédiaire; EM, Ebro massif; IM, Iberian massif; LDB, Le Danois block; NCR, Navarre-Cantabrian ramp; SB, Santander sub-basin; SVB, Sare Vera sub-basin; UB, Ulzama sub-basin; ZB, Zumaya sub-basin; 1, Rio Miera flexure; 2, Las Losas fault; 3, Pamplona fault. The Ulzama sub-basin (UB) is an eastward extension on the NCR and the Santander sub-basin (SB) is a westward extension of the NCR toward the Santander coastal domain.

distinguished in the BCB at this time: (1) a flysch and pelagic depocentre in the NE on the plate boundary (Basque trough), bounded to the south by a continental slope and split into several smaller depocentres; and (2) a carbonate ramp in the south and east, on the north Iberian shelf. The carbonate ramp can be subdivided into three domains: (1) a highly subsiding outer or distally steepened part of the carbonate ramp (the Navarre-Cantabrian ramp (NCR); Floquet 1991), which remained marine during Cenomanian to Santonian times, and records the deposition of up to 4000 m of outer shelf sediments; (2) an outer, less subsiding ramp to the NW in the Santander coastal domain (formed by the Santander sub-basin (SB) and Le Danois block (LDB), and separated from the more subsiding NCR in the SE by the palaeogeographic feature of 'Rio Miera flexure'; Feuillée & Rat 1971; Fig. 12.20); (3) the inner or proximal part of the carbonate ramp (the Castilian ramp *sensu stricto* (CR); Fig. 12.20; Floquet 1998), which was not constantly flooded during the Cenomanian to Santonian interval and records environments ranging from mid-shelf to terrestrial environments, though dominated by shallow-marine carbonates.

Northeastward, the deep flysch Basque trough that developed at the plate boundary became split into a series of smaller shallows and more subsiding sub-basins, such as the Plencia sub-basin (in the Biscay synclinorium area), the Ulzama sub-basin (on the northern margin of the Iberian plate) and the Zumaya, Sare Vera and dépression intermédiaire sub-basins (on the southern margin of the European plate; Fig. 12.20). As a result of the progressive northeastward widening and deepening of this flysch trough, Late Cretaceous deep-water pelagic and flysch sediments were deposited on the previously emergent Basque massifs. Nowadays, the Late Cretaceous sedimentary cover of the Basque massifs is preserved only in the Sare Vera, dépression intermédiaire, Nappe des Marbres and Ulzama sub-basins. These sub-basins show a succession of sedimentary facies made of rudist carbonates and bioclastic limestones succeeded by several metres of pelagic carbonates, that in turn are overlain firstly with calcareous flysch, and then with siliciclastic flysch, thereby terminating the Cretaceous history of sedimentation. Stratigraphic boundaries between these sedimentary units are highly diachronous across the different sub-basins (Schwentke & Kuhnt 1992).

Northwestward, in the high block of Le Danois (Fig. 12.20), sediments are mainly Early Cretaceous in age, with only very thin Late Cretaceous deposits (Malod *et al.* 1982). However, in the borehole Mar Cantabrica H-1X, situated on the southern margin of the Le Danois block, a 446 m thick succession of Cenomanian biomicrites, sandstones and marlstones was drilled (IGME 1987).

Beginning in late Albian times, marine gulfs encroached upon the Iberian plate from both the NNW and SSE, between the Iberian massif in the west and the Ebro massif in the east (Alonso *et al.* 1993). These gulfs developed to form an Atlantic ramp on the northern side of the plate (where the NCR, SB and CR were located) and a Tethyan ramp on its southeastern side. The growth extension of the Atlantic ramp was the result of the southeastward movement of Iberia and the opening of the Bay of Biscay. The ramp gradually disappeared in the latest Cretaceous to early Palaeogene interval due to the build-up of compressive forces as a result of the reversed northward movement of Iberia.

Late Cretaceous sediments in the carbonate ramp of the north Iberian shelf (CR and NCR) can be viewed as a series of five megasequences, each deposited over 5 to 10 Ma and separated by tectonically enhanced unconformities (Fig. 12.21). These megasequences are Mena (late Albian to late Cenomanian), Angulo (late Cenomanian to earliest Coniacian), Losa (early Coniacian to late Santonian), Sedano (Campanian) and Urbasa (Late Campanian to Maastrichtian). Late Cretaceous sedimentary cycles recorded by these five megasequences reflect an interplay between eustatic sea-level changes, inherited palaeotopography, tectonics, climate and sediment supply. This cyclicity was superimposed on a long ranging tectonosedimentary cycle of 35 Ma that embraces the whole of Late Cretaceous time. Local plate tectonic processes played a role in both long-term and shorter-term relative sea-level changes, so that, for example, a decoupling of maximum flooding and maximum water depth in Turonian and Santonian times has its causes in plate tilt and palaeotopography.

Another regional influence was an anomalously high heat flow in the basin, which was responsible for a certain amount of thermotectonic subsidence additional to that solely related to crustal extension. Lateral salt movement is also responsible for a part of the tectonic subsidence. Judging from the shape of the areas with highest subsidence, it is clear that there was also some influence from oblique-slip faults south of Europe/Iberia.

Climate was important in the development of depositional cycles in the BCB. The subtropical position of Iberia provided generally warm and humid conditions suitable for high biogenic production and a high growth potential of the carbonate ramp. On the other hand, pelagic areas seem to be rather oligotrophic compared to more highly productive pelagic realms in the western Tethys (Scaglia facies) and NW Europe (grey and white chalk facies). Wetter conditions on the western border of the BCB (Floquet 1998) probably favoured deposition of marginal terrigenous deposits there, whereas more arid conditions along the eastern border favoured deposition of sabkha facies around the Ebro massif. The effects of climatic cooling (Barrera & Savin 1999) may have enhanced a Campanian–Maastrichtian regression.

Mena megasequence. This megasequence is characterized by the increasing landward onlap of hemipelagic to pelagic deposits composed of marl–limestone alternations (e.g. Arceniega Fm; Fig. 12.21). In deep water sections of the NCR and the Basque Trough (Leioa, Galarreta, Gordoa, borehole Urbasa-2; Gräfe & Wiedmann 1993; Rodríguez-Lázaro *et al.* 1998; Gräfe 2000), the Cenomanian succession comprises some tens of metres of calcareous claystones and marls (Gordoa Fm; Fig. 12.21; Ciordia Member of the Arceniega Fm; Amiot 1982), or calcareous flysch with intercalated basaltic volcanics or tuffites (Elgueta Fm; Fig. 12.20; Mathey 1987). The marginal facies comprises subtidal to intertidal carbonate ramp deposits (Santa Maria de las Hoyas and Dosante formations; Fig. 12.21). The sequence is overlain by an unconformity in the marginal areas, and by prograding Pithonella Limestones (Cabezas Fm) in the hemipelagic areas (Fig. 12.21; Floquet 1998).

Angulo megasequence. During and after the deposition of the 130 m thick Cabezas Fm (Fig. 12.21) renewed flooding of the ramp resulted in the deposition of ammonite-foraminiferal claystones and marlstones (Puentedey, Santa Cruz del Tozo, Picofrentes and Monterde formations; Fig. 12.21). In the northern Castilian ramp, maximum flooding occurred synchronously with maximum deepening, whereas on the southern Castilian ramp this appears not to be the case (Floquet 1998).

Basinward, the Turonian succession comprises up to 1000 m of foraminiferal marls (proximal; Hornillatorre Fm), and marl–limestone alternations (distal; Valle de Mena Fm), with only a few tens of metres

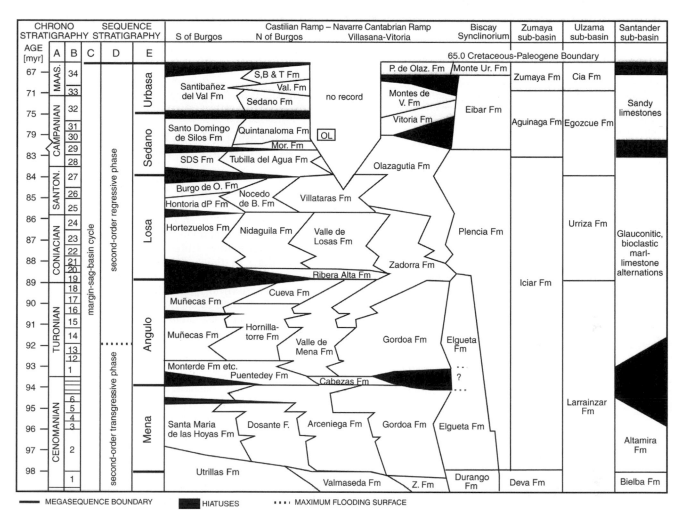

— MEGASEQUENCE BOUNDARY   ■ HIATUSES   ···· MAXIMUM FLOODING SURFACE

Fig. 12.21. Stratigraphy of the Upper Cretaceous succession in the Basque-Cantabrian basin, including the Castilian ramp, Navarre-Cantabrian ramp, Zumaya sub-basin, Ulzama sub-basin and Santander sub-basin (see Figure 12.24 for location of depositional domains and sub-basins). Chronostratigraphy and megasequences (left) are based on data of the Navarre-Cantabrian ramp and modified after Gräfe (1999). For source of data see Gräfe (1994) and Wiedmann in Gischler *et al.* (1994). Timescale after Gradstein *et al.* (1995). To the right, lithostratigraphic units are displayed according their palaeogeographic position in the BCB. Lithostratigraphy after Gräfe (1994) modified from Amiot (1982), Floquet *et al.* (1982) and Mathey (1982). Only the most important hiatuses are shown in black (the number of hiatuses is much larger, increasing in the proximal areas of the BCB). Key: column headers: A, stages; B, biozones; C, major second-order cycles (supercycles); D, transgressive–regressive facies cycles; E, megasequences. Formations: Burgo de O. Fm, Burgo de Osma Fm; Hontoria dP Fm, Hontoria del Pinar Fm; Monte Ur. Fm, Monte Urco Fm; Montes de V. Fm, Montes de Vitoria Fm; Monterde Fm etc., Monterde, Santa Cruz del Tozo and Picofrentes formations; Mor. Fm, Moradillo de Sedano Fm; Nocedo de B. Fm, Nocedo de Burgos Fm; OL, Oro Limestone; P. de Olaz. Fm, Puerto de Olazagutia Fm; SDS Fm, Santo Domingo de Silos Fm; S, B & T Fm, Sobrepeña, Bozoo and Torme formations; Val Fm, Valdenoceda Fm; Z. Fm, Zufia Fm. Biozones: 1, *Graysonites* sp. and *Hypoturrilites mantelli* zones; 2, *M. mantelli* and *M. dixoni* zones; 3, *Turrilites costatus* Zone; 4, *Cunningtoniceras cunningtoni* Zone; 5, *Eucalycoceras spathi* Zone; 6, *Calycoceras naviculare* Zone; 7, *Metoicoceras muelleri* Zone; 8, *Metoicoceras geslinianum* Zone; 9, *Vascoceras gamai* Zone; 10, *Spathites subconciliatus* Zone; 11, *Leoniceras discoidale* and *Watinoceras* sp. zones; 12, *Mammites nodosoides* Zone; 13, *Wrightoceras munieri* and *Leoniceras fleuriausianum* zones; 14, *Collignoniceras woollgari* and *Romaniceras inerme* zones; 15, *Romaniceras ornatissimum* Zone; 16, *Romaniceras deverianum* and *Colligoniceras carolinum* zones; 17, *Subprionocyclus neptuni* Zone; 18, *Germariceras germari* Zone; 19, *Forresteria petrocoriensis* Zone; 20, *Tissotioides haplophyllus* Zone; 21, *Romaniceras hispanicum* Zone; 22, *Gauthiericeras margae* Zone; 23, *Hemitissotia turzoi* Zone; 24, *Hemitissotia lenticeratiformis* Zone; 25, *Texanites gallicus* Zone; 26, *Texanites hourcqi* Zone; 27, *Texanites presoutoni* Zone; 28, *Bevahites subquadratus* Zone; 29, *Scaphites hippocrepis* and *Neocrioceras riosi* zones; 30, *Hoplitoplacenticeras marroti* and *Trachyscaphites spiniger* zones; 31, *Bostrychoceras polyplocum* Zone; 32, *Nostoceras hyatti* and *Pseudokossmaticeras tercense* zones; 33, *Pachydiscus neubergicus* Zone; 34, *Pachydiscus* spp. Zone.

of pelagic foraminiferal claystones being deposited on the deep water edge of the NCR (Gordoa Fm; Fig. 12.21). Calcareous flysch and tuffites occur in the Basque trough, including the Zumaya, Sare Vera and Ulzama sub-basins (Plencia, Elgueta, Iciar and Larrainzar formations; Fig. 12.21; Schwentke & Kuhnt 1992). In the Santander sub-basin (SB), glauconite-rich marlstones and bioclastic limestones were deposited in a facies similar to those in the Hornillatorre Fm (Wiese & Wilmsen 1999).

In the latest middle to late Turonian interval, a tendency towards marine regression led to the progradation of shallow-marine carbonates to the north (Muñecas Fm; Fig. 12.21). In late Turonian times the whole outer proximal carbonate ramp became covered by prograding rudist carbonates (Cueva Fm; Fig. 12.21).

Losa megasequence. The base of the Losa megasequence shows 300 m of shallow marine carbonates (Ribera Alta Fm) pinching out towards the proximal ramp, which was emergent during this time (Fig. 12.21). This emergence is recorded by a weak angular unconformity on the Castilian ramp (Floquet 1991; Gräfe & Wiedmann 1993), whereas more basinward the Ribera Alta Fm interfingers with calciturbidites (Fig. 12.21). Subsequent initially increasing relative sea level (Coniacian) produced a variety of calcareous facies belts (Hortezuelos Fm; Fig. 12.21; Floquet 1998), the inner ramp part of which was later subaerially exposed. Between these marginal facies and the deep marine realms, ammonite- and echinoid-rich claystones and bioclastic limestones were deposited (Nidaguila and Valle de Losas formations; Fig. 12.21). Deep marine Coniacian deposits comprise distal calciturbidites and planktonic foraminiferal marl–limestone alternations (Zadorra, Olazagutia, and Plencia formations; Fig. 12.21). The deposition of calciturbidites is also a characteristic feature in the Ulzama and Zumaya sub-basins (Iciar and Urriza formations; Fig. 12.21; Schwentke & Kuhnt 1992).

Early Santonian flooding of the Castilian ramp occurred from north to south, leading to the deposition of planktic foraminiferal marlstones (Olazagutia Fm), micritic ammonite-rich limestones and marls (Villataras Fm) and open marine limestones of the Nocedo de Burgos Fm (Fig. 12.21). Marginal areas show sabkha and intertidal deposits (Hontoria del Pinar and Burgo de Osma formations; Fig. 12.21). Late Santonian infilling of the Iberian basin resulted in shallow marine facies belts prograding northwards (Floquet 1998), and led ultimately to subaerial exposure and karstification of the whole proximal ramp (Fig. 12.21).

Sedano megasequence. The Santonian–Campanian boundary marks the onset of strong compressive tectonism in parts of the basin, and the consequent enhancement of siliciclastic facies conditions over carbonate facies deposition. Deep marine deposition was dominated by thick turbiditic wedges (Mathey 1987). In deeper basinal realms, and in the Zumaya and Ulzama sub-basins, the Sedano megasequence shows a transition from calcareous turbidites (Plencia and Iciar formations) to siliciclastic turbidites (Eibar, Aguinaga and Egozcue formations; Fig. 12.21). In the Santander sub-basin (SB), a Campanian hiatus separates glauconitic marlstones and bioclastic limestones below from sandy limestones and sandstones above. On the distal shelf of the NCR, the marl–limestone alternations of the Vitoria Fm are rapidly prograded by deltaic sandstones of the Montes de Vitoria Fm (Fig. 12.21). In the central NCR, no sediments younger than Santonian are preserved except on the top of salt diapirs, such as at Murguia, where bioclastic limestones (Oro Limestone) record Campanian carbonate facies deposition.

Renewed Campanian transgressive pulses resulted in the deposition of the Tubilla del Agua Fm (rich in *Lacazina*), followed by the prograding sandstones of the Moradillo de Sedano Fm and finally by three prograding rudist biostromes of the Quintanaloma Fm (Fig. 12.21). The base of the Quintanaloma Fm has been dated with rudists as early Campanian (Floquet 1991). Marginal facies belts are represented by the Santo Domingo de Silos Fm (Floquet 1998; Fig. 12.21).

Urbasa megasequence. This final megasequence comprises siliciclastic sediments (Sedano and Sobrepeña formations), shallow marine rudist (Valdenoceda Fm) and orbitoidal carbonates (Torme and Puerto de Olazagutia formations; Fig. 12.21). Basinwards, thick flysch wedges occur in the north of the NCR (Biscay synclinorium), as well as in the more northerly Zumaya sub-basin (Eibar and Zumaya formations). Towards the Cretaceous–Tertiary boundary, widespread regression caused a basin-wide unconformity which separates the Cretaceous succession from Tertiary rocks. The unconformity is most pronounced towards the south where it embraces both the late Maastrichtian and early Palaeocene intervals (Fig. 12.21). It is less pronounced and even almost conformable towards the north of the NCR (Biscay synclinorium) and in the Zumaya sub-basin, where marlstones and flysch deposition occurred (Zumaya section; Küchler & Kutz 1989; Schwentke & Kuhnt 1992; Gräfe 1994). Bathyal flysch deposits are restricted to early Maastrichtian times (lower member of Zumaya Fm), whereas the upper Maastrichtian succession is characterized by hemipelagic foraminiferal-rich red marls (upper member of Zumaya Fm, Cia Fm and Monte Urco Fm; Fig. 12.21). Latest Maastrichtian pelagic claystones and Danian limestones exhibit a well developed orbital cyclicity that allows precise dating of the

Cretaceous–Palaeogene boundary in Zumaya (Ten Kate & Sprenger 1993). An important fact, that has palaeogeographic implications for the whole of Late Cretaceous time, is that Cenomanian–Santonian flysch in the Zumaya sub-basin and around the plate boundary (Biscay synclinorium), as well as the thick siliciclastic flysch of Campanian–Maastrichtian age, were derived mainly from the east to SE, in contrast to a northern provenance for the Albian–early Cenomanian 'black flysch' (Deva Fm).

The Iberian basin (RM, MS, RS, MAFN, JLS, AG, MNM, BC, JFG-H, FO, JG, JM-Ch)

The Iberian basin was a wide intracratonic basin located in the northeastern part of Iberia (Fig. 12.1) during a period of Mesozoic crustal thinning. It was later inverted during Palaeogene Alpine convergence between Europe and Africa, to produce the present-day Iberian and Catalonian Coastal Ranges and parts of the surrounding Ebro, Duero and Tajo basins. The Mesozoic rifting of the basin mostly occurred during two episodes: Late Permian–Triassic and late Oxfordian–Early Cretaceous. These rifting phases were each followed by post-rift periods dominated by thermal subsidence that lasted for most of Jurassic and Late Cretaceous time (Salas *et al.* 2001).

The following account of Cretaceous Iberian basin evolution starts with an Early Cretaceous continental rifting phase, during which four strongly subsident sub-basins were generated (Cameros, Maestrat, Columbrets and South Iberian sub-basins) and thick successions of continental to shallow-marine carbonates and clastics were deposited. Special attention is given to the exceptional Lower Cretaceous fossil site of Las Hoyas, in Cuenca province, which has provided one of the world's best preserved records of late Barremian vertebrates and plants. This is followed by an examination of the Late Cretaceous post-rift interval (late Albian to Santonian), when the Iberian basin formed a wide but shallow seaway between the north and the south Iberian continental margins (Fig 12.1), and carbonate platforms reached their maximum extent across the basin. Finally, latest Cretaceous times are considered, when the basin experienced a broad regression related to tectonic movements that initiated the Alpine orogeny in Spain.

Lower Cretaceous of the Iberian basin (RM, RS)

The Lower Cretaceous rocks of the Iberian basin were deposited during a prolonged phase of intracontinental rifting which spanned latest Oxfordian to middle Albian times. This phase coincided with northward propagation of rifting from the central Atlantic and the gradual opening of the North Atlantic oceanic basin. Transtensional rifting in the Bay of Biscay culminated in middle Aptian crustal separation and the onset of sea-floor spreading (Ziegler 1988*a*; Le Vot *et al.* 1996; Vergés & Garcia-Senz 2001). In the Iberian Ranges this rifting phase led to the progressive destruction of the Middle and Late Jurassic carbonate platforms and the development of a new system of extensional depocentres that included the Cameros, Maestrat, Columbrets and South Iberian sub-basins (Fig. 12.22). These basins are discordantly superimposed on the Triassic Iberian rift and contain a sedimentary and structural record of three main phases of tectonic subsidence (Salas *et al.* 2001) and 13 depositional sequences (J.9, J.10 and K1.1 to K1.10, Fig. 12.22).

Synrift subsidence commenced at different times across the area: latest Oxfordian in the Maestrat sub-basin, early Tithonian in the Cameros sub-basin, and Berriasian in the South

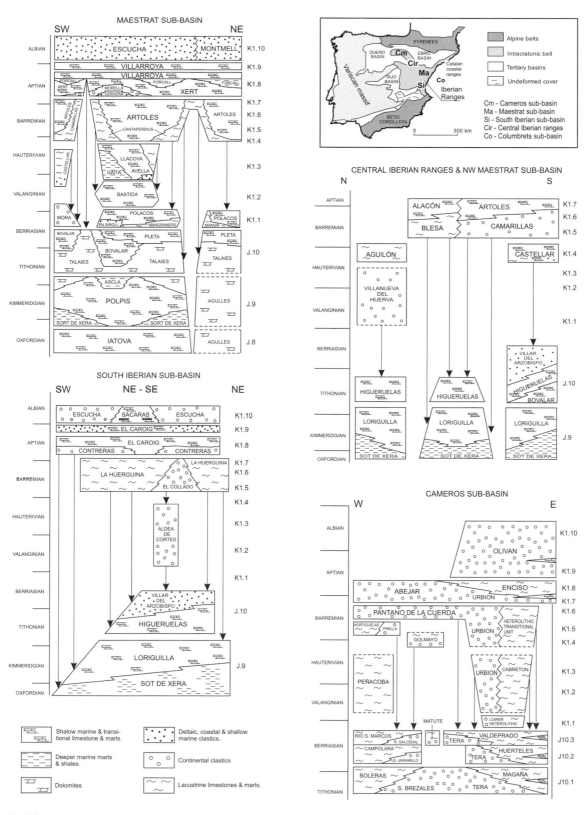

Fig. 12.22. Chrono-lithostratigraphic chart of the four main sectors of the Iberian basin during the Late Jurassic–Early Cretaceous rifting stage, showing the depositional environments, depositional sequences and lithostratigraphic units (formations). Based on Salas *et al.* (2001). Top right corner: simplified tectonic map of the Iberian peninsula showing the Iberian Ranges and the present location of the Late Jurassic–Early Cretaceous rifted sub-basins.

Iberian sub-basin (and probably the Columbrets sub-basin; Fig. 12.22). The synrift sedimentary fill of these depocentres is up to 5 km thick and bounded by two significant unconformities. The stratigraphic gap associated with the basal unconformity spans early Tithonian to Barremian times in the Cameros sub-basin (Mas *et al.* 1993), whereas in the Maestrat sub-basin it is much smaller and affects only the upper part of the latest Oxfordian *Planula* biozone (Aurell 1991). The upper boundary of the synrift succession is formed by an intra-Albian unconformity which preceded the onset of the Late Cretaceous post-rift thermal subsidence stage during which a large carbonate platform developed in northeastern Iberia.

The palaeogeographic evolution of northeastern Iberia during the Late Jurassic–Early Cretaceous rifting cycle is presented in Figs 12.22 and 12.23. During this rifting cycle, pre-existing fracture systems, which were partly inherited from Triassic rifting and the late Variscan wrench faulting, were reactivated. At the end of the previous post-rift period (early Oxfordian post-rift stage 1 of Salas *et al.* 2001), the Soria seaway extended from southeastern Iberia along the trend of the Triassic Iberian rift to the margins of the Bay of Biscay rift, thereby connecting the Atlantic with the Tethys sea (Vera 2001). This seaway was bounded to the SW by the Iberian meseta and to the NE by the Ebro high. These highs were accentuated during the Late Jurassic–Early Cretaceous rifting cycle and acted as significant palaeotopographic thresholds. Towards the southern margin of the evolving rift system, a smaller palaeohigh appeared during Early Cretaceous times and separated the Maestrat and South Iberian sub-basins (Fig. 12.23).

The Late Jurassic–Early Cretaceous evolution of the Iberian rift system can be divided into the following three stages: latest Oxfordian–late Hauterivian (146.5–117.5 Ma), latest Hauterivian–early Albian (117.5–103 Ma), and middle Albian (103–98 Ma). At the onset of the latest Oxfordian rifting, tectonically induced marine regression closed the Soria seaway (Fig. 12.23). During early Kimmeridgian times, tectonic activity increased, leading to the re-opening of the Soria seaway in response to a relative rise in sea level and the subsidence of tilted extensional fault blocks. As a consequence, the rapid drowning of the sponge-rich late Oxfordian carbonate platform occurred in the southeastern parts of the evolving rift, seen in the Maestrat sub-basin. During Tithonian times, the Soria seaway was interrupted again by the early synrift sedimentation in the Cameros sub-basin, which formed part of a set of smaller extensional sub-basins. During Tithonian to late Hauterivian times, rapid subsidence of basins forming part of the Iberian rift system led to significant palaeogeographic changes, particularly in its northwestern parts. Overall, this time interval was a regressive episode that was only interrupted by minor late Berriasian and early Valanginian transgressive pulses from the Tethys ocean. The South Iberian sub-basin developed during Berriasian times into a NW–SE trending narrow trough which became filled with terrigenous sediments, whereas the northwestern basins of the Iberian rift system coalesced and drained into the Bay of Biscay rift.

The latest Hauterivian to early Albian rifting phase was associated with a major transgression, and accelerated tectonic subsidence facilitated the transgression of the Atlantic and Tethys seas, resulting in a retreat of continental facies towards the SE and NW, respectively. This tendency towards flooding was interrupted by a short-lived early Aptian regression and terminated by a middle Albian regression (Fig. 12.23). In the Maestrat and South Iberian sub-basins, these regressions resulted in the progradation of deltaic complexes. The middle Albian renewal of crustal extension produced rapid subsidence of fault-controlled troughs and the accumulation of thick deltaic and lacustrine successions such as the coal-bearing Escucha Fm in the Maestrat sub-basin.

Maestrat sub-basin. Late Jurassic–Early Cretaceous rifting produced four main fault zones in the Maestrat sub-basin, with the faults partly soling out within a succession (>1000 m thick) of middle Muschelkalk shales and salts (Salas & Guimerà 1996). In areas where these Muschelkalk sediments are thin, the fault system is rooted in a deeper Palaeozoic detachment horizon (Roca *et al.* 1994). Rapid subsidence of fault-tilted blocks, combined with a eustatic rise in sea level, resulted in the rapid drowning of an Oxfordian sponge-rich carbonate ramp (J.8, Fig. 12.22) and the accumulation of Kimmeridgian deeper water carbonate sediments (Salas & Casas 1993; J.9). Thin bedded calcareous mudstones and sponge build-ups capped the crests of fault blocks. These facies passed vertically into anoxic basinal marls (Ascla Fm) in the evolving hanging-wall basins. The Tithonian to Berriasian sequence (J.10) is composed of assorted platform carbonates, characterized by tidal flats and fringing oolitic-bioclastic shoals, which graded basinwards into hemipelagic *Calpionella* limestones up to 1000 m thick. The Valanginian to Barremian sequence (K1.1–K1.7) is up to 1500 m thick and characterized along the basin margins by estuarine shallow-water carbonate platforms (with large freshwater discharges) where carbonate production was dominated by molluscs and calcareous algae, and oolitic bioclastic shoals and coralgal boundstones were abundant. The earliest Aptian succession (lower part of K1.8) consists of an up to 100 m thick tidally dominated delta complex, the upper delta plain deposits of which are rich in dinosaur remains (Morella Fm). During Aptian times (K1.8–K1.9) marine conditions prevailed again, producing the widespread deposition of up to 1100 m of rapidly prograding shallow-water carbonate platform sediments, characterized by orbitolinids, calcareous algae and rudists. The lower to middle Albian succession (K1.10) comprises very extensive spreads of tidally influenced deltaic sediments, up to 500 m thick, with rich coal layers (Escucha Fm; Querol *et al.* 1992).

Cameros sub-basin. This basin (Fig. 12.22) differs fundamentally from the Maestrat sub-basin in both its geometry and its sedimentary infill, consisting predominantly of alluvial and lacustrine deposits, with only very rare marine incursions (Gómez-Fernández 1992; Alonso & Mas 1993). It displays a pre-inversion synclinal geometry and is not bounded by major basement-involving faults. During its evolution, depocentres migrated progressively northwards, shown by northward onlap of its sediments over the Jurassic pre-rift series. Palinspastic reconstructions indicate that the evolution of this basin was controlled by a south-dipping ramp along a major extensional fault that soles out in the Variscan basement and displaces the hanging-wall by about 33 km to the south (Guimerà *et al.* 1995). The sedimentary succession of the basin can be divided into six unconformity-bounded main depositional sequences which correlate with the J.10–K1.10 sequences of the marine basins (Fig. 12.22; Mas *et al.* 1993). The Tithonian-Berriasian initial syn-rift sequence (J.10) consists predominantly of alluvial and lacustrine deposits that reach a maximum thickness of up to 3000 m in the eastern part of the basin, where they rest on Kimmeridgian marine strata. Further west the sequence overlies progressively older Jurassic strata. Two minor discontinuities indicate that the sequence can be subdivided into one Tithonian (J10.1) and two Berriasian (J10.2 and J10.3) successions characterized by lacustrine deposits, including shallow freshwater carbonates, which show only occasional marine influences. During the deposition of J10.3, an evaporitic playa-lake developed in the eastern part of the basin.

The Berriasian–Valanginian sequence (K1.1) occurs only at the eastern end of the basin, where it attains a thickness of up to 200 m of fluvial and lacustrine sediments. The Valanginian–Hauterivian sequences (K1.2, K1.3, K1.4) are absent in the central part of the basin. In the west part they comprise lacustrine carbonates (100 m thick), whereas in the east they are siliciclastic and mixed siliciclastic–carbonate fluvial and lacustrine sediments (500 m thick). The K1.4 sequence is only represented in the southernmost part of the basin

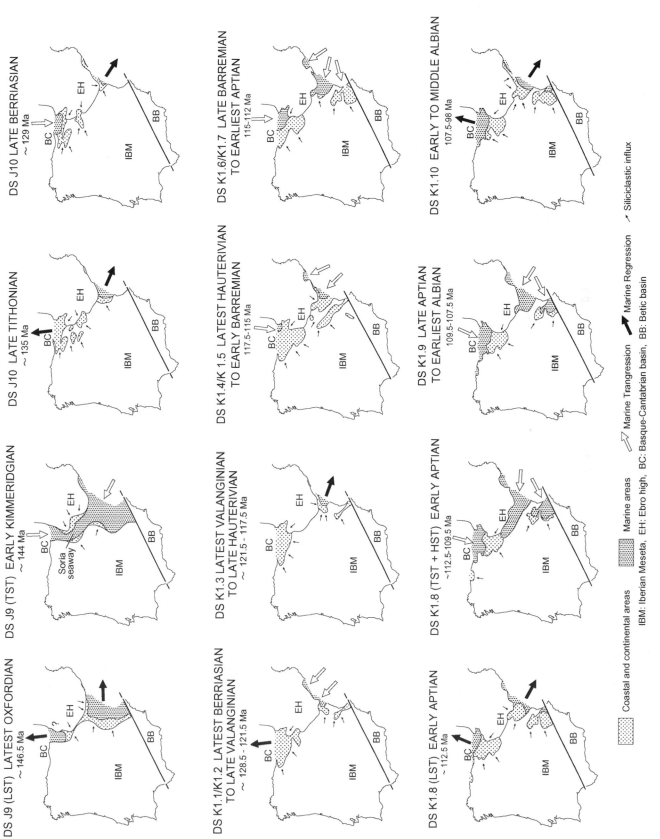

Fig. 12.23. Palaeogeographic maps showing evolution of the Iberian rift system during the latest Oxfordian to middle Albian times.

(Soria sector) in an area that underwent rapid subsidence which resulted in a succession up to 800 m thick.

The overlying Barremian succession is fluvio-lacustrine in origin, and was initially deposited in two separate depocentres to east and west (K1.5, late Hauterivian?–early Barremian). A westward expansion of sedimentation from the eastern area (Fig. 12.23) produced the fluvial sediments of the K1.6 subsequence, and a coalescence of the two depocentres, a process that was completed during the late Barremian–early Aptian interval (K1.7, K1.8). Subsidence rates across the area were different, with 1900 m of clastic fluvial deposits in the east grading NE into 1100 m of lacustrine carbonates, whereas in the west only 800 m of fluvial sediments were deposited. A marked marine influence (lagoonal deposits) can be detected in this part of the succession. Finally, during the deposition of the final late Aptian–middle Albian synrift sequences (K1.9, K1.10), sedimentation was restricted to the eastern depocentre where up to 1500 m of alluvial clastics containing rare and thin lacustrine carbonate intercalations are preserved. An important intra-Albian erosional unconformity forms the upper boundary of this sequence.

South Iberian sub-basin. This NW–SE trending, 300 km long basin is located to the south of the Maestrat sub-basin (Fig. 12.22) from which it is separated by the Valencia high. It developed during the Berriasian to middle Albian rifting phase and contains more than 2000 m of synrift sediments. The lower part of this succession is mainly developed in continental and lacustrine facies whereas the upper part consists essentially of shallow marine carbonates (Vilas *et al.* 1983). The Tithonian–Berriasian sequence (J.10), associated with the beginning of faulting and tilting of the previous carbonated ramps (sequence J.9), is dominated by shallow subtidal carbonate bars (Higueruelas Fm) and tidal flats and deltaic siliciclastic facies (Villar del Arzobispo Fm) which are up to 600 m thick in the depocentre.

The Valanginian to Hauterivian succession (K1.2–K1.3) lies unconformably over the previously described units (J.10), and was deposited in a very narrow and elongated fault-bounded trough. It comprises up to 200 m of mixed carbonate-clastic to clastic deposits, representing lagoonal, tidal flats and coastal alluvial plain environments. After this episode the basin extended laterally and the Barremian sequences (K1.5–K1.7) onlapped unconformably over the basement. These Barremian sequences locally reach up to 700 m thick, and consist of siliciclastic alluvial deposits (El Collado Fm) and shallow lacustrine carbonate facies (La Huérguina Fm) with some marine incursions from the SE. As in the Maestrat sub-basin, marine conditions prevailed during Aptian times (K1.8–K1.9) when shallow-water carbonate platform sediment (El Caroig Fm) up to 500 m thick laterally interfingered with continental siliciclastic units (Contreras Fm). The overlying early to middle Albian sequence (K1.10) is siliciclastic and comprises up to 150 m of coastal alluvial and deltaic deposits (Escucha Fm) and peri-tidal mixed siliciclastic-carbonate facies (Sácaras Fm).

Columbrets sub-basin. This depocentre is located offshore beneath the Valencia Trough (Fig. 12.22) and apparently developed mainly during the Late Jurassic–Early Cretaceous rift cycle. At present it corresponds to a NE–SW orientated syncline which resulted from its partial inversion in Palaeogene times, and shows >8 km of Mesozoic and Palaeocene rocks (Roca 1996). The sedimentary fill of this sub-basin is poorly known and only defined by seismic reflection data calibrated by a few wells. It consists of more than 8 km of Mesozoic and Palaeocene rocks (Roca 1996).

Las Hoyas: an exceptional Lower Cretaceous fossil site (MAF-M, JLS, MNM, FO)

The Las Hoyas fossil site (Fig. 12.24) is a world famous 'Konservat-lagerstätte' located in the southern part of the Serranía de Cuenca (southwestern Iberian Ranges). It belongs within the La Huérguina Limestone Fm and formed in a small basin (50 km²) linked to the early Cretaceous intracontinental rifting. This Las Hoyas basin was filled with 400 m of upper Barremian alluvial and lacustrine sediments arranged in four stratigraphic units separated by local unconformities (Frege-

nal Martínez 1998; Fregenal Martínez & Meléndez 2000). The fossil site occurs within the second of these units, which is entirely composed of carbonate lacustrine facies deposited in an extensive, perennial, shallow-water carbonate lake system that underwent strong water-level oscillations and was fed by groundwaters and ephemeral storm-generated streams in the surrounding palustrine plain. The fossiliferous facies are laminated limestones that show a rhythmic alternation of millimetre-thick dark and light laminae and were deposited in the basinal areas of the lake under permanent subaqueous and bottom anaerobic conditions. Basinal carbonate deposits resulted from bioinduced precipitation, allochthonous influxes and reworking of detrital particles that reached the bottom as dilute turbidites, as well as from microbial mats growing on the lake floor. These microbial mats, the rapid sedimentation and the anoxic to anaerobic conditions of the lake bottom are the main factors invoked to explain the exceptional fossil preservation and their associated taphonomic features (Sanz *et al.* 2001*a,b*).

The present-day knowledge of Las Hoyas biota includes 103 different organic forms, both plants and animals. The terrestrial floral assemblage is mainly composed of conifers, ferns, cycads and Bennetitales, although some evidence for both angiosperms and gnetals has also emerged. The aquatic floral assemblage is mainly made up of charophytes, with Las Hoyas exposures yielding exceptional specimens, which can include remains of the vegetative apparatus. The most abundant floral remains belong to the enigmatic genus *Montsechia*, which was probably an aquatic angiosperm.

The faunal assemblage includes both terrestrial and aquatic organisms, with invertebrate forms of the latter including molluscs (bivalves and gastropods) and arthropods (very abundant crayfish remains, some peracarids, ostracods and insects, mainly Hemiptera and Coleoptera). Fishes are represented by 19 different forms belonging to Semionotiformes, Pycnodontiformes, Amiiformes, coelacanths and primitive teleosteans, and, just recently, two tiny shark specimens have been found.

The invertebrate terrestrial faunal assemblage is composed of rare myriapods and several insect orders (Ephemeroptera, Hemiptera, Hymenoptera, Odonata, Coleoptera, Diptera, Mecoptera, Isoptera, Neuroptera, Orthoptera and Blattoidea). Terrestrial vertebrates include frogs, salamanders, albanerpetontids, turtles, lizards, crocodiles, non-avian dinosaurs and birds. The research on some of these groups has yielded significant conclusions on the evolutionary history of some tetrapod lineages (McGowan & Evans 1995; Pérez-Moreno *et al.* 1994; Sanz *et al.* 1988, 1996, 2001*a,b*)

Upper Cretaceous of the Iberian basin (MS, AG, BC, JFG-H, JG)

During Late Cretaceous times, progressive waning of tectonic activity favoured the development of a wide and shallow seaway along an intracratonic basin (Upper Cretaceous Iberian basin, UCIB) that linked the northern and southern continental margins of Iberia. The UCIB was bordered by two emergent areas, the Iberian (Hesperian) massif to the west and the Ebro massif to the NE, and was initially coincident with the present Iberian Ranges where there are excellent exposures. During the Late Cretaceous epoch, however, marine transgression extended westwards so that the UCIB sediments onlapped the Iberian massif. These sediments are mostly presently buried under the Cenozoic Duero and Tajo basins,

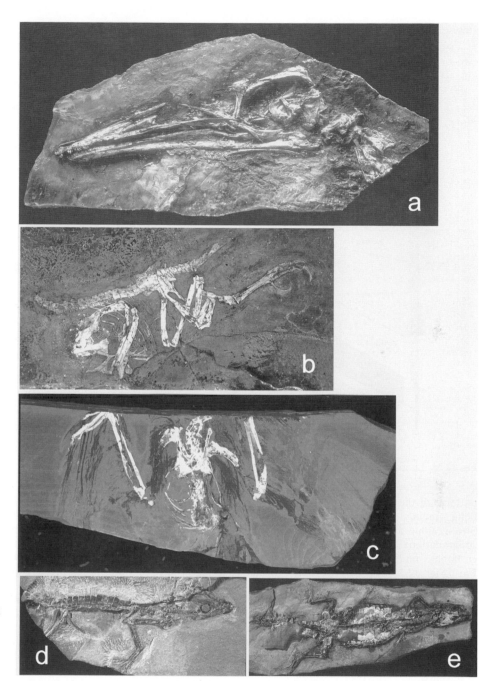

Fig. 12.24. Some vertebrate fossils from Las Hoyas fossil site (Lower Cretaceous, Iberian Ranges). **(a)** *Pelecanimimus polyodon* (Dinosauria, Ornithomimosauria). Transferred skull. **(b)** *Iberomesornis romerali* (Dinosauria, Aves). Fluorescence-induced photograph by UV light. **(c)** *Eoalulavis hoyasi* (Dinosauria, Aves). Fluorescence-induced photograph by UV light. **(d)** Las Hoyas crocodyliform (Crocodylomorpha, Crocodyliformes). **(e)** Las Hoyas Neosuchia (Crocodylomorpha, Neosuchia).

but they can be studied both along the basin margins (Cantabrian Mountains, Central Range and Montes de Toledo) as well as in oil-well data.

Sedimentation within the UCIB was controlled by both eustasy and regional tectonics. The Late Cretaceous epoch was a period of tectonic quiescence, mainly characterized by relatively slow thermal subsidence during which a thick sedimentary pile (up to 800 m) was deposited on the UCIB (Fig. 12.25). Sedimentation was thus primarily a consequence of the well documented worldwide mid-Cretaceous eustatic sea-level rise (García *et al.* 1989*b*, 1996). This important transgression enhanced the more local effects of thermal subsidence and sediment loading. The eustatic sea-level rise

reached a maximum rate during the Cenomanian stage and the sea attained its highest level in early Turonian times (Haq *et al.* 1988), when shallow seas covered vast continental areas. Regional tectonic effects were never so important as to obliterate these background eustatic changes which can, in many cases, be traced throughout the entire basin (García *et al.* 1993, 1996; Alonso *et al.* 1993) and correlated with other basins of Iberia (Martín-Chivelet 1992; Gräfe 1999).

Eustatic sea-level changes (low and high order) produced both deepening- and shallowing-upward sequences (from second to fifth order) that spread along the entire UCIB (Fig. 12.26). Sequence boundaries are easily recognized at the basin margins due to minor erosion, coastal onlap and toplap

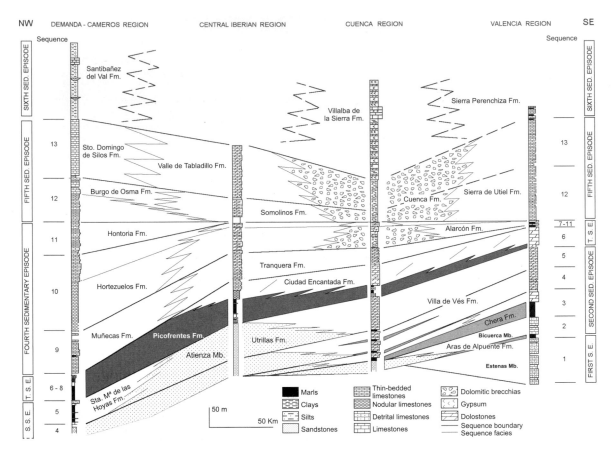

Fig. 12.25. Lithostratigraphic units in the Late Cretaceous Iberian basin (UCIB).

patterns of the high-order sequences, and shifts in the facies belts (mainly basinwards in relation to sea-level falls). Stratigraphic architecture is mainly aggradational, but two progradational events are recorded by the presence of thick carbonate units bearing clinoforms: the Ciudad Encantada Fm (Fig. 12.25) of early–middle Turonian age, and the Hortezuelos–Hontoria formations (Fig. 12.25) of late Coniacian–Santonian age.

The lower part of the Upper Cretaceous succession (late Albian to early Campanian) is mainly composed of shallow marine limestones with intercalated sandstones and marls (Fig. 12.25). Diagenetic processes related to the sequence boundaries, however, induced pervasive dolomitization of many of the limestone units. Sandstones are usually located both at the base of the succession and at the margins of the basin (Figs 12.25 and 12.26); they are currently interpreted as landward stepping coastal clastic wedges, mainly along the Iberian Massif margin of the basin (García *et al.* 1989*b*), and have complex indentations with the carbonate units (Fig. 12.25). On the other hand, marly units are either shallow marls with amalgamated palaeosols (e.g. Chera marls; Fig. 12.25), or hemipelagic deposits with ammonites (Picofrentes and Hortezuelos marls; Fig. 12.25) recording the deepest events in the UCIB, which occurred at the Cenomanian-Turonian boundary (Picofrentes marls) and during the Coniacian stage (Hortezuelos marls). The upper part of the Upper Cretaceous succession (early Campanian to Maastrichtian) consists of mudstones and evaporites with minor limestone intercalations (Fig. 12.25).

The Upper Cretaceous succession is organized into a large number of formal lithostratigraphic units (Fig. 12.25; Canerot 1982; Floquet *et al.* 1982; Vilas *et al.* 1982; García *et al.* 1989*b*; Floquet 1991; Segura *et al.* 1999), and has a maximum thickness of about 800 m, although with important variations across the basin. Biostratigraphic data are poor, except for some intervals, mainly the deep-water marls of the Picofrentes and Hortezuelos formations. Correlations are mainly based on lithostratigraphy and sequences (Fig. 12.26). Many sequence boundaries can be traced along the UCIB (Fig. 12.26), allowing the use of fossiliferous data to constrain sequence ages. Several fossil groups have been used for chronostratigraphical purposes, including ammonites (Wiedmann 1964; Floquet 1991), benthic foraminifera (Fourcade & García 1982; Calonge 1989; Giménez 1989; Floquet 1991; Schroeder *et al.* 1993; Gischler *et al.* 1994; Calonge *et al.* 1996) and rudists (Floquet 1991; Segura *et al.* 2000).

In the UCIB the stratigraphic record can be divided into six sedimentary episodes, typically bounded by unconformities as detailed below (Fig. 12.26). The mostly unconformable boundaries of these sequences can be traced across the basin and were produced by both eustatic and tectonic events. Tectonism was an especially important influence on palaeogeographic changes in the basin during four time intervals: (i) Late Albian, which represents the cessation of the previous rift stage and the start of the post-rift interval; (ii) Middle to late Cenomanian, when the Iberian plate tilted northwards (e.g. Giménez 1989; Segura *et al.* 1993*a,b*; Alonso *et al.* 1993) in response to the opening of the Bay of Biscay and the

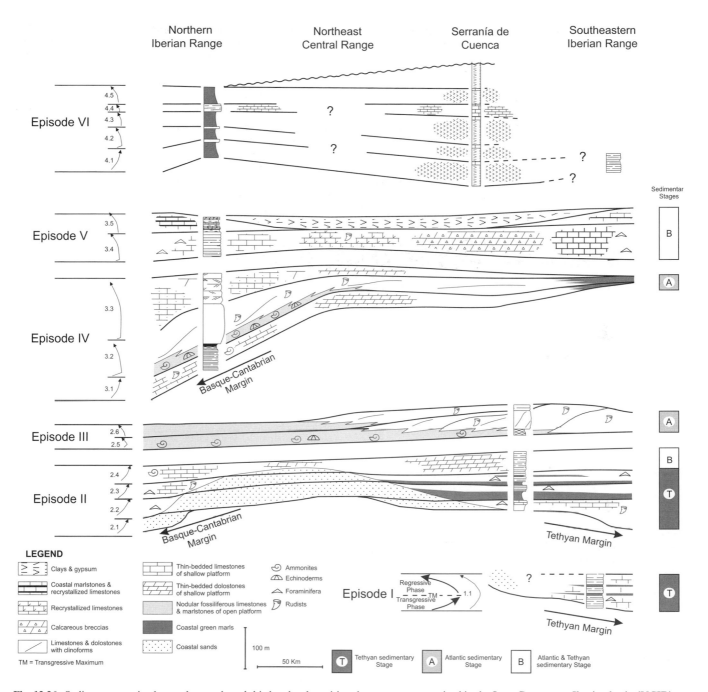

Fig. 12.26. Sedimentary episodes, and second- and third-order depositional sequences recognized in the Late Cretaceous Iberian basin (UCIB).

anticlockwise rotation of the Iberian plate; (iii) Santonian, when a new phase of tilting occurred, but this time towards the south; (iv) Early Campanian, when basin inversion began in response to the first Alpine compressive movements.

First sedimentary episode (late Albian). This episode (Fig. 12.26) was restricted to the southeastern areas of the Iberian Ranges. Sedimentation reflects the late Albian sea-level rise and it is represented by just one sequence, deposited under coastal conditions (Fig. 12.26). The base of the episode is a major sedimentary break and rests on either the Escucha Fm, in the Maestrat sub-basin, or the Sácaras Fm, in the South Iberian sub-basin. The sequence is mainly composed of brown, sandy limestones facies (Estenas Mb; e.g. Carenas *et al.* 1994) grading upwards to low-energy marls (Losilla Mb; e.g. García *et al.* 1993).

Second sedimentary episode (late Albian–late Cenomanian). During this episode (Fig. 12.26) sedimentation started mainly on the Tethyan margin of the basin (Aras de Alpuente Fm), but subsequent gradual, multiepisodic transgression over the UCIB took place at both its northern and southern margins (early and middle Cenomanian), with both margins eventually becoming connected by the end of this episode (late Cenomanian; Fig. 12.26). Mixed terrigenous–carbonate shallow platform facies initially overlap continental–coastal silici-clastic sediments in the emergent areas. Sea-level rises and falls produced an alternation of tabular, open-platform shallow-water carbonate bodies (Alatoz Fm and the lower part of the Villa de Vés Fm; Fig. 12.25), deposited during highstands, and wide, thin low-energy coastal marls (Losilla, Chera, Pinarueco, Poveda and Pozuel marls deposited during lowstands; Fig. 12.25; e.g. Carenas *et al.* 1989, 1994). Towards the emergent areas, mainly towards the Iberian

Massif, the marls grade laterally into coastal clastic wedges (part of the Utrillas Fm; Fig. 12.25), which were also deposited during lowstands indicating source area reactivation. In the northern part of the UCIB, where links existed with the Atlantic Ocean, coeval successions are represented by the Utrillas and Santa María de la Hoyas formations. In the late Cenomanian interval the basin tilted northwards, causing basin levelling. Transgressions connected the Atlantic and Tethyan margins and extended farther west and NE, towards the emergent Iberian and Ebro massifs, to produce a broad and shallow carbonate platform (upper part of the Villa de Vés Fm; Fig. 12.25) with a narrow, terrigenous coastal fringe restricted to the Iberian margin.

Third sedimentary episode (late Cenomanian–middle Turonian). During this episode (Fig. 12.26) renewed northward tilting of the basin produced a major palaeogeographic change. The evolution of the former platform was abruptly interrupted and almost the entire basin was invaded by a rapid Atlantic transgression. Sedimentation was mainly related to the Atlantic margin, with communication with the Tethyan margin remaining restricted (Fig. 12.26). This major change took place at the same time as an important sea-level rise (García *et al.* 1989*a*), and the basin became very deep in its northern part, whereas the southern areas showed uplift or very low subsidence (Martín-Chivelet & Giménez 1993). Sedimentation started with the strongly transgressive deposits of the Picofrentes Fm (Fig. 12.25), a unit composed of ammonite-bearing hemipelagic marls that locally rest on shallow-water coastal sediments (Utrillas Fm) of the previous episode. The Picofrentes marls grade both laterally and upwards into a strongly prograding carbonate platform (Ciudad Encantada Fm; Fig. 12.25), which was deposited in the southern part of the basin and records rapid northward progradation linked to early–middle Turonian sea-level fall (Segura *et al.* 1989, 1993*a*; García-Hidalgo *et al.* 1997). The Ciudad Encantada Fm grades farther south into a thin unit of marly limestones with palaeosols (Alarcón Fm; Giménez 1989; Martín-Chivelet & Giménez 1993; Segura *et al.* 1993*a*). This episode ends with basinward progradation of a thin coastal clastic wedge (Somolinos sands; Fig. 12.25).

Fourth sedimentary episode (mid Turonian–Santonian). The palaeogeography of this episode (Fig. 12.26) was similar to the previous one, with the basin being open to the Atlantic margin but with restricted or blocked access to the Tethyan margin (Fig. 12.26). Sedimentation began with the deposition of shallow-water carbonate rocks of the Tranquera and Muñecas formations (Fig. 12.25), linked to a new sea-level rise. Deepening of the basin induced the deposition of deeper water nodular limestones and marls with abundant ammonites, echinoderms and pelecypods (lower part of the Hortezuelos Fm; Fig. 12.25; Floquet 1991). These marls grade upwards and laterally southwards into thick-bedded, massive, poorly stratified limestones (upper part of the Hortezuelos Fm; Fig. 12.25) and thick-bedded limestones with clinoforms (Hontoria Fm; Fig. 12.25) that are currently interpreted as rapid progradation deposits related to a sea-level stillstand and fall.

These units thin southwards and southeastwards grading laterally into a thin marly unit with palaeosols (Alarcón marls; Martín-Chivelet & Giménez 1993). This episode ends with a major palaeogeographic change in the basin.

Fifth sedimentary episode (late Santonian–earliest Campanian). Southward tilting induced a restructuring of the basin (Fig. 12.26) and was accompanied by a general increase in the subsidence rate which, in conjunction with a global eustatic rise (Haq *et al.* 1988), reestablished a marine connection between the northern and southern margins of the UCIB (Fig. 12.26). Thick shallow-water carbonate units were deposited in response to these changes, and were characterized by a predominance of inner-platform, tidal and sabkha deposits (Cañadilla, Burgo de Osma, Cuenca and Sierra de Utiel formations; Fig. 12.25) facies. This new palaeogeography persisted until latest Santonian or earliest Campanian times, when the first Pyrenean compressive events took place. The onset of compression subdivided the UCIB into several highs (some of which became important emergent areas) and troughs, each with an increasingly separate sedimentary record, which makes correlation difficult (Fortanete, Santo Domingo de Silos, Valle de Tabladillo, Sierra Perenchiza formations; Fig. 12.25), so that stratigraphic relationships between these formations remain poorly understood.

Sixth sedimentary episode (early Campanian–Maastrichtian). This episode (Fig. 12.26) started with an important marine regression both towards the north and south (e.g. Alonso *et al.* 1987), linked to an end-Cretaceous relative sea-level fall so that continental deposits became increasingly important. In the central area of the basin this final Cretaceous succession comprises clays and marls with gypsum intercalations (Fig. 12.25), and minor limestone and dolostone beds (Villalba de la Sierra Fm; Meléndez 1971). In the Torralba, Tribaldos and Santa Bárbara oil wells, located in the Tertiary Tajo basin immediately west of the Iberian Ranges, five gypsum–limestone or dolostone cycles can be recognized, with a total thickness of about 200 m. In the northern Iberian basin coeval deposits comprise clays and limestones deposited in coastal or continental environments (Santibañez del Val Fm; Floquet 1991), whereas in the southern Iberian basin a thick unit of lacustrine limestones and red clays were deposited. At the end of the Cretaceous period, shallow marine deposits remained only in the southernmost and northernmost areas of the basin, in the transitions to the Iberian continental margins.

The editors W. Gibbons and T. Moreno are gratefully acknowledged for their invitation to participate in this book and their careful help in the editing of the manuscript. T. Simó and D. Sanders are thanked for their critical reviews, which greatly helped in improving the original manuscript. J. López-Gómez, M. Losantos, E. Pi and four other anonymous colleagues are thanked for their comments and suggestions on the earlier versions of different parts of the chapter. Also, X. B., E. C. and C. P. want to thank O. Ballespi and D. Maynou for their help in the preparation of some of the figures.

13 Tertiary

ANA M. ALONSO-ZARZA, ILDEFONSO ARMENTEROS, JUAN C. BRAGA,
ARSENIO MUÑOZ, VICTORIANO PUJALTE, EMILIO RAMOS (coordinators),
JULIO AGUIRRE, GASPAR ALONSO-GAVILÁN, CONCHA ARENAS,
JUAN IGNACIO BACETA, JESÚS CARBALLEIRA, JOSÉ P. CALVO,
ANGEL CORROCHANO, JOAN J. FORNÓS, ANGEL GONZÁLEZ,
ARÁNZAZU LUZÓN, JOSÉ M. MARTÍN, GONZALO PARDO, AITOR PAYROS,
ANTONIO PÉREZ, LUIS POMAR, JUAN MANUEL RODRIGUEZ &
JOAQUÍN VILLENA

Tertiary (Palaeogene and Neogene) deposits crop out widely across both the Iberian peninsula and the Balearic Islands (Fig. 13.1), and record a dramatic sequence of events during plate convergence. The anticlockwise rotation of an initially isolated Mesozoic Iberian plate was followed by late Cretaceous–Cenozoic interaction with both the European and African plates. This ultimately created two great Alpine mountain belts (Pyrenean-Basque-Cantabrian and Betic-Balearic) (Fig. 13.1), each of which generated major Cenozoic foreland basins (Ebro and Guadalquivir). Away from these mountain belts, two large Cenozoic intraplate depressions (Duero and Tajo basins) flank a central horst (Central Range). Another important group of depocentres occurs within a string of Neogene grabens situated along the eastern side of mainland Spain (Fig. 13.1), forming part of a long-lived and still-active extensional system linking the Valencia trough with the Rhine and Rhone grabens in Germany and France. Further SE, Neogene extension propagated from the Valencian trough into the southern Betic orogen and created a series of basins from Alicante to Granada and beyond. Tertiary sedimentary rocks in Spain were thus deposited during and after Alpine compression in the Iberian area. This chapter summarizes the main characteristics of these sediments, moving broadly from north to south, a direction reflecting the diachronous shift in Cenozoic Alpine deformation from the Pyrenees to the Betic-Balearic region.

Western Pyrenees and Basque-Cantabrian region (VP, JIB, APa)

The Palaeogene and Neogene periods are represented in the western Pyrenees and Basque-Cantabrian region by a wide variety of sedimentary rocks, from coarse grained alluvial conglomerates to deep water marine hemipelagites. Deposition of these rocks took place during the long interval of compressional tectonism caused by the diachronous collision and partial subduction of the Iberian plate under the European plate (ECORS-Pyrenean Team 1988), a process that brought about a change from generally marine to mostly terrestrial conditions. Throughout early Palaeogene times (early compressional phase) the Pyrenean domain was an east–west trending elongate interplate embayment, opening westwards into the palaeo-Bay of Biscay, and surrounded elsewhere by shallow shelf areas (Fig. 13.2A; Plaziat 1981). The deepest part of that embayment (i.e. the so-called Basque basin) was created in early Campanian time, at the beginning of the

Pyrenean convergence, by enlargement and coalescence of precursor sub-basins (Pujalte *et al.* 1998). However, further convergence during Eocene times led to progressive emergence of the Pyrenean axial zone, with subdivision of the former single basin into northern Pyrenean and southern Pyrenean foreland basins, and to expansion of terrestrial conditions (Fig. 13.2B). Additional thrusting and uplift during late Eocene and Miocene times (late compressional phase) further reduced the marine areas and eventually resulted in basin inversion with concurrent removal of an important fraction of the original strata (Fig. 13.2C and D). Remaining Cenozoic deposits are exposed today in several disconnected outcrops, the largest of which are depicted in Figure 13.3. Early Palaeogene deep water deposits of the Basque basin are preserved in the Biscay syncline and in the Gipuzkoa monocline (Fig. 13.3). Coeval shallow water, slope and base-of-slope deposits of the southern Iberian margin, along with late Palaeogene and Neogene deposits, are preserved in the Urbasa-Andía, Miranda, Villarcayo, San Román and San Vicente de la Barquera synclines (Fig. 13.3). Palaeogene deposits of the northern margin of the Basque basin, and post-tectonic Oligocene and Neogene accumulations, are known to occur offshore in the Bay of Biscay (Fig. 13.3; Soler *et al.* 1981), but these are still poorly documented and will not be discussed here.

The Palaeogene and Neogene succession of the western Pyrenees and Basque-Cantabrian region can be subdivided into a number of packages, each recording a particular set of tectonic conditions. These packages are usually (but not always) separated from each other by stratigraphic discontinuities (particularly prominent within terrestrial and shallow marine strata), linked to variations of the rate of tectonic shortening and, in some cases, to major sea-level changes (Fig. 13.4). An outline of the Tertiary evolution of the study area, and a summary of the features of each of these stratigraphic packages, are given below.

Palaeocene

The Palaeocene epoch was a time of overall transgression in the study area. Major transgression had begun in latest Maastrichtian times, reversing a previous regressive trend spanning the Campanian to early Maastrichtian interval. Along the Iberian margin (e.g. Urbasa-Andia syncline), relative rise of sea level initially led to the encroachment of carbonate ramps (latest Cretaceous and early Danian) which

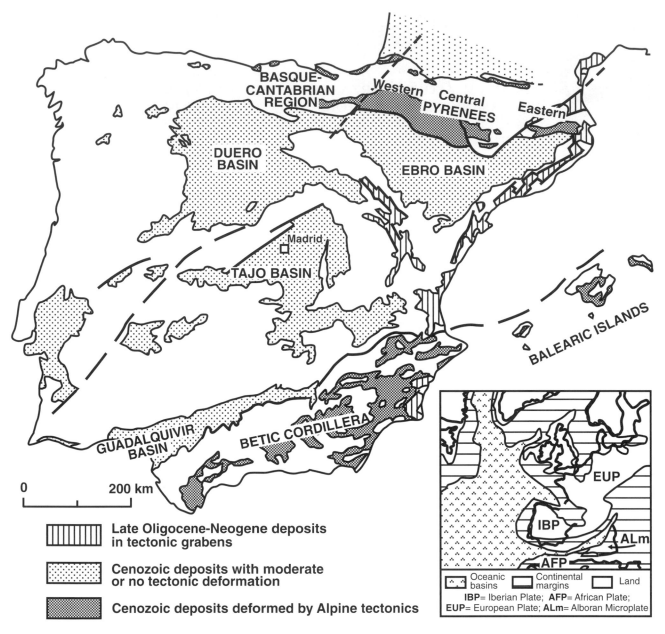

Fig. 13.1. Map showing Cenozoic outcrops in mainland Spain, the Balearic Islands and neighbouring areas. Inset: Palaeocene palaeogeography of the North Atlantic domain (modified from Zeigler 1990).

became transformed into a rimmed carbonate platform following the growth of a coralgal reef belt during late Danian times. This reef belt separated the contemporary shallow and deep water domains (Fig. 13.4), persisting as a distally steepened ramp even after continued transgression forced a backstep during Thanetian time. A number of relative sea-level falls, superimposed on the overall transgression, have allowed subdivision of the uppermost Cretaceous–Palaeocene succession into eight depositional sequences (Pujalte *et al.* 2000). The most significant of these sea-level falls, occurring between Danian and Thanetian times, led to a long subaerial exposure of the Danian platform, creating a prominent stratigraphic boundary (i.e. the 'Mid-Palaeocene Unconformity'; MPU in Fig. 13.4) and inducing extensive diagenetic alteration of the Danian carbonates (Baceta *et al.* 2001).

Seaward from the palaeoshelf edge, Palaeocene successions are typified by stacks of carbonate breccias and turbidites (5 in Figs 13.4 and 13.5) that record accumulation in base-of-slope aprons of materials resedimented from the shallow ramps and platforms. Deposition of breccias has been interpreted as having only been important during lowstand intervals (particularly during the middle Palaeocene sea-level fall), whereas shedding of carbonate turbidites seems to have taken place during both lowstand and highstand periods. As a rule, shallow water and base-of-slope successions are separated by a narrow belt (about 2–4 km wide) where the main Palaeocene deposits are channelized accumulations of coarse grained carbonate breccias. This belt is interpreted as a gullied upper slope (Fig. 13.5). Base-of-slope breccias and turbidites pinch out basinwards between the hemipelagic

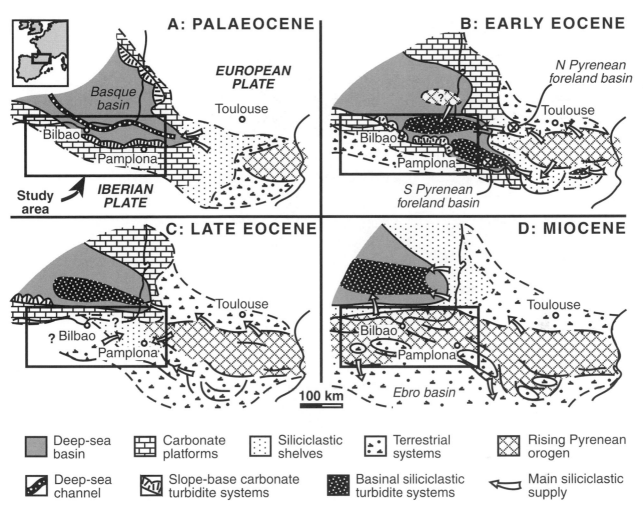

Fig. 13.2. Palaeogeographic evolution of the Pyrenean and Basque-Cantabrian region during the Tertiary era (based on Plaziat (1981) and our own data).

alternations of limestones and marls that typified the basin floor setting (6 in Figs 13.4 and 13.5).

In the deep water Basque basin the following succession is observed: (i) a 1700 m thick Campanian–lower Maastrichtian siliciclastic flysch unit (2 in Fig. 13.4), interpreted as deposited during a preceding phase of overall regression; (ii) a comparatively thin upper Maastrichtian accumulation of hemipelagic marls, limestones and thin-bedded turbidites (up to 135 m; 3 in Fig. 13.4), interpreted as laid down during the initial phase of overall transgression; and (iii) an even thinner Palaeocene succession of alternating hemipelagic limestones and marls (130 m; 6 in Fig. 13.4) that is coeval with the main phase of transgression. In addition to a strong reduction of siliciclastic input, these lithological variations reflect a sharp drop in sedimentation rates, from about 11 cm/ka (compacted) during the Late Cretaceous regressive phase, to 4 cm/ka during initial transgression, and just 1.3 cm/ka during the main Palaeocene transgression. Sedimentation rates were particularly low (around 0.8 cm/ka) during Danian times, when the Basque basin is considered to have been starved of terrigeneous sediments (Pujalte et al. 1998).

Coincidentally with the time of maximum starvation, an erosional deep-sea channel was excavated along the axis of the

Basque basin (Fig. 13.2). Concentrated and confined turbidite currents flowed through this channel and deposited coarse grained turbiditic deposits of both carbonate and siliciclastic nature (7 in Fig. 13.4). Field data demonstrate that the channel stretched along the entire length of the basin, being at least 200 km long, about 8–10 km wide, and up to 300 m deeper than the surrounding 1000 m deep sea floor. Once created, the deep-sea channel persisted as a prominent geomorphologic feature of the Basque basin until the end of the Palaeocene epoch.

Earliest Ypresian (early Ilerdian)

The overall Palaeocene transgression peaked during earliest Ypresian (early Ilerdian) times, causing the drowning of the former platform margin, pushing the carbonate factory landward, and prompting an increase in carbonate production. As a consequence, the early Ilerdian interval is represented in inner platform settings by a widespread facies of bioclastic grainstones and packstones rich in larger benthic foraminifers, notably alveolinids, which is recognizable in most of the Pyrenean domain (i.e. the Alveolina Limestone Formation (Fm); 8 in Fig. 13.4). Towards the outer platform this unit

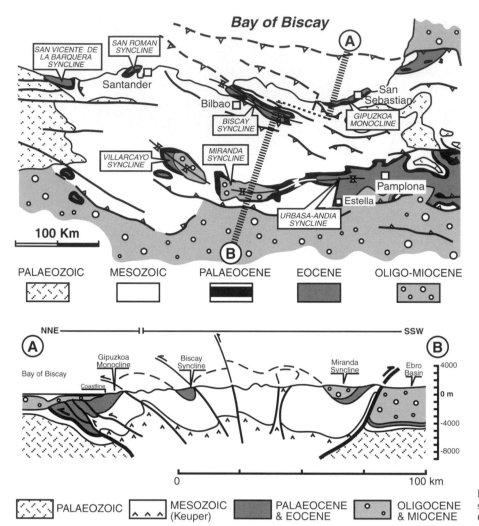

Fig. 13.3. Simplified outcrop map and cross-section of the western Pyrenees and Basque-Cantabrian region.

grades to nummulitid-rich grainstones and, further seawards, to alternations of marls and limestones with abundant planktonic foraminifers which drape slope and base-of-slope settings (Unanu Fm; 9a in Fig. 13.4). Coeval deeper water successions, in addition to marl–limestone alternations, contain numerous intercalations of thin-bedded turbidites. In base-of-slope settings, these turbidites are exclusively of a bioclastic carbonate nature (often including nummulitids and alveolinids), and thus were clearly shed from the flanking carbonate platforms. In basinal settings, however, turbidites are predominantly of siliciclastic composition, probably transported by WNW-flowing axial currents coming from incipient reliefs on the eastern Pyrenees (Hondarribia Fm; 9b in Fig. 13.4). Increase in both hemipelagic production and turbiditic input greatly enlarged deep-sea sedimentation rates in early Ypresian time. This resulted in a rapid smoothing out of the submarine topography that had typified the Basque basin during most of Palaeocene time, including the burying of its conspicuous deep-sea channel.

Different lines of evidence suggest that tectonism was subdued in the study area during the Palaeocene and earliest Ypresian interval. For instance, eight Palaeocene depositional sequences can be recognized with similar thickness and facies arrangement over most of the Pyrenean domain (Pujalte *et al.* 2000; Payros *et al.* 2000*b*). Also, the position of the platform

margin, the base-of-slope apron and the deep-sea channel remained essentially stable throughout the whole interval (Fig. 13.5). Finally, independent evidence afforded by Vergés *et al.* (1995) indicates that the Palaeocene–earliest Ypresian interval was characterized in the Pyrenees by a very low rate of tectonic shortening. Interestingly, oxygen isotope data compiled by Abreu *et al.* (1998) from Deep Sea Drilling Project/Ocean Drilling Program sites and from well-dated outcrop profiles show a distinctive trend towards lighter isotopic values from late Maastrichtian to early Ypresian time, a clear indication of a climatic warming. It is considered likely, therefore, that the coeval transgression described above was essentially driven by a long-term eustatic sea level rise.

Early Ypresian–mid-Lutetian

Compressional movements accelerated sharply during early Ypresian and remained high until at least middle Lutetian times (Vergés *et al.* 1995) forcing dramatic modifications in the palaeogeographic scenario (Figs 13.2B and 13.5). However, due to the oblique convergence of the Iberian and European plates, the Pyrenean orogeny was a diachronous east-to-west process, so that these palaeogeographic changes were introduced gradually, as tectonism propagated westwards (Fig. 13.5).

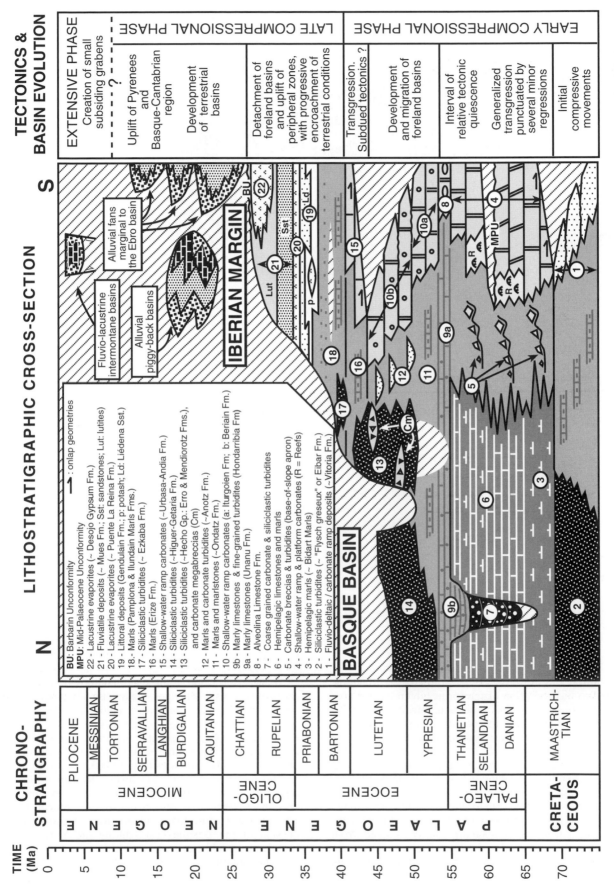

Fig. 13.4. Chronostratigraphic framework of the latest Cretaceous-Tertiary succession of the study area, indicating the main lithostratigraphic units mentioned in the text (encircled numbers).

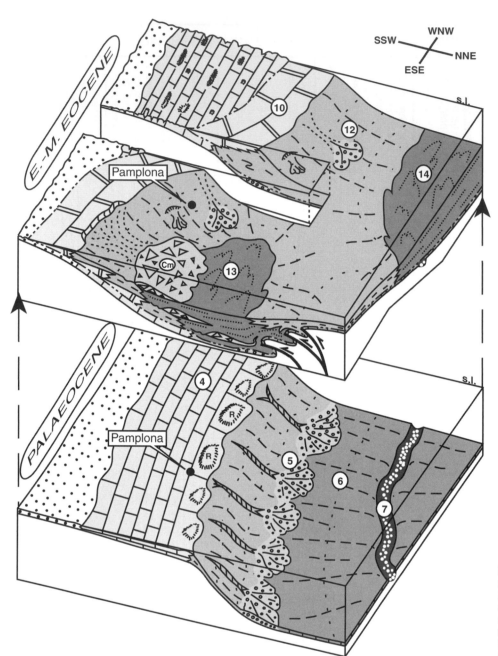

Fig. 13.5. Three-dimensional reconstruction of a segment of the Iberian margin and adjacent basinal areas during Palaeocene and early–middle Eocene times (not to scale). Encircled numbers refer to lithostratigraphic units in Figure 13.4. Cm, carbonate megabreccias; R, reefs.

In the central and western Pyrenees, the increased tectonic activity led to incorporation of the former deep basinal area into the orogenic wedge. This process involved uplift of the future Pyrenean axial zone, the development of northward- and southward-directed thrust systems, and the creation of foreland basins on both sides of the growing orogen (i.e. the north Pyrenean and south Pyrenean foreland basins; Fig. 13.2B). In each of these foreland basins, two zones can be differentiated, a deep water foredeep and a flanking shallow water shelf. The south Pyrenean foredeep was infilled with the Hecho Group (Gp) (13 in Figs 13.4 and 13.5), a thick wedge of siliciclastic turbidites (*c.* 4000 m) clearly sourced from fluviodeltaic systems located to the east. The north Pyrenean foredeep was also a site of active turbiditic accumulation, represented by the Higuer-Getaria Fm (14 in Figs 13.4 and

13.5). This is a >2500 m thick unit composed mostly of siliciclastic turbidites, well exposed along the Gipuzkoa coast (NE part of the Basque-Cantabrian region; Fig. 13.3). Detailed analysis by several authors (e.g. Van Vliet 1978; Crimes 1976; Pujalte *et al.* 2000) has demonstrated that these turbidites were in part derived from source areas situated to the north and NE of the present-day outcrops (Fig. 13.2B). Therefore, although roughly the same age as the Hecho Gp, the Higuer-Getaria Fm represents a turbiditic system different from, and largely unconnected with, the one in the south Pyrenean foredeep. This provides additional evidence for the progressive subdivision of the Basque basin during the Ypresian–middle Lutetian interval (Figs 13.2B and 13.5).

Carbonate ramps developed on the Iberian cratonic margin of the south Pyrenean foreland basin, whose constructional dynamics was also largely controlled by the coeval tectonism. In the area situated to the east of Pamplona, propagation of loading linked to the growing orogenic wedge determined a southward migration of the foredeep depocentre and a concurrent flexural subsidence of the Iberian margin. Migration of the foredeep is recorded by the onlap of the Hecho Gp turbiditic wedge to the south (Labaume et al. 1985), whereas the margin subsidence is reflected in the stepped retreat of the shallow carbonate ramps (Fig. 13.5; Barnolas & Teixell 1994). This evolution was episodic, as demonstrated by the alternating periods of ramp growth and drowning (Barnolas & Teixel 1994; Payros et al. 1999). Retreat of the margin was also punctuated by large-scale collapses of the carbonate ramps, probably triggered by high-magnitude earthquakes. Such resedimentation events are recorded by spectacular accumulations of carbonate megabreccia sheets intercalated within the Hecho Gp (Cm in Figs 13.4 and 13.5), individual examples of which can be 200 m thick and extend for >150 km. Precise age dating has demonstrated that these carbonate megabreccias are temporarily clustered into four distinctive time intervals (Payros et al. 1999), providing further evidence of the episodic nature of the southward displacement of the foreland basin.

Flexural subsidence of the Iberian margin diminished westwards and, eventually, it became zero along a hinge zone situated in the Urbasa-Andia syncline, to the west of Pamplona (Figs 13.2 and 13.5). Landward of this hinge zone, the sea floor was progressively uplifted, leading first to its gradual shoaling and later to its subaerial exposure, causing the karstification of Palaeocene–lowermost Ypresian carbonates. Seaward of the hinge zone, however, the margin was gently tilted towards the basin, and the development of carbonate ramps was not interrupted (10 in Fig. 13.5).

Gentle tilting of the margin during most of Ypresian time resulted in just mild progradation recorded by the Iturgoien and Ondatz formations (10a and 11 in Fig. 13.4), which represent respectively the proximal and oversteepened distal parts of bioclastic carbonate ramps. An increased tilting rate during late Ypresian to middle Lutetian times reduced the production of shallow water carbonates to a comparatively narrow belt which rapidly prograded seawards. Such progradation is best seen in the bioclastic carbonates of the Beriain Fm (10b in Fig. 13.4), a unit typified by a low-angle, offlapping clinoformal geometry denoting a rapid regression. Inspection of such geometry also reveals several angular unconformities, interpreted as indicative of a pulsating tectonic uplift and progradation.

The late Ypresian to middle Lutetian regression/progradation is also reflected in the deep ramp/slope deposits of the Anotz Fm, a unit mainly comprising thick (and commonly slumped) accumulation of marls and marlstones rich in planktonic foraminifers (12 in Figs 13.4 and 13.5). The Anotz Fm also contains four discrete lithological units made up of stacked bioclastic grainstones rich in nummulitids and other shallow water fossils, but which also display abundant graded bedding and Bouma sequences, demonstrating deposition from turbiditic currents. Despite their fossiliferous content, these units are therefore interpreted as carbonate turbiditic lobes accumulated at the base of the slope. Abundant channelling and cut-and-fill features indicate that they are the result of small-scale but frequent resedimentation events, while palaeocurrents attest to provenance from the proximal ramp carbonates of the Beriain Fm. Interestingly, the age of

the turbiditic lobes matches those of the time intervals in which the south Pyrenean carbonate megabreccias are clustered (Payros et al. 1999), and also seems to be coeval with the intraformational unconformities of the Beriain Fm. It is likely, therefore, that the whole area was influenced by the same episodes of active tectonism, although east and west of Pamplona the results were quite different (Fig. 13.5).

Late Lutetian

The so-called 'Biarritzian transgression' (Plaziat 1981) took place in the Pyrenees and Basque-Cantabrian area during late Lutetian time. In the study area, its most obvious result was a dramatic onlap of the Urbasa-Andia Fm (15 in Fig. 13.4) onto tilted and progressively older karstified carbonates. This onlap, best seen in the type area of the unit (the Urbasa-Andia syncline; Fig. 13.3), is interpreted as recording the marine reflooding of the area uplifted during the late Ypresian–middle Lutetian episode, and eventually brought about the superposition of upper Lutetian carbonates directly on Cretaceous deposits. At the same time, resedimentation of clastic materials into basinal areas was reduced. For instance, the Erize Fm (16 in Fig. 13.4), which is laterally equivalent to the Urbasa-Andia Fm, is represented by a homogeneous accumulation of marls with only thin and scattered intercalations of carbonate turbidites. Also, siliciclastic turbidites of the Hecho Gp became progressively less frequent, eventually disappearing during latest Lutetian times (Fig. 13.4). Finally, in the San Vicente de la Barquera syncline (NW of the Basque-Cantabrian region; Fig. 13.3), middle Lutetian shallow ramp carbonates rich in larger foraminifers are overlain by upper Lutetian ('Biarritzian') marls with abundant planktic foraminifers, providing clear evidence of a rapid deepening in depositional conditions readily attributable to coeval transgression. The eustatic and/or tectonic origin of this 'Biarritzian transgression' has not yet been analysed. Nevertheless, the seeming absence of internal unconformities in shallow water deposits of this age, and the rarity of resedimentation events in deeper water settings, suggests that late Lutetian tectonism was more subdued than during the previous early Ypresian–middle Lutetian interval.

Bartonian–Priabonian

Active tectonics resumed in the study area during the Bartonian age. As a result, the south Pyrenean foreland basin became gradually incorporated into the orogenic thrust sheets, while the western part of the Pyrenean axial zone, and parts of the Iberian margin, emerged to form subaerial conditions (Fig. 13.2C). Rivers flowing from the newly created land delivered increasing amounts of sediments to the remaining marine areas via prograding deltaic systems, causing progressive shallowing. Thus, a 1500 m thick Bartonian–Priabonian succession accumulated in the Pamplona area, recording the overfilling of the south Pyrenean foreland basin. The lower part of this shallowing-up succession is represented by the Ezkaba Fm (17 in Fig. 13.4), a channel-levee, siliciclastic turbiditic system fed directly from the Pyrenean axial zone (Payros et al. 1997). The Ezkaba Fm is intercalated within, and overlain by, the Pamplona Marls Fm (18 in Fig. 13.4), a marly unit still recording comparatively deep marine conditions. However, these marls are followed by the Ilundain Marls Fm, a unit deposited in shallower marine conditions, as evidenced by its fossil content and by abundant sandy tempestite beds. These two marly units are locally separated by an angular unconformity, indicative of intra-Bartonian tectonic movements.

Tectonic activity further increased during Priabonian time, leading first to subdivision of the foreland basin and later on to its detachment and conversion into a piggy-back basin. Growth folds began to develop in the Pamplona area, probably related to the inception of east–west trending blind thrust sheets. Seawater circulation became restricted in the subsiding synclines, favouring deposition of euxinic marls and evaporitic deposits (e.g. potash member of the Gendulain Fm; 19 in Fig. 13.4), whereas erosion occurred in the crests of growing anticlines. The late Priabonian Liédena Member (Mb) is the youngest marine unit of the Pamplona area. This unit accumulated in nearshore environments and is famous for its numerous and well-preserved bird tracks (Payros *et al.* 2000*a*). The Liédena Mb also contains conglomerate clasts of Lutetian turbidites, clear proof that the Hecho Gp deposits were already unroofed, and implying that by Priabonian time the axial zone had been uplifted at least 1500 m (Payros *et al.* 2000*a*).

Bartonian–Priabonian deposits have rarely been positively identified in the rest of the study area. An interesting outcrop has been recently recognized by Astibia *et al.* (2000) in the southern limb of the Miranda syncline (Fig. 13.3). There, a middle Priabonian alluviolacustrine succession with abundant vertebrate remains (notably mammals) rests unconformably on tilted late Cretaceous–Palaeocene rocks. However, in an isolated outcrop in the San Vicente de la Barquera syncline, a near complete Bartonian–Priabonian succession of marine carbonates has been preserved (Fig. 13.3). This succession therefore provides clear evidence that marine conditions persisted in the NW part of the Basque-Cantabrian region until at least early Oligocene time (Fig. 13.2C).

Early–middle Oligocene

Except for the isolated outcrop near San Vicente de la Barquera, Oligocene deposits of the study area are almost exclusively of terrestrial facies. These deposits attest therefore to the near-total emergence and continentalization of the western Pyrenees and Basque-Cantabrian region, and to the inversion of the earlier Palaeogene marine embayment. Typical of that interval was the development of internally draining, fluviolacustrine basins, commonly containing evaporites, recording episodes of semi-arid climatic conditions.

The most complete and best known Oligocene terrestrial succession outcrops around the village of Estella, in the southern limb of the Urbasa-Andia syncline (Fig. 13.3). This succession begins with the Puente la Reina Fm (20 in Fig. 13.4), a lacustrine evaporitic unit that marks the Eocene–Oligocene transition (Salvany 1989, 1997). These evaporites define a 40 km wide sub-basin connected laterally to the east with fluvial complexes of the Jaca basin. The Puente la Reina evaporites are overlain by the Mues Sandstone (21 in Fig. 13.4), a red bed unit accumulated by rivers flowing to the NE from the newly uplifted Iberian Ranges. After this fluviatile interlude, evaporite deposition resumed in the central part of a new, large lacustrine basin (Desojo Gypsum Fm; 22 in Fig. 13.4). Marginal zones of that basin were typified by red lutites with intercalations of sandstones, charophyte-bearing limestones

and minor evaporites (upper member of the Mues Fm; 21 in Fig. 13.4).

Late Oligocene, Miocene and Pliocene

A final period of active compressional tectonics took place in the western Pyrenees and Basque-Cantabrian region during late Oligocene times, intensifying folding/thrusting rates and leading locally to diapiric extrusions of Keuper evaporites. In the Estella area, this tectonic deformation caused a progressive southward tilting of lower Oligocene and younger deposits, their subsequent erosion being recorded by a prominent intra-Oligocene unconformity (Barbarin unconformity; Riba 1992; BU in Figs 13.4 and 13.6). The relief of uplifted areas also increased, enhancing denudation rates and average grain sizes of products being exported to the Bay of Biscay or to the Ebro basin. Near Estella, in the northern margin of the Ebro basin, late Oligocene and Miocene sedimentation took place in large-scale alluvial fan systems, which evolved southwards into fluvial systems flowing into ephemeral lacustrine basins (Fig. 13.6; Salvany 1989, 1997). Alluvial fans also developed radially around diapiric highs, such as the Estella diapir. Cycles of progradation–retrogradation, separated by minor angular unconformities, are recognizable in all these alluvial fans, indicating alternate episodes of tectonism and quiescence (Fig. 13.6; Baceta *et al.* in press).

A part of the upper Oligocene–Miocene sedimentary detritus was stored in piggy-back basins developed on actively deforming thrust sheets, such as those located in the Villarcayo and Miranda synclines (Figs 13.3 and 13.6). These two basins exhibit a similar architectural pattern, clearly imposed by the geometry of the basal decollement (Fig. 13.6). For instance, their fills have a broadly synclinal arrangement, being thicker and more complete in the basin axis (Riba 1976). Basin depocentres migrated northwards, producing progressive onlap and more pronounced basal unconformities on the northern side of both basins (Fig. 13.6).

Deposits within these syntectonic piggy-back basins typically are composed of proximal and distal alluvial fan clastics and of fresh-water carbonates. Proximal alluvial fan deposits, mostly composed of calcareous conglomerates, occur near the basin margins. As a rule, they are much better developed in the northern margins than in the southern ones, a clear reflection of the relative areal extent and relief of the respective catchment areas (Fig. 13.6). These marginal conglomeratic accumulations exhibit several angular unconformities, probably created by tectonic movements linked to displacement along the basal decollement (Fig. 13.6). Distal alluvial fan deposits are the most abundant type of sediment in these piggy-back basins, and are represented by mudstones and siltstones with intercalations of channelized conglomerates and sandstones. Fresh-water carbonates consist of metre-thick accumulations of marls and limestones, with charophyta remains and thin-shelled gastropods. They often contain calcified root traces (rhizoconcretions) that indicate very shallow lacustrine–palustrine conditions, but no evaporites have so far been reported.

Compressional movements ceased some time during mid- to late Miocene times, and were replaced during the late Miocene and/or early Pliocene intervals by a phase of extensional tectonism during which a number of isolated basins were created (e.g. near the villages of Ocio, Bernedo, Campezo and Murgia in Alava province; near Acedo, Nazar

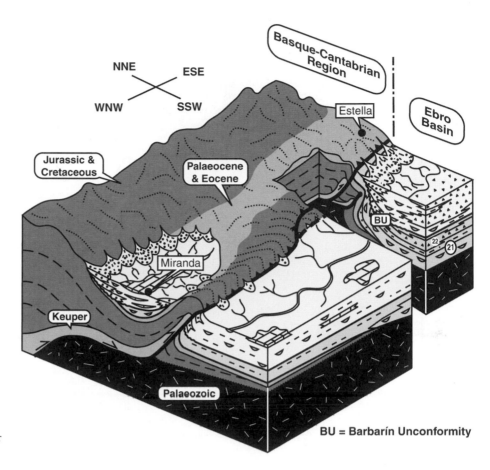

Fig. 13.6. Three-dimensional reconstruction of a segment of the southern part of the Basque-Cantabrian region during the Miocene epoch (not to scale). Encircled numbers (21 and 22) refer to lithostratigraphic units in Figure 13.4.

and Oco in Navarra province). These basins are located in tectonic grabens, created in most cases by fault reactivation, and are infilled with alluviolacustrine deposits. Although small (0.5–3 km wide and up to 3 km long) compared to coeval extensional basins in eastern Spain (Fig. 13.1), these grabens can contain remarkably thick sedimentary sequences (up to 300 m).

Ebro basin (northeastern Spain) (AM, CA, AG, AL, GP, APe, JV)

The broadly triangular Ebro basin is bounded by the Pyrenean Range to the north, the Iberian Ranges to the south and the Catalan Coastal Ranges to the east (Fig. 13.7). It is generally viewed as the southern foreland basin of the Pyrenees, with an asymmetrical Tertiary sedimentary fill thickening northwards (Quirantes 1978; Riba *et al.* 1983). However, in its western region it is considered to be a symmetrical foreland basin (Muñoz-Jiménez & Casas-Sainz 1997), since both the Pyrenees and Iberian Ranges have produced thrust sheets with a throw of more than 20 km, reducing the basin width by about 70% (Jurado & Riba 1996). The geodynamic evolution of the basin was thus closely linked to the structural development of the Pyrenean orogen and the Iberian and Catalan Coastal Ranges during Alpine orogeny (Puigdefàbregas *et al.* 1986; Guimerà & Álvaro 1990; Zoetemeijer *et al.* 1990; Casas-Sainz 1992). As a consequence, the Ebro basin basically acquired its final structure during the late Oligocene interval, coinciding with the last main Pyrenean thrusting phases and tectonic activity on the other two bounding margins.

Stratigraphy and sedimentology

Detailed studies of the Tertiary stratigraphic record in the Ebro basin (Pérez 1989; González 1989; Muñoz 1992; Arenas 1993; Villena *et al.* 1996a,b; Angulo *et al.* 2000; Luzón 2001) have recognized, in the outcropping autochthonous deposits eight tectosedimentary units (TSU) bounded by sedimentary breaks of basinal extent (after definition of Pardo *et al.* 1989). Correlation of these units has been established by mapping, by the similarity in evolution of their vertical sequences, and by palaeontological data. The ages of the units have been determined using fossil sites which, although few and scattered and not always allowing accurate biostratigraphic correlation, sometimes provide critical age evidence. In addition, in some sectors of the basin, ages have been constrained more precisely by magnetostratigraphic studies (Agustí *et al.* 1994b; Gomis *et al.* 1999; Barberà 1999; Pérez-Rivarés, pers. comm. 2001). Based on these data, the first three TSUs are regarded here as Palaeogene, the fourth extends across the Palaeogene–Neogene transition and the other four are Neogene (Fig. 13.8). Each unit has a generally complex but characteristic sequence, related to variations in subsidence, sediment supply and base level in different parts of the basin. Many lithostratigraphic units have been named (e.g. Riba *et al.* 1983; Villena *et al.* 1992), but details of these are avoided in this work in order to review the stratigraphy in a simple and clear way. However, some names of the main marine and lacustrine formations are introduced below for reference. Most of the Ebro basin sediments are continental and include a wide variety of alluvial deposits, as well as both fresh-water lacustrine

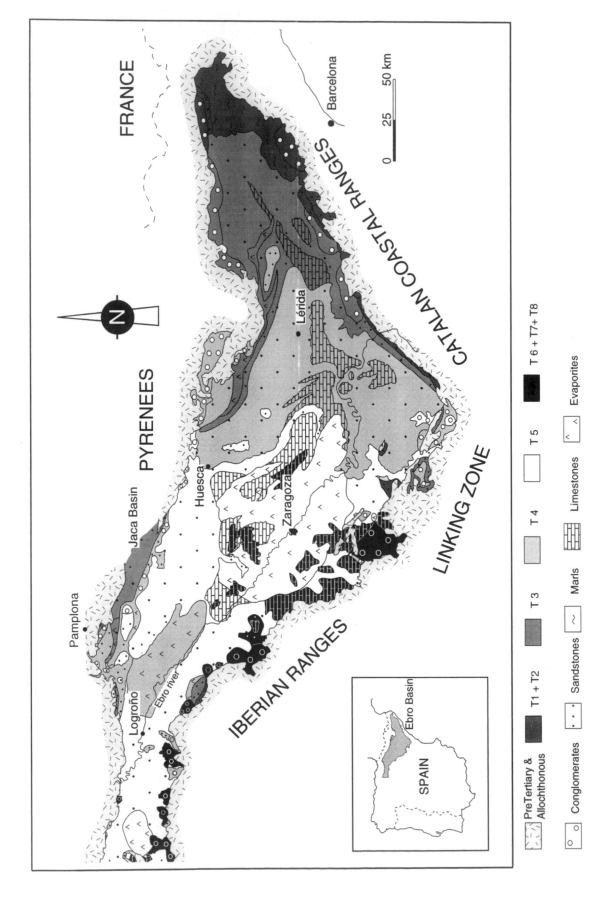

Fig. 13.7. Geological map of the Ebro basin tectosedimentary units (T1–T8) with lithofacies (mappable facies associations) distributions for every unit.

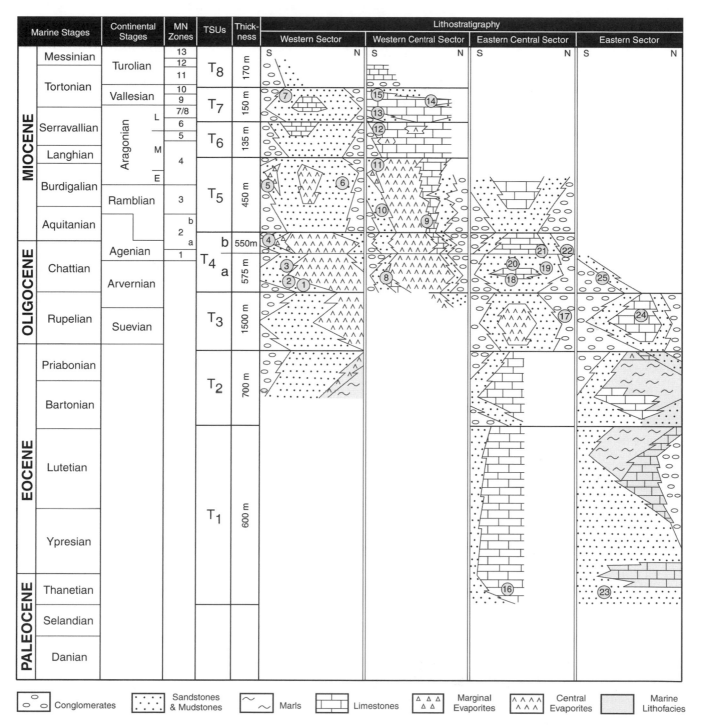

Fig. 13.8. Chronostratigraphy, lithological composition (marine lithofacies are shaded grey) and sequence evolution of the Ebro basin tectosedimentary units (TSUs). MN Zones, Mammal Neogene zones. Approximate positions of the main fossil sites are indicated by numbers in the lithostratigraphic columns: **1**, Bergasa and Arnedo; **2**, Autol; **3**, Quel and Carretil; **4**, Islallana and Fuenmayor-2; **5**, Los Agudos; **6**, Miranda de Arga; **7**, Cellórigo; **8**, Las Torcas; **9**, Barranco Foyas, Ereta de las Monjas, Paridera del Cura, San Juan and La Galocha; **10**, Tudela I and II; **11**, Tarazona and Monteagudo; **12**, Villanueva de Huerva; **13**, Moyuela; **14**, San Caprasio; **15**, El Buste; **16**, Vidaliella gerundensis; **17**, Peraltilla; **18**, Mina del Pilar; **19**, Torrente de Cinca 4, 7 and 18; Fraga 4 and 6; **20**, Fraga 7, Velilla de Cinca 5, and Ballobar 12; **21**, Fraga 11; Torrente de Cinca 68; Clara, Ontiñena, and Ballobar 21; **22**, Santa Cilia; **23**, Vidaliella gerundensis; **24**, Tárrega and Calaf; **25**, Gandesa. Chronostratigraphy after Friend (1996).

carbonates and saline playa lake deposits. Marine sediments are all pre-Oligocene in age (Fig. 13.8) and include shelf carbonates and marls, coastal and deltaic detrital facies, and evaporites. Alluvial deposits exist in most sequences and were deposited both in relatively localized alluvial fans *sensu stricto* (<15 km long), and in larger systems (up to 100 km long) with higher transport efficiency (cf. Colombo 1989). Details of each TSU are given below.

Palaeogene units. The oldest of the three TSUs deposited during the Palaeogene interval (T1) is Thanetian–Bartonian in age (Anadón *et al.* 1979; Colombo 1986; Pérez 1989; Villena *et al.* 1992) and crops out only in the east and SE (Figs 13.7–13.9). It typically displays a markedly asymmetrical cyclic pattern, initially fining upwards followed by a much thicker coarsening-upward sequence. Its lower boundary is a mappable unconformity above a basement of different Palaeozoic and Mesozoic rocks. In the southernmost eastern sector this unit is solely composed of continental facies deposited in alluvial plains associated with carbonate and, less commonly, evaporitic lacustrine areas. To the NE greater lithological variety is present, with easterly derived fluvial, and northerly derived fluviodeltaic sediments interfingering with marine platform carbonates (Alveoline limestones, Orpí Fm, in the lower part of the unit, and Tavertet Fm in the upper part; Fig. 13.9).

The second unit (T2) is Bartonian–Priabonian in age, and crops out in the eastern sector (Figs 13.7 and 13.8), with another fining-upward, then coarsening-upward succession. In the west only the upper part of the unit crops out (Fig. 13.8). Its lower boundary is a sedimentary break marking a change from coarsening-upward to fining-upward and is locally represented by a mappable unconformity. In the western sector, alluvial facies represent deposits of high transport efficiency fans that originated along the southern basin margin and are laterally associated to the north with marine carbonate and evaporitic environments (Pamplona Marls and Pamplona Saline formations; Figs 13.8 and 13.9). To the east the unit displays a wide variety of facies and depositional environments, ranging from entirely continental in the SE to mostly marine in the NE. The continental sediments include both high and low transport efficiency alluvial fan deposits of southern and eastern provenance, and in the basin centre are correlated with carbonate lacustrine deposits. The marine facies further north are represented by carbonate platform (Igualada Fm), evaporitic (Cardona Fm) and detrital coastal deposits (Figs 13.7–13.9), but alluvial and fan delta deposits of northern and eastern provenance are also present. The chronostratigraphic assignment of this unit is based on marine fossil associations (Palli 1972; Reguant 1967; Gich *et al.* 1967; Ferrer *et al.* 1968; Ferrer 1971; Anadón *et al.* 1979), relationships between fossil-bearing marine and continental facies, and fossil associations in continental facies (Anadón 1978; Anadón *et al.* 1979, 1983*a*; Colombo 1980).

Unit T3 is late Priabonian–Rupelian in age, shows another fining-upward/coarsening-upward sequence and crops out mainly, but not exclusively, in the eastern sector (Figs 13.7 and 13.8). Its lower boundary is a sedimentary break, locally showing syntectonic unconformities (seen at localities such as San Miguel de Montclar in the east and Gatún in the west). This unit was deposited in continental environments disconnected from the sea: both high and low transport efficiency alluvial fans discharged from the three basin margins into central evaporitic (Barbastro Fm to the east and Puente La Reina Fm to the west; Figs 13.7–13.9) and carbonate (Tárrega Fm in the east) lacustrine systems. The chronostratigraphic assignment of this unit is based on fossils within the Ebro basin (Anadón & Feist 1981; Anadón *et al.* 1987, 1992; Agustí *et al.* 1987; Álvarez Sierra *et al.* 1987, 1990).

Palaeogene–Neogene transition unit. The fourth unit (T4) is Arvernian–Agenian in age, crops out widely throughout the basin, and has been divided into two subunits (T4a and T4b). Differences in T4 sequences from one sector in the basin to another can be attributed to the activity of local tectonic structures, causing modifications to drainage patterns. Its lower boundary is a sedimentary break, locally with syntectonic unconformities (seen at localities such as Anguiano and Préjano in the western sector), as is the boundary between T4a and T4b along the Pyrenean margin (seen at Barbarín and Agüero). Unit T4 records deposition from both low and high transport efficiency alluvial fans from all three basin margins, feeding into a large central lacustrine system with carbonate depositional environments to the east (Mequinenza Fm; Figs 13.7–13.9) and saline, mostly sulphate depositional environments in the central and western sectors (Zaragoza Fm in the centre, and Falces and Lerín formations to the west; Figs 13.7–13.9). The chronostratigraphy of this unit (and overlying Neogene units T5, T6, T7 and T8) is based on the presence of vertebrate faunas from numerous sites within the basin (Fig. 13.8; Crusafont *et al.* 1966*b*; Santafé *et al.* 1982; Cabrera 1983; Cuenca 1983; Agustí *et al.* 1987, 1988, 1994*a*; Martínez-Salanova 1987; Álvarez

Sierra 1987; Álvarez Sierra *et al.* 1987, 1990; Cuenca & Canudo 1991; ITGE 1991; Cuenca *et al.* 1989, 1992; Muñoz 1992) and is discussed further by Villena *et al.* (1992).

Neogene units. The Neogene stratigraphic record is divided into four TSUs, with the oldest (T5) being Agenian–middle Aragonian in age, and having the largest outcrop and much lateral variation across the basin (Figs 13.7 and 13.8). The lower boundary of unit T5 is marked by syntectonic unconformities along the basin margins (e.g. unconformity of Calcón in the eastern central sector) and a sudden influx of coarser sediment in the basin centre (Fig. 13.9). This unit resulted from deposition by high and low transport efficiency alluvial fans from both northern and southern margins. Distal parts of the fans interfingered with carbonate (Alcubierre Fm) and saline (Zaragoza and Cerezo de Río Tirón formations) lacustrine systems (Figs 13.7–13.9). In the western sector a trunk fluvial system flowed from west to east and was connected to an important playa lake complex (Zaragoza Fm) through wide alluvial plains and carbonate lake environments (Tudela Fm). In areas located between alluvial fans of southern provenance, gypsum facies developed close to the southern basin margin (e.g. Ablitas and Ribafrecha Gypsum; Figs 13.8 and 13.9).

The second Neogene unit (T6) is middle-late Aragonian in age and crops out in the central and western regions of the basin. Its vertical sequence fines upward and its lower boundary is a sedimentary break (syntectonic unconformity at Ledesma de la Cogolla in the southwestern sector). Facies of this unit were again deposited by high and low transport efficiency alluvial fans that coalesced in the west into a trunk fluvial system, feeding shallow, carbonate lacustrine areas that occupied the lowest parts of the basin (Puerto de la Brújula Fm in the west, and Alcubierre Fm in the central region; Figs 13.7–13.9).

Unit T7 is late Aragonian-Vallesian in age and crops out only in the western half of the basin. Like T6, in the western central sector this unit shows onlap at the southern basin margin, and has a fining-upward, then a thicker coarsening-upward sequence. In some places the basal sedimentary break is an angular unconformity, above which there is a sharp increase in grain size. Extensive progradation of fluvial systems of northern provenance is recorded by T7 (Arenas & Pardo 1991, 2000), with the distal parts of these systems reaching areas close to the southern basin margin, where a related carbonate lacustrine system was strongly wave-dominated (Alcubierre Fm; Figs 13.7–13.9). In the western sector of the basin, low transport efficiency alluvial fans continued to develop.

The youngest unit (T8) is Turolian in age and has a very limited outcrop only present in the southwestern part of the basin. Its lower boundary is locally an unconformity (e.g. at Yerga), and coincides with a sharp change in grain size. Facies of this unit again represent deposits of high and low transport efficiency alluvial fans, the distal parts of which were connected to carbonate lacustrine systems. Thick oncolitic deposits accumulated during this time (Muela de Borja Fm; Figs 13.7–13.9).

Palaeogeography

The Tertiary palaeogeographic evolution of the Ebro basin was mainly controlled by the tectonic regime in which the basin developed. Tectonics conditioned the extent and topography of the sediment source areas in the orogenic wedge and foreland, the geographic location and orientation of the basin margins and depositional areas, and the subsidence and sedimentation rates through space and time. Latest Cretaceous to late Miocene sedimentation took place during north–south convergence between the Iberian and European plates. Superimposed upon this compressional context, an east–west extensional regime that produced widespread rifting from the Rhine graben to the Alboran sea began in the Miocene in the Catalan region, and extended westward during Miocene times.

Ebro basin stratigraphy suggests that tectonic activity was the dominant control on the sedimentary evolution of the basin, influencing variations in the sediment supply/subsidence ratio (SS/Sub). During periods of increasing SS/Sub,

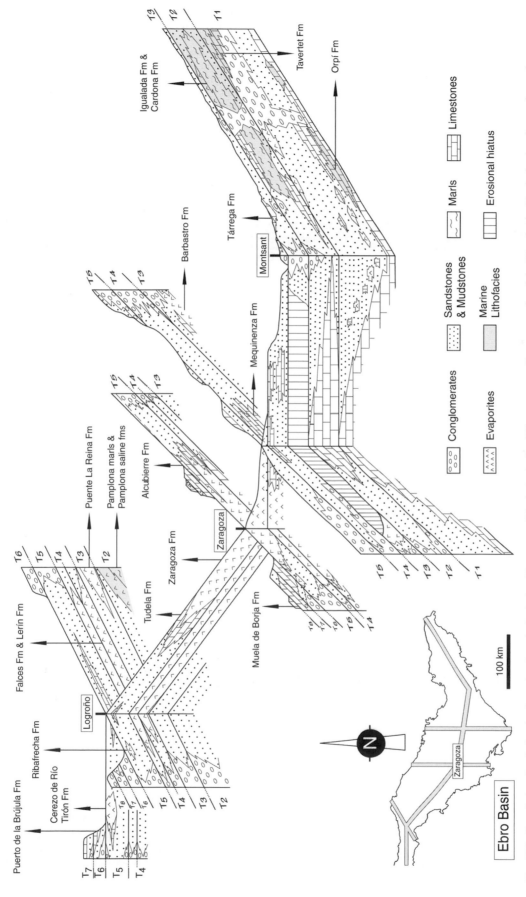

Fig. 13.9. Correlation sketch showing lateral and vertical relationships among lithofacies in the different tectosedimentary units of the Ebro basin. Vertical axis refers to timescale.

progradation of the alluvial fans basinward and, in most cases, shrinkage of the lacustrine systems are recorded. In contrast, during periods of decreasing SS/Sub, retrogradation of the alluvial fans towards the basin margins and expansion of the lacustrine systems are recorded. Additionally, in lacustrine depositional areas, base-level variations due to climatic changes played a relevant role, and might have modified the overall evolution of the sedimentary systems from one sector to another (Figs 13.8 and 13.10). Figure 13.11 provides an interpretation of Ebro basin palaeogeography from late Eocene to late Miocene times. Each timeframe is taken to represent a period when the SS/Sub was at a minimum, so that from Oligocene times onward, central lacustrine systems were at their maximum development.

Albian to Santonian sinistral transcurrent motion and anti-clockwise rotation of Iberia with respect to Europe resulted in the opening of the Bay of Biscay. During this time a complex, strike-slip fault-related flysch basin developed in the Pyrenean area to the north of the axial zone. At that time the present south Pyrenean zone (including the axial zone), along with the present Ebro basin, constituted part of the northern margin of the Iberian plate, and sometimes was inundated by marine transgressions from the flysch basin (e.g. Cenomanian and Santonian transgressions). Then, from late Santonian times onwards, convergence between the European and Iberian plates began the development of the Pyrenean thrust-and-fold belt. At the end of the Cretaceous period, continental sedimentation (Garumnian facies) occupied the eastern and

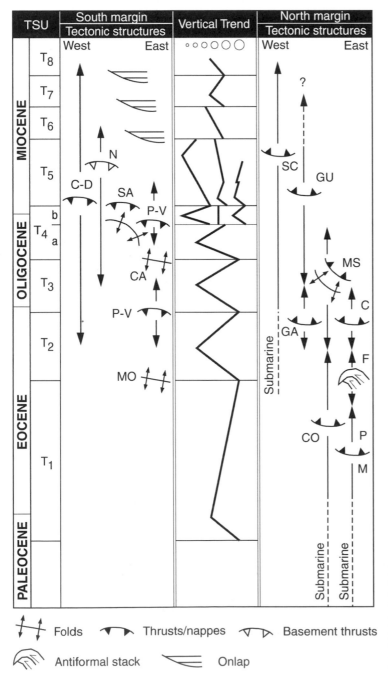

Fig. 13.10. Table summarizing the tectonic–sedimentation relationships for both the northern and southern margins of the Ebro basin. The main active tectonic structures and the vertical trend of the tectosedimentary units are indicated. Key: **C**, Cadí; **CA**, Calanda; **C-D**, Cameros-Demanda; **CO**, Cotiella; **F**, Freser; **GA**, Gavarnie; **GU**, Guarga; **M**, Montgrí; **MO**, Montalbán; **MS**, Marginal Sierras; **N**, Nájera, Baños and Arnedo; **P**, Pedraforca; **P-V**, Portalrubio-Vandellós; **SA**, Sierra de Arcos; **SC**, Sierra de Cantabria.

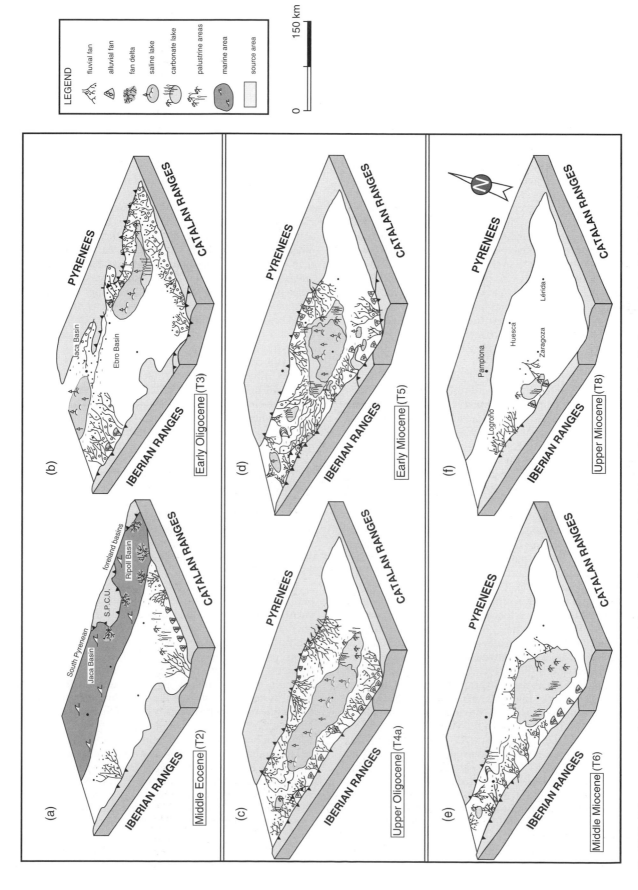

Fig. 13.11. Palaeogeographic sketches for times of minimum sediment supply/subsidence ratios, illustrating the Tertiary evolution of the Ebro basin.

southern sector of the south Pyrenean margin, and the submarine emplacement of the Boixols and Cotiella thrust sheets took place in the south Pyrenean central unit (SPCU), along with the Montgrí thrust sheet in the east.

Deposition of unit T1 began during Thanetian times when continental sedimentation in the east, represented by fluvial and lacustrine systems, connected westward to marine environments. At the beginning of Ypresian (Ilerdian) times continental environments shifted southward as a consequence of a transgressive stage that affected the region, whereas carbonate platform depositional environments (Orpí Fm) were present to the north (Figs 13.8 and 13.9). Subsequent Ypresian (Cuisian) regression restricted marine facies to areas further north and west, and the progradation of alluvial fans and fan deltas took place from the south, east and in some cases from the north, as a consequence of emergence of the Cotiella, Pedraforca and Montgrí thrust sheets. As a result of markedly decreasing SS/Sub due to tectonic loading, the south Pyrenean margin progressively deepened to form east–west orientated, turbiditic troughs to the west (Jaca basin) and east (Ripoll basin) of the SPCU. A phase of increasing SS/Sub during early Lutetian times caused the filling of these turbiditic troughs as alluvial systems prograded from the south, east and north. However, during this period, and due to the movement of the sedimentary trough southward, a marine input from the east is recorded (Figs 13.8 and 13.9; Tavertet Fm).

The passage from unit T1 to T2 took place during an increase in the SS/Sub, when large thrust units finished their emplacement (e.g. Cotiella nappe and Freser antiformal stack; Fig. 13.10). This affected the Pyrenean troughs and led to deposition of extensive and thick deltaic platform and fan delta lithofacies. A subsequent decrease in SS/Sub was coincident with a second transgressive stage (Figs 13.8 and 13.11a). To the south, depositional environments were continental, with alluvial systems feeding from both the Iberian and Catalan basin margins. Alluvial fans from the south were of high transport efficiency, whereas those from the east were predominantly of low transport efficiency (fluvial and alluvial fans, respectively, in Fig. 13.11a). The distal parts of both systems were occupied by palustrine carbonate areas.

Deposition of the upper part of unit T2 took place in the NE during the initial stage of the Cadí nappe emplacement (Fig. 13.10), inducing the isolation of the Ripoll basin, along with a change from platform to restricted, evaporitic environments (Cardona Fm). During this time, progradation of the alluvial systems sourced from the basin margins overwhelmed the previously palustrine areas and shifted the marine depositional environments westwards. With regard to tectonic activity over this period, the Marginal Sierras thrust sheets were active at the southern extremity of the SPCU, the Gavarnie thrust sheet was emplaced southward in the western central sector of the basin, while in the Iberian margin the Portalrubio-Vandellós and Cameros-Demanda thrusts started to develop. In this context, high transport efficiency alluvial fans built out from both the south and east (SPCU) and were connected to marine transitional environments (Pamplona Saline Fm).

During the deposition of unit T3 (early Oligocene) the south Pyrenean foreland basins finally became fully continental. To the east, the emergent thrust front of the Cadí nappe (Vallfogona thrust) formed the northern margin of the youngest south Pyrenean foreland basin (Ebro basin). To the west, the Guarga thrust sheet initiated the formation of the External Sierras, separating the piggy-back Jaca basin from the foreland Ebro basin. Within this context, high and low transport efficiency alluvial fans developed from the three basin margins. During the deposition of the middle part of the unit, these alluvial systems retrograded (decreasing SS/Sub) and favoured the development of evaporitic lacustrine systems in the east (Barbastro Fm) and west (Puente La Reina Fm). Differences between these two areas include the fact that the western lacustrine basin was fed by westward-flowing rivers in the Jaca and Ebro basins, and the eastern lacustrine system displayed a carbonate fringe (Tárrega Fm) around its eastern part (Fig. 13.11b).

The transition from unit T3 to T4 is interpreted to have taken place during a new period of increasing SS/Sub. During this time, progradation of the alluvial systems caused the eastern lacustrine system to disappear and the western lacustrine system to shift SE. Along the Pyrenean margin, the southward movement of the Guarga thrust sheet uplifted the older Jaca basin to the north into an area of erosion that from then on became the sediment source for the depositional environments within the Ebro basin. High and low transport efficiency alluvial systems formed along the three basin margins discharging into an evaporitic lacustrine system in the basin centre. During the time of deposition of the middle part of unit T4, and interpreted as a result of decreasing SS/Sub, the evaporitic lacustrine system extended over a large part of the basin centre (e.g. Falces, Lerín and Zaragoza Fms) (Fig. 13.11c). Shallow carbonate lacustrine environments were restricted to the western and eastern extremes of the basin. The eastern ones formed a fringe around the central saline system, whereas the western ones (La Bureba area) remained isolated by an east-flowing trunk fluvial system that reached its maximum development during unit T5 deposition (Fig. 13.11d).

Coinciding with another period of thrust activity along the northern (Guarga and Sierra de Cantabria thrust) and southern (Portalrubio-Vandellós and Sierra de Arcos thrusts) basin margins, an increase in the SS/Sub caused progradation of the alluvial systems basinward and substantial shrinkage of the lacustrine system, which disappeared altogether in the east. This context characterized the transition from unit T4 to T5. During the subsequent period of decreasing SS/Sub (Fig. 13.11d) an evaporitic lacustrine system occupied the central part of the basin (Zaragoza Fm), with a shallow carbonate lacustrine fringe (Alcubierre and Tudela formations) developed around its northern, eastern and western margins. This central lacustrine system had a distinctive sedimentary evolution, probably controlled by climate, and was fed by high transport efficiency alluvial systems from the northern and southern margins, and by an important, mostly meandering fluvial system flowing from the west. In the western region of the basin, an evaporitic lacustrine system with similar characteristics existed, but its extent was much smaller (Cerezo de Río Tirón Fm). Along the margin of the Iberian Ranges, small, isolated saline lakes developed within areas between alluvial fans (e.g. Ribafrecha, Ablitas and Borja Gypsum; Fig. 13.11d). During the sedimentation of unit T5, the Nájera, Baños and Arnedo basement thrusts (Fig. 13.10), which had started to develop in Rupelian times, reached their maximum activity and caused strong variations (160 to 820 m) in sediment thickness.

The transition between units T5 and T6 (middle Aragonian) occurred during another increase in the SS/Sub. A progradation of the alluvial systems and a general shrinkage of the central lacustrine system related to renewed movements on the Sierra de Cantabria and Cameros-Demanda thrust sheets. By this time the three basin margins had established the same position as seen today. From middle Aragonian times onwards (unit T6) a significant palaeogeographic change took place, during which the evaporitic, mostly sulphate lacustrine systems that had occupied the basin centre since the basin became isolated from the sea, were replaced with mostly carbonate lacustrine systems that extended throughout the centre and western part of the basin. This system extended further south with respect to the evaporitic system developed during unit T5, and the alluvial fans from the south had a correspondingly smaller extent (Fig. 13.11e). Deposits of this unit onlapped the Iberian margin of the basin, which evolved as a passive margin from this time, with the exception of the western part of the basin where thrusting along both the southern and northern margins created a symmetrical foreland basin (Muñoz-Jiménez & Casas-Sainz 1997).

During latest Aragonian times, a new period of increasing SS/Sub, related to the movement of the Cameros-Demanda and Sierra de Cantabria thrusts, took place at the beginning of unit T7 deposition. Extensive progradation of alluvial fans from the south and north caused a significant shrinkage of the lacustrine system. This system became reduced to small carbonate ponded areas within distal plains at the beginning

of unit T7, although it later expanded when the SS/Sub diminished. During the Vallesian interval, distal alluvial deposits from the north reached areas close to the central part of the Iberian margin, and the lacustrine system was pushed further south than it had ever been.

During the Vallesian–Turolian transition (base of unit T8), a final period of increasing SS/Sub caused progradation of the alluvial systems, at least in the southern part of the western and central sectors of the basin. During Turolian times (Fig. 13.11f), depositional environments in the SW comprised low transport efficiency alluvial fans derived from the south, and distal sectors of siliciclastic fluvial systems presumably flowing from the north. Carbonate fluvial and shallow lacustrine environments included areas in which abundant oncolitic facies accumulated. Deposits of this unit represent the last stages of sedimentary trough displacement from north to south, that is, from areas that at present correspond to the Pyrenean Range to areas close to the Iberian Ranges.

Duero basin (northern Spain) (IA, AC, GA-G, JC, JMR)

The continental Tertiary Duero basin is located in the north-western region of the Iberian Peninsula (Fig. 13.12). Throughout its history from early Eocene to late Miocene times, this intraplate basin underwent a multistage evolution controlled

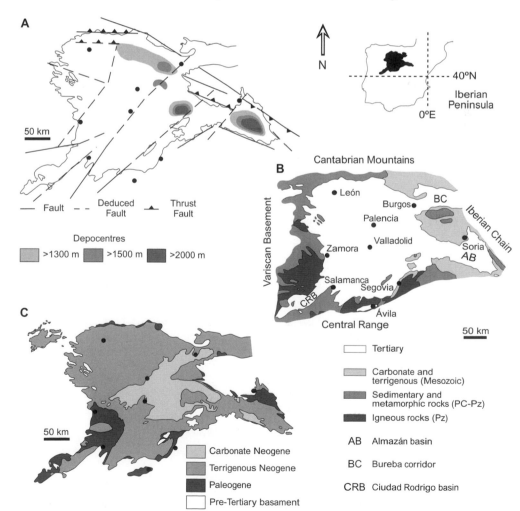

Fig. 13.12. The Tertiary Duero basin. **(A)** Structural framework and main depocentres. **(B)** Rim of pre-Tertiary rocks. **(C)** Schematic geological map of Tertiary rocks. Key: PC, Precambrian; Pz, Palaeozoic.

by faults, some of which were reactivated strike-slip late Variscan structures, during Alpine orogenic events (Fig. 13.12A). The basin is markedly asymmetric, with the Tertiary successions ranging up to only 400 m thick in the west and south, but reaching thickness of over 3000 m in the maximum subsidence axis running parallel to the northern and eastern borders (Fig. 13.12). The palaeogeographical evolution of the basin throughout Tertiary times was mainly controlled by: (a) convergence between the European and Iberian plates; (b) collision between the Iberian and African plates; (c) Tertiary inversion of the Mesozoic Iberian and Basque-Pyrenean basins during compressional phases; and (d) uplift of the mountain belt forming the southern margin of the basin (Central Range; Fig. 13.12B).

The Duero basin fill comprises siliciclastic and detrital carbonate alluvial sediments, and carbonate and evaporite lacustrine sediments deposited during Palaeogene (middle Eocene–early Oligocene) and Neogene times. Palaeogene outcrops are restricted to the basin margins by the much more extensive Neogene cover (Fig. 13.12C). Different compositions of the source area, combined with climatic changes through time have influenced the compositions of the terrigenous Tertiary facies and have led to a complex evolution of the Duero sedimentary record. Along the western margin, the basin is surrounded by a Variscan basement (Fig. 13.12B) composed of Precambrian and Palaeozoic greenschist facies and pre-Variscan and Variscan igneous rocks. The Central Range to the south trends ENE–WSW and consists of Palaeozoic metamorphic rocks and abundant granitoid rocks. To the north the basin is bounded by the Cantabrian Mountains (Fig. 13. 12B), most of which comprise Palaeozoic terrigenous and carbonate sedimentary rocks in the west and Mesozoic carbonate and terrigenous facies in the east. Finally, to the east, the Duero basin borders on the Iberian Ranges which is divided into two branches separated by the Almazán basin (Fig. 13.12B) and

is made up of terrigenous and carbonate Mesozoic sedimentary rocks.

With regard to climatic conditions, the latitudinal change of the Iberian peninsula from nearly 30° at the beginning of Tertiary times to nearly 40° at present, together with the formation of mountainous barriers, have both influenced basin sedimentation. The climate evolved from Late Cretaceous–early Eocene humid-tropical conditions to a middle Eocene–early Miocene subtropical climate with arid phases, and then to mild subtropical conditions with dry-warm alternations throughout the rest of the Miocene epoch.

It is useful to overview the Duero basin in terms of four geographical domains surrounding a central area (Fig. 13.13). Major geological influences on each of these domains include: (i) reactivation of Variscan basement faults in the west; (ii) N–S compression and southward-directed thrusting during the uplift of the Cantabrian mountains in the north; (iii) inversion of the Mesozoic Iberian basin in the east; (iv) uplift of the Central Range in the south.

Seven main stages of lithostratigraphic and palaeogeographic evolution can be identified from the study of Duero sediments (Portero *et al.* 1982; Corrochano & Armenteros 1989; Armenteros & Corrochano 1994; Pérez González *et al.* 1994; Santisteban *et al.* 1996; unpublished work by the authors). A preliminary, latest Mesozoic stage records sedimentation between the Basque-Cantabrian basin to the north and the Iberian basin to the SE (Fig. 13.14A). Renewed sedimentation in early Eocene times (stage 2) was a response to extension and the development of grabens and half-grabens (Fig. 13.14B). Compression initiated in late Eocene times (Fig. 13.14C) produced more widespread sedimentation (stage 3) followed by a major intra-Oligocene hiatus (Fig. 13.13). A return to sedimentation in late Oligocene through early Miocene times (stage 4) finally produced a unified Duero basin with a single main depocentre (Fig. 13.14D). During the middle–late Miocene interval (stage 5; Fig. 13.14E), following

Fig. 13.13. Chronostratigraphic correlation across the different domains of the Duero basin. Ages are as quoted in Harland *et al.* (1990). Column a: standard Tertiary series and stages. Column b: mammal ages, on the left and Mammal Palaeogene (MP, 7–30) and Mammal Neogene (MN, 1–17) zones. Column c: continental stages and local zones (W–I). Fossiliferous sites: AR, Armuña; AZ, Ariza; B, Babilafuente; BV, Villavieja del Cerro; C, Cabrerizos; CA; Cetina de Aragón; D, Dueñas; DE, Deza; G, El Guijo; MA, Mazaterón; MO, Montejo de la Vega; MDP, Molino del Pico; PI, Piquera de San Esteban; QC, Quintanilla del Coco; S, Sanzoles; SC, Sta Clara; SI, Simancas; TA, Tariego; TM1, Torremormojón; TM6, Torremormojón; VA, Valladolid; VC, Villavieja del Cerro; VF, Los Valles de Fuentidueña; ZA, Zamora. Most representative Tertiary units: 1, siliceous Sandstones; 2, Lychnus Limestone Fm; 3, Corrales Fm; 4, Sta Cecilia Limestone Fm; 5, Mazaterón-Cihuela Limestones; 6, Vegaquemada Fm; 7, Adearrubia Fm; 8, Deza Limestone; 9, Candanedo Fm; 10, Villalba de Adaja/'Series Rojas'; 11, Dueñas; 12, Tierra de Campos; 13, 'Cuestas'; 14, 'Páramo Inferior' Limestone; 15, 'Páramo Superior' Limestone; 16, Siliceous Conglomerates ('Rañas').

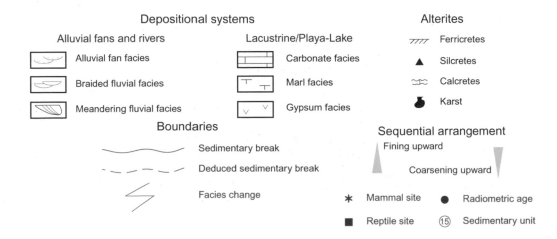

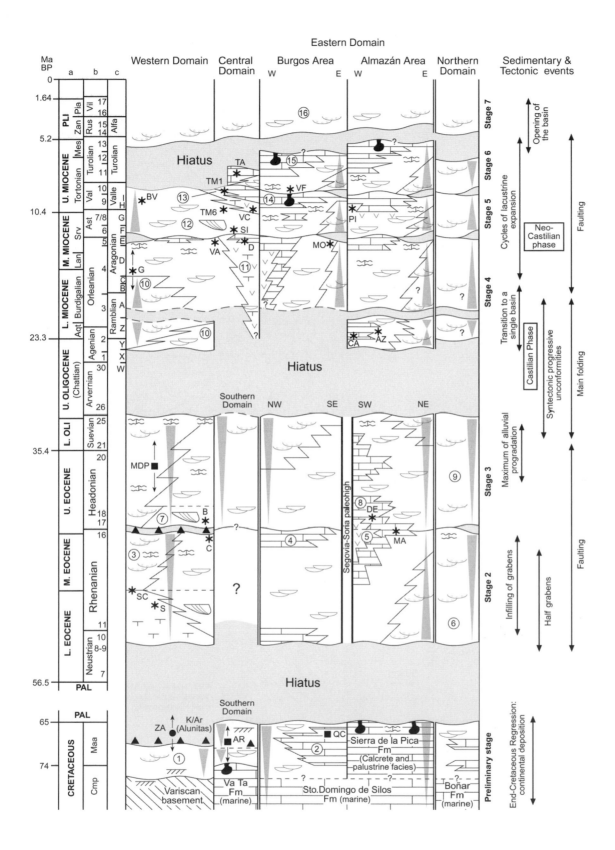

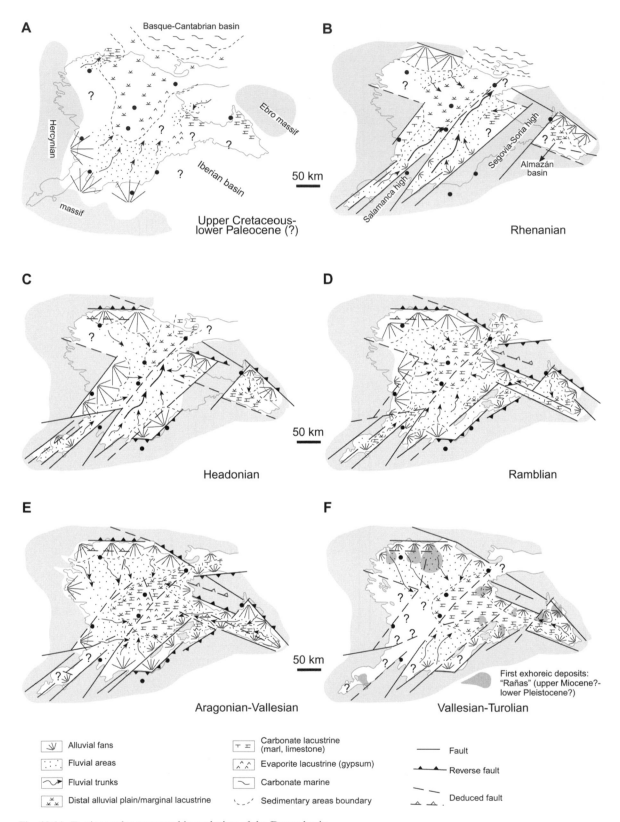

Fig. 13.14. Tertiary palaeogeographic evolution of the Duero basin.

an initial phase of tectonic activity and renewed sediment supply, relatively stable conditions set in and there was a major expansion of lake environments. Another tectonic pulse (late Miocene: beginning of stage 6) induced a final phase of basin infilling (Fig. 13.14F) before the area became open to the Atlantic and achieved its present-day configuration as an external drainage basin (stage 7). Further details are provided below. Palaeogeographic reconstructions for the four first stages are based on wireline logs, cores and, in some cases, geophysical data, and are integrated in a coherent model supported by surface information from the basin margins.

Stage 1 (Late Cretaceous–?early Palaeocene). In the west, fluvial and alluvial fan successions are discordant on the Variscan basement (Corrochano 1977, 1982; Alonso-Gavilán 1981; Fig. 13.13), whereas in the east fluvial and especially shallow carbonate lacustrine systems extend around the Ebro massif (Alonso et al. 1993; Fig. 13.14). The basin was open to the north, where marine sedimentation continued in the Basque-Cantabrian basin (Fig. 13.14A). A tropical climate prevailed, with relatively dry periods that allowed silcrete formation (Corrochano 1977; Bustillo & Martín-Serrano 1980; Alonso-Gavilán 1981; López Martínez 1989), consistent with the position of the northern Iberian plate at 30° latitude.

In the western domain, the mainly Cretaceous succession is represented by a ferruginous, siliciclastic sequence of conglomerates, sandstones and mudstones (Corrochano & Pena dos Reis 1986) derived from a deeply weathered Variscan basement that shows ferruginous alteration (ferricretes) and an upward increase in kaolinite content (Bustillo & Martín Serrano 1980; Molina et al. 1997). The siliciclastic successions, exhibiting a coarsening-upward trend, were deposited in braided fluvial systems and associated progradational alluvial fans. They contain thick mottled silcretes, formed by processes that were probably repeated over millions of years on a stable basement. Each silicification episode seems to have been developed as a duricrust at the end of individual pulses of sedimentation. The presence of K-bearing alunite in opal-rich silicified layers has allowed K-Ar radiometric dating that indicates genesis between 58 and 67 Ma ago (Blanco et al. 1982).

In the rest of the basin, continental sediments, most of which can be attributed to the Maastrichtian (Floquet & Meléndez 1982; Alonso et al. 1993), lie above an upward-shallowing dolomitic marine sequence corresponding to the Sto Domingo de Silos Fm and equivalent units (Fig. 13.13). In the southern domain, a fluvial ferruginous formation (Fernández Garcia et al. 1989; Olmo Sanz & Martínez Salanova 1989), similar to that in the west, was deposited over a karstified surface developed at the top of a Campanian regressive marine carbonate sequence (Va Ta Fm; Fig. 13.13). In the eastern domain, the Iberian basin was mainly occupied by shallow carbonate lacustrine environments (Sierra de la Pica Fm and equivalent units). These lacustrine sediments overlie an almost imperceptible sedimentary break at the top of the last Cretaceous marine formations and record a Late Cretaceous marine regression. In the Burgos area (Pol & Carballeira 1986) and the northern domain (Colmenero et al. 1982; Corrochano 1989) a transition to fluvial deposits can be observed. The presence of the gastropod *Lychnus* and dinosaur remains in the Burgos area succession allows us to attribute these continental deposits to a Maastrichtian age (Pol et al. 1992).

Stage 2 (early–middle Eocene). The sediments deposited during this stage are bounded by unconformities, with the lower one representing a major hiatus (Fig. 13.13). Sedimentation was initiated by a regime of intraplate extension induced by the subduction of Bay of Biscay oceanic lithosphere southwards under the Iberian plate during Palaeocene and Eocene times (Boillot & Malod 1988). The area became compartmentalized by NW–SE, NE–SW and NNE–SSW fault systems that created basins and the intervening horst palaeohighs of Salamanca and Segovia-Soria (Fig. 13.14B). The sedimentary break situated at the top of the stage correlates with an important phase of compressive activity in the Betic margin (Vera 1988).

This structural setting produced a complex palaeogeography: in the west grabens and half-grabens opened to a marine area located to the NE, whereas the Almazán basin became isolated (Fig. 13.14B). The successions show variable thicknesses due to the different degrees of subsidence, and differences in topographic level between the grabens. Despite this local variation, to the west of the Segovia-Soria high (Fig. 13.14B) a typical succession begins with carbonate (marls and limestones) lacustrine and palustrine systems that developed into axial fluvial systems running parallel to the graben axes. The latter evolved from high- to low-sinuosity fluvial systems, which with time were replaced by transverse progradational fan systems (Corrales Fm) sourced from the graben margins (Corrochano 1977; Jiménez et al. 1983). The coarse-grained compositions are affected by the nearby source areas. In the western domain, arkose and sublitharenite sandstones are derived from the low-grade metamorphic and igneous rocks of the Variscan basement. In the northern domain polymictic facies are sourced from the Cantabrian Palaeozoic rocks; whereas sublitharenite sandstones derived from the old Ebro Massif are widespread in the Burgos area. Finally, carbonate conglomerates and litharenites coming from Upper Cretaceous rocks belonging to the incipient relief of the Iberian Ranges are predominant in the Almazán basin. In the latter area, a fining-upward succession evolved from retrogradational alluvial fans at the base to overlying carbonate and evaporite lacustrine–palustrine systems (Armenteros & Bustillo 1996). In the Burgos area, a similar sequence developed, producing sandy fluvial deposits associated with carbonate palustrine systems (Pol & Carballeira 1986).

Data supplied by mammal sites (S, SC, C and MA in Fig. 13.13) indicate that this compartmentalization stage and its infilling was mainly developed during Rhenanian times, since the fossiliferous sites range from 10 to 17 MP (Mammal Palaeogene zone) (Peláez-Campomanes et al. 1989; synthesis in Jiménez 1992). This stage follows a hiatus comprising most of the Palaeocene and the early Eocene intervals and representing a dramatic palaeogeographic change during the initiation of the Duero basin. The top of the succession in the west is marked by silcretes that record persistent exposure, in contrast to the maximum expansion of lacustrine environments in the more rapidly subsiding eastern areas. Climatic conditions were subtropical, with the existence of long dry periods recorded by the presence of calcrete, silcrete, magnesian clay and evaporites (Armenteros & Bustillo 1996) as well as by faunal indicators (Jiménez 1974; López Martínez 1989).

Stage 3 (late Eocene–early Oligocene). Broadly speaking, the main palaeogeographic trends were inherited from the previous stage. Widespread alluvial fan progradation took place from virtually all surrounding basin margins and, simultaneously, intervening palaeohighs were worn down by erosion. A more structured drainage network evolved and flowed towards and into an extensive lacustrine area in the NE (Fig. 13.14C). The sedimentary sequence shows a broad coarsening-upward pattern, starting with axial fluvial meandering/braided systems that were replaced upwards by large progradational alluvial fan systems (Fig. 13.13). Lacustrine systems were represented by shallow carbonate lakes surrounded by broad, periodically flooded, marginal plains (palustrine system). In the Almazán basin, which continued to be isolated, the lacustrine depocentre was displaced to the south (Armenteros et al. 1989; Bond 1996). As in the previous stage, coarse lithofacies closely reflect the nature of the source area: siliciclastic (litharenites and arkoses) in the western domain, and polymictic conglomerates and litharenites in the rest of the region.

Sedimentation during this stage was coincident with increasing regional tectonic compression. This compression produced southward thrust faulting along the Cantabrian border (Pulgar et al. 1999), and initiated uplift of the Central Range as stress was transmitted towards the centre of the Iberian plate from both the Pyrenean (NNE) and Betic (SE) plate margins. Along the southern border of the Duero basin, southward dipping reverse faults were produced by the reactivation of SW–NE Variscan structures (De Vicente et al. 1996d). In the eastern domain, the Iberian Ranges were uplifted, with the main structures following the NW–SE Iberian trend. From this stage to middle Miocene times, the Iberian Ranges were subjected to the effects of convergence at the northern (Pyrenean) and southern (Betic) margins of the Iberian plate, so that its formation was a result of the superposition of Pyrenean and Betic palaeostresses (Muñoz Martín & de Vicente 1996). Inside the Duero basin, the effects of this compression are revealed by fault alignments, facies changes and depocentre locations. This compressive regime in the Duero basin correlates with a coeval tectonic event known as the Castilian phase in the Madrid (Tajo) basin (Aguirre *et al.* 1976). This tectonic event represented a significant point in the history of the basin since from this time onwards the tectonic regime was reversed, changing from the strongly compressional late Palaeogene–early Miocene phase to an extensional regime characterized by normal and strike-slip faults.

The formation of a thick phreatic calcrete at the end of stage 3 (well exposed in the west; Corrochano 1977; Alonso Gavilán 1984, 1986) together with evidence of sclerophyllous subtropical forests (López Martínez 1989) indicate increasing aridity compared with the previous stage. Mammal fossils indicate an early Headonian age for the base of the sedimentary sequence (Peláez-Campomanes *et al.* 1989), while the age of the upper boundary is less well constrained but is probably early Oligocene (Suevian). In the Almazán basin, the overlying stage 4 sediments containing Oligocene–Miocene fauna rest unconformably on the stage 3 succession, indicating the presence of a significant time gap at the end of stage 3.

Stage 4 (late Oligocene–middle Miocene). This stage followed a major intra-Oligocene hiatus related to the Castilian compressional phase mentioned above. Two sequences, separated by a sedimentary break, are recognized. The oldest is Agenian–Ramblian in age and the other extends from Ramblian to middle Aragonian (Fig. 13.13). Lacustrine depocentres coincide with a maximum subsidence axis along a fringe located next to, and running parallel to, the northern and eastern borders (Fig. 13.12A). The first sequence is coarsening-upward in almost all the basin margins and reflects a compressional late Ramblian (?) phase coeval with maximum deformation due to compression in the Betic margin (Burdigalian; Vera 1988). This latter event gave rise to widespread tectonic instability and finally led to the discontinuity surface that caps this sequence and that is seen as a disconformity in the western border, an unconformity in the Almazán basin, and probably a paraconformity in the central domain (Fig. 13.13). In both the Almazán basin and the central domain (known from drilling) the lower sequence, dated by mammalian fossils (Calvo *et al.* 1993), shows an expansion of carbonate palustrine and evaporite environments at the base (Varas *et al.* 1999).

The Ramblian–Aragonian sequence records a major expansion of the lacustrine environments and a concomitant retrogradation of the alluvial and fluvial systems, producing an overall fining-upward sequence. Sedimentation occurred in shallow fresh-water carbonate lakes and palustrine environments that alternated with precipitation in evaporitic playa-lake systems. The top of the sequence has been dated as middle Aragonian using mammalian fossils (López Martínez *et al.*. 1986; Mazo *et al.* 1998; Fig. 13.13). Lake sedimentation was widespread in the central domain to the NE of Valladolid, although only the upper part of the lacustrine succession crops out here (Dueñas unit; 11 in Fig. 13.13). In the Almazán basin, lacustrine facies produced palustrine carbonates associated with calcretes. In the Bureba corridor that links the Duero and Ebro basins, an evaporite depocentre was developed during Agenian and Ramblian times, with a marginal association of marls and gypsum and, in the inner zone, marls, glauberite and anhydrite in a succession up to 200 m thick (Fig 13.14D). With this palaeogeography, the Duero basin attained an almost single-basin configuration for the first time, and the previously prominent Palaeogene fault-block compartmentalization nearly vanished. The main depocentre for this newly unified Duero basin lay in the area of Valladolid, with smaller ones persisting at the southeastern tip of the Almazán basin and Bureba corridor (Fig. 13.14D).

In all sequences, coarse-grained rocks appear as siliciclastic facies in the western domain, and petromictic conglomerates and litharenites elsewhere. Lacustrine successions consist of carbonates, marls and evaporites (interstitial and laminated sedimentary gypsum) with associated calcretes. During this stage, relatively arid conditions persisted, as indicated by evidence of open forests inhabited by herds of herbivorous animals (López Martínez 1989), calcrete development and evaporite precipitation in the central domains of the basin. The predominant presence of the 'Series Rojas' (10 in Fig. 13.13; Corrochano 1977; Alonso Gavilán 1981; Gracia *et al.* 1981), mostly along the western margin, is due to an in situ reddening weathering process consistent with a Mediterranean-type climate (Blanco *et al.* 1989).

Stage 5 (middle–late Miocene). During middle Aragonian times, a further period of instability gave rise to a disconformity that is traceable throughout almost the whole of the basin and dated using mammalian fossils (Fig. 13.12). This disconformity reflects a change in the stress field acting on the basin from extensional back to a compressional regime. It has been attributed to the Guadarrama phase in the Central Range (Capote *et al.* 1990) which may be correlated with the Neo-Castilian tectonic phase of the Madrid (Tajo) basin (Aguirre *et al.* 1976). This tectonic activity resulted in the development of SW–NE trending reverse faults, and the incision of deep valleys around the basin margins (Pol & Carballeira 1982). These dramatic changes induced the growth of marginal alluvial fans which fed well-defined and large fluvial systems whose influence during the onset of this stage reached the central part of the basin, where the former lacustrine environments were at times replaced by mud flats and palustrine areas in the Valladolid–Palencia sector. During this time, the area attained its maximum development as a single basin, although differential subsidence induced local variations in sedimentary environments within the central domain (Corrochano *et al.* 1991; Fig. 13.14E).

Coarse-grained rocks of the alluvial fans again reflect nearby source area compositions, whereas the fluvial sandy facies are chiefly immature sandstones, litharenites and arkoses, these latter mainly occurring in the southwestern and western margins (Corrales *et al.* 1978, 1986; Corrochano et al. 1983). Fluvial sedimentation, represented by the 'Arcillas de la Tierra de Campos' in the central basin (12 in Fig. 13.13), progressively receded and gave way, both up the succession and in the basin centre, to carbonate and gypsiferous sedimentation (known as the 'Cuestas' unit; 13 in Fig. 13.13) that represents a marginal fluvial-lacustrine zone. This is especially well developed along the western, northwestern and southwestern borders, where terrigenous-carbonate deposition in a distal alluvial plain–paludal–shallow lake system developed (Carballeira & Pol 1986; Corrochano *et al.* 1986; Corrochano & Armenteros 1989). On the southeastern and eastern borders, this belt is represented by a distal alluvial plain/mud flat–shallow carbonate lacustrine system, characterized by flood-related carbonate sequences that become more abundant upwards. Facies changes show a transition from brown sandy mudstones to calcretes and finally to palustrine carbonates and/or pond carbonates, palaeosols (calcretes) and shallow pond/palustrine carbonates (Armenteros 1986). In the basin centre, a chemical facies association consisting of marls, carbonates (limestones and dolostones) and gypsum (interstitial, sedimentary precipitated, and detrital) dominates (Fig. 13.14E). This association was deposited in a dry/saline mud flat–ephemeral saline lake system, alternating in time and space with shallow fresh-water carbonate lakes (Mediavilla 1986/87; Armenteros 1991). The stage ends with an extensive and expansive limestone level (known as 'Páramo Inferior' unit; 14 in Fig.13.13) which, despite its slightly diachronous nature, provides the most reliable marker horizon for correlation in the upper Miocene (Armenteros 1986). This widespread carbonate level represents shallow lacustrine sedimentation in the central basin, passing outwards to carbonate palustrine environments that in turn grade out into a receding distal alluvial plain, with abundant palaeosols (calcretes) that are especially well developed in the east. Overall, this carbonate unit records a period of tectonic stability. The lithological indicators of palaeoclimatic conditions, such as the widespread development of evaporites, calcretes and some paragenetic mineral associations linked to confined basins (Armenteros 1986; Mediavilla 1986/87; Armenteros *et al.* 1995), indicate significant dry periods during this stage. The flora of the area points to a seasonal temperate-warm climate, with cold and wet periods alternating with hot and dry ones (Rivas *et al.* 1994). Dry periods were especially well developed during the main evaporite depositional episode ('Cuestas' unit).

Stage 6 (late Miocene). During late Vallesian times, another disconformity records renewed tectonism, interpreted as corresponding to the Torrelaguna phase (Capote *et al.* 1990), and resulting in the movement of strike-slip and normal faults (Fig. 13.14F). This break in sedimentation resulted in the development of a karst surface on the underlying carbonates of the 'Páramo Inferior' (Corrochano & Armenteros 1989). The sedimentary sequence produced during this stage began with a sharp progradation of the alluvial fan and fluvial systems that locally reached the basin centre (Fig. 13.14F). The succession, which is broadly similar to that of stage 5 (minus evaporites), is exposed only in the eastern half of the basin. The thickest successions are found in the Burgos area and along a NW–SE fringe next to the southern border of the Almazán area, where subsidence persisted in depocentres parallel to the Iberian Ranges (Fig. 13.12A). The sequence shows several limestone units that today form plateaux (locally called 'páramos') at different levels (Fig. 13.13) which become successively younger, higher, and better preserved eastwards (Armenteros 1986), a result of the progressive west-to-east opening of the basin to the Atlantic ocean.

Mammalian ages have yielded a Vallesian age in the SE (VF on Fig. 13.13) and a Turolian age near the base of the sequence in the centre of the basin (TA on Fig. 13.13). Based on its Vallesian age, the VF fossiliferous site has been attributed to the previous stage (Calvo *et al.* 1993). However, a correlation based on the underlying 'Páramo Inferior' marker would suggest rather that it belongs to the last Miocene succession (stage 6; Armenteros 1986, 1991; Corrochano & Armenteros 1989). The almost complete absence of evaporites and prevailing open lacustrine sedimentation throughout the succession suggest less arid conditions than in the previous stage, an interpretation consistent with an increase in taxon diversity due to the establishment of relatively colder and more humid conditions (Rivas 1991).

Stage 7 (?late Miocene–?early Pleistocene). The establishment of an external drainage system induced a dramatic sedimentary break and the cessation of Tertiary sedimentation. The basin opened progressively by river capture from west to east, until its present configuration was attained (Martín Serrano 1991). The present-day river network has therefore inherited its configuration from the Miocene fluvial systems described above. The oldest deposits within this palaeogeographic setting are brown-yellow siliceous gravels associated with red sandy muds. They are known by the Spanish name of 'Rañas' (16 in Fig. 13.13), and overlie the disconformity surface on the last Neogene sequence (see Chapter 14). They are still partially preserved on the highest pediment platforms, appearing all around the borders of the basin, and are developed on both Tertiary and pre-Tertiary basement (Fig. 13.14F). The sediments and surfaces of the Rañas are related to the beginning of the present river system and represent the link between the end of the filling and the beginning of the opening of the basin. A diachronous age for the Rañas has been attributed to the progressive nature of the basin-opening process, and to the development of the pediments and their deposits (Martín Serrano 1991). Tentatively, the ages of these processes are suggested to be late Miocene–Pliocene in the west and late Pliocene–earliest Pleistocene in the east, and climatic conditions were wetter than previously. Finally, Quaternary erosion and sedimentation produced a fluvial terrace system and alluvial deposits related to the evolution of the current drainage network.

Summary and overview

From its initiation over 65 Ma ago the Duero basin has had a multistage evolution and its terrestrial infilling was controlled by tectonic compression from both the southern and northern margins of the Iberian peninsula, as well as by palaeoclimatic changes. The lithostratigraphic record has been divided into seven tectonosedimentary units separated by discontinuities, although from a broader perspective it can be grouped into two composite sequences, one Palaeogene and the other Neogene, separated by a long intra-Oligocene hiatus (Fig. 13.13).

Gradual Late Cretaceous marine regression brought about the progressive eastward establishment of terrestrial sedimentation in the region (Fig. 13.14A). Later, during the Palaeocene and Eocene epochs, compression in the north due to the convergence of the European and Iberian plates produced fault-controlled basins with intervening horsts (Fig. 13.14B). Rhenanian sedimentation in the basins produced coarsening-upward sequences in the west where there was a lower subsidence rate, and fining-upward sequences in the east where subsidence was greater.

By late Eocene (Headonian) times, widespread alluvial fan

progradation took place throughout the basin as a consequence of compressional stresses coming from both the Pyrenean and Betic plate margins (Fig. 13.14C). This tectonic episode brought about the uplift of the Cantabrian mountains, the Central Range and the Iberian Ranges. A general coarsening-upward succession resulted in an increasingly organized fluvial drainage system. A major tectonic event (Castilian phase) put an end to this stage and induced major unconformity.

Renewed sedimentation in late Oligocene times was linked to a sharp palaeogeographic change characterized by the establishment of lacustrine systems in the central domain and Almazán basin. Subsequently a widespread early Miocene (Ramblian) alluvial progradation occurred, coinciding with climactic tectonism related to strong compression along the Betic margin (Burdigalian–Ramblian). The stage culminated (middle Miocene) in the development of widespread lacustrine environments and for the first time the basin reached an almost single configuration (Fig. 13.14D). Subsequent middle Miocene tectonic activity, correlated with Neocastilian tectonism (middle Aragonian), initiated a new phase of sedimentation that was to continue into late Miocene times (Vallesian). A fining-upward sequence was deposited as carbonate and evaporite lacustrine environments became increasingly established across a wide area in the centre of the basin (Fig. 13.14E). Alluvial fans and fluvial systems were centripetally arranged around the central basin, and, by the middle of the stage, had become dramatically restricted, especially in the eastern domain. Finally, renewed tectonic activity in late Miocene times (Vallesian) induced unconformity and initiated the last depositional sequence in the basin infill, whose architecture was similar to that of the previous sequence except for the absence of evaporites (Fig. 13.14F).

Infilling of the basin ceased from around late Neogene times as it progressively began to open towards the Atlantic ocean. The process advanced from west to east, so that both fluvial siliciclastic deposits and the coeval pediments around the basin margins show a diachroneity that did not reach the east until Pleistocene times. This stage represents the link between the end of Tertiary infilling and the present erosional state of the basin (Fig. 13.14F).

With regard to palaeoclimatic evolution, the Tertiary succession records changes due both to the latitudinal variation (from 30°N in the Cretaceous to 40°N in the final Neogene) and to the effects of mountain uplift. The late Cretaceous climate was hot-humid tropical, passing through warm-humid subtropical conditions during Palaeocene–early Eocene times, to more arid subtropical conditions during the middle Eocene–Oligocene interval. This aridity persisted through early Miocene times (as recorded by evaporites and sclerophyllous forests), favoured by uplift of the surrounding mountains. The middle to upper Miocene climate was characterized by alternating climate, with two especially dry late Miocene intervals occurring around the Aragonian–Vallesian boundary (maximum development of gypsum facies) and during Turolian times. Relatively more humid and colder conditions then became established, lasting up to the end of Neogene times.

Tajo basin (AMA-Z, JPC)

Tertiary sedimentation in central Spain was focused into the Tajo basin, an intracratonic depocentre bounded by the

Central Range in the north, the Montes de Toledo in the south, and the Iberian Ranges in the east (Fig. 13.15). The structure and kinematics of these borders reflect differences in the transmission of stresses from the Iberian plate boundaries during Alpine tectonism (Alvaro *et al.* 1979). Variscan granitoid and metamorphic rocks of the Central Range form a pop-up structure rooted, at a depth of 11 ± 2 km, in a basal detachment that dips gently southward (De Vicente *et al.* 1996c). The contact between this mountainous uplift and the basin sediments to the south is a NE–SW trending reverse fault with a throw of more than 2000 m, which was active from Palaeogene into middle Miocene times. The Montes de Toledo in the south are composed mainly of granites and high grade metamorphic rocks that have been faulted over the sediments of the Madrid basin on an east–west trending reverse fault that dips 40–50° southwards. To the east, the Iberian and Altomira ranges are formed mostly by Mesozoic sedimentary rocks. The western part of the Iberian Ranges (the Castilian branch) forms the NE margin of the Tajo basin and consists of compressional and extensional duplexes associated with a group of anastomosing, NW–SE trending, strike-slip faults. The Altomira Range is a narrow, east-dipping fold-and-thrust belt trending north–south, emplaced during the latest Palaeogene to early Miocene interval. The presence of the Altomira Range in Neogene times created a barrier to sediment transport, subdividing the area into the Madrid basin and the much smaller Loranca basin ('intermediate depression') in the SE (Fig. 13.15).

Loranca basin (intermediate depression)

The Loranca basin is located between the Iberian and Altomira ranges (Fig. 13.15). This area has been characterized as a foreland basin produced by a westward-moving Iberian fold-thrust belt (Gómez *et al.* 1996). The basin is filled by 1000–1400 m of Eocene to Quaternary terrestrial sediments. Several authors have described the general stratigraphy of the basin in terms of five major successions, although complete agreement on the name, age and composition of each unit is still to be reached, especially for the Neogene deposits (Díaz-Molina & López-Martínez 1979; Torres & Zapata 1986; Gómez *et al.* 1996; Arribas *et al.* 1997; Torres *et al.* 1997).

During Palaeogene times two main sedimentary successions were deposited. The first succession (Eocene to upper Oligocene) consists of sandstones, gravels, mudstones and limestones deposited within a fluvial system that flowed from south to north. The second succession (upper Oligocene to lower Miocene) comprises sediments deposited in humid alluvial systems with the wide development of palustrine areas. Three subsequent Neogene successions have been recognized by Torres *et al.* (1997). The first of these is partially coincident with unit III of Gómez *et al.* (1996), is Ramblian in age, and is characterized by sandstones, gravels, mudstones, limestones and bioturbated gypsum deposited during an expansion of lacustrine evaporite environments. For the middle Miocene succession, the stratigraphic schemes proposed by Torres *et al.* (1997) and Gómez *et al.* (1996) show strong discrepancies.

Fig. 13.15. Geological setting of the Tertiary Tajo basin, the area of which today includes the Madrid basin, the smaller Loranca basin ('intermediate depression') and the intervening Altomira Range. Numbers indicate lithologies of the basin margins and main Tertiary units: 1, plutonic rocks; 2, shales, marbles, quartzites and gneisses; 3, shales and metagreywackes; 4, shales, quartzites and metavolcanic rocks; 5, Mesozoic basement (mainly carbonates); 6, Palaeogene (terrigenous and carbonates); 7, lower to upper Miocene terrigenous sediments; 8, lower Miocene unit; 9, intermediate Miocene unit; 10, upper Miocene unit; 11, Pliocene; 12, Quaternary.

According to the latter authors, there is a considerable time gap covering most of Aragonian and even Vallesian times. On the contrary, Torres *et al.* (1997) do not recognize such an unconformity, and consider that their second Neogene succession is Aragonian to lower Vallesian in age and formed by yellow sandstones, red mudstones, limestones and bioturbated gypsum. Finally, the third and last Neogene succession of Torres *et al.* (1997) includes clastic deposits and lacustrine carbonates that record upper Miocene sedimentation and are probably coincident with unit V of Gomez *et al.* (1996).

Madrid basin

The broadly triangular Madrid basin occupies >10 000 km^2, and is filled by Tertiary terrestrial deposits which range in thickness from a maximum of 3500 m in the west, to approximately 2000 m in other parts of the basin. This thick succession has been recognized from deep boreholes and seismic profiles (Junco & Calvo 1983). Palaeogene strata constitute a majority of the total thickness of the Tertiary deposits in the central parts of the basin. The Neogene sediments are about 800 m thick and have been divided into three Miocene stratigraphic units (Alberdi *et al.* 1984; Calvo *et al.* 1989) overlain by a thin Pliocene succession (Fig. 13.16). This distinction is based upon observable differences in lithologies in vertical sections as well as on the recognition of sedimentary discontinuities (palaeokarstic surfaces, erosive and/or minor angular disconformities) between the units. The chronology of these units has been determined mainly by macro- and micromammal faunas present at different stratigraphic levels throughout the Neogene succession (Calvo *et al.* 1990).

The Palaeogene rocks of the Madrid basin crop out in the northernmost part of the basin and also form a narrow belt fringing the east of the Altomira Range (Fig. 13.15). Two main stratigraphic units are distinguished (Portero & Olivé 1984), the lower being the *c.* 1100 m thick Torrelaguna-Uceda unit, which comprises Upper Cretaceous to upper Eocene red mudstones, gypsum and conglomerates. It was deposited in proximal to distal alluvial fan systems grading to evaporitic lakes. The upper unit is the *c.* 900 m thick Beleña de Sorbe–Torremocha de Jadraque unit, which has been subdivided by Arribas (1994) into an upper Eocene carbonate succession overlain by a lower Oligocene detrital succession. Overall, this upper unit represents the progradation of a large alluvial fan system over previous lacustrine environments.

The oldest of the three Miocene successions (lower unit) is Ramblian to middle Aragonian in age (Fig. 13.16). The lower limit of the unit has been recognized in seismic profiles whilst its upper limit has been dated from micro- and macromammal faunas from the Torrijos locality (Aguirre *et al.* 1982). The overlying succession (intermediate unit: Aragonian to lower Vallesian) contains abundant mammal localities, especially in the outskirts of the city of Madrid and in the northeastern part of the basin (Calvo *et al.* 1990). Both the lower and upper limits of the unit are defined by palaeokarstic surfaces, developed in gypsum (Rodríguez-Aranda & Calvo 1997) and carbonate strata (Cañaveras *et al.* 1996), respectively (Fig. 13.16). The base of youngest sequence (upper unit: upper Vallesian to Turolian) is marked by terrigenous deposits, which disconformably overlie a palaeokarstic surface in the central parts of the basin. Its upper limit is defined by a palaeokarst developed in the carbonate strata that form the top of the unit (Sanz 1996). The Neogene sedimentary fill of the basin ends with a relatively thin package of terrigenous and carbonate (mainly calcrete) sediments of Pliocene age (Fig. 13.16).

These Neogene deposits record the development of a mosaic of continental environments from alluvial fan to evaporite lakes (Fig. 13.17), whose evolution in space and time mainly reflects the influences of climate, tectonism and source rocks. Palaeoclimatic curves have been obtained through the study of the mammal associations (Calvo *et al.* 1993) (Fig. 13.16) and show two more humid periods during Ramblian and Vallesian times. The wetter conditions that prevailed during the Vallesian stage easily explains the wide development of shallow carbonate lakes at this time. On the other hand, during Ramblian times a wide development of evaporite deposits seems to be in contrast with the prevailing climate, but can be explained if one takes into account the uplift of older evaporite deposits from Altomira during that time. The main tectonic pulses in the basin were induced by fault movements along the various margins (De Vicente *et al.* 1996*a*) and were responsible for sometimes clearly disconformable boundaries between the different stratigraphic units. Further details on the lower, intermediate and upper units that comprise the Miocene succession of the Madrid basin are provided below.

Lower unit. In the central part of the basin, the Miocene lower (or Saline) unit (García del Cura 1979) comprises a thick (*c.* 500 m) evaporite succession, which is composed predominantly of anhydrite (commonly transformed to gypsum in outcrops), halite, clays, and non-skeletal carbonates (magnesite, dolomite), as well as important deposits of sodium sulphates (García del Cura *et al.* 1996). The evaporite facies grades laterally into reddish-green mudstones containing anhydrite and/or gypsum nodules, and then into coarser clastics deposited in alluvial environments (Ordóñez *et al.* 1991; Fig. 13.16). A concentric arrangement of these facies in the centre, south and east is deduced from mapping of the lacustrine and alluvial sequences in the basin, which suggests a geomorphic and hydrologically closed basin during sedimentation (Fig. 13.17). In the west, however, the deposits consist mainly of arkoses and associated sandy clays, which were supplied from the granitic and high-grade metamorphic rocks of the Central and Toledo ranges. Alluvial facies derived from both uplands coalesced distally in this part of the basin, the most distal areas of the alluvial systems being represented by mudstones with associated calcretes and/or carbonate pond deposits (Fig. 13.16).

A lacustrine system, developed in the central part of the basin, has been interpreted as a shallow but perennial saline lake that experienced significant changes in water level, and was periodically desiccated (Ordóñez *et al.* 1991). Glauberite–anhydrite and anhydrite–magnesite couplets, together with halite beds and clays, form most of the saline lake strata. Glauberite occurs mainly as euhedral, discoidal and bipyramidal crystals that are interpreted as having nucleated in the brine, subsequently sank to the bottom, and probably continued growing after burial (Ordóñez & García del Cura 1994). Besides this occurrence, glauberite is also associated with mudstones, forming cyclic sequences in which it is typically nodular and/or banded and probably representative of marginal lake areas during shallowing-upward trends (Ortí 2000).

Anhydrite beds display nodular/enterolithic and laminar fabrics, providing evidence that their formation took place under both intrasedimentary and subaqueous conditions. Magnesite is present as centimetre-thick layers between the sulphate deposits, although its origin, either as a primary precipitate or an early diagenetic product from hydromagnesite and/or huntite, is not yet clearly established. Halite is present as centimetre- to decimetre-thick beds showing bottom-nucleated crystals (mainly chevron structures); poikilotopic clear halite and halite cements are also common (Ordóñez & García del Cura 1994). Besides these evaporite minerals, polyhalite displaying spherulitic fabrics has been occasionally recognized in some sedimentary sequences. Finally, thenardite, an anhydrous sodium

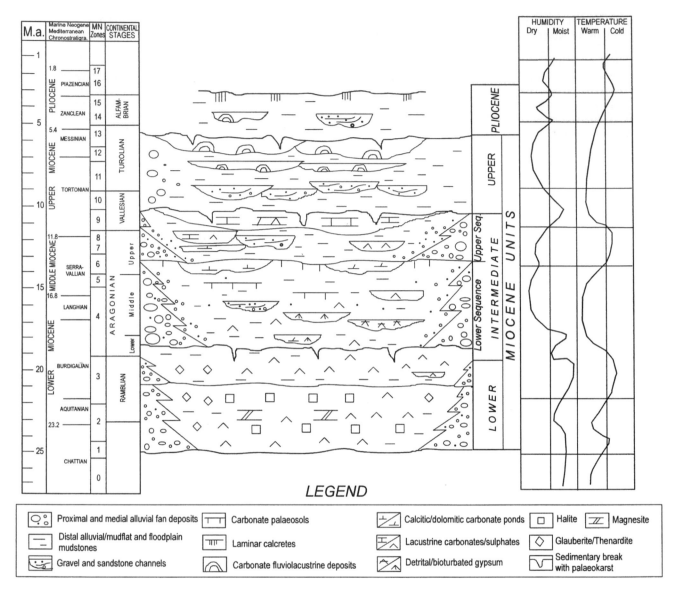

Fig. 13.16. Miocene lithostratigraphy integrated with sedimentary and climatic record of the Tajo basin.

sulphate, forms relatively thick deposits in the uppermost part of the lower unit. These thenardite deposits have been interpreted as a result of erosion and incongruent dissolution of the underlying glauberite deposits, leading to the formation of sodium- and sulphate-enriched brines that contributed to the precipitation of almost pure thenardite (Ordóñez & García del Cura 1994).

Gypsum is widely present in outcrops of the lower unit. However, most of the gypsum comprises secondary gypsum fabrics due to hydration of precursor sulphates, especially anhydrite and glauberite. In contrast, primary gypsum is not common, with the exception of some deposits located near the eastern margin of the basin (Altomira Range), where very distinctive gypsum fabrics, the so-called 'Christmas-tree gypsum' (Rodríguez-Aranda *et al.* 1995), form beds up to 2.5 m thick. The deposits are interpreted as having accumulated in restricted lakes away from the main evaporite lake basin (Fig. 13.17). Brines in these marginal lakes reached only moderate concentrations, probably because of the more permanent supply of fresh water from the nearby uplands. As a result, only gypsum precipitated subaqueously and grew up from the lake floor, whereas other saline phases characteristic of more concentrated saline conditions are absent.

Lacustrine saline deposits of the lower unit of the Madrid basin are interpreted to have been sourced by extensive recycling of Mesozoic and Palaeogene evaporite formations that were uplifted in eastern and northern parts of the basin (Altomira and Iberian ranges). This conclusion is based on ^{34}S data (Ordóñez *et al.* 1991) that provide evidence of the isotopic relationship between the older and newly formed evaporites. Besides the existence of a large source for solutes, precipitation of evaporites was also facilitated by the fact that the Madrid basin lay in a local 'rain shadow' region in central Iberia during the early and middle Miocene intervals (Calvo *et al.* 1996).

Intermediate unit. This succession has the greatest outcrop area, and its thickness varies from 200 m in the NE to 50 m in central and southern areas. These variations in thickness have been interpreted as indicative of the influence of a NW–SE trending strike-slip fault zone, active during the compression of the Iberian Ranges (De Vicente *et al.* 1996*a*). The unit spans the middle Aragonian to early Vallesian interval (Fig. 13.16) and has a sharp base, marked by an extensive palaeokarst developed on the underlying lower unit and clearly visible in central areas of the basin (Calvo *et al.* 1984). In the NE the base of the unit shows clastic deposits that have prograded over

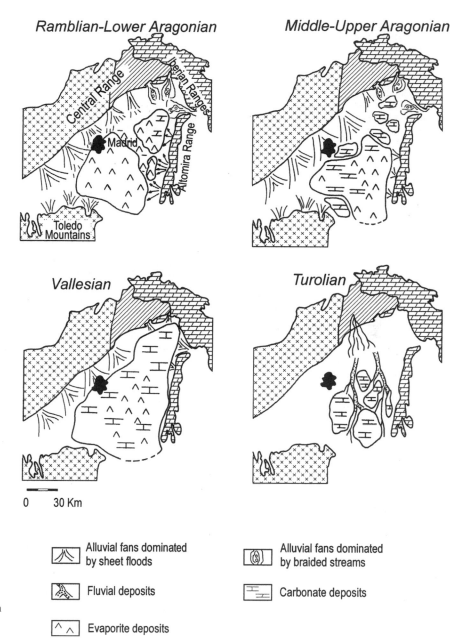

Fig. 13.17. Miocene palaeogeographic evolution of sedimentary environments in the Madrid basin.

the lacustrine carbonates of the lower unit (Alonso-Zarza *et al.* 1990). The intermediate unit comprises two sequences (lower and upper) each showing clastic alluvial sediments at the base passing gradually upwards to lacustrine carbonates and/or evaporites. The boundary between the two sequences is defined by an abrupt progradation of clastic deposits on lacustrine sediments that occurred in upper Aragonian times (Fig. 13.15).

Alluvial fan deposits form a discontinuous fringe around the basin margins and their width varies from hundreds of metres to 30 km. In the north the most proximal alluvial fan facies have coarse conglomerates full of granite boulders, although more typical sediments are thick, stacked coarse arkosic sands deposited by ephemeral floods. Distal alluvial fan facies comprise fine arkoses, clays (smectites, illite, sepiolite) and calcretes–dolocretes. In the NE wide alluvial fans developed, entrenched within structurally controlled palaeovalleys (Alonso-Zarza *et al.* 1993). Braided streams, occasionally meandering, deposited coarse gravel beds and dominated these fans. Within regular but steep slopes of the margins, small alluvial fans and scree slope deposits were dominated by debris flows. In the south, proximal alluvial facies are rarely

preserved and sheet-like sandstones grading distally into red mudstones are more typical (Sanz 1996).

Extensive palaeosol development and the presence of carbonate ponds are particularly characteristic of this unit (Sanz *et al.* 1995). Calcretes have been studied mainly in the north and NE (Wright & Alonso-Zarza 1990; Alonso-Zarza *et al.* 1992b) where these soils have been analysed as an integral part of the sedimentary record and the palaeolandscape. Pedostratigraphic studies have shown that in the NE the more mature soils occur in the medial fan areas, where the common effect of decreasing sedimentation rates down-fan, and the coarse-grained size of the deposits favoured more mature soil development. In floodplains, palaeosol maturity increases with the distance to the main channels, and so a pedofacies relationship (Bown & Kraus 1987) is observed (Alonso-Zarza *et al.* 1992b). In contrast, in the north continuous sedimentation of arkosic fans did not favour soil development, resulting in the more mature soils being located in the most distal fan fringe, where calcretes and dolocretes display calcified root cells interpreted as *Microcodium* b (Alonso-Zarza *et al.* 1998a). In these areas, silcretes are also present (Bustillo & Bustillo 2000).

Carbonate ponds developed in areas relatively close to the basin margins, either in floodplain environments (NE) or in mudflat (south) settings (Sanz *et al.* 1995). Ponds formed during periods of reduced clastic input as the alluvial systems were retreating towards the margins under periods of tectonic quiescence. This led to a modification of the previous drainage systems that became more widely spaced, allowing the formation of unconnected, low relief and wet depressions. These depressions were favourable sites for the precipitation of very shallow lake carbonates (ponds), either micrites or dolomicrites with rare bioclasts, with variable amounts of detrital grains and abundant evidence for desiccation and bioturbation.

In the lower sequence of the intermediate unit, lacustrine deposits consist of carbonates and evaporites that interfinger with mudstones (locally Mg-rich clays) at lake margins (Ordoñez *et al.* 1991). Evaporite deposits occur in the south and east of the basin, forming in shallow lakes of moderate salinity (Rodríguez-Aranda & Calvo 1998). In these lakes, chemical precipitates included selenite and lenticular gypsum, which were reworked by sheet floods to produce extensive detrital gypsum deposits (Sanz *et al.* 1994). The bioturbation of massive chemical gypsum beds and intercalated gypsiferous marls by invertebrates and plants gave rise to a highly characteristic lithofacies, the bioturbated gypsum deposits, which is very common in continental Tertiary basins of Spain (Rodríguez-Aranda & Calvo 1998). In the northern areas of the basin, around Madrid, lacustrine deposits are mostly dolomitic and include nodular and massive bedded dolostones, stromatolites, bedded limestones and dolostones with gypsum moulds. These together represent a slightly evaporitic carbonate environment (Calvo *et al.* 1995) and contrast with the freshwater carbonates found in the NE (Alonso-Zarza *et al.* 1990).

In the upper sequence of the intermediate unit, lacustrine deposits are mostly carbonates, with evaporites being more restricted to areas in the south and east. The topmost part of the intermediate unit records a wide expansion of these lacustrine areas (Fig.13.16). Palustrine carbonates are characteristic of this stage (Alonso-Zarza *et al.* 1992a), but locally deeper lakes developed (Bellanca *et al.* 1992).

Upper unit. This unit comprises up to 50 m of terrigenous and carbonate sediments which occur mainly in the central and eastern part of the Madrid basin. This distribution contrasts with the concentric facies arrangement shown by the underlying units (Fig. 13.17), and has been interpreted as resulting from a change in late Miocene times from compressive to extensional regional stresses (De Vicente *et al.* 1996a). The lower part of the unit comprises mainly terrigenous deposits that accumulated in north–south trending fluvial depositional systems supplied by granitic and high-grade metamorphic rocks derived from the Central Range. The fluvial facies consists of superposed fining-upward sequences, which can be interpreted as deposits of an anastomosing alluvial system. River distribution was also controlled by geomorphological constraints derived from the evolution of the karstic landscape sculpted at the top of the underlying Miocene unit (Cañaveras *et al.* 1996). Carbonate sediments form the upper part of the unit and are mainly fresh-water tufas and fossiliferous micrites, containing variable amount of oncoids, gastropods and charophyte stems and gyrogonites. They are interpreted as having accumulated in a mosaic of shallow, fresh-water lakes subject to water level fluctuations, favouring the development of a strong palustrine imprint in the carbonates. This effect, together with extensive meteoric diagenesis related to karstification, contributed to a high degree of induration in the carbonates despite the fact that they did not undergo significant burial (Sanz 1996; Wright *et al.* 1997).

Pliocene outcrops are sparsely distributed within the Madrid basin. In the south and east two different Pliocene units are recognized. The lower unit, named the 'red series' by Pérez-González (1982) or 'Pliocene detrital unit' by Sanz (1996), consists of 0–40 m of red mudstones, sandstones and conglomerates. This unit crops out following a north–south trend, which may indicate that sedimentation occurred in the tectonic depression known as the Tajo syncline (Capote & Fernández-Casals 1976). The unit was deposited by fluvial systems in which many of the channels were dominated by carbonate sedimentation including oncoliths, stromatolites,

tufas and ooids (Ordoñez & García del Cura 1983). Channels filled by siliciclastic grains, floodplain deposits, lacustrine carbonates and calcretes are also present in this unit (Sanz 1996). The upper Pliocene unit consists of a multistorey, 1–5 m thick laminar calcrete (Pérez-González 1982; Sanz 1996) that shows a lower 'chalky' part and an upper part formed by amalgamated carbonate laminae. Detailed studies of this laminar calcrete (Sanz 1996) indicate that it formed by different erosion–sedimentation–calcrete formation periods, strongly influenced by organic activity, during a prolonged period of tectonic quiescence.

Southern Spain (JCB, JMM, JA)

The geology of southern Spain is dominated by the Betic Cordillera and a series of Neogene basins that developed and filled while the Betic mountains uplifted (Fig. 13.18). The Betic Cordillera is the westernmost segment of the European part of the Alpine orogenic belt and is the result of convergence of the passive southern margin of the Palaeozoic Iberian Massif (the External Zones) and a continental crustal wedge mostly composed of metamorphic rock complexes (the Internal Zones or Alborán Domain). A series of nappes of marine sediments (the Flysch Units or Campo de Gibraltar Complex) occur along the contact between the External Zones and the Internal Zones (Fig. 13.18). The Palaeogene to middle Miocene sedimentary record in these three major crustal domains will be treated here separately. To a large extent, the Palaeogene to early Miocene sedimentary evolution in each domain was a continuation of its Mesozoic depositional history. Middle Miocene sedimentation in the External Zones occurred during compressional stacking and displacement of nappes whereas within the Internal Zones it was coeval with extensional tectonism. This latter phase of crustal extension controlled the palaeogeographic evolution of the Betic region which formed the northern margin of the Alboran basin in the western Mediterranean Sea. From late Miocene times onwards, tectonism within the Betic Cordillera produced discrete basins (the Neogene basins) that evolved in connection either with the Mediterranean sea or with the Atlantic ocean.

Palaeogene to middle Miocene in the External Zones

The External Zones represent the Mesozoic-to-middle Miocene southern continental margin of the Iberian Massif. This passive margin had been divided by Jurassic rifting into Prebetic and Subbetic domains (García-Hernández *et al.* 1980; Vera 1988; Chapter 11). The Prebetic constituted the more proximal domain where continental and shallow marine sedimentation prevailed, whereas to the south the Subbetic domain was a pelagic basin from the end of Early Jurassic times. Individual rifted blocks caused local variations in Subbetic sedimentation (Vera 1988), but sediments in this domain generally record an overall change from Triassic to lowermost Jurassic continental and shallow-platform marine environments to pelagic settings from Pliensbachian times onwards (García-Hernández *et al.* 1980). The Palaeogene sequences in the Prebetic domain are mainly controlled by tectonics, interpreted as far-field responses to compression during the Pyrenean collision (Geel *et al.* 1998; Geel 2000). In addition, eustatic sea-level changes may have contributed to cyclicity during late Eocene and Oligocene times. Compressional tectonics led to the emplacement of Subbetic nappes upon the

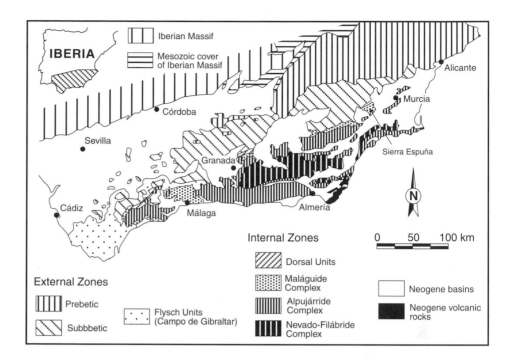

Fig. 13.18. Geological scheme of the Betic Cordillera in southern Spain.

Prebetic area so that the original transition from platform to slope and basin deposits can now only be observed locally (Geel *et al.* 1998).

Prebetic. The Palaeogene deposits of the Prebetic domain comprise continental and shallow marine sediments to the north and NW and deeper shelf and slope facies to the south and SE. These sediments are poorly known and their descriptions are almost entirely limited to unpublished PhD theses, geological maps from the early 1970s and internal reports of the Instituto Geológico y Minero de España (Jerez-Mir 1981). Partially dolomitized Danian white bioclastic shelf carbonates rich in coral and algal fragments and other invertebrate debris overlie an erosion surface on Upper Cretaceous rocks. Thanetian limestones intercalating with, and changing southeastwards to, marly limestones and marls with planktonic foraminifers complete the reported Prebetic Palaeocene succession. The limestones are mostly micritic, include small coral patches and large foraminifers, and probably represent an open-platform environment.

Eocene to Oligocene deposits have been intensively studied in the eastern Prebetic (Alicante region), whereas in the western sector available information is again restricted to geological maps and unpublished reports. Fourteen sequences separated by unconformities have been recognized in the Eocene platform deposits of the Alicante region (Geel *et al.* 1998; Fig. 13.19). Most sections, however, only contain part of these sequences, probably due to tectonism. Using planktonic and loose large benthic foraminifers, six of the sequences can be attributed to the early Eocene interval, five to the middle Eocene, and three to the late Eocene (Geel 2000).

Lower Eocene successions mainly consist of limestones formed on a ramp and grading southeastwards into slope marls and marly limestones. Inner-ramp carbonates contain miliolids and other imperforate foraminifers such as alveolinids. Middle- and outer-ramp packstones are rich in large perforate foraminifers (nummulitids and *Discocyclina*; Geel 2000). Middle and upper Eocene successions in the platform also comprise carbonates (limestones and dolostones) but have intercalated beds of quartz sandstones in their most proximal sections (Fig. 13.19). Retrograding ramp carbonates in the lower part of the sequences show imperforate foraminifers characterizing their innermost facies, with perforate nummulites and *Discocyclina* being the most common components in the packstones of the middle and outer ramp. In contrast, the prograding carbonates that complete the sequences were deposited on shelves rimmed by coral reefs. Miliolids and alveolinids are the main bioclasts in the protected shelf area,

whereas perforate foraminifers and coralline red algae are common in the outer shelf facies (Geel 2000). Laterally equivalent slope deposits that crop out to the south and SE comprise mass-flows and turbidites with north and northwestern provenances included in marls with planktonic foraminifers (Geel *et al.* 1998; Geel 2000).

In the Oligocene platform deposits, four sequences separated by erosional surfaces have been distinguished (Fig. 13.19). All four sequences are attributable to the lower Oligocene (Rupelian) and are interpreted as recording transgressions and sea-level highstands, with the unconformities between being thought to correspond to sea-level lowstands (Geel 2000). During the transgressions, the inner platform facies was characterized by grainstones with miliolids and small lepidocyclinids together with coralline red algae. Coral reefs grew in the mid-shelf and backstepped to the NW. Perforate large foraminifers are the most common components in the outer platform packstones. During sea-level highstands, the prograding limestones formed on a shelf rimmed by coral reefs (Geel 2000). As in the Eocene sequences, protected-shelf facies are rich in miliolids and alveolinids and other imperforates. Quartz sandstones and silts occur in the most proximal sections and are interbedded with, and prograde on top of, the miliolid grainstones (Jerez-Mir 1981; Geel 2000). Rotaliids and *Lepidocyclina* are frequent in the outer-shelf packstones and grainstones. In the distal sections of the Alicante Prebetic, the lower Oligocene deposits consist of marls intercalating mass-flows and turbidites with current directions indicating a source area to the north and NW. Upper Oligocene, nodular pelagic limestones with no coeval rocks on the platform can be locally identified (Fig. 13.19).

A phase of faulting at the end of the Oligocene epoch divided the Prebetic domain into blocks that were to have different sedimentary evolutions during the early and middle Miocene intervals (Geel *et al.* 1992). The former Palaeogene platform to the north and NW of the domain was partitioned into smaller segments, while new relief emerged in the SE. A narrow, NE–SW orientated strip of the Prebetic became part of the foreland trough of the Betics, the so-called north Betic straits.

Three sedimentary sequences have been recognized in lower to middle Miocene deposits of the Alicante region and were probably controlled by tectonism related to plate movements in the western Mediterranean (Geel *et al.* 1992; Fig. 13.19). The lowermost, Aquitanian in age, comprises platform carbonates (limestones with miogypsinids and *Lepidocyclina*) that change laterally to marls and submarine-fan detrital sediments. Overlying an unconformity on top of the previous unit, the Burdigalian–lower Langhian succession

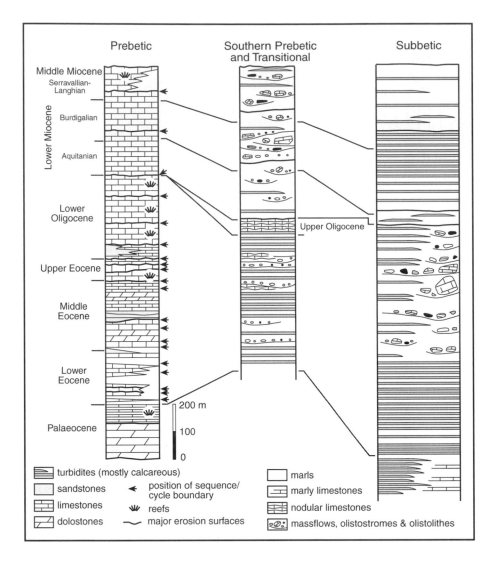

Fig. 13.19. Composite, simplified columns of the Palaeogene to middle Miocene deposits in the Prebetic and Subbetic domains, External Zones of the Betic Cordillera. Data for the Prebetic from the Alicante region after Geel *et al.* (1992, 1998) and Geel (2000). Data for the Subbetic from Comas (1978) and Soria (1993).

consists of platform limestones rich in coralline algae that pass basinwards to marls including olistoliths and turbidites. The southern emergent areas became the main source of detrital material. The upper Langhian–Serravallian sequence also overlies an erosional unconformity and comprises deposits similar to those in the underlying sequence, although with a different spatial distribution of the facies belts.

Subbetic. As mentioned above, the Subbetic area was the Triassic to Miocene distal domain of the passive southern Iberian margin, with marine pelagic sedimentation prevailing from Early Jurassic times onwards. Particularly characteristic was the widespread deposition of alternating micritic limestones and marls (or marly limestones) rich in nannoplankton and planktonic foraminifers from the Late Cretaceous epoch to the end of the middle Eocene interval (Vera & Molina 1999). These limestone and marl alternations change laterally to localized bodies of turbidite and mass-flow deposits (Fig. 13.19). Palaeocene turbidites mainly consist of calcarenites rich in reworked *Microcodium*. Lower to middle Eocene turbiditic calcarenites, calcirudites and mass-flows with clasts and blocks of Mesozoic rocks change laterally to marls and micritic limestones.

In the northern sections of the Subbetic domain, upper Eocene, Oligocene and Aquitanian deposits comprise turbidites and massflows, whereas the southern areas are characterized by very thick mass-flow accumulations and huge olistoliths (more than 3 km in outcrop width) included in autochthonous marls (Comas 1978; Fig. 13.19). The Eocene to Aquitanian turbidites in the Subbetic are

mostly calcarenites and calcirudites rich in large benthic foraminifers, coralline algae and other bioclasts derived from the platform. The turbidite facies indicate the occurrence of submarine fans at the toe of the slope that descended basinwards from the Prebetic platform of the Iberian margin. Rocks from the Mesozoic sedimentary pile of this margin were exposed in submarine cliffs and canyons and fed the boulders and blocks in the mass-flows. Such exposure and collapse of Mesozoic materials was particularly common in the southern part of the domain, pointing to the existence of tectonically active submarine highs that delimited the southern margin of the Subbetic trough (Comas 1978). Deep-water pelagic sedimentation continued in most of the Subbetic region during the rest of early and middle Miocene times. Sedimentation was coeval with ongoing compression within the External Zones after their early Miocene collision with the Internal Zones. The Subbetic became stacked into a series of nappes that were active at least until Tortonian times. This Miocene syntectonic sedimentary evolution of the Subbetic domain is poorly known, and both data and hypotheses on the depositional environments and age of the successive units are few. Several units separated by unconformities have been recognized in lower to middle Miocene rocks of the southern Subbetic domain (Fig. 13.19). They mainly comprise marls intercalated with turbidite sandstones and conglomerates that locally change to calcarenites and sandstones formed on platforms around submarine highs or emergent areas (Soria 1993). The lower to middle Miocene deposits in the northern Subbetic area and in its transition to the Prebetic domain mostly consist of marls, turbidite and mass-flow deposits that include large olistoliths of Mesozoic and Cenozoic rocks derived from the Subbetic sedimentary pile (Fig. 13.19). They are interpreted as deep-water deposits accumulated in the north Betic straits that acted as the foreland basin of the Subbetic nappes. These chaotic deposits were subsequently involved in the displacement of the Subbetic nappes and accumulated as exotic elements in the Tortonian marls of the Guadalquivir basin (Vera 2000).

Palaeogene–lower Miocene 'flysch units' (Campo de Gibraltar complex)

The so-called 'flysch units' (Campo de Gibraltar complex) occur as a series of tectonic units in a narrow belt along the contact between the External Zones and Internal Zones of the Betic Cordillera. This belt widens in the province of Cádiz, where these units crop out extensively on the northern side of the Straits of Gibraltar (Fig. 13.18). In southern Spain, the 'flysch units' mainly comprise Palaeogene and lower Miocene rocks, but several tectonic units made up mostly of Mesozoic (Cretaceous) rocks occur as well in small, dispersed outcrops (Martín-Algarra 1987). Similar rocks in laterally equivalent tectonic settings can be mapped throughout the Rif Cordillera in northern Morocco and into the Kabylia Mountains of Algeria and Tunisia, on the southern side of the Mediterranean sea. The Cenozoic deposits of the Campo de Gibraltar complex can be subdivided into two major types (Algeciras and Aljibe units) distinguished by their sedimentary characteristics and tectonic position, although several units intermediate in character have also been recognized (Pendón 1978). These deposits are all deep-water marine sediments that can be dated with planktonic foraminifer assemblages.

The Algeciras unit and its equivalents comprise Palaeocene to Oligocene clays, silts and silty marls intercalated with thin layers of turbidite sandstones and conglomerates, which in many cases are of carbonate composition (Fig. 13.20). As in other Palaeogene sequences in southern Spain, reworked *Microcodium* can locally be a major component of the turbidite calcarenites. In the excellent exposures along the coast south of Algeciras, this unit consists of a monotonous rhythmite (up to 1000 m in thickness) of turbidite sandstones, silts and marls Oligocene to Aquitanian in age (Fig. 13.20).

The Aljibe unit and its equivalents, collectively known as the Numidian units or Numidian nappe, tectonically overlie the Algeciras-type ones. The lower part of the Numidian units is up to 100 m thick and consists of clays and marls and intercalated thin beds of turbidite calcarenites and calcareous breccias, locally rich in derived *Microcodium*. The upper part of these units is made up of a very thick (>1000 m) succession of frequently channelled and amalgamated beds of quartz sandstones with thin beds of brownish to grey silts (Pendón 1978; Fig. 13.20). The age of these deposits probably ranges from Oligocene to Aquitanian (Martín-Algarra 1987). Some 10 m of clays and marls with thin beds of turbidite sandstones and siliceous layers locally occur on top of the Numidian sandstones (Fig. 13.20). These fine-grained deposits contain lower Burdigalian planktonic foraminifers (Didon *et al.* 1984).

Brownish to grey clays, probably lower Burdigalian in age, occur as a mappable unit along with other Campo de Gibraltar units. These clays have been described by many local names and collectively ascribed to the 'Neonumidian flysch' (Bourgeois 1978) or the 'Numidoid' formation (Olivier 1984). The Numidoid clays are interbedded with Numidian-type quartz sandstones and siliceous deposits, and engulf blocks and huge olistoliths of rocks with a wide range of ages (Mesozoic to early Miocene) and diverse provenances from the Internal Zones and from other units of the Campo de Gibraltar complex.

The lower clays and marls of the Algeciras-type units have been interpreted as deposits of a submarine plain locally reached by distal submarine fan sediments. They were covered during Oligocene times by prograding outer submarine-fan lobes in which rhythmic turbidite sandstones, silts and marls formed (Pendón 1978). The nature of the lithoclasts in the sandstones and the scarce conglomerates and

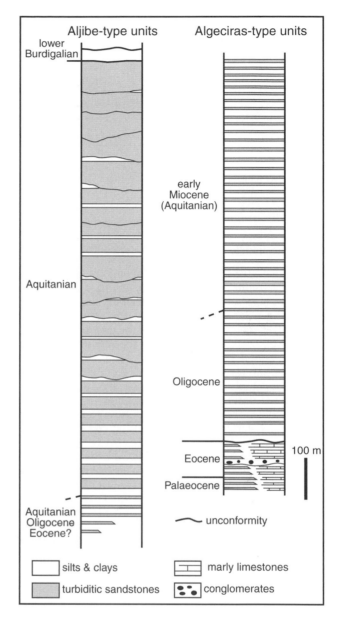

Fig. 13.20. Simplified columns of the two major types of 'flysch units' (Campo de Gibraltar complex), after Martín-Algarra (1987).

breccias in these units suggest that the Internal Betic Zones, in particular the Maláguide complex, were the source area of the detrital materials (Peyre 1974). The Aljibe-type units also reflect an upward change from submarine-plain to submarine-fan deposits. The composition and morphology of quartz grains, together with other components in the Numidian sandstones, indicate that North Africa was the source area for the sediments (Guerrera & Puglisi 1983). The two sets of submarine-fan deposits are thought to have formed in a long trough (the 'flysch trough') separating the Internal Zones from the African plate to the south and the Iberian plate to the west (Martín-Algarra 1987; Fig. 13.21). The trough was closed by the convergence of all these crustal segments and the flysch units were squeezed and tectonically superposed on both the Internal Zones and the External Zones. The Numidoid clays probably formed during the last period of trough evolution in deep areas locally covered by the Numidian submarine fans, onto which blocks and mass-flows descended from adjoining, tectonically active submarine reliefs.

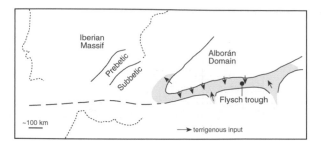

Fig. 13.21. Palaeogeographic sketch of the location of the 'flysch trough' between the African plate and the Alborán domain during the Aquitanian. The Algeciras-type units were fed from the Alborán domain whereas the Aljibe-type units were fed from North Africa (modified after Martín-Algarra 1987).

Palaeogene–middle Miocene of the Internal Zones

The Internal Zones constitute a crustal wedge emplaced against the southern Iberian margin during early Miocene times. They consist of three stacked tectonic complexes which in ascending order are the Nevado-Filábride, the Alpujárride and the Maláguide (Fig. 13.18). Only the uppermost complex contains well-developed Palaeogene deposits since the lower two were buried in deep crustal locations and underwent metamorphism for at least part of the Palaeogene. The Alpujárride complex was exhumed and cooled to near-surface temperatures during early Miocene times (*c.* 21–19 Ma BP; Platt & Whitehouse 1999) and the Nevado-Filábride was unroofed and started to be eroded at the end of the middle Miocene interval (Johnson *et al.* 1997).

Palaeogene–lower Miocene deposits overlying the Mesozoic series in the Maláguide complex have been recognized in all the major outcrops of the complex. Geel (1973) first made a systematic description of the Maláguide Palaeogene in the northern part of the Almería province and defined several stratigraphic units, but the most recent study on the Cenozoic materials of this complex was carried out by Martín-Martín (1996) in Sierra Espuña, where the most complete and best-exposed sections occur.

Five major sedimentary units separated by unconformities can be distinguished in the Palaeogene–lower Miocene succession in Sierra Espuña (Fig. 13.22). The lowermost unit, Palaeocene in age according to its planktonic foraminifer and nannoplankton assemblages, consists of alternating silty marls and calcarenites, calcareous sandstones and conglomerates. The carbonate beds contain *Microcodium*, marine invertebrates, foraminifers (including large foraminifers) and dasyclad and coralline algal fragments. The deposits of the overlying lower-middle Eocene (upper Ypresian–lower Lutetian) unit formed in shallow marine and transitional settings. The shallowest deposits occur to the SE now and consist of marls, marly limestones rich in gastropods and lignite beds. These deposits change laterally to thick (up to 150 m) inner-platform limestones with imperforate foraminifers (miliolids and alveolinids) that in turn pass northwestwards to open-platform limestones rich in nummulitids and coralline algae. Sandstones and conglomerates are mixed with the platform carbonates in the lower part of the sequence (Martín-Martín 1996; Fig. 13.22). The middle-upper Eocene (upper Lutetian–Priabonian) unit includes materials deposited in sedimentary environments similar to those of the underlying sequence, and which pass basinwards to marls intercalated with calcarenite beds. Internal cyclicity can be recognized within a general deepening trend of the facies in this unit (Fig. 13.22). The overlying lower Oligocene sediments are more restricted in thickness (up to 15 m) and outcrop due to erosion (Fig. 13.22). The shallowest deposits are marls with gastropods formed in backshore environments, and these change laterally to inner-platform calcareous

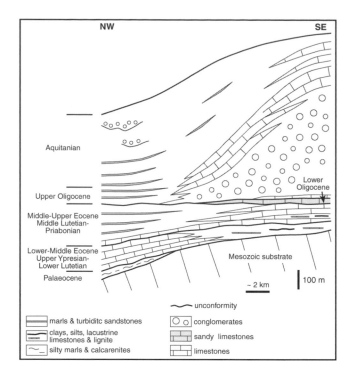

Fig. 13.22. Stratigraphy of the Palaeogene–Lower Miocene deposits from the Maláguide complex in Sierra Espuña (data compiled from Martín-Martín 1996).

sandstones and conglomerates and then to open-platform limestones with quartz grains and *Lepidocyclina*. The last unit is late Oligocene–Aquitanian in age and includes algal limestones with nummulitids and *Lepidocyclina* formed on a marine platform on which fan-deltas developed locally. Thick conglomerates (up to 600 m) with calcareous clasts eroded from the Maláguide Mesozoic substrate and underlying Palaeogene units accumulated on these fan-deltas. The algal limestones change laterally to marls interbedded with turbiditic calcarenites, sandstones and conglomerates formed in submarine fans (Fig. 13.22).

Palaeogene to lower Miocene sediments, mostly deep-water marine marls and calcareous turbidites, unconformably overlie Mesozoic rocks in very small, dispersed outcrops of some tectonic units palaeogeographically related to the Alpujárride domain and collectively included in the 'dorsal units' shown on Figure 13.18 (Martín-Algarra 1987). A polymictic breccia of angular clasts derived from the underlying Palaeozoic and Mesozoic rocks is associated with the pelagic marls and the Mesozoic substrate in several of these 'dorsal units'.

Younger Miocene marine sediments unconformably overlie different tectonic units of the Alpujárride and Maláguide complexes. They comprise polymictic breccias and conglomerates, turbidite sandstones and calcarenites, siliceous beds, marly limestones and marls containing nannoplankton and planktonic foraminifers dated as latest Aquitanian–early Burdigalian (Sanz de Galdeano *et al.* 1993). These mostly deep-water marine sediments have various local names and usually appear in small scattered outcrops at the base of the Miocene sequences in intermontane Neogene basins resting on Internal Zone basement (Figs 13.23 and 13.24). They represent the first marine sediments to cover the exhumed metamorphic rocks of the Alpujárride complex, which, along with the Maláguide complex, provided the source for clasts in the breccias and conglomerates.

Langhian shallow-water conglomerates, sandstones and reef carbonates unconformably overlie the older Miocene successions as well as Alpujárride and Maláguide rocks, and record the uplift and emergence of the precursor topography to the present-day Sierra Nevada–Sierra de los Filabres mountains (Braga *et al.* 1996a; Fig. 13.24). This mountain chain is built by the complexes of the Internal Zones and includes the highest mountains in the Iberian peninsula,

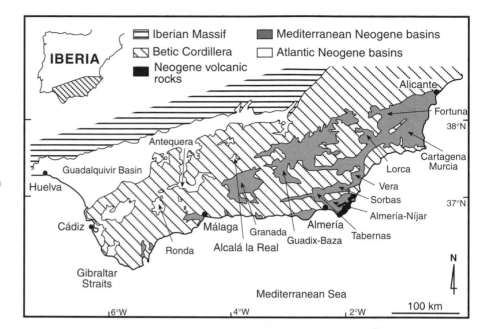

Fig. 13.23. The Neogene basins in southern Spain. Fortuna, Lorca, Guadix-Baza and Granada are inner Mediterranean-linked basins, while Vera, Sorbas, Tabernas, Almería-Níjar and Cartagena-Murcia are outer Mediterranean-linked basins. The inner basins were isolated from the Mediterranean sea and became continental during latest Tortonian or early Messinian times, whereas the outer basins remained connected to the Mediterranean sea during the rest of the Miocene epoch.

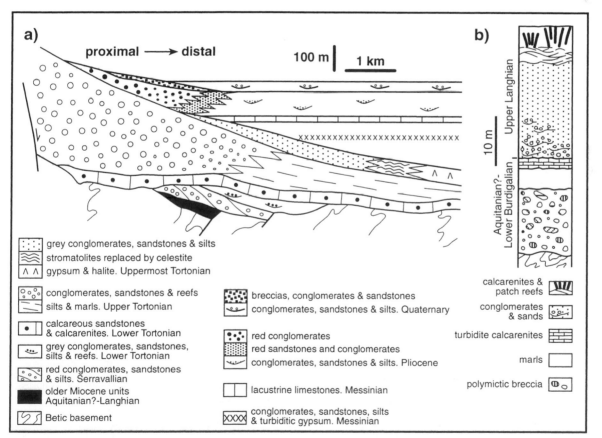

Fig. 13.24. (a) Neogene stratigraphy of the southeastern margin of the Granada basin, representative of the inner, Mediterranean-linked Neogene basins. (b) Detailed column of the older Miocene units (Aquitanian?–Langhian) (after Braga *et al.* 1990, 1996*a*).

with several peaks above 3000 m. The Langhian shallow-water deposits grade outwards from the axis of the Sierra Nevada–Sierra de los Filabres precursor into deeper-water marls and turbidite sandstones.

Finally, continental red conglomerates, sandstones and siltstones spread over older Neogene deposits and the Alpujárride-Maláguide

basement in small outcrops dispersed over a large area of the Internal Zones (Fig. 13.24). Locally they include micromammal fossil assemblages that indicate a Serravallian age (Martín-Suárez *et al.* 1993). They pass northwards and southwards (i.e. away from the precursors of the major mountain chain) into beach and deeper-marine facies (Serrano 1990; Soria 1993).

Neogene basins

A series of sedimentary basins, the so-called Neogene basins, developed as discrete entities both on and between the External Zones and the Internal Zones at the beginning of late Miocene times (Fig. 13.23). The palaeogeography and extent of these intermontane, late Neogene basins has strongly conditioned the modern landscape of southern Spain, since major present-day depressions correspond to the areas occupied by the basins. At the same time, a major foreland basin, the Guadalquivir basin, developed along the northwestern front of the Betic Cordillera (Vera 2000; Fig. 13.23).

The intermontane basins located on the External Zones are either continental (Prebetic basins) or marine. The continental basins are of restricted occurrence and include lacustrine conglomerates, sandstones and mudstones passing upwards into diatomites of late Tortonian–Messinian age (Calvo *et al.* 1978; Elizaga & Calvo 1988). With regard to the marine depocentres (e.g. the Ronda basin), these were in fact marginal embayments of the Guadalquivir basin and as such linked to the Atlantic ocean. The majority of the Betic intermontane basins opened directly to the Mediterranean sea, although some maintained links with the Atlantic ocean during late Miocene times (Soria *et al.* 1999).

Guadalquivir basin and Atlantic-linked intermontane basins. The Guadalquivir basin is the foreland basin of the Subbetic nappes and it opens directly to the Atlantic in the WSW, in the Gulf of Cádiz area (Sanz de Galdeano & Vera 1992). Its upper Miocene–Pliocene sedimentary infill comprises five units, interpreted by Sierro *et al.* (1996) as unconformity-bounded sequences. From bottom to top these are early–middle Tortonian, middle–late Tortonian, latest Tortonian–earliest Messinian, late Messinian–early Pliocene, and early Pliocene in age, according to their planktonic foraminifer and nannoplankton fossil assemblages. Maximum thicknesses are normally around 300–400 m for each of the units, although they may reach up to 700 m in places (Roldán-García 1995). All units have bioclastic sandstones deposited on a platform at the eastern margin of the basin and prograded to the WSW on top of silty clays and marls deposited on the platform slopes. Basinal deposits consist of clays and marls with locally developed thick bodies of sandstone turbidites. Alluvial and fluvial red sandstones of uncertain age (late Pliocene–Pleistocene?) overlie the marine deposits in Huelva (Zazo 1979). At the southern margin of the basin, olistostrome deposits and olistoliths, located at, but detached from, the frontal part of the Subbetic nappes, are intercalated within upper Tortonian clays and marls. Messinian clays and sandstones, up to 450 m thick, locally with calcarenites, occur on top of the olistostromic units (Riaza & Martínez del Olmo 1996).

South of Cádiz, marine sedimentation continued up to the end of the Pliocene epoch and three units separated by unconformities can be recognized (Aguirre 1995). These range in age from latest Messinian to late Pliocene according to their planktonic foraminifer assemblages. The lower unit is made up of inner-platform bioclastic sandstones that change gradually to fine-grained sediments deposited on an outer platform. The second unit is made up of beach and backshore calcirudites, sandstones and siltstones. The last unit in the Bay of Cádiz is a shallowing-upward sequence of wave- and tidal-dominated delta deposits prograding to the SSW. Fine-grained sandstones and siltstones formed in small interdistributary bays where oyster bioconstructions developed (Aguirre 1998a). In the southern areas protected from terrigenous influx, rhodolith-dominated carbonates were deposited in a small sheltered bay, whereas coeval lacustrine sediments were deposited in inland areas (Aguirre *et al.* 1993, 1995). Finally, the Ronda, Antequera, Alcalá la Real and several other minor basins formed large late Miocene embayments along the southern edge of the Guadalquivir basin and became filled mostly by platform temperate carbonates and mixed siliciclastic–carbonate deposits interfingering laterally with basinal marls.

Mediterranean-linked intermontane basins. Two main types of Mediterranean-linked basins can be distinguished. (a) Inner basins (distant from the present-day Mediterranean sea) mainly occur at the contact between the External Zones and Internal Zones. The Fortuna, Lorca, Guadix-Baza and Granada basins belong to this type, with the Granada basin being our choice of the most representative example (Figs 13.23 and 13.24). (b) Outer basins (near to the present-day Mediterranean), such as the Vera, Sorbas, Tabernas, Almería and Cartagena-Murcia basins, are located on the Internal Zones. The Sorbas basin has been selected as the representative example (Figs 13.23 and 13.26).

Both types of basin have a similar sedimentary evolution up to late Tortonian times. During the latest Tortonian–early Messinian interval the inner basins were isolated from the Mediterranean sea and became continental. The outer basins, however, remained connected to the Mediterranean sea during the rest of Miocene times, this continuing in some cases through the Pliocene epoch, except during the short-lived Messinian salinity crisis (Hsü *et al.* 1977; Riding *et al.* 1998).

Lower Tortonian sedimentary rocks typically consist of temperate marine platform bioclastic calcarenites up to 50 m thick, with abundant remains of bryozoans, coralline algae and bivalves (Brachert *et al.* 1996; Esteban *et al.* 1996). They also contain a variable amount of siliciclastic deposits (sandstones and conglomerates), which can be locally dominant. Cross-bedding stratification is the most conspicuous internal sedimentary structure (Rodríguez-Fernández 1982; Betzler *et al.* 1997).

Huge conglomerate–sandstone bodies up to several hundred metres thick formed at the foot of the main emergent areas during late Tortonian times (Dabrio *et al.* 1978; Braga *et al.* 1990). These bodies consist mostly of debris-flow conglomerates, with individual clasts up to several metres in size, and sandstones deposited in fan deltas that locally include coral reefs (Martín *et al.* 1989; Braga *et al.* 1990; Fig. 13.24). These coastal deposits change laterally to basinal, fine-grained sediments (silts and marls) containing planktonic nannofossils and foraminifers. Thick (up to 500 m) turbidite sandstone bodies locally formed in channels and lobes of submarine-fan systems (i.e. in the Tabernas basin; Kleverlaan 1989) and interfinger with the basinal marls.

As mentioned above, the inner and outer Mediterranean-linked basins evolved independently from latest Tortonian-early Messinian times onwards. Inner basins such as the Granada, Lorca and Fortuna basins became isolated from the sea and desiccated (Fig. 13.25). Except for the Guadix-Baza basin, significant evaporite deposits (up to several hundred metres in thickness) accumulated in these basins. They mainly consist of gypsum and, in some cases such as in the Granada (Dabrio *et al.* 1982; Martín *et al.* 1984) and Lorca (Playà *et al.* 2000) basins, included salt (halite) deposits. Younger sediments overlying the evaporites are in all cases of continental origin.

The most complete continental sedimentary record, up to 500 m thick, occurs in the Guadix-Baza basin. Alluvial and fluviatile sediments (conglomerates, sandstones and siltstones) pass laterally into lacustrine deposits (silts and marls, limestones and gypsum). Several units, separated by unconformities, can be distinguished, and the sediments range in age from Turolian to Pleistocene (García-Aguilar & Martín 2000). The Granada basin exhibits a similar continental sedimentary record (Fernández *et al.* 1996; Fig. 13.24), although it also contains coal layers, a few metres thick, intercalated with marginal lacustrine deposits of Turolian age.

Bioclastic calcarenites with abundant remains of bryozoans, bivalves and coralline algae, and a variable siliciclastic (sandstone/microconglomerate) content, accumulated in the outer, Mediterranean-linked, intermontane basins during latest Tortonian–earliest Messinian times (Martín *et al.* 1996; Braga *et al.* 2001; Fig. 13.26). These sediments are temperate, carbonate-platform deposits with a thickness of around 40 m, changing basinwards to marls containing planktonic foraminifers (Martín & Braga 1994; Martín *et al.* 1999; Sánchez-Almazo *et al.* in press). Two Messinian reef units, each around 50 m thick, overlie the temperate carbonates (Fig. 13.26). The lower one consists of coral (*Porites* and *Tarbellastraea*) and

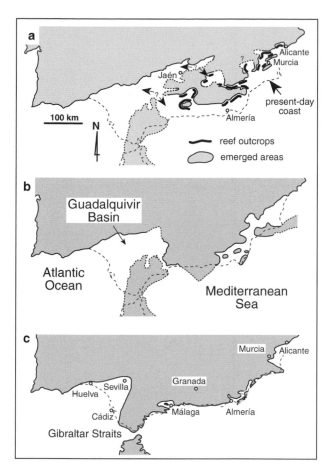

Fig. 13.25. Palaeogeographic evolution of southern Spain from the late Tortonian to early Pliocene interval (after Esteban *et al.* (1996) for the Miocene, and Montenat (1977), Sierro *et al.* (1996) and Aguirre (1998*b*) for the Pliocene). **(A)** Late Tortonian; **(B)** Early Messinian; **(C)** Early Pliocene.

Halimeda bioherms occurring within calcarenites and/or silts and marls (Braga *et al.* 1996*b*; Martín *et al.* 1997). The upper unit comprises fringing reefs composed almost exclusively of *Porites* heavily encrusted by microbial, micritic carbonates (Riding *et al.* 1991*b*). These fringing reefs extend more than 20 km laterally and prograde seawards for more than 1 km (Dabrio *et al.* 1981; Braga & Martín 1996). Reef carbonates change laterally basinwards to marls with planktonic foraminifers and diatomites.

A clearly marked erosional surface occurs on top of the fringing reefs (Dabrio *et al.* 1981; Riding *et al.* 1991*b*) and their laterally equivalent basinal marls (Riding *et al.* 1998, 1999). This exposure surface developed during the Messinian desiccation event that resulted in the precipitation of huge masses of salt in the centre of the Mediterranean (Hsü *et al.* 1977). Thick (>100 m) selenite–gypsum deposits (Dronkert 1977) directly overlie this erosional surface in the outer basins (Fig. 13.26). The gypsum beds onlap the underlying Messinian marls and fringing-reef talus-slopes (Riding *et al.* 1999). These beds are marine evaporites according to their strontium, oxygen and sulphur isotope ratios (Playà *et al.* 2000). Silt and marl marine layers containing planktonic foraminifers and calcareous nanno-plankton are intercalated within the gypsum, which has been inter-preted as deposited during repeated episodes of evaporitic conditions as the sea re-invaded some of these tectonically barred, marginal basins immediately after the main Mediterranean desiccation event (Riding *et al.* 1998).

The unit on top of the gypsum is also Messinian in age and consists of coastal sandstones (locally conglomerates), up to 70 m thick,

changing laterally to basinal silts and marls containing planktonic foraminifers and nannoplankton (Riding *et al.* 1998). Microbial (stromatolite and thrombolite) domes are ubiquitous within the unit (Riding *et al.* 1991*a*; Martín *et al.* 1993; Calvet *et al.* 1996; Braga & Martín 2000). Small *Porites* patch-reefs, together with coralline algae and bivalves also occur locally in the sandstones/conglomerates (Fig. 13.26).

Continental deposits of late Messinian–early Pliocene age (Martín-Suárez *et al.* 2000) overlie the Messinian marine sequences in the Sorbas basin, although a final marine invasion during the early Pliocene interval is recorded by thin bioclastic sandstones. In the basins closest to the Mediterranean, Pliocene deposits accumulated mainly in marine settings (e.g. Elche, Vera, Almería-Níjar and Málaga basins; Fig. 13.25). A marked unconformity divides the Pliocene sedi-mentary record into two units (Montenat 1977, 1990; Boorsma 1992; Aguirre 1998*b*). The lower unit is an upward-shallowing sequence, early Pliocene to earliest late Pliocene in age (Montenat 1977, 1990; Aguirre 1998*b*) and can be recognized from Catalonia to Cádiz (Martinell 1988; Aguirre 1998*b*). The unit comprises outer platform silts and clays at the base and shallow nearshore deposits at the top. The shallower facies include siliciclastic and mixed siliciclastic–carbonate deposits formed on prograding deltas (Völk 1966; Postma 1983; Aguirre 1998*b*), and carbonate sediments deposited in starved areas beyond the influence of terrigenous input.

The lower unit sedimentation was terminated by a phase of intra-Pliocene deformation that caused a major palaeogeographic reorganization of the basins. The palaeocoast shifted to a position close to the present-day coastline with marine sediments preserved only in small palaeoembayments. In the Almería-Níjar basin, the sediments belonging to the Pliocene upper unit consist of fan-delta deposits (conglomerates, sandstones and silts) that prograded in a sheltered bay where ahermatypic coral banks grew (Aguirre & Jiménez 1998). The unit can be dated as late Pliocene by its planktonic foraminifer assemblages (Aguirre 1998*b*). Gilbert-type deltas developed in the southern margin of the Serrata de Níjar (SE Almería-Níjar basin) and in the Vera basin (Völk 1966; Postma & Roep 1985; Boorsma 1992).

Balearic islands (JJF, LP, ER)

Mallorca, Menorca, Eivissa, Formentera, Cabrera and other minor islands (Fig. 13.27) compose the Balearic archipelago. These islands are the emergent parts of the Balearic promon-tory, a thickened continental crustal unit forming the NE continuation of the Alpine Betic thrust and fold-belt. The major topographic highs of these islands are horsts formed during post-middle Miocene extension and expose deformed Palaeozoic to middle Miocene rocks. The intervening grabens are flat areas, filled by upper Miocene and younger sequences that can exceed 1000 m in thickness (Benedicto *et al.* 1993). We have split the Tertiary rocks into nine sequences, each bounded by regional unconformities and representing pre-, syn- and post-orogenic sedimentation. These sequences can comprise just one, or a number of, formal or informal litho-stratigraphic units. The pre-orogenic sequences (I and II) are composed of Palaeogene rocks, the synorogenic sequences (III, IV and V) are early to middle Miocene in age, and the post-orogenic sequences (VI, VII, VIII and IX) were deposited in late Miocene to late Pliocene times.

Eocene–Oligocene: the pre-orogenic sequences

These rocks crop out on Mallorca, Menorca and Cabrera islands, and range in age from Lutetian (middle Eocene) to Chattian (late Oligocene; Ramos 1988; Ramos *et al.* 1989). They lie unconformably on a Mesozoic basement that is mostly Early Cretaceous on Mallorca and Menorca and Jurassic–Cretaceous in age on Cabrera. The pre-orogenic

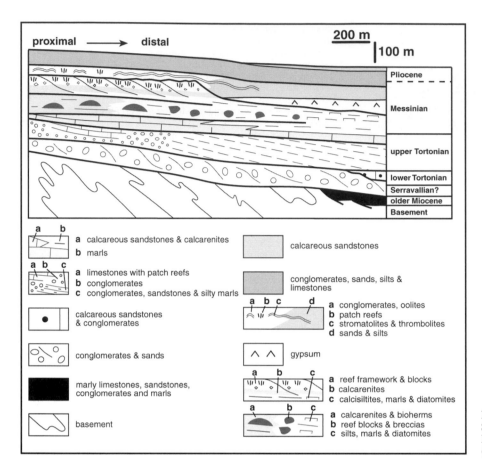

Fig.13.26. Neogene stratigraphy of the Sorbas basin, representative of the outer, Mediterranean-linked Neogene basins (modified from Martín & Braga 1994).

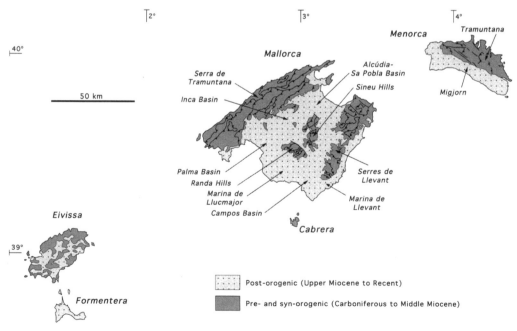

Fig. 13.27. Geological sketch of the Balearic islands showing the major morphostructural domains cited in the text.

record from these Teriary rocks is incomplete, so that palaeo-geographical and palaeoclimatological reconstructions of the Balearic basins during the Palaeogene are difficult to make. Despite this, Ramos *et al.* (2001) have been able to reconstruct a middle Eocene to Oligocene palaeogeography for the Mallorca area, which begins (middle–late Eocene) with

marine shelf carbonates in the SE (SE Mallorca and Cabrera) isolated from fresh-water lake carbonates in the NW (NW Mallorca and probably north of Menorca; Fig. 13.29A). During late Eocene to earliest Oligocene times, relative sea-level rise produced transgressive onlap of marine strata over the highlands towards the NW. These transgressive deposits have been documented only in Mallorca where coral patches and nummulitic banks occurred. Subsequent uplift of the NW highlands later in the early Oligocene interval reversed this trend, and an alluvial system prograded towards the south and SE (Fig. 13.29B). Thus, to the NW of the Balearic Islands there was a subtropical–tropical continental area prior to the opening of the Valencia trough, with mammal assemblages indicating no geographical barriers between Balearic and Iberian populations during Oligocene times (Ramos *et al.* 2001).

Sequence I. This includes both marine and non-marine sediments, deposited in two separate palaeogeographic domains: SE (Cabrera and SE of Mallorca) and NW (north of Menorca and NW of Mallorca). The SE domain contains littoral and shallow marine deposits, whereas the NW domain comprises lacustrine and palustrine carbonates that are thickest in Mallorca but are thin to locally absent in Menorca.

In the SE domain, the oldest rocks are marine shelf deposits of the S'Envestida Calcarenites Fm (S'E in Fig. 13.28) which crops out at the SW end of the Serres de Llevant on Mallorca, and on Cabrera. The lower part of the sequence is composed of 75 m thick marls, interbedded with siltstones, skeletal packstones and grainstones. Large foraminifera (*Nummulites*, *Operculina*, *Discocyclina*, *Assilina*, *Alveolina* and *Orbitolites*; Ramos 1988, 1993), ahermatipic corals (Álvarez *et al.* 1989), echinoderms, rhodoliths and *Ophiomorpha* burrows are abundant. Locally, packstones form in situ accumulations of monospecific *Nummulites* A and B forms (*Nummulites* banks). This lower succession gradually changes upward into a 20 m thick shallow marine, white limestone rich in *Nummulites* and miliolids. The S'Envestida Calcarenites Fm has been given a latest Lutetian–Bartonian age on the basis of its *Nummulites* and *Alveolina* content (Ramos 1988; Ramos *et al.* 1989).

In the NW domain, the Peguera Limestone Fm (Pe in Fig. 13.28) is 140 m thick along the SE edge of the Serra de Tramuntana (Mallorca) but only a few decimetres thick in the north of Menorca. These limestones contain thin layers of fine-grained red beds, organic-rich marls, and coal, and represent lacustrine and palustrine sequences. Ramos *et al.* (2001) distinguished three facies assemblages in the Peguera Limestone Fm, recording a lacustrine basin evolution from initiation and widening and deepening of the lake to lake-margin progradation and final basin infill.

In the marginal palustrine carbonates, root traces, brecciation,

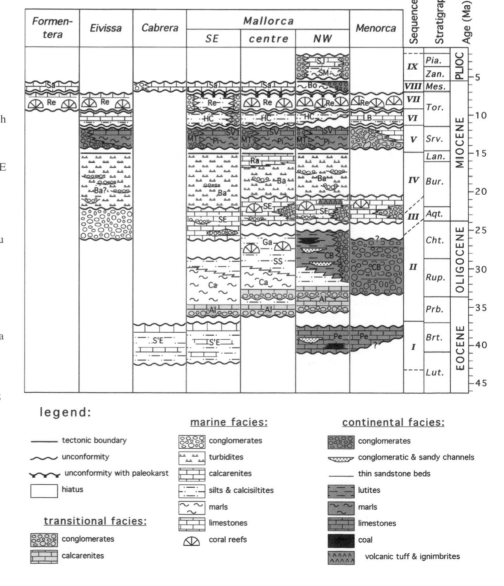

Fig. 13.28. Simplified stratigraphic sketch for the Tertiary successions of the Balearic islands. The column for Mallorca has been divided into three domains: SE, centre and NW. Column SE refers to the Serres de Llevant for the pre- and synorogenic sequences and the Campos basin, as well as Marina de Llevant and Marina de Llucmajor areas for the post-orogenic sequences. The centre column refers to Randa and Sineu Hills for the pre- and post-orogenic sequences and the northern part of Campos and the southern part of the Alcudia basins for the post-orogenic sequences. Column NW refers to the Serra de Tramuntana for the pre- and synorogenic sequences and the northwestern sides of Palma, Inca and Sa Pobla basins for the post-orogenic sequences. Key: **Al**, Alaró Calcarenites Fm; **Ba**, Banyalbufar Turbiditic Fm; **Bo**, Bonanova marls unit; **Ca**, Calvari Marls Fm; **CB**, Cala Blanca Detrital Fm; **Ga**, Galdent Limestones Fm; **HC**, *Heterostegina* calcisiltites unit; **LB**, lower bar unit; **MT**, marginal terrigenous complex; **Pe**, Peguera Limestones Fm; **Pi**, Pina marls unit; **Ra**, Randa calcarenites unit; **Re**, reefal complex; **Sa**, Santanyí limestone unit (terminal complex); **SE**, Sant Elm Calcarenites Fm; **S'E**, S'Envestida Calcarenites Fm; **SJ**, Sant Jordi calcarenites unit; **SM**, Son Mir calcisiltites unit; **SS**, Son Santre Lutites and Sandstones Fm; **SV**, Son Verdera limestones unit.

Microcodium and small-sized oncolites are widespread, and tufa and travertine precipitation occurred. In the depocentres the lacustrine sedimentation was characterized by rhythmic coal and organic-rich laminated limestone accumulation, evolving upwards into massive or laminated limestone. The coals, which have long been mined in Mallorca, are classified as sub-bituminous C/B type, whereas the scattered organic matter indicates a type III kerogene (Ramos *et al.* 2001). The Peguera Limestone Fm includes abundant charophyta (Ramos 1988), gastropods (Vidal 1917), mammals (De Bruijn *et al.* 1978; Hugueney & Adrover 1982), crocodile and turtle (Ramos 1988), and plant remains (Bauzá 1961). On the basis of their charophyte and mammal content, Ramos (1988) attributed a Bartonian age to the formation.

Sequence II. This sequence crops out on Mallorca and Menorca and includes marine and non-marine deposits organized in a transgressive–regressive succession. The transgressive and highstand deposits comprise littoral and shallow marine carbonates (Al, Ca, SS and Ga in Fig. 13.28) and onlap, toward the NW, both the older Palaeogene rocks (sequence I) and the Mesozoic basement. The regressive deposits comprise fan-delta and alluvial fan lithofacies, prograding towards the SE.

The Alaró Calcarenites and the Calvari Marls formations represent the transgressive suite of this sequence (Fig. 13.28). Basal conglomerates are overlain by calcarenites containing a typically shallow-marine fauna of complex porcelaneous larger foraminifera such as *Peneroplis evolutus* Henson, *Austrotrillina howchini* (Schlumberger) and *Praerhapydionina delicata* Henson (Colom 1929, 1975a; Ramos 1988), as well as *Nummulites* and *Operculina* (Ramos *et al.* 1989). The Alaró Fm passes upwards and basinward (SE) into the finer-grained Calvari Marls Fm. The latter crops out mostly in the SE of Mallorca (Serres de Llevant), as well as locally in the centre of the island where it is overlain by the terrigenous Son Sastre Lutites and Sandstones Fm and Galdent Limestone Fm, composed of coral patch-reefs (Fig. 13.28). The marine deposits of the Galdent and Son Sastre formations are rich in corals (Álvarez *et al.* 1989) and larger foraminifera are also widespread. Locally, the littoral facies yields abundant open-marine gastropods (Vidal 1905) although opportunistic forms, characteristic of brackish-water environments, usually dominate (Ramos & Martinell 1985). On the basis of their larger foraminifera content, Ramos *et al.* (1989) attributed a Priabonian (late Eocene) to Rupelian (early Oligocene) age to these formations.

The non-marine Cala Blanca Detrital Fm (CB in Fig. 13.28) crops out on Mallorca (Serra de Tramuntana and Central zone) and on the northern part of Menorca (Fig. 13.28). Alluvial-fan and fan-delta conglomerates and sandstones dominate in the NW, but pass southeastwards into fine-grained sandstones and lutites, representing alluvial mud flats, and then to limestones and coal-bearing sequences related to lacustrine basins (Ramos 1988; Ramos *et al.* 2001). Fluvial and lacustrine deposits contain a significant amount of microbialites, mainly stromatolites and oncolites (Colom 1961; Ramos 1988; Ramos *et al.* 2001). Palaeocurrent structures suggest that terrigenous sediment was supplied from the denudation of limestone mountains (mainly Jurassic and Cretaceous limestone) located towards the NW of the present-day position of the island.

The fossil content of the proximal alluvial facies of the Cala Blanca Fm is negligible, whereas its distal-alluvial to lacustrine facies has yielded fossil mammals (Hugueney & Adrover 1989–90; Hugueney 1997) at several localities in Mallorca. Other fossil finds include gastropods (Esu 1984), charophytes from Mallorca (Ramos 1988) and Menorca (Bourrouilh 1983; Ramos 1988), and Álvarez-Ramis *et al.* (1987) have studied the overall macro- and microflora in Mallorca. On the basis of these finds, the Cala Blanca Detrital Fm has been attributed to the late Rupelian to Chattian interval (Oligocene).

Lower-middle Miocene: the synorogenic sequences

These successions accumulated during the Alpine orogeny which affected the Balearic area during early and middle Miocene times. The sediments, which onlap both Palaeogene and Mesozoic rocks, have been involved in folding and thrusting. The initial compressive stage produced imbricated

NW-directed thrust sheets, but later (post-Langhian) there evolved an extensional regime that produced a series of horsts and grabens bounded by NE–SW and NW–SE striking vertical faults. The most representative exposures of synorogenic sequences occur in Mallorca (albeit in rather restricted outcrops), although they are also found on Menorca and Eivissa, but are absent on Cabrera and Formentera. According to Ramos *et al.* (1989), the synorogenic Tertiary succession can be split into three unconformity-bounded sequences (III to V in Fig. 13.28).

During earliest Miocene times, shallow marine carbonates with some terrigenous influence onlapped an irregular palaeotopography (Fig. 13.29C) which to the north emerged as islands where lacustrine and pyroclastic rocks were deposited. Lacustrine palaeobotanical data suggest warm and humid conditions in the northern highlands (Arènes & Depape 1956), with volcanic eruptions being related to the opening of the Valencia trough (Mitjavila *et al.* 1990).

The emplacement of thrust sheets towards the NW during Burdigalian–Langhian times, produced a series of foreland basins in which southerly derived coeval turbiditic sedimentation occurred (Fig. 13.29D). Proximal to distal turbiditic deposition patterns suggest progressive deepening of the basin, probably related both to tectonic subsidence and a general middle Miocene sea-level rise. In some areas, such as the central zone of Mallorca, the foreland basins were infilled enough to induce shallow marine conditions in a carbonate-dominated environment.

The final (Serravallian) compressional episodes produced a general emergence and erosion of the uplifted areas, leading to rapid resedimentation of the older turbiditic marls into small, fault-controlled basins (Fig. 13.29E). In these intramontane basins, palustrine (Pina Marls unit) and mixed lacustrine–palustrine (Son Verdera limestones) environments predominated (Simó & Ramón 1986; Ramos *et al.* 2000). The rate of tectonic subsidence in these basins was rapid (Benedicto *et al.* 1993), leading to thick accumulation of non-marine successions, and the development of small alluvial fans along the fault-bounded basin margins (the marginal terrigenous complex).

Micromammal faunas in the middle Miocene deposits suggest that the opening of the Valencia trough, and the Algerian basin, already constituted efficient biological barriers hampering connections with both the Iberian and African plates (Mein & Adrover 1982; Adrover *et al.* 1983–84). This indicates the insular nature of the Balearic domain during middle Miocene times, but the limits of these ancestral Balearic islands are as yet poorly defined.

Sequence III. This unit consists of marine limestones, littoral deposits, and alluvial fan sediments (including palaeosols) that rest with angular unconformity on a Mesozoic and Palaeogene basement. It typically reaches a thickness of several tens of metres, and was first defined by Rodríguez-Perea (1984) as the Sant Elm Calcarenitic Fm (SE in Fig. 13.28). It crops out mainly on Mallorca, but also on Menorca (Bourrouilh 1983) and in the north of Eivissa (Rangheard 1971). The most common lithologies comprise carbonate lithoclastic and bioclastic calcarenites, with coral patch reefs, red algae biostromes and seaweed deposits recording a shallow shelf environment. At some localities, the presence of planktonic foraminifers with sponges record sedimentation in a deeper (aphotic) shelf. Both the shallower and deeper limestones are interlayered with conglomerates with a calcarenitic marine matrix, mostly in the lower parts of the succession. Additionally, in the Serra de Tramuntana, sequence III includes a basal interval of brackish-water lacustrine deposits (Oliveros *et al.* 1960; Colom 1975) which yielded a palaeofloral assemblage (Arènes & Depape 1956).

TERTIARY 331

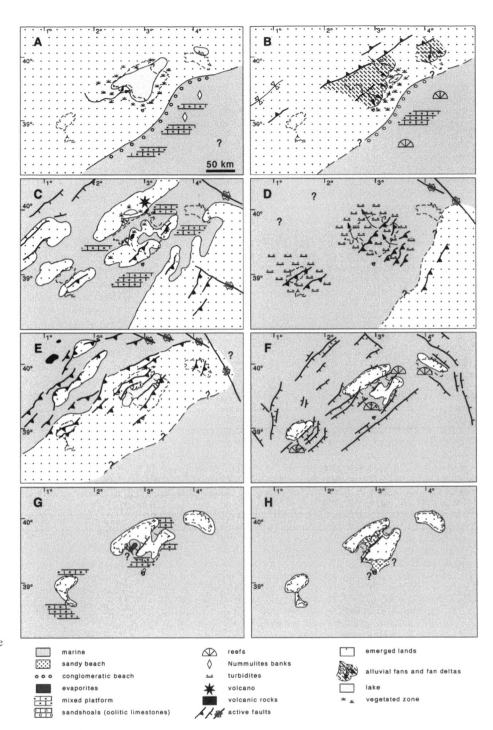

Fig. 13.29. Palaeogeographic reconstructions and Tertiary evolution of the Balearic area. **(A)** Bartonian (sequence I). **(B)** Oligocene (upper part of sequence II). **(C)** Aquitanian (sequence III). **(D)** Burdigalian (sequence IV). **(E)** Serravalian (sequence V). **(F)** Tortonian (sequence VI). **(G)** Messinian (sequence VIII). **(H)** Pliocene (sequence IX). Based on data of Biju-Duval *et al.* (1977), Dercourt *et al.* (1986) and Roca (1992).

Calc-alkaline rhyodacitic tuffs have been described from the north and NW of Mallorca (Wadsworth & Adams 1989; Mitjavila *et al.* 1990), with K-Ar dates by Mitjavila *et al.* (1990) indicating a Burdigalian age (18.6 to 19 Ma BP). This is broadly consistent with an early Miocene (Aquitanian or Burdigalian) age given by foraminifera in the northern outcrops of the Sant Elm Calcarenitic Fm (Rodríguez-Perea 1984), as opposed to late Oligocene ages which have been found in the SE of Mallorca (Anglada & Serra-Kiel 1986).

On Menorca, sequence III crops out in the NE (Fig. 13.28) and comprises nearly 10 m of reefal limestones interfingered with well-rounded conglomerates (Obrador 1970; Fornós 1987). On Eivissa it crops out in the NE (Fig. 13.28) and comprises up to 50 m of

conglomerates made of pebbles of black dolomite and volcanic lithologies, derived from a source area exposing Triassic rocks (Rangheard 1971).

Sequence IV. This sequence is dominated by marine turbiditic calcarenites and grey marls, with minor breccia, conglomerates and chert beds, and commonly preserves olistostomic deposits in its lower part. It crops out widely across Mallorca and Eivissa (reaching a maximum thickness of 450 m in NW Mallorca), but it is not present on Menorca, Formentera and Cabrera. It includes the Banyalbufar Turbiditic Fm (Ba in Fig. 13.28; Rodríguez-Perea 1984) and the informally named Randa Calcarenites unit (Ra in Fig. 13.28). The sequence onlaps both Mesozoic and older Tertiary rocks, on a basal erosive surface (sometimes a poorly developed hardground) which locally is an angular unconformity. The upper boundary is frequently truncated by thrusting. According to Rodríguez-Perea & Pomar (1983) and Rodríguez-Perea (1984), the sediments of the Banyalbufar Turbiditic Fm record the infilling of small foreland basins related to the emplacement of thrust sheets. The turbidites show a deepening-upward trend, interpreted as evolving from proximal deep-sea fan environments to basin plain. Sole marks are rare but they show a NNE-directed palaeoflow (Rodríguez-Perea 1984). Facies variations within the outcrops across the Balearic Islands reveal marl-dominated sequences in SE and central Mallorca, calcarenites and conglomerates interbedded within marls in NW Mallorca (Serra de Tramuntana) and a dominance of conglomerates and breccias interbedded within sandy marls and limestones on Eivissa, suggesting more proximal settings to the west and NW (Rangheard 1971). Based on planktonic foraminifera from the NW and central zone of Mallorca, the age of the Banyalbufar Turbiditic Fm has been established as latest Aquitanian–Burdigalian (González-Donoso *et al.* 1982).

The Randa Calcarenites unit is a 130 m thick succession which crops out on the Randa hills, in the central zone of Mallorca, but has been also recognized in boreholes in the central zone. These bioclastic calcarenites are mainly composed of red algae fragments and benthonic foraminifera. They overlie distal facies of the Banyalbufar turbidites. In their lower part, slumps and olistostromes are frequent while towards the top of the Randa succession, erosion gullies, large-scale cross-bedding, and swaley and hummocky cross-lamination are characteristic. The unit has been interpreted as a shallowing-upward succession, evolving from distal turbiditic sedimentation into a carbonate ramp, on which red algae and larger foraminifera thrived in a wave- and storm-dominated environment.

Sequence V. This sequence is widespread on Mallorca, Menorca and Eivissa, where it has been recognized both in outcrop and in the subsurface. It was deposited in small basins during the final episodes of Miocene tectonic compression, resting with angular unconformity on older pre- and synorogenic rocks. On Mallorca and Eivissa, three informal lithostratigraphic units have been differentiated: the Pina marls, the Son Verdera limestones and the Marginal terrigenous complex (Pi, SV and MT respectively in Fig. 13.28).

The Pina marls unit varies in thickness from a few metres up to more than 500 m in the depocentres. It is mainly composed of massive grey gypsiferous marls that locally are interbedded with sandy layers. The upper part of this unit contains pedogenic structures and some centimetre-thick coal beds. Fossils in the Pina marls consist of rare charophyta, ostracod and mammal remains (Oliveros *et al.* 1960; Mein & Adrover 1982; Adrover *et al.* 1983–84) mixed with abundant resedimented Burdigalian foraminifera (Colom 1975). The unit has been interpreted as recording deposition in evaporitic mud flats, swamps and brackish shallow-water lacustrine environments during Serravallian times (Oliveros *et al.* 1960).

The Son Verdera limestones unit conformably overlies the Pina marls, ranges from 10 m to more than 70 m thick (Ramos *et al.* 2000), and is mainly composed of limestones and fine-grained terrigenous deposits organized into two facies associations. The first of these comprises alternating massive grey marls, organic-matter-bearing silt-stones, and well-stratified brownish limestones with chert nodules, abundant gastropods and ostracodes, and lesser amounts of charophytes, fish otholites and mammals. According to Ramos *et al.* (2000), the facies records deposition in shallow brackish lacustrine environments. The second facies association comprises massive greenish marls with abundant carbonate- and iron-rich nodules, and thin limestone layers with stromatolitic laminations. They contain plant debris and minor gastropod and mammal remains, and display mud cracks and root structures. According to Ramos *et al.* (2000) this facies assemblage records mixed deposition in a palustrine–marginal brackish lacustrine environment.

The two units described above grade into the marginal terrigenous complex toward the periphery of the basins. This complex is mainly composed of lenticular to tabular stratified reddish sandstone and conglomerate beds containing subangular pebbles. The succession has a poorly defined coarsening-upward sequence and is interlayered with terrigenous mudstones that locally contain pedogenic structures. They

are interpreted as recording coarse terrigenous sedimentation on small alluvial fans that interfinger with muddy floodplain deposits.

On Menorca, sequence V records the accumulation of alluvial fans and fan-delta systems in active fault-controlled basins. The succession comprises boulder-sized conglomerates that interfinger, offshore, with calcareous marine sands.

Upper Miocene–Pliocene: the post-orogenic sequences

Post-orogenic Tertiary sequences crop out in all the Balearic islands, but are best represented on Mallorca, Menorca and Formentera. They comprise flat-lying carbonate platforms onlapping horst structures, and basinal deposits that infill subsiding basins (half-grabens) between the horsts. These sequences have undergone only slight tilting and flexure associated with normal and strike-slip faulting during late Neogene to middle Pleistocene times. The palaeogeographic evolution of the area during this interval started with early Tortonian carbonate ramps (sequence VI) prograding around palaeoislands (horst structures) produced during previous (Serravallian) tectonism. Subsequently (late Tortonian–early Messinian), extensive coral reefs (sequence VII) prograded over these ramp carbonates around the same palaeoislands. The area of present-day Mallorca comprised two palaeo-islands separated by a narrow seaway (Pomar 1979), Menorca was a small island (only the Tramuntana area), and Eivissa was similar in size to today, but with platforms extending towards the south and occupying the area of present-day Formentera (Fig. 13.29F).

Although deposited under similar conditions of high-frequency sea-level fluctuations, these two carbonate sequences exhibit different internal facies architectures and distribution of heterogeneities (Pomar 2001). During early Tortonian times, carbonate production occurred both in the shallow euphotic zone (foramol) and in the deeper oligophotic zone (rhodalgal). During the late Tortonian to early Messinian interval, carbonate production mainly occurred in the euphotic zone where framework-producing biota (coral reefs) formed a rimmed platform. The change of effective accommodation space to form two successive depositional sequences is interpreted to have resulted from an ecological change rather than significant relative sea-level change, namely by the shift from grain- to framework-producing biota. While an increase of surface-water temperature cannot be excluded, a trophic change seems to have been fundamental in controlling this biotic change (Pomar 2001). This trophic change in seawater could have been the result of the climatic change from humid to arid that occurred around the early–late Tortonian transition (Calvo *et al.* 1993). Thus these late Miocene carbonate platforms are thought to illustrate how changes in intrabasinal environmental conditions (nutrients and/or temperature) produce changes in stratal patterns and facies architecture if they affect the biological system (Pomar 2001).

Subsequent deposition of late Miocene gypsiferous marls within the Palma basin records shallow- and restricted-water deposition in the centre of the basins, now rimmed by the previous reef complexes, during a major drop of sea level that occurred at the end of the reef complex progradation (Pomar *et al.* 1996). Finally, during Pliocene times (Fig. 13.29H), carbonate production became more significant, contributing to the partial infilling of the basins remaining from latest Messinian times, between the main ranges of Mallorca (Serra de Tramuntana and Serres de Llevant) and southern Menorca.

Several lithostratigraphic units have been described in Mallorca and Menorca by Obrador *et al.* (1983, 1992), Pomar (1991, 1993), Pomar & Ward (1994, 1995, 1999) and Pomar *et al.* (1996), and in Eivissa and Formentera by Simó & Giner (1983). These lithostratigraphic units can be assembled into four unconformity-bounded sequences (sequences VI to IX; Fig. 13.28). The lowermost boundary is an angular unconformity over the older pre- and synorogenic sequences, and the upper limit corresponds to a Pleistocene unconformity.

Sequence VI. On Mallorca, the lowest sequence corresponds to the 200 m thick *Heterostegina* calcisiltites unit (Pomar 1979; Pomar *et al.* 1996; HC in Fig. 13.28) and crops out in the SW of Mallorca (although it is also known from boreholes). The unit unconformably overlies a folded and thrusted basement, whereas its upper limit is both conformable (basinward) and, in some places, unconformable (e.g. SW Palma Bay where it is overlain by the upper Miocene reef complex). It comprises fine-grained bioturbated packstones, locally interbedded with coarser grainstone layers, containing the larger foraminifera *Heterostegina* and planktonic foraminifera, scaphopods, rhodoliths, echinoids and large oysters. No coral reefs are known in this unit (Pomar *et al.* 1996). On Mallorca, the *Heterostegina* calcisiltite unit has been attributed to the early Tortonian by Pomar *et al.* (1996).

On Menorca, the sequence is up to 500 m thick in the subsurface, and it was defined as the Lower Bar unit (Obrador *et al.* 1983; LB in Fig. 13.28) from outcrops in the south and west of the island. This sequence is composed of foreshore deposits which unconformably onlap and backstep onto older units and by a prograding distally steepened ramp (Obrador *et al.* 1992; Pomar 2001). The upper boundary is an erosional surface in a proximal setting that becomes conformable distally, where it is overlain by a 1–2 m thick *Heterostegina*–packstone bed and, locally, by a thin phosphate-iron crust. The lithostratigraphic succession contains four facies assemblages. (1) Cross-bedded pebbly sandstone (beach deposits) at the base, grading basinward to (2) highly bioturbated, fine- to medium-grained, skeletal dolopackstones containing foraminifera and mollusc remains. These interfinger basinward with coarser cross-bedded grainstones (subaqueous dunes) with rhodoliths and larger foraminifera (*Heterostegina* and *Amphistegina*). (3) Rhodolithic rudstones, floatstones and grainstones with abundant red-algae fragments and benthic foraminifers, and minor echinoids, bryozoans and molluscs. Red-algae genera include *Lithothamnion*, *Mesophyllum*, *Lithoporella*, *Titanoderma*, *Sporolithon* and *Spongites*. This lithofacies form large-scale clinobeds and correspond to prograding ramp-slope deposits (Pomar 2001). (4) Fine-grained wackestone–packstone laminated deposits with planktonic foraminifera that, in proximal settings, are interbedded with 0.5–2 m thick graded beds (turbiditic deposits), channelled rhodolithic rudstones (debris-flow deposits) and large-scale cross-bedded grainstone units (toe-of-slope sand drift deposits).

On Eivissa, the lowermost post-orogenic sequences are less than 50 m thick, crop out in the NE, and comprise white marls and limestones with molluscs, foraminifera and red algae (Rangheard 1971).

Sequence VII. This sequence, known as the 'reef complex' (Re in Fig. 13.28), is characterized by the extensive progradation of reef-rimmed platforms in all the Balearic islands. On the basin margins the lower boundary is an erosion surface on either sequence VI or folded basement rocks. Basinward, the sequence lies conformably on sequence VI. Regional considerations, foraminiferal assemblages (Bizon *et al.* 1973; Alvaro *et al.* 1984*b*), and K-Ar dates (Pomar *et al.* 1996) indicate a late Tortonian–early Messinian age for the 'reef complex'.

Many recent publications on the Mallorca 'reef complex' from both outcrops and borehole data (Pomar 1991, 1993; Bosence *et al.* 1994; Pomar & Ward 1994, 1995, 1999; Pomar *et al.* 1996) have emphasized the genetic processes controlling stratigraphic architecture, chiefly high-frequency cyclicity of sea level, based on high-resolution sequence stratigraphy analysis. In these reef-rimmed platforms, heterogeneity in the lithofacies architecture was controlled by changes in carbonate production during relative sea-level fluctuations in a system dominated by framework-producing biota (Pomar & Ward

1995, 1999). The Mallorcan reef complexes built up on shallow marine shelves adjacent to basins. The most extensive accumulation of these progradational reefal carbonates was on the Llucmajor platform, a 20 km wide carbonate platform on the southeastern side of the Palma basin. The 'reef complex' shows the following facies associations.

(i) *Off-reef open shelf lithofacies.* This comprises two main lithologies. The first of these comprises coarse-grained red-algal grainstones–packstones with rhodoliths at the base of the reefal platforms. The second forms fine-grained, highly burrowed, flat-lying packstone–wackestone beds containing planktonic foraminifera, ostracods, and fine-grained detritus of oysters, other bivalves, echinoids and red algae.

(ii) *Forereef-slope lithofacies.* This shows basinward-dipping clinobeds composed of fine-grained red algae–mollusc packstone–grainstone with rhodoliths passing landward to steeply dipping (10–30°) layers of skeletal and intraclastic grainstone, packstone, rudstone and floatstone. The rocks are formed from fragments of red algae, corals, bivalves, gastropods, echinoids, bryozoans and green algae, and interfinger landward with coral reefs.

(iii) *Reef framework and reefal rudstone (reef core) facies.* This is mainly composed of two coral genera, *Tarbellastraea* and *Porites*, although *Siderastraea* also occurs. Secondary framework components are encrustations of red algae, foraminifera, bryozoans, worm tubes and vermetid gastropods. The reefs are vertically zoned according to depth-controlled growth morphology of the corals. Deeper-water coral colonies are platy with finger-like vertical projections, intermediate colonies are predominantly branching, and shallow-water colonies are hemispheroidal to columnar or even domal in the reef-crest position (Pomar 1991).

(iv) *Outer-lagoonal lithofacies.* This comprises skeletal grainstone–packstone with coral patch-reefs and contain abundant red-algae fragments, echinoids, molluscs, benthic foraminifera and coral fragments. Inner-lagoonal lithofacies are composed of grainstones, packstones and wackestones–mudstones with miliolids, thin bivalves, peloids and cerithid gastropods. On the basin margins, especially near the Serra de Tramuntana, these lithofacies interfinger with terrigenous calcareous conglomerates and sandstones related to small fan-deltas (marginal complex).

On Menorca and Eivissa, outcrops are not as extensive as on Mallorca, and the quality of exposures is lower, whereas on Formentera accessibility is the main problem. Nevertheless, facies associations and reefal architecture similar to those seen on Mallorca have been recognized in several places (Rangheard 1971; Simó & Giner 1983; Obrador *et al.* 1983, 1992).

Sequence VIII. This sequence is recorded only in Mallorca, Cabrera and the west of Formentera. It is composed of three informal stratigraphic units: the Bonanova marls, the gypsum and grey marls and the Santanyí limestones units. The lower boundary is an erosion surface, locally with karstic dissolution caves (Fornós 1999). The upper boundary, also an unconformity, is overlain by Pliocene deposits (Pomar *et al.* 1990).

The Bonanova marls unit (Bo in Fig. 13.28) crops out only in the NW margin of the Palma basin on Mallorca, where it is 35 m thick. It is composed of marls with abundant oysters, gastropods, corals and scallops, which pass upwards into conglomerates, sandstones and silty red beds. It represents the progradation of a fan delta system from the basin margin.

The Santanyí limestones unit (the former 'terminal carbonate complex' of Esteban 1979–80; Sa in Fig. 13.28) crops out in the Marinas de Llevant and Llucmajor, Palma and Campos basins, in the south and SE of Mallorca, and in the west of Formentera (Fornós & Pomar 1984). The lower boundary is an erosion surface over the 'reef complex', whereas the upper boundary is not exposed. The unit is up to 30 m thick and comprises four major lithofacies, which are from base to top: (1) miliolid packstones and grainstones with vertical root traces (mangrove swamps); (2) stromatolitic boundstones and mudstones (intertidal–subtidal); (3) cross-bedded oolitic grainstones with giant subtidal thrombolites and stromatolites; and (4) skeletal grainstones and stromatolites (restricted subtidal). Near Palma, giant thrombolitic and stromatolitic domes several metres in diameter, are capped by metre-scale cyanobacteria domes and worm 'reefs', which, in turn, are overlain by cross-bedded oolitic grainstone.

The gypsum and grey marls unit occurs only within the Palma basin (where it was first described by Fornós & Pomar (1984) as a lithofacies within the Santanyí limestones) and can be correlated with massive gypsum deposits cored in the centre of the Palma basin. The unit is about 10 m thick in outcrop and is composed of dolomite, grey marls with stromatolites, marine molluscs and fish debris. Alvaro *et al.* (1984*b*) and Simó & Ramón (1986) have described the presence of laminated marls with diatomites, fishes, fresh- to brackish-water ostracodes and *Chara*.

The stratigraphical relationship between these three units remains doubtful because of a lack of chronostratigraphical data and correlation problems. The gypsum and grey marls unit comprises restricted shallow marine deposits which conformably overlie the deeper-water deposits of the 'reef complex'. The Bonanova marls represent a regressive succession on top of the shallow reefal platform and may be coeval with the gypsum and grey marls unit. The Santanyí limestones represent a shallow open marine transgressive unit, overlying an erosional surface on top of the shallow-water facies (reef core and lagoon) of the 'reef complex' (Fig. 13.29G). It remains uncertain whether this unit was deposited before (Pomar *et al.* 1996) or after (Esteban 1996) the Mediterranean salinity crisis.

Sequence IX. The outcrop of this Pliocene sequence is restricted to a small area in Mallorca, although it has been widely recognized in boreholes from the Palma, Inca and Sa Pobla basins. It forms a regressive succession, up to 370 m thick in the Palma basin, that overlies a major erosion surface on top of the gypsum and grey marls unit (sequence VIII) and the forereef-slope lithofacies of the 'reef complex' (sequence VII). Quaternary deposits unconformably overlie it.

Depositional environments recorded by this sequence range from shallow marine (Son Mir calcisiltites) at the base, to aeolian (Sant Jordi calcarenites) at the top (SM and SJ respectively in Fig. 13.28). Both units represent the final infilling of remaining basins rimmed by the upper Miocene reefs, and both are interbedded, at the basin margins, with conglomerates and sandstones interpreted as fan-delta deposits. The Son Mir calcisiltites unit (Barón & Pomar 1985) is up to 300 m thick, and contains pyritized foraminifers and abundant molluscs, bryozoans, a characteristic pectinid (*Ammussium*), and benthonic and planktonic foraminifera. The foraminiferal assemblage dates this unit as Zanclean (early Pliocene; Colom 1985). A grey marly interval exists at the base, containing abundant ostracods and foraminifers. The Sant Jordi calcarenites unit (Barón & Pomar 1985) is up to 70 m thick, attributed to the Plasencian (late Pliocene; Colom

1985) and comprises large-scale cross-bedded skeletal calcarenites containing an abundance of molluscs and benthonic foraminifera. These rocks have been interpreted as shoreface, foreshore and aeolian deposits.

V.P., J.I.B., A.Pa. (Basque-Cantabrian): The editing efforts of W. Gibbons and T. Moreno, and the reviews of C. Dabrio and P. Burgess greatly clarified our first version, and are much appreciated. Funds for fieldwork were generously provided by the Basque Country University through Research Projects UPV 121.310-BE233/93 and UPV 121.310-EB191/98.

A.M., C.A., A.G., A.L., G.P., A.P., and J.V. (Ebro): This contribution summarizes the results of several projects financed by the Spanish Government since 1985, as well as several scholarships. At present our work is supported by Project PB97-0882-C03-02 of the Dirección General de Enseñanza Superior e Investigación Científica. We would like to thank P. Noone for supervising the English grammar. P. Burgess, an anonymous reviewer, T. Moreno and W. Gibbons are thanked for their valuable suggestions.

I.A., A.C., G.A.G., J.C., J.M.R. (Duero): We acknowledge financial support from the DGICYT Project PB98-0668-C02-02 and the help of the Central Language Office of the University of Salamanca in reviewing the first English version. We wish to thank P. Burgess and G. Pardo for their constructive criticism and the editors for their suggestions and invaluable help in the final English version.

A.M.A.-Z., J.P.C.-S. (Tajo): We sincerely thank V.P. Wright and P. Burgess for their reviews of this work. W. Gibbons and T. Moreno are thanked for their efforts on carrying out this project. This work forms part of Project PB-98-0691-C03-03.

J.C.B., J.M.M., J.A. (southern Spain): We sincerely thank W. Gibbons and T. Moreno for their kind invitation to contribute to this chapter and for their editing work. Constructive comments by C. Betzler and P. Burgess are greatly appreciated. We are grateful to C. Laurin for correcting the English text. This work was supported by 'Dirección General de Enseñanza Superior e Investigación Científica' Project PB97-0809 and the 'Junta de Andalucía'.

J.J.F., L.P., E.R. (Balearic Islands): We would like to thank P. Burgess, W. Gibbons and J. A. Vera for their constructive criticism, which enabled us to considerably improve this paper. Financial support was provided by the Spanish DGICYT project Fs PB97-0882-C03-01 and PB98-0132 and the Comissionat per Universitats i Recerca de la Generalitat de Catalunya, Research Group of Geodynamics and Basin Analysis (SGR 1997-00073).

14 Quaternary

MATEO GUTIÉRREZ-ELORZA (coordinator), JOSE MARÍA GARCÍA-RUIZ,
JOSÉ LUIS GOY, F. JAVIER GRACIA,
FRANCISCO GUTIÉRREZ-SANTOLALLA, CARLOS MARTÍ,
ANGEL MARTÍN-SERRANO, ALFREDO PÉREZ-GONZÁLEZ &
CARIDAD ZAZO, with a section on human fossils by EMILIANO AGUIRRE

Spanish Quaternary sediments and landforms record glacial, alluvial, fluvial, lacustrine, aeolian, coastal and volcanic environments. Within these environments numerous processes acted on different lithologies and structures during changing climate, neotectonic activity, and anthropogenic influence. Consequently, the current landscape of Spain comprises a complex palimpsest of inherited and active landforms.

Spain has a mountainous relief, with an average height of 660 m for the whole Iberian peninsula. This high relief is related to the presence of extensive plateaux surrounded by mountain ranges (Cantabrian mountains, Pyrenees, Catalonian Coastal Ranges, Iberian Ranges and Betic Cordillera). The highest elevation in the Iberian peninsula is found in the Betic Cordillera (Mulhacén, 3481 m), although in the Pyrenees there are many peaks above 3000 m.

The hydrographic network of the Iberian peninsula has a main divide separating the basins draining to the Mediterranean sea from those draining into the Atlantic ocean. Most of the major rivers drain to the Atlantic and run for a significant part of their course through Tertiary depressions such as those of the Duero, Tajo and Guadalquivir basins. The Ebro river is the main Mediterranean fluvial system and flows through the

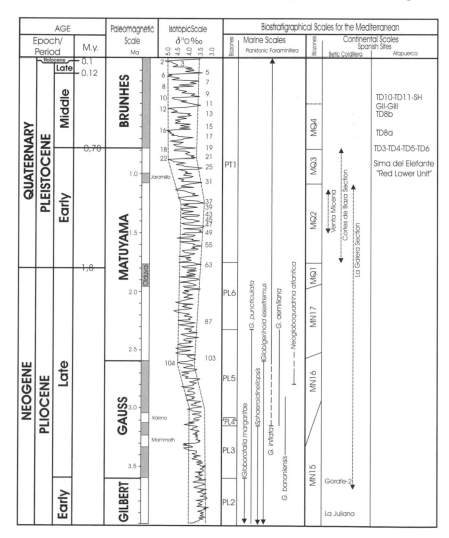

Fig. 14.1. Chronostratigraphic scale for the Mediterranean and Atlantic (modified after Bardají 1999). Palaeomagnetic scale after Cande and Kent (1995). Isotopic scale corresponding to ODP-677 (0–2 Ma) and ODP-846 (2–4 Ma), after Shackleton (1997). Marine scales after Berggren *et al.* (1995), Iaccarino (1985). Continental biozones after Mein (1975). Betic sites after Agustí *et al.* (1997), Agustí & Moyá Solá (1998). Atapuerca sites after Cuenca Bescós *et al.* (2001).

Ebro depression. These Iberian rivers drain a landscape affected by contrasting climatic belts, with a humid zone in the north, a semi-arid zone situated in the SE and in the large Tertiary depressions, and semi-humid conditions elsewhere.

Iberian Massif

Spanish Quaternary geology should be viewed within the broad context of Iberian geological history since the emergence of a Variscan basement in Late Carboniferous times. Subsequent peneplanation of an initially mountainous Variscan land surface left only residual resistant ridges (e.g. Ordovician quartzites) that protruded above extensive spreads of Mesozoic sediments. Cenozoic orogenesis, produced by the interactions between the Iberian and African plates, rejuvenated the landscape and produced the mosaic of synorogenic and late-orogenic basins described in previous chapters. At the beginning of Neogene times, many of these basins had undergone a transition from marine to continental conditions, and the Iberian peninsula became dominated by internal fluvial drainage. The transition from Pliocene to Pleistocene times is considered to mark the change from internal to external drainage in inland Spain (Pérez-González 1982; Bardají *et al.* 2000).

Raña

Raña is the name given to extensive, generally thin, clast-supported siliciclastic conglomerates of fluvial origin. They rest on Tertiary or older bedrock across inland Spain. These deposits are found at different topographic levels, and are always higher than the oldest river terraces (Martín-Serrano 1991). They are typically strongly weathered with partially disintegrated pebbles and intense alteration of the argillaceous material, giving place to well developed soils (ultisols/oxisols according to Espejo (1985) and Molina (1991). These deposits are generally considered as Pliocene–Quaternary in age (Vaudour 1979; Pérez-González & Gallardo 1987), although the exact ages of these presumably diachronous deposits are poorly constrained (Martín-Serrano 1991). The Raña deposits are common in the Montes de Toledo to the south of Madrid, where they partially fill old valleys preserving ancient landscapes, as seen in Los Yebenes, Retuerta and Robledo. Further south, Raña deposits are spread across broad, gently north-dipping, 600–800 m high tablelands, whereas to the west, in Extremadura, the Raña has a lower height and continuity but gains in textural maturity.

The Raña deposits, in addition to covering the old peneplain surfaces of central and southern Spain, are also found as alluvial platforms related to the incision of the major rivers like the Guadalquivir, Tajo (Jaraicejo, 680–480 m) and Guadiana (Sierra de San Pedro, 300–400 m; Mérida, 300–350 m). They also occur in valleys within the Sierra Morena (eg. Rumblar, 500–450 m; Cárdena-Montoro, 460–370 m), and as remnants of alluvial fans draining the Central Range in Riaza, Ojos Albos, Somosierra and the Ciudad Rodrigo basin. Further north, in the Duero basin, Raña deposits form spectacular alluvial fans fed by the Cantabrian mountains, such as the 52 km long fan around Guardo (Palencia province). In Zamora and León provinces the Raña deposits are found at a height of 1000–1300 m (the 'haute surface alluvial' of Herail (1984)), as seen at Omanas, Valduerna, Eria, Tera and Aliste.

Fluvial network

The origins of the modern Spanish fluvial network go back into pre-Quaternary times, and are closely linked to Cenozoic tectonism (Martín-Serrano 1991). Several works specifically have studied the fluvial systems draining Variscan basement massifs in Iberia (Herail 1984; Cantano 1996; Martín-Serrano *et al.* 1998; Vidal *et al.* 1998; Sánchez & Blanco 1999). In the north of Spain, the fluvial network draining the Cantabrian mountains was initiated during Oligocene–Miocene uplift, so that the well known Cantabrian gorges and deep valleys, such as Deva, Cares and Sella, were partially formed during Cenozoic times. The drainage is remarkably asymmetric. The south-draining rivers flowing into the 1000 m high Castilian northern meseta have their vertical incision limited by Neogene base levels. In contrast, northward flowing Cantabrian rivers show steep gradients (2000 m in 30–70 km), deep incision, and well developed terraces and floodplains in their lower courses. As an example, the Nalón river has erosion surfaces preserved at 90–100 m (late Pliocene), 80–85 m and 50–60 m (early Pleistocene), 24–32 m (middle Pleistocene) and 15–20 m (late Pleistocene or Holocene; Hoyos 1989).

Galician rivers may have an even more ancient origin, with Pagés (2000) recognizing a latest-Mesozoic fluvial network contemporaneous with Atlantic opening. Remnants of this old network include Atlantic coastal valleys such as Grande, Castro, Tambre and Xallas, as well as inland valleys such as Narla, Modeo and the upper Miño. By Quaternary times the fluvial systems in Galicia had long been established and underwent little modification (60–80 m maximum incision), since recent Afro-Iberian plate interactions had only a limited effect on this part of Spain. The evolution of the Ebro, Duero and Tajo fluvial systems are analysed in more detail below, as is the Quaternary geology of the Betics, Pyrenees and Iberian ranges.

Quaternary glaciation

The Quaternary glaciation has been more important in the northern sector due to a combination of several factors including relief, latitude and precipitation. Although it has been generally accepted that there has been one glacial phase 50–40 ka BP, recent work carried out by Vidal *et al.* (1999) provides evidence for a younger glaciation (30 ka BP). Compared with the evidence for actual glaciation, signs of periglacial activity such as nivation forms, rock glaciers and solifluction phenomena are rare and commonly difficult to interpret.

In the Cantabrian mountains there is evidence for glaciation at heights ranging from 900 to 1200 m, Peña Urbina being the highest glaciated area. In the Picos de Europa the glacial processes are linked with karstic phenomena (Martín-Serrano 1994). Here, glacial erosional landforms are widespread although deposits are very rare, with the exception of the Enol and Ercina tills, which lie at a height of 1300 m. To the north of the Picos de Europa the ice flowed down to 650 m, whereas the glacial landforms and fluvioglacial deposits descended to only 800 m in the south-facing slopes. Further to the west, in Galicia and León, the glacial activity affected a few areas above 1500 m, such as in the Segundera, Cabrera and Teleno mountains, where a glacier covered a plateau-like summit. The best known examples of moraines occur in the southeastern slopes of the Sanabria mountains, at a height of 1000 m. In the Central Range, Sierra de Gredos was the most intensively glaciated area, although evidence for glacial activity has also been found in the Bejar, Serrota, Guadarrama and Somosierra areas (Pedraza 1994). Hanging cirques were carved in both the north- and south-facing slopes of the Sierra de Gredos. Glacial tills are found in this sierra above 1380 m and 1500 m in the northern and southern slopes respectively.

Betic Cordillera

The Betic Cordillera extends for nearly 600 km along the southern and southeastern coasts of the Iberian peninsula, and then re-emerges in the Balearic islands (Fig. 14.2). The Cordillera has been traditionally subdivided into internal zones, external zones and Neogene basins (Chapters 13 and 16). The sedimentary filling of the latter basins is firstly of marine character but becomes progressively continental, giving rise to important accumulations of Quaternary deposits.

One of the wider Neogene basins is the Guadalquivir depression, located to the NW, between the Subbetics and the Iberian Massif. This basin is almost 350 km long and mostly around 30 km wide, although at its southernmost termination it widens to nearly 70 km. The sedimentary filling of this foreland basin comprises deposits of both Neogene (Chapter 13) and Quaternary age, which together can reach thicknesses of around 2000 m. Another important area of Quaternary

deposition has been within the 'Eastern Betics left-lateral shear zone' (Montenat et al. 1987), a long and narrow depression of sigmoidal geometry that extends in a SW–NE direction (Fig. 14.3) from Almería to Alicante (Silva et al. 1993). Other important basins are those located in the littoral zone of the Cadiz Bay–Gibraltar Strait to the SW, and in the southern Valencia littoral zone to the NE.

In the Betic Cordillera there is a close relationship between Quaternary deposits and three young fault systems running broadly ENE, NW and NNE. Those trending ENE–WSW, parallel to the orogenic strike, include the longest young faults in the Betics (Sanz de Galdeano 1983). The fault set trending NW–SE involves mostly dextral displacement and has influenced the growth of important littoral lagoons such as those of La Mata, Torrevieja, and Mar Menor, all lying to the south of Alicante (Fig. 14.2). Finally, the NNE–SSW trending fault set shows sinistral offsets of the drainage network and commonly controls both the coastal outline and the distribution of Quaternary marine and terrestrial deposits. More important examples include the Palomares, Carboneras, and the Alhama de Murcia faults, the latter controlling the deposition of alluvial fans in the Guadalentín corridor, SW of Murcia (Fig. 14.3). The description of the Quaternary deposits given below is subdivided on the basis of the environments under which they formed: (1) weathering, pedogenic and karstic; (2) slope;

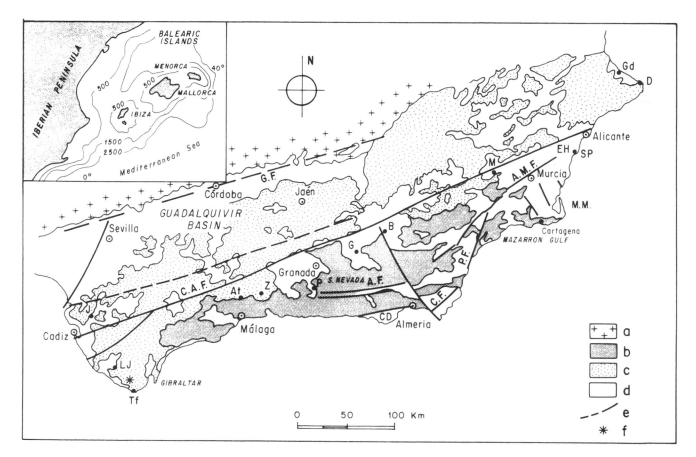

Fig. 14.2. Schematic map of the Betic Cordillera showing the main structural domains: a, Iberian Massif; b, internal zones and Gibraltar flysch units; c, external zones; d, Neogene and Quaternary Basins; e, main faults and inferred faults; f, landslides controlled by seismotectonics. Faults: AF, Alpujarras; AMF, Alhama-Murcia; CAF, Cadiz-Alicante; CF, Carboneras; GF, Guadalquivir; PF, Palomares. Localities: At, Antequera; B, Baza; CD, Campo de Dalias; D, Denia; EH, El Hondo; G, Guadix; Gd, Gandia; J, Jerez; LJ, La Janda; M, Mula; MM, Mar Menor; P, Padul; SP, Santa Pola; Tf, Tarifa; Z, Zafarraya.

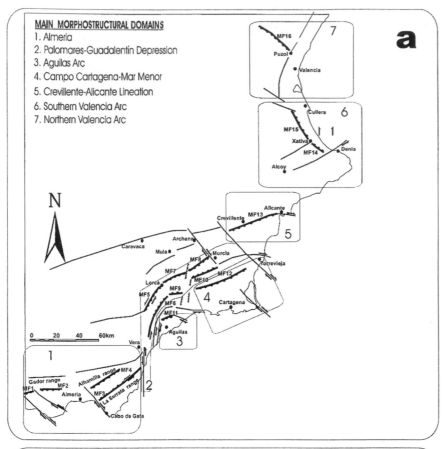

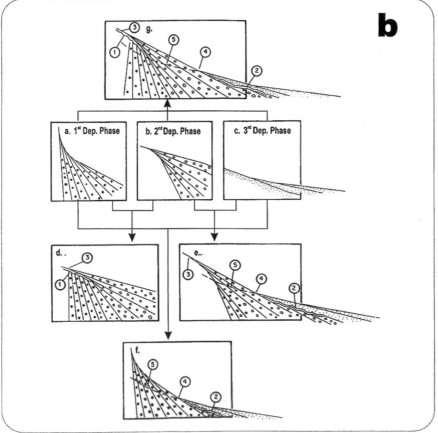

Fig. 14.3. **(a)** Main active mountain fronts distributed along the Eastern Betics shear zone (modified after Silva *et al.* 1992*b*, 2002). **(b)** Depositional phases, sedimentary style and geomorphologic expression of alluvial fan systems in the Guadalentin depression (after Silva *et al.* 1992*a*).

(3) glacial and periglacial; (4) lacustrine and palustrine; (5) fluvial; (6) littoral and pre-littoral (including both aeolian and marine).

Weathering, pedogenic and karstic deposits. The most characteristic example of this kind of deposit is the kaolinitic-ferruginous 'Bonares sands' in the Guadalquivir basin, the most probable age of which is late Pliocene (Huelva) to Quaternary (Cádiz). The maximum altitude of the Bonares sands ranges between 180 and 130 m, while the depth of weathering reaches 14 m. The weathering profile, with abundant quartz grains and kaolinite, shows a very characteristic upper horizon of ferruginous pisoliths that resembles a ferricrete. This kind of weathering requires a humid tropical climate, with significant precipitation (1000–2300 mm) and a marked dry season, as well as climatic stability, good drainage, smooth relief, high groundwater table close to the surface, and dense plant cover.

Calcareous accumulations up to 3 m thick are very common in the Betic Cordilleras and formed by leaching of carbonates from red soils (pedogenic) or during later diagenesis. They have been used to identify ancient surfaces mainly associated with deposits of early and middle Pleistocene age (Dumas 1977). The best examples of pedogenic calcretes are found in the terrace sequences of the main rivers such as the Guadalquivir, Guadalete, Segura, and Júcar, whereas those of diagenetic–pedogenic origin on alluvial fan sequences typically developed at the foot of the main mountain fronts (Alonso Zarza et al. 1998b).

Karst landforms are widely represented in the calcareous areas of the Betic Cordillera, with good examples of karrens (e.g. Torcal de Antequera, Málaga), dolines, poljes and travertines. The most significant karstic feature in the Betics is the Zafarraya Polje (Granada), which preserves >70 m of infilled detrital deposits and decalcified clays (Lhenaff 1986). Travertines are abundant on the slopes of calcareous ranges, and seem to be linked to the alternation of colder and warmer climatic episodes (Delannoy et al. 1989). Early–middle Pleistocene (>350 ka BP), middle-late Pleistocene (150–40 ka BP) and late Pleistocene–Holocene (<20 ka BP) sequences have been differentiated, by means of isotopic dating, in the travertinic platforms of Talox (Horcajos river, Málaga). Durán et al. (1988) suggest that the travertines developed in the southern border of Mijas Range (Málaga) correlate with the warmer (interstadial) period IS7, whereas those developed in Benalmádena (Málaga) are IS5 in age, and those from Torremolinos (Málaga) belong to IS2 (Würm).

Slope deposits. Hillslope deposits may be subdivided into colluvium and landslides. Although the effects of climatic change can be locally important in producing colluvium deposits, in the Betic area both colluvium and landslide deposits are in most cases related to fault lines, emphasizing the importance of neotectonic activity in the development of the modern landscape (Baena et al. 2002). Landslides are usually associated with lineaments across which important lithological changes occur, in most cases related to the outline of ancient Betic thrust faults. One of the most outstanding examples is provided by the landslides located along the contact between the Almodovar river and La Momia Range, where at least three sliding phases can be recognized in the area of Tarifa–El Algar (Cadiz), indicating important Quaternary seismotectonic activity along the main structural lineaments running NE–SW and NW–SE (Goy et al. 1996).

Glacial and periglacial deposits. The Sierra Nevada (Fig. 14.2) preserves the southernmost glacial landforms in Europe, and is the only place in the Betics where glacial and periglacial deposits are found. The greatest concentration of ancient glacier systems developed on the northern side of the mountain range (e.g. Dilar, Monachil, Genil, Maitena), determined by an orientation that favoured the snow input. Despite this, it is on the steep southern faces where glacial deposits are best preserved (e.g. Lanjarón, Toril, Poqueira, Trévelez), being found in both valley glacier systems with important lateral moraines (e.g. Genil, Poqueira), as well as associated with cirque glaciers with shorter lateral and terminal moraines. Glacial tills usually preserve their original morphology especially when they developed on less steep slopes such as valley, intermediate (slope) and cirque glaciers. This is the case for the Lanjarón glacier system, where up to five glacial phases, located between 1700 and 2800 m, can easily

be recognized, and where till remains at 1400 m possibly indicate a former glaciation (Gómez Ortiz et al. 1998). The same glacial phases are recognized in the glacier systems of 'Siete Lagunas', between 2400 and 3130 m, and Poqueira, between 1980 and 3150 m.

In general, and according to sedimentological criteria, two glacial stages with moraine development can be identified in the Sierra Nevada (Messerli 1965), one from the Riss (moraine at 1340 m) and the other from the Würm (moraine at 1700 m). Remains from the Late Glacial stage are recorded at altitudes higher than 2800 m, being represented by small lateral and terminal moraines. Holocene glaciation was reduced to rare rock glaciers developed during colder stages such as the Little Ice Age (Goy & Zazo 1989b; Goy et al. 1994; Gómez Ortiz et al. 1998).

During glacial stadials, as well as to some extent during interstadials, these high mountain areas were subjected to seasonal freeze–thaw processes that reshaped previous glacial landforms and generated new periglacial ones. Some of the most outstanding of these landforms are seen in the 'summit high plains' (cryoplanation terraces, patterned ground, stone fields) in the cirques (rock glaciers, névé moraines, gelifluction lobes), and in the slopes (debris-mantled slopes, grèzes litées). Whereas today periglacial conditions exist only above around 2700 m, during the Late Glacial stage the periglacial area in the eastern Alpujarras (Granada) reached down to 1000 m.

The most representative periglacial landforms in the Sierra Nevada form rock glaciers located higher than 2700 m, where perennial snow remained during the Late Glacial stage. In the Veleta Cirque, rock glaciers are located below the melting snow niches of Picacho del Veleta and in the declivity of Cerro de los Machos, where they are characterized by large angular blocks without any matrix (Gómez Ortiz et al. 1998). Other well known periglacial landforms are those from Sierra de la Tramontana in Mallorca, where shallow nivo-karstic processes gave place to deposits such as block flows and solifluction lobes (Grimalt Gelabert & Rodríguez Perea 1994).

Lacustrine and palustrine deposits. These deposits are found both inland (the 'inner basins') and along the coast. Examples from the inner basins include the Guadix-Baza basin (Chapter 13), the centre of which shows marly lacustrine Pliocene–Pleistocene sediments grading out into limestones and marls bioturbated by rootlets. Cerastoderma bearing sandy-silty sediments overlie these facies, with thin discontinuous carbonate levels and sporadic channel-fill facies. This general sequence is capped by shallow water micritic palustrine limestones of middle Pleistocene age in the area of Caniles. In contrast, the Durcal-Padul depression (Granada), also an inner basin, preserves peat deposits associated with the activity of boundary faults which favoured the accumulation of more than 100 m of organic sediments. Pollen analyses show the top 72 m of these sediments belong to middle and late Pleistocene and Holocene times (Menéndez Amor & Florschütz 1964; Pons & Reille 1988).

In the basins distributed along the Betic coastline, some of the best examples of lacustrine deposits are seen at El Hondo (Alicante) and La Janda (Cadiz), the last lagoonal areas of which were drained in the twentieth century (Luque et al. 1999). Upper Holocene deposits at these localities mostly comprise clayey-silty and clayey-sandy materials very rich in organic matter. In the Cartagena-Mar Menor basin (Murcia), as well as in the central part of the Guadalentín depression (Lorca-Totana, Murcia), former playa lake environments linked to alluvial fan systems (Goy & Zazo 1989a; Somoza 1993; Silva et al. 1996) are also nowadays not functional basins, mostly as a result of human activity. Other deposits found near the coast include those in the Mula basin (Murcia) where seismic activity has promoted landslides that impeded the drainage, creating palustrine areas during late Pleistocene times (Mather et al. 1995). Other lagoonal littoral deposits are those associated with brackish paralic environments, such as all the coastal lagoons distributed along the Betic littoral from Valencia to Cádiz and Huelva.

Fluvial deposits. Fluvial deposits in the Betics represent piedmont, alluvial cones and fans, terraces, and alluvial plain environments (Table 14.1). Deposits lacking channelized flows tend to relate to mountain fronts and were mainly generated by sheet flow, whereas channelized sediments are typically associated with valley river beds and alluvial plains. Within the Betic inner basins, such as the Guadix-Baza basin, the Pliocene–Pleistocene sedimentary record shows two

Table 14.1. Relationship between type of deposit and slope angle in fluvial systems related to mountain fronts (after Goy 1984)

Deposit	Examples of slope angles							
Colluvium	Alcira 2°	Jaraco 4°31′	Cuartell 5°01′19″	Llombay 5°16′26″	N. Puzol 5°03′41″	Tabernes 7°07′30″	Casinos 7°35′41″	Mongo 10°08′
Piedmont	Cuartell 0°41′	W. Picasent 1°41′5″	N. Perenchiza 1°47′24″	NW Villavieja 1°51′	Perenchiza 2°28′48″	Canals 3°22′	NW Castellón 3°39′	N. Puzol 5°03′41″
Alluvial cone	Alcira 0°48′0″	W. Puzol 1°32′	N. Villavieja 1°37′16″	S. Villavieja 2°01′17″	S. Canals 2°01′55″	W. Gandía 3°17′2″		
Pediment cover	P. Larga 0°44′21″	W. Benifayó 0°53′17″	Casinos 1°10′03″	Pego 1°21′50″	Jaraco 1°28′14″			
Alluvial fan	Gorgos 0°15′22″	Serpis 0°21′6″	Cañoles 0°33′26″	Seco 0°43′12″	R. Gallinera 0°48′15″	Girona 0°51′48″		

facies: dominantly alluvial fans (Guadix Formation (Fm)), and dominantly lacustrine sediments (Baza Fm). These two formations pass laterally into each other, and their age spans from late Miocene (Messinian) up to middle Pleistocene (Vera 1970; Peña 1979, 1985; Goy *et al.* 1989; Viseras 1991; Guerra-Merchán 1993).

Further towards the coast, fluvial deposits are typically alluvial fans and cones associated with faulted mountain fronts (Figs 14.3 and 14.4). In Almería (Campo Dalías and La Serrata–Campo de Níjar) (Fig. 14.3a, 1) seven and nine alluvial fan generations have been differentiated respectively at the mountain fronts of Sierra de Gador and Sierra Alhamilla, with a staircase geometry for the most ancient (early Pleistocene), and overlapping and offlapping geometries for the more recent generations (middle and late Pleistocene). The Campo de Níjar sector is bounded towards the south by the hills of La Serrata, which is made up of Miocene and Pliocene volcanic and sedimentary materials. This range is bounded by a system of N40°–45°E parallel faults, with sinistral strike-slip drainage offsets and, towards the south, a vertical component of movement.

In the Guadalentín depression (Fig. 14.3a, 2) recent activity along the Lorca-Alhama, Palomares and North Carrascoy faults produced important progressive discordances in the alluvial fan deposits, which in many cases surpass 200 m in thickness (Silva *et al.* 1992a). Three depositional phases with different sedimentary style and geomorphologic expression have been defined in this area (Silva *et al.* 1992b, 2001), and suggest a progressive decrease in regional tectonic activity and an increase in aridity during the Quaternary period (Harvey 1990). An initial syntectonic depositional phase is characterized by debris flow processes, higher fan gradient and a depositional style dominated by proximal offlap aggradation (Fig. 14.3b, a). A second depositional phase was dominated by proximal onlap aggradation culminating with a general backfilling episode that surpasses the outline of the fault. The sedimentary facies change from debris flow at the bottom to sheet flow and fluvial processes at the top (Fig. 14.3b, b). Finally, a third, post-tectonic, depositional phase produced dissection in the proximal area, resulting in trenching below the upper surfaces and a shift of deposition towards distal sites. The alluvial fans present a smooth gradient and are dominated by distal offlap aggradation (Fig. 14.3b, c). Combinations of these depositional styles offer four different models for fan development (Fig. 14.3b, b, d, e, f).

Alluvial deposits have also been studied in the Valencian coastal area (Fig. 14.3a, 6, 7; Goy 1984), the Cartagena-Mar Menor basin (Fig. 14.3a, 4; Somoza 1993), and the Elche depression (Fig. 14.3a, 5; Goy & Zazo 1989a). In addition to the important influences of neotectonics in these southeastern areas, the effects of climate and sea-level changes are also well recorded (Dumas 1977; Goy & Zazo 1989b; Somoza 1993; Harvey *et al.* 1999; Stokes & Mather 2000), with the most important aggradation impulses coinciding with Pleistocene cold phases.

The most complete sequence of fluvial terraces is found in the Guadalquivir river valley, between Sevilla and Córdoba, where 14 terrace levels are distributed between +215 m above the thalweg (T-1) and the present alluvial plain at +6 m (T-14). Baena & Díaz del Olmo (1997) grouped these terraces from highest to lowest into those aged >800 ka (T1–T3), 800–300 ka (T4–T9), 300–80 ka (T10–T12),

and a lowest group (T12–T14) representing Late Glacial and Holocene time (Fig. 14.5).

Abandoned ancient alluvial plains are located in tectonically active areas and preserved by fluvial capture processes. Such is the case of the lower Guadalquivir river basin, where a former north–south branch flowed into the Cadiz Bay during early Pleistocene times, its morphology and deposits being observable in the depression of Llanos de Caulina (east of Jerez). Another outstanding example is found in the Guadalentín river, which initially flowed into the Mazarrón Gulf through the Murcia–Las Moreras Rambla before changing its course towards the Mar Menor through the Fuente Alamo Rambla. It was finally captured by the Sangonera river (river Segura drainage basin), probably during middle-late Pleistocene times, leaving fluvial deposits >60 m thick to the SE of Totana (Silva *et al.* 1996). Other examples of fluvial capture are described by Calvache & Viseras (1995, 1997).

Aeolian deposits. The aeolian record in the Betic area is represented by two main kinds of deposits: aeolian silts and dune systems. The aeolian silts are wind-generated deposits derived from the coastal platform during lowstands of sea level. Between the Serpis river (Gandía, Valencia) and Rambla Gallinera (Denia, Alicante) these deposits are found on top of alluvial fan sediments, showing two aeolian sequences separated by a black- brown soil which probably records a change of climate correlated with the Würm interstadial (Goy 1984).

Dune systems develop along the littoral zone extending towards inner areas. One of the most outstanding examples is provided by the oolithic and terrigenous deposits that crop out between Santa Pola and Los Arenales del Sol (Alicante), where, on top of late Pliocene lagoonal deposits, several consolidated dune sequences, separated by red palaeosols, record early-middle Pleistocene sedimentation (Goy & Zazo 1989b). The thickest and most extensive fossil dune systems occur in the Balearic islands, specifically in the Palma, Alcudia and Campos basins on Mallorca, as well as in the SE of the island where Butzer (1975) described two fossil dune systems separated by palaeosols in Es Bancassos, close to Cala Figuereta (nine aeolian horizons), and in Sa Plana, close to Cala Marmols (seven horizons). In the former, the aeolian deposits reach 130–140 m above sea level, recording sedimentation from Pliocene–Pleistocene times up to the present. The most complete early Pleistocene sequence is found at Bahía Azul where 11 dune generations, separated by palaeosols, have been recognized (Goy *et al.* 1994; González-Hernández *et al.* 1999). In addition, a very complete late Pleistocene dune sequence, comprising at least five generations, is recorded in this area, with three Holocene dune generations occurring to the south in Campos Bay.

Pyrenees and Basque-Cantabrian mountains

The Quaternary record in the Pyrenees and Basque-Cantabrian mountains has been well studied (Fig. 14.6). Abundant Pyrenean glacial deposits have been studied since the end of the nineteenth century (Chueca *et al.* 1998) and

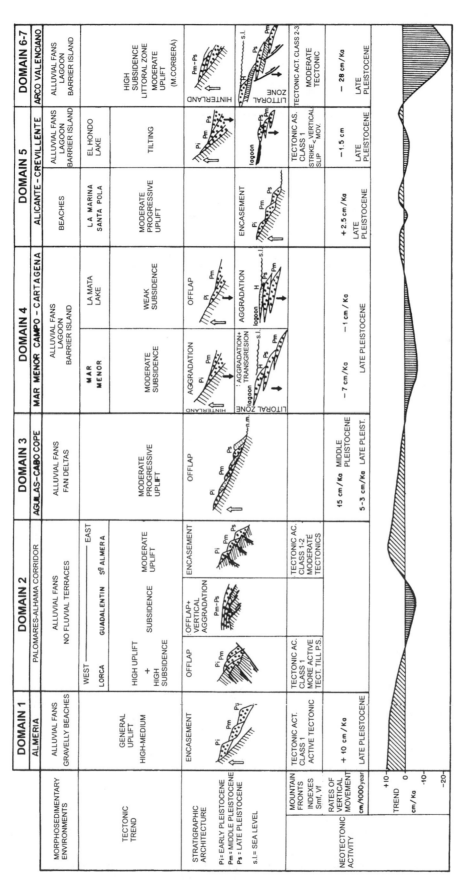

Fig. 14.4. Relationship between neotectonics, geomorphology and Quaternary sedimentation in eastern and southeastern Spain. For domains 1 to 7, see Figure 14.3a.

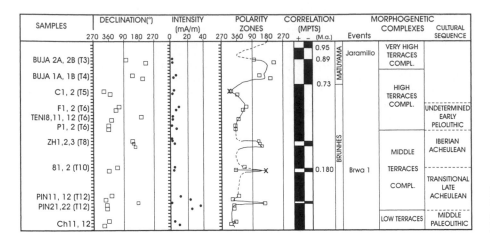

Fig. 14.5. Chronosequence of the Guadalquivir fluvial terraces based on magnetometry and cultural data (after Baena & Diaz del Olmo 1997).

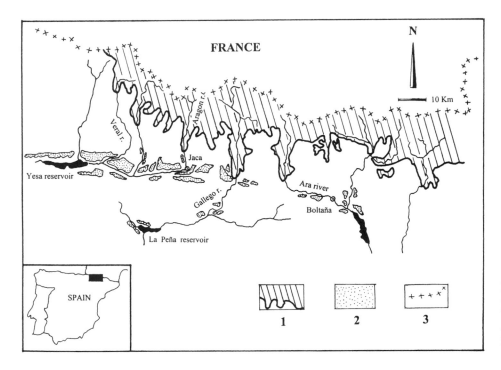

Fig. 14.6. Maximum extent of glaciers and middle Pleistocene terraces in the western Pyrenees. 1, Maximum extent of glaciers; 2, Middle Pleistocene terraces and covered pediments; 3, French border.

have become the subject of intense discussion concerning the number of glaciations that affected the Pyrenees. Additionally, the presence of large basins within the range (Canal de Berdún, Val Ancha, Conca de Tremp, la Cerdaña) has favoured the deposition of major fluvial successions. Relationships between glacial and fluvial deposits are complex, but help us to assess the glacial periods. In the Basque-Cantabrian mountains the Quaternary period is represented by fluvial terraces, much less well developed than in the Pyrenees, slope deposits (screes), and by small glacial deposits close to the highest divides. Finally, both the Pyrenees and the Basque-Cantabrian mountains have been affected by a number of large landslides.

Glacial deposits and stages. There is clear evidence for the discontinuous presence of upper Pleistocene glaciers across a wide area stretching from the Castro-Valnera Massif (1707 m) in the Cantabrian mountains, to the Canigó Massif (2785 m) in the eastern Pyrenees. In most of the Basque-Cantabrian mountains only a few remains of glacial deposits and landforms can be found, isolated around the highest peaks. Eastwards from the Ori peak, in the western Pyrenees, glacial deposits are common, both in the main and secondary valleys, provided that the altitude exceeds 2000 m. The most important glaciated valleys are, from west to east, Aragón Subordán, Aragón, Gállego, Ara, Cinca, Cinqueta, Esera, Noguera Ribagorzana, Noguera de Tor, Garona, Noguera Pallaresa, Noguera de Cardós, Valira and Carol. Some of these glaciers exceeded 30 km in length and 500 m in thickness (800 m in some overexcavated basins; Bordonau 1992).

The most detailed studies, including isotopic dates, confirm that the majority of the glacial remains belong to the Würm period. Nevertheless the presence of deposits corresponding to older stages is increasingly evident. Thus, in the Gállego valley deeply weathered granitic boulders of glacial origin have been found far away from the main valley (Serrano 1998). In the Noguera Ribagorzana valley, Vilaplana (1983) describes a fluvioglacial terrace 180 m above the present channel, as well as morainic boulders incorporated within a periglacial deposit covered by a till from the last glacial maximum. In the Aragón valley, where eight frontal moraines are spread across some 3 km, the problem is very complex, though there is evidence for at least two different glacial periods (Martí-Bono 1973; Vidal-Bardán & Sánchez-Carpintero 1990). Finally, in the Carol valley (Upper Segre valley) the Puigcerdá moraines suggest the occurrence of several glacial periods (Calvet 1998), one of them being very old and possibly related to the isotopic stades (IS) 16 or 32.

The Würm glaciation is very well represented everywhere in the Pyrenees. Studies by Vilaplana (1983), Martínez de Pisón (1989), Bordonau (1992), Martí-Bono (1996), Serrano (1998) and Chueca *et al.* (1998) have established different episodes within the Würm, from the glacial maximum to the Little Ice Age. The glacial maximum occurred between 50 000 and 70 000 years ago, that is, prior to the period of maximum cold (around 20 000 years BP). Lateral morainic deposits show the existence of three glacial pulses in many valleys (Aragón Subordán, Aragón, Gállego, Ara and Esera). During the maximum extent of the ice, only the highest peaks stood above an almost continuous ice mass that interconnected the different valleys through the mountain passes. The glacial tongues of the main valleys extended down to 800 and 900 m. Most of the frontal moraines of this stage have disappeared as a result of fluvial erosion, but good examples have been preserved in the Aragón, Gállego, Noguera Ribagorzana and Carol valleys.

After this glacial maximum, the Pyrenean glaciers underwent a strong spatial shrinkage. Montserrat (1992) demonstrated, from the study of sediments in the Tramacastilla lake (Gállego valley), that 30 ka BP ago the ice was restricted to the head area. A later advance called the 'valley-glacier stage' (Bordonau 1992) took place some 25–20 ka BP (Martínez de Pisón 1989; Serrano 1998). In the Gállego valley, sediment accumulated in a doline after the glacial retreat has given an age of 20.8 ka BP (Montserrat 1992). After this secondary but important advance, three other less important stages have been identified: (i) 'valley-glacier stage at altitude' (Bordonau 1992; 16–15 ka BP); (ii) 'cirque-glacier stage' (14–13 ka BP) with glaciers restricted to small ice masses around 4 to 8 km in length, sometimes covered by blocks and boulders ('debris-covered glaciers'), as in the Escarra and Lana Mayor valleys; (iii) 'wall-glaciers stage' (11–10 ka BP) with moraines located very close to the cirque backwalls and frequent development of rock glaciers. There are no data for the possible existence of glaciers during Holocene times. However, an important advance took place during the Little Ice Age in the massifs above 3000 m with sedimentation of very fresh moraines (Aneto-Maladeta, Posets, Monte Perdido and Infierno-Balaitús massifs). Finally, in addition to the moraines, many valleys preserve glacio-lacustrine and glaciotorrential deposits, these having been mainly studied in the Gállego (Serrano 1998), Ara (Serrat *et al.* 1983), Esera and Noguera Ribagorzana (Bordonau 1992) valleys.

In the Basque-Cantabrian mountains the presence of small glaciated areas has been detected in the Cinco Villas massif and in the Aralar sierra, with small frontal moraines located at a distance of 1 to 2 km from the cirque. The most important glacial massif was Castro-Valnera, in the Cantabrian mountains (Martínez de Pisón & Arenillas 1979), where the glacial tongues reached down to 425 m in the Asón valley, to 600 m in the Miera valley, and to 750 m on the southern slopes of the Trueba valley where the former glacial front shows the remains of four morainic arcs (Serrano 1996). Eastward, in Peña Lusa (1562 m), another glacial valley has been identified in the Gándara valley where glaciers descended further than normal due to the influence of Atlantic-derived precipitation. Finally, in the Pas valley González-Díez *et al.* (1999) have identified several generations of glacial cirques ranging in age from around 55–45 ka to 11–10 ka.

Fluvial deposits and pediments. The main Pyrenean rivers run from north to south, thus crossing the geological structure and developing generally narrow valleys. Consequently, along most of their course the rivers show small, young terraces. In the Aragón Subordán valley three levels of terracing can be distinguished at 40 m, 20 m, and 8 m, restricted to a narrow strip that laterally connects with slope colluvium. In other valleys it is difficult to find remains of more than one terrace level due to lateral fluvial erosion.

Superimposed on the north–south drainage pattern, the presence of easily erodible materials (Eocene marls, Keuper clays) has favoured the development of wide east–west trending valleys (Pamplona Basin, Canal de Berdún, Val Ancha, Basa and Ara valleys) and intramontane basins (Cuenca de Campo, Conca de Tremp). These valleys and depressions are several kilometres in width and preserve several terrace levels in the Aragón, Gállego, Ara, Cinca, Esera and Noguera Ribagorzana valleys. In addition, the headwaters of the river Segre

have eroded a large structural depression in which there are widespread Quaternary deposits.

It is difficult to correlate the terrace levels found in the different Pyrenean rivers as the evolution of each river depends not only on the general climatic conditions but also on the local lithology and structure. However, study of the terraces in the main rivers suggests that a general sequence can be identified in the Pyrenees. Thus, in most of the main rivers between four and six terrace levels have been found (Peña 1994), with a very recent low level (2 m) being flooded during low frequency discharges, and higher levels located at 10, 20, 35–40, 60 and 90–120 m. Some of these levels are occasionally absent, for example the 40 m level in the Aragón river, or the 60 m terrace in the Segre river, but, in general, all the main Pyrenean rivers seem to have been subjected to the same stages of incision–aggradation. The highest level (90–120 m, if indeed it represents just one level) is very poorly preserved, and its age dates back to the middle (or even early) Quaternary. In contrast, the 60 m level is widely developed and is covered by red soils which pre-date the Würm moraines of the Aragón valley (Vidal-Bardán & Sánchez-Carpintero 1990). In the wider valleys there is also a complex system of alluvial pediments connecting the slopes and the terrace systems and dating back to possible Pliocene–Quaternary times in places such as the Conca de Tremp (Peña 1983). Finally, the most recent level was constructed during the Holocene and is widespread, being related to anthropogenic erosion in the hillslopes.

In the Basque-Cantabrian mountains fluvial deposits are found along all the rivers flowing into the Cantabrian Sea, but they are isolated, mostly recent remains of terraces (Hazera 1968). An exception to this is provided by the existence of at least eight levels in the Pas river basin (González-Díez *et al.* 1999). In the basins draining towards the Ebro river, the existence of large structural depressions (Vitoria basin, Miranda-Treviño depression) favoured fluvial sedimentation and the preservation of the terraces (e.g. the middle stretch of the Zadorra river).

Slope deposits. Talus located at the foot of rockwalls (especially limestones) is covered by thick screes basically developed during the cold stages of upper Pleistocene times. In the Pyrenees and Basque-Cantabrian mountains different types of slope deposits have been identified. Many of them were produced under past climatic conditions, are located at relatively low altitudes, and cover or connect with the lowest levels of fluvial terraces. The highest ones (>2200 m) are still active due to frequent, intense freeze–thaw cycles.

Most of the slope deposits can be considered as amorphous colluvium not necessarily linked to very cold environments. Others are the consequence of geomorphic processes derived from cold, total or partially periglacial temperatures, and show the typical structure of stratified screes. Their study has allowed the reconstruction of palaeo-environmental conditions over the last 25 000 years. García-Ruiz *et al.* (2000) have demonstrated the existence of two stages in the development of stratified screes. The first one coincides with the upper Pleistocene coldest period (25–20 ka BP) and the second one with the Late Glacial stage 15–13 ka BP. Those corresponding to the upper Pleistocene coldest stages have been developed mainly by means of stone-banked solifluction lobes with clear stratification of openwork and matrix-rich layers, vertical clast-size classification in the openwork layers, and lateral continuity of the beds. The occasional presence of *grèzes litées* emphasizes the intensity of freeze–thaw cycles. In places, stratified screes show frontal festoons, discontinuities in the layers and an almost total absence of vertical grading, thus suggesting that debris flows were of major importance. Most of the stratified screes in the central Pyrenees are located on east- or SE-facing slopes, at the foot of limestone rockwalls. Westward, especially in the Basque-Cantabrian mountains, their presence is exceptional, due to less severe climatic conditions. In some cases the slope deposits show palaeosols with abundant ashes corresponding to wildfires which occurred around 35–33 ka BP, above which new massive debris was probably deposited during the Little Ice Age.

Mass movements are very abundant both in the Pyrenees and in the Basque-Cantabrian mountains. Many of these mass movements are deep-seated, with a rotational plane (sometimes accompanied by flows) and have great geomorphological consequences. The spatial distribution of deep-seated mass movements reveals the importance of lithology. Most of them are located in Palaeozoic shales and schists (e.g. in the Gállego valley), as well as in the Eocene flysch sector, occuring at very different altitudes and aspects, and sometimes induced by recent channel incision. In many cases the mass movements are located on deglaciated slopes.

González-Díaz *et al.* (1999) have obtained a number of dates from landslides of the Cantabrian mountains, ranging from over 120 ka BP to the present day. Dated examples from the Pyrenees include those in the Gállego valley mentioned above, where a large landslide has been dated as 20 120 ± 150 years BP, although a similar one, near Tramacastilla lake, seems to belong to the middle Holocene. The large slump of Biescas-Arguisal, also in the Gállego valley, is probably very old, and perhaps linked to deglaciation. The most important triggering factors have been fluvial incision, tectonic activity and temporal increases in precipitation. Corominas & Moya (1999) conclude that deep-seated landslides typically occur after a period of moderate rainfall (>200 mm over several weeks).

Iberian and Catalonian Coastal ranges

The Iberian Ranges today form an important fluvial divide in the Iberian peninsula (Fig. 14.7), separating rivers flowing to the Ebro basin and Catalan coastal plains (Mediterranean drainage) from those flowing to the Duero and Tajo basins (Atlantic drainage). The present geomorphology of the Iberian Ranges was created by late Pliocene tectonism. Extensive Neogene erosional surfaces were deformed and displaced from their original positions. This tectonic phase was responsible for the present distribution and altitude of the high sierras

(from north to south: Demanda, Urbión, Cebollera, Moncayo, Albarracín, Gúdar and Javalambre; Fig. 14.7). At the same time, new tectonic depressions were formed (Jiloca graben and the Catalan coastal grabens: Baix Ebre, El Camp, Vallés-Penedés), while other pre-existing ones were reactivated (Calatayud graben, Alfambra-Teruel graben). This pulse interrupted calcareous sedimentation in the Neogene lacustrine basins, producing a palaeoenvironmental change to alluvial plains, and leading to the generation of wide mantled pediments and alluvial sheets during middle to late Villafranchian times. At the same time, the topographic uplift of the sierras favoured incision and most grabens were fluvially captured. The tectonic pulses have continued, although with less intensity, through Quaternary times up to the present (Gutiérrez & Gracia 1997). Much of the recent geological evolution of the range is recorded in the Quaternary sediments which, although not very widespread, show a rich variety of glacial, periglacial, slope, pediment, fluvial, lacustrine, tufa, aeolian and karstic deposits.

Glacial and periglacial deposits. Quaternary glacial deposits only appear in Demanda, Urbión, Cebollera and Moncayo sierras (Fig. 14.7), in the form of small morainic ridges, generated under a single glacial phase. In the Urbión mountains, Thornes (1968) differentiated between upper and lower cirque moraines: the former lack any soil cover and coarsen upwards, whereas the latter are covered by brown-reddish soils and incipient podzols. Moraine thicknesses reach 30–40 m, and several lacustrine clay deposits up to 3 m thick

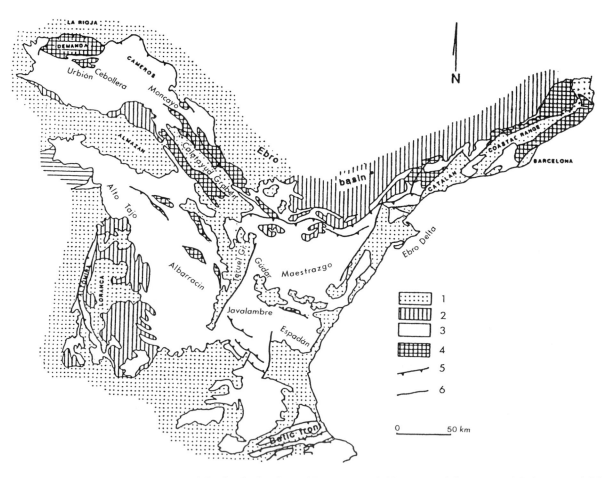

Fig. 14.7. Map of the Iberian Ranges and the Catalonian Coastal Ranges. Key: 1, Neogene and Quaternary; 2, Palaeogene; 3, Mesozoic; 4, Palaeozoic; 5, thrust; 6, fault (modified from Santanach 1997*a*).

develop between the moraine ridges. Several fluvioglacial terraces appear downslope, in the Laguna Negra, a lake of glacial origin. In the Sierra Demanda, García Ruiz (1979) described chaotic moraine deposits 2–3 m high. In the Sierra Moncayo and Sierra Cebollera there are similar deposits, although in the latter the moraines are much thicker.

In high zones of the Iberian Ranges there are many periglacial features and deposits inherited from the last Pleistocene cold stage. The most important ones consist of block-streams, gelifluction lobes, screes and other debris deposits. In the Sierra Cebollera there are rock glaciers up to 600 m long, with huge blocks, and several protalus ramparts (Arnáez & García Ruiz 2000). In the Sierra Moncayo slope and block-streams, solifluction lobes and *grèzes litées* are common, with heterometric clasts, clay matrix and podzolic soils (Pellicer 1980).

In the Sierra Albarracín spectacular block-streams about 2 km long develop in some valleys, fed by blocky slopes that sustain peat deposits, and are now completely stabilized (Gutiérrez & Peña 1989a). In these mountains there also appear several *grèzes litées* and other small relict periglacial deposits in many other high parts of the ranges (Alto Tajo, Molina Highplains, Javalambre, Gúdar, Maestrazgo).

Slope and pediment deposits. In the Iberian Ranges the most important slope deposits are linked to periglacial processes. The older ones are formed by compacted breccias, very common in the Mediterranean coastal zone and in some inner sierras. In the Sierra Maestrazgo they are coeval with a middle Pleistocene fluvial terrace, so they are clearly older than the nearby periglacial deposits (Calvo 1986).

An accumulative solifluction episode is recognized in the ranges. In some places, as in the Sierra Espadán (Castellón), slope deposits show several reworking phases associated with Pleistocene cold-climate denudation processes, revealing a very complex morphogenetic evolution (Butzer & Mateu 1999). Nevertheless, all these deposits have since been affected by fluvial dissection and anthropogenic erosion, and many of them contain Late Bronze Age–Early Iron Age pottery. A new although less important slope sedimentation episode has also been identified, containing archaeological remains from the Middle Ages (Gutiérrez & Peña 1998).

Pediment deposits appear in the most important grabens and inner depressions, especially in the Catalan coastal zone. They are mainly represented by mantled pediments, generally controlled by fluvial base levels, and are formed by angular clasts of local origin. The older ones are represented by alluvial fans deposited during the Pliocene–Pleistocene transition, as a consequence of the tectonic rise of the sierras in late Pliocene times. They usually show reddish deposits with an important development of a calcareous duricrust (calcrete) in the upper levels. They are very extensive in La Puebla de Valverde–Sarrión depression, where they are more than 80 m thick, covering former Villafranchian levels (Gutiérrez & Peña 1989a). Purely Quaternary pediments are less widespread and generally rather thin within the Iberian Ranges, and connect with fluvial terraces in the main valleys. In the Catalonian coastal grabens they are much more widespread, connecting with marine terraces. In this zone they were dated by U/Th-series and thermoluminescence by Villamarín et al. (1999), yielding ages ranging between 300 and 100 ka. Both the older and the more recent pediments are sometimes deformed by Quaternary tectonics, expressed as soft fracturing or tilting.

Fluvial deposits. The transition of the internally drained interior basins to outwardly directed drainage was not coeval everywhere but diachronous between, and even within, basins (Gutiérrez et al. 1996). The capture of the Calatayud graben (Fig. 14.7) by the Jalón river, which flows into the Ebro basin, was probably triggered by the tectonic movement of a NE–SW fault during early Pliocene times (Gutiérrez 1998). A second capture took place during the late Pliocene interval, this time by the Huerva river. Something similar happened in the Daroca and Jiloca grabens, which were progressively captured by the Jiloca river (Jalón basin). The Teruel graben (Fig.14.8) was captured during early Pliocene times by the Turia river, which flows directly to the Mediterranean sea (Gutiérrez et al. 1996). In the Catalonian coastal grabens, a late Pliocene alluvial fan records

a transition to the exterior drainage of these depressions. As in the Iberian ranges, the late Neogene reactivation of transverse faults (NW–SE) controlled the location of the main fluvial valleys that drain the coastal grabens into the Mediterranean sea (Sala 1994). All these captures, and the new valleys formed, probably used ancient fluvial depressions inherited from the Neogene network. The stream captures and the following divide migrations resulted in a dynamic geomorphic disequilibrium, where many drainage basins were enlarged at the expense of adjacent basins (Miller 1980). There remain, however, some depressions that still have not been captured by the fluvial network (southern Jiloca graben, Gallocanta graben).

The rivers of the Iberian Ranges usually form straight and deep valleys so that fluvial sediments are rare and located low down, close to the main channels. Only when the rivers flow through the wider inner basins do they develop significant fluvial deposits, distributed across several levels of stepped terraces. The best known terrace system is the Alfambra–Guadalaviar–Turia basin, which crosses the Alfambra–Teruel graben, with the river flowing towards the Mediterranean sea. In this graben the river develops ten terrace levels at 1–2, 4–5, 9–15, 20–23, 32–35, 38–44, 50, 55–60, 75 and 100 m above the modern channel (Gutiérrez 1998). Intermediate levels were dated as middle Pleistocene (Riss) by macro- and micromammals (Moissenet 1985). The deposits are formed by cross-laminated polygenetic gravels, with the higher levels being strongly cemented by a calcrete, as is very common in all the fluvial systems of the range. The thickness of the fluvial deposits is only a few metres, although some higher levels reach >60 m in places due to synsedimentary karstic subsidence by dissolution of the evaporitic Miocene substratum.

In the Calatayud graben the Jalón river also develops ten terrace levels, at 3–5, 20–25, 30–35, 45, 50–55, 60–65, 70–75, 85–90, 100–105 and 115 m above the modern channel, with the four oldest levels all being early Pleistocene in age (Gutiérrez 1998). As in Teruel, the Jalón river terraces are affected by both ductile and brittle deformation, as well as important thickening (locally >100 m) due to karstic subsidence (Gutiérrez 1996). Other minor rivers, such as Huerva, Martín, Guadalope, show small and dispersed terrace levels when flowing over soft materials of the inner basins.

In the Catalonian Coastal Ranges most rivers are sourced from the Pyrenees (Ter, Llobregat and, mainly, the Ebro river). The heights of their terrace levels are similar to their equivalents in the Ebro basin: 4–6 m, 10–12 m, 30–40 m and 80–90 m. They pass into submerged fluvial terraces, under the Ebro deltaic system. Although the Ebro delta is the most important one in this coastal zone, some other Catalonian rivers have also developed large deltas during the Holocene (Llobregat and Tordera rivers, among others). All these deltas have shown a spectacular historic growth, due to increasing input of detritus linked to deforestation and human desertification. The present erosive trend is related to the high number of dams constructed in the basins, which have considerably reduced the arrival of sediments to the coast (Serrat 1989).

Lacustrine deposits. Lacustrine deposits in the Iberian Ranges are represented mainly by those of Gallocanta lake (14 km^2), located in the central sector of the ranges. This lake was generated during the final stages of deepening of a karstic polje, due to the existence of a basal insoluble and impervious level (Upper Triassic clays). Gallocanta lake belongs to a greater Quaternary interior basin, in which other minor lakes and ponds broadly reflect the past extent of former polje bottoms (Gracia et al. 1999). Quaternary deposits in the basin are represented by stepped lacustrine terraces formed of laminated clastic sediments that again reflect the greater extent of the lake in the past (Fig. 14.9), with heights fluctuating between 0.5 and 8 m above the present maximum lake level (Gracia 1995). The sediments covering the centre of the lake consist of carbonate and sulphate muds, and during dry periods the shore is covered by a discontinuous salt crust. Some core profiles along the lake coast show that lacustrine sediments are only a few metres thick below the water body, and are formed by alternating clays, carbonates and detrital elements. Their geochemical and mineralogical analysis has provided valuable palaeo-environmental and sedimentological information about the Holocene evolution of the lake (Schütt 1997).

Outside the Iberian Ranges there is another important lake located in a zone between the Pyrenees and the Catalonian Coastal Ranges,

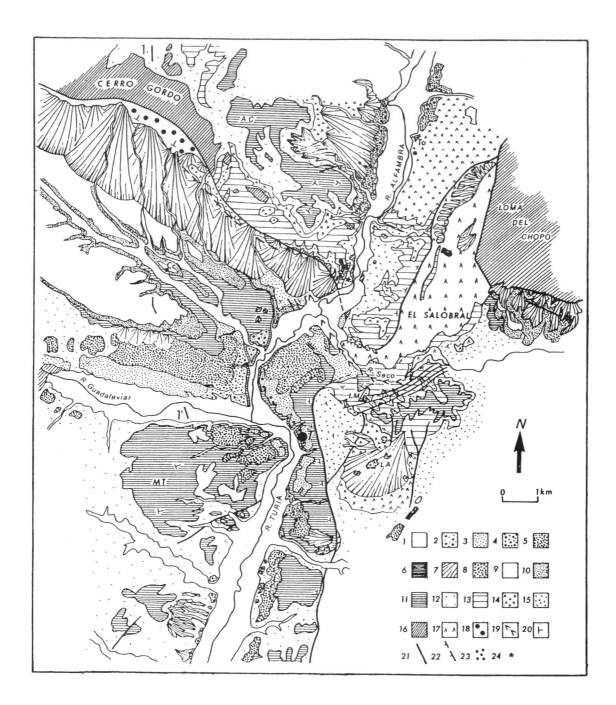

Fig. 14.8. Quaternary units of the central Teruel graben: geomorphological map and north–south cross-section. Key: 1, floodplain; 2, decoupled lower fluvial terrace; 3, lower terrace; 4, decoupled intermediate terrace; 5, intermediate terrace; 6, travertines; 7, lower and intermediate mantled pediments; 8, Villafranchian conglomerates; 9, Gea alluvial fan (lower Villafranchian); 10, Gea detrital red unit; 11, Pliocene limestones; 12, Celadas Miocene–Pliocene detrital red unit; 13, Turolian limestones; 14, Miocene gypsum; 15, Vallesian detritic red unit; 16, Jurassic and Cretaceous units; 17, Upper Triassic evaporites ('diapiric Triassic'); 18, Lower and Middle Triassic; 19, Río Seco valley; 20, dominant dip of strata; 21, fault; 22, fault supposed or fossilized; 23, solution dolines; 24, Baños mining railway station. Locations: Ac, Mesa of Celadas; Cc, Concud; LA, Los Aljezares; LM, Los Mansuetos; MT, Mesa of Teruel; T, Teruel; To, Tortajada; Vi, Villalba Baja (after Moissenet 1985).

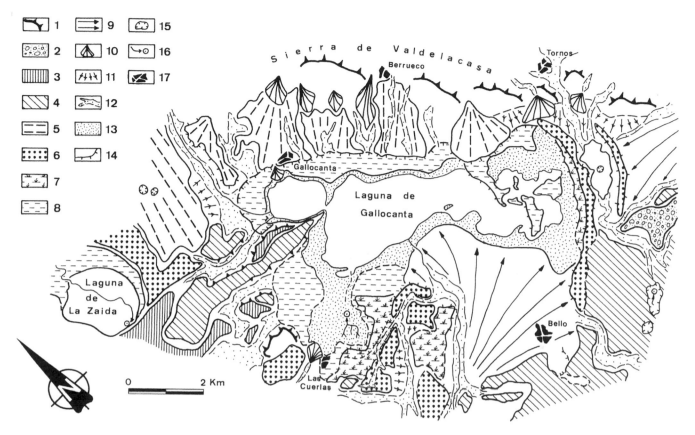

Fig. 14.9. Geomorphological map of Gallocanta lake and surroundings. Key: 1, structural scarpment; 2, Neogene; 3 and 4, stepped karstic corrosion surfaces; 5, Quaternary pediment; 6, higher lacustrine terrace; 7, middle lacustrine terrace; 8, lower lacustrine terrace; 9, recent alluvial fan; 10, debris cone; 11, slope; 12, flat-bottomed valley; 13, lacustrine plain of occasional flooding; 14, scarp in Quaternary deposits; 15, doline; 16, water sink; 17, village (after Gracia *et al.* 1999).

near the Ampurdan basin. This is the Banyoles-Besalú lake, with Quaternary carbonate deposits that developed during ancient stages of lacustrine evolution. These deposits range in age from early Pleistocene to present and include many archaeological and palaeontological remains, such as a human mandible of *Homo neanderthalensis* (Julià & Bischoff 1991). Detailed palaeoclimatic reconstructions have been made using palynological analysis of the lacustrine record (Pérez-Obiol & Julià 1994).

Other deposits. Numerous tufa deposits appear in the Iberian Ranges, commonly related to fluvial valleys and springs, such as at Tajuña, Gallo, Mesa, Piedra, Jiloca, Júcar, Guadalaviar, Mijares, and Palancia. Spectacular stepped tufa barriers and terraces can be seen in the Ruidera lacustrine system, along the southern border of the range, with ages ranging between 10 and 140 ka (Martínez *et al.* 1988). In Catalonia (Serrat 1989), at the base of the Anoia tufa (Capellades zone), there is a level with *Paralephas trogontheri nesti*, characteristic of early Pleistocene times. In recent decades several radiometric methods have been used for dating tufa deposits in the range, indicating some degree of correlation with Quaternary warm periods.

In the Alfambra–Teruel graben, near Escorihuela, there is an aeolian deposit fossilized by other Pleistocene alluvial units. It is formed by up to 10 m of well sorted brown-yellowish sands, with aeolian ripples and spectacular cross-lamination. Thin levels of fluvial sands and fine gravels are also included in some places, and the clasts are affected by aeolian processes, indicating an arid to semi-arid palaeoenvironment (Gutiérrez & Peña 1976).

Finally, karstic infillings at archaeological and palaeontological cave sites are relatively abundant, especially in the Catalonian Coastal Ranges (e.g. abric Romaní in Barcelona and Serinyà in Gerona). Pollen and macromammal analysis revealed a clear link with Pyrenean palaeoclimatic evolution (Butzer & Freeman 1968). Cold phases at 50–45, 26–16 and 16–13 ka BP have been identified from many prehistoric settlements. Other karstic deposits are represented by fluviokarstic accumulations in poljes, formed by a mixture of red clays (terras rossas) and fan deposits with calcareous clasts. Some of these ancient karstic deposits include important Pleistocene palaeontological and archaeological remains, as at Layna and Torralba, in the western Iberian Ranges (Howell *et al.* 1995). However, the most important Pleistocene palaeontological setting in the Iberian Ranges is Atapuerca (Burgos), in the NW part of the area. At this site, several clastic deposits infill former karstic depressions to depths of tens of metres. They record a wide time interval, from middle Pleistocene to Holocene, and contain thousands of palaeomammal remains, as well as six remains of the new Hominid specimen *Homo antecessor*, a precursor to the Neanderthals (Bermúdez de Castro *et al.* 1997). The richness of the fauna and the high number and complexity of the deposits makes this site the most important Quaternary locality in the Iberian peninsula (see further details below).

The Ebro depression

The Ebro depression was one of the largest Tertiary basins of the Iberian peninsula (Chapter 13). At the end of the Tertiary period, the basin changed from a closed, interior basin to one drained by the Ebro river. The incision of the Tertiary sediments by externally draining river systems generated the main present-day geomorphological features of the Ebro depression. About one-third of this depression is today covered by Quaternary deposits (Gutiérrez & Peña 1989b, 1994b; Fig. 14.10), with the most widespread sediments being covered pediments and fluvial terrace deposits commonly crowned by

Fig. 14.10. Simplified map of the Quaternary deposits in the Ebro depression (modified from Perez-González *et al.* 1989).

Table 14.2. Heights in metres of the terraces of the main fluvial systems in the Ebro depression (after Julián 1996)

	Ebro Rioja	Ebro Navarra	Ebro Zaragoza	Ebro Caspe	Oja	Najerilla	Iregua	Aragón	Arga	Ega	Gállego	Jalón	Huerva	Cinca	Segre	Noguera Ribagorz.
T1	2–4	3–5	3–6	2–4	0.5–1	1–2	5	2–5	3–6	2	2–9	2–3	1–3	2–3	1.5–2	1.5–8
T2	6–12	8–11	10–14	7–11	8–10	10–18	10–12	8–10	15–20	10–15	12–27	7.5–9	8–12	10	10	10
T3	15–20	17–20	29–34	15–20	20–25	20–30	20–25	15–20	25–31	25–30	21–35	25–30	17–22	20	18–20	20
T4	30–40	23–31	64–73	30–40	30	40–45	30–35	23–35	40–50	35–45	34–48	42–49	30–35	45	35–40	35–40
T5	50–60	45–60	106–115	45–60	40	60	40–50	40–45	56–60	50–70	55–78	58–63	53–60	60	60	50–65
T6	65–80	50–55	127–138	70–80	50	75–80	60	90	70–80	95–110	80–104	72–84	68–74	85–90	80–90	90–100
T7	90–110	70–80	156–168	90–110	70	120–150	100	100–115	95–100	140–160	121–126	95–110	82–87	100–105		115–120
T8	130–150	85–98	198–220	135–145	80	160–170	165–170	130–140			150–175	155–160	106–111	115–120		150
T9	170–180	100–115		180	90	190	235	170					160	150		
T10		120–140			112	245								190		
T11		185												200		

calcretes. There are also infilled valleys, talus flatiron sequences, lacustrine saline deposits and aeolian sediments.

Mantled pediment deposits. Structural platforms, locally known as 'muelas' or 'planas', are capped by horizontal limestones which locally record the end of Tertiary internal drainage sedimentation. In the central sector of the depression, to the south of the Ebro river, La Plana and La Muela de Zaragoza structural platforms are locally covered by thin detrital deposits unconformably overlying a deformed Miocene calcareous bedrock. These deposits are made up of clasts from the Iberian Ranges, and correspond to covered pediments inclined towards the Ebro river (Zuidam 1976). Similar deposits occur to the north of the Ebro river in the highest mantled pediments of La Plana Negra and Montes de Castejón. These pediments are, however, inclined towards the south, and the clasts have a Pyrenean origin. Alluvial supply was clearly both from the Iberian Ranges and the Pyrenees towards the axis of the depression where an ancient Ebro river was located (Gutiérrez & Peña 1994b).

Later progressive incision of the fluvial systems and selective erosion of the Tertiary sediments led to the generation of the main morphological features of the depression. Extensive covered pediment levels were developed at the foot of the structural reliefs and surrounding ranges, and coeval stepped sequences of fluvial terraces were deposited on the valley flanks. Depending on their relative chronology and position the covered pediments have been differentiated into two groups: Pliocene–Pleistocene and Quaternary. The Pliocene–Pleistocene mantled pediments are located on topographic divides and are more widespread in the Pyrenean piedmont (Alberto et al. 1984) than in the Iberian piedmont (Fig. 14.10). These covered pediments have a gentle slope and are capped by calcretes. The Quaternary pediments are linked to the base levels of the evolving fluvial systems and form stepped sequences at the foot of the upland areas. These alluvial deposits comprise poorly bedded sequences with angular clasts, and are generally less than 10 m thick. The pediments commonly show a rock-cut surface in the proximal area that gives way to an aggradation surface towards the base level. The deposits of the pediments in the distal area are locally interfingered with the coeval fluvial terrace sediments. The amount of calcium carbonate increases with the relative age of the pediment surface and deposit so that thick, well developed calcified horizons are found on the oldest and highest pediment levels (Sancho & Meléndez 1992).

Fluvial terrace deposits. Stepped sequences of Quaternary fluvial deposits <10 m thick are widespread in the Ebro depression, being more common along the Pyrenean tributaries of the Ebro river than in the Iberian tributaries (Fig. 14.10 and Table 14.2; Julián 1996). In La Almunia de Doña Godina area (Zaragoza Province), at the Iberian margin of the Ebro basin, some terraces of the Jalón river are locally thickened and deformed due to Pleistocene synsedimentary tectonic subsidence. The deposits of numerous fluvial systems are also affected by synsedimentary subsidence due to underlying dissolution of evaporitic Tertiary bedrock made up of Ca-sulphates (gypsum and anhydrite), halite and Na-sulphates (glauberite and thenardite) (Benito et al. 1998; Gutiérrez et al. 2001; see also below). Some palaeontological remains have been found in Quaternary terrace deposits, such as Elephas sp. in Logroño (Ebro river) and in the outskirts of Zaragoza city (Ebro and Gállego rivers) and Bos sp. and Equus sp. in Los Llanos de Urgell (Lérida). These fossils were found in coarse-grained sediments and do not provide precise chronological information due mainly to their wide chronostratigraphical distribution. Archaeological remains found in the fluvial valleys of the Ebro depression also do not have chronological value for dating the fluvial deposits, since they have been collected from the surface of the terraces. In the Guadalope river valley, Macklin & Passmore (1995), applying the infra-red stimulated luminescence technique, have obtained ages of 115 ± 17 ka for the 23 m terrace and 28 ± 4 ka for the 12 m terrace. In the terraces of the lower reaches of the Gállego river affected by this evaporite dissolution subsidence, Benito et al. (1998) differentiate two chronological groups of terraces using palaeomagnetic reversals: Matuyama age (pre-780 ka) from T_1 to T_6 terraces (175 to 85 m) and Brunhes age (post-780 ka) from T_7 to T_{12} (70 to 2 m). Sancho et al. (2000) have magnetostratigraphically dated the terraces T_5 (60 m), T_6 (85–90 m) and T_8 (100–105 m) of the Cinca river as Brunhes (post-780 ka).

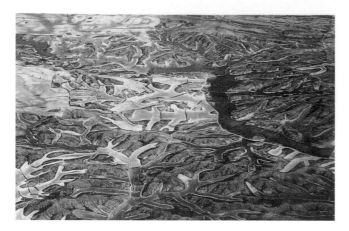

Fig. 14.11. Aerial view of infilled valleys developed in Neogene evaporitic formations in the central sector of the Ebro depression (outskirts of Zaragoza city).

In the central sector of the Ebro depression there is a distinctive dense network of flat-bottomed infilled valleys (Fig. 14.11) generally developed in evaporitic and clay lithofacies by alluvial processes (Gutiérrez & Arauzo 1997; Gutiérrez & Gutiérrez 1998). In some areas these sediments contain abundant archaeological remains that allow dating of late Holocene accumulations (Zuidam 1976; Burillo *et al.* 1985). Dates covering most of the Holocene have been obtained by radiocarbon dating in different units of the infilled valleys (Peña *et al.* 1993). In the Alcañiz area (Teruel), Stevenson *et al.* (1991) and Macklin *et al.* (1994) differentiate three Holocene accumulation stages (dated as 3849 to 500 BC, 410 BC to AD390 and post-Roman) separated by incision stages induced by changes in vegetation cover.

Slope deposits. Talus 'flatirons' are relatively common around the scarped structural reliefs of the Ebro depression. They are also known as tripartite slopes or triangular slope facets and are characteristic landforms of arid and semi-arid zones. Their origin corresponds initially to slopes covered by debris, which later on become incised by water erosion so that some relict slopes become isolated and separated from the scarp (Fig. 14.12). In their distal parts they grade into pediments, and finally into fluvial or lacustrine terraces. The facets are concave and covered by poorly sorted detritus <8 m thick. For a facet to be generated a stage in which accumulation is predominant must be followed by other stages in which erosion processes prevail. The repetition of several of those alternating stages leads to the generation of polyphase talus flatiron sequences, with the most ancient ones located furthest from the scarp.

The generally accepted model of origin is related to climatic changes. The stages with more vegetation cover correspond to relatively colder periods with dominant debris accumulation on the slope. A decrease in the vegetation cover during relatively warm periods can trigger the incision of the slope, creating the facets. A critical point is considered to be a vegetation cover corresponding to 70% and a mean annual precipitation of 400–500 mm, below which erosion increases considerably. In recent times, human activities, including destruction of the vegetation cover, can be the main cause of the incision of the talus slopes (Gutiérrez & Peña 1998*a*; Sancho *et al.* 1988).

The absolute date of the most recent facet obtained in Mezalocha (Zaragoza Province) by radiocarbon dating (2930 + 60 years BP) correlates with the Cold Period of the Middle Bronze Age. The AMS dates of the two oldest facets (27 862 + 444 years BP and 35 570 + 490 years BP) can be correlated with the H_3 and H_4 Heinrich events that correspond to global cold periods (Gutiérrez *et al.* 1998*a*). In this region, using palaeotopographic reconstructions based on talus flatiron profiles, scarp retreat rates ranging from 0.9 to 1 mm/year have been estimated (Gutiérrez *et al.* 1998*b*).

Lacustrine deposits. In the central sector of the Ebro depression there are several areas of interior drainage with ephemeral lakes (Fig. 14.13). Some of these lakes have evaporitic sedimentation that has been exploited in historical times. Mineralogical studies indicate surface precipitation of chlorides (halite) and sulphates (gypsum, mirabilite-thenardite and bloedite) together with small amounts of carbonates (Pueyo 1978–79). Despite the numerous boreholes drilled by different research groups, there are no absolute dates to correlate the sediments between the different lacustrine basins. In El Pito and La Jabonera saline lakes (Bujaraloz area) a change from humid conditions in early Holocene times to the current arid conditions with a humid phase in between have been inferred (Schütt & Baumhauer 1996). In the Salada de Mediana, according to Valero-Garcés *et al.* (2000), the lower sections correspond to the Late Glacial phase with very humid periods separated by more arid periods. From the fourteenth century, in Salada de Mediana as in Salada de Chiprana, there was a rise in the water level of the lake due to irrigation practices. In the Alcañiz area, Macklin *et al.* (1994) infer alternating arid and humid periods and the subsequent influence of human activity due to irrigation.

Evaporite dissolution subsidence and Quaternary alluvial deposits

In numerous areas within the Ebro depression, currently active subsidence related to the subsurface dissolution of evaporites constitutes a geohazard of great economic and social impact. It affects buildings, communication routes, irrigation networks and agriculture, causing large financial losses and numerous inconveniences (Gutiérrez & Gutiérrez 1998; Gutiérrez *et al.* 2001). Quaternary alluvial deposits

Fig. 14.12. Talus flatirons in Mezalocha (Zaragoza province).

Fig. 14.13. Salada de Mediana saline lake (central sector of the Ebro depression).

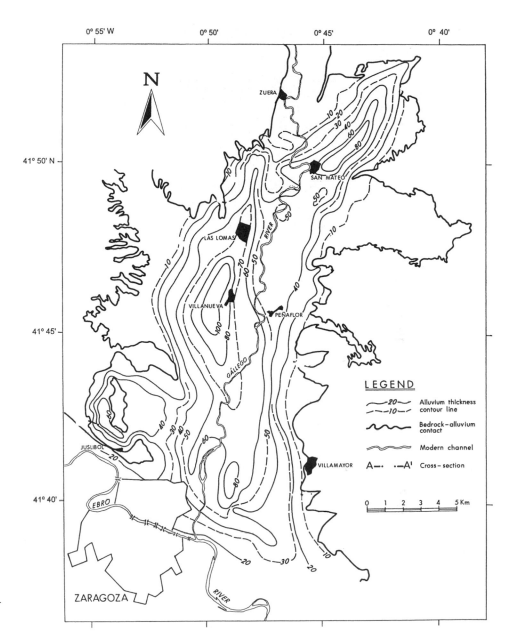

Fig. 14.14. Alluvium isopach map of the lower reach of the Gállego river. The Quaternary terrace deposits are thickened due to evaporite dissolution-induced synsedimentary subsidence (after Benito *et al.* 1998).

above highly soluble evaporite sediments may change in a short distance from <10 m to >110 m in thickness (Fig. 14.14). The boundary between the alluvium and evaporite bedrock generally shows a very irregular geometry, and the alluvial cover can be affected by abundant ductile and brittle deformation (Fig. 14.15). This contrasts with the underlying evaporites that frequently remain undeformed, a fact that disproves a tectonic origin for this type of thickening and deformation. In some cases, the deposits of different terrace levels are superimposed and bounded by either angular unconformities or disconformities.

These anomalous features observed in alluvial deposits above evaporitic formations in several Spanish Tertiary basins have been explained by an evolutionary model of synsedimentary karstic subsidence (Gutiérrez 1996; Gutiérrez & Arauzo 1997; Benito *et al.* 1998, 2000; Gutiérrez & Gutiérrez 1998; Gutiérrez *et al.* 2001). As the evaporite substratum is removed by suballuvial dissolution, the alluvial cover subsides

either in a ductile way by sagging (Fig. 14.15), or in a brittle way by collapse. The differential karstification of the bedrock and the consequent subsidence of the alluvial cover produce closed depressions (sinkholes, dolines) of variable size and geometry. In these closed depressions the water table may be close to, or above, the surface leading to the development of ponds and palustrine environments. The alluvial system responds by aggrading in the subsiding areas, and tends to reach a condition of dynamic equilibrium in which the subsidence rate is balanced by the aggradation rate. This synsedimentary karstic subsidence causes localized thickening and deformation in the alluvial deposits. The sedimentary fill of these 'dissolution-induced basins' has a typical basin structure with cumulative wedge-out systems at the margins and convergent dips towards the core. A map of alluvium isopachs in the Gállego river valley shows that these Quaternary deposits can reach up to 110 m thick and fill several dissolution-induced basins and troughs in the halite and

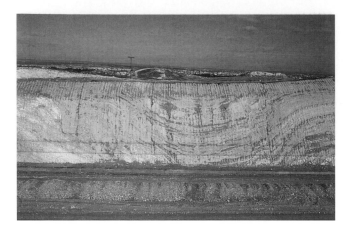

Fig. 14.15. Quaternary mantled pediment deposit overlying Tertiary evaporites in a trench dug for the Madrid–Barcelona high-speed train (Zaragoza city). The local thickening and synformal deformation of the alluvial cover records a synsedimentary subsidence caused by the subsurface dissolution of the evaporites. The arrow indicates a rucksack 60 cm high.

Na-sulphate-bearing Zaragoza Gypsum Fm (Fig. 14.14). In this fluvial system two thickening periods related to greater subsidence and karstification have been dated by palaeomagnetic reversals as Matuyama (pre-78 ka BP) and Brunhes (post-78 ka BP), and linked to intervals of high water supply (Benito *et al.* 1998). The deposits belonging to some terrace levels of the Jalón and Huerva rivers also show thickenings of at least 60 m. The Ebro and Aragón river terrace deposits are affected by deformation and anomalous thickenings where they overlie Oligocene evaporite formations in Navarra (Benito *et al.* 2000). These thickened alluvial deposits are economically important since they constitute valuable water reservoirs and rich sources of aggregates.

Synsedimentary subsidence induced by suballuvial karstification of evaporites has also been studied in infilled valleys that drain into the main Ebro river valley. In some of the infilled valleys thickened and deformed sediments include a high proportion of carbonate and carbonaceous facies originally deposited in palustrine environments developed in dissolution-induced depressions (Gutiérrez & Arauzo 1997). Several outcrops show the superposition of different sedimentary units separated by angular unconformities.

The Duero depression

The Duero basin is another of the large inland depocentres of the Iberian peninsula that originated in Tertiary times (Chapter 13). Known as the northern meseta (Fig. 14.16) it covers approximately 50 000 km² and has a mean altitude of 850 m, some 150 m higher than the southern (Tajo) meseta which lies to the south of the Central Range. The basin has mountainous margins: the Cantabrians to the north, the Asturian-Leonese region to the NW, the Iberian Ranges to the east and SE, the Central Range to the south, and the Central Iberian region, or Salmantino-Zamorana peneplain (mean height 700 to 800 m), to the west. A particular feature of this depression is that it is almost entirely drained by one major river, the Duero and its tributaries, with the main exception of its SE sector, the Almazán sub-basin, which is drained by the Jalón river, a tributary of the Ebro. Most authors argue that

the direction of rivers draining the Duero basin is structurally controlled by Alpine and late Variscan reactivated fault systems (Pérez-González *et al.* 1994).

From the perspective of Pleistocene geology, this area is not as well known as the Tajo basin, mainly due to a notable lack of vertebrate fauna and palaeolithic archaeological records in stratigraphic settings. Exceptions are the archaeological sites of La Maya (Salamanca) (Santonja & Pérez González 1984), and the Atapuerca (Burgos) archaeopalaeontological sites, which will be discussed below.

Figure 14.16 shows the characteristic Duero deposits: piedmonts and fluvial terraces in the northern region; fluvial terraces and erosion surfaces (with or without a coating of superficial deposits) to the south of the main river, sandy aeolian deposits in the SE; and erosive depositional structural surfaces formed on late Neogene carbonate lacustrine deposits of the *páramos castellanos* in the east and central area. At least two erosional surfaces associated with the latter have been described and mapped, including numerous dolines, terra-rosa relicts and limestone crusts of Pliocene age (Molina & Armenteros 1986).

Fluvial deposits. Fluvial deposits (Fig. 14.16) are undoubtedly the most conspicuous Quaternary sediments of the Duero basin. The most ancient sediments, of probable late Pliocene age, are piedmont deposits up to 10 m thick comprising siliceous gravel and quartzite conglomerates eroded from the Montes de León (Martín-Serrano 1991). This high alluvial plain lies at 1200–1250 m, is called the Brañuelas surface (Birot & Solé 1954), and is equivalent to the Guardo Raña (1080 m) preserved on the eastern margin of the basin. Soil profiles (Herail 1984) are characterized by thick A horizons, low pH (4.6 to 4.8), and the appearance of thin discontinuous ferruginous pans, 3 to 7 m deep. It is thus a highly developed soil of the ultisol type. Terrace sequences are always stepped systems with commonly a large number of levels, e.g. up to 18 or 20 terraces in the Carrión and Arlanzón rivers (Zazo *et al.* 1983), with the lowest (youngest) being 4–7 m above the present alluvial plain, and the highest (oldest) lying at a relative height of 110 m. In the central sector of the Duero valley, the highest terrace preserved in the Tordesilla-Medina del Campo profile is found at 144 m above the river (Pérez-González *et al.* 1989). Terrace sequence lithofacies differ somewhat from central northern to central southern sectors of the Duero basin. The northern sector shows extensive regional development of terraces related to the Cantabrian and Galaico-Leoneses rivers (e.g. Órbigo, Esla, Cea, Carrión, Tera, Duerna; Fig. 14.16). Terrace sediments are quartzite-rich gravels with only limited development of sandy facies, with overbank muds, silts and sands being better preserved in low terraces and in the present floodplain. Terrace sediments in the southern rivers (Zapardiel, Adaja, Eresma and Cega) are again mostly quartzitic gravels, but sandy lithofacies may also be important (Tortosa *et al.* 1997). These southern terraces are associated with extensive erosion surfaces on which a detrital cover of sandy and pebbly deposits sometimes reaches >10 m thick. The latter arise from continuous reorganization of fluvial networks by capture and avulsion (Fernández 1988; Fernández & Garzón 1994; Pérez-González *et al.* 1994) in a cold, dry climate.

Estimates for the ages of the terrace systems of the Duero basin are hindered by a lack of preserved palaeofaunal and lithic artefacts. However, an exception can be found in the palaeolithic sequence of La Maya in the river Tormes (Santonja & Pérez-González 1984), with industries dating from the middle Palaeolithic (level at 8 m) to the lower Palaeolithic (levels at 14 m, 34 m and 50 m). These, along with data on the relative altitudes of terraces, have been compared with fluvial sequences in the Madrid basin, which serve as a chronological reference.

Aeolian deposits. Substantial aeolian sand deposits, dunes and cover sands, have been identified and mapped SE of the Duero river (Fig. 14.16), in the region known as Tierra de Pinares. The following aeolian features have been described: deflation hollows scoured in Tertiary detrital materials of middle Pleistocene age (Pérez-González *et al.* 1994), dunes and cover sands (Hernández-Pacheco 1923; Alcalá del

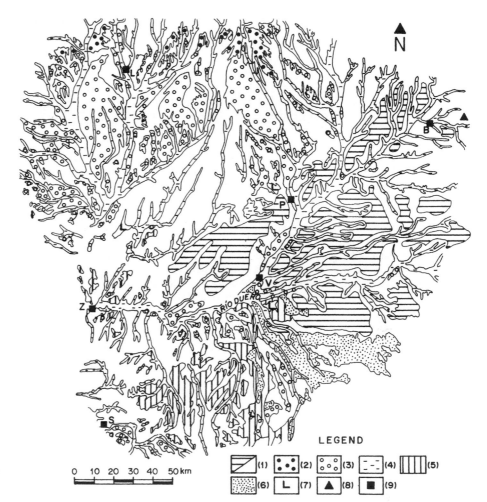

Fig. 14.16. Quaternary deposits of the Duero basin (modification from the Quaternary Map, 1989, and the Geological Map of Spain, 1994, both 1:1 000 000, ITGE). Legend: (1) continental Tertiary, uppermost deposits are lacustrine páramo limestones; (2) Raña piedmont; (3) fluvial terraces; (4) floodplains; (5) erosion surfaces with or without detritic cover; (6) aeolian sands; (7) endorheic depressions; (8) Atapuerca sites; (9) cities: B, Burgos; L, León; P, Palencia; S, Salamanca; V, Valladolid; Z, Zamora.

Olmo 1972; Portero & Olmo 1982; Olivé & Gutierrez Elorza 1982) overlying the extensive fluvial terraces described previously, and aeolian deposits interstratified with fluvial facies of the rivers draining the area. These aeolian deposits are texturally fine and medium quartz sands with a <5% silt and clay fraction. They were derived from both the Quaternary deposits and Tertiary detrital bedrock. The dunes are almost all parabolic in form, although blow-out dunes, climbing dunes and transverse forms of alignment perpendicular to dominant SW winds also occur. Crest height does not usually exceed 4 m, although the thickness of the cover sands may reach over 10 m. Recent thermoluminescence dates (Bateman & Diez Herrero 1999) indicate a dominant sand accumulation period from 12.5 to 11 ka BP, corresponding to the cold, dry phase of the Younger Dryas. Temiño *et al.* (1997) described three Holocene aeolian reactivation stages from 10.2 to 2.2 ka BP (^{14}C), separated by phases of stability related to a high water table (wetland–dune systems) at 10.2, 6.1 and 2.2–1.5 ka BP.

Atapuerca archaeological sites

The archaeopalaeontological sites of the Sierra de Atapuerca are located to the east of Burgos (Fig. 14.16) on the NE margin of the Duero basin. The region comprises ancient karst developed over partly exposed Mesozoic limestones that form a NNW–SSE anticline. The Edelweiss pot-holing team (Martín-Merino *et al.* 1981) has mapped over 3 km of caves in the Sierra de Atapuerca karst. La Sima de los Huesos, located in one of these (Cueva Mayor), is where Torres discovered the remains of human fossils in 1976, a find that included a complete jaw bone studied by Aguirre & Lumley

(1977). These human remains, alongside those of *Ursus deningeri*, were found in brown mud. At present, over 3000 fossils corresponding to some 30 hominids of both sexes and different ages have been found. These hominids are considered ancestors of the Neanderthals, and have been dated to 300 ka BP by uranium series, electron spin resonance (ESR) and palaeomagnetic techniques (Arsuaga *et al.* 1997; Parés *et al.* 2000). On the south side of the Atapuerca anticline, an abandoned railway cutting has exposed several caves overtopped by deposits of boulders, gravels and brown or red muds from 15 to 20 m thick. These deposits contain rich faunal associations of micro- and macromammals, birds and reptiles. In level TD6 of the Dolina (or Gran Dolina) cavity in the 'aurora' stratum (20–25 cm thick, composed of mudstones with limestone clasts), lithic artefacts and almost a hundred fossil remains belonging to at least six individuals (associated with the micromammal *Mimomys savini*) have been recovered (Carbonell *et al.* 1995). These fossils have been palaeomagnetically dated as belonging to the lower Pleistocene (Matuyama Chron >0.78 Ma BP) by Parés & Pérez-González (1995) and Falguères *et al.* (1999), using a combination of ESR and U-series applied to fossilized ungulate tooth enamel. Analysis of the human fossils has led to the proposal of a new species of the genus *Homo*, *H. antecessor*, as an ancestor of *H. neanderthalensis* and *H. sapiens* (Bermúdez de Castro *et al.* 1997). Further details of the remarkable Atapuerca site are provided at the end of this chapter.

The Tajo basin

The area considered here once comprised one wide Cenozoic depocentre, but the subsequent uplift of the Sierra de Altomira has subdivided it into the Madrid basin, the Manchega plain ('Llanura manchega') and the western sector of the Júcar basin (sometimes referred to as the 'intermediate depression'; Fig. 14.17). Broadly speaking, the Tajo basin (or southern meseta) is bordered to the north and west by Precambrian and Palaeozoic mountains composed of igneous rocks, slate and quartzite, and to the south and east by relatively low-lying hills of Mesozoic carbonates. These marked lithological and topographic contrasts have strongly influenced Quaternary sedimentation.

Fluvial terraces and piedmonts, together with aeolian deposits, travertine and tufa, are characteristic Quaternary deposits across this southern meseta. Quaternary materials also include early Pleistocene alkaline and ultra-alkaline volcanic rocks in Campo de Calatrava (Ancochea 1982). Moreover, fauna including *Mammuthus meridionalis* among other mammalian species, have been described underlying basaltic lapilli in a terrace of the Guadiana river, NE of Valverde de Calatrava. Equally characteristic forms and deposits may be seen in the small playa lakes in which detrital and saline sedimentation was studied by Peña & Marfil (1986), and more recently by Schütt (2000) in the seasonal lakes of Cucharaz y Sancho Gómez in Campo de Calatrava and the Manchega plain. The Madrid basin preserves deposits of terrarossa and calcified soils overlying the Páramo limestones of La Alcarria (Vaudour 1979). Finally, recent periglacial deposits of

grézes litêes have been described (González Martín 1986) on the slopes of the valleys that dissect La Alcarria.

Fluvial deposits. The Raña piedmont alluvial plain is well developed to the NE of Madrid and is formed by materials derived from the quartzite and slate Somosierra, Ayllón and Alto Rey ranges, east of the Central Range. It is similarly well represented in Tertiary subbasins west of Ciudad Real and in the Júcar valley where it is known as Aluviones de Casas Ibañez or Plataforma de Casasimarro (Cabra *et al.* 1988). In each case, this piedmont was established before the formation of the first terrace of the current fluvial system, and is composed of clast-supported gravels mainly of quartzite, quartz and slates, with largest grain sizes corresponding to the boulder or cobble fractions. Its thickness does not exceed 10 m and it supports highly developed soils classified as Calcic Palexeralf or Ultic Palexeralf (Espejo 1985; Gallardo *et al.* 1987). Its age has not yet been clearly established but is about 2.5 million years old in the Madrid basin (Pérez-González 1982). To the east and west of the Manchega plain, the end of Tertiary sedimentation and subsequent unconformable deposition of the Raña is more precisely dated as having occurred at the beginning of the Matuyama Chron, some 2 Ma BP (Gallardo-Millán & Pérez-González 2000).

River terraces, alluvial deposits of the east and central Manchega plain, and deposits of alluvial fans, are also sediments of fluvial origin that are common in areas of the southern meseta. The valleys of the river Tajo and its northern bank tributaries, along with the Guadiana and Júcar rivers, have been the subject of several studies since the midnineteenth century because of the presence of lithic industries and mammalian fauna, particularly in the Manzanares, Jarama and Tajo rivers (Fig. 14.17). Syntheses of special interest include those by Royo Gómez (1929), Vaudour (1979), Alférez (1977), Pérez-González (1982), Aguirre (1989), Santonja & Villa (1990), and Sesé & Soto (2000). These works have led to a reasonable understanding of the construction of these valleys and their deposits, and of the fauna and

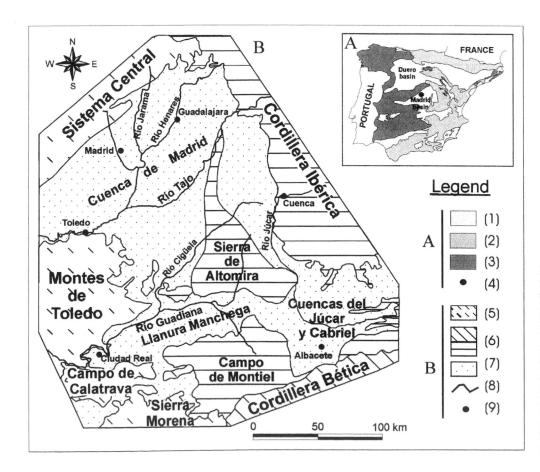

Fig. 14.17. Main morphostructural units of the south submeseta and their location in the Iberian peninsula (note: cuenca = basin). **(A)** Legend: (1) Tertiary basins; (2) Alpine chains; (3) Iberian Massif; (4) Madrid. **(B)** Legend: (5) Precambrian–Palaeozoic Montes de Toledo, Sierra Morena and Central Range; (6) Mesozoic and Palaeogene of Iberian Ranges, Campo de Montiel and Betic Cordillera; (7) continental Tertiary basins; (8) main rivers; (9) towns.

archaeological artefacts they contain. A first consideration is that the geometry and number of terrace facies are conditioned not only by climate (a factor not yet well understood) but also by tectonics and the lithology of the substrate (Alía 1960; Pérez-González 1982; Giner & Vicente 1995; Silva *et al.* 1997).

Archaeological or palaeontological sites are relatively abundant in the lower reaches of the Manzanares river, in the Jarama valley to the east of Madrid, and in the terraces of the Tajo river, near Toledo. Faunal associations of vertebrates indicate a temperate or warm-temperate climate for the Middle and Lower Pleistocene and fauna of the Upper Pleistocene such as *Mammuthus primigenius* and the hairy rhinocerous (*Coelodonta antiquitatis*) have been identified, indicating a cold climate (Aguirre 1989; Sesé & Soto 2000; Sesé *et al.* 2000).

Chronological determinations have so far taken into account only faunal or archaeological contents, and have correlated +60 m terraces with the lower–middle Pleistocene boundary based on *Mammuthus meridionalis*. Terraces from +60 m to +18–20 m are considered to represent the middle Pleistocene based on acheulian artefacts (Pinedo, Aridos etc.) and *Palaeoloxodon antiquus*. Terraces at +12–15 m and +7–9 m have been dated as upper Pleistocene. The Holocene interval is represented by alluvial plains found at +3–5 m.

Further fluvial deposits are those found associated with cover pediments and alluvial fans, these ranging in age from the early Pleistocene to the present day. At times these deposits are of great thickness such as the muds, sands and gravels that cover the river terraces of the river Tajo around the city of Toledo. The best developed sandy gravel alluvial fans are those that encircle Campo de Montiel and the eastern Palaeozoic relief of the Montes de Toledo. The longest fans extend for 18 km, as seen in the alluvial fan of the Alto Guadiana in the central plain of La Mancha. Massive carbonates and calcified soils on these fans are up to 15 m thick and show different stages of formation indicating intermittent activity over a prolonged period.

Travertine and tufa deposits. These deposits are always related to limestone landscapes and are especially abundant in the Madrid basin, the Júcar valley in Albacete, the Campo de Montiel, and on the western margin of the Iberian Ranges along the course of the river Tajuña as it flows through La Alcarría. Travertine terraces (Torres *et al.* 1995; Fernández Fernández *et al.* 2000) are composed of gravels (locally with boulders) overlain by biogenic carbonate. These terraces are found at different relative heights above the river channels with thicknesses often exceeding 20 m. Available chronological data (U-series and Th/U) ascribe the different levels to the oxygen-isotope stages 7, 5 and 1. Tufa deposits, many of which are Holocene in age (Pedley *et al.* 1996; García del Cura *et al.* 2000), are controlled to some extent by surface and subterranean flow, as well as by the mineral content of the water.

Sand and clay dunes. These dunes, occupying central and eastern areas of the Manchega plain, were first described by Pérez-González (1982). The main dune forms identified in eastern areas of the Manchega plain are parabolic dunes up to 7 m high, formed by westerly winds, and sourced from the local sediments of the river Júcar alluvial system on which they lie. The alluvial plain of San Juan in the central region of the Manchega plain contains clay dunes (>10% of clay), sand dunes and sand-sheets covering areas up to 30 km × 7 km in extent. Dunes do not generally exceed 3–4 m in height (although locally they may reach 8 m), and include ovoid, parabolic, transverse and longitudinal types formed by winds from the west, NW and SW. Thermoluminescence and optical stimulated luminescence studies (Rendell *et al.* 1994) indicate an initial phase of aeolian activity around 28–20 ka BP, one around 10 ka BP that would correspond to the Young Dryas, and some reactivated aeolian activity in the Late Holocene (4000–300 years BP).

Coastal geology

The Spanish coastline shows tremendous variation, including as it does the tectonically active Betic Cordillera and the volcanically active Canaries shorelines, the warm salty Mediterranean (tidal range <0.5 m) and the cooler Atlantic (tidal range 4 m) coasts. Although the present coastal outline

records the response to the most recent transgression that has occurred during the present interglacial (Holocene), something like 100 m of sea-level change has occurred during late Pleistocene times and left its mark on the Spanish shoreline.

The Pliocene–Pleistocene boundary

On the southern Atlantic coast, the Pliocene–Pleistocene boundary is represented by the Bonares Sands Fm ('Arenas de Bonares') or Red Sands ('Arenas Rojas'). These comprise siliciclastic sands that become more conglomeratic upwards, record deltaic conditions and are overlain by a palaeosol. On the northern Atlantic coasts, the Pliocene–Pleistocene boundary has traditionally been linked to a marine erosion surface, called 'La Rasa', located at +100–120 m above present mean sea level (Fig. 14.18). In the Canary Islands, palaeomagnetic and volcanic data seem to indicate a late Pliocene–early Pleistocene age for an 80 m marine terrace (Agaete, Gran Canaria; Klug 1968). The faunal content of the sediments on this terrace suggests similar oceanographic conditions to those at present. Finally, recent work carried out along the Mediterranean coastline shows that the Pliocene–Pleistocene boundary does not record any important change either in geodynamic or in climatic conditions (Bardají *et al.* 2000). The extinction of the warm fauna with *Strombus coronatus*, which was traditionally assumed as representing the signal for Quaternary cooling, has became an obsolete criterion since palaeomagnetic data showed that this species crossed the Pliocene–Pleistocene boundary. The cold fauna assemblage recorded in the Pliocene–Pleistocene boundary stratotype in Vrica (Italy) has not yet been found in the Spanish coastal zone.

Mediterranean coastline: Spanish mainland

The eastern and SE coasts of Spain present a great richness and diversity of Quaternary deposits, especially those associated with the Betic domain where important neotectonic activity has promoted the subaerial exposure of very complete marine sequences in upfaulted areas. The main deposits are marine terraces, alluvial fans, aeolian dunes, and sediments associated with deltas, spit bar and barrier island–lagoonal systems.

Marine terraces. Gigout *et al.* (1955) described two *Strombus bubonius*-bearing coastal marine terraces and assigned them to the Flandrian (2 m) and Tyrrhenian (5 m) transgressions, and two older levels (20–25 m and 80 m) that were associated with the more ancient Tyrrhenian and Sicilian transgressions respectively. The work of Lhenaff (1977) on the Malaga coast, and Dumas (1977) on the Murcia and Alicante coasts, together with a study done by Ovejero & Zazo (1971) on the Almería coast, were the first studies to consider both the tectonic setting as well as correlations between coeval marine and terrestrial deposits.

In SE Spain, marine terrraces are well preserved in Almeria (Campo Dalías: highest terrace at 150 m; Goy & Zazo 1986), Murcia (Aguilas: highest terrace at 80 m; Bardají *et al.* 1999), and Alicante (El Molar Range–La Marina: highest terrace at 65 m; Goy & Zazo 1988; Somoza 1993). Palaeomagnetic analyses carried out in these areas have allowed differentiation between the early and middle Pleistocene levels. A higher number of terraces is recorded during middle Pleistocene times, each one representing different interglacials. Along the coast of Alicante (Somoza 1993) and Murcia (Bardají *et al.* 1999) early-middle Pleistocene alluvial fan deposits are interleaved with marine terraces, indicating that in addition to neotectonic effects, glacio-eustatic factors influenced sea-level changes in these areas. As a general rule, the better-developed terraces are those corresponding to marine isotopic stages (MIS) 11 and 9.

At the end of middle Pleistocene times, during MIS 7, the entry of equatorial warm fauna (Senegalese fauna) into the Mediterranean is commonly recorded, although *Strombus bubonius* has only been found in the coast of Almería (Zazo & Goy 1989), Murcia and Alicante (Zazo *et al.* 2001). However, it was not until the late Pleistocene (MIS 5) that this warmer fauna invaded the whole Mediterranean mainland coastal area. Along the Betic coast, all the marine levels corresponding to the different substages of the last interglacial (MIS 5e, 5c and 5a) bear *S. bubonius*, although MIS 5a can be clearly defined only along the Almería coast.

The climatic and eustatic variability during MIS 5e (Zazo & Goy 1989; Zazo *et al.* 1993) is recorded in the most complete sequences by the development of three marine units separated by marked erosion surfaces, terrestrial deposits or palaeosols. Stearns & Thurber (1967) made the first isotopic measurements (Th/U) on the coasts of Almería, Murcia and Alicante, and were later followed by the works of Dumas (1977), Bernat *et al.* (1978, 1982), Brückner & Radtke (1986), Hillaire-Marcel *et al.* (1986), Brückner (1986) and McLaren & Rowe (1996). A synthesis and discussion on the age of the middle and late Pleistocene marine terraces, based on field criteria and Th/U measurements was published by Causse *et al.* (1993) and Goy *et al.* (1993), deducing mean ages of *c.* 128 ka for MIS 5e, *c.* 98 ka for MIS 5c, and 85–80 ka for MIS 5a. Finally amino-acid racemization analyses have been carried out by Hearty *et al.* (1987) particularly on the last interglacial (MIS 5e) marine terraces of Almeria and Alicante.

Alluvial fans. In those coastal zones where the relationships between marine terraces and alluvial fans can be observed, and studied independently from neotectonic effects, data suggest that major changes in the evolution of the alluvial fans were linked to climatic and sea-level changes. Bardají *et al.* (1987) and Somoza (1993) found a close relationship between sea-level lowstands (glacial stages) and fan development in the Alicante and Murcia areas. Harvey *et al.* (1999) suggest that at least in the Cabo de Gata area (Almería), major sediment pulses coincide with global glacials and dissection occurs during the intervening global interglacials. The most complete and best dated littoral sequence of alluvial fans is located in the area of Campoamor (Alicante), where the distal facies of the fans crop out along the cliff. The development of red palaeosols, with well developed Bt and BCa horizons, together with palaeomagnetic analyses, has allowed the differentiation of at least six sequences of alluvial fans developed during middle Pleistocene times along this coast (Somoza *et al.* 1989).

Aeolian deposits. Aeolian deposits began to develop during the last interglacial along the southeastern coastline of Almería, Murcia and Alicante, with the only known exception being a dune from the middle–late Pleistocene transition (Goy & Zazo 1986) in Almería. Although the spatial development of these dunes is not very important, their sometimes oolithic character, usually related to MIS 5e, gives them great stratigraphic value (Goy *et al.* 1993) and climatic significance. Much of the last interglacial (MIS 5e) was characterized by warmer-than-present superficial marine waters, and scarce detrital supply to the coast under a relatively arid climate. During Holocene times, the most important dune development is recorded in the coasts of Alicante (Guardamar del Segura) and Almería (Cabo de Gata). The relationship between these aeolian units and spit bar systems indicates that it was not until 2500 years ago that dunes began to develop in response to the activity of SW winds (Goy *et al.* 1998).

Deltaic deposits. The most important deltas are located along the eastern (Ebro river, Tarragona) and southeastern coasts (Adra and Andarax rivers in Almería, Guadalhorce and Velez rivers in Málaga, and Jucar and Turia rivers in Valencia). Deltaic conditions acquired great importance during Holocene times, being studied with the help of drilling and isotopic dating (Ebro delta) as well as by historical data (Adra, Velez, and Jucar-Turia deltas). The development of the Ebro delta began during the highstand of the Holocene transgression around 7000 years ago, and has continued to recent times, with two pulses of sedimentation related to relative sea-level changes occurring during the last millenium (Somoza *et al.* 1998). Along the southern

coast, the most important infilling and progradation on the coastal alluvial plains of the ancient rivers flowing towards the Alborán sea, have taken place since the sixteenth century partly in response to human activity (Hoffman & Schultz 1988; Lario *et al.* 1995).

Spit bar and barrier island–lagoon systems. Barrier island–lagoon systems have acquired great importance during the Holocene interval, being almost geographically coincident with those developed during the last interglacial (Alicante: Torrevieja and La Mata Lagoons; Murcia: Mar Menor; Almería: Guardias Viejas). The most outstanding examples of spit bar systems are those closing the Valencia, Albufera, Santa Pola lagoons (Alicante) and the Cabo de Gata lagoon (Almería), although the most complete Holocene sequence is found cropping out in Roquetas (Almería). This system started to develop during the maximum of the Holocene transgression around 7000 years ago, and analyses of the beach ridges that compose each spit unit within the system suggest that relative sea-level oscillations, related to short more arid periods, caused the most important phases of coastal progradation during relative fall in sea level (Goy *et al.* in press).

Mediterranean coastline: Balearic islands

Following the excellent synthesis of Cuerda (1989) it has become clear that the most complete and best dated marine and continental Quaternary sequences in the Balearic islands are the marine terraces, aeolian dunes and alluvial fans in Mallorca.

Marine terraces. A sequence of seven staircased marine terraces has developed along the northern margin of Palma Bay in Mallorca, rising up to 60 m, with the four oldest being early and middle Pleistocene in age. U-series measurements and amino-acid racemization were carried out by Hearty (1987) who stated that multiple minor oscillations occur during MIS 5, and two mid-Pleistocene sea levels, one lower than present and the other at about 14 m above present sea level. Later Hillaire-Marcel *et al.* (1996) using U-series measurements (thermal infrared multispectral scanner) at the Campo de Tiro (Palma Bay) section, concluded that the last interglacial is represented by three marine units (MIS 5e) developed between 135 and 117 ka BP, and one marine unit (MIS 5c) aged *c.* 100 ka. Stable isotope and U-series studies of overgrowths on speleothems (Vesica *et al.* 2000) in littoral caves reveal a record of Quaternary interglacials corresponding to MIS 9 (or older), 7, 5e, 5c and 5a. During the Holocene transgression (Goy *et al.* 1997) small fluviomarine terraces and wide barrier island–lagoon systems developed in the southern and northeastern part of Mallorca. Carbon-14 dating has allowed the reconstruction of the changes occurring since the transgressive maximum (*c.* 65 years BP). In the other islands of the archipelago, outcrops of Quaternary marine deposits are poor, discontinous, and date back only to the last interglacial.

Aeolian deposits. The most complete and best dated sequences are again observed on Mallorca. In Palma Bay nine dune systems separated by erosional surfaces or red palaeosols are preserved on top of a marine terrace considered to be of early Pleistocene age on the basis of its faunal content (Cuerda & Sacarés 1966). Palaeomagnetic analyses (González-Hernández *et al.* 1999) indicate that the marine terrace plus the three first dune systems were deposited under reverse polarity (Matuyama epoch). More recent sequences, from the middle and late Pleistocene and Holocene, have been studied to the south and NE of Palma (Rose *et al.* 1999) and indicate enhanced aeolian activity during MIS 6 and 2. Finally the most notable aeolian deposit on the other Balearic islands is the oolithic sand associated with the last interglacial in Formentera.

Alluvial fans. Western Mallorca is characterized by 30 m high cliffs made up almost entirely of broad alluvial fan systems, originating from the southern slope of the Northern Range. At least three important sequences can be identified due to the presence of alternating erosional surfaces or well-developed red palaeosols. They are assumed to be early–mid Pleistocene in age by their relationship with the Last Interglacial marine deposits.

Atlantic coastline: Gulf of Cádiz

The geomorphology of the coast around Huelva has been strongly influenced by changes in the courses of the major rivers that currently flow south, namely the Guadiana, Tinto-Odiel, Gaudalquivir, and Guadalete rivers. During early Pleistocene times the palaeo-Tinto river flowed towards the Portuguese coast with a NE–SW direction (Rodriguez Vidal *et al.* 1991), while the Guadiana river flowed westwards to the coast south of Lisbon (Portugal). Middle to late Pleistocene tectonically induced catchment promoted the displacement of this fluvial network towards the SE. Meanwhile in Cadiz, early Pleistocene estuaries of the Guadalquivir and Guadalete rivers were connected by the existence of a channel located eastwards of Jerez de la Frontera. Deposits corresponding to an ancient delta have been located in the northeastern part of the present Gulf of Cádiz. During early–middle Pleistocene times, barrier island systems developed along the bay, separating lagoons and marshlands from the open sea. Cádiz and San Fernando were at this time two small islands emerging within a wide bay.

In the eastern part of the Gulf of Cádiz, several marine terraces record at least four early to middle Pleistocene highstands. Numerous U-series dates on the last interglacial units reveal a 0.15 mm/year uplift of the central part of the Straits of Gibraltar over the last 130 ka (Zazo *et al.* 1999*a*). During the last glacial period a 100–120 m drop in sea level induced the incision of wide valleys by the main rivers. Holocene transgression promoted the drowning of these valleys, which became wide estuaries. The sedimentary filling of these valleys (Zazo *et al.* 1999*b*; Dabrio *et al.* 2000) suggests that the maximum of the Flandrian transgression occurred here around 6500 years ago.

At present most of the estuaries are silted up and systems of spit bars and marshes have developed at the mouths of the main rivers. Radiometric dating together with archaeological and historical data point to coastal progradation phases over the last 4000 years, mainly driven by climatic changes and, over the last 500 years, by human activity (Zazo *et al.* 1994; Lario *et al.* 1995; Rodríguez Ramírez *et al.* 1996). Finally, the most important aeolian accumulation in the entire Spanish mainland exists on the coast of Huelva, where the thickness of the dune systems from the last glacial period and present interglacial reaches 100 m (Zazo *et al.* 1999*c*).

Atlantic coastline: Cantabria and Galicia

The northwestern and northern coasts of Spain are the least well understood due to a general paucity of Pleistocene deposits. Several erosional or depositional marine levels record ancient sea-level positions, the most complete sequence occurring in the central part of Asturias (Fig. 14.18), where four marine levels range in height from 60 to 5 m and are considered as early and middle Pleistocene in age (Hoyos 1989). The most continuous marine terrace, both on the Cantabrian coast and in Galicia, is a raised beach belonging to the last interglacial, and ranging in height from present sea level up to 2.5 m (Alonso & Pagés 2000; Martínez Graña *et al.* 2000). Inland, fluvial terrace deposits are associated with steep and short rivers, such as the Nalón river where the following terraces have been distinguished (Hoyos 1989): 80–85 m, 50–60 m (early Pleistocene); 40–35 m, 32–24 m, 20–15 m (middle Pleistocene); 5 m and 2 m (late Pleistocene and Holocene).

Coastal sediments from the last glacial and Holocene are probably the best studied, and include solifluction deposits, colluvial deposits and dunes, separated in many cases by palaeosols and eroded by alluvial sediments (Alonso & Pagés 2000). The present interglacial is represented at the surface by rare peat levels and aeolian dunes, while Flandrian infilling of ancient estuaries along the easternmost Cantabrian coast have been studied in detail by Cearreta (1998) and Cearreta & Murray (2000), and supported by numerous radiometric dates. The estuaries began to develop around 8000 years ago and their recent infilling, depositing a shallowing-upward sequence, commenced around 3000 years ago.

Fig. 14.18. Image of 'La Rasa', a marine erosional surface considered as the geomorphologic indicator of the Pliocene–Pleistocene boundary in the northern Cantabrian coast of Spain (Cabo Vidio, Asturias).

Canary islands

The best and most complete Quaternary chronology in the Canaries has been obtained from sequences located along the coasts of Fuerteventura and Lanzarote. Apart from volcanic layers, these sequences consist mainly of marine terraces and dune-palaeosols. In the other islands, with the exception of Gran Canaria where urban expansion has destroyed most of the outcrops, there are discontinuous exposures of marine terraces from the last interglacial (Meco 1987).

Many papers in the 1960s examined raised marine terraces in the volcanic Canarian archipelago, especially on Lanzarote and Fuerteventura. Driscoll *et al.* (1965), Tinkler (1966), Crofts (1967), Lecointre *et al.* (1967) and Hernández Pacheco (1969) cited Quaternary marine terraces and described several levels at topographic elevations ranging from +55 to 0 m. Crofts (1967) recognized seven terraces in Fuerteventura whereas Driscoll *et al.* (1965) and Tinkler (1966) registered six in Lanzarote at very similar elevations. As the faunal content of many of these layers included *Strombus bubonius* and other warm species, Hernández Pacheco (1969) concluded that *Strombus bubonius* occurred in all the Quaternary marine terraces. However, a faunal review of these marine terraces induced Meco (1977, 1987) and Meco *et al.* (1997) to conclude that with the exception of those terraces located below 7 m above sea level (a.s.l.), all the marine levels previously dated as Quaternary constituted in fact one marine terrace (from 70 to 7 m in elevation) of Miocene–early Pleistocene age and characterized by the warm faunal assemblage *Strombus coronatus*, *Nerita emiliana* and *Gryphaea virleti*. According to this interpretation the last interglacial would be represented by one terrace (0 to +5 m a.s.l.) with warm fauna characterized by the presence of *Strombus bubonius*, and the Holocene also represented by one terrace with a fauna similar to that of the present day. Later works (Zazo *et al.* 1997, 2000a) based on morphosedimentary analyses and mapping, as well as on U-series and ^{14}C measurements, view at least seven of the 12 marine terraces developed between 70 m and 0 m to be of Quaternary age, all with a warm fauna (except that from the Holocene). Two sea-level highstands are recorded during MIS 5e, and Holocene sea-level oscillations in the last 4500 years are related to the intensification of trade winds (Zazo *et al.* 1997).

Finally, with regard to aeolian deposits, numerous dune sediments (with the oldest being capped with a thick calcareous crust) alternating with palaeosols (with abundant land snail shells and insects) were deposited during Quaternary times. The four most ancient dune systems are related to the early and middle Pleistocene marine terraces. In late Pleistocene times, several dune sequences developed during the more arid periods, alternating with humid episodes, one of which, dated as around 31 ka BP (Hillaire-Marcel *et al.* 1995), seems to be the most prominent. Large mobile dune accumulations have developed in the northeastern part of Lanzarote and northeastern and southern Fuerteventura.

Human fossils (EA)

The oldest supposedly human fossil reported from Spain is a controversial cranial fragment of an infant from Orce, Granada (Figs 14.19 and 14.20; Agustí & Moyà Solà 1987; Gibert *et al.* 1998). Palaeomagnetic evidence and faunal correlations suggest an age between 1.4 and 1.2 Ma (Agustí *et al.* 1986). The site, Venta Micena, crops out on the margin of the Baza basin within the Betic Cordillera (Chapter 13; Fig.14.19), and is affected by neotectonic disturbance. At that time there was a lake with fertile surroundings and high faunal diversity, under a mild but somewhat arid climate. Mammal bone accumulations over a large area are interpreted as hyena dens (Arribas *et al.* 1996). Also near Orce, the Fuentenueva 3 site contains archaeological evidence without fossil humans. The faunal assemblage, with early microtines and the evolved, hypsodont variety of *Mammuthus meridionalis*, suggests an age of around 1 Ma, close to the Jaramillo magnetic episode.

A phalanx from the second digit was collected in Cueva Victoria, Murcia (Figs 14.19 and 14.20), and described as human (Gibert & Pérez-Pérez 1989). If the original deposit is the old bone breccia filling the major cavities of Cueva Victoria, as suggested, its age would be no younger than 1.1 Ma, with the bone accumulation in the site being due to hyenas. The presence of typical *Mammuthus meridionalis meridionalis* and other mammal species (Agustí *et al.* 1986) suggests a correlation with Solilhac (France) and Tell'Ubeidiya (Israel).

By far the most relevant hominid fossil site in Spain, however, is at the Sierra de Atapuerca in the province of Burgos in northern Spain (Figs 14.20 and 14.21). The Sierra de Atapuerca karst system, which at present includes three sites with human fossils (Fig. 14.23), originated in late Pliocene times and evolved as a juvenile karst until maturation, with a number of lateral openings, in the later part of early Pleistocene time (Aguirre 1998). Fossils were found in a deep, infilled cavity known as the Gran Dolina, in a nearby cave known as the Galería, and in a sloping chamber below a vertical shaft known as the Sima de los Huesos (Fig. 14.24). Human fossils from the 'aurora bed' in the Gran Dolina represent two children, two adolescents and two adults, totalling six individuals. Taphonomic studies demonstrated that they were eaten in the same fashion as were other mammals such as red deer, roe deer, megacerines, horses, bison and rhinoceros, and were, therefore, an early case of dietary cannibalism (Fernández Jalvo *et al.* 1999). They have been described as a new species *Homo antecessor* (Bermúdez de Castro *et al.* 1997) and are an outstanding representation of human occupants in Eurasia during the poorly known time span of late Early and early Middle Pleistocene. In contrast, the human fossil record from the Sima de los Huesos is much younger, but has yielded the largest number of hominid fossils in Europe: more than 3000 fragments from a minimum of 27 individuals. The human population recorded at Sima de los Huesos is classified with other late Middle Pleistocene European hominids (similar to Mauer, Swanscombe, Arago, Steinheim, and Petralona: Fig. 14.20) as *Homo heidelbergensis*, or *H. sapiens heidelbergensis*, and interpreted to be ancestral to the Neanderthals (Aguirre *et al.* 1977; Arsuaga *et al.* 1997). The origin of this population is to be found most probably in Africa: the resemblances between Sima de los Huesos (SH) mandibles and some African fossils (Aguirre & Lumley 1977; Aguirre *et al.* 1980) are reinforced by the strong similarities in facial skulls of Bodo, Petralona and SH specimens (Rightmire 1996).

The size of the Atapuerca SH human fossil sample allows studies of overall variability, ontogenetic variation (Rosas 1998), and palaeodemography (Bermúdez de Castro & Nicolás 1997). Palaeodemographic inferences depend on the scenario of site formation. If the carcasses were deposited discontinuously, either by their relatives after natural death or by carnivores, the mean age may represent the average life

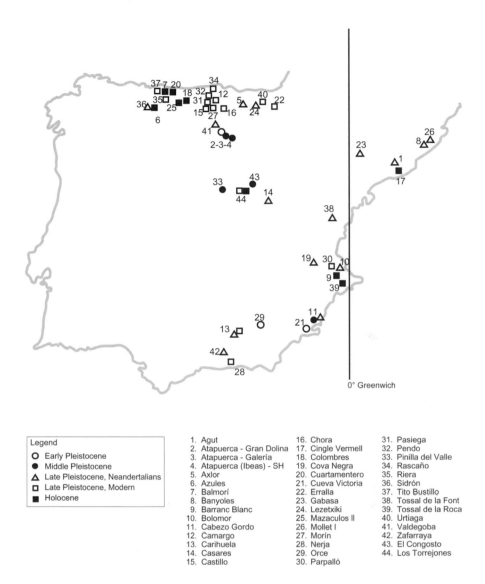

Fig. 14.19. Sites with human fossils in Spain.

Legend

○ Early Pleistocene
● Middle Pleistocene
△ Late Pleistocene, Neandertalians
□ Late Pleistocene, Modern
■ Holocene

1. Agut
2. Atapuerca - Gran Dolina
3. Atapuerca - Galería
4. Atapuerca (Ibeas) - SH
5. Axlor
6. Azules
7. Balmorí
8. Banyoles
9. Barranc Blanc
10. Bolomor
11. Cabezo Gordo
12. Camargo
13. Carihuela
14. Casares
15. Castillo

16. Chora
17. Cingle Vermell
18. Colombres
19. Cova Negra
20. Cuartamentero
21. Cueva Victoria
22. Erralla
23. Gabasa
24. Lezetxiki
25. Mazaculos II
26. Mollet I
27. Morín
28. Nerja
29. Orce
30. Parpalló

31. Pasiega
32. Pendo
33. Pinilla del Valle
34. Rascaño
35. Riera
36. Sidrón
37. Tito Bustillo
38. Tossal de la Font
39. Tossal de la Roca
40. Urtiaga
41. Valdegoba
42. Zafarraya
43. El Congosto
44. Los Torrejones

expectancy. If sudden death overcame a group taking refuge in a cave entrance, possibly as the result of a landslide caused by heavy rain, the observed ages would represent a living population (Aguirre 2000*b*; Andrews & Fernández-Jalvo 1997). Taphonomic, zooarchaeological and palaeoecological studies in the Galería and Gran Dolina outcrops suggest overall a varied nutrition with a major vegetarian component, and only occasional meat consumption. Opportunistic use of dying animals that fell into sinkholes as natural traps has been demonstrated at several successive levels in the Galería (Díez *et al.* 1999*a,b*). Selected and transported animal pieces were consumed at the cave entrance, and in Gran Dolina (unit TD10) there is evidence for the consumption of whole ungulates, as well as for tool-making activity (Díez & Rosell 1998)

Other human fossils assigned to a middle Pleistocene age in Spain (Fig. 14.20) include two teeth (upper M1; upper M3) from Pinilla del Valle, Madrid province, with associated fauna, flora and lower palaeolithic complex. This find lies below a fallen calcareous shelter in the Somosierra mountains, and ESR and U-series dating methods provide ages of nearly 200 ka (Alférez 1985). Also recorded is a mandible and other pieces in a deep infilled pit at Cabezo Gordo, Torre-Pacheco

village, Murcia province, which also yields Neanderthals in younger deposits (Walker *et al.* 1998).

Atapuerca-Gran Dolina. The Gran Dolina filled with sediments from late Early to early Late Pleistocene time. Its entrance became sealed in the early Middle Pleistocene interval, and a new, higher opening was created by a combination of hillside erosion and roof collapse. The faunal assemblage represented in the lower deposits, before the closure of the entrance, is the one known in Europe as early Cromerian, or early Galerian, with *Mimomys savini*, *Panthera gombaszoegensis*, *Stephanorhinus etruscus*, an early bison. Later units contain a younger palaeofauna, with *Arvicola sapidus*, *Microfus agrestis*, *Panthera leo*, *Stephanorhinus hemitoechus*, known in Europe from late middle Pleistocene deposits (Aguirre 1995).

The Gran Dolina has a long, well calibrated, and diverse sedimentological, palaeoecological, and archaeological record (Hoyos & Aguirre 1995). Eleven stratigraphic members (TD1–11) have been defined, which from bottom to top are as follows. Mb TD1 consists of silts of endokarstic origin. Mb TD2 shows a transition to exokarstic conditions, with fallen blocks and speleothemes, a few lenses of coarse waterlain sediments, and a hard, calcareous capping. The pollen record is poor, but *Artemisia* and *Cupressaceae* may indicate relatively cold conditions similar to modern times. Mb TD3 is a thin bed containing both small and large mammalian fossils, indicative of open cave, temperate climate and increased humidity conditions (possibly

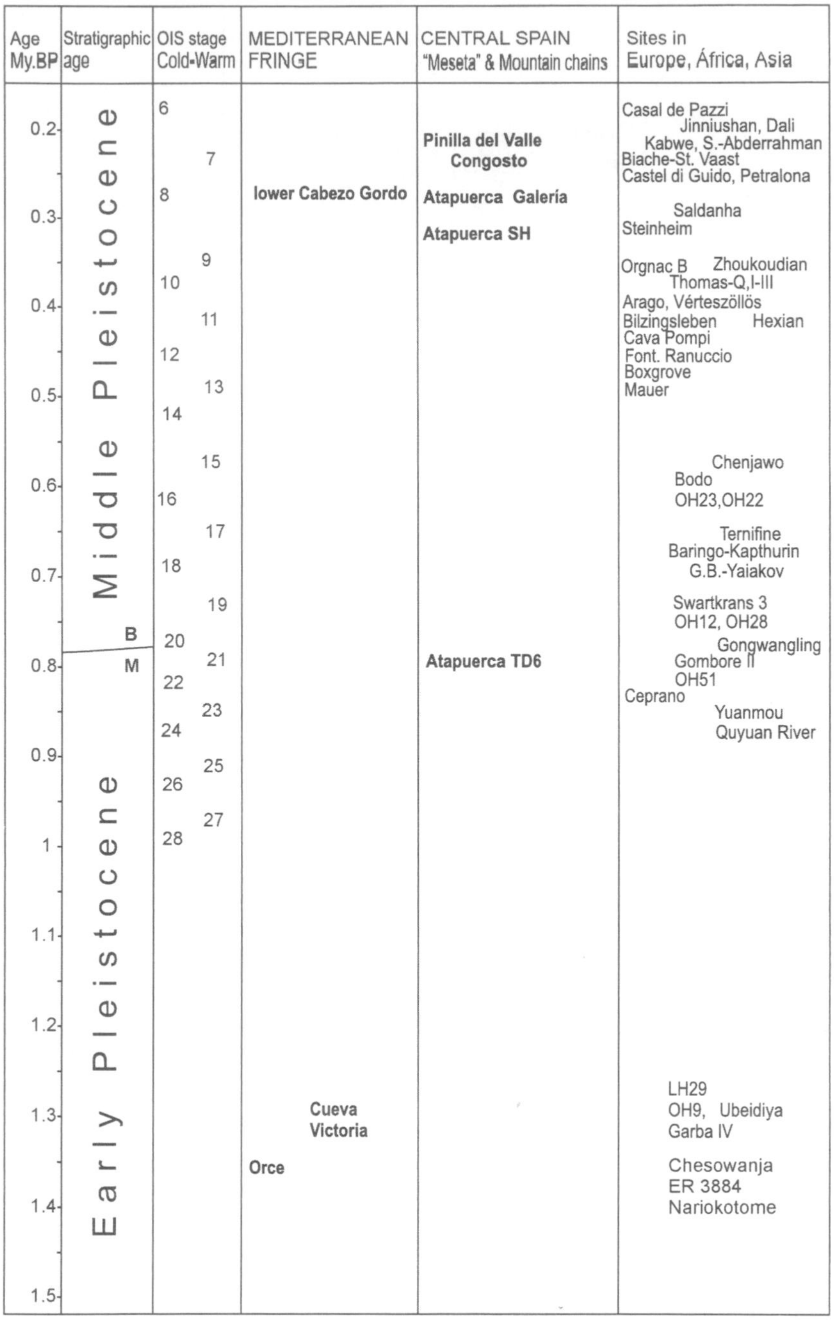

Age My.BP	Stratigraphic age	OIS stage Cold-Warm	MEDITERRANEAN FRINGE	CENTRAL SPAIN "Meseta" & Mountain chains	Sites in Europe, África, Asia
0.2-	Middle Pleistocene	6			Casal de Pazzi
		7		Pinilla del Valle Congosto	Jinniushan, Dali Kabwe, S.-Abderrahman Biache-St. Vaast Castel di Guido, Petralona
0.3-		8	lower Cabezo Gordo	Atapuerca Galería	Saldanha Steinheim
		9		Atapuerca SH	
0.4-		10			Orgnac B Zhoukoudian Thomas-Q,I-III Arago, Vérteszöllös Bilzingsleben Hexian Cava Pompi Font. Ranuccio Boxgrove Mauer
		11			
		12			
0.5-		13			
		14			
0.6-		15			Chenjawo Bodo OH23,OH22
		16			
		17			Ternifine Baringo-Kapthurin G.B.-Yaiakov
0.7-		18			
		19			Swartkrans 3 OH12, OH28
	B	20			Gongwangling Gombore II OH51
0.8-	M	21		Atapuerca TD6	
		22			Ceprano
		23			Yuanmou Quyuan River
0.9-	Early Pleistocene	24			
		25			
		26			
		27			
1-		28			
1.1-					
1.2-					
1.3-			Cueva Victoria		LH29 OH9, Ubeidiya Garba IV
			Orce		
1.4-					Chesowanja ER 3884 Nariokotome
1.5-					

Fig. 14.20. Localities in Spain with published human fossils of early and middle Pleistocene age, with reference to human fossil sites from Africa and Eurasia in the same timespan (1.5–0.2 Ma BP).

oxygen isotope stage (OIS) 25 or 27). Mb TD4 starts with sedimentary evidence for two cold maxima and a final transition to a wetter, warmer episode, possibly OIS 23/25. In its upper part, pollen indicates a transition from Atlantic temperate to Mediterranean drier vegetation, although contamination cannot be excluded. Fossils of large mammals may have fallen in from the old entrance located 10 m above an adjacent wall. Near the top of Mb TD4, a few quartzite technical Mode 1 tools and cores testify to human presence at the entrance. There are also mammal bones with traces of human activity (Rosell et al. 1998). Mb TD5 starts with the collapse, slip and erosion of underlying deposits. Sedimentary and pollen data show two phases of temperate to warm conditions sandwiching a cooler, but not too dry, phase. Following an unconformity, there is evidence for a return to cooler conditions at the top of the deposits. Microvertebrates, together with stone artefacts, were collected from the section exposed by the railway cutting (Sesé & Gil 1987). Mb TD6 shows increasingly

cold conditions but then, in its upper third, records a change to a mild, temperate and humid climate. Xeric vegetation is replaced by more dense deciduous forest (García-Antón 1995), stone artifacts are common, and faunal diversity increases. Magnetic polarity was reversed (Parés & Pérez González 1999), and the change in climate probably correlates with OIS 22–21 dated as c. 795 ka BP (Berggren et al. 1995b). At that time the upper part of Mb TD6 reached the level of the cave entrance, forming a 30 cm thick red clayish horizon named the 'aurora bed'. This unit, in a test pit less than 7 m², yielded 268 artifacts, 4104 animal remains and 92 human fossils. The lithic assemblage is classified as evolved Mode 1 (Carbonell et al. 1999), and the faunal assemblage is known in other late early Pleistocene European sites (Cuenca Bescós et al. 1999; Made 1999). Ages of 676 ± 101, 762 ± 114, 770 ± 116 ka were obtained with combined ESR and U-series methods for the TD6 'aurora bed' (Falguères et al. 1999). Mb TD7 lies unconformably over TD6, represents a warm stage with seasonal,

Fig. 14.21. The 'Gran Dolina' fill, stratified deposits in Sierra de Atapuerca (Burgos), dissected by the railway cutting. Excavation at Mb TD10 (Equipo de Investigadores de Atapuerca, Servicio de Fotografía, Museo Nacional de Ciencias Naturales, Madrid).

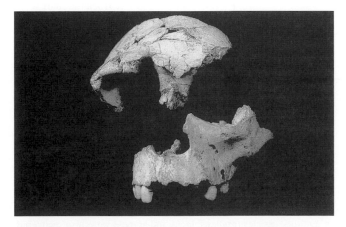

Fig. 14.22. Cranial fragments ATD6-15 (above) and ATD6-69 (below), from the 'aurora bed', Gran Dolina, Atapuerca. Anterior views (Equipo de Investigadores de Atapuerca, Servicio de Fotografía, Museo Nacional de Ciencias Naturales, Madrid).

decreasing rainfall, and is poor in fossil mammals. Two pollen samples indicate an increased presence of Mediterranean taxa and semi-arid conditions, with arboreal pollen decreasing near the top. Its upper third contains the Brunhes–Matuyama magnetic reversal (Parés & Pérez González 1999). Mb TD8 includes three brecciated subunits with unconformities, erosion, collapse and evidence for an occluded entrance. At least two cold phases can be distinguished, plus a temperate phase near the end of the period represented by this member (Fernández-Jalvo & Andrews 1992; Hoyos & Aguirre 1995). Fossils are rare, but *Hippopotamus amphibius* is represented. These deposits are cemented and are overlain by a stalagmitic crust. Mb TD9 is a thin bed of fine deposits which thicken towards the interior, have abundant bat dung, and only partially fill an empty, eroded space below the encrusted top of Mb TD8. New cave entrances opened during the relatively warm period represented by this member, as erosion affected both the cave roof and the hillside. Mb TD10 starts with cold oscillations, first dry then moist, and its upper half reflects temperate warm oscillations with seasonal heavy rains turning to cold and dry conditions producing unconformities. Roof collapses and mud flows are evident, and both small and large mammal fossils are abundant, as are (in the upper half) artifacts with traces of lasting human occupation and tool-making, with work on both skin and wood having been demonstrated (Márquez *et al.* 2001). Tools belonging to an evolved Mode 2 (Acheulean) technology are recognized, along with the presence of the Levallois technique. The mammal fauna is typical of the European upper Middle Pleistocene, and a correlation of this level with OIS episode 9 is most likely, with a weighted U-series/ESR age of 372 ± 33 ka having recently been obtained by Falgueres *et al.* 1999. Finally Mb TD11 includes a double cold oscillation in its lower part and a temperate interval. Stone artifacts attest to the transition to Mode 3 technology. The terminal beds, with gravels, fill up the old cavity to the irregular roof and are strongly cemented with carbonates.

The preserved human bones in the 'aurora bed' include: five facial and eight cranial vault fragments, part of a mandible, seven vertebrae or vertebral fragments, eight ribs or rib portions, three clavicles, two radii, one incomplete femur, two patellae, two carpal bones, two metacarpal bones, 11 hand phalanges, two metatarsals, five foot phalanges, and 14 isolated teeth. Postcranial bones display morphological similarities to modern humans and not to the Neanderthals (Carretero *et al.* 1999; Lorenzo *et al.* 1999). The mandibular fragment is morphologically similar to Ternifine III, to Zhoukoudian, and to *Homo ergaster*, and Dmanisi (Rosas & Bermúdez de Castro 1999*a*). Primitive features are seen in the dental morphology, and a 'modern' pattern in dental eruption is indicative of an increased duration of growth and development (Bermúdez de Castro *et al.* 1999). Observed features of the facial skeleton include the contour of the lower margin of the zygomatic process of maxilla, the presence of post-canine fossa, an arched zygomatic process, an infero-lateral orbital margin, and a short face with vertical profile, all absent in Neanderthals, in their ancestors known in Europe, but present in Zhoukoudian (Arsuaga *et al.* 1999) and Dali (Fig. 14.22). The Atapuerca occupants at the time of TD6 'aurora bed' are considered likely to be related to ancestors dispersed over Eurasia 1 Ma ago, but the evolution of very early human populations in Eurasia remains controversial and poorly understood.

Atapuerca-Galería. A cranial fragment was found under an arch connecting the Galería with more exterior, now infilled cavities of a small karstic system called the Tres Simas complex. Its age is probably early OIS episode 8. A mandibular fragment was collected by T.J. Torres in 1976, in the floor, not far from the position of the skull fragment (Arsuaga *et al.* 1999*a*; Rosas & Bermúdez de Castro 1999*b*). The Galería infilling was excavated with excellent results revealing fauna, flora, artifacts and records of human behaviour in late middle Pleistocene times (Díez *et al.* 1999*b*).

Atapuerca-Sima de los Huesos. The Atapuerca SH hominid fossil sample lies more than 400 m inside Cueva Mayor, Ibeas de Juarros (Fig.14.23), below the vertical shaft called the Sima de los Huesos (Figs 14.24 and 14.25; Bischoff *et al.* 1997). The SH site is not yet fully excavated. The SH individuals were all less than 40 years old, with the commonest age classes being adolescent (eight) and young adults (nine). Ten females and eight males have been identified, along with

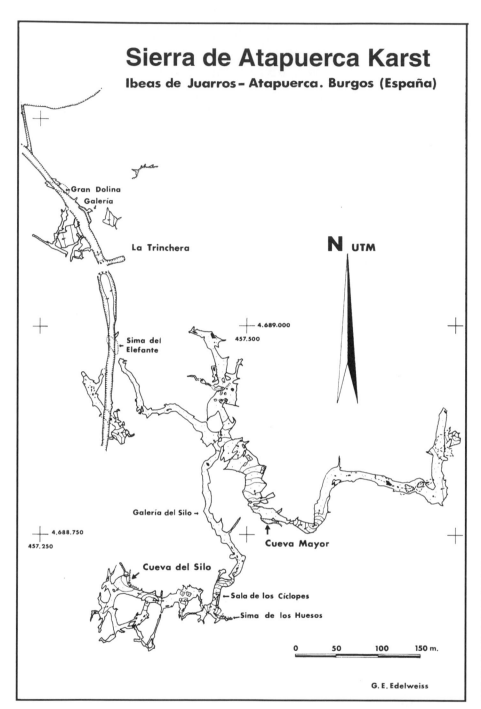

Sierra de Atapuerca Karst

Ibeas de Juarros – Atapuerca. Burgos (España)

Gran Dolina
Galería

La Trinchera

N UTM

4.689.000
457.500

Sima del
Elefante

Galería del Silo →

Cueva Mayor

4.688.750
457.250

Cueva del Silo

← Sala de los Cíclopes

← Sima de los Huesos

0   50   100   150 m.

G. E. Edelweiss

Fig. 14.23. Projection of the Cueva Mayor–Cueva del Silo karst subsystem on surface map with the railway cutting, presently remaining cavities and the exposed sections of Gran Dolina, Galería and Sima del Elefante (Ibeas de Juarros and Atapuerca, Burgos) (Equipo de Investigadores de Atapuerca, Servicio de Fotografía, Museo Nacional de Ciencias Naturales, Madrid, and Grupo Espeleológico, Edelweiss, Diputación Provincial de Burgos).

a further nine individuals of undetermined sex, and all parts of the human skeleton are more or less represented, although they were found mixed and broken. Almost all were collected in situ, within a pit with red silty-clayish matrix, and only a very few were mixed with bear fossils (*Ursus s. deningeri*) at the contact with the overlying bed. Different authors and different analyses converge on *c.* 320 ka as the most probable age for the SH deposits (Yokoyama 1989; Bischoff *et al.* 1997). Two hundred of these fossils were recovered from accumulations discarded years ago by cavers in search of bear teeth.

Nine crania have been described plus 19 temporal and some occipital, parietal, frontal bones and portions of the facial skeleton (Arsuaga *et al.* 1997; Martínez & Arsuaga 1997). Three mandibles are complete, five are incomplete, and there are several significant

mandibular fragments (Rosas 1997). By 1995, 205 isolated teeth had been found in addition to 91 in place in the jaws (Bermúdez de Castro & Nicolás 1997). Fifteen clavicles or clavicular fragments, one complete scapula (plus 16 fragments), six humeri (plus 27 humeral fragments), and all bones of the postcranial skeleton are represented, including hand and foot phalanges. The SH record of long bones of the locomotor system represents more than 80% of the global record for the whole Middle Pleistocene interval. Two hyoid bones and a set of auditory ossicles (malleus, incus and stapes) were also recovered. Temporo-mandibular arthropaty is common, but cranial osteoma and cranial trauma rare. Cranium 4 exhibits symptoms of ear hyperostosis and spondylartrosis. Cranium 5 has invasive alveolar infection from apical abscesses in both maxillar and mandiblar teeth. Cribra orbitalia

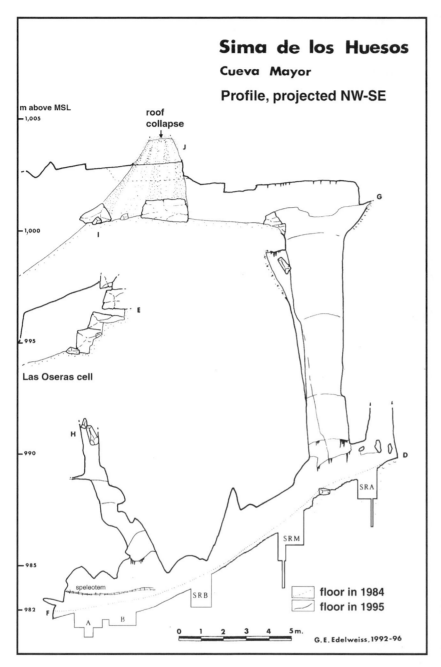

Sima de los Huesos

Cueva Mayor

Profile, projected NW-SE

m above MSL

roof collapse

Las Oseras cell

············· floor in 1984

floor in 1995

0 1 2 3 4 5m.

G. E. Edelweiss, 1992-96

Fig. 14.24. Vertical cut projection of Sima de los Huesos and connecting cavities, Cueva Mayor. The major concentration of human fossils has been found in Pit B (Grupo Espeleológico, Edelweiss, Diputación Provincial de Burgos).

is common, enamel hypoplasia is observed in 40% of individuals (generally mild, acute in three cases), and a case of meningioma has been described (Pérez *et al.* 1997; Pérez & Gracia 1998).

Neanderthals

The Neanderthal race, or species (*Homo sapiens neanderthalensis*, or *H. neanderthalensis*) is represented at various sites in the Iberian peninsula although, with rare exceptions, the remains are poorly preserved. All sites record relatively young ages, with that of Boquete de Zafarraya in Málaga province in particular preserving some of the last representatives of the Neanderthal race before its extinction.

The complete mandible from Bañolas (Banyoles), Gerona (Girona) province (Figs 14.19 and 14.26), of an aged individual was recovered by Roura and Alsius from a travertine quarry in 1887 (Cazurro 1909) and described as *H. neanderthalensis* (Hernández-Pacheco & Obermaier 1915), a classification that is contested. Lumley (1972) assigned it to an 'ante-neandertalian'. Wear produced deeply abraded surfaces on the teeth with a very strong inclination outwards and downwards, and the mandibular body is deformed. There are no associated tools. Assigned ages vary from 17.6 ka (^{14}C) to 111.0 ± 73.0 ka (U-series), with the most reliable date being *c.* 45 ka (Juliá & Bischoff 1993). There is still a lake in Bañolas, in a tectonic depression, and not far from here a right upper molar was collected with 'archaic musterian', in the Mollet I site at Reclau (Maroto *et al.* 1987). The Abric Agut site is a rock shelter near Capellades, Barcelona province (Figs 14.19 and 14.26), and in 1912 yielded four teeth associated with a Mousterian (Mode 3) tool assemblage (Lumley 1973).

A Neanderthal tooth and a phalanx are reported from the Moros de Gabasa cave in Huesca province (Lorenzo 1992), with a Mode 3 tool assemblage, dated as <45 ka (Figs 14.19 and 14.26). Lezetxiki

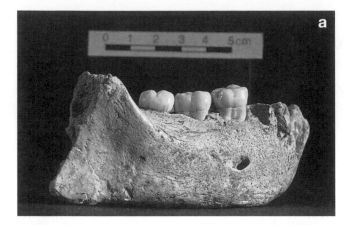

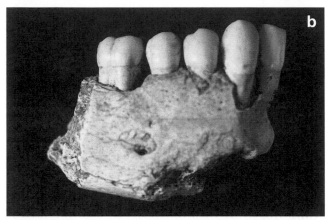

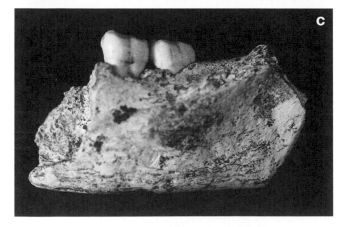

Fig. 14.25. Human mandibles from Sima de los Huesos, Cueva Mayor. (**A**) AT-1 of individual I, found in 1976; (**B**) AT-2, of individual II, 1976, with some of the teeth fitted later; (**C**) AT-75, individual VI, the first to be found in situ, 1984.

Cave, Guipuzcoa province (Fig. 14.19), yielded a right humerus of an adult female, and two Neanderthal teeth (Basabe 1966, 1970). The latter are assigned to an early late Pleistocene cold phase, and are associated with a Mode 3 lithic tool assemblage. The humerus was associated with a faunal assemblage characteristic of late middle and early late Pleistocene times (Altuna 1972; Altuna *et al.* 1982). An incomplete upper dental series comes from the Axlor shelter (Basabe 1973, 1982) near Bilbao, Vizcaya province (Fig. 14.19) associated with a Mousterian/Mode 3 assemblage.

Parts of three skeletons and two incomplete mandibles, totalling 120 pieces but with no skulls, were informally extracted in 1994

from Cueva del Sidrón, Piloña, Asturias (Fig. 14.19), and have been studied by J. Egocheaga and co-workers in Oviedo University since late 1998. A preliminary morphological and statistical study of the mandibular remains suggests that they are Neanderthals (Rosas & Aguirre 1999). Their closest similarities are with Krapina, Regourdou, Kebara and various specimens of the Atapuerca SH population. This may suggest that they belong to either an early variety or a conservative deme of Neanderthals.

Valdegoba Cave near Huérmeces in Burgos province (Fig. 14.19), yielded an adult mandible and ten deciduous teeth associated with Mode 3 tools and faunal remains. These are assigned to a mid-late Pleistocene age (Díez 1991). The Cueva de los Casares at Riba de Saelices in Guadalajara province (Fig. 14.19) provided an adult V metacarpal from bed 12, described by Basabe, and an early Late Pleistocene fauna, studied by J. Altuna (both in Barandiarán 1973), along with a Mousterian tool industry. The site Tossal de la Font (Fig. 14.19) is a cave fill of last interglacial (earliest late Pleistocene) age near Vilafamés in Castellón province. It contains Mousterian tools (Gusi *et al.* 1984) in addition to which half of a humerus and almost half of an innominate were described and classified as Neanderthal (Arsuaga & Bermúdez de Castro 1984).

Cova Negra de Bellús (Fig. 14.19), near Xátiva in Valencia province, was discovered in 1933 by the priest G. Viñes (1942). The sedimentary sequence, studied by P. Fumanal (1986), records several climate fluctuations from basal late Pleistocene times, OIS episode 5e, to mid-late OIS episode 3. A parietal bone of a Neanderthal (Fusté 1953, 1958), Cova Negra 1, is correlated with OIS 5a to OIS 4, and is associated with Mousterian tool technology. An immature mandibular fragment plus an isolated milk molar, Cova Negra 2, from a lower horizon, probably OIS 5b, was described later (Arsuaga *et al.* 1989), as was an upper central incisor, Cova Negra 3, from an upper bed with Charentian tool technology.

The La Carihuela cave (Fig.14.19), opening over an intramontane basin in the Betic Cordillera near Piñar, Granada province, includes a small network of cavities with deposits containing records of successive human occupation from early late Pleistocene to middle Holocene times. In the lower units 32 Mousterian horizons were found, with Bronze Age remains on top. Two incomplete adult parietals, Piñar 1 and Piñar 2, came from beds 6 and 7 of Spahni respectively, equivalent to beds V and VI (Vega-Toscano *et al.* 1988). Piñar 1 is associated with a late Mousterian assemblage with denticulates and is from a cold phase, probably OIS 3b. Piñar 2 is associated with typical Mousterian and a temperate period, OIS 3a. The frontal bone of a Neanderthal child, Piñar 3 (García-Sánchez 1960; Lumley & García-Sánchez 1971) was found in Bed 9/VIII, representing a mild period, OIS 5a, between two cold phases. Piñar 3 was associated with a lithic assemblage classified as 'local Mousterian' (Vega-Toscano *et al.* 1988). It was originally found with a frontal bone of a woolly rhinoceros (*Coelodonta antiquitatis*) and a handful of ochre powder. Fossils of modern humans also came from La Carihuela upper beds (see below; other references in Aguirre & Bermúdez de Castro 1991). Finally, the Boquete de Zafarraya site at Alcaucín in Málaga province (Fig.14.19), is a cave in which the right femur of a Neanderthal was initially found (Zafarraya 1), followed by a complete mandible associated with Mode 3 artifacts (Barroso *et al.* 1984; Hublin *et al.* 1995). Although the age of Zafarraya remains under discussion, as stated above, the occupants can be considered among the last representatives of the Neanderthal race before its extinction.

The well-preserved cranium of a female more than 40 years old was discovered at Forbe's Quarry (Gibraltar) in 1848, shortly after the child skull of Engis (Belgium, 1930) but before the cranium of the Neander valley (Germany) was described as a new species, *Homo neanderthalensis*. Another famous find of a neanderthalian in Gibraltar is a fragmentary skull of a 4-year old infant in Devil's Tower, Gibraltar, by Dorothy Garrod in 1926. The study of the latter has recently been renewed (Zollikofer *et al.* 1995; Stringer 2000) but the precise age of these two significant fossils is still unknown, as is that of a human tooth excavated from Gorham's cave in Gibraltar. Another Iberian tooth known as neanderthalian was found at Gruta Nova da Columbeira in Portugal. Finally, there is the recent discovery of a 4-year old child who lived nearly 25 000 years ago at Lagar Velho (Lapedo, Portugal) and has been described as sharing Neanderthal and modern traits (Duarte *et al.* 1999).

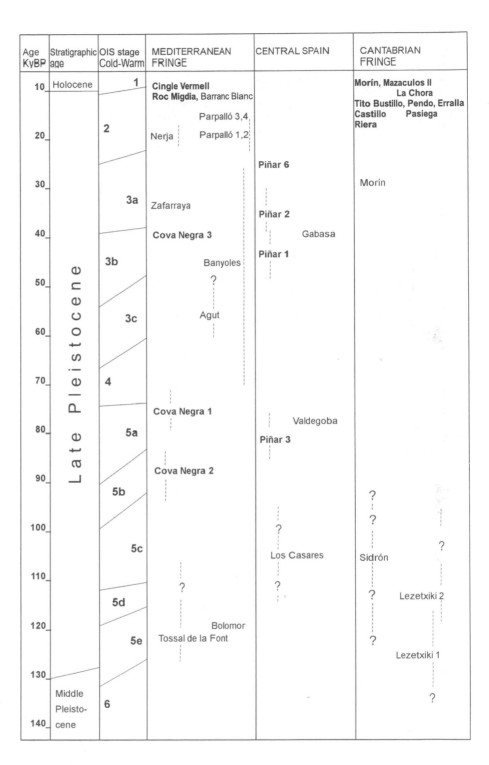

Fig. 14.26. Sites with human fossils, in three major geographical divisions of Spanish territory, of late Pleistocene and the transition to Recent (Holocene) age.

'Modern' humans

Fossils representing anatomically modern humans have been found in Iberian sediments deposited over the last third of late Pleistocene time. A tibia from La Carihuela, Piñar 6, described as obviously 'modern' was found at the base of Bed III close to a change to cold climate, and was associated with a Mousterian, or Epi-Mousterian tool assemblage. Fragmentary cranial and mandibular remains interpreted as of earliest Aurignacian (bed 18G, *c.* 38.5 ka BP; bed 18C, *c.* 40 ka BP) were found in Bed 18 of El Castillo cave, Santander province (Garralda 1989a; Cabrera & Bernaldo de Quirós 1996). A skull was found by the priest Lorenzo Sierra at Camargo in 1909, and was also associated with Aurignacian tools (Obermaier 1924, 1925). Two clay forms found in an Aurignacian layer in Cueva Morín (Fig. 14.19), dated as 27.3 ka BP, have been interpreted as natural casts of human corpses (Freeman & González Echegaray 1970).

In Spain Solutrean assemblages are dated between 21 and 17 ka BP (Straus & Clark 1986). Three fragmentary human remains associated with Solutrean lithic assemblages have been described from La Riera cave in Cantabria. In addition there is a distal humerus and cranial fragments from a lower bed in Parpalló cave in the Spanish Levant, and four skeletons in Nerja cave, Málaga province (Fig. 14.19; Turbón *et al.* 1994). Successive changes in the Nerja palaeoenvironment and resources during late upper Pleistocene times have been studied by Jordá (1986) and Aura-Tortosa *et al.* (1993).

After four millennia the Solutrean mode was replaced by Magdalenian technologies. In northern Spain, two human teeth, aged 16.2 ka, with associated fauna and lower Magdalenian technology were found in Erralla, Guipuzcoa province (Fig. 14.19). Two frontal bones from El Castillo (Fig. 14.19) were described as ritual objects (Obermaier 1925), with an additional half-skull being added, and the cultural interpretation revised (Garralda 1989*b*).

Two human teeth related to late Magdalenian times were found in Tito Bustillo cave in Asturias (Garralda 1976), with dates between 13.0 and 13.5 ka BP. Also in Asturias, a 14 000 year old cranial vault and a tooth have been described from Rascaño (Fig. 14.19). In Santander province, maxillary and mandibular remains with teeth are known from La Chora (González-Echegaray *et al.* 1962), as is part of an upper maxilla in La Pasiega (Fig. 14.19). Further SE, a middle to upper Magdalenian mandibular fragment and four teeth have been described from an upper level of Parpalló cave in Valencia province (Garralda 1992). Remains of a skull from Urtiaga, Guipuzcoa province, dated as 10.28 ± 0.19 ka BP, and a fragmentary

infantile skull from El Pendo, Santander (Fig. 14.19), discovered by Jesús Carballo represent populations with final Magdalenian technology, near the end of the last glaciation in latest Pleistocene times.

The presently known human fossil record from the last, dramatic and rapid change from a glacial to a warm climate during the transition from latest Pleistocene to Holocene times (OIS 2 and 1), is derived from several sites in northern and eastern Spain. In Asturias (Fig. 14.19) the record includes a mandible from Balmorí (with 'Asturiense' lithic tools), an incomplete skeleton of an adult male from Los Azules (with Azzilian tools: 9.54–9.43 ka BP), a fragmentary mandible of a female from Mazaculos II (9.89 ka BP), a partial female skeleton from Colombres (9.3–6.5 ka BP), and a skull from Cuartamentero (with 'Asturian' mesolithic tools). Further east and SE, several incomplete skeletons with Epi-Palaeolithic tools (9.76 ka BP) are known from Cingle Vermell, Barcelona. In addition, there are two cranial fragments from Barranc Blanc, near Rótova, Valencia, one of which (BB-2) is associated with microlamellar Epi-Gravettian tools and roughly dated between 13.47 and 8.45 ka BP (Bernaldo de Quirós & Moure 1978). Finally, three isolated teeth are known from Tossal de la Roca, Alicante, and were again found with geometric Epi-Palaeolithic tools (Garralda 1989*c*, 1991). In Portugal, one complete mandible was recovered from a Solutrean bed in Gruta do Correio-Mór (Loures). Other localities with human fossils associated with upper Palaeolithic assemblages in Portugal are Salemas and Gruta da Casa Moura. The kjökkenmödding Moita do Sebastiâo furnished a large sample of a population with Epi-Palaeolithic tools and a date of 7350 ± 50 years BP (Caria Mendes 1985).

15 Alpine tectonics I: the Alpine system north of the Betic Cordillera

RAMÓN CAPOTE, JOSEP ANTON MUÑOZ, JOSÉ LUIS SIMÓN (coordinators),
CARLOS L. LIESA & LUIS E. ARLEGUI

The current structure and geomorphology of the Iberian peninsula are, for the most part, a direct consequence of Mesozoic and Cenozoic tectonic activity. North of the Betic Cordillera, in its foreland, this post-Palaeozoic tectonic evolution and the resulting Alpine structure is rather complex compared to other European foreland areas. This is a consequence of the Iberian peninsula being a small continental lithospheric plate, which, after the Variscan orogeny, moved relatively independently of its two great neighbours, the European and African plates. The existence of two plate boundaries, following the King's trough–Azores–Biscay rise–north Spanish trough in the north and the Azores–Gibraltar line in the south, and the relatively small size of the Iberian plate, explain why so much tectonic activity was transmitted to the interior of the Iberian peninsula, especially during Alpine collision.

Tectonic setting (RC)

Several mountain ranges, Tertiary sedimentary basins of different types, and various less deformed platform areas all record Mesozoic and Cenozoic tectonic activity north of the Betic Cordillera (Fig. 15.1). The relatively more stable core of the peninsula, the Iberian Massif, forms a Variscan basement now exposed in the central and western part of the Iberian plate, where it is hardly deformed by post-Variscan events. The western limit of the Iberian Massif defines a passive continental margin that was formed by Mesozoic extension during the opening of the North Atlantic ocean. To the north of the Iberian Massif, a chain of mountains runs for 950 km from Galicia and Leon through Cantabria and into the Pyrenees. Further north still, offshore around the base of the north Spanish continental shelf, there is an accretionary prism that marks the position of a subduction zone (Fig. 15.1). The eastern boundary of Iberia is defined by the Cenozoic extensional basin known as the Valencia trough, most of which lies below sea level in the western Mediterranean (Fig. 15.1). Across the centre of the Iberian plate run two compressional mountain belts, the Iberian Ranges, which continue into the Catalonian Coastal Ranges, and the Central Range (Fig. 15.1). Finally, Cenozoic basins with different tectonic origins are

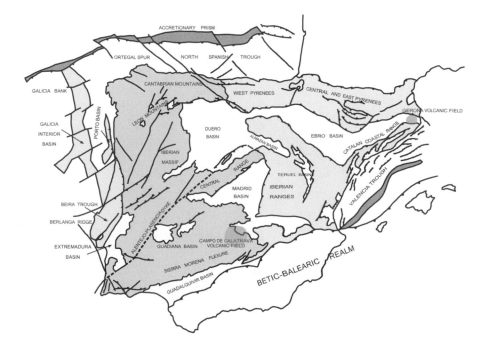

Fig. 15.1. The main tectono-geomorphological units of the Iberian peninsula north of the Betic Cordillera.

distributed all over the peninsula. The Ebro basin, for example, is a foreland basin associated with the Pyrenees, whereas the Duero and Madrid basins are compressive basins forming the high inland mesetas of Spain on either side of the Central Range.

The lithosphere beneath the peninsula has a thickness of 110 ± 5 km, which diminishes towards the boundaries to just 60 km at the Valencia trough. In the stable interior, the crust has a thickness of 31–32 km and a typical continental structure. In the Alpine ranges the crust is thicker (Banda 1996), up to 50 km in the Pyrenees, 35 km in the Iberian Ranges, and 33 to 34 km under the Central Range (Suriñach & Vegas 1988). This crust thins towards the Iberian margin, extensional tectonics again being most evident in the Valencia trough, which has a crust <14 km thick. Whereas there is moderate seismicity concentrated in the Pyrenees, within the interior the seismicity is much lower, with only local exceptions in some parts of the Iberian Ranges. Of additional relevance to the understanding of Alpine tectonism in central and northern Spain is the presence of two Cenozoic volcanic intraplate provinces, the Gerona and Campos de Calatrava volcanic zones (López-Ruiz *et al.* 1993), and magmatism is also known offshore beneath the Valencia trough (Fig. 15.1). The heat flow (Fernández *et al.* 1998) is on the average moderate, with a mean value of 65 ± 10 mW/m^2 in the continental areas but with great lateral variations, reaching maximum values towards the Mediterranean coast (70–90 mW/m^2).

Relative plate motions and the evolution of tectonic regimes

Tectonics both in the interior and at the margins of Iberia were controlled by the drift of the European and African plates away from the North American plate. As the Atlantic ocean was opening, this continental drift determined relative motions of the Iberian plate with respect to the two much larger contiguous plates (Fig. 15.2). Detailed reconstruction of these relative motions has been possible using Atlantic magnetic anomalies, from the classical models of Dewey *et al.* (1973), Biju-Duval *et al.* (1977) and Tapponier (1977), to the more refined models of Srivastava *et al.* (1990), Roest &

Srivastava (1991), and Malod & Mauffret (1990). According to these models the post-Variscan Iberian plate acted rather independently until around 84 Ma BP (chron 34) when it joined the African plate (Malot 1989). The main active plate boundary then became the King's trough–Azores–Biscay rise–north Spanish trough line lying to the north of Iberia (Srivastava *et al.* 1990). Another key change took place around 42 Ma BP (chron 18), when the active plate boundary switched to the Azores–Gibraltar fracture zone. From chron 18 to 6 (24 Ma BP) the Iberian plate moved independently but from 24 Ma BP to the present day Iberia has been joined with the Eurasian plate. The reconstruction of the plate positions shows mainly Mesozoic anticlockwise rotation of Iberia with respect to Eurasia, in accordance with the results of palaeomagnetic studies (Van der Voo 1969; Galdeano *et al.* 1989). For much of Mesozoic time the path of the Iberian plate shows lateral motions relative to the other two plates, thereby producing appropriate conditions for an extensional or transtensional tectonic setting, under which the main Mesozoic sedimentary basins formed. Two rifting events have been detected in the sedimentary record, the first one during the Early to Middle Triassic interval and the other one of Late Jurassic–Early Cretaceous age. However, around 84 Ma BP (chron 34), convergent motions began (Figs 15.2 and 15.3), and initiated subduction along both northern and southern plate boundaries, producing the inversion and compressive deformation of the Mesozoic basins, and the rise of the Alpine ranges, such as the Pyrenees. At the end of Palaeocene times a continental collision process started to transmit this compression to the interior of the Iberian foreland, and so generate the Iberian Ranges and the Central Range.

Mesozoic tectonic evolution

A first question to be resolved is: When did the Alpine tectonic regime become established? Some sedimentary basins, now fragmented and dispersed, started to form during Permian times, at the end of the Variscan orogeny, but it seems that they should not be related to the beginning of Alpine evolution. These Permian basins were created in an extensional

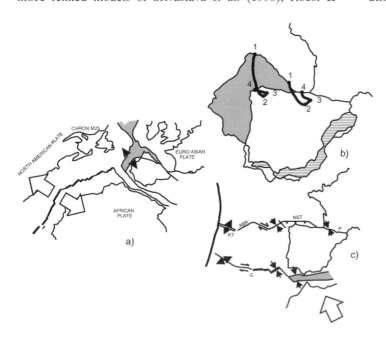

Fig. 15.2. (a) Relative position of the Iberian plate with respect to North American, Euro-Asian and African plates at chron M25 (Late Jurassic), during north Atlantic active spreading (after Srivastava *et al.* 1990). (b) Iberian motion and rotation with respect to the Eurasian plate, between chron M21 (Late Jurassic) and chron 6 (early Miocene). The points along the path are positions in chrons M21, M0, 31 and 6 (after Vegas *et al.* 1996). (c) Active plate margins during the convergent stage and alpine compressive stage. Key: ABR, Azores–Biscay rise; G, Azores–Gibraltar fault; KT, King's trough; NST, north Spanish trough. Black arrows show tectonic regime at plate boundaries. White arrows show movement direction of plates.

regime during large displacements along brittle faults, with associated calc-alkaline volcanism. The latter is found both at the base of Mesozoic clastic sedimentary deposits as well as in emergent areas away from the basins (Doblas *et al.* 1994*b*). Instead, it seems that the beginning of the Alpine cycle should be located at the beginning of a rifting regime during which the various Mesozoic basins took shape (Alvaro *et al.* 1979). A good record of these initial stages is found in the lower part of the sedimentary series in the intraplate mountain ranges of the interior of Iberia, in particular in the Iberian Ranges (Fig. 15.3) and the Catalonian Coastal Ranges. Tectonic subsidence produced grabens in which there were accumulations of red fluvial and littoral sandstones and conglomerates. Several transgressive–regressive megasequences became deposited (Arche & López-Gómez 1992), including formations of marine carbonates and evaporites. The main basins (Iberian Ranges and Pyrenees) record graben subsidence controlled by the motion of blocks along very active normal faults (Alvaro *et al.* 1979; Vegas & Banda 1982). The values of β calculated for the different basins show a relatively moderate total stretching for the Iberian Ranges (Alvaro 1987) and a larger amount of extension for sectors of the Pyrenees. The change from tectonic to thermal subsidence took place during Late Triassic times, when large evaporitic, clastic and dolomitic deposits were deposited under a regressive sedimentary regime. The sedimentary succession also includes basic (tholeiitic) igneous materials. This stage of thermal subsidence continues through the Jurassic period, when shallow water marine carbonates formed the dominant facies in all basins. The situation is clearly extensional and it is in this context, while the Mesozoic basins were in the thermal stage, that a prominent dolerite dyke (the Alentejo-Plasencia dyke) was emplaced across the emergent part of the Iberian Massif (Fig. 15.1). This great dyke has been dated as Late Triassic to Middle Jurassic in age (Schermerhorn *et al.* 1978; Dunn *et al.* 1998).

All evidence then points to a new significant rifting stage in all basins (Salas & Casas 1993), starting at the end of Late Jurassic times and continuing through the Early Cretaceous epoch. Extensional tectonism ('Neokimmeric' and 'Austric' tectonic movements) is recorded by erosion, continental sedimentation in several fault-controlled important basins (Purbeck and Weald facies), and a series of prominent unconformities in the Cretaceous sequence. This second rifting episode was related to the above-mentioned independent rotation of the Iberian plate prior to its interaction with the European and African plates. A final marine transgression across the interior of Iberia during the Late Cretaceous epoch records a new stage of thermal subsidence, depositing marine carbonates across wide areas. In addition to this overall picture, a refined study using six wells that cover a large part of the Iberian Ranges has allowed the detection of several other minor rifting episodes during Mesozoic time (Van Wees *et al.* 1998).

Tectonic inversion and compressive deformation of basins

Compressive Alpine deformation seems to have started during Late Cretaceous time, and was related to the new relative motion of the lithospheric plates, which changed to a convergent regime that produced tectonic inversion of the Mesozoic basins. The tectonic stress fields of this stage have been studied using fault population analysis in the Alpine ranges as well as in the less deformed Tertiary basins and platforms. At least two main directions of shortening show a change from north–south (Pyrenean compression) to NNW–SSE (Betic compression), with time (Fig. 15.2).

The most important Alpine range, the Pyrenees, was formed along the northern limit of the Iberian plate, with a near north–south shortening direction (Guimerá 1984, 1996). It is very important to realize that the Pyrenees between Spain and France represent only the eastern and central part of an orogen that continues towards the west as the Basque-Cantabrian mountains (Fig. 15.1), which are rightly viewed as the western sector of the Pyrenees (Muñoz 1992). Even farther west the orogenic belt continues into the Palaeozoic Cantabrian and Leon mountains and the Galicia highlands (Santanach 1994), in the NW corner of the Iberian plate.

The formation of this system was related to subduction processes that culminated in continental collision starting in the Eocene epoch. The subduction of oceanic crust of the Bay of Biscay under the north Spanish margin started probably about 83 Ma ago and developed an oceanic accretionary prism at the base of the continental border. Seismic reflection profiles, such as the ECORS through the Pyrenees (Choukroune *et al.* 1989) and the ESCI Project (Santanach 1997*b*), across several mountain ranges in the peninsula, have revealed the deep Alpine structure of northern Spain. The structure of the Pyrenees involves both the Variscan basement and the Mesozoic–Palaeocene sedimentary prism deposited on the passive margin of the Bay of Biscay, an extensional structure that was formed mostly during the rotation of the peninsula which started during Late Jurassic–Early Cretaceous time. In the southern part (south Pyrenean zone) several thrust sheets were formed and displaced towards the Ebro foreland basin. In the north Pyrenean zone the structure is formed by folds and high-angle reverse faults directed towards the north.

The stresses transmitted to the central zones of the interior of the Iberian plate, along a direction close to north–south or

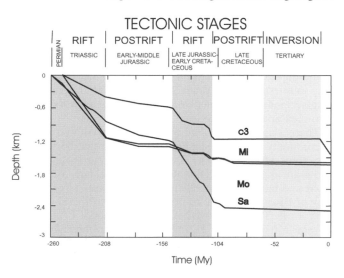

Fig. 15.3. Examples of total subsidence curves for four wells in the Maestrazgo region (eastern Iberian Ranges), showing the two main rifting stages and the corresponding post-rift thermal periods. During Tertiary times the tectonic inversion of basins was produced by a compressive tectonic regime. Key: C3, Amposta well; Mi, Mirambell well; Mo, Maestrazgo well; Sa, Salzedella well (after Salas & Casas 1993).

NNE–SSW, produced a relatively important compressive deformation. This Palaeogene and Miocene deformation inverted the Iberian basin to form the Iberian Ranges (Alvaro *et al.* 1979), an intraplate mountain belt with a moderately folded and thrusted Mesozoic–Cenozoic cover. Further west, the compression reactivated late Variscan faults to uplift a number of basement blocks without Mesozoic cover (or in some cases with just a thin veneer of Upper Cretaceous sediments) and form the Central Range. The latter mountains form a complex NE–SW orientated pop-up structure and separate the Madrid and Duero basins (Fig. 15.1). The direction of shortening was in this case NW–SE and the most important uplift phase started as recently as around 10 Ma BP and continues to the present day, as revealed by fission-track data (Sell *et al.* 1995; De Bruijne & Andriessen 2000). The seismic refraction profiles in the Central Range (Suriñach & Vegas 1988) show that the continental crust is slightly thickened, by about 2 km, compared to the normal thickness of the Variscan Iberian Massif.

Syntectonic and post-tectonic Tertiary continental sediments were deposited in several basins bordering the uplifted massifs. The deepest of these basins has resulted from strong rifting in the eastern and northeastern part of Iberia. This has been associated with the creation of the Valencia trough (Fig. 15.1), produced by a progressive change of the compressive stress field (Simón 1986*b*) into an extensional one, heralding Neotectonic events in the eastern part of the Iberian plate. The crust was thinned during this extensional tectonism not only under the Valencia trough but also in the SE part of the Iberian Range. In central areas, the controlling faults continue to be active at present, with a NW–SE shortening direction and moderate seismic activity. Most earthquakes, however, remain concentrated in the Pyrenees, marking the old plate boundary.

The Pyrenees (JAM)

The Pyrenees form a doubly vergent collisional mountain belt which resulted from Mesozoic–Cenozoic interaction between the Afro-Iberian and European plates (Roest & Srivastava 1991; Olivet 1996). The area is part of a larger orogen extending for some 1500 km from the eastern Alps along the Mediterranean coast to the Atlantic ocean NW of the Iberian peninsula. The Pyrenean mountains are flanked by two main foreland basins, the Aquitanian basin in the north and the Ebro basin to the south (Fig. 15.4), and they display different characteristics along strike. In the east the Pyrenean mountains have been overprinted by Neogene extensional features related to the opening of the Gulf of Lyon and the drift of the Corso-Sardinian block. In contrast, the main part of the mountain range between France and Spain forms a continental collisional orogen without this late extensional overprint. Here, the orogen developed over a previously thinned continental crust but without intervening oceanic crust between the two plates. Further west still, the oceanic crust of the Bay of Biscay became involved in the Pyrenean orogen but, unlike most oceanic lithosphere, was only moderately subducted. Instead, deformation was mainly concentrated in the previously thinned continental crust south of a Bay of Biscay oceanic zone.

The Pyrenees is a mountain range produced by tectonic inversion of Triassic–Cretaceous extensional to transtensional rift systems. This rifting was associated with the fragmentation of southern Variscan Europe and western Tethys as a result of the break-up of Pangaea, as well as with the opening of the central Atlantic ocean and the Bay of Biscay, and the resulting rotation of Iberia (Roest & Srivastava 1991; Olivet 1996). Convergence occurred from late Santonian to middle Miocene times as the Afro-Iberian plates moved generally northward against Europe. As a result, the earlier extensional structures were firstly inverted, then incorporated into the thrust system.

The size of this orogen, the quality of the exposures, the unusually good preservation of the synorogenic strata and the knowledge of its lithospheric structure together make the Pyrenees an excellent natural laboratory for investigating orogenic processes and foreland basin formation mechanisms.

Main geological features and structural units

The Pyrenees has been classically divided into structural units across strike (Séguret 1972; Muñoz *et al.* 1986), although they also show significant structural differences along strike. The along-strike division takes into account the longitudinal variation of the geometry of the thrust system on a crustal scale and the resulting asymmetry of the double orogenic wedge, with structural differences mostly having been controlled by the inherited geometry of an Early Cretaceous extensional fault system. According to these criteria, the Pyrenean range can be divided into two main parts: the Aragonese-Catalan Pyrenees (subdivided into eastern, central and west-central Pyrenees) and the Basque-Cantabrian or western Pyrenees (subdivided into the Basque and Cantabrian Pyrenees; Fig. 15.4). The boundary between both parts corresponds to an inverted Early Cretaceous transfer zone referred to in the literature as the Pamplona fault.

The Aragonese-Catalan Pyrenees forms the main part of the mountain belt and defines the Franco-Iberian border. It is characterized by a thrust system that displays an asymmetric double wedge of upper crustal rocks. This double wedge divides the mountains into the northern and the southern Pyrenees (Figs. 15.5 and 15.12). The northern wedge is formed by a northward directed imbricate stack (north Pyrenean thrust system) involving basement and cover rocks. The southern wedge is southward directed and consists of an imbricate stack involving cover rocks (south Pyrenean thrust system) and a duplex of basement rocks (known as the 'axial zone'). The southern wedge is wider than the northern one, and both displacement and cumulative shortening are correspondingly also greater.

There is no doubt that the Basque-Cantabrian Pyrenees are in continuity with the main Pyrenean range regardless of their structural differences. The Basque-Cantabrian Pyrenees also show an upper crustal double wedge. The northern thrust front and most of the northern wedge are submerged in the Bay of Biscay along the Cantabrian or north Iberian margin (Fig. 15.4). This fact, together with the absence of the north Pyrenean fault, makes it difficult to continue westwards the longitudinal structural subdivisions proposed for the Aragonese-Catalan Pyrenees (Séguret 1972; Muñoz *et al.* 1986). Another characteristic of the Basque-Cantabrian Pyrenees is the existence of numerous diapirs made of Upper Triassic salt and evaporites (Serrano *et al.* 1989). Most of the diapirs are related to Early Cretaceous extensional structures although they have been reactivated during the Cenozoic era.

Northern Pyrenees. These form a narrow thrust-and-fold belt in the Aragonese-Catalan Pyrenees separated on land from the southern

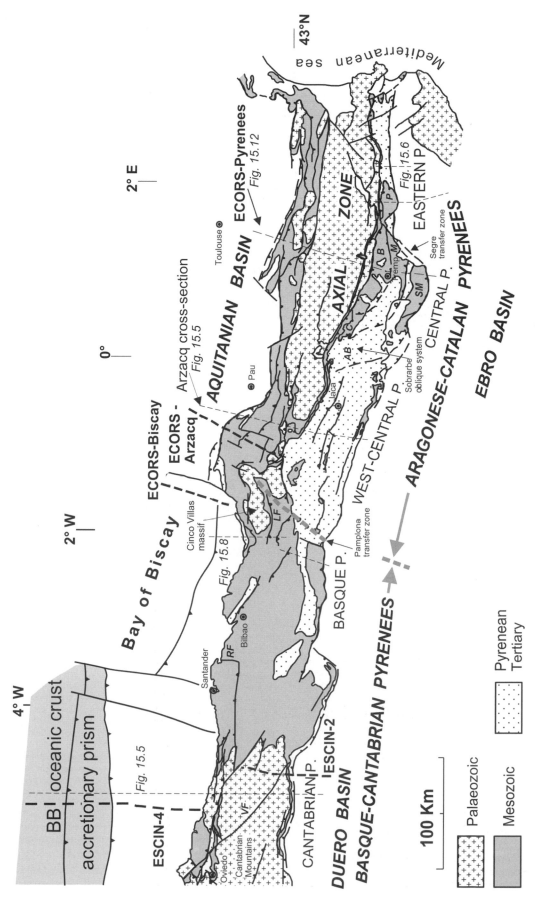

Fig. 15.4. Structural map of the Pyrenees. The northern Pyrenees are located north of the axial zone in the central and eastern Pyrenees and north of a structural divergence axis in the west-central Pyrenees (see Fig. 15.5). Key: AB, Ainsa basin. Thrust sheets: B, Bóixols thrust sheet; C, Cotiella thrust sheet; M, Montsec thrust sheet; P, Pedraforca thrust sheet; SM, Serres Marginals thrust sheet. Faults: LF, Leiza fault; RF, Ramales fault; VF, Ventaniella fault. Modified from Teixell (1996).

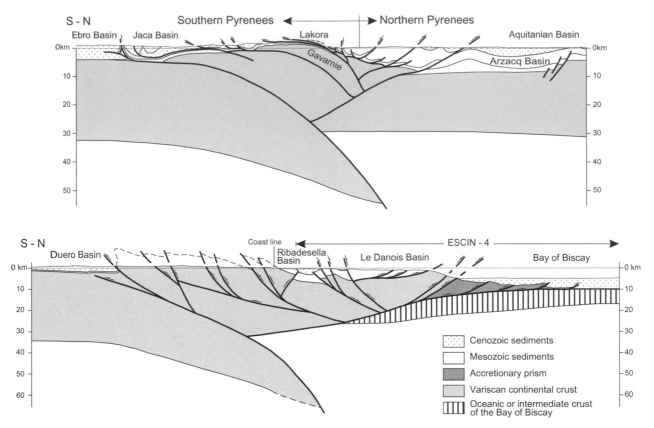

Fig. 15.5. Crustal cross-sections of the west-central Pyrenees (top) and the Cantabrian Pyrenees (bottom). The west-central cross-section has been modified from the Arzacq cross-section of Teixell (1998). The Cantabrian Pyrenees cross-section is based on published geophysical data (Pulgar *et al.* 1996; Álvarez-Marrón *et al.* 1997) and seismic data of the Cantabrian platform (Sánchez 1991). Compare the crustal structure with the ECORS Pyrenees cross-section of Figure 15.12. See Figure 15.4 for location.

Pyrenees by the north Pyrenean fault. This fault apparently terminates at the western end of the axial zone where it is truncated by thrust structures (Hall & Johnson 1986). In the west-central Pyrenees the boundary corresponds with a zone of divergence (Teixell 1998; Figs 15.4 and 15.5).

The north Pyrenean fault is a major strike-slip fault which developed during the Middle Cretaceous sinistral displacement of Iberia (Choukroune 1976). The fault evolved from an initial Albo-Cenomanian transtensional regime (Debroas 1990; Goldberg & Maluski 1988), to a later transpressional regime during the onset of convergence in early Senonian time (Puigdefàbregas & Souquet 1986; Debroas 1990). During the transtensional regime pull-apart basins were formed and thermal metamorphism developed. Lower crustal granulites and ultramafic upper mantle rocks (lherzolites) are embedded in Jurassic and Lower Cretaceous metamorphic rocks along a narrow strip parallel to the north Pyrenean fault (Choukroune 1976; Vielzeuf & Kornprobst 1984). These rocks were carried to upper crustal levels during the strike-slip faulting. Neither post-Variscan metamorphic rocks nor lower crustal rocks crop out anywhere else in the Pyrenees.

North of the north Pyrenean fault, north-directed thrusts involve basement and cover rocks, and the Variscan basement forms culminations known as the north Pyrenean massifs. Thick successions of Upper Cretaceous turbidites unconformably overlie this basement. The non-metamorphic and weakly deformed character of these turbidites contrasts with the strongly deformed metamorphic rocks cropping out in the north Pyrenean fault zone. The structural style of the north Pyrenean thrust sheets is strongly controlled by the inversion of the Early Cretaceous extensional faults.

Southern Pyrenees. Two main units form the southern Pyrenees in the Aragonese-Catalan Pyrenees: the upper and lower thrust sheets (Muñoz *et al.* 1986). The upper thrust sheets consist of Mesozoic and

Palaeogene rocks which have been detached from their basement along Late Triassic evaporites and thrust on top of Palaeogene rocks of the south Pyrenean foreland basin (Figs 15.5–15.7 and 15.12). Mesozoic deposits are only tens of metres thick in the southernmost units and progressively thicken to up to 7 km northwards. They form a sedimentary wedge that results from the erosion of the early Mesozoic units as well as the southward thinning and pinching out of the Cretaceous stratigraphic units (pre- and synorogenic). The geometry of this wedge is imposed by Early Cretaceous extensional faults and their inversion during the initial stages of convergence (Late Cretaceous). Inversion tectonics is a structural feature of the upper thrust sheets, which also show numerous oblique and lateral structures, some of them related to the original Mesozoic basin configuration. From these structures an approximately north–south transport direction can be deduced.

To the north, and below the upper thrust sheets, basement rocks of the lower thrust sheets constitute a duplex (axial zone). This duplex involves only upper and mid-crustal rocks, its sole thrust being located at about 15 km below the top of the basement. In the eastern and central Pyrenees the basement duplex is characterized by an antiformal stack (Figs 15.4, 15.5 and 15.12).

The upper unit of the lower thrust sheets is characterized by an incomplete and reduced Mesozoic succession overlain by a Palaeogene foreland basin platform and turbiditic sequences. These cover sequences were deposited in the footwall of the upper thrust sheets synchronously with their emplacement, and they unconformably overlie Palaeozoic basement rocks. In the eastern Pyrenees (Cadí thrust sheet) a Mesozoic succession is absent, whereas in the central and west-central Pyrenees (Gavarnie thrust sheet) it is mostly represented by Upper Cretaceous rocks.

Eastern Pyrenees. These are located east of the river Segre. North of the south Pyrenean main frontal thrust (Vallfogona thrust), south

Pyrenean thrust sheets are narrower in a north–south direction than in the central Pyrenees (Fig. 15.4). Provided that the amount of shortening is similar in both parts of the chain, a change in the structural geometry must occur. In fact, the eastern Pyrenean thrust sheets are mainly piled one on top of the other instead of imbricated as observed in the central Pyrenees. Each one of the stacked thrust sheets involving cover rocks displays a different stratigraphy (Vergés 1993; Vergés et al. 1995). They have been grouped into two main structural units: the Pedraforca and the Cadí thrust sheets (Fig. 15.6).

The Pedraforca thrust sheet consists of several units of Mesozoic and Palaeogene rocks (Guerin-Desjardins & Latreille 1962; Séguret 1972). The uppermost unit represents an inverted Early Cretaceous basin (Vergés 1993) and contains the most complete and thickest Mesozoic succession. It was the first to be emplaced (Late Cretaceous), unlike the lowermost unit which has the thinnest series and was the last to be emplaced (middle Eocene). Below the Pedraforca thrust sheet, the Cadí thrust sheet is characterized by a thick succession of synorogenic Palaeocene–Eocene sediments unconformably overlying Palaeozoic rocks or a very reduced Mesozoic section (Fig. 15.6, Muñoz et al. 1986; Vergés 1993). The Eocene succession contains turbiditic and slope sediments deposited in a narrow trough in front of the Pedraforca thrust sheet. The above-described thrust sheets were tilted and deformed by the stacking of basement thrust sheets (Muñoz 1985). Both the Cadí thrust sheet and the basement piggy-back antiformal stack developed from middle Eocene to middle Oligocene times (Puigdefàbregas et al. 1986).

South of the Vallfogona thrust, the eastern Pyrenees are characterized by a wide detached thrust-and-fold belt involving the Ebro foreland basin sediments (Fig. 15.4). The geometry of these structures is controlled by the stratigraphic position and cartographic distribution of the three main salt detachment Eocene horizons of the foreland succession (Vergés et al. 1992). The eastern Pyrenees deformation front is located at the southern tip of these evaporitic basins and it shows a characteristic thrust wedge or triangle zone geometry (Sans et al. 1996).

Central Pyrenees. These are characterized by a widespread regional outcrop of the upper thrust sheets (the south Pyrenean central unit of Séguret 1972). They form three main imbricated thrust sheets detached over Upper Triassic evaporites, and from top (north) to bottom (south) are: Bóixols, Montsec and Serres Marginals (Figs 15.4 and 15.7). Seismic and well data show that, at present, these units lie on top of the autochthonous Palaeogene rocks of the Ebro foreland basin (Fig. 15.7; Cámara & Klimowitz 1985; Teixell & Muñoz 2000). The upper thrust sheets of the central Pyrenees are the westward continuation of the three main stacked Pedraforca thrust sheets along the Segre oblique thrust system (Vergés 1993). The structural style of the central Pyrenees thrust sheets largely depends on the geometry of the Early Cretaceous extensional system and the thickness of the Mesozoic series.

The Bóixols thrust sheet consists of a thick (over 5000 m), mainly Lower Cretaceous, Mesozoic sequence. It developed during late Santonian to late Maastrichtian time by the inversion of Early Cretaceous extensional basins (Garrido 1973; Berástegui et al. 1990; Bond & McClay 1995). The thrust front is largely unexposed as it is unconformably overlapped by the Maastrichtian Arén Formation (Fm) (Souquet 1967; Garrido & Ríos 1972). It is characterized by thrust and fold structures developed in the footwall of the inverted extensional faults.

South of Bóixols, the Montsec thrust sheet shows a simple structure. It consists of a broad syncline that supports the piggy-back Tremp-Graus basin fill of Palaeocene to upper Eocene calcareous and clastic sediments (Figs 15.4 and 15.7; Nijman 1998). This thrust sheet includes a complete Mesozoic sequence about 3000 m thick, comprising mainly Upper Cretaceous synorogenic limestones. These limestones represent the marginal facies of the foreland basin as they were deposited synchronously with the turbiditic trough at the Bóixols thrust front. The Montsec thrust sheet is bounded to the south by the Montsec thrust that displays an arcuate map pattern. Its western oblique zone is unconformably overlain by the middle-upper Eocene sediments of the Tremp-Graus Basin (Fig. 15.4; Puigdefàbregas et al. 1992). This confirms the early Eocene age of the Montsec thrust, as is also indicated by syntectonic sediments at the thrust front in the Ager and Ainsa basins. The Montsec thrust sheet continues to the NW into

the Cotiella thrust sheet where it appears as a spectacular shallow-dipping thrust on top of the Eocene turbiditic rocks of the Ainsa basin (Séguret 1972; Garrido 1973; Muñoz et al. 1994).

The Serres Marginals thrust sheets are located between the south Pyrenean frontal thrust to the south and the Montsec thrust to the north (Pocoví 1978b; Martínez-Peña & Pocoví 1988). They consist of several small imbricate units characterized by Mesozoic successions that are reduced to the south but are thicker and more complete to the north (Pocoví 1978b). These units are overlain by a lower Eocene succession and by upper Eocene–lower Oligocene conglomerates, sandstones and gypsums which were deposited synchronously with thrust development. This succession shows spectacular linked relationships between sedimentation and tectonics (Vergés & Muñoz 1990; Meigs 1997; Teixell & Muñoz 2000).

To the north and below the imbricated cover of the upper thrust sheets, basement rocks form an antiformal stack (axial zone). The contact between the upper thrust sheets cover and the antiformal stack basement corresponds to a passive-roof backthrust (Morreres backthrust). During the development and southward displacement of the basement antiformal stack the cover units became wedged northwards over the top of the basement, as similarly described in other orogenic belts. The antiformal stack comprises three main structural units, which from top to bottom are the Nogueres, Orri and Rialp thrust sheets (Fig. 15.7; Muñoz 1992). The frontal tip of the Nogueres thrust sheet is preserved in the southern limb of the antiformal stack and is known as the Nogueres zone (Dalloni 1930do). Nogueres zone thrusts involve Triassic and Palaeozoic rocks (Stephano-Permian post-Variscan sequences and Silurian–Devonian–Lower Carboniferous Variscan rocks). Thrusts have been steepened to a subvertical orientation by underthrusting, so that hanging-wall anticlines display a downward-facing fold geometry (Fig. 15.7). The Orri thrust sheet is characterized by distinct Devonian and Stephano-Permian successions (Zwart 1979). The lowermost Rialp thrust sheet crops out only in the Rialp tectonic window, where Triassic rocks are exposed underneath the Cambro-Ordovician and Devonian rocks of the Orri thrust sheet.

West-central Pyrenees. These are dominated by a wide syncline of Palaeogene sediments known as the Jaca basin (Figs 15.4, 15.5 and 15.7; Puigdefàbregas 1975). Mesozoic rocks crop out on both sides of the syncline, along thrust and fold belts in high relief areas. The northern belt, adjacent to the basement rocks of the axial zone, is referred to as the internal sierras ('sierras interiores'; Teixell 1996). The southern belt, along the south Pyrenean thrust front, is known as the external sierras ('sierras exteriores'; Alastrué et al. 1957) and is the western continuation of the marginal sierras ('serres marginals') of the central Pyrenees (Fig. 15.4; Millán et al. 2000).

The boundary between the west-central Pyrenees and the central Pyrenees is a fold and thrust system of north–south to NW–SE trending structures in the footwall of the Montsec-Cotiella thrust sheet (Muñoz et al. 1994), here referred as the Sobrarbe oblique fold and thrust system (Fig. 15.4). These structures developed during middle and upper Eocene times (Millán et al. 1994; Poblet et al. 1998) and have experienced significant clockwise rotation (Pueyo et al. 1997; Pueyo 2000; Dinarés & Muñoz, unpublished data) during the southward displacement of the South Pyrenean central unit.

In the west-central Pyrenees, the basement thrust sheets are not piled one on top of the other as they are in the central Pyrenees. Instead, they constitute a piggy-back imbricate stack underthrust below the cover units (Fig. 15.7; Cámara & Klimowitz 1985; Teixell 1996). As a result, the structural relief of the axial zone antiformal stack decreases westwards, the structures plunge in the same direction at its western extremity and the Palaeozoic outcrops terminate. The basement units of the imbricate stack constitute upper crustal ramp anticlines with flat tops at different elevations, separated by monoclinal frontal limbs (Teixell 1996, 1998; Figs 15.5 and 15.7). This along-strike change of the tectonic style in the central Pyrenees is the result of the along-strike variations of the inherited geometry of the intracrustal weak zones with respect to the position of the cover detachments.

Upper Cretaceous and lower Palaeogene rocks of the internal sierras overlie the basement rocks of the axial zone (Gavarnie thrust sheet; Figs 15.5 and 15.7). They are deformed by an imbricate stack of several slices and by fault-propagation folds (Alonso & Teixell 1992; Teixell 1996). North of the western end of the axial zone, the floor

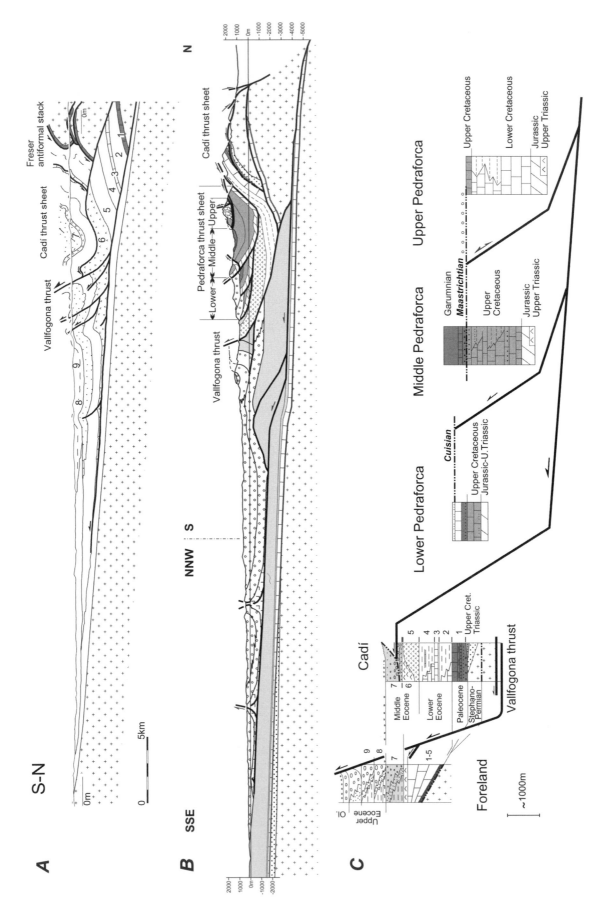

Fig. 15.6. Cross-sections through the eastern Pyrenees. The cross-section across the Pedraforca thrust sheet is from Vergés (1993). See Figure 15.4 for location.

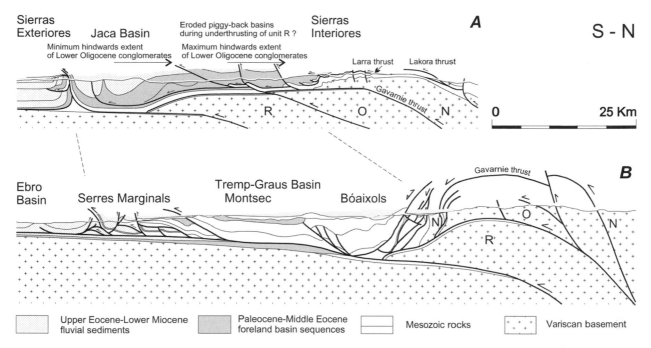

Fig. 15.7. A comparision between a cross-section through the west-central Pyrenees (A) and a cross-section of the central Pyrenees located along the ECORS profile (B) illustrates the differences in the tectonic style and in the uplift and exhumation partitioning of the south Pyrenean thrust sheets. The supposed base of late syntectonic conglomerates has been projected above the present topography. The upper cross-section has been modified from Teixell (1996) and its frontal part has been projected from the east following Millán (1996). It corresponds with the southern part of the section of Figure 15.5. See Figure 15.4 for location. From Muñoz *et al.* (1997).

thrust of the internal sierras imbricate stack (Larra thrust) branches into the Lakora thrust, a major south-directed thrust which carried basement and Mesozoic rocks on top of the Upper Cretaceous sediments (Figs 15.5 and 15.7; Teixell 1998).

The Jaca basin can be subdivided into northern and southern sectors. The northern part comprises lower and middle Eocene turbidites deposited in a foreland basin turbiditic trough then deformed by a thrust system and related folds. The southern Jaca basin is dominated by a thick succession of upper Eocene–Oligocene continental sediments displaying a synclinorium affected by detachment folds (Puigdefàbregas 1975). These sediments were deposited during the emergence of the south Pyrenean frontal thrust in the external sierras that were thrust over a lower-middle Eocene carbonate platform, which was coeval with the northern Jaca turbidites. As a consequence the Jaca basin is viewed as a piggy-back basin active during synorogenic continental sedimentation.

The external sierras are characterized by the superposition of two distinct sets of structures. The first of these was a middle Eocene–lower Oligocene oblique system of folds, trending north–south to NW–SE and continuing to the east into the Sobrarbe thrust-and-fold oblique system. The second set of structures was a late Oligocene–early Miocene thrust system and a frontal detachment fold, trending WNW–ESE (Millán *et al.* 1994, 1995*b*, 2000).

The geometry of the thrust front of the west-central Pyrenees differs from that of the eastern transects. In the external sierras the foreland is not detached because of the absence of salt horizons. Moreover, the thrust front is strongly emergent and pinned at the southern pinch-out of the Triassic detachment level (Fig. 15.7). Thrust pinning resulted in a dextral rotation of the external sierras (Pueyo 2000) and the development of a lift-off frontal anticline several kilometres high (Soler & Puigdefàbregas 1970; Millán 1996) which folded previous frontal thrust structures.

Basque Pyrenees. These are characterized by both north- and south-directed thrusting, although with a predominance of north-vergent structures (Fig. 15.8). South-directed structures are restricted to the south Pyrenean frontal thrust and a south-vergent synform (Villarcayo and Miranda-Treviño synclines) which supports a piggy-back

basin of Palaeogene and lower Miocene sediments (Figs 15.4 and 15.8). The north-directed thrust wedge consists of several thrust sheets involving a very thick Cretaceous succession. Tectonic style is largely controlled by the inversion of Early Cretaceous extensional and strike-slip faults (Cuevas *et al.* 1999). North-directed thrusts and related structures of the Basque Pyrenees are referred to as the Basque arc (Feuillée & Rat 1971) because of their concave geometry in map view (Fig. 15.4). This arc is delineated by three main structures that from north to south are named the Bilbao anticlinorium, Vizcaya synclinorium and Vizcaya anticlinorium (Fig. 15.8). The northernmost Basque thrust sheets are thrust on top of the Tertiary rocks of the Landes plateau. This latter plateau was an uplifted area during Early Cretaceous times between the Parentis basin in the north and the Basque basins to the south, presently inverted and incorporated into the Basque Pyrenean thrust sheets (Sánchez 1991).

The significant change of the vergence of the structures from the west-central Pyrenees to the Basque Pyrenees occurs across a deformation zone trending NNE–SSW which has been referred to as the Pamplona fault. This fault has been interpreted as a major crustal strike-slip fault across the Pyrenees with a sinistral (Rat 1988), dextral (Muller & Roger 1977; Turner 1996), or combined (Martínez-Torres 1989) movement. The Pamplona fault may be considered as a transfer zone of the thrust system inherited from a previous transfer zone of the Early Cretaceous extensional system. The latter interpretation is supported by differences between Lower Cretaceous successions on both sides of the fault, as well as the absence of significant palaeomagnetic rotations (Larrasoaña 2000). The Pamplona fault has been recently interpreted as a hanging-wall drop fault resulting from accommodation of the different thicknesses of Mesozoic sequences during southward displacement above the south Pyrenean frontal thrust (Larrasoaña 2000). The diapirs aligned along the Pamplona transfer zone would have been developed during the Early Cretaceous epoch and reactivated during Cenozoic deformation.

The Basque massifs, exposing Palaeozoic rocks, crop out on both sides of the Pamplona transfer zone. The Cinco Villas massif is the westernmost and largest of the Basque massifs (Fig. 15.4). It is involved in the northern-directed thrust sheets of the Basque Pyrenees and shows a pop-up structure between inverted Early

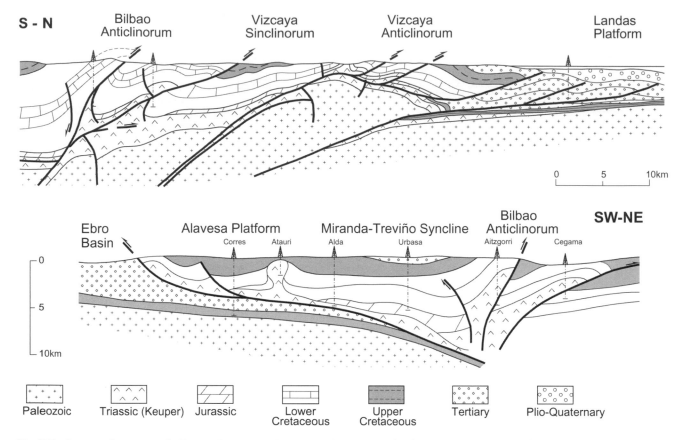

Fig. 15.8. Cross-sections across the Basque Pyrenees redrawn from Cámara 1989 (top) and Serrano *et al.* 1989 (bottom). The two sections form a complete transect from the Ebro foreland basin to the Aquitaine foreland at the Landas platform. See Figure 15.4 for location.

Cretaceous basins (Fig. 15.8). South of the Cinco Villas massif, the Leiza fault follows a narrow strip of lower crustal and mantle rocks as well as metamorphic Mesozoic rocks (Martínez-Torres 1989). It has been considered as the western continuation of the north Pyrenean fault into the Basque Pyrenees (Choukroune & Mattauer 1978; Rat 1988). Its significance as the boundary between the Iberian and the European plates during Early Cretaceous sinistral strike-slip motion of Iberia has been emphasized (Olivet 1996). Nevertheless, correlation between the Leiza and the north Pyrenean faults has recently been ruled out (Larrasoaña 2000). The Leiza fault and its westward continuation into an Early Cretaceous strike-slip fault system (García-Mondéjar *et al.* 1996) could be part of a wide deformation zone that includes the Palaeozoic Basque massifs and which has accommodated the sinistral motion of Iberia and distributed the strike-slip displacement between different faults (Larrasoaña 2000).

Cantabrian Pyrenees. These are characterized by the basement uplift of the Cantabrian mountains and the subduction of the Bay of Biscay oceanic crust (Fig. 15.5). The asymmetry of the Cantabrian Pyrenees changes again with respect to the Basque Pyrenees. Southward-directed structures predominate onshore, whereas northward-directed ones are mainly restricted to the accretionary prism. The divergence axis is located within the Danois basin, at the northern extremity of the Cantabrian platform (Fig. 15.5). The transition from the Basque Pyrenees to the Cantabrian Pyrenees also corresponds with a north–south trending transfer and accommodation zone inherited from the Early Cretaceous fault system. Offshore this transfer zone is aligned with the eastern extremity of the Bay of Biscay oceanic crust (Fig. 15.11) and its subduction below the Cantabrian platform. Onshore, the north–south trending Ramales fault represents the boundary between the north-vergent structures of the Bilbao anticlinorium and the south-vergent structures of the Cantabrian Pyrenees at Santander (Fig. 15.4, Cámara 1989). Southwards, some diapirs occur and the frontal south Pyrenean thrust shows a salient.

The Variscan basement of the Cantabrian mountains has been uplifted and thrust on top of the Tertiary sediments of the Duero basin, with a southward displacement estimated to be about 25 km (Alonso *et al.* 1996). East of the Cantabrian mountains the Mesozoic cover rests on top of the basement in apparent continuity, although the structures observed further east still suggest a detachment of the cover succession on top of Upper Triassic evaporites (Hernaiz & Solé 2000). Below this detachment, thrust structures affect the Variscan basement and have been interpreted to merge downwards into an intracrustal detachment (Fig. 15.5; Alonso *et al.* 1996; Hernaiz & Solé 2000).

The Cantabrian Pyrenees show a distinct structural unit at their southern front known as the 'Burgos platform and folded band' (Hernaiz & Solé 2000). It consists of several thrust sheets of Mesozoic rocks thrust over Tertiary sediments of the Duero basin. The eastern extremity of the Burgos platform is an oblique set of thrust structures which represents the boundary between the Ebro and the Duero foreland basins (Fig. 15.4).

Along the coast, between Santander and Oviedo, southward-directed thrusts involve basement and cover rocks and result from the reactivation of extensional Early Cretaceous faults. These thrusts are bracketed between the NW–SE trending Ventaniella fault westwards and the north–south trending Ramales fault eastwards (Fig. 15.4; Cámara 1989). This south-directed thrust system continues into the Cantabrian platform where it incorporates the Tertiary Ribadesella basin (Fig. 15.5; Sánchez 1991). The edge of the Cantabrian platform is characterized by north-vergent thrusts and folds affecting the Mesozoic succession of the Danois basin and basement rocks on the continental slope (Boillot *et al.* 1979; Álvarez-Marrón *et al.* 1995). In front of the Cantabrian continental slope there is an accretionary prism deforming Tertiary and Mesozoic sediments of the Bay of Biscay oceanic basin (Fig. 15.5; Deregnaucourt & Boillot 1982). The accretionary prism is 60 km wide in the east (Fig. 15.5) but reduces in size westwards (20 km) changing its internal structure and morphology (Álvarez-Marrón *et al.* 1997).

Foreland basins

Since the initial stages of the Pyrenean collision in Late Santonian–Campanian times two foreland basins developed, one on each side of a central double-wedge of basement rocks (Figs 15.4 and 15.5). The northern foreland (Aquitanian) basin consists of a thick (a few kilometres) succession of Upper Cretaceous turbidites overlain by a Palaeogene sequence up to 4 km thick. Most of the latter is represented by continental deposits as marine platform sediments of only early Ypresian age are observed (Buis & Rey 1975). The Aquitanian basin mainly developed in the footwall of the north Pyrenean frontal thrust and was not greatly involved in the north Pyrenean thrust system.

The south Pyrenean foreland basin is wider and has a thicker composite succession than the Aquitanian basin. Its filling is characterized by alternating marine sediments of Late Cretaceous and early-middle Eocene age, and continental deposits of Palaeocene and late Eocene–Miocene age. The first synorogenic deposits comprise Upper Cretaceous turbidites and marls which grade upward in the central and eastern Pyrenees into an uppermost Cretaceous–Palaeocene shallow water and continental succession. The Eocene series above this is characterized by thick, mainly marine sediments with strong lateral variations in facies and thickness. This variability stems from basin partitioning and the development of piggy-back basins (Tremp-Graus and Ager basins) during the southward displacement of the upper thrust sheets (Puigdefàbregas *et al.* 1986, 1992; Nijman 1998). The piggy-back basins contain lower Eocene platform and continental terrigenous facies that, in the footwall of the upper thrust sheets, grade into deeper marine turbidites and pro-delta marls. These marine Eocene successions were incorporated into the south Pyrenean thrust system and were overthrust on top of initially marginal facies, so that new piggy-back basins developed below the previous ones (Jaca basin in the west-central Pyrenees and Ripoll basin in the eastern Pyrenees). The autochthonous part of the south Pyrenean foreland basin, southward from the south Pyrenean frontal thrust, is known as the Ebro basin in the Aragonese-Catalan and Basque Pyrenees and as the Duero basin in the Cantabrian Pyrenees (Fig. 15.4). It is mainly filled by continental sediments after early Priabonian evaporites which record the closure and isolation of the Ebro Basin. Upper Eocene–lower Miocene continental clastics filled the enclosed basin and progressively backfilled and buried the south Pyrenean thrust system during its late stages of development (Coney *et al.* 1996).

Timing of deformation

The Pyrenean orogen is characterized by the preservation of synorogenic strata that closely constrain the age of the structures. Growth strata produced in the initial stages of thrusting are preserved only in the southern Aragonese-Catalan Pyrenees where timing of deformation is best constrained. Other criteria such as thermochronology based on apatite fission tracks, and indirect criteria from regional geology, are also available to infer deformation ages. Deformation of the Pyrenean double-wedge migrated outwards, although synchronous thrusting and break-back sequences coexisted with a piggy-back thrusting mode (Fig. 15.9; Martínez *et al.* 1988; Vergés & Muñoz 1990).

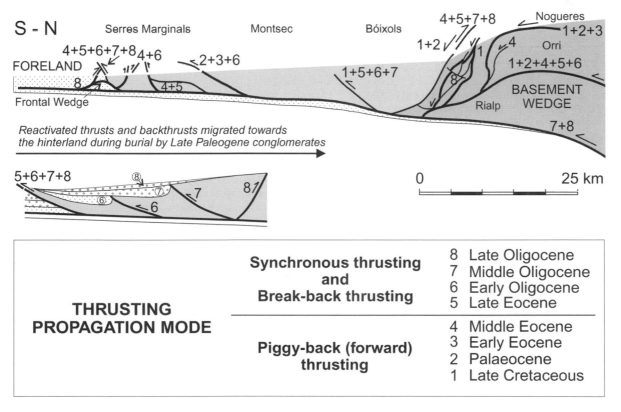

Fig. 15.9. Timing of deformation and thrust sequence in the southern central Pyrenees along the ECORS cross-section. Compare with Figures 15.7 and 15.12 for location of structures.

After a period of late Albian to early Santonian thermal subsidence and transpression (Puigdefàbregas & Souquet 1986), convergence started in the Aragonese-Catalan Pyrenees in late Santonian times. The onset of the convergence is recorded by a significant unconformity at the bottom of a turbiditic succession in the internal parts of the south Pyrenean and Aquitaine foreland basins, and below the platform sequence in the marginal areas (Garrido 1973; Garrido & Ríos 1972). Late Cretaceous contractional structures and growth strata are spectacularly preserved in the southern central Pyrenees and were produced by the inversion of older (Early Cretaceous) extensional faults (Bond & McClay 1995). Late Cretaceous structures are also observed or inferred in the Aquitanian basin and the west-central Pyrenees (Teixell 1996). However, there is no evidence of these early structures in the Basque-Cantabrian Pyrenees. As a result, it has been considered that the onset of the convergence occurred later in the Basque-Cantabrian Pyrenees, starting as early as the middle-late Eocene interval (Choukroune 1973). Sedimentation of late Santonian to middle Eocene age has been interpreted to have been controlled by thermal subsidence in a continental marginal setting (Rat 1988). It is probably not reasonable to suggest such a difference in age for the earliest contractional structures on both sides of the Pamplona transfer zone. South-directed Late Cretaceous thrusting of the Eaux Chaudes–Lakora–Pic d'Orhy system (De Luca *et al.* 1985; Teixell 1996) would have been transferred into north-directed thrusting north of Alduides and in the Cinco Villas massif west of the Pamplona transfer zone. In the Cantabrian Pyrenees, Late Cretaceous contractional structures would have been restricted to the accretionary prism as suggested by Boillot (1984). Synchronous onset of convergence in Late Cretaceous times along the entire Pyrenean belt fits better with the kinematics of the Iberian plate (Olivet 1996).

After this Late Cretaceous tectonic inversion of the Early Cretaceous extensional basins, the subsequent deformation events may be grouped into different periods. Firstly, the Palaeocene was a period of relative tectonic quiescence. As a result, there is little evidence of tectonic activity (Biure and Montsec thrust sheets in the eastern and central Pyrenees respectively; Pujadas *et al.* 1989). During early Eocene times an increase of the shortening rate has been deduced (Vergés 1993; Vergés *et al.* 1995), and induced the emplacement of the Biure, Middle Pedraforca, Montsec and Cotiella thrust sheets. Ages are very well recorded by marine strata deposited both in the shallow water piggy-back basins and in the foreland turbiditic troughs (Soler & Garrido 1970; Puigdefàbregas *et al.* 1992; Muñoz *et al.* 1994; Nijman 1998). In the Basque-Cantabrian Pyrenees no evidence of these structures is observed onshore because only the southern margin of the turbiditic trough and its platform marginal equivalents are preserved (Pujalte *et al.* 2000). Nevertheless, the processes that occurred at the platform margin are equivalent to those described in the west-central Pyrenees (Pujalte *et al.* 2000) where they have been related to thrust evolution (Labaume *et al.* 1985; Barnolas & Teixell 1994).

During middle-late Eocene times deformation propagated forwards to incorporate the foreland turbiditic troughs (Ripoll and Jaca) into the south Pyrenean thrust system. These troughs became piggy-back basins and were filled up by deltaic and continental sediments that record the emplacement of the Lower Pedraforca and Cadí thrust sheets in the eastern

Pyrenees (Vergés *et al.* 1998) and the growth of an oblique fold system in the west-central Pyrenees (Millán *et al.* 1994; Poblet *et al.* 1998). These folds show dextral rotation around a vertical axis during the piggy-back displacement of the Montsec thrust sheet on top of the marginal sierras thrust sheets which were developed during this period. All these frontal cover structures emerged from the basement-involved Gavarnie, Nogueres and Cadí thrusts. In the Cantabrian Pyrenees it is accepted that subduction in the Bay of Biscay was active from at least middle Eocene times, whatever the timing uncertanties of its onset (Álvarez-Marrón *et al.* 1997). This is based on the age of a major unconformity between the middle and upper Eocene times in the Bay of Biscay and the Cantabrian platform (Sibuet *et al.* 1971; Boillot *et al.* 1971). It has been considered historically, but incorrectly, that this period represents the main tectonic event in the Pyrenees and the onset of the deformation in the Basque-Cantabrian Pyrenees.

The forward propagation of deformation in the southern Pyrenees was modified during the last stage of the evolution of the thrust belt (from late Eocene times) by a break-back reactivation of the older thrusts, and by the development of new, minor out-of-sequence thrusts. The younger thrust structures of the southern Pyrenees are very well recorded by the continental sediments of the south Pyrenean foreland basin once it was cut off from the Atlantic ocean in early Priabonian times to form the internally draining Ebro foreland basin (Coney *et al.* 1996). The progressive burial of the Pyrenean thrust front changed the thrust kinematics from a major forward-thrusting propagation mode to a synchronous thrusting mode (coeval forward and hindward thrusting; Fig. 15.9). In the hanging-wall of the frontal thrust, older thrusts were reactivated towards the hinterland as syntectonic conglomerates filled the foreland basin and progressively buried the thrust wedge. Buried thrusts became inactive and the subsequent active thrusts stepped backwards to the maximum extent of the conglomerates, resulting in break-back thrust sequences (Fig. 15.15; Vergés & Muñoz 1990; Burbank *et al.* 1992b; Meigs 1997; Muñoz *et al.* 1997) which coexisted with a continuously active thrust front (synchronous thrusting).

Ages of the younger Pyrenean structures have been historically based on mammal biostratigraphy of the continental syntectonic sediments (Crusafont *et al.* 1966a) with quantitative data provided by magnetostratigraphic surveys (Hogan & Burbank 1996; Meigs & Burbank 1997) and thermochronological studies (Fitzgerald *et al.* 1999). The end of the contractional deformation in the Pyrenees was diachronous. It ended by middle Oligocene times in the eastern Pyrenees and continued until the middle Miocene interval in the Cantabrian Pyrenees (Vergés 1993). Thrust activity was probably inhibited in the east by the onset of extension in the Valencia trough. In the central Pyrenees, thrust deformation has been dated magnetostratigraphically to have been as young as late Oligocene (Meigs & Burbank 1997). In the internal parts, uplift of the Orri basement thrust sheet during underthrusting of the lower and younger unit (Rialp) has been dated as middle Oligocene (Fitzgerald *et al.* 1999). In the west-central Pyrenees the youngest frontal structures occurred during sedimentation of lower Miocene conglomerates (Arenas *et al.* 2001). In the Basque-Cantabrian Pyrenees deformation in the accretionary prism ended during Burdigalian times (Álvarez-Marrón *et al.* 1997) synchronously with the end of thrust activity along the southern front (Hernaiz & Solé 2000).

Geophysical data and crustal structure

The crustal and lithospheric structures of the Pyrenees have been constrained by different geophysical techniques (deep reflection and refraction seismic profiles, gravity, magnetotellurics, magnetic anomalies, tomography, heat flow). It is difficult to find another orogen with such an amount and quality of geophysical data as is available in the Pyrenees. The data that best constrain the Pyrenean crustal structure are from the deep seismic reflection profiles, mainly the ECORS Pyrenees profile (Fig. 15.10; Choukroune et al. 1989). The ESCIN programme completed another transect across the Cantabrian Pyrenees from the Bay of Biscay to the Duero basin (Fig. 15.4; Álvarez-Marrón et al. 1995; Pulgar et al. 1995). The profiles of this programme together with the IAM-12 and ECORS Bay of Biscay profiles (Pinet et al. 1987; Marillier et al. 1988; Álvarez-Marrón et al. 1997) complete a crustal image of the Basque-Cantabrian Pyrenees. More recently, a deep seismic reflection survey (LISA) has been conducted in the eastern extremity of the Pyrenees (Nercessian et al. 2001; Gallart et al. 2001).

An extensive seismic refraction–wide angle reflection survey has also been completed to fill the gap of crustal structure information between the Cantabrian and the central Pyrenees seismic reflection profiles (Fernández-Viejo et al. 1998, 2000). All the above-mentioned seismic data together with the refraction data (Daignières et al. 1982) and the ECORS Arzacq deep reflection profile (Grandejan 1992; Daignières et al. 1994) have been integrated to compile the map of Figure 15.11 in which the main features of the Pyrenean crustal structure are depicted.

The main result of these seismic surveys is the demonstration of the subduction of the Iberian plate below the European plate. This was already shown by the ECORS profile in the central Pyrenees (Fig. 15.10) and has been confirmed also to exist in the Basque-Cantabrian Pyrenees by the ESCIN-2 reflection profile (Pulgar et al. 1996) and the wide angle refraction surveys (Fernández-Viejo et al. 2000).

The relationships between the Moho offset produced by the subduction of the Iberian plate and the north Pyrenean fault have been a matter of debate among Pyrenean geologists and geophysicists (Daignières et al. 1982; Deramond et al. 1985; Choukroune et al. 1989; Mattauer 1990; Muñoz 1992; Vacher & Souriau 2001). Geophysical data confirming the existence of subducted continental Iberian crust in the Basque-Cantabrian Pyrenees, west of the termination of the north Pyrenean fault and north of the Leiza fault (considered to be the continuation of the north Pyrenean fault into the Basque Pyrenees; Choukroune 1976; Martínez-Torres 1989) rules out any causal relationships between these faults and the observed Moho offset. The intersection of the European Moho by the subduction plane (cut-off line) may be traced all along the Pyrenees (Fig. 15.11). Northwards, the cut-off line of the Iberian Moho is not so well defined seismically, because of the steep attitude of both the subduction plane and the Iberian Moho. Its minimum northward extent has been represented on the map of the Figure 15.11. The area between the two cut-off lines represents the zone where the Iberian and European crusts are superposed at depth. This area correponds with the location of the highest gradients of the Bouguer gravity anomaly (Casas et al. 1997).

The seismic structure of the eastern extremity of the Pyrenees is marked by continuous crustal thinning towards the Mediterranean (Nercessian et al. 2001). As a result, the crustal structure is destroyed and a new structural grain parallel to the Mediterranean coast is superimposed (Fig. 15.11). In the easternmost Pyrenees the subducted Iberian crust is not observed (Gallart et al. 2001).

In the Basque Pyrenees less crustal thickening is observed with respect to the other parts of the chain although the subduction of the Iberian crust is still clear (Fig. 15.11) (Gallart et al. 1999; Fernández-Viejo et al. 2000). This apparent difference of crustal thickening could be the result of the shortening of a previously thinner crust in the Basque domain or alternatively some of the crustal shortening could have been transferred into the Iberian Ranges, as suggested by a deeper Moho southwards from the Basque Pyrenees (Fig. 15.11). In the Cantabrian Pyrenees an abrupt crustal thickening is observed from the Variscan crust with a 30–32 km thickness (Pérez-Estaún et al. 1994) to the deformed and subducted Iberian crust eastwards (Fernández-Viejo et al. 2000). This abrupt transition trends NW–SE and coincides with oblique faults at the surface (Fig. 15.11).

Several different interpretations of the crustal structure of the Pyrenees have been given on the basis of the combined geological and geophysical data (Roure et al. 1989; Mattauer 1990; Muñoz 1992; Teixell 1998). I argue that an explanation in which the orogenic double-wedge involves only upper crustal rocks provides the best geometry in which to integrate all these data (Fig. 15.12). The crust appears to have been decoupled, with the lower crust (below the upper crustal double-wedge) being subducted together with the lithospheric mantle. This inferred crustal subduction is compatible with other geophysical data as well as the absence of post-Variscan metamorphic rocks or lower crustal rocks in the Pyrenean orogenic double-wedge. Subducted Iberian lower crust has been detected down to a depth of 80–100 km in the Aragonese-Catalan Pyrenees from a magnetotelluric survey (Pous et al. 1995; Ledo et al. 2000) as well as from seismic tomography (Souriau & Granet 1995). Conductivity anomalies have been explained as due to either partial melting of the subducted crust (Fig. 15.10; Glover et al. 2000) or eclogitic metamorphism with water release, the latter idea fitting better with a recently proposed three-dimensional density model of the Pyrenean deep structure (Vacher & Souriau 2001). The deformation style of the Pyrenean orogen being due to simple crustal subduction is consistent with numerical modelling by Beaumont et al. (2000). In the west-central Pyrenees subduction of the Iberian lower crust has been related to the indentation of the European crust (Teixell 1998). This crustal indent geometry has been modified to fit better with the crustal structure proposed for the central Pyrenees (Figs 15.5 and 15.12).

In the Basque-Cantabrian Pyrenees subduction of the European plate southwards, underneath the Iberian plate, was proposed because of its oceanic nature in the Bay of Biscay (Boillot & Capdevila 1977). The change in the polarity of subduction with respect to the Aragonese-Catalan Pyrenees was interpreted to occur across a flip zone coinciding with the Pamplona transfer zone (Engeser & Schwentke 1986). A crustal cross-section through the Cantabrian Pyrenees based on the seismic profiles shows a structure that could be interpreted as a double subduction with a subducted Iberian crust southwards and a subducted Bay of Biscay crust northwards (Fig. 15.5; Álvarez-Marrón et al. 1997). However, a problem arises with the nature and southward extent of the Bay of Biscay crust. Álvarez-Marrón et al. (1997) interpreted the high

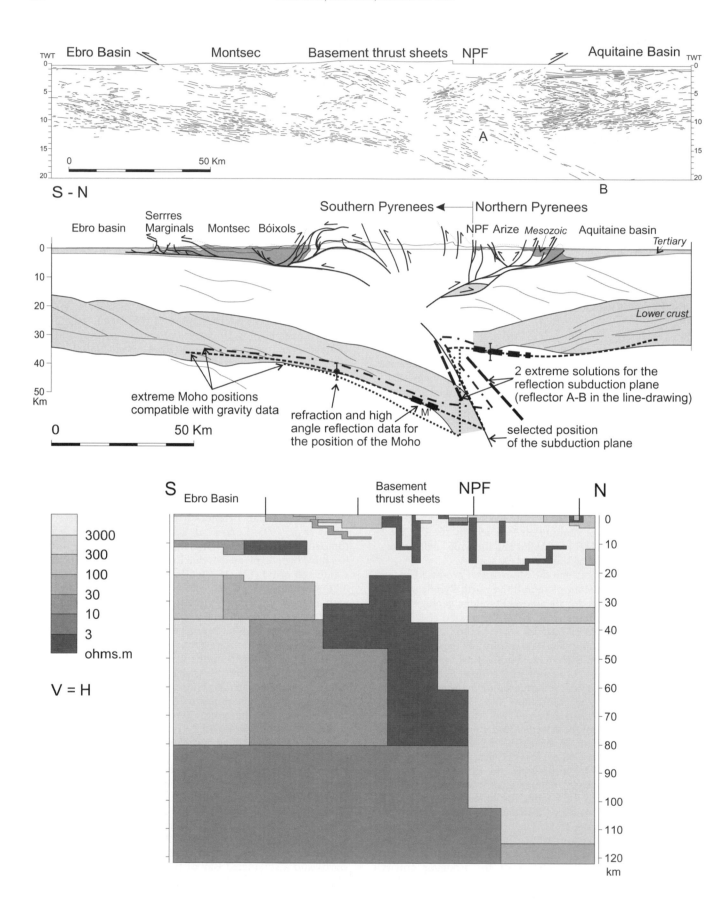

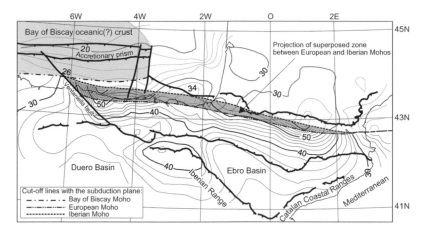

Fig. 15.11. Structural sketch showing the main features of the crustal structure of the Pyrenees. Depth to the Moho has been represented for both the Iberian and the European plates (depth lines each 2 km). The narrow grey strip between the cut-off lines of the Moho represents the zone where the Iberian and European Mohos are superposed vertically. The cut-off line of the European Moho is reasonably well defined seismically, but that of the Iberian Moho is not, owing to its steep attitude. Its minimum northward extent has been represented in the map. The maximum extent of the oceanic crust of the Bay of Biscay has been depicted in dark grey, coinciding with the location of the abyssal plain. However, the nature of its eastern part is not clear (see text for further discussion). The position of the cut-off line of the Bay of Biscay crust has been based on the interpretation of the IAM-12 and ESCIN-4 seismic profiles by Álvarez-Marrón *et al.* (1997). Moho depth data are based on all the available deep reflection and refraction seismic data (Daignières *et al.* 1982, 1994; Pinet *et al.* 1987; Choukroune *et al.* 1989; Álvarez-Marrón *et al.* 1995; Pulgar *et al.* 1995, 1996; Fernández-Viejo *et al.* 1998, 2000; Gallart *et al.* 1999, 2001; Nercessian *et al.* 2001).

velocity crust observed below and southwards of the accretionary prism (Pulgar *et al.* 1996) as a subducted oceanic slab. Its southern tip has been represented in the map of Figure 15.11. On the contrary, Fernández-Viejo *et al.* (1998, 2000) considered that the eastern part of the crust below the abyssal plain of the Bay of Biscay was probably a thinned continental crust instead of a truly oceanic one. This configuration would have favoured the indentation of the Bay of Biscay crust below the Cantabrian platform and the delamination and subduction of the Iberian lower crust northwards below the European plate (Fig. 15.5). In contrast, oceanic crust in the western part of the Bay of Biscay would have been subducted southwards and prevented significant deformation of the Iberian Variscan crust as observed west of the Ventaniella fault (Fig. 15.11). Independently of this interpretation, a flip in the subduction polarity presumably must have occurred in the Cantabrian Pyrenees.

Balanced and restored cross-sections

Balanced and restored cross-sections were constructed not only to integrate geophysical and geological data but also to estimate the amount of orogenic contraction (Fig. 15.12). A geometrical solution of a crustal cross-section of the central Pyrenees along the ECORS profile gave a total shortening of 147 km (Muñoz 1992). However, this value increases to 165 km if the internal deformation of the crust below the sole thrust of the Pyrenean thrust system is restored (Beaumont *et al.* 2000). Other cross-section restorations of the central Pyrenees have estimated shortening values of over 100 km

(Williams & Fischer 1984; Deramond *et al.* 1985; Roure *et al.* 1989). A shortening calculation for a crustal cross-section in the eastern Pyrenees yielded a shortening estimate of about 125 km (Vergés *et al.* 1995).

A recent study of the kinematics of the Iberian plate has revealed that shortening in the central Pyrenees cannot be less than 150 km (Olivet 1996). Different proposed models for reconstruction of the Iberian plate as well as cross-sections west of the ECORS transect, show that shortening decreases westwards down to values of 100 km (Olivet 1996). Teixell (1998) calculated a shortening value of 80 km for the west-central Pyrenees. The estimated duration of convergence in the central Pyrenees is about 60 Ma, which gives a mean shortening rate of 2.5 mm/year. A similar shortening rate has also been deduced in the eastern Pyrenees during a shorter period of convergence (Vergés *et al.* 1995).

The thrust transport direction was consistently north–south to NNE–SSW throughout most of the tectonic evolution as deduced by the map pattern of the structures, kinematic criteria along thrust planes, and the absence of significant rotation around a vertical axis in the central Pyrenees (Dinarès *et al.* 1992). This implies a near-normal convergence throughout the main orogenic phase.

The restored cross-section gives an estimate of the geometry of the crust before the Pyrenean collision (Fig. 15.12). The geometry of the inherited structures (Variscan cleavage and thrusts, late Variscan extensional faults and Early Cretaceous extensional system) displays a mostly north-dipping listric geometry over layered lower crust. This geometry has been observed in the undeformed part of the ECORS Pyrenees

Fig. 15.10. Geophysical data along the ECORS Pyrenees profile. Top: reflection seismic time section after Choukroune *et al.* (1989) and Berástegui *et al.* (1993). Centre: interpreted crustal structure based on geophysical and geological data; gravity constraints from Torné *et al.* (1989); high-angle reflection and refraction data from Daignières *et al.* (1989); geological data from Muñoz (1992) and Berástegui *et al.* (1993). Main reflectors of the ECORS seismic profile have been depth-converted, gravity and refraction data taken into account, and then incorporated into the crustal cross-section. NPF, north Pyrenean fault. Bottom: two-dimensional electrical resistivity model from a magnetotelluric survey from Pous *et al.* (1995) located a few kilometres east of the ECORS profile.

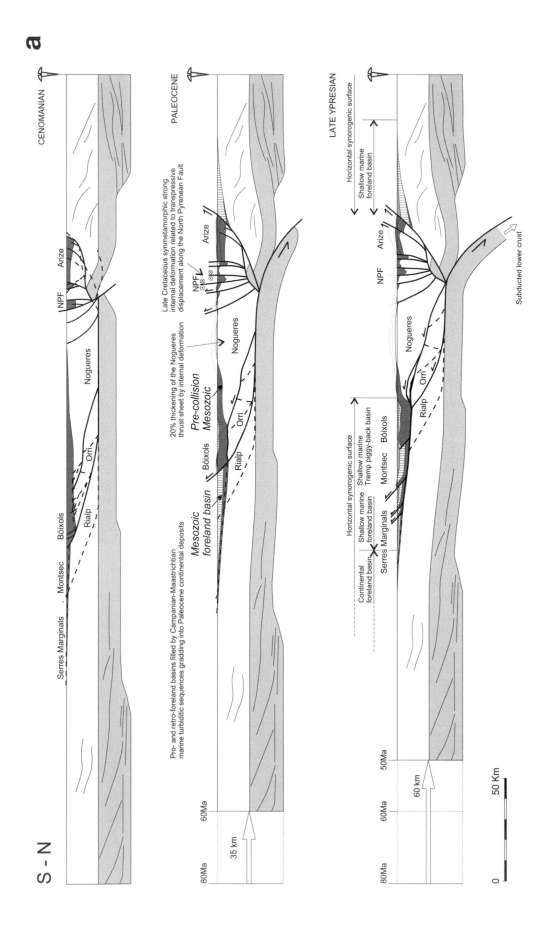

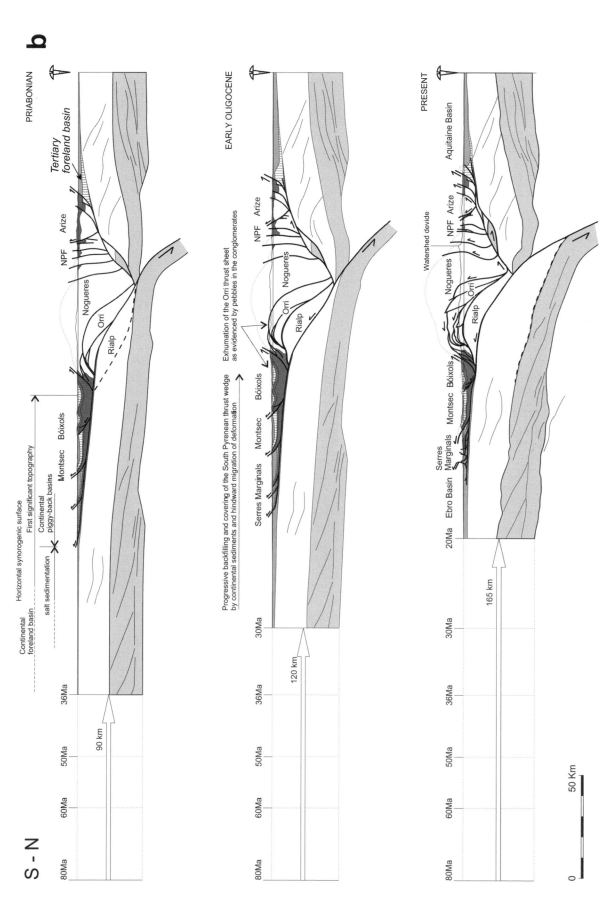

Fig. 15.12. Balanced, partially restored and fully restored cross-sections of the crust along the ECORS profile. **(a)** Sequential restoration from the stage prior to convergence (in late Cenomanian–Turonian times) to the early Eocene. Note the geometry of the Early Cretaceous extensional system and the north Pyrenean strike-slip fault zone. **(b)** Late Eocene and early Oligocene stages and the present deformed state. Note the deformation of the Iberian upper crust and the significant amount of uplift and exhumation since early Oligocene times as evidenced by thermochronology data (Fitzgerald *et al.* 1999). Reference frame holds the European plate and arrows indicate estimated total convergence at each stage. Upper and lower crust are shown in white and pale grey, respectively. From Beaumont *et al.* (2000).

profile, deduced after the restoration, or made by comparison with other areas. The restored crustal geometry is consistent with the areas around the Pyrenees that were affected by the Mesozoic extensional events but were not subsequently deformed by contractional structures, e.g. in the Aquitaine foreland (Le Pichon & Barbier 1987; Pinet *et al.* 1987; Boillot & Malod 1988; Marillier *et al.* 1988). A similar geometry is observed in most of the BIRPS deep reflection profiles across the Mesozoic extensional basins of northwestern Europe where the lower layered crust is not intersected by upper crustal extensional faults (Cheadle *et al.* 1987). The location of these discontinuities favoured the delamination of the crust, the upper part forming an orogenic double wedge. The crust below this intracrustal detachment was then subducted beneath the European crust.

Geodynamic evolution

In the Pyrenees the geometry of the thrust structures, as well as their ages, are constrained by the exceptionally well preserved synorogenic strata that allow the construction of a series of cross-sections as orogeny progressed (Fig. 15.12). Restored cross-sections across the central Pyrenees have been made using an area mass balance method and taking into account the shortening partitioning between the two sides of the orogenic double-wedge, the extent and depth of the foreland basins and palaeotopography where preserved (Muñoz 1992; Vergés *et al.* 1995). These partially restored cross-sections integrate the crustal structure deduced from geophysical data and summarize the available information on the geodynamic evolution of the Pyrenees from the geology. They form the basis for comparison with numerical models to provide insight into the fundamental processes of orogenic growth and foreland basin development (Millán *et al.* 1995a; Beaumont *et al.* 2000).

The restored cross-section shows an Early Cretaceous extensional system which evolved, in late Albian–early Cenomanian times, into a combined sinistral strike-slip and extensional dip-slip fault system driven by the north Pyrenean fault. This fault penetrated the crust and probably the whole lithosphere. After a period of Cenomanian to late Santonian transpression along the north Pyrenean fault (Puigdefàbregas & Souquet 1986), the first stage of north–south convergence (late Santonian–Maastrichtian) inverted the previous Early Cretaceous main extensional faults. The geometry of this system, as well as the location of the intracrustal weak detachment zones, not only determined the structure at the surface but also on a crustal scale. Inversion tectonic features are dominant in the northern Pyrenees and in the northern upper thrust sheets of the southern Pyrenees. They are also spectacular in the Basque-Cantabrian Pyrenees where the reactivation of the Early Cretaceous extensional faults was coeval with diapiric flow of thick Triassic evaporites. Numerical geodynamical modelling corroborates the idea that the tectonic style of the Pyrenees and partitioning of the deformation between both sides of the orogenic double-wedge is strongly influenced by the inversion of previous extensional features (Beaumont *et al.* 2000). This conclusion can also be inferred from the variations of structural pattern along the strike of the chain. Strongly subsiding troughs filled by turbidites developed in the footwall of the inverted faults. These deep marine foreland basins of the initial stages were superimposed on previous marine post-rift basins developed during a thermal subsidence phase (Brunet 1986).

The Palaeocene epoch was a time with relatively low rates of plate convergence between Europe and Africa (Roest & Srivastava 1991). In the eastern and central Pyrenees, the Early Cretaceous extensional faults were completely inverted so that the stretched upper crust recovered its initial pre-Cretaceous length, and probably the whole crust also attained its pre-Cretaceous crustal thickness. The eastern and central parts of the foreland basins changed from marine to continental as topography developed and the amount of eroded material was sufficient to fill the basins. In the Basque-Cantabrian Pyrenees the crust would have recovered its initial thickness later than in the central and eastern Pyrenees as a result of greater stretching of the Pyrenean crust during Early Cretaceous times, coupled with a younger and lesser amount of convergence. In this area Palaeocene rocks are represented by deep-water carbonate and siliciclastic sediments deposited conformably on the Upper Cretaceous turbidites (Pujalte *et al.* 2000).

During early–middle Eocene times the thrusting rate increased. Both foreland basins experienced a deepening which produced the widest expansion of marine deposits seen in the Pyrenean foreland basins (Puigdefàbregas *et al.* 1992; Burbank *et al.* 1992a). The thrust front in the southern side of the central Pyrenees advanced rapidly due to the presence of a weak detachment level (Triassic evaporites) below the Mesozoic deforming wedge. Shallow marine deposits were deposited in the foreland as well as in piggy-back basins which demonstrate a subhorizontal mean topography over the southern frontal wedge. Strongly subsident troughs filled by turbidite sequences were developed south of the uplifted basement in the footwall of the upper thrust sheets. Some relief existed towards the hinterland as deduced from the north–south river systems supplying basement-derived clastics and from the geometry and location of proximal alluvial fans. A maximum topography of 1–2 km has been calculated based on palaeotopographic reconstructions and flexural modelling (Vergés *et al.* 1995; Millán *et al.* 1995a). Flexure of the Iberian and European plates was produced by a combination of topographic loading and subduction loading (Millán *et al.* 1995a; Beaumont *et al.* 2000). The northern frontal thrust was mainly pinned and no piggy-back basins developed in the Aquitanian basin.

The last stage (late Eocene–middle Miocene) of Pyrenean orogenic growth is characterized by a change of deformational style (Fig. 15.12). Iberian upper crustal units were underthrust below the previously southward-displaced basement and cover units. A basement antiformal stack developed in the middle of the chain synchronously with further southward overthrusting of the southern lower cover thrust sheets on top of the foreland. The geometry of the emergent thrust front as well as the foreland structure of the southern Pyrenees were strongly controlled by evaporitic horizons deposited in the foreland basin succession (Vergés *et al.* 1992; Sans *et al.* 1996). In the northern Pyrenees the frontal thrust migrated 6 km into the foreland. At this time both foreland basins were filled by continental deposits, and relief and erosion rates increased, as deduced from the increase of the exhumation rates of the basement thrust sheets (Fitzgerald *et al.* 1999). The Ebro basin became closed and separated from the Atlantic as a result of the growth of tectonic relief during inversion of the Early Cretaceous basins in the Basque-Cantabrian Pyrenees. Erosional debris from the Pyrenees and other highlands surrounding the Ebro basin (Iberian and Catalonian Coastal ranges) progressively filled the basin and then backfilled, burying the flanking thrust belts on its margins (Coney *et al.* 1996). This progressive backfilling forced deformation to migrate towards the hinterland and as a consequence reactivation of previously developed thrust structures and break-back thrust sequences occurred (Vergés & Muñoz 1990). The southern central Pyrenees were almost completely buried by early–middle Miocene times.

Finally, an abrupt subsequent Miocene–Pliocene re-excavation of the southern Pyrenees and the Ebro basin began the development of the present fluvial system and was due to some combination of Miocene rifting of the western Mediterranean and the Messinian salinity crisis. River incision has also been favoured by ongoing post-orogenic isostatic rebound during thermal re-equilibration of the lithospheric root and the subducted lower crust (Pous *et al.* 1995; Fitzgerald *et al.* 1999).

The Iberian Ranges (JLS, CLL, LEA)

The Iberian Ranges ('Sistema Iberico', Iberian Cordillera, Iberian Chains) form a NW–SE striking intraplate fold belt located in eastern central Spain (Fig. 15.13). This isolated, underpopulated highland area is characterized by broad hills and mountains that rarely rise above 2000 m, and exposures show much less overall deformation than seen in the great mountain belts along the Iberian plate margins to the north (Pyrenees) and south (Betics). It represents an uplifted ('inverted') Mesozoic basin, and preserves spectacular examples of folded and faulted Jurassic and Cretaceous sediments.

Mesozoic extensional tectonics

The Mesozoic sedimentary successions of the Iberian Ranges were deposited in an extensional intraplate basin roughly

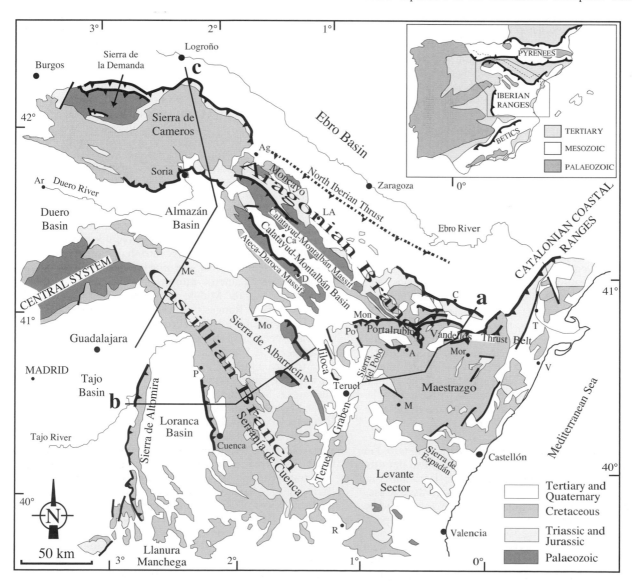

Fig. 15.13. Schematic map showing the general structure of the Iberian Ranges and its sectors. Inset: location of the Iberian Ranges in the Iberian peninsula. Localities: A, Aliaga; Ag, Ágreda; Al, Albarracín; Ar, Aranda de Duero; C, Calanda; Ca, Calatayud; D, Daroca; LA, La Almunia de Doña Godina; M, Mora de Rubielos; Me, Medinaceli; Mo, Molina de Aragón; Mor, Morella; P, Priego; Po, Portalrubio; R, Requena; T, Tortosa; V, Vinaroz. Cross-sections a, b, and c are shown in Figures 15.16 and 15.17.

coinciding with the present-day fold belt. This basin evolved as a complex graben, oblique to the Betic–Balearic passive margin, and underwent compressive deformation during Tertiary times. Subsidence began in Early Permian times as a consequence of the collapse of the Variscan orogen. The crust in central Spain was overthickened (up to 50–60 km) during Variscan compression, and overheated by granitic intrusions. This caused a critical decrease in viscosity and resistance some 10–20 Ma after orogenic compression ceased, giving rise to gravitational instability and extensional collapse (Doblas *et al.* 1994b). At the same time, a network of large wrench faults developed within the right-lateral megashear zone linking the Appalachians and the Urals (Arthaud & Matte 1977). The Iberian plate was bounded to the north by the Pyrenean–Bay of Biscay fault system, and within the plate two main fault sets striking NW–SE and NE–SW propagated across the Variscan orogen (Arthaud & Matte 1975; Capote 1978, 1983a).

Rifting began after mid-Late Permian times, as a consequence of the continental break-up of Pangaea (Arche & López-Gómez 1996). Situated between North America, Europe and Africa, the Iberian plate was submitted to the double influence of Atlantic opening and westward propagation of Tethys. Late Variscan wrench faults became reactivated as normal faults, and new faults developed, producing a fracture pattern that was to dominate Mesozoic tectonic evolution.

Sequence stratigraphy and subsidence analysis of the Mesozoic successions show a large number of pulsating phases of rapid tectonic subsidence followed by slower subsidence periods, well correlated throughout the Iberian basin (Van Wees *et al.* 1998). Two main evolutionary cycles can be distinguished, Late Permian–Middle Jurassic and Late Jurassic–Late Cretaceous, each one including rift and post-rift stages (Alvaro 1987; Sánchez-Moya *et al.* 1992; Salas & Casas 1993; Salas *et al.* 2001). The first cycle involved Late Permian–Hettangian rifting and Sinemurian–Oxfordian post-rifting phases, and the second involved Kimmeridgian–middle Albian rifting and late Albian-Maastrichtian post-rifting phases.

The formation and evolution of the Iberian and Pyrenean basins during the Mesozoic era correlate well with the evolution of the north Atlantic marginal basins. Triassic rift and Jurassic post-rift phases are related to the opening and spreading of the Tethys toward the west. Late Jurassic–Early Cretaceous rifting and Late Cretaceous post-rifting phases occurred as a consequence of the spreading of the central Atlantic and then of the north Atlantic, accompanied by the opening of the Bay of Biscay and the counterclockwise rotation of Iberia (Salas & Casas 1993).

Late Permian-Middle Jurassic cycle. Rift subsidence started in Permian times and was especially active during two periods, Late Permian–Early Triassic and Late Triassic–Early Jurassic, each one giving rise to a regional unconformity (Salas & Casas 1993). Subsidence rate decreased during the Early–Middle Triassic interval, in what has been interpreted as a short post-rift event (Sánchez-Moya *et al.* 1992).

Previous NW–SE and NE–SW striking faults and anisotropies, inherited from Late Variscan orogeny, conditioned the development of extensional structures (Fig. 15.14a). During the first stages, only a few narrow, isolated half-grabens formed close to the SE boundary, controlled by a normal, NW–SE striking, listric fault dipping to the NE (Serranía de Cuenca fault; Arche & López-Gómez 1996). The

infilling sediments onlap the Palaeozoic basement, decreasing in thickness to the NE. During Late Permian times they evolved into wider half-grabens which finally joined to produce an Early Triassic nearly symmetric graben (Sopeña *et al.* 1988). Extension propagated to the NE with time, with the deformation being accommodated by roll-over of the hanging-wall and development of subsidiary, antithetic southwestward-dipping faults (Molina–Teruel–Espadán fault zone). To the NE, further extension led to formation of the Ateca–Montalbán high, the Serranía de Cuenca fault becoming inactive and sedimentation being shifted to the northeastern margin of the Iberian basin and the newly created Triassic Ebro basin (Arche & López-Gómez 1996; Fig. 15.14b).

Extension rates within the Iberian basin increased from NW (thinning factor β = 1.05–1.15) to SE (β = 1.3), according to calculations made by Alvaro (1987), Sánchez-Moya *et al.* (1992) and Arche & López-Gómez (1996). Differences in extension rates were accommodated by contemporaneous NNE–SSW striking transfer faults. The most important were located at the eastern sector (Requena–Mora d'Ebre and Castellón–Vinaroz faults; see Fig. 15.14a), and they became more active during Middle–Late Triassic times.

During the Triassic–Jurassic transition, reactivation of listric faults involving block tilting, half-graben formation and erosion occurred. This gave rise to a regional unconformity recognized in many areas throughout the ranges (San Román & Aurell 1992; Roca *et al.* 1994). In some cases erosion reached down to Lower Triassic strata. Sedimentary breccias within the Cortes de Tajuña Fm, derived from erosion of tilted blocks and half-grabens, are a consequence of extensional break-up of the Upper Triassic shallow carbonate ramp (San Román & Aurell 1992).

By that time, progressive stretching and thinning of the crust gave rise to intrusion of alkaline magmas from the mantle through deep extensional faults. Large basaltic sills, controlled by NW–SE structural trends, intruded Upper Triassic to Hettangian sediments of the northeastern basin boundary (Moncayo and Cameros areas). Subvolcanic intrusions are also abundant in the Middle and Upper Triassic rocks of the eastern Iberian Ranges, showing two magmatic affinities: alkaline in the Teruel–Castellón area, and subalkaline (with tholeiitic tendency) to transitional in the Valencia area (Martínez *et al.* 1997). This magmatic activity has been attributed to partial melting of the base of the lithosphere induced by a mantle plume, accompanied by further decompressional melting as lithospheric stretching progressed (Alvaro *et al.* 1979; Martínez *et al.* 1997; Salas *et al.* 2001).

Thermal subsidence and volcanic activity mainly dominated the Sinemurian to Oxfordian post-rift stage (Salas & Casas 1993). During this stage submarine volcanism, mainly pyroclastic, occurred in the eastern Iberian Ranges, with basaltic and trachybasaltic emissions of Pliensbachian, Toarcian and Bajocian ages (Gautier 1968; Ortí & Vaquer 1980). This magmatism shows little differentiation and is less alkaline than the previous, Triassic magmatism, indicating that magmas ascended quickly from shallower levels, through a thinned lithosphere (Martínez *et al.* 1997). Outcrop patterns of igneous rocks suggest that the volcanism follows the two major Variscan fracture directions (NW–SE and NE–SW; Ortí & Vaquer 1980).

Late Jurassic-Late Cretaceous cycle. Late Jurassic-Early Cretaceous rift episodes fragmented the Iberian basin to

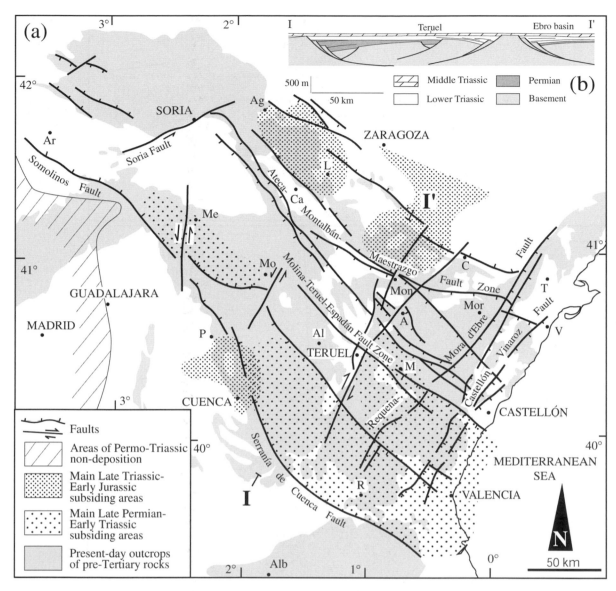

Fig. 15.14. (a) Main subsiding areas and active faults during the two main rifting periods (Late Permian–Early Triassic and Late Triassic–Early Jurassic) of the Late Permian–Middle Jurassic cycle. Oblique lines represent areas of non-deposition during the Late Permian–Triassic interval. **(b)** Cross-section I–I′ (modified from Arche & López-Gómez 1999) shows a schematic view of the Late Permian–Early Triassic grabens across the Iberian basin (see location in a). Based on Capote (1978), Platt (1990), San Roman & Aurell (1992), Arche & López-Gómez (1996, 1999). Localities as in Figure 15.13.

produce four important palaeogeographic domains (Fig. 15.15a): northwestern (Cameros), central, southwestern and eastern (Maestrazgo) domains (Soria *et al.* 2000). Each domain is divided into several smaller depositional basins, controlled by local tectonic structures. The northwestern domain includes the Cameros basin. The central domain comprises two areas of Lower Cretaceous sedimentation, namely the Aguilón and Oliete basins. The southwestern domain consists of the Serranía de Cuenca and Valencia basins. Finally, the eastern domain consists of six basins: Penyagolosa, La Salzedella, Morella, El Perelló, Las Parras and Galve. The most important depocentres during this rift stage were placed in the Cameros and Maestrazgo basins, with synrift deposits reaching thicknesses of 8000 m and 4000 m, respectively.

The overall geometry of the subsiding areas was controlled

by NW–SE and NE–SW late- and post-Variscan faults, reactivated as essentially normal dip-slip faults (Canérot 1974; Alvaro *et al.* 1979; Capote 1983*b*; Guiraud & Séguret 1984). These faults propagated into the Jurassic cover, giving rise to several minor faults parallel to them. The most important ones are La Rambla, Las Parras, Herbers, Segre, Turmell, Miravete and Cedrillas faults in the eastern domain (Salas & Guimerà 1996; Soria 1997; Soria *et al.* 1997*b*; Liesa *et al.* 2000*a,b*), and the Soria, San Leonardo, Quintanilla-Hortigüela and northern Cameros faults in the northwestern domain (Guiraud & Séguret 1984; Platt 1990; Casas 1993). There is much evidence for the uplift and erosion of local highs, tilting of the pre- and synrift sequence, and angular unconformities related to this extensional tectonic activity. At the same time, a general emergence of the central and western parts of the basin

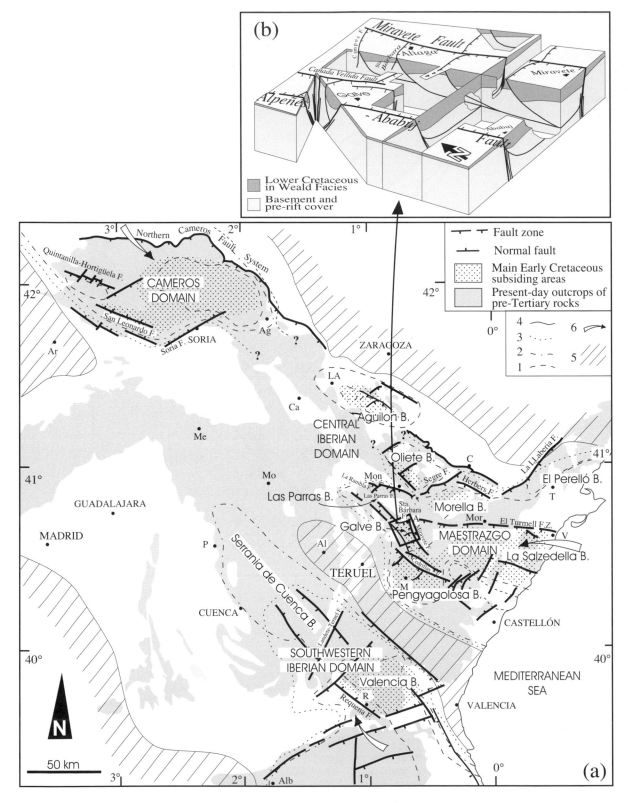

Fig. 15.15. (a) Main subsiding areas, palaeogeographic domains, basins and active faults during the Late Jurassic–Early Cretaceous rifting stage. Legend: lines 1, 2, 3 and 4 limit the areas of minimum extension of the Lower Cretaceous sedimentation: 1, Berriasian–Valanginian; 2, Hauterivian–Barremian; 3, Aptian; 4, Albian; 5, areas of Early Cretaceous non-deposition; 6, marine transgressions. Localities as in Figure 15.13. **(b)** Tectonic model proposed by Soria (1997) for the Lower Cretaceous of the Galve basin. Based on Canérot (1974), Arias *et al.* (1979), García (1982), Mas *et al.* (1982), Rincón (1982), Salomón (1982), Vilas *et al.* (1982), Rincón *et al.* (1983), Fregenal & Meléndez (1993), Platt (1990), Liesa *et al.* (1996, 2000*a,b*), Salas & Guimerà (1996), A. M. Casas *et al.* (1997), Soria (1997), Soria *et al.* (1997*a,b*, 2000), Casas & Gil (1998), Cortés *et al.* (1999), Salas *et al.* (2001).

occurred. The listric geometry of the faults explains the tilting of some blocks, and the fan-shaped geometry of the sub-basins (Querol *et al.* 1992; Soria *et al.* 1997*a*; Liesa *et al.* 1996; Salas & Guimerà 1997; Fig. 15.15b). Usually they were detached either along Middle–Upper Triassic strata (where these units are thick) or within the basement (Roca *et al.* 1994; Salas & Guimerà 1997). Some minor extensional systems use local shallower detachment levels, such as lowermost Cretaceous mudstones (Simón *et al.* 1999*a*).

According to the analysis of subsidence curves constructed for the Maestrazgo domain (Salas & Casas 1993) and the Cameros domain (Mas *et al.* 1993), two main rifting episodes can be distinguished: Kimmeridgian–Berriasian and Barremian–early Albian. In the classic regional literature, these tectonic events have been referred to as Neokimmerian and Austrian phases, respectively. However, the tectonic evolution is more complex than this because a much larger number of episodes of extensional synsedimentary faulting, together with related unconformities, are registered within the Upper Jurassic–Lower Cretaceous series. The beginning of rifting was not coeval throughout the Iberian basin. Synrift subsidence started in the Maestrazgo domain during latest Oxfordian times, but was delayed until early Tithonian and Berriasian times in the Cameros basin and south Iberian domain respectively (Salas *et al.* 2001). Towards the end of Early Cretaceous times tectonic subsidence decreased considerably across the Iberian basin, and both thermal subsidence and uplift played a more important role (Salas & Casas 1993). Late Albian emergence of the region above sea level led to widespread fluvial sedimentation (Utrillas Fm) across an important erosion surface. During the subsequent Late Cretaceous expansion of wide carbonate platforms across Iberia, only a few small synsedimentary normal faults appear to have been active, although some of them in the Cameros and Maestrazgo basins attain offsets of several hundred metres (Salas *et al.* 2001).

The role of NW–SE and NE–SW basement structures on the origin and development of the Lower Cretaceous Iberian basin is different in each domain (Liesa 2000). According to Platt (1990), NW–SE faults of the Cameros basin acted as extensional faults, while the NE–SW faults are transverse faults in the sense of Gibbs (1984). At marginal areas of the Maestrazgo domain (Las Parras and Galve basins) the situation is the reverse: the NW–SE to NNW–SSE Variscan faults are transfer faults of the extensional imbricate system formed by NE–SW to ENE–WSW south-dipping, listric normal faults (Soria 1997; Soria *et al.* 2000). Detailed analysis of Wealden facies in several basins within the Maestrazgo domain (Peñagolosa, Galve, Las Parras and Oliete; Soria 1997) indicates a progressive expansion of the deposition area towards the NW (Fig. 15.15a). This occurred by successive activation of new NW–SE and NE–SW striking faults (Cedrillas, Miravete, Santa Bárbara, Las Parras, La Rambla faults).

Movement on these fault systems during Late Jurassic and Early Cretaceous times induced significant extensional deformation and crustal thinning superimposed on the structures of Late Permian–Early Triassic age. The central domain of the Iberian basin underwent more intense crustal thinning (β = 1.13 for the Cretaceous; β = 1.26 for the whole Mesozoic) than the southwestern domain (β = 1.05 for the Cretaceous; β = 1.12 for the whole Mesozoic; Alvaro 1987). Higher values (β = 1.15 to 1.35) have been obtained from the geometry of the Cameros basin during the Tithonian–early Albian extensional episode (Guimerà *et al.* 1995; Casas & Gil 1998). Salas *et al.* (2001) calculate β values of 1.1–1.2 at two wells within the Maestrazgo basin. An average β value of about 1.3 and a cumulative absolute extension of 40–47 km across the Iberian basin may be estimated for the whole Mesozoic era (Guimerà *et al.* 1996, 2000; Salas *et al.* 2001).

Associated with this Late Jurassic–Early Cretaceous crustal extension a pervasive fracture pattern, consisting of normal faults and

joints, developed in the Jurassic limestones. This fracture pattern consists of two pairs of mutually orthogonal fracture sets (Liesa 2000): NW–SE (or NNW–SSE) and NE–SW; east–west to ESE–WNW and north–south to NNE–SSW. Palaeostress analysis from micro- and mesofractures (Capote *et al.* 1982; Aranda & Simón 1993; Liesa *et al.* 1996, Liesa 2000) suggests a tectonic regime close to multidirectional extension ($\sigma_2 \approx \sigma_3$). This extensional stress field inside the Iberian plate was probably induced by the superposed effect of rifting at both the eastern Iberian (east–west to ESE–WNW extension) and Pyrenean-Cantabrian (NNE extension) plate margins. Local stress directions were also conditioned by the influence of inherited post-Variscan, NW–SE and NE–SW striking fractures (Liesa 2000), so that σ_3 trajectories usually trend normal to them. Differences in location with respect to plate margins and trending of the main inherited structures justify the different models proposed by Platt (1990) and Soria (1997) for the Cameros domain and the marginal basins of the Maestrazgo domain, respectively.

Local folding, angular unconformities, and other features related to Neokimmerian and Austrian phases were interpreted as due to weak compression by Riba (1959), Meléndez (1971) and Viallard (1973). Other researchers (Villena 1971; Alvaro *et al.* 1979) linked local folding to movement of large NW–SE strike-slip faults. Nevertheless, it is difficult to accept that a compressional regime existed throughout the Iberian Ranges since compressional evidence is rather local and doubtful (Capote 1983*b*). In more recent papers, local folding is related to extensional faults and they can be explained as rollover or drag folds on listric faults showing flat-ramp geometry (Soria 1997; Cortés *et al.* 1999).

The only clear exception is the Cameros basin, which underwent a compressional stage at the end of Early Cretaceous times (Casas & Gil 1998). Immediately after Early Cretaceous basin formation, shortening brought about folding with development of axial planar cleavage throughout the eastern Cameros basin. The orientation of these folds was strongly controlled by the initial geometry of the basin. The central part of the Cameros basin records low-grade metamorphism (Guiraud & Séguret 1984 zone; $T \geq 420°C$, P = 1–3 kbar), related to the lithostatic load created by several thousands of metres of alluvial and lacustrine sediments. Its thermal peak has been dated between 108 and 86 Ma BP (Goldberg *et al.* 1988; Casquet *et al.* 1992; Mantilla 1999). This metamorphic peak is younger than cleavage, so that the cleavage-related folding must have occurred between early Albian (top age of the involved sediments) and Cenomanian times (Casas & Gil 1998). Cleavage and metamorphism lessen away from the Cameros basin, both towards the east, in Sierra del Moncayo (Gil & Pocoví 1994), and towards the west in Sierra de la Demanda.

Tertiary orogeny: positive inversion of the Iberian basin

By the beginning of the Cenozoic era, the tectonic regime within the Iberian basin switched from extensional to compressional. Alpine compression affected the Variscan basement, the Mesozoic cover and the syntectonic Tertiary deposits, and gave rise to a double verging intraplate belt with a sharp contrast between basement and cover deformation (Fig. 15.16). According to the presence of several internal Tertiary basins and to differences in the deformation style (as a result of several factors such as thickness of the cover and detachment levels, and geometry of previous fractures), it is possible to identify five structural units within the chain (Fig. 15.13): Demanda–Cameros (NW), Aragonian branch (central NE), Maestrazgo (SE), Castilian branch (central SW) and Sierra de Altomira (SW).

Demanda–Cameros. This comprises an allochthonous unit showing an overall asymmetric pop-up structure (Fig. 15.17). It is displaced northward about 20–35 km over the western side of the Tertiary Ebro basin (the latter being filled with a 4000–5000 m thick sequence of Tertiary deposits) and some 2–3 km southward over the Almazán basin (Casas 1990, 1993; Guimerà & Alvaro 1990; Platt 1990; Guimerà *et al.* 1995). Southward-dipping thrusts, with strikes changing from east–west (western sector, Sierra de la Demanda) to ENE–WSW

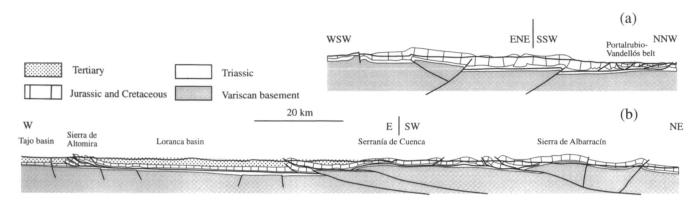

Fig. 15.16. General cross-sections across the Iberian basin (sections a and b in Fig. 15.13). Modified from Guimerà & Álvaro (1990) and Salas *et al.* (2001).

(central sector) and NW–SE (eastern sector), make up the northern boundary of the Demanda–Cameros massif. At Sierra de la Demanda, the detachment level is located in Precambrian slates, whereas it climbs up to Upper Triassic sediments in Cameros by means of a lateral ramp (Casas 1990; Guimerà & Alvaro 1990). The hanging-wall of the Cameros thrust comprises the thick (up to 8000 m) Upper Jurassic–Lower Cretaceous succession previously deposited in the Cameros basin and now affected by several large folds.

Aragonian branch. This comprises two NW–SE trending basement uplifts: Ateca-Daroca and Calatayud-Montalbán massifs. Alpine tectonics reactivated some previous Variscan and late Variscan faults

as strike-slip faults with a northward-directed thrust component. Borehole data indicate that the boundary with the Ebro basin is a thrust involving basement rocks (north Iberian thrust; Sánchez *et al.* 1990; Salas *et al.* 2001), although Neogene post-tectonic deposits now overlie this structure.

Maestrazgo. The central and southern sectors of this unit show a thick Mesozoic cover (up to 6000 m in the SE corner) and an almost flat-lying structure, with a few gentle, NW–SE trending folds (Canérot 1969, 1974; Simón 1982). To the north, the cover thins and shows a thrust-and-fold belt in the Portalrubio–Vandellós area (Canérot 1974; Guimerà 1988; Fig. 15.16a). This belt shows several changes in trend,

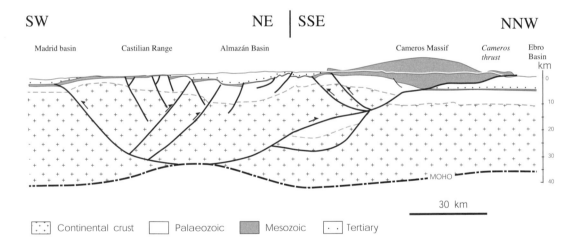

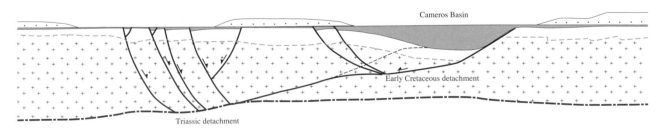

Fig. 15.17. Crustal-scale cross-section of the Cameros massif and Almazán basin (section c in Fig. 15.13) and restored section before the positive tectonic inversion. After Casas *et al.* (2000*b*).

with three major inflections where the structure veers from NW–SE to NE–SW; the easternmost inflection represents the link with the Catalonian Coastal Ranges. The Mesozoic cover is generally detached over the Middle–Upper Triassic succession, although to the south thrust surfaces probably cut down into the Variscan basement (Salas *et al.* 2001). The maximum single offset is 5–8 km at the Montalbán thrust (Guimerà 1988; Casas *et al.* 2000*a*).

Castilian branch. This unit is separated from the Aragonian branch by the Tertiary Almazán and Calatayud–Montalbán basins. In the central sector (Sierra de Albarracín) the Mesozoic cover is thin and mostly consists of Triassic and Jurassic units. Deformation of the cover is limited to NW–SE trending, NE-verging basement-involved folds (Riba 1959; Capote *et al.* 1982; Fig. 15.16b). At the southwestern part of the Castilian branch (Serranía de Cuenca), the cover includes a thick Cretaceous sequence and is detached from the basement, with fold axes again trending NW–SE but verging to the SE (Viallard 1973). The Levante sector is the southeastern end of the Castilian branch, with a similar style of deformation and separated from it by the Neogene Teruel graben.

Sierra de Altomira. This is a north–south trending fold-and-thrust belt affecting a thin cover, mainly Cretaceous in age, detached on Middle–Upper Triassic strata, and thrust over the Tajo basin (Muñoz & De Vicente 1998*a*; Fig. 15.16b). Displacement on the belt becomes smaller to the south, where the structures are covered by Neogene post-orogenic deposits in the Llanura Manchega area. The Tertiary Loranca basin separates this sector from the rest of the Iberian Ranges.

Tertiary inversion: the role of structural heritage. Tectonic inversion of the Iberian basin basically took place along the large normal faults that were active during the Mesozoic era, most of them inherited from Variscan and post-Variscan times. In some cases, Tertiary compression did not produce total inversion of the Mesozoic basins, but extensional structures account for the nucleation of major Tertiary folds, since the faults created an obstacle impeding the movement of the cover over the regional detachment level.

The reactivation of steeply dipping inherited fault surfaces as reverse faults is mechanically difficult. Those oriented nearly orthogonal to the regional compression did not undergo reactivation. In these cases stress concentration produced important contractional structures (tight folds, cleavage) by buttressing against the fault planes (Cortés *et al.* 1999; Liesa 2000). Faults oblique to the maximum stress trajectories moved as strike-slip faults. A number of steep basement faults (especially those belonging to the NE–SW set) were reactivated as strike-slip faults triggering en echelon folds and thrusts as well as inducing sharp rotations of contractional structures.

The most active basins during the Late Jurassic–Early Cretaceous rifting (Cameros and Maestrazgo domains) underwent total positive inversions. Basement faults whose displacement gave rise to the development of the Cameros basin controlled the different directions of the Cameros thrust front (Casas 1990, 1993; Fig. 15.17). After an early phase of slight compression in mid-Cretaceous times (Casas & Gil 1998; Gil 1999), the main positive inversion took place during the Tertiary period, when faults moved with a reverse component, recovering the former normal offset and raising the Cameros block to its present position (Casas 1990, 1993). Guimerà *et al.* (1995) interpreted the Cameros basin as an extensional-ramp basin developed over a south-dipping ramp within a deep extensional fault inside the Variscan basement. According to this model, the thrust that accommodated Tertiary inversion was a new fault formed within the weak Upper Triassic horizons.

In the Maestrazgo sector, the positive inversion of the southward-dipping Early Cretaceous faults bounding the Maestrazgo basin produced the Portalrubio–Vandellós thrust belt. The trend of the thrust belt varies between NW–SE and NE–SW, these orientations being parallel to the main normal faults (i.e. Las Parras and La Rambla faults) active during the Early Cretaceous rifting (Liesa *et al.* 1996, 2000*b*; Soria 1997; see Figs 15.13 and 15.15a). Within the internal parts of the Maestrazgo basin, inversion was conspicuous in the western zone, whereas many extensional faults were not inverted at the easternmost sector (Salas & Guimerà 1997). Some north–south to NW–SE striking faults (i.e. Cañada Vellida and Miravete faults) and ENE-WSW faults (i.e. Campos and Santa Bárbara faults) were reactivated as reverse faults recovering hectometric to kilometric normal throws, and gave rise to the development of large-amplitude folds (Guimerà & Salas 1996; Salas & Guimerà 1997; Soria 1997; Simón *et al.* 1998; Liesa *et al.* 2000*a*).

Basement and cover deformation. The compressional architecture of the Iberian Ranges is strongly conditioned by the superposition of two structural levels showing a different behaviour: basement (Variscan basement, Permian and Lower Triassic) and cover (Jurassic, Cretaceous and Tertiary). Both are separated by a regional detachment that constitutes, in many areas, the sole thrust of the cover (Guimerà & Alvaro 1990). This detachment is generally located within the Upper Triassic marls and evaporites, though in the eastern Iberian Ranges it also includes Middle Triassic sediments (Viallard 1973; Guimerà 1988). In some cases, thick incompetent formations interbedded within the Mesozoic cover (Purbeck–Weald facies, Escucha and Utrillas formations) behave as local detachment levels giving rise to very disharmonic folding (Viallard 1973, 1983; Simón 1980). On the other hand, thrusting involving the Variscan basement suggests the existence of deeper detachments. These may be located either in slate units within the Palaeozoic succession (Cortés & Casas 1996; Liesa *et al.* 2000*a,b*) or in a mid-crustal low-velocity level (Guimerà & Alvaro 1990) such as that identified by Banda *et al.* (1981) at a depth of 7–11 km.

The magnitude of crustal shortening for the whole chain has been estimated as up to 57 km (from geometric calculations on large folds and thrusts along structural cross-sections; Salas *et al.* 2001) or 75 km (by comparing the present-day crustal profile with that restored to Late Cretaceous time; Guimerà *et al.* 1996). Partial values of shortening are about 15 km in the Maestrazgo sector (Salas *et al.* 2001), 16–18 km in the Sierra de Altomira–Loranca basin–Sierra de Bascuñana (Muñoz 1997; Muñoz & De Vicente 1998*a*), 38 km for the whole Cameros pop-up structure (Guimerà *et al.* 1995), and 20 km for the Aragonian branch (Salas *et al.* 2001).

Different tectonic styles may be found depending on the thickness of the cover and detachment levels (Alvaro 1995). First, examples of folds and thrusts involving the basement (thick-skinned tectonics) are found in the Aragonian branch and Sierra de Albarracín–Serranía de Cuenca. Secondly, tight folds and thrusts involving only the detached cover (thin-skinned tectonics) are found in northern Maestrazgo, and in the Sierra de Altomira–Loranca basin, where a thin cover overlies the detachment level. Finally, major thrusting with only gentle buckling is found in the Maestrazgo and Cameros sectors where the cover is thick.

Basement deformation was partially guided by the late-Variscan strike-slip faults. Kinematic indicators and map relationships generally show dextral movement on the NW–SE striking planes and sinistral movement on the NE–SW faults (Guimerà 1988). Long NW–SE transpressive dextral-reverse faults crop out in the Palaeozoic materials of the Aragonian branch (Ateca and Daroca massifs). In the Daroca fault small shallow-dipping thrusts probably become vertical and join the master fault at depth (Colomer & Santanach 1988). Some of these thrusts cut across the Tertiary deposits of the Calatayud–Montalbán basin. In the Ateca massif, transpression resulted in a prominent box-shaped fold involving both basement and cover, with the related fault probably detached at a depth of about 8–12 km (Cortés & Casas 1996; Casas *et al.* 1998). NW–SE striking basement faults in the Castilian branch are not so large, and they are associated with high-amplitude periclines whose Variscan cores overthrust Mesozoic materials towards the NE (Sierra de Albarracín; Riba 1959) or the SW (Serranía de Cuenca; Viallard 1973; Liesa & Casas 1994). NE–SW sinistral faults do not crop out within the Iberian Ranges but they can be inferred from cover deformation. In this way, the large inflections of the Portalrubio–Vandellós fold-and-thrust belt are controlled by two inferred NNE–SSW left-lateral wrench faults in the Variscan basement (Simón Gómez 1981; Guimerà 1988). Finally, a few east–west striking basement faults have been reactivated as pure reverse (Sierra de la Demanda thrust; Casas 1993; Liesa & Casas Sainz 1994) or dextral-reverse faults (northern Maestrazgo).

The Variscan–Lower Triassic basement did not behave only as an ensemble of rigid blocks bounded by strike-slip faults. East–west orientated Variscan folds were reactivated in Sierra de la Demanda (Liesa & Casas Sainz 1994) and the northeastern boundary of the Almazán basin (Casas *et al.* 2000b). The core of the Sierra de Espadán anticlinorium shows well developed axial planar cleavage in Lower Triassic sandstones and siltstones (Gutiérrez & Pedraza 1974; Simón 1982) that becomes a crenulation cleavage cutting the Variscan slaty cleavage in the underlying Palaeozoic rocks (Simón 1986a).

Cover deformation mainly consists of folds and thrusts whose density is controlled by the thickness of the cover and the presence of basement faults (Fig. 15.16). Their dominant trend is parallel to the NW–SE trend of the Iberian Ranges, although both gradual changes in orientation (e.g. Serranía de Cuenca and Sierra de Altomira show a change from NW–SE to north–south) and sudden bends (northern Maestrazgo, change from NW–SE to east–west and NE–SW) occur. In other areas, especially in the southern part of the Iberian Ranges, folds of different orientations are found together. Interference structures showing transversal (east–west to ENE–WSW) folds superposed on longitudinal (NW–SE to NNW–SSE) folds have been described in several areas, notably in the southeastern sector (close to the Betic Cordillera; Champetier 1972), the Aliaga region (in the northern Maestrazgo area; Simón 1980; Guimerà 1988; Simón *et al.* 1998), in the Sierra del Pobo (Liesa 2000), and in the Almazán basin (Simón 1991a). In a few cases east–west to ENE–WSW folds are older than the NW–SE folds (Gómez & Babín 1973; Capote *et al.* 1982). Local coeval development of different fold trends as a result of constrictional deformation has also been proposed at the core of the central inflection of the northern Maestrazgo sector (Guimerà 1988).

Major cover thrusts appear at the northern boundaries of the inverted Cameros and Maestrazgo Cretaceous basins. The Cameros–Demanda thrust is the most important in the Iberian Ranges, showing a displacement of 20–35 km (Casas 1990, 1993; Guimerà & Alvaro 1990; Fig. 15.17). The latest direction of displacement, as inferred from kinematic indicators, was 340–345°, oblique to the north to NNE inferred regional compression (Casas 1990; Casas *et al.* 1992). According to the model proposed by Casas & Simón (1992), the eastern ramp of the Cameros thrust (striking 155°) underwent a right-lateral movement with a small reverse component, which gave rise to rigid displacement of the Cameros block towards the NNW and produced a 'guided movement' on the western, ENE–WSW striking frontal ramp. In other Tertiary deformation events a northward displacement might also have taken place (Casas 1993).

The curving thrust belt that forms the northern border of the Maestrazgo sector and links into the Catalonian Coastal Ranges is an imbricate system whose sole fault lies in the mostly mid-Triassic regional detachment level (Guimerà & Alvaro 1990). In contrast, thrusts forming the western sector of the belt, close to the Palaeozoic

massifs of the Aragonian branch (the Montalbán and Portalrubio thrusts), are single structures involving the basement (Casas *et al.* 2000a; Liesa *et al.* 2000b). The Montalbán thrust has 5–8 km of displacement (Guimerà 1988; González & Guimerà 1993; Casas *et al.* 2000a) and shows a strongly deformed hanging-wall, with folds of several orientations. Transport directions show two main events: an early displacement towards the NNE and a late movement towards the NNW (Liesa 1999; Fig. 15.18). Finally, at Sierra de Altomira and in the southwestern part of the Castilian branch, displacement took place towards the west and SW, with structures verging in the same direction. The boundary between these contrasting vergences is located along the 'Hesperian line' (Stille 1931), a central axis running through the Castilian branch (Fig. 15.18).

Tectonosedimentary relationships: the age of deformation. Several Tertiary sedimentary basins are directly related to the Iberian Ranges. The Ebro, Duero, and Tajo basins all lie adjacent to, but outside, the Ranges, whereas the internal Calatayud-Montalbán, Loranca, Almazán, and northern Maestrazgo depocentres lie within them and were filled with intramontane syntectonic deposits. Some of these depocentres are piggy-back basins, formed and filled in the hanging-wall of thrust sheets. These basins include thin-skinned types, such as Loranca and Aliaga, related to the shallow thrusts of Sierra de Altomira and Montalbán respectively (Calvo *et al.* 1990; González & Guimerà 1993). Others are thick-skinned piggy-back basins, such as the Almazán basin, transported over the deep-rooted Demanda-Cameros thrust (Guimerà *et al.* 1995; Casas *et al.* 2000b).

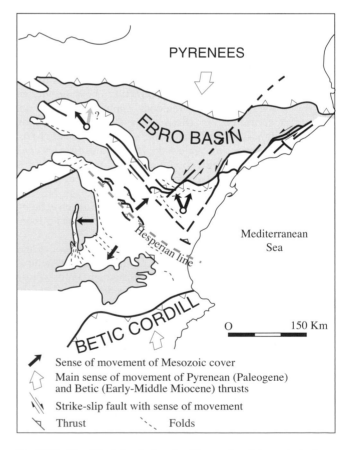

Fig. 15.18. Simplified sketch showing kinematics of the main faults and sense of movement of the Mesozoic–Tertiary cover during Alpine compression. Modified from Guimerà & Álvaro (1990).

A number of unconformities have been identified within the Tertiary deposits, allowing establishment of the timing of folds and thrusts. Lack of palaeontological dating makes it difficult to know the exact age of deposits and hence that of compressive deformations, although their age can be inferred by correlation of unconformity-bounded units or tectonosedimentary units (TSU; González 1989; Pérez 1989; Pardo et al. 1989).

Compression is likely to have started by early-middle Eocene times in the Iberian Ranges, as opposed to the Catalonian Coastal Ranges where the main folds and thrusts were already propagating by then (Guimerà 1984) and the first unconformities were appearing in internal and adjacent basins. The earliest structures in the fold-and-thrust belt linking the Iberian and Catalonian Coastal ranges probably formed around this time, although they have not been clearly identified (González 1989; González & Guimerà 1993; Villena et al. 1996a; Casas et al. 2000a). Development of the main structures in the Cameros–Demanda massif and Almazán basin probably started by the Eocene–Oligocene boundary (Muñoz & Casas 1997; Casas et al. 2000a).

The deformation peak may be placed within late Oligocene times, when compressive structures of different orientations developed, especially NNW–SSE to WNW–ESE folds and thrusts. These are seen in the Aragonian branch–Almazán basin (Simón 1991b; Casas et al. 2000a), the northern Maestrazgo, Montalbán and Aliaga areas (González 1989; Pérez 1989; González & Guimerà 1993; Simón et al. 1998), and the Sierra de Altomira–Loranca basin (Díaz Molina & López-Martínez 1979; Calvo et al. 1996). This climactic deformation event ended with the largest and most important unconformity, Agenian in age, close to the Oligocene–Miocene boundary (González et al. 1988a,b; Calvo et al. 1993, 1996; Villena et al. 1996b).

Throughout early to middle Miocene times, folding and thrusting were still developing in several areas of the chain, though at a lesser rate. The main structure still active by the beginning of the late Miocene is the Cameros–Demanda thrust (Casas 1990; Muñoz & Casas 1997). The smaller Daroca thrust has allowed accurate dating of rocks affected by this structure (Aragonian, middle Miocene). Deposits with an age attributed to the middle Miocene and early part of the late Miocene are also involved in thrusts and folds within the Loranca basin (Gómez et al. 1996). Other compressional macrostructures have been considered to be coeval using regional correlation (Viallard 1973; Simón & Paricio 1988; González 1989; Pérez 1989; Guimerà & Alvaro 1990). Many of them correspond to transversal, east–west to ENE–WSW trending fold axes and thrusts, eventually superimposed on longitudinal (NNW–SSE to ENE–WSW) folds (Simón 1980; Simón et al. 1998).

In general, compressive deformation progressed westwards through the Iberian Ranges. This diachroneity applies to the beginning of folding and thrusting as well as to the transition to an extensional regime that becomes progressively younger to the west. Tectonosedimentary analysis along the boundary between the Iberian Ranges and Ebro basin (Villena et al. 1996a) corroborates this evolutionary pattern.

Evolution of intraplate compressive stress fields. Several compression directions coexisting within the Iberian Ranges have been deduced from microstructural analysis of stylolites, tension gashes, and dynamic analysis of fault populations. In general, these directions are compatible with the trends of large structures such as folds and thrusts. In recent decades, two types of models have been proposed to explain these results (Simón 1990). (1) The first model proposes the occurrence of several successive compressive stress directions created by different geodynamic forces external to the Iberian Ranges (Alvaro 1975; Alvaro et al. 1979; Capote et al. 1982; Capote 1983b; Simón 1982; De Vicente 1988; De Vicente et al. 1996c; Giner 1996; Muñoz 1997; Muñoz & De Vicente 1998b; Cortés 1999). (2) The second model proposes the existence of just one NNE directed compression, with the other stress directions obtained from brittle microstructural analysis being deviations caused by the presence of basement and cover discontinuities (Guimerà 1988; Casas 1990).

In our opinion, a single direction of regional compression cannot explain the variety of macrostructures, superposed folds and palaeostress directions found in the Iberian Ranges. Major faults in the Aragonian branch record both dextral and sinistral strike-slip movements (Calvo 1993) which are not compatible with small-scale stress perturbations. Local palaeostress directions show several maxima that are homogeneous throughout the region and coexist at many sites, which suggests that they constitute independent events (Liesa 2000). The tectonic complexity is a consequence of combined time (successive tectonic phases) and space (single mechanism) inhomogeneities (Simón 1990). In this way, the concept of a regional, perturbed stress field needs to be used as an essential tool to explain the tectonic evolution of the Iberian Ranges within its geodynamic setting. This concept has been applied in numerous sectors of the Iberian Ranges during the 1990s (Casas et al. 1992; Liesa & Simón 1994; Casas & Maestro 1996; Cortés et al. 1996; Muñoz 1997; Maestro 1999; Liesa 2000).

Recently, Liesa (2000) compiled 1523 local σ_1 directions in the northeastern Iberian peninsula from analysis of brittle microstructures (with 1289 data points in the Iberian Ranges), and distinguished three mean intraplate compressive stress fields (Fig. 15.19): an Iberian stress field (σ_1 NE–SW), a Betic (sensu lato) stress field (average σ_1 NW–SE) and a Pyrenean stress field (σ_1 N to NNE). Three successive stages with slightly different compression directions (ESE, SE and SSE or Guadarrama) have been differentiated within the Betic stress field, representing a progressive clockwise rotation of the compression direction. These compressive stress fields played a different role in the development of structures within the Iberian Ranges and the rest of the northeastern Iberian peninsula. The Iberian compression (average direction NE–SW) is well represented across the whole Iberian Ranges, being responsible for its main macrostructures. The Betic compression, showing a clockwise rotation of σ_1 from ESE through SSE, initiated compressive structures in the Catalonian Coastal Ranges, then produced folds and thrusts in the Iberian Ranges which are transverse to the main NW–SE structural trend. The Pyrenean compression (average direction NNE–SSW) is responsible for the final structure of the Pyrenean orogenic belt, as well as for the Sierra de Cameros and Sierra de la Demanda thrusts in the northwestern Iberian Ranges.

The stress fields show significant spatial variations of σ_1 directions and stress ratio ($R = (\sigma_2 - \sigma_3)/(\sigma_1 - \sigma_3)$), which have been recognized from outcrop scale to very large distances (Fig. 15.19a, b, c). They are associated with strike-slip faults, and locally with thrusts, following the models proposed by Anderson (1951), Auzias (1995) or Homberg et al. (1997).

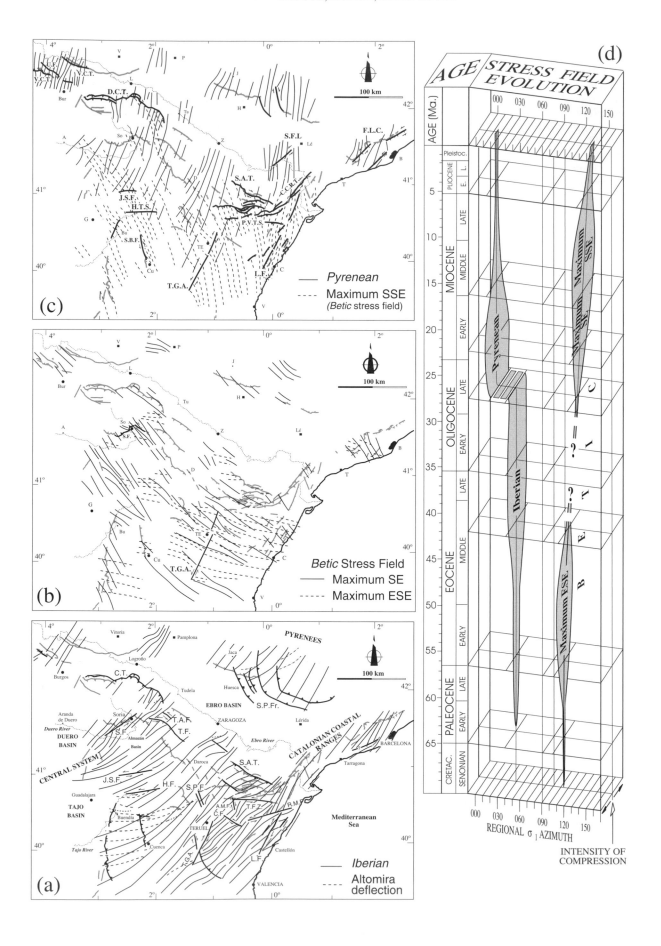

Nevertheless, clockwise or anticlockwise deviations of σ_1 trajectories usually do not exceed 20° . Therefore, they should not bring about confusion in the assignment of a particular local stress state to its genetically related regional stress field (Liesa 2000).

Owing to the inhomogeneous spatial distribution of the different stress fields, their chronological relationships show a complex evolution that greatly depends on the geographic location within the Iberian Ranges. These relationships have been constrained from outcrop observations of cross-cutting relationships between microstructures, as well as from dating of macrostructures. According to Liesa (2000), the following sequence of compressive stress fields can be inferred (Fig. 15.19d): (1) ESE maximum of the Betic stress field (early–middle Eocene); (2) Iberian stress field (middle Eocene–late Oligocene); (3) SE and SSE maxima (Guadarrama) of the Betic stress field (late Oligocene–Miocene); and (4) Pyrenean stress field (Miocene). The Betic and Pyrenean stress fields have remained active until the present day although usually they are not superposed in a given area, owing to their inhomogeneous spatial distribution.

The intraplate stress fields recorded in the Iberian Ranges were transmitted from the plate boundaries, as a consequence of relative motions of Africa, Iberia and Europe. The Iberian and Pyrenean stress fields arose from the convergence between Iberia and Europe, and they probably correlate with the two main mountain-building stages of the Pyrenees (emplacement of upper and lower thrust sheets, respectively; Liesa 2000). The three stages of the Betic stress field (regional compressive directions ESE, SE, and SSE) are related to the complex evolution of the plate boundary between Iberia and Africa. Indentation of the Alborán block induced westward propagation of the Betic orogen up to its present location. This explains both the systematic vertical-axis rotation described by Lonergan & White (1997) for the kinematic evolution of the Betic Cordillera, and the clockwise rotation of the regional maximum stress (σ_1). In fact, every stress field results from a combination of collisional resistance forces at the Pyrenean and Betic margins and the push force of the Atlantic ridge. This explains why regional σ_1 trajectories, especially those of the Iberian stress field, deviate up to a nearly east–west trend (Altomira deflection) as they approach the western boundary of the Iberian Ranges (Serranía de Cuenca-Sierra de Altomira).

Neogene-Quaternary extension

Extensional macrostructures. A large network of post-middle Miocene extensional faults, most of them oblique to the compressive folds and thrusts, appears in the Iberian Ranges (Fig. 15.20). These faults developed after the compressive structure of the chain had been established and as the eastern margin of the Iberian peninsula became dominated by the influence of rifting in the Valencia trough (Alvaro et al. 1979; Simón 1982; Simón & Paricio 1988). During Oligocene and Miocene times, rifting propagated southward (Vegas et al. 1979), so that the Iberian Ranges were affected later than some other regions of the Alpine belt such as Languedoc (middle–upper Oligocene;

Mattauer 1973) or Catalonia (early Miocene; Cabrera 1981), but earlier than the Betic Cordillera (upper Miocene; Montenat 1977).

The two main directions of faults inherited from late-Variscan and Mesozoic times, which moved as reverse and strike-slip faults during the Palaeogene compression, have been reactivated again as Neogene–Quaternary normal faults. Due to the dominant WNW–ESE σ_3 trajectories related to rifting, Neogene extension was mainly accommodated by NNE–SSW faults. Seismic data and deep structural reconstructions indicate that these faults have a listric geometry and are detached on a middle crustal level 13–15 km deep (Roca & Guimerà 1992). The incompatibility between the upper crustal thinning inferred from this geometry ($\beta = 1.4$–1.5) and from geophysical data ($\beta = 1.8$) suggests that present-day crustal thinning may be partially inherited from Mesozoic extension (Roca & Guimerà 1992).

The movement of the main normal faults created graben, filled with Neogene–Quaternary sediments, both parallel and oblique to the compressive structures of the chain. The Calatayud basin is a NW–SE trending basin whose structure and evolution are not well known. It probably originated during late Oligocene or early Miocene times under a compressive regime, its southwestern boundary (Daroca area) being a shallow-dipping thrust (Colomer & Santanach 1988), but it then underwent post-Miocene extensional reactivation (Gracia et al. 1988). As a consequence of this history, it exhibits the most complete sedimentary series among the internal Neogene basins of the Iberian Ranges (early Miocene to Quaternary). The Teruel and Maestrazgo grabens are controlled by NNE–SSW striking normal faults, parallel to the structural trend of the Valencia trough. They represent the onshore deformation of the eastern Iberian Neogene rift (Vegas et al. 1979; Simón 1982; Roca & Guimerà 1992). The Jiloca graben shows a NNW–SSE trend, although its boundaries coincide with NW–SE faults showing an en echelon pattern (Fig. 15.20).

Extension propagated westward from the inner parts of the Valencia trough. The filling of the offshore trough was initiated by early Miocene times, with rifting being interrupted in the middle Miocene interval (Alvarez & Meléndez 1994). The Maestrazgo graben probably developed during the early to middle Miocene interval (Anadón & Moissenet 1996), whereas infilling of the Teruel graben started at the beginning of late Miocene times (Simón & Paricio 1988; Garcés et al. 1997; Alcalá et al. 2000; Alonso-Zarza & Calvo 2000). Still later, subsidence of the Jiloca graben occurred during the late Pliocene interval (Simón 1989), and local reactivation of normal faults occurred during Pleistocene times. Several NNE–SSW striking faults at the boundaries of the Teruel and Maestrazgo graben had undergone deca- to hectometric displacements by early Pleistocene times (Simón 1982, 1983; Simón et al. 1983). NW–SE striking faults at the eastern limit of the Jiloca graben also show metric to decametric offsets affecting middle and upper Pleistocene sediments (Simón & Soriano 1989).

The Ebro basin was also affected by Neogene extensional deformation. Rocks of the central Ebro basin show a nearly flat-lying structure, although a number of mild deformation events can be recognized. The main macrostructure is the Ebro syncline (Quirantes 1978; Arlegui 1996), a wide NW–SE trending fold with gently dipping limbs (up to 4–6°) which extends over some 200 km along the Ebro valley. Another interesting macrostructural feature, also following the dominant NW–SE structural trend of the basin, is a dense cluster of tectonic lineaments identified on aerial photographs and satellite images (Arlegui & Soriano 1998). They have been interpreted, from both field observations and seismic profiles, as normal faults showing metre-scale offsets. At depth, seismic and well information compiled by Lanaja (1987) and IGME (1990) reveal NW–SE to WNW–ESE striking faults in the basement under the Ebro syncline, which could have induced the generation of cover structures. At the Bárdenas

Fig. 15.19. Trajectories of σ_1 and active structures for the different compressive stress fields related to the tectonic building of the Iberian Ranges. (**a**) Iberian stress field. (**b**) Maxima ESE and SE of the Betic stress field. (**c**) Pyrenean stress field and maximum SSE (Guadarrama) of the Betic stress field. (**d**) Evolution on time of stress fields. After Liesa (2000).

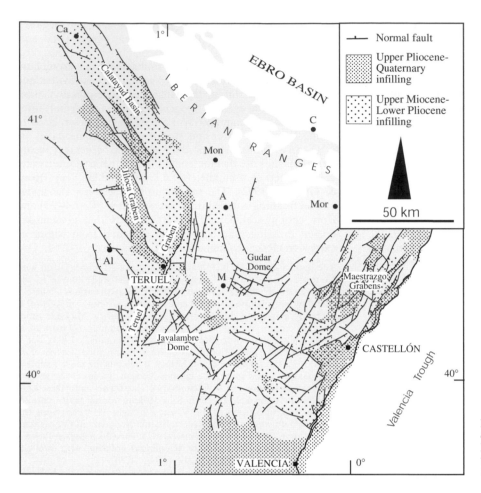

Fig. 15.20. Extensional structures and deposition areas for the late Miocene–early Pliocene and late Pliocene–Quaternary in the eastern Iberian Ranges (after Simón 1982, 1989).

region (western Ebro basin), the main structures follow a different trend. There, early to middle Miocene rocks are affected by a dense swarm of north–south trending normal faults showing metre- to decametre-scale offsets (Gracia & Simón 1986; Arlegui & Simón 1997).

The Tertiary rocks of the Ebro basin are also affected by several sets of fractures at the outcrop scale (Simón 1989; Hancock & Engelder 1989; Hancock 1991; Arlegui 1996; Arlegui & Simón 2001). Three regional joint sets and several sets of small-scale faults are present, closely mirroring the macrostructural trends (NW–SE and north–south). All of them have been widely used for palaeostress analysis (Arlegui & Simón 1998, 2001; Simón *et al.* 1999*b*).

Evolution of deformation. Two main episodes (late Miocene and Pliocene–Pleistocene in age) have been distinguished within the extensional process (Viallard 1973; Simón 1982). The first one produced the main grabens in the eastern Iberian Ranges (Teruel, Maestrazgo), whose NNE trend is parallel to the axis of rifting. This implies a dominant ESE extension direction that is also evident at a microstructural scale. This extensional state was attained through a gradual transition from the late (either NNW–SSE or NNE–SSW) compression (Simón 1982, 1986*b*; Guimerà 1988).

The whole Iberian Ranges and Spanish meseta were planed by erosion during Neogene times, resulting in either several nested levels or only one polyphase and polygenic erosion surface, depending on the tectonic activity of each area (Gracia *et al.* 1988). The sculpting of the largest surface (the 'fundamental erosion surface' of Solé Sabarís (1978); Peña

et al. 1984) ended by late Miocene or early Pliocene times (Simón 1982; Gracia *et al.* 1988), during a period of relative tectonic quiescence between both extensional episodes. Pliocene–Pleistocene extension then deformed this surface, giving rise to the main morphotectonic features of the region. The age of related deposits suggests that this second extensional episode started at the beginning of the late Pliocene, around 3 Ma BP (Simón 1982, 1989). Contour maps of the fundamental erosion surface have been used to show the geometry of Pliocene–Quaternary structures (Simón 1982, 1989), which is characterized by large faulted domes (Gúdar, Javalambre) and grabens (Calatayud, Jiloca, Teruel, Maestrazgo). Some of the grabens reactivated previous faults active during the Miocene epoch (Teruel, Maestrazgo, Calatayud), whereas the Jiloca graben is an entirely new structure filled with sediments only of Pliocene–Quaternary age.

Stress fields. The Neogene–Quaternary extensional stress field within the Iberian Ranges and Ebro basin was inhomogeneous in both space and time, and involved the superimposition of two primary stress systems, one compressional and one extensional, caused by different geodynamic mechanisms (Simón 1986*a*, 1989, 1990). North–south compression, caused by the convergence between Europe, Iberia and Africa, was probably a combination of NNE directed Pyrenean compression and NNW directed Betic compression, related to convergence at the northern and southern margin of Iberia, respectively (Cortés *et al.* 1996; Cortés 1999; Liesa 2000). NW–SE to

WNW–ESE extension was related to rifting at the eastern margin of Iberia and opening of the Valencia trough. This superposition resulted in an extensional stress field with σ_2 trajectories trending approximately north–south, with local compressive stress states registered where and when the extensional component was not intense enough (Simón 1989). This stress field has been active up to the present day, as indicated by stress tensors inferred from focal mechanisms (Herraiz et al. 2000).

The extensional stresses exhibited well-defined WNW–ESE trajectories during the Miocene epoch in both the Iberian Ranges and Ebro basin (Simón 1982, 1986b; Arlegui & Simón 1993; Arlegui 1996; Liesa 2000). On the contrary, onshore fracturing during Pliocene–Quaternary times developed within a near-multidirectional extension regime, according to palaeostress analysis achieved from fault populations (Simón 1982, 1989; Arlegui 1996; Arlegui & Simón 1998; Cortés 1999; Liesa 2000). Multidirectional extension is also consistent with the style of macrostructures (domes, reactivated faults of varied directions) and has been interpreted as a consequence of crustal doming related to rifting at the eastern Iberian Ranges (Simón 1982, 1989).

Finally, a NNE–SSW extensional phase related to isostatic rebound of the Pyrenees might also be superimposed on the former stress systems within the central Ebro basin. This may be responsible for the development of a multiscale regional fracture set orientated WNW–ESE, including map-scale normal faults (tectolineaments) and penetrative jointing. The former are superimposed on a previous north–south regional joint set which reflects the roughly north–south orientated intraplate stress field (Arlegui 1996; Arlegui & Simón 2001).

The Central Range (RC)

The Central Range is a NE–SW orientated mountain range, in places rising to a height of over 2500 m, in which the Variscan basement has been uplifted by reverse faults during Alpine compression (Fig. 15.21). Unlike the Iberian Ranges, the Central Range has no Mesozoic sediments of significant thickness and the compressive structure includes only basement blocks (Figs 15.21 and 15.22), so that it was not produced by the inversion of a previous extensional Mesozoic basin. This distinctive mountain range has an average width of 80 km, and runs for 480 km southwestwards from the Iberian Ranges into Portugal. The interpretation of the structure has been much debated and several models have been proposed (Warburton & Alvarez 1989; Banks & Warburton 1991; Vegas et al. 1990; Ribeiro et al. 1990a). The isolated position of the Central Range in the centre of the peninsula, and the structure and kinematics of this unit can only be explained by considering

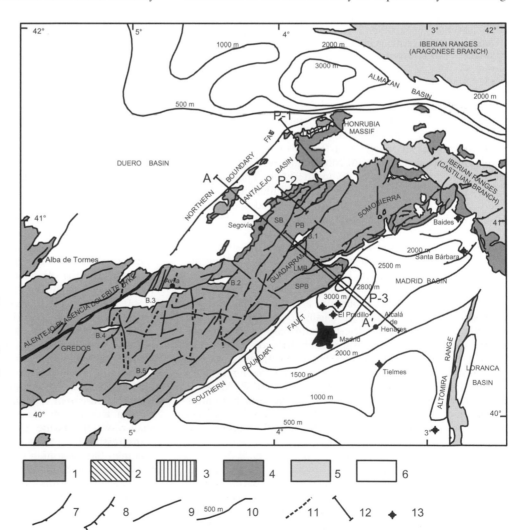

Fig. 15.21. Main features of the Central Range and neighbouring Tertiary basins. Legend: 1, Variscan basement; 2, Triassic sediments; 3, Jurassic formations; 4, Upper Cretaceous limestones; 5, Iberian Ranges units; 6, Tertiary sediments; 7, reverse fault; 8, normal fault; 9, undifferentiated fault; 10, basement depth contour; 11, alkaline Mesozoic dyke; 12, structural profile; 13, exploration oil well. Cross-section A–A′ is shown in Figure 15.23. P-1, P-2 and P-3 are partial profiles making up Figure 15.22a. Basins: B.1, Lozoya tertiary basin; B.2, Campo Azálvaro basin; B.3, Ambles basin; B.4, high Alberche basin; B.5, Tietar basin. Blocks: LMB, La Maliciosa block; PB, Peñalara block; SB, Segovia block; SPB, San Pedro block.

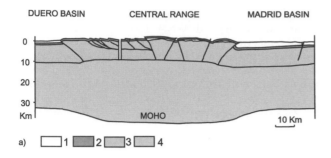

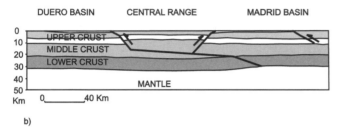

Fig. 15.22. Two interpretative structural profiles in the Central Range. (**a**) Faulted basement blocks and draped folds in the upper crust with a thickened middle and lower crust. Interpretation from Gonzalez Casado & De Vicente (1996) 1 = Tertiary sediments; 2 = Cretaceous marine sedimentary rocks; 3 = Brittle upper crust; 4 = Ductile lower crust. (**b**) Gravity model of Central Range from Tejero *et al.* (1996), with the thickened middle and lower crust and a lithospheric elastic layer supporting the extra weight.

how Alpine compressive stress was transmitted to the interior of the continent. These compressive stresses continue to act at the present day, with the basal detachment presumably located within or beneath the brittle upper crust.

On both sides of the Central Range there are two large sedimentary basins filled with continental Tertiary materials derived from the erosion of the uplifted belt. These are the Duero and Madrid basins in whose sedimentary successions can be found, in the form of clastic wedges and unconformities, a stratigraphic record of the tectonic events that uplifted the Central Range. The Duero basin record is the more complex because although it can be viewed as a subcircular intracratonic basin within the Iberian Massif, along its northern boundary and in its eastern extension into the Almazán Basin (Fig. 15.21) it is more like two foreland basins related to the Cantabrian mountains and the Iberian Ranges. The Madrid basin (really a sub-basin of the Tajo basin: see Chapter 13) is an asymmetric foreland basin, with a depocentre running along the southern border of the Central Range, and contains up to 3500 m of Tertiary sediments. Within the Central Range there are several smaller Tertiary basins (Lozoya, Campo Azálvaro, Amblés, High Alberche and Tietar) that are filled with clastic sediments and bounded by reverse faults. These smaller basins provide valuable information on the uplift history of the mountain range.

Pre-Tertiary events

The Central Range basement is part of the central Iberian zone of the Iberian Massif, and comprises Palaeozoic metasediments and orthogneisses that were metamorphosed and deformed in the Variscan orogeny and intruded by large

granitoid bodies. In the western parts of the Central Range these granites constitute most of the outcrop (Bellido *et al.* 1981). In addition to several dyke swarms linked to both the granitoid emplacement and late Variscan faulting (Capote *et al.* 1987; Doblas *et al.* 1994b), there is a group of younger, Mesozoic dykes. These include north–south striking alkaline dykes, mainly syenites and lamprophyres, in the middle part of the Central Range (Avila region; Nuez *et al.* 1981; Villaseca & Nuez 1986). This dyke swarm is linked to the same phase of extension that produced the early grabens in the Mesozoic basins of the Pyrenees and the Iberian Ranges (Alvaro *et al.* 1979). In addition, there is a series of NE–SW basic dykes, the most prominent of which is the Alentejo–Plasencia doleritic dyke. This is a great dyke (200–300 m thick) with a total length of 500 km, running from the SW Portuguese coast to the Central Range, crossing it from Plasencia to near Avila. Locally, such as in the area west of Avila, the dyke is coarse grained enough to produce gabbroic textures (De Figuerola *et al.* 1974). K-Ar and ^{40}Ar/^{39}Ar data have yielded discrepant ages of 168–148 Ma (Middle Jurassic) and 203 ± 2 Ma (Early Jurassic) respectively (Schermerhorn *et al.* 1978; Dunn *et al.* 1998). The emplacement of this great dyke is linked to the early opening of the Atlantic ocean (Schermerhorn *et al.* 1978; Vegas 2000), along a line later reactivated during early and middle Miocene times when the dyke outcrop became associated with a series of sinistral strike-slip faults creating small pull-apart Tertiary basins (Capote *et al.* 1996).

There is not much information on pre-Upper Cretaceous sedimentation in the Central Range area. We know that in the eastern boundary a number of faults separate an Iberian zone to the east (with Triassic and Jurassic rocks), from a Central Range zone in which only Triassic sediments are preserved under Cretaceous formations (Capote *et al.* 1982). Jurassic exposures around the Honrubia Massif (Fig. 15.21) show that the Central Range was once partially covered by shallow-water Jurassic limestones, which have now disappeared due to erosion. This erosive period was related to Late Jurassic and Early Cretaceous tectonic events known as the Neokimmeran and Austrian movements, with erosion of Jurassic strata occurring on uplifted fault blocks reactivating both NE–SW and NW–SE fault sets (Salas & Casas 1993). Fission track analysis (Sell *et al.* 1995; Bruijne & Andriessen 2000; Bruijne 2001) on apatites from Central Range rocks shows two Mesozoic accelerated cooling events, one at 120 ± 20 Ma BP and the other at 88 ± 20 Ma BP, which correspond to two stages of uplift and erosion in the Central Range. Uplift models point to a cooling rate of 20°C/Ma and to an erosion of 3 km of Triassic and Jurassic sedimentary cover (Bruijne & Andriessen 2000). Hydrothermal alteration in basement dykes, dated at 100 Ma BP (Caballero *et al.* 1992), is interpreted to have been a consequence of these Cretaceous tectonic phases.

Upper Cretaceous sedimentary formations, transgressive over the basement, are found along the boundaries of the eastern half of the Central Range. These sandstones and shallow-water marine carbonates were deposited in two transgressive–regressive cycles, one Cenomanian–Turonian and the other Senonian in age (Alonso 1981; Gil *et al.* 1993). Oil exploration drilling has proved similar Upper Cretaceous thin carbonate formations in the Madrid basin, lying directly on the Variscan basement (Querol 1989).

Tertiary sedimentation

Latest-Cretaceous marine regression across the central region of the Iberian peninsula resulted in continental sedimentation, the record of which shows it to have been organized in pre-, syn- and post-orogenic sequences. In the Duero basin (Santisteban *et al.* 1996*a,b*; Martín-Serrano *et al.* 1996) the first of these sequences (pre-orogenic complex: Late Cretaceous to Palaeocene), comprises siliciclastic, carbonate and evaporitic continental sediments in the SW sector and some marine deposits towards the east. The second sequence (synorogenic complex: Eocene to Oligocene), is formed by three clastic sequences separated by unconformities, and shows a general coarsening-upward sequence. The youngest of these tectonosedimentary units (post-orogenic complex: Miocene to Quaternary) comprises sandstones, limestones and evaporites in a generally fining-upward sequence (Mediavilla *et al.* 1996).

The Madrid basin has a triangular shape with its southern boundary defined by the uplifted Montes de Toledo basement block. The Madrid basin is limited to the NW and east by two compressive units, each with very different characteristics. Along its eastern boundary the frontal thrust-and-fold belt of Sierra de Altomira is faulted over the Tertiary sediments of the Madrid basin, producing a piggy-back displacement of the Loranca basin, a Tertiary foreland basin of the Iberian Ranges. The Central Range forms the northwestern boundary which is defined by a long and complex reverse fault line known as the southern boundary fault. The western sector of the Madrid basin can be considered as a foreland basin of the Central Range, with a deep depocentre where Tertiary sediments reach up to 3500 m thick close to the southern boundary fault, and with a forebulge in the central zone, to the east of the Tielmes oil exploration well (Figs 15.21 and 15.23). This geometry is related to the depression of the lithosphere under the weight of the pop-up unit and the clastic wedge produced by erosion of the rising blocks.

Exploratory drilling for oil in the centre and at the western margin of the Madrid basin (El Pradillo, Tielmes, Baides and Santa Bárbara wells; Figs 15.21 and 15.23) revealed a two-unit pre-orogenic succession under Miocene strata. A lower unit is formed by clay with anhydrite, with a sediment transport direction from south to north. Above this is an unconformity overlain by an upper, coarser-grained unit comprising a clastic

marginal facies and anhydrites interbedded with clays in the central areas, with a change in sediment transport direction which becomes centripetal into the basin. The overlying Miocene formations have been divided into three units (Calvo *et al.* 1996). A lower unit (upper Agenian–lower Aragonian) shows conglomerates and sandstones at the basin margins, grading laterally towards the centre into mudstones, carbonates and finally evaporites. An intermediate unit (middle Aragonian–lower Vallesian) is separated form the latter by an unconformity at the boundaries and by a prominent disconformity in the central areas (Capote & Carro 1968; Calvo *et al.* 1996). The upper unit (Turolian), comprises mainly lacustrine carbonates underlain by fluvial clastic sediments lying on the disconformity.

Finally, within the Central Range itself are various smaller basins that include both pre-orogenic and synorogenic sediments (Fig. 15.21). The former belong to the so-called 'pre-arkosic cycle' (Upper Cretaceous–Palaeocene), and are siliceous detrital sediments with high iron content (siderolitic materials). The younger, synorogenic sediments are sandstones and conglomerates of the so-called synorogenic 'arkosic cycle' (Eocene–Oligocene).

Stress fields and Cenozoic tectonic episodes

The first studies intending to establish the stress fields in the centre of the peninsula during Alpine compression were based on the assessment of the orientation of stylolites formed in carbonate formations of Early to Middle Jurassic age (Alvaro & Capote 1973; Alvaro 1975; Capote *et al.* 1982). The orientation of the stylolitic peaks in the Iberian Ranges was used to define several shortening directions that have been confirmed by studies of calcite twins (González Casado & García Cuevas 1999). A new phase of the study was the application of fault population analysis methods to the central regions of the peninsula in order to define the stress tensors (Simón 1984, 1986*b*; De Vicente 1988; De Vicente *et al.* 1996*b*; Muñoz & De Vicente 1996, 1998*a*; Liesa 2000). The application of these methods in the Central Range resulted in the elucidation of three distinct tectonic episodes (Capote *et al.* 1982, 1990). The first of these (Iberian stage or Castilian phase; Aguirre *et al.* 1976) was Oligocene–early Miocene in age and had a NE–SW to NNE–SSW shortening direction. The second episode (Guadarrama stage or Neocastilian phase) was middle-late

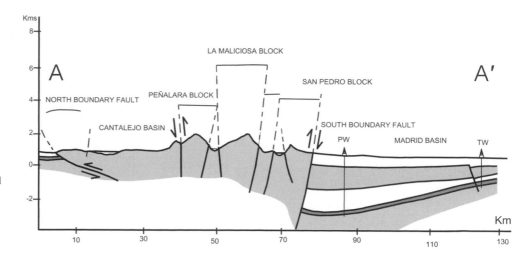

Fig. 15.23. Interpretation of basement block uplifting in Central Range from fission track data (location of profile shown in Fig. 15.21). Key: PW, Pradillo oil well; TW, Tielmes oil well. Modified from Bruijn & Andriessen (2000).

Miocene (intra-Aragonian) in age and characterized by NW–SE shortening. This latter stage induced reverse movements on NE–SW faults, and is the best represented stress field in the Central Range. Finally, the late Miocene Torrelaguna stage can be considered as a weakened continuation of the previous stress field, with a compression direction varying from NW–SE to north–south. It affects Miocene deposits and appears to have been active from latest-Miocene to present times. Studies made during a high-speed railway line (AVE) construction project through the Guadarrama Range have confirmed the current existence of a NW–SE compressive stress. The first and last of these three tectonic episodes are recorded by fission track data from the Guadarrama Range (Sell *et al.* 1995; De Bruijne & Andriessen 2000; De Bruijne 2001).

Structure and kinematics

There is widespread agreement that faults active during Alpine compression included late-Variscan structures reactivated in Tertiary times by new stress fields. In the Guadarrama–Somosierra sector, the general structure is formed of blocks uplifted by reverse NE–SW trending faults, with those in the centre of the range being the highest. Between Segovia and Honrubia (Fig. 15.23) the faults which control the blocks have small relative motions and the whole area is not very deeply eroded, with Cretaceous strata forming monoclines which in some cases have been interpreted as fault-propagation folds (Gómez Ortiz & Babín 1996, 1998). The geometry of the structures can be very well controlled and it is in this sector that the best profiles have been performed and the shortening has been confidently calculated (Fig. 15.22). To the SE the structure is somewhat more complex, although the presence of faulted and folded Cretaceous and Palaeogene strata again provides useful information on the deformation history (Sanchez Serrano *et al.* 1993). An initial model proposed for the deep structure argued for NW directed crustal thrusts rooting in a detachment in the base of the upper crust. The model envisaged imbricated backthrusts in the zone of the Sierra de Guadarrama and other detachments in the thin sedimentary cover along the northern boundary, giving rise to a stacked repetition of Cretaceous formations (Warburton & Alvarez 1989; Banks & Warburton 1991). However, such a model does not fit the field data provided by well exposed sections (De Vicente *et al.* 1996*c*) and an interpretation is preferred in which high-angle reverse faults in the basement go down to a middle and lower crust which was thickened by ductile deformation (Fig. 15.22a). In the Segovia sector, however, a shallower (6–7 km deep) main detachment surface is more likely.

The central blocks, forming the highest part of the sierras, have suffered strong erosion, eliminating the Cretaceous deposits so that structural profiles lack control. Only in some interior basins is it possible to control the geometry of the structures because Cretaceous and Tertiary sediments are preserved, as seen in the Lozoya basin (Guadarrama area), where these materials are clearly overthrust by the crystalline basement (Goicoechea *et al.* 1991). However, fission track analysis from the Guadarrama samples (De Bruijne 2001) has yielded data that help establish the uplift rate, and how much erosion has taken place. Uplift was greater in the centre and southern boundary of the Guadarrama area, so that the southern boundary fault is at a higher angle than the northern boundary fault, resulting in an asymmetric pop-up geometry. The thickness of eroded material in the uplifted blocks varies between 3.3 and 5.9 km, the latter figure corresponding to the central blocks (Fig. 15.23). The highest uplift and erosion rates occurred during the most recent phases of tectonism, starting 10 Ma ago, and have been continuing with an accelerating rate over the last 5 Ma (De Bruijne 2001). Along the SE boundary block the thermal history shows a slight heating because during late Eocene–early Oligocene times it was covered by at least 2 km of sediments. These sediments probably represent the clastic marginal facies that are missing in the Madrid basin, suggesting that the boundary of the Central Range migrated towards the basin during the Oligocene epoch.

Seismic refraction profiles and gravimetric models across the Central Range (Suriñach & Vegas 1988; Paulsen & Viser 1993; Tejero *et al.* 1996) show that the crust is slightly thickened, by about 2–4 km (Fig. 15.22b). The crust of central Spain typically comprises a 12 km thick upper crust (Banda *et al.* 1981) with three layers: an upper layer of sediments (0–3 km; 4.5 km/s P-wave velocity), a basement (3–7 km; 6.0 km/s), and an underlying low-velocity layer (7–12 km; 5.6 km/s). The middle crust lies at a depth of 12–22 km (6.4 km/s), and the lower crust lies between 22 and 31 km (6.8 km/s). This typical crustal structure is modified in the Central Range, which has a root about 3 km thicker across a zone 70 to 120 km wide. Furthermore, the upper limit of the middle crust is about 2–3 km higher than the level of the surrounding zones, indicating that it is here that thickening has occurred.

The thermal flux in the Central Range is about 70 mW/m^2 and in the Tertiary basins it is about 60 mW/m^2, to a large degree due to the radiogenic heat produced by the late Variscan granitoids and by the arkosic sediments in the Tertiary basins (Fernández *et al.* 1998). Crustal rheological profiles and gravity studies (Tejero & Ruiz 2000) both suggest that an effective elastic layer of 15 km thickness exists below the Central Range (Gómez-Ortiz 2001). This accords with a downwarping of the Moho (shown by seismic profiles) caused by the weight of the Central Range relief, which is supported by the elastic part of the lithosphere.

J.A.M. thanks O. Fernández and J. Gallart for their collaboration and gratefully acknowledges financial support from DGSIC (PB97-0882-CO3-03 project) and Dirección General de Investigación (MCYT, REN 2001-1734-COS-03 project).

16 Alpine tectonics II: Betic Cordillera and Balearic Islands

JOSÉ MIGUEL AZAÑÓN, JESÚS GALINDO-ZALDÍVAR,
VICTOR GARCÍA-DUEÑAS & ANTONIO JABALOY

The Betic and Rif cordilleras, lying to the north and south of the Alborán sea, form an arc-shaped mountain belt joining across the Straits of Gibraltar. The arc developed during, and partly in response to, late Mesozoic to Cenozoic convergence between Africa and Iberia. Three main pre-Miocene tectonic domains have been identified within the arc (Fig. 16.1a). The first of these represents the palaeomargins of the southern part of the Iberian plate and the northern (Maghrebian) part of the African plate. Both palaeomargins comprise autochthonous, parautochthonous, and/or allochthonous non-metamorphic Mesozoic and Tertiary cover overlying a Variscan basement. These palaeomargins were deformed in response to Alpine events and now form the External Zones of the two cordilleras. The second major tectonic domain comprises deformed Cretaceous to Miocene deep-water 'flysch' sediments located in the western Betics and along the northern part of Africa from the Gibraltar Strait to the Kabylies (Fig. 16.1; e.g. Durand-Delga 1980). These 'flysch' sediments are thought to have been deposited in a basin located between the palaeomargins of Iberia and Africa and the rocks that form the internal part of the mountain belt (Balanyá & García-Dueñas 1987). The third major tectonic unit is known as the Alborán domain or Internal Zones (Balanyá & García-Dueñas 1987), and mainly comprises three nappe complexes of variable metamorphic grade, which are, from bottom to top, the Nevado-Filábride, the Alpujárride and the low-grade Maláguide complexes (Fig. 16.1). In addition, sedimentary rocks known as the 'dorsal' and 'pre-dorsal' units of Triassic to lower Neogene age are considered as part of the Internal Zones, and crop out structurally between the flysch units and the Malaguide complex.

The Guadalquivir basin is located to the north of the Betic Cordillera (Fig. 16.1). This basin is filled by Neogene to Quaternary rocks and is traditionally interpreted as the Betic foreland basin. Northwards, Guadalquivir basin sediments lie unconformably over pre-Palaeozoic and Palaeozoic rocks of the Iberian Massif, which rise from the orogen-parallel hills of the Sierra Morena, interpreted as the result of the flexure of the lithosphere under the weight of the Alpine orogen (van der Beek & Cloetingh 1992).

Early Miocene collision of the Alborán domain and flysch units with the south Iberian and Maghrebian palaeomargins induced the development of a thin-skinned External Zone foreland fold-and-thrust belt (Balanyá & García-Dueñas 1987). At the same time, rifting and crustal extension affected the whole Internal Zones, and extensional fault systems developed (Platt & Vissers 1989; Comas et al. 1992; García-Dueñas et al. 1992; Jabaloy et al. 1992; Watts et al. 1993). Subsequently, from late Tortonian to Pliocene times, the Alborán region underwent north–south to NW–SE compression, which produced the emergence of part of the Miocene Alborán basin and both folding (mainly open kilometre-scale east–west striking folds) and faulting of the previous extensional systems (Weijermars et al. 1985; Comas et al. 1992; García-Dueñas et al. 1992; Rodríguez-Fernández & Martín-Penela 1993).

Geophysical data

Since the 1970s, the deep structure of the Betic Cordillera has been investigated by various geophysical techniques. Seismic refraction profiles for the Betic Cordillera and the Alborán sea indicate sharp differences between the crusts of the two regions. The continental crust of the Alborán sea has a minimum thickness of 16 km and lies over an anomalous mantle with a reduced propagation velocity (Hatzfeld 1976; WGDSSA 1978; Suriñach & Vegas 1993). In contrast, the continental crust of the central Betic Cordillera lies over a normal mantle and reaches thicknesses of up to 38 km in the central part (WGDSSS 1977; Banda et al. 1993) and up to 35 km in the Straits of Gibraltar (WGDSSA 1978). Towards the eastern Betic Cordillera there is a decrease in crustal thickness (Banda & Ansorge 1980; Banda et al. 1993) that is also confirmed by magnetotelluric data (Pous et al. 1999). This change in thickness can be abrupt (Banda & Ansorge 1980) or progressive (Banda et al. 1993). Towards the Iberian Massif, the crust gradually diminishes in thickness down to 30–32 km (Suriñach & Vegas 1988; ILIHA DSS Group 1993).

Gravimetric studies (Hatzfeld 1976; Casas & Carbó 1990) have confirmed the distribution of the crustal thicknesses deduced from the seismic data. In addition, more recent works (Torné & Banda 1992; Galindo-Zaldívar et al. 1997; Torné et al. 2000) have reported that the variation in crustal thickness between the Betic Cordillera and the Alborán sea is very abrupt in areas near the current coastline.

Most of the large Neogene basins of the Betic Cordillera and Alborán sea have been studied by conventional seismic profiling focusing mainly on their sedimentary infill. In addition, the ESCIBETICAS deep seismic reflection profiles (García-Dueñas et al. 1994) reveal that the crust in the central cordillera has a nearly transparent upper level (up to approximately 5 seconds two-way travel-time (5 s TWT), corresponding to 15–20 km depth, depending on the seismic velocities used). In contrast, the lower crust has several reflectors ending in a prominent band around 11 s TWT, which is interpreted as the Moho, and the upper mantle in the region is also reflective. In spite of large, kilometric folds at the surface (Weijermars 1985), the reflectors in the lower crust and the Moho are subhorizontal, indicating a crustal detachment at the base of the compensation level of the folds, probably at around 10–15 km (Galindo-Zaldívar et al. 1997; Martínez-Martínez et al. 1998). This detachment probably coincides with the contact at 10 km determined by seismic refraction profiles in the central and eastern Betic Internal Zones (Banda et al. 1993). Studies of earthquakes in the Granada basin (Morales et al. 1997) indicate that seismicity concentrates at that depth, probably also coinciding with the transition between brittle and ductile deformation conditions.

Aeromagnetic anomalies within the Iberian Massif (Ardizone et al. 1989) extend southwards beneath the Guadalquivir basin and External Zones (Galindo-Zaldívar et al. 1997), suggesting at least partial continuity of the Iberian basement below the Betic Cordillera. Seismic tomography research recently performed in the Betic Cordillera and the Alborán sea has led to the proposal of a model involving continental subduction of the Iberian Massif beneath the Betic Cordillera (Serrano et al. 1998; Morales et al. 1999) and the consequent generation of intermediate seismicity (30–110 km deep) in the Málaga region (Fig. 16.2). These data indicate that the deep structure of the cordillera is also heterogeneous: whereas in the western cordillera the subducting plate dips 45° SSE and is currently active, in the central cordillera the continental crust of the Iberian Massif is directly in contact with Internal Zones rocks (upper plate), as suggested by the deep seismic reflection profiles.

From tomographic studies within the mantle, Blanco & Spakman (1993b) identify a high-velocity ('cold') body below the Betic Cordillera, which has associated deep seismicity at depths of 640–670 km. The origin of this anomaly remains controversial. Blanco & Spakman (1993b), Lonergan & White (1997) and Bijwaard et al. (1998) proposed that it is due to a detached slab of subducted oceanic lithosphere. In contrast, Calvert et al. (2000) suggest that it represents delaminated upper mantle, as previously suggested by Seber et al. (1996a).

External Zones (South Iberian domain)

The northern, least metamorphosed, most 'external' part of the Betic Cordillera can be divided broadly into the Prebetic and Subbetic zones (Fig. 16.1). The Prebetic zone crops out mainly in the eastern and, to a lesser extent, in the central sectors of the cordillera. The rocks are mainly Triassic to lower Tortonian platform carbonates, although there are some continental siliciclastic deposits. The Subbetic zone crops out further south, as an ENE–WSW orientated belt comprising rocks of Triassic to early Burdigalian age, which are mostly carbonates with some basalt. Whereas the rocks of Triassic to Early Jurassic age show continental or shallow marine facies, those of Middle to Late Jurassic age show shallow-platform or pelagic facies (Chapter 11; García-Hernández et al. 1980).

In addition to the stratigraphical and palaeogeographical interpretation of the Subbetic zone in terms of a series of fault block highs and pelagic basins (Chapter 11), any structural analysis must focus on the fact that approximately half of its outcrops are olistolithic fragments of different Mesozoic and Cenozoic rocks of varying size embedded in a matrix. The matrix consists principally of middle Miocene marls in the western sector and around the Guadalquivir basin, and of Triassic and Keuper facies rocks in the central and eastern sectors of the cordillera. In some areas near Antequera, this matrix also includes fragments of metamorphic rocks and peridotite (Morata 1993). In addition to these abundant olistostromes, there are also outcrops in which the units are more stratigraphically coherent, such as around Ronda in the western sector of the cordillera, north of Granada and Loja in the central sector, and most of the eastern sector.

Mesozoic and Palaeogene evolution

Several, superposed rifting stages controlled the Mesozoic evolution of the External Zones (south Iberian domain). Basic volcanism in the Keuper facies of the cordillera suggests that extension started as early as Triassic (Morata 1993). The first definite rifting stage began in the middle Carixian (lower part of the Pleinsbachian stage) and caused the break-up of the shallow carbonate platform into a series of fault blocks separated by deeper basin areas. The original orientation of these subsiding basins seems to have been north–south, a direction

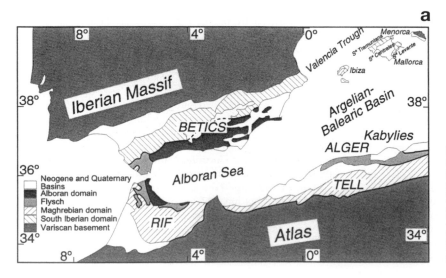

a

Fig. 16.1. (a) Geological setting of the Betic Cordillera in the western Mediterranean. (b) Geological sketch map of the Betic Cordillera. Legend: a, Iberian Massif; b, Subbetic zone; c, Prerif and Mesorif; d, Prebetic zone; e, olistostromes and breccias; f, flysch; g, Nevado-Filábride complex; h, Alpujárride complex (p, peridotites); i, Malaguide; j, flysch; k, Neogene basins; dots, volcanic rocks. Lines labelled 3, 6A, 6B show locations of sections in Figures 16.3 and 16.6.

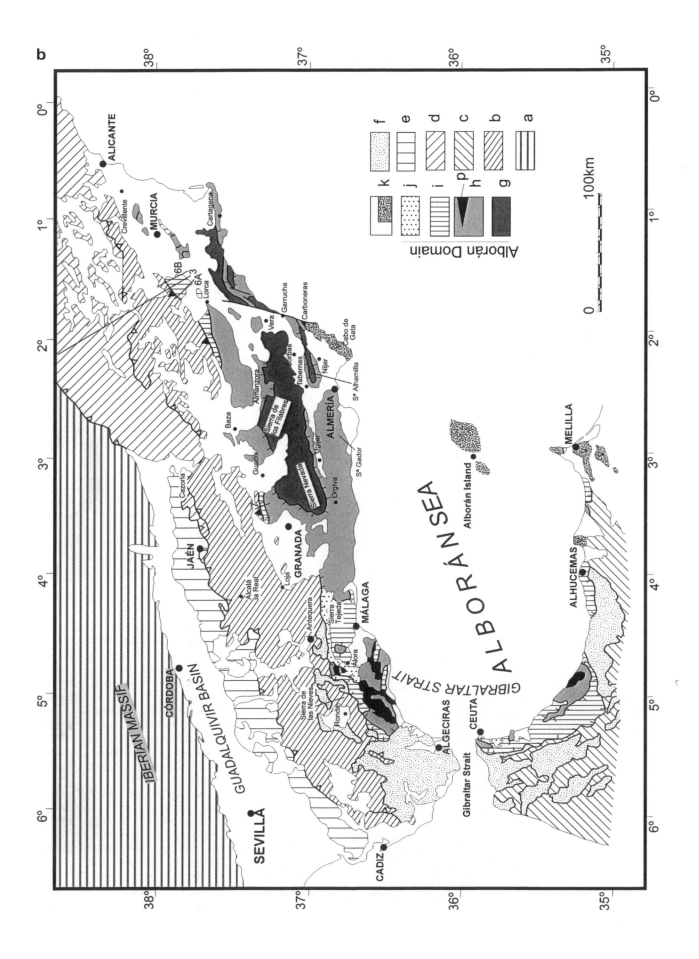

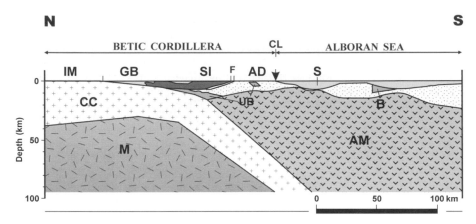

Fig. 16.2. Main features of the deep structure of the Betic Cordillera and Alborán sea along a north–south cross-section through the Málaga meridian. Modified from Morales *et al.* (1999). Key: AD, Alborán domain; AM, anomalous mantle; B, basic rocks; CC, continental crust; CL, coastline; F, flysch; GB, Guadalquivir basin; IM, Iberian Massif; M, mantle; S, Neogene sediments of the Alborán sea; SI, south Iberian domain; UB, ultrabasic rocks.

still preserved by facies changes in those areas not affected by subsequent vertical axis rotation (e.g. Osete *et al.* 1988; Platzman 1992). Few data exist, however, on the original structural controls of the zone, except in the east, where Rey (1993) has reported some high-angle synsedimentary normal faults of this age. This extension stage seems to have been active up until Middle Jurassic time.

The next rifting stage in the eastern sector of the cordillera took place during the Late Jurassic to Late Cretaceous interval (Vilas *et al.* 2001). This rifting stage produced high-angle normal faults that gave rise to half-graben basins. In the Prebetic zone, these normal faults have a NNE–SSW trend (Vilas *et al.* 2001). In the Subbetic zone, the normal faults generated in this stage have two sets of directions, east–west to NNW–SSE, which, after correcting for the vertical axis rotations, would have had an approximate original direction of north–south to ENE–WSW (E. Fernández-Fernández pers. comm.). This extensional stage produced tectonic subsidence that placed the sea floor near the carbonate compensation level during Aptian-Albian times (E. Fernández-Fernández pers. comm.), as indicated by the Fardes Formation (Fm) deposits (Reicherter 1994). It is likely that this rifting stage is responsible for the formation of the listric fault that separated the Prebetic from Intermediate units during this period and which has been described in the seismic profiles of the eastern region by Banks & Warburton (1991) (LF in Fig. 16.3). This listric fault in places allowed the deposition of nearly 3 km of sediments during Early Cretaceous times.

During Jurassic times, the central Subbetic zone also underwent several stages of volcanism and intrusion of basic igneous rocks with an intraplate alkaline affinity localized along an ENE–WSW belt that is about 100 km long and just a few kilometres wide (Morata 1993). The age of an associated prehnite–pumpellite facies metamorphism remains unknown, although it might be broadly linked to high heat flow during crustal extension. It has also been shown that, during these rifting stages, the evaporitic Keuper facies Triassic rocks underwent diapirism, intruding overlying Jurassic–Cretaceous deposits (Nieto *et al.* 2001). These diapirs occasionally reached the sea floor, producing rock deposits with Triassic facies intercalated with Cretaceous rocks (Sanz de Galdeano 1973; Nieto *et al.* 2001). The Palaeogene evolution of this margin is very poorly known. Contractional deformation has been proposed, but no associated structures have been reported.

Neogene evolution of the Subbetic zone

The overall structure of the Subbetic zone (Figs 16.1b and 16.3) includes several tectonic units that thrust to the NW on the northern side and to the SE on the southern side, forming a broadly synformal structure (Banks & Warburton 1991). Most studies (i.e. Banks & Warburton 1991; Allerton *et al.* 1993; Galindo-Zaldívar *et al.* 1997) contend that the thrusts in the northern limb continue beneath the cordillera's Internal Zones (Fig. 16.3). On the other hand, De Smet's model (1984) proposes that the Subbetic zone is a large flower structure associated with the movement of a strike-slip fault in the basement.

The general structure of the western sector of the Subbetics is a NW-vergent fold-and-thrust belt (Kirker & Platt 1998; Crespo-Blanc & Campos 2001) that deforms sediments of early Burdigalian age. The folds have a predominantly NE–SW trend and usually verge toward the NW, though several box folds can be observed (Kirker & Platt 1998; Crespo-Blanc & Campos 2001). The geometry of these folds is highly dependent on the rock rheology: the massive Jurassic limestones and dolostones define folds with 3–5 km wavelengths, while the marly Cretaceous to Palaeogene rocks define decimetric-scale folds with which is associated a cleavage (Crespo-Blanc & Campos 2001). These folds are cut by a thrust system with a sole thrust in the Keuper facies of the Triassic rocks. Kinematic criteria indicate a top-to-the-NW sense of movement for this system. Several out-of-sequence thrusts cut both fold limbs (Crespo-Blanc & Campos 2001). Later deformations include normal faults and late folding that produced very open NE–SW folds probably during the late Tortonian to late Messinian interval (Crespo-Blanc & Campos 2001). An important characteristic of this region is that the Internal Zones lie above the External Zones on a low-angle fault zone whose kinematics are under debate (Balanyá *et al.* 1997; Kirker & Platt 1998).

In the eastern sector, the southern Subbetic units were thrust SE over the flysch and the Internal Zones during middle Burdigalian times (Lonergan 1993). The structures are very similar to those of the western sector of the cordillera, but the vergence is opposite, with a SE-vergent fold-and-thrust system that is early to middle Burdigalian in age (Lonergan 1993). Fold trends describe an arc from NNE–SSW to ENE–WSW. They verge towards the SSE to SE, and feature a cleavage related with the rock rheology in the Cretaceous to

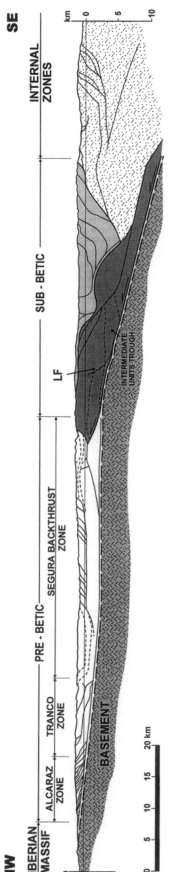

Fig. 16.3. Balanced cross-section of the eastern Betic Cordillera. Location shown in Figure 16.1. Modified from Banks & Warburton (1991). Subbetic units are grey; intermediate units are dark grey.

Palaeogene marls. Thrusts have a top-to-the-SE sense of movement (Lonergan 1993) and cut the previous folds.

The record of the most recent deformation episodes is well observed in the central sector of the cordillera where there are folds of Burdigalian age. These folds generally verge southwards in the southern half of the central Subbetic zone, and northwards in the northern part. In the intermediate zones where there is a change in vergence, folds with several hinges (mushroom-shaped folds) can be found (García-Dueñas 1967). WSW-directed faulting related to extensional detachments in the northern limb of the Sierra Tejeda antiform cuts both limbs of these Burdigalian folds (Galindo-Zaldívar *et al.* 2000). The fault surfaces are also cut by later reverse faults of Tortonian age, with a NNW sense of movement of the hanging-wall (Galindo-Zaldívar *et al.* 2000).

At the front of the orogenic belt, Subbetic breccias overlie early Tortonian Guadalquivir basin sediments on a system of trailing imbricate fan thrusts (Blankenship 1992; Roldán-García 1995), in turn overlain by late Tortonian to Messinian sediments (Roldán-García 1995). Normal faults in this area have a wide range of strikes and may indicate late radial extension.

Across the whole Subbetic Zone the hinges of the Burdigalian folds define several arcuate structures in plan view. A number of palaeomagnetic studies of the Subbetic Zone have focused on these rotations (Allerton *et al.* 1993; Osete *et al.* 1988, 1989; Platzman 1992, 1994). In the Subbetics most of the folds trend WSW–ENE, parallel to the major alignment of basalt in the middle Subbetics, but Osete *et al.* (1988, 1989) show that the basalts and the folds have undergone clockwise vertical axis rotation of approximately 60°. In areas where the folds have roughly north–south directions, however, the palaeomagnetic studies reveal small vertical axis rotations (Platzman 1994) or counterclockwise rotations around 10° (Allerton *et al.* 1993). These data suggest that the Burdigalian folds were generated with a roughly north–south trend (probably NNE–SSW) and were later rotated to their present-day orientations.

Balearic Islands

The Betic Cordillera continues northeastward towards the Balearic Islands. Its Mesozoic and younger rocks may be correlated with those of the Betic External Zones (South Iberian domain), and were deposited on the southeastern margin of the Iberian Massif. The present-day structure of the region is the result of overprinted extensional and compressional structures (Sabat *et al.* 1995; Gelabert-Ferrer 1998). In early and middle Miocene times, the Balearic Islands were separated from the Iberian peninsula during the development of the Valencia trough, while at the same time compressional structures in the Balearic promontory were active (Sabat *et al.* 1995). Since late Miocene times, the development of extensional structures in the southeastern margin of the Balearic promontory has been related to the development of the Algerian-Balearic basin, the margin of which has steep slopes and is known as the Emile Baudot scarp. The northeastern boundary of the Balearic promontory is defined by the margin of the Ligur-Provençal basin.

The Balearic promontory is a NE–SW orientated block of continental crust with several highs that crop out as the islands, the main highs forming, from SW to NE, Ibiza, Mallorca and Menorca. Crustal thickness is quite variable in the region (Gallart *et al.* 1995; Sabat *et al.* 1995). While the

Valencia trough is formed by thinned continental crust with a minimum thickness of 14–15 km, in the Balearic promontory the crustal thickness reaches 30 km (Gallart *et al.* 1995). To the SE there is crustal thinning towards the Algerian-Balearic basin, which is probably floored by oceanic crust (Gallart *et al.* 1995; Sabat *et al.* 1995).

Palaeozoic rocks (Silurian to Permian) crop out mainly in Menorca, and only locally in Mallorca. They most likely constitute a fragment of the Iberian Massif, being lithologically similar to the Palaeozoic rocks of Cerdeña, and bear no evident relation with the rocks that belong to the Internal Zones of the Betic Cordilleras. Most of the Balearic Islands are formed by Mesozoic and Cenozoic limestones and marls with interlayered siliciclastic beds, typical of the Betic External Zones. The Triassic rocks are of German-Andalusian facies, as also seen in the External Zones.

The main tectonic stages responsible for the present-day structure of the Balearic Islands started after a long period of predominantly extensional tectonics during the Mesozoic era. The first alpine deformation was a compressional stage that affected rocks up to late Oligocene–Langhian age, followed by an extensional episode (Sabat *et al.* 1988; Ramos-Guerrero 1989; Gelabert-Ferrer 1998). The alpine compression produced ENE–WSW orientated thrusting with a top-to-the-NW sense of movement, and the development of coeval strike-slip faults orientated NW–SE (Ramos-Guerrero 1989; Gelabert-Ferrer 1998).

Each of the Balearic Islands has its local tectonic characteristics. The structure of Mallorca features several NE–SW orientated horsts and grabens, with the main horsts being located in the Sierra de la Tramuntana, on the NW side of the island, in the lower relief Sierras Centrales, and in the Sierra de Levante, which marks the SE edge of the island. Although Hermite (1879) produced an early sketch of the structure of the Mallorca, the first important geological studies are those by Fallot (1922), who described the lithology and some tectonic aspects of the Sierra de la Tramuntana, and the research by Darder (1925) on the Sierra de Levante. The recent tectonic studies by Gelabert-Ferrer (1998) indicate that Mallorca was formed by the superposition of several thrust sheets of Mesozoic to Palaeogene rocks with NW vergence, and has ramp and flat structures typical of the Betic External Zones. Some of these thrust sheets may contain basement rocks of Palaeozoic age, although the Triassic rocks of Keuper facies act as a detachment level. The deformation progresses northeastward, as the structures of the Sierra de Levante are older than those of the Sierra de la Tramuntana, and the thrust sequence is essentially piggy-back. These thrust structures are cut by normal fault systems of post-Langhian age, some of them with listric geometry, that dip mainly towards the SE. In the Mallorca cross-section, the shortening associated with alpine compression was around 44%, and the subsequent extension was about 5% (Gelabert-Ferrer 1998).

Ibiza shows a tectonic style similar to that of Mallorca: it is made up of several superposed tectonic units (Fallot 1922; Rangheard 1971) favoured by detachment levels located in the Triassic rocks of Keuper facies. The general orientation of the thrusts and folds is NE–SW, and the vergence of the structures is towards the NW.

Menorca, meanwhile, is geologically very different from the other islands. It is formed mainly by a Mesozoic and Cenozoic cover detached from the basement at the Triassic rocks of Keuper facies. The Mesozoic rocks are deformed by thrusts and folds mainly verging towards the SW. This island is affected by a dense array of fractures, active at different stages during the Alpine orogeny, which were probably inherited from Variscan deformations (Bourrouilh 1973).

Campo de Gibraltar flysch units

Several tectonic units containing sedimentary sequences with turbiditic deposits ranging in age from Early Cretaceous to the Lower Miocene exist between the External and Internal zones. They are known as the Campo de Gibraltar flysch in Spain and as the Maghrebian flysch in the Rif, Kabylians and Sicily. Another name for them, the 'flysch trough domain', derives from palaeogeographic reconstructions supposing an origin in a deep basin.

In the Rif and Kabylias mountains, the sequences of the flysch units are traditionally divided into Massylian and Mauritanian units using palaeogeographic reconstructions. The Mauritanian units were located near the Internal Zones whereas the Massylian units were located near North Africa and External Zones (South Iberian domain). Whereas the Massylian sequences contain mostly Cretaceous turbidites, the Mauritanian units comprise a thick Oligocene turbidite sequence overlain by a siliciclastic 'Numidian flysch' that reaches up to Upper Burdigalian in age. The basements to each of these sequences crop out in the Rif and Sicily and comprise Jurassic E-MORB with pillow basalts indicating that the flysch sediments were deposited on an oceanic crust (Durand-Delgá *et al.* 2000). In the Betic Cordillera, however, there are no basement outcrops of the flysch domain and only detached sedimentary sequences can be observed within the main outcrop occurring in the west around Algeciras (Fig. 16.1b; Pendón 1978; Martín-Algarra 1987 and references herein).

The main tectonic unit in this western outcrop is the Aljibe unit (with rocks ranging from late Senonian to early Miocene). The lower part of this unit comprises green and red siltstones and calcareous sandstones alternating with siltstones. The upper part comprises sandstones and siltstones of the Aljibe Fm (equivalent to the Numidian flysch; Luján *et al.* 2000). The Aljibe unit has an internal structure characterized by folds associated with a thrust system with a top-to-the-west or -WNW sense of movement. Back-thrusts also developed with a top-to-the-east or -ESE sense of movement (Luján *et al.* 2000).

Three other tectonic units, called the Almarchal, Bolonia and Algeciras units, have been defined in this western outcrop of the flysch deposits. The Almarchal and Bolonia units are the continuation in the Betics of the well-developed Tanger and Tala Lakrah units respectively in the Rif. The two units have structures, both folds and thrusts, facing west or NW. In the Bolonia unit these structures are recumbent folds whereas in the Almarchal unit they are reverse faults and thrusts. However, the geometry of these units and their relationships with the Aljibe unit are currently poorly known.

Further NE, flysch units also crop out near to Ronda and Antequera in a narrow band between the Internal and External zones (Fig. 16.1b). They are usually located below Subbetic zone rocks and thrust over Internal Zone rocks. Additionally, the flysch units can also form part of olistostromes within a sedimentary complex known as the Alozaina complex (Balanyá & García-Dueñas 1987). This complex is upper Burdigalian in age, and is superposed over the Internal Zones by NW-directed mass sliding (Balanyá & García-Dueñas 1987; González-Lodeiro *et al.* 1996).

Internal Zones

Nevado-Filábride complex

Although the Nevado-Filábride complex constitutes the structurally lowermost metamorphic complex of the Betic Cordillera, the rocks are exposed at the highest present-day elevations within the mountain belt (reaching 3482 m at Mulhacén in the Sierra Nevada) as a consequence of young folding and uplift that has occurred since Late Miocene times. The Nevado-Filábride complex crops out mainly within large antiformal structures that form the Sierra Nevada, Sierra de los Filabres and Sierra Alhamilla (Figs 16.4 and 16.5). In the eastern Betic Cordillera, Nevado-Filábride rocks are found in isolated blocks which have been deformed by Neogene to Quaternary strike-slip faults.

The first geological research on the Nevado-Filábride rocks was done by Von Drasche (1879), who studied the upper slopes of Sierra Nevada. Subsequently Brouwer (1926) distinguished two main tectonic groups of rocks, the lower being the crystalline rocks of Sierra Nevada, comprising graphite-bearing schists and quartzites, and the upper being the 'Mischungzone', comprising a great variety of rocks including schists, marbles, gneiss and serpentinites. The term Nevado-Filábride was proposed by Egeler (1963), taking into account the similarities of the rocks that crop out in both the Sierra Nevada and in the Sierra de los Filabres (Fig. 16.4).

The Nevado-Filábride rocks can be divided into two lower and upper lithological sequences that are poorly dated. The lower sequence has been attributed to the Palaeozoic, based on the finds of Middle Devonian Chaetetes in the eastern sector of the cordillera (Lafuste & Pavillon 1976), and on Permian–Triassic Rb-Sr dates obtained for orthogneiss in Sierra de los Filabres (Priem et al. 1966). Most studies suggest a Permian–Mesozoic age for the upper sequence (Puga et al. 1975; Gómez-Pugnaire & Franz 1988; Galindo-Zaldívar et al. 1989), although recently Gómez-Pugnaire et al. (2000) have suggested a Palaeozoic age for the upper lithological units of the Nevado-Filábride complex based on metamorphic data.

Two members form the lower sequence, the lower of these being rich in graphite-bearing metapelites, whereas the upper member mainly comprises metasandstones, although rare outcrops of marbles and orthogneisses also occur. The

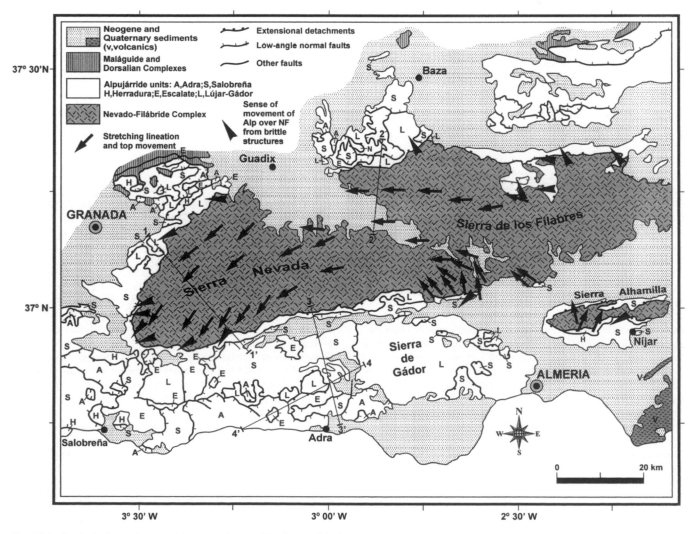

Fig. 16.4. Geological sketch map of the central part of the internal Betic Cordillera. Lines 1–1', 2–2', 3–3', 4–4' show locations of sections in Figure 16.5. Data from various sources cited in the text.

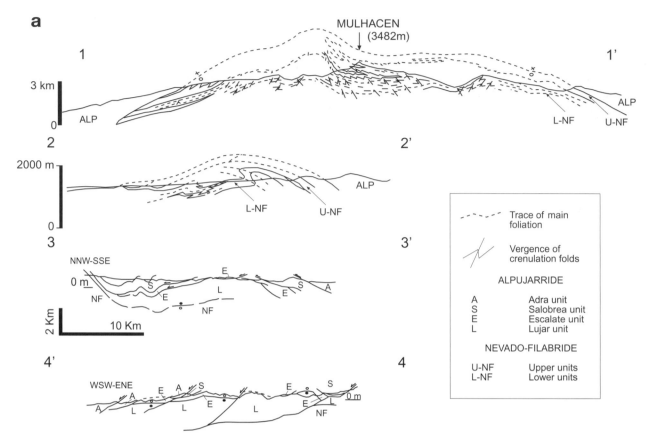

Fig. 16.5. (a) Representative cross-sections of the Nevado-Filábride complex structure in Sierra Nevada (1–1') and Sierra de los Filabres (2–2'), and of the Alpujárride complex west of Sierra de Gádor (3–3' and 4–4'). Locations are shown in Figure 16.4. **(b)** Lithostratigraphic sequences of the Alpujárride units in the central sector of the Cordillera, with representative mineral assemblages.

graphite-bearing schists of the lower member form the oldest exposed rocks in the Nevado-Filábride complex and are more than 2000 m thick, with an unexposed base. The lower part of this lower member is of greenschist to lower amphibolite facies, whereas the upper part records higher temperatures. The typical association of the rocks forming the lower part is: quartz + white mica + garnet (mainly almandine) ± plagioclase (albite and Na-oligoclase) ± chloritoid (Fe) ± chlorite ± epidote. There is no agreement on whether the metamorphism of these rocks is Alpine (Puga *et al.* 1975) or Variscan (Gómez-Pugnaire & Franz 1988). These rocks may have reached maximum temperatures of 400–550°C.

In the upper part of the lower member, alpine associations such as quartz + white mica + garnet (pyrope–glossularite) + kyanite + chloritoid (Fe-Mg) ± staurolite ± chlorite have been recognized (Gómez-Pugnaire & Franz 1988; De Jong 1993). Locally, in this upper part older high-temperature/low-pressure relict minerals can be found, with thermodynamic features incompatible with the above-described assemblage. These minerals are large, have been partially replaced, and consist of garnet (mainly almandine) + chloritoid (Fe) + andalusite + staurolite (Fe) + biotite ± cordierite (pseudomorphs). There is general agreement that this older assemblage is of Variscan age, although partially transformed by the alpine metamorphism (Puga 1976; Gómez-Pugnaire 1979; Gómez-Pugnaire & Sassi 1983).

The upper member of the lower sequence comprises alternating metapsammites, black slates and fine-grained micaschists, with rare intercalations of graphite-bearing marbles, and it has a maximum thickness of about 1200 m.

The upper sequence comprises the Las Casas marbles, overlain by the Tahal schists, and lies unconformably over the lower sequence (Nijhuis 1964), with deformed basal conglomerates preserved in some areas (Gómez-Pugnaire *et al.* 1981; Jabaloy 1991). This sequence can reach more than 1500 m in thickness and comprises light-coloured quartzites and conglomerates (more abundant at the base), and micaschists (predominant in the upper part). The sequence continues with the 800 m of Las Casas marbles, comprising intercalations of calcschists, schist and metabasites. In addition, metamorphosed igneous rocks (orthogneisses, eclogites, amphibolites, serpentinites and harzburgites) are observed, mainly in the upper part of the sequence. Metamorphic study of these rocks reveals a high-pressure overprint reaching up to 14–16 kbar and 600–650°C in metapelites (Gómez-Pugnaire & Camara 1990) and even more in the eclogites. The metamorphic conditions subsequently evolved to intermediate pressure (7 kbar) and temperature (550–600°C). The age of the metamorphism has been dated as 48.4 ± 2.2 Ma BP for the first event and 24.6 ± 3.6 Ma BP for the second one using the Ar^{39}/Ar^{40} method (Monié *et al.* 1991).

There has been a great deal of discussion about the

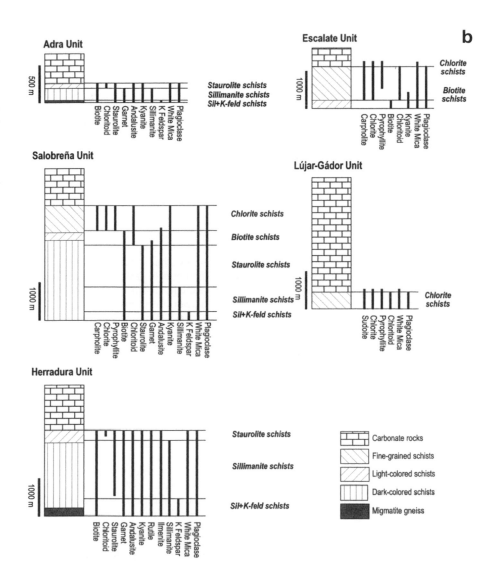

interpretation of the metabasic rocks, including the suggestion that they represent magmatism in a thinned continental crust (Gómez-Pugnaire 1979; Bakker *et al.* 1989) or the last remnants of an ophiolitic sequence (Bodinier *et al.* 1987). Whatever their origin, these rocks, dated as Middle to Late Jurassic (Hebeda *et al.* 1980), record a stage of intense extension and crustal thinning in the region.

Microstructural studies of the Nevado-Filábride rocks in the Sierra Nevada and Sierra de los Filabres (Martínez-Martínez 1984; García-Dueñas *et al.* 1988; Galindo-Zaldívar *et al.* 1989; Jabaloy *et al.* 1992, 1993; De Jong 1993), in Sierra Alhamilla (Platt & Behrman 1986) or in the eastern part of the cordillera (Alvarez 1987), among others, identify a main deformation stage. This deformation produced penetrative foliations, sometimes with stretching lineations, that post-date the metamorphic events. Structures older than this penetrative deformation have been mostly obliterated but are locally preserved as relict foliations and folds within metamorphic minerals. In the lower structural levels of the Nevado-Filábride complex, earlier structures consisting of folds with a related slaty cleavage have been preserved, probably because of the lower intensity of the main deformation phase (Jabaloy *et al.* 1993).

The main foliation probably formed with a slight eastward dip and was deformed by later folds. Stretching lineations show trends that rotate from NW–SE in the eastern regions of the cordillera to east–west in the Sierra de los Filabres and the central part of Sierra Nevada, and NE–SW in the western Sierra Nevada (Fig. 16.4; Platt & Behrman 1986; Frizon de Lamotte *et al.* 1991). The intensity of the fabric generally increased upwards, affecting a thickness of some 800 m of rocks in the eastern part of the cordillera, to more than 2000 m in Sierra Nevada. S-C ductile structures, quartz c-axis fabrics, asymmetric pressure shadows and mica fish indicate a top-to-the-west component for the non-coaxial strain of this fabric (Platt & Behrman 1986; García-Dueñas *et al.* 1988; Galindo-Zaldívar *et al.* 1989; Frizon de Lamotte *et al.* 1991; Jabaloy *et al.* 1993; Martínez-Martínez & Azañón 1997; Fig. 16.4). Downwards in the Nevado-Filábride complex, the fabric becomes more coaxial and stretching lineations cannot be recognized. This fabric is axial planar to recumbent folds which sometimes show sheath geometry and range in size from millimetric to kilometric, such as those observed in Sierra Nevada and Sierra de los Filabres (Fig. 16.5). The large folds generally have straight hinge lines and northward or northwestward vergences (García-Dueñas *et al.* 1988; Galindo-Zaldívar *et al.* 1989). The main deformation stage occurred in conditions of retrograde greenschist facies metamorphism, and radiometric dating suggests an age of 16–17 Ma for the end of this event (Monié *et al.* 1991).

The main foliation and related structures were subsequently deformed by folds with crenulation cleavages (with hinge lines generally subparallel to the previous stretching lineations) to produce hook-shaped interference figures. These folds can be up to kilometre-sized in Sierra de los Filabres (Fig. 16.5). Additionally, in the upper part of the Nevado-Filábride complex, there are well-developed ductile–brittle and brittle extensional crenulation cleavages (Platt & Vissers 1980) that mainly indicate a top-to-the-west sense of movement. These structures are more intense and brittle upwards, towards the contact with the Alpujárride complex.

All previous structures are cut by brittle fault gouges at the contact between the Nevado-Filábride and the Alpujárride complexes, which is generally known as the Mecina fault (González-Lodeiro *et al.* 1984). This fault has been interpreted as a low-angle normal fault with a top-to-the-west sense of movement (Figs 16.4 and 16.5; González-Lodeiro *et al.* 1984; García-Dueñas & Martínez-Martínez 1988; Galindo-Zaldívar *et al.* 1989; Jabaloy *et al.* 1993; Martínez-Martínez & Azañón 1997). In addition, tensional joints have developed with a main set whose strike is orthogonal to the slip motion of the Mecina Fault. The activity of this fault has produced the exhumation of the Nevado-Filábride rocks, first in the eastern Betic Cordillera (Serravallian) and later in the central regions (Tortonian; Johnson *et al.* 1997). Tortonian sediments in eastern Sierra Nevada overlie this fault. Overall, the deformation history of the Nevado-Filábride rocks may be related to the evolution of a crustal-scale shear zone located at the contact with the Alpujárride complex, with early ductile conditions becoming progressively more brittle with time.

The difference in metamorphic conditions, mainly in temperature, between the lower and the upper rocks of the Nevado-Filábride complex, has led several authors to propose that the major structure consists of two or three thrust nappes (Puga *et al.* 1975; Martínez-Martínez 1984; García-Dueñas *et al.* 1988). In the lower part of the Nevado-Filábride complex, where the main deformation is least intense, it is possible to recognize sharp contacts that can be attributed to thrust structures (Jabaloy *et al.* 1993). However, in the upper structural levels there is no agreement on the supporting evidence for the thrust nappe interpretation since all the rocks exhibit a mylonitic fabric younger than the metamorphic climax and the deformation intensity does not seem to be related to the contacts.

Alpujárride complex

The Alpujárride complex comprises several tectonic units that underwent stacking and subsequent extension, as indicated by the alpine metamorphic record (Torres Roldán 1981; Vissers *et al.* 1995; Azañón *et al.* 1997; Balanyá *et al.* 1997). The lithostratigraphic sequence within each of the units is similar for the entire Alpujárride complex. The Alpujárride type sequence includes, from bottom to top: (1) a dark-coloured metapelitic formation, attributed to the Palaeozoic, made up of schists, graphitic micaschists, and quartzschists, with migmatite gneisses appearing towards the bottom as the metamorphic grade increases downwards in the sequence; (2) a light-coloured

metapelitic-metapsammitic formation, also attributed to the Palaeozoic, comprising schists, quartzschists, and metaquartzites; (3) a Permo-Triassic (?) metapelitic-metapsammitic formation consisting of fine-grained schists and minor metaquartzites with calcschists in the uppermost part, transitional with the overlying carbonate rock formation; and (4) a carbonate rock formation, Middle and Late Triassic in age (Braga & Martín 1987*a*, and references therein), made up of calcareous and dolomitic marbles.

One of the most remarkable features of this complex is the outcropping of ultramafic rocks in the Ronda massif, to the west of Málaga (Fig. 16.1). The entire Ronda peridotite can now be considered according to both geological (Lundeen 1978; Tubía & Cuevas 1986) and gravimetric data (Torné *et al.* 1992) as a large segmented slab included within the nappe stack of the Alpujárride complex. Furthermore, recent detailed structural maps show that within the ultramafic massifs it is possible to distinguish several tectonic slices bounded by shear zones (Sánchez-Gómez *et al.* 1995). The contacts between the Ronda peridotite and the surrounding Alpujarrides units are shear zones (Balanyá *et al.* 1997; Tubía *et al.* 1997) that trend parallel to the contacts. The sheared lherzolites exhibit pervasive mylonitic foliation and NE or ENE stretching lineation (relative movement of the hanging-wall towards the NE), mainly defined by orthopyroxene porphyroclasts (Balanyá *et al.* 1997; Tubía & Cuevas 1987; Tubía *et al.* 1997). The peridotite can thus be identified as an independent tectonic element composed mainly of lherzolite and minor harzburgite and mafic layers (Dickey 1970) that reach a thickness of 4.5 km in the Sierra Bermeja. The temperature conditions of shearing have been determined as 828–881°C on the basis of the Ca/Al solubility versus temperature in orthopyroxenes (van der Wal 1993). The pressure conditions attained by the garnet lherzolite during shearing are poorly constrained owing to the difficulties involved in applying barometers and the controversy surrounding the significance of different lherzolite facies.

The thermobarometric history of the Ronda peridotite indicates an evolution from high pressure–high temperature to low pressure with a very moderate decrease of temperature. Emplacement models proposed for the Ronda peridotites reveal important discrepancies over the mode of exhumation of these rocks from a depth of something like 180 km. There is also controversy about: (a) the lithospheric, asthenospheric or mixed source area for these rocks (van der Wal & Vissers 1996; Tubía & Cuevas 1987; Vauchez & Garrido 2001); (b) the interpretation of the lower shear zone as of an extensional (van der Wal & Vissers 1996) or compressional nature (Tubía *et al.* 1997); (c) the age of the different stages of the exhumation process.

The main fabric within the rocks of each Alpujárride unit is a synmetamorphic foliation, which we define as S_2. This regional foliation is thought to be synchronous for the entire Alpujárride complex, as it appears to be the only deformation phase that developed between the same two characteristic metamorphic stages recorded in these rocks: the first under high pressure–low temperature conditions and the second characterized by much lower pressure conditions. A previous S_1 foliation is recognizable only in small, lens-shaped domains bounded by S_2 foliation planes or as inclusions in porphyroblasts such as plagioclase or garnet (where present) in the metapelitic rocks. Small-scale isoclinal folds with an S_1 relict foliation or the S_2 main foliation parallel to the axial plane are only rarely observed, and the vergence of neither F_1 nor F_2 folds has yet been established. An L_2 mineral lineation, defined by quartz ribbons in quartz-rich bands within the metapelitic formations or by calcite in the carbonate rocks, displays no consistent orientation.

High pressure–low temperature metamorphic assemblages have been recognized within most of the Alpujárride units of the eastern (e.g. Bakker *et al.* 1989; De Jong 1991), central (e.g. Goffé *et al.* 1989; García-Casco & Torres Roldán 1996; Azañón *et al.* 1997, 1998), and western Betics (Westerhoff 1977; Tubía & Gil-Ibarguchi 1991; Balanyá *et al.* 1997; Tubía *et al.* 1997). These assemblages record the oldest metamorphic event to affect the Alpujárride complex (event D1). This event was followed by an almost isothermal pressure decrease (e.g. Westerhoff 1977; Goffé *et al.* 1989; Tubía & Gil-Ibarguchi 1991; García-Casco & Torres Roldán 1996; Azañón *et al.* 1997, 1998; Balanyá *et al.* 1997), although local heating at low pressure has been reported (Platt *et al.* 1998; Soto & Platt 1999). During the drop in pressure the main regional foliation (S_2) developed and the metamorphic zones were brought closer to parallelism with the main foliation (Westerhoff 1977; Aldaya *et al.* 1979; Torres Roldán 1981; Balanyá *et al.* 1993, 1997; García-Casco & Torres Roldán 1996; Azañón *et al.* 1997; Argles *et al.* 1999). The next event is registered by north-vergent F_3 folds, low grade metamorphism, and final, post-metamorphic nappe stacking (Azañón *et al.* 1997, 1998). U/Pb ages of zircon rims (closure temperature around 800°C) in Palaeozoic rocks show that high-grade metamorphism is early Miocene in age (Sánchez-Rodríguez 1998; Platt & Whitehouse 1999), although older ages have been obtained from the core of complex polyphase zircons (e.g. Zeck & Whitehouse 1999). Moreover, the structure and metamorphic evolution of Permian–Triassic and Palaeozoic rocks from the metapelitic sequence are similar (e.g. Tubía *et al.* 1992; Azañón *et al.* 1997, 1998).

Figure 16.5b illustrates the distribution of the metamorphic mineral assemblages within the metapelitic sequence of each of the Alpujárride units cropping out in the central Betics. These assemblages also appear in the lower part of the overlying carbonate sequence when impure levels are present. The mineral index was used in order to distinguish chlorite, biotite, staurolite, sillimanite and sillimanite + K-feldspar mineral zones. The following observations deserve special attention: (1) within each unit the metamorphic grade increases systematically downwards in the sequence, that is, perpendicular to the S_2 main foliation; (2) the metamorphic zones are closely spaced; (3) the mineral zones associated with the S_2 main foliation trend parallel to the lithological boundaries; and (4) within the column of metapelitic rocks, the position of a given metamorphic zone varies from one unit to the next, and in consequence, so does the metamorphic grade of each unit. After the chain metamorphic stage, the next fold phase to affect the Alpujárride units produced kilometre-scale, north-vergent F_3 folds with associated small-scale chevron-type folds (Fig. 16.5) and locally developed S_3 crenulation cleavage, pervasive only in F_3 hinges. The axes of F_3 folds strike east–west to ENE–WSW and are generally sub-horizontal or gently plunging towards the west. The Alpujárride nappe stack was subsequently strongly thinned (D4) during the Miocene rifting that resulted in the opening of the Alborán basin (Comas *et al.* 1992; García-Dueñas *et al.* 1992; Vissers *et al.* 1995). In the central Betics, the geometric distribution of the Alpujárride units is attributed to the inter-ference of successive low-angle normal faults and detachment systems with different extensional directions (García-Dueñas *et al.* 1992; Crespo-Blanc *et al.* 1994; Crespo-Blanc 1995). Thus, the present-day Alpujárride units are extensional tectonic units, bounded by brittle faults.

The large-scale extensional geometry of these faults can be seen in a SSW–NNE directed cross-section (Fig. 16.5a, cross-sections 3 and 4). The remarkably severe thinning or even omission of units in the northern part of the section is due to a fan of listric normal faults with a NE-NNE sense of movement of the hanging-wall. These faults commonly tend to coalesce into a single fault, and the whole extensional system clearly cuts through the F3 folds. The relationships with a later sedimentary cover constrain the age of the basement brittle faults: Serravallian deposits clearly post-date the top-to-the NE-NNE extensional fault system and Tortonian rocks cover the top-to-the-WSW fault system (Crespo-Blanc *et al.* 1994).

The D3 Alpujárride nappe stack thinned during event D4 was finally folded (mainly kilometre-scale folds with east–west trending axes) and faulted (high-angle normal and reverse faults and conjugate systems of strike-slip faults) during D5. The lowermost complex of the Internal Zones, the Nevado-Filábride complex, appears in the core of an east–west to ENE–WNW striking kilometre-scale fold, the Sierra Nevada antiform. South of Sierra Nevada, a second antiform is defined by undulating low-angle normal faults with a northerly directed hanging-wall. This second antiform, with an approximate east–west trend, continues to the east as far as the Sierra Alhamilla region, where Weijermars *et al.* (1985) dated it as latest Tortonian. The north–south to NW–SE compression that began during the late Tortonian is still underway (e.g. Comas *et al.* 1992; Rodríguez-Fernández & Martín-Penela 1993). This D5 compressive event caused the final uplift of part of the Miocene Alborán Basin and its basement.

Maláguide and associated complexes

In the Betic Cordillera, the Maláguide complex is so termed because its largest outcrop lies north of the city of Málaga. ENE of this outcrop are several others separated by Neogene basins forming a discontinuous WSW–ENE belt that crops out NE of Granada, and continues to the east near Lorca (Fig. 16.1). WSW of Málaga city there is a set of discontinuous outcrops along the coastline and other outcrops in the northern part of the Internal Zones (Fig. 16.1). In addition, several small klippen of Maláguide rocks lying over Alpujárride rocks have also been reported. Around the northern margin of the Internal Zones, this complex is associated with Mesozoic–Cenozoic rocks with several tectonic units that Durand Delga & Foucault (1968) and Didon *et al.* (1973) termed the 'dorsal' and 'pre-dorsal' complexes, respectively. These terms had a strong palaeogeographic connotation and referred to those units that, during the Mesozoic and Cenozoic eras, were located at the transition between the Internal Zones and the domain where the flysch successions were being deposited.

The Maláguide complex. The Maláguide sequence comprises a large, lower set of Palaeozoic (and older?) rocks called the Piar Group (Gp) (Geel 1973; Martín-Algarra 1987), followed by a Mesozoic–Palaeogene carbonate succession, and an uppermost set of siliciclastic rocks of late Oligocene–early Miocene age. This complete Maláguide sequence can be observed in only a few places within the cordillera as in many places the Mesozoic and Cenozoic cover is detached from the Palaeozoic basement. The largest Palaeozoic outcrops are found in the north and west of Málaga, along the coastline SW of Málaga, and in the Chirivel–Lorca region to the east of Granada. The largest outcrop with a coherent, although also deformed, Mesozoic–Cenozoic sucession of Maláguide rocks is the Sierra Espuña (NW of Lorca; Paquet 1969; Lonergan 1991, 1993).

The Palaeozoic base of the Piar Gp succession begins with the Morales Fm, dated at the top as Silurian and at the base as Ordovician and possibly older. This formation consists of slates and greywackes with occasional levels of conglomerates, cherts and limestones. The base of the sequence contains bands of phyllite and schist with garnet and andalusite. Above phyllite and schist lies the Devonian Sancti Petri Fm, colloquially called 'Calizas Alabeadas' ('warped limestones'). It comprises clastic, turbiditic limestones corresponding to a carbonate flysch separated by several layers of slaty calcareous mudstones. Above this formation, the Falcoña Fm comprises levels of siliceous slates and rocks and is in turn overlain by the Almogia Fm which contains conglomerates, sandstones and slaty mudstones with Culm facies that include olistostromes of carbonate rocks. Both the Falcoña and Almogia formations are Lower Carboniferous in age, and the Piar Gp is overlain unconformably by the Upper Carboniferous Marbella Conglomerate.

The Mesozoic part of the Maláguide sequence begins at the base with the Saladilla Fm, which contains Triassic fossils and is of continental facies: conglomerates, sandstones and red mudstones, together with lesser amounts of dolomites and gypsum that are found mostly in the upper part. Above this formation are carbonate rocks with primarily platform facies ranging from Mesozoic to Palaeogene in age and showing numerous changes in facies and thickness. The best-developed rock sequence begins with the Castillón Fm, which contains Jurassic limestones, and is overlain by the Vélez-Rubio and Belmonte formations which comprise Cretaceous limestones, marls and marly limestones. The overlying Xiquena Fm (Geel 1973) consists of marls and limestones with nummulites of early Eocene age, above which lie calcarenites and marls of the middle Eocene Jardín Fm.

The uppermost part of the Maláguide complex comprises sediments that unconformably overstep various structures generated by the Alpine orogeny (Fig. 16.6). Above the earliest thrusts there is a sequence of conglomerates containing pebbles derived from older Maláguide complex formations (Paquet 1969); it has been dated as late Eocene by Lonergan (1991) and as Oligocene by Martín-Martín *et al.* (1996). Lonergan (1993) termed this sequence the Late Eocene unit, whereas Martín-Martín *et al.* (1996) called it the As Fm. The sequence is unconformably overlain by the Granada City Gp ('Grupo Ciudad de Granada'), a succession of late Oligocene–Aquitanian conglomerates, sandstones and mudstones.

Most of the Maláguide Palaeozoic rocks record several phases of Variscan deformation and have been studied in greater detail in rocks from the same complex in the Rif (Chalouan & Michard 1990). These authors distinguished three Variscan deformational phases affecting Devonian rocks. A first generation (D1) of minor folds with associated axial plane foliation (S_1) are deformed by a second generation (D2) of recumbent folds with crenulated foliation (S_2) and associated shear zones with NE–SW stretching lineations. The

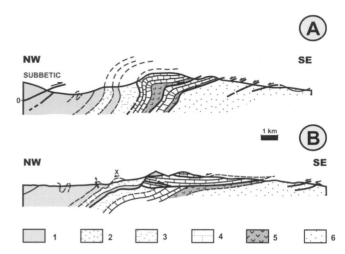

Fig. 16.6. Cross-sections of the Maláguide complex WNW of Lorca (southern Spain). Locations are shown in Figure 16.1b. Legend: 1, Upper Eocene–Lower Miocene limestones, conglomerates, sandstones and marls; 2, Oligocene conglomerates; 3, Lower Eocene nummulitic limestones; 4, Jurassic and Cretaceous limestones and dolostones; 5, Upper Triassic gypsum and marls; 6, Triassic red beds and dolostones. Modified from Lonergan (1991).

Lower Carboniferous rocks show just one deformational phase (D3), with NW–SE to east–west folds and a vertical axial plane with an associated axial plane foliation (S_3). None of these structures are observed in the Triassic rocks. The last phase of Variscan deformation (D3) was accompanied by greenschist to anchizonal metamorphism. In the Betic Cordillera, similar structures are recognized deforming the Palaeozoic rocks, although only two deformational phases have been distinguished affecting Lower Carboniferous or older rocks (Balanyá 1991).

Attempts at reconstructing the sequence of alpine deformations of the Maláguide complex are faced with the problem of the fragmentary nature of its outcrops. Despite this, a sequence of deformational stages can be proposed, beginning with a set of compressional deformations during latest Palaeogene times. These deformations produced the detachment of part of the Maláguide succession, which was deformed into a thin-skinned thrust system resulting in the Sierra Espuña thrusts (Lonergan 1993; Fig. 16.6). The detachment level is located within the Triassic Saladilla Fm, the basal part of which remains attached to the Palaeozoic basement. The kinematic structures observed on these thrust surfaces indicate a NNW sense of movement of the hanging-wall (Lonergan 1991). This thrust system is deformed by later folds and thrusts, the age of which is currently under debate, with proposals ranging from the middle Eocene–middle Oligocene (Lonergan 1991) to the late Oligocene (Martín-Martín *et al.* 1996). In the Rif, the oldest alpine thrusts involve the Palaeozoic basement, are essentially brittle, and are Eocene to Chatian in age (Chalouan & Michard 1990). However, at the base of the lowest unit in the Rif there is a low-grade metamorphism dated at 25 Ma BP (Chalouan & Michard 1990). The front of this anchizonal Alpine metamorphism seems to rise towards the east, since it affects Triassic rocks (Saladilla Fm) in the eastern Betics (Nieto *et al.* 1994; Lonergan & Platt 1995).

The next deformational stage, although not well documented, corresponds to widespread extension event during Aquitanian times. This extension produced the present-day contact between the Maláguide and Alpujárride complexes in the central and eastern sectors of the Betic Cordillera. It is a brittle extensional detachment with an ENE sense of movement of the hanging-wall, and lies above a ductile shear zone in the footwall (Alpujárride) with the same kinematics (Aldaya et al. 1991; Lonergan & Platt 1995). However, the low-angle brittle normal faults that developed in the Maláguide sequence during this stage generally have a northerly sense of movement of the hanging-wall (González-Lodeiro et al. 1996). Relationships with unconformable formations dated as Aquitanian (Granada City Gp) and latest Aquitanian–Burdigalian (Viñuela Gp), as well as fission-track data for the extensional detachment (Johnson 1993), indicate that the extension took place during the early Miocene interval. Associated with this extension during the same period (22 to 23 Ma BP), a phase of basic magmatism with island-arc tholeiite affinities intruded east–west and north–south families of dykes (Torres Roldán et al. 1986). All of these structures seem to indicate nearly radial extension of the rocks of the complex. At the end of this stage, the Internal Zones became a subsiding basin in which marine facies rocks from the Viñuela Gp were deposited (Sanz de Galdeano et al. 1993).

The third phase of alpine deformation in the Maláguide complex corresponds to the compression associated with the convergence of the Internal Zones with the Betic Cordillera External Zones. In the central and eastern regions, this deformation produced NE–SW folds verging SE in the deposits unconformably overlying the Maláguide complex (Viñuela Fm; González-Lodeiro et al. 1996), which has an associated incipient axial plane cleavage. This deformation also created thrusts with a SE sense of movement of the hanging-wall, thrusting Subbetic rocks over the Maláguide sequence in the central and eastern Betics (Paquet 1969; Lonergan 1993). In the western sector, this compression gave rise to north–south axis folds verging both west and east (Balanyá 1991).

The next deformation phase affecting the Maláguide complex was related to a middle Miocene extension that produced most of the low-angle extensional detachments seen in the Internal Zones. These faults, with a SW and south sense of movement of the hanging-wall, greatly thinned the Maláguide sequence. Three of these detachments stand out: the Alcaucín extensional detachment, which eliminated all of the Maláguide in the central sector of the cordillera (Fernández-Fernández et al. 1992); the extensional detachment that crops out east of Málaga, cutting out the base of the Maláguide complex and thus preventing the observation of its original relationship with the Alpujárride complex (González-Lodeiro et al. 1996); and the Piedras Recias fault, in the extreme west of the Internal Zones (Balanyá et al. 1993). More recent deformations have folded these faults over north–south and east–west folds that interfere in domes such as in Sancti Petri, west of Málaga.

The 'dorsal' and 'pre-dorsal' complexes. The Mesozoic–Cenozoic rocks of the 'dorsal' complex show different sequences that allow the differentiation of 'external dorsal' units, now located closer to the Iberian Massif, and 'internal dorsal' units, now located closer to the Internal Zones. The internal dorsal units have a sequence very similar to that seen in the upper part of the Maláguide succession and the two may be related directly (e.g. Wildi 1983). However, the external dorsal units show carbonate-dominated sequences (sometimes metamorphosed to marbles), which typically show Alpujárride-like Upper Triassic carbonate formations overlain by Lower Jurassic carbonate platform facies rocks, followed by young pelagic limestones, and Oligocene carbonate turbidites (Horca Fm).

The pre-dorsal complex consists of Miocene olistostromes that contain a mixture of diverse reworked Mesozoic and Cenozoic sedimentary rocks. Kirker & Platt (1998) have termed these rocks the synorogenic Lower Miocene and they most likely correspond to the synorogenic sediments deposited during the convergence of the Internal Zone (Alborán domain) rocks with Iberia. The pre-dorsal complex is separated from underlying rocks by a detachment which has been interpreted as a normal fault (Balanyá 1991) or a thrust (Kirker & Platt 1998).

In outcrops where the dorsal complex is relatively simple in structure (e.g. Sierra de las Nieves) one can recognize a tight NE–SW syncline verging NW, with axial plane foliation in the overturned limb (Dürr 1967). This fold is associated with thrust surfaces with a NW sense of movement of the hanging-wall (Kirker & Platt 1998). The fold-and-thrust surfaces affect early Aquitanian rocks and have associated synorogenic deposits (Brecha de la Nava) also attributed to the Aquitanian (Martín-Algarra 1987).

In outcrops with a more complicated structure, one can observe several very thin units thrust over one another and then folded (generally into a synform) with a hinge parallel to the margin of the Internal Zones and a vergence towards the interior of the domain (Fig. 16.7). The initial thrusting affected the Oligocene Horca Fm and it is therefore given an Aquitanian age. The later folding has not been dated with any precision, but it is thought to be early-middle Miocene (Balanyá 1991). Still later structures affecting these rocks correspond to the development of normal faults and associated folds that have deformed the Internal Zones since middle Miocene times.

Neogene basins

The Neogene–Quaternary basins located in the Betic Cordillera show a wide variety of structures and tectonic settings. During Tortonian times, the Alborán sea covered most of the cordillera, and subsequent uplift produced a series of emergent areas separated by sedimentary basins. The Alborán sea (Fig. 16.1), lying above rocks of the Internal Zones, is the largest of these basins and is currently still below sea level. Further north, however, are several examples of once-marine basins now uplifted above sea level. Tortonian marine rocks are now exposed at >1500 m near Granada, and >2000 m in Sierra de Gádor (Fig. 16.4). The sediments in these uplifted basins across the orogen are commonly faulted and sometimes folded. The basins located in the intramontane depressions of the central and western cordillera, which developed in the hanging-wall of a crustal detachment, show intense deformations and have been considered as piggy-back basins (Jabaloy et al. 1992).

The basins located in the western cordillera (Gulf of Cádiz, Algeciras, Ronda and Alora) are generally affected by tilting and normal faulting. However, a reverse fault in the Gulf of Cádiz has deformed up to the most recent Quaternary

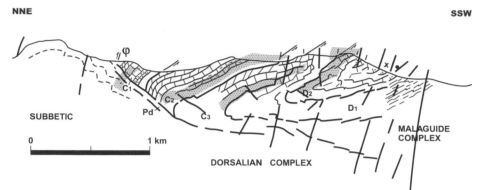

Fig. 16.7. Cross-section of the dorsal complex NE of Granada. Key: C_1, C_2, C_3, external dorsal units; D_1, D_2, internal dorsal units; Pd, predorsalian complex. Modified from Balanyá (1991).

sediments, and is probably the surface expression of active deep faults accommodating ongoing NW–SE shortening in the Betic Cordillera (Rodero 1999).

In the central cordillera, the basins generally develop on open synforms, locally bounded by faults. The Granada basin is a large half-graben with a sedimentary wedge thickening northwards and bounded to the north and east by normal faults dipping southwards and with ENE–WSW and NW–SE strikes (Ruano *et al.* 2000). The Alcalá la Real, Almanzora, Orgiva, Ugíjar and Sorbas-Tabernas basins (Fig. 16.1), among others, are examples of basins located on synformal structures. In some of these basins, such as Alcalá la Real, the internal structure records the overprinting of open folds with different trends (NW–SE and NE–SW) and ages. Other basins are deformed by an even more complex overprinting of folds, normal and reverse faults, such as seen in the Ugíjar basin (Galindo-Zaldívar 1986), implying a local alternation of NW–SE compressional and extensional deformations. The development of folds in the Sierra Nevada during Pliocene times (Johnson 1997), and in the Sierra de los Filabres, has controlled the deep structure of the Guadix-Baza basin, together with faulting that deforms parts of its northwestern border (Estévez & Sanz de Galdeano 1983). However, this deep structure is poorly known due to the lack of deep outcrops and geophysical data.

Most of the basins of the eastern cordillera are controlled by strike-slip faults. The Sorbas-Tabernas basin is asymmetrical, thickening southwards towards the east–west transpressive dextral strike-slip fault located along the northern border of Sierra Alhamilla (Ott d'Estevou 1980). The Vera-Garrucha basin is deformed by the NNE–SSW sinistral Palomares fault that has produced synsedimentary deformation along its border (Montenat *et al.* 1987). The Níjar-Carboneras basin also shows folds and faults related to the NE–SW sinistral Carboneras fault that continues to the Alborán sea. Futher east, other basins (e.g. Lorca, Totana) also developed in the same context, with important wrench faults. The strike-slip faults of these regions indicate, in general, north–south to NW–SE compression simultaneous with orthogonal extension.

Finally, the Guadalquivir basin represents the foreland basin of the Betic orogen and shows an asymmetrical geometry, with sediment thickness increasing southeastwards. The most interesting tectonic feature is the emplacement of olistostromes derived from the Subbetic Zone during Miocene (up to middle Tortonian) times (Roldán-García 1995).

Betic Cordillera tectonic evolution and models

The Betic Cordillera is composed of rocks ranging from Palaeozoic or even Precambrian age (including rocks affected by the Variscan orogeny) up to the present day. It belongs to the western part of the Alpine orogen, and together with the Rif Cordillera and the Alborán sea, records the evolution of part of the Eurasian-African plate boundary. The region has been the subject of controversy, with numerous, and at times contradictory, models being proposed, ranging from old ideas involving mantle diapirs (Weijermars 1985) to more recent debates over the roles of subduction-driven extension versus the influence of lithospheric root detachment (see below).

Many workers have viewed the Internal Zone rocks as belonging to a mobile 'Alborán domain' that was displaced westwards, a model first proposed by Andrieux *et al.* (1971), who considered that an Alborán microplate collided with Europe and Africa during its westward displacement, producing the Betic and Rif cordilleras. A similar model was proposed by Balanyá & García-Dueñas (1987), in which the Alborán crustal domain also moved westwards, thrusting over the south Iberian and north African margins, and undergoing internal deformation. Other interpretations view the build-up of the cordillera as a consequence of oceanic crust subduction, emphasizing the models for the opening of the western Mediterranean proposed by Bijú-Duval *et al.* (1978) and Rehault *et al.* (1984). These authors envisage westward migration of the Alborán domain in the hanging-wall of a subduction zone that ran from the Rif through the Betic, the Kabylias and the Calabria chains, to the east of the Iberian peninsula (Jabaloy *et al.* 1992; Lonergan & White 1997). There have been various suggestions for the polarity of this subduction zone. De Jong (1991) and Morley (1993) proposed a model with a northward-dipping subduction located along the African margin, a view sustained by Zeck *et al.* (1992) and Zeck (1996). Alternatively, a double system with oceanic crust being subducted both north and south of the Alborán sea, ending in late Miocene times, was considered by Torres Roldán *et al.* (1986). Such models provide a ready explanation for Miocene extension and extensional collapse of the Betic Cordillera driven by slab rollback and detachment.

Other models consider that the present-day setting of the orogen is a consequence of gravitational collapse of thickened crustal lithosphere located approximately in the Alborán sea during the Alpine orogeny. Removal of the lithospheric root to this thickened crust caused collapse and radial extension in the Internal Zones, driving compressional orogeny in the

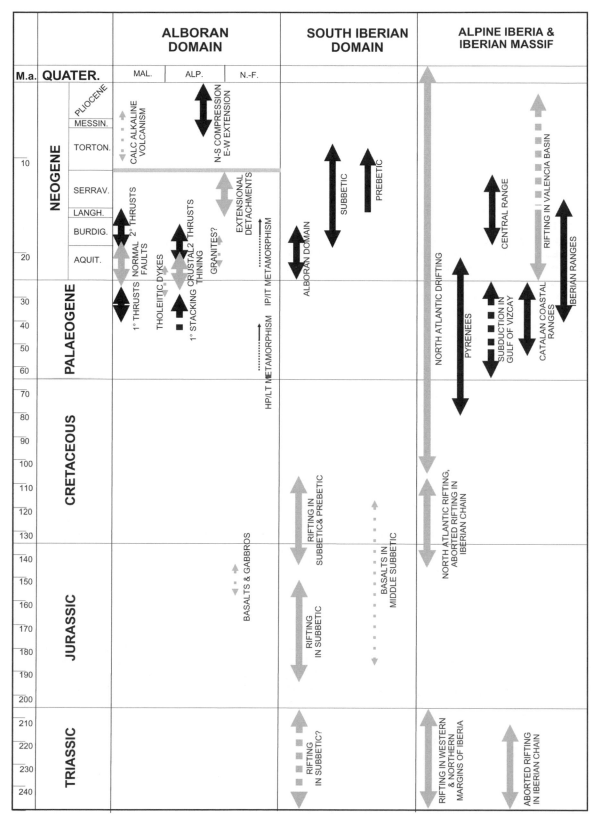

Fig. 16.8. Main features of the tectonic evolution of the Betic Cordillera (Alborán and south Iberian domains), alpine Iberia and Iberian Massif. Black arrows, compressive deformation events. Grey arrows, extensional deformation events.

External Zones of the south Iberian and north African margins (e.g. Watts *et al.* 1993; Platt *et al.* 1998). The removal of the orogen root may have been caused by mantle convection beneath a fixed depocentre (Platt & Vissers 1989; Platt *et al.* 1998) or by mantle delamination beneath an eastward-migrating depocentre (García-Dueñas *et al.* 1992; Docherty & Banda 1995; Seber *et al.* 1996*a*; Calvert *et al.* 2000). Seismic tomography data for crust and upper mantle (Serrano *et al.* 1998; Morales *et al.* 1999; Calvert *et al.* 2000), combined with the determination of stress states in the region (Morales *et al.* 1999), reveal the existence of active continental subduction under the Betic Cordillera (Fig. 16.2) and this needs to be taken into account in any model trying to explain the evolution of the region.

The post-Variscan evolution of the Betic area began during the Mesozoic era (Fig. 16.8), when the south Iberian domain constituted the passive margin of the Iberian Massif and underwent mainly extensional tectonics associated with deformation of the Tethyan ocean. Most palaeogeographic reconstructions (e.g. Dercourt *et al.* 1993) indicate that the maximum distance between Iberia and Africa was less than 200 km and that a large transform fault related to the opening of the Tethys and the Atlantic oceans separated the two. Continental thinning under this tectonic regime may have progressed up to incipient oceanic spreading in pull-apart basins. The Prebetic was located near the Iberian margin while the Subbetic was located basinwards, and was broken up by normal faults in a basin-and-swell structure that determined the differences seen in the subsequent sedimentary record. The Alborán domain (Internal Zones) was probably located eastward of its present-day position, in the Algerian-Balearic basin. Whereas in the Nevado-Filábride and Alpujárride complexes there are no sediments more recent than Triassic, in the Maláguide complex Mesozoic sediments are similar to those of the south Iberian domain (External Zones).

The Alpine orogeny in the region probably started in the Late Cretaceous epoch and continues up to the present day (Fig. 16.8). In the Internal Zones, metamorphic complexes were stacked by thrusting and affected by high pressure–low temperature metamorphic events, first in the Nevado-Filábride complex, located at present in the lowermost structural position, and later in the Alpujárride complex (Azañón *et al.* 1997). Between the Internal and External zones the flysch units were deposited. During the approach of the Alborán (Internal) and south Iberian (External) domains, produced by the westward migration of the Alborán domain in Palaeogene–Lower Miocene times, these flysch units were deformed by fold-and-thrusts, developing an internal structure similar to those of accretionary prisms. Continued closure folded and thrust the uppermost crust of the south Iberian domain (External Zones), probably with an east–west to NW–SE shortening direction (García-Dueñas 1969), and most of the different tectonic units underwent clockwise rotation (Osete *et al.* 1988; Platzman 1992). At the same time, however,

the lower crust of the south Iberian domain, which corresponds to the Iberian Massif, was subducted below the rocks of the Internal Zones and finally under the Alborán sea, as is evidenced by the geophysical data (Fig. 16.2; e.g. Galindo-Zaldívar *et al.* 1997; Morales *et al.* 1999).

The south Iberian domain (External Zones) corresponds to an orogenic wedge that shows two active thrust fronts. One of these faces towards the foreland and is capped by upper Tortonian rocks of the Guadalquivir basin, although there are out-of-sequence deformations deforming sediments as young as Quaternary. The other front faces towards the Internal Zones and was covered by sediments during middle Burdigalian times.

During early Miocene times large, low-angle top-to-the-east normal faults existed at the Alpujárride–Maláguide contact (Aldaya *et al.* 1991; Lonergan & Platt 1995). Later, low-angle faults also became active in the Alpujárride complex, showing a top-to-the-north sense of movement, although there is some disagreement as to whether they are normal or reverse in nature (Cuevas *et al.* 1986; Crespo-Blanc *et al.* 1994). One of the major low-angle normal faults located in the Nevado-Filábride–Alpujárride contact remained active up until the Tortonian stage, with a top-to-the-west sense of movement (García-Dueñas & Martínez-Martínez 1988; Galindo-Zaldívar *et al.* 1989; Jabaloy *et al.* 1993). This fault developed simultaneously with thrust structures in the south Iberian domain, also with a top-to-the-west sense of movement (Galindo-Zaldívar *et al.* 2000). The most recent compressional deformations in the External Zones correspond to top-to-the-NW thrust structures, which affect rocks up to Tortonian in age.

The structures active since the late Miocene are responsible for the present-day features of the cordillera. The most important active deep structure corresponds to the continental subduction of the Iberian Massif beneath the Betic Cordillera. The continental lithosphere in places reaches a depth of 100 km, although its features vary in different sectors of the cordillera. The collision of the Alborán domain and the subduction of the Iberian Massif have produced large folds forming most of the present-day high ground of the Betic Cordillera during principally NW–SE shortening, together with the uplift of the entire region. This uplift of the Betic Cordillera has led to the progressive reduction in size of the Alborán sea, which is the largest Neogene basin of the region. Simultaneously with the shortening, recent NE–SW extension occurred, as evidenced by normal and strike-slip faulting. Overall, these neotectonic events have shaped the present-day landscape of southern Spain during continuing NW–SE convergence between the African and Eurasian plates.

We thank N. Fry, L. Lonergan and the editors, W. Gibbons and T. Moreno for the comments that have improved the quality of this chapter. We are grateful to C. Laurin and J. L. Sanders for the English revision.

17 Cenozoic volcanism I: the Iberian peninsula

JOSÉ LÓPEZ-RUIZ, JOSÉ MARÍA CEBRIÁ & MIGUEL DOBLAS

Cenozoic volcanic rocks in mainland Spain are located in the NE, east, SE and southern central parts of the peninsula (Fig. 17.1), and are mostly Neogene in age, although there was minor activity in both late Oligocene and Quaternary times (Fig. 17.2). Four main provinces can be recognized (Fig. 17.1): the SE volcanic province (SEVP, Cabo de Gata–Mazarrón–Cartagena); the Calatrava volcanic province in south central Spain (CVP); the NE volcanic province (NEVP, Ampurdán–Selva–Garrotxa); and the Gulf of Valencia volcanic province (GVVP), with both onshore and offshore emissions. The SEVP is the most heterogeneous and complex area with calc-alkaline (CA), high-K calc-alkaline (KCA), shoshonitic (SH) and ultrapotassic (UP) rocks, and alkaline basalts (AB). The CVP and NEVP are characterized by alkaline basalts, with occasional leucitites in the CVP. The GVVP contains CA rocks and AB, although the few available data on the offshore volcanism do not exclude a greater lithological variation.

Other outcrops of igneous rocks also exist but are either volumetrically insignificant, incompletely studied or of unconstrained age: the Málaga tholeiitic dyke swarms in the western Betics (Torres Roldán *et al.* 1986; Turner *et al.* 1999); the tholeiites of Alborán island (Bellon & Brousse 1977; Aparicio *et al.* 1991); the volcaniclastic layer of the Lanaja–Peñalba area in the Ebro basin (Odin *et al.* 1997); and the basaltic neck of Nuévalos in the Iberian Ranges (Ancochea *et al.* 1987; Hoyos *et al.* 1998). Of these, only the tholeiitic magmatism will be examined further, as this is key to the understanding of the Betic orogen.

In this chapter we first review currently available geological, petrological and geochemical data, as well as the petrogenetic models (including the compositional signature of the respective mantle sources) proposed for the volcanism in the SEVP, CVP and NEVP. The scarce data from the GVVP, which are limited to AB rocks in the Columbretes islands, Cofrentes and Picassent (e.g. Martí *et al.* 1992; Aparicio & García 1995), do not allow a detailed petrogenetic analysis of this province. We then discuss the different geodynamic models that have been previously suggested for Cenozoic volcanism in Iberia. Finally, combining the geochemical results

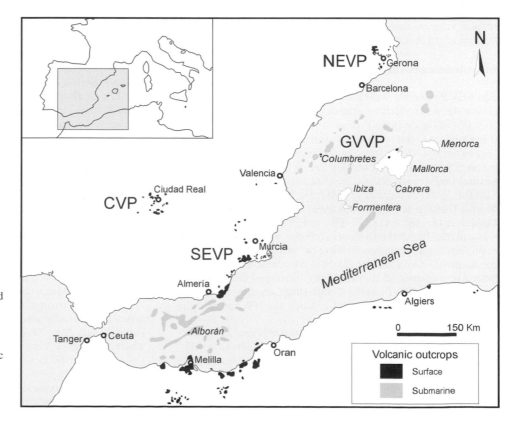

Fig. 17.1. Map showing the location of the four major Cenozoic volcanic provinces of the Iberian peninsula and related regions: SEVP, SE volcanic province; CVP, Calatrava volcanic province; GVVP, Gulf of Valencia volcanic province; NEVP, NE volcanic province. Submarine volcanic outcrops or subcrops in the Alborán sea after Comas *et al.* (1992) and in the Valencia trough from Martí *et al.* (1992).

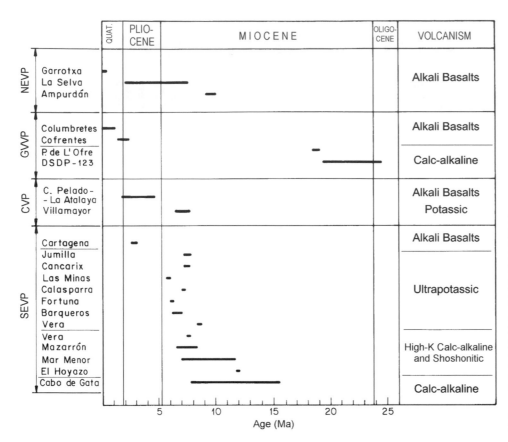

Fig. 17.2. Absolute ages and petrological affinities of the four major Cenozoic volcanic provinces of the Iberian peninsula. Data compiled from: Bellon & Brousse (1977), Nobel *et al.* (1981), Bellon *et al.* (1983), Di Battistini *et al.* (1987) for the SE volcanic province (SEVP); Ancochea *et al.* (1979), Bonadonna & Villa (1986) for the Calatrava volcanic province (CVP); Ryan *et al.* (1972), Saenz & López-Marinas (1975), Rivière *et al.* (1981), Mitjavila *et al.* (1990), Aparicio *et al.* (1991), Marti *et al.* (1992) for the Gulf of Valencia volcanic province (GVVP); and Donville (1973*a,b*), Araña *et al.* (1983) for the NE volcanic province (NEVP).

with other available geological, geodynamic and geophysical data, we propose a new tectonomagmatic scenario which accounts for the genesis of the different Cenozoic magmatic episodes in the Iberian peninsula with implications to other related regions such as the European volcanic province or northwestern Africa.

SE volcanic province

The SEVP covers *c.* 9000 km² in the SE corner of the Iberian peninsula, where Betic metamorphic rocks (Alpujarride and Nevado-Filábride units) form a basement to a series of transtensional, transpressional and extensional Neogene to Quaternary sedimentary basins (continental to shallow marine) bounded by a complex set of fractures (Fig. 17.3a). Several fault trends decisively influenced the evolution of the SEVP (e.g. Hernández *et al.* 1987; De Larouzière *et al.* 1988; Martín-Escorza & López-Ruiz 1988; Sanz de Galdeano 1990; Doblas *et al.* 1991): (1) NE–SW to ENE–WSW normal and transcurrent (dextral and sinistral) faults which constitute the main tectonic elements controlling the sedimentary basins and the Cabo de Gata sector; (2) NW–SE dextral and sinistral faults cross-cutting SE Iberia; (3) NS oriented sinistral faults disrupting the SEVP; and (4) late-stage arc-shaped thrusts (convex to the NW and vergent towards the NW) around the costal area of Aguilas.

The SEVP comprises the following series (Fig. 17.3b): calc-alkaline (CA), high-K calc-alkaline (KCA), shoshonitic (SH), ultrapotassic (UP), and alkaline basalts (AB) (López-Ruiz & Rodríguez-Badiola 1980). Volcanism in the SEVP developed in two stages, the first represented by the eruption of CA, KCA, SH and UP rocks, and the second characterized by the

extrusion of small volumes of alkaline basalts to the NW of Cartagena. The first stage occurred during the Langhian–Messinian interval, over a time span of 9 Ma (Fig. 17.2), beginning with the CA series in the SE coastal area of Cabo de Gata (15 to 7 Ma BP), followed inland by the KCA (El Hoyazo, Mazarrón and Mar Menor) and the SH series (Vera, Mazarrón and Cartagena; 12 to 6 Ma BP) and, finally, the UP series in Vera, Mazarrón and the northern sector of the province (8 to 6 Ma BP; Bellon & Brousse 1977; Nobel *et al.* 1981; Bellon *et al.* 1983; Di Battistini *et al.* 1987). After a gap of *c.* 4 Ma, the second stage saw the eruption of alkaline basalts near Cartagena around 2.8–2.6 Ma BP (Bellon *et al.* 1983).

CA volcanism is restricted to Cabo de Gata (I on Fig. 17.3b) where magmas erupted during several pulses separated by periods of inactivity. Highly explosive eruptions with ash-flow tuffs and breccias alternate with other less explosive phenomena triggering lavas and domes. Bioclastic marine limestones rich in fragmented foraminifers, bryozoa, bivalves and algae were deposited during periods of inactivity (e.g. Coello & Castañón 1965; Fúster *et al.* 1965; Páez & Sánchez Soria 1965; León 1967; Sánchez-Cela 1968; Cunningham *et al.* 1990; Rytuba *et al.* 1990; Arribas 1993).

KCA and SH volcanism is located in the sector of El Hoyazo–Vera–Mazarrón–Cartagena–Mar Menor (II on Fig. 17.3b). In Mazarrón–Cartagena most of the outcrops of both series are spatially associated. However, the outcrops of El Hoyazo and Mar Menor are representative of the KCA, whereas those of Vera (except the ultrapotassic verites) are typical of the SH. Both magmatic types are characterized by domes and dykes with only rare lavas and pyroclastic deposits. The domes are typically made up of fragments of volcanic rocks embedded in a matrix of the same composition. They

a

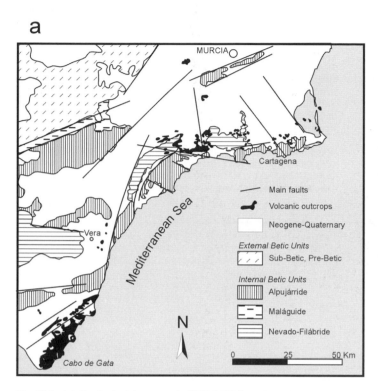

b

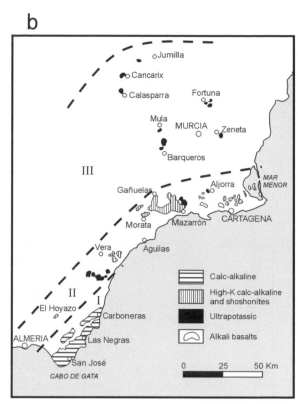

Fig. 17.3. (a) Geological framework (ITGE 1994) and **(b)** distribution of calc-alkaline, high-K calc-alkaline, shoshonitic, ultrapotassic rocks, and alkaline basalts (López-Ruiz & Rodríguez-Badiola 1980) in the SE volcanic province. Three different volcanic zones are shown in (b): calc-alkaline (I), high-K calc-alkaline and shoshonitic (II), and ultrapotassic (III).

locally show brecciated or massive structures with frequent columnar jointing.

The ultrapotassic (lamproitic) volcanism formed necks (Cancarix, Las Minas, Barqueros, Aljorra, Mazarrón and Vera), dykes (Fortuna and Mula) and, more rarely, lavas (Jumilla). With only a few exceptions (Vera, Mazarrón, Aljorra), UP rocks are located in the northern sector of this volcanic region (III on Fig. 17.3b). The necks have an external facies showing breccias made up of volcanics and host rocks. The internal facies is vitreous or very finely crystalline, becoming holocrystalline in the central portion of the neck. This variation suggests a rapid and violent ascent of the magma, resulting in strong explosive eruptions of pyroclastic material, followed by lava flows. Remains of pyroclastic deposits that constructed volcanic edifices over the ascent conduit can be found in Barqueros and Cancarix, while relics of the lava flows can be observed in Barqueros and Vera (Fúster *et al.* 1967).

Basaltic volcanism is restricted to a small area to the NW of Cartagena (Fig. 17.3b) with pyroclastic rocks and small lava flows. Despite their young age, the erupted materials are highly weathered and the volcanic vents are strongly eroded. The lavas contain abundant and metasomatized peridotite xenoliths (Sagredo 1972, 1973; Boivin 1982; Dupuy *et al.* 1986; Capedri *et al.* 1989), as well as inclusions of quartz-feldspar granulites and albitic schists (Navarro 1973). The ultramafic xenoliths are interpreted as fragments of the lithospheric mantle entrained by the magma during its ascent, whereas the others are fragments of the lower and upper crust, respectively.

Petrology and geochemistry. The SEVP has been studied in detail by many authors (e.g. López-Ruiz & Rodríguez-Badiola 1980; Molin 1980; Munksgaard 1984; Nixon *et al.* 1984; Venturelli *et al.* 1984, 1988, 1991, 1993; Toscani *et al.* 1990, 1995; López-Ruiz & Wasserman 1991; Zeck 1992; Benito 1993; Salvioli & Venturelli 1996; Benito *et al.* 1999; Turner *et al.* 1999; Zeck *et al.* 1999). A detailed petrological and mineralogical description of these rocks was undertaken by López-Ruiz & Rodríguez-Badiola (1980). A representative selection of major, trace elements and isotopic ratios of these rocks can be found in Benito *et al.* (1999) and Turner *et al.* (1999).

CA rocks comprise basaltic andesites, pyroxene- and amphibole-bearing andesites and dacites. The basaltic andesites and the andesites consist of plagioclase (An_{95-68}), orthopyroxene (En_{79-55} Fs_{18-43} Wo_{3-2}), clinopyroxene (En_{50-41} Fs_{9-17} Wo_{41-42}), magnetite (Usp_{29-11}) and ilmenite (Ilm_{80-74}), accompanied by amphibole and biotite in the most silica-rich samples. Dacites are constituted by plagioclase (An_{90-67}), orthopyroxene (En_{58-55} Fs_{40-43} Wo_2), amphibole (hornblende-cummingtonite), biotite, magnetite (Usp_{26-21}), ilmenite (Ilm_{80-73}) and quartz.

KCA rocks are predominantly corundum-normative high-K andesites and dacites. Although their mineralogy is similar to their equivalents in CA series, amphibole is never present and biotite is the main ferromagnesian mineral.

SH rocks are represented by banakites and latites. The banakites consist of plagioclase (richer in albite and orthoclase than their equivalents in CA and KCA series), orthopyroxene (En_{52-51} Fs_{46-47} Wo_2), clinopyroxene (En_{51-46} Fs_{6-10} Wo_{43-44}), sanidine (Or_{75-64}), biotite and quartz. In latites, plagioclase (An_{53-43}) and sanidine (Or_{75}) coexist, clinopyroxene is absent, orthopyroxene (En_{85-62} Fs_{14-37} Wo_1) is scarce, and biotite and quartz show higher abundances than in banakites.

KCA and SH rocks contain two groups of inclusions: metapelitic xenoliths; and quartz-diorite, basaltoid and quartz-gabbroic inclusions. The first group is the more abundant, representing two-thirds of the total volume of xenoliths. The petrology and chemical composition of both groups have been studied by Zeck (1970, 1992) and Molin (1980). Metapelitic xenoliths range from a few millimetres to 50 cm, although most of them are <10 cm, and contain biotite and sillimanite with variable amounts of cordierite, almandine, spinel and quartz. Three main types of metapelitic xenoliths can be distinguished according to the relative abundances of the following minerals: biotite–sillimanite–almandine, spinel–cordierite and cordierite–quartz. These represent end-members with a complete transition between them (Zeck 1970). Although their distribution is irregular within the volcanic suite, some types predominate in specific areas. Biotite–sillimanite–almandine parageneses with variable cordierite are the most abundant at El Hoyazo, while spinel–cordierite ones predominate at Mar Menor. The textures and mineralogy of the metapelitic xenoliths suggest that they represent refractory residues produced after partial melting of gneisses that were initially made up of biotite, garnet, sillimanite, feldspars and quartz (Zeck 1970; Cesare et al. 1997; Cesare & Gómez-Pugnaire 2001). The disaggregation of these xenoliths generated anhedral cordierite, plagioclase, spinel, zircon and graphite, euhedral garnets, and corroded quartz. The composition and zoning of the garnet xenocrysts have been studied by López-Ruiz et al. (1977), Molin (1980) and Munksgaard (1985).

UP rocks are olivine-diopside-richterite madupitic lamproites (jumillites), enstatite-sanidine-phlogopite lamproites (cancalites), hyalo-enstatite-phlogopite lamproites (fortunites) and hyalo-olivine-diopside-phlogopite lamproites (verites). The matrix of the vitreous samples is rich in SiO_2, Al_2O_3 and K_2O while, in the holocrystalline varieties, the glass is replaced by potassic richterite and iron-rich sanidine. Apatite, calcite, rutile and spinels are the most important accessory minerals.

CA rocks are poor in K_2O, TiO_2, P_2O_5 and incompatible and compatible trace elements relative to KCA and SH, whereas UP rocks are richer in K_2O and incompatible trace elements, as well as in MgO, Ni and Cr, relative to the other groups (Figs 17.4 and 17.5). All series show high large-ion lithophile element (LILE)/light rare

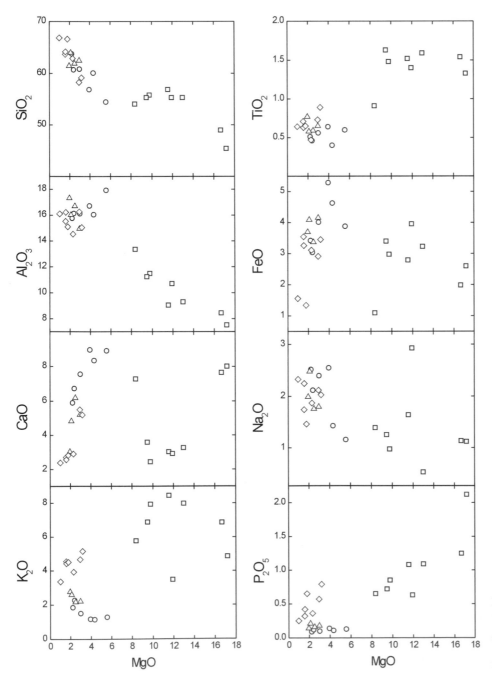

Fig. 17.4. MgO–major element diagrams (values in wt%) for the calc-alkaline (circles), high-K calc-alkaline (triangles), shoshonitic (rhombs) and ultrapotassic (squares) lavas of the SE volcanic province. Data from Benito et al. (1999).

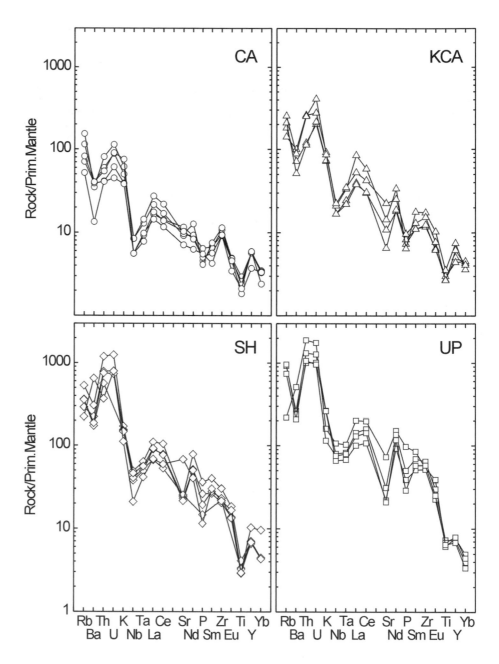

Fig. 17.5. Multielement diagrams normalized to primitive mantle (after Sun & McDonough 1989) for the calc-alkaline (CA), high-K calc-alkaline (KCA), shoshonitic (SH) and ultrapotassic (UP) lavas of the SE volcanic province. Data from Benito *et al.* (1999).

earth element (LREE), LILE/high field strength element (HFSE) and B/Be ratios, a small and persistent negative Eu anomaly, as well as a strong correlation between $^{87}Sr/^{86}Sr$, B and 1/Sr (Benito *et al.* 1999).

CA rocks exhibit a relatively wide range of SiO_2 and their abundance in major elements is similar to equivalent rocks of other areas. They are characterized by high contents of Al_2O_3 and CaO, moderate values of Na_2O, and low percentages of TiO_2, K_2O and P_2O_5 (Fig. 17.4). This series displays a typical enrichment in LILE (higher in Rb and Th than in Ba) and LREE, and negative anomalies of Sr, Nb, Ta and Ti (Fig. 17.5). With increasing SiO_2, a progressive enrichment of K_2O, Rb, Th, Zr and REE is observed, as well as a decrease in $Fe_2O_3^*$, MgO, CaO, TiO_2, P_2O_5, Co, Ni and V.

KCA rocks show a narrow variation range in SiO_2 relative to CA series, they are richer in TiO_2, K_2O and P_2O_5, poorer in FeO and CaO, and Al_2O_3 displays similar values (Fig. 17.4). They are also enriched in LILE and LREE, and have negative anomalies of Nb, Ta, Sr, P and Ti (Fig. 17.5). The abundance of all these elements, as well as Ni, Cr, V and Co, is higher than in CA rocks (with similar SiO_2 content), but their Ba/La, La/Nb, Th/Nb and Zr/Nb ratios are

equivalent. With increasing SiO_2, Al_2O_3 remains virtually constant, while an increase in K_2O and P_2O_5, and a decrease in MgO and CaO is observed.

SH rocks are characterized by higher contents of TiO_2, K_2O and P_2O_5 (Fig. 17.4). They are also strongly enriched in LILE (in this case the higher values correspond to Th and not to Rb, as in the two other series) and LREE relative to the other incompatible trace elements (Fig. 17.5). In contrast, their abundance in heavy rare earth elements (HREE) and compatible elements falls within the range of KCA rocks. As a result of the enrichment in highly incompatible elements relative to the moderately incompatible elements, the LILE/LREE and LILE/HFSE ratios are higher than those of CA and KCA rocks. Al_2O_3, MnO, MgO, CaO, TiO_2 and P_2O_5 decrease as silica increases.

UP rocks display much higher contents of MgO, Ni and Cr, as well as K_2O, TiO_2, P_2O_5, LILE (particularly Th) and LREE, than the other groups (Figs 17.4 and 17.5). On the other hand, they have moderate to low contents of Al_2O_3, CaO, Sc and V.

Regarding the isotopic data, the $\delta^{18}O$ values of CA, KCA, SH and UP rocks lie between +9.5 and +20.3‰ (Munksgaard 1984;

López-Ruiz & Wasserman 1991). There is no good correspondence between these values and the abundance of SiO_2, MgO or incompatible elements. However, there is a strong positive correlation ($r = 0.92$) between $\delta^{18}O$ and the loss-on-ignition values. This correlation, as well as the evidence for extensive devitrification of the matrix in many of the samples (>50% of the rock volume), suggests that the measured $\delta^{18}O$ values are not primary. For this reason, López-Ruiz & Wasserman (1991) corrected the measured $\delta^{18}O$ values following the method of Ferrara et al. (1985) to account for the alteration effects. The calculated primary values display a more restricted range (+8.6 to +11.8‰) with lower $\delta^{18}O$ values for CA rocks (+8.6 to +10.2‰) and the highest $\delta^{18}O$ values for UP lavas (+10.2 to +11.8‰).

The $^{87}Sr/^{86}Sr$ and $^{143}Nd/^{144}Nd$ ratios in CA, KCA, SH and UP rocks lie in the ranges 0.70802–0.72288 and 0.511980–0.512483, respectively (Powell & Bell 1970; Munksgaard 1984; Nelson et al. 1986; Toscani et al. 1990; Venturelli et al. 1991; Benito et al. 1999; Turner et al. 1999; Zeck et al. 1999; and additional unpublished data), showing a generalized increase in radiogenic Sr from CA to UP (Fig. 17.6). In contrast to the oxygen isotopes, the $^{87}Sr/^{86}Sr$ ratios have not been significantly modified by alteration processes. This is demonstrated by the Sr isotope analyses of Toscani et al. (1990) in CA whole rock, plagioclase and groundmass, which show very small differences, usually within analytical error. Furthermore, this is supported by the good correlation of the Sr isotopes relative to $^{87}Rb/^{86}Sr$ (Munksgaard 1984).

Lead isotopic ratios are high in the four series, and there are no significant differences among them (Fig. 17.7). $^{206}Pb/^{204}Pb$, $^{207}Pb/^{204}Pb$ and $^{208}Pb/^{204}Pb$ ratios range respectively from 18.72–18.91, 15.59–15.70 and 38.58–39.02 in CA lavas, to 18.66–18.84, 15.63–15.74 and 38.83–39.20 in UP rocks (Hertogen et al. 1985, 1988; Nelson et al. 1986; Arribas & Tosdal 1994; Turner et al. 1999; and additional unpublished data).

The AB (basanites and olivine basalts) contain olivine (Fo_{90-80}), clinopyroxene (En_{42-36} Fs_{13} Wo_{45-51}), plagioclase (An_{63-53}), magnetite and analcime. SiO_2 and Al_2O_3 have low abundances, whereas Na_2O, K_2O, TiO_2 and P_2O_5 contents are high. In contrast to the other series, there is a strong enrichment in HFSE and conspicuous negative anomalies in Rb and K (Fig. 17.8). Their Sr, Nd and Pb isotopic ratios are very homogeneous (0.70437–0.70560, 0.51258–0.51259 and 18.71–18.83, 15.52–15.56 and 38.42–38.64, respectively; Capedri et al. 1989; Turner et al. 1999), significantly lower than those observed in CA to UP lavas (see Figs 17.6 and 17.7).

Petrogenetic model

The temporal and spatial close relationship of CA, KCA, SH and UP rocks, and their comparable geochemical signatures (e.g. high LILE/HFSE, LREE/HFSE and B/Be ratios, high Sr, Pb and O isotopic ratios, and low Nd isotopic ratios) suggest a somehow linked petrogenetic origin. On the other hand, KCA and SH lavas entrain abundant gneissic xenoliths whose mineral and chemical compositions indicate extraction of an anatectic fraction, thus suggesting that KCA and SH magmas interacted with liquids derived from the upper crust.

Following the classic works by Osann (1906) and Jérémine & Fallot (1928) on the UP rocks, the origin of the magmas from the SEVP has been intensively studied. Most of the authors focused on a specific group of rocks: Toscani et al. (1990), Fernández-Soler (1996) and Zeck et al. (1999) considered only CA rocks; Zeck (1970, 1992) and Munksgaard (1984) restricted their work to CA and SH rocks; Nixon et al. (1984) and Venturelli et al. (1984, 1988) focused on the UP rocks. In contrast, López-Ruiz & Rodríguez-Badiola (1980), López-Ruiz & Wasserman (1991), and Benito et al. (1999) studied the genesis of the whole SEVP magmatism. Several petrogenetic hypotheses have been proposed, including: anatexis of pelitic metamorphic rocks (e.g. Zeck 1970; Munksgaard 1984); anatexis of crustal rocks of different lithologies and ages (Zeck et al. 1999); mixing of anatectic and mantle-derived magmas (e.g. Zeck 1992); assimilation of pelitic material with magmas similar to CA rocks (e.g. Bellon et al. 1983); partial melting of a subduction-enriched mantle source (e.g. Venturelli et al. 1984) that had lost a magmatic component prior to the metasomatic event (Venturelli et al. 1988); partial melting of a mantle source metasomatized by fluids derived from sediments (e.g. Nelson et al. 1986; López-Ruiz & Wasserman 1991); and partial melting of a mantle metasomatized by sediment-derived fluids, later interacting with crustal liquids (KCA, SH rocks; Benito et al. 1999).

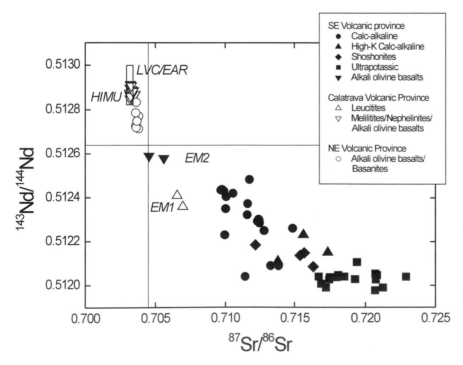

Fig. 17.6. $^{87}Sr/^{86}Sr$–$^{143}Nd/^{144}Nd$ diagram for the lavas of the SE volcanic province (data from Nelson et al. 1986; Turner et al. 1999; Zeck et al. 1999; and unpublished data), the Calatrava volcanic province (unpublished data) and the NE volcanic province (data from Cebriá et al. 2000). The values of the low velocity component/European asthenospheric reservoir (LVC/EAR) are taken from Hoernle et al. (1995) and Cebriá & Wilson (1995). Mantle components HIMU, EM1 and EM2 are from Zindler & Hart (1986) and Hart (1988).

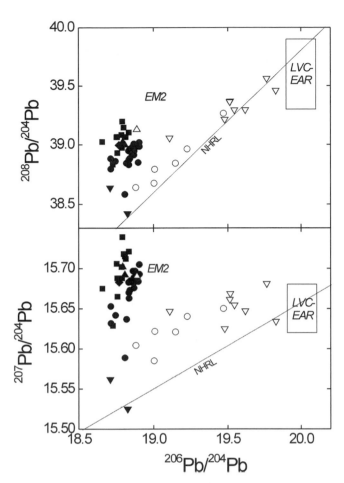

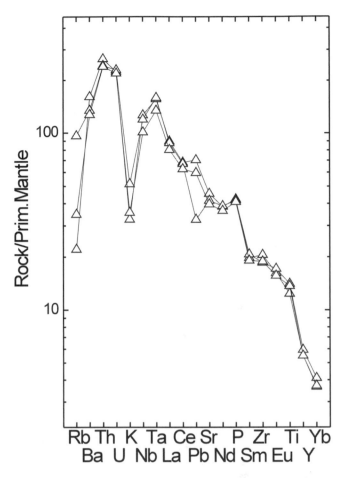

Fig. 17.7. $^{206}Pb/^{204}Pb-^{207}Pb/^{204}Pb$ and $^{206}Pb/^{204}Pb-^{208}Pb/^{204}Pb$ diagrams for the lavas of the SE volcanic province (data from Nelson *et al.* 1986; Arribas & Tosdal 1994; Turner *et al.* 1999; and unpublished data), the Calatrava volcanic province (unpublished data) and the NE volcanic province (data from Cebriá *et al.* 2000). Symbols and data sources for the LVC/EAR and mantle components as in Figure 17.6. NHRL is the northern hemisphere reference line of Hart (1984).

Fig. 17.8. Multielement diagrams normalized to primitive mantle (after Sun & McDonough 1989) for the alkali olivine basalts of the SE volcanic province (after López-Ruiz 1999). In contrast with the calc-alkaline, high-K calc-alkaline, shoshonitic and ultrapotassic rocks of the previous volcanic stage (see Fig. 17.5), the alkali olivine basalts of this province show an enrichment in HFSE and negative anomalies in Rb and K.

The obvious geochemical similarity between the groups implies that the genetic model should apply to all of them. With this in mind, it is argued that the petrogenetic hypotheses that best explain the geochemical characteristics of CA, KCA, SH and UP lavas as a whole are those of López-Ruiz & Wasserman (1991) and Benito *et al.* (1999). The latter authors propose a petrogenetic model involving two mixing processes (Fig. 17.9). During a first episode, source contamination (metasomatism) takes place within the lithospheric mantle by fluids derived from subducted pelagic sediments (between 1 and 5%). This process would yield a heterogeneous source enriched in incompatible elements (especially LILE and LREE) and $^{87}Sr/^{86}Sr$, with only minor modification of the $\delta^{18}O$ values. The partial melting of this metasomatized mantle gives rise to CA, KCA, SH and UP magmas highly enriched in LILE and LREE, which inherit the isotopic signature of their source. During their ascent to the surface, KCA and SH magmas interact with crustal liquids from the Betics basement. This mixing process produced magmas enriched in $\delta^{18}O$ and, to a lesser extent, $^{87}Sr/^{86}Sr$. In contrast, although the crustal

contribution is high (>18% at Mar Menor, >33% at Vera and >35% at El Hoyazo), the effect on the incompatible trace element abundances is minor because of the similar concentrations of these elements in both crustal- and mantle-derived liquids.

Regarding the younger alkaline basalts (AB), their geochemistry suggests derivation from a sublithospheric mantle source containing a K-bearing phase such as phlogopite. These basalts could have interacted to a small extent with the lithospheric mantle (which was probably almost sterile after the generation of the CA to UP magmas). As we will argue in the geodynamic scenario, this sublithospheric component can be found beneath a huge region, from the eastern central Atlantic to Europe (Hoernle *et al.* 1995). It may even have been associated with a long-lived Triassic–Jurassic, central Atlantic superplume (Oyarzun *et al.* 1997).

To summarize, CA and UP magmas are interpreted as generated from a lithospheric mantle intensely and heterogeneously metasomatized by fluids derived from pelagic sediments. CA magmas melted from less enriched mantle

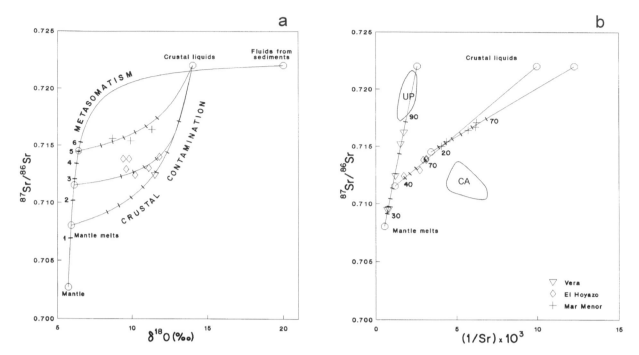

Fig. 17.9. $^{87}Sr/^{86}Sr$–$\delta^{18}O$ (a) and $^{87}Sr/^{86}Sr$–$(1/Sr) \times 10^3$ (b) diagrams for the KCA and SH rocks of the SE volcanic province, modified from Benito *et al.* (1999). **(a)** Sr and O isotopic ratios of the metasomatized mantle source and the proportion of sediment-derived fluids that pervaded the lithospheric mantle. **(b)** Concentration of Sr in the melts derived from the continental crust and the metasomatized mantle, as well as the proportion in which the crustal liquids have participated in the generation of KCA and SH magmas. The Sr and O isotopic ratios and the Sr concentrations of the unmodified lithospheric mantle and the fluids derived from pelagic sediments are assumed to correspond to the average composition of the depleted lithospheric mantle and marine sediments, taking also into account the mobility of Sr during subduction (López-Ruiz & Wasserman 1991; Benito *et al.* 1999). The corresponding ratios in the crustal liquids are considered identical to the restitic xenoliths entrained in the KCA and SH rocks. The CA and UP rocks (whose fields are included for comparison) do not fit this model as they did not undergo crustal contamination.

domains and so they have lower incompatible trace element abundances (especially LILE and LREE) than the UP lavas. In contrast, KCA and SH, although also produced from this same mantle source, subsequently interacted with crustal liquids. In the CA/UP and KCA/SH magmas, the proportion of sediment-derived fluids dominates the abundance of both incompatible elements and the isotopic ratios of Sr and Pb in the melt. This sedimentary component is also held responsible for the depletion in HFSE and for a slight Eu negative anomaly, as these geochemical characteristics are typical of most oceanic sediments (e.g. Ben Othman *et al.* 1989; Lin 1992; Cousens *et al.* 1994). On the contrary, the concentration of major elements is related to the conditions of melting and the subsequent fractional crystallization processes (including assimilation in KCA and SH magmas) which give rise to the different petrological types. This decoupling of trace elements from major elements, and the presence of sediment-derived fluids in the mantle source of the four magma types, explain the similar trace element patterns of CA, KCA, SH and UP rocks (Fig. 17.5). Finally, the relative geochemical simplicity of the late AB magmas, as compared to the other volcanic series, is explained by invoking the participation of the same sub-lithospheric mantle source that can be found beneath Europe and the eastern central Atlantic.

Calatrava volcanic province

The CVP outcrops in south central Spain within Variscan rocks locally covered by late Cenozoic sediments (Fig. 17.10).

The Palaeozoic basement rocks are Lower Ordovician quartzites, Upper Ordovician limestones and Ordovician–Silurian slates, all of which have been variably affected by Variscan folds and thrusts and Alpine brittle deformations. This basement is unconformably overlain by upper Miocene to Quaternary fluvial and lacustrine sediments deposited in a series of fault-bounded basins (*c*. 200 km²) that formed as a result of extensional tectonics imposed since late Miocene time. A complex fracture pattern shaped the geometry of the basins (east–west to ENE–WSW, NW–SE and NE–SW), and normal faults bounding NW–SE orientated grabens (López-Ruiz *et al.* 1993) controlled the facies and thicknesses of the Cenozoic continental sediments, giving the region a basin-and-range geomorphology. According to K-Ar age determinations (Ancochea *et al.* 1979; Bonadonna & Villa 1986), volcanism had started by late Miocene times (7.7–6.4 Ma BP) when olivine leucitites were erupted, followed after a gap of 1.7 Ma by alkaline basaltic magmas which were extruded until Quaternary times (1.75 Ma BP; see Fig. 17.2).

The CVP is made up of more than 200 volcanic outcrops generated by strombolian and hydromagmatic eruptions. The strombolian eruptions produced relatively small monogenetic cinder cones and lava flows, whereas hydromagmatism created explosive craters and maar structures, the latter being the most distinctive volcanic structures of the CVP. Although there are more than 50 maars reported in the region, most are heavily eroded and less than 20 preserve the original ring structure. Their average diameters are usually of *c*. 600 m (internal ring) and *c*. 1000 m (external ring), and they resulted

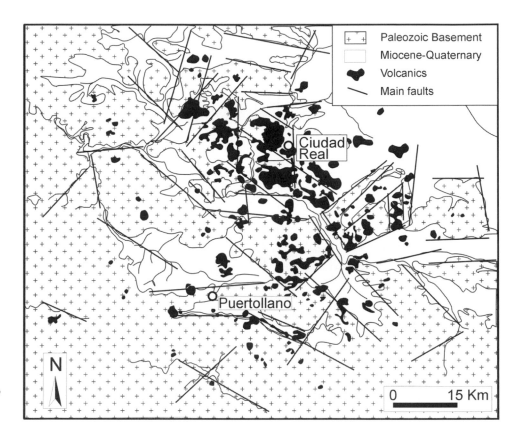

Fig. 17.10. Simplified geological map of the Calatrava volcanic province (Cebriá & López-Ruiz 1995).

from highly explosive phreatic and phreatomagmatic eruptions caused by the interaction of magma (either directly or acting as a heat source) with groundwater. Thus, most maars are found in sedimentary basins, although some are also located within the Variscan basement, these commonly producing lateral blasts and non-circular shaped tuff rings.

On a regional scale the CVP defines a subcircular shaped region (Fig. 17.10), although the emission vents are statistically distributed along predominantly NW–SE trends (Ancochea & Brändle 1982). Volcanism and Pliocene–Quaternary sedimentation were coeval, producing alternations of interbedded volcanic and sedimentary deposits (see fig. 2 of López-Ruiz *et al.* 1993).

Petrology and geochemistry. The CVP rocks have received less attention than those of the SEVP and the NEVP, although a detailed petrological and geochemical study of this volcanism has been carried out by López-Ruiz *et al.* (1993) and Cebriá & López-Ruiz (1995, 1996). The mafic volcanic rocks of the CVP contain olivine, clinopyroxene and Fe-Ti oxides. The first two phases occur both as phenocrysts and microcrysts, whereas the third one appears only in the matrix and as inclusions in phenocrysts. Melilite, feldspathoids (nepheline, leucite or sodalite), plagioclase (never associated with feldspathoids), biotite and other minor phases (apatite, perovskite, analcime) are found only in some groups, mostly as microphenocrysts. Carbonates and zeolites form secondary infillings in vesicles and fractures. In all the groups, olivine shows a uniform composition around Fo_{86}, and clinopyroxenes range from augite to Ti-diopside, usually with normal zoning. Green Fe-rich cores to clinopyroxene phenocrysts are commonly surrounded by normal zoning. High-Ti biotites, which have been interpreted as mantle xenocrysts, are also found in these basalts. The remaining mineral phases such as plagioclase (An_{65}) and feldspathoids have uniform compositions (Cebriá & López-Ruiz 1995).

The volcanic rocks of the CVP can be classified on the basis of their

modal mineralogy into: olivine melilitites (OLM; with melilite), olivine nephelinites (OLN; with nepheline as the only leucocratic phase), alkaline olivine basalts (AOB; with plagioclase) and olivine leucitites (LEU; with leucite). However, this modal classification cannot be applied to rocks with a cryptocrystalline or vitreous matrix. For this reason, it is better to use a chemical/normative classification which is consistent with the modal classification: OLM (with Lc + Cs), OLN (with Or ± Lc and <5% Ab), AOB (with Or and >5% Ab) and LEU (with $K_2O > Na_2O$) (Cebriá & López-Ruiz 1995).

The CVP volcanics have major element compositions analogous to those of primitive alkaline basalts and leucitites from intraplate continental zones, such as the Neogene to present volcanic province of western and central Europe (Wilson & Downes 1991). Concerning the variations between groups, OLM, OLN and AOB define a suite characterized by a progressive enrichment in SiO_2, Al_2O_3 and FeO and a depletion in the other major elements (Fig. 17.11).

The trace element abundances of the CVP volcanics are also typical of those observed in continental intraplate alkaline mafic magmas. Figure 17.12 displays representative analyses of the CVP rocks on primitive mantle normalized trace element diagrams. All groups are characterized by highly fractionated patterns and strong enrichments with respect to the primitive mantle values, and the abundance in trace elements is higher than those observed in typical OIB (e.g. Clague & Frey 1982; Sun & McDonough 1989).

OLM, OLN and AOB display very similar trace element patterns (Fig. 17.12) with a tendency towards higher enrichment from Yb to Nb relative to the primitive mantle and a negative anomaly in Rb and K (Cebriá & López-Ruiz 1995). In the OLM–OLN–AOB sequence there is a lower enrichment in all the elements, and this is more prominent in the highly incompatible elements than in the moderately incompatible ones (average $[La/Yb]_N$ ratios = 27.2 in OLM, 21.8 in OLN, and 18.3 in AOB). Additionally, the negative K and Rb anomalies also decrease in the OLM–OLN–AOB sequence.

LEU display a very different trace element pattern from the associated basaltic rocks (Fig. 17.12). Many of their characteristics are typical of ultrapotassic rocks, but they cannot be assigned to a specific group according to the classification of Foley *et al.* (1987). The trace

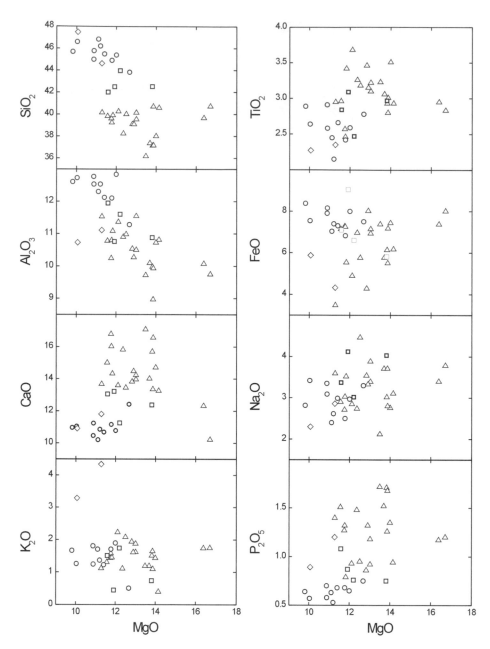

Fig. 17.11. MgO–major element diagrams (values in wt%) for olivine melilitites (triangles), olivine nephelinites (squares), alkali olivine basalts (circles) and leucitites (rhombs) of the Calatrava volcanic province. Data from Cebriá & López-Ruiz (1995).

element abundances and the [La/Yb]$_N$ ratio (28.1) of LEU are similar to OLM. However, LEU are poorer in Nb, Sr, P, Ti and Y, richer in K, Nd, Zr, and strongly enriched in Rb (Cebriá & López-Ruiz 1995). Another significant feature is a negative Nb anomaly, which is typical of lithospheric mantle sources modified by crustal-derived fluids (e.g. Bradshaw *et al.*1993).

Sr-Nd isotope determinations for OLM, OLN and AOB fall in the ranges 0.70314–0.70356 and 0.512849–0.512914, respectively, plotting close to the HIMU component (Fig. 17.6). Similarly, their $^{206}Pb/^{204}Pb$, $^{207}Pb/^{204}Pb$ and $^{208}Pb/^{204}Pb$ ratios are in the ranges 19.11–19.83, 15.62–15.68, 39.06–39.56, respectively (see Fig. 17.7). In contrast, Sr-Nd isotope determinations for LEU show higher $^{87}Sr/^{86}Sr$ values (0.70655–0.70696) and lower $^{143}Nd/^{144}Nd$ ratios (0.512359–0.512409) than the related basaltic rocks, plotting near the EM components, and the Pb isotopic ratios are 18.89, 15.69 and 39.13 (see Figs 17.6 and 17.7). These values are typical of equivalent continental magmatic associations (e.g. Eifel region; Wilson & Downes 1991). The primary δ^{18}O ratios corrected according to the method of Ferrara *et al.* (1985) are in the range 5.76–6.13‰ for OLM, OLN and BAS and are >7.16‰

for the leucitites (Cebriá & López-Ruiz 1995), which are also similar to primitive basalts and leucitites from intraplate settings (Kyser 1986).

Petrogenetic model

Most of the CVP lavas represent relatively primitive magma compositions (e.g. Ni > 200 ppm, Mg# ≥ 68) and host peridotite xenoliths. The homogeneity of the Sr–Nd–Pb isotopic ratios and the fact that the trace element patterns of the primary OLM, OLN and AOB are very similar, indicate that this basaltic suite results from variable degrees of partial melting (F) of a relatively homogeneous mantle source (Cebriá & López-Ruiz 1995). The variations observed in highly incompatible elements (e.g. LREE, Zr, P) suggest that the degree of melting has increased from OLM to OLN and

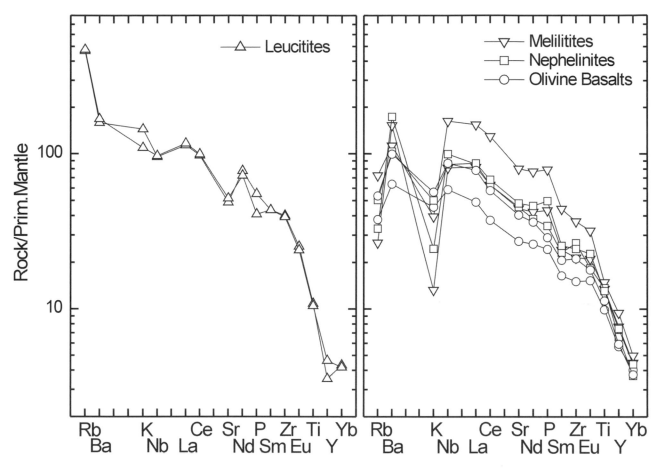

Fig. 17.12. Multielement diagrams normalized to primitive mantle (after Sun & McDonough 1989) for the lavas of the Calatrava volcanic province. Data from Cebriá & López-Ruiz (1995).

AOB. This also explains the variation in the degree of fractionation (as shown by $[La/Yb]_N$).

Some of the LEU also have geochemical characteristics typical of primary magmas. However, their distinctive trace element patterns and their different isotopic signatures compared to the associated basalts, suggest that they may be derived from a compositionally distinct source. The hypothesis that the basaltic suite and LEU derive from different sources has been tested by quantitative trace element modelling (Cebriá & López-Ruiz 1995). The results of this model for the basaltic suite are shown in Table 17.1. This set of parameters reproduces their observed abundances in incompatible trace elements by degrees of melting between 5.0% and 17.6%. These results also suggest that the mantle source of the CVP rocks was enriched in incompatible trace elements with respect to primitive mantle abundances, and that the degree of enrichment is roughly proportional to the degree of incompatibility of the elements considered: 6 to 10 times for the highly incompatible elements (La, Ce, Nb) and 1.2 times for the moderately incompatible ones (Yb, Lu, Y). Additionally, this model supports the idea that the basaltic suite can be derived from a virtually homogeneous source.

The trace element characteristics of LEU cannot be reproduced using the calculated mantle source composition shown in Table 17.1, thus confirming the likelihood that their source must be compositionally different. Unfortunately, the absence

of a series of magmas produced by variable degrees of melting does not allow similar inferences to those used for OLM, OLN and AOB. Cebriá & López-Ruiz (1995) calculated a possible F value of c. 3.7% for the leucitites assuming the D_0 and P_L values of Table 17.1.

Calculated bulk distribution coefficients for the CVP basalts were used by Cebriá & López-Ruiz (1995) to infer the presence of some minerals in the mantle source and to evaluate their maximum contribution to the melt. For example, the values of D_0 and P_L for Yb, Lu and Y, as well as the low abundances and narrow variation ranges of Yb and Lu, can only be explained if garnet was present in the source and contributed in a low proportion to the melt ($<2\%$ if $D_{Gt/liq}$ for Yb and Lu = c. 11.5). On the other hand, the variations in K_2O and Rb require the presence of a potassic mineral in the residue. In the CVP, Cebriá & López-Ruiz (1995) used Sr to discriminate between phlogopite and K-amphibole: as Sr behaves as a highly incompatible element, phlogopite is the most likely potassic mineral in the source for the basaltic rocks, for which D^{Sr} is very small relative to K-amphibole (Arth 1976; Roden 1981). The parameters calculated above can also be used to infer the absence of certain accessory phases. For example, the highly incompatible behaviour of P_2O_5, Zr and LREE suggests that apatite, zircon and sphene are absent in the residue of the basaltic suite. Finally, for the LEU, the negative Sr and P spikes in the normalized trace element diagrams

Table 17.1. Parameters calculated for the partial melting model of the primitive lavas from Calatrava

	D_0	P_L	C_0
P₂O₅	0.014	0.041	423.308
Sr	0.028	0.071	110.878
La	0.009	0.049	5.926
Ce	0.003	0.014	11.709
Nd	0.015	0.021	6.667
Sm	0.026	0.055	1.404
Eu	0.043	0.092	0.521
Gd	0.055	0.121	1.615
Dy	0.098	0.201	1.479
Yb	0.143	0.221	0.554
Lu	0.151	0.153	0.081
Y	0.085	0.182	5.954
Zr	0.043	0.101	33.993
Nb	0.011	0.023	7.683

After Cebriá & López-Ruiz (1995).
D_0, bulk distribution coefficients of the elements for the source paragenesis; P_L, bulk distribution coefficients of the elements for the minerals contributing to the liquid; C_0, concentration of the elements in the source (in ppm).

imply that apatite might be a residual phase in their source. However, the participation of a K-bearing phase in the liquid is higher, as suggested by the stronger enrichment in K, Rb and Ba. The most likely phase is again phlogopite because of its higher contents in Rb and K relative to amphibole.

According to the petrogenetic model of Cebriá & López-Ruiz (1995), OLM, OLN and AOB constitute a suite generated by variable degrees of partial melting (F = 5–17%) of a nearly homogeneous, enriched mantle source, i.e. up to 10 times the primitive mantle values for the highly incompatible elements, and about 1.2 times the moderately incompatible ones. This mantle source, composed of Ol + Opx + Cpx + Gt + Phl, is depleted in ^{87}Sr and enriched in ^{143}Nd relative to the primitive mantle, and its signature is highly similar to the sublithospheric reservoir that can be traced all across Europe and the eastern central Atlantic (Hoernle *et al.* 1995). This means that the sodic lavas from the CVP represent nearly 'pure' European sublithospheric liquids that have not interacted with the lithosphere during their ascent. The leucitites (LEU) seem to be derived by low degrees of partial melting (c. 4%) from a different mantle source, with high Rb, Ba and K contents, and high ^{87}Sr values. This source would correspond to a lithospheric mantle of lherzolite composition containing phlogopite and apatite.

In summary, therefore, geochemical data suggest the participation of two different components in the genesis of the CVP rocks: (1) a ^{87}Sr-depleted and ^{143}Nd-enriched source relative to primitive mantle in OLM, OLN and AOB, which would represent a HIMU-like reservoir; and (2) a ^{87}Sr-enriched and ^{143}Nd-depleted source for the LEU. If we accept the hypothesis of Hoernle *et al.* (1995) that western/central Europe is underlain by a homogeneous sublithospheric reservoir with a common isotopic composition (Hoernle *et al.* 1995; Cebriá & Wilson 1995), the first component corresponds to this reservoir. The second component may then be assigned to an enriched portion of the continental mantle lithosphere.

NE volcanic province

This volcanic province is usually subdivided into three main areas: Garrotxa (also known as Olot), Ampurdán and Selva.

It is located in the NE corner of the Iberian peninsula (Fig. 17.13) at the intersection of the eastern Pyrenees (north), the Ebro basin (west), and the Catalonian Coastal Ranges (south). Two major sets of fractures have decisively influenced the generation and evolution of the NEVP: (1) NW–SE directed transtensional faults that transect the whole area SE from the Pyrenean front, reaching the Mediterranean as the north Balearic fault system (e.g. Doblas & Oyarzun 1990; Martí *et al.* 1992); and (2) NE–SW orientated transtensional faults (to the south and east) related to Cenozoic extension in the Gulf of Valencia and Catalonian Coastal Ranges.

K-Ar ages (Donville 1973*a,b*; Araña *et al.* 1983) suggest that volcanism started shortly after the opening of the Neogene basins in Ampurdán (10–9 Ma BP), continued in Selva (7–2 Ma BP) and finally ended up in Garrotxa (0.7–0.11 Ma BP; Figs 17.2 and 17.13). Although the general evolution of the volcanism as deduced from K-Ar dating is roughly valid, the time-span for the activity in each area is likely to be wider. For example, thermoluminiscence-based plagioclase ages for Garrotxa (Guerín *et al.* 1986) indicate a lower age limit of 11500 years. Such a young age for the Garrotxa volcanism is consistent with the high number of well-preserved outcrops in the region (c. 100; 50 of them cinder cones).

The NEVP consists of some 200 volcanic outcrops irregularly distributed across c. 2500 km² (Fig. 17.13). Volcanism is represented by some well-preserved strombolian cones and lava flows, although some also show sporadic hydromagmatic activity. Explosive craters are small (usually less than 300 m) and there is a wide variety of deposits produced by hydromagmatic eruptions (Martí & Mallarach 1987), including both dry and wet pyroclastic surges. The erosion of the older edifices allows the observation of numerous outcrops of pyroclasts, basaltic lavas and necks.

Petrology and geochemistry. On the basis of their modal and normative mineralogy, the volcanic rocks of the NEVP can be divided into four groups (López-Ruiz & Rodríguez-Badiola 1985): leucite basanites (LBAS), nepheline basanites (NBAS), alkaline olivine basalts (AOB) and trachytes (TR). The latter are restricted to a few outcrops in the Ampurdán area that might represent volcanic necks, and show variable degrees of alteration. Their mineralogy is mainly represented by phenocrysts of plagioclase (andesine-oligoclase) and occasionally phlogopite, amphibole and apatite, embedded in a groundmass of feldspar and Fe-Ti oxides (López-Ruiz & Rodríguez-Badiola 1985; Díaz *et al.* 1996).

LBAS, NBAS and AOB are found in the three areas of the NEVP, although LBAS crop out mainly in Garrotxa. The three groups have very similar mineralogies, all being porphyritic, with olivine (Fo$_{92-74}$) and clinopyroxene (En$_{53-30}$ Fs$_{4-23}$ Wo$_{43-47}$) as phenocrysts, and with a groundmass of plagioclase (An$_{71-58}$, also as microphenocrysts), titanomagnetite, leucite and nepheline (López-Ruiz & Rodríguez-Badiola 1985). The presence of mantle xenoliths is a common feature of all groups. The only significant differences are the higher abundance of leucite in LBAS and the presence of analcime in the groundmass of NBAS. The mineral chemistry is similar in all groups and the compositional variations observed in Garrotxa cover the full range for the NEVP (López-Ruiz & Rodríguez-Badiola 1985). The only regional variations can be attributed to differences in the degree of alteration, which is usually greater in the older rocks (i.e. in Ampurdán and Selva). For these reasons, the geochemical discussion is focused on the rocks from Garrotxa, which are representative of the whole region, excluding TR (López-Ruiz & Rodríguez-Badiola 1985).

Most lavas from Garrotxa have Na₂O + K₂O > 5%, high proportions of MgO, TiO₂ and P₂O₅, and Mg# = 57.7–68.3, typical of relatively primitive alkaline basalts. This is supported by the occasional presence of peridotite xenoliths. LBAS are characterized by higher K₂O values (>1.9%) relative to NBAS and AOB (<1.76%), which

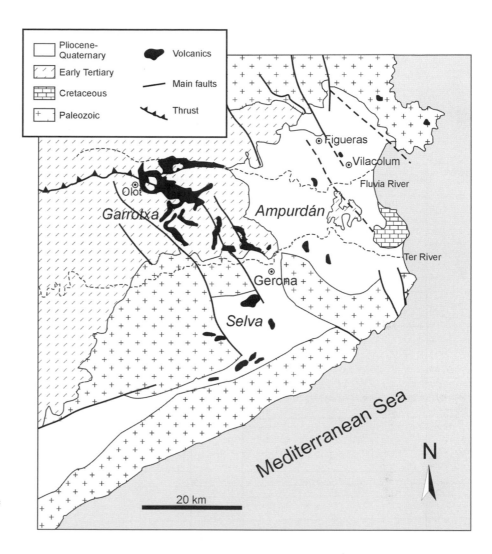

Fig. 17.13. Simplified geological map of the NE volcanic province (modified from Guerin *et al.* 1986).

cannot be readily distinguished on the basis of their major element composition, except for a slightly higher SiO_2 abundance in AOB.

Major elements versus MgO diagrams (Fig. 17.14) show that the three groups crudely follow coherent patterns of depletion in SiO_2 and Al_2O_3 and enrichment in P_2O_5, with increasing MgO, and a wide variation in TiO_2. The apparent scattering of Na_2O and CaO in those diagrams is due to the relatively wide range of LBAS compositions, which is different from the usual trends of enrichment in CaO and depletion in Na_2O relative to MgO.

Normalized trace element diagrams for these lavas are presented in Figure 17.15. All groups show similar patterns, characterized by variable degrees of enrichment relative to primitive mantle concentrations, lowest for Y and Yb (4–5 times) and highest for Ba and Nb-Ta (which display positive anomalies; Cebriá *et al.* 2000). There are no significant compositional differences between the primary and the few evolved lavas, indicating that the latter underwent only a small degree of fractional crystallization. LBAS show the highest enrichment in all elements, whereas the NBAS and AOB display nearly identical patterns and levels of abundance. The only significant variations between the more enriched samples (represented by LBAS) and the less enriched ones (NBAS and AOB) is the progressively larger negative anomaly in Rb and Th. Most samples show a small positive Zr anomaly, although this is very weak or absent in NBAS. The normalized diagram in Figure 17.15 shows that, although the subdivision between NBAS and AOB is valid when considering their petrography and their major element compositions, it has no significance when considering trace elements (and isotopic ratios, as we will show).

These trace element patterns are also similar to those of other Cenozoic lavas of Europe (e.g. Chauvel & Jahn 1984; Wilson & Downes 1991; Wedepohl *et al.* 1994; Cebriá & López-Ruiz 1995) which have HIMU-like characteristics such as positive Nb-Ta anomalies (e.g. St. Helena, Bouvet, and Ascension islands; Weaver 1991) and negative Th anomalies (e.g. Tubuai island; Chauvel *et al.* 1992). However, the strong positive Ba anomaly, which is not usual in HIMU patterns, is typical of some EM1 lavas (e.g. Gough island; Weaver 1991).

The Sr–Nd–Pb isotopic characteristics of Garrotxa are shown in Figures 17.6 and 17.7. Their $^{87}Sr/^{86}Sr$ and $^{143}Nd/^{144}Nd$ ratios plot in a narrow field with average values of 0.70365 ± 7 and 0.51274 ± 3, respectively. The $^{206}Pb/^{204}Pb$, $^{207}Pb/^{204}Pb$, $^{208}Pb/^{204}Pb$ ratios are in the ranges 18.88–19.47, 15.59–15.68, 38.64–39.26, respectively (Cebriá *et al.* 2000). Thus, the Garrotxa lavas have an intermediate signature between HIMU and EM1, similar to contemporaneous basalts from other parts of Europe (Wilson & Downes 1991). This interpretation agrees with the relatively low $\delta^{18}O$ values (*c.* 6.17‰; Benito *et al.* 1992), which fall within the range of typical intracontinental primary lavas (e.g. Kyser 1986; Harmon & Hoefs 1995; Hoefs 1997). Although this might represent the signature of the mantle source, its slightly higher value as compared to the nearly constant ratio of MORBs (5.7 ± 0.2‰) suggests the participation of a relatively enriched source (Hoefs 1997).

Despite the lack of a sharp linear relationship between $^{87}Sr/^{86}Sr$ and $^{143}Nd/^{144}Nd$, Pb isotopes show excellent correlations ($r = 0.83$ for $^{206}Pb/^{204}Pb$–$^{207}Pb/^{204}Pb$ and $r = 0.98$ for $^{206}Pb/^{204}Pb$–$^{208}Pb/^{204}Pb$) nearly parallel to the NHRL, even if they are relatively more enriched in $^{207}Pb/^{204}Pb$ and slightly less in $^{208}Pb/^{204}Pb$ (Cebriá *et al.* 2000). These ratios fall also in intermediate positions between HIMU and EM1.

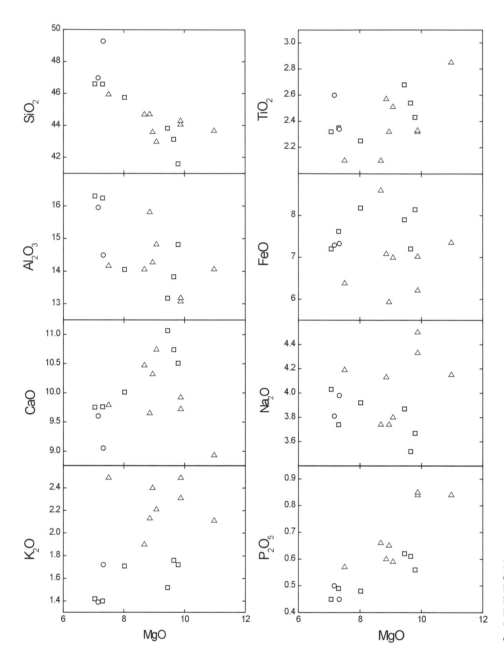

Fig. 17.14. MgO–major element diagrams (values in wt%) for the leucite basanites (triangles), nepheline basanites (squares), and alkali olivine basalts (circles) for the lavas of the NE volcanic province. Data from Cebriá *et al.* (2000).

Although this variation could be interpreted as a secondary isochron of 1.57 Ga (κ = 3.706), we will show that it represents the interaction of two compositionally distinct components.

Petrogenetic model

The origin of the NEVP lavas has been previously studied by Araña *et al.* (1983), López-Ruiz & Rodríguez-Badiola (1985), López-Ruiz *et al.* (1986), and, more comprehensively by Cebriá *et al.* (2000). The homogeneity of the trace element patterns of the Garrotxa lavas, together with other geochemical characteristics (e.g. mantle-like isotopic signature, presence of mantle xenoliths) suggest that they are the result of partial melting from a single enriched mantle source. This hypothesis was tested for the Garrotxa primary melts using the quantitative modelling of Cebriá & López-Ruiz (1996) by

Cebriá *et al.* (2000). The results of this modelling suggest melting percentages between 4% and 16% and an enriched mantle source with a K-bearing phase. The calculated distribution coefficients (Table 17.2) allowed Cebriá *et al.* (2000) to establish several constraints on the mineralogy of the mantle source. For example, the bulk distribution coefficients calculated for Yb confirm that garnet was present during melting, but that its proportion in the source and its contribution to the liquid should be relatively small (<2% and <8% respectively, with $D_{Gt/liq}^{Yb} = 6.4$; Halliday *et al.* 1995). Concerning the inferred K-bearing phase, the bulk distribution coefficients calculated for Rb, Ba and K can easily be reproduced with assemblages including phlogopite as the only potassic phase, or phlogopite + kaersutite. However, the partition coefficients calculated for Sr are easier to attain in the latter case. Finally, the low values of the bulk distribution coefficients for K–Rb–Ba are in

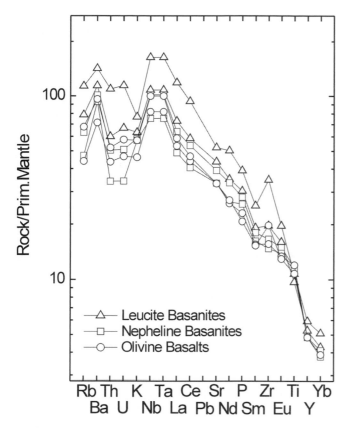

Fig. 17.15. Multielement diagram normalized to primitive mantle (after Sun & McDonough 1989) for the lavas of the NE volcanic province. Data from Cebriá *et al.* (2000).

phase that can fractionate Rb and that $D_{Phl/liq}^{Rb} = 1.7$ (Halliday *et al.* 1995), the source would have *c.* 2% of phlogopite entering the liquid in a proportion <12%.

These characteristics indicate that the magmas were derived from enriched portions of the mantle lithosphere. However, the isotopically heterogeneous character of this source (as suggested by Pb isotopes) may be related to the interaction of several reservoirs. According to Cebriá *et al.* (2000), this apparent contradiction can be explained in terms of the common geochemical features of the European volcanic province (Hoernle *et al.* 1995; Cebriá & Wilson 1995). This hypothesis proposes that the geochemically similar sodic lavas of Europe (melilitites, nephelinites, basanites, alkali olivine basalts) were derived from a common HIMU-like sublithospheric source (the low velocity component (LVC) of Hoernle *et al.* (1995) or the European asthenospheric reservoir (EAR) of Cebriá & Wilson (1995)), whereas the potassic lavas (leucitites, and some leucite basanites) originated from enriched portions of the mantle lithosphere. A good example of these two extreme compositions can be found in the CVP, as shown in the previous section.

The lavas of Garrotxa do not represent 'pure' European sublithospheric liquids (see Figures 17.6 and 17.7). The trace element and isotopic signature of these rocks are not significantly different from those of the Massif Central, which resulted from the interaction between the sublithospheric reservoir and a lithospheric component (Cebriá & Wilson 1995). Thus, the geochemical characteristics of the Garrotxa mantle source may also be the result of interacting sublithospheric and lithospheric components. If this is correct, the partial melting curve of the Garrotxa lavas as plotted in a trace element C^i/C^j–C^i diagram should end up (F = 100%) at the composition of the hybrid source, which is located along a mixing curve between two components (Fig. 17.16; Sims & DePaolo 1997). In Garrotxa, the melting curve and the composition of the hybrid source can be obtained from the modelling of Cebriá *et al.* (2000). For the mixing curve that defines the source, the hypothesis of Cebriá *et al.* (2000) implies that the composition of one component should be represented by sublithospheric LVC/EAR-derived liquids (see Figs 17.6 and

agreement with the observations made for the normalized trace element diagrams, indicating that the proportion of phlogopite ± kaersutite in the source has been small, and that these phases entered the melt in percentages allowing them to remain in the residue. For example, if we consider the parameters in Table 17.2, and assuming that phlogopite is the only

Fig. 17.16. Log(La/Nb)–log(La) diagram for the primary lavas of Garrotxa (symbols as in Fig. 17.14) and calculated partial melting and mantle metasomatism (binary mixing) curves modified after Cebriá *et al.* (2000). LVC/EAR: low velocity component/European asthenospheric reservoir. The trace element composition of the LVC/EAR liquids is assumed to correspond to the average concentrations of those samples from the CVP whose isotopic ratios match those proposed for the LVC/EAR. The assumed composition of the lithospheric component (La = 2.5 ppm and La/Nb = 0.579) is similar to the lithospheric spinel peridotites reported by McDonough (1990).

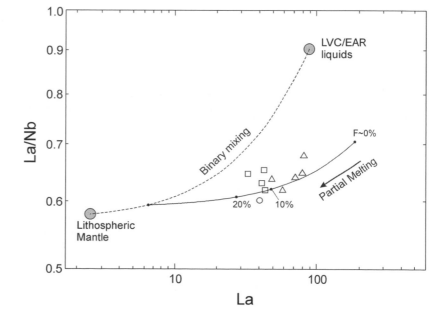

Table 17.2. Parameters calculated for the partial melting model of the primitive lavas from Garrotxa

	D_0	P_L	C_0
K_2O	0.109	0.316	0.390
P_2O_5	0.070	0.217	0.115
Rb	0.032	0.103	5.594
Ba	0.118	0.340	150.859
Sr	0.118	0.340	163.605
La	0.029	0.184	5.372
Ce	0.035	0.100	13.678
Nd	0.057	0.184	7.070
Sm	0.090	0.270	1.553
Eu	0.117	0.336	0.541
Tb	0.160	0.425	0.251
Yb	0.149	0.502	0.428
Lu	0.151	0.151	0.069
Y	0.152	0.411	5.576
Zr	0.038	0.119	31.311
Hf	0.067	0.206	0.856
Nb	0.037	0.129	9.723
Ta	0.045	0.136	0.599
Th	0.011	0.071	0.465

After Cebriá *et al.* (2000).
D_0, bulk distribution coefficients of the elements for the source paragenesis; P_L, bulk distribution coefficients of the elements for the minerals contributing to the liquid; C_0, concentration of the elements in the source (in ppm, except K_2O and P_2O_5, in %).

17.7). The composition of the lithospheric component is more difficult to establish as there are no available data and the mantle lithosphere displays wide ranges (e.g. McDonough 1990). The La/Nb–La diagram shows that, as expected, the sublithospheric composition is not the source of the Garrotxa lavas, and that the composition of the lithospheric component should have La < 6.3 ppm and La/Nb < 0.594 (Fig. 17.16). These conditions are reached with La = 2.5 ppm and La/Nb = 0.579, which are very close to the average of McDonough (1990) for lithospheric spinel peridotites (La = 2.6 ppm and La/Nb = 0.54). This implies a *c.* 5% participation of LVC/EAR-derived liquids in a lithospheric mantle source, in which the required high concentration in Nb also suggests an enriched source similar to some K-rich fertile lherzolites (e.g. sample 20 of Hartmann & Wedepohl 1990). Furthermore, the Pb isotopic ratios observed in the Garrotxa lavas can also be explained by this enrichment event: the addition of *c.* 4–5% of sublithospheric liquids ($^{206}Pb/^{204}Pb = 20.0$, $^{207}Pb/^{204}Pb = 15.65$, $^{208}Pb/^{204}Pb = 39.75$, Pb = 5 ppm; Wilson & Downes 1991; Wedepohl *et al.* 1994; Hegner *et al.* 1995; Hoernle *et al.* 1995; Wilson *et al.* 1995) into an enriched lithospheric source similar to the high-K peridotites mentioned above (e.g. $^{206}Pb/^{204}Pb = 18.28$, $^{207}Pb/^{204}Pb = 15.56$, $^{208}Pb/^{204}Pb = 38.43$, Pb = 0.16 ppm; Rosenbaum & Wilson 1997) produces a hybrid source with the characteristics required for the Garrotxa lavas. This supports the idea that the relatively high Pb ratios of these lavas are not the consequence of a long-lived evolution of liquids stored in the lithosphere by an old enrichment episode. They rather are inherited from a metasomatic event involving small amounts of sublithospheric liquids.

To summarize, the petrogenetic model proposed by Cebriá *et al.* (2000) involves at least two enrichment episodes of the lithospheric mantle before melting produced the Garrotxa lavas. The first event (which could be related to Late Variscan extension) generated a series of domains similar to those

identified in other European volcanic regions as the source of high-K lavas (leucitites, leucite basanites). Following the emplacement of the LVC/EAR component under the NEVP, a second enrichment episode facilitated the infiltration of sublithospheric liquids stored in the mantle lithosphere. Finally, melting involved both the stored sublithospheric liquids and the old enriched portions of the mantle lithosphere. The lavas produced from this hybrid source show trace element variations indicative of partial melting from a single mantle source with a relatively heterogenous isotopic signature. An additional consequence is that the inferred presence of garnet in the source might not be a feature of the mantle lithosphere, and could represent a geochemical relic of a second enrichment event produced by deeper sublithospheric liquids. This has been suggested in the Rhön area of Germany to account for the presence of spinel–clinopyroxene clusters that substitute garnet in mantle lithosphere xenoliths as a result of the destabilization produced by mantle uplifting (Witt-Eickschen & Kramm 1997). The strong enrichment in Rb, Ba and the inferred presence of a K-bearing phase can be attributed to both enrichment events. K-bearing phases are common in enriched portions of the lithosphere and they have also been reported in the mantle source of sublithospheric-derived liquids (e.g. Wilson & Downes 1991; Cebriá & López-Ruiz 1995). Finally, this model also accounts for the presence in the Carpathian-Pannonian region of pervasive sublithospheric veins in mantle xenoliths (Rosenbaum & Wilson 1997), suggesting that the infiltration/storage of sublithospheric-derived melts in the lithosphere is a common characteristic of the European domain.

Geodynamic setting

In this section we review the different geodynamic models that have been suggested for Cenozoic volcanism in the Iberian peninsula, and we propose a new tectonomagmatic model that integrates the available geological, geophysical and geochemical data. Few geodynamic scenarios have been proposed for tholeiitic volcanism in the western Betics and Alborán island, and for alkaline basalts in the CVP and NEVP. By contrast, many, often conflicting, tectonomagmatic models have been suggested for calc-alkaline and related volcanism of the SEVP and GVVP. The geodynamic model proposed here argues that Cenozoic volcanism in Iberia and related regions can best be understood in terms of four stages: tholeiitic western Betics/proto-Alborán stage; calc-alkaline Valencia Basin stage; calc-alkaline to ultrapotassic Betics/Rif, Alborán stage; and, alkaline basaltic trans-Moroccan, western Mediterranean, European stage.

Previous models

Different models have been proposed for the western Betics/proto-Alborán, SEVP, GVVP, CVP and NEVP domains. Some of these models have implications regarding neighbouring Cenozoic volcanic areas, e.g. Alborán Sea, northwestern Africa, and western/central Europe. With regard to the earliest magmatism, conflicting absolute ages exist for tholeiitic east–west orientated dyke swarms in Málaga (23–22 Ma BP; Torres Roldán *et al.* 1986; and 30–27 Ma BP; Turner *et al.* 1999) and volcanic rocks in Alborán Island (25–20 Ma BP; Bellon & Brousse 1977; and 18–7 Ma BP, Aparicio *et al.* 1991). One geodynamic scenario for this magmatism suggests an early Miocene back-arc basin model involving two centripetal (dipping towards the Alborán region) subduction zones (Torres Roldán *et al.* 1986). As we will argue in the following section, this scheme can be ruled out because subduction processes ceased around the Cretaceous–Tertiary boundary. In contrast, Turner *et al.* (1999) suggest that this magmatism might be ascribed to the extensional collapse of the Betics–Rif orogenic system, involving the convective removal of three lithospheric bodies. However, the existence of several detached lithospheric mantle roots beneath this area is incompatible with the geophysical data showing only one removed body (e.g. Seber *et al.* 1996b).

Even more controversial has been the interpretation of the SEVP and GVVP. Geodynamic models suggested for volcanism in these two provinces can be roughly subdivided into six groups: compressional, extensional, subduction, detached lithospheric roots, transtensional, and mantle upwellings. Early models emphasizing compressional tectonics (e.g. Andrieux *et al.* 1971; Araña & Vegas 1974) were superceded by a recognition that extensional tectonics played a key role in the evolution of the Betics (e.g. Doblas & Oyarzun 1989a,b; Platt & Vissers 1989; Van der Wal & Vissers 1993). Initial extensional models explained the volcanism in terms of leaking high-angle normal faults (Torrés Roldán 1979; Puga 1980) but, not having access to later ideas on the extensional collapse of orogens, they were unable to account for the whole picture. The extensional model by Doblas & Oyarzun (1989a,b) envisaged the development of a 'basin-and-range' type province with widespread low-angle normal faulting, upwelling phenomena, and associated lithospheric melting and extrusion of volcanic rocks, a scenario resulting directly from the collapse of the overthickened Betics–Rif orogen. The older ideas on compressional tectonics (e.g. Araña & Vegas 1974) became further weakened by an increasing acceptance that there is no clearly defined 'Alborán microplate', as demonstrated by the deformation and seismic characteristics of the Alborán sea and its uppermost mantle (Vissers *et al.* 1995; Seber *et al.* 1996b; Lonergan & White 1997).

With regard to the many subduction models proposed to explain Betic magmatism, there are essentially three major categories: (1) intracontinental subduction with northward and southward polarity (Maury *et al.* 2000); (2) static oceanic subduction with either a single north-dipping (Araña & Vegas 1974) or south-dipping (Torres Roldán 1979) Benioff zone, or with double polarity (López-Ruiz & Rodríguez-Badiola 1980; Torres Roldán *et al.* 1986); (3) retreating oceanic subduction rollback migrating towards the eastern Mediterranean (Doglioni *et al.* 1997; Carminati *et al.* 1998), towards the west (Gibraltar strait and the eastern Atlantic passive margin; Royden 1993), or in both directions at the same time (Lonergan & White 1997; Wilson & Bianchini 1999). Calc-alkaline volcanism of this sector has, within this context, been misinterpreted by some in terms of classical 'orogenic' magmatism associated with an Andean-type or back-arc-type active margin (e.g. Banda & Channell 1979; Doglioni *et al.* 1997; Wilson & Bianchini 1999).

There are abundant arguments that indicate that there was no active Neogene subduction within this southwestern European domain: (1) no deep oceanic trenches, ophiolites, or continent-directed metamorphic polarities are observed (De Larouzière *et al.* 1988; Doblas & Oyarzun 1990; Oyarzun *et al.* 1995); (2) the SE Spanish margin is too narrow (150 × 25 km)

and thin (*c.* 20 km) to explain the complex volcanic suite in terms of active subduction; (3) the Alborán sea is too small to allow double north- and south-directed subduction (Doblas & Oyarzun 1989a); (4) timing of the magmatic events is post-collisional (López-Ruiz & Rodríguez-Badiola 1980; Doblas & Oyarzun 1989a; Benito *et al.* 1999; Maury *et al.* 2000); (5) the geochemical and geochronological polarities observed outwards from the Alborán sea (NW-directed in Spain and south-directed in Morocco) are incompatible with a single subduction scenario (Doblas & Oyarzun 1989a); (6) the thermal history of the metamorphic rocks of the Alborán sea does not support the rollback subduction model (Platt *et al.* 1998); (7) there is no geophysical evidence for the hypothetical Neogene to present existence of active subduction (Vissers *et al.* 1995; Seber *et al.* 1996b; Mezcua & Rueda 1997). However, we should point out that a north-dipping subduction system probably existed during Late Cretaceous to early Tertiary convergence and collision in the Betics–Rif orogenic belt (Doblas *et al.* 1991; Turner *et al.* 1999; Benito *et al.* 1999).

The concept of removal of lithospheric roots in overthickened orogens has recently received increased attention as the ultimate cause of late-orogenic extensional collapse of mountain belts such as the Betics–Rif system, where continued convergence exists between the African and Eurasian plates (e.g. Seber *et al.* 1996b; Burg & Ford 1997; Platt *et al.* 1998). Two main mechanisms have been proposed (see review in Burg & Ford 1997; Platt *et al.* 1998): asymmetric delamination of lithosphere (Docherty & Banda 1995; Mezcua & Rueda 1997; Platt *et al.* 1998; Calvert *et al.* 2000); and symmetric convective removal of the lithospheric roots (Vissers *et al.* 1995; Seber *et al.* 1996b). The two hypotheses (delamination and symmetric removal) are both capable of explaining the gross geological features of the Betics–Rif orogenic system, and it is commonly difficult to discriminate between them (Vissers *et al.* 1995; Seber *et al.* 1996b; Lonergan & White 1997; Platt *et al.* 1998): both depend on postulates about the rheology and long-term behaviour of the mantle lithosphere in a region of active convergence, neither of which are easy to measure or observe directly (Platt *et al.* 1998). Some data argue against the delamination model: the observed symmetry both in surface geology and subsurface structures (Seber *et al.* 1996b); the absence of widespread and massive crustal melting (Turner *et al.* 1999); theoretical physical and rheological constraints (Turner *et al.* 1999); the supposedly oblique trend of the Alborán–Valencia rift system with respect to the Betics–Rif orogenic belt (Doglioni *et al.* 1997); and the absence of either east- or west-directed age or deformation polarities outwards from the Valencia basin (Turner *et al.* 1999). Other data argue against the symmetrical convective removal model. For example, the high metamorphic temperatures deduced for the Alborán basement rocks and the shallow depths at which they attained them, require that virtually all the lithospheric mantle had been removed (Platt *et al.* 1998). Also, such supposed symmetry is not supported by the asymmetric geometry of the uppermost mantle inferred from intermediate and very deep seismicity (Seber *et al.* 1996b; Mezcua & Rueda 1997). Lonergan & White (1997) point out that neither of the two models (delamination and symmetric removal) can explain the large vertical-axis block rotations. Finally, a complex delamination model involving a subducted slab has also been suggested (Blanco & Spakman 1993a; Zeck 1999), as well as a model proposing the existence of three different bodies of detached lithospheric mantle (Turner *et al.*

1999). According to Calvert *et al.* (2000) a modified delamination model, involving a subducted oceanic slab (Blanco & Spakman 1993*a*; Zeck 1999), should be ruled out as it would require any subducted lithosphere to remain hanging in the mantle for 60 Ma before finally detaching and descending into the mantle. The modified symmetric removal scenario of Turner *et al.* (1999), involving several detached lithospheric roots, appears to us incompatible with the geophysical data showing a single major symmetrical body (e.g. Seber *et al.* 1996*b*) and with the NE-trend of the alkaline basaltic stage.

A transtensional model involving a SW–NE trending 'trans-Alborán' sinistral shear zone has been suggested by Hernández *et al.* (1987) and De Larouzière *et al.* (1988) to account for Cenozoic volcanism in SE Spain and the Alborán sea. Although it seems doubtful whether this fault could deliver enough shear heating to create the magmatism, it could have acted as a conduit for mantle-derived magmas (Zeck 1996). According to Maury *et al.* (2000), however, this transtensional model cannot account for the east–west orientated Cenozoic magmatic events of the Mediterranean Maghreb margin. Neither does it explain the following: the crude reverse geochemical and geochronological polarities observed in SE Spain and Morocco; the geochemical signatures of the calc-alkaline to ultrapotassic volcanic suites of the SEVP; the east–west orientation of the Mazarrón–Cartagena volcanic field; the strong crustal thinning observed towards the Alborán realm; and the generalized extensional collapse of the region with unroofing of metamorphic core complexes.

Mantle upwelling models evolved from the early ideas of Van Bemmelen (1966) and Ritsema (1970), through Weijermars (1988), who suggested that the Alborán sea formed as a result of an ascending mantle diapir, to Doblas & Oyarzun (1989*a,b*, 1990) who argued that the extensional disruption of the Betics–Rif orogen and the Valencia basin were characterized by low-angle detachment tectonics, and upwelling/exhumation of deep-seated core complexes. Although the idea of mantle upwelling became unpopular in the early days of plate tectonics theory, it is now widely recognized that both horizontal and vertical movements occur. In fact, most of the currently competing models in Iberia involve some form of mantle upwelling process, particularly the detached lithospheric root model, the subduction rollback scheme, and the extensional scenarios. Moreover, a recent model by Oyarzun *et al.* (1997) argues that large-scale Late Triassic to present asymmetric mantle upwelling phenomena were active within a gigantic NE-directed province (eastern central Atlantic, western Africa and Europe), ultimately related to the impingement of a Triassic–Jurassic superplume in the proto-central Atlantic.

With regard to the CVP, this was originally interpreted in terms of unspecified hot-spot processes or alpine compression (Ancochea 1982). Latter ideas suggested a more complex model in terms of NW-directed indentation of the eastern Betic Cordillera, triggering NW–SE normal faults in its foreland and the ascent of a mantle diapir (Doblas *et al.* 1991;

López-Ruiz *et al.* 1993). Several authors have suggested that large-scale transtension can be recognized in a similar NE-directed trend from Morocco up to northern Europe, constituting a gigantic Cenozoic tectonic province (e.g. Tapponnier 1977; Sanz de Galdeano 1990).

The alkali basalts of the NEVP were interpreted by Martí *et al.* (1992) as related to NW–SE normal faults which would be the onshore continuation of the north Balearic fault zone, the whole associated with the late opening stage of the Gulf of Valencia. These authors suggest that two extensional events favoured the interaction of different mantle sources. Recently, Cebriá *et al.* (2000) have argued that late indentation phenomena in the eastern Pyrenees might have triggered NW–SE normal faults in the foreland of the NEVP. The geochemical characteristics of this volcanic province indicate two mantle enrichment episodes, the second one related to the ascent of a mantle diapir from a so-called low-velocity component or European common sublithospheric reservoir (e.g. Hoernle *et al.* 1995; Cebriá *et al.* 2000).

A new tectonomagmatic model

We argue that Cenozoic volcanism in Iberia might be understood in terms of four major tectonomagmatic stages, within different areas, commonly bearing contrasted geological/geochemical characteristics and ages: (1) Oligocene to earliest Miocene tholeiitic stage in the western Betics–proto-Alborán domain; (2) Late Oligocene to early Miocene calc-alkaline stage in the GVVP; (3) Middle to late Miocene calc-alkaline to ultrapotassic stage in the SEVP; and (4) Late Miocene to present alkaline basaltic stage in the SEVP, CVP, GVVP, and NEVP.

Tholeiitic western Betics–proto-Alborán stage. Initial synorogenic extension began during the Oligocene and earliest Miocene interval in the western Betics–proto-Alboran realm, leading to the collapse of its gravitationally unstable belt (e.g. Doblas & Oyarzun 1989*b*). The localized extensional disruptions triggered crustal thinning, partial detachment of the overthickened lithospheric roots, and upwelling of hot mantle (Fig. 17.17c). This induced the generation of tholeiitic magmatism along leaking normal faults, cropping out today as the Málaga dyke swarms and the Alborán Island volcanics.

Calc-alkaline Valencia trough stage. The major opening phase of the Valencia trough occurred in late Oligocene–early Miocene times, involving the SE-directed clockwise rotation of the Balearic rise (Fig. 17.18; Doblas & Oyarzun 1990). Generalized asymmetric extensional conditions prevailed, producing the NE–SW orientated and SE-dipping Valencia-Catalonian low-angle detachment surface, accompanied by NE–SW 'basin-and-range' type grabens and horsts bounded by high-angle normal faults (Fig. 17.18). This strong crustal thinning triggered upwelling of sublithospheric material,

---

Fig. 17.17. Tectonomagmatic scenario proposed for the evolution of the Betics–Alborán–Rif realm. (**a**) Late Cretaceous convergence between Africa and Iberia with a north-dipping subduction zone (AS, asthenosphere; CC, continental crust; LI, lithosphere; OC, oceanic crust; MML, metasomatized mantle lithosphere). (**b**) Late Cretaceous to early Palaeogene collision between Africa and Iberia, generating the doubly vergent Betics–Rif orogenic belt. (**c**) Oligocene to earliest Miocene synorogenic extensional phenomena in the western Betics/proto-Alborán realm and associated tholeiitic magmatism (PAI, proto-Alborán island; MD, Málaga dykes). (**d**) Middle to late Miocene extensional collapse of the Betics–Rif orogenic belt and generation of the Alborán sea and Gibraltar arc. Volcanism: CA, calc-alkaline lavas; KCA, high-K calc-alkaline lavas; SH, shoshonites; UP, ultrapotassic rocks. BD, Betics detachment; BT, Betics frontal thrusts; MD, Moroccan detachment; MT, Moroccan frontal thrusts.

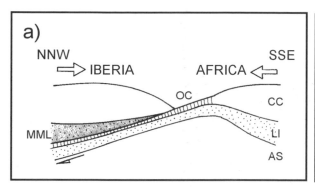

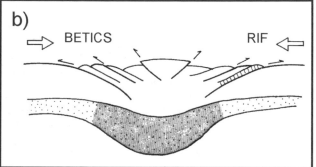

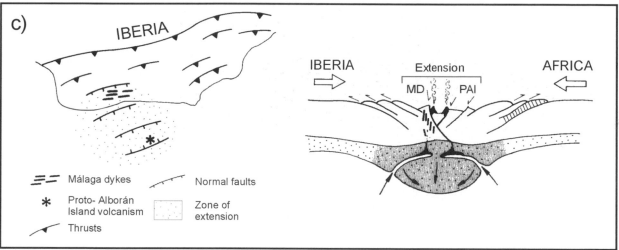

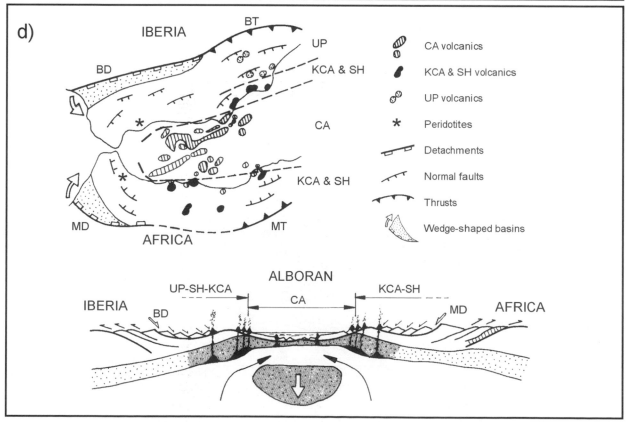

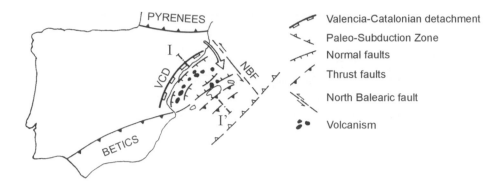

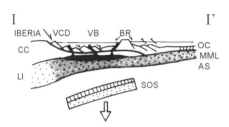

Fig. 17.18. Tectonomagmatic scenario proposed for the late Oligocene to early Miocene opening of the Valencia trough (VB) and asssociated calc-alkaline volcanism. Key: AS, asthenosphere; BR, Balearic rise; CC, continental crust; LI, lithosphere; MML, metasomatized mantle lithosphere; NBF, north Balearic fault; OC, oceanic crust; SOS, subducted oceanic slab; VCD, Valencia-Catalonian detachment.

melting of the lithospheric mantle, and, ultimately, extrusion of calc-alkaline volcanics along leaking high-angle normal faults pervading the Valencia trough (Doblas & Oyarzun 1990). The NW–SE orientated, sinistral north Balearic transform/transfer fault zone formed the northern boundary of the rotating Valencia trough (Fig. 17.18). A Late Cretaceous to earliest Tertiary subduction zone dipping towards the NW probably existed to the east (Fig. 17.18). However, as there are no available isotopic data for the CA volcanics of the Valencia trough, the existence of a subduction-related

metasomatized lithospheric mantle beneath this realm remains conjectural.

Calc-alkaline to ultrapotassic Betics–Rif–Alborán stage. The tectonomagmatic history of this whole region can be regarded as resulting from the overprinting of the Late Cretaceous to early Palaeogene Betics–Rif alpine compressional orogenic system by late Palaeogene to Neogene extension (Fig. 17.17; Doblas & Oyarzun 1989a,b). The convergence between Africa and Eurasia was characterized by northward-directed

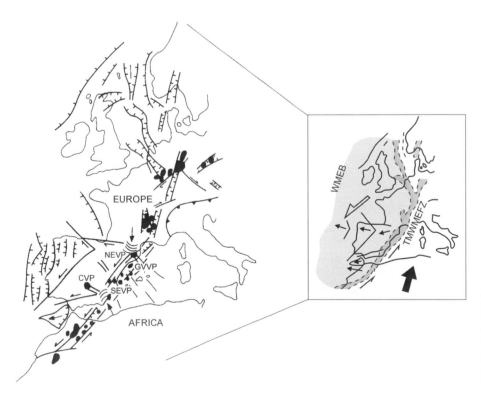

Fig. 17.19. Tectonomagmatic scenario proposed for the latest Miocene to present alkaline basaltic volcanism associated to a complex transtensional megafault (TMWMEFZ, trans-Moroccan, western mediterranean, European fault zone). WMEB, western Mediterranean–European block.

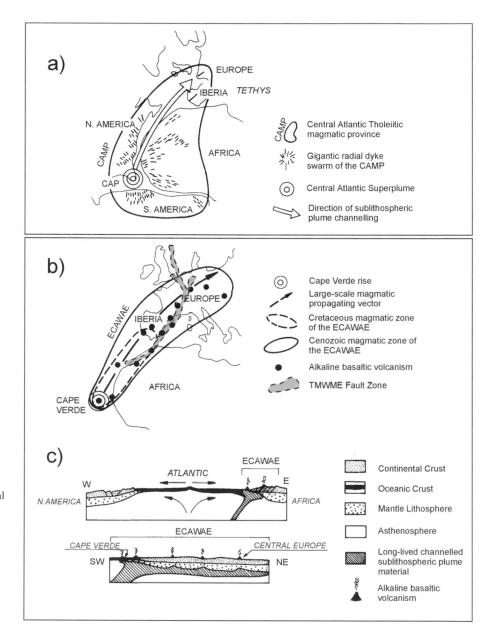

Fig. 17.20. Sketches depicting the Late Triassic to present tectonomagmatic evolution of the NE-directed eastern central Atlantic, western Africa and Europe (ECAWAE) domain. From Oyarzun *et al.* (1997). **(a)** Pre-drift Triassic–Jurassic configuration. CAMP, central Atlantic tholeiitic magmatic province; CAP, central Atlantic superplume. **(b)** Cretaceous to present passive margin scenario of the ECAWAE. **(c)** Schematic cross-sections highlighting asymmetric mantle upwelling phenomena in the Cenozoic ECAWAE.

subduction (Fig. 17.17a) culminating in Late Cretaceous times with collision/obduction between Iberia and North Africa (Fig. 17.17b). This alpine collision is interpreted as having built a symmetrical orogenic edifice with north- and south-vergent nappe stacking (Betics and Rif, respectively; Fig. 17.17b).

Synorogenic extensional disruption of the Betics–Rif mountain belt started in the west during Oligocene to earliest Miocene times, triggering the tholeiitic magmatism of the previous stage (Fig. 17.17c). The main phase of post-orogenic extension took place during Serravalian to Tortonian times in the east, and continued until Messinian times in the west (Doblas & Oyarzun 1989a,b). Meanwhile, the eastern external front of the Betic and Rif mountains underwent final gravitational-related thrusting (Fig. 17.17d). Coeval volcanism pervaded SE Spain (SEVP), the Alborán sea, and northern Africa (Fig. 17.17d). A crude symmetrical distribution of the different magmatic suites and ages of emplacement exists northwards and southwards from Alborán (Fig. 17.17d). This

defines a central region with the older calc-alkaline volcanics (SE tip of Spain; Alborán sea; and coastal sector of northern Africa), two peripheral belts (narrow in SE Spain and wider in northern Africa) with slightly younger K-rich calc-alkaline and shoshonitic rocks, and an outer belt of the youngest volcanism (ultrapotassic) in the SEVP (Bellon & Brousse 1977; López-Ruiz & Rodríguez-Badiola 1980; Doblas & Oyarzun 1989a; Zeck *et al.* 1999). The geochronological polarity is not perfect as the ages partially overlap in some of the belts. The geochemical variations in the magmas are interpreted in terms of different depths of melting and contrasted degrees of partial melting from a heterogeneously metasomatized mantle, above a descending lithospheric root that was detached during the extensional events (Fig. 17.17d).

Alkaline basaltic trans-Moroccan, western Mediterranean, European stage. The last tectonomagmatic stage is interpreted to have resulted from an increased rate of convergence between Africa and Eurasia from latest Miocene times to the present (Tapponnier 1977; Sanz de Galdeano 1990; Doblas *et al.* 1991; Ziegler 1994). North–south directed compression became predominant and a gigantic realm from Morocco up to central Europe was disrupted by a complex megafault zone (trans-Moroccan, western Mediterranean, European; TMWME), bounding a west-directed escaping western Mediterranean European block (WMEB; Fig. 17.19). This fault system is expressed as a complex zone linking the following sectors: an ENE-orientated transpressional sinistral southernmost sector in Morocco; a huge NE-directed predominantly sinistral sector running from northern Morocco, through the Alborán sea, eastern Iberia, to the Valencia and Lyon basins; a NNE- to north-directed predominantly extensional zone crossing France and Germany; and a north- to NNW-directed extensional zone in northern Europe and the North Sea.

The TMWME megafault zone was pervaded by Cenozoic alkaline basaltic volcanism (Morocco, Alborán, SEVP, GVVP, NEVP, Massif Central, and Rhenish Massif), and its influence also triggered magmatism outside its main trend in the south-central part of Iberia (CVP) and the Bohemian Massif in northern Europe. Within this broad context, the CVP is viewed as resulting from NW-directed indentation in the eastern Betics, generating NW–SE normal faults that reused old Variscan trends (Fig. 17.19). Similarly, the NEVP

is interpreted as resulting from SE-directed indentation in the eastern Pyrenees, generating NW–SE normal faults that reused the previous trend of the north Balearic transform/transfer fault zone. In contrast, the alkaline basalts of the GVVP (Cofrentes, Picassent, and Valencia basin) are more directly related to the sinistral transtensional displacement of the TMWME.

We argue that a key factor in the generation of the TMWME-related alkaline basaltic volcanism is an association with a huge and long-lived (Late Triassic to present) area of asymmetric mantle upwelling (Oyarzun *et al.* 1997). This area is postulated to extend northeastward through the eastern central Atlantic, western Africa and Europe, and closely follows the trends of the TMWME megafault zone (Fig. 17.20). This mantle upwelling is thought to relate to the impingement of a Triassic–Jurassic superplume head at the former triple junction between North America, South America and Africa (Oyarzun *et al.* 1997). The plume head material was subsequently channelled northeastward along the transtensional TMWME megafault zone which provided a giant conduit for the leakage of alkaline basalts to the Earth's surface (Oyarzun *et al.* 1997).

This work was financially supported by the Spanish Dirección General de Investigación through Project PB98-0507. We wish to thank Julian Pearce and the editors for helpful and constructive reviews.

18 Cenozoic volcanism II: the Canary Islands

JUAN CARLOS CARRACEDO (coordinator),
FRANCISCO JOSÉ PÉREZ TORRADO, EUMENIO ANCOCHEA,
JOAQUÍN MECO, FRANCISCO HERNÁN, CARMEN ROSA CUBAS,
RAMÓN CASILLAS, EDUARDO RODRÍGUEZ BADIOLA &
AGUSTINA AHIJADO

The Canarian archipelago comprises seven main volcanic islands and several islets that form a chain extending for *c.* 500 km across the eastern Atlantic, with its eastern edge only 100 km from the NW African coast (Fig. 18.1). The islands have had a very long volcanic history, with formations over 20 million years old cropping out in the eastern Canaries. Thus all stages of the volcanic evolution of oceanic islands, including the submarine stage as well as the deep structure of the volcanoes, can be readily observed. Rainfall and vegetation cover are relatively low, with the exception of the island of La Palma, favouring both geological observation and rock preservation. Furthermore, the absence of surface water has promoted groundwater mining by means of up to 3000 km of subhorizontal tunnels (locally known as 'galerías'). These galerías are especially numerous in Tenerife, La Palma and El Hierro, and allow the direct observation and sampling of the deep structure of the island volcanoes without requiring expensive and indirect geophysical methods (Carracedo 1994, 1996a,b).

Since the early work of famous naturalists such as Leopold von Buch, Charles Lyell, and Georg Hartung, the Canaries have been viewed as a 'special' volcanic island group and their origin has been closely related to African continental tectonics (Fúster *et al.* 1968a,b,c,d; McFarlane & Ridley 1969; Anguita & Hernán 1975; Grunau *et al.* 1975). However, a wealth of geological data made available in recent years, especially on the western islands, has led to the conclusion that the Canaries (and the Cape Verde islands) are similar in many aspects to hotspot-induced oceanic island volcanoes, such as the Hawaiian archipelago (for recent maps and aerial photographs see info@grafcan.rcanaria.es, ign@ign.es, igme@igme.es, info@grafcan.rcanaria.es). However, although the Canarian and Hawaiian volcanoes show common constructional and structural features (rift zones, multiple gravitational

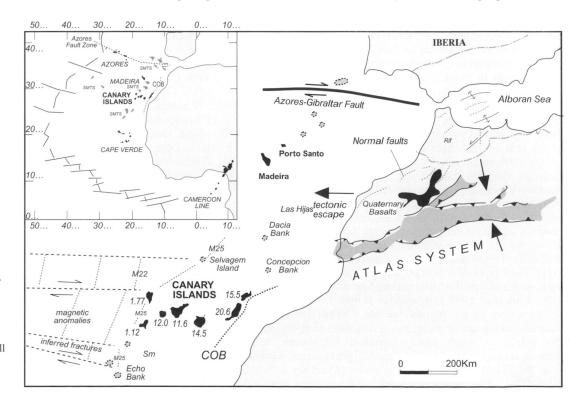

Fig. 18.1. Geographic and geodynamic framework of the NW African continental margin volcanic groups. Ages of the Canaries from Cantagrel *et al.* (1984), Ancochea *et al.* (1996), Coello *et al.* (1992), Guillou *et al.* (1996, 2001), McDougall & Schmincke 1976 (modified from Carracedo *et al.* 1998).

collapses), they also show some interesting differences in the geochemical evolution of their magmas and the amount of subsidence (Schmincke 1973, 1976, 1982; Carracedo 1984, 1999; Carracedo *et al.* 1998).

Tectonomagmatic setting, origin and evolution (JCC)

The Canary Islands developed in a geodynamic setting characterized by old (Jurassic) oceanic lithosphere lying close to a passive continental margin, and on a very slow-moving tectonic plate (the African plate). The absolute easterly motion of the African plate over fixed hotspots has been estimated (O'Connor & Duncan 1990) for the Walvis Ridge at *c.* 7° in latitude and 34° in longitude for the last 60 Ma. In the region of the Canaries these values may be as low as 2.4° and 5° (*c.* 9 mm/year), respectively, for this period.

Another potentially very important difference between the Canaries and most other oceanic island groups is that the Canaries are located adjacent to a region of intense active deformation, comprising the Atlas and Rif mountains, Alboran sea and Betic Cordillera provinces of the alpine orogenic belt. Some authors postulate the extension to the Canaries of an offshore branch of the trans-Agadir fault, associated with the Atlas system (Anguita & Hernán 1975, 1986, 2000), although there is no obvious geological or geophysical evidence for this (e.g. Dillon & Sougy 1974). Neither the Canarian archipelago as a whole, nor the individual islands and their volcanic centres and rifts follow the postulated extension of the Atlas fault to the Canaries. In fact, the islands of Fuerteventura and Lanzarote are parallel to the continental margin, whereas most of the remaining islands of the archipelago follow a general east–west trend, and the dual line of La Palma and El Hierro forms a north–south trend. Rifts in the western islands are radial and do not relate to the Atlas trend (Carracedo 1994). Rihm *et al.* (1998) demonstrate the presence of a group of apparently young seamounts (Las Hijas seamounts) located 70 km SE of El Hierro and probably destined to become the next Canarian islands. Their dyke and rift orientations are similar to those of the existing Canaries, and their location is consistent with the age-progression trend of volcanism in the Canarian archipelago and the average spacing of these islands.

The propagation of continental Atlas structures into oceanic lithosphere is likely to be mechanically unfeasible (e.g. Vink *et al.* 1984; Steckler & ten Brink 1986; ten Brink 1991). It is evident that the >150 Ma old oceanic lithosphere at the African margin in the Atlas region is considerably stronger than the continental lithosphere, precluding any fracture propagation from the Atlas towards the Canaries. Also relevant is the fact that the volcanic trend defined by the islands of Fuerteventura and Lanzarote, and an associated chain of abundant seamounts off cape Juby (Dillon & Sougy 1974), run along the continental–oceanic boundary (Figs 18.1 and 18.2). This boundary is characterized by the presence of a 10 km thick layer of sediments, and provides a zone of relative weakness and a preferential pathway for magmas (Schmincke 1982; Vink *et al.* 1984; Carracedo *et al.* 1998).

The Cape Verde islands, located 500 km off the west African continental margin, are very similar to the Canaries but exhibit a much more prominent lithospheric swell, estimated to be between 400 km and 1500 km across (Grunau *et al.* 1975) and 1500 m high at its centre (Courtney & White 1986). The apparent lack of a Canarian lithospheric swell was used by Filmer & McNutt (1988) as an argument against the presence of a hotspot in the Canaries and has also been noted by other authors (Hoernle & Schmincke 1993; Watts 1994). However, Canales & Dañobeitia (1998) analysed a number of seismic lines in the vicinity of the archipelago and demonstrated the existence of a subdued (*c.* 500 m maximum elevation) lithospheric depth anomaly around the Canary Islands. These authors proposed that this anomaly could be related to a swell that was otherwise obscured by the weight and perhaps also mechanical effects of the thick sedimentary cover along the NW African continental margin and by the weight of the volcanic rocks of the islands themselves. Swell building and magmatism for hotspots interacting with slowly moving plates are different from those seen in Hawaiian-type, fast-moving, open-ocean lithosphere, as noted by Morgan & Price (1995). Monnereau & Cazenave (1990) found that the Cape Verde, Madeira and Canary archipelagos have a much smaller island platform relief in proportion to their swell relief when compared with the Hawaiian hotspot. Satellite and surface gravity data from the Canary and Cape Verde islands (Liu 1980) show them to be the only regions of convection-generated tensional stress fields in NW Africa.

Twentieth century Atlantic intraplate seismicity (Wysession *et al.* 1995) within the African plate has been concentrated in two areas: within the Cape Verde swell and in a region of similar extent around the Canaries. The authors interpret this seismicity as being related to plume-generated magmatic activity. As suggested by Schmincke (1979), the persistence of volcanism for very long periods (>20 Ma in some islands) will require the concentration of mantle melting in a small area over a long period, an idea compatible with a mantle plume beneath the archipelago. Analysis of the seismic wave attenuation in the Canary Islands indicates zones with a high degree of attenuation and a dominance of intrinsic absorption over scattering attenuation, which points to a strong asthenosphere in the archipelago, probably hotspot-rejuvenated crust. Zones of higher values of lithospheric inelasticity are found precisely beneath the islands of Tenerife, La Palma and El Hierro (Canas *et al.* 1994, 1998). The absence of significant subsidence in the Canaries over very long time intervals (Carracedo *et al.* 1998; Carracedo 1999) may be related to slow swell evolution of the Canarian hotspot on a slow-motion plate.

Analysis of isotopic variations with distance and time in the Canaries has provided further evidence for a mantle plume origin. Hoernle *et al.* (1991) reported isotopic systematics of lavas from Gran Canaria that appear to have a plume-like composition, with high $^{238}U/^{204}Pb$. According to these authors, the plume was located to the west of Gran Canaria during the Pliocene–Recent epochs. Hoernle & Schmincke (1993) analysed major and trace elements in the island of Gran Canaria, concluding that mafic magmas were probably formed by decompression melting in an upwelling column of asthenospheric material. More recently, Hoernle *et al.* (1995) found evidence from seismic tomography and isotope geochemistry of a large region of upwelling in the upper mantle extending from the eastern Atlantic to the western Mediterranean.

The volume and distribution of the island volcanoes built above this Canarian hotspot provide interesting information about the evolution of the archipelago. The islands rest on an oceanic floor that deepens progressively westwards, reaching a depth of 4000 m in the area of La Palma and El Hierro (Fig. 18.2A). Shaded-relief images (Fig. 18.2B and C), which give

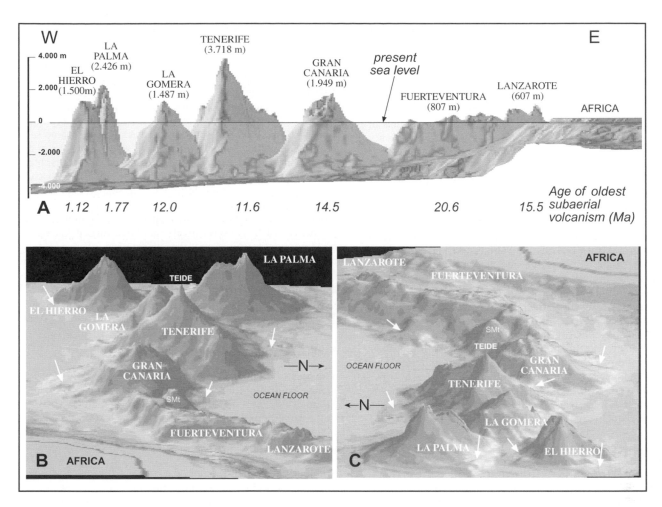

Fig. 18.2. (**A**) Shaded relief cross-section (east–west) of the Canaries showing the subaerial/submarine volumes of the island edifices and the corresponding oldest age of the subaerial volcanism for each island (from Carracedo 1999). (**B**) Shaded relief view of the Canaries from the east. (**C**) Shaded relief view of the Canaries from the west. Arrows indicate debris avalanche deposits from giant landslides (modified from Carracedo *et al.* 1998).

an 'empty ocean' view of the Canarian chain, clearly show that the elevation and emerged volume of the islands increase as their age decreases, with a generally westward trend. Three different groups of islands can be recognized by comparing the volume and aspect ratio of the islands with their relative ages (Fig. 18.2A): (1) Lanzarote, Fuerteventura, Gran Canaria and La Gomera are clearly older islands whose subaerial edifices have largely been mass-wasted by erosion; (2) Tenerife, the highest and most voluminous island, is probably at the peak of volcanic construction; (3) La Palma and El Hierro are both still in a very juvenile stage of growth.

Destructive processes eroding the Canaries are important in shaping island geomorphology and provide a major contribution to sea-floor sediments both surrounding the islands and on the Atlantic abyssal plains to the north. Recent deep-sea surveys and ocean drilling programmes have revealed several debris avalanche deposits around the islands (Holcomb & Searle 1991; Weaver 1991; Weaver *et al.* 1992, 1998; Watts & Masson 1995; Masson 1996; Urgelés *et al.* 1997, 1998, 1999; Teide Group 1997; Funk & Schmincke 1998). These works and related studies onshore (Carracedo 1994, 1999; Carracedo *et al.* 1998, 1999a,b; Ancochea *et al.* 1994; Guillou *et al.* 1996,

2001; Stillman 1999) have shown giant landslides to be a dominant feature of the islands' erosion.

Large-scale distribution and age progression in most oceanic island chains are well explained by the steady movement of lithospheric plates over fixed mantle plumes, yielding chains of consecutive discrete volcanoes. In this model, a new island starts to form when the bulk of the previous one has already developed, the inter-island distance being governed by lithospheric thickness and rigidity (Vogt 1974; ten Brink 1991). At the initiation of an oceanic island chain the first magma intrudes through unloaded pristine lithosphere. As the first island develops the volcanic load grows and increasing compressive flexural stresses eventually exceed the tensile stresses of the mantle plume, blocking magma pathways to the surface. This precludes the initiation of new eruptions in the vicinity of any new island volcano (ten Brink 1991; Carracedo *et al.* 1997, 1998; Hieronymus & Bercovici 1999).

An interesting feature of the Canarian island chain is the fact that the islands of La Palma and El Hierro are growing simultaneously and form a north–south trending dual line of island volcanoes, perpendicular to the general trend of the archipelago (Fig. 18.2B and C; see also Fig 18.14A and B).

Dual-line volcanoes, such as the Kea and Loa trends in the Hawaiian islands, have been associated with changes in plate motion, resulting in the location of a volcanic load off the hotspot axis. Compressive stresses related to the off-axis volcano block the formation of the next island and split the single line of volcanoes into a dual line of alternating positions of volcanoes (Hieronymus & Bercovici 1999).

In the Canarian chain, the association of a dual line of volcanoes with a change in direction of the African plate is not clear since such a change took place between the formation of Gran Canaria and La Gomera (14.5 to 12 Ma BP). A possible model to explain the splitting of the Canaries in a dual line may be related to the anomalous position of La Gomera, apparently older than Tenerife but ahead in the hotspot pathway. In fact, as pointed out above, the islands of Lanzarote, Fuerteventura, Gran Canaria and La Gomera (old Canaries) seem to form a different group from Tenerife, La Palma and El Hierro (young Canaries). Some perturbation in plate motion may have resulted in the initiation of volcanism off the previous hotspot pathway, forming the island of Tenerife NE from La Gomera, compressive stresses from which may have forced the splitting of the archipelago to form a dual line of island volcanoes (Carracedo *et al.* 1997, 1998; Carracedo 1999).

In summary, the timing of eruptive activity in the islands, their morphological and structural features, seismic signature and geochemical evolution (Schmincke 1973, 1976, 1982; Carracedo 1975, 1999; Hoernle *et al.* 1991, 1995; Hoernle & Schmincke 1993; Carracedo *et al.* 1997, 1998; Canas *et al.* 1994, 1998) clearly converge on a slow-moving hotspot model.

Fuerteventura (RC, AA)

Fuerteventura is the easternmost and second largest (1677 km^2 including the Isla de Lobos) of the Canary Islands, and the closest (100 km) to the African coast (Figs 18.1 and 18.2). Together with Lanzarote and the Conception Bank, it lies along a SSW–NNE volcanic line (East Canary Ridge) that runs parallel to the continental margin, and is built on a 3000 m deep ocean floor. Fuerteventura and Lanzarote are not geologically different islands *sensu stricto* since they are separated by only a narrow stretch of sea (La Bocaina) which is shallower than the lowest sea level at glacial maxima (Fig. 18.2).

The topography of Fuerteventura is characteristic of a post-erosional island with abundant Quaternary deposits (beach sand, dunes, scree, alluvium) and minor rejuvenation volcanism (Fig. 18.3). It is predominantly smooth, with broad valleys separated by sharp erosive interfluves (locally known as 'cuchillos' or knives). The bulk of the island is predominantly of low altitude (<500 m), except in the southern peninsula of Jandía, where the maximum elevation of the island, Pico de La Zarza, reaches 807 m. The leeward (western) coastline is low, with common shell sand beaches and dunes, in contrast to high cliffs on the windward (eastward) side of the island.

The geological history of Fuerteventura is the most complex and prolonged of the Canary Islands (Fúster *et al.* 1968b, 1980; Stillman *et al.* 1975; Le Bas *et al.* 1986; Coello *et al.* 1992; Ancochea *et al.* 1996; Steiner *et al.* 1998; Balogh *et al.* 1999). The island is located on thick (>11 km), strongly reworked oceanic crust, and four main lithological units are exposed (Fig. 18.3): Mesozoic oceanic crust, submarine volcanic complexes, Miocene subaerial volcanic complexes, and Pliocene–Quaternary sedimentary and volcanic rocks.

Mesozoic oceanic crust

This fragment of Mesozoic oceanic crust comprises a thick (*c.* 1600 m) sedimentary sequence that rests on tholeiitic N-MORB basalts of Early Jurassic age (Robertson & Stillman 1979a; Robertson & Bemouilli 1982; Steiner *et al.* 1998). The basaltic rocks form sheet flows, pillow lavas and breccias and represent the only outcropping Mesozoic oceanic basement described so far in the central Atlantic. The Mesozoic sedimentary sequence is Early Jurassic to Late Cretaceous in age and consists of terrigenous quartzitic clastic sediments, black shales, redeposited limestones, marls and chalks with chert nodules. The succession is part of a deep-sea fan deposited on the west African continental margin (Fúster *et al.* 1968b; Robertson & Stillman 1979a; Robertson & Bemouilli 1982; Steiner *et al.* 1998) and is subvertical to overturned. Cleavage–bedding relationships and the orientation of minor folds indicate the presence of a major NE facing reclined fold (Robertson & Stillman 1979a). This fold has been interpreted as having been generated by dextral motion along a N-S orientated transcurrent shear zone (Robertson & Stillman 1979a) or northward-directed thrust faulting (Gutiérrez 2000).

Submarine volcanic complexes

The Mesozoic succession is unconformably overlain by a submarine volcanic complex of uncertain age (Robertson & Stillman 1979b; Le Bas *et al.* 1986; Stillman 1987, 1999; Gutiérrez 2000). While some authors suggest that the earliest pillow lavas are probably Palaeocene to lower Eocene, and that the build-up of the island continued up to early Miocene times (Robertson & Stillman 1979b; Le Bas *et al.* 1986; Stillman 1987, 1999; Balogh *et al.* 1999), others consider that the main period of submarine construction of the island was concentrated in Oligocene times (Fúster *et al.* 1980; Cantagrel *et al.* 1993; Sagredo *et al.* 1996).

This tectonically uplifted submarine volcanic complex is dominated by pillow lavas and hyaloclastites of basaltic and trachybasaltic composition, exposed in the western part of the island, but deeply eroded and variably dipping. Near the Ajuy valley the lowest volcaniclastic sediments and volcanic breccias are inverted or steeply dipping, in general conformity with the stratigraphically underlying Mesozoic sedimentary rocks (Robertson & Stillman 1979b; Le Bas *et al.* 1986). In contrast, to the north near El Valle, the middle and upper parts of the submarine sequence are only gently inclined to the west or to the east (Fúster *et al.* 1968b; Robertson & Stillman 1979b; Le Bas *et al.* 1986; Gutiérrez 2000). The submarine volcanic rocks are unconformably overlain by littoral and shallow-water marine deposits (reefal bioclastic sediments, beach sandstones and conglomerates), gently inclined to the west and representing the transition from submarine to subaerial activity (Fúster *et al.* 1968b; Robertson & Stillman 1979b; Le Bas *et al.* 1986; Gutiérrez 2000).

Associated with the submarine volcanic complex is a sequence of plutonic and hypabyssal intrusions that have been subdivided into an early syenite–ultramafic series, and later syenite–carbonatite complexes. The former has been called the Tierra Mala Formation (Fm) by Le Bas *et al.* (1986), or the A1 rock group by Balogh *et al.* (1999). It comprises ultramafic and mafic rocks (alkali pyroxenites, hornblendites and amphibole gabbros) intruded by *c.* 65 Ma (^{39}Ar/^{40}Ar; Balogh *et al.* 1999) syenites (Fúster *et al.* 1980; Le Bas *et al.* 1986; Ahijado

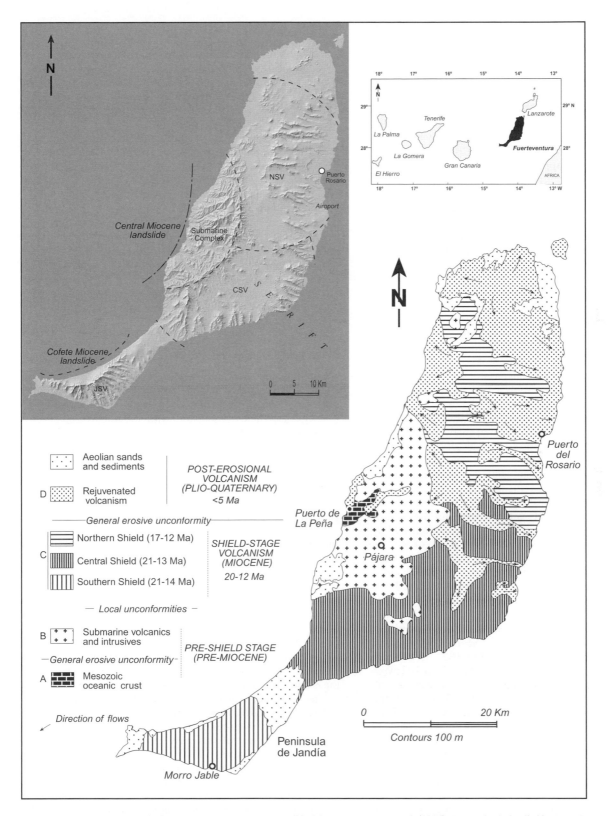

Fig. 18.3. Simplified geological map of Fuerteventura. Modified from Ancochea *et al.* (1996). Inset: shaded relief image of Fuerteventura with indication of the main geomorphological and tectonic features (image GRAFCAN). Giant landslides after Stillman (1999). Shield volcanoes: CSV, central shield volcano; SSV, southern shield volcano; NSV, northern shield volcano (after Ancochea *et al.* (1996)).

1999), and is exposed along the western coast of the island (Tostón Cotillo, Ajuy and Punta del Peñon Blanco). The later syenite–carbonatite rocks form three complexes which from north to south are: Esquinzo, Ajui-Solapa and Punta del Peñon Blanco (Fúster *et al.* 1980; Barrera *et al.* 1981; Le Bas *et al.* 1986; Ahijado 1999; Balogh *et al.* 1999). Their emplacement took place during a relatively short period of time around 25 Ma BP (Le Bas *et al.* 1986; Cantagrel *et al.* 1993; Sagredo *et al.* 1996; Ahijado 1999; Balogh *et al.* 1999). Ijolite crops out in Esquinzo and Caleta de la Cruz, and younger nephelinite dykes also appear. The emplacement of these rocks is coeval with the occurrence of several upper Oligocene–lower Miocene brittle–ductile to ductile shear zones (Casillas *et al.* 1994; Fernández *et al.* 1997a) arranged in a nearly orthorhombic pattern, with kinematic criteria indicating east–west horizontal extension. Shear zone activity took place after earlier deformation had affected the submarine sequence.

Miocene subaerial volcanic complexes

The early stages of subaerial growth of the island are the result of the formation of three different adjacent large basaltic volcanic complexes (Fig. 18.3 inset): the southern (SVC), central (CVC) and northern (NVC) edifices of Ancochea *et al.* (1992, 1996). They comprise mainly alkali basaltic and trachybasaltic lavas, with minor trachytic differentiates, and interbedded pyroclasts with abundant subvertical dykes. The limited amount of pyroclastics and abundant feeder dykes indicate the mostly low eruptive explosivity typical of shield volcanoes.

Each complex has its own prolonged history that may be longer than 10 Ma, during which time several periods of activity alternated with gaps characterized by major erosive episodes and giant landslides (Ancochea *et al.* 1996; Stillman 1999). The ages of these volcanic complexes (Ancochea *et al.* 1996) differ slightly. The CVC is the oldest with the bulk of construction occurring between 22 and 18 Ma BP, followed by a significant pause, and then a later smaller phase of activity from 17.5 to 13 Ma BP. In the NVC, the main activity took place between 17 and 12 Ma BP, and in the SVC volcanism occurred between 21 and 14 Ma BP. It has been estimated that the CVC volcano may have reached an elevation of over 3000 m (Javoy *et al.* 1986; Stillman 1999).

The deeply eroded remnants of these volcanic complexes crop out over much of the central E and SE part of the island and in the Jandía peninsula to the SW (Fig. 18.3). In the core of both the NVC and the CVC there are plutonic rocks (pyroxenites, gabbros and syenites) and a basaltic-trachybasaltic dyke swarm which represent the hypabyssal roots of the evolving subaerial volcanic complexes (Ancochea *et al.* 1996; Balogh *et al.* 1999). The dykes are extremely abundant, mostly orientated NNE–SSW (although some are NE–SW and others NW–SE), and seem to have involved crustal stretching of around 30 km (Fúster *et al.* 1968b; López Ruiz 1970; Stillman 1987; Ahijado 1999). The plutonic rocks can be subdivided into earlier gabbros and pyroxenites, and later gabbro–syenite–trachysyenite complexes. The former occur as several NNE–SSW elongated bodies (Gastesi 1969) that produce high-grade thermal metamorphic effects on the host rocks (Stillman *et al.* 1975; Hobson *et al.* 1998; Ahijado 1999), and have been dated (K-Ar) at around 20–21 Ma BP (Sagredo *et al.* 1996; Balogh *et al.* 1999). The later plutons include concentric intrusions of gabbros and syenites which form the Vega de Río

y Palmas ring complex, dated at 18.4–20.8 Ma BP (K-Ar; Le Bas *et al.* 1986; Cantagrel *et al.* 1993). These intrusions were emplaced along conical fractures associated with the relaxation of the stress field that produced the dyke swarm. Other late plutonic bodies include syenites and trachysyenites of the Betancuria subaerial complex (13–14 Ma BP; Cantagrel *et al.* 1993; Muñoz & Sagredo 1996) and the Morro del Sol and Morro Negro gabbros.

Metamorphism has affected most of the rocks which form the Mesozoic oceanic crust, the submarine volcanic complex, the lower part of the subaerial volcanic complexes, and the plutonic bodies and dyke swarms related to these complexes. An intense hydrothermal metamorphism in epidote–albite greenschist facies was probably produced by the massive intrusion of dyke swarms (Fúster *et al.* 1968b; Stillman *et al.* 1975; Robertson & Stillman 1979b; Javoy *et al.* 1986). Additionally, contact metamorphism affected the host rocks of the plutons related to the subaerial volcanic complexes. In some places, near the plutonic bodies, this metamorphism reached pyroxene hornfels facies and led to the partial melting of the host rocks (Stillman *et al.* 1975; Muñoz & Sagredo 1994; Hobson *et al.* 1998).

Plio-Quaternary sedimentary and volcanic rocks

After Miocene magmatism, the subaerial volcanoes were deeply eroded, losing a large part of their original volume. Magmatic activity recommenced with the formation of Pliocene lava fields associated with basaltic volcanic vents aligned along fractures, and remaining active in prehistoric times (Cendrero 1966; Coello *et al.* 1992). Pliocene–Quaternary sedimentation produced littoral and shallow-water marine deposits overlain by aeolian deposits with intercalations of alluvial fans and palaeosols (Meco 1991).

Lanzarote (JCC, ERB)

The island of Lanzarote lies at the NE edge of the Canarian chain just 140 km from the African coast (Fig. 18.1). Elongated in a NE–SW trend, parallel to the continental margin, it is 60 km long and 20 km wide and covers 862 km² (905 km² including the small northern islets of Graciosa, Mña. Clara and Alegranza). The topography is characteristic of post-erosional islands (Fig. 18.4 inset), with deeply eroded volcanoes, broad valleys ('barrancos'), precipitous cliffs, and wide lowlands covered with aeolian sands locally called 'jables'. The highest elevation is 670 m (Peñas del Chache, in the northern part of the island) and, unlike neighbouring Fuerteventura, the island has seen volcanic eruptions in historical times.

The spectacular lava fields of the 1730–1736 eruption and the presence of high-temperature fumaroles (<600°C) were the subject of most of the early geological reports (Hausen 1959; Hernández Pacheco 1960; Bravo 1964). Marine deposits and raised beaches were studied by Meco (1977), and the first general geological study, including a geological map of the island, was published by Fúster *et al.* (1968c). Radiometric dating and geomagnetic stratigraphy were published by Abdel-Monem *et al.* (1971), Coello *et al.* (1992), Carracedo & Soler (1992) and Carracedo & Rodríguez Badiola (1993). More recently the historical eruptions (1730 and 1824) were studied by Carracedo & Rodríguez Badiola (1991) and Carracedo *et al.* (1992).

The geology of Lanzarote and Fuerteventura is very similar,

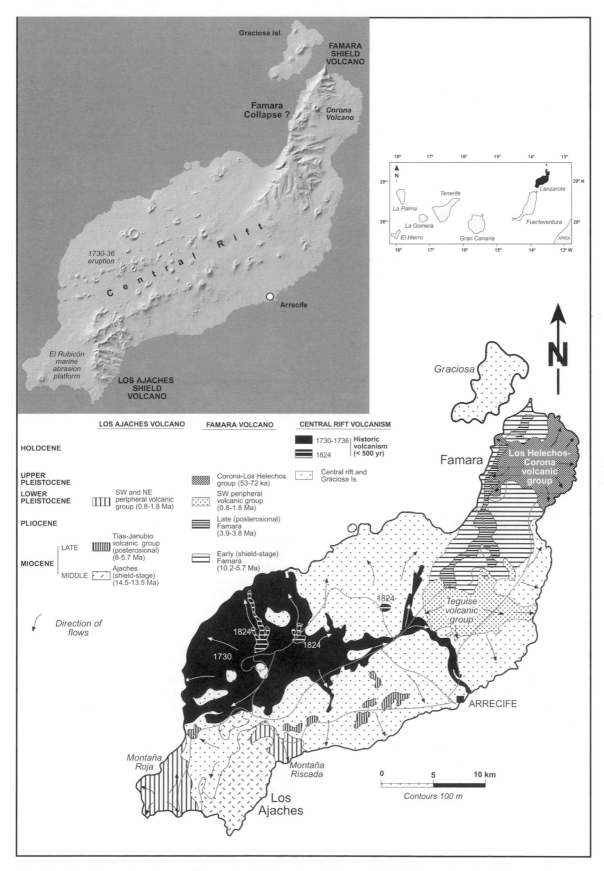

Fig. 18.4. Simplified geological map of Lanzarote. Modified from Fúster *et al.* (1968*c*). Inset: shaded relief image of Lanzarote with indication of the main geomorphological and tectonic features (image GRAFCAN).

both islands being in a very advanced post-erosional stage of development. Only the lack of significant subsidence in the Canaries (Carracedo 1999) explains why these islands are still emergent, although they have lost most of their original bulk due to catastrophic mass wasting and marine and meteoric erosion. In Lanzarote the geology is dominated by two main shield volcanoes (the Ajaches and Famara volcanoes) which developed as independent island volcanoes (Fúster *et al.* 1968c; Carracedo & Rodríguez Badiola 1993) and were subsequently connected by eruptive products from a central rift (Fig. 18.4). The remnants of the Ajaches volcano are preserved as basanite–alkali basalt–trachyte lavas located in the south of the island, and have been dated (K-Ar) as 19.0–6.1 Ma BP (Abdel-Monem *et al.* 1971) and 15.5–6.6 Ma BP (Coello *et al.* 1992). Comparison of the geomagnetic polarities with the established GPTS (Carracedo & Rodríguez Badiola 1993) suggests that the main shield-building stage of this volcano extends from *c.* 14.5 to 13.5 Ma BP, with volcanism from *c.* 8.0 to 5.7 Ma BP corresponding to the main post-erosional stage of the volcano, with minor later rejuvenation volcanism (Fig. 18.4).

The Famara shield volcano forms a 24 km long, NE–SW trending elongate edifice located at the northernmost part of the island (Fig. 18.4). The western half of the volcano is bound by a 600 m cliff, most probably representing the retrograded scarp of a giant gravitational collapse that mass wasted that portion of the volcano. Published radiometric (K-Ar) ages range from 10.0 to 5.3 Ma (Abdel-Monem *et al.* 1971) or 10.2 to 3.8 Ma (Coello *et al.* 1992). Geomagnetic stratigraphy suggests a main shield-building stage from 10.2 to 5.7 Ma BP and a main post-erosional stage from 3.9 to 3.8 Ma BP, with some Quaternary residual rejuvenation activity (Carracedo & Soler 1992; Carracedo & Rodríguez Badiola 1993). This trend suggests an alternation of the main eruptive phases of both the Famara and Ajaches volcanoes, similar to that proposed for the islands of La Palma and El Hierro (Carracedo *et al.* 1999a,b). The composition of the Famara volcano lavas is predominantly basanitic, with minor gradation to alkali basalts.

Recent volcanism

Quaternary volcanism is represented by peripheral eruptions around the main shields, producing the Mña. Roja and Caldera Riscada volcanoes in the Ajaches shield and the Teguise and Corona volcanic groups in the Famara shield (Fig. 18.4). The Corona lavas advanced along a wave-cut platform located 70 m below the present sea level. This eruption produced one of the largest known lava tubes in the world, 6.8 km long with sections reaching 25 m in diameter, 2 km of which are at present submerged (−80 m) following sea-level rise after the last glaciation.

Holocene eruptions are very few, probably limited to the historical 1730 and 1824 events. The 1730–1736 eruption (Fig. 18.4) is the second largest basaltic fissure eruption recorded in historical time (after the 1783 Laki eruption in Iceland). It lasted for 68 months and produced a total lava volume of *c.* 700 m^3 × 10^6, compared with less than three months and 66 m^3 × 10^6 for the second largest historical eruption in the Canaries. Over 30 volcanic vents were formed in five main multi-event eruptive phases, aligned along a 14 km long, N80°E-trending fissure (Carracedo *et al.* 1992). Precise reconstruction of this eruption was greatly facilitated by detailed eye witness accounts, such as the report of the parish priest of

Yaiza (included in the work of Buch 1825), and particularly the official reports of the local authorities to the Royal Court of Justice, found on file in the Spanish Archivo General de Simancas (Carracedo *et al.* 1990, 1992; Carracedo & Rodríguez Badiola 1991).

An interesting feature of the 1730 activity was the eruption of transitional lavas from olivine melanephelinites through basanites and alkali basalts to tholeiites, suggesting that the lava compositions represent nearly unmodified primary melts from a heterogeneous mantle source (Carracedo *et al.* 1990; Carracedo & Rodríguez Badiola 1991, 1993). Isotopic variability found in these lavas may be partly explained as well by the melting or melt-mixing of this composite mantle source (Sigmarsson *et al.* 1998). Finally, in 1824, a short eruption from three vents aligned along a N70°E, 13 km long fissure, produced small amounts of basaltic lavas (Fig. 18.4).

Gran Canaria (FJPT, JCC)

Gran Canaria lies at the centre of the Canarian chain (Fig. 18.1) and is the third largest island (1532 km^2) of the archipelago, 45 km in diameter with a maximum elevation of 1950 m (Pico de las Nieves). It is a typical post-erosional island with important rejuvenation volcanism, and has a geomorphology that changes greatly from the NE (predominantly post-erosive volcanism) to the SW (exclusively shield-stage volcanism). A dense radial network of deep barrancos dissects the island, forming a rugged topography. The coastal landforms vary considerably, reflecting the above-mentioned differences in age, with sheer, vertical, high cliffs in the western part and coastal platforms and wide beaches and dunes in the southern and eastern regions (Fig. 18.5).

Numerous geological studies have been carried out on Gran Canaria since the eighteenth century (see Bourcart & Jeremine 1937; Hausen 1962; Fúster *et al.* 1968d; Schmincke 1968, 1976, 1993; and references therein), and geological maps of the island have recently been published (ITGE 1990, 1992). Rocks belonging to the seamount stage do not crop out, but oceanographic work on the volcaniclastic apron around the island was carried out during the Ocean Drilling Program (ODP) Leg 157 (Weaver *et al.* 1998). Seismic and bathymetric profiles indicate that the submarine stage formed at least 90% of the bulk volume of the island (Schmincke & Sumita 1998). Boreholes up to 300 m deep reveal graded hyaloclastite tuffs and debris flow deposits, interpreted as being derived from shallow submarine eruptions (Schmincke & Segschneider 1998). No apparent unconformity separates these submarine deposits from those of the subaerial volcanism, the latter commonly being interbedded with the former. A similarity in geochemical composition suggests a common magmatic source for both submarine and subaerial volcanic products (Schmincke & Segschneider 1998), the sole difference being in eruptive mechanisms. Once subaerial volcanism had become fully established, the onland geology of Gran Canaria records the growth of a shield volcano (evolving through shield to caldera and post-caldera stages), followed by *c.* 3 Ma of erosion, after which there was renewed magmatic activity.

Shield growth

Activity during this stage produced a complex shield volcano, 60 km in diameter, over 2000 m high, with a volume of over 1000 km^3, and occupying both the present extent of the island

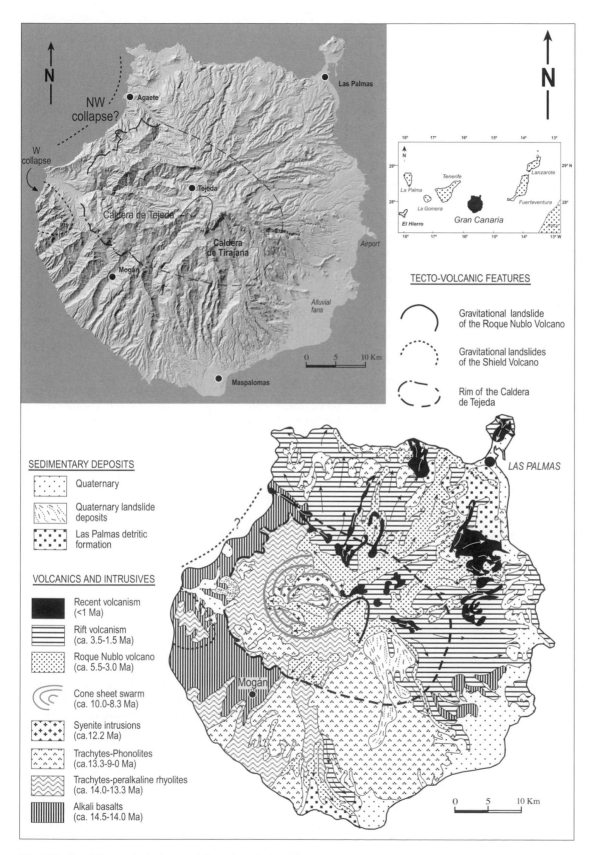

Fig. 18.5. Simplified geological map of Gran Canaria (modified from ITGE 1992). Inset: shaded relief image of Gran Canaria indicating the main geomorphological and tectonic features (image GRAFCAN).

as well as several kilometres offshore to the west (Schmincke 1976, 1993; ITGE 1990, 1992). At present, shield-stage rocks are exposed mainly in the western and southwestern coastal cliffs, where they reach thicknesses of up to 1000 m (Fig. 18.5). Volcanic activity was characterized by Hawaiian-type eruptive styles, involving sustained eruption of effusive lavas (alkali-basalts to mugearites) with minor interlayered pyroclastics and a swarm of feeder dykes (Fúster *et al.* 1968*d*; Schmincke 1976, 1993). The growth period for this stage is very short, constrained to between 14.5 to 14 Ma BP according to K-Ar and ^{39}Ar/^{40}Ar ages from Bogaard *et al.* (1988) and Bogaard & Schmincke (1998). This short time of growth and the sparse occurrence of pyroclastic rocks strongly contrasts with the shield stages of the other Canaries, where shield stages exceeded at least 1 Ma and produced predominantly pyroclastic facies from more explosive eruptions.

A clear intraformational unconformity at the SW of the island suggests the presence of a gravitational collapse embayment, with thick, interbedded debris-avalanche deposits and infilling lavas of slightly younger ages mass wasting the southwestern flank of the shield (Schmincke 1976, 1993). The arcuate outline of the NW coast (San Nicolás to Agaete) has also been attributed to a gravitational collapse. The original shape and extent of the volcanoes forming the shield stage of Gran Canaria are unknown, although Schmincke (1976, 1993) defined three volcanoes: the northwest volcano, centred close to Agaete; the north volcano, near the Aldea de San Nicolás, and the southeast volcano, with a centre close to Agüimes. Schmincke & Sumita (1998) postulated a fourth volcano to the north of Arucas, and suggested an east to west migration of volcanic activity in the shield stage of Gran Canaria, with the southeast volcano being the youngest. Analysis of dyke distribution and orientation carried out during the geological mapping of the island (ITGE 1990, 1992) suggests the presence of a main shield volcano centred in the Mesa del Junquillo (at the northern edge of the cone sheet west of Tejeda). The volcanoes defined by Schmincke (1976, 1993) and Schmincke & Sumita (1998) may have functioned as rift zones associated with this main shield volcano, which subsequently collapsed to form the Caldera de Tejeda.

Caldera stage volcanism

A shallow (*c.* 4–5 km) rhyolitic magmatic chamber was emplaced towards the final stages of shield development, periodically fed from a deeper (sublithospheric *c.* 14 km) basaltic source (Freundt & Schmincke 1992). These rhyolitic magmas produced the first highly explosive eruptions and, subsequently, highly welded ignimbrites. Ignimbritic eruptions and high eruptive discharges characteristic of the shield stage may have caused the abrupt emptying of the magmatic chamber, finally collapsing the summit of the volcano and creating the Caldera de Tejeda (Schmincke 1967; Hernán 1976). This elliptical, NW–SE trending, 20 × 17 km, 1000 m deep caldera is the most prominent geomorphological feature of Gran Canaria (Fig. 18.5). The caldera rim, with contacts usually dipping 45° towards the centre of the island, is observable at present along its western half, and is marked by bright green-blue hydrothermal alteration layers (locally known as 'azulejos'). Simultaneously with the formation of the caldera, the rhyolitic chamber was replenished with basaltic magmas. The weight of the cauldron block forced the violent emission of *c.* 80 km^3 of magma through the caldera rim fissures

(Freundt & Schmincke 1992), producing a single, 30 m thick ignimbrite named P1 by Schmincke (1976, 1993) and dated at 14 Ma BP (Bogaard *et al.* 1988; Bogaard & Schmincke 1998). The P1 ignimbrite covers >400 km^2 of the shield-stage basalts around the caldera and constitutes a prominent marker bed recording the creation of the Caldera de Tejeda.

Post-caldera stage volcanism

This stage is characterized by the eruption of large volumes (*c.* 1000 km^3) of silicic lavas and ignimbrites from the caldera rim (ring fractures). The presence of 'fiamme' and rheomorphic folds suggests boiling-over-type eruptive mechanisms during expulsion of high temperature ignimbrites. Two magmatic phases, comprising extra- and intra-caldera domains, can be defined: a predominantly peralkaline rhyolite–trachyte initial phase (Mogán Group (Gp) in the terminology of Schmincke (1976, 1993)) and a later trachyte–phonolite phase (Fataga Gp). Both phases were fed from a shallow magma reservoir, periodically replenished with deeper basaltic magmas, later differentiated by crystal fractionation (Schmincke 1976, 1993). A peralkaline rhyolite–trachyte initial phase (*c.* 14–13.3 Ma BP; volume estimates of about 300–500 km^3) was heralded by the eruption of the P1 ignimbrite. The extra-caldera deposits, composed of 15 cooling units up to 300 m thick (Schmincke 1976, 1993), mantle most of the shield volcano. Flow direction and periclinal slopes dipping at 7–9° relate these eruptions to the caldera rim (Schmincke & Swanson 1967). The lack of interbedded epiclastic deposits suggests high eruptive rates and short periods of emission. The intra-caldera volcanic deposits filled most of the caldera, but later intrusions have hindered stratigraphic correlation with the extra-caldera volcanics. However, it seems evident that both were erupted from the same vents located at the edge of the caldera.

During the trachyte–phonolite second eruptive phase (*c.* 13.3–8.3 Ma BP; volume estimates >500 km^3), extra-caldera deposits formed sequences up to 1000 m thick of successive ignimbrites with interlayered lava flows that become more abundant towards the top. Interbedded epiclastic deposits indicate frequent interruptions of eruptive activity and episodic mass wasting processes, determined by Bogaard & Schmincke (1998) to have occurred between 12.33–12.07, 11.36–10.97 and 9.85–8.84 Ma BP. Eruptions continued from vents along the caldera rim and, probably, from a stratovolcano located near the Cruz Grande, in the SE of the island (Schmincke 1976, 1993).

Intra-caldera activity in this second phase is mainly intrusive, consisting of three main events (Schmincke 1967, 1976, 1993; Hernán 1976): alkali syenites (*c.* 12.2 Ma BP), trachytic-phonolitic cone sheets (*c.* 10–8.3 Ma BP), and phonolitic-nephelinitic plugs (*c.* 8.5 Ma BP). The syenites form small stocks located at altitudes up to 1200 m in the centre of the caldera, whereas the phonolitic-nephelinitic plugs define a circular trend at the edge of the 12 km diameter cone sheet outcrop. Trachytic and phonolitic intrusions cut the complete sequence, producing a 3 km doming of the region (Hernán & Vélez 1980). Dyke density increases sharply northwestwards to form >90% of the outcrop, and dyke dips increase from about 30° on the periphery to 50° in the centre. In cross-section, the cone-sheet distribution points to a common focus located at a depth of about 2 km (Hernán & Vélez 1980). The similarity in age and composition of all these intrusives with the extra-caldera trachyte-phonolite ignimbrites and lavas suggests that they represent a subvolcanic facies.

Erosional stage

Following the shield and caldera stages the island entered a long period (c. 3 Ma) of erosion, with eruptions limited to minor phonolitic events on the northern slopes (Pérez Torrado et al. 2000). A radial pattern of canyons cut deep into the island, reaching the basaltic shield volcano. This palaeotopography controlled the distribution of the subsequent posterosional volcanism (Schmincke 1976, 1993).

Sediments originated in this erosional period were deposited in alluvial fans predominantly on the northeastern, eastern, and southern coastal platforms (Fig. 18.5), forming a Lower Member (Mb) of the Las Palmas Detrital Fm (LPDF; ITGE 1990, 1992). The onset of renewed volcanism (see below) coincides in Gran Canaria with an important transgression (Lietz & Schmincke 1975), during which a Middle Mb of the LPDF (ITGE 1990, 1992) was formed. These marine sediments, with abundant Pliocene fauna, now crop out in the NNE at heights of 50–110 m (ITGE 1990).

Post-erosional magmatism

Renewed volcanism (5.5 Ma BP to present), has shown three main phases: Roque Nublo, post-Roque Nublo and recent volcanism (Pérez Torrado 2000). Roque Nublo volcanism involved strombolian eruptions of basanitic and nephelinitic magmas localized in the central and southern slopes of the island, initially forming a NW–SE aligned series of vents. Later eruptions (4.6 Ma BP) became focused in the centre of the island, building the large and complex Roque Nublo stratovolcano over a period of >1.5 Ma (Pérez Torrado et al. 1995). Volcanic activity began with the emission of large amounts of lavas (basanites–alkali basalts to trachytes–phonolites) that were immediately channelled into the network of canyons excavated during the previous erosional stage. Some of the first mafic lavas flowed up to 20 km towards the NNE coastal areas, where they formed a thick (>20 m) sequence of pillow lava flows, with minor pillow breccia and hyaloclastites, above the marine deposits of the Middle Mb of the LPDF. When the magma changed to trachytic-phonolitic compositions (at about 3.9 Ma BP), explosive volcanic activity occurred in the summit area of the stratovolcano, producing breccia-type ignimbritic deposits (Pérez Torrado et al. 1997). Eruptive activity of the volcano ended by around 3 Ma BP with the intrusion of phonolitic plugs (see Fig. 18.18B).

The distribution and geometry of the Roque Nublo volcanics (c. 200 km³) suggest that this volcano may have exceeded 2500 m in height, with asymmetric flanks defined by extended and gentle slopes in the north and short and steep sides in the south (Pérez Torrado et al. 1995). Several gravitational collapses mass wasted the Roque Nublo stratovolcano, generating 25 km long debris avalanche deposits (García Cacho et al. 1994; Mehl & Schmincke 1999). These deposits have been identified in the submarine ODP Leg 157 boreholes (Schmincke & Sumita 1998).

The second main magmatic phase (Post-Roque Nublo) has been thought by some authors to have started after a >0.5 Ma gap in volcanic activity (McDougall & Schmincke 1976; Schmincke 1976, 1993; ITGE 1990, 1992). However, the Roque Nublo terminal phonolitic intrusions seem coeval with the initial basanitic eruptions of the post-Roque Nublo phase (Pérez Torrado et al. 1995) and the epiclastic deposits considered representative of this erosive gap have been subsequently reassigned to different stratigraphic units of the island. The volcanic activity of this stage is characterized by strombolian vents along a NW–SE rift, with basanite–nephelinite and trachybasaltic lavas forming a 500 m thick sequence (about 10 km³) that covers large areas on the northern slopes of the island. The ages of these lavas suggest the main eruptive period was from 3 to 1.7 Ma BP (McDougall & Schmincke 1976; ITGE 1990, 1992).

Finally, the most recent volcanism has involved rare, dispersed minor eruptions of highly alkalic magmas (basanites, nephelinites). These eruptions produced phreatomagmatic tuff cones and calderas such as the Caldera de Bandama, and strombolian vents such as the Montañón Negro volcano, dated at 3.5 ka BP (Nogales & Schmincke 1969), one of the youngest eruptions known on Gran Canaria.

La Gomera (FH, CRC)

La Gomera is intermediate in size (c. 380 km²) between the two other western islands, La Palma and El Hierro (Fig. 18.1). The island reaches a maximum height of 1487 m in the centre (Alto de Garajonay) above a small (40 km²) plateau ('meseta') from which several radial, deep, and narrow ravines run down to the coast (Fig. 18. 6 inset). These ravines interrupt steep coastal cliffs, 50 to 850 m high, and those in the south are separated by gently sloping flat-topped ridges, which form a distinct feature of the island. Other noteworthy features of La Gomera landscape include the common presence of the so-called 'roques' or 'fortalezas', remains of eroded felsic lava plugs and domes, and wall-like subvertical dykes which stand out from the host rock because of preferential erosion. By contrast, due to the lack of Quaternary volcanic activity, recent cinder cones and lava fields, very common on all the other islands, are absent on La Gomera. The only well-preserved volcanic cone is La Caldera, near the southern coast (Fig. 18.6).

The geology of this island can be subdivided into an old basement of submarine and intrusive origin, an overlying trachytic-phonolitic complex, older and younger basalts, and a series of late felsic domes (Fig. 18.6). Previous studies include those of Gagel (1925), Blumenthal (1961), Bravo (1964) and Hausen (1971), with more specific publications dealing with the old basement (Cendrero 1971), geochronology (Abdel-Monem et al. 1971, 1972; Féraud 1981; Cantagrel et al. 1984), felsic domes (Cubas 1978), and trachytic-phonolitic rocks (Rodríguez 1988).

The oldest rocks exposed on La Gomera are known as the basal complex and represent the basement developed during the pre-shield stage of the island (Cendrero 1971). As in Fuerteventura and La Palma, submarine (basaltic and trachytic) pillow lavas and tuffs and very thinly bedded (pelitic, silicic, and carbonate) sediments crop out on La Gomera, but they are restricted to a few small exposures and some screens in the north, and so there is a lack of palaeontological age data. The main rocks forming this La Gomera basal complex are mafic (gabbros and olivine gabbros) and ultramafic (wehrlites and pyroxenites) in composition, and as extremely dense dyke swarm which represents up to 90% of the unit. Radiometric (K-Ar) ages for intrusive lithologies range between 20 and 14 Ma BP, and probably record late stages of cooling and

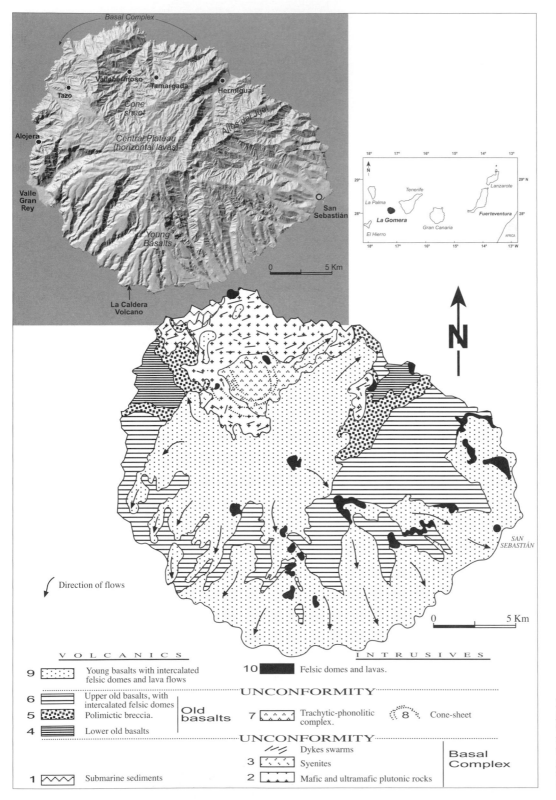

Fig. 18.6. Simplified geological map of La Gomera. Inset: shaded relief image of La Gomera with indication of the main geomorphological and tectonic features (image GRAFCAN).

emplacement. The only outcrop of syenite in the island (usually considered to be part of the basal complex) is to be found in Tamargada. These fine-grained rocks are clearly the youngest among the plutonics and therefore yielded a much younger K-Ar age of *c.* 9 Ma (Cantagrel *et al.* 1984). Most of the dykes are basaltic in composition, but show a great variety of aphanitic and porphyritic types that give rise to a complicated, irregular network. Several felsic (trachytic to phonolitic) dykes are also present, but these are related to the overlying trachytic-phonolitic series.

The trachyte–phonolite complex, recording the next phase in the magmatic evolution of the island, crops out in a deeply eroded, well drained area between Las Rosas and Valle-hermoso. It comprises fractured and highly weathered trachytic to phonolitic massive lavas, breccias, domes and dykes that seem to lie unconformably on the basement, although the contact relationship is rarely exposed. The dykes, which are abundant but not so numerous as in the basal complex, have been grouped into two radial swarms (Huertas et al. 2000) and one cone sheet complex (Rodríguez 1987; Hernán et al. 2000). The earlier radial swarm, the axis of which is centred at Tamargada, is most likely related to the syenites exposed in the same area, whilst the later one, centred south of Vallehermoso, seems to be associated with the cone sheets which were fed from a dome-shaped magma body some 1350 m below sea level (Hernán et al. 2000).

A group of basaltic rocks known as the 'lower old basalts' lies unconformably on the basal complex, but, unfortunately, a lack of critical exposures where these basalts are observed in direct contact with the trachytic-phonolitic rocks has led to questions about their relative stratigraphic position. Some authors have considered the trachytic-phonolitic complex to be the oldest subaerial unit, whereas others contend that the lower old basalts represent the first subaerial activity. A few felsic dykes, probably belonging to the trachyte–phonolite complex, cut through these lower old basalts. If the latter interpretation is accepted, then these old basalts mark the onset of the island shield stage. The unit consists mainly of thin pahoehoe lava flows, and subordinate interlayered pyroclastic horizons, cut by a dense dyke swarm, with abundant sills. Most lava flows are porphyritic ankaramites and plagioclase basalts that reach a maximum thickness of 250 m in the single large area where they are exposed, between Tazo and Alojera in the NW. Smaller outcrops include one near the town of Hermigua, where Cubas et al. (1994) identified submarine levels at the base of the sequence, and one or two more restricted spots in the south where the barrancos have excavated deeper. These outcrops represent remnants of a larger edifice that once occupied the northern onland and offshore areas of the present island. The initial subaerial basaltic shield may have been active between at least 11 and 9 Ma BP, as suggested by the K-Ar age from a dyke (Cantagrel et al. 1984) and two ages from lava flows near Hermigua (Cubas et al. 1994).

A polymictic breccia up to 300 m thick lies unconformably over the lower old basalts. This breccia includes heterogeneous, subrounded and angular clasts up to 50 cm in diameter and surrounded by a cemented fine-grained groundmass. Its origin has been interpreted as due to either highly explosive or laharic activity. Occasionally, where the lower old basalts are lacking, the breccia rests directly over the basal complex rocks. The breccia is presumably a remnant of an almost entirely removed initial shield volcano. Near its top, the breccia is conformably interlayered with new basaltic flows known as the 'upper old basalts'. These new basalts are spread over the entire island and are better preserved than the lower old basalts. They were erupted effusively from fissure vents and built a large new shield, which has since been capped by younger rocks. The remains of this shield are at present exposed by erosion at the bottom and in the walls of the main barrancos, with the maximum visible thickness (over 500 m) being preserved in Altos de Juel. The pile consists mainly of thin lava flows (ankaramites, plagioclase basalts, and aphanitic basalts), some pyroclastic layers and a few buried pyroclastic cones, with thick trachybasalt flows near the top of the sequence. According to Cubas (1978), a few of the trachytic

domes cutting only the upper old basalts but not the young basalts (see below) could also be a part of this unit. However, other authors (Bravo 1964; Cendrero 1971; Rodríguez 1988) distinguished a single, more recent episode of felsic domes. The numerous dykes intruding the old basalts, more abundant in the lower than in the upper ones, acted as feeders for the different subaerial units. The most consistent radiometric (K-Ar and $^{39}Ar/^{40}Ar$) ages obtained by several authors (Abdel-Monem et al. 1971; Féraud 1981; Cantagrel et al. 1984) place the upper old basalts activity between about 9 and 6.5 Ma BP. The available age data suggest that either an interval of quiescence accompanied by gradual denudation or a violent rapid destruction of the lower basaltic shield took place between approximately 10 and 9 Ma BP.

The most recent basaltic unit is separated from the upper older basalts by a marked unconformity representing c. 1.5 Ma and clearly visible only in the western area. The unit comprises a thick (500 to 1000 m) pile of mostly basaltic lava flows in which olivine and augite-bearing porphyritic basalts, aphanitic basalts and trachybasalts are the most representative types. Plagioclase basalts, common in earlier episodes, were only rarely erupted during this younger activity. The individual flows are very thick (5–10 m), only rarely cut by dykes, and generally well preserved, exhibiting frequent columnar, and sometimes spherical jointing. Whilst in the central sector the flows are characteristically flat, they dip gently seawards in the southern peripheral sector. Basaltic pyroclasts are fairly well represented in the form of interlayered sheets and buried cones, and felsic pyroclastic layers, lava flows and, most of all, phonolitic domes also occur. The previously mentioned cone, La Caldera, belongs to this phase of activity, as does another, less well-preserved one almost completely covered by vegetation in the central sector. K-Ar and $^{39}Ar/^{40}Ar$ ages determined from basaltic flows and felsic domes indicate concentrated activity approximately between 5 and 4 Ma BP, although a single datum from a lava at the top of the section in Arure would extend the activity up to 2.8 Ma BP. The emission of the latest, more heterogeneous materials must have accentuated the island relief and enlarged its surface considerably. How large and high the island became after the construction of this third and most complex shield is not known. High coastal cliffs around the island, and especially along the north and west coasts, suggest that the present island is only a small fraction of the size it once was.

Tenerife (EA)

Tenerife is the largest (2058 km^2) and highest (3718 m) island in the Canaries and has a complex volcanic history. The oldest visible materials (the Old Basaltic Series; Fúster et al. 1968a) are preserved in three isolated and deeply eroded massifs (Fig. 18.7): Anaga (to the NE), Teno (to the NW) and Roque del Conde (to the south). Each has been formed in late Miocene and early Pliocene times as a result of several independent volcanic cycles interspersed with long pauses in activity (Ancochea et al. 1990).

After the Old Basaltic Series, the volcanic activity became concentrated within two large edifices: the central composite volcano of Las Cañadas and the 'cordillera dorsal', a SW–NE volcanic ridge linking Las Cañadas and the Anaga massif (Fig. 18.7). On each side of the cordillera dorsal, two large trapezoidal depressions have been formed, the so-called 'valleys' of Güímar and La Orotava.

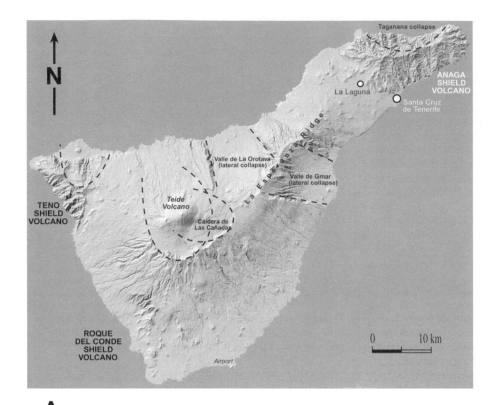

A

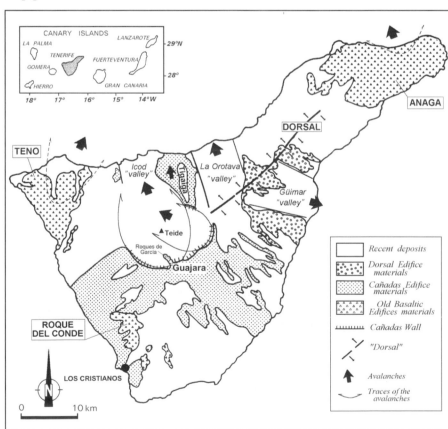

B

Fig. 18.7. (**A**) Shaded relief image of Tenerife with indication of the main geomorphological and tectonic features (image GRAFCAN). (**B**) General sketch map of Tenerife (modified from Cantagrel *et al.* 1999 and Ancochea *et al.* 2000).

On the upper part of the Las Cañadas volcanic edifice, there is a semi-elliptic depression with a NE trending axis 16 km long, known as the 'Caldera de Las Cañadas'. This depression is partially surrounded by a southeastern cliff, Las Cañadas Wall, over 25 km long and reaching a maximum altitude (2712 m) and height (500 m) at Guajara (Fig. 18.7). The most recent activity is represented by many monogenetic basaltic centres scattered across the island and by the Teide–Pico Viejo edifice, volcanic products from which partially fill the Las Cañadas caldera. Some basaltic eruptions have taken place in historical times (last 500 years). Further details and additional references on the geology of the island can be found in Fúster *et al.* (1968*a*), Carracedo (1979) and Ancochea *et al.* (1990, 1999).

With regard to the oldest massifs, that of Anaga comprises a complex sequence of basaltic lava flows and volcaniclastic levels intruded by basaltic and phonolitic dykes and domes (Fig. 18.8a). Three units have been identified (Fig. 18.9), the oldest of which (Lower Anaga) appears in the north in a large arcuate landform called Arco de Taganana and consists of volcaniclastic rocks and debris flow deposits cut by abundant dykes. The age of this unit is not well defined, due to intense alteration and the fact that a K-Ar age of 16.1 Ma from an ankaramite of this unit (Abdel-Monem *et al.* 1972) is much older than any other for Tenerife and so must be viewed with caution (Ancochea *et al.* 1990). The overlying Middle Anaga unit has a total thickness of about 1000 m, is separated from the former by an unconformity, and dips towards the south. It comprises basaltic pyroclastic and lava flows, with subordinate felsic plugs and lava flows. The ages obtained from this unit vary between 6.5 Ma and 4.5 Ma (Carracedo 1975; Féraud 1981; Féraud *et al.* 1985; Ancochea *et al.* 1990). The youngest unit (Upper Anaga: 3.3–3.7 Ma BP) is basaltic to phonolitic in

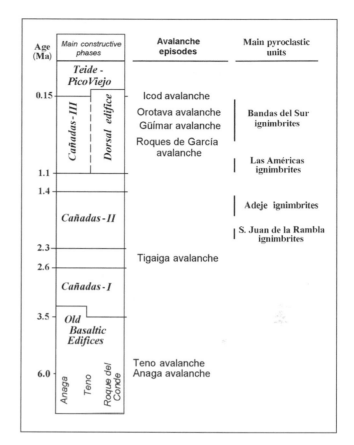

Fig. 18.9. Simplified volcanic stratigraphy, avalanche episodes and main pyroclastic units of Tenerife (modifies from Cantagrel *et al.* 1999).

composition, appears in the western part of the massif, and lies above a clear unconformity.

The Teno edifice is made up of only two units: a lower sequence (Lower Teno) of basaltic pyroclastics and lava flows, dipping seawards, covered unconformably by an upper sequence (Upper Teno) of subhorizontal (700 m) basaltic lava flows, with some trachytes, and cut by abundant dykes (Fig. 18.8b). The upper part of the Lower Teno unit has been dated by Ancochea *et al.* (1990) at 6.7–6.1 Ma BP, the Upper Teno unit at 5.6–5.0 Ma BP, and the youngest age yet obtained is from a 4.5 Ma phonolitic plug. In the lower part of the Upper Teno unit a debris flow deposit (*c.* 6 Ma BP; Cantagrel *et al.* 1999) represents the subaerial remains of a major collapse (Teno avalanche; Fig. 18.8b). Finally, the Roque del Conde edifice comprises 1000 m of basaltic lava flows (Fig. 18.8c), with ages mainly between 8.5 and 6.4 Ma (Ancochea *et al.* 1990).

Over the last 3.5 Ma the central part of Tenerife has been occupied by shield or central composite volcanoes that at times have reached more than 3000 m in height (Ancochea *et al.* 1999, 2000). Four main phases have been recognized: Cañadas I, II, III, and the presently active Teide–Pico Viejo complex ('Cañadas IV'; Fig. 18.9). Initial volcanic activity (Cañadas I volcano) took place in the western part of the island, between *c.* 3.5 Ma and 2.6 Ma BP, and produced mainly basalts, trachybasalts and trachytes. The remains of this phase crop out in the Cañadas wall and at the bottom of several

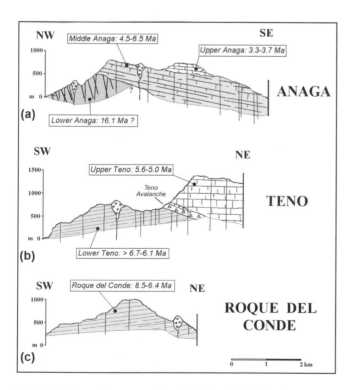

Fig. 18.8. Schematic cross-sections of old basaltic edifices in Tenerife.

radial ravines. Its main emission centre (3000 m high) was located in the central part of what is now the Cañadas 'caldera' (Ancochea *et al.* 1999). This edifice underwent partial destruction by failure and flank collapse to the north (Ibarrola *et al.* 1993; Cantagrel *et al.* 1999) forming a debris avalanche deposit (Tigaiga breccia) on which a second volcano was built (Cañadas II volcano) between 2.4 and 1.4 Ma BP. Towards the end of this second period of activity, major explosive eruptions took place forming ignimbrites, pyroclastic flows, and ash fall deposits of trachytic composition, now exposed in the SW lower slopes.

The next constructional phase, from 1.1 to 0.15 Ma, built a new volcano (Cañadas III volcano) which produced trachybasaltic lava flows and abundant phonolitic products. The later stages (0.7–0.15 Ma BP) are again strongly marked by major explosive eruptions, producing the Bandas del Sur ignimbrites from an eastern eruptive centre, whose pyroclastic products are widely exposed across the SE lower slopes of Tenerife. Several of these pyroclastic eruptions had sufficient volume to trigger the formation of a complex vertical collapse caldera(s) in the summit region (Martí *et al.* 1994, 1997; Bryan *et al.* 1998). However, their geometry and location within the Cañadas II and III edifices is poorly defined because several debris avalanche events later broke through these earlier structures.

The formation mechanisms of the present Las Cañadas 'caldera' have long been discussed. Models for erosion, collapse, explosion and avalanche have been proposed (e.g. Hausen 1949, 1956, 1961; Bravo 1962; McFarlane & Ridley 1969; Araña 1971; Coello 1973; Booth 1973; Ancochea *et al.* 1990, 1998; Carracedo 1994; Martí *et al.* 1994; Watts & Masson 1995). In recent years, the discussion has focused on the alternatives of vertical

collapse versus lateral collapse, or a combination of both. Martí *et al.* (1997) suggested that Las Cañadas is the result of a complex sequence of vertical collapse events. These vertical collapses may have played a major role in triggering lateral collapses. For Cantagrel *et al.* (1999) the present shape of the wall at Las Cañadas is not related to vertical collapses, but is instead a product of repeated flank failures.

Numerous pyroclastic trachytic and phonolitic eruptions occurred during the construction of Las Cañadas, the early studies of which provided a landmark in the international development of tephrochronology (Walker 1973; Booth 1973). This pyroclastic activity showed an eastward migration through time (Martí *et al.* 1994; Ancochea *et al.* 1999) and became increasingly common, so that pyroclastic deposits of all types represent a large proportion of the volcanic products that make up the Cañadas volcanic edifice (Ancochea *et al.* 1990, 1999; Martí *et al.* 1994). Pyroclastic rocks are especially well exposed on the SE lower slopes (Bandas del Sur) of the island, where they have been extensively studied (Alonso 1989; Bryan *et al.* 1998), but are also present in various stratigraphic positions in all the other sectors of the Cañadas edifice. According to their geographic position and isotopic ages, four main pyroclastic phases may be identified in the Cañadas edifice over the last 2 Ma (Figs 18.9 and 18.10). The first 'San Juan de la Rambla' phase (*c.* 2 Ma BP), whose outcrops are presently restricted to the north of Tenerife, occurred during the first period of construction of the Cañadas II edifice. The second 'Adeje' phase (1.8–1.5 Ma BP) is much more important in volume and erupted in successive distinct pyroclastic events during the second part of construction of the Cañadas II edifice. Widespread in the SW, the pyroclastic

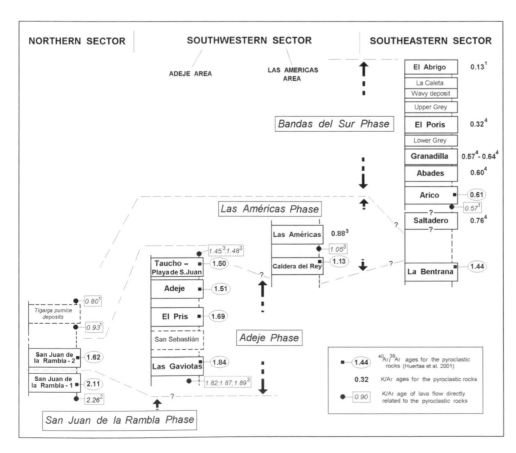

Fig. 18.10. Schematic stratigraphic colums of the main pyroclastic deposits showing possible correlations. K-Ar ages (in Ma) from the following references: [1]Ancochea *et al.* (1990); [2]Ibarrola *et al.* (1993); [3]Fúser *et al.* (1994); [4]Bryan *et al.* (1998).

deposits of this phase have also been recognized in the Tigaiga massif in the north and in the deepest levels of the Barranco de la Bentrana in the SE (Bandas del Sur). The third 'Las Américas' phase (1.1–0.9 Ma BP) is presently recognized around the town of Los Cristianos, in the south of the island, and in the Tigaiga massif in the north, although some pyroclastic units stratigraphically located between the La Bentrana and Arico ignimbrites could also belong to this phase (Fig. 18.10). The fourth 'Bandas del Sur' phase (Wolff 1985; Alonso 1989; Bryan et al. 1998; 0.7–0.15 Ma BP) is by far the best exposed and also appears to be the most important in volume. Widespread across the SE slopes of Tenerife, it is known in the north at the top of the Tigaiga massif but has not yet been described in the SW. Several tens of pyroclastic units have been identified but their lateral extent is commonly limited so that no unique stratigraphic column can be drawn, although a simplified representation is presented in Fig. 18.10 (see Bryan 1995 and Bryan et al. 1998 for details). The coeval eruption of the dorsal edifice (Ancochea et al. 1999) produced a basaltic NE-trending volcanic ridge, the westernmost flows of which are interbedded with felsic rocks from the Cañadas III central edifice. The so-called 'valleys' of La Orotava and Güímar, transversal to the ridge axis (Fig. 18.7) also formed during this period. Finally, the most recent volcanic activity in Tenerife is represented by the Teide–Pico Viejo complex, formed by basalts, trachytes and phonolites, which partly fill Las Cañadas 'caldera' (e.g. Ablay et al. 1998), and by many small scattered basaltic monogenetic volcanoes, mainly on the dorsal ridge. Six such basaltic volcanoes have erupted in historical times (see Fig. 18.21) and the Teide–Pico Viejo complex has also shown historical activity.

Much attention has recently focused on very large debris avalanche deposits found off the north coast of Tenerife (Ancochea et al. 1990, 1998; Carracedo 1994; Watts & Masson 1995; Teide Group 1997). On the SE flank of the dorsal ridge, a south-directed, large landslide formed the Güímar valley later than 0.80 Ma BP. Another six, north-directed debris avalanche events have been identified in Tenerife (Cantagrel et al. 1999): the Anaga and Teno (c. 6 Ma BP) in the Old Basaltic edifices, followed by the Tigaiga event in Cañadas I volcano (>2.3 Ma BP), Los Roques de García (<1.4 Ma BP); La Orotava avalanche (c. 0.6 Ma BP) that formed the Orotava valley; and the <0.15 Ma BP Icod avalanche (Fig. 18.10).

El Hierro (JCC, FJPT, ERB)

The island of El Hierro, SW of La Palma (Fig. 18.1), is the smallest (287 km^2) and least populated of the archipelago. It is also one of the steepest oceanic shield volcanoes in the world, with slopes that frequently exceed 30°, with the emergent summit, Malpaso, rising to 1501 m from a depth below sea level of 3700–4000 m. The characteristic trilobate morphology of the island is the result of the development of a regular (120°), three-branched rift zone (Carracedo 1994, 1996a) that forms the ridges of the volcano (Fig. 18.11). The sectors between the rifts form arcuate embayments, originating by catastrophic mass wasting during successive giant landslides. El Hierro and La Palma form a dual line of island volcanoes, with volcanism apparently alternating between the two islands (Carracedo et al. 1999a), and only La Palma showing important eruptive activity during Holocene times.

Published reports on the geology of El Hierro are relatively scarce prior to 1996 and focused on the general geology and petrology of the island (Hausen 1964), and on the geochronology (Abdel-Monem et al. 1972; Fúster et al. 1993). However, after 1996, much work has been published on the geochronology (Guillou et al. 1996), geology and tectonics (Day et al. 1997; Carracedo et al. 1999b) and submarine geology (Masson 1996; Urgelés et al. 1997, 1998). Recent mapping with K-Ar dated geomagnetic reversals has made possible stratigraphic correlation between selected sections, permitting the reconstruction of the geological evolution of the island and the definition of three main volcanic edifices (Fig. 18.11): the Tiñor and El Golfo volcanoes, and the subsequent development of a triple rift system.

Tiñor volcano

This constitutes the first stage of subaerial growth of El Hierro and its present outcrop is confined to the NE flank of the island and the interior of the Las Playas embayment (Fig. 18.11). Tiñor volcano lavas are basaltic and characterized geochemically by relatively primitive picrobasaltic to hawaiitic-tephritic compositions. According to the K-Ar ages and magnetostratigraphy, Tiñor volcano developed very rapidly and continuously between c. 1.2–0.8 Ma BP. There is no consistent compositional variation with time that can be mapped in the field, but it is possible to recognize early, intermediate and late stage magmatic phases. The early stage has produced a basal unit of thin flows on steep slopes, with more pyroclastic rocks and dykes than in younger basalts. The intermediate stage shows thicker lavas with dips that become subhorizontal in the centre of the edifice, probably reflecting the mature stage of growth of the volcano. Late stage activity produced more explosive, wide-cratered vents (the Ventejís volcano group) and associated xenolith-rich lavas. This explosive stage seems to have immediately preceded the collapse of the NW flank of the Tiñor volcano at c. 0.88 Ma BP (Figs 18.11 and 18.12).

El Golfo

This represents a new volcanic edifice developed over the remnants of the Tiñor volcano, filling its collapse embayment and totally burying it. Continued volcanism from c. 0.55 to 0.17 Ma BP built a 2000 m high shield volcano, 20 km in diameter, with a basal, mostly pyroclastic unit overlain by an upper unit composed mostly of lava flows. The basal unit was produced mainly by strombolian and surtseyan eruptions, with abundant cinder cones, tuff rings and subordinate lava flows. It is densely intruded by NE-, ESE- and WNW-trending swarms of dykes and sills that match the present volcanic vent systems and indicate that a triple rift system may have been an important feature of the El Golfo edifice. The lavas of the upper unit, in contrast, show a relatively small number of exposed feeder dykes, probably because they were mainly erupted from the summit of the volcano. These lavas show petrographic characteristics similar to (but becoming more evolved upwards than) those of the Tiñor volcano: olivine–augite basalts at the base, grading upward to microcrystalline lavas with scarce clinopyroxene phenocrysts. The more evolved trachytic rocks higher in the El Golfo edifice show abundant phenocrysts of alkali feldspar, aegirine–augite and opaque minerals in a trachytic matrix forming a network of

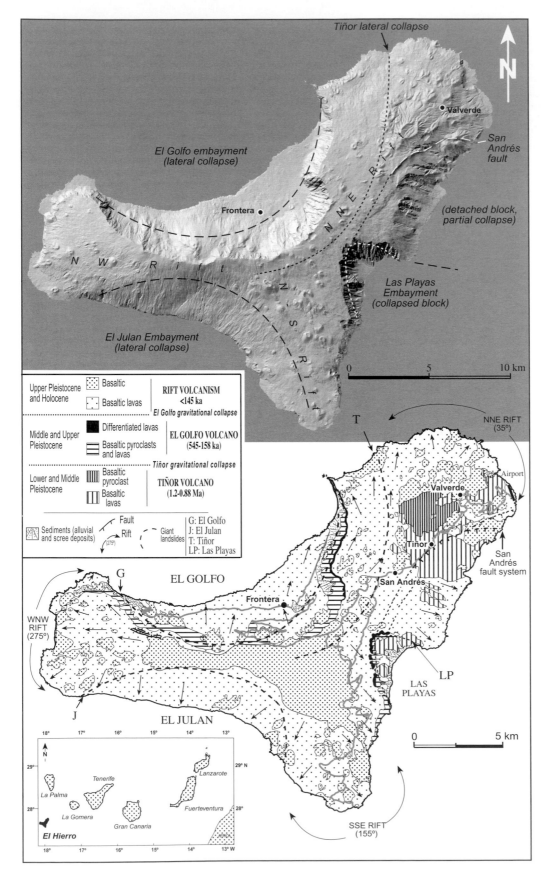

Fig. 18.11. Simplified geological map of El Hierro (from Carracedo *et al.* 1999*b*). Inset: shaded relief image of El Hierro with indication of the main geomorphological and tectonic features (image GRAFCAN).

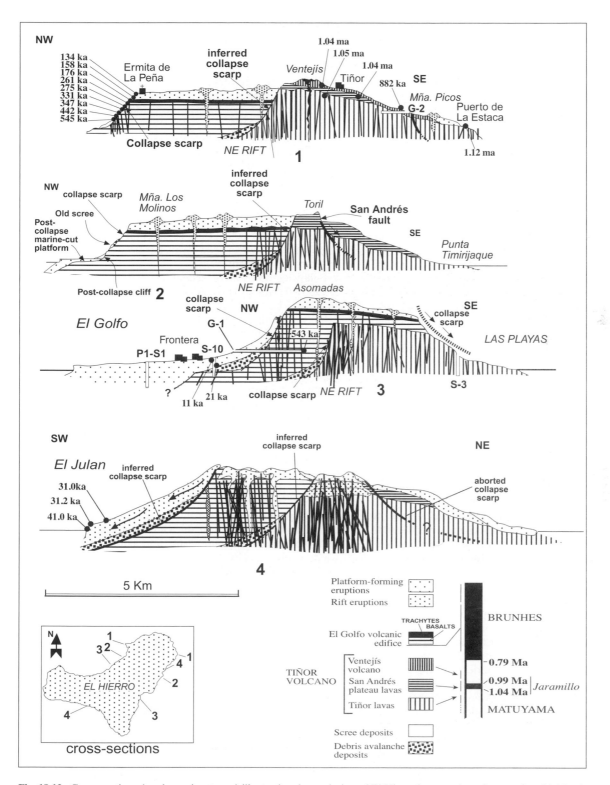

Fig. 18.12. Cross-sections (see lower inset map) illustrating the evolution of El Hierro by accretion of successive shield volcanoes (modified from Carracedo *et al.* 1999*b*).

feldspars and clinopyroxenes. The youngest rocks are formed by several differentiated lava flows (trachybasalts, trachytes) and block-and-ash deposits that probably correspond to the terminal stages of activity of the volcano, which was subsequently modified by lateral collapse. The El Julan collapse, on the SW flank of the volcano, is older than 158 ka, the age of the rift lavas that later partially fill this embayment. The SE flank underwent an aborted collapse between *c.* 261 and 176 ka BP, during which the San Andrés fault system (Fig. 18.11) was formed (Day *et al.* 1997). The youngest lateral collapse in El Hierro formed the northern El Golfo embayment and took place 15 ka BP ago according to Masson (1996) or between 21 and 134 ka BP according to Carracedo *et al.* (1999*b*).

Rift volcanism

The latest stage of growth of the island saw volcanism reorganized into a three-branched rift system, in which all the rifts were simultaneously active without the development of a central vent complex. Rift lavas show a slightly more alkaline and silica-undersaturated compositional trend, with alkaline picrobasalts, basanites and tephrites being the predominant petrographic types. The maximum age of these eruptions is constrained by the differentiated lavas topping El Golfo volcano (176 ka). K-Ar ages from 145 to 134 ka have been obtained at the top of El Golfo cliff (Fig. 18.12, 1). After this time the north flank of the island collapsed producing the present El Golfo embayment (Masson 1996; Carracedo *et al.* 1997, 1999*b*; Urgelés *et al.* 1997). Since this collapse there has been only moderate eruptive activity with lavas post-dating the last glacial maximum, filling the El Golfo embayment and forming coastal platforms. The last dated eruption (2.5 ka BP, Guillou *et al.* 1996) is from a small vent at the NE rift. However, an eruptive vent at the westernmost part of the island has been associated with the seismic crisis of 1793 (Hernández Pacheco 1982).

La Palma (JCC, ERB, FJPT)

The island of La Palma lies at the northwestern edge of the Canarian chain (Fig. 18.1), occupies a 706 km^2 triangular area, and forms a volcanic edifice that reaches a maximum elevation of 2426 m a.s.l. (Roque de Los Muchachos), resting on a 4000 m deep ocean floor. La Palma is formed mainly by two volcanoes: a circular, 25 km diameter shield volcano in the north (northern shield in Fig. 18.13 inset) and a north–south elongated, 20 km long Cumbre Vieja rift in the south (CV in Fig. 18.13 inset). The former, extinct for the last 400 ka, is deeply eroded, with a radial network of deep barrancos and a 6 km diameter erosional depression (Caldera de Taburiente) on its SW flank. The Cumbre Vieja volcano has been highly active since *c.* 120 ka BP, with six eruptions in the last 500 years, two of them in the twentieth century (1949 and 1971).

Early reports on the geology of La Palma were mainly focused on the spectacular Caldera de Taburiente and the submarine lavas cropping out in its interior, starting with the works of Buch (1825), Lyell (1865), Reiss (1861), Gagel (1908) and Sapper (1906). The submarine volcanic rocks have been studied by Middelmost (1970), Hernández Pacheco (1973), Staudigel (1981) and Staudigel & Schmincke (1984), among others. Geochronological, stratigraphic and structural studies have been carried out by Abdel-Monem *et al.* (1972), Ancochea *et al.* (1994), Guillou *et al.* (1998), Carracedo *et al.* (1999*a*) and Guillou *et al.* (2001), among others. Swath bathymetry and side-scan sonar studies of the marine flanks of La Palma (see Fig. 18.20B) have been published by Holcomb & Searle (1991), Weaver *et al.* (1992) and Urgelés *et al.* (1999).

The geological evolution of La Palma is characterized by the growth of three main volcanoes: a submarine volcano (seamount), the now extinct northern shield volcano, and the still active southern Cumbre Vieja rift (Fig. 18.13). Initial submarine magmatism produced alkali basaltic to trachytic

pillow lavas and hyaloclastites, basaltic feeder dyke swarms, and gabbroic, trachytic and phonolitic intrusions. The entire complex has been affected by hydrothermal metamorphism, grading from zeolite to albite–epidote–hornfels facies, probably corresponding to a metamorphic gradient of *c.* 200–300°C/km (Staudigel & Schmincke 1984), and crops out on the floor and lower walls of the Caldera de Taburiente (CT in Fig. 18.13 inset).

The 20 km diameter submarine volcano has been dated at 4–3 Ma BP (Staudigel *et al.* 1986), and is separated from overlying subaerial volcanics by an angular unconformity. The angularity of this unconformity, and its present height of 1500 m, is due to uplift, tilting and erosion during the relatively long period required for the subaerial volcanism to become established, when soft, easily eroded pyroclastics were being produced. A 400–600 m thick sedimentary unit made of breccias, agglomerates and sediments lies between the submarine and subaerial volcanoes.

Following the emergence of La Palma above sea level around 1.77 Ma BP, two contiguous volcanic edifices were constructed. The earlier of these two subaerial volcanic centres forms the northern half of La Palma (Fig. 18.13) and provides an excellent illustration of a typical shield-stage Canarian volcano (Ancochea *et al.* 1994; Navarro & Coello 1994; Carracedo *et al.* 2001). The well-exposed sections provided by the Caldera de Taburiente walls allow the observation of the magmatic history from its initial stages, and over 100 water tunnels or 'galerías' permit the study and sampling of the internal structure of the volcanic edifice. Geomagnetic reversal data and precise radiometric dating (59 K-Ar and Ar^{40}/Ar^{39} ages) show that this shield is built of two subaerial volcanoes, Garafía and Taburiente volcanoes, developed from 1.77 to 0.4 Ma BP (Guillou *et al.* 2001).

The Garafía volcano developed discordantly over the seamount, increasing the island in areal extent and elevation, but subsequently became completely covered by the Taburiente volcano. Outcrops of the Garafía volcano (GV in Fig. 18.13 inset) are therefore rare and confined to erosional windows provided by the deep barrancos in the northern flank of the shield (Fig. 18.13). Continuous volcanic activity from 1.77 to 1.20 Ma BP built an edifice with steep flanks (slopes of 30–35°), 23 km in diameter and probably reaching 3000 m in height. Lava sequences around 400 m thick and comprising olivine–pyroxene and pyroxene–plagioclase basalts, with minor trachybasalts, constructed a 315 km^3 volcano with eruptive rates of 0.6 km^3/ka. Overgrowth and increasing instability triggered a giant lateral collapse towards the SW at *c.* 1.2 Ma BP (Figs 18.13 and 18.20). Breccias and sediments inside the Caldera de Taburiente (Fig. 18.13) may represent the avalanche deposits and post-collapse sediments associated with this first catastrophic tectonic event on La Palma.

The Taburiente volcano (TV in Fig. 18.13 inset) is the result of the resumption of the eruptive activity in the northern shield after the collapse of the Garafía volcano. From *c.* 1.1 to 0.4 Ma BP, a sequence of lavas over 1000 m thick completely covered the previous edifice and considerably enlarged the island in area and height. The first post-collapse lavas filled the collapse embayment and then spilled over the flanks of the Garafía volcano. Lavas flowing against the back scarp of the collapse formed a 400 m sequence of horizontal lavas which, with later differential erosion, subsequently formed a central plateau (CP in Fig. 18.13 inset) perched on the top of the shield. Volcanism then resumed at *c.* 0.8 Ma BP, changing

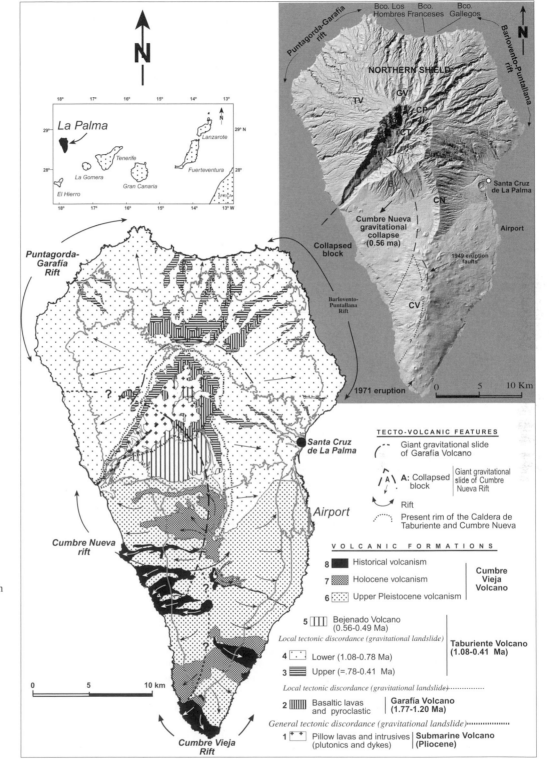

Fig. 18.13. Simplified geological map of La Palma (modified from Carracedo *et al.* 2001). Inset: shaded relief image of La Palma with indication of the main geomorphological and tectonic features (image GRAFCAN). Key: BV, Bejenado volcano; CP, central plateau (horizontal lavas filling the Garafía collapse embayment); CT, Caldera de Taburiente; Cumbre Nueva, Cumbre Nueva ridge; Cumbre Vieja, Cumbre Vieja volcano; GV, Garafía volcano; TV, Taburiente volcano.

from eruptions through dispersed vents in the shield to vents along well-defined rift zones (Figs 18.13 and 18.18A). In the later stages, a 3000 m high central volcano with differentiated lavas and explosive eruptions developed at the centre of the shield, together with rift-zone eruptions up until *c.* 0.4 Ma BP.

Volcanic activity may have already started to migrate southwards focusing predominantly in the southern Cumbre Nueva rift (CN in Fig. 18.13), which became progressively destabilized and collapsed towards the SW at *c.* 0.5 Ma BP. This collapse formed a prominent embayment (Fig. 18.13 inset) and initiated the 15 km long, 7 km wide, 2 km deep, *c.* 100 km³ Caldera de Taburiente, first considered by Buch (1825) to be a typical 'uplifted crater', then an erosion caldera by Lyell (1865), and finally an avalanche caldera by Ancochea *et al.*

(1994). Studies by Carracedo *et al.* (1994, 1999*a,b*, 2001) and Paris & Carracedo (2001) concluded that this is an erosion caldera, initiated by a tectonic event, the Cumbre Nueva giant landslide. The main part of the depression developed by subsequent erosional retreat of the embayment walls. Linear incision rates in the caldera are often in excess of 1 m/ka and headward erosion rates exceed 3 m/ka, with the highest values at the boundaries of the collapse. Higher erosion rates in the western valley (Barranco de Las Angustias), coalescence of valleys and damming by lava flows of the Bejenado volcano led to the formation of a single circular depression. The Bejenado volcano (BV in Fig. 18.13 inset) developed inside this Cumbre Nueva collapse, representing a continuation of eruptive activity at the Cumbre Nueva rift, simultaneously with eruptions along other rifts of the shield. Remnants of the collapse embayment wall may still be present in slide blocks (Torevas) inside the caldera and underneath the Bejenado lavas.

The northern shield ceased volcanic activity at *c.* 0.4 Ma BP. However, apparently younger adventive vents on the Bejenado volcano may represent an eruptive interval between the northern shield and the Cumbre Vieja rift activity. Lavas of the Taburiente volcano show higher degrees of magmatic evolution, from primary basanites and basalts (olivine–pyroxene and amphibole) to differentiated rocks such as tephrites, phonolites and trachytes. The most differentiated lavas were erupted during the latest stages of the evolution of the volcano, and associated with more explosive eruptive mechanisms. Continuing southward migration of volcanism in La Palma ultimately resulted in the extinction of the northern shield *c.* 0.4 Ma BP and the formation in the last 130 ka of the Cumbre Vieja rift (CV in Fig. 18.13 inset). Eruptions from Cumbre Vieja occurred in two main pulses, one from *c.* 125 to 80 ka BP, and another from *c.* 20 ka BP to present (Carracedo *et al.* 1999*a*), building a 20 km long, 1949 m high ridge (Fig. 18.13). Reorganization of the rift from *c.* 7 ka BP apparently increased the instability of the volcano (Day *et al.* 1999), although the previous volcanoes of La Palma took considerably longer periods (0.5 and 0.8 Ma) to reach their instability threshold. The Cumbre Vieja volcano may collapse in the geological future or may evolve to a stable configuration.

Historical (<500 years) eruptions are located along the Cumbre Vieja rift (Figs 18.13 and 18.21), the most recent one (Teneguía volcano, 1971) being at its southernmost tip, with submarine vents extending the rift further to the south. An association of recent eruptions with phonolitic plugs is clearly evident, probably because these fractured plugs provide an easy pathway for the ascending magmas. Juvenile phonolites have been extruded in several of these recent eruptions, changing the characteristic effusive (strombolian) eruptions to more explosive mechanisms, with abundant block-and-ash deposits and magma mixing. Interaction with groundwater is frequent, and most of the recent eruptions show phreatomagmatic features (Kluegel *et al.* 1999).

Canarian geochronology, stratigraphy and evolution: an overview (JCC, FJPT)

Since the early K-Ar ages of Abdel-Monem *et al.* (1971, 1972), extensive geochronological studies have been carried out in the Canary Islands and more than 450 radiometric (K-Ar and ^{39}Ar/^{40}Ar) ages from volcanic rocks of the different islands have been published. At least 105 of these ages, from volcanic

lithologies in the islands of La Palma and El Hierro, have been obtained with stringent requirements: sampling from well-controlled stratigraphic sections, using only microcrystalline groundmass, replicated analyses, combined use of K-Ar and ^{39}Ar/^{40}Ar methods and systematic comparison of the palaeomagnetic polarities of the samples with the currently accepted geomagnetic reversal timescales (Guillou *et al.* 1996, 1998, 2001). A plot of the published radiometric ages from the Canaries (Fig. 18.14A) shows three groups of islands: (1) Lanzarote, Fuerteventura and Gran Canaria, with subaerial volcanism 14.5 Ma or older and two main stages of volcanic growth separated by long periods of inactivity (erosional gap in Fig. 18.14A); (2) La Gomera, with subaerial volcanism not older than 12 Ma and only the pre-erosional gap stage of growth; and (3) Tenerife, La Palma and El Hierro, with subaerial volcanism younger than *c.* 7.5 Ma and only the juvenile shield stage of growth.

An interruption in the volcanic activity (i.e. erosional gap) of individual islands is a common feature of hotspot oceanic island groups. This feature was used in the Hawaiian islands to separate two main volcanostratigraphic units: the shield stage and the post-erosional or rejuvenated stage (Clague & Dalrymple 1987; Walker 1990). The application to the Canaries of this distinction (Carracedo *et al.* 1998; Carracedo 1999) solves many of the problems raised by the original use of the term 'series' in the volcanostratigraphy of the islands (Fúster *et al.* 1968*a,b,c,d*), a term these days used for geological units formed during the same timespan and with synchronous boundaries. The use of terms such as 'old' and 'recent' series led to considerable confusion, since the 'old' series of La Palma or El Hierro is considerably younger than the 'recent' series of Fuerteventura, Lanzarote or Gran Canaria. This confusion is avoided when the concept of shield-stage and post-erosional or rejuvenation volcanism is applied to the Canaries, which can then be separated accordingly (Fig. 18.14B).

There are, however, some peculiarities in the stratigraphic relationship of the Canaries when compared to other oceanic groups such as the Hawaiian archipelago. In fact, a comparison of the two archipelagoes reveals significant differences: (a) in the age of the oldest islands (Fuerteventura is about four times older than Kauai); and (b) two islands in the Canarian archipelago are still in the shield stage, whereas in the Hawaiian islands only the active volcanoes on the island of Hawaii are at present in this stage of growth. Another important difference between the Canarian and the Hawaiian volcanoes is the extinction sequence (Stearns 1946; Clague & Dalrymple 1987). In the Hawaiian islands, the main stage of growth (typically lasting *c.* 1 Ma according to Langenheim & Clague 1987) of each island volcano is nearly completed before the next one emerges, whereas in the Canaries, three islands spanning at least 7.5 Ma are at present in their shield stage of growth.

In Fuerteventura, La Gomera and La Palma the products of shield volcanism rest upon variably deformed and uplifted sequences of submarine sediments, volcanic rocks (mainly pillow basalts), dyke swarms and plutonic intrusions which form the cores of these islands. These formations (termed the 'basal complex' by Bravo in 1964) are consistently separated from the subaerial volcanism by a major unconformity. Because of subsidence these pre-shield formations do not crop out in most of the volcanic ocean islands, except the Canaries and Cape Verde islands, notably on Maio (Stillman *et al.* 1982). Early interpretations linked these formations to uplifted

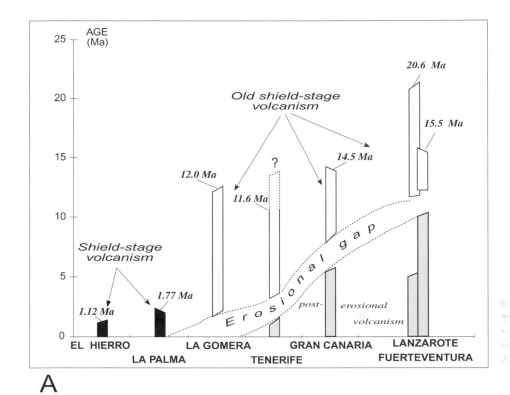

Fig. 18.14. (**A**) Published K-Ar ages from lavas of the Canary Islands. The presence of a gap in the eruptive activity allows the separation of two main stratigraphic units: shield-stage and post-erosional-stage volcanism. Canarian ages from Abdel-Monem *et al.* (1971, 1972), McDougall & Schmincke (1976), Carracedo (1979), Ancochea *et al.* (1990, 1994, 1996), Coello *et al.* (1992), Pérez Torrado *et al.* (1995), Guillou *et al.* (1996, 2001) X-axis not to scale. (**B**) Oldest ages of the subaerial volcanism of the Canary Islands. Sources of ages as in (A). (modified from Carracedo 1999).

blocks of 'oceanic basement' in the pre-plate tectonic sense (Hausen 1958; Fúster *et al.* 1968*b,c*). However, this interpretation proved to be inconsistent with the fact that the igneous rocks are younger than the oceanic sedimentary sequences (Robertson & Stillman 1979*a,b*). Studies in the Caldera de Taburiente in La Palma (Staudigel & Schmincke 1984) demonstrated that the 'basal complex' represents the seamount stage of the growth of these islands and, as anticipated in the Hawaiian group, in oceanic islands in general. Similar conclusions had been reached for the 'basal complex' of Fuerteventura (Stillman 1987).

Detailed geological mapping inside the Caldera de Taburiente (Carracedo *et al.* 2001) showed that several

formations previously included in the 'basal complex' of La Palma could be assigned to younger subaerial stratigraphic units. When these units were excluded, the remaining formations conformed to the seamount described by Staudigel & Schmincke (1984). Similar circumstances are present in the geology of Fuerteventura and La Gomera, in which the rocks of the basal seamount are discordant with the subaerial volcanic rocks. We therefore propose that the term 'basal complex' be discarded and replaced by the general term 'seamount' or 'submarine volcanic edifice'. However, the latter terms have a genetic meaning, and would require, as in the subaerial volcanism, the definition of stratigraphic units and intrusive chronology.

Subsidence history

An important difference between the Canaries and other oceanic island groups (such as the Hawaiian islands) is the comparative absence of Canarian subsidence (Carracedo *et al.* 1998; Carracedo 1999). Individual islands in the Hawaiian group subside and eventually become seamounts. The amount, age and rate of subsidence have been derived from studies of submarine canyons, submerged coral reefs and coastal terraces of lava deltas (Moore & Fornari 1984; Moore 1987; Moore & Campbell 1987). The lack of significant vertical movements of the Canary Islands in their post-seamount stages becomes evident from the observation of the position of contemporary sea levels, in the form of marine abrasion platforms, littoral and beach sedimentary deposits, coastal volcanic deposits (hyaloclastite-based lava deltas and surtseyan tuff rings), and erosional palaeocliffs, all widespread in the Canary Islands. These features consistently occur close to present sea level, within the range of eustatic sea-level changes (Carracedo 1999). Marine abrasion platforms and marine and beach deposits interbedded with volcanic formations of different ages up to several million years old also appear consistently close to present-day sea level in Fuerteventura and Lanzarote (Meco & Stearns 1981; Carracedo & Rodríguez Badiola 1993). Surtseyan tuff rings appear at present sea level in the coastlands of the majority of the Canaries, as well as shallow pillow lavas of ages ranging from the Miocene to present day (Ibarrola *et al.* 1991; Carracedo & Rodríguez Badiola 1993).

Near-horizontal seismic reflectors observed in the volcanic apron of Gran Canaria (Funk & Schmincke 1998) show this island to have been stable at least since the late stages of shield building around 14 Ma ago. These reflectors reach the south flank of Tenerife, where they interbed with the volcanic aprons, providing evidence of the stability of this island during its entire volcanic history. It seems, therefore, that the islands of the Canarian archipelago, located close to the African continental margin, have been extremely stable, undergoing neither subsidence nor uplift after emergence.

The lack of post-emergence uplift, in contrast to the major uplift implied by the occurrence of the seamount formations, is also of interest since it implies that prior to emergence large intrusive complexes grew within the volcanoes. Conversely, post-emergence, endogenous growth of the islands was limited. This is consistent with the many geochemical and petrological data sets for subaerial volcanic suites in the Canary Islands, which indicate that these suites are for the most part fed by magma reservoirs in the underlying oceanic crust and/or oceanic lithosphere. Only the more evolved suites of rocks show some evidence for the presence of shallow magma reservoirs (Kluegel *et al.* 1999).

Different aspect ratios of the shield-stage and post-erosional-stage islands can be readily observed in the images of Figure 18.2. However, this is not a constructional feature since Fuerteventura and Gran Canaria may have reached similar or even greater elevations (Pérez Torrado *et al.* 1995; Stillman 1999), nor is it a consequence of subsidence, as discussed above, but of different stages of erosion. The Canaries (and the Cape Verde islands) apparently remain emergent for long periods of time, even exceeding 25 Ma, until completely mass wasted through gravitational collapses, relatively frequent in the juvenile stages of growth, and erosion.

Magma production rates and eruptive frequency

Another difference between Canarian volcanoes and the Hawaiian shields is that the latter involve much greater volumes and a higher frequency of eruptions (Walker 1990). The total volume of erupted magma and production rates are difficult or impossible to evaluate in the Canaries, especially for the western islands, in spite of the high quality and amount of age data available. The discontinuous character of volcanism, in which eruptive gaps, inherently difficult to date, may predominate over periods of activity, makes true evaluation of magma production rates unreliable unless large time intervals are compared. Most of the island edifices include non-volcanic sedimentary parts and the boundaries between islands are difficult to define. Furthermore, giant lateral collapses repeatedly removed large fractions of the mass of an island, especially during the shield-building stage, and redistributed them offshore over distances of hundreds or even thousands of kilometres. Several megaturbidites deposited in the Madeira abyssal plain within the past 1 Ma have been shown to have originated in the Canarian archipelago (Weaver *et al.* 1992). The rate of encroachment of volcaniclastic deposits onto the apron of Gran Canaria has been estimated to be of the order of >110 m/Ma in the shield stage, 55–110 m/Ma in the post-caldera, 22 m/Ma in the erosional gap stage and *c.* 66 m/Ma in the post-erosional stage (Schneider *et al.* 1998).

Despite these uncertainties, Schmincke (1982) computed some estimates of the total volume of the islands in the Canarian archipelago, including the products of submarine and subaerial volcanism, intrusions and sedimentary materials, but did not consider materials removed by mass wasting and by gravitational collapses. He obtained a range of estimated volumes of islands as follows: 5.3×10^3 km³ for the smallest island of El Hierro to 23.8×10^3 km³ for Gran Canaria and 30.6×10^3 km³ for Fuerteventura. As pointed out by Schmincke (1982), the average total volume is remarkably similar to that of many shields in the Hawaiian islands (*c.* 20×10^3 km³). This may imply that volumes of *c.* 20×10^3 km³ are optimum values for the maximum growth of these oceanic islands, at least for the Canaries and the Hawaiian islands. However, actual values would require computing magma-production estimates using the combined volumes of the islands plus the offshore deposits.

Eruption rates during the subaerial shield-building stage, which are two to three orders of magnitude greater than those during the post-erosional stage, are therefore implied by considering a period of the order of tens of thousands of years. But even this may not be a sufficient averaging period because of switching activity between shield-stage islands on timescales of the order of hundreds of thousands of years. A further complication is the occurrence of episodes of relatively intense post-erosional volcanism such as that which produced the Roque Nublo volcano on Gran Canaria (Pérez Torrado *et al.* 1995).

The eruptive histories of the islands of El Hierro (Guillou *et al.* 1996) and La Palma (Guillou *et al.* 1998; Carracedo *et al.* 1999*a,b*; Guillou *et al.* 2001) are probably geochronologically the best constrained of any of the Canary Islands. The uncomplicated development of these islands, which are still in their juvenile stage of shield growth, together with the abundant and accurate K-Ar ages and magnetic stratigraphy allow the closest possible approach to the reconstruction of the entire emerged volcanic history of any of the Canaries. The present emerged volume of El Hierro, of c. 150 km³, has been produced in the last 1.5 Ma, giving an apparent average magma production rate of 0.1 km³/ka. However, rates increase significantly if we take into consideration the three known consecutive giant lateral collapses that affected the island, each clearly exceeding 100 km³.

In La Palma, the northern Garafía and Taburiente volcanoes form a shield of c. 235 km³ resting unconformably over the uplifted submarine edifice (Carracedo et al. 1999a,b; Guillou et al. 2001). In the southern half of the island, the Cumbre Vieja volcano has a volume of c. 130 km³. Both have been constructed without important interruptions in their eruptive activity, the former between 1.7 and 0.4 Ma BP, and the latter between 0.15 Ma BP and the present. The eruptive rates for these volcanoes are, therefore, 0.18 and 0.86 km³/ka, respectively. In comparison, average magma supply rates during the entire history of the island of Hawaii have been estimated to be of the order of 20 km³/ka (Moore & Clague 1992). A similar evaluation of shield-stage magma production rates in the presently post-erosional islands is highly problematic. This is because it is impossible to evaluate the volume removed by lateral collapses (Stillman 1999). It is difficult to determine even the number of collapses in these deeply eroded islands, let alone the volumes of individual collapses.

Eruptive frequency and volume vary considerably, as observed for the historical eruptions over the last 500 years. The 1730–1736 eruption of Lanzarote is the largest to occur in the archipelago in this period, involving an eruptive volume as much as an order of magnitude larger than any other from historical eruptions in other Canarian islands (Carracedo et al. 1992). However, the previous eruption in Lanzarote may be that of the corona volcano, dated at 53 ka BP (Guillou, unpublished data). In the same period, as many as 100–1000 smaller eruptions may have taken place in the shield-building-stage islands of El Hierro, La Palma and Tenerife.

Petrology and geochemistry: an overview (EA)

The study of the rock composition of the Canary archipelago has been the subject of numerous works, especially by Fúster and co-workers and by Schmincke and collaborators (see Fúster 1975; Schmincke 1982). Over the last ten years, studies have focused on the isotopic composition of the magmas (e.g. Hoernle & Tilton 1991; Hoernle et al. 1991; Hoernle & Schmincke 1993; Marcantonio et al. 1995; Thirlwall et al. 1997; Thomas et al. 1999).

The plots of Fig. 18.15 show that the great majority of analyses of volcanic rocks fall in the alkaline, silica-undersaturated field of the total alkali versus silica (TAS) diagram, showing the typical variation trend of the alkali basalt kindred. There is a generally bimodal grouping into basalt–basanite and trachyte–phonolite compositions. A similar variation is found in the plutonic lithologies, pyroxenites, gabbros, alkali gabbros, ijolites and syenites and alkali syenites. On the other hand, with the exception of the Cape Verde islands, Fuerteventura is the only known oceanic island with exposed carbonatites (aegirinic, feldspathic and biotitic alvikites (Barrera et al. 1981), and sövites with sanidine, aegirine–augite, biotite, apatite, pyrochlore and magnetite (Demény et al. 1998)).

Around 1500 analyses of fresh volcanic rocks ($H_2O < 2\%$; $CO_2 < 1\%$) belonging to all these islands have been plotted on the TAS diagram (IUGS; Fig. 18.15), showing that most of them correspond to moderately alkaline (alkali basalts–trachytes) or highly alkaline (basanite–phonolite) rocks. Rocks of tholeiitic affinity (hypersthene-normative, but with no Ca-poor pyroxene) have been recognized only in the oldest units of Gran Canaria and in the most recent lavas of Lanzarote (Ibarrola 1969; Fúster 1975; Schmincke 1982; Carracedo et al. 1992). Ultra-alkaline rocks appear mainly in Gran Canaria (olivine melilitites and olivine nephelinites) and Fuerteventura (olivine nephelinites; Ancochea et al. 1996), corresponding in both these islands to the latest phases of their activity.

There are notable differences in the alkalinity and abundance of rock types between the islands (Fig. 18.15). Gran Canaria has rocks embracing all compositions from the most to the least alkaline, whereas in the other islands the alkalinity of lavas is more homogeneous. The most alkaline is La Palma, whereas Tenerife is less so (on the boundary between highly and moderately alkaline), and El Hierro, La Gomera, Fuerteventura and Lanzarote show the least overall alkalinity (all moderately alkaline).

The islands with the least felsic rocks (< 1%) are Lanzarote and Hierro, with even intermediate compositions being also very rare in Lanzarote. Fuerteventura is mainly formed of basaltic and trachybasaltic rocks, with c. 4% intermediate rocks and only around 1% trachytic rocks being present. La Palma and La Gomera have a somewhat higher overall abundance of felsic rocks (c. 3%), and in the central islands (Gran Canaria and Tenerife) evolved rocks are at their most important (> 10%).

Mg variation diagrams (Fig. 18.16) show similar trends in all islands, with decreasing MgO in basic rocks (basalts, basanites, nephelinites: MgO > 6%), both the CaO and the FeO contents decrease slightly, while Al_2O_3 rises considerably and TiO_2 has a tendency to increase slightly. These variations record the major role played by olivine and, to a lesser extent, clinopyroxene in the differentiation history of the magmas. Similarly, a change in the slope of the curves (decreasing in FeO, CaO and TiO_2) in rocks with <6% MgO (Fig. 18.16) records the influence of clinopyroxene crystallization control in the fractionation of tephrites, trachybasalts and basaltic trachyandesites. A further change of slope in the $MgO–Al_2O_3$ diagram shown by trachytic-phonolitic rocks (MgO < 1%) reflects the importance of plagioclase in the final stages of fractionation (Fig. 18.16). Trace element data confirm the previous trends, with the most incompatible elements (Fig. 18.17b) increasing from basalts to trachybasalts and basaltic trachyandesites (but not Ti). A change takes place in trachyandesites, with an important decrease in Ti and P, possibly due to Fe–Ti oxide and apatite fractionation. The decrease in Ti and P in trachytes and phonolites is accompanied by a Ba, Sr and Eu decrease, also observed in the rare earth element (REE) diagrams (Fig. 18.17c), in which trachybasalt and trachyandesites are enriched in REE content relative to the basalts, while the trachytic and phonolitic rocks show a negative Eu anomaly.

With respect to the primitive mantle, the rocks are enriched in a similar way in all the Canary Islands (Fig. 18.17a). A low amount of K and Rb suggests that a phase such as phlogopite or amphibole remains in the residuum at low degrees of partial melting (Hoernle & Schmincke 1993). The contents in trace element and radiogenic isotopes are characteristic of HIMU OIBs (Sun & McDonough 1989; Weaver 1991), although with some variations both between and within depending on the age of the various units (Hoernle et al. 1991; Thirlwall et al. 1997).

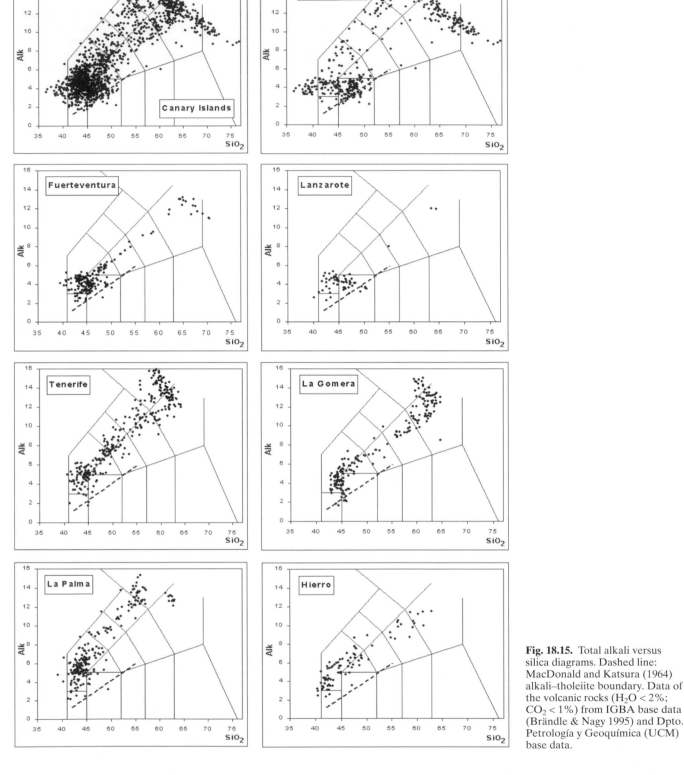

Fig. 18.15. Total alkali versus silica diagrams. Dashed line: MacDonald and Katsura (1964) alkali–tholeiite boundary. Data of the volcanic rocks ($H_2O < 2\%$; $CO_2 < 1\%$) from IGBA base data (Brändle & Nagy 1995) and Dpto. Petrología y Geoquímica (UCM) base data.

Canarian volcanic structure: an overview (JCC, FJPT)

The western and eastern Canaries show clear differences in structure and other important volcanic characteristics. The western islands display frequent, small-volume eruptions, high-aspect-ratio island edifices, well-defined, multibranched, long-lasting rifts and frequent massive flank collapses. Conversely, in the eastern islands volcanism is scarce and scattered, the islands have low aspect ratios, and rifts and giant landslides seem to be absent. Early interpretations related

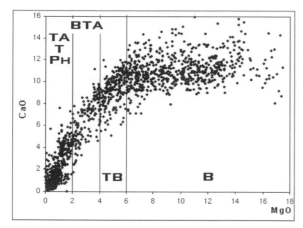

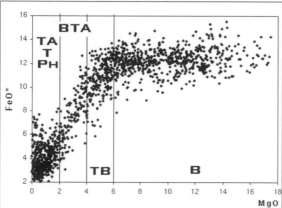

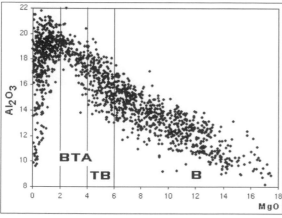

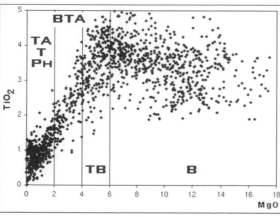

these apparent contrasts to different geological and geophysical characteristics between the 'eastern Canaries', possibly underlain by continental crust, and the 'western Canaries', resting on oceanic crust (Rothe & Schmincke 1968; Dash & Bosshard 1969). However, subsequent studies have clearly shown the presence of Mesozoic oceanic crust beneath the entire Canarian archipelago (Schmincke *et al.* 1998; Steiner *et al.* 1998). These structural differences therefore instead reflect the different stages of evolution of the islands, with structural features readily observed in the young islands becoming removed by erosion in the older ones (Carracedo *et al.* 1998; Carracedo 1999).

The form, structure and landscape of the Canaries are characterized by four main features: (1) shield volcanoes, such as the Garafía-Taburiente in La Palma; (2) central strato-volcanoes, such as the Roque Nublo in Gran Canaria or the Teide volcano in Tenerife; (3) rift zones, locally known as 'dorsals', such as the Cumbre Vieja volcano in La Palma; and (4) collapse structures: vertical collapse calderas such as the Caldera de Tejeda in Gran Canaria, and gravitational collapse scarps and embayments such as the Caldera de Taburiente in La Palma or the Caldera de Las Cañadas in Tenerife.

Shield volcanoes

Several adjoining (or superimposed) shield volcanoes frequently form the bulk of the subaerial island edifices in the Canaries, as is also the case for the Hawaiian islands. Ancochea *et al.* (1996) showed that the island of Fuerteventura is composed of three shields (northern, central and southern), aligned in a NE–SW trend. This alignment continues in the adjacent island of Lanzarote with another two shield volcanoes (Los Ajaches and Famara volcanoes). The dimensions of these volcanoes vary from 30 to 47 km in diameter and they originally may have reached heights of up to 3 km above sea level (Stillman 1999). According to Ancochea *et al.* (1996), the duration of the volcanic activity in these shields varied from 3 to *c.* 7 Ma, although recent work shows that the ages of the volcanoes are considerably reduced when palaeomagnetic reversals and radiometric dating are applied with stringent requirements

A 50 km diameter, *c.* 2000 m high, shield volcano of *c.* 1000 km³ developed in Gran Canaria apparently in a very short time (<500 ka; McDougall & Schmincke 1976; Bogaard *et al.* 1988; Bogaard & Schmincke 1998), building a predominantly effusive volcano, in which most of the dykes and vents are clustered in three convergent rift zones. The Gran Canaria shield-stage volcanism continued for more than 6 Ma, with a complex succession of processes including the formation of a collapse caldera and a cone-sheet swarm, and the emission of large volumes (*c.* 1000 km³) of differentiated rocks (peralkalic rhyolites, trachytes and phonolites) in highly explosive eruptions.

In Tenerife, the ages of the two main shield volcanoes (Teno-Centro and Anaga) have been bracketed using geomagnetic reversals and radiometric dating (Abdel-Monem *et al.* 1972; Carracedo 1975, 1979). The Teno-Centro shield, 48 km in diameter, was constructed between *c.* 7.2 and 5.3 Ma BP,

Fig. 18.16. Plots of MgO versus major oxides from Canary Islands volcanic rocks. Key: B, basaltic rocks; BTA, basaltic trachyandesites and phonotephrites; PH, phonolites; T, trachytes; TA, trachyandesites and tephriphonolites; TB, trachybasalts and tephrites.

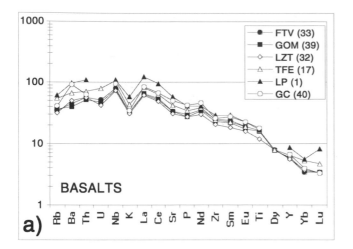

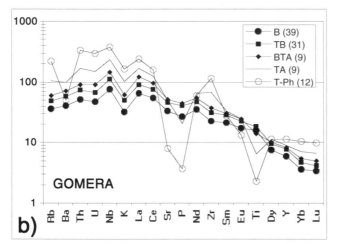

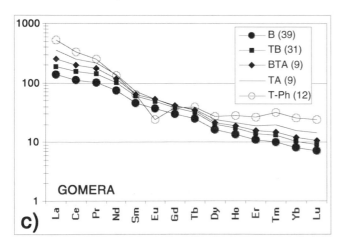

Fig. 18.17. Plots of incompatible trace elements normalized to primitive mantle (after Sun & McDonough 1989). **(a)** Basaltic rock averages (MgO > 6%). FTV, Fuerteventura; GC, Gran Canaria; GOM, Gomera; LP, La Palma; LZT, Lanzarote; TFE, Tenerife. **(b)** Rock averages from Gomera. **(c)** REE averages from Gomera normalized to chondrite (after Nakamura 1974). Abbreviations for rock types as in Figure 18.16. Gomera data from Dpto. Petrología y Geoquímica UCM (unpublished).

whereas the Anaga shield, 20 km in diameter, grew between 5.3 and 3.0 Ma BP. In a more recent work, Ancochea *et al.* (1990) defined three shields in Tenerife: Teno, Roque del Conde and Anaga, with time spans of 6.7–5.0, 8.5–6.4 and 6.5–4.5 Ma BP, respectively.

The Taburiente volcano in La Palma provides the best opportunity to define the characteristics (duration, volume, eruptive rates, magmatic evolution, etc.) of a well-preserved, entire shield. A similar volcanic history is observed in El Hierro, where near-continuous volcanic activity constructed the island from *c.* 1.12 Ma BP (Guillou *et al.* 1996). Eruptions produced a conspicuous three-branched, regular rift system and up to four consecutive lateral collapses (Carracedo *et al.* 1999b), one of them aborted (Day *et al.* 1997) with magmas evolving from basalts to trachybasalts and trachytes.

A duration of *c.* 1.5–2 Ma and an evolution from basaltic to mafic phonolites and trachytes seem to be characteristic of the western Canaries shield volcanoes. In the eastern Canaries, however, shield volume and duration may be considerably greater, as mentioned above, with diameters >30 km and growth time >3 Ma. Additional precise ages of these shield volcanoes are needed to obtain an improved understanding of their volcanic histories comparable to those of the shields in the western Canaries.

An important contrasting feature of the Canarian shields when compared with the Hawaiian shield volcanoes is that in the latter the volcanism in the most productive stage is typically tholeiitic, with minor volumes of alkali basalts and differentiated magmas. Silica-poor magmas (alkali basalts, basanites and nephelinites) predominate only in the rejuvenated volcanism of the post-erosional stages. In contrast, in the Canarian shields this difference is not evident and magmas of varied composition, from alkali basalts to phonolites and trachytes, are present in both stages.

Stratovolcanoes

Stratovolcanoes were built in different stages of the central Canaries: the Roque Nublo stratovolcano in Gran Canaria and the Teide–Pico Viejo volcano in Tenerife. The formation of stratovolcanoes requires the presence of shallow magmatic chambers and prolonged periods of repose to allow the differentiation of magmas. The Canarian stratovolcanoes have higher elevations and slopes (3718 m and 25–30% for Teide–Pico Viejo volcano in Tenerife, and 3000 m and 15–30% for Roque Nublo volcano in Gran Canaria) than the shields, but their basal diameters are considerably lower (*c.* 5.5 km and 5–10 km for the Teide–Pico Viejo and the Roque Nublo volcanoes respectively). Stratovolcanoes are, therefore, very unstable and gravitational collapses are more frequent than in shield volcanoes, although volumes are lower.

Parasitic vents and intrusions in the Canarian stratovolcanoes show a combination of radial and concentric trends (Martí & Mitjavila 1995; Pérez Torrado *et al.* 1995), whereas radial or three-armed rift zones predominate in shield volcanoes (Carracedo 1994, 1999). The compositions of lavas show variations from basic (alkali basalt–basanite) to silicic (trachyte–phonolite) compositions, the latter accounting for 20% of the total volume of the volcanoes. Crystal fractionation is the predominant process involved, although magma mixing and gaseous transfer processes have also been identified in the Teide–Pico Viejo and Roque Nublo volcanoes.

Eruptive mechanisms in these Canarian stratovolcanoes are

significantly more explosive than in the shield volcanoes, with subplinian, plinian and vulcanian eruptions having been identified respectively in the Las Cañadas, Teide–Pico Viejo and Roque Nublo stratovolcanoes. Eruptive rates of Canarian stratovolcanoes have been calculated at 0.75 km³/ka for Teide–Pico Viejo (Ancochea *et al.* 1990) and 0.1 km³/ka for Roque Nublo (Pérez Torrado *et al.* 1995). A comparison of the volcanic history of the Roque Nublo stratovolcano and the Taburiente shield shows a much longer duration for the former (Fig. 18.18), involving *c.* 2 Ma, a period which includes 12 different magnetozones corresponding to the Gilbert–Gauss chrons. In contrast the Taburiente shield only lasted 1.4 Ma, and was erupted over six magnetozones in the Matuyama–Brunhes chrons. However, while volcanic activity occurred in the Taburiente shield during all the afore-mentioned magnetozones (Fig. 18.18A), there is evidence of volcanism in the Roque Nublo stratovolcano in only two of the 12 consecutive magnetozones (Fig. 18.18B). This clearly suggests long periods of continued volcanism in the Canarian shields, contrasting with short intervals of volcanism separated by long periods of repose in the stratovolcanoes.

Rift zones

Rift zones characterized by concentrations of vents and dense dyke swarms have been described in the majority of oceanic volcanic islands (Walker 1992). In the Canaries such zones typically form a tight cluster of eruptive vents aligned along narrow ridges, known locally as 'dorsals' (Fig. 18.19). These active volcanic structures show some characteristics of Hawaiian rifts (Tilling & Dvorak 1993), such as forceful injection of dykes, but they lack others, notably a direct connection with a shallow underlying summit magma chamber and caldera. Rift zones can be observed in exceptional detail in the Canaries, because the volcanoes are crossed at different depths and altitudes by many accessible water tunnels totalling >3000 km, excavated for mining groundwater (Carracedo 1994, 1996*a,b*). Their deep structure is characterized by numerous densely packed and closely parallel feeder dykes (Fig. 18.19A), with the swarm density increasing rapidly from the margins towards the axis of the rift zones and with depth. In outcrop, they form narrow zones of clustered eruptive vents, similarly increasing in density towards the axis of the rift zone and usually widening like a fan towards the lower flanks of the volcano (Fig.

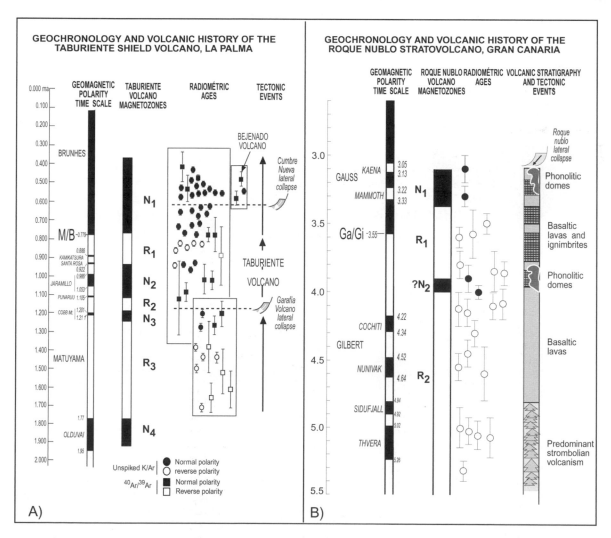

Fig. 18.18. (**A**) Volcanic and tectonic history of a typical Canarian shield volcano, the northern shield of La Palma (modified from Guillou *et al.* 2001). (**B**) Volcanic and tectonic history of a typical Canarian stratovolcano, the Roque Nublo volcano in Gran Canaria (modified from Pérez Torrado *et al.* 1995).

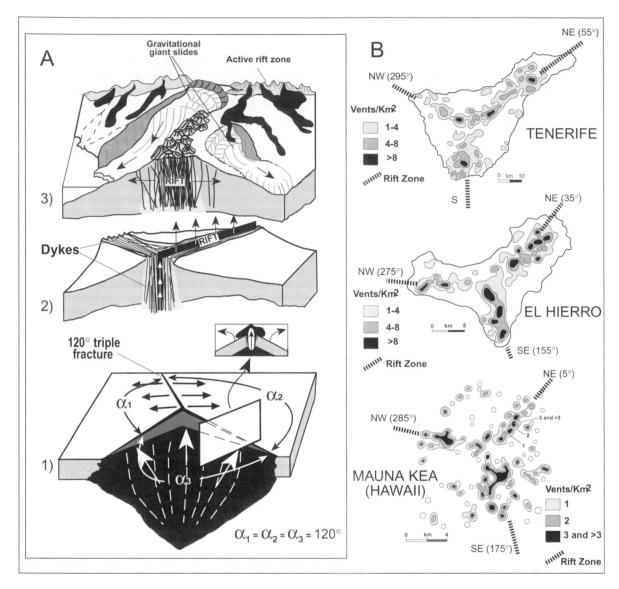

Fig. 18.19. (**A**) Model combining the formation of rifts and giant landslides in oceanic islands (modified from Carracedo 1994). (**B**) Distribution of eruptive vents in three-armed rifts in the Canaries. Mauna Kea vents from Porter (1972) (modified from Carracedo 1994).

18.19B). Walker (1992) proposed that rift zones in Hawaii contain about 65% of dykes and have a high bulk density (c. 2.8–2.9 mg/m³), much greater than the host vesicular lavas (bulk density c. 2 mg/m³). The surface rift zone geometry in the Canaries is characterized by three converging rifts, with angles of 120°, as in oceanic island volcanoes (Wentworth & MacDonald 1953; MacDonald 1972). Such a geometry is clearly observed in the distribution of Mauna Kea vents (Fig. 18.19B) mapped by Porter (1972), as seen in Tenerife and El Hierro in the Canary Islands (Carracedo 1994, 1999). However, the original regular geometry of rift zones is commonly lost during the evolution of the volcano, especially if buttressed among older volcanoes, as frequently happens in the very active initial shield-stage phases, such as seen in the island of Hawaii or in La Palma.

The presence of tripartite rift zones in the Canaries was first inferred by gravity measurements carried out by McFarlane & Ridley (1969). These authors observed a local Bouguer anomaly showing a three-pointed star shape coinciding with topographic ridges. They interpreted this feature as reflecting very high concentrations of dykes at depth, intruding along fissure zones, an interpretation that proved to be correct when the core of the rifts was observed in the 'galerías'. A simple model explaining the genesis of these features and their regular geometry was proposed by Carracedo (1994), in which three-armed rifts with angles of 120° resulted from least-effort fracturing due to magmatic intrusion (Fig. 18.19A, 1). Active rift zones in the Canaries are presently well developed only on those islands in the juvenile phases of shield building, with important recent eruptive activity (i.e. Tenerife, El Hierro and La Palma; Figs 18.19 and 18.21). This may reflect the observation of Vogt & Smoot (1984) that magmatism has to be frequent and intense for the rift zones to stay hot enough (thermal memory) to concentrate successive injections. Repeated magma injection swell deformation, and eventual rupture and resulting emplacement of blade-like dykes constitute a repetitive process that, by progressively increasing anisotropy, forces the new dykes to inject parallel to the main structure (Fig. 18.19A, 2; Carracedo 1994, 1996a,b). According to Walker (1992) there may be a limit to the growth of rift zones (65% in the Koolau dyke complex in Oahu, Hawaiian islands), with some mechanism restraining entry of magma into the complex after a certain intensity of intrusion and attendant widening are attained.

Collapse structures

Destructive collapse events, producing substantial changes in geomorphology and drainage systems, are the most conspicuous features of the Canarian landscape. Two genetic types can be defined: vertical collapse calderas, related to the abrupt emptying of magmatic chambers during highly explosive eruptions, and lateral collapse calderas and valleys associated with giant landslides. The best (and probably unique) observable vertical collapse caldera in the Canary Islands is the Caldera de Tejeda, in Gran Canaria, which formed some 14 Ma ago. This NW–SE orientated elliptical (20 × 17 km) caldera is located in what may have been the original summit of the basaltic shield of Gran Canaria. It has a fault system that separates at least three different blocks, each around 3 km wide, probably subsiding along listric faults into the caldera (Schmincke 1993). The vertical displacement of this fault system, with fault planes dipping 45° towards the caldera interior, has been estimated at c. 1000 m. Hydrothermal alteration of the intracaldera pyroclastic rocks through these fractures can be readily observed. From studies by Schmincke (1967), Hernán (1976) and Freundt & Schmincke (1992), the genesis of this collapse caldera may be related to the rapid growth of the basaltic shield of Gran Canaria, and the later abrupt emptying and collapse of a shallow rhyolitic magma chamber. The volume of ignimbrites associated with the caldera collapse has been calculated at c. 80 km³, covering 400 km² of the SW flanks of the shield. After collapse, the caldera underwent a resurgence (Smith & Bailey 1968) due to the intrusion of a conical dyke swarm.

The lateral collapse events in the Canaries owe their origin to the fact that the volcanoes typically grow to a considerable height (Fig. 18.2), with gravitational stresses and dyke injections both tending to progressively increase the mechanical instability of the edifice, especially in the most active shield-stage phases of growth. The development of triple-branched rifts promotes this edifice overgrowth, with steep, unstable flanks and concentrations of dykes that destabilize the flanks through magma overpressure during emplacement (Swanson et al. 1976). These rifts induce mechanical and thermal pressurization of pore fluids, while extensional stresses develop in the axial zones of the rifts by dyke wedging, eventually exceeding the stability threshold and triggering massive flank failures. Slide blocks consistently form perpendicular to, or between, two branches of the rift system, with the remaining rift acting as a buttress (Carracedo 1994, 1996a,b). These collapses represent processes that oceanic volcanoes require in order to restore equilibrium and reduce altitudes that may restrain or suppress the internal plumbing system. They adequately explain the origin of morphological scarps, valleys and calderas (e.g. Caldera de Taburiente, Las Cañadas, valleys of La Orotava, Güímar) that were difficult to rationalize by erosive processes alone. The most recent known such event in the Canaries was the El Golfo landslide in El Hierro (8 in Fig.

18.20A) which took place around 120 ka ago (Carracedo et al. 1999b).

At least 11 of these giant collapses have so far been identified in the Canaries (Fig. 18.20A), both onshore (Ancochea et al. 1994, 1999; Carracedo 1994, 1996a,b, 1999; Carracedo et al. 1998; Stillman 1999) and offshore (Holcomb & Searle 1991; Weaver et al. 1992; Masson & Watts 1995; Watts & Masson 1995; Masson 1996; Urgelés et al. 1997, 1998, 1999). On occasions they can abort, as did the San Andrés flank collapse in El Hierro (Day et al. 1997), and probably one block may have detached from the Cumbre Nueva collapse in La Palma (see Fig. 18.13). Other examples may take place slowly, such as the gradual collapse of the southern flank of Kilauea volcano (Swanson et al. 1976; Smith et al. 1999). However, there is little doubt that these extremely low-probability events can produce disastrous effects. Two giant, and probably highly destructive, landslides occurred at c. 1.2 and 0.5 Ma BP, from the fast-developing Taburiente shield in La Palma (see Figs 18.18A and 18.20A, B). The average heights and time span required for the growth of these volcanoes to reach the instability threshold to trigger such a giant collapse seem to be c. 2500–3000 m and 0.5 Ma, respectively. In contrast, the Cumbre Vieja in southern La Palma is only 0.12–0.15 Ma old and <2000 m high.

Recent climatic changes (JM)

Post-Miocene erosion and sedimentation were influenced by palaeoclimate, so their analysis allows an attempt to reconstruct the climatic history of the Canary Islands over the last 6 Ma. Geoclimatic variations and related sea-level changes are especially well recorded by marine deposits, aeolian sands and calcretes in the eastern Canaries. Miocene–Pliocene marine deposits crop out in the islands of Lanzarote, Fuerteventura and Gran Canaria at +25 m to +55 m, +10 to +80 m and +70 m, respectively. They were originally interpreted as Quaternary in Fuerteventura and Lanzarote, and Miocene in Gran Canaria (Lyell 1865), but later palaeontological and geomorphological studies suggested a Messinian or early Pliocene age (Meco 1977; Meco et al. 1997). On Fuerteventura, pillow lavas with K-Ar ages of c. 5.8 and 5.0 Ma (Meco & Stearns 1981; Coello et al. 1992) overlie these marine deposits. In Lanzarote (Meco 1977) they are overlain by lavas with a K-Ar age of 6.7 Ma (Coello et al. 1992) and in Gran Canaria by pillowed lava with a K-Ar age of c. 4.37 and c. 4.25 Ma (Lietz & Schmincke 1975).

The presence of lower Pleistocene raised marine deposits in northwestern Gran Canaria at +85 m a.s.l. was first reported by Denizot (1934). Molluscs indicate an age on the Pliocene–Pleistocene boundary during a climatic interglacial similar to that of the present day. Middle Pleistocene marine terraces crop out along the northern coast of Gran Canaria at +35 m a.s.l., and are again interpreted as interglacial. Finally, upper Pleistocene marine deposits at +12 m a.s.l. . on the NE coast of Gran Canaria, first reported by Lyell (1865) and Rothpletz & Simonelli (1890), are similar in age and faunal composition to the last interglacial marine deposits of the south coast of Fuerteventura and Lanzarote at c. +5 m a.s.l. (Meco et al. 1997).

During post-Messinian regression, coinciding with the establishment of the cold Canary current, Pliocene bioclastic aeolian sands entirely covered the islands of Fuerteventura, Lanzarote and the NE of Gran Canaria to elevations of

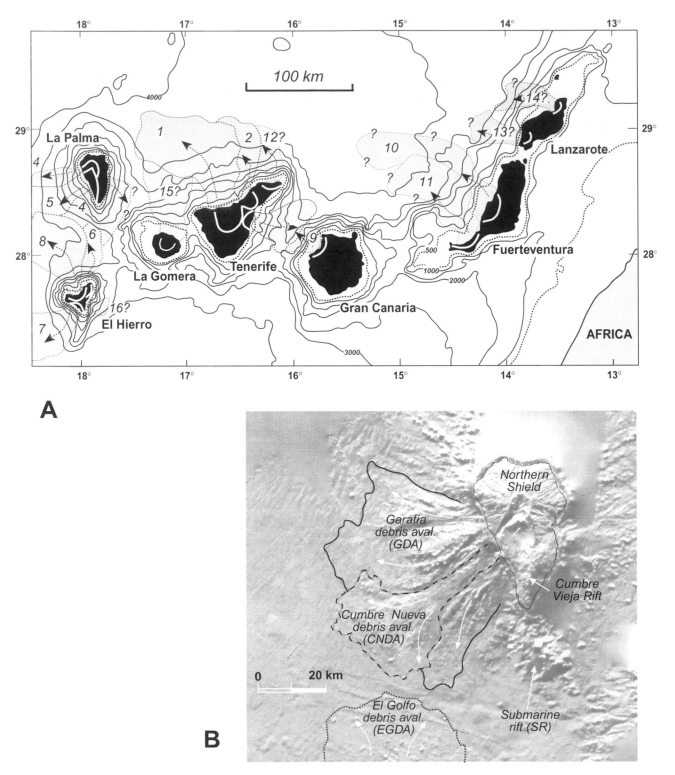

Fig. 18.20. (A) Documented and inferred giant landslides in the Canaries (modified from Urgelés *et al.* 1995, 1997; Carracedo *et al.* 1998). **(B)** Shaded relief images of bathymetric and topographic data of La Palma showing debris avalanches and related giant landslides (modified from Urgelés *et al.* 1999).

c. 400 m. These biocalcarenites include terrestrial gastropods, tortoises, eggshells of large birds and alluvial horizons with angular volcanic clasts, fragmented Messinian marine fossils and vertical roots. After the installation of a drainage system, middle Pleistocene–lower Holocene loessic palaeosols

interbedded between aeolian sands record humid spells in a generally arid regime (Meco & Petit-Maire 1986; Meco *et al.* 1997).

Calcrete on Gran Canaria was first described by Buch (1825), and on Lanzarote and Fuerteventura by Hartung

(1857). The calcrete is frequently several metres thick and was formed on top of the aeolian sands by dissolution of carbonate fragments of the marine fossil flora and fauna. This process took place during the climatic change from wet and warm to arid and cold at the beginning of the first Quaternary glaciation. Wind reworking of the underlying sands through erosion windows in the calcrete formed the middle Pleistocene to present dunes and beaches of the Fuerteventura and Lanzarote islands, although Pleistocene dunes are also covered by very thin calcretes.

Holocene volcanism and modern hazards

Holocene volcanic eruptions have occurred across the entire Canarian group with the exception of La Gomera. However, the intensity of Holocene volcanism is not constant throughout the archipelago, with the shield-stage islands having eruptive frequencies and volumes at least 10–100 times greater than those for the post-erosional islands (Fig. 18.21). Ten to 100 eruptions have been identified in La Palma (Carracedo *et al.* 1999*a*; Guillou *et al.* 2001), El Hierro (Guillou *et al.* 1996) and Tenerife (Fúster *et al.* 1968*a*), whereas less than ten eruptions took place in Gran Canaria (Fúster *et al.* 1968*d*; ITGE 1992), Fuerteventura (Fúster *et al.* 1968*b*) and Lanzarote (Fúster *et al.* 1968*c*; Carracedo *et al.* 1992). The occurrence of a relatively very long and voluminous eruption in Lanzarote (1730–1736) has prompted the erroneous idea of similar levels of volcanic activity along the entire Canarian chain. However, this idea is inconsistent with the fact that,

prior to the 1730 eruption, the only eruptions of note on Lanzarote were those of the Corona volcano (53 ka) and Los Helechos volcanic group (72 ka).

Geochronological control for the late Pleistocene–Holocene volcanism has still been poorly addressed, with the exception of the island of La Palma. Volcanism preserving youthful morphology has generally been grouped and mapped as 'Recent', 'Quaternary' or 'Basaltic Series III and IV' (Fúster *et al.* 1968*a,b,c,d*). However, differences in age are significant, probably due to important variations in erosion rates in relation to changes in climatic conditions and/or geographic location. The use of precise radiometric dating and sea-level changes related to glaciations allowed the definition of several stratigraphic subunits in the upper Pleistocene–Holocene successions of La Palma (Carracedo *et al.* 1999*a*), defining more precisely the stratigraphy and mapping of the volcanic activity. The Cumbre Vieja volcano, previously considered older than 620 ka (Abdel-Monem *et al.* 1972) or 540 ka (Ancochea *et al.* 1994), has been constrained to *c.* 120 ka (Guillou *et al.* 1998; Carracedo *et al.* 1999*b*).

Historical eruptions during the last 500 years that occurred in the Canary Islands are shown in Figure 18.21. These eruptions have happened when the islands had a low population or, later, in places with very few inhabitants. Only the 1706 eruption of Garachico on Tenerife, the anomalously long 1730 eruption of Lanzarote and the most recent eruptions of La Palma (1949 and 1971) posed any significant threat to the people and the economy of the region. Notwithstanding, the spectacular increase in the Canary Islands' population (1.8

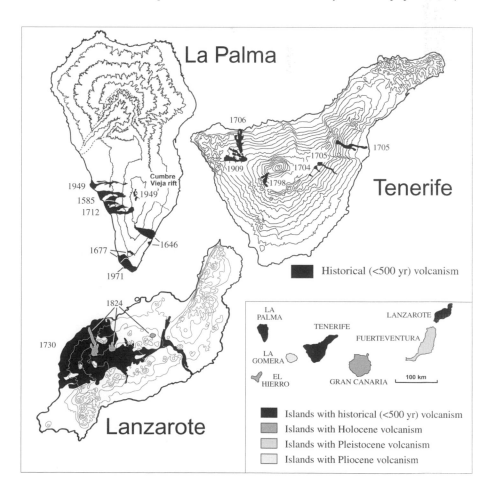

Fig. 18.21. Recent (inset) and historic (<500 years) volcanism in the Canary Islands (modified from Carracedo 1994). Note that most of the historic flows reach the coast in La Palma (juvenile island) and not in Tenerife (post-erosive).

million inhabitants and 10–11 million visitors annually) has accordingly increased the risks.

The first eyewitness account of a volcanic eruption in the Canaries is that of Torriani on the 1585 Jedey eruption in La Palma (Torriani 1592). Important errors in the identification of which vents were active and which lava flows were produced in the different eruptive events are common in the geological literature, even in cases of historical volcanism with contemporary descriptions, such as the 1585 and 1677 eruptions of La Palma (Santiago 1960; Carracedo *et al.* 1996). Modern eruptions on La Palma and Tenerife are clearly related to the rift zones, whereas in Lanzarote they are located in the central, NE–SW tectonovolcanic feature. Historical eruptions in the Canaries are characteristically basaltic fissure eruptions, with the Jedey or Tahuya eruption (1585, La Palma) producing both basalts and juvenile phonolites, and the 1730 eruption of Lanzarote producing tholeiites (Carracedo & Rodríguez Badiola 1991; Carracedo *et al.* 1992). Eruption duration varies from eight days to more than six years, but is typically one to three months. Volumes range from 0.2×10^6 to *c.* 700×10^6 m^3, most commonly being 10×10^6 to 40×10^6 m^3. The area covered by lavas during an eruption varies from 0.2×10^6 to 150×10^6 m^2, with values generally between 3×10^6 and 10×10^6 m^2. All historical eruptions showed seismic precursors, from less than one year to only a few hours prior to eruption onset.

Although El Hierro is geologically the youngest island of the Canaries, most of the recent eruptions have occurred in La Palma where at least ten eruptive events have occurred over the last 2500 years, compared with only one eruption in El Hierro in the same period (Mña. Chamuscada, ^{14}C age of 2500 ± 70 years BP; Guillou *et al.* 1996). Recent felsic, explosive volcanism is limited to the Teide–Pico Viejo volcanic complex in Tenerife. However, according to Barberi (1989), the occurrence of several basaltic eruptions adventive to this complex suggests that its magmatic chamber is very reduced in size or inactive, an observation in agreement with the lack of explosive eruptions parallel to these basaltic events.

The most likely potential volcanic hazards in modern times are related to basaltic fissure eruptions in the shield-stage islands, particularly in the Cumbre Vieja volcano of southern La Palma, where the remote possibility of a catastrophic collapse (with resulting tsunami destruction) also exists. However, the absence of any detectable displacement of the unstable western flank (Moss *et al.* 1999) and the lack of seismicity seem to exclude any immediate hazard. From available information for older giant collapse in the Canaries, it may take tens, or even hundreds of thousands of years for the Cumbre Vieja volcano to become critically unstable, assuming that the volcano does not evolve to stable configurations in the geological future. Furthermore, whereas the Garafía and Cumbre Nueva volcanoes developed for 570 and 640 ka respectively before collapsing (Carracedo *et al.* 1999*a,b*; Guillou *et al.* 2001), the age of the Cumbre Vieja volcano is only *c.* 120 ka.

The authors thank C. Stillman, R. Tilling and A. Kerr for long hours dedicated to the careful revision of this chapter and for their very helpful comments.

19 Economic and environmental geology

ROSARIO LUNAR, TERESA MORENO (coordinators),
MANUEL LOMBARDERO, MANUEL REGUEIRO,
FERNANDO LÓPEZ-VERA, WENCESLAO MARTÍNEZ DEL OLMO,
JUAN M. MALLO GARCÍA, JOSÉ A. SAENZ DE SANTA MARIA,
FÉLIX GARCÍA-PALOMERO, PABLO HIGUERAS, LORENA ORTEGA &
RAMÓN CAPOTE

Spain has a great variety of metallic and industrial rock and mineral deposits, as well as important energy and water resources. Within the European Union it has a pre-eminent position, being the country with the highest level of production of raw materials for its own use (Table 19.1). Spain is a first-rank producer of several non-metallic minerals such as celestite, sodium sulphate, magnesite, potassium and sepiolite, and ornamental rocks such as granite and marble. There are huge quarrying operations currently active in gypsum, clays, slate and aggregate. Spanish ores include examples of world-class deposits such as Almadén, by far the largest mercury deposit in the world, and the Iberian Pyritic Belt with its giant and supergiant massive sulphide deposits that include the world's largest at Rio Tinto. Exploration programmes developed in the 1990s have resulted in the discovery of new

deposits, both in already active mining districts (e.g. Migollas, Aguas Teñidas, Las Cruces and Los Frailes in the Pyritic Belt, and new mercury reserves in Almadén) and in new areas (e.g. El Valle-Carlés for gold, Aguablanca for nickel). However, despite the large reserves of metallic minerals that exist in the country, mining is only currently active for copper, mercury, gold and zinc (Table 19.2). One legacy of the long mineral exploitation history in Spain results from the fact that, before the 1980s, environmental damage was considered to be an inevitable consequence of the extractive Spanish mining industry. This has produced many environmental problems associated with pollution in water, air and soils, with infamous examples including Portman Bay (linked to La Unión mines in Murcia), and the disposal pond break in Aznalcóllar (Huelva). However, increasing environmental concern over

Table 19.1. Main Spanish mineral resources

Substance	Main deposits	Production/Reserves
Energetic resources		
Coal	*Higher rank coals*: Central Coal Basin, Ciñera-Matallana. *Brown coal*: Andorra, As Pontes	36% Spanish needs
Oil & Gas	Casablanca (Tarragona), Albatros (Cantabrian Sea)	2% Spanish needs
Uranium	Mina Fe (Salamanca)	Total reserves 37 500 t
Metallic resources		
Hg	Almadén (Ciudad Real)	No. 1 world producer (now closed)
Au	Iberian Pyritic Belt (Huelva, Sevilla), El Valle and Carles (Asturias)	No. 1 EU producer
Ni	Aguablanca (Huelva)	Total reserves 30 Mt, 0.7% Ni, 0.5% Cu, 0.75 g/t Pt+Pd+Au
Pb-Zn	Iberian Pyritic Belt (Huelva, Sevilla)	Total production 280 t: 45% S, 0.7–2% Cu, 0.4–2% Pb, 1–12% Zn, 0.5–1 g/t
	Reocín (Santander), La Troya (Guipúzcoa)	21 Mt & 3.7 Mt
Industrial minerals and dimension stones		
Celestite	Montevives-Escuzar (Granada)	No. 2 world producer, only EU producer
Glauberite, Thenardite	Cerezo del Río Tirón (Burgos), Villarrubia de Santiago (Toledo), Villaconejos (Madrid)	No. 1 EU producer
Gypsum	Sorbas (Almeria)	No. 1 EU producer
Sepiolite	Madrid	No. 1 world producer
Feldspar	Carrascal del Río (Segovia)	No. 1 EU producer
Fluorspar	Asturias	No. 2 EU producer
Slates	Orense, León	No. 1 world producer
Granite	Several sites	No. 1 EU producer
Marble	Murcia, Almeria, Alicante	No. 2 world producer

Data from ITGE (1998).

Table 19.2. Production of metallic, energy and non-metallic minerals in period 1995–1999

	1995	1996	1997	1998	1999
Energy minerals (kt, unless otherwise stated)					
Anthracite	6 274	6 440	6 603	5 857	5 436
Coal	7 387	7 195	7 200	6 128	6 295
Black lignite	4 037	4 071	4 012	3 925	3 704
Brown lignite	10 775	9 585	8 462	9 749	8 831
Oil	652	516	513	379	306
Natural gas (million m^3)	416	458	466	174	112
Uranium (t U$_3$O$_8$)	0.3	0.3	0.3	0.3	0.3
Metallic minerals (kt, unless otherwise stated)					
Iron	965.9	695	58	52	30
Pyrite	403.3	1 042	993	868	–
Copper*	22.9	38.4	38.4	37.2	3.5
Zinc*	172.4	145	147	128	153
Lead	30.1	24	23	19	28
Gold (kg)*	3.3	2.7	1.8	3.3	5.0
Silver (t)*	101.9	103	66	25	96
Mercury (t)*	1.5	861	0.39	0.67	0.43
Tin (t)*	2	2	4	5	7
Non-metallic minerals					
Special clays (t)	1 003 123	1 054 282	1 120 000	894 512	1 030 000
Sulphur from pyrites (t S)	405 673	439 010	445 201	405 373	276 135
Baryte (t BaSO$_4$)	26 141	70 878	90 176	70 299	62 000
Kaolin (t)	316 074	345 680	352 000	380 000	400 000
Celestite (t SrSO$_4$)	106 000	107 538	88 874	111 122	128 457
Kyanite (t)	1 000	50	23	46	–
Diatomite (t)	44 623	39 115	43 207	56 145	55 000
Feldspar (t)	413 284	440 252	398 000	429 000	430 000
Fluorite (t CaF$_2$)	113 700	112 319	120 039	130 411	107 381
Sodium sulphate (t NaSO$_4$)	813 242	854 900	923 100	1 001 104	885 974
Lithium (t lepidolite)	10 000	8 250	9 124	7 676	10 000
Raw magnesite (t MgO)	252 200	245 862	170 618	201 505	183 000
Mica (t)	2 628	2 507	3 273	2 602	3 000
Pumice (t)	677 414	600 372	640 653	737 054	750 000
Potash (t K$_2$O)	808 580	717 064	770 114	504 867	507 942
Salt (t)	3 685 333	3 436 647	3 593 765	3 699 697	3 620 000
Industrial silica (t)	4 608 705	4 801 954	5 800 000	5 900 000	6 000 000
Talc (t)	89 000	109 756	118 355	110 150	110 430
Common clay (kt)	27 885	30 879	35 680	36 200	36 400
Aggregates (kt)	227 000	225 200	230 000	248 000	282 000
Calcium carbonate (kt)	1 500	1 500	1 750	1 800	1 800
Dolomite (kt)	1 500	1 130	1 200	1 350	1 500
Dunite (kt)	979	803	869	751	800
Cement Raw materials (kt)	40 290	39 112	43 400	50 700	51 000
Gypsum (kt)	9 000	8 500	8 300	8 986	8 500
Granite (kt)	1 375	1 295	1 575	1 625	–
Marble (kt)	2 151	2 347	2 880	3 120	–
Roofing slate (kt)	551	705	555	615	–

Source: Dirección General de Minas, Ministerio de Industria y Energía and Reguerio *et al.* (2000*b*).
*Metal content.

recent years has resulted in the imposition (since 1986) of strict laws controlling emissions as well as the remediation of damaged areas.

Regarding energy resources, the country has a great coal mining tradition going back over 150 years, and is still (in 2001) the third largest coal producer in Europe, with abundant coal reserves still remaining. Uranium reserves are also considerable and production has supplied much of the internal market demand (figures up to 2000). By contrast, production of oil and gas, which mainly comes from Mediterranean and Cantabrian fields, is very limited and Spain depends mostly upon an external supply of hydrocarbons. Finally, there is a group of small mines where gem-type and other specialist

minerals occur, although these are of only limited economic interest.

This chapter deals firstly with metallic then non-metallic deposits, before moving on to examine Spanish non-renewable energy resources (gas, oil and coal). The next section deals with hydrogeology, a critical issue in Spain, where almost three-quarters of the land is subhumid or semi-arid. Intensive use of groundwater and superficial waters in Spain is producing serious environmental problems, most notably soil erosion, the contamination of riverbeds and aquifers, and alteration to coastal ecosystems and littoral sedimentation. Finally, and continuing with the environmental theme, the chapter provides a brief overview of seismic hazard in Spain.

Although not a country widely recognized as presenting significant seismic risk, southern Spain lies on the active African-European plate boundary and has a history of highly destructive earthquakes.

Metallic resources (RL, TM, FGP, PH, LO)

Metallic mineral deposits in Spain include those in both plutonic and volcanic settings, hydrothermal deposits (vein, skarn and shear-zone related) and stratabound ores of many different types. This wide spectrum of mineralizations has for the most part been produced, or influenced, by at least one of the two main orogenic events that have

affected Spain so that Spanish metallic ores are typically classified in terms of whether they are 'Variscan' (i.e. rocks affected, or produced by the Variscan orogeny) or 'Alpine' (i.e. rocks affected by Afro-Iberian-European plate collision).

Pre-Variscan and Variscan (Palaeozoic) ore deposits

Mercury. The Iberian peninsula contains not just the most famous mercury mine in the world (Almadén) but also includes an unusually wide range of cinnabar deposits with various host rocks of Palaeozoic to Mesozoic age (Fig. 19.1). These include barite-related stratabound deposits in

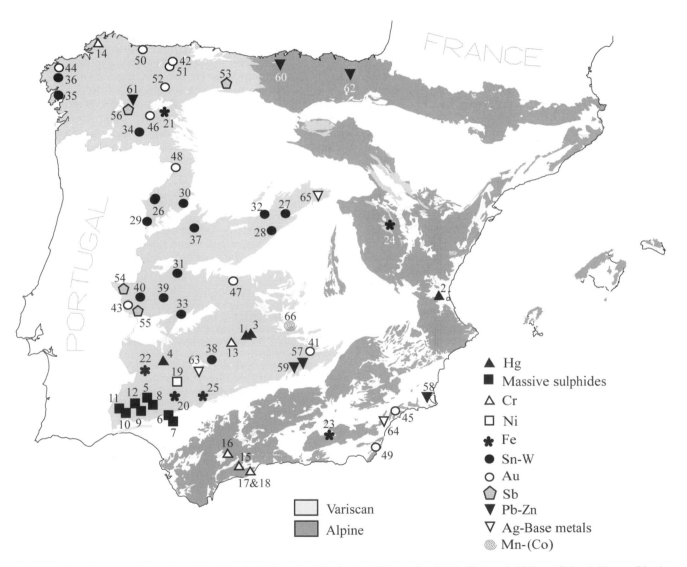

Fig. 19.1. Location map of the main ore deposits in Spain referred in the text. **Hg**: 1, Almadén; 2, Chóvar; 3, El Entredicho; 4, Usagre. **Massive sulphides**: 5, Aguas Teñidas; 6, Aznalcóllar–Los Frailes; 7, Las Cruces; 8, Rio Tinto; 9, Sotiel–Migollas–Masa Valverde; 10, Tharsis; 11, Vallejín; 12, La Zarza. **Cr**: 13, Calzadilla de los Barros (+PGE); 14, Cabo Ortegal (+PGE); 15, La Gallega (+Ni); 16, Jarales (Carratraca); 17, Marbella (+graphite); 18, Ronda (+PGE). **Ni**: 19, Aguablanca. **Fe**: 20, Cala; 21, Cotos Wagner–Vivaldi; 22, Monchi–La Berrona–Burgillos–San Guillermo; 23, Marquesado–Alquife–Las Piletas; 24, Ojos Negros; 25, El Pedroso. **Sn-W**: 26, Barruecopardo; 27, Bustarviejo; 28, Cabeza Lijar; 29, La Fregeneda; 30, Golpejas; 31, Mina Teba; 32, Otero de Herreros; 33, La Parrilla; 34, Penouta; 35, Sant Finx; 36, Santa Comba; 37, Los Santos; 38, Oropesa; 39, El Trasquilón–Las Navas; 40, Tres Arroyos. **Au**: 41, Almuradiel; 42, Carles; 43, La Codosera; 44, Corcoesto–Albores; 45, Lomo de Bas (+Ag-Sn); 46, Las Médulas; 47, Nava de Ricomalillo; 48, Pino; 49, Rodalquilar (+Sn); 50, Salave; 51, El Valle–Boinas; 52, Villamanín. **Sb**: 53, Burón; 54, Mari Rosa (Au); 55, San Antonio (W, Hg); 56, Villarbacú. **Pb-Zn**: 57, La Carolina; 58, La Crisoleja–La Unión–Mazarrón (+Ag-Sn); 59, Linares; 60, Reocín–La Florida–Novales; 61, Rubiales; 62, La Troya. **Ag-base metals**: 63, Guadalcanal; 64, Herrerías–Sierra Almagrera; 65, Hiendelaencina–Atienza District. **Mn-Co**: 66, Calatrava.

Cambrian limestones and dolomites at Usagre (Badajoz; Tornos & Locutura 1989), epigenetic and stratabound deposits associated with Cambrian and Carboniferous rocks in Asturias (Martínez García 1983*b*; Luque *et al.* 1990), vein infillings in Triassic dolomite in the Iberian Ranges (Tritlla & Cardellach 1993, 1997), and in Buntsandstein sandstones (Chóvar, in Castellón; Tritlla 1994). All these, however, are eclipsed by the Almadén deposit, since their reserves and historic production have been relatively very low (in the order of thousands of flasks; 1 flask = 34.5 kg) compared to the millions of flasks produced in the Almadén district. The following account will focus upon this Almadén deposit from which something like one-third of all the world's mercury has been mined.

Almadén lies within the Palaeozoic succession of the Central Iberian Zone and is dominated by a relatively monotonous sequence of marine shelf siliciclastic sediments, with minor carbonates and mafic magmatic rocks (Fig. 19.2). The mercury mining here is especially notable for four main reasons: (1) the Almadén area has been continuously mined since pre-Roman days; (2) the huge amount of Hg involved, with estimated original reserves of over 8 million flasks (8 MF. 1 flask = 34.5 kg); (3) the clear link between mantle-derived magmatism and mineralization; (4) the richness and monometallic nature of the mineralizations, with little sign of geochemically related elements such as gold, silver, arsenic or antimony.

Uses of the mineral and the metal have changed drastically with time: pre-Roman (possibly even prehistoric) peoples used cinnabar for painting whereas Romans exploited it for vermilion, the tint for the crimson togas of senators and principals. During medieval times, alchemists tried to transmute mercury into gold. A huge increase in the demand for the metal occurred during the conquest of Central and South America, linked to the use of mercury to recover gold and silver by means of amalgamation. Later, other applications, such as in electrolytic pools, or batteries, supported demand for the metal. The greatest crisis for the global use of mercury came with the Minamata disaster in Japan, when coastal spillage from a chemical factory caused extensive poisoning of coastal communities by highly toxic methyl mercury. Other similar pollution events, combined with the misuse of seeds dressed with Hg-fungicides in Iran, have led to a catastrophic downturn in the global market for mercury (Baakir *et al.* 1973;

Kudo & Turner 1999) and alternatives are sought wherever possible.

In the Almadén area, the main environmental problem associated with mercury mining has been the progressive poisoning of workers with jobs in direct contact with mercury vapours (mostly underground miners and metallurgical plant workers). Such problems have been minimized by the introduction of reduced time schedules for this type of work, typically to a maximum of only six hours in eight days per month. Another potential problem is that around the Almadén mines the mercury levels in soils and water are abnormally high. This is a result of human mining activity combined with primary and secondary dispersion of mercury away from the natural outcrop of the mercury deposits. Such enhanced levels have presumably persisted in the area from pre-human times, and so far have not been shown to have caused any abnormal effects on the local flora and fauna (Romero & Oliveros 1986). This perhaps surprising fact is probably due to the relative scarcity of pyrite in the Hg deposits, together with an abundance of carbonates, which has prevented the formation of acid-mine drainage. The pH of waters and soils in the area is thus typically neutral or even slightly alkaline (Urbina *et al.* 2000), inhibiting the solution, transport and reconcentration of heavy metals in general (Smith & Huyck 1999) and of mercury in particular (Grey 1997; Grey *et al.* in press).

Investigations on the possible influence of the central Iberian Hg-mineralized areas on the contamination of the Mediterranean basin (nearly 400 km away) have revealed little or no effect (Ferrera *et al.* 1997). Of more relevance to this question are two mercury-bearing mineralized areas, of late Miocene age, which occur in southern Spain (Almería), namely Las Herrerías and Valle del Azoque (Martínez Frías *et al.* 1998).

The main mines within the Almadén area are Nuevo Entredicho and Nueva Concepción (in Ordovician and Silurian host rock), Almadén itself, El Entredicho, Las Cuevas, Vieja Concepción, Pilar de la Legua (all in Silurian host rock), Corchuelo, Guadalperal, and Burcio-Tres Hermanas (all in Devonian host rock; Fig. 19.3). Of these mines Almadén has been by far the most productive (7.5 MF), followed in turn by El Entredicho (0.35 MF), Nueva Concepción (0.185 MF), Las Cuevas (0.15 MF) and Vieja Concepción (0.1 MF). The host rocks for the mineralization are both sedimentary and igneous: quartzite in Almadén, El Entredicho, Vieja Concepción, Burcio-Tres Hermanas and Pilar de la Legua; pyroclastic and detrital rocks in Las Cuevas, Nueva Concepción and Nuevo Entredicho; and non-pyroclastic

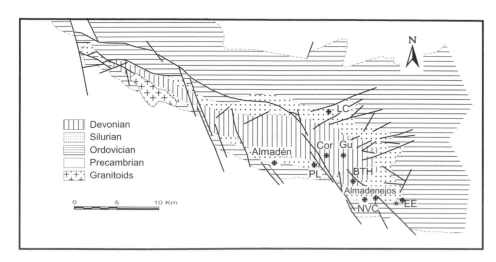

Fig. 19.2. Geological sketch map of the Almadén syncline. Abbreviations for mercury deposits: BTH, Burcio-Tres Hermanas; Cor, Corchuelo; EE, El Entredicho; Gu, Guadalperal; LC, Las Cuevas; NVC, Nueva- and Vieja Concepción; PL, Pilar de la Legua.

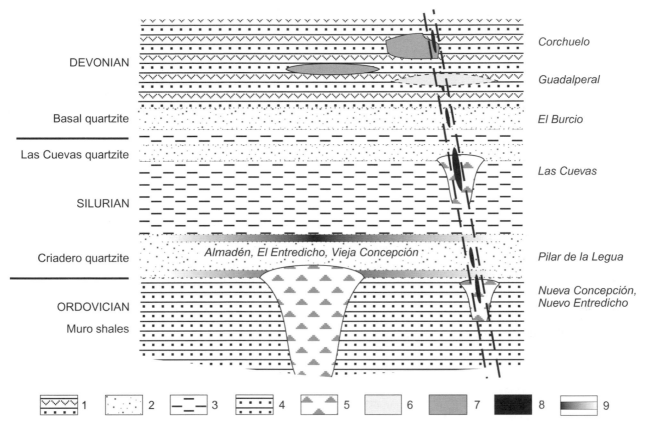

Fig. 19.3. Scheme of the different relationships between Hg mineralizations and host rocks in the Almadén district. Names on the different Hg deposits shown in italics. Legend: 1, quartzites, shales and volcanic rocks; 2, quartzites; 3, shales; 4, shales and quartzites; 5, basic pyroclastic rocks (Frailesca); 6, basic lava flows; 7, diabases; 8, Hg discordant mineralizations. 9, stratiform Hg mineralization in Criadero quartzite, with decreasing Hg content away from the frailesca source.

igneous rocks in Corchuelo, Guadalperal, and in part at Nueva Concepción (Borrero & Higueras 1990).

Magmatic rocks are unusually common at Almadén, in places forming the dominant rock type. They occur most commonly within Silurian and Devonian rocks, and are closely linked to the mercury mineralization. There are four main lithologies: (1) alkaline porphyritic rocks of continental affinity ranging from basanites/nephelinites through olivine basalts, trachybasalts, trachyandesites, trachytes and locally rhyolites; (2) alkaline pyroclastic diatreme-like bodies with a patchy, fragmental texture that has been given the local name of 'frailesca' after the robes of Franciscan monks; (3) tholeiitic dolerites common throughout the Palaeozoic sequence and typically altered to a prehnite–pumpellyite–actinolite–epidote-bearing low-grade metamorphic assemblage; (4) highly altered and sometimes fuchsitic ultramafic rocks present as xenoliths in the porphyritic and pyroclastic lithologies (Higueras & Munhá 1993; Higueras et al. 2000).

The Hg mineralization is not only directly linked to the igneous rocks, but has been strongly influenced by pervasive low-grade alteration. A so-called 'regional' alteration, affecting the alkaline rocks (porphyritic and pyroclastic), has produced carbonates (calcite–dolomite–ankerite–magnesite–siderite), chlorite and illite (including Cr-illite/fuchsite; Morata et al. 2002). This is a propilitic alteration caused by CO_2 metasomatism and depletion in SiO_2, Na_2O and the more mobile trace elements (Higueras 1993). The alteration displays a zoned pattern around the Hg stratabound deposits, with the maximum carbonatization (Fe-Mg carbonates) occurring close in, and the minimum carbonatization (Ca-Mg carbonates) occurring more distally. Laser Ar/Ar dating of the Cr-illite/fuchsite (≤420 Ma BP; Hall et al. 1997) indicates that this alteration could be a post-magmatic event. In addition to this 'regional' event, a 'local argillitic alteration' has also been recognized. This has been produced by an acidic alteration of the magmatic host rocks by sulphur-rich fluids (Higueras et al.

1999; Hernández et al. 2000). Relative depletion of MgO, CaO, Na_2O and corresponding enhancement of Al_2O_3 and SiO_2 has favoured zonal growth of secondary minerals such as pyrophyllite, kaolinite, illite and pyrite. Laser Ar/Ar dating of this illite (360 ± 2 Ma; Hall et al. 1997) indicates that this process is clearly unrelated to the formation of the stratabound deposits, and may be related to the high thermal gradient associated with the magmatic activity.

The deposits are stratabound in Almadén, El Entredicho and Vieja Concepción and discordant, vein-related in the other mines. The stratabound deposits lie within the Lower Silurian Criadero quartzite close to a 'frailesca' vent. From a cut-off minimum grade of 0.1%, Hg values reach 10–12% in specific, highly concentrated areas such as in the San Pedro, San Francisco, and San Nicholas ore bodies. The Hg mineralization is clearly post-sedimentary, pre-diagenetic and probably related to the 'regional' alteration event. Cinnabar can be seen growing around original surfaces of detrital quartz grains then being itself overgrown by diagenetic quartz (Saupé 1967, 1990). The vein deposits are of various types, with different sizes, grades and host rocks, usually being richest when they are associated with 'frailesca'. A well studied, typical example of this kind of mineralization is given by Las Cuevas (Higueras et al. 1999) where Hg is concentrated in subvertical volumes 25 m wide and up to 150 m high, with grades reaching 25%. Cinnabar appears as vein fillings and semi-massive replacements of the igneous host rock. Unlike the stratabound deposits, the vein Hg is commonly associated with abundant pyrite, produced by the 'local argillitic alteration'.

The Almadén mine has a mining history going back to pre-Roman times. Since the Romans its production has ceased only for mining problems, the most famous of such events being the fires of 1639 and 1755 (lasting three years), and the flood of 1766. In the late twentieth century production declined and then stopped altogether in 1991 in response to the crisis of demand created by increasing recognition of the environmental problems linked to mercury use. At present (2001) Almadén mine is once again operating, but only on a relatively small scale, recovering residual reserves. The other mines have shorter histories: El Entredicho was mined during Arabic rule, then again during the seventeenth and eighteenth centuries, and finally reopened and exhausted between 1981 and 1997. Las Cuevas shows evidence of important Roman exploitation, was rediscovered in 1983, and actively mined from 1990 to 1999. Nueva Concepción was discovered in 1779 and active between 1795 and 1861. Currently, therefore, only Almadén mine is active, since the recent closures of El Entredicho (September 1997) and Las Cuevas (September 2000).

Massive sulphides: the Iberian Pyritic Belt. Like Almadén, the Iberian Pyritic Belt (IPB) is a world-class metallogenic province. It extends across an area in SW Iberia measuring 250 by 40 km, from the Atlantic coast in Portugal to the Guadalquivir basin in Spain. It is characterized by the presence of abundant (>250) giant and supergiant massive sulphide deposits and small associated Mn deposits. These massive sulphide deposits have been known and exploited since pre-Roman times, although the main discoveries were made in the 1960s and include major deposits in both Spain and Portugal such as Rio Tinto, Aznalcóllar, Tharsis, La Zarza, Sotiel, Aljustrel and Neves-Corvo (Fig. 19.4). Three-hundred megatonnes (300 Mt) have been produced in the last hundred years and reserves amount to between 700 and 1000 Mt of massive sulphide ores with an average content of 46% S, 42% Fe and 2–4% Cu + Pb + Zn (Strauss & Madel 1974; Barriga 1990). Although some of the older mines are now declining, estimates of sulphide reserves are continuously increasing with new discoveries such as Migollas, Aguas Teñidas, Las Cruces, Los Frailes, Masa Valverde and Vallejín (Sáez *et al.* 1999*b*).

The IPB deposits show similar features to other volcano-sedimentary provinces across the world, although the sheer amount of massive sulphides deposited in the region makes the area unique. Lydon (1988) and Sáez *et al.* (1999*b*) have pointed out that the deposition of such large sulphide masses cannot be explained by simple physicochemical changes in the deposition environment. In this sense, Sáez *et al.* (1999*b*) propose that the IPB has an intermediate character between volcanic-hosted (VHMS) and sedimentary-hosted massive sulphide (SHMS or SEDEX) deposits and highlight the key role of both volcanic-hydrothermal systems and sediment-related processes in the generation of this exceptional metallogenic province.

The first known mining activity in the IPB dates back 5000 years with discoveries being focused on rare malachite impregnations in volcanic rocks (Blanco & Rothemberg 1981). Later the gossan outcrops were mined to obtain Au and Ag, as well as secondary minerals in the Cu cementation area. In both cases abundant slag material was produced and concentrated in areas such as Tharsis, Rio Tinto and others (up to 6–8 Mt during Roman times; Rothemberg & García Palomero 1986). This metalliferous waste material produced the first known environmental problem associated with the IPB mineralizations. The next intensive exploitation occurred in the second half of the nineteenth century, focusing on high Cu levels in the cemented sulphides. Minerals with Cu contents up to 2–6% were accumulated in large areas, directly or by previous calcination ('teleras'), for leaching and later Cu recovery (Pinedo Vara 1963). These partially leached materials produced atmospheric pollution during calcinations, combined with fluvial pollution by acid metalliferous waters, especially in the Tinto and Odiel rivers. It was not until the beginning of the twentieth century that the big and numerous open pits for sulphides (pyrites) were opened. These pits, which continued operating until the 1970s, produced sulphur as the major element, and Cu, Pb, Zn, Au and Ag as residual

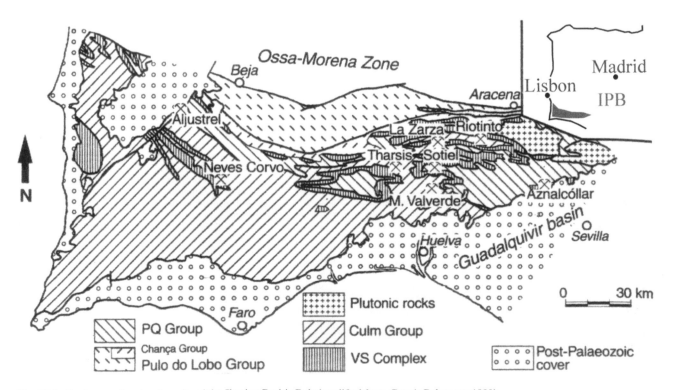

Fig. 19.4. Geology and major deposits of the Iberian Pyritic Belt (modified from García Palomero 1990).

elements (from the calcined pyrites). In Rio Tinto concentrates with 1–2% Cu were obtained by flotation from stockworks (García Palomero 1990). All exploitations from this time produced big sulphide-rich waste tips, creating a huge visual impact and increasing river pollution.

During 1970–2000, exploitations have concentrated on metal recovery using sulphide flotation in the Cu-rich areas (0.6–3% in stockwork and 3–7% in massive sulphides) and polymetallic-rich areas (Cu + Pb + Zn + Ag + Au, 4–7% in massive sulphides). Other precious metals have been recovered from leached gossan (1–2 ppm Au and 40–50 ppm Ag). Mining activity during this time generated the same environmental problems as before, plus a new risk related to the accumulation of several millions of tons of tailings with abundant metallic sulphides in tailing dumps. Following disastrous outbreaks of polluted tailings into the fluvial system, such as happened at Aznalcóllar (García Guinea et al. 1998), urgent work is currently taking place to manage the disposal and control of effluent leakage. Within this context, the highly polluted Rio Tinto, with pH 2–2.5 and high levels of metallic cations, has developed a strange ecosystem suited to such extreme conditions (Amils et al. 1997), and is the subject of study by those interested in the early development of life on Earth (López Archilla & Amils 1999). Despite the extreme surface damage, aquifer pollution is not a huge problem in this area due to the impermeability of the IPB rocks, with the exception of some mines located near Miocene–Quaternary sedimentary sequences.

Regional geology. The geology and metallogeny of the IPB have been the subjects of many studies (e.g. Williams 1962; Strauss & Madel 1974; Williams et al. 1975; Carvalho et al. 1976b; Strauss et al. 1977; Carvalho 1979; García Palomero 1980; Solomon et al. 1980; Barriga & Carvahlo 1983; Barriga 1990; Saéz et al. 1999b). The IPB is located in the south Portuguese zone of the Iberian Massif and comprises an Upper Devonian–Lower Carboniferous succession with three well-defined lithostratigraphic units, the Slate-Quartzite Group (Gp) (Fammenian), the volcano-sedimentary complex (Tournasian), and the Culm Gp (Visean) (Schermerhorn 1970). The Slate-Quartzite Gp (PQ on Fig. 19.4) is a >500 m thick monotonous sequence of slates and quartzites, with lenses of limestones towards the top. The overlying volcano-sedimentary complex is a >1000 m thick unit hosting massive sulphide mineralizations and composed mainly of bimodal acid and mafic volcanic rocks (Munhá 1983; Sáez et al. 1996), interbedded with slates, cherts, jaspers and greywackes. The Culm Gp is a turbiditic unit made up of slates and greywackes, with thickness varying from 500 to 1000 m. All three successions have been overprinted by chlorite-grade regional metamorphism, intensely folded and locally overturned during the Variscan orogeny (Lécolle 1972).

Ore geology. The IPB massive sulphide and Mn deposits are always located within the volcano-sedimentary complex and occur associated with the final stages of felsic volcanic cycles. Three main styles of mineralization occur (massive sulphide stratiform orebodies, stockworks with sulphides, and Mn lenses) all three of which are closely related to each other and to the volcanism.

The massive sulphide stratiform orebodies occur at the top of acid pyroclastic sequences and in laterally equivalent more distal facies comprising acid volcanics and interbedded sediments (Carvalho 1979; García Palomero 1980; Barriga 1983). The stockworks usually occur in acid volcanics and are underlain by sulphide masses hosted by pyroclastic rocks (e.g. Rio Tinto, La Zarza, Neves Corvo) although sometimes the spatial relationship between them is not evident. Such stockworks are generally absent under orebodies hosted by sediments or by distal volcanic facies (Barriga 1990). The Rio Tinto stockworks and sulphide masses at San Dionisio and Planes-San Antonio deposits provide some of the best examples showing the connection between the sulphide orebodies and their stockwork roots that act as channels for hydrothermal fluid upwelling (Williams et al. 1975).

The massive sulphide orebodies are stratiform lenticular masses within pyroclastic-sedimentary units comprising mostly pyrite (48–50% S) with low average contents of other ore minerals, such as chalcopyrite (1% Cu), sphalerite (2–3% Zn), galena (1–2% Pb) and very low contents of a large number of trace-element-bearing and secondary minerals (Rambaud 1969; Sierra 1984). These massive lenses can yield from 1 Mt to 400–500 Mt of ore (Cerro Colorado; García Palomero 1990), with lenticular bodies of 30–50 Mt being particularly common (e.g. in Neves Corvo, Aljustrel, Aznalcollar, Sotiel, Rio Tinto, Tharsis). They exhibit sedimentary internal structures such as layering of sulphides and pyroclastics, slumping and synsedimentary breccias with fragments of both rocks and sulphides, suggesting their formation on the marine floor (García Palomero 1980). Although the sulphides are now recrystallized, primary features such as coloform and framboidal textures can be recognized (Read 1967). The stockworks comprise three-dimensional networks of sulphide veinlets and stringers, 1 mm to 1 m thick, cross-cutting the upper parts of the felsic volcanic sequences and producing a strong hydrothermal alteration (sericitic, silicic and chloritic). The main sulphide is pyrite, although chalcopyrite, sphalerite and galena, among others, can be locally abundant (García Palomero 1980; Halsall 1989; Silveira 1996). These stockworks can be very large in size, up to 500 m thick and extending over several square kilometres, and economically valuable as illustrated by the Cerro Colorado stockwork at Rio Tinto, which has 200 Mt of Cu ore. Finally, the Mn lenses consist of pyrolusite impregnations associated with tuffites, jaspers and cherts, both in mafic and felsic volcanic rocks (Rambaud 1969; Lécolle 1972; García Palomero 1980; Halsall 1989; Silveira 1996). They occur both with productive and non-productive volcanic episodes (Barriga 1990).

Hydrothermal alteration. The volcano-sedimentary complex shows strongly hydrothermal alteration on both regional and local scales. The regional alteration consists of a general hydrothermal overprint produced by heated seawater circulating through both mafic and felsic volcanics (Munhá 1979, 1990; Munhá & Kerrich 1980; Barriga 1983, 1990; Barriga & Kerrich 1984). The local alteration is typically related to ore deposition and results from the interaction between mineralizing hydrothermal fluids and the host rocks. It is focused within stockwork areas or footwall rocks and develops in a roughly concentric zonal pattern with an inner chloritic zone and a peripheral sericitic zone, accompanied by other alterations that include silicification, sulphidization and carbonatization (García Palomero 1980; Sáez et al. 1999b; and references herein). Hanging-wall alteration has also been observed indicating that hydrothermal activity persisted after main ore deposition (Barriga 1990).

Ore genesis. The IPB massive sulphide deposits were formed at or near the sea floor by sulphide precipitation from hydrothermal fluids. The stockworks represent the paths for fluid circulation and acted as feeders for the stratiform sulphide masses (Fig. 19.5). They developed mostly at the end of the main felsic volcanic cycle at temperatures ranging from 400°C at the base of the stockwork to 100°C at the top (data from mineral equilibria, fluid inclusions and stable isotopes; García Palomero 1980; Eastoe et al. 1986; Mitsuno et al. 1986; Munhá et al. 1986a; Halsall 1989; Silveira 1996; Almodóvar et al. 1998). The stratiform sulphide masses formed either above the stockworks, or in both proximal and distal euxinic depressions on the sea floor by deposition from hydrothermal fluids expelled by the feeders and/or by replacement of soft sediments, subsequently transformed into pyritic muds (Schemerhorn 1971; Almodóvar et al. 1998). The Mn lenses were formed under oxidizing conditions away from these depressions. The $\delta^{18}O$ and δD isotopic data (Barriga & Kerrich 1984; Munhá et al. 1986a) indicate that both the fluids producing hydrothermal regional alteration and the ore-forming fluids are essentially seawater, although there may be some degree of mixing with deep-seated waters of different origin. The $\delta^{34}S$ signature (ranging from −15‰ to +10‰, with lighter values in the stockworks than in the sulphide masses) points to a sulphur derived from sea water, with some sulphur supplied by bacterial reduction (Rambaud 1969; Eastoe et al. 1986; Mitsuno et al. 1986, Kase et al. 1990; Velasco et al. 1998). Lead isotopes indicate a homogeneous and crustal source for metals (Marcoux et al. 1992; Leistel et al. 1994; Thhiéblemont et al. 1994; Marcoux 1998).

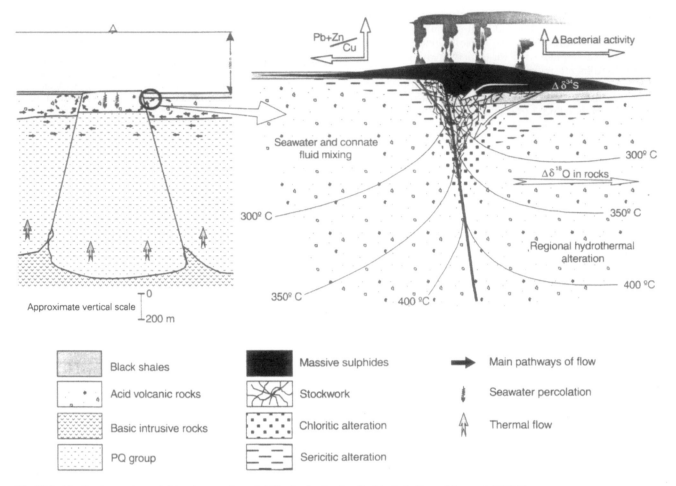

Fig. 19.5. Idealized genetic model for the massive sulphides in the Iberian Pyritic Belt. From Sáez *et al.* (1999*b*).

After the volcanism and deposition of the Culm sequence, the IPB rocks and mineralizations were folded by the Variscan orogeny and the sulphides recrystallized. Later uplift and erosion exposed the ores which then underwent weathering processes and developed gossans with goethite and limonite and underlying supergene enrichment zones with calcocite and covellite (Rambaud 1969; García Palomero 1980, 1990; García Palomero *et al.* 1986; Amoros *et al.* 1981; Martín 1981; Kosakevitch *et al.* 1993).

Cr and platinum-group elements. Although there are no chromium mines in Spain, chromite, rich in platinum-group elements (PGE), of potential economic interest occurs in mafic and ultramafic rocks in Cabo Ortegal (NW Spain) and Calzadilla de los Barros (SW Spain, Fig. 19.1).

Cabo Ortegal. The PGE-rich chromites occur in ultramafic rocks which were thrust over high-pressure granulites during Variscan collision (Arenas *et al.* 1986; Monterrubio *et al.* 1990; Monterrubio 1991; Monterrubio & Lunar 1992; Martínez Catalán *et al.* 1996). These ultramafics comprise a basal tectonized mantle harzburgite and an overlying lower dunite–pyroxenite–upper dunite (*c.* 375 m, *c.* 350 m, *c.* 225 m, respectively) unit known as the Herbeira Layered Complex (HLC) Moreno 1999; Moreno *et al.* 2001). The HLC is interpreted as a layered intrusion crystallized at the mantle–crust interface. Disseminated spinels with low PGE values are found throughout the mantle harzburgites, which locally contain chromite-rich, but Pt-Pd-depleted, dunitic pods. Within the dunites of the HLC, both chromite and PGE are more abundant than within the harzburgites. This enrichment is most extreme in the higher part of the upper dunites which contains thick (>10 cm) and abundant chromitite layers

comprising 90% massive Fe^{3+}-Ti chromite. PGE enrichment upwards through the upper dunites has produced spectacularly high PGE values (>13 000 ppb) and preferential fractionation of Pt and Pd which reach values of 10 900 ppb (Moreno *et al.* 2001).

The ultramafic massifs of Cabo Ortegal present a great variety of platinum-group minerals (PGM) interpreted as mostly secondary in origin rather than primary magmatic (Moreno *et al.* 1999). They include, in order of abundance: Pt-Pd sulphides; Pt-, Rh base metal alloys; Pt-, Pd-, Rh-bearing As, Bi, Sb and Te; Os-, Ir-, Ru-bearing PGM; Au-Hg-Pb-bearing PGM; and PGE-bearing oxides. The majority of them are composite grains with irregular morphologies, very small in size (<10 μm), and located in the altered silicates interstitial to chromite grains. Other common locations are inside chromites, in sulphides (pentlandite, chalcopyrite and millerite), or at the contacts between chromites and silicates.

Calzadilla de los Barros. The Malcocinado Fm in the Ossa Moreno Zone includes three chromitiferous serpentinized massifs covering an area of *c.* 14 km² near Calzadilla de los Barros (Badajoz; Monterrubio 1991). These serpentinites are hosted in Precambrian volcanic rocks and interpreted as olistoliths within synorogenic sequences (Quesada *et al.* 1991). Chromitites are present in the massifs together with pyroxenites, chloritites, tremolitites, talc-carbonated rocks, amphibolites and basic dykes (Monterrubio 1991). Although ultrabasic rocks are almost completely serpentinized it is still possible to distinguish very altered dunites and harzburgites. Chromite mineralization occurs mainly in one 30 × 1.5 m thick body in Cabeza Gorda and in non-in-situ chromitic samples in Cerro Cabrera. In both cases it is possible to observe disseminated (1–2 mm) and massive chromites, all very fractured, altered and with increasing iron contents towards the rims. Compositionally they are Al-rich chromites (Cr/(Cr+Al) < 0.6) and they probably originated in an ophiolitic complex (Monterrubio 1991).

Nickel. The most interesting nickel deposit in Spain is the Aguablanca Ni-Cu-PGE mineralization, discovered in 1995. Aguablanca (Fig. 19.1) lies at the southern border of the Ossa Morena Zone and comprises magmatic Ni-Cu sulphides and PGM hosted by mafic and ultramafic igneous rocks. The estimated reserves in 1998 were 30–35 Mt (0.7% Ni, 0.6% Cu, 0.02% Co, 0.3 ppm Pt, 0.3 ppm Pd, 0.15 ppm Au) and the deposit may be exploited in the near future. This finding is important not only from an economic point of view (Ortega *et al.* 2000), but also because it is the first mineralization of this type found in the Iberian peninsula (Lunar *et al.* 1997).

The ore deposit is located in the northern part of the Aguablanca sill, a pre-Variscan folded igneous body consisting of mineralized gabbros and gabbroic cumulates, with subvertical layering, that grade southwards into amphibole–biotite diorites. To the north the Aguablanca stock intrudes the Bodonal-Cala complex, a metasedimentary and metavolcanic sequence (Neoproterozoic–Lower Cambrian) folded into a D1 Variscan overturned anticline. To the south, the Aguablanca diorites are cut by the Santa Olalla pluton, a Variscan tonalitic intrusion with flat-lying foliation (Casquet *et al.* 1998).

The mineralization occurs in two elongated subvertical bodies hosted by gabbros, norites and subvertical layers of pyroxenite and peridotite. Three types of ore are recognized: disseminated (Ni/Cu = 1–1.5), massive (Ni/Cu = 2–5) and breccia ores. The disseminated sulphides comprise most of the mineralization and grade into folded pods of semi-massive and massive sulphides with local breccias in the core and northern part of the main body. The massive and brecciated ores account for 15% and 0.7% respectively of the total mineralized rocks. The primary sulphide mineralogy is simple and includes magmatic pyrrhotite, pentlandite and chalcopyrite. Platinum group minerals, mostly consisting of Pt and Pd tellurides occurring within the sulphides, were initially exsolved during the magmatic stage and lately remobilized during the metamorphism of the ore. Pt and Pd contents in 100% sulphides are higher in the disseminated ore. Late hydrothermal pyrite is superimposed on the previous assemblages. Igneous host rocks show generalized alteration of pyroxene into amphibole, local serpentinization and talc-carbonate alteration and a pervasive low-grade alteration (calcite–chlorite–epidote–sericite). The mineralogical, textural and chemical features indicate that the original magmatic mineralization was formed by fractional crystallization of a sulphide melt exsolved by the Aguablanca mafic–ultramafic magma. The magmatic ore was reworked by deformation and metamorphism during the Variscan orogeny and overprinted by later hydrothermal events probably related to the intrusion of the Santa Olalla Variscan pluton.

Iron. Although iron has been extracted in Spain since pre-Roman times, the main phase of activity took place around the turn of the twentieth century when Spanish iron mining accounted for up to 10% of world production. In 1918 there were 430 operative mines, 210 in 1960, only 35 in 1973, and since 1980 production decline has been continuous.

The most important Palaeozoic iron ores are oolitic sedimentary deposits in the NW (Cotos Wagner, Vivaldi) and the skarn of Cala in the SW (Huelva). Other iron-containing skarns such as San Guillermo, Monchi and La Berrona in Badajoz and El Pedroso in Sevilla are of relatively minor importance (Lunar 1991). The most important exploitations in the Iberian Range of eastern Spain are located in Sierra Menera (Ojos Negros; Fig. 19.1) and contain oxides (hematites and goethite) and carbonates hosted in Ashgillian dolomites.

Oolitic iron deposits. There are many Spanish sedimentary oolitic iron deposits of Arenig–Caradoc age, the most important ones being located in the West Asturian-Leonese Zone (Lunar 1977; Lunar & Amoros 1979). Minor deposits occur in the Iberian Ranges, the Cantabrian Zone, the northern Central Iberian Zone, and the Ossa Morena Zone (Lunar 1991). The mineralization comprises oxides (magnetite), carbonates (siderite) and silicates (chlorite). Important mineralogical and chemical variations are found in those cases where mineralizations have been affected by contact metamorphism.

Stratigraphically, the most important iron-rich levels are related to a thick Llanvirn–Llandeilo pelitic unit that lies above a quartzitic sandy formation equivalent to the Armorican quartzite (Arenig). Iron horizons are located preferentially in the lower and middle part of the pelitic unit and in the transition to the quartzite formation. Up to seven discontinuous iron-rich horizons, 0.5–17 m thick, can be observed in the same unit. The most modern units also contain oolitic iron mineralizations, usually related to the sub-Caradoc disconformity.

Iron copper skarns. Calcic iron skarns are frequently developed in the Ossa Morena Zone, and several contain magnetite deposits of economic interest such as Cala, Monchi, San Guillermo, La Berrona, Burguillos and El Pedroso (Ruiz 1976; Velasco 1976; Casquet & Velasco 1978; Casquet 1980; Velasco & Amigo 1981; Locutura *et al.* 1990). Total reserves exceed 100 Mt with 25–35% Fe in Cala, La Berrona and San Guillermo mines, and appreciable Cu grades in some mines (e.g. up to 0.4% Cu in Cala).

The skarns are developed at the contact between Variscan plutonic rocks and marbles of Precambrian or Early Cambrian age, and are mainly composed of andradite, hedenburgite, actinolite, epidote and chlorite. Ore mineralogy includes magnetite chalcopyrite, hematite, pyrite, pyrrhotite, traces of gold and arsenopyrite, and local enrichment in U and rare earth element (REE) minerals such as uraninite. Such ore deposits are typical of the Fe-(Cu-Au) skarn type of Einaudi *et al.* (1981) and Einaudi & Burt (1982), and form a well defined group geotectonically related to early plutonism along an active continental margin (Locutura *et al.* 1990).

Sn-W. This is one of the most characteristic Variscan mineralizations in Spain, being related to post-orogenic granites as seen elsewhere in Europe such as in Cornwall and Brittany. Although tin mining is one of the oldest in Spain (together with gold, silver, copper and lead), tungsten mining is much more recent, reaching its maximum development in the 1940s during the so-called 'tungsten fever'. In the mid-1970s both tin and tungsten mining were active, but since the beginning of the 1980s production has declined dramatically as first tin and then tungsten production went into crisis. The last Spanish mine (San Finx) closed in 1991, although in terms of future potential the quantity and variety of Iberian Sn-W deposits remain some of the best in Europe.

Most of the Spanish Sn-W deposits are concentrated within the Central Iberian Zone of the Iberian Massif (Fig. 19.1), where there are abundant S-type granites generated by anatexis of continental crust (Hutchinson 1983). Different types of Sn-W deposits are found in Spain (Gumiel 1984), although only vein-type and disseminations in granitic cupolas are of economic interest. The recognized types are disseminations and related veins, pegmatites, skarns, vein-type, and disseminations and veinlets within hydrothermally altered siliciclastic rocks.

Disseminations and veins contain cassiterite, columbo-tantalite and minor sulphides confined to the roof of greisenized granitic cupolas, examples including Golpejas, Penouta and El Trasquilón (Fig. 19.1). *Pegmatites* show disseminations of cassiterite, columbite and tantalite, some being complex pegmatites with rare metals (e.g. Li, Be, Rb, Cs), examples including La Fregeneda, Tres Arroyos and Las Navas (Fig. 19.1). *Skarns and skarnoids* exhibit scheelite-sulphide (e.g. arsenopyrite, chalcopyrite) as seen for example in Los Santos and Otero de

Herreros. *Vein-type deposits* display veins and stockworks hosted in country rocks surrounding granitic cupolas. They are quartz-filled veins with (a) wolframite and/or cassiterite and (b) scheelite–cassiterite, accompanied by sulphides that may include arsenopyrite, molybdenite, chalcopyrite, sphalerite, stannite and Ag-bearing galena. Wall-rock alteration typically involves greisenization, tourmalization and silicification. Examples include La Parrilla, Mina Teba, Barruecopardo, Santa Comba, San Finx, Bustarviejo and Cabeza Lijar (Fig. 19.1). *Disseminations and veinlets within hydrothermally altered siliciclastic rocks* contain cassiterite and base metal sulphides and include examples such as Oropesa (Cordoba, discovered in the 1980s).

The fluids associated with the greisenized cupolas, the pegmatites and the vein-type deposits were hot: $CO_2 \pm CH_4$ bearing brines circulating at temperatures between 200 and 500°C and at pressures up to 2 kbar (Mangas & Arribas 1987; Noronha *et al.* 1999; Vindel *et al.* 2000). Disseminations, then pegmatites and finally veins record progressively decreasing pressure–temperature conditions with two main stages of mineralization recognized for veins: a high-temperature stage (>350°C) responsible for wolframite–cassiterite–arsenopyrite deposition; and a lower-temperature stage (350–200°C) related to sulphide precipitation.

Antimony. More than 60 antimony occurrences and deposits are found in Spain, most of them within the Variscan domain, notably in the Central Iberian Zone. The San Antonio mine (Badajoz; Fig. 19.1) is the most important from an economic point of view as it was active until the end of the 1980s. The geology and metallogeny of the Iberian antimony deposits have been studied by Gumiel & Arribas (1987), Ortega & Vindel (1995) and Ortega *et al.* (1996). Three main types of deposits are found: (i) Sb-W-Hg stratabound deposits associated with Palaeozoic shales and limestones with interbedded volcanics such as San Antonio (Badajoz) and Villarbacú (Lugo); (ii) Sb-Au vein type deposits hosted by Precambrian and Palaeozoic metasedimentary rocks and related to late Variscan granitoids, such as Mari Rosa (Cáceres); (iii) Sb deposits in late Variscan volcanic feeder dykes cross-cutting Carboniferous rocks, such as Burón (León; Paniagua *et al.* 1987), and several Sb occurrences in La Carolina lead district.

Gold. Spain has been producing precious metals since ancient times. The main zones are the Pyritic Belt (both gossan and complex sulphides) and the NW part of the peninsula which has been considered an auriferous area since at least Roman times (it was the principal gold mining area in the Roman empire). Romans worked in different types of gold deposits and produced more than 200 t from more than 600 mines. After this, interest in this large gold province waned considerably until 1973 when increasing prices stimulated several companies such as RTZ, Consolidated Gold Fields and Rio Narcea Gold Mines (RNGM), to initiate exploration works that have resulted in the identification of deposits such as Salave, Carles and El Valle in Asturias (Fig. 19.1). RNGM research has provided the main impetus in recent gold exploitation and is a key player in the enlargement of known reserves and ensuring the survival of gold mining activity in Spain.

Four types of Variscan gold deposits are recognized (Castroviejo 1995, 1998). (i) *Au-sulphides vein deposits in regional shear zones.* The gold ore is mainly asssociated with arsenopyrite and generally shows intense hydrothermal alteration. Examples of this type of deposit include Albores (Coruña), Pino (Zamora), La Codosera (Badajoz) and Nava de Ricomalillo (Toledo; Fig. 19.1). (ii) *Hydrothermal Au-sulphide deposits in granitoids.* This ore deposit is related to granodiorites and quartzdiorites, with well developed hydrothermal

alteration (chloritic, sericitic, propilitic, albitic and 'hongorock' – the local name describing the 'hongo' mushroom shape of the altered zone) being linked to the mineralization. A notable example is the Salave mine where unusual hydrothermal concentrations of Au (up to 2.5 g/t) are found with 16.5 Mt reserves (Harris 1980). (iii) *Skarn mineralizations (Carles type) superimposed by epithermal alteration (El Valle type).* These mineralizations appear in the Rio Narcea area, the easternmost of the Asturian gold belts (Fig. 19.6). This gold belt includes one open pit mine at El Valle, one mine under development at Carlés, two more known gold deposits and three prospect areas under various stages of exploration with estimated gold reserves of 1.8 million ounces (Clifford 2000). El Valle (Fig. 19.1) is the largest of the Rio Narcea deposits and consists of a NE-orientated mineralized system with five major zones of gold mineralization around the Boinas laccolith. El Valle ore is the result of gold deposition in two stages (Martin Izard *et al.* 1997, 1998). The first of these involves the development of a copper–gold skarn at the contact between the Boinas granodiorite and limestone and dolomite of the Larnaca Fm (Cambrian). The ore assemblage is characterized by chalcopyrite, bornite, magnetite and electrum. This skarn-type mineralization was overprinted by two later epithermal alteration events of the calcite–adularia type associated with the emplacement of porphyritic dykes. The low temperature ore assemblage is dominant in the deposit and includes electrum, native copper, cuprite, chalcocite, hematite, pyrite and gold-bearing jasperoids. The Carlés deposit is located 10 km north of El Valle and consists of a pyroxene and garnet skarn developed along the contact between Rañeces carbonates (Devonian) and a granodiorite intrusion. The mineralization contains gold associated with arsenopyrite, chalcopyrite, pyrrhotite and magnetite. The main environmental problem related to the exploitation of the gold ores at Rio Narcea derives from the cyanide leaching method used for extracting gold. Waste material in disposal ponds is likely still to be impregnated with dilute cyanide solution. As part of a remediation programme, RNGM has incorporated cyanide destruction on the tailings (Inco SO_2 system) and is planning to construct systems for cyanide regeneration and on-site generation of SO_2 (Clifford 2000). (iv) *Sb (As)-Au veins.* This type includes traces and veins of only minor economic importance at present, and includes examples such as Almuradiel and Villamanin.

Pb-Zn. Lead–zinc deposits in Spain are numerous and varied (Fig. 19.1), occur in very different geological settings, and are mostly either vein or stratabound ores. Vein ores include a large number of deposits mostly grouped around Linares-La Carolina (Rios 1977) and Alcudia valley area (south central Iberian zone; Palero *et al.* 1992), and around Córdoba and Azuaga (Ossa Morena Zone). Stratabound ores are hosted in carbonates (e.g. Rubiales, Lugo), although their origin does not seem to be sedimentary (Tornos *et al.* 1997), and none are currently being exploited.

Rubiales deposit. This was discovered in 1968, with exploitation starting in 1977 and mining an estimated reserve volume of 20 Mt with average grades of 1.3% Pb, 7.3% Zn and 400 g Ag/t. Rubiales is a 1200 × 600 × 30 m lenticular hydrothermal deposit located in a shear zone (Arias 1992). The Lower Cambrian host rock is named the 'transition series', and is located within the Western Asturian-Leonese Zone. The mineralization is controlled both structurally (shear zone) and lithologically (calcareous levels in the transition series between slate levels). The mineralogy is simple: sphalerite and galena, with pyrite, chalcopyrite and pyrrhotite as accessories.

Linares–La Carolina District. This is the main Pb-Zn vein field in Spain, and is a world-class mineral deposit. It includes a total of 1300 deposits mined mainly during Roman times and in the nineteenth to early twentieth century, when between 1875 and 1920 it was the leading world producer of lead ore. The mineralization consists of subvertical veins of metric thickness and up to 10 km long that are hosted in Ordovician–Lower Carboniferous metasediments and the Santa Elena granite. These are hydrothermal deposits of mostly argentiferous galena (300–500 g Ag/t Pb) filling late-Variscan extensional faults, typically trending WNW–ESE, west–east and SW–NE, and associated with the Pedroches batholith (Lillo 1992).

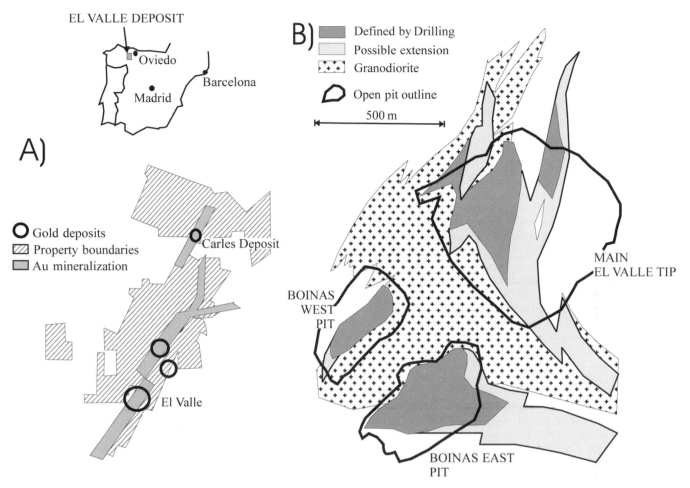

Fig. 19.6. Location of El Valle Au deposit showing **(A)** where the mineralization is, and **(B)** a map of the mined area (modified from Rio Narcea Gold Mines Ltd; web page http://www.rionarcea.com).

Ag–base metal deposits. Late Variscan hydrothermal activity resulted in important epithermal silver–metal (Pb-Zn-Cu) vein formation in the eastern part of the Central Range (Martínez Frías 1986, 1992, 1996; Martínez Frías *et al.* 1992*b*) as well as minor Ba-F–metal vein deposits hosted in granitoids (Lillo *et al.* 1992). During the Variscan orogeny, the central Iberian crust was thickened by compressional tectonics, heated, weakened, and subsequently overthickened by massive late Variscan granitic intrusions. Extensional collapse of this crust led to the formation of a basin-and-range-type province within which andesitic volcanism and hydrothermal activity induced epithermal systems that leached, transported and precipitated silver and base metals in fractures cutting metamorphic core complexes (Concha *et al.* 1992; Doblas *et al.* 1988). The main districts are the Hiendelaencina–Atienza district and the Guadalcanal district (Seville) mines.

The Hiendelaencina–Atienza district. These deposits (Hiendelaencina, La Bodera and Congostrina) present a rather special case in which Palaeozoic shallow-seated epithermal deposits escaped Mesozoic–Cenozoic erosion by rapid sedimentation of, and burial by, Permo-Triassic sediments. This contrasts with the more typical situation in Andean type mineralization within Cenozoic orogenic belts, where uplift–erosion processes tend to erode shallow-seated

mineralization rapidly (Martínez Frías 1992; Oyarzun *et al.* 1998). This explains why Hiendelaencina, a Palaeozoic silver epithermal deposit of shallow emplacement, has been preserved until today.

The district was mined extensively during the period 1844–1925, when ore with grades of up to 200 kg/t Ag was extracted, mostly from the Rico vein. This vein strikes for about 5 km and displays an oxidized top with barite, silver chlorides and bromides. With increasing depth the assemblage grades into complex silver sulphosalts (pyrargyrite, freislebenite, stephanite, miargyrite, freibergite) and ruby silvers, with increasing contents of base metal sulphides, such as galena, sphalerite, chalcopyrite and arsenopyrite. The veins underwent several mineralization episodes, giving rise to breccia-like textures. Hydrothermal alteration at Hiendelaencina includes quartz–sericite, propilitization and adularization, while at Atienza the andesites display pervasive propilitization, giving the rocks a typical green colour in the field. Fluid inclusions in quartz and fluorite samples from veins have homogenization temperatures of 100–160°C in fluorite and 90–120°C in quartz, which clearly indicates mineral precipitation under epithermal conditions.

The Guadalcanal (Sevilla) mines. Guadalcanal was discovered in 1555 and exploited irregularly until the end of the nineteenth century (Enadimsa 1986). The paragenesis, virtually restricted to the area called Pozo Rico–Guadalcanal consists of native silver, miargirite and stephanite and pirargirite, and is similar to other examples in Norway (Kongsber) and Canada (Cobalt). The ore is a 0.5 m thick vein with an average grade of 10 kg/t producing up to 70 000 kg of Ag during the mid-fourteenth century.

Alpine (Mesozoic to Recent) ore deposits

Stratabound Fe deposits. These are very numerous in the Nevado–Filábride and Alpujárride complexes in the Betic Cordillera of southern Spain (Fig. 19.1), the most important district being Marquesado del Zenete in Granada (Torres 1992). The mineralization typically occurs as stratiform and irregular bodies hosted in Triassic marbles. They are considered to have an early sedimentary–diagenetic origin, with clear stratigraphic, lithologic and palaeogeographic controls, and were formed in a transitional continental–marine lagoonal sedimentary facies. The main minerals are hematite, siderite, ankerite, geothite and magnetite with sulphides as accessories. There are also numerous iron mineralizations known from the Subbetic zone to the north (see Chapter 16), the most important of which is the magnetite mine of Cehegin (Murcia). These are located at the contact between basic intrusions and Triassic limestones and dolomites.

Cr–Ni and graphite mineralization. The small volumes of chromite, niqueline and Fe-Ni-Cu sulphide deposits associated with ultramafic rocks in the Betics of southern Spain are more a mineralogical curiosity than of any real economic importance. However, some of them had important exploitation episodes in the past, such as La Gallega in Ojén, the Jarales district in Carratraca or the graphite mines in Marbella. Mineralization occurs within the peridotitic massifs of Ronda (Gervilla 1992; Garrido 1995; Gutierrez Narvona 1999), with equivalent deposits being found around the Betic–Rif orogen in the Beeni Bousera massif (Morocco) and in Algeria (Gervilla & Leblanc 1990). There are two types of mineralization: (i) chromite and Ni-arsenides with pyroxenes and/or cordierite in an unusual and geochemically anomalous association that shows either Cr+Ni or only Cr mineralization; (ii) Fe-Ni-Cu sulphides with variable amounts of graphite and minor proportions of chromite, pyroxenes, plagioclase and phlogopite. PGE content in the sulphides is extremely low (<0.8 ppm) by comparison with contents in magmatic sulphides, but there is some Pt and Pd enrichment.

Gold concentration in Neogene and Recent detrital sediments. Gold of alluvial origin is found in both northeastern and southern Spain, occurring as fine, millimetre- to centimetre-sized plates. This type of gold deposit was of particular strategic importance in Roman times with up to 600 mines in operation (Pérez & Sánchez 1992). The Darro mine in Granada province has been panned in the Genil and Darro rivers since Roman times, and was also mined in the conglomerates of the Alhambra Fm. The host rock is a Pliocene alluvial-fan conglomerate up to 200 m thick. In Las Médulas (Caracedo) gold was worked by the Romans from the first century AD using the 'ruina montium' method, excavating the mountain with hydraulic power and running the resulting product along channels where the gold was extracted by gravity processes. During the first half of the twentieth century Oligocene–Quaternary gold placers such as those in the Orbigo, Duerma, Omañas and Sil rivers in the NW of Spain were once again exploited. At the moment, however, such mines are of only historic and archaeological interest.

Pb-Zn mineralization. Pb-Zn deposits occur both in the Basque-Cantabrian basin in the north, and in the Betic Cordillera to the south of Spain. In the Basque-Cantabrian basin there are stratiform mineralizations in Lower Cretaceous (Aptian–Albian transition in Urgonian facies) carbonates and mudstones, related to Mississippi Valley or Sedex type deposits. Examples of Mississippi Valley type Zn-Pb deposits include the Reocín, Novales and La Florida deposits (Fig. 19.1). The mineralizations typically display a simple paragenesis of sphalerite, galena (rare), pyrite, marcasite (in Reocín) or barite (La Florida; Herrero 1989; Seebol *et al.* 1992). Coloform textures are very common, with dolomites and ankerite alternating with sphalerite. Reocín, hosted within iron-rich dolomites, is the most important Zn deposit in Spain (and the second in the EU) having been mined for more than 100 years, both underground and in open pits. The main Sedex type Zn-Pb deposit in Spain is La Troya and consists of a subhorizontal stratiform north–south striking body cut by several faults. It contains three mineralizations: massive sulphides, silicified limestones and sideritic limestones, and is comparable to Tynagh and Silvermines in Ireland (Fernández Martínez *et al.* 1992). Finally, in the Betic Cordillera F-Pb-Zn stratiform, palaeokarst and vein deposits of minor economic value are present (Fenoll *et al.* 1987), hosted mainly in the Middle–Upper Triassic carbonate sequence in the Alpujárride complex (Lujar, Gador and Baza Sierras). The mineral paragenesis is simple, with sphalerite, galena, pyrite, marcasite, fluorite and barite.

Late Alpine mineralizations: the Cartagena–Almeria volcanic belt

The SE corner of Spain has been intensively exploited for Au, Ag and base metals since pre-Roman times. The main mining districts in this area include Rodalquilar (Au), Herrerías-Sierra Almagrera (Ag), La Unión (Pb-Zn-Ag-Sn) and La Crisoleja (Sn-Pb-Ag). The mineralization is linked to the Cartagena–Almería volcanic belt, a short, narrow (150 × 25 km) NE–SW trending zone of middle to late Miocene age. Volcanism changes over time from Au-rich, Sn-poor calc-alkaline to Sn-rich, Au-poor high-K calc-alkaline, shoshonitic and, finally, lamproitic series. The mineral deposits (Au, Ag, Pb-Zn-Ag, Sn, Mn and Sb) are mostly epithermal and associated with caldera-type structures and dome complexes (Manteca & Ovejero 1992; Rodríguez 1992; Morales *et al.* 1995, 2000; Oyarzun *et al.* 1995, 1998). The six main districts are: (i) Rodalquilar (Au-Sn), a caldera-related Au epithermal deposit; (ii) Herrerias–Sierra Almagrera (Ag); (iii) Lomo de Bas (Au-Ag-Sn), a porphyry-type mineralization; (iv) Mazarrón (Pb-Zn-(Ag)) vein-type ores with acid-sulphate alteration; (v) La Unión (Pb-Zn-(Ag-Sn)) stratabound sulphide deposits in carbonate sequences in the Nevado–Filábride and Alpujárride complexes; (vi) La Crisoleja (Sn-(Pb-Ag)), related to subvolcanic domes.

Rodalquilar. This is an epithermal acid-sulphate Au district (Arribas 1992, 1993; Arribas *et al.* 1995) related to an 8 km wide dominantly rhyolitic caldera (Fig. 19.1). Volcanic activity took place 11 Ma ago (Tortonian) and began with the eruption of ash-flow tuffs (Cinto unit), followed by the formation of a second caldera nested in the centre of the older structure (Lomilla caldera) and accompanied by the eruption of more ash-flow tuffs (Lázaras unit). Volcanism ended with the production of andesite flows and intrusions. This episode induced widespread fracturing and hydrothermal activity that led to pervasive advanced argillic alteration (alunite) and gold deposition in the higher levels of the hydrothermal system. Contemporaneously, Pb-Zn-Ag-Au vein-type deposits formed distally (1.5 to 3 km away). Both types of mineralization were related to magmatic hydrothermal processes following the emplacement of andesitic magma during the late stages of the formation of Rodalquilar caldera. Pyrite is generally the most important sulphide phase, with other minerals including gold, calaverite, sphalerite and galena. Sn has proved a useful pathfinder for gold mineralization (Pineda 1984), with background levels of 53 ppm/1.4 ppm Au and anomalous values of 385 ppm Sn/ 7.6 ppm Au.

Herrerías–Sierra Almagrera. Herrerías is a stratabound sulphide deposit that includes a field of well preserved fossil marine fumaroles within Miocene sandy marls belonging to the Herrerías trough (Martínez Frías *et al.* 1992a; López Gutierrez *et al.* 1993, 1995). The fumarole field is about 130 m by 10 m and contains more than 30 vents with hydrothermal Fe-Mn oxides, disseminated base-metal sulphides, native silver, barite, jasper, chalcedony and gypsum (rare) within strongly altered (silicification, sericitization) marine sediments. Veined epithermal mineralization in Sierra Almagrera is spatially and genetically related to Herrerías (Martínez Frías 1991, 1998; Morales *et al.* 1995, 2000). During the nineteenth century the Herrerías–Sierra Almagrera and Hiendelaencina districts were the most important silver producers in Europe. At present Herrerías is the only open pit mining barite in the Iberian peninsula.

La Unión. This Pb-Zn-(Ag-Sn) district provides an excellent example of the interplay between basin formation, normal faulting, volcanism, hydrothermal activity and mineral deposition (Lopez Garcia *et al.* 1988; Manteca & Ovejero 1992), all synergistically combining to create exceptional concentrations of metals (7 Mt Zn + Pb, 65 Mt Fe). Hydrothermal activity related to middle Miocene subvolcanic magmatism was focused along normal faults bounding a graben. The mineralization occurs (i) within a strongly altered zone above the Miocene footwall (probably a preferred horizon for hydrothermal fluid circulation; Lunar et al. 1982; Amoros et al. 1984); (ii) in pebbly mudstone beds where the hydrothermal activity led to dissolution, void formation and mineral deposition; and (iii) in fault breccias along the normal faults that bound the Miocene sediments.

La Crisoleja. This is a peculiar deposit located within La Unión district and associated with a dacitic dome of late Miocene age that developed a stockwork and intense hydrothermal alteration. The mineralization consists of cassiterite, hematite, pyrite, Pb-jarosite and goethite, with the presence of silver having also been reported (Manteca & Ovejero 1992).

Pleistocene–Quaternary Co-rich Mn mineralization in central Spain

The Calatrava region in Ciudad Real (central Spain) hosts a series of Co-Mn hot spring mineralizations (Fig. 19.1), linked to late Miocene–Pliocene volcanism in the southern central part of the Iberian peninsula (Crespo & Lunar 1997; Chapter 17). The deposits are Pliocene in age and display unusual geochemical and mineralogical characteristics, including very high Co contents, low contents of As, Sb and other related trace elements, and a manganese oxide mineralogy constituted by lithiophorite, birnessite, todorokite and criptomelane. These mineralizations display a variety of morphologies: spring aprons and feeders, pisolitic beds, wad beds and tufa-like replacements of plant debris. The spring aprons are found along or near to normal faults bounding small basins and topographic highs. Tufa-like Mn deposits are found near to the spring sources, while both pisolitic and wad beds are clearly distal facies occurring within the Pliocene basins. The deposits were mined in the past (1880–1963), and more recently (1985–1991) have received renewed interest regarding the possibilities of Co extraction from the manganese ores.

Industrial minerals and rocks (ML, MR)

As previous chapters have shown, Spain has a varied and complex geology in which almost all sedimentary, metamorphic and igneous settings are represented. This geological variety has produced many occurrences of useful materials, some exploited as far back as pre-Roman times. The earliest known indications of organized mining date from the Tartessan civilization, which inhabited the Guadalquivir basin around 3000 years ago. Although ancient mining mainly exploited metallic ores, it is well known that the Phoenicians started salt exploitation in Iberia, a practice that continued uninterrupted over many centuries. In the Carthaginian Carthago Nova (today Cartagena in SE Spain) there were important marine brine exploitations for the production of 'garum', a fish preserved in brine (the caviar of those times), which persisted after the Carthaginian defeat of the city by Escipion the African, a conquest that ended the war between Romans and Carthaginians. Romans used Spanish marble widely as a monumental stone, not just the famous Macael (Almería) deposit, but also some other less known outcrops such as those near Talavera de la Reina (Toledo; Urbina *et al.* 1997). Indeed the mineral wealth of Iberia provided a key reason for the Roman invasion.

After at least two millennia of mining history, Spain currently produces around 600 Mt of industrial minerals and rocks (IM&R) each year, worth ex-works 2728 million Euros and constituting over 66% of the national mining production. Regueiro & Marchán (2000) have recently revised the principal figures of this industry: Spain is the third world producer of dimension stone after China and Italy, the first world producer of roofing slate, the second world producer of marble and the first European producer of granite. Among industrial minerals, Spain is European leader in the production of celestite (second world producer after Mexico), sodium sulphate (sole European producer), fluorite (second European producer), gypsum (third world producer and first in Europe), feldspar (biggest feldspathic sand reserves in Europe) and sepiolite (70% of world reserves). Most of the IM&R production is related to the construction industry. Additionally, Spain is currently the top world producer of glazes and a leader in ceramic tile production (Table 19.2).

Despite the current social trend in which three main factors (difficulties of land access, restrictive environmental protection and lengthy administrative procedures) combine to impede the development of this industry, Spain has so far maintained its position as a world-class producer of IM&R (Table 19.1). This production has shown a growth in almost all commodities, and is the only subsector of a generally declining mining industry which is currently still increasing output.

Mineral resources, in their widest sense, are usually subdivided into energy resources, metallic ores, industrial minerals (or non-metallic minerals), rocks and water (Lorenz 1997). The differences between water resources, energy resources and the rest are clear, although some radioactive minerals can be considered both metallic ores and nuclear fuels. However, the classification of a particular substance as a metallic ore or an IM&R is sometimes not so obvious. Even the 'industrial rock' or 'industrial mineral' concepts are not clearly defined. For example, the industrial minerals magnesite and dolomite are involved to a limited extent in the extraction of the metal Mg. Similarly, chromite is used both as an industrial mineral in refractories, and as a metallic ore, it being the main source of chromium.

One of the most useful definitions was established by the National Spanish Mining Plan in 1971 (Ministerio de Industria 1971): 'Industrial minerals & rocks are those mineral substances used directly (or after adequate treatment) in industrial processes for their physical and chemical properties and not for the substances that can potentially be extracted from them or their energy'. In the following subsections we provide an alphabeticlly ordered summary of the main Spanish IM&R, covering both exploited deposits and those that may be potentially exploitable.

Aggregates

Aggregates provide the leading non-energetic mining sector in Spain, with a production in 2000 of 400 Mt, worth around 1900 million Euros. It is a very dynamic sector although widely disseminated across Spain, with some 2065 exploitations of which only 132 produce more than 0.25 million tonnes per annum (Mtpa). The operations are run by more than 1800 different companies, although only 20% of companies represent 75% of the total production. The Spanish aggregate per capita consumption is currently 7.8 tonnes per annum (tpa), about average for Europe, as compared to 1993 when per capita consumption was 4.9 tpa, one of the lowest in Europe. Such tremendous growth is related to the development of the National Infrastructures Master Plan 1993–2007 and by a seemingly unstoppable demand for housing (more than 500 000 in 1999).

Fifty-four per cent of the national aggregate consumption is of calcareous rocks, 30% sand and gravel and 16% igneous and metamorphic rocks. The main materials used as construction aggregates are limestone, dolomite, and sand and gravel. Quartzite, quartz, gneiss (mainly orthogneisses) and granitoid rocks are second in importance, being used especially in Galicia and other western Spanish regions where no calcareous deposits are easily available. Schist, slate and porphyry are commonly used, but usually as low-quality aggregates. High-quality aggregates for the upper layer of road pavements and high-speed railway ballast are obtained from quartzite formations (in the widely outcropping Armorican quartzite), some orthogneisses, dolerites, peridotites and hornfels. In the Canary Islands aggregates are produced mainly from volcanic rocks, with phonolite and basalt being widely used, and for sand-size materials the most popular aggregate is the 'picón', a loose pyroclastic basaltic pumice. Spain has also a considerable aggregate production as by-products of other mining activities (kaolin, quartz, granite).

Alabaster

Alabaster has been widely used throughout the history of Spain in the construction of monuments, particularly in Navarra, La Rioja, Castilla and Aragón. The famous statue of the Siguenza´s squire in the cathedral of Siguenza (Guadalajara), the magnificent plateresque frontispiece of the Colegiata of Santa Mª de Calatayud (Zaragoza) and the main altarpiece of the Pilar Basilica in Zaragoza are some examples of monuments made out of alabaster (Díaz-Rodriguez 1991).

The annual production of alabaster is around 40 000 tpa which comes from two areas in Aragón: Fuentes del Ebro-La Zaida (Zaragoza) and Azaila (Teruel). The alabaster is worked to produce decorative objects in local workshops and is also exported, mainly to Italy. The Spanish alabaster deposits are located in the evaporitic formations of the continental Tertiary basins of the rivers Tagus and Ebro: Palaeocene of Jadraque–Cogolludo (Guadalajara), Miocene of Albitas (Navarra), Oligocene and Miocene of Aragón and Oligocene of Camarasa (Lérida). There are also occurrences in other Tertiary basins (Guadix basin, east of Granada) and in the continental Permo-Triassic and Triassic rocks of the Betic Cordillera (Huercal-Overa near Almería, Lanjarón south of Granada, and La Paca in Murcia; IGME 1985).

The main deposits, located in Aragón, were formed in a playa-lake mud-flat environment (Díaz-Rodriguez 1990). The alabaster appears in nodules and rounded blocks up to 1 m in diameter isolated within gypsum beds. The origin of such alabaster is quite complex. The primary evaporitic gypsum is transformed by diagenesis to anhydrite, which later rehydrates during exhumation, forming gypsum with the typical microcrystalline texture of alabaster (Murray 1964).

Aluminium silicates (andalusite, kyanite, sillimanite)

The main industrial use of this group of minerals is in the manufacture of synthetic mullite, providing alumina to refractory products. In Spain these minerals have never been exploited intensively. Only artisan exploitations of kyanite in the El Pino-Touro (La Coruña) area (Marchán 1998), with a production under 1000 tpa, and sillimanite in A Toxiza (Lugo) with only anecdotal production, have been reported.

Deposits of these minerals are scattered through the metamorphic terrains of the Variscan Massif. The El Pino-Touro deposit is sedimentary, however, comprising Quaternary fluvial sands with kyanite clasts from the surrounding metamorphic areas (Toyos 1989; Lombardero 1998a). The andalusite deposits are related to the contact metamorphic aureole of Variscan granite bodies intruded into pelites of low metamorphic grade, e.g. Boal (Asturias), Riello (León), Salamanca, Baños de la Encina (Jaén). Andalusite has also been reported in metamorphic segregation veins in Goyán (Pontevedra), El Cardoso de la Sierra (Madrid) and Pinares (Ávila). None of these has ever been exploited although some have been investigated, such as the metamorphic aureole of the Forgoselo (La Coruña) plutonic body (Marchán 1998) and some other zones in Galicia (Toyos & Ferrero 1990), particularly in O Rosal (Pontevedra; Ferrero *et al.* 1992) and Verín (Orense; Ferrero *et al.* 1993).

With regards to sillimanite, although a common mineral in the Variscan Massif (the northern Madrid mountains have yielded abundant archaeological artifacts proving the use of sillimanite in hammer heads by its ancient dwellers), it usually appears in its fibrolite variety, mixed with mica and other minerals thus making mineral processing expensive and difficult.

Barite

Spanish production of barite (see Table 19.2) is now not as important as it was in the 1960s and 1970s (Fig. 19.7). Most of the production comes from Las Herrerías mine (Cuevas de Almanzora, Almería). Production is also reported in Tobed (Zaragoza) and Caldas de Besaya (Cantabria). The barite mineralizations are included in a wide variety of geological environments (Urbano 1998). A brief summary by geological zones is given below.

Cantabrian and West Asturian-Leonese zones: (i) stratabound mineralizations in Cambrian and Carboniferous limestones (Babia, León); (ii) vein mineralizations in Precambrian and Palaeozoic (Cangas de Narcea, Salas, Asturias); (iii) exhalative-sedimentary mineralizations in Neogene basins overlying Variscan basement; (iv) infill of karstic cavities in carbonate rocks with barite cobbles (Nieves Mine, Caldas de Besaya, Cantabria). *Central Iberian Zone*: (i) stratabound mineralizations linked to Silurian volcano-sedimentary series (Alcañizes Syncline, Zamora); (ii) vein mineralizations in plutonic and metamorphic rocks (Alcaracejos (Córdoba), Sotillo de la Adrada (Madrid) and Almorox (Toledo)); (iii) Ba-Cu mineralizations linked to palaeosols developed at the Palaeozoic–Triassic unconformity, and linked to Alpine-related fractures (La Carolina, Aldeaquemada and Baños de la Encina (Jaén)). *Ossa-Morena Zone*: (i) vein mineralizations in plutonic and metamorphic rocks of the well known old mining district of Villaviciosa (Córdoba); (ii) infill of karstic cavities in Palaeozoic carbonate rocks with barite cobbles (Llerena, Badajoz). *Pyrenees*: (i) vein mineralizations in plutonic or metamorphic rocks (Bellmunt, Lérida); (ii) infill of karstic cavities in Palaeozoic carbonate rocks with barite cobbles (Rocabruna, Gerona). *Catalonian Coastal Ranges*: (i) stratabound and vein mineralizations in detrital Permo-Triassic rocks (Escornalbou, Barcelona). *Iberian*

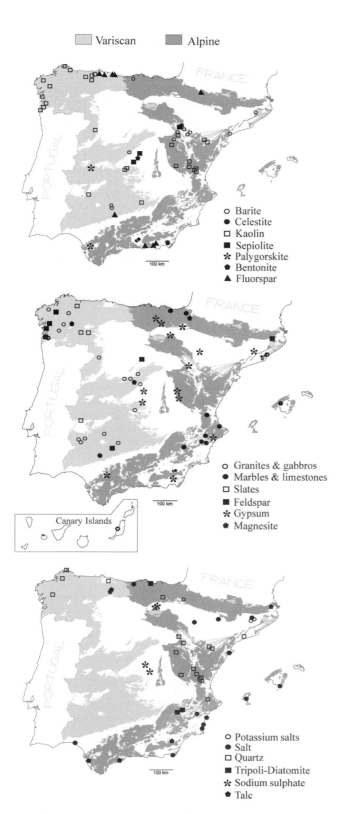

Fig. 19.7. Location map of the main industrial minerals and rocks deposits in Spain.

Ranges: (i) vein mineralizations in Palaeozoic detrital rocks (Tobed mine, Zaragoza); (ii) vein mineralizations in Permo-Triassic detrital rocks (Sierra del Espadán). *Betic Cordillera*: (i) vein mineralizations in Triassic carbonate rocks (Sierra de Baza, Granada); (ii)

stratabound and karstic infills in carbonate Triassic rocks (Sierra de Baza, Granada); (iii) exhalative sedimentary mineralizations in Neogene basins (Cuevas de Almanzora, Almería).

Calcium carbonate

Three companies control 80% of the ground calcium carbonate (GCC) production in Spain and are also leaders in technological development, with facilities in Albox and Purchena (Almería), Tarragona and Zaragoza. The main production zone in Spain is located north of Tarragona. Precipitated calcium carbonate (PCC) has gradually lost market as these synthetic products are being substituted by natural ultramicronized GCC which due to technological improvements are now finer and present better properties at lower costs. Saleable Spanish micronized GCC production is estimated to be 1.5 Mtpa. Another 1.5 Mtpa of non-micronized GCC is used in grit, animal feedstock, lime and other fillers. The market is dominated by filler uses in paper, paints and plastics, for animal feedstock, and in ceramics and glass where it is used as a chemically active agent. One of the biggest consumers is the soda factory (Solvay process) of Torrelavega (Cantabria), which produces GCC for internal use and also produces a short tonnage of PCC. GCC is obtained by milling and micronizing from calcareous lithologies. In eastern and southern Spain many sedimentary calcareous formations of different ages crop out, so it is not surprising that the raw limestones for GCC are varied. Examples include the loose calcareous deposits of the Miocene of northern Tarragona province (Arboç and Castellet i la Gornal), the Upper Jurassic limestone of Belchite (Zaragoza) in the Iberian Ranges, the calcarous marble of Macael (Almería), and the travertine limestone of Albox. Whiteness and chemical purity are the main factors to be studied and controlled.

Celestites

Spain is the second world producer of celestite and the only European producer. There are currently two operations in production at Escuzar and Montevives, Granada (SE Spain; Fig. 19.7). Celestite production, after a drop in 1997, recovered in 1998 and rocketed to 128 500 t in 1999, with an ex-works value worth nearly 6 million Euros. Spanish deposits represent the biggest known world reserves of celestite (Regueiro 1998).

The Escuzar deposit is exploited by a multinational company with a production in 1999 of 88 457 t and reserves estimated at 4 Mt. Montevives is handled by a Spanish company with a production in 1999 of about 40 000 t and estimated reserves of 8 Mt of several commercial grades. The production boom is related with the new 30 million Euro strontium carbonate plant inaugurated in January 2000 in Cartagena (Murcia) with a production capacity of 35 000 tpa, 11% of the world consumption.

Both celestite deposits (Montevives and Escuzar) are located in the Tertiary rocks of the Granada basin, an intramontane depocentre lying on the boundary between the internal and external zones of the Betic Cordillera. The primary mineralization is located in a lower marine evaporitic unit almost 90 m thick. In Montevives, celestite replaces calcite in two stromatolitic limestone layers with thickness varying between 0.2 and 1 m, both layers being exploited with an average grade of over 86%. Close to the current exploitation there are occurrences of celestite in the gypsiferous upper part of the evaporitic lower unit. Replacement of calcite and infilling of fissures and cavities by celestite seem to be synsedimentary, induced by salinity changes due to the mixture of sulphate salt waters and fresh waters during late Miocene times. In contrast, the Escuzar deposit is secondary. Boulders of mineralized limestone with strontium, celestite and dolomite filling the cavities of a karst developed in gypsum on the lower evaporitic unit (Ortega-Huertas 1992).

Many other occurrences of celestite are known elsewhere in Spain. In the Betic Ranges there are hydrothermal occurrences in Huercal-Overa (Almería) and in Hortichuela there are replacements of dolomite by celestite. The mineralization of Saltador in Lorca (Murcia) is probably synsedimentary or early diagenetic and was exploited in the 1960s. In Alicante there is diagenetic celestite in limestone and marls of the Cretaceous succession. In Jaén there are several occurrences in Keuper and Lower Triassic rocks. Finally, in Vic (Gerona) there are indications of celestite in Eocene sediments (Ortega-Huertas 1992).

Clay (common clays)

Within this wide group we include red clays for bricks, roof and ceramic tiles, ball clays used in wall and floor tiles, and 'sericitic' white clays used in ceramics and cement. The Spanish tile industry has undergone several major changes linked to the use of the homogenization plants and spray-dryers needed for modern floor and wall tile preparation.

Red clays. Two main ceramic production sectors consume common clays: brick and roof tiles, and wall and floor tiles. Spain consumes around 26 Mtpa of clays for brickmaking, with exploitations being located all over the country close to the factories as the low value of the raw material does not allow for lengthy transport. Toledo, Barcelona and Valencia are the main producing provinces with over 2 Mtpa each, followed by Alicante, Jaén and La Rioja which produce over 1 Mtpa. There are over 400 companies operating in this sector, most being family concerns, although some national and multinational industrial groups have an involvement. Ten of the family-owned companies are big enough to consume over 115 000 tpa of red clays (Regueiro *et al.* 2000*b*).

Clays for brickmaking and roof tiles do not have very strict industrial specifications, being considered suitable if they have an adequate plasticity to be extruded and a vitrifying temperature of around 1000°C (González-Díaz 1992). Specifications for clays to be used in wall and floor tiles are much more stringent, requiring in particular an absence of organic material, easy deflocculation, good pressing behaviour and the ability to gresificate (sinterize) at not very high temperature. These clays are mainly composed of quartz and illite with variable quantities of smectites, kaolinite and other clay minerals. For floor tiles, non-calcareous clays are commonly used (usually from continental formations) whereas wall tiles use marly clays (mostly marine). The main geological suppliers of red clay in Spain belong to the continental Triassic (Asturias, Iberian Ranges, Betics), Purbeck facies of the Jurassic (Asturias, Iberian Ranges, Balearic islands), Weald and Utrillas facies (fluvial and lateritic sediments) of Cretaceous age (Cantabria, Basque Country, Iberian Ranges) and several formations in the continental Tertiary basins (Duero, Ebro, Tagus, Calatayud, Guadix, Extremadura and Galicia) and the marine Tertiary of the Guadalquivir basin and eastern Spain. Residual clays from Quaternary terraces are also used locally (González-Díaz 1992).

Red clays for wall and floor tiles are extracted mainly in the Valencia region, in the villages of Villar del Arzobispo and Chulilla (Utrillas and Weald facies), San Juan de Moró (Buntsandstein fluvial facies), Alcora (marine Tertiary), and with a small portion coming from the Utrillas facies of Galve (Teruel). Extraction, micronization and trading of such clays involves a few important mining companies, and even some multinationals. Only a few wall and floor tiles producers have their own clay quarries.

Ball clays. Ball clays are used by several ceramic sectors with white body tiles being the most important product, although speciality ball clays are also used in glazes, tableware, porcelain and sanitaryware. Ball clays used in Spain are mainly domestic (75%) although a certain amount (25%) is imported from Great Britain, Germany, France and Ukraine. Total ball clay consumption in Spain is around 1.2 Mtpa.

Spanish ball clays are mainly found in the Cretaceous estuarine-deltaic fluvial deposits of the Iberian Ranges (Utrillas and Escucha facies), which are rather similar to the estuarine Palaeogene deposits

in the UK (Bristow & Robson 1994). They are currently being mined in Teruel (La Cañada de Verich, Estercuél, Ariño-Oliete area) and are mainly composed of poorly ordered, very fine (<2 µm) kaolinite or kaolinite–illite, quartz and minor amounts of illite, smectites and other clay minerals (Barba *et al.* 1997). Usually they contain high alumina and low carbonates, alkali and iron, so they are refractory and white or pale cream after firing. Cretaceous lacustrine and fluvial ball clays are also found in Asturias, Cantabria and Burgos. The Tertiary continental basins of Galicia, particularly those associated with lignite deposits, also contain fire clay and ball clay deposits, and some are exploited.

Sericitic slates. This name is used in Extremadura and northern Andalucia for certain white clays formed by supergene and hydrothermal weathering of Palaeozoic slates of the Iberian Massif. Usually these deposits are linear, following the traces of major late Variscan or Alpine faults. They are typically very irregular, with weathered zones alternating with zones where the slate is very little altered or not transformed at all. The most common paragenesis is illite, pyrophyllite and low kaolinite (Mesa 1992). Commercially they are similar to ball clays although their properties are slightly different. Most of the old exploitations are abandoned although recently one company has reopened its sericite operation from hydrothermally altered low grade slates in La Zarza de Alange (Badajoz) and in 1999 sold 50 000 t for ceramics and cement.

Clay (special clays)

We include in this group kaolin and kaolinitic sands, sepiolite, paligorskite and bentonite. These minerals are more valuable than common ceramic clays and are employed in several different industries to manufacture grouts and refractories, fillers, absorbents, sealants, drilling muds, and ionic exchangers. Kaolin is also widely used in sanitaryware, porcelains and in the paper industry as a coating agent.

Kaolin. Ceramic kaolin has seen a considerable production increase in the last few years due to the rapid growth of the ceramic industry in Spain (Fig. 19.7). Almost 57% of total Spanish kaolin production (400 000 tpa) is consumed by the domestic and international ceramic industry. Estimated reserves of kaolin exceed 100 Mt, suggesting a promising future for a possible expansion of production, particularly given the current improvement in consuming sectors. Twenty-two per cent of the national production is exported for the paper, ceramics, fibreglass, paints and rubber industries.

Three types of kaolin deposit have been reported in Spain: sedimentary, hydrothermal, and residual (Doval *et al.* 1991; Mesa 1992).

(i) *Sedimentary kaolin.* The main Spanish kaolin resources fall into this type of deposit, which is divided into Iberian Ranges, Asturias and Pontevedra subtypes. Iberian Ranges kaolin is found in the Weald, Utrillas and Escucha facies (Lower Cretaceous) and originated during Jurassic tropical alteration of granitic and acid metamorphic rocks, erosion of the lateritic soil generated, and redeposition by rivers during Cretaceous times as kaolinic sand and kaolin bodies. Kaolinization is thought to have been produced in all these stages. This kaolin is medium-ordered, has less than 20% kaolinite crystals under 20 µm, and sometimes minor amounts of allophane and hydroxides (remnants of their lateritic origin). Deposits of this kaolin are widely exploited (producing around 170 000 tpa) in Guadalajara, Cuenca, Teruel, Zaragoza, and Valencia, usually in big opencast operations. The Asturias subtype is a flint-clay kaolin (tonstein), originating as a diagenetic alteration of a narrow, very continuous, acid volcanic ash layer in the Ordovician Cuarcita de Barrios Fm (Armorican quartzite) of the Cantabrian Zone (Iberian Massif; Cembreros & Cossio 1990). It has a high kaolinite content (>90%) and conchoidal fracture. It is mined underground for grouts and refractory uses (about 40 000 tpa). Similar kaolin layers have been reported in Sierra Morena and Córdoba (Doval *et al.* 1991). The Pontevedra subtype is a Neogene to Quaternary proximal kaolin deposit, generated by nearby weathering of acid plutonic and metamorphic rocks of the Iberian Massif. Some of them are mined, but with only limited production.

(ii) *Hydrothermal kaolin.* This has been generated by hydrothermal alteration of acid Variscan igneous rocks. The genetic processes are not yet well understood, but are probably similar to those described for the kaolinization of Cornwall granites (Manning *et al.* 1996). Most deposits are developed on low-pressure, epi-metamorphic emplaced granitoids, as seen in both Cornwall and Galicia. There are deposits of this type along the northern coast of La Coruña and Lugo, as well as in Toledo and Segovia. The kaolin is mainly composed of well-ordered and very fine kaolinite and halloysite (Mesa 1992). A special subtype is the Burela (Lugo) deposit, which originated by meta-morphism and weathering of an albitite (probably an old acid volcanic layer) in the Cambrian succession of the Western Asturian-Leonese Zone of the Iberian Massif (Doval *et al.* 1991) and produced the highest quality kaolin in Spain (used for porcelain and paper coating). Production of hydrothermal kaolin in La Coruña and Lugo is currently around 100 000 tpa (Marchán 1998).

(iii) *Residual kaolin* deposits are in-situ weathered acid igneous rocks which have not experienced transport (soils). Paraño (Lugo) and Sayago (Zamora) are the main deposits, both currently mined, with a production of 25 000 tpa.

Sepiolite. Spain is the world leader in sepiolite production, with its most important deposits located in the Tagus basin (Fig. 19.7). Madrid hosts the world's biggest sepiolite deposit, with reserves of an extra-ordinarily high quality estimated at over 15 Mt. Sepiolite production is about 0.8 Mt and is currently on the increase, with Madrid, Toledo and Zaragoza being the main production zones. Sepiolite and paligorkite are both Mg-bearing phyllosilicates having specific industrial properties due to the peculiar acicular habit of their crystals, with sheets forming microscopic tubes and an effective pore size of 0.15 μm for sepiolite and 3 μm for paligorkite (Galán 1992). The unique absorptive, rheological and catalytic properties of both minerals result in many industrial applications such as molecular filtering, grease and insecticide absorption, cat litter and food additives.

Although marine sepiolite occurrences have been described, the main industrial deposits are located in continental lacustrine deposits. The Madrid basin, part of the continental Tajo Tertiary basin, hosts the most important deposits known in the world. There are two layers more than 1 m thick and with great lateral continuity. They were deposited in lakes at the distal end of alluvial fans during the arid climate that characterized inland Iberia during most of Miocene times (Castillo 1991; Galán 1992). The Vicálcaro–Barajas deposits are extensively exploited by a Spanish multinational and some other national companies, most of the material being exported. Some other levels of minor interest but also exploited occur in Valdemoro and Parla (Madrid), Yunquillos, Esquivias and Cabañas de la Sagra (Toledo). This latter deposit holds the purest sepiolite deposit of the whole basin. Another important deposit currently exploited is located in Orera in the Calatayud basin (Zaragoza), and corresponds to a marginal deposit of an alluvial fan which during late Miocene times closed a valley producing a lake in which sepiolite was formed (Gonzalo-Corral pers. comm.; Castillo 1991).

Sepiolitic clays and sepiolite–paligorskite clays also occur in the Cuesta facies of the Duero Tertiary basin, in Lebrija (Sevilla) where they have been exploited to filter sherry (so it is locally called 'tierra del vino'), and in both the Guadix basin (Granada) and the Tertiary continental basin of Galicia (Castillo 1991; Galán 1992).

Palygorskite (attapulgite). Spanish production of this mineral (some 160 000 tpa) is not as important as that of sepiolite, but it is enough to supply the domestic industry and even to export a small tonnage. Paligorskite is being produced in Cáceres and Segovia by two big (multinational) and a few other Spanish mining companies (Fig. 19.7). SW Spain deposits (Sevilla and Cádiz) are of coastal origin, having formed in small slightly saline water lakes where mixed siliciclastic–carbonate deposition occurred. Palygorskite and palygorskite–sepiolite layers are interbedded with illitic clays and marls. Palaeoclimatic evidence indicates arid to subhumid weathering during the genesis of these deposits during the Pliocene epoch (Castillo 1991; Galán 1992). In Torrejón el Rubio (Cáceres), a small (250 km^2) pull-apart Tertiary basin on the Esquisto Grauváquico Fm (Palaeozoic slates) hosts the biggest known palygorskite deposit in Spain. Its genesis is related to the alteration of chlorite slates along the Palaeozoic–Tertiary unconformity and later erosion and redeposition of this layer by rivers (Castillo 1991). The currently exploited deposit

of Maderuelo–Bercimuel (Segovia) was probably generated by a Mg enrichment of an illitic hardpan or calcrete soil, in an alluvial-fan sedimentary environment.

Bentonite and Fuller's earth. Bentonite is the name used for plastic clays formed by minerals of the smectite group, and includes six different di- and trioctahedral phyllosilicates. Most bentonite deposits result from the alteration of felsic or intermediate volcanic ashes in lakes or coastal zones, although late igneous (deuteric) and hydrother-mal bentonites have also been cited. The name Fuller´s earth is sometimes used synonymously for calcic bentonites, although it is occasionally also used for bentonites containing carbonates, opal and other clays such as paligorskite and sepiolite. The main properties of bentonites include a very fine particle size (<2 μm), and the capacity to absorb and lose cations located in the crystalline structure but only loosely joined with it. Water suspensions of some bentonites are thixotropic and very viscous, due to the formation of gels. Interactions between smectites and organic compounds are complex so that smectites have both absorption and catalytic properties.

The main industrial applications of bentonite and Fuller´s earth are foundry moulds, to provide viscosity and transportation capability to drilling muds, in filtering and clearing of wines and juices, as an industrial absorbent and as a filler, and in water treatment. Other uses include chemical and pharmaceutical industries, ceramics (as a plasti-fier) and as a sealant of stored radioactive waste. By a simple treatment with soda ash, calcitic bentonites can be activated and trans-formed into sodic bentonites with a higher ionic exchange capacity. Today bentonites are starting to displace sepiolite as cat litter due to its easier handling and removability. Bentonite production in Spain has never been high (around 0.15 Mtpa) and there are only four producing companies. The traditional bentonite source in Spain has been Nijar (Almería) which is of hydrothermal origin (40–70°C) by alteration of andesites and rhyolites of the Cabo de Gata Neogene volcanic complex (Fig. 19.7). These bentonites have a great chemical and mineralogical variability which has been studied by several authors (Doval 1992). In the Tertiary rocks of the Madrid basin there are sedimentary calcic bentonites, originating by diagenesis of newly formed smectites in a continental mud-flat basin (Doval 1992). They are currently extracted in Yuncos (Toledo), Pinto and Valdemoro (Madrid).

Dimension stone

With a production of over 7 Mtpa, Spain is currently the third global stone producer after China and Italy. National natural stone production in 2000 reached 7.6 Mt, out of which 25% was granite, 61% marble and 10% roofing slate (Table 19.2) with other rocks such as sandstone, phonolite and ignimbrite also being quarried.

Granite. Granite quarries are mainly located in low-pressure, late Variscan plutons. Most commercial varieties belong to the 'shallow undeformed' type (Bellido *et al.* 1987), these being frequently pinkish, cream or brownish granites. The colour of granite (usually due to the feldspar colour and biotite content) is thought to be related to some extent to the emplacement depth and the chemistry of the body (i.e. white and pinkish = shallow, grey or dark = deep). In addition to the Variscan basement a few other granites are extracted in the axial zone of the Pyrenees (including a wide, coarse-grained pegmatite dyke of high commercial value) and in the Catalonian Coastal Ranges. A review of the general geology and technological properties of Spanish dimension stone granites is described in Lombardero & Quereda (1992). Mineralogical composition, grain size, the presence of magmatic flow and/or deformation fabrics, primary (miarolitic) and secondary (episienitization) porosity and alteration, are the main geological features that control the quality of a plutonic body for orna-mental purposes. Dark greenish dolerites are also included in the commercial term 'granite'.

Most quarries are in Galicia (Fig. 19.7), where about a dozen varieties (included the globally exported variety known as Rosa Porriño) are intensively exploited (more than 100 000 m^3/year). Other big quarries (more than 10 000 m^3/year each) are located in Madrid (Blanco Cristal, Blanco Aurora, Crema Champán), and a few coloured varieties in Extremadura, where there are areas with groups of small quarries. Isolated quarries can be found in western Andalucia (Negro Santa Olalla, a tonalite), Salamanca, Zamora, Avila, Toledo and Lérida (the Azul Arán pegmatite).

Marble. Marble *sensi stricto* crops out mainly in the Betic Cordillera interbedded with micaschists, calcschists and recrystallized limestones and dolomites (Fig. 19.7). Marble beds are tightly folded and very irregular, with variations in thickness, number of beds, and lateral continuity. Nevertheless, they are exploited in the Sierra de los Filabres area (Almería) where the famous Blanco Macael and other white, grey, yellow and cream varieties are mined in many big quarries. The Cambrian carbonates of the Iberian Massif include potentially valuable marbles in many places, but there are as yet only a few small quarries in them (Alconera in Badajoz, Aroche in Huelva, Guijuelo in Salamanca).

The most important ornamental stone potential in Spain is provided by limestone which, although not marble *sensu stricto* is commonly marketed under that name. Jurassic limestones are intensively quarried in southern Spain, white-cream limestones are mined in Granada, Sevilla and Córdoba, red limestones are quarried in Murcia, Alicante, Granada and Málaga, and brecciated dolostones (the valuable brown varieties Beige Serpiente and Marrón Imperial) are exploited in Valencia. Other limestones noted for their value as ornamental stone include Tertiary (Crema Marfil) pink, brown and cream Lower Jurassic and Upper Cretaceous limestones in Valencia and Tarragona, and black, grey, cream and red limestones in the Cretaceous Basque-Cantabrian basin (including the very famous Negro Marquina). The horizontal Pliocene lacustrine limestone of the Duero and Tagus basins, although only a few metres thick, is also quarried in Valladolid, Burgos and Madrid, thanks to its minimal overburden (thus making exploitation easy), polishing quality and, sometimes, a travertine texture (prescribed by architects for many indoor uses; Lombardero & Quereda 1992).

Slate. Spain produces and controls about 80% of the international market of the world's roofing slate. The economic value of this activity is especially important in SE Galicia (Valdeorras, Orense) and western León (La Cabrera), where a few big companies plus many small companies produce more than 0.7 Mtpa of high-quality slate tiles for roofs (Fig. 19.7). The best roofing slate is quarried from massive metapelitic formations of lower greenschist facies metamorphic grade (Roberts *et al.* 1990; García-Guinea *et al.* 1997). Many geological features affect the quality of slate for roofing. Slight differences in the grain size and mineralogical composition result in noticeable changes in the splitting ability of the rock. Pyrite and other Fe-sulphide contents, slaty cleavage development, minor structures such as crenulation cleavage or kink-bands, folding style, metamorphic grade and many other factors control the exploitability of slate bodies.

Nearly all roofing slate quarries in Spain are located in three metapelitic Ordovician formations originally deposited on the Gondwanan marine shelf (like similar examples elsewhere in western Europe, North America and Argentina; Lombardero & Reile 1997). The Pizarras de Luarca Fm in the Central Iberian Zone is characterized by massive black slates and is intensively quarried in the Truchas (Orense and León) and Alcañices (Zamora) synclines. The Casayo and Rozadais formations of the Truchas syncline (Barros 1989) are also intensively exploited, but for paler light to dark grey slates that crop out in metre- to decimetre-thick beds. Thinner beds are only exploitable in fold hinges, where they have been thickened by similar folding. The black Tremadocian slate of the Sierra del Caurel (Lugo) and the green Cambrian slate of the Pizarras de Cándana Fm, both in the Western Asturian-Leonese Zone, are also exploited. Isolated quarries are located in the Devonian slates of La Codosera (Badajoz) syncline, in the pre-Ordovician silty slate of Bernardos (Segovia), and in the Carboniferous of Sotiel (Huelva).

Dolomite

Spanish dolomite production for all uses is estimated at 1.5 Mtpa. The major dolomite producer in Spain (and the only refractory grade dolomite producer) currently produces 285 000 tpa from three major dolomite quarries, located in Bueras (in a Cretaceous formation of the Basque-Cantabrian basin) and Peñas Negras (Castellón). The same group also produces 75 000 tpa of dolomitic lime at their facilities in Santullán (Cantabria). Total refractory grade dolomite production is 25 000 tpa of sinter dolomite produced at their plant in Santoña (Cantabria). The other main company produces 0.5 Mtpa from two marble quarries located in the Triassic Unidad Blanca of the Internal Betic Zone at Coin (Málaga) and has three processing plants. Most of the production is sold for chemical uses (28%), filler (39%) and aggregates (8%). Forty-five per cent of the production is exported through their own loading facilities at the port of Málaga. Spain holds huge resources of high grade Jurassic and Cretaceous dolomites that have yet to be investigated. Many of these deposits are large, very fractured or brecciated secondary dolomite massifs, allowing easy exploitation with low explosive consumption.

Feldspar

Spanish feldspar production has experienced a sharp increase in the last few years, especially in 1998 and 1999, and although official statistics indicate a lessening of production, total production is thought to currently be around 0.65 Mtpa of feldspars, feldspathic sands and kaolin plant by-product. Most of this production is consumed by the ceramic (sanitary- and tableware, tiles, glazes, porcelain, refractories) and glass industries, with only 2% going to other industries (paints, plastics, rubber). Economic Spanish reserves of feldspars are estimated at around 40 Mt (2.5 Mt sodium feldspar and 37.5 Mt potassium feldspar). Total national resources of all feldspar types could reach as much as 600 Mt (Regueiro *et al.* 2000*b*). The biggest Spanish producer (about 0.2 Mtpa) is a subsidiary of a huge glass company. There are also three medium-size companies (individual production about 0.1 Mtpa) and a few small producers, one of them extracting pegmatitic feldspar with Li-mica (lepidolite; La Fregeneda, Salamanca), marketed under the name of 'lithium feldspar'. There is in fact great confusion regarding the commercial term 'feldspar' as it includes both high purity feldspars and feldspathoids, as well as milled feldspathic rocks containing significant amounts of quartz, mica and other impurities. The main feldspar deposits are veins or dykes (aplites–pegmatites), plutonic rocks (nephelinic syenites, alaskites, granitoids), volcanic rocks (rhyolites, phonolites), sedimentary rocks (feldspathic sands, arkoses) and soils (resulting from weathering of granites with megacrysts of feldspar). An excellent synthesis of Spanish feldspar deposits has been published by Sánchez-Muñoz & García-Guinea (1992).

Pegmatitic feldspars are exploited in Muras (Lugo), El Vellón (Madrid) and Llansá (Gerona) and have been exploited in many places in the past, particularly in the Iberian Massif (Galicia, Zamora, Salamanca, Ávila, Segovia, Madrid, Extremadura and Córdoba). Most pegmatites are associated with felsic plutonic bodies, but some are of metamorphic segregation origin, as seen in Sierra Albarrana (Azor 1997) and Fuenteobejuna (both in Córdoba province). There are a few other occurrences that were previously exploited, such as the episyenites of Valderrodrigo (Salamanca). In El Vellón (Madrid) there is an opencast operation extracting a sodium-feldspar-rich aplite. Albitized and kaolinized granites are mined for feldspar at Cazalla de la Sierra (Sevilla). There are also operations exploiting albitites (tabular bodies of former volcanic tuffs interbedded within the Candana Fm of the Cambrian succession) in Barreiros (Lugo), and there is a similar but as yet unexploited deposit near Cazalla de la Sierra (Sevilla). Most of the industrial feldspar in Spain, however, comes from feldspathic sands of fluvial origin but reworked later by the wind, forming continental dunes now located in Segovia where more than 100 Mt have been evaluated (Sánchez-Muñoz & García-Guinea 1992). The most important feldspathic sand resource to be developed in the future is in the Tertiary arkoses covering large areas on either side of the Central Range: exploration works are currently being carried out in Salamanca and Toledo (Fig. 19.7). Finally, concentrations of loose feldspar crystals in soils over megacrystalline granites have been investigated in Puerto de Villatoro (Ávila) and are currently exploited in Ceclavín (Cáceres).

Fluorspar

National fluorspar production was very important in the 1960s and 1970s, but currently it has decreased to about 0.1 Mt and is obtained from only two deposits located in Asturias (Villabona-Arlós and Caravias-Berbes districts; Fig. 19.7). Spain nevertheless remains Europe's second biggest producer after France (Regueiro et al. 2000a), with reserves estimated at 5 Mt. The Asturian deposits belong to the Cantabrian Zone of the Iberian Massif, have been very well studied (García-Iglesias & Loredo 1992), and are found associated with the Permo-Triassic unconformity as well as within the Palaeozoic basement. Their morphology is vein-like (intruded through fractures in the basement) or as coatings associated with the unconformity. The most frequent paragenesis is fluorite–calcite–silica–(pyrite–barite). Away from Asturias, other (currently unexploited) deposits occur in four quite different areas: (i) the Ossa-Morena Zone of the SW Iberian Massif has vein-type fluorspar associated with Variscan granitic plutonism (Cerro Muriano and El Cabril, Córdoba); (ii) the Coastal Catalonian Ranges have vein mineralizations with a paragenesis of fluorite–calcite–silica–(barite–sulphides); (iii) the axial zone of the Pyrenees has sedax type mineralizations in Palaeozoic carbonate rocks; (iv) the Betic Cordillera has stratabound deposits in Triassic dolomites (Sierra de Baza, Sierra de Gador, Sierra de Lújar; Urbano 1998).

Gypsum

Spain is one of the most important world producers of gypsum, probably third after the USA and China. In Europe, Spain is a clear leader in production and consumption and is the biggest exporter on the continent. National gypsum production is estimated to be 8.5 Mtpa and resources are estimated in the range of 60 Mm^3. The Spanish gypsum sector is in the hands of European multinational groups, the largest of which produces more than 3 Mtpa and controls 65% of the production through its subsidiaries. This group mines the Sorbas deposit in Almería (Fig. 19.7), the biggest European gypsum operation, from where more than 1.5 Mt of gypsum are exported through the Garrucha and Carboneras ports. A second industrial group with interests in many branches of prefabricated construction products (such as fibre cement, gypsum, plaster) has many production units in Spain and holds 25% of national production. There are also a few small producers, usually family-owned quarries and factories, mainly for local or regional consumption. Most of the total production is fired to obtain plaster and other construction products, but gypsum and anhydrite are also used in agriculture, the chemical industry, coal mining and pharmacy.

Formations bearing gypsum and anhydrite are found scattered all over the outcrop of Mesozoic and Cenozoic rocks in Spain, and many of them are exploited. The main marine formations occur in the Upper Triassic–Lower Jurassic successions in central Spain, the Jurassic and Cretaceous rocks of the Iberian Ranges, the Eocene foreland basin of the Pyrenees, and the Neogene evaporites of the Mediterranean coast (from Catalonia to the Betic Cordillera). Continental gypsum formations are located in the Tertiary successions of the Ebro, Duero and Tagus basins. Small Neogene intramontane basins in the Iberian and Catalonian Coastal ranges and Betic Cordillera also hold important deposits (Regueiro et al. 1997).

Iron oxides (pigments)

Spain is a traditional producer of iron oxide pigments, a basic component in the manufacture of ceramic glazes, and red oxide from hematite and other iron minerals is known worldwide as 'Spanish red'. Production zones are located in Zaragoza, Vizcaya, Jaén, Granada and Málaga. Total production is in the order of 80 000 tpa with a production value exworks of around 21 million Euros. The main Spanish producer operates an underground mine at Tierga (Zaragoza) and a processing plant in Sopuerta (Vizcaya). The main ore is located between the Jalón beds and the Ribota dolomite (Cambrian of the Aragonese branch of the Iberian Ranges). Other producers also operate exploitations in Zaragoza and in Granada, with four micronizing plants operating in Jaén, Málaga and Almería.

Magnesite

The estimated Spanish raw magnesite production is 0.6 Mtpa (20 million Euros) from which 140 000 t of caustic magnesite and 62 000 t of sinter (dead burned) magnesite are produced. The main uses are as refractory bricks and concrete, animal food, and in the chemical industry and agriculture. The major Spanish deposits are located in Navarra and Lugo (Fig. 19.7). The Eugui (Navarra) deposit is mined in an open pit and consists of interstratified coarse-grained spar magnesite and dolomite lenses and slate layers, lying concordantly with Carboniferous (Namurian) slates and dolomites of the Quinto Real Palaeozoic massif (west Pyrenees axial zone). Many hypotheses have been proposed to explain the magnesite genesis, but a diagenetic replacement of dolomite by magnesite seems to be generally accepted. Resources have been estimated to be 380 Mt (Olmedo et al. 1992), with the company currently producing 0.4 Mtpa in a plant located at Zubiri (Navarra) for use in the refractory industry.

The Lugo deposit is near the village of Rubián and is operated by a national company belonging to a coal mining group. It produces 200 000 tpa of magnesite which is treated at their own plant at Monte Castelo to produce caustic magnesite, used in agriculture and refractories. The deposit represents a lateral change of facies of the Cándana limestone (Cambrian of the West Asturian-Leonese Zone) which has two main magnesite layers with a thickness of about 2 m (upper) and 14 m (lower), with the latter level being currently exploited by underground mining. The origin of these magnesites is primary (sedimentary in a lagoonal marine environment), recrystallized during later diagenesis and low-grade metamorphism (Doval et al. 1977).

There are a few other magnesite deposits scattered across Spain, such as the now-abandoned one in Puerto de la Cruz Verde (Madrid), formed in a metamorphic setting, and evaporitic deposits in the Tertiary succession of the Ebro and Tagus basins. There is also an arid-climate magnesite enrichment in Quaternary soils developed over dolomitic rocks in Sierra de Gador (Almería).

Potassium salts

Since 1997 potassium salts have been exploited only in Barcelona (Catalonian potash basin), from two deposits in Suria and Sallent (Fig. 19.7). The former is operated by a multinational company that has enlarged production of silvinite and carnalite in its Suria underground mine to reach 656 000 t (K_2O content). The potash deposit in Sallent produces potassium chloride and sodium chloride (as a by-product) from a treatment plant and is used for fertilizers. The Catalonian potash deposits lie within the Ebro basin, a former wide marine basin covering southern France and NE Spain during Eocene times. The uplift of the Pyrenees divided the basin into the Aquitanian and Ebro gulfs, the latter being separated from the open sea by a reef barrier. Two productive potash layers are located in the Saline Member (Mb) of the Cardona Fm (upper Eocene), with thicknesses of 3.5 m (lower) and 0.8 m (upper). Anhydrite, very pure halite layers and marls are also common lithologies of this formation. The internal structure of the basin is quite complicated, as it has been subjected to multiple folding and faulting events since Eocene times, with diapiric uplift of salt bodies still being active. The reserves in this basin have been estimated as 300 Mt of equivalent K_2O (Rubio 1997), with a good summary of the Catalonian potash mineralization having been provided by Ramírez (1991).

Silvinite deposits are also known from Navarra where they are located in a similar sedimentary environment to those of the Ebro Basin. Here the stratigraphic column also includes anhydrite, gypsum, halite, marls and sandstones. The Subiza mine, operated until 1997, is in a wide anticline, with a main sylvinite layer 2 to 2.5 m thick. Resources in this basin reach 8.5 Mt of equivalent K_2O (Rubio 1997).

Salt

Salt is exploited in Spain in coastal and inland salines (Fig. 19.7), by underground mining and in wells by underground dissolution. Total production is about 3.6 Mtpa, this comprising 1.9 Mt of halite mines, 0.3 Mt of a potassium salts mining by-product, 1.3 Mt of coastal salines and 0.1 Mt of inland salines (Rubio 1997). The main uses of sodium chloride are in the chemical industry (soda production by Solvay process), ice control on roads, water treatment, and as a food preservative. Rock salt deposits are located in the Triassic succession (Keuper) which is widely exposed in many areas of the north (Cantabria, Basque provinces) and east (Pyrenees, Iberian Ranges, Betic Cordillera) of Spain. The plastic behaviour of the Upper Triassic evaporites produced diapirs and decollement surfaces during the Alpine orogeny, so that many halite deposits are located in strongly tectonized areas. Many inland salines evaporate salts from springs, fountains or wells on Triassic evaporite terrains. Cenozoic salt deposits are located in saline units of the Ebro, Duero and Tagus basins. The underground halite mine in Remolinos (Zaragoza), halite and

silvinite in Suria and Sallent (Barcelona) and Subiza (Navarra) are all located in the Ebro basin. There are also some saline formations in intramontane and marginal basins of the Betic Cordillera in southern Spain.

Silica minerals and rocks

The industrial usage of silica raw materials is very wide and includes abrasives, foundry moulds, refractory bricks and mortars, sand filters, glass fibre, ceramics, chemical industry (sodium silicate), cement, glass and in construction, mainly as aggregates for mortar and concrete. The Spanish production of siliceous materials is currently around 7.5 Mtpa (excluding diatomite and aggregates). Generally speaking the industry demands raw materials with a high silica content (>80%), although for some uses (glass, ferrosilica, refractories, etc.) higher purities (over 95%) are demanded, even up to 99.9% for special uses.

Silica sands and quartzites. Silica sand production has experienced a considerable increase in the last few years, almost doubling from 1996 to 1999, and is currently estimated as being around 6 Mt from more than 20 operations all over Spain. Most of the production comes from specific silica sand mining (70%), but it also results as a by-product of feldspathic sand processing in Segovia, kaolin extraction in Valencia and sandstone and quartzite quarrying. Seventy per cent of Spanish production capacity is in the hands of only two companies. Total silica sand resources in Spain could reach 500 Mt, taking into account many deposits yet to be fully investigated. Metallurgy (foundry sands) and glass industries are the main silica sand consumers, followed by abrasives, refractory bricks, cement and sand filters. Each use has its own specifications, sometimes related to the chemistry of the raw material, sometimes to the grain size and shape, or with the impurity content.

Many sandy formations of the Mesozoic and Cenozoic successions of Spain are quarried as silica raw material. Lower Cretaceous (Weald and Utrillas facies) sandstones crop out widely in the Cantabrian, Basque-Cantabrian and Iberian Ranges areas, and are intensively exploited. Pliocene, Miocene and Oligocene sands are also quarried in dozens of localities in Cantabria, Tarragona, Castellón, Valencia, Albacete and Ciudad Real, and Tertiary and Quaternary sands are quarried in Galicia and Andalucía. One of the biggest Spanish operations is located in the Quaternary sediments of the upper Ebro basin, in Arija (Burgos). The deposit has resulted from the erosion of Cretaceous (Utrillas facies) silica sand washed and purified by rain and fluvial waters.

Quartzite is also exploited in Spain, mainly for glass and aggregates. There are dozens of quarries in different Palaeozoic formations, many of them in the Armorican quartzite (Lower Ordovician of the Iberian Massif). Massive pure quartzite is hard and expensive to exploit, so the quarries are often in brecciated or very fractured rock massifs, as in Boñar (León), Caldas de Besaya (Cantabria) and Atienza (Guadalajara).

Quartz. Quartz is mainly mined in Galicia (about 1.5 Mtpa) for refractories, ferrosilica and metallic-Si production and high-value uses such as electronic and optical quartz. Most exploitations are in La Coruña and Lugo (Fig. 19.7), although in Gerona quartz is a by-product of feldspar mining (Marchán 1998). Spanish quartz appears in the form of veins of pegmatitic, metamorphic segregation and hydrothermal origins. The most important dykes are located along the northern coast of La Coruña (O Barquero dyke, 30–50 m thick and 10 km long; Crabifosse *et al.* 1989; and Pico Sacro dyke, 100–300 m thick and 3 km long; Muñoz *et al.* 1998). These are exploited mainly for ferroalloys, and more than half the production is exported. Other main uses are in the production of metallic silicon and the manufacture of silicon carbide. Small tonnages of quartz crystals with high value are used in the electronic industry as oscillators or piezoelectric quartz, as nucleii to manufacture synthetic quartz and in optics. Other important pegmatitic deposits occur in La Fregeneda (Salamanca), Sierra Albarrana (Córdoba) and Llansá (Gerona).

Tripoli. Spanish production of tripoli is very small, almost 5000 tpa, and comes from several small operations located near Castro-Urdiales (Cantabria), handled by a local company (Fig. 19.7), and in abrasives and polishing. Grain size, colour and oil absorption capacity are the main properties that control the commercial grade of tripoli. The deposits are of karstic origin, the opaline sand being mixed with clays infilling cavities of an exhumed palaeokarst developed on limestones containing small diagenetic grains of chalcedony and microcrystalline quartz. This limestone belongs to the Supraurgonian complex (Cretaceous) of the Basque-Cantabrian basin.

Diatomite. Only three companies (one multinational and two domestic) exploit diatomite in Spain, with an average production of about 55 000 tpa. Its main uses are as fillers and filters, pozzolanic additives for cements, and in foodstock. Low-grade diatomite is also used in the manufacture of insulating bricks at a plant in Andujar (Jaén). Purity, porosity, specific surface, grain size and state (breakage) of the diatom shells are the main factors controlling the commercial grade of diatomite.

The main Spanish diatomite deposits are located in shallow marine, lacustrine, and continental formations of Miocene and Pliocene age. Marine diatomites of the Guadalquivir basin are locally named 'moronitas' (from Morón, Cádiz) or 'albarizas' (from latin 'alba', white). The marly diatomite-bearing formation is up to 200 m thick and of Langhian–Tortonian age. The deposits are irregular, with their shape and quality controlled by upwelling marine palaeofloods surging from the Atlantic into the basin. They are quarried on a small scale in Martos (Jaén) and used for insulating bricks. The main producing areas in Spain are located in the continental diatomitic marls of the Neogene basins of Albacete, Murcia and Almería (Regueiro *et al.* 1993). Some individual layers yield over 90% opaline silica. Small deposits in the Miocene of Madrid and La Cerdaña (Lérida) basins have also been reported.

Sodium sulphate

Sodium sulphate minerals have been exploited in Spain since the nineteenth century (Ordoñez & García del Cura 1992). Very old, abandoned mines exist in gypsum–glauberite–halite–thenardite deposits south of Madrid (Fig. 19.7). Today Spain is the sole European and top world producer of sodium sulphate. In 1999 production reached 0.88 Mt, with exploitations in Madrid, Toledo and Burgos. The exploitations in Madrid and Burgos extract glauberite by opencast dissolution, whereas in Toledo the mineral thenardite is extracted by underground chamber and pillar mining. Estimated reserves reach 730 Mt (Regueiro *et al.* 2000*a*). The industrial grade of sodium sulphate is controlled by the purity and mineral type. Different grades are used for paper, glass, detergents, chemical industry and other minor uses.

The Tertiary Madrid basin hosts important deposits of glauberite and thenardite. Another area of Spain yielding major deposits is that linking the Duero and Ebro Tertiary basins (Cerezo de Rio Tirón–Belorado; Burgos and La Rioja). This latter deposit was formed in a continental sub-basin, isolated by the detrital sediments of a fan-delta, generating saline deposits of anhydrite and glauberite with dolomite and clay interbeds. In Rio Tirón there are glauberite beds almost 20 m thick with a grade over 90%. In the Madrid basin the mineralization is located in the lower saline unit (upper Oligocene–lower Miocene) and includes glauberite, thenardite, magnesite, gypsum, halite and polyhalite. A good summary of the geology of the sodium sulphate deposits in this zone can be found in Ordoñez & García del Cura (1992). These authors propose a genetic model with two salt precipitation stages, one with a paragenesis of anhydrite, glauberite, thenardite and halite (polyhalite) deposited in a great permanent salt lake, and another with thenardite, glauberite and halite (anhydrite, magnesite) deposited in a minor non-permanent lake, with frequent salinity changes due to changes in the inflow of fresh water during the rainy season. There are occurrences of sodium sulphate in many other Tertiary basins, such as Alcanadre (La Rioja), Remolinos and Calatayud (Zaragoza).

Talc

Spanish talc deposits are located in León and Málaga (Fig. 19.7) and currently just one multinational company controls all Spanish production. Quarries are located at Puebla de Lillo (León) and Málaga, the materials are treated in plants located in Boñar (León) and Mijas (Málaga) and total production capacity is 120 000 tpa. The talc has multiple industrial applications, such as in ceramics, refractories, sealants, paper, industrial absorbents, olive oil production, polishing, pharmaceutical and cosmetics.

Spanish talc deposits belong to two main genetic groups: (i) related to metasomatic or hydrothermal alteration of dolomites; and (ii) associated with the alteration of magnesium-rich ultramafic rocks. The main deposits are located in Puebla de Lillo (León), at a high altitude (1700–2000 m), and are exploited opencast only in the summer, as weather conditions are poor in winter. Local geology is complicated by Variscan thrusts and late-Variscan and Alpine faults juxtaposing quartzites and other siliceous rocks against Palaeozoic dolomites. Talc deposits were once also exploited in the eastern Pyrenees, occurring in Palaeozoic rocks of the axial zone of Gerona (Massanet de Cabrenys and La Bajol) and being genetically similar to the talc of the French Pyrenees (Rodas & Luque 1992). Their origin is related to hydrothermal processes acting on micaschists and marbles along tectonic contacts and in the presence of a nearby granodiorite. In Somontín (Almería) talc mineralization appears also to be associated with hydrothermal processes acting on dolomitic marbles in the schists of the Alpujarrides and on ultrabasic rocks in the Nevado-Filábrides. The main examples of talc mineralization related to ultrabasic rocks, however, are located in the serpentinized peridotites of the Ronda Massif (Málaga), where they are currently exploited mainly in the production of olive oil. Serpentinization occurred prior to the talc-forming process which involved the circulation of SiO_2- and CO_2-rich hydrothermal fluids through a great fracture in serpentine, generating a 4–5 m thick band of very pure talc which passes laterally to mixed talc and unaltered serpentine (Rodas & Luque 1992).

Decorative rocks and minerals

The Lower Cretaceous Weald facies rocks of Navajún (Soria) host one of the most amazing deposits of very perfect cubic crystals of pyrite. Samples of them are found in public and private mineral collections all over the world. One opencast mine exploits these perfect crystals, some of which reach sizes of up to 20 cm. The pink quartz dyke of Oliva de Plasencia (Cáceres) is similarly exploited for collections and decorative purposes. Another decorative material formerly exploited in Asturias is jet, used in the traditional handmade silver jewellery of Santiago de Compostela (La Coruña). The origin of jet is related to the transportation of Carboniferous coal fragments and their redeposition in Cretaceous sediments (Muñoz *et al.* 1986).

Dunite

There is currently only one dunite exploitation in Spain, located in Landoi (northern coast of La Coruña), and producing about 0.7 Mtpa of steel-grade dunite. The quarry is within the basic–ultrabasic complex of Cabo Ortegal in the Galicia–Tras os Montes Zone of the Iberian Massif.

Garnet

There have been at least two attempts to exploit garnet in Spain. Between 1995 and 1997 secondary concentrations of garnet (almandine) in soils of the El Hoyazo volcano (Nijar, Almería) were exploited. The original garnet occurs as xenocrysts in dacitic rocks of the Cabo de Gata volcanic complex (Benito *et al.* 1998; Marchán 1998; Lunar *et al.* 1999; Muñoz Espadas *et al.* 2000). Today (2001) another company is developing a grossular deposit in Badenas (Teruel) and plans to install a production plant to produce 10 000 tpa of dense aggregate and abrasive-grade garnet. The installation, if successful, will be the only garnet producer in southern Europe.

Graphite

The existence of graphite deposits in Spain has been known since ancient times. William Bowles (an Irish geologist, naturalist and engineer serving the Spanish Crown) in the eighteenth century mentions the graphite occurrences of Real de Monasterio (Seville) and Serranía de Ronda (Málaga; Bowles 1775). There was sporadic mining activity in Spain from at least 1700 to 1963, when the Guadamur mines in Toledo closed. The geological setting of the Spanish graphites is very varied. In many cases graphite is related to slates or marbles with carbonaceous material affected by regional or contact metamorphism (Sierra de Aracena in Huelva, Axial Pyrenees in Huesca, Guadamur and La Puebla de Montalban in Toledo, La Bastida in Salamanca). In other occurrences the origin of graphite is tectonic, associated with faults in a compressive regime (Villarbacú in Lugo, El Muyo in Segovia). Finally, some occurrences are associated with dykes and veins in the ultrabasic peridotites of Ronda (Málaga; Luque & Rodas 1992; Lasheras *et al.* 1995).

Peat and leonardite

Both substances are currently exploited in Spain for agricultural uses (soil conditioner and fulvic and humic acid production). Peat deposits currently undergoing exploitation include both mountain (Burgos, Cantabria, Lugo, Granada) and coastal (Castellón) types. Leonardite is exploited in the oxidized near-surface zones of the Tertiary lignite layers of Ariño and Calanda (Teruel).

Vermiculite

Vermiculite was exploited in Spain only from 1951 to 1959 and then in very small amounts. Known deposits are associated with mafic and ultramafic rocks (Sierra de Aracena in Huelva and the ultrabasic massif in Ronda), gabbro-dioritic plutons generating a skarn in carbonate rocks (Santa Olalla Massif in Huelva), and even granitic intrusions. In all cases the genesis points to a supergenic alteration (either by low-temperature hydrothermal fluids or by infiltrating meteoric waters) of preexisting phlogopite or biotite micas (Luque *et al.* 1992).

Wollastonite

There are many occurrences of wollastonite in the Iberian Massif, most of them related to siliceous limestones affected by contact metamorphism in a low-pressure tectonic setting. Notable examples are those in Colmenar Viejo (Madrid), Mérida (Badajoz), Aroche (Huelva) and several in Salamanca of which only one (in Aldea del Obispo) is currently under exploitation, but even this has a very low production. The deposit has developed in the contact aureole of a Variscan granite intruded into Vendian low-grade metasediments, which includes a sandy marble limestone (Lombardero 1998*b*). The intended use for this short-fibre wollastonite is in the ceramic industry of Castellón. The Aroche deposit was investigated in 1988, estimated at 1.5 Mt of long-fibre wollastonite, and is located in Cambrian marbles currently exploited as dimension stone.

Non-renewable energy resources: oil and gas (WMO, JM)

Since the first exploratory well was drilled in Spain in 1894 until the beginning of what can be viewed as the first stage of modern hydrocarbon exploration around 1957, only some 45 shallow wells (<1000 m deep) were drilled (Lanaja 1987). Most of them were located close to surficial oil shows or in structures identified by surface geology aided by generally poor analogue seismic data. None of these wells resulted in a discovery. From the late 1950s until 1968, when the first offshore well was drilled, 93 exploratory wells were drilled and the first two commercial findings in Spain discovered: the Castillo gas field in 1960, and Ayoluengo oil and gas field in 1964, both in the Cantabrian basin.

Excluding appraisals, development wells and stratigraphic bores, by the end of 2000 a total of 464 wells (305 onshore and 159 offshore) had been drilled throughout the different Spanish basins. These efforts resulted in the discovery of one oil field and 21 gas fields onshore while ten oil fields and four gas fields were tapped in Spanish waters. In addition to these commercial findings, some 15 other oil and gas discoveries also deserve a mention, although in most cases their resources could not be commercially developed at the time because the required technologies were not yet available.

Total figures for aggregated volumes of recoverable hydrocarbons discovered in Spain are 300 million barrels of oil and 0.6 trillion (10^{12}) cubic feet of gas, the vast majority of which has already been produced so that the fields are consequently depleted and abandoned (Table 19.3). Production levels saw a maximum of 65 000 barrels of oil equivalent per day (BOEPD) in 1986, having declined since then, reaching 12 750 BOEPD in December 2000. The distribution of the 464 wells (Fig. 19.8A) is geographically uneven, most of them being concentrated on those areas with earlier discoveries.

As of today, five sedimentary basins are known to host well developed petroleum systems, although perhaps surprisingly only three of them (Gulf of Valencia, Guadalquivir–Gulf of Cádiz and Cantabrian) have been investigated to what can be called a mature level of exploration and even these have not been fully explored. In spite of this, since 1987 only

Table 19.3. Recoverable hydrocarbons discovered in Spain

Year	No. exploratory wells	Field/discovery	Hydrocarbon type	Reserves oil: Bbls × 10⁶ gas: Bcf	Reservoir age	Source reservoir age	Basin name
1961	15	Castillo	Dry gas	1.3	Late Cretaceous	Cretaceous ?	Onshore Cantábrica
1964	121	Ayoluengo	Oil & gas	17.2	U. Jur. & Cret.	Toarcian	Onshore Cantábrica
1970	202	Amposta	Oil	55.6	Barremian	Burdigalian	Gulf of Valencia
1973	224	G. Cádiz B1	Biogenic gas	?	Oligocene	Tortonian	Gulf of Cádiz
1965	140	Riudaura	Wet gas	?	Lutetian	Low. Eocene	East Pyrenees
1967	145	Hontomin	Oil	?	Dogger	Toarcian	Onshore Cantábrica
	157	Centenera	Dry gas	?	Santonian	?	Central Pyrenees
	172	Caldones	Dry gas		?	Carboniferous	Asturias
1975	255	Dorada	Oil	16.6	Jurassic	Burdigalian	Gulf of Valencia
	257	Tarragona D	Heavy oil	?	Triassic	Burdigalian	Gulf of Valencia
	262	Casablanca	Oil	149.0	Jurassic	Burdigalian	Gulf of Valencia
	264	Mc. C2	Oil & gas	?	Cret. & Eocene	Toarcian	Offshore Cantábrico
1976	264	Cajigar	Wet gas	?	Jurassic	Toarcian	Central Pyrenees
	267	Tarraco	Oil	14.5	Jurassic	Burdigalian	Gulf of Valencia
1978	285	Montanazo C	Oil	?	Jurassic	Burdigalian	Gulf of Valencia
	291	Serrablo	Dry gas	45.5	Lutetian	Ypressian	Central Pyrenees
	295	G. Cádiz B2	Biogenic gas	<3.0	Messinian	Tortonian	Gulf of Cádiz
	297	G. Cádiz B3	Biogenic gas	40.0	Messinian	Tortonian	Gulf of Cádiz
	301	G. Cádiz C2	Biogenic gas	6.6	Messinian	Tortonian	Gulf of Cádiz
	305	Delta E1	Oil	2.3	Jurassic	Burdigalian	Gulf of Valencia
1979	313	Mc-M1	Heavy oil	?	Aptian	Toarcian ?	Offshore Cantábrica
	315	G. Cádiz B4	Biogenic gas	13.4	Messinian	Tortonian	Gulf of Cádiz
	316	Atlantida 2-1	Biogenic gas	45.9	Messinian	Tortonian	Gulf of Cádiz
	321	G. Cádiz B5	Biogenic gas	12.7	Messinian	Tortonian	Gulf of Cádiz
1980	326	Barcelona A	Heavy oil	?	Barremian	Burdigalian	Gulf of Valencia
	338	Gaviota	Wet gas	305.2	U. Cretaceous	Sthephanian	Gulf of Vizcaya
1981	345	Albatross	Wet gas	14.5	U. Cretaceous	Sthephanian	Gulf of Vizcaya
1982	369	Palancares	Biogenic gas	4.1	Tortonian	Tortonian	Guadalquivir
1982	383	Neptuno 2	Wet gas	?	Messinian	?	Gulf of Cádiz
1983	392	Ciervo	Dry gas	1.2	Tortonian	Carboniferous ?	Guadalquivir
	394	Rincón	Biogenic gas	4.2	Tortonian	Tortonian	Guadalquivir
	399	Atlaántida 2-2	Biogenic gas	8.9	Messinian	Tortonian	Gulf of Cádiz
	400	G. Cádiz C4	Biogenic gas	7.2	Messinian	Tortonian	Gulf of Cádiz
1984	401	Cadialso	Wet gas	?	Jurassic	Toarcian ?	Onshore Cantábrica
	402	Salmonete	Oil	>1.5	Jurassic	Burdigalian	Gulf of Valencia
	403	Angula	Oil	>0.6	Jurassic	Burdigalian	Gulf of Valencia
	405	Rosas 2-1	Oil	?	?	?	Gulf of Valencia
	407	R. Segura	Wet gas	?	Neocamian	Neocomian	Betics
	408	G. Cádiz C5	Biogenic gas	>3.2	Messinian	Tortonian	Gulf of Cádiz
1985	415	Marismas 3	Biogenic gas	5.1	Tortonian	Tortonian	Guadalquivir
	416	Marismas 4	Biogenic gas	5.0	Tortonian	Tortonian	Guadalquivir
	417	Sevilla 1	Biogenic gas	0.6	Tortonian	Tortonian	Guadalquivir
	419	Sevilla 3	Biogenic gas	1.8	Tortonian	Tortonian	Guadalquivir
	420	S. Carlos 3-3	Oil	2.1	Barremian	Burdigalian	Gulf of Valencia
1986	421	Marismas 3	Biogenic gas	7.1	Tortonian	Tortonian	Guadalquivir
	430	Córdoba B2	Biogenic gas	0.3	Tortonian	Tortonian	Guadalquivir
	433	Córdoba C1	Biogenic gas	0.2	Tortonian	Tortonian	Guadalquivir
	443	S. Juan V1	Biogenic gas	1.1	Tortonian	Tortonian	Guadalquivir
	444	S. Juan V6	Biogenic gas	2.2	Tortonian	Tortonian	Guadalquivir
	447	S. Juan R1	Biogenic gas	2.9	Tortonian	Tortonian	Guadalquivir
1992	450	Saladillo	Biogenic gas	5.4	Tortonian	Tortonian	Guadalquivir
	451	Asperillo	Biogenic gas	2.1	Tortonian	Tortonian	Guadalquivir
1995	456	Rodaballo	Oil	2.1	Jurassic	Burdigalian	Gulf of Valencia
1996	459	Boquerón	Oil	4.2	Jurassic	Burdigalian	Gulf of Valencia
1997	459	Tarajales	Biogenic gas	2.2	Tortonian	Tortonian	Guadalquivir
1998	460	Sta. Clara	Biogenic gas	4.1	Tortonian	Tortonian	Guadalquivir
1999	463	Chipirón	Oil	9.1	L. Cret.	Burdigalian	Gulf of Valencia
2000	464	Barracuda	Oil	1.1	Jurassic	Burdigalian	Gulf of Valencia

two of these basins (Guadalquivir–Gulf of Cádiz and Gulf of Valencia) still register current exploratory activity (Alvarez 1994). Exploration of these basins is typically of the kind where four or five wells are drilled annually and result in two or three small discoveries (1 to 10 million barrels of oil equivalent) that are developed almost instantaneously. This recent concentration of exploratory activities has induced a change from the previously low rate of technical success in Spain (one in sixteen) to the current much higher rate of two in three. This greater success has been aided by improved seismic data and better knowledge of the geological models and the petroleum systems involved.

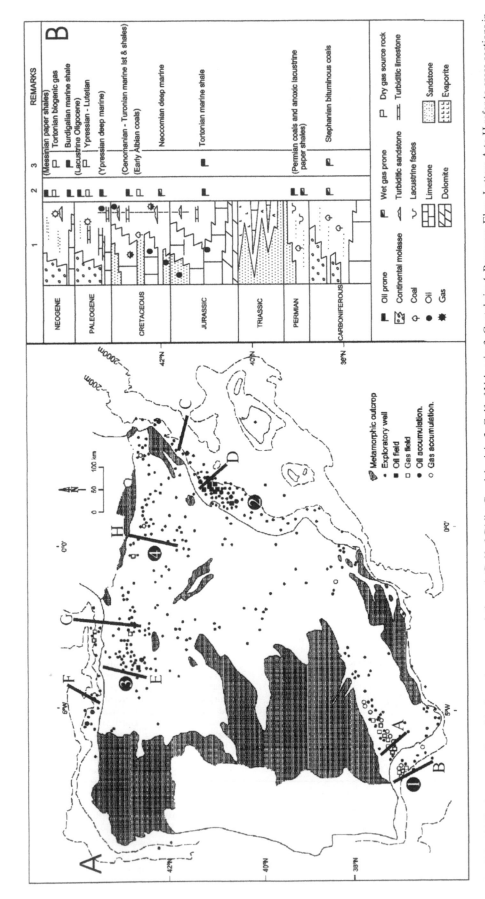

Fig. 19.8. **(A)** Exploratory wells and hydrocarbon accumulations: 1, Gulf of Cádiz–Guadalquivir; 2, Gulf of Valencia; 3, Cantabria; 4, Pyrenees–Ebro. Letters A to H refer to cross-sections in Figure 19.9. **(B)** Summary of source rocks and reservoirs.

Exploration in the rest of the basins has remained practically inactive since 1987 and, as a consequence, little or no benefits have been obtained there from the use of modern exploration technologies, such as 3D seismic data. Thus, although Spain obviously does not belong to the privileged group of major oil-producing countries, there is plenty of room for improvement in the amount and quality of exploration data in several basins within the country, and much exploratory potential remains.

A lithostratigraphic sketch (Fig. 19.8B) showing the most significant links between oil–gas potential and the palaeogeographies of the Spanish sedimentary basins has been produced by integrating subsurface and field data. Petroleum source rocks deposited during Stephanian, Domerian (Liass), Neocomian, Lutetian (middle Eocene), Burdigalian and Tortonian times have been identified as responsible for commercial accumulations discovered so far in Spain. In addition to these well known petroleum systems, there are a number of source rocks of different ages (Permian, Upper Jurassic, Upper Cretaceous, Eocene, Oligocene and Miocene) that are known to be linked to oil and gas (Permanyer et al. 1988; Martínez Rius et al. 1996; Ardevol et al. 2000) but have not yet been associated with commercial production. The following additional information is relevant to each of the stratigraphic units.

Carboniferous (Namurian to Stephanian). These rocks contain a number of levels with high generative potential for gas-condensate and reservoirs in Namurian limestones and dolomites and in Westphalian and Stephanian sandstones and fractured conglomerates. Although regionally they are still inside the gas expulsion window, Stephanian lithologies show in some areas a much higher maturation level than that in the Westphalian strata for reasons still unknown (hydrothermal activity and rapidly subsiding local troughs have been invoked). The late Variscan unconformity is overlain in many Spanish basins by Triassic basal clastics or by Albian–Cenomanian carbonates. The petroleum system comprising Carboniferous source rocks and Permo-Triassic reservoirs, so common in many other European basins, remains almost unexplored in the vast majority of the Spanish basins due to the lack of sufficiently good seismic results. It has to be stressed, however, that in some basins Carboniferous and Albian–Cenomanian strata are proven elements of a well characterized petroleum system.

Permian. These contain coaly source rocks (Rio Guadiato basin) as well as paper shales deposited in small anoxic lakes (Iberian Ranges). The reservoir potential, however, seems limited.

Triassic. Rift zones host clastic reservoirs close to the basin margins while carbonates are widely present on the external palaeogeographies of all the Spanish basins. Up to three different salt levels recognized within the Triassic represent very efficient seals.

Jurassic. The Domerian–Toarcian eustatic rise in sea level was responsible for one of the more widespread source rocks in Spain (Quesada et al. 1993). This source rock (marly Lower Jurassic) and several reservoirs within Jurassic, Cretaceous and Palaeogene strata are the main elements of this petroleum system.

Cretaceous. Neocomian and Albian source rocks contain type III kerogens but have sourced only small gas and gas-condensate accumulations in the Cantabrian, Pyrenean, Rioja and Betic basins. The Cenomanian–Turonian anoxic event did not enable the deposition of organic-rich facies over the large but shallow shelves, as they were very oxygenated and high-energy realms. Thus, the Cretaceous subcrop is evaluated as having a high reservoir potential and a fair to good generative potential for gas-condensate, especially in Lower Cretaceous depocentres located on slopes with high progradation and sedimentation rates.

Palaeogene. Palaeogene sedimentation in Spain is mostly characterized by red and evaporitic continental facies coeval with Alpine compression, uplift and erosion. Important exceptions to this occur in the Betics and Pyrenees, with the Pyrenean flysch constituting another commercial petroleum system in Spain (Cámara & Klimowitz 1985).

Neogene. In the foreland basins related to the Betic Cordillera, the marine Miocene succession contains the two youngest, and so far most prolific, petroleum systems in Spain. The first is located in the Gulf of Valencia and is based on an upper Burdigalian marine source rock (Watson 1979; Soler y José et al. 1983; Albaigés et al. 1986) that fossilized an angular unconformity associated with Alpine compression and uplift. The second was developed on the Guadalquivir–Gulf of Cádiz basin and is closely related to the generation of biogenic gas from upper Tortonian hemipelagic shales and the deposition of Tortonian–Messinian sand reservoirs (Suarez et al. 1989; Riaza & Martinez del Olmo 1996; Martinez del Olmo et al. 1996).

Review of productive basins

To help with the visualization of those basins where commercial discoveries have been made, eight conceptual traverses have been drawn projecting the actual position of several wells (Fig. 19.9). Locations of these cross-sections are shown on Figure 19.8A; they show hydrocarbon fields in the SW (Guadalquivir and Gulf of Cádiz), east (Valencia), north (Cantabria and Gulf of Vizcaya), and NE (south Pyrenees). The complex geology of these areas is not the ideal environment in which to conduct efficient exploration activities without using the most accurate geochemical and seismic data that modern technologies can provide. However, in common with many other examples worldwide, this was the situation in which more than 75% of the exploratory wells were drilled in Spain. Given the new opportunities that modern technology provides, combined with continuously growing markets, several areas in Spain (the central and eastern Pyrenees, Betic Cordillera, southern Gulf of Valencia and the deep waters of Cantabrian, Mediterranean and Alboran seas and the Gulf of Cádiz) have all become attractive targets for a renewed exploratory effort.

Rio Guadalquivir and Gulf of Cádiz basins (Fig. 19.9A, B). These basins represent the SW segment of the Betic Cordillera Miocene foreland and are characterized by the presence of a huge and chaotic blanket of olistostromic masses with thicknesses in excess of 3000 m. These masses constitute an impenetrable barrier for seismic reflection techniques, and so have forced hydrocarbon exploration to concentrate on the more external olistostrome-free zones. All the accumulations discovered in these basins have been sourced from biogenic gas originating within an upper Tortonian deep marine source rock (average total organic carbon (TOC) of only 0.5%), possibly fixed originally as gas hydrates and later mobilized and trapped within the surrounding sands.

The producing reservoirs, with porosities ranging from 20 to 25%, belong to two different turbiditic systems: (a) onshore upper Tortonian–Messinian turbidites with a high progradational rate, reduced thickness and large depositional area; (b) offshore, thicker but more restricted turbidites deposited during Messinian–early Pliocene times (Suarez et al. 1989; Riaza & Martínez del Olmo 1996). All the accumulations except one, controlled by a fault over a diapir crest, are pure stratigraphic traps or subtle closures generated by differential compaction. Modern exploration on these basins started in 1973, and since then improvements in seismic data and, more especially, our understanding of turbidite depositional systems, have maintained the success rate above 50%, with discovered accumulations ranging from 5 to 40 billion cubic feet.

Gulf of Valencia basin (Fig. 19.9C, D). This basin represents the easternmost segment of the Betic Cordillera Miocene foreland (Martínez del Olmo 1993, 1996) and, as in the Guadalquivir–Gulf of Cádiz basin, exploration has focused on the northernmost fringe where it is olistostrome-free. All the commercial discoveries originated from an upper Burdigalian–lower Langhian marine source rock

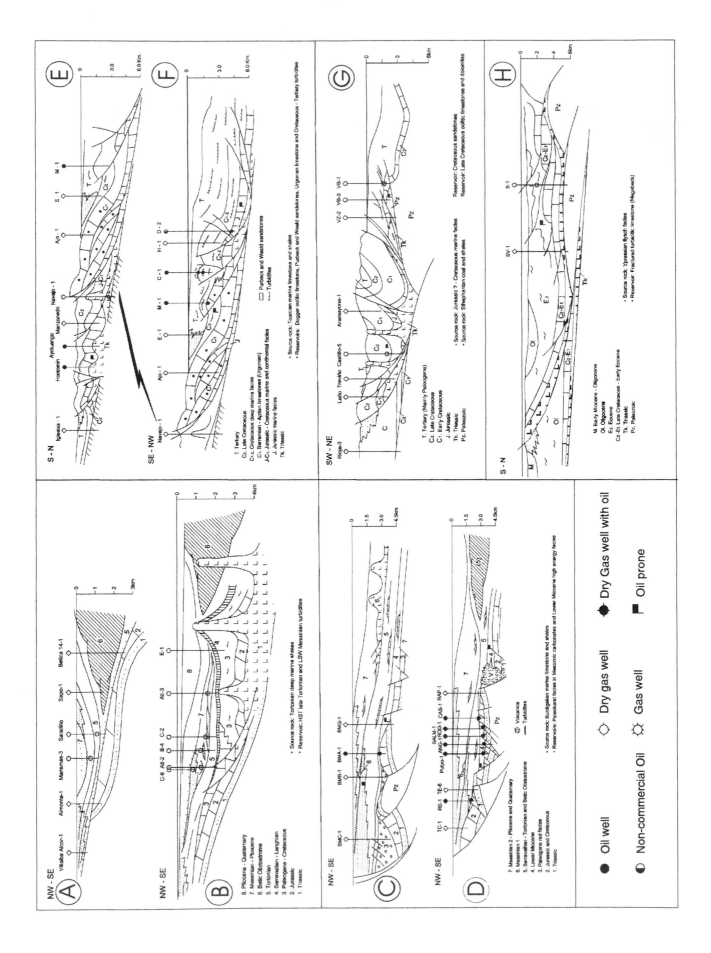

with a TOC of 3% (Watson 1979; Albaigés *et al.* 1986; Clavell 1991). Under an overburden of 3000–3300 m it reached peak maturity as recently as in late Pliocene–Pleistocene times. Short to mid-range distance migration occurred efficiently along the angular and erosive unconformity that separates Miocene sediments from its complex subcrop of Palaeozoic, Mesozoic and Palaeogene rocks. All the accumulations developed in reservoirs intimately related to this unconformity, the most significant ones being generated during Palaeogene times by subaerial karstic processes acting on Mesozoic carbonates (Watson 1979; Martínez del Olmo & Esteban 1983; Stampfli & Höcker 1989; Seemann *et al.* 1990; Zacharakis *et al.* 1996), aided by later (Messinian) hot dolomitization. The reservoirs mostly have a matrix-plus-fracture average porosity in the range of 8 to 10%, while permeability is extremely high (it is not uncommon to find wells with daily production rates in excess of 10 000 barrels of oil) due to the combined effects of dissolution plus a late tectonic fracturing. Immediately above the unconformity, Miocene high-energy transgressive facies, conglomerates and reefal carbonates also provide reservoirs, although they are restricted and less important economically. Productive traps may be described under two different scales of observation: (a) very rough palaeo-reliefs related to subaerial karstification processes taking place under a tropical climate; and (b) traps controlled by up-to-the basin faults (Martínez del Olmo 1993, 1996) that rejuvenated the unconformity mainly during early and middle Miocene times.

Exploration in the Gulf of Valencia started in 1968, with the third well drilled making the first commercial discovery. By 1976, with just 22 exploratory wells drilled, most of the hydrocarbon had been found (Amposta, Dorada, Montanazo and Tarraco oil fields). Since 1994, 3D seismic started to be used as an exploration tool and, consequently, a high rate of success has been achieved and new accumulations discovered, which, in spite of their small sizes (3 to 10 million barrels of oil), have been immediately developed. The olistostrome-free deep waters of the Gulf of Valencia have not yet seen any real exploration, this being also the case for several deep grabens in which Miocene volcanic layers are, for the moment, inhibiting collection of the necessary seismic data.

Cantabrian and Gulf of Vizcaya basins (Fig. 19.9E, F, G). The Cantabrian basin is located on the central segment of the sedimentary margin established on the northern edge of the Iberian plate since Triassic times. Three different petroleum systems have been identified on both sectors (onshore and offshore) of the basin. The first of them has a marine Domerian–Toarcian source rock with TOC values >5% (Soler y José *et al.* 1981; Quesada *et al.* 1993). The large size of its depositional area, together with the sedimentary and structural complexities of the basin (Riaza 1996), caused the current levels of maturity to range from immature to overmature. Regionally, it seems that the critical factor to enable the finding of the Lias source rock still inside the oil window is to have had an appropriate thickness of onshore Cretaceous (Quesada *et al.* 1996) and offshore Palaeogene deposits. During the two main expulsion phases (Late Cretaceous and Oligocene) migration pathways are assumed to have been mainly vertical, and to a lesser extent lateral. Vertical migration through fractures explains why there is not a unique reservoir, as several lithologies have proven oil-bearing: Middle Jurassic oolitic limestones, Upper Jurassic to Lower Cretaceous (Purbeck to Weald) high-sinuosity clastic rocks (Alvarez del Buergo & García 1996), Lower Cretaceous (Urgonian) limestones and Upper Cretaceous and Eocene sandy and calcareous turbidites in Hontomin, Ayoluengo and Mar Cantabrico M-1 and C-1 (Fig. 19.9E).

The Castillo gas field (Fig. 19.9F) is interpreted as relating to a high degree of maturation reached by the Jurassic source rock, perhaps supplemented by other coaly Cretaceous sources identified in a southern depocentre of the onshore Cantabrian basin (Serrano & Martínez del Olmo 1990). This small discovery, made in 1960, together with both the abundant gas shows recorded in all the wells drilled in this subbasin, and the absence at that time of modern

geochemical concepts and tools, explains why such a big exploratory effort was made between 1960 and 1980 in this difficult and over-mature basin.

A second petroleum system has been identified in the Gulf of Vizcaya basin, which is located in the easternmost segment of the Cantabrian sea. Its main elements are a Stephanian source rock feeding Upper Cretaceous calcareous reservoirs in which gas-condensates were produced. These accumulations share a similar north Pyrenean geological context with the marine Aquitaine foreland basin by showing: (i) northern tectonic vergences; (ii) calcareous shelf deepening southwards; (iii) Palaeogene marine molasse deposits prograding northwards. The Stephanian coals and bituminous shales still remain within the oil window (oil resistivity ranges from 0.75 to 1.0) and have probably been that way since Oligocene–Miocene times. Migration is interpreted to be vertical using the available fracture network, while lateral short-distance migration came into play in those areas where the Variscan unconformity was fossilized by Late Cretaceous sedimentation. Reservoirs lie within a complex sedimentary system involving a high-energy external shelf edge (oolitic bars, apron facies, micrites, etc.) that was affected by processes of secondary fracturing and dolomitization. The productive traps are complex folds created by north-vergent thrusts that rejuvenated the Variscan unconformity during the late Eocene Alpine compressional phases (Fig. 19.9F). In spite of the poor quality of seismic data, this being the main exploratory limitation, and the structural complexities present in such a tectonic environment, the exploration has resulted in a success rate close to 50% since it began in 1980.

Pyrenean basin (Fig. 19.9H). The Pyrenean Cordillera and its foreland (Puigdefabregas *et al.* 1986) was one of the first regions in Spain to be explored for hydrocarbons. Discoveries in the Pyrenean basins triggered a high level of activity between 1960 and 1975. Just when almost all this early exploration had ceased due to a lack of good results, the first commercial discovery was made in 1979 (Serrablo gas field) in a new and unexplored play. This small dry gas accumulation is interpreted to have been sourced from a Cuisian humic source rock with low TOC values but having an average thickness of over 240 m. The high maturity level (oil resistivity ranges from 1.0 to 1.5) found at moderate depths, suggests the erosion of much of the upper Eocene–Oligocene overburden originally deposited in frontal syntectonic synclinal basins developed in the southern Pyrenees. Although difficult to prove, this context fits well with a model involving short-distance vertical migration. The reservoir of this petroleum system comprises middle and upper Eocene calcareous turbidites (Cámara & Klimowitz 1985). These are quite continuous regionally, very compact, and affected by a patchy network of fractures, apparently controlled by the geometry of south-verging thrust anticlines, that give the reservoir a high permeability in spite of its low porosity.

Non-renewable energy resources: coal (JASS)

Coal is the only abundant non-renewable energy source in Spain, and over the last 150 years the mining of 59 Carboniferous and Cretaceous coal and lignite basins has contributed hugely to the economic development of the country. Even today (2001), coal accounts for 35.6% of primary energy production and remains the second most important energy source. Although huge reserves remain, however, many Spanish coals are difficult and expensive to extract, so that most coal is currently imported and the industry is in decline.

Pre-Westphalian coal basins

The oldest coals, all of minimal economic interest, are located in SW Spain (Ossa Morena Zone) in a group of small Dinantian, Visean and lower Namurian basins located within

Fig. 19.9. Schematic cross-sections: (**A, B**) the Guadalquivir and Cádiz basins (biogenic gas system); (**C, D**) the Gulf of Valencia basin (oil petroleum system); (**E, F**) western Cantabrian basin (oil petroleum system); (**G**) eastern Cantabrian and Gulf of Vizcaya basins (dry and wet gas systems); (**H**) Pyrenean (Jaca) basin (dry gas system). Locations of cross-sections are shown in Figure 19.8A.

Badajoz (Extremadura), Sevilla and Córdoba (Andalucia) provinces. The basins include (Fig. 19.10) Valdeinfierno, Los Santos De Maimona, Berlanga, El Couce and Benajarafe, Bienvenida and Casas De La Reina. The coals in these basins are all bituminous, and form part of a lithologically varied transgressive–regressive sequence (<800 m) that rests unconformably on Palaeozoic basement.

Westphalian coal basins

These are present in SW (Ossa Morena Zone) and northern (Cantabrian Zone) Spain and have a wider outcrop, a greater sediment thickness, and contain the more economically valuable mines.

Southwest Spain. (1) The Guadiato basin (Peñarroya-Belmez-Espiel; Fig. 19.10) includes three different coal bands of which the most northerly contains the most important bituminous and anthracite coal locations in SW Spain. The coals are Westphalian B in age and lie within a sequence of basal conglomerates, limestones, shales, and

sandstones all folded into a north-verging overturned syncline. (2) The Villanueva del Rio y Minas basin defines another asymmetric syncline with Westphalian sediments (conglomerates, sandstones, shales and coal) lying unconformably upon a Famenian–upper Visean basement.

Cantabrians. The Westphalian basins in the north of Spain are characterized by their wide outcrop, the number of exploitable coal seams, and their coal quality (bituminous coals with <40% of volatile matter (VM)). The most important are the Quirós, Teverga and La Camocha basins (Westphalian A or B) and, especially, the central coal basin (CCB). These basins originally formed part of a single large basin subsequently deformed and dismembered into several outcrops due to Variscan tectonism. The more important basins are (Fig. 19.10) Teverga, Quirós, Central Basin, Naranco, Santo Firme–Ferroñes, La Camocha, Viñón, Fresnedo–Libardón, Cofiño–Ribadesella, La Marea –Caleao–San Isidro, Villamayor–Beleño, Llerandi–Sebarga, Riaño, Villamanin–Carmenes and San Emiliano.

The CCB is the biggest (1400 km^2) and most economically important coal outcrop in Spain. It comprises 6000 m of Visean to upper Westphalian D alternating limestones, shales, sandstones and coals, the latter being most common in the upper 2800 m. The sedimentary environment comprised a wide and stable coastal area

Fig. 19.10. Locations of main Spanish coal basins (A, anthracite; B, bituminous; S, sub-bituminous; BC, brown coal). 1, Tineo (A); 2, Cangas (A); 3, Carballo (A); 4, Rengos (A); 5, Tormaleo-Ibias (A); 6, Degaña, Cerredo and Villablino (B & A); 7, San Emiliano (B); 8, Fabero–Toreno–Bembibre (El Bierzo) (B & A); 9, Cuenca Central, Quirós, Teverga and Marginales (B); 10, Riaño, Cármenes–Villamanín (B); 11, La Magdalena (B); 12, Ciñera–Matallana (B); 13, Sabero (B); 14, Guardo (B & A); 15, Barruelo–San Cebrian–Casavegas (B & A); 16, Villamayor–La Marea–Caleao–San Isidro–Beleño (B); 17, Llerandi and Sebarga (B); 18, Gamonedo–Cabrales (B); 19, Cofiño–Ribadesella (B); 20, Fresnedo–Libardón (B); 21, Viñon (B); 22, La Camocha (B); 23, Arnao (B); 24, Santo Firme and Ferroñes (B); 25, Naranco (B); 26, Meirama (BC); 27, Puentes de Garcia Rodríguez (BC); 28, Vera de Bidasoa (B); 29, Sallent de Gallego (A); 30, Tamajón (B); 31, Pont de Suert (BC); 32, Seo de Urgell (BC); 33, San Juan de las Abadesas (BC); 34, Bisaurri (S); 35, Cajigar (S); 36, Capella; Laguarres (S); 37, Pobla de Segur (S); 38, Tremp (S); 39, Calaf (S); 40, Prats Alp (S); 41, Berga and Tuixent (S); 42, La Demanda (S); 43, Mequinenza (S); 44, Olite–Andorra–Estercuel (S); 45, Utrillas–Aliaga (S); 46, Castellote–Cuevas de Cañart (S); 47, Foz–Calanda (S); 48, La Ginebrosa–Beceite (S); 49, Los Santos de Maimona (B); 50, Bienvenida; Casas de Reina; Berlanga (B); 51, Guadalcanal; Fuente del Arco; Alanis; San Nicolas del Puerto (B); 52, Valdeinfierno (B); 53, El Couce; Benajarafe (B); 54, Guadiato (Peñarroya–Belmez–Espiel) (B & A); 55, Villanueva del Rio y Minas; Viar (B); 56, Puertollano (B); 57, Arenas del Rey (BC); 58, Henarejos (B); 59, Baleares (S).

bordered by swamps and forests. Cyclic sedimentation produced almost 100 coal seams of variable thickness (0.25–3 m). Within the main outcrop area (the Caudal-Nalón unit of Luque & Sáenz de Santa María 1993), 11 formations have been defined: Levinco, Llanón, Tendeyón, Caleras, Generales, San Antonio, María Luisa, Sotón, Entrerregueras, Sorriego and Modest–Oscura (Garcia – Loygorri et al. 1971). The three lowest formations (Levinco, Llanón and Tendeyón; Westphalian A to lower Westphalian D) are together around 2200 m thick, collectively referred to as 'Subhullero', and of low economic value. Most of the minor coal workings known from this area have been won from the Levinco and Llanón formations. The 1000 m thick Tendeyón Fm is dominated by calcareous shales and limestones, and culminates in a 10 m thick white quartzitic marker horizon found throughout the entire basin and known as 'La Cuarcita de la Cruz'.

Above La Cuarcita de la Cruz the Caleras Fm marks the beginning of both the upper Westphalian D and the productive coal measures, known as the 'Hullero'. Near the base of the overlying Generalas Fm (265 to 325 m) up to four thick coal seams have been mined across the whole basin from San Antonio mine in the south to Lieres mine in the north. The overlying San Antonio Fm (240–365 m) has a strong marine influence and therefore few coals (only three workable coal seams have been found). The formation culminates with a thick (100 m), continuous sandstone marker (Arenisca de La Voz) which underlies the María Luisa Fm (270–310 m). This, together with the overlying Sotón Fm (380–480 m), contains the best and most abundant coals (up to 21 exploitable horizons). Above the Sotón Fm the Entrerregueras (340 m), Sorriego (300 m) and Modest–Oscura (575 m visible) formations show an increasingly continental influence upwards. These higher formations only exist in the Nalón valley, where they are preserved in synclinal cores and typically show thick zones of thinly bedded coals and carbonaceous shales, producing wide but dirty coal-rich horizons. The Modest–Oscura Fm extends from late Westphalian D to Stephanian A in age. The coals in this area are mostly bituminous but grade increases southwards from 35% VM in the north (e.g. Lieres and Pumarabule shafts) to semi-anthracites (5–10% VM) and even anthracites near the León fault.

The coal-bearing sedimentary sequences of the many other smaller basins preserved around the CCB vary in local stratigraphic detail but will not be described here. All of these basins, including the CCB, lie in synclines that have been deformed by two near-perpendicular fold systems (Fuente & Sáenz de Santa María 1999). The first deformation was syn-Westphalian and produced folds and thrusts with typically Variscan NE–SW axial trends, whereas the second was Stephanian and has east–west orientated fold axes. These structures produce dome-and-basin fold interference patterns with tight synclines and steep to overturned limbs. Later polyphase faulting and reactivation episodes have created very complex geology that has proved a considerable impediment to mining.

Late Westphalian-Stephanian coal basins

Cantabrians. In the eastern outcrop of the Cantabrian coal belt there are several late Variscan basins (such as the Guardo-Barruelo and Cabrales basins) that contain sediments deposited in intramontane settings and preserve small amounts of anthracitic coals. A series of still younger 'Post-Variscan' coal basins are also found in the Cantabrian–West Asturian-Leonese area. These are typically Stephanian in age, small in extent, contain anthracitic coals associated with continental sediments, and show a relatively simple structure. Good examples occur in the west (Tineo, Cangas, Carballo and Rengos basins), the north (Ferroñes, Arnao, Puerto Ventana basins), the south (La Magdalena, Ciñera-Matallana, Sabero and several small basins aligned along the León fault), and the east (Sebarga and Gamonedo-Cabrales basins).

Pyrenees. A series of coal basins ranging in age from late Westphalian to Stephanian occur in the Variscan core of the Pyrenees (Fig. 19.10). The most important of these are the Pont de Suert and Seo de Urgell basins which are located near Aguiró and Seo de Urgell. They are small, elongate, east–west orientated, and up to about 15 km long and 500 to 800 m wide. The sequence comprises 2300 m of conglomerates, shales, and coal seams up to 3 m thick (e.g. Pont de Suert) with

interbedded volcanic rocks. Other, more minor, basins with Palaeozoic coal seams include those of San Juan de las Abadesas and Vera de Bidasoa.

South central Spain. The Puertollano coal basin (upper Stephanian B or C) extends in an east–west direction for 11 km with a maximum width of 4 km (Fig. 19.10). The coal-bearing strata lie unconformably on an Ordovician basement and are covered by Miocene and Quaternary sediments. The 500 m thick succession is dominated by bituminous black shales, sandstones and coal. The sediments are variously lacustrine (shales), marshy (coal) and fluvial (channelled sandstones). Sedimentation was accompanied by acid volcanism with frequent ash emissions. The workable coal seams reach up to 4 m in thickness and are bituminous with 36–40% ash content and 4500 kcal/kg PCS (Poder Calorifico superior).

Stephanian to Permian basins

This group includes the Fuente del Arco, Guadalcanal, Alanís and San Nicolás Del Puerto basins (Stephanian–Autunian), which represent the remains of a larger basin dissected by erosion (Fig. 19.10). They show horizontal strata resting unconformably on a Lower Cambrian basement. The sequences show basal conglomerates and sandstones grading up into shales, sandstones and occasional coals. The most southerly of these basins in SW Spain is the Viar basin (Fig. 19.10) which is entirely Permian in age. Sediment thickness in this basin can reach up to 1000 m, with abundant conglomerates and basalts interbedded with sandstones, shales and coal seams. The sequence has been folded around an asymmetric syncline with generally gentle dips except along the northeastern fault-bounded margin.

Mesozoic coal basins

Mesozoic coal basins in Spain were formed in Aptian–Albian times. Regional latest Aptian regression produced swampy coastlines (paralic facies) well suited to coal generation. These basins contain the most important sub-bituminous coals (black lignite) in Spain, with the best examples being located in Teruel, in the SW of Tarragona province, among the Ebro massif, and in the Vinaroz marine basin. On a large scale most of these outcrops can be viewed as part of the same east–west orientated sedimentary system subsequently isolated by tectonic events. On a smaller scale, there are small basins and indications of lignite in many parts of eastern Spain that have not been investigated in detail, but which are clearly associated with the Utrillas facies (e.g. western Soria and the Castilian branch of the Iberian Ranges). Lignite indications at this stratigraphic level are also recognized in Castellón, Alicante and Valencia.

There are two coal-bearing facies called Escucha and Utrillas separated by a disconformity. The best coals are associated with lagoonal deposits: seams with poorer quality, but commonly more laterally continuous, were formed in deltaic plains with lacustrine or marshy facies. The Andorra–Oliete–Estercuel basin (Fig. 19.10) has the widest outcrop of all the Mesozoic coal basins (600 km^2). Its mining activity has been very important, with several seams called O (0.30–2 m), P (1.00 m), Q (3.00 m), R (1.00 m) and S (0.50 m) yielding sub-bituminous coal (black lignite) with 3000 to 5000 kcal/kg PCS.

The Utrillas–Aliaga basin (Fig. 19.10) covers 350 km^2 to the SE of the previous one and is separated from it by a fault called Umbral de Montalbán. The coal seams are cleaner in this basin and have a higher heating power. In the lower part of the Escucha Fm there are six coal seams, two being worked (4th and 6th) with thicknesses of up to 2.5 m. Above this in the Escucha Fm many other coal seams exist, notably the 1st (1.8 m) and the 2nd (1.2 m) seams. The

Castellote–Cuevas De Cañart, Foz–Calanda and La Ginebrosa–Beceite basins (Fig. 19.10) are relatively small outcrops with a very similar geology to the basins immediately to the west. Finally, the Mequinenza basin is the most northeasterly of the Mesozoic coal outcrops and covers around 300 km². It exposes a diverse range of facies: lagoonal, deltaic plain and lacustrine coastal. Although coal seams are typically lenticular, there are more than 100 of them, the most important occurring in the Tercero (ten coal seams >50 cm thick) and Cuarto formations (five coal seams up to 35 cm). The coal is black lignite (sub-bituminous coal), with a heating power between 4000 and 6000 kcal/kg PCS.

Cretaceous-Tertiary coal basins

NE Spain and the Balearics. After the deposition of the Utrillas coals, conditions suitable for coal formation and preservation did not occur again until latest Cretaceous times, when several carbonaceous sedimentary units were deposited. These are generally referred to as the 'Garumniense facies' and are best developed in the Berga–Tuixent basin in the south Pyrenees area (Fig. 19.10). Located north of Barcelona, this outcrop extends for about 25 km in an east–west direction and is around 8 km wide. The sediments are both continental and lacustrine, and were deposited from Maastrichtian until Palaeocene times. The coal is black lignite (sub-bituminous coals) of 6000–6500 kcal/kg PCS. Four formations, called Cero, Primeras, Segundas and Terceras, have been distinguished but the irregular character of the layers makes them difficult to work. Structurally, three sub-basins exist, with only one of these (the Fígols–Vallcebre sub-basin) being big and rich enough to be worth mining. Alpine deformation of these sediments has, in their northern outcrop, tilted them vertically.

Black lignite deposits have also been mined in lowest Tertiary rocks in Mallorca. They occur in lacustrine sediments (Oligocene) and are best developed in the vicinity of Alaró where they rest on a Lower Cretaceous basement and where the productive units (45 m thick) have several coal seams 3–20 m thick (3500 to 5500 kcal/kg PCS: sub-bituminous coals). In general, these are poor quality lignites, rich in ash and sulphur, and for this reason the mines have been closed.

Lower Oligocene black lignites occur within the Calaf basin (160 km²), 60 km NW of Barcelona (Fig. 19.10). In this basin lacustrine materials were deposited on an Eocene basement and are themselves overlain by 350 m of detrital molassic sediments that pass laterally into coal-bearing horizons (4500 to 5500 kcal/kg PCS). Another localized lignite-bearing continental basin occurs further north: the Prats–Alp basin of the Central Pyrenees (Fig. 19.10). This is located in the NE part of the Cerdaña graben, between Lérida and Gerona, in a sedimentary sequence that includes clays and thin lignite layers of Miocene age, overlain by clays and pale sands (Pontian age) with lignite levels of 3000 to 4000 kcal/kg PCS.

NW Spain. The most important lignite-bearing Neogene outcrops in Spain are located in the NW on the Palaeozoic Iberian Massif. Faulting associated with Alpine tectonism in Galicia produced localized continental depressions that became filled with sediments and lignite from Aquitanian times onwards. There are many such small depocentres but only two have economic importance, these being the Meirama and Puentes de García Rodriguez basins. The NW–SE orientated Meirama basin is located about 25 km SW of La Coruña and is only 3.5 km by 1 km in size (Fig. 19.10) but has lignite-bearing units up to 300 m thick, yielding brown coals of low quality (48% humidity, 15.8% ashes, 20.5% VM, 1.3% S, 15.3% fixed carbon, 1740 kcal/kg PCI (Poder Calorifico inferior) and 2100 kcal/kg PCS). The basin developed along a Variscan strike-slip fault reactivated during the Miocene epoch, when subsidence created a lacustrine environment against the fault. The basin continued to fill until Quaternary times. The Puentes de García Rodriguez basin is a NW–SE orientated Miocene basin measuring 8 km by 2.5 km with a narrow, shallow central section that divides the basin into 'Campo Este' and 'Campo Oeste' (Fig. 19.10). The basin is very similar to that at Meirama, being controlled by Variscan faults. Sands, clays and even conglomerates were deposited with the lignite which has been subdivided into 19

seams (each 0.5–25 m thick) with a total thickness of 117 m. The average quality of the lignite is 41.1% humidity, 22.8% ashes, 21.4% VM, 2.5% S and 2181 kcal/kg PCS.

Betic Cordillera. Lignite deposits in the Betics of SE Spain are linked to Alpine tectonic events, as seen in NW Spain. The only area containing significant accumulations of lignite is the Arenas del Rey basin (Fig. 19.10), located at the southern limit of the Miocene Granada basin which lies unconformably on a Triassic and Palaeozoic basement syncline. Sedimentation in this structural graben during late Neogene times took place in an actively subsiding trough under humid climatic conditions under which organic matter rapidly accumulated. Numerous lignite indications exist in other parts of the widespread Tertiary outcrop of eastern Spain, such as in mountainous lake basins in Castellón, Valencia and Alicante. These minor basins have deposited sands, muds and clays with lignite levels that are thin and lack economic potential.

Quaternary peat deposits

Peat deposits occur in many parts of Spain, but have only rarely been of significant economic value. Peat-bog workings occurs in Llanes (Asturias) where black peat (with a heating power of 6043 kcal/kg) and brown peat (5893 kcal/kg) are both extracted. Examples of peat bogs that were worked in the past include deposits in the Castellón, Valencia and Alicante areas.

Future developments

Coal mining in the European Union shows a continuing decline in favour of cheaper imports. The forecasts for 2001 show a European production of 131 Mt (66% of the 197 Mt mined in 1990), with imports increasing by 47% from the 61 Mt of 1990 to 90 Mt in 2001. According to the Spanish national energy plan, the use of coal should remain practically constant although the percentage contribution of national coal will depend on government subsidy.

Coal production in Spain is dedicated almost completely to electric power generation with only a small part being used in the domestic sector. Whereas the production of black lignite held up during the 1990s, brown lignite mining declined progressively due to its high sulphur content and the need to mix it with imported coal low in ash. The demand for both bituminous coal and anthracite diminished gradually by 2% annually during the 1990s, and this decline is predicted to continue. Production costs are higher than the price of imported coals and only one-third of Spanish mines are likely to be able to compete without subsidy. It seems inevitable, therefore, that production will continue to fall over future years. Brown lignite production fell to 10 Mt in 2001 (16.4 Mt in 1990), with sub-bituminous, bituminous and anthracitic coals not exceeding 14 Mt (19.5 Mt in 1990).

Coal for the steel industry was expected to fall to 3 Mt in 2001 from 4.5 Mt in 1990, and a similar decline in coal consumption is expected within the cement sector due to an increasing use of petroleum coke. However, in electricity power generation, coal consumption should increase if the demand grows according to European patterns (*c.* 2.7% increase annually up to 2005). The use of imported coals will increase, with the importation of low-sulphur-content coals contributing, by means of mixtures, to the consumption of national high-sulphur-lignite reserves. In 2010, coal consumption in Spain is predicted to be 43 Mtpa (of which 21 Mt will be imported), as compared to the 46.3 Mt consumed in 1990.

Coal gasification. Underground coal gasification relies on a partial oxidation of coal by means of oxygen and water vapour to produce a gas mixture of CO, H_2 and CH_4 that can be extracted and burnt to produce electricity. Several projects are currently ongoing in Spain, such as one in Teruel, financed by a European grouping (Spain, Belgium and United Kingdom) aided by the EU-funded THERMIE programme. This is currently being operated to a depth of 500 m, but extrapolation to 1000 m depth is planned if the project is successful.

Coal bed methane. Coal bed methane is obtained by drilling wells to various depths (between 700 and 1500 m), fracturing the coal seams by means of high-pressure fluid injection, and recovering methane still preserved in the coal seams. The typical gas content of a coal seam varies between 8 and 14 m^3/ton. In 1992, Unión Texas and Hunosa España began an investigation in the Sama and El Entrego synclines (CCB in Asturias). Two holes drilled to 2100 m (Central Asturias #1 and Modest #1) obtained strong gas indications when cutting different coal seams in the Sotón and María Luisa formations. Exploitation has not yet been carried out, due to drilling and other costs, but the potential for coal gas fields remain high.

Environmental consequences. Coal mining and combustion create environmental problems on global, regional and local scales. The latter continue, and are even enhanced, after mine closure when problems such as acid mine water, gas accumulation, subsidence, and waste tip stability need monitoring and remediation.

Acid drainage. In general, waters circulating through coal mines are acid and contain high concentrations of sulphates, iron, and heavy metals such as zinc, copper and arsenic. The contamination problems posed by such mine water flows are complex, with each basin presenting a specific array of problems.

Surface gas flow. Once the mine is closed, residual coal continuously desorbs methane gas that tends to work its way to the surface. Gas in buildings located above the disused mine can accumulate to explosive concentrations and create disasters in the absence of monitoring and remediation. In some cases, the use of this gas is possible, as has been the case in certain coal basins in France (e.g. Nord Pas de Calais and Lorena basins).

Subsidence. This is common in mining areas and the danger does not disappear after mine closure. Subsidence above long tunnels can lead to a characteristically linear topographic expression (called 'minados' in Spain). In a single basin there can be 100 old mine workings, each posing a potential subsidence risk.

Waste tips. Piles of mining debris are not inert and need to be placed under surveillance and control in order to prevent potentially disastrous events such as slides or spontaneous combustion. The environmental control and remediation of abandoned coal mining areas will remain important for many years. Such work will be carried out by multidisciplinary teams involving environmental geoscientists, chemists, biologists and engineers, and is likely to require major funding by the EU and the Spanish government.

Water resources (FLV)

One of the factors that has allowed Spain to become a modern country is the regulation of its water resources. If Spain did not use dams only 9% of the water demands for industry and watering could be supplied, in contrast with an average figure of 40% for most other European countries. Similarly, most European countries do not need to irrigate the land, whereas in Spain more than 3×10^6 ha are watered, which requires 80% of the total consumed water. In order to satisfy these necessities Spain currently has 1200 big dams, some of them built by the Romans 2000 years ago and still operative. Thus Spain has the third highest number of dams in Europe, and the fifth highest number in the world.

The use of groundwater is no less intense than for superficial waters. There are 2500-year-old wheels preserved in Menorca and Cádiz, which were built with a similar technology to that used now by some populations in the Sahara area. The intensive use of groundwater is originally attributed to the Muslim population who inhabited Spain between the seventh and fourteenth centuries, using kanaq technology. This involves the use of long tunnels excavated in alluvial fans, a technique coming originally from Iraq. This same technology and infrastructure were used to supply the city of Madrid until the mid-nineteenth century, with the tunnel network extending up to 127 km in length (Oliver Asins 1958). Today 12.2 million Spanish people, 31% of the population, are supplied by groundwater. Groundwaters also supply one-third of the irrigated lands in Spain (700 000 ha) plus 300 000 ha using mixed irrigation with superficial water.

Eighty per cent of Spain's water resources is used in agriculture, 12% is consumed directly by the country's 40 million inhabitants and 60 million tourists, and the rest is used by industry. The distribution of water resources in Spain reflects the extreme geographical and meteorological diversity found across the country. The northernmost strip of the peninsula is characterized by its mild climate, with low-pressure systems of Atlantic origin affecting it throughout the year. Along much of the Mediterranean coast, and in part of inland Andalucia (the Guadalquivir basin), the weather is mild with dry summers and warm winters. Across most of the rest of the peninsula, the predominant climate is continental, being characterized by dry summers and cold winters. In the Canary Islands and the Mediterranean coastal strip between Almería and Murcia, the climate is dry, with scarce precipitation, mild winters and warm summers. Thus, whereas the average annual rainfall in Spain is 648 mm/year (equivalent to 346 km^3/year of water), this figure conceals an extremely irregular distribution, ranging from over 1600 mm/year in the north, to a mere 300 mm/year in large areas of the SE and less than 200 mm/year in some parts of the Canary Islands (MIMAM 2000). On a monthly scale, the rainiest month tends to be December and the driest June. The part of mainland Spain with the lowest rainfall is the Segura basin with scarcely 50 mm/year, which is about 20 times less than Galicia in the NW, five times less than the national average, and three times less than the average per day in Spain.

Although snowfalls are frequent in mountainous regions, snow retention never lasts more than four months, except in some snowfields in the Pyrenees. The average rate of evapotranspiration across Spain is 464 mm/year (figures for 55-year period, 1941/42–1995/96), although once again this marks great regional differences, with the maximum values being found in the southern half of the peninsula, the Canaries and the middle reaches of the Ebro valley (MIMAM 1998). The difference between precipitation and evapotranspiration is the fraction of water (known as 'useful rain') that may be used as a result of superficial and subterranean runoff. The average annual figure for water available from direct runoff and groundwater is 220 mm, or about 111 000 hm³/year (74% runoff, 26% groundwater). This all originates on Spanish territory except for 300 hm³/year which come from Andorra and France (on the headwaters of the river Segre, a tributary of the Ebro).

Surface water

Spain's fluvial network is asymmetrical, with 69% of the peninsula (400 839 km²) draining into the Atlantic and the remaining 31% (18 661 km²) draining into the Mediterranean. The most important basins are those of the Tajo (Tagus) (1007 km long and 81 947 km² in area), the Ebro (910 km and 85 997 km²), the Duero (895 km and 98 375 km²), the Guadiana (778 km and 67 500 km²) and the Guadalquivir (667 km and 57 442 km²) (Fig. 19.11). The administrative organizations responsible for water management in these basins are the Organismos de Cuenca defined in the Water Act of 1985, which include the old Confederaciones Hidrográficas, created in 1926, and the Comisarías del Agua. Rainfall patterns are measured by the Instituto Nacional de Meteorología (national meteorological institute, INM) which currently runs 5080 weather stations (80 of them automatic). River levels are monitored and flood warnings given by the Sistema Automático de Información Hidrológica (automatic hydrologic information system, SAIH). Another network, the Red Oficial de Estaciones de Aforo (official network of capacity stations, ROEA) measures river flow rates, reservoir and canal levels. All these data are stored in the HIDRO database of the Centro Estudios Hidrográficos (centre for hydrographic studies).

Many parts of Spain are prone to both drought and flood conditions. In the three most serious recent droughts, 1941–1944, 1979–1982 and 1990–1994, the decrease in average annual rainfall was as much as 35%. These recurrent droughts occurred over wide areas of the peninsula and caused major crises of irrigation and supply, leading to emergency situations. At the other end of the scale are floods caused by freak precipitation, when the rainfall of a few hours can exceed the annual average, a problem especially recurrent in Mediterranean continental areas. Over the last 500 years, there has been an average of five major floods per year (MOPT 1993), with average annual material losses amounting to about 500 million Euros, not to mention environmental damage and the loss of human life. There are over 1000 risk areas, of which 70 are officially considered high risk (potential loss of human lives), 250 medium risk and the remainder low risk. The geomorphology of the affected areas, human deforestation activity and building work in the basins, among other factors, can exacerbate the effects of rivers flooding. Today such events are combated under the LINDE plan (Villaroya Aldea 1997), which defines the limits of the areas in basins prone to flooding and establishes mitigation procedures.

Groundwater

Aquifers, geological formations capable of storing and transmit economic quantities of water, are irregularly distributed in Spain. There are various types of aquifers, broadly grouped into karstic, detrital, volcanic, plutonic and metamorphic, although the latter two are of only minor importance. The total storage capacity for these Spanish aquifers (MIMAM 1998), i.e. water stored to a depth of 200 m, is 125 000 hm³ (120 000 hm³ on the peninsula, 2500 hm³ each in the Canary and Balearic islands). Average annual recharge is about 30 km³, and about 19% of this (6000 hm³) is extracted from waterwheels. They

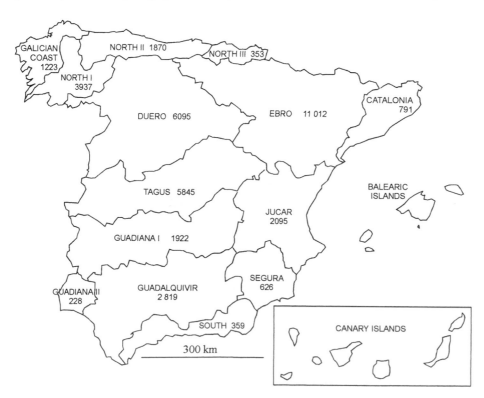

Fig. 19.11. Territorial distribution of basin organization showing total estimated water reserves in hm³/year (from MIMAM 2000).

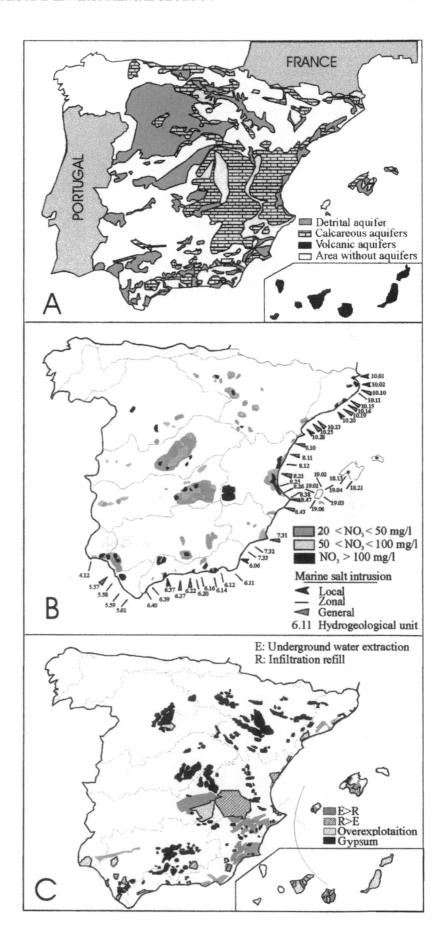

Fig. 19.12. (A) Distribution of detrital; karstic and volcanic aquifers. 'Areas without aquifers' are in fact areas of low permeability with very small aquifers (MINER-MOPTMA 1994). **(B)** Pollution by nitrates and marine incursion into groundwater (modified from MINER-MOPTMA 1994). **(C)** Overexploitation of aquifers and gypsum salinating groundwater (modified from MINER-MOPTMA 1994).

are currently monitored by a national network of 1910 sites spread over 160 000 km^2 (almost 90% of the permeable surface), with an average area per piezometer of 85 km^2 (DGOH 1992). Since 1996, the DGOHCA (Dirección General de Obras Hidráulicas y Calidad de las Aguas) of the MIMAM (Ministerio de Medio Ambiente) has been responsible for the publication of data concerning the measurement networks of surface and groundwater.

Karstic aquifers occur in all limestone areas of Spain (Fig. 19.12A), notably concentrated across the Mesozoic carbonate formations and, to a lesser degree, in Cenozoic deposits. These formations occupy over one-fifth of the peninsula and the Balearic islands. As karstified rocks, they readily take in rainwater and runoff, forming important internal karstic networks and copious springs. In the Cantabrian mountains they occupy wide stretches of land, with many springs discharging large quantities of water into the Sella, Cares, Deva, Miera and Asón rivers and even the Ebro in its headwaters. They also occur in the Pyrenees, particularly in the higher parts of the basins of the Cinca, Segre and Ter rivers.

Between the Iberian Ranges and the Cazorla and Segura ranges there are large and continuous aquifers of limestone and dolomite feeding the headwaters of the Tagus, Jiloca, Mijares, Guadalaviar, Gabriel and Júcar rivers (in the Iberian Ranges). The Guadiana river surfaces amid marshy lakes in the high plateau of Montiel and then disappears by infiltration into the great La Mancha aquifer of Ciudad Real, which naturally discharges into Los Ojos and Las Tablas de Daimiel, one of Spain's most important wetland areas.

The Cazorla and Segura ranges are drained by springs which give rise to the Segura, Mundo and Guadalquivir rivers. From Cádiz to Alicante and to Ibiza and Mallorca, the Betic Cordillera lack widespread hydrologic continuity. Those areas with high rainfall get more infiltration, which they discharge into rivers that are less extensive (due to the proximity to the coast) but have a considerable flow rate, such as the Serpis, Algar, Genil, Adra, Guadalhorce, Guadalfeo, Guadiaro and Guadalete.

Detrital aquifers are mainly located within Cenozoic sediments on which the wide valleys of the large peninsula rivers such as the Duero, Tajo (Tagus) and lower Guadalquivir have been formed. These aquifers are the largest in Spain, occupying around 100 000 km^2 and occurring in large, mainly clayey, outcrops that are more or less consolidated, with low permeability but containing interspersed levels of sand and gravel, which, using suitable technology and wells 400 or 500 m deep, can give a flow of >100 l/s.

Also of great hydrogeological importance are the Quaternary alluvial deposits along rivers and on the coastal plains. These deposits have a high permeability and storage capacity, which, as they are connected to rivers, are readily exploited by means of wells. Noteworthy among these aquifers are those of the Ebro river and the lower reaches of its Pyrenean tributaries, the floodplain of the Guadiana river in Badajoz, as well as those of the Guadalquivir, Segura and Llobregat, the coastal plains of Castellón, Valencia and Gandía-Denia and the Palma plain.

Volcanic aquifers are very rare in the peninsula (Fig. 19.12A) but they are very common in the Canary Islands. The hydrogeological structure of these islands is very complex and consists of numerous small aquifers, which are not interconnected due to many discontinuities. Except on the islands of Lanzarote and Fuerteventura, which have scant precipitation, and the small island of El Hierro, water flowing from the areas below the peaks used to collect in deep ravines. Nowadays this natural phenomenon has almost disappeared because of the intensive exploitation of groundwater. The volcanic aquifers of the Canaries, together with other smaller ones in Campo de Calatrava in the province of Ciudad Real and Olot in Girona, have a surface area of 8000 km^2.

Groundwater is commonly used to supplement surface reservoir supply. The most common strategy is alternating use, to guarantee the supply of water in drought periods. This strategy has been used to supply Madrid, in the Mijares system, in the Castellón plain, in the upper and middle reaches of the Segura, in Motril, Vélez and in the lower Guadalhorce, while mixed surface and groundwater have also been used for irrigation (Sánchez & Alvarez 1994).

A complementary practice to this joint use is the artificial recharge of aquifers to store surface runoff or recycled water

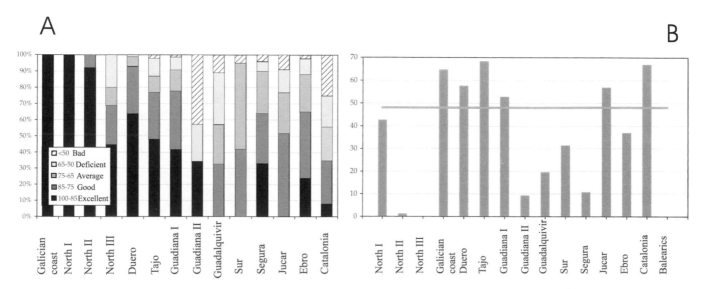

Fig. 19.13. (**A**) Situation of water quality expressed as a percentage of the length of the river network according to the Índice General de Calidad (MIMAM 2000). (**B**) Volume of degraded water in lakes and reservoirs due to eutrophication (MIMAM 1998).

from local treatment plants. This improves aquifer management, particularly where water pressures are falling or where there are problems with marine incursion, as at Campo de Dalias (Almería). Although this has been done many times throughout Spain using various techniques, it still is not general practice. The longest experience so far has been gained from the recharging of wells with treated sewage water (up to 20 hm³ per year) in the alluvial plain of the Llobregat river.

Desalination of sea water collected by means of wells in coastal areas and of brackish continental water contributes 222 hm³/year, which is put to urban-tourist, industrial and agricultural uses. This gives Spain the highest number of desalination plants in Europe (30% in 1998) with new plants constantly being built.

Water quality and pollution

Water quality has been systematically controlled since 1962 when the Control Oficial de la Calidad del Agua (official quality control for water, COCA) network was created. The data from the COCA network are the most extensive and have been used as a basis to define the Índice General de Calidad (quality general index, ICG) and the suitability of water for human consumption. The ICG affords a measurement of the quality of the water in Spanish rivers, based on an aggregation formula with 23 parameters. According to this index, a value of 100 corresponds to water of excellent quality suitable for all uses, a value below 65 indicates unsuitability for most uses, and one below 50 means water of poor quality (Fig. 19.13A). In addition to the COCA network, an Ictiofauna (ichthyofauna) network was established to test the suitability of water for fish life. From 1993 on, both networks were integrated into the Red Integrada de Calidad de las Aguas (water quality integrated network) with the aim of sampling stretches of rivers

with the frequency and intensity that their increasing use required. Today this network also includes the Estaciones Automáticas de Alerta (automatic warning stations) that analyse and transmit data on chosen chemical parameters in real time. In addition to this network there are others, such as the Red Nacional de Control de la Radiactividad Ambiental (national network for environmental radioactivity control).

Regarding groundwater, the Red de Observación de Calidad de Aguas Subterráneas (groundwater quality observation network, ROCAS) has been progressively enlarged since its establishment in 1970 to its present number of 1650 stations, which are complemented by the 798 stations of the ROI (Red de Observación Marina) network, which deals with marine incursions in coastal aquifers. Potential degradation of groundwater depends on the vulnerability of aquifers to polluting activities. In 1989 the Instituto Geológico y Minero de España (IGME) assessed Spain in terms of high, medium and low risk areas depending on the hydrogeological characteristics of the terrain and the existence of polluting activities. Findings showed 28% of the country to be at high risk, 34% at medium risk and 38% at low risk.

Because of the large areas affected, the two most serious problems of groundwater pollution are the presence of nitrates due to the use of fertilizers in agriculture, and marine infiltration due to the exploitation of coastal aquifers (Fig. 19.12B). Another important form of pollution is the salination of water due to the overexploitation of aquifers and the occurrence of gypsiferous formations (Fig. 19.12C). A fourth problem with water pollution is created by livestock waste, which is organic and bacteriological. Finally, pesticides and heavy metals have been detected locally due to urban waste as well as mining and industrial activities (MIMAM 1998).

With regard to lakes and reservoirs, the main cause of pollution of large bodies of standing water is the dumping of

Table 19.4. Destructive historic earthquakes in Spain from the fourteenth century, with intensity IX or greater at epicentre

Number	Year	Epicentre	Intensity	Magnitude	Comments
1	1396	Tavernes (Valencia)	IX		Fortress and bridges collapsed, hundreds of houses destroyed, rock fall, ground crack, new springs
2	1428	Queralps (Gerona)	IX–X		800 deaths. Important damages in churches and castles, number of houses damaged, big ground cracks
3	1431	Atarfe (Granada)	IX		Towers and mosques cracked, wall of Alhambra partially collapsed
4	1504	Carmona (Sevilla)	IX		100 deaths. Destruction of walls, collapse of church vaults, damages in a great number of houses, landslides, ground cracks, changes in ground water regimen
5	1518	Vera (Almería)	IX–X		Destruction all around the village, relocation of the city, castles damaged
6	1522	Almería	IX		More than 2500 casualties. Almería completely destroyed, 80 villages devastated, harbour ruined, tsunami, considered as important as Lisbon earthquake
7	1680	Málaga	IX	6.8–7.4	200 deaths and 250 injured in Málaga alone. Muslim main castle destroyed, 852 houses totally ruined and 1250 damaged
8	1804	Dalías (Almería)	IX		150–200 deaths. Churches and castle towers fallen down and several hundreds of houses destroyed, aftershocks during seven months
9	1829	Torrevieja (Almería)	X	6.9	399 deaths, 388 injured. Over 2900 houses destroyed and more than 2000 damaged. Aftershocks for months, ground cracks and change in water regime
10	1884	Arenas de Rey (Granada)	IX	6.5–6.7	750–900 deaths, around 2000 injured. Over 1000 houses destroyed and 17 000 damaged. Landslides, ground cracks, liquefaction of soils, disturbances in water regime, big aftershocks last one year

urban or agricultural sewage, which is organic and causes eutrophication of the water. This is a fertilization process with nutritive substances, especially nitrogen and phosphorus in forms that are easily assimilated by aquatic vegetation, which cause an increase in algae and productivity at all levels in the food chain, together with a deterioration of the original physical and chemical characteristics of the water. Depending on chlorophyll concentration, water is classified with decreasing pollution as hypereutrophic, eutrophic, mesotrophic, oligotrophic and ultraoligotrophic. Figure 19.13B shows the proportions by volume of degraded water (eutrophic and hypereutrophic) according to the total volume in each basin. It will be noticed that the most degraded water is that of the Tajo (Tagus) river (68%) and the inland basins of Catalonia (67%).

Final statement

Given current trends towards climate warming, there is a likelihood of future decreases in the average flow rate of Spanish rivers (MIMAM 1998). Models suggest that an increase of 1°C in average annual temperature would decrease the average flow of rivers by 5%, with extreme values of 11% in the Segura and Guadiana basins. Some evidence already exists for geomorphological change induced by a drop in the amount of riverborne sediment, such as shrinking of the Ebro delta and sandbars in the Guadiana river. Another change that is clearly evident is the progressive drying of wetland areas. Over the last ten years, the total surface area of wetlands in Spain has decreased by some 25% (from an initial area of 1567 km²; MOPT 1991). Most (*c.* 300 km²) of these dried wetlands are

connected to aquifers, such as the Gallocanta lakes in Zaragoza, the Tablas de Daimiel in La Mancha (Serna & Gaviria 1995), and the Guadalquivir marshes. Other environmental problems brought about by the overexploitation of aquifers include subsidence (e.g. in the city of Murcia) and deterioration of water quality.

Generally speaking, two opposing tendencies can be observed in Spanish society regarding water conservation. Some advocate improvement in the management of hydrogeological resources, combined with ecologically sound conservation measures (Bentabol 1990). Others retain concepts deeply rooted in the national water culture, for example water discharged into the sea by rivers is considered wasted. There is continuing pressure to redistribute water resources by interbasin transfers to compensate natural climatic imbalance. Given the likehood that the current level of water use is reaching the limit of economic and environmental sustainability, and the possibility of significant future climate change, the good management of Spanish water resources is becoming ever more critically important.

Seismic hazard (RC)

Compared with some other Mediterranean regions, such as Greece or Italy, Spain shows only relatively moderate seismic activity, and there is consequently only a low awareness of the potential seismic hazard that exists across the country. In reality, at least ten earthquakes have produced massive destruction and significant loss of life since the fourteenth century, these reaching intensities of IX or X (Table 19.4). Moreover, the earthquake considered to be one of the greatest

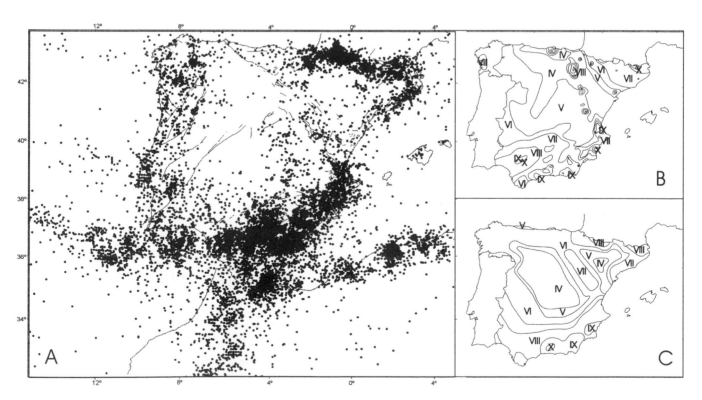

Fig. 19.14. (A) Distribution of seismicity in the Ibero-Maghrebian region. Data from National Geographic Institute (IGN). **(B)** Maps of seismic hazard in Spain from earthquake data, showing maximum intensity felt. **(C)** As (B) with intensities for a return period of 1000 years. Data from IGN (1992*b*).

in recorded history, the Lisbon earthquake of 1 November 1755, whose magnitude has been estimated as 8.5, originated in the Gulf of Cádiz. This earthquake was felt all over Spain, with intensities which varied from VII in the westernmost areas, adjacent to the Portuguese border, to IV in Catalonia.

The record of Spanish earthquakes starts in 330 BC and includes over 18 000 events, most of which are instrumental. Historical earthquakes are recorded from the references of ancient authors, the chronicles of medieval kingdoms, the civil registers of towns and villages all over the country, and from the news in the press and a variety of other reports in more recent times. The assignment of seismological parameters to historical earthquakes is full of uncertainties. For the period beginning AD 1300, only intensities larger than VIII at the epicentre are well assigned. The instrumental period starts in 1920 and from then on errors have diminished as the national seismic network has been improved. In 1980 the deployment of a more modern network dramatically diminished the epicentral error and allowed an accurate measurement of other seismological parameters as well as the determination of the focal mechanisms of some of the earthquakes. The quality of the data obtained with the current network (and future planned improvements) allows establishment of clear links between the seismicity and the geology, the identification of seismological structures and, in some cases, the analysis of the fracture processes during the seismic event.

Distribution of earthquakes

A map of earthquake epicentres (Fig. 19.14A) shows that Spanish seismic activity is clearly concentrated within certain zones (Mezcua et al. 1991). The most important region is the southern and southeastern Iberian peninsula (the Ibero-Maghrebian zone), which represents the diffuse but actively convergent plate boundary between Eurasia and Africa. This seismically active region includes the Gulf of Cádiz, the Alborán sea and the Betic Cordillera. It is in these areas where most of the seismic activity is concentrated, but there is another region of important activity within the Pyrenees, the second largest orogenic range in the country. The concentration of seismicity in this region occurs along what was once the northern limit of the Iberian plate, which has at times behaved independently from the rest of Eurasia and Africa. Another earthquake area, albeit much less active, is found inland, associated with the Iberian Ranges. Finally, there is yet another isolated region of relatively minor activity in the NW of the peninsula, in the region of Galicia. The central parts of the Iberian peninsula, that is, the Central Range, the Duero and Tajo Tertiary basins, and the Palaeozoic Iberian Massif in Extremadura, have very little seismicity.

Nearly all recorded earthquakes are of superficial focus, depths ranging from 4 to 50 km. The only recorded deep earthquakes, with depths of up to 640 km, have taken place in the Betic region and in the Gulf of Cádiz (Buforn et al. 1991). Although current seismological activity in Spain is closely related to the regional geodynamic framework of Eurasian-African plate interaction, the neotectonic picture is in detail relatively complex. Whereas parts of the country remain in a state of compression, other areas are currently in extension. This is particularly the case for much of the southern and eastern areas of the peninsula, where extension has been related to the collapse of the Betic orogen, giving rise to the Alborán basin (Dewey et al. 1973), as well as to the creation and propagation of the rift system in the Valencia trough. The greater concentration of earthquakes in the south and SW has to do with the active boundary between the converging lithospheric plates of Eurasia and Africa.

Evaluation of seismic hazard and active faults

Because the last destructive earthquake took place as long ago as 1895, the seismic risk has been rather forgotten in Spain. Traditionally the evaluation of seismological hazard in Spain has been based on past records, on the basis of which the country has been divided into seismological provinces (Martín Martín 1984), based mainly on Alpine tectonic units and a few neotectonic provinces. Both deterministic and probabilistic criteria have been applied in the seismological and tectonic studies undertaken for nuclear plants, large reservoirs and future storage of radioactive wastes. At the scale of the Iberian peninsula maps have been produced to illustrate (IGN 1992a,b) maximum felt intensity and most likely maximum intensity for different return periods (Fig. 19.14b,c). The maximum intensities felt are in the regions of Granada and Murcia, with values of IX and X. As for the probabilistic calculations for different return periods, the areas of maximum intensity essentially match the areas where epicentres have been concentrated. At the other extreme, some interior regions appear which are considered to be extremely stable with very low expected intensities. However, in all these hazard evaluations the characteristics of the seismic sources (active faults) are not considered in detail.

The convergence of the Eurasian and African plates at the Ibero-Maghrebian zone gives rise to a shortening of 4 mm/year (Argus et al. 1989). This deformation is transmitted to a network of faults in the interior of the peninsula whose velocities are moderate, being greatest in the areas close to the contact region between the plates. The tectonic stress field in the continent has been determined from the studies of earthquake focal mechanisms (Buforn et al. 1988a,b) and from population analysis of active Quaternary faults. If the epicentre distribution map is superimposed over the fracture network it becomes clear that the seismological activity is controlled by sliding along active Quaternary faults. The earthquake return period in each fault must be relatively long because no two destructive earthquakes in the last 2000 years have taken place along the same fault. Ongoing studies in neotectonics, active tectonics and palaeoseismicity are beginning to shed new light over the study of seismological hazard in Spain by extending the observation period back on a geological timescale.

In the Betic Cordillera the fault network showing neotectonic and Quaternary activity controls the seismicity, and includes NW–SE faults, which act as normal faults, ENE–WSW faults which act as reverse faults and east–west faults with a mainly transcurrent component of movement (Sanz de Galdeano 1983). This seismotectonic picture is compatible with a NW-directed regional shortening (Capote & De Vicente 1989), as indicated by earthquake focal mechanisms around the Arc of Gibraltar. The link between earthquake distribution and this fault network allows determination of the seismic regional sources (Sanz de Galdeano & López Casado 1988). Some of the more active faults are hundreds of kilometres long and show signs of a higher than normal sliding rate. Among these, the Alhama–Murcia fault has been investigated by trenching and at least two events of coseismic slid 15 000 years have been detected (Martínez-Díaz *et al.* in press). The mean velocity estimated for this fault is 0.2–0.4 mm/year, although it is a little slower for the Holocene (about 0.1 mm/year).

In Cataluña a number of faults in the Vallés–Penedés Tertiary basin have been investigated, with no seismological activity greater than magnitude 4. Palaeoseismological activity studies along the El Camp fault give a sliding rate of 0.12–0.16 mm/year since the start of the Miocene epoch and of 0.02 mm/year for the last 125 000 years. The faults are probably segmented into lengths ranging from 14 to 25 km and the magnitudes of the detected palaeo-earthquakes reach values between 6.4 and 6.7. Finally, even in the relatively stable areas within Spain there are several faults with associated seismological activity or evidence of tectonic motions

during Quaternary times. In Portugal mean velocities of 0.2–0.5 mm/year have been determined for the Manteigas–Bragança transcurrent fault (Cabral 1989, 1995) and from 0.03 to 0.1 mm/year for the Ponsul reverse fault (Dias & Cabral 1989). At the Alentejo–Plasencia fault, which runs along a great Mesozoic basic dyke, geological evidence allows the detection of Quaternary deformation and an estimated mean sliding rate of around 0.01 mm/year (Villamor *et al.* 1996). These fault velocities imply return periods of between 1000 and 10 000 years, or even 100 000 years, for earthquakes in the most stable areas of the peninsula.

The authors would like to thank S. Dominy, W. Gibbons, J. Martínez-Frías and T. Ramsay for their very useful and detailed reviews of the chapter.

References

Abbreviations used in reference list

AAPG   American Association of Petroleum Geologists
AGGEP   Asociación de Geólogos y Geofísicos Españoles del Petróleo
BRGM   Bureau de Recherches Géologiques et Minières
CNRS   Centre Nationale de la Recherche Scientifique
CSIC   Consejo Superior de Investigaciones Científicas
DGOH   Dirección General de Obras Hidráulicas
IAS   International Association of Sedimentologists
IGCP   International Geological Correlation Project
IGME   Instituto Geológico y Minero de España
IGN   Instituto Geográfico Nacional
INQUA   International Quaternary Association
ITGE   Instituto Tecnológico y Geominero de España
IUGS   International Union of Geological Sciences
SEPM   Society of Economic Paleontologists and Mineralogists
USGS   United States Geological Survey

ABAD, A. 1989. El Cámbrico Inferior de Terrandes (Gerona). Estratigrafía, facies y paleontología. *Batalleria*, **2**, 47–56.

ÁBALOS, B. 1990. *Cinemática y mecanismos de la deformación en régimen de transpresión. Evolución estructural y metamórfica de la zona de cizalla de Badajoz–Córdoba.* PhD Thesis, University of Pais Vasco.

ÁBALOS, B. 1992. Variscan shear-zone deformation of late Precambrian basement in SW Iberia, implications for circum-Atlantic pre-Mesozoic tectonics. *Journal of Structural Geology*, **14**, 807–823.

ÁBALOS, B. 1997. Omphacite fabric variation in the Cabo Ortegal eclogite (NW Spain): relationships with strain symmetry during high-pressure deformation. *Journal of Structural Geology*, **19**, 621–637.

ÁBALOS, B. 2001. Nuevos datos microestructurales sobre la existencia de deformaciones precámbricas en la Sierra de la Demanda (Cordillera Ibérica). *Geogaceta*, **30**, 3–6.

ÁBALOS, B. & DÍAZ CUSÍ, J. 1995. Correlation between seismic anisotropy and major geological structures in SW Iberia: a case study on continental lithosphere deformation. *Tectonics*, **14**, 1021–1040.

ÁBALOS, B., GIL IBARGUCHI, I. & EGUÍLUZ, L. 1991*a*. Structural and metamorphic evolution of the Almadén de la Plata Core (Seville, Spain) in relation to syn-metamorphic shear between the Ossa-Morena and South-Portuguese zones of the Iberian Variscan Fold Belt. *Tectonophysics*, **191**, 365–387.

ÁBALOS, B., GIL IBARGUCHI, I. & EGUÍLUZ, L. 1991*b*. Cadomian Subduction/Collision and Variscan Transpression in the Badajoz-Córdoba Shear Belt (SW Spain). *Tectonophysics*, **199**, 51–72.

ÁBALOS, B., GIL IBARGUCHI, J. I. & EGUÍLUZ, L. 1993. A reply to 'Cadomian subduction/collision and Variscan transpression in the Badajoz-Córdoba shear belt, southwest Spain. A discussion on the age of the main tectonometamorphic events', by A. AZOR, F. GONZÁLEZ LODEIRO & J. F. SIMANCAS. *Tectonophysics*, **217**, 347–353.

ÁBALOS, B., MENDIA, M. S. & GIL IBARGUCHI, J. I. 1994. Structure of the Cabo Ortegal eclogite-facies zone (NW Iberia). *Comptes Rendus de l'Académie des Sciences de Paris*, **319**, 1231–1238.

ABATI, J. 2000. *Petrología metamórfica y geocronología de la unidad culminante del Complejo de Órdenes en la región de Carvallo (Galicia, NW del Macizo Ibérico).* PhD Thesis, Complutense University, Madrid.

ABATI, J., DUNNING, G. R., ARENAS, R., DÍAZ GARCÍA, F., GONZÁLEZ CUADRA, P., MARTÍNEZ CATALÁN, J. R. & ANDONAEGUI, P. 1999. Early Ordovician orogenic event in Galicia (NW Spain): evidence from U-Pb ages in the uppermost unit of the Ordenes Complex. *Earth and Planetary Science Letters*, **165**, 213–228.

ABDEL-MONEM, A., WATKINS, N. D. & GAST, P. W. 1971. Potassium-argon ages, volcanic stratigraphy and geomagnetic polarity history of the Canary Islands: Lanzarote, Fuerteventura, Gran Canaria and La Gomera. *American Journal of Science*, **271**, 490–521.

ABDEL-MONEM, A., WATKINS, N. D. & GAST, P. W. 1972. Potassium-argon ages, volcanic stratigraphy and geomagnetic polarity history of the Canary Islands: Tenerife, La Palma and Hierro. *American Journal of Science*, **272**, 805–825.

ABLAY, G. J., CARROLL, M. R., PALMER, M. R., MARTÍ, J. & SPARKS, S. J. 1998. Basanite-phonolite lineages of the Teide-Pico Viejo volcanic complex, Tenerife, Canary Islands. *Journal of Petrology*, **39**, 905–936.

ABREU, V. S., HARDENBOL, J., HADDAD, G. A., BAUN, G. R., DROXLER, A. W. & VAIL, P. R. 1998. Oxygen isotope synthesis: a Cretaceous ice-house? *In*: GRACIANSKY P-CH. DE, HARDENBOL, J., JACQUIN, T. & VAIL, P. R. (eds) *Mesozoic and Cenozoic Sequence Stratigraphy of European Basins.* SEPM, Special Publications, **60**, 75–80.

ACEÑOLAZA, G. F. & GUTIÉRREZ-MARCO, J. C. 1998*a*. Estructuras de fijación de pelmatozoos (equinodermos) en el Ordovícico Medio de la Zona Centroibérica española. *Coloquios de Paleontología*, **49**, 23–40.

ACEÑOLAZA, G. F. & GUTIÉRREZ-MARCO, J. C. 1998*b*. *Helminthopsis abeli* Ksiazkiewicz, un icnofósil del Ordovícico Superior de la Zona Centroibérica española. *Geogaceta*, **24**, 7–10.

ACEÑOLAZA, G. F. & GUTIÉRREZ-MARCO, J. C. 1999. Icnofósiles del Ordovícico terminal (Pizarras Chavera, Pizarras de Orea: Hirnantiense) de algunas localidades españolas. *Boletín Geológico y Minero*, **110**, 123–134.

ACOSTA, A. 1998. *Estudio de los fenómenos de fusión cortical y generación de granitoides asociados a las peridotitas de Ronda.* PhD Thesis, University of Granada.

ACOSTA, P. & GARCÍA-HERNÁNDEZ, M. 1988. Las facies de plataforma carbonatada del Jurásico inferior y medio en la Sierra de Cazorla (Zona Prebética). *Geogaceta*, **5**, 39–41.

ADROVER, R., AGUSTÍ, J., MOYÁ, S. & PONS, J. 1983–84. Nueva localidad de micromamíferos insulares del Mioceno Medio en las proximidades de San Lorenzo en la isla de Mallorca. *Paleontología i Evoluciò*, **18**, 121–129.

AGIRREZABALA, L. M. 1996. *El Aptiense-Albiense del Anticlinorio Nor-Vizcaino entre Gernika y Azpeitia.* PhD Thesis, University of País Vasco.

AGIRREZABALA, L. M. & GARCÍA-MONDÉJAR, J. 1989. La serie de talud urgoniano de Ea (Bizkaia): caracteres sedimentológicos e implicaciones paleogeográficas. *In*: *Libro Homenaje a Rafael Soler*, AGGEP, Madrid, 15–25.

AGUADO, R. 1992. *Nannofósiles del Cretácico de la Cordillera Bética (sur de España): Bioestratigrafía.* PhD Thesis, University of Granada.

AGUADO, R., O'DOGHERTY, L., REY, J. & VERA, J. A. 1991. Turbiditas calcáreas del Cretácico al norte de Vélez Blanco (Zona Subbética): Bioestratigrafía y génesis. *Revista de la Sociedad Geológica de España*, **4**, 271–304.

AGUADO, R., MOLINA, J. M. & O'DOGHERTY, L. 1993. Bioestratigrafía y litoestratigrafía de la Formación Carbonero (Barremiense-Aptiense?) en la transición del Subbético externo-Subbético medio (sur de Jaén). *Cuadernos de Geología Ibérica*, **17**, 325–344.

AGUADO, R., DE GEA, G. A. & RUIZ-ORTIZ, P. A. 1996. Datos bioestratigráficos sobre las formaciones cretácicas del Dominio Intermedio en el corte tipo (sur de Jaén). Zonas Externas de las Cordilleras Béticas. *Geogaceta*, **20**, 197–200.

AGUADO, R., CASTRO, J. M., COMPANY, M. & DE GEA, G. 1999. Aptian bioevents, an integrated biostratigraphic analysis of the Almadich Formation (Inner Prebetic Domain, SE. Spain). *Cretaceous Research*, **20**, 663–683.

ÁGUEDA, J. A., BAHAMONDE, J. R., BARBA, F. J. *et al.* 1991. Depositional environments in Westphalian coal-bearing successions of the Cantabrian Mountains, northwest Spain. *Bulletin de la Société Géologique de France*, **162**(2), 325–337.

AGUIRRE, E. 1989. Vertebrados del Pleistoceno continental. *In*: PÉREZ-GONZÁLEZ, A., CABRA, P. & MARTÍN-SERRANO, A. (eds) *Mapa del Cuaternario de España a escala 1:1.000.000 y Memoria.* ITGE, Madrid, 47–70.

AGUIRRE, E. 1995. Registro faunístico Pleistoceno antiguo de Atapuerca. *Trabajos de Prehistoria*, **52**, 47–60.

AGUIRRE, E. 1998. El proyecto de Atapuerca. Propósito, estrategia y primeros resultados. *In*: AGUIRRE, E. (ed.) *Atapuerca y la evolución humana.* Fundación Ramón Areces, Madrid, 15–48.

AGUIRRE, E. 2000*a*. *Evolución humana. Debates actuales y vías abiertas.* Real Academia de Ciencias Exactas, Físicas y Naturales, Madrid, 11–115.

AGUIRRE, E. 2000*b*. Sima de los Huesos. Escenarios de la formación del yacimiento, crítica y sesgo demográfico, *In*: CARO, L., RODRÍGUEZ, H., SÁNCHEZ, E., LÓPEZ, B. & BLANCO, M. J. (eds) *Tendencias Actuales de Investigación en la Antropología Física Española.* University of León, 31–42.

AGUIRRE, E. & BERMÚDEZ DE CASTRO, J. M. 1991. Agut. Axlor. Banyoles. Carihuela. Cova Negra. Cueva Victoria. Ibeas (Atapuerca). Lezetxiki. Mollet I. Orce. Pinilla del Valle. Tossal de la Font. Valdegoba. Zafarraya. *In*: ORBAN, R. (ed.) *Hominid remains – No. 4. Spain.* Université Libre de Bruxelles.

AGUIRRE, E. & LUMLEY, M. A. 1977. Fossil man from Atapuerca, Spain: Their bearing on Human Evolution in the Middle Pleistocene. *Journal of Human Evolution*, **6**, 681–688.

AGUIRRE, E., DÍAZ MOLINA, M. & PÉREZ GONZÁLEZ, A. 1976. Datos paleomastológicos y fases tectónicas en el Neógeno de la meseta sur española. CSIC. *Trabajos sobre el Neógeno-Cuaternario*, **5**, 7–29.

AGUIRRE, E., LUMLEY, M. A., BASABE, J. M. & BOTELLA, M. 1980. Affinities between the mandibles of Atapuerca and l'Arago, and some East African Fossil Hominids. *In*: LEAKEY, R. A. & OGOT, B. A. (eds) *Proceedings of the VIIIth PCPQS*, Nairobi, 5–10 Sept. 1977. ILLMIAP, Nairobi, 171–174.

AGUIRRE, E., ALBERDI, M. T., JIMÉNEZ, E., MARTÍN ESCORZA, C., MORALES, J., SESÉ, C. & SORIA, D. 1982. Torrijos: nueva fauna con Hispanotherium de la cuenca media del Tajo. *Acta Geológica Hispánica*, **17**, 39–61.

AGUIRRE, J. 1995. Implicaciones estratigráficas y paleogeográficas de dos discontinuidades estratigráficas en los depósitos pliocenos de Cádiz (SW de España). *Revista Sociedad Geológica de España*, **8**, 153–166.

AGUIRRE, J. 1998*a*. Bioconstrucciones de *Saccostrea cuccullata* Born, 1778 en el Plioceno superior de Cádiz (SW de España): Implicaciones paleoambientales y paleoclimáticas. *Revista Española de Paleontología*, **13**, 27–36.

AGUIRRE, J. 1998*b*. El Plioceno del SE de la Península Ibérica (provincia de Almería). Síntesis estratigráfica, sedimentaria, bioestratigráfica y paleogeográfica. *Revista Sociedad Geológica de España*, **11**, 297–315.

AGUIRRE, J. & JIMÉNEZ, A. P. 1998. Fossil analogues of the present-day ahermatypic *Cladocora caespitosa* coral banks: sedimentary setting, dwelling community, and taphonomy (Late Pliocene, W Mediterranean). *Coral Reefs*, **17**, 203–213.

AGUIRRE, J., BRAGA, J. C. & MARTÍN, J. M. 1993. Algal nodules in the upper Pliocene deposits at the coast of Cádiz (S Spain). *In*: BARATTOLO, F., DE CASTRO P. & PARENTE, M. (eds) *Studies on Fossil Benthic Algae.* Muchi, Modena, 1–7.

AGUIRRE, J., CASTILLO, C., FERRIZ, F. J., AGUSTÍ, J. & OMS, O. 1995. Marine-continental magnetobiostratigraphic correlation of the Dolomys subzone (middle of Late Ruscinian): Implications for the Late Ruscinian age. *Palaeogeography, Palaeoclimatology, Palaeoecology*, **117**, 139–152.

AGUSTÍ, J. & MOYÁ-SOLÁ, S. 1987. Sobre la identidad del fragmento craneal atribuido a *Homo* sp. en venta Micena (Orce, Granada). *Estudios Geológicos*, **43**, 535–538.

AGUSTÍ, J. & MOYÁ-SOLÁ, S. 1998. The Early Pleistocene mammal turnover in Spain: Evidence against an 'End-Villafranquian' event. *In*: KOLFSCHOTEN, T. V. & GIBBARD, P. L. (eds) *The Dawn of the Quaternary* (Proceedings of the SEQS-Euroman Symposium 1996). Medelingen Nederlands Institut voor Toegepaste Geowetenschappen, Harlem, 513–519.

AGUSTÍ, J., MOYÁ-SOLÁ, S. & PONS-MOYÁ, J. 1986. Venta Micena (Guadix-Baza basin, South-Eastern Spain): its place in the Plio-Pleistocene mammal succession in Europe. *Geologica Romana*, **25**, 33–62.

AGUSTÍ, J., ANADÓN, P., ARBIOL, S., CABRERA, L., COLOMBO, F. & SÁEZ, A. 1987. Biostratigraphical characteristics of the Oligocene sequences of North-Eastern Spain (Ebro and Campins Basins). *Muenchner Geowissenschaftliche Abhandlungen, Reihe A: Geologie und Palaeontologie*, **10**, 34–42.

AGUSTÍ, J., CABRERA, L., ANADÓN, P. & ARBIOL, S. 1988. A Late Oligocene-Early Miocene rodent biozonation from the SE Ebro Basin (NE Spain): a potential mammal stage stratotype. *Newsletters on Stratigraphy*, **18**(2), 81–97.

AGUSTÍ, J., ARENAS, C., CABRERA, L. & PARDO, G. 1994*a*. Characterisation of the Latest Aragonian-Early Vallesian (Late Miocene) in the central Ebro Basin (NE Spain). *Scripta Geologica*, **106**, 1–10.

AGUSTÍ, J., BARBERÀ, X., CABRERA, L., PARÉS, J. M. & LLENAS, M. 1994*b*. Magnetobiostratigraphy of the Oligocene-Miocene transition in the Ebro Basin (Eastern Spain): state of the art. *Muenchner Geowissenschaftliche Abhandlungen, Reihe A: Geologie und Palaeontologie*, **26**, 161–172.

AGUSTÍ, J., OMS, O., GARCÉS, M. & PARÉS, J. M. 1997. Calibration of the Late Pliocene–Early Pleistocene transition in the continental beds of the Guadix-Baza Basin (Southeastern Spain). *Quaternary International*, **40**, 93–100.

AHIJADO, A. 1999. *Las intrusiones plutónicas e hipoabisales del sector meridional del Complejo Basal de Fuerteventura.* PhD Thesis, Complutense University, Madrid.

AHLBERG, P. 1984. Lower Cambrian trilobites and biostratigraphy of Scandinavia. *Lund Publications in Geology*, **22**, 1–38.

AKKERMAN, J. H., MAIER, G. & SIMON, O. J. 1980. On the Geology of the Alpujarride Complex in the western Sierra de las Estancias (Betic Cordilleras, SE Spain*). Geologie en Mijnbouw*, **59**, 363–374.

ALASTRUÉ, E., ALMELA, A. & RÍOS, J. M. 1957. *Explicación al Mapa Geológico de la Provincia de Huesca, E. 1:200.000.* IGME, Madrid.

ALBAIGÉS, J., ALGABA, J., CLAVELL, E. & GRIMALT, J. 1986. Petroleum Geochemistry of the Tarragona Basin (Spanish Mediterranean offshore). *Organic Geochemistry*, **10**, 441–450.

ALBAREDE, F. & MICHARD-VITRAC, A. 1978. Age and significance of the north-Pyrenean metamorphism. *Earth and Planetary Science Letters*, **40**, 327–332.

ALBERDI, M. T., HOYOS, M., JUNCO, F., LÓPEZ-MARTÍNEZ, N., MORALES, J., SESÉ, C. & SORIA, M. D. 1984. Biostratigraphy and sedimentary evolution of the continental Neogene in the Madrid area. *Paléobiologie Continentale*, Montpellier, **14**, 47–68.

ALBERTO, F., GUTIÉRREZ, M., IBÁÑEZ, M. J., MACHÍN, J., PEÑA, J. L., POCOVÍ, A. & RODRÍGUEZ, J. 1984. *El Cuaternario de la Depresión del Ebro en la región aragonesa. Cartografía y síntesis de los conocimientos existentes.* University of Zaragoza and Estación Experimental Aula Dei, Zaragoza.

ALCALÁ, L., ALONSO-ZARZA, A. M., ÁLVAREZ SIERRA, M. A. *et al.* 2000. El registro sedimentario y faunístico de las cuencas de Calatayud-Daroca y Teruel. Evolución paleoambiental y paleoclimática durante el Neógeno. *Revista de la Sociedad Geológica de España*, **13**(2), 323–343.

ALCALÁ DEL OLMO, L. 1972. Estudio sedimentológico de los arenales de Cuéllar (Segovia). *Estudios Geológicos*, **28**, 345–358.

ALDAYA, F. 1969. *Los Mantos Alpujárrides al Sur de Sierra Nevada.* PhD Thesis, University of Granada.

ALDAYA, F., CARLS, P., MARTÍNEZ GARCÍA, E. & QUIROGA, J. L. 1976. Nouvelles précisions sur la série de San Vitero (Zamora, nord-ouest de l'Espagne). *Comptes Rendus de l'Académie des Sciences, Paris (D)*, **283**, 881–883.

ALDAYA, F., GARCÍA-DUEÑAS, V. & NAVARRO VILA, F. 1979. Los Mantos Alpujárrides del tercio central de las Cordilleras Béticas.

Ensayo de correlación tectónica de los Alpujárrides. *Acta Geológica Hispánica*, **14**, 154–166.

ALDAYA, F., ALVAREZ, F., GALINDO-ZALDÍVAR, J., GONZÁLEZ-LODEIRO, F., JABALOY, A. & NAVARRO-VILÁ, F. 1991. The Malaguide-Alpujarride contact (Betic Cordilleras, Spain): a brittle extensional detachment. *Comptes Rendus de l'Académie des Sciences de Paris*, **313**, 1447–1453.

ALEKSEEV, A. S., KONONOVA, L. I. & NIKISHIN, A. M. 1996. The Devonian and Carboniferous of the Moscow Syneclise (Russian Platform): stratigraphy and sea-level changes. *Tectonophysics*, **268**, 149–168.

ALFÉREZ, F. 1977. Estudio del sistema de terrazas del río Tajo al W de Toledo. *Estudios Geológicos*, **33**, 223–250.

ALFÉREZ, F. 1985. Dos molares humanos procedentes del yacimiento del Pleistoceno medio de Pinilla del Valle (Madrid). *Trabajos de Antropología*, **19**, 303.

ALFONSO, P. & MELGAREJO, J. C. 2000. Boron vs. Phosphorous in granitic pegmatites: the Cap de Creus case (Catalonia, Spain). *Journal of the Czech Geological Society*, **45**(1–2), 131–141.

ALFONSO, P., CORBELLA, M. & MELGAREJO, J. C. 1995. Nb-Ta-Minerals from the Cap de Creus pegmatite field, eastern Pyrenees: distribution and geochemical trends. *Mineralogy and Petrology*, **55**, 53–69.

ALÍA, M. 1960. Sobre la tectónica profunda de la fosa del Toja. *Notas y Comunicaciones del IGME*, **58**, 125–162.

ALIBERT, C., DEBON. F. & TERNET, Y. 1988. Le pluton à structure concentrique du Néouvielle (Hautes Pyrénées): typologie chimique, âge et genèse. *Comptes Rendues de l'Académie des Sciences de Paris*, **306**, 49–54.

ALLEN, P. A. & ALLEN, J. R. 1990. *Basin Analysis*. Blackwell, Oxford.

ALLER, J. & BASTIDA, F. 1993. Anatomy of the Mondoñedo Nappe basal shear zone (NW Spain). *Journal of Structural Geology*, **15**, 1405–1419.

ALLERTON, S., LONERGAN, L., PLATT, J. P., PLATZMAN, E. S. & MCCLELLAND, E. 1993. Paleomagnetic rotations in the eastern Betic Cordillera, Southern Spain. *Earth and Planetary Science Letters*, **119**, 225–241.

ALMERA, J. 1891*a*. Caracterización del Muschelkalk en Gavá, Begas y Pallejá. *Crónica Científica*, **14**, 132–136, 276–281.

ALMERA, J. 1891*b*. Importancia del descubrimiento del '*Monograptus priodon*' cerca S. Vicens dels Horts. *Crónica Científica*, **14**(321), 116–118.

ALMERA, J. 1891*c*. Mapa geológico y topográfico de la provincia de Barcelona. Región Primera o de contornos de la capital. E. 1:400.000. Diputación Provincial Barcelona.

ALMERA, J. 1914. *Mapa geológico y topográfico de la provincia de Barcelona. Región Quinta o del Montseny. E. 1:40.000.* Diputación Provincial Barcelona.

ALMODÓVAR, G. R., SÁEZ, R., PONS, J. M., MAESTRE, A., TOSCANO, M. & PASCUAL, E. 1998. Geology and genesis of the Aznalcóllar massive sulphide deposits, Iberian Pyrite Belt, Spain. *Mineralium Deposita*, **33**, 111–136.

ALONSO, A. 1981. El Cretácico de la Provincia de Segovia (Borde N. Del Sistema Central). *Seminarios de Estratigrafía, Universidad Complutense de Madrid, Serie Monografías*, **7**, 271 pp.

ALONSO, A. & MAS, J. R. 1990. El Jurásico superior en el sector Demanda-Cameros (La Rioja-Soria). *Cuadernos de Geología Ibérica*, **14**, 173–198.

ALONSO, A. & MAS, R. 1993. Control tectónico e influencia del eustatismo en la sedimentación del Cretácico inferior de la cuenca de los Cameros, España. *Cuadernos de Geología Ibérica*, **17**, 285–310.

ALONSO, A. & PAGÉS, J. L. 2000. El registro sedimentario del final del Cuaternario en el litoral Noroeste de la Península Ibérica. Márgenes Cantábrico y Atlántico. *Revista de la Sociedad Geológica de España*, **13**(1), 17–29.

ALONSO, A., FLOQUET, M., MAS, J. R., MELÉNDEZ, A., MELÉNDEZ, N., SALOMÓN, J. & VADOT, J. P. 1987. Modalités de la régression marine sur le detroit ibérique (Espagne) à la fin du Cretacé. *Memoires Geologiques de la Universite de Dijon*, **11**, 91–112.

ALONSO, A., FLOQUET, M., MAS, R. & MELÉNDEZ, A. 1993. Late Cretaceous carbonate platforms, origin and evolution, Iberian Range, Spain. *In*: SIMÓ, J. A.T., SCOTT, R. W. & MASSE,

J.-P. (eds) *Cretaceous Carbonate Platforms*. AAPG Memoir, **56**, 297–313.

ALONSO, J. J. 1989. Estudio volcanoestratigráfico y volcanológico de los piroclastos sálicos del sur de Tenerife. *Col. Investigación Universidad de La Laguna*.

ALONSO, J. L. 1985. *Estructura y Evolución Tectonoestratigráfica del Manto del Esla (Zona Cantábrica, NW de España)*. Diputación Provincial de León, Instituto Fray Bernardino de Sahagún.

ALONSO, J. L. 1989. Fold reactivation involving angular unconformable sequences: theoretical analysis and natural examples from the Cantabrian Zone (Northwest Spain). *Tectonophysics*, **170**, 57–77.

ALONSO, J. L & TEIXELL, A. 1992. Forelimb deformation in some natural examples of fault-propagation folds. *In*: MCCLAY, K. (ed.) *Thrust Tectonics*. Chapman & Hall, London, 175–180.

ALONSO, J. L., ÁLVAREZ MARRÓN, J., ALLER, J. *et al.* 1992. Estructura de la Zona Cantábrica. *In*: GUTIÉRREZ MARCO, J., SAAVEDRA, J. & RÁBANO, I. (eds) *Paleozoico Inferior de Ibero-América*. University of Extremadura, Badajoz, 423–434.

ALONSO, J. L., PULGAR, J. A., GARCÍA-RAMOS, J. C. & BARBA, P. 1996. Tertiary basins and Alpine tectonics in the Cantabrian Mountains (NW Spain). *In*: FRIEND, P. F. & DABRIO, C. (eds) *Tertiary Basins of Spain*. Cambridge University Press, Cambridge, 214–227.

ALONSO-GAVILÁN, G. 1981. *Estratigrafía y sedimentología del Paleógeno en el borde suroccidental de la cuenca del Duero (provincia de Salamanca)*. PhD Thesis, University of Salamanca.

ALONSO-GAVILÁN, G. 1984. Evolución del sistema fluvial de la Formación Areniscas de Aldearrubia (Paleógeno superior) (provincia de Salamanca). *Mediterranea, Serie Geología*, **3**, 107–130.

ALONSO-GAVILÁN, G. 1986. Paleogeografía del Eoceno superior-Oligoceno en el SO de la Cuenca del Duero (España). *Studia Geológica Salmanticensia*, Ediciones Universidad Salamanca, **22**, 71–92.

ALONSO OLAZABAL, A., CARRACEDO, M. & ARANGUREN, A. 1999. Petrology, magnetic fabric and emplacement in a strike–slip regime of a zoned peraluminous granite: the Campanario–La Haba pluton, Spain. *In*: CASTRO, A., FERNÁNDEZ, C. & VIGNERESSE, J. L. (eds) *Understanding Granites. Integrating New and Classical Techniques*. Geological Society, London, 177–191.

ALONSO-ZARZA, A. M. & CALVO, J. P. 2000. Palustrine sedimentation in a episodically subsiding basin: the Miocene of the northern Teruel Graben (Spain). *Palaeogeography, Paleoclimatology, Palaeoecology*, **160**, 1–21.

ALONSO-ZARZA, A. M., CALVO, J. P. & GARCÍA DEL CURA, M. A. 1990. Litoestratigrafía y evolución paleogeográfica del Mioceno del borde noreste de la Cuenca de Madrid (prov. Guadalajara). *Estudios Geológicos*, **46**, 415–432.

ALONSO-ZARZA, A. M., CALVO, J. P. & GARCÍA DEL CURA, M. A. 1992*a*. Palustrine sedimentation and associated features – graainification and pseudo-microkarst- in the Middle Miocene (Intermediate Unit) of the Madrid Basin, Spain. *Sedimentary Geology*, **76**, 43–61.

ALONSO-ZARZA, A. M., WRIGHT, V. P., CALVO, J. P. & GARCÍA DEL CURA, M. A. 1992*b*. Soil-landscape relationships in the middle Miocene of the Madrid Basin. *Sedimentology*, **39**, 17–35.

ALONSO-ZARZA, A. M., CALVO, J. P. & GARCÍA DEL CURA, M. A. 1993. Palaeogeomorphological controls on the distribution and sedimentary styles of alluvial systems, Neogene of the NE of the Madrid Basin (Central Spain). *In*: MARZO, M. & PUIGDEFÁBREGAS, C. (eds) *Alluvial Sedimentation*. Special Publications, IAS, **17**, 277–292.

ALONSO-ZARZA, A. M., SANZ, M. E., CALVO, J. P. & ESTÉVEZ, P. 1998*a*. Calcified root cells in Miocene pedogenic carbonates of the Madrid Basin: evidence for the origin of *Microcodium* b. *Sedimentary Geology*, **116**, 81–97.

ALONSO-ZARZA, A. M., SILVA, P. G., GOY, J. L. & ZAZO, C. 1998*b*. Fan-surface dynamics and biogenic calcrete development: Interactions during ultimate phases of fan evolution in the semiarid SE Spain (Murcia). *Geomorphology*, **24**, 147–167.

ALTUNA, J. 1972. Fauna de mamíferos de los yacimientos prehistóricos de Guipuzcoa. *Munibe*, **24**, 1–464.

ALTUNA, J., MARIEZKURENA, K., ARMENDÁRIZ, A., DEL BARRIO, L.,

UGALDE, T. & PEÑALVER, J. 1982. Carta arqueológica de Guipuzcoa. *Munibe*, **34**, 1–231.

ALVARADO, A. & HERNÁNDEZ-SAMPELAYO, A. 1945. Zona occidental de la cuenca del Rubagón. *Boletín IGME*, **58**, 1–44.

ALVAREZ, C. 1994. Hydrocarbons in Spain: exploration and production. *First Break*, **12**, 43–46.

ÁLVAREZ, C., DOUBINGER, J. & DIÉGUEZ, C. 1969. Estudio paleo-obotánico de la flora de Ogassa. *Estudios Geológicos*, **27**, 267–277.

ÁLVAREZ, E. & MELÉNDEZ, F. 1994. Características generales de las subcuencas del margen peninsular mediterráneo ('Rift' del Surco de Valencia). *Acta Geológica Hispánica*, **29**(1), 67–79 (published 1996).

ALVAREZ, F. 1987. *La tectónica de la Zona Bética en la región de Aguilas*. PhD Thesis, University of Salamanca.

ÁLVAREZ, G., BUSQUETS, P., VILAPLANA, M. & RAMOS, E. 1989. Fauna coralina paleógena de las Islas Baleares (Mallorca y Cabrera), España. *Batalleria*, **3**, 61–68.

ALVAREZ DE BUERGO, E. & GARCÍA, A. 1996. Cálculo de reservas remanentes de hidrocarburos en zonas estructuralmente complejas: aplicación al campo de Ayoluengo IV. *Geogaceta*, **20**, 161–168.

ÁLVAREZ-MARRÓN, J., PULGAR, J. A., DAÑOBEITIA, J. J. *et al.* 1995. Results from the ESCI-N4 marine deep seismic profile in the northern Iberian Margin. *Revista de la Sociedad Geológica de España*, **8**(4), 355–363.

ÁLVAREZ-MARRÓN, J., RUBIO, E. & TORNÉ, M. 1997. Subduction-related structures in the North Iberian Margin. *Journal of Geophysical Research*, **102**, 22497–22511.

ALVAREZ NAVA, H., GARCIA CASQUERO, J. L., GIL, A. *et al.* 1988. Unidades litoestratigráficas de los materiales precámbrico-cámbricos de la mitad suroriental de la Zona Centro Ibérica. *II Congreso Geológico de España*, **1**, 49–22.

ÁLVAREZ-RAMIS, C. 1980. Sur la macroflore du Crétacé continantal de l'Espagne. *Mémoires de la Société Geologique de France*, **139**, LIX, 5–9.

ÁLVAREZ-RAMIS, C., RAMOS, E. & FERNÁNDEZ-MARRÓN, T. 1987. Estudio paleobotánico del Cenozoico de la zona central de Mallorca: Yacimiento de Son Ferragut. *Boletín Geológico y Minero*, **98**, 349–356.

ÁLVAREZ-SIERRA, M. A. 1987. Estudio sistemático y bioestratigráfico de los Eomyidae (Rodentia) del Oligoceno superior y Mioceno inferior español. *Scripta Geologica*, **86**, 1–207.

ÁLVAREZ-SIERRA, M. A., DAAMS, R., LACOMBA, J. I., LÓPEZ-MARTÍNEZ, N. & SACRISTÁN-MARTÍN, M. A. 1987. Succession of micro-mammal faunas in the Oligocene of Spain. *Muenchner Geowissenschaftliche Abhandlungen, Reihe A: Geologie und Palaeontologie*, **10**, 43–48.

ÁLVAREZ-SIERRA, M. A., DAAMS, R., LACOMBA, J. L., LÓPEZ MARTÍNEZ, N., VAN DER MEULEN, A. J., SESÉ, C. & DE VISSER, J. 1990. Paleontology and biostratigraphy (micromammals) of the continental Oligocene–Miocene deposits of the North-Central Ebro Basin (Huesca, Spain). *Scripta Geologica*, **94**, 1–77.

ÁLVARO, J. J. 1995. Propuesta de una nueva unidad litoestratigráfica para el Cámbrico Medio-Superior de las Cadenas Ibéricas (NE España): El Grupo Acón. *Boletín de la Real Sociedad Española de Historia Natural (Geología)*, **90**, 95–106.

ÁLVARO, J. J. & LIÑÁN, E. 1997. Nuevos datos acerca del Bilbiliense (Cámbrico Inferior terminal) en las Cadenas Ibéricas y su correlación con otras áreas. *Revista Española de Paleontología*, **12**, 277–280.

ÁLVARO, J. J. & VENNIN, E. 1998. Stratigraphic signature of a terminal Early Cambrian regressive event in the Iberian Peninsula. *Canadian Journal of Earth Sciences*, **35**, 402–411.

ÁLVARO, J. J. & VIZCAÏNO, D. 1998. Révision biostratigraphique du Cambrien moyen du versant méridional de la Montagne Noire (Languedoc, France). *Bulletin de la Société Géologique de France*, **169**, 233–242.

ÁLVARO, J. J., LIÑÁN, E., GOZALO, R. & GÁMEZ-VINTANED, J. A. 1993a. Estratigrafía del tránsito Cordubiense-Ovetiense (Cámbrico Inferior) en la Cadena Ibérica Occidental (España). *Cuadernos do Laboratorio Xeolóxico de Laxe*, **18**, 147–162.

ÁLVARO, J. J., LIÑÁN, E., GOZALO, R. & SDZUY, K. 1993b. The palaeo-geography of the northern Iberia at the Lower-Middle Cambrian transition. *Bulletin de la Société Géologique de France*, **164**, 843–850.

ÁLVARO, J. J., LIÑÁN, E., VENNIN, E. & GOZALO, R. 1995. Palaeogeo-graphical evolution within a passive margin with syndepositional faulting: the Marianian deposits (Lower Cambrian) of the Iberian Chains (NE Spain). *Neues Jahrbuch Geologie und Paläontologie, Mh.*, **1995**, 521–540.

ÁLVARO, J. J., LIÑÁN, E. & VIZCAÏNO, D. 1998. Biostratigraphical significance of the genus *Ferralsia* (Lower Cambrian, Trilobita). *Geobios*, **31**, 499–504.

ÁLVARO, J. J., ROUCHY, J. M., BECHSTÄDT, T. *et al.* 2000. Evaporitic constraints on the southward drifting of the western Gondwana margin during Early Cambrian times. *Palaeogeography, Palaeo-climatology, Palaeoecology*, **160**, 105–122.

ÁLVARO, M. 1975. Estilolitos tectónicos y fases de plegamiento en el área de Sigüenza (borde del Sistema Central y la Cordillera Ibérica). *Estudios Geológicos*, **31**, 241–247.

ÁLVARO, M. 1987. La subsidencia tectónica en la Cordillera Ibérica durante el Mesozoico. *Geogaceta*, **3**, 34–37.

ÁLVARO, M. 1995. La Cadena Ibérica. *In*: MELÉNDEZ, A. & GUTIERREZ, M. (eds) *XXIX Curso de Geología Práctica*, Teruel, 1–28.

ÁLVARO, M. & CAPOTE, R. 1973. Las estructuras menores de las calizas jurásicas de un anticlinal de la Sierra de Altomira (Cuenca, España). *Estudios Geológicos*, **24**, 467–478.

ÁLVARO, M. & DEL OLMO, P. 1992. *Palma*, Spanish Geological Survey, Report and Map, Scale 1:50 000. 1–64.

ÁLVARO, M., CAPOTE, R. & VEGAS, R. 1979. Un modelo de evolución geotectónica para la Cadena Celtibérica. *Acta Geológica Hispánica*, **14**, 172–177.

ÁLVARO, M., BARNOLAS, A., DEL OLMO, P., RAMÍREZ DEL POZO, J. & SIMÓ, A. 1984a. Estratigrafía del Jurásico. *In*: ÁLVARO, M., BARNOLAS, A., DEL OLMO, P., RAMÍREZ DEL POZO, J. & SIMÓ, A. (eds) *Sedimentología del Jurásico de Mallorca*. GEM-IGME-CGS, 43–71.

ÁLVARO, M., BARNOLAS, A., DEL OLMO, P., RAMÍREZ DEL POZO, J. & SIMÓ, A. 1984b. El Neógeno de Mallorca: Caracterización sedi-mentológica y biostratigráfica. *Boletín Geológico y Minero*, **95**, 3–25.

ÁLVARO, M., BARNOLAS, A., CABRA, A. *et al.* 1989. El Jurásico de Mallorca (Islas Baleares). *Cuadernos de Geología Ibérica*, **13**, 67–120.

ÁLVARO, M., DEL OLMO, P. & BATLLE, A. 1992. *Sóller*, Spanish Geological Survey, Report and Map, Scale 1:50 000.

AMBROSE, T. 1972. *The stratigraphy and structure of the pre-Carbon-iferous rocks North-Est of Cervera de Pisuerga. Cantabrian Mountains, Spain*. PhD Thesis, University of Sheffield.

AMEGLIO, L., VIGNERESSE, J. L. & BOUCHEZ, J. L. 1997. Granite pluton geometry and emplacement mode inferred from combined fabric and gravity data. *In*: BOUCHEZ, J. L., STEPHENS, W. E. & HUTTON, D. H. W. (eds) *Granite: from Segregation of Melt to Emplacement Fabrics*. Kluwer, Dordrecht, 200–214.

AMILS, R., LÓPEZ-ARCHILLA, A. I. & MARÍN, I. 1997. Modelos de vida en condiciones límite como base para la exobiología. *Academia Ciencias Exactas, Fisicas y Naturales*, **91**, 87–99.

AMIOT, M. 1982. El Cretácico superior de la Región Navarro-Cantabra. *In*: *El Cretácico de España*. Complutense University, Madrid, 88–111.

AMOROS, J. L., LUNAR, R. & TAVIRA, P. 1981. Jarosite: A silver bearing mineral of the gossan of Río Tínto (Huelva) and La Unión (Cartagena). *Mineralium Deposita*, **16**, 205–213.

AMOROS, J. L, LOPEZ, J. A., LUNAR, R., MARTINEZ, J., SIERRA, J. & VINDEL, E. 1984. Chalcopyrite-sphalerite textures in some Spanish syngenetic and epigenetic deposits: Guadarrama Mountain, Aznalcollar and La Union. *In*: WAUSCHKUHN *et al.* (eds) *Syngenesis and Epigenesis in the Formation of Minerals Deposits*, Springer-Verlag, Berlin, 18–27.

ANADÓN, P. 1978. *El Paleógeno continental anterior a la transgresión Biarritziense (Eoceno Medio) entre los ríos Gaià y Ripoll (Provin-cias de Tarragona y Barcelona)*. PhD Thesis, University of Barcelona.

ANADÓN, P. & FEIST, M. 1981. Charophytes et biostratigraphie du Paléogène inférieur du bassin de l'Èbre oriental. *Palaeonto-graphica, B*, **178**, 143–168.

ANADÓN, P. & MOISSENET, E. 1996. Neogene basins in the Eastern Iberian Range. *In*: FRIEND, P. F. & DABRIO, C. J. (eds) *Tertiary Basins of Spain. The Stratigraphic Record of Crustal Kinematics*. Cambridge University Press, Cambridge, 68–76.

ANADÓN, P., COLOMBO, F., ESTEBAN, M., MARZO, M., ROBLES, S., SANTANACH, P. & SOLÉ SUGRAÑES, L. 1979. Evolución tectonoestratigráfica de los Catalánides. *Acta Geológica Hispánica*, **14**, 242–270.

ANADÓN, P., FEIST, M., HARTENBERGER, J. L., MULLER, C. & VILLALTA-COMELLA, J. 1983a. Un exemple de corrélation biostratigraphique entre échelles marines et continentales dans l'Éocène: la coupe de Pontils (Bassin de l'Ebre, Espagne). *Bulletin de la Société Géologique de France*, **25**(7), 747–755.

ANADÓN, P., JULIVERT, M. & SÁEZ, A. 1983b. VIII El Carbonífero de las Cadenas Costeras Catalanas. *In*: MARTÍNEZ DÍAZ, C. (ed.) *Carbonífero y Pérmico de España*. IGME, Madrid, 331–336.

ANADÓN, P., VIANEY-LIAUD, M., CABRERA, L. & HARTENBERGER, J. L. 1987. Gisements à vertébrés du Paléogène de la zona orientale du bassin de l'Ebre et leur apport à la stratigraphie. *Paleontologia i Evolució*, **21**, 117–131.

ANADÓN, P., CABRERA, L., CHOI, S. J., COLOMBO, F., FEIST, M. & SÁEZ, A. 1992. Biozonación del Paleógeno continental de la zona oriental de la Cuenca del Ebro mediante caráfitas: implicaciones en la biozonación general de caráfitas de Europa occidental. Homenaje a Oriol Riba Arderiu. *Acta Geológica Hispánica*, **27**, 69–94.

ANCOCHEA, E. 1983. *Evolución espacial y temporal del volcanismo reciente de España Central*. Complutense University, Madrid, **203/83**.

ANCOCHEA, E. & BRANDLE, J. L. 1982. Alineaciones de volcanes en la región volcánica central española. *Revista de Materiales y Procesos Geológicos*, **2**,115–133.

ANCOCHEA, E., GIULIANI, A. & VILLA, I. 1979. Edades radiométricas K-Ar del vulcanismo de la región central española. *Estudios Geológicos*, **35**, 131–135.

ANCOCHEA, E., PERNI, A. & HERNÁN, F. 1980. Caracterización geoquímica del vulcanismo del área de Atienza (provincia de Guadalajara. España). *Estudios Geológicos*, **36**, 327–337.

ANCOCHEA, E., MUÑOZ, M. & SAGREDO, J. 1987. Las rocas volcánicas neógenas de Nuévalos (provincia de Zaragoza). *Geogaceta*, **3**, 7–10.

ANCOCHEA, E., ARENAS, R., BRANDLE, J. L., PEINADO, M. & SAGREDO, J. 1988. Caracterización de las rocas metavolcánicas silúricas del noroeste del Macizo Ibérico. *Geociencias*, **8**, 23–34.

ANCOCHEA, E., FÚSTER, J. M., IBARROLA, E. *et al*. 1990. Volcanic evolution of the island of Tenerife (Canary Islands) in the light of new K-Ar data. *Journal of Volcanology and Geothermal Research*, **44**, 231–249.

ANCOCHEA, E., CASQUET, C., JAMOND, C., DÍAZ DE TÉRAN, J. R. & CENDRERO, A. 1992. Evolution of the eastern volcanic ridge of the Canary Islands based on new K-Ar data *Journal of Volcanology and Geothermal Research*, **53**, 251–274.

ANCOCHEA, E., HERNÁN F., CENDRERO, A., CANTAGREL, J. M., FÚSTER, J. M., IBARROLA, E. & COELLO, J. 1994. Constructive and destructive episodes in the building of a young oceanic island: La Palma, Canary Islands, and genesis of the Caldera de Taburiente. *Journal of Volcanology and Geothermal Research*, **60**, 3–4, 243–262.

ANCOCHEA, E., BRANDLE, J. L., CUBAS, C. R., HERNAN, F. & HUERTAS, M. J. 1996. Volcanic complexes in the Eastern Ridge of the Canary Islands: the Miocene activity of the island of Fuerteventura. *Journal of Volcanology and Geothermal Research*, **70**, 183–204.

ANCOCHEA, E., CANTAGREL, J. M., FÚSTER, J. M., HUERTAS, M. J. & ARNAUD, N. O. 1998. Vertical and lateral collapses on Tenerife (Canary Islands) and other volcanic ocean islands: comment. *Geology*, **26**, 861–862.

ANCOCHEA, E., HUERTAS, M. J., CANTAGREL, J. M., COELLO, J., FÚSTER, J. M., ARNAUD, N. O. & IBARROLA, E. 1999. Evolution of the Cañadas Edifice and its implications for the origin of the Cañadas Caldera (Tenerife, Canary Islands). *Journal of Volcanology and Geothermal Research*, **88**, 177–199.

ANCOCHEA, E., HUERTAS, M. J., CANTAGREL, J. M., FÚSTER, J. M. & ARNAUD, N. O. 2000. Cronología y evolución del edificio Cañadas, Tenerife, Islas Canarias. *Boletín Geológico y Minero*, **111**, 3–16.

ANDERSON, J. P. 1951. *The Dynamics of Faulting and Dyke Formation with Application to Britain*. Oliver & Boyd, Edinburgh.

ANDONAEGUI, P. 1990. *Geoquímica y geocronología de los granitoides del Sur de Toledo*. PhD Thesis, Complutense University, Madrid.

ANDREIS, R. R. & WAGNER, R. 1983. Estudios de abanicos aluviales en el borde norte de la cuenca Westfaliense B de Peñarroya-Bélmez. *In*: LEMOS DE SOUSA, M. J. (ed.) *Contributions to the Carboniferous Geology and Palaeontology of the Iberian Peninsula*. Universidad do Porto, 171–227.

ANDREWS, J. R., ARBIZU, M. A., BARROUQUÈRE, G. *et al*. 1996. Dévonien-Carbonifère Inférieur. *In*: *Synthèse géologique et géophysique des Pyrénées*. BRGM and ITGE, Orleans and Madrid, **1**, 235–301.

ANDREWS, P. & FERNÁNDEZ JALVO, Y. 1997. Surface modifications of the Atapuerca fossil humans. *Journal of Human Evolution*, **33**, 191–217.

ANDRIEUX, J., FONTBOTÉ, J. M. & MATTAUER, M. 1971. Sur un modèle explicatif de l'Arc de Gibraltar. *Earth and Planetary Science Letters*, **12**, 191–198.

ANDRIEUX, J., FRIZON DE LAMOTTE, D. & BRAUD, J. 1989. A structural scheme for the Western Mediterranean area in Jurassic and Early Cretaceous times. *Geodinamica Acta*, **3**(1), 5–15.

ANGLADA, E. & SERRA-KIEL, J. 1986. El Paleógeno y tránsito al Neógeno en el área del macizo de Randa (Mallorca). *Boletín Geológico y Minero*, **97**, 580–589.

ANGUITA, F. & HERNÁN, F. 1975. A propagating fracture model versus a hot spot origin for the Canary Islands. *Earth and Planetary Science Letters*, **27**, 1, 11–19.

ANGUITA, F. & HERNÁN, F. 1986. Geochronology of some Canarian dyke swarms: contribution to the volcano-tectonic evolution of the Archipelago. *Journal of Volcanology and Geothermal Research*, **30**, 155–162.

ANGUITA, F. & HERNÁN, F. 2000. The Canary Islands origin: a unifying model. *Journal of Volcanology and Geothermal Research*, **103**, 1–26.

ANGULO, A., BRACERO, C. & MUÑOZ, A. 2000. Caracterización de las Unidades Tectosedimentarias de la Comarca de La Bureba (Burgos, España) y su correlación con las cuencas terciarias del Ebro y Duero. *Geotemas*, **1**(2), 19–23.

ANSTED, D. T. 1860. On the geology of Málaga and the southern part of Andalucia. *Quarterly Journal of the Geological Society of London*, **15** (suppl.), 585–604.

ANTHONIOZ, P. M. 1972. *Les complexes polymétamorphiques précambriens de Morais et Bragança (NE du Portugal): étude pétrographique et structurale*. Memórias dos Serviços Geológicos de Portugal, **20**, 1–112.

APALATEGUI, O., BORRERO, J, DELGADO, M., ROLDÁN, F. J. & SÁNCHEZ, R. 1985. *Mapa Geológico de España a E. 1:50.000 no. **901** (Villaviciosa de Córdoba)*. 2ª serie-MAGNA. IGME, Madrid.

APALATEGUI, O., EGUILUZ, L. & QUESADA, C. 1990. Ossa Morena Zone, Structure. *In*: DALLMEYER, R. D. & MARTÍNEZ GARCÍA, E. (eds) *Pre-Mesozoic Geology of Iberia*. Springer-Verlag, Berlin, 280–291.

APARICIO, A. & GARCÍA, R. 1995. El volcanismo de las Islas Columbretes (Mediterráneo Occidental). Quimismo y mineralogía. *Boletín Geológico y Minero*, **106**, 468–488.

APARICIO, A., MITJAVILA, J. M., ARAÑA, V. & VILLA, I. M. 1991. La edad del volcanismo de las islas Columbrete Grande y Alborán (Mediterráneo occidental). *Boletín Geológico y Minero*, **102**, 562–570.

APRAIZ, A. 1998. *Geología de los macizos de Lora del Río y Valuengo (Zona de Ossa-Morena). Evolución tectonometamórfica y significado geodinámico*. PhD Thesis, University of Pais Vasco.

ARAMBURU, C. 1989. *El Cambro-Ordovícico de la Zona Cantábrica (NO de España)*. PhD Thesis, University of Oviedo.

ARAMBURU, C. & GARCÍA-RAMOS, J. C. 1988. Presencia de la discontinuidad sárdica en la Zona Cantábrica. *Geogaceta*, **5**, 11–13.

ARAMBURU, C. & GARCÍA-RAMOS, J. C. 1993. La sedimentación cambro-ordovícica en la Zona Cantábrica (NO de España). *Trabajos de Geología, Universidad de Oviedo*, **19**, 45–73.

ARAMBURU, C., TRUYOLS, J., ARBIZU, M. *et al*. 1992. El Paleozoico Inferior en la Zona Cantábrica. *In*: GUTIÉRREZ MARCO, J. C.,

SAAVEDRA, J. & RÁBANO, I. (eds) *Paleozoico Inferior de Ibero-America.* University of Extremadura, Badajoz, 397–421.

ARAMBURU, C., ARBIZU, M., GUTIÉRREZ-MARCO, J. C., MÉNDEZ-BEDIA, I., RÁBANO, I. & TRUYOLS, J. 1996. Primera identificación de materiales del Ordovícico Medio en le sección de Los Barrios de Luna (Zona Cantábrica, noroeste de España). *Geogaceta*, **20**, 7–10.

ARAÑA, V. 1971. Litología y estructura del edificio Cañadas, Tenerife (Islas Canarias). *Estudios Geológicos*, **27**, 95–137.

ARAÑA, V. & VEGAS, R. 1974. Plate tectonics and volcanism in the Gibraltar Arc. *Tectonophysics*, **24**, 197–211.

ARAÑA, V., APARICIO, A., MARTÍN-ESCORZA, C. *et al.* 1983. El volcanismo neógeno-cuaternario de Catalunya: caracteres estructurales, petrológicos y geodinámicos. *Acta Geológica Hispánica*, **18**, 1–17.

ARANBURU, A. 1998. *El Aptiense-Albiense de Trucíos-Güeñes (oeste de Bizkaia).* PhD Thesis, University of País Vasco.

ARANBURU, A., FERNÁNDEZ-MENDIOLA, P. A. & GRACÍA-MONDÉJAR, J. 1992. Contrasting styles of paleokarst infill in a block-faulted carbonate ramp (Lower Albian, Trucios, N. Spain). *Geogaceta*, **11**, 42–44.

ARANDA, M. & SIMÓN, J. L. 1993. Aspectos de la tectónica cretácica y terciaria en la Cuenca de Utrillas (Teruel) a partir de los datos de minería del interior. *Revista de Sociedad Geológica de España*, **6**(1–2), 123–129.

ARANGUREN, A. 1997. Magnetic fabric and 3D geometry of the Hombreiro-Sta. Eulalia pluton: implications for the Variscan structures of eastern Galicia, NW Spain. *Tectonophysics*, **273**, 329–344.

ARANGUREN, A. & TUBÍA, J. M. 1992. Structural evidence for the relationship between thrusts, extensional faults and granite intrusion in the Variscan belt of Galicia (Spain). *Journal of Structural Geology*, **14**, 1229–1237.

ARANGUREN, A., LARREA, F. J., CARRACEDO, M., CUEVAS, J. & TUBÍA, J. M. 1997. The Los Pedroches batholith (southern Spain): polyphase interplay between shear zones in transtension and setting of granites. *In*: BOUCHEZ, J. L., STEPHENS, W. E. & HUTTON, D. H. W. (eds) *Granite: from Segregation of Melt to Emplacement Fabrics.* Kluwer, Dordrecht, 215–229.

ARBIN, P., HAVLÍCEK, V. & TAMAIN, G. 1978. La 'Formation d'Enevrio' de l'Ordovicien de la Sierra Morena (Espagne), et sa faune à *Drabovia praedux* nov.sp. (Brachiopoda). *Bulletin de la Société Géologique de France* [7], **20**, 29–37.

ARBIZU M., GUTIÉRREZ-MARCO, J. C., LIÑÁN, E. & RÁBANO I. 1997. Fósiles del Paleozoico Inferior del Manto de Mondoñedo (Lugo). *In*: GRANDAL D'ANGLADE, A., GUTIÉRREZ-MARCO, J. C. & SANTOS FIDALGO, L. (eds) *XIII Jornadas de Paleontología*, A Coruña, Libro de Resúmenes y Excursiones, 333–352.

ARBOLEYA, M. L. 1981. La estructura del Manto del Esla (Cordillera Cantábrica, León). *Boletín Geológico y Minero*, **92**, 19–40.

ARBOLEYA, M. L. 1989. Fault rocks of the Esla Thrust (Cantabrian Mountains, N Spain): an example of foliated cataclasites. *Annales Tectonicae*, **3**, 99–109.

ARBONA, J., FONTBOTÉ, J.-M., GONZÁLEZ-DONOSO, J. M. *et al.* 1984–85. Precisiones biosestratigráficas y aspectos sedimentológicos del Jurásico-Cretácico basal de la isla de Cabrera (Baleares). *Cuadernos de Geología, Universidad de Granada*, **12**, 169–186

ARBUÉS, P., PI, E. & BERÁSTEGUI, X. 1996. Relaciones entre la evolución sedimentaria del Grupo de Areny y el cabalgamiento de Bóixols. *Geogaceta*, **20**, 446–449.

ARCHE, A. & LÓPEZ-GÓMEZ, J. 1992. Una nueva hipótesis sobre las primeras etapas de la evolución tectonosedimentaria de la cuenca permotriásica del SE de la Cordillera Ibérica. *Cuadernos de Geología Ibérica*, **16**, 115–143.

ARCHE, A. & LÓPEZ-GÓMEZ, J. 1996. Origin of the Permian-Triassic Iberian Basin, central-eastern Spain. *Tectonophysics*, **266**, 443–464.

ARCHE, A. & LÓPEZ-GÓMEZ, J. 1999. Tectonic and geomorphic controls on the fluvial styles of the Eslida Formation, Middle Triassic, Eastern Spain. *Tectonophysics*, **315**, 187–207.

ARCHE, A., COMAS-RENGIFO, M. J., GÓMEZ, J. J. & GOY, A. 1977. Evolución vertical de los sedimentos carbonatados del Lías medio y superior en Sierra Palomera (Teruel). *Estudios Geológicos*, **33**, 571–574.

ARCHE, A., RAMOS, A. & SOPEÑA, A. 1983. El Pérmico de la Cordillera Ibérica y bordes del Sistema Central. *In*: MARTÍNEZ GARCÍA, E. (ed.) *Carbonífero y Pérmico de España*. Ministerio de Industria y Energía, 423–438.

ARCHE, A., LÓPEZ-GÓMEZ, J. & GARCÍA-HIDALGO, J. F. 2001. Depósitos regresivos del Carniense (Triásico Superior) en el SE de la Cordillera Ibérica. Controles climáticos, tectónicos y eustáticos. *Geotemas*, **3**(1), 71–74.

ARDEVOL, L., KLIMOWITZ, J., MALAGÓN, J. & NAGTEGAAL, P. J. C. 2000. Depositional sequence response to foreland deformation in the Upper Cretaceous of the Southern Pyrenees, Spain. *AAPG Bulletin*, **84**, 566–587.

ARDIZONE, J., MEZCUA, J. & SOCIAS, I. 1989. *Mapa aeromagnético de España peninsular*. Instituto Geográfico Nacional, MOPU, Madrid.

ARENAS, C. 1993. *Sedimentología y paleogeografía del Terciario del margen pirenaico y sector central de la Cuenca del Ebro (zona aragonesa occidental)*. PhD Thesis, University of Zaragoza.

ARENAS, C. & PARDO, G. 1991. Significado de la ruptura sedimentaria entre las Unidades Tectosedimentarias N2 y N3 en el centro de la Cuenca del Ebro. *Geogaceta*, **9**, 67–70.

ARENAS, C. & PARDO, G. 2000. Neogene lacustrine deposits of the north-central Ebro Basin, northeastern Spain. *In*: GIERLOWSKI-KORDESCH, E. & KELTS, K. (eds) *Lake Basins Through Space and Time*. AAPG Studies in Geology, **46**, 395–406.

ARENAS, C., MILLÁN, H., PARDO, G. & POCOVÍ, A. 2001. Ebro basin continental sedimentation associated with late Pyrenean compressional Pyrenean tectonics (north-eastern Iberia): controls on basin margin fans and alluvial systems. *Basin Research*, **13**, 65–89.

ARENAS, R. 1985. *Evolución petrológica y geoquímica de la Unidad Alóctona inferior del Complejo Metamórfico básico-ultrabásico de Cabo Ortegal (Unidad de Moeche) y del Silúrico parautóctono. Cadena Hercínica Ibérica (NW de España)*. PhD Thesis, Complutense University, Madrid.

ARENAS, R. 1991. Opposite P-T-t paths of Hercynian metamorphism between the upper units of the Cabo Ortegal Complex and their substratum (northwest of Iberian Massif). *Tectonophysics*, **191**, 347–364.

ARENAS, R., CASQUET, C. & PEINADO, M. 1980. El metamorfismo en el sector de Riaza (Somosierra, Sistema Central Español). Implicaciones Geoquímicas y Petrológicas. *Cuadernos del Laboratorio Geológico de Laxe*, **1**, 117–146.

ARENAS, R., GIL IBARGUCHI, J. I., GONZALEZ LODEIRO, F. *et al.* 1986. Tectonostratigraphic units in the Complexes with mafic and related rocks of the NW of the Iberian Massif. *Hercynica*, **2**, 87–110.

ARENAS, R., RUBIO PASCUAL, F. J., DÍAZ GARCÍA, F. & MARTÍNEZ CATALÁN, J. R. 1995. High-pressure micro-inclusions and development of an inverted metamorphic gradient in the Santiago Schists (Ordenes Complex, NW Iberian Massif, Spain): evidence of subduction and syncollisional decompression. *Journal of Metamorphic Geology*, **13**, 141–164.

ARENAS, R., ABATI, J., MARTÍNEZ CATALÁN, J. R., DÍAZ GARCÍA, F. & RUBIO PASCUAL, F. J. 1997. P-T evolution of eclogites from the Agualada Unit (Ordenes Complex, northwest Iberian Massif, Spain): implications for crustal subduction. *Lithos*, **40**, 221–242.

ARÈNES, J. & DEPAPE, G. 1956. La flore Burdigalienne des iles Baléares (Majorque). *Revue Générale de Botanique*, **63**, 347–390.

ARGLES, T. W., PLATT, J. P. & WATER, D. J. 1999. Attenuation and excision of a crustal section during extensional exhumation: the Carratraca Massif, Betic Cordillera, Southern Spain. *Journal of the Geological Society, London*, **156**, 149–162.

ARGUS, D. F., GORDON, R. G., DEMETS, C. & STEIN, S. 1989. Closure of the Africa-Eurasia-North America plate motion circuit and tectonics of the Gloria fault. *Journal of Geophysical Research*, **94**, 5585–5602.

ARGYRIADIS, I., DE GRACIANSKY, P. C., MARCOUX, J. & RICOU, L. E. 1980. The opening of the Mesozoic Tethys between Eurasia and Arabia-Africa. *In*: AUBOIN, J., DEBELMAS, J. & LATREILLE, M. (eds) *Geology of the Alpine chains born of the Tethys*. Mémoires du Bureau de Recherches Géologiques et Minières, **115**, 200–213.

ARIAS, C. 1978. *Estratigrafía y paleogeografía del Jurásico superior y*

Cretácico inferior del nordeste de la provincia de Albacete. Seminarios de Estratigrafía, Serie Monografías, **3**.

ARIAS, C., MAS, R., GARCÍA, A., ALONSO, A, VILAS, L., RINCÓN R. & MELÉNDEZ, N. 1979. Les facies urgoniens et leurs variations pendant la transgression aptienne occidentale de la Chaîne Ibérique (Espagne). *Geobios Mémoire Spécial*, **3**, 11–23.

ARIAS, P. 1992. Geoquímica y mineralogía del yacimiento de Pb-Zn de Rubiales (Lugo). *In*: GARCÍA GUINEA, J. & MARTÍNEZ-FRÍAS, J. (eds) *Recursos Minerales de España*. CSIC, Colección Textos Universitarios **15**, 964–985.

ARLEGUI, L. E. 1996. *Diaclasas, fallas y campo de esfuerzos en el sector central de la cuenca del Ebro.* PhD Thesis, University of Zaragoza.

ARLEGUI, L. E. & SIMÓN, J. L. 1993. El sistema de diaclasas N-S en el sector central de la Cuenca del Ebro. Relación con el campo de esfuerzos neógeno. *Revista de la Sociedad Geológica de España*, **6**(1–2), 115–122.

ARLEGUI, L. E. & SIMÓN, J. L. 1997. El sistema de fallas normales de Las Bárdenas (Navarra) en el marco evolutivo del campo de esfuerzos neógeno. *In*: CALVO, J. P. & MORALES, J. (eds) *Avances en el conocimiento del Terciario Ibérico*. Universidad Complutense de Madrid & Museo Nacional de Ciencias Naturales, Madrid, 29–32.

ARLEGUI, L. E. & SIMÓN, J. L. 1998. Reliability of paleostress analysis from fault striations in near multidirectional extension stress fields. Example from the Ebro Basin, Spain. *Journal of Structural Geology*, **20**(7), 827–840.

ARLEGUI, L. E. & SIMÓN, J. L. 2001. Geometry and distribution of regional joint sets in a non-homogeneous stress field: case study in the Ebro basin (Spain). *Journal of Structural Geology*, **23**(3–4), 297–313.

ARLEGUI, L. E. & SORIANO, M. A. 1998. Characterizing lineaments from satellite images and field studies in the central Ebro basin (NE Spain). *International Journal of Remote Sensing*, **19**(16), 3169–3185.

ARMENTEROS, I. 1986. *Estratigrafía y sedimentología del Neógeno del sector suroriental de la Depresión del Duero.* Ediciones Diputación de Salamanca, Serie Castilla y León, **1**.

ARMENTEROS, I. 1991. Contribución al conocimiento del Mioceno lacustre de la Cuenca terciaria del Duero (sector centro-oriental, Valladolid-Peñafiel-Sacramenia-Cuéllar). *Acta Geologica Hispanica*, **26**, 97–131.

ARMENTEROS, I. & BUSTILLO, M. A. 1996. Sedimentología, paleoalteraciones y diagénesis en la unidad Carbonática de Cihuela (Eoceno superior de la Cuenca de Almazán, Soria). *Geogaceta*, **20**(2), 266–269.

ARMENTEROS, I. & CORROCHANO, A. 1994. Lacustrine record in the continental Tertiary Duero basin (northern Spain). *In*: GIERLOWSKI E. & KELTS, K. (eds) *Global Geological Record of Lacustrine Basins*. Vol. I. Cambridge University Press, Cambridge, 47–52.

ARMENTEROS, I., DABRIO, C. J., GUISADO, R. & SÁNCHEZ DE VEGA, A. 1989. Megasecuencias sedimentarias del Terciario del borde oriental de la Cuenca de Almazán (Soria-Zaragoza). *Studia Geológica Salmanticensis*, Ediciones Universidad Salamanca, **5**, 107–127.

ARMENTEROS, I., BUSTILLO, M. A. & BLANCO, J. A. 1995. Pedogenic and groundwater processes in closed Miocene basin (northern Spain). *Sedimentary Geology*, **99**, 17–36.

ARNÁEZ, J. & GARCÍA RUIZ, J. M. 2000. El periglaciarismo en el Sistema Ibérrico noroccidental. *In*: PEÑA, J. L., SÁNCHEZ-FABRE M. & LOZANO M. V. (eds) *Procesos y formas periglaciares en la montaña mediterránea*. Instituto de Estudios Turolenses, Teruel, 113–126.

ARNAIZ, I., ROBLES, S. & PUJALTE, V. 1991. Correlación entre registros de sondeos y series de superficie del Aptiense-Albiense continental del extremo SW de la Cuenca Vascocantábrica y su aplicación a la identificación de zonas lignitíferas. *Geogaceta*, **10**, 65–68.

ARRANZ, E. 1997. *Petrología del macizo granítico de La Maladeta (Huesca-Lérida): estructura, mineralogía, geoquímica y petrogénesis.* PhD Thesis, University of Zaragoza.

ARRIBAS, A., PALMQVIST, P. & MARTÍNEZ-NAVARRO, B. 1996. Estudio tafonómico cuantitativo de la asociación de macromamíferos de Venta Micena. *In*: MELÉNDEZ, G., BLASCO, M. F. & PÉREZ, J. (eds) *II Reunión de Tafonomía y Fosilización, Zaragoza, 13–15 Junio 1996*. Institución Fernando el Católico, Zaragoza, 27–38.

ARRIBAS, A. Jr. 1992. Los yacimientos del sureste peninsular. *In*: GARCÍA GUINEA, J. & MARTÍNEZ-FRÍAS, J. (eds) *Recursos Minerales de España*. CSIC, Colección Textos Universitarios **15**, 875–893.

ARRIBAS A. Jr. 1993. *Mapa geológico del distrito minero de Rodalquilar, Almería, escala 1:25.000*. Instituto Tecnológico Geominero de España, Madrid.

ARRIBAS A. Jr. & TOSDAL, R. M. 1994. Isotopic composition of Pb in ore deposits of the Betic Cordillera, Spain: origin and relationship to other European deposits. *Economic Geology*, **89**, 1074–1093.

ARRIBAS, A. Jr, CUNNINGHAM, C. G., RYTUBA, J. J. *et al.* 1995. Geology, geochronology, fluid inclusions and isotope geochemistry of the Rodalquilar gold alunite deposit, Spain. *Economic Geology*, **90**, 795–822.

ARRIBAS, J. 1984. *Sedimentología y diagénesis del Buntsandstein y Muschelkalk de la Rama Aragonesa de la Cordillera Ibérica (Provincias de Soria y Zaragoza)*. PhD Thesis, Complutense University, Madrid.

ARRIBAS, J. 1985. Base litoestratigráfica de las facies Buntsandstein y Muschelkalk de la Rama Aragonesa de la Cordillera Ibérica (Zona Norte). *Estudios Geológicos*, **41**, 47–57.

ARRIBAS, J. 1987. Las facies superiores del Muschelkalk en el borde de la Rama Aragonesa de la Cordillera Ibérica. *Cuadernos de Geología Ibérica*, **11**, 557–574.

ARRIBAS, J. & SORIANO, J. 1984. La porosidad en las areniscas triásicas (Rama Aragonesa de la Cordillera Ibérica). *Estudios Geológicos*, **40**, 341–353.

ARRIBAS, J., MARFIL, R. & DE LA PEÑA, J. A. 1985. Provenance of Triassic feldspatic sandstones in the Iberian Range (Spain): Significance of quartz types. *Journal of Sedimentary Petrology*, **55**, 864–868.

ARRIBAS, J., GÓMEZ-GRAS, D., ROSELL, J. & TORTOSA, A. 1990. Estudio comparativo entre las areniscas Paleozoicas y Triásicas de la isla de Menorca: Evidencias de procesos de reciclado. *Revista de la Sociedad Geológica de España*, **3**, 105–116.

ARRIBAS, J., DÍAZ-MOLINA, M., GÓMEZ, J. J., MORALES, J., TORTOSA, A. & AZANZA, B. 1997. Terciario entre la Sierra de Altomira y la Serranía de Cuenca. Reconstrucción tridimensional de almacenes sedimentarios en cinturones de meandros. Puntos de Interés geológico. Cuenca de Loranca. *In*: ALCALÁ, L. & ALONSO-ZARZA, A. M. (eds) *Itinerarios geológicos en el Terciario del centro y este de la Península Ibérica*. HC multimedia, Madrid, 41–56.

ARRIBAS, M. E. 1994. Paleogene of the Madrid Basin (northeast sector), Spain. *In*: GIERLOWSKI-KORDESCH, A. & KELTS, K. (eds) *Global Geological Record of Lake Basins. Vol. I.* Cambridge University Press, Cambridge, 255–260.

ARSUAGA, J. L. & BERMÚDEZ DE CASTRO, J. M. 1984. Estudio de los restos humanos del yacimiento de la Cova del Tossal de la Font (Vilafamés, Castellón). *Cuadernos de Prehistoria y Arqueología Castellonenses*, **10**, 19–34.

ARSUAGA, J. L., GRACIA, A., MARTÍNEZ, I., BERMÚDEZ DE CASTRO, J. M., ROSAS, A., VILLAVERDE, V. & FUMANAL, M. P. 1989. The human remains from Cova Negra (Valencia, Spain) and their place in European Pleistocene human evolution. *Journal of Human Evolution*, **18**, 55–92.

ARSUAGA, J. L., MARTÍNEZ, I., GRACIA, A. & LORENZO, C. 1997. The Sima de los Huesos crania (Sierra de Atapuerca, Spain). A comparative study. *Journal of Human Evolution*, **33**, 219–281.

ARSUAGA, J. L., GRACIA, A., LORENZO, C., MARTÍNEZ, I. & PÉREZ, P. J. 1999a. Resto craneal humano de Galería/Cueva de los Zarpazos (Sierra de Atapuerca, Burgos). *In*: CARBONELL, E., ROSAS, A. & DÍEZ, J. C. (eds) *Atapuerca. Ocupaciones humanas y paleoecología del yacimiento de Galería*. Junta de Castilla y León, Valladolid, Arqueología en Castilla y León, Memorias, **7**, 233–235.

ARSUAGA, J. L., MARTÍNEZ, I., LORENZO, C., GRACIA, A., MUÑOZ, A., ALONSO, O. & GALLEGO, J. 1999b. The human cranial remains from Gran Dolina Lower Pleistocene site (Sierra de Atapuerca, Spain). *Journal of Human Evolution*, **37**, 431–457.

ARTH, J. G. 1976. Behavior of trace elements during magmatic processes. A summary of theoretical models and their applications. *Journal of Research of the US. Geological Survey*, **4**, 41–47.

ARTH, J. G. 1979. Some trace elements in trondhjemites – their implications to magma genesis and paleotectonic setting. *In*: BARKER, F. (ed.) *Trondhjemites, Dacites and Related Rocks*. Developments in Petrology, **6**, Elsevier, Amsterdam, 123–132.

ARTHAUD, F. & MATTE, PH. 1975. Les décrochements tardi-herciniens du Sud-Ouest de l'Europe: Géometrie et essai de reconstruction des conditions de la deformation. *Tectonophysics*, **25**, 139–171.

ARTHAUD, F. & MATTE, PH. 1977. Late Paleozoic strike-slip faulting in southern Europe and northern Africa: Result of a right lateral shear zone between the Appalachian and the Urals. *Geological Society of America Bulletin*, **88**, 1305–1320.

ARZ, J. A. & ARENILLAS, I. 1996. Discusión de los modelos de extinción para los foraminíferos planctónicos del límite Cretácico/Terciario en el corte de Agost (Cordilleras Béticas). *Actas de la Real Sociedad Española de Historia Natural*, **XII Bienal**, 281–285.

ASHAUER, H. & TEICHMULLER, R. 1935. Origen y desarrollo de las Cordilleras variscas y alpídicas de Cataluña. *Publicaciones extranjeras de Geología de España*, **3**, 7–102. Madrid, 1946.

ASTIBIA, H., ARANBURU, A., PEREDA SUBERBIOLA, X. *et al.* 2000. Un nouveau site à vertébrés continentaux de l'Eocène supérieur de Zambrana (Bassin de Miranda-Treviño, Alava, Pays Basque). *Geobios*, **32**, 233–248.

AURA-TORTOSA, J. E., FERNÁNDEZ-PERIS, J. & FUMANAL, M. P. 1993. Medio físico y corredores naturales: notas sobre el poblamiento paleolítico del País Valenciano. *Recerques del Museu d'Alcoi*, **2**, 89–107.

AURELL, M. 1991. Identification of systems tracts in low-angle carbonate ramps: examples from the Upper Jurassic of the Iberian Chain (Spain). *Sedimentary Geology*, **73**, 101–115.

AURELL, M. & BÁDENAS, B. 1997. The pinnacle reefs of Jabaloyas (Late Kimmeridgian, NE Spain): vertical zonation and associated facies related to sea level changes. *Cuadernos de Geología Ibérica*, **22**, 37–64.

AURELL, M. & MELÉNDEZ, A. 1993. Sedimentary evolution and sequence stratigraphy of the Upper Jurassic in the central Iberian Chain, northeast Spain. IAS, Special Publication, **18**, 343–368.

AURELL, M., FERNÁNDEZ LÓPEZ, S. & MELÉNDEZ, G. 1994a. The Middle-Upper Jurassic oolitic ironstone bed in the Iberian Range (Spain). Eustatic implications. *Geobios, Special Memoir*, **17**, 549–561.

AURELL, M., MAS, R., MELÉNDEZ, A. & SALAS, R. 1994b. El tránsito Jurásico-Cretácico en la Cordillera Ibérica: relación tectónica-sedimentación y evolución paleogeográfica. *Cuadernos de Geología Ibérica*, **18**, 369–396.

AURELL, M., BOSENCE, D. W. J. & WALTHAM, D. A. 1995. Carbonate ramp depositional systems from a late Jurassic epeiric platform (Iberian basin, Spain. a combined computer modelling and outcrop analysis. *Sedimentology*, **42**, 75–94.

AURELL, M., PÉREZ-URRESTI, I., RAMAJO, J., MELÉNDEZ, G. & BÁDENAS, B. 1997. La discordancia de Moyuela (Zaragoza): precisiones sobre la tectónica extensional en el límite Oxfordiense-Kimmeridgiense en la Cuenca Ibérica. *Geogaceta*, **22**, 21–24.

AURELL, M., BÁDENAS, B., BOSENCE, D. W. J. & WALTHAM, D. A. 1998. Carbonate production and offshore transport on a Late Jurassic carbonate ramp (kimmeridgian, Iberian basin, NE Spain): evidence from outcrops and computer modelling. Geological Society, London, Special Publication, **149**, 137–161.

AURELL, M., BÁDENAS, B., BELLO, J., DELVENE, G., MELÉNDEZ, G., PÉREZ-URRESTI, I. & RAMAJO, J. 1999a. El Calloviense y el Jurásico superior en la Cordillera Ibérica nororiental y la Zona de Enlace con la Cordillera Costero-Catalana, en los sectores de Sierra de Arcos, Calanda y Xerta-Paüls. *Cuadernos de Geología Ibérica*, **25**, 111–137.

AURELL, M., BÁDENAS, B. & BORDONABA, A. P. 1999b. El Bathoniense-Kimmeridgiense (Jurásico medio-superior) en la región de Obón-Torre de las Arcas (Teruel). *Geogaceta*, **25**, 19–22.

AURELL, M., MELÉNDEZ, G., BÁDENAS, B., PÉREZ-URRESTI, I. & RAMAJO, J. 2000. Sequence Stratigraphy of the Callovian-Berriasian (Middle Jurassic-Lower Cretaceous) of the Iberian basin (NE Spain). *GeoResearch Forum*, **5**, 281–292.

AUTRAN, A. 1980. Les Granites des Pyrénées. *In*: AUTRAN, A. & DERCOURT, J. (coords) *Evolutions Géologiques de la France*. BRGM, Memoirs **107**, 71–76.

AUTRAN, A. 1996a. Magmatisme Hercynien: Pluton de Montlouis-Andorra. *In*: BARNOLAS, A. & CHIRON, J. C. (eds) *Synthèse géologique et géophysique des Pyrénées. 1- Cycle Hercynien*. BRGM-ITGE, Orléans and Madrid, 441–445.

AUTRAN, A. 1996b. Magmatisme Hercynien: Pluton de Costabonne. *In*: BARNOLAS, A. & CHIRON, J. C. (eds) *Synthèse géologique et géophysique des Pyrénées. 1- Cycle Hercynien*. BRGM-ITGE, Orléans and Madrid, 447–448.

AUTRAN, A. 1996c. Magmatisme Hercynien: Massifs de granites alumineux du Canigou et d'Ax-les-Thermes. *In*: BARNOLAS, A. & CHIRON, J. C. (eds) *Synthèse géologique et géophysique des Pyrénées. 1- Cycle Hercynien*. BRGM-ITGE, Orléans and Madrid, 446–447.

AUTRAN, A. 1996d. Magmatisme Hercynien: Petits plutons de granitoides et de gabbrodiorites à faciès charnockitique, mis en place dans le socle granulitique Hercynien des massifs nord-Pyreneens. *In*: BARNOLAS, A. & CHIRON, J. C. (eds) *Synthèse géologique et géophysique des Pyrénées. 1- Cycle Hercynien*. BRGM-ITGE, Orléans and Madrid, 446–447.

AUTRAN, A. & COCHERIE, A. 1996. Magmatisme Hercynien: Pluton de Saint Laurent de Cerdans- La Jonquera et Pluton associé de Batère. *In*: BARNOLAS, A. & CHIRON, J. C. (eds) *Synthèse géologique et géophysique des Pyrénées. 1- Cycle Hercynien*, BRGM-ITGE, Orléans and Madrid, 448–452.

AUTRAN, A. FONTEILLES, M. & GUITARD, G. 1970. Relations entre les intrusions de granitoïdes, l'anatexie et le métamorphisme régional considerées principalement du point de vue de l'eau: cas de la Chaîne Hercynienne des Pyrénées Orientales. *Bulletin Société Geologique de France*, **12**, 673–731.

AUZIAS, V. 1995. *Contribution a la caractérisation tectonique des réservoirs fracturés. I: Modélisation photoélasticimétrique des perturbations de contrainte au voisinage des faille et de la fracturation associeé: aplication pétrolière. II: Mécanismes de développement en 3D des diaclases dans un analogue de réservoir, le Dévonien tabulaire du Caithness (Ecosse)*. PhD Thesis, University of Montpellier II.

AZAMBRE, B. & RAVIER, J. 1978. Les écailles de gneiss du faciès granulite du Port de Saleix et de la région de Lherz (Ariège). Nouveaux témoins du socle profond des Pyrénées. *Bulletin Société Geologique de France*, **20**, 3, 221–228.

AZAMBRE, B. & ROSSY, M. 1976. Le magmatisme alcalin d'âge crétacé dans les Pyrénées occidentales et l'arc basque; ses relations avec le métamorphisme et la tectonique. *Bulletin de la Société Géologique de France*, **18**, 1725–1728.

AZAÑÓN, J. M. & GOFFÉ, B. 1997. Ferro- and magnesiocarpholite assemblages as record of high-P, low-T metamorphism in the Central Alpujárrides, Betic Cordillera (SE Spain). *European Journal of Mineralogy*, **9**, 1035–1051.

AZAÑÓN, J. M., CRESPO-BLANC, A. & GARCÍA-DUEÑAS, V. 1997. Continental collision, crustal thinning and nappe forming during the pre-Miocene evolution of the Alpujarride Complex (Alboran Domain, Betic). *Journal of Structural Geology*, **19**, 1055–1071.

AZAÑÓN, J. M., GARCÍA-DUEÑAS, V. & GOFFÉ, B. 1998. Exhumation of high-pressure metapelites and coeval crustal extension in the Alpujarride Complex (Betic Cordillera). *Tectonophysics*, **285**, 231–252.

AZCÁRRAGA, J. 2000. *Evolución tectónica y metamórfica de los mantos inferiores de grado alto y alta presión del complejo de Cabo Ortegal*. PhD Thesis, University of País Vasco.

AZÉMA, J. 1977. *Étude géologique des zones externes des Cordilléres Bétiques aux confins des Provinces d'Alicante et de Murcie (Espagne)*. PhD Thesis, Université de París.

AZÉMA, J., CHAMPETIER, Y., FOUCAULT, A., FOURCADE, E. & RANGHEARD, J. 1971. Le Jurassique dans le partie oriental des Zones Externes des Cordillères Bétiques: essai de coordination. *Cuadernos de Geología Ibérica*, **2**, 91–110.

AZÉMA, J., FOUCAULT, A., FOURCADE, E. *et al.* 1979. *Las microfacies del Jurásico y Cretácico de las Zonas Externas de las Cordilleras Béticas*. University of Granada.

AZOR, A. 1997. *Evolución Tectonometamórfica del Límite entre las zonas Centroibérica y de Ossa–Morena (Cordillera Varisca, SO de España)*. PhD Thesis, University of Granada.

AZOR, A., GONZÁLEZ LODEIRO, F., HACAR RODRÍGUEZ, M., MARTÍN PARRA, L. M., MARTÍNEZ CATALÁN, J. R. & PÉREZ-ESTAÚN, A. 1992. Estratigrafía y estructura del Paleozoico en el Dominio del Ollo de Sapo. *In*: GUTIÉRREZ-MARCO, J. C., SAAVEDRA, J. & RÁBANO, I. (eds) *Paleozoico Inferior de Ibero-América*. University of Extremadura, 469–483.

AZOR, A., GONZÁLEZ LODEIRO, F. & SIMANCAS, J. F. 1994. Tectonic evolution of the boundary between the Central Iberian and Ossa Morena Zones (Variscan belt, southwest Spain). *Tectonics*, **13**, 45–61.

AZOR, A., BEA, F., GONZÁLEZ LODEIRO, F. & SIMANCAS, J. F. 1995. Geochronological constraints on the evolution of a suture: the Ossa Morena/Central Iberian contact (Variscan belt, southwest Iberian Peninsula). *Geologische Rundschau*, **84**, 375–383.

BAAKIR, F., DAMLUJI, S. F., AMIN-ZAKI, L. *et al.* 1973. Methyl mercury poisoning in Iraq. *Science*, **181**, 230–241.

BABIN, C. & GUTIÉRREZ-MARCO, J. C. 1985. Un nouveau cycloconchide (Mollusca, Bivalvia) du Llanvirn inférieur (Ordovicien) de Monts de Tolède (Espagne). *Geobios*, **18**, 609–616.

BABIN, C. & GUTIÉRREZ-MARCO, J. C. 1991. Middle Ordovician bivalves from Spain and their phyletic and palaeogeographic significance. *Palaeontology*, **34**, 109–147.

BABIN, C. & GUTIÉRREZ-MARCO, J. C. 1992. Intéret paléobiogeographique de la presence du genre *Trocholites* (Cephalopoda, Nautiloidea) dans le Dobrotivá (Llandeilo) inférieur d'Espagne. *Neues Jahrbuch für Geologie und Paläontologie Monatshefte*, **1992**, 519–541.

BABIN, C. & HAMMANN, W. 2001. Une nouvelle espèce de *Modiolopsis* (Bivalvia) dans l'Arenig (Ordovicien Inférieur) de Daroca (Aragón, Espagne): réflexions sur la denture des bivalves primitifs. *Revista Española de Paleontología*, **16**, 269–282.

BACETA, J. I., WRIGHT, V. P. & PUJALTE, V. 2001. Palaeo-mixing zone karst features from Palaeocene carbonates of north Spain: criteria for recognizing a potentially widespread but rarely documented diagenetic system. *Sedimentary Geology*, **139**, 205–216.

BACETA, J. I., GÓMEZ, I. & HERNÁNDEZ SAMANIEGO, A. (in press). *Mapa Geológico de Navarra a escala 1:25.000, Hoja 172–I (Arroniz)*. Departamento de Obras Públicas, Transporte y Comunicaciones del Gobierno de Navarra.

BÁDENAS, B. 1996. El Jurásico superior de la Sierra de Aralar (Guipuzcoa y Navarra): caracterización sedimentológica y paleogeográfica. *Estudios Geológicos*, **52**, 147–170.

BADENAS, B. 1999. *La sedimentación en las rampas carbonatadas del Kimmeridgiense en las cuencas del este de la placa Iberica*. PhD Thesis, University of Zaragoza.

BÁDENAS, B. 2001. Análisis sedimentológico y tectónica sinsedimentaria del Kimmeridgiense inferior en un sector marginal de la Cuenca Ibérica (Sierra Menera-Sierra Palomera, Teruel). *Revista de la Sociedad Geológica de España*, **14**, 101–112.

BÁDENAS, B. & AURELL, M. 1997. El Kimmeridgiense de los Montes Universales (Teruel): distribución de facies y variaciones del nivel del mar. *Cuadernos de Geología Ibérica*, **22**, 15–36

BÁDENAS, B. & AURELL, M. 2001a. Proximal-distal facies relationship and sedimentary processes in a storm dominated carbonate ramp (Kimmeridgian, northwest of the Iberian Ranges, Spain). *Sedimentary Geology*, **139**, 319–342.

BÁDENAS, B. & AURELL, M. 2001b. Kimmeridgian palaeogeography and basin evolution of northeastern Iberia. *Palaeogeography, Palaeoclimatology, Palaeoecology*, **168**, 291–310.

BÁDENAS, B., AURELL, M. & MELÉNDEZ, A. 1993. Estratigrafía Secuencial y Sedimentología del Jurásico superior del Noreste de la provincia de Albacete. *Estudios Geológicos*, **49**, 253–366.

BÁDENAS, B., AURELL, M., FONTANA, B., GALLEGO, M. R. & MELÉNDEZ, G. 1997. Estratigrafía y evolución sedimentaria del Jurásico en la Cordillera Vasco-Cantábrica Oriental (Navarra y Guipuzcoa). *IV Congreso de Jurásico de España, libro de comunicaciones*, 41–43.

BAENA, J. & JEREZ, L. 1982. *Síntesis para un ensayo paleogeográfico entre la Meseta y la Zona Bética (s.s.)*. Colección Informe, IGME, Madrid.

BAENA, J., MORENO, F., NOZAL, F. *et al.* 2001. *Mapa Neotectónico y Sismotectónico de España a Escala 1:1.000.000*. ITGE and ENRESA.

BAENA, R. & DÍAZ DEL OLMO, F. 1997. Resultados paleomagnéticos de la Raña del Hespérico Meridional (Montoro, Córdoba). *Geogaceta*, **21**, 31–34.

BAEZA-ROJANO, L. J., RUIZ GARCÍA, C., RUIZ MONTES, M. & SÁNCHEZ, A. 1981. Mineralización exhalativo-sedimentaria de sulfuros polimetálicos en la Sierra Morena cordobesa (España). *Boletín Geológico y Minero*, **92**, 203–216.

BAHAMONDE, J. R. 1990. *Estratigrafía y Sedimentología del Carbonífero medio y superior de la Región del Manto del Ponga (Zona Cantábrica)*. PhD Thesis, University of Oviedo.

BAHAMONDE, J. R. & COLMENERO, J. R. 1993. Análisis estratigráfico del Carbonífero medio y superior del Manto del Ponga (Zona Cantábrica). *Trabajos de Geología, Oviedo*, **19**, 155–193.

BAHAMONDE, J. R. & NUÑO, C. 1991. Características geológicas del sinclinal de Santa María de Redondo (Zona Cantábrica, Palencia). *Boletín Geológico y Minero*, **102**, 219–239.

BAHAMONDE, J. R., COLMENERO, J. R. & VERA, C. 1997. Growth and demise of late Carboniferous carbonate platforms in the eastern Cantabrian Zone, Asturias, northwestern Spain. *Sedimentary Geology*, **110**, 99–122.

BAHAMONDE, J. R., VERA, C. & COLMENERO, J. R. 2000. A steep-fronted Carboniferous carbonate platform: clinoformal geometry and lithofacies (Picos de Europa Region, NW Spain). *Sedimentology*, **47**, 645–664.

BAKKER, H. E., DE JONG, K., HELMERS, H. & BIERMAN, C. 1989. The geodynamic evolution of the Internal Zone of the Betic Cordilleras (South-East Spain): a model based on structural analysis and geothermobarometry. *Journal of Metamorphic Geology*, **7**, 359–381.

BALANYÁ, J. C. 1991. *Estructura del Dominio de Alborán en la parte Norte del Arco de Gibraltar*. PhD Thesis, University of Granada.

BALANYÁ, J. C. & GARCÍA-DUEÑAS, V. 1987. Les directions structurales dans le Domaine d'Alborán de part et d'autre du Détroit de Gibraltar. *Comptes Rendus de l'Académie des Sciences de Paris*, **304**, 929–932.

BALANYÁ, J. C., AZAÑÓN, J. M., SÁNCHEZ-GÓMEZ, M. & GARCÍA-DUEÑAS, V. 1993. Pervasive ductile extension, isothermal decompression and thinning of the Jubrique unit in the Paleogene (Alpujárride Complex, western Betics Spain). *Comptes Rendus de l'Académie des Sciences de Paris*, **316**, 1595–1601.

BALANYÁ, J. C., GARCÍA-DUEÑAS, V., AZAÑÓN, J. M. & SÁNCHEZ-GÓMEZ, M. 1997. Alternating contractional and extensional events in the Alpujarride nappes of the Alboran Domain (Betics, Gibraltar Arc). *Tectonics*, **16**, 226–238.

BALDWIN, C. T. 1977. The stratigraphy and facies associations of trace fossils in some Cambrian and Ordovician rocks of north western Spain. *In*: CRIMES, T. P. & HARPER, J. C. (eds) *Trace Fossils 2*. Geological Journal, Special Issue, **9**, 9–40.

BALOGH, K., AHIJADO, A., CASILLAS, R. & FERNÁNDEZ, C. 1999. Contributions to the chronology of the Basal Complex of Fuerteventura, Canary Islands. *Journal of Volcanology and Geothermal Research*, **90**(1–2), 81–102.

BANDA, E. 1996. Deep crustal expression of Tertiary basins in Spain. *In*: FRIEND, P. F. & DABRIO, C. J. (eds) *Tertiary Basins of Spain. The Stratigraphic Record of Crustal Kinematics*. World and Regional Geology, **6**. Cambridge University Press, Cambridge, 15–19.

BANDA, E. & ANSORGE, J. 1980. Crustal structure under the central and eastern part of the Betic Cordillera. *Geophysical Journal of the Royal Astronomical Society*, **63**, 515–532.

BANDA, E. & CHANNELL, J. 1979. Evidencia geofísica para un modelo de evolución de las cuencas del Mediterráneo occidental. *Estudios Geológicos*, **35**, 5–14.

BANDA, E. & SANTANACH, P. 1992. Structure of the crust and upper mantle beneath the Balearic Islands (Western Mediterranean). *Earth and Planetary Science Letters*, **49**, 219–230.

BANDA, E., SURIÑACH, E., APARICIO, A., SIERRA, J. & RUIZ DE LA PARTE, E. 1981. Crust and upper mantle structure of the central Iberian Meseta (Spain). *Geophysics Journal of Royal Astronomy Society*, **67**, 779–789.

BANDA, E., GALLART, J., GARCÍA-DUEÑAS, V., DAÑOBEITIA, J. J. &

MAKRIS, J. 1993. Lateral variation of the crust in the Iberian peninsula: new evidence from the Betic Cordillera. *Tectonophysics*, **221**, 53–66.

BANDRÉS, A., EGUILUZ, L., GONZALO, J. C. & CARRACEDO, M. 1999. El Macizo de Mérida, un arco volcánico cadomiense reactivado en el hercínico. *Geogaceta*, **25**, 27–30.

BANDRÉS, A., EGUÍLUZ, L., MENÉNDEZ, M., ORTEGA, L. A. & GIL IBARGUCHI, J. I. 2000. El macizo precámbrico de Mérida (SW de España): petrografía, geoquímica, geocronología y significado geodinámico. *Cuadernos del Laboratorio Xeológico de Laxe*, **25**, 159–163.

BANKS, C. J. & WARBURTON, J. 1991. Mid crustal detachment in the Betic system of southeast Spain. *Tectonophysics*, **191**, 275–289.

BARANDIARÁN, I. 1973. *La Cueva de los Casares (en Riba de Saelices, Guadalajara)*. Dirección General de Bellas Artes, Madrid, Excavaciones Arqueológicas en España, **76**.

BARBA, A., FELIÚ, C., GARCÍA-TEN, J., GINÉS, F., SANCHEZ, E. & SANZ, V. 1997. *Materias primas para la fabricación de soportes de baldosas cerámicas*. Instituto de Tecnología Cerámica, Castellón de la Plana (ISBN 84-923176-04).

BARBA, P. 1991. *Estratigrafía y Sedimentología de la sucesión westfaliense del borde Sureste de la Cuenca Carbonífera Central*. PhD Thesis, University of Oviedo.

BARBA, P. & COLMENERO, J. R. 1994. Estratigrafía y Sedimentología de la sucesión Westfaliense del borde sureste de la Cuenca Carbonífera Central (Zona Cantábrica, N de España). *Studia Geológica Salmanticiensia, Salamanca*, **XXX**, 139–204.

BARBERÀ, X. 1999. *Magnetoestratigrafia del Oligocé del sector sudoriental de la Conca de l'Ebre: implicacions magnetocrnologiques i secuencials*. PhD Thesis, University of Barcelona.

BARBERI, F. 1989. Riesgo volcánico y vigilancia. *In*: Araña, V. & Coello, J. (eds) *Los volcanes y la caldera del Parque Nacional del Teide (Tenerife, Islas Canarias)*. ICONA (Instituto Conservación de la Naturaleza), Madrid, 387–396.

BARBERO, L. 1995. Granulite facies metamorphism in the Anatectic Complex of Toledo, Spain: Late Hercynian tectonic evolution by crustal extension. *Journal of the Geological Society, London*, **152**, 365–382.

BARBERO, L. & VILLASECA, C. 2000. Eclogite facies relics in metabasites from the Sierra de Guadarrama (Spanish Central System): P-T estimations and implications for the Hercynian evolution. *Mineralogical Magazine*, **64**, 815–836.

BARBERO, L., VILLASECA, C., ROGERS, G. & BROWN, P. E. 1995. Geochemical and isotopic disequilibrium in crustal melting: An insight from the anatectic granitoids from Toledo, Spain. *Journal of Geophysical Research*, **100**(B8), 15745–15765.

BARD, J. P. 1969. *Le Métamorphisme régional progressif des Sierras d'Aracena en Andalousie Occidentale (Espagne): sa place dans le segment hercynien sud-Ibérique*. PhD Thesis, Scientific and Technical University, Montpellier.

BARD, J. P., BURG, J. P., MATTE, PH. & RIBEIRO, A. 1980. La Chaine Hercynienne d'Europe occidentale en termes de Tectonique des Plaques. In: Géologie de l'Europe, du Précambrien aux basins sédimentaires post-Hercyniennes. *Mémoires du Bureau de Recherches Géologiques et Minières*, **108**, 233–246.

BARDAJÍ, T. 1999. *Evolución Geomorfológica durante el Cuaternario de las Cuencas Neógenas Litorales del Sur de Murcia y Norte de Almería*. PhD Thesis, University of Complutense, Madrid.

BARDAJÍ, T., DABRIO, C. J., GOY, J. L., SOMOZA, L. & ZAZO, C. 1987. Sedimentologic features related to Pleistocene sea level changes in the SE Spain. *Trabajos sobre Neógeno-Cuaternario*, **10**, 79–93.

BARDAJÍ, T., SILVA, P. G., GOY, J. L., ZAZO, C., DABRIO, C. J. & CIVIS, J. 1999. Recent evolution of the Aguilas Arc Basins (SE Spain): Sea-level record and neotectonics. *INQUA Mediterranean Sea-level Subcommission Newsletter*, **21**, 21–26.

BARDAJÍ, T., GOY, J. L. & ZAZO, C. 2000. El límite Plio-Pleistoceno: un debate todavía abierto. *Cuaternario y Geomorfología*, **14**(1–2), 77–92.

BARNOLAS, A. & CHIRON, J. C. 1996. *Synthèse géologique et géophysique des Pyrénées. Tome 1: Introduction. Géophysique. Cycle Hercynien*. BRGM and IGME, Orléans and Madrid.

BARNOLAS, A. & GARCÍA-SANSEGUNDO, J. 1992. Caracterización estratigráfica y estructural del Paleozoico de Les Gavarres

(Cadenas Costero Catalanas, NE de España). *Boletín Geológico y Minero*, **103**, 94–108.

BARNOLAS, A. & SIMÓ, A. 1984. Sedimentología. *In*: ÁLVARO, M., BARNOLAS, A., DEL OLMO, P., RAMÍREZ DEL POZO, J. & SIMÓ, A. (eds) *Sedimentología del Jurásico de Mallorca*. GEM-IGME-CGS, 73–119.

BARNOLAS, A. & TEIXEL, A. 1994. Platform sedimentation and collapse in a carbonate-dominated margin of a foreland basin (Jaca basin, Eocene, southern Pyrenees). *Geology*, **22**, 1107–1110.

BARNOLAS, A., GARCÍA VÉLEZ, A. & SOUBRIER, J. 1980. Sobre la presencia del Caradoc en Las Gavarres. *Acta Geologica Hispanica*, **15**, 1–13.

BARÓN, A. & POMAR, L. 1985. Stratigraphic correlation tables: area 2c Balearic Depression, *In*: STEININGER, F. F., SENES, J., KLEEMANN, K. & ROG, F. (eds) *Neogene of the Mediterranean, Tethys and Paratethys*. Institute of Paleontology, University of Vienna, 1: 17 and 2: 17.

BARRERA, E. & SAVIN, S. M. 1999. Evolution of late Campanian-Maastrichtian marine climates and oceans. *In*: BARRERA, E. & JOHNSON, C. C. (eds) *Evolution of the Cretaceous Ocean–Climate System*. Geological Society of America, Special Papers, **332**, 245–282.

BARRERA, J. L., FERNANDEZ SANTIN, S., FÚSTER, J. M. & IBARROLA, E. 1981. Ijolitas-Sienitas-Carbonatitas de los Macizos del Norte del del Complejo Plutónico Basal de Fuerteventura (Islas Canarias). *Boletín Geológico y Minero*, **92**, 4, 309–321.

BARRERA, J. L., BELLIDO, F., PABLO, J. G. & ARPS, C. E. S. 1982. Evolución petrológico geoquímica de los granitoides hercínicos del NO gallego. *Cuadernos do Laboratorio Xeolóxico de Laxe*, **3**, 21–52.

BARRERA, J. L., FARIAS, P., GONZÁLEZ LODEIRO, F. *et al.* 1989. *Mapa y Memoria de la Hoja, no. 17/27 (Orense-Verín) del Mapa Geológico de España a escala 1:200.000*. IGME, Madrid.

BARRIGA, F. 1983. *Hydrothermal metamorphism and ore genesis at Aljustrel, Portugal*. PhD Thesis, University of Western Ontario.

BARRIGA, F. 1990. Metallogenesis in the Iberian Pyrite Belt. *In*: DALLMEYER R. D. & MARTÍNEZ GARCÍA E. (eds) *Pre-Mesozoic Geology of Iberia*. Springer-Verlag, Berlin, 369–379.

BARRIGA, F. & CARVAHLO, D. 1983. Carboniferous volcanogenic sulphide mineralisations in South Portugal (Iberian Pyirite Belt). *Memorias dos Servicos Geológicos de Portugal*, **29**, 99–113.

BARRIGA, F. & KERRICH, R. 1984. Extreme ^{18}O-enriched volcanics and ^{18}O-evolved marine water, Aljustrel, Iberian Pyrite Belt: transition from high to low Rayleigh number vonvective regimes. *Geochimica et Cosmochimica Acta*, **48**, 1021–1031.

BARROIS, Ch. 1882. Recherches sur les terrains anciens des Asturies et de la Galice. *Mémoires Société Géologique Nord*, **2**, 1–630.

BARROIS, Ch. 1892. Observations sur le terrain dévonien de la Catalogne. *Annales Société Géologique Nord*, **19**, 61–73.

BARRÓN, E., COMAS-RENGIFO, M. J. & TRINCAO, P. 1999. Estudio palinológico del tránsito Pliensbachiano/Toarciense en la Rambla del Salto (Sierra Palomera, Teruel, España). *Cuadernos de Geología Ibérica*, **25**, 171–187.

BARRON, E. J. & MOORE G. T. 1994. *Climate model application in paleoenvironmental analysis*. SEPM, Short Course, **33**.

BARROS, J. C. 1989. Nuevos datos geológicos y cartográficos sobre el flanco sur del Sinclinorio de Truchas. *Cuadernos del Laboratorio Geológico de Laxe*, **14**, 93–116.

BARROSO, C., MEDINA, P., SANCHIDRIÁN, J. L., GARCÍA-SÁNCHEZ, M. & RUIZ-BUSTOS, A. 1984. Le gisement moustérien de la grotte du 'Boquete de Zafarraya' (Alcaucín, Andalousie). *L'Anthropologie*, **88**, 133–134.

BARTHEL, K. W., CEDIEL, F. & GEYER, O. F. 1966. Der Subetische Jura von Cehegín (Prov. Murcia, Spanien). *Mitteilungen der Bayerischen Staatssammlung für Paläontologie und historische Geologie*, **6**, 167–211.

BASABE, J. M. 1966. El húmero premusteriense de Lezetxiki (Guipuzcoa). *Munibe*, **18**, 13–22

BASABE, J. M. 1970. Dientes humanos del Paleolítico de Lezetxiki (Mondragón). *Munibe*, **22**, 113–124.

BASABE, J. M. 1973. Dientes humanos del Musteriense de Axlor (Dima, Vizcaya). *Trabajos de Antropología*, **16**, 187–202.

BASABE, J. M. 1982. Restos fósiles humanos de la región vasco-cantábrica. *Cuadernos de la Sección de Antropología y Etnología. Sociedad de Estudios Vascos*, **1**, 69–83.

BASSETT, M. G. & OWENS, R. M. 1996. Discussion on a revision of Ordovician Series and Stage divisions from the historical type area (with a reply by R. A. FORTEY, D. A. T. HARPER, J. K. INGHAM, A. W. OWEN & A. W. A. RUSHTON). *Geological Magazine*, **133**, 767–772.

BASTIDA, F. 1980. *Las estructuras de la primera fase de deformación herciniana en la Zona Asturoccidental-leonesa (Costa Cantábrica, N. W. de España)*. PhD Thesis, University of Oviedo.

BASTIDA, F. & PULGAR, J. A. 1978. La estructura del manto de Mondoñedo entre Burela y Tapia de Casariego (Costa Cantábrica, NW de España). *Trabajos Geológicos*, **10**, 75–124.

BASTIDA, F., MARCOS, A., MARQUÍNEZ, J., MARTÍNEZ CATALÁN, J. R., PÉREZ ESTAÚN, A. & PULGAR, J. A. 1984. *Mapa y Memoria de la Hoja no. 1 (La Coruña) del Mapa Geológico de España a escala 1:200.000*. IGME, Madrid.

BASTIDA, F., MARTÍNEZ CATALÁN, J. R. & PULGAR, J. A. 1986. Structural, metamorphic and magmatic history of the Mondoñedo nappe (Hefcynian belt, NW Spain). *Journal of Structural Geology*, **8**, 415–430.

BASTIDA, J., BESTEIRO, J., REVENTOS, M., LAGO, M. & POCOVI, A. 1989. Los basaltos alcalinos subvolcánicos espilitizados de Arándiga (provincia de Zaragoza): estudio mineralógico y geoquímico. *Acta Geológica Hispánica*, **24**, 115–130.

BASTIDA, F., BRIME, C., GARCÍA-LÓPEZ, S. AND SARMIENTO, G. N. 1999. Tectonothermal evolution in a region with thin-skinned tectonics: the western nappes in the Cantabrian Zone (Variscan belt of NW Spain). *International Journal of Earth Sciences*, **88**, 38–48.

BATALLER, J. R. 1933. El Triasic Català. *Boletín de la Sociedad de Ciencias Naturales (Club Muntanyenc)*, **13**, 3–12.

BATEMAN, M. D. & DÍEZ-HERRERO, A. 1999. Thermoluminescence dates and palaeoenvironmental information of the late Quaternary sand deposits, Tierra de Pinares, Central Spain. *Catena*, **4**, 277–291.

BATEMAN, R., MARTÍ, H. P. & CASTRO, A. 1992. Mixing of cordierite granitoids and piroxene gabbro, and fractionation in the Santa Olalla tonalite (Andalucia) *Lithos*, **28**, 111–131.

BAUDELOT, S. & TAUGOURDEAU-LANZ, J. 1986. Decouverte d'une microflore dans les Pyrenees catalanes atribuable au Norien-Rhetien. *Review of Paleobiologie*, **5**, 5–9.

BAULUZ, B., MAYAYO, M. J., FERNÁNDEZ-NIETO, C. & GONZÁLEZ LÓPEZ, J. M. 2000. Geochemistry of Precambrian and Paleozoic siliciclastic rocks from the Iberian Range (NE Spain): implications for source-area weathering, sorting, provenance and tectonic setting. *Chemical Geology*, **168**, 135–150.

BAUZA, F. 1876. Breve reseña geológica de las provincias de Tarragona y Lérida. *Boletín de la Comisión del Mapa Geológico de España*, **III**, 115–123.

BAUZÁ, J. 1961. Contribución al conocimiento de la flora fósil de Mallorca. *Estudios Geológicos*, **14**, 43–44.

BAYER, R. & MATTE, Ph. 1979. Is the mafic/ultramafic massif of Cabo Ortegal (northern Spain) a nappe emplaced during a Variscan obduction? – A new gravity interpretation. *Tectonophysics*, **57**, T9–T18.

BEA, F. 1982. Sobre el significado de la cordierita en los granitoides del batolito de Avila (Sistema Central español). *Boletín Geológico y Minero*, **67**, 59–67.

BEA, F. 1985. Los granitoides hercínicos de la mitad occidental del batolito de Avila (sector de Gredos). Aproximación mediante el concepto de Superfacies. *Revista de la Real Academia de las Ciencias Exactas, Naturales y Física de Madrid*, **LXXIX**, 549–572.

BEA, F. & MORENO-VENTAS, I. 1985. Estudio petrológico de los granitoides del área centro-norte de la Sierra de Gredos (Batolito de Avila; Sistema Central Español). *Studia Geológica Salmantinensis*, **XX**, 137–174.

BEA, F. & PEREIRA, D. 1990. Estudio petrológico del complejo anatéctico de la Peña Negra (Batolito de Avila, España Central). *Revista de la Sociedad Geológica de España*, **3**(1–2), 87–103.

BEA, F., SÁNCHEZ, J. G. & SERRANO-PINTO, M. 1987. Una compilación geoquímica (elementos mayores) de los granitoides del Macizo Hespérico. *In*: BEA, F., CARNICERO, A., GONZALO, J. C., LÓPEZ-

PLAZA, M. & RODRÍGUEZ, M. D. (eds) *Geología de los Granitoides y Rocas Básicas Asociadas del Macizo Hespérico*. Rueda, Madrid, 87–194.

BEA, F., FERSHTATER, G. B. & CORRETGÉ, L. G. 1992. The geochemistry of phosphorous in granite rocks and the effect of aluminium. *Lithos*, **29**, 43–56.

BEA, F., MONTERO, P. & MOLINA, J. F. 1999. Mafic precursors, peraluminous granitoids, and lamprophyres in the Avila batholith: A model for the generation of Variscan batholiths in Iberia. *Journal of Geology*, **107**, 411–419.

BEAUMONT, C., MUÑOZ, J. A., HAMILTON, J. & FULLSACK, P. 2000. Factors controlling the Alpine Evolution of the Central Pyrenees inferred from a comparison of observations and geodynamic models. *Journal of Geophysical Research*, **105**, 8121–8145.

BEETSMA, J. J. 1995. *The late Proterozoic/Paleozoic and Hercynian crustal evolution of the Iberian Massif, N Portugal: as traced by geochemistry, and Sr-Nd-Pb isotope systematics of pre-Hercynian terrigenous sediments and Hercynian granitoids*. PhD Thesis, Vrije University, Amsterdam.

BEGE, V. 1970. *Der Armoricanische Quartzit in Spanien. Paläogeographie, Fazies, Sedimentation des frühen Ordovizium*. Dissertation, University of Heidelberg.

BEHMEL, H. 1970. Beiträge zur Stratigraphie und Paläontologie des Juras von Ostspanien. V. Stratigraphie un Fazies im präebetischen Jura von Albacete und Nord-Murcia. *Neues Jarbuch für Geologie und Paläontologie. Abhandlungen*, **137**(1), 1–102.

BELLANCA, A., CALVO, J. P., CENSI, P., NERI, R. & POZO, M. 1992. Recognition of lake-level changes in Miocene lacustrine units, Madrid Basin, Spain. Evidence from facies analysis, isotope geochemistry and clay mineralogy. *Sedimentary Geology*, **76**, 135–153.

BELLIDO, F. 1996. *Geoquímica del magmatismo de la Faja Pirítica*. Open file report, Instituto Tecnológico Geo-Minero de España, 1–89.

BELLIDO, F., CAPOTE, R., CASQUET, C., FÚSTER, J. M., NAVIDAD, M., PEINADO, M. & VILLASECA, C. 1981. Características generales del cinturón hercínico en el sector oriental del Sistema Central Español. *Cuadernos de Geología Ibérica*, **7**, 15–51.

BELLIDO, F., GONZÁLEZ-LODEIRO, F., KLEIN, E., MARTÍNEZ-CATALÁN, J. R. & PABLO-MACIÁ, J. G. 1987. *Las rocas graníticas hercínicas del norte de Galicia y del occidente de Asturias*. IGME, Madrid, Memorias, **101**.

BELLIDO, F., BRANDLE, J. L., LASALA, M. & REYES, J. 1992. Consideraciones petrológicas y cronológicas sobre las rocas graníticas hercínicas. *Cuadernos del Laboratorio Geológico de Laxe*, **17**, 241–261.

BELLON, H. & BROUSSE, R. 1977. Le magmatisme périméditerranéen occidental. Essai de synthèse. *Bulletin de la Société Géologique de France*, **19**, 469–480.

BELLON, H., BLACHÈRE, H., CROUSILLES, M. *et al.* 1979. Radiochronologie, évolution tectono-magmatique et implications métallogéniques dans les Cadomo-variscides du Sud-Est Hespérique. *Bulletin de la Société Géologique de France*, **21**, 113–120.

BELLON, H., BORDET, P. & MONTENAT, C. 1983. Chronologie du magmatisme néogène des Cordillères Bétiques (Espagne méridionale). *Bulletin de la Société Géologique de France*, **25**, 205–217.

BEN OTHMAN, D., WHITE, W. M. & PATCHETT, J. 1989. The geochemistry of marine sediments, island arc magma genesis, and crust-mantle recycling. *Earth and Planetary Science Letters*, **94**, 1–21.

BENDER, J. F., HANSON, G. N. & BENCE, A. E. 1982. The Cortlandt complex: evidence for large-scale liquid inmiscibility involving granodiorite and diorite magmas. *Earth and Planetary Science Letters*, **58**, 330–334.

BENEDICTO, A., RAMOS, E., CASAS, A., SÀBAT, F. & BARÓN, A. 1993. Evolución tectonosedimentaria de la cubeta neógena de Inca (Mallorca). *Revista de la Sociedad Geológica de España*, **6**, 167–176.

BENITO, G., PÉREZ-GONZÁLEZ, A., GUTIÉRREZ, F. & MACHADO, M. J. 1998. River response to Quaternary large-scale subsidence due to evaporite slution (Gálllego River, Ebro Basin, Spain). *Geomorphology*, **22**, 243–263.

BENITO, G., GUTIÉRREZ, F., PÉREZ-GONZÁLEZ, A. & MACHADO, M. J.

2000. Geomorphological and sedimentological features in Quaternary fluvial systems affected by solution-induced subsidence (Ebro Basin, NE-Spain). *Geomorphology*, **33**, 209–224.

BENITO, R. 1993. *Hibridación del manto y asimilación de corteza continental en el magmatismo calco-alcalino y shoshonítico del SE de España*. PhD Thesis, University of Autónoma, Madrid.

BENITO, R., LÓPEZ-RUIZ, J. & TURI, B. 1992. Valores de dO¹⁸ de los basaltos alcalinos de la región volcánica de Olot, La Garrotxa, Cataluña. *Estudios Geológicos*, **48**, 43–46.

BENITO, R., MARTINEZ FRIAS, J., LUNAR, R. & WOLF D. 1998. El Hoyazo: a unique garnet-rich volcanic complex in southeast Spain. *Transactions of the Institution of Mining and Metallurgy*, **107**, 158–164.

BENITO, R., LÓPEZ-RUIZ, J., CEBRIÁ, J. M., HERTOGEN, J., DOBLAS, M., OYARZUN, R. & DEMAIFFE, D. 1999. Sr and O isotope constraints on source and crustal contamination in the high-K calc-alkaline and shoshonitic Neogene volcanic rocks of the SE Spain. *Lithos*, **46**, 773–802.

BENKE, K., DÜRKOOP, A., ERRENST, C. & MENSINK, H. 1981. Die korallenkalke im Ober-Jura der nord-westlichen Iberischen Ketten (Spanien). *Facies*, **4**, 27–94.

BENTABOL, H. 1990. *Las aguas de España y Portugal*. Estudio Tipo de la Viuda e Hijos de M. Tello, Madrid.

BERÁSTEGUI, X., GARCÍA, J. M. & LOSANTOS, M. 1990. Structure and sedimentary evolution of the Organyà basin (Central South Pyrenean Unit, Spain) during the Lower Cretaceous. *Bulletin de la Société Géologique de France*, **8**, 251–264.

BERÁSTEGUI, X., LOSANTOS, M., MUÑOZ, J. A. & PUIGDEFÀBREGAS, C. 1993. *Tall geològic del Pirineu Central 1/200.000*. Institut Cartogràfic de Catalunya, Barcelona.

BERGER, E., KAUFMANN, E. U. & SACHER, L. 1968. Sedimentologische untersuchungen im Jungpaläozoikum der Ostlichen Iberischen Ketten (Spanien). *Geologische Rundschau*, **57**, 472–483.

BERGGREN, W. A., HILGEN, F. J., LANGEREIS, C. G. *et al.* 1995a. Late Neogene Chronology: New perspectives in high resolution stratigraphy. *GSA Bulletin*, **107**, 1272–1287.

BERGGREN, W. A., KENT, D. V., SWISHER, C. & AUBRY, M. P. 1995b. A revised Cenozoic geochronology and chronostratigraphy. *In*: BERGGREN, W. A., KENT, D. V., AUBRY, M. P., HARDENBOL, J. (eds) *Geochronology, Time-Scales and Global Stratigraphic Correlation*. SEPM, Tulsa, Special Publications, **54**, 129–212.

BERGSTRÖM, J. & AHLBERG, P. 1981. Uppermost Lower Cambrian biostratigraphy in Scania, Sweden. *Geologiska Föreningens i Stockholm Förhandlingar*, **103**, 193–214.

BERGSTRÖM, S. M. & MASSA, D. 1992. Stratigraphic and biogeographic significance of Upper Ordovician conodonts from northwestern Libya. *In*: SALEM, M. J., HAMMUDA, O. S. & ELIAGOUBI, B. A. (eds) *The Geology of Libya*, Vol. 4, Elsevier, Amsterdam, 1323–1342.

BERMÚDEZ DE CASTRO, J. M. & NICOLÁS, M. E. 1997. Palaeodemography of the Atapuerca-SH Middle Pleistocene hominid sample. *Journal of Human Evolution*, **33**, 333–355.

BERMÚDEZ DE CASTRO, J. M., ARSUAGA, J. L., CARBONELL, E., ROSAS, A., MARTÍNEZ, I. & MOSQUERA, M. 1997. A hominid from the Lower Pleistocene of Atapuerca, Spain: possible ancestor to Neandertals and modern humans. *Science*, **276**, 1392–1395.

BERMÚDEZ DE CASTRO, J. M., ROSAS, A., CARBONELL, E., NICOLÁS, M. E., RODRÍGUEZ, J. & ARSUAGA, J. L. 1999. A modern human pattern of dental development in Lower Pleistocene hominids from Atapuerca-TD6 (Spain). *Proceedings of the National Academy of Sciences USA*, **96**, 4210–4213.

BERNALDO DE QUIRÓS, F. & MOURE, J. A. 1978. Cronología del Paleolítico y el Epipaleolítico Peninsulares. *C-14 y Prehistoria de la Península Ibérica*. Fundación J. March, Madrid, Serie Unversidad, **77**, 17–35.

BERNARD-GRIFFITHS, J., PEUCAT, J. J., CORNICHET, J., IGLESIAS PONCE DE LEÓN, M. & GIL IBARGUCHI, J. I. 1985. U-Pb, Nd isotope and REE geochemistry in eclogites from the Cabo Ortegal Complex, Galicia, Spain: an example of REE inmobility conserving MORB-like patterns during high-grade metamorphism. *Chemical Geology*, **52**, 217–225.

BERNAT, M., BOUSQUET, J. C.& DARS, R. 1978. I₀-U dating of the

Ouljian stage from Torre García (Southern, Spain). *Nature*, **275**, 302–303.

BERNAT, M., ECHAILLER, J. VC. & BUSQUET, J. C. 1982. Nouvelles datations Io-U sur des Strombes du Dernier Interglaciaire en Mediterranée. *Comptes Rendues de la Académie des Sciences*, Paris, **295**(II), 1023–1026.

BERNAUS, J. M. 2000. L'Urgonien du bassin d'Organyà, NE Espagne: Micropaléontologie, sédimentologie et stratigraphie séquentielle. *Géologie Alpine*. Mémoire hors de série, **33**, 25–26.

BERNOULLI, D. 1972. North Atlantic and Mediterranean Mesozoic Facies: A comparison. *In*: HOLLISTER, C. D., EWING, J. L. *et al.* *Initial Reports of the Deep Sea Drilling Project*, Volume XI. US Government Printing Office, Washington, 801–871.

BERNOULLI, D. & JENKYNS, H. C. 1974. Alpine, Mediterranean and central Atlantic Mesozoic facies in relation to the early evolution of the Tethys. *In*: DOTT, R. H. & SHAVER, R. H. (eds) *Modern and Ancient Geosynclinal Sedimentation*. SEPM, Tulsa, Special Publications, **19**, 129–160.

BERNOULLI, D. & LEMOINE, M. 1980. Birth and early evolution of the Tethys: the overall situation. *In*: AUBOIN, J., DEBELMAS, J. & LATREILLE, M. (eds) *Geology of the Alpine Chains Born of the Tethys*. Mémoires du Bureau de Recherches Géologiques et Minières, **115**, 168–179.

BERTRAND, M. & KILIAN, W. 1889. Etudes sur les terrains secondaires et tertiares dans les provinces de Grenade et Malaga. In Mission d'Andalousie. *Mémoires Académie des Sciences Paris*, **30**, 377–379.

BESEMS, R. E. 1981. Aspects of middle and late Triassic palynology. 1. Palynostratigraphical data from the Chiclana de Segura Formation of the Linares-Alcaraz region (southeastern Spain) and correlation with palynological assemblages from the Iberian Peninsula. *Review of Palaeobotany and Palynology*, **32**, 257–273.

BETZLER, C., BRACHERT, T. C., BRAGA, J. C. & MARTÍN, J. M. 1997. Nearshore, temperate, carbonate depositional systems (lower Tortonian, Agua Amarga Basin, southern Spain): implications for carbonate sequence stratigraphy. *Sedimentary Geology*, **113**, 27–53.

BEUF, S., BIJU-DUVAL, B., CHARPAL DE, O., ROGNON, P., GARIEL, O. & BENNACEF, A. 1971. *Les grès du Paléozoïque inférieur au Sahara. Sédimentation et discontinuités, évolution structurale d'un craton*. Publications de l'Institut Français du Pétrole, **18**.

BICKLE, M. J., WICKHAM, S. M., CHAPMAN, H. J. & TAYLOR, H. P. 1988. A strontium, neodymium and oxigen isotope study of hydrothermal metamorphism and crustal anatexis in the Trois Seigneurs Massif, Pyrenees, France. *Contributions to Mineralogy and Petrology*, **100**, 399–417.

BIJU-DUVAL, B., DECOURT, J. & LE PINCHON, X. 1977. From the Tethys ocean to the Mediterranean seas: a plate tectonic model of the evolution of the western Alpine system. *In*: BIJU-DUVAL, B. & MONTADERT, I. (eds) *Structural History of the Mediterranean Basin*. Technip, Paris, 143–164.

BIJU-DUVAL, B., LE TOUZEY, J. & MONTADERT, L. 1978. Structure and evolution of the Mediterranean Basins. *Initial Reports Deep Sea Drilling Project*, **42**, 951–984.

BIJWAARD, H., SPAKMAN, W. & ENGDAHL, E. R. 1998. Closing the gap between regional and global travel-time tomography. *Journal of Geophysical Research*, **103**, 30055–30078.

BILOTTE, M. 1985. Le Crétacé Supérieur des plates-formes est-Pyrénéennes. *Strata*, Série 2, Mémoires, **1**, Toulouse.

BINNEKAMP, J. G. 1965. Lower Devonian brachiopods and stratigraphy of North Palencia (Cantabrian Mountains, Spain). *Leidse Geologische Mededelingen*, **33**, 1–62.

BIROT, P. & SOLÉ, L. 1954. Recherches morphologiques dans le Nord-Ouest de la Peninsule Ibérique. *Memoires et Documents du Centre National de la Recherche Scientifique*, Paris, **4**, 9–61.

BISCHOFF, J. L., FITZPATRICK, J. A., LEÓN, L., ARSUAGA, J. L., FALGUÈRES, C., BAHAIN, J. J. & BULLEN, T. 1997. Geology and preliminary dating of hominid-bearing sedimentary fill of the Sima de los Huesos Chamber, Cueva Mayor of the Sierra de Atapuerca, Burgos, Spain. *Journal of Human Evolution*, **33**, 109–154.

BISCHOFF, L. 1974. Ein neues Silur-Vorkommen bei Serracin in der östlichen Sierra de Guadarrama, Zentralspanien. *Neues Jahrbuch für Geologie und Paläontologie, Abhandlungen*, **147**, 218–235.

BISCHOFF, L., LENZ, H., MULLER, P. & SCHMIDT, K. 1978. Geochemische und geochronologische Untersuchungen and Metavulkaniten und Orthogniessen der Ostlichen Sierra de Guadarrama (Spanien). *Neues Jahrbuch für Geologie und Palaeontologie. Abhandlungen*, **155**, 275–298.

BIXEL, F. 1984. *Le volcanisme stéphano-permien des Pyrénées*. PhD Thesis, Paul Sabatier University, Toulouse.

BIXEL, F. 1988. Le volcanisme stéphano-permien des Pyrénées Atlantiques. *Bulletin des Centres de Recherches Exploration-Production*, **12**, 661–706.

BIXEL, F., MULLER, J. & ROGER, P. 1985. *Carte geologique du Pic du midi d'Ossau et haut bassin du rio Gallego (1/25000)*. Institute Géodynamique Université de Bordeaux, **3**.

BIZON, G., BIZON, J. J., BOURROUILH, R. & MASSA, D. 1973. Présence aux Îlles Baléares (Méditerranée Occidentale) de sédiments 'messiniens' déposés dans une mer ouverte á salinité normale. *Comptes Rendus des seánces de l'Académie des Sciences de Paris*, **277**, 985–988.

BLANCO, A. & ROTHEMBERG, B. 1981. *Exploración Arqueometalúrgica de Huelva*. Labor, Barcelona.

BLANCO, C. G., VALENZUELA, M., SUÁREZ DE CENTI, C. & FERNÁNDEZ-PELLO LOIS, M. 1996. Características geoquímicas preliminares de azabaches artesanales del Kimmeridgiense de Asturias. *Geogaceta*, **20**(3), 677–680.

BLANCO, J. A., CORROCHANO, A., MONTIGNY, R. & THUIZAT, R. 1982. Sur l'âge du début de la sédimentation dans le bassin tertiaire du Duero (Espagne). Attribution au Paléocène par datation isotopique des alunites de l'unité inférieure. *Comptes Rendus de l'Académie des Sciences*, **295**, 259–262.

BLANCO, J. A., CANTANO, M., ARMENTEROS, I., FERNÁNDEZ MACARRO, B. & SÁNCHEZ MACÍAS, S. 1989. Superposición de procesos de alteración en la serie roja miocena de la fosa de Ciudad Rodrigo *Studia Geológica Salmanticensia*, Universidad Salamanca, **5**, 223–238.

BLANCO, M. J. & SPAKMAN, W. 1993*a*. Delay time tomography of the Iberian peninsula. *In*: MEZCUA, J. & CARREÑO, E. (eds) *Iberian Lithosphere Heterogeneity and Anisotropy (ILIHA)*. Monographs, **10**, IGN, 321–329.

BLANCO, M. J. & SPAKMAN, W. 1993*b*. The P-wave velocity structure of the mantle below the Iberian Península: evidence for subducted lithosphere below southern Spain. *Tectonophysics*, **221**, 13–34.

BLANKENSHIP, C. L. 1992. Structure and paleogeography of the External Betic Cordillera, southern Spain. *Marine and Petroleum Geology*, **9**, 256–264.

BLATRIX, P. & BURG, J. P. 1981. [40]Ar/[39]Ar dates from Sierra Morena (southern Spain). Variscan metamorphism and Cadomian orogeny. *Neues Jahrbuch für Mineralogie Monatshefte*, **10**, 470–478.

BLUMENTHAL, M. 1927. Versuch einer tektonischen gliederug der betischen cordilleren von Central, und Sud-West Andalusien. *Eclogae Geologicae Helveticae*, **20**, 487–592.

BLUMENTHAL, M. 1949. Estudio geológico de las cadenas costeras al oeste de Málaga, entre el Río Guadalorce y el Río Verde. *Boletín del IGME*, **62**, 11–203.

BLUMENTHAL, M. M. 1961. Rasgos principales de la geología de las Islas Canarias con datos sobre Madeira. *Boletín Geológico y Minero*, **72**, 1–30.

BODIN, J. & LEDRU, P. 1986. Nappes hercyniennes précoces à matériel dévonien hétérotopique dans les Pyrénées ariégoises. *Comptes Rendus de l'Académie des Sciences de Paris*, **302**, 969–974.

BODINIER, J. L., MORTEN, L., PUGA, E. & DÍAZ DE FEDERICO, A. 1987. Geochemistry of metabasites from the Nevado-Filábride Complex, Betic Cordilleras, Spain: relics of a dismembered ophiolitic secuence. *Lithos*, **20**, 235–245.

BOERSMA, K. T. 1973. Devonian and Lower Carboniferous conodont biostratigraphy, Central Spanish Pyrenees. *Leidse Geologische Mededelingen*, **49**(2), 303–377.

BOGAARD, P. & SCHMINCKE, H. U. 1998. Chronostratigraphy of Gran Canaria. *In*: WEAVER, P. P. E., SCHMINCKE, H. U., FIRTH, J. V. &

DUFFIELD, W. (eds) *Proceedings of the Ocean Drilling Program, Science Results*, **157**, 127–140.

BOGAARD, P., SCHMINCKE, H. U. & FREUNDT, A. 1988. Eruption ages and magma supply rates during the Miocene evolution of Gran Canaria. Single-crystal 39Ar/40Ar laser ages. *Naturwissenschaften*, **75**, 616–617.

BOGOLEPOVA, O. K., GUTIÉRREZ-MARCO, J. C. & ROBARDET, M. 1998. A brief account on the Upper Silurian cephalopods from the Valle syncline, province of Seville (Ossa Morena Zone, southern Spain). *Temas Geológico-Mineros ITGE*, **23**, 63–66.

BOILLOT, G. 1984. Le Golfe de Gascogne et les Pyrenees. *In*: BOILLOT, G. (ed.) *Les marges continentales actualles et fossiles autour de la France*. Masson, Paris, 5–81.

BOILLOT, G. & CAPDEVILA, R. 1977. The Pyrenees: subduction and collision? *Earth and Planetary Science Letters*, **35**, 151–160.

BOILLOT, G. & MALOD, J. 1988. The north and north-west spanish continental margin: a review. *Revista Sociedad Geológica de España*, **1**, 295–316.

BOILLOT, G., DUPEUBLE, P. A., LAMBOY, M., D'OZOUVILLE, L. & SIBUET, J. C. 1971. Structure et histoire geologique de la marge continentale au Nord de l'Espagne (entre 4º et 9ºW). *In*: DEBYSIER, J., LE PICHON, X. & MONTARDET, M. (eds) *Histoire structurale du Golfe de Gascogne*, Vol. 6. Technip, Paris, 1–52.

BOILLOT, G. DEPEUBLE, P. A. & MALOD, J. 1979. Subduction and tectonics on the continental margin of Northern Spain. *Marine Geology*, **32**, 53–70.

BOISSONNAS, J. 1974. *Carte géologique de la France a 1/50 000, feuille pic de Maubermé*. BRGM, Orléans.

BOIVIN, P. A. 1982. *Interactions entre magmas basaltiques et manteau supérieur: arguments apportés par les enclaves basiques des basaltes alcalins. Exemples du Deves (Massif-Central Français) et du volcanisme Quaternaire de la région de Carthagene (Espagne)*. PhD Thesis, University of Clermont-Ferrand.

BONADONNA, F. P. & VILLA, I. 1986. Estudio geocronológico del volcanismo de Las Higueruelas. *Actas Castilla-La Mancha: Espacio y Sociedad*, **3**, 249–253.

BOND, J. 1996. Tectono-sedimentary evolution of the Almazán Basin, NE Spain. *In*: Friend, F. & Dabrio, C. (eds) *Tertiary Basins of Spain: the Stratigraphic Record of Crustal Kinematics*. World and Regional Geology, **6**, Cambridge University Press, Cambridge 203–213.

BOND, R. M.G. & McCLAY, K. R. 1995. Inversion of a Lower Cretaceous extensional basin, south central Pyrenees, Spain. *In*: BUCHANAN, J. G. & BUCHANAN, P. G. (eds) *Basin Inversion*, Geological Society, London, Special Publication, **88**, 415–431.

BONJOUR, J. L. & ODIN, G-S. 1989. Recherche sur les volcanoclastites des Séries Rouges Initiales en presqu'île de Crozon: premier âge radiométrique de l'Arénig. *Géologie de la France*, **4**, 3–8.

BONJOUR, J. L., PEUCAT, J. J., CHAUVEL, J. J., PARIS, F. & CORNICHET 1988. U-Pb zircon dating of the Early Palaeozoic (Arenigian) transgression in Western Brittany (France) ; a new constraint for the Lower Palaeozoic time-scale. *Chemical Geology (Isotope Geoscience section)*, **72**, 329–336.

BONNATTI, E. 1985. Punctiform initiation of sea floor spreading in the Red Sea during transition from a continental to an oceanic rift. *Nature*, **315**, 33–37.

BOOGAARD, M. VAN DEN. 1963. Conodonts of Upper Devonian and Lower Carboniferous age from Southern Portugal. *Geologie Mijnbouw*, **42**(8), 248–259.

BOOGAARD, M. VAN DEN & SCHERMERHORN, L. J. G. 1980. Conodont faunas from Portugal and southwestern Spain. Part 4 – A Famennian conodont fauna near Nerva (Rio Tinto). *Scripta Geologica*, **56**, 1–14.

BOOGAARD, M. VAN DEN & SCHERMERHORN, L. J. G. 1981. *Conodont faunas from Portugal and southwestern Spain. Part 6. A lower Famennian conodont fauna at Monte Forno da Cal (South Portugal)*. Rijksmuseum van Geologie en Mineralogie, Leiden, Scripta Geologica, **63**, 1–16.

BOOGAARD, M. VAN DEN & VÁZQUEZ GUZMÁN, F. 1981. *Conodont faunas from Portugal and southwestern Spain. Part 5. Lower Carboniferous conodonts at Santa Olalla de Cala (Spain)*. Rijksmuseum van Geologie en Mineralogie, Leiden, Scripta Geologica, **61**.

BOORSMA, L. J. 1992. Syn-tectonic sedimentation in a Neogene strike-slip basin containing a stacked Gilbert-type delta (SE Spain). *Sedimentary Geology*, **81**, 105–123.

BOOTH, B. 1973. The Granadilla pumice deposit of southern Tenerife, Canary Islands. *Proceedings of the Geologists' Association*, **84**, 353–369.

BORDONABA, A. P. & AURELL, M. 2001. El Hettangiense-Sinemuriense (Jurásico inferior) en el sector Montalbán-Oliete (Teruel): análisis de facies y evolución sedimentaria. *Revista de la Sociedad Geológica de España*, **14**, 135–146.

BORDONABA, A. P., AURELL, M. & CASAS, A. 1999. Control tectónico y distribución de las facies en el tránsito Triásico-Jurásico en el sector de Oliete (Teruel). *Geogaceta*, **25**, 43–46.

BORDONABA, A. P., AURELL, M. & BÁDENAS, B. 2000. La Unidad Almonacid de la Cuba en el sector de Obón-Castel de Cabra. *Geotemas*, **1**, 163–167.

BORDONAU, J. 1992. *Els complexes glàccio-lacustres relacionats ams el darrer cicle glacial als Pirineus.* Geoforma Ediciones, Logroño, 251 pp.

BORN, A. 1918. Die *Calymene tristani*-Stufe (mittleres Untersilur) bei Almaden, ihre Fauna, Gliederung und Verbreitung. *Abhandlungen der senckenbergischen naturforschenden Gesellschaft*, **36**, 309–358.

BORREGO, A. G., HAGEMANN, H. V., BLANCO, C. G., VALENZUELA, M., & SUÁREZ DE CENTI, C. 1996. The Pliensbachian (Early Jurassic) 'anoxic' event in Asturias, northern Spain: Santa Mera Member, Rodiles Formation. *Organic Geochemistry*, **25**(5–7), 295–309.

BORRERO, J. & HIGUERAS, P. 1990. Nuevos conocimientos sobre la geología y metalogénesis de los yacimientos de mercurio de Almadén (Ciudad Real). *Boletín Geológico y Minero*, **101**, 48–65.

BOSENCE, D. W. J., POMAR, L., WALTHAM, D. A. & LANKASTER, H. G. 1994. Computer modelling of a Miocene carbonate platform, Mallorca, Spain. *AAPG Bulletin*, **78**, 247–266.

BOUCHEZ, J. L. & GLEIZES, G. 1995. Two-stage deformation of the Mont Louis-Andorra granite pluton (Variscan Pyrenees) inferred from magnetic susceptibility anisotropy. *Journal of the Geological Society*. London, **152**, 669–679.

BOULIN, J. L. 1968. *Les zones internes des Cordillères Bétiques de Málaga à Motril.* PhD Thesis, University of Paris.

BOULOUARD, C. & VIALLARD, P. 1971. Identification du Permien dans la Chaîne Ibérique. *Comptes Rendus de l'Académie des Sciences de Paris*, **272**, 2441–2444.

BOULOUARD, C. & VIALLARD, P. 1982. Reduction ou lacune du Trias inférieur sur la bordure méditerrannée de la Chaîne Ibérique: arguments palynologiques. *Comptes Rendus de l'Académie des Sciences de Paris*, **295**, 803–808.

BOULTER, C. A. 1993. Comparison of Rio Tinto, Spain, and Guaymas Basin, Gulf of California: An explanation of a supergiant massive sulfide deposit in an ancient sill-sediment complex. *Geology*, **21**, 801–804.

BOURCART, J. & JEREMINE, E. 1937. La Grande Canarie. Etude géologique et lithologique. *Bulletin Volcanologique*, **2**, 3–77.

BOURGEOIS, J. 1978. *La transversale de Ronda, Cordillères Bétiques, Espagne. Donnes gèologiques pour un modèle d'èvolution de l'Arc de Gibraltar.* PhD Thesis, University of Besançon.

BOURGOIS, J. 1980. Pre-Triassic fit and alpine tectonics of continental blocks in the western Mediterranean: Discussions and reply. *Geological Society of America Bulletin*, Part I, **91**, 632–634.

BOURROILH, R. 1973. *Stratigraphie, sédimentologie et tectonique de l'île de Minorque et du Nord-East de Majorque (Baléares). La terminación nord-orientale des Cordillères Bètiques en Mediterranée occidentale.* PhD Thesis, University of Pierre and Marie Curie, Paris.

BOURROILH, R. 1983. *Estratigrafía, sedimentología y tectónica de la isla de Menorca y del Noroeste de Mallorca (Baleares).* IGME, Madrid, Memórias, **99**.

BOURROILH, R. & GORSLINE, D. S. 1979. Pre-Triassic fit and alpine tectonics of continental blocks in the western Mediterranean. *Geological Society of America Bulletin*, Part I, **90**, 1074–1083.

BOURROILH, R. & GORSLINE, D. S. 1980. Pre-Triassic fit and alpine tectonics of continental blocks in the western Mediterranean: Discussions and reply. *Geological Society of America Bulletin*, Part I, **91**, 634–636.

BOUTET, C., RANGHEARD, Y., ROSENTHAL, P., VISSCHER, H., DURAND-DELGA, M. 1982. Découverte d'une microflore d'age Norien dans la Sierra Norte de Majorque (Baleares, España). *Comptes Rendus de l'Académie des Sciences de Paris*, **294**, 1267–1270.

BOUYX, E. 1970. Contribution a l'étude des Formations Anteordoviciennes de la Meseta Méridionale (Ciudad Real et Badajoz). IGME, Madrid, Memorias, **73**.

BOUYX, E., BLAISE, J., BRICE, D. *et al.* 1992. Implications paléogéographiques des affinités nord-gondwaniennes et rhénanes des faunes dévoniennes de la zone de Meguma (Appalaches septentrionales). *Comptes rendus de l'Académie des Sciences de Paris*, **315**(2), 337–343.

BOUYX, E., BLAISE, J., BRICE, D. *et al.* 1997 Biostratigraphie et paléobiogéographie du Siluro-Devonien de la zone de Meguma (Nouvelle-Ecosse, Canada). *Canadian Journal Earth Sciences*, **34**, 1295–1309.

BOWLES, W. 1775. *Introducción a la Historia Natural y a la Geografía Física de España.* D. Francisco Manuel de Mena, Madrid.

BOWMAN, M. B. 1985. The sedimentology and paleogeographic setting of late Namurian-Westfalian. A basin-fill succession in the San Emiliano and Cármenes areas of NW León, Cantabrian Mountains, NW Spain. *In*: LEMOS DE SOUSA, H. J. & WAGNER, R. H. (eds) *Papers on the Carboniferous of the Iberian Peninsula (Sedimentology, Stratigraphy, Paleontology, Tectonics and Geochemistry).* Annales da Faculdade de Ciências, Universidade do Porto, Special Supplement to Vol. **64** (1983), 117–168.

BOWN, T. M. & KRAUS, M. J. 1987. Integration of channel and floodplain suites in aggrading alluvial systems. I. Developmental sequences and lateral relations of lower Eocene alluvial palaeosols, Willwood formation, Bighorn basin, Wyoming. *Journal of Sedimentary Petrology*, **57**, 589–601.

BOWRING, S. A. & ERWIN, D. H. 1998. A new look at evolutionary rates in deep time: Uniting paleontology and high-precision geochronology. *GSA Today*, **8**(9), 1–8.

BOYER, F., KRYLATOV, S. & STOPPEL, D. 1974. Sur le problème de l'existence d'une lacune sous les lydiennes à nodules phosphatés du Dinantien des Pyrénées et de la Montagne Noire (France, Espagne). *Geologische Jahrbuch*, B, **9**, 1–60.

BOYER, S. E. & ELLIOTT, D. 1982. Thrust systems. *AAPG Bulletin*, **66**, 1196–1230.

BOYNTON, W. V. 1984. Geochemistry of the rare earth elements meteorite studies. *In*: HENDERSON (ed.) *Rare Earth Element Geochemistry.* Elsevier, Amsterdam, 65–114.

BRACHERT, T. C., BETZLER, C., BRAGA, J. C. & MARTÍN, J. M. 1996. Record of climatic change in neritic carbonates: turnover in biogenic associations and depositional modes (late Miocene, southern Spain). *Geologische Rundschau*, **85**, 327–337.

BRADSHAW, T. K., HAWKESWORTH, C. J. & GALLAGHER, K. 1993. Basaltic volcanism in the Southern Basin and Range: no role for a mantle plume. *Earth and Planetary Science Letters*, **116**, 45–62.

BRAGA, J. C. 1983. *Ammonites del Domerense de la Zona Subbetica (Cordilleras Beticas, Sur de España).* PhD Thesis, University of Granada.

BRAGA, J. C. & MARTÍN, J. M. 1987a. Distribución de las algas Dasycladaceas en el Trias alpujárride. *Cuadernos de Geología Ibérica*, **11**, 475–489.

BRAGA, J. C. & MARTÍN, J. M. 1987b. Sedimentación cíclica lagunar y bioconstrucciones asociadas en el Trías superior alpujárride. *Cuadernos de Geología Ibérica*, **11**, 459–473.

BRAGA, J. C. & MARTÍN, J. M. 1996. Geometries of reef advance in response to relative sea-level changes in a Messinian (uppermost Miocene) fringing reef (Cariatiz reef, Sorbas Basin, SE Spain). *Sedimentary Geology*, **107**, 61–81.

BRAGA, J. C. & MARTÍN, J. M. 2000. Subaqueous siliciclastic stromatolites: a case history from Late Miocene beach deposits in the Sorbas basin of SE Spain. *In*: RIDING, R. & AWRAMIK, S. M. (eds) *Microbial Sediments.* Springer-Verlag, Berlin, 226–232.

BRAGA, J. C., GARCÍA-GÓMEZ, R., JIMENEZ, A. P. & RIVAS, P. 1981. *Correlaciones en el Lias de las Cordilleras Béticas.* Curso de Conferencias sobre el Programa Internacional de Correlación Geológica. Real Academia de Ciencias Exactas, Físicas y Naturales, Madrid, 161–181.

BRAGA, J. C., COMAS-RENGIFO, M. J., GOY, A. & RIVAS, P. 1984a. The

Pliensbachian of Spain: Ammonite succesions, boundaries and correlations. *In*: MICHELSEN, O. & ZEISS, A. (eds) *International Symposium on Jurassic Stratigraphy*, Vol. 1 Erlangen, 160–176.

BRAGA, J. C., MARTÍN-ALGARRA, A. & RIVAS, P. 1984b. Biostratigraphic sketch of the Lower Liassic in the Betic Cordillera. *In*: MICHELSEN, O. & ZEISS, A. (eds) *International Symposium on Jurassic Stratigraphy*, Vol. 1 Erlangen, 178–190.

BRAGA, J. C., COMAS-RENGIFO, M. J., GOY, A. & RIVAS, P. 1985. Le Pliensbachien dans la chaine cantabrique orientale entre Castillo Pedroso et Reinosa (Santander, Espagne). *Les Cahiers de l'Institut Catholique de Lyon*, **14**, 69–83.

BRAGA, J. C., COMAS-RENGIFO, M. J., GOY, A., RIVAS, P. & YÉBENES, A. 1988. El Lías inferior y medio en la zona central de la Cuenca Vasco-Cantábrica (Camino, Santander). *III Coloquio de Estratigrafía y Paleogeografía del Jurásico de España, libro guía de las excursiones, Ciencias de la Tierra (Instituto de Estudios Riojanos)*, **11**, 17–45.

BRAGA, J. C., MARTÍN, J. M. & ALCALÁ, B. 1990. Coral reefs in coarse terrigenous sedimentary environments (upper Tortonian, Granada basin, S. Spain). *Sedimentary Geology*, **66**, 135–150.

BRAGA, J. C., JIMENEZ, A. P., MARTÍN, J. M. & RIVAS, P. 1996a. Middle Miocene, coral-oyster reefs (Murchas, Granada, southern Spain). *In*: FRANSEEN, E. K., ESTEBAN, M., WARD, W. C & ROUCHY, J. M. (eds) *Models for Carbonate Stratigraphy from Miocene Reef Complexes of Mediterranean Regions*. Concepts in Sedimentology and Paleontology Series, SEPM, Tulsa, **5**, 131–139.

BRAGA, J. C., MARTÍN, J. M. & RIDING, R. 1996b. Internal structure of segment reefs: *Halimeda* algal mounds in the Mediterranean Miocene. *Geology*, **24**, 35–38.

BRAGA, J. C., MARTÍN, J. M. & WOOD, J. L. 2001. Submarine lobes and feeder channels of redeposited, temperate carbonate and mixed siliciclastic-carbonate platform-deposits (Vera Basin, Almería, southern Spain). *Sedimentology*, **48**, 99–116.

BRANDEBOURGER, E., CASQUET, C., DEBON, F. *et al.* 1983. Nota previa sobre la petrografía y la geoquímica de los granitoides de la Sierra de Guadarrama (España). *Studia Geológica Salmantinensis*, **XVIII**, 251–264.

BRÄNDLE, J. L. & NAGY, G. 1995. The state of the 5th version of IGBA: Igneous Petrological Data Base. *Computers & Geosciences*, **21**, 425–432

BRASIER, M. D. 1985. Evolutionary and geological events across the Precambrian–Cambrian boundary. *Geology Today*, **1**, 141–146.

BRASIER, M. D. 1995. The basal Cambrian transition and Cambrian bio-events (from terminal Proterozoic extinctions to Cambrian biomeres). *In*: WALLISER, O. H. (ed.) *Global Events and Event Stratigraphy in the Phanerozoic*. Springer-Verlag, Berlin, 113–138.

BRASIER, M. D., PEREJÓN, A. & SAN JOSÉ, M. A. 1979. Discovery of an important fossiliferous Precambrian–Cambrian sequence in Spain. *Estudios Geológicos*, **35**, 379–383.

BRAVO, T. 1962. El circo de Las Cañadas y sus dependencias. *Boletín de La Real Sociedad Española de Historia Natural (Sec. Geol.)*, **60**, 93–108.

BRAVO, T. 1964. Estudio geológico y petrográfico de la isla de La Gomera. *Estudios Geológicos*, **20**, 1–56.

BRENCHLEY, P. J., ROMANO, M. & GUTIÉRREZ-MARCO, J. C. 1986. Proximal and distal hummocky cross-stratified facies on a wide Ordovician Shelf in Iberia. *In*: KNIGHT, R. J. & MCLEAN, J. R. (eds) *Shelf Sands and Sandstones*, Canadian Society for Petroleum Geologists, Memoir **11**, 241–255.

BRENCHLEY, P. J., ROMANO, M., YOUNG, T. P. & STORCH, P. 1991. Hirnantian glaciomarine diamictites – evidence for the spread of glaciation and its effect on Upper Ordovician faunas. *In*: BARNES, C. R. & WILLIAMS, S. H. (eds) *Advances in Ordovician Geology*. Geological Survey of Canada, Paper **90–9**, 325–336.

BRINKMANN, R. 1931. Betikum und Keltiberikum in Südöstspanien. *Abhandlungen der Gesellschaft der Wissenschaften zu Göttingen (Mathematisch-Physikalische Klasse)*, **3**, 749–856 (translated into Spanish by Gómez de Llarena, J. 1948. Las cadenas béticas y celtibéricas del sureste de España. *Publicaciones Extranjeras sobre Geología de España*, **4**, 305–434).

BRIQUEU, L. & INNOCENT, C. 1993. Datation U/Pb sur zircon et géochimie isotopique Sr et Nd du volcanisme permien des

Pyrénées occidentales (Ossau et Anayet). *Comptes Rendues de l'Académie des Sciences de Paris*, **316**, 623–628.

BRISTOW, C. & ROBSON, J. L. 1994. Paleogene basin development in Devon. *Transactions of the Institution of Mining and Metallurgy*, **103**, 163–174.

BROUTIN, J. 1977. Nouvelles données sur la flore des bassins autunostephaniens des environs de Gudalcanal (Province de Seville, Espagne). *Cuadernos de Geología Ibérica*, **4**, 91–98.

BROUTIN, J. 1981. *Étude paléobotanique et palynologique du passage Carbonifère-Permien dans les bassins continentaux du Sud-Est de la zone d'Ossa-Morena (environs de Guadalcanal, Espagne du Sud). Implications paléogéographiques et stratigraphiques*. PhD Thesis, University of Paris VI.

BROUTIN, J., DOUBINGER, J., GISBERT, J. & SATTA-PASSINI, S. 1988. Premieres datations palynologiques dans le facies Buntsandstein des Pyrenees catalanes espagnoles. *Comptes Rendus de l'Académie des Sciences*, **306**, 169–163.

BROUTIN, J., FERRER, J., GISBERT, J. & NMILA, A. 1992. Prémiere découverte d'une microflore thuringienne dans le facies saxonien de l'île de Minorque (Baleares, Espagne). *Comptes Rendus de l'Académie des Sciences*, **315**, 117–122.

BROUTIN, J., CHATEAUNEUF, J., GALTIER, J. & RONCHI, A. 2000. L'Autunien d'Autun reste-t-il une reference pour les depots continentaux du Permien inferieur d'Europe? *Geologie de la France*, **2**, 17–31.

BROUWER, A. 1964. Deux faciès dans le Dévonien des Montagnes Cantabriques Méridionales. *Breviora Geologica Asturica*, **8**(1–4), 3–10.

BROUWER, H. A. 1926. Zur geologie der Sierra Nevada. *Geologische Rundschau*, **17**, 118–137.

BRÜCKNER, H. 1986. Stratigraphy evolution and age of Quaternary marine terraces in Morocco and Spain. *Zeitscrift für Geomorphologie*. N.F., **62**, 83–101.

BRÜCKNER, H. & RADTKE, U. 1986. Paleoclimatic implications derived from profiles along the spanish mediterranean coast. *In*: LÓPEZ VERA, F. (ed.) *Quaternary Climate in the Western Mediterranean*. Universidad Autónoma de Madrid, Madrid, 467–486.

BRUIJNE, C. H. 2001. *Denudation, intraplate tectonics and far field effects. An integrated fission track study in central Spain*. PhD Thesis, Vrije University, Amsterdam.

BRUIJNE, C. H. & ANDRIESSEN, P. A. M. 2000. Interplay of intraplate tectonics and surface processes in the Sierra de Guadarrama (central Spain) assessed by apatite fission track analysis. *Physics and Chemistry of the Earth*, **25**, 555–563.

BRUN, J. P. & PONS, J. 1981. Strain patterns of pluton emplacement in a crust undergoing non-coaxial deformation, Sierra Morena, Southern Spain. *Journal of Structural Geology*, **3**, 219–229.

BRUNER, K. R. & SMOSNA, R. 2000. Stratigraphic-tectonic relations in Spain's Cantabrian Mountains: fan delta meets carbonate shelf. *Journal of Sedimentary Research*, **70**, 1302–1314.

BRUNER, K. R., SMOSNA, R. & MARTÍNEZ GARCÍA, E. 1998. Comparative analysis of fan-delta facies from the Carboniferous of northwestern Spain. *Revista de la Sociedad Geológica de España*, **11**, 181–194.

BRUNET, M. F. 1986. The influence of the evolution of the Pyrenees on adjacent basins. *Tectonophysics*, **129**, 343–354.

BRYAN, S. 1995. Bandas del Sur pyroclastics, southern Tenerife. *In*: MARTÍ J. & MITJAVILA J. (eds) *A Field Guide to the Central Volcanic Complex of Tenerife (Canary Islands)*. Serie Casa de Los Volcanes, **4**, Cabildo de Lanzarote.

BRYAN, S., MARTÍ, J. & CAS, R. A. F. 1998. Stratigraphy of the Bandas del Sur Formation: an extracaldera record of Quaternary phonolitic explosive eruptions from the Las Cañadas edifice, Tenerife (Canary Islands). *Geological Magazine*, **135**, 605–636.

BUCH, L. VON 1825. *Physikalische Beschreibung der Canarischen Inseln*. Berlin.

BUDDINGTON, A. F. 1959. Granite emplacement with special reference to North America. *Geological Society of American Bulletin*, **70**, 671–747.

BUDUROV, K., CALVET, F., GOY, A., MÁRQUEZ-ALIAGA, A., TRIFONOVA, E. & ARCHE, A. 1993. Middle Triassic stratigraphy and correlation in parts of the Tethys realm (Bulgaria and Spain). *In*: HAGDORM, H. & SEILACHER, A. (eds) *Proceedings Muschelkalk Schöntaler*

Symposium, 1991. Sonderbände der Gesellschaft für Naturkunde in Wüttemberg **2**, 157–164.

BUFORN, E., UDÍAS, A. & COLOMBÁS, M. A. 1988*a*. Seismicity, source mechanism and tectonics of the Azores-Gibraltar plate boundary. *Tectonophysics*, **152**, 89–118.

BUFORN, E., UDÍAS, A. & MEZCUA, J. 1988*b*. Seismicity and focal mechanisms in South Spain. *Bulletin Seismological Society of America*, **78**, 2008–2024.

BUFORN, E., UDÍAS, A. & MEZCUA, J. & MADARIAGA, R. 1991. A deep earthquake under South Spain, 8 March 1990. *Bulletin of the Seismological Society of America*, **81**, 1–5

BUGGISCH, W., MEIBURG, P. & SCHUMANN, D. 1982. Facies, paleogeography and intra-Devonian stratigraphic gaps of the Asturo-Leonese Basin (Cantabrian Mts./Spain), *In*: KULLMANN, J., SCHÖNENBERG, R. & WIEDMANN, J. (eds) *Subsidenzentwicklung im Kantabrischen Variszikum und an passiven kontinentalrändern der Kreide: Teil 1. Variszikum. Neues Jahrbuch Geologie Paläontologie Abhandlungen*, **163**(2), 212–230.

BUIS, M. G. & REY, J. 1975. Une évolution sédimentaire de type deltaique: le passage du Tertiaire marin au Tertiaire continental entre l'Ariège et le Douctouyre (Pyrénées Ariégeoises) *Bull. Soc. Hist. Nat. Toulouse*, **111**, 80–95.

BULARD, P. F. 1972. *Le Jurassique moyen et supérieur de la Chaine Ibérique sur la bordure du bassin de l'Ebre (Espagne).* PhD Thesis, University of Nice.

BULARD, P. F., GÓMEZ, J. J., THIERRY, J., TINTANT, H. & VIALLARD, P. 1974. La discontinuite entre Jurassique Moyen et Jurassique Superieur dans les Chaines Ibériques. *Comte Rendu de l'Académie des Sciences de Paris*, **278**, 2107–2110.

BULARD, P. F., FEUILLÉE, P. & FLOQUET, M. 1979. Le limite Jurassique moyen-Jurassique supérieur dans la Sierra d'Aralar (Pyrénnées basques espagnoles*). Cuadernos Geología Ibérica*, **10**, 179–196.

BULTYNCK, P. 1971. Le Silurien Supérieur et le Dévonien Inférieur de la Sierra de Guadarrama (Espagne centrale). Deuxième partie: Assemblages de conodontes à *Spathognathodus. Bulletin de l'Institut royal des Sciences naturelles de Belgique*, **47**(3), 1–43.

BULTYNCK, P. & SOERS, E. 1971. Le Silurien supérieur et le Dévonien inférieur de la Sierra de Guadarrama (Espagne centrale). Première partie: stratigraphie et tectonique. *Bulletin de l'Institut royal des Sciences naturelles de Belgique, Sciences de la Terre*, **47**(1), 1–22.

BUNTFUSS, J. 1970. Die Geologie der Kustenketten zwischen dem Río Verde und dem Campo de Gibraltar (Westliche Betische Kordilleren, Südspanien). *Geologie Jahrbuch*, **88**, 373–420.

BURBANK, D., PUIGDEFABREGAS, C. & MUÑOZ, J. A. 1992*a*. The chronology of the Eocene tectonic and stratigraphic development of the eastern Pyrenean foreland basin, northeast Spain. *Geological Society of America Bulletin*, **104**, 1101–1120.

BURBANK, D., VERGÉS, J., MUÑOZ, J. A. & BENTHAM, P. 1992*b*. Coeval hindward- and forward-imbricating thrusting in the central southern Pyrenees, Spain: timing and rates of shortening. *Geological Society of America Bulletin*, **104**, 3–17

BURG, J. P. & FORD, M. 1997. Orogeny through time: an overview. *In*: Burg, J. P. & Ford, M. (eds) *Orogeny Through Time*. Geological Society, London, Special Publications, **121**, 1–17.

BURG, J. P., IGLESIAS, M., LAURENT, PH., MATTE, PH. & RIBEIRO, A. 1981. Variscan intracontinental deformation: the Coimbra-Córdoba shear zone (SW Iberian Peninsula). *Tectonophysics*, **78**, 161–177.

BURG, J. P., BALÉ, P., BRUN, J. P. & GIRARDEAU, J. 1987. Stretching lineations and transport direction in the Ibero-Armorican Arc during the Siluro-Devonian collision. *Geodinamica Acta*, **1**, 71–81.

BURGOS, J. C. & PASCUAL, E. 1976. El stock básico del Norte de Villaviciosa de Córdoba (Complejo Los Ojuelos-La Coronada). Sierra Morena (España). *Cuadernos Geológicos*, **7**, 69–122.

BURILLO, F., GUTIÉRREZ, M. & PEÑA, J. L. 1985. Las acumulaciones holocenas y su datación arqueológica en Mediana de Aragón (Zaragoza). *Cuadernos de Investigación Geográfica*, **11**, 193–207.

BUSNARDO, R., MOUTERDE, R. & LINARES, A. 1966. Découverte de l'Hettangien dans la coupe de Alhama de Granada (Andalousie). *Comptes Rendus de l'Académie des Sciences, Paris*, **263**, 1036–1039.

BUSTILLO, M. A. & BUSTILLO, M. 2000. Miocene silcrete in argillaceous playa deposits, Madrid Basin, Spain: petrological and sedimentological features. *Sedimentology*, **47**, 1023–1037.

BUSTILLO, M. A. & MARTÍN-SERRANO, A. 1980. Caracterización y significado de las rocas silíceas y ferruginosas del Paleoceno de Zamora. *Tecniterrae* (Madrid), **36**, 14–29.

BUTENWEG, P. 1968. Geologische Untersuchungen im Östteil der Sierra Morena nordöstlich von La Carolina (Provinz Jaén, Spanien). *Münstersche Forschungen zur Geologie und Paläontologie*, **6**, 1–126.

BUTZER, K. W. 1975. Pleistocene littoral sedimentary cycles of the Mediterranean Basin: A Mallorquin view. *In*: BUTZER, K. W. & ISAAC, G. (eds) *After the Australopithecines*. Mouton Press, The Hague, 25–71.

BUTZER, K. W. & FREEMAN, L. G. 1968. Pollen analysis at the Cueva del Toll, Catalonia: a critical re-appraisal. *Geologie Mijnbouw*, **47**, 116–120.

BUTZER, K. W. & MATEU, J. F. 1999. Pleistocene versus Holocene: Geomorphological change in a small but steep Watershed of Mediterranean Spain. *In*: ROSSELLÓ, V. (ed.) *Geoarqueologia i Quaternari litoral. Memorial M. P. Fumanal*. University of Valencia, 97–111.

CABALLERO, J. M., CASQUET, C., GALINDO, C., GONZÁLEZ-CASADO, J. M., SNELLING, N. & TORNOS, F. 1992. Dating of hydrothermal events in the Sierra de Guadarrama, Iberian Hercynian Belt, Spain. *Geogaceta*, **11**, 18–22.

CABANIS, B. & LE FUR-BALOUET, S. 1989. Les magmatismes stéphanopermiens des Pyrénées marqueurs de l'évolution géodynamique de la chaîne. *Bulletin des Centres de Recherches Exploration-Production Elf-Aquitaine*, **13**, 105–130.

CABANIS, B. & LE FUR-BALOUET, S. 1990. Le magmatisme Crétacé des Pyrénées – apport de la géochimie des éléments en traces – conséquences chronologiques et géodynamiques. *Bulletin des Centres de Recherches Exploration-Production Elf-Aquitaine*, **14**, 155–184.

CABRA, P., DÍAZ DE NEIRA, A., ENRILE, A., LÓPEZ OLMEDO, F. & PÉREZ-GONZÁLEZ, A. 1988. *Cartografía geológica y memoria de las hojas geológicas a escala 1:50.000 de La Roda (742), La Gineta (765) y Quintanar del Rey (717).* ITGE MINER, Madrid.

CABRAL, J. 1989. An example of intraplate neotectonic activity, Vilariça basin, Northeast Portugal. *Tectonics*, **8**, 285–303.

CABRAL, J. 1995. *Neotectónica en Portugal continental*. Memorias Insituto Geologico y Mineiro, **31**.

CABRERA, L. 1981. Influencia de la tectónica en la sedimentación continental de la cuenca del Vallés-Penedés (provincia de Barcelona, España) durante el Mioceno inferior. *Acta Geológica Hispánica*, **16**(3), 165–171.

CABRERA, L. 1983. *Estratigrafía y Sedimentología de las formaciones lacustres del tránsito Oligoceno-Mioceno del SE de la Cuenca del Ebro.* PhD Thesis, University of Barcelona.

CABRERA, V. & BERNALDO DE QUIRÓS, F. 1996. The origins of the Upper Palaeolithic: a Cantabrian perspective. *In*: CARBONELL, E. & VAQUERO, M. (eds) *The Last Neandertals. The First Anatomically Modern Humans*. Fundació catalana per a la Recerca, Barcelona, 251–265.

CADILLAC, H., CANEROT, J. & FAUREÉ, PH. 1981. Le Jurassique inférieur aux confins des Iberides et des Catalanides (Espagne). *Estudios Geológicos*, **37**, 187–198.

CALONGE, A. 1989. *Bioestratigrafía del Cenomaniense de la Cordillera Ibérica por Foraminíferos bentónicos.* PhD Thesis, Complutense University, Madrid.

CALONGE, A., GARCÍA, A. & SEGURA, M. 1996. Middle Cretaceous biostratigraphic units in the Iberian Ranges (Spain) based on Alveolinids. *Mitteilungen aus dem Geologisch-Paläontologischen*, Institut der Universität Hamburg, **77**, 149–158.

CALVACHE, M. I. & VISERAS, C. 1995. Consecuencias geomorfológica derivadas de un proceso de captura fluvial. *Geogaceta*, **18**, 93–96.

CALVACHE, M. I. & VISERAS, C. 1997. Long-term control mechanisms of stream piracy processes in Southeast Spain. *Earth Surface Processes and Landform*, **22**, 93–106.

CALVERT, A., SANDVOL, E., SEBER, D., BARAZANGI, M., ROECKER, S., MOURABIT, T., VIDAL, F., ALGUACIL, G. & JABOUR, N. 2000. Geodynamic evolution of the lithosphere and upper mantle beneath the Alborán region of the western Mediterranean:

constraints from travel time tomography. *Journal of Geophysical Research*, **105**, 10871–10898.

CALVET, F. & MARZO, M. 1994. El Triásico de las Cordilleras Costero Catalanas: Estratigrafía, Sedimentología y Análisis Secuencia. *Field Guide, III Coloquio de Estratigrafía y Sedimentología del Triásico y Pérmico de España*, 1–53.

CALVET, F & RAMÓN, X. 1987. Estratigrafía, sedimentología y diagénesis del Muschelkalk inferior de los Catalánides. *Cuadernos de Geología Ibérica*, **11**, 141–169.

CALVET, F. & TUCKER, M. 1988. Outer ramp cycles in the upper Muschelkalk of the Catalan basin, northeast Spain. *Sedimentary Geology*, **57**, 185–198.

CALVET, F. & TUCKER, M. 1995. Triassic Reef-mound complexes (Ladinian, upper Muschelkalk) Catalan Ranges, Spain. *In*: MONTY, C., BOSENCE, D. BRIDGES, P & PRATT, B. (eds) *Mud Mounds: Origin and Evolution*. IAS, Special Publications, **23**, 311–333.

CALVET, F., MARCH, M. & PEDROSA, A. 1987. Estratigrafía, sedimentología, y diagénesis del Muschelkalk superior de los Catalánides. *Cuadernos de Geología Ibérica*, **11**, 171–198.

CALVET, F., TUCKER, M. & HENTON, J. 1990. Middle Triassic carbonate ramp systems in the Catalan Basin, northeast Spain: facies, systems tracks, sequences and controls. *In*: TUCKER, M., WILSON, J., CREVELLO, P., SARG, J. & READ, J. (eds) *Carbonate Platforms*. International Association of Sedimentologists, Special Publications, **9**, 79–108.

CALVET, F., SOLÉ DE PORTA, N. & SALVANY, J. M. 1993. Cronoestratigrafía (Palinología) del Triásico surpirenaico y del Pirineo Vasco-Cantábrico. *Acta Geológica Hispánica*, **28**, 33–48.

CALVET, F., ZAMARREÑO, I. & VALLÈS, D. 1996. Late Miocene reefs of the Alicante-Elche basin, southeast Spain. *In*: FRANSEEN, E. K., ESTEBAN, M., WARD, W. C & ROUCHY, J. M. (eds) *Models for Carbonate Stratigraphy from Miocene Reef Complexes of Mediterranean Regions*. Concepts in Sedimentology and Paleontology Series, **5**, SEPM, Tulsa, 177–190.

CALVET, M. 1998. Los complejos fluvioglaciares de Cerdanya-Capcir (Pirineos orientales) y sus enseñanzas. *In*: GÓMEZ ORTIZ, A. & PÉREZ ALBERTI, A. (eds) *Las huellas glaciares de las montañas españolas*. University of Santiago de Compostela, 263–290.

CALVO, A. 1986. *Geomorfología de laderas en la montaña del País Valenciano*. PhD Thesis, University of Valencia.

CALVO, J. M. 1993. *Cinemática de las fallas discontinuas en el sector central de la Cordillera Ibérica*. PhD Thesis, University of Zaragoza.

CALVO, J. P., ELIZAGA, E., LÓPEZ-MARTÍNEZ, N., ROBLES, F. & USERA, J. 1978. El Mioceno superior continental del Prebético externo: Evolución del estrecho nord-bético. *Boletín Geológico y Minero*, **89**, 9–28.

CALVO, J. P., HOYOS, M., GARCÍA DEL CURA, M. A. & ORDÓÑEZ, S. 1984. Caracterización sedimentológica de la Unidad Intermedia del Mioceno en la zona sur de Madrid. *Revista Materiales y Procesos Geológicos*, **2**, 145–176.

CALVO, J. P., ALONSO ZARZA, A. M. & GARCÍA DEL CURA, M. A. 1989. Models of Miocene marginal lacustrine sedimentation in response to varied source areas and depositional regimes in the Madrid Basin, central Spain. *Palaeogeography, Palaeoclimatology, Palaeoecology*, **90**, 199–214.

CALVO, J. P., HOYOS, M., MORALES, J. & ORDÓÑEZ, S. 1990. Neogene stratigraphy, sedimentology and raw materials of the Madrid Basin. *Paleontologia i Evolució, Mèmoria Especial*, **2**, 63–95.

CALVO, J. P., DAAMS, R., MORALES, J. *et al.* 1993. Up-to-date Spanish continental Neogene synthesis and paleoclimatic interpretation. *Revista de la Sociedad Geológica de España*, **6** (3–4), 29–40.

CALVO, J. P., JONES, B. F., BUSTILLO, M. FORT, ALONSO-ZARZA, A. M. & KENDALL, C. 1995. Sedimentology and geochemistry of carbonates from lacustrine sequences in the Madrid Basin, central Spain. *Chemical Geology*, **123**, 173–191.

CALVO, J. P., ALONSO ZARZA, A. M., GARCÍA DEL CURA, M. A., ORDÓÑEZ, S., RODRÍGUEZ-ARANDA, J. P. & SANZ MONTERO, M. E. 1996. Sedimentary evolution of lake systems through the Miocene of the Madrid Basin: paleoclimatic and paleohydrological constraints. *In*: FRIEND, P. F. & DABRIO, C. J. (eds) *Tertiary*

Basins of Spain: the Stratigraphic Record of Crustal Kinematics. Cambridge University Press, Cambridge, 272–277.

CÁMARA, P. 1989. La terminación estructural occidental de la Cuenca Vasco-Cantábrica. *In*: *Libro Homenaje a Rafael Soler*. Asociación de Geólogos y Geofísicos Españoles del Petróleo, 27–35.

CÁMARA, P. & KLIMOWITZ, J. 1985. Interpretación geodinámica de la vertiente centro-occidental surpirenaica (cuencas de Jaca y Tremp). *Estudios Geológicos*, **41**, 391–404.

CAMPOS, J. 1979. Estudio geológico del Pirineo Vasco al W del Bidasoa. *Munibe*, **31**, 3–139.

CAMPOS, S., AURELL, M. & CASAS, A. 1996. Origen de las brechas de la base del Jurásico en Morata de Jalón (Zaragoza). *Geogaceta*, **20**, 887–890.

CANALES, J. P. & DAÑOBEITIA, J. 1998. The Canary Islands swell: a coherence analysis of bathymetry and gravity. *Geophysical Journal International*, **132**, 479–488.

CANALES, M. L., GOY, A., HERRERO, C. & URETA, S. 1993. Foraminíferos del Aaliense en el sector suroccidental de la Cuenca Vasco-Cantábrica. *Treballs del museu de geologia de Barcelona*, **3**, 19–39.

CANAS, J. A., PUJADES, L. G., BLANCO, M. J., SOLER, V. & CARRACEDO, J. C. 1994. Coda-Q distribution in the Canary Islands. *Tectonophysics*, **246**, 245–261.

CANAS, J. A., UGALDE, A., PUJADES, L. G., CARRACEDO, J. C., SOLER, V. & BLANCO, M. J. 1998. Intrinsic and scattering seismic wave attenuation in the Canary Islands. *Journal of Geophysical Research*, **103**(B7), 15037–15050.

CAÑAVERAS. J. C., CALVO, J. P., HOYOS, M. & ORDÓÑEZ, S. 1996. Paleomorphologic features of an intra-Vallesian paleokarst, Tertiary Madrid Basin: significance of paleokarstic surfaces in continental basin analysis. *In*: FRIEND, P. F. & DABRIO, C. J. (eds) *Tertiary Basins of Spain: the Stratigraphic Record of Crustal Kinematics*. Cambridge University Press, Cambridge, 278–284.

CANDE, S. C. & KENT, D. V. 1995. Revised calibration of the geomagnetic polarity time scale for the late Cretaceous and Cenozoic. *Journal of Geophysical Research*, **100**(B4), 6093–6095.

CANÉROT, J. 1969. Observations géologiques dans la région de Montalban, Aliaga et Alcorisa (province de Teruel, Espagne). *Bulletin de la Société Géologique de France*, **7**(11), 854–861.

CANEROT, J. 1971. Le Jurassique dans le partie meridionale du Maestrazgo (prov. de Castellón): stratigraphie et paleogeographie. *Cuadernos de Geología Ibérica*, **2**, 323–332.

CANÉROT, J. 1974. *Recherches géologiques aux confins des chaînes Ibérique et Catalane*. PhD Thesis, ENADIMSA (Empresa Nacional ADARO (ENADIMSA)).

CANEROT, J. 1982. Ibérica central – Maestrazgo. *In*: *El Cretácico de España*. University of Complutense, Madrid, 273–344.

CANO ALONSO, R., PASCUAL GARCÍA, J. & PÉREZ SILVA, F. 1958. Localización del Gothlandiense en la Hoja de Sallent (Huesca). *Notas y Comunicaciones del IGME*, **49**, 53–63.

CANTAGREL, J. M., CENDRERO, A., FÚSTER J. M., IBARROLA, E. & JAMOND, C. 1984. K-Ar Chronology of the volcanic eruptions in the Canarian Archipelago: Island of La Gomera. *Bulletin Volcanologique*, **47**(3), 597–609.

CANTAGREL, J. M., FÚSTER, J. M., PIN, C., RENAUD, U. & IBARROLA, E. 1993. Age miocene inferieur des carbonatites de Fuerteventure (23 Ma:U-Pb zircon) et le magmatisme précoce d'une ile océanique (Iles Canaries). *Comptes Rendus de L'Académie des Sciences de Paris*, **316**, 1147–1153.

CANTAGREL, J. M., ARNAUD, N. O., ANCOCHEA, E., FÚSTER, J. M. & HUERTAS, M. J. 1999. Repeated debris avalanches on Tenerife and genesis of Las Cañadas caldera wall (Canary Islands). *Geology*, **27**, 739–742.

CANTANO, M. 1996. *Evolución morfodinámica del sector suroccidental de la Cuenca de Ciudad Rodrigo, Salamanca*. PhD Thesis, University of Huelva.

CANUDO, J. I., KELLER, G. & MOLINA, E. 1991. Cretaceous-Tertiary boundary extinction pattern and faunal turnover at Agost and Caravaca, SE Spain. *Marine Micropaleontology*, **17**, 319–341.

CAPDEVILA, R. 1966. Sur la présence de sills basiques et ultrabasiques métamorphisés dans la region de Villalba (Lugo, Espagne). *Comptes Rendus de l'Académie des Sciences de Paris*, **262**, 2193–2196.

CAPDEVILA, R. 1969. *Le métamorphisme régional progresif et les granites dans le segment hercynien de Galice nord-oriental (NW de l'Espagne)*. PhD Thesis, Scientific and Technical University, Montpellier.

CAPDEVILA, R. & VIALETTE, Y. 1970. Estimation radiométrique de l'âge de la deuxième phase tectonique en Galice moyenne (Nord-Ouest de l'Espagne). *Comptes Rendus de l'Académie des Sciences de Paris*, **270**, 2527–2530.

CAPDEVILA, R., CORRETGÉ, L. G. & FLOOR, P. 1973. Les granitoïdes varisques de la Meseta Ibérique. *Bulletin de la Société Géologique de France*, **15**, 209–228.

CAPEDRI, S., VENTURELLI, G., SALVIOLI-MARIANI, E., CRAWFORD, A. J. & BARBIERI, M. 1989. Upper-mantle xenoliths and megacrysts in an alkali basalt from Tallante, South-eastern Spain. *European Journal of Mineralogy*, **1**, 685–699.

CAPOTE, R. 1978. Tectónica española. *Seminario Criterios Sísmicos aplicados a Instalaciones Nucleares y Obras Públicas*. Asociación Española de Ingeniería Sísmica, Madrid, 1–30.

CAPOTE, R. 1983*a*. La fracturación subsecuente a la orogenia hercínica. *In*: COMBA, J. A. (coord.) *Geología de España. Libro Jubilar J. M. Ríos*, Vol. II. IGME, Madrid, 17–25.

CAPOTE, R. 1983*b*. La tectónica de la Cordillera Ibérica. *In*: Comba, J. A. (coord.) *Geología de España. Libro Jubilar J. M. Ríos*, Vol. II. IGME, Madrid, 109–120.

CAPOTE, R. & CARRO, S. 1968. Existencia de una red fluvial intramio-cena en la depresión del Tajo. *Estudios Geológicos*, **24**, 91–95.

CAPOTE, R. & DE VICENTE, G. 1989. El marco geológico y tectónico. *In*: *Mapa del Cuaternario de España, escala 1:1.000.000*. Memorias del Instituto Tecnológico y Geominero de España, 9–19.

CAPOTE, R. & FERNÁNDEZ-CASALS, M. J. 1976. La tectónica postmio-cena del sector central de la Depresión del Tajo. *Boletín Geológico y Minero*, **89**, 114–122.

CAPOTE, R. & GONZÁLEZ LODEIRO, F. 1983. *La estructura herciniana en los afloramientos paleozoicos de la Cordillera Ibérica. In*: COMBA, J. A. (coord.) *Libro Jubilar J. M. Ríos*, Vol. I. IGME, Madrid, 513–529.

CAPOTE, R., DÍAZ, M., GABALDÓN, V. *et al.* 1982. Evolución sedimen-tológica y tectónica del ciclo alpino en el tercio noroccidental de la Rama Castellana de la Cordillera Ibérica. *Temas Geológico Mineros*, **5**.

CAPOTE, R., GONZÁLEZ-CASADO, J. M. & DE VICENTE, G. 1987. Análisis poblacional de la fracturación tardihercínica en el sector central del Sistema Central Ibérico. *Cuadernos del Laboratorio Xeoloxico de Laxe*, **11**, 305–314.

CAPOTE, R., DE VICENTE, G. & GONZÁLEZ CASADO, J. M. 1990. Evolución de las deformaciones alpinas en el Sistema Central español (S. C. E.). *Geogaceta*, **7**, 20–22.

CAPOTE, R., VILLAMOR, P. & TSIGE, M. 1996. La tectónica alpina de la Falla de Alentejo-Plasencia (Macizo Hespérico). *Geogaceta*, **20**, 921–924.

CARACUEL, J. E. 1996. *Asociaciones de megainvertebrados, evolución ecosedimentaria e interpretaciones ecoestratigráficas en umbrales epioceánicos del Tethys Occidental (Jurásico superior)*. PhD Thesis, University of Granada.

CARACUEL, J. E. & OLÓRIZ, F. 1998. Revisión Estratigráfica del jurásico Superior de la Sierra Norte (Mallorca). *Revista de la Sociedad Geológica de España*, **11**(3–4), 345–353.

CARACUEL, J. E. & OLÓRIZ, F. 1999. Recent data on the Kimmeridgian-Tithonian boundary in the Sierra Norte of Mallorca (Spain), with notes on the genus *Hybonoticeras* breistroffer. *Geobios*, **32**, 575–591.

CARACUEL, J. E., OLÓRIZ, F. & SANDOVAL, J. 1994. Datos preliminares sobre el Bajociense superior en el sector de Coll Baix (Península de Alcudia, Mallorca). *Geogaceta*, **16**, 67–69.

CARACUEL, J. E., EL KADIRI, K. & OLÓRIZ, F. 1995. Les Radiolarites d'âge Callovien de la 'Fm. Puig d'en Parè' (Sierra Norte, Majorque). *Geobios*, **28**(6), 675–681.

CARACUEL, J. E., OLÓRIZ, F. & RODRÍGUEZ-TOVAR, F. J. 1998. Inter-pretaciones ecoestratigráficas en el estudio del Oxfordiense terminal y Kimmeridgiense basal (Jurásico superior) de la Cordillera Bética. *Cuadernos de Geología Ibérica*, **24**, 43–68.

CARACUEL, J. E., OLÓRIZ, F. & RODRÍGUEZ-TOVAR, F. J. 1999. Ammonite biostratigraphy from the Planula and Platynota Zones in the Lugar Section (External Subbetic, southern Spain). *Profil*, **16**, 107–120.

CARACUEL, J. E., OLÓRIZ, F. & RODRÍGUEZ-TOVAR, F. J. 2000. Oxfordian biostratigraphy from the Lugar Section (External Subbetic, southern Spain). *GeoResearch Forum*, **6**, 55–64.

CARBALLEIRA, J. & POL, C. 1986. Características y evolución de los sedimentos lacustres miocenos de la región de Tordesillas (Facies de las Cuestas) en el sector central de la cuenca del Duero. *Studia Geológica Salmanticensia*, Universidad Salamanca, **22**, 213–246.

CARBONELL, E., BERMÚDEZ DE CASTRO, J. M. *et al.* 1995. Lower Pleistocene hominids and artifacts from Atapuerca-TD6 (Spain). *Science*, **269**, 826–830.

CARBONELL, E., GARCÍA-ANTÓN, M. D., MALLOL, C. *et al.* 1999. The TD6 level lithic industry from Gran Dolina, Atapuerca (Burgos, Spain): production and use. *Journal of Human Evolution*, **37**, 653–693.

CARENAS, B., GARCÍA, A., CALONGE, A., PEREZ. P. & SEGURA, M. 1989. Middle Cretaceous (Upper Albian-Turonian) in the central sector of the Iberian Ranges (Spain). *In*: WIEDMANN, J. (ed.) *Cretaceous of the Western Tethys*. E. Schweizerbart'sche Verlags-buchandlung, Stuttgart, 265–279.

CARENAS, B., SEGURA, M., GARCÍA, A., GARCÍA-HIDALGO, J., RUIZ, G. & BRAVO, C. 1994. La Fm. Calizas de Aras de Alpuente (Vilas *et al.* 1982) en la región Norte de Valencia. *Cuadernos de Geología Ibérica*, **18**, 241–269.

CARIA MENDES, J. 1985. *As origens do Homen*. Fundaçao Calouste Gulbenkian, Lisboa.

CARIOU, E., MELÉNDEZ, G., SEQUEIROS, L. & THIERRY, J. 1988. Biochronologie du Callovien de la province d'ammonites subméditerranéenne: Reconnaissance dans le chaînes Ibériques de subdivisions fines distinguées dans le centre-ouest de la France. *Proceedings of 2nd International Symposium on Jurassic Stratigraphy*, **1**, 395–406.

CARLS, P. 1969. Die Conodonten des tieferen Unter-Devons der Guadarrama (Mittel-Spanien) und die Stellung des Grenzbere-iches Lochkovium/Pragium nach der rheinischen Gliederung. *Senckenbergiana lethaea*, **50**, 303–355.

CARLS, P. 1971. Stratigraphische bereinstimmungen im hochsten Silur und tieferen Unter-Devon zwischen Keltiberien (Spanien) und Bretagne (Frankreich) und das Alter des Grès de Gdoumont (Belgien). *Neues Jahrbuch Geologie Paläontologie, Monatshefte*, **1971**, 195–212.

CARLS, P. 1974. Die Proschizoporiinae (Brachiopoda, Silurium-Devon) der Östlichen Iberischen Ketten (Spanien). *Senckenber-giana lethaea*, **55**, 153–227.

CARLS, P. 1975. The Ordovician of the eastern Iberian chains near Fombuena and Luesma (Prov. Zaragoza, Spain). *Neues Jahrbuch für Geologie und Paläontologie, Abhandlungen*, **150**, 127–146.

CARLS, P. 1977. The Silurian–Devonian boundary in northeastern and central Spain. *In*: *The Silurian–Devonian boundary*. International Union of Geological Sciences, Series A, **5**, 143–158.

CARLS, P. 1983. La Zona asturoccidental-leonesa en Aragón y el Macizo del Ebro como prolongación del Macizo cantábrico. *In*: COMBA, J. A. (coord.) *Geología de España, Libro Jubilar J. M. Ríos*, Vol. 3. IGME, Madrid, 11–32.

CARLS, P. 1988. The Devonian of Celtiberia (Spain) and Devonian paleogeography of SW Europe. *In*: MCMILLAN, N. J., EMBRY, A. F. & GLASS, D. J. (eds) *Devonian of the World*. Canadian Society Petroleum Geologists Calgary, Memoirs, **14**(1), 421–466.

CARLS, P. & LAGES, R. 1983. Givetium und Oberdevon in den Östlichen Iberischen Ketten (Spanien). *Zeitschrift deutschen geologischen Gesellschaft*, **134**, 119–142.

CARLS, P. & VALENZUELA-RÍOS, J. I. 1998. The ancestry of the Rhenish Middle Siegenian brachiopod fauna in the Iberian Ranges and its palaeozoogeography (Early Devonian). *Revista Española Paleontología*, Special volume, Homenaje Prof. Gonzalo Vidal, 123–142.

CARLS, P. & VALENZUELA-RÍOS, J. I. 1999. Similitudes y diferencias estratigráficas entre el Pridoliense-Praguiense celtibérico y armoricano. *Revista Española Paleontología*, **14**(2), 279–292.

CARLS, P. & VALENZUELA-RÍOS, J. I. 2000. *Faunas, correlations, various boundaries, and opinions concerning the early Emsian* sensu lato *and a* Cancellata *Boundary.* Report to the Emsian Working Party of the Subcommission on Devonian Stratigraphy, Macquarie University.

CARLS, P., GANDL, J., GROOS-UFFENORDE, H. *et al.* 1972. Neue Daten zur Grenze Unter-/Mittel-Devon. *Newsletter in Stratigraphy*, **2**, 115–147.

CARMINATI, E., WORTEL, M. J. R., SPAKMAN, W. & SABADINI, R. 1998. The role of slab detachment processes in the opening of the western-central Mediterranean basins: some geological and geophysical evidence. *Earth and Planetary Science Letters*, **160**, 651–665.

CARNICERO, A. 1980. *Estudio petrológico del metamorfismo y los granitoides entre Cipérez y Aldea del Obispo (oeste de la provincia de Salamanca).* PhD Thesis, University of Salamanca.

CARNICERO, A. & CASTRO, A. 1982. El Complejo Básico de Barcarrota: Su petrología y estructura. *Boletín Geológico y Minero*, **93**, 165–171.

CARRACEDO, J. C. 1975. *Estudio Paleomagnético de la Isla de Tenerife.* PhD Thesis, Complutense University, Madrid.

CARRACEDO, J. C. 1979. Paleomagnetismo e historia volcánica de Tenerife. *Aula de Cultura de Tenerife.*

CARRACEDO, J. C. 1984. Geología de las Islas Canarias. *In*: *Geografía Física de Canarias.* Editorial Interinsular Canaria, Santa Cruz de Tenerife.

CARRACEDO, J. C. 1994. The Canary Islands: an example of structural control on the growth of large oceanic-island volcanoes. *Journal of Volcanology and Geothermal Research*, **60**, 225–241.

CARRACEDO, J. C. 1996a. Morphological and structural evolution of the western Canary Islands: Hotspot-induced three-armed rifts or regional tectonic trends? *Journal of Volcanology and Geothermal Research*, **72**, 151–162.

CARRACEDO, J. C. 1996b. A simple model for the genesis of large gravitational landslide hazards in the Canary Islands. *In*: MCGUIRE, W. J., JONES, A. P. & NEUBERG, J. (eds) *Volcano Instability on the Earth and Other Planets.* Geological Society, London, Special Publication, **110**, 125–135.

CARRACEDO, J. C. 1999. Growth, structure, instability and collapse of Canarian volcanoes and comparison with Hawaiian volcanoes. *Journal of Volcanology and Geothermal Research*, **94**, 1–4, 1–19.

CARRACEDO, J. C. & RODRÍGUEZ BADIOLA, E. 1991. *Lanzarote: La erupción volcánica de 1730* (with a colour, 1/25.000 geological map of the eruption). Editorial MAE, Madrid.

CARRACEDO, J. C. & RODRÍGUEZ BADIOLA, E. 1993. Evolución geológica y magmática de la isla de Lanzarote (Islas Canarias). *Revista de la Academia Canaria de Ciencias*, **4**, 25–58.

CARRACEDO, J. C. & SOLER, V. 1992. Anomalously shallow palaeomagnetic inclinations and the question of the age of the Canarian Archipelago. *Geophysical Journal International*, **122**, 393–406.

CARRACEDO, J. C., RODRÍGUEZ BADIOLA, E. & SOLER, V. 1990. Aspectos volcanológicos y estructurales, evolución petrológica e implicaciones en riesgo volcánico de la erupción de 1730 en Lanzarote, Islas Canarias. *Estudios Geológicos*, **46**, 25–55.

CARRACEDO, J. C., RODRÍGUEZ BADIOLA, E. & SOLER, V. 1992. The 1730–1736 eruption of Lanzarote, Canary Islands: A long, high-magnitude basaltic fissure eruption. *Journal of Volcanology and Geothermal Research*, **53**, 1–4, 239–250.

CARRACEDO, J. C., DAY, S., GUILLOU, H. & RODRÍGUEZ BADIOLA, E. 1996. The 1677 eruption of La Palma, Canary Islands. *Estudios Geológicos*, **52**, 345–357.

CARRACEDO, J. C., DAY, S. J., GUILLOU, H., RODRÍGUEZ BADIOLA, E., CANAS, J. A. & PÉREZ TORRADO, F. J. 1997. Origen y evolución del volcanismo de las Islas Canarias. *In*: *Ciencia y Cultura en Canarias.* Museo de la Ciencia y el Cosmos, Santa Cruz de Tenerife, 67–89.

CARRACEDO, J. C., DAY, S. J., GUILLOU, H., RODRÍGUEZ BADIOLA, E., CANAS, J. A. & PÉREZ TORRADO, F. J. 1998. Hotspot volcanism close to a passive continental margin: the Canary Islands. *Geological Magazine*, **135**(5), 591–604.

CARRACEDO, J. C., DAY, S. J., GUILLOU, H. & GRAVESTOCK, P. 1999a. The later stages of the volcanic and structural evolution of La Palma, Canary Islands: The Cumbre Nueva giant collapse and the Cumbre Vieja volcano. *Geological Society of America Bulletin*, **11**(5), 755–768.

CARRACEDO, J. C., DAY, S. J., GUILLOU, H. & PÉREZ TORRADO, F. J. 1999b. Giant Quaternary landslides in the evolution of La Palma and El Hierro, Canary Islands. *Journal of Volcanology and Geothermal Research*, **94**, 1–4, 169–190.

CARRACEDO, J. C., GUILLOU, H., RODRÍGUEZ BADIOLA, E. *et al.* 2001a. *Mapa Geológico Nacional (MAGNA) de La Palma*, (Hojas geológicas y memorias). ITGE, Madrid.

CARRACEDO, J. C., RODRÍGUEZ BADIOLA, E., GUILLOU, S., NUEZ, H. J. DE LA & PÉREZ TORRADO, F. J. 2001b. Geology and Volcanology of the Western Canaries: La Palma and El Hierro. *Estudios Geológicos*, **57**, 171–295.

CARRÉ, D., HENRY, J.-L., POUPON, G. & TAMAIN, G. 1970. Les quartzites Botella et leur faune trilobitique. Le problème de la limite Llandeilien-Caradocien en Sierra Morena. *Bulletin de la Société Géologique de France [7]*, **12**, 774–785.

CARRERAS, F. J., RAMIREZ DEL POZO, J., GIANNINI, G., PORTERO, J., DEL OLMO, P. DEL & AGUILAR, M. 1978. *Reinosa.* Spanish Geological Survey, Report and Map, Scale 1:50 000, **38**.

CARRERAS, F. J., PORTERO, J. M. & DEL OLMO, P. DEL 1979. *Los corrales de Buelna.* Spanish Geological Survey, Report and Map, Scale 1:50 000, **58**.

CARRERAS, J. 1975. Las deformaciones tardi-hercínicas en el litoral septentrional de la península del Cabo de Creus (Prov. Gerona, España): La génesis de las bandas miloníticas. *Acta Geológica Hipanica*, **10**, 109–115.

CARRERAS, J. & CAPELLA, I. 1994. Tectonic levels in the Palaeozoic basement of the Pyrenees: a review and a new interpretation. *Journal of Structural Geology*, **16**, 1509–1524.

CARRERAS, J. & DEBAT, P. 1996. La Tectonique Hercynienne. *In*: BARNOLAS, A. & CHIRON, J. C. (eds) *Synthèse géologique et géophysique des Pyrénées. Vol. 1: Introduction. Géophysique. Cycle Hercynien.* BRGM and IGME, Orléans and Madrid, 585–677.

CARRERAS, J. & DRUGUET, E. 1994. Structural zonation as a result of inhomogeneous non-coaxial deformation and its control on syntectonic intrusions: an example from the Cap de Creus area, eastern Pyrenees. *Journal of Structural Geology*, **16**, 1525–1534.

CARRERAS, J. & LOS SANTOS, M. 1982. Geological setting of the Roses granodiorite (E–Pyrenees, Spain). *Acta Geológica Hipanica*, **17**, 211–217.

CARRETERO, J. M., LORENZO, C. & ARSUAGA, J. L. 1999. Axial and appendicular skeleton of *Homo antecessor. Journal of Human Evolution*, **37**, 459–499.

CARRINGTON DA COSTA, J. 1950. *Noticia sobre a carta geológica de Buçaco, de Nery Delgado.* Special Publication, Ediciones Serviços Geológicos de Portugal, Lisboa, 1–27.

CARVALHO, D. 1979. Geology, metallogeny and prospecting of South Portugal massive sulphide deposits. *Comunicacoes dos Servicos Geológicos de Portugal*, **65**, 169–191.

CARVALHO, D., CORREIA, H. A. C. & INVERNO, C. 1976a. Contribuição para o conhecimento geológico do Grupo de Ferreira-Ficalho. Suas relações com a Faixa Piritosa e o Grupo do Pulo do Lobo. *Mémorias e Noticias*, **82**, 145–169.

CARVALHO, D., CORREIA, H. A. C. & INVERNO, C. M. C. 1976b. Livro guía das excursoes geologicas na Faixa Piritosa Iberica. *Comunicacoes dos Servicos Geológicos de Portugal*, **60**, 271–315.

CASAS, A., KEAREY, P., RIVERO, L. & ADAM, C. R. 1997. Gravity anomaly map of the Pyrenean region and comparison of the deep geological structure of the western and eastern Pyrenees. *Earth and Planetary Science Letters*, **150**, 65–78.

CASAS, A. M. 1990. *El Frente Norte de las Sierras de Cameros: Estructuras cabalgantes y campo de esfuerzos.* PhD Thesis, University of Zaragoza (Zubía, Monográfico no. 4, Instituto de Estudios Riojanos (CSIC), Logroño, 1992).

CASAS, A. M. 1993. Tectonic inversion and basement thrusting in the Cameros Massif (Northern Spain). *Geodinamica Acta*, **6**, 202–216.

CASAS, A. M. & GIL, A. 1998. Extensional subsidence, contractional folding andthrust inversion of the Eastern Cameros Basin (NW Spain). *Geologische Rundschau*, **86**, 802–818.

CASAS, A. M. & MAESTRO, A. 1996. Deflection of a compressional stress field by large-scale basement faults. A case study from the Tertiary Almazán Basin (Spain). *Tectonophysics*, **255**, 135–156.

CASAS, A. M. & SIMÓN, J. L. 1992. Stress field and thrust kinematics: a model for the tectonic inversion of the Cameros Massif (Spain). *Journal of Structural Geology*, **14**, 521–530.

CASAS, A. M., SIMÓN, J. L. & SERÓN, F. J. 1992. Stress deflection in a tectonic compressional field: a model for the northwestern Iberian Chain, Spain. *Journal of Geophysical Research*, **97**(B5), 7183–7192.

CASAS, A. M., CORTÉS, A. L., LIESA, C. L., MELÉNDEZ, A. & SORIA, A. R. 1997. The structure of the northern margin of Iberian Range between the Sierra de Arcos and the Montalbán anticline. *Cuadernos de Geología Ibérica*, **23**, 243–268.

CASAS, A. M., CORTÉS, A. L., GAPAIS, D., NALPAS, T. & ROMÁN, T. 1998. Modelización analógica de estructuras asociadas a compresión oblicua y transpresión, Ejemplos del NE de peninsular. *Revista de la Sociedad Geológica de España*, **11**(3–4), 137–150.

CASAS, A. M., CASAS, A., PÉREZ, A., TENA, S., BARRIER, L., GAPAIS, D. & NALPAS, TH. 2000a. Syn-tectonic sedimentation and thrust-and-fold kinematics at the intra-mountain Montalbán Basin (northern Iberian Chain, Spain). *Geodinamica Acta*, **1**, 1–17.

CASAS, A. M., CORTÉS, A. & MAESTRO, A. 2000b. Intra-plate deformation and basin formation during the Tertiary within the Northern Iberian plate: origin an evolution of the Almazán Basin. *Tectonics*, **19**(2), 258–289.

CASAS, J. & CARBÓ, D. 1990. Deep structure of the Betic Cordillera derived from the interpretation of a complete Bouguer anomaly map. *Journal of Geodynamics*, **12**, 137–147.

CASAS-SAINZ, A. M. 1992. El frente Norte de las Sierras de Cameros: Estructuras cabalgantes y campo de esfuerzos. *Zubía*. Monográfico **4**, Instituto de Estudios Riojanos, Longroño.

CASILLAS, R. 1989. *Las asociaciones plutónicas tardihercínicas del Sector Occidental de la Sierra de Guadarrama – Sistema Central Español (Las Navas del Marqués – S. Martín de Valdeiglesias). Petrología, Geoquímica, Génesis y Evolución*. PhD Thesis, Complutense University, Madrid.

CASILLAS, R., VIALETTE, I., PEINADO, M., DUTHOU, JL. & PIN, CH. 1991. Ages et caractèristiques isotopiques (Sr-Nd) des granitoïdes de la Sierra de Guadarrama occidentale (Espagne). Séance spécialisée de la Soc Géol France a la mémoire de Jean Lameyre. *Granites océaniques et continentaux*.

CASILLAS, R., AHIJADO, A. & HERNÁNDEZ-PACHECO, A. 1994. Zonas de cizalla ductil en el complejo basal de Fuerteventura. *Geogaceta*, **15**, 117–120.

CASQUET, C. 1980. *Fenómenos de endomorfismo, metamorfismo y metasomatismo de contacto en los mármoles de Ribera de Cala (Huelva)*. PhD Thesis, Complutense University, Madrid.

CASQUET, C. & NAVIDAD, M. 1985. El metamorfismo en el Sistema Central español. Comparación entre el sector central y el oriental en base al zonado del granate. *Revista de la Real Academia de Ciencias Exactas, Físicas y Naturales*, **79**, 523–548.

CASQUET, C. & VELASCO, F. 1978. Contribución a la geología de los skarn calcicos en torno a Santa Olalla de Cala (Huelva-Badajoz). *Estudios Geológicos*, **43**, 399–405.

CASQUET, C., GALINDO, C., GONZÁLEZ CASADO, J. M. *et al.* 1992. El metamorfismo en la Cuenca de Cameros. Geocronología e implicaciones tectónicas. *Geogaceta*, **11**, 18–22.

CASQUET, C., EGUILUZ, L., GALINDO, C., TORNOS, F. & VELASCO, F. 1998. The Aguablanca Cu-Ni-(PGE) intraplutonic ore deposit (Extremadura, Spain). Isotope (Sr, Nd, S) constraints on the source and evolution of magmas and sulfides. *Geogaceta*, **24**, 71–74.

CASTAÑARES, L. M., ROBLES, S. & VICENTE BRAVO, J. C. 1997. Distribución estratigráfica de los episodios volcánicos submarinos del Albiense-Santoniense de la Cuenca Vasca (sector Gernika-Plentzia, Bizkaia). *Geogaceta*, **22**, 43–46.

CASTAÑARES, L. M., ROBLES, S., GIMENO, D. & VICENTE BRAVO, J. C. 2001. The submarine volcanic system of the Errigoiti Formation (Albian-Santonian of the Basque-Cantabrian basin, northern Spain): Stratigraphic framework, facies and sequences. *Journal of Sedimentary Research*, **71**, 318–333.

CASTELLTORT, F. X. 1986. *Estratigrafía del Muschelkalk mitjà dels Catalànids I Sedimentologia de les seves unitats detrìtiques*. PhD Thesis, University of Barcelona.

CASTILLO, A. 1991. Geología de los yacimientos de minerales del grupo de la Palygorskita-Sepiolita. *In*: LUNAR, R. & OYARZUN, R.

(eds) *Yacimientos Minerales*. Editorial Centro de Estudios Ramón Areces, Madrid (ISBN 84-87191-74-6).

CASTILLO HERRADOR, F. 1974. Le Trias évaporitique des bassins de la Vallée de l'Ebre et de Cuenca. *Bulletin de la Société Géologique de France*, **XVI**(7), 49–63.

CASTILLO-HERRADOR, F. 1974. Le Trias eváporitique des bassins de la Vallée de l'Ebre et de Cuenca. *Bulletin de la Société Geologique du France*, **16**, 49–63.

CASTRO, A. 1981. *Estudio petrológico del área de Barcarrota-Higuera de Vargas (Badajoz, Sierra Morena Occidental)*. MSc Thesis, University of Salamanca.

CASTRO, A. 1984. *Los Granitoides y la Estructura Hercínica en Extremadura Central*. PhD Thesis, University of Salamanca.

CASTRO, A. 1986. Structural pattern and ascent model in the Central Extremadura batholith, Hercynian belt, Spain. *Journal of Structural Geology*, **8**, 633–645.

CASTRO, A. 1987. Implicaciones de la zona Ossa-Morena y dominios equivalentes en el modelo geodinámico de la Cadena Hercínica europea. *Estudios Geológicos*, **43**, 249–260.

CASTRO, A. 2001. Plagioclase morphologies in assimilation experiments. Implications for disequilibrium melting in the generation of granodiorite rocks. *Mineralogy and Petrology*, **71**, 31–49.

CASTRO, A. & FERNÁNDEZ, C. 1998. Granite intrusion by externally induced growth and deformation of the magma reservoir, the example of the Plasenzuela pluton, Spain. *Journal of Structural Geology*, **20**, 1219–1228.

CASTRO, A., FERNÁNDEZ, C., DE LA ROSA, J. D. *et al.* 1996a. Triple-junction migration during Paleozoic plate convergence: the Aracena metamorphic belt, Hercynian massif, Spain. *Geologische Rundschau*, **85**, 180–185.

CASTRO, A., FERNÁNDEZ, C., DE LA ROSA, J. D., MORENO VENTAS, I. & ROGERS, G. 1996b. Significance of MORB-derived amphibolites from the Aracena metamorphic belt, Soutwest Spain. *Journal of Petrology*, **37**, 235–260.

CASTRO, A., FERNÁNDEZ, C., EL-HMIDI, H., EL-BIAD, M., DÍAZ, M., DE LA ROSA, J. D. & STUART, F. 1999a. Age constraints to the relationships between magmatism, metamorphism and tectonism in the Aracena metamorphic belt, southern Spain. *International Journal of Earth Sciences*, **88**, 26–37.

CASTRO, A., PATIÑO DOUCE, A. E., CORRETGÉ, L. G., DE LA ROSA, J. D., EL-BIAD, M. & EL-HMIDI, H. 1999b. Origin of peraluminous granites and granodiorites, Iberian massif, Spain. An experimental test of granite petrogenesis. *Contributions to Mineralogy and Petrology*, **135**, 255–276.

CASTRO, A., CORRETGÉ, L. G., EL-BIAD, M., EL-HMIDI, H., FERNÁNDEZ, C. & PATIÑO DOUCE, A. E. 2000. Experimental constraints on Hercynian Anatexis in the Iberian Massif, Spain. *Journal of Petrology*, **41**, 1471–1488.

CASTRO, J. M. 1998. *Las plataformas del Valanginiense superior al Albiense superior en el Prebético de Alicante*. PhD Thesis, University of Granada.

CASTRO, J. M. & RUIZ-ORTIZ, P. A. 1991. Nivel condensado con estromatolitos pelágicos en el Cretácico de la Sierra de Estepa (Subbético Externo, prov. Sevilla). *Revista de la Sociedad Geológica de España*, **4**, 305–319.

CASTRO, J. M., CHECA, A. & RUÍZ-ORTIZ, P. A. 1990. Cavidades kársticas con relleno de Calloviense Superior y Oxfordiense Inferior (Subbético Externo; Sierra de Estepa, provincia de Sevilla). *Geogaceta*, **7**, 61–63.

CASTROVIEJO, R. 1995. A typological classification of Spanish precious metals deposits. *Cuadernos Laboratorio Xeoloxico de Laxe*, **20**, 253–279

CASTROVIEJO, R. 1998. Nuevas aportaciones a la tipología de metales preciosos para exploración en España. *Boletín Geológico y Minero*, **109**, 77–101.

CAUS, E., GARCÍA-SENZ, J., RODÉS, D. & SIMÓ, A. 1990. Stratigraphy of the Lower Cretaceous (Berriasian–Barremian) sediments in the Organyà Basin, Pyrenees, Spain. *Cretaceous Research*, **11**, 313–320.

CAUS, E., GÓMEZ-GARRIDO, A., SIMÓ, A. & SORIANO, K. 1993. Cenomanian–Turonian platform to basin integrated stratigraphy in the South Pyrenees (Spain). *Cretaceous Research*, **14**, 531–555.

CAUS, E., TEIXELL, A. & BERNAUS, J. M. 1997. Depositional model of a

Cenomanian–Turonian extensional basin (Sopeira Basin) NE Spain: Interplay between tectonics, eustasy and biological productivity. *Paleogeogrpahy, Palaeoclimatology, Palaeoecology*, **129**, 23–36.

CAUSSE, C., GOY, J. L., ZAZO, C. & HILLAIRE-MARCEL, C. 1993. Potential chronologique (Th/U) des faunes Pléistocènes méditerranéenes: Example des terrasses marines des régions de Murcie et Alicante (Sud-Est de l'Espagne). *Geodinamica Acta*, **6**(2), 121–134.

CAVET, P. 1957. Le Paléozoique de la zone axiale des Pyrénées orientales françaises entre la Rousillon et l'Andorre (étude stratigraphique et paléontologique). *Bulletin Service de la Carte Géologique de France*, **55**(254), 1–216.

CAZURRO, M. 1909. Las cuevas de Serinyá y otras estaciones prehistóricas del NE de Cataluña. *Anuari del Institut d'Estudis Catalans*, **2**, 24–25.

CEARRETA, A. 1998. Holocene sea-level change in the Bilbao estuary (north Spain): foraminiferal evidence. *Micropaleontology*, **44**(3), 265–276.

CEARRETA, A. & MURRAY. 2000. AMS ^{14}C dating of Holocene estuarine deposits: consequences of high-energy and reworked foraminifera. *The Holocene*, **10**(1), 155–159.

CEBRIÁ, J. M. & LÓPEZ-RUIZ, J. 1995. Alkali basalts and leucitites in an extensional intracontinental plate setting: The Late Cenozoic Calatrava Volcanic Province (Central Spain). *Lithos*, **35**, 27–46.

CEBRIÁ, J. M. & LÓPEZ-RUIZ, J. 1996. A refined method for trace element modelling of nonmodal batch partial melting processes: The Cenozoic continental volcanism of Calatrava, Central Spain. *Geochimica et Cosmochimica Acta*, **60**, 1355–1366.

CEBRIÁ, J. M. & WILSON, M. 1995. Cenozoic mafic magmatism in Western/Central Europe: a common european asthenospheric reservoir? *Terranova Abstracts Supplements*, **7**, 162.

CEBRIÁ, J. M., LÓPEZ-RUIZ, J., DOBLAS, M., OYARZUN, R., HERTOGEN, J. & BENITO, R. 2000. Geochemistry of the Quaternary alkali basalts of Garrotxa (NE volcanic province, Spain): a case of double enrichment of the mantle lithosphere. *Journal of Volcanology and Geothermal Research*, **102**, 217–235.

CECCA, F., FOURCADE, E. & AZÉMA, J. 1992. The disappearance of the 'Ammonitico Rosso'. *Palaeogeography, Palaeoclimatology, Palaeoecology*, **99**, 55–70.

CEMBREROS, V. M. & COSSIO, J. R. 1990. *Estudio integral del caolín en cuatro áreas selecionadas de Asturias*. IGME, Madrid, Report **11–236**.

CENDRERO, A. 1966. Los volcanes recientes de Fuerteventura (Islas Canarias). *Estudios Geológicos*, **22**, 201–226.

CENDRERO, A. 1971. The Volcano-plutonic Complex of La Gomera (Canary Islands). *Bulletin Volcanologique*, **34**, 537–561.

CESARE, B. & GÓMEZ-PUGNAIRE, M. T. 2001. Crustal melting in the Alboran domain: Constraints from xenoliths of the Neogene volcanic province. *Physics and Chemistry of the Earth*, **26**, 255–260.

CESARE, B., SALVIOLI, E. & VENTURELLI, G. 1997. Crustal anatexis and melt extraction during deformation in the restite xenoliths at El Joyazo (SE Spain). *Mineralogical Magazine*, **61**, 15–27.

CHACÓN, B. & MARTÍN-CHIVELET, J. 1999. El Cretácico terminal y Paleoceno de la Sierra del Carche (dominio prebético. Jumilla). Caracterización estratigráfica y sedimentológica. *Geogaceta*, **26**, 11–14.

CHACÓN, B. & MARTÍN-CHIVELET, J. 2001*a*. Implicaciones tectosedimentarias de la discontinuidad estratigráfica del Maastrichtiense medio en Aspe (Prebético de Alicante). *Revista de la Sociedad Geológica de España*, **14**, 123–133.

CHACÓN, B. & MARTÍN-CHIVELET, J. 2001*b*. Discontinuidades y conformidades correlativas en las series hemipelágicas del final del Cretácico en el Prebético. Caracterización biocronoestratigráfica. *Geotemas*, **3**(2), 177–180.

CHACÓN, J., FERNÁNDEZ, J., MITROFANOV, F. & TIMOFEEV, B. V. 1984. Primeras dataciones microfitopaleontológicas en el sector de Valverde de Burguillos-Jerez de los Caballeros (Anticlinorio de Olivenza-Monesterio). *Cuadernos do Laboratorio Xeologico de Laxe*, **8**, 211–220.

CHALOUAN, A. 1986. *Les nappes ghomarides (Rif septentrional,*

Maroc). Un terrain Varisque dans la chaîne alpine. PhD Thesis, University of Strasbourg.

CHALOUAN, A. & MICHARD, A. 1990. The Ghomarides nappes, Rif coastal range, Morocco, a Variscan chip in the Alpine belt. *Tectonics*, **9**, 1565–1583.

CHAMPETIER, Y. 1972. *Le Prébétique et l'Ibérique cotiers dans le Sud de la Province de Valence et le Nord de la Province d'Alicante (Espagne)*. PhD Thesis, University of Nancy.

CHANTRAINE, J., CHAUVEL, J. J., BALÉ, P., DENIS, E. & RABU, D. 1988. Le Briovérien (Protérozoïque supérieur à terminal) et l'orogenèse cadomienne en Bretagne (France). *Bulletin de la Société Géologique de France*, **8**, 821–829.

CHAPPELL, B. W. & WHITE, A. J. R. 1974. Two contrasting granite types Pacific. *Geology*, **8**, 173–174.

CHARLET, J. M. 1979. Le massif granitique de La Maladeta (Pyrénées Centrales Espagnoles). Synthese des données geologiques. *Annales de la Société Géologique de Belgique*, **102**, 313–323.

CHARLET, J. M. 1982. Les grainds traits géologiques du massif de La Maladeta (Pyrénées Centrales Espagnoles). *Pirineos*, **116**, 57–66.

CHARPENTIER, J. L., LETHIERS, F. & TAMAIN, G. 1976. Les 'Schistes Aquisgrana' à ostracodes du Dévonien supérieur-terminal en Sierra Morena Orientale (Espagne). *Annales Société Géologique Nord*, **36**, 353–362.

CHAUVEL, C. & JAHN, B. M. 1984. Nd-Sr isotope and REE geochemistry of alkali basalts from the Massif Central, France. *Geochimica et Cosmochimica Acta*, **48**, 93–110.

CHAUVEL, C., HOFMANN, A. W. & VIDAL, PH. 1992. HIMU-EM: The French Polynesian connection. *Earth and Planetary Science Letters*, **110**, 99–119.

CHAUVEL, J. 1973. Les echinodermes Cystoïdes de l'Ordovicien de Cabo de Peñas (Asturies). *Breviora Geologica Asturica*, **17**, 30–32.

CHAUVEL, J. & LE MENN, J. 1979. Sur quelques Echinodermes (Cystoïdes et Crinoïdes) de l'Ashgill d'Aragon (Espagne). *Geobios*, **12**, 549–587.

CHAUVEL, J. & MELÉNDEZ, B. 1978. Les Echinodermes (Cystoïdes, Asterozoaires, Homalozoaires) de l'Ordovicien moyen des Monts de Tolède (Espagne). *Estudios Geológicos*, **34**, 75–87.

CHAUVEL, J. & MELÉNDEZ, B. 1986. Note complementaire sur les Echinodermes ordoviciens de Sierra Morena. *Estudios Geológicos*, **42**, 451–459.

CHAUVEL, J. & TRUYOLS, J. 1977. Sur la présence du genre *Destombesia* Chauvel (Echinoderme Cystoïde) dans l'Ordovicien des Asturies (Espagne). *Breviora Geologica Asturica*, **21**, 37–40.

CHAUVEL, J., DROT, J., PILLET, J. & TAMAIN, G. 1969. Précisions sur l'Ordovicien moyen et supérieur de la 'série-type' du Centenillo (Sierra Morena orientale, Espagne). *Bulletin de la Société Géologique de France [7]*, **11**, 613–626.

CHAUVEL, J., MELÉNDEZ, B. & LE MENN, J. 1975. Les Echinodermes (Cystoïdes et Crinoïdes) de l'Ordovicien supérieur de Luesma (Sud de l'Aragon, Espagne). *Estudios Geológicos*, **31**, 351–364.

CHEADLE, M. J., MCGEARY, S., WARNER, M. R. & MATTHEWS, D. H. 1987. Extensional structures on the western UK continental shelf, a review of evidence from deep seismic profiling. *In*: COWARD, M. P., DEWEY, J. F. & HANCOCK, P. L. (eds) *Continental Extensional Tectonics*. Geological Society, London, Special Publications, **28**, 445–465.

CHLUPAC, I., LUKES, P., PARIS, F. & SCHÖNLAUB, H. P. 1985. The Lochkovian-Pragian boundary in the Lower Devonian of the Barrandian area (Czechoslovakia). *Jahrbuch Geologischen Bundesanstalt*, **128**, 9–41.

CHLUPÁC, I., FERRER, E., MAGRANS, J., MAÑÉ, R. & SANZ, J. 1997. Early Devonian eurypterids with Bohemian affinities from Catalonia (NE Spain). *Batalleria*, **7**, 9–21.

CHLUPÁC, I., HAVLÍCEK, V., KRÍZ, J., KUKAL, Z. & STORCH, P. 1998. *Palaeozoic of the Barrandian (Cambrian to Devonian)*. Czech Geological Survey, Prague.

CHOUKROUNE, P. 1973. Phase teconique d'âge variable dans les Pyrénées: évolution du domaine plissé pyrénéen au cours du Tertiaire. *Comptes Rendus de l'Académie des Sciences de Paris*, **276**, 909–912.

CHOUKROUNE, P. 1976. *Structure et évolution téctonique de la Zone Nord-Pyrénéenne. Analyse de la deformation dans une portion de la chaîne à schistosité subverticale.* Mémoires de la Société géologique de France, **55**.

CHOUKROUNE, P. 1992. Tectonic evolution of Pyrenees. *Annual Review of Earth and Planetary Sciences*, **20**, 143–158.

CHOUKROUNE, P. & ECORS TEAM. 1989. The ECORS Pyrenean deep seismic profile reflection data and the overall structure of an orogenic belt. *Tectonics*, **8**, 23–39.

CHOUKROUNE, P. & MATTAUER, M. 1978. Tectonique des plaques et Pyrénées: sur le fonctionnement de la faille transformante nord-pyréneenne.; comparaison avec les modéles actuales. *Bulletin de la Société Géologique de France*, **5**, 689–700.

CHOUKROUNE, P., ROURE, F., PINET, B. & ECORS TEAM. 1990. Main results of the ECORS Pyrenees profile. *Tectonophysics*, **173**, 411–423.

CHUECA, J., PEÑA, J. L., LAMPRE, F., GARCÍA-RUIZ, J. M. & MARTÍ-BONO, C. 1998. *Los glaciares del Pirineo aragonés: Estudio de su evolución y extensión actual.* University of Zaragoza.

CLAGUE, D. A. & DALRYMPLE, G. B. 1987. The Hawaiian-Emperor volcanic chain. Part I: Geologic evolution. *In*: DECKER, W., WRIGHT, T. L. & STAUFFER, P. H. (eds) *Volcanism in Hawaii*. USGS, Professional Papers, **1350**(1), 5–54.

CLAGUE, D. A. & FREY, F. A. 1982. Petrology and trace element geochemistry of the Honolulu volcanics, Oahu: Implications for the oceanic mantle below Hawaii. *Journal of Petrology*, **23**, 447–504.

CLAVELL, E. 1991. *Geología del petroli de les conques terciáries de Catalunya.* PhD Thesis, University of Barcelona.

CLIFFORD, D. 2000. Rio Narcea's ongoing growth. *Mining Magazine*, February, 82–86.

CLIN, M. 1959. *Étude géologique de la Haute Chaîne des Pyrénées centrales entre le Cirque de Troumouse et le Cirque du Lys.* PhD Thesis, University of Nancy.

CLIN, M. & DEBON, F. 1996. Magmatisme Hercynien: Leucogranites de Tramezaigues et Bossost. *In*: BARNOLAS, A. & CHIRON, C. J. (eds) *Synthèse géologique et géophysique des Pyrénées. 1 – Cycle Hercynien.* BRGM and ITGE, Orléans and Madrid, 427.

CLOSAS, M. J. & FONT-ALTABA, M. 1960. Estudio de un yacimiento de bauxita en el Paleozoico de León. *Estudios Geológicos*, **16**(3), 157–161.

COCHERIE, A. 1978. *Géochemie des terres rares dans les granitoïdes.* Thèse 3ème cycle, Rennes.

COCHERIE, A. 1985. *Interaction Manteau-Croûte: son rôle dans la genèse d'associations plutoniques calco-alcalines, contraintes géochimiques (élements en traces et isotopes du strontium et de l'oxigène).* PhD Thesis, University of Rennes (Documents du BRGM **90**).

COCHERIE, A., WICKHAM, S. M. & AUTRAN, A. 1996. Géochimie des isotopes stables et radioactifs.-géochronologie. *In*: BARNOLAS, A. & CHIRON, C. J. (eds) *Synthèse géologique et géophysique des Pyrénées. 1 – Cycle Hercynien.* BRGM and ITGE, Orléans and Madrid, 476–480.

COCKS, L. R. M. 2000. The Early Palaeozoic geography of Europe. *Journal of the Geological Society, London*, **157**, 1–10.

COCKS, L. R. M. 2001. Ordovician and Silurian global geography. *Journal of the Geological Society, London*, **158**, 197–210.

COCKS, L. R. M. & FORTEY, R. A. 1990. Biogeography of Ordovician and Silurian faunas. *In*: MCKERROW, W. S. & SCOTESE, C. R. (eds) *Palaeozoic Palaeogeography and Biogeography*, Geological Society, London, Memoirs, **12**, 97–104.

COCKS, L. R. M. & MCKERROW, W. S. 1993. A reassessment of the early Ordovician 'Celtic' brachiopod province. *Journal of the Geological Society, London*, **150**, 1039–1042.

COCKS, L. R. M., MCKERROW, W. S. & VAN STAAL, C. R. 1997. The margins of Avalonia. *Geological Magazine*, **134**, 627–636.

COELLO, J. 1973. Las series volcánicas en subsuelos de Tenerife. *Estudios Geológicos*, **29**, 491–512.

COELLO, J. & CASTAÑÓN, A. 1965. Las sucesiones volcánicas de la zona de Carboneras (Almería). *Estudios Geológicos*, **21**, 145–166.

COELLO, J., CANTAGREL, J. M., HERNÁN, F. *et al.* 1992. Evolution of the Eastern volcanic ridge of the Canary Islands based on new K-Ar data. *Journal of Volcanology and Geothermal Research*, **53**, 251–274.

COLCHEN, M. 1971. Les Formations carbonifères de la Sierra de la

Demanda: comparison avec celles de l'ensemble cantabro-asturien. *Trabajos de Geología, Oviedo*, **3**, 53–68.

COLCHEN, M. 1974. *Géologie de la Sierra de la Demanda, Burgos-Logroño (Espagne).* IGME, Madrid, Memorias, **85**.

COLMENERO, J. R. & BAHAMONDE, J. R. 1986. Análisis estratigráfico y sedimentológico de la cuenca Estefaniense de Sebarga (Región de Mantos, Zona Cantábrica). *Trabajos de Geología, Oviedo*, **16**, 103–119.

COLMENERO, J. R. & PRADO, J. G. 1993. Coal basins in the Cantabrian Mountains, northwestern Spain. *International Journal of Coal Geology*, **23**, 215–229.

COLMENERO, J. R., GARCÍA-RAMOS, J. C., MANJÓN, M. & VARGAS, I. 1982. Evolución de la sedimentación terciaria en el borde N de la Cuenca del Duero entre los valles del Torío y Pisuerga (León-Palencia). *Temas Geológico Mineros, IGME*, **6**(I), 171–181.

COLMENERO, J. R., ÁGUEDA, J. A., FERNÁNDEZ, L. P., SALVADOR, C. I., BAHAMONDE, J. R. & BARBA, P. 1988. Fan-delta systems related to the Carboniferous evolution of the Cantabrian Zone, north-western Spain. *In*: NEMEC, W. & STEEL, R. J. (eds) *Fan Deltas: Sedimentology and Tectonics Settings.* Blackie & Son, London, 267–285.

COLMENERO, J. R., BAHAMONDE, J. R. & BARBA, P. 1996. Las facies aluviales asociadas a los depósitos de carbón en las cuencas este-fanienses de León (borde sur de la Cordillera Cantábrica). *Cuadernos de Geología Ibérica*, **21**, 71–92.

COLOM, G. 1929. Nota sobre las calizas con miliolas del Estampiense de Mallorca. *Memórias de la Real Sociedad Española de Historia Natural*, **15**, 237–241.

COLOM, G. 1961. La paléoécologie des lacs du Ludien-Stampien Inférieur de l'ile de Majorque. *Revue de Micropaléontologie*, **4**, 17–29.

COLOM, G. 1975. *Geología de Mallorca*, Vol. 1. Instituto de Estudios Baleáricos, Palma de Mallorca, 40–84.

COLOM, G. 1980. Estudio sobre las microfacies y micropaleontología de la isla de Cabrera. *Revista Española de Micropaleontología*, **12**(1), 47–64.

COLOM, G. 1985. Estratigrafía y sedimentología del Andaluciense y del Plioceno de Mallorca (Baleares). *Boletín Geológico y Minero*, **96**, 235–302.

COLOM, G. & ESCANDELL, B. 1960–62. L'évolution du géosinclinal baléar. *Mémoires h.-série. Société géologique de France*, **I**, 125–136 (Livre à la mémoire du Professeur P. Fallot).

COLOMBO, F. 1980. *Estratigrafía y Sedimentología del Terciario inferior continental de los Catalánides.* PhD Thesis, University of Barcelona.

COLOMBO, F. 1986. Estratigrafía y sedimentología del Paleoceno continental del borde meridional occidental de Los Catalánides (Provincia de Tarragona, España). *Cuadernos de Geología Ibérica*, **10**, 295–334.

COLOMBO, F. 1989. Abanicos aluviales. *In*: Arche, A. (coord.) *Sedimentología*. Colección Nuevas Tendencias, CSIC, Madrid, **1**, 143–218.

COLOMER, M. & SANTANACH, P. 1988. Estructura y evolución del borde suroccidental de la Fosa de Calatayud-Daroca. *Geogaceta*, **4**, 29–31.

COMAS, M. C. 1978. *Sobre la geología de los Montes Orientales. Sedimentación y evolución paleogeográfica desde el Jurásico al Mioceno inferior (Zona Subbética, Andalucía).* PhD Thesis, University of Bilbao.

COMAS, M. C., OLÓRIZ, F. & TAVERA, J. M. 1981. The red nodular lime-stones (Ammonitico Rosso) and associated facies: A key for settling slopes or swell areas in the Subbetic Upper Jurassic submarine topography (southern Spain). *In*: FARINACCI, A. & ELMI, S. (eds) *Rosso Ammonitico Symposium Proceedings*, Roma, 144–136.

COMAS, M. C., RUÍZ-ORTIZ, P. A. & VERA, J. A. 1982. El Cretácico de las Unidades Intermedias y Zona Subbética. *In*: GARCÍA, A. (coord.) *El Cretácico de España.* Complutense University, Madrid, 570–603.

COMAS. M. C., GARCÍA DUEÑAS, V. & JURADO, M. J. 1992. Neogene tectonic evolution of the Alborán basin from MCS data. *Geo-Marine Letters*, **12**, 157–164.

COMAS, M. C., PLATT, J. P., SOTO, J. I. & WATTS, A. B. 1999. The origin and tectonic history of the Alborán Basin: Insights from Leg 161. *In*: ZAHN, R., COMAS, M. C. & KLAUS, A. (eds) *ODP Proceedings, Scientific Results*, **161**, 555–579.

COMAS-RENGIFO, M. J. 1985. *El Pliensbachiense de la Cordillera Ibérica.* PhD Thesis, Complutense University, Madrid.

COMAS-RENGIFO, M. J., GOY, A. & YÉBENES, A. 1988. El Lías en el sector suroccidental de la Sierra de la Demanda (Castrovido, Burgos). *III Coloquio de Estratigrafía y Paleogeografía del Jurásico de España, libro guía de las excursiones, Ciencias de la Tierra (Instituto de Estudios Riojanos)*, **11**, 119–142.

COMAS-RENGIFO, M. J., GOY, A., MELÉNDEZ, G., MILLÁN, H., NAVARRO, J. J. & POCOVÍ, A. 1989. Caracterización bioestratigráfica del Lías de San Felices (Prepirineo Meridional, extremo occidental de las Sierras Exteriores). *Cuadernos de Geología Ibérica*, **13**, 175–184.

COMAS-RENGIFO, M. J., GÓMEZ, J. J., GOY, A. & RODRIGO, A. 1998. El Sinemuriense y Pliensbachiense en la sección de Alfara, Cordillera Costero Catalana (Tarragona). *Cuadernos de Geología Ibérica*, **24**, 173–184.

COMAS-RENGIFO, M. J., GÓMEZ, J. J., GOY, A., HERRERO, C., PERILLI, N. & RODRIGO, A. 1999. El Jurásico inferior en la sección de Almonacid de la Cuba (Sector central de la Cordillera Ibérica, Zaragoza). *Cuadernos de Geología Ibérica*, **25**, 27–58.

COMBA, J. A., ALVARADO, M., CAPOTE, R. *et al.* (coords) 1983. *Geología de España. Libro jubilar de J. M. Ríos.* IGME, Madrid.

COMBES, P. J. 1969. *Recherches sur la genèse des bauxites dans le NE de l'Espagne, le Languedocien et l'Ariège (France).* PhD Thesis, University of Montpellier (Mémoires du Centre d'Études et Recherches Hidrogéologiques **3–4**).

COMPANY, M. 1987. *Los Ammonites del Valanginiense del sector Oriental de las Cordilleras Béticas (SE de España).* PhD Thesis, University of Granada.

COMPANY, M., GARCÍA-HERNÁNDEZ, M., LÓPEZ GARRIDO, A. C., VERA, J. A. & WILKE, H. 1982. Interpretación genética y paleogeográfica de las turbiditas y materiales redepositados del Senoniense superior en la Sierra de Aixorta (Prebético Interno, Provincia de Alicante). *Cuadernos de Geología Ibérica*, **8**, 449–463.

COMTE, P. 1937. La série cambrienne et silurienne du León. *Comptes Rendus de l'Académie des Sciences Paris*, **202**, 337–341.

COMTE, P. 1959. *Recherches sur les terrains anciens de la Cordillère Cantabrique.* IGME, Madrid, Memorias **60**.

CONCHA, A., OYARZUN, R., LUNAR, R., SIERRA, J., DOBLAS, M. & LILLO, J. 1992. The Hiendelaencina epithermal silver-base metal district, Central Spain: tectonic and mineralizing processes. *Mineralium Deposita*, **27**, 83–89.

CONEY, P., JONES, D. L. & MONGER, J. W. H. 1980. Cordilleran suspect terranes. *Nature*, **288**, 329–333.

CONEY, P., MUÑOZ, J. A., MCCLAY, K. & EVENCHICK, C. 1996. Synctectonic burial and post-tectonic exhumation of the southern Pyrenees foreland fold-thrust belt. *Journal of the Geological Society, London*, **153**, 9–16.

CONTE, J., GASCÓN, F., LAGO, M. & CARLS, P. 1987. Materiales stephano-pérmicos en la fosa de Fombuena (provincia de Zaragoza). *Boletín Geológico y Minero*, **48**, 460–470.

COO, J. M. C., DEELMAN, J. C. & BAAN, V. VAN DER 1971. Carbonate facies of the Santa Lucia Formation (Emsian-Couvinian) in Leon and Asturias, Spain. *Geologie Mijnbouw*, **50**(3), 359–366.

COOPER, R. A., FORTEY, R. A. & LINDHOLM, K. 1991. Latitudinal and depth zonation of early Ordovician graptolites. *Lethaia*, **24**, 199–218.

COOPER, R. A., NOWLAN, G. S. & WILLIAMS, S. H. 2001. Global stratotype section and point for base of the Ordovician System. *Episodes*, **24**, 19–28.

COPPONEX, J. P. 1958. Observations géologiques sur les Alpujárrides occidentales (Cordillères Bétiques, Espagne). *Boletín Geológico y Minero*, **70**, 79–203.

CORBALÁN, F. & MELÉNDEZ, G. 1986. Nuevos datos bioestratigráficos sobre el Jurásico superior de la rama Castellana de la Cordillera Ibérica. *Acta Geológica Hispánica*, **21–22**, 555–560.

COROMINAS, J. & MOYA, J. 1999. Reconstructing recent landslide activity in relation to rainfall in the Llobregat River basin, Eastern Pyrenees, Spain. *Geomorphology*, **30**, 79–93.

CORRALES, I. 1971. La sedimentación durante el Estefaniense B-C en Cangas de Narcea, Rengos y Villablino (NW de España). *Trabajos de Geología*, **3**, 69–73.

CORRALES, I., MANJON, M. & VALLADARES, I. 1974. La serie carbonatada de Navarredonda de la Rinconada (Salamanca, España). *Studia Geologica*, **8**, 85–91.

CORRALES, I., CARBALLEIRA, J., CORROCHANO, A., POL, C. & ARMENTEROS, I. 1978. Las facies miocenas del sector sur de la Cuenca del Duero. *Publicaciones Departamento Estratigrafía, University of Salmanca*, **9**, 7–15.

CORRALES, I., CARBALLEIRA, FLOR, G., POL, C. & CORROCHANO, A. I. 1986. Alluvial systems in the northwestern part of the Duero Basin (Spain). *Sedimentary Geology*, **47**, 149–166.

CORRETGÉ, L. G. 1971. *Estudio petrológico del batolito de Cabeza de Araya (Cáceres).* PhD Thesis, University of Salamanca.

CORRETGÉ, L. G. & SUÁREZ, O. 1990. Igneous rocks. Cantabrian and Palentian Zones. *In*: DALLMEYER, R. D. & MARTÍNEZ GARCÍA, E. (eds) *Pre-Mesozoic Geology of Iberia*. Springer-Verlag, Berlin, 72–79.

CORRETGÉ, L. G., UGIDOS, J. M. & MARTÍNEZ, F. J. 1977. Les series granitiques du secteur Centre–Occidental Espagnol. *La chaine varisque d'Europe moyenne et occidentale. Colloques Internationaux du Centre National de la Reserche Scientifique*, **243**, 453–461.

CORRETGÉ, L. G., BEA, F. & SUAREZ, O. 1985. Las características geoquímicas del batolito de Cabeza de Araya (Cáceres, España): implicaciones petrogenéticas. *Trabajos de Geología*, **15**, 219–238.

CORRETGÉ, L. G., SUÁREZ, O. & GALÁN, G. 1990. West Asturien-Leonese Zone. Igneous rocks. *In*: DALLMEYER, R. D. & MARTÍNEZ GARCÍA, E. (eds) *Pre-Mesozoic Geology of Iberia*. Springer-Verlag, Berlin, 115–128.

CORRETGÉ, P. 1969. El complejo ortoneísico de Pola de Allande. *Boletín Geológico y Minero*, **80**, 289–306.

CORROCHANO, A. 1977. *Estratigrafía y sedimentología del Paleógeno en la provincia de Zamora.* PhD Thesis, University of Salamanca.

CORROCHANO, A. 1982. Costra ferralítica y Unidad basal del Paleógeno en Zamora. *Temas Geológico Mineros, IGME*, **6**(I), 802–806.

CORROCHANO, A. 1989. Facies del Cretácico terminal y arquitectura secuencial de los abanicos aluviales terciarios del borde norte de la Depresión del Duero (valle de las Arrimadas, León). *Studia Geológica Salmanticensia, Ediciones Universidad Salamanca*, **5**, 89–106.

CORROCHANO, A. & ARMENTEROS, I. 1989. Los sistemas lacustres de la Cuenca terciaria del Duero. *Acta Geologica Hispanica*, **24**(3–4), 259–279.

CORROCHANO, A. & PENA DOS REIS, R. 1986. Analogías y diferencias en la evolución sedimentaria de las cuencas del Duero, Occidental Portuguesa y Lousa (Península Ibérica). *Studia Geológica Salmanticensia, Ediciones Universidad Salamanca*, **22**, 309–326.

CORROCHANO, A., CARBALLEIRA, J., POL, C. & CORRALES, I. 1983. Los sistemas deposicionales terciarios de la depresión de Peñaranda-Alba y sus relaciones con la fracturación. *Studia Geológica Salmanticensia, Ediciones Universidad Salamanca*, **19**, 187–199.

CORROCHANO, A., FERNÁNDEZ MACARRO, B., RECIO, C., BLANCO, J. A. & VALLADARES, I. 1986. Modelo sedimentario de los lagos neógenos de la Cuenca del Duero. Sector Centro-Occidental. *Studia Geológica Salmanticensia, Ediciones Universidad Salamanca*, **22**, 93–110.

CORROCHANO, A., ARMENTEROS, I., PÉREZ, C., SÁNCHEZ DE VEGA, A. & SAN DIMAS, L. F. 1991. Distribución de arcillas en la Unidad de Cuestas (Neógeno de la Cuenca del Duero). *Geogaceta*, **10**, 22–24.

CORTÉS, A. L. 1999. *Evolución tectónica reciente de la Cordillera Ibérica, Cuenca del Ebro y Pirineo centro-ocidental.* PhD Thesis, University of Zaragoza.

CORTÉS, A. L. & CASAS, A. M. 1996. Deformación alpina del zócalo y cobertera en el borde norte de la Cubeta de Azuara (Cordillera Ibérica). *Revista de la Sociedad Geológica de España*, **9**(1–2), 51–66.

CORTÉS, A. L., LIESA, C. L., SIMÓN, J. L., CASAS, A. M., MAESTRO, A. & ARLEGUI, L. 1996. El campo de esfuerzos compresivo neógeno en el NE de la Península Ibérica. *Geogaceta*, **20**(4), 806–809.

CORTÉS, A. L., LIESA, C. L., SORIA, A. R., & MELÉNDEZ, A. 1999. Role of the extensional structures in the location of folds and thrusts during tectonic inversion (Northern Iberian Chain, Spain). *Geodinámica Acta*, **12**(2), 113–132.

COURTNAY, R. C. & WHITE, R. S. 1986. Anomalous heat flow and geoid across the Cape Verde Rise: evidence for dynamic support from a thermal plume in the mantle. *Geophysical Journal of the Royal Astrophysical Society*, **87**, 815–867.

COUSENS, B. L., ALLAN, J. F. & GORTON, M. P. 1994. Subduction-modified pelagic sediments as the enriched component in back-arc basalts from the Japan Sea: Ocean Drilling Program Sites 797 and 794. *Contributions to Mineralogy and Petrology*, **117**, 421–434.

COUTO, H., PIÇARRA, J. M. & GUTIÉRREZ-MARCO, J. C. 1997. El Paleozoico del anticlinal de Valongo (Portugal). *In*: GRANDAL D'ANGLADE, A., GUTIÉRREZ-MARCO, J. C. & SANTOS FIDALGO, L. (eds) *XIII Jornadas de Paleontología y V Reunión Internacional Proyecto 351 PICG*. A Coruña, Libro de Resúmenes y Excursiones, 270–290 (ISBN 84-605-6825-3).

CRABIFOSSE, S., FERRERO, A. & MONGE, C. 1989. Aportación al conocimineto del cuarzo en Galicia. *Cuadernos del Laboratorio Geológico de Laxe*, **14**, 225–236.

CRAMER-DÍEZ, F. H. & DÍEZ, M. C. 1978. Iberian chitinozoans. I, Introduction and summary of pre-Devonian data. *Palinología*, no. extraord. **1**, 149–201.

CRAMER-DÍEZ, F. H., JULIVERT, M. & DÍEZ, M. C. 1972. Llandeilian chitinozoans from Rioseco, Asturias, Spain. Preliminary note. *Breviora Geologica Asturica*, **16**, 23–25.

CREMADES, J. & LINARES, A. 1982. Contribución al conocimiento del Albense superior – Cenomanense del Prebético de la Provincia de Alicante. *Cuadernos de Geología Ibérica*, **8**, 721–738.

CRESPO, A. & LUNAR, R. 1997. Terrestrial hot-spring Co-rich Mn mineralization in the Pliocene-Quaternary Calatrava Region (Central Spain). *In*: NICHOLSON, K., HEIN, J. R., BÜHN, B. & DASGUPTA, S. (eds) *Manganese Mineralization: Geochemistry and Mineralogy of Terrestrial and Marine Deposits*. The Geological Society, London, Special Publications, **119**, 253–264.

CRESPO BLANC, A. 1989. *Evolución geotectónica del contacto entre la Zona de Ossa-Morena y la Zona Surportuguesa en las Sierras de aracena y Aroche (Macizo Ibérico Meridional): un contacto mayor en la Cadena Hercínica Europea*. PhD Thesis, University of Sevilla.

CRESPO BLANC, A. 1995. Interference pattern of extensional fault systems: a case study of the Miocene rifting of the Alboran basement (North of Sierra Nevada, Betic Chain). *Journal of Structural Geology*, **17**, 1559–1569.

CRESPO BLANC, A. & CAMPOS, J. 2001. Structure and kinematics of the South Iberian paleomargin and its relationship with the Flysch Trough units: Extensional tectonics within the Gibraltar Arc fold-and-thrust belt (western Betics). *Journal of Structural Geology*, **23**, 1615–1630.

CRESPO BLANC, A. & OROZCO, M. 1991. The boundary between the Ossa-Morena and Southportuguese Zones (Southern Iberian Massif): a major suture in the European Hercynian Chain. *Geologische Rundschau*, **80**, 691–702.

CRESPO BLANC, A., OROZCO, M. & GARCÍA-DUEÑAS, V. 1994. Extension versus compression during the Miocene tectonic evolution of the Betic chain. Late folding of normal fault systems. *Tectonics*, **13**, 78–88.

CRIMES, T. P. 1976. Sand fans, turbidites, slumps and the origin of the bay of Biscay: a facies analysis of the Guipuzcoan flysch. *Palaeogeography, Palaeoclimatology, Palaeoecology*, **19**, 1–15.

CRIMES, T. P. 1994. The period of early evolutionary failure and the dawn of evolutionary success: the record of biotic changes across the Precambrian–Cambrian boundary. *In*: DONOVAN, S. K. (ed.) *The Palaeobiology of Trace Fossils*. John Wiley & Sons, Chichester, 105–133.

CRIMES, T. P. & MARCOS, A. 1976. Trilobite traces and the age of the lowest part of the Ordovician reference section for NW Spain. *Geological Magazine*, **113**, 350–356.

CRIMES, T. P., MARCOS, A. & PÉREZ-ESTAÚN, A. 1974. Upper Ordovician turbidites in Western Asturias: a facies analysis with particular reference to vertical and lateral variations. *Palaeogeography, Palaeoclimatology, Palaeoecology*, **15**, 169–184.

CRIMES, T. P., LEGG, I., MARCOS, A. & ARBOLEYA, M. 1977. ?Late

Precambrian– low Lower Cambrian trace fossils from Spain. *In*: CRIMES, T. P. & HARPER, J. C. (eds) *Trace Fossils 2. Geological Journal* Special Issue, **9**, 91–138.

CRITTENDEN, M. D., JR., CONEY, P. J. & DAVIS, G. H. (eds) 1980. *Cordilleran metamorphic core complexes*. Geological Society of America, Memoirs, **153**.

CROFTS, R. 1967. Raised beaches and chronology in north west Fuerteventura. Canary Islands. *Quaternaria*, **9**, 247–260.

CRUSAFONT, M., RIBA, O. & VILLENA, J. 1966a. Nota preliminar sobre un nuevo yacimiento de vertebrados aquitanienses en Santa Cilia (Río Formiga, Provincia de Huesca) y sus consecuencias geológicas. *Notas y Comunicaciones del IGME*, **83**, 7–14.

CRUSAFONT, M., TRUYOLS, J. & RIBA, O. 1966b. Contribución al conocimiento de la estratigrafía del Terciario continental de Navarra y Rioja. *Notas y Comunicaciones del IGME*, **90**, 53–76.

CRUZ-SANJULIÁN, J., OLÓRIZ, F. & SEQUEIROS, L. 1973. El Jurásico superior entre el Torcal de Antequera y Cañete la Real (Cordilleras Béticas, región occidental). *Cuadernos de Geología de la Universidad de Granada*, **4**, 15–25.

CUBAS, C. R. 1978. Estudio de los domos sálicos de la Isla de La Gomera (Islas Canarias). I. Vulcanología. *Estudios Geológicos*, **34**, 53–70.

CUBAS, C. R., HERNÁN, F., ANCOCHEA, E., BRÄNDLE, J. L. & HUERTAS, M. J. 1994. Serie Basáltica Antigua Inferior en el sector de Hermigua. Isla de La Gomera. *Geogaceta*, **16**, 15–18.

CUENCA, G. 1983. Nuevo yacimiento de vertebrados del Mioceno inferior del borde meridional de la Cuenca del Ebro. *Estudios Geológicos*, **39**, 217–224.

CUENCA, G. & CANUDO, J. I. 1991. El límite Oligoceno-Mioceno con roedores fósiles en la Cuenca del Ebro: Fraga y Ballobar, Provincia de Huesca. *Azara*, **3**, 35–51.

CUENCA, G., AZANZA, B., CANUDO, J. I. & FUERTES, V. 1989. Los micro-mamíferos del Mioceno inferior de Peñalba (Huesca). Implicaciones bioestratigráficas. *Geogaceta*, **6**, 75–77.

CUENCA, G., CANUDO, J. I., LAPLANA, C. & ANDRÉS, J. A. 1992. Bio y cronoestratigrafía con mamíferos en la Cuenca Terciaria del Ebro: ensayo de síntesis. *Acta Geológica Hispánica*, **27**, 127–143 (Homenaje a Oriol Riba Arderiu).

CUENCA BESCÓS, G., LAPLANA, C. & CANUDO, J. L. 1999. Biochronological implications of the Arvicolidae (Rodentia, Mammalia) from the Lower Pleistocene hominid-bearing level of Trinchera Dolina 6 (TD6, Atapuerca, Spain). *Journal of Human Evolution*, **37**, 353–373.

CUENCA BESCÓS, G., CANUDO, I. & LAPLANA, C. 2001. La séquence des rongeurs (*Mammalia*) des sites du Pléistocène inférieur et moyen d'Atapuerca (Burgos, Espagne). *L'Anthropologie*, **105**(1), 115–130.

CUERDA, J. 1989. *Los tiempos cuaternarios en Baleares*. Dirección General de Cultura, Gobierno Balear.

CUERDA, J. & SACARÉS, J. 1966. Nueva contribución al estudio del Pleistoceno marino del término de Lluchmajor (Mallorca). *Boletín de la Real Sociedad Española de Historia Natural*, **12**, 63–100.

CUETO, L. A., EGUILUZ, L., LLAMAS, F. J. & QUESADA, C. 1983. La granodiorita de Pallarés, un intrusivo Precámbrico en la alineación Olivenza-Monesterio (Zona de Ossa Morena). *Memórias dos Serviços Geológicos de Portugal*, **LXIX**, 219–226.

CUEVAS, J. 1990. *Microtectónica y metamorfismo de los Mantos Alpujárrides del tercio central de las Cordilleras Béticas (entre Motril y Adra). Boletín Geológico y Minero de España*, **101**, 1–129 (PhD Thesis, University of País Vasco, 1988).

CUEVAS, J., ALDAYA, F., NAVARRO-VILA, F. & TUBÍA, J. M. 1986. Caractérisation de deux étapes de charriage principales dans les nappes Alpujarrides centrales (Cordillères Bétiques, Espagne). *Comptes Rendus de l'Académie des Sciences de Paris*, **302**, 1177–1180.

CUEVAS, J., ARANGUREN, A., BADILLO, J. M. & TUBÍA, J. 1999. Estudio estructural del sector central del Arco Vasco (cuenca Vasco-Cantábrica). *Boletín Geológico y Minero*, **110**, 3–18.

CUNNINGHAM, C. G., ARRIBAS JR., A., RYTUBA, J. J. & ARRIBAS, A. 1990. Mineralized and unmineralized calderas in Spain; Part I, evolution of the Los Frailes Caldera. *Mineralium Deposita*, **25**, S21–S28.

DABRIO, C. J., FERNÁNDEZ, J., PEÑA, J. A., RUIZ-BUSTOS, A. & SANZ DE

GALDEANO, C. 1978. Rasgos sedimentarios de los conglomerados miocénicos del borde noreste de la Depresión de Granada. *Estudios Geológicos*, **34**, 89–97.

DABRIO, C. J., ESTEBAN, M. & MARTÍN, J. M. 1981. The coral reef of Níjar, Messinian (uppermost Miocene), Almería province, S.E. Spain. *Journal of Sedimentary Petrology*, **51**, 521–539.

DABRIO, C. J., MARTÍN, J. M. & MEGÍAS, A. G. 1982. Signification sédimentaire des évaporites de la dépression de Grenade (Espagne). *Bulletin Societè Géologique France*, **24**, 705–710.

DABRIO, C. J., ZAZO, C., GOY, J. L. *et al.* 2000. Depositional history of estuaire infill during the last postglacial transgression (Gulf of Cádiz, Southern Spain). *Marine Geology*, **162**, 381–404.

DAHM, H. 1966. Stratigraphie und Paleogeographie in Kantabrischen Jura (Spanien). *Beih. Geol. Jb.*, **44**, 13–54.

DAIGNIÈRES, M., GALLART, J., BANDA, E., HIRN, A. 1982. Implications of the seismic structure for the orogenic evolution of the Pyrenean range. *Earth and Planetary Science Letters*, **57**, 88–100.

DAIGNIÈRES, M., DE CABISSOLE, B., GALLART, J., HIRN, A., SURIÑACH, E. & TORNÉ, M. 1989. Geophysical constraints on the deep structure along the ECORS Pyrenees line. *Tectonics*, **8**, 1051–1058.

DAIGNIÈRES, M., SÉGURET, M., SPECHT, M. AND ECORS TEAM. 1994. The Arzaq-Western Pyrenees ECORS Deep Seismic Profile. *In*: MASCLE A. (ed.) *Hydrocarbon and Petroleum Geology of France*. European Association of Petroleum Geologists, Special Publications, Springer-Verlag, Berlin, **4**, 199–208.

DALLMEYER, R. D. & GIL IBARGUCHI, J. I. 1990. Age of the amphibolitic metamorphism in the Ophiolitic Unit of the Morais allochthon (Portugal): implications for early Hercynian orogenesis in the Iberian massif. *Journal of the Geological Society, London*, **147**, 873–878.

DALLMEYER, R. D. & MARTÍNEZ GARCÍA, E. (eds) 1990. *Pre-Mesozoic Geology of Iberia*. Springer-Verlag, Berlin.

DALLMEYER, R. D. & QUESADA, C. 1992. Cadomian vs. Variscan evolution of the Ossa-Morena zone (SW Iberia): field and $^{40}Ar/^{39}Ar$ mineral age constraints. *Tectonophysics*, **216**, 339–364.

DALLMEYER, R. D. & TUCKER, R. D. 1993. U-Pb zircon age for the Lagoa augen gneiss, Morais Complex, Portugal: tectonic implications. *Journal of the Geological Society, London*, **150**, 405–410.

DALLMEYER, R. D., RIBEIRO, A. & MARQUES, F. 1991. Polyphase Variscan emplacement of exotic terranes (Morais and Bragança Massifs) onto Iberian successions: evidence from $^{40}Ar/^{39}Ar$ mineral ages. *Lithos*, **27**, 133–144.

DALLMEYER, R. D., MARTÍNEZ CATALÁN, J. R., ARENAS, R. *et al.* 1997. Diachronous Variscan tectonothermal activity in the NW Iberian Massif: evidence from $^{40}Ar/^{39}Ar$ dating of regional fabrics. *Tectonophysics*, **277**, 307–337.

DALLONI, M. M. 1910. Etude géologique des Pyrénées de l'Aragón. *Annales des Faculté des Sciences du Marseille*, **19**.

DALLONI, M. 1930. Etude géologique des Pyrénées de Catalogne. *Annales des Faculté des Sciences du Marseille*, **26**, 140–182.

DAÑOBEITIA, J. J., ARGUEDAS, M., GALLART, J. & BANDA, E. 1992. Deep crustal configuration of the Valencia though and its Iberian and Balearic borders from extensive refraction and wide-angle reflection profiling. *Tectonophysics*, **208**, 37–55.

DARDER, B. 1914. El Triásico de Mallorca. *Trabajos del Museo Nacional de Ciencias Naturales*, **7**, 1–85.

DARDER, B. 1925. La tectonique de la region orientale de l'île de Majorque. *Bulletin de la Société géologique de la France*, **25**, 245–278.

DASH, B. P. & BOSSHARD, E. 1969. Seismic and gravity investigations around the western Canary Islands. *Earth and Planetary Science Letters*, **7**, 169–177.

DAY, S. J., CARRACEDO, J. C. & GUILLOU, H. 1997. Age and geometry of an aborted rift flank collapse: The San Andrés fault, El Hierro, Canary Islands. *Geological Magazine*, **134**(4), 523–537.

DAY, S. J., CARRACEDO, J. C., GUILLOU, H. & GRAVESTOCK, P. 1999. Recent structural evolution of the Cumbre Vieja Volcano, La Palma, Canary Islands: Volcanic rift zone reconfiguration as a precursor to volcanic flank instability? *Journal of Volcanology and Geothermal Research*, **94**, 135–167.

DE BRUIJNE, C. H. & ANDRIESSEN, P. A. V. 2000. Interplay of Intraplate Tectonics and Surface Processes in the Sierra le Guadarrama (Central Spain) Assessed by Apatite Fission track Analysis. *Physics and Chemistry of the Earth*, **25**, 555–563.

DE BRUIJN, H., SONDAAR, P. Y. & SANDERS, E. A. C. 1978. On a new species of Pseudoltinomys (Theridomydae, Rodentia) from the Paleogene of Mallorca. *Proceedings of the Koninklijke Nederlandse Akademie van Wetenschappen*, **82**, 1–10.

DE FIGUEROLA, L. C. G., CORRETGE, L. G. & BEA, F. 1974. El dique de Alentejo-Plasencia y haces de diques básicos de Extremadura. (Estudio comparativo). *Boletín Geológico y Minero*, **85**, 40–69.

DE JONG, G., DALSTRA, H., BOORDER, H. & SAVAGE, J. F. 1991. Blue amphiboles, Variscan deformation and plate tectonics in the Beja Massif, South Portugal. *Comunicações dos Serviços Geológicos de Portugal*, **77**, 59–64.

DE JONG, J. D. 1971. Molasse and clastic-wedge sediments of the Southern Cantabrian Mountains (NW Spain) as geomorphological and environmental indicators. *Geologie Mijnbouw*, **50**, 399–416.

DE JONG, K. 1990. Alpine tectonics and rotation pole evolution of Iberia. *Tectonophysics*, **184**, 279–296

DE JONG, K. 1991. *Tectono-metamorphic studies and radiometric dating in the Betic Cordilleras (SE Spain), with implications for the dynamics of extension and compression in the western Mediterranean area*. PhD Thesis, University of Amsterdam.

DE JONG, K. 1993. Large-scale polyphase deformation of a coherent HP/LT metamorphic unit: the Mulhacen Complex in the eastern Sierra de los Filabres. *Geologische en Mijnbouw*, **71**, 327–336.

DE LA PEÑA, J. A., FONOLLÁ, F., RAMOS, J. & MARFIL, R. 1977. Identificación del Autuniense en la Rama Aragonesa de la Cordillera Ibérica (provincia de Soria). *Cuadernos de Geología Ibérica*, **4**, 123–134.

DE LA ROCHE, H. 1964. Sur l'expresion graphique des relations entre la composition chimique et la composition minèralogique quantitative del roches cristalines. Presentation d'un diagrame destiné à l'etude chimico-mineralogique des massifs granitiques ou granodioritiques. Aplication aux Vosgues cristalines. *Science de la Terre*, **IX**, 293–337.

DE LA ROSA, J. D. 1992. *Petrología de las rocas básicas y granitoides del batolito de la Sierra Norte de Sevilla, zona Surportuguesa, Macizo Ibérico*. PhD Thesis, University of Sevilla.

DE LA ROSA, J. D., ROGERS, G. & CASTRO, A. 1993. Relaciones $^{87}Sr/^{86}Sr$ de Rocas Básicas y Granitoides del batolito de la Sierra Norte de Sevilla. *Revista de la Sociedad Geológica de España*, **6**, 141–149.

DE LA ROSA, J. D., JENNER, G., CASTRO, A., VALVERDE, P. & TUBRET, M. 1998. Aplicación de análisis de isótopos de U-Th-Pb mediante LAM-ICP-MS en el estudio de herencias de circones de granitoides del Macizo Ibérico. *Boletín de la Sociedad Española de Mineralogía*, **21–A**, 184–185.

DE LA ROSA, J. D., CASTRO, A. & JENNER, G. 1999. Age and relationships between the deep continental crust of South Portuguese and Ossa-Morena zones in the Aracena metamorphic belt revealed by zircon inherited. *XV Reunión de Geología del Oeste Peninsular (International Meeting on Cadomian Orogens)*, Badajoz, Spain. Extended Abstracts, 81–88.

DE LAROUZIÈRE, F., BOLZE, J., BORDET, P., HERNÁNDEZ, J., MONTENAT, CH. & OTT D'ESTEVOU, PH. 1988. The Betic segment of the lithospheric Trans-Alborán shear zone during the late Miocene. *Tectonophysics*, **152**, 41–52.

DE LUCA, P., DUÉE, G. & HERVOUET, Y. 1985. Évolution et déformations du bassin de flysch du Crétacé supérieur de la haute chaîne (Pyrénées Basco-béarnaises – région du Pic d'Orhy). *Bulletin de la Société Géologique de France*, **2**, 249–262.

DE RUIG, M. J. 1992. *Tectono-sedimentary evolution of the Prebetic fold belt of Alicante (SE Spain)*. PhD Thesis, Vrije University, Amsterdam.

DE SITTER, L. U. 1961. Le Précambrien dans la Chaîne cantabrique. *Comptes Rendus Sommaire des Séances de la Société Géologique de France*, **9**, 1–253.

DE SITTER, L. U. & ZWART, H. J. 1960. Tectonic development in supra and infra-structures of a mountain chain. *Proceedings of the 21st International Geological Congress*, Copenhagen, **18**, 248–256.

DE SMET, M. E. M. 1984. Wrenching in the external zone of the Betic Cordilleras, southern Spain. *Tectonophysics*, **107**, 57–79.

DE VICENTE, G. 1988. *Análisis poblacional de fallas. El sector de enlace*

Sistema Central-Cordillera Ibérica. PhD Thesis, Complutense University, Madrid.

DE VICENTE, G., CALVO, J. P. & MUÑOZ-MARTÍN, A. 1996a. Neogene tectono-sedimentary review of the Madrid Basin. *In*: FRIEND, P. F. & DABRIO, C. J. (eds) *Tertiary Basins of Spain: the Stratigraphic Record of Crustal Kinematics.* Cambridge University Press, Cambridge, 268–271.

DE VICENTE, G., GINER, J. L., MUÑOZ-MARTÍN, A., GONZÁLEZ-CASADO, J. M. & LINDO, R. 1996b. Determination of the present-day stress tensor and neoetctonic interval in the Spanish Central Range and Madrid Basin, central Spain. *Tectonophysics*, **266**, 405–424.

DE VICENTE G., GONZÁLEZ-CASADO, J. M., MUÑOZ-MARTÍN, GINER, J. & RODRÍGUEZ-PASCUA, M. A.1996c. Structure and Tertiary evolution of the Madrid Basin. *In*: FRIEND, P. F. & DABRIO, C. J. (eds) *Tertiary Basins of Spain: the Stratigraphic Record of Crustal Kinematics.* Cambridge University Press, Cambridge, 263–267.

DE VICENTE, G., HERRAIZ, M., GINER, J. L., LINDO, R., CABAÑAS, L. & RAMÍREZ, M. 1996d. Características de los esfuerzos activos inter-placa en la Península Ibérica. *Geogaceta*, **20**(4), 909–912.

DE VRIES, W. C. & ZWAAN, K. B. 1967. Alpujárride succesion in the Central Sierra de las Estancias, Province of Almería, SE Spain. *Koninklijke Nederlandse Akademie Van Wetenschappen*, **70**, 443–453.

DEAN, W. T. & MONOD, O. 1997. Cambrian development of the Gond-wanaland margin in southeastern Turkey. *Turkish Association of Petroleum Geologists, Special Publication*, **3**, 61–74.

DEBON, F. 1975. *Les massifs granitoïdes à structure concentrique de Cauterets-Panticosa (Pyrénées occidentales) et leurs enclaves.* PhD Thesis, University of Nancy (*Sciences de la Terre, Mémoire*, **33**).

DEBON, F. 1980. Genesis of the three concentrically-zoned granitoid plutons of Cauterets-Panticosa. *Geologische Rundschau*, **69**(1), 107–130.

DEBON, F. & AUTRAN, A. 1996. Magmatisme Hercynien: Mise en place. *In*: BARNOLAS, A. & CHIRON, J. C. (eds) *Synthèse géologique et géophysique des Pyrénées.* 1- *Cycle Hercynien.* BRGM-ITGE, Orléans and Madrid, 466–467.

DEBON, F. & GUITARD, A. 1996. Magmatisme Hercynien: Métamor-phisme et Plutonisme hercyniens. Carte de synthèse. *In*: BARNOLAS, A. & CHIRON, C. J. (eds) *Synthèse géologique et géophysique des Pyrénées.* 1. *Cycle Hercynien.* BRGM-ITGE, Orléans and Madrid.

DEBON, F. & LE FORT, P. 1983. A chemical-mineralogical classification of common plutonic rocks and associations. *Transactions of the Royal Society of Edinburgh. Earth Sciences*, **73**, 135–149.

DEBON, F. & ZIMMERMANN, J. L. 1988. Le pluton hercynien de Bassiès (Pyrénées, Zone Axiale): typologie chimique, âge et remaniements isotopiques. *Comptes Rendues de l'Académie des Sciences de Paris*, **306**, 897–902.

DEBON, F., ENRIQUE, P. & AUTRON, A. 1995. Magnetisme hercynien. *In*: BARNOLAS, A. & CHIRON, J. C. (eds) *Synthese Geologique ey Geophysique des Pyrenees.* 1. *Cycle Hercynien.* BRGM-ITGE, Orléans and Madrid, 361–499.

DEBRENNE, F. & LOTZE, F. 1963. Die Archaeocyatha des spanischen Kambriums. *Akademie der Wissenschaft und der Literatur Abhandlungen der Naturswissenshaftligen Klase*, **1963**(2), 107–143.

DEBRENNE, F. & ZAMARREÑO, I. 1970. Sur la découverte d'Archéocy-athes dans le Cambrien du NW de l'Espagne. *Brevoria Geológica Astúrica*, **14**, 1–11.

DEBROAS, E. J. 1990. Le Flysch noir albo-cénomanien témoin de la structuration albienne à sénonienne de la zone nord-pyrénéenne en Bigorre (Hautes Pyrénées, France). *Bulletin de la Société Géologique de France*, **8**, VI(2), 273–285.

DÉGARDIN, J. M. 1979. Découverte du genre *Phyllograptus* (Grapto-lites) dans l'Ordovicien des Pyrénées Atlantiques: conséquences stratigraphiques. *Geobios*, **12**, 321–329.

DÉGARDIN, J. M. 1984. Graptolites du Silurien de la région de Camprodon, Province de Gérone, Pyrénées orientales espag-noles. *Annales de la Société Géologique du Nord*, **103**, 57–74.

DÉGARDIN, J. M. 1988. Le Silurien des Pyrénées, biostratigraphie, paléogéographie. *Publications de la Société Géologique du Nord*, **15**, 1–525.

DÉGARDIN, J. M. 1990. The Silurian of the Pyrenees. *Journal of the Geological Society, London*, **147**, 687–692.

DÉGARDIN, J. M. (coord.) 1995. Ordovicien Supérieur – Silurien. *In*: BARNOLAS A. & CHIRON J. C. (eds) *Synthèse géologique et géophysique des Pyrénées, 1, Cycle Hercynien.* BRGM-ITGE, Orléans and Madrid, 211–233.

DÉGARDIN, J. M. & DE WEVER, P. 1985. Radiolaires siluriens dans les Pyénées centrales espagnoles. *Annales de la Société Géologique du Nord*, **104**, 121–125.

DÉGARDIN, J. M. & LETHIERS, F. 1982. Une microfaune (Conodonta, Ostracoda) dans le Silurien terminal des Pyrénées centrales espagnoles. *Revista Española de Micropaleontología*, **14**, 335–358.

DÉGARDIN, J. M. & PARIS, F. 1978. Présence de chitinozoaires dans les calcaires siluro-dévoniens de la Sierra Negra (Pyrénées centrales espagnoles). *Géobios*, **11**, 769–777.

DÉGARDIN, J. M. & PILLET, J. 1983. Nouveaux Trilobites du Silurien des Pyrénées centrales espagnoles. *Annales de la Société Géologique du Nord*, **103**, 83–92.

DEL MORO, A. & ENRIQUE, P. 1996. Edad Rb-Sr mediante isocrona de minerales de las tonalitas biotitico-hornbléndicas del Macizo del Montnegre (Cordilleras Costeras Catalanas). *Geogaceta*, **20**(2), 491–494.

DEL OLMO, P. DEL, OLIVÉ, A, GUTIERREZ, M, AGUILAR, M. LEAL, M. & PORTERO, J. M. 1983. *Paniza.* Spanish Geological Survey, Report and Map. Scale 1:50 000, **438**.

DELANNOY, J. J., DÍAZ DEL OLMO, F. & PULIDO, A. 1989. *Reunión franco-espagnole sur les karsts mediterraneéns d'Andalousie occidentale.* Livret-Guide Bétiques, **89**.

DELGADO, F., ESTEVEZ, A., MARTÍN, J. M. & MARTÍN ALGARRA, A. 1981 Observaciones sobre la estratigrafía de la formación carbonatada de los mantos alpujárrides (Cordilleras béticas). *Estudios Geológicos*, **37**, 45–57.

DELGADO, M. 1971. Esquema geológico de la Hoja número 878 de Azuaga (Badajoz). *Boletín Geológico y Minero*, **82**, 277–286.

DELGADO, M., LIÑÁN, E., PASCUAL, E. & PÉREZ LORENTE, F. 1977. Criterios para la diferenciación en dominios en Sierra Morena Central. *Studia Geologica*, **12**, 75–90.

DELGADO-QUESADA, M. 1971. Esquema geológico de la hoja número 878 de Azuaga (Badajoz). *Boletin Geológico y Minero*, **82**, 277–286.

DELVOLVÉ, J. J. 1987. *Un bassin synorogénique varisque. Le Culm des Pyrénées centro-occidentales.* PhD Thesis, University of Toulouse.

DELVOLVÉ, J. J. (coord.) 1995. Carbonifère à faciès Culm. *In*: BARNOLAS, A., CHIRON, J. C. & GUÉRANGÉ, B. (eds) *Synthèse Géologique et Géophysique des Pyrénées – Volume 1: Intro-duction. Géophysique. Cycle hercynien.* BRGM and IGME, Orléans and Madrid, 303–338.

DEMÉNY, A., AHIJADO, A., CASILLAS, R. & VENNEMANN, T. W. 1998. Crustal contamination and fluid/rock interaction in the carbon-atites of Fuerteventura (Canary Islands, Spain): a C, O, H isotope study. *Lithos*, **44**, 101–115

DENIZOT, G. 1934. Sur la structure des lles Canaries considerée dans ses rapports avec le probleme de l'Atlantide. *Comptes Rendus de La Académie des Sciences de Paris*, **199**, 372–373.

DEPAOLO, D. J., KYTE, F. T., MARSHALL, B. D., O'NEIL, J. R. & SMIT, J. 1983. Rb-Sr, Sm-Nd, K-Ca, O and H isotopic study of Cretaceous– Tertiary boundary sediments, Caravaca, Spain: Evidence for an oceanic impact site. *Earth and Planetary Science Letters*, **64**, 356–373.

DERAMOND, J., GRAHAM, R. H. HOSSACK, J. R. BABY, P. & CROUZET, G. 1985. Nouveau modéle de la chaîne des Pyrénées. *Comptes Rendus de l'Académie des Sciences, Paris*, **301**, 1213–1216.

DERCOURT, J., ZONENSHAIN, L. P., RICOU, L. E. *et al.* 1985. Présentation de 9 cartes paléogéographiques au 1 20 000 000ᵉ s'étendant de l'Atlantique au Pamir pour la période du Lias à l'Actuel. *Bulletin de la Société Géologique de France*, **8**, I (5), 637–652.

DERCOURT, J., ZONENSHAIN, L. P., RICOU, L. E. *et al.* 1986. Geological evolution of the Tethys belt from the Atlantic to the Pamir since Lias. *Tectonophysics*, **123**, 241–315.

DERCOURT, J., RICOU, L. E., & VRIELYNICK, B. (eds) 1993. *Atlas Tethys Palaeoenvironmental Maps.* Gauthier-Villars, Paris.

DERCOURT, J., FOURCADE, E., CECCA, F., AZÉMA, J., ENAY, R., BASSOULLET, J. P. & COTTEREAU, N. 1994. Palaeoenvironment of the Jurassic system in the western and central Tethys (Toarcian, Callovian, Kimmeridgian, Tithonian): An overview. *Geobios*, Mémoire Spéciale **17**, 625–644.

DERCOURT, J., GAETANI, M., VRIELYNCK, B. *et al.* (eds) 2000. *Atlas Peri-Tethys, Palaeogeographical Maps*. CCGM/CGMW, Paris.

DEREGNAUCOURT, D. & BOILLOT, G. 1982. Structure géologique du Golfe de Gascogne. *Bulletin BRGM*, **2**, 149–178.

DEREIMS, A. 1898. Recherches géologiques dans le sud de l'Aragón. *Annales de Stratigraphie et Paléontologie du Laboratoire de Géologie de la Facultait de Sciences de Paris. Annales Hébert*, **II**, 1–198.

DESPARMET, R., MONROSE, H. & SCHMITZ, U. 1972. Zur altersstyellum der eruptiv-Gesteine und tuffite in Spanien. *Münsfersche Forschungen zur Geologie und Palaöntologie*, **4**(24), 3–16.

DESTOMBES, J. P. 1950. Le Trias d'Amelie-les-Bains (Pyrenees Orientales*). Bulletin Société Histoire Naturel*, **85**, 251–253.

DEWEY, J. F., PITMAN, W. C., RYAN, W. B. F. & BONNIN, J. 1973. Plate tectonics and the evolution of the Alpine system. *Geological Society of American Bulletin*, **84**, 3137–3180.

DEWEY, J. F., HELMAN, M. L., TURCO, E., HUTTON, D. H. W. & KNOTT, S. D. 1989. Kinematics of western Mediterranean. *In*: COWARD, M. P., DIETRICH, D. & PARK, R. G. (eds) *Alpine Tectonics*. Geological Society, London, Special Publications, **45**, 265–283.

DEYNOUX, M. 1980. Les formations glaciaires du Précambrien terminal et de la fin de l'Ordovicien en Afrique de l'Ouest. Deux exemples de glaciations d'inlandsis sur une plateforme stable. *Travaux du Laboratoire des Sciences de la Terre, Saint Jérome, Marseille (B)*, **17**, 1–554.

DGOH. (Dirección general de Obras Hidráulicas). 1992. *Establecimiento y explotación de redes oficiales de control de aguas subterráneas*. Informe **2745**, Madrid.

DI BATTISTINI, G., TOSCANI, L., IACCARINO, S. & VILLA, I. M. 1987. K/Ar ages and the geological setting of calc-alkaline volcanic rocks from Sierra de Gata, SE Spain. *Neues Jahrbuch für Mineralogie*, **8**, 369–383.

DIAS, R. P. & CABRAL, J. 1989. Neogene and Quaternary Reactivation of the Ponsul Fault in Portugal. *Comunicagôes Servico Geologico de Portugal*, **75**, 3–28.

DÍAZ, N., GIMENO, D., LOSANTOS, M. & SEGURA, C. 1996. Las traquitas de Arenys d'Empordà (Alt Empordà, NE de la Península Ibérica): características generales. *Geogaceta*, **20**, 572–575.

DÍAZ GARCÍA, F. 1986. La Unidad de Agualada. Borde W del Complejo de Órdenes (NW de España). *Trabajos de Geología*, **16**, 3–14.

DÍAZ GARCÍA, F., ARENAS, R., MARTÍNEZ CATALÁN, J. R., GONZÁLEZ DEL TÁNAGO, J. & DUNNING, G. 1999a. Tectonic evolution of the Careón ophiolite (Northwest Spain): a remnant of oceanic lithosphere in the Variscan belt. *Journal of Geology*, **107**, 587–605.

DÍAZ GARCÍA, F., MARTÍNEZ CATALÁN, J. R., ARENAS, R. & GONZÁLEZ CUADRA, P. 1999b. Structural and kinematic analysis of the Corredoiras detachment: evidence for early Variscan syn-convergent extension. *International Journal of Earth Sciences*, **88**, 337–351.

DÍAZ MOLINA, M. & LÓPEZ-MARTÍNEZ, N. 1979. El Terciario continental de la Depresión intermedia (Cuenca). Bioestratigrafía y Paleogeografía. *Estudios Geológicos*, **39**, 149–167.

DÍAZ-RODRIGUEZ, L. A. 1990. *Estimación del potencial de alabastro en el Valle del Ebro*. Unpublished report **11–258**, IGME, Servicio de Documentación, Madrid.

DÍAZ-RODRIGUEZ, L. A. 1991. El alabastro: un enigmatico mineral industrial ornamental. Criterios para su reconocomiento. *Boletín del Museo Arqueológico Nacional*, **9**, 101–112.

DICKEY, J. S. JR. 1970. Partial fusion products in alpine-type peridotites: serrania de la Ronda and other examples. *In*: *Fiftieth Anniversary Symposia, Mineralogy and Petrology of the Upper Mantle*. Mineralogical Society of America, Special Publications, **3**, 33–49.

DIDON, H., DURAND-DELGA, M., ESTERAS, M., FREINBERG, H., MAGNÉ, J. & SUTTER, G. 1984. La Formation des Grès numidiens de l'arc de Gibraltar s'intercale stratigraphiquement entre des argiles oligocènes et de marnes burdigaliennes. *Comptes Rendus de l'Académie des Sciences de Paris*, **299**, 121–128.

DIDON, W. P., DURAND-DELGA, M. & KORNPROBST, J. 1973. Homologies géologiques entre les deux rives du détroit de Gigraltar. *Bulletin de la Société Géologique de la France*, **7**, 77–105.

DÍEZ, C. 1991. La grotte de Valdegoba (Huérmeces, Burgos, Espagne). Un gisement du Paléoolithique moyen avec des restes humains. *L'Anthropologie*, **95**, 329–330.

DÍEZ, J. C. & ROSELL, J. 1998. Estrategias de subsistencia de los Homínidos de la Sierra de Atapuerca. *In*: AGUIRRE, E. (ed.) *Atapuerca y la evolución humana*. Fundación Ramón Areces, Madrid, 361–390.

DÍEZ, J. C., FERNÁNDEZ-JALVO, Y., ROSELL, J. & CÁCERES, I. 1999a. Zooarchaeology and taphonomy of Aurora Stratum (Gran Dolina, Sierra de Atapuerca, Spain). *Journal of Human Evolution*, **37**, 623–652.

DÍEZ, J. C., MORENO, V., RODRÍGUEZ, J., ROSELL, J., CÁCERES, I. & HUGUET, R. 1999b. Estudio arqueológico de los restos de macrovertebrados de la unidad GIII de Galería (Sierra de Atapuerca). *In*: CARBONELL, E., ROSAS, A. & DÍEZ, J. C. (eds) *Atapuerca: ocupaciones humanas y paleoecología del yacimiento de Galería*. Junta de Castilla y León, Arqueología en Castilla y León Memorias, **7**, 265–281.

DÍEZ, J. B., BROUTIN, J., FERRER, J., GISBERT, J. & LIÑÁN, E. 1994. Estudio paleobotánico de los afloramientos triásicos de la localidad de Rodanas (Epila, Zaragoza), Rama Aragonesa de la Cordillera Ibérica. *Cuadernos de Geología Ibérica*, **20**, 205–214.

DÍEZ, J. B., GRAUVOGEL-STAMM, L., BROUTIN, J., FERRER, J., GISBERT, J. & LIÑÁN, E. 1996. Première découverte d'une paléoflore anisienne dans le faciès 'Buntsandstein' de la Branche Aragonaise de la Cordillère Ibérique (Espagne). *Comptes Rendus de l'Académie des Sciences de Paris*, **323**, 341–347.

DÍEZ BALDA, M. A. 1980. La sucesión estratigráfica del Complejo esquisto-grauváquico al Sur de Salamanca. *Estudios Geológicos*, **36**, 131–138.

DÍEZ BALDA, M. A. 1986. El complejo Esquisto-Grauváquico, las series paleozoicas y la estructura hercínica al Sur de Salamanca. *Acta Salmanticensia, Sección de Ciencias*, **52**, 1–162.

DÍEZ BALDA, M. A. & FOURNIER VIÑAS, C. 1981. Hallazgo de acritarcos en el complejo Esquistograuváquico al Sur de Salamanca. *Acta Geológica Hispánica*, **16**, 131–134.

DÍEZ BALDA, M. A. & VEGAS, R. 1992. La estructura del dominio de los pliegues verticales de la Zona Centro Ibérica. *In*: GUTIÉRREZ MARCO, J., SAAVEDRA, J. & RÁBANO, I. (eds) *Paleozoico Inferior de Ibero-América*. University of Extremadura, Badajoz, 523–534.

DÍEZ BALDA, M. A., VEGAS, R. & GONZÁLEZ-LODEIRO, F. 1990. Central-Iberian Zone. Autochtonous sequences. Structure. *In*: DALLMEYER, R. D. & MARTÍNEZ-GARCÍA, E. (eds) *Pre-Mesozoic Geology of Iberia*. Springer-Verlag, Berlin, 172–188.

DÍEZ BALDA, M. A., MARTÍNEZ CATALÁN, J. R. & AYARZA, P. 1995. Syncollisional extensional collapse parallel to the orogenic trend in a domain of steep tectonics: the Salamanca Detachment Zone (Central Iberian Zone, Spain). *Journal of Structural Geology*, **17**, 163–182.

DILLON, W. P. & SOUGY, M. A. 1974. Geology of West Africa and Canary and Cape Verde Islands. *In*: NAIRN, A. E. M. & STEHLI, F. G. (eds) *The Ocean Basins and Margins, Vol. 2: The North Atlantic*. Plenum Press, New York, 315–390.

DINARÉS, J., MCCLELLAND, E. & SANTANACH, P. 1992. Contrasting rotations within thrust sheets and kinematics of thrust-tectonics as derived from palaeomagnetic data: an example from the southern Pyrenees. *In*: MCCLAY, K. (ed.) *Thrust Tectonics*. Chapman & Hall, London, 265–275.

DISSLER, E., DORÉ, F., DUPRET, L., GRESSELIN, F. & LE GALL, J. 1988. L'évolution géodynamique cadomienne du Nord-Est du Massif armoricain. *Bulletin de la Société Géologique de France*, **8**, 810–814.

DOBLAS, M. 1991. Late Hercynian extensional and transcurrent tectonics in Central Iberia. *Tectonophysics*, **191**, 325–334.

DOBLAS M. & OYARZUN, R. 1989a. Mantle core complexes and Neogene extensional detachment tectonics in the western Betic Cordilleras, Spain: An alternative model for the emplacement of the Ronda peridotite. *Earth and Planetary Science Letters*, **93**, 76–84.

DOBLAS, M. & OYARZUN, R. 1989b. Neogene extensional collapse in the western Mediterranean (Betic-Rif alpine orogenic belt): Implications for the genesis of the Gibraltar Arc and magmatic activity. *Geology*, **17**, 430–433.

DOBLAS, M. & OYARZUN, R. 1990. The late Oligocene–Miocene opening of the North Balearic Sea (Valencia basin, western Mediterranean): A working hypothesis involving mantle upwelling and extensional detachment tectonics. *Marine Geology*, **94**, 155–163.

DOBLAS, M, OYARZUN, R., LUNAR, R., MAYOR, N. & MARTÍNEZ FRIAS, J. 1988. Detachment faulting and late-paleozoic epithermal Ag-base metal mineralisations in the Spanish Central System. *Geology*, **16**, 800–803.

DOBLAS, M., LÓPEZ-RUIZ, J., CEBRIÁ, J. M., HOYOS, M. & MARTÍN-ESCORZA, C. 1991. Late Cenozoic indentation/escape tectonics in the eastern Betic Cordilleras and its consequences on the Iberian foreland. *Estudios Geológicos*, **47**, 193–205.

DOBLAS, M., LÓPEZ RUIZ, J., OYARZUN, R. *et al.* 1994*a*. Extensional tectonics in the central Iberian Peninsula during the Variscan to Alpine transition. *Tectonophysics*, **238**, 95–116.

DOBLAS, M., OYARZUN, R., SOPEÑA, A. *et al.* 1994*b*. Variscan–late variscan-early Alpine progressive extensional collapse of central Spain. *Geodinamica Acta*, **7**, 1–14.

DOCHERTY, C. & BANDA, E. 1995. Evidence for the eastward migration of the Alboran Sea based on regional subsidence analysis: A case for basin formation by delamination of the subcrustal lithosphere? *Tectonics*, **14**, 804–818.

DOGLIONI, C., GUEGUEN, E., SÁBAT, F. & FERNÁNDEZ, M. 1997. The western Mediterranean extensional basins and the Alpine orogen. *Terra Nova*, **9**, 109–112.

DOMÍNGUEZ, P. & GUTIÉRREZ, J. C. 1990. Primeros representantes ibéricos del género *Anatifopsis* (Homalozoa, Stylophora) y su posición sistemática. *Acta Salmanticensia*, **68**, 121–131.

DONAIRE, T. 1995. *Petrología y geoquímica de rocas granitoides y enclaves asociados del batolito de Los Pedroches (Macizo Ibérico)*. PhD Thesis, University of Huelva.

DONAIRE, T., SÁEZ, E. & PASCUAL, E. 1998. Evidencias petrográficas de interacción entre el magma félsico y un nivel sedimentario rico en radiolarios en la Faja Pirítica Ibérica. *Geogaceta*, **24**, 111–114.

DONAIRE, T., PASCUAL, E., PIN, C. & DUTHOU, J. L. 1999. Two-stage granitoid-forming event from an isotopically homogeneous crustal source: The Los Pedroches batholith, Iberian Massif, Spain. *Geological Society of American Bulletin*, **111**, 1897–1906.

DONAYRE, F. 1873. Bosquejo de una descripción física y geológica de la provincia de Zaragoza. *Memoria de la Comisión del Mapa Geológico de España*, **1**, 1–125.

DONVILLE, B. 1973*a*. Ages potassium-argon des vulcanites du Bas-Ampurdan (Nord-Est de l'Espagne). *Comptes Rendus de l'Académie des Sciences de Paris*, **276**, 3253–3256.

DONVILLE, B. 1973*b*. Ages potassium-argon des roches volcaniques de la dépression de La Selva (Nord-Est de l'Espagne). *Comptes Rendus de l'Académie des Sciences de Paris*, **277**, 1–4.

DOUBINGER, J., ADLOFF, M. C., RAMOS, A., SOPEÑA, A. & HERNANDO, S. 1977. Primeros estudios palinológicos en el Pérmico y Triásico del Sistema Ibérico y bordes del Sistema Central. *Revista de Palinología*, **1**, 27–33.

DOUBINGER, J., LÓPEZ-GÓMEZ, J. & ARCHE, A. 1990. Pollen and spores from the Permian and Triassic sediments of the southeastern Iberian Ranges, Cueva de Hierro (Cuenca) to Chelva-Manzanera (Valencia-Teruel) region, Spain. *Review of Paleobotany and Palynology*, **66**, 25–45.

DOVAL, M. 1992. Bentonitas. *In*: García-Guinea, J. & Martínez-Frías, J. (eds) *Recursos Minerales de España*. Consejo Superior de Investigaciones Científicas, Madrid, Colección Textos Universitarios, **15**, 45–70.

DOVAL, M., BRELL, M. & GALÁN, E. 1977. El yacimiento de magnesita de Incio, Lugo, España. *Boletín Geológico y Minero*, **88–1**, 50–64.

DOVAL, M., GARCÍA-ROMERO, E., LUQUE, J., MARTÍN VIVALDI, J. L. & RODAS, M. 1991. Arcillas industriales: yacimientos y aplicaciones. *In*: LUNAR, R. & OYARZUN, R. (eds) *Yacimientos Minerales*. Editorial Centro de Estudios Ramón Areces, Madrid (ISBN 84-87191-74-6).

DRISCOLL, E. M., HENDRY, G. L. & TINKLER, K. J. 1965. The geology and geomorphology of Los Ajaches, Lanzarote. *Geological Journal*, **4**, 321–334.

DRONKERT, H. 1977. The evaporites of the Sorbas basin. *Revista Instituto de Investigaciones Geológicas Diputación Provincial-Universidad de Barcelona*, **32**, 55–76.

DROT, J. & MATTE, P. 1967. Sobre la presencia de capas del Devoniano en el límite de Galicia y León (NW de España). *Notas y Comunicaciones del IGME*, **93**, 87–92.

DRUGUET, E. 1997. *The structure of the NE Cap de Creus peninsula. Relationships with metamorphism and magmatism*. PhD Thesis, University of Barcelona.

DRUGUET, E. 2001. Development of high thermal gradients by coeval transpression and magmatism during the Variscan orogeny: insights from the Cap de Creus (Eastern Pyrenees). *Tectonophysics*, **332**, 275–293.

DRUGUET, E., ENRIQUE, P. & GALÁN, G. 1995. Tipologia de granitoides y rocas asociadas del complejo migmatítico de la Punta dels Farallons (Cap de Creus, Pirineo Oriental). *Geogaceta*, **18**, 46–49.

DUARTE, C., MAURICIO, J., PETTIT, P. B., SOUTO, P., TRINKAUS, E., VAN DER PLICHT, H. & ZILHAO, J. 1999. The early Upper Paleolithic human skeleton from the Abrigo do Lagar Velho (Portugal) and modern human emergence in Iberia. *Proceedings, National Academy of Sciences, USA*, **96**, 7604–7609.

DUBAR, G., MOUTERDE, R. & LLOPIS, N. 1963. Première récolte d'une Ammonite de l'Hettangien inférieur dans les calcaires dolomitiques de la région de Aviles (Asturies, Espagne du Nord). *Comptes Rendus de l'Académie des Sciences, Paris*, **257**, 2306–2308.

DUCASSE, L. & VELASQUE, P. C. 1988. *Géotraverse dans la partie occidentale des Pyrénées de l'avant-pays Aquitain au bassin de l'Ebre*. PhD Thesis, University of Aix-Marseille.

DUMAS, B. 1977. *Le Levant espagnol, la gènese du relief*. PhD Thesis, University of Paris XII.

DUNN, A. M., REYNOLDS, P. H., CLARKE, D. B. & UGIDOS, J. M. 1998. A comparison of the age and composition of the Shelburne dyke, Nova Scotia, and the Messejana dyke, Spain. *Canadian Journal of Earth Sciences*, **35**, 1110–1115.

DUPONT, R. 1979. *Cadre géologique et metalogenese des gisements de fer du sud de la province de Badajoz (Sierra Morena occidentale, Espagne)*. PhD Thesis, University of Nancy.

DUPONT, R. & VEGAS, R. 1978. Le Cambrien inferieur du Sud de la province de Badajoz. Distribution des series sedimentaires et volcaniques asocies. *Comptes Rendues de l'Académie des Sciences de Paris*, **286**, 447–450.

DUPONT, R., LINARES, E. & PONS, J. 1981. Premières datations radiométriques par el méthode potassium-argon des granitoides de la Sierra Morena Occidentale (Province de Badajoz, Espagne): conséquencesgéologiques et métallogéniques. *Boletin Geologico y Minero*, **92**, 370–374.

DUPUY, C., DOSTAL, J. & BOIVIN, P. A. 1986. Geochemistry of ultramafic xenoliths and their host alkali basalts from Tallante, southern Spain. *Mineralogical Magazine*, **50**, 231–239.

DURÁN, H. 1990. El Paleozoico de Les Guilleries. *Acta Geológica Hispánica*, **25**(1/2), 83–103.

DURAN, M. 1985. *El Paleozoico de les Guilleries*. PhD Thesis, Autónoma University, Barcelona.

DURAN, M., GIL IBARGUCHI, J. I., JULIVERT, M. & UBACH, S. 1984. Early Paleozoic and volcanism in the Catalonian Coastal Ranges (North Western Mediteranean). *PICG 5, Newsletter*, **6**, 33–43.

DURAND-DELGA, M. 1980. La Méditerranée occidentale: étapes de sa genèse et problèmes structuraux liés à celle-ci. *Mémoire hors série de la Société géologique de France*, **10**, 203–224.

DURAND-DELGA, M. & FOUCAULT, A. 1968. La Dorsale bétique, nouvel élément paléogéographique et structural des Cordillères bétiques au bord Sud de la Sierra Arana (prov. de Grenade, Espagne). *Bulletin de la Société géologique de la France*, **7**, 723–728.

DURAND-DELGA, M., ROSSI P., OLIVIER, P. & PUGLISI, D. 2000. Situation structurale et nature ophiolitique de roches basiques jurassiques associées aux flyschs maghrébins du Rif (Maroc) et de Sicile (Italie). *Comptes Rendus de l'Académie des Sciences de Paris*, **331**, 29–38.

DURÁN, J. J., GRUN, R. & SORIA, J. M. 1988. Edad de las formaciones travertínicas del flanco meridional de la sierra de Mijas (provincia de Málaga, Cordilleras Béticas). *Geogaceta*, **5**, 61–63.

DÜRR, S. 1967. Geologie der Serranía de Ronda und irher sudwestlichen Ausläufer (Andalousien). *Geologica Romana*, **6**, 1–73.

EASTOE, C. J., SOLOMON, M. A. & GARCÍA PALOMERO, F. 1986. Sulphur isotope study of massive and stockwork pyrite deposits at Rio Tinto, Spain. *Transactions of the Institution of Mining and Metallurgy*, **95**, 201–207.

EBINGER, C. 1989. Tectonic development of the western branch of the East African rift system. *Geological Society of America Bulletin*, **101**, 885–903.

ECORS-PYRENEAN TEAM. 1988. The ECORS deep reflection seismic survey across the Pyrenees. *Nature*, **331**, 508–511.

EDEL, J. B. & WEBER, K. 1995. Cadomian terranes, wrench faulting and thrusting in the central Europe Variscides: geophysical and geological evidence. *Geologische Rundschau*, **84**, 412–432.

EDEN, C. P. 1991. *Tectonostratigraphic analysis of the northern extent of the Oceanic Exotic Terrane, Northwestern Huelva Province, Spain*. PhD Thesis, University of Southampton.

EGELER, C. G. 1963. On the tectonics of the eastern Betic Cordilleras (SE Spain). *Geologische Rundschau*, **53**, 260–269.

EGELER, C. G. & SIMON, O. J. 1969. Orogenic evolution of the Betic Zone (Betic Cordilleras, Spain), with emphasis on the nappe structures. *Geologie en Mijnbouw*, **48**, 296–305.

EGUÍLUZ, L. 1988. *Petrogénesis de rocas ígneas y metamórficas en el Anticlinorio Burguillos-Monesterio, Macizo Ibérico Meridional*. PhD Thesis, University of País Vasco.

EGUÍLUZ, L. & ÁBALOS, B. 1992. Tectonic setting of Cadomian low-pressure metamorphism in the central Ossa-Morena Zone (Iberian Massif, SW Spain). *Precambrian Research*, **56**, 113–137.

EGUILUZ, L., CARRACEDO, M. & APALATEGUI, O. 1989. Stock de Santa Olalla de Cala (Zona de Ossa–Morena, España) *Studia Geológica Salmanticensia*, **4**, 145–157.

EGUÍLUZ, L., ABALOS, B. & GIL IBARGUCHI, J. I. 1990. Eclogitas de la Banda de Cizalla Badajoz-Córdoba (Suroeste de España). Datos petrográficos y significado geodinámico. *Geogaceta*, **7**, 28–31.

EGUÍLUZ, L., APRAIZ, A., ÁBALOS, B. & MARTÍNEZ TORRES, L. M. 1995. Évolution de la zone d'Ossa Morena (Espagne) au course du Protérozöique supérieur: corrélations avec l'orogène cadomien nord armoricain: *Géologie de France*, **3**, 35–47.

EGUÍLUZ, L., APRAIZ, A., MARTÍNEZ TORRES, L. M. & PALACIOS, T. 1997. Estructura del sector de Zafra: implicaciones en la subdivisión de unidades cámbricas en la Zona de Ossa-Morena (ZOM). *Geogaceta*, **22**, 59–62.

EGUÍLUZ, L., CASAS, A., BANDRÉZ, A. & PINTO, V. 1999. Alegrete-San Pedro de Mérida thrust: The boundary between the Ossa Morena and Central Iberian Zones. *XV Reunión de Geología del Oeste Peninsular (International Meeting on Cadomian Orogens). Journal of Conference Abstracts*, **4**(3), 1009.

EGUÍLUZ, L., GIL IBARGUCHI, J. I., ÁBALOS, B. & APRAIZ, A. 2000. Superposed Hercynian and Cadomian orogenic cycles in the Ossa-Morena zone and related areas of the Iberian Massif. *Geological Society of America Bulletin*, **112**, 1398–1413.

EICHMÜLLER, K. 1985. Die Valdeteja Fm: Aufbau und Geschichte einer oberkarbonischen Karbonatplattform (Kantabrisches Gebirge, Nordspanien). *Facies*, **13**, 45–155.

EICHMÜLLER, K. & SEIBERT, P. 1984. Faziesentwicklung zwischen Tournai und Westfal D im Kantabrischen Gebirge (NW-Spanien). *Zeitschrift der Deutschen Geologischen Gesellschaft*, **135**, 163–191.

EINAUDI, M. & BURT, D. M. 1982. Introduction, terminology, classification and composition of skarn deposits. *In*: EINAUDI, M. & BURT, D. (eds) *Special Issue Devoted to Skarn Deposits, Economic Geology* and *Bulletin of the Society of Economic Geologists*, **77**, 745–754.

EINAUDI, M., MEINERT, L & NEWBERRY, R. 1981. Skarn deposits. *Economic Geology*, **75**, 317–391.

EL-HMIDI, H. 2000. *Petrología y geoquímica de los sistemas andesíticos ricos en Mg: estudio petrológico y experimental de las noritas de la Banda Metamórfica de Aracena, SO de España*. PhD Thesis, University of Huelva.

EL-KHAYAL, A. A. & ROMANO, M. 1985. Lower Ordovician trilobites from the Hanadir Shale of Saudi Arabia. *Palaeontology*, **28**, 401–412.

ELIAS, J. 1934. Inclinació i inflexions dels estrats paleozoics i triàsica dels voltants de Terrassa. *Boletín del Instituto Catalán de Historia Natural*, **34**, 53–56.

ELICKI, O. 1997. Biostratigraphic data of the German Cambrian – present state of knowledge. *Freiberger Forschungsheft*, **C466**, 155–165.

ELÍZAGA, E. 1980. Los sedimentos terrígenos del Cretácico Medio del Sur de la Meseta y el Norte del Prebético Externo. Hipótesis sedimentológicas. *Boletín Geológico y Minero*, **XCI-V**, 619–638.

ELIZAGA, E. & CALVO, J. P. 1988. Evolución sedimentaria de las cuencas lacustres neógenas de la zona prebética (Albacete, España). Relación, posición y efectos del vulcanismo durante la evolución. Interés minero. *Boletín Geológico y Minero*, **99**, 837–846.

ELMI, S., GOY, A., MOUTERDE, R., RIVAS, P. & ROCHA, R. 1989. Correlaciones biostratigráficas en el Toarciense de la Península Ibérica. *Cuadernos de Geología Ibérica*, **13**, 265–277.

EMIG, C. C. & GUTIÉRREZ-MARCO, J. C. 1997. Signification des niveaux à lingulidés à la limite supérieure du Grès Armoricain (Ordovicien, Arenig, Sud-Ouest de l'Europe). *Geobios*, **30**, 481–495.

ENADIMSA. 1986. *La mineria Andaluza*. Junta de Andalucía, Consejería de Economía y Fomento, Dirección General de Industria y Minas, Libro Blanco.

ENGESER, T. & SCHWENTKE, W. 1986. Towards a new concept of the tectonogenesis of the Pyrenees. *Tectonophysics*, **129**, 233–242.

ENGESER, T., REITNER, J., SCHWENTKE, W. & WIEDMANN, J. 1984. Die kretazisch-alttertiäre Tektogenese des Basko-Kantabrischen Beckens (Nordspanien). *Zeitschrift der deutschen geologischen Gesellschaft*, **135**, 243–268.

ENRIQUE, P. 1984. The hercynian post-tectonic plutonic and hypabyssal rocks of the Montnegre Massif, Catalonian Coastal Ranges (NE Spain). *In*: SASSI, F. P. & JULIVERT, M. (eds) *IGCP no. 5, Newsletter*, **3**, 45–55.

ENRIQUE, P. 1985. *La asociación plutónica tardi-Herciniana del Macizo del Montnegre, Catalánides Septentrionales (Barcelona)*. PhD Thesis, University of Barcelona.

ENRIQUE, P. 1989. Caracterización geoquímica mediante elementos mayores de los granitoides de la vertiente meridional del Pirineo Central. *Studia Geológica Salmanticensia*, **4**, 41–60.

ENRIQUE, P. 1990. The hercynian intrusive rocks of the Catalonian Coastal Ranges (NE Spain). *Acta Geològica Hispànica*, **25**, 39–64.

ENRIQUE, P. 1995. Una nueva interpretación sobre las relaciones entre el emplazamiento de granitoides y el metamorfismo regional hercinianos en el Pirineo Oriental. *Geogaceta*, **18**, 203–206.

ENRIQUE, P & DEBON, F. 1987. Le pluton calcoalcalin de Montnègre (Chaînes Côtières Catalanes, Espagne): étude isotopique Rb-Sr et comparaison avec les granites hercyniens des Pyrénées, Sardaigne et Corse. *Comptes Rendues de l'Académie des Sciences de Paris*, **305**, 1157–1162.

ENRIQUE, P., ANDREICHEV, & SOLÉ, J. 1999. Rb-Sr dating of a multi-intrusive plutonic sequence from the Catalan Coastal Batholith. EUG (European Union of Geosciences), **10**, Strasbourg, 809.

ERDTMANN, B.-D. 1998. Neoproterozoic to Ordovician/Silurian Baltica and Laurentia interaction with (Proto-) Gondwana: Critical review of macro- and microplate transfer models. *Acta Universitatis Carolinae, Geologica*, **42**, 409–418.

ERDTMANN, B.-D., MALETZ, J. & GUTIÉRREZ-MARCO, J. C. 1987. The new Early Ordovician (Hunneberg Stage) graptolite genus *Paradelograptus* (Kinnegraptidae), its phylogeny and biostratigraphy. *Paläontologische Zeitschrift*, **61**, 109–131.

ERRENST, CH. 1990. Das korallenführende Kimmeridgium der nordwestlichen Iberischen ketten und agrenzender Gebiete (Fazies, Pallaogeographie und Beschreibung der korallenfauna). *Palaeontographica Abt. A*, **214**, 121–207.

ESCUDER, J., INDARES, A. & ARENAS, R. 1996. Termobarometría núcleo-borde en granates con zonado difusional: El ejemplo del Domo Gneísico del Tormes, NO de Salamanca. *Geogaceta*, **20**, 617–620.

ESCUDER, J., HERNÁIZ, P. P., VALVERDE-VAQUERO, P., RODRÍGUEZ FENÁNDEZ, R. & DUNNING, G. 1998. Variscan syncollisional extension in the Iberian Massif: structural metamorphic and geochronolgical evidence from the Somosierra sector of the Sierra de Guadarrama (Central Iberian Zone, Spain). *Tectonophysics*, **290**, 87–109.

ESCUDER VIRUETE, J. 1998. Relationships between structural units in the Tormes Gneiss Dome (NW Iberian Massif, Spain): geometry, structure and kinematics of contractional and extensional Variscan deformation. *Geologische Rundschau*, **87**, 165–179.

ESCUDER VIRUETE, J. 2000. P-T paths derived from garnet growth zoning in an extensional setting: an example from the Tormes gneiss dome (Iberian Massif, Spain). *Journal of Petrology*, **41**, 1489–1515.

ESCUDER VIRUETE, J., ARENAS, R. & MARTÍNEZ CATALÁN, J. R. 1994. Tectonothermal evolution associated with Variscan crustal extension in the Tormes Gneissic Dome (NW Salamanca Iberian Massif, Spain). *Tectonophysics*, **238**, 117–138

ESCUDER VIRUETE, J., VILLAR, P., RODRÍGUEZ FERNÁNDEZ L. R., MONTESERÍN, V. & SANTIESTEBAN, J. I. 1995. Evolucion tectonotérmica del área metamórfica del SO de Salamanca (Zona Centro-Ibérica, O de España). *Boletín Geológico y Minero*, **106**, 303–315.

ESCUDER VIRUETE, J., INDARES, A. & ARENAS, R. 1997. P-T path determinations in the Tormes Gneissic Dome, NW Iberian Massif, Spain. *Journal of Metamorphic Geology*, **15**, 645–663.

ESCUDER VIRUETE, J., HERNÁIZ HUERTA, P. P., VALVERDE VAQUERO, P., RODRÍGUEZ FERNÁNDEZ, R. & DUNNING, G. 1998. Variscan syncollisional extension in the Iberian massif: structural, metamorphic and geochronological evidence from the Somosierra sector of the Sierra de Guadarrama (Central Iberian Zone, Spain). *Tectonophysics*, **242**, 56–82.

ESCUDER VIRUETE, J., INDARES, A. & ARENAS, R. 2000. P-T paths derived from garnet growth zoning in an extensional setting: an example from the Tormes Gneissic Dome (Iberian Massif, Spain). *Journal of Petrology*, **41**, 1488–1518.

ESPEJO, R. 1985. The ages and soils of two levels of Raña sufaces in Central Spain. *Geoderma*, **35**, 223–239.

ESTEBAN, M. 1979–80. Significance of the Upper Miocene reefs of the Western Mediterranean. *Palaeogeography, Palaeoclimatology, Palaeoecology*, **29**, 169–188.

ESTEBAN, M. 1996. An overview of Miocene reefs from Mediterranean areas: General trends and facies models. *In*: FRANSEEN, E. K., ESTEBAN, M., WARD, W. C. & ROUCHY, J.-M. (eds) *Models for Carbonate Stratigraphy from Miocene Reef Complexes of Mediterranean Regions*. SEPM, Tulsa, Concepts in Sedimentology and Paleontology, **5**, 3–53.

ESTEBAN, M. & JULIÁ, R. 1973. Discordancias erosivas intrajurásicas en las Catalánides. *Acta Geológica Hispánica*, **8**, 153–157.

ESTEBAN, M., POMAR, L., MARZO, M. & ANADÓN, P. 1977. Naturaleza del contacto entre el Muschelkalk inferior y el Muschelkalk medio en la zona de Aiguafreda. *Cuadernos de Geología Ibérica*, **4**, 201–210.

ESTEBAN, M., BRAGA, J. C., MARTÍN, J. M. & SANTISTEBAN, C. 1996. Western Mediterranean reef complexes. *In*: FRANSEEN, E. K., ESTEBAN, M., WARD, W. C & ROUCHY, J. M. (eds) *Models for Carbonate Stratigraphy from Miocene Reef Complexes of Mediterranean Regions*. Concepts in Sedimentology and Paleontology Series, **5**, SEPM, Tulsa, 55–72.

ESTÉVEZ, A. & SANZ DE GALDEANO, C. 1983. Néotectonique du secteur central des Chaînes Bétiques (Bassin du Guadix-Baza et de Grenade). *Revue de Géologie Dynamique et Géographie Physique*, **21**, 23–34.

ESU, D. 1984. Gasteropodi dei bacini continentali Terziari Eocenico-Oligocenici dell'isola di Maiorca (Baleari). *Thalassia Salentina*, **14**, 85–99.

EVANS, D. H. 2000. A cephalopod fauna from the Middle Ordovician of Saudi Arabia. *Palaeontology*, **43**, 573–589.

EVE. 1988. *Mapa Geológico del País Vasco. 64–II, San Sebastian, E: 1/25.000*. Ente Vasco de la Energía (ed.), Bilbao.

EVE. 1991. *Mapa Geológico del País Vasco. 64–III, Villabona, E: 1/25.000*. Ente Vasco de la Energía (ed.), Bilbao.

FALGUÈRES, C., BAHAIN, J.-J., YOKOYAMA, Y. *et al*. 1999. Earliest humans in Europe: the age of TD6 Gran Dolina, Atapuerca, Spain. *Journal of Human Evolution*, **37**, 343–352.

FALLOT, P. 1922. *Étude geologique de la Sierra de Majorque*. PhD Thesis, University of Paris.

FALLOT, P. 1943. *El sistema Cretácico en las Cordilleras Béticas*. Memorias del Instituto Lucas Mallada, CSIC, Madrid.

FALLOT, P., FAURE-MURET, A, FONTBOTÉ, J. M. & SOLÉ SABARIS, L.

1960. Estudios sobre las series de Sierra Nevada y de la llamada Mischungzone. *Boletín Geológico y Minero*, **71**, 345–557.

FÄRBER, A. & JARITZ, W. 1964. Die Geologie des westasturichen Kustengebietes zwischen San Esteban de Pravia und Ribadeo (NW Spanien). *Geologische Jahrbuch*, **81**, 679–738.

FARIAS, P. 1982. La estructura del sector central de Picos de Europa. *Trabajos de Geología*, **12**, 63–72.

FARIAS, P. 1989. *La Geología de la región del sinforme de Verín (Cordillera Hercinica, NW de España)*. PhD Thesis, University of Oviedo (*Serie Nova Terra*, 1990, **2**, 1–201).

FARIAS, P. 1992. El Paleozoico Inferior de la Zona de Galicia-Tràs-os-Montes (Cordillera Hercinica, NW de España). *In*: GUTIÉRREZ MARCO, J., SAAVEDRA, J. & RÁBANO, I. (eds) *Paleozoico Inferior de Ibero-América*. University of Extremadura, Badajoz, 495–504.

FARIAS, P., GALLASTEGUI, G., GONZÁLEZ LODEIRO, F. *et al*. 1987. Aportaciones al conocimiento de la litoestratigrafía y estructura de la Galicia Central. *Memórias da Faculdade de Ciências, Universidade do Porto*, **1**, 411–431.

FARIAS ARQUER, P. 1992. El Paleozoico Inferior de la Zona de Galicia-Tras os Montes (Cordillera Hercinica, NW de España). *In*: GUTIÉRREZ-MARCO, J. C., SAAVEDRA, J. & RÁBANO, I. (eds) *Paleozoico Inferior de Ibero-América*, University of Extremadura, Badajoz, 495–504.

FATKA, O., KRAFT, J., KRAFT, P., MERGL, M., MIKULAS, R. & STORCH, P. 1995. Ordovician of the Prague Basin: stratigraphy and development. *In*: COOPER, C., DROSER, M. L. & FINNEY, S. (eds) *Ordovician Odyssey*. Society for Sedimentary Geology, Pacific Section, Fullerton, **77**, 171–176 (ISBN 1-878861-70-0).

FAURÉ, P. 1984. Le Lias de la partie centro-oriental des Pyrénées espagnoles (provinces de Huesca, Lérida et Barcelona). *Bulletin de la Société d'Histoire Naturelle de Toulouse*, **121**, 23–37.

FEDONKIN, M., LIÑÁN, E. & PEREJÓN, A. 1985. Icnofósiles de las rocas precámbrico-cámbricas de la Sierra de Córdoba, España. *Boletín de la Real Sociedad Española de Historia Natural (Geología)*, **81** (year 1983), 5–14.

FENOLL, P., DELGADO, F., FONTBOTÉ, L. *et al*. 1987. *Los yacimientos de fluorita, plomo, cinc y bario del sector central de la Cordillera Bética*. University of Granada.

FÉRAUD, G. 1981. *Datations des réseaux de dykes et de roches volcaniques sous-marines par les méthodes K-Ar et 40Ar-39Ar. Utilisation des dykes comme marqueurs de Paleocontraintes*. PhD. Thesis, University of Nice.

FÉRAUD, G., GIANNERINI, G., CAMPREDON, R. & STILLMAN, C. J. 1985. Geochronology of some Canarian dyke swarms: contribution to the volcano-tectonic evolution of the archipelago. *Journal of Volcanology and Geothermal Research*, **25**, 29–52.

FERNÁNDEZ, A., CHAUVEL, J. J. & MORO, M. C. 1998. Comparative study of the Lower Ordovician ironstones of the Iberian Massif (Zamora, Spain) and of the Armorican Massif (Central Brittany, France). *Journal of Sedimentary Research*, **68**, 53–62.

FERNÁNDEZ, C. & CASTRO, A. 1999. Pluton accommodation at high strain rates in the upper continental crust. The example of the Central Extremadura batholith, Spain. *Journal of Structural Geology*, **21**, 1143–1149.

FERNÁNDEZ, C., CASILLAS, R., AHIJADO, A., PERELLO, V. & HERNÁNDEZ-PACHECO, A. 1997a. Shear zones as result of intraplate tectonics in oceanic crust: an example of the Basal Complex of Fuerteventura (Canary Islands). *Journal of Structural Geology*, **19**, 41–57.

FERNÁNDEZ, C., CASTRO, A., DE LA ROSA, J. D. & MORENO-VENTAS, I. 1997b. Rheological aspects of magma transport inferred from rock structures. *In*: BOUCHEZ, J. L., STEPHENS, W. E. & HUTTON, D. H. W. (eds) *Granite: from Segregation of Melt to Emplacement Fabrics*. Kluwer Academic Publishers, Dordrecht, 75–94.

FERNÁNDEZ, J. 1984. Capas rojas triásicas del borde sureste de la Meseta, síntesis estratigráfica y sedimentológica. *Mediterránea, Serie de Estudios Geológicos*, **3**, 89–105.

FERNÁNDEZ, J. & DABRIO, C. J. 1985. Fluvial architecture of the Buntsandstein facies red beds in the Middle to Upper Triassic (Ladinian-Norian) of the southeastern edge of the Iberian Meseta (Southern Spain). *In*: MADER, D. (ed.) *Aspects of Fluvial Sedimentation in the Lower Triassic Buntsandstein of Europe*.

Lecture Notes in Earth Sciences, **4**, Springer-Verlag, Berlin, 411–435.

FERNÁNDEZ, J. & GIL, A. 1989. Interpretación sedimentaria de los materiales triásicos de facies Buntsandstein en las zonas Externas de las Cordilleras Béticas, y en la Cobertera Tabular de la Meseta. *Revista Sociedad Geológica España*, **2**, 113–124.

FERNÁNDEZ, J., VISERAS, C. & SORIA, J. M. 1996. Pliocene–Pleistocene continental infilling of the Granada and Guadix basins (Betic Cordillera, Spain): the influence of allocyclic and autocyclic processes on the resultant stratigraphic organization. *In*: FRIEND, P. F. & DABRIO, C. J. (eds) *Tertiary Basins of Spain: the Stratigraphic Record of Crustal Kinematics*. Cambridge University Press, Cambridge, 366–371.

FERNÁNDEZ, L. P. 1990. *Estratigrafía, Sedimentología y Paleogeografía de la Región de Riosa, Quirós y Teverga-San Emiliano*. PhD Thesis, University of Oviedo.

FERNÁNDEZ, L. P. 1993. La Formación San Emiliano (Carbonífero de la Zona Cantábrica, no. de España): Estratigrafía y extensión lateral. Algunas implicaciones paleogeográficas. *Trabajos de Geología*, **19**, 97–122.

FERNÁNDEZ, L. P. 1995. *El Carbonífero*. *In*: Aramburu, C. & Bastida, F. (eds) *Geología de Asturias*. Trea S. L., Gijón, 63–80.

FERNÁNDEZ, L. P., ÁGUEDA, J. A., COLMENERO, J. R., SALVADOR, C. I. & BARBA, P. 1988. A coal-bearing fan-delta complex in the Westfalian D of the Central Coal Basin, Cantabrian Mountains, northwestern Spain: implications for the recognition of humid-type fan deltas, *In*: NEMEC, W. & STEEL, R. J. (eds) *Fan Deltas: Sedimentology and Tectonic Settings*. Blackie & Son, London, 286–302.

FERNÁNDEZ, L. P., FERNÁNDEZ-MARTINEZ, E., MENDEZ-BEDIA, I., RODRIGUEZ, S. & SOJO, F. 1995. Devonian and Carboniferous reefal facies from the Cantabrian Zone (NW Spain). Guide Book of the Field Trip A: VII International Symposium on Fossil Cnidaria and Porifera. *Gráficas Barona*, 1–76.

FERNÁNDEZ, L. P., FERNÁNDEZ-MARTINEZ, E., MENDEZ-BEDIA, I., RODRIGUEZ, S. & SOJO, F. 1997. A sequential approach to the study of reefal facies in the Candas and Portilla Formations (Middle Devonian) of the Cantabrian Zone (NW Spain). *Boletin Real Sociedad Española Historia Natural (Sección Geológica)*, **92**(1–4), 23–33.

FERNÁNDEZ, M., MARZÁN, I., CORREIA, A. & RAMALHO, E. 1998. Heat flow, heat production and lithospheric thermal regime in the Iberian Peninsula. *Tectonophysics*, **291**, 29–53.

FERNÁNDEZ, P. 1988. Evolución cuaternaria y sistemas de terrazas en la subfosa terciaria de Valverde del Majano y el Macizo de Sta Mª. Real de Nieva (Segovia). *Boletín de la Real Sociedad Española de Historia Natural (Geol.)*, **84**, 69–83.

FERNÁNDEZ, P. & GARZÓN, G. 1994. Ajustes en la red de drenaje y morfoestructura en los ríoos del centro-sur de la cuenca del Duero. *In*: ARNÁEZ, J., GARCÍA RUIZ, J. M. & GÓMEZ VILLAR, A. (eds) *Geomorfología de España*. Sociedad Española de Geomorfología, Logroño, 471–484.

FERNÁNDEZ CASALS, M. J. & GUTIÉRREZ-MARCO, J. C. 1985. Aspectos estratigráficos de la Cadena Hercínica en el Sistema Central. *Revista de la Real Academia de Ciencias Exactas, Físicas y Naturales de Madrid*, **79**, 487–509.

FERNÁNDEZ FERNÁNDEZ, A., GARCÍA DEL CURA, M. A., GONZÁLEZ MARTÍN, J. A. & ORDOÑEZ, S. 2000. Morfogénesis y sedimentación carbonática pleistocena en el valle del Júcar (Albacete). *Geotemas*, **1**, 353–357.

FERNÁNDEZ-FERNÁNDEZ, E., CAMPOS, J. & GONZÁLEZ-LODEIRO, F. 1992. Estructuras extensionales en los materiales Alpujárrides al E de Málaga (Sierra Tejeda, Cordilleras Béticas). *Geogaceta*, **12**, 13–15.

FERNÁNDEZ GARCÍA, P., MAS, R., RODAS, M., LUQUE DEL VILLAR, F. J. & GARZÓN, M. G. 1989. Los depósitos aluviales del Paleógeno basal en el sector suroriental de la cuenca del Duero (provincia de Segovia): evolución y minerales de la arcilla característicos. *Estudios Geológicos*, **45**, 27–44.

FERNANDEZ JALVO, Y. & ANDREWS, P. 1992. Small Mammal Taphonomy of Gran Dolina, Atapuerca (Burgos), Spain. *Journal of Archaeological Science*, **19**, 407–428.

FERNÁNDEZ-JALVO, Y., DÍEZ, J. C., CÁCERES, I. & ROSELL, J. 1999. Human cannibalism in the Early Pleistocene of Europe (Gran Dolina, Sierra de Atapuerca, Burgos, Spain). *Journal of Human Evolution*, **37**, 591–622.

FERNÁNDEZ-LÓPEZ, S. 1985. *El Bajociense en la Cordillera Ibérica*. PhD Thesis, Complutense University, Madrid.

FERNÁNDEZ-LÓPEZ, S. 1997. Ammonites, ciclos tafonómicos, y ciclos estratigráficos en plataformas epicontinentales carbonáticas. *Revista Española de Paleontología*, **12**, 151–174.

FERNÁNDEZ-LÓPEZ, S. & GÓMEZ, J. J. 1990a. Evolution tectonosedimentaire et genese des associations d'Ammonites dans le secteur central du bassin Iberique (Espagne) pendant l'Aalenien. *Cahiers de la Université Catholique de Lyon, Sér. Sciences*, **4**, 39–52.

FERNÁNDEZ-LÓPEZ, S. & GÓMEZ, J. J. 1990b. Facies aalenienses y bajocienses, con evidencias de emersión y carstificación, en el sector central de la Cuenca Ibérica. *Cuadernos de Geología Ibérica*, **14**, 67–111.

FERNÁNDEZ-LÓPEZ, S. & MELÉNDEZ, G. 1994. Abrasion surfaces on internal moulds of ammonites as palaeobathymetric indicators. *Palaeogeography, Palaeoclimatology, Palaeoecology*, **110**, 29–42.

FERNÁNDEZ-LÓPEZ, S. & MELÉNDEZ, G. 1995. Taphonomic gradients in Middle Jurassic ammonites of the Iberian Range (Spain). *Geobios*, **18**, 155–165.

FERNÁNDEZ-LÓPEZ, S. & SUÁREZ-VEGA, L. C. 1979. Estudio bioestratigráfico del Aaleniense y Bajociense en Asturias. *Estudios Geológicos*, **35**, 231–239.

FERNÁNDEZ-LÓPEZ, S., MELÉNDEZ, G. & SUÁREZ-VEGA, L. C. 1978. El Dogger y Malm en Moscardón (Teruel). *Grupo Español del Mesozoico, Guía de Excursiones al Jurásico de la Cordillera Ibérica*, **VI**, 1–20.

FERNÁNDEZ-LÓPEZ, S., MELÉNDEZ, G. & SEQUEIROS, L. 1985. Le Dogger et Malm de la Sierra Palomera (Teruel). *Strata*, **2**, 142–153.

FERNÁNDEZ-LÓPEZ, S., GÓMEZ, A. & URETA, M. S. 1988a. Características de la plataforma carbonatada del Dogger en el sector meridional de la Sierra de la Demanda (Soria). *III Coloquio de Estratigrafía y Paleogeografía del Jurásico de España, libro guía de las excursiones, Ciencias de la Tierra (Instituto de Estudios Riojanos)*, **11**, 119–142.

FERNÁNDEZ-LÓPEZ, S., GOY, A. & URETA, M. S. 1988b. El Toarciense superior, Aaleniense y Bajociense en Camino (Santander). Precisiones bioestratigráficas. *III Coloquio de Estratigrafía y Paleogeografía del Jurásico de España, libro guía de las excursiones, Ciencias de la Tierra (Instituto de Estudios Riojanos)*, **11**, 47–62.

FERNÁNDEZ-LÓPEZ, S., AURELL, M., GARCÍA-JORAL, F. *et al.* 1996. El Dogger de los catalánides: unidades litoestratigráficas y elementos paleogeográficos. *Revista Española de Paleontología*, número extraordinario, 122–139.

FERNÁNDEZ-LÓPEZ, S., AURELL, M., GARCÍA-JORAL, F. *et al.* 1998a. La plataforma de Tortosa (Cuenca Catalana) durante el Jurásico Medio: unidades litoestratigráficas, paleogeografía y ciclos ambientales. *Cuadernos de Geología Ibérica*, **24**, 185–221.

FERNÁNDEZ-LÓPEZ, S., GARCÍA-JORAL, F., GÓMEZ, J. J., HENRIQUES, M. H. P. & MARTÍNEZ, G. 1998b. La diferenciación paleogeográfica de la Cuenca Catalana al principio del Jurásico medio. *Revista de la Sociedad Geológica de España*, **11**, 3–22.

FERNÁNDEZ-MARTÍN, M. L., SÁEZ, R. & PASCUAL, E. 1996. Piroxenos relictos: una clave para la interpretación del vulcanismo en la Faja Pirítica Ibérica. *Geogaceta*, **20**, 568–571.

FERNÁNDEZ MARTINEZ, J., FANO ARDANAZ, H. & OJENBARRENA SAN MARTÍN, L. 1992. La mineralización de Pb-Zn de mina Troya (Guipuzcoa). *In*: GARCÍA GUINEA, J. & MARTÍNEZ-FRÍAS, J. (eds) *Recursos Minerales de España*. CSIC, Colección Textos Universitarios, **15**, 985–999.

FERNÁNDEZ-MENDIOLA, P. A. & GARCÍA-MONDÉJAR, J. 1995. Volcanoclastic sediments in the early Albian Mañaria carbonate platform (northern Spain). *Cretaceous Research*, **16**, 451–463.

FERNÁNDEZ-MENDIOLA, P. A., GÓMEZ-PÉREZ, I. & GARCÍA-MONDÉJAR, J. 1993. Aptian-Albian carbonate platforms: central Basque-Cantabrian Basin, northern Spain. *In*: SIMO, J. A., SCOTT, R. W. & MASSE, J. P. (eds) *Cretaceous Carbonate Platforms*. AAPG Memoir, **56**, 315–324.

FERNÁNDEZ REMOLAR, D. C. 1999. Las calizas fosforíticas del Ovetiense Inferior de la Sierra de Córdoba, España. *Boletín de la*

Real Sociedad Española de Historia Natural (Geología), **95**, 15–45.

FERNÁNDEZ-SOLER, J. M. 1996. *El volcanismo calco-alcalino en el Parque Natural de Cabo de Gata-Níjar (Almería). Estudio volcanológico y petrológico.* Sociedad Almeriense de Historia Natural, Almería.

FERNÁNDEZ-SUÁREZ, J. 1994. *Petrología de los granitos peralumínicos y metamorfismo de la banda Boal-Los Ancares.* PhD Thesis, University of Oviedo.

FERNÁNDEZ SUÁREZ, J., GUTIÉRREZ ALONSO, G., JENNER, J. A. & JACKSON, S. E. 1998. Geochronology and geochemistry of the Pola de Allande granitoids (northern Spain): their bearing on the Cadomian-Avalonian evolution of northwest Iberia. *Canadian Journal of Earth Sciences*, **35**, 1439–1453.

FERNÁNDEZ SUÁREZ, J., GUTIÉRREZ ALONSO, G., JENNER, G. A. & TUBRETT, M. N. 1999. Crustal sources in lower Paleozoic rocks from NW Iberia: insights from laser ablation U-Pb ages of detrital zircons. *Journal of the Geological Society, London*, **156**, 1065–1068.

FERNÁNDEZ SUÁREZ, J., DUNNING, G., JENNER, G. A. & GUTIÉRREZ ALONSO, G. 2000a. Variscan collisional magmatism and deformation in NW Iberia: constraints from U-Pb geochronology of granitoids. *Journal of the Geological Society London*, **157**, 565–576.

FERNÁNDEZ SUÁREZ, J., GUTIÉRREZ ALONSO, G., JENNER, G. A. & TUBRETT, M. N. 2000b. New ideas on the Proterozoic–Early Paleozoic evolution of NW Iberia: insights from U-Pb detrital zircon ages. *Precambrian Research*, **102**, 185–206.

FERNÁNDEZ-VIEJO, G., GALLART, J., PULGAR, J. A., CÓRDOBA, D. & DAÑOBEITIA, J. J. 1998. Crustal transition between continental and oceanic domains along the North Iberian margin from wide angle seismic and gravity data. *Geophysical Research Letters*, **25**(23), 4249–4252.

FERNÁNDEZ-VIEJO, G., GALLART, J., PULGAR, J. A., CÓRDOBA, D. & DAÑOBEITIA, J. J. 2000. Seismic signature of Variscan and Alpine tectonics in NW Iberia: Crustal structure of the Cantabrian Mountains and Duero basin. *Journal of Geophysical Research*, **105**(B2), 3001–3018.

FERNEX, F. 1968. *Tectonique et paléogéographie du Bétique et du pénibétique orientaux. Transversale de la Paca-Lorca-Aguilas (Cordillères Bétiques, Espagne Méridionale).* PhD Thesis, University of Paris.

FERRARA, G., LAURENZI, M. A., TAYLOR, H. P., TONARINI, S. & TURI, B. 1985. Oxygen and strontium isotope studies of K-rich volcanic rocks from the Alban Hills, Italy. *Earth and Planetary Science Letters*, **75**, 13–28.

FERREIRA, E. 1983. Rochas granitoides hercínicas post-tectónicas del area de Satao. *Memorias Nota Publicacoes Museu Geología*, **96**, 39–74.

FERREIRA, N., IGLESIAS, M., NORONHA, F., PEREIRA, E., RIBEIRO, A. & RIBEIRO, M. L. 1987. Granitoides da zona Centro–Ibérica e seu enquadramento geodinâmico. *In*: BEA, F., CARNICERO, A., GONZALO, J. C., LÓPEZ PLAZA, M. & RODRÍGUEZ ALONSO, M. D. (eds) *Geología de los granitoides y rocas asociadas del Macizo Hespérico. Libro Homenaje a L. C. García Figuerola.* Rueda, Madrid, 37–53.

FERRER, E., MAGRANS, J. & MAÑÉ, R. 1992. Euriptèrids (merostomats) del Devonià inferior de Bruguers (Gava) i Santa Creu d'Olorda (Sant Feliu de Llobregat). *I Trobada d'Estudiosos de Garraf. Monografies Diputació de Barcelona*, **19**, 33–38.

FERRER, J. 1971. El Paleoceno y Eoceno del borde suroriental de la Depresión del Ebro (Cataluña). *Memoires Suisses de Paleontologie*, **90**, 1–70.

FERRER, J., ROSELL, J. & REGUANT, S. 1968. Síntesis litoestratigráfica del Paleógeno del borde oriental de la depresión del Ebro. *Acta Geológica Hispánica*, **3**(3), 54–56.

FERRERA, R., MASERTI, B. E., ANDERSON, M., EDNER, H., RAGNARSON, P. & SVANBERG, H. 1997. Mercury degassing rate from mineralized areas in the Mediterranean basin. *Water, Air and Soil Pollution*, **93**, 53–66.

FERRERO, A., RUIZ, J. E. & TOYOS, J. M. 1992. *Estudio de concentración de andalucitas en O Rosal.* Unpublished report **11–327**, IGME, Madrid.

FERRERO, A., ROEL, J. & TOYOS J. M. 1993. *Investigación de andalucita en el área de Verín (Orense).* Unpublished report **11–340**, IGME, Madrid.

FERRÉS, M. 1998. *Le complexe granitique alcalin du massif du Cadiretes (Chaînes Côtières Catalanes, NE de l'Espagne): étude pétrologique et géocronologie 40Ar/39Ar et Rb-Sr.* PhD Thesis, University of Genève (*Terre & Environnement*, **13**).

FERRÉS, M. & ENRIQUE, P. 1994. Granitos de afinidad alcalina en el batolito calcoalcalino de las Cadenas Costeras Catalanas (NE de la Península Ibérica). *Boletin de la Sociedad Española de Mineralogía*, **17-1**, 75–76.

FERRÉS, M. & ENRIQUE, P. 1996. El complejo leucogranítico tardihercínico de afinidad alcalina de Tossa de Mar (Cadenas Costeras catalanas, NE de España). *Geogaceta*, **20**(3), 601–604.

FERRES, M., ENRIQUE, P., DELALOYE, M. & SINGER, B. S. 1997. Magmatic and thermal history of the central Catalan Coastal Batholith (NE Spain): New Constraints from 40Ar/39Ar incremental-heating studies. *Terra Nova*, **9**, 503.

FEUILLÉE, P. 1983. The Flysch Noir. *In*: *Vue sur le Crétacé bascocantabrique et nord-ibérique. Une marge et son arrière-pays, ses environnements sédimentaires.* Mémoires Géologiques de la Université de Dijon, **9**, 79–81.

FEUILLÉE, P. & RAT, P. 1971. Structures et paléogéographies Pyrénéo-Cantabriques. *In*: DEBYSER, J., LE PICHON, X. & MONTARDET, L. (eds) *Histoire Structurale du Golfe de Gascogne.* Publications de l'Institute Français du Pétrole, Collection Colloques et Séminaires, **22**, Technip, Paris, 1–48.

FEZER, R. 1988. Die oberjurassische karbonatische Regressionsfazies im südwestlichen Keltiberikum zwischen Griegos und Aras de Alpuente (Prov. Teruel, Cuenca, Valencia, Spanien). *Arb. Institut fur Geologie und Paläontologie, Univ. Stuttgart*, **84**, 1–119.

FIEGE, K. 1936. Stratonomische Beobachtungen in der Grauwackenfazies des Harzer Kulms. *In*: *Geburtstag von Hans Stille, Festschrift zum* **60**, 44–64.

FILMER, P. E. & McNUTT, M. K. 1988. Geoid anomalies over the Canary Islands group. *Marine Geophysical Research*, **11**, 77–87.

FITZGERALD, P. G., MUÑOZ, J. A., CONEY, P. J. & BALDWIN, S. L. 1999. Asymmetric exhumation across the central Pyrenees: implications for the tectonic evolution of a collisional orogen. *Earth and Planetary Science Letters*, **173**, 157–170.

FLOQUET, M. 1991. *La plate-forme nord-castillane au Crétacé supérieur (Espagne). Arrière-pays ibérique de la marge passive bascocantabrique. Sédimentation et Vie.* Mémoires Géologiques de la Université de Dijon, **14**.

FLOQUET, M. 1998. Outcrop cycle stratigraphy of shallow ramp deposits, the Late Cretaceous Series on the Castilian ramp (northern Spain). *In*: GRACIANSKY, P. C. DE, HARDENBOL, J., JACQUIN, T. & VAIL, P. R. (eds) *Mesozoic and Cenozoic Sequence Stratigraphy of European Basins.* SEPM, Tulsa, Special Publications, **60**, 343–361.

FLOQUET, M. & MELÉNDEZ, A. 1982. Características sedimentarias y paleogeográficas de la regresión finicretácica en el sector central de la Cordillera Ibérica. *Cuadernos de Geología Ibérica*, **8**, 237–257.

FLOQUET, M. & RAT, P. 1975. Un exemple de interrelation entre socle, paléogéographie et structure dans l'arc pyrénéen basque: la Sierra de Aralar. *Revue de Géographie physique et de Géologie dynamique*, **17**, 497–512.

FLOQUET, M., ALONSO, A. & MELÉNDEZ, A. 1982. El Cretácico superior de la Región Cameros-Castilla. *In*: *El Cretácico de España.* Complutense University, Madrid, 387–456.

FOLEY, S. F., VENTURELLI, G., GREEN, D. H. & TOSCANI, L. 1987. The ultrapotassic rocks: characteristics, classification, and constraints for petrogenetic models. *Earth Science Reviews*, **24**, 81–134.

FOMBELLA, M. A. 1978. Acritarcos de la Formación Oville, edad Cámbrico Medio-Tremadoc, Provincia de León, España. *Palinología*, núm. extra., 245–261.

FOMBELLA BLANCO, M. A. 1984. Age palynologique du Blastomilonitic Grabben, zone occidentale de la Galice. *Revue de Micropaléontologie*, **27**, 113–117.

FONOLLÁ, R., TALENS, J., COY, A., MELÉNDEZ, F. & ROBLES, F. 1972. *Mapa Geológico de España a E. 1:50.000 no. 665 (Mira).* 2ª Serie-MAGNA. IGME, Madrid.

FONSECA, P. 1996. Domínios meridionais da zona de Ossa-Morena e limites com a zona Sul Portuguesa: metamorfismo de alta pressão relacionado com a Sutura Varisca Ibérica. *In*: ARAÚJO, A. A. & PEREIRA, M. F. (eds) *Estudo sobre a Geología da Zona de Ossa-Morena (Maciço Iberico)*. Livro de Homenagem ao Profesor Francisco Gonçalves. University of Evora, 133–168.

FONSECA, P. & RIBEIRO, A. 1993. Tectonics of the Beja-Acebuches ophiolite: a major suture in the Iberian Variscan Foldbelt. *Geologische Rundschau*, **82**, 440–447.

FONTANA, B., GALLEGO, M. R., JURADO, M. J. & MELÉNDEZ, G. 1994*a*. A correlation of subsurface and surface data of the middle-upper Jurassic between the Ebro basin and the central Iberian chain. *Geobios, Special Memoir*, **17**, 563–574.

FONTANA, B., GALLEGO, M. R., MELÉNDEZ, G., AURELL, M. & BÁDENAS, B. 1994*b*. Las calizas con esponjas del Bajociense de la Cordillera Vasco Cantábrica-Oriental, Navarra. *Geogaceta*, **15**, 30–33.

FONTEILLES, M. 1970. Géologie des terrains métamorphiques et granitiques du massif hercynien de l'Agly (Pyrénées orientales). *Bulletin du BRGM*, **3**, 21–72.

FORGHANI, A. H. 1965. Sur la structure annulaire du massif éruptive de Bordères (Hautes Pyrénées). *Comptes Rendues de l'Académie des Sciences de Paris*, **260**, 6943–6945.

FORNÓS, J. J. 1987. *Les plataformes carbonatades de les Balears*. PhD Thesis, University of Barcelona.

FORNOS, J. J. (ed.) 1998. *Aspectes Geològics de les Balears (Mallorca, Menorca i Ibiza)*. Univesity of the Balearic Isles.

FORNÓS, J. J. 1999. Karst collapse phenomena in the Upper Miocene of Mallorca (Balearic Islands, Western Mediterranean). *Acta Geologica Hungarica*, **42**, 237–250.

FORNÓS, J. J. & POMAR, L. 1984. Facies, ambientes y secuencias de plataforma carbonatada somera (Formación Calizas de Santanyí) en el Mioceno terminal de Mallorca (Islas Baleares). *Publicaciones de Geología, U. A. B.* (Universitat Autònoma de Barcelona), **20**, 319–338.

FORNOS, J. J., RODRÍGUEZ-PEREA, A. & SABAT, F. 1988. Shelf facies of the Middle-Upper Jurassic, Arta Caves ('Serres de Llevant', Mallorca, Spain). *Congreso Geológico de España, Granada, Comunicaciones*, **1**, 75–78.

FORTEY, R. A. & MORRIS, S. F. 1982. The Ordovician trilobite *Neseuretus* from Saudi Arabia, and the palaeogeography of the *Neseuretus* fauna related to Gondwanaland in the Earlier Ordovician. *Bulletin of the British Museum (Natural History), Geology series*, **36**, 63–75.

FORTEY, R. A., HARPER, D. A. T., INGHAM, J. K., OWEN, A. W. & RUSHTON, A. W. A. 1995. A revision of Ordovician series and stages from the historical type area. *Geological Magazine*, **132**, 15–30.

FORTEY, R. A., HARPER, D. A. T., INGHAM, J. K., OWEN, A. W., PARKES, M. A., RUSHTON, A. W. A. & WOODCOCK, N. H. 2000. *A Revised Correlation of Ordovician Rocks in the British Isles*. Geological Society, London, Special Reports, **24**.

FORTUIN, A. R., 1984. Late Ordovician glaciomarine deposits (Orea Shale) in the Sierra de Albarracin, Spain. *Palaeogeography, Palaeoclimatology, Palaeoecology*, **48**, 245–261.

FOUCAULT, A. 1971. *Etude géologique des environs des sources du Guadalquivir (provinces de Jaen et de Grenade, Espagne méridionales)*. PhD Thesis, University of Paris.

FOUCAULT, A. & PAQUET, J. 1970. L'estructure de l'ouest de la Sierra Harana (prov. de Grenade, Espagne). *Comptes Rendus de l'Académie des Sciences de Paris*, **272**, 2756–2758.

FOURCADE, E. 1970. *Le Jurassique et le Crátacé aux confins des Chaînes Bétiques et Ibériques (Sud-Est de l'Espagne)*. PhD Thesis, University of Paris.

FOURCADE, E. & GARCÍA, A. 1982. El Albiense superior y el Cenomaniense con Foraminíferos bentónicos del Sur de la Cordillera Ibérica (provincias de Cuenca y Valencia). *Cuadernos de Geología Ibérica*, **8**, 369–389.

FOURCADE, E., AZÉMA, J., CHABRIER, G., CHAUVE, P., FOUCAULT, A. & RANGHEARD, Y. 1977. Liaisons paléogéographiques au Mésozoique entre les Zones Externes Bétiques, Baléares, Corso-Sardes et Alpines. *Revue de Géographie Physique et Géologie Dynamique (2)*, **XIX**, 377–388.

FOURCADE, E., CHAUVE, P. & CHABRIER, G. 1982. Stratigraphie et tectonique de l'île d'Ibiza, témoin du prolongement de la nappe subbétique aux Baleares (Espagne). *Eclogae geologicae Helvetiae*, **75**(2), 415–436.

FOURCADE, S. & ALLÈGRE, C. J. 1981. Trace element behaviour in granite genesis. The calc-alkaline plutonic association from the Quérigut Complex (Pyrenees, France). *Contributions to Mineralogy Petrology*, **76**, 177–195.

FOURCADE, S. & JAVOY, M. 1991. Sr-Nd-O isotopic features of mafic microgranular enclaves and host granitoids from the Pyrenees, France: Evidence for their hybrid nature and inference on their origin. *In*: DIDIER, J. & BARBARIN, B. (eds) *Enclaves and Granite Petrology*. Elsevier, Amsterdam, 345–364.

FOURCADE, S. PEUCAT, J., MARTINEAU, F., CUESTA, A., CORRETGÉ, L. G. & GIL IBARGUCHI J. I. 1991. Análisis de isótopos de oxígeno y edad Rb-Sr del plutón zonado de Caldas de Reyes (Galicia, España). *Geogaceta*, **6**, 7–9.

FRANCO, M. P. & GARCÍA DE FIGUEROLA, L. C. 1986. Las rocas básicas y ultrabásicas en el extremo occidental de la Sierra de Avila (provincias de Avila y Salamanca). *Studia Geológica Salmanticensia*, **XXIII**, 193–219.

FRANKE, W. & ENGEL, W. 1986. Synorogenic sedimentation in the Variscan Belt of Europe. *Bulletin de la Société Géologique de France*, **28**, 25–33.

FRANKE, W. & ONCKEN, O. 1995. Zur pre-devonischen Geschichte des Rhenohercynischen Beckens. *Nova Acta Leopoldina, Neue Folge*, **71**, 53–72.

FRANKE, W., DALLMEYER, D. & WEBER, K. 1995. Geodynamic evolution. *In*: DALLMEYER, R. D., FRANKE, W. & WEBER, K. (eds) *Pre-Permian Geology of Central and Western Europe*. Springer-Verlag, Berlin, 579–593.

FRANKENFELD, H. 1982. Das Ende der Devonischen Riff-Fazies im nordspanischen Variszikum. *Neues Jahrbuch für Geologie und Paläontologie Abhandlungen*, **163**, 238–241.

FRANKENFELD, H. 1983. El manto del Montó-Arauz: interpretación estructural de la Región del Pisuerga-Carrión (Zona Cantábrica, España). *Trabajos de Geología*, **13**, 37–47.

FREEMAN, L. G. & GONZÁLEZ-ECHEGARAY, J. 1970. Aurignacian structural features and burials at Cueva Morín (Santander, Spain). *Nature*, **26**, 722–726.

FREGENAL, M. A. & MELÉNDEZ, N. 1993. Sedimentología y evolución paleogeográfica de la Cubeta de Las Hoyas (Cretácico inferior, Serranía de Cuenca). *Cuadernos de Geología Ibérica*, **17**, 231–256.

FREGENAL MARTÍNEZ, M. A. 1998. *Análisis de la cubeta sedimentaria de Las Hoyas y su entorno paleogeográfico (Cretácico Inferior, Serranía de Cuenca). Sedimentología y aspectos tafonómicos del yacimiento de Las Hoyas*. PhD Thesis, Complutense University, Madrid.

FREGENAL MARTÍNEZ, M. A. & MELÉNDEZ, N. 2000. The lacustrine fossiliferous deposits of the Las Hoyas subbasin (Lower Cretaceous, Serranía de Cuenca, Iberian Ranges, Spain). *In*: GIERLOWSKI-KORDESCH, E. H. & KELTS, K. (eds) *Lake Basins through Space and Time*. AAPG, Studies on Geology, **46**, 303–314.

FREUNDT, A. & SCHMINCKE, H. U. 1992. Mixing of rhyolite, trachyte and basalt magma erupted from a vertically and laterally zoned reservoir, composite flow P1, Gran Canaria. *Contributions to Mineralogy and Petrology*, **112**, 1–19.

FRÈYCHENGES, M. & PEYBERNES, B. 1991. Association de Foraminiferes benthiques dans le Trias carbonaté des Pyrenees Espagnoles. *Acta Geologica Hispánica*, **26**, 67–73.

FRICKE, W. 1941. *Die Geologie des Grenzgebietes Zwischen Nordöstlicher Sierra Morena und Extremadura*. PhD Thesis, University of Berlin.

FRIEND, P. F. 1996. Tertiary stages and ages, and some distinctive stratigraphic approaches. *In*: FRIEND, P. & DABRIO, C. (eds) *Tertiary Basins of Spain*. World and Regional Geology 6, Cambridge University Press, Cambridge, 3–5.

FRIZON DE LAMOTTE, D., ANDRIEUX, J. & GUÉZOU, J. C. 1991. Cinématique des chevauchements néogènes dans l'Arc bético-rifain: discussion sur les modèles géodynamiques. *Bulletin de la Société Géologique de la France*, **162**, 611–626.

FRYDA, J. & GUTIÉRREZ-MARCO, J. C. 1996. An unusual new Sinuitid mollusc (Bellerophontoidea, Gastropoda) from the Ordovician of Spain. *Journal of Paleontology*, **70**, 602–609.

FRYDA, J., ROHR, D. M., ROBARDET, M. & GUTIÉRREZ-MARCO, J. C. 2001. New microdomatid genus (Gastropoda) from Ashgill limestones of Seville (Ossa Morena Zone, Spain) with a revision of Ordovician Microdomatidea. *Alcheringa*, **25**, 117–127.

FUENTE, P. & SÁENZ DE SANTA MARÍA, J. A. 1999. Tectónica y microtectónica de la Cuenca Carbonífera Central de Asturias. *Trabajos de Geología*, **21**, 121–140.

FUMANAL, M. P. 1986. *Sedimentología y clima en el País Valenciano. Las Cuevas habitadas en el Cuaternario reciente.* Servicio de Investigación Prehistórica, Diputación Provincial de Valencia, Trabajos Varios, **83**.

FUNK, T. & SCHMINCKE, H. U. 1998. Growth and destruction of Gran Canaria deduced from seismic reflection and bathymetric data. *Journal of Geophysical Research*, **103**, 15393–15407.

FUSTÉ, M. 1953. Parietal neandertalense de Cova Negra (Játtiva). *Trabajos Varios del S.I.P.*, **37**, 1–32.

FUSTÉ, M. 1958. Endokranialer Ausguss des Neandertaler Parietale von Cova Negra. *Anthropologisches Anzeigen*, **21**, 268–273.

FÚSTER, J. M. 1975. Las Islas Canarias: un ejemplo de evolución espacial y temporal del vulcanismo oceánico. *Estudios Geológicos*, **31**, 439–463.

FÚSTER, J. M., AGUILAR, M. & GARCÍA, A. 1965. Las sucesiones volcánicas en la zona del Pozo de los Frailes dentro del vulcanismo cenozoico del Cabo de Gata (Almería). *Estudios Geológicos*, **21**, 199–222.

FÚSTER, J. M., GASTESI, P., SAGREDO, J. & FERMOSO, M. L. 1967. Las rocas lamproíticas del sureste de España. *Estudios Geológicos*, **22**, 35–69.

FÚSTER, J. M., ARAÑA, V., BRÄNDLE, J. L., NAVARRO, J. M., ALONSO, V. & APARICIO, A. 1968a. *Geología y volcanología de las Islas Canarias: Tenerife.* Instituto Lucas Mallada (CSIC), Madrid.

FÚSTER, J. M., CENDRERO, A., GASTESI, P., IBARROLA, E .& LÓPEZ RUIZ, J. L. 1968b. *Geología y volcanología de las Islas Canarias. Fuerteventura.* Instituto Lucas Mallada (CSIC), Madrid.

FÚSTER, J. M., FERNÁNDEZ SANTÍN, S. & SAGREDO, J. 1968c. *Geología y volcanología de las Islas Canarias: Lanzarote.* Instituto Lucas Mallada (CSIC), Madrid.

FÚSTER, J. M., HERNÁNDEZ PACHECO, A., MUÑOZ, M., RODRÍGUEZ BADIOLA, E. & GARCÍA CACHO, L. 1968d. *Geología y volcanología de las Islas Canarias: Gran Canaria.* Instituto Lucas Mallada (CSIC), Madrid.

FÚSTER, J. M., MUÑOZ, M., SAGREDO, J., YEBENES, A., BRAVO, T. & HERNÁNDEZ-PACHECO, A. 1980. Excursión no. 121 A + c del 26° I. G. C. a las Islas Canarias. *Boletín del IGME*, **XCII-II**, 351–390.

FÚSTER, J. M., HERNÁN, F., CENDRERO, A., COELLO, J., CANTAGREL, J. M., ANCOCHEA, E. & IBARROLA, E. 1993. Geocronología de la isla de El Hierro (Islas Canarias *Boletín de La Real Sociedad Española de Historia Natural (Sec. Geol.)*, **88**(1–4), 85–97.

FÚSTER, J. M., IBARROLA, E., SNELLING, N. J., CANTAGREL, J. M., HUERTAS, M. J., COELLO, J. & ANCOCHEA, E. 1994. Cronología K-Ar de la Formación Cañadas en el sector suroeste de Tenerife: implicaciones de los episodios piroclásticos en la evolución volcánica. *Boletín de La Real Sociedad Española de Historia Natural (Sección Geología)*, **89**, 25–41.

GABALDÓN, V. 1989. *Plataformas siliciclásticas externas: facies y su distribución areal. (Plataformas dominadas por tormentas).* PhD Thesis, Autònoma University, Barcelona.

GABALDÓN, V. & QUESADA, C. 1986. Examples de bassins houillers limniques du sud-ouest de la péninsule Ibérique: évolution sédimentaire et contrôle structural. *Memoire de la Société géologique de France*, **149**, 27–36.

GABALDÓN, V., GARROTE, A. & QUESADA, C. 1983. Geología del Carbonífero inferior del Norte de la Zona de Ossa Morena. Introducción a la excursión. 5ª Reunión Grupo Ossa Morena. *Temas Geológico-Mineros, IGME, Madrid*, **7**, 101–137.

GAERTNER H. R. VON. 1930. Obersilurische Faunen aus des spanischen Pyrenäen. *Nachrichten von der Gesellschaft der Wissenschaften zu Göttingen, Mathematisch-Physicalische Klasse*, **4**, 179–188.

GAGEL, C. 1908. Die caldera von La Palma. *Zeitschr Deutschen Gesellschaft Erdk.*, Berlin, 168–186, 222–250.

GAGEL, C. 1925. Begleitworte zu der Karte von La Gomera mit einen Anhang uber die Calderafrage. *Z. Deutsch Geologische Gesellschaft A. Abhandlungen*, **77**, 551–575.

GALÁN, E. 1992. Palygorskita y Sepiolita. *In*: GARCÍA-GUINEA, J. & MARTÍNEZ-FRÍAS, J. (eds) *Recursos Minerales de España*. Consejo Superior de Investigaciones Científicas. Colección Textos Universitarios, **15**, 71–94.

GALÁN, G. 1984. *Las rocas graníticas del sector norte del Macizo de Vivero (Lugo, no. de España)*. PhD Thesis, University of Oviedo.

GALÁN, G., PIN, C. & DUTHON, J. 1996. Sr–Nd isotopic record of multistage interactions between mantle-derived magmas and crustal components in a collision contex – the ultramafic- granitoids association from Vivero (Hercynian belt, NW Spain). *Chemical Geology*, **131**, 67–91.

GALDEANO, A., MOREAU, M. G., POZZI, J. P., BRETHOU, P. Y. & MALOD, J. A. 1989. New paleomagnetic results from Cretaceous sediments near Lisboa (Portugal) and implications for the rotation of Iberia. *Earth and Planetary Sciences Letters*, **92**, 95–106.

GALERA-FERNÁNDEZ, J. M. 1987. *Estudio del Devoniano del Pirineo Central español*. PhD Thesis, Polytechnic University, Madrid.

GALINDO, C. & PORTUGAL FERREIRA, M. R. 1988. *Analisis radiometricos (K/Ar) del Macizo de Almendral (Zona de Ossa-Morena, Badajoz)*. Reunión del oeste Peninsular, Bragança.

GALINDO, C., PORTUGAL FERREIRA, M., CASQUET, C. & PRIEM, H. 1990. Dataciones Rb/Sr en el complejo plutónico Taliga-Barcarrota (CPTB) (Badajoz). *Geogaceta*, **8**, 7–10.

GALINDO-ZALDÍVAR, J. 1986. Etapas de fallamiento neógenas en la mitad occidental de la depresión de Ugíjar (Cordilleras Béticas). *Estudios Geológicos*, **42**, 1, 1–10.

GALINDO-ZALDÍVAR, J., GONZÁLEZ-LODEIRO, F. & JABALOY, A. 1989. Progressive extensional shear structures in a detachment contact in the Western Sierra Nevada (Betic Cordilleras, Spain). *Geodinamica Acta*, **3**, 73–85.

GALINDO-ZALDÍVAR, J., JABALOY, A., GONZÁLEZ-LODEIRO, F. & ALDAYA, F. 1997. Crustal structure of the central sector of the Betic Cordillera (SE Spain). *Tectonics*, **16**, 18–37.

GALINDO-ZALDÍVAR, J., RUANO, P., JABALOY, A. & LÓPEZ-CHICANO, M. 2000. Kinematics of faults between Subbetic Units during the Miocene (central sector of the Betic Cordillera). *Comptes Rendus de l'Académie des sciences de Paris*, **331**, 811–816.

GALLARDO, J., PÉREZ-GONZÁLEZ, A. & BENAYAS, J. 1987. Paleosuelos de los piedemonte villafranquienses de las terrazas pleistocenas de las región del valle del Henares-Alto Jarama. *Boletín Geológico y Minero*, **98**, 27–39.

GALLARDO-MILLÁN, J. L. & PÉREZ-GONZÁLEZ, A. 2000. Magnetoestratigrafía del relleno neógeno en las cuencas del Campo de Calatrava (Ciudad Real). *Geotemas*, **1**, 101–104.

GALLART, J., FERNÁNDEZ-VIEJO, G., DIAZ, J., VIDAL, N. & PULGAR, J. A. 1995. Deep structure of the transition between the Cantabrian Mountains and the north Iberian margin from wide-angle ESCI-N data. *Revista de la Sociedad Geológica de España*, **8**, 365–382.

GALLART, J., PULGAR, J. A., PEDREIRA, D., GALLASTEGUI, J., DÍAZ, J. & CARBONELL, R. 1999. Wide-angle seismic measurements from the Pyrenees to the Cantabrian Mountains and the North Iberian margin: evidence for the lateral extent of an Alpine crustal thickening. *Communications of the Dublin Institute for Advanced Studies, Geophysical Bulletin*, **49**, 77–80.

GALLART, J., DÍAZ, J., NERCESSIAN, A., MAUFFRET, A. & DOS REIS, T. (in press). The eastern end of the Pyrenees: Seismic features at the transition to the NW Mediterranean. *Geophysical Research Letters*.

GALLASTEGUI, G. 1993. *Petrología del Macizo Granodiorítico de Bayo-Vigo*. PhD Thesis, University of Oviedo.

GALLASTEGUI, G., HEREDIA, N., RODRÍGUEZ FERNÁNDEZ, L. R. & CUESTA, A. 1990. El stock de Peña Prieta en el contexto del magmatismo de la Unidad del Pisuerga-Carrión (Zona Cantábrica, N de España). *Cuadernos del Laboratorio Geológico de Laxe*, **15**, 203–217.

GALLASTEGUI, G., ARAMBURU, C., BARBA, P., FERNÁNDEZ, L. P. & CUESTA, A. 1992. Vulcanismo del Paleozoico Inferior en la Zona Cantábrica (NO de España). *In*: GUTIÉRREZ-MARCO, J. C., SAAVEDRA, J. & RÁBANO, I. (eds) *Paleozoico Inferior de Ibero-América*. University of Extremadura, 435–452.

GALLASTEGUI, J., PULGAR, J. A. & ÁLVAREZ-MARRÓN, J. 1997. 2D seismic modeling of the Variscan foreland thrust and fold belt crust in NW Spain from ESCIN-1 deep seismic reflection data. *Tectonophysics*, **269**, 21–32.

GALLEGO, M. R. & MELÉNDEZ, G. 1997. Síntesis estratigráfica y paleogeográfica del Jurásico en el sector meridional de la cuenca Vasco-Cantábrica (provincia de Álava, N. de España). *IV Congreso de Jurásico de España, libro de comunicaciones*, 73–76.

GALLEGO, M. R., AURELL, M., BÁDENAS, B., FONTANA, B. & MELÉNDEZ, G. 1994. Origen de las brechas de la base del Jurásico de Leitza (Cordillera Vasco Cantábrica-Oriental, Navarra). *Geogaceta*, **15**, 26–29.

GALLEMÍ, J. MARTINEZ, R & PONS, J. M. 1983. Coniacian-Maastrichtian of the Tremp area (south Central Pyrenees). *Newsletter Stratigraphy*, **12**(1), 1–17.

GÁMEZ, J. A., FERNÁNDEZ-NIETO, C., GOZALO, R., LIÑÁN, E., MANDADO, J. & PALACIOS, T. 1991. Bioestratigrafía y evolución ambiental del Cámbrico de Borobia (Provincia de Soria, Cadena Ibérica Oriental). *Cuadernos do Laboratorio Xeolóxico de Laxe*, **16**, 251–271.

GÁMEZ-VINTANED, J. A. & LIÑÁN, E. 1996. Significant ichnologial data during the Neoproterozoic-early Cambrian transition in Iberia Series. *In*: LIÑÁN, E., GÁMEZ VINTANED, J. A. & GOZALO, R. (eds) *II Field Conference of the Cambrian Stage Subdivision Working Groups. International Subcommission on Cambrian Stratigraphy*. University of Zaragoza, 101–102.

GAND, G., KERP, H., PARSONS, C. & MARTÍNEZ-GARCÍA, E. 1997. Palaeoenvironmental and stratigraphic aspects of animal traces and plant remains in spanish Permian red beds (Peña Sagra, Cantabrian Mountains, Spain). *Geobios*, **30**, **2**, 295–318.

GANDL, J. 1972. Die Acastavinae und Asteropyginae (Trilobita) Keltiberiens (NE-Spanien). *Abhandlungen der Senckenbergischen Naturforschenden Gesellschaft*, **530**, 1–184.

GARCÉS, M., KRIJGSMAN, W., VAN DAM, J., CALVO, J. P., ALCALÁ, L. & ALONSO-ZARZA, A. M. 1997. Late Miocene alluvial sediments from the Teruel area: Magnetostratigraphy, magnetic susceptibility, and facies organization. *Acta Geológica Hispánica*, **32**(3–4), 171–184.

GARCÍA, A. (Coord.). 1982. *El Cretácico de España*. Complutense University, Madrid.

GARCÍA, A., CARENAS, B., PEREZ, P., SEGURA, M. & CALONGE, A. 1989a. Les cycles sédimentaires dans les faciès de plate-forme nord-Téthysienne de la Chaîne Ibérique centrale de l'Albien supérieur au le Cénomanien moyen dans. *Geobios, memoire spéciale*, **11**, 151–160.

GARCÍA, A., SEGURA, M., CALONGE, A. & CARENAS, B. 1989b. Unidades estratigráficas para la organización de la sucesión sedimentaria de la Plataforma del Albiense-Cenomaniense de la Cordillera Ibérica. *Revista de la Sociedad Geológica de España*, **2**(3–4), 303–333.

GARCÍA, A., SEGURA, M., GARCÍA-HIDALGO, J. F. & CARENAS, B. 1993. Mixed siliciclastic and carbonate platform of Albian-Cenomanian age from the Iberian basin (Spain). *In*: SIMO, J. A. T., SCOTT, B. W. & MASSE, J. P. (eds) *Cretaceous Carbonate Platforms*. AAPG, Memoirs, **56**, 255–269.

GARCÍA, A., SEGURA, M. & GARCÍA-HIDALGO, J. F. 1996. Cycles and hiatuses in the upper Albian–Cenomanian of the Iberian Ranges (Spain): a Cyclostratigraphic approach. *Sedimentary Geology*, **103**, 175–200.

GARCÍA-AGUILAR, J. M. & MARTÍN, J. M. 2000. Late Neogene to Recent continental history and evolution of the Guadix-Baza basin (SE Spain). *Revista Sociedad Geológica de España*, **13**, 65–77.

GARCÍA-ALCALDE, J. L. 1992. El Devónico de Santa María del Mar (Castrillón, Asturias, España). *Revista Española Paleontología*, **7**(1), 53–79.

GARCÍA-ALCALDE, J. L. 1995. L'évolution paléogéographique prévarisque du la Zone Cantabrique septentrionale (Espagne). *Revista Española Paleontología*, **10**(1), 9–29.

GARCÍA-ALCALDE, J. L. 1996. El Devónico Astur-Leonés en la Zona Cantábrica (N de España). *Revista Española Paleontología*, Madrid, extra volume, 58–71.

GARCÍA-ALCALDE, J. L. 1997. North Gondwanan Emsian events. *Episodes*, **20**(4), 241–246.

GARCÍA-ALCALDE, J. L. 1998. Devonian events in northern Spain. *Newsletter Stratigraphy*, **36**(2/39), 157–175.

GARCÍA-ALCALDE, J. L., MONTESINOS, J. R., TRUYOLS-MASSONI, M. *et al.* 1988. El Silúrico y el Devónico del Dominio Palentino (NO de España). *Revista Sociedad Geológica España*, **1**(1–2), 7–13.

GARCÍA-ALCALDE, J. L., MONTESINOS, J. R., TRUYOLS-MASSONI, M., GARCÍA-LÓPEZ, S., ARBIZU, M. A. & SOTO, F. 1990. The Palentine Domain (Palentian Zone). *In*: DALLMEYER, R. D. & MARTÍNEZ GARCÍA, E. (eds) *Pre-Mesozoic Geology of Iberia*. Springer-Verlag, Berlin, 20–23.

GARCÍA-ANTÓN, M. 1995. Paleovegetación del Pleistoceno Medio de Atapuerca a través del análisis polínico. *In*: BERMÚDEZ DE CASTRO, J. M., ARSUAGA, J. L. & CARBONELL, E. (eds) *Evolución humana en Europa y los yacimientos de la Sierra de Atapuerca, I*. Junta de Castilla y León, 147–165.

GARCÍA CACHO, L., DÍEZ-GIL, J. L & ARAÑA, V. 1994. A large volcanic debris avalanche in the Pliocene Roque Nublo stratovolcano, Gran Canaria, Canary Islands. *Journal of Volcanology and Geothermal Research*, **63**, 217–229.

GARCÍA CASCO, A. 1993. *Evolución metamórfica del complejo gneísico de Torrox y series adyacentes (Alpujárrides Centrales)*. PhD Thesis, University of Granada.

GARCÍA CASCO, A. & TORRES ROLDÁN, R. L. 1996. Disequilibrium induced by fast decompression in St-Bt-Grt-Ky-Sill-And metapelites from the Betic Belt (Southern Spain). *Journal of Petrology*, **37**, 1207–1239.

GARCÍA CASQUERO, J. L., BOELRIJK, N. A. I. M., CHACÓN, J. & PRIEM, H. N. A. 1985. Rb-Sr evidence for the presence of Ordovician granites in the deformed basement of the Badajoz-Córdoba Belt, SW Spain. *Geologische Rundschau*, **74**, 379–384.

GARCÍA DE CORTÁZAR, A. & PUJALTE, V. 1982: Litoestratigrafía y facies del Grupo Cabuérniga (Malm-Valanginiense inferior) al S. de Cantabria y NE. de Palencia. *Cuadernos de Geología Ibérica*, **8**, 5–21.

GARCÍA DE FIGUEROLA, F. L., CORRETGÉ, L. G. & BEA, F. 1974. El dique de Alentejo-Plasencia y haces de diques básicos de Extremadura. (Estudio comparativo). *Boletín Geológico y Minero*, **337**, 308–337.

GARCÍA DE FIGUEROLA, L. C. & MARTÍNEZ GARCÍA, E. 1972. El Cámbrico Inferior de la Rinconada (Salamanca, España Central). *Studia Geologica*, **3**, 33–41.

GARCÍA DEL CURA, M. A. 1979. *Las sales sódicas, calcosódicas y magnésicas de la Cuenca del Tajo*. Serie Universitaria, **109**, Fundación Juan March, Madrid.

GARCÍA DEL CURA, M. A., DABRIO, J. C. & ORDÓÑEZ, S. 1996. Mineral resources of the Tertiary deposits of Spain. *In*: FRIEND, P. F. & DABRÍO, C. J. (eds) *Tertiary Basins of Spain: the Stratigraphic Record of Crustal Kinematics*. Cambridge University Press, Cambridge, 26–40.

GARCÍA DEL CURA, M. A., PEDLEY, H. M., ORDOÑEZ, S. & GONZÁLEZ MARTÍN, J. A. 2000. Petrology of a barrage tufa system (Pleistocene to recent) in the Ruidera Lakes Natural Park (Central Spain). *Geotemas*, **1**, 359–363.

GARCÍA-DUEÑAS, V. 1967a. La Zona Subbética al norte de Granada. *Notas y Comunicaciones del IGME*, **101–102**, 73–100. PhD Thesis, University of Granada.

GARCIA DUEÑAS, V. 1969. Les Unités allochtones de la Zone Subbetique, dans la transversale de Grenade (Cordillères Bétiques, Espagne). *Revue de Géographie Physique et de Géologie Dynamique*, **11**, 211–222.

GARCÍA DUEÑAS, V. & BALANYÁ, J. C. 1986. Estructura y naturaleza del Arco de Gibraltar. *Malleo Boletim Informativo da Sociedade Geologica de Portugal*, **2**, 23.

GARCÍA-DUEÑAS, V. & MARTÍNEZ-MARTÍNEZ, J. M. 1988. Sobre el adelgazamiento mioceno del Dominio Cortical de Alborán, el Despegue Extensional de Filabres (Béticas orientales). *Geogaceta*, **5**, 53–55.

GARCÍA-DUEÑAS, V., MARTÍNEZ-MARTÍNEZ, J. M., OROZCO, M. & SOTO, J. 1988. Plis-nappes, cisaillements syn- à post-métamorphiques et cisaillements ductiles-fragiles en distension dans les Nevado-Filabrides (Cordillères bétiques, Espagne). *Comptes Rendus de l'Académie des Sciences de Paris*, **307**, 1389–1395.

GARCÍA-DUEÑAS, V., BALANYÁ, J. C. & MARTÍNEZ-MARTÍNEZ, J. M.

1992. Miocene extensional detachments in the outcropping basement of the Northern Alboran Basin (Betics) and their tectonic implications. *Geo-Marine Letters*, **12**, 88–95.

GARCÍA-DUEÑAS, V., BANDA E., TORNÉ M., CÓRDOBA D. & ESCI-BÉTICAS WORKING GROUP. 1994. A deep seismic reflection survey across the Betic Chain (southern Spain): first results. *Tectonophysics*, **232**, 77–89.

GARCÍA-ESPINA, R. 1997. *La estructura y evolución tectonoestratigráfica del borde occidental de la Cuenca Vasco-Cantábrica (Cordillera Cantábrica, NO de España)*. PhD Thesis, University of Oviedo.

GARCÍA GARMILLA, F. 1987. *Las formaciones terrígenas del Wealdense y del Aptiense inferior en los anticlinorios de Bilbao y Ventoso (Vizcaya, Cantabria): Estratigrafía y Sedimentación*. PhD Thesis, University of País Vasco.

GARCÍA GARZÓN, L., DE PABLO MACIÁ, J. G. & DE LLAMAS, J. 1981. Edades absolutas obtenidas mediante el método Rb/Sr en dos cuerpos de ortogheises en Galicia occidental. *Boletín Geológico y Minero*, **92–94**, 463–466.

GARCÍA-GIL, S. 1991. Las unidades litoestratigráficas del Muschelkalk en el NW de la Cordillera Ibérica. *Boletín de la Real Sociedad Española de Historia Natural, Sección Geología*, **86**, 21–51.

GARCÍA-GIL, S. 1995. Evolución sedimentaria de la zona de enlace Cordillera Ibérica-Sistema Central, margen occidental de la cuenca del Tethys durante el Triásico Medio. *Cuadernos de Geología Ibérica*, **19**, 99–128.

GARCÍA GUINEA, J., LOMBARDERO, M., ROBERTS, B. & TABOADA, J. 1997. Spanish roofing slate deposits. *Transactions of the Institution of Mining and Metallurgy*, **106**, 205–214.

GARCÍA GUINEA, J., MARTINEZ-FRIAS, J. & HARFFY, M. 1998. The Aznalcollar tailings dam burst and its ecological impact in S Spain. *Nature & Resources*, **34**(4), 45–48.

GARCÍA-HERNÁNDEZ, M. 1978. *El Jurásico terminal y el Cretácico inferior en las sierras de Cazorla y Segura (zona prebética)*. PhD Thesis, University of Granada.

GARCÍA-HERNÁNDEZ, M. 1981. Biozonation du Crétacé infèrieur à l'aide des Foraminifères benthiques et des Algues Dasycladacees dans le Prebetique occidental (Cordillères Betiques, Espagne). *Géobios*, **14–2**, 261–267.

GARCÍA HERNÁNDEZ, M. & LIÑÁN, E. 1983. Estromatolitos y facies asociadas en la Formación Santo Domingo (Cámbrico inferior de la Sierra de Córdoba). *In*: COMBA, J. A. (coord.) *Libro Jubilar de J. M. Ríos*, Vol. 3. IGME, Madrid, 125–132.

GARCÍA-HERNÁNDEZ, M. & LÓPEZ-GARRIDO, A. C. 1979. El tránsito Jurásico-Cretácico en la Zona Prebética. *Cuadernos de Geología de la Universidad de Granada*, **10**, 535–544.

GARCÍA-HERNÁNDEZ, M. & LÓPEZ-GARRIDO, A. C. 1988. The Prebetic platform during the Jurassic: a sedimentary evolution upon a distensive margin. *Second International Symposium on Jurassic Stratigraphy*, **II**, 1017–1030.

GARCÍA-HERNÁNDEZ, M., LÓPEZ-GARRIDO, A. C. & OLÓRIZ, F. 1979. El Oxfordense y el Kimmeridgiense inferior en la Zona Prebética. *Cuadernos de Geología de la Universidad de Granada*, **10**, 527–533.

GARCÍA-HERNÁNDEZ, M., LÓPEZ-GARRIDO, A. C., RIVAS, P., SANZ DE GALDEANO, C. & VERA, J. A. 1980. Mesozoic palaeogeographic evolution of the External Zones of the Betic Cordillera. *Geologie en Mijnbouw*, **59**, 155–168.

GARCÍA-HERNÁNDEZ, M., LÓPEZ-GARRIDO, A. C. & OLÓRIZ, F. 1981. Etude des calcaires noduleux du Jurassique supérieur de la Zone Prébétique (Cordillèrees Bétiques, SE de l'Espagne). *In*: FARINACCI, A. & ELMI, S. (eds) *Rosso Ammonitico Symposium Proceedings*, Roma, 419–434.

GARCÍA-HERNÁNDEZ, M., LÓPEZ-GARRIDO, A. C. & VERA, J. E. 1982. El Cretácico de la zona Prebética. *In*: *El Cretácico de España*. Complutense University, Madrid, 526–570.

GARCÍA-HERNÁNDEZ, M., LUPIANI, E. & VERA, J. A. 1986–87. Discontinuidades estratigráficas en el Jurásico de Sierra Gorda (Subbético interno, Provincia de Granada). *Acta Geológica Hispánica*, **21–22**, 339–349.

GARCÍA-HERNÁNDEZ, M., LÓPEZ-GARRIDO, A. C., MARTÍN-ALGARRA, A., MOLINA, J. M., RUÍZ-ORTIZ, P. A. & VERA, J. A. 1989. Las discontinuidades mayores del Jurásico de las Zonas Externas de

las Cordilleras Béticas: análisis e interpretación de los ciclos sedimentarios. *Cuadernos de Geología Ibérica*, **13**, 35–52.

GARCÍA-HIDALGO, J. F. 1993a. Las pistas fósiles del Alcudiense Superior en el Anticlinal de Ibor. Consideraciones cronoestratigráficas. *Geogaceta*, **13**, 33–35.

GARCÍA-HIDALGO, J. F. 1993b. Las pistas fósiles de los anticlinales de Alcudia y Abenójar (Zona Centroibérica). Edad de las series. *Geogaceta*, **14**, 57–59.

GARCÍA-HIDALGO, J. F. 1993c. Pistas fósiles en la 'Serie de Carrascalejo'. Implicaciones cronoestratigráficas. *Geogaceta*, **13**, 36–37.

GARCÍA-HIDALGO, J. F, SEGURA, M. & GARCÍA, A. 1997. El Cretácico del borde septentrional de la Rama Castellana de la Cordillera Ibérica. *Revista de la Sociedad Geológica de España*, **10**, 39–53.

GARCÍA-IGLESIAS, J. & LOREDO, J. 1992. Yacimientos de fluorita en Asturias. *In*: GARCÍA-GUINEA, J. & MARTÍNEZ-FRÍAS, J. (eds) *Recursos Minerales de España*. Consejo Superior de Investigaciones Científicas, Colección Textos Universitarios, **15**, 487–500.

GARCÍA JORAL, F. & GOY, A. 2000. Stratigraphic distributions of Toarcian brachiopods from the Iberian Range (Spain) and its relation to deposicional sequences. *GeoResearch Forum*, **6**, 381–386.

GARCÍA-LÓPEZ, S., JULIVERT, M., SOLDEVILA, J., TRUYOLS-MASSONI, M. & ZAMARREÑO, I. 1990. Bioestratigrafía y facies de la sucesión carbonatada del Silúrico Superior y Devónico Inferior de Santa Creu d'Olorda (Cadenas Costeras Catalanas, NE de España). *Acta Geologica Hispanica*, **25**, 141–168.

GARCÍA-LÓPEZ, S., GARCÍA-SANSEGUNDO, J. & ARBIZU, M. 1991. Devonian of the Arán Valley Synclinorium, Central Pyrenees, Spain: Stratigraphical and Paleontological data. *Acta Geológica Hispánica*, Barcelona, **26**(1), 55–66.

GARCÍA-LÓPEZ, S., RODRÍGUEZ-CAÑERO, R., SANZ-LÓPEZ, J., SARMIENTO, G. & VALENZUELA-RÍOS, J. J. 1996. Conodontos y episodios carbonatados en el Silúrico de la Cadena Hercínica meridional y el Dominio Sahariano. *Revista Española de Paleontología*, extra volume, 1996, 33–57.

GARCÍA-LÓPEZ, S., BRIME, C., BASTIDA, F. & SARMIENTO, G. 1997. Simultaneous use of thermal indicators to analyze the transition from diagenesis to metamorphism: an example from the Variscan Belt of Northwest Spain. *Geological Magazine*, **134**, 323–334.

GARCÍA-LÓPEZ, S., SANZ LÓPEZ, J. & PARDO ALONSO, M. V. 1999. Conodontos (bioestratigrafía, biofacies y paleotemperaturas) de los sinclinales de Almadén y Guadalmez (Devónico-Carbonífero Inferior), Zona Centroibérica meridional, España. *Revista Española Paleontología*, Oviedo, extra volume, Homenaje Prof. J. Truyols, 161–172.

GARCIA-LOYGORRI, A., ORTUÑO, G., CARIDE DE LIÑAN, C., GERVILLA, M., GREBER, CH. & FEYS, R. 1971. El Carbonífero de la Cuenca Central Asturiana. *Trabajos de Geología*, **3**, 101–150.

GARCÍA-MONDÉJAR, J. 1982. Aptiense y Albiense. Región Vasco-Cantábrica y Pirineo Navarro. *In*: *El Cretácico de España*. Complutense University, Madrid, 63–84.

GARCÍA-MONDÉJAR, J. 1985. Aptian Albian reefs (Urgonian) in the Asón- Soba area. *In*: MILÁ, M. D. & ROSELL, J. (eds) *Excursion Guide Book*, 6th European Regional Meeting of Sedimentology. IAS, Lleida, 329–352.

GARCÍA-MONDÉJAR, J. 1990. The Aptian–Albian carbonate episode of the Basque-Cantabrian basin (northern Spain): general characteristics, controls and evolution. *In*: TUCKER, M. E., WILSON, J. L., CREVELLO, P. D., SARG, J. F. & READ, J. F. (eds) *Carbonate Platforms: Facies, Sequences and Evolution*. International Association of Sedimentologists, Special Publications, **9**, 257–290.

GARCÍA-MONDÉJAR, J. 1996. Plate reconstruction of the Bay of Biscay. *Geology*, **24**, 635–638.

GARCÍA-MONDÉJAR, J. & FERNÁNDEZ-MENDIOLA, P. A. 1992. Incised valley fills at a lower Albian sequence boundary (western Basque-Cantabrian basin, north Spain). *Geogaceta*, **11**, 105–107.

GARCÍA-MONDÉJAR, J. & FERNÁNDEZ-MENDIOLA, P. A. 1995. Albian carbonate mounds: comparative study in the context of sea-level variations (Soba, northern Spain). *In*: MONTY, C. L. U., BOSENCE, D. W. J., BRIDGES, P. H. & PRATT, B. R. (eds) *Carbonate mud mounds: their origin and evolution*. International Association of Sedimentologists, Special Publications, **23**, 359–384.

GARCÍA-MONDÉJAR, J. & PUJALTE, V. 1975. Contemporaneous

tectonics in the Early Cretaceous of Central Santander province, North Spain. *IX International Congress on Sedimentology. Tectonics and Sedimentation. Nice*, **IV**, 131–137.

GARCÍA-MONDÉJAR, J. & PUJALTE, V. 1983. Origen, karstificación y enterramiento de unos materiales carbonatados albienses (Punta del Castillo, Górliz, Vizcaya). *X Congreso Nacional de Sedimentología, Menorca, Abstracts*, 3.9–3.12.

GARCÍA-MONDÉJAR, J. & ROBADOR, A. 1986–87. Sedimentación y paleogeografía del Complejo Urgoniano (Aptiense-Albiense) en el área de Bermeo (región Vasco-Cantábrica septentrional). *Acta Geologica Hispanica*, **21–22**, 411–418.

GARCÍA-MONDÉJAR, J. PUJALTE, V. & ROBLES, S. 1986. Características sedimentológicas secuenciales y tectoestratigráficas del Triásico de Cantabria y norte de Palencia. *Cuadernos de Geología Ibérica*, **10**, 151–172.

GARCÍA-MONDÉJAR, J., PUJALTE, V., ROBLES, S., CASTRO, J. & VALLÉS, J. 1989. Sistemas deposicionales, facies y evolución tectonoestratigráfica de la cubeta pérmica de Peña Labra-Peña Sagra (borde occidental de la cuenca Vasco-Cantábrica. Cantabria y Palencia). *In: Libro Homenaje a Rafael Soler*, AGGEP, Madrid, 53–65.

GARCÍA-MONDÉJAR, J., ARANBURU, A., ROSALES, I. & FERNÁNDEZ-MENDIOLA, P. A. 1993. El surco paleogeográfico de Trucíos-Somorrostro (Albiense inferior, W de Vizcaya). *Geogaceta*, **13**, 38–42.

GARCÍA-MONDÉJAR, J., AGIRREZABALA, L. M., ARANBURU, A., FERNÁNDEZ-MENDIOLA, P. A., GÓMEZ-PÉREZ, I., LÓPEZ-HORGUE, M. & ROSALES, I. 1996. Aptian–Albian tectonic pattern of the Basque-Cantabrian Basin (northern Spain). *Geological Journal*, **31**, 13–45.

GARCÍA NAVARRO, E. 2001. *Análisis de poblaciones de fallas en el extremo suroccidental de la península ibérica.* PhD Thesis, University of Huelva.

GARCÍA PALACIOS, A. & RÁBANO, I. 1996. Hallazgo de trilobites en pizarras negras graptolíticas del Silúrico inferior (Telychiense, Llandovery) de la Zona Centroibérica (España). *Geogaceta*, **20**, 239–241.

GARCÍA PALACIOS, A., GUTIÉRREZ-MARCO, J. C. & HERRANZ ARAÚJO, P. 1996. Edad y correlación de la 'Cuarcita de Criadero' y otras unidades cuarcíticas del límite Ordovícico-Silúrico en la Zona Centroibérica meridional (España y Portugal). *Geogaceta*, **20**, 19–22.

GARCÍA PALOMERO, F. 1980. *Caracteres geológicos y relaciones morfológicas y genéticas de los yacimientos del Anticlinal de Riotinto.* PhD Thesis, Excelentísima Diputacion Provincial de Huelva, Huelva.

GARCÍA PALOMERO, F. 1990. *Rio Tinto Deposits – Geology and Geological Models for their Exploration and Ore-reserve Evaluation.* The Institute of Mining and Metallurgy, Sulphide Deposits, London.

GARCÍA PALOMERO, F., BEDIA, J. L., GARCÍA MAGARIÑO, M. & SIDES, E. J. 1986. Nuevas investigaciones y trabajos de evaluación de reservas de gossan en Minas de Rio Tinto. *Boletín Geológico y Minero*, **97**, 82–102.

GARCÍA-RAMOS, J. C. & GUTIÉRREZ-CLAVEROL, M. 1995. La Geología de Franja Costera Oriental y de la Depresión Prelitoral de Oviedo-Cangas de Onís. *In:* ARAMBURU, C. & BASTIDA, F. (eds) *Geología de Asturias.* Editorial Trea, Gijón, 247–258.

GARCÍA-RAMOS, J. C. & ROBARDET, M. 1992. Hierros oolíticos ordovícicos de la Zona de Ossa Morena. *Publicaciones del Museo de Geología de Extremadura*, **3**, 123–132.

GARCÍA-RAMOS, J. C., SUÁREZ DE CENTI, C., PANIAGUA, A. & VALENZUELA, M. 1987. Los depósitos de hierro oolítico de Asturias y León: ambiente de depósito y relación con el vulcanismo. *Geogaceta*, **2**, 38–40.

GARCÍA-RAMOS, J. C., VALENZUELA, M., & SUÁREZ DE CENTI, C. 1992. Icnofósiles, procesos sedimentarios y facies en una rampa carbonatada del Jurásico de Asturias. *Sociedad Geológica de España, Reunión monográfica sobre biosedimentación.* University of Oviedo, 5–89.

GARCÍA-RAMOS, J. C. M. 1978. Estudio e interpretación de las principales facies sedimentarias comprendidas en las formaciones Naranco y Huergas (Devónico Medio) en la Cordillera Cantábrica. *Trabajos Geología Universidad Oviedo*, **10**, 195–247.

GARCÍA-RAMOS, J. C. M. & COLMENERO, J. R. 1981. Evolución sedi-

mentaria y paleogeográfica durante el Devónico en la Cordillera Cantábrica. *Real Academia Ciencias Exactas, Físicas Naturales (PICG)*, **2**, 61–76.

GARCÍA-ROYO, C. & ARCHE, A. 1987. El Triásico de la región Nuévalos-Cubel (Zaragoza). Sedimentación en un sector del borde de la cuenca del surco Molina-Valencia. *Cuadernos de Geología Ibérica*, **11**, 575–605.

GARCÍA RUIZ, J. M. 1979. El glaciarismo cuaternario en la Sierra de la Demanda (prov. de Logroño y Burgos, España). *Cuadernos de Investigación*, **5**, 3–25.

GARCÍA-RUIZ, J. M., MARTÍ-BONO, C., VALERO, B., GONZÁLEZ-SAMPÉRIZ, P., LORENTE, A. & BEGUERÍA, S. 2000. Derrubios de ladera en el Pirineo central español. significación cronológica y paleoclimática. *In:* PEÑA, J. L., SÁNCHEZ-FABRE, M. & LOZANO-TENA, M. V. (eds) *Procesos y formas periglaciares en la montaña mediterránea.* Instituto de Estudios Turolenses, Teruel, 63–80.

GARCÍA-SÁNCHEZ, M. 1960. Restos humanos del Paleolítico medio y superior y del Neo-Eneolítico de Piñar (Granada). *Trabajos del Instituto Benardino Sahagún de Antropología y Etnología*, **15**, 17–22.

GARCÍA SANSEGUNDO, J. 1992. *Estratigrafía y estructura de la Zona Axial pirenaica en la transversal del Valle de Arán y de la Alta Ribagorça.* Boletín Geológico y Minero, ITGE, Madrid.

GARRALDA, M. D. 1976. Dientes humanos del Magdaleniense de Tito Bustillo. *In:* MOURE, J. A. & CANO, M. (eds) *Excavaciones en la Cueva de Tito Bustillo (Asturias).* Instituto de Estudios Asturianos, Oviedo, 195–200.

GARRALDA, M. D. 1989*a*. Upper Paleolithic human remains from El Castillo cave (Santander, Spain). *In:* GIACOBINI, G. (ed.) *Hominidae.* Jaca Book, Milan, 479–482.

GARRALDA, M. D. 1989*b*. Lower Magdalenian Human remains from 'El Castillo' cave (Santander, Spain). *In:* EIBEN, O. G. (ed.) *European Populations in Past, Present and Future. Humanbiologie*, Budapest, **19**, 21–25.

GARRALDA, M. D. 1989*c*. Les populations post-paléotithiques de'Espagne: morphologie, culture et écologie. *In:* HERSHKOVITZ, J. (ed.) *People and Culture in Change.* BAR International Series, **508**(i), 505–516.

GARRALDA, M. D. 1991. Azules. Balmorí. Barranc Blanc. Camargo. Casares. Castillo. Chora. Cingle Vermell. Colombres. Cuartamentero. Erralla. Mazaculos II. Marín. Nerja. Parpalló. Pasiega. Pendo. Riera. Roc del Migdía. Tito Bustillo. Tossal de la Roca. *In:* ORBAN, E. (ed.) *Hominid Remains. No. 4. Spain.* Université Libre de Bruxelles.

GARRALDA, M. D. 1992. Les Magdaléniens en Espagne. Anthropologie et contexte paléoécologique. *In: Le Peuplement Magdalénien. Paléogéographie Physique et Humaine.* CTHS, Paris, 63–70.

GARRIDO, A. 1973. *Estudio geológico y relación entre tectónica y sedimentación del Terciario y Secundario de la vertiente meridional pirenaica en su zona central (prov. Huesca y Lérida).* PhD Thesis, University of Granada.

GARRIDO, A. & RÍOS, L. M. 1972. Síntesis geológica del Terciario y Secundario entre los ríos Cinca y Segre (Pirineo central de la vertiente surpirenaica, provincias de Huesca y Lérida). *Boletín Geológico y Minero*, **83**, 1–47.

GARRIDO, C. J. 1995. *Estudio geoquímico de las capas máficas del Macizo ultramáfico de Ronda.* PhD Thesis, University of Granada.

GASTESI, P. 1969. Petrology of the ultramafic and basic rocks of Betancuria massif, Fuerteventura Island (Canarian Archipelago). *Bulletin Volcanologique*, **33**, 1008–1038.

GAUTIER, F. 1968. Sur l'existence et l'âge d'un paléovulcanisme dans le Jurassique sud-aragonais (Espagne). *Comptes Rendues Sommaires Société Géologique Française*, **3**, 74.

GEBAUER, D. & MARTÍNEZ-GARCÍA, E. 1993. Geodynamica significance, age and origin of the Ollo de Sapo augengneiss (NW Iberian Massif, Spain). *In: Geological Society of America Annual Meeting, Boston, Mass. Abstracts with programs*, A342.

GEEL, T. 1973. The geology of the Betic Malaga, the subbetic and the region between these two units in the Vélez Rubio area (SE Spain), *GUA Papers of Geology*, **5**, 1–181.

GEEL, T. 2000. Recognition of stratigraphic sequences in carbonate platform and slope deposits: empirical models based on microfacies analysis of Palaeogene deposits in southeastern Spain. *Palaeogeography, Palaeoclimatology, Palaeoecology*, **155**, 211–238.

GEEL, T., ROEP, TH. B., KATE, W. TEN & SMIT, J. 1992. Early-Middle Miocene stratigraphic turning points in the Alicante region (SE Spain): reflections of Western Mediterranean plate-tectonic reorganization. *Sedimentary Geology*, **75**, 223–239.

GEEL, T., ROEP, TH. B., VAIL, P. R. & VAN HINTE, J. E. 1998. Eocene tectono-sedimentary patterns in the Alicante region (Southeastern Spain). *In*: DE GRACIANSKY, P. C., HARDENBOL, J., JACQUIN, T. & VAIL, P. R. (eds) *Mesozoic and Cenozoic Sequence Stratigraphy of European Basins*. SEPM, Special Publications, **60**, 289–302.

GELABERT, B., SABAT, F. & RODRIGUEZ-PEREA, A. 1992. A structural outline of the Serra de la Tramuntana of Mallorca, Balearic Islands. *Tectonophysics*, **203**, 167–183.

GELABERT-FERRER, B. 1998. *La estructura geológica de la mitad occidental de la Isla de Mallorca*. PhD Thesis, University of Barcelona.

GERVILLA, F. 1992. Depósitos de cromita-arseniuros de Ni-(Au-EGP) y de sulfuros de Fe-Ni-Cu y grafito, asociados a las rocas ultramáficas del sur de España. *In*: GARCÍA GUINEA, J. & MARTÍNEZ-FRÍAS, J. (eds) *Recursos Minerales de España*. Textos Universitarios, **15**, CSIC, 275–291.

GERVILLA, F. & LEBLANC, M. 1990. Magmatic ores in high-temperature alpine-type lherzolite massifs (Ronda, Spain and Beni-Bousera, Morocco). *Economic Geology*, **85**, 112–132.

GERVILLA, M., BEROIZ, C., BARÓN, A. & RAMIREZ DEL POZO, J. 1978. *Mapa Geológico de España, E: 1:50.000, Hoja 29, Oviedo, Segunda serie*. Instituto Geológico Minero, 1–64.

GESSA, S., TRUYOLS MASSONI, M. & ROBARDET, M. 1994. Quantitative analysis of *Homoctenowakia bohemica bohemica* (Tentaculitoids) from the Lochkovian of the Valle syncline, Ossa-Morena Zone (SW Spain). *Revista Española de Paleontología*, **9**, 203–210.

GEYER, G. 1990. Proposal of formal lithostratigraphical units for the Terminal Proterozoic to early Middle Cambrian of southern Morocco. *Newsletters on Stratigraphy*, **22**(2/3), 87–109.

GEYER, G. & ELICKI, O. 1995. The Lower Cambrian trilobites from the Görlitz Synclinorium (Germany) – review and new results. *Paläontologische Zeitschrift*, **69**, 87–119.

GEYER, G. & LANDING, E. 1995. The Cambrian of the Moroccan Atlas regions. *Beringeria*, Special Issue **2**, 7–46.

GEYER, G. & SHERGOLD, J. H. 2000. The quest for internationally recognized divisions of Cambrian time. *Episodes*, **23**, 188–195.

GEYER, O. F. & PELLEDUHN, R. 1979. Sobre la Estratigrafía y la facies espongiolítica del Kimmeridgiense de Calanda (provincia de Teruel). *Cuadernos de Geología, Universidad de Granada*, **10**, 67–72.

GIBBONS, W. & HARRIS, A. L. 1994. *A revised correlation of Precambrian rocks in the British Isles*. Geological Society, London, Special Report, **22**.

GIBBS, A. D. 1984. Structural evolution of extensional basin margins. *Journal of the Geological Society, London*, **141**, 609–620.

GIBERT, J. & PÉREZ-PÉREZ, A. 1989. A human phallanx from the Lower Palaeolithic site of Cueva Victoria (Murcia, Spain). *Human Evolution*, **4**, 307–316.

GIBERT, J., CAMPILLO, D., ARQUÉS, J. M., GARCÍA-OLIVARES, E., BORJA, C. & LOWENSTEIN, J. 1998. Hominid status of the Orce cranial fragment reasserted. *Journal of Human Evolution*, **34**, 203–217.

GIBSON, R. L. 1989. The relationship between deformation and metamorphism in the Canigou massif, Pyrenees: a case study. *Geologie en Mijnbouw*, **68**, 345–356

GICH, M., ROSELL, J., REGUANT, S. & CLAVELL, E. 1967. Estratigrafía del Paleógeno en la zona de tránsito entre la Cordillera Prelitoral Catalana y el Prepirineo. *Acta Geológica Hispánica*, **2**(6), 13–18.

GIESE, U., REITZ, E. & WALTER, R. 1986. Contributions to the stratigraphy of the Pulo do Lobo succession in Southwest Spain. *Comunicações dos Serviços Geológicos de Portugal*, **74**, 79–84.

GIESE, U., HOEGEN, R. VON, HOLLMANN, G. & WALTER, R. 1994. Geology of the southwestern Iberian Meseta I. The Palaeozoic of the Ossa Morena Zone north and south of the Olivenza-Mones-

terio Anticline (Huelva province, SW Spain). *Neues Jahrbuch für Geologie und Paläontologie, Abhandlungen*, **192**, 293–331.

GIGOUT, M., SOLÉ SABARÍS, L. & SOLÉ, N. 1955. Sur le Quaternaire méditerranéen d'Andalousie. *Compte Rendu Sommaire de la Société Géologique Française*, **9–10**, 177–179.

GIL, A. 1999. *La estructura de la Sierra de Cameros: Deformación dúctil y su significado a escala cortical*. PhD Thesis, University of Zaragoza.

GIL, A. & POCOVI, A. 1994. La esquistosidad alpina del extremo NW de la Cadena Ibérica Oriental (Sierra del Moncayo): distribución, génesis y significado tectónico. *Revista de la Sociedad Geológica de España*, **7**(1–2), 91–113.

GIL, J., GRACÍA, A. & SEGURA, M. 1993. Secuencias deposicionales del Cretácico en el flanco sur del Sistema Central. *Geogaceta*, **13**, 43–45.

GIL CID, M. D. 1981. Los trilobites Agnóstidos del Cámbrico Inferior y Medio de España. *Boletín Geológico y Minero*, **92**, 111–126.

GIL CID, M. D. & JAGO, J. B. 1989. New data on the Lower Cambrian Trilobites of Cortijos de Malagón (Spain). *Estudios Geológicos*, **45**, 91–99.

GIL CID, M. D., PEREJÓN, A. & SAN JOSÉ, M. A. 1976. Estratigrafía y Paleontología de las calizas cámbricas de los Navalucillos (Toledo). *Tecniterrae*, **13**, 1–19.

GIL CID, M. D., DOMÍNGUEZ, P., CRUZ, M. C. & ESCRIBANO, M. 1996*a*. Primera cita de un Blastoideo Coronado en el Ordovícico Superior de Sierra Morena oriental. *Revista de la Sociedad Geológica de España*, **9**, 253–267.

GIL CID, M. D., DOMÍNGUEZ, P., ESCRIBANO, M. & SILVÁN, E. 1996*b*. Un nuevo Rombífero, *Homocystites geyeri* n. sp., en el Ordovícico de El Viso del Marqués (Ciudad Real). *Geogaceta*, **20**, 216–219.

GIL CID, M. D., DOMÍNGUEZ ALONSO, P., CRUZ GONZÁLEZ, M. C. & ESCRIBANO RÓDENAS, M. 1996*c*. Nuevo macrocystellidae (Echinodermata, Cystoidea Rhombifera) para el Ordovícico español. *Estudios Geológicos*, **52**, 175–183.

GIL CID, M. D., DOMÍNGUEZ ALONSO, P. & SILVÁN POBES, E. 1996*d*. Reconstrucción y modo de vida de *Heviacrinus melendezi* nov.gen. nov.sp. (Disparida Iocrinidae), primer crinoide descrito para el Ordovícico medio de los Montes de Toledo (España). *Revista de la Sociedad Geológica de España*, **9**, 19–27.

GIL CID, M. D., DOMÍNGUEZ ALONSO, P., SILVÁN POBES, E. & ESCRIBANO RÓDENAS, M. 1996*e*. *Bohemiaecystis jefferiesi* n. sp.; primer Cornuta para el Ordovícico español. *Estudios Geológicos*, **52**, 313–326.

GIL CID, M. D., DOMÍNGUEZ ALONSO, P. & SILVÁN POBES, E. 1998. *Coralcrinus sarachagae* gen. nov. sp. nov., primer crinoide (Disparida, Inadunata) descrito en el Ordovícico medio de Sierra Morena. *Coloquios de Paleontología*, **49**, 115–128.

GIL CID, M. D., DOMÍNGUEZ, P., TORRES, M. & JIMÉNEZ, I. 1999. A mathematical tool to analyze radially symmetrical organisms and its application to a new camerate from Upper Ordovician of South Western Spain. *Geobios*, **32**, 861–867.

GIL, E. 1987. *Taxonomía y bioestratigrafía de micromamíferos del Pleistoceno medio, especialmente roedores, de los rellenos cársticos de la trinchera del ferrocarril de la Sierra de Atapuerca (Burgos)*. Tesis Doctoral, Universidad de Zaragoza.

GIL IBARGUCHI, J. I. & ARENAS, R. 1990. Metamorphic evolution of the Allochthonous Complexes from the northwest of the Iberian Peninsula. *In*: DALLMEYER, R. D. & MARTÍNEZ GARCÍA, E. (eds) *Pre-Mesozoic Geology of Iberia*. Springer-Verlag, Berlin, 237–246.

GIL IBARGUCHI, J. I. & DALLMEYER, R. D. 1991. Hercynian blueschist metamorphism in north Portugal: tectonothermal implications. *Journal of Metamorphic Geology*, **9**, 539–549.

GIL IBARGUCHI, J. I. & JULIVERT, M. 1988. Petrología de la aureola metamórfica de la granodiorita de Barcelona en la Sierra de Collcerola (Tibidabo). *Estudios Geológicos*, **44**, 353–374.

GIL IBARGUCHI, J. I. & MARTÍNEZ, F. J. 1982. Petrology of garnet-cordierite-sillimanite gneisses from the El Tormes Thermal Dome, Iberian Hercynian Foldbelt (NW Spain). *Contributions to Mineralogy and Petrology*, **80**, 14–24.

GIL IBARGUCHI, J. I. & ORTEGA GIRONÉS, E. 1985. Petrology, structure and geotectonic implications of glaucophane-bearing eclogites and related rocks from Malpica-Tuy (MT) Unit, Galicia, Northwest Spain. *Chemical Geology*, **50**, 145–162.

GIL IBARGUCHI, J. I., MENDIA, M., GIRARDEAU, J. & PEUCAT, J. J. 1990. Petrology of eclogites and clinopyroxene-garnet metabasites from the Cabo Ortegal Complex (northwestern Spain). *Lithos*, **25**, 133–162.

GIL IBARGUCHI, J. I., ABALOS, B., AZCÁRRAGA, J. & PUELLES, P. 1999. Deformation, high-pressure metamorphism and exhumation of ultramafites in a deep subduction/collision setting (Cabo Ortegal, NW Spain). *Journal of Metamorphic Geology*, **17**, 747–764.

GIL-PEÑA, I., SANZ-LÓPEZ, J., BARNOLAS, A. & CLARIANA, P. 2000. Secuencia sedimentaria del Ordovícico Superior en el margen occidental del Domo de Orri (Pirineos centrales). *Geotemas*, **1**, 187–190.

GILBERT, J. S., BICKLE, M. J. & CHAPMANN, H. J. 1994. The origin of Pyrenean Hercynian volcanic rocks (France-Spain): REE and Sm-Nd constraints. *Chemical Geology*, **111**, 207–226.

GIMÉNEZ, R. 1989. La megaséquence transgressive – regressive du Cénomanien supérieur dans la région meridionale de la Chaîne Ibérique (provinces de Valence et d'Albacete, Espagne. *Geobios, Memoire Speciale*, **11**, 43–52.

GIMÉNEZ, R., MARTÍN-CHIVELET, J. & VILAS, L. 1993. Upper Albian to Middle Cenomanian carbonate platforms of Betic and Ibérian basins (Spain). *In*: SIMÓ, J. A. T., SCOTT, R. W. & MASSE, J. P. (eds) *Cretaceous Carbonate Platforms*. AAPG, Memoirs, **56**, 271–281.

GINER, J. 1978. Origen y significado de las brechas del Lías de la Mesa de Prades (Tarragona). *Estudios Geológicos*, **34**, 529–533.

GINER, J. 1980. *Estudio sedimentológico y diagenético de las facies carbonatadas del Jurásico de las Catalánides, Maestrazgo y Rama Aragonesa de la Cordillera Ibérica*. PhD Thesis, University of Barcelona.

GINER, J. & BARNOLAS, A. 1979. Las construcciones arrecifales del Jurásico superior de la Sierra de Albarracín. *Cuadernos de Geología, Universidad de Granada*, **10**, 73–82.

GINER, J. & BARNOLAS, A. 1980. Los biohermes de espongiarios del Bajociense superior de Moscardón (Teruel). *Acta Geológica Hispánica*, **15**, 105–108.

GINER, J. L. 1996. *Análisis neotectónico y sismotectónico en la parte centro-oriental de la cuenca del Tajo*. PhD Thesis, Complutense University, Madrid.

GINER, J. L. & VICENTE, de G. 1995. Crisis tectónicas recientes en el sector central de la cuenca de Madrid. *In*: ALEIXANDRE, T. & PÉREZ-GONZÁLEZ, A. (eds) *Reconstrucción de paleoambientes y cambios climáticos durante el Cuaternario*. Monografias CCMA, CSIC, **3**, Madrid, 141–161.

GINKEL, A. C. VAN & VILLA, E. 1991. Some fusulinids from the Moscovian-Kasimovian transition in the Carboniferous of the Cantabrian Mountains (NW Spain). *Proceedings Koninklijke Nedeerlandse Akademie van Wetenschappen*, **94**, 299–359.

GINKEL, A. C. VAN & VILLA, E. 1999. Late fusulinellids and early schwagerinid foraminifera: relationships and occurrences in the Las Llacerias section (Moscovian/Kasimovian), Cantabrian Mountains, Spain). *Journal of Foraminifera Research*, **29**, 263–290.

GIRARDEAU, J. & GIL IBARGUCHI, J. I. 1991. Pyroxenite-rich peridotites of the Cabo Ortegal Complex (Northwestern Spain): evidence for large-scale upper-mantle heterogeneity. *Journal of Petrology*, Special Lherzolites Issue, 135–154.

GISBERT, J. 1981. *Estudio geológico-petrológico del Estefaniense-Pérmico de la Sierra del Cadí (Pirineo de Lérida): diagénesis y sedimentología*. PhD Thesis, University of Zaragoza.

GISBERT, J. 1983. El Pérmico de los Pirineos españoles. *In*: MARTÍNEZ GARCÍA, E. (ed.) *Carbonífero y Pérmico de España*. Ministerio de Industria y Energía, Madrid, 405–420.

GISBERT, J. 1984. Les molasses post-hercyniennes dans le Haut Urgell et la Cerdagne occidentale (Pyrenees Catalanes, Espagne). *Comptes Rendus de l'Académie des Sciences, Paris*, **298**, 883–888.

GISBERT, J., MARTÍ, J. & GASCÓN, F. 1985. *Guía de la Excursión al Estefaniense, Pérmico y Triásico inferior del Pirineo Catalán*. Institut d'Estudis Illerdencs, 1–79.

GISCHLER, E., GRÄFE, K. U. & WIEDMANN, J. 1994. The Upper Cretaceous *Lacazina* Limestone in the Basco-Cantabrian and

Iberian Basins of Northern Spanien: Cold-water grain associations in warm-water environments. *Facies*, **30**, 209–246.

GLEIZES, G., NÉDÉLEC, A., BOUCHEZ, J. L., AUTRAN, A. & ROCHETTE, P. 1993. Magnetic susceptibility of the Mont Louis-Andorra ilmenite-type granite (Pyrenees): A new tool for the petrographic characterization and regional mapping of zoned granite plutons. *Journal of Geophysical Research*, **98**, 4317–4331.

GLEIZES, G., LEBLANC, D. & BOUCHEZ, J. L. 1997. Variscan granites of the Pyrenees revisited: their role as syntectonic markers of the orogen. *Terra Nova*, **9**, 38–41.

GLEIZES, G., LEBLANC, D. & BOUCHEZ, J. L. 1998a. The main phase of the Hercynian orogeny in the Pyrenees is a dextral transpression. *In*: HOLDSWORTH, R. E., STRACHAN, R. A. & DEWEY, J. F. (eds) *Continental Trasnpressional and Transtensional Tectonics*. Geological Society of London, Special Publications, **135**, 267–273.

GLEIZES, G., LEBLANC, D., SANTANA, V., OLIVIER, PH. & BOUCHEZ, J. L. 1998b. Sigmoidal structures featuring dextral shear during emplacement of the Hercynian granite complex of Cauterets-Panticosa (Pyrenees). *Journal of Structural Geology*, **20**, 1229–1245.

GLOVER, P., POUS, J., QUERALT, P., MUÑOZ, J. A., LIESA, M. & HOLE, M. J. 2000. Integrated two dimensional lithospheric conductivity modelling in the Pyrenees using field-scale and laboratory measurements. *Earth and Planetary Science Letters*, **178**, 59–72.

GOFFÉ, B., MICHARD, A., GARCÍA-DUEÑAS, V. *et al.* 1989. First evidence of high-pressure, low-temperature metamorphism in the Alpujarride nappes, Betic Cordillera (SE Spain). *European Journal of Mineralogy*, **1**, 139–142.

GOICOECHEA, P., DOBLAS, M., HERNÁNDEZ ENRILE, J. L. & UBANELL, A. G. 1991. Estudio cinemático de las fallas alpinas que delimitan la fosa tectónica del Lozoya (Sistema Central). *Geogaceta*, **9**, 24–27.

GOLDBERG, J. M. & MALUSKI, H. 1988. Données nouvelles et mise au point sur l'âge du métamorphisme pyrénéen. *Comptes Rendus de l'Académie des Sciences, Paris*, **306**(II), 429–435.

GOLDBERG, J. M., GUIRAUD, M., MALUSKI, H. & SEGURET, M. 1988. Caractéres pétrologiques et âge du métamorphisme en contexte distensif du bassin sur décrochement de Soria (Crétacé inférieur, Nord Espagne). *Comptes Rendues de l'Académie des Sciences de Paris*, **307**(II), 521–527.

GOMBAU, I. 1877. Reseña fisico-geológica de la província de Tarragona. *Boletín de la Comisión del Mapa Geológico de España*, **4**, 181–250.

GÓMEZ, J. J. 1979. *El Jurásico en facies carbonatadas del sector levantino de la Cordillera Ibérica*. PhD Thesis, Complutense University, Madrid (Seminarios de Estratigrafía, Serie Monografías, **4**).

GÓMEZ, J. J. 1991. Jurásico. *In*: IGME (ed.) *Memoria del Mapa Geológico de España, 1:200.000, Daroca*. 31–82.

GÓMEZ, J. J. & BABÍN, R. 1973. Evidencia de tres generaciones de pliegues en el anticlinal de Sot (Cordillera Ibérica; provincia de Valencia). *Estudios Geológicos*, **29**, 381–388

GÓMEZ, J. J. & GOY, A. 1979. Las unidades litoestratigráficas del Jurásico medio y superior, en facies carbonatadas del sector levantino de la Cordillera Ibérica. *Estudios Geológicos*, **35**, 569–598.

GÓMEZ, J. J. & GOY, A. 1997a. El Jurásico de la Cordillera Ibérica: Estratigrafía Secuencial y Paleogeografía. *IV Congreso de Jurásico de España, libro de comunicaciones*, 15–17.

GÓMEZ, J. J. & GOY, A. 1997b. El tránsito Triásico-Jurásico en la sección de Decantadero (Lécera, Zaragoza). *Publicaciones del Seminario de Paleontología de Zaragoza*, **3**, 21–30.

GÓMEZ, J. J. & GOY, A. 1998. Las unidades litoestratigráficas del tránsito Triásico-Jurásico en la región de Lécera (Zaragoza). *Geogaceta*, **23**, 63–66.

GÓMEZ, J. J. & GOY, A. 1999. Las unidades carbonatadas y evaporíticas del tránsito Triásico-Jurásico en la región de Lécera (Zaragoza, España). *Cuadernos de Geología Ibérica*, **25**, 13–23.

GÓMEZ, J. J. & GOY, A. 2000. Definition and organization of limestone-marl cycles in the Toarcian of the northern and east-central part of the Iberian Subplate (Spain). *GeoResearch Forum*, **5**, 301–310.

GÓMEZ, J. J., DÍAZ-MOLINA, M. & LENDÍNEZ, A. 1996. Tectono-sedimentary analysis of the Loranca Basin (Upper Oligocene-Miocene, Central Spain): a 'non-sequenced' foreland basin. *In*: FRIEND, P. F. & DABRIO, C. J. (eds) *Tertiary Basins of Spain: the*

Stratigraphic Record of Crustal Kinematics. Cambridge University Press, Cambridge, 285–294.

GÓMEZ FERNANDEZ, J. C. 1992. *Análisis de la cuenca sedimentaria de Los Cameros durante sus etapas iniciales de relleno en relación con su evolución paleogeográfica*. PhD Thesis, Complutense University, Madrid.

GÓMEZ-GARRIDO, A. 1989. Biostratigrafía (Foraminíferos Planctónicos) del Cretácico Superior del Surpirineo Central. *Revista Española de Micropaleontología*. **21**(1), 145–171.

GÓMEZ-GRAS, D. 1993a. El Permotrías de la Cordillera Costero Catalana: Facies y petrología sedimentaria (parte I). *Boletín Geológico y Minero*, **104**, 37–56.

GÓMEZ-GRAS, D. 1993b. El Permotrías de las Baleares y de la vertiente mediterránea de la Cordillera Ibérica y del Maestrat: Facies y Petrología Sedimentaria (Parte II). *Boletín del IGME*, **104**, 467–515

GOMEZ-LLUECA. 1920. Sur la géologie de Cabrera, Conejera et autres îles voisines. *Comptes Rendus de l'Académie des Sciences, Paris*, **171**, 1158.

GÓMEZ-ORTIZ, A., SALVADOR-FRANCH, F., SÁNCHEZ-GÓMEZ, S. & SIMÓN-TORRES, M. 1998. Morfología de Cumbres de Sierra Nevada. Una aproximación a la dinámica glaciar y periglaciar. *In*: GÓMEZ-ORTIZ, A., SALVADOR-FRANCH, F., SHULTZ, L., & GARCÍA NAVARRO (eds) *Itinerarios geomorfológicos por Andalucía Oriental*. University of Barcelona, 37–63.

GÓMEZ ORTÍZ, D. 2001. *La estructura de la corteza en la zona central de la Península Ibérica*. PhD Thesis, Complutense University, Madrid.

GÓMEZ ORTIZ, D. & BABÍN, R. 1996. Los pliegues de propagación de falla de la región centro oriental del Sistema Central Español. Análisis geométrico. *Revista de la Sociedad Geológica de España*, **9**, 297–309.

GÓMEZ ORTÍZ, D. & BABÍN, R. 1998. Geometría de las fallas inversas de la zona de Sepúlveda (borde norte del Sistema Central) a partir de modelos de pliegues de propagación de falla. *Geogaceta*, **23**, 67–70.

GÓMEZ-PÉREZ, I. 1994. *El modelo de plataforma carbonatada-cuenca de Gorbea (Aptiense superior-Albiense, Bizkaia)*. PhD Thesis, University of País Vasco.

GÓMEZ-PÉREZ, I., ARANBURU, A., FERNÁNDEZ-MENDIOLA, P. A. & GARCÍA-MONDÉJAR, J. 1994. Un sistema de paleoalto carbonatado-surco siliciclástico (Gorbea-Artzentales, Albiense medio, Bizkaia). *Geogaceta*, **16**, 74–77.

GÓMEZ-PÉREZ, I., FERNÁNDEZ-MENDIOLA, P. A. & GARCÍA-MONDÉJAR, J. 1998. Platform margin paleokarst development from the Albian of Gorbea (Basque-Cantabrian region, N Iberia). *Acta Geologica Hungarica*, **41/1**, 3–22.

GÓMEZ-PUGNAIRE, M. T. 1979. *La evolución del metamorfismo alpino en el Complejo Nevado-Filábride de la Sierra de Baza (Cordilleras Béticas, España)*. PhD Thesis, University of Granada.

GÓMEZ PUGNAIRE, M. T. 1981. *La evolución del metamorfismo alpino en el Complejo Nevado-Filábride de la Sierra de Baza (Cordilleras Béticas, España)*. Tecniterrae, **4**, 1–130.

GÓMEZ PUGNAIRE, M. T. 1992. El Paleozoico de las Cordilleras Béticas. *In*: GUTIÉRREZ-MARCO, J. C., SAAVEDRA, J. & RÁBANO, I. (eds) *Paleozoico Inferior de Ibero-América*. University of Extremadura, 593–606.

GÓMEZ-PUGNAIRE, M. T. & CÁMARA, F. 1990. La asociación de alta presión distena+talco+fengita coexistente con escapolita en metapelitas de origen evaporítico (Complejo Nevado-Filábride, Cordilleras Béticas). *Revista de la Sociedad Geológica de España*, **3**, 373–384.

GÓMEZ-PUGNAIRE, M. T. & FRANZ, G. 1988. Metamorphic evolution of the Palaeozoic series of the Betic Cordilleras (Nevado-Filabride complex, SE Spain) and its relationship with the Alpine orogeny. *Geologische Rundschau*, **77**, 619–640.

GÓMEZ-PUGNAIRE, M. T. & SASSI, F. P. 1983. Pre-alpine metamorphic features and alpine overprints in some parts of the Nevado-Filabride basement (Betic Cordilleras, Spain). *Memorie degli Instituti di Geología e Mineralogia dell'Universitá di Padova*, **36**, 49–72.

GÓMEZ-PUGNAIRE, M. T., FONTBOTÉ J. M. & SASSI, J. P. 1981. On the occurrence of a metaconglomerate in the Sierra de Baza (Nevado-Filabride Complex, Betic Cordilleras, Spain). *Neues Jahrbuch für Geologie und Paläontologie Monatsheften*, **7**, 405–418.

GÓMEZ PUGNAIRE, M-T., CHACÓN, J., MITROFANOV, F. & TIMOFEEV, V. 1982. First report on pre-Cambrian rocks in the graphite-bearing series of the Nevado-Filábride Complex (Betic Cordilleras, Spain). *Neues Jahrbuch fuer Geologie und Palaeontologie, Monatshefte*, **3**, 176–180.

GÓMEZ-PUGNAIRE, M. T., BRAGA, J. C., MARTÍN, J. M., SASSI, F. P. & DEL MORO, A. 2000. Regional implications of a Palaeozoic age for the Nevado-Filábride Cover of the Betic Cordillera, Spain. *Schweizerische Mineralogische und Petrographische Mitteilungen*, **80**, 45–52.

GOMIS, E., PARÉS, J. M. & CABRERA, L. 1999. Nuevos datos magnetoestratigráficos del tránsito Oligoceno-Mioceno en el sector SE de la Cuenca del Ebro (Provincias de Lleida, Zaragoza y Huesca, NE de España). *Acta Geológica Hispánica*, **32**(3–4), 185–199.

GONZÁLEZ, A. 1989. *Análisis tectosedimentario del terciario del borde SE de la Depresión del Ebro (sector bajoaragonés) y cubetas ibéricas marginales*. PhD Thesis, University of Zaragoza.

GONZÁLEZ, A. & GUIMERÀ, J. 1993. Sedimentación sintectónica en una cuenca transportada sobre una lámina de cabalgamiento: la cubeta terciaria de Aliaga. *Revista de la Sociedad Geológica de España*, **6**, 151–165.

GONZÁLEZ, A., GUTIÉRREZ, M. & LERANOZ, B. 1988a. Las superficies de erosion neógenas en el sector central de la Cordillera Ibérica. *Revista de la Sociedad Geológica de España*, **1**(1–2), 135–142.

GONZÁLEZ, A., PARDO, G. & VILLENA, J. 1988b. El análisis tectosedimentario como instrumento de correlación entre cuencas. *Actas II Congreso Nacional Geología*, Granada, Sociedad Geológica de España, vol. Simposiums, 175–184.

GONZÁLEZ, F., MORENO, C., RODRÍGUEZ, R. M. & SÁEZ, R. 2000. Las pizarras negras del sinclinal de Las Herrerías, Faja Pirítica Ibérica. *Geogaceta*, **29**, 71–74.

GONZÁLEZ CASADO, J. M. 1987. *Estudio geológico de la zona de cizalla de Berzosa-Honrubia (Sistema Central Español)*. PhD Thesis, Complutense University, Madrid.

GONZALEZ CASADO, J. M. & DE VICENTE, G. 1996. Evolución alpina del Sistema Central Español. *In*: SEGURA, M., DE BUSTAMANTE, I. & BARDAJI, T. (eds) *Itinerarios Geológicos desde Alcalá de Henares*. Acalá de Henares, 141–151.

GONZÁLEZ CASADO, J. M & GARCÍA CUEVAS, C. 1999. Calcite twins from microveins as indicators of deformation history. *Journal of Structural Geology*, **21**, 875–889.

GONZÁLEZ-CASADO, J. M., CABALLERO, J. M., CASQUET, C., GALINDO, G. & TORNOS, F. 1996. Palaeostress and geotectonic interpretation of the Alpine cycle onset in the Sierra de Guadarrama (eastern Iberia-Central System) based on evidence from epysyenites. *Tectonophysics*, **262**, 213–219.

GONZÁLEZ CLAVIJO, E. 1997. *La Geología del Sinforme de Alcañices, Oeste de Zamora*. PhD Thesis, University of Salamanca.

GONZÁLEZ-DÍAZ, I., REMONDO, A., DÍAZ DE TERÁN, J. R. & CENDRERO, A. 1999. A methodological approach for the analysis of the temporal occurance and triggering factors of landslides. *Geomorphology*, **30**, 95–114.

GONZÁLEZ-DÍAZ, I. 1992. Arcillas comunes. *In*: GARCÍA-GUINEA, J. & MARTÍNEZ-FRÍAS, J. (eds) *Recursos Minerales de España*. Consejo Superior de Investigaciones Científicas. Colección Textos Universitarios, **15**, 95–114.

GONZÁLEZ-DONOSO, J. M., LINARES, D., PASCUAL, I. & SERRANO, F. 1982. Datos sobre la edad de las secciones del Mioceno Inferior de Es Port d'es Canonge y de Randa (Mallorca). *Boletín Sociedad Historia Natural Balears*, **26**, 229–232.

GONZÁLEZ-ECHEGARAY, J., GARCÍA-GUINEA, M. A., BEGINES, A. & MADARIAGA, B. 1962. *La Cueva de La Chora (Santander)*. Excavaciones Arqueológicas en España, **26**, Dirección General de Bellas Artes, Madrid.

GONZÁLEZ-HERNÁNDEZ, E. M., MÖRNER, N. A., GOY, J. L., ZAZO, C. & SILVA, P. G. 1999. Resultados paleomagnéticos de los depósitos Plio-Pleistocenos de la Cuenca de Palma (Mallorca, España). *Estudios Geológicos*, **56**, 163–173.

GONZÁLEZ LASTRA, J. 1978. Facies salinas en la Caliza de Montaña (Cordillera Cantábrica) *Trabajos de Geología*, **10**, 249–265

GONZÁLEZ LODEIRO, F. 1980. *Estudio geológico estructural de la terminación oriental de la Sierra del Guadarrama.* PhD Thesis, University of Salamanca.

GONZÁLEZ-LODEIRO, F., OROZCO M., CAMPOS J. & GARCÍA-DUEÑAS V. 1984. Cizallas dúctiles y estructuras asociadas en los Mantos del Mulhacén y Veleta, primeros resultados sobre Sierra Nevada y Sierra de los Filabres. *In: El borde mediterráneo español, evolución del Orógeno bético y Geodinámica de las depresiones neógenas,* Granada, 5–8.

GONZÁLEZ-LODEIRO, F., ALDAYA, F., GALINDO-ZALDÍVAR, J. & JABALOY, A. 1996. Superimposition of extensional detachments during the Neogene in the internal zones of the Betic cordilleras. *Geologische Rundschau,* **85,** 350–362.

GONZÁLEZ MARTÍN, J. A. 1986. Las manifestaciones frías mediterráneas en la cuenca baja del Tajuña, durante el Cuaternario reciente. *In: Actas de Geomorfología.* Alianza Editorial, Madrid, 229–238.

GONZALO, J. C. 1988. El plutonismo hercínico en el área de Mérida (Extremadura Central, España). *In:* BEA, F., CARNICERO, A., GONZALO, J. C., LÓPEZ PLAZA, M. & RODRÍGUEZ ALONSO, M. D. (eds) *Geología de los granitoides y rocas asociadas del Macizo Hespérico. Libro Homenaje a L. C. García Figuerola.* Rueda, Madrid, 345–356.

GOURVENNEC, R. 1990. Un genre nouveau de Cyrtiacea (Brachiopoda) du Silurien supérieur d'Espagne. *Géobios,* **23,** 141–147.

GOURVENNEC, R., BOUYX, E., BRICE, D. & LE MENN, J. 1997. The Siluro-Devonian palaeobiogeography of the Meguma terrane and its relationships with North Gondwana. *In:* PIRES, C. C., GOMES, M. E. P. & COKE, C. (coords) *Comunicaçoes XIV Reuniao de Geología do Oeste Peninsular. Evoluçao geologica do Maciço Iberico e seu enquadramento continental,* Vila Real, 65–70.

GOY, A. 1995. Ammonoideos del Triásico Medio de España: Bioestratigrafía y correlaciones. *Cuadernos de Geología Ibérica,* **19,** 21–60.

GOY, A. & Márquez-Aliaga, A. 1998. Bivalvos del Triásico Superior en la Formación Imón (Cordillera Ibérica, España). *Boletín de la Real Sociedad Española de Historia Natural, Sección Geología,* **94,** 77–91.

GOY, A. & MARTÍNEZ, G. 1990. Biozonación del Toarciense en el área de La Almunia de Doña Godina-Ricla (Sector central de la Cordillera Ibérica). *Cuadernos de Geología Ibérica,* **14,** 11–54.

GOY, A. & PÉREZ-LÓPEZ, A. 1996. Presencia de cefalópodos del tránsito Anisiense-Ladiniense en las facies Muschelkalk de la Zona Subbética (Cordillera Bética). *Geogaceta,* **20,**183–186.

GOY, A. & URETA, M. S. 1986. Leioceratinae (Ammonitina) del Aaleniense inferior de Fuentelsaz (Cordillera Ibérica, españa). *Bolletino de la Societa Paleontologica Italiana,* **3**(25), 213–216.

GOY, A. & URETA, M. S. 1990. El Aaleniense en la Cordillera Ibérica. *Les Cahiers de l'Université Catholique de Lyon, Série Sciences,* **4,** 73–87.

GOY, A. & URETA, S. 1988. Ammonitina del Toarciense superior en la Sierra Norte de Mallorca (España). *Boletín de la Real Sociedad Española de Historia Natural (Geología),* **84** (1–2), 19–38.

GOY, A., GÓMEZ, J. J. & YÉBENES, A. 1976. El Jurásico de la Rama Castellana de la Cordillera Ibérica (mitad norte). I Unidades litoestratigráficas. *Estudios Geológicos,* **32,** 391–423.

GOY, A., MELÉNDEZ, G., SEQUEIROS, L. & VILLENA, J. 1979. El Jurásico Superior del sector comprendido entre Molina de Aragón y Monreal del Campo (Cordillera Ibérica). *Cuadernos de Geología, Universidad de Granada,* **10,** 95–106.

GOY, A., JIMENEZ, A., MARTÍNEZ, G. & RIVAS, P. 1988. Difficulties in correlating the Toarcian ammonite succesion of the Iberian and Betic Cordilleras. *In:* Rocha, R. B. & Soares, A. F. (eds) *Second International Symposium on Jurassic Stratigraphy,* Lisboa, Vol. I, 155–178.

GOY, A., MARTÍNEZ, G. & URETA, S. 1994. The Toarcian in the Pozazal-Reinosa region (Cantabrian Mountains, Spain). *Coloquios de Paleontología,* **46,** 93–127.

GOY, A., MARTÍNEZ, G. & URETA, M. S. 1995. Ammonitina (Hammatoceratidae) of the Toarcian and Aalenian in the Serra de Llevant (Isle of Mallorca, Spain). *Hantkeniana,* **1,** 97–104.

GOY, J. L. 1984. Cartografía y Memoria del Cuaternario. *In: Mapa Geológico de Valencia 1/200.000.* Diputación Provincial Universidad de Valencia, IGME, 39–47.

GOY, J. L. & ZAZO, C. 1986. Western Almería (Spain) coastline changes since the Last Interglacial. *Journal of Coastal Research,* **1,** 89–93.

GOY, J. L. & ZAZO, C. 1988. Sequences of the Quaternary marine levels in Elche Basin (Eastern Betic Cordillera, Spain). *Palaeogeography, Palaeoclimatology, Palaeoecology,* **68,** 301–310.

GOY, J. L. & ZAZO, C. 1989*a.* The role of neotectonics in the morphologic distribution of the Quaternary marine and continental deposits of the Elche basin, southeast Spain. *Tectonophysics,* **163,** 219–225.

GOY, J. L. & ZAZO, C. 1989*b.* Cordilleras Bética y Baleares. *In: Territorio y Sociedad en España, I. Geografía Física.* Taurus, Madrid, 81–98.

GOY, J. L., ZAZO, C., DABRIO, C. J., HOYOS, M. & CIVIS, J. 1989. Geomorfología y evolución dinámica del Sector Suroriental de la Cuenca de Guadix-Baza (árrea Baza-Caniles). *Trabajos sobre Neógeno-Cuaternario,* **11,** 97–112.

GOY, J. L., ZAZO, C., BARDAJÍ, T., SOMOZA, L., CAUSSE, C. & HILLAIRE-MARCEL, C. 1993. Elémments d'une chronostratigraphie du Tyrrhénnien des réggions d'Alicante-Murcia, Sud-Est de l'Espagne. *Geodinamica Acta,* **6,** 103–119.

GOY, J. L., ZAZO, C. & RODRÍGUEZ-VIDAL, J. 1994. Cordilleras Beticas-Islas Baleares. *In:* GUTIÉRREZ ELORZA, M. (ed.) *Geomorfología de España.* Rueda, Madrid, 123–157.

GOY, J. L., ZAZO, C., SILVA, P. G., LARIO, J., BARDAJÍ, T. & SOMOZA, L. 1996. Evaluación geomorfológica del comportamiento neotectónico del Estrecho de Gibraltar (Zona Norte) durante el Cuaternario. *In: IV Coloquio Internacional sobre el Enlace Fijo del Estrecho de Gibraltar. El Medio Físsico,* Vol. II. SECEG, Madrid, 56–69.

GOY, J. L., ZAZO, C. & CUERDA, J. 1997. Evolución de las áreas margino-litorales de la Costa de Mallorca (I. Baleares) durante el Ultimo y Presente Interglacial: Nivel del mar Holoceno y clima. *Boletín Geológico y Minero,* **108,** 127–135.

GOY, J. L., ZAZO, C., DABRIO, C. J., BAENA, J., HARVEY, A., SILVA, P. G., GONZALO, F. & LARIO, J. 1998. Sea level and climate changes in the Cabo de Gata lagoon (Almería) during the last 6,500 yrBP. *INQUA Mediterranean sea-level Subcommission Newsletter,* **20,** 11–18.

GOY, J. L., ZAZO, C., DABRIO, C. J. (in press). A beach-ridge progradation complex reflecting periodical sea-level and climate variabilityduring the Holocene (Gulf of Almeria, Western-Mediterranean). *Geomorphology.*

GOZALO, R. 1994. Geología y Paleontología (ostrácodos) del Devónico Superior de Tabuenca (NE de la Cadena Ibérica Oriental). *Memorias Museo Paleontológico Universidad Zaragoza,* **6,** 1–291.

GOZALO, R. 1995. El Cámbrico de las Cadenas Ibéricas. *In:* GÁMEZ VINTANED, J. A. & LIÑÁN E. (eds) *Memorias de las IV Jornadas Aragonesas de Paleontología: 'La expansión de la vida en el Cámbrico'. Libro homenaje al Prof. Klaus Sdzuy.* Institución 'Fernando el Católico', Zaragoza, 137–167.

GOZALO, R. & LIÑÁN, E. 1988. Los materiales hercínicos de la Cordillera Ibérica en el contexto del Macizo Ibérico. *Estudios Geológicos,* **44,** 399–404.

GOZALO, R. & LIÑÁN, E. 1995. Leonian (early Middle Cambrian) *Paradoxides* biostratigraphy. *Beringeria,* Special Issue **2,** 169–171.

GOZALO, R. & LIÑÁN, E. 1999. Middle Cambrian correlation in the Mediterranean Subprovince and its palaeogeographical consequences. *Journal of Conference Abstracts,* **4,** Annual meeting of IGCP Project 376, 1012.

GOZALO, R., ÁLVARO, J. J., LIÑÁN, E., SDZUY, K. & TRUYOLS, J. 1993. La distribución de *Paradoxides (Acadoparadoxides) mureroensis* SDZUY, 1958 (Cámbrico Medio basal) y sus implicaciones paleobiogeográficas. *Cuadernos do Laboratorio Xeolóxico de Laxe,* **18,** 217–230.

GOZALO, R., LIÑÁN, E. & ÁLVARO, J. 1994. Trilobites de la familia Solenopleuropsidae Thoral, 1947 del Cámbrico Medio de la Unidad de Alconera (zona de Ossa Morena, SO España). *Boletín de la Real Sociedad Española de Historia Natural (Geología),* **89,** 43–54.

GRAAFF, W. J. E. VAN DE 1971. Facies distribution and basin configuration in the Pisuerga area before the Leonian Phase. *Trabajos de Geología*, **3**, 161–177.

GRACIA, A. S., GARCÍA MARCOS, J. J. & JIMÉNEZ, E. 1981. Las fallas de El Cubito: geometría, funcionamiento y sus implicaciones cronoestratigráficas en el Terciario de Salamanca. *Boletín Geológico y Minero de España*, **92**(4), 267–273.

GRACIA, F. J. 1995. Shoreline forms and deposits in Gallocanta Lake (NE Spain). *Geomorphology*, **11**, 323–335.

GRACIA, F. J. & SIMÓN, J. L. 1986. El campo de fallas miocenas de la Bardena Negra (Provs. de Navarra y Zaragoza). *Boletín Geológico y Minero*, **97**, 693–703.

GRACIA, F. J., GUTIÉRREZ, F. & GUTIÉRREZ, M. 1999. Evolución geomorfológica del polje de Gallocanta (Cordillera Ibérica). *Revista Sociedad Geológica de España*, **12**, 351–368.

GRADSTEIN, F. M. & OGG, J. 1996. A Phanerozoic time scale. *Episodes*, **19**(1–2), 3–5.

GRADSTEIN, F. M., AGTERBERG, F. P., OGG, J. G., HARDENBOL, J., VAN VEEN, P., THIERRY, J. & HUANG, Z. 1995. A Triassic, Jurassic and Cretaceous time scale. *In*: BERGGREN, W. A., KENT, D. V., AUBRY, M. P. & HARDENBOL, J. (eds) *Geochronology, Time Scales and Global Stratigraphic Correlation*. SEPM, Tulsa, Special Publications, **54**, 95–126.

GRÄFE, K.-U. 1994. *Sequence Stratigraphy in the Cretaceous and Paleogene (Aptian to Eocene) of the Basco-Cantabrian Basin (N. Spain)*. Tübinger geowissenschaftliche Arbeiten, (A) **18**.

GRÄFE, K.-U. 1999. Sedimentary cycles, burial history and foraminiferal indicators for systems tracts and sequence boundaries in the Cretaceous of the Basco-Cantabrian Basin (Northern Spain). *Neues Jahrbuch für Geologie und Paläontologie, Abhandlungen*, **212**, 85–130.

GRÄFE, K.-U. 2000. Cenomanian planktic foraminiferal stratigraphy in a deeper-water section (Galarreta, Basco-Cantabrian Basin, Northern Spain). *Newsletters on Stratigraphy*, **38**, 101–115.

GRÄFE, K.-U. & WIEDMANN, J. 1993. Sequence stratigraphy in the Upper Cretaceous of the Basco-Cantabrian Basin (Northern Spain). *Geologische Rundschau*, **82**, 327–361.

GRANDJEAN, G. 1992. *Mise en evidence des structures crustales dans une portion de chaine et de leur relation avec les bassins sedimentaires. Application au Pyrenees occidentales au travers du projet ECORS-Arzacq Pyrenees*. PhD Thesis, University of Montpellier II.

GREILING, L. & PUSCHMANN, H. 1965. Die Wende Silurium/Devon am St. Creu d'Olorde bei Barcelona (Katalonien). *Senckenbergiana lethaea*, **46**(4/6), 453–457.

GREY, J. E. 1997. Environmental geochemistry, mercury speciation, and effects to surrounding ecosystems of a belt of mercury deposits and abandoned mercury mines in southwestern Alaska, USA. *In*: PAPUNEN, M. (ed.) *Mineral Deposits*. Balkema, Rotterdam, 899–902.

GREY, J. E., THEODORAKOS, P. M., BAILEY, E. A. & TURENER, R. R. 2000. Distribution, speciation and transport of mercury in stream-sediments, stream-water, and fish collected near abandoned mercury mines in south-western Alaska, USA. *The Science of the Total Environment*, **260**, 21–33.

GRIMALT GELABERT, M. & RODRÍGUEZ PEREA, A. 1994. El modelado periglaciar en Baleares. Estado de la cuestión. *In*: GÓMEZ ORTIZ, A., SIMÓN TORRES, M. & SALVADOR FRANCH, F. (eds) *Periglaciarismo en la Península Ibérica, Canarias y Baleares*. Monografías de la SEG, **7**, 198–201.

GRIMAUD, S., BOILLOT, G., COLLETTE, B. J., MAUFFRET, A., MILES, P. R. & ROBERTS, D. B. 1982. Western extension of the Iberian-European plate boundary during the early Cenozoic (Pyrenean) convergence: a new model. *Marine Geology*, **45**, 63–67.

GROOT, J. J., DE JONGE, R. B. G., LANGEREIS, C. G., TEN KATE, W. G. H. G. & SMIT, J. 1989. Magnetostratigraphy of the Cretaceous–Tertiary boundary at Agost (Spain). *Earth and Planetary Science Letters*, **94**, 385–397.

GROSS, G. 1966. Palaozoikum und Tertiar am Puig Moreno (prov. Terual, Spanien). *Neues Jahrbuch für Geologie und Paläontologie Abhandlungen*, **9**, 554–562.

GRUNAU, H. R., LEHNER, P., CLEINTUAR, M. R., ALLENBACH, P. & BAKKER, G. 1975. New radiometric ages and seismic data from Fuerteventura (Canary Islands), Maio (Cape Verde) Islands), and Sao Tomé (Gulf of Guinea). *In*: BORRADAILLE, G. J. *et al.* (eds) *Progress in Geodynamics*. Royal Netherlands Academy of Arts and Sciences, 90–118.

GUÉRIN, G., BENHAMOU, G. & MALLARACH, J. M. 1986. Un exemple de fusió parcial en medi continental. El vulcanisme quaternari de Catalunya. *Vitrina*, **1**, 20–26.

GUERIN-DESJARDINS, B. & LATREILLE, M. 1962. Estudio geológico de los Pirineos españoles entre los ríos Segre y Llobregat (prov. de Lérida). *Boletín Geológico y Minero*, **73**, 329–371.

GUERRA-MERCHÁN, A. 1993. *Origen y relleno sedimentario de la cuenca neógena del Corredor de Almanzora y áreas limítrofes (Cordilleras Béticas)*. PhD Thesis, University of Málaga.

GUERRERA, F. & PUGLISI, D. 1983. Le Arenarie di Yesoma in Somalia: un possibile equivalente meridionale delle piu'note 'Nubian Sandstones'. *Rendiconti Società Geologica Italiana*, **6**, 42–47.

GUERRERA, F., MARTÍN-ALGARRA, A. & PERRONE, V. 1993. Late Oligocene–Miocene syn-/-late-orogenic successions in Western and Central Mediterranean Chains from the Betic Cordillera to the Southern Apennines. *Terra Nova*, **5**, 525–544.

GUILLOCHEAU, F. 1991. Modalités d'empilement des séquences génétiques dans un bassin de plate-forme (Dévonien armoricain): Nature et distorsion des différents ordres de séquences de dépots emboitées. *Bulletin Centres Recherche Exploration-Production Elf Aquitaine*, **15**(2), 383–410.

GUILLOU, H., CARRACEDO, J. C., PÉREZ TORRADO, F. & RODRÍGUEZ BADIOLA, E. 1996. K-Ar ages and magnetic stratigraphy of a hotspot-induced, fast grown oceanic island: El Hierro, Canary Islands. *Journal of Volcanology and Geothermal Research*, **73**, 141–155.

GUILLOU, H., CARRACEDO, J. C & DAY, S. J. 1998. Dating of the Upper Pleistocene-Holocene volcanic activity of La Palma using the unspiked K-Ar technique. *Journal of Volcanology and Geothermal Research*, **86**, 137–149.

GUILLOU, H., CARRACEDO, J. C & DUNCAN, R. 2001. K-Ar, 39Ar/40Ar Ages and Magnetostratigraphy of Brunhes and Matuyama Lava Sequences from La Palma Island. *Journal of Volcanology and Geothermal Research*, **106**, 175–194.

GUIMERÁ, J. 1984. Paleogene evolution of deformation in the north-eastern Iberian Peninsula. *Geological Magazine*, **121**, 413–420.

GUIMERÁ, J. 1988. *Estudi estructural de l'enllaç entre la Serralada Iberica y la Serralada Costanera Catalana*. PhD Thesis, University of Barcelona.

GUIMERÁ, J. 1996. Cenozoic evolution of eastern Iberia: Structural data and dynamic model. *Acta Geologica Hispánica*, **29**, 57–66.

GUIMERÀ, J. & ÁLVARO, M. 1990. Structure et évolution de la compression alpine dans la Chaîne ibérique et la Chaîne côtiere catalane (Espagne). *Bulletin de la Société Géologique de France*, **VI**, 339–348.

GUIMERÀ, J. & SALAS, R. 1996. Inversión terciaria de la falla normal mesozoica que limitaba la subcuenca de Galve. *Geogaceta*, **20**(7), 1701–1703.

GUIMERÀ, J., ALONSO, A. & MAS, R. 1995. Inversion of an extensional-ramp basin by a newly formed thrust: the Cameros basin (N Spain). *In*: J. G. BUCHANAN & P. G. BUCHANAN (eds) *Basin Inversion*. Geological Society, London, Special Publication, **88**, 433–453.

GUIMERÀ, J., SALAS, R., VERGES, J. & CASAS, A. 1996. Mesozoic extension and Tertiary compressive inversion in the Iberian Chain: Results from the analysis of a gravity profile. *Geogaceta*, **20**(7), 1691–1694.

GUIMERÀ, J., SALAS, R., MAS, R., MARTÍN-CLOSAS, C., MELÉNDEZ, A. & ALONSO, A. 2000. The Iberian Chain: tertiary inversion of a mesozoic intraplate basin. *Geotemas*, **1**(1), 67–69.

GUIRAUD, M. & SEGURET, M. 1984. Releasing solitary overstep model for the Late Jurassic-Early Cretaceous (Wealdian) Soria Strike-Slip Basin (North Spain). *In*: BIDDLE, K. T. & CHRISTIE-BLICK, N. (eds) *Strike-Slip Deformation, Basin Formation and Sedimentation*. SEPM, Tulsa, Special Publications, **37**, 159–175.

GUITARD, G. 1960. Sur la présence et l'âge d'un granite d'affinité charnockitique dans le massif de l'Agly (Pyrénées orientales).

Comptes Rendues de l'Académie des Sciences de Paris, **251**, 2554–2555.

GUITARD, G. 1964. Un exemple de structure en nappe de style pennique dans la chaîne hercynienne: les gneiss stratoïdes du Canigou (Pyrénées Orientales). *Comptes Rendus de l'Académie des Sciences de Paris*, **258**, 4597–4599.

GUITARD, G. 1970. *Le métamorphisme hercynien mésozonal et les gneiss oeillés du massif du Canigou, (Pyrénées Orientales).* BRGM, Mémoires, **63**.

GUITARD, G., GEYSSANT, J. & LAUMONIER, B. 1984. Les plissements hercyniens tardifs dans le Paléozoïque inférieur du versant nord du Canigou. 1ère partie: analyse géométrique et chronologie des phases superposées. Relations avec le granite de Mont-Louis et le métamorphisme régional. *Géologie de la France*, **4**, 95–125.

GUITARD, G., AUTRAN, A. & FONTEILLES, M. 1996. Le substratum Précambrien du Paléozoïque. *In*: BARNOLAS, A. & CHIRON, J. C. (eds) *Synthèse géologique et géophysique des Pyrénées. 1: Introduction. Géophysique. Cycle Hercynien.* BRGM and IGME, Orléans and Madrid, 137–155.

GUMIEL, P. 1984. *Tipología de los yacimientos de estaño y wolframio del Macizo Ibérico.* I Congreso Español Geología, Segovia, 183–216.

GUMIEL, P. & ARRIBAS, A. 1987. Antimony deposits in the Iberian Peninsula. *Economic Geology*, **82**, 1453–1463.

GUSI, F., GIBERT, J., AGUSTÍ, J. & PÉREZ-CUEVA, A. 1984. Nuevos datos del yacimiento Cova del Tossal de la Font (Vilafamés, Castellón). *Cuadernos de Prehistoria y Arqueología Castellonenses*, **10**, 7–18.

GUTIÉRREZ, F. 1996. Gypsum karstification induced subsidence: effects on alluvial systems and derived geohazards (Calatayud Graben, Iberian Range, Spain). *Geomorphology*, **16**, 277–293.

GUTIÉRREZ, F. 1998. *Fenómenos de subsidencia por disolución de formaciones evaporíticas en las fosas neógenas de Teruel y Calatayud (Cordillera Ibérica).* PhD Thesis, University of Zaragoza.

GUTIÉRREZ, F. & ARAUZO, T. 1997. Subsidencia kárstica sinsedimentaria en un sistema aluvial efímero: El Barranco de Torrecilla (Depresión del Ebro, Zaragoza). *Cuadernos de Geología Ibérica*, **22**, 349–372.

GUTIÉRREZ, F., GRACIA, F. J. & GUTIÉRREZ, M. 1996. Consideraciones sobre el final del relleno endorreico de las fosas de Calatayud y Teruel y su paso al exorreísmo. Implicaciones morfoestratigráficas y estructurales. *In*: GRANDAL, A. & PAGÉS, J. (eds) *IV Reunión de Geomorfología.* SEG, A Coruña, 23–43.

GUTIÉRREZ, F., ORTÍ, F., GUTIÉRREZ, M., PÉREZ-GONZÁLEZ, A., BENITO, G., GRACIA, J. & DURÁN, J. J. 2001. The stratigraphical record and activity of evaporite dissolution subsidence in Spain. *Carbonates and Evaporites*, **16**(1), 46–70.

GUTIÉRREZ, M. 2000. *Estudio petrológico, geoquímico y estructural de la serie volcánica submarina del Complejo Basal de Fuerteventura (Islas Canarias): caracterización del crecimiento submarino y de la emersión de la isla.* PhD Thesis, University of La Laguna.

GUTIÉRREZ, M. & GRACIA, F. J. 1997. Environmental interpretation and evolution of the Tertiary erosion surfaces in the Iberian Range (Spain). *In*: WIDDOWSON, M. (ed.) *Palaeosurfaces: Recognition, Reconstruction and Palaeoenvironmental Interpretation.* Geological Society, London, Special Publications, **120**, 147–158.

GUTIÉRREZ, M. & GUTIÉRREZ, F. 1998. Geomorphology of Tertiary gypsum formations in the Ebro Depression. *Geoderma*, **87**, 1–29.

GUTIÉRREZ, M. & PEDRAZA, J. 1974. Existencia de pizarrosidad alpina en la Cordillera Ibérica. *Boletín Geológico y Minero*, **85**, 269–270.

GUTIÉRREZ, M. & PEÑA, J. L. 1976. Glacis y terrazas en el curso medio del Río Alfambra (provincia de Teruel). *Boletín Geológico y Minero*, **87**, 561–570.

GUTIÉRREZ, M. & PEÑA, J. L. 1989a. La Cordillera Ibérica. *In*: PÉREZ GONZÁLEZ, A., CABRA, P. & MARTÍN-SERRANO, A. (eds) *Mapa del Cuaternario de España.* ITGE, Madrid, 141–151.

GUTIÉRREZ, M. & PEÑA, J. L. 1989b. Depresión del Ebro. *In*: PÉREZ-GONZÁLEZ, A., CABRA, R. & MARTÍN-SERRANO, A. (eds) *Mapa del Cuaternario de España.* ITGE, Madrid, 129–139.

GUTIÉRREZ, M. & PEÑA, J. L. 1994. Depresión del Ebro. *In*: GUTIÉRREZ, M. (ed.) *Geomorfología de España.* Rueda, Madrid, 305–349.

GUTIÉRREZ, M. & PEÑA, J. L. 1998. Geomorphology and late Holocene climatic change in Northeastern Spain. *Geomorphology*, **23**, 205–217.

GUTIÉRREZ, M., SANCHO, C. & ARAUZO, T. 1998a. Scarp retreat rates in semiarid environments from talus flatirons (Ebro Basin, NE Spain). *Geomorphology*, **25**, 11–21.

GUTIÉRREZ, M., SANCHO, C., ARAUZO, T. & PEÑA, J. L. 1998b. Evolution and paleoclimatic meaning of the talus flatirons in the Ebro Basin, NE of Spain. *In*: ALSHARHAN, A. S., GLENNIE, K. W. & WITTLE, G. L. (eds) *Quaternary Deserts and Climate Change.* Balkema, Amsterdam, 593–599.

GUTIÉRREZ ALONSO, G. 1996. Strain partitioning in the footwall of the Somiedo Nappe: structural evolution of the Narcea Tectonic Window, NW Spain. *Journal of Structural Geology*, **18**, 1217–1229.

GUTIERREZ-CAVEROL, M. & MANJÓN, M. 1984. Los depósitos evaporíticos del tránsito Permotrías-Lías en Asturias (Mina Felisa). *Revista de Minas*, **4**, 37–49.

GUTIÉRREZ-MARCO, J. C. 1982. Descubrimiento de nuevos niveles con graptolitos ordovícicos en la unidad 'Pizarras con *Didymograptus*' – Schneider 1939 – (Prov. Huelva, SW. de España). *Comunicações dos Serviços Geológicos de Portugal*, **68**, 241–246.

GUTIÉRREZ-MARCO, J. C. 1984. Una interesante señal de actividad biológica en el Ordovícico de los Montes de Toledo. *Coloquios de Paleontología*, **39**, 17–25.

GUTIÉRREZ-MARCO, J. C. 1986a. *Graptolitos del Ordovícico español.* PhD Thesis, Complutense University, Madrid.

GUTIÉRREZ-MARCO, J. C. 1986b. Notas sobre el desarrollo y estructura proximal del rhabdosoma en algunos graptolitos ordovícicos del SO de Europa. *Paleontogía i Evolució*, **20**, 191–201.

GUTIÉRREZ-MARCO, J. C. 1997. *Tolmachovia babini* nov. sp., nuevo ribeirioide (Mollusca, Rostroconchia) del Ordovícico Medio de la Zona Centroibérica Española. *Geobios-Mémoire Spécial*, **20**, 291–298.

GUTIÉRREZ-MARCO, J. C. 1999. Algunos ejemplos inusuales de nematulario en graptolitos silúricos del noreste de España. *Temas Geológico-Mineros ITGE*, **26**, 239–243.

GUTIÉRREZ-MARCO, J. C. 2000. Revisión taxonómica de '*Echinosphaerites*' *murchisoni* Verneuil y Barrande, 1855 (Echinodermata, Diploporita) del Ordovícico Medio centroibérico (España). *Geogaceta*, **27**, 121–124.

GUTIÉRREZ-MARCO, J. C. 2001. Cistoideos rombíferos (Echinodermata) de la Caliza Urbana (Ordovícico Superior) de la Zona Centroibérica, España. *Coloquios de Paleontología*, **52**, 91–100.

GUTIÉRREZ-MARCO, J. C. & ACEÑOLAZA, F. G. 1987. *Araneograptus murrayi* (Hall) (Graptoloidea, Anisograptidae): su identidad con *D. yaconense* Turner y distribución en España y Sudamérica. *Actas X Congreso Geológico Argentino*, Tucumán, **1**, 321–334.

GUTIÉRREZ-MARCO, J. C. & ACEÑOLAZA, G. F. 1999. *Calix inornatus* (Meléndez, 1958) (Echinodermata, Diploporita, Ordovícico): morfología de la región oral de la teca y revisión bioestratigráfica. *Temas Geológico-Mineros ITGE*, **26**, 557–565.

GUTIÉRREZ-MARCO, J. C. & BABIN, C. 1999. *Lyrodesma* et autres mollusques bivalves des Quartzites Botella (Ordovicien moyen) de la Zone Centre-Ibérique (Espagne). *Revista Española de Paleontología*, special volume, homenaje prof. J. Truyols, 229–238.

GUTIÉRREZ-MARCO, J. C. & BAEZA CHICO, E. 1996. Descubrimiento de *Aristocystites metroi* Parsley y Prokop, 1990 (Echinodermata, Diploporita) en el Ordovícico Medio centroibérico (España). *Geogaceta*, **20**, 225–227.

GUTIÉRREZ-MARCO, J. C. & LENZ, A. C. 1998. Graptolite synrhabdosomes: Biological or taphonomic entities? *Paleobiology*, **24**, 37–48.

GUTIÉRREZ-MARCO, J. C. & MARTÍN SÁNCHEZ, J. 1983. Estudio de los Monoplacóforos (Mollusca) del Ordovícico de los Montes de Toledo (España central). *Estudios Geológicos*, **39**, 379–385.

GUTIÉRREZ-MARCO, J. C. & MELÉNDEZ, B. 1987. Nuevos hallazgos de Estilóforos (Homalozoos) en los materiales ordovícicos de la Zona Centroibérica. *Coloquios de Paleontología*, **41**, 41–50.

GUTIÉRREZ-MARCO, J. C. & PINEDA VELASCO, A. 1988. Datos bioestratigráficos sobre los materiales silúricos del subsuelo de El

Centenillo (Jaén). *Comunicaciones II Congreso Geológico de España*, Granada, **1**, 91–94.

GUTIÉRREZ-MARCO, J. C. & RÁBANO, I. 1987a. Trilobites y graptolitos de las lumaquelas terminales de los 'Bancos Mixtos' (Ordovícico superior de la zona Centroibérica meridional). *Boletín Geológico y Minero*, **98**, 647–669.

GUTIÉRREZ-MARCO, J. C. & RÁBANO, I. 1987b. Paleobiogeographical aspects of the Ordovician mediterranean faunas. *Geogaceta*, **2**, 24–26.

GUTIÉRREZ-MARCO, J. C. & RÁBANO, I. 1996. First Iberian representatives of the genus *Hanadirella* (Problematica, Ordovician) and reevaluation of its biostratigraphic significance. *In*: BALDIS, B. & ACEÑOLAZA, F. G. (eds) *Early Paleozoic Evolution in NW Gondwana*. Serie Correlación Geológica, **12**, Tucumán, 271–272.

GUTIÉRREZ-MARCO, J. C. & RÁBANO, I. 1997. Los materiales del Ordovícico y Silúrico de la región limítrofe entre los Dominios de Caurel-Peñalba (Zona Asturoccidental-leonesa) y Truchas (flanco Norte del antiforme del Ollo de Sapo: Zona Centroibérica). *In*: GRANDAL D'ANGLADE, A., GUTIÉRREZ-MARCO, J. C. & SANTOS FIDALGO, L. (eds) *XIII Jornadas de Paleontología y V Reunión Internacional Proyecto 351 PICG*. A Coruña, Libro de Resúmenes y Excursiones, 298–313 (ISBN 84-605-6825-3).

GUTIÉRREZ-MARCO, J. C. & ROBARDET, M. 1991. Découverte de la zone à *Parakidograptus acuminatus* (base du Llandovery) dans le Silurien du synclinorium de Truchas (Zone asturo-léonaise, no. de l'Espagne): conséquences stratigraphiques et paléogéographiques au passage Ordovicien-Silurien. *Comptes Rendus de l'Académie des Sciences, Paris, (2)*, **312**, 729–734.

GUTIÉRREZ-MARCO, J. C. & RODRÍGUEZ, L. 1987. Descubrimiento de graptolitos arenigienses en la Escama de Rioseco (zona Cantábrica, N de España). *Cuadernos do Laboratório Xeolóxico de Laxe*, **11**, 209–220.

GUTIÉRREZ-MARCO, J. C. & STORCH, P. 1997. Graptolitos silúricos del sinclinal de Castrillo (Zona Asturoccidental-Leonesa, NO de España): revisión del yacimiento descubierto por Casiano de Prado en 1855. *Geogaceta*, **22**, 89–92.

GUTIÉRREZ-MARCO, J. C. & STORCH, P. 1998. Graptolite biostratigraphy of the Lower Silurian (Llandovery) shelf deposits of the Western Iberian Cordillera, Spain. *Geological Magazine*, **135**, 71–92.

GUTIÉRREZ-MARCO, J. C., CHAUVEL, J., MELÉNDEZ, B, & SMITH, A. B. 1984a. Los equinodermos (Cystoidea, Homalozoa, Stelleroidea, Crinoidea) del Paleozoico inferior de los Montes de Toledo y Sierra Morena (España). *Estudios Geológicos*, **40**, 421–453.

GUTIÉRREZ-MARCO, J. C., LUNAR, R. & AMORÓS, J. L. 1984b. Los depósitos de hierro oolítico en el Ordovícico de España. Significado paleogeográfico. *I Congreso Español de Geología, Segovia*, **2**, 501–525 (ISBN 84-500-9834-3).

GUTIÉRREZ MARCO, J.C., RÁBANO, I., PRIETO, M. & MARTÍN, J. 1984c. Estudio bioestratigráfico del Llanvirn y Llandeilo (Dobrotiviense) en la parte meridional de la Zona Centroibérica (España). *Cuadernos de Geología Ibérica*, **9**, 289–321.

GUTIÉRREZ-MARCO, J.C., RÁBANO, I. & ROBARDET, M. 1984d. Estudio bioestratigráfico del Ordovícico en el sinclinal del Valle (provincia de Sevilla, SO. de España). *Memórias e Notícias, Coimbra*, **97**, 12–37.

GUTIÉRREZ-MARCO, J. C., SAN JOSÉ, M. A. & PIEREN, A. P. 1990. Central-Iberian Zone, Autochthonous sequences: post-Cambrian Palaeozoic Stratigraphy. *In*: DALLMEYER, R. D. & MARTÍNEZ GARCÍA, E. (eds) *Pre-Mesozoic Geology of Iberia*. Springer-Verlag, Berlin, 160–171.

GUTIÉRREZ-MARCO, J.C., SAAVEDRA, J. & RÁBANO, I. (eds) 1992. *Paleozoico Inferior de Ibero-América*. University of Extremadura.

GUTIÉRREZ-MARCO, J. C., RÁBANO, I., SAN JOSÉ, M. A., HERRANZ, P. & SARMIENTO, G. N. 1995. Oretanian and Dobrotivian stages vs. 'Llanvirn-Landeilo' Series in the Ordovician of the Iberian Peninsula. *In*: COOPER, J. D., DROSER, M. L. & FINNEY, S. C. (eds) *Ordovician Odyssey: short papers for the 7th ISOS, Las Vegas*. Society for Sedimentary Geology, Pacific Section, Fullerton, **77**, 55–59.

GUTIÉRREZ-MARCO, J. C., ALBANI, R., ARAMBURU, C. *et al.* 1996a. Bioestratigrafía de la Formación Pizarras del Sueve (Ordovícico

Medio) en el sector septentrional de la Escama de Laviana-Sueve (Zona Cantábrica, N de España). *Revista Española de Paleontología*, **11**, 48–74.

GUTIÉRREZ-MARCO, J. C., ARAMBURU, C., ARBIZU, M., MÉNDEZ-BEDIA, I., RÁBANO, I. & VILLAS, E. 1996b. Rasgos estratigráficos de la sucesión del Ordovícico Superior en Portilla de Luna (Zona Cantábrica, noroeste de España). *Geogaceta*, **20**, 11–18.

GUTIÉRREZ-MARCO, J.C., CHAUVEL, J. & MELÉNDEZ, B. 1996c. Nuevos equinodermos (cistideos y blastozoos) del Ordovícico de la Cordillera Ibérica (NE de España). *Revista Española de Paleontología*, **11**, 100–119.

GUTIÉRREZ-MARCO, J.C., LENZ, A. C., ROBARDET, M. & PIÇARRA, J.M. 1996d. Wenlock–Ludlow graptolite biostratigraphy and extinction: a reassessment from the southwestern Iberian Peninsula (Spain and Portugal). *Canadian Journal of Earth Sciences*, **33**, 656–663.

GUTIÉRREZ-MARCO, J. C., ARAMBURU, C., ARBIZU, M., MÉNDEZ-BEDIA, I., RÁBANO, I., TRUYOLS, J. & VILLAS, E. 1997a. Caracterización estratigráfica del Ordovícico Superior en el Manto de Mondoñedo (Zona Asturoccidental-leonesa, no. de España): primeras dataciones paleontológicas y correlación. *In*: GRANDAL D'ANGLADE, A., GUTIÉRREZ-MARCO, J. C. & SANTOS FIDALGO, L. (eds) *XIII Jornadas de Paleontología y V Reunión Internacional Proyecto 351 PICG*. A Coruña, Libro de Resúmenes y Excursiones, 33–37 (ISBN 84-607-6825-3).

GUTIÉRREZ-MARCO, J. C., BABIN, C. & PORRO MAYO, T. 1997b. Moluscos bivalvos de las facies cuarcíticas del Ordovícico Inferior centroibérico. *Geogaceta*, **22**, 85–88.

GUTIÉRREZ-MARCO, J. C., RÁBANO, I. & STORCH, P. 1997c. Fósiles ordovícico-silúricos de Galicia. *In*: GRANDAL D'ANGLADE, A., GUTIÉRREZ-MARCO, J. C. & SANTOS FIDALGO, L. (eds) *XIII Jornadas de Paleontología y V Reunión Internacional Proyecto 351 PICG*. A Coruña, Libro de Resúmenes y Excursiones, 10–16.

GUTIÉRREZ-MARCO, J. C., PORRO MAYO, T., HERRANZ ARAÚJO, P. & GARCÍA PALACIOS, A. 1997d. Dos nuevos yacimientos con graptolitos silúricos en la región de Alange (Badajoz). *Geogaceta*, **21**, 131–133.

GUTIÉRREZ-MARCO, J. C., ROBARDET, M. & PIÇARRA, J. M. 1998. Silurian stratigraphy and paleogeography of the Iberian Peninsula (Spain and Portugal). *Temas Geológico-Mineros ITGE*, **23**, 13–44.

GUTIÉRREZ-MARCO, J. C., ARAMBURU, C., ARBIZU, M., ROBARDET, M. & ROQUÉBERNAL, J. 1999a. Revisión bioestratigráfica de las pizarras del Ordovícico Medio en el noroeste de España (Zonas Cantábrica, Asturoccidental-leonesa y Centroibérica septentrional). *Acta Geologica Hispanica*, **34**, 3–87.

GUTIÉRREZ MARCO, J. C., FERRER, E., ROBARDET, M. & ROQUE-BERNAL, J. 1999b. Graptolitos multirramosos del Devónico de las Cordilleras Costero Catalanas (noreste de España). *Temas Geológico-Mineros ITGE*, **26**, 610–617.

GUTIÉRREZ-MARCO, J. C., RÁBANO, I., SARMIENTO, G. N. *et al.* 1999c. Faunal dynamics between Iberia and Bohemia during the Oretanian and Dobrotivian (late Middle–earliest Upper Ordovician), and biogeographic relations with Avalonia and Baltica. *Acta Universitatis Carolinae, Geologica*, **43**, 487–490.

GUTIÉRREZ-MARCO, J.C., ROQUÉ BERNAL, J., ROBARDET, M.& IBÁÑEZ SOTILLOS, R. 1999d. Graptolitos de la Biozona de *Coronograptus cyphus* (Rhuddaniense: Silúrico inferior) en el área del Montseny (Cadenas Costeras Catalanas, noreste de España). *Temas Geológico-Mineros ITGE*, **26**, 618–622.

GUTIÉRREZ-MARCO, J.C., ACEÑOLAZA, G. F. & ACEÑOLAZA, F. G. 2000. Primer registro del gasterópodo *Lesueurilla* y euomphalomorfos afines en el Ordovícico Inferior de Argentina y España. Su interés paleobiogeográfico. *Boletín Geológico y Minero*, **111**, 85–94.

GUTIÉRREZ-MARCO, J. C., SARMIENTO, G. N., ROBARDET, M., RÁBANO, I. & VANEK, J. 2001. Upper Silurian fossils of Bohemian type from NW Spain and their palaeogeographical significance. *Journal of the Czech Geological Society*, **46**(3), 161–172.

GUTIÉRREZ NARVONA, R. 1999. *Implicaciones metalogenéticas (cromo y elementos del grupo del platino) de los magmas/fluidos residuales de un proceso de precolación a gran escala en los macizos ultramáficos de Ronda y Ojén (Béticas, Sur de España)*. PhD Thesis, University of Granada.

HABERFELNER, E. 1931. Eine revision der graptolithen der Sierra Morena (Spanien). *Abhandlungen der Senckenbergischen Naturforschenden Gesellschaft*, **43**, 19–66.

HAFENRICHTER, M. 1979. Paläontologish-Ökologische und lithofazielle Untersuchungen des 'Ashgill-Kalkes' (Jungordovizium) in Spanien. *Arbeiten aus dem Paläontologischen Institut Würzburg*, **3**, 1–139.

HAFENRICHTER, M. 1980. The lower and upper boundary of the Ordovician system of some selected regions (Celtiberia, Eastern Sierra Morena) in Spain. Part II: The Ordovician–Silurian boundary in Spain. *Neues Jahrbuch für Geologie und Paläontologie, Abhandlungen*, **160**, 138–148.

HALL, C. A. & JOHNSON, J. A. 1986. Apparent western termination of the north Pyrenean fault and tectonostratigraphic units of the western north Pyrenees, France and Spain. *Tectonics*, **5**, 607–627.

HALL, C. M., HIGUERAS, P., KESLER, S. E., LUNAR, R., DONG, A. N. & HALLIDAY, A. N. 1997. Laser $^{40}Ar/^{39}Ar$ dating of mercury mineralisation, Almadén district, Spain. *Earth and Planetary Science Letters*, **148**, 287–298.

HALLIDAY, A. N., LEE, D. C., TOMMASINI, S., DAVIES, G. R., PASLICK, C. R., FITTON, J. G. & JAMES, D. E. 1995. Incompatible trace elements in OIB and MORB and source enrichment in the sub-oceanic mantle. *Earth and Planetary Science Letters*, **133**, 379–395.

HALSALL, C. E. 1989. *Intrusive Magmatism, Volcanism, and Massive Sulphide Mineralisation at Rio Tinto, Spain*. PhD Thesis, University of London.

HAMMANN, W. 1974. Phacopina und Cheirurina (Trilobita) aus dem Ordovizium von Spanien. *Senckenbergiana lethaea*, **53**, 1–151.

HAMMANN, W. 1976*a*. Trilobiten aus dem oberen Caradoc der östlichen Sierra Morena (Spanien). *Senckenbergiana lethaea*, **57**, 35–85.

HAMMANN, W. 1976*b*. The Ordovician of the Iberian Peninsula – A review. *In*: BASSETT, M. G. (ed.) *The Ordovician System: Proceedings of a Palaeontological Association Symposium, Birmingham 1974*. University of Wales Press and National Museum of Wales, Cardiff, 387–409.

HAMMANN, W. 1983. Calymenacea (Trilobita) aus dem Ordovizium von Spanien; ihre Biostratigraphie, Ökologie und Systematik. *Abhandlungen der senckenbergischen naturforschenden Gesellschaft*, **542**, 1–177.

HAMMANN, W. 1992. The Ordovician trilobites from the Iberian Chains in the province of Aragón, NE-Spain. I. The trilobites of the Cystoid Limestone (Ashgill Series). *Beringeria*, **6**, 3–219.

HAMMANN, W. & HENRY, J.-L. 1978. Quelques espèces de *Calymenella, Eohomalonotus* et *Kerfornella* (Trilobita, Ptychopariida) de l'Ordovicien du Massif Armoricain et de la Péninsule Ibérique. *Senckenbergiana lethaea*, **59**, 401–429.

HAMMANN, W. & LEONE, F. 1997. Trilobites of the post-Sardic (Upper Ordovician) sequence of southern Sardinia. Part I. *Beringeria*, **20**, 3–217.

HAMMANN, W. & RÁBANO, I. 1987. Morphologie und Lebensweise der Gattung *Selenopeltis* (Hawle & Corda, 1847) und ihre Vorkomenn im Ordovizium von Spanien. *Senckenbergiana lethaea*, **68**, 91–137.

HAMMANN, W. & SCHMINCKE, S. 1986. Depositional environment and systematics of a new ophiuroid, *Taeniaster ibericus* n.sp., from the Middle ordovician of Spain. *Neues Jahrbuch für Geologie und Paläontologie Abhandlungen*, **173**, 47–74.

HAMMANN, W., ROBARDET, M. & ROMANO M. 1982. *The Ordovician System in southwestern Europe (France, Spain and Portugal). Correlation Chart and Explanatory Notes*. IUGS, Publication **11**.

HANCOCK, P. L. 1991. Determining contemporary stress directions from neotectonic joint system. *Philosophical Transactions of the Royal Society of London*, **A337**, 29–40

HANCOCK, P. L. & ENGELDER, T. 1989. Neotectonic joints. *Geological Society of America Bulletin*, **101**, 1197–1208.

HANKOCK, J. T. & KAUFFMAN, E. G. 1979. The great transgressions of the late Cretaceous. *Journal of the Geological Society London* **136**, 175–186.

HAQ, B. U., HANDERBOL, J. & VAIL, P. R. 1988. Mesozoic and Cenocoic chronostratigraphy and cycles of sea level change. *In*: *Sea Level Changes. An Integrated Approach*. SEPM, Special Publications, **42**, 71–78.

HARLAND, W. B., ARMSTRONG, R. L., COX, A. V., CRAIG, L. E., SMITH, A. G. & SMITH, D. G. 1990. *A Geologic Time Scale 1989*. Cambridge University Press, Cambridge.

HARMON, R. S. & HOEFS, J. 1995. Oxygen isotope heterogeneity of the mantle deduced from global ^{18}O systematics of basalts from different geotectonic settings. *Contributions to Mineralogy and Petrology*, **120**, 95–114.

HARRIS, M. 1980. Gold mineralisation at the Salave gold prospect, northwest Spain. *Transactions Institute of Mining and Metallurgy*, **Feb**, 1–4.

HART, S. R. 1984. A large-scale isotope anomaly in the Southern Hemisphere mantle. *Nature*, **309**, 753–757.

HART, S. R. 1988. Heterogeneous mantle domains: signatures, genesis and mixing chronologies. *Earth and Planetary Science Letters*, **90**, 273–296.

HARTEVELT, J. J. A. 1970. Geology of the upper Segre and Valira Valleys, Central Pyrenees, Andorra/Spain. *Leidse Geologische Mededelingen*, **45**, 167–236.

HARTMANN, G. & WEDEPOHL, K. H. 1990. Metasomatically altered peridotite xenoliths from the Hessian Depression (Northwest Germany). *Geochimica et Cosmochimica Acta*, **54**, 71–86.

HARTUNG, G. 1857. Die geologischen Verhaltnisse der Inseln Lanzarote und Fuerteventura. *Neue Denkschrift Allgemeine Schweizerische Gesellschaft Naturwissenschaften*, **15**, 1–168.

HARVEY, A. M. 1990. Factors influencing Quaternary alluvial fan development in Southeast Spain. *In*: ROCHOCKI, A. H., CHURCH, M. (eds) *Alluvial Fans: a Field Approach*. John Wiley, Chichester, 247–269.

HARVEY, A. M., SILVA, P. G., MATHER, A. E., GOY, J. L., SOTOKES, M. & ZAZO, C. 1999. The impact of Quaternary sea-level and climatic change on coastal alluvial fans in the Cabo de Gata ranges, southeast Spain. *Geomorphology*, **28**, 1–22.

HATZFELD, D. 1976. Etude sismologique et gravimétique de la structure profonde de la mer d'Alboran, mise en évidence d'un manteau anormal. *Comptes Rendus de l'Académie des Sciences de Paris*, **283**, 1021–1024.

HAUDE, R. 1992. Scyphocrinoiden, die Bojen-Seelilien im höhem Silur-tiefen Devon. *Palaeontographica*, Abt. A, **222**, 141–187.

HAUSEN, H. 1949. Om Calderabildningar med saärskild hänsyn till Kanarieöarna. *Societas Scientiarum Fennica, Commentationes Physico-Mathematicae*, **28**, B, 2.

HAUSEN, H. 1956. Contributions to the geology of Tenerife (Canary Islands). *Societas Scientiarum Fennica, Commentationes Physico-Mathematicae*, **18**, 1–254.

HAUSEN, H. 1958. Contribución al conocimiento de las formaciones sedimentarias de Fuerteventura (Islas Canarias). *Anuario de Estudios Atlánticos*, **4**, 37–84.

HAUSEN, H. 1959. On the geology of Lanzarote, Graciosa and the isletas (Canarian Archipelago). *Societas Scientiarum Fennica, Commentationes Physico-Mathematicae*, **23**, 4.

HAUSEN, H. 1961. Canarian calderas; a short review based on personal impressions, 1947–1957. *Bulletin de la Commission Geologique de Finlande*, **196**, 33, 179–213.

HAUSEN, H. 1962. New contributions to the geology of Gran Canary (Gran Canaria, Canary Islands). *Societas Scientiarum Fennica, Commentationes Physico-Mathematicae*, **27**, 1, 1–418.

HAUSEN, H. 1964. Rasgos generales de la isla de El Hierro. *Anuario de Estudios Atlánticos*, **10**, 547–593.

HAUSEN, H. 1971. Outlines of the Geology of Gomera. *Societas Scientiarum Fennica, Commentationes Physico-Mathematicae*, **41**, 1–53.

HAVLÍCEK, V. 1961. Paleozoikum. *In*: Cepek, L. *et al.* (eds) *Vysvetlivky k prehledné geologické mape CSSR 1:200,000, M-33-XX Plzen*. Praha.

HAVLÍCEK, V. 1976. Evolution of Ordovician brachiopod communities in the Mediterranean Province. *In*: BASSETT, M. G. (ed.) *The Ordovician System, Proceedings of a Palaeontological Association Symposium, Birmingham 1974*. University of Wales Press and National Museum of Wales, Cardiff, 349–358.

HAVLÍCEK, V. & FATKA, O. 1992. Ordovician of the Prague Basin (Barrandian area, Czechoslovakia). *In*: WEBBY, B. D. & LAURIE, J. R. (eds) *Global Perspectives on Ordovician Geology*. Balkema, Rotterdam, 461–472.

HAVLÍCEK, V. & JOSOPAIT, V. 1972. Articulate brachiopods from the Iberian Chains, Northeast Spain (Middle cambrian-Upper Cambrian-Tremadoc). *Neues Jahrbuch für Geologie und Paläontologie Abhandlungen*, **140**, 238–353.

HAVLÍCEK, V. & MAREK, L. 1973. Bohemian Ordovician and its international correlation. *Casopis pro mineralogii a geologii*, **18**, 225–232.

HAVLÍCEK, V. & VANEK, J. 1966. The biostratigraphy of the Ordovician of Bohemia. *Sbornik Geologických Ved, Paleontologie*, **8**, 7–69.

HAVLÍCEK, V. & VANEK, J. 1996. Dobrotivian/Berounian boundary interval in the Prague Basin with a special emphasis on the deepest part of the trough (Ordovican, Czech Republic). *Vestník Ceského geologického ústavu*, **71**, 225–243.

HAZERA, J. 1968. La région de Bilbao et son arrière pays. *Munibe*, **20**, 1–354.

HEARTY, P. J. 1987. New data on the Pleistocene of Mallorca. *Quaternary Science Reviews*, **6**, 245–257.

HEARTY, P. J., HOLLIN, J. T. & DUMAS, B. 1987. Geochronology of Pleistocene littoral deposits on the Alicante and Almería coasts of Spain. *Trabajos sobre Neógeno-Cuaternario*, **10**, 95–107.

HEBEDA, E. M., BOELRIJK, N. A. I. M., PRIEM, H. N. A. & VENDURMEN, R. H. 1980. Excess radiogenic Ar and undisturbed Rb-Sr systems in basic intrusives subjected to Alpine metamorphism in SE Spain. *Earth and Planetary Science Letters*, **47**, 81–90.

HEDDEBAUT, C. 1973. *Etudes gélogiques dans les Massifs paléozoiques basques*. PhD Thesis, University of Lille.

HEDDEBAUT, C. 1975. Etudes géologiques dans les massifs paléozoïques basques. *Bulletin du BRGM* (2ème série), section IV, **1**, 5–30.

HEGNER, E., WALTER, H. J. & SATIR, M. 1995. Pb-Sr-Nd isotopic compositions and trace element geochemistry of megacrysts and melilitites from the Tertiary Urach volcanic field: source composition of small volume melts under SW Germany. *Contributions to Mineralogy and Petrology*, **122**, 322–335.

HEINZ, W. 1984. *Kartierung altpaläozoischer Schichten und Beschreibung des phreatomagmatischen Vulkanismus im Ordovizium des südlichen Kantabrischen Gebirges (Provin León, Spanien)*. Diploma Thesis, University of Tübingen.

HEINZ, W., LOESCHKE, J. & VAVRA, G. 1985. Phreatomagmatic volcanism during the Ordovician of the Cantabrian Mountains (NW Spain). *Geologische Rundschau*, **74**, 623–639.

HELMIG, H. M. 1965. The geology of the Valderrueda, Tejerina, Ocejo and Sabero coal basins (Cantabrian Mountains, Spain). *Leidse Geologische Mededelingen*, **32**, 75–149.

HEMLEBEN, CH. & REUTHER, C.-D. 1980. Allodapic limestones of the Barcaliente Fm (Namurian A) between Luna and Cea Rivers (Southern Cantabrian Mountains, Spain). *Neues Jahrbuch für Geologie und Paläontologie Abhandlungen*, **159**, 225–255.

HENN, A. & JAHNKE, H. 1984. Die palentinische Faziesentwicklung im Devon des Kantabrischen Gebirges. *Zeitschrift deutschen geologischen Gesellschaft*, **135**, 131–147.

HENNINGSEN, D. 1982. Zusammensetzung und Herkunft der sandigen gesteine des Devons und Karbons von Menorca (Balearen, Mittlemeer). *Neues Jahrbuch für Geologie und Paläontologie Monatshefte*, **11**, 736–746.

HERAIL, G. 1984. *Géomorphologie et gîtologie de l'or detritique. Piémonts et bassins intramontagneux du Nord Ouest de l'Espagne*. Centre National de la Recherche Scientifique, Toulouse.

HERBIG, H.-G. 1983. El Carbonífero de las Cordilleras Béticas. *In*: MARTÍNEZ DÍAZ, C. (ed.) *Carbonífero y Pérmico de España*. IGME, Madrid, 343–356.

HERBIG, H. G. 1985. An Upper Devonian limestone slide block near Marbella (Betic Cordillera, Southern Spain) and the palaeogeographic relations between Malaguides and Menorca. *Acta Geológica Hispánica*, **20**(2), 155–178.

HEREDIA, N. 1991. *Estructura geológica de la Región del Mampodre y áreas adyacentes (Zona Cantábrica)*. PhD Thesis, University of Oviedo.

HEREDIA, N., RODRÍGUEZ FERNÁNDEZ, L. R. & WAGNER, R. H. 1990. Carboniferous of the Palentian Zone. *In*: DALLMEYER, R. D. & MARTÍNEZ GARCÍA, E. (eds) *Pre-Mesozoic Geology of Iberia*. Springer-Verlag, Berlin, 34–38.

HERMITE, H. 1879. *Etudes géologiques sur les îles Baléares; premier partie Majoruqe et Minorque*. PhD Thesis, University of Paris.

HERNAIZ, P. P. & SOLÉ, J. 2000. Las estructuras del diapiro de Salinas de Rosío y del alto de San Pedro-Iglesias y sus implicaciones en la evolución tectónica de la transversal burgalesa de la Cordillera Vascocantábrica-Cuenca del Duero. *Revista de la Sociedad Geológica de España*, **13**, 471–486.

HERNÁN, F. 1976. Estudio petrológico y estructural del complejo traquítico-sienítico de Gran Canaria. *Estudios Geológicos*, **32**, 279–324.

HERNÁN, F. & VÉLEZ, R. 1980. El sistema de diques cónicos de Gran Canaria y la estimación estadística de sus características. *Estudios Geológicos*, **36**, 65–73.

HERNAN, F., PERNI, A. & ANCOECHEA, E. 1981. El vulcanismo del área de Atienza. Estudio petrológico. *Estudios Geológicos*, **37**, 13–25.

HERNÁN, F., CUBAS, C. R., HUERTAS, M. J., BRÄNDLE, J. L. & ANCOCHEA, E. 2000. Geometría del enjambre de diques cónicos de Vallehermoso. La Gomera (Islas Canarias). *Geogaceta*, **27**, 87–90.

HERNÁNDEZ, A., JÉBRAK, M., HIGUERAS, P., OYARZUN, R., MORATA, D. & MUNHÁ, J. 2000. The Almadén mercury mining district, Spain. *Mineralium Deposita*, **34**, 539–548.

HERNÁNDEZ, J., DE LAROUZIÈRE, F. D., BOLZE, J. & BORDET, P. 1987. Le magmatisme néogene bético-rifain et le couloir de décrochement 'trans-Alborán'. *Bulletin de la Société Géologique de France*, **3**, 257–267.

HERNÁNDEZ, J. M. 2000. *Sedimentología, paleogeografía y relaciones tectonica/ sedimentación de los sistemas fluviales, aluviales y palustres de la cuenca rift de Aguilar (Grupo Campóo, Jurásico superior-Cretácico inferior de Palencia, Burgos, Cantabria)*. PhD Thesis, University of País Vasco.

HERNÁNDEZ, J. M., PUJALTE, V., ROBLES, S. & MARTÍN-CLOSAS, C. 1999. Propuesta de una nueva división estratigráfica para el Grupo Campóo en su area tipo (Jurásico superior-Cretácico basal del N de Burgos y S de Cantabria). *Revista de la Sociedad Geológica de España*, **12**, 277–296

HERNÁNDEZ ENRILE, J. L. 1971. Las rocas porfiroides del límite Cámbrico-Precámbrico en el flanco meridional del anticlinal Olivenza-Monesterio. *Boletín Geológico y Minero*, **82**(3–4), 143–154.

HERNÁNDEZ ENRILE, J. L. 1991. Extensional tectonics of the Toledo ductile-brittle shear zone, central Iberian Massif. *Tectonophysics*, **191**, 311–324.

HERNÁNDEZ ENRILE, J. L. & GUTIÉRREZ ELORZA, M. 1968. Movimientos caledónicos (fases salaírica, sárdica y érica) en la Sierra Morena occidental. *Boletín de la Real Sociedad Española de Historia Natural (Geología)*, **66**, 21–28.

HERNÁNDEZ-MOLINA, F. J., SANDOVAL, J., AGUADO, R., O'DOGHERTY, L., COMAS, M. C. & LINARES, A. 1991. Olistoliths from the Middle Jurassic in Cretaceous materials of the Fardes formation, Biostratigraphy (Subbetic Zone, Betic Cordillera). *Revista de la Sociedad Geológica de España*, **4**, 79–104.

HERNÁNDEZ-PACHECO, A. 1973. Sobre la supuesta existencia de unas 'rocas graníticas' en la isla de La Palma (Canarias). *Estudios Geológicos*, **19**, 107–109.

HERNÁNDEZ-PACHECO, A. 1982. Sobre una posible erupción en 1793 en la isla de El Hierro (Canarias). *Estudios Geológicos*, **38**, 15–25.

HERNÁNDEZ PACHECO, E. 1907. Los martillos de piedra y las piedras con cazoletas de las antiguas minas de cobre de la Sierra de Córdoba. *Boletín de la Real Sociedad Española de Historia Natural*, **7**, 279–292.

HERNÁNDEZ-PACHECO, E. 1960. En relación con las grandes erupciones volcánicas del siglo XVIII y 1824 en Lanzarote. *El Museo Canario Las Palmas de Gran Canaria*, 239–254.

HERNÁNDEZ-PACHECO, E. & OBERMAIER, H. 1915. *La mandíbula neandertaloide de Bañolas*. Museo Nacional de Ciencias Naturales, Madrid. Publicaciones de la Comisión de Investigaciones Paleontológicas y Prehistóricas, **6**.

HERNÁNDEZ-PACHECO, F. 1923. Las arenas voladoras de la provincia de Segovia. *Boletín de la Real Sociedad Española de Historia Natural (Geol.)*, **23**, 211–216.

HERNÁNDEZ-PACHECO, F. 1969. Los niveles de playas cuaternarias de

Lanzarote. *Revista de la Real Academia de Ciencias Exactas, Físsicas y Naturales de Madrid*, **63**, 903–961.

HERNÁNDEZ SAMPELAYO, P. 1922. *Hierros de Galicia*. Memoria del IGME, **4**.

HERNÁNDEZ SAMPELAYO, P. 1926. Yacimientos de graptolítidos en la zona de Almadén. *Boletín de la Real Sociedad Española de Historia Natural*, **26**, 335–338.

HERNÁNDEZ SAMPELAYO, P. 1942. *Explicación del Nuevo Mapa Geológico de España. Tomo II. El Sistema Siluriano*. Memorias del IGME, **45**(1–2).

HERNÁNDEZ SAMPELAYO, P. 1944. De la fauna Gotlandiense. *Dalmanites batalleri* SAMP. – corrección de *Phacops longicaudatus* MURCH. – *Dalmanites longicaudatus*, enmienda de Font y Sagüé. *Notas y Comunicaciones del IGME*, **13**, 4–8.

HERNÁNDEZ SAMPELAYO, P. 1948. '*Pradoceras (Kotoceras) kobayashi*' n.sp. del Ordoviciense de Ciudad Real. *Boletín del IGME*, **61**, 49–53.

HERNÁNDEZ SAMPELAYO, P. 1960. Graptolítidos españoles, recopilados por Rafael Fernández Rubio. *Notas y Comunicaciones del IGME*, **57**, 3–78.

HERNANDO, S. 1973. El Pérmico de la región Atienza-Somolinos (Provincia de Guadalajara). *Boletin Geológico y Minero*, **LXXXIV**, 231–235.

HERNANDO, S. 1977. Pérmico y Triásico de la región de Ayllón-Atienza (provincias de Segovia, Soria y Guadalajara). *Seminarios de Estratigrafía*, **2**, 1–408.

HERNANDO, S., SCHOTT, J. J., THUIZAT, T. & MONTIGNY, R. 1980. Age des andesites et des sediments interstratifiés de la region d'Atienza (Espagne): étude stratigraphique, géochronologique et paléomagnetique. *Sciences Geologiques Bulletin*, **32**, 119–128.

HERRAIZ, M., DE VICENTE, G., LINDO-ÑAUPARI, R. *et al.* 2000. The recent (upper Miocene to Quaternary) and present tectonic stress distributions in the Iberian Peninsula. *Tectonics*, **19**(4), 762–786.

HERRANZ, P., SAN JOSÉ, M. A. & VILAS, L. 1977. Ensayo de correlación del Precámbrico entre los Montes de Toledo y el Valle del Matachel. *Estudios Geológicos*, **33**, 327–342.

HERRANZ, P., PIEREN, A. P. & SAN JOSÉ, M. A. 1999. El área 'Lusitano-Mariánica' como una nueva zona del Macizo Hespérico. Argumentos estratigráficos. *In*: GÁMEZ-VINTANED, J. A., EGUÍLUZ, L. & PALACIOS, T. (eds) *XV Reunión de Geología del Oeste Peninsular & International Meeting on Cambrian Orogens IGCP 376*, Diputación de Badajoz, 133–139 (English abstract published in *Journal of Conference Abstracts*, **4**, 1012–1013, for 2000).

HERRANZ ARAÚJO, P. 1985. El Precámbrico y su cobertera paleozoica en la región centro-oriental de la provincia de Badajoz. *Seminarios Estratigrafia*, **10**(2), 1–833.

HERRANZ ARAÚJO, P. 1994. Las discontinuidades estratigráficas principales en el sector central del NE de 'Ossa Morena': Rango y significado tectosedimentario. Memorias e Noticias. *Publicaçaes Museu Laboratorio Mineralógico Geológico Universidade Coimbra*, **97**, 51–80.

HERRANZ ARAÚJO, P., PIEREN, A. P. & SAN JOSÉ, M. A. DE 1999. El área 'Lusitano-Mariánica' como una nueva zona del Macizo Hespérico. Argumentos estratigráficos. *XV Reunión de Geología del Oeste Peninsular (International Meeting on Cadomian Orogens)*. *Journal of Conference Abstracts*, **4**(3), 1009.

HERRERO, J. M. 1989. *Las mineralizaciones de Zn, Pb, F en el sector occidental de Vizcaya: mineralogía, geoquímica y metalogenia*. PhD Thesis, University of Pais Vasco.

HERTOGEN, J., LÓPEZ-RUIZ, J., RODRÍGUEZ-BADIOLA, E., DEMAIFFE, D. & WEIS, D. 1985. Petrogenesis of ultrapotassic volcanic rocks from S.E. Spain: Trace elements and Sr-Pb isotopes. *Terra Cognita*, **5**, 215–216.

HERTOGEN, J., LÓPEZ-RUIZ, J., DEMAIFFE, D. & WEIS, D. 1988. Modelling of source enrichment and melting processes for the calcalkaline-shoshonite-lamproite suite from S.E. Spain. *Chemical Geology*, **70**, 153.

HEWARD, A. 1978a. Alluvial fan sequence and lacustrine sediments from the Stephanian A and B (La Magdalena, Ciñera y Sabero) coalfields, northern Spain. *Sedimentology*, **25**, 451–488.

HEWARD, A. 1978b. Alluvian fan sequence and megasequence models: with examples from Westphalian D–Stephanian B coalfields, Northern Spain. *In*: MIALL, A. (ed.) *Fluvial Sedimentology*.

Canadian Society of Petroleum Geologists, Memoir **5**, 669–702.

HEWARD, A. & READING, H. G. 1980. Deposits associated with a Variscan continental strike-slip system, Cantabrian Mountains, Northern Spain. *In*: BALLANCE, P. F. & READING, H. G. (eds) *Sedimentation in Oblique-slip Mobile Zones*. IAS, Special Publications, **4**, 105–125.

HIERONYMUS, C. F. & BERCOVICI, D. 1999. Discrete alternating hotspot islands formed by interaction of magma transport and lithospheric flexure. *Nature*, **397**, 604–607.

HIGUERAS, P. 1993. Alteration of basic igneous rocks from the Almaden mercury mining district. *In*: FENOLL, P., GERVILLA, F. & TORRES, J. (eds) *Current Research on Geology Applied to Mineral Deposits*. University of Granada, 131–134.

HIGUERAS, P. & MUNHÁ, J. 1993. Geochemical constraints on the petrogenesis of mafic magmas in the Almadén mercury mining district. *Terra Abstracts*, **6**, 12–13.

HIGUERAS, P., OYARZUN, R., LUNAR, R., SIERRA, J. & PARRAS, J. 1999. The Las Cuevas deposit, Almadén district (Spain): An unusual case of deep-seated advanced argillic alteration related to mercury mineralisation. *Mineralium Deposita*, **34**, 211–214.

HIGUERAS, P., OYARZUN, R., MUNHÁ, J. & MORATA, D. 2000. The Almadén metallogenic cluster (Ciudad Real, Spain): alkaline magmatism leading to mineralisation process at an intraplate tectonic setting. *Revista de la Sociedad Geológica de España* **13–1**, 105–119.

HILLAIRE-MARCEL, C., CARRO, O., CAUSSE, C., GOY, J. L. & ZAZO, C. 1986. Th/U dating of Strombus bubonius-bearing marine terraces in southeastern Spain. *Geology*, **14**, 613–616.

HILLAIRE-MARCEL, C., GHALEB, B., GARIEPY, C., ZAZO, C., HOYOS, M. & GOY, J. L. 1995. U-series dating by the TIMS Technique of land snails from paleosols in the Canary Islands. *Quaternary Research*, **44**, 276–282.

HILLAIRE-MARCEL, C., GARIEPY, C., GHALEB, B., GOY, J. L., ZAZO, C. & CUERDA, J. 1996. U-series measurements in Tyrrhenian deposits from Mallorca. Further evidence for two last Interglacials high sea-levels in the Balearic islands. *Quaternary Science Reviews*, **15**, 53–62.

HINKELBEIN, K. 1969. El Triásico de los alrededores de Albarracín. *Teruel*, **41**, 35–75.

HIRSCH, F. 1966. Sobre la presencia de conodontos en el Muschelkalk superior de los Catalánides. *Notas de la Comisión del IGME*, **90**, 85–92.

HIRSCH, F., MÁRQUEZ-ALIAGA, A. & SANTISTEBAN, C. 1987. Distribución de moluscos y conodontos del tramo superior del Muschelkalk en el sector occidental de la provincia sefardí. *Cuadernos Geología Ibérica*, **11**, 799–814.

HIRT, A. M., LOWRIE, W., JULIVERT, M. & ARBOLEYA, M. L. 1992. Paleomagnetic results in support of a model for the origin of the Asturian arc. *Tectonophysics*, **213**, 321–339.

HOBSON, A., BUSSY, F. & HERNANDEZ, J. 1998. Shallow-level magmatization of gabbros in a metamorphic contact aureole, Fuerteventura Basal Complex, Canary Islands. *Journal of Petrology*, **39**, 125–137.

HOEDEMAEKER, PH. J. 1973. *Olisthostromes and other delapsional deposits, and their occurrence in the region of Moratalla (Prov. of Murcia, Spain)*. Scripta Geologica, **19**.

HOEFS, J. 1997. *Stable Isotope Geochemistry* (fourth edition). Springer-Verlag, New York.

HOERNLE, K. & SCHMINCKE, H. U. 1993. The role of partial melting in the 15–Ma geochemical evolution of Gran Canaria: a blob model for the Canarian hotspot. *Journal of Petrology*, **34**, 599–626.

HOERNLE, K. & TILTON, G. 1991. Sr-Nd-Pb isotope data for Fuerteventura (Canary Islands) basal complex and subaerial volcanics: applications to magma genesis and evolution. *Schweizerische Mineralogische und Petrographische Mitteilungen*, **71**, 3–18.

HOERNLE, K., TILTON, G. & SCHMINCKE, H. U. 1991. Sr-Nd-Pb isotopic evolution of Gran Canaria: evidence for shallow enriched mantle beneath the Canary Islands. *Earth and Planetary Science Letters*, **106**, 44–63.

HOERNLE, K., ZHANG, Y. S. & GRAHAM, D. 1995. Seismic and geochemical evidence for large-scale mantle upwelling beneath the Eastern Atlantic and Western and Central Europe. *Nature*, **374**, 34–39.

HOFFMANN, G. & SCHULTZ, H. D. 1988. Coastline shifts and Holocene stratigraphy on the Mediterranean coast of Andalucia (Southeastern Spain). *In*: *Archeology of Coastal Changes, Proceedings of the First International Symposium 'Cities on the Sea. Past and Present, Israel'.* Auner Raban, 53–70.

HOGAN, P. J. & BURBANK, D. W. 1996. Evolution of the Jaca piggy-back basin and emergence of the External Sierra, southern Pyrenees. *In*: FRIEND, P. F. & DABRIO, C. J. (eds) *Tertiary Basins of Spain, the Stratigraphic Record of Crustal Kinematics.* Cambridge University Press, 153–160.

HOLCOMB, R. T. & SEARLE, R. C. 1991. Large landslides from oceanic volcanoes. *Marine Geotechnology*, **10**, 19–32.

HOLTZ, F. 1987. *Evolution strusturale, métamorphique et géoquimique des granitoïdes hercyniens et de leur encaissant dans la région de Montalegre, Tras–os–Montes (Portugal).* PhD Thesis, University of Nancy I.

HOMBERG, C., HU, J. C., ANGELIER, J., BERGERAT, F. & LACOMBE, O. 1997. Characterization of stress perturbations near major fault zones: Insights from 2–D distinct-element numerical modelling and field studies (Jura mountains). *Journal of Structural Geology*, **19**, 703–718.

HOTTINGER, L., DROBNE, K. & CAUS, E. 1989. Late Cretaceous, larger, complex miliolids (Foraminifera) endemic in the Pyrenean Faunal Province. *Facies*, **21**, 99–134.

HOWELL, F. C., BUTZER, K. W., FREEMAN, L. G. & KLEIN, R. G. 1995. Observations on the Acheulean occupation site of Ambrona (Soria province, Spain) with particular reference to recent investigations (1980–1983) and the lower occupation. *Jabrbuch des Römisch-Germanischen*. Zentralmuseum Mainz, **38**, 33–82.

HOYOS, M. 1989. La Cornisa Cantábrica. *In*: PÉREZ-GONZÁLEZ, A., CABRA, R. & MARTÍN-SERRANO, A. (eds) *Mapa del Cuaternario de España.* ITGE, Madrid, 105–119.

HOYOS, M. & AGUIRRE, E. 1995. El registro paleoclimático pleistoceno en la evolución del karst de Atapuerca (Burgos): el corte de Gran Dolina. *Trabajos de Prehistoria*, **52**, 31–45.

HOYOS, M., GILLOT, P. Y., SANZ, E., SOLER, V., SÁNCHEZ-MORAL, S. & CAÑAVERAS, J. C. 1998. El volcanismo de Nuévalos (Zaragoza): Situación morfoestructural y edad. *Estudios Geológicos*, **54**, 103–107.

HSÜ, K. J., MONTADERT, L., BERNOULLI, D. *et al.* 1977. History of the Messinian salinity crisis. *Nature*, **267**, 399–403.

HUBBARD, R. PAPE, J. & ROBERTS, D. G. 1985. Depositional sequence mapping to illustrate the evolution of a passive continental margin. *In*: BERG, O. R. & WOOLVERTON, D. G. (eds) *Seismic Stratigraphy II. An Integrated Approach.* AAPG, Memoir **39**, 93–115.

HUBLIN, J.-J., BARROSO, C., MEDINA, P., FONTUGNE, M. & REYES, J. L. 1995. The musterian site of Zafarraya (Andalucía, Spain): dating and implications on the paleolithic peopling of Western Europe. *Comptes Rendus de l'Académie des Sciences de Paris.*, série IIa, **321**, 931–937.

HUERTAS, M. J. & VILLASECA, C. 1994. Les derniers cycles magmatiques posthercyniens du système central espagnol: less essaims filoniens calco-alcalins. *Schweizerische Mineralogische und Petrographische Mitteilungen*, **74**, 383–401.

HUERTAS, M. J., BRÄNDLE, J. L., ANCOCHEA, E., HERNÁN, F. & CUBAS, C. R. 2000. Distribución de los diques sálicos del Norte de La Gomera. *Geogaceta*, **27**, 91–94.

HUGUENEY, M. 1997. La faune de Gliridés (Rodentia, Mammalia) de Paguera (Majorque, Espagne): particularisme dans l'Oligocène majorquin. *Geobios, Mém. Spéciale*, **20**, 299–305.

HUGUENEY, M. & ADROVER, R. 1982. Le peuplement des Baléares (Espagne) au Paléogène. *Geobios, Mémoire Spéciale*, **6**, 439–449.

HUGUENEY, M. & ADROVER, R. 1989–90. Rongeurs (Rodentia, Mammalia) de l'Oligocène de Sineu (Baleares, Espagne). *Paleontologia i Evolució*, **23**, 157–169.

HUTCHINSON, C. S. 1983. *Economic Deposits and their Tectonic Setting.* MacMillan, London.

IACCARINO, S. 1985. Mediterranean Miocene and Pliocene planktic foraminifera. *In*: BOLLI, H. M., SAUNDERS, J. S. & PERCH-NIELSEN, K. (eds) *Plankton Stratigraphy.* Cambridge University Press, 283–314.

IBARROLA, E. 1969. Variation trends in basaltic rocks of the Canary Islands. *Bulletin Volcanologique*, **33**, 729–777.

IBARROLA, E., VILLASECA, C., VIALETTE, Y., FÚSTER, J. M., NAVIDAD, M., PEINADO, M. & CASQUET, C. 1987. Dating of Hercynian granitesin the Sierra de Guadarrama (Spanish Central System). *In*: BEA, F., CARNICERO, A., GONZALO, J. C., LÓPEZ PLAZA, M. & RODRÍGUEZ ALONSO, M. D. (eds) *Geología de los granitoides y rocas asociadas del Macizo Hespérico. Libro Homenaje a L. C. García Figuerola.* Rueda, Madrid, 377–383.

IBARROLA, E., ANCOCHEA, E. & HUERTAS, M. J. 1991. Rocas volcánicas submarinas en la base de la Formación Cañadas. Macizo de Tigaiga (N. de Tenerife). *Geogaceta*, **9**, 17–20.

IBARROLA, E., ANCOCHEA, E., FÚSTER, J. M., CANTAGREL, J. M., COELLO, J., SNELLING, N. J. & HUERTAS, M. J. 1993. Cronoestratigrafía del Macizo de Tigaiga: evolución de un sector del Edificio Cañadas (Tenerife, Islas Canarias). *Boletín de La Real Sociedad Española de Historia Natural (Sec. Geol)*, **88**, 57–72.

IGLESIAS, M. & ROBARDET, M. 1980. El Silúrico de Galicia Media (central). Su importancia en la paleogeografía varisca. *Cuadernos do Laboratorio Xeolóxico de Laxe*, **1**, 99–115.

IGLESIAS, M., RIBEIRO, M. L. & RIBEIRO, A. 1983. *La interpretación aloctonista de la estructura del Noroeste Peninsular. Libro Jubilar J. M. Ríos 'Geología de España, Volumen I'.* IGME, Madrid, 459–467.

IGME. 1982. *Síntesis geológica de la Faja Pirítica del SO de España.* IGME, Colección Memorias, **98**.

IGME. 1985. *Proyecto de investigación de alabastros en España.* Report **127**, IGME, Madrid.

IGME 1987. *Contribucion de la exploración petrolífera al conocimiento de la Geología de España.* IGME, Madrid.

IGME. 1990. *Documentos sobre la Geología del subsuelo de España.* Tomo VI (Cuenca del Ebro y Pirineo). IGME, Madrid.

IGN. 1992a. *Análisis Sismotectónico de la Península Ibérica, Baleares y Canarias.* Publicaciones Tecnológicas IGN, **22**.

IGN. 1992b. *Atlas Nacional de España. Sección II: Geofísica.* Publucaciones del MOP 8.1–8.20.

ILIHA DSS GROUP. 1993. A deep seismic sounding investigation of lithospheric heterogeneity and anisotropy beneath the Iberian Peninsula. *Tectonophysics*, **221**, 35–51.

INNOCENT, C., BRIQUEU, L. & CABANIS, B. 1994. Sr-Nd isotope and trace element geochemistry of late Variscan volcanism in the Pyrenees: magmatism in post-orogenic extension. *Tectonophysics*, **238**, 161–181.

INSTITUT CARTOGRAFIC DE CATALUNYA. 1994–97. *Mapa Geològic de Catalunya 1:25.000. 297-1-1 (Cala Montgó); 297-1-2 (L'Estartit); 296-2-1 (L'Escala); 296-2-2 (Torroella de Montgrí); 258-1-1 (Figueres).*

IRVINE, T. N. & BARAGAR, W. R. A. 1971. A guide to the chemical classification of the common volcanic rocks. *Canadian Journal of Earth Science*, **8**, 523–548.

ITGE. 1989. *Las Aguas Subterráneas en España.* Informe de Síntesis, Madrid.

ITGE. 1990. *Geological maps (1:25.000) of Gran Canaria. Proyecto MAGNA.* ITGE, Madrid.

ITGE. 1991. *Mapa Geológico de España. Escala 1:200.000. Hoja no. 40 (Daroca).* Ministerio de Industria, Madrid.

ITGE. 1992. *Proyecto MAGNA, geological map (1:100.000) of Gran Canaria.* ITGE, Madrid.

ITGE. 1994. *Mapa Geológico de la Península Ibérica, Baleares y Canarias.* ITGE, Madrid.

ITGE. 1998. *Guía para la Investigación de los Recursos Minerales de España.* ITGE, Madrid.

IUGS. 2000. *International Stratigraphic Chart & Explanatory Note*, compiled by J. Remane and ICS Subcommissions. Division of Earth Sciences, UNESCO, Paris.

IWANIW, I. 1984. Lower Cantabrian basin margin deposits in NW Spain-a model for valley-fill sedimentation in a tectonically active humic climatic setting. *Sedimentology*, **31**, 91–110

IWANIW, I. 1985. The sedimentology of Lower Cantabrian basin margin deposits in NE León, Spain. *In*: LEMOS DE SOUSA, M. J. & WAGNER, R. H. (eds) *Papers on the Carboniferous of the Iberian Peninsula (Sedimentology, Stratigraphy, Paleontology, Tectonics and Geochemistry). Annales da Faculdade de Ciências, Universidade do Porto*, Special Supplement, **64** (1983), 49–115.

JABALOY, A. 1991. *La estructura de la región occidental de la Sierra de los Filabres*. PhD Thesis, University of Granada.

JABALOY, A., GALINDO-ZALDÍVAR, J. & GONZÁLEZ-LODEIRO, F. 1992. The Mecina Extensional System: its relation with the Post-Aquitanian piggy-back basins and the paleostresses (Betic Cordilleras, Spain). *Geo-Marine Letters*, **12**, 96–103.

JABALOY, A., GALINDO-ZALDÍVAR, J. & GONZÁLEZ-LODEIRO, F. 1993. The Alpujarride Nevado-Filábride extensional shear zone, Betic-Cordillera, SE Spain. *Journal of Structural Geology*, **15**, 555–569.

JAEGER, H. 1976. Das Silur und Unterdevon vom thüringischen Typ in Sardinien und seine regionalgeologische Beteudung. *Nova Acta Leopoldina*, 224, **45**, 263–299.

JAEGER, H. 1977. Thuringia and Sardinia. *In*: The Silurian–Devonian Boundary. IUGS, Stuttgart, Series A, **5**, 117–125.

JAEGER, H. 1986. Graptolithen als Tentakulitenfalle. *Zeitschrift für geologische Wissenschaften*, **14**, 669–671.

JAEGER, H. & ROBARDET, M. 1979. Le Silurien et le Dévonien basal dans le Nord de la Province de Séville (Espagne). *Geobios*, **12**, 687–714.

JAHNKE, H., HENN, A., MADER, H. & SCHWEINEBERG, J. 1983. Silur und Devon im Arauz-Gebiet (Prov. Palencia, N-Spanien). *Newsletters on Stratigraphy*, **13**, 40–66.

JANSA, L., BUJAL, J. & WILLIAMS, G. 1980. Upper Triassic salt deposits of the Western North Atlantic. *Canadian Journal of Earth Sciences*, **17**, 547–559.

JAVOY, M., STILLMAN, C & PINEAU, C. 1986. Oxygen and hydrogen isotope studies on the basal complexes of the Canary Islands: implications on the conditions of their genesis. *Contributions to Mineralogy and Petrology*, **92**, 225–235.

JENKYNS, H. C, & CLAYTON, C. J. 1986. Black shales and carbon isotopes in pelagic sediments from the tethyan lower Jurassic. *Sedimentology*, **33**, 87–106.

JÉRÉMINE, E. & FALLOT, P. 1928. Sur la présence d'une variété de jumillite aux environs de Calasparra (Murcia). *Comptes Rendus de l'Académie des Sciences de Paris*, **188**, 800.

JEREZ-MIR, L. 1973. *Geología de la Zona Prebética, en la transversal de Elche de la Sierra y sectores adyacentes (provincias de Albacete y Murcia)*. PhD Thesis, University of Granada.

JEREZ-MIR, L. 1981. *Estudio geológico, geotectónico y tectonosedimentario de la Zona Prebética en relación con las demás cadenas béticas e ibérica*. Informe interno IGME, Madrid.

JIMÉNEZ, A. P. 1986. *Estudio paleontológico de los ammonites del Toarciense inferior y medio de las Cordilleras Béticas (Dactylioceratidae e Hildoceratidae)*. PhD Thesis, University of Granada.

JIMENEZ, A. P., JIMENEZ DE CISNEROS, C., RIVAS, P. & VERA, J. A. 1996. The Early Toarcian anoxic event in the westernmost Tethys (Subbetic): Paleogeographic and paleobiogeographic significance. *Journal of Geology*, **104**, 399–416.

JIMÉNEZ, E. 1974. Iniciación al estudio de la paleoclimatología del Paleógeno de la cuenca del Duero y su posible relación con el resto de la Península Ibérica. *Boletín Geológico y Minero de España*, **85**(5), 6–12.

JIMÉNEZ, E. 1992. Las dataciones del Paleógeno de Castilla y León. Junta de Castilla y León. *In*: Jiménez, E. (coord.) *Vertebrados fósiles de Castilla y León*. Consejería de Cultura y Turismo, 39–41.

JIMÉNEZ, E., CORROCHANO, A. & ALONSO GAVILÁN, G. 1983. El Paleógeno en la cuenca del Duero. *In*: *Geología de España*. IGME, Madrid, **2**, 489–494.

JIMÉNEZ ONTIVEROS, P. & HERNÁNDEZ ENRILE, J. L. 1983. Rocas miloníticas indicadoras de la deformación progresiva en la zona de cizalla hercínica de Juzbado-Penalva do Castelo. *Studia Geológica Salmanticensis*, **18**, 139–158.

JOHNSON, C. 1993. Contrasted thermal histories of different nappe complexes in SE Spain: evidence for complex crustal extension. *In*: SÉRANNE, M. M. J. (ed.) *Late Orogenic Extension in Mountain Belts*. BRGM, Orleans.

JOHNSON, C. 1997. Resolving denudational histories in orogenic belts with apatite fission-track thermochronology and structural data: an example from southern Spain. *Geology*, **25**, 623–626.

JOHNSON, C., HARBURY, N. & HURFORD, A. J. 1997. The role of extension in the Miocene denudation of the Nevado-Filábride Complex, Betic Cordillera (SE Spain). *Tectonics*, **16**, 189–204.

JONES, P. J. 1995. *Timescales. 5: Carboniferous. Australian phanerozoic timescales. Biostratigraphic charts and explanatory notes*. Second Series, Australian Geological Survey Organisation.

JORDÁ, J. F. 1986. *La Prehistoria de la Cueva de Nerja (Mállaga)*. Patronato de la Cueva de Nerja, Mállaga.

JOSEPH, J. 1973. *Le Paléozoique de la nappe de Gavarnie entre le cirque de Troumouse et la gave de Pau*. PhD Thesis, University of Toulouse.

JOSEPH, J., BRICE, D. & MOURAVIEFF, N. 1980. Données paléontologiques nouvelles sur le Frasnien des Pyrénées centrales et occidentales: implications paléogeographiques. *Bulletin Société Histoire Naturelle Toulouse*, **116**(1–2), 16–41.

JOSOPAIT, V. 1972. Das Kambrium und das Tremadoc von Ateca (Westliche Iberische Ketten, NE-Spanien). *Münstersche Forschungen zur Geologie und Paläontologie*, **23**, 1–121.

JOSOPAIT, V. & SCHMITZ, U. 1971. Beitrag zur Stratigraphie im Unter und Mittel-Kambrium der Sierra de la Demanda (NE-Spanien). *Münstersche Forschungen zur Geologie und Paläontologie*, **19**, 85–89.

JUCH, D. & SCHÄFER, D. 1974. L'Hercynien de Maya et de la vallée d'Arizakun dans la partie orientale du massif de Cinco-Villas (Pyrénées Occidentales d'Espagne). *Pirineos*, **111**, 41–58, Jaca

JULIÁ, R. & BISCHOFF, J. L. 1991. Radiometric dating of Quaternary deposits and the hominid mandible of Lake Banyolas, Spain. *Journal of Archaeological Science*, **18**, 707–722.

JULIÁ, R. & BISCHOFF, J. L. 1993. Datación radiométrica de los depósitos cuaternarios y de la mandíbula humana del lago de Banyoles. *In*: MAROTO, J. (ed.) *La mandíbula de Banyoles en el context dels fossils humans del Pleistocè*. Centre d'Investigacions Arqueològiques, Girona, 91–101.

JULIÁN, A. 1996. *Cartografía y Correlación General de la Acumulaciones Cuaternarias de la Depresión del Ebro*. PhD Thesis, University of Zaragoza.

JULIVERT, M. 1971. Décollement tectonics in the Variscan Cordillera of the northwest Spain. *American Journal of Sciences*, **270**, 1–29.

JULIVERT, M. 1978. Hercynian orogeny and Carboniferous paleogeography in Northwestern Spain: a model of deformation-sedimentation relationships. *Zeitschrift der Deutschen geologischen Gesellschaft*, **129**, 565–592.

JULIVERT, M. 1979. A cross section through the northern part of the Iberian Massif: its position within the Hercynian fold belt. *Krystallinikum*, **60**, 107–128.

JULIVERT, M. 1981. A cross section through the northern part of the Iberian Massif. *Geologie in Minjbow*, **14**, 51–67.

JULIVERT, M. 1983a. *La estructura de la Zona Cantábrica. Libro Jubilar J. M. Ríos 'Geología de España, Volumen I'*. IGME, Madrid, 339–381.

JULIVERT, M. 1983b. *La Estructura de la Zona Asturoccidental-Leonesa. Libro Jubilar J. M. Ríos 'Geología de España, Volumen I'*. IGME, Madrid, 381–408.

JULIVERT, M. & ARBOLEYA, M. L. 1984a. A geometrical and kinematic approach to the nappe structure in an arcuate fold belt: the Cantabrian nappes (Hercynian chain, NW Spain). *Journal of Structural Geology*, **6**, 499–519.

JULIVERT, M. & ARBOLEYA, M. L. 1984b. Curvature increase and structural evolution of the core (Cantabrian Zone) of the Ibero-Armorican Arc. *Bulletin des Sciences Géologiques*, **37**, 5–11.

JULIVERT, M. & ARBOLEYA, M. L. 1986. Areal balancing and estimate of areal reduction in a thin-skinned fold-and-thrust belt (Cantabrian Zone, NW Spain): constraints on its emplacement mechanism. *Journal of Structural Geology*, **8**, 407–414.

JULIVERT, M. & DURÁN, H. 1990a. Paleozoic stratigraphy of the Central and Northern part of the Catalonian Coastal Ranges (NE Spain). *Acta Geologica Hispanica*, **25**, 3–12.

JULIVERT, M. & DURÁN, H. 1990b. The Hercynian structure of the Catalonian Coastal Ranges (NE Spain). *Acta Geológica Hispánica*, **25**, 13–21.

JULIVERT, M. & DURÁN, H. 1992. El Paleozoico Inferior de las Cadenas Costeras Catalanas. *In*: GUTIÉRREZ MARCO, J., SAAVEDRA, J. & RÁBANO, I. (eds) *Paleozoico Inferior de Ibero-América*. University of Extremadura, Badajoz, 607–613.

JULIVERT, M. & MARCOS, A. 1973. Superimposed folding under

flexural conditions in the Cantabrian zone (Hercynian Cordillera, Northwest Spain). *American Journal of Science*, **273**, 353–375.

JULIVERT, M. & MARTÍNEZ, F. J. 1980. The Paleozoic of the Catalonian Coastal Ranges (Northwestern Mediterranean). *In*: Sassi, F. P. (ed.) *IGCP 5 Newlsetter*, **2**, 124–128.

JULIVERT, M. & MARTÍNEZ, F. J. 1983. *El Paleozoico de las Cadenas Costeras Catalanas*. *Libro Jubilar J. M. Ríos 'Geología de España, Volumen I'*. IGME, Madrid, 529–536.

JULIVERT, M. & MARTÍNEZ, F. J. 1987. The structure and evolution of the Hercynian Fold Belt in the Iberian Peninsula. *In*: SCHAER, J. P. & RODGERS, J. (eds) *The Anatomy of Mountain Belts*. Princeton University Press, Princeton, 65–103.

JULIVERT, M. & MARTÍNEZ GARCÍA, E. 1967. Sobre el contacto entre el Precámbrico y el Cámbrico en la parte meridional de la cordillera Cantábrica y el papel del Precámbrico en la orogénesis Herciniana. *Acta Geológica Hispánica*, **2**, 107–110.

JULIVERT, M. & TRUYOLS, J. 1983. El Ordovícico en el Macizo Ibérico. *In*: COMBA, J. A. (coord.) *Libro Jubilar J. M. Ríos, Geología de España*, Vol. 1. IGME, Madrid, 192–246.

JULIVERT, M., FONTBOTÉ, J. M., RIBEIRO, A. & CONDE, L. 1972a. *Mapa Tectónico de la Península Ibérica y Baleares a escala 1:1.000.000 y Memoria Explicativa*. IGME, Madrid.

JULIVERT, M., MARCOS, A. & TRUYOLS, J. 1972b. L'évolution paléogéographique du nord-ouest de l'Espagne pendant l'Ordovicien-Silurien. *Bulletin de la Société Géologique et Minéralogique de Bretagne [C]*, **4**, 1–7.

JULIVERT, M., FONTBOTÉ, J. M., RIBEIRO, A., NABAIS-CONDE, L. E. 1974. *Mapa tectónica de la Péninsula Ibérica y Baleares, escala 1:1.000.000*. IGME, Madrid.

JULIVERT, M., TRUYOLS, J. & VERGÉS, J. 1983a. El Devónico en el Macizo Ibérico. *In*: COMBA, J. A. (coord.), *Geología de España, Libro Jubilar J. M. Ríos*, Vol. 1. IGME, Madrid, 265–311.

JULIVERT, M., VEGAS, R., ROIZ, J. M. & MARTÍNEZ RÍUS, A. 1983b. La estructura de la extensión sureste de la Zona Centro-Ibérica con metamorfismo de bajo grado. *In*: COMBA, J. A. (coord.) *Libro Jubilar J. M. Ríos 'Geología de España, Volumen I'*. IGME, Madrid, 477–490.

JULIVERT, M., DURÁN, H., RICKARDS, R. B. & CHAPMAN, A. J. 1985. Siluro-Devonian graptolite stratigraphy of the Catalonian Coastal Ranges. *Acta Geologica Hispanica*, **20**, 199–207.

JULIVERT, M., DURAN, N., GARCÍA-LÓPEZ, S., GIL IBARGUCHI, J. I., TRUYOLS, M. & VILLAS, E. 1987. Pre-Carboniferous rocks in the Catalonian coastal Ranges: volcanism, stratigraphic sequence and fossil content. *In*: FLÜGEL, H. W., SASSI, F. P. & GRECULA, P. (eds) *Pre-Variscan Events in the Alpine Mediterranean Mountain Belts*. Mineralia Slovaca, 313–322.

JUNCO, F. & CALVO, J. P. 1983. Cuenca de Madrid. In: *Geología de España*, II. IGME, Madrid, 534–542.

JURADO, M. 1990. El Triásico y el Liásico basal evaporíticos del subsuelo de la cuenca del Ebro. *In*: ORTÍ, F. & SALVANY, J. (eds) *Formaciones evaporíticas de la Cuenca del Ebro y cadenas periféricas y de la zona de Levante*. Enresa, Madrid, 21–28.

JURADO, M. J. & RIBA, O. 1996. The Rioja Area (westernmost Ebro basin): a ramp valley with neighbouring piggybacks. *In*: FRIEND, P. & DABRIO, C. (eds) *Tertiary Basins of Spain*. World and Regional Geology **6**, Cambridge University Press, 173–179.

KAIHO, K. & LAMOLDA, M. A. 1999. Catastrophic extinction of planktonic foraminifera at the Cretaceous–Tertiary boundary evidenced by stable isotopes and foraminiferal abundance at Caravaca, Spain. *Geology*, **27**, 355–358.

KANIS, J. 1956. Geology of the eastern zone of the Sierra del Brezo (Palencia, Spain). *Leidse Geologisches Mededelingen*, **21**(2), 377–445.

KASE, K., YAMAMOTO, M., NAKAMURA, T. & MITSUNO, C. 1990. Ore mineralogy and sulfur isotope study of the massive sulfide deposit of Filon Norte, Tharsis Mine, Spain. *Mineralium Deposita*, **25**, 289–296.

KEASBERRY, E. 1979. An interpretation model of semi-circular Bouguer anomalies found over the peripheral belt of the Ordenes Complex (NW Spain). *Geologie en Mijnbouw*, **58**, 65–70.

KEGEL, W. 1929. Das Gotlandium in den Kantabrischen Ketten Nordspaniens. *Zeitschrift der deutschen geologischen Gesellschaft*, **81**, 35–62.

KENTER, J. A. M., REYNER, J. J. G., VAN DER STRAATEN, H. C. & PEPER, T. 1990. Facies patterns and subsidence history of the Jumilla-Cieza-Region (Southeastern Spain). *Sedimentary Geology*, **67**, 263–280.

KENTER, J. A. M., VAN HOEFLAKEN, F., BAHAMONDE, J. R., BRACCO GARTNER, G. L. & KLEIM, L. 2002. Anatomy and lithofacies of an intact and seismic-scale carboniferous carbonate platform (Asturias, NW Spain). *In*: ZEMPLOLICH, W. & COOK, H. E. (eds) *Paleozoic Carbonates of the Commonwealth of Independent States (CIS): Subsurface Reservoirs and Outcrops Analogues*. SEPM, Special Publication, **74**, 185–207.

KEPPIE, J. D. 1993. Synthesis of Paleozoic deformational events and terrane accretion in the Canadian Appalachians. *Geologische Rundschau*, **82**, 381–431.

KEPPIE, J. D. & DALLMEYER, R. D. 1987. Dating transcurrent terrane accretion: an example from the Meguma and Avalon composite terranes in the northern Appalachians. *Tectonics*, **6**, 831–847.

KIRKER, A. I. & PLATT, J. P. 1998. Unidirectional slip vectors in the western Betic Cordillera: implications for the formation of the Gibraltar arc. *Journal of the Geological Society of London*, **155**, 193–207.

KLAPPER, G. 1977. Conodonts. *In*: *The Silurian–Devonian Boundary*. IUGS, Stuttgart, Series A, **5**, 318–319.

KLEINSMIEDE, W. F. 1960. Geology of the Valley de Aran (Central Pyrenees). *Leidse Geologische Mededelingen*, **25**, 129–147.

KLEVERLAAN, K. 1989. *Tabernas fan complex. A study of a Tortonian fan complex in a Neogene basin, Tabernas, Province of Almería, SE Spain*. PhD Thesis, University of Amsterdam.

KLUEGEL, A., SCHMINCKE, H. U., WHITE, J. D. L. & HOERNLE, K. A. 1999. Chronology and volcanology of the 1949 muti-vent rift-zone eruption of La Palma (Canary Islands). *Journal of Volcanology and Geothermal Research*, **94**(1–4), 267–282.

KLUG, H. 1968. Morphologischen Studien auf den Kanarischen Inseln. Biträge zur Küsstenentwickelung und Talbindung auf einem vulkanischen archipel. *Geographische Institut Universitat Kiel Schriften*, **24/3**.

KNIGHT, J. A. 1971. The sequence and stratigraphy of the eastern end of the Sabero Coalfield (León, NW Spain). *Trabajos de Geología*, **3**, 193–229.

KNIGHT, J. A., BURGER, K. & BIEG, G. 2000. The pyroclastic tonsteins of the Sabero Coalfield, north-western Spain, and their relationship to the stratigraphy and structural geology. *International Journal of Coal Geology*, **44**, 187–226.

KOCKEL, F. & STOPPEL, D. 1963. Nuevos hallazgos de conodontos y algunos cortes en el Paleozoico de Málaga (Sur de España). *Boletín Geológico y Minero*, **68**, 133–170.

KOLB, S. & WOLF, R. 1979. Distribution of *Cruziana* in the Lower Ordovician sequence of Celtiberia (NE Spain) with a revision of the *Cruziana rugosa*-group. *Neues Jahrbuch für Geologie und Paläontologie Monatshefte*, **1979**, 457–474.

KOLLMEIER, J. M., VAN DER PLUIJM, B. A. & VAN DER VOO, R. 2000. Analysis of Variscan dynamics; early bending of Cantabria-Asturias Arc, northern Spain. *Earth and Planetary Science Letters*, **181**, 203–216.

KOOPMANS, B. N. 1962. The sedimentary and structural history of the Valsurvio Dome. Cantabrian Mountains, Spain. *Leidse Geologische Mededelingen*, **26**, 121–232.

KOREN, T. N., LENZ, A. C., LOYDELL, D. K., MELCHIN, M. J., STORCH, P. & TELLER, L. 1996. Generalized graptolite zonal sequence defining Silurian time intervals for global paleogeographic studies. *Lethaia*, **29**, 59–60.

KORN, D. 1997. The Palaeozoic ammonoids of the South Portuguese Zone. *Memorias Instituto Geológico Mineiro Portugal*, **33**, 1–32

KORNPROBST, J. 1971. *Contribution à l'étude pétrographique et structurale de la zone interne du Rif*. PhD Thesis, University of Paris.

KOSAKEVITCH, A., PALOMERO, F., LECA, X., LEISTEL, J. M., LENATRE, N. & SABOL, F. 1993. Controles climatique et geomorphologique de la cocentration de L'or dans les chapeaux de Fer de Rio Tinto. *Comptes Rendus de l'Académie des Sciences de. Paris*, **316**, 85–90.

KOSSMAT, F. 1927. Gliederung des Variszischen Gebirgebaues. *Abhandlungen des Sächsisches Geologisches Landesamtes*, **1**, 1–39.

KOZUR, H. & SIMON, O. P. 1972. Contribution to the Triassic micro-fauna and stratigraphy of the Betic Zone (Southern Spain). *Revista Española de Micropaleontología*, Special Anniversary Volume, **30**, 143–158.

KOZUR, H., MULDER-BLANKEN, C. W. & SIMON, O. 1985. On the Triassic of the Betic Cordilleras (southern Spain), with special emphasis on holothurian sclerites. *Proceedings, Koninklijke Nederlandse Académie van Wetenschappen Series B*, **88**, 83–110.

KRANS, T. F., GUIT, F. A. & OFWEGEN, L. P. VAN. 1982. Facies-patterns in the Lower Devonian carbonates of the Lebanza Formation (Cantabrian Mountains, Province of Palencia, NW Spain). *In*: KULLMANN, J., SCHÖNENBERG, R. & WIEDMANN, J. (eds) *Subsidenz-entwicklung im Kantabrischen Variszikum und an passiven kontinentalrändern der kreide. Teil 1. Variszikum. Tübingen, Neues Jahrbuch Geologie Paläontologie Abhandlungen*, Stuttgart, **163**(2), 192–211.

KRIEGSMAN, L. M. 1989. Structural geology of the Lys-Caillouas massif, Central Pyrenees. Evidence for a large scale recumbent fold of late variscan age. *Geodynamica Acta*, **3**, 163–170.

KRIZ, J., JAEGER, H., PARIS, F. & SCHÖNLAUB, H. P. 1986. Pridoli – the Fourth Subdivision of the Silurian. *Jahrbuch der Geologischen Bundesanstalt*, **129**, 291–360.

KÜCHLER, T. & KUTZ, A. 1989. Biostratigraphie des Campan bis Unter-Maastricht der E-Barranca und des Urdiroz/Imiscoz-Gebietes (Navarra, N-Spanien). *In*: WIEDMANN, J. (ed.) *Cretaceous of the Western Tethys*, Schweizerbart, Stuttgart, 191–213.

KUDO, A. & TURNER, R. 1999. Mercury contamination of Minamata Bay: Historical overview and progress towards recovery. *In*: EBINGHAUS, R., TURNER, R., DE LACERDA, L. D., VASILIEV, O. & SALOMONS, W. (eds) *Mercury Contaminated Sites*. Springer-Verlag, Berlin, 143–158.

KUHNT, W. & OBERT, D. 1991. Evolution crétacée de la marge tellienne. *Bulletin de la Société Géologique de la France*, **162**, 515–522.

KUIJPER, R. P., PRIEM, H. N. A. & DEN TEX, E. 1982. Late Archean–early Proterozoic sources ages of zircons in rocks from the Paleozoic orogen of Western Galicia, NW Spain. *Precambrian Research*, **19**, 1–29.

KULLMANN, J. & CALZADA, S. 1982. Goniatiten (Cephal.) aus hercynischen Unter-Devon der Ost-Pyrenäen. *Neues Jahrbuch Geologie Paläontologie Monatshefte*, **10**, 593–599.

KYSER, T. K. 1986. Stable isotope variations in the mantle. *In*: VALLEY, J. W., TAYLOR, H. P. & O'NEIL, J. R. (eds) *Stable Isotopes in High Temperature Geological Processes*. Mineralogical Society of America, Washington DC, 141–164.

LABAUME, P., SÉGURET, M. & SEYVE, C. 1985. Evolution of a turbiditic foreland basin and analogy with an accretionary prism: example of the Eocene South-Pyrenean Basin. *Tectonics*, **4**, 661–685.

LAFUSTE, J. & PAVILLON, M. J. 1976. Mise en évidence d'Eifelien daté au sein des terrains métamorphiques des zones internes des Cordillères bétiques. Intérêt de ce nouveau repère stratigraphique. *Comptes Rendus de l'Académie des Sciences de Paris*, **283**, 1015–1018.

LAGO, M. & POCOVI, A. 1984. Le vulcanisme calco-alcalin d'age Stephanien-Permien dans La Chaine Iberique (Est de l'Espagne). *Supplement Bulletin de Mineralogie*, Paris, **110**, 42–44.

LAGO, M. & POCOVI, A. 1991. Petrología y Geoquímica. *In*: *Hoja Magna, 1:50.000, no. 380 (Borobia)*. ITGE, Madrid, 91–101.

LAGO, M., POCOVI, A., ZACHMANN, D., ARRANZ, E., CARLS, P., TORRES, J. A. & VAQUER, R. 1991. Comparación preliminar de las manifestaciones magmáticas, calco-alcalinas y stephaniense-pérmicas, de la Cadena Ibérica. *Cuaderno do Laboratorio Xeoloxico de Laxe*, **16**, 95–107.

LAGO, M., ÁLVARO, J., ARRANZ, E., POCOVI, A. & VAQUER, R. 1992. Condiciones del emplazamiento, petrología y geoquímica de las riolitas calco-alcalinas Stephaniense-Pérmicas en las Cadenas Ibéricas. *Cuadernos Laboratorio Xeologico de Laxe*, **17**, 187–198.

LAGO, M., AUQUÉ, L., ARRANZ, E., GIL, A. & POCOVI, A. 1993. Caracteres de la fosa de Bronchales (Stephaniense-Pérmico) y de la turmalinización asociada a las riolitas calco-alcalinas (Provincia de Teruel). *Cuaderno do Laboratorio Xeoloxico de Laxe*, **18**, 65–80.

LAGO, M., ARRANZ, E., POCOVI, A., VAQUER, R. & GIL, A. 1994.

Petrología y geoquímica de los basaltos calco-alcalinos, Autuniense, de Ojos Negros (Cadena Ibérica oriental, Teruel). *Boletin Geologico y Minero*, **105**, 73–81.

LAGO, M., GIL, A., ARRANZ, E., BASTIDA, J. & POCOVI, A. 1995. Emplazamiento, petrología y geoquímica del complejo volcanoclástico de Orea (Guadalajara, Cadena Ibérica Occidental). *Cuaderno Do Laboratorio Xeoloxico De Laxe*, **20**, 195–212.

LAGO, M., GIL, A., POCOVI, A., ARRANZ, E., BASTIDA, J., AUQUÉ, L. & LAPUENTE, M. P. 1996. Rasgos geológicos del magmatismo Autuniense en la Sierra de Albarracín (Cadena Ibérica occidental). *Cuadernos de Geología Ibérica*, **20**, 139–157.

LAKE, P. A. 1991. *The biostratigraphy and structure of the Pulo do Lobo Domain and Iberian Pyrite Belt Domin within Huelva province, southwest Spain*. PhD Thesis, Southampton University.

LAKE, P. A., OSWIN, W. N. & MARSHALL, J. E. A. 1988. A palynological approach to terrane analysis in the South Portuguese Zone. *Trabajos de Geología*, **17**, 125–131.

LAMOLDA, M. A., MATHEY, B., ROSSY, M. & SIGAL, J. 1983. La edad del volcanismo Cretácico de Vizcaya y Guipuzcoa. *Estudios geológicos*, **39**, 151–155.

LANAJA, J. M. 1987. *Contribución de la exploración petrolífera al conocimiento de la Geología de España*. IGME (Instituto Geológico y Minero de España), Madrid.

LANCELOT, J. R. & ALLEGRET, A. 1982. Radiochronologie U/Pb de l'orthogneiss alcalin de Pedroso (Alto Alentejo, Portugal) et évolution anté-hercynienne de l'Europe occidentale. *Neues Jahrbuch für Mineralogie Monatshefte*, **9**, 385–394.

LANCELOT, J. R., ALLEGRET, A. & IGLESIAS PONCE DE LEÓN, M. 1985. Outline of Upper Precambrian and Lower Paleozoic evolution of the Iberian Peninsula according to U-Pb dating of zircons. *Earth & Planetary Science Letters*, **74**, 325–337.

LANDING, E. 1994. Precambrian–Cambrian boundary global stratotype ratified and a new perspective of Cambrian time. *Geology*, **22**, 179–182.

LANGENHEIM, V. A. M. & CLAGUE, D. A. 1987. The Hawaiian-Emperor volcanic chain. Part. II: Stratigraphic framework of volcanic rocks of the Hawaiian Islands. *In*: DECKER, W., WRIGHT, T. L. & STAUFFER, P. H. (eds) *Volcanism in Hawaii*. USGS, Professional Paper, **1350**, 1, 55–84.

LARDIÉS, M. D. 1990. Observaciones bioestratigráficas y sedimentológicas sobre el Calloviense en la Provincia de Zaragoza. *Cuadernos de Geología Ibérica*, **14**, 157–172.

LARIO, J., ZAZO, C., DABRIO, C. J., SOMOZA, L., GOY, J. L., BARDAJI, T. & SILVA, P. G. 1995. Record of recent Holocene sediment imput on spit bars and deltas of Southern Spain. *In*: CORE, B. (ed.) *Holocene Cyclic Pulses and Sedimentations. Journal of Coastal Research*, Special Issue, **17**, 241–245.

LARRASOAÑA, J. C. 2000. *Estudio magnetotectónico de la zona de transición entre el Pirineo central y occidental; implicaciones estructurales y geodinámicas*. PhD Thesis, University of Zaragoza.

LASHERAS, F., LOMBARDERO, M. & RODRÍGUEZ, J. I. 1995. *Exploración del grafito en España*. Report **11–415**, IGME, Madrid.

LASHERAS, E., LAGO, M., GARCÍA, J. & ARRANZ, E. 1999a. Emplazamiento de sills y diques del Pérmico superior en el Macizo de Cinco Villas (Pirineo navarro). *Geogaceta*, **25**, 123–126.

LASHERAS, E., LAGO, M., GARCÍA, J. & ARRANZ, E. 1999b. Geoquímica del magmatismo, Pérmico superior, del Macizo de Cinco Villas (Pirineo navarro). *Geogaceta*, **25**, 119–122.

LASHERAS, E., LAGO, M., GARCÍA, J. & ARRANZ, E. 1999c. Petrología de diques doleríticos y basaltos, Pérmico superior, en el Macizo de Cinco Villas (Pirineo navarro). *Geogaceta*, **25**, 115–118.

LAUBSCHER, H. & BERNOULLI, D. 1977. Mediterranean and Tethys. *In*: NAIRN, A. E. M., KANES, W. H. & STHELI, F. G. (eds) *The Ocean Basins and Margins*, Vol. 4A. Plenum Press, New York, 1–28.

LAUMONIER, B. 1988. Les groupes de Canaveilles et de Jujols ('Paléozoïque inférieur') des Pyrénées Orientales. Arguments en faveur de l'âge essentiellement Cambrien de ces séries. *Hercynica*, **4**, 25–38.

LAUMONIER, B. (coord.) 1996. Cambro – Ordovicien. *In*: BARNOLAS, A. & CHIRON J. C. (eds) *Synthèse Géologique et Géophysique des Pyrénées*. BRGM and ITGE, Orleans and Madrid, 157–209.

LAUMONIER, B. 1998. Les Pyrénées centrales et orientales au début du Paléozoïque (Cambrien s.l.): évolution paléogéographique et géodynamique. *Geodinamica Acta*, **11**, 1–11.

LE BAS, M. J., REX, D. C. & STILLMAN, C. J. 1986. The Early magmatic chronology of Fuerteventura, Canary Islands. *Geological Magazine*, **123**(3), 287–298.

LE BAS, M. J., LE MAITRE, R. W., STRECKEISEN, A. & ZANETTIN, B. 1986. A chemical classification of volcanic rocks based on the total alkali-silica diagram. *Journal of Petrology*, **27**, 745–750

LE PICHON, X. & BARBIER, F. 1987. Passive margin formation by low-angle faulting within the upper crust, the Northern Bay of Biscay margin. *Tectonics*, **6**, 133–150.

LE PICHON, X., SIBUET, J.-C. & BONNIN, J. 1970. La faille nord-pyrénéenne, faille transformante liée à l'ouverture du Golfe de Gascogne. *Comptes Rendus de l'Académie des Sciences de Paris*, **271**, 1941–1944.

LE VOT, M., BITEAU, J. J. & MASSET, J. M. 1996. The Aquitaine basin: oil and gas producing in the foreland of the Pyrenean fold-and-thrust belt. New exploration perspectives. In: ZIEGLER, P. A. & HORVÀTH, F. (eds) *Peri-Tethys Memoir 2: Structure and Prospects of Alpine Basins and Forelands*. Mémoires du Muséum national d'Histoire naturelle, **170**, 159–171.

LEAL, N., PEDRO, J., MOITA, P., FONSECA, P., ARAÚJO, A. & MUNHÁ, J. 1997. Metamorfismo nos sectores meridionais da zona de Ossa-Morena: actualização de conhecimentos. In: ARAÚJO, A. A. & PEREIRA, M. F. (eds) *Estudo sobre a Geología da Zona de Ossa-Morena (Maciço Iberico)*. Livro de Homenagem ao Profesor Francisco Gonçalves, Universidade de Evora, 119–132.

LEBLANC, D., GLEIZES, G., ROUX, L. & BOUCHEZ, J. L. 1996. Variscan dextral transpression in the French Pyrenees: new data from the Pic des Trois-Seigneurs granodiorite and its country rocks. *Tectonophysics*, **261**, 331–345.

LEBLANC, M., MORALES, J. A., BORREGO, J. & ELBAZ-POOLICHET, J. 2000. 4.500 year old mining pollution in southern Spain: Long-term implications for modern mining polluting. *Economic Geology*, **95**, 655–662.

LECLERC, J. 1971. *Etude geologique du massif du Maigmo et de ses abords (Province d'Alicante, Espagne)*. PhD Thesis, University of Paris.

LECLERC, J. & AZÉMA, J. 1976. Le crétace dans la region d'Agost (Province d'Alicante, Espagne) et ses accidents sedimentaires. *Cuadernos de Geología. Universidad de Granada*, **7**, 35–51.

LECOINTRE, G., TINKLER, K. J. & RICHARDS, G. 1967. The marine Quaternary of the Canary Islands. *Academy of Natural Science of Philadelphia Proceedings*, **119**, 325–344.

LÉCOLLE, M. 1972. Successions lithologique et stratigraphique dans la Province d'Huelva (Espagne); positions des mineralisations manganesiferes et pyrituses. *Comptes Rendus de l'Académie des Sciences de Paris*, **274**, 505–508.

LÉCOLLE, M. 1977. *La ceinture sud-ibérique: un exemple de province à amas sulfurés volcano-sédimentaires (tectonique, métamorphisme, stratigraphie, volcanisme, paléogéographie et metallogénie)*. PhD Thesis, University of Pierre and Marie Curie, Paris.

LEDO, J., AYALA, C., POUS, J., QUERALT, P., MARCUELLO, A. & MUÑOZ, J. A. 2000. New geophysical constraints on the deep structure of the Pyrenees. *Geophysical Research Letters*, **27**, 1037–1040.

LEFÈBVRE, B. & GUTIÉRREZ-MARCO, J. C. 2002. New Ordovician mitrocystitidan mitrates (Echinodermata, Stylophora) from the Central Iberian zone (Spain). *Neues Jahrbuch für Geologie und Paläontologie Abhandlungen*, **224**.

LEFORT, J. P. 1983. A new geophysical criterion to correlate the Acadian and Hercynian orogenies of western Europe and eastern North America. In: HATCHER, R. D., WILLIAMS, H. & ZIETZ, I. (eds) *Contributions to the Tectonics and Geophysics of Mountain Chains*. Geological Society of America, Memoir, **158**, 3–18.

LEFORT, J. P. & HAWORTH, R. T. 1979. The age and origin of deepest correlative structures recognized in Canada and Europe. *Tectonophysics*, **59**, 139–150.

LEFORT, J. P. & RIBEIRO, A. 1980. La faille Porto-Badajoz-Cordoue a-t-elle controlé l'évolution de l'océan paléozoique sud-armoricain? *Bulletin de la Société Géologique de France*, **22**, 455–462.

LEINFELDER, R. R. 1993. Upper Jurassic reef types and controlling factors. *Profil*, **5**, 1–45.

LEISTEL, J. M., BONIJOLY, D., BRAUX, C. et al. 1994. *The massive sulphide deposits of the South Iberian Pyrite province: geological setting and exploration criteria*. Documents du BRGM **234**.

LEISTEL, J. M., MARCOUX, E., THIÉBLEMONT, D. et al. 1998a. The volcanic-hosted massive sulphide deposit of the Iberian Pyrite Belt. Review and preface to the Thematic Issue. *Mineralium Deposita*, **33**, 82–97.

LEISTEL, J. M., MARCOUX, E., THIÉBLEMONT, D. et al. 1998b. The volcanic-hosted massive sulphide deposits of the Iberian Pyrite Belt. *Mineralium Deposita*, **33**, 2–30.

LEMOINE, M., TRICART, P. & BOILLOT, G. 1987. Ultramafic and gabbroic ocean floor of the Ligurian Tethys (Alps, Corsica, Apennines): In search of a genetic model. *Geology*, **15**, 622–625.

LEMOS DE SOUSA, M. J. & OLIVEIRA, J. T. (eds) 1983. *The Carboniferous of Portugal*. Serviços Geológicos de Portugal, Lisboa, Memórias, **29**.

LENDÍNEZ, A., RUIZ FERNÁNDEZ DE LA LOPA, V. & CARLS, P. 1989. *Moyuela*. Spanish Geological Survey, Report and Map, Scale 1:50 000, **466**.

LENZ, A. C., ROBARDET, M., GUTIÉRREZ-MARCO, J. C. & PIÇARRA, J. M. 1997. Devonian graptolites from southwestern Europe: a review with new data. *Geological Journal*, **31**, 349–358.

LEÓN, C. 1967. Las formaciones volcánicas del Cerro de los Lobos (Almería, sureste de España). *Estudios Geológicos*, **23**, 15–28.

LEONE, F., HAMMANN, W., LASKE, R., SERPAGLI, E. & VILLAS, E. 1991. Lithostratigraphic units and biostratigraphy of the post-sardic Ordovician sequence in south-west Sardinia. *Bollettino della Società Paleontologica Italiana*, **30**, 201–235.

LEPVRIER, C. & MARTÍNEZ-GARCÍA, E. 1990. Fault development and stress evolution of the post-Hercynian Asturian Basin (Asturias and Cantabria, northwestern Spain). *Tectonophysics*, **184**, 345–356.

LEPVRIER, C., MARTINEZ-GARCÍA, E., ROBLES, S. & VICENTE-BRAVO, J. C. 1992. Kinematics of Aptian–Albian synsedimentary faults in the Basque Basin (N. Spain), as related to the rifting of the Bay of Biscay. *14e Réunion des Sciences de la Terre, Toulouse, Résumes*, 98.

LERET, G., CÁMARA, P. & LERET, I. 1982. Aportación al conocimiento estratigráfico y sedimentológico del Cretácico en la Zona Prebética oriental (transversal de Villena-Alicante). *Cuadernos de Geología Ibérica*, **8**, 465–481.

LETERRIER, J. 1972. *Étude pétrographique et géochimique du Massif granitique de Quérigut (Ariège)*. CNRS, Nancy, Memoir **23**.

LETHIERS, F. 1983. Paléobiogéographie des faunes d'ostracodes au Dévonien supérieur. *Lethaia*, **16**, 39–49.

LEUTWEIN, F., SAUPE, F., SONET, J. & BOUYX, E. 1970. Première mesure géochronologique en Sierra Morena. La granodiorite de Fontannosas (Ciudad Real, Espagne). *Geologie en Mijnbouw*, **49**, 297–304.

LHENAFF, R. 1977. *Recherches géomorphologiques sur les Cordillères Bétiques centre-occidentales (Espagne)*. PhD Thesis, University of Paris-Sorbonne.

LHENAFF, R. 1986. Les grans poljés des Cordilleres Bétiques andalouses et leurs rapports avec l'organisation endokarstique. *Karstologia Mémoires*, **1**, 101–112.

LIESA, C. L. 2000. *Fracturación y campos de esfuerzos compresivos alpinos en la Cordillera Ibérica y el NE peninsular*. PhD Thesis, Zaragoza University.

LIESA, C. L. & CASAS SAINZ, A. M. 1994. Reactivación alpina de pliegues y fallas del zócalo hercínico de la Cordillera Ibérica: ejemplos de la Sierra de la Demanda y la Serranía de Cuenca. *Cuadernos del Laboratorio Xeologico de Laxe*, **19**, 119–135.

LIESA, C. L. & SIMÓN, J. L. 1994. Fracturación a distintas escalas y campos de esfuerzos durante la tectogénesis alpina en el area de Mosqueruela (Teruel). *Estudios Geológicos*, **50**(1–2), 47–57.

LIESA, C. L., SORIA, A. R. & MELÉNDEZ, A. 1996. Estudio preliminar sobre la tectónica sinsedimentaria del Cretácico inferior en el borde septentrional de la Cubeta de Aliaga (Cordillera Ibérica). *Geogaceta*, **20**(7), 1707–1710.

LIESA, C. L., SORIA, A. R. & MELÉNDEZ, A. 2000a. Lacustrine evolution in a basin controlled by extensional faults: the Galve subbasin, Teruel, Spain. In: GIERLOWSKI-KORDESCH, E. H. & KELTS, K. R. (eds) *Lake Basins through Space and Time*. AAPG Studies in Geology, **46**, 295–302.

LIESA, C. L., SORIA, A. R. & MELÉNDEZ, A. 2000b. Estructura extensiva cretácica e inversión terciaria del margen noroccidental de la subcuenca de Las Parras (Cordillera Ibérica, España). *Geotemas*, **1**(2), 231–234.

LIESA CARRERA, C. L. 1999. Estructura y cinemática del arco de cabalgamientos Portalrubio-Vandellós en el sector de Castellote (Teruel). *Revista Mas de las Matas*, **18**, 9–37.

LIESA, M. 1988. *El metamorphisme del versant sud del massis del Roc de Frausa (Pireneu oriental)*. PhD Thesis, University of Barcelona.

LIETZ, J. & SCHMINCKE, H. U. 1975. Miocene-Pliocene sea level changes and volcanic episodes on Gran Canaria (Canary Islands) in the light of new K-Ar ages. *Palaeogeography, Palaeoclimatology and Palaeoecology*, **18**, 213–239.

LILLO, J. 1992. *Geology and geochemestry of Linares-La Carolina Pb. ore field (southeastern border of the hesperian massif*. PhD Thesis, University of Leeds.

LILLO, J., OYARZUN, R., LUNAR, R., DOBLAS, M., GONZALEZ, A. & MAYOR, N. 1992. Geological and metallogenic aspects of late Variscan Ba- (F)-(base-metal) vein deposits of Spanish Central System. *Transactions of the Institute of Mining and Metallurgy*, **101**, 24–32.

LIN, P. N. 1992. Trace element and isotopic characteristics of western Pacific pelagic sediments: Implications for the petrogenesis of Mariana Arc magmas. *Geochimica et Cosmochimica Acta*, **56**, 1641–1654.

LIÑÁN, E. 1974. Las formaciones cámbricas del norte de Córdoba. *Acta Geológica Hispánica*, **9**, 15–20.

LIÑÁN, E. 1978. *Bioestratigrafía de la Sierra de Córdoba*. PhD Thesis, University of Granada.

LIÑÁN, E. 1984a. Introducción al problema de la Paleogeografía del Cámbrico de Ossa Morena. *Cuadernos do Laboratorio Xeolóxico de Laxe*, **8**, 283–314.

LIÑÁN, E. 1984b. Los icnofósiles de la Formación Torreárboles (¿Precámbrico?-Cámbrico Inferior) en los alrededores de Fuente de Cantos, Badajoz. *Cuadernos do Laboratorio Xeolóxico de Laxe*, **8**, 47–74.

LIÑÁN, E. & FERNÁNDEZ-CARRASCO, J. 1984. La Formación Torreárboles y la paleogeografía del límite Precámbrico-Cámbrico en Ossa Morena (Flanco Norte de la alineación Olivenza-Monesterio). *Cuadernos do Laboratorio Xeolóxico de Laxe*, **8**, 315–328.

LIÑÁN, E. & GÁMEZ-VINTANED J. A. 1993. Lower Cambrian palaeogeography of the Iberian Peninsula and its relations with some neighbouring European areas. *Bulletin de la Société Géologique de France*, **164**, 831–842.

LIÑÁN, E. & GOZALO, R. 1986. Trilobites del Cámbrico inferior y medio de Murero (Cordillera Ibérica). *Memorias del Museo Paleontológico de la Universidad de Zaragoza*, **2**, 1–104.

LIÑÁN, E. & PALACIOS, T. 1983. Aportaciones micropaleontológicas para el conocimiento del límite Precámbrico-Cámbrico en la Sierra de Córdoba, España. *Comunicaçoes dos Serviços Geológicos de Portugal*, **69**, 227–234.

LIÑÁN, E. & PALACIOS, T. 1987. Asociaciones de pistas fósiles y microorganismos de pared orgánica del Proterozoico, en las facies esquisto-grauváquicas del norte de Cáceres. Consecuencias regionales. *Boletín de la Real Sociedad Española de Historia Natural*, **82**, 211–232.

LIÑÁN, E. & PEREJÓN, A. 1981. El Cámbrico inferior de la 'Unidad de Alconera', Badajoz (SW de España). *Boletín de la Real Sociedad Española de Historia Natural (Sección Geológica)*, **79**, 125–148.

LIÑÁN, E. & QUESADA, C. 1990. Ossa Morena Zone. Stratigraphy. Rift Phase (Cambrian). *In*: DALLMEYER, R. D. & MARTÍNEZ GARCÍA, E. (eds) *Pre-Mesozoic Geology of Iberia*. Springer-Verlag, Berlin, 259–266.

LIÑÁN, E. & SDZUY, K. 1978. A trilobite from the Lower Cambrian of Córdoba (Spain) and its stratigraphical significance. *Senckenbergiana Lethaea*, **59**, 387–399.

LIÑÁN, E. & TEJERO, R. 1988. Las formaciones precámbricas del antiforme de Paracuellos (Cadenas Ibéricas). *Boletín de la Real Sociedad Española de Historia Natural*, **84**, 39–49.

LIÑÁN, E., MORENO-EIRIS, E., PEREJÓN, A. & SCHMITT, M. 1982. Fossils from the basal levels of the Pedroche Formation, Lower Cambrian (Sierra Morena, Córdoba, Spain). *Boletín de la Real Sociedad Española de Historia Natural (Geología)*, **79** (year 1981), 277–286.

LIÑÁN, E., PALACIOS, F. & PEREJÓN, A. 1984. Precambrian–Cambrian boundary and correlation from southwestern and central part of Spain. *Geological Magazine*, **121**, 221–228.

LIÑÁN, E., FERNÁNDEZ-NIETO, C., GÁMEZ, J. A. *et al.* 1993a. Problemática del límite Cámbrico Inferior-Medio en Murero (Cadenas Ibéricas, España). *Revista Española de Paleontología*, No. Extraordinario, 26–39.

LIÑÁN, E., PALACIOS, T., VILLAFAINA, M., GOZALO, R. & ÁLVARO, J. J. 1993b. Middle Cambrian acritarchs in the levels with *Solenopleuropsis* and *Sao* (Trilobita) from Zafra (Badajoz province, Spain). Biostratigraphical consequences. *Terra Nova*, **6**, 3–4.

LIÑÁN, E., PEREJÓN, A. & SDZUY, K. 1993c. The Lower-Middle Cambrian stages and stratotypes from the Iberian Peninsula: a revision. *Geological Magazine*, **130**, 817–833.

LIÑÁN, E., GRANT, S. & NAVARRO, D. 1994. El Precámbrico de Codos (Zaragoza). Cadena Ibérica Oriental. *Reunión de Xeología e Minería de NO Peninsular, Resúmenes, Laxe, España*, 3.

LIÑÁN, E., ÁLVARO, J. J., GOZALO, R., GÁMEZ-VINTANED, J. A. & PALACIOS, T. 1995. El Cámbrico Medio de la Sierra de Córdoba (Ossa Morena, S de España): trilobites y paleoicnología. Implicaciones bioestratigráficas y paleoambientales. *Revista Española de Paleontología*, **10**, 219–238.

LIÑÁN, E., VILLAS, E., GÁMEZ VINTANED, J. A., ÁLVARO, J., GOZALO, R., PALACIOS, T. & SDZUY, K. 1996. Síntesis paleontológica del Cámbrico y Ordovícico del Sistema Ibérico (Cadenas Ibéricas y Cadenas Hespéricas). *Revista Española de Paleontología*, No. extraord., 21–32.

LIÑÁN, E., GONÇALVES, F., GÁMEZ VINTANED, J. A. & GOZALO, R. 1997. Evolución paleogeográfica del Cámbrico de la Zona de Ossa Morena basada en el registro fósil. *In*: ARAÚJO, A. A. & PEREIRA, M. F. (eds) *Estudo sobre a Geología da Zona de Ossa Morena (Maciço Ibérico), Livro de Homenagem ao Professor Francisco Gonçalves*. University of Évora, 1–26.

LINARES, A. & CREMADES, J. 1988. Graysonites (Mantelliceratinae Ammonitina) from the lowermost Cenomanian of the betic cordillera (Spain). *Geobios*, **21**, 307–317.

LINARES, A. & SANDOVAL, J. 1993. El Aaleniense de la Cordillera Bética (Sur de España): análisis bioestratigráfico y caracterización paleogeográfica. *Revista de la Sociedad Geológica de España*, **6**, 177–206.

LINARES, A., MOUTERDE, R. & RIVAS, P. 1971. El Lias en el Sector Central de la Zona Subbética (vista de conjunto). *In*: *I Coloquio de Estratigrafía y Paleogeografía del Jurásico de España*, Vitoria, Vol. 2, 227–236.

LINNEMANN, U. 1998. Relative sea-level fluctuations and tectonic setting of the Ordovician type section of Thuringia (Saxo-Thuringian Terrane, Central European Variscides, Germany). *Acta Universitatis Carolinae, Geologica*, **42**, 465–471.

LINNEMANN, U. & HEUSE, T. 2000. The Ordovician of the Schwarzburg Anticline: Geotectonic setting, biostratigraphy and sequence stratigraphy (Saxo-Thuringian Terrane, Germany). *Zeitschrift der Deutschen Geologischen Gesellschaft*, **151**, 471–491.

LIU, H. S. 1980. Convection generated stress field and intraplate volcanism. *Tectonophysics*, **65**, 225–244.

LLOMPART, C. 1979. Aportaciones a la paleontología del Lías de Menorca. *Boletín de la Sociedad de Historia Natural de Baleares*, **23**, 87–116.

LLOMPART, C. 1980. Nuevo afloramiento del Lías fosilífero menorquín. *Boletín de la Sociedad de Historia Natural de Baleares*, **24**, 85–88.

LLOMPART, C., PALLI, L. & ROSELL, J. 1984. Aportaciones al conocimiento del Mesozoico de la provincia de Girona: Jurásico. *Publicaciones de Geología, Universidad Autonoma Barcelona*, **20**, 339–354.

LLOMPART, C., ROSELL, J. & MÁRQUEZ-ALIAGA, A. 1987. El Muschelkalk de la isla de Menorca. *Cuadernos Geología Ibérica*, **11**, 323–335.

LLOPIS, N. 1961. Estudio geológico de la región de Cabo de Peñas (Asturias). *Boletín del Instituto Geológico y Minero de Asturias*, **72**, 237–248.

LLOPIS LLADÓ, N. 1965. Sur la paléogéographie du Dévonien du Nord de l'Espagne. *Compte Rendu Sommaire des séances de la Société Géologique de France*, **9**, 290–292.

LLOPIS LLADÓ, N. 1966. Sobre la estratigrafía del Silúrico de Andorra y el límite Silúrico-Devónico. V Congreso Internacional de Estratigrafía del Pirineo, Jaca-Pamplona, *Pirineos*, **81–82**, 79–86.

LLOPIS LLADÓ, N. & ROSELL, J. 1968. Algunas aportaciones a la estratigrafía del Silúrico-Devónico de 'Las Nogueras' al E de Gerri de la Sal (Lérida). *Acta Geológica Hispánica*, **3**, **5**, 113–116.

LOCUTURA, J, TORNOS, F., FLORIDO, P. & BAENA, J. 1990. Ossa Morena Zone: Metallogeny. *In*: DALLMEYER, R. D. & MARTINEZ, E. (eds) *Premesozoic Geology of Iberia*. Springer-Verlag, Berlin, 321–330.

LOESCHKE, J. 1983. Igneous and pyroclastic rocks in Devonian and Lower Carboniferous strata of the Cantabrian Mountains (NW Spain). *Neues Jahrbuch fuer Geologie und Palaeontologie. Abhandlungen*, **8**, 495–504.

LOESCHKE, J. & ZEIDLER, N. 1982. Early Palaeozoic sills in the Cantabrian Mountains (Spain) and their geotectonic environment. *Neues Jahrbuch fuer Geologie und Palaeontologie. Abhandlungen*, **7**, 419–439.

LOEVEZIJN, G. B. S. VAN 1989. Extinction pattern for the Middle-Upper Devonian stromatoporoid-coral reefs; a case of study from the Cantabrian Mountains. *Proceedings Koninklijke Nederlandse Akademie Wetenscappen*, **92**(1), 61–74.

LOMBARDERO, M. 1998*a*. Andalucita, Cianita y Sillimanita. *In*: Urbano, R. (ed.) *Guía para la investigación de los Recursos Minerales en España*. IGME, Madrid, 110–111.

LOMBARDERO, M. 1998*b*. Wollastonita. *In*: Urbano, R. (ed.) *Guía para la investigación de los Recursos Minerales en España*. IGME, Madrid, 112.

LOMBARDERO, M. & QUEREDA, J. M. 1992. La Piedra Natural para la construcción. *In*: GARCÍA-GUINEA, J. & MARTÍNEZ-FRÍAS, J. (eds) *Recursos Minerales de España*. Consejo Superior de Investigaciones Científicas, Colección Textos Universitarios, **15**, 1115–1152.

LOMBARDERO, M. & REILE, E. 1997. Natural Stone of San Luis (Argentina): Geological potential and business opportunity. *Roc-Maquina*, **24**, 47–54.

LONERGAN, L. 1991. *Structural Evolution of the Sierra Espuña, Betic Cordillera, SE Spain*. PhD Thesis, Oxford University.

LONERGAN, L. 1993. Timing and Kinematics of deformation in the Malaguide Complex, Internal Zone of the Betic Cordillera, Southeast Spain. *Tectonics*, **12**, 460–476.

LONERGAN, L. & PLATT, J. 1995. The Malaguide–Alpujarride boundary: a major extensional contact in the Internal Zone of the eastern Betic Cordillera, SE Spain. *Journal of Structural Geology*, **17**, 1665–1671.

LONERGAN, L. & WHITE, N. 1997. Origin of the Betic-Rif mountain belt. *Tectonics*, **16**, 504–522.

LOON, A. J. VAN 1972. A prograding deltaic complex in the Upper Carboniferous of the Cantabrian Mountains (Spain): the Prioro-Tejerina Basin. *Leidse Geologische Mededelingen*, **48**, 1–81.

LÓPEZ-ARCHILLA, A. I. & AMILS, R. 1999. A comparative ecological study of two acidic rivers in Southwestern Spain. *Microbial Ecologi*, **38**, 146–156.

LÓPEZ DÍAZ, F. 1992. *Evolución Estructural de la Antiforma de Navalpino (Zona Centroibérica)*. PhD Thesis, University of Oviedo.

LÓPEZ DÍAZ, F. 1995. Late Precambrian series and structures in the Navalpino variscan anticline (Central Iberian Peninsula). *Geologische Rundschau*, **84**, 151–163.

LÓPEZ GARCÍA, J., LUNAR, R. & OYARZUN, R. 1988. Silver and lead mineralogy in gossan type deposits of Sierra de Cartagena (S. E. Spain). *Transactions of the Institute of Mininig and Metallurgy*, **97**, 82–88.

LÓPEZ-GARRIDO, A. C. 1969. Primeros datos sobre la estratigrafía de la región Chiclana de Segura- Río Madera (Zona Prebética, provincia de Jaén). *Acta Geológica Hispánica*, **4**, 84–90.

LÓPEZ-GARRIDO, A. C., PÉREZ-LÓPEZ, A. & SANZ DE GALDEANO, C. 1997. Présence de faciès Muschelkalk dans des unités alpujarrides de la région de Murcie (Cordillère bétique, sud-est de l'Espagne) et implications paléogéographiques. *Comptes Rendus de l'Académie des Sciences Paris, série IIa*, **324**, 647–654.

LÓPEZ-GÓMEZ, J. 1985. Sedimentología y estratigrafía de los materiales pérmicos y triásicos del sector SE de la Rama Castellana de la Cordillera Ibérica entre Cueva de Hierro y Chelva (provincias de Cuenca y Valencia). *Seminarios de Estratigrafía, serie monografías*, **11**, 1–344.

LÓPEZ-GÓMEZ, J. & ARCHE A. 1986. Estratigrafía del Pérmico y Triásico en facies Buntsandstein y Muschelkalk en el sector SE de la Rama Castellana de la Cordillera Ibérica. *Estudios Geológicos*, **42**, 259–270.

LÓPEZ-GÓMEZ, J. & ARCHE A. 1992. Paleogeographical significance of the Röt (Anisian, Triassic) Facies (Marines Formation) in the Iberian Ranges, Eastern Spain. *Palaeogeography, Palaeoclimatology, Palaeoecology*, **103**, 347–361.

LÓPEZ-GÓMEZ, J. & ARCHE, A. 1993. Sequence stratigraphy analysis and paleogeographic interpretation of the Buntsandstein and Muschelkalk facies (Permo-Triassic) in the SE Iberian Ranges, eastern Spain. *Palaeogeography Palaeoclimatology Palaeoecology*, **103**, 179–201.

LÓPEZ-GÓMEZ, J., MAS, R. & ARCHE, A. 1993. The evolution of the Middle Triassic (Muschelkalk) carbonate ramp in the SE Iberian Ranges, eastern Spain: sequence stratigraphy, dolomitization processes and dynamic controls. *Sedimentary Geology*, **87**, 165–193.

LÓPEZ-GÓMEZ, J., ARCHE, A., CALVET, F. & GOY, A. 1998. Epicontinental marine carbonate sediments of the Middle and Upper Triassic in the westernmost part of the Tethys Sea, Iberian Peninsula. *In*: BACHMANN, G. H. & LERCHE, I. (eds) *Epicontinental Triassic, Zentralblatt für Geologie und Paläontologie*, (1), **9–10**, 1033–1084.

LÓPEZ GUTIERREZ, J., MARTÍNEZ FRIAS, J., LUNAR, R. & LÓPEZ, J. A. 1993. El rombohorst mineralizado de Las Herrerías: Un caso de 'doming' e hidrotermalismo submarino mioceno en el S. E. Ibérico. *Estudios Geológicos*, **49**,13–19.

LÓPEZ GUTIERREZ, J., MARTÍNEZ FRIAS, J. & LUNAR, R. 1995. El area mineralizada de Vera- Garrucha como base de un modelo tectonometalogenético de tipo 'Basin and Range' en el SE ibérico. *In*: MARTINEZ FRIAS, J., REY, J. & LUNAR, R. (eds) *Geología y Metalogenia en Ambientes Oceánicos: Depósitos hidrotermales submarinos*. Instituto Español de Oceanografía, **18**, 89–95.

LÓPEZ-HORGUE, M., FERNÁNDEZ-MENDIOLA, P. A. & GARCÍA-MONDÉJAR, J. 1994. El modelo de plataforma carbonatada del Albiense superior de Sopeña (Karrantza, Bizkaia). *Geogaceta*, **16**, 78–81.

LÓPEZ-HORGUE, M. A., ARANBURU, A., FERNÁNDEZ-MENDIOLA, P. A. & GARCÍA-MONDÉJAR, J. 2000. Existencia de una discordancia angular con laguna de Albiense medio en el Complejo Urgoniano de Ranero (Ramales-Karrantza, región vasco-cantábrica). *Geogaceta*, **28**, 89–92.

LÓPEZ MARTÍNEZ, N. 1989. Tendencias en Paleobiogeografía: el futuro de la biogeografía del pasado. *In*: Aguirre, E. (coord.) *Nuevas Tendencias, Paleontología*, **10**, 271–296.

LÓPEZ MARTÍNEZ, N., GARCÍA MORENO, E. & ÁLVAREZ SIERRA, A. 1986. Paleontología y bioestratigrafía (micromamíferos) del Mioceno medio y superior del sector central de la cuenca del Duero. *Studia Geológica Salmanticensia, Ediciones Universidad Salamanca*, **22**, 191–212.

LÓPEZ MORO, F. J. 2000. *Las rocas plutónicas calcoalcalinas y shoshoníticas del Domo Varisco del Tormes (Centro-Oeste español)*. PhD Thesis, University of Salamanca.

LÓPEZ PLAZA, M. 1982. *Contribución al conocimiento de la dinámica de los cuerpos graníticos en la penillanura salmantino–zamorana*. PhD Thesis, University of Salamanca.

LÓPEZ PLAZA, M. & CARNICERO, A. 1987. El plutonismo Hercínico de la penillanura salmantino–zamorana (Centro Oeste de España). Visión de conjunto en el contexto geológico regional. *In*: BEA, F., CARNICERO, A., GONZALO, J. C., LÓPEZ PLAZA, M. & RODRÍGUEZ ALONSO, M. D. (eds) *Geología de los granitoides y rocas asociadas del Macizo Hespérico. Libro Homenaje a L. C. García Figuerola*. Rueda, Madrid, 53–69.

LÓPEZ PLAZA, M. & GONZALO, J. C. 1993. Caracterización geoquímica de las anatexitas del Domo del Tormes (provincias de Salamanca y Zamora). *Revista de la Sociedad Geológica de España*, **6**, 3–24.

LÓPEZ PLAZA, M. & MARTÍNEZ CATALAN, J. R. 1987. Síntesis estructural de los granitoides del Macizo Hespérico. *In*: BEA, F., CARNICERO, A., GONZALO, J. C., LÓPEZ PLAZA, M. & RODRÍGUEZ ALONSO, M. D. (eds) *Geología de los Granitoides y rocas* asociadas del Macizo Hespérico. *Libro Homenaje a L. C. García Figuerola*. Rueda, Madrid, 195–210.

LÓPEZ-RUIZ, J. 1999. El campo volcánico Neógeno del SE de España. *Enseñanza de las Ciencias de la Tierra*, **7**, 244–253.

LÓPEZ-RUIZ, J. & RODRÍGUEZ-BADIOLA, E. 1980. La región volcánica neógena del sureste de España. *Estudios Geológicos*, **36**, 5–63.

LÓPEZ-RUIZ, J. & RODRÍGUEZ-BADIOLA, E. 1985. La región volcánica Mio-Pleistocena del NE de España. *Estudios Geológicos*, **41**, 105–126.

LÓPEZ-RUIZ, J. & WASSERMAN, M. D. 1991. Relación entre la hidratación/desvitrificación y el d^{18}O en las rocas volcánicas neógenas del SE de España. *Estudios Geológicos*, **47**, 3–11.

LÓPEZ-RUIZ, J., RODRÍGUEZ-BADIOLA, E. & GARCÍA-CACHO, L. 1977. Origine des grenats des roches calco-alcalines du Sud-Est de l'Espagne. *Bulletin Volcanologique*, **40**, 1–12.

LÓPEZ-RUIZ, J., RODRÍGUEZ-BADIOLA, E. & CEBRIÁ, J. M. 1986. Petrogénesis de los basaltos alcalinos de la Garrotxa, región volcánica del NE de España. *Geogaceta*, **1**, 28–31.

LÓPEZ-RUIZ, J., CEBRIÁ, J. M., DOBLAS, M., OYARZUN, R., HOYOS, M. & MARTÍN, C. 1993. The late Cenozoic alkaline volcanism of the central Iberian Peninsula (Calatrava Volcanic Province, Spain): intraplate volcanism related to extensional tectonics. *Journal of the Geological Society, London*, **150**, 915–922.

LÓPEZ RUIZ, J. L. 1970. Estudio petrografico y geoquimico del Complejo filoniano de Fuerteventura (Islas Canarias). *Estudios Geológicos*, **26**, 173–208.

LORENZ, W. 1997. Criteria for the assesment of non-metallic mineral deposits. *Geologisches Jahrbuch*, **127**, 299–326.

LORENZO, C., ARSUAGA, J. L. & CARRETERO, J. M. 1999. Hand and foot remains from the Gran Dolina Early Pleistocene site (Sierra de Atapuerca, Spain). *Journal of Human Evolution*, **37**, 501–522.

LORENZO, J. I. 1992. El origen del homnbre y la Paleoantropología en Aragón. *In*: MOLINA, E. (ed.) *Origen y evolución del hombre*. University of Zaragoza, SIUZ Cuadernos Interdisciplinares, **2**, 105–134.

LOSANTOS, M., PALAU, J. & SANZ, J. 1986. Considerations about Hercynian thrusting in the Marimanya Massif (Central Pyrenees). *Tectonophysics*, **129**, 71–79.

LOSANTOS, M., BERÁSTEGUI, X., MUÑOZ, J. A. & PUIGDEFABREGAS, C. 1988. Corte geológico cortical del Pirineo Central (Perfil ECORS): Evolución geodinámica de la Cordillera Pirenaica. *Simposio sobre: Cinturones orogénicos. II Congreso Geológico de España*. Sociedad Geológica de España, 7–16.

LOTZE, F. 1929. Stratigraphie und Tektonik des Keltiberischen Grundgebirges (Spanien) *Abhandlungen der Gesellschaft der Wissenschaften zu Göttingen Mathematisch-Physikalische Klasse Neue Folge*, **14**(2).

LOTZE, F. 1945. Zur Gliederung des Varisciden der Iberischen Meseta. *Geotektonische Forschungen*, **6**, 78–92.

LOTZE, F. 1956a. Das Präkambrium Spaniens. *Neues Jahrbuch für Geologie und Paläontologie. Monatschefte*, **8**, 373–380.

LOTZE, F. 1956b. Über sardischen Bewegungen im Spanien und ihre Beziehungen zur assynthischen Faltung. *In: Geotektonische Symposium zu Ehren von Hans Stille*, Stuttgart, 128–139.

LOTZE, F. 1957a. Zum Alter norddwestspanischer Quarzit-Sandstein-Folgen. *Neues Jahrbuch für Geologie und Paläontologie, Mh.*, **1957**, 464–471.

LOTZE, F. 1957b. Zum Alter nordwestpanischer Quarzit-Sandstein-Folgen. *Neues Jahrbuch für Geologie und Paläontologie Monatshefte*, **10**, 128–139.

LOTZE, F. 1958. Zur Stratigraphie des spanischen Kambriums. *Geologie*, **7**, 727–750.

LOTZE, F. 1961a. *Das Kambrium Spaniens. Teil 1: Stratigraphie Traducido por J. Gómez de Llarena*. Memorias del IGME, **75**.

LOTZE, F. 1961b. Das Kambrium Spaniens. Teil I: Stratigraphie. *Abhandlungen der Gesellschaft der Wissenschaften zu Göttingen (Mathematisch-Physikalische Klasse)*, **6**, 283–501 (translated to Spanish by Gómez de Llarena, J., 1970, El Cámbrico de España. *Memorias del IGME*, **70**).

LOTZE, F. 1961c. Das Kambrium Spaniens. Teil I: Stratigraphie. *Akademie der Wissenschaften und der Literatur, Abhandlungen der mathematisch-naturwissenschaftlichen Klasse*, **1961**(6), 1–216.

LUCAS, C. & GISBERT-AGUILAR, J. (coords). 1995. Carbonifère Supérieur-Permien. *In*: BARNOLAS, A., CHIRON, J. C. & GUÉRANGÉ, B. (eds) *Synthèse Géologique et Géophysique des Pyrénées – Volume 1: Introduction. Géophysique. Cycle hercynien*. BRGM and IGME, Orléans and Madrid, 339–359.

LUCAS, C., TAUGOURDEAU-LANZ, J. & TEFIANI, M. 1980. Un repere palynologique dans le Trias des Corbieres (France). *Comptes Rendus de l'Académie des Sciences*, **294**, 111–116.

LUJÁN, M., BALANYÁ, J. C. & CRESPO-BLANC, A. 2000. Contractional and extensional tectonics in Flysch and Prebetic units (Gibraltar Arc, SE Spain): new constraints on the emplacement mechanisms. *Comptes Rendus de l'Académie des sciences de Paris*, **330**, 631–637

LUMLEY, M.-A. DE 1972. La mandíbula de Bañolas. *Ampurias*, **33–34**, 1–91.

LUMLEY, M.-A. DE 1973. Anténéandertaliens et Néandertaliens du bassin méditerranéen européen. *Études Quaternaires*, **2**, 551–558.

LUMLEY, M.-A. DE & GARCÍA-SÁNCHEZ, M. 1971. L'enfant néandertalien de Carigüela à Piñar (Andalousie). *L'Anthropologie*, **75**, 29–56.

LUNAR, R. 1977. *Mineralogénesis de los Yacimientos de Hierro del NW de la Península*. Memorias del IGME, **90**.

LUNAR, R. 1991. Yacimientos Sedimentarios de Hierro. *In*: LUNAR, R. & OYARZUN, R. (eds) *Yacimientos Minerales*. Editoral Ceura, 352–374.

LUNAR, R. 1992. Yacimientos de Hierro Oolítico en el Ordovícico de España. *In*: GARCÍA GUINEA, J. & MARTÍNEZ-FRÍAS, J. (eds) *Recursos Minerales de España*. CSIC, Colección Textos Universitarios, **15**, 557–568.

LUNAR, R. & AMORÓS, J. L. 1979. Mineralogy of the oolitic iron deposits of the Ponferrada-Astorga Zone, northwestern Spain. *Economic Geology*, **74**, 751–762.

LUNAR, R., MANTECA, J. I., RODRIGUEZ, P. & AMOROS, J. L. 1982. Estudio mineralógico y geoquímico del gossan de los depósitos de Fe- Pb- Zn de la Unión (Sierra de Cartagena). *Boletín del IGME*, **93**, 244–253.

LUNAR, R., ORTEGA, L., SIERRA, J., GARCÍA PALOMERO, F., MORENO, T. & PRICHARD, H. M. 1997. Ni-Cu-(PGM) mineralisation associated with mafic and ultramafica rocks: the recently discovered Aguablanca ore deposit, SW Spain. *In*: PAPUNEN, H. (ed.) *Mineral Deposits*. Balkema, Rotterdam.

LUNAR, R., MARTINEZ-FRIAS, J., BENITO, R. & WOLF, D. 1999. Garnets in Europe: geology and beneficiation of the only economic ore deposit of Spain. *Geotimes*, **44–1**, 23–28.

LUNDEEN, M. T. 1978. Emplacement of the Ronda peridotite, Sierra Bermeja, Spain. *Geological Society of America Bulletin*, **89**, 172–180.

LUQUE C. & SÁENZ DE SANTA MARÍA J. A. 1993. Hunosa: una geologia difícil. *Tierra y Tecnología*, **3**, 1–8.

LUQUE C., MARTÍNEZ-GARCÍA, E. & RUIZ, F. 1990. Metallogenesis (of the Cantabrian Zone). *In*: DALLMEYER, R. D. & MARTÍNEZ GARCÍA, E. (eds) *Pre-Mesozoic Geology of Iberia*. Springer-Verlag, Berlín, 80–87.

LUQUE, F. J. & RODAS, M. 1992. Yacimientos españoles de grafito. *In*: GARCÍA-GUINEA, J. & MARTÍNEZ-FRÍAS, J. (eds) *Recursos Minerales de España*. CSIC, Colección Textos Universitarios, **15**, 513–526.

LUQUE, F. J., RODAS, M. & JUSTO-ERBEZ, A. 1992. Yacimientos españoles de vermiculita. *In*: GARCÍA-GUINEA, J. & MARTÍNEZ-FRÍAS, J. (eds) *Recursos Minerales de España*. CSIC, Colección Textos Universitarios, **15**, 1431–1442.

LUQUE, L., ZAZO, C., RECIO, J. M. *et al.* 1999. Evaluación sedimentaria de la Laguna de la Janda (Cádiz) durante el Holoceno. *Cuaternario y Geomorfología*, **13**(3–4), 43–50.

LUX, D. R., DE YOREO, J. J., GUIDOTTI, C. V. & DECKER, E. R. 1986. The role of plutonism in low pressure/high-temperature metamorphic belt formation. *Nature*, **323**, 794–797.

LUZÓN, A. 2001. *Análisis tectosedimentario de los materiales terciarios continentales del sector central de la Cuenca del Ebro (Provincias de Huesca y Zaragoza)*. PhD Thesis, University of Zaragoza.

LYDON, J. W. 1988. Ore deposits models 14, volcanogenic massive sulphide deposits. Part 2: Genetic models. *Geoscience Canadian*, **15**, 43–65.

LYELL, C. 1865. *Elements of Geology* (sixth edition). London.

MACDONALD, G. A. 1972. *Volcanoes*. Prentice-Hall, Englewood Cliffs.

MACDONALD, G. A. & KATSURA, T. 1964. Chemical composition of Hawaiian lavas. *Journal of Petrology*, **5**, 82–133.

McDONOUGH, W. F. 1990. Constraints on the composition of the continental lithosphere mantle. *Earth and Planetary Science Letters*, **101**, 1–18.

McDOUGALL, I. & SCHMINCKE, H. U. 1976. Geochronology of Gran Canaria, Canary Islands: Age of shield building volcanism and other magmatic phases. *Bulletin of Volcanology*, **40**, 1–21.

McFARLANE, D. I. & RIDLEY, W. I. 1969. An interpretation of gravity data for Tenerife, Canary Islands. *Earth and Planetary Science Letters*, **4**, 481–486.

McGOWAN, G. & EVANS, S. E. 1995. Albanerpetontid amphibians from the Cretaceous of Spain. *Nature*, **373**, 143–145.

McGUILLAVRY, H. G., ROEP, T. B. & GEEL, T. 1960. Notes on the Betic of Málaga near Vélez Rubio (SE Spain). *Koninklijke Nederlandse Akademie Wetenschappen*, Amsterdam, **63**(5), 623–626.

McLAREN, S. J. & ROWE, P. J. 1996. The reliability of Uranium-series mollusc dates from the Western Mediterranean basin. *Quaternary Science Reviews*, **15**, 709–717.

McPHIE, J., DOYLE, M. & ALLEN, R. 1993. *Volcanic Textures: A Guide To the Interpretation of Textures in Volcanic Rocks*. University of Tasmania, Australia.

MAAS, K. 1974. The geology of Liébana. Cantabrian Mountains, Spain; flysh area. *Leidse Geologische Mededelingen*, **49**, 379–465.

MACAYA, J., GONZÁLEZ LODEIRO, F., MARTÍNEZ CATALÁN, J. R. & ÁLVAREZ, F. 1991. Continuous deformation ductile thrusting and backfolding of cover and basement in the Sierra de Guadarrama Hercynian orogen of central Spain. *Tectonophysics*, **191**, 291–309.

MACEDO, C. A. R. 1988. *Granitoides do Complexo Xisto-Grauváquico e Ordovícico na Regiao entre Trancoso e Pinhel (Portugal Central)*. PhD Thesis, University of Coimbra.

MÄCKEL, G. H. 1985. The geology of the Maláguide Complex and its bearing on the geodynamic evolution of the Betic-Rif Orogen (southern Spain and northern Morocco). *GUA Papers of Geology*, **22**, 1–263.

MACKLIN, M. G. & PASSMORE, D. G. 1995. Pleistocene environmental change in the Guadalope Basin, northeast Spain: fluvial and archaeological records. *In*: LEWIN, J., MACKLIN, M. G. & WOODSWARD, J. C. (eds) *Mediterranean Quaternary River Environments*. Balkema, Rotterdam, 103–113.

MACKLIN, M. G., PASSMORE, D., STEVENSON, A. C. & DAVIS, B. A. 1994. Responses of rivers and lakes to Holocene environmental change in the Alcañiz Region, Teruel, North-East Spain. *In*: MILLINGTON, A. C. & PYE, K. (eds) *Environmental Change in Drylands: Biogeographical and Geomorphological Perspectives*. Wiley, Chichester, 113–130.

MADE, J. VAN DER 1999. Ungulates from Atapuerca TD6. *Journal of Human Evolution*, **37**, 389–414.

MAESTRO, A. 1999. *Estructura y evolución alpina de la Cuenca de Almazán (Cordillera Ibérica)*. PhD Thesis, University of Zaragoza.

MAESTRO-MAIDEU, E., ESTRADA, R. & REMACHA, E. 1998. La sección del Carbonífero en el Priorat Central (Prov. de Tarragona). *Geogaceta*, **23**, 91–94.

MAJESTÉ-MENJOULAS, C. & RÍOS, L. M. (coords). 1995. Dévonien-Carbonifère Inferiéur. *In*: BARNOLAS, A., CHIRON, J. C. & GUÉRANGÉ, B. (eds) *Synthèse Géologique et Géophysique des Pyrénées – Volume 1: Introduction. Géophysique. Cycle hercynien*. BRGM and IGME, Orléans and Madrid, 235–301.

MAJOOR, F. J. M. 1988. *A Geochronological Study of the Axial Zone of the Central Pyrenees, with Emphasis on Variscan Events and Alpine Resseting*. Laboratorium Voor Isotopen-Geologie, Amsterdam, Verhandeling No. **6**.

MALETZ, J. 1997. Graptolites from the *Nicholsonograptus fasciculatus* and *Pterograptus elegans* Zones (Abereiddian, Ordovician) of the Oslo region, Norway. *Greifswalder Geowissenschaftliche Beiträge*, **4**, 5–98.

MALLADA, L. 1890. Reconocimiento geográfico y geológico de la provincia de Tarragona. *Boletín de la Comisión del Mapa Geológico de España*, **XVI**, 1–175.

MALLADA, L. 1895. Explicación del Mapa Geológico de España. Tomo II. Sistemas Cambriano y Siluriano. *Memorias de la Comisión del Mapa Geológico de España*, **20**, 1–515.

MALLÓ, A., FONTAN, F., MELGAREJO, J. C. & MATA, J. M. 1995. The Albera zoned pegmatite field, Eastern Pyrenees, France. *Mineralogy & Petrology*, **55**, 103–116.

MALOD, J. A. & MAUFFRET, A. 1990. Iberian plate motions during the Mesozoic. *Tectonophysics*, **184**, 261–278.

MALOD, J.-A., BOILLOT, G., CAPDEVILA, R. *et al.* 1982. Subduction and tectonics on the continental margin off northern Spain, observations with the submersible Cyana. *In*: LEGGETT, J. K. (ed.) *Trench – Forearc Geology*. Geological Society, London, Special Publications, **10**, 309–315.

MALOT, J. A. 1989. Ibérides et plaque ibérique. *Bulletin Société géologique France*, **5**, 927–934.

MAMET, B. & MARTÍNEZ GARCÍA, E. 1995. Permian Microcodiaceans (Algae, Incertae Sedis) Sotres Limestones, Asturias. *Revista Española de Micropaleontología*, **27**, 107–106.

MANGAS, J. & ARRIBAS, A. 1987. Fluid inclusion study in different types of tin deposits associated with the hercynian granites of Western Spain. *Chemical Geology*, **61**, 193–208.

MANJÓN, M., GUTIERREZ CLAVEROL, M. & MARTÍNEZ GARCÍA, E. 1992. La sucesión posthercínica preliásica del área de Villabona (Asturias, N de España). *Actas del III Congreso de Geología de España*, **2**, 107–111.

MANNING, D. A. C., HILL, P. I. & HOWE, J. H. 1996. Primary lithological variation in the kaolinized St. Austell Granite, Cornwall, England. *Journal of the Geological Society*, **153**, 827–838.

MANTECA, J. I & OVEJERO, G. 1992. Los yacimientos de Zn, Ag, Fe del distrito Minero de la Union – Cartagena (Betica oriental). *In*: GARCÍA GUINEA, J. & MARTÍNEZ-FRÍAS, J. (eds) *Recursos Minerales de España*. CSIC, Colección Textos Universitarios, **15**, 1085–1103.

MANTILLA, L. C. 1999. *El metamorfismo hidrotermal de la Sierra de Cameros (La Rioja-España): petrología, geoquímica, geocronología y contexto estructural de los procesos de interacción fluido-roca*. PhD Thesis, Complutense University, Madrid.

MARCANTONIO, F., ZINDLER, A., ELLIOT, T. & STAUDIGEL, H. 1995. Os isotope systematics of La Palma, Canary Islands: Evidence for recycled crust in the mantle source of HIMU ocean islands. *Earth and Planetary Science Letters*, **133**, 397–410.

MARCH, M. 1991. *Los conodontos del Triásico Medio (Facies Muschelkalk) del Noroeste de la Península Ibérica y de Menorca*. PhD Thesis, University of Valencia.

MARCHÁN, C. 1998. *Panorama Minero 1996*. IGME, Madrid.

MARCOS, A. 1970. Sobre la presencia de un flysch del Ordovícico superior en el occidente de Asturias. *Breviora Geologica Asturica*, **14**, 13–28.

MARCOS, A. 1973. Las Series del Paleozoico Inferior y la estructura herciniana del Occidente de Asturias (NW de España). *Trabajos de Geología, Universidad de Oviedo*, **6**, 1–113.

MARCOS, A. & MARQUÍNEZ, J. 1984. La estructura de la Unidad Gildar-Montó (Cordillera Cantábrica). *Trabajos de Geología*, **14**, 53–64.

MARCOS, A. & PÉREZ-ESTAÚN, A. 1981. La estratigrafía de la Serie de los Cabos en la Zona de Vegadeo (Zona Asturoccidental-Leonesa, NW de España). *Trabajos de Geología*, Oviedo, **11**, 89–94.

MARCOS, A. & PHILIPPOT, A. 1972. Nota sobre el Silúrico del occidente de Asturias (NW de España). *Breviora Geologica Asturica*, **16**, 39–42.

MARCOS, A. & PULGAR, F. J. 1982. An approach to the tectonostratigraphic evolution of Cantabrian thrust and fold belt, Variscan Cordillera of NW Spain. *Neues Jahrbuch für Geologie und Paläontologie Abhandlungen*, **163**, 256–260.

MARCOUX, E. 1998. Lead isotope systematics of the giant massive sulphide deposits in the Iberian Pyrite Belt. *Mineralium Deposita*, **33**, 45–58.

MARCOUX, E., LEISTEL, J. M., SOBOL, F., MILÉSI, J., LESCUYER J. L. & LECA, X. 1992. Signature isotopique du plomb des amas sulfurés de la province de Huelva, Espagne. Conséquences métallogéniques et géodynamiques. *Comptes Rendus de l'Académie des Sciences de Paris*, **314**, 1469–1476.

MARILLIER, F. TOMASSINO, A., PATRIAT, PH. & PINET, B. 1988. Deep structure of the Aquitaine Shelf: constraints from expanding spread profiles on the ECORS Bay of Biscay transect. *Marine and Petroleoum Geology*, **5**, 65–74.

MARIN, P. 1974. *Stratigraphie et évolution paléogéographique post-hercynienne de la Chaîne Celtibérique orientale aux confins de l'Aragón et du Haut-Maestrazgo (provinces de Teruel et Castellón de la Plane, Espagne). I. Le socle paléozoique et la couverture Pemo?-Triasique.* PhD Thesis, Claude-Bernard University, Lyon.

MARIN, P. & PLUSQUELLEC, Y. 1973. Sur des '*Combophyllum*' (Tetra-coralliaires) du Dévonién de Montalbán (Province de Teruel, Espagne). *Annales Société Géologique Nord*, **93**, 39–54.

MAROTO, J., SOLER, N. & MIR, A. 1987. La cueva de Mollet I (Serinyá, Gerona). *Cypsela*, **6**, 101–110.

MARQUES, B., OLÓRIZ, F. & RODRÍGUEZ-TOVAR, F. J. 1991. Interactions between tectonics and eustasy during the upper Jurassic and lowermost Cretaceous. Examples from the South of Iberia. *Bulletin de la Société Géologique de France*, **126**(6), 1109–1124.

MARQUES, F. G., RIBEIRO, A. & PEREIRA, E. 1992. Tectonic evolution of the deep crust: Variscan reactivation by extension and thrusting of Precambrian basement in the Bragança and Morais Massifs (Tras-os-Montes, NE Portugal). *Geodinamica Acta*, **5**, 135–151.

MARQUES, F. O., RIBEIRO, A. & MUNHÁ, J. 1996. Geodynamic evolution of the Continental Allochthonous Terrane (CAT) of the Bragança Nappe Complex, NE Portugal. *Tectonics*, **15**, 747–762.

MÁRQUEZ, B., OLLÉ, A., SALA, R. & VERGÉS, J. M. 2001. Perspectives méthodologiques de l'analyse fonctionnelle des ensembles lithiques du Pléistocene inférieur et moyen d'Atapuerca (Burgos, Espagne). *L'Anthropologie*, **105**, 281–299.

MÁRQUEZ, L. & TRIFONOVA, E. 1990. *Ammobacilites hiberensis* sp. nov. (foraminifera) from the upper Muschelkalk of Catalonian Ranges (Spain). *Revista Española de Paleontología*, **5**, 77–80.

MÁRQUEZ, L., LÓPEZ-GÓMEZ, J. & TRIFONOVA, E. 1994. Datación (foraminíferos) y ambientes sedimentarios de la Formación Dolomías de Landete, Anisiense, Facies Muschelkalk, provincia de Cuenca. *Boletín de la Real Sociedad Española de Historia Natural (Sección Geología)*, **89**, 99–107.

MÁRQUEZ-ALIAGA, A. 1985. *Bivalvos del Triásico Medio del sector meridional de la Cordillera Ibérica y de los Catalánides.* Complutense University, Madrid, **40**.

MARQUÍNEZ, J. 1978. Estudio geológico del sector SE de los Picos de Europa (Cordillera Cantábrica, NW de España). *Trabajos de Geología*, **10**, 295–315.

MARQUÍNEZ, J. 1989. Mapa Geológico de la Región del Cuera y Picos de Europa (Cordillera Cantábrica, NW de España). *Trabajos de Geología*, **18**, 137–144.

MARQUÍNEZ, J. & MARCOS, A. 1984. La estructura de la Unidad Gildar-Montó (Cordillera Cantábrica). *Trabajos de Geología*, **14**, 53–64.

MARQUÍNEZ, J., MÉNDEZ, C. A., MENÉNDEZ-ÁLVAREZ, J. R., SÁNCHEZ DE POSADA, L. C. & VILLA, E. 1982. Datos bioestratigráficos de la sucesión carbonífera (Tournaisiense-Kasimoviense) de las Llacerías, Picos de Europa, Norte de España. *Trabajos de Geología*, **12**, 187–193.

MARQUÍNEZ, J. L. 1984. *La Geología del área esquistosa de Galicia Central (Cordillera Herciniana, NW de España).* Memorias del IGME, Madrid, **100**.

MARRE, J. 1973. *Le Complexe éruptif de Quérigut. Petrologie, structurologie, cinematique de mise en place.* PhD Thesis, University of Toulouse.

MARTÍ, J. 1986. *El volcanisme explosiu tardihercinià del Pirineu Català.* PhD Thesis, University of Barcelona.

MARTÍ, J. 1991. Caldera-like structures related to Permo-Carboniferous volcanism of the Catalan Pyrenees (NE Spain). *Journal of Volcanology and Geothermal Research*, **45**, 173–186.

MARTÍ, J. & MALLARACH, J. M. 1987. Erupciones hidromagmáticas en el volcanismo cuaternario de Olot (Girona). *Estudios Geológicos*, **43**, 31–40.

MARTÍ, J. & MITJAVILA, J. (eds) 1995. A field guide to the central volcanic complex of Tenerife (Canary Islands). Cabildo de Lanzarote. *Serie Casa de los Volcanes*, **4**.

MARTÍ, J., MUÑOZ, J. A. & VAQUER, R. 1986. Les roches volcaniques de l'Ordovicien Supérieur de la région de Ribes de Freser-Rocabruna (Pyrénées Catalanes): caractères et signification. *Comptes Rendues de l'Académie des Sciences de Paris*, **302**, 1237–1242.

MARTÍ, J., MITJAVILA, J., ROCA, E. & APARICIO, A. 1992. Cenozoic magmatism of the Valencia trough (western Mediterranean): relationship between structural evolution and volcanism. *Tectonophysics*, **203**, 145–165.

MARTÍ, J., MITJAVILA, J. & ARAÑA, V. 1994. Stratigraphy, structure and geochronology of the Las Cañadas caldera (Tenerife, Canary Islands). *Geological Magazine*, **131**, 715–727.

MARTÍ, J., HURLIMANN, M., ABLAY, G. J. & GUDMUNDSSON, A. 1997. Vertical and lateral collapses on Tenerife (Canary Islands) and other volcanic ocean islands. *Geology*, **25**, 879–882.

MARTÍ-BONO, C. 1973. Nota sobre los sedimentos morrénicos del río Aragón. *Pirineos*, **107**, 39–46.

MARTÍ-BONO, C. 1996. *El glaciarismo cuaternario en el Alto Aragón Occidental.* PhD Thesis, University of Barcelona.

MARTÍN, E. 1981. *Mineralogía y génesis del Cerro Colorado (Rio Tinto).* PhD Thesis, Complutense University, Madrid.

MARTÍN, J. M. & BRAGA, J. C. 1987a. Alpujárride carbonate deposits (southern Spain)-marine sedimentation in a Triassic Atlantic. *Palaeogeogreography, Palaeoclimatology and Palaeoecology*, **59**, 243–260.

MARTÍN, J. M. & BRAGA, J. C. 1987b. Bioconstrucciones del Anisiense-Ladiniense en el Trias Alpujárride. *Cuadernos Geología Ibérica*, **11**, 421–444.

MARTÍN, J. M. & BRAGA, J. C. 1994. Messinian events in the Sorbas Basin in southeastern Spain and their implications in the recent history of the Mediterranean. *Sedimentary Geology*, **90**, 257–268.

MARTÍN, J. M., ORTEGA HUERTAS, M. & TORRES RUIZ, J. 1984. Genesis and evolution of strontium deposits of the Granada Basin (southeastern Spain): evidence of diagenetic replacement of a stromatolite belt. *Sedimentary Geology*, **39**, 281–298.

MARTÍN, J. M., BRAGA, J. C. & RIVAS, P. 1989. Coral successions in Upper Tortonian reefs in SE Spain. *Lethaia*, **22**, 271–286.

MARTÍN, J. M., BRAGA, J. C. & RIDING, R. 1993. Siliciclastic stromatolites and thrombolites, late Miocene, S.E. Spain. *Journal of Sedimentary Petrology*, **63**, 131–139.

MARTÍN, J. M., BRAGA, J. C., BETZLER, C. & BRACHERT, T. C. 1996. Sedimentary model and high-frequency cyclicity in a Mediterranean, shallow-shelf, temperate-carbonate environment (uppermost Miocene, Agua Amarga Basin, Southern Spain). *Sedimentology*, **43**, 263–277.

MARTÍN, J. M., BRAGA, J. C. & RIDING, R. 1997. Late Miocene *Halimeda* alga-microbial segment reefs in the marginal Mediterranean Sorbas Basin, Spain. *Sedimentology*, **44**, 441–456.

MARTÍN, J. M., BRAGA, J. C. & SÁNCHEZ-ALMAZO, I. M. 1999. The Messinian record of the outcropping marginal Alborán Basin deposits: significance and implications. *Proceedings ODP Scientific Results*, **161**, 543–551.

MARTÍN-ALGARRA, A. 1987. *Evolución geologica alpina del contacto entre las zonas internas y las zonas externas de la Cordillera Bética (sector central y occidental).* PhD Thesis, University of Granada.

MARTÍN-ALGARRA, A. 1995. El Triásico del Maláguide-Gomáride (Formación Saladilla, Cordillera Bética Occidental y Rif Septentrional): Nuevos datos sobre su estratigrafía y significado paleogeográfico. *Cuadernos de Geología Ibérica*, **19**, 249–278.

MARTÍN-ALGARRA, A., CHECA, A., OLÓRIZ, F. & VERA, J. A. 1983. Un modelo de sedimentación pelágica en cavidades kársticas: La Almola (Cordillera Bética). *In:* OBRADOR, A. (ed.) *X Congreso Nacional de Sedimentología, Menorca. Comunicaciones*, 3.21–3.24.

MARTÍN-ALGARRA, A., RUIZ-ORTIZ, P. A. & VERA, J. A. 1992. Factors controlling Cretaceous turbidite deposition in the Betic Cordillera. *Revista de la Sociedad geológica de España*, **5**, 53–80.

MARTÍN-ALGARRA, A., SOLÉ DE PORTA, N. & MARQUEZ-ALIAGA, A. 1995. Nuevos datos sobre la estratigrafía, paleontología y procedencia paleogeográfica del Triásico de las escamas del Corredor del Boyar (Cordillera Bética Occidental). *Cuadernos de Geología Ibérica*, **19**, 279–307.

MARTÍN-ALGARRA, A., O'DOGHERTY, L., AGUADO, R. & GURSKY, H.-J. 1998. Stratigraphy, petrography and palaeogeographical significance of australpine-type jurassic radiolarites of the Nieves Unit (Parauta Formation, Rondaides, Western Betic Cordillera). *Geogaceta*, **24**, 211–214.

MARTÍN-CHIVELET, J. 1990. La transgresion del Cenomaniense superior en el Prebético. *Geogaceta*, **8**, 86–88.

MARTÍN-CHIVELET, J. 1991. Sedimentación lacustre finicretácica en el Prebético de Murcia: caracterización estratigráfica. *Geogaceta*, **9**, 70–73.

MARTÍN-CHIVELET, J. 1992. *Las plataformas carbonatadas del Cretácico superior de la Margen Bética (Altiplano de Jumilla Yecla, Murcia)*. PhD Thesis, Complutense University, Madrid.

MARTÍN-CHIVELET, J. 1994. Litoestratigrafía del Cretácico superior del Altiplano de Jumilla-Yecla (Zona Prebética). *Cuadernos de Geología Ibérica*, **18**, 117–173.

MARTÍN-CHIVELET, J. 1995. Sequence stratigraphy of mixed carbonate siliciclastic platforms developed in a tectonically active setting: the Upper Cretaceous of the Betic continental margin (Spain). *Journal of Sedimentary Research*, **65b**, 235–254.

MARTÍN-CHIVELET, J. 1996. Late Cretaceous stratigraphic patterns and subsidence history of the Betic continental margin (Jumilla-Yecla region, SE Spain). *Tectonophysics*, **265**, 191–211.

MARTÍN CHIVELET, J. 1999. *Cambios Climáticos. Una aproximación al Sistema Tierra*. Libertarias-Prodhufi, Madrid.

MARTÍN-CHIVELET, J. & GIMÉNEZ, R. 1992. Paleosols in microtidal sequences: Sierra de Utiel Formation, Upper Cretaceous, SE Spain. *Sedimentary Geology*, **81**, 125–145.

MARTÍN-CHIVELET, J. & GIMÉNEZ, R. 1993. Évolutions sédimentaires et tectoniques des plates-formes du sud-est de l'Espagne au cours du Cénomanien superieur – Coniacien inferieur. *Cretaceous Research*, **14**, 509–518.

MARTÍN-CHIVELET, J., PHILIP, J. & TRONCHETTI, G. 1990. Les Formations à rudistes du Crétacé supérieur (Cénomanien moyen – Sénonien inférieur) du domaine prebetique (Sierra du Cuchillo, Region de Yecla, Espagne). *Géologie Méditerranéenne*, **17**, 139–151.

MARTÍN-CHIVELET, J., RAMÍREZ DEL POZO, J., TRONCHETTI, G. & BABINOT, J. F. 1995. Palaeoenvironments and evolution of the upper Maastrichtian platform in the Betic continental margin, SE Spain. *Palaeogeography, Palaeoclimatology, Palaeoecology*, **119**, 169–186.

MARTÍN-CHIVELET, J., GIMÉNEZ, R. & LUPERTO-SINNI, E. 1997. La discontinuidad del Campaniense basal en el Prebético ¿Inicio de la convergencia alpina en la Margen Bética? *Geogaceta*, **22**, 121–124.

MARTÍN-CLOSAS, C. 1989. *Els carofits del Cretaci inferior de les conques periferiques del bloc de l'Ebre*. PhD Thesis, University of Barcelona.

MARTÍN ESCORZA, C. 1976. Las 'Capas de transición', Cámbrico Inferior y otras series preordovícicas (¿Cámbrico superior?) en los Montes de Toledo surorientales: sus implicaciones geotectónicas. *Estudios Geológicos*, **32**, 591–613.

MARTÍN-ESCORZA, C. & LÓPEZ-RUIZ, J. 1988. Un modelo geodinámico para el volcanismo Neógeno del sureste Ibérico. *Estudios Geológicos*, **44**, 243–251.

MARTÍN IZARD, A., CEPEDAL, M. A., RODRÍGUEZ-PEVIDA, L., SPIERING, E., GONZÁLEZ, S., VARELA, A. & MALDONADO, C. 1997. The El Valle deposit: an example of prophyry-related copper-gold skarn mineralisation overprinted by late epithermal events, Cantabrian mountains, Spain. *In*: PAPUNEN (ed.) *Mineral Deposits*. Balkema, Rotterdam.

MARTÍN IZARD, A., CEPEDAL, M. A., FUERTES-FUENTE, M. *et al.* 1998. Los yacimientos de oro-cobre del Cinturón del Rio Narcea, Asturias, España. *Boletín Geológico y Minero*, **109**, 479–496.

MARTÍN MARTÍN, A. J. 1984. *Riesgo Sísmico en la Península Ibérica*. PhD Thesis, IGN, Spain.

MARTÍN-MARTÍN, M. 1996. *El Terciario del dominio Maláguide en Sierra Espuña (Cordillera Bética oriental, SE de España): Estratigrafía y evolución paleogeográfica*. PhD Thesis, University of Granada.

MARTÍN-MARTÍN, M., EL-MAMOUNE, B., MARTÍN-ALGARRA, A. & MARÍN-PÉREZ, J. A. 1996. The Internal-External zone boundary in the eastern Betic Cordillera (SE Spain): Discussion. *Journal of Structural Geology*, **18**, 523–524.

MARTÍN MERINO, M. A., DOMINGO MENA, S. & ANTÓN PALACIOS, J.

1981. Estudio de las cavidades de la zona BU-IV. A. (Sierra de Atapuerca). *Kaite, Estudios de Espeleología Burgalesa*, **2**, 41–76.

MARTÍN-SERRANO, A. 1991. La definición y el encajamiento de la red fluvial actual sobre el Macizo Hespérico en el marco de su geodinámica alpina. *Revista de la Sociedad Geológica de España*, **4**, 337–351.

MARTÍN-SERRANO, A. 1994. Macizo Hespérico Septentrional. *In*: GUTIÉRREZ M. (ed.) *Geomorfología de España*. Rueda, Madrid, 25–62.

MARTÍN-SERRANO, A., SANTISTEBAN, J. I. & MEDIAVILLA, R. 1996. Tertiary of Central System basins. *In*: FRIEND, P. & DABRIO, C. (eds) *Tertiary Basins of Spain: The Stratigraphic Record of Crustal Kinematics*, Cambridge University Press, 255–260.

MARTÍN-SERRANO, A., CANTANO, M., CARRAL, P., RUBIO, F. & MEDIAVILLA, R. 1998. La degradación cuaternaria del piedemonte del río Yeltes (Salamanca). *Cuaternario y Geomorfología*, **12**, 5–17.

MARTÍN-SUÁREZ, E., FREUDENTHAL, M. & AGUSTI, J. 1993. Micromammals from the Middle Miocene of the Granada Basin (Spain). *Geobios*, **26**, 377–387.

MARTÍN-SUÁREZ, E., FREUDENTHAL, M., KRIJGSMAN, W. & RUTGER-FORTUIN, A. 2000. On the age of the continental deposits of the Zorreras Member (Sorbas basin, SE Spain). *Geobios*, **33**, 505–512.

MARTINELL, J. 1988. An overview of the marine Pliocene of N.E. Spain. *Géologie Méditerranée*, **15**, 227–233.

MARTÍNEZ, A., VERGÉS, J. & MUÑOZ, J. A. 1988. Secuencias de propagación del sistema de cabalgamientos de la terminación oriental del manto del Pedraforca y relación con los conglomerados sinorogénicos. *Acta Geológica Hispánica*, **23**, 119–127.

MARTÍNEZ, F. J. 1974. *Estudio del área metamórfica y granítica de los Arribes del Duero (Provincias de Salamanca y Zamora)*. PhD Thesis, University of Salamanca.

MARTÍNEZ, F. J. & RÖLET, J. 1988. Late Palaeozoic metamorphism in the northwestern Iberian Peninsula, Brittany and related areas in SW Europe. *In*: HARRIS, A. L. & FETTES, D. J. (ed.) *The Caledonian – Appalachian Orogen*. Geological Society, London, Special Publications, **38**, 611–620.

MARTÍNEZ, F. J., JULIVERT, M., SEBASTIÁN, A., ARBOLEYA, M. L. & GIL IBARGUCHI, J. I. 1988. Structural and thermal evolution of high-grade areas in the northwestern parts of the Iberian Massif. *American Journal of Science*, **288**, 969–996.

MARTÍNEZ, F. J., CARRERAS, J., ARBOLEYA, M. L. & DIETSCH, C. 1996. Structural and metamorphic evidence of local extension along the Vivero fault coeval with bulk crustal shortening in the Variscan chain (NW Spain). *Journal of Structural Geology*, **18**, 61–73.

MARTÍNEZ, G., MELÉNDEZ, G. & SEQUEIROS, L. 1997. Excursión al Jurásico de Aguilón y Ricla: aspectos didácticos y geológicos. *In*: GÁMEZ, J. I. & LIÑAN, E. (eds) *Vida y ambientes del Jurásico (Homenaje científico a la Profesora Asunción Linares)*. Instituto Fernándo El Católico, **1863**, 91–132

MARTÍNEZ, I. & ARSUAGA, J. L. 1997. The temporal bones from Sima de los Huesos Middle Pleistocene Site (Sierra de Atapuerca, Spain). A phylogenetic approach. *Journal of Human Evolution*, **33**, 283–318.

MARTÍNEZ, J., HENTZSCH, B., LÓPEZ, F. & MARTÍNEZ, J. 1988. Edad de las terrazas y diques travertínicos de las Lagunas de Ruidera y sus implicaciones paleoclimáticas. *Estudios Geológicos*, **44**, 75–81.

MARTÍNEZ, M. B., 1982. Influencia del sustrato en las estructuras de la cobertera deslizada de las Sierras Marginales, Huesca. *Acta Geológica Hispánica*, **17**, 235–240.

MARTÍNEZ, R. 1982. *Ammonoideos cretácicos del pre-Pirineo de la provincia de Lleida*. Autónoma University, Barcelona, Publicaciones de Geología, **17**.

MARTÍNEZ, R. M., LAGO, M., VALENZUELA, J. I., VAQUER , R., SALAS, R. & DUMITRESCU, R. 1997. El volcanismo triásico y jurásico del sector SE de la Cadena Ibérica y su relación con los estadios de rift mesozoicos. *Boletín Geológico y Minero*, **108**(4), 367–376.

MARTÍNEZ-ÁLVAREZ, J. A., TORRES ALONSO, M. & GUTIÉRREZ CLAVEROL, M. 1975. *Mapa Geológico de España*. Scale 1:50.000. Foz. IGME, Madrid.

MARTÍNEZ CATALÁN, J. R. 1980. L'apparition du chevauchement basal de la nappe de Mondoñedo dans le Dôme de Lugo (Galice, Espagne). *Comptes Rendus de l'Académie des Sciences, Paris*, **290**, 179–282.

MARTÍNEZ CATALÁN, J. R. 1985. *Estratigrafía y estructura del Domo de Lugo (sector Oeste de la Zona Asturoccidental-Leonesa)*. Corpus Geologicum Gallaeciae, **2** (PhD Thesis, University of Salamanca, 1981).

MARTÍNEZ CATALÁN, J. R. 1990a. A non-cylindrical model for the northwest Iberian allochthonous terranes and their equivalents in the Hercynian belt of Western Europe. *Tectonophysics*, **179**, 253–272.

MARTÍNEZ CATALÁN, J. R. 1990b. West Asturian-Leonese Zone: Introduction. *In*: DALLMEYER, R. D. & MARTÍNEZ GARCÍA, E. (eds) *Pre-Mesozoic Geology of Iberia*. Springer-Verlag, Berlin, 91.

MARTÍNEZ CATALÁN, J. R. & ARENAS, R. 1992. Deformación extensional de las unidades alóctonas superiores de la parte oriental del Complejo de Órdenes (Galicia). *Geogaceta*, **11**, 108–111.

MARTÍNEZ CATALÁN, J. R., PÉREZ ESTAÚN, A., BASTIDA, F., PULGAR, J. A. & MARCOS, A. 1990. Structure: West-Asturian-Leonese Zone (Part III). *In*: DALLMEYER, R. D. & MARTÍNEZ GARCÍA, E. (eds) *Pre-Mesozoic Geology of Iberia*. Springer-Verlag, Berlin, 103–114.

MARTÍNEZ CATALÁN, J. R., HACAR RODRÍGUEZ, M. P., VILLAR ALONSO, P., PÉREZ-ESTAÚN, A. & GONZÁLEZ LODEIRO, F. 1992a. Lower Paleozoic extensional tectonics in the limit between the West Asturian-Leonese and Central Iberian Zones of the Variscan Fold-Belt in NW Spain. *Geologische Rundschau*, **81**, 545–560.

MARTÍNEZ CATALÁN, J. R., PÉREZ ESTAÚN, A., BASTIDA, F., PULGAR, J. A. & MARCOS, A. 1992b. La zona asturoccidental-leonesa: estructura. *In*: GUTIÉRREZ MARCO, J., SAAVEDRA, J. & RÁBANO, I. (eds) *Paleozoico Inferior de Ibero-América*. University of Extremadura, Badajoz, 463–468.

MARTÍNEZ CATALÁN, J. R., AYARZA ARRIBAS, P., PULGAR, J. A. *et al.* 1995. Results from the ESCI-N3.3 marine deep seismic profile along the Cantabrian continental margin. *Revista de la Sociedad Geológica de España*, **8**, 341–354.

MARTÍNEZ CATALÁN, J. R., ARENAS, R., DIAZ GARCIA, F., RUBIO PASCUAL, F., ABATI, J. & MARQUINEZ, J. 1996. Variscan exhumation of a subducted Paleozoic continental margin: the basal units of the Ordenes Complex, Galicia, NW Spain. *Tectonics*, **15**, 106–121.

MARTÍNEZ CATALÁN, J. R., ARENAS, R., DÍAZ GARCÍA, F. & ABATI, J. 1997. Variscan accretionary complex of northwest Iberia: Terrane correlation and succession of tectonothermal events. *Geology*, **25**, 1103–1106.

MARTÍNEZ-CATALÁN, J. R., ARENAS, R., DÍAZ GARCÍA, F. & ABATI, J. 1999. Allocthonous units in the Variscan Belt of Iberia, terranes and accrectional history. *In*: SINHA, A. K. (ed.) *Proceedings of the Thirteenth International Conference on Basement Tectonics, Blacksburg, Virginia, June 1997*. Kluwer, Dordrecht. *Basement Tectonics*, **13**, 65–84.

MARTÍNEZ DE PISÓN, E. 1989. Morfología glaciar del valle de Benasque (Pirineo aragonés). *Ería*, **18**, 51–64.

MARTÍNEZ DE PISÓN, E. & ARENILLAS, M. 1979. Algunos problemas de morfología glaciar en la España atlántica. *Acta Geológica Hispánica*, **14**, 445–450.

MARTÍNEZ DEL OLMO, W. 1993. Depositional sequences in the Gulf of Valencia Tertiary basin. *In*: FRIEND, P. & DABRIO, C. (eds) *Tertiary Basins of Spain: the Stratigraphic Record of Crustal Kinematics*. Cambridge University Press, 55–67.

MARTÍNEZ DEL OLMO, W. 1996. *Secuencias de Depósito y estructuración diapírica en Mesozoico y Neógeno del Prebético y Golfo de Valencia desde sondeos y líneas sísmicas*. PhD Thesis, Complutense University, Madrid.

MARTÍNEZ DEL OLMO, W. & ESTEBAN, M. 1983. Paleokarst development (Western Mediterranean). *In*: SHOLLE, P. A., BEHOUT, D. G. & MOORE, C. M. (eds) *AAPG*, Memoir **33**, 93–95.

MARTÍNEZ DEL OLMO, W., LERET, G. & MEGÍAS, A. G. 1982. El límite de la plataforma carbonatada del Cretácico Superior en la zona Prebética. *Cuadernos Geología Ibérica*, **8**, 597–614.

MARTÍNEZ DEL OLMO, W., RIAZA MOLINA, C. & TORRESCUSA, S. 1996.

Descenso Eustático messiniense en una cuenca atlántica. El cañón submarino del Río Guadalquivir (SO de España). *Geogaceta*, **20**, 138–142.

MARTÍNEZ-DÍAZ, J. J., MASANA, E., HERNÁNDEZ-ENRILE, J. L. & SANTANACH, P. 2001. Evidence for co-seismic events of recurrent prehistoric deformation along the Alhama de Murcia fault, southeastern Spain. *In*: MASANA, E. & SANTANAC, P. (eds) *Paleosismicidad en España*. Monografía de Acta Geologica Hispanica, **36**, 315–327.

MARTÍNEZ FRIAS, J. 1986. *Mineralogía y metalogenia de las mineralizaciones de plata del Sector oriental del Sistema Central*. Tesis Doctoral. Universidad Complutense de Madrid, España.

MARTÍNEZ FRIAS, J. 1991. Sulphide and sulphosalt mineralogy and paragenesis from Sierra Almagrera (Betic Cordillera). *Estudios Geológicos*, **47**, 271–279.

MARTÍNEZ FRIAS, J. 1992. The Hiendelaencina mining district (Guadalajara, Spain). *Mineralium Deposita*, **27**, 206–212.

MARTÍNEZ FRIAS, J. 1996. The Congostrina mine: a unique case of Bi- and Sb-rich silver sulphosalt association in the Hiendelaencina Mining District (Spain). *Neues Jahrbuch fur Mineralogie*, **8**, 377–383.

MARTÍNEZ FRIAS, J. 1998. An ancient Ba-Sb-Ag-Fe-Hg- bearing hydrothermal system in SE Spain. *Episodes*, **21–4**, 248–252.

MARTÍNEZ FRIAS, J., GARCÍA GUINEA, J., LOPEZ RUIZ, J. & REYNOLS, G. 1992a. Discovery of fossil fumaroles in Spain. *Economic Geology*, **7**, 444–446.

MARTÍNEZ FRIAS, J., LUNAR, R. & VINDEL, E. 1992b. Los yacimientos de plata del Sistema Central Español. Características y Modelos. *In*: GARCÍA GUINEA, J. & MARTÍNEZ FRÍAS, J. (eds) *Recursos Minerales de España*. CSIC, Colección Textos Universitarios, **15**, 921–935.

MARTÍNEZ FRIAS, J., NAVARRO, A., LUNAR, R. & GARCÍA-GUINEA, J. 1998. Mercury pollution in the marine environment: A natural venting system in the SW Mediterranean margin. *Nature & Resources*, **34**, 3, 9–15.

MARTÍNEZ GARCÍA, E. 1981. El Paleozoico de la zona Cantábrica Oriental (Noroeste de España). *Trabajos de Geología de la Universidad de Oviedo*, **11**, 97–127.

MARTÍNEZ GARCÍA, E. 1983a. El Pérmico de la región Cantábrica. *In*: MARTÍNEZ DÍAZ, C. (ed.) *Carbonífero y Pérmico de España*. X Congreso Internacional de Estratigrafía y Geología del Carbonífero. IGME, Madrid, 389–402.

MARTÍNEZ GARCÍA, E. 1983b. Permian mineralisations in the Cantabrian Mountains (North-West Spain). *In*: SCHNEIDER, H.-J. (eds) *Mineral Deposits of the Alps and the Alpine Epoch in Europe*. Springer-Verlag, Berlin, 259–274.

MARTÍNEZ-GARCÍA, E. 1991. Hercynian syn-orogenic and post-orogenic successions in the Cantabrian and the Palentian zones (NW Spain). Comparison with other western European occurrences. *Giornale di Geologia*, **53**, 209–228.

MARTÍNEZ GARCÍA, E. 1999. Orogénesis y sedimentación a finales del Paleozoico en el NW del Macizo Ibérico (Asturias, Cantábria y Palencia). *Libro Homenaje a José Ramírez del Pozo*. AGGEP, Madrid, 167–174.

MARTÍNEZ GARCÍA, E. & FOMBELLA, M. A. 1997. Datación palinológica de la Unidad de Curbeiro (Pontevedra, España) y su importancia en la geología de Galicia. *In*: GRANDAL D'ANGLADE, A., GUTIÉRREZ-MARCO, J. C. & SANTOS FIDALGO, L. (eds) *XIII Jornadas de Paleontología y V Reunión Internacional Proyecto 351 PICG*, A Coruña, Libro de Resúmenes y Excursiones, 32–33 (ISBN 84-605-6825-3).

MARTÍNEZ GARCÍA, E. & VILLA, E. 1998. El desarrollo estratigráfico de las unidades alóctonas del área de Gamonedo-Cabrales (Picos de Europa, Asturias, NW de España). *Geogaceta*, **24**, 219–222.

MÁRTINEZ GARCÍA, E., LUQUE, C., BURKHARDT, R & GUTIERREZ CLAVEROL, M. 1991. Hozarco: un ejemplo de mineralización de Pb-Zn-Hg de edad Pérmica (Cordillera Cantábrica, NW de España). *Boletín de la Sociedad Española de Mineralogía*, **14**, 107–106.

MARTÍNEZ GARCÍA, E., LUQUE, C., BURKHARDT, R. & GUTIERREZ CLAVEROL, M. 1994. Primer descubrimeiento de huellas de tetrápodos, invertebrados y plantas en el Pérmico de la Cordillera Cantábrica (Peña Sagra, Cantábria). *Proccedings III Coloquio de*

Estratigrafía y Paleogeografía del Pérmico y Triásico de España, Cuenca, 77–78.

MARTÍNEZ GARCÍA, E., COQUEL, R., GUTIERREZ CLAVEROL, M. & QUIROGA, J. L. 1998. Edad del 'Tramo de transición' entre el Pérmico y el Jurásico en el área de Gijón (Asturias, NW de España). *Geogaceta*, **24**, 215–222.

MARTÍNEZ-GRAÑA, A., GOY, J. L. & ZAZO, C. 2000. Actividad tectónica en el Noroeste Peninsular, en base a los registros de los depósitos costeros de los últimos 130.000 años (Ria Arosa-Pontevedra, Galicia). *Geotemas*, **1**, 263–266.

MARTÍNEZ-MARTÍNEZ, J. M. 1984. *Evolución tectono-metamórfica del Complejo Nevado-Filábride en el sector entre Sierra Nevada y Sierra de los Filabres (Cordilleras Béticas)*. PhD Thesis, University of Granada.

MARTÍNEZ MARTÍNEZ, J. M. 1984–85. Las sucesiones nevado-filábrides en la Sierra de los Filabres y Sierra Nevada. Correlaciones. *Cuadernos de Geología Universidad de Granada*, **12**, 127–144.

MARTINEZ-MARTINEZ, J. M. & AZAÑON, J. M. 1997. Mode of extensional tectonics in the southeastern Betics (SE Spain): Implications for the tectonic evolution of the peri-Alborán orogenic system. *Tectonics*, **16**, 205–225.

MARTÍNEZ-MARTÍNEZ, J. M., SOTO, J. I. & BALANYÁ, J. C. 1998. Crustal decoupling and intracrustal flow beneath domal exhumed core complexes, Betic (SE Spain). *Terra Nova*, **9**, 223–227.

MARTÍNEZ-PEÑA, M. B. & POCOVÍ, A. 1988. El amortiguamiento frontal de la estructura de la cobertera surpirenaica y su relación con el anticlinal de Barbastro-Balaguer. *Acta Geológica Hispánica*, **23**, 81–94.

MARTÍNEZ-POYATOS, D., SIMANCAS, J. F., AZOR, A. & GONZÁLES LODEIRO, F. 1998. Evolution of a Carboniferous piggyback basin in the southern Central Iberian Zone (Variscan Belt, SW Spain). *Bulletin de la Société Géologique de France*, **169**, 573–578.

MARTÍNEZ RIUS, A., RIVERO, L. & CASAS, A. 1996. Interpretación de subsuelo en la zona del Ripollés (Pirineo oriental) y su aplicación en la prospección petrolífera. *Geogaceta*, **20**, 157–160.

MARTÍNEZ-RUIZ, F. 1994. *Geoquímica y mineralogía del tránsito Cretácico-Terciario en las Cordilleras Béticas y en la cuenca Vasco-Cantábrica*. PhD Thesis, University of Granada.

MARTÍNEZ-RUIZ, F., ORTEGA, M. & PALOMO, I. 1999. Positive Eu anomaly development during diagenesis of the K/T boundary ejecta layer in the Agost section (SE Spain): implications for the trace-element remobilization. *Terra Nova*, **11**, 290–296.

MARTÍNEZ-SALANOVA, J. 1987. *Estudio paleontológico de los micro-mamíferos del Mioceno inferior de Fuenmayor (La Rioja)*. Ciencias de la Tierra, **10**, Instituto de Estudios Riojanos, Logroño.

MARTÍNEZ-TORRES, L. M. 1989. *El Manto de los Mármoles (Pirineo Occidental): geología estructural y evolución geodinámica*. PhD Thesis, University of País Vasco.

MARZO, M. 1980. *El Buntsandstein de los Catalánides: Estratigrafía y procesos de sedimentación*. PhD Thesis, University of Barcelona.

MARZO, M. 1986. Secuencias fluvio-eólicas en el Buntsandstein del macizo de Garraf (provincia de Barcelona). *Cuadernos de Geología Ibérica*, **10**, 207–233.

MARZO, M. & ANADÓN, P. 1977. Evolución y características sedimentológicas de las facies fluviales basales del Buntsandstein de Olesa de Montserrat (provincia de Barcelona). *Cuadernos de Geología Ibérica*, **4**, 211–222.

MARZO, M. & CALVET, F. 1985. El Triásico de los Catalánides. *II Coloquio Estratigráfico y Paleogeográfico del Pérmico y Triásico de España*. Institut d'Estudis Ilerdencs, Lleida.

MAS, R, ALONSO, A. & MELÉNDEZ, N. 1982. El Cretácico basal 'Weald' de la Cordillera Ibérica suroccidental (NW de la provincia de Valencia y E de la de Cuenca). *Cuadernos de Geología Ibérica*, **8**, 309–335.

MAS, R., ALONSO, A. & MELÉNDEZ, N. 1984. La Formación Villar del Arzobispo: un ejemplo de llanura de mareas siliciclástica asociada a plataformas carbonatadas. Jurásico terminal (NW de Valencia y E de Cuenca). *Publicaciones de Geología, Univ. Autonoma Barcelona*, **20**, 175–188.

MAS, R., ALONSO, A. & GUIMERÀ, J. 1993. Evolución tectonosedimentaria de una cuenca extensional intraplaca: la cuenca finijurásica-eocretácica de Los Cameros (La Rioja-Soria). *Revista de la Sociedad Geológica de España*, **6**, 129–144.

MASCLE, G. H. 1980. Pre-Triassic fit and alpine tectonics of continental blocks in the western Mediterranean: Discussions and reply. *Geological Society of America Bulletin*, Part I, **91**, 631–634.

MASSA, D. & BOURROUILH, R. 2000. Upper Ordovician carbonate mud-mound of northern Ghadamis Basin, Libya: a review. *Sedimentary Basins of Libya. Abstracts Second Symposium Geology of Northwest Libya*, Tripoli. Gutenberg Press, Malta, 63.

MASSE, J. P., ARIAS, C. & VILAS, L. 1992. Stratigraphy and biozonation of a reference Aptian-Albian p.p. Tethyan carbonate platform succession: The Sierra del Carche series (oriental Prebetic zone-Murcia, Spain). *In: New Aspects on Tethyan Cretaceous Fossil Assemblages*. Schriftenreihe der erdwissenschaftlichen Kommission der Österreichischen Akademie der Wissenschaften, **9**, 201–221.

MASSE, J. P., ARIAS, C. & VILAS, L. 1993. Caracterización litoestratigráfica y bioestratigráfica del Valanginiense superior-Hauteriviense inferior en el Prebético de la zona septentrional de Murcia. *Revista Española de Micropaleontología*, **XXV-3**, 123–136.

MASSE, J. P., ARIAS, C. & VILAS, L. 1998. Lower Cretaceous Rudist faunas of Southeast Spain: An Overview. *Geobios*, Mémoire spécial, **22**, 193–210.

MASSON, D. G. 1996. Catastrophic collapse of the volcanic island of Hierro 15 Kaago and the history of landslides in the Canary Islands. *Geology*, **24**, 3, 231–234.

MASSON, D. G. & WATTS, A. 1995. Slope failures and debris avalanches on the flanks of volcanic oceanic islands: The Canary Islands, off northwest Africa. *Landslide News*, **9**, 21–24.

MATA, J. & MUNHÁ, J. 1986. Geodynamic significance of high grade metamorphic rocks from Degolados-Campo Maior (Tomar-Badajoz-Córdoba Shear Zone). *Maleo*, **2**, 28.

MATA, J. & MUNHÁ, J. 1990. Magmatogénese de Metavulcanitos Câmbricos do Nordeste Alentejano: os estádios iniciais de 'rifting' continental. *Comunicações dos Serviços Geológicos de Portugal*, **76**, 61–89.

MATHER, A. E., SILVA, P. G., GOY, J. L., HARVEY, A. M. & ZAZO, C. 1995. Tectonics versus climate: An example from late Quaternary aggradational and dissectional sequences of the Mula basin, southeast Spain. *In*: LEWIN, S., MACKLIN, M. G., WOODWARD, J. C. (eds) *Mediterranean Quaternary River Environments*. Balkema, Rotterdam, 77–97.

MATHEY, B. 1982. El Cretácico superior del Arco Vasco. *In*: *El Cretácico de España*. Complutense University, Madrid, 111–136.

MATHEY, B. 1986. *Les Flyschs Crétacé Supérieur des Pyrénées Basques: age, anatomie, origine du matériel, milieu de dépôt et la relation avec l'ouverture du Golfe de Gascogne*. PhD Thesis, University of Bourgogne.

MATHEY, B. 1987. *Les flyschs Crétacé supérieur des Pyrénées Basques*. Mémoires Géologiques de l'Université de Dijon, **12**.

MATTAUER, M. 1973. *Les déformations des matériaux de l'écorce terrestre*. Hermann, París.

MATTAUER, M. 1990. Vue autre interprétation du profil ECORS Pyrénées. *Bulletin de la Société Géologique de France*, **VI(2)**, 307–311.

MATTE, P. 1967. Le Précambrien supérieur schisto-gréseux de l'Ouest des Asturies (Nord-Est de l'Espagne) et ses relations avec les séries précambriennes plus internes de l'arc galicien. *Comptes Rendus de l'Académie des Sciences de Paris*, **264**, 1769–1772.

MATTE, P. 1968. La structure de la virgation hercynienne de Galice (Espagne). *Revue de Géologie Alpine*, **44**, 155–280.

MATTE, P. 1976. Raccord des segments hercyniens d'Europe sud-occidental. *Nova Acta Leopoldina*, **224**, 239–262.

MATTE, P. 1986. Tectonics and plate tectonics model for the Variscan belt of Europe. *Tectonophysics*, **126**, 329–374.

MATTE, P. 1991. Accretionary history and crustal evolution of the Variscan belt in Western Europe. *In*: HATCHER, R. D. JR. & ZONENSHAIN, L. (eds) *Accretionary Tectonics and Composite Continents. Tectonophysics*, **196**, 309–337.

MATTE, P. 2001. The Variscan collage and orogeny (480–290 Ma) and the tectonic definition of the Armorican microplate: a review. *Terra Nova*, **13**, 122–128.

MATTE, P. & RIBEIRO, A. 1967. Les rapports tectoniques entre le Précambrienn ancien et le Paléozoïque dans le Nord-Ouest de la

Péninsule Ibérique: grandes nappes ou extrusions?. *Comptes Rendus de l'Académie des Sciences, Paris*, **264**, 2268–2271.

MAURY, R. C., FOURCADE, S., COULON, C. *et al.* 2000. Post-collisional Neogene magmatism of the Mediterranean Maghreb margin: a consequence of slab breakoff. *Comptes Rendus de l'Académie des Sciences de Paris*, **331**, 159–173.

MAYORAL, E. 1991. Actividad bioerosiva de briozoos ctenostomados en el Ordovícico superior de la Zona Cantábrica del Macizo Hespérico (Cabo Vidrias, Oviedo). *Revista Española de Paleontología*, **6**, 27–36.

MAYORAL, E., GUTIÉRREZ-MARCO, J. C. & MARTINELL, J. 1994. Primeras evidencias de briozoos perforantes (Ctenostomata) en braquiópodos ordovícicos de los Montes de Toledo (zona Centroibérica meridional, España). *Revista Española de Paleontología*, **9**, 185–194.

MAZÍN, J. M. & MARTÍN, M. 1983. Interet stratigraphique des microrestes vertebrés mesozoyques. Exemple du Trias d'Amelie-Les-Basins. *Bulletin Société Geologiqué France*, **25**, 785–787.

MAZO, A. V., MADE, J. VAN DER, JORDÁ, J. F., HERRÁEZ, E. & ARMENTEROS, I. 1998. Fauna y bioestratigrafía del yacimiento Aragoniense de Montejo de la Vega de la Serrezuela (Segovia). Montejo de la Vega. *Estudios Geológicos*, **54**, 231–248.

MECO, J. 1977. *Los Strombus neogenos y cuaternarios del Atlántico euroafricano. Taxonomía, bioestratigrafia y Paleoecología.* PhD Thesis, Complutense University, Madrid.

MECO, J. 1987. Islas Canarias. In: *Mapa del Cuaternario de España. E: 1.000.000.* ITGE, Madrid, 233–243.

MECO, J. 1991. *Los depósitos del inicio del Plioceno y sus fósiles (Fuerteventura).* Cabildo Insular de Fuerteventura, Casa Museo de Betancuria.

MECO, J. & PETIT-MAIRE, N. 1986. *El Cuaternario reciente de Canarias.* Las Palmas, Marseille.

MECO, J. & STEARNS, C. E. 1981. Emergent littoral deposits in the Eastern Canary Islands. *Quaternary Research*, **15**, 199–208.

MECO, J., PETIT-MAIRE, N. FONTUGNE, M., SHIMMIELD, G., HARROP, P. & RAMOS, A. J. 1997. The Quaternary deposits in Lanzarote and Fuerteventura (Eastern Canary Islands, Spain. An overview. *In*: MECO, J. & PETIT-MAIRE, N. (eds) *Climates of the past.* University of Las Palmas, Gran Canaria, 123–140.

MEDIAVILLA, R., DABRIO, C. J., MARTÍN-SERRANO, A & SANTISTEBAN, J. I. 1996. Lacustrine Neogene systems of the Duero Basin: evolution and controls. *In*: FRIEND, P. & DABRIO, C. (eds) *Tertiary Basins of Spain: The Stratigraphic Record of Crustal Kinematics.* Cambridge University Press, Cambridge, 228–236.

MEDIAVILLA, R. M. 1986/87. Sedimentología de los yesos del sector central de la depresión del Duero. *Acta Geologica Hispanica*, **21–22**, 35–44.

MEHL, K. W. & SCHMINCKE, H. U. 1999. Structure and emplacement of the Pliocene Roque Nublo debris avalanche deposit, Gran Canaria, Spain. *Journal of Volcanology and Geothermal Research*, **94**, 105–134.

MEIGS, A. J. 1997. Sequential development of selected Pyrenean thrust faults. *Journal of Structural Geology*, **19**, 481–502.

MEIGS, A. J. & BURBANK, D. W. 1997. Growth of the South Pyrenean wedge. *Tectonics*, **16**, 239–258.

MEIN, P. 1975. Rapport d'activité du Groupe de travail 'Vertebrés': mise à jour de la biostratigraphie du Néogéne basée sur les mammiféres. *Annals de Geólogie de Pays Hellénique*, **3**, 1367–1372.

MEIN, P. & ADROVER, R. 1982. Une faunule de Mammifères insulaires dans le Miocène moyen de Majorque (Iles Baléares). *Geobios, Mémoire Spéciale*, **6**, 451–463.

MEISSNER, R. 1989. Rupture, creep, lamellae and crocodiles: happenings in the continental crust. *Terra Nova*, **1**, 17–28.

MELÉNDEZ, A., AURELL, M., BÁDENAS, B. & SORIA, A. R. 1995. Las rampas carbonatadas del Triásico Medio en el sector central de la Cordillera Ibérica. *Cuadernos de Geología Ibérica*, **19**, 173–199.

MELÉNDEZ, B. 1951. Sobre un notable Cistideo del Silúrico español, *Echinosphaerites murchisoni* de Vern. y Barr. *Libro Jubilar (1849–1949) del IGME*, **2**, 1–15.

MELÉNDEZ, B. 1958. Nuevo Cistideo del Ordoviciense de los Montes de Toledo. *Notas y Comunicaciones del IGME*, **50**, 323–329.

MELÉNDEZ, B. 1959. Los *Echinosphaerites* del Silúrico de Luesma (Zaragoza). *Estudios Geológicos*, **15**, 269–276.

MELÉNDEZ, B. & CHAUVEL, J. 1982. Sur quelques Cystoidées cités par les Drs. J. Almera et M. Faura dans l'Ordovicien de Barcelone. *Acta Geologica Hispanica*, **14** (1979), 318–321.

MELÉNDEZ, B. & HEVIA, I. 1947. La fauna ashgilliense del Silúrico aragonés. *Boletín de la Universidad de Granada*, **19**, 247–259.

MELÉNDEZ, B., TALENS, J., FONOLLÁ & ÁLVAREZ-RAMIS, C. 1983. Las cuencas carboníferas del sector central de la Cordillera Ibérica (Henarejos y Montalbán). *In*: MARTÍNEZ DÍAZ, C. (ed.) *Carbonífero y Pérmico de España.* IGME, Madrid, 209–220.

MELÉNDEZ, F. 1971. *Estudio geológico de la Serranía de Cuenca.* PhD Thesis, Complutense University, Madrid.

MELÉNDEZ, F. & RAMÍREZ DEL POZO, J. 1972. El Jurásico de la Serranía de Cuenca. *Boletín Geológico y Minero*, **83–84**, 313–342.

MELÉNDEZ, G. 1989. *El Oxfordiense en el sector central de la Cordillera Ibérica (provincias de Zaragoza y Teruel).* PhD Thesis, Instituto Fernándo el Católico-Instituto Estudios Turolenses.

MELÉNDEZ, G. & FONTANA, B. 1993. Biostratigraphic correlation of the Middle Oxfordian sediments in the Iberian Chain, eastern Spain. *Acta Geologica Polonica*, **43**, 192–211.

MELÉNDEZ, G. & LARDIÉS, M. D. 1988. El Calloviense y Oxfordiense de Ricla (Prov. Zaragoza). *III Coloquio de Estratigrafía y Paleogeografía del Jurásico de España, libro guía de las excursiones, Ciencias de la Tierra (Instituto de Estudios Riojanos)*, **11**, 265–282.

MELÉNDEZ, G., SEQUEIROS L. & BROCHWIC-LEWINSKI, W. 1983. Lower Oxfordian in the Iberian Chain, Spain, Part I: Biostratigraphy and nature of gaps. *Bull. Acad. Pol. des Sc. Série des Sciences de la Terre*, **30**, 157–172.

MELÉNDEZ, G., AURELL, M. & ATROPS, F. 1990. Las unidades del Jurásico superior en el sector nororidental de la Cordillera Ibérica: nuevas subdivisiones litoestratigráficas. *Cuadernos de Geología Ibérica*, **14**, 225–245.

MELÉNDEZ, G., BELLO, J., DELVENE, G. & PÉREZ-URRESTI , I. 1997. El Jurásico medio y superior (Calloviense-Kimmeridgiense) en Ventas de San Pedro. Reconstrucción paleogeográfica y análisis tafonómico. *Cuadernos de Geología Ibérica*, **23**, 269–300.

MELÉNDEZ, G., BELLO, J., DELVENE, G., PÉREZ-URRESTI , I., RAMAJO, J. & ATROPS, F. 1999. Middle and Upper Jurassic at the Calanda-Mas de las Matas area, in the region of River Guadalope (NE Iberian Chain, E Spain). *Profil*, **16**, 275–296.

MELÉNDEZ HEVIA, F., OLÓRIZ SÁEZ, F. & POMAR GOMÁ, L. 1988. Relaciones entre las Baleares y las Béticas. *Symposium on the Geology of the Pyrinees and Betics. Abstracts*, 7–8.

MÉLOU, M. 1973. Le genre *Aegiromena* (Brachiopode Strophomenida) dans l'Ordovicien du Massif Armoricain (France). *Annales de la Société Géologique du Nord*, **93**, 253–264.

MÉLOU, M. 1975. Le genre *Heterorthina* (Brachiopoda, Orthida) dans la formation des Schistes de Postolonnec (Ordovicien), Finistère, France. *Geobios*, **8**, 191–208.

MÉLOU, M. 1976. Orthida (Brachiopoda) de la Formation de Postolonnec (Ordovicien), Finistère, France. *Geobios*, **9**, 693–717.

MÉLOU, M., OULEBSIR, L. & PARIS, F. 1999. Brachiopodes et chitinozoaires ordoviciens dans le NE du Sahara algérien: implications stratigraphiques et paléogeographiques. *Geobios*, **32**, 823–839.

MENDES, F. 1968. Contribution à l'étude géochronologique para la méthode au Strontium desformations cristallines du Portugal. *Boleios do museu e laboratori minero geológico do facultade de ciencias de Lisboa*, **11**, 1–150.

MENDES, F., FUSTER, J. M., IBARROLA, E. & FERNÁNDEZ SANTÍN, S. 1972. L'age de quelques granites de la Sierra de Guadarrama (Systeme Central Espagnol). *Revista Faculta Ciencias de Lisboa*, **17**(1), 345–365.

MÉNDEZ-BEDIA, I. 1976. Biofacies y litofacies de la Formación Moniello-Santa Lucia (Devónico de la Cordillera Cantábrica, NW de España). *Trabajos Geología Universidad Oviedo*, **9**, 1–93.

MÉNDEZ-BEDIA, I. 1984. Primera nota sobre los estromatoporoideos de la Formación Moniello (Devónico de la Cordillera Cantábrica, NW de España). *Trabajos Geología Universidad Oviedo*, **14**, 151–159.

MÉNDEZ-BEDIA, I. & SOTO, F. 1984. Paleoecological succession in a Devonian organic buildup (Moniello Formation, Cantabrian Mountains, NW Spain). *Geobios*, **8**, 151–158.

MÉNDEZ-BEDIA, I., SOTO, F. & FERNÁNDEZ-MARTINEZ, E. 1994. Devonian reef types in the Cantabrian Mountains (NW Spain) and their faunal composition. *Courier Forschungsinstitut Senckenberg*, **172**, 161–183.

MENDIA, M. S. 2000. *Petrología de la unidad eclogítica del Complejo de Cabo Ortegal (NW de España)*. PhD Thesis, University of País Vasco (Serie Nova Terra, **16**).

MENÉNDEZ AMOR, S. & FLORSCHÜTZ, F. 1964. Results of the preliminary palynological investigation of samples from a 50 m boring in southern Spain. *Boletín Real Sociedad Española de Historia Natural (Geol.)*, **62**, 251–255.

MENSINK, H. & MERTMANN, D. 1984. Diskontinuitaten im Unter-Callovium der nordwestlichen Keltiberischen Ketten (Spanien). *Neues Jahrbuch für Geologie und Paläontologie, Abhandlungen*, **162**, 189–223.

MERCIER, A., ROUX, L. & WICKHAM, S. M. 1996. Magmatisme Hercynien: Granites du Massif des Trois Seigneurs. *In*: BARNOLAS, A. & CHIRON, C. J. (eds) *Synthèse géologique et géophysique des Pyrénées. 1- Cycle Hercynien*. BRGM and ITGE, Orléans and Madrid, 454–457.

MERGL, M. & LIÑÁN, E. 1986. Some Cambrian Brachiopoda of the Cordillera Iberica and their biostratigraphical significance. *In*: VILLAS, E. (coord.) *Memorias I Jornadas de Paleontología. Zaragoza*. Diputación General de Aragón, Zaragoza, 159–179.

MESA, J. M. 1992. Caolín y arcillas caoliníferas. *In*: GARCÍA-GUINEA, J. & MARTÍNEZ-FRÍAS, J. (eds) *Recursos Minerales de España*. CSIC, Colección Textos Universitarios, **15**, 27–44.

MESCHEDE, M. 1985. The geochemical character of volcanic rocks of the Basco-Cantabrian Basin, Northeastern Spain. *Neues Jahrbuch für Geologie und Paläontologie, Monatshefte*, **1985**(2), 115–128.

MESSERLI, B. 1965. *Beiträge zur Geomorphologie der Sierra Nevada (Andalusien)*. Juris Verlag, Zurich.

METTE, W. 1989. Acritarchs from Lower Paleozoic rocks of western Sierra Morena, SW-Spain and biostratigraphic results. *Geologica et Palaeontologica*, **23**, 1–19.

MEY, P. H. W. 1967. The Geology of the upper Ribagorzana and Baliera Valleys, Central Pyrenees, Spain. *Leidse Geologische Mededelingen*, **41**, 153–220.

MEZCUA, J. & RUEDA, J. 1997. Seismological evidence for a delamination process in the lithosphere under the Alborán Sea. *Geophysical Journal International*, **129**, F1–F8.

MEZCUA, J., RUEDA, J. & MARTÍNEZ, S. 1991. Seismicity of the Ibero-Maghrebian Region. *In*: MEZCUA, J. & UDÍAS, A. (eds) *Seismicity, Seismotectonics and Seismic Risk of the Ibero-Maghrebian Region*. IGN Monografía, **8**, 17–28.

MIDDLEMOST, E. A. K. 1970. San Miguel de La Palma, a volcanic island in section. *Bulletin Volcanologique*, **34**, 216–239.

MILLÁN, H. 1996. *Estructura y cinemática del frente de cabalgamiento surpirenaico en las Sierras Exteriores Aragonesas*. PhD Thesis, University of Zaragoza.

MILLÁN, H., AURELL, M. & MELÉNDEZ, M. 1994. Synchronous detachment folds and coeval sedimentation in the Pyrenean External Sierras (Spain). A case study for a tectonic origin of sequences and tracts. *Sedimentology*, **41**, 1001–1024.

MILLÁN, H., DEN BEZEMER, T., VERGÉS, J. *et al.* 1995a. Palaeo-elevation and effective elastic thickness evolution at mountain ranges: inferences from flexural modelling in the Eastern Pyrenees and Ebro Basin. *Marine and Petroleum Geology*, **12**, 917–928.

MILLÁN, H., POCOVÍ, A. & CASAS, A. 1995b. El frente de cabalgamiento surpirenaico en el extremo occidental de las Sierras Exteriores: sistemas imbricados y pliegues de despegue. *Revista de la Sociedad Geológica de España*, **8**, 73–90.

MILLÁN, H., PUEYO, E., AURELL, M., LUZÓN, A., OLIVA, B., MARTÍNEZ-PEÑA, M. B. & POCOVÍ, A. 2000. Actividad tectónica registrada en los depósitos terciarios del frente meridional del Pirineo central. *Revista de la Sociedad Geológica de España*, **13**, 279–300.

MILLER, B. V., SAMSON, S. D. & D'LEMOS, R. S. 1999. Time span of plutonism, fabric development & cooling in a Neoproterozoic magmatic arc segment: U-Pb age constraints from syn-tectonic plutons, Sark, Channel Islands, UK. *Tectonophysics*, **312**, 79–95.

MILLER, V. C. 1980. Divide migration and stream capture in the Oliete area, Spain. *ITC (International Institute for Aerial Survey and Earth Sciences) Journal*, **1980-2**, 328–348.

MIMAM (Ministerio de Medio Ambiente). 1998. *Calidad y contaminación de las aguas subterráneas en España. Propuestas de protección*. Secretaria de Estado de Aguas y Costas, Madrid, **144**.

MIMAM (Ministerio de Medio Ambiente). 2000. *Libro Blanco del Agua en España*, **637**, Madrid.

MINER – MOPTMA. 1994. *Libro blanco de las aguas subterráneas*. Serie Monografías. **135**. Madrid.

MINISTERIO DE INDUSTRIA. 1971. *Plan Nacional de la Minería*. Report, Dirección General de Minas, Madrid.

MIROUSE, R. 1962. *Recherches geologiques dans la partie occidentale de la zone axiale primaire des Pyrenees*. PhD Thesis, University of Toulouse.

MIROUSE, R. 1966. *Recherches geologiques dans la partie occidentale de la zone primaire axiale des Pyrénées*. Memoires Carte Géologique France.

MITJAVILA, J. & MARTI, J. 1986. El volcanismo triásico del sur de Catalunya. *Revista de Investigacions Geologiques*, **42–43**, 89–103.

MITJAVILA, J., RAMOS, E. & MARTÍ, J. 1990. Les ignimbrites del Puig de l'Ofre (Serra de Tramuntana, Mallorca): Nouvelles precisions geologiques sur leur position et datation radiometrique. *Comptes Rendus des séances de l'Académie des Sciences de Paris*, **311**, 687–692.

MITJAVILA, J., MARTÍ, J. & SORIANO, C. 1997. Magmatic evolution and tectonic setting of the Iberian Pyrite Belt volcanism. *Journal Petrology*, **38**, 727–755.

MITSUNO, C., NAKAMURA, T., KANEHIRA, K. *et al.* 1986. *Geological studies of the Iberian Pyrite Belt with special reference to its genetical correlation of the Yanahara ore deposits and others in the inner zone of southwest Japan*. University of Okayama, Japan.

MOCZYDŁOWSKA, M. 1999. The Lower-Middle Cambrian boundary recognized by acritarchs in Baltica and at the margin of Gondwana. *Bolletino della Società Paleontologica Italiana*, **38**, 207–225.

MOISSENET, E. 1985. Le Quaternaire moyen alluvial du fossé de Teruel (Espagne). *Physio-Geo*, **14/15**, 61–78.

MOLIN, D. 1980. *Le volcanisme miocène du Sud-Est de l'Espagne (Provinces de Murcia et d'Almeria)*. PhD Thesis, University of Paris VI.

MOLINA, E. 1991. *Geomorfología y geoquímica del paisaje. Dos ejemplos en el interior de la Meseta Ibérica*. Acta Salmanticensia, University of Salamanca (Biblioteca de las Ciencias, **72**).

MOLINA, E. & ARMENTEROS, I. 1986. Los arrasamientos pliocenos y plio-pleistocenos en el sector suroriental de la cuenca del Duero. *Studia Geológica Salmanticensia*, **22**, 293–307.

MOLINA, E., ARENILLAS, I. & ARZ, J. A. 1996. The Cretaceous/Tertiary boundary mass extinction in planktic foraminifera at Agost (Spain). *Revue de Micropaleontology*, **39**, 225–243.

MOLINA, E., GARCÍA TALEGÓN, J. & VICENTE, M. A. 1997. Palaeo-weathering profiles develop on the Iberian Hercynian Basement and their relationship to the oldest Tertiary surface in central and western Spain. *In*: WIDDOWSON, M. (ed.) *Palaeosurfaces: Recognition, Reconstruction and Palaeoenvironmental Reconstruction*. Geological Society, London, Special Publications, **120**, 175–185.

MOLINA, J. M. 1987. *Análisis de facies del Mesozoico en el Subbético Externo (Provincias de Córdoba y Sur de Jaén)*. PhD Thesis, University of Granada.

MOLINA, J. M. & RUÍZ-ORTÍZ, P. A. 1990. Nuevos datos y modelo genético sobre brechas jurásicas generadas en relación con fallas transcurrentes (Subbético Externo. Provincia de Córdoba). *Geogaceta*, **7**, 56–59.

MOLINA, J. M. & VERA, J. A. 1996. La Formación Milanos en el Subbético Medio (Jurásico Superior): Definición y Descripción. *Geogaceta*, **20**, 1, 39–42.

MOLINA, J. M., VERA, J. A. & GEA, G. A. DE 1998. Vulcanismo submarino del Santoniense en el Subbético. Datación con nanofósiles e interpretación (Formación Capas Rojas, Alamedilla, provincia de Granada). *Estudios Geológicos*, **54**, 191–197.

MOLINA, J. M., O'DOGHERTY, L., SANDOVAL, J. & VERA, J. A. 1999a. Jurassic radiolarites in a Tethyan continental margin (Subbetic, southern Spain): palaeobathymetric and biostratigraphic considerations. *Palaeogeography, Palaeoclimatology, Palaeoecology*, **150**, 309–330.

MOLINA, J. M., RUÍZ-ORTÍZ, P. A. & VERA, J. A. 1999b. A review of polyphase karstification in extensional tectonic regimes: Jurassic and Cretaceous examples, Betic Cordillera, southern Spain. *Sedimentary Geology*, **129**, 71–84.

MOLLAT, H. 1968. Schichtfolge und tektonischer Bau der Sierra Blanca und ihrer Umgebung (Betische Kordilleren, Südspanien). *Geologie Jahrbuch*, **86**, 471–532.

MON, R. 1971. Estudio Geológico del extremo occidental de los Montes de Málaga y de la Sierra de Cartama (prov. de Málaga). *Boletín Geológico Minero*, **82**, 132–146.

MONIÉ, P., GALINDO-ZALDÍVAR, J., GONZÁLEZ-LODEIRO, F., GOFFÉ, B. & JABALOY, A. 1991. $^{40}Ar/^{39}Ar$ geochronology of Alpine tectonism in the Betic Cordilleras (southern Spain). *Journal of the Geological Society of London*, **148**, 288–297.

MONNEREAU, M. & CAZENAVE, A. 1990. Depth and geoid anomalies over oceanic hotspots swells: a global survey. *Journal of Geophysical Research*, **95**, 15429–15438.

MONTADERT, L., ROBERTS, D. G., DE CHARPAL, O. & GUENNOC, P. 1979. Rifting and subsidence of the northern continental margin of the Bay of Biscay. *Initial Reports of the Deep Sea Drilling Project*, **48**, 1025–1059.

MONTENAT, C. 1977. Les bassins Néogènes du Levant d'Alicante et de Murcia (Cordillères bétiques orientales – Espagne). Stratigraphie, paleogeographie et evolution dynamique. *Documents Laboratoire Géologie Faculté Sciences Lyon*, **69**, 1–345.

MONTENAT, C. (ed.) 1990. Les Bassins Néogènes du Domaine Bétique Oriental (Espagne). *Documents et Travaux IGAL*, **12–13**, 392 pp.

MONTENAT, C., OTT D'ESTEVOU, P. & MASSE, P. 1987. Tectonic-sedimentary characters of the Betic Neogene basins evolving in a crustal transcurrent shear zone (SE Spain). *Bulletin des Centres de Recherche Exploration-Production Elf-Aquitaine*, **11**, 1–22.

MONTERRUBIO, S. 1991. *Mineralizaciones asociadas a rocas ultrabasicas en el Hercinico español*. PhD Thesis, Complutense University, Madrid.

MONTERRUBIO, S., & LUNAR, R. 1992. Mineralizaciones de Cr-EGP en el complejo de Cabo Ortegal (NW de España). *In*: GARCÍA-GUINEA, J. & MARTÍNEZ-FRÍAS, J. (eds) *Recursos Minerales de España*. CSIC, Colección Textos Universitarios, **15**, 291–317.

MONTERRUBIO, S., LUNAR, R., OYARZUN, R. & SÁENZ, J. 1990. NW Spain: A potential for Cr and PGM mineralisations. *Mining Magazine*, **163**, 106–109.

MONTESERÍN, V., PÉREZ ROJAS, A., SAN JOSÉ, M. A. *et al.* 1987. *Mapa Geológico de España a escala 1:50000. Hoja no. 652 (Jaraicejo)*. IGME, Madrid, 1–46.

MONTESINOS, J. R. & SANZ-LÓPEZ, J. 1999a. *Falcitornoceras* and *Cheiloceras* (Ammonoidea, Goniatitida) from the lower Famennian of the Iberian Peninsula and their biostratigraphic applications. *Newsletters Stratigraphie*, **37**(3), 163–175.

MONTESINOS, J. R. & SANZ-LÓPEZ, J. 1999b. Ammonoideos del Devónico Inferior y Medio en el Pirineo Oriental y Central. Antecedentes históricos y nuevos hallazgos. *Revista Española Paleontología*, Madrid, No. extra. Homenaje Prof. J. Truyols, 97–108.

MONTESINOS, J. R. & TRUYOLS-MASSONI, M. 1987. La Fauna de Anetoceras y el límite Zlichoviense-Dalejiense en el Dominio Palentino (NO de España). *Cuadernos Laboratorio Xeoloxico Laxe*, **11**, 191–208.

MONTSERRAT, J. 1992. *Evolución glaciar y postglaciar del clima y la vegetación en la vertiente sur del Pirineo: estudio palinológico*. Instituto Pirenaico de Ecología, Zaragoza.

MOORE, J. G. 1987. Subsidence in the Hawaiian Ridge. *In*: DECKER, W., WRIGHT, T. L. & STAUFFER, P. H. (eds) *Volcanism in Hawaii*. USGS, Professional Papers, **1350**, 1, 85–100.

MOORE, J. G. & CAMPBELL, J. F. 1987. Age of tilted reefs, Hawaii. *Journal of Geophysical Research*, **92**, 2641–2646.

MOORE, J. G. & CLAGUE, D. A. 1992. Volcano growth and evolution of the island of Hawaii. *Geological Society of America Bulletin*, **104**, 1471–1484.

MOORE, J. G. & FORNARI, D. J. 1984. Drowned reefs as indicators of the rate of subsidence of the islands of Hawaii. *Journal of Geology*, **92**, 752–759.

MOORE, M. 1964. Giant submarine landslides on the Hawaiian Ridge. *USGS Professional Paper*, **501**, D, 95–98.

MOPT (Ministerio de Obras Públicas y Transporte). 1991. *Estudio de las zonas húmedas de la España península*. Inventario y tipificación, Madrid.

MOPT. 1993. *Plan Hidrológico Nacional*. Memoria y anteproyecto, **253**, Madrid.

MORAD, S., AL-ASAM I., LONGSTAFF, F., MARFIL, R., JOHANSEN, H. & MARZO, M. 1995. Diagenesis of a mixed siliciclastic/evaporitic sequence of the middle Muschelkalk (Middle Triassic), the Catalan Coastal Range, NE Spain, *Sedimentology*, **42**, 749–768.

MORALES, J., SERRANO, I., VIDAL, F. & TORCAL, F. 1997. The depth of the earthquake activity in the Central Betics (Southern Spain). *Geophysical Research Letters*, **24**, 3289–3292.

MORALES, J., SERRANO, I., JABALOY, A. *et al.* 1999. Active continental subduction beneath the Betic Cordillera and the Alborán Sea. *Geology*, **27**, 735–738.

MORALES, S., FENOLL HACH-ALI, P. & BOTH, R. 1995. Mineralogy, paragenesis and geochemistry of base metal ore deposits of the Aguilas-Sierra Almagrera, Spain. *In*: PASAVA, J. (ed.) *Mineral Deposits: From their Origin to their Enviromental Impacts*. Balkema, Rotterdam, 369–372.

MORALES, S., CARRILLO, F., FENOLLI, P., DE LA FUENTE, F. & CONTRERAS, E. 2000. Epithermal Cu-Au mineralisation in the Palai-Islica deposit, Almeria, southeastern Spain: fluid inclusion evidence for mixing of fluids as a guide to gold mineralisation. *Canadian Mineralogist*, **38**, 553–565.

MORATA, D. 1993. *Petrología y geoquímica de las ofitas de las Zonas Externas de las Cordilleras Béticas*. PhD Thesis, University of Granada.

MORATA, D., HIGUERAS, P., DOMÍNGUES-BELLA, S., PARRAS, J., VELASCO, F. & APARCICO, P. 2001. Fuchsite and other Cr-rich phyllosilicates in ultramafic enclaves from the Almadén mercury mining district. *Clay Minerals*, **36**, 34–354.

MORENO, C. 1987. *Las facies Culm del anticlinorio de la Puebla de Guzmán (Huelva, España)*. PhD Thesis, University of Granada.

MORENO, C. 1988. Dispositivos turbidíticos sincrónicos en el Carbonífero Inferior de la Faja Pirítica Ibérica (Zona Sur-Portuguesa). *Estudios Geológicos*, **44**, 233–242.

MORENO, C. 1993. Postvolcanic paleozoic of the Iberian Pyrite Belt: an example of basin morphologic control on sediment in a turbidite basin. *Journal of Sedimentary Petrology*, **63**, 1118–1128.

MORENO, C. & SÁEZ, R. 1989. Petrología y procedencia de las areniscas del Culm de la parte occidental de Faja Pirítica Ibérica (Zona Sur-Portuguesa). *Boletín Geológico y Minero*, **100–I**, 134–147.

MORENO, C. & SEQUEIROS, L. 1989. The Basal Shaly Fm of Iberian Pyrite Belt (South Portuguese Zone): Early Carboniferous bituminous deposits. *Paleogeography, Paleoclimatology, Paleoecology*, **73**, 233–241.

MORENO, C., SIERRA, S., & SÁEZ, R. 1995. Mega-debris flows en el tránsito Devónico-Carbonífero de la Faja Pirítica Ibérica. *Geogaceta*, **17**, 9–11.

MORENO, C., SIERRA, S. & SÁEZ, R. 1996. Evidence for catastrophism at the Famenian-Dinantian boundary in the Iberian Pyrite Belt. Recent Advances. *In*: STROGEN, P., SOMERVILLE, D. & JONES, G. LL. (eds) *Lower Carboniferous Geology*. Geological Society, London, Special Publications, **107**, 153–162.

MORENO, F. 1977. Tectónica y sedimentación de las Series de Tránsito (Precámbrico Terminal) en el Anticlinal de Valdelacasa y el Valle de Alcudia. Ausencia de Cámbrico. *Studia Geológica Salmanticensia*, **12**, 123–136.

MORENO, F., VEGAS, R. & MARCOS, A. 1976. Sobre la edad de las series ordovícicas y y cámbricas relacionadas con la discordancia 'Sárdica' en el anticlinal de Valdelacasa (Montes de Toledo, España). *Breviora Geologica Asturica*, **20**, 8–16.

MORENO, T. 1999. *Platinum-group elements and chromite mineralisation in ultramafic rocks: a case study from the Cabo Ortegal Complex, Northwest Spain*. PhD Thesis, Cardiff University.

MORENO, T., PRICHARD, H. M., LUNAR, R., MONTERRUBIO, S. & FISHER, P. 1999. Formation of a secondary platinum-group mineral assemblage in chromitites from the Herbeira ultramafic massif in Cabo Ortegal, NW Spain. *European Journal of Mineralogy*, **11**, 363–378.

MORENO, T., GIBBONS, W., PRICHARD, H. & LUNAR, R. 2001. Platiniferous chromitite and the tectonic setting of ultramafic rocks in Cabo Ortegal (north West Spain). *Journal of the Geological Society*, **158**, 601–614.

MORENO-ÉIRIS, E. 1987. Los montículos arrecifales de Algas y Arqueociatos del Cámbrico Inferior de Sierra Morena. *Publicaciones especiales del Boletín Geológico y Minero*, 1–127.

MORENO-VENTAS, I. 1991. *Petrología de los granitoides y rocas básicas de la sierra de Gredos*. PhD Thesis, University of Sevilla.

MORENO-VENTAS, I., ROGERS, G. & CASTRO, A. 1995. The role of hybridization in the genesis of Hercynian granitoids in the Gredos massif Inferences from Sm/Nd isotopes. *Contributions to Mineralogy and Petrology*, **120**, 137–149.

MORGAN, W. J. & PRICE, E. 1995. Hotspot melting generates both hotspot volcanism and a hotspot swell? *Journal of Geophysical Research*, **100**(B5), 8045–8062.

MORILLO, M. J. & MELÉNDEZ, F. 1979. El Jurásico de la Alcarria-La Mancha. *Cuadernos de Geología, Universidad de Granada*, **10**, 149–166.

MORLEY, C. K. 1993. Discussion of origins of hinterland basins to the Rif-Betic Cordillera and Carpathians. *Tectonophysics*, **226**, 359–376.

MORZADEC, P., PARIS, F., PLUSQUELLEC, Y., RACHEBOEUF, P. & WEYANT, M. 1988. Devonian stratigraphy and paleogeography of the Armorican Massif (western France). *In*: MCMILLAN, N. J., EMBRY, A. F. & GLASS, D. J. (eds) *Devonian of the World*. Canadian Society of Petroleum Geologists, 401–420.

MOSS, J., MCGUIRE, W. J. & PAGE, D. 1999. Ground deformation monitoring of a potential landslide at La Palma, Canary Islands. *Journal of Volcanology and Geothermal Research*, **94**, 1–4, 251–265.

MOUTERDE, R., FERNÁNDEZ-LÓPEZ, S. GOY, A., LINARES, A., RIVAS, P., RUGET, CH. & SUÁREZ-VEGA, L. C. 1978. El Jurásico en la región de Obón (Teruel). *Grupo Español del Mesozoico, Guía de Excursiones al Jurásico de la Cordillera Ibérica*, II.1–II.13.

MULLER, D. 1969. *Perm und Trias in Valle del Batzán, Spanische West-Pyrenees*. PhD Thesis, Clausthal University.

MULLER, D. 1973. Perm und Trias in Valle del Batzán, Spanische West-Pyrenees. *Neues Jahrbuch für Geologie und Paläontologie. Abhandlungen*, **142**, 30–43.

MULLER, J. & ROGER, P. 1977. L'evolution des Pyrénées (domaine central et occidental). Le segment hercynien, la chaine de fond alpine. *Geologie Alpine*, **53**, 141–191.

MUNHÁ, J. 1979. Blue anphiboles, metamorphic regime and plate tectonic modelling in the Iberian Pyrite Belt. *Contributions to Mineralogy and Petrology*, **69**, 279–289.

MUNHÁ, J. 1983. Variscan magmatism in the Iberian Pyrite Belt. *In*: LEMOS DE SOUSA, M. J. & OLIVEIRA, J. T. (eds) *The Carboniferous of Portugal*. Serviços Geológicos de Portugal, Lisboa, Memórias, **29**, 39–81.

MUNHÁ, J. 1990. Metamorphic evolution of the South Portuguese/Pulo do Lobo Zone. *In*: DALLMEYER, R. D. & MARTÍNEZ GARCÍA, E. (eds) *Pre-Mesozoic Geology of Iberia*. Springer-Verlag, Berlin, 363–368.

MUNHÁ, J. & KERRICH, R. 1980. Sea water–basalt interaction in spilites from the Iberian Pyrite Belt. *Contributions to Mineralogy and Petrology*, **73**, 191–200.

MUNHÁ, J., BARRIGA, F. & KERRICH, R. 1986a. High ^{18}O ore-forming fluids in volcanic hosted base metal massive sulphide deposits: geologic $^{18}O/^{16}O$ and D/H evidence for the Iberian Pyrite Belt; Crandon, Wisconsin; and Blue Hill, Maine. *Economic Geology*, **81**, 530–552.

MUNHÁ, J., OLIVEIRA, J. T., RIBEIRO, A., OLIVEIRA, V., QUESADA, C. & KERRICH, R. 1986b. Beja-Acebuches Ophiolite: characterization and geodynamic significance. *Maleo, Boletím Informativo da Sociedade Geológica de Portugal*, **2/13**, 31.

MUNKSGAARD, N. C. 1984. High d^{18}O and possible pre-eruptional Rb-Sr isochrons in cordierite-bearing Neogene volcanics from SE Spain. *Contributions to Mineralogy and Petrology*, **87**, 351–358.

MUNKSGAARD, N. C. 1985. A non-magmatic origin for compositionally zoned euhedral garnets in silicic Neogene volcanics from SE Spain. *Neues Jahrbuch für Mineralogie*, **2**, 73–82.

MUÑOZ, A. 1992. *Análisis tectosedimentario del Terciario del sector occidental de la Cuenca del Ebro (Comunidad de La Rioja)*. Instituto de Estudios Riojanos, Logroño, Ciencias de la Tierra, **15**.

MUÑOZ, A. 1997. *Evolución geodinámica del borde oriental de la Cuenca del Tajo desde el Oligoceno hasta la actualidad*. PhD Thesis, Complutense University, Madrid.

MUÑOZ, A. & CASAS, A. M. 1997. The Rioja trough: tecto-sedimentary evolution of a foreland symmetric basin. *Basin Research*, **9**, 65–85.

MUÑOZ, A. & DE VICENTE, G. 1996. Campos de paleoesfuerzos terciarios en el borde oriental de la cuenca del Tajo (España central). *Geogaceta*, **20**, 913–916.

MUÑOZ, A. & DE VICENTE, G. 1998a. Origen y relación entre las deformaciones y esfuerzos alpinos de la zona centro-oriental de ll Península Ibérica. *Revista de la Sociedad Geológica de España*, **11**, 57–70.

MUÑOZ, A. & DE VICENTE, G. 1998b. Cuantificación del acortamiento alpino y estructura en profundidad del extremo sur-occidental de la Cordillera Ibérica (Sierras de Altomira y Bascuñana). *Revista de la Sociedad Geológica de España*, **11**, 233–252.

MUÑOZ, J. A. 1985. *Estructura alpina i herciniana a la vora sud de la zona axial del Pirineu oriental*. PhD Thesis, University of Barcelona.

MUÑOZ, J. A. 1992. Evolution of a continental collision belt: ECORS-Pyrenees crustal balanced cross-section. *In*: MCCLAY, K. (ed.) *Thrust Tectonics*. Chapman & Hall, London, 235–246.

MUÑOZ, J. A., PUIGDEFÁBREGAS, C. & FONTBOTÉ, J. M. 1983. El Pirineo *In*: COMBA (ed.) *Libro Jubilar J. M. Ríos, 'Geologíc de España'*, Vol. 2. IGME, Madrid, 161–205.

MUÑOZ, J. A., MARTÍNEZ, A. & VERGÉS, J. 1986. Thrust sequences in the eastern Spanish Pyrenees. *Journal of Structural Geology*, **8**, 399–405.

MUÑOZ, J. A., MCCLAY, K. & POBLET, J. 1994. Synchronous extension and contraction in frontal thrust sheets of the Spanish Pyrenees. *Geology*, **22**, 921–924.

MUÑOZ, J. A., CONEY, P., MCCLAY, K. & EVENCHICK, C. 1997. Discussion on syntectonic burial and post-tectonic exhumation of the southern Pyrenees foreland fold-thrust belt. *Journal of the Geological Society, London*, **154**, 361–365.

MUÑOZ, M. & SAGREDO, J. 1994. Reajustes mineralógicos y geoquímicos producidos durante el metamorfismo de contacto de diques basálticos (Fuerteventura, Islas Canarias). *Boletín Sociedad Española de Mineralogía*, **17**(1), 86–87.

MUÑOZ, M. & SAGREDO, J. 1996. Evidencias de mezcla de magmas en el edificio volcánico- subvolcánico de Betancuria (Islas Canarias). *Geogaceta*, **20**(3), 554–557.

MUÑOZ, P., COSSIO, J., CEMBREROS, V. & BAHAMONDE, J. 1986. *Posibilidades de azabaches en Asturias*. Report **11–137**, IGME, Madrid.

MUÑOZ, P., AIZPURÚA, F. J., GARCÍA, E., GÓMEZ, G. & NAVARRO, J. V. 1998. *Mapa de Rocas y Minerales Industriales*, ***hoja 8***. IGME, Madrid (ISBN 86-7840-354-X).

MUÑOZ ESPADAS, M. J, LUNAR, R, & MARTINEZ FRIAS, J. 2000. The garnet placer deposit from SE of Spain: industrial recovery and geochemical features. *Episodes*, **23–4**, 266– 270.

MUÑOZ-JIMÉNEZ, A & CASAS-SAINZ, A. M. 1997. The Rioja Trough (N Spain): tectosedimentary evolution of a symmetric foreland basin. *Basin Research*, **9**, 65–85.

MUÑOZ MARTÍN, A. & DE VICENTE, G. 1996. Campos de paleoesfuerzos terciarios en el borde oriental de la cuenca del Tajo (España Central). *Geogaceta*, **20**(4), 913–916.

MURPHY, J. B. & NANCE, R. D. 1991. Supercontinent model for the contrasting character of Late Proterozoic orogenic belts. *Geology*, **19**, 469–472.

MURPHY, J. B. KEPPIE, J. D., DOSTAL, J. & HYNES, A. J. 1990. Late Precambrian Georgeville group: A volcanic arc rift succession in the Avalon Terrane of Nova Scotia. *In*: D'LEMOS, R. S., STRACHAN, R. A. & TOPLEY, C. G. (eds) *The Cadomian Orogeny*. Geological Society, London, Special Publication, **51**, 383–393.

MURRAY, R. C. 1964. Origin and diagenesis of gypsum and anhidrite. *Journal of Sedimentary Petrology*, **34**, 512–523.

MUTTI, E. 1985. Turbidite systems and their relations to depositional sequences. *In*: ZUFFA, G. C. (ed.) *Reading Provenance from Arenites*. NATO-ASI Series, Reidel, New York, 65–93.

NÄGLER, T. F., SCHÄFER, H. J. & GEBAUER, D. 1995. Evolution of the Western European continental crust: implications from Nd and Pb isotopes in Iberian sediments. *Chemical Geology*, **121**, 345–357.

NAGTEGAAL, P. J. C. 1969. Sedimentology, paleoclimatology and diagenesis of Post-Hercynian continental deposits in South-Central Pyrenees, Spain. *Leidse Geologische Mededelingen*, **42**, 143–238.

NAKAMURA, N. 197. Determination of REE, Ba, Fe, Mg, Na and K in carbonaceous and ordinary chrondites. *Geochim. Cosmochim. Acta*, **38**, 757–773.

NANCE, R. D., MURPHY, J. B., STRACHAN, R. A., D'LEMOS, R. S. & TAYLOR, G. K. 1991. Late Proterozoic tectonostratigraphic evolution of the Avalonian and Cadomian terranes. *Precambrian Research*, **53**, 41–78.

NAVARRO, J. M. & COELLO, J. J. 1994. *Mapa geológico del P. N. de Taburiente*. ICONA.

NAVARRO, L. F. 1973. Enclaves metamórficos localizados en las rocas basálticas del noroeste de Cartagena (provincia de Murcia). *Estudios Geológicos*, **29**, 77–81.

NAVARRO VILÁ, F. 1976. *Los Mantos Alpujárrides y Maláguides al N de Sierra Nevada*. PhD Thesis, University of Bilbao.

NAVIDAD, M. 1996. Magmatisme Pré-Hercinien: Volcanisme. *In*: BARNOLAS, A. & CHIRON, J. C. (eds) *Synthèse Géologique et Géophysique des Pyrénées. Tome 1-Cycle Hercynien*. BRGM and ITGE, Orléans and Madrid, 364–376.

NAVIDAD, M. & ÁLVARO, M. 1985. El vulcanismo alcalino del Triásico Superior de Mallorca. *Boletín del IGME*, **96**, 10–22.

NAVIDAD, M. & BARNOLAS, A. 1991. El magmatismo (ortogneises y volcanismo del Ordovícico Superior) del Paleozoico de los Cataládines. *Boletín Geológico y Minero*, **102**, 187–202.

NAVIDAD, M. & CARRERAS, J. 1995. Pre-Hercynian magmatism in the eastern Pyrenees (Cap de Creus and Albera massif) and its geochemical seeting. *Geologie en Mijnbouw*, **74**, 65–77.

NAVIDAD, M. & PEINADO, M. 1976. Facies volcanosedimentarias en el Guadarrama Central (Sistema Central Español). *Studia Geológica*, **12**, 137–159.

NAVIDAD, M., PEINADO, M. & CASILLAS, R. 1992. El magmatismo pre-Hercínico del Centro Peninsular (Sistema Central Español). *In*: GUTIÉRREZ MARCO, J., SAAVEDRA, J. & RÁBANO, I. (eds) *Paleozoico Inferior de Ibero-América*. University of Extremadura, Badajoz, 485–494.

NELSON, D. R., MCCULLOCH, M. T. & SUN, S. 1986. The origins of ultra-potassic rocks as inferred from Sr, Nd and Pb isotopes. *Geochimica et Cosmochimica Acta*, **50**, 231–245.

NERCESSIAN, A., MAUFFRET, A., DOS REIS, A. T., VIDAL, R., GALLART, J. & DIAZ, J. 2001. Deep reflection seismic images of the crustal thinning in the eastern Pyrenees and western Gulf of Lion. *Journal of Geodynamics*, **31**, 211–225.

NESBITT, R. W., PASCUAL, E., FANNING, C. M., TOSCANO, M., SÁEZ, R. & ALMODOVAR, G. R. 1999. U-Pb dating of stock-work zircons from the eastern Iberian Pyrite Belt. *Journal of the Geological Society, London*, **156**, 7–10.

NEUWEILER, F. 1993. Development of Albian microbialites and microbialite reefs at marginal platform areas of the Vasco-Cantabrian basin (Soba reef area, Cantabria, N. Spain). *Facies*, **29**, 231–250.

NEUWEILER, F., GAUTRET, P., THIEL, V., LANGES, R., MICHAELIS, W. & REITNER, J. 1999. Petrology of Lower Cretaceous carbonate mud mounds (Albian, N. Spain): insights into organomineralic deposits of the geological record. *Sedimentology*, **46**, 837–859.

NIETO, F., VELILLA N., PEACOR, D. R. &. ORTEGA-HUERTAS. M. 1994. Regional retrograde alteration of sub-greenschist facies chlorite to smectite. *Contributions to Mineralogy and Petrology*, **115**, 243–252.

NIETO, J. M., PUGA, E., MONIÉ, P., DÍAZ DE FEDERICO, A. & JAGOUTZ, E. 1997. High-pressure metamorphism in metagranites and orthogneiss from the Mulhacén Complex (Betic Cordillera, Spain). *Terra Nova*, **9**, 22–23.

NIETO, L. M. 1996. *La cuenca subbética mesozoica en el sector oriental de las Cordilleras Béticas*. PhD Thesis, University of Granada.

NIETO, L. M., DE GEA, G. A., AGUADO, R., MOLINA, J. M. & RUIZ-ORTIZ,

P. A. 2001. Procesos sedimentarios y tectónicos en el tránsito Jurásico/Cretácico: precisiones bioestratigráficas (Unidad del Ventisquero, Zona Subbética). *Revista de la Sociedad Geológica de España*, **14**, 35–46.

NIJHUIS, H. 1964a. On the stratigraphy of the Nevado-Filabride units as exposed in the eastern Sierra de los Filabres (SE Spain). *Geologie en Mijnbouw*, **43**, 321–325.

NIJHUIS, H. 1964b. Plurifacial alpine metamorphism in the south-eastern Sierra de los Filabres, south of Lubrin. PhD Thesis, University of Amsterdam.

NIJMAN, W. 1998. Cyclicity and basin axis shift in a piggy-back basin: towards modelling of the Eocene Tremp-Ager Basin. *In*: MASCLE, A., PUIGDEFÀBREGAS, C., LUTERBACHER, H. P. & FERNÁNDEZ, M. (eds) *Cenozoic Foreland Basins of Western Europe*. Geological Society, London, Special Publications, **134**, 135–162.

NIJMAN, W. & SAVAGE, J. F. 1989. Persistent basement wrenching as controlling mechanism of Variscan thin-skinned thrusting and sedimentation, Cantabrian Mountains, Spain. *Tectonophysics*, **169**, 281–302.

NIXON, P. H., THIRWALL, M. F., BUCKLEY, F. & DAVIES, C. J. 1984. Spanish and Western Australian lamproites: aspects of whole rock geochemistry. *In*: KORNPROBST, J. (ed.) *Kimberlites and Related Rocks*. Elsevier, Amsterdam, 285–296.

NOBEL, F. A., ANDRIESSEN, P. A. M., HEBEDA, E. H., PRIEM, H. N. A. & RONDEEL, H. E. 1981. Isotopic dating of the post-Alpine Neogene volcanism in the Betic Cordilleras, southern Spain. *Geologie en Mijnbouw*, **60**, 209–214.

NOGALES, J. & SCHMINCKE, H. U. 1969. El pino enterrado de la Cañada de las Arenas (Gran Canaria). *Cuadernos de Botánica Canaria*, **5**, 23–25.

NORONHA, F., VINDEL, E., LÓPEZ, J. A., DÓRIA, A., GARCÍA, E. & CATHELINEAU, M. 1999. Fluids related to tungsten ore deposits in northern Portugal and Spanish Central System: a comparative study. *Revista de la Sociedad Geológica de España*, **12**, 397–403.

NOZAL MARTÍN, F., GARCÍA CASQUERO, J. L. & PICART BOIRA, J. 1988. Discordancia Intraprecámbrica y series sedimentarias en el sector sur-oriental de los Montes de Toledo. *Boletín Geológico y Minero*, **99**, 473–489.

NUEZ, J. DE LA., UBANELL, A. & VILLASECA, C. 1981. Diques lamprofídicos norteados con facies brechoidales eruptivas en la región de la Paramera de Avila (Sistema Central Español). *Cuadernos del Laboratorio Xeolóxico de Laxe*, **3**, 53–73.

O'CONNOR, J. M. & DUNCAN, R. A. 1990. Evolution of the Walvis Ridge-Rio Grande Rise hotspot system: implications for African and South American plate motions over plumes. *Journal of Geophysical Research*, **95**, 17475–17502.

O'DOGHERTY, L. 1994. *Biochronology and paleontology of Mid-Cretaceous radiolarians of mid Cretaceous from Northern Apennines (Italy) and Betic Cordillera (Spain)*. Mémoires de Géologie (Lausanne), **21**.

O'DOGHERTY, L., MOLINA, J. M., RUÍZ-ORTIZ, P. A., SANDOVAL, J. & VERA, J. A. 1997. La Formación Radiolarítica Jarropa: Definición y significado en el Jurásico subbético (Cordillera Bética). *Estudios Geológicos*, **53**, 145–157.

O'DOGHERTY, L., SANDOVAL, J., & VERA, J. A. 2000. Ammonite faunal turnover tracing sea level changes during the Jurassic (Betic Cordillera, southern Spain). *Journal of the Geological Society, London*, **157**, 723–736.

O'DOGHERTY, L., MARTÍN-ALGARRA, A., GURSKY, H. J. & AGUADO, R. 2001. The Middle Jurassic radiolarites and pelagic limestones of the Nieves Unit (Rondaide Complex, Betic Cordillera): Basin starvation in a rifted marginal slope of the western Tethys. *Geologische Rundschau*, **90**, 831–846.

OBATA, M. 1980. The Ronda peridotites: Garnet-, Spinel-, and Plagioclase-Lherzolite facies and the P-T trajectories of a high-temperature mantle intrusion. *Journal of Petrology*, **21**, 533–572.

OBERMAIER, H. 1924. *Fossil Man in Spain*. Yale University Press, New Haven.

OBERMAIER, H. 1925. *El Hombre fósil*. Museo Nacional de Ciencias Naturales, Madrid. Publicaciones de la Comisión de Investigaciones Paleontológicas y Prehistóricas, **9**.

OBRADOR, A. 1970. Estudio estratigráfico y sedimentológico de los materiales miocénicos de la isla de Menorca. *Acta Geológica Hispánica*, **5**, 19–23.

OBRADOR, A., POMAR, L., RODRÍGUEZ-PEREA, A. & JURADO, M. J. 1983. Unidades deposicionales del Neógeno menorquín. *Acta Geológica Hispánica*, **18**, 87–97.

OBRADOR, A., POMAR, L. & TABERNER, C. 1992. Late Miocene megabreccia of Menorca (Balearic Islands): A basis for the representation of a megabreccia ramp deposit. *Sedimentary Geology*, **79**, 203–223.

OCHSNER, A. 1993. *U-Pb Geochronology of the Upper Proterozoic-Lower Paleozoic geodynamic evolution in the Ossa-Morena Zone (SW Iberia): Constraints on the timing of the cadomian orogeny.* PhD Thesis, Eidgenössische Technische Hoschschule, Zürich, **10**.392.

ODIN, G. S., CUENCA, G., CANUDO, J. I., COSCA, M. & LAGO, M. 1997. Biostratigraphy and geochronology of a Miocene continental volcaniclastic layer from the Ebro Basin. *In*: MONTANARI, A., ODIN, G. S. & COCCIONI, R. (eds) *Miocene Stratigraphy: An Integrated Approach.* Elsevier, Amsterdam, 297–310.

OEN, I. S. 1958. The geology, petrology and ore deposits of the Viseu region, northern Portugal. *Comunicações dos Servicios Geológicos de Portugal*, **XLI**, 5–199.

OEN, I. S. 1970. Granite intrusion, folding and metamorphism in central Portugal. *Boletin Geológico y Minero*, **81**, 271–298.

OGG, J., ROBERTSON, A. H. F. & JANSA, L. F. 1983. Jurassic sedimentation history of site 534 (western North Atlantic) and of the Atlantic-Tethys Seaway. *In*: SHERIDAN, R. E., GRANDSTEIN, F. M. *et al. Initial Reports of the Deep Sea Drilling Project*, Volume LXXVI. US Government Printing Office, Washington, 829–884.

OLIVÉ, A. & GUTIÉRREZ ELORZA, M. 1982. *Cartografía geomorfológica a escala 1:50.000 de la hoja de Arévalo (455).* Serie MAGNA, IGME. Madrid.

OLIVEIRA, J. T. 1983. The marine Carboniferous of south Portugal: A stratigraphic and sedimentological approach. *In*: Lemos de Sousa, M. J. & Oliveira, J. T. (eds) *The Carboniferous of Portugal.* Serviços Geológicos de Portugal, Lisboa, Memórias, **29**, 3–37.

OLIVEIRA, J. T. 1990. Stratigraphy and syn-sedimentary tectonism in the South Portuguese Zone. *In*: DALLMEYER, R. D. & MARTÍNEZ GARCÍA, E. (eds) *Pre-Mesozoic Geology of Iberia.* Springer-Verlag, Berlin, 334–347.

OLIVEIRA, J. T. & QUESADA, C. 1998. A comparison of stratigraphy, structure and paleogeography of the South Portuguese Zone and Southwest England, European Variscides. Geoscience in Southwest England. *Proceedings of the Ussher Society*, **9**(3), 141–150.

OLIVEIRA, J. T. & WAGNER-GENTIS, C. H. T. 1983. The Mértola and Mira Fm boundary betwen Dugueno and Almada de Ouro, marine Carboniferous of south Portugal. *In*: LEMOS DE SOUSA, M. J. (ed.) *Contributions to the Carboniferous Geology and Palaeontology of the Iberian Peninsula.* Universidade do Porto, 1–39.

OLIVEIRA, J. T., HORN, M. & PAPROTH, E. 1979. Preliminary note on the stratigraphy of the Baixo Alentejo Flysch Gp, Carboniferous of southern Portugal and on the palaeogeographic developments, compared to corresponding units in northwest Germany. *Comunicaçoes dos Serviços Geológicos de Portugal, Lisboa*, **65**, 151–168.

OLIVEIRA, J. T., CUNHA, T., STREEL, M. & VANGUESTEINE, M. 1986a. Dating the Horta da Torre Formation, a new lithostratigraphic unit of the Ferreira-Ficalho Group, South Portuguese Zone: geological consequences. *Comunicações dos Serviços Geológicos de Portugal*, **72**, 26–34.

OLIVEIRA, J.T., GARCÍA-ALCALDE, J. L., LIÑÁN, E. & TRUYOLS, J. 1986b. The Famennian of the Iberian Peninsula. *Annales Société Géologique Belgique*, **109**, 159–174.

OLIVER ASINS, J. 1958. *Historia del nombre Madrid.* CSIC, Instituto Miguel Asin, **412**.

OLIVEROS, J. M., ESCANDELL, B. & COLOM, G. 1960. El Burdigaliense superior salobre-lacustre en Mallorca. *Memorias del IGME*, **61**, 265–348.

OLIVET, J. L. 1996. La cinématique de la plaque ibérique. *Bulletin des Centres de Recherches Exploration-Production Elf-Aquitaine*, **20**, 131–195.

OLIVIER, P. 1984. *Evolution de la limite entre Zones Internes et Zones*

Externes dans l'arc de Gibraltar (Maroc-Espagne). PhD Thesis, Paul Sebatier University, Toulouse.

OLMEDO, F., YUSTA, I., PESQUERA, Y. & VELASCO, F. 1992. El yacimiento de magnesita de Eugui (Navarra). *In*: GARCÍA-GUINEA, J. & MARTÍNEZ-FRÍAS, J. (eds) *Recursos Minerales de España.* CSIC, Colección Textos Universitarios, **15**, 637–648.

OLMO SANZ, A. & MARTÍNEZ SALANOVA, J. 1989. El tránsito Cretácico-Terciario en la Sierra del Guadarrama y áreas próximas de las Cuencas del Duero y Tajo. *Studia Geológica Salmanticensia, Ediciones Universidad Salamanca*, **5**, 55–69.

OLÓRIZ, F. 1978. *Kimmeridgiense-Tithónico inferior en el sector central de las Cordilleras Béticas (Zona Subbética). Paleontología. Bioestratigrafía.* PhD Thesis, University of Granada.

OLÓRIZ, F. 2000. Time-averaging and long-term palaeoecology in macroinvertebrate fossil assemblages with ammonites (Upper Jurassic). *Revue de Paléobiologie*, Volume Spécial **8**, 123–140.

OLÓRIZ, F. & RODRÍGUEZ-TOVAR, F. J. 1993a. Lower Kimmeridgian biostratigraphy in the Central Prebetic Southern Spain. Cazorla and Segura de la Sierra sectors). *Neues Jarbuch für Geologie und Paläontologie. Monafshefte*, **3**, 150–170.

OLÓRIZ, F. & RODRÍGUEZ-TOVAR, F. J. 1993b. Reconsideración del límite Oxfordiense-Kimmeridgiense en el perfil de Puerto Lorente (Prebético Externo). *Geogaceta*, **13**, 92–94.

OLÓRIZ, F. & RODRÍGUEZ-TOVAR, F. J. 1999. Biostratigraphic review and ecostratigraphic analysis of the Fuente Alamo profile (External Prebetic, Betic Cordillera, Spain). *Profil*, **16**, 73–81.

OLÓRIZ, F. & RODRÍGUEZ-TOVAR, F. J. 2000. *Diplocraterion*: A useful marker for sequence stratigraphy and correlation in the Kimmeridgian, Jurassic (Prebetic Zone, Betic Cordillera, southern Spain). *Palaios*, **15**, 544–550.

OLÓRIZ, F. & RUÍZ-ORTIZ, P. A. 1980. Precisiones estratigráficas sobre el Malm de la Unidad del Jabalcuz-San Cristobal. Consideraciones sobre la edad de los sedimentos redepositados. *Cuadernos de Geología, Universidad de Granada*, **11**, 187–193.

OLÓRIZ, F., TAVERA, J. M. & CRUZ-SANJULIÁN, J. 1979. Precisiones sobre la bioestratigrafía del Jurásico superior del Subbético Interno en el sector de Teba (Provincia de Málaga, Cordilleras Béticas). *Cuadernos de Geología, Universidad de Granada*, **10**, 299–305.

OLÓRIZ, F., VALENZUELA, M., GARCÍA-RAMOS, J. C. & SUÁREZ DE CENTI, C. 1988. The first record of the Genus *Eurasenia* (Ammonitina) from the Upper Jurassic of Asturias (Northern Spain). *Geobios*, **21**, 741–748.

OLÓRIZ, F., MARQUES, B. & RODRÍGUEZ-TOVAR, F. J. 1991. Eustatism and faunal associations. Examples from the South Iberian Margin during the Late Jurassic (Oxfordian-Kimmeridgian). *Eclogae geologicae Helvetiae*, **84**(1), 83–106.

OLÓRIZ, F., RODRÍGUEZ-TOVAR, F. J., CHICA-OLMO, M. & PARDO, E. 1992. The marl-limestone rhythmites from the Lower Kimmeridgian (Platynota Zone) of the central Prebetic and their relationship with variations in orbital parameters. *Earth and Planetary Science Letters*, **111**, 407–424.

OLÓRIZ, F., RODRÍGUEZ-TOVAR, F. J., MARQUES, B. & CARACUEL, J. E. 1993. Ecostratigraphy and sequence stratigraphy in the high frequency sea-level fluctuations: Example from Jurassic macro-invertebrate assemblages. *Palaeogeography, Palaeoclimatology, Palaeoecology*, **101**, 131–145.

OLÓRIZ, F., RODRÍGUEZ-TOVAR, F. J. & MARQUES, B. 1994. Macroin-vertebrate assemblages and ecostratigraphic structuration within a Highstand System Tract. An example from the Lower Kimmeridgian in southern Iberia. *Geobios*, Memoire Special **17**, 605–614.

OLÓRIZ, F., CARACUEL, J. E., MARQUES, B. & RODRÍGUEZ-TOVAR, F. J. 1995a. Asociaciones de tintinnoides en facies ammitico rosso de la Sierra Norte (Mallorca). *Revista Española de Paleontología*, No. Homenaje al Dr. Guillermo Colom, 77–93.

OLÓRIZ, F., CARACUEL, J. E. & RODRÍGUEZ-TOVAR, F. J. 1995b. Using ecostratigraphic trends in sequence stratigraphy. *In*: HAQ, B. U. (ed.) *Sequence Stratigraphy and Depositional Response to Eustatic, Tectonic and Climatic Forcing*, Kluwer, New York, 59–85.

OLÓRIZ, F., CARACUEL, J. E., RUÍZ-HERAS, J. J., RODRÍGUEZ-TOVAR, F. J. & MARQUES, B. 1996. Ecostratigraphic approaches, sequence stratigraphy proposals and block tectonics: examples from epioceanic swell areas in south and east Iberia. *Palaeogeography, Palaeoclimatology, Palaeoecology*, **121**, 273–295.

OLÓRIZ, F., MARQUES, B. & CARACUEL, J. E. 1998. The Middle-Upper Oxfordian of Central Sierra Norte (Mallorca, Spain), and progressing ecostratigraphic approach in western Tethys. *Geobios*, **31**(3), 319–336.

OLÓRIZ, F., MOLINA-MORALES, J. M. & SERNA-BARQUERO, A. 1999*a*. Revisión estratigráfica del intervalo Kimmeridgiense medio-Tithonico basal en el perfil G_{10} del sector de Venta Quesada (Sierra Gorda, provincia de Granada). *Geogaceta*, **26**, 67–70.

OLÓRIZ, F., REOLID, M. & RODRÍGUEZ-TOVAR, F. J. 1999*b*. Fine-resolution ammonite biostratigraphy at the Río Gazas-Chorro II section in Sierra de Cazorla (Prebetic Zone, Jaén province, southern Spain). *Profil*, **16**, 83–94.

OLÓRIZ, F., VILLASEÑOR, A. B. & GONZÁLEZ.ARREOLA, C. (in press). Major lithostratigraphic units in land-outcrops in North-Central Mexico and the subsurface of the Northern Rim of the Gulf of Mexico Basin (Upper Jurassic-lowermost Cretaceous): a proposal for correlation of tectono-eustatic sequences. *The Journal of South American Earth Sciences*.

ONÉZIME, J. 2001. *Environnement structural et géodynamique des minéralisations de la Ceinture Pyriteuse Sud-Ibèrique: leur place dans l'évolution hercynienne.* PhD Thesis, University of Orleans.

ÖPIK, A. A. 1966. The Ordian stage of the Cambrian and its Australian Metadoxidae. *Bureau of Mineral Resources, Australia, Bulletin*, **92**, 133–169.

ORDÓÑEZ, B. 1998. *Geochronological studies of the Pre-Mesozoic basement of the Iberian Massif: the Ossa Morena zone and the Allochthonous Complexes within the Central Iberian zone.* PhD Thesis, Eidgenössische Technische Hoschschule, Zürich, **12.940**.

ORDÓÑEZ, B., GEBAUER, D. & EGUÍLUZ, L. 1997*a*. Late Cadomian formation of two anatectic gneiss domes in the Ossa Morena zone: Monesterio and Mina Afortunada. *XIV Reunião de Geología do Oeste Peninsular Vila Real, Abstracts*, 161–164.

ORDÓÑEZ, B., GEBAUER, D. & EGUÍLUZ, L. 1997*b*. SHRIMP zircon ages dating protolith formation of orthogneisses and their migmatization: Results from the Coimbra-Badajoz-Córdoba shear belt. *XIV Reunião de Geología do Oeste Peninsular Vila Real, Abstracts*, 165–168.

ORDÓÑEZ, B., GEBAUER, D. & EGUÍLUZ, L. 1998. SHRIMP age-constraints for the calc-alkaline volcanism in the Olivenza-Monesterio antiform (Ossa-Morena, SW Spain). *Goldschmidt Conference, Toulouse*, 35.

ORDÓÑEZ, B., GEBAUER, D., SCHÄFER, H.-J., GIL IBARGUCHI, J. I. & PEUCAT, J. J. 2001. A single Devonian subduction event for the HP/HT metamorphism of the Cabo Ortegal Complex within the Iberian Massif. *Tectonophysics*, **232**, 359–385.

ORDÓÑEZ, S. & GARCÍA DEL CURA, M. A. 1983. Recent and ancient fluvial carbonates in Central Spain. *In*: COLLINSON, J. D. & LEWIN, J. (eds) *Modern and Ancient Fluvial Systems.* IAS, Special Publication, **6**, 485–497.

ORDÓÑEZ, S. & GARCÍA DEL CURA, M. A. 1992. El sulfato sódico natural en España: Las sales sódicas de la cuenca de Madrid. *In*: GARCÍA-GUINEA, J. & MARTÍNEZ-FRÍAS, J. (eds) *Recursos Minerales de España.* CSIC, Colección Textos Universitarios, **15**, 1129–1250.

ORDÓÑEZ, S. & GARCÍA DEL CURA, M. A. 1994. Deposition and diagenesis of sodium-calcium sulfate salts in the Tertiary saline lakes of the Madrid Basin, Spain. *In*: RENAUT R. W. & LAST, W. M. (eds) *Sedimentology and Geochemistry of Modern and Ancient Saline Lakes.* SEPM, Special Publication, **50**, 229–238.

ORDÓÑEZ, S., CALVO, J. P., GARCÍA DEL CURA, M. A., ALONSO-ZARZA, A. M. & HOYOS, M. 1991. Sedimentology of sodium sulphate deposits and special clays from the Tertiary Madrid Basin (Spain). *In*: ANADÓN, P., CABRERA, L. & KELTS, K. (eds) *Lacustrine Facies Analysis.* IAS, Special Publication, **13**, 39–55.

ORTEGA, E. & GONZÁLEZ LODEIRO, F. 1986. La discordancia intra-Alcudiense en el dominio meridional de la Zona Centroibérica. *Breviora Geológica Astúrica*, **3**, 27–32.

ORTEGA, L. & VINDEL, E. 1995. Evolution of ore-forming fluids associated with late Hercynian antimony deposits in Central/Western Spain: case study of Mari Rosa and El Juncalón. *European Journal of Mineralogy*, **7**, 655–673.

ORTEGA, L., OYARZUN, R. & GALLEGO, M. 1996. The Mari Rosa late Hercynian Sb-Au deposit, western Spain. Geology and geochemistry of the mineralizing processes. *Mineralium Deposita*, **31**, 172–187.

ORTEGA, L., PRICHARD, H. M., LUNAR, R., GARCÍA PALOMERO, F., MORENO, T. & FISHER, P. C. 2000. The Aguablanca discovery. *Mining Magazine*, February, 78–80.

ORTEGA, L. A. 1998. *Estudio petrológico del granito sincinemático de dos micas de A Espenuca (A Coruña).* PhD Thesis, Laboratorio Xeoloxico de Laxe (Serie Nova Terra **14**).

ORTEGA, L. A. & GIL IBARGUCHI, I. 1990. The genesis of late Hercynian granitoids from Galicia (NW Spain): inferences from REE studies. *Journal of Geology*, **98**, 189–211.

ORTEGA, L. A., CARRACEDO, M., LARREA, F. J. & GIL IBARGUCHI, J. I. 1996. Geochemistry and tectonic environment of volcanosedimentary rocks from the Ollo de Sapo Formation (Iberian Massif, Spain). *In*: DEMAIFFE, D. (ed.) *Petrology and Geochemistry of Magmatic Suites of Rocks in the Continental and Oceanic Crusts.* University of Brussels, 277–290.

ORTEGA, L. A., ARANGUREN, A., MENÉNDEZ, M. & GIL IBARGUCHI, J. I. 2000. Petrogénesis, edad y emplazamiento del granito tardi-Hercínico de Veiga (antiforme del Ollo de Sapo, Noroeste de España). *Cadernos do Laboratorio Xeoloxico de Laxe*, **25**, 265–268.

ORTEGA-HUERTAS, M. 1992. Yacimientos de Estroncio en España. *In*: GARCÍA-GUINEA, J. & MARTÍNEZ-FRÍAS, J. (eds) *Recursos Minerales de España.* CSIC, Colección Textos Universitarios, **15**, 429–440.

ORTÍ, F. 1974. El Keuper del Levante Español. Litoestratigrafía, Petrología y Paleogeografía de la cuenca. *Estudios Geológicos*, **30**, 7–46.

ORTÍ, F., 1987. Aspectos sedimentológicos de las evaporitas del Triásico y del Liásico inferior en el Este de la Península Ibérica. *Cuadernos de Geología Ibérica*, **11**, 837–858.

ORTÍ, F. 1990. Introducción a las evaporitas de la Cuenca Terciaria del Ebro. *In*: Ortí, F. & Salvany, J. (eds) *Formaciones evaporíticas de la Cuenca del Ebro y cadenas periféricas y de la zona de Levante.* Enresa, Madrid, 62–66.

ORTÍ, F. 2000. Unidades glaberíticas del Terciario Ibérico: Nuevas aportaciones. *Revista de la. Sociedad Geológica de España*, **13**, 227–249.

ORTÍ, F. & BAYO, A. 1977. Características litoestratigráficas del Triásico Medio y Superior en el Baix Ebre. *Cuadernos de Geología Ibérica*, **4**, 223–238.

ORTÍ, F. & SALVANY, J. 1990. *Formaciones evaporíticas de la Cuenca del Ebro y cadenas periféricas, y de la zona de Levante.* Enresa, Madrid.

ORTÍ, F. & VAQUER, R. 1980. Volcanismo jurásico del sector valenciano de la Cordillera Ibérica. Distribución y trama estructural. *Acta Geológica Hispánica*, **15**(5), 127–130.

ORTÍ, F., GARCÍA-VEIGAS, J., ROSELL, J., JURADO, M. & UTRILLA, R. 1996. Formaciones salinas de las cuencas triásicas de la Península Ibérica: caracterización petrológica y geoquímica. *Cuadernos de Geología Ibérica*, **20**, 13–356.

OSANN, A. 1906. Ueber einige Alkaligesteine aus Spanien. *In*: Wuelfing, E. A. (ed.) *Festschrift H. Rosenbusch Gewidmet von seinen Schülern zum siebzigsten Geburtstag*, 263–310.

OSETE, M. L., FREEMAN, R. & VEGAS, R. 1988. Preliminary palaeomagnetic results from the Subbetic Zone (Betic Cordillera, southern Spain); kinematic and structural implications. *Physics of the Earth and Planetary Interiors*, **57**, 283–300.

OSETE, M. L., FREEMAN, R. & VEGAS, R. 1989. Investigaciones paleomagnéticas en la Zona Subbética. *Cuadernos de Geología Ibérica*, **12**, 39–58.

OSETE, M. L., VILLALAÍN, J. J., OSETE, C. & GIALANELLA, P. R. 2000. Evolución de Iberia durante el Jurásico a partir de datos paleomagnéticos. *Geotemas*, **1**, 116–119.

OTT D'ESTEVOU, P. 1980. *Evolution dynamique du basin néogène de Sorbas (Cordillères bétiques orientales, Espagne).* PhD Thesis, Documents et travaux de l'IGAL, Paris, **1**.

OULEBSIR, L. & PARIS, F. 1995. Chitinozoaires ordoviciens du Sahara algérien: biostratigraphie et affinités paléogéographiques. *Review of Palaeobotany and Palynology*, **86**, 49–68.

OVEJERO, G. & ZAZO, C. 1971. Niveles marinos pleistocenos en Almería (SE de España). *Quaternaria* **XV**, 141–158.

OVTRACHT, A. & TAMAIN, G. 1970. Tectonique en Sierra Morena (España). *Comptes Rendus de l'Académie des Sciences*, **270**, 2634–2636.

OWENS, R. M. & HAMMANN, W. 1990. Proetide trilobites from the Cystoid Limestone (Ashgill) of NW Spain, and the suprageneric classification of related forms. *Paläontologische Zeitschrift*, **64**, 221–244.

OYARZUN, R., MARQUEZ, A, ORTEGA, L., LUNAR, R. & OYARZUN, J. 1995. A late Miocene metallogenic province in southeast Spain: atypical Andean-type processes on a smaller scale. *Transactions of the Institute of Mining and Metallurgy*, **104**, 197–202.

OYARZUN, R., DOBLAS, M., LÓPEZ-RUIZ, J. & CEBRIÁ, J. M. 1997. Opening of the central Atlantic and asymmetric mantle upwelling phenomena: Implications for long-lived magmatism in western North-Africa and Europe. *Geology*, **25**, 727–730.

OYARZUN, R., DOBLAS, M. & LUNAR, R. 1998. Exploration concepts based on contrasting tectonic settings of epithermal precious- and base metal deposits: some Chilean and Spanish examples. *Transactions of the Institute of Mining and Metallurgy*, **107**, 99–108.

PÁEZ, A. & SÁNCHEZ SORIA, P. 1965. Vulcanología del Cabo de Gata, entre San José y Vela Blanca. *Estudios Geológicos*, **21**, 223–246.

PAGÉS, J. 2000. Origen y evolución geomorfológica de las rías atlánticas de Galicia. *Revista de la Sociedad Geológica de España*, **13**(3–4), 393–403.

PALACIOS, T. 1982. *El Cámbrico entre Viniegra de Abajo y Mansilla (Sierra de la Demanda, Logroño). Trilobites e icnofósiles.* Instituto de Estudios Riojanos, Logroño.

PALACIOS, T. 1989. *Microfósiles de pared orgánica del Proterozoico Superior (región central de la Península Ibérica).* Memorias del Museo Paleontológico de la Universidad de Zaragoza, 3(2).

PALACIOS, T. 1993. Acritarchs from the volcanosedimentary group Playon beds. Lower-Upper Cambrian, Sierra Morena, Southern Spain. *Terra Nova*, **6**, 3.

PALACIOS, T. 1997. Acritarcos del Cámbrico superior de Borobia, Soria: implicaciones bioestratigráficas. *In*: GRANDAL D'ANGLADE, A., GUTIÉRREZ-MARCO, J. C. & SANTOS FIDALGO, L. (eds) *XIII Jornadas de Paleontología. Libro de Resúmenes y excursiones.* University of La Coruña, 90–91.

PALACIOS, T. & DELGADO, D. 1999. Acritarch assemblages at the Lower-Middle Cambrian boundary in the Iberian Peninsula and their utility in global correlation. *Journal of Conference Abstracts*, **4**, Annual meeting of IGCP Project 376, 1017.

PALACIOS, T. & MOCZYDLOWSKA, M. 1998. Acritarch biostratigraphy of the Lower-Middle Cambrian boundary in the Iberian Chains, province of Soria, northeastern Spain. *Revista Española de Paleontología*, no. extr. Homenaje al Prof. Gonzalo Vidal, 65–82.

PALACIOS, T. & VIDAL, G. 1992. Lower Cambrian acritarchs from Northern Spain: the Precambrian-Cambrian boundary and biostratigraphic implications. *Geological Magazine*, **129**, 421–436.

PALACIOS, T., GÁMEZ VINTANED, J. A., FERNÁNDEZ-REMOLAR, D. & LIÑÁN, E. 1999a. The Lowermost Cambrian in the Valdelacasa anticline (central Spain): Some new palaeontological data. *Journal of Conference Abstracts*, **4**, Annual meeting of IGCP Project 376, 1017.

PALACIOS, T., GÁMEZ VINTANED, J. A., FERNÁNDEZ-REMOLAR, D. & LIÑÁN, E. 1999b. New palaeontological data at the Vendian-Cambrian transition in central Spain. *Journal of Conference Abstracts*, **4**, 265.

PALAU, J. 1995. *El plutó de Marimanya i el seu encaixant.* PhD Thesis, University of Barcelona.

PALAU, J. 1998. *El magmatisme calcoalcalí del massís de Marimanya i les mineralitzacions As-Au-W associades.* Institut Cartogràfic de Catalunya. Generalitat de Catalunya, Monografíes técniques no. **4**.

PALAU, J. & SANZ, J. 1989. The Devonian units of the Marimanya massif and their relationship with the Pyrenean Devonian facies areas. *Geodinamica Acta*, **3**(2), 171–182.

PALERO, F., BOTH, R A., MARTÍN IZARD, A., MANGAS, J. & REGILON, R. 1992. Metalogénesis de los yacimientos de la region del Valle de Alcudia (Sierra Morena Occidental). *In*: GARCÍA GUINEA, J. & MARTÍNEZ-FRÍAS, J. (eds) *Recursos Minerales de España.* CSIC, Colección Textos Universitarios, **15**, 1027–1069.

PALERO, F. J. 1993. Tectónica pre-hercínica de las series infraordovícicas del anticlinal de Alcudia y la discordancia intraprecámbrica en su parte oriental (Sector meridional de la Zona Centroibérica). *Boletín Geológico y Minero*, **104**, 227–242.

PALLI, L. 1972. *Estratigrafía del Paleógeno del Empordá y zonas limítrofes.* Publicaciones de Geología, Universidad Autónoma de Barcelona, **1**.

PALMER, A. R. & JAMES, N. P. 1980. The Hawke Bay Event: a circum-Iapetus regression near the Lower-Middle Cambrian Boundary. *In*: WONES, D. R. (ed.) *The Caledonides in U.S.A.* Department of Geological Sciences, Virginia Polytechnic Institute and State University, Memoir **2**, 15–18.

PANIAGUA, A., RODRÍGUEZ, S. & GUTIÉRREZ, J. 1987. Mineralizaciones de As-Sb-Au asociadas a rocas ígneas filonianas del NE de León: las minas de Burón. *Boletín de la Sociedad Española de Mineralogía*, **10**, 25–26.

PAPA, F. 1964. *De Bunter tussen Cervera en Brañosera (N. Palencia).* Internal report, Institut of Geology, Leiden University, The Netherlands.

PAQUET, J. 1969. *Etude Géologique de l'Ouest de la province de Murcie (Espagne).* Mémoires de la Société Géologique de la France, **48 (111)**.

PAQUET, J. 1974. Tectonique éocene dans les Cordillères Bétiques: vers une nouvelle conception de la paléogéographie Méditerranée occidentale. *Bulletin de la Société Géologique de France*, **16**, 58–73.

PARDO, A., ORTÍZ, N. & KELLER, G. 1996. Latest Maastrichtian and Cretaceous–Tertiary boundary foraminiferal turnover and environmental changes at Agost, Spain. *In*: MACLEOD, N. & KELLER, G. (eds) *The Cretaceous-Tertiary Mass Extinction: Biotic and Environmental Events.* W. W. Norton, New York, 139–171.

PARDO, G., VILLENA, J. & GONZALEZ, A. 1989. Contribución a los conceptos y a la aplicación del análisis tectosedimentario. Rupturas y unidades tectosedimentarias como fundamento de correlaciones estratigráficas. *Revista de la Sociedad Geológica de España*, **2**, 199–221.

PARDO, M. V. & GARCÍA-ALCALDE J. L. 1984. Biostratigrafía del Devónico de la región de Almadén (Ciudad Real, España). *Trabajos de Geología Universidad Oviedo*, **14**, 79–120.

PARDO ALONSO, M. V. 1998. Update on the Silurian–Devonian transition in the Almadén area, Central-Iberian Zone, Spain. *Temas Geológico-Mineros, ITGE*, **23**, 110–114.

PARDO ALONSO, M. V. 1999a. The southern boundary of the Central-Iberian Zone (Spain), a new proposal: the Castuera-San Benito fault. *XV Reunión Geología Oeste Peninsular (International Meeting on Cadomian Orogens).* Journal of Conference Abstracts, **4**(3), 1017.

PARDO ALONSO, M. V. 1999b. El límite inferior de la laguna estratigráfica intradevónica en Herrera del Duque (Badajoz). *Trabajos Geología Universidad Oviedo*, volumen extraordinario homenaje J. Truyols, **21**, 253–263.

PARDO ALONSO, M. V. & GARCÍA ALCALDE, J. L. 1996. El Devónico de la Zona Centroibérica. *Revista Española de Paleontología*, no. extraordina, 72–81.

PARDO ALONSO, M. V. & GOZALO GUTIÉRREZ, R. 1999. Historia de los estudios paleontológicos en el Devónico de la región de Almadén (Zona Centroibérica, España): periodo 1834–1990. *Revista Española Paleontología*, Madrid, no. extraordina. Homenaje Prof. J. Truyols, 217–228.

PARÉS, J. M. & PÉREZ-GONZÁLEZ, A. 1995. Paleomagnetic age for hominid fossils at Atapuerca Archaeological site, Spain. *Science*, **269**, 830–832.

PARÉS, J. M. & PÉREZ-GONZÁLEZ, A. 1999. Magnetochronology and stratigraphy at Gran Dolina section, Atapuerca (Burgos, Spain). *Journal of Human Evolution*, **37**, 325–342.

PARÈS, J. M. & VAN DER VOO, R. 1992. Paleozoic paleomagnetism of Almadén, Spain: a cautionary note. *Journal of Geophysical Research*, **97**(B6), 9353–9356; **97**(B10), 14245.

PARÉS, J. M., VAN DER VOO, R., STAMAKATOS, J. & PÉREZ ESTAÚN, A. 1994. Remagnetizations and postfolding oroclinal rotations in the Cantabrian/Asturian arc, northern Spain. *Tectonics*, **13**, 1461–1471.

PARÉS, J. M., PÉREZ-GONZÁLEZ, A., WEIL, A. B. & ARSUAGA, J. L. 2000. On the age of the hominid fossils at the Sima de los Huesos, Sierra de Atapuerca, Spain: Paleomagnetic evidence. *American Journal of Physical Anthopology*, **111**, 451–461.

PARGA-PONDAL, I. & GÓMEZ DE LLARENA, J. 1963. Yacimientos fosilíferos en las pizarras metamórficas de Guntín (Lugo, Galicia). *Boletín de la Real Sociedad Española de Historia Natural (Geología)*, **61**, 83–88.

PARGA PONDAL, I., MATTE, P. & CAPDEVILA, R. 1964. Introduction a la geologie de l'Ollo de sapo, formation porphyroide antesilurienne du nord-ouest de l'Espagne. *Notas y Cominicaciones del IGME*, **76**, 119–154.

PARGA PONDAL, I., VEGAS, R., & MARCOS, A. 1983. *Mapa Geológico del Macizo Hespérico – escala 1:500.000 y Memoria*. Publicacions da Area de Xeoloxía e Minería do Seminario do Estudos Galegos, Ediciós do Castro, La Coruña.

PARIS, F. 1981. Les Chitinozoaires dans le Paléozoïque du Sud-Ouest de l'Europe (Cadre géologique-Etude systématique-Biostratigraphie. *Mémoires de la Société Géologique et Minéralogique de Bretagne*, **26**, 1–496.

PARIS, F. 1990. The Ordovician chitinozoan biozones of the Northern Gondwana Domain. *Review of Palaeobotany and Palynology*, **66**, 181–209.

PARIS, F. 1992. Application of chitinozoans in long-distance Ordovician correlations. *In*: WEBBY, B. D. & LAURIE, J. R. (eds) *Global Perspectives on Ordovician Geology*. Balkema, Rotterdam, 23–33.

PARIS, F. 1993. Evolution paléogéographique de l'Europe au Paléozoïque inférieur: le test des Chitinozoaires. *Comptes Rendus de l'Académie des Sciences, Paris*, **316**, sér. II, 273–280.

PARIS, F. 1998. Early Palaeozoic palaeogeography of northern Gondwana regions. *Acta Universitatis Carolinae, Geologica*, **42**, 473–483.

PARIS, F. 1999. Palaeobiodiversification of Ordovician chitinozoans from northern Gondwana. *Acta Universitatis Carolinae, Geologica*, **43**, 283–286.

PARIS, F. & LE POCHAT, G. 1994. The Aquitaine Basin. *In*: KEPPIE, J. D. (ed.) *Pre-Mesozoic Geology in France and Related Areas*. Springer-Verlag, Berlin, 405–415.

PARIS, F. & ROBARDET, M. 1977. Paléogéographie et relations ibéro-armoricaines au Paléozoïque anté-Carbonifère. *Bulletin de la Société Géologique de France* (7), **19**, 1121–1126.

PARIS, F. & ROBARDET, M. 1990. Early Palaeozoic palaeobiogeography of the Variscan regions. *Tectonophysics*, **177**, 193–213.

PARIS, F., ELAOUAD-DEBBAJ, Z., JAGLIN, J. C., MASSA, D. & OULEBSIR, L. 1995. Chitinozoans and Late Ordovician Glacial events on Gondwana. *In*: COOPER, J. D., DROSSER, M. L. & FINNEY, S. C. (eds) *Ordovician Odyssey*. The Pacific Section, Society for Sedimentary Geology, Fullerton, Book **77**, 171–176 (ISBN 1-878861-70-0).

PARIS, F., DEYNOUX, M. & GHIENNE, J. F. 1998. Découverte de Chitinozoaires à la limite Ordovicien-Silurien en Mauritanie; implications paléogéographiques. *Comptes Rendus de l'Académie des Sciences, Paris, Sciences de la terre et des planètes*, **326**, 499–504.

PARIS, F., BOURAHROUH, A. & HÉRISSÉ, A. L. 2000*a*. The effects of the final stages of the Late Ordovician glaciation on marine palynomorphs (chitinozoans, acritarchs, leiospheres) in well Nl-2 (NE Algerian Sahara). *Review of Palaeobotany and Palynology*, **113**, 87–104.

PARIS, F., WINCHESTER-SEETO, T., BOUMENDJEL, K. & GRAHN, Y. 2000*b*. Toward a global biozonation of Devonian chitinozoans. *Courier Forschungsinstitut Senckenberg*, **220**, 39–55.

PARIS, R. & CARRACEDO, J. C. 2001. Formation d'une caldera d'érosion et instabilité récurrente d'une île de point chaud: la Caldera de Taburiente, La Palma, Iles Canaries. *Géomorphologie* 2001(2), 93–105.

PASCAL, A. 1985. *Les systémes biosédimentaires urgoniens (Aptien-Albien) sur la marge nord-Ibérique*. Mémorie Géologique de l'Université de Dijon, **10**.

PASCUAL, E. 1981. *Investigaciones geológicas en el sector Córdoba-Villaviciosa de Córdoba*. PhD Thesis, University of Granada.

PASCUAL, E. & PÉREZ-LORENTE, F. 1987. La alineación o eje magmático de Villaviciosa de Córdoba-La Coronada. *In*: BEA, F.,

CARNICERO, A., GONZALO, J. C., LÓPEZ PLAZA, M. & RODRÍGUEZ ALONSO, M. D. (eds) *Geología de los granitoides y rocas asociadas del Macizo Hespérico. Libro Homenaje a L. C. García Figuerola*. Rueda, Madrid, 365–376.

PASCUAL, E., RUIZ DE ALMODÓVAR, G., SÁEZ, R., TOSCANO, M. & DONAIRE, T. 1994. Petrología y geoquímica de tobas vítreas del área de Aznalcóllar (Faja Pirítica Ibérica). *Boletín de la Sociedad Española de Mineralogía*, **17**, 155–156.

PATAC, I. 1920. La formación uraliense asturiana. Estudios de cuencas carboníferas. *Compañía Asturiana de Artes Gráficas*, 1–5.

PATIÑO DOUCE, A. E. 1995. Experimental generation of hybrid silicic melts by reaction of high–al basalt with metamorphic rocks. *Journal of Geophysical Research*, **100**, 15623–15639.

PATIÑO DOUCE, A. E. 1999. What do experiments tell us about the relative contributions of crust and mantle to the origin of granitic magmas? *In*: CASTRO, A., FERNÁNDEZ, C. & VIGNERESSE, J. L. (eds) *Understanding Granites. Integrating New and Classical Techniques*. Geological Society, London, 55–75.

PAULSEN, H. & VISSER, J. 1993. The crustal structure in Iberia inferred from P-wave coda. *In*: Mezcua & Carreño (eds) *Iberian Lithosphere Heterogeneity and Anisotropy ILIHA*. IGN, Serie Monografía, **10**, 3–18.

PAYROS, A., PUJALTE, V., ORUE-ETXEBARRIA, X. & BACETA, J. I. 1997. Un sistema turbidítico Bartoniense de tipo 'channel-levee' en la Cuenca de Pamplona: implicaciones tectónicas y paleogeográficas. *Geogaceta*, **22**, 145–148

PAYROS, A., PUJALTE, V. & ORUE-ETXEBARRIA, X. 1999. The South Pyrenean Eocene Carbonate Megabreccias revisited: new interpretation based on evidence from the Pamplona Basin. *Sedimentary Geology*, **125**, 165–194.

PAYROS, A., ASTIBIA, H., CEARRETA, A., PEREDA-SUBERBIOLA, X., MURELAGA, X. & BADIOLA, A. 2000*a*. The Upper Eocene South Pyrenean coastal deposits (Liedena Sandstone, Navarre): sedimentary facies, benthic forainifera and avian ichnology. *Facies*, **42**, 107–132.

PAYROS, A., PUJALTE, V. BACETA, J. I. *et al.* 2000*b*. Lithostratigraphy and sequence stratigraphy of the upper Thanetian to middle Ilerdian strata of the Campo section (southern Pyrenees, Spain): Revision and new data. *Revista de la Sociedad Geológica de España*, **13**, 213–226.

PEDLEY, H. M., ANDREWS, J. E., ORDOÑEZ, S., GONZÁLEZ MARTÍN, J. A., GARCÍA DEL CURA, M. A. & TAYLOR, D. M. 1996. Climatically controled fabrics in freshwater carbonates: a comparative study of barrage tufas from Spain and Britain. *Palaeo*, **121**, 239–257.

PEDRAZA, J. 1994. El Sistema Central. *In*: GUTIÉRREZ M. (ed.) *Geomorfología de España*. Rueda, Madrid, 63–100.

PEDRO, J. C., FONSECA, P., LEAL, N. & MUNHÁ, J. 1995. Estudo petrológico e estrutural do evento tectono-metamórfico varisco de alta pressão no sector de Safira-Santiago do Escoural (SW da Zona de Ossa-Morena). *Memórias do Museo e Laboratorio Mineralogico e Geología, Universidade do Porto*, **4**, 781–786.

PELÁEZ, J. R., GARCÍA HIDALGO, J. F., HERRANZ, P., PIEREN, A. P., VILAS, L. & SAN JOSÉ, M. A. 1989. Upper Proterozoic in central Spain. *Abstracts 28 International Geological Congress, Washington*, **2**, 590–591.

PELÁEZ-CAMPOMANES, P., DE LA PEÑA, A. & LÓPEZ MARTÍNEZ, N. 1989. Primeras faunas de micromamíferos del Paleógeno de la Cuenca del Duero. *Studia Geológica Salmanticensia, Ediciones Universidad Salamanca*, **5**, 135–157.

PELLICER, F. 1980. El periglaciarismo del Moncayo. *Geographicalia*, **7/8**, 3–25.

PEÑA, J. A. 1979. *La depresión de Guadix-Baza. Estratigrafía del Plioceno-Pleistoceno*. PhD Thesis, University of Granada.

PEÑA, J. A. 1985. La Depresión de Guadix-Baza. *Estudios Geológicos*, **41**, 33–46.

PEÑA, J. A. & MARFÍL, R. 1986. La sedimentación salina actual en las lagunas de La Mancha: una síntesis. *Cuadernos de Geología Ibérica*, **10**, 235–270.

PEÑA, J. L. 1983. *La Conca de Tremp y las Sierras Prepirenaicas comprendidas entre los ríos Segre y Noguera Ribagorzana*. Instituto de Estudios Ilerdenses, Lérida.

PEÑA, J. L. 1994. Cordillera Pirenaica. *In*: GUTIÉRREZ ELORZA, M. (ed.) *Geomorfología de España*, Rueda, Madrid, 159–225.

PEÑA, J. L., GUTIÉRREZ ELORZA, M., IBÁÑEZ, M. et al. 1984. Geomorfología de la provincia de Teruel. Publ. Instituto de Estudios Turolenses, Teruel.

PEÑA, J. L., ECHEVERRÍA, M. T., PETIT-MAIRE, N. & LAFONT, R. 1993. Cronología e interpretación de las acumulaciones holocenas de la Val de las Lenas (Depresión del Ebro, Zaragoza). Geographicalia, 30, 321–332.

PENDÓN, J. G. 1978. Sedimentación turbidítica en las Unidades del Campo de Gibraltar. PhD Thesis, University of Granada.

PENHA, H. M. & ARRIBAS, A. 1974. Datación gecronológica de algunos granitos uraníferos españoles. Boletin Geológico y Minero, 85, 271–273.

PERDIGAO, J. C. 1967. Descoberta de Mesodevónico em Portugal (Portalegre). Comunicaçoes Serviços Geológicos Portugal, 52, 27–48.

PERDIGAO, J. C. 1973. A fauna dos gres e quartzitos silúrico-devónicos de Portalegre e a sua posiçao estratigráfica. Comunicaçoes Serviços Geológicos Portugal, 56, 5–32.

PERDIGAO, J. C. 1974. O Devónico de Portalegre. Comunicaçoes Serviços Geológicos Portugal, 57, 203–228.

PERDIGAO, J. C. 1979. O Devónico de Dornes (Paleontologia e estratigrafia). Comunicaçoes Serviços Geológicos Portugal, 65, 193–199.

PERDIGAO, J., OLIVEIRA, J. T. & RIBEIRO, A. 1982. Noticia explicativa da folha 44-B (Barrancos). Serviços Geológicos de Portugal, Lisboa, LX.

PEREIRA, M. D., RONKIN, Y. & BEA, F. 1992. Dataciones Rb/Sr en el Complejo Anatéctico de la Peña Negra (Batolito de Avila, España central): Evidencias de magmatismo Pre-Hercínico. Revista de la Sociedad Geológica de España, 5, 129–134.

PEREIRA, Z. 1999. Palinoestratigrafia do Sector Sudoeste da Zona Sul Portuguesa. Comunicações do Instituto Geológico e Mineiro de Portugal, 86, 1, 25–57.

PEREIRA, Z., SÁEZ, R., PONS, J. M., OLIVEIRA, J. T. & MORENO, C. 1996. Edad devónica (Estruniense) de las mineralizaciones de Aznalcóllar (Faja Pirítica Ibérica) en base a palinología. Geogaceta, 20, 1609–1612.

PEREIRA, Z., MEIRELES, C. & PEREIRA, E. 1999. Upper Devonian palynomorphs of NE sector of Tras-os-Montes (Central-Iberian Zone). XV Reunión Geología Oeste Peninsular (International Meeting on Cadomian Orogens). Journal of Conference Abstracts, 4(3), 1018.

PEREJÓN, A. 1972. Primer descubrimiento y descripción de Arqueociatos en la provincia de Salamanca. Studia Geologica, 4, 143–149.

PEREJÓN, A. 1973. Contribución al conocimiento de los Arqueociatidos de los yacimientos de Alconera (Badajoz). Estudios Geológicos, 29, 179–206.

PEREJÓN, A. 1975a. Arqueociatos de lo subórdenes Monocyathina y Dokidocyathina. Boletín de la Real Sociedad Española de Historia Natural (Geología), 73, 125–145.

PEREJÓN, A. 1975b. Arqueociatos Regulares del Cámbrico inferior de Sierra Morena (SW de España). Boletín de la Real Sociedad Española de Historia Natural (Geología), 73, 147–193.

PEREJÓN, A. 1976. Nuevos datos sobre los Arqueociatos de Sierra Morena. Estudios Geológicos, 32, 5–33.

PEREJÓN, A. 1986. Bioestratigrafía de los Arqueociatos en España. Cuadernos de Geología Ibérica, 9 (year 1984), 213–316.

PEREJÓN, A. 1989. Arqueociatos del Ovetiense en la sección del Arroyo Pedroche, Sierra de Córdoba, España. Boletín de la Real Sociedad Española de Historia Natural (Geología), 84, 143–247.

PEREJÓN, A. 1994. Palaeogeographic and biostratigraphic distribution of Archaeocyatha in Spain. Courier Forschungsinstitut Senckenberg, 172, 341–354.

PEREJÓN, A., MORENO-EIRIS, E. & ABAD, A. 1994. Montículos de arqueociatos y calcimicrobios del Cámbrico inferior de Terrades, Gerona (Pirineo oriental, España). Boletín de la Real Sociedad Española de Historia Natural (Geología), 89, 55–95.

PEREJÓN, A., FRÖHLER, M., BECHSTÄDT, T., MORENO-EIRIS, E. & BONI, M. 2000. Archaeocyathan assemblages from the Gonnesa Group, Lower Cambrian (Sardinia Italy) and their sedimentologic context. Bolletino della Società Paleontologica Italiana, 39, 257–291.

PÉREZ, A. 1989. Estratigrafía y sedimentología del Terciario del borde meridional de la Depresión del Ebro (sector riojano y aragonés) y cubetas de Muniesa y Montalbán. PhD Thesis, University of Zaragoza.

PÉREZ, L. C. & SANCHEZ, F. J. 1992. Los yacimientos de oro de las medulas de Carucedo (Leon). In: GARCÍA GUINEA, J. & MARTÍNEZ-FRÍAS, J. (eds) Recursos Minerales de España. CSIC, Colección Textos Universitarios, 15, 861–873.

PÉREZ, P. J. & GRACIA, A. 1998. Los homínidos de Atapuerca: información sobre modos de vida a partir de datos paleoepidemiológicos. In: AGUIRRE, E. (ed.) Atapuerca y la evolución humana. Fundación Ramón Areces, Madrid, 333–360.

PÉREZ, P. J., GRACIA, A., MARTÍNEZ, I. & ARSUAGA, J. L. 1997. Paleopathological evidence of the cranial remains of the Sima de los Huesos Middle Pleistocene site (Sierra de Atapuerca, Spain). Description and preliminary inferences. Journal of Human Evolution, 33, 409–421.

PÉREZ-ARLUCEA, M. 1985. Estratigrafía y sedimentología del Pérmico y Triásico en el sector Molina de Aragón-Albarracín (provincias de Guadalajara y Teruel). PhD Thesis, Complutense University, Madrid.

PÉREZ-ARLUCEA, M. & SOPEÑA, A. 1985. Estratigrafía del Pérmico y Triásico en el sector central de la Rama Castellana de la Cordillera Ibérica (provincias de Guadalajara y Teruel). Estudios Geológicos, 41, 207–222.

PÉREZ-ARLUCEA, M & TRIFONOVA, E. 1993. Stratigraphy of the Middle Triassic in part of the Iberian Ranges (Spain) based on foraminifera data. Geologica Balcanica, 23, 23–33.

PÉREZ ESTAÚN, A. 1973. Datos sobre la sucesión estratigráfica del Precámbrico y la estructura del extremo sur del Antiforme del Narcea (NW de España). Breviora Geológica Astúrica, 17, 5–16.

PÉREZ ESTAÚN, A. 1974a. Aportaciones al conocimiento del Carbonífero de San Clodio (Prov. de Lugo). Breviora Geologica Asturica, 18, 3–8.

PÉREZ-ESTAÚN, A. 1974b. Algunas precisiones sobre la sucesión ordovícica y silúrica de la región de Truchas. Breviora Geologica Asturica, 18, 23–25.

PÉREZ ESTAÚN, A. 1978. Estratigrafía y estructura de la rama Sur de la Zona Asturoccidental-Leonesa. Memorias del IGME, 92.

PÉREZ ESTAÚN, A. & BASTIDA, F. 1990. Structure: Cantabrian Zone. In: DALLMEYER, R. D. & MARTÍNEZ GARCÍA, E. (eds) Pre-Mesozoic Geology of Iberia. Springer-Verlag, Berlin, 55–69.

PÉREZ-ESTAÚN, A. & MARCOS, A. 1981. La Formación Agüeira en el sinclinorio de Vega de Espinareda: aproximación al modelo de sedimentación durante el Ordovícico superior en la zona Asturoccidental-leonesa (NW de España). Trabajos de Geología, Oviedo, 11, 135–145.

PÉREZ ESTAÚN, A. & MARTÍNEZ, F. J. 1978. El Precámbrico del Antiforme del Narcea en el sector de Tineo-Cangas de Narcea (NW de España). Trabajos de Geología, 10, 367–387.

PÉREZ-ESTAÚN, A., BASTIDA, F., ALONSO, J. L. et al. 1988. A thin-skinned tectonics model for an arcuate fold and thrust belt: The Cantabrian Zone (Variscan Ibero-Armorican Arc). Tectonics, 7, 517–537.

PÉREZ-ESTAÚN, A., BASTIDA, F., MARTÍNEZ CATALÁN, J. R., GUTIÉRREZ MARCO, J. C., MARCOS, A. & PULGAR, J. A. 1990. West Asturian-Leonese Zone. Stratigraphy. In: DALLMEYER, R. D. & MARTÍNEZ GARCÍA, E. (eds) Pre-Mesozoic Geology of Iberia. Springer-Verlag, Berlin, Heidelberg, 92–102.

PÉREZ ESTAÚN, A., MARTÍNEZ CATALÁN, J. R. & BASTIDA, F. 1991. Crustal thickening and deformation sequence in the footwall to the suture of the Variscan belt of northwest Spain. Tectonophysics, 191, 243–253.

PÉREZ-ESTAÚN, A., MARCOS, A., MARTÍNEZ CATALÁN, J. R., BASTIDA, F. & PULGAR, J. A., 1992. Estratigrafía de la Zona Asturoccidental-Leonesa. In: GUTIÉRREZ-MARCO, J. C., SAAVEDRA, J. & RÁBANO, I. (eds) Paleozoico Inferior de Ibero-América, University of Extremadura, Badajoz, 453–461.

PÉREZ ESTAÚN, A., PULGAR, J. A., BANDA, E., ALVAREZ MARRÓN, J. & ESCI-N RESEARCH GROUP. 1994. Crustal structure of the external variscides in northwest Spain from deep seismic reflection profiling. Tectonophysics, 232, 91–118.

PÉREZ-ESTAÚN, A., PULGAR, J. A., ALVAREZ-MARRÓN, J. & ESCI-N

GROUP. 1995. Crustal structure of the Cantabrian Zone: seismic image of a Variscan foreland thrust and fold belt (NW Spain). *Revista de la Sociedad Geológica de España*, **8**, 307–321.

PÉREZ-GARCÍA, A., ROBLES, S. & VICENTE BRAVO, J. C. 1993. Modelo genético de las secuencias arenosas de plataforma de la Formación Balmaseda (Albiense de la Cuenca Vascocantábrica), N de España. *Geogaceta*, **14**, 76–79.

PÉREZ-GARCÍA, A., ROBLES, S. & VICENTE BRAVO, J. C. 1997. Arquitectura estratigráfica del sistema de plataforma dominada por tormentas de la Fm. Valmaseda (Albiense sup.-Cenomaniense inf. de la cuenca Vascocantábrica). *Geogaceta*, **22**, 153–156.

PÉREZ-GONZÁLEZ, A. 1982. *Neógeno y Cuaternario de la llanura manchega y sus relaciones con la Cuenca del Tajo*. PhD Thesis, Complutense University, Madrid.

PÉREZ-GONZÁLEZ, A. & GALLARDO, J. 1987. La Raña al sur de Somosierra y Sierra Ayllón: un piedemonte escalonado del Villafranquiense medio. *Geogaceta*, **2**, 29–32.

PÉREZ-GONZÁLEZ, A., CABRA, P. & MARTÍN-SERRANO, A. 1989. *Mapa del Cuaternario de España*. ITGE, Madrid.

PÉREZ GONZÁLEZ, A., MARTÍN-SERRANO, A. & POL, C. 1994. Depresión del Duero. *In*: GUTIÉRREZ ELORZA, M. (ed.) *Geomorfología de España*. Rueda, Madrid, 351–388.

PÉREZ-LÓPEZ, A. 1991. *El Trías de facies germánica del sector Central de la Cordillera Bética*. PhD Thesis, Granada University.

PÉREZ-LÓPEZ, A. 1996. Sequence model for coastal-plain depositional systems of Upper Triassic (Betic Cordillera, Southern Spain). *Sedimentary Geology*, **101**, 99–117.

PÉREZ-LÓPEZ, A. 1998. Epicontinental Triassic of the southern Iberian Continental Margin (Betic Cordillera). *Zentralblatt für Geologie und Paläontologie*, **I**, 9–10, 1009–1031.

PÉREZ-LÓPEZ, A. & SANZ DE GALDEANO, C. 1994. Tectónica de los materiales triásicos en el sector central de la Zona Subbética (Cordillera Bética). *Revista Sociedad Geológica de España*, **7**, 141–153.

PÉREZ-LÓPEZ, A., SOLÉ DE PORTA, N., MÁRQUEZ SANZ, L. & MÁRQUEZ-ALIAGA, A. 1992. Caracterización y datación de una unidad carbonática de edad Noriense (Fm. Zamoranos) en el Trías de la Zona Subbética. *Revista Sociedad Geológica de España*, **5**, 113–127.

PÉREZ LORENTE, F. 1979. *Geología de la Zona Ossa Morena al norte de Córdoba (Pozoblanco-Bélmez-Villaviciosa de Córdoba)*. PhD Thesis, University of Granada.

PÉREZ-MORENO, B. P., SANZ, J. L., BUSCALIONI, A. D., MORATALLA, J. J., ORTEGA, F. & RASSKIN-GUTMAN, D. 1994. A unique multitoothed ornithomimosaur from the Lower Cretaceous of Spain. *Nature*, **370**, 363–367.

PÉREZ-OBIOL, R. & JULIÀ, R. 1994. Climatic change on the Iberian Peninsula recorded in a 30.000 Yr pollen record from Lake Banyoles. *Quaternary Research*, **41**, 91–98.

PÉREZ TORRADO, F. J. 2000. *Volcanoestratigrafía del Grupo Roque Nublo, Gran Canaria*. Univ. Las Palmas de Gran Canaria. Cabildo, Gran Canaria.

PÉREZ TORRADO, F. J., CARRACEDO, J. C. & MANGAS, J. 1995. Geochronology and stratigraphy of the Roque Nublo Cycle, Gran Canaria, Canary Islands. *Journal of the Geological Society of London*, **152**, 807–818.

PÉREZ TORRADO, F. J., MARTÍ, J., MANGAS, J. & DAY, S. J. 1997. Ignimbrites of the Roque Nublo group, Gran Canaria, Canary Islands. *Bulletin Volcanologique*, **58**, 647–654.

PÉREZ TORRADO, F. J., SCHNEIDER, J. L., GIMENO, D., WASSMER, P. & CABRERA, M. C. 2000. Mecanismos de transporte y emplazamiento de depósitos volcanoclásticos en el litoral NE de Gran Canaria (Islas Canarias). *Geotemas*, **1**(3), 329–333.

PÉREZ-VALERA, F., SOLÉ DE PORTA, N. & PÉREZ-LÓPEZ, A. 2000. Presencia de facies Buntsandstein (Anisiense-Ladiniense?) en el Triásico de Calasparra, Murcia. *Geotemas*, **1**, 209–212.

PERILLI, N. 1999. Calcareous nannofossil biostratigraphy of Toarcian-Aalenian transition at Fuentelsalz section (Iberian Range, East Spain). *Cuadernos de Geología Ibérica*, **25**, 189–212.

PERMANYER, A., VALLES, D. & DORRONSORO, C. 1988. Source rock potential of an Eocene Carbonate slope: the Armancies Formation of the South Pyrenean Basin, Northeast Spain. *AAPG Bulletin*, **72/8**, 1019.

PERRET, M. F. 1993. Recherches micropaléontologiques et biostratigrafiques (conodontes-foraminifères) dans le Carbonifère Pyreneen. *Strata*, **2**(21), 1–597.

PERROUD, H. 1982. Contribution à l'étude paléomagnétique de l'Arc ibéro-armoricain. *Bulletin de la Société géologique et minéralogique de Bretagne [C]*, **14**, 1–114.

PERROUD, H. 1983. Palaeomagnetism of Palaeozoic rocks from the Cabo de Peñas, Asturia, Spain. *Geophysical Journal of the Royal Astronomical Society*, **75**, 201–215.

PERROUD, H. & BONHOMMET, N. 1981. Palaeomagnetism of the Ibero-Armorican arc and the Hercynian orogeny in Western Europe. *Nature*, **292**(5822), 445–448.

PERROUD, H. & BONHOMMET, N. 1984. A Devonian palaeomagnetic pole for Armorica. *Geophysical Journal of the Royal Astronomical Society*, **77**, 839–845.

PERROUD, H., CALZA, F. & KHATTACH, D. 1991. Paleomagnetism of the Silurian volcanism at Almadén, southern Spain. *Journal of Geophysical Research*, **96**, 1949–1962.

PESQUERA, A. 1985. *Contribución a la mineralogía, petrología y metalogenia del macizo paleozoico de Cinco Villas (Pirineos Vascos)*. PhD Thesis, University of Pais Vasco.

PETFORD, N., CRUDEN, A. R., MCCAFFREY, K. J. W. & VIGNERESSE, J. L. 2000. Granite magma formation, transport and emplacement in the Earth's crust. *Nature*, **408**, 669–673.

PEUCAT, J. J., BERNARD-GRIFFITHS, J., GIL IBARGUCHI, J. I., DALLMEYER, R. D., MENOT, R. P., CORNICHET, J. & IGLESIAS PONCE DE LEÓN, M. 1990. Geochemical and geochronological cross section of the deep variscan crust: the Cabo Ortegal high-pressure nappe (NW Spain). *In*: MATTE, P. (ed.) *Terranes in the Variscan Belt of Europe and Circum-Atlantic Paleozoic Orogens*. *Tectonophysics*, **177**, 263–292.

PEYBERNES, B. 1976. *Le Jurassique et le Crétacé Inférieur des Pyrénées Franco-Espagnoles entre la Garonne et la Méditerranée*. PhD Thesis, University of Toulouse.

PEYBERNES, B. 1991. Les séquences de dépôt du Dogger des Pyrénées Centrales et Orientales Franco-Espagnoles. *Comptes Rendus de l'Académie des Sciences de Paris*, **313**(II), 209–214.

PEYBERNÈS, B. & PELISSIÉ, T. 1985. Essai de reconstruction de la paléogéographie des dépôts contemporains de la fin du rifting téthysien avant la transgression bathonienne sur le 'Haut-fond Occitan' (SW de la France). *Comptes Rendus de l'Académie des Sciences de Paris*, **301**(II), 533–538.

PEYRE, Y. 1974. *Géologie d'Antequera et de sa région (Cordillères Bétiques, Espagne)*. PhD Thesis, University of Paris.

PFEFFERKORN, H. 1968. *Geologie des Gebietes zwischen Serpa und Mértola (Baixo Alentejo, Portugal)*. Münster Forschungen für Geologie und Paläontologie, **9**.

PHILIP, J. 1983. Le Campanien et le Maastrichtien á rudistes et grands Foraminifères de Quatretonda (Province de Valence, Espagne): une clef pour la biozonation et les correlations stratigraphiques dans le domaine mésogéen. *Géologie Mediterranéenne*, **3–4**, 87–98.

PHILIP, J. 1985. Sur les relations des marges téthysiennes au Campanien et au Maastrichtien déduites de la distribution des rudistes. *Bulletin de la Société Géologique de France*, **8**(1,5), 723–732.

PIÇARRA J. M., GUTIÉRREZ-MARCO J. C., OLIVEIRA J. T., ROBARDET M. & JAEGER, H. 1992. Bioestratigrafia do Silurico da Zona de Ossa Morena (Portugal-Espanha): revisão critica dos dados existentes. *Publicaciones del Museo de Geología de Extremadura*, **1**, 118–119.

PIÇARRA, J. M., GUTIÉRREZ-MARCO, J. C., LENZ, A. C. & ROBARDET, M. 1998. Pridoli graptolites from the Iberian Peninsula: a review of previous data and new records. *Canadian Journal of Earth Sciences*, **35**, 65–75.

PICKERILL, R. K., ROMANO, M. & MELÉNDEZ, B. 1984. Arenig trace fossils from the Salamanca area, western Spain. *Geological Journal*, **19**, 249–269.

PIEREN, A., ARECES, J., TORAÑO, J. & MARTÍNEZ-GARCÍA, E. 1995. Estratigrafía y estructura de los materiales permotriásicos del sector Gijón-La Camocha (Asturias). *Cuadernos de Geología Ibérica*, **19**, 309–335.

PIEREN, A. P., HERRANZ ARAÚJO, P. & GARCÍA GIL, S. 1991. Evolución de los depósitos continentales del proterozoico superior en 'La Serena', Badajoz (Zona Centro-Ibérica). *Cuaderno do Laboratorio Xeolóxico de Laxe*, **16**, 179–191.

PIEREN-PIDAL, A. & GUTIÉRREZ-MARCO, J. C. 1990. Datos bioestratigráficos de los materiales silúricos del sinclinal de Herrera del Duque (Badajoz). *Geogaceta*, **8**, 58–61.

PILLOLA, G. L. 1998. The 'Sardic Unconformity'. *In*: SERPAGLI, E. (ed.) *Sardinia Guide-book, ECOS VII. Giornale di Geología [3]*, **60**, 175.

PILLOLA, G. L. & GUTIÉRREZ-MARCO, J. C. 1988. Graptolites du Tremadoc du Sud-ouest de la Sardaigne (Italie): Paléoécologie et contexte tecto-sédimentaire. *Geobios*, **21**, 553–565.

PILLOLA, G. L., GÁMEZ-VINTANED, J. A., DABARD, M. P., LEONE, F., LIÑÁN, E. & CHAUVEL, J.-J. 1994. The Lower Cambrian ichnospecies *Astropolichnus hispanicus*: palaeoenvironment and palaeogeographical significance. *Bolletino Societa Paleontologica Italiana*, special volume **2**, 253–267.

PILLOLA, G. L., LEONE, F. & LOI, A. 1995. The Lower Cambrian Nebida Group of Sardinia. *Rediconti del Seminario della Facoltà di Scienze dell'Universtà di Cagliari*, supplementary volume **65**, 27–60.

PIN, C. 1989. *Essai sur la chronologie et l'évolution géodynamique de la chaîne hercinienne d'Europe*. PhD Thesis, Blaise Pascal University, Clermont Ferrant.

PIN, C., ORTEGA, L. A. & GIL IBARGUCHI, J. I. 1992. Mantle-derived, early Paleozoic A-type metagranitoids from the NW Iberian massif: Nd isotope and trace-element constraints. *Bulletin de la Société Géologique de France*, **163**, 483–494.

PIN, C., LIÑÁN, E., PASCUAL, E., DONAIRE, T. & VALENZUELA, A. 1999. Late Proterozoic crustal growth in Ossa Morena: Nd isotope and trace element evidence from the Sierra de Córdoba volcanics. *International Meeting Cadomian Orogens Annual Meeting of IGCP Project 376. Journal of Conference Abstracts*, **4**, 1019.

PINARELLI, L. & ROTTURA, A. 1995. Sr and Nd isotopic study and Rb-Sr geochronology of the Bejar granites, Iberian Massif, Spain. *European Journal of Mineralogy*, **7**, 577–589.

PINEDA, A. 1984. Las mineralizaciones metálicas y su contexto geológico en el area volcánica neógena de Cabo de Gata (Almería, SE de España). *Boletín Geológico y Minero*, **95–96**, 569–592

PINEDA, A. 1987. La Caliza Urbana (Ordovícico Superior) y sus tramos volcanoclásticos en el subsuelo del norte de El Centenillo (Jaén). *Boletín Geológico y Minero*, **98**, 780–793.

PINEDO VARA, I. 1963. *Piritas de Huelva*. Edito. SUMMA, Madrid.

PINET, B., MONTARDET, L. & ECORS SCIENTIFIC PARTY 1987. Deep seismic reflection and refraction profiling along the Aquitaine shelf (Bay of Biscay). *Geophysical Journal of the Royal Astronomical Society*, **89**, 305–312.

PINTO, M. S. 1983. Geochronology of portuguese granitoids: a contribution. *Studia Geológica Salmanticensia*, **18**, 277–306.

PINTO, M. S., CASQUET, C., IBARROLA, E., CORRETGÉ, L. G. & FERREIRA, M. 1987. Síntesis geochronológica dos granitóides do Maciço Hespérico. *In*: BEA, F., CARNICERO, E., GONZALO, J. C., LÓPEZ PLAZA, M. & RODRÍGUEZ, M. D. (eds) *Geología de los granitoides y rocas asociadas del Macizo Hespérico. Libro homenaje a L. C. García de Figuerola*. Rueda, Madrid, 69–86.

PITCHER, W. S. 1997. *The Nature and Origin of Granite*. Chapman & Hall, Glasgow.

PITCHER, W. S. & BERGER, A. R. 1972. *The Geology of Donegal: A Study of Granite Emplacement and Unroofing*. John Wiley, Chichester.

PLATT, J. P. & BEHRMANN, J. H. 1986. Structures and fabrics in a crustal-scale shear zone, Betic Cordillera, SE Spain. *Journal of Structural Geology*, **8**, 15–33.

PLATT, J. P. & VISSERS, R. L. M. 1980. Extensional structures in anisotropic rocks. *Journal of Structural Geology*, **2**, 379–410.

PLATT, J. P. & VISSERS, R. L. M. 1989. Extensional collapse of thickened continental lithosphere: A working hypothesis for the Alboran Sea and Gibraltar arc. *Geology*, **17**, 540–543.

PLATT, J. P. & WHITEHOUSE M. J. 1999. Early Miocene high-temperature metamorphism and rapid exhumation in the Betic Cordillera (Spain): evidence from U-Pb zircon ages. *Earth and Planetary Science Letters*, **171**, 591–605.

PLATT, J. P., SOTO, J. I., WHITEHOUSE, M. J., HURFORD, A. J. & KELLEY, S. P. 1998. Thermal evolution, rate of exhumation, and tectonic significance of metamorphic rocks from the floor of the Alborán extensional basin, western Mediterranean. *Tectonics*, **17**, 671–689.

PLATT, N. H. 1990. Basin evolution and fault reactivation in the Wester Cameros Basin, Northern Spain. *Journal of the Geological Society of London*, **147**, 165–175.

PLATZMAN, E. S. 1992. Paleomagnetic rotations and the kinematics of the Gibraltar Arc. *Geology*, **20**, 311–314.

PLATZMAN, E. S. 1994. East-west thrusting and anomalous magnetic declinations in the Sierra Gorda, Betic Cordillera, southern Spain. *Journal of Structural Geology*, **16**, 11–20.

PLATZMAN, E., PLATT, J. P., KELLEY, S. P. & ALLERTON, S. 2000. Large clockwise rotations in an extensional allochthon, Alborán Domain (southern Spain). *Journal of the Geological Society, London*, **157**, 1187–1197.

PLAYÀ, E., ORTÍ, F. & ROSELL, L. 2000. Marine to non-marine sedimentation in the upper Miocene evaporites of the Eastern Betics, SE Spain: sedimentological and geochemical evidence. *Sedimentary Geology*, **133**, 135–166.

PLAZIAT J. C. 1981. Late Cretaceous to late Eocene paleogeography evolution of southwest Europe. *Palaeogeography, Palaeoclimatology, Palaeoecology*, **36**, 263–320.

PLUSQUELLEC, Y. 1987. Révision de *Michelinia transitoria* Knod, 1908 (Tabulata, Dévonien de Bolivie). *Annales Société Géologique Nord*, **105**, 249–252.

POBLET, J., MUÑOZ, J. A., TRAVE, A. & SERRA-KIEL, J. 1998. Quantifying the kinematics of detachment folds using the three-dimensional geometry: application to the Mediano anticline (Pyrenees, Spain). *Geological Society of America Bulletin*, **110**, 111–125.

POCOVI, A. 1978a. Estudio geológico de las Sierras Marginales catalanas. *Acta Geológica Hispánica*, **13**, 73–79.

POCOVÍ, A. 1978b. *Estudio geológico de las Sierras Marginales Catalanas (Prepirineo de Lérida)*. PhD Thesis, University of Barcelona.

POL, C. & CARBALLEIRA, J. 1982. Las facies conglomeráticas terciarias de la región de Covarrubias (Burgos). *Temas Geológico Mineros, IGME*, **6(II)**, 509–525.

POL, C. & CARBALLEIRA, J. 1986. El sinclinal de Santo Domingo de Silos: estratigrafía y paleogeografía de los sedimentos continentales (borde este de la Cuenca del Duero). *Studia Geológica Salmanticensia, Ediciones Universidad Salamanca*, **22**, 7–36.

POL, C., BUSCALIONI, A. D., CARBALLEIRA, J. *et al.* 1992. Reptiles and mammals from the Late Cretaceous new locality Quintanilla del Coco (Burgos province, Spain). *Neues Jahrbuch für Geologie und Palaontologie. Abhandlungen*, **184**(3), 279–314.

POMAR, L. 1976. Tectónica de gravedad en los depósitos Mesozoicos, Paleógenos y Neógenos de Mallorca (España). *Boletín de la Sociedad de Historia Natural de Baleares*, **21**, 159–175.

POMAR, L. 1979. La evolución tectosedimentaria de las Baleares: análisis crítico. *Acta Geológica Hispànica*, **14**, 293–310.

POMAR, L. 1991. Reef geometries, erosion surfaces and high-frequency sealevel changes, Upper Miocene Reef Complex, Mallorca, Spain. *Sedimentology*, **38**, 243–269.

POMAR, L. 1993. High-resolution sequence stratigraphy in prograding Miocene carbonates: application to seismic interpretation. *In*: LOUCKS, R. G. & SARG, J. F. (eds) *Carbonate Sequence Stratigraphy, Recent Developments and Applications*. AAPG, Memoirs, **57**, 389–407.

POMAR, L. 2001. Ecological control of sedimentary accommodation: evolution from a carbonate ramp to rimmed shelf, Upper Miocene, Balearic Islands, *Palaeogeography, Palaeoclimatology, Palaeoecology*, **175**, 249–272.

POMAR, L. & WARD, W. C. 1994. Response of a Miocene carbonate platform to high-frequency eustasy. *Geology*, **22**, 131–134.

POMAR, L. & WARD, W. C. 1995. Sea-level changes, carbonate production and platform architecture: The Llucmajor Platform, Mallorca, Spain. *In*: HAQ, B. U. (ed.) *Sequence Stratigraphy and Depositional Response to Eustatic, Tectonic and Climatic Forcing*. Kluwer, Dordrecht, 87–112.

POMAR, L. & WARD, W. C. 1999. Reservoir-scale heterogeneity in depositional packages and diagenetic patterns on a reef-rimmed platform, Upper Miocene, Mallorca, Spain. *AAPG Bulletin*, **83**, 1759–1773.

POMAR, L., RODRÍGUEZ-PEREA, A., SÀBAT, F. & FORNÓS, J. J. 1990. Neogene stratigraphy of Mallorca Island. *Paleontologia i Evolució, Mem. Espec.*, **2**, 271–320.

POMAR, L., WARD, W. C. & GREEN, D. G. 1996. Upper Miocene reef complex of the Llucmajor area, Mallorca, Spain. *In*: FRANSEEN, E., ESTEBAN, M., WARD, W. C. & ROUCKY, J. M. (eds) *Models for carbonate stratigraphy from Miocene Reef complexes of the Mediterranean regions*. SEPM, Concepts in Sedimentology and Paleontology, **5**, 191–225.

PONS, A. & REILLE, M. 1988. The Holocene and Upper Pleistocene pollen record from Padul (Granada, Spain): a new study. *Palaeogeography, Palaeoclimatology, Palaeoecology*, **66**, 243–263.

PONS, J. 1982. *Un modèle d'évolution des complexes plutoniques: gabbros et granitoides de la serra Morena occidentale*. PhD Thesis, Paul Sabatier University.

PONS, J. M., GALLEMÍ, J., HÖFLING, R. & MOUSSARIAN, E. 1994. Los Hippurites del Barranc del Racó, microfacies y fauna asociada (Maastrichtiense superior, sur de la Provincia de Valencia). *Cuadernos de Geología Ibérica*, **18**, 273–307.

PORTER, S. C. 1972. Distribution, morphology and size frequency of cinder cones on Mauna Kea. *Geological Society of America Bulletin*, **84**, 607–612.

PORTERO, J. M. & OLIVÉ, A. 1984. El Terciario del borde meridional del Guadarrama y Somosierra. In: *Geología de España*. II IGME, Madrid, 527–543.

PORTERO, J. M. & DEL OLMO, P. 1982. *Mapa Geológico de España a escala 1:50.000*. Hoja de Portillo (400). IGME, Madrid.

PORTERO, J. M., DEL OLMO, P. & RAMÍREZ DEL POZO, I. 1982. Síntesis del Terciario continental de la Cuenca del Duero. *Temas Geológico Mineros, IGME*, **6**(I), 11–37.

POSTAIRE, B. 1982. *Systématique Pb commun et U-Pb sur zircons*. PhD Thesis, University of Rennes.

POSTMA, G. 1983. Water-scape structures in the context of a depositional model of a mass-flow dominated conglomeratic fan-delta (Abrioja Formation, Pliocene, Almería basin, SE Spain). *Sedimentology*, **30**, 91–103.

POSTMA, G. & ROEP, T. B. 1985. Resedimented conglomerates in the bottomsets of Gilbert-type gravel deltas. *Journal of Sedimentary Petrology*, **6**, 874–885.

POUGET, P., LAMOUROUX, C. & DEBAT, P. 1988. Le dôme de Bosost (Pyrenees centrales): réinterpretation majeure de sa forme et de son évolution tectonométamorphique. *Comptes Rendues de l'Académie des Sciences de Paris*, **307**(2), 949–955.

POUGET, P., LAMOUROUX, C., DAHMANI, A. *et al.* 1989. Typologie et mode de mise en place des roches magmatiques dans les Pyrénéess hercyniennes. *Geologische Rundschau*, **78**, 537–554.

POUS, J., MUÑOZ, J. A., LEDO, J., LIESA, M. 1995. Partial melting of the subducted continental lower crust in the Pyrenees. *Journal of the Geological Society, London*, **152**, 217–220.

POUS, J., QUERALT, P., LEDO J. & ROCA, E. 1999. A high electrical conductive zone at lower crustal depth beneath the Betic Chain (Spain). *Earth and Planetary Sciences Letters*, **167**, 35–45.

POWELL, J. L. & BELL, K. 1970. Strontium isotopic studies of alkalic rocks. Localities from Australia, Spain and the western United States. *Contributions to Mineralogy and Petrology*, **27**, 1–10.

PRADO, C. DE 1855. Mémoire sur la Géologie d'Almadén, d'une partie de la Sierra Morena et des Montagnes de Tolède. *Bulletin de la Société Géologique de France [2]*, **12**, 182–204.

PRADO, C. DE 1856. *Mapa Geológico de la Provincia de Palencia (Escala 1:400.000)*. Boletin Comisión Mapa Geológico España, Madrid.

PRADO, C. DE 1864. *Descripción física y geológica de la provincia de Madrid*. Junta General de Estadística, Madrid.

PRADO, C. DE & VERNEUIL, E. DE 1850. Note géologique sur les terrains de Sabero et de ses environs dans les montagnes de Leon (Espagne) (C. de Prado), suivie d'une description des fossiles de ces terrains (E. de Verneuil). *Bulletin Société Géologique France*, **7**(2), 137–186.

PRADO, C. DE, VERNEUIL, E. DE & BARRANDE, J. 1855. Mémoire sur la géologie d'Almaden, d'une partie de la Sierra Morena et des montagnes de Tolède (C. de Prado), suivi d'une description des fossiles qui s'y rencontrent (E. de Verneuil & J. Barrande). *Bulletin Société Géologique France*, **12**(2), 1–86.

PRADO, J. 1972. Nota sobre la petrología de la zona de Viñón (Asturias). *Studia Geológica*, **3**, 7–32.

PRESCOTT, D. M. 1988. The geochemistry and palaeoenvironmental significance of iron pisoliths and ferromanganese crusts from the Jurassic of Mallorca, Spain. *Eclogae geologicae Helvetiae*, **81**(2), 387–414.

PRIEM, H. N., BOELRUK, N. A., HEBEDA, E. H. & VERSCHUREN, R. H. 1966. Isotopic age determinations on tourmaline granite- gneiss (South-Eastern Sierra de los Filabres). *Geologie en Mijnbouw*, **45**, 184–187.

PRIEM, H. N. A., BOELRIJK, N. A. I. M., VERSCHURE, R. H., HEBEDA, E. H. & FLOOR, P. 1966. Isotopic evidence for Upper-Cambrian or Lower-Ordovician granite emplacement in the Vigo area, North-Western Spain. *Geologie en Mijnbouw*, **45**, 36–40.

PRIEM, H. N. A., BOELRIJK, N. A. I. M., VERSCHURE, R. H., HEBEDA, E. H. & VERDURMEN, E. A. TH. 1970. Dating events of acid plutonism through the Paleozoic of the western Península. *Eclogae Geologie Helvetiae*, **63**, 255–274.

PRIEWALDER, H. 1997. SEM-Revision of a Chitinozoan Assemblage from the Uppermost San Pedro Fm (Pridoli), Cantabrian Mountains (Spain). *Jahrbuch der Geologischen Bundesanstalt*, **140**, 73–93.

PUEYO, E. 2000. *Rotaciones paleomagnéticas en sistemas de pliegues y cabalgamientos. Tipos, causas, significado y aplicaciones*. PhD Thesis, University of Zaragoza.

PUEYO, E., MILLÁN, H., POCOVÍ, J. & PARÉS, J. M. 1997. Cinemática rotacional del cabalgamiento basal surpirenaico en las Sierras Exteriores Aragonesas: Datos magnetotectónicos. *Acta Geológica Hispánica*, **32**, 237–256.

PUEYO, J. J. 1978–79. La precipitación evaporítica actual en las lagunas saladas del área: Bujaraloz-Sástago, Caspe, Alcañiz y Calanda (provs. de Zaragoza y Teruel). *Revista del Instituto de Investigaciones Geológicas*, **33**, 5–56.

PUGA, E. 1976. *Investigaciones petrológicas en Sierra nevada occidental (Cordilleras Béticas)*. PhD Thesis, Granada University.

PUGA, E. 1977. Sur l'existence dans le complexe de la Sierra Nevada (Cordillères Bétiques, Espagne du Sud) d'éclogites et sur leur origine probable à partir d'une croute océanique mésozoïque. *Comptes Rendus de l'Académie des Sciences, Paris*, **285**(16), 1379–1382.

PUGA, E. 1980. Hypothèses sur la genèse des magmatismes calcoalcalins, intraorogéniques et postorogéniques alpins dans les Cordillères Bétiques. *Bulletin de la Société Géologique de France*, **22**, 243–250.

PUGA, E. & DÍAZ DE FEDERICO, A. 1976. Pre-Alpine metamorphism in the Sierra Nevada Complex (Betic Cordilleras, Spain). *Cuadernos de Geología*, **7**, 161–171.

PUGA, E. DÍAZ DE FEDERICO, A. & FONTBOTÉ, J. M. 1974. Sobre la individualización y sistematización de las unidades profundas de la Zona Bética. *Estudios Geológicos*, **30**, 543–548.

PUGA, E., FONTBOTÉ, J. M. & MARTÍN-VIVALDI, J. L. 1975. Kyanite pseudomorphs after andalusite in polymetamorphic rocks of Sierra Nevada (Betic Cordillera, Southern Spain). *Schweizerische Mineralogische und Petrographische Mitteilungen*, **55**, 227–241.

PUGA, E., DÍAZ DE FEDERICO, A., MOLINA-PALMA, J. F., NIETO, J. M. & TENDERO-SEGOVIA, J. A. 1993. Fiel trip to the Nevado-Filabride Complex (Betic Cordilleras, SE Spain). *Ofioliti*, **18**(1), 37–60.

PUGA, E., DÍAZ DE FEDERICO, A. & DEMANT, A. 1995. The eclogitized pillows of the Betic Ophiolitic Association: relics of the Tethys Ocean floor incorporated in the Alpine chain after subduction. *Terra Nova*, **7**(1), 31–43.

PUIGDEFÀBREGAS, C. 1975. La sedimentación molásica en la cuenca de Jaca. *Pirineos*, **104**, 1–188.

PUIGDEFABREGAS, C. & SOUQUET, P. 1986. Tectono-sedimentary cycles and depositional sequences of the Mesozoic and Tertiary from the Pyrenees. *Tectonophysics*, **129**, 172–203.

PUIGDEFABREGAS, C., MUÑOZ, J. A. & MARZO, M. 1986. Thrust belt development in the Eastern Pyrenees and related depositional sequences in the southern foreland basin. *In*: ALLEN P. A. & HOMEWOOD P. (eds) *Foreland Basins*, IAS, Special Publications, **8**, 229–246.

PUIGDEFÀBREGAS, C., MUÑOZ, J. A. & VERGÉS, J. 1992. Thrusting and foreland basin evolution in the Southern Pyrenees. *In*: McCLAY, K. (ed.) *Thrust Tectonics*. Chapman & Hall, London, 247–254.

PUJADAS, J., CASAS, J. M., MUÑOZ, J. A. & SÀBAT, F. 1989. Thrust tectonics and Paleogene syntectonic sedimentation in the Empordà area (Southeastern Pyrenees). *Geodinamica Acta*, **3**, 195–206.

PUJALTE, V. 1977. *El complejo Purbeck-Weald de Santander. Estratigrafía y sedimentación*. PhD Thesis, University of Bilbao.

PUJALTE, V. 1981. Sedimentary succession and palaeoenvironments within a fault-controlled basin: the Wealden of the Santander area, Northern Spain. *Sedimentary Geology*, **28**, 293–325.

PUJALTE, V. 1982a. El tránsito Jurásico–Cretácico, Berriasiense, Valanginiense, Hauteriviense y Barremiense de la región Vasco-Cantábrica. *In: El Cretácico en España*. Complutense University, Madrid. 49–61.

PUJALTE, V. 1982b. La evolución paleogeográfica de la cuenca 'Wealdense' de Cantabria. *Cuadernos de Geología Ibérica*, **8**, 65–83.

PUJALTE, V. 1985. The Wealden basin of Santander. *In*: MILÁ, M. D. & ROSELL, J. (eds) *6th European Regional Meeting of Sedimentology, IAS, Lleida. Excursion guide book*. 351–371.

PUJALTE, V. 1989a. Ensayo de correlación de las sucesiones del Oxfordiense-Barremiense de la región Vasco-Cantábrica basado en macrosecuencias deposicionales: implicaciones paleogeográficas. *Cuadernos de Geología Ibérica*, **13**, 199–216.

PUJALTE, V. 1989b. Macrosecuencias deposicionales del Oxfordiense-Barremiense de la región Vasco-Cantábrica: Implicaciones estratigráficas y paleogeográficas. *In: Libro Homenaje a Rafael Soler*. AGGEP, Madrid, 105–114.

PUJALTE, V. & MONGE, C. 1985. A tide dominated delta system in a rapidly subsiding basin: the Middle Albian–Lower Cenomanian Valmaseda Formation of the Basque-Cantabrian region, northern Spain. *6th European Regional Meeting of Sedimentology, IAS, Lleida. Abstracts*, 381–384.

PUJALTE, V. & ROBLES S. 1989. Las cuencas aluvio-lacustres del Malm de la parte occidental de la región Vascocantábrica: Facies y significado tectonoestratigráfico. *Geogaceta*, **6**, 100.

PUJALTE, V., ROBLES, S. & GARCÍA-MONDÉJAR, J. 1986–87. Características sedimentológicas y paleogeográficas del fan-delta albiense de la Formación Monte Grande y sus relaciones con el Flysch Negro (Arminza-Górliz, Vizcaya). *Acta Geologica Hispanica*, **21–22**, 141–150.

PUJALTE, V., ROBLES, S. & VALLES, J. C. 1988. El Jurásico marino de las zonas de alto sedimentario relativo del borde SW de la Cuenca Vasco-Cantábrica (Rebolledo de la Torre, Palencia). *II Coloquio de Estratigrafía y Paleogeografía del Jurásico de España, libro guía de las excursiones, Ciencias de la Tierra (Instituto de Estudios Riojanos)*, **11**, 85–94.

PUJALTE, V., ROBLES, S. & HERNÁNDEZ J. M. 1996. La sedimentación continental del grupo Campóo (Malm-Cretácico basal de Cantabria, Burgos y Palencia): testimonio de un reajuste hidrográfico al inicio de una fase rift. *Cuadernos de Geología Ibérica*, **21**, 227–251.

PUJALTE, V., BACETA, J. I., ORUE-ETXEBARRIA, X. & PAYROS, A. 1998. Paleocene strata of the Basque Country, W Pyrenees, N Spain: facies and sequence development in a deep-water, starved basin. *In*: GRACIANSKY P-CH. DE, HARDENBOL, J., JACQUIN, T. & VAIL, P. R. (eds) *Mesozoic and Cenozoic Sequence Stratigraphy of European Basins*. SEPM, Special Publications, **60**, 311–325.

PUJALTE, V., ROBLES, S., ORUE-ETXEBARRIA, X. BACETA, J. I., PAYROS, A. & LARRUZEA, I. F. 2000. Uppermost Cretaceous–Middle Eocene strata of the Basque-Cantabrian region and Western Pyrenees: a sequence stratigraphic perspective. *Revista de la Sociedad Geológica de España*, **13**, 191–211.

PULGAR, J. 1980. *Análisis e interpretación de las estructuras originadas durante las fases de replegamiento en la Zona Astur-Occidental-Leonesa (Cordillera Herciniana, NW de España)*. PhD Thesis, University of Oviedo.

PULGAR, J. A., BASTIDA, F., MARCOS, A., PÉREZ-ESTAÚN, A., VARGAS, I. & RUIZ, F. 1981. *Memoria explicativa de la Hoja no. 100 (Degaña) del Mapa Geológico de España E. 1:50,000 (Segunda Serie)*. IGME, Madrid, 1–35.

PULGAR, J., PÉREZ-ESTAÚN, A., GALLART, J., ÁLVAREZ-MARRÓN, J., GALLASTEGUI, J., ALONSO, J. L. & ESCIN GROUP. 1995. The ESCI-N2 deep seismic reflection profile: a traverse across the Cantabrian Mountains and adjacent Duero basin. *Revista de la Sociedad Geológica de España*, **8**(4), 383–394.

PULGAR, J. A., GALLART, J., FERNANDEZ-VIEJO, G., PÉREZ-ESTAÚN, A., ÁLVAREZ-MARRÓN, J. & ESCIN GROUP. 1996. Seismic image of the Cantabrian Mountains in the western extension of the Pyrenean belt from integrated reflection and refraction data. *Tectonophysics*, **264**, 1–19.

PULGAR, J. A., ALONSO, J. L., ESPINA, R. G. & MARÍN, J. A. 1999. La deformación alpina en el basamento varisco de la Zona Cantábrica. *Trabajos de Geología, University of Oviedo*, **21**, 283–294.

PUSCHMANN, H. 1967. Zum Problem der Schichtlücken im Devon der Sierra Morena (Spanien). *Geologische Rundschau*, **56**, 528–542.

PUSCHMANN, H. 1968a. Stratigraphische Untersuchungen im Paläozoikum des Montseny (Katalonien/Spanien). *Geologische Rundschau*, **57**, 1066–1088.

PUSCHMANN, H. 1968b. La série paléozoïque du Montseny (Catalogne, Espagne du Nord-Est). *Comptes Rendus de l'Académie des Sciences, Paris*, **266**, 657–659.

PUSCHMANN, H. 1970. Das Paläozoikum der nördlichen Sierra Morena am Beispiel der Mulde von Herrera del Duque (Spanien). *Geologie*, Berlin, **19**, 309–329.

QUARCH, H. 1975. Stratigraphie und Tektonik des Jungpaläozoikums im Sattel von Montalbán (Östliche Iberische Ketten, NE-Spanien). *Geologisches Jahrbuch*, **B16**, 3–43.

QUEROL, R. 1989. Geología del subsuelo de la Cuenca del Tajo. ITGE, Madrid.

QUEROL, X., SALAS, R., PARDO, G. & ARDEVOL, L. 1992. Albian coal-bearing deposits of the Iberian Range in northeastern Spain. *In*: McCABE, J. P. & PARRISH, J. T. (eds) *Controls and Distribution and Quality of Cretaceous Coals: Boulder, Colorado*. Geological Society of America, Special Papers, **267**, 193–208.

QUESADA, C. 1983. El Carbonífero de Sierra Morena. *In*: MARTÍNEZ DÍAZ, C. (ed.) *Carbonífero y Pérmico de España*. IGME, Madrid, 243–277.

QUESADA, C. 1990a. Precambrian successions in SW Iberia: their relationship to Cadomian orogenic events. *In*: D'LEMOS, R. S., STRACHAN, R. A. & TOPLEY, C. G. (eds) *The Cadomian Orogeny*. Geological Society, London, Special Publications, **51**, 353–362.

QUESADA, C. 1990b. Geological constraints on the Paleozoic tectonic evolution of tectonostratigraphic terranes in the Iberian Massif. *Tectonophysics*, **185**, 225–245.

QUESADA, C. 1990c. Precambrian terranes in the Iberian Variscan Foldbelt. *In*: STRACHAN, R. A. & TAYLOR, G. K. (eds) *Avalonian and Cadomian Geology of the North Atlantic*. Blackie, New York, 109–133.

QUESADA, C. 1992. Evolución tectónica del Macizo Ibérico. *In*: GUTIÉRREZ MARCO, J. C., SAAVEDRA, J. & RÁBANO, I. (eds) *Paleozoico Inferior de Ibero-America*. University of Extremadura, Badajoz, 173–190.

QUESADA, C. 1996. Estructura del sector español de la Faja Pirítica: implicaciones para la exploración de yacimientos. *Boletín Geológico y Minero*, **107**, 265–278.

QUESADA, C. 1997. Evolución geodinámica de la Zona de Ossa-Morena durante el ciclo Cadomiense. *In*: ARAÚJO, A. A. & PEREIRA, M. F. (eds) *Estudo sobre a Geología da Zona de Ossa-Morena (Maciço Iberico)*. Livro de Homenagem ao Profesor Francisco Gonçalves. University of Evora, 205–230.

QUESADA, C. 1998. A reappraisal of the structure of the Spanish segment of the Iberian Pyrite Belt. *Mineralium Deposita*, **33**, 31–44.

QUESADA, C. & CUETO, L. A. 1994. *Memoria explicativa de la Hoja no. 895 (Encinasola) del Mapa Geológico de España E. 1:50,000 (Segunda Serie)*. IGME, Madrid, 1–90.

QUESADA, C., APALATEGUI, O., EGUÍLUZ, L., LIÑÁN, E. & PALACIOS, T. 1990a. Ossa-Morena Zone. 2. Stratigraphy. 2.1. Precambrian. *In*: DALLMEYER, R. D. & MARTÍNEZ GARCÍA, E. (eds) *Pre-Mesozoic Geology of Iberia*. Springer-Verlag, Berlin, 252–258.

QUESADA, C., ROBARDET, M. & GABALDÓN, V. 1990b. Ossa Morena

Zone. Stratigraphy: synorogenic phase (Upper Devonian–Carboniferous–Lower Permian). *In*: DALLMEYER, R. D. & MARTÍNEZ GARCÍA, E. (eds) *Pre-Mesozoic Geology of Iberia.* Springer-Verlag, Berlin, 273–279.

QUESADA, C., BELLIDO, F., DALLMEYER, R. D. *et al.* 1991. Terranes within the Iberian Massif: correlations with West African sequences. *In*: DALLMEYER, R. D. & LECORCHÉ, J. P. (eds) *The West African Orogens and Circum-Atlantic Correlations.* Springer, Berlin, 267–293.

QUESADA, C., FONSECA, P. E., MUNHA, J., OLIVEIRA, J. T. & RIBEIRO, A. 1994. The Beja-Acebuches Ophiolite (Southern Iberia Variscan fold belt): Geological characterization and geoddynamic significance. *Boletin Geológico y Minero*, **105**, 3–49.

QUESADA, S. & ROBLES, S. 1995. Distribution of Organic Facies in the Liassic Carbonate Ramps of the Western Basque-Cantabrian Basin (Northern Spain). *17th International Meeting on Organic Geochemistry, Field Trip Guidebook.*

QUESADA, S., PUJALTE, V., ROBLES, S. & VICENTE, J. C. 1990. Las formaciones esponjiolíticas del Dogger de la región Vasco-Cantábrica: Características y posibilidades petrolíferas. *Geogaceta*, **7**, 26–28.

QUESADA, S., ROBLES, S. & PUJALTE, V. 1991. Correlación secuencial y sedimentológica entre registros de sondeos y series de superficie de Jurásico Marino de la Cuenca de Santander (Cantabria, Palencia y Burgos). *Geogaceta*, **10**, 3–6.

QUESADA, S., ROBLES, S. & PUJALTE, V. 1993. El Jurásico Marino del margen suroccidental de la Cuenca Vasco-Cantábrica y su relación con la exploración de hidrocarburos. *Geogaceta*, **13**, 92–96.

QUESADA, S., ROBLES, S. & DORRONSORO, C. 1996. Caracterización de la roca madre del Lias y su correlacción con el petróleo de campo de Ayoluengo en base al analisis de cromatográfia de gases e isótopos de carbono (Cuenca Vasco-Cantábrica, España). *Geogaceta*, **20**, 176–179.

QUESADA, S., DORRONSORO, C. ROBLES, S., CHALER, R. & GRIMALT, J. O. 1997. Geochemical correlation of oil from the Ayoluengo field to Liassic black shale units in the southwestern Basque-Cantabrian Basin (northern Spain). *Organic Geochemistry*, **27**, 25–40.

QUINTERO, I. 1962. Graptolites en la provincia de Lugo. *Notas y Comunicaciones del IGME*, **65**, 61–82.

QUIRANTES, J. 1978. *Estudio sedimentológico y estratigráfico del Terciario continental de los Monegros.* PhD Thesis, Instituto Fernando el Católico (CSIC), Zaragoza.

QUIROGA, J. L. 1982. Estudio geológico del Paleozoico del W de Zamora. *Trabajos Geología Universidad Oviedo*, **12**, 205–226.

RÁBANO, I. 1981. Phacopina (Trilobita) del Ordovícico de Horcajo de los Montes (Ciudad Real, España). *Estudios Geológicos*, **37**, 269–283.

RÁBANO, I. 1983. The Ordovician trilobite *Hungioides* Kobayashi, 1936 (Asaphina, Dikelokephalinidae) from Spain. *Geobios*, **16**, 431–441.

RÁBANO, I. 1984. Nuevas observaciones sobre *Placoparia (Placoparia) cambriensis* Hicks, 1875 (Trilobita, Cheirurina) en el Llanvirn centroibérico. *Coloquios de Paleontología*, **39**, 7–16.

RÁBANO, I. 1989*a*. El género *Uralichas* Delgado, 1892 (Trilobita, Lichaida) en el Ordovícico de la Península Ibérica. *Boletín Geológico y Minero*, **100**, 21–47.

RÁBANO, I. 1989*b*. Trilobites del Ordovícico Medio del sector meridional de la Zona centroibérica española. Part 1. *Boletín Geológico y Minero*, **100**, 307–338.

RÁBANO, I. 1989*c*. Trilobites del Ordovícico Medio del sector meridional de la Zona centroibérica española. Part 2. *Boletín Geológico y Minero*, **100**, 541–609.

RÁBANO, I. 1989*d*. Trilobites del Ordovícico Medio del sector meridional de la Zona centroibérica española. Part 3. *Boletín Geológico y Minero*, **100**, 767–841.

RÁBANO, I. 1989*e*. Trilobites del Ordovícico Medio del sector meridional de la Zona centroibérica española. Part 4. *Boletín Geológico y Minero*, **100**, 971–1032.

RÁBANO, I. & GUTIÉRREZ-MARCO, J. C. 1983. Revisión del género *Ectillaenus* Salter, 1867 (Trilobita, Illaenina) en el Ordovícico de la Península Ibérica. *Boletín de la Real Sociedad Española de Historia Natural (Geología)*, **81**, 225–246.

RÁBANO, I., PEK, I. & VANEK, J. 1985. New Agnostina (Trilobita) from the Llanvirn (Ordovician) of Spain. *Estudios Geológicos*, **4**, 439–445.

RÁBANO, I., GUTIÉRREZ-MARCO, J. C. & ROBARDET, M. 1993. Upper Silurian Trilobites of Bohemian affinities from the West Asturian-Leonese Zone (NW Spain). *Géobios*, **26**, 361–376.

RACHEBOEUF, P. R. & ROBARDET, M. 1986. Le Pridoli et le Dévonien inférieur de la Zone d'Ossa Morena (sud-ouest de la Peninsule Iberique. Etude des brachiopodes). *Geologica Palaeontologica*, **20**, 11–37.

RACHEBOEUF, P. R., LETHIERS, F., BABIN, C., ROLFE, W. D. I. & MAREZ, E. DE. 1986. Les faunes du Dévonien supérieur d'Alange (province de Badajoz, Sud-Ouest de l'Espagne). *Géologie Méditerranéenne*, **12–13**(1–2), 37–47.

RACHEBOEUF, P. R., FERRER BATET, E. & MAGRANS, J. 1993. Un nouvel assemblage faunique du Dévonien inférieur de Catalogne (NE de l'spagne). *Treballs del Museu de Geología de Barcelona*, **3**, 5–18.

RADIG, F. 1964. Die Lebensspur *Tomaculum problematicum* Groom 1902, in Llandeilo der Iberischen Halbinsel. *Neues Jahrbuch für Geologie und Paläontologie Abhandlungen*, **119**, 12–18.

RADIG, F. 1966. Eine Oberdevon-Fauna aus dem östlichen Asturien (Spanien) und die Schichtlücke unter den Knollenkalken des Visé. *Zeitschrift deutschen geologischen Gesellschaft*, **115**, 515–523.

RAGUIN, E. 1977. Le massif de l'Aston dans les Pyrénées de l'Ariège. *Bulletin BRGM*, **1**, 89–119.

RAMAJO, J. & AURELL, M. 1997. Análisis sedimentológico de las discontinuidades y depósitos asociados del Calloviense superior-Oxfordiense medio en la Cordillera Ibérica Noroccidental. *Cuadernos de Geología Ibérica*, **22**, 213–236.

RAMAJO, J., AURELL, M., BÁDENAS, B., BELLO, J., DELVENE, G., MELÉNDEZ, G. & PÉREZ-URRESTI, I. 1999. Síntesis del Oxfordiense de la Cuenca Ibérica Nororidental y su correlación con la Cuenca Catalana. *Cuadernos de Geología Ibérica*, **25**, 111–137.

RAMBAUD, F. 1978. Distribución de focos volcánicos y yacimientos en la banda pirítica de Huelva. *Boletín Geológico y Minero*, **89**, 223–233.

RAMBAUD PÉREZ, F. 1969. *El Sinclinal Carbonífero de Río Tinto (Huelva) y sus Mineralizaciones Asociadas.* IGME, Memorias, **71**.

RAMÍREZ, A. 1991. Yacimientos potásicos. *In*: LUNAR, R. & OYARZUN, R. (eds) *Yacimientos Minerales.* Editorial Centro de Estudios Ramón Areces, Madrid (ISBN 84-87191-74-6).

RAMÍREZ DEL POZO, J. 1969. Bioestratigrafía y paleogeografía del Jurásico de la costa asturiana. *Boletín Geológico y Minero*, **80**, 307–332.

RAMÍREZ DEL POZO, J. 1971. *Bioestratigrafía y Microfacies del Jurásico y Cretácico del Norte de España (Región Cantábrica).* Memoria del IGME, **78**.

RAMÍREZ DEL POZO, J. & MARTÍN-CHIVELET, J. 1994. Bioestratigrafía y Cronoestratigrafía del Coniaciense – Maastrichtiense en el sector prebético de Jumilla – Yecla. *Cuadernos de Geología Ibérica*, **18**, 83–116.

RAMÓN, X. 1989. Análisis secuencial y sedimentología del Lías en los Pirineos centrales. *Cuadernos de Geología Ibérica*, **13**, 159–173.

RAMON, X. & CALVET, F. 1987. Estratigrafía y sedimentología del Muschelkalk inferior del dominio Montseny-Llobregat. *Estudios Geológicos*, **43**, 471–487.

RAMÓN, X., AURELL, M. & MELÉNDEZ, G. 1992. Stratigraphy and associated unconformities in the Middle to Upper Jurassic on the south central Pyrenees, Spain. *II Congreso Geológico de España, Simposios*, **2**, 161–167.

RAMOS, A. 1979. *Estratigrafía y paleogeografía del Pérmico y Triásico al oeste de Molina de Aragón (Província de Guadalajara).* Seminarios de Estratigrafía. Serie Monografías, **6**.

RAMOS, A. 1995. Transition from alluvial to coastal deposits (Permian-Triassic) on the Island of Mallorca, western Mediterranean. *Geological Magazine*, **132**, 435–447.

RAMOS, A. & DOUBINGER, J. 1989. Prémieres datations palinologiques dans le faciès Buntsandstein de l'île de Majorque (Baleares, Espagne). *Comptes Rendus de l'Académie des Sciences de Paris*, **309**, 1089–1094.

RAMOS, A. & RODRIGUEZ-PEREIRA, A. 1985. Découverte d'un

affleurement de terrains paléozoïques dans l'Ile de Majorque (Baleares, Espagne). *Comptes Rendus de l'Académie des Sciences de Paris*, **301**, 1205–1207.

RAMOS, A., DOUBINGER, J. & VIRGILI, C. 1976. El Pérmico inferior de Rillo de Gallo (Guadalajara). *Acta Geológica Hispánica*, **XI**, 65–70.

RAMOS, E. 1988. *El Paleógeno de las Baleares. Estratigrafía y Sedimentología*. PhD Thesis, University of Barcelona.

RAMOS, E. 1993. El Paleogen. *In*: Alcover, J. A., Ballesteros, E. & Fornós, J. J. (eds) *Història Natural de l'Arxipèlag de Cabrera*. Monografias de la Societat Historia Natural de Balears, **2**, 87–103.

RAMOS, E. & MARTINELL, J. 1985. Datos preliminares sobre la malacofauna marina del Oligoceno de Mallorca. *Iberus*, **5**, 1–9.

RAMOS, E., RODRÍGUEZ-PEREA, A., SÀBAT, F. & SERRA-KIEL, J. 1989. Cenozoic tectosedimentary evolution of Majorca Island. *Geodinàmica Acta*, **3**, 53–72.

RAMOS, E., BERRIO, I., FORNÓS, J. J. & MORAGUES, LL. 2000. The Middle Miocene Son Verdera Lacustrine-Palustrine System (Santa Margalida Basin, Mallorca). *In*: GIERLOWSKI-KORDESCH, E. H. & KELTS, K. (eds) *Lake Basins through Space and Time*. AAPG, Studies in Geology, **46**, 441–448.

RAMOS, E., CABRERA, L., HAGEMANN, H. W., PICKEL, W. & ZAMARREÑO, I. 2001. Paleogene lacustrine record in Mallorca (NW Mediterranean, Spain): Depositional, palaeogeographic and palaeoclimatic implications for the ancient southeastern Iberian margin. *Palaeogeography, Palaeoclimatology, Palaeoecology*, **172**, 1–37.

RAMOS-GUERRERO, E. 1989. *El Paleógeno de las Baleares, Estratigrafia y sedimentología*. PhD Thesis, University of the Balearic Isles, University of Barcelona.

RAMOS-GUERRERO, E., RODRÍGUEZ-PEREA, A., SABAT, F. & SERRA-KIEL, J. 1989. Cenozoic evolution of Mallorca island. *Geodinamica Acta*, **3**(1), 53–72.

RANGHEARD, Y. 1970. Principales donnés stratigraphiques et tectoniques des îles d'Ibiza et de Formentera (Baleares); situation paléogéographique et structurale de ces îles dans les Cordilleres bétiques. *Comptes Rendus de l'Académie des Sciences, Paris*, Série D, **270**, 1227–1230.

RANGHEARD, Y. 1971. *Étude géologique des îles d'Ibiza et de Formentera (Baléares)*. IGME, Memorias, **82**.

RAT, P. 1959. *Les pays crétacés basco-cantabriques (Espagne)*. PhD Thesis, Publications de l'Université de Dijon, **18**.

RAT, P. 1962. Contribution à l'étude stratigraphique du Purbeckien-Wealdien de la région de Santander (Espagne). *Bulletin de la Société Géologique de France*, **IV**, 3–12.

RAT, P. 1988. The Basque-Cantabrian Basin between the Iberian and European plates: some facts but still many problems. *Revista de la Sociedad Geóogica de España*, **1**, 327–348.

RATCLIFFE, N. M., ARMSTRONG, R. L., MOSE, D. G., SENESCHAL, R. WILLIAMS, N. & BAIAMONTE, M. J. 1982. Emplacement history and tectonic significance of the Cortlandt Complex, related plutons, and dike swarms in the taconide zone of southeastern New York based on K-Ar and Rb-Sr investigations. *American Journal of Science*, **282**, 358–390.

RAVEN, J. G. M. 1983. Conodont biostratigraphy and depositional history of the Middle Devonian to Lower Carboniferous in the Cantabrian Zone (Cantabrian Mountains, Spain). *Leidse Geologische Mededelingen*, **52**(2), 265–339.

RAVIER, J. 1959. Le métamorphisme des terrains secondaires des Pyrénées. *Mémoires de la Société Géologique de France*, **38**, 86–250.

RAZIN, P. 1989. *Évolution tecto-sédimentaire alpine des Pyrénées Basques a lóuest de la transformante de Pamplona (province du Labourd)*. PhD Thesis, University of Bordeaux.

READ, R. A. 1967. *Deformation and metamorphism of the pyritic San Dionisio ore body, Rio Tinto, Spain*. PhD Thesis, Imperial College, London.

RECHE, J., MARTÍNEZ, F. J., ARBOLEYA, M. L., DIETSCH, C. & BRIGGS, W. D. 1998. Evolution of a kyanite-bearing belt within a HT-LP orogen: the case of NW Variscan Iberia. *Journal of Metamorphic Geology*, **16**, 379–394.

REGUANT, S. 1967. *El Eoceno marino de Vich*. Memorias del IGME, **LXVIII**.

REGUEIRO, M. 1998. Strontium the Global Club: from sweet refinement to the WWW. *In*: O'DRISCOLL, M. (ed.) *Proceedings of the 13th Industrial Minerals International Congress*. Industrial Minerals Information Ltd, Kuala Lumpur, Malaysia, 58–87.

REGUEIRO, M. & MARCHÁN, C. 2000. Industrial Minerals of Spain. Europe's mining stronghold. *Industrial Minerals*, **394**, 53–65.

REGUEIRO, M., CALVO, J. P., ELÍZAGA, E. & CALDERÓN, V. 1993. Spanish diatomite: Geology and economics. *Industrial Minerals*, **306**, 57–67.

REGUEIRO, M., MARTÍNEZ-DURÁN, P. & GONZALO-CORRAL, F. 2000a. Rocas y Minerales Industriales de España. *In*: CALVO, B. (ed.) *Rocas y Minerales Industriales de Iberoamérica*. ITGE, Madrid, 295–308.

REGUEIRO, M., SÁNCHEZ, E., SANZ, V. & CRIADO, E. 2000b. Cerámica industrial en España. *Boletín de la Sociedad Española De Cerámica y Vidrio*, **39**, 5–30.

REHAULT, J. P., MASCLE, J. & BOILLOT, G. 1984. Evolution géodynamique de la Méditerranée depuis l'Oligocène. *Memorias Societa Geologica Italiana*, **27**, 85–96.

REICHERTER, K. 1994. *The Mesozoic Tectono-sedimentary evolution of the central Betic seaway (External Betic Cordillera, Southern Spain)*. PhD Thesis, University of Tübingen.

REICHERTER, K., WIEDMANN, J. & HERBIN, J. P. 1996. Distribution of organic-rich sediments in Subbetic sections during the Aptian-Turonian (Betic Cordillera, Southern Spain). *Revista de la Sociedad geológica de España*, **9**, 75–88.

REICHERTER, K. R. & PLETSCH, T. K. 2000. Evidence for a synchronous circum-Iberian subsidence event and its relation to the African – Iberian plate convergence in the Late Cretaceous. *Terra Nova*, **12**, 141–147.

REIJERS, T. J. A. 1972. Facies and diagenesis of the Devonian Portilla Limestone Formation between the river Esla and the Embalse de la Luna, Cantabrian Mountains, Spain. *Leidse Geologische Mededelingen*, Leiden, **47**, 163–249.

REISS, W. 1861. *Die Diabas und lavenformation del Insel Palma*. Kreidel, Weisbaden, 11–20.

REITNER, J. 1982. Die Entwicklung von Inselplattformen und Diapir-Atollen im Alb des Basko-Kantabrikums (Nordspanien). *Neues Jahrbuch für Geologie und Paläontologie, Abhandlungen*, **165**, 87–101.

REITNER, J. 1986. A comparative study of the diagenesis in diapir-influenced reef atolls and a fault block reef platform in the Late Albian of the Vasco-Cantabrian basin (northern Spain). *In*: SCHROEDER, J. H. & PURSER, B. H. (eds) *reef Diagenesis*, Springer-Verlag, Berlin, 186–209.

RENDELL, H. M., CALDERÓN, T., PÉREZ-GONZÁLEZ, A., GALLARDO, J., MILLÁN, A. & TOWNSEND, P. 1994. Thermoluminescence and optically stimulated luminescence dating of Spanish dunes. *Quaternary Geochronology (Quaternary Science Reviews)*, **13**, 429–432.

REQUADT, H. 1972. Zur stratigraphie und Fazies des Unter-und Mitteldevons in den spanischen Westpyrenäen. *Clausthaler Geologische Abhandlungen*, **13**, 1–113.

REQUADT, H. 1974. Aperçu sur la stratigraphie et le facies du Dévonien inférieur et moyen dans les Pyrénées Occidentales d'Espagne. *Pirineos*, **111**, 109–127.

REQUADT, H. 1979. Geologie des Unter- und Mitteldevons im Südwestlichen Aldudes-Quinto Real Massiv (Spanische Westpyrenäen). *Clausthaler Geologische Abhandlungen*, **12**, 75–88.

RESPAUT, J. P. & LANCELOT 1983. U-Pb dating on zircons and monazites of the synmetamorphic emplacement of the Ansignan charnockite (Agly Massif, France). *Neues Jahrbuch fuer Mineralogie, Abhandlungen*, **147**, 21–34.

REUTHER, C.-D. 1977. Das Namur im südlichen Kantabrischen Gebirge (Nord-spanien)-Krustenbewegungen und Faziesdifferenzierung im Übergang Geosynklinale-Orogen. *Clausthaler Geologische Abhandlungen, Clausthal-Zellerfeld*, **28**, 1–122.

REY, J. 1993. *Análisis de la cuenca subbética durante el Jurásico y Cretácico en la transversal Caravaca Vélez-Rubio*. PhD Thesis, University of Granada.

REY, D. & RAMOS, A. 1991. Estratigrafía y sedimentología del Pérmico y Triásico del sector Deza-Castejón (Soria). *Revista de la Sociedad Geológica de España*, **4**(1–2), 105–125.

REY, J., ANDREO, B., GARCÍA-HERNÁNDEZ, M., MARTÍN-ALGARRA, A.

& VERA, J. A. 1990. The Liassic 'Lithiotis' facies north of Vélez Rubio (Subbetic Zone). *Revista de la Sociedad Geológica de España*, **3**(1–2), 199–212.

RIAZA, C. 1996. Inversión estructural en la cuenca mesozoica del offshore asturiano. Revisión de un modelo exploratorio. *Geogaceta*, **20**, 169–171.

RIAZA, C. & MARTÍNEZ DEL OLMO, W. 1996. Depositional model of the Guadalquivir–Gulf of Cádiz Tertiary basin. *In*: FRIEND, P. F. & DABRIO, C. J. (eds) *Tertiary Basins of Spain: the Stratigraphic Record of Crustal Kinematics*. World and Regional Geology, **6**, Cambridge University Press, Cambridge, 330–338.

RIBA, O. 1959. *Estudio geológico de la Sierra de Albarracín*. Monografía 16, Instituto Lucas Mallada Investgaciones Geológicas.

RIBA, O. 1976. Tectogenèse et sédimentation: deux modèles de discordances syntectoniques pyrénéennes. *Bulletin du Bureau de Recherches Géologiques et Minières*, **4**, 383–401.

RIBA, O. 1992. Las secuencias oblicuas en el borde Norte de la Depresión del Ebro en Navarra y la Discordancia de Barbarín. *Acta Geológica Hispánica*, **27**, 55–68.

RIBA, O., REGUANT, S. & VILLENA, J. 1983. Ensayo de síntesis estratigráfica y evolutiva de la cuenca terciaria del Ebro. *In*: COMBA, J. A. (coord.) *Libro Jubilar J. M. Ríos. Geología de España*, vol. 2. IGME. Madrid, 131–159.

RIBEIRO, A. 1974. Contribution à l'étude tectonique de Trás-os-Montes Oriental. *Memórias dos Serviços Geológicos de Portugal*, **24**, 1–168.

RIBEIRO, A. 1990. Central-Iberian Zone: Introduction. *In*: DALLMEYER, R. D. & MARTÍNEZ GARCÍA, E. (eds) *Pre-Mesozoic Geology of Iberia*. Springer Verlag, Berlin, 143–144.

RIBEIRO, A. & SILVA, J. B. 1983. Structure of the South Portuguese Zone. *In*: LEMOS DE SOUSA, M. & OLIVEIRA, J. T. (eds) *The Carboniferous of Portugal*. Memórias dos Serviços Geológicos de Portugal, **29**, 83–89.

RIBEIRO, A., QUESADA, C. & DALLMEYER, R. D. 1987. Tectonostratigraphic terranes and the geodynamic evolution of the Iberian Variscan Foldbelt. *Conference on Plate Tectonics and Deformation, Gijón-Oviedo*, 60–61.

RIBEIRO, A., KULLBERG, M. C., KULLBERG, J. C., MANUPELLA, G. & PHIPPS, S. 1990a. A review of Alpine tectnics in Portugal: Foreland detachment in basement and cover rocks. *Tectonophysics*, **184**, 357–366.

RIBEIRO, A., PEREIRA, E. & DIAS, R. 1990b. Central Iberian Zone. Allochthonous sequences. Structure in the northwest of the Iberian Peninsula. *In*: DALLMEYER, R. D. & MARTÍNEZ GARCÍA, E. (eds) *Pre-Mesozoic Geology of Iberia*. Springer Verlag, Berlin, 220–236.

RIBEIRO, A., QUESADA, C. & DALLMEYER, R. D. 1990c. Geodynamic evolution of the Iberian Massif. *In*: DALLMEYER, R. D. & MARTÍNEZ GARCÍA, E. (eds) *Pre-Mesozoic Geology of Iberia*. Springer Verlag, Berlin, 399–409.

RIBEIRO, M. L. 1986. *Geología e petrología da regiao a SW de Macedo de Cavaleiros (Tras-os-Montes oriental)*. PhD Thesis, University of Lisbon.

RIBEIRO, M. L. & FLOOR, P. 1987. Magmatismo peralcalino no Maciço Hespérico: sua distribuçâo e significado geodinâmico. *In*: BEA, F., CARNICERO, A., GONZALO, J. C., LÓPEZ PLAZA & RODRÍGUEZ ALONSO, M. D. (eds) *Geología de los granitoides y rocas asociadas del Macizo Hespérico. Libro Homenaje a L. C. García de Figuerona*. Rueda, Madrid, 211–222.

RICHARDSON J. B., RODRÍGUEZ GONZÁLEZ, R. M. & SUTHERLAND, S. J. E. 2000. Palynology and recognition of the Silurian/Devonian boundary in some British terrestrial sediments by correlation with Cantabrian and other European marine sequences – a progress report. *Courier Forschungsinstitut Senckenberg*, **220**, 1–7.

RICHARDSON J. B., RODRÍGUEZ GONZÁLEZ, R. M. & SUTHERLAND, S. J. E. 2001. Palynological zonation of mid-Palaeozoic sequences from the Cantabrian Mountains, NW Spain: implications for inter-regional and inter-facies correlation of the Ludford/Pridoli and Silurian/Devonian boundaries, and plant dispersal patterns. *Bulletin of the Natural History Museum, Geology Series*, **57**(2), 115–162.

RICHTER, G. 1930. Die Iberischen Ketten zwischen Jalón und Demanda. *Beiträge zur Geologie der westlichen Mediterrangebiete Abhandeungen der Gesselschaft der Wissenchaften zu Gottingen. Mathematik-Physik Klasse*, (5), **16**, 3.

RICHTER, G. & TEICHMÜLLER, R. 1933. Die entwiicklung der Keltiberischen Ketten. *Beiträge zur Geologie der westlichen Mediterrangebiete Abhandeungen der Gesselschaft der Wissenchaften zu Gottingen. Mathematik-Physik Klasse*, (9–11) **7**, 1–118.

RICOU, L.-E. 1987. The tethyan oceanic gates: A tectonic approach to major sedimentary changes within Tethys. *Geodinamica Acta*, **1**(4/5), 225–232.

RIDING, R., BRAGA, J. C. & MARTÍN, J. M. 1991a. Oolite stromatolites and thrombolites, Miocene, Spain: analogues of Recent giant Bahamian examples. *Sedimentary Geology*, **71**, 121–127.

RIDING, R., MARTÍN, J. M. & BRAGA, J. C. 1991b. Coral-stromatolite reef framework, Upper Miocene, Almería, Spain. *Sedimentology*, **38**, 799–818.

RIDING, R., BRAGA, J. C., MARTÍN, J. M. & SÁNCHEZ-ALMAZO, I. M. 1998. Mediterranean Messinian Salinity Crisis: constraints from a coeval marginal basin, Sorbas, SE Spain. *Marine Geology*, **146**, 1–20.

RIDING, R., BRAGA, J. C. & MARTÍN, J. M. 1999. Late Miocene Mediterranean desiccation: topography and significance of the 'Salinity Crisis' erosion surface on-land in southeast Spain. *Sedimentary Geology*, **123**, 1–7.

RIES, A. & SHACKLETON, R. M. 1971. Catazonal complexes of northwestern Spain and north Portugal; remmants of a hercynian thrust plate. *Natural and Physical Science*, **234**, 65–69.

RIES, A. C. 1979. Variscan metamorphism and K-Ar dates in the Variscan Fold Belt of S Brittany and Nw Spain. *Journal of the Geological Society London*, **136**, 89–103.

RIGBY, J. K., GUTIÉRREZ-MARCO, J. C., ROBARDET, M. & PIÇARRA, J. M. 1997. First articulated Silurian sponges from the Iberian Peninsula, Spain and Portugal. *Journal of Paleontology*, **71**, 554–563.

RIGHTMIRE, G. P. 1996. The human cranium from Bodo, Ethiopia: evidence for speciation in the Middle Pleistocene? *Journal of Human Evolution*, **31**, 21–39.

RIHM, R., JACOBS, C. L., KRASTEL, S., SCHMINCKE, H. U. & ALIBÉS, B. 1998. Las Hijas Seamounts, the next Canary Islands? *Terra Nova*, **10**, 121–125.

RINCÓN, R. 1982. Minerales pesados en las facies detríticas del Cretácico inferior de la Cordillera Ibérica suroccidental. *Cuadernos de Geología Ibérica*, **8**, 259–265.

RINCÓN, R., VILAS, L., ARIAS, C., GARCÍA, A., MAS, R., ALONSO, A. & MELÉNDEZ, N. 1983. El Cretácico de las cordilleras intermedias y borde de la Meseta. *In*: COMBA, J. A. (coord.) *Geología de España. Libro Jubilar J. M. Ríos*, vol. II. IGME, Madrid, 79–103.

RÍOS, J. & ALMELA, A. 1954. El Triásico de Santa Perpetua (Tarragona). *Boletín de la Real Sociedad Española de Historia Natural*, Volumen Homenaje a E. Hernández Pacheco, 567–570.

RÍOS, J. M., BELTRÁN, F. J., LANAJA, J. M. & MARIN, F. J. 1979. Contribución a la geología de la Zona Axial Pirenaica, valles del Cinca y Esera, Provincia de Huesca. *Acta Geológica Hispánica*, **14**, 271–279.

RIOS, S. 1977. *Estudio geológico del metalotecto plumbífero del Ordoviciense (La Carolina – Santa Elena, Sierra Morena Oriental)*. PhD Thesis, Polytechnic University of Madrid.

RITSEMA, A. R. 1970. On the origin of the western Mediterranean sea basins. *Tectonophysics*, **10**, 609–623.

RIVAS, M. R. 1991. The development of vegetation and climate during the Miocene in the south-eastern sector of the Duero Basin (Spain). *Review of Palaeobotany and Palynology*, **67**, 341–351.

RIVAS, M. R., ALONSO-GAVILÁN, G., VALLE, M. F. & CIVIS, J. 1994. Miocene palynology of central sector of the Duero basin (Spain) in relation to paleogeography and palaeoenvironment. *Review of Palaeobotany Palynology*, **82**, 251–264.

RIVAS, P. 1973. *Estudio paleontológico estratigráfico del Lias (Sector central de las Cordilleras Béticas)*. PhD Thesis, University of Granada.

RIVIÉRE, M., BELLON, H. & BONNOT-COURTOIS, CH. 1981. Aspects géochimiques et géochronologiques du volcanisme pyroclastique dans le Golfe de Valence: Site 123 DSDP, Leg 13 (Espagne)-Conséquences géodynamiques. *Marine Geology*, **41**, 295–307.

ROBARDET, M. 1976. L'originalité du segment hercynien sud-ibérique au Paléozoïque inférieur: Ordovicien, Silurien et Dévonien dans le nord de la province de Séville (Espagne). *Comptes Rendus de l'Académie des Sciences, Paris*, série D, **283**, 999–1002.

ROBARDET, M. 1981. Late Ordovician tillites in the Iberian Peninsula. *In*: HAMBREY, M. J. & HARLAND, W. B. (eds) *Earth Pre-Pleistocene Glacial Record*. Cambridge University Press, 585–589.

ROBARDET, M. 1982. The Silurian-earliest Devonian succession in South Spain (Ossa Morena Zone) and its paleogeographical signification. *IGCP No. 5 Newsletter*, **4**, 72–77.

ROBARDET, M. & DORÉ, F. 1988. The late Ordovician diamictic formations from southwestern Europe: north-Gondwana glaciomarine deposits. *Palaeogeography, Palaeoclimatology, Palaeoecology*, **66**, 19–31.

ROBARDET, M. & GUTIÉRREZ-MARCO, J. C. 1990a. Ossa Morena Zone. Stratigraphy. Passive margin phase (Ordovician – Silurian – Devonian). *In*: DALLMEYER R. D. & MARTÍNEZ GARCÍA, E. (eds) *Pre-Mesozoic Geology of Iberia*. Springer-Verlag, Berlin, 267–272.

ROBARDET, M. & GUTIÉRREZ-MARCO, J. C. 1990b. Sedimentary and faunal domains in the Iberian Peninsula during Lower Paleozoic times. *In*: DALLMEYER R. D. & MARTÍNEZ GARCÍA E. (eds) *Pre-Mesozoic Geology of Iberia*. Springer-Verlag, Berlin, 383–395.

ROBARDET, M., VEGAS, R. & PARIS, F. 1980. El techo del Ordovícico en el centro de la Península Ibérica. *Studia Geológica Salmanticensia*, **16**, 103–121.

ROBARDET, M., WEYANT, M., LAVEINE, J. P. & RACHEBOEUF, P. R. 1986. Le Carbonifère inférieur du synclinal du Cerrón del Hornillo (province de Seville, Espagne). *Revue de Paléobiologie*, **5**, 71–90.

ROBARDET, M., WEYANT, M., BRICE, D. & RACHEBOEUF, P. R. 1988. Dévonien supérieur et Carbonifère inférieur dans le nord de la province de Seville (Espagne). Age et importance de la prèmiere phase hercynienne dans la zone d'Ossa-Morena. *Comptes Rendus de l'Académie des Sciences*, **307**, 1091–1095.

ROBARDET, M., PARIS, F. & RACHEBOEUF, P. R. 1990. Palaeogeographic evolution of southwestern Europe during Early Palaeozoic times. *In*: McKERROW, W. S. & SCOTESE, C. R. (eds) *Palaeozoic Palaeogeography and Biogeography*. Geological Society, London, Memoirs, **12**, 411–419.

ROBARDET, M., GROSS-UFFENORDE, H., GANDL, J. & RACHEBOEUF, P. R. 1991. Trilobites et Ostracodes du Dévonien inférieur de la Zone d'Ossa Morena (Espagne). *Géobios*, **24**, 333–348.

ROBARDET, M., BLAISE, J., BOUYX, E. *et al.* 1993. Paléogéographie de l'Europe occidentale de l'Ordovicien au Dévonien. *Bulletin de la Société Géologique de France*, **164**, 683–695.

ROBARDET, M., BONJOUR, J. L., PARIS, F., MORZADEC, P. & RACHEBOEUF, P. R. 1994a. Ordovician, Silurian, and Devonian of the Medio-North-Armorican Domain. *In*: KEPPIE, J. D. (ed.) *Pre-Mesozoic Geology in France and Related Areas*. Springer-Verlag, Berlin, 142–151.

ROBARDET, M., VERNIERS, J., FEIST, R. & PARIS, F. 1994b. Le Paléozoïque de France, contexte paléogéographique et géodynamique. *Géologie de la France*, **3**, 3–31.

ROBARDET, M., PIÇARRA, J. M., STORCH, P., GUTIÉRREZ-MARCO, J. C. & SARMIENTO, G. N. 1998. Ordovician and Silurian stratigraphy and faunas (graptolites and conodonts) in the Ossa Morena Zone of the SW Iberian Peninsula (Portugal and Spain). *Temas Geológico-Mineros ITGE*, **23**, 289–318.

ROBARDET, M., RÁBANO, I., GUTIÉRREZ-MARCO, J. C., SARMIENTO, G. N. & VANEK, J. 2000. La 'Caliza de Scyphocrinites' (Silúrico Superior) del Norte de Sevilla: avance de resultados paleontológicos y bioestratigráficos. *1 Congreso Ibérico de Paleontología – XVI Jornadas de la Sociedad española de Paleontología – VIII International Meeting of IGCP 421*, Evora, Libro de Resúmenes, 254–255 (ISBN 972-778-026-1).

ROBARDET, M., PARIS, F. & PLUSQUELLEC, Y. 2001. Comment on 'New Early Devonian paleomagnetic data from NW France: Paleogeography and implications for the Armorican microplate hypothesis' by J. Tait. *Journal of Geophysical Research*, **106**(B7), 13,307–13,310.

ROBERT, J. 1980. *Etude geologique et metellogenetique de la Val de*

Rivas, Pyrenees Espagnoles. PhD Thesis, University of Franche-Compté.

ROBERTS, B., MORRISON, C. & HIRONS, S. 1990. Low grade metamorphism of the Manx Group: A comparative study of white mica crystallinity tchniques. *Journal of the Geological Society London*, **147**, 271–277.

ROBERTSON, A. H. F. & BERNOUILLI, D. 1982. Stratigraphy, facies and significance of Late Mesozoic and Early Tertiary Sedimentary rocks of Fuerteventura (Canary Islands) and Maio (Cape Verde Islands). *In*: VON RAD, HIAZ SAMTHEIN AND SEIBOLD (eds) *Geology of the Northwest African Continental Margin*, 498–525.

ROBERTSON, A. H. F. & STILLMAN, C. J. 1979a. Late Mesozoic sedimentary rocks of Fuerteventura, Canary Islands. Implications for West Africa continental margin evolution. *Journal of the Geological Society of London*, **136**, 47–60.

ROBERTSON, A. H. F. & STILLMAN, C. J. 1979b. Submarine volcanic and associated sedimentary rocks of the Fuerteventura Basal Complex, Canary Islands. *Geological Magazine*, **116**, 203–214.

ROBLES, R. & ALVAREZ NAVA, H. 1988. Los materiales Precámbrico-Cámbricos del Domo de las Hurdes: existencia de tres series sedimentarias separadas por discordancias, SO de Salamnca (Zona Centro Ibérica). *II Congreso Geológico de España*, **1**, 185–188.

ROBLES, S. & QUESADA, S. 1995. La rampa dominada por tempestades del Lias inferior de la zona occidental de la Cuenca Vasco-Cantábrica. *XIII Congreso Español de Sedimentología, libro de comunicaciones*, 109–110.

ROBLES, S., GARCÍA-MONDÉJAR, J. & PUJALTE, V. 1987. Sistemas aluviales pérmicos del área de Peña Labra-Peña Sagra (Cantabria y Palencia). *Cuadernos de Geología Ibérica*, **11**, 5–21.

ROBLES, S., GARCÍA-MONDÉJAR, J. & PUJALTE, V. 1988a. A retreating fan-delta system in the Albian of Biscay, northern Spain: facies analysis and palaeotectonic implications. *In*: NEMEC, W. & STEEL, R. J. (eds) *Fan Deltas: Sedimentology and Tectonic Settings*. Blackie, Edinburgh, 187–211.

ROBLES, S., PUJALTE, V. & GARCÍA-MONDÉJAR, J. 1988b. Evolución de los sistemas sedimentarios del margen continental cantábrico durante el Albiense y Cenomaniense, en la transversal del litoral vizcaíno. *Revista Sociedad Geológica de España*, **1**(3–4), 409–441.

ROBLES, S., PUJALTE, V. & VALLES, J. C. 1989. Sistemas sedimentarios del Jurásico de la parte occidental de la Cuenca Vasco-Cantábrica. *Cuadernos de Geología Ibérica*, **13**, 185–198.

ROBLES, S., PUJALTE, V., HERNÁNDEZ, J. M. & QUESADA, S. 1996. La sedimentación aluvio-lacustre de la Cuenca de Cires (Júrasico sup-Berriasiense de Cantabria): un modelo evolutivo de las cuencas lacustres ligadas a la etapa temprana del rift Nord-Ibérico. *Cuadernos de Geología Ibérica*, **21**, 227–251.

ROCA, E. 1992. *L'estructura de la conca Catalano-Balear: paper de la compressió i de la distensió en la seva gènesi*. PhD Thesis, University of Barcelona.

ROCA, E. 1996. La evolución geodinámica de la cuenca Catalano-Balear y áreas adyacentes desde el Mesozoico hasta la actualidad. *Acta Geologica Hispanica*, **29**(1), 3–25.

ROCA, E. & GUIMERÀ, J. 1992. The Neogene structure of the eastern Iberian margin: structural constraints on the crustal evolution of the Valencia trough (western Mediterranean). *Tectonophysics*, **203**, 203–218.

ROCA, E., GUIMERÀ, J. & SALAS, R. 1994. Mesozoic extensional tectonics in the southeast Iberian Chain. *Geological Magazine*, **131**(2), 155–168.

ROCA, E., SANS, M., CABRERA, LL. & MARZO, M. 1999. Modelo tectosedimentario del sector central y septentrional del margen catalán sumergido (cubetas de Barcelona, Sant Feliu, Begur y Riumors-Roses). *Libro Homenaje a José Ramírez del Pozo*. AGGEP, Madrid, 99–217.

RODAS, M. & LUQUE, F. J. 1992. Yacimientos españoles de talco. *In*: GARCÍA-GUINEA, J. & MARTÍNEZ-FRÍAS, J. (eds) *Recursos Minerales de España*. CSIC, Colección Textos Universitarios, **15**, 1387–1401.

RODEN, M. F. 1981. Origin of coexisting minette and ultra-mafic breccia, Navajo volcanic field. *Contributions to Mineralogy and Petrology*, **77**, 195–206.

RODERO, J. 1999. *Dinámica sedimentaria y modelo evolutivo del margen continental sudoriental del Golfo de Cadiz durante el*

Cuaternario superior (Pleistoceno medio- Holoceno). PhD Thesis, University of Granada.

RODRÍGUEZ, J. A. 1987. Un complejo de diques cónicos en la isla de La Gomera, Islas Canarias. *Estudios Geológicos*, **43**, 41–45.

RODRÍGUEZ, J. A. 1988. *El complejo traquítico fonolítico de La Gomera. Islas Canarias*. PhD Thesis, Complutense University, Madrid.

RODRÍGUEZ, P. 1992. Distrito de Mazarrón Zn + Pb +Ag. Mineralizaciones, potencial y trabajos de evaluación. *Tierra y Tecnología*, **3**, 28–33.

RODRÍGUEZ, S. 1978. Corales rugosos del Devónico de la Sierra del Pedroso. *Estudios Geológicos*, **34**, 331–350.

RODRÍGUEZ, S. 1992. Evolución de la cuenca. In: *Análisis Paleontológico y Sedimentológico de la Cuenca Carbonífera de Los Santos de Maimona (Badajoz)*. Coloquios de Paleontología, Editorial Complutense, Madrid, 249–255.

RODRÍGUEZ, S. 1996. Development of coral reef-facies during the Visean at Los Santos de Maimona, SW Spain. *In*: STRONGEN, P., SOMERVILLE, P. & JONES, G. L. (eds) *Recent Advances in Lower Carboniferous Geology*. Geological Society, London, Special Publications, **107**, 145–152.

RODRÍGUEZ, S. & SOTO, F. 1979. Nuevos datos sobre los corales rugosos del Devónico de la Sierra del Pedroso. *Estudios Geológicos*, **35**, 345–354.

RODRÍGUEZ ALLER, J., COSCA, M., GIL IBARGUCHI, J. I. & DALLMEYER, R. D. 1997. Eo-Hercynian HP metamorphism of a subducted continental crust (Malpica-Tui allochthon, NW Spain): new petrological and age constraints. *Fifth International Eclogite Conference, Ascona*. Terra Nova, **9**, 29.

RODRÍGUEZ ALONSO, M. D. 1985. *El complejo esquisto-grauváquico y el Paleozoico en el centro-oeste español*. Acta Salmanticensia Ciencias, Universidad de Salamanca, **51**.

RODRÍGUEZ-ARANDA, J. P. & CALVO, J. P. 1997. Desarrollo de paleokarstificación en facies yesíferas del Mioceno de la Cuenca de Madrid. Implicaciones en el análisis evolutivo de sucesiones lacustres evaporíticas. *Boletín Geológico Minero*, **108**, 377–392.

RODRÍGUEZ-ARANDA, J. P. & CALVO, J. P. 1998. Trace fossils and rhizoliths as a tool for sedimentological and palaeoenvironmental analysis of ancient continental evaporite successions. *Palaeogeography, Palaeoclimatology, Palaeoecology*, **140**, 383–399.

RODRÍGUEZ-ARANDA, J. P., ROUCHY, J. M., CALVO, J. P., ORDÓÑEZ, S. & GARCÍA DEL CURA, M. A. 1995. Unusual twinning features in large primary gypsum crystals formed in salt lake conditions, Middle Miocene, Madrid Basin, Spain – palaeoenvironmental implications. *Sedimentary Geology*, **95**, 123–132.

RODRÍGUEZ-CAÑERO, R. 1993. *Contribución al estudio de los conodontos del Paleozoico del Complejo Maláguide (Cordillera Bética)*. PhD Thesis, University of Málaga.

RODRÍGUEZ ESTRELLA, T. 1977. Síntesis geológica del Prebético de la provincia de Alicante. 1: Estratigrafía. *Boletín Geológico y Minero*, **88**(3), 183–214.

RODRÍGUEZ ESTRELLA, T. 1979. *Geología e Hidrogeología del Sector de Alcaraz – Lietor-Yeste (Provincia de Albacete)*. Memoria IGME, **97**.

RODRÍGUEZ-FERNÁNDEZ, J. 1982. *El Mioceno del sector central de las Cordilleras Béticas*. PhD Thesis, University of Granada.

RODRÍGUEZ-FERNÁNDEZ, J. & MARTÍN-PENELA, A. J. 1993. Neogene evolution of the Campo-de-Dalias and the surrounding offshore areas – (northeastern Alboran Sea). *Geodinamica Acta*, **6**, 255–270.

RODRÍGUEZ FERNÁNDEZ, L. R. 1994. *Estratigrafía y estructura de la región de Fuentes Carrionas y áreas adyacentes (Cordillera Herciniana, NO de España)*. Laboratorio Xeolóxico de Laxe, Sada (La Coruña), Nova Terra, **9**.

RODRÍGUEZ FERNÁNDEZ, L. R. & HEREDIA, N. 1987. La estratigrafía del Carbonífero y la estructura de la unidad del Pisuerga-Carrión. NO de España. *Cuadernos do Laboratorio Xeolóxico de Laxe, Sada (La Coruña)*, **12**, 207–229.

RODRÍGUEZ FERNÁNDEZ, L. R. & HEREDIA, N. 1990. Structure: Palentian Zone. *In*: DALLMEYER, R. D. & MARTÍNEZ GARCÍA, E. (eds) *Pre-Mesozoic Geology of Iberia*. Springer-Verlag, Berlin, 69–71.

RODRÍGUEZ FERNÁNDEZ, L. R., HEREDIA, N., LOBATO, L. & VELANDO, F. 1985. *Mapa y Memoria de la Hoja no. 106 (Camporredondo)*

del Mapa Geológico de España a escala 1:50.000. Serie MAGNA. IGME, Madrid.

RODRÍGUEZ GONZÁLEZ, R. M. 1983. *Palinología de las formaciones del Silúrico superior-Devónico inferior de la Cordillera Cantábrica, Noroeste de España*. Institución Fray Bernardino de Sahagún-Servicio de Publicaciones de la Universidad de León.

RODRIGUEZ-LÁZARO, J., PASCUAL, A. & ELORZA, J. 1998. Cenomanian events in the deep western Basque Basin, the Leioa section. *Cretaceous Research*, **19**, 673–700.

RODRÍGUEZ NÚÑEZ, V. M., GUTIÉRREZ-MARCO, J. C. & SARMIENTO, G. 1989. Rasgos bioestratigráficos de la sucesión silúrica del Sinclinal de Guadarranque (provincias de Cáceres, Badajoz y Ciudad Real). *Coloquios de Paleontología*, **42**, 83–106.

RODRÍGUEZ-PEREA, A. 1984. *El Mioceno de la Serra Nord de Mallorca. Estratigrafía, sedimentología e implicaciones estructurales*. PhD Thesis, University of Barcelona.

RODRÍGUEZ-PEREA, A. & POMAR, L. 1983. El Mioceno de la Sierra Norte de Mallorca (sector centro-occidental). *Acta Geológica Hispánica*, **18**, 105–116.

RODRÍGUEZ-PEREA, A., RAMOS-GUERRERO, E., POMAR, L., PANIELLO, X., OBRADOR, A. & MARTÍ, J. 1987. El Triásico de las Baleares. *Cuadernos Geología de Ibérica*, **11**, 295–321.

RODRÍGUEZ RAMÍREZ, A., RODRÍGUEZ VIDAL, J., CÁCERES, L. *et al.* 1996. Recent coastal evolution of the Doñana National Park (SW Spain). *Quaternary Science Reviews*, **15**, 803–809.

RODRÍGUEZ-TOVAR, F. J. 1993. *Evolución sedimentaria y ecoestratigráfica en plataformas epicontinentales del margen Sudibérico durante el Kimmeridgiense inferior*. PhD Thesis, University of Granada.

RODRÍGUEZ VIDAL, J., CÁCERES, L. & RODRÍGUEZ RAMÍREZ, A. 1991. La red fluvial cuaternaria en el piedemonte de Sierra Morena Occidental. *Cuadernos de Investigación Geográfica, Universidad de la Rioja*, **17**, 37–45.

ROEP, T. B. 1972. Stratigraphy of the Permo-Triassic Saladilla formation and its tectonic setting in the Betic of Malaga (Vélez Rubio region, SE Spain). *Koninklijke Nederlandse Académie van Wetenschappen Series B*, **75**, 223–247.

ROEP, TH. B. 1980. Condensed Cretaceous limestones in a section near Xiquena, Betic of Malaga, SE Spain. *Proceedings of the Koninklijke Nederlandse Akademie Van Wetenschappen*, **93**, 183–200.

ROEST, W. R. & SRIVASTAVA, S. P. 1991. Kinematics of the plate boundaries between Eurasia, Iberia and Africa in the North Atlantic from the Late Cretaceous to the present. *Geology*, **19**, 613–616.

ROGER, P. 1970. Note preliminaire sur l'etude du 'gres rouge' de l'extremité occidentale de la zone des Nogueres (Huesca). *Comptes Rendus Société Geologiqué France*, **3**, 109–110.

ROLDÁN-GARCÍA, F. J. 1995. *Evolución Neógena de la Cuenca del Guadalquivir*. PhD Thesis, University of Granada.

RÖLZ, P. 1975. Beiträge zum Aufban des jungpräkambrischen und altpaläozoischen Grundgebirges in den Provinzen Salamanca und Cáceres (Sierra de Tamames, Sierra de Francia und östliche Sierra de Gato), Spanien. *Münstersche Forschungen zur Geologie und Paläontologie*, **36**, 1–68.

ROMANO, M. 1982. The Cambrian–Ordovician boundary in Spain, Portugal, and north-west France. *In*: BASSET, M. G. & DEAN, W. T. (eds) *The Cambrian-Ordovician Boundary: Sections, Fossil Distributions, and Correlations*. National Museum of Wales, Cardiff, Geological Series, **3**, 71–75.

ROMANO, M. 1991. Lower to Middle Ordovician trace fossils from the Central Iberian Zone of Portugal and Spain. *In*: BARNES, C. R. & WILLIAMS, S. H. (eds) *Advances in Ordovician Geology*. Geological Survey of Canada, Paper **90–9**, 191–204.

ROMÃO, J. M., GUTIÉRREZ-MARCO, J. C., RÁBANO, I., OLIVEIRA, J. T. & MARQUES GUEDES, A. 1995. A Formação de Cabeço do Peão (Ordovícico Superior) no sinforma Amêndoa-Carvoeiro (SW da ZCI) e a sua correlaçâo estratigráfica na província mediterrânica. *Memórias do Museu e Laboratório Mineralógico e Geológico da Universidade do Porto*, **4**, 121–126.

ROMARIZ, C. 1962. Graptolitos do Silúrico Português. *Revista de Faculdade de Ciências, Universidade de Lisboa, Série C*, **10**, 115–312.

ROMARIZ, C. 1963. Graptolitos da colecçâo de Nery Delgado provenientes de jacidas espanholas. *Boletim do Museu e Laboratório*

Mineralógico e Geológico da Faculdade de Ciências, Lisboa, **9**, 131–134.

ROMARIZ, C. 1969. Graptolitos silúricos do Noroeste Peninsular. *Comunicações dos Serviços Geológicos de Portugal*, **53**, 107–155.

ROMER, R. L. & SOLER, A. 1995. U-Pb age and lead isotopic characterization of Au-bearing skarn related to the Andorra granite (Pyrenees, Spain). *Mineralium Deposita*, **30**, 374–383.

ROMERO, J. M. & OLIVEROS, J. M. 1986. El mercurio. *Papeles de economía española*, **29**, 282–303.

ROQUÉ, J. 1999. La Biozona *ascensus-acuminatus* en el Silúrico de las Cadenas Costeras Catalanas (NE de España). *Temas Geológico-Mineros ITGE*, **26**, 632–637.

ROSALES, I. 1995. *La plataforma carbonatada de Castro Urdiales (Aptiense/Albiense, Cantabria)*. PhD Thesis, University of País Vasco.

ROSALES, I. 1999. Controls on carbonate platform evolution on active fault blocks: the Lower Cretaceous Castro Urdiales platform (Aptian-Albian, northern Spain). *Journal of Sedimentary Research*, **69**, 447–465.

ROSALES, I., ARANBURU, A. & GARCÍA-MONDÉJAR, J. 1989. Estratigrafía del Urgoniano entre San Pedro de Galdames y Zalla (Aptiense/Albiense medio, Bizkaia, Euskal Herria). *Kobie*, **18**, 87–96.

ROSALES, I., FERNÁNDEZ-MENDIOLA, P. A. & GARCÍA-MONDÉJAR, J. 1994. Carbonate depositional sequence development on active fault blocks: the Albian in the Castro Urdiales area, northern Spain. *Sedimentology*, **41**, 861–882.

ROSALES, I., MEHL, D., FERNÁNDEZ-MENDIOLA, P. A. & GARCÍA-MONDÉJAR, J. 1995. An unusual poriferan community in the Albian of Islares (north Spain): Palaeoenvironmental and tectonic implications. *Palaeogeography, Palaeoclimatology, Palaeoecology*, **119**, 47–61.

ROSALES, I., QUESADA, S & ROBLES, S. 2001. Primary and diagenetic isotopic signal in fossils and hemipelagic carbonates: the lower Jurassic of northern Spain. *Sedimentology*, **48**, 5, 1149–1169.

ROSAS, A. 1997. A gradient of size and shape for the Atapuerca sample and Middle Pleistocene hominid variability. *Journal of Human Evolution*, **33**, 319–331.

ROSAS, A. 1998. Modelos de crecimiento en mandíbulas fósiles de homínidos. Atapuerca un nuevo paradigma. *In*: AGUIRRE, E. (ed.) *Atapuerca y la evolución humana*. Fundación Ramón Areces, Madrid, 239–275.

ROSAS, A. & AGUIRRE, E. 1999. Restos humanos neandertales de la Cueva del Sidrón, Piloña, Asturias. Nota preliminr. *Estudios Geológicos*, **55**, 181–190.

ROSAS, A. & BERMÚDEZ DE CASTRO, J. M. 1999a. The ATD-5 mandibular specimen from Gran Dolina (Atapuerca, Spain). Morphological study and phylogenetic implications. *Journal of Human Evolution*, **37**, 567–590.

ROSAS, A. & BERMÚDEZ DE CASTRO, J. M. 1999b. Descripción y posición evolutiva de la mandíbula AT76–T1H del yacimiento de Galería (Sierra de Atapuerca). *In*: CARBONELL, E., ROSAS, A. & DÍEZ, J. C. (eds) *Atapuerca: Ocupacines humanas y paleoecología del yacimiento de Galería*. Junta de Castilla y León, Arqueología en Castilla y León, Memorias, **7**, 237–243.

ROSE, J., MENG, X. & WATSON, C. 1999. Palaeoclimate and palaeoenvironmental responses in the western Mediterranean over the last 140 Ka: evidence from Mallorca, Spain. *Journal of the Geological Society, London*, **156**, 435–448.

ROSELL, J. 1967. Estudio geológico del sector del pre-Pirineo comprendido entre los ríos Segre y Noguera Ribagorzana (prov. de Lérida). *Revista Pirineos*, **21**, 9–214.

ROSELL, J. & ELÍZAGA, E. 1989. Evolución tectosedimentaria del Paleozoico de la Isla de Menorca. *Boletín Geológico y Minero*, **100**, 193–204.

ROSELL, J. CÁCERES, I. & HUGUET, R. 1998. Systèmes d'occupation anthropique pendant le Pleistocène Inférieur et Moyen à la Sierra de Atapuerca (Burgos, Espagne). *Quaternaire*, **9**, 355–360.

ROSENBAUM, J. M. & WILSON, M. 1997. Multiple enrichment of the Carpathian-Pannonian mantle: Pb-Sr-Nd isotope and trace element constraints. *Journal of Geophysical Research*, **102**, 14947–14961.

ROTHE, P. & SCHMINCKE, H. U. 1968. Contrasting origins of the eastern and western islands of the Canarian Archipelago. *Nature*, **218**, 1152–1154.

ROTHEMBERG, B. & GARCÍA PALOMERO, F. 1986. *The Rio Tinto Enigma No More*. IAMS (Institute for Archeo-Metallurgica Studies) no. 8. London.

ROTHPLETZ, A. & SIMONELLI, V. 1980. Die marinen Ablagerungen auf Gran Canaria. *Zeitschrift der deustschen geologischen Gesellschaft*, **42**, 677–736.

ROURE, F., CHOUKROUNE, P., BERÁSTEGUI, X. *et al.* 1989. ECORS Deep Seismic data and balanced cross-sections, geometric constraints to trace the evolution of the Pyrenees. *Tectonics*, **8**, 41–50.

ROUTHIER, P., AYE, F., BOYER, C., LÉCOLLE, M., MOLIÈRE, P., PICOT, P. & ROGER, G. 1978. *La ceinture sud-ibérique à amas sulfurés dans sa partie espagnole médiane. Tableau géologique et métallogénique. Synthèse sur le type amas sulfurés volcano-sédimentaires*. BRGM, *Mémoires*, **94**.

ROUX, L. 1977. *L'évolution des roches des faciès granulite et le problème des ultramafitites dans le massif de Castillon*. PhD Thesis, Paul Sabatier University, Toulouse.

ROWLAND, S. M. & GANGLOFF, R. A. 1988. Structure and palaeoecology of Lower Cambrian reefs. *Palaios*, **3**, 111–135.

ROYDEN, L. H. 1993. Evolution of retreating subduction boundaries formed during continental collision. *Tectonophysics*, **12**, 629–638.

ROYO GÓMEZ, J. 1929. *Cartografía geológica y memoria de la hoja geológica a escala 1:1.000.000 de Madrid (559). 1ª Edición*. ITME, Madrid.

RUA, C. DE LA 1985. Restos humanos de Erralla. *Munibe*, **37**, 195–198.

RUANO, P., GALINDO-ZALDÍVAR, J. & JABALOY, A. 2000. Evolución geológica desde el Mioceno del sector noroccidental de la depresión de Granada (Cordilleras Béticas). *Revista de la Sociedad Geológica de España*, **13**, 148–155.

RUBIO, J. 1997. *Inventario Nacional de recursos de cloruro sódico y sales potásicas*. IGME. Madrid (ISBN 84-7840-288-8).

RUIZ, C. 1976. Génesis de los depósitos de hierro de SW de la provincia de Badajoz: mina Monchi. *Boletín Geológico y Minero*, **87**, 15–31.

RUIZ-LÓPEZ, J. L., LEÓN-COULLAUT, J., SOLER, M., BABIANO, F., FERNÁNDEZ, J. & APALATEGUI, O. 1979. *Mapa geológico de España a E. 1:50.000 no. 896 (Higuera la Real)*. 2ª serie-MAGNA. IGME, Madrid.

RUIZ-ORTIZ, P. A. 1980. *Análisis de facies del Mesozoico de las Unidades Intermedias (entre Castril – provincia de Granada- y Jaén)*. PhD Thesis, University of Granada.

RUIZ-ORTIZ, P. A. & CASTRO, J. M. 1998. Carbonate depositional sequences in shallow to hemipelagic platform deposits, Aptian, Prebetic of Alicante (SE Spain). *Bulletin de la Société Géologique de France*, **169**, 21–33.

RUIZ-ORTIZ, P. A., MOLINA, J. M. & NIETO, L. M. 1996. Turbiditas calcáreas y otros fenómenos de resedimentación en el Jurásico Superior – Cretácico Inferior de la unidad de Huelma (Jaén). Zonas Externas de las Cordilleras Béticas. *Geogaceta*, **20**, 323–384.

RUIZ-ORTIZ, P. A., GEA, G. A. DE & AGUADO, R. 2001. Cañón submarino del Cretácico Inferior en un área de pendiente del paleomargen sudibérico. Unidad de Huelma (Jaén). *Revista de la Sociedad Geológica de España*, **14**(3–4), 175–188.

RUSHTON, A. W. A. & POWELL, J. H. 1998. A review of the stratigraphy and trilobite faunas from the Cambrian Burj Formation in Jordan. *Bulletin of The Natural History Museum (Geology Series)*, **54**, 131–146.

RUSSO, A. & BECHSTÄDT, T. 1994. Evolución sedimentológica y paleogeográfica de la Formación Vegadeo (Cámbrico inferior-medio) en la zona entre Visuña y Piedrafita do Caurel (Lugo, NO de España). *Revista de la Sociedad Geológica de España*, **7**, 299–310.

RYAN, W. B. F., HSÜ, K. J., HONNOREZ, J. *et al.* 1972. Petrology and geochemistry of the Valencia trough volcanic rocks. *Initial Reports of the Deep Sea Drilling Project*, **13**, 767–773.

RYTUBA, J. J., ARRIBAS JR., A., CUNNINGHAM, C. G. *et al.* 1990. Mineralized and unmineralized calderas in Spain; Part II, evolution of the Rodalquilar caldera complex and associated gold-alunite deposits. *Mineralium Deposita*, **25**, S29–S35.

SÀBAT, F., MUÑOZ, J. A. & SANTANACH, P. 1988. Transversal and oblique structures at the Serres de LLevant thrust belt (Mallorca Island). *Geologische Rundschau*, **77**, 529–538.

SÀBAT, F., ROCA, E., MUÑOZ, J. A. *et al.* 1995. Role of extension and compression in the evolution of the eastern margin of Iberia: the ESCI-València Trough seismic profile. *Revista de la Sociedad Geológica de España*, **8**, 431–448.

SABATIER, H. 1991. Vaugnerites, special lamprophyre-derived mafic enclaves in some Hercynian granites from Western and Central Europe. *In*: DIDIER, J. & BABARIN, B. (eds) *Enclaves and Granite Petrology*. Elsevier, Amsterdam. 63–81.

SACHER, L. 1966. Uber Karbonische Sedimente bei Montalbán in den Östlichen Iberischen Ketten (Spanien). *Neues Jahrbuch für Geologie und Paläontologie Monatshefte*, **7**, 437–443.

SÁENZ, C. & LÓPEZ-MARINAS, J. M. 1975. La edad del vulcanismo de Cofrentes (Valencia). *Tecniterrae*, **6**, 8–14.

SÁEZ, R. & MORENO, C. 1997. Geology of the Puebla de Guzmán Anticlinorium. *In*: BARRIGA, F. & CARVALHO, D. (eds) *Geology and VMS Deposits of the Iberian Pyrite Belt*. Society of Economic Geologists, Guidebook Series, Littleton (Colorado), **27**, 131–136.

SÁEZ, R., ALMODÓVAR, G. R. & PASCUAL, E. 1988. Mineralizaciones estratoligadas de scheelita en la Faja Pirítica del Suroeste Ibérico. *Boletin de la Sociedad Española de Mineralogía*, **1**, 135–141.

SÁEZ, R., ALMODÓVAR, G. R. & PASCUAL, E. 1996. Geological constraints on massive sulphide genesis in the Iberian Pyrite Belt. *Ore Geology Reviews*, **11**, 429–451.

SÁEZ, R., ALMODÓVAR, G. R. & PASCUAL, E. 1999a. Reply to the comments by C. H. Boulter on 'Geological constraints on massive sulphide genesis in the Iberian Pyrite Belt'. *Ore Geology Reviews*, **14**, 151–153.

SÁEZ, R. PASCUAL. E., TOSCANO, M. & ALMODÓVAR, G. R. 1999b. The Iberian type of volcano-sedimentary massive sulphide deposits. *Mineralium Deposita*, **34**, 549–570.

SAGREDO, J. 1972. Enclaves peridotíticos encontrados en los afloramientos basálticos al noroeste de Cartagena (provincia de Murcia). *Estudios Geológicos*, **28**, 119–135.

SAGREDO, J. 1973. Estudio de las inclusiones de rocas ultramáficas con anfíbol que aparecen en los basaltos al noroeste de Cartagena (provincia de Murcia). *Estudios Geológicos*, **29**, 53–62.

SAGREDO, J. & PEINADO, M. 1992. Vulcanismo Cámbrico de la Zona de Ossa-Morena. *Paleozoico Inferior de Ibero-América*. University of Extremadura, 567–576.

SAGREDO, J., MUÑOZ, M. & GALINDO, C. 1996. Características petrológicas y edad K-Ar de las sienitas nefelínicas del Morro del Recogedero (Fuerteventura, Islas Canarias). *Geogaceta*, **20**(2), 506–509.

SALA, M. 1994. Sistema Costero Catalán. *In*: GUTIÉRREZ, M. (ed.) *Geomorfología de España*. Rueda, Madrid, 287–303.

SALAS, R. 1989. Evolución estratigráfica secuencial y tipos de plataformas de carbonatos del intervalo Oxfordiense-Berriasiense en las Cordilleras Ibérica Oriental y Costero Catalana Meridional. *Cuadernos de Geología Ibérica*, **13**, 121–157.

SALAS, R. & CASAS, A. 1993. Mesozoic extensional tectonics, stratigraphy, and crustal evolution during the Alpine cycle of the eastern Iberian basin. *Tectonophysics*, **228**, 33–55.

SALAS, R. & GUIMERÀ, J. 1996. Main structural features of the lower Cretaceous Maestrat basin (Eastern Iberian Range). *Geogaceta*, **20**(7), 1704–1706.

SALAS, R. & GUIMERÀ, J. 1997. Estructura y estratigrafía secuencial de la cuenca de Maestrazgo durante la etapa de rift jurásica superior-cretácica inferior (Cordillera Ibérica oriental). *Boletín Geológico y Minero*, **108**(4–5), 393–402.

SALAS, R., GUIMERÀ, J., MAS, R., MARTÍN-CLOSAS, C., MELÉNDEZ, A. & ALONSO, A. 2001. Evolution of the Mesozoic Central Iberian Rift System and its Cainozoic inversion (Iberian Chain). *In*: ZIEGLER, P. A., CAVAZZA, W., ROBERTSON, A. F. H. & CRASQUIN-SOLEAU, S. (eds) *Peri-Tethys Memoir 6: Peri-Tethyan Rift/Wrench Basins and Passive Margins*. Mémoires du Muséum national d'Histoire naturelle, **186**, 145–185.

SALDAÑA, M. J. 1993. *Análisis estratigráfico y sedimentológico de la cuenca carbonífera de Guardo-Cervera*. PhD Thesis, University of Oviedo.

SALOMÓN, J. 1982. El Cretácico inferior. *In*: García, A. (coord.) *El Cretácico de España*. Complutense University, Madrid, 345–387.

SALPETEUR, I. 1976. *Etude structurales et pétrographique de la zone de Paymogo (Nord de la Province d'Huelva). Métallogenèse des amas sulfurés asocies*. PhD Thesis, University of Nancy.

SALVADOR, C. I. 1989. *Estratigrafía y Sedimentología del Norte de la Cuenca Carbonífera Central Asturiana*. PhD Thesis, University of Oviedo.

SALVADOR, C. I. 1993. La sedimentación durante el Westfaliense en una cuenca de antepaís (Cuenca Carbonífera Central de Asturias, N de España). *Trabajos de Geología*, **19**, 195–264.

SALVANY, J. M. 1986. *El Keuper dels Catalànids. Petrologia i sedimentologia*. Internal Report, University of Barcelona.

SALVANY, J. M. 1989. Los sistemas lacustres evaporíticos del sector navarro-riojano de la Cuenca del Ebro durante el Oligoceno y Mioceno inferior. *Acta Geológica Hispánica*, **24**, 231–241.

SALVANY, J. M. 1990. Introducción a las evaporitas triásicas de las cadenas perioféricas de la cuenca del Ebro: Catalánides, Pirineo y región Cantábrica. *In*: ORTÍ, F. & SALVANY, J. (eds) *Formaciones evaporíticas de la Cuenca del Ebro y cadenas periféricas y de la zona de Levante*. Enresa, Madrid, 9–20.

SALVANY, J. M. 1997. Continental evaporitic sedimentation in Navarra during the Oligocene to Lower Miocene: Falces and Lerín Formations. *In*: BUSSON, G. & SCHEIBER, B. C. (eds) *Sedimentary Deposition in Rift and Foreland Basins in France and Spain*. Columbia University Press, New York, 397–410.

SALVANY, J. M. & ORTÍ, F. 1987. El Keuper de los Catalánides. *Cuadernos de Geología Ibérica*, **11**, 215–236.

SALVIOLI, E. & VENTURELLI, G. 1996. Temperature of crystallization and evolution of the Jumilla and Cancarix lamproites (SE Spain) as suggested by melt and solid inclusions in minerals. *European Journal of Mineralogy*, **8**, 1027–1039.

SAMSON, S. D. & D'LEMOS, R. S. 1999. A precise late Neoproterozoic U-Pb zircon age for the syntectonic Perelle quartz diorite, Guernsey, Channel Islands, UK. *Journal of the Geological Society, London*, **156**, 47–54.

SAN JOSÉ, M. A. 1983. El complejo sedimentario pelítico-grauváquico. *In*: Comba, J. A. (ed.) *Libro Jubilar JM Ríos*, Vol. 1. IGME, Madrid, 91–99.

SAN JOSÉ, M. A. 1984. Los materiales anteordovícicos del anticlinal de Navalpino. *Cuadernos de Geología Ibérica*, **9**, 81–117.

SAN JOSÉ, M. A., PELÁEZ PRUNEDA, J. R., VILAS MINONDO, L. & HERRANZ ARAÚJO, P. 1974. Las series ordovícicas y preordovícicas del sector central de los Montes de Toledo. *Boletín Geológico y Minero*, **85**, 21–31.

SAN JOSÉ, M. A., PIEREN, A. P., GARCÍA HIDALGO, J. F., VILAS, L., HERRANZ, P., PELÁEZ, J. R. & PEREJÓN, A. 1990. Autochthonous sequences in the Central Iberian Zone: ante-Ordovician stratigraphy. *In*: DALLMEYER, R. D. & MARTÍNEZ GARCÍA, E. (eds) *Pre-Mesozoic Geology of Iberia*. Springer Verlag, Berlin, 147–159.

SAN JOSÉ, M. A., RÁBANO, I., HERRANZ, P. & GUTIÉRREZ-MARCO, J. C. 1992. El Paleozoico Inferior de la Zona Centroibérica meridional. *In*: GUTIÉRREZ-MARCO, J. C., SAAVEDRA, J. & RÁBANO, I. (eds) *Paleozoico Inferior de Ibero-América*. University of Extremadura, 505–521.

SAN MIGUEL, A. 1966. Estudio petrológico de los diques de pórfidos graníticos de la garganta del Ter entre el Pasteral y Susqueda (Gerona). *Instituto Investigación Geología Diputación de Barcelona*, **20**, 73–81.

SAN MIGUEL DE LA CÁMARA, M. 1936. *Estudio de las rocas eruptivas de España*. Memorias de la Academia de Ciencias de Madrid, Serie Ciencias Naturales, **VI**.

SAN ROMÁN, J. & AURELL, M. 1992. Palaeogeographical significance of the Triassic-Jurassic unconformity in the north Iberian basin (Sierra del Moncayo, Spain). *Palaeogeography, Palaeoclimatology, Palaeoecology*, **99**, 101–117.

SÁNCHEZ, A. & ALVAREZ, J. J. 1994. *Identificación de los reagaíos con aguas subterráneas en España*. Symposium Nacional: Presente y Futuro de los regadios españoles. CEDE-COIA (Centro de Experimentación de Obras Públicas), Madrid.

SÁNCHEZ, A. & BLANCO, J. A. 1999. La depresión terciaria de Zarza de Granadilla y sus bordes. *Studia Geológica Salmanticensia*, **7**, 101–120.

SÁNCHEZ, F. 1991. *Evolución estructural post-kimmérica de la plataforma continental vasco-cantábrica.* PhD Thesis, Polytechnic University of Madrid.

SÁNCHEZ, J. A., SAN ROMÁN SALDAÑA, J., DE MIGUEL CABEZA , J. L. & MARTÍNEZ GIL, F. J. 1990. El drenaje subterráneo de la Cordillera Ibérica en la depresión del Ebro: aspectos geológicos. *Geogaceta*, **8**, 115–118.

SÁNCHEZ-ALMAZO, I. M., SPIRO, B, BRAGA, J. C. & MARTÍN, J. M. 2001. Constraints of stable isotope signatures on the depositional palaeoenvironments of upper Miocene reef and temperate carbonates in the Sorbas Basin, SE Spain. *Palaeogeography, Palaeoclimatology, Palaeoecology*, **175**, 153–172.

SÁNCHEZ-CARRETERO, R., CARRACEDO, M., EGUILUZ, L., GARROTE, A. & APALATEGUI, O. 1989a. El magmatismo calcoalcalino del Precámbrico terminal en la Zona de Ossa-Morena (Macizo Ibérico). *Revista de la Sociedad Geológica de España*, **2**, 7–21.

SANCHEZ-CARRETERO, R., CARRACEDO, M., GIL IBARGUCHI, J. I., ORTEGA, L. A. & CUESTA, A. 1989b. Unidades y datos geoquímicos del magmatismo hercínico de la 'Alineación de Villaviciosa de Córdoba-La Coronada' (Ossa-Morena Oriental). *Studia Geológica Salmanticense*, **4**, 105–130.

SÁNCHEZ-CARRETERO, R., EGUÍLUZ, L., PASCUAL, E. & CARRACEDO, M. 1990. Igneous rocks of the Ossa Morena Zone. *In*: DALLMEYER, R. D. & MARTÍNEZ GARCÍA, E. (eds) *Pre-Mesozoic Geology of Iberia*. Springer-Verlag, Berlin, 292–313.

SÁNCHEZ-CARRETERO, R., CARRACEDO, M., EGUILUZ, L. & ALONSO OLAZABAL, A. 1999. Magmatismo alcalino tardicadomiense en la zona de Ossa–Morena (Macizo Ibérico): Cartografía, petrografía y geoquímica preliminar del Macizo de Almendral. *Geogaceta*, **26**, 87–90.

SÁNCHEZ-CELA, V. 1968. Estudio petrológico de las sucesiones volcánicas del sector central de la formación del Cabo de Gata (Almería). *Estudios Geológicos*, **24**, 1–38.

SÁNCHEZ DE LA TORRE, L., ÁGUEDA, J., COLMENERO, J. & MANJÓN, M. 1977. La serie permotriásica en la región de Villaviciosa (Asturias). *Cuadernos de Geología Ibérica*, **4**, 329–338.

SÁNCHEZ DE LA TORRE, L., ÁGUEDA-VILLAR, J. A., COLMENERO-NAVARRO, J. R., GONZÁLEZ-LASTRA, J. A. & MARTÍN-LLANEZA, J. 1981. Emplazamiento de deltas progradantes y facies asociadas en el Westfaliense superior del borde oriental de la Cuenca Carbonífera Central (Asturias). *Trabajos de Geología*, **11**, 191–201.

SÁNCHEZ DE LA TORRE, L., ÁGUEDA, J., COLMENERO, J. R., GARCÍA-RAMOS, J. C. & GONZÁLEZ LASTRA, J. 1983. Evolución sedimentaria y paleogeográfica del Carbonífero en la Cordillera Cantábrica. *In*: MARTÍNEZ DÍAZ, C. (ed.) *Carbonífero y Pérmico de España*. IGME, Madrid, 133–150.

SÁNCHEZ DE POSADA, L. C., MARTÍNEZ CHACÓN, M. L., MÉNDEZ FERNÁNDEZ, C., MENÉNDEZ ALVAREZ, J. R., TRUYOLS, J. & VILLA, E. 1990. Carboniferous Pre-Stephanian rocks of the Asturian-Leonese Domain (Cantabrian Zone). *In*: DALLMEYER, R. D. & MARTÍNEZ GARCÍA, E. (eds) *Pre-Mesozoic Geology of Iberia*. Springer-Verlag, Berlin, 24–33.

SÁNCHEZ DE POSADA, L. C., VILLA, E., MARTÍNEZ CHACÓN, M. L., RODRÍGUEZ, R. M., RODRÍGUEZ, S. & COQUEL, R. 1999. Contenido paleontológico y edad de la sucesión de Demúes (Carbonífero, Zona Cantábrica). *Trabajos de Geología*, **21**, 339–352.

SÁNCHEZ FERRER, F. 1991. *Evolución estructural post-Kimmérica de la plataforma continental Vasco-Cantábrica.* PhD Thesis, Polytechnic University of Madrid.

SÁNCHEZ-GÓMEZ, M., GARCÍA-DUEÑAS, V. & MUÑOZ, M. 1995. Relations structurales entre les Péridotites de Sierra Bermeja et les unités alpujarrides sous-jacentes (Benahavís, Ronda, Espagne). *Comptes Rendus de l'Académie des Sciences de Paris*, **321**, 885–892.

SÁNCHEZ-MOYA, Y. 1992. *Evolución sedimentológica y controles estructurales de un borde de cuenca extensional: Comienzo del Mesozoico en un sector del margen occidental de la Cordillera Ibérica.* PhD Thesis, Complutense University, Madrid.

SÁNCHEZ-MOYA, Y., SOPEÑA, A., MUÑOZ, A. & RAMOS, A. 1992. Consideraciones teóricas sobre el análisis de la subsidencia: aplicaciones a un caso real en el borde de la cuenca triásica Ibérica. *Revista de la Sociedad Geológica de España*, **5**(3–4), 21–39.

SÁNCHEZ-MUÑOZ, L. & GARCÍA-GUINEA, J. 1992. Feldespatos: Yacimientos, Mineralogía y Aplicaciones. *In*: GARCÍA-GUINEA, J. & MARTÍNEZ-FRÍAS, J. (eds) *Recursos Minerales de España*. CSIC, Colección Textos Universitarios, **15**, 441–470.

SÁNCHEZ-RODRÍGUEZ, L. 1998. *Pre-Alpine and Alpine evolution of the Ronda Ultramafic Complex and its country-rocks (Betic chain, southern Spain): U-Pb SHRIMP zircon and fission-track dating.* PhD Thesis, Swiss Federal Institute of Technology, Zürich.

SÁNCHEZ SERRANO, F., DE VICENTE MUÑOZ, G. & GONZÁLEZ CASADO, J. M. 1993. Cortes compensados para la deformación principal alpina en el borde sur oriental del Sistema Central español (Zona de Tamajón, Guadalajara). *Revista de la Sociedad Geológica de España*, **6**, 7–14.

SANCHO, C. & MELÉNDEZ, A. 1992. Génesis y significado ambiental de los caliches pleistocenos de la región del Cinca (Depresión del Ebro). *Revista de la Sociedad Geológica de España*, **5**, 81–93.

SANCHO, C., GUTIÉRREZ, M., PEÑA, J. L. & BURILLO, F. 1988. A quantitative approach to scarp retreat starting from triangular slope facets, central Ebro Basin, Spain. *Catena Supplement*, **13**, 139–146.

SANCHO, C., LEWIS, C. J., MCDONALD, E. V. & PEÑA, J. L. 2000. Primeros datos sobre la relación entre el levantamiento postorogénico de los Pirineos y el patrón de encajamiento fluvial del río Cinca durante el Cuaternario (Depresión del Ebro). *VI Reunión Nacional de Geomorfología*. Madrid, 49.

SANDOVAL, J. 1983. *Bioestratigrafía y Paleontología (Stephanocerataceae y Perisphinctaceae) del Bajocense y Bathonense de las Cordilleras Béticas.* PhD Thesis, University of Granada.

SANDOVAL, J. 1990. A revision of the Bajocian divisions in the Subbetic Domain (southern Spain). *Memorie Descrittive della Carta Geologica d'Italia*, **40**, 141–162.

SANDOVAL, J. 1994. The Bajocian stage in the Island of Majorca: Biostratigraphy and ammonite assemblages. *Miscellania del Servizio Geológico Nazionale*, **5**, 203–215.

SANDOVAL, J., HENRIQUES, M. H., URETA, S., GOY, A. & RIVAS, P. 2001. The Lias/Dogger boundary in Iberia: Betic and Iberian Cordilleras and Lusitanian basin. *Bulletin de la Société Geologique de France*, **172**, 385–387.

SANS, M., MUÑOZ, J. A. & VERGÉS, J. 1996. Triangle zone and thrust wedge geometries related to evaporitic horizons (southern Pyrenees). *Bulletin of Canadian Petroleum Geology*, **44**, 375–384.

SANTAFÉ, J. V., CASANOVAS, L. & ALFÉREZ, E. 1982. Presencia del Vallesiense en el Mioceno continental de la Depresión del Ebro. *Revista de la Real Academia de Ciencias Exactas, Físicas y Naturales*, **76**, 277–284.

SANTAMARÍA, J. 1995. *Los yacimientos de fosfato sedimentario en el límite Precámbrico-Cámbrico del anticlinal de Valdelacasa (Zona Centro Ibérica).* PhD Thesis, Autónoma University, Barcelona.

SANTANACH, P. 1994. Las Cuencas Terciarias gallegas en la terminación occidental de los relieves pirenaicos. *Cuadernos del Laboratorio Xeoloxico de Laxe*, **19**, 57–71.

SANTANACH, P. 1997a. The Ebro Basin in the structural framework of the Iberian Plate. *In*: BUSSON, J. & SCHREIBER, B. C. (eds) *Sedimentary Deposition in Rift and Foreland Basins in France and Spain (Palaeogene and Lower Neogene)*. Columbia University Press, New York, 304–318.

SANTANACH, P. (ed.) 1997b. ESCI. Estudios Sísmicos de la Corteza Ibérica. *Revista de la Sociedad Geológica de España*, **8**, 301–543.

SANTANTONIO, M. 1993. Facies associations and evolution of pelagic carbonate platform/basin systems: examples from the Italian Jurassic. *Sedimentology*, **40**, 1039–1067.

SANTIAGO, M. 1960. Los volcanes de La Palma (Islas Canarias). *El Museo Canario*, **75–76**, 281–346.

SANTIESTEBAN, C. & TABERNER, C. 1987. Depósitos evaporíticos de ambiente sabkha preservados como pseudomorfos en dolomta, en los materiales superiores de la facies Muschelkalk de la Sierra de Prades. *Cuadernos de Geología Ibérica*, **11**, 199–214.

SANTISTEBAN, J. I., MEDIAVILLA, R. & MARTÍN-SERRANO, A. 1996a. Alpine tectonic framework of south-western Duero basin. *In*: FRIEND, F. & DABRIO, C. (eds) *Tertiary Basins of Spain: The Stratigraphic Record of Crustal Kinematics*. World and Regional Geology, **6**, Cambridge University Press, 188–195.

SANTISTEBAN, J. I., MEDIAVILLA, R., MARTÍN-SERRANO, A. & DABRIO, C. J. 1996b. The Duero Basin: a general overview. *In*: FRIEND, F. & DABRIO, C. (eds) *Tertiary Basins of Spain: The Stratigraphic Record of Crustal Kinematics*. World and Regional Geology, **6**, Cambridge University Press, 183–187.

SANTONJA, M. & PÉREZ GONZÁLEZ, A. 1984. Las industrias paleolíticas de la Maya I en su ámbito regional. *Excavaciones Arqueológicas en España, Ministerio de Cultura*, **135**, 5–347.

SANTONJA, M. & VILLA, P. 1990. The Lower Paleolithic of Spain and Portugal. *Journal of World Prehistory*, **4**, 45–94.

SANTOS, J. F., MATA, J., GONÇALVES, F. & MUNHÁ, J. 1987. Contribuição para o conhecimento geológico-petrológico da região de Santa Susana: O complexo vulcanosedimentar da Toca da Moura. *Comunicações dos Serviços Geológicos de Portugal*, **73**, 29–48.

SANTOS, J. F., ANDRADE, A. A. S. & MUNHÁ, J. 1990. Magmatismo orogénico varisco no limite meridional da Zona de Ossa-Morena. *Comunicações dos Serviços Geológicos de Portugal*, **76**, 91–124.

SANTOS, J. F., MARQUES, F. O., MUNHÁ, J. & TASSINARI, C. 1995. A new isotopic tale for Bragança eclogites. *Terra Nova*, **7**, Abstracts Supplement 1, 108.

SANTOS, J. F., MARQUES, F. O., MUNHÁ, J., RIBEIRO, A. & TASSINARI, C. 1997. First dating of a Precambrian (1.0 to 1.1 Ga) HP/HT metamorphic event in the uppermost allochthonous unit of the Bragança Massif (Iberian variscan chain, northern Portugal). *Terra Nova*, **9**, 497.

SANTOS, J. F., SCHARER, U., GIL IBARGUCHI, J. I. & GIRARDEAU, J. 2002. Genesis of Pyroxenite-rich peridotite at Cabo Ortegal (NW Spain): Geochemical and Pb-Sr-Nd isotope Data. *Journal of Petrology*, **43**, 17–43.

SANTOS ZALDUEGUI, J. F., SCHÄRER, U. & GIL IBARGUCHI, J. I. 1996. Isotope constraints on the age and origin of magmatism and metamorphism in the Malpica-Tuy Allochthon. *Chemical Geology*, **121**, 91–103.

SANTOS ZALDUEGUI, J. F., SCHÄRER, U., GIL IBARGUCHI, J. I. & GIRARDEAU, J. J. 1997. Origin and evolution of the Paleozoic Cabo Ortegal ultramafic-mafic complex (NW Spain): U-Pb, Rb-Sr and Pb-Pb isotope data. *Chemical Geology*, **129**, 281–304.

SANZ, J. L., BONAPARTE, J. F. & LACASA, A. 1988. Unusual Early Cretaceous bird from Spain. *Nature*, **331**, 433–435.

SANZ, J. L., CHIAPPE, L. M., PÉREZ-MORENO, B. P., BUSCALIONI, A. D., MORATALLA, J. J., ORTEGA, F. & POYATO-ARIZA, F. J. 1996. An Early Cretaceous bird from Spain and its implications for the evolution of avian flight. *Nature*, **382**, 442–445.

SANZ, J. L., CHIAPPE, L. M., FERNÁNDEZ-JALVO, Y., ORTEGA, F., SANCHÉZ-CHILLÓN, B., POYATO-ARIZA, F. J. & PÉREZ-MORENO, B. P. 2001a. An Early Cretaceous pellet. *Nature*, **409**, 998–1000.

SANZ, J. L., FREGENAL MARTÍNEZ, M. A., MELÉNDEZ, N. & ORTEGA, F. 2001b. Las Hoyas Lake. *In*: BRIGGS, D. E. G. & CROWTHER, P. R. (eds) *Palaeobiology II*. Blackwell, Oxford.

SANZ, M. E. 1996. *Sedimentología de las Formaciones Neógenas del Sur de la Cuenca de Madrid*. Cedex, Madrid.

SANZ, M. E., RODRÍGUEZ-ARANDA, J. P., CALVO, J. P. & ORDÓÑEZ, S. 1994. Tertiary detrital gypsum in the Madrid Basin, Spain: criteria for interpreting detrital gypsum in continental evaporitic sequences. *In*: RENAUT, R. W. & LAST, W. M. (eds) *Sedimentology and Geochemistry of Modern and Ancient Saline Lakes*. SEPM, Special Publications, **50**, 217–228.

SANZ, M. E., ALONSO-ZARZA, A. M. & CALVO, J. P. 1995. Carbonate pond deposits related to semi-arid alluvial systems: examples from the Tertiary Madrid Basin, Spain. *Sedimentology*, **42**, 437–452.

SANZ DE GALDEANO, C. 1973. *Geología de la transversal Jaén-Frailes (Provincia de Jaén)*. PhD Thesis, University of Granada.

SANZ DE GALDEANO, C. 1983. Los accidentes y fracturas principales de las Cordilleras Béticas. *Estudios Geológicos*, **39**, 157–165.

SANZ DE GALDEANO, C. 1990. La prolongación hacia el sur de las fosas y desgarres del norte y centro de Europa: una propuesta de interpretación. *Revista de la Sociedad Geológica de España*, **3**, 231–241.

SANZ DE GALDEANO, C. 1997. *La Zona Interna Bético-Rifeña (antecedentes, unidades tectónicas, correlaciones y bosquejo de reconstrucción paleogeográfica)*. Monográfica Tierras del Sur, University of Granada, **18**.

SANZ DE GALDEANO, C. & LÓPEZ CASADO, C. 1988. Fuentes sísmicas en el ámbito Bético-Rifeño. *Revista Geofísica*, **44**, 175–198.

SANZ DE GALDEANO, C. & VERA, J. A. 1992. Stratigraphic record and palaeogeographical context of the Neogene basins in the Betic Cordillera, Spain. *Basin Research*, **4**, 21–36.

SANZ DE GALDEANO, C., SERRANO, F., LÓPEZ-GARRIDO, A. C. & MARTÍN-PÉREZ, J. A. 1993. Palaeogeography of the Late Aquitanian-Early Burdigalian Basin in the western Betic Internal Zone. *Geobios*, **26**, 43–55.

SANZ DE GALDEANO, C., ANDREO, B., GARCÍA-TORTOSA, F. J. & LÓPEZ GARRIDO, A. C. 2001. The Triassic palaeogeographic transition between the Alpujarride and Malaguide complexes, Betic-Rif Internal Zone (S Spain, N Marocco). *Palaeogeography, Palaeoclimatology, Palaeoecology*, **167**, 157–173.

SANZ LÓPEZ, J. 1995. *Estratigrafía y bioestratigrafía (Conodontos) del Silúrico superior-Carbonífero inferior del Pirineo oriental y central*. PhD Thesis, University of Barcelona.

SANZ LÓPEZ, J. & SARMIENTO, G. N. 1995. Asociaciones de conodontos del Ashgill y Llandovery en horizontes carbonatados del valle del Freser (Girona). *In*: LÓPEZ, G., OBRADOR, A. & VICENS, E. (eds) *Comunicaciones XI Jornadas de Paleontología*. Tremp, 157–160 (ISBN 84-600-9248-8).

SANZ-LÓPEZ, J., BARNOLAS, A. & GARCÍA-SANSEGUNDO, J. 1998. Le Silurien supérieur et le Dévonien inférieur à facies carbonatés du massif des Gavarres (chaînes côtières catalanes, Nord-Est de l'Espagne): stratigraphie et relation structurale avec les séries sous-jacentes. *Comptes Rendus de l'Académie des Sciences, Paris*, **326**, 893–900.

SANZ-LÓPEZ, J., VALENZUELA-RÍOS, J. I., GARCÍA-LÓPEZ, S., GIL PEÑA, I. & ROBADOR, A. 1999. Nota preliminar sobre la estratigrafía y el contenido en conodontos del Pridoli-Lochkoviense inferior en la unidad de els Castells (Pirineo central). *Temas Geológico-Mineros ITGE*, **26**, 638–642.

SANZ-LÓPEZ, J., MELGAREJO, J.-C. & CRIMES, T. P. 2000. Stratigraphy of Lower Cambrian and unconformable Lower Carboniferous beds from the Valls unit (Catalonian Coastal Ranges). *Comptes Rendus de l'Académie des Sciences de Paris*, sér. IIa, **330**, 147–153.

SAPPER, K. 1906. Beiträge zur kenntnis von Palma und Lanzarote. *Petermanns Geographische Mitteilungen*, **7**, 145–153.

SARMIENTO, G. N. 1990. Conodontos de la Zona Ordovicicus (Ashgill) en la Caliza Urbana, Corral de Calatrava (Ciudad Real). *Geogaceta*, **7**, 54–46.

SARMIENTO, G. N. 1993. *Conodontos ordovícicos de Sierra Morena (Macizo Hespérico meridional)*. PhD Thesis, Complutense University, Madrid.

SARMIENTO, G. N. & GARCÍA PALACIOS, A. 1996. Conodontos silúricos (Telychiense-Sheinwoodiense) en las facies sapropelíticas negras de Corral de Calatrava (Ciudad Real), España. *In*: PALACIOS, T. & GOZALO, R. (eds) *Comunicaciones XII Jornadas de Paleontología*, Badajoz, 109–111 (ISBN 84-7723-262-8).

SARMIENTO, G. N. & GUTIÉRREZ-MARCO, J. C. 1999. Microfósiles ordovícicos en olistolitos carboníferos de la Cuenca del Guadiato, Adamuz (Córdoba). *Temas Geológico-Mineros ITGE*, **26**, 580–584.

SARMIENTO, G. N. & RODRÍGUEZ NÚÑEZ, V. M. 1991. Conodontos telychienses (Silúrico inferior) del Sinclinal del Guadarranque (Zona Centroibérica, Macizo Hespérico). *Revista Española de Paleontología*, no. extraordinario, 151–156.

SARMIENTO, G. N., MÉNDEZ-BEDIA, I., ARAMBURU, C., ARBIZU, M. & TRUYOLS, J. 1994. Early Silurian conodonts from the Cantabrian Zone, NW Spain. *Geobios*, **27**, 507–522.

SARMIENTO, G. N., GUTIÉRREZ-MARCO, J. C. & RÁBANO, I. 1995a. A biostratigraphical approach to the Middle Ordovician conodonts from Spain. *In*: COOPER, J. D., DROSER, M. L. & FINNEY, S. C. (eds) *Ordovician Odyssey*. Pacific Section, Society for Sedimentary Geology, Fullerton, Book **77**, 61–64 (ISBN 1-878861-70-0).

SARMIENTO, G. N., SANZ LÓPEZ, J. & BARNOLAS, A. 1995b. Conodontos del Ashgill en las Calizas de Madremanya, Les Gavarres (Girona). *In*: LÓPEZ, G., OBRADOR, A. & VICENS, E. (eds) *Comunicaciones XI Jornadas de Paleontología*. Tremp, 161–163 (ISBN 1-84-600-9248-8).

SARMIENTO, G. N., SANZ-LÓPEZ, J. & GARCÍA-LÓPEZ, S. 1998. Silurian conodonts from the Iberian Peninsula – an update. *Temas Geológico-Mineros ITGE*, **23**, 119–124.

SARMIENTO, G. N., GARCÍA-LÓPEZ, S. & BASTIDA, F. 1999a. Conodont colour alteration indices (CAI) of Upper Ordovician limestones from the Iberian Peninsula. *Geologie en Mijnbouw*, **77**, 77–91.

SARMIENTO, G. N., GÜL, M. A., KOZLU, H. & GÖNCÜOGLU, M. C. 1999b. Darriwilian conodonts from the Taurus Mountains, southern Turkey. *Acta Universitatis Carolinae, Geologica*, **43**, 37–40.

SARMIENTO, G. N., GUTIÉRREZ-MARCO, J. C. & ROBARDET, M. 1999c. Conodontos ordovícicos del Noroeste de España. Aplicación al modelo de sedimentación de la región limítrofe entre las Zonas Asturoccidental-leonesa y Centroibérica durante el Ordovícico Superior. *Revista de la Sociedad Geológica de España*, **12**, 477–500.

SARMIENTO, G. N., PIÇARRA, J. M., REBELO, J. A., ROBARDET, M., GUTIÉRREZ-MARCO, J. C., STORCH, P. & RÁBANO, I. 1999d. Le Silurien du synclinal de Moncorvo (NE du Portugal): biostratigraphie et importance paléogéographique. *Géobios*, **32**, 749–767.

SAUPÉ, F. 1967. Note préliminaire concernant la genèse du gisement de mercure d'Almadén (Province de Ciudad Real, Espagne). *Mineralium Deposita*, **2**, 6–33.

SAUPÉ, F. 1971a. Stratigraphie et pétrographie du 'Quartzite du Criadero' (Valentien) à Almadén (province de Ciudad Real, Espagne). *Mémoires du BRGM*, **73**, 139–147.

SAUPÉ, F. 1971b. La série ordovicienne et silurienne d'Almadén (Province de Ciudad Real, Espagne); point des connaissances actuelles. *Mémoires du BRGM*, **73**, 355–365.

SAUPÉ, F. 1990. Geology of the Almadén Mercury Deposit, Province of Ciudad Real, Spain. *Economic Geology*, **85**, 482–510.

SAVAGE, J. F. 1979. The Hercynian orogeny in the Cantabrian Mountains, N Spain. *Krystallinikum*, **14**, 91–108.

SAVAGE, J. F. 1981. Geotectonic cross section through the Cantabrian Mountains, Northern Spain. *Geologie in Minjbow*, **81**, 3–5.

SAVOSTIN, L. A., SIBUET, J. C., ZONENSHAIN, L. P., LE PICHON, X. & ROULET, M. J. 1986. Kinematic evolution of the Tethys belt from the Atlantic Ocean to the Pamirs since the Triassic. *Tectonophysics*, **228**, 33–55.

SAVOY, L. E. 1992. Environmental record of Devonian-Mississippian carbonate and low-oxygen facies transitions, southernmost Canadian Rocky Mountains and northwesternmost Montana. *Geological Society of America Bulletin*, **104**, 1412–1432.

SAWKINS, L. J. G. 1990. *Metal Deposits in Relation to Plate Tectonics*, (second edition). Springer-Verlag, Berlin.

SCHÄFER, H. J. 1990. *Geochronological investigations in the Ossa-Morena Zone, SW Spain*. PhD dissertation, Eidgenössische Technische Hoschschule, Zürich, **9.246**.

SCHÄFER, H. J., GEBAUER, D. & NÄGLER, T. F. 1991. Evidence for Silurian eclogite- and granulite-facies metamorphism in the Badajoz-Córdoba shear belt, SW Spain. *Terra Nova*, **6**, 11.

SCHÄFER, H. J., GEBAUER, D., NÄGLER, T. F. & EGUÍLUZ, L. 1993. Conventional and ion-microprobe U-Pb dating of detrital zircons of the Tentudía Group (Serie Negra, SW Spain): implications for zircon systematics, stratigraphy, tectonics and Precambrian/Cambrian boundary. *Contributions to Mineralogy and Petrology*, **113**, 289–299.

SCHERMERHOLM, L., PRIEM, H., BOELRIJK, N. & HEBEDA, E. 1978. Age and origin of the Mesejana dolerite fault-dyke system (Portugal and Spain). *Journal of Geology*, **86**, 299–309.

SCHERMERHORN, L. J. G. 1959. *Igneous, metamorphic and ore geology of the Castro Daire–Sao Pedro do Sul–Sátao Region (Northern Portugal)*. Geological Institute, University of Amsterdam.

SCHERMERHORN, L. J. G. 1971. An outline stratigraphy of the Iberian Pyrite Belt. *Boletín Geológico y Minero*, **82**, 239–268.

SCHERMERHORN, L. J. G., PRIEM, H. N. A., BOELRIJK, N. A. I. M., HEBEDA, E. H., VERDUMEN, E. A. & VERSCHURE, R. H. 1978. Age and origin of the Messejana dolerite fault-dike system (Portugal and Spain) in the light of the opening of the North Atlantic ocean. *Journal of Geology*, **86**, 299–309.

SCHMIDT, H. 1931. Das Paläozoikum der spanischen pyrenaen. *Arbeitet Gesselschaft Göttingen, Mathematik-Physik*, **3**(5,8), 981–1065.

SCHMIDT, M. 1933. Beobachtungen über die Trias von Olesa de Montserrat und Vallirana in Katalonien um den Keuper von Alicante. *Geologie de la Mediterranee Occidentale*, **23**, 1–7.

SCHMIDT-THOMÉ, M. 1973. Beiträge zur Feinstratrigraphie des Unterkambriums in den Iberischen Ketten (Nordost-Spanien). *Geologisches Jahrbuch, Reihe B*, **7**, 3–43.

SCHMINCKE, H. U. 1967. Cone sheet swarm, resurgence of Tejeda Caldera, and the early geologic history of Gran Canaria. *Bulletin Volcanologique*, **31**(1), 53–162.

SCHMINCKE, H. U. 1968. Faulting versus erosion and the reconstruction of the Mid-Miocene shield volcano of Gran Canaria. *Geologische Mitteilungen*, **8**, 23–50.

SCHMINCKE, H. U. 1973. Magmatic evolution and tectonic regime in the Canary, Madeira, and Azores Island groups. *Geological Society of America Bulletin*, **84**, 633–648.

SCHMINCKE, H. U. 1976. Geology of the Canary Islands. *In*: KUNKEL, G. (eds) *Biogeography and Ecology in the Canary Islands*. W. Junk, The Hague, 67–184.

SCHMINCKE, H. U. 1979. Age and crustal structure of the Canary Islands. *Journal of Geophysics*, **46**, 217–224.

SCHMINCKE, H. U. 1982. Volcanic and chemical evolution of the Canary Islands. *In*: VON RAD, U., HINZ, K., SARNTHEIN, M. & SEIBOLD, E. (eds) *Geology of the Northwest African Margin*, 273–306.

SCHMINCKE, H. U. 1993. *Geological Field Guide of Gran Canaria* (sixth edition). Pluto-Press, Kiel (Germany).

SCHMINCKE, H. U. & SEGSCHNEIDER, B. 1998. Shallow submarine to emergent basaltic shield volcanism of Gran Canaria: evidence from drilling into the volcanic apron. *In*: WEAVER, P. P. E., SCHMINCKE, H. U., FIRTH, J. V. & DUFFIELD, W. (eds) *Proceedings ODP, Scientific Results, 157: College Station, TX (Ocean Drilling Program)*, 141–181.

SCHMINCKE, H. U. & SUMITA, M. 1998. Volcanic evolution of Gran Canaria reconstructed from apron sediments: synthesis of VICAP project drilling. *Proceedings ODP, Scientific Results, 157: College Station, TX (Ocean Drilling Program)*, 443–469.

SCHMINCKE, H. U. & SWANSON, D. A. 1967. Laminar viscous flowage structures in ash-flow tuffs from Gran Canaria, Canary Islands. *Journal of Geology*, **75**, 641–664.

SCHMINCKE, H. U., KLÜEGEL, A., HASTEEN, T. H., HOERNLE, K. & VAN DEN BOGAARD, P. 1998. Samples from the Jurassic ocean crust beneath Gran Canaria, La Palma and Lanzarote (Canary Islands). *Earth and Planetary Science Letters*, **163**, 343–360.

SCHMITZ, U. 1971. Stratigraphie und Sedimentologie im Kambrium und Tremadoc der Westlichen Iberischen Ketten nördlich Ateca (Zaragoza), NE-Spanien. *Münstersche Forschungen zur Geologie und Paläontologie*, **22**, 1–123.

SCHNEIDER, H. 1939. *Altpaläozoikum bei Cala in der westlichen Sierra Morena (Spanien)*. Dissertation, University of Berlin.

SCHNEIDER, H. 1951. Das Paläozoikum im Westteil der Sierra Morena (Spanien). *Zeitschrift Deutschen Geologischen Gesellschaft*, **103**, 134–135.

SCHNEIDER, J. L., BRUNNER, C. A. & KUTTNER, S. 1998. Epiclastic sedimentation during the upper Miocene–lower Pliocene volcanic hiatus of Gran Canaria: evidence from Sites 953 and 954. *Proceedings ODP, Scientific Results, 157: College Station, TX (Ocean Drilling Program)*, 293–314.

SCHOTT, J. J. & PERES, A. 1987. Paleomagnetism of the Lower Cretaceous redbeds from northern spain, evidence for a multistage acquisition of magnetization. *Tectonophysics*, **139**, 239–253.

SCHRIEL, W. 1929. Der geologischen Bau des Katalonischen Küstengebirges zwischen Ebromündung und Ampurdán. *Sitzung zur Geologie der west Mediterrangebietes Abhandlungen der Gesselschaft der Wissenfchsften zu Göttingen*, **2**, 1–79.

SCHRIEL, W. 1930. Die Sierra de la Demanda und die Montes Obarenes. *Abhandlungen der Gesellschaft der Wissenschaften zu Göttingen Matematisch-Physikalische Klasse*, **16**, 1–105.

SCHROEDER, R., GARCÍA, A., CERCHI, A. & SEGURA, M. 1993. El Albense- Cenomanense del Pto. del Remolcador (Cordillera Ibérica, Maestrat, Castelló): secuencias deposicionales y biozonación con grandes foraminíferos. *Geogaceta*, **14**, 69–72.

SCHUDACK, M. 1987. Charophyten flora und Fazielle entwiklung der Grenzschichten marine Jura/ Wealden in den Nordwest Iberischen Ketten (Mit Wergleichen zu Asturien und Kantabrien). *Paleontographica Abt. B*, **204**, 1–180.

SCHULZE, R. 1982. *Deckenbau und Flysch sedimentation im Variszikum des Masivs von Mouthoumet (Süd Frankreich)*. PhD Thesis, University of Göttingen.

SCHÜTT, B. 1997. Reconstruction of Holocene paleoenvironments in the endorheic basin of Laguna de Gallocanta, Central Spain, by investigation of mineralogical and geochemical characters from lacustrine sediments. *Journal of Paleolimnology*, **20**, 217–234.

SCHÜTT, B. 2000. Holocene paleohydrology of playa lakes in northern and central Spain: a reconstruction based on the mineral composition of lacustrine sediments. *In*: ZOLITSCHLLA, B., WUELF, S. & NEGENDANK, J. F. W. (guest eds) *Mediterranean Lacustrine Records: a Contribution to the ELDP. Quaternary International*, **73/74**, 7–27.

SCHÜTT, B. & BAUMHAUER, R. 1996. Playa sedimente aus dem Zentralen Ebrobecken/Spanien als Indikatoren fúrr holozäne klimaschwankungen-ein vorläufige Bericht. *Petermans Geographische Mitteilungen*, **140**, 33–42.

SCHWEINEBERG, J. 1987. Silurische Chitinozoen aus der Provinz Palencia (Kantabrisches Gebirge, N-Spanien). *Göttinger Arbeiten für Geologie und Paläontologie*, **33**, 1–94.

SCHWENTKE, W. 1990. *Upper Cretaceous Tectono – Sedimentary and facies evolution of the Basque Pyrenees (Spain)*. Tübinger geowissenschaftliche Arbeiten, (A) **7**.

SCHWENTKE, W. & KUHNT, W. 1992. Subsidence history and continental margin evolution of the Western Pyrenean and Basque Basins. *Palaeogeography, Palaeoclimatology, Palaeoecology*, **95**, 297–318.

SCOTESE, C. R. & BARRETT, S. F. 1990. Gondwana's movement over the South Pole during the Palaeozoic: evidence from lithological indicators of climate. *In*: MCKERROW, W. S. & SCOTESE, C. R. (eds) *Palaeozoic Palaeogeography and Biogeography*. Geological Society, London, Memoirs, **12**, 75–85.

SDZUY, K. 1961. Das Kambrium Spaniens. Teil II: Trilobiten. *Akademie der Wissenschaften und der Literatur, Abhandlungen der mathematisch-naturwissenschaftlichen Klasse*, **1961** (7–8), 499–690 (217–408).

SDZUY, K. 1962. Trilobiten aus dem Unter-Kambrium der Sierra Morena (S. Spanien). *Senckenbergiana lethaea*, **43**, 181–229.

SDZUY, K., 1968. Trilobites del Cámbrico Medio de Asturias. *Trabajos de Geología, Universidad de Oviedo*, **1**, 77–133.

SDZUY, K. 1969. Bioestratigrafía de la Griotte Cámbrica de los Barrios de Luna (León) y de otras sucesiones comparables. *Trabajos de Geología, Universidad de Oviedo*, **2**, 45–58.

SDZUY, K. 1971a. Acerca de la correlación del Cámbrico inferior de la Península Ibérica. *I Congreso Hispano-Luso-Americano de Geología Económica, Sección 1 Geología*, **2**, 753–768.

SDZUY, K. 1971b. La subdivisión bioestratigráfica y la correlación del Cámbrico Medio de España. *I Congreso Hispano-Luso-Americano de Geología Económica, Sección 1 Geología*, **2**, 769–782.

SDZUY, K. 1972. Das Kambrium der acadobaltischen Faunenprovinz. *Zentralblatt für Geologie und Paläontologie*, Teil II, **1972**, 1–91.

SDZUY, K. 1987. Trilobites de la Base de la formación Jalón (Cámbrico inferior) de Aragón. *Revista Española de Paleontología*, **2**, 3–8.

SDZUY, K. 1995. Acerca del conocimiento actual del Sistema Cámbrico y del Límite Cámbrico Inferior-Cámbrico Medio. *In*: GÁMEZ VINTANED, J. A. & LIÑÁN E. (eds) *Memorias de las IV Jornadas Aragonesas de Paleontología: 'La expansión de la vida en el Cámbrico'. Libro homenaje al Prof. Klaus Sdzuy*. Institución 'Fernando el Católico', Zaragoza, 253–263.

SDZUY, K. & LIÑÁN, E. 1993. Rasgos Paleogeográficos del Cámbrico Inferior y Medio del Norte de España. *Cuadernos do Laboratorio Xeolóxico de Laxe*, **18**, 189–215.

SDZUY, K., LIÑÁN, E. & GOZALO, R. 1999. The Leonian Stage (early Middle Cambrian): a unit for Cambrian correlation in the Mediterranean subprovince. *Geological Magazine*, **136**, 39–48.

SEBASTIÁN, A., RECHE, J. & DURÁN, H. 1990. Hercynian metamorphism in the Catalonian Coastal Ranges. *Acta Geológica Hispánica*, **25**, 31–38.

SEBER, D., BARAZANGI, M., IBENBRAHIM A. & DEMNATI, A. 1996a. Geophysical evidence for lithospheric delamination beneath the Alboran Sea and Rif-Betic mountains. *Nature*, **379**, 785–790.

SEBER, D., BARAZANGI, M., TADILI, B. A., RAMDANI, M., IBENBRAHIM, A. & SARI, D. B. 1996b. Three-dimensional upper mantle structure beneath the intraplate Atlas and interplate Rif mountains of Morocco. *Journal of Geophysical Research*, **101**, 3125–3138.

SEEBOL, I., FERNÁNDEZ, G., REINOSO, J., ALONSO, J. A., ESCAYO, M. A. & GOMEZ, M. 1992. Yacimientos estratoligados de blenda, galena y marcasita en dolomias. Mina de Reocín (Cantabria). *In*: GARCÍA GUINEA, J. & MARTÍNEZ-FRÍAS, J. (eds) *Recursos Minerales de España*. CSIC, Colección Textos Universitarios, **15**, 947–964.

SEEMANN, U., PUMPIN, V. F. & CASSON, N. 1990. *Amposta oil field. Structurals Traps I: Traps Associated With Tectonic Faulting*. AAPG, Special Publication, 1–20.

SEGURA, M., PEREZ, P., CARENAS, B., GARCÍA, A. & CALONGE, A. 1989. Le Cénomanien supérieur-Turonien dans la zone centrale de la Chaîne Ibérique (Espagne): une étape sédimentaire trés particuliére dans l'evolution de la plate-forme Crétacé. *Geobios*, **11**, 161–167.

SEGURA, M., GARCÍA, A., GARCÍA-HIDALGO, F. J. & CARENAS, B. 1993a. The Cenomanian–Turonian transgression in the Iberian Ranges (Spain): depositional sequences and the Cenomanian–Turonian boundary. *Cretaceous Research*, **14**, 519–529.

SEGURA, M., GARCÍA-HIDALGO, J. F., CARENAS, B. & GARCÍA, A. 1993b. Upper Cenomanian Platform from central Estern Iberian (Spain). *In*: SIMO, J. A. T., SCOTT, B. W & MASSE J. P. (eds) *Cretaceous Carbonate Platforms*. AAPG, Memoirs, **56**, 283–296.

SEGURA, M., GARCÍA-HIDALGO, J. F., GARCÍA, A., RUIZ, G. & CARENAS, B. 1999. El Cretácico de la zona de intersección del Sistema Central con la Cordillera Ibérica: unidades litoestratigráficas y secuencias deposicionales. Libro Homenaje a José Ramirez del Pozo. *Asociación Española de Géologos y Geofísicos de Petróleo*. 129–139.

SEGURA, M., GIL, J. & PONS, J. M. 2000. Identificación de *Bournonia gardónica* (Toucas, 1907) en el Cretácico superior del Barranco de las Cuevas (Patones, Madrid). *Geotemas*, **1**(2), 321–324.

SÉGURET, M. 1972. *Étude tectonique des nappes et séries décollées de la partie centrale du versant sud des Pyrénées*. USTELA, Série Géologie Structurale **2**. Montpellier.

SEILACHER, A. 1970. *Cruziana* stratigraphy of 'non-fossiliferous' Palaeozoic sandstones. *In*: CRIMES, T. P. & HARPER, J. C. (eds) *Trace Fossils*. Seel House Press, Liverpool, 447–476 (*Geological Journal Special Issue*, **3**).

SELL, I., POUPEAU, G., CASQUET, C., GALINDO, C. & GONZÁLEZ-CASADO, J. M. 1995. Exhumación alpina del bloque meorfotectónico Pedriza-La Cabrera (Sierra de Guadarrama, Sistema Central Español): potencialidad de la termocronología por trazas de fisión en apatitos. *Geogaceta*, **18**, 23–26.

SEQUEIROS, L. 1974. *Paleobiogeografía del Calloviense y Oxfordense en el Sector Central de la Zona Subbética*. PhD Thesis, University of Granada.

SEQUEIROS, L. 1987. Caracterización cuali-cuantitativa del Calloviense de Cabra (Cordillera Bética, España). *Boletín de la Sociedad Española de Historia Natural*, **83**, 25–46.

SEQUEIROS, L. & CARIOU, E. 1984. Síntesis bioestratigráfica del Calloviense de Ricla (Zaragoza, Cordillera Ibérica). *Estudios Geológicos*, **40**, 411–419.

SEQUEIROS, L. & MELÉNDEZ, G. 1979. Nuevos datos bioestratigráficos del Calloviense y Oxfordiense de Aguilón (Cordillera Ibérica, Zaragoza). *Cuadernos de Geología, Universidad de Granada*, **10**(1981), 167–178.

SERNA, J. & GAVIRIA, M. 1995. *La quimera del agua*. Siglo veintiuno (eds) **444**.

SERRA, M. 1990. Geología de los materiales paleozoicos del área de Capellades (Prov. de Barcelona). *Acta Geológica Hispanica*, **25**, 123–132.

SERRANO, A. & MARTÍNEZ DEL OLMO, W. 1990. Tectónica salina en el Dominio Cantabro-Navarro: evolución, edad y origen de las estructuras salinas. *In*: ORTI, F. & SALVANY, J. M. (eds) Formaciones evaporíticas de las Cuenca del Ebro y cadenas periféricas y de la zona de Levante. Enresa, Madrid, 39–53.

SERRANO, A., MARTÍNEZ DEL OLMO, W. & CÁMARA, P. 1989. Diapirismo del Trías salino en el dominio Cántabro-Pirenaico. *In*: *Libro Homenaje a Rafael Soler*. AGGEP, Madrid, 115–121.

SERRANO, E. 1996. El complejo morrénico frontal del valle de Trueba (Espinosa de los Monteros, Burgos). *Cadernos do Laboratorio Xeolóxico de Laxe*, **21**, 505–517.

SERRANO, E. 1998. *Geomorfología del Alto Gállego, Pirineo aragonés*. Institución Fernando El Católico, Zaragoza.

SERRANO, F. 1990. El Mioceno Medio en el área de Níjar (Almería, España). *Revista Sociedad Geológica de España*, **3**, 65–77.

SERRANO, I., MORALES, J., ZHAO, D., TORCAL, F. & VIDAL, F. 1998. P-wave tomographic images in the Central Betics-Alboran Sea (South Spain) using local earthquakes: contribution for continental collision. *Geophysical Research Letters*, **25**, 4031–4034.

SERRANO PINTO, M. & GIL IBARGUCHI, J. I. 1987. Revisión de datos geocronológicos e isotópicos de granitoides hercínicos y ante-hercínicos de la Region Galaico-Castellana. *Memorias Museo Laboratorio de Mineralogia y Geología, Universidade do Porto*, **1**, 171–186.

SERRAT, D. 1989. Sistema Costero Catalán. *In*: PÉREZ GONZÁLEZ, A., CABRA, P. & MARTÍN-SERRANO, A. (eds) *Mapa del Cuaternario de España*. ITGE, Madrid, 125–128.

SERRAT, D., VILAPLANA, J. M. & MARTÍ-BONO, C. 1983. Some depositional models of glaciolacustrine environments in Southern Pyrenees. *In*: EVENSON, E., SCHLUCHTER, CH. & RABASSA, J. (eds) *Tills and Related Deposits*. Balkema, Rotterdam, 231–244.

SERVAIS, T. & METTE, W. 2000. The *messaoudensis-trifidum* acritarch assemblage (Ordovician: late Tremadoc-early Arenig) of the Barriga Shale Formation, Sierra Morena (SW Spain). *Review of Palaeobotany and Palynology*, **113**, 145–163.

SESÉ, C. & GIL, E. 1987. Los micromaniferos del Pleistoceno Medio del complejo cárstico de Atapuerca (Burgos). *In*: AGUIRRE, E., CARBONELL, E. & BERMÚDEZ DE CASTRO, J. M. (eds) *El hombre fosíl de Ibeas y el Pleistoceno de la Sierra Atapuerca*. Junta de Castilla y León, Valladolid, 47–54.

SESÉ, C. & SOTO, E. 2000. Vertebrados del Pleistoceno de Madrid. *In*: MORALES, J., NIETO, M., AMEZUA, L., FRAILE, S. *et al.* (eds) *Patrimonio Paleontológico de la Comunidad de Madrid. Arqueología, Paleontología y Etnografía*. Monográfico, **6**, Madrid, 216–243.

SESÉ, C., SOTO, E. & PÉREZ-GONZÁLEZ, A. 2000. Mamíferos de las terrazas del valle del Tajo: primeros datos de micromamíferos del Pleistoceno en Toledo (España central). *Geogaceta*, **28**, 137–140.

SEYFRIED, H. 1978. Der Subbetische Jura von Murcia (Súdost-Spanien). *Geologisches Jahrbuch*, **29**, 3–204.

SEYFRIED, H. 1979. Ensayo sobre el significado paleogeográfico de los sedimentos del Jurásico de las Cordilleras Béticas Orientales. *Cuadernos de Geología, Universidad de Granada*, **10**, 317–348.

SHACKLETON, N. J. 1997. The deep-sea sediment record and the Pliocene–Pleistocene boundary. *Quaternary International*, **40**, 33–35.

SHELLEY, D. & BOSSIÈRE, G. 2000. A new model for the Hercynian Orogen of Gondwanan France and Iberia. *Journal of Structural Geology*, **22**, 757–776.

SHERGOLD, J. H. & SDZUY, K. 1991. Late Cambrian trilobites from the Iberian Mountains, Zaragoza Province, Spain. *Beringeria*, **4**, 193–235.

SHERGOLD, J. H., LIÑÁN, E. & PALACIOS, T. 1983. Late Cambrian trilobites from the Najerilla Formation, north-eastern Spain. *Palaeontology*, **26**, 71–92.

SHERGOLD, J. H., FEIST, R. & VIZCAINO, D. 2000. Early late Cambrian trilobites of Australo-Sinian aspect from the Montagne Noire, southern France. *Palaeontology*, **43**, 599–632.

SIBUET, J. C., PAUTOT, G. & LE PICHON, X. 1971. Interprétation structurale du golfe de Gascogne à partir des profils sismiques. *In*: DEBYSIER, J., LE PICHON, X. & MONTARDET, M. (eds) *Histoire Structurale du Golfe de Gascogne*. Technip, **6**, 10-1–10-32.

SIERRA, J. 1984. Geología, mineralogía y metalogenia del yacimiento de Aznalcóllar. Litoestratigrafia y tectónica. *Boletín Geológico y Minero*, **95**, 553–568.

SIERRA, S. & MORENO, C. 1997. La cuenca pérmica del río Viar, SO de España: Análisis petrográfico de las areniscas. *Cuadernos de Geología Ibérica*, **22**, 447–472.

SIERRA, S. & MORENO, C. 1998. Arquitectura fluvial de la cuenca pérmica del Viar (SO de España). *Revista de la Sociedad Geológica de España*, **11**, 197–212.

SIERRA, S., MORENO, C. & GONZÁLEZ, F. 1999. Los abanicos aluviales de la Cuenca pérmica del Viar (SO de España): caracterización sedimentológica y petrográfica. *Geogaceta*, **25**, 195–198.

SIERRA, S., MORENO, C. & GONZÁLEZ, F. 2000. El vulcanismo pérmico de la cuenca del Viar (SO de España): Caracterización de la Secuencia Volcanoclástica Gris. *Geogaceta*, **27**, 159–162.

SIERRO, F. J., GONZÁLEZ-DELGADO, J. A., DABRIO, C. J., FLORES, J. A. & CIVIS, J. 1996. Late Neogene depositional sequences in the foreland basin of Guadalquivir (SW Spain). *In*: FRIEND, P. F. & DABRIO, C. J. (eds) *Tertiary Basins of Spain: The Stratigraphic Record of Crustal Kinematics*. Cambridge University Press, Cambridge, 339–345.

SIGMARSSON, O., CARN, S. A. & CARRACEDO, J. C. 1998. Systematics of U-series nuclides in primitive lavas from the 1730–36 eruption on Lanzarote, Canary Islands, and implications for the role of garnet pyroxenites during oceanic basalt formations. *Earth Planet. Sci. Lett.*, **162**, 137–151.

SILVA, J. B. 1989. *Estrutura de uma geotransversal da Faixa Piritosa: Zona do Vale do Guadiana*. PhD Thesis, University of Lisbon.

SILVA, J. B., OLIVEIRA, J. T. & RIBEIRO, A. 1990. Structural outline of the South Portuguese Zone. *In*: DALLMEYER, R. D. & MARTÍNEZ GARCÍA, E. (eds) *Pre-Mesozoic Geology of Iberia*. Springer-Verlag, Berlin, 348–362.

SILVA, P. G., GOY, J. L. & ZAZO, C. 1992a. Discordancias progresivas y expresión geomorfológica de los abanicos aluviales cuaternarios de la Depresión tectónica del Guadalentín (Murcia, SE España). *Geogaceta*, **11**, 57–70.

SILVA, P. G., HARVEY, A. M., ZAZO, C. & GOY, J. L. 1992b. Geomorphology, depositional style and morphometric relationships of Quaternary alluvial fans in the Guadalentín depression (Murcia, Southeast Spain). *Zeitschrift für Geomorphologie N.F.*, **35**, 325–341.

SILVA, P. G., GOY, J. L., SOMOZA, L., ZAZO, C. & BARDAJÍ, T. 1993. Landscape response to strike-slip faulting linked to collisional settings Quaternary Tectonics and basin formation in the Eastern Betics (SE Spain). *Tectonophysics*, **224**, 289–303.

SILVA, P. G., GOY, J. L., ZAZO, C. & BARDAJÍ, T. 1996. Evolución reciente del drenaje en la depresión del Guadalentín (Murcia, SE España). *Geogaceta*, **20**(5), 89–91.

SILVA, P. G., CAÑAVERAS, J. C., ZAZO, C., SÁNCHEZ-MORAL, S., LARIO, J. & SANZ, E. 1997. 3D soft-sediment deformation structures: existence for Quaternary seismicity in the Madrid basin, Spain. *Terra Nova*, **9**, 208–212.

SILVA, P. G., GOY, J. L., ZAZO, C. & BARDAJÍ, T. (in press). Fault-generated mountain fronts in Southeast Spain: geomorphologic assesment of tectonic and earthquake activity. *Geomorphology* (special issue on 'Landscape development in Southern Spain').

SILVEIRA, I. M. 1996. *Efeitos Mineralogicos e Geoquimicos da Alteracao em Rochas Vulcanicas Felsicas de Rio Tinto F. P. I*. PhD Thesis, Lisbon University.

SIMANCAS, F. 1980. Evolución tardihercínica de un área situada al NW de la provincia de Sevilla. *Temas Geológico-Mineros*, **4**, 237–360.

SIMANCAS, J. F. 1983. *Geología de la extremidad oriental de la Zona Sudportuguesa*. PhD Thesis, University of Granada.

SIMANCAS, J. F. 1985. Estudio estratigráfico de la cuenca del Viar. *Temas Geológico-Mineros*, **5**, 7–17.

SIMANCAS, J. F. & RODRIGUEZ GORDILLO, J. F. 1980. Magmatismo basaltico hercínico tardío en el NW de Sevilla. *Cuadernos Geológicos de la Universidad de Granada*, **11**, 49–60.

SIMANCAS, J. F., MARTÍNEZ POYATOS, D., EXPÓSITO, I., AZOR, A. & GONZÁLEZ LODEIRO, F. 2001. The structure of a major suture zone in the SW Iberian Massif: the Ossa-Morena/Central Iberian contact. *Tectonophysics*, **332**, 295–308.

SIMÓ, A. 1986. Carbonate platform depositional sequences, Upper Cretaceous, south-central Pyrenees (Spain). *Tectonophysics*, **129**, 205–231.

SIMÓ, A. & GINER, J. 1983. El Neógeno de Ibiza y Formentera (Islas Baleares). *Revista de Investigaciones Geológicas*, **36**, 67–81.

SIMÓ, A. & RAMON, X. 1986. Análisis sedimentológico y descripción de las secuencias deposicionales del Neógeno postorogénico de Mallorca. *Boletín Geológico y Minero*, **157**, 445–472.

SIMÓN, J. L. 1980. Estructuras de superposición de plegamientos en el borde NE de la Cadena Ibérica. *Acta Geológica Hispánica*, **15**(5), 137–140.

SIMÓN, J. L. 1983. Tectónica y neotectónica del sistema de fosas de Teruel. *Teruel*, **69**, 21–97.

SIMÓN, J. L. 1984. *Compresión y distensión alpinas en la Cadena*

Ibérica Oriental. PhD Thesis, University of Zaragoza (Publicaciones del Instituto de Estudios Turolenses, Teruel, Spain).

SIMÓN, J. L.1986a. Sobre las deformaciones del paleozoico en el macizo de l Desierto de Las Palmas (Castellón). *Estudios Geológicos*, **42**, 407–414.

SIMÓN, J. L. 1986b. Analysis of a gradual change in stress regime (example from eastern Iberian Chain, Spain). *Tectonophysics*, **124**, 37–53.

SIMÓN, J. L. 1989. Late Cenozoic stress field and fracturing in the Iberian Chain and Ebro Basin (Spain). *Journal of Structural Geology*, **11**(3), 285–294.

SIMÓN, J. L. 1990. Algunas reflexiones sobre los modelos tectónicos aplicados a la Cordillera Ibérica. *Geogaceta*, **8**, 123–130.

SIMÓN, J. L. 1991a. Tectónica Alpina. *In: Mapa Geológico de España a escala 1:50.000 de Alhama de Aragon (hoja nº 346)*. ITGE, Madrid, 42–48.

SIMÓN, J. L. 1991b. Tectónica Alpina. *In: Mapa Geológico de España a escala 1:50.000 de Torrijo de la Cañada (hoja nº 408)*. ITGE, Madrid, 54–62.

SIMÓN, J. L. & PARICIO, J. 1988. Sobre la compresión neógena en la cordillera ibérica (algunas precisiones a proposito de los comentarios de J. Guimerá al trabajo 'Aportaciones al conocimiento de la compresión tardía en la Cordillera Iberica centro-oriental: la cuenca neógena inferior de Mijares. Teruel-Castellón'). *Estudios Geológicos*, **44**, 271–283.

SIMÓN, J. L. & SORIANO, M. A. 1989. La falla de Concud (Teruel): Actividad cuaternaria y régimen de esfuerzos asociado. *In: El Cuaternario de España y Portugal*, Vol. 2. ITGE, Madrid, 729–737 (published 1993).

SIMÓN, J. L., PÉREZ CUEVA, A. & CALVO, A. 1983. Morfogénesis y neotectónica en el sistema de fosas del Maestrat (Provincia de Castellón). *Estudios Geológicos*, **39**, 167–177.

SIMÓN, J. L., ARENAS, C., ARLEGUI, L. E. *et al.* 1998. *Guía del Parque Geológico de Aliaga*, Univeridad de Zaragoza, Zaragoza.

SIMÓN, J. L., LIESA, C. L. & SORIA, A. R. 1999a. Un sistema de fallas normales sinsedimentarias en las unidades de facies Urgon de Aliaga (Teruel, Cordillera Ibérica). *Geogaceta*, **24**, 291–294.

SIMÓN, J. L., ARLEGUI, L. E., LIESA, C. L. & MAESTRO, A. 1999b. Stress perturbations registered by jointing near strike-slip, normal and reverse faults: Examples from de Ebro Basin, Spain. *Journal of Geophysical Research*, **B104**(7), 15141–15153.

SIMON, W. 1951. Untersuchungen im Paläozoikum von Sevilla (Sierra Morena, Spanien). *Abhandlungen der senckenbergischen naturforschenden Gesellschaft*, **485**, 31–62.

SIMÓN GOMEZ, J. L. 1981. Reactivación alpina del desgarre del segre en el borde NE de la Cadena Ibérica. *Tervel*, **65**, 195–209.

SIMS, K. W. W. & DEPAOLO, D. J. 1997. Inferences about mantle magma sources from incompatible element concentration ratios in oceanic basalts. *Geochimica et Cosmochimica Acta*, **61**, 765–784.

SJERP, N. 1967. The geology of the San Isidro-Porma area (Cantabrian Mountains, Spain). *Leidse Geologische Mededelingen*, **39**, 55–128.

SKEVINGTON, D. 1974. Graptolite faunas from central and northwestern Spain. *Breviora Geologica Asturica*, **18**, 61–64.

SLOSS, L. L. 1963. Sequences in the cratonic interior of North America. *Geological Society of America Bulletin*, **74**, 93–103

SMIT, J. 1990. Meteorite impact, extinctions and the Cretaceous–Tertiary boundary. *Geologie in Mijnbouw*, **69**, 187–204.

SMIT, P. W. 1966. *Sedimentpetrografie en milieu analyse van de Permo-Trias afzettingen langs de ooste lijke rand van het Cantabrische gebergte (Spanje)*. Internal Report, Leiden University, The Netherlands.

SMITH, J. R., MALAHOFF, A. & SHOR, A. N. 1999. Submarine geology of the Hilina slump and morpho-structural evolution of Kilauea Volcano, Hawaii. *Journal of Volcanology and Geothermal Research*, **94**(1–4), 59–88.

SMITH, K. S. & HUYCK, H. L. O. 1999. An overview of the abundance, relative mobility, bioavailability, and human toxicity of metals. *In*: PLUMLEE, G. S. & LOGSDON, M. J. (eds) *The Environmental Geochemistry of Mineral Deposits*. Reviews in Economic Geology, **6A**, Mineralogical Society of America, Michigan, 29–70.

SMITH, R. L. & BAILEY, R. A. 1968. Resurgent cauldrons. *Geological Society of America*, **116**, 613–662.

SOEDIONO, H. 1971. *Geological investigations in the Chirivel area (provincie of Almería, southeastern Spain)*. PhD Thesis, Amsterdam University.

SOERS, E. 1971. Stratigraphie et Géologie structurale de la partie orientale de la Sierra de Guadarrama (Espagne centrale). *Studia Geológica Salamanca*, **4**, 7–94.

SOLDEVILA BARTOLÍ, J. 1992a. La sucesión paleozoica en el sinforme de La Codosera-Puebla de Obando (Provincias de Cáceres y Badajoz, SO de España). *Estudios Geológicos*, **48**, 353–362.

SOLDEVILA BARTOLÍ, J. 1992b. La sucesión paleozoica en el sinforme de la Sierra de San Pedro (Provincias de Cáceres y Badajoz, SO de España). *Estudios Geológicos*, **48**, 363–379.

SOLÉ, J. 1993. *Le massif granitique du Montnegre (sud de la Costa Brava, Catalogne). Etude pétrologique, géochimique et géochronologique*. PhD Thesis, University of Genève.

SOLÉ, J. & VILLALOBOS, L. 1974. Maya de Baztán. Spanish Geological Survey, Report and map, Scale 1:50.000, **66**.

SOLÉ, J., DELALOYE, M. & ENRIQUE, P. 1998. K-Ar ages in biotites and K-feldspars from the Catalan Coastal batholith: Evidence of a post-Hercynian overprinting. *Eclogae Geologicae Helvetiae*, **91**, 139–148.

SOLÉ DE PORTA, N., CALVET, F. & TORRENTÓ, L. 1987. Análisis palinológico del Triásico de los Catalánides (NE de España). *Cuadernos de Geología Ibérica*, **11**, 237–254.

SOLÉ SABARÍS, L. 1978. Los rebordes Oriental y Meridional de la Meseta: Cordillera Ibérica y Sierra Morena. *In*: M. de Teran *et al.* (eds) *Geografía General de España*. Ariel, Madrid, 74–85.

SOLÉ SABARIS, L., VIRGILI, C. & JULIVERT, M. 1956. Características estratigráficas del Trías en la zona limítrofe entre las provincias de Barcelona y Tarragona. *Estudios Geológicos*, **31–32**, 287–300.

SOLER, E. 1980. Splites et métallogénie. La province pyrito-cuprifère de Huelva (SW Espagne). *Mémoires de Sciences de la Terre*, **39**, 1–461.

SOLER, M. & GARRIDO, A. 1970. La terminación occidental del manto del Cotiella. *Pirineos*, **98**, 5–12.

SOLER, M. & PUIGDEFÀBREGAS, C. 1970. Líneas generales de la geología del Alto Aragón Occidental. *Pirineos*, **96**, 5–20.

SOLER, R., LÓPEZ VILCHEZ, J. & RIAZA, C. 1981. Petroleum geology of the Bay of Biscay. *In: Petroleum Geology of the Continental Shelf of North-West Europe*. Institute of Petroleum, London, 474–482.

SOLER Y JOSÉ, R. 1971. El Jurásico marino de la Sierra de Aralar (Cuenca Cantábrica oriental): Los problemas postkimméricos. *Cuadernos de Geología Ibérica*, **2**, 509–532.

SOLER Y JOSÉ, R. 1972a. Las series Jurasicas del 'purbekiense' neocomiense de Guernica. *Boletín Geológico y Minero*, **83**, 221–230.

SOLER Y JOSÉ, R. 1972b. El Jurásico y Cretáceo inferior de Leiza y Tolosa (Cuenca Cantabrica Oriental): los problemas postkimméricos. *Boletín Geológico y Minero*, **83**, 582–594.

SOLER Y JOSÉ, R., LÓPEZ-VILCHEZ, J. & RIAZA, C. 1981. Petroleum geology of the Bay of Biscay. *In:* ILLING L. V. & HOBSON G. D. (eds) *Petroleum Geology of the Continental Shelf of North-west Europe*. Heyden, London, 474–482.

SOLER Y JOSÉ, R., MARTÍNEZ DEL OLMO, W., MEGÍAS, A. G. & ABEGER, J. A. 1983. Rasgos básicos del Neógeno del Mediterráneo Español. *Mediterránea Servicios Geológicos*, **1**, 71–82.

SOLOMON, M., WALSHE, J. L. & GARCÍA PALOMERO, F. 1980. Formation of massive sulphide deposits at Rio Tinto, Spain. *Transactions of the Institute of Mining and Metallurgy*, **89**, 16–24.

SOMMER, W. 1965. *Stratigraphie und Tektonik im Östl. Guadarrama-Gebirge (Spanien)*. Arbeiten Geologisch-Paläontologischen Institut Westf. Wilhelms-Universität Münster, **1**.

SOMOZA, L. 1993. *Estudio del Cuaternario litoral entre Cabo de Palos y Guardamar (Murcia-Alicante). Las variaciones del nivel del mar en relación con el contexto geodinámico*. Instituto Español de Oceanografía, Madrid, Special Publications, **12**.

SOMOZA, L., ZAZO, C., GOY, J. L. & MÖRNER, N. A. 1989. Estudio geomorfológico de secuencias de abanicos aluviales cuaternarios (Alicante-Murcia, España). *Cuaternario y Geomorfología*, **3**(1–4), 73–82.

SOMOZA, L., BARNOLAS, A., ARASA, A., MAESTRO, A., REES, J. G. & HERNÁNDEZ-MOLINA, F. J. 1998. Architectural stacking patterns of the Ebro delta controlled by Holocene high-frequency eustatic fluctuations delta lobe seitching and subsidence processes. *Sedimentary Geology*, **117**, 11–32.

SOPEÑA, A. 1979. Estratigrafía del Pérmico y Triásico del NO de la provincia de Guadalajara. *Seminarios de Estratigrafía, Serie Monografías*, **5**, 1–329.

SOPEÑA, A., VIRGILI, C., ARCHE, A., RAMOS, A. & HERNANDO, S. 1985. El Triásico *In*: COMBA, J. A. (ed.) *Libro Jubilar J. M. Rios. Geología de España*, 2. IGME, Madrid, 47–62.

SOPEÑA, A., LÓPEZ, J., ARCHE, A., PÉREZ-ARLUCEA, M., RAMOS, A., VIRGILI, C. & HERNANDO, S. 1988. Permian and Triassic rift basins of the Iberian Peninsula. *In*: MANSPEIZER, W. (ed.) *Triassic–Jurassic Rifting*. Elsevier, Amsterdam, Developments in Geotectonics, **22**, 757–786.

SOPEÑA, A. DOUBINGER, J, RAMOS, A. & PÉREZ-ARLUCEA, M. 1995. Palynologie du Permian et du Trias dans le Centre de la Péninsule Ibérique. *Science Geologique Bulletin*, **48**, 119–157.

SORIA, A. R. 1983. Le decollement de la couverture dans la Chaîne Ibêrique méridionale: effect de raccourcissements différentiels entre substratum et couverture. *Bulletin de la Société Géologique de France*, **25**(3), 379–387.

SORIA, A. R. 1997. *La sedimentación en las cuencas marginales del surco Ibérico durante el Cretácico Inferior y su control tectónico*. PhD Thesis, University of Zaragoza.

SORIA, A. R., LIESA, C. L. & MELÉNDEZ, A. 1997a. Tectónica extensional cretácica en la subcuenca de Oliete (Cordillera Ibérica central). *Geogaceta*, **22**, 203–206.

SORIA, A. R., MELÉNDEZ, A., MELÉNDEZ, M. N. & LIESA, C. L. 1997b. Evolución de dos sistemas continentales en la Cubeta de Aguilón (Cretácico Inferior): Interrelación sedimentaria entre depósitos aluviales y lacustres y su control tectónico. *Cuadernos de Geología Ibérica*, **22**, 473–507.

SORIA, A. R., MELÉNDEZ, M. N., MELÉNDEZ, A., LIESA, C. L., AURELL, M. & GÓMEZ-FERNÁNDEZ, J. C. 2000. The Early Cretaceous of the Iberian Basin (Northeastern Spain). *In*: GIERLOWSKI-KORDESCH, E. H. & KELTS, K. R. (eds) *Lake Basins through Space and Time*. AAPG, Studies in Geology, **46**, 285–294.

SORIA, J. M. 1993. *La sedimentación neógena entre Sierra Arana y el río Guadiana Menor. Evolución desde un margen continental hasta una cuenca intramontañosa*. PhD Thesis, University of Granada.

SORIA, J. M., FERNÁNDEZ, J. & VISERAS, C. 1999. Late Miocene stratigraphy and palaeogeographic evolution of the intramontane Guadix Basin (Central Betic Cordillera, Spain): implications for an Atlantic–Mediterranean connection. *Palaeogeography, Palaeoclimatology, Palaeoecology*, **151**, 255–266.

SORIANO, C. & MARTÍ, J. 1999. Facies Analysis of Volcano-Sedimentary Successions Hosting Massive Sulfide Deposits in the Iberian Pyrite Belt, Spain. *Economic Geology*, **94**, 867–882.

SOS, V. 1933. Los fósiles triásicos españoles del Museo de Ciencias Naturales de Madrid. *Boletín de la Sociedad Española de Historia Natural*, **33**, 287–302.

SOTO, F. 1982. Synaptophyllum (Rugosa) aus dem Unterdevon des kantabrischen Gebirges (Colle, Prov. Leon). *Neues Jahrbuch Geologischen Paläontologischen Abhandlungen*, **163**(2), 236–238.

SOTO, J. I. & PLATT, J. P. 1999. Petrological and structural evolution of high-grade metamorphic rocks from the floor of the Alboran Sea Basin, Western Mediterranean. *Journal of Petrology*, **40**, 21–60

SOULA, J. C. 1982. Characteristics and mode of emplacement of gneiss domes and plutonic domes in central-eastern Pyrenees. *Journal of Structural Geology*, **4**, 313–342.

SOULA, J. C., DEBAT, J., DERAMOND, J. & POUGET, P. 1986. A dynamic model of the structural evolution of the Hercynian Pyrenees. *Tectonophysics*, **129**, 29–51.

SOUQUET, P. 1967. *Le Cretacé superieur sud-pyrénéenne en Catalogne, Aragon et Navarre*. PhD Thesis, University of Toulouse.

SOUQUET, P. 1986a. Els Pirineus (Jurasic). In: Santanach, P. (coord.) *Historia Natural del Paisos Catalans, I. Geología*. Enciclopedia Catalana, Barcelona, 307–320.

SOUQUET, P. 1986b. Els Pirineus. *In*: FOLCH, R. (ed.) *Historia Natural del Paisos Catalans, 1*. Geología, Barcelona, 280–284.

SOUQUET, P. & PEYBERNÉS, B. 1991. Stratigraphie séquentielle du cycle Albien dans les Pyrénées franco-espagnoles. *Bull. Centres Rech. Explor.-Prod. Elf-Aquitaine*, **15**, 195–213.

SOUQUET, P., DEBROAS, E. J., BOIRIE, J. M. *et al.* 1985. Le Groupe du Flysch Noir (Albo-Cénomanien) dans les Pyrénées. *Bulletin des Centres de Recherches Exploration-Production Elf-Aquitaine*, **9**, 183–252.

SOURIAU, A. & GRANET, M. 1995. A tomographic study of the lithosphere beneath the Pyrenees from local and teleseismic data. *Journal of Geophysical Research*, **100**(B9), 18117–18134.

SPJELDNAES, N. 1961. Ordovician climatic zones. *Norsk Geologisk Tidsskrift*, **41**, 45–77.

SPJELDNAES, N. 1967. The palaeogeography of the Tethyan Region during the Ordovician. *In*: ADAMS, C. J. & AGER, D. V. (eds) *Aspects of Tethyan Biogeography*. Systematics Association Publications, London, **7**, 45–57.

SRIVASTAVA, S. P., ROEST, W. R., KOVACS, L. C., OAKEY, G., LÉVESQUE, S., VERHOEF, J. & MACNAB, R. 1990. Motion of Iberia since the Late Jurassic: Results from detailed aeromagnetic measurements in the Newfoundland Basin. *Tectonophysics*, **184**, 229–260.

STAMPFLI, G. M. & HÖCKER, C. F. W. 1989. Messinian paleorelief from 3-D seismic survey in the Tarraco concession area (Spanish Mediterranean Sea). *Geologie Minjbouw*, **68**, 201–210.

STAUDIGEL, H. 1981. *Der basale Komplex von La Palma. Submarine vulkanische prozesse, Petrologie, Geochemie und sekundare prozesse mi herausgehobenen, submarinen Teil einer ozeanischen Inseln*. PhD Thesis, University of Bochum, Germany.

STAUDIGEL, H. & SCHMINCKE, H. U. 1984. The Pliocene seamount series of La Palma, Canary Islands. *Journal of Geophysical Research*, **89**(B-13), 11190–11215.

STAUDIGEL, H., FÉRAUD, G. & GIANNERINI, G. 1986. The history of intrusive activity of the island of La Palma (Canary Islands). *Journal of Volcanology and Geothermal Research*, **27**, 299–322.

STEARNS, CH. & THURBER, D. 1967. 230 Th/234 U dates of Pleistocene marine fossils from the Mediterranean and Morocan littorals. *Progress in Oceanography, New York*, **4**, 293–305.

STEARNS, H. T 1946. Geology of the Hawaiian Islands. *Hawaii Division of Hydrography Bulletin*, **8**, 1–105.

STECKLER, M. S. & TEN BRINK, U. S 1986. Lithospheric strength variations as a control on new plate boundaries: examples from the northern Red Sea. *Earth and Planetary Science Letters*, **79**, 120–132.

STEINER, C., HOBSON, A., FAVRE, P. & STAMPLI, G. M. 1998. Early Jurassic sea-floor spreading in the central Atlantic – the Jurassic sequence of Fuerteventura (Canary Islands). *Geological Society of America Bulletin*, **110**(10), 1304–1317.

STEL, J. H. 1975. The influence of hurricanes upon the quiet depositional conditions in the Lower Emsian La Vid Shales of Colle (NW Spain). *Leidse Geologischen Mededelingen*, **49**(3), 475–486.

STEVENSON, A. C., MACKLIN, M. G., BENAVENTE, J. A., NAVARRO, C., PASSMORE, D. & DAVIS, B. A. 1991. Cambios ambientales durante el Holoceno en el valle medio del Ebro: sus implicaciones arqueológicas. *Cuaternario y Geomorfología*, **5**, 149–164.

STEWART, S. A. 1995. Paleomagnetic analysis of fold kinematics and implications for geological models of the Cantabrian/Asturian arc, north Spain. *Journal of Geophysical Research*, **100**, 20079–20094.

STILLE, H. 1931. Die Keltiberische Scheitelung. *Nachr. Ges. Wiss. Göttingen, Mat-Phys.*, **10**, 138–164.

STILLE, H. 1939. Bemerkungen betreffend die 'Sardische' Faltung und den Ausdruck 'ophiolithisch'. *Zeitschrift der deutschen geologischen Gesellschaft*, **91**, 771–773.

STILLMAN, C. J. 1987. A Canary Islands dyke swarm: implication for the formation of oceanic islands by extensional fissural volcanism. *In*: HALLS, H. C. & FAHRIG, W. F. (eds) *Mafic Dyke Swarms*. Geological Association of Canada, Special Papers, **34**, 243–255.

STILLMAN, C. J. 1999. Giant Miocene landslides and the evolution of Fuerteventura, Canary Islands. *Journal of Volcanology and Geothermal Research*, **94**(1–4), 89–104.

STILLMAN, C. J., FÚSTER, J. M., BENNELL-BAKER, M., MUÑOZ, M., MEWING, S. & SAGREDO, J. 1975. Basal complex of Fuerteventura is an oceanic intrusive complex with rift-system affinities. *Nature*, **257**, 469–471.

STILLMAN, C. J., FURNES, H., LE BAS, M. J., ROBERTSON, A. H. F. & ZIELONKA, J. 1982. The geological history of Maio, Cape Verde islands. *Journal of the Geological Society of London*, **139**, 347–361.

STOKES, M. & MATHER, A. E. 2000. Response of Plio-Pleistocene alluvial systems to tectonically induced base-level change, Vera. Basin, SE Spain. *Journal of the Geological Society*, **157**, 303–316.

STOLL, H. M. & SCHRAG, D. P. 2000. High-resolution stable isotope records from the Upper Cretaceous rocks of Italy and Spain: Glacial episodes in a greenhouse planet? *Geological Society of America Bulletin*, **112**(2), 308–319.

STORCH, P. 1983. The genus *Diplograptus* (Graptolithina) from the lowermost Silurian of Bohemia. *Vestník Ustredniho ústavu geologického*, **58**, 150–178.

STORCH, P. 1998. New data on Telychian (Upper Llandovery, Silurian) graptolites from Spain. *Journal of the Czech Geological Society*, **43**, 113–141.

STORCH, P. & GUTIÉRREZ-MARCO, J. C. 1998. Silurian sections of the Castilian Branch of the Iberian Cordillera (provinces of Guadalajara and Teruel. *Temas Geológico-Mineros ITGE*, **23**, 326–335.

STORCH, P., GUTIÉRREZ-MARCO, J. C., SARMIENTO, G. N. & RÁBANO, I. 1998. Upper Ordovician and Lower Silurian of Corral de Calatrava, southern part of the Central Iberian Zone. *Temas Geológico-Mineros ITGE*, **23**, 319–325.

STRAUS, L. G. & CLARK, G. A. 1986. *La Riera Cave. Stone Age Hunter-Gatherer Adaptations in Northern Spain*. Arizona State University, Anthropological Research Papers, **36**.

STRAUSS, G. K. 1970. *Sobre la geología de la provincia piritífera del suroeste de la Península Ibérica y de sus yacimientos, en especial sobre la mina de pirita de Lousal (Portugal)*. IGME, Madrid, Memorias, **77**.

STRAUSS, G. K. & MADEL, J. 1974. Geology of massive sulphide deposits in the Spanish–Portuguese pyrite belt. *Geologische Rundschau*, **63**, 191–211.

STRAUSS, G. K. & MADEL, J. 1977. Exploration practice for strata-bound volcanogenic sulphide deposits in the Spanish-Portuguese Pyrite Belt. *In*: KLEMM, D. D. & SCHNEIDER, H. J. (eds) *Time and Strata-bound Ore Deposits*. Springer-Verlag, Berlin, 55–93.

STRECKEISEN, A. 1979. Classification and nomenclature of volcanic rocks, lamprophyres, carbonatites, and melilitic rocks: Recommendations and suggestions of the IUGS Subcommission on the Systematics of Igneous Rocks. *Geology*, **7**, 331–335.

STRINGER, C. 2000. Gibraltar and the Neanderthals. In: FINLAYSON, C., FINLAYSON, G. & FA, D. (eds) *Gibraltar during the Quaternary*. Gibraltar Government Heritage Publications, Gibraltar, Monographs, **1**, 197–200.

SUÁREZ, J., MARTÍNEZ DEL OLMO, W., SERRANO, A. & LERET, G. 1989. Estructura del sistema turbidítico de la Formación Arenas del Guadalquivir. *In*: *Libro Homenaje a R. Soler y José*. AGGEP, Madrid, 123–136.

SUÁREZ, O., RUIZ, F., GALÁN, J. & VARGAS, I. 1978. Edades Rb-Sr de granitoides del occidente de Asturias (Nw de España). *Trabajos de Gelogia*, **10**, 437–442.

SUÁREZ, O., CORRETGÉ, L. G. & MARTÍNEZ, F. J. 1990. Distribution and characteristics of Hercynian metamorphism: West-Asturian-Leonese Zone (Part V). *In*: DALLMEYER, R. D. & MARTÍNEZ GARCÍA, E. (eds) *Pre-Mesozoic Geology of Iberia*. Springer-Verlag, Berlin, 129–133.

SUÁREZ DE CENTI, C., GARCÍA RAMOS, J. C. & VALENZUELA, M. 1989. Icnofósiles del Silúrico de la Zona Cantábrica (NO de España). *Boletín Geológico y Minero*, **100**, 339–394.

SUÁREZ DEL RÍO, L. M. & SUÁREZ, O. 1976. Estudio petrológico de los porfiroides precámbricos en la zona de Cudillero (Asturias). *Estudios Geológicos*, **32**, 53–59.

SUÁREZ RODRÍGUEZ, A. 1988. Estructura del área de Villaviciosa-Libardón (Asturias, Cordillera Cantábrica). *Trabajos de Geología de la Universidad de Oviedo*, **17**, 87–98.

SUÁREZ-RUIZ, I. 1987. *Caracterización, clasificación y estudio de la evolución de la materia orgánica dispersa en el Jurásico de Asturias y Cantabria*. PhD Thesis, University of Oviedo.

SUÁREZ-VEGA, L. C. 1974. Estratigrafía del Jurásico en Asturias. *Cuadernos de Geología Ibérica*, **3**, 1–368.

SUN, S. S. & MCDONOUGH, A. 1989. Chemical and isopotic systematics of oceanic basalts: implications for mantle composition and processes, *In*: SAUNDERS, A. D. & NORRY, M. J. (eds) *Magmatism in the Oceanic Basins*. Geological Society, London, Special Publications, **42**, 313–345.

SUÑER COMA, E. 1957. Los Graptolítidos del Silúrico superior de la Cordillera Costera Catalana. I, Santa Creu d'Olorde (Can Farrés). *Estudios Geológicos*, **33**, 45–84.

SURIÑACH, E. & VEGAS, R. 1988. Lateral inhomogeneities of the Hercynian crust in central Spain. *Physics of Earth and Planetary Interiors*, **51**, 226–234.

SURIÑACH, E. & VEGAS, R. 1993. Estructura general de la corteza en una transversal del Mar de Alborán a partir de datos de sísmica de refracción-reflexión de gran ángulo. Interpretación geo-dinámica. *Geogaceta*, **14**, 126–128.

SUTCLIFFE, O. E., DOWDESWELL, J. A., WHITTINGTON, R. J., THERON, J. N. & CRAIG, J. 2000. Calibrating the Late Ordovician glaciation and mass extinction by the eccentricity cycles of Earth's orbit. *Geology*, **28**, 967–970.

SWANSON, D. A., DUFFIELD, W. A. & FISKE, R. S. 1976. *Displacement of the South flank of Kilauea Volcano: the result of forceful intrusion of magma in the rift zones*. USGS Professional Papers, **963**.

SWEET, W. C. & BERGSTRÖM, S. M. 1984. Conodont provinces and biofacies of the Late Ordovician. *Geological Society of America, Special Paper* **196**, 69–86.

TAIT, J. A., BACHTADSE, V., FRANKE, W. & SOFFEL, H. C. 1997. Geodynamic evolution of the European Variscan fold belt: palaeomagnetic and geological constraints. *Geologische Rundschau*, **86**, 585–598.

TAMAIN, G. 1967. El Centenillo, zone de référence pour l'étude de l'Ordovicien de la Sierra Morena orientale (Espagne). *Comptes Rendus de l'Académie des Sciences, Paris, [D]*, **265**, 389–392.

TAMAIN, G. 1971*a*. El Alcudiense y la Orogénesis Cadomiense en el sur de la Meseta Ibérica (España). *Primer Centenario de la Real Sociedad Española de Historia Natural*, 437–464.

TAMAIN, G. 1971*b*. L'Ordovicien est-marianique (Espagne). Sa place dans la province méditerranéenne. *Mémoires du BRGM*, **73**, 403–416.

TAPPONNIER, P. 1977. Evolution téctonique du système alpin en Méditerranée: Poinçement et écrasement rigide-plastique. *Bulletin Société Géologique de France*, **7**, 437–460.

TAVERA, J. M. 1985. *Los ammonites del Tithónico superior-Berriasense de la Zona Subbética (Cordilleras Béticas)*. PhD Thesis, University of Granada.

TAVERA, J. M., AGUADO, R., COMPANY, M. & OLÓRIZ, F. 1994. Integrated biostratigraphy of the Durangites and Jacobi Zones (J/K boundary) at the Puerto Escaño section in southern Spain (province of Cordoba). *Geobios*, Mémoire Spécial **17**, 469–476.

TAYLOR, S. R. & MCLENNAN, S. M. 1985. *The Continental Crust: its Composition and Evolution*. Blackwell, London.

TEIDE GROUP. 1997. Morphometric interpretation of the northwest and southeast slopes of Tenerife, Canary Islands. *Journal of Geophysical Research*, **102**, 20325–20342.

TEIXELL, A. 1992. *Estructura alpina en la transversal de la terminación occidental de la Zona Axial Pirenaica*. PhD Thesis, University of Barcelona.

TEIXELL, A. 1996. The Ansó transect of the southern Pyrenees: basement and cover thrust geometries. *Journal of the Geological Society, London*, **153**, 301–310.

TEIXELL, A. 1998. Crustal structure and orogenic material budget in the West-Central Pyrenees. *Tectonics*, **17**, 395–406.

TEIXELL, A. & MUÑOZ, J. A. 2000. Evolución tectono-sedimentaria del Pirineo meridional durante el Terciario: una síntesis basada en la transversal del río Noguera Ribagorçana. *Revista de la Sociedad Geológica de España*, **13**, 251–264.

TEJERO, R & RUIZ, J. 2000. Brittle–ductile transition in Central Iberian Peninsula Crust. *Geogaceta*, **27**, 163–166.

TEJERO, R., PERUCHA, M. A., RIVAS, A. & BERGAMÍN, J. F. 1996. Gravity and structural models of Spanish Central System. *Geogaceta*, **20**, 947–950.

TEMIÑO, J., GARCÍA-HIDALGO, J. F. & SEGURA, M. 1997. Caracterización y evolución geológica del sistema dunas-humedales de Cantalejo (Segovia). *Estudios Geológicos*, **53**, 135–143.

TEN BRINK, U. S. 1991. Volcano spacing and plate rigidity. *Geology*, **19**, 397–400.

TEN KATE, W. G. H. Z. & SPRENGER, A. 1993. Orbital cyclicities above and below the Cretaceous/Paleogene boundary at Zumaya

(N Spain), Agost and Relleu (SE Spain). *Sedimentary Geology*, **87**, 69–101.

THIÉBLEMONT, D., MARCOUX, E., TEGYEY, M. & LEISTER, J. M. 1994. Mise en place de la ceinture pyriteuse sud-ibérique dans unpaléoprisme dacrétion? Arguments pétrologiques. *Bulletin Société Géologique de France*, **164**, 407–423.

THIEBLÉMONT, D., PASCUAL, E. & STEIN, G. 1998. Magmatism in the Iberian Pyrite Belt: petrological constraints on a metallogenic model. *Mineralium Deposita*, **33**, 98–110.

THIRWALL, M. F., JENKINS, C., VROON, P. C. & MATTEY, D. P. 1997. Crustal interaction during construction of oceanic islands: Pb-Sr-Nd-O isotope geochemistry of the shield basalts of Gran Canaria, Canary Islands. *Chemical Geology*, **135**, 233–262.

THOMAS, L. E., HAWKESWORTH, C. J., VAN CALSTERN, P., TURNER, S. P. & ROGERS, N. W. 1999. Melt generation beneath ocean islands: A U-Th-Ra isotope study from Lanzarote in the Canary Islands. *Geochimica et Cosmochimica Acta*, **63**, 4081–4099.

THORNES, J. 1968. Glacial and periglacial features in the Urbion Mountains, Spain. *Estudios Geológicos*, **24**, 249–258.

TILLING, R. I. & DVORAK, J. J, 1993. Anatomy of a basaltic volcano. *Nature*, **363**, 125–133.

TINKLER, K. J. 1966. Volcanic chronology of Lanzarote (Canary Islands). *Nature*, **209**, 1122–1123.

TORNÉ, M. & BANDA, E. 1992. Crustal thinning from the Betic Cordillera to the Alboran sea. *Geo-Marine Letters*, **12**, 76–81.

TORNÉ, M., DE CABISSOLE, B., BAYER, R., CASAS, A., DAIGNIERES, M. & RIVERO, A. 1989. Gravity constraints in the deep structure of the Pyrenean Belt along the ECORS profile. *Tectonophysics*, **165**, 105–116.

TORNÉ, M., BANDA, E., GARCÍA DUEÑAS, V. & BALANYÁ, J. C. 1992. Mantle-lithosphere bodies in the Alborán crustal domain (Ronda peridotites, Betic-Rif orogenic belt). *Earth and Planetary Science Letters*, **110**, 163–171.

TORNÉ, M., FERNÁNDEZ, M., COMAS, M. C. & SOTO, J. I. 2000. Lithospheric structure beneath the Alboran Basin – Results from 3D gravity modeling and tectonic relevance. *Journal of Geophysical Research*, **105**, 3209–3228.

TORNOS, F. & LOCUTURA, J. 1989. Mineralizaciones epitermales de Hg en Ossa Morena (Usagre, Badajoz). *Boletín de la Sociedad Española de Mineralogía*, **12**, 363–374.

TORNOS, F., RIBERA, F., ARIAS, D., LOREDO, J. & GALINDO, C. 1997. The carbonate-hosted Zn-Pb deposits of NW Spain; stratabound and discordant deposits related to the Variscan deformation. *In*: SANGSTER, D. F. (ed.) *Carbonate-Hosted Lead-Zinc Deposits*. Society of Economic Geologists, Special Publications, **4**, 195–203.

TORRES, J. 1992. Los yacimientos estratoligados de hierro de las cordilleras Beticas: el distrito del marquesado del Zenete (Granada). *In*: GARCÍA GUINEA, J. & MARTÍNEZ-FRÍAS, J. (eds) *Recursos Minerales de España*. CSIC, Colección Textos Universitarios, **15**, 569–584.

TORRES, T. 1990. Primeros resultados de una datación palinológica en el Keuper de la Rama Castellana de la Cordillera Ibérica y el Subbético frontal. *In*: ORTÍ, F & SALVANY, J. (eds) *Formaciones evaporíticas de la cuenca del Ebro y cadenas periféricas, y de la zona de Levante*. Enresa, Madrid, 219–223.

TORRES, T. & ZAPATA, J. L. 1986. Caracterización de dos sistemas de abanicos aluviales húmedos en el Terciario de la Depresión Intermedia (Cuenca-Guadalajara). *Actas Geología Hispanica*, **21–22**, 45–53.

TORRES, T., CANOIRA, L., COELLO, F. J. *et al.* 1995. Datación e interpretación paleoambiental de los travertinos de Priego (Cuenca) y río Blanco (Soria). Sector central de la Cordillera Ibérica. *In*: ALEIXANDRE, T. & PÉREZ-GONZÁLEZ, A. (eds) *Reconstrucción de paleoambientes y cambios climáticos durante el Cuaternario*. CCMA CSIC, Madrid, Monografias **3**, 113–124.

TORRES, T., GARCÍA ALONSO, P., NESTARES, T. & ORTÍZ, J. E. 1997. Terciario entre la Sierra de Altomira y la Serranía de Cuenca. Aspectos básicos de la paleogeografía cenozoica de la Depresión Intermedia. *In*: ALCALÁ, L. & ALONSO-ZARZA, A. M. (eds) *Itinerarios geológicos en el Terciario del centro y este de la Península Ibérica*. HC Multimedia, Madrid, 57–70.

TORRES ROLDÁN, R. L. 1979. The tectonic subdivision of the Betic Zone (Betic Cordilleras, southern Spain): its significance and one possible geotectonic scenario for the westernmost alpine belt. *American Journal of Science*, **279**, 19–51.

TORRES ROLDÁN, R. L. 1981. Plurifacial metamorphic evolution of the Sierra Bermeja Peridotite aureole (Southern Spain). *Estudios Geológicos*, **37**, 115–133.

TORRES ROLDÁN, R. L., POLI, G. & PECCERILLO, A. 1986. An early Miocene arc-tholeiitic magmatic dike event from the Alborán Sea – Evidence for precollisional subduction and back-arc crustal extension in the westernmost Mediterranean. *Geologische Rundschau*, **75**, 219–234.

TORRIANI, L. 1592. *Descripcion e historia del reino de las Islas Canarias*. Wölfel, 1940 (Spanish translation by A. Cioranescu, Edit. Goya, 1978, with the map of the 1585 eruption made by Torriani).

TORRUBIA, J. 1754. *Aparato para la Historia Natural Española. Tomo Primero*. Madrid, Imprenta de los Herederos de Don Agustín de Gordejuela y Sierra.

TORTOSA, A., ARRIBAS, J., GARZÓN, G., FERNÁNDEZ, P. & PALOMARES, M. 1997. Análisis petrológico de depósitos de terrazas aplicado al estudio de los procesos de captura en los ríos Adaja, Voltoya y Eresma (provincias de Segovia y Valladolid). *Revista de la Sociedad Geológica de España*, **10**, 131–145.

TOSCANI, L., VENTURELLI, G., BARBIERI, M., CAPEDRI, S., FERNÁNDEZ, J. M. & ODDONE, M. 1990. Geochemistry and petrogenesis of two-pyroxene andesites from Sierra de Gata (SE Spain). *Mineralogy and Petrology*, **41**, 199–213.

TOSCANI, L., CONTINI, S. & FERRARINI, M. 1995. Lamproitic rocks from Cabezo Negro de Zeneta: brown micas as a record of magma mixing. *Mineralogy and Petrology*, **55**, 281–292.

TOYOS, J. M. 1993. Minerales silicoaluminosas: Situación actual, tipos de yacimientos y posibilidades de explotación en Galicia. *Cuadernos del Laboratorio Geológico de Laxe*, **14**, 237–246.

TOYOS, J. M. & FERRERO, A. 1990. *Investigación minera de minerales silicoaluminosos en Galicia*. Report **11–275**, IGME, Madrid.

TRICALINOS, J. 1928. Untersuchungen Uber den bau der Celtiberischen Ketten der Nordostlichen Spaniens. *Deutschen Geologischen Gesellschaft*, **80**, 409–483.

TRITLLA, J. 1994. *Geología y metalogenia de las mineralizaciones de Ba-Hg de la Sierra del Espadán (Provincia de Castellón)*. PhD Thesis, University of Barcelona.

TRITLLA, J. & CARDELLACH, E. 1993. Origin of the carbonate-hosted mercury veins from the Espadán mountains (Iberian ranges, eastern Spain): evidence from fluid inclusions and stable isotopes. *In*: FENOLL, P., TORRES, J. & GERVILLA, F. (eds) *Current Research in Geology Applied to Mineral Deposits*. University of Granada, 265–268.

TRITLLA, J. & CARDELLACH, E. 1997. Fluid inclusion systematics in pre-ore minerals from carbonate-hosted mercury veins from the Espadán Ranges (East of Spain). *Chemical Geology*, **137**, 91–106.

TRUYOLS, J. & JULIVERT, M. 1983. El Silúrico en el Macizo Ibérico. *In*: COMBA, J. A. (coord.) *Libro Jubilar J. M. Ríos, Geología de España, 1*. IGME, Madrid, 246–265.

TRUYOLS, J., PHILIPPOT, A. & JULIVERT, M. 1974. Les Formations siluriennes de la Zone Cantabrique et leurs faunes. *Bulletin de la Société Géologique de France*, **16**, 23–35.

TRUYOLS, J., ARBIZU, M. A., GARCÍA ALCALDE, J. L. GARCÍA LÓPEZ, S. MÉNDEZ BEDIA, I., SOTO, F. & TRUYOLS MASSONI, M. 1990. Cantabrian and Palentian Zones; Stratigraphy: The Asturian-Leonese Domain (Cantabrian Zone). *In*: DALLMEYER, R. D. & MARTÍNEZ GARCÍA, E. (eds) *Pre-Mesozoic Geology of Iberia*, Springer Verlag, Berlin, 10–19.

TRUYOLS, J., ARAMBURU, C., ARBIZU, M. *et al.* 1996. La Formación vulcanosedimentaria del Castro (Ordovícico-Silúrico) en el Cabo Peñas (Zona Cantábrica, NO España). *Geogaceta*, **20**, 15–18.

TRUYOLS-MASSONI, M. 1986. *Nowakia acuaria* (Praguiense, Devónico inferior) de la Zona Asturoccidental-Leonesa (NO de España). *Breviora Geologica Asturica*, **27**, 12–16.

TRUYOLS-MASSONI, M. & QUIROGA, J. L. 1981. Tentaculites dacrioconáridos en el sinforme de Alcañices (prov. de Zamora). *Cuadernos Laboratorio Xeolóxico Laxe*, **2**, 171–173.

TUBÍA, J. M. 1988. Estructura de los Alpujárrides occidentales: Cinemática y condiciones de emplazamiento de las peridotitas de Ronda. *Publicaciones especiales del Boletín Geológico y Minero de España*, **99**, 165–212 (PhD Thesis, University of País Vasco, 1986).

TUBÍA, J. M. & CUEVAS, J. 1986. High-temperature emplacement of the Los Reales peridotite nappe (Betic Cordillera, Spain). *Journal of Structural Geology*, **8**, 473–482.

TUBÍA, J. M. & CUEVAS, J. 1987. Structures et cinématique liées à la mise en place des péridotites de Ronda (Cordillères Béetiques, Espagne). *Geodinámica Acta*, **1**, 59–69.

TUBÍA, J. M. & GIL-IBARGUCHI, J. I. 1991. Eclogites of the Ojén nappe: a record of subduction in the Alpujárride complex (Betic Cordilleras, southern Spain). *Journal of the Geological Society of London*, **148**, 801–804.

TUBÍA, J. M., CUEVAS, J., NAVARRO-VILÁ, F., ALVAREZ, F. & ALDAYA, F. 1992. Tectonic evolution of the Alpujárride Complex (Betic Cordillera, Southern Spain). *Journal of Structural Geology*, **14**, 193–203.

TUBÍA, J. M., CUEVAS, J. & GIL-IBARGUCHI, J. I. 1997. Sequential development of the metamorphic aureole beneath the Ronda peridotites and its bearing on the tectonic evolution of the Betic Cordillera. *Tectonophysics*, **279**, 227–252.

TURBÓN, D., PÉREZ-PÉREZ, A. & LALUEZA, C. 1994. Los restos humanos del nivel solutrense de la cueva de Nerja (Málaga). *In*: BERNIS, C. *et al.* (eds) *Biología de poblaciones humanas: problemas metodológicos e interpretación ecológica*. Autónoma University, Madrid, 51–62.

TURNER, J. A. 1996. Switches in subduction direction and the lateral termination of mountain belts, Pyrenees-Cantabrian transition, Spain. *Journal of the Geological Society of London*, **153**, 563–571.

TURNER, S. P., PLATT, J. P., GEORGE, R. M. M., KELLEY, S. P., PEARSON, D. G. & NOWELL, G. M. 1999. Magmatism associated with orogenic collapse of the Betic-Alborán domain, SE Spain. *Journal of Petrology*, **40**, 1011–1036.

UBACH, J. 1990. Geología de los materiales paleozoicos de las escamas de la Cordillera prelitoral catalana al este del río Llobregat. *Acta Geológica Hispanica*, **25**, 113–121.

UBANELL, A. G. 1981. Significado tectónico de los principales sistemas de diques en un sector del sistema central español. *Cuadernos de Geología Ibérica*, **7**, 607–622.

UGIDOS, J. M. 1973. *Estudio Petrológico del área Bejar-Plasencia (Salamanca Cáceres)*. PhD Thesis, University of Salamanca.

UGIDOS, J. M. 1990. Granites as a Paradigm of Genetic processes of granitic rocks: I-types vs S-types. *In*: DALLMEYER, R. D. & MARTÍNEZ GARCÍA, E. (eds) *Pre-Mesozoic Geology of Iberia*. Springer-Verlag, Berlin, 189–206.

UGIDOS, J. M., ARMENTEROS, I., BARBA, P., VALLADARES, M. I. & COLMENERO, J. R. 1997*a*. Geochemistry and petrology of recycled orogen-derived sediments: a case study from Upper Precambrian siliciclastic rocks of the Central Iberian Zone, Iberian Massif, Spain. *Precambrian Research*, **84**, 163–180.

UGIDOS, J. M., VALLADARES, M. I., RECIO, C., ROGERS, G., FALLICK, A. E. & STEPHENS, W. E. 1997*b*. Provenance of Upper Precambrian/Lower Cambrian shales in the Central Iberian Zone, Spain: evidence from a chemical and isotopic study. *Chemical Geology*, **136**, 55–70.

UGIDOS, J. M., VALLADARES, M. I., BARBA, P., ARMENTEROS, I. & COLMENERO, J. R. 1999. Geochemistry and Sm-Nd isotope systematics on the Upper Precambrian–Lower Cambrian sedimentary successions in the Central Iberian Zone, Spain. *Journal of Conference Abstracts*, **4**, Annual meeting of IGCP Project 376, 1021.

UNDERWOOD, C. J., DEYNOUX, M. & GHIENNE, J.-F. 1998. High palaeolatitude (Hodh, Mauritania) recovery of graptolite faunas after the Hirnantian (end Ordovician) extinction event. *Palaeogeography, Palaeoclimatology, Palaeoecology*, **142**, 91–105.

URBANO, R. 1998. *Guía para la investigación de los Recursos Minerales en España*. IGME. Madrid (ISBN 84-7840-335-3).

URBINA, D., SÁNCHEZ, A., LOMBARDERO, M. & REGUEIRO, M. 1997. Mármoles romanos y canteras en Talavera de La Reina. *Zephyrvs*, **50**, 273–287.

URBINA, M., BIESTER, H., HIGUERAS, P. & LORENZO, S. 2000. Utilización de la técnica de descomposición térmica de muestras con contenidos en mercurio para la identificación de problemas medioambientales producidos por este elemento. Datos preliminares referidos al distrito de Almadén. *Cuadernos Laboratorio Xeolóxico de Laxe*, **25**, 365–367.

URETA, M. S. 1985. *Bioestratigrafía y Paleontología (Ammonitina) del Aaleniense en el sector noroccidental de la Cordillera Ibérica*. PhD Thesis, Complutense University, Madrid (Colección Tesis Doctorales, no. **158/85**).

URETA, M. S. 1988. El Aaleniense en el borde nororiental de la Sierra de los Cameros (Muro de Aguas, Logroño). *II Coloquio de Estratigrafía y Paleogeografía del Jurásico de España, libro guía de las excursiones, Ciencias de la Tierra (Instituto de Estudios Riojanos)*, **11**, 299–308.

URETA, M. S., GOY, A., GÓMEZ, J. J. & MARTÍNEZ, G. 1999. El Toarciense superior y Aaleniense en la sección de Moyuela. *Cuadernos de Geología Ibérica*, **25**, 59–72.

URGELÉS, R., CANALS, M., BARAZA, J., ALONSO, B. & MASSON, D. G. 1997. The last major megalandslides in the Canary Islands: The El Golfo debris avalanche and the Canary debris flow, west Hierro Island. *Journal of Geophysical Research*, **102**, 20305–20323.

URGELÉS, R., CANALS, M., BARAZA, J. & ALONSO, B. 1998. Seismostratigraphy of the western flank of El Hierro and La Palma (Canary Islands): a record of the Canary volcanism. *Marine Geology*, **146**, 225–241.

URGELÉS, R., MASSON, D. G., CANALS, M., WATTS, A. B. & LE BAS, T. 1999. Recurrent large-scale landsliding on the west flank of La Palma, Canary Islands. *Journal of Geophysical Research*, **104**, 25331–25348.

VACAS, J. M. & MARTÍNEZ CATALÁN, J. R. 1987. El Sinforme de Alcañices en la transversal de Manzanal del Barco. *Studia Geológica Salmanticensia*, **24**, 151–175.

VACHARD, D., COLIN, J. P., ROSELL, J., HOCHULI, P. 1989. Incursions de microfaunes alpines dans le Trias des Isles Baléares et des Pyrénées espagnoles. *Comptes Rendus de l'Académie des Sciences*, **308**, 947–952.

VACHARD, D., COLIN, J. P., HOCHULI, R. A. & ROSELL, J. 1990. Bostratigraphie, Foraminifères, palynoflores et ostracodes du Rhetien du Bac Grillera (Pyrenees orientales espagnoles). *Geobios*, **23**, 521–537.

VACHER, P. & SOURIAU, A. 2001. A three-dimensional model of the Pyrenean deep structure based on gravity modelling, seismic images and petrological constraints. *Geophysical Journal International*, **145**, 460–470.

VAIL, P. R., AUDEMARD, F., BOWMAN, S. A., EISNER, P. N. & PÉREZ-CRUZ, C. 1991. The stratigraphic signatures of tectonics, eustasy and sedimentology – an overview. *In*: EINSELE, G., RICKEN, W. & SEILACHER, A. (eds) *Cycles and Events in Stratigraphy*. Springer-Verlag, Berlin, 619–659.

VALENZUELA, M. 1988. *Estratigrafía, Sedimentología y Paleogeografía del Jurásico de Asturias*. PhD Thesis, University of Oviedo.

VALENZUELA, M., GARCÍA-RAMOS, J. C. GONZÁLEZ LASTRA, J. A. & SUÁREZ DE CENTI, C. 1985. Sedimentación cíclica margo-calcárea de plataforma en el Lías de Asturias. *Trabajos de Geología, Universidad Oviedo*, **15**, 45–52.

VALENZUELA, M., GARCÍA-RAMOS, J. C. & SUÁREZ DE CENTI, C. 1986. The Jurassic sedimentation in Asturias (N Spain). *Trabajos de Geología, Universidad Oviedo*, **16**, 121–132.

VALENZUELA, M., GARCÍA-RAMOS, J. C. & SUÁREZ DE CENTI, C. 1989. La sedimentación en una rampa carbonatada dominada por tempestades, ensayos de correlación de ciclos y eventos de la ritmita margo-calcárea del Jurásico de Asturias. *Cuadernos de Geología Ibérica*, **13**, 217–235.

VALENZUELA, M., GARCÍA-RAMOS, J. C. & SUÁREZ DE CENTI, C. 1992. *Hardgrounds* a techo de ciclos de somerización y ralentización en una rampa carbonatada del Lías de Asturias. *Geogaceta*, **11**, 70–73.

VALENZUELA, M., DÍAZ GONZÁLEZ, T. E., GUTIÉRREZ VILLARÍAS, M. I. & SUÁREZ DE CENTI, C. 1998. La Formación Lastres del Kimmeridgiense de Asturias: Sedimentología y estudio paleobotánico inicial. *Cuadernos de Geología Ibérica*, **24**, 141–171.

VALENZUELA-RÍOS, J. I. 1984. *Estudio geológico de un sector de las Cadenas Ibéricas Orientales entre Minas Tiergas, Mesones y Nigüella(Zaragoza)*. PhD Thesis, University of Zaragoza.

VALENZUELA-RÍOS, J. I. 1994. Conodontos del Lochkoviense y Praguiense (Devónico Inferior) del Pirineo Central Español. *Memorias Museo Paleontológico Universidad Zaragoza*, **5**, 1–178.

VALENZUELA-RÍOS, J. I. 1996. Conodontos del Wenlock y Ludlow (Silúrico) del Valle de Tena (Pirineos Aragoneses). *Geogaceta*, **19**, 91–93.

VALENZUELA-RÍOS, J. I. 1997. Can *Polygnathus pireneae* be the index of a standard conodont Zone? *Newsletters in Stratigraphy*, **35**(3), 173–179.

VALENZUELA-RÍOS, J. I. & CARLS, P. 1996. Conodontos e Invertebrados del Devónico Medio del valle de Tena (Huesca, Pirineo aragonés). *Coloquios Paleontología*, **46**, 43–59, 196–199.

VALERO, B. 1991. *Los sistemas lacustres carbonatados del Stephaniense y Pérmico en el Pirineo central y occidental.* PhD Thesis, University of Zaragoza.

VALERO, B. L. 1993. Lacustrine depositional and related volcanism in a transtensional tectonic setting: Upper Stephanian–Lower Autunian in the Aragon-Bearn Basin, Western Pyrenees (Spain-France). *Sedimentary Geology*, **83**, 133–160.

VALERO, B. & GISBERT, J. 1992. Shallow carbonate lacustrine facies models in the Permian of the Aragón-Bearn Basin (W Spanish-French Pyrenees). *Carbonates and Evaporites*, **7**, 94–107.

VALERO-GARCÉS, B. L., GONZÁLEZ-SAMPÉRIZ, P., DELGADO-HUERTAS, A., NAVAS, A., MACHÍN, J. & KOLTS, K. 2000. Late glacial and Late Holocene environmental and vegetational change in Salada Mediana, central Ebro Basin, Spain. *Quaternary International*, **73/74**, 29–46.

VALLADARES, I., UGIDOS, J. M. & RECIO, C. 1993. Criterios geoquímicos de correlación y posible área fuente de las pelitas del Precambrico Superior-Cámbrico Inferior en la Zona Centro Ibérica (Macizo Ibérico. España). *Revista de la Sociedad Geológica de España*, **6**, 37–45.

VALLADARES, M. I. 1995. Siliciclastic-carbonate slope apron in an immature tensional margin (Upper Precambrian–Lower Cambrian), Central Iberian Zone, Salamanca, Spain. *Sedimentary Geology*, **94**, 165–186.

VALLADARES, M. I. & RODRÍGUEZ ALONSO, M. D. 1988. Depositional processes of submarine channel-fill carbonate (Late Precambrian, Salamanca, Spain). *Revista de la Sociedad Geológica de España*, **1**, 165–175.

VALLADARES, M. I., BARBA, P., COLMENERO, J. R., ARMENTEROS, I. & UGIDOS, J. M. 1998. La sucesión sedimentaria del Precámbrico Superior-Cámbrico Inferior en el sector central de la Zona Centro Ibérico: Litoestratigrafía, geoquímica y facies sedimentarias. *Revista de la Sociedad Geologica de España*, **11**, 271–283.

VALLADARES, M. I., UGIDOS, J. M., BARBA, P., ARMENTEROS, I. & COLMENERO, J. R. 1999. Upper Proterozoic–Lower Cambrian shales in the Central Iberian Zone: Chemical features and implications for other peri-Gondwanan areas. *Journal of Conference Abstracts*, **4**, Annual meeting of IGCP Project 376, 1021–1022.

VALLADARES, M. I., BARBA, P., UGIDOS, J. M., COLMENERO, J. R. & ARMENTEROS, I. 2000. Upper Neoproterozoic–Lower Cambrian sedimentary successions in the Central Iberian Zone (Spain): sequence stratigraphy, petrology and chemostratigraphy. Implications for other European zones. *International Journal of Earth Science*, **89**, 2–20.

VALLADARES, M. I., UGIDOS, J. M., BARBA, P. & COLMENERO, J. R. 2002. Contrasting geochemical features of the Central Iberian Zone shales (Iberian Massif, Spain): Implications for the evolution of Upper Proterozoic–Lower Cambrian sediments and their sources in other Peri-Gondwana areas *Tectonophysics*, **352**, 121–132.

VALVERDE VAQUERO, P. 1997. *An integrated field, geochemical and U-Pb geochronological study of the southwest Hermitage Flexure (Newfoundland Appalachians, Canada) and the Sierra de Guadarrama (Iberian Massif, Central Spain): a contribution to the understanding of the geological evolution of circum-Atlantic peri-Gondwana.* PhD Thesis, Memorial University of Newfoundland, St. John's.

VALVERDE VAQUERO, P. & DUNNING, G. 2000. New U-Pb ages for Early Ordovician magmatism in Central Spain. *Journal of the Geological Society, London*, **157**, 15–26.

VALVERDE-VAQUERO, P. & FERNÁNDEZ, F. J. 1996. Edad de enfriamiento U/Pb en rutilos del Gneis de Chimparra (Cabo Ortegal, NO de España). *Geogaceta*, **20**, 475–478.

VALVERDE-VAQUERO, P., CUESTA FERNÁNDEZ, A., GALLASTEGUI, G., SUÁREZ, O., CORRETGÉ, L. G. & DUNNING, G. R. 1999. U-Pb dating of late Variscan Magmatism in the Cantabrian zone (Northern Spain). *Journal of Conference, EUG* (European Union of Geosciences) 10.

VAN BEMMELEN, R. W. 1966. Origin of the Western Mediterranean Sea. *Verhand. Koning. Nederlanden Geologisches Mijnbouw*, **26**, 13–52.

VAN CALSTEREN, P. W. C. 1977. Geochronological, geochemical and geophysical investigations in the Hercynian basement of Galicia (NW Spain). *Leidse Geologische Mededelingen*, **51**, 57–61.

VAN CALSTEREN, P. W. C. & DEN TEX, E. 1978. An early Paleozoic continental rift system in Galicia (NW Spain). *In*: RAMBERG, J. B. & NEUMAN, E. R. (eds) *Tectonics and Geophysics of Continental Rifts*. Reidel, Dordrecht, 125–132.

VAN CALSTEREN, P. W. C., BOELRIJK, N. A. I. M., HEBEDA, E. H., PRIEM, H. N. A., DEN TEX, E., VERDURMEN, E. A. T. & VERSCHURE, R. H. 1979. Isotopic dating of older elements (including the Cabo Ortegal mafic-ultramafic complex) in the hercynian orogen of NW Spain: manifestations of a presumed early Palaeozoic mantle-plume. *Chemical Geology*, **24**, 35–56.

VAN DEN EECKHOUT, B. & ZWART, H. J. 1988. Hercynian crustal-scale extensional shear zone in the Pyrenees. *Geology*, **16**, 135–138.

VAN DER BEEK, P. A. & CLOETINGH, S. 1992. Lithospheric flexure and the tectonic evolution of the Betic Cordilleras (SE Spain). *Tectonophysics*, **203**, 325–344.

VAN DER VOO, R. 1969. Paleomagnetic evidence for the rotation of the Iberian Peninsula. *Tectonophysics*, **7**, 5–56.

VAN DER VOO, R. 1993. *Paleomagnetism of the Atlantic, Tethys and Iapetus Oceans.* Cambridge University Press.

VAN DER VOO, R., STAKAMATOS, J. A. & PARÉS, J. M. 1997. Kinematic constraints on thrust-belt curvature from syndeformational magnetizations in the Lagos del Valle syncline in the Cantabrian Arc, Spain. *Journal of Geophysical Research*, **102**, 10105–10119.

VAN DER WAL, D. 1993. *Deformation processes in mantle peridotites.* PhD Thesis, University of Utrecht (Geológica Ultraiectina, **102**).

VAN DER WAL, D. & VISSERS, R. L. M. 1993. Uplift and emplacement of upper mantle rocks in the western Mediterranean. *Geology*, **21**, 1119–1122.

VAN DER WAL, D. & VISSERS, R. L. M. 1996. Structural petrology of the Ronda Peridotite, SW Spain; deformation history. *Journal of Petrology*, **37**, 23–43.

VAN OVERMEEREN, R. A. 1975. A gravity investigation of the catazonal rock complex at Cabo Ortegal (NW Spain). *Tectonophysics*, **26**, 293–307.

VAN VLIET, A. 1978. Early Tertiary deep-water fans of Guipúzcoa, northern Spain. *In*: Stanley, D. J. & Kelling, G. (eds) *Sedimentation in Submarine Canyons, Fans and Trenches*. Dowden, Hutchinson and Ross, Stroudsburg, Pennsylvania, 190–209.

VAN WEES, J. D., ARCHE, A., BEIJDORFF, C. G., LÓPEZ-GÓMEZ, J. & CLOETING, S. A. P. L. 1998. Temporal and spatial variations in tectonic subsidence in the Iberian Basin (eastern Spain): inferences from automated forward modelling of high-resolution stratigraphy (Permian-Mesozoic). *Tectonophysics*, **300**, 285–310.

VAN ZUUREN, A. 1969. Structural petrology of an area near Santiago de Compostela (NW Spain). *Leidse Geologische Mededelingen*, **45**, 1–71.

VANDENBERG, J. 1980. New paleomagnetic data from the Iberian Peninsula. *Geologie en Mijnbouw*, **59**, 49–60.

VANEK, J. & VOKÁK, V. 1997. Trilobites of the Bohdalec Formation (Upper Berounian, Ordovician, Prague Basin, Czech Republic). *Palaeontologia Bohemiae*, **3**, 20–50.

VANNIER, J. 1986a. Ostracodes Palaeocopa de l'Ordovicien (Arenig-Caradoc) ibéro-armoricain. *Palaeontographica* Abt A, **193**, 77–143.

VANNIER, J. 1986b. Ostracodes Binodicopa de l'Ordovicien (Arenig-Caradoc) ibéro-armoricain. *Palaeontographica* Abt A, **193**, 145–218.

VANNIER, J. 1987. Le genre *Ceratopsis* (Ostracoda, Palaeocopa) dans l'Ordovicien de l'Europe et de l'Amérique du Nord: phylogenèse, paléoécologie et implications paléobiogéographiques. *Geobios*, **20**, 725–755.

VANNIER, J. & BABIN, C. 1995. *Hanadirella* El-Khayal, 1985, un arthropode (?) énigmatique de l'Ordovicien nord-gondwanien. *Geobios*, **28**, 473–485.

VARAS, M. J., BARBA, P. & ARMENTEROS, I. 1999. Estratigrafía del Mioceno en el sector suroriental de la Cuenca de Almazán (Zaragoza). *Revista de la Sociedad Geológica de España*, **12**(1), 63–75.

VAUCHEZ, A. 1975. Tectoniques tangentielles superposées dans le segment hercynien sud-ibérique: les nappes et plis couchés de la region d'Alconchel-Fregenal de la Sierra (Badajoz). *Boletín Geológico y Minero*, **86**, 573–580.

VAUCHEZ, A. & GARRIDO, C. 2001. Seismic properties of an asthenospherized lithospheric mantle: constrains from lattice preferred orientations in peridotite from the Ronda massif. *Earth and Planetary Science Letters*, **192**, 235–249.

VAUCHEZ, A. & NICOLAS, A. 1991. Mountain building: strike-parallel motion and mantle anisotropy. *Tectonophysics*, **185**, 183–201.

VAUDOUR, J. 1979. *La región de Madrid*. Ophrys, Aix-en-Provence.

VÁZQUEZ, J. M., MORENO, F., GARCÍA DE MIGUEL, J. M., RUIZ, C. & GONZÁLEZ, A. P. 1992. *Mapa y Memoria de la Hoja no. 685 (Los Yébenes) del Mapa Geológico de España a escala 1:50.000*. Serie MAGNA. IGME, Madrid.

VEEN, J. VAN 1965. The tectonic and stratigraphic history of the Cardaño Area, Cantabrian Mountains, Northwest Spain. *Leidse Geologische Mededelingen*, **35**, 45–104.

VEGAS, R. 1971. Precisiones sobre el Cámbrico del centro y sur de España. El problema de la existencia de Cámbrico en el valle de Alcudia y en las sierras de Cáceres y N de Badajoz. *Estudios Geológicos*, **27**, 419–425

VEGAS, R. 2000. The intrusion of the Plasencia (Messejana) dike as part of the Circum-Atlantic early Jurassic magmatism: Tectonic implications in the southwestern Iberian peninsula. *Geogaceta*, **27**, 175–178.

VEGAS, R & BANDA, E. 1982. Tectonic framework and Alpine evolution of the Iberian Peninsula. *Earth Evolution Sciences*, **4**, 320–343.

VEGAS, R., FONTBOTÉ, J. M. & BANDA, E. 1979. Widespread neogene rifting superimposed on alpine regions of the Iberian Peninsula. *Proceedings of the Symposium on Evolution and Tectonism of the Western Mediterranean and Surrounding Areas*, EGS, Viena. IGN, Special Publications, **201**, 109–128.

VEGAS, R., VAZQUEZ, J. T., SURINACH, E. & MARCOS, A. 1990. Model of distributed deformation, block rotation and crustal thickening for the formation of the Spanish Central System. *Tectonophysics*, **184**, 367–378.

VEGAS, R., JUAREZ, M. T. & KÄLIN, O. 1996. Tectonic and geodynamic significance of paleomagnetic rotations in the Iberian Chain, Spain. *Geogaceta*, **19**, 11–14.

VEGA-TOSCANO, L. G., HOYOS, M., RUIZ-BUSTOS, A. & LAVILLE, H. 1988. La séquence de la Grotte de la Carihuela (Piñar, Grenade): chronostratigraphie et paléoécologie du Pleistocène supérieur au sud de la Péninsule Ibérique. *In*: OTTE, M. (ed.) *L'Homme de Néandertal 2. L'environnement*. ERAUL, Liège, **29**, 169–180.

VELASCO, F. 1976. *Mineralogía y metalogenia de los skarns de Santa Olalla (Huelva)*. PhD Thesis, University of Lejona, PaisVasco.

VELASCO, F. & AMIGO, J. M. 1981. Mineralogy and origin of the skarn from Cala (Huelva, Spain). *Economic Geology*, **76**, 719–727.

VELASCO, F., SÁNCHEZ-ESPAÑA, J., BOYCE, A. J., FALLICK, A. E., SÁEZ, R. & ALMODÓVAR, G. R. 1998. A new sulfur isotopic study of some IPB deposits: evidence of a textural control on sulfur isotope composition. *Mineralium Deposita*, **34**, 4–18.

VENNIN, E., ÁLVARO, J. J. & VILLAS, E. 1998. High-latitude pelmatozoan-bryozoan mud-mounds from the late Ordovician northern Gondwana platform. *Geological Journal*, **33**, 121–140.

VENTURELLI, G., CAPEDRI, S., DI BATTISTINI, G., CRAWFORD, A., KOGARKO, L. N. & CELESTINI, S. 1984. The ultrapotassic rocks from southeastern Spain. *Lithos*, **17**, 37–54.

VENTURELLI, G., MARIANI, E. S., FOLEY, S. F., CAPEDRI, S. & CRAWFORD, A. J. 1988. Petrogenesis and conditions of crystallization of Spanish lamproitic rocks. *Canadian Mineralogist*, **26**, 67–79.

VENTURELLI, G., CAPEDRI, S., BARBIERI, M., TOSCANI, L., SALVIOLI, E. & ZERBI, M. 1991. The Jumilla lamproite revisited: a petrological oddity. *European Journal of Mineralogy*, **3**, 123–145.

VENTURELLI, G., SALVIOLI, E., TOSCANI, L., BARBIERI, M. & GORGONI, C. 1993. Post-magmatic apatite+hematite+carbonate assemblage in the Jumilla lamproites. A fluid inclusion and isotope study. *Lithos*, **30**,139–150.

VERA, J. A. 1970. Estudio estratigráfico de la Depresión de Guadix-Baza. *Boletín Geológico y Minero*, **81**, 429–462.

VERA, J. A. 1988. Evolución de los sistemas de depósito en el margen ibérico de la Cordillera Bética. *Revista de la Sociedad Geológica de España*, **1**, 373–391.

VERA, J. A. 1998. El Jurásico de la Cordillera Bética: Estado actual de conocimientos y problemas pendientes. *Cuadernos de Geología Ibérica*, **24**, 17–42.

VERA, J. A. 2000. El Terciario de la Cordillera Bética: estado actual de conocimientos. *Revista Sociedad Geológica de España*, **13**, 345–373.

VERA, J. A. 2001. Evolution of the Southern Iberian Continental Margin. *In*: ZIEGLER, P. A., CAVAZZA, W., ROBERTSON A. F. H. & CRASQUIN-SOLEAU, S. (eds) *Peri-Tethys Memoir 6: Peri-Tethyan Rift/Wrench Basins and Passive Margins*. Mémoires du Muséum National d'Histoire naturelle, **186**, 109–143.

VERA, J. A. & MARTÍN-ALGARRA, A. 1994. Mesozoic stratigraphic breaks and pelagic stromatolites in the Betic Cordillera, Southern Spain. *In*: BERTRAND-SARFATI, J. & MONTY, C. (eds) *Phanerozoic Stromatolites II*. Kluwer, Dordrecht, 319–344.

VERA, J. A. & MOLINA, J. M. 1998. Shallowing-upward cycles in pelagic troughs (Upper Jurassic, Subbetic, Southern Spain). *Sedimentary Geology*, **119**, 103–121.

VERA, J. A. & MOLINA, J. M. 1999. La Formación Capas Rojas: caracterización y génesis. *Estudios Geológicos*, **55**, 45–66.

VERA, J. A., GARCÍA-HERNÁNDEZ, M., LÓPEZ-GARRIDO, A. C., COMAS, M. C., RUIZ-ORTIZ, P. A. & MARTÍN-ALGARRA, A. 1982. El Cretácico de la Cordillera Bética. *In*: GARCÍA, A. (ed.) *El Cretácico de España*. Complutense University, Madrid, 515–632.

VERGÉS, J. 1993. *Estudi tectònic del vessant sud del Pirineu central i oriental: evolució en 3D*. PhD Thesis, University of Barcelona.

VERGÉS, J. 1999. *Estudi geològic del vessant sud del Pirineu oriental i central. Evolució cinemàtica en 3-D*. Institut Cartogràfic de Catalunya, Monografies tècniques **7**.

VERGES, J. & GARCIA-SENZ, J. 2001. Mesozoic evolution and Cainozoic inversion of the Pyrenean Rift. *In*: ZIEGLER, P. A., CAVAZZA, W., ROBERTSON, A. F. H. & CRASQUIN-SOLEAU, S. (eds) *Peri-Tethys Memoir 6: Peri-Tethyan Rift/Wrench Basins and Passive Margins*. Mémoires du Muséum national d'Histoire naturelle, **186**, 187–212.

VERGÉS, J. & MUÑOZ, J. A. 1990. Thrust sequences in the Southern Central Pyrenees. *Bulletin de la Société Géologique de France*, **8**, **VI(2)**, 265–271.

VERGÉS, J., MUÑOZ, J. A. & MARTÍNEZ, A. 1992. South Pyrenean fold-and-thrust belt: role of foreland evaporitic levels in thrust geometry. *In*: MCCLAY, K. (ed.) *Thrust Tectonics*. Chapman & Hall, London, 255–264.

VERGÉS, J., MILLAN, H., ROCA, E. *et al.* 1995. Eastern Pyrenees and related foreland basins: pre-, syn- and post-collisional crustal-scale cross-sections. *Marine and Petroleum Geology*, **12**, 893–915.

VERGÉS, J., MARZO, M., SANTAEULARIA, T., SERRA-KIEL, J., BURBANK, D. W., MUÑOZ, J. A. & GIMÉNEZ-MONTSANT, J. 1998. Quantified vertical motions and tectonic evolution of the SE Pyrenean foreland basin. *In*: MASCLE, A., PUIGDEFÀBREGAS, C., LUTER-BACHER, H. P. & FERNÁNDEZ, M. (eds) *Cenozoic Foreland Basins of Western Europe*. Geological Society, London, Special Publications, **134**, 107–134.

VERHOEFF, P. N. W., VISSERS, R. L. M. & ZWART, H. J. 1984. A new interpretation of the structural and metamorphic history of the western Aston massif (central Pyrenees, France). *Geologie en Mijnbouw*, **63**, 399–410.

VERNEUIL, E. DE & ARCHIAC, A. D' 1845. Note sur les fossiles du terrain paléozoique des Asturies. *Bulletin Société Géologique France*, Paris, **2**(2), 458–480.

VERNEUIL, E. DE & BARRANDE, J. 1855. Description des fossiles trouvés dans les terrains Silurien et Dévonien d'Almadén, d'une partie de la Sierra Morena et des Montagnes de Tolède. *Bulletin de la Société Géologique de France [2]*, **12**, 964–1025.

VERNEUIL, E. DE & COLLOMB, E. 1853. Coup d'oeil sur la constituion géologique de quelques provinces de l'Espagne. *Bulletin Société Géologique France*, **10**(2), 61–176.

VERNEUIL, E. DE & LARTET, L. 1863. Sur le calcaire à *Lichnus* des environs de Segura (Aragón). *Bulletin Société Géologique France*, **20**(2), 684–698.

VERNEUIL, E. DE & LORIÈRE, G. 1854. Aperçu d'un voyage géologique et tableau des altitudes prises en Espagne pendant l'été de 1853. *Bulletin Société Géologique France*, **11**(2), 661–711.

VERNIERS, J., NESTOR, V., PARIS, F. DUFKA, P., SUTHERLAND, S. & VAN GROOTEL, G. 1995. A global Chitinozoa biozonation for the Silurian. *Geological Magazine*, **132**, 651–666.

VESICA, P. L., TUCCIMEI, P., TURI, B., FORNOS, J. J., GINÉS, A. & GINÉS, J. 2000. Late Pleistocene paleoclimates and sea-level change in the Mediterranean as inferred from stable isotope and U-series studies of overgrowths on speleothems, Mallorca, Spain. *Quaternary Science Reviews*, **19**, 865–879.

VIA BOADA, L., VILLALTA, J. & ESTEBAN, M. 1977. Paleontología y paleoecología de los yacimientos fosilíferos del Muschelkalk superior entre Alcover y Mont-Ral (Montañas de Prades, provincia de Tarragona). *Cuadernos de Geología Ibérica*, **4**, 247–256.

VIALETTE, Y., CASQUET, C., FUSTER, J. M., IBARROLA, E., NAVIDAD, M., PEINADO, M. & VILLASECA, C. 1986. Orogenic granitic magmatism of pre-Hercynian age in the Spanish Central System (S. C. S.). *Terra Cognita*, **6**, 143.

VIALETTE, Y., CASQUET, C., FÚSTER, J. M., IBARROLA, E., NAVIDAD, M., PEINADO, M. & VILLASECA, C. 1987. Geochronological study of orthogneisses from the Sierra de Guadarrama (Spanish Central System). *Neues Jahrbuch für Mineralogie Monatshefte*, **10**, 465–479.

VIALLARD, P. 1973. *Recherches sur le cycle Alpin dans la Chaine Ibérique sud-occidentale*. PhD Thesis, Paul Sabatier University, Toulouse.

VICENTE-BRAVO, J. C. & ROBLES, S. 1991. Geometría y modelo deposicional de la secuencia Sollube del Flysch Negro (Albiense medio, norte de Bizkaia). *Geogaceta*, **10**, 69–72.

VICENTE-BRAVO, J. C. & ROBLES, S. 1995. Large-scale mesotopographic bedforms from the Albian Black Flysch, northern Spain: characterization, setting and comparison with recent analogues. *In*: PICKERING, K. T., HISCOTT, R. N., KENYON, N. H., RICCI LUCCHI, F. & SMITH, R. D. A. (eds) *Atlas of Deep Water Environments: Architectural Style in Turbidite Systems*. Chapman & Hall, London, 216–226.

VIDAL, G., JENSEN, S. & PALACIOS, T. 1994a. Neoproterozoic (Vendian) ichnofossils from Lower Alcudian strata in central Spain. *Geological Magazine*, **131**, 169–179.

VIDAL, G., PALACIOS, T., GÁMEZ VINTANED, J. A., DÍEZ BALDA, M. A. & GRANT, S. W. F. 1994b. Neoproterozoic-early Cambrian geology and paleontology of Iberia. *Geological Magazine*, **131**, 729–765.

VIDAL, G., PALACIOS, T., MOCZYDLOWSKA, M. & GUBANOV, A. P. 1999. Age contraints from small shelly fossils on the early Cambrian terminal Cadomian Phase in Iberia. *Geologiska Föreningens i Stockholm Förhandlingar*, **121**, 137–143.

VIDAL, J. R., YEPES, J. & RODRÍGUEZ, R. 1998. Evolución geomorfológica del Macizo Hespérrico Peninsular. Estudio de un sector comprendido entre las provincias de Lugo y Ourense (Galicia, NW de España). *Cuadernos Laboratorio Xeolóxico de Laxe*, **23**, 165–199.

VIDAL, J. R., FERNÁNDEZ, D., MARTÍ, K. & FERREIRA, A. B. 1999. Nuevos datos para la cronología glaciar pleistocena en el NW de la Península Ibérica. *Cuadernos Laboratorio Xeolóxico de Laxe*, **24**, 7–29.

VIDAL, L. M. 1905. Note sur l'Oligocéne de Majorque. *Bulletin de la Société Géologique de France*, **5**, 651–654.

VIDAL, L. M. 1914. Nota paleontológica sobre el Silúrico superior del Pirineo catalán. *Memorias de la Real Academia de Ciencias y Artes de Barcelona* (3), **11**(19), 307–313.

VIDAL, L. M. 1917. Edad geológica de los lignitos de Selva y Binisalem (Mallorca). *Memórias de la Real Sociedad Española de Historia Natural*, **10**, 343–358.

VIDAL, M. 1998. Le modèle des biofaciès à Trilobites: un test dans l'Ordovicien inférieur de l'Anti-Atlas, Maroc. *Comptes Rendus de l'Académie des Sciences, Paris, Sciences de la terre et des planètes*, **327**, 327–333.

VIDAL-BARDÁN, M. & SÁNCHEZ-CARPINTERO, L. 1990. Análisis e interpretación de algunas cuestiones que plantea el complejo de morrenas y terrazas del río Aragón (Huesca). *Cuaternario y Geomorfología*, **4**, 107–118.

VIELZEUF, D. 1984. *Relations de phases dans le faciès granulite et implications géodynamiques. L'exemple des granulites des Pyrénées*. PhD Thesis, University of Clermont-Ferrand.

VIELZEUF, D. & KORNPROBST, J. 1984. Crustal splitting and the emplacement of Pyrenean lherzolites and granulites. *Earth and Planetary Science Letters*, **67**, 383–386.

VIGNERESSE, J. L. & BOUCHEZ, J. L. 1997. Succesive granitic magma batche during pluton emplacement: the case of Cabeza de Araya (Spain). *Journal of Petrology*, **38**, 1767–1776.

VILAPLANA, J. M. 1983. *Estudi del glaciarisme quaternari de les alts valls de la Ribagorça*. PhD Thesis, University of Barcelona.

VILAS, L. & QUEROL, R. 1999. El límite septentrional de la extensión prebética en el sector de Murcia. *In*: *Libro Homenaje a José Ramírez del Pozo*. AGGEP, Madrid, 219–226.

VILAS, L. & SAN JOSÉ, M. A. 1990. Autochthonous sequences in the Central Iberian Zone: Stratigraphy. *In*: DALLMEYER, R. D. & MARTÍNEZ GARCÍA, E. (eds) *Pre-Mesozoic Geology of Iberia*. Springer Verlag, Berlin, 145–146.

VILAS, L., HERNÁNDO, S., GARCÍA-QUINTANA, A., RINCÓN, R. & ARCHE, A. 1977. El Triásico de la región de Monterde-Alhama de Aragón (provincia de Zaragoza). *Cuadernos de Geología Ibérica*, **4**, 467–484.

VILAS, L., MAS, R., GARCÍA, A., ARIAS, C., ALONSO, A., MELÉNDEZ, N. & RINCÓN, R. 1982. Ibérica Suroccidental. *In*: GARCÍA, A. (coord.) *El Cretácico de España*. Complutense University, Madrid, 457–515.

VILAS, L., ALONSO A., ARIAS, C., GARCÍA, A., MÁS, J. R., RINCÓN, R. & MELÉNDEZ, N. 1983. The Cretaceous of the Southwestern Iberian Ranges (Spain). *Zitteliana*, **10**, 245–254.

VILAS, L., MASSE, J. P. & ARIAS, C. 1993. Aptian Mixed Terrigenous and Carbonate Platforms from Iberic and Prebetic Regions, Spain. *In*: SIMÓ, J. A. T., SCOTT, R. W. & MASSE. J. P. (eds) *Cretaceous Carbonate Platforms*. AAPG, Memoirs, **56**, 243–253.

VILAS, L, DABRIO, C. J. PELÁEZ, J. R. & GRACÍA-HERNÁNDEZ M. 2001. Dominios sedimentarios generados durante el período extensional Cretácio Inferior sobre Cazorla y Hellín (Béticas externas). Su implicación en la estructura actual. *Revista de la Sociedad Geológica de España*, **14**, 113–122.

VILASECA, S. 1920. Contribució a léstudi dels terrenys triàsics de la provincia de Tarragona. *Treballs Museu Siènces Naturals de Barcelona*, **VIII**, 1–66.

VILLA, E. & BAHAMONDE, J. R. 2001. Accumulations of Ferganites (Fusulinacea) in shallow turbidite deposits from the Carboniferous of Spain. *Journal of Foraminifera Research*, **31**, 173–190.

VILLA, E. & HEREDIA, N. 1991. Aportaciones al conocimiento del Carbonífero de la Región de Mantos y de la Cuenca Carbonífera Central (Cordillera Cantábrica, NO de España), *Boletín Geológico y Minero*, **99**, 757–769.

VILLA, E. & MARTÍNEZ GARCÍA, E. 1989. El Carbonífero superior marino de Dobros (Picos de Europa, Asturias, NW de España). *Trabajos de Geología*, **18**, 77–93.

VILLA, E., ESCUDER, J. & GINKEL, A. C. VAN. 1996. Fusulináceos y edad de los afloramientos carboníferos de Puig Moreno (Cordillera Ibérica, Teruel, España). *Revista Española de Paleontología*, **11**, 207–215.

VILLALOBOS, L. 1975. *Sumilla*. Spanish Geological Survey. Report and map, scale 1:50.000, **90**.

VILLALTA, J. F. DE & ROSELL, J. 1969. Nuevas aportaciones al conocimiento de la estratigrafía del Devónico de Gerri de la Sal (Lérida). *Acta Geológica Hispánica*, **4**, 108–111.

VILLAMARÍN, J. A., MASANA, E., CALDERÓN, T., JULIÀ, R. & SANTANACH, P. 1999. Abanicos aluviales cuaternarios del Baix Camp (provincia de Tarragona): resultados de dataciones radiométricas. *Geogaceta*, **25**, 211–214.

VILLAMOR, P., CAPOTE, R. & TSIGE, M. 1996. Neotectonic activity of the Alentejo-Plasencia fault in the Extremadura (Hesperian Massif). *Geogaceta*, **20**, 925–928.

VILLAROYA ALDEA, C. 1997. *La delimitacion del dominio público hidráulico y de sus zonas inundables.* El proyecto LLINDE, Madrid.

VILLAS, E. 1983. Las formaciones del Ordovícico medio y superior de las Cadenas Ibéricas y su fauna de braquiópodos. *Estudios Geológicos*, **39**, 359–377.

VILLAS, E. 1985. Braquiópodos del Ordovícico medio y superior de las Cadenas Ibéricas Orientales. *Memorias del Museo Paleontológico de la Universidad de Zaragoza*, **1**, 1–156.

VILLAS, E. 1992. New Caradoc brachiopods from the Iberian Chains (Northeastern Spain) and their stratigraphic significance. *Journal of Paleontology*, **66**, 772–793.

VILLAS, E. 1995. Caradoc through Early Ashgill brachiopods from the Central-Iberian Zone (Central Spain). *Geobios*, **28**, 49–84.

VILLAS, E. & COCKS, L. R. M. 1996. The first Early Silurian brachiopod fauna from the Iberian Peninsula. *Journal of Paleontology*, **70**(4), 571–588.

VILLAS, E., DURÁN, H. & JULIVERT, M. 1987. The Upper Ordovician clastic sequence of the Catalonian Coastal Ranges and its brachiopod fauna. *Neues Jahrbuch für Geologie und Paläontologie Abhandlungen*, **174**, 55–74.

VILLAS, E., GISBERT, J. & MONTESINOS, R. 1989. Brachiopods from volcanoclastic Middle and Upper Ordovician of Asturias (Northern Spain). *Journal of Paleontology*, **63**, 554–565.

VILLAS, E., ARBIZU, M., BERNÁRDEZ, E., MÉNDEZ-BEDIA, I. & ARAMBURU, C. 1995a. *Protambonites primigenius* (Brachiopoda, Clitambonitidina) y el límite Cámbrico-Ordovícico en la Serie de Los Cabos (Zona Asturoccidental-leonesa, NO de España). *Revista Española de Paleontología*, **10**, 140–150.

VILLAS, E., HARPER, D. A. T., MÉLOU, M. & VIZCAÏNO, D. 1995b. Stratigraphical significance of the *Sbovodaina* species (Brachiopoda, Heterorthidae) range in the Upper Ordovician of South-Western Europe. *In*: COOPER, J. D., DROSSER, M. L. & FINNEY, S. C. (eds) *Ordovician Odyssey: Short Papers from the Seventh International Symposium on the Ordovician System.* Pacific Section, Society for Sedimentary Geology, Fullerton, Book **77**, 97–98 (ISBN 1-878861-70-0).

VILLAS, E., LORENZO, S. & GUTIÉRREZ-MARCO, J. C. 1999. First record of a Hirnantia Fauna from Spain, and its contribution to the Late Ordovician palaeogeography of northern Gondwana. *Transactions of the Royal Society of Edinburgh: Earth Sciences*, **89**, 187–197.

VILLASECA, C., BARBERO, L., HUERTAS, M. J., ANDONAEGUI, P. & BELLIDO, F. 1993. A cross-section through Hercynian granites of Central Iberian Zone. *Excursion Guide*. CSIC, Madrid.

VILLASECA, C., EUGERCIOS, L., SNELLING, N., HUERTAS, MJ. & CASTELLÓN, T. 1995. Nuevos datos Geocronológicos (Rb-Sr, K-Ar) de granitoides hercínicos de la Sierra de Guadarrama. *Revista de la Sociedad Geologica de España*, **8**, 137–148.

VILLASECA, C., BARBERO, L. & ROGERS, G. 1998. Crustal origin of Hercynian peraluminous granitic batholiths of Central Spain: petrological, geochemical and isotopic (Sr, Nd) constraints. *Lithos*, **43**, 55–79.

VILLASECA, C., DOWNES, H., PIN, C. & BARBERO L. 1999. Nature and composition of the lower continental crust in Central Spain and the granulite–granite linkage: inferences from granulitic xenoliths. *Journal of Petrology* **40**, 1465–1496.

VILLENA, J. 1971. *Estudio geológico del sector de las Cadenas Ibéricas comprendido entre Molina de Aragón y Monreal. del Campo (Prov. de Guadalajara y Teruel)*. PhD Thesis, University of Granada.

VILLENA, J. & PARDO, G. 1983. El Carbonífero de la Cordillera Ibérica. *In*: Martínez Díaz, C., (ed.) *Carbonífero y Pérmico de España.* IGME, Madrid, 191–206.

VILLENA, J., RAMIREZ DEL POZO, J., LINARES, A. y RIBA, O. 1971. Características estratigráficas del Jurásico de la región de Molina de Aragón. *Cuadernos de Geología Ibérica*, **2**, 355–374.

VILLENA, J., GONZÁLEZ, A., MUÑOZ, A., PARDO, G. & PÉREZ, A. 1992. Síntesis estratigráfica del Terciario del borde Sur de la Cuenca del Ebro: unidades genéticas. *Acta Geológica Hispánica*, Homenaje a Oriol Riba Arderiu, **27**, 225–245.

VILLENA, J., PARDO, G., PÉREZ, A., MUÑOZ, A. & GONZÁLEZ, A. 1996a. The Tertiary of the Iberian margin of the Ebro basin: paleogeography and tectonic control. *In*: FRIEND, P. & DABRIO, C. (eds)

Tertiary Basins of Spain: the Stratigraphic Record of Crustal Kinematics. World and Regional Geology, **6**, Cambridge University Press, 83–88.

VILLENA, J., PARDO, G., PÉREZ, A., MUÑOZ, A. & GONZÁLEZ, A. 1996b. The Tertiary of the Iberian margin of the Ebro basin: sequence stratigraphy. *In*: FRIEND, P. & DABRIO, C. (eds) *Tertiary Basins of Spain: the Stratigraphic Record of Crustal Kinematics.* World and Regional Geology **6**, Cambridge University Press, 77–82.

VINDEL, E., LÓPEZ, J. A., MARTÍN CRESPO, T. & GARCÍA, E. 2000. Fluid evolution and hydrothermal processes of the Spanish Central System. *Journal of Geochemical Exploration*, **69–70**, 359–362.

VIÑES, G. 1942. Cova Negra de Bellúss. I. Notas sobre las excavaciones practicadas. *Trabajos Varios del SIP, Valencia*, **6**, 7–13.

VINK, G. E., MORGAN, W. J. & ZHAO, W. L. 1984. Preferential rifting of continents: a source of displaced terranes. *Journal of Geophysical Research*, **89**, 10072–10076.

VIRGILI, C. 1952. Hallazgo de nuevos *Ceratites* en el Triásico Mallorquín. *Memorias de la Comisión del Instituto Geológico Provincial de Barcelona*, **9**, 19–38.

VIRGILI, C. 1955. El tramo rojo intermedio del Muschelkalk de las Catalánides. *Memorias y Comisión del Instituto Geológico provincial, Barcelona*, **XIII**, 37–78.

VIRGILI, C. 1958. El Triásico de los Catalánides. *Boletín del IGME*, **LXIX**, 1–856.

VIRGILI, C. 1963. Le Trias du Nordest de l'Espagne. *Bulletin BRGM*, **15**, 469–481.

VIRGILI, C. & JULIVERT, M. 1954. El Triásico de la Sierra de Prades. *Estudios Geológicos*, **10**, 216–242.

VIRGILI, C., HERNÁNDO, S., RAMOS, A. & SOPEÑA, A. 1976. Le Permian en Espagne. *In*: FALKE, H. (ed.) *The Continental Permian in the Central West and South Europa.* Brill, Leiden, 91–109.

VIRGILI, C., SOPEÑA, A., RAMOS, A. & HERNÁNDO, S. 1983. El relleno posthercínico y la sedimentación mesozoica. *In*: COMBA, J A. (ed.) *Geología de España, 2.* IGME, Madrid, 25–36.

VISERAS, C. 1991. *Estratigrafía y sedimentología del relleno aluvial de la Cuenca de Guadix (Cordilleras Béticas)*. PhD Thesis, University of Granada.

VISSCHER, H., BRUGMAN, W. & LÓPEZ-GÓMEZ, J. 1982. Nota sobre la presencia de una palinoflora triásica en el supuesto Pérmico de Cueva de Hierro (Serranía de Cuenca), España. *Revista Española de Micropaleontología*, **14**, 315–322.

VISSERS, R. L. M. 1981. A structural study of the central Sierra de los Filabres (Betic Zone, SE Spain) with emphasis on deformational processes and their relation to the Alpine metamorphism. *GUA Paper of Geology*, **15**, 1–154.

VISSERS, R. L. M. 1992. Variscan extension in the Pyrenees. *Tectonics*, **11**, 1369–1384.

VISSERS, R. L. M., PLATT, J. P. & VAN DER WAL, D. 1995. Late orogenic extension of the Betic Cordillera and the Alborán domain: A lithospheric view. *Tectonics*, **14**, 786–803.

VITRAC-MICHARD, A. & ALLÈGRE, C. J. 1975. A study of the formation and history of a piece of continental crust by 87Rb-87Sr method: the case of the French Oriental Pyrenees. *Contribution to Mineralogy and Petrology*, **50**, 257–285.

VITRAC-MICHARD, A., ALBAREDE, F. DUPUIS, C. & TAYLOR, H. P. 1980. The genesis of Variscan (Hercinian) plutonic rocks: inferences from Sr, Pb, and O studies on the Maladeta Igneous Complex, Central Pyrenees (Spain). *Contribution to Mineralogy and Petrology*, **72**, 57–72.

VOGEL, D. E. 1967. Petrology of eclogite- and pirigarnite-bearing polymetamorphic rock complex at Cabo Ortegal, NW Spain. *Leidse Geologische Mededelingen*, **40**, 121–213.

VOGEL, D. E. 1984. Cabo Ortegal, mantle plume or double klippe. *Geologie en Mijnbouw*, **63**, 131–140.

VOGEL, D. E. & ABDEL MONEM, A. A. 1971. Radiometric evidence for a Precambrian metamorphic event in NW Spain. *Geologie in Minjbouw*, **50**, 749–750.

VOGT, P. R. 1974. Volcano spacing, fractures and thickness of the lithosphere. *Earth and Planetary Science Letters*, **21**, 235–252.

VOGT, P. R. & SMOOT, N. C. 1984. The Geisha guyots: multibeam bathymetry and morphometric interpretation. *Journal of Geophysical Research*, **89**, 11085–11107.

Völk, H. R. 1966. Aggradational directions and biofacies in the youngest postorogenic deposits of southeastern Spain. A contribution to the determination of the age of the Mediterranean coast of Spain. *Palaeogeography, Palaeoclimatology, Palaeoecology*, **2**, 313–331.

Von Drasche, R. 1879. Bosquejo geológico de la zona superior de Sierra Nevada. *Boletín del Comité del Mapa Geológico de España*, **6**, 353–388.

Wadsworth, W. J. & Adams, A. E. 1989. Miocene volcanic rocks from Mallorca. *Proceedings Geological Association*, **100**, 107–112.

Wagner, R. H. 1959. Sur la presence d'une nouvelle phase tectonique 'leonienne' d'âge Westphalien D dans le nord-ouest de l'Espagne. *Comptes Rendues de l'Académie des Sciences*, **249**, 2804–2806

Wagner, R. H. 1960. Middle westphalian floras from the northern Palencia (Spain) in relation with the Curavacas phase of folding. *Estudios Geológicos*, **XVI**, 55–92

Wagner, R. H. 1971*a*. Account of the International Field Meeting on the Carboniferous of the Cordillera Cantábrica, 19–26 Sept. 1970. *Trabajos de Geología*, **3**, 1–39.

Wagner, R. H. 1971*b*. The stratigraphy and structure of the Ciñera-Matallana coalfield (prov. León, N. W. Spain). *Trabajos de Geología*, **4**, 385–429.

Wagner, R. H. 1971*c*. Carboniferous nappe structures in north-western Palencia (Spain). *Trabajos de Geología*, **4**, 431–459.

Wagner, R. H. 1999. Peñarroya, a strike-slip controlled basin of early Westphalian age in southwest Spain. *Bulletin of the Czech Geological Survey*, **74**, 87–108.

Wagner, R. & Martínez García, E. 1982. Description of an early Permian flora from Asturias and comments on similar occurrences in the Iberian Peninsula. *Trabajos de Geología de la Universidad de Oviedo*, **12**, 273–287.

Wagner, R. H. & Martínez García, E. 1998. Floral remains from the highest Valdeón Fm, a marine Stephanian unit South of the Picos de Europa, and comparisons with the eastern Asturias, NW Spain. *Revista Española de Paleontología*, **13**, 93–106.

Wagner, R. H. & Wagner-Gentis, C. H. T. 1963. Summary of the stratigraphy of Upper Palaeozoic rocks in NE Palencia, Spain. *Proceeding Koninklijke Nederlandse Akademie van Wetenschappen (b)*, **LXVI**(2), 149–163

Wagner, R. H. & Winkler Prins, C. F. 1985. The Cantabrian and Barruelian stratotypes: A summary of basin development and biostratigraphic inFm. *In*: Lemos de Sousa, M. J. & Wagner, R. H. (eds) *Papers on the Carboniferous of the Iberian Peninsula (Sedimentology, Stratigraphy, Paleontology, Tectonics and Geochemistry). Annales da Faculdade de Ciências, Universidade do Porto*, **64**, Special Supplement (1983), 359–410.

Wagner, R. H. & Winkler-Prins, C. F. 1999. Carboniferous stratigraphy of the Sierra ddel Brezo in northern Palencia: evidence of major uplifts. *Trabajos de Geología, Universidad de Oviedo*, **21**, 385–403.

Wagner, R. H., Winkler Prins, C. J. & Riding, E. 1971. Lithostratigraphic units of the lower part of the Carboniferous in northern León, Spain. *Trabajos de Geología*, **4**, 603–663.

Wagner, R. H., Carballeira, Ambrose, T. & Martínez Garcia, E. 1984. *Memoria de la Hoja Geológica E. 1:50.000 n° 107 (17–7), Barruelo de Santullán*. IGME, Madrid.

Wagner, R. H., Talens, J. & Meléndez, B. 1985. Upper Stephanian stratigraphy and megaflora of Henarejos (province of Cuenca) in the Cordillera Ibérica, Central Spain. *In*: Lemos de Sousa, M. J. & Wagner, R. H. (eds) *Papers on the Carboniferous of the Iberian Peninsula (Sedimentology, Stratigraphy, Paleontology, Tectonics and Geochemistry). Annales da Faculdade de Ciências, Universidade do Porto*, **64**, Special Supplement (1983), 445–480.

Walker, G. P. L. 1973. Explosive volcanic eruptions: A new classification scheme. *Geologische Rundschau*, **62**, 4322–4346.

Walker, G. P. L. 1990. Geology and volcanology of the Hawaiian Islands. *Pacific Sciences*, **44**, 315–347.

Walker, G. P. L. 1992. Coherent intrusion complexes in large basaltic volcanoes – a new structural model. *Journal of Volcanology and Geothermal Research*, **50**, 41–54.

Walker, M. J., Gibert, J., Sánnchez, F. *et al*. 1998. Two SE Spanish middle palaeolithic remains. Sima de las Palomas del Cabezo Gordo and Cueva Negra del Estrecho del Río Quípar (Murcia

province). *Internet Archaeology*, **5** <http://intarch.ac.uk/journal/issue5/walker-index.html>.

Wallis, R. J. 1983. Early evolution and sedimentation in the Puertollano basin (Ciudad Real, central Spain). *Annales da Faculdade de Ciências, Universidade do Porto*, **64**, 269–282.

Walliser, O. 1964. Conodonten des Silurs. *Abhandlungen Hessischen Landesamtes Bodenforschung*, **41**, 1–106.

Walliser, O. 1996. Patterns and causes of global events. *In*: Walliser, O. H. (ed.) *Global Events and Event Stratigraphy*. Springer-Verlag, Berlin, 7–19.

Walter, R. 1963. Beitrag zur Stratigraphie das Kambiums im Galicien (Nord-west-Spanien). *Neues Jahrbuch für Geologie und Paläontologie, Abh.*, **117**, 360–371.

Walter, R. 1968. Die Geologie in der nordöstlichen Provinz Lugo (Nordwest Spanien). *Geotektonische Forschungen*, **27**, 3–70.

Walter, R. 1969. Das Silurium Spaniens und Portugal. *Zentralblatt für Geologie und Paläontologie, Teil I, Stuttgart*, **1969**(5), 857–902.

Walter, R. 1977. Zwei geologische Traversen durch die südliche Iberische Meseta, Spanien. *Münstersche Forschungen zur Geologie und Paläontologie*, **42**, 1–55.

Wang, R. 1991. Conodonten-Biofazies und -Biostratigraphie der Moyuela-Formation (Eifelium) in Keltiberien, Spanien. *Neues Jahrbuch Geologie Paläontologie, Monatshefte*, **1991**, 177–189.

Warburton, J & Alvarez, C. 1989. A thrust tectonic interpretation of the Guadarraña Mountains, Spanish Central System. *In*: *Libro Homenaje a Rafael Soler*. AGGEP, Madrid, 147–155.

Waterlot, M. 1983. VII El Carbonífero de los Pirineos. *In*: Martínez Díaz, C. (ed.) *Carbonífero y Pérmico de España*. IGME, Madrid, 281–328.

Watson, H. J. 1979. *Casablanca Field Offshore Spain. A Paleomorphic Trap*. AAPG, Memoirs, **32**.

Watts, A. B. 1994. Crustal structure, gravity anomalies and flexure of the lithosphere in the vicinity of the Canary Islands. *Geophysical Journal International*, **119**, 648–666.

Watts, A. B. & Masson, D. G. 1995. A giant landslide on the north flank of Tenerife, Canary Islands. *Journal of Geophysical Research*, **100**(B12), 24487–24498.

Watts, A. B., Platt, J. P. & Buhl, P. 1993. Tectonic evolution of the Alboran Sea basin. *Basin Research*, **5**, 153–177.

Weaver, B. L. 1991. The origin of ocean island basalt end-member compositions: trace element and isotopic constraints. *Earth and Planetary Science Letters*, **104**, 381–397.

Weaver, P. P. E., Rothwell, R. G., Ebbing, J., Gunn, D. E. & Hunter, P. M. 1992. Correlation, frequency of emplacement and source directions of megaturbidites on the Madeira Abyssal Plain. *Marine Geology*, **109**, 1–20.

Weaver, P. P. E., Schmincke, H. U., Firth, J. V. & Duffield, W. (eds) 1998. *Proceedings ODP, Scientific Results*, **157**: College Station, TX (Ocean Drilling Program).

Webby, B. D. 1998. Steps toward a global standard for Ordovician stratigraphy. *Newsletters on Stratigraphy*, **36**, 1–33.

Wedepohl, K. H., Gohn, E. & Hartmann, G. 1994. Cenozoic alkali basaltic magmas of western Germany and their products of differentiation. *Contributions to Mineralogy and Petrology*, **115**, 253–278.

Weijermars, R. 1985. Uplift and subsidence history of the Alboran Basin and a profile of the Alboran Diapir (W-Mediterranean). *Geologie en Mijnbouw*, **64**, 349–356.

Weijermars, R. 1988. Neogene tectonics in the Western Mediterranean may have caused the Messinian Salinity Crisis and associated glacial event. *Tectonophysics*, **148**, 211–219.

Weijermars, R., Roep, T. B., Van den Eeckhout, B., Postma, G. & Kleverlaan, K. 1985. Uplift history of a Betic fold nappe inferred from Neogene–Quaternary sedimentation and tectonics (in the Sierra Alhamilla and Almería, Sorbas and Tabernas Basins of the Betic Cordilleras, SE Spain). *Geologie en Mijnbouw*, **64**, 397–411.

Wendt, J. & Aigner, 1985. Facies patterns and depositional environments of Palaeozoic cephalopod limestones. *Sedimentary Geology*, **44**, 263–300.

Wentworth, C. K. & MacDonald, G. A. 1953. Structures and forms of basaltic rocks in Hawaii. *USGS Bulletin*, **994**.

Westerhoff, A. B. 1977. On the contact relations of high-temperature peridotites in the serrania de Ronda, southern Spain. *Tectonophysics*, **39**, 579–591.

WEYANT, M., BRICE, D., RACHBOEUF, P. R., BABIN, C. & ROBARDET, M. 1988. Le Dévonien supérieur du synclinal du Valle (province de Seville, Espagne). *Revue Paléobiologie*, **7**(1), 233–260.

WGDSSA. 1978 (WORKING GROUP FOR DEEP SEISMIC SOUNDING IN ALBORAN 1974–1975). Crustal seismic profiles in the Alboran Sea. Preliminary results. *Pure and Applied Geophysics*, **116**, 166–180.

WGDSSS. 1977 (WORKING GROUP FOR DEEP SEISMIC SOUNDING IN SPAIN 1974–1975). Deep seismic soundings in southern Spain. *Pure and Applied Geophysics*, **115**, 721–735.

WHITTINGTON, H. B. & HUGHES, C. P. 1972. Ordovician geography and faunal provinces deduced from trilobite distribution. *Philosophical Transactions of the Royal Society, London, B, Biological Sciences*, **263**, 235–278.

WICKHAM, S. M. 1987. Crustal anatexis and granite petrogenesis during low pressure regional metamorphism; the Trois Seigneurs Massif, Pyrenees, France. *Journal of Petrology*, **28**, 127–169.

WICKHAM, S. M. & OXBURGH, E. R. 1986. A rifted tectonic setting for Hercynian high-thermal gradient metamorphism in the Pyrenees. *Tectonophysics*, **129**, 53–69.

WIEDMANN, J. 1964. Le Crétacé supérieur de l'Espagne et du Portugal et ses Céphalopodes. *Estudios Geológicos*, **20**, 107–148.

WIEDMANN, J. & BOESS, J. 1984. Ammonitenfunde aus der Biskaya-Synkline (Nordspanien) – Kreidegliederung und Alter des Kreide-Vulkanismus. *Eclogae Geologica Helvetica*, **77**, 483–510.

WIESE, F. & WILMSEN, M. 1999. Sequence stratigraphy in the Cenomanian to Campanian of the North Cantabrian Basin (Cantabria, Spain*). Neues Jahrbuch für Geologie und Paläontologie, Abhandlungen*, **212**, 131–173.

WILDBERG, H. G. H., BISCHOFF, L. & BAUMANN, A. 1989. U-Pb ages of zircons from meta-igneous and meta-sedimentary rocks of Sierra de Guadarrama: implications for the Central Iberian crustal evolution. *Contributions to Mineralogy and Petrology*, **103**, 253–262.

WILDE, S. 1990. The Bathonian and Callovian of the Northwest Iberian Range: stages and facies palaeogeographical differentiation on an epicontinental platform. *Cuadernos de Geología Ibérica*, **14**, 113–142.

WILDI, W. 1983. La chaîne tello-rifaine (Algérie, Maroc, Tunisie): structure, stratigraphie et évolution du Trias au Miocène. *Revue de Géologie Dynamique et de Géographie Physique*, **24**, 201–297.

WILLIAMS, D. 1962. Further reflections on the origin of the Poephyries and Ores of Rio Tinto. Spain. *Transactions of the Institution of Mining and Metallurgy*, **71**, 265–266.

WILLIAMS, D., STANTON, R. L. & RAMBAUD, F. 1975. The Planes – San Antonio pyritic deposit of Riotinto, Spain: its nature, environment and genesis. *Transactions of the Institute of Mining and Metallurgy*, **84**, 73–82.

WILLIAMS, G. D. & FISCHER, M. W. 1984. A balanced section across the Pyrenean orogenic belt. *Tectonics*, **3**, 773–780.

WILMSEN, M. 2000. Evolution and demise of a mid-Cretaceous carbonate shelf, the Altamira Limestones (Cenomanian) of northern Cantabria (Spain). *Sedimentary Geology*, **133**, 195–226.

WILSON, M. & BIANCHINI, G. 1999. Tertiary–Quaternary magmatism within the Mediterranean and surrounding regions. *In*: DURAND, B., JOLIVET, L., HORVATH, F. & SERANNE, M. (eds) *The Mediterranean Basins: Tertiary Extension within the Alpine Orogen*. Geological Society, London, Special Publications, **156**, 141–168.

WILSON, M. & DOWNES, H. 1991. Tertiary–Quaternary extension-related alkaline magmatism in western and central Europe. *Journal of Petrology*, **32**, 811–849.

WILSON, M., ROSENBAUM, J. M. & DUNWORTH, E. A. 1995. Melilitites: partial melts of the thermal boundary layer? *Contributions to Mineralogy and Petrology*, **119**, 181–196.

WITT-EICKSCHEN, G. & KRAMM, U. 1997. Mantle upwelling and metasomatism beneath Central Europe: Geochemical and isotopic constraints from mantle xenoliths from the Rhön (Germany). *Journal of Petrology*, **38**, 479–493.

WOLF, R. 1980*a. Lithology and acritarchs of the Lower Ordovician formations of Celtiberia (NE Spain) with stratigraphic and palaeoenvironmental implications*. Dissertation, University of Würzburg.

WOLF, R. 1980*b*. The lower and upper boundary of the Ordovician System of some selected regions (Celtiberia, Eastern Sierra Morena) in Spain. Part I: The Lower Ordovician sequence of Celtiberia. *Neues Jahrbuch für Geologie und Paläontologie Abhandlungen*, **160**, 118–137.

WOLFF, J. A. 1985. Zonation mixing and eruptions of silica-undersaturated alkaline magma: a case study from Tenerife, Canary Islands. *Geological Magazine*, **122**, 623–640.

WRIGHT, V. P. & ALONSO-ZARZA, A. M. 1990. Pedostratigraphic models for alluvial fan deposits: a tool for interpreting ancient sequences. *Journal of the Geological Society*, **147**, 8–10.

WRIGHT, V. P., ALONSO-ZARZA, A. M., SANZ, M. E. & CALVO, J. P. 1997. Diagenesis of Late Miocene micritic lacustrine carbonates, Madrid Basin, Spain. *Sedimentary Geology*, **114**, 81–95.

WU, X. 1990. The evolution of humankind in China. *Acta Anthropologica Sinica*, **9**(4), 321 (Chinese, with English abstract).

WÜRM, A. 1911. Untersuchungen über den geologischen bau der Trias von Aragonien. *Zeitschift der Deutchen. Geologischen. Gesellschaft*, **63**, 38–175.

WÜRM, A. 1919. Beiträge zur Kenntnis der Trias von Katalonien. *Zeitschift der Deutchen. Geologischen. Gesellschaft*, **71**, 153–160.

WYSESSION, M. E., WILSON, J., BARTKÓ, L. & SAKATA, R. 1995. Intraplate seismicity in the Atlantic Ocean Basin: A teleseismic catalog. *Bulletin of the Seismological Society of America*, **85**, 755–774.

YÉBENES, A., COMAS-RENGIFO, M. J., GÓMEZ, J. J. & GOY, A. 1988. Unidades tectosedimentarias en el Lias de la Cordillera Ibérica. *III Congreso de Estratigrafía y Paleogeografía del Jurásico de España, abstracts*, 108–109.

YENES, M., GURIÉREZ-ALONSO, G. & ALVAREZ, F. 1996. Dataciones K-Ar de los granitoides del área La Alberca-Béjar (Sistema Central Español). *Geogaceta*, **20**, 479–482.

YOKOYAMA, Y. 1989. Direct gamma-ray spectrometric dating of ante-neandertalian and neandertalian human remains. *In*: GIACOBINI, G. (ed.) *Hominidae*. Jaca, Milan, 387–390.

YOUNG, T. P. 1988. The lithostratigraphy of the upper Ordovician of central Portugal. *Journal of the Geological Society, London*, **145**, 377–392.

YOUNG, T. P. 1989. Eustatically controlled ooidal ironstone deposition: facies relationships of the Ordovician open-shelf ironstones of Western Europe. *In*: YOUNG, T. P. & TAYLOR, W. E. G. (eds) *Phanerozoic Ironstones*. Geological Society, London, Special Publication, **46**, 51–63.

YOUNG, T. P. 1990. Ordovician sedimentary facies and faunas of Southwest Europe: palaeogeographic and tectonic implications. *In*: MCKERROW, W. S. & SCOTESE, C. R. (eds) *Palaeozoic Palaeogeography and Biogeography*. Geological Society, London, Memoirs, **12**, 421–430.

YOUNG, T. P. 1992. Ooidal ironstones from Ordovician Gondwana: a review. *Palaeogeography, Palaeoclimatology, Palaeoecology*, **99**, 321–347.

ZACHARAKIS, T. G., NAVARRO COMET, J., ARRIETA MURILLO, A. & ESTEBAN, M. 1996. Determination of porosity and facies trends in a complex carbonate reservoir, by using 3–D seismic, borehole tools, and outcrop geology. *AAPG Bulletin*, **80**, 1347.

ZAMARREÑO, I. 1972. Las litofacies carbonatadas del Cámbrico de la Zona Cantábrica (NW de España) y su distribución paleogeográfica. *Trabajos de Geología, Universidad de Oviedo*, **5**, 1–118.

ZAMARREÑO, I. 1975. Peritidal origin of Cambrian carbonates in Northwest Spain. *In*: GINSBURG, R. N. (ed.) *Tidal Deposits: a Case Book of Recent Examples and Fossil Counterparts*. Springer Verlag, Berlin, 323–332.

ZAMARREÑO, I. 1976. Depósitos carbonatados de tipo 'tidal flat' en el Devónico inferior del NW de España: la Dolomía de Bañugues. *Trabajos Geología Universidad Oviedo*, **8**, 59–85.

ZAMARREÑO, I. 1983. El Cámbrico del Macizo Ibérico. *In*: COMBA, J. A. (coord.) *Libro Jubilar J. M. Rios, 1*. IGME, Madrid, 117–192.

ZAMARREÑO, I. & DEBRENNE, F. 1977. Sédimentologie et biologie des constructions organogènes du Cambrien inférieur du Sud de l'Espagne. *BRGM Mémoires*, **89**, 49–61.

ZAMARREÑO, I. & JULIVERT, M. 1968. Estratigrafía del Cámbrico del Oriente de Asturias y estudio petrográfico de las facies carbonatadas. *Trabajos de Geología, Universidad de Oviedo*, **1**, 135–163.

ZAMARREÑO, I. & PEREJÓN, A. 1976. El nivel carbonatado del Cámbrico de Piedrafita (Zona Asturoccidental-Leonesa, NW de España): tipos de facies y faunas de Arqueociatos. *Breviora Geológica Astúrica*, **20**, 17–32.

ZAMARREÑO, I., HERMOSA, J. L., BELLAMY, J. & RABU, D. 1975. Litofacies de nivel carbonatado del Cámbrico de la región de Ponferrada (Zona Asturoccidental-Leonesa, NW de España). *Breviora Geológica Astúrica*, **19**, 40–48.

ZAMARREÑO, I., VEGAS, R. & MORENO, F. 1976. El nivel carbonatado de los Navalucillos y su posición en la sucesión Cámbrica de los Montes de Toledo occidentales (Centro de España). *Breviora Geológica Astúrica*, **20**, 56–64.

ZAZO, C. 1979. El problema del límite Plio-Pleistoceno en el litoral S y SE de España. *Trabajos Neógeno-Cuaternario*, **9**, 65–72.

ZAZO, C. & GOY, J. L. 1989. Sea level changes in the Iberian Peninsula during the last 200,000 years. *In*: SCOTT, D., PIRAZZOLI, P. & HONING, G. (eds) *Late Quaternary Correlations and Applications*. Kluwer, Devente, **256**, 27–39.

ZAZO, C., GOY, J. L. & HOYOS, M. 1983. Estudio geomorfológico de los alrededores de la Sierra de Atapuerca (Burgos). *Estudios Geológicos*, **39**, 179–185.

ZAZO, C., GOY, J. L., DABRIO, C. J., BARDAJÍ, T., SOMOZA, L. & SILVA, P. G. 1993. The Last Interglacial in the Mediterranean as a model for the Present Interglacial. *Global Planetary Change*, **7**, 109–117.

ZAZO, C., GOY, J. L., SOMOZA, L. *et al*. 1994. Holocene sequence of sea-level fluctuations in relation to climatic trends in the Atlantic-Mediterranean linkage coast. *Journal of Coastal Research*, **10**, 933–945.

ZAZO, C., HILLAIRE-MARCEL, C., GOY, J. L., GHALEB, B. & HOYOS, M. 1997. Cambios del nivel del mar-clima en los últimos 250 Ka (Canarias Orientales). *Boletín Geológico y Minero*, **108**, 159–169.

ZAZO, C., DABRIO, C. J., BORJA, F. *et al*. 1999a. Pleistocene and Holocene aeolian facies on the Huelva coast (Southern Spain): climatic and neotectonic implications. *Tectonophysics*, **77**, 209–224.

ZAZO, C., DABRIO, C. J., GONZÁLEZ, A. *et al*. 1999b. The record of the latter glacial and interglacial periods in the Guadalquivir marshlands (Mari López drilling, S.W. Spain). *Geologie Mijnbouw*, **26**, 123–126.

ZAZO, C., SILVA, P. G., GOY, J. L. *et al*. 1999c. Coastal uplift in continental collision plate boundaries: data from the Last Interglacial marine terraces of the Gibraltar Strait area (South Spain). *Tectonophysics*, **301**, 95–109.

ZAZO, C., GOY, J. L., HOYOS, M. *et al*. 2000. Raised marine terraces recording oceanic warm conditions and uplifiting trend during Pliocene and Quaternary (Eastern Canary Islands, Spain). *Quaternary International*, **63–64**, 202.

ZAZO, C., GOY, J. L., DABRIO, C. J. *et al*. (in press). Pleistocene raised marine teraces of Spanish Mediterranean and Atlantic coasts:record of coastal uplift, sea-level highstands and climate changes. *Marine Geology*.

ZECK, H. P. 1970. An Erupted Migmatite from Cerro del Hoyazo, SE Spain. *Contributions to Mineralogy and Petrology*, **26**, 225–246.

ZECK, H. P. 1992. Restite-melt and mafic-felsic magma mixing and mingling in an S-type dacite, Cerro del Hoyazo, southeastern Spain. *Transactions Royal Society of Edinburgh: Earth Sciences*, **83**, 139–144.

ZECK, H. P. 1996. Betic-Rif orogeny: subduction of Mesozoic Tethys lithosphere under eastward drifting Iberia, slab detachment shortly before 22 Ma, and subsequent uplift and extensional tectonics. *Tectonophysics*, **254**, 1–16.

ZECK, H. P. 1999. Alpine plate kinematics in the western Mediterranean: a westward-directed subduction regime followed by slab roll-back and slab detachment. *In*: DURAND, B., JOLIVET, L.,

HORVATH, F. & SERANNE, M. (eds) *The Mediterranean Basins: Tertiary Extension Within the Alpine Orogen*. Geological Society, London, Special Publications, **156**, 109–120.

ZECK, H. P. & WHITEHOUSE, M. J. 1999. Hercynian, Pan-African, Proterozoic and Archean ion-microprobe zircon ages for a Betic core complex, Alpine Belt, W Mediterranean – consequences for its P-T-t path. *Contributions to Mineralogy and Petrology*, **134**, 134–149.

ZECK, H. P., MONIÉ, P., VILLA, I. M. & HANSEN, B. T. 1992. Very high rates of cooling and uplift in the Alpine belt of the Betic Cordilleras, southern Spain. *Geology*, **20**, 79–82.

ZECK, H. P., KRISTENSEN, A. B. & NAKAMURA, E. 1999. Inherited Palaeozoic and Mesozoic Rb-Sr isotopic signatures in Neogene calc-alkaline volcanics, Alborán volcanic province, SE Spain. *Journal of Petrology*, **40**, 511–524.

ZHOU, Z. & DEAN, W. T. 1989. Trilobite evidence for Gondwanaland in east Asia during the Ordovician. *Journal of Southeastern Asia Earth Sciences*, **3**, 131–140.

ZIEGLER, P. 1988. Laurussia – the Old Red Continent. *In*: MCMILLAN, N. J., EMBRY, A. F. & GLASS, D. J. (eds) *Devonian of the World*. Canadian Society of Petroleum Geologists, Calgary, Memoirs, **14**(1), 15–48.

ZIEGLER, P. A. 1988a. *Evolution of the Arctic-North Atlantic and the Western Tethys*. AAPG, Memoirs, **43**.

ZIEGLER, P. A. 1988b. Post-Hercynian plate reorganization in the Tethys and Arctic-North Atlantic Domains. *In*: MANSPEIZER, W. (ed.) *Triassic–Jurassic Rifting. Continental Breakup and the Origin of the Atlantic Ocean and Passive Margins, Part B, III Related Mesozoic Atlantic Rift Basins, Western Europe*. Elsevier, Amsterdam, 711–756.

ZIEGLER, P. A. 1990. *Geological Atlas of Western and Central Europe*. Shell Internationale Petroleum Maatschappij BV, International Lithosphere Program, **148**.

ZIEGLER, P. A. 1994. Cenozoic rift system of western and central Europe: an overview. *Geologie en Mijnbouw*, **73**, 99–127.

ZINDLER, A. & HART, S. 1986. Chemical geodynamics. *Annual Reviews of Earth and Planetary Sciences*, **14**, 493–571.

ZOETEMEIJER, R., DESEGAULX, P., CLOETINGH, S., ROURE, F. & MORETTI, I. 1990. Lithospheric dinamics and tectonic-stratigraphic evolution of the Ebro Basin. *Journal of Geophysical Research*, **95**, 2701–2711.

ZOLLIKOFER, C., PONCE DE LEÓN, M., MARTÍN, R. & STUCKI, P. 1995. Neanderthal computer skulls. *Nature*, **375**, 283–285.

ZUIDAM, R. A. van. 1976. *Geomorphological development of the Zaragoza region, Spain*. ITC (*International Institute for Aerial Survey and Earth Sciences*) *Journal*, Enschede.

ZULUAGA, M. C., AROSTEGUI, J., GARCÍA-GARMILLA, F. & VELASCO, F. 1992. Mineralogía de arcillas y diagénesis en la sección tipo de la Formación Gordexola (Albiense inferior a medio, flanco sur del anticlinorio de Bilbao). *Boletín de la Sociedad Española de Mineralogía*, **15**, 113–122.

ZWART, H. J. 1962. On the determination of polymetamorphic mineral associations, and its application to Bosost area (Central Pyrenees). *Geologische Rundschau*, **52**, 38–65.

ZWART, H. J. 1963a. Metamorphic history of the Central Pyrenees. *Leidse Geologische Mededelingen*, **28**, 321–376.

ZWART, H. J. 1963b. The structural evolution of the Paleozoic of the Pyrenees. *Geologische Rundschau*, **53**, 170–205.

ZWART, H. J. 1965. Geological map of the Paleozoic of the Central Pyrenees. Part II, Valle de Aran, sheet 4. *Leidse Geologische Mededelingen*, **28**, 321–376.

ZWART, H. J. 1968. The Paleozoic crystalline rocks of the Pyrenees in their structural setting. *Krystalinikum*, **6**, 125–140

ZWART, H. J. 1979. The geology of the central Pyrenees. *Leidse Geologische Mededelingen*, **50**, 1–74.

Index

Page numbers in *italic* refer to figures and those in **bold** to tables.